T5-CVD-780

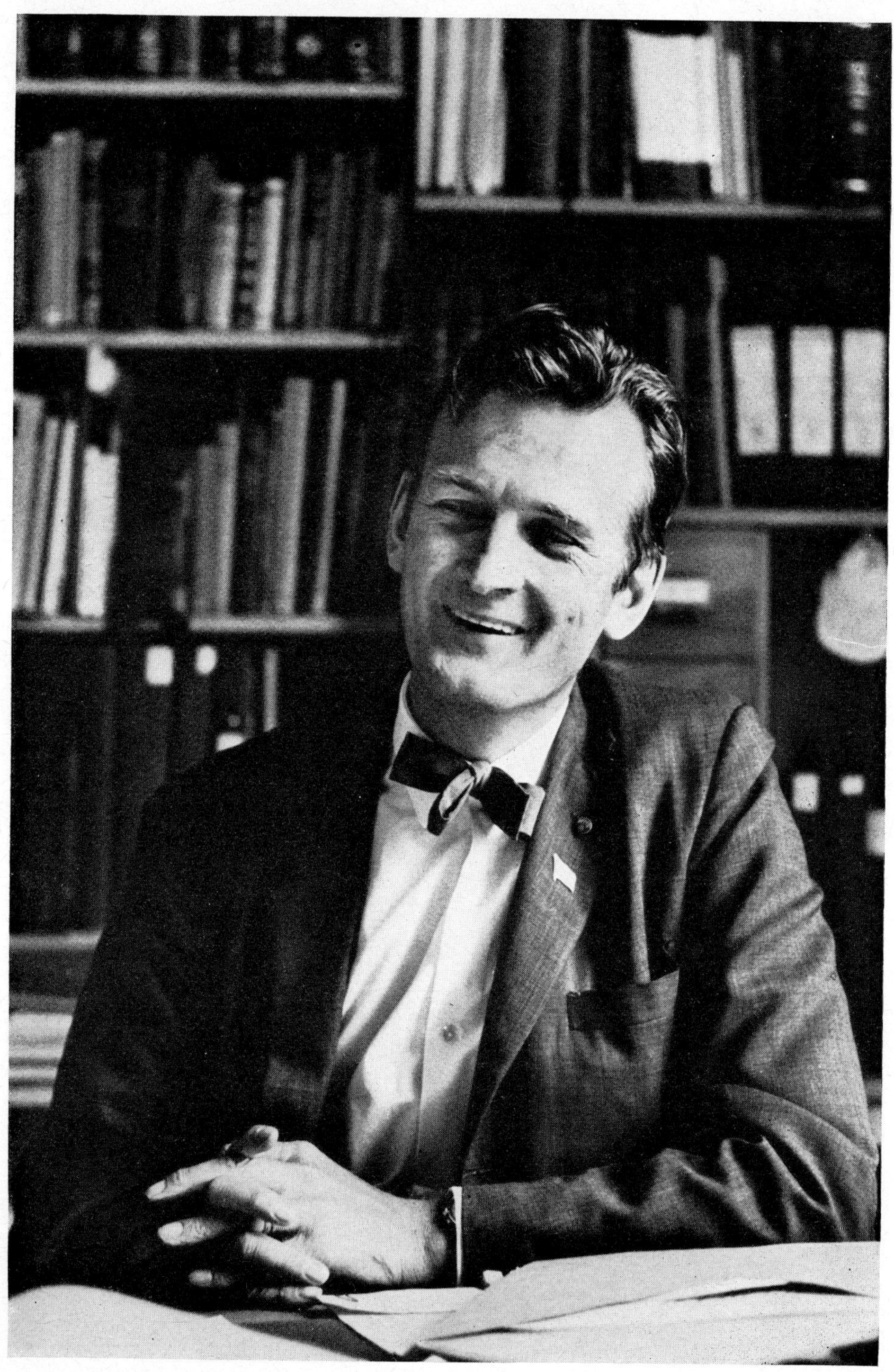

Lars Gunnar Sillén

STABILITY CONSTANTS OF METAL-ION COMPLEXES

Supplement No 1

Dedicated to the Memory of Lars Gunnar Sillén

PART I

INORGANIC LIGANDS

Compiled by Lars Gunnar Sillén

PART II

ORGANIC INCLUDING MACROMOLECULE LIGANDS

Compiled by Arthur E Martell

SPECIAL PUBLICATION No 25

SUPPLEMENT No 1 TO SPECIAL PUBLICATION No 17

THE CHEMICAL SOCIETY
BURLINGTON HOUSE, LONDON, W1V 0BN

This tabulation, like the first and second editions, has been prepared by the editors under the auspices of the Commission on Equilibrium Data of the Analytical Chemistry Division of the International Union of Pure and Applied Chemistry.

SBN 85186 019 2

Printed in Great Britain by Alden & Mowbray Ltd
at the Alden Press, Oxford

LARS GUNNAR SILLÉN

11th July 1916–23rd July 1970

The untimely death of Professor Lars Gunnar Sillén at the age of only 54 is a blow to the field of co-ordination chemistry. His passing will be mourned by the large number of co-ordination chemists all over the world who knew him and his work. Few Swedish chemists have participated so actively in various international congresses and conferences, and in the IUPAC. His effectiveness in these meetings was facilitated by his excellent command of, and ability to converse in, several languages such as Russian, English, German, and French.

Lars Gunnar Sillén began his scientific career as a crystallographer, but soon transferred his interests to solution chemistry. His main contributions to co-ordination chemistry are his extensive studies of the hydrolysis of metal ions in solution. In these systems polynuclear complexes are the rule rather than the exception, and he was highly successful in developing methods for the treatment of the complicated equilibria encountered under varying solution conditions. The methods of equilibrium analysis that he applied to polynuclear hydroxometal complexes were carried out first with graphical methods and more recently with computer techniques. The results of these studies provided new information and important new concepts on the nature and mechanism of formation of complex polynuclear species in solution.

In recent years Sillén embarked on a new research programme, which consisted of the application of his method of equilibrium analysis to sea water. Through this approach he was able to give a simplified but clear description of the most important processes responsible for the present composition of sea water.

Lars Gunnar Sillén was a generous and noble person—a lover of nature and of his fellow man. He was dedicated to the principle that all nations and all political systems should live together in harmony, and he worked continuously toward that objective.

Lars Gunnar Sillén was a very prolific scientist. He actively participated in the production of more than 500 scientific papers, although his name appears on only a fraction of them. Co-ordination chemists all over the world are indebted and grateful to Professor Sillén for his compilation of data for the present Supplement and former issues of 'Stability Constants'. The quality of this work is so high that it is doubted that it can be ever equalled by those who follow in this endeavour.

During the last year of his life, when he completed the Inorganic Ligands Tables of this Supplement, his wife Birgit helped him in many ways, including the carrying of books and references to and from his hospital room. In this manner the final proofing of the Inorganic Ligands Tables was completed, just before his death on 23rd July 1970.

We, who both have been associated with Lars Gunnar Sillén in the production of the Tables and in other professional activities, are grateful for the opportunity we had to know him and to learn from him his sense of purpose, high ideals, and high standards of quality in all phases of his professional and personal life.

ERIK HÖGFELDT
ARTHUR E. MARTELL

PREFACE

THE present volume is the first supplement to the second edition of 'Stability Constants of Metal-ion Complexes' (Chem. Soc. Special Publication no 17, published in 1964), henceforth referred to as the "1964 Tables". The two volumes are intended to be used together for a rapid survey of the literature on a particular topic.

An attempt has been made to cover the pertinent literature up to the end of 1968, including earlier papers that were omitted from the 1964 Tables, and to correct errors in the 1964 Tables. In the inorganic part, the completeness of the search has suffered somewhat from the long illness of the main compiler, but it is hoped that much important material has not been missed. In the organic part, papers were earlier omitted when the reaction conditions were not well defined, or when there were other reasons to doubt the results. On the request of many users, the organic part (like the inorganic part) now tries to cover all literature data and to make critical comments where the results seem to be incomplete or dubious. In spite of all effort, the compilers may well have misunderstood a paper, and in such case comments from the authors would be welcome.

The section 'How to use the tables' has been brought up to date. We are very grateful to Professor K.W. Sykes of Queen Mary College, London E1, for the revision included in this volume. There are very few changes from the 1964 Tables: a few methods are added. When data are given for compounds of 'noble gases', these are treated as group 8B. In the organic part the value of m in protonation constants K_{mn} is omitted since it always comes out as 1; in the inorganic part, however, it has been thought necessary to observe the difference between, say (in the case of $F^- - H^+$ complexes), K_2 $(HF + F^- \rightleftharpoons HF_2^-)$ and K_{12} $(H^+ + HF \rightleftharpoons H_2F^+)$.

The style of the two sections is, as before, slightly different but we do not think this will inconvenience the reader.

Limits of error (±) are now sometimes given in the organic part, and have a range of meaning; many times they are the standard deviations given by the author, in others they represent the limits of the range of values given for the same constant. A new feature of the organic section is a Functional Group Index, including also the 1964 Tables.

(References to Russian journals in the inorganic part refer to the Russian originals, in the organic part to the translation; in the latter case, the page number of the originals are added in parentheses.)

We are feeling embarrassed still to give ΔH and ΔS values in calorie units. However, it would have been quite impractical to switch over to SI units in the supplement.

The compilers wish to express their thanks and appreciation to those who assisted in the work. Charles Saunders, Eileen Lewis and Professor Miloš Bartušek (organic part) have helped with literature searches. The members of the secretarial staffs at KTH and TAMU have very competently typed the manuscript. Professors S. Misumi and K. Yamasaki have provided us with equilibrium data published in Japanese journals. The following have helped us by reviewing sections of the manuscript and offering comments, corrections and suggestions about missing data: S. Ahrland, P.J. Antikainen, C.V. Banks, R.L. Benoit, M.T. Beck, G. Biedermann, R.L. Gustafson, I. Leden, Y. Marcus, R. Motekaitis, R. Näsänen, R. Österberg, J.F. Powell, H. Sigel, F.A. Snavely, L. Sommer, L.M. Venanzi and K.B. Yatsimirski. Many others have sent us data in reprints or private communications.

We would appreciate continued comments from users and authors concerning omissions, errors and possible ways of improvement. The comments should be sent to 'Professor E. Högfeldt, Dept. of Inorganic Chemistry, KTH, S-100 44 Stockholm 70, Sweden' (inorganic part) and to 'Professor A.E. Martell, Dept. of Chemistry, Texas A & M University, College Station, Texas 77843, USA' (organic part). This kind of continued review will, we hope, increase the usefulness of 'Stability Constants' to all those concerned.

LARS GUNNAR SILLÉN
ERIK HÖGFELDT

ARTHUR E. MARTELL
ROBERT M. SMITH

CONTENTS

TABLES

INDEXES

HOW TO USE THE TABLES

1. GENERAL ARRANGEMENT. This supplement contains three types of data: (i) data published between the completion of the 1964 Tables and the end of 1968; (ii) data published before the completion of the 1964 Tables but omitted from them; (iii) corrections to entries in the 1964 Tables. The general arrangement of all this material follows that used in the 1964 Tables.

Each table summarises the appropriate data up to 1968 for the association of one particular ligand with all the metallic ions which have been studied in conjunction with it, as far as the compilers located them. Method of measurement, composition, and temperature of the media to which the data refer, appropriate stability constants for the various complexes formed, and references to the original papers are given for each ligand-metal pair. Acid dissociation constants of the ligands are recorded by including the hydrogen ion among the metals as one of the cations with which the ligands associate. Redox equilibria are represented by including the electron as a ligand, and hydrolysis of the metallic ions is described by regarding the hydroxyl ion as one of the ligands. The tables are divided into two sections, the first dealing with inorganic ligands and the second with organic ligands. Ligands listed in the 1964 edition generally have the same table numbers as before; new ligands have been inserted in the appropriate places by the use of decimal numbering, *e.g.* 141·1 and 141·2 between 141 and 142. In the following description of the arrangement and presentation of the data, all statements refer to both inorganic and organic sections unless the contrary is specifically stated.

2. ORDER OF LIGANDS AND METAL IONS—INORGANIC SECTION. The inorganic ligands are placed in the following order, which is based on the Periodic System:

electron:	redox equilibria.
hydroxide:	
A-groups:	vanadate, niobates and tantalates, chromate, molybdate, wolframate (tungstate), manganates, rhenates, cyanoferrate (II), cyanoferrate (III), cyanocobaltate (III), other ligands with A-group elements.
group 3B:	borate.
group 4B:	cyanide, cyanate, thiocyanate, selenocyanate, other ligands with CN and B group elements, carbon monoxide, carbonate, other ligands with C, silicate, germanate, stannate, other ligands with 4B elements.
group 5B:	ammonia; amide, imide and nitride; hydrazine, hydroxylamine, azide, nitric oxide, nitrite, nitrate, other ligands with N, hypophosphite, phosphite, phosphate, diphosphate, trimetaphosphate, triphosphate, tetrametaphosphate, polyphosphates, amidophosphates, other ligands with P, arsenite, arsenate, other ligands with As, ligands with Sb or Bi.
group 6B:	water, peroxide, sulphide, thiosulphate, sulphite, sulphate, amidosulphate, other ligands with S, selenide, selenite, selenate, telluride, tellurite, tellurate.
group 7B:	fluoride, chloride, hypochlorite, chlorite, chlorate, perchlorate, bromide, hypobromite, bromate, iodide, hypoiodite, iodate, periodate, other ligands with 7B elements, halides: mixed and comparative.
group 8B:	ligands with 8B elements ('noble gases').

Metal ions are placed down the first column of each inorganic ligand table in the order of the following form of the Periodic System. Where an element has several valencies, a lower oxidation state precedes a higher one.

group 1A:	H, Li, Na, K, Rb, Cs, Fr.
group 2A:	Be, Mg, Ca, Sr, Ba, Ra.
group 3A+4*f*:	Sc, Y, La, Ce, Pr, Nd, Pm, Sm, Eu, Gd, Tb, Dy, Ho, Er, Tm, Yb, Lu, Ac.
group 4A:	Ti, Zr, Hf, Th.
group 5A:	V, Nb, Ta, Pa.
group 6A+5*f*:	Cr, Mo, W, U, Np, Pu, Am, Cm, Bk, Cf, Es, Fm, Md, No, Lw.
group 7A:	Mn, Tc, Re.
group 8A:	Fe, Co, Ni, Ru, Rh, Pd, Os, Ir, Pt.
group 1B:	Cu, Ag, Au.
group 2B:	Zn, Cd, Hg.
group 3B:	B, Al, Ga, In, Tl.
group 4B:	C, Si, Ge, Sn, Pb.

group 5B:	N, P, As, Sb, Bi.
group 6B:	O, S, Se, Te, Po.
group 7B:	F, Cl, Br, I, At.
group 8B:	He, Ne, Ar, Kr, Xe, Rn.

Many inorganic tables contain headings such as 'Group $2A^{2+}$ or Group $4f^{3+}$', where stability constants of related ions are compared; these entries are not repeated under the separate ions. Mixed complexes, *e.g.* those which contain one metal ion and two different ligands or one ligand and two different metal ions, are usually listed only once in the tables; sometimes a cross-reference is given, but readers looking for mixed complexes are advised to search in all possible places. Table 80, 'Halogenides, mixed and comparative', contains many equilibrium constants which are not listed under the separate halogenide ions.

3. ORDER OF LIGANDS AND METALLIC IONS—ORGANIC SECTION. Organic ligands, except the macromolecules which are listed at the end in the alphabetical order of their names, are arranged according to Beilstein's system, a brief description of which is given below. Metallic ions are placed in the alphabetical order of their international symbols, except the hydrogen ion which is placed first in each table.

Beilstein's system

(*a*) The formulæ are written with the elements in the order:

C H O N Cl Br I F S P

followed by the other elements in the alphabetical order of their symbols.

(*b*) All formulæ containing p carbon atoms precede those containing $(p+1)$ carbon atoms where p is any integer.

(*c*) For a given value of p, all formulæ containing q different elements other than carbon precede those containing $(q+1)$ other elements, where q is any integer.

(*d*) For given values of p and q, formulæ containing different combinations of elements are arranged as follows:

(i) the first element is always carbon;

(ii) the second element is chosen in turn from the elements after carbon given in the list in (*a*);

(iii) for a given choice of the second element (X), the third is selected in turn from those after X in the list given in (*a*);

(iv) in general, for a given choice of the first r elements, the rth of which is Z, the $(r+1)$th is selected in turn from those after Z in the list given in (*a*).

(*e*) The above rules may be illustrated by compounds containing only C, H, O, N, and Cl. Provided the number of carbon atoms is constant, the following order holds irrespective of how many atoms of H, O, N, or Cl the molecule contains:

$q = 1$: CH, CO, CN, CCl

$q = 2$: CHO, CHN, CHCl, CON, COCl, CNCl

$q = 3$: CHON, CHOCl, CHNCl, CONCl

(*f*) Formulæ with the same combination of elements are ordered by reference to the numbers of the various atoms in the molecule. Let X and Y be any pair of successive elements in a molecule $C_p \ldots X_i Y_j \ldots$, composed of a given combination of elements; i and j are any integers other than zero; then:

(i) all formulæ containing i atoms of X precede those containing $(i+1)$ atoms of X, irrespective of j (the number of atoms of Y);

(ii) for a given value of i, all formulæ containing j atoms of Y precede those containing $(j+1)$ atoms of Y.

(*g*) The rules in (*f*) may be illustrated by the order of the eight formulæ of the type CHON which can be constructed from 1C, 2H, 2O, and 2N:

$CHON$ $CHON_2$ CHO_2N CHO_2N_2 CH_2ON CH_2ON_2 CH_2O_2N $CH_2O_2N_2$

(*h*) The rules that determine the order of isomers will not be discussed here. Since the Tables contain relatively few isomers of any given formula, readers should be able to find isomeric ligands by trial and error without much difficulty.

4. METHOD OF MEASUREMENT. The method by which the stability constants were measured is shown in the second column of each Table by the following abbreviations:

act	activity coefficient, not specified.
aix	anion exchange.

ana	chemical analysis.
bp	boiling point.
cal	calorimetry.
cfu	centrifuge or ultracentrifuge.
cix	cation exchange.
col	colorimetry (often for measuring pH).
con	conductivity.
dis	distribution between two phases.
E	electromotive force, not specified.
esr	electron spin resonance.
est	estimated.
fp	freezing point.
gl	glass electrode.
hyp	from some hypothesis.
iE	current–voltage studies.
ir	infrared spectra.
ix	ion exchange.
kin	rate of reaction.
lit	from critical survey of literature data.
mag	magnetic susceptibility.
MHg	e.m.f. with amalgam electrode.
nmr	nuclear magnetic resonance.
oth	other methods, specified in 'results' column.
pH_2O, pL	partial pressure of substance indicated.
pH	pH method, not specified.
pol	polarography.
pre	preparative work.
qh	quinhydrone electrode.
Ram	Raman spectra.
red	e.m.f. with redox electrode.
sol	solubility.
sp	spectrophotometry.
tp	electrical migration or transference number.
tyn	tyndallometry, nephelometry.
X	*X*-ray diffraction.
*X*e	electron diffraction.
*X*n	neutron diffraction.
var	several methods, specified in 'results' column.
ΔG	combination of thermodynamic data.
Ag, Cd, etc.	e.m.f. with electrode of metal stated.
H, O, etc.	e.m.f. with electrode of gas stated.
conJ, spJ, etc.	conductimetry, spectrophotometry, etc., by 'Job's method' of continuous variation.
glΔ, HΔ, etc.	e.m.f. by differential method (minimum buffer capacity).
?	method not known to compilers.

5. TEMPERATURE. The third column of each table gives the temperature in °C. 'rt' denotes room temperature, and '?' is used if the value was not found by the compilers. When results for several temperatures are available, only the upper and the lower limits of the range are stated in the third column. The fifth column then gives either the individual temperatures in parentheses after the corresponding values of the equilibrium constant, or an empirical expression for the variation of the equilibrium constant with temperature. In such expressions, '*t*' denotes °C and '*T*' K.

6. MEDIUM. The nature of the medium to which the equilibrium constants refer is specified in the fourth column of each table, as indicated by the examples below. In general, concentrations are expressed in mole litre^{-1} (M), but the relatively few which are reported in mole kg.$^{-1}$ (m) are not distinguished; consequently neither the symbol M nor m is used in this column. Unless otherwise stated, the solvent is water.

→0	constants extrapolated to zero ionic strength.
0 corr	constants corrected to zero ionic strength by the application of some theoretical or empirical formula; this procedure is not always sharply distinguished from extrapolation.
0·06	an ionic strength of 0·06 mole litre^{-1}.
3 $NaClO_4$	constant concentration of the substance stated (3 mole litre^{-1} $NaClO_4$).
2($NaClO_4$)	ionic strength held constant at the value stated (2 mole litre^{-1}) by addition of the inert salt shown in parentheses.
I($NaClO_4$)	measurements made at a series of ionic strengths (I) with sodium perchlorate as the inert salt; the individual values are given in the fifth column in parentheses placed after the corresponding equilibrium constants.
2(Na)ClO_4	concentration of the anion (ClO_4^-) held constant at the value stated (2 mole litre^{-1}) with the ion shown in parentheses as the inert cation.
I(NO_3^-)	measurements made at a series of nitrate-ion concentrations, the individual values of which are given in the fifth column in parentheses placed after the corresponding equilibrium constants.
dil	dilute solution, concentration usually not more than 0·01 mole litre^{-1}.
var	ionic medium varied, and in some cases no special attempt was made to control the ionic strength.
KCl var	the medium was mainly aqueous KCl at various concentrations.
EtOH	ethanol as solvent.
org	various organic solvents.
50% MeOH	the solvent consists of the stated percentage of the component specified, water being the other component. Percentage compositions are given by weight unless otherwise stated.

7. RESULTS. The fifth column of the tables records the logarithm to base 10 (log K) of the equilibrium constants with the convention that 'log' and '=' are always omitted; *e.g.* 'K_2 2·35' means 'log K_2 = 2·35' and 'K_4< 0·3' means 'log K_4< 0·3'. Concentrations ([ ]) are mostly expressed in mole litre^{-1} and should be assumed to be in these units unless otherwise stated, though the occasional mole kg^{-1} may not be distinguished if the solvent is water. Where explicit reference is made to a particular concentration scale, this is done by referring to M units (molarity, mole litre^{-1}), *m* units (molality, mole kg^{-1}), or *x* units (mole fraction). All pressures (*p*) are in atmospheres unless otherwise stated.

A formula of the type H_pL or L is given in the heading of each ligand table except those for organic macromolecules, in order to define the entity L in terms of which the equilibrium constants are expressed. In the organic section, most of the data are represented by means of a few standard types of equilibrium constant which are explained in paragraph 8, and all other equilibria are defined by giving the chemical equation after each constant. The inorganic section, however, contains a much wider variety of equilibrium constants, which are defined in paragraph 9 in terms of a systematic notation of subscripts and superscripts.

Some other comments apply to both the inorganic and organic sections. Ligand formulæ such as BO_2^-, $H_2BO_3^-$, and $B(OH)_4^-$, which differ only in the number of molecules of solvent they contain, or complex formulæ which differ only in the number of medium ions associated with them, cannot be distinguished by equilibrium methods. The same equilibrium constant may sometimes be expressed in different ways, so occasionally different symbols may be used for the same or analogous constants. Particular formulæ for which there is good structural evidence are usually given, but otherwise an arbitrary formula is used in the conventional manner. No attempt is made to show structural features of complex formation which cannot strictly be deduced from equilibrium considerations, but in many cases the reader can easily select the appropriate structure. Thus lysine (2,6-diaminohexanoic acid, HL) may acquire an extra proton to become H_2L^+ which then reacts with a metal ion M^{n+} to give the complex MHL^{n+}. The equilibrium formulation does not indicate which of the two amino-groups has lost its proton, but it is very probably the one in the 2-position. Throughout the tables, as a rule, only the minimum number of constants required to represent the authors' results are given. Constants for many other equilibria can be calculated by combining the tabulated values.

In defining the activity scale for hydrogen ions, it is usual to make the activity equal to the concentration in infinitely dilute aqueous solution. When other media are used, it would be logical to refer the activity scale to an infinitely dilute solution of hydrogen ions in that medium, but all workers have not been consistent in this respect. If it is important to know which convention has been used, reference should be made to the original literature.

8. EQUILIBRIUM CONSTANTS—ORGANIC SECTION.

(a) *Protonation of ligands*

$$H^+ + L \rightleftharpoons HL \qquad K_1 = \frac{[HL]}{[H^+][L]}$$

$$H^+ + H_{n-1}L \rightleftharpoons H_nL \qquad K_n = \frac{[H_nL]}{[H^+][H_{n-1}L]}$$

Example: $C_3H_7O_2N$ 2-Aminopropanoic acid (α-alanine) HL$-$H$^+$ (table 171) $K_1(25°)$ 9·89, $K_2(25°)$ 2·37.

$$K_1 = \frac{[CH_3{\cdot}CH(NH_3^+){\cdot}CO_2^-]}{[H^+][CH_3{\cdot}CH(NH_2){\cdot}CO_2^-]} = 10^{9.89} \text{ mole}^{-1} \text{ litre at } 25°C$$

$$K_2 = \frac{[CH_3{\cdot}CH(NH_3^+){\cdot}CO_2H]}{[H^+][CH_3{\cdot}CH(NH_3^+){\cdot}CO_2^-]} = 10^{2.37} \text{ mole}^{-1} \text{ litre at } 25°C$$

Note the change of practice from the 1964 Tables, where acidity constants for $n \geqslant 2$ were denoted by K_{1n} to emphasise that the complex H_nL contains only one ligand molecule. The acidity constants listed at the beginning of each table may refer either to values obtained in original investigations of acidity constants or to values used in the determination of various metal-ion stability constants; in this latter case the reference is to the paper in which the metal-ion results are given. If the reader wishes to find whether the authors themselves measured the acidity constants, or whether they took the values from the literature, then the original paper should be consulted. In any calculations on the equilibria, inconsistencies may be avoided in part by combining the proton–ligand and metal-ion–ligand stability constants reported by the same authors.

(b) *Combination of ligands with metal ions*

(i) Consecutive or stepwise constants, *K*, are used as follows whenever possible:

$$ML_{n-1} + L \rightleftharpoons ML_n \qquad K_n = \frac{[ML_n]}{[ML_{n-1}][L]}$$

Example: $C_3H_7O_2N$ 2-Aminopropanoic acid (α-alanine) HL$-$Cu^{2+} (table 171) $K_1(20°)$ 8·66, $K_2(20°)$ 7·13.

$$K_1 = \frac{[Cu(CH_3{\cdot}CH(NH_2){\cdot}CO_2)^+]}{[Cu^{2+}][CH_3{\cdot}CH(NH_2){\cdot}CO_2^-]} = 10^{8.66} \text{ mole}^{-1} \text{ litre}$$

$$K_2 = \frac{[Cu(CH_3{\cdot}CH(NH_2){\cdot}CO_2)_2]}{[Cu(CH_3{\cdot}CH(NH_2){\cdot}CO_2)^+][CH_3{\cdot}CH(NH_2){\cdot}CO_2^-]} = 10^{7.13} \text{ mole}^{-1} \text{ litre}$$

(ii) Cumulative or gross constants, β_n, are used if they are the only quantities which are determined or if the sequence of stepwise constants is incomplete.

$$M + nL \rightleftharpoons ML_n \qquad \beta_n = \frac{[ML_n]}{[M][L]^n} = \prod_{i=1}^{n} K_i$$

Example: $C_3H_4N_2$ 1,3-Diazole (imidazole) L$-$Cu^{2+} (table 144) β_4 12·98

$$\beta_4 = \frac{[Cu(C_3H_4N_2)_4{}^{2+}]}{[Cu^{2+}][C_3H_4N_2]^4} = 10^{12.98} \text{ mole}^{-4} \text{ litre}^4$$

(iii) If the authors report their data in the form of free-energy changes and give no equilibrium constants, only free-energy changes are listed in the tables. $\Delta G_n°$ is the standard Gibbs free-energy change for the reaction of equilibrium constant K_n, and is related to the latter by the equation $\Delta G_n° = -RT \ln K_n$, where R is the gas constant (1·987 cal. deg.$^{-1}$ mole^{-1}) and T is the temperature in K. Alternatively the chemical equation to which the free-energy change refers may be given in parentheses.

(iv) All equilibria other than those detailed in (i), (ii), and (iii) above, and in (v) below, are dealt with by writing the chemical equation in parentheses after the symbol K.

Examples: $C_4H_6O_6$ 2,3 Dihydroxybutanedioic acid (tartaric acid) (Table 203)

H_2L-Cu^{2+} $K[Cu^{2+}+2OH^-+2L^{2-}\rightleftharpoons Cu(OH)_2L_2{}^{4-}]$ 9·85

H_2L-Fe^{3+} $K[2\,Fe^{3+}+2L^{2-}\rightleftharpoons Fe_2(H_{-1}L)_2+2H^+]$ 11·87

$$K = \frac{[Cu(OH)_2(O_2C\cdot CH(OH)\cdot CH(OH)\cdot CO_2)_2{}^{4-}]}{[Cu^{2+}][OH^-]^2[O_2C\cdot CH(OH)\cdot CH(OH)\cdot CO_2{}^{2-}]^2} = 10^{9.85}\text{ mole}^{-4}\text{ litre}^4$$

$$K = \frac{[Fe_2(O_2C\cdot CH(O)\cdot CH(OH)\cdot CO_2)_2][H^+]^2}{[Fe^{3+}]^2[O_2C\cdot CH(OH)\cdot CH(OH)\cdot CO_2{}^{2-}]^2} = 10^{11.87}\text{ mole}^{-1}\text{ litre}$$

As far as equilibrium considerations are concerned, the first reaction could equally well be formulated as one in which each hydroxyl ion, instead of being added to the complex by substitution for a water molecule in hydration shell of the cupric ion, removes a proton from a hydroxyl group of the tartrate ion, L^{2-}. Such reactions differ only in the elements of the solvent, water. Where the authors have suggested that the latter alternative is followed, as in the second example, a negative subscript is used to show that a further proton is removed from L^{2-}.

(v) Current practices in the publication of metal complex equilibria involving macromolecules are varied and frequently widely different from the conventions used for small organic ligands. The equilibrium constant for the interaction of a given metal-ion with a particular ligand group in a macromolecule varies with the electrostatic free energy needed to change the electric charge of the macromolecule, which is itself a function of all the ionic equilibria in which the various types of ligand group in the macromolecule take part. Semi-empirical expressions, based on a theoretical analysis due to Lindestrøm-Lang (C.R. trav. lab. Carlsberg, 1924, **15,** no. 7) and Scatchard (*Ann. N.Y. Acad. Sci.*, 1949, **51,** 660), are used to represent the variation with macromolecular charge of an 'intrinsic' constant for each type of ligand group present. To improve the clarity and usefulness of the macromolecular data, all terms and symbols are defined in the remarks section. The tables give not only the 'intrinsic' constants, but also the parameters involved in their calculation, so that the data can be used to calculate the concentrations and activities of species in solution under various experimental conditions.

9. EQUILIBRIUM CONSTANTS—INORGANIC SECTION.

(a) *Consecutive or stepwise constants: K*

(i) Addition of ligand (L)

K_n is used for the stepwise addition of ligands to metallic ions as explained in the section 8(b)(i), where an example is given.

(ii) Addition of protonated ligand (HL) with elimination of proton

$$ML_{n-1}+HL\rightleftharpoons ML_n+H^+ \qquad {}^*K_n = \frac{[ML_n][H^+]}{[ML_{n-1}][HL]}$$

Example: $SO_4{}^{2-}-Sc^{3+}$ (table 56) $^*K_1$ 1·49 [$K_1(H^+)$ 1·08]

$$K_1 = \frac{[ScSO_4{}^+][H^+]}{[Sc^{3+}][HSO_4{}^-]} = 10^{1.49}$$

In deriving this result, it was assumed that

$$K_1(SO_4{}^{2-}-H^+) = \frac{[HSO_4{}^-]}{[H^+][SO_4{}^{2-}]} = 10^{1.08}\text{ mole}^{-1}\text{ litre}$$

(iii) Addition of protonated ligand (H_pL)

$$M(H_pL)_{n-1}+H_pL\rightleftharpoons M(H_pL)_n \qquad H_pL: K_n = \frac{[M(H_pL)_n]}{[M(H_pL)_{n-1}][H_pL]}$$

Example: $CO_3{}^{2-}-Nd^{3+}$ (table 21) HL^-: K_3 2·71, K_4 1·80

$$HL^-: K_3 = \frac{[Nd(HCO_3)_3]}{[Nd(HCO_3)_2{}^+][HCO_3{}^-]} = 10^{2.71}\text{ mole}^{-1}\text{ litre}$$

$$HL^-: K_4 = \frac{[Nd(HCO_3)_4^-]}{[Nd(HCO_3)_3][HCO_3^-]} = 10^{1.80} \text{ mole}^{-1} \text{ litre}$$

(iv) Addition of gaseous ligand [L(g)]

$$ML_{n-1} + L(g) \rightleftharpoons ML_n \qquad K_{pn} = \frac{[ML_n]}{[ML_{n-1}]p_L}$$

This constant does not appear to be used in this supplement.

(v) Inner-sphere and outer-sphere complexes

Experimental methods which respond to both types of complex give the sum $K_n = K_n^{in} + K_n^{out}$, where K_n^{in} and K_n^{out} are the separate constants for equilibria in which the ligand displaces solvent molecules from the inner and outer co-ordination spheres of the central atom respectively. This notation is only used if the authors distinguished between the two types of complex. If the equilibrium between inner and outer complexes has been studied, it is expressed by $K_n^{is} = K_n^{out}/K_n^{in}$; there is no change from the 1964 Tables where the definitions of K_n^{is} on pages xii, xiii and 241 are in error and should be as above.

Example: $SCN^- - Co[NH_3]_5^{3+}$ (table 17) K_1^{out} 0·65

$$K_1^{out} = \frac{[\{Co(NH_3)_5H_2O\}^{3+}SCN^-]}{[Co(NH_3)_5H_2O^{3+}][SCN^-]} = 10^{0.65} \text{ mole}^{-1} \text{ litre}$$

(vi) Addition of central atom (M)

$$M + L \rightleftharpoons ML \qquad K_1 = \frac{[ML]}{[M][L]}$$

$$M + M_{n-1}L \rightleftharpoons M_nL \ (n \geqslant 2) \qquad K_{1n} = \frac{[M_nL]}{[M][M_{n-1}L]}$$

These constants are used for the addition of either metal ions or protons to a ligand.

Example: $CO_3^{2-} - H^+$ (table 21) K_1 10·32 (25°), K_{12} 6·35 (25°)

$$K_1 = \frac{[HCO_3^-]}{[H^+][CO_3^{2-}]} = 10^{10.32} \text{ mole}^{-1} \text{ litre at 25°C}$$

$$K_{12} = \frac{[H_2CO_3]}{[H^+][HCO_3^-]} = 10^{6.35} \text{ mole}^{-1} \text{ litre at 25°C}$$

(b) *Cumulative or gross constants* : β

In β_{nm} and $^*\beta_{nm}$ the subscripts n and m denote the composition of the complex M_mL_n formed; when $m = 1$, the second subscript (=1) is omitted.

(i) addition of central atoms (M) and ligands (L)

$$mM + nL \rightleftharpoons M_mL_n \qquad \beta_{nm} = \frac{[M_mL_n]}{[M]^m[L]^n}$$

Examples: NH_3–Cu^{2+} (table 27), β_4 12·63

$$\beta_4 = \frac{[Cu(NH_3)_4^{2+}]}{[Cu^{2+}][NH_3]^4} = 10^{12.63} \text{ mole}^{-4} \text{ litre}^4$$

I^-–Ag^+ (table 75), β_{13} 18·96

$$\beta_{13} = \frac{[Ag_3I^{2+}]}{[Ag^+]^3[I^-]} = 10^{18.96} \text{ mole}^{-3} \text{ litre}^3$$

(ii) addition of central atoms (M) and protonated ligands (HL) with elimination of protons

$$mM + nHL \rightleftharpoons M_mL_n + nH^+ \qquad ^*\beta_{nm} = \frac{[M_mL_n][H^+]^n}{[M]^m[HL]^n}$$

Example: $OH^- - Be^{2+}$ (table 2) $*\beta_{96} - 34{\cdot}5$

$$*\beta_{96} = \frac{[Be_6(OH)_9{}^{3+}][H^+]^9}{[Be^{2+}]^6} = 10^{-34.5} \text{ mole}^4 \text{ litre}^{-4}$$

Note that if $L = OH^-$, $HL = H_2O$.

(c) *Solubility constants* : K_s

(i) Solid M_aL_b in equilibrium with free ions in solution

$M_aL_b(s) \rightleftharpoons aM + bL \qquad K_{s0} = [M]^a[L]^b$

Examples: I^-–Ag^+ (table 75), K_{s0} $-12{\cdot}0$

$K_{s0} = [Ag^+][I^-] = 10^{-12.0}$ mole^2 litre^{-2} (solid = AgI)

$S^{2-} - Ag^+$ (table 53) $K_{s0}[Ag_2S(s) \rightleftharpoons 2Ag^+ + S^{2-}]$ $-49{\cdot}7$

$K_{s0} = [Ag^+]^2[S^{2-}] = 10^{-49.7}$ mole^3 litre^{-3} (solid = Ag_2S)

K_{s0} is the conventional solubility product and the subscript zero indicates that the equilibrium of the solid with the simple (uncomplexed) species M and L is considered. Unless otherwise stated, the solid M_aL_b is the stable substance normally formed from the central atom M and the ligand L, *e.g.*, AgI and Ag_2S in the above illustrations.

(ii) Solid M_aL_b in equilibrium with complex M_mL_n and ligand L in solution.

$$\frac{m}{a} M_aL_b(s) \rightleftharpoons M_mL_n + \left(\frac{mb}{a} - n\right)L \qquad K_{snm} = [M_mL_n][L]^{(mb/a-n)}$$

In K_{snm} the subscripts n and m denote the composition of the complex M_mL_n formed in solution. When $m = 1$, the second subscript (=1) is omitted. The ligand L may occur on either the right- or the left-hand side of the chemical equation, or be absent, depending on whether $\left(\frac{mb}{a} - n\right)$ is positive, negative, or zero respectively.

Examples: $I^- - Cu^+$ (table 75) K_{s4} $-2{\cdot}52$

$$K_{s4} = \frac{[CuI_4{}^{3-}]}{[I^-]^3} = 10^{-2.52} \text{ mole}^{-2} \text{ litre}^2 \; [CuI(s) + 3I^- \rightleftharpoons CuI_4{}^{3-}]$$

$Cl^- - Ag^+$ (table 67) $K_{s12}[AgCl(s) + Ag^+ \rightleftharpoons Ag_2Cl^+]$ $-1{\cdot}96$ in $NaNO_3(1)$, x units

$$K_{s12} = \frac{xAg_2Cl^+}{xAg^+} = 10^{-1.96}$$

where x denotes mole fraction in liquid $NaNO_3$ and the reaction is that given in the square brackets above.

(iii) Protonated ligand reacts with elimination of proton

$$\frac{m}{a} M_aL_b(s) + \left(\frac{mb}{a} - n\right)H^+ \rightleftharpoons M_mL_n + \left(\frac{mb}{a} - n\right)HL; \; *K_{snm} = \frac{[M_mL_n][HL]^{(mb/a-n)}}{[H^+]^{(mb/a-n)}}$$

Example: $OH^- - Cd^{2+}$ (table 2) $*K_{s0}$ $14{\cdot}03$ [β–$Cd(OH)_2$]

$$*K_{s0} = \frac{[Cd^{2+}]}{[H^+]^2} = 10^{14.03} \text{ mole}^{-1} \text{ litre } [\beta\text{–}Cd(OH)_2(s) + 2H^+ \rightleftharpoons Cd^{2+} + 2H_2O]$$

(d) *Acidic and basic constants*

(i) When L is hydroxide (OH^-), HL is water (H_2O) and $*K_n$ is the nth acid dissociation constant for the hydrolysis of a metal ion.

Example: OH^-–Fe^{3+} (table 2) $*K_1$ $-3{\cdot}00$

$$*K_1 = \frac{[FeOH^{2+}][H^+]}{[Fe^{3+}]} = 10^{-3.00} \text{ mole litre}^{-1}$$

(ii) See 9(a)(vi) for the use of H^+ as the central atom to represent protolytic constants.

(iii) Other acidic constants are denoted by K_a, followed by parentheses enclosing the formula of the species donating the proton.

Example: $OH^- - Pa^{5+}$ (table 2) $K_a[Pa(OH)_4^+]$ $-1{\cdot}05$

$$K_a[Pa(OH)_4^+] = \frac{[PaO(OH)_3][H^+]}{[Pa(OH)_4^+]} = 10^{-1{\cdot}05} \text{ mole litre}^{-1}$$

(iv) Basic constants for the transfer of a proton from the solvent to the ligand are denoted by K_b followed, if necessary, by parentheses enclosing the formula of the species accepting the proton.

Example: NH_3–H^+ (table 27) K_b $-4{\cdot}75$

$$K_b = \frac{[NH_4^+][OH^-]}{[NH_3]} = 10^{-4{\cdot}75} \text{ mole litre}^{-1}$$

(e) *Special constants*

(i) K (equation)

The equation defines the reaction to which K refers, as in 8 (b) (iv).

(ii) $\dagger K_{nm}$, $\dagger K_{snm}$

The corresponding reaction is given as a footnote to the table, or in parentheses after the constant when the latter is first used in the same table.

(iii) β [formula]

The formula gives the composition of the complex in terms of the species from which it is formed. Species with negative subscripts are eliminated in the formation of the complex.

Examples: $F^- - Pa^{5+}$ (table 66) $\beta[Pa(OH)_3^{2+}(HF)_nH^+_{1-n}]$ $3{\cdot}65(n = 1)$, $7{\cdot}65(n = 2)$

$$\beta[Pa(OH)_3^{2+}(HF)] = \frac{[Pa(OH_2)(OH)_2F^{2+}]}{[Pa(OH)_3^{2+}][HF]} = 10^{3{\cdot}65} \text{ mole}^{-1} \text{ litre}$$

$$\beta[Pa(OH)_3^{2+}(HF)_2H_{-1}^+] = \frac{[Pa(OH_2)(OH)_2F_2^+][H^+]}{[Pa(OH)_3^{2+}][HF]^2} = 10^{7{\cdot}65} \text{ mole}^{-1} \text{ litre}$$

(iv) K_s [formula]

The formula gives the composition of the solid phase in terms of the species with which it is in equilibrium in solution. Species with negative subscripts are eliminated in the formation of the solid.

Example: $PO_4^{3-} - Mg^{2+}$ (table 38) $K_s[Mg^{2+}NH_4^+L^{3-}(H_2O)_6]$ $-13{\cdot}15$

$K_s[Mg^{2+}NH_4^+L^{3-}(H_2O)_6] = [Mg^{2+}][NH_4^+][PO_4^{3-}] = 10^{-13{\cdot}15} \text{ mole}^3 \text{ litre}^{-3}$

$[MgNH_4PO_4{\cdot}6H_2O(s) \rightleftharpoons Mg^{2+} + NH_4^+ + PO_4^{3-}]$

(v) K_{pnm}, $\dagger K_{pnm}$

Equilibria involving a gas are in general denoted by K_p, and the composition of the complex formed is indicated by the figures given after p in the subscript. For consecutive addition of gaseous ligands, see 9(*a*)(iv). Otherwise the corresponding reactions are usually given in the appropriate ligand table; for reactions involving carbon dioxide, reference to the bottom of page 135 and the top of page 136 in the 1964 Tables may also prove helpful.

Examples: $CO_3^{2-} - H^+$ (table 21) $K_{p0}K_1^2$ (2·20)

$$K_{p0}K_1^2 = \frac{[HCO_3^-]^2}{[CO_3^{2-}]p_{CO_2}} = 10^{2{\cdot}20} \text{ mole litre}^{-1} \text{ atm}^{-1}$$

$CO_2(g) + CO_3^{2-} + H_2O \rightleftharpoons 2HCO_3^-$ i.e. $CO_2(g) + H_2O \rightleftharpoons 2H^+ + CO_3^{2-}$
$+ 2\{CO_3^{2-} + H^+ \rightleftharpoons HCO_3^-\}$

$CO_3^{2-} - H^+$ (table 21) $K_{p12}[CO_2(g) + H_2O \rightleftharpoons H_2CO_3]$ $-1{\cdot}99$ m units

$$K_{p12} = \frac{[H_2CO_3]}{p_{CO_2}} = 10^{-1{\cdot}99} \text{ mole kg}^{-1} \text{ atm}^{-1}$$

$CO_3^{2-} - Mg^{2+}$ (table 21) $\dagger K_{ps0}$ $-5{\cdot}20$ (magnesite)

$$†K_{ps0} \text{ (magnesite)} = \frac{[Mg^{2+}][HCO_3^-]^2}{p_{CO_2}} = 10^{-5\cdot20} \text{ mole}^3 \text{ litre}^{-3} \text{ atm}^{-1}$$

$$MgCO_3 \text{ (s, magnesite)} + CO_2(g) + H_2O \rightleftharpoons Mg^{2+} + 2HCO_3^-$$

(vi) K_D [equation]

Equilibria involving distribution between two liquid phases, such as water and an organic solvent, are denoted by K_D followed by the appropriate equation in brackets. All species not labelled as being in the organic phase may be assumed to be in aqueous solution.

Example: $NO_3^- - UO_2^{2+}$ (table 34)

$K_D[UO_2^{2+} + 2L^- + 2T \text{ (in } CCl_4) \rightleftharpoons UO_2L_2T_2 \text{ (in } CCl_4)]$ 1·77

$[T = (BuO)_3PO]$

$$K_D = \frac{[UO_2(NO_3)_2\{(BuO)_3PO\}_2(\text{in } CCl_4)]}{[UO_2^{2+}][NO_3^-]^2[(BuO)_3PO(\text{in } CCl_4)]^2} = 10^{1\cdot77} \text{ mole}^{-4} \text{ litre}^4$$

(f) *Redox equilibria*

Results are given in table 1 of the inorganic section in two forms: as standard electrode potentials, and as equilibrium constants for the addition of the electron as a ligand.

According to the I.U.P.A.C. 1953 convention,* the electrode potential (E_1) of, say, the Fe^{2+}–Fe^{3+} system on the hydrogen scale is the potential of the right-hand electrode less that of the left-hand electrode in the cell

$$Pt,H_2/H^+ \qquad Fe^{2+},Fe^{3+}/Pt$$

when hydrogen gas is at unit pressure and hydrogen ions are at unit activity. The electrode potential has its standard value $E°$ when the ratio of the activities of Fe^{2+} and Fe^{3+} is unity. The electrode potential is also related by the equation

$$-\Delta G_1 = zFE_1$$

to the Gibbs free-energy change (ΔG_1) of the reaction

$$Fe^{3+} + \tfrac{1}{2}H_2 \text{ (unit pressure)} \longrightarrow Fe^{2+} + H^+ \text{ (unit activity)}, \qquad (1)$$

which takes place when one faraday (F) of positive charge ($z = 1$) passes through the cell from left to right. The standard electrode potential $E_1°$ is similarly related to the standard free-energy change $\Delta G_1°$. Thus

$$\Delta G_1 = \Delta G_1° + \frac{RT}{F} \ln \frac{\{Fe^{2+}\}}{\{Fe^{3+}\}}$$

$$E_1 = E_1° + \frac{RT}{F} \ln \frac{\{Fe^{3+}\}}{\{Fe^{2+}\}},$$

where { } denotes activity.

The overall chemical reaction (1) may be represented formally by two stages in which electrons take part:

$$Fe^{3+} + e^- \rightleftharpoons Fe^{2+} \qquad (2)$$

$$\tfrac{1}{2}H_2 \rightleftharpoons H^+ + e^- \qquad (3)$$

An electron-activity scale may be established by taking the equilibrium constant of the second stage to be unity for all temperatures and all ionic media:

$$i.e., K_3 = \frac{\{H^+\}\{e^-\}}{p_{H_2}^{1/2}} = 1 \text{ (activities and fugacity are equilibrium values)}$$

and $\Delta G_3° = \Delta H_3° = \Delta S_3° = 0$.

With the electron activity defined, an equilibrium constant may be written for reaction (2), which may be regarded as the addition of the electron as a ligand:

$$K_2 = \frac{\{Fe^{2+}\}}{\{Fe^{3+}\}\{e^-\}} \text{ (activities are equilibrium values).}$$

But $K_2 = K_1/K_3 = K_1$,

$$-\Delta G_1° = zFE_1° = RT \ln K_1,$$

and thus

$$\log_{10} K_2 = zE_1°\left(\frac{RT}{F} \ln 10\right)^{-1}$$

* I.U.P.A.C. Manual of Physio-Chemical Symbols and Terminology, Butterworths, London, 1959, p. 4.

For a reaction involving electrons, the fifth column of table 1 gives K, the equilibrium constant analogous to K_2 above, followed in parentheses by the chemical equation written as a reduction with z electrons on the left-hand side, the logarithm to base 10 of K, and then in parentheses again the standard electrode potential and the units in which it is expressed; if electrons are not involved, as in reaction (1) above, the standard potential is omitted.

Potentials reported before 1948 are given in international volts (1 international volt = 1·00034 absolute volt), and subsequent ones in absolute volts. If in the original paper $E°$ was calculated from $\Delta G°$, the equilibrium constant has often been calculated directly from the free-energy change rather than from the potential. The factor RTF^{-1} ln 10 has the following values in millivolts (absolute): 54·195 (0°C), 59·155 (25°C), 64·115 (50°C), 74·035 (100°C).

It is hoped that the equilibrium constants will prove more useful than the potentials, since concentrations can be calculated from them more readily.

Example (25°C, zero ionic strength):

The equilibrium constant

$$K = \frac{[Eu^{3+}]p_{H_2}^{\frac{1}{2}}}{[Eu^{2+}][H^+]}$$

for the reaction

$$Eu^{2+} + H^+ \rightleftharpoons Eu^{3+} + \tfrac{1}{2}H_2$$

can be calculated from the data for the constituent reactions:

$$K(Eu^{3+} + e^- \rightleftharpoons Eu^{2+}) - 9{\cdot}3\ (-0{\cdot}55V)$$

$$K(H^+ + e^- \rightleftharpoons \tfrac{1}{2}H_2(g)0\ \text{(definition)}$$

In terms of the equilibrium constants:

$$K = 10^{-(-9\cdot3)+0} = 10^{9\cdot3}\ \text{mole}^{-1}\ \text{litre atm}^{\frac{1}{2}}$$

and in terms of the potentials:

$$\log_{10} K = \left(\frac{zFE°}{2{\cdot}303RT}\right) = \frac{-(-0{\cdot}55)+0}{0{\cdot}059155} = 9{\cdot}3\ (K \text{ in mole}^{-1}\ \text{litre atm}^{\frac{1}{2}}).$$

Many types of equilibrium constant can be calculated by combining the appropriate redox data, as illustrated on pages xvi and xvii of the 1964 Tables, where there is also a discussion of a quantity pE, which represents the electron activity in a manner analogous to that in which pH represents hydrogen ion activity.

10. ENTHALPY AND ENTROPY CHANGES. These are included in the fifth column and, unless otherwise stated, ΔH is in kilocalories and ΔS in calories degree^{-1}, each for the numbers of moles in the corresponding chemical equation. Where the symbol K is used for the equilibrium constant, H or S is given the same superscript or subscript as the corresponding K : *e.g.*, $K_n\ \Delta H_n$; $K_{s0}\ \Delta H_{s0}$; $\dagger K_{nm}\ \Delta\dagger H_{nm}$; etc. For inorganic equilibria which are represented by a cumulative constant, β, the following notation is used: $\beta_n\ \Delta H_{\beta n}$; $\beta_{nm}\ \Delta H_{\beta nm}$; $*\beta_{nm}\ \Delta * H_{\beta nm}$; etc.

Readers making rough estimates of heat-content changes from temperature coefficients of equilibrium constants, or *vice versa*, may find the following numerical values useful. In the equation

$$\frac{d \log K}{dT} = \frac{\Delta H}{2{\cdot}303RT^2} = \text{constant} \times \Delta H$$

the constant is 0·00293 (0°C), 0·00246 (25°C), and 0·00209 (50°C) if ΔH is expressed in kcal.

11. REMARKS. The following abbreviations are also used in the fifth column, chiefly in the inorganic section:

ev M_mL_n	evidence for the existence of the complex M_mL_n.
cpx	complex.
cat	cationic (positive).
ani	anionic (negative).
unch	uncharged.
polyn	polynuclear

?	author's doubt expressed in the reference given.
(?)	compiler's doubt.
(K_1 3·71)	the value log K_1 = 3·71 was assumed (*e.g.*, from earlier work).

For a convention used in the organic section for references to acidity constant determinations, see Section 8(*a*).

12. REFERENCES. The symbols in the last column of each table relate to the references which are listed at the end of each table. Each reference is characterised by two figures denoting the year of publication followed by a capital letter denoting the name of the author, or of the first author when there are several. All years up to and including 1900 are given as 00, and to distinguish between references with the same year and capital letter, a lower-case letter is added as in 64S, 64Sa, 64Sb, etc.

In the inorganic section, the plus sign (+), the oblique stroke (/), and the signs for equality (=) and approximate equality (≈) are used to give additional information by means of the conventions illustrated by the examples below. References placed after the oblique stroke used mainly, or exclusively, the data of other workers, and the plus sign denotes that additional data were taken from the references following it.

68M	conclusions of 68M from data in 68M.
67R+66Aa	conclusions of 67R from data in 67R and in 66Aa.
65Sb+	conclusions of 65Sb from data in 65Sb and in numerous or unspecified other references.
64S≈63S	conclusions and data of 64S are substantially the same as those of 63S.
64Sa = 63S	conclusions and data of 64Sa are identical with those of 63S.
52R/66S	conclusions of 66S from data in 52R.
65Ba/	conclusions of compiler from data in 65Ba.
/67Bc	conclusions of 67Bc from data in numerous or unspecified other references.

If a critical survey is claimed, 'lit' is used.

13. INDEXES. The following indexes of table numbers are provided. For the 1964 tables and this supplement: inorganic ligands alphabetically by name; organic ligands and macromolecules alphabetically by name; metals alphabetically by international symbol for both inorganic and organic ligands; organic functional groups alphabetically by name. It is hoped that this new type of index, covering both editions, will be useful to readers searching for data on the interaction of a given functional group with a series of metal ions, or the effect of its environment within the ligand on the interaction of a given functional group with a given metal ion.

K. W. SYKES.
QUEEN MARY COLLEGE,
UNIVERSITY OF LONDON.

PART I—INORGANIC LIGANDS

TABLES

1 Redox equilibria, e^-

Metal	*Method*	*Temp*	*Medium*	*Log of equilibrium constant, remarks*	*Ref*
e^-aq	kin	25	dil	$K(e^- \rightleftharpoons e^-aq) = K[\frac{1}{2}H_2(g) \rightleftharpoons H^+ + e^-aq]$ −45·3(−2·68 v) $K(H^+ + e^-aq \rightleftharpoons Haq)$ 9·6, e^-aq = hydrated electron	66Ha
D	ΔG	0—75	0	$K(D^+ + e^- \rightleftharpoons \frac{1}{2}D_2)$ −0·129(0°; −6·96 mv), −0·0985(25°; −5·828 mv), −0·0633(50°; −4·061 mv), −0·0241(75°; −1·666 mv)	68M
Group 1A$^+$	*E*	25	MeNHCHO	$K[M^+ + Cl^- + Ag(s) \rightleftharpoons M(s) + AgCl(s)]$ −52·806 (M = Li; −3·1237 v), −47·447 (M = Na; −2·8067 v), −51·073 (M = K; −3·0212 v), −50·49(M = Cs; −2·987 v), $K[Na^+ + Br^- + Ag(s) \rightleftharpoons Na(s) + AgBr(s)]$ −45·871 (−2713·5 mv) in MeHNCHO(l), *m* units	66Lb
Li	*E*	25	0	$K[Li^+ + e^- \rightleftharpoons Li(x, \text{in Hg})]$ −36·99(−2188 mv)	/67Bb
	E	25	0	$K[Li(s) \rightleftharpoons Li(x, \text{in Hg})]$ 14.26(843·8 mv)	68C
	E	25	0 corr	$K[Li^+ + e^- \rightleftharpoons Li(x, \text{in Hg})]$ −37·13(−2196·3 mv) $K[Li^+ + e^- \rightleftharpoons Li(s)]$ −51·39(−3040·1 mv)	68H
Na	*E*	25	0	$K[Na^+ + e^- \rightleftharpoons Na(x, \text{in Hg})]$ −33·10(−1958 mv)	/67Bb
K	*E*	25	0	$K[K^+ + e^- \rightleftharpoons K(x, \text{in Hg})]$ −33.32(−1971 mv)	/67Bb
	E	25	EtOH	$K[K^+ + e^- \rightleftharpoons K(s)]$ −48·43(−2865 mv) in EtOH(l) → 0, also values for 6 H_2O–EtOH mixt	68D
		25	→ 0	$K[K^+ + e^- \rightleftharpoons K(s)]$ −49·41(−2923·0 mv)	
Be	cal, ΔG	25	0 corr	$K[Be^{2+} + 2e^- \rightleftharpoons Be(s)]$ −66·5 to −66·8	65Ba/
Ca	*E*	25	0	$K[Ca^{2+} + 2e^- \rightleftharpoons Ca(x, \text{in Hg})]$ −67·42(−1994 mv)	/67Bc
	lit	25	0 corr	$K[Ca^{2+} + 2e^- \rightleftharpoons Ca(x, \text{in Hg})]$ −67·48(−1996 mv) $K[Ca^{2+} + 2e^- \rightleftharpoons Ca(s)]$ ⩽ −94·27(⩽ −2809 mv)	/68B
Eu	*E*	25	0	$K(Eu^{3+} + e^- \rightleftharpoons Eu^{2+})$ −5.9(−0·35 v)	36M/65Mc
	E	25	0 corr	$K(Eu^{3+} + e^- \rightleftharpoons Eu^{2+})$ −9·3(−0·55 v)	63A
	cal	25	0 corr	$\Delta H(Eu^{3+} + e^- \rightleftharpoons Eu^{2+})$ = 19·6, ΔS = 32·4, $\Delta H[Eu^{2+} + 2e^- \rightleftharpoons Eu(s)]$ = 120.9	65Sb+
	kin	25	2($NaClO_4$), 0·5 H^+	$K(Eu^{3+} + Cr^{2+} \rightleftharpoons Eu^{2+} + Cr^{3+})$ ≈ −0·1 $K(Eu^{3+} + Cr^{2+} + Cl^- \rightleftharpoons Eu^{2+} + CrCl^{2+})$ −0·36	66A
Tm	sp	25		$K(Tm^{3+} + e^- \rightleftharpoons Tm^{2+})$, −42(−2.5 v, est)	64Ba
Ti	est	25	0	$K(Ti^{3+} + e^- \rightleftharpoons Ti^{2+})$ −34(−2 v)	63O
	sp		H^+ var	ev cpx 1Ti^{IV} + 1Fe^{2+}	65Rb
Th	sp, est	25		$K(Th^{4+} + e^- \rightleftharpoons Th^{3+})$ −41(−2·4 v)	65M
V	cal	25	0 corr	$\Delta H(Fe^{2+} + VO_2^+ + 2H^+ \rightleftharpoons Fe^{3+} + VO^{2+} + H_2O)$ = −19·7	66B
	kin	25	$I(LiClO_4)$	$K(VO^{2+} + Cu^+ + 2H^+ \rightleftharpoons V^{3+} + Cu^{2+} + H_2O)$ 3·04(I = 3), 2·85(I = 1)	68Sc

continued overleaf

1 e^- (contd)

Metal	*Method*	*Temp*	*Medium*	*Log of equilibrium constant, remarks*	*Ref*
Nb	pol	25?	12 HCl	$K(Nb^{V}+e^- \rightleftharpoons Nb^{IV})$ −3·60(242−455 = −213 mv)	54C
	pol, pre	25	1 $KClO_4$	$K(Nb_6Cl_{12}^{4+}+2e^- \rightleftharpoons Nb_6Cl_{12}^{2+})$ 14·40(426 mv)	65Ma
Ta	kin, sp	15—22	0·1 ($HClO_4$)	$K(Ta_6Cl_{12}^{3+}+Fe^{3+} \rightleftharpoons Ta_6Cl_{12}^{4+}+Fe^{2+})$ −1·74(15°, kin), −1·80(22°, sp)	66E
Mo	*E*	25?	cH_2SO_4	$K[Mo(CN)_6^{3-}+e^- \rightleftharpoons Mo(CN)_6^{4-}]$ 14·64(c = 4; 866 mv), 13·79(c = 0·5; 816 mv) also for some other c values, 12·26(→ 0; 725 mv)	63M
	sp	25	8·6 HBr	$K[Mo^{VI}_2 \rightleftharpoons (MoOBr_4^-)_2+Br_3^-]$ 0·4	64A
W	*E*	18?	4 H_2SO_4	$K[W(CN)_6^{3-}+e^- \rightleftharpoons W(CN)_6^{4-}]$ 8·59(496 mv)	67M
U	pCl_2, sp	650	MCl(l)	$K[2UO_2^{2+}+2Cl^- \rightleftharpoons 2UO_2^{+}+Cl_2(p\ atm)]$ −5·92 in (Li, K)Cl(1, eut)	64W

1964 *Erratum*: p. 6, *in ref.* 61 Sd, *for* "328·8 v" *read* "328·8 mv"

Metal	*Method*	*Temp*	*Medium*	*Log of equilibrium constant, remarks*	*Ref*
Cm	sp, est	25		$K(Cm^{4+}+e^- \rightleftharpoons Cm^{3+})$ 55(3·25 v)	65M
Cf	pre		var	ev Cf^{2+}	68Ca
Md	est	25	0	$K(Md^{3+}+e^- \rightleftharpoons Md^{2+})$ −1·7(−0·10 v)	67Ma
Mn	*E*	25	0	$K(MnO_4^-+e^- \rightleftharpoons MnO_4^{2-})$ 9.43(558 mv)	56Ca
	E	25	6—12 KOH	$K(MnO_4^{2-}+e^- \rightleftharpoons MnO_4^{3-})$ 4.14(285 mv)	
	E	10	0 corr	$K(MnO_4^{2-}+e^- \rightleftharpoons MnO_4^{3-})$ 4.6(0·26 v)	67T
				$K[MnO_2(s, x)+H^++e^- \rightleftharpoons MnO_{1.5}(s, 1-x)+\frac{1}{2}H_2O]$ (single solid solution phase); same equilibrium described in other ways, see refs 66Ka, 66Va	
	E	25(?)	0	25·7($x \approx$ 0·95, 1·52 v), 17·8(x = 0·75, 1·05 v)	64B
	E	22	0	23·3(x = 1·0), 21·3(x = 0·9), 19·3(x = 0·8), 17·7(x = 0·7), 17·0(x = 0·6), activity coefficients given for solid solutions	65N
	E	25	9 KOH	$K\{Mn^{4+}(x_4$, in solid) $+e^-(+H_2O) \rightleftharpoons Mn^{3+}(x_3$, in solid) $[+H^+$(in solid)$+OH^-]\}$ 1·74(103mv), x units	66Ka
	ana, *E*	20	0·1 SO_4^{2-}	data on equilibria of solid solution γ-MnO_2H_{4-2n}, equilibrium value of $[Mn^{2+}][OH^-]^2$ −22·96 (n = 1·84), −23·85(n = 1·86), −24·74(n = 1·88), −25·41(n = 1·90), −26·27(n = 1·92), −27·14(n = 1·94), $x = 2n-3$	66Va
	E	18	→ 0	$K(Mn^{IV}+2e^- \rightleftharpoons Mn^{2+})$ 52·3(1·51 v) (extrapolation from 1·5 M-HCl)	67L
	pol	25	$c\ H_2SO_4$	$K(Mn^{II}+Mn^{IV} \rightleftharpoons 2Mn^{III})$ 2·67(c = 12), 2·30(c = 10·5), 2·02(c = 9), 1·86(c = 7·5), 1·63(c = 6), 1·30(c = 4·5) in c M-H_2SO_4	68P
	E	25?	?	$K[Mn(CN)_5NO^{2-}+e^- \rightleftharpoons Mn(CN)_5NO^{3-}]$ 11·09 (656 mv)	64J
	E	25	0 corr	$K[Mn(CN)_5NO^{2-}+e^- \rightleftharpoons Mn(CN)_5NO^{3-}]$ 10·09 (597 mv)	64Jb
	ΔG	25	0	$K[Mn_3O_4(s)+8H^++2e^- \rightleftharpoons 3Mn^{2+}+4H_2O]$ 59·34	52R/66S
	ΔG	25	0	$K[Mn(OH)_3(s)+3H^++e^- \rightleftharpoons Mn^{2+}+3H_2O]$ 31·12	52R/66S
	E	15	0	$K[Mn^{2+}+2e^- \rightleftharpoons Mn(s)]$ −40·86(−1168 mv)	63Ja/64Jc
Re	cal, ΔG	25	0	$K(ReO_4^-+8H^++6Cl^-+3e^- \rightleftharpoons ReCl_6^{2-}+4H_2O)$ 9·6(0·19 v)	66Bc+
				$K[ReCl_6^{2-}+4e^- \rightleftharpoons Re(s)+6Cl^-]$ 34(0·50 v)	

continued

Metal	*Method*	*Temp*	*Medium*	*Log of equilibrium constant, remarks*	*Ref*
Fe	ΔG	25	0 corr	$\Delta H(Fe^{3+} + e^- \rightleftharpoons Fe^{2+}) = -10$	68Lb+
	cal	25 ?	0 corr	$\Delta H(2Fe^{2+} + H_2O_2 + 2H^+ \rightleftharpoons 2Fe^{3+} + 2H_2O) = -69{\cdot}8$	68Sb

1964 *Erratum*: p. 11, *line* 4 *from end for* "½ $Fe_3(CO)_{12}$" *read* "⅓ $Fe_3(CO)_{12}$"

Metal	*Method*	*Temp*	*Medium*	*Log of equilibrium constant, remarks*	*Ref*
	E	25	0 corr	$K[FeOOH(s) + 3H^+ + e^- \rightleftharpoons Fe^{2+} + 2H_2O]$	58L/66S
				15·35(908 mv)	
		25	0 corr	15·87(939 mv)	58L+60Ma/66S
	ΔG	25	0	$K[Fe_3O_4(s) + 8H^+ + 2e^- \rightleftharpoons 3Fe^{2+} + 4H_2O]$ 33·21 ?	52R/66S
	E	25	0 corr	$K[Fe_3O_4(s) + 8H^+ + 2e^- \rightleftharpoons 3Fe^{2+} + 4H_2O]$	58L/66S
				40·77(1206 mv)	
		25	0 corr	41·58(1230 mv)	58L+60Ma/66S
	ΔG	25	0 corr	$K[Fe^{2+} + 2e^- \rightleftharpoons Fe(s)]$ −15·99(−473 mv),	68Lb+
				$\Delta H = 22{\cdot}1$	
	E	25	→ 0	$K[Fe(CN)_6^{3-} + e^- \rightleftharpoons Fe(CN)_6^{4-}]$ 6·160(364·4 mv)	65L
		20—35	→ 0	6·160+0·026(25−*t*)[*t*°; 364·4−2·8 (*t*−25) mv]	
	E	25	0 corr	$K[Fe(CN)_6^{3-} + e^- \rightleftharpoons Fe(CN)_6^{4-}]$ 6·26(370·4 mv)	66R
	E	25	0 corr	$K[Fe(CN)_6^{3-} + e^- \rightleftharpoons Fe(CN)_6^{4-}]$ 6·00(355 mv)	67Ha
				$\Delta H = -26{\cdot}7$, $\Delta S = -62{\cdot}1$ (from data 15—30°)	
	E	25	→ 0	$K[Fe(CN)_6^{3-} + e^- \rightleftharpoons Fe(CN)_6^{4-}]$ 6·103(361·0 mv)	68Ma
	E	25 ?	var	$K[Fe(CN)_5H_2O^{2-} + e^- \rightleftharpoons Fe(CN)_5H_2O^{3-}]$	66Mb
				6·96(412 mv)	
Co	cal	15	4 $HClO_4$	$\Delta H(Co^{3+} + Fe^{2+} \rightarrow Co^{2+} + Fe^{3+}) = -26{\cdot}3$	64Ja
	ΔG	25	0	$K(Co^{3+} + e^- \rightleftharpoons Co^{2+})$ 33(1·95 v)	64Ja+
	E	2—25	5·6 $HClO_4$	$K(Co^{3+} + e^- \rightleftharpoons Co^{2+})$ 24·7(2°; 1·35 v),	67Bd
				23·3(→ 25°; 1·38 v)	
	cal, ΔG	25	0 corr	$K[Co^{2+} + 2e^- \rightleftharpoons Co(s)]$ −9·8(−0·29 v)	66G+
	ΔG	25	0 corr	$K[Co^{2+} + 2e^- \rightleftharpoons Co(s)]$ −9·70(−287 mv), $\Delta H = 14{\cdot}0$	68Lb+
	dep	18—22	0·5 $NaClO_4$	$K[Co^{2+} + 2e^- \rightleftharpoons Co(s)]$ −7·05(−205 mv)	68Z
			$SO_4 \rightarrow 0$	$K[Co^{2+} + 2e^- \rightleftharpoons Co(s)]$ −9·80(−285 mv)	
			NO_3 var	$K[CoNO_3^+ + 2e^- \rightleftharpoons Co(s) + NO_3^-]$ −4·50(−131 mv)(?)	
			$I^- \rightarrow 0$	$K[Co^{2+} + 2e^- \rightleftharpoons Co(s)]$ −8·15(−237 mv)	
	E	25	→ 0	$K[Co_2Fe(CN)_6(s) + 4e^- \rightleftharpoons 2Co(s) + Fe(CN)_6^{4-}]$	66Mc
				−18·20(−269·1 mv)	
Ni	*E*	225—250	85% KOH	$K[Ni^{IV} + \frac{1}{2}H_2(g) \rightleftharpoons Ni^{III}]$ 12·12 (225°; 1198 mv),	68V
				11·49(250°; 1193 mv) in 85% KOH	
				$K[Ni^{III} + \frac{1}{2}H_2(g) \rightleftharpoons Ni^{II}]$ 11·82(225°; 1168 mv),	
				11·31(250°; 1174 mv) in 85% KOH	
	ΔG	25	0 corr	$K[Ni^{2+} + 2e^- \rightleftharpoons Ni(s)]$ −7·71(−228 mv),	68Lb+
				$\Delta H = 12{\cdot}8$	
Ru	ΔG	25	0 corr	$K(RuO_4 + e^- \rightleftharpoons RuO_4^-)$ 1·67(0·99 v)	57C
	ana	25	0 corr	$K[4RuO_4^- + 4H^+ \rightleftharpoons 3RuO_4 + RuO_2(H_2O)_x(s)]$ 27·4	57C
	sp	14—32	NaOH var	$K(RuO_4^- + MnO_4^{2-} \rightleftharpoons RuO_4^{2-} + MnO_4^-)$	66L
				0·64(20°)	
				$\Delta H = -4{\cdot}0$, $\Delta S = -11$	
		20	0 corr	$K(RuO_4^- + e^- = RuO_4^{2-})$ 10·23(595 mv)	66L+56Ca
	E, sp	25	1 H_2SO_4	ev slow reactions, species with fractional oxidation	67Ba
				numbers *e.g.* 3·5 and ≈ 3·65	
	pol	25 ?	dil	$K(Ru^{3+} + e^- \rightleftharpoons Ru^{2+})$ 3·7(0·22 v) (in	65Mb
				$H^+CH_3C_6H_4SO_3^-$, dil)	

continued overleaf

1 e^- (contd)

Metal	*Method*	*Temp*	*Medium*	*Log of equilibrium constant, remarks*	*Ref*
Ru (*contd*)	*E*	25	0 corr	$K(Ru^{3+}+e^- \rightleftharpoons Ru^{2+})$ 4·204(248·7 mv), ΔH = 10·1, ΔS = 53 (sign misprinted), values for various *I* and 15—25°	66Bd
		25	1 H_2SO_4	$K(Ru^{3+}+e^- \rightleftharpoons Ru^{2+})$ 3·0(0·18 v)	67Ba
	ana	25	→ 0	$K[Ru(NH_3)_6{}^{3+}+e^- \rightleftharpoons Ru(NH_3)_6{}^{2+}]$ 3·62 (214 mv)	65E
	E	→ 25	0 corr	4·1(0·24 v)	
	E	→ 25	0 corr	$K[Ru(NH_3)_5{}^{3+}+e^- \rightleftharpoons Ru(NH_3)_5{}^{2+}]$ 3·4(0·20 v)	65E
	E	25	0 corr	$K[Ru(NH_3)_5OH_2{}^{3+}+e^- \rightleftharpoons Ru(NH_3)_5OH_2{}^{2+}]$ 2·7(0·16 v)	68Md
	E	25	0 corr	$K[Ru(NH_3]_6{}^{3+}+e^- \rightleftharpoons Ru(NH_3)_6{}^{2+}]$ 1·7(0·10 v)	68Md
	E	25	0 corr	$K(Ru\,en_3{}^{3+}+e^- \rightleftharpoons Ru\,en_3{}^{2+})$ 3·6(0·21 v)	68Md
	pol	25	var	$K(RuCl_2{}^{+}+e^- \rightleftharpoons RuCl_2)$ −0.2(−0·01 v) $K(RuCl_3+e^- \rightleftharpoons RuCl_3{}^-)$ −1·7(−0·10 v)	66Bd
Pd	*E*	25	?	$K(Pd\,en_2X_2{}^{2-}+2e^- \rightleftharpoons Pd\,en_2{}^{2+}+2X^-)$ 38·21(X = Cl; 1130 mv); 23·40 (X = Br; 692 mv), 21.13(X = I; 625 mv)	65B
	E	25	0 corr	$K[Pd^{2+}+2e^- \rightleftharpoons Pd(s)]$ 30·9(915 mv)	67I
	cal	25	0·1 NaI	$\Delta H[Pd^{2+}+3I^- \rightleftharpoons Pd(s)+I_3{}^-]$ −24·90	
	E	25	*I*($HClO_4$)	$K[Pd^{2+}+2e^- \rightleftharpoons Pd(s)]$ 33·67(*I* = 4·87; 996 mv), 32·19(*I* = 3·46; 952 mv), 31·25(*I* = 2·22; 924·4 mv) 31·11(*I* = 1·06; 920 mv)	68La
Os	sp, kin		1 KOH	$K(Os^{VIII}+Os^{VI} \rightleftharpoons 2\,Os^{VII})$ 0·89 $Os^{VI} = OsO_4{}^{2-}$, $Os^{VIII} = OsO_6{}^{4-}$	68Ba
	E	25	0 corr	$K[Os(bipy)_3{}^{3+}+e^- \rightleftharpoons Os(bipy)_3{}^{2+}]$ 14·96(884·7 mv) $K[Os(bipy)_2pyCl^{2+}+e^- \rightleftharpoons Os(bipy)_2pyCl^{+}]$ 8·18(483·6 mv) and 17 other similar reactions	66B
Ir	*E*	20	→ 0	$K(IrCl_6{}^{2-}+e^- \rightleftharpoons IrCl_6{}^{3-})$ 17·49(1017 mv)	44D
	E	25	1 HCl 1 NaCl	$K(IrCl_6{}^{2-}+e^- \rightleftharpoons IrCl_6{}^{3-})$ 15·76 (932 mv), 15·77(933 mV)	64K
	E	25	0·2 HNO_3	$K(IrCl_5{}^{-}+e^- \rightleftharpoons IrCl_5{}^{2-})$ 17(1.0 v)	65C
	E	25	0·4 HNO_3	$K(IrCl_4+e^- \rightleftharpoons IrCl_4{}^{-})$ 20(1·2 v)	
	E	25	0·4 $HClO_4$	$K[1,2,3,IrCl_3(OH_2)_3{}^{+}+e^- \rightleftharpoons 1,2,3\text{-}IrCl_3(OH_2)_3]$ 22·0(1.30 v)	67E
	E	25	0·4 $HClO_4$, 0·2 Na_2HPO_4	$K[trans\text{-}IrCl_4(OH_2)_2+e^- \rightleftharpoons trans\text{-}IrCl_4(OH_2)_2{}^-]$ 20·6(1·22 v)	
Pt	ana	60	3 M(Cl, ClO_4)	$K[Pt(s)+PtCl_6{}^{2-}+2Cl^- \rightleftharpoons 2\,PtCl_4{}^{2-}]$ −1·70(M = H), −1·35(M = Na)	67G+
		60	3 M(Cl, ClO_4)	$K[PtCl_6{}^{2-}+2Ag(s) \rightleftharpoons PtCl_4{}^{2-}+2AgCl(s)]$ 15·82(M = H; 523 mv), 16·01(M = Na; 529 mv), standard potentials against Cl^-/AgCl, Ag given	68G+
	E	25	dil	$K(PtA_4Br_2{}^{2+}+2e^- \rightleftharpoons PtA_4{}^{2+}+2Br^-)$ (A = NH_3) 19·71(583 mv) $K(PtA_2Br_4{}^{0}+2e^- \rightleftharpoons PtA_2Br_2{}^{0}+2Br^-)$ 20·18(*cis*; 597 mv), 20·29(*trans*; 600 mv) $K(PtA_4I_2{}^{2+}+2e^- \rightleftharpoons PtA_4{}^{2+}+2I^-)$ 14·57 (431 mv) $K(PtA_2I_4{}^{0}+2e^- \rightleftharpoons PtA_2I_2{}^{0}+2I^-)$ 13·19(*cis*; 390 mv), 12·95(*trans*; 383 mV)	49G
	E	25	→ 0	A = NH_3, L = $NO_2{}^-$ $K[(AL)_2X_2Pt+2e^- \rightleftharpoons (AL)_2Pt+2X^-]$ 26·17(X = Cl; 759 mv), 26·97(X = Br; 782 mv)	66C

continued

Metal	*Method*	*Temp*	*Medium*	*Log of equilibrium constant, remarks*	*Ref*
Pt (*contd*)				$K(A_2L_2X_2Pt + 2e^- \rightleftharpoons A_2L_2Pt + 2X^-)$ 26·31(X = Cl; 763 mv), 27·10(X = Br; 786 mv)	
				$K[A_2(LCl)_2Pt + 2e^- \rightleftharpoons A_2LClPt + L^- + Cl^-]$ 24·90(722 mv)	
				$K(ALAXLXPt + 2e^- \rightleftharpoons ALAXPt + L^- + X^-)$ 26·83(X = Cl; 778 mv), 27·83(X = Br; 807 mv)	
	E	25	1 MCl	$K(Pt\,en_2Cl_2^{2+} + 2e^- \rightleftharpoons Pt\,en_2^{2+} + 2Cl^-)$ 19·58(M = Na; 579·0 mv), 19·14(M = H; 566·0 mv)	68Ga
			1 MCl	$K[Pt(RNH_2)_4Cl_2^{2+} + 2e^- \rightleftharpoons Pt(RNH_2)_4^{2+} + 2Cl^-]$ R = Me: 20·85(M = Na; 616·8 mv), 21·03(M = H; 622·0 mv), R = Et: 23·02(M = Na; 681 mv), 23·39(M = H; 691·9 mv)	68Ga
			1 M_2SO_4	R = Et: 22·92(M = Na; 678·0 mv), 27·50(M = H; 813·4 mv), R = Pr: 24·06(M = Na; 711·6 mv), 28·15(M = H; 832·7 mv) R = Bu: 25·06(M = Na; 741·3 mv), 28·33(M = H; 838·0 mv)	

1964 *Erratum*: p. 16, $PtCl_6^{2-} + 2e^- \rightleftharpoons PtCl_4^{2-} + 2Cl^-$, *last line* "24·8(40°; 770 mv) *belongs to* ref. 37G, *not to* 61Y; *Next formula should be* "$PtBr_6^{2-} + 2e^- \rightleftharpoons PtBr_4^{2-} + 2Br^-$"

Metal	*Method*	*Temp*	*Medium*	*Log of equilibrium constant, remarks*	*Ref*
Cu	ana	25	0	$K[Cu^{2+} + Cu(s) \rightleftharpoons 2Cu^+]$ −5·94	65E
	ana, kin	25	0·2 H_2SO_4	$K[Cu^{2+} + Cu(s) \rightleftharpoons 2Cu^+]$ −6·24	68T
	ΔG	25	0 corr	$K[Cu^{2+} + 2e^- \rightleftharpoons Cu(s)]$ 11·66(345 mv), $\Delta H = -15{\cdot}7$	68Lb+
	E	25	2 $Na(NO_3)$	$K[CuBr_2^- + e^- \rightleftharpoons Cu(s) + 2Br^-]$ 3·3(≈ 195 mv)	67P/
	E	25	HF(l)	$K[CuF_2(s) + 2H^+ + 2e^- \rightleftharpoons Cu(s) + 2HF]$ 9·33(276 mv) in HF(l)	68Bh
Ag	kin	25	3·5 $HClO_4$	$K(Co^{2+} + Ag^{2+} \rightleftharpoons Co^{3+} + Ag^+)$ 0·12	63K
	kin	26·2	6·18 HNO_3	$K(Ce^{III} + Ag^{II} \rightleftharpoons Ce^{IV} + Ag^I)$ 6·58	66Z
	E	26·2	6·18 HNO_3	5·37	
	kin	25	4 $HClO_4$	$K(Co^{II} + Ag^{II} \rightleftharpoons Co^{III} + Ag^+)$ 1·64	67H
	E	25	*I*(Na)ClO_4	$K[Ag^+ + e^- \rightleftharpoons Ag(s)]$ 13·106(*I* = 3; 775·3 mv), 13·253(*I* = 2; 784·0 mv) 13·395(*I* = 1; 792·4 mv)	67K
	E	15—25	NH_2CHO	$K[Ag^+ + e^- \rightleftharpoons Ag(s)]$ 12·272(15°; 701·6 mv), 11·668(25°; 690·2 mv) in NH_2CHO(l), → 0, *m* units	68Bf
	E	25	0 corr	$K[Ag_4Fe(CN)_6(s) + 4e^- \rightleftharpoons 4Ag(s) + Fe(CN)_6^{4-}]$ 9·99(147·8 mv)	64R
	E	10—50	→ 0	$K[Ag_2C_2O_4(s) + 2e^- \rightleftharpoons 2Ag(s) + C_2O_4^{2-}]$ 17·04(10°; 478·7 mv), 15·71(25°; 464·7 mv) 13·76(50°; 441·2 mv), also values for intermediate temp	68Mb
	E	25	0 corr	$K[Ag_2S(s) + H^+ + 2e^- \rightleftharpoons 2Ag(s) + SH^-]$ −9·20(−272 mv)	63Cb
				$K[AgCl(s) + e^- \rightleftharpoons Ag(s) + Cl^-]$	
	E	25	*w*% $MeCO_2H$	3·558(*w* = 10; 210·5 mv), 1·885(*w* = 60; 111·5 mv), in *w*% $MeCO_2H$(→ 0), also data for *w* = 20, 30, 40, 50	32O/39Ha
	E	5—60	D_2O	4·082(5°; 225·28 mv), 3·836(15°; 219·31 mv), 3·595(25°; 212·66 mv), 3·358(35°; 205·32 mv), 3·126(45°; 197·33 mv), 3·012(50°; 193·10 mv), in D_2O, → 0	64G

continued overleaf

1 e^- (contd)

Metal	Method	Temp	Medium	Log of equilibrium constant, remarks	Ref
Ag (*contd*)	ΔG	0—100	0 corr	4·367(0°; 236·68 mv), 3·759(25°; 222·38 mv), 3·189(50°; 204·46 mv), 2·551(80°; 178·74 mv), 2·143(100°; 158·63 mv), also values for intermediate temp	64Ac+
	E	25—225	→0	3·753(25°; 222·0 mv), 2·985(60°; 197·3 mv), 2·358(90°; 169·9 mv), 1·662(125°; 131·3 mv), 1·189(150°; 99·8 mv), 0·731(175°; 65·0 mv), 0·292(200°; 27·4 mv), −0·133(225°; −13·1 mv)	60G/64L
	E	25	I(Na)ClO_4	3·442(I = 3; 203·6 mv), 3·728(I = 2; 220·5 mv), 3·929(I = 1; 232·4 mv)	67K
	E	25—225	D_2O→0	3·540(25°; 209·4 mv), 2·766(60°; 182·8 mv), 2·125(90°; 153·1 mv), 1·406(125°; 111·1 mv), 0·912(150°; 76·6 mv), 0·434(175°; 38·6 mv), −0·029(200°; −2·7 mv), −0·477(225°; −47·1 mv), m units, in D_2O	64L
	E	25	dioxan+MeOH	−1·22(70%; −72 mv), −1·49(65%; −88 mv), −1·72(60%; −102 mv), −1·89(54%; −112 mv), −2·50(50%; −148 mv), −2·74(45%; −162 mv), −3·35(40%; −198 mv) in w% MeOH+dioxan, → 0	64Sa
	E	35—70	MeCONHMe	3·465(35°; 211·87 mv), 3·311(40°; 205·73 mv), 3·183(45°; 200·91 mv), 3·035(50°; 194·56 mv), 2·914(55°; 189·72 mv), 2·777(60°; 183·57 mv), 2·525(70°; 171·94 mv) in MeCONHMe(l)	65D
	E	25	w% MeOH	3·759(w = 0; 222·34 mv), 3·643(w = 10; 215·5 mv), 3·535(w = 20; 209·1 mv), 3·433(w = 30; 203·1 mv), 3·327(w = 40; 196·8 mv), 3·222(w = 50; 190·6 mv), 3·073(w = 60; 181·8 mv), 2·845(w = 70; 168·3 mv), 2·522(w = 80; 149·2 mv), −0·171(w = 100; −10·1 mv), in w% MeOH, → 0	65Pb+
	E	25	H_2NCHO	3·357(198·6 mv) in H_2NCHO(l)	66Aa
	E	40	org	2·699(w = 29·98; 167·72 mv), 2·260(w = 43·1; 140·41 mv), 0·439(w = 73·85; 27·27 mv) in w% dioxan+MeCONHMe(l)	66D
	E	25—55	H_2NCHO	3·357(25°; 198·6 mv), 3·220(30°; 193·7 mv), 3·088(35°; 188·8 mv), 2·982(40°; 185·3 mv), 2·853(45°; 180·1 mv), 2·734(50°; 175·3 mv), 2·634(55°; 171·5 mv), ΔH = −46·1 kJ(25°) ΔS = −90·5 J K^{-1}, in H_2NCHO(l, → 0)	67A
	E	5—45	$C_2H_4(OH)_2$	0·834(5°; 46·0 mv), 0·610(15°; 34·9 mv) 0·397(25°; 23·5 mv), 0·193(35°; 11·8 mv), −0·006(45°; −0·4 mv) in $C_2H_4(OH)_2$(l), also values for glycol–water mixtures, m units	67S
	E	5—25	NH_2CHO	4·012(5°; 221·4 mv), 3·673(15°; 210·0 mv), 3·384(25°; 200·2 mv), in NH_2CHO(l), m units, → 0	68Bf
	E	25	R_2O	−3·11(in Et_2O; −184 mv), −3·45(in Pr_2O; −204 mv), −3·67(in Bu_2O; −217 mv), −3·89[in $(C_5H_{11})_2O$; −230 mv]	68I
	E	25	org	values given for mixtures dioxan–EtOH and hexane–EtOH	68Sa
	E	25	Me_2NCOMe	2·536(150·0 mv) in Me_2NCOMe(l) (→ 0)	68Sd
				$K[AgBr(s)+e^- \rightleftharpoons Ag(s)+Br^-]$	
	E	25	w% MeOH	1·125(w = 10; 66·55 mv), 0·947(w = 43·12; 56·0 mv) in w% MeOH (→ 0)	63F

continued

Metal	Method	Temp	Medium	Log of equilibrium constant, remarks	Ref
Ag (contd)	*E*	25	*w*% dioxan	1·013(*w* = 20; 59·91 mv), 0·538(*w* = 45; 31·83 mv) in *w*% dioxan, → 0	65F
	E	25	*w*% $MeCO_2H$	1·025 (*w* = 10; 60·65 mv), −0·652(*w* = 60; −38·55 mv) in *w*% $MeCO_2H$(→ 0), data also for *w* = 20 and 40	66Ba
	E	25	MeOH	−2·345(−138·7 mv), in MeOH(l), −0·130(−7.7 mv) in 87·68 wt% MeOH	67Aa
	E	15—45	*w*% glycol	values given for mixtures glycol–H_2O (0 corr)	67Be
	E	25	*w*% MeOH	values given for 20, 33, 50, 68, and 90% MeOH	67F
	E	25	H_2NCHO	1·635(96·7 mv) in H_2NCHO(l), → 0	68Mc
				$K[AgI(s) + e^- \rightleftharpoons Ag(s) + I^-]$	
	E	0—50	→ 0	−2·701(0°; −146·37 mv), −2·667(5°; −147·19 mv), −2·638(10°; −148·22 mv), −2·614(15°; 149·42 mv), −2·593(20°; −150·81 mv), −2·577(25°; −152·44 mv), −2·561(30°; −154·05 mv), −2·550(35°; −155·90 mv), −2·541(40°; −157·88 mv), −2·534(45°; −159·98 mv), −2·530(50°; −162·19 mv)	64H
	E	20—200	0	−2·587(20°; −150·48 mv), −2·570(25°; −152·04 mv), −2·521(50°; −161·65 mv), −2·548(75°; −176·0 mv), −2·601(100°; −192·6 mv), −2·690(125°; −212·5 mv), −2·829(150°; −237·5 mv), −3·037(175°; −270·0 mv), −3·296(200°; −309·4 mv)	65K
	E	25	*w*% MeOH	−2·601(*w* = 10; −153·85 mv), −2·517(*w* = 43·12; −148·9 mv), in *w*% MeOH (→ 0)	63F
	E	25	*w*% dioxan	−2·559(*w* = 20; −151·36 mv) −2·709(*w* = 45; −160·25 mv) in *w*% dioxan, → 0	65F
	E	25	*w*% $MeCO_2H$	−2·721(*w* = 10; −160·95 mv), −4·166(*w* = 60; −246·45 mv) in *w*% $MeCO_2H$(→ 0), data also for *w* = 20 and 40	66Ba
	E	25	MeOH	−5·369(−317·6 mv), in MeOH(l), −3·187(−188·5 mv) in 87·68 wt% MeOH	67Aa
	E	25	*w*% MeOH	values given for 20, 33, 50, 68, and 90% MeOH	67F
	E	25—45	→ 0	−2·574(25°; −152·24 mv) −2·549(35°; −155·86 mv), −2·534(45°; −159·94 mv), $\Delta H = -0{\cdot}21$, $\Delta S = -12{\cdot}43$	68Me
			100% MeOH	−5·373(25°; −317·86 mv), −5·414(35°; −331·01 mv), −5·533(45°; −349·26 mv), $\Delta H = -3{\cdot}46$, $\Delta S = -36{\cdot}01$. Data also for 15, 30, 45, 60, 75, 90, and 99% MeOH	
Au	*E*	25	0	$K[Au(SCN)_4^- + 2e^- \rightleftharpoons Au(SCN)_2^- + 2SCN^-]$ 21·06(623 mv)	66Pa
				$K[Au(SCN)_2^- + e^- \rightleftharpoons Au(s) + 2SCN^-]$ 11·19(662 mv)	
				$K[Au(SCN)_4^- + 3e^- \rightleftharpoons Au(s) + 4SCN^-]$ 32·25(636 mv)	66Pa
	E	25	var	T = $(NH_2)_2CS$; $K(AuT_2^+ + e^- \rightleftharpoons Au(s) + 2T)$ 6·4(0·38 v)	64Ka
	E	25	0 corr	$K(AuCl_4^- + 2e^- \rightleftharpoons AuCl_2^- + 2Cl^-)$ 31·14(921 mv)	65P
				$K[AuCl_4^- + 3e^- \rightleftharpoons Au(s) + 4Cl^-]$ 50·46(995 mv)	
				$K[AuCl_2^- + e^- \rightleftharpoons Au(s) + 2Cl^-]$ 19·41(1148 mv)	
				$K[AuCl_4^- + 2Au(s) + 2Cl^- \rightleftharpoons 3AuCl_2^-]$ −7·68	
	ΔG	25	0	$\Delta H[AuCl_4^- + 3e^- \rightleftharpoons Au(s) + 4Cl^-]$ = −87·0	66P

continued overleaf

1 e^- (contd)

Metal	*Method*	*Temp*	*Medium*	*Log of equilibrium constant, remarks*	*Ref*
Au	ana	25	0 corr	$K[AuBr_4^- + 2Au(s) + 2Br^- \rightleftharpoons 3AuBr_2^-]$ −5·34	63E
(*contd*)	*E*	25	0 corr	$K[AuBr_2^- + e^- \rightleftharpoons Au(s) + 2Br^-]$ 16·21(959 mv)	63E
	E, ana	25	0 corr	$K(AuBr_4^- + 2e^- \rightleftharpoons AuBr_2^- + 2Br^-)$ 27·12(802 mv)	
				$K[AuBr_4^- + 3e^- \rightleftharpoons Au(s) + 4Br^-]$ 43·31(854 mv)	
	E	25	0 corr	$K(AuBr_4^- + 2e^- \rightleftharpoons AuBr_2^- + 2Br^-)$ 26·98(798 mv)	65Pa
				$K[AuBr_4^- + 3e^- \rightleftharpoons Au(s) + 4Br^-]$ 43·21(852 mv)	
				$K[AuBr_2^- + e^- \rightleftharpoons Au(s) + 2Br^-]$ 16·23(960 mv)	
				$K[AuBr_4^- + 2Au(s) + 2Br^- \rightleftharpoons 3AuBr_2^-]$ −5·48	
	ΔG	25	0	$\Delta H[AuBr_4^- + 3e^- \rightleftharpoons Au(s) + 4Br^-] = -69{\cdot}2$	66P
Zn	ΔG	25	0 corr	$K[Zn^{2+} + 2e^- \rightleftharpoons Zn(s)]$ −25·83(−764 mv),	68Lb+
				$\Delta H = 36{\cdot}9$	
	var	500	$ZnCl_2$(l)	ev Zn_2^{2+} in $ZnCl_2$(l) (sol, sp, chronopotentiometry)	66V
	sol	702—777	$ZnCl_2$(l)	$K[Zn(a) + Zn^{2+} \rightleftharpoons 2Zn^+]$ −4·4(702°), −4·15(777°);	66Be
				$K[Zn(a) + Zn^{2+} \rightleftharpoons Zn_2^{2+}]$ −2(702°), −1·7(777°)	
				in $ZnCl_2$(l), *x* units	
Cd	*E*	25	→ 0	$K[Cd^{2+} + 2e^- \rightleftharpoons Cd(s)]$ −13·64(−403·5 mv)	66Bb
	E	25	*I*(Na)ClO_4	$K[Cd^{2+} + 2e^- \rightleftharpoons$ Cd(in 2-phase amalgam)]	66K
				−12·192(*I* = 3; −360·6 mv),	
				−12·195(*I* = 1; −360·7 mv)	
	E	25	*I*(Na)ClO_4	$K[Cd^{2+} + 2e^- \rightleftharpoons Cd(s)]$ −13·899(*I* = 3; −411·1 mv),	67K
				−13·889(*I* = 2; −410·8 mv),	
				−13·903(*I* = 1; −411·2 mv)	
	ΔG	25	0 corr	$K[Cd^{2+} + 2e^- \rightleftharpoons Cd(s)]$ −13·63(−403 mv),	68Lb+
				$\Delta H = 18{\cdot}1$	
	E	25	H_2NCHO	$K[CdCl_2(s) + 2e^- \rightleftharpoons Cd(s) + 2Cl^-]$	67R+66Aa
				−20·86(−616·9 mv) in H_2NCHO(l)	
	E	5—25	H_2NCHO	$K[CdCl_2(s) + 2e^- \rightleftharpoons Cd(s) + 2Cl^-]$	68Bg
				−21·78(5°; −601 mv), −21·20(15°; −606 mv),	
				−20·69(25°; −612 mv) in H_2NCHO(l), *m* units	
	red	580	$CdCl_2$(l)	ev Cd_2^{2+} or Cd° in $CdCl_2$(l), no ev Cd^+	63T
	red	277	$NaAlCl_4$(l)	ev Cd_2^{2+} in $NaAlCl_4$(l)	66M
	sp	350	melt	$K[Cd(l) + Cd^{2+} \rightleftharpoons Cd_2^{2+}]$ 0·23 in $NaAlCl_4$(l)	68Pa
Hg	*E*	20?	Me_2NCHO	$K[Hg^{2+} + Hg(l) \rightleftharpoons Hg_2^{2+}]$ 0·7 in Me_2NCHO(l),	67Ka
				0·1M-$HClO_4$	
				$K[Hg_2SO_4(s) + 2e^- \rightleftharpoons 2Hg(l) + SO_4^{2-}]$	
	E	25	→ 0	20·71(612·5 mv)	65Ca
	E	25	*m* M_2SO_4	$E°$ values for 0·5—1·9*m*-Na_2SO_4, and	65S
				0·5—4·0*m* Li_2SO_4 as medium	
	E	25	→ 0	20·811(615·53 mv)	
				$K[Hg_2Cl_2(s) + 2e^- \rightleftharpoons 2Hg(l) + 2Cl^-]$	
	E	5—45	→ 0	9·890(5°; 272·89 mv), 9·681(10°; 271·94 mv),	63G
				9·474(15°; 270·83 mv), 9·270(20°; 269·57 mv),	
				9·066(25°; 268·16 mv), 8·865(30°; 266·60 mv),	
				8·665(35°; 264·89 mv), 8·467(40°; 263·04 mv),	
				8·271(45°; 261·04 mv)	
		25	0	$\Delta H = -8{\cdot}23$, $\Delta S = -6{\cdot}85$, $\Delta C_p = -41$	
	ΔG	0—100	0 corr	10·092(0°; 273·47 mv), 9·066(25°; 268·155 mv),	/64Ac+
				8·075(50°; 258·85 mv), 6·951(80°; 243·52 mv),	
				6·211(100°; 229·92 mv)	
	E	15—35	*m* MCl	$E°$ values for 15, 25, 35°, *m* = 0·5, 1·0, 2·0, 4·0,	65S
				M = Li, Na, K, *m* units	

continued

Metal	Method	Temp	Medium	Log of equilibrium constant, remarks	Ref
Hg (contd)		15—35	→ 0	9·476(15°; 270·88 mv), 9·068(25°; 268·22 mv), 8·663(35°; 264·81 mv)	
	E	25	→ 0	9·059 to 9·062(267·93 to 268·04 mv)	67C
	E	25	H_2NCHO	8·287(245·1 mv) in H_2NCHO(l)	67R+66Aa
	E	25	Me_2CO	−0·42(−12·5 mv) in 95 weight% Me_2CO, *m* units, also values for other Me_2CO–H_2O mixtures	65Sa
				$K[Hg_2Br_2(s)+2e^- \rightleftharpoons 2Hg(l)+2Br^-]$	
	E	5—45	→ 0	5·110(5°; 141·00 mv), 5·014(10°; 140·84 mv), 4·914(15°; 140·48 mv), 4·812(20°; 139·93 mv), 4·707(25°; 139·23 mv), 4·602(30°; 138·39 mv), 4·495(35°; 137·41 mv), 4·389(40°; 136·36 mv), 4·285(45°; 135·24 mv)	63G
		25	0	$\Delta H = -4{\cdot}25$, $\Delta S = -3{\cdot}60$, $\Delta C_p = -38{\cdot}1$	
	E	25	→ 0	4·713(139·40 mv)	64S
		25	EtOH, *w*%	0·784(23·20 mv), in 95 weight % EtOH, → 0, also values for intermediate *w* values	
	E	15—35	→ 0	4·921(15°; 140·67 mv), 4·713(25°; 139·40 mv), 4·501(35°; 137·60 mv)	66Sa
		15—35	MeOH	−0·375(15°; −10·71 mv), −0·701(25°; −20·72 mv), −1·024(35°; −31·31 mv) in 99·5 wt% MeOH, also values for mixtures MeOH–H_2O	
	sol	25	0	$K[Hg(l) \rightleftharpoons Hg(aq)]$ −6·55(25°), $\Delta H = 15{\cdot}9$, also *K* values for 16 organic solvents	68S
Ga	*E*	25?	0 corr	$K[Ga^{3+}+3e^- \rightleftharpoons Ga(s)]$ −67·25(−1326 mv)	67Va
	ana	20—90	0·06 $GaCl_3$	$K[Ga^{3+}+2Ga(s) \rightleftharpoons 3Ga^+]$ −11·7(20°), −10·1(30°), −9·75(35°), −9·5(45°), −9·0(60°), −9·8?(75°), −10·0?(90°)	68K
In	*E*	15—60	→ 0	$K[In^{3+}+3e^- \rightleftharpoons In(s)]$ −17·82(15°; −339·5 mv), −17·15(25°; −338·2 mv), −16·58(35°; −337·9 mv), −16·08(45°; −338·3 mv), −15·35(60°; −338·1 mv)	63Ca
	E, pol	20	0·7 $HClO_4$	$K[In^++e^- \rightleftharpoons In(s)]$ −2·17(−126 mv)	65V
		20	0·7 $HClO_4$	$K[In^{3+}+2In(s) \rightleftharpoons 3In^+]$ −10·89	65V+
	ana	24	MeCN	$K[In^{3+}+2In(inHg) \rightleftharpoons 3In^+]$ −0·54, M units, in MeCN(l)	67Hb
Tl	*E*	25	3 H^+,4($NaClO_4$)	$K(Tl^{3+}+2I^- \rightleftharpoons Tl^++I_2)$ 25·34	66J
	iE		1 $HClO_4$	ev Tl^{2+}, $K(2Tl^{2+} \rightleftharpoons Tl^{3+}+Tl^+) \gg 0$	63J
	E	25	3 (Na)ClO_4	$K[Tl^++e^- \rightleftharpoons Tl$(in Hg, saturated)] −6·606(−390·8 mv)	66Ga
	E	25	*I*(Na)ClO_4	$K[Tl^++e^- \rightleftharpoons Tl(s)]$ −6·649(*I* = 3; −393·3 mv), −6·336(*I* = 2; −374·8 mv), −6·038(*I* = 1; −357·2 mv)	67K+
	E	25	*I*(Na)ClO_4	$K[TlCl(s)+e^- \rightleftharpoons Tl(s)+Cl^-]$ −9·72(*I* = 3; −575 mv), −9·43(*I* = 2; −558 mv), −9·23(*I* = 1; −546 mv)	67K+
	E	0—80	→ 0	$K[TlCl(s)+e^- \rightleftharpoons Tl(x$, in Hg$)+Cl^-]$ −9·352(0°; −506·8 mv), −8·796(25°; −520·3 mv), −8·372(50°; −536·8 mv), −7·992(80°; −560·0 mv), also values for intermediate temp and for 7 *x* values	65Md
C	*E*	25	0	$K[C_6H_4O_2(p\text{-quinone})+2H^++2e^- \rightleftharpoons C_6H_4(OH)_2]$ 23·657(699·72 mv) ("quinhydrone electrode")	64Sb
	E	25	*I*(Na)ClO_4	$K[C_6H_4O_2+2H^++2e^- \rightleftharpoons C_6H_4(OH)_2]$ 22·92(*I* = 3, 677·8 mv), 23·15(*I* = 2, 684·6 mv), 23·34(*I* = 1, 690·2 mv)	67K

continued overleaf

1 e^- (contd)

Metal	*Method*	*Temp*	*Medium*	*Log of equilibrium constant, remarks*	*Ref*
Ge	*iE*, kin	25 ?	1 H_2SO_4	$K(Ge^{IV}+2e^- \rightleftharpoons Ge^{2+})$ 0(0 v)	65R
			1 H_2SO_4	$K[Ge^{2+}+2e^- \rightleftharpoons Ge(s)]$ 8·1(0·24 v)	65R
	iE	25 ?	1 H_2SO_4	$K[Ge^{IV}+4e^- \rightleftharpoons Ge(s)]$ 8·38(124 mv)	65Ra+65R
Pb	*E*	25	→ 0	$K[PbO_2(s)+4H^++SO_4^{2-}+2e^- \rightleftharpoons PbSO_4(s)+2H_2O]$ 57·142(1690·1 mv)	65Ca
	red	580—700	PbX_2(l)	ev Pb_2^{2+} or Pb°, no ev Pb^+ in $PbCl_2$(l), PbI_2(l)	63T
	red	277	$NaAlCl_4$(l)	ev Pb^+ [or $Pb_n^{(2n-1)+}$] in $NaAlCl_4$(l)	66M
	var	518—530	$PbCl_2$(l)	ev Pb_2^{2+} in $PbCl_2$(l) (chronopotentiometry, sol, sp)	66V
N	*E*	25	0 corr	$K[HNO_2+H^++e^- \rightleftharpoons NO(g)+H_2O]$ 16·62(983 mv) $K[NO^++e^- \rightleftharpoons NO(g)]$ 24·7(1·46 v)	66Sb = 67Sa
P	*E*	22	var	$K(H_4P_2S_2O_6+2e^-+2H^+ \rightleftharpoons 2H_3PO_3S)$ 12·0(0·35 v)	65Na
Sb	fp	≈ 660	NaI(l)	ev Sb^{3-} and Sb_3^{3-} in NaI(l)	68O

1964 *Addendum*: p. 27, *for ref.* "62 Hb" *read* "62 Hb = 63H"

Metal	*Method*	*Temp*	*Medium*	*Log of equilibrium constant, remarks*	*Ref*
Bi	Bi	264	$BiCl_3$(l)	$K(4Bi^+ \rightleftharpoons Bi_4^{4+})$ 6·43 in $BiCl_3$(l), M units	61Ta/63B
	sp	264	$BiCl_3$(l)	$K(4Bi^+ \rightleftharpoons Bi_4^{4+})$ 6·58 in $BiCl_3$(l), M units	63B
	sp, sol		MCl(l)	ev Bi^+, Bi_5^{3+} and one species with <0·6+ per Bi in NaCl–$AlCl_3$(l) and KCl–$ZnCl_2$(l)	65Bb
	sp	130—190	MCl(l)	$K(6Bi^+ \rightleftharpoons Bi^{3+}+Bi_5^{3+})$ 8·68(130°), 5·60(190°) in $Na_{0.37}Al_{0.63}Cl_{2.26}$(l, eut), M units	67B
	sp	380	MCl(l)	$K(6Bi^+ \rightleftharpoons Bi^{3+}+Bi_5^{3+})$ 10·23, $K[2Bi(l)+Bi^{3+} \rightleftharpoons 3Bi^+]$ −5·85, $K[4Bi(l)+Bi^{3+} \rightleftharpoons Bi_5^{3+}]$ −1·51, M units, in $K_{0.28}Zn_{0.72}Cl_{1.72}$(l, eut)	
	sp	250	$NaAlCl_4$(l)	ev Bi_8^{2+} in $NaAlCl_4$(l)	67Bf
	fp	650—680	melt	ev Bi^{3-} in NaI(l), KI(l) *etc.*	68O
O	*E*	263—356	MNO_3(l)	$K[2Ag^++O^{2-} \rightleftharpoons 2Ag(s)+\frac{1}{2}O_2(g)]$ 12·02(536K; 639·0 mv), 9·67(639K; 612·8 mv), in $Na_{0.5}K_{0.5}NO_3$(l), also *K* for intermediate temp	65Ka
	E	25	0	$K[2H^++\frac{1}{2}O_2(g)+2e^- \rightleftharpoons H_2O]$ 41·6(1·23 v), also discussion of other redox reactions on Pt electrodes	64Wa
S	*E*	25	0 corr	$K[nS(s)+2e^- \rightleftharpoons S_n^{2-}]$ −15·70(n = 1; −464·44 mv), −16·6(n = 2, est; −492 mv), −15·4(n = 3, est; −455 mv), −11·93(n = 4; −353 mv), −11·70 (n = 5; −346 mv), −12·00(n = 6; −355 mv)	63Cb
Se	con	280	Se(l)	ev Se_n^+ in Se(l)	66La = 68L
	E	25	→ 0	$K[H_2SeO_3+4H^++4e^- \rightleftharpoons Se(s)+3H_2O]$ 50·3(744 mv)	66O
	var		H_2SO_4	ev Se_4^{2+} and Se_8^{2+} in H_2SO_4(l), HSO_3F(l), and $H_2S_2O_7$(l) (con, fp, sp, mag)	68Bb
	var		var	ev Se_4^{2+} in $Se_4(SO_3F)_2$(s) and other solids, and in H_2SO_4 (l) (con, fp, nmr, pre)	68Bc
Te	pol	25	var	$K[Te(s)+2H^++2e^- \rightleftharpoons H_2Te]$ −17·2(−0·51 v)	63Pa
	pol	25	var	$K[Te(s)+2e^- \rightleftharpoons Te^{2-}]$ −32·1(−0·95 v)	63Pa
	pol	25	var	$K[Te(s)+Te^{2-} \rightleftharpoons Te_2^{2-}]$ 3·7	63Pa+23Ka
	E	25 ?	var	$K[2Te(s)+2e^- \rightleftharpoons Te_2^{2-}]$ −28·57(−845 mv)	64P
	sp		MCl(l)	ev Te_{2n}^{n+}, probably Te_4^{2+} in NaCl–$AlCl_3$(l)	68Bd
	var		HSO_3F	ev Te_4^{2+} in HSO_3F(l) (pre, sp, mag, fp)	68Be

continued

Metal	Method	Temp	Medium	Log of equilibrium constant, remarks	Ref
Po	dis, kin	25?	2 HCl	$K[Po^{IV}+e^- \rightleftharpoons Po^{III}(PoCl_6^{2-}, PoCl_5^{2-})]$ 6·71(397 mv)	66H
				$K[Po^{IV}+2e^- \rightleftharpoons Po^{II}(PoCl_4^{2-})]$ 12·27(363 mv);	
				$K(Po^{III}+e^- \rightleftharpoons Po^{II})$ 5·58(330 mv);	
				$K[Po^{II}+2e^- \rightleftharpoons Po(s)]$ 24·7(0·73 v)	66H+56B
	E	25	0 corr	$K[PoCl_4^{2-}+2e^- \rightleftharpoons Po(s)+4Cl^-]$ 17·2(510 mv)	65Ea
				$K(PoCl_6^{2-}+2e^- \rightleftharpoons PoCl_4^{2-}+2Cl^-)$ 24·1(712 mv)	
F	*E*	0	HF(l)	$K[\frac{1}{2}F_2(g)+e^- \rightleftharpoons HF\text{–}H^+]$ [in HF(l)] 51·07(2768 mv)	29F
Cl	*E*	25	I(Na)ClO$_4$	$K[\frac{1}{2}Cl_2(g)+e^- \rightleftharpoons Cl^-]$ 22·671(I = 3; 1341·1 mv),	67K
				22·957(I = 2; 1358·0 mv), 23·158(I = 1; 1369·9 mv)	
Br	*E*	0—50	0 corr	$K(\frac{1}{2}Br_2+e^- \rightleftharpoons Br^-)$ 20·257(0°, 1097·8 mv),	66Ma
				19·479(10°; 1094·3 mv), 18·741(20°; 1090·0 mv),	
				18·381(25°; 1087·3 mv), 18·033(30°; 1084·6 mv),	
				17·354(40°; 1078·2 mv), 16·700(50°; 1070·7 mv)	
	X		solid	ev Br_2^+ in $Br_2Sb_3F_{16}$(s)	68E
	var		HSO_3F	ev Br_2^+ and Br_3^+ in HSO_3F(l) (pre, sp, Ram)	68Gb
	E	25	c H$^+$	$K(BrCl+2e^- \rightleftharpoons Br^-+Cl^-)$ 39·04(6M-H$^+$;	27F
				1154·8 mv), 41·28(4M-H$^+$; 1221 mv),	
				$K(Br_2+Cl_2 \rightleftharpoons 2BrCl)$ 3·49 (6 or 4M-H$^+$);	
				$K(Br_2+2Cl^- \rightleftharpoons Cl_2+2Br^-)$ −11·52(6M-H$^+$),	
				−11·25(4M-H$^+$)	
I	cal	25	0 corr	$\Delta H[\frac{1}{2}I_2(s)+e^- \rightleftharpoons I^-] = -13{\cdot}45$	67V+
	var		IF_5(l)	ev I^+ in IF_5(l) (con, sp, mag)	63Aa
	sp, pre		HSO_3F(l)	ev I^+, I_3^+ in HSO_3F(l)	65A
	fp, con		H_2SO_4	ev I_3^+, IO^+ (no I^+), I_5^+ *etc.* in H_2SO_4(l)	65G
	var		HSO_3F(l)	ev I_2^+, I_3^+, I_5^+ in HSO_3F(l) (con, fp, mag, sp)	66Gb
	var	−70 to −90	HSO_3F(l)	$K(2I_2^+ \rightleftharpoons I_4^{2+})$ 1·36(−70°), 2·23(−86·5°)	68G
				ΔH = 10(sp);	
				K 2·48(fp) m units in HSO_3F(l); agrees with (mag, con)	
Xe	pre, est	25	0	$K(H_4XeO_6+2H^++2e^- \rightleftharpoons XeO_3+3H_2O)$ 100(3·0 v)	64Aa
				$K[XeO_3+6H^++6e^- \rightleftharpoons Xe(g)+3H_2O]$ 180(1·8 v)	
	est	25	0	$K[XeF_2+2H^++2e^- \rightleftharpoons Xe(g)+2HF]$ 74(2·2 v)	64Ab

27F G. S. Forbes and R. M. Fuoss, *J. Amer. Chem. Soc.*, 1927, **49**, 142
29F K. Fredenhagen and O. T. Krefft, *Z. Elektrochem.*, 1929, **35**, 670
32O B. B. Owen, *J. Amer. Chem. Soc.*, 1932, **54**, 1758
39Ha H. S. Harned and B. B. Owen, *Chem. Rev.*, 1939, **25**, 31
44D F. P. Dwyer, H. A. McKenzie, and R. S. Nyholm, *J. Proc. Roy. Soc. New South Wales*, 1944, **78**, 260
49G A. A. Grinberg and B. Z. Orlova, *Zhur. priklad. Khim.*, 1949, **22**, 441
52R F. D. Rossini, D. D. Wagman, W. H. Evans, S. Levine, and I. Jaffe, *Nat. Bur. Stand., Circular* 500, 1952
54C D. Cozzi and S. Vivarelli, *Z. Elektrochem.*, 1954, **58**, 177
56Ca A. Carrington and M. C. R. Symons, *J. Chem. Soc.*, 1956, 3373
57C R. E. Connick and C. R. Hurley, *J. Phys. Chem.*, 1957, **61**, 1018
58L O. N. Lapteva, *Zhur. priklad. Khim.*, 1958, **31**, 1210
60Ma B. N. Mattoo, *Zhur. priklad. Khim.*, 1960, **33**, 2015
61Ta L. E. Topol, S. J. Yosim, and R. A. Osteryoung, *J. Phys. Chem.*, 1961, **65**, 1511
63A L. B. Anderson and D. J. Macero, *J. Phys. Chem.*, 1963, **67**, 1942
63Aa E. E. Aynsley, N. N. Greenwood, and D. H. W. Wharmby, *J. Chem. Soc.*, 1963, 5369
63B C. R. Boston, G. P. Smith, and L. C. Howick, *J. Phys. Chem.*, 1963, **67**, 1849
63Ca A. K. Covington, M. A. Hakeem, and W. F. K. Wynne-Jones, *J. Chem. Soc.*, 1963, 4394
63Cb P. L. Cloke, *Geochim. Cosmochim. Acta*, 1963, **27**, 1265
63E D. H. Evans and J. J. Lingane, *J. Electroanalyt. Chem.*, 1963, **6**, 1
63F D. Feakins and P. Watson, *J. Chem. Soc.*, 1963, 4686.
63G S. R. Gupta, G. J. Hills, and D. J. G. Ives, *Trans. Faraday Soc.*, 1963, **59**, 1874 (Cl), 1886 (Br)
63H A. Hershaft and J. D. Corbett, *Inorg. Chem.*, 1963, **2**, 979
63J J. Jordan and H. A. Catherino, *J. Phys. Chem.*, 1963, **67**, 2241
63Ja G. Jangg and H. Kirchmayr, *Z. Chem.*, 1963, **3**, 47

continued overleaf

1 e^- (contd)

63K J. B. Kirwin, F. D. Peat, P. J. Proll, and L. H. Sutcliffe, *J. Phys. Chem.*, 1963, **67**, 2288
63M W. U. Malik and S. I. Ali, *Indian. J. Chem.*, 1963, **1**, 374
63O J. W. Olver and J. W. Ross, jun., *J. Amer. Chem. Soc.*, 1963, **85**, 2565
63Pa A. J. Panson, *J. Phys. Chem.*, 1963, **67**, 2177
63T L. E. Topol, *J. Phys. Chem.*, 1963, **67**, 2222
64A J. F. Allen and H. M. Neumann, *Inorg. Chem.*, 1964, **3**, 1612
64Aa E. H. Appelman and J. G. Malm, *J. Amer. Chem. Soc.*, 1964, **86**, 2141
64Ab E. H. Appelman and J. G. Malm, *J. Amer. Chem. Soc.*, 1964, **86**, 2297
64Ac J. C. Ahluwalia and J. W. Cobble, *J. Amer. Chem. Soc.*, 1964, **86**, 5381
64B H. Bode and A. Schmier, *Ber. Bunsengesellschaft Phys. Chem.*, 1964, **68**, 954
64Ba J. C. Barnes and P. Day, *J. Chem. Soc.*, 1964, 3886
64G R. Gary, R. G. Bates, and R. A. Robinson, *J. Phys. Chem.*, 1964, **68**, 1186
64H H. B. Hetzer, R. A. Robinson, and R. G. Bates, *J. Phys. Chem.*, 1964, **68**, 1929
64J W. Jakób, A. Gołębiewski, and T. Senkowski, in "Theory and structure of complex compounds" (symposium papers, Wrocław 1962), Pergamon, Warsaw, 1964, p. 381
64Ja D. A. Johnson and A. G. Sharpe, *J. Chem. Soc.*, 1964, 3490
64Jb W. Jakób and T. Senkowski, *Roczniki Chem.*, 1964, **38**, 1751
64Jc G. Jangg, *Monatsh.*, 1964, **95**, 1103
64K V. I. Kvartsov and G. M. Petrova, *Zhur. neorg. Khim.*, 1964, **9**, 1010
64Ka V. P. Kazakov, A. I. Lapshin, and B. I. Peshchevitskii, *Zhur. neorg. Khim.*, 1964, **9**, 1299
64L M. H. Lietzke and R. W. Stoughton, *J. Phys. Chem.*, 1964, **68**, 3043
64P A. J. Panson, *J. Phys. Chem.*, 1964, **68**, 1721
64R P. A. Rock and R. E. Powell, *Inorg. Chem.*, 1964, **3**, 1593
64S K. Schwabe, R. Urlass, and A. Ferse, *Ber. Bunsengesellschaft Phys. Chem.*, 1964, **68**, 46
64Sa A. M. Shkodin, T. P. Sogoyan, L. I. Karkuzaki, and L. I. Kozynyuk, *Ukrain. khim. Zhur.*, 1964, **30**, 237
64Sb J.-P. Schwing, *J. Chim. phys.*, 1964, **61**, 491
64W D. A. Wenz, M. D. Adams, and R. K. Steudenberg, *Inorg. Chem.*, 1964, **3**, 989
64Wa N. Watanabe and M. A. V. Devanathan, *J. Electrochem. Soc.*, 1964, **111**, 615
65A F. Aubke and G. H. Cady, *Inorg. Chem.*, 1965, **4**, 269
65B A. V. Babaeva and E. Ya. Khananova, *Zhur. neorg. Khim.*, 1965, **10**, 2579
65Ba I. J. Bear and A. G. Turnbull, *J. Phys. Chem.*, 1965, **69**, 2828
65Bb N. J. Bjerrum, C. R. Boston, G. P. Smith, and H. L. Davis, *Inorg. Nuclear Chem. Letters*, 1965, **1**, 141
65C J. C. Chang and C. S. Garner, *Inorg. Chem.*, 1965, **4**, 209
65Ca A. K. Covington, J. V. Dobson, and Lord Wynne-Jones, *Trans. Faraday Soc.*, 1965, **61**, 2050
65D L. R. Dawson, W. H. Zuber, jun., and H. C. Eckstrom, *J. Phys. Chem.*, 1965, **69**, 1335
65E J. F. Endicott and H. Taube, *Inorg. Chem.*, 1965, **4**, 437
65Ea J. F. Eichelberger, G. R. Grove, and L. V. Jones, *Mound Laboratory Report MLM*-1250, 1965
65F D. Feakins and D. J. Turner, *J. Chem. Soc.*, 1965, 4986
65G R. A. Garrett, R. J. Gillespie, and J. B. Senior, *Inorg. Chem.*, 1965, **4**, 563
65K G. Kortüm and W. Häussermann, *Ber Bunsengesellschaft Phys. Chem.*, 1965, **69**, 594
65Ka R. N. Kust, *J. Phys. Chem.*, 1965, **69**, 3662
65L J. Lin and W. G. Breck, *Canad. J. Chem.*, 1965, **43**, 766
65M J. H. Miles, *J. Inorg. Nuclear Chem.*, 1965, **27**, 1595
65Ma R. E. McCarley, B. G. Hughes, F. A. Cotton, and R. Zimmerman, *Inorg. Chem.*, 1965, **4**, 1491
65Mb E. E. Mercer and R. R. Buckley, *Inorg. Chem.*, 1965, **4**, 1692
65Mc D. J. Macero, L. B. Anderson, and P. Malachesky, *J. Electroanalyt. Chem.*, 1965, **10**, 76
65Md T. Mussini and P. Longhi, *Ricerca sci.*, 1965, **A**, **8**, 1352
65N K. Neumann and E. von Roda, *Ber. Bunsengesellschaft Phys. Chem.*, 1965, **69**, 347
65Na H. Neumann, I. Z. Steinberg, and E. Katchalski, *J. Amer. Chem. Soc.*, 1965, **87**, 3841
65P J. Pouradier, M.-C. Gadet, and H. Chateau, *J. Chim. phys.*, 1965, **62**, 203
65Pa J. Pouradier and M.-C. Gadet, *J. Chim. phys.*, 1965, **62**, 1181
65Pb M. Paabo, R. G. Bates, and R. A. Robinson, *Analyt. Chem.*, 1965, **37**, 462
65R W. E. Reid, jun., *J. Phys. Chem.*, 1965, **69**, 2269
65Ra W. E Reid, jun., *J. Phys. Chem.*, 1965, **69**, 3168
65Rb M. L. Reynolds, *J. Chem. Soc.*, 1965, 2993
65S K. Schwabe and E. Ferse, *Ber. Bunsengesellschaft Phys. Chem.*, 1965, **69**, 383
65Sa K. Schwabe and K. Wankmüller, *Ber. Bunsengesellchaft Phys. Chem.*, 1965, **69**, 528
65Sb C. T. Stubblefield, J. L. Rutledge, and R. Phillips, *J. Phys. Chem.*, 1965, **69**, 991
65V R. E. Visco, *J. Phys. Chem.*, 1965, **69**, 202
66A A. Adin and A. G. Sykes, *J. Chem. Soc. (A)*, 1966, 1230
66Aa R. K. Agarwal and B. Nayak, *J. Phys. Chem.*, 1966, **70**, 2568
66B D. A. Buckingham, F. P. Dwyer, and A. M. Sargeson, *Inorg. Chem.*, 1966, **5**, 1243
66Ba H. P. Bennetto, D. Feakins, and D. J. Turner, *J. Chem. Soc. (A)*, 1966, 1211
66Bb J. L. Burnett and M. H. Zirin, *J. Inorg. Nuclear Chem.*, 1966, **28**, 902
66Bc R. H. Busey, K. H. Gayer, R. A. Gilbert, and R. B. Bevan, jun., *J. Phys. Chem.*, 1966, **70**, 2609
66Bd R. R. Buckley and E. E. Mercer, *J. Phys. Chem.*, 1966, **70**, 3103
66Be G. G. Bombi, G. A. Sacchetto, and M. Fiorani, *Chem. Comm.*, 1966, 563

continued

66C I. I. Chernyaev, G. S. Muraveiskaya, and L. S. Korablina, *Zhur. neorg. Khim.*, 1966, **11**, 1339
66D L. R. Dawson, K. H. Kim, and H. C. Eckstrom, *J. Phys. Chem.*, 1966, **70**, 775
66E J. H. Espenson and R. E. McCarley, *J. Amer. Chem. Soc.*, 1966, **88**, 1063
66G R. N. Goldberg, R. G. Riddell, M. R. Wingard, H. P. Hopkins, C. A. Wulff, and L. G. Hepler, *J. Phys. Chem.*, 1966, **70**, 706
66Ga M. Gamsjäger, W. Kraft, and W. Rainer, *Monatsh.* 1966, **97**, 833
66Gb R. J. Gillespie, and J. B. Milne, *Inorg. Chem.*, 1966, **5**, 1577
66H H. Haïssinsky and E. Pluchet, *J. Inorg. Nuclear Chem.*, 1966, **28**, 2861
66Ha E. J. Hart, S. Gordon, and E. M. Fielden, *J. Phys. Chem.*, 1966, **70**, 150
66J L. Johansson, *Acta Chem. Scand.*, 1966, **20**, 2156
66K W. Kraft, H. Gamsjäger, and E. Schwarz-Bergkampf, *Monatsh.*, 1966, **97**, 1134
66Ka A. Kozawa and R. A. Powers, *J. Electrochem. Soc.*, 1966, **113**, 870
66L E. V. Luoma and C. H. Brubaker, jun., *Inorg. Chem.*, 1966, **5**, 1637
66La M. Lundkvist and L. G. Sillén, *Acta Chem. Scand.*, 1966, **20**, 1723
66Lb E. Luksha and C. M. Criss, *J. Phys. Chem.*, 1966, **70**, 1496
66M T. C. F. Munday and J. D. Corbett, *Inorg. Chem.*, 1966, **5**, 1263
66Ma T. Mussini and G. Fàita, *Ricerca sci.*, 1966, **36**, 175
66Mb W. U. Malik and H. Om, *Indian J. Chem.*, 1966, **4**, 106
66Mc W. U. Malik and A. Das, *Indian J. Chem.*, 1966, **4**, 203
66O Sh. D. Osman-Zade and A. T. Vagramyan, *Elektrokhimiya*, 1966, **2**, 85
66P J. Pouradier, *J. Chim. phys.*, 1966, **63**, 694
66Pa J. Pouradier and M.-C. Gadet, *J. Chim. phys.*, 1966, **63**, 1467
66R P. A. Rock, *J. Phys. Chem.*, 1966, **70**, 576
66S L. G. Sillén, *Arkiv Kemi*, 1966, **25**, 159
66Sa K. Schwabe and E. Ferse, *Ber. Bunsengesellschaft Phys. Chem.*, 1966, **70**, 849
66Sb G. Schmid and U. Neumann, *Ber. Bunsengesellschaft Phys. Chem.*, 1966, **70**, 1165 (short)
66V J. D. Van Norman, J. S. Bookless, and J. J. Egan, *J. Phys. Chem.*, 1966, **70**, 1276
66Va K. J. Vetter and N. Jaeger, *Electrochim. Acta*, 1966, **11**, 401
66Z E. Zeltmann, preliminary data quoted by R. W. Dundon and J. W. Gryder *Inorg. Chem.*, 1966, **5**, 989
67A R. K. Agarwal and B. Nayak, *J. Phys. Chem.*, 1967, **71**, 2062
67Aa M. Alfenaar, C. L. de Ligny, and A. G. Remijnse, *Rec. Trav. chim.*, 1967, **86**, 555
67B N. J. Bjerrum, C. R. Boston, and G. P. Smith, *Inorg. Chem.*, 1967, **6**, 1162
67Ba F. Brito and D. Lewis, *Arkiv Kemi*, 1967, **26**, 401
67Bb J. N. Butler, R. Huston, and P. T. Hsu, *J. Phys. Chem.*, 1967, **71**, 3294
67Bc J. N. Butler and R. Huston, *J. Phys. Chem.*, 1967, **71**, 4479
67Bd L. M. Borisova and A. L. Rotinyan, *Elektrokhimiya*, 1967, **3**, 1163
67Be S. K. Banerjee, K. K. Kundu, and M. N. Das, *J. Chem. Soc. (A)*, 1967, 161
67Bf N. J. Bjerrum and G. P. Smith, *Inorg. Nuclear Chem. Letters*, 1967, **3**, 165; *Inorg. Chem.*, 1967, **6**, 1968
67C A. K. Covington, J. V. Dobson, and Lord Wynne-Jones, *Electrochim. Acta*, 1967, **12**, 513, 525
67E A. A. El-Awady, E. J. Bounsall, and C. S. Garner, *Inorg. Chem.*, 1967, **6**, 79
67F D. Feakins and R. P. T. Tomkins, *J. Chem. Soc., A*, 1967, 1458
67G O. Ginstrup and I. Leden, *Acta Chem. Scand.*, 1967, **21**, 2689
67H D. H. Hutchital, N. Sutin, and B. Warnqvist, *Inorg. Chem.*, 1967, **6**, 838
67Ha G. I. H. Hanania, D. H. Irvine, W. A. Eaton, and P. George, *J. Phys. Chem.*, 1967, **71**, 2022
67Hb J. B. Headridge and D. Pletcher, *Inorg. Nuclear Chem. Letters*, 1967, **3**, 475
67I R. M. Izatt, D. Eatough, and J. J. Christensen, *J. Chem. Soc.*, *(A)*, 1967, 1301; R. M. Izatt, G. D. Watt, D. Eatough, and J. J. Christensen, *ibid.*, p. 1304
67K W. Kraft, *Monatsh.*, 1967, **98**, 1978
67Ka N. V. Kiet and M. Bréant, *Compt. rend.*, 1967, **264**, *C*, 1042.
67L V. N. Lisov, *Ukrain. khim. Zhur.*, 1967, **33**, 849
67M W. U. Malik, J. P. Jain, and J. K. Khandelwal, *Indian J. Chem.*, 1967, **5**, 275
67Ma J. Malý and B. B. Cunningham, *Inorg. Nuclear Chem. Letters*, 1967, **3**, 445
67P D. G. Peters and R. L. Caldwell, *Inorg. Chem.*, 1967, **6**, 1478
67R M. de Rossi, G. Pecci, and B. Scrosati, *Ricerca sci.*, 1967, **37**, 342
67S U. Sen, K. K. Kundu, and M. N. Das, *J. Phys. Chem.*, 1967, **71**, 3665
67Sa G. Schmid and U. Neumann, *Z. phys. Chem. (Frankfurt)*, 1967, **54**, 150
67T R. Thiele and R. Landsberg, *Z. phys. Chem. (Leipzig)*, 1967, **236**, 95
67V A. F. Vorob'ev, A. F. Broier, and S. M. Skuratov, *Doklady Akad. Nauk. S.S.S.R.*, 1967, **173**, 385
67Va A. T. Vagramjan and T. I. Leshawa, *Z. phys. Chem. (Leipzig)*, 1967, **234**, 57
68B J. N. Butler, *J. Electroanalyt. Chem.*, 1968, **17**, 309
68Ba J.-P. Beaufils, M. Hellin, and F. Coussemant, *Compt. rend.*, 1968, **266**, *C*, 496
68Bb J. Barr, R. J. Gillespie, R. Kapoor, and K. C. Malhotra, *Canad. J. Chem.*, 1968, **46**, 149
68Bc J. Barr, D. B. Crump, R. J. Gillespie, R. Kapoor, and P. K. Ummat, *Canad. J. Chem.*, 1968, **46**, 3607
68Bd N. J. Bjerrum and G. P. Smith, *J. Amer. Chem. Soc.*, 1968, **90**, 4472
68Be J. Barr, R. J. Gillespie, R. Kapoor, and G. P. Pez, *J. Amer. Chem. Soc.*, 1968, **90**, 6855
68Bf R. W. C. Broadbank, S. Dhabanandana, K. W. Morcom, and B. L. Muju, *Trans. Faraday Soc.*, 1968, **64**, 3311
68Bg R. W. C. Broadbank, B. L. Muju, and K. W. Morcom, *Trans. Faraday Soc.*, 1968, **64**, 3318

continued overleaf

1 e⁻ (contd)

68Bh B. Burrows and R. Jasinski, *J. Electrochem. Soc.*, 1968, **115**, 348
68C D. R. Cogley and J. N. Butler, *J. Phys. Chem.*, 1968, **72**, 1017
68Ca L. H. Cohen, A. H. W. Aten, jun., and J. Kooi, *Inorg. Nuclear. Chem. Letters*, 1968, **4**, 429; criticized by S. Fried and D. Cohen, *ibid.*, p. 611
68D A. J. Dill, L. M. Itzkowitz, and O. Popovych, *J. Phys. Chem.*, 1968, **90**, 4580
68E A. J. Edwards, G. R. Jones, and R. J. C. Sills, *Chem. Comm.*, 1968, 1527
68G O. Ginstrup and I. Leden, *Acta Chem. Scand.*, 1968, **22**, 1163
68Ga A. A. Grinberg, S. Ch. Dkhara, and M. I. Gel'fman, *Zhur. neorg. Khim.*, 1968, **13**, 2199
68Gb R. J. Gillespie and M. J. Morton, *Chem. Comm.*, 1968, 1565
68Gc R. J. Gillespie, J. B. Milne, and M. J. Morton, *Inorg. Chem.*, 1968, **7**, 2221
68H R. Huston and J. N. Butler, *J. Phys. Chem.*, 1968, **72**, 4263
68I M. V. Ionin and V. G. Shverina, *Zhur. fiz. Khim.*, 1968, **42**, 1819
68K L. F. Kozin and A. G. Egorova, *Izvest. Akad. Nauk. Kazakh. S.S.R., Ser. khim.*, 1968, no. 3, 6
68L M. Lundkvist, *Acta Chem. Scand.*, 1968, **22**, 281
68La O. G. Levanda, I. I. Moiseev, and M. N. Vargaftig, *Izvest. Akad. Nauk. S.S.S.R., Ser. khim.*, **1968**, 2368
68Lb J. W. Larson, P. Cerutti, H. K. Garber, and L. G. Hepler, *J. Phys. Chem.*, 1968, **72**, 2902
68M J. D. E. McIntyre and M. Salomon, *J. Phys. Chem.*, 1968, **72**, 2431
68Ma R. C. Murray, jun., and P. A. Rock, *Electrochim. Acta*, 1968, **13**, 969
68Mb P. B. Mathur and S. M. A. Naqvi, *Indian J. Chem.*, 1968, **6**, 311
68Mc K. W. Morcom and B. L. Muju, *Nature*, 1968, **217**, 1046
68Md T. J. Meyer and H. Taube, *Inorg. Chem.*, 1968, **7**, 2369
68Me J. M. McIntyre and E. S. Amis, *J. Chem. and Eng. Data*, 1968, **13**, 371
68O M. Okada, R. A. Guidotti, and J. D. Corbett, *Inorg. Chem.*, 1968, **7**, 2118
68P G. Piccardi and R. Guidelli, *Ricerca sci.*, 1968, **38**, 46
68Pa R. A. Potts, R. D. Barnes, and J. D. Corbett, *Inorg. Chem.*, 1968, **7**, 2558
68S J. N. Spencer and A. F. Voigt, *J. Phys. Chem.*, 1968, **72**, 464, 471
68Sa A. M. Shkodin, T. P. Sogoyan, and L. I. Kozynyuk, *Ukrain. khim. Zhur.*, 1968, **34**, 135; A. M. Shkodin and T. P. Sogoyan, *ibid.*, p. 1234
68Sb A. Sousa-Alonso, I. Chadwick and R. J. Irving, *J. Chem. Soc. (A)*, 1968, 2364
68Sc K. Shaw and J. H. Espenson, *J. Amer. Chem. Soc.*, 1968, **90**, 6622
68Sd B. Scrosati, G. Pecci, and G. Pistoia, *J. Electrochem. Soc.*, 1968, **115**, 506
68T G. W. Tindall and S. Bruckenstein, *Analyt. Chem.*, 1968, **40**, 1402
68V W. M. Vogel, *Electrochim. Acta*, 1968, **13**, 1815
68Z D. M. Ziv, I. S. Kirin, V. A. Ishina, and A. F. Ivanchenko, *Radiokhimiya*, 1968, **10**, 329, 335, 339; D. M. Ziv, I. S. Kirin, G. M. Sukhodolov, and A. F. Ivanchenko, *ibid.*, p. 423

2 Hydroxide, OH⁻

Metal	*Method*	*Temp*	*Medium*	*Log of equilibrium constant, remarks*	*Ref*
H^+	gl	25	1—2000 atm	$K_1(p\text{ atm})/K_1(1\text{ atm})$ $-0{\cdot}17(p=500)$, $-0{\cdot}33(p=1000)$, $-0{\cdot}47(p=1500)$, $-0{\cdot}60(p=2000)$, $\Delta V_1 = 20{\cdot}4$ cm³	63Ha
	cal	25	0 corr	$\Delta H_1 = -13{\cdot}335$	63Hb
	cal	25	→ 0	$\Delta H_1 = -13{\cdot}337$	63Hc
	con	−17	ice	K_1 20·1 in ice, M units	63Lb
	cal	25	0 corr	$\Delta H_1 = -13{\cdot}336$	63Va
	H	25	I(KCl)	K_1 14·34(I = 3·72), 14·17(I = 2·99), 13·96(I = 2), 13·80(I = 1), 13·76(I = 0·5), 13·78(I = 0·1)	64Fb ≈ 33H
		25	c $BaCl_2$	K_1 13·61(c = 1·5), 13·56(c = 1), 13·55(c = 0·5), 13·58(c = 0·25)	
	H	25	$I(NaClO_4)$	K_1 14·42(I = 3·8), 14·20(I = 2·99), 13·97(I = 2), 13·80(I = 1), 13·74(I = 0·5), 13·78(I = 0·1)	64Fb
		25	c Na_2SO_4	K_1 12·77(c = 1·52), 12·78(c = 1), 12·88(c = 0·5), 13·01(c = 0·25), 13·24(c = 0·1)	
	gl	25	H_2O–D_2O	$K_1(x_D)$ diagram given for H_2O–D_2O mixt	64Se
	H	25	mNaCl	K_1 14·22(m = 6), 13·88(m = 3), 13·62(m = 1), m units	65Fa
	gl	25	2 (Na)NO_3	$K_1 \approx 13{\cdot}86$	65H

continued

Metal	*Method*	*Temp*	*Medium*	*Log of equilibrium constant, remarks*	*Ref*
H^+ (*contd*)	gl	25	3 ($LiClO_4$)	K_1 13·55(?)	65K
	gl	25	3 ($NaClO_4$)	K_1 14·00(?)	
	oth	25	→ 0	ΔV_1 = 22·11 ml	65Hf+/65D
	cal	25	EtOH var	ΔH_1 = −13·3(x = 0), −13·6(0·1), −12·6(0·2), −11·3(0·3), −10·0(0·4), −8·7(0·5), −7·4(0·6), −6·2(0·7), −5·2(0·8), −5·1(x = 0·9), in EtOH (mole fraction x)+H_2O(l)	66Bf
	con	500—1000	0	K_1 3·8(500°; 40 kbar), 2·8(750°; 52 kbar), 2·1(1000°; 64 kbar), for ρ^{-1} = 0·75 cm^3g^{-1} K_1 2·6(500°; 70 kbar), 1·7(750°; 80 kbar), 1·0(1000°; 104 kbar) for ρ^{-1} = 0·65 cm^3g^{-1}	66Ha
	cal	25	3 (Na)ClO_4	ΔH_1 = −13·05	67Aa
	cal	25	3 (Li)ClO_4	ΔH_1 = −13·5	67Ab
	H	25	0·1	K_1 13·96 (in 0·1M-KCl, Me_4NCl, or KNO_3)	67Ac
			1 (KCl)	K_1 13·96	
			1 ($NaClO_4$)	K_1 13·95	
			1 (Me_4NCl)	K_1 14·11	
	H, gl	25	var	K_1 13·96[3 M Na(ClO_4)], 13·88[2 M K(Cl)], 13·96[2 M (K)Cl]	67Bf
		35	1 Na(ClO_4)	K_1 13·86	
	gl, Ag	25	→ 0	K_1 13·994, m scale	67Gc
	ΔG	25—306	0 corr	K_1 13·995(25°), 13·27(50°), 12·27(100°), 11·58(156°), 11·19(218°), 11·46(306°) ΔH_1 = −13·335(125°), ΔS_1 = 19·31(25°)	/67H
	H	−60	NH_3(l)	K_1 16 in NH_3(l), 0·1M-KI	67Hb
	H?	25?	I CsCl	K_1 15·36(I = 6·9), 14·87(I = 5·34), 14·46(I = 4), 14·18(I = 3), 13·97(I = 2), 13·85(I = 1)	67La
			I KBr	K_1 14·53(I = 4·2), 14·16(I = 3)	
			I KI	K_1 15·05(I = 5·1), 14·68(I = 4), 14·29(I = 3), 14·02(I = 2), 13·83(I = 1)	
			I Me_4NCl	K_1 15·80(I = 4·6), 15·38(I = 4), 14·72(I = 3), 14·28(I = 2), 13·88(I = 1)	
			I Et_4NCl	K_1 15·65(I = 3·18), 15·35(I = 2·9), 15·00(I = 2·5), 14·78(I = 2·2), 14·22(I = 1·5)	
	H	20?	C_4H_8O	K_1 32 in C_4H_8O(THF)(l), 0·1M-Bu_4NClO_4	67Pa
	gl, H	25	7·75 Bu_4NBr	K_1 15·8(gl), 15·72(H)	67Sd
	cal	0—70	0 corr	ΔH_1 = −14·62(0°), −14·14(10°), −13·78(18°), −13·48(25°), −12·81(40°), −12·24(55°), −11·60(70°)	67V
		25	I($NaNO_3$)	ΔH_1 = −13·62(I = 2·98), −13·61(I = 1·00), −13·60(I = 0·51), −13·49(I = 0·042), −13·48(0 corr), also values for these I values and other temp	
	cal	25	→ 0	ΔH_1 = −13·350	68Gf
D^+	oth	13·5	$HClO_4$ var	K(H_2O+H_2DO^+ ⇌ HDO+H_3O^+) −0·02 (fractionation D/H, vapour/solution)	64Hc
	pHDO	13·5	NaOH var	K(H_2O+OD^- ⇌ HDO+OH^-) 0·62	64Hd
	D, Ag	5—50	D_2O, 0 corr	K_1 15·740(5°), 15·526 (10°), 15·326 (15°), 15·136(20°), 14·955(25°), 14·784(30°), 14·622(35°), 14·468(40°), 14·322(45°), 14·182(50°), in D_2O, m units; also K_1 values with M and x units.	66Cc

continued overleaf

2 OH⁻ (contd)

Metal	Method	Temp	Medium	Log of equilibrium constant, remarks	Ref
D^+ (contd)				$\Delta H_1 = -15{\cdot}37(5°), -14{\cdot}31(25°), -12{\cdot}88(50°)$, $\Delta S_1 = 16{\cdot}76(5°), 20{\cdot}43(25°), 25{\cdot}03(50°)$, $\Delta C_p = 54{\cdot}8(25°)$ *m* units	
	ΔG	5—45	0	$K(2D_3O^+ + 3H_2O \rightleftharpoons 2H_3O^+ + 3D_2O)$ 1·01(5°), 0·98(15°), 0·95(25°), 0·93(35°), 0·90(45°)	/66Sa
	gl, Ag	25	$D_2O, \to 0$	K_1 14·856, *m* scale, in D_2O, also values for D_2O–H_2O mixt	67Gc
	cal	25	$D_2O, \to 0$	ΔH_1 $(D^+ + OD^- \to D_2O) = -14{\cdot}488$ in D_2O (l), $\to 0$	68Gf
Li^+	H	25	3($NaClO_4$)	K_1 −0·18	64O
	act	25?	0 corr	K_1 −0·1	/66Mb
	H	20?	C_4H_8O	K_1 11 in C_4H_8O(THF)(l), 0·1M-Bu_4NClO_4	67Pa
Na^+	H	20?	C_4H_8O	K_1 8 in C_4H_8O(THF)(l), 0·1M-Bu_4NClO_4	67Pa
Group $2A^{2+}$	sol	700—900	MCl(l)	K_s $(M^{2+}O^{2-})$ in (Na, K)Cl(l) given for M = Ca, Sr, Ba	66N
Be^{2+}	H, gl, qh	25	3 (Na)ClO_4	$*\beta_{33}$ −8·664, $*\beta_{12}$ −3·22, $*\beta_2$ −10.87, no ev (2,2) or (4,4) cpx	61Ca+56Ka/64Ha
	oth	20	0·1 (KCl)	$*K_1$ −5·68, $*K_2 \leqslant -6{\cdot}7$	64W
		20	0·1 ($NaClO_4$)	$*K_1$ −5·71, $*K_2 \leqslant -6{\cdot}7$ (gl, rapid flow)	
	gl	?	0·5 ($NaClO_4$)	$*\beta_{33}$ −8·81, $*\beta_{12}$ −3·24, $*\beta_2$ −11·0	65Ba
	dis, gl	25	dil	$*\beta_2$ −13·65, $*\beta_3$ −24·11, no ev $BeOH^+$	65Ga
	oth	25	dil	ev $Be_3(OH)_3{}^{3+}$ (coagulation method)	65M
	sol	240—510	$M(NO_3)$(l)	$\Delta H[BeO(s) + H_2O + 2OH^- \rightleftharpoons Be(OH)_4{}^{2-}] = 4{\cdot}8$ $\Delta S = 7{\cdot}5$, ev $Be(OH)_4{}^{2-}$ possibly $Be_4O(OH)_{12}{}^{6-}$ in $K_{0.4}Na_{0.6}(NO_3,OH)$(l)	65Sf
	gl	25	0·5 $NaClO_4$	$*\beta_{33}$ −8·81, $*\beta_{12}$ −3·20, $*\beta_2$ −11·0	67Be
	fp	32	Na_2SO_4 sat	ev $Be_3(OH)_3{}^{3+}$	
	H, qh	0—60	1 NaCl	$*\beta_{33}$ −10·08(0°), −8·91(25°), −7·67(60°), $\Delta H = 16{\cdot}0, \Delta S = 15{\cdot}3$; $*\beta_{12}$ −3·64(0°), −3·43(25°), −2·93(60°), $\Delta H = 5{\cdot}0, \Delta S = 1{\cdot}4$; $*\beta_{75}$ −28·66(0°), −25·33(25°), −22·11(60°), $\Delta H = 45{\cdot}3, \Delta S = 35{\cdot}2$	67Ma

1964 *Erratum*: p. 41, *in ref.* 29P *delete* "27P" *from the second line*

Metal	Method	Temp	Medium	Log of equilibrium constant, remarks	Ref
					67O
	gl	25	dioxan–H_2O	$*K_1 \leqslant -6$, $*\beta_{12}$ −3·66, $*\beta_2$ −10·84, $*\beta_{22}$ −7·15, $*\beta_{33}$ −8·75 [in 0·2 dioxan + 0·8 H_2O, 3M-(Li)ClO_4]	
	gl	25	35% dioxan	$*\beta_{33}$ −8·65, $*\beta_{12}$ −3·29, $*\beta_2$ −11·5 in 35% dioxan (10 mol %), I = 3($LiClO_4$)	67Oa
	gl	25	3 ($LiClO_4$)	$*\beta_{33}$ −8·74, $*\beta_{12}$ −3·27, $*K_1 \leqslant -5{\cdot}4$, $*\beta_2$ −11·5	
	gl	25	2 (K)NO_3	$*\beta_{12}$ −3·28, $*\beta_{33}$ −8·90, $*\beta_{43}$ −16, $*\beta_{86}$ −27·5, $*\beta_{96}$ −34·5	68L
	gl	25	3 KCl	$*\beta_{12}$ −3·18, $*\beta_{33}$ −8·91	68Pa

1964 *Addendum*: p. 41, *for ref.* "13K" *read* "13K ≈ 04Ka"

Metal	Method	Temp	Medium	Log of equilibrium constant, remarks	Ref
Mg^{2+}	sol, gl	25	0 corr	K_{s0} −11·15, K_1 2·60	63Hd
	X		solid	ev$(Et_2OMg)_4Br_6O$ (Mg_4O in centre) in solid	64Si
Ca^{2+}	gl	75	$Ca(NO_3)_2(H_2O)_4$	$*K_1$ −10·3, $K_s[\tfrac{1}{2}Ca(OH)_2(s)(+\tfrac{1}{2}Ca^{2+}) \rightleftharpoons CaOH^+]$ −1 to −1·3 in $Ca(NO_3)_2(H_2O)_4$(l)	65C

continued

Metal	Method	Temp	Medium	Log of equilibrium constant, remarks	Ref
Ca^{2+}	cal	25	0 corr	$\Delta H_{s0} = -4{\cdot}36$, $\Delta H_1 = 2{\cdot}0$	65Hd
(*contd*)		25	0 corr	$\Delta H_{s0} = -4{\cdot}29$, $\Delta S_{s0} = -38{\cdot}2$, $\Delta S_1 = 12{\cdot}3$	65Hd+59Be
	sol	0—350	0 corr	K_{s0} −4·88(0°), −5·03(25°), −5·25(50°), −5·52(75°), −5·83(100°), −6·57(150°), −7·43(200°), −8·37(250°), −9·38(300°), −10·45(→ 350°); $\Delta H_{s0} = -3{\cdot}04(25°), -31{\cdot}4(300°)$; $\Delta S_{s0} = -33{\cdot}2(25°), -97{\cdot}7(300°)$; $\Delta C_p = -62{\cdot}4(25°), -144(300°)$; values also for intermediate temp	67Y
Sr^{2+}	H, Ag	25	0	$\Delta H_1 = 1{\cdot}16$, $\Delta S_1 = 7{\cdot}6$, $\Delta C_p = 31$	37Ha/67H
Ba^{2+}	cal	25	0 corr	$\Delta H_{s0} = 13{\cdot}7$, $\Delta S_{s0} = 29$, $\Delta H_1 = 1{\cdot}1$, $\Delta S_1 = 14$ (K_1 2·2, agrees better than $K_1 \sim 0{\cdot}6$)	65He+
	H, Ag	25	0	$\Delta H_1 = 1{\cdot}23$, $\Delta S_1 = 7{\cdot}1$, $\Delta C_p = 74$	37Ha/67H
Sc^{3+}	gl	25	1 ($NaClO_4$)	$*K_1$ −5·14, $*\beta_{22}$ −6·12, $*\beta_{43}$ −14·18, $*\beta_{53}$ −17·48	54Ka/66Aa
	lit	25	0 corr	K_{s0} −29·7	/62Ae
	sol, gl	25	0 corr	K_{s0} −29·70	/63Ab
	sol, gl	20	0 corr	K_{s0} −29·0 to −29·1(K_w −14·00)	63Kb
	sol, gl	25	1 (Na)ClO_4	$*K_{s0}$ 10·50	63S+
	gl		1 KNO_3	ev $Sc_2(OH)_2{}^{4+}$	64Fa
	fp		KNO_3 sat	ev $Sc_2(OH)_2{}^{4+}$	
	gl	25	1 ($NaClO_4$)	$*K_1$ −5·11, $*\beta_{22}$ −6·14, $*\beta_{43}$ −13·00, $*\beta_{53}$ −17·47	66Aa
	cfu	25	1 ($NaClO_4$)	no ev cpx with > 3 Sc	
	sp, gl	20	0·1	$*K_1$ −4·90, $*K_2$ −5·78, $*K_3$ −6·58	68Ad
	sol, ir	25	2—14 KOH	ev ani cpx, ev $K_2Sc(OH)_5(H_2O)_4(s)$	68I
Y^{3+}	H	25	0 corr	K_{s0} −24·5	60E
	qh, gl	25	3 (Li)ClO_4	$*K_1$ −9·1, $*\beta_{22}$ −14·30, $*\beta_{53}$ −33·8	64Bb
	X		var	ev polyn cpx in solution	65V
	gl	25	0·3 ($NaClO_4$)	$*K_1$ −8·34[$K_1(H^+)$ 13·73]	66Fa
	tyn	20	dil	K_{s0} −25·7	66Oa
Group $4f^{3+}$	hyp	25	3 ($LiClO_4$)	K_1 4·3(Ce), 4·7(Eu), 4·5(Gd), 4·9(Lu) *etc.* [4·1(La)]	61B/65Ka
	gl	25	0·3 ($NaClO_4$)	$*K_1$ −9·06(La), −8·55(Pr), −8·43(Nd), −8·34(Sm), −8·31(Eu), −8·35(Gd), −8·16(Tb), −8·10(Dy), −8·04(Ho), −7·99(Er), −7·95(Tm), −7·92(Yb), −7·90(Lu), [$K_1(H^+)$ 13·73]	66Fa
	tyn	20	dil	K_{s0} −20·1(La), −22·3(Pr), −23·9(Nd), −24·5(Sm), −24·2(Gd), −23·6(Dy), −23·9(Er), −25·1(Yb), −25·3(Lu)	66Oa
	tyn	rt	"10% sea water"	K_{s0} −27·7(La), −27·8(Pr), −27·1(Nd, Sm), −27·5(Gd, Dy), −28·4(Yb), −27·5(Lu), −27·4(Y); K_{s3} −5·3(La), −4·9(Pr), −5·1(Nd), −4·9(Sm, Eu, Gd, Dy, Yb), −5·1(Lu), −4·8(Y) hence β_3 22·4(La) *etc.*, also values for "25, 50, and 75% sea water"	67P
La^{3+}	gl	?	→ 0	K_{s0} −22·60	57Mf
	H	25	0 corr	K_{s0} −21·5	60E = 59S
	gl	25	3 (Li)ClO_4	$*K_1$ −10·12, $*\beta_{12}$ −9·98, $*\beta_{95}$ −71·43	61B/64L
	sol, qh	20	→ 0	K_{s0} −27·71 [$K_1(H^+)$ 14·07]	64K
	sol		0·38 $NaNO_3$	K_{s0} −19·19 (fresh), −20·89 (fresh, 0 corr)	65Z
	sol	25	NaOH var	no ev ani cpx	66Ia

continued overleaf

2 OH^- (contd)

Metal	*Method*	*Temp*	*Medium*	*Log of equilibrium constant, remarks*	*Ref*
Ce^{3+}	gl	?	$\rightarrow 0$	K_{s0} −24·40	57Mf
	gl	25	3 (Li)ClO_4	$^*\beta_{53}$ −35·75	64Bc
		25	0·005 (KNO_3)	$^*K_1$ −4·2(?) (K_w −13·83)	65S
	gl	tp	dil	$^*K_1$ −9·29 (one solution)	67S
Ce^{4+}	red	25	$I(ClO_4^-)$	ev $Ce_2(OH)_4^{4+}$	65Hb
	sp	25	I(Na)ClO_4	$^*K_1$ −0·7(I = 0·9 to 1·7), no ev polynuclear cpx	66O
	gl, red	25	3 (Na)NO_3	$^*\beta_{32}$ −1·68, $^*\beta_{42}$ −2·29, ev cpx with > 2Ce, best agreement with $^*\beta_{12.6}$ −1·98 (misprint in synopsis)	67D
	sp	25	$MeCO_2H$ var	ev trimer such as $Ce_3O_3(MeCO_2H)^{6+}$ in $MeCO_2H$, equil constants given	68W
Pr^{3+}	var	25	0 corr	K_{s0} −22·08(sol, gl, Ag)	63Aa = 62A
	sol, gl	25	0 corr	K_{s0} −22·08	/63Ab
Nd^{3+}	H	25	0 corr	K_{s0} −23·26	60E = 62Ae
	sol, gl	25	0 corr	K_{s0} −23·89	/63Ab
	gl, pol	20	0 corr	K_{s0} −23·92	67Ad
Sm^{3+}	H	25	0 corr	K_{s0} −23·89	60E
	pol, gl	20	0 corr	K_{s0} −25·20	66Ka
Gd^{3+}	sol, gl	25	0 corr	K_{s0} −26·89	/63Ab
	sol	25	NaOH var	$K_{s3} \approx -4$, ev weak ani cpx?	66Ia/
Dy^{3+}	sp, sol	20	0 corr	K_{s0} −27·90	64Kc
	sol	25	NaOH var	$K_{s3} \approx -2{\cdot}7$, ev weak ani cpx	66Ib/
	pol, gl	20	0 corr	K_{s0} −25·00	66Ka
Er^{3+}	sol	25	NaOH var	$K_{s3} \approx -2{\cdot}9$, ev ani cpx for $[OH^-]$ > 12M	66Ib/
	sp, sol	20	0 corr	K_{s0} −26·71	66K
Yb^{3+}	sol	25	NaOH var	K_{s3} −4·4, ev ani cpx for $[OH^-]$ > 8M	66Ia
	gl, pol	20	0 corr	K_{s0} −25·06	67Ad
Lu^{3+}	sol, gl	25	0 corr	K_{s0} −27·00	/63Ab
	sol	25	NaOH var	$K_{s3} \approx -3{\cdot}2$, ev ani cpx for $[OH^-]$ > 6M	66Ib/
	pol, gl	20	0 corr	K_{s0} −26·85	68Kb
Ac^{3+}	sol		0 corr	K_{s0} −18·68 (fresh), −20·89(aged)	65Z
Ti^{3+}	gl	25	3 KBr	$^*K_1$ −2·55, $^*\beta_{22}$ −3·30	68Pb
Ti^{4+}	gl		$\rightarrow 0$	K_{s0} −53·10	57Mf
	X		solid	ev tetrahedral TiO_4^{4-} in Ba_2TiO_4(s)	61Bf
	sol	18	0·1 ($NaClO_4$)	K_{s4} −5·5	64Nd
	cix	18	0 corr	K_1 18·0, K_2 17·2, K_3 12·5, K_4 11·0(K_w ?)	
	cix	18	$TiCl_4$ var	ev TiO^{2+} at pH > 0·65, Ti^{4+} at pH < 0·32	64Z
Zr^{4+}	dis, pH	25	0 corr	K_1 14·59, β_2 29·31, β_3 43·64(K_w −13·90) [0 corr from 1 M(Na)ClO_4]	61Pa/66Sb
	tyn	20	sea water	K_{s4} −4·6 in sea water, various dilutions	66Ba
	tyn	20	1 $NaClO_4$	$^*K_{s0}$ [$Zr(OH)_4(s) + 4H^+ \rightleftharpoons Zr^{4+} + 4H_2O$] ≈ 3·8 K_{s4} −4·36, ev(+1)cation, *e.g.* $^*K_{s3}$ 0·2	66Bd
		20—40	dil	$^*K_{s0} \approx$ 4·6(20°), 5·05(40°), K_{s4} −3·9(20°, 40°) *e.g.* K_{s3} 0·27(20°), 0·06(40°)	
	tyn	20	"50% sea water"	K_{s4} −4·6	66Bh
	X		solid	ev $Zr_2(OH)_2^{6+}$ groups in $Zr_2(OH)_2(SO_4)_3(H_2O)_4$(s)	66M
	dis	25	0 corr	K_1 14·58, β_2 29·38, β_3 43·72, β_4 57·85 (K_w −13·90) (?) [0 corr from 1M-(Li)ClO_4]	66Sb
		25	0 corr	$^*K_1$ 0·68, $^*K_2$ 0·90, $^*K_3$ 1·13, $^*K_4$ 1·13	66Sb+/67Sb
	fp	−3	KNO_3 sat	ev cpx with, on the average, 7—20 Zr	68Kc

continued

Metal	*Method*	*Temp*	*Medium*	*Log of equilibrium constant, remarks*	*Ref*
Hf^{4+}	dis	25	HNO_3 var	ev Hf_2 cpx and Hf_n chains, $K(Hf_n+Hf \rightleftharpoons Hf_{n+1}) \approx 1{\cdot}5$	64La
	oth	25	3 (Na)Cl	ev slow reactions, probably cpx with ⩽ 4 Hf and fine-grained HfO_2 (light scattering)	68Cb
Th^{4+}	H, qh	25	1 (Na)ClO_4	$*\beta_1$ −3·71, $*\beta_{22}$ −4·44 (misplaced in table), $*\beta_{84}$ −18·78, $*\beta_{15.6}$ −36·42	54H/68H
	gl	25	1 (Na)ClO_4	$*\beta_1$ −4·12, $*\beta_2$ −7·81, $*\beta_{22}$—4·61, $*\beta_{84}$ −19·01, $*\beta_{15.6}$ −36·76, *m* units	54K/65Bc
	gl, H	25	3 (Na)Cl	$*\beta_{21} \approx$ −9·1, $*\beta_{12} \approx$ −2·65, $*\beta_{22}$ −4·70, $*\beta_{32}$ −8.83, $*\beta_{14,6}$ −36·53, $*\beta_{15,6}$ −40·37, no ev cpx (8, 4), (12, 5), (17, 7), *etc.*	64Hb
	sol, gl	17	0·1 (Na)ClO_4	K_{s4} −6·32, K_1 9·4, K_2 8·9, K_3 8·1, K_4 8·1, K_{s0} −41·14 (K_w −14·00 ?)	64Nc
	qh, gl	0	1 (Na)ClO_4	$*\beta_1$ −4·32, $*\beta_2$ −8·48, $*\beta_{22}$ −5·60, $*\beta_{84}$ −22·79, $*\beta_{15,6}$ −43·84, *m* units	65Bc
	gl, H	95	1 (Na)ClO_4	$*\beta_1$ −2·29, $*\beta_2$ −4·50, $*\beta_{22}$ −2·55, $*\beta_{84}$ −10·49, $*\beta_{15,6}$ −20·63, *m* units	
	sol	95	1 (Na)ClO_4	$*K_{s0}$ [$ThO_2(s)+4H^+ \rightleftharpoons Th^{4+}+2H_2O$] 4·26, $*\beta_1$ −2·26, $*\beta_2$ −4·54, *m* units	
		25	1 (Na)ClO_4	$\Delta^*H_1 = 5{\cdot}9$, $\Delta^*H_{\beta 2} = 13{\cdot}9$, $\Delta^*H_{\beta 22} = 14{\cdot}8$, $\Delta^*H_{\beta 84} = 57{\cdot}7$, $\Delta^*H_{\beta 15,6} = 108{\cdot}5$, $\Delta^*S_1 = 0{\cdot}9$, $\Delta^*S_{\beta 2} = 11{\cdot}0$, $\Delta^*S_{\beta 22} = 28{\cdot}4$, $\Delta^*S_{\beta 84} = 106{\cdot}6$, $\Delta^*S_{\beta 15,6} = 195{\cdot}7$	
		0	1 (Na)ClO_4	$*\beta_1$ −4·31, $*\beta_2$ −8·46, $*\beta_{22}$ −5·59, $*\beta_{84}$ −22·80, $*\beta_{15,6}$ −43·81	65Bc/68H
		25	1 (Na)ClO_4	$*\beta_1$ −4·23, $*\beta_2$ −7·69, $*\beta_{22}$ −4·61, $*\beta_{84}$ −19·16, $*\beta_{15,6}$ −37·02	
		95	1 (Na)ClO_4	$*\beta_1$ −2·25, $*\beta_2$ −4·51, $*\beta_{22}$ −2·59, $*\beta_{84}$ −10·44, $*\beta_{15,6}$ −20·61, no attempt to test other models	
	oth	25	1 $NaClO_4$	ev $ThOH(ClO_4)_2^+$, species with 2—3 Th(1·6—1·9 OH/Th) and 140—150 Th(3 OH/Th) (light scattering)	65Ha
	tyn	20	sea water	K_{s4} −5 in sea water, various dilution	66Ba
	tyn	20	"50% sea water"	K_{s4} −5	66Bh
	cfu	25	1 (Na)ClO_4	ev polyn cpx, slow reactions for > 2·5 OH^-/Th	66H
	cfu	25	1 (Na)Cl	ev polyn cpx, slow reactions for > 1 OH^-/Th	
	fp	−3	NH_4ClO_4 sat	ev cpx with up to ≈ 7Th	66Hb
	cix, gl	25	0·5 ($NaClO_4$)	K_1 11·64, K_2 10·80, K_3 10·62, K_4 10·45 (K_w ?)	67Bc
	qh	25	4 (Na)NO_3	$*\beta_{22}$ −5·5, $*\beta_{63}$ −17·92, $*\beta_{12,4}$ −37·2,	68Da
	qh	25	4 (Na)NO_3	$*\beta_{12}$ −2·72, $*\beta_{53}$ −12·42, $*\beta_{42}$-10·49, $*\beta_{84}$ −19·2, $*\beta_{14,6}$ −36·2	
	H, gl	25	3 (Na)Cl	ev cpx with 6Th, best agreement with ("IVB") $*\beta_1$ −5·0, $*\beta_{22}$ −4·76, $*\beta_{32}$ −8·94, $*\beta_{52}$ −16·99, $*\beta_{13}$ −1·36, $*\beta_{33}$ −6·83, $*\beta_{84}$ −21·11, $*\beta_{14,6}$ −36·58, $*\beta_{25,10}$ −65·35	68H
	X		solid	ev binuclear $[(H_2O)_3(NO_3)_3Th]_2(OH)_2$ units in $ThOH(NO_3)_3(H_2O)_4(s)$	68J
	X		2 Th(ClO_4)	ev Th_2 cpx for up to 1 OH/Th, ev cpx with ⩾ 4Th for higher OH/Th ratios	68Ja
V^{2+}	pol	25	0·2 ($NaClO_4$)	ev V_2O^{2+} predom at 0·1M-H^+ (?)	63Bg
	H	15—35	VSO_4 var	$*K_1$ −6·85(15°), −6·49(25°), −6·10(35°), $\Delta^*H_1 = 15{\cdot}5$	64Pb
V^{3+}	gl	22·5	→ 0	$*K_1$ −2·4, $*K_2$ −3·85	63Pc
	gl	20	3 (Na)Cl	$*K_1$ −3·15, $*\beta_2$ −7·3, $*\beta_{22}$ −4.1	63Pd
	gl	20	1 (Na)Cl	$*K_1$ −2·85, $*\beta_2$ −6·7, $*\beta_{22}$ −3·9	
	sp	23	1 ($LiClO_4$)	ev $V_2(OH)_2^{4+}$	64Nb

continued overleaf

2 OH^- (contd)

Metal	Method	Temp	Medium	Log of equilibrium constant, remarks	Ref
V^{3+}	sp		1 (Na)Cl	ev VOH^{2+}, $V_2(OH)_2^{4+}$, other polyn cpx	65P+
(contd)	H, gl	25	3 (K)Cl	$^*K_1$ −3·07, $^*\beta_{22}$ −3·96, $^*\beta_2 \approx$ −7·5, $^*\beta_{32} \approx$ −8·7	66B
	X		solid	ev $V(OH)_6^{3-}$ in $Sr_3[V(OH)_6]_2(s)$	66Mc
	sp	20	1 (Na)Cl	agrees with 63Pc, 63Pd, ev larger cpx, $V_3(OH)_4^{5+}$?	66P
	mag	20	3 (Na)Cl	ev$[aqV(OH)_2Vaq]^{4+}$, not $(aqVOVaq)^{4+}$	66Pa
	H	25	3 (K)Cl	$^*K_1$ −3·07, $^*\beta_{22}$ −3·93, $^*\beta_{32}$ −8·0	68D
VO^{2+}	nmr		H^+ var	ev VOH^+ for $[H^+] \geqslant 1$ M	63Ra
	X		solid	ev VO_4^{4-} in $Ba_3VO_5(s)$	65Ma
Pa^{4+}	dis	25	3 (Li)ClO_4	$^*K_1$ −0·14, $^*\beta_2$ −0·52, $^*\beta_3$ −1·77, ev Pa_3 cpx	65Gc=66G
	dis	25	3 (Li)ClO_4	$^*K_1$ −0·14, $^*\beta_2$ −0·52, $^*\beta_3$ −1·6 to −2·0, $^*\beta_4$ −5·3, ev polyn cpx	68Gc
Pa^{5+}	dis	25	3 (Li)ClO_4	ev $Pa(OH)_3^{2+}$ (or PaO_2H^{2+}) dominant for 1—3 M H^+ ev $Pa(OH)_4^+$ or PaO_2^+ dominant for 0·001—0·003 M-H^+	64Gd
	dis	25	3 (Li)ClO_4	$K_a[Pa(OH)_4^+]$ −1·05	65Gc
	dis, sp	25	3 (Li)ClO_4	ev $PaOOH^{2+}$ (unstable), $Pa(OH)_3^{2+}$ (stable), ev polyn cpx with charge 1+/Pa but not with 2+/Pa, $K[M^+ + M_{n-1}{}^{(n-1)+} \rightleftharpoons M_n{}^{n+}]$ 7·2, $M^+ = Pa(OH)_4^+$, ev slow irreversible reaction	65Gd
	dis	20	HNO_3 var	ev slow formation of inert polyn cpx	65Sg
	dis, sp	25	3 (Li)ClO_4	PaO_2H^{2+}: $^*K_1$ −1·05, $^*\beta_2$ −5·55, no ev $(PaO_2H^{2+})_n$, ev $[PaO(OH)_2^+]_n$, $n > 1$	66G
	dis	25	3 (Li)ClO_4	$K_a[PaO_2(OH)_2^+]$ −5·55$[K_a(PaO_2H^{2+})$ −1·05]	68Gd
Cr^{3+}	oth	25	var	ev $(H_2O)_4Cr(OH)_2Cr(H_2O)_4^{4+}$($^{18}O$ exchange)	64Kb
	fp		5·3 $HClO_4$	ev Cr_3–OH cpx	64Lb
	var		var	ev Cr_2OH^{5+},$Cr_2(OH)_2^{4+}$,$Cr_3(OH)_4^{5+}$ (green, linear) (sp, mag, fp, gl, esr, cix)	64Ta
	gl, cix	37—68	1($NaClO_4$)	$^*K_1$ −3·60(37·5°), —3·42(50°), −3·09(67·5°); $^*\beta_{22}$ −4·54(37·5°), −4·24(50°), −3·72(67·5°), $\Delta^*H_{\beta 22}$ = 12·5; $^*\beta_{43}$ −7·54(37·5°), −6·73(50°), −5·9(67·5°), $\Delta^*H_{\beta 43}$ = 24·9	
	oth	20	0·1 ($NaClO_4$)	$^*K_1$ −4.21, $^*K_2 \leqslant$ −5·8 (gl, rapid flow)	64W
	sp, qh	25	2 ($NaClO_4$)	$^*\beta_{22}$ −3·35, $^*\beta_{44}$ −5·11, $^*\beta_{64}$ −10·91 (misprints in eq. 15—17) ev mixed cpx Cr–OH–Al	67Sc
	X		solid	ev $Cr_4(OH)_6(NH_3)_{12}^{6+} = R^{6+}$ in $RCl_6(H_2O)_4(s)$	68B
	sp, gl	25	1 ($NaClO_4$)	$^*\beta_{22}$ −3·46, ev $Cr_{n+1}(OH)_{2n}^{3+n}$	68M
Cr en_3^{2+}	sp	25	*w*% dioxan	K_1 0·8($w \rightarrow 0$), 1·41(w = 20), 1·85(w = 30), 2·48(w = 40) in *w*% dioxan	67Cb
$Cr(NH_3)_5H_2O^{3+}$					
	gl	20	0·1 ($NaNO_3$)	$^*K_1$ −5·10(H_2O), −5·26(D_2O), −5·28(20% dioxan)	68Cc
$Cr(H_2O)_5Cl^{2+}$			?	$^*K_1$ −5·2 (not found in ref 64W as quoted)	/62Sd
Cr^{4+}	*X*		solid	ev CrO_4^{4-} in $Ba_3CrO_5(s)$	65Ma
Cr^{5+}	*X*		solid	ev CrO_4^{3-} in $Li_3CrO_4(s)$ and several other solids	65W
Mo^{4+}	var		var	ev $MoOOH^+$ (pre, pol, gl, fp)	66Se
Mo^{5+}	fp	32	Na_2SO_4 sat	ev $Mo_4O_8^{4+}$	65Sj
	var		var	ev $[HOOCl_2Mo(OH)_2MoCl_2OOH]^{2-}$ (paramagnetic) $(H_2O)_2HOOMoO_2MoOOH(H_2O)_2$ (diamagnetic) (mag, pre, sp)	65Wa

continued

Metal	Method	Temp	Medium	Log of equilibrium constant, remarks	Ref
U^{4+}	sp	25	0 corr	$*K_1$ $-1{\cdot}11$	50K/67Sb
	sp	20	1 (Li)ClO_4	$*K_1$ $-1{\cdot}57$	64M
	ana	700	MCl(l)	$K[UO_2(s)+UCl_4 \rightleftharpoons (UOCl_2)_2]$ $-0{\cdot}88$ in $Na_{\frac{1}{2}}K_{\frac{1}{2}}Cl(l)$, x units	64Na
UO_2^+	sp	0—25	2(H,Li)ClO_4	$K(UO_2^+ + UO_2^{2+} \rightarrow U_2O_4^{3+})$ 1.20 (25°) $\Delta H = -1{\cdot}9$, $\Delta S = -1$	65N
UO_2^{2+}	X		solid	ev UO_6^{6-} units in $Ba_3UO_6(s)$	54Ra

1964 *Erratum*: p. 50, *line ending with* "49A/62Rb" *should come immediately below line ending with* "49A/56R"

Metal	Method	Temp	Medium	Log of equilibrium constant, remarks	Ref
	cal	25	3 (Na)ClO_4	$\Delta^*H_{\beta 22} = 9{\cdot}5$, $\Delta^*H_{\beta 53} = 24{\cdot}4$, $\Delta^*S_{\beta 22} = 4{\cdot}3$, $\Delta^*S_{\beta 53} = 6$	62S+/68Ac
	cal	25	3 (Na)ClO_4	$*\beta_{22}$ $-6{\cdot}02$, $*\beta_{53}$ $-16{\cdot}54$	62S/68Ac
	sp, gl	25	0·1 ($NaClO_4$)	$*\beta_{22}$ $-6{\cdot}09$(gl), $-6{\cdot}28$(sp), $*\beta_{12}$ $-2{\cdot}5$(gl), $-1{\cdot}9$(sp)	64Ba
	sp	22—25	w% EtOH	$*\beta_{22}$ $-6{\cdot}2$ (in 30% EtOH), $-4{\cdot}8$ (in 50% EtOH), 0·1M-($NaClO_4$)	
	cal	25	6 HNO_3	$\Delta H[UO_3(s)+2H^+ \rightleftharpoons UO_2^{2+}+H_2O] = -18{\cdot}61(\alpha\text{-}UO_3)$, $-17{\cdot}89(\beta\text{-}UO_3)$, $-17{\cdot}04(\gamma\text{-}UO_3)$, $-18{\cdot}65(\varepsilon\text{-}UO_3)$, $-20{\cdot}8(A\text{-}UO_3)$, $-11{\cdot}15[\alpha\text{-}UO_3(H_2O)_{0.85}]$, $-11{\cdot}40[\beta\text{-}UO_2(OH)_2]$, $-11{\cdot}97[\varepsilon\text{-}UO_2(OH)_2]$	64Cc
	var	rt	1 $NaClO_4$	ev only (?) $(UO_2)_2(OH)_2^{2+}$ and $(UO_2)_2(OH)_3^+$ (sp, lit, pol, gl)	65Bb
	pH	?	var	equil constants for $(UO_2)_2(OH)_2^{2+}$, $(UO_2)_4(OH)_6^{2+}$, $UO_2(OH)_2$, $UO_2(OH)_3^-$, and $UO_2(OH)_4^{2-}$ calculated from a single titration curve (?)	65Ia
	sol, gl	rt	1 $NaClO_4$	$K_s[Na^+_{0.14}UO_2^{2+}(OH^-)_{2.14}]$ $-23{\cdot}92$	66Bb
	tyn, gl	rt	dil	$K_s[K^+_2(UO_2^{2+})_7(OH^-)_{16}]$ $-7\times 23{\cdot}5$	66Ta
	var		var	ev$(UO_2)_6(OH)_{10}(H_2O)_{14}^{2+}$ (pre, con?, fp?, hyp)	66Wa
	sol, gl	20—50	0 corr	K_{s0} $(UO_2^{2+}OH^-{}_2)$ $-21{\cdot}12(20°)$, $-20{\cdot}87(25°)$, $-20{\cdot}63(30°)$, $-20{\cdot}25(40°)$, $-20{\cdot}06(50°)$, also $K_{s1}(UO_2OH^+OH^-)$ values (?, polynuclear species neglected). Same $K_{s0}(\pm 0{\cdot}1)$ for $UO_3(s)$, $UO_2(OH)_2(s)$, $UO_2(OH)_2H_2O(s)$	67Ga
	fp	32	Na_2SO_4 sat	ev cpx with up to 8 U	67J
	gl	25	5 (Mg)NO_3	$*K_1$ $-5{\cdot}53$, $*\beta_{22}$ $-6{\cdot}52$, $*\beta_{53}$ $-17{\cdot}76$	68S
		25	3 (Mg)NO_3	$*K_1$ $-5{\cdot}38$, $*\beta_{22}$ $-6{\cdot}34$, $*\beta_{53}$ $-17{\cdot}37$	
Np^{4+}	oth	25	dil	ev cat and unch cpx such as $NpOH^{3+}$ and $Np(OH)_4$ (coagulation method)	66Wc
NpO_2^+	sp	50	5 (Y)ClO_4,2H^+	$K(NpO_2^+ + Cr^{3+} \rightleftharpoons Np^VCr^{III})$ 0·33	64Sc
		25—50	8 (Mg,ClO_4)2H^+	$K(NpO_2^+ + Cr^{3+} \rightleftharpoons Np^VCr^{III})$ 0·4(25°), 0·43(35°), 0·30(50°) $\Delta H = -3{\cdot}3$, $\Delta S = -9{\cdot}0(25°)$	
	sp	25—50	8 (Mg,ClO_4)	$K(NpO_2^+ + Rh^{III} \rightleftharpoons Np^VRh^{III})$ 0·52(25°), 0·37(35°), 0·33(50°), $\Delta H = -3{\cdot}6$, $\Delta S = -10$	67M
NpO_2^{3+}	pre, sp		var	ev NpO_5^{3-} in solids and alkaline solution ev NpO_2^{3+} in acidic solution	68K
	sol	20	1 NaOH	$K_s[Co(NH_3)_6^{3+}NpO_5^{3-}]$ $-7{\cdot}29$	
Pu^{4+}	sol, gl	24	0 corr	K_{s0} $-56{\cdot}3$ to $-47{\cdot}3$ ($\rightarrow$ 0)	65Pa
PuO_2^+	sp		NO_3 var	no ev cpx at pH = 2—6	64Gb = 65Gb
	var		NO_3 var	$K_{s0}(PuO_2^+OH^-)$ $-9{\cdot}3$, ev ani cpx at pH $\approx$ 8 (sol, gl, sp, tp)	68Z

continued overleaf

2 OH^- (contd)

Metal	*Method*	*Temp*	*Medium*	*Log of equilibrium constant, remarks*	*Ref*
PuO_2^{2+}	sol, gl	24	OH^- var	ev ani cpx, $(PuO_2)_2(OH)_5^-$?	65Pa
Pu^{7+}	pre, sp		var	ev PuO_5^{3-} in solids and solution (1—5M-NaOH)	68Ka
Mn^{2+}					

1964 *Addendum* p. 52, *for ref* "13K" *read* "13K ≈ 04Ka"

Metal	*Method*	*Temp*	*Medium*	*Log of equilibrium constant, remarks*	*Ref*
	H, gl	25	1 (Na)SO_4	$^*K_1$ −10·5, $^*\beta_{12}$ −9·9, $^*\beta_{32}$ −25·4	68F
Mn^{3+}	sp	23	6 (Na)ClO_4	$^*K_1 \approx$ 0·6	64D
	sp	23	5—6 ($HClO_4$)	$^*K_1$ 0·2	64F
	sp	25	4 (Mn^{2+},$H^+ClO_4^-$)	$^*K_1$ −0·06	65Wc
	sp	1—35		Δ^*H_1 = 4·8, Δ^*S_1 = 15·7	
	sp	25	4 [$Mn(ClO_4)_2$]	$^*K_1$ −0·03, Δ^*H_1 = 4·8, Δ^*S_1 = 15·7	67W
	kin	12·4	3 ($NaClO_4$)	$^*K_1$ −0·20 (Δ^*H_1 = 4·8)	68R
Mn^{4+}	*X*		solid	ev MnO_4^{4-} in Ba_3MnO_5(s)	65Ma
Fe^{2+}	gl	25	1 Na(ClO_4)	$^*K_1$ −6·51, $^*\beta_2$ −11·5	53Ha/63K
	gl	20	0 corr	K_{s0} −15·82, K_{s1} −9·43 [$K_1(H^+)$ 14·15]	63Dc
	kin, gl	25	1 ($NaClO_4$)	$^*K_1$ −3·32 (?)	65Wb
	gl	0—35	0·01 ($NaClO_4$)	$^*K_1$ −4·6(0°), −4·0(10°), −3·8(15°), −3·3(25°), −2·8(30°), −2·0(35°) (?)	68Wa
	gl	25	1 ($NaClO_4$)	$^*K_1$ −3·3 (?)	
	kin	25	*I*($NaClO_4$)	$^*\beta_2$ −7·96(*I* = 4), −6·35(*I* = 1), −5·85(*I* = 0), −5·89(*I* = 0·01) (?)	

1964 *Erratum*: p. 52, *in ref* 61Be, *for the entry read*

Metal	*Method*	*Temp*	*Medium*	*Log of equilibrium constant, remarks*	*Ref*
	"gl	12—30	1 (Na)ClO_4	$^*K_1$ −8·50(12°), −8·03(15°), −7·15(20°), −6·8(25°), −5·95(30°) (?)	61Be"
Fe^{3+}	gl		→ 0	K_{s0} −36·85	57Mf
	sp	−60 to 20	80% MeOH	$^*K_1$ −2·49(−59°), −2·45(−50°), −2·09(−42°), −2·04(−25°), −1·49(0°), −1·44(20°), Δ^*H_1 = 3·9 in 80% MeOH, *I* = 0·15	61Bg
	cal	25	3 (Na)ClO_4	Δ^*H_1 = 11·0, $\Delta^*H_{\beta22}$ = 10·0, $\Delta^*H_{\beta43}$ = 14·3, Δ^*S_1 = 23, $\Delta^*S_{\beta22}$ = 20·0, $\Delta^*S_{\beta43}$ = 22	62Sa/68Ab
	tyn	25	0 corr	K_{s0} −38·6 (fresh precipitate, $K_1(H^+)$14·00)	64Pd
	X		solid	ev FeO_4^{5-} in Ba_2LaFeO_5(s)	65Ma
	red, gl	25	0·5 (Na)Cl	no ev soluble OH–Fe cpx in 0·5 M-NaCl, for solid "β-FeOOH", see Cl^- table, ref 66B	66Be
	var		var	ev stable colloid, $Fe(OH)_{3-x}(NO_3)_x$ (*x* = 0·5—0·7) (gl, cfu, dialysis)	66Sd
	sol, gl	25	NO_3^- var	K_{s0} −38·2(fresh), −39·2(after 48 h), [$K_1(H^+)$ 14·22]	68Gg
	red, gl	25—45	3 ($NaClO_4$)	$^*K_1$ −3·00(25°), −2·73(35°), −2·52(45°), Δ^*H_1 = 10·2, Δ^*S_1 = 20·4; $^*\beta_{22}$ −2·36(25°), −2·17(35°), −2·03(45°); $\Delta^*H_{\beta22}$ = 11·14, $\Delta^*S_{\beta22}$ = 27·2	68Za
Fe^{4+}	*X*		solid	ev FeO_4^{4-} in Ba_3FeO_5(s)	65Ma
$Fe(CN)_5NO^{2-}$	pol	20	?	β_2 4·87 (summary) or 4·90 (text)	54Z
	sp	15—35	1 (KCl)	β_2 [$Fe(CN)_5NO^{2-} + 2\,OH^- \rightleftharpoons Fe(CN)_5NO_2^{4-} + H_2O$] 6·62(15°), 6·18(25°), 5·84(35°); $\Delta H_{\beta2}$ = −16·2, $\Delta S_{\beta2}$ = −26·1, see also Table 33, NO_2^-	66S
	sp	25	0·5 ($NaNO_2$)	K_1 0·50 ?, β_2 4·50	68Me

continued

Metal	*Method*	*Temp*	*Medium*	*Log of equilibrium constant, remarks*	*Ref*
Co^{2+}					
964 *Addendum*: p. 54, *for ref* "13K" *read* "13K ≈ 04Ka"					
	gl	15—40	0 corr	$*K_1$ −9·96(15°), −9·85(25°), −9·62(35°), −9·50(40°), $\Delta *H_1$ = 8·2	63Bf
		25	0	$*K_2$ −8·9(K_{s0} −15·7, K_{s2} −6·4, K_w −14)	63Bf+
	gl	28	1 ($NaClO_4$)	$*K_1$ −9·82, no ev polynuclear cpx for 0·10M-Co^{2+}	63Sg
	var		solid	ev $Co(OH)_6{}^{4-}$ in $Ba_2Co(OH)_6$(s) (ir, mag, sp)	65Hc
	gl	25	3 (Ba)ClO_4	$*K_1$ −9·75, $*\beta_{12}$ −9·44, $*\beta_{66}$ −42·55 (best fit) or $*K_1$ −10·20, $*\beta_{12}$ −9·37, $*\beta_{44}$ −29·30	67Ca
Co^{3+}	*X*		solid	ev $CoO_4{}^{5-}$ in Ba_2LaCeO_5(s)	65Ma
	kin	25	3 ($NaClO_4$)	$*K_1$ −0·66	66C
	kin	22—24	var	ev dimer, *e.g.* $Co_2(OH)_2{}^{4+}$, and Co^{3+}	67Ae
	kin	7	0·25	$*K_1$ −1·3	68Hc
$Co(NH_3)_5H_2O^{3+}$					
	gl	20	0·1 ($NaNO_3$)	$*K_1$ −6·18(H_2O), −6·35(D_2O), −6·33(20% dioxan)	68Cc
	gl, sp	25	0·3 (Na)ClO_4	$*K_1$ −6·22	68Se
			D_2O	$*K_1$ −6·81 in D_2O, 0·3M-(Na)ClO_4	
$Co(ND_3)_5OD_2{}^{3+}$					
	gl, sp	25	D_2O	$*K_1$ −6·70 in D_2O, 0·3M-(Na)ClO_4	68Se
Co $en_3{}^{3+}$	oth, sp	25	*I*($NaClO_4$)	K_1 1·7(dilatometry, *I* = 0·02—0·09), 1·42 (sp, *I* = 0·01)	66Ca
	sp	15—35	0·01	K_1 1·40(15°), 1·42(25°), 1·44(35°)	67Cb
		25	*w*% dioxan	K_1 1·66(*w* = 10), 2·01(*w* = 20), 2·46(*w* = 30), 3·10(*w* = 40) in *w*% dioxan	
Co $en_2ClOH_2{}^{2+}$					
	kin, est	25—30	dil	$*K_1$ ≈ −8(25°), ≈ −7·4(30°) (*cis*)	56Pd+
		25		$*K_1$ −7·2 (*trans*)	60Bh
	kin	25	dil	$*K_1$ −6·7(*cis*), −5·7(*trans*)	63Ca
	?	25	dil	$*K_1$ −6·8(*cis*), −5·8(*trans*)	63Cb
	gl	10—20	0·1($NaClO_4$)	Co $en_2ClH_2O^{2+}$, *cis*: $*K_1$ −7·47(10°), −7·13(20°); *trans*; $*K_1$ −6·37(10°), −6·11(20°);	64Hg
Co^{3+}cpx	?	25	?	Co $en_2NO_2H_2O^{2+}$; $*K_1$ −6·4(*trans*), −6·3 (*cis*)	56Aa/60Tb
	kin	50—60	0·1 ($NaClO_4$)	*trans*-Co $en_2NH_3H_2O^{3+}$; $*K_1$ −5·1(50°), −5·2(60°)	62Md
	gl	18	dil	*trans*-$Co(NH_3)_4CNH_2O^{2+}$; $*K_1$ −8·6 *trans*-$Co(NH_3)_4NO_2H_2O^{2+}$; $*K_1$ −9·0	64B
	gl	10—20	0·1 ($NaClO_4$)	Co trien$(H_2O)_2{}^{3+}$, *cis*-α; $*K_1$ −5·8(10°), −5·4(20°), $*K_2$ −7·8(10°), −7·3(20°); *cis*-β: $*K_1$ −5·6(10°), −5·3(20°), $*K_2$ −7·9(10°), −7·8(20°); Co trienClH_2O^{2+}, *cis*-α: $*K_1$ −7·46(10°), −7·11(20°); *cis*-β: $*K_1$ −6·2(10°), −5·8(20°); *cis*-$Co(NH_3)_4ClH_2O^{2+}$; $*K_1$ −6·6(20°)	64Hg
	gl	0	var	Co $en_2SO_4H_2O^+$; $*K_1$ −6·3	65Bd
	sp	25	dil?	Co $en_2(NH_3)RNH_2{}^{3+}$; K_1 1·79(R = H), 1·73(R = OH), 1·70(R = Et)	66Ca
	gl	3—44	1 (Na)ClO_4	Co $en_2(H_2O)_2{}^{3+}$; $*K_1$ −6·34(3°), −5·98(25°), −5·93(28°, *cis*); −5·02(2°), −4·55(25°), −4·49(28°, *trans*);	66Cd

continued overleaf

2 OH⁻ (contd)

Metal	*Method*	*Temp*	*Medium*	*Log of equilibrium constant, remarks*	*Ref*
Co^{3+}cpx (*contd*)				Co $en_2NH_3H_2O^{3+}$; $*K_1$ −6·55(1°), −6·05(25°), 5·95(29°), −5·73(44°, *cis*); −6·25(1°), −5·70(25°), −5·65(28°), −5·38(44°, *trans*)	
	sp	30	1 ($NaClO_4$)	$K(en_2CoOHOH_2^{2+}$, *cis* ⇌ *trans*) −0·10	68La
	var		var	ev$\{(NH_3)_4Co[\mu\text{-}(OH)_2]\}_2Co(CN)_2^{3+}$ ev$[(NH_3)_4Co(\mu\text{-}OH, \mu\text{-}NO_2)]_2Co(CN)_2^{3+}$, (pre, sp, ir)	68Sh
	sp, kin	25—70	1·0	$Co(NH_3)_5C_2O_4H^{2+}$; $*K_1$ −2·06(25°), sp, −1·77 (70°, kin)	66A
	kin	0—25	0·1 ($NaClO_4$)	$Co(NH_3)_5Cl^{2+}$; K_1 0·61(0°), 0·57(25°)	66Cb
		0·8	0·1 ($NaClO_4$)	Co $en_2NH_3Cl^{2+}$; K_1 0·52	
		0·8	0·1 ($NaClO_4$)	Co trienNH_3Cl^{2+}; K_1 0·32	
	gl?	20	0·1	$Co(NH_3)_5CH_3OH^{3+}$; $*K_1$ −5·58	66Ja
	X		solid	ev A^{3+} = $(NH_3)_3Co(OH)_3Co(NH_3)_3^{3+}$ in ABr_3(s) and AI_3(s)	67A
	sp	25	1 (H, NaCl)	$Co(NH_3)_5OCONH_3^+$; $*K_1$ −0·83	67B
Co^{4+}	*X*		solid	ev CoO_4^{4-} in Ba_3CoO_5(s)	65Ma
Ni^{2+}					

1964 *Addendum*: p. 56, *for ref* "13K" *read* "13K ≈ 04Ka"

Metal	*Method*	*Temp*	*Medium*	*Log of equilibrium constant, remarks*	*Ref*
	gl	28	1 ($NaClO_4$)	$*K_1$ −10·01, no ev polynuclear cpx for < 0·09 M-Ni^{2+}	63Sg
	gl	25—50	1 (Na)ClO_4	$*K_1$ −9·76(25°), −9·65(30°), −9·52(40°), −9·3(50°), $\Delta *H_1$ = 7·76	63Bd
	gl	20	*I* KNO_3	$*K_1$ −10·18(I = 1·5), −10·26(I = 1·0), also for 9 lower *I* values, −10·05($I \to 0$)	64Pa
		15—42	0 corr	$*K_1$ −10·22(15°), −10·05(20°), −9·86(25°), −9·75(30°), −9·58(36°), −9·43(42°) $\Delta *H_1$ = 12·4, $\Delta *S_1$ = 3(25°)	
	gl	25	3 (Na)ClO_4	$*\beta_{44}$ −27·37, $*\beta_{12}$ ≈ −10?	65B
	gl	25	3 (Na)Cl	$*\beta_{44}$ −28·42, $*\beta_{12}$ −9·3	65Be
	sp, ir		solid	ev $Ni(OH)_6^{4-}$ in $Ba_2Ni(OH)_6$(s)	65Hc
	gl	25	3 (Na)NO_3	$*\beta_{33}$ −21·58, $*\beta_{12}$ −9·6	66Bg
	cal	25	3 (Na)Cl	$\Delta *H_{\beta 44}$ = 43·0, $\Delta *S_{\beta 44}$ = 14	68Aa+
Ru^{3+}	sol, gl		var	K_{s0} −38 [$K_1(H^+)$ 14]	68Ba
$Ru^{3,4+}$	gl	25	1 (Na)SO_4	ev polyn cpx, ≈ $[Ru^{III}Ru^{IV}(OH)_6^+]_x$	67Ba
Ru^{4+}	oth	25	0·1 $HClO_4$	ev ion of charge 4+, not 2+ (Donnan equil)	64Wa
	sol, gl		var	K_{s0} −49 [$K_1(H^+)$14]	68Ba
Ru^{8+}	*p* RuO_4	10—90	dil	$K[RuO_4(aq) \rightleftharpoons RuO_4(g, p$ in atm)], −1·01(25°), 8·72—2900T^{-1}(10—25°), 6·16—2138T^{-1}(25—90°)	64N
$Ru(NH_3)_5H_2O^{3+}$	gl	25	dil	$*K_1$ −4·2	64Bd

1964 *Errata*: p. 56, *the "metal"* $RuO_4(H_2O)_n$ *refers to refs.* 50S, 54Mb, *and* 63Sd: *in ref* 54Mb, for "$HRuO_5^+$" *read* "$HRuO_4^+$"

Metal	*Method*	*Temp*	*Medium*	*Log of equilibrium constant, remarks*	*Ref*
Rh^{3+}	cix, fp		var	ev only Rh^{3+} in acidic solutions	64Cd
	gl	25		$*K_1$ −3·2(25°), −2·4(est, 64·4°)	64Pc
	gl	75—85	2·5 ($NaClO_4$)	$*K_1$ −3·40(25°), −3·20(45°), −3·08(60°), −2·96(→ 75°), −2·92(→ 80°), −2·89(→ 85°), $\Delta *H_1$ = 4·3	66Sf

continued

Metal	*Method*	*Temp*	*Medium*	*Log of equilibrium constant, remarks*	*Ref*
$Rh(NH_3)_5OH_2^{3+}$					
	gl	9—35	0·2 ($NaClO_4$)	$*K_1$ −6·63(9·4°), −6·24(35°), $\Delta^*H_1 = 6, \Delta^*S_1 = -9$	67Pb
trans-Rh $en_2X(OH_2)_2^{2+}$					
	gl	20	dil	$*K_1$ −6·44(X = Cl), −6·07(X = Br), −6·55(X = I)	67Bb
Pd^{2+}	sp, gl	25	0 corr	K_1 13·0(gl), 12·4(sp); β_2 25·8(gl), 26·5(sp)	67I
	sol	25		K_{s2} −2·65	
$PdCl_2(H_2O)_2$	sp, gl		var	$*K_1$ −2	66Wb
Os^{4+}	pol	25	NaOH var	ev $OsO(OH)_3^-$, $Os_2O_2(OH)_{10}^{6-}$, $OsO(OH)_5^{3-}$	68C
Os^{6+}	ir		solid	ev $OsO_2X_4^{2-}$(X = Cl, CN, OH), $OsO_2(OH)_2X_2^{2-}$ (X = NO_2^-, Cl^-, *etc.*) in K^+ salts	64Ga
	pol	25	NaOH var	ev $OsO(OH)_5^-$, $OsOOH^{3+}$, $OsO(OH)_3^+$?	68C
$OsO_2(OH)_4$	ir		solid	ev $OsO_4(OH)_2^{2-}$ in K^+ salt	64Ga
	gl	20	0·25 Na_2SO_4	$*K_1$ −7·24	67Bd
	sp	20	dil	$*K_1$ −7·2, $*K_2$ −12·2, $*K_3$ −13·95, $*K_4 = K_a(OsO_5OH^{3-})$ −14·17 (K_w −14)	
Os en_3^{3+}	gl	25?	dil	$*K_1$ −5·10	66W
Pt^{2+}cpx	gl	25	0·1 $NaNO_3$	*cis*-$PtA_2(H_2O)_2^{2+}$, same $*K_1$ and $*K_2$ for $A_2 = (NH_3)_2$ (ref 39J), $(MeNH_2)_2$, $(EtNH_2)_2$, $C_2H_4(NH_2)_2$, $C_3H_6(NH_2)_2$; *trans*-$PtA_2(H_2O)_2^{2+}$, same $*K_1$ and $*K_2$ for $A_2 = (NH_3)_2$ (ref 39J), $(MeNH_2)_2$, $(EtNH_2)_2$; Pt dipy$(H_2O)_2^{2+}$; $*K_1$ −4·7, $*K_2$ −5·7; *trans*-$Pt(NH_2OH)_2(H_2O)_2^{2+}$; $*K_1$ −5·6, $*K_2$ −7·7; $PtNH_3$ py$(H_2O)_2^{2+}$; $*K_1$ −5·2, $*K_2$ −6·85 (*trans*); $*K_1$ −4·1, $*K_2$ −6·7 (*cis*); Pt $py_2(H_2O)_2^{2+}$; $*K_1$ −4·5, $*K_2$ −6·3 (*trans*); $*K_1$ −3·9, $*K_2$ −6·4 (*cis*)	63Gb
	gl	25	0·3 $NaNO_3$	$*K_1$ and $*K_2$ for 10 mixed cpx of Pt^{2+} with $MeNH_2$, $EtNH_2$, NH_2OMe, NH_2OH, py; see also NH_2OH table	66Ga
	gl	25	0·3 $NaNO_3$	Pt en$(H_2NOH)_2^{2+}$; $*K_1$ −7·68, $*K_2$ −10·7 also $*K_n$ values for Pt^{II}–oxime cpx	68G
	gl?	25?	dil?	$Pt(NH_3)_2(H_2O)_2^{2+}$; $*K_1$ −5·63, $*K_2$ −9·25 (*cis*) $*K_1$ −4·23, $*K_2$ −7·30 (*trans*)	68P
Pt^{4+}cpx	sol	20	0·1 K(OH, Cl)	$Pt(CN)_4(NH_3)_2^0$; $*K_1$ −12·12	63C
	gl	20	dil	Pt en$(CN)_2NH_3Cl^+$; $*K_1$ −8·6	64Ca
	sol	20	var	$Pt(NH_3)_2(CN)_3X^0$; $*K_1$ −12·7(X = Cl), −12·8(Br), −13·0 (I) $Pt(CH_3NH_2)_2(CN)_3X^0$; $*K_1$ −12·9(X = Cl, Br), −13·3(I) $Pt(CH_3NH_2)_2(CN)_2I_2^0$; $*K_1$ −14·05 $Pt(NH_3)_2(CN)_2NO_2NO_3^0$; $*K_1$ −12·8	64Cb
	gl	25	dil	$Pt(NH_3)_{4-n}py_nCl_2^{2+}$; $*K_1$ −11·17(*n* = 0), −10·00(*n* = 1), −9·95(*n* = 2, *trans*), −9·39 (*n* = 2, *cis*), −8·62(*n* = 3); with *n* = 4, cpx decomposed by OH^-	66Gb

continued overleaf

2 OH^- (contd)

Metal	*Method*	*Temp*	*Medium*	*Log of equilibrium constant, remarks*	*Ref*
Pt^{4+}cpx	gl	25?	dil	$Pt(NH_3)_2py_2Cl_2^{2+}$; $^*K_1$ −9·96 (*trans*), −9·39 (*cis*)	67Gb
(*contd*)	gl	25	dil	$Pt(NH_3)_5Cl^{3+}$; $^*K_1$ −8·05, $^*K_2$ −10·72	68Ga
				$Pt(NH_3)_3pyNH_3Cl^{3+}$; $^*K_1$ −6·92, $^*K_2$ −10·52	
				$Pt(NH_3py)_2NH_3Cl^{3+}$; $^*K_1$ −5·74, $^*K_2$ −10·12	
Me_3Pt^+	*X*		solid	ev tetrahedral $(Me_3PtOH)_4$ in $Me_3PtOH(s)$	68Sf
Cu^+	*X*		solid	ev $Cu_4O_4^{4-}$ rings in KCuO(s)	68Hd
Cu^{2+}					

1964 *Addendum*: p. 59, *for ref* "13K" *read* "13K ≈ 04Ka"

Metal	*Method*	*Temp*	*Medium*	*Log of equilibrium constant, remarks*	*Ref*
	gl	25	0 corr	$^*K_1$ −7·34, $^*\beta_{22}$ −10·57	64A
	gl, sol	25?	0 corr	K_{s0} −19·32[$Cu(OH)_2(s)$], −18·7[$Cu(OH)_2(H_2O)_3(s)$]	64Gc
	oth	20	0·1 ($NaClO_4$)	$^*\beta_{22}$ −10·78 (gl, rapid flow)	64W
	sol, gl	25	0·2 (Na)ClO_4	$^*K_{s0}$ 8·92[$Cu(OH)_2$], 7·89(CuO)	65Se
		25	0 corr	$^*K_{s0}$ 8·68, K_{s0} −19·32[$Cu(OH)_2$], $^*K_{s0}$ 7·65, K_{s0} −20·35(CuO)	
	X		solid	ev[$(MeNH_2)_2H_2OCu(OH)_2Cu(NH_2Me)_2]_2^{4+}$ in $Cu_2(MeNH_2)_4(OH)_2SO_4H_2O(s)$	66I
	X		solid	ev $(C_5H_5NCu)_4Cl_6O$ (Cu_4O in centre) in solid	67K
	cal	25	3 (Na)ClO_4	$\Delta^*H_{\beta 22} = 15{\cdot}8$, $\Delta^*S_{\beta 22} = 4{\cdot}4$	68A
	gl	25	55% dioxan	$^*K_1$ −7·60, $^*\beta_{22}$ −10·95 in 55% dioxan, 3M-(Li)ClO_4	68O
	sol	25	0 corr	K_1 6·0, K_2 7·18, K_3 1·24, K_4 0·14, K_{s0} −19·89	68Sb
Cu^{3+}	pre		var	$Cu(OH)_4^-$ decomposes very rapidly	65Mb
Ag^+	sol	25	1 ($NaClO_4$)	K_{s0} −8·17, K_1 3·02, β_2 4·69 [$K_1(H^+)$ 14·00]	67Gd
	Ag	350—425	KNO_3(l)	K_s ($Ag^+{}_2O^{2-}$) −15·6(350°), −15·1(375°), −14·8(400°), −14·47(425°); $\Delta H_s = 28{\cdot}5$, $\Delta S_s = -25{\cdot}3$ in KNO_3(l)	68Sa
	Ag	250—350	MNO_3(l)	K_s ($Ag^+{}_2O^{2-}$) −13.8(250°), −13·3(275°), −13·0(300°), −12·7(325°), −12·6(350°); $\Delta H_s = 14{\cdot}8$, $\Delta S_s = -33{\cdot}5$ in $Na_{0.5}K_{0.5}NO_3$(l), *m* units	
	X		solid	ev $Ag_4O_4^{4-}$ rings in KAgO(s), CsAgO(s)	68Sg
Ag^{3+}	var	27	K(NO_3,OH)var	ev $Ag(OH)_4^-$ (pre, pol, sp)	68Cd
Au^+	*X*		solid	ev $Au_4O_4^{4-}$ rings in CsAuO(s)	68Wb
Au^{3+}cpx	gl	25	0·5 (Na)ClO_4	Au dien X^{2+}; $^*K_1$ −4·0(X = Cl), −4·5(X = Br), −5·8(X = OH),	63Be
				Au en_2^{3+}; $^*K_1$ −6·3	
			0·5 (Na)Cl	Au dien Cl^{2+}; $^*K_1$ −4·7	
				Au en_2^{3+}; $^*K_1$ −7·2	
			0·5 (Na)Br	Au dien Br^{2+}; $^*K_1$ −5·2	
Me_2Au^+	*X*, pC_6H_6		var	ev rings $(Me_2AuOH)_4$ in solid and C_6H_6(l)	68Ge
	gl		dil	ev cpx with up to 3 OH^-/Au	
Zn^{2+}					

1964 *Errata*: p. 61, *for ref* "13K" *read* "13K ~ 04Ka": p. 63, *in ref* 61Sc, *for* "$H^+{}_{-2}$" *read* "$H^+{}_{-1}$"

Metal	*Method*	*Temp*	*Medium*	*Log of equilibrium constant, remarks*	*Ref*
	gl	25	2 KCl	$^*K_1$ −9·01, $^*\beta_{12}$ −7·20	63Sh
	gl	25	2 NaCl	$^*K_1$ −9·11, $^*\beta_{12}$ −7·49	
	gl	25	2 (K)Cl	$^*K_1$ −9·01, $^*\beta_{12}$ −7·20	64Sa
		25	2 (Na)Cl	$^*K_1$ −9·12, $^*\beta_{12}$ −7·48	

continued

Metal	Method	Temp	Medium	Log of equilibrium constant, remarks	Ref
Zn^{2+}	MHg	25	3 NaCl	β_4 13·35, β_{62} 26·77	64Sb
(*contd*)	var	25	0·2 (KNO_3 or $NaClO_4$)	$^*K_{s0}$ 12·70(am), 11·98(β_1), 12·02(β_2), 11·96(γ), 12·07(δ), 11·75(ε), 11·56(active ZnO), 11·39(inactive ZnO) [$K_1(H^+)$ 13·70]	64Sd = 63S
		25	0 corr	K_{s0} 12·48(am), 11·76(β_1), 11·80(β_2), 11·74(γ), 11·85(δ), 11·53(ε), 11·34(active ZnO), 11·17(inactive ZnO), (sol, gl, MHg)	
	dis	25	3 ($NaClO_4$)	$^*\beta_2$ −20·10, $^*\beta_3$ −28·83, $^*\beta_4$ −38·72, no ev $ZnOH^+$ ev $Zn(OH)_2L_2$ in C_6H_6(l), L = R_2NH, R = branched alkyl group	65Sb
	gl	25	3 (Na)Cl	$^*K_1$ −9·25, $^*\beta_{12}$ −7·50	65Sc
		25	3 (K)Cl	$^*K_1$ −9·26, $^*\beta_{12}$ −7·47, $^*\beta_{24}$ −13·32?	
	sol	25	1 ($NaClO_4$)	K_{s0} −16·76, K_1 6·31, β_2 11·19, β_3 14·31, β_4 17·70	67Gd
	sol, gl	25	1 ($NaClO_4$)	K_{s0} −16·76 [$K_1(H^+)$ 14·00]	68Gb
	Zn	20	KOH var	ev $Zn(OH)_4^{2-}$ in 2—17M-KOH	68Mb
	gl, MHg	25	0 corr	K_s [$Zn^{2+}(OH^-)_2(H_2O)_{0\cdot1}$] −17·33, valid also in EtOH–H_2O mixt	68Sc
Cd^{2+}					

1964 *Addendum*: p. 63, *for ref* "13K" *read* "13K ≈ 04Ka"

Metal	Method	Temp	Medium	Log of equilibrium constant, remarks	Ref
	sol, gl	25	3 (Na)ClO_4	$^*K_{s0}$ 14·03[β-$Cd(OH)_2$], 14·22[γ-$Cd(OH)_2$]	63S
	var	25	0 corr	K_{s0} −14·09, $^*K_1$ −7·92, $^*K_2$ −11·38, $^*K_3$ −14·32, $^*K_4$ −13·96, $^*K_5$ −14·64, $^*K_6$ −14·88 (?) (MHg, gl, sol) [$K_1(H^+)$ 14·00]	64S
	sol	25	1 ($NaClO_4$)	K_1 17·76 (?)	64Sh
	sol	24·5	*I*Na(ClO_4)	K_{s1} −8·5, K_{s2} −6·0, K_{s0} −5·9, K_{s4} −5·5 (I = 1 or 7)	65R
	cal	25	3 (Li)ClO_4	Δ^*H_1 = 13·1, $\Delta^*H_{\beta12}$ = 10·9, $\Delta^*H_{\beta44}$ = 40·6, Δ^*S_1 = −3, $\Delta^*S_{\beta12}$ = −5, $\Delta^*S_{\beta44}$ = −9	67Ab + 62Ba
Hg^{2+}	gl, red	25	0·5 Na(ClO_4)	$^*K_1$ −3·68, $^*K_2$ −2·57	52H/63K
	X		solid	ev HgO_2^{2-} in Na_2HgO_2(s) *etc.*	64H
	cal	25	3 (Na)ClO_4	Δ^*H_1 = 7·23, Δ^*H_2 = 2·61, $\Delta^*H_{\beta12}$ = 3·06, $\Delta^*H_{\beta22}$ = 12·7, Δ^*S_1 = 7·9, Δ^*S_2 = −3·4, $\Delta^*S_{\beta12}$ = −2·1, $\Delta^*S_{\beta22}$ = 18·8	67Aa + 62A

1964 *Erratum*: p. 65, *ref.* 63F, *for* "K_{s4}−4·5" *read* "K_{s3}−4·5"

Metal	Method	Temp	Medium	Log of equilibrium constant, remarks	Ref
$MeHg^+$	gl, cal	20	0·1 (KNO_3)	$^*K_1$ −4·59, K_{12}($MOH + M^+ \rightleftharpoons M_2OH^+$) 2·37[$K_1(H^+)$ 13·96], ΔH_1 = −8·5, ΔS_1 = 13·7	63Sf = 65Sd
RHg^+	gl	25	0·1 (KNO_3)	K_1 9·00(Me), 8·80(Et), 8·66(Pr), 8·61(Bu) [$K_1(H^+)$ 13·78]	68Zb
		20	0·1 (KNO_3)	K_1 9·32 (Me)	
Al^{3+}	kin	100	0·01 $AlCl_3$	$^*K_1$ −2·78 ?, K_1 9·36 ?	04Ka ≈ 13K
		100	$AlCl_3$ var	$^*\beta_2$ −6·25, β_2 18·63	
	pre, gl		var	ev "$Al_2O_3H^+$aq" [≈ $Al_{2n}(OH)_{5n}^{n+}$], reacts slowly with H	47Sa
	gl		→ 0	K_{s0} −33·45	57Mf
	gl	25	3 Na(ClO_4)	$^*\beta_{17,7}$ −48·8, $^*\beta_{34,13}$ −97·6	64Be
	oth	25	dil	ev $Al_8(OH)_{20}^{4+}$ (coagulation method)	64Ma +
	Al	25?	NaOH var	ev $Al_n(OH)_{4n+2}^{(n+2)-}$ at pH > 12·4, n = 40—100 (?)	64P

continued overleaf

2 OH^- (contd)

Metal	Method	Temp	Medium	Log of equilibrium constant, remarks	Ref
Al^{3+} (contd)	X		$Al(NO_3)$ var	ev cpx in solution of same size as $Al_{13}O_4(OH)_{24}(H_2O)_{12}^{7+}$ (ref 60J) (small-angle X-ray scattering)	64R
	pre		var	ev $Al_4O_5^{2+}$ (?)	64Sg+
	var	20	0·1 (KCl)	$Al^{3+}+OH^-$: several fast and slow reactions (gl, rapid flow)	64W
				$[Al(OH)_2^+]_x$, average $x = 6$—8 (cfu), no rapidly "protonable" group (gl, rapid flow)	
	gl	25	2 (Na)ClO_4	$*\beta_{22}$ −7·07, $*\beta_{32,13}$ −104·5,	65Aa
	cfu			ev cpx with ≈ 13Al	
	gl	30—40	2 $NaNO_3$	ev $Al_4(OH)_8^{4+}$, $Al_7(OH)_{16}^{5+}$ (also ir)	65F
	var		OH^- var	ev only $Al(OH)_4^-$ in alkaline solutions [gl, (sp), fp (lit), diffusion]	65Ja+
	con	30–70	NaOH var	ev anions of charge −1 and more negative	65L
	X		solid	ev AlO_4^{5-} in Ba_2LaAlO_5(s)	65Ma
	con, gl	25	→ 0	$*K_1$ −4·9	65Na
	X		solid	ev $(HO)_3AlOAl(OH)_3^{2-}$ in $K_2AlO(OH)_6$(s)	66J
	X, ir		solid	ev $Al(OH)_6^{3-}$ in $Ca_3Al_2(OH)_{12}$(s)	67C
	qh	25	2 ($NaClO_4$)	ev polyn cpx	67Sc
	oth	25	$AlCl_3$ dil	$*K_1$ −4·5 (dissociation field effect relaxation)	68Ha
	kin	25	1 ($NaClO_4$)	$*K_1$ −4·31	68Sd
Ga^{3+}	dis, gl	20	1 (NaCl)	$*K_1$ −1·40, $*K_2$ −1·75, $*K_3$ −2·12	65A
	X		solid	ev GaO_4^{5-} in Ba_2LaGaO_5(s)	65Ma
	dis	20	0·1 ($NaClO_4$)	$*K_1$ −0·44, $*K_2$ −0·80, $*K_3$ −1·05	65Si/66Sc
	sol, pH	25	dil	ev polyn cat cpx	66Bc
	sol, gl	60	3 (Na)ClO_4	$*K_{s0}$ [α-GaOOH(s)+3H$^+$ ⇌ Ga^{3+}] 3·66, ev polyn cpx	67G
	gl	25	3 (K)Cl	$\beta_{5n,2n}$−213, (n ≈ 20), ev $(Ga^{3+})_x$	67Ha
	Ga	25?	0?	β_4 38·7	67Va+
	aix		NaOH var	ev only $Ga(OH)_4^-$	68Ca
	sp, gl	25	0·1	$*K_1$ −2·9, $*K_2$ −3·7, $*K_3$ −4·4	68N
Me_2Ga^+	gl	25	0·3 ClO_4^-	$*K_1 \leqslant$ −5·38, $*K_2 \leqslant$ −8·82	68Hb
	var		var	ev $Me_2Ga(OH)_2^-$, $(Me_2GaOH)_4$ and $Me_2Ga(OH_2)_2^+$ (ir, Ram, nmr, pre)	68T
In^{3+}	dis	20	1·0	$*K_1$ −2·11, $*K_2$ −2·45, $*K_3$ −2·68 [$K_1(H^+)$ 14·00]	65Sh
Tl^{3+}	red	25	3 (Na)ClO_4	$*K_1$ −1·14, $*K_2$ −1·43	53Ba/63K
	red	25	3 (Li)ClO_4	$*K_1$ −1·18, $*K_2$ −1·42	64Ka
	sol	25	NaOH var	ev ani cpx	65I
	gl, red	25	10 NH_4NO_3	K_1 15·7, β_2 30·7 (K_w −14·00?) (?, see Fig. 3)	67L
			2 $Mg(NO_3)_2$	K_1 15·40, β_2 28·66 (K_w −14·00?)	
			2 $enH_2(NO_3)_2$	K_1 15·45, β_2 28·64 (K_w −14·00?)	
Me_2Tl^+	kin	25	0 corr	K_1 1·04, K[2MOH ⇌ $M_2(OH)_2$] ≈ 0·3	59Ld
Ge^{2+}	sol	25	$HClO_4$ var	$*K_{s1}$ [$Ge(OH)_2$(s)+H$^+$ ⇌ $GeOH^+$+H_2O] −1·26	64G
				K_{s2} ≈ −1·7, no ev Ge^{2+} [white $Ge(OH)_2$(s)]	
		25	NaOH var	ev reddish colloid [brown GeO(s) ?]	
Sn^{2+}	MHg, gl	25	3 (Na)ClO_4	$*\beta_{43}$ −6·85, $*\beta_{32}$ ≈ −6·7, $*\beta_{22}$ ≈ −4·6	58Ta/64L
	sol, gl	25	1 ($NaClO_4$)	$*K_{s1}$ [SnO(s)+H$^+$ ⇌ $SnOH^+$] −0·28	66Ma
R_2Sn^{2+}	gl	25	3 (Na)ClO_4	$*K_1$ −3·54(Me), −3·40(Et), −2·92(Pr)	66T
				K[MOH^+ ⇌ $M_2(OH)_2^{2+}$] 2·48(Me), 2·43(Et), 2·27(Pr)	
Me_2Sn^{2+}	gl	24·8	dil	$*K_1$ −3·11, ev $Me_2Sn(OH)_4^{2-}$	53R

continued

Metal	Method	Temp	Medium	Log of equilibrium constant, remarks	Ref
Me_2Sn^{2+} (*contd*)	gl	25	3 (Na)ClO_4	$*K_1$ −3·55, $*\beta_2$ −9·00, $*\beta_{22}$ −4·52, $*\beta_{64}$ −16.14 or $*K_1$ −3·54, $*\beta_2$ −8·98, $*\beta_{22}$ −4·60, $*\beta_{32}$ −9·76, $*\beta_{43}$ −10·40	62T/64Tb
	gl	25	3 (Na)ClO_4,H_2O	$*K_1$ −3·55, $*\beta_2$ −9·00, $*\beta_{22}$ −4·52, $*\beta_{64}$ −16·14; or $*K_1$ −3·54, $*\beta_2$ −8·98, $*\beta_{22}$ −4·60, $*\beta_{32}$ −9·76, $*\beta_{43}$ −10·40	64T
			3 (Na)ClO_4,D_2O	$*K_1$ −4·06, $*\beta_2$ −10·16, $*\beta_{22}$ −4·22, $*\beta_{64}$ −18·53; or $*K_1$ −3·95, $*\beta_2$ −10·06, $*\beta_{22}$ −5·34, $*\beta_{32}$ −11·07, $*\beta_{43}$ −11·96	
	gl	25	0·1 (Na)Cl	$*K_1$ −3·245, $*\beta_2$ −8·52, $*\beta_{22}$ −5·00, $*\beta_{64}$ −16·85; or $*K_1$ −3·25, $*\beta_2$ −8·535, $*\beta_{22}$ −5·05, $*\beta_{32}$ −9·81, $*\beta_{43}$ −11·52	64Tb
	H	25	0·1 (Na)Cl	$*K_1$ −3·25, $*K_2$ −5·27, (ref 64T), $*K_3$ −11·5, $*K_4$ −11·7	65T
Et_2Sn^{2+}	gl	25	dil	$*K_1$ −2·40, $*K_2$ −4·58, $*\beta_{22}$ −3·40, $*\beta_{32}$ −6·70 [$K_1(H^+)$ 13·85]	68Ae
			1 K(Cl)	$*K_1$ −2·65, $*K_2$ −4·84, $*\beta_{22}$ −4·00, $*\beta_{32}$ −7·60 [$K_1(H^+)$ 13·90]	
R_3Sn^+	gl	25	3 (Na)ClO_4	$*K_1$ −6·60 (R = Me), −6·81 (R = Et)	66T
Me_3Sn^+	gl	25	dil	$*K_1$ −6·16, $*\beta_{12}$ −5·40, $*\beta_{22}$ −13·50 [$K_1(H^+)$ 13·84]	68Ae
			2 K(Cl)	$*K_1$ −6·40, $*\beta_{12}$ −5·45, $*\beta_{22}$ −13·85 [$K_1(H^+)$ 13·94]	
Et_3Sn^+	gl	25	3 (Na)ClO_4	$*K_1$ −6·81 (in H_2O), −7·50 (in D_2O)	64T
Ph_3Sn^+	dis	30	0·1 ($NaNO_3$)	K_1 9·2, K_D[$Ph_3SnOH \rightleftharpoons Ph_3SnOH(org)$] 4·0 (org = C_6H_6), 2·8(org = iso-BuCOMe)	65Sa
Pb^{2+}					

1964 *Addendum*: p. 69, *for ref* "13K" *read* "13K ~ 04Ka"

Metal	Method	Temp	Medium	Log of equilibrium constant, remarks	Ref
	MHg, gl	25	*I* (Na)ClO_4	good agreement with model given by ref 60Oa	60Oa/65E
	X		var	ev tetrahedral $Pb_4(OH)_4^{4+}$ in solution	62Eb
	gl	25	2 (Na)ClO_4	$*K_1$ −7·93, $*\beta_{44}$ −19·35	64He
	oth	25	1 $NaClO_4$	ev $Pb_4(OH)_4(ClO_4)_2^{2+}$, $Pb_6(OH)_8(ClO_4)_3^+$ no ev cpx with > 6Pb (light scattering)	64Hf
	cfu		var	agrees with model of ref 60Oa	65E
	gl	25	2 (Na)NO_3	$*K_1$ −8·84, $*\beta_{12}$ −7·11, $*\beta_{44}$ −21·72	65H
	Ram		var	ev tetrahedral $Pb_4(OH)_4^{4+}$ and $Pb_6(OH)_8^{4+}$ (octahedral?)	67Mb
	MHg	25	3 Na(Cl)	β_2 7·78, β_3 9·962	67Sa
	X		1—4 ClO_4	ev tetrahedral Pb_4 cpx and irregular Pb_6 cpx in solution	68Jb+
	Ram		var	ev octahedral $Pb_6(OH)_8^{4+}$, tetrahedral $Pb_4(OH)_4^{4+}$	68Md
$Pb^{2+,4+}$	sol	25?	0 corr	K_s [$PbO_{1.57}(s) + 1{\cdot}57OH^- \rightleftharpoons 0{\cdot}57PbO_2(OH)_2^{2-} + 0{\cdot}43PbOOH^-$] −3·83; K_s [$PbO_{1.33}(s) + 1{\cdot}33OH^- \rightleftharpoons 0{\cdot}33PbO_2(OH)_2^{2-} + 0{\cdot}67PbOOH^-$] −3·44	67Cc
Pb^{4+}	sol	25?	0 corr	K_s [$PbO_2(\beta, s) + 2OH^- \rightleftharpoons PbO_2(OH)_2^{2-}$] −4·13	67Cc
Me_2Pb^{2+}	gl	25	3 (Na)ClO_4	$*\beta_2$ −15·54, $*\beta_3$ −28·52, $*\beta_{22}$ −10·83, $*\beta_{43}$ −24·31	66F
Ph_3Pb^+	dis	30	0·1 ($NaNO_3$)	K_1 7·7, K_D [$Ph_3PbOH \rightleftharpoons Ph_3PbOH(org)$] 2·9(org = $CHCl_3$), 2·2(org = iso-BuCOMe)	65Sa

continued overleaf

2 OH^- (contd)

Metal	Method	Temp	Medium	Log of equilibrium constant, remarks	Ref
R_4N^+	H	20?	C_4H_8O	K_1 10(R = Me), 7(Et), 0(Bu), in C_4H_8O(l), 0·1M-Bu_4NClO_4	67Pa
Sb^{3+}	pol	25	NaOH var	ev monomer and dimer, $Sb(OH)_4^-$ and $Sb_2O(OH)_4$?	63Ma
	var		NaOH var	ev $SbO_2(H_2O)_n^-$, no ev polyn anion (diffusion, sp, pre, pol)	65J
	var		$HClO_4$ var	ev $SbO(H_2O)_n^+$, no ev polyn cation (aix, cix, tp, sp, diffusion)	
	sp	23	$HClO_4$ var	$*K_3$ [$SbO^+ + 2H_2O \rightleftharpoons Sb(OH)_3 + H^+$] −1·42	68Ma
Bi^{3+}	*X*		solid	ev $Bi_2O_2^{2+}$ or $Bi_2(OH)_2^{4+}$ in $BiOHCrO_4$ (s, monoclinic)	64Aa
	X		solid	ev Bi_2OOH^{3+} in $Bi_2O(OH)_2XO_4$(s), X = S or Se; ev $Bi_2(OH)_2^{4+}$ in $BiOHXO_4H_2O$(s), X = S, or Se	64Ab
	X		solid	ev Bi_6 or Bi_{12} units in $BiONO_3(H_2O)_{0.5}$(s)	68Kd
	Ram		var	ev octahedral $Bi_6(OH)_{12}^{6+}$	68Mc

1964 *Erratum*: p. 70, *in ref* 54F, *for* "$Bi_4O_4^+$" *read* "$Bi_4O_4^{4+}$".

Metal	Method	Temp	Medium	Log of equilibrium constant, remarks	Ref
Po^{4+}	cix	21	0·035 ($NaClO_4$)	$*K_1$ −3·4, $*\beta_2$ −8·15	63Ka
	dis	18—22	0·1 (Na)ClO_4	$*K_1$ −1·10, $*\beta_2$ −2·20, $*\beta_3$ −3·06, $*\beta_4$ −4·80	64Sf
	dis	?	3 (Li)ClO_4	$*K_1$ −0·14, $*\beta_2$ −0·52, $*\beta_3$ −1·77	65G
Xe^{8+}	*X*		solid	ev XeO_6^{4-} in $Na_4XeO_6(H_2O)_8$(s)	64I
	X		solid	ev XeO_6^{4-} in $Na_4XeO_6(H_2O)_6$(s)	64Za

04Ka C. Kullgren, dissertation, Uppsala, 1904.

1964 *Addendum*
33Ja add: discussion, E. Laue and H. L. Johnston *et al.*, *J. Amer. Chem. Soc.*, 1934, **56**, 1249
47Sa M. E. Shishniashvili, V. A. Kargin, and A. L. Batsanadze, *Zhur. fiz. Khim.*, 1947, **21**, 391
53R E. G. Rochow and D. Seyferth, *J. Amer. Chem. Soc.*, 1953, **75**, 2877
54Ra W. Rüdorff and F. Pfitzer, *Z. Naturforsch.*, 1954, **9b**, 568
54Z P. Zuman and M. Kabát, *Chem. listy*, 1954, **48**, 368

1964 *Addendum*
55Mb add: C. L. Rulfs and R. J. Meyer, *J. Amer. Chem. Soc.*, 1955, 77, 4505
56Aa S. Ašperger and C. K. Ingold, *J. Chem. Soc.*, 1956, 2862 (compiler cannot find equilibrium data quoted by ref. 60Tb)
56Pd R. G. Pearson, R. E. Meeker, and F. Basolo, *J. Amer. Chem. Soc.*, 1956, **78**, 2673
57Mf N. N. Mironov and A. I. Odnosevtsev, *Zhur. neorg. Khim.*, 1957, **2**, 2202
59Ld J. K. Lawrence and J. E. Prue, *Chem. Soc. Special Publ. No.* 13, 1959, 186
60Bh M. E. Baldwin, private communication, quoted by ref 60Tb
60E V. I. Ermolenko, dissertation, Kiev, 1960, quoted by ref 62Ea
60Tb M. L. Tobe, *Sci. Progr.*, 1960, **48**, 483

1964 *Erratum:* p. 82, *in ref* 61Be, *for* "A. Bolzan" *read* "J. A. Bolzán"

61Bf J. A. Bland, *Acta Cryst.*, 1961, **14**, 875
61Bg E. J. Bowers and H. D. Weaver, jun., *Proc. Indiana Acad. Sci.*, 1961, **71**, 101
62Ae N. V. Aksel'rud, in 'Khimiya rastvorov redkozomel'nykh elementov', *Izd. ANUSSR*, Kiev 1962, vol. 3, p. 3
62Ea V. I. Ermolenko, in 'Khimiya rastvorov redkozomel'nykh elementov', *Izd. ANUSSR*, Kiev 1962, vol. 3, p. 148
62Eb O. E. Esval, Ph.D. Thesis, North Carolina, 1962, quoted by ref 67Mb
62Md D. F. Martin and M. L. Tobe, *J. Chem. Soc.*, 1962, 1388
63Aa N. V. Aksel'rud, in 'Redkozemel'nye elementy', Izd. Nauka, Moscow, 1963, p. 175
63Ab N. V. Aksel'rud, *Uspekhi Khim.*, 1963, **32**, 800
63Bd J. A. Bolzán, E. A. Jáuregui, and A. J. Arvia, *Electrochim. Acta*, 1963, **8**, 841
63Be W. H. Baddley, F. Basolo, H. B. Gray, C. Nölting, and A. J. Poë, *Inorg. Chem.*, 1963, **2**, 921
63Bf J. A. Bolzán, J. J. Podesta, and A. J. Arvia, *Anales Asoc. Quím Argentina*, 1963, **51**, 43
63Bg W. J. Biermann and W.-K. Wong, *Canad. J. Chem.*, 1963, **41**, 2510
63C I. I. Chernyaev, A. V. Babkov, and N. N. Zheligovskaya, *Zhur. neorg. Khim.*, 1963, **8**, 2441
63Ca S. C. Chan, *J. Chem. Soc.*, 1963, 5137
63Cb S. C. Chan and M. L. Tobe, unpubl, quoted by ref. 63Ca

continued

63Dc J. Dauphin, S. Dauphin, D. Chatonier, and G. Andraud, *Bull. Soc. chim. France*, 1963, 2751; J. Dauphin, S. Dauphin, D. Chatonier, and M.-T. Vialatte, *ibid.*, p. 2754
63Gb A. A. Grinberg, Kh. I. Gil'dengershel', and E. P. Panteleeva, *Zhur. neorg. Khim.*, 1963, **8**, 2226
63Ha S. D. Hamann, *J. Phys. Chem.*, 1963, **67**, 2233
63Hb J. D. Hale, R. M. Izatt, and J. J. Christensen, *J. Phys. Chem.*, 1963, **67**, 2605
63Hc J. D. Hale, R. M. Izatt, and J. J. Christensen, *Proc. Chem. Soc.*, 1963, 240
63Hd P. B. Hostetler, *Amer. J. Sci.*, 1963, **261**, 238
63K N. P. Komar', *Uchenye Zapiski Khar'kov Univ.*, 1963, **133** (*Trudy Khim. Fak. KhGU.*, **19**) 189
63Ka H. Koch and H. Schmidt, *Z. Naturforsch.*, 1963, **18b**, 936
63Kb P. N. Kovalenko and K. N. Bagdasarov, *Izvest. V.U.Z. Khim.*, 1963, **6**, 546
63Lb L. Levi, O. Milman, and E. Suraski, *Trans. Faraday Soc.*, 1963, **59**, 2064
63Ma L. Meites and R. H. Schlossel, *J. Phys. Chem.*, 1963, **67**, 2397
63Pc L. Pajdowski, *Roczniki Chem.*, 1963, **37**, 1351
63Pd L. Pajdowski, *Roczniki Chem.*, 1963, **37**, 1363
63Ra A. I. Rivkind, *Zhur. strukt. Khim.*, 1963, **4**, 664
63S P. Schindler, *Chimia* (*Switz.*) 1963, **17**, 313
63Sf M. Schellenberg, dissertation, Eidgenössische Technische Hochschule, Zürich, 1963
63Sg J. Shankar and B. C. De Souza, *Austral. J. Chem.*, 1963, **16**, 1119
63Sh G. Schorsch and J. Byé, *Compt. rend.*, 1963, **257**, 2833
63Va C. E. Vanderzee and J. A. Swanson, *J. Phys. Chem.*, 1963, **67**, 2608
64A F. Achenza, *Ann Chim.* (*Italy*) 1964, **54**, 240
64Aa B. Aurivillius and A. Löwenhielm, *Acta Chem. Scand.*, 1964, **18**, 1937
64Ab B. Aurivillius, *Acta Chem. Scand.*, 1964, **18**, 2375
64B I. B. Baranovskii and A. V. Babaeva, *Zhur. neorg. Khim.*, 1964, **9**, 2163
64Ba M. Bartušek and L. Sommer, *Z. phys. Chem.* (*Leipzig*), 1964, **226**, 309
64Bb G. Biedermann and L. Ciavatta, *Arkiv Kemi*, 1964, **22**, 253
64Bc G. Biedermann and L. Newman, *Arkiv Kemi*, 1964, **22**, 303
64Bd J. A. Broomhead, F. Basolo, and R. G. Pearson, *Inorg. Chem.*, 1964, **3**, 826
64Be G. Biedermann, *Svensk kem. Tidskr.*, 1964, **76**, 362
64C I. I. Chernyaev, N. N. Zheligovskaya, and T. N. Leonova, *Zhur. neorg. Khim.*, 1964, **9**, **347**
64Ca I. I. Chernyaev, A. V. Babkov, and N. N. Zheligovskaya. *Zhur. neorg. Khim.*, 1964, **9**, 576
64Cb I. I. Chernyaev and A. V. Babkov, *Zhur. neorg. Khim.*, 1964, **9**, 2307
64Cc E. H. P. Cordfunke, *J. Phys. Chem.*, 1964, **68**, 3353
64Cd M. Cola and A. Perotti, *Gazzetta*, 1964, **94**, 191
64D H. Diebler and N. Sutin, *J. Phys. Chem.*, 1964, **68**, 174
64F J. P. Fackler, jun. and I. D. Chawla, *Inorg. Chem.*, 1964, **3**, 1130
64Fa F. Fromage and J. Faucherre, *Compt. rend.*, 1964, **259**, 3274
64Fb R. Fischer and J. Byé, *Bull. Soc. chim. France*, 1964, 2920
64G K. H. Gayer and O. T. Zajicek, *J. Inorg. Nuclear Chem.*, 1964, **26**, 2123
64Ga W. P. Griffith, *J. Chem. Soc.*, 1964, 245
64Gb A. D. Gel'man and V. P. Zaitseva, *Doklady Akad. Nauk. S.S.S.R.*, 1964, **157**, 1403
64Gc E. Sh. Ganelina, *Zhur. priklad. Khim.*, 1964, **37**, 1358
64Gd R. Guillaumont and G. Boussières, *Bull. Soc. chim. France*, 1964, 2098
64H R. Hoppe and H. J. Röhrborn, *Z. anorg. Chem.*, 1964, **329**, 110
64Ha S. Hietanen and L. G. Sillén, *Acta Chem. Scand.*, 1964, **18**, 1015
64Hb S. Hietanen and L. G. Sillén, *Acta Chem. Scand.*, 1964, **18**, 1018
64Hc K. Heinzinger and R. E. Weston, jun., *J. Phys. Chem.*, 1964, **68**, 744
64Hd K. Heinzinger and R. E. Weston, jun., *J. Phys. Chem.*, 1964, **68**, 2179
64He R. Hugel, *Bull. Soc. chim. France*, 1964, 1462
64Hf F. C. Hentz, jun., and S. Y. Tyree jun., *Inorg. Chem.*, 1964, **3**, 844
64Hg C. J. Hawkins, A. M. Sargeson, and G. H. Searle, *Austral. J. Chem.*, 1964, **17**, 598
64I J. A. Ibers, W. C. Hamilton, and D. R. MacKenzie, *Inorg. Chem.*, 1964, **3**, 1412
64K P. N. Kovalenko and K. N. Bagdasarov, *Zhur. neorg. Khim.*, 1964, **9**, 534
64Ka F. Ya. Kul'ba, Yu. B. Yakovlev, and V. E. Mironov, *Zhur. neorg. Khim.*, 1964, **9**, 2573
64Kb R. B. Kolaczkowski and R. A. Plane, *Inorg. Chem.*, 1964, **3**, 322
64Kc P. N. Kovalenko and K. N. Bagdasarov, *Zhur. priklad. Khim.*, 1964, **37**, 735
64L Ts.-ch. Lyan and Yu.-M. Du, *Zhur. neorg. Khim.*, 1964, **9**, 1333
64La E. N. Lebedeva, S. S. Korovin, and A. M. Rozen, *Zhur. neorg. Khim.*, 1964, **9**, 1744
64Lb J. Levitan and M. Ardon, *Israel J. Chem.*, 1964, **2**, 269
64M H. A. C. McKay and J. L. Woodhead, *J. Chem. Soc.*, 1964, 717
64Ma E. Matijević, G. E. Janauer, and M. Kerker, *J. Colloid. Sci.*, 1964, **19**, 333
64N A. B. Nikol'skii, *Zhur. neorg. Khim.*, 1964, **9**, 2511
64Na D. Naumann, G. Tschirne, and W. Burk, *Z. anorg. Chem.*, 1964, **332**, 63
64Nb T. W. Newton and F. B. Baker, *Inorg. Chem.*, 1964, **3**, 569
64Nc B. I. Nabivanets and L. N. Kudritskaya, *Ukrain. khim. Zhur.*, 1964, **30**, 891
64Nd B. I. Nabivanets and V. V. Lukachina, *Ukrain. khim. Zhur.*, 1964, **30**, 1123
64O H. Ohtaki, *Acta Chem. Scand.*, 1964, **18**, 521

continued overleaf

2 OH^- (contd)

64P R. C. Plumb and J. W. Swaine, jun., *J. Phys. Chem.*, 1964, **68**, 2057
64Pa D. D. Perrin, *J. Chem. Soc.*, 1964, 3644
64Pb J. Podlaha and J. Podlahová, *Coll. Czech. Chem. Comm.*, 1964, **29**, 3164
64Pc W. Plumb and G. M. Harris, *Inorg. Chem.*, 1964, **3**, 542
64Pd R. F. Platford, *Canad. J. Chem.*, 1964, **42**, 181
64R W. V. Rausch and H. D. Bale, *J. Chem. Phys.*, 1964, **40**, 3391
64S V. B. Spivakovskii and L. P. Moisa, *Zhur. neorg. Khim.*, 1964, **9**, 2287
64Sa G. Schorsch, *Bull. Soc. chim. France*, 1964, 1449
64Sb G. Schorsch, *Bull. Soc. chim. France*, 1964, 1456
64Sc J. C. Sullivan, *Inorg. Chem.*, 1964, **3**, 315
64Sd P. Schindler, H. Althaus, and W. Feitknecht, *Helv. Chim. Acta*, 1964, **47**, 982
64Se P. Salomaa, L. L. Schaleger, and F. A. Long, *J. Amer. Chem. Soc.*, 1964, **86**, 1
64Sf I. E. Starik, N. I. Ampelogova, and B. S. Kuznetsov, *Radiokhimiya* 1964, **6**, 519
64Sg B. Siegel and R. L. Johnson, *Nature*, 1964, **204**, 375
64Sh J. Ste-Marie, A. E. Torma, and A. O. Gübeli, *Canad. J. Chem.*, 1964, **42**, 662
64Si G. Stucky and R. E. Rundle, *J. Amer. Chem. Soc.*, 1964, **86**, 4821
64T R. S. Tobias and M. Yasuda, *J. Phys. Chem.*, 1964, **68**, 1820
64Ta G. Thompson, Thesis, U. of Cal., Berkeley, 1964, UCRL-11410
64Tb R. S. Tobias and M. Yasuda, *Canad. J. Chem.*, 1964, **42**, 781
64W H. Wenger, dissertation, Eidgenössische Technische Hochschule, Zürich, 1964
64Wa R. M. Wallace, *J. Phys. Chem.*, 1964, **68**, 2418
64Z A. I. Zhukov and A. S. Nasarov, *Zhur. neorg. Khim.*, 1964, **9**, 1465
64Za A. Zalkin, J. D. Forrester, and D. H. Templeton, *Inorg. Chem.*, 1964, **3**, 1417
65A I. P. Alimarin, Sh. A. Khamid, and I. V. Puzdrenkova, *Zhur. neorg. Khim.*, 1965, **10**, 389
65Aa J. Aveston, *J. Chem. Soc.*, 1965, 4438
65B K. A. Burkov, L. S. Lilič, and L. G. Sillén, *Acta Chem. Scand.*, 1965, **19**, 14
65Ba F. Bertin, G. Thomas, and J.-C. Merlin, *Compt. rend.*, 1965, **260**, 1670
65Bb V. Baran, *Z. Chem.*, 1965, **5**, 56
65Bc C. F. Baes, jun., N. J. Meyer, and C. E. Roberts, *Inorg. Chem.*, 1965, **4**, 518
65Bd C. G. Barraclough and R. S. Murray, *J. Chem. Soc.*, 1965, 7047
65Be K. A. Burkov and L. S. Lilich, *Vestnik Leningrad. Univ.*, 1965, no. 10 (*Fiz. Khim.*, no. 1), 103
65C R.-P. Courgnaud and B. Trémillon, *Bull. Soc. chim. France*, 1965, 752
65D L. A. Dunn, R. H. Stokes, and L. G. Hepler, *J. Phys. Chem.*, 1965, **69**, 2808
65E O. E. Esval and J. S. Johnson, jun., *J. Phys. Chem.*, 1965, **69**, 959
65F J. J. Fripiat, F. Van Cauwelaert, and H. Bosmans, *J. Phys. Chem.*, 1965, **69**, 2458
65Fa A. Ferse, *Z. phys. Chem. (Leipzig)*, 1965, **229**, 51
65G R. Guillaumont, *Compt. rend.*, 1965, **260**, 1416
65Ga R. W. Green and P. W. Alexander, *Austral. J. Chem.*, 1965, **18**, 651
65Gb A. D. Gel'man and V. P. Zaitseva, *Radiokhimiya*, 1965, **7**, 49
65Gc R. Guillaumont, *Bull. Soc. chim. France*, 1965, 135
65Gd R. Guillaumont, *Bull. Soc. chim. France*, 1965, 2106
65H R. Hugel, *Bull. Soc. chim. France*, 1965, 968
65Ha F. C. Hentz, jun. and S. Y. Tyree, jun., *Inorg. Chem.*, 1965, **4**, 873
65Hb H. Holzapfel and K. Dittrich, *Z. Chem.*, 1965, **5**, 314
65Hc W. E. Hatfield, J. F. Anders, and L. J. Rivela, *Inorg. Chem.*, 1965, **4**, 1088
65Hd H. P. Hopkins, jun. and C. A. Wulff, *J. Phys. Chem.*, 1965, **69**, 6
65He H. P. Hopkins, jun. and C. A. Wulff, *J. Phys. Chem.*, 1965, **69**, 1980
65Hf L. G. Hepler, J. M. Stokes, and R. H. Stokes, *Trans. Faraday Soc.*, 1965, **61**, 20
65I B. N. Ivanov-Emin, V. I. Rybina, and V. I. Kornev, *Zhur. neorg. Khim.*, 1965, **10**, 1005
65Ia Y. J. Israéli, *Bull. Soc. chim. France*, 1965, 193
65J G. Jander and H. J. Hartmann, *Z. anorg. Chem.*, 1965, **339**, 239, 248, 256
65Ja K. F. Jahr and I. Pernoll, *Ber. Bunsengesellschaft Phys. Chem.*, 1965, **69**, 221, 226
65K F. Ya. Kul'ba, Yu. B. Yakovlev, and V. E. Mironov, *Zhur. neorg. Khim.*, 1965, **10**, 1624
65Ka V. N. Kumok and V. V. Serebrennikov, *Zhur. neorg. Khim.*, 1965, **10**, 2011
65L R. V. Lundquist, U.S. Bureau of Mines Rept. Invest. no. 6582, 1965
65M E. Matijević, *J. Colloid Sci.*, 1965, **20**, 322
65Ma M. Mansmann, *Z. anorg. Chem.*, 1965, **339**, 52
65Mb J. S. Magee, jun. and R. H. Wood, *Canad. J. Chem.*, 1965, **43**, 1234
65N T. W. Newton and F. B. Baker, *Inorg. Chem.*, 1965, **4**, 1166
65Na T. Nishide and R. Tsuchiya, *Bull. Chem. Soc. Japan*, 1965, **38**, 1398
65P L. Pajdowski, *Chem. Zvesti*, 1965, **19**, 192
65Pa J. A. Pérez-Bustamante, *Radiochim. Acta*, 1965, **4**, 67
65R D. E. Ryan, J. R. Dean, and R. M. Cassidy, *Canad. J. Chem.*, 1965, **43**, 999
65S A. V. Stepanov and V. P. Shvedov, *Zhur. neorg. Khim.*, 1965, **10**, 1000
65Sa G. K. Schweitzer and S. W. McCarty, *J. Inorg. Nuclear Chem.*, 1965, **27**, 191
65Sb T. Sekine, *Acta Chem. Scand.*, 1965, **19**, 1526
65Sc G. Schorsch, *Bull. Soc. chim. France*, 1965, 988

continued

65Sd G. Schwarzenbach and M. Schellenberg, *Helv. Chim. Acta*, 1965, **48**, 28
65Se P. Schindler, H. Althaus, F. Hofer, and W. Minder, *Helv. Chim. Acta*, 1965, **48**, 1204
65Sf M. E. Shying, J. Aggett, and R. B. Temple, *Austral. J. Chem.*, 1965, **18**, 1719
65T R. S. Tobias and C. E. Freidline, *Inorg. Chem.*, 1965, **4**, 215
65Sg V. I. Spitsyn, R. A. D'yachkova, and V. P. Khlebnikov, *Radiokhimiya*, 1965, **7**, 257
65Sh A. Kh. Sherif, I. P. Alimarin, and I. V. Puzdrenkova, *Vestnik Moskov. Univ.*, 1965, *Khim.*, 3, 71
65Si A. P. Savostin, *Zhur. neorg. Khim.*, 1965, **10**, 2565
65Sj P. Souchay, M. Cadiot, and M. Duhameux, *Compt. rend.*, 1965, **260**, 186
65V R. C. Vickery, *Nature*, 1965, **208**, 678
65W K. A. Wilhelmi and O. Jonsson, *Acta Chem. Scand.*, 1965, **19**, 177
65Wa W. Wojciechowski, B. Jeżowska-Trzebiatowska, and N. Rudolf, *Chem. Zvesti*, 1965, **19**, 229
65Wb C. F. Wells and M. A. Salam, *Nature*, 1965, **205**, 690
65Wc C. F. Wells and G. Davies, *Nature*, 1965, **205**, 692
65Z D. M. Ziv and I. A. Shestakova, *Radiokhimiya*, 1965, **7**, 175
66A C. Andrade and H. Taube, *Inorg. Chem.*, 1966, **5**, 1087
66Aa J. Aveston, *J. Chem. Soc. (A)*, 1966, 1599
66B F. Brito, *Anales de Quim.*, 1966, **62**, 193
66Ba H. Bilinski and M. Branica, *Croat. Chim. Acta*, 1966, **38**, 263
66Bb V. Baran and M. Tympl, *J. Inorg. Nuclear Chem.*, 1966, **28**, 89
66Bc F. M. Brewer, P. L. Goggin, and G. S. Reddy, *J. Inorg. Nuclear Chem.*, 1966, **28**, 361
66Bd H. Bilinski, M. Branica, and L. G. Sillén, *Acta Chem. Scand.*, 1966, **20**, 853
66Be G. Biedermann and J. T. Chow, *Acta Chem. Scand.*, 1966, **20**, 1376
66Bf G. L. Bertrand, F. J. Millero, C.-H. Wu, and L. G. Hepler, *J. Phys. Chem.*, 1966, **70**, 699
66Bg K. A. Burkov and L. V. Ivanova, *Vestnik Leningrad Univ.*, 1966, no. 16 (*Fiz. Khim.*, no. 3), 120
66Bh H. Bilinski and M. Branica, *Croat. Chim. Acta*, 1966, **38**, 263
66C T. J. Conocchioli, G. H. Nancollas, and N. Sutin, *Inorg. Chem.*, 1966, **5**, 1
66Ca S. C. Chan and F. Leh, *J. Chem. Soc. (A)*, 1966, 129
66Cb S. C. Chan, *J. Chem. Soc. (A)*, 1966, 1124
66Cc A. K. Covington, R. A. Robinson, and R. G. Bates, *J. Phys. Chem.*, 1966, **70**, 3820
66Cd R. D. Cannon and J. E. Earley, *J. Amer. Chem. Soc.*, 1966, **88**, 1872
66F C. E. Freidline and R. S. Tobias, *Inorg. Chem.*, 1966, **5**, 354
66Fa U. K. Frolova, V. N. Kumok, and V. V. Serebrennikov, *Izvest. V.U.Z. Khim.*, 1966, **9**, 176
66G R. Guillaumont, *Rev. Chim. minérale*, 1966, **3**, 339
66Ga A. A. Grinberg, A. I. Stetsenko, and N. D. Mitkinova, *Zhur. neorg. Khim.*, 1966, **11**, 2075
66Gb A. A. Grinberg, Kh. I. Gil'dengershel', and V. F. Budanova, *Zhur. neorg. Khim.*, 1966, **11**, 2519
66H F. C. Hentz, jun., and J. S. Johnson, *Inorg. Chem.*, 1966, **5**, 1337
66Ha W. Holzapfel and E. U. Franck, *Ber. Bunsengesellschaft Phys. Chem.*, 1966, **70**, 1105
66Hb K. Hahnefeld, dissertation, Freie Univ., Berlin, 1966
66I Y. Iitaka, K. Shimizu, and T. Kwan, *Acta Cryst.*, 1966, **20**, 803
66Ia B. N. Ivanov-Emin, E. N. Siforova, M. Mekes Fisher, and V. Mel'yado Kampos, *Zhur. neorg. Khim.*, 1966, **11**, 475
66Ib B. N. Ivanov-Emin, E. N. Siforova, V. Mel'yado Kampos, and E. Balestr Lafert, *Zhur. neorg. Khim.*, 1966, **11**, 1975
66J G. Johansson, *Acta Chem. Scand.*, 1966, **20**, 505
66Ja R. B. Jordan, A. M. Sargeson, and H. Taube, *Inorg. Chem.*, 1966, **5**, 1091
66K P. N. Kovalenko and G. G. Shchemeleva, *Zhur. priklad. Khim.*, 1966, **39**, 2440
66Ka P. N. Kovalenko, L. T. Azhipa, and M. M. Evstifeev, *Zhur. neorg. Khim.*, 1966, **11**, 2689
66M D. B. McWhan and G. Lundgren, *Inorg. Chem.*, 1966, **5**, 284
66Ma R. E. Mesmer and R. R. Irani, *J. Inorg. Nuclear Chem.*, 1966, **28**, 493
66Mb W. L. Masterton and L. H. Berka, *J. Phys. Chem.*, 1966, **70**, 1924
66Mc P. Müller, dissertation, Technische Hochschule, Karlsruhe, 1966
66N D. Naumann and G. Reinhard, *Z. anorg. Chem.*, 1966, **343**, 165
66O H. G. Offner and D. A. Skoog, *Analyt. Chem.*, 1966, **38**, 1520
66Oa Z. Orhanović, B. Pokrić, H. Füredi, and M. Branica, *Croat. Chem. Acta*, 1966, **38**, 269
66P L. Pajdowski, *J. Inorg. Nuclear Chem.*, 1966, **28**, 433
66Pa L. Pajdowski and B. Jeżowska-Trzebiatowska, *J. Inorg. Nuclear Chem.*, 1966, **28**, 443
66S J. H. Swinehart and P. A. Rock, *Inorg. Chem.*, 1966, **5**, 573
66Sa P. Salomaa and V. Aalto, *Acta Chem. Scand.*, 1966, **20**, 2035
66Sb A. S. Solovkin and A. I. Ivantsov, *Zhur. neorg. Khim.*, 1966, **11**, 1897
66Sc A. P. Savostin, *Zhur. neorg. Khim.*, 1966, **11**, 2817
66Sd T. G. Spiro, S. E. Allerton, *et al.*, *J. Amer.Chem. Soc.*, 1966, **88**, 2721
66Se P. Souchay, M. Cadiot, and M. Duhameaux, *Compt. rend.*, 1966, *C*, **262**, 1524
66Sf K. Swaminathan and G. M. Harris, *J. Amer. Chem. Soc.*, 1966, **88**, 4411
66T R. S. Tobias, H. N. Farrer, M. B. Hughes, and B. A. Nevett, *Inorg. Chem.*, 1966, **5**, 2052
66Ta B. Tomažić and M. Branica, *Croat. Chim. Acta*, 1966, **38**, 249
66W G. W. Watt, J. T. Summers, E. M. Potrafke, and E. R. Birnbaum, *Inorg. Chem.*, 1966, **5**, 857
66Wa J. L. Woodhead, A. M. Deane, A. C. Fox, and J. M. Fletcher, *J. Inorg. Nuclear Chem.*, 1966, **28**, 2175
66Wb R. Wyatt, *Chem. Weekblad*, 1966, **62**, 310
66Wc J. O. Wear and E. Matijević, *J. Phys. Chem.*, 1966, **70**, 3825

continued overleaf

2 OH^- (contd)

67A P. Andersen, *Acta Chem. Scand.*, 1967, **21**, 243
67Aa R. Arnek and W. Kąkołowicz, *Acta Chem. Scand.*, 1967, **21**, 1449
67Ab R. Arnek and W. Kąkołowicz, *Acta Chem. Scand.*, 1967, **21**, 2180
67Ac G. Anderegg, *Helv. Chim. Acta*, 1967, **50**, 2333
67Ad L. T. Azhipa, P. N. Kovalenko, and M. M. Evstifeev, *Zhur. neorg. Khim.*, 1967, **12**, 1138
67Ae M. Anbar and I. Pecht, *J. Amer. Chem. Soc*, 1967, **89**, 2553
67B D. A. Buckingham, I. I. Olsen, A. M. Sargeson, and H. Satrapa, *Inorg. Chem.*, 1967, **6**, 1027
67Ba F. Brito and D. Lewis, *Arkiv Kemi*, 1967, **26**, 391
67Bb H. L. Bott and A. J. Poë, *J. Chem. Soc.*, (*A*), 1967, 205
67Bc M. Beran, *Coll. Czech. Chem. Comm.*, 1967, **32**, 1368
67Bd J.-C. Bavay, G. Nowogrocki, and G. Tridot, *Bull. Soc. chim. France*, 1967, 2026
67Be F. Bertin, G. Thomas, and J.-C. Merlin, *Bull. Soc. chim. France*, 1967, 2393
67Bf M. Barrès, *Rev. Chim. minerale*, 1967, **4**, 803
67C C. Cohen-Addad, P. Ducros, and E. F. Bertaut, *Acta Cryst.*, 1967, **23**, 220
67Ca M. P. Collados, F. Brito, and R. Díaz Cadavieco, *Anales Fiz. Quim.*, 1967, *B*, 843
67Cb S. C. Chan, *J. Chem. Soc.* (*A*), 1967, 2103
67Cc P. Chartier, *Bull. Soc. chim. France*, 1967, 2706
67D P. R. Danesi, *Acta Chem. Scand.*, 1967, **21**, 143; private communication on misprint
67G H. Gamsjäger and P. Schindler, *Helv. Chim. Acta*, 1967, **50**, 2053
67Ga Yu. I. Gryzin and K. Z. Koryttsev, *Zhur. neorg. Khim.*, 1967, **12**, 101
67Gb Kh. I. Gil'dengershel', *Zhur. neorg. Khim.*, 1967, **12**, 389
67Gc V. Gold and B. M. Lowe, *J. Chem. Soc.* (*A*), 1967, 936
67Gd A. O. Gubeli and J. Ste-Marie, *Canad. J. Chem.*, 1967, **45**, 827
67H H. C. Helgeson, *J. Phys. Chem.*, 1967, **71**, 3121
67Ha J. Haladjian and G. Carpéni, *J. Chim. phys.*, 1967, **64**, 1338
67Hb M. Herlem, *Bull. Soc. chim., France*, 1967, 1687
67I R. M. Izatt, D. Eatough, and J. J. Christensen, *J. Chem. Soc.* (*A*), 1967, 1301
67J D. V. S. Jain, C. M. Jain, and R. N. Vaid, *J. Chem. Soc.* (*A*), 1967, 915
67K B. T. Kilbourn and J. D. Dunitz, *Inorg. Chim. Acta*, 1967, **1**, 209
67L B. I. Lobov, F. Ya. Kul'ba, and V. E. Mironov, *Zhur. neorg. Khim.*, 1967, **12**, 334, 341
67La M. Lucas, *Bull. Soc. chim. France*, 1967, 3842
67M R. K. Murmann and J. C. Sullivan, *Inorg. Chem.*, 1967, **6**, 892
67Ma R. E. Mesmer and C. F. Baes, jun., *Inorg. Chem.*, 1967, **6**, 1951
67Mb V. A. Maroni and T. G. Spiro, *J. Amer. Chem. Soc.*, 1967, **89**, 45
67O H. Ohtaki, *Inorg. Chem.*, 1967, **6**, 808
67Oa H. Ohtaki and H. Kato, *Inorg. Chem.*, 1967, **6**, 1935
67P B. Pokrić and M. Branica, *Croat. Chem. Acta*, 1967, **39**, 11
67Pa J. Perichon and R. Buvet, *Bull. Soc. chim. France*, 1967, 3697
67Pb A. J. Poë, K. Shaw, and M. J. Wendt, *Inorg. Chim. Acta*, 1967, **1**, 371
67S P. L. Sarma and M. S. Davis, *J. Inorg. Nuclear Chem.*, 1967, **29**, 2607
67Sa G. Schorsch and N. Ingri, *Acta Chem. Scand.*, 1967, **21**, 2727
67Sb A. S. Solovkin, Z. N. Tsvetkova, and A. I. Ivantsov, *Zhur. neorg. Khim.*, 1967, **12**, 626
67Sc Yu. I. Sannikov, E. I. Krylov, and V. M. Vinogradov, *Zhur. neorg. Khim.*, 1967, **12**, 2651
67Sd J. Steigman and D. Sussman, *J. Amer. Chem. Soc.*, 1967, **89**, 6400
67V V. P. Vasil'ev and G. A. Lobanov, *Zhur. fiz. Khim.*, 1967, **41**, 838
67Va A. T. Vagramjan and T. I. Leshawa, *Z. phys. Chem.* (*Leipzig*), 1967, **234**, 57
67W C. F. Wells and G. Davies, *J. Chem. Soc.* (*A*), 1967, 1858
67Y L. B. Yeatts and W. L. Marshall, *J. Phys. Chem.*, 1967, **71**, 2641
68A R. Arnek and C. C. Patel, *Acta Chem. Scand.*, 1968, **22**, 1097
68Aa R. Arnek, *Acta Chem. Scand.*, 1968, **22**, 1102
68Ab R. Arnek and K. Schlyter, *Acta Chem. Scand.*, 1968, **22**, 1327
68Ac R. Arnek and K. Schlyter, *Acta Chem. Scand.*, 1968, **22**, 1331
68Ad V. P. Antonovich and V. A. Nazarenko, *Zhur. neorg. Khim.*, 1968, **13**, 1805
68Ae M. Asso and G. Carpéni, *Canad. J. Chem.*, 1968, **46**, 1795
68B E. Bang, *Acta Chem. Scand.*, 1968, **22**, 2671
68Ba C. Brémard, G. Nowogrocki, and G. Tridot, *Bull.Soc. chim. France*, 1968, 1961
68C J. G. Connery and R.E. Cover, *Analyt. Chem.*, 1968, **40**, 87
68Ca A. Chrétien and D. Bizot, *Compt. rend.*, 1968, **266**, *C*, 1688
68Cb D. B. Copley and S. Y. Tyree, jun., *Inorg. Chem.*, 1968, **7**, 1472
68Cc S. C. Chan and K. Y. Hui, *Austral. J. Chem.*, 1968, **21**, 3061
68Cd G. L. Cohen and G. Atkinson, *J. Electrochem. Soc.*, 1968, **115**, 1236
68D S. M. Dumpiérrez and F. Brito, *Anales de Quim*, 1968, **64**, 115
68Da P. R. Daris, M. Magini, S. Margherita, and G. D. Alessandro, *Energia Nucleare*, 1968, **15**, 335
68F S. Fontana and F. Brito, *Inorg. Chim. Acta*, 1968, **2**, 179
68G A. A. Grinberg, A. I. Stetsenko, and S. G. Strelin, *Zhur. neorg. Khim.*, 1968, **13**, 814, 1089
68Ga A. A. Grinberg, Kh. I. Gil'dengershel', and V. F. Budanov, *Zhur. neorg. Khim.*, 1968, **13**, 1365
68Gb A. O. Gubeli and J. Ste-Marie, *Canad. J. Chem.*, 1968, **46**, 1707

continued

68Gc R. Guillaumont, *Bull. Soc. chim. France*, 1968, 162
68Gd R. Guillaumont, *Bull. Soc. chim. France*, 1968, 168
68Ge G. E. Glass, J. H. Konnert, M. G. Miles, D. Britton, and R. S. Tobias, *J. Amer. Chem. Soc.*, 1968, **90**, 1131
68Gf R. N. Goldberg and L. G. Hepler, *J. Phys. Chem.*, 1968, **72**, 4654
68Gg A. A. van der Giessen, dissertation, Technische Hogeschool, Eindhoven, 1968
68H S. Hietanen and L. G. Sillén, *Acta Chem. Scand.*, 1968, **22**, 265
68Ha L. P. Holmes, D. L. Cole, and E. M. Eyring, *J. Phys. Chem.*, 1968, **72**, 301
68Hb S. J. Harris, unpublished, quoted by ref 68T
68Hc J. Hill and A. McAuley, *J. Chem. Soc.* (*A*), 1968, 2405
68Hd K. Hestermann and R. Hoppe, *Z. anorg. Chem.*, 1968, **360**, 113
68I B. I. Ivanov-Emin, L. D. Borzova, B. E. Zaitsev, and M. Mekes Fisher, *Zhur. neorg. Khim.*, 1968, **13**, 2238
68J G. Johansson, *Acta Chem. Scand.*, 1968, **22**, 389
68Ja G. Johansson, *Acta Chem. Scand.*, 1968, **22**, 399
68Jb G. Johansson and Å. Olin, *Acta Chem. Scand.*, 1968, **22**, 3197
68K N. N. Krot, M. P. Mefod'eva, T. V. Smirnova, and A. D. Gel'man, *Radiokhimiya*, 1968, **10**, 412; N. N. Krot, M. P. Mefod'eva, F. A. Zakharova, T. V. Smirnova, and A. D. Gel'man, *ibid.*, p. 630; N. N. Krot, M. P. Mefod'eva, and A. D. Gel'man, *ibid.*, p. 634
68Ka Yu. A. Komkov, N. N. Krot, and A. D. Gel'man, *Radiokhimiya*, 1968, **10**, 625
68Kb P. N. Kovalenko, L. T. Azhipa, and M. M. Evstifeev, *Zhur. priklad. Khim.*, 1968, **41**, 198
68Kc E. V. Komarov, V. N. Krylov, and M. F. Pushlenkov, *Zhur. neorg. Khim.*, 1968, **13**, 467
68Kd G. Kiel and G. Gattow, *Naturwiss*, 1968, **55**, 389
68L E. Lanza and G. Carpéni, *Electrochim. Acta*, 1968, **13**, 519
68La S. F. Lincoln and D. R. Stranks, *Austral. J. Chem.*, 1968, **21**, 1745
68M J. I. Morrow and J. Levy, *J. Phys. Chem.*, 1968, **72**, 885
68Ma S. K. Mishra and Y. K. Gupta, *Indian J. Chem.*, 1968, **6**, 757
68Mb V. Ya. Mishin and N. A. Parpiev, *Uzbek. khim. Zhur.*, 1968, no. 2, 72
68Mc V. A. Maroni and T. G. Spiro, *Inorg. Chem.*, 1968, **7**, 183
68Md V. A. Maroni and T. G. Spiro, *Inorg. Chem.*, 1968, **7**, 188
68Me J. Mašek and J. Dempír, *Inorg. Chim. Acta*, 1968, **2**, 443
68N V. A. Nazarenko, V. P. Antonovich, and E. M. Nevskaya, *Zhur. neorg. Khim.*, 1968, **13**, 1574
68O H. Ohtaki, *Inorg. Chem.*, 1968, **7**, 1205
68P J. R. Perumareddi and A. W. Adamson, *J. Phys. Chem.*, 1968, **72**, 414
68Pa M. R. Pâris and Cl. Gregoire, *Analyt. Chim. Acta*, 1968, **42**, 431
68Pb M. R. Pâris and Cl. Gregoire, *Analyt. Chim. Acta*, 1968, **42**, 439
68R D. R. Rosseinsky and M. J. Nicol, *J. Chem. Soc.* (*A*), 1968, 1022
68S U. Schedin and M. Frydman, *Acta Chem. Scand.*, 1968, **22**, 115
68Sa A. M. Shams el Din, T. Gouda, and A. A. El Hosary, *J. Electroanalyt. Chem.*, 1968, **17**, 137
68Sb V. B. Spivakovskii and G. V. Makovskaya, *Zhur. neorg. Khim.*, 1968, **13**, 1555
68Sc V. B. Spivakovskii and G. V. Makovskaya, *Zhur. neorg. Khim.*, 1968, **13**, 2764
68Sd K. Srinivasan and G. A. Rechnitz, *Analyt. Chem.*, 1968, **40**, 1818
68Se R. C. Splinter, S. J. Harris, and R. S. Tobias, *Inorg. Chem.*, 1968, **7**, 897
68Sf T. G. Spiro, D. H. Templeton, and A. Zalkin, *Inorg. Chem.*, 1968, **7**, 2165
68Sg H. Sabrowsky and R. Hoppe, *Z. anorg. Chem.*, 1968, **358**, 241
68Sh H. Siebert and R. Schiedermaier, *Z. anorg. Chem.*, 1968, **361**, 169
68T R. S. Tobias, M. J. Sprague, and G. E. Glass, *Inorg. Chem.*, 1968, **7**, 1714
68W K. B. Wiberg and P. C. Ford, *Inorg. Chem.*, 1968, **7**, 369
68Wa C. F. Wells and M. A. Salam, *J. Chem. Soc.* (*A*), 1968, 24
68Wb H.-D. Wasel-Nielen and R. Hoppe, *Z. anorg. Chem.*, 1968, **359**, 36
68Z V. P. Zaitseva, D. P. Alekseeva, and A. D. Gel'man, *Radiokhimiya*, 1968, **10**, 537
68Za O. E. Zvyagintsev and S. B. Lyakhmanov, *Zhur. neorg. Khim.*, 1968, **13**, 1230
68Zb P. Zanella, G. Plazzogna, and G. Tagliavini, *Inorg. Chim. Acta*, 1968, **2**, 340

Vanadate, $L^{2-} = VO_2(OH)_3^{2-} = HVO_4^{2-}$ 3

Metal	*Method*	*Temp*	*Medium*	*Log of equilibrium constant, remarks*	*Ref*
				Complexes with at most 1 H^+ *per* L^{2-}, *thus charge per* V $\leqslant -1$	
H^+		25	0·5 Na(Cl)	$K(L^{2-} \rightleftharpoons HL^- + OH^-)$ −5·89 $K(2L^{2-} \rightleftharpoons HL_2^{3-} + OH^-)$ −3·16, $K(3L^{2-} \rightleftharpoons H_3L_3^{3-} + 3OH^-)$ −10·56 $K(4L^{2-} \rightleftharpoons H_4L_4^{4-} + 4OH^-)$ −13·55,	59I/64B

continued overleaf

3 $L^{2-} = VO_2(OH)_3^{2-} = HVO_4^{2-}$ (contd)

Metal	*Method*	*Temp*	*Medium*	*Log of equilibrium constant, remarks*	*Ref*
H^+				$K(4L^{2-} \rightleftharpoons H_2L_4^{6-} + 2OH^-)$ −4·70	
(*contd*)				$K(2L^{2-} \rightleftharpoons L_2^{4-})$ 0·44	

1964 *Addendum*: p. 86, *for ref* "59Jd ≈ 59Jc" *read* "59Jd ≈ 59Jc ≈ 63F"

Metal	*Method*	*Temp*	*Medium*	*Log of equilibrium constant, remarks*	*Ref*
	fp	−1	Na_2SO_4 sat	ev $H_4L_4^{4-}$ ($V_4O_{12}^{4-}$)	63Ja
	cfu			ev $H_4L_4^{4-}$ ($V_4O_{12}^{4-}$)	63Sb
	Cu	350	KNO_3(l)	$K(V_2O_7^{4-} + O^{2-} \rightleftharpoons 2VO_4^{3-})$ 6·65 in KNO_3(l)	63Sc
	fp	33	Na_2SO_4 sat	ev $H_4L_4^{4-}$	64N
	oth	25	var	ev $V_4O_{12}^{4-}$ (diffusion method)	64S
	Hg		0·1 ($NaClO_4$)	ev $H_3L_3^{3-}$($V_3O_9^{3-}$), not $H_4L_4^{4-}$($V_4O_{12}^{4-}$) at log [V] ≈ −4 to −2·3	64Sa
	O	350	KNO_3(l)	$K(V_2O_7^{4-} + O^{2-} \rightleftharpoons 2VO_4^{3-})$ 6·65 in KNO_3(l)	64Sc
	oth	25	NaHL var	$H_4L_4^{4-}$ predom for 0·08 to 0·2M-V, perhaps smaller cpx at lower [V]	65C
	oth	25	Na_2L	L_2^{4-} predom, for 0·08 to 0·2M-V (light scattering)	
	nmr		var	ev six species, tentative assignment	65H
	nmr, gl	23	1 ($NaClO_4$)	ev VO_4^{3-}, HVO_4^{2-}, $V_2O_7^{4-}$, $HV_2O_7^{3-}$, VO_3^-, $V_3O_9^{3-}$, or $V_4O_{12}^{4-}$ (agrees "broadly" with ref 59I)	65Ha
	Ram, ir		var	ev VO_4^{3-}, HVO_4^{2-}, $V_2O_7^{4-}$(L_2^{4-}), $HV_2O_7^{3-}$(HL_2^{3-}), $(VO_3)_n^{n-}$($H_nL_n^{n-}$)	66G
	var	32	var	ev cpx with 0, 1, and 1·6 H^+ per L^{2-} such as L_2^{4-}, $H_4L_4^{4-}$, and $H_{16}L_{10}^{4-}$ (gl, con, conJ)	66Sa
	oth	25	0·5 (NaCl)	*T*-jump relaxation data consistent with HL^-, L^{2-}, HL_2^{3-}, and $H_3L_3^{3-}$, equil constants of ref 59I	66W
	X		solid	ev $V_2O_7^{4-}$ in $Cd_2V_2O_7$(s)	67A
	gl	40	0·5 Na(Cl)	$K(3HL^- \rightleftharpoons H_3L_3^{3-})$ 7·14 $K(4HL^- \rightleftharpoons H_4L_4^{4-})$ 10·10	67B
	X		solid	ev $V_2O_7^{4-}$ in $Mn_2V_2O_7$(s)	67D
	Ram, ir		OH^-	ev tetrahedral VO_4^{3-} at high [OH^-]	67M

Complexes with more than 1 H^+ *per* L^{2-}, *charge per* V > −1

Metal	*Method*	*Temp*	*Medium*	*Log of equilibrium constant, remarks*	*Ref*
	gl, red	20?	0 corr	$K(HL^- + 2H^+ \rightleftharpoons VO_2^+)$ 5·26	64D
	dis	25	0·5 Na(ClO_4)	$K_{13}(H_2L + H^+ \rightleftharpoons VO_2^+)$ 3·20, $K_{12}(HL^- + H^+ \rightleftharpoons H_2L)$ 3·78 $K_D[H_2L(aq) \rightleftharpoons H_2L(hexol)]$ 1·99	64Da
	X		solid	ev $V_{10}O_{28}^{6-}$ in $Ca_3V_{10}O_{28}(H_2O)_{16}$ (s, pascovite) and $K_2Zn_2V_{10}O_{28}(H_2O)_{16}$ (s, hummerite)	64E
	gl, fp	33	Na_2SO_4 sat	$*K(10VO_2^+ \rightleftharpoons H_{16}L_{10}^{4-} + 14H^+)$ −2·4 $K_a(H_{16}L_{10}^{4-})$ −2·19, no ev species with 6 V	64N
	sol, pH		dil	ev $HV_4O_7^{7+}$, $V_8O_{11}^{18+}$, $V_3O_8^-$(?), K_s values(?)	64R
	kin	25	0·06	$K_{13}(H_2L + H^+ \rightleftharpoons VO_2^+)$ 3·21, $K(H_3L_2^- + H^+ \rightleftharpoons 2H_2L)$ −3·55	64Y
	fp	32	Na_2SO_4 sat	ev $H_4L_4^{4-}$, $H_7L_5^{3-}$($V_5O_{14}^{3-}$), $H_9L_6^{3-}$($HV_6O_{17}^{3-}$) and $H_{16}L_{10}^{4-}$ ($H_2V_{10}O_{28}^{4-}$)	65B+
	nmr, gl	23	1 ($NaClO_4$)	ev $V_{10}O_{28}^{6-}$, $HV_{10}O_{28}^{5-}$, $H_2V_{10}O_{28}^{4-}$, VO_2^+	65Ha
	oth	20	0·1	$\beta_{13}(3H^+ + L^{2-} \rightleftharpoons VO_2^+)$ 16·17 (gl, sp from equilibria with EDTA)	65P
	gl	40	0·5 Na(Cl)	$K(3HL^- \rightleftharpoons H_3L_3^{3-})$ 7·14, $K(4HL^- \rightleftharpoons H_4L_4^{4-})$ 10·10, $K(5H^+ + 10HL^- \rightleftharpoons H_{15}L_{10}^{5-})$ 57·42, $K(6H^+ + 10HL^- \rightleftharpoons H_{16}L_{10}^{4-})$ 61·81,	66B

continued

Metal	*Method*	*Temp*	*Medium*	*Log of equilibrium constant, remarks*	*Ref*
H^+				$K(3H^+ + 6HL^- \rightleftharpoons H_9L_6^{3-})$ 33·04	
(*contd*)	cal	25	→ 0	Δ^*H_s [$\frac{1}{2}V_2O_5(s) + H^+ \rightleftharpoons VO_2^+ + \frac{1}{2}H_2O$] = −4·2	66Ba
	X		solid	ev $V_{10}O_{28}^{6-}(H_{14}L_{10}^{6-})$ in $K_2Zn_2V_{10}O_{28}(H_2O)_{16}(s)$	66E
	sp	19	1·01 ($NaClO_4$)	$^*K(10VO_2^+ \rightleftharpoons H_{16}L_{10}^{4-} + 14H^+)$ −6·93	66I
	gl	40	0·5 Na(Cl)	$K(5H^+ + 10HL^- \rightleftharpoons H_{15}L_{10}^{5-})$ 57·42, $K(6H^+ + 10HL^- \rightleftharpoons H_{16}L_{10}^{4-})$ 61·81 $K(3H^+ + 6HL^- \rightleftharpoons H_9L_6^{3-})$ 33·04	67B
	dis, gl	25	0·1 ($NaClO_4$)	K_{12} 3·45, K_{13} 3·35	67T
Ca^{2+}	*X*		solid	ev $[(H_2O)_5Ca]_2V_{10}O_{28}^{2-}$ in $Ca_3V_{10}O_{28}(H_2O)_{17}$ (s, pascoite)	66S
V^{4+}	pre, sp		var	ev $V_{10}O_{26}^{4-}$, $HV_{10}O_{26}^{4-}$ (stable), $HV_{10}O_{24}^{4-}$ (stable), (with 2, 3, 7 V^{4+})	63O = 64Oa
	pre, sp		var	ev $V_{10}O_{24}^{4-}$, $HV_{10}O_{25}^{4-}$, $V_{10}O_{25}^{4-}$ (with 6, 5, 4 V^{4+})	63Oa = 64Oa
	var		var	ev $HV_6O_{15}^{3-}$ (with 4 V^{4+}), (pre, sp, pol)	64O
	gl, sp		0·1 NaCl	$K(2HVO_4^{2-} + V_4O_9^{2-} + H_2O \rightleftharpoons HV_6O_{15}^{3-} + 3OH^-)$ −8·4	
	gl, sp	45	1 NaCl	$K(0{\cdot}7H_2V_{10}O_{28}^{4-} + 3VO^{2+} + 7{\cdot}2OH^- \rightleftharpoons HV_{10}O_{26}^{4-} + 3{\cdot}8H_2O)$ 71	64Oa
		rt	1 NaCl	$K(0{\cdot}3H_2V_{10}O_{28}^{4-} + 7VO^{2+} + 16{\cdot}8OH^- \rightleftharpoons HV_{10}O_{24}^{4-} + 8{\cdot}2H_2O)$ 143	
		rt	1 NaCl	$K(0{\cdot}8H_2V_{10}O_{28}^{4-} + 2VO^{2+} + 4{\cdot}8OH^- \rightleftharpoons V_{10}O_{26}^{4-} + 3{\cdot}2H_2O)$ 49	
		15	1 NaCl	$K(0{\cdot}4H_2V_{10}O_{28}^{4-} + 6VO^{2+} + 14{\cdot}4OH^- \rightleftharpoons V_{10}O_{24}^{4-} + 7{\cdot}6H_2O)$ 140	
		0	1 NaCl	$K(0{\cdot}5H_2V_{10}O_{28}^{4-} + 5VO^{2+} + 12OH^- \rightleftharpoons HV_{10}O_{25}^{4-} + 6H_2O)$ 104	
		0	1 NaCl	$K(0{\cdot}6H_2V_{10}O_{28}^{4-} + 4VO^{2+} + 9{\cdot}6OH^- \rightleftharpoons V_{10}O_{25}^{4-} + 5{\cdot}4H_2O)$ 103	
	var		var	ev $Q^{4-} = H_4V_{10}O_{28}^{4-}$ in $(Et_4N)_4Q(s)$, MeCN(l) and H_2O (pre, esr, ir, con, sp)	68H
	pre			ev $V_{10}O_{24}^{6-}$ in "$Na_3V_5O_{12}(H_2O)_8$" and K salt	68Ha
Cu^{2+}	pre		var	ev cpx, probably $Cu(OH)_4V_{10}O_{28}^{8-}$, in solids and solution	66J
Tl^+	sol	28	dil	$K_s(Tl^+VO_3^-)$ −8·26, $K_s(Tl^+{}_4V_2O_7^{4-})$ −18·59	64Sb
P^{5+}	spJ		var	ev cpx PV_2 and P_3V, apparent equil constants given	63Ma
	oth		NaCl sat	$\beta[P^V(V_6)_2]$ (?) 6·08, by paper chromatography	67R

63F J. Fuchs and K. F. Jahr, *Chem. Ber.*, 1963, **96**, 2460
63Ja K. F. Jahr, H. Schroth, and J. Fuchs, *Z. Naturforsch.*, 1963, **18b**, 1133
63Ma S. P. Moulik, B. N. Ghosh, and D. K. Mullick, *J. Indian Chem. Soc.*, 1963, **40**, 743
63O S. Ostrowetsky, *Compt. rend.*, 1963, **257**, 1937
63Oa S. Ostrowetsky and P. Souchay, *Compt. rend.*, 1963, **257**, 2036
63Sb H. P. Stock and K. F. Jahr, *Z. Naturforsch.*, 1963, **18b**, 1134
63Sc A. M. Shams el Din and A. A. A. Gerges, *J. Inorg. Nuclear Chem.*, 1963, **25**, 1537
64B F. Brito, N. Ingri, and L. G. Sillén, *Acta Chem. Scand.*, 1964, **18**, 1557
64D F. Z. Dzhabarov and S. V. Gorbachev, *Zhur. neorg. Khim.*, 1964, **9**, 2399
64Da D. Dyrssen and T. Sekine, *J. Inorg. Nuclear Chem.*, 1964, **26**, 981
64E H. T. Evans, jun., A. G. Swallow, and W. H. Barnes, *J. Amer. Chem. Soc.*, 1964, **86**, 4209
64N A. W. Naumann and C. J. Hallada, *Inorg. Chem.*, 1964, **3**, 70
64O S. Ostrowetsky, *Bull. Soc. chim. France*, 1964, 1012
64Oa S. Ostrowetsky, *Bull. Soc. chim. France*, 1964, 1018
64R L. B. Reznik and P. N. Kovalenko, *Ukrain. khim. Zhur.*, 1964, **30**, 514
64S J. Simon and K. F. Jahr, *Z. Naturforsch.*, 1964, **19b**, 165
64Sa G. Schwarzenbach, Proceedings of the Symposium on Co-ordination Chemistry, Tihany, 1964 (Akad. Kiadó, Budapest 1965) p. 145

continued overleaf

3 $L^{2-} = VO_2(OH_3)^{2-} = HVO_4^{2-}$ (contd)

64Sb R. S. Saxena and M. L. Mittal, *Indian J. Chem.*, 1964, **2**, 332
64Sc A. M. Shams el Din and A. A. Gerges, *Electrochim. Acta*, 1964, **9**, 613
64Y K. B. Yatsimirskii and V. E. Kalinina, *Zhur. neorg. Khim.*, 1964, **9**, 1117
65B J. Beltrán and C. Guillem, *Anales de Quim*, 1965, **61**, 919
65C D. B. Copley, A. K. Banerjee, jun., and S. Y. Tyree, jun., *Inorg. Chem.*, 1965, **4**, 1480
65H J. V. Hatton, Y. Saito, and W. G. Schneider, *Canad. J. Chem.*, 1965, **43**, 47
65Ha O. W. Howarth and R. E. Richards, *J. Chem. Soc.*, 1965, 864
65P L. Przyborowski, G. Schwarzenbach, and T. Zimmermann, *Helv. Chim. Acta*, 1965, **48**, 1556
66B F. Brito, *Anales de Quím.*, 1966, **62**, 123
66Ba G. L. Bertrand, G. W. Stapleton, C. A. Wulff, and L. G. Hepler, *Inorg. Chem.* 1966, **5**, 1283
66E H. T. Evans, jun., *Inorg. Chem.*, 1966, **5**, 967
66G W. P. Griffith and T. D. Wickins, *J. Chem. Soc. (A)*, 1966, 1087
66I A. A. Ivakin, *Zhur. priklad. Khim.*, 1966, **39**, 277
66J K. F. Jahr and F. Preuss, *Chem. Ber.*, 1966, **99**, 1602
66S A. G. Swallow, F. R. Ahmed, and W. H. Barnes, *Acta Cryst.* 1966, **21**, 397
66Sa R. S. Saxena and O. P. Sharma, *J. Inorg. Nuclear Chem.* 1966, **28**, 1881
66W M. P. Whittaker, J. Asay, and E. M. Eyring, *J. Phys. Chem.*, 1966, **70**, 1005
67A P. K. L. Au and C. Calvo, *Canad. J. Chem.*, 1967, **45**, 2297
67B F. Brito, *Acta Chem. Scand.*, 1967, **21**, 1968
67D E. Dorm and B.-O. Marinder, *Acta Chem. Scand.*, 1967, **21**, 590
67M A. Müller, B. Krebs, W. Rittner, and M. Stockburger, *Ber. Bunsengesellschaft Phys. Chem.*, 1967, **71**, 182
67R R. Ripan, A. Duca, and V. Cordiş, *Rev. Roumaine Chim.*, 1967, **12**, 375
67T M. Tanaka and I. Kojima, *J. Inorg. Nuclear Chem.*, 1967, **29**, 1769
68H C. Heitner-Wirguin and J. Selbin, *J. Inorg. Nuclear Chem.*, 1968, **30**, 3181
68Ha E. Hayek and U. Pallasser, *Monatsh.*, 1968, **99**, 2494

4 Niobate and Tantalate

Metal	*Method*	*Temp*	*Medium*	*Log of equilibrium constant, remarks*	*Ref*
				Niobates	
H^+	pre		org	ev $Nb_6O_{19}^{8-}$, $Nb_{24}O_{65}^{10-}$, $H_xNb_{12}O_{37}^{(14-x)-}$ $x = 6$—9	63F
	var	25	1 (KCl)	ev predom cpx, probably 6 Nb and some K^+ from medium, charge ≈ -2, [cfu, light scattering, (ir)]	64Nc
	var	25	var	ev $Nb_6O_{19}^{8-}$ $HNb_6O_{19}^{7-}$, $H_2Nb_6O_{19}^{6-}$, and larger cpx, no ev $Nb_5O_{16}^{7-}$ (con, pre, gl)	65D = 63L/65D
	Ram		solid, soln	same cpx ($Nb_6O_{19}^{8-}$) in solution as in crystals $K_7HNb_6O_{19}(H_2O)_{13}(s)$	65T
	gl, con	25	var	ev $H_xNb_6O_{19}^{(8-x)-}$, $x = 0, 1, 2$, and larger (probably 3)	66D = 67D
	Ram		12 KOH	no ev NbO_4^{3-}, only polynuclear cpx	66G
	aix		KCl var	ev anion of charge 7− (agrees with ref 64N)	67S
Ce^{3+}	pre, con		var	ev $CeNb_2O_6(OH)_4^{3-}$	64L
Mn^{4+}	var		var	ev $H_{2n}MnNb_{12}O_{37+n}^{10-}$ (pre, cix, sp, esr)	67Da
Co^{2+}	pre, con		var	ev $CoNb_2O_6(OH)_2^{2-}$	64L
Cu^{2+}	pre, con		var	ev $CuNb_4O_{12}(OH)_2^{4-}$	64L
				Tantalates	
H^+	var	25	1 KCl	ev $Ta_6O_{19}^{8-}$ for pH = 10—13 (gl, sp, ir, light scattering)	63N
	cfu, Ram	25	1 K(Cl)	ev $Ta_6O_{19}^{8-}$	64A
	cfu	25	1 KCl	ev $Ta_6O_{19}^{8-}$, single anion	64Nb

continued

NbTa (contd) 4

Metal	Method	Temp	Medium	Log of equilibrium constant, remarks	Ref
H^{2} (contd)	tp		H^+ var	ev ani, unch and cat cpx in 2—8M-HCl, 0·1—8M-HNO_3; smaller cpx on addn of HCl or H_2SO_4	64Na
	con, pH		dil	$K_8(Ba^{2+}{}_4Ta_6O_{19}{}^{8-})$ −34·55, ev $H_nTa_6O_{19}{}^{(8-n)-}$, $n = 0$ and two positive values, no ev $Ta_5O_{16}{}^{7-}$	65Da
	Ram		var	same cpx ($Ta_6O_{19}{}^{8-}$) in solution as in crystals $K_8Ta_6O_{19}(H_2O)_{16}(s)$	65T
	pre		solid	ev $H_xTa_{12}O_{37}{}^{(14-x)-}$ ($x = 5, 8, 9$), $H_xTa_{18}O_{55}{}^{(20-x)-}$ ($x = 12, 13$)	66P
	aix		KCl var	ev anion of charge 7− at pH = 12	67S

63F J. Fuchs, K. F. Jahr, and G. Heller, *Chem. Ber.*, 1963, **96**, 2472
63N W. H. Nelson and R. S. Tobias, *Inorg. Chem.*, 1963, **2**, 985
64A J. Aveston and J. S. Johnson, *Inorg. Chem.*, 1964, **3**, 1051
64L A. V. Lapitskii, L. G. Vlasov, and V. I. Bezrukov, *Doklady Akad. Nauk S.S.S.R.*, 1964, **154**, 868
64Na B. I. Nabivanets, *Zhur. neorg. Khim.*, 1964, **9**, 1079
64Nb W. H. Nelson and R. S. Tobias, *Inorg. Chem.*, 1964, **3**, 653
64Nc W. H. Nelson and R. S. Tobias, *Canad. J. Chem.*, 1964, **42**, 731
65D Y. Dartiguenave, M. Lehne, and R. Rohmer, *Bull. Soc. chim. France*, 1965, 62
65Da Y. Dartiguenave, M. Dartiguenave, M. Lehné, and R. Rohmer, *Bull. Soc. chim. France*, 1965, 3173
65T R. S. Tobias, *Canad. J. Chem.*, 1965, **43**, 1222
66D Y. Dartiguenave, M. Dartiguenave, M. Lehne, and R. Rohmer, *Bull. Soc. chim. France*, 1966, 1679
66G W. P. Griffith and T. D. Wickins, *J. Chem. Soc.(A)*, 1966, 1087
66P W. Plötz, dissertation, Freie Univ., Berlin, 1966
67D Y. Dartiguenave, *Bull. Soc. chim. France*, 1967, 381
67Da B. W. Dale and M. T. Pope, *Chem. Comm.*, 1967, 792
67S F. W. Smith, *J. Inorg. Nuclear Chem.*, 1967, **29**, 1161

Chromate, $CrO_4{}^{2-}$ 5

Metal	Method	Temp	Medium	Log of equilibrium constant, remarks	Ref
				see also Table 56, $SO_4{}^{2-}$ and Table 67, Cl^- at metal Cr^{6+}	
H^+	*X*		solid	ev $Cr_2O_7{}^{2-}$ in $(NH_4)_2Cr_2O_7(s)$	51B
	sp	25	1 ($LiClO_4$)	K_{12} −0·62, †K_{22} 1·99, no ev $HCr_2O_7{}^-$ ($H_3L_2{}^-$)	53T/64T
	dis	?	1 (Na)ClO_4	K_{12} −0·21	60S/66T
	sp	25	I($NaClO_4$)	K_{12} ($H^+ + HL^- \rightleftharpoons H_2L$) −0·7($I$ = 3), −0·8(I = 1)	64H
	gl	25	1·5 (KNO_3)	K_1 5·76, †K_{22} 2·14	64M
	Cu	350	KNO_3(l)	$K(Cr_2O_7{}^{2-} + O^{2-} \rightleftharpoons 2CrO_4{}^{2-})$ 12·26 in KNO_3(l)	64Sa = 63S
	sp	20	I($NaClO_4$)	†K_{22} ($2HCrO_4{}^- \rightleftharpoons Cr_2O_7{}^{2-} + H_2O$) 2·00($I$ = 1·0), 1·90(I = 0·5)	65L
	dis	20	I(Na_2SO_4)	K_1 5·90(I = 0·4), 6·12 to 6·19(I = 0·1), 6·56 to 6·63 (→ 0)	66H
	sp	15—35	1 ($LiClO_4$)	K_{12} −0·83(15°), −0·61(25°), −0·42(35°) ΔH_{12} = 8·5, ΔS_{12} = 26	66T
	fp	32·3	Na_2SO_4 sat	K_1 4·96, †K_{22} 2·36	67J
	O	271—342	MNO_3(l)	$K(Cr_2O_7{}^{2-} + O^{2-} \rightleftharpoons 2CrO_4{}^{2-})$ 7·92(271°), 7·62(314°), 7·48(342°) in $Na_{0.5}K_{0.5}NO_3$(l), also values for intermediate temp ΔH = −10, ΔS = −17(330°)	67K
	sp	25	C_1 ($NaClO_4$)	†K_{22} 1·89+0·083C_1 (C_1 from 1 to 7)	67L
			C_2 ($NaNO_3$)	†K_{22} 1·87+0·047C_2 (C_2 from 1 to 7)	
			$NaClO_4$+$NaNO_3$	†K_{22} 1·89−0·02C_2/(C_1+C_2)+0·083C_1+0·047C_2	

continued overleaf

5 CrO_4^{2-} (contd)

Metal	*Method*	*Temp*	*Medium*	*Log of equilibrium constant, remarks*	*Ref*
H^+	sp	14·5	$I(NaClO_4)$	$\dagger K_{22}$ 2·26(I = 3), 2·17(I = 1·5), 2·08(I = 0·9) *etc.*	68H
(*contd*)		6—35	0·1 ($NaClO_4$)	$\dagger K_{22}$ 1·92(6·4°), 1·85(14·5°), 1·81(20°), 1·74(25°), 1·68(34·5°)	
		6—35	0 corr	$\dagger K_{22}$ 1·68(6·4°), 1·64(14·5°), 1·58(20°), 1·51(25°), 1·46(34·5°)	
	sp	15—45	0 corr	$\dagger K_{22}$ 1·64(15°), 1·53(25°), 1·40(35°), 1·30(45°),	68L
		25		$\Delta\dagger H_{22} = -4{\cdot}7$, $\Delta\dagger S_{22} = -8{\cdot}8$	
	gl	5—60	0 corr	K_1 6·44(5°), 6·48(15°), 6·49(25°), 6·52(35°), 6·57(45°), 6·64(60°)	68La
	sp	15—45	0 corr	K_1 6·47(15°), 6·50(25°), 6·53(35°), 6·59(45°)	
		25	0 corr	$\Delta H_1 = 1{\cdot}3$, $\Delta S_1 = 34$, $\Delta C_p = -35$	
	sp	25	→ 0	$\dagger K_{22}$ ($2HCrO_4 \rightleftharpoons Cr_2O_7^{2-} + H_2O$) 1·36, $\Delta\dagger H_{22} = -3{\cdot}8$, also graphs $\dagger K_{22}(I)$ for solutions of KNO_3, KCl, NaCl, K_2SO_4, Na_2SO_4, Li_2SO_4, $CaCl_2$, $MgCl_2$, $MgSO_4$ and $AlCl_3$	68P
Group 1A$^+$	act	25?	0 corr	K_1 0·7(Li^+), 0·7(Na^+), 0·8(K^+)	/66M
Y^{3+}	var		var	ev YL_3^{3-} (gl, high-frequency titration, aix, cix, sol)	66R
La^{3+}	con	25	0 corr	K_1 3·92, K_2 (β_2?) 3·29	66E
Er^{3+}	var		var	ev ErL_3^{3-} (gl, high-frequency titration, aix, cix, sol)	66R
Fe^{3+}	sp	0	0·084	$*K_1$ 0·15	63E
$Co(NH_3)_5^{3+}$	sp, gl	25	→ 0	$*K_1$ [$Co(NH_3)_5^{3+} + HL^- \rightleftharpoons Co(NH_3)_5L^+ + H^+$] −1·01, also $*K_1$ for I = 0·0025—0·488 ($LiClO_4$) and 10—35°	64S
Ph_3Sn^+	dis	30	0·1 ($NaNO_3$)	K_D [$Ph_3SnOH(org) + H^+ + HL^- \rightleftharpoons Ph_3SnHL(org) + H_2O$] −6·7(org = C_6H_6)	65S

51B A. Byström and K. A. Wilhelmi, *Acta Chem. Scand.*, 1951, **5**, 1003
63E J. H. Espenson and E. L. King, *J. Amer. Chem. Soc.*, 1963, **85**, 3328
63S A. M. Shams el Din and A. A. A. Gerges, *J. Inorg. Nuclear Chem.*, 1963, **25**, 1537
64H G. P. Haight, jun., D. C. Richardson, and N. H. Coburn, *Inorg. Chem.*, 1964, **3**, 1777
64M J. G. Mason and A. D. Kowalak, *Inorg. Chem.*, 1964, **3**, 1248
64S J. C. Sullivan and J. E. French, *Inorg. Chem.*, 1964, **3**, 832
64Sa A. M. Shams el Din and A. A. A. Gerges, *Electrochim. Acta*, 1964, **9**, 613
64T J. Y. Tong, *Inorg. Chem.*, 1964, **3**, 1804
65L O. Lukkari, *Suomen Kem.*, 1965, **38**, *B*, 121
65S G. K. Schweitzer and S. W. McCarty, *J. Inorg. Nuclear Chem.*, 1965, **27**, 191
66E V. I. Ermolenko, *Dopovidi Akad. Nauk Ukrain. R.S.R.*, 1966, 85
66H J. Hála, O. Navrátil, and V. Nechuta, *J. Inorg. Nuclear Chem.*, 1966, **28**, 553
66M W. L. Masterton and L. H. Berka, *J. Phys. Chem.*, 1966, **70**, 1924
66R D. I. Ryabchikov and N. S. Vagina, *Zhur. neorg. Khim.*, 1966, **11**, 1038
66T J. Y. Tong and R. L. Johnson, *Inorg. Chem.*, 1966, **5**, 1902
67J D. V. S. Jain and C. M. Jain, *J. Chem. Soc.* (*A*), 1967, 1541
67K R. N. Kust, *Inorg. Chem.*, 1967, **6**, 2239
67L J. Lagrange and J. Byé, *Bull. Soc. chim. France*, 1967, 1490
68H K. E. Howlett and S. Sarsfield, *J. Chem. Soc.* (*A*), 1968, 683
68L H. G. Linge and A. L. Jones, *Austral. J. Chem.*, 1968, **21**, 1445
68La H. G. Linge and A. L. Jones, *Austral. J. Chem.*, 1968, **21**, 2189
68P B. Perlmutter-Hayman and Y. Weissmann, *Israel J. Chem.*, 1968, **6**, 17

Molybdate, MoO_4^{2-} 6

1964 *Errata*: p. 92, *in ref* 45B = 54B, *for* "$Mo_6O_{13}^{2-n}$" *read* "$Mo_4O_{13}^{2-n}$": *for ref* "61S" *read* "61S = 64Sg"

Metal	*Method*	*Temp*	*Medium*	*Log of equilibrium constant, remarks*	*Ref*
H^+	*X*		solid	ev $Mo_8O_{26}^{4-}$ in $(NH_4)_4Mo_8O_{26}(H_2O)_5(s)$	50Lc
	con		var	ev cpx with ratio H^+/L^{2-} = 0·67($Mo_3O_{11}^{4-}$?), 1·0($Mo_4O_{14}^{4-}$?), 1·14($Mo_7O_{24}^{6-}$), 1·5($Mo_4O_{13}^{2-}$) and 1·75($HMo_4O_{13}^{-}$?)	63B
	var	25	HNO_3 var, $HClO_4$ var	ev mononuclear cation (MoO_2^{2+} ?) and polynuclear cation (with 2Mo ?) (diffusion, sp, dis)	63J
	sp	20	0 corr	K_1 4·24, K_{12} 4·00	63R
	sol	26	0 corr	$K[MoO_2(OH)_2(s)+OH^- \rightleftharpoons HMoO_4^- + H_2O] \sim -1{\cdot}8$ (? added $[OH^-]$ used in K) $K[MoO_2(OH)_2(s)+H^+ \rightleftharpoons MoO_2OH^+ + H_2O] \sim -0{\cdot}3$ (? added $[H^+]$ used in K) $K_s\,[H_2MoO_4(s) \rightleftharpoons H_2MoO_4]$ −1·1 to −2·1	63Sa
	cfu	25	1	ev species with on the average > 6 and < 9 Mo for < 1·5 H^+ bound/MoO_4^{2-}; larger species for > 1·5 H^+/Mo(1M-NaCl, LiCl, $NaClO_4$ medium)	64A
	Ram		var, solid	ev $Mo_7O_{24}^{6-}$ in 3·7M-$Li_{0.86}H_{1.14}MoO_4$ ev $Mo_8O_{26}^{4-}$? in 2M-$Li_{0.5}H_{1.5}MoO_4$	
	gl	25	1 Na(Cl)	several schemes give acceptable agreement, such as K_1 3·55, β_{12} 7·20, β_{78} 52·81, β_{79} 57·39, $\beta_{7,10}$ 61·02, $\beta_{7,11}$ 63·40 (best fit) or K_1 3·53, β_{12} 7·26, β_{78} 52·80, β_{79} 57·42, $\beta_{7,10}$ 60·84, $\beta_{8,12}$ 71·56; or K_1 3·69, β_{12} 7·29, $\beta_{8,9}$ 59·92, $\beta_{8,10}$ 64·85, $\beta_{8,11}$ 68·75, $\beta_{8,12}$ 71·75	64A
	var		var	ev only $Mo_4O_{13}^{2-}$ and $HMo_4O_{13}^{-}$, no ev other cpx (?) (aix, sp, gl)	64H
	gl, qh	25	3 Na(ClO_4)	K_1 3·89, β_{12} 7·50, β_{78} 57·74, β_{79} 62·14, $\beta_{7,10}$ 65·68, $\beta_{7,11}$ 68·21, ev large cpx, *e.g.* $\beta_{19,34}$ 196·3 ?, ev cation of charge +1 *e.g.* $\beta_{5,2}$ 19 ?	64S ~ 63S
	qh	25	3 Na(Cl)	ev $Mo_7O_{24}^{6-}$, $HMo_6O_{20}^{3-}$, $Mo_6O_{20}^{4-}$, no ev $HMo_6O_{21}^{5-}$	64Sc
	tp	22	0·1 (Na)ClO_4	K_1 3·57, β_{12} 8·32, β_{13} 9·11	65C
	tp	22	0·1 (Na)NO_3	K_1 3·52, β_{12} 8·36, β_{13} 8·69 values in ref 63Ca withdrawn	
	qh	25	dil	K_1 3·87, β_{12} 8·06, β values for 28 species with 2—7 Mo	65Ca
	var	20	0·2 NaCl	ev species with 7—8 Mo($Mo_7O_{24}^{6-}$, $Mo_8O_{26}^{4-}$) (sp, pH, cfu)	65G
	O	350	KNO_3(l)	$K(Mo_3O_{10}^{2-} + 2O^{2-} \rightleftharpoons 3MoO_4^{2-})$ 23·0 in KNO_3(l)	65Sb
	fp	0	$(NH_4)_6Mo_7O_{24}$ var	ev $Mo_4O_{13}^{2-}$, $Mo_3O_{11}^{4-}$, $NH_4Mo_4O_{13}^{-}$ (?)	66G
	gl		var	ev $H_2Mo_4O_{13}$ (?)	
	oth	18—25	0·3 $NaNO_3$	ev species with two different numbers of Mo for ⩽ 1·5 H^+ per L^{2-}, species with one number of Mo (supposedly 8) for > 1·5 H^+ per L^{2-}, ev $Mo_4O_{10}(OH)_6^{2-}$ as short-lived intermediate (*T*-jump relaxation)	66Ga
	var		var	ev Mo_4 cpx, $Mo_7O_{24}^{6-}$, $Mo_8O_{26}^{4-}$, and other cpx, no ev Mo_2 or Mo_3 cpx (cfu, gl, sp, ir)	66H
	kin	25	0 corr	$\beta[H^+(HL^-)_2]$ 8·25, $\beta[H^+_3(HL^-)_6]$ 25·19 (?)	67A
	qh	25	3 Na(Cl)	β_{78} 56·36, β_{67} 52·70, ev $HMo_6O_{20}^{3-}$	67H
			2 Na(Cl)	β_{78} 54·66, β_{67} 51·21, ev $HMo_6O_{21}^{5-}$?	

continued overleaf

6 MoO_4^{2-} (contd)

Metal	*Method*	*Temp*	*Medium*	*Log of equilibrium constant, remarks*	*Ref*
H^+ (*contd*)			1 Na(Cl)	β_{78} 53·11, β_{67} 54·41, ev $HMo_6O_{21}^{5-}$?	
			0·1 Na(Cl)	β_{67} 50·49	
	O	260—344	MNO_3(l)	$K(Mo_2O_7^{2-}+O^{2-} \rightleftharpoons 2MoO_4^{2-})$ 7·60(260°), 7·07(310°), 6·68(342°), $\Delta H = -17$, $\Delta S = 4$ in $Na_{0.5}K_{0.5}NO_3$(l)	67Ka
	pol	25	Na(Cl)	ev species with 1·14, 1·33, 1·43 and 1·50 H^+/L^{2-} ($Mo_7O_{24}^{6-}$, $Mo_6O_{20}^{4-}$, $H_2Mo_7O_{24}^{4-}$, $HMo_6O_{20}^{3-}$)	67L
	sp	20	I($NaClO_4$)	$K_{14}(MoO_2OH^+ + H^+ \rightleftharpoons MoO_2^{2+})$ 0·08(I = 2), 0·39(I = 1), 0·62(I = 0·5); $K_{13}K_{14}$ 0·19(I = 2), 1·11(I = 1), 1·36(I = 0·5); K_{12} 0·43(I = 2), 0·85(I = 1), 0·96(I = 0·5)	67V
	dis, sp	20	1 HNO_3	$\beta[(MoO_2^{2+})_n]$ 1·98(n = 2), 5·0(n = 3), 8·09(n = 4), $K(2MoO_2^{2+} + H_2O \rightleftharpoons Mo_2O_5^{2+} + 2H^+)$ 3·22	67Va
			1 $HClO_4$	$\beta[(MoO_2^{2+})_n]$ 2·72(n = 2), 5·96(n = 3), also values for some (Li, H)NO_3 mixtures	
	cal	25	3 Na(ClO_4)	$\Delta H_1 = 14$, $\Delta H_{\beta 78} = -56{\cdot}0$, $\Delta S_1 = 65$, $\Delta S_{\beta 78} = 76$; $H^+ + P^{6-} \rightleftharpoons HP^{5-}$, $\Delta H = 2{\cdot}6$, $\Delta S = 29$ ($P^{6-} = Mo_7O_{24}^{6-}$); $H^+ + HP^{5-} \rightleftharpoons H_2P^{4-}$, $\Delta H = 0{\cdot}8$, $\Delta S = 19$; $H^+ + H_2P^{4-} \rightleftharpoons H_3P^{3-}$, $\Delta H = -0{\cdot}6$, $\Delta S = 10$	68A+64S
	oth		var	ev mainly mononuclear cpx for [Mo] < 0·001 M, cations with ⩽ 3 Mo in 1—10 M-HNO_3, anions for pH > 2 (dialysis, electrodialysis)	68B
	cal	800	melt	$K[Na_2MoO_4(l) + 2MoO_3(l) \rightleftharpoons Na_2Mo_3O_{10}(l)]$ 0·0 in melt $Na_2MoO_4 + MoO_3$, x units	68Ba
	X		solid	ev $Mo_7O_{24}^{6-}$ in $(NH_4)_6Mo_7O_{24}(H_2O)_4$(s)	68E
	pre, ir		var	ev $Mo_6O_{19}^{2-}$ in solid and in Me_2CO(l)	68F
	X		solid	ev $Mo_7O_{24}^{6-}$ in $K_6Mo_7O_{24}(H_2O)_4$(s)	68G
	var	25	3 Na(ClO_4)	ev mainly mono- and ≈ hepta-nuclear species, one larger anion and cations (final β's to follow) (gl, qh, sol). Survey of earlier work	68Sa
Group $4f^{3+}$	con	25	0·0012	K_1 4·22(La), 4·42(Ce), 4·42(Pr), 4·74(Nd), 4·85(Sm), 4·74(Eu), 4·40(Gd), 4·51(Tb), 4·49(Dy), 4·32(Ho), 4·26(Er), 4·23(Yb)	68D
La^{3+}	sol		dil	$K_{s0}(La^{3+}{}_2L^{2-}{}_3)$ −20·66, K_1 3·68	68D
Ce^{4+}	X		solid	ev $CeMo_{12}O_{42}^{8-}$ in $(NH_4)_2H_6CeMo_{12}O_{42}(H_2O)_{12}$(s)	68Da
Ti^{4+}	gl		var	ev $TiMo_{12}O_{40}^{4-}$, no cpx of higher charge	68S
Cr^{3+}	X		solid	ev $H_6CrMo_6O_{24}^{3-}$ in $Na_3H_6CrMo_6O_{24}(H_2O)_8$(s)	66Pa
	sp, gl		constant I	β_6 18·33; $\beta[(CrL_6)_m(H_xPO_4)L^{2-}{}_{n-6m}]$ 11(m = 1, n = 8), 19(m = 2, n = 16), 24(m = 3, n = 24), 32(m = 4, n = 30), 35(m = 5, n = 36), 39(m = 6, n = 42), ?	67K
Mo^{5+}	pre, sp		var	ev brown $Mo_6O_{18}^{2-}$, red $HMo_6O_{18}^{2-}$, yellow-brown $HMo_6O_{17}^{-}$	64O
Mn^{2+}	var		solid	ev $MnH_6Mo_6O_{24}^{4-}$ (pre, ir, X powder)	68L
Mn^{4+}	var		Na_2SO_4 sat	ev $MnMo_9O_{32}^{6-}$ and $H_6MnMo_9O_{32}$ (fp, gl, sp)	66B
Fe^{3+}	sol	?	var	K_1 7·90	64Z
	oth		NaCl sat	$\beta(Fe^{3+}Mo_6O_{21}^{6-})$ (?) 4·26 by paper chromatography	66Da
Co^{2+}	var		solid	ev $CoH_6Mo_6O_{24}^{4-}$ (pre, ir, X powder)	68L

continued

Metal	*Method*	*Temp*	*Medium*	*Log of equilibrium constant, remarks*	*Ref*
Co^{3+}	oth		NaCl sat	$\beta(Co^{3+}Mo_6O_{21}^{6-})$ (?) 3·96 by paper chromatography	67D
Ni^{2+}	var		solid	ev $NiH_6Mo_6O_{24}^{4-}$ (pre, ir, X powder)	68L
Ni^{4+}	fp, sp		Na_2SO_4 sat	ev $NiMo_9O_{32}^{6-}$	66B
Cu^{2+}	var		solid	ev $CuH_6Mo_6O_{24}^{4-}$ (pre, ir, X powder)	68L
Zn^{2+}	var		var	ev cpx $ZnMo_6$ (sol, sp, con)	66L
	var		solid	ev $ZnH_6Mo_6O_{24}^{4-}$ (pre, ir, X powder)	68L
Cd^{2+}	var		var	ev cpx $CdMo_{\approx 6}$ (sol, sp, con)	66L
	var		var	ev cpx $CdMo_3$ at pH ≈ 3 to 4, β_3 5(con), 4·4 to 4·2(cix), 4·6(sol)	67La
Al^{3+}	oth		NaCl sat	$\beta(Al^{3+}Mo_6O_{21}^{6-})$ (?) 5·2, by paper chromatography	66Da
	var		var	ev $AlMo_6O_{21}^{3-}$ (con, gl, cal)	67Wb
	sp		dil	$\beta(Al^{3+}Mo^{VI}_6)$ 20·0 at pH ≈ 2·5	68La
Si^{4+}	dis	20	var	ev several cpx	64Se
	dis, sp		var	ex cpx $SiMo_{12}$ at pH = 1·8	65Sa
Si^{4+}, Ti^{4+}	sp		0·3 (Na_2SO_4)	$\beta\{[TiO(OH)_2^0]_2Si(OH)_4^0(Mo_2O_7^{2-})_4\}$ 22·10 (?)	67B
Si^{4+}, Mo^{5+}	sp		var	ev cpx $SiMo^V{}_4Mo^{VI}{}_8$	68Aa
Si^{4+}, Co^{2+}	pre, sp		var	ev $SiCo^{II}Mo_{11}O_{40}H_2^{6-}$	67W
P^{5+}	dis, sp	20	var	ev several cpx, anions of charge (−1) and larger	64Se
	sp		0·25 (NaCl)	ev cpx PMo_{12}, PMo_{10}, PMo_{20}, PMo_{24}, apparent β given, *e.g.* $\beta[P^V(Mo_3O_{10}^{2-})_4]$ −11·1 at pH = 3·3 (?)	65K
	dis, sp		var	ev cpx PMo_{12} and PMo_{18} at pH = 1·8	65Sa
	oth		var	$\beta[P^V(Mo_6)_2]$ (?) 8·1(NaCl sat), 7·7(0·03M-$HClO_4$), 4·85(dil, pH = 5·6); by paper chromatography	66D
	var		var	ev cpx with 9, 11, 12, and 16 Mo/P (cal, gl, con, sp)	66Sd
P^{5+}, Ce^{4+}	sp		var	ev mixed cpx	64Sd
P^{5+}, $4A^{4+}$	dis		1 $HClO_4$	Q = PMo_{12}-cpx, $\beta(TiO^{2+}Q)$ 3·73, $\beta(Zr^{4+}Q)$ 5·72, $\beta(Hf^{4+}Q)$ 4·87	66Sc
P^{5+}, Ti^{4+}	oth		NaCl sat	$\beta[Ti^{IV}P^V(Mo_6)_2]$ (?) 12·55, by paper chromatography	67D
P^{5+}, Zr^{4+}	dis		var	ev cpx $PZrMo_{12}$, charge 4−	66Sb
P^{5+}, Hf^{4+}	sp, dis		var	ev mixed cpx	64Sf
	dis, sp		var	ev cpx $PHfMo_{12}$	66S
P^{5+}, Th^{4+}	sp		var	ev mixed cpx	64Sd
P^{5+}, V^{5+}	pre, sp		var	ev cpx $PVMo_{11}$, PV_2Mo_{10}, PV_3Mo_9	64C
	var		var	ev $PVMo_{11}O_{40}^{4-}$, $PV_2Mo_{10}O_{40}^{5-}$, $HPV_3Mo_9O_{40}^{5-}$, $PV_3Mo_6O_{30}^{4-}$; $K(PVMo_{11}O_{40}^{4-} \rightleftharpoons 0{\cdot}5PV_2Mo_{10}O_{40}^{5-} + 0{\cdot}5PMo_{12}O_{40}^{3-})$ −1·26(sp, fp, tp, gl)	67C
	con	25	→ 0	} ev $PVMo_{11}O_{40}^{4-}$, $PV_2Mo_{10}O_{40}^{5-}$, no ev cpx of these	68H
	gl	25	3 $NaClO_4$	} ions with H^+	
	dis, gl		1 $NaClO_4$	$Q^{4-} = PVMo_{11}O_{40}^{4-}$, $K_a(H_4Q) \approx K_a(H_3Q^-) \approx K_a(H_2Q^{2-}) \approx K_a(HQ^{3-})$ −0·75	68Sb
	var		var	ev $PVMo_{11}O_{40}^{4-}$, $PV_2Mo_{10}O_{40}^{5-}$, $PV_3Mo_9O_{40}^{6-}$, (pre, gl, sp, ir)	68T
P^{5+}, Nb^{5+}	pre, sp		3 H_2SO_4	$\beta(P^VNb^VMo^{VI}{}_{10})$ 30·7 in 3M-H_2SO_4	65B
P^{5+}, Mo^{5+}	var	25	var	($Mo_{18}O_{62}$ = R) ev P_2R^{6-}, $H_2P_2R^{6-}$, $H_4P_2R^{6-}$, $H_6P_2R^{6-}$ (pre, red, pol)	67P

continued overleaf

6 MoO_4^{2-} (contd)

Metal	*Method*	*Temp*	*Medium*	*Log of equilibrium constant, remarks*	*Ref*
P^{5+}, Mo^{5+}	sp		var	ev several cpx, one with $PMo^{V}{}_3Mo^{VI}{}_8$ or $PMo^{V}{}_2Mo^{VI}{}_{10}$	68Aa
(*contd*)	pol	?	1 ($NaClO_4$)	$\beta(H^+PMo_{12}O_{40}{}^{n-})$; α-series: 2·1($n$ = 5), 6(n = 6), 8·8(n = 7), 10·7(n = 8), 12·2(n = 9) β-series: 2·7(n = 5), 6·8(n = 6), 9·5(n = 7), 10·8(n = 8)	68Fa
	gl	?	1 ($NaClO_4$)	$\beta[H^+, H_nPMo_{12}O_{40}{}^{(7-n)-}]$ 9·3(n = 0), 6·9(n = 1), 4·5(n = 2)	
P^{5+}, Fe^{3+}	sp		var	ev cpx $FePMo_{12}$	65S
As^{5+}, Mo^{5+}	var	25	var	($Mo_{18}O_{62}$ = R) ev As_2R^{6-}, $H_2As_2R^{6-}$, $H_4As_2R^{6-}$, $H_6As_2R^{6-}$ (pre, red, pol)	67P
Te^{4+}	var		var	ev $TeMo_6O_{24}{}^{8-}$ (pre, cix, qh)	64Sa
Te^{6+}	fp	< 0	Na_2SO_4 sat	ev cpx $TeMo_6$, $TeMo_7$, $TeMo_{12}$	64Sb
	sp		0·025	$\beta[Te^{VI}(Mo^{VI}{}_6)]$ − 3·7 at pH = 2—3	66Sa
	var		var	ev $TeMo_6O_{24}{}^{6-}$, $TeMo_7O_{27}OH^{7-}$, $TeMo_{12}O_{45}{}^{12-}$ (con, gl, cal)	67Wa+
	X		solid	ev $TeMo_6O_{24}{}^{6-}$ in $(NH_4)_6TeMo_6O_{24}Te(OH)_6(H_2O)_7$(s)	68E
Te^{6+}, V^{5+}	cix, pre			ev$(TeV_6Mo_4O_{36}{}^{12-}aq)_n$	66P
I^{7+}	var		var	ev $IMo_6O_{24}{}^{5-}$(= Q^{5-}), HQ^{4-}, H_2Q^{3-}, H_3Q^{2-}, H_4Q^- (con, gl, cal)	66J

50Lc I. Lindqvist, *Arkiv Kemi*, 1950, **2**, 349
63B J. Beltrán and F. Puerta, *Anales de Quim.*, 1963, **59**, 271
63J I. L. Jenkins and A. G. Wain, *J. Appl. Chem.*, 1963, **13**, 561
63R E. F. C. H. Rohwer and J. J. Cruywagen, *J. S. African Chem. Inst.*, 1963, **16**, 26
63Sa U. V. Seshaiah and S. N. Banerji, *Proc. Nat. Acad. Sci., India*, 1963, **33**, 61
64A J. Aveston, E. W. Anacker, and J. S. Johnson, *Inorg. Chem.*, 1964, **3**, 735
64C P. Courtin, F. Chauveau, and P. Souchay, *Compt. rend.*, 1964, **258**, 1247
64H C. Heitner-Wirguin and R. Cohen, *J. Inorg. Nuclear Chem.*, 1964, **26**, 161
64O S. Ostrowetsky, *Bull. Soc. chim. France*, 1964, 1003
64S Y. Sasaki and L. G. Sillén, *Acta Chem. Scand.*, 1964, **18**, 1014
64Sa Z. F. Shakhova and T. L. Ku, *Zhur. neorg. Khim.*, 1964, **9**, 1848
64Sb H. Schroth, dissertation, Freie Univ., Berlin, 1964
64Sc J. P. Schwing, *J. Amer. Chem. Soc.*, 1964, **86**, 1879
64Sd Yu. F. Shkaravskii, *Ukrain. khim. Zhur.*, 1964, **30**, 241
64Se Yu. F. Shkaravskii, *Ukrain. khim. Zhur.*, 1964, **30**, 670
64Sf Yu. F. Shkaravskii, *Ukrain. khim. Zhur.*, 1964, **30**, 1170
64Sg J. P. Schwing, *J. Chim. phys.*, 1964, **61**, 508
64Z N. A. Zhukova and L. I. Lebedeva, *Zhur. neorg. Khim.*, 1964, **9**, 480
65B V. F. Barkovskii and M. I. Zaboeva, *Zhur. neorg. Khim.*, 1965, **10**, 900
65C J. Chojnacka, *Roczniki Chem.*, 1965, **39**, 161
65Ca J. Chojnacki and B. Oleksyn, *Roczniki Chem.*, 1965, **39**, 1141
65G O. Glemser, W. Holznagel, and S. I. Ali, *Z. Naturforsch*, 1965, **20b**, 192
65K W. Kemula and S. Rosołowski, *Roczniki Chem.*, 1965, **39**, 123
65S Z. F. Shakhova and E. N. Dorokhova, *Zhur. neorg. Khim.*, 1965, **10**, 2060
65Sa Yu. Shkaravskii, *Ukrain. khim. Zhur.*, 1965, **31**, 94
65Sb A. M. Shams el Din and A. A. El Hosary, *J. Electroanalyt. Chem.*, 1965, **9**, 349
66B L. C. W. Baker and T. J. R. Weakley, *J. Inorg. Nuclear Chem.*, 1966. **28**, 447
66D A. Duca and T. Budiu, *Rev. Roumaine Chim.*, 1966, **11**, 585
66Da A. Duca and T. Budiu, *Rev. Roumaine Chim.*, 1966, **11**, 817
66G J. Guignard, *J. Chim. phys.*, 1966, **63**, 569
66Ga O. Glemser and W. Höltje, *Angew Chem.*, 1966, **78**, 756
66H W. Höltje, dissertation, Univ. Göttingen, 1966
66J K. F. Jahr, J. Fuchs, and G. Wiese, *Z. Naturforsch.*, 1966, **21b**. 11
66L L. I. Lebedeva, *Vestnik Leningrad. Univ.* 1966, no. 10 (*Fiz. Khim.* no. 2), 144
66P S. Prasad and K. C. Pathak, *Bull. Chem. Soc. Japan*, 1966, **39**, 2666
66Pa A. Perloff, thesis, Georgetown Univ., Washington D.C., 1966
66S Yu. F. Shkaravskii and Ya. F. Ometsinskaya, *Ukrain. khim. Zhur.*, 1966, **32**, 1023
66Sa Z. F. Shakhova, T. L. Ku and F. P. Sudakov, *Vestnik Moskov. Univ., Khim.* 1966, no. 2, 69

continued

MoO_4^{2-} (contd) 6

66Sb Yu. F. Shkaravskii, *Zhur. neorg. Khim.*, 1966, **11**, 120
66Sc Yu. F. Shkaravskii, *Zhur. neorg. Khim.*, 1966, **11**, 797
66Sd H. Stark, dissertation, Freie Univ., Berlin, 1966
67A I. I. Alekseeva, *Zhur. neorg. Khim.*, 1967, **12**, 1840
67B V. F. Barkovskii and T. L. Radovskaya, *Zhur. neorg. Khim.*, 1967, **12**, 991
67C P. Courtin and F. Chauveau, *Bull. Soc. chim. France*, 1967, 2461
67D A. Duca and T. Budiu, *Rev. Roumaine Chim.*, 1967, **12**, 479
67H M. Haeringer and J. P. Schwing, *Bull. Soc. chim. France*, 1967, 708
67K K. Kyong and L. I. Lebedeva, *Vestnik Leningrad. Univ.* 1967, no. 16 (*Fiz. Khim.* no. 3), 156
67Ka R. N. Kust, *Inorg. Chem.*, 1967, **6**, 2239
67L P. Lagrange and J.-P. Schwing, *Bull. Soc. chim. France*, 1967, 718
67La L. I. Lebedeva and G. V. Kuznetsova, *Vestnik Leningrad. Univ.* 1967, no. 4 (*Fiz. Khim.* no. 1), 106
67P E. Papaconstantinou and M. T. Pope, *Inorg. Chem.*, 1967, **6**, 1152
67V S. P. Vorob'ev, I. P. Davydov, and I. V. Shilin, *Zhur. neorg. Khim.*, 1967, **12**, 2142
67Va S. P. Vorob'ev, I. P. Davydov and I. V. Shilin, *Zhur. neorg. Khim.*, 1967, **12**, 2665
67W T. J. R. Weakley and S. A. Malik, *J. Inorg. Nuclear Chem.*, 1967, **29**, 2935
67Wa G. Wiese and J. Fuchs, *Z. Naturforsch.*, 1967, **22b**, 124
67Wb G. Wiese and J. Fuchs, *Z. Naturforsch.*, 1967, **22b**, 469
68A R. Arnek and I. Szilard, *Acta Chem. Scand.*, 1968, **22**, 1334
68Aa N. A. Alikina, V. F. Barkovskii, T. L. Radovskaya, and V. S. Shvarev, *Zhur. neorg. Khim.*, 1968, **13**, 1880
68B A. K. Babko and G. I. Gridchina, *Zhur. neorg. Khim.*, 1968, **13**, 123
68Ba A. Brenner and W. H. Metzger, jun., *J. Electrochem. Soc.*, 1968, **115**, 258
68D N. K. Davidenko, G. A. Komashko, and K. B. Yatsimirskii, *Zhur. neorg. Khim.*, 1968, **13**, 117
68Da D. D. Dexter and J. V. Silverton, *J. Amer. Chem. Soc.*, 1968, **90**, 3589
68E H. T. Evans, jun., *J. Amer. Chem. Soc.*, 1968, **90**, 3275
68F J. Fuchs and K. F. Jahr, *Z. Naturforsch.*, 1968, **23b**, 1380
68Fa J.-M. Fruchart and P. Souchay, *Compt. rend.*, 1968, **266**, *C*, 1571
68G B. M. Gatehouse and P. Leverett, *Chem. Comm.*, 1968, 901
68H C. J. Hallada, G. A. Tsigdinos, and B. S. Hudson, *J. Phys. Chem.*, 1968, **72**, 4304
68L A. La Ginestra, F. Giannetta and P. Fiorucci, *Gazzetta*, 1968, **98**, 1197
68La L. I. Lebedeva and K. Kyong, *Vestnik Leningrad. Univ.* 1968, no. 4 (*Fiz. Khim.* no. 1), 127
68S Z. F. Shakova and E. N. Semenovskaya, *Zhur. neorg. Khim.*, 1968, **13**, 1887
68Sa Y. Sasaki and L. G. Sillén, *Arkiv Kemi*, 1968, **29**, 253
68Sb P. Souchay, F. Chauveau and P. Courtin, *Bull. Soc. chim. France*, 1968, 2384
68T G. A. Tsigdinos and C. J. Hallada, *Inorg. Chem.*, 1968, **7**, 437

Wolframate (tungstate) WO_4^{2-} 7

Metal	*Method*	*Temp*	*Medium*	*Log of equilibrium constant, remarks*	*Ref*
H^+	*X*		solid	ev $H_2W_{12}O_{42}^{12-}$ in $Na_{10}W_{12}O_{41}(H_2O)_{28}(s)$	52L/65L
	X		solid	ev WO_6^{6-} in $Sr_3WO_6(s)$, *etc.*	63B
	gl	50	3 Li(Cl)	β_{67} 53·98, $\beta_{12,14}$ 110·03, $\beta_{12,18}$ 132·51	64A
	cfu, Ram	35, 50	var	ev $W_{12}O_{41}^{10-}$, $W_{12}O_{39}^{6-}$, $HW_6O_{21}^{5-}$	
	sol, pH	22?	dil	$K_s(H^+{}_2L^{2-}) \approx -15{\cdot}5$ (authors give $-21{\cdot}6$, reaction?)	64Ra/
	var		var	ev cpx with ratio $H^+/L^{2-} \approx 8:7$ (*para* $W_7O_{24}^{6-}$?), $\approx 3:2$(*meta*, $W_4O_{13}^{2-}$?) (gl, con, conJ)	64S
	kin	25	var	$K_1 \approx 3{\cdot}5$, $K_{12} \approx 2{\cdot}3$, $K_{13}(H_2L+H^+ \rightleftharpoons H_3L^+)$ 0	64Ya
	oth	25, 50	2 ($NaClO_4$)	in solutions with 1·5 H^+/L^{2-} ev cpx $\approx$ 16 W (decomposes very slowly), W_{12} cpx (stable) (light scattering)	65C
	ir, nmr		solid	ev $W_{12}O_{38}(OH)_2^{6-}$ (*meta*), $W_{12}O_{36}(OH)_{10}^{10-}$ (*para B*), $W_{24}O_{72}(OH)_{12}^{12-}$ (*ψ meta*)	65Ga+
	gl		var	β_{67}(*para A*) 57·5, $\beta_{12,18}$(*meta*) 142, also *T*-jump relaxation	65Ga
	cal	25	0·84 NaOH	$\Delta H[WO_3(s)+2\,OH^- \rightleftharpoons WO_4^{2-}+H_2O] = -13{\cdot}80$; $\Delta H[H_2WO_4(s)+2\,OH^- \rightleftharpoons WO_4^{2-}+2H_2O] = -13{\cdot}05$	65S
	O	350	$KNO_3(1)$	$K(WO_3+O^{2-} \rightleftharpoons WO_4^{2-})$ 14·0 in $KNO_3(1)$, M units	65Sa

continued overleaf

7 WO_4^{2-} (contd)

Metal	Method	Temp	Medium	Log of equilibrium constant, remarks	Ref
H^+	gl, fp	32	Na_2SO_4 sat	$\beta_{12,14}(14H^+ + 12L^{2-} \rightleftharpoons H_{10}W_{12}O_{46}^{10-})$ 114·7	65T
(*contd*)	kin	25	0·003	K_1 3·70, K_{12} 2·20	65Y
	fp	32	Na_2SO_4 sat	ev species with 1, 3, and 6 W	66J
	cal		dil	$\Delta H[W_{12}O_{39}^{6-}(\text{meta}) + 4OH^- \rightleftharpoons W_{12}O_{41}^{10-} + 2H_2O] = -22{\cdot}2$ $\Delta H(W_{12}O_{39}^{6-} + 18OH^- \rightleftharpoons 12WO_4^{2-} + 9H_2O) = -102$	66S
	nmr		var	ev two central H in metatungstate, $H_2W_{12}O_{40}^{6-}$	66Pa
	O	316—370	MNO_3(l)	$K(2WO_4^{2-} \rightleftharpoons W_2O_7^{2-} + O^{2-})$ −2·48(316°), −2·12(348°), −1·74(370°), also *K* for intermediate *T*, ΔH = 21·0, ΔS = 24·8 in $Na_{0.5}K_{0.5}NO_3$(l), *m* units	67K
	var		var	ev $W_6O_{19}^{2-}$ in solid and in Me_2CO(l) (pre, cfu, ir)	68F
Li^+	*X*		solid	ev $LiW_4O_{16}^{7-}$ in $Li_2WO_4(H_2O)_{4/7}$(s) = $Li^+_{13}(LiW_4O_{16}^{7-})(WO_4^{2-})_3(H_2O)_4$	66H
Ce^{3+}	pre		var	ev $CeW_8O_{28}^{5-}$	66R = 64I
Ce^{4+}	pre		var	ev $CeW_8O_{28}^{4-}$	66R = 64I
Mo^{6+}	var	20	dil	ev $MoO_4HWO_4^{3-}$ (spJ, kin, gl)	65Ya
W^{5+}	pol	25	1 SO_4^{2-}	(Q = $W_{12}O_{40}$) ev H_2Q^{7-}, H_2Q^{8-}, H_4Q^{6-}, H_6Q^{6-}	66P
	X, pre		solid	ev $H_2W_{12}O_{40}^{8-}$ (with $2W^V$) and $H_2W_{12}O_{40}^{12-}$ (with $6W^V$) units in K^+ salts	66T ≈ 68T
$W^{5+,3+}$	var		var	ev $H_2W_{12}O_{40}^{8-}$, $H_2W_{12}O_{40}^{12-}$, $H_2W_{12}O_{16}(OH)_{24}^{3-}$?, $H_2W_{12}O_4(OH)_{36}^{4+}$, and $H_2W_{12}O_4(OH)_{28}(OH_2)_8^{2+}$, or $H_2W_{12}O_4(OH)_{24}(OH_2)_{12}^{6+}$ (pre, sp, pol, gl, con)	67T
Fe^{3+}, W^{5+}	pol	25	1 SO_4^{2-}	(Q = $W_{12}O_{40}$) ev FeQ^{5-}, FeQ^{6-}, FeQ^{7-}, FeQ^{8-}, H_4FeQ^{6-}	66P
Fe^{3+}, Ni^{2+}	*X*		solid	ev $H_3FeNiW_{11}O_{40}^{6-}$, not $FeNiW_{12}O_{42}^{7-}$	68R
Co, W^{5+}	pol, red	25, 30	1 SO_4^{2-}	(Q = $W_{12}O_{40}$) ev CoQ^{6-}, CoQ^{7-}, CoQ^{8-}, H_2CoQ^{6-}, H_4CoQ^{6-}	66P
$Co^{2+,3+}$	var		var	ev $H_xXMW_{11}O_{40}$ anions (X in tetrahedral centre, M octahedral like W) with (X, M) = (Si^{4+}, Co^{2+}), (H^+_2, Co^{3+}), (Co^{2+}, Co^{2+}), (Co^{3+}, Co^{2+}), ref 59B withdrawn (pre, *X*, fp, gl)	66B
Ni^{2+}, Mo^{6+}					63M

1964 *Erratum*: p. 98, *in ref* 63M, *for* "$NiMo_{6-n}W_{24}O_nH_6^{4-}$" *read* "$NiMo_{6-n}W_nO_{24}H_6^{4-}$"

Metal	Method	Temp	Medium	Log of equilibrium constant, remarks	Ref
Zn^{2+}	*X*		solid	ev cpx $ZnMW_{11}$(8−, M^{II} = Ni, Co, Cu, Zn, M^{IV} = V), (7−, M^{III} = Mn), (6−, M^{II} = Mn)	68T
B^{3+}	kin	25	0·2 (KCl)	no ev cpx	64Y
	X		solid	ev cpx BW_{11}(9−), $BM^{II}W_{11}$(7−, M = Mn, Co, Ni, Cu, Zn) in solids	68T
Ga^{3+}	var		var	ev anion ($H_2GaW_{11}O_{40}^{9-}$) (pre, *X*)	66B
Si^{4+}	*X*		$H_4SiW_{12}O_{40}$ sat	ev $SiW_{12}O_{40}^{4-}$ group in soln	62Be
Si^{4+}, W^{5+}	pol, red	25, 30	1 SO_4^{2-}	(Q = $W_{12}O_{40}$) ev SiQ^{4-}, SiQ^{5-}, SiQ^{6-}, H_2SiQ^{6-}, SiQ^{8-} ?	66P
Si^{4+}, Mn^{3+}	*X*		solid	ev cpx SiW_{11}(8−), ev $SiMn^{III}W_{11}O_{40}^{7-}$ in K salt	68T
Si^{4+}, 8A	pre, sp		var	ev $SiCo^{III}W_{11}O_{40}H_2^{5-}$, $SiNi^{II}W_{11}O_{40}H_2^{6-}$	67W
Ge^{4+}	*X*		solid	ev cpx GeW_{11}(8−) in K salt	68T

continue

Metal	*Method*	*Temp*	*Medium*	*Log of equilibrium constant, remarks*	*Ref*
Ge^{4+}, 8A	pre, sp		var	ev $GeCo^{II}W_{11}O_{40}H_2^{6-}$, $GeCo^{III}W_{11}O_{40}H_2^{5-}$, $GeNi^{II}W_{11}O_{40}H_2^{6-}$	67W
P^{5+}	*X*		$H_3PW_{12}O_{40}$ sat	ev $PW_{12}O_{40}^{3-}$ group in soln	62Be
	kin	25	0·2 (KCl)	ev cpx PW apparent K_1 1·2(0·2M-H^+), 0·9(0·05M-H^+)	64Y
P^{5+}, $M^{2+,3+}$	var		var	ev $P_2MW_{17}O_{62}H_2^{8-}$($M = Mn^{2+}, Co^{2+}, Ni^{2+}, Zn^{2+}$) $P_2MW_{17}O_{62}H_2^{7-}$($M = Mn^{3+}, Co^{3+}$) (pre, tp, cix, sp, mag)	68M = 67M
	X		solid	ev $PW_{11}O_{39}^{7-}$, $PMn^{II}W_{11}O_{39}OH_2^{5-}$ in K salts ev $P_2MW_{17}O_{61}OH_2$(8−, M^{II} = Mn, Co, Ni, Cu, Zn), (7−, M^{III} = Cr, Mn, Fe, Ga)	68T
P^{5+}, V^{5+}	sp, pre			ev $PVW_{11}O_{40}^{4-}$, $PV_2W_{10}O_{40}^{5-}$, $PV_3W_6O_{31}H^{5-}$ $\beta(H^+{}_4PVW_{11}O_{40}^{4-})$ 2·20, $\beta(H^+{}_5PV_2W_{10}O_{40}^{5-})$ 1·89, $\beta(H^+{}_3PW_{12}O_{40}^{3-})$ 2·74	68C
P^{5+}, W^{5+}	pol, red	25, 30	1 SO_4^{2-}	ev PQ^{3-}, PQ^{4-}, PQ^{5-}, HPQ^{6-} ($Q = W_{12}O_{40}$)	66P
	pol	25	var	ev PQ^{n-}, n = 3, 4, 5, 6, 7, 8, 9 ($Q = W_{12}O_{40}$)	67P
	red, pol	25	var	($R = W_{18}O_{62}$) ev P_2R^{6-}, P_2R^{7-}, P_2R^{8-}, P_2R^{10-} (red, pol); P_2R^{9-}, P_2R^{11-}, P_2R^{12-}(pol); (separate series from A-P_2R^{6-} and B-P_2R^{6-})	
P^{5+}, Ni^{2+}	pre, sp		var	ev $PNiW_{11}O_{40}H_2^{5-}$	67W
As^{5+}	kin	25	0·2 (KCl)	ev cpx AsW, apparent K_1 1·5(0·2M-H^+), 2·4(0·05M-H^+)	64Y
As^{5+}, $M^{2+,3+}$	var		var	ev $As_2MW_{17}O_{62}H_2^{8-}$($M = Mn^{2+}, Co^{2+}, Ni^{2+}$), $As_2MW_{17}O_{62}H_2^{7-}$($M = Mn^{3+}, Co^{3+}$) (pre, cix, tp, sp, mag)	68M = 67M
As^{5+}, W^{5+}	red, pol	25	var	($R = W_{18}O_{62}$) ev As_2R^{6-}, As_2R^{7-}, As_2R^{8-}, As_2R^{10-} (red, pol), As_2R^{9-}, As_2R^{11-}, As_2R^{12-}(pol)	67P
Se^{4+}, V^{5+}	pre, gl		var	ev $SeV_2W_{10}O_{40}^{6-}$	65P
Te^{4+}	sp, con		H_2SO_4 var, HCl var	ev cpx TeW_6, TeW_{12}, TeW_{14}, TeW_{18}	65G

62Be A. A. Babad-Zakhryapin and N. S. Gorbunov, *Izvest. Akad. Nauk. S.S.S.R., Ser. khim.*, 1962, 1870
63B I. N. Belyaev, V. S. Filip'ev, and E. G. Fesenko, *Zhur. strukt. Khim.*, 1963, **4**, 719
64A J. Aveston, *Inorg. Chem.*, 1964, **3**, 98
64R R. Ripan and I. Todorut, *Roczniki Chem.*, 1964, 38, 1587
64Ra L. B. Reznik and P. N. Kovalenko, *Ukrain. khim. Zhur.*, 1964, **30**, 514
64S R. S. Saxena and O. P. Sharma, *Z. anorg. Chem.*, 1964, 333, 154
64Y K. B. Yatsimirskii and K. E. Prik, *Zhur. neorg. Khim.*, 1964, **9**, 178
64Ya K. B. Yatsimirskii and K. E. Prik, *Zhur. neorg. Khim.*, 1964, **9**, 1838
65C H. R. Craig and S. Y. Tyree, jun., *Inorg. Chem.*, 1965, **4**, 997
65G E. Sh. Ganelina and N. I. Nerevyatkina, *Zhur. neorg. Khim.*, 1965, **10**, 894
65Ga O. Glemser, W. Holznagel, W. Höltje, and E. Schwarzmann, *Z. Naturforsch.*, 1965, **20b**, 725
65L W. N. Lipscomb, *Inorg. Chem.*, 1965, **4**, 132
65P N. A. Polotebnova, *Zhur. neorg. Khim.*, 1965, **10**, 1498
65S V. I. Spitsyn and N. N. Patsukova, *Zhur. neorg. Khim.*, 1965, **10**, 2396
65Sa A. M. Shams el Din and A. A. El Hosary, *J. Electroanalyt. Chem.*, 1965, **9**, 345
65T A.-P. Tai and T.-W. Wang, *Scientia Sinica*, 1965, **14**, 568; *Acta Sci. Nat. Univ. Nankin*, 1964, **8**, 395
65Y K. B. Yatsimirskii and V. F. Romanov, *Zhur. neorg. Khim.*, 1965, **10**, 1607
65Ya K. B. Yatsimirskii and V. F. Romanov, *Zhur. neorg. Khim.*, 1965, **10**, 1613
66B L. C. W. Baker, V. S. Baker, K. Eriks, *et al. J. Amer. Chem. Soc.*, 1966, **88**, 2329
66H A. Hüllen, *Ber. Bunsengesellschaft Phys. Chem.*, 1966, **70**, 598
66J D. V. S. Jain and S. K. Dogra, *J. Chem. Soc. (A)*, 1966, 284
66P M. T. Pope and G. M. Varga, jun., *Inorg. Chem.*, 1966, **5**, 1249
66Pa M. T. Pope and G. M. Varga, jun., *Chem. Comm.*, 1966, 653

continued overleaf

7 WO_4^{2-} (contd)

66R R. Ripan and I. Todoruţ, *Rev. Roumaine Chim.*, 1966, **11**, 691, 1279
66S V. I. Spitsyn and G. V. Kosmodem'yanskaya, *Zhur. neorg. Khim.*, 1966, **11**, 1397
66T C. M. Tourné and P. Souchay, *Compt. rend.*, 1966, **263**, *C*, 1142
67K R. N. Kust, *Inorg. Chem.*, 1967, **6**, 157
67M S. A. Malik and T. J. R. Weakley, *Chem. Comm.*, 1967, 1094
67P M. T. Pope and E. Papaconstantinou, *Inorg. Chem.*, 1967, **6**, 1147
67T C. Tourné, *Bull. Soc. chim. France*, 1967, 3196, 3199, 3214
67W T. J. R. Weakley and S. A. Malik, *J. Inorg. Nuclear Chem.*, 1967, **29**, 2935
68C M. P. Courtin, *Bull. Soc. chim. France*, 1968, 4799
68F J. Fuchs and K. F. Jahr, *Z. Naturforsch.*, 1968, **23b**, 1380
68M S. A. Malik and T. J. R. Weakley, *J. Chem. Soc.* (*A*), 1968, 2647
68R R. Ripan and M. Puşcaşiu, *Z. anorg. Chem.*, 1968, **358**, 82
68T C. M. Tourné and G. Tourné, *Compt. rend.*, 1968, **266**, *C*, 1363

8 Manganates, MnO_4^{2-}, MnO_4^-

Metal	*Method*	*Temp*	*Medium*	*Log of equilibrium constant, remarks*	*Ref*
				Manganate(VI), MnO_4^{2-}	
H^+	est	20	var	K_1 12·5	67H
				Manganate(VII), *permanganate*, MnO_4^-	
H^+	con, dis		var	$K_D[H^+ + L^- \rightleftharpoons H^+ \text{(in TBP)} + L^- \text{(in TBP)}] \approx -0{\cdot}5$, $K_D[H^+ + L \rightleftharpoons HL \text{(in TBP)}] \approx 1$, K_1 1·5 in TBP	68R

67H K.-H. Heckner and R. Landsberg, *J. Inorg. Nuclear Chem.*, 1967, **29**, 423
68R A. M. Rozen and A. S. Skotnikov, *Doklady Akad. Nauk S.S.S.R.*, 1968, **182**, 1098

9 Rhenate(VII) (per-rhenate) ReO_4^-

Metal	*Method*	*Temp*	*Medium*	*Log of equilibrium constant, remarks*	*Ref*
H^+	tp		HCl var	ev cat cpx in 6M-HCl	66G
	sp, kin		H_2SO_4 var	ev Re_2O_7 predom in > 12·5M-H_2SO_4, $\beta(H^+L^-) \approx \beta(H^+Br^-)$, perhaps −8 to −9 (kin)	67R
	dis		H_2SO_4 var	$K_D[H^+ + L^- + 2T\text{(org)} \rightleftharpoons HLT_2\text{(org)}]$ 0·44, T = TBP, org = kerosene, M units	67Z
	var		$HReO_4$ var	ev cpx, perhaps polyn, in > 4·5M-HL(nmr, ir, Ram)	67U
Group 1A$^+$	tp, con	18	0 corr	no ev cpx (Na^+), K_1 1·2(K^+), 1·0(Cs^+)	63S
$Co(NH_3)_5^{3+}$	var		var	ev $Co(NH_3)_5OReO_3^{2+}$ (pre, kin, ir)	68L
Ph_4P^+	sol	10—50	dil	K_{s0} −8·96(10°), −8·80(15°), −8·69(20°), −8·56(25°), −8·40(30°), −8·08(40°), −7·74(50°), ΔH_{s0} = 11·0(20°)	66O
Ph_4As^+	sol	10—50	dil	K_{s0} −8·58(10°), −8·52(15°), −8·43(20°), −8·32(25°), −8·22(30°), −8·06(40°), −7·81(50°), ΔH_{s0} = 7·2(20°)	66O

references on facing pag

63S V. P. Shvedov and K. Kotegov, *Radiokhimiya*, 1963, **5**, 374
66G D. S. Gaibakyan and M. V. Darbinyan, *Armyan. khim. Zhur.*, 1966, **19**, 27
66O T. Ôkubo and F. Aoki, *J. Chem. Soc. Japan*, 1966, **87**, 1103
67R C. L. Rulfs, R. A. Pacer, and R. F. Hirsch, *J. Inorg. Nuclear Chem.*, 1967, **29**, 681
67U K. Ulbricht, R. Radeglia and H. Kriegsmann, *Z. anorg. Chem.*, 1967, **356**, 22
67Z A. N. Zelikman and L. Dregan, *Zhur. neorg. Khim.*, 1967, **12**, 261
68L E. Lenz and R. K. Murmann, *Inorg. Chem.*, 1968, **7**, 1880

Cyanoferrate(II), $Fe(CN)_6^{4-}$ 10

Metal	*Method*	*Temp*	*Medium*	*Log of equilibrium, constant remarks*	*Ref*
H^+	red	25	0 corr	K_1 4·28, K_{12} 2·3	67H
	cal	25	dil	$\Delta H_1 = 0{\cdot}5, \Delta S_1 = 21; \Delta H_{12} = 1, \Delta S_{12} = 14$	
	sp	27	1 KCl	K_1 2·33, K_{12} 0·49	68L
Group 1A$^+$	tp	25	0 corr	K_1 1·78(Li^+), 2·08(Na^+), 2·42—2·65(Rb^+), 2·85(Cs^+) [K_1(K^+) 2·30]	66S
K^+	*E*	25—39	0·1 (Me_4NCl)	K_1 1·5(25°), 1·6(39°), (K^+ glass electrode)	66C
			0 corr	K_1 2·2(25°), 2·4(39°), $\Delta H = 4, \Delta S = 23$	
	ir		var	no ev inner ("water-free") cpx	66T
	gl	25	C K_4L	K_1 1·78(C = 0·004), 2·13(C = 0·0004), 2·35 (0 corr), (K^+-sensitive glass electrode)	67E
	gl	10—45	0·001 K_4L	K_1 1·95(10°), 2·00(25°), 2·01(40°) $\Delta H_1 = 0{\cdot}6$	
	cal	25	dil	$\Delta H_1 = 1{\cdot}0, \Delta S_1 = 13$	
Ca^{2+}	tp	25	0 corr	K_1 3·59	66N
	ir		var	no ev inner ("water-free") cpx	66T
Sr^{2+}	tp	25	0 corr	K_1 3·59	66N
Cr^{3+}	oth		var	ev $Cr^{3+}L^{4-}$ (from ultrasonic velocity by "Job's method")	66M
Mo^{5+}	sp		HCl var	ev cpx	67T
Co^{2+}	Co	25	0 corr	$K_{s0}(Co^{2+}{}_2L^{4-})$ −37·32	66Ma
Co en_3^{3+}	oth	20	2 Na(ClO_4)	K_1 0·4, β_2 1·62 (polarimetry)	67L
			0·21 ($NaClO_4$)	K_1 2·0, β_2 3·8 (circular dichroism)	
Cu^+	Cu	25	0 corr	$K_s(K^+{}_2Cu^+{}_2L^{4-})$ −26·66	64R
Ag^+	Ag	25	0 corr	$K_{s0}(Ag^+{}_4L^{4-})$ −44·07	64R
Zn^{2+}	MHg	25	0 corr	$K_{s0}(Zn^{2+}{}_2L^{4-})$ −15·68, $K_s(K^+Zn^{2+}{}_{1.5}L^{4-})$ −21·09	64R
Cd^{2+}	MHg	25	0 corr	$K_{s0}(Cd^{2+}{}_2L^{4-})$ −17·38, $K_s(K^+{}_2Cd^{2+}L^{4-})$ −17·09	64R
Tl^+	cal, MHg	25	3 ($LiClO_4$)	K_1 0·82, $\Delta H_1 = -1{\cdot}78, \Delta S_1 = 2{\cdot}2$	67M
	sol			K_1 0·5?	
Pb^{2+}	MHg	25	0 corr	$K_{s0}(Pb^{2+}{}_2L^{4-})$ −18·02	64R

64R P. A. Rock and R. E. Powell, *Inorg. Chem.*, 1964, **3**, 1593
66C R. W. Chlebek and M. W. Lister, *Canad. J. Chem.*, 1966, **44**, 437
66M W. U. Malik and J. Singh, *Bull. Chem. Soc. Japan*, 1966, **39**, 2541
66Ma W. U. Malik and A. Das, *Indian J. Chem.*, 1966, **4**, 203
66N G. F. Nichugovskii and V. P. Shvedov, *Radiokhimiya*, 1966, **8**, 118
66S V. P. Shvedov and G. F. Nichugovskii, *Radiokhimiya*, 1966, **8**, 66
66T N. Tanaka, Y. Kobayashi, and M. Kamada, *Bull. Chem. Soc. Japan*, 1966, **39**, 2187

continued overleaf

10 $Fe^{II}(CN)_6^{4-}$ (contd)

67E W. A. Eaton, P. George, and G. I. H. Hanania, *J. Phys. Chem.*, 1967, **71**, 2016
67H G. I. H. Hanania, D. H. Irvine, W. A. Eaton, and P. George, *J. Phys. Chem.*, 1967, **71**, 2022
67L R. Larsson, *Acta Chem. Scand.*, 1967, **21**, 257
67M I. F. Mavrin, F. Ya. Kul'ba, and V. E. Mironov, *Zhur. fiz. Khim.*, 1967, **41**, 1659
67T V. M. Tarayan and S. V. Vartanyan, *Armyan. khim. Zhur.*, 1967, **20**, 179
68L S. A. Levison and R. A. Marcus, *J. Phys. Chem.*, 1968, **72**, 358

11 Cyanoferrate(III), $Fe(CN)_6^{3-}$

Metal	*Method*	*Temp*	*Medium*	*Log of equilibrium constant, remarks*	*Ref*
Group $1A^+$	sol	25	3 (Li)NO_3	K_1 −0·3(Na^+), 0·30(K^+), 0·52(Cs^+)	67R
Na^+	sol, cal	25	3 (LiCl)	K_1 −0·77, ΔH_1 = −4·0, ΔS_1 = −17	66Ma
K^+	*E*	25—39	0·1 (Me_4NCl)	K_1 0·85(25°), 0·9(39°), (K^+-glass electrode)	66C
			0 corr	K_1 1·4(25°), 1·5(39°), ΔH_1 = 3, ΔS_1 = 15	
	sol, cal	25	3 (LiCl)	K_1 −0·42, ΔH_1 = −5·4, ΔS_1 = −20	66Ma
	ir		var	no ev inner ("water-free") cpx	66T
	gl	25	*c* K_3L	K_1 1·03(*c* = 0·005), 1·29(*c* = 0·001), 1·46(0 corr), (K^+-sensitive glass electrode)	67E
		10—45	0·00125 K_3L	K_1 1·21(10°), 1·25(25°), 1·28(45°), ΔH_1 = 0·5	
	cal	25	dil	ΔH_1 = 0·5, ΔS_1 = 8	
	sol	25	*I*(Li)NO_3	K_1 0·30(*I* = 3), 0·18(*I* = 2), −0·24(*I* = 0·5), 0·90(*I* = 0)	67R
	sol	25	*I*(Li)Cl	K_1 −0·42(*I* = 3), −0·49(*I* = 2), −0·27(*I* = 1), −0·22(*I* = 0·5), 0·90(*I* = 0)	
Cs^+	sol	25	3 (LiCl)	K_1 −0·26	66Ma
Group $2A^{2+}$	ir		var	no ev inner ("water-free") cpx for Mg, Ca, Ba	66T
	sol	25	3 ($LiNO_3$)	K_1 0·04(Mg^{2+}), 0·15(Ca^{2+}), 0·23(Sr^{2+})	67R
Mg^{2+}	sol, cal	25	3 (LiCl)	K_1 −1·03, ΔH_1 = −3·4, ΔS_1 = −16	66Ma
Sr^{2+}	sol	25	3 (LiCl)	K_1 −0·77	66Ma
Ba^{2+}	sol, cal	25	3 (LiCl)	K_1 −0·60, ΔH_1 = −3·7, ΔS_1 = −15	66Ma
La^{3+}	con	25	→ 0	K_1 3·74	63D
	con	25	→ 0, *p* atm	K_1 3·58(2000 atm), 3·62(1495 atm), 3·68(985 atm), 3·74(480 atm), 3·80(1 atm); $\Delta V_1 \approx$ 8 ml(1 atm), 4·6 ml(2000 atm)	64H
	con	25	H_2NCHO	K_1 2·40 in H_2NCHO(l)	65J
	var		0 corr	K_1 varies with interionic distance assumed	/68M
Nd^{3+}	con	25	→ 0	K_1 3·74	63D
Gd^{3+}	con	25	→ 0	K_1 3·74	63D
Cr^{3+}	oth		var	ev $Cr^{3+}{}_3L^{3-}{}_2$ (from ultrasonic velocity by "Job's method")	66M
Fe^{3+}	kin, sp	9—25	0·5 $HClO_4$	$K_1{}^{out}$ 0·79(9°), 1·54(25°) $K(Fe^{3+}L^{3-} \rightleftharpoons FeL) \approx 0·5$	67S
CoY^{2-}	sp, kin	25	0·6 ($NaMeCO_2$)	H_4Y = EDTA; K_1 3·18(kin), 3·20(sp), $K[CoYL^{5-} + L^{3-} \rightleftharpoons Co^{III}YL^{4-} + Fe(CN)_6^{4-}]$ −1·15	67H
$Co(NH_3)_6{}^{3+}$	con	25	H_2NCHO	K_1 2·63 in H_2NCHO(l)	65J
Co $en_3{}^{3+}$	con	25	H_2NCHO	K_1 2·66 in H_2NCHO(l)	65J
R_4N^+	nmr	25	var	K_1 1·3(R = Me), 1(R = Et, Pr, *etc.*)	65L

references on facing pa

63D H. S. Dunsmore, T. R. Kelly, and G. H. Nancollas, *Trans. Faraday Soc.*, 1963, **59**, 2606
64H S. D. Hamann, P. J. Pearce, and W. Strauss, *J. Phys. Chem.*, 1964, **68**, 375
65J G. P. Johari and P. H. Tewari, *J. Phys. Chem.*, 1965, **69**, 2862
65L D. W. Larsen and A. C. Wahl, *Inorg. Chem.*, 1965, **4**, 1281
66C R. W. Chlebek and M. W. Lister, *Canad. J. Chem.*, 1966, **44**, 437
66M W. U. Malik and J. Singh, *Bull. Chem. Soc. Japan*, 1966, **39**, 2541
66Ma V. E. Mironov and Yu. I. Rutkovskii, *Zhur. neorg. Khim.*, 1966, **11**, 1792
66T N. Tanaka, Y. Kobayashi, and M. Kamada, *Bull. Chem. Soc. Japan*, 1966, **39**, 2187
67E W. A. Eaton, P. George, and G. I. H. Hanania, *J. Phys. Chem.*, 1967, **71**, 2016
67H D. H. Huchital and R. G. Wilkins, *Inorg. Chem.*, 1967, **6**, 1022
67R Yu. I. Rutkovskii and V. E. Mironov, *Zhur. neorg. Khim.*, 1967, **12**, 3287
67S D. L. Singleton and J. H. Swinehart, *Inorg. Chem.*, 1967, **6**, 1536
68M R. A. Matheson, *J. Phys. Chem.*, 1968, **72**, 3330

Cyanocobaltate(III), $Co(CN)_6^{3-}$ 12

Metal	*Method*	*Temp*	*Medium*	*Log of equilibrium constant, remarks*	*Ref*
La^{3+}	con	25	$\rightarrow 0$	K_1 3·74	63D
	con	25	w% dioxan	K_1 4·09(w = 10), 4·60(w = 20 in w% dioxan, $\rightarrow 0$	63D
	con	25	H_2NCHO	K_1 2·39 in H_2NCHO(l)	65J
Co en_3^{3+}	con	25	H_2NCHO	K_1 2·78 in H_2NCHO(l)	65J
Ag^+	Ag	25	0 corr	$K_{s0}(Ag^+{}_3L^{3-})$ −25·41	65R
Cd^{2+}	con	25	0 corr	K_1 4·17	64R
Hg_2^{2+}	Hg	25	0 corr	$K_{s0}[(Hg_2^{2+})_3L^{3-}{}_2]$ −36·72	65R
$MeHg^+$	gl	20	0·1 (KNO_3)	K_1 4·15, $\beta_{12}(M^+{}_2L^{3-})$ 7·65	63S = 65S

63D H. S. Dunsmore, T. R. Kelly, and G. H. Nancollas, *Trans. Faraday Soc.*, 1963, **59**, 2606
63S M. Schellenberg, dissertation, Eidgenössische Technische Hochschule, Zürich, 1963
64R T. P. Radhakrishnan, S. C. Saraiya, and A. K. Sundaram, *J. Inorg. Nuclear Chem.*, 1964, **26**, 382
65J G. P. Johari and P. H. Tewari, *J. Phys. Chem.*, 1965, **69**, 2862
65R P. A. Rock, *Inorg. Chem.*, 1965, **4**, 1667
65S G. Schwarzenbach and M. Schellenberg, *Helv. Chim. Acta*, 1965, **48**, 28

Other ligands with A-group elements 13

Metal	*Method*	*Temp*	*Medium*	*Log of equilibrium constant, remarks*	*Ref*
				Pertechnetate TcO_4^-	
H^+	dis	25	1 (Na, H)NO_3	$K_D[H^+ + L^- \rightleftharpoons$ HL (in cyclohexanol)] 1·66	60B/64B
				$K_D[Na^+ + L^- \rightleftharpoons$ NaL (in cyclohexanol)] −0·56	
	sp, kin		H_2SO_4 var	ev Tc_2O_7 predom in > 9·5M-H_2SO_4(sp), $\beta(H^+L^-) \approx \beta(H^+Br^-)$ perhaps −8 to −9 (kin)	67R
Group 1A$^+$	tp, con	18	0 corr	no ev cpx (Na^+), K_1 0·91(K^+), 0·64(Cs^+)	63S
K^+	lit	25	0 corr	K_{s0} −0·89	/53C

53C J. W. Cobble, W. T. Smith, jun., and G. E. Boyd, *J. Amer. Chem. Soc.*, 1953, **75**, 5777
60B G. E. Boyd and Q. V. Larsen, *J. Phys. Chem.*, 1960, **64**, 988 (figure 7, p. 995)
63S V. P. Shvedov and K. V. Kotegov, *Radiokhimiya*, 1963, **5**, 374
64B A. Beck, D. Dyrssen, and S. Ekberg, *Acta Chem. Scand.*, 1964, **18**, 1695
67R C. L. Rulfs, R. A. Pacer, and R. F. Hirsch, *J. Inorg. Nuclear Chem.*, 1967, **29**, 681

14 Borate, $B(OH)_4^-$

Metal	Method	Temp	Medium	Log of equilibrium constant, remarks	Ref
H^+	*X*		solid	ev BO_3^{3-} in MBO_3(s), M = Sc, La, In	32G
	X		solid	ev $B_3O_6^{3-}$ in KBO_2(s)	37Za

1964 *Erratum*: p. 106, *in ref* 63I, 63Ia: "$\dagger\beta_{11}$" *should be* "$\dagger\beta_{10}(HL+OH^- \rightleftharpoons L^- + H_2O)$" *throughout*

Metal	Method	Temp	Medium	Log of equilibrium constant, remarks	Ref
	X		solid	ev $Q^{3-} = B_5O_6(OH)_6^{3-}$ in ulexite, $NaCaQ(H_2O)_5$(s)	64C
	X		solid	ev $B_3O_3(OH)_3$ rings in HBO_2(s, *o*-rh)	64P = 40T
	gl, Ag	25?	org	K_1 13·23 in 50% C_6H_6+50% MeOH, 11·36 in 47% C_6H_6+47% MeOH+6% H_2O	65A
	H	25	dioxan	K_1 10·045 in 20% dioxan, 11·406 in 45% dioxan, ($\rightarrow$ 0)	65F
	X		solid	ev BO_4^{4-} in Fe_3BO_6(s)	65W
	Ram	750	Li-B-O(l)	ev BO_3^{3-}, $B_2O_5^{4-}$, $B_3O_7^{5-}$ and larger polyions in melts with various ratios Li_2O/B_2O_3	66B
	gl	25	HL var	$*\beta_{54}(5HL \rightleftharpoons H_4L_5^- + H^+)$ −6·9	66K
	fp		melts	ev BO_4^{5-}, BO_3^{3-}, BO_2^- and polyn cpx in KCl(l) and KBr(l)	66M
	dis, H, gl	25	2 K(Cl)	$*\beta_{pq} = K[pHL \rightleftharpoons (p-q)H^+ + H_qL_p^{(p-q)-}]$ K_1 8·87, $*\beta_{32}$ −6·90, $*\beta_{31}$ −16·30, $*\beta_{54}$ −6·55, $*\beta_{53}$ −13·80, (K_w −13·88), K_D[HL(aq) $\rightleftharpoons$ HL (in octanol-1)] −0·74	67B
		25	3 Na(ClO)$_4$	K_1 8·77, $*\beta_{32}$ −6·42, $*\beta_{31}$ −15·70, (K_w −13·96), K_D −0·56	
		35	3 Na(ClO$_4$)	K_1 8·80, $*\beta_{32}$ −6·70, $*\beta_{31}$ −15·80, $*\beta_{54}$ −6·10, $*\beta_{53}$ −13·47, (K_w −13·86), K_D −0·55	
		25	K^+ var	K_1 8·92, $*\beta_{43}$ −6·70, $*\beta_{42}$ −15·25, $*\beta_{54}$ −6·57, $*\beta_{53}$ −14, (K_w −13·81)	
		25	dil	K_D −0·76	
	nmr	25	NaH_4L_5 var	$K(L^- + 2HL \rightleftharpoons H_2L_3^-)$ 1·9, $K(L^- + 4HL \rightleftharpoons H_4L_5^-)$ 2·05, pH values from ref 56K	67M+
			$NaHL_2$ var	$K(HL+OH^- \rightleftharpoons L^- + H_2O)$ 5·0, ev several polyn species, good agreement with ref 63I, 63Ia, 63Ib	
	MHg	25	3 Na(Cl)	no ev monoborate of charge < −1 (such as HBO_3^{2-} or BO_3^{3-}) for [OH^-] up to 0·5M	67S
	dis	25	dil	K_D[HL $\rightleftharpoons$ HL(org)] −1·5[org = $(C_8H_{17})_3N$], −0·89($C_{12}H_{25}OH$), −0·38(C_4H_9OH), −0·55[$(C_4H_9O)_3PO$], −0·28[DAMF = $(AmO)_2MePO$], K_D for 14 other "org"; no HL in $C_xH_yCl_z$ solvents	67V
	gl	25	D_2O	K_1 9·74 in D_2O, 9·22 in H_2O, $K_1/K_1(H_2O)$, $0·399x+0·131x^2-0·014x^3$ in $D_{2x}H_{2-2x}O$(l), 0 corr	68G
	gl		var	ev mainly H_2B_2 and B_2^{2-} in dilute soln (?)	68N
Rb^+	oth	910—1350	gas	ev $RbBO_2$(g) (transpiration method)	66A
Fe^{3+}	var		var	K_1 8·5(sp, red), β_2 15·6(hyp), β_3 20·6(hyp), 20·3(red)	64B
Cu^{2+}	sol, pH	22	var	$K_s(Cu^{2+}OH^-L^-)$ −17·3, β_3 15·2—15·7 $K_s(Cu^{2+}L^-{}_2)$ −14·5, K_1 7·13, K_2 5·32, K_3 2·72	65S

32G V. M. Goldschmidt and H. Hauptmann, *Nachr. Ges. Wiss. Göttingen*, 1932, **39**, 53
37Za W. H. Zachariasen, *J. Chem. Phys.*, 1937, **5**, 919
40T H. Tazaki, *J. Sci. Hiroshima Univ.*, 1940, **10**, *A*, 55
56K P. H. Kemp, "The Chemistry of Borates", Borax Consolidated Ltd., London, 1956
64B N. B. Burchinskaya, *Ukrain. khim. Zhur.*, 1964, **30**, 177
64C J. R. Clark and D. E. Appleman, *Science*, 1964, **145**, 1295
64P C. R. Peters and M. E. Milberg, *Acta Cryst.*, 1964, **17**, 229
65A V. V. Aleksandrov, L. L. Spivak, and L. K. Zakharchenko, *Zhur. fiz. Khim.*, 1965, **39**, 58

continued

65F D. Feakins and D. J. Turner, *J. Chem. Soc.*, 1965, 4986
65S M. B. Shchigol', *Zhur. neorg. Khim.*, 1965, **10**, 2097
65W J. G. White, A. Miller, and R. E. Nielsen, *Acta Cryst*, 1965, **19**, 1060
66A C. E. Adams and J. T. Quan, *J. Phys. Chem.*, 1966, **70**, 331
66B W. Bues, G. Förster, and R. Schmitt, *Z. anorg. Chem.*, 1966, **344**, 148
66K L. I. Katzin and E. Gulyas, *J. Amer. Chem. Soc.*, 1966, **88**, 5209
66M A. Mitchell, *Trans. Faraday Soc.*, 1966, **62**, 1296
67B M. Barrès, *Rev. Chim. minérale*, 1967, **4**, 803
67M R. K. Momii and N. H. Nachtrieb, *Inorg. Chem.*, 1967, **6**, 1189
67S G. Schorsch and N. Ingri, *Acta Chem. Scand.*, 1967, **21**, 2727
67V E. E. Vinogradov and L. A. Azarova, *Zhur. neorg. Khim.*, 1967, **12**, 1624
68G V. Gold and B. M. Lowe, *J. Chem. Soc. (A)*, 1968, 1923
68N R. F. Nickerson, *J. Inorg. Nuclear Chem.*, 1968, **30**, 1447

Cyanide, CN^- 15

Metal	*Method*	*Temp*	*Medium*	*Log of equilibrium constant, remarks*	*Ref*
H^+	pol	30	2 ($NaNO_3$)	K_1 8·52	56N
	gl	20—33	→0	K_1 9·36(20°), 9·19(26°), 9·05(33°), $\Delta H_1 = -9{\cdot}9$, $\Delta S_1 = 9{\cdot}2$	66Ba
V^{2+}	cal	25·4	var	$\Delta H_{\beta 6} = -47{\cdot}0$	64G
Cr^{2+}	cal	25·4	var	$\Delta H_{\beta 6} = -63{\cdot}2$	64G
Cr^{3+}	var		var	ev $CrCN^{2+}$ and $CrNC^{2+}$ (kin, cix, sp)	65E
$CrNO^{2+}$	kin	27	var	$\beta[H^+CrNOL_2(H_2O)_3{}^0]$ 1·2, $\beta[H^+CrNOL(H_2O)_4{}^+]$ 0·7	68Bb
Mo^{4+}	*X*		solid	ev $Mo(CN)_8{}^{4-}$ in $K_4Mo(CN)_8(H_2O)_2(s)$	39H
	var		var	ev $MoO_2L_4{}^{4-}$ not $Mo(OH)_4L_4{}^{4-}$ (*X*, ir, pre)	67La
	X		solid	ev $MoO_2(CN)_4{}^{4-}$ and $MoO_2H(CN)_4{}^{3-}$ rather than $Mo(OH)_4(CN)_4{}^{4-}$ and $Mo(OH)_3H_2O(CN)_4{}^{3-}$	67Lb
	spJ	25	var	$\beta[VO^{2+}MoL_4(OH)_3H_2O^{3-}]$ 4·86	68D
	X		solid	ev $MoO_2L_4{}^{4-}$ and not $Mo(OH)_4L_4{}^{4-}$ in $NaK_3MoO_2L_4(H_2O)_6(s)$	68Db
	gl	25	0 corr	$\beta(H^+MoO_2L_4{}^{4-})$ 12·62, $\beta(H^+MoOOHL_4{}^{3-})$ 9·98	68Pa
W^{4+}	var		var	ev $WO_2L_4{}^{4-}$, not $W(OH)_4L_4{}^{4-}$ (X, ir, pre)	67La
Mn^{2+}	cal	25·4	var	$\Delta H_{\beta 6} = -34{\cdot}5$	64G
Mn^{3+}	pre, *X*		solid	ev $MnL_6{}^{3-}$ in solids	64F
Re^{2+}	gl	30	0·5 KCl	$\beta(H^+ReL_5{}^{3-})$ 10·5, $\beta(H^+HReL_5{}^{2-})$ 1·57	64Sa
	var		var	ev $ReL_6{}^{4-}$ (mag, pre, sp)	65Sa
Re^{3+}	pre, mag		var	ev $ReL_6{}^{3-}$	60C
Re^{5+}	gl	27	0·5 (KNO_3)	$\beta(H^+ReO_2L_4{}^{3-})$ 4·2, $\beta(H^+HReO_2L_4{}^{2-})$ 1·4, no ev $H_3ReO_2L_4$	64Cb
Fe^{2+}	pol	10	dil	$K(2FeL_5H_2O^{3-} \rightleftharpoons Fe_2L_{10}{}^{6-} + 2H_2O) \approx 2$	64E
	cal	25·4	var	$\Delta H_{\beta 6} = -73{\cdot}7$	64G
	ir		solid	ev $Fe(LH)_4L_2$ (*trans*) in $H_4FeL_6(s)$	65G
	cal	25	0 corr	$\Delta H_{\beta 6} = -85{\cdot}77$	65W
	ΔG	25	0	β_6 35·4	65W+
	ΔG	25	0	β_6 36·9, $\Delta S_{\beta 6} = -119{\cdot}0$	65W+/65B
	X		solid	ev $FeL_6{}^{4-}$ in $H_4FeL_6(s)$	66P

continued overleaf

15 CN^- (contd)

Metal	*Method*	*Temp*	*Medium*	*Log of equilibrium constant, remarks*	*Ref*
Fe^{3+}	cal	25·4	var	$\Delta H_{\beta 6} \approx -68$	64G
	cal	25	0 corr	$\Delta H_{\beta 6} = -70{\cdot}14$	65W
	ΔG	25	0	β_6 43·6	65W+
	ΔG	25	0	β_6 43·9, $\Delta S_{\beta 6} = -34{\cdot}3$	65W+/65B
	sp, con		Me_2SO	ev FeL_2^+, FeL_3, FeL_4^- in Me_2SO(l)	67C
Co^{2+}	cal	25·4	var	$\Delta H_{\beta 5}? = -74{\cdot}4$	64G
	sp		KCN var	ev CoL_3^-, CoL_4^{2-}, CoL_5^{3-}	66Sa
	sp		var	$K_6 \ll 0$, ev $CoL_5H_2O^{3-}$ and some cpx with less L	67P
	sp	20	3 ($NaNO_3$)	$\beta(Rb^+CoL_5^{3-})$ 0·57	
	var		$MeNO_2$	ev CoL_2, CoL_4^{2-}, CoL_5^{3-} in $MeNO_2$(l) (sp, con, Ag)	68G
	cal	25	0	$\Delta H_{\beta 5}(Co^{2+} + 5L^- \rightleftharpoons CoL_5^{3-}) = -61{\cdot}5$	68I
Co^{2+}, Co^{3+}	sp	25	0·51 (KCl)	$K[CoL_5^{3-} + \frac{1}{2}H_2(aq) \rightleftharpoons HCoL_5^{3-}]$ 2·59	67B
	cal	25	0	$\Delta H(2CoL_5^{3-} + H_3O^+ \rightarrow CoL_5H_2O^{2-} + HCoL_5^{3-}) = -32{\cdot}0$	68I
Co^{3+}	kin	40	1 ($NaClO_4$)	$K_a(CoL_5H_2O^{2-})$ −9·7 (= Table 2, ref. 62Ha)	65Ha
	ir, *X*		solid	ev CoL_6^{3-} in Ag_3CoL_6(s), H_3CoL_6(s)	67L
	var		var	ev $HCoL_5^{3-}$(hydride H) in solid and solution and probably $H_2CoL_4^{3-}$ in soln (pre, sp, ir)	68Bc
	gl, sp	25	0	$K_a(CoL_5H_2O^{2-})$ −9·7	68I
	MHg	30	KCN var	β_6 17·7 (?, MHg electrode measures $[Co^{2+}]$ if anything)	68K
Ni^0	pre		solid	ev NiL_4^{4-} in $K_4Ni(CN)_4$(s)	42E
	pre		var	ev NiL_4^{4-} and $Ni_2L_6^{6-}$ in salts	65M
Ni^{2+}	con		ROH	ev NiL_3^- in MeOH(l), EtOH(l)	38G
	sp		var	ev NiL_4SCN^{3-}, NiL_4I^{3-}, NiL_4Br^{3-}	63Ba
	X		solid	ev $Ni(CN)_4^{2-}$ in $K_2Ni(CN)_4$(s)	64V
	sp	23	2·5 ($NaClO_4$)	K_5 −0·69	64Va
	nmr		5 NH_3	ev NiL_4^{2-}, no ev NiL_6^{4-}, results of ref 59Bb dubious	
	sp, ir	25	4 K(F)	K_5 0·03, *m* units, no ev NiL_6^{2-} or NiL_5F^{2-}	65C
	pre		solid	ev NiL_5^{3-} in $Cr(NH_3)_6Ni(CN)_5(H_2O)_2$(s)	66R
	var		org	ev NiL_2 and NiL_4^{2-} in Me_2SO(l), NiL_4^{2-} in Me_2NCOMe(l), $C_3H_6O_2CO$(l) (sp, con, Ag)	68Ga
	spJ	190	KSCN(l)	β_4 19·5 in KSCN(l), M units	68H
	MHg	30	KCN var	β_4 22·2	68K
	sp, gl	25	0·1 ($NaClO_4$)	β_4 30·5, $\beta(H^+NiL_4^{2-})$ 5·4, $\beta(H^+HNiL_4^-)$ 4·5 $[K_1(H^+)$ 9·0]	68Ka
	kin	25	0·1 ($NaClO_4$) or $HClO_4$ var	$\beta(H^+H_2NiL_4)$ 2·6	
	X		solid	ev two forms of NiL_5^{3-} (square pyramids, trigonal bipyramids) in Cr $en_3NiL_5(H_2O)_{1.5}$(s)	68Ra
Ni^{2+}cpx	sp	22	0 corr	$\beta(NiA^{2+}L^-)$ 1·4, A = cyclic tetramine	65Cb
Ru^{2+}	ir		solid	ev $Ru(LH)_4L_2$(*trans*) in H_4RuL_6(s)	65G
Rh^{3+}	var		var	ev $HRhL_4H_2O^{2-}$ with H–Rh (not $HRhL_5$, ref 59G) (pre, nmr, ir)	66L
Pd^{2+}	Pd	25	KCN var	β_4 51·6, $\{K[Pd^{2+} + 2e^- \rightleftharpoons Pd(s)]$ 33·4}, also β_4 for 10—60° assuming same *K* and same RTF^{-1} as for 25°C.	65F
	gl, Pd	25	0 corr	β_4 42·4, K_5 2·9	67Ia
	cal	25	0 corr	$\Delta H_{\beta 4} = -92{\cdot}3$, $\Delta H_5 = -0{\cdot}2$, $\Delta S_{\beta 4} = -116$, $\Delta S_5 = 8$	
	cal	25	0·1 Na^+	$\Delta H(PdBr_4^{2-} + 4L^- \rightleftharpoons PdL_4^{2-} + 4Br^-) = -78{\cdot}8$	

continued

Metal	*Method*	*Temp*	*Medium*	*Log of equilibrium constant, remarks*	*Ref*
Os^{2+}	ir		solid	ev $Os(LH)_4L_2$ (*trans*) in $H_4O_sL_6(s)$	65G
$Os^{2+\text{ to }6+}$	pol		var	ev $OsO_2L_4{}^{2-}$, $OsL_6{}^{3-}$ or other Os^{III} cpx, $OsL_6{}^{4-}$	57M
Ir	ir, var		var	ev $IrHL_5{}^{3-}$ in soln and solid (pre, ir, nmr)	67K
Pt^{4+}	var		dil	ev $PtL_6{}^{2-}$ (pre, gl, con)	64Ca
	ir, Ram		var	ev $PtL_4X_2{}^{2-}$, X = Cl, Br, I	65J
Cu^+	cal		Me_2SO	ev $CuL_2{}^-$, $CuL_4{}^{3-}$, in $Me_2SO(l)$	64D
	cal	25	≈ 0·6 (Na)$CuL_{>2}$	K_3 5·0, K_4 2·64	65Bb
	Ram		var	ev $CuL_2{}^-$, $CuL_3{}^{2-}$, $CuL_4{}^{3-}$, no ev mixed cpx with Br^- or I^-	66C
	pol	195	KSCN(l)	β_2 5·70 in KSCN(l)	67E
	gl	25	0 corr	K_3 5·30, K_4 1·5 [$K_1(H^+)$ 9·21]	67I
	cal	25	0 corr	$\Delta H_3 = -11{\cdot}1$, $\Delta H_4 = -11{\cdot}2$, $\Delta S_3 = -13{\cdot}4$, $\Delta S_4 = -31$ $\Delta H_{\beta 2} = -29{\cdot}1\{\Delta H[Cu^+ + Cl^- \rightleftharpoons CuCl(s)] = -4{\cdot}58\}$ $\Delta S_{\beta 2} = -12$, (β_2 23·94, from ref 50V)	
	var	25	aix resin	ev $CuL_2{}^-$, $CuL_3{}^{2-}$, $CuL_4{}^{3-}$ in anion exchange resin, estimates of K_n in resin (aix, gl, ir)	68C
	X		solid	ev tetrahedral $CuL_4{}^{3-}$ in $K_3CuL_4(s)$	68R
	var		var	ev $CuL_4{}^-$, probably $(NC)_2CuNCCN$ in $NaCuL_4(H_2O)_4(s)$ and solution (pre, ir, sp)	68T
Cu^{2+}	con		ROH	ev $CuL_3{}^-$, $CuL_4{}^{2-}$ in MeOH(l), EtOH(l)	38G
	MHg	−45	60% MeOH	β_4 26·7 in 60 *w*% MeOH	65P
Cu^{2+}cpx	sp	22	0 corr	$\beta(CuA^{2+}L^-)$ 2·4—2·95, A = four-cyclic tetramine	65Cb
Ag^+	Ag	150	$MNO_3(l)$	K_{s0} −10·45, β_2 13·50 in $Li_{0.43}K_{0.57}NO_3$(l, eut), *m* units	65Ba
	Ag	248	MNO_3 (l)	K_1 4·04, β_2 8·04 in $Na_{0.5}K_{0.5}NO_3(l)$	65T
	Ag, gl	20·5	0 corr	β_2 20·85, K_3 0·95, $K_{s0}K_{s2}$ −11·35	65Z
	Ag	178—204	KSCN(l)	$K_1 \approx$ 3·2 in KSCN(l) (ion fraction units)	66B
	Ag	20	NH_3	K_1 6·45, K_2 5·45 in $NH_4NO_3(NH_3)_{1.3}(l)$	66T
	Ag	30	1 (KNO_3)	β_3 20·30, β_4 20·79	67A
	Ag	25	1	β_4 20·37	67Ba
	ir, nmr		HF	ev $AgHL^+$ and $Ag(HL)_2{}^+$ in HF(l)	67D
				$K[2AgHL^+ \rightleftharpoons Ag^+ + Ag(HL)_2{}^+]$ 0·66 in HF(l)	67D/
	pol	195	KSCN(l)	β_2 3·20 in KSCN(l)	67E
	Ag	21	4·25 ($NaClO_4$)	β_2 20·23, $\beta(Ag^+L^-OH^-)$ 13·23	67Z
	Ag	30	KCN var	β_2 21·1	68K
	Ag	370	MCl(l)	K_1 1·16, β_2 1·79; or K_1 1·11, β_2 1·85, β_{12} 2·6 ? in $Li_{0.59}K_{0.41}Cl(l)$, *m* units	68W
Ag^+, 8A	sol	20	0·1	$\beta\{Ag^+[Pt(NH_3)_2L_4]^\circ\}$ 1·37	63Ca
	pre, con	20	dil	ev Pt en$L_2X_2Ag^+$ (X = OH, Cl, NO_3)	64C
	var		var	ev $AgQ_2{}^{5+}$, $Q^{2+} = Co(NH_3)_5L^{2+}$ (ir, pre, sp)	64S
Au^+	pol	195	KSCN(l)	β_2 11·23 in KSCN(l)	67E
Au^{3+}	var		var	ev $AuL_2X_2{}^-$, X = Cl, Br, and I (pre, ir, Ram)	64J
	ir, pre		var	ev $AuL_4{}^-$, AuL_3Cl^-, no ev $AuL_5{}^{2-}$ or $AuL_6{}^{3-}$ up to 10M-L^-	65S
	X		solid	ev $Au(CN)_4{}^-$ in $KAu(CN)_4H_2O(s)$	68B

continued overleaf

15 CN^- (contd)

Metal	Method	Temp	Medium	Log of equilibrium constant, remarks	Ref
Zn^{2+}	cal	25·4	var	$\Delta H_{\beta 4} = -27{\cdot}7$	64G
	gl	25	0 corr	β_2 11·07, K_3 4·98, K_4 3·57, β_4 19·62, no ev ZnL^+ [$K_1(H^+)$ 9·21]	65I
	cal	25	0	$\Delta H_{\beta 2} = -10{\cdot}8$, $\Delta H_3 = -8{\cdot}4$, $\Delta H_4 = -8{\cdot}6$	
	X		solid	ev tetrahedral ZnL_4^{2-} in K_2ZnL_4(s)	66S
	pol	195	KSCN(l)	β_2 5·1, β_3 6·6 (6·1 ?), β_4 8·65 in KSCN(l)	67E
Cd^{2+}	fp, Ram		var	ev CdL_3^-, CdL_4^{2-}	64Gb
	dis	30	0·1 ($NaClO_4$)	K_1 5·8, β_2 11·1	65H
	cal	25	3 ($NaClO_4$)	$\Delta H_1 = -7{\cdot}4$, $\Delta H_2 = -7{\cdot}7$, $\Delta H_3 = -7{\cdot}1$, $\Delta H_4 = -7{\cdot}0$; $\Delta S_1 = 0{\cdot}3$, $\Delta S_2 = -2{\cdot}5$, $\Delta S_3 = -3{\cdot}1$, $\Delta S_4 = -9{\cdot}1$	66G+43I
	pol	195	KSCN(l)	K_1 1·8, β_2 3·9, β_3 5·15 in KSCN(l)	67E
Hg^{2+}					

1964 *Addenda*: p. 112, *for ref* "58N" *read* "58N = 56N"; *for ref* "59N" *read* "59N = 56N"; *for ref* "61Na" *read* "61Na = 56N"; *for ref* "59N+58N" *read* "59N+58N = 56N".

Metal	Method	Temp	Medium	Log of equilibrium constant, remarks	Ref
	Hg	25	var	β_2 36·57, β_3 40·64, β_4 43·54, β_5 44·40, β_6 45·19	64A
	sp	25	dil	$K(HgL_2 + HgX_2 \rightleftharpoons 2HgXL)$ 0·93(X = Cl), 0·29(Br), −0·96(I)	64B = 63B = 64B
			dioxan	$K(HgL_2 + HgX_2 \rightleftharpoons 2HgXL)$ 1·08(Cl), 0·24(Br), −0·7(I) in dioxan(l)	64Bb = 63B
	fp, Ram		var	ev HgL_2, HgL_3^-, HgL_4^{2-}	64Gb
	pre, con		var	ev HgL_2Cl^-, HgL_2Br^-, HgL_2SCN^-	65A
	Hg, red	10—40	0 corr	K_1 17·97(10°), 17·00(25°), 16·26(40°), K_2 16·74(10°), 15·75(25°), 15·02(40°)	65Ca
	gl			K_3 3·81(10°), 3·56(25°), 3·37(40°) K_4 2·81(10°), 2·66(25°), 2·42(40°)	
	cal	25	0 corr	$\Delta H_1 = -23{\cdot}0$, $\Delta H_2 = -25{\cdot}5$, $\Delta H_3 = -7{\cdot}6$, $\Delta H_4 = -7{\cdot}2$; $\Delta S_1 = 0{\cdot}7$, $\Delta S_2 = -13{\cdot}4$, $\Delta S_3 = -9{\cdot}0$, $\Delta S_4 = -12{\cdot}1$	
	pol	195	KSCN(l)	β_2 6·50 in KSCN(l)	67E
	Hg	30	3 (KNO_3)	β_5 40·5, β_6 42·6, β_7 38·2 (?)	68A
	Ram	25	5 NaI	ev HgL_3I^{2-}, $HgL_2I_x^{x-}$, $HgLI_3^{2-}$, apparent K_1 2·27, β_2 5·99	68Ca
	sp		HgL_2 var	$K(HgI_2 + HgL_2 \rightleftharpoons 2HgLI)$ −0·89	
	sol	25	0·4 HgI_2	$K(HgI_2 + HgL_2 \rightleftharpoons 2HgLI)$ −0·85	
	Hg, pol	25 ?	*w*% C_5H_5N	K_1 22·9, β_2 26·5, β_3 28·4(w = 20), β_2 28·6(w = 80) in *w*% C_5H_5N, also values for w = 10, 40, 60	68P
Hg^{2+}, Cr^{3+}	sp	25	1 ($LiClO_4$)	$\beta(CrL^{2+}Hg^{2+}) \geqslant 5$	68Ba
Hg^{2+}, Co^{3+}	var		var	ev HgQ_3^{8+}, HgQ_4^{10+}, Q = $Co(NH_3)_5L^{2+}$ (ir, pre, sp)	64S
$MeHg^+$	gl, cal	20	0·1 (KNO_3)	K_1 14·0 or 14·1 {$\beta[Hg^{2+}(CN^-)_2]$ 34·71]} $K_2 < 0{\cdot}5$, $K_{12} < 0{\cdot}7$, $\Delta H_1 = -22{\cdot}1$, $\Delta S_1 = -11{\cdot}4$	63Sa = 63Sb
Tl^{3+}	cix	8—32	0·5 $H(ClO_4)$	$K_n < 11$ (n = 1, 2)	58H
Pb^{2+}	Pb	180—185	KSCN(l)	$K_1 \approx 2{\cdot}4$ in KSCN(l), ion fraction units	66B
Te^0	pre, ir		var	no ev $TeCN^-$	64Ga
	pre, ir		var	ev $TeCN^-$ in Me_2CO(l), Me_2NCHO(l), and Et_4N^+ salt	68Da

references on facing page

38G W. L. German, *J. Chem. Soc.*, 1938, 1027
39H J. L. Hoard and H. H. Nordsieck, *J. Amer. Chem. Soc.*, 1939, **61**, 2853
42E J. W. Eastes and W. M. Burgess, *J. Amer. Chem. Soc.*, 1942, **64**, 1187
56N L. Newman, Ph.D. thesis, Massachussetts Institute of Technology, 1956
57M L. Meites, *J. Amer. Chem. Soc.*, 1957, **69**, 4631
58H R. A. Horne, *J. Inorg. Nuclear Chem.*, 1958, **6**, 338
63Ba M. Beck and J. Bjerrum, *Magyar Kém. Folyóirat*, 1963, **69**, 455
63Bb M. T. Beck and F. Gaizer, *Magyar Kém. Folyóirat*, 1963, **69**, 555, 560
63Ca I. I. Chernyaev, A. V. Babkov, and N. N. Zheligovskaya, *Zhur. neorg. Khim.*, 1963, **8**, 2441
63Sa M. Schellenberg, dissertation, Eidgenössische Technische Hochschule, Zürich, 1963
64A A. M. Azzam and I. A. W. Shimi, *Z. anorg. Chem.*, 1964, **327**, 89
64B M. T. Beck and F. Gaizer, *J. Inorg. Nuclear Chem.*, 1964, **26**, 1755
64Ba M. T. Beck and F. Gaizer, *Acta Chim. Acad. Sci. Hung.*, 1964, **41**, 423
64C I. I. Chernyaev, A. V. Babkov, and N. N. Zheligovskaya, *Zhur. neorg. Khim.*, 1964, **9**, 576
64Ca I. I. Chernyaev and A. V. Babkov, *Zhur. neorg. Khim.*, 1964, **9**, 2253
64Cb M. C. Chakravorti, *J. Indian Chem. Soc.*, 1964, **41**, 477
64D R. E. Dodd and R. P. H. Gasser, *Proc. Chem. Soc.*, 1964, 415
64E G. Emschwiller, *Compt. rend.*, 1964, **259**, 4281
64F A. Ferrari, E. Morisi, and M. E. Tani, *Acta Cryst.*, 1964, **17**, 311
64G F. H. Guzzetta and W. B. Hadley, *Inorg. Chem.*, 1964, **3**, 259
64Ga N. N. Greenwood, R. Little, and M. J. Sprague, *J. Chem. Soc.*, 1964, 1292
64Gb W. P. Griffith, *J. Chem. Soc.*, 1964, 4070
64J L. H. Jones, Proceedings of the Symposium on Co-ordination Chemistry, Tihany, 1964 (Akad. Kiadó, Budapest, 1965), p. 349; *Inorg. Chem.*, 1964, **3**, 1581
64S H. Siebert, *Z. anorg. Chem.*, 1964, **327**, 63
64Sa S. Sen, *Z. anorg. Chem.*, 1964, **333**, 160
64V N.-G. Vannerberg, *Acta Chem. Scand.*, 1964, **18**, 2385
64Va A. L. Van Geet and D. N. Hume, *Inorg. Chem.*, 1964, **3**, 523
65A A. K. Agrawal and R. C. Mehrotra, *Z. anorg. Chem.*, 1965, **336**, 66
65B R. H. Busey, *J. Phys. Chem.*, 1965, **69**, 3179
65Ba G. G. Bombi, M. Fiorani, and G. A. Mazzocchin, *J. Electroanalyt. Chem.*, 1965, **9**, 457
65Bb A. Brenner, *J. Electrochem. Soc.*, 1965, **112**, 611
65C J. S. Coleman, H. Petersen, jun., and R. A. Penneman, *Inorg. Chem.*, 1965, **4**, 135
65Ca J. J. Christensen, R. M. Izatt, and D. Eatough, *Inorg. Chem.*, 1965, **4**, 1278
65Cb Y. M. Curtis and N. F. Curtis, *Austral. J. Chem.*, 1965, **18**, 1933
65E J. H. Espenson and J. P. Birk, *J. Amer. Chem. Soc.*, 1965, **87**, 3280
65F A. B. Fasman, G. G. Kutyukov, and D. V. Sokol'skii, *Zhur. neorg. Khim.*, 1965, **10**, 1338
65G A. P. Ginsberg and E. Koubek, *Inorg. Chem.*, 1965, **4**, 1186
65H H. E. Hellwege and G. K. Schweitzer, *J. Inorg. Nuclear Chem.*, 1965, **27**, 99
65Ha A. Haim, R. J. Grassi, and W. K. Wilmarth, *Adv. Chem. Series*, 1965, **49**, 31
65I R. M. Izatt, J. J. Christensen, J. W. Hansen, and G. D. Watt, *Inorg. Chem.*, 1965, **4**, 718
65J L. H. Jones and J. M. Smith, *Inorg. Chem.*, 1965, **4**, 1677
65M A. Müller, dissertation, Technische Hochschule, Munich, 1965
65P R. Paterson and J. Bjerrum, *Acta Chem. Scand.*, 1965, **19**, 729
65S J. M. Smith, L. H. Jones, I. K. Kressin, and R. A. Penneman, *Inorg. Chem.*, 1965, **4**, 369
65Sa S. Sen, *Z. anorg. Chem.*, 1965, **340**, 82
65Sb G. Schwarzenbach and M. Schellenberg, *Helv. Chim. Acta*, 1965, **48**, 28
65T H. T. Tien, *J. Phys. Chem.*, 1965, **69**, 3763
65W G. D. Watt, J. J. Christensen and R. M. Izatt, *Inorg. Chem.*, 1965, **4**, 220
65Z J. Zsakó and E. Petri, *Rev. Roumaine Chim.*, 1965, **10**, 571
66B W. E. Bennett and W. P. Jensen, *J. Inorg. Nuclear Chem.*, 1966, **28**, 1829
66Ba J. H. Boughton and R. N. Keller, *J. Inorg. Nuclear Chem.*, 1966, **28**, 2851
66C D. Cooper and R. A. Plane, *Inorg. Chem.*, 1966, **5**, 16
66G P. Gerding, *Acta Chem. Scand.*, 1966, **20**, 2771
66L D. N. Lawson, M. J. Mays, and G. Wilkinson, *J. Chem. Soc. (A)*, 1966, 52
66P M. Pierrot, R. Kern, and R. Weiss, *Acta Cryst.*, 1966, **20**, 425
66R K. N. Raymond and F. Basolo, *Inorg. Chem.*, 1966, **5**, 949
66S A. Sequeira and R. Chidambaram, *Acta Cryst*, 1966, **20**, 910
66Sa T. Suzuki and T. Kwan, *J. Chem. Soc. Japan*, 1966, **87**, 395
66T A. Thiébault, M. Herlem, and J. Badoz-Lambling, *Rev. Chim. minérale*, 1966, **3**, 1005
67A I. A. Ammar, S. Darwish, and K. Rizk, *Electrochim. Acta*, 1967, **12**, 647
67B M. G. Burnett, P. J. Connolly, and C. Kemball, *J. Chem. Soc. (A)*, 1967, 800
67Ba G. A. Boos and A. A. Popel', *Zhur. neorg. Khim.*, 1967, **12**, 2086
67C B. Csiszar, V. Gutmann, and E. Wychera, *Monatsh*, 1967, **98**, 12
67D M. F. A. Dove and J. G. Hallett, *Chem. Comm.*, 1967, 571
67E A. Eluard and B. Tremillon, *J. Electroanalyt. Chem.*, 1967, **13**, 208
67I R. M. Izatt, H. D. Johnston, G. D. Watt, and J. J. Christensen, *Inorg. Chem.*, 1967, **6**, 132

continued overleaf

15 CN⁻ (contd)

67Ia R. M. Izatt, G. D. Watt, D. Eatough, and J. J. Christensen, *J. Chem. Soc.* (*A*), 1967, 1304
67K K. Krogmann and W. Binder, *Angew. Chem.*, 1967, **79**, 902
67L A. Ludi, H. O. Güdel, and V. Dvořák, *Helv. Chim. Acta*, 1967, **50**, 2035
67La S. J. Lippard and B. J. Russ, *Inorg. Chem.*, 1967, **6**, 1943
67Lb S. J. Lippard, H. Nozaki and B. J. Russ, *Chem. Comm.*, 1967, 118
67P J. M. Pratt and R. J. P. Williams, *J. Chem. Soc.* (*A*), 1967, 1291
67Z I. Zsakó and E. Fekete, *Studia Univ. Babeş-Bolyai, Ser. Chem.*, 1967, 45
68A I. A. Ammar, S. Darwish, and K. Rizk, *Electrochim. Acta*, 1968, **13**, 797
68B C. Bertinotti and A. Bertinotti, *Compt. rend.*, 1968, **267**, *B*, 1227
68Ba J. P. Birk and J. H. Espenson, *Inorg. Chem.*, 1968, **7**, 991
68Bb J. Burgess, B. A. Goodman, and J. B. Raynor, *J. Chem. Soc.* (*A*), 1968, 501
68Bc R. G. S. Banks and J. M. Pratt, *J. Chem. Soc.* (*A*), 1968, 854
68C J. S. Coleman, R. George, L. Allaman, and L. H. Jones, *J. Phys. Chem.*, 1968, **72**, 2605
68Ca J. S. Coleman, R. A. Penneman, L. H. Jones, and I. K. Kressin, *Inorg. Chem.*, 1968, **7**, 1174
68D K.-ud-Din and M. A. Beg, *J. Indian Chem. Soc.*, 1968, **45**, 455
68Da A. W. Downs, *Chem. Comm.*, 1968, 1290
68Db V. W. Day and J. L. Hoard, *J. Amer. Chem. Soc.*, 1968, **90**, 3374
68G V. Gutmann and K. H. Wegleitner, *Monatsh.*, 1968, **99**, 368
68Ga V. Gutmann and H. Bardy, *Z. anorg. Chem.*, 1968, **361**, 213
68H J. Hennion, J. Nicole, and G. Tridot, *Compt. rend.*, 1968, **267**, *C*, 831
68I R. M. Izatt, G. D. Watt, C. H. Bartholomew, and J. J. Christensen, *Inorg. Chem.*, 1968, **7**, 2236
68K D. K. Kalani, *Lab. Practice*, 1968, **17**, 188
68Ka G. B. Kolski and D. W. Margerum, *Inorg. Chem.*, 1968, **7**, 2239
68P N. Payen-Baldy and M. Machtinger, *Bull. Soc. chim. France*, 1968, 2221
68Pa J. van de Poel and H. M. Neumann, *Inorg. Chem.*, 1968, **7**, 2086
68R R. B. Roof, jun., A. C. Larson, and D. T. Cromer, *Acta Cryst.*, 1968, **B24**, 269
68Ra K. N. Raymond, P. W. R. Corfield, and J. A. Ibers, *Inorg. Chem.*, 1968, **7**, 1362
68T S. K. Tobia and M. F. El-Shahat, *J. Chem. Soc.* (*A*), 1968, 2444
68W S. H. White, D. Inman, and B. Jones, *Trans. Faraday Soc.*, 1968, **64**, 2841

16 Cyanate, OCN⁻ (+fulminate, CNO⁻)

Metal	*Method*	*Temp*	*Medium*	*Log of equilibrium constant, remarks*	*Ref*
				Cyanate, OCN^-	
general	hyp			K_1 values for various metal ions deduced from electronegativities	67L
H^+	gl	20—33	→ 0	K_1 3·70(20°), 3·57(26°), 3·69(33°), $\Delta H_1 \approx -1$, $\Delta S_1 \approx 10$	66B
Mn^{2+}	var		var	ev tetrahedral $Mn(NCO)_4{}^{2-}$ (con, ir, sp, X)	64Fa = 65F
Fe^{2+}	ir		var	ev tetrahedral $Fe(NCO)_4{}^{2-}$	65F
Fe^{3+}	sp		var	β_2 3·72	66L
	sp		var	K_1 2·04	66Lb
			0·7 ($NaClO_4$)	K_1 2·15, β_2 2·56	
Co^{2+}	var		var	ev tetrahedral $Co(NCO)_4{}^{2-}$ (pre, sp, mag) (ir, ref 65F)	61C = 65F
	sp	27	1·5 ($NaNO_3$)	β_4 2·67	66C
	sp		var	K_1 1·80, K_2 1·26, K_3 1·04, K_4 0·90(0·84?)	66La
Ni^{2+}	var	≈ 25	solid, org	ev tetrahedral $NiL_4{}^{2-}$ in solids, MeCN(l), Me_2SO(l), $PhNO_2$(l) (ir, mag, sp, con, pre, X)	64F = 64Fa
	sp		var	ev $NiL_4{}^{2-}$	66Lc
	sp		var	K_1 1·97, K_2 1·56, K_3 1·37, K_4 1·3	67La
Pd^{2+}	ir, pre		var	ev $Pd(NCO)_4{}^{2-}$	65Fa

continued

OCN^- (contd) 16

Metal	Method	Temp	Medium	Log of equilibrium constant, remarks	Ref
Cu^{2+}	var		var	ev $Cu(NCO)_4^{2-}$ (con, mag, sp, *X*, ir)	64Fa = 65F
	sp		var	ev CuL_4^{2-}	66Lc
	sp		var	K_1 2·70, K_2 2·01, K_3 1·43, K_4 1·31	67Lb
	sp		MeOH	$\beta(Cu^{2+}L^-_{1.5})$ 5·1 (?) in MeOH(l)	67Q
Zn^{2+}	var		$MeNO_2$	ev tetrahedral ZnL_4^{2-} in solids and $MeNO_2$(l) (con, ir, Ram, X)	67F = 64Fa
Cd^{2+}	ir		var	ev tetrahedral $Cd(NCO)_4^{2-}$	65F
Sn^{4+}	ir, pre		var	ev $Sn(NCO)_6^{2-}$	65Fa

Fulminate, CNO^-

Metal	Method	Temp	Medium	Log of equilibrium constant, remarks	Ref
Fe^{2+}	pre		var	ev FeL_6^{4-}	00N
	var		var	ev $Fe(CN)_5L^{4-}$ (pre, mag, ir)	64B
	pre, ir		var	ev FeL_6^{4-}	64S
Co^{3+}	pre, ir		var	ev CoL_6^{3-}	64S
Ni^{2+}	pre		var	ev NiL_4^{2-}	64S
Ru^{2+}	pre, ir		var	ev RuL_6^{4-}	64S
Pd^{2+}	pre, sp		solid	ev $Pd(CNO)_4^{2-}$ in solid	66F
OsO_2^{2+}	pre		var	ev $OsO_2(CNO)_4^{2-}$	66F
Pt^{2+}	pre, sp		solid	ev $Pt(CNO)_4^{2-}$ in solid	66F
Ag^+	X		solid	ev Ag_6L_6 rings in rhombohedral AgCNO(s)	65B
Hg^{2+}	pre, ir		var	ev HgL_4^{2-}	64S

00N J. U. Nef, *Annalen*, 1894, **280**, 291
61C F. A. Cotton and M. Goodgame, *J. Amer. Chem. Soc.*, 1961, **83**, 1777
64B W. Beck, *Z. anorg. Chem.*, 1964, **333**, 115
64F J. P. Fackler, jun., G. E. Dolbear, and D. Coucouvanis, *J. Inorg. Nuclear Chem.*, 1964, **26**, 2035
64Fa D. Forster and D. M. L. Goodgame, *J. Chem. Soc.*, 1964, 2790
64S E. Schuierer, dissertation, Technische Hochschule, Munich, 1964
65B D. Britton and J. D. Dunitz, *Acta Cryst*, 1965, **19**, 662
65F D. Forster and D. M. L. Goodgame, *J. Chem. Soc.*, 1965, 262
65Fa D. Forster and D. M. L. Goodgame, *J. Chem. Soc.*, 1965, 1286
66B J. H. Boughton and R. N. Keller, *J. Inorg. Nuclear Chem.*, 1966, **28**, 2851
66C R. Cohen-Ahad and P. Veys, *Bull. Soc. chim. France*, 1966, 1740
66F K. Feldl, dissertation, Technische Hochschule, Munich, 1966
66L A. Łodzińska and H. Puzanowska, *Roczniki Chem.*, 1966, **40**, 1365
66La A. Łodzińska, *Roczniki Chem.*, 1966, **40**, 1585
66Lb A. Łodzińska, *Roczniki Chem.*, 1966, **40**, 1595
66Lc A. Łodzińska and D. Jagodzińska, *Roczniki Chem.*, 1966, **40**, 2011
67F D. Forster and W. D. Horrocks, jun., *Inorg. Chem.*, 1967, **6**, 339
67L A. Łodzińska, *Roczniki Chem.*, 1967, **41**, 1007
67La A. Łodzińska, *Roczniki Chem.*, 1967, **41**, 1155
67Lb A. Łodzińska, *Roczniki Chem.*, 1967, **41**, 1437
67Q M. Quastlerová and Z. Valtr, *Chem. Zvesti*, 1967, **21**, 894

Thiocyanate, SCN^- 17

Metal	Method	Temp	Medium	Log of equilibrium constant, remarks	Ref
general	hyp			K_1 values for various metal ions deduced from electronegativities	67L

continued overleaf

17 SCN⁻ (contd)

Metal	Method	Temp	Medium	Log of equilibrium constant, remarks	Ref
H^+	dis	25	0 corr	K_1 −1·84	65Mb
			7·2 $NaClO_4$	K_D[HL ⇌ HL(org)] 0·31(CCl_4), −0·22($CHCl_3$), 0·59(C_6H_{14}), −0·30(mesitylene), −0·41(C_6H_6), −1·30(Bu_2O)	
	sp		0 corr	$K_1 \approx -1{\cdot}4$?	
	dis	25?	var	K_D[$H^+ + L^- + T(org) \rightleftharpoons HLT(org)$] 1·26 T = DAMFK = (iso-$C_5H_{11}O)_2CH_3PO$, org = *m*-xylene	65S
	gl	20—33	→ 0	K_1 1·0(20°), 1·0(26°), 0·5(33°), $\Delta H_1 = -13$ (?), $\Delta S_1 = -40$ (?)	66B
	pol	?	3 ($NaClO_4$)	K_1 −0·7	66Tb
	dis	20	0 corr	K_D[HL ⇌ HL(in 2-methylpentanone)] 1·97 [$K_1(H^+)$ −1·85]	67D
	dis	23	HL var	K_D[$H^+ + L^- + T(org) = HLT(org)$] −0·60, T = THF($C_4H_4O$), org = C_6H_6	68R
K^+	con	25	org	K_1 1·3 in $NC(CH_2)_4CN$(l), 0 corr	67Sa
Be^{2+}	aix, pre		var	ev ani cpx	66N
	dis		var	K_D[$Be^{2+} + 2L^- \rightleftharpoons BeL_2(org)$] −0·5, org = EtCOMe	67Bb
Sc^{3+}	aix		2 NH_4L	ev ani cpx	64Ha
	sp	20	0·6 ($HClO_4$)	K_1 0·20 [$K_1(Fe^{3+})$ 2·15]	64Ka
	sp	20	3·65 $HClO_4$	K_1 0·7—1·0	67Ka
Y^{3+}	sp	20	0·6 ($HClO_4$)	K_1 −0·07 [$K_1(Fe^{3+})$ 2·15]	64Ka
Group 4*f* $^{3+}$	sp	20	0·6 ($HClO_4$)	K_1 < −0·1(La^{3+}), −0·2(Nd^{3+}), 0·09(Sm^{3+}), 0·21(Gd^{3+}), 0·12(Dy^{3+}), 0·16(Er^{3+}) [$K_1(Fe^{3+})$ 2·15]	64Ka
	dis		2 Na(Cl)	K_D[$M^{3+} + 3L^- \rightleftharpoons ML_3$ (in TBP)] 2·4(M = La), 2·6(Ce), 2·4(Pr), 2·3(Nd) K_D[$M^{3+} + H^+ + 4L^- \rightleftharpoons HML_4$ (in TBP)] 3·4(La), 3·85(Ce), 3·3(Pr), 3·7(Nd), "ML_3" = ML_3T_4, "HML_4" = HML_4T_5, T = TBP = $(BuO)_3PO$	65G
	X, con		var	ev $M(NCS)_6^{3-}$ in salts and in $PhNO_2$(l), $MeNO_2$(l) (M = Pr, Nd, Eu, Sm, Ho, Er, Tm, Yb, Y)	68Mf
La^{3+}	dis	25	5 Na(ClO_4)	K_1 0·24, β_2 < −0·62, β_3 −0·73	64S
	dis	25	5 Na(ClO_4)	K_1 0·24, β_2 < −0·6, β_3 −0·73	65Sa
	dis	27	1 ($NaClO_4$)	K_1 0·06, β_2 0·05	68Ra
Nd^{3+}	dis	25—55	1 ($NaClO_4$)	K_1 0·81(25°), 0·61(40°), 0·47(55°), $\Delta H_1 = -5{\cdot}47$, $\Delta S_1 = -14{\cdot}7$; K_2 0·11(25°)	65C
Sm^{3+}, Eu^{3+}	dis		var	K_D[$M^{3+} + 3L^- + 5T(org) = ML_3T_5(org)$] ≈ 2 (unit?) (M = Sm or Eu), T = TBP, org = C_6H_6	67G
Eu^{3+}	sp	25	0 corr	$K_1 \approx 0{\cdot}7$	64B
	dis	25	5 Na(ClO_4)	K_1 0·32, $\beta_2 \approx -0{\cdot}1$, β_3 −0·36	64S = 65Sa
	dis	25	1 ($NaClO_4$)	K_1 0·70, K_2 0·13	65C
	dis	25	5 Na(ClO_4)	K_2K_3 −0·4, K_3 0·5, K_4 −0·47 K_D[$EuL_3 \rightleftharpoons EuL_3(org)$] 2·05 org = 5% TBP in hexane	65Sc
Lu^{3+}	dis	25	5 Na(ClO_4)	K_1 0·45, $\beta_2 \approx -1{\cdot}3$, β_3 −0·14	64S = 65Sa
Ac^{3+}	dis	27	1 ($NaClO_4$)	K_1 0·05, β_2 −0·09	68Ra
Ti^{4+}	pre		var	ev TiL_6^{2-} in K_2TiL_6(s), H_2O, MeCN, NH_3(l)	64Sb

continued

Metal	*Method*	*Temp*	*Medium*	*Log of equilibrium constant, remarks*	*Ref*
Zr^{4+}	dis		2 or 3 $HClO_4$	ev ZrL_2^{2+}, ZrL_4, ZrL_6^{2-}	64Va
	sp		I?($NaClO_4$)	(0·8M-H^+): K_1 3·8, β_2 7·3, β_3 10·8, β_4 14·0, β_5 17·1, β_6 21·0, β_7 22·9, β_8 25·6; (0·1M-H^+): K_1 2·0, β_2 3·4, β_3 4·7, β_4 5·8, β_5 6·9, β_6 7·9, β_7 8·9, β_8 9·9	66Gd
	dis	20	org	ev ZrL_4 in 2-methylpentanone(l)	67D
Hf^{4+}	dis		PhCOMe	ev HfL_4, HfL_6^{2-} in PhCOMe(l)	65O
	sp		I?($NaClO_4$)	0·8M-H^+: K_1 2·6, β_2 4·9, β_3 7·1, β_4 9·2, β_5 11·1, β_6 12·9, β_7 14·7, β_8 16·5; 0·1M-H^+: K_1 2·0, β_2 4·0, β_3 5·7, β_4 7·2, β_5 8·7, β_6 10·0, β_7 11·1, β_8 12·2	66Gd
Th^{4+}	con, pre	25	MeOH	ev ThL_8^{4-}, $ThL_7H_2O^{3-}$, $ThL_5(H_2O)_3^-$, $ThL_4(H_2O)_4$, $ThL_5OH^-(H_2O)_{2-3}$ in MeOH(l)	64M
	sp		Me_2CO	K_1 3·5, β_3 9·57, β_4 12·55 in Me_2CO(l)	66Gc
	sp		MeOH	K_1 3·37, β_2 6·66, β_3 9·82, β_4 12·89 in MeOH(l)	
	sp, tp		Me_2NCHO	K_1 3·20, β_2 6·28, β_3 9·26, β_4 12·12, β_5 14·92, β_6 17·7 in Me_2NCHO(l)	
V^{2+}	kin	24	1 ($NaClO_4$)	K_1 1·04 (T-jump method)	68Ka
	sp	11—45	0·84 ($LiClO_4$), 0·5 H^+	K_1 1·61(11°), 1·43(25°), 1·18(45°), $\Delta H_1 = -5{\cdot}2$, $\Delta S_1 = -11$	68Me
	sp	25	1 ($NaClO_4$)	K_1 1·43	68O
V^{3+}	sp	23—25	MeOH	β_6 15—16 in MeOH(l)	63Gf
	sp	5—37	1 ($NaClO_4$)	K_1 2·16(5°), 2·14(12°), 2·07(25°), 1·94(37°), $\Delta H_1 = -3{\cdot}5$, $\Delta S_1 = -2{\cdot}2$	67Bc
	kin	23	1 ($NaClO_4$)	K_1 2·18 (T-jump method)	68Ka
VO^{2+}	X		solid	ev $VOL_4H_2O^{2-}$ in $(NH_4)_2VOL_4(H_2O)_5$(s)	63Ha = 64H
	Ag	10—40	0 corr	K_1 2·48(10°), 2·32(25°), 2·15(40°), $\Delta H_1 = -4{\cdot}1$, $\Delta S_1 = -2{\cdot}8$; $K_2 \approx$ 1·4(10°), 1·36(25°), 1·32(40°)	68S
Nb^{5+}	var		org	K_1 3·58, β_2 6·74, β_3 9·23 in MeOH(l); K_1 4·37, β_2 8·58, β_4 16·92 in BuOH(l); K_1 3·08, β_2 6·11, β_3 8·92, β_4 11·55, β_5 14·15, β_6 16·72, β_7 19·28 in Me_2NCHO(l) (Nb added as $NbCl_5$) [sp, pre, (con)]	64Ga
	dis		org	ev NbL_6^- in TBP(l)	66Gb
	var		var	ev $Nb(NCS)_6^-$ in $C_2H_4Cl_2$(l), MeCN(l), and salts (pre, ir, con)	67Bf
Ta^{5+}	sp, pre		MeOH	K_1 3·12, β_2 5·48, β_3 7·77 in MeOH(l)	64Gc
	con, sp		BuOH	K_1 3·68, β_2 7·05, β_3 11·42 in BuOH(l)	
	sp, pre		Me_2NCHO	K_1 3·15, β_2 5·92, β_3 8·55, β_4 11·06, β_5 13·52, β_6 15·96 in Me_2NCHO(l)	
	sp J		MeOH	ev cpx with 8,2 and 0·5 L per Ta in MeOH(l)	66D
	dis		org	ev TaL_6^- in TBP(l)	66Gb
	var		var	ev $Ta(NCS)_6^-$ in $C_2H_4Cl_2$(l), MeCN(l), and salts (pre, ir, con)	67Bf
Cr^{3+}	X		solid	ev CrL_6^{3-} in $K_3Cr(SCN)_6(H_2O)_4$(s)	53Z
	con, spJ		org	ev CrL_3, CrL_4^-, CrL_6^{2-} in Me_2CO(l) and MeOH(l)	63Gd
	sp		org	β_3 5·6—6·0, $\beta_6 \approx$ 8·3 in Me_2CO(l); β_3 4·3—5·5 in MeOH(l)	

continued overleaf

17 SCN⁻ (contd)

Metal	Method	Temp	Medium	Log of equilibrium constant, remarks	Ref
Cr^{3+}	kin	25	1 $HClO_4$	$K(CrSCN^{2+} \rightleftharpoons CrNCS^{2+})$ 5·5	65Hb
(contd)	var		solid, org	ev CrL_6^{3-} (pre, con, ir)	67S
$Cr(NH_3)_5^{3+}$	sp, kin	60	1·71 (NaBr)	K_1 0·6	67Da
		30—60	0·106 ($NaClO_4$)	K_1 2(30°), 1·8(45°), 1·7(60°), $\Delta H_1 = -5, \Delta S_1 = -6$	
		23	0·16 ($NaClO_4$)	K_1 1·0	

1964 *Erratum*: p. 118, *line* 7, *in ref* 59Ba, *for temperature* "rt" *read* "20".

Metal	Method	Temp	Medium	Log of equilibrium constant, remarks	Ref
Mo^{3+}	var		solid, org	ev MoL_6^{3-} (pre, con, ir)	67S
	X		solid	ev $Mo(NCS)_6^{3-}$ in $K_3MoL_6H_2OMeCO_2H(s)$	68K
Mo^{5+}	var		H^+ var	ev $MoO_2L_2^-$, $MoOL_4^-$ (sp, dis, pre), ev $MoOL_5^{2-}$ (only in solid, pre)	64T
	sp, pol		var	ev $Mo_4O_4(OH)_8L^{3+}$, $Mo_4O_4(OH)_6L_3^{3+}$, $Mo_2O_2OHL_3^{2+}$	65D
	sp		org	K_1 3·85, ev MoL_2^{3+} in MeOH(l) K_1 5·0, K_2 4·4, K_3 4·0, K_4 3·4 in Me_2CO(l) ("Mo^{5+}" = $MoCl_5$)	65U
	esr	23	2 H(ClO_4)	$K(MoOL_4^- + nT \rightleftharpoons MoOL_{4-n}T_n^{n-1} + nL^-)$ −1·64(n = 1), −3·24(n = 2), −6·19(n = 3), T = $(NH_2)_2CS$	68Mc
Mo^{6+}	sp, hyp		1 ? H_2SO_4	$K_1 \approx$ 2·5(sp), K_2(hyp, misprint) K_3 0·96(hyp), K_4 0·08(hyp)	63Ma
W^{5+}	var		var	ev WOL_5^{2-}, $WO(OH)_2L_2^-$, $WO_2L_3^{2-}$, $W_2O_3L_8^{4-}$ (mag, pol, pre)	65Ba
W^{6+}	sp, con		org	ev cpx with 1—6 L/W K_1 3·5, β_2 7·2 in Me_2CO(l), K_1 3·1, β_2 6·3, β_4 13·25 in MeCOEt(l) β_6 20·6 in cyclohexanone(l), W added as WCl_5	67U
UO_2^{2+}	sp	20	1 ($NaClO_4$)	K_1 0·73, β_2 0·96, β_3 0·8	49A/68Sa
	sp	22—23	I($NaNO_3$)	K_1 0·71, β_2 0·72 (I = 4·0) K_1 0·72, β_2 0·70 (I = 2·5) K_1 1·5, β_2 1·9 (0 corr)	64Vb
	X		solid	ev $UO_2(NCS)_5^{3-}$ in $Cs_3UO_2L_5(s)$	66Aa
	ir		4 ($NaNO_3$)	calculated $\bar{n}$ agrees with ref 49A	68L
Pu^{3+}	dis	25	1 ($NaClO_4$)	K_1 0·46, K_2 0·29	65C
	aix, tp	25	var	ev ani cpx	66C
	cix	25	3 Na(ClO_4)	K_1 0·04, K_2 −0·14, K_3 −0·6	
Am^{3+}	dis	25	5 Na(ClO_4)	K_1 0·85, β_3 0·55, β_4 0·00	64S = 65S
	dis	25—55	1 ($NaClO_4$)	K_1 0·50(25°), 0·40(40°), 0·19(55°), $\Delta H_1 = -4·36, \Delta S_1 = -12·3$; K_2 0·34(25°)	65C
	dis	25	5 Na(ClO_4)	K_2K_3 −0·2, K_4 −0·13, $K_D[AmL_3 \rightleftharpoons AmL_3(org)]$ 2·5, org = 5% TBP in hexane	65Sc
Cm^{3+}	dis	25	1 ($NaClO_4$)	K_1 0·43, K_2 0·42	65C
Cf^{3+}	dis	25	1 ($NaClO_4$)	K_1 0·49	65C
Mn^{2+}	dis	20	1·5 ($NaClO_4$)	$K_D[Mn^{2+} + 2L^- \rightleftharpoons MnL_2(org)]$ −0·07, org = $MeCOBu^i$, K_1 0·73, β_2 1·30	64Tb
	var		var	ev MnL_4^{2-} in org, solid, MnL_6^{4-} in solid (con, ir, mag, pre, sp)	65Fb
	cal	25	0 corr	$\Delta H_1 = -0·92, \Delta S_1 = 2·5$ (K_1 1·23)	67N
	E	35	→ 0	K_1 1·57	68Pa

continu

Metal	*Method*	*Temp*	*Medium*	*Log of equilibrium constant, remarks*	*Ref*
Tc^{4+}	dis		var	ev TcL_6^{2-} [joined with $(C_8H_{17})_3NH^+$] in cyclohexane	64O
$Re^{2,3+}$	pol	25	MeCN	ev $Re_2L_8^{3-}$, $Re_2L_8^{4-}$ in MeCN(l)	67Ca
Re^{3+}	var		var	ev $Re_2L_8^{2-}$ (pre, con, ir, mag, sp)	67C
Re^{4+}	sp, aix		HCl var	ev cpx with 1 L and 2 L (ani, probably $ReO_2L_2^{2-}$), K_2 3·64	64Ta
				K_1 3·64	64Ta/66T
	sp		4 ? H_2SO_4	ev cpx with 1—4 L per Re, K_1 3·7	66T
	var		var	ev ReL_6^{2-} (pre, ir, sp, mag)	67B
	var		var	ev $HRe_2OL_{10}^{3-}$, $H_2Re_2OL_{10}^{2-}$ (ir, pre, mag, sp)	67W
	var		var	ev $Re_3(OH)_2L_{14}^{4-}$ (ir, mag, pre, sp)	67Wa
$Re^{4,5+}$	pre, sp		var	ev $Re_2O_2L_8^{3-}$	65Oa
Re^{5+}	pre, sp		var	ev $ReOL_5^{2-}$	65Oa
	sp, aix		var	ev cpx with 1, 2, 3 L per Re^V	65Mc
	pre, ir		var	ev $Re(SCN)_6^-$	67Bd
Fe^{2+}	var		var	ev $Fe(NCS)_4^{2-}$ in org, solid (con, ir, mag, pre, sp)	65Fb
	kin, sp	25	3 (Mg, ClO_4) 1·8 H^+	K_1 0·85(kin), 0·81(sp)	67Cb
Fe^{2+}cpx	sp, gl		KL var	Fe(dimethylglyoxime)$_2^{2+}$: K_1 0·35, K_2 0·95	67Ba
Fe^{3+}	oth	rt	1·4 ($NaClO_4$)	K_1 2·35, K_2 1·55, K_3 1·31, K_4 0·66, K_5 0·21 (paper electrophoresis)	64J
	sp	20	$I(HClO_4)$	K_1 2·15(I = 0·6), 2·20(I = 0·3), 2·27(I = 0·15)	64Ka
	sp	22—23	4 ($NaNO_3$)	K_1 2·1, K_2 1·3, K_3 0·5, K_4 0, K_5 −0·1, K_6 −0·1, β_6 3·7	64V
			0 corr	K_1 3·1, K_2 2·2, K_3 0·9	
	var		var	ev $Fe(NCS)_6^{3-}$ in org, solid (con, ir, mag, pre, sp)	65Fb
	dis	25	3 ($LiClO_4$), 0·2 H^+	K_1 2·18, K_2 1·42, K_3 1·40, K_4 1·30, K_5 −0·07, K_6 −0·09, β_6 6·14	65M
				$K_D[FeL_4^-(aq) \rightleftharpoons FeL_4^-(Et_2O)]$ 1·5	
		25	3 ($LiNO_3$), 0·2 H^+	$\beta(M^+FeL_6^{3-})$ −1.1(Na^+), −0·25(K^+), −0·1(Rb^+), 0·1(Cs^+), −0·1(NH_4^+)	
	sp	25	3 ($LiClO_4$), 0·2 H^+	K_1 2·19, K_2 1·48, K_5 0·00, K_6 −0·035	65Ma
	sp	25	$I(LiNO_3)$, 0·1 H^+	K_1 2·26(I = 6·1), 2·08(I = 3·1), 1·91(I = 1·1), 2·02(I = 0·6), 2·17(I = 0·1); β_2 3·93(I = 6·1), 2·99(I = 1·1), 3·60(I = 0·1) ΔH_1 = −2·2(I = 6·1), −1·5(I = 1·1), −2·2(I = 0·1) ΔS_1 = 3(I = 6·1), 4(I = 1·1), 2(I = 0·1) $\Delta H_{\beta 2}$ = −5·8(I = 6·1), −5·2(I = 1·1), −5·0(I = 0·1) $\Delta S_{\beta 2}$ = −2(I = 6·1), −3(I = 1·1), −0·3(I = 0·1), also K_1 and β_2 for 35° and 45°	
	sp		$I(KNO_3)$	K_1 1·96(I = 2), 1·97(I = 1·5), 2·00(I = 1), 2·08(I = 0·5), 2·11(I = 0·4), 3·09(0 corr)	66V
	kin, sp	25	3 ($MgClO_4$), 1·8 H^+	K_1 2·26(kin), 2·24(sp)	67Cb
	ir, pre		solid	ev FeL_6^{3-}	67S
	sp	25	Me_2SO	K_1 3·07 in Me_2SO(l), I = 0·024($NaNO_3$)	68Lc
	sp	20	1·2 ($NaClO_4$), 0·2 H^+	K_1 2·12	68Ma
	red	25	1 ($NaClO_4$)	K_1 2·10, β_2 3·14	68P
Fe^{3+} cpx	var		var	$\beta(FeY^-H^+L^-)$ −1·10, H_4Y = EDTA, (dis, sp, gl)	67La
$Fe(C_5H_5)_2^+$	red	25	0 corr	K_1 1·0	67Na

continued overleaf

17 SCN^- (contd)

Metal	*Method*	*Temp*	*Medium*	*Log of equilibrium constant, remarks*	*Ref*
Co^{2+}	*X*		solid	ev $Co(NCS)_4^{2-}$ in $Na_2CoL_4(H_2O)_8$(s)	53P
	dis	25	3 $Na(ClO_4)$	K_1 −0·45, K_2 −0·62, $K_D[CoL_2 \rightleftharpoons CoL_2(\text{in S})]$ 3·51, $K_D[CoL_2+2S(\text{in S}') \rightleftharpoons CoL_2S_2(\text{in S}')]$ 1·12, S = iso-BuCOMe = hexone, S′ = iso-BuCHOHMe = hexol	63D
	sp		0 corr	$K[CoL_4(H_2O)_2^{2-}$(octahedral) $\rightleftharpoons CoL_4^{2-}$(tetrahedral) $+2H_2O]$ −0·33	63Va
	sp	23	3 $K(NO_3)$	K_1 0·63, β_3 −0·38, no ev CoL_2	64K
	var	var	→ 0	K_1 1·75 (from data of ref 51Ka, 51S, 58Sa, 60Ta, 61D), 1·37 ? (from data of ref 58Ya)	/64K
	sp	20	$I(HClO_4)$	K_1 1·10(I = 0·6), 1·18(I = 0·3), 1·28(I = 0·15) [$K_1(Fe^{3+})$ from ref 64Ka]	64Ka
	X		solid	ev $Co(NCS)_4^{2-}$ in $Hg[SC(NH_2)_2]_4CoL_4$(s)	65Ka
	sol	?	$I(KNO_3)$	$K_s[Co^{2+}L^-_2(C_5H_5N)_4]$ −12·67(I = 0·5), −12·87(I = 0·2), −13·08(I = 0·05), −13·11(I → 0)	65P
	dis		20% NH_4L	$K_D[Co^{2+}+6L^- \rightleftharpoons CoL_6^{4-}$(org in ion pair)] 6·8—6·9, org = $C_2H_4Cl_2$(l), polyethyleneglycol added. Estimates of $K[CoL_{n-1}$(org ?) $+ L^- = CoL_n$(org ?)]	65R
	cal	25	0 corr	ΔH_1 = −1·63, ΔS_1 = 2·2 (K_1 1·72)	67N
	E	25	→ 0	K_1 1·77	68Pa
Co^{2+}cpx	sp, gl		KL var	Co(dimethylglyoxime)$_2^{2+}$; K_1 4·9, K_2 2·9	67Ba
$Co(NH_3)_5^{3+}$	sp	25	0·5	K_1^{out} 0·65	63H
	sp	45	1 $(NaClO_4)$	K_1^{out} −0·37	67Lb
Co^{3+} cpx	kin	70	sulpholan	*trans*-Co $en_2NO_2Cl^+$; K_1^{out} 0·91 in sulpholan(l)	68A
Ni^{2+}	*X*		solid	ev NiL_6^{4-} in $K_2Ni(NCS)_6(H_2O)_4$(s)	53Z
	dis	20	3 $(NaClO_4)$	K_1 1·19, β_2 1·68, β_3 1·30, β_4 1·54	64Tb
			1·5 $(NaClO_4)$	$K_D[Ni^{2+}+2L^- \rightleftharpoons NiL_2(\text{org})]$ −0·14, org = $MeCOBu^i$ K_1 1·14, β_2 1·75, β_3 1·70, β_4 2·04	
	var		solid	ev NiL_4^{2-} (ir, mag, sp)	65Fa
	cal	25	0 corr	ΔH_1 = −2·26, ΔS_1 = 0·5 (K_1 1·76)	67N
	var		org	ev NiL^+ in Me_2NCOMe(l), NiL_2, NiL_3^-, NiL_4^{2-} in Me_2CO(l), MeCN(l) *etc.*	68Gd
	pol	15—35	≈ 0·65 $(HClO_4)$	K_1 1·34(15°), 1·24(25°), 1·17(35°), ΔH_1 = −3·45, ΔS_1 = −5·8	68Mb
	E	35	→ 0	K_1 1·85	68Pa
Ru^{3+}	dis, sp	25	var	ev $RuOL_n^{1-n}$, n = 0, 1, 2, 3	66Oa
	var		solid, org	ev RuL_6^{3-} (pre, sp, fp, ir)	66S
	var		solid, org	ev RuL_6^{3-} (pre, con, ir)	67S
Rh^{3+}	var		solid, org	ev RhL_6^{3-} (pre, con, ir)	67S
Pd^{2+}	Pd	18—20	var	β_4 27·6{$K[Pd^{2+}+2e^- \rightleftharpoons Pd(s)]$ 33·4}	63Gg
	var	20—25	dil	β_2 8·4(sp), β_4 19·46(Ag, 25°), ev PdL^+(sp), K_{s4} 1·63(sol, 20°), K_{s0} −17·8	64Gb
	sol	20	var	$K_s[PdI_2(s)+L^- \rightleftharpoons PdI_2L^-]$ −0·47	
	Pd	25	KI var	β_4 26{$K[Pd^{2+}+2e^- \rightleftharpoons Pd(s)]$ 33·4}, also β_4 values for 10—60°, assuming same K and same RTF^{-1} as for 25°!	65F
	dis		C_6H_6	ev $H_4OPdL_4A_4$ (A = $C_5H_{10}OH$, BuOH), $(H_3O)_2PdL_4A_6$ (A = cyclohexanone) in C_6H_6(l)	65Ga

continued

Metal	*Method*	*Temp*	*Medium*	*Log of equilibrium constant, remarks*	*Ref*
Pd^{2+} (*contd*)	sp	25	1 Na^+, 0·1 H^+	β_4 28·22, $\beta(Pd^{2+}Br^-{}_nL^-{}_{4-n})$ 25·85(n = 1), 22·25(n = 2), 18·15(n = 3), $\beta(Pd^{2+}Cl^-L^-{}_3)$ 25·19[$\beta(Pd^{2+}Br^-{}_4)$ 13·05]	66Ba
	sp	25	1 NaCl, 0·1 H^+	$K(PdCl_{5-n}L_{n-1}{}^{2-}+L^- \rightleftharpoons PdCl_{4-n}L_n{}^{2-}+Cl^-)$ 6·03(n = 1), 4·90(n = 2), 3·59(n = 3), 3·03(n = 4), β_4 28·67[$\beta_4(Cl^-)$ 11·12]	67Be
	con	24	org	ev $PdL_4{}^{2-}$ in Me_2NCHO(l)	67S
Os^{3+}	var		var	ev $OsL_6{}^{3-}$ in solid and org (pre, con, ir)	67S
Ir^{3+}	var		var	ev $IrL_6{}^{3-}$ in solid and org (pre, con, ir)	67S
Pt^{4+}	*X*		solid	ev $PtL_6{}^{2-}$ in $PbPt(SCN)_6$(s)	53Z
	sp	35	1·1 (Na)ClO_4	$Pt(NH_3)_4{}^{4+} = Q^{4+}$; $K(QCl_2{}^{2+}+L^- \rightleftharpoons QClL^{2+}+Cl^-)$ 2·55	67M
		35	0·2 $H(ClO_4)$	$K(QClL^{2+}+L^- \rightleftharpoons QL_2{}^{2+}+Cl^-)$ 1·08	
	var		var	ev $PtL_6{}^{2-}$ in solid and org (pre, con, ir)	67S
Pt $en_3{}^{4+}$	sp	10—40	0 corr	no ev cpx	63Gc
Cu^+	sol	20	2 $Na(NO_3)$	K_{s0} −11·32, $\beta(Cu^+L^-Br^-)$ 7·76, $\beta(Cu^+L^-NO_2{}^-)$ 8·37	66Sa
		20	4 $Na(NO_3)$	K_{s0} −11·15, $\beta(Cu^+L^-Cl^-)$ 7·31	
	sol	20	4·4 (NH_4NO_3)	K_{s4} −2·71	68Gc
Cu^{2+}	var		solid	ev $CuL_4{}^{2-}$ (ir, mag, sp)	65Fa
	X		solid	ev $CuL_2(NH_3)_2$ in $Cu(SCN)_2(NH_3)_2$(s)	65H
	cal	25	0 corr	$\Delta H_1 = -3{\cdot}00$, $\Delta S_1 = 0{\cdot}6$ (K_1 2·33)	67N
	sp	30	0 corr	K_1 2·39	68D
Ag^+	Ag	25	org	K_{s0} −13·9 in MeOH, −9·9 in $HCONH_2$, −11·5 in Me_2NCHO, −10·5 in Me_2NCOMe, −6·5 in Me_2SO, −9·6 in MeCN, −8·5 in $(Me_2N)_3PO$; β_2 11·9 in Me_2NCHO, 11·4 in Me_2NCOMe, 7·4 in Me_2SO, 9·7 in $(Me_2N)_3PO$	67A
	Ag	30	sulpholan	K_{s0} −16·27, β_2 16·04 in sulpholan(l), I = 0·1 (Et_4NClO_4)	68Da
	sol	20	2 (KNO_3)	$K_{s63}[3AgL(s)+3L^- \rightleftharpoons Ag_3L_6{}^{3-}]$ −1·35	68Gb
	ir		C_5H_5N	K_1 2·55, β_2 4·44, β_3 4·49, β_{12} 0·15, β_{22} 0·42 in C_5H_5N(l)	68La
	sol	24	Me_2SO	$\beta_{12}(Ag^+{}_2L^-)$ 6·40 in Me_2SO(l), I = 0·5(Et_4NClO_4)	68Lb
	Ag	25	C_5H_5N	K_1 2·24, β_2 3·83, β_3 3·69, β_{12} 2·34 in C_5H_5N(l), 1·5M-py$H^+ClO_4{}^-$; K_1 2·54, β_2 4·6 in C_5H_5N(l), no py$HClO_4$ added	68M

1964 *Erratum*: p. 122, *line* 3 *from end in ref* 56K *for* "ΔH_{s0} = 9·84" *read* "ΔH_{s0} = 19·6 (misprint 9·84)"

Metal	*Method*	*Temp*	*Medium*	*Log of equilibrium constant, remarks*	*Ref*
Au^+	Au	25	3 $(NaClO_4)$	K_1 15·27, β_2 16·98{$K[Au^++e^- \rightleftharpoons Au(s)]$ 28·4(1·68V)}	66K
Zn^{2+}	MHg	25	4 $Li(ClO_4)$	K_1 1·11, β_2 1·81, β_3 2·81, β_4 2·80	66M
			4 $(LiClO_4)$	$\beta(M^+ZnL_3{}^-)$ −1·89(Na^+), −1·38(K^+), −1·24($NH_4{}^+$, Rb^+), −0·74(Cs^+); $\beta(M^+ZnL_4{}^{2-})$ −1·28(Na^+), −0·77(K^+), −0·64($NH_4{}^+$, Rb^+), −0·14(Cs^+)	
	ir, Ram		$MeNO_2$	ev tetrahedral $ZnL_4{}^{2-}$ in $MeNO_2$(l)	67F
	X		solid	ev $Zn(NCS)_4{}^{2-}$ in $Cd[SC(NH_2)_2]_2ZnL_4$(s)	67K
	cal	25	0 corr	$\Delta H_1 = 0{\cdot}2$, $\Delta S_1 = 9{\cdot}2$ (K_1 1·85)	67N
	aix	?	0 corr	K_1 1·57, β_2 1·56, β_3 1·51, β_4 3·02	68Na
	E	35	→0	K_1 1·42	68Pa

continued overleaf

17 SCN⁻ (contd)

Metal	Method	Temp	Medium	Log of equilibrium constant, remarks	Ref
Cd^{2+}	MHg	20	MeOH	K_1 3·0, β_2 5·5, β_3 5·9, β_4 6·2 in MeOH(l), I = 1·6M($NaClO_4$)	64G
			Me_2NCHO	K_1 3·0, β_2 4·5, β_3 6·3, β_4 6·7 in Me_2NCHO(l); I = 1·2M($NaClO_4$)	
			MeCN	β_4 13·33, β_5 14·0, in MeCN(l), I = 1·6M($NaClO_4$); K_1 2·3, β_2 3·5, β_3 4·3, β_4 5·2 in 70 vol% MeCN; I = 1·6M($NaClO_4$)	
	sp	20	I($HClO_4$)	K_1 1·32(I = 0·6), 1·44(I = 0·3), 1·53(I = 0·15) [K_1(Fe^{3+}) from ref 64Ka]	64Ka
	dis	20	3 ($NaClO_4$)	K_D[$Cd^{2+} + 2L^- \rightleftharpoons CdL_2$(org)] 2·79, org = MeCOBu^i K_1 1·60, β_2 2·60, β_3 2·90	64Tb
			1·5 ($NaClO_4$)	K_D 1·60, K_1 1·32, β_2 1·98, β_3 2·55	
	dis	30	1 ($NaClO_4$)	K_1 0·7, β_2 1·5	65Ha
	aix	rt	0 corr	K_1 1·74, K_2 0·66, K_3 −1·0, K_3K_4 0·51, β_4 2·91	66A
	cal	25	3 ($NaClO_4$)	ΔH_1 = −1·94, ΔH_2 = −1·73, ΔH_3 = −1·57, ΔH_4 = −1·0; ΔS_1 = 0·0, ΔS_2 = −2·0, ΔS_3 = −4·2, ΔS_4 = −3·5 (K_1 1·42, β_2 2·24, β_3 2·48, β_4 2·48)	66G+
	pol	30	2 (KNO_3)	K_1 1·00, β_2 1·74, β_3 0·85, β_4 1·64	67H
	cal	25	0 corr	ΔH_1 = −0·70, ΔS_1 = 9·2 (K_1 3·51)	67N
	MHg	25	3 ($NaClO_4$)	K_1 1·41, β_2 2·24, β_3 2·48, β_4 2·48	68G
			2 ($NaClO_4$)	K_1 1·34, β_2 2·05, β_3 2·25, β_4 2·03	
			1 ($NaClO_4$)	K_1 1·32, β_2 1·99, β_3 2·03, β_4 1·88	
			0·5 ($NaClO_4$)	K_1 1·35, β_2 2·04, β_3 2·08, β_4 1·98	
			0·25 ($NaClO_4$)	K_1 1·43, β_2 2·10, β_3 2·30	
	cal		2 ($NaClO_4$)	ΔH_1 = −2·22, ΔH_2 = −1·85, ΔH_3 = −2·2 ΔS_1 = −1·3, ΔS_2 = −3·0, ΔS_3 = −6·5	68Ga
			1 ($NaClO_4$)	ΔH_1 = −2·29, ΔH_2 = −1·93, ΔH_3 = −2·2, ΔS_1 = −1·6, ΔS_2 = −3·4, ΔS_3 = −7·2	
			0·5 ($NaClO_4$)	ΔH_1 = −2·24, ΔH_2 = −2·45, ΔS_1 = −1·3, ΔS_2 = −5·0	
			0·25 ($NaClO_4$)	ΔH_1 = −2·24, ΔH_2 = −2·7, ΔS_1 = −1·0, ΔS_2 = −6·0	
	E	35	→ 0	K_1 2·08	68Pa
Hg^{2+}	sol, sp	25	1 ($NaClO_4$)	K[$HgI_2(s) + 2L^- \rightleftharpoons HgI_2L_2^{2-}$] −0·27 β($Hg^{2+}I^-{}_2L^-{}_2$) 27·92, [K_s($Hg^{2+}I^-{}_2$) −28·19]	64C
	pol	30	var	β_2 17·60, β_3 20·40, β_4 21·23	64Sa
	X		solid	ev HgL_4^{2-} in $CuHgL_4$(s)	64Kb
	sol	20	4·6 (K, Ca, NO_3)	K_{s2} −1·83, K_s[$HgL_2(s) + Hg^{2+} \rightleftharpoons Hg_2L_2^{2+}$] −0·70 ev $Hg_2L_2^{2+}$ rather than HgL^+ ([Hg^{II}] from 0·06 to 0·73 M)	66Ga
	sol	20	4·6 [$Ca(NO_3)_2$]	K_s[$HgL_2(s) + Hg^{2+} \rightleftharpoons Hg_2L_2^{2+}$] −0·81 ([$Hg^{II}$] = 0·4—1·0)	68Gb
	Hg, pol	25?	w% C_5H_5N	β_2 7·6, β_3 9·2, β_4 10(w = 20), β_3 9·7(w = 80) in w% C_5H_5N, also values for w = 10, 40, 60	68Pb
	sol	20	2 Na(NO_3)	K_s[$HgL_2(s) + Br^- \rightleftharpoons HgL_2Br^-$] 0·56 K_s[$HgL_2(s) + 2NO_2^- = HgL_2(NO_2)_2^{2-}$] −1·03	68Y
RHg^+	sp	25?	1 (KNO_3)?	K_1 5·82(Me), 5·85(Et), 5·96(Pr), 6·0(Pr^i, Bu, Bu^i), [K_1(Fe^{3+}) 2·96]	65T
	sol	25	1 (KNO_3)	K_{s1}[$RHgL(s) \rightleftharpoons RHgL$] 1·37(Me, I = 3), 0·56(Et), 0·38(Pr), 0·17(Pr^i), −0·54(Bu), 0·10(Bu^i), K_2 0·02(Me), 0·15(Et), 0·19(Pr), 0·20(Pr^i), 0·22(Bu), 0·30(Bu^i)	

continued

Metal	*Method*	*Temp*	*Medium*	*Log of equilibrium constant, remarks*	*Ref*
$MeHg^+$	gl, cal	20	0·1 (KNO_3)	K_1 6·05, $K_{12}(ML+M^+ \rightleftharpoons M_2L^+)$ 1·65 $\Delta H_1 = -11{\cdot}2, \Delta S_1 = -10{\cdot}6$	63Sa = 65Sb
	gl	20	0·1 ($NaClO_4$)	$\beta[Co(NH_3)_5NCS^{2+}MeHg^+]$ 3·20	
$EtHg^+$	aix	25	KL var	$K_2 \sim 0{\cdot}4, K_3 \sim 0{\cdot}4$	65B
	sol	25	1 Na(ClO_4)	K_{s1} −2·48, K_2 −0·10, K_3 0·20	65Bb
	sol	25	50% MeOH	K_{s1} −1·76, K_2 0·24, K_3 −0·25 in 50 vol% MeOH, 1M-Na(ClO_4),	
$BuHg^+$	sol, oth	25	50% MeOH	K_{s1} −1·89, K_2 0·36, K_3 −0·25 in 50 vol% MeOH, 1M-Na(ClO_4), fit also optical rotation data	65Bb
HgY^{2-}	spJ		0·10 ($KClO_4$)	K_1 5·9, H_4Y = EDTA	68N
Al^{3+}	pre		var	ev AlL_6^{3-} in K_3AlL_6(s), MeCN(l), NH_3(l), ev AlL_3Cl^- in MeCN(l)	64Sc
	sp	30	var	no ev cpx	68D
Ga^{3+}	sp	20	0·6 ($HClO_4$)	K_1 1·18 [$K_1(Fe^{3+})$ 2·15]	64Ka
	sp, Ag	30—35	0 corr	K_1 2·15(30°, sp), 2·26(35°, Ag)	68D
	aix		6 ($NaClO_4$)	ev ani cpx	68Md
In^{3+}	sp	20	0·6 ($HClO_4$)	K_1 2·34 [$K_1(Fe^{3+})$ 2·15]	64Ka
	pol	25	2 ($NaClO_4$)	K_1 1·7, β_2 2·3, β_3 2·08, β_4 3·22	65N
	sp	30—35	0 corr	K_1 3·15 (30°, sp), 3·26 (35°, Ag)	68D
Tl^+	sol	25	dil	K_{s0} −3·77	58M
	sol	25	4 ($LiClO_4$)	K_1 0·20, β_2 −0·05, β_3 −0·58, β_4 −0·80 $\beta(K^+TlL_4^{3-})$ −0·1	65K
	pol	25	2·5 (KNO_3)	K_1 0·17—0·21, β_2 −0·07—0·01, β_3 −0·44	66O
Sn^{2+}	MHg	?	1·6 ($NaClO_4$)	K_1 1·02, β_2 1·54, β_3 1·46	63Ge
		?	MeOH	K_1 3·7, β_2 5·6, β_3 6·4, β_4 6·55 in MeOH(l), I = 1·6($NaClO_4$)	
		?	Me_2NCHO	K_1 2·04, β_2 3·70, β_3 4·58 in Me_2NCHO(l), I = 1·2($NaClO_4$)	
		?	MeCN	β_4 16·82 in MeCN(l), I = 1·6($NaClO_4$)	
$Me_nSn^{(4-n)+}$	pre, tp		var	ev ani cpx $Me_3SnL_2^-$, $Me_2SnL_4^{2-}$, $MeSnL_5^{2-}$	65Ca
$MeSn^{3+}$	red	25	1 ($NaClO_4$)	K_1 1·48, β_2 2·20, β_3 3·32	68P
Me_2Sn^{2+}	red	25	1 ($NaClO_4$)	K_1 0·43, β_2 1	68P
Pb^{2+}	cal	25	0 corr	$\Delta H_1 = 0{\cdot}3, \Delta S_1 = 6{\cdot}0$ (K_1 1·09)	67N
	MHg	25	Me_2NCHO	K_1 1·30, β_2 1·80, β_3 1·90, β_4 2·04 in Me_2NCHO(l), I = 1·2($NaClO_4$)	68Sb
R_4N^+	con	25	$PhNO_2$	K_1 1·57 in $PhNO_2$(l), → 0, R = C_5H_{11}	66La
NO^+	sp	0—20	0 corr	$K(HNO_2+H^++L^- \rightleftharpoons NOL+H_2O)$ 1·67(0°), 1·51(20°), $\Delta H = -2{\cdot}9$	63Sb
Bi^{3+}	sol	25	2 ($NaClO_4$), 1 H^+	$\beta(Bi^{3+}Cl^-L^-)$ 4·71, $\beta(Bi^{3+}Cl^-L^-_2)$ 5·67, $\beta(Bi^{3+}Br^-L^-)$ 4·08, $\beta(Bi^{3+}Br^-L^-_2)$ 5·30	65J
	sol, sp	20	2 ($NaClO_4$), 1 H^+	$\beta(Bi^{3+}Cl^-L^-)$ 4·71, ev $BiClL_2$	65L
	sp		3·9 Na^+, 0·25 H^+ (ClO_4)	$K(BiL_5Br^{3-}+L^- \rightleftharpoons BiL_6^{3-}+Br^-)$ −2·03, $K(BiL_3Br_3^{3-}+2L^- \rightleftharpoons BiL_5Br^{3-}+2Br^-)$ −2·77, $\beta(Bi^{3+}L^-{}_5Br^-)$ 6·21, $\beta(Bi^{3+}L^-{}_3Br^-{}_3)$ 9·0, (β_6 4·18)	66J
I_2	dis		var	$K_D[M^++L^-+2I_2(org) \rightleftharpoons M^+(org)+LI_4^-(org)]$ 0·65(M = Na), 3·94(M = Cs), org = $PhNO_2$	66Ta
I^+	sp	1—25	1·2 ($KClO_4$), 1·0 H^+	ev IL_2, $K(IL_2^- \rightleftharpoons I^-+L_2)$ −5·7(1°), −5·3(4°), −5·34(6·6°), −5·0(9°), −4·96(12°), −4·1(→ 25°)	66L

references overleaf

17 SCN⁻ (contd)

53P A. Preisinger, *Tschermaks mineralog. petrog. Mitt.*, 1953, **3**, 376
53Z G. S. Zhdanov, Z. V. Zvonkova, and V. P. Glushkova, *Zhur. fiz. Khim.*, 1953, **27**, 106
58M V. E. Mironov, dissertation, 1958, Leningrad Tekh. Inst., quoted by ref 63K
63D G. D'Amore, V. Chiantella, and F. Corigliano, *Ann. Chim. (Italy)*, 1963, **53**, 1466
63Gc D. C. Giedt and C. J. Nyman, *J. Phys. Chem.*, 1963, **67**, 2491
63Gd A. M. Golub and R. A. Kostrova, *Ukrain. khim. Zhur.*, 1963, **29**, 784
63Ge A. M. Golub and V. M. Samoilenko, *Ukrain. khim. Zhur.*, 1963, **29**, 789
63Gf A. M. Golub and R. A. Kostrova, *Dopovidi Akad. Nauk Ukrain. R.S.R.*, 1963, 1061
63Gg A. A. Grinberg, N. V. Kiseleva, and M. I. Gel'fman, *Doklady Akad. Nauk S.S.S.R.*, 1963, **153**, 1327
63H A. Haim and H. Taube, *Inorg. Chem.*, 1963, **2**, 1199
63Ha A. C. Hazell, *J. Chem. Soc.*, 1963, 5745
63K F. Ya. Kul'ba and V. E. Mironov, "Khimiya Talliya" (Chemistry of Thallium), Goskhimizdat, Leningrad, 1963, p. 67
63Ma M. A. Mubayadzhyan, *Izvest. Akad. Nauk Armyan S.S.R., Khim.*, 1963, **16**, 229
63Sa M. Schellenberg, dissertation, Eidgenössische Technische Hochschule, Zürich, 1963
63Sb G. Stedman and P. A. E. Whincup, *J. Chem. Soc.*, 1963, 5796
63Va V. P. Vasil'ev, V. N. Vasil'eva, F. I. Ivanova, and O. N. Morozova, *Izvest. V.U.Z., Khim.*, 1963, **6**, 712
64B J. C. Barnes, *J. Chem. Soc.*, 1964, 3880
64C D. M. Czakis-Sulikowska, *Roczniki Chem.*, 1964, **38**, 1435
64G A. M. Golub and V. M. Samoilenko, *Zhur. neorg. Khim.*, 1964, **9**, 70
64Ga A. M. Golub and A. M. Sych, *Zhur. neorg. Khim.*, 1964, **9**, 1085
64Gb A. M. Golub and G. B. Pomerants, *Zhur. neorg. Khim.*, 1964, **9**, 1624
64Gc A. M. Golub and A. M. Sych, *Izvest. Akad. Nauk Latv. S.S.R., Khim.*, 1964, 387
64H A. C. Hazell, *Acta Cryst.*, 1964, **17**, 1155
64Ha H. Hamaguchi, N. Onuma, M. Kishi, and R. Kuroda, *Talanta*, 1964, **11**, 495
64J E. Jósefowicz and J. Masłowska, in "Theory and structure of complex compounds" (symposium papers, Wrocław 1962), Pergamon, Warszawa, 1964, p. 525
64K V. N. Kumok, *Zhur. neorg. Khim.*, 1964, **9**, 362
64Ka V. N. Kumok and V. V. Serebrennikov, *Zhur. neorg. Khim.*, 1964, **9**, 2148
64Kb A. Korczyński, in "Theory and structure of complex compounds" (symposium papers, Wrocław 1962), Pergamon, Warszawa, 1964, p. 405
64M A. K. Molodkin and G. A. Skotnikova, *Zhur. neorg. Khim.*, 1964, **9**, 60, 1493
64O Y. Oka and T. Kato, *J. Chem. Soc. Japan*, 1964, **85**, 784
64S T. Sekine, *J. Inorg. Nuclear Chem.*, 1964, **26**, 1463
64Sa T. D. Seth and R. C. Kapoor, *J. Polarog. Soc.*, 1964, **10**, 17
64Sb O. Schmitz-DuMont and B. Ross, *Angew Chem.*, 1964, **76**, 304
64Sc O. Schmitz-DuMont and B. Ross, *Angew Chem.*, 1964, **76**, 647
64T M. M. Tananaiko, *Zhur. neorg. Khim.*, 1964, **9**, 608
64Ta V. M. Tarayan and L. G. Mushegyan, *Izvest Akad. Nauk. Armyan. S.S.R., Khim.*, 1964, **17**, 46
64Tb S. Tribalat and J.-M. Calderó, *Compt. rend.*, 1964, **258**, 2828
64V V. P. Vasil'ev and P. S. Mukhina, *Zhur. neorg. Khim.*, 1964, **9**, 1134
64Va I. V. Vinarov, A. I. Orlova, G. I. Byk, and N. F. Kislitsa, *Ukrain. khim. Zhur.*, 1964, **30**, 758
64Vb V. P. Vasil'ev and P. S. Mukhina, *Izvest. V.U.Z., Khim.*, 1964, **7**, 711
65B R. Barbieri, M. Giustiniani, and E. Cervo, *J. Inorg. Nuclear Chem.*, 1965, **27**, 1325
65Ba H. Böhland and E. Niemann, *Z. anorg. Chem.*, 1965, **336**, 225
65Bb R. Barbieri and J. Bjerrum, *Acta Chem. Scand.*, 1965, **19**, 469
65C G. R. Choppin and J. Ketels, *J. Inorg. Nuclear Chem.*, 1965, **27**, 1335
65Ca A. Cassol, R. Portanova, and R. Barbieri, *J. Inorg. Nuclear Chem.*, 1965, **27**, 2275
65D M. Duhameux and M. Cadiot, *Compt. rend.*, 1965, **260**, 1153
65F A. B. Fasman, G. G. Kutyukov, and D. V. Sokol'skii, *Zhur. neorg. Khim.*, 1965, **10**, 1338
65Fa D. Forster and D. M. L. Goodgame, *Inorg. Chem.*, 1965, **4**, 823
65Fb D. Forster and D. M. L. Goodgame, *J. Chem. Soc.*, 1965, 268
65G A. M. Golub, M. I. Olevinskii, and E. F. Lutsenko, *Ukrain. khim. Zhur.*, 1965, **31**, 12
65Ga A. M. Golub and G. V. Pomerants, *Ukrain. khim. Zhur.*, 1965, **31**, 104
65H C.-L. Huan and Y.-Z. Chen, *Scientia Sinica*, 1965, **14**, 924
65Ha H. E. Hellwege and G. K. Schweitzer, *J. Inorg. Nuclear Chem.*, 1965, **27**, 99
65Hb A. Haim and N. Sutin, *J. Amer. Chem. Soc.*, 1965, **87**, 4210
65J E. Jósefowicz and H. Ladzińska-Kulinska, *Roczniki Chem.*, 1965, **39**, 1175
65K F. Ya. Kul'ba, V. E. Mironov, and G. Mrnyakova, *Zhur. neorg. Khim.*, 1965, **10**, 1393
65Ka A. Korczyński and M. A. Porai-Koshits, *Roczniki Chem.*, 1965, **39**, 1567
65L H. Ladzińska-Kulińska, *Roczniki Chem.*, 1965, **39**, 1137
65M V. E. Mironov and Yu. I. Rutkovskii, *Zhur. neorg. Khim.*, 1965, **10**, 1069
65Ma V. E. Mironov and Yu. I. Rutkovskii, *Zhur. neorg. Khim.*, 1965, **10**, 2670
65Mb T. D. B. Morgan, G. Stedman, and P. A. E. Whincup, *J. Chem. Soc.*, 1965, 4813
65Mc L. G. Mushegyan and V. M. Tarayan, *Izvest. Akad. Nauk Armyan. S.S.R., Khim.*, 1965, **18**, 118
65N Y. Narusawa, J. Hashimoto, and H. Hamaguchi, *Bull. Chem. Soc. Japan*, 1965, **38**, 234
65O A. I. Orlova, I. V. Vinarov, and L. M. Burtnenko, *Ukrain. khim. Zhur.*, 1965, **31**, 775
65Oa T. Ôkubo, *J. Chem. Soc. Japan*, 1965, **86**, 1142

continu

65P I. V. Pyatnitskii and M. Durdyev, *Ukrain. khim. Zhur.*, 1965, **31**, 1247
65R W. Rittner, dissertation, Univ. Göttingen, 1965
65S O. A. Sinegribova and G. A. Yagodin, *Zhur. neorg. Khim.*, 1965, **10**, 1250
65Sa T. Sekine, *Acta Chem. Scand.*, 1965, **19**, 1519
65Sb G. Schwarzenbach and M. Schellenberg, *Helv. Chim. Acta.*, 1965, **48**, 28
65Sc T. Sekine, *Bull. Chem. Soc. Japan*, 1965, **38**, 1972
65T V. F. Toropova and M. K. Saikina, *Zhur. neorg. Khim.*, 1965, **10**, 1166
65U N. V. Ul'ko, *Ukrain. khim. Zhur.*, 1965, **31**, 887
66A G. Alexandrides and C. Cummiskey, *J. Inorg. Nuclear Chem.*, 1966, **28**, 2025
66Aa E. G. Arutyunyan and M. A. Porai-Koshits, *Zhur. strukt. Khim.*, 1966, **7**, 393
66B J. H. Boughton and R. N. Keller, *J. Inorg. Nuclear Chem.*, 1966, **28**, 2851
66Ba A. A. Biryukov, V. I. Shlenskaya, and I. P. Alimarin, *Izvest. Akad. Nauk. S.S.S.R., Khim.*, 1966, 3
66C A. Cassol and L. Magon, *Ricerca sci.*, 1966, **36**, 1194
66D J. Dehand, R. Dorschner, and R. Rohmer, *Compt. rend.*, 1966, **262**, *C*, 1882
66G P. Gerding, *Acta Chem. Scand.*, 1966, **20**, 2771
66Ga E. A. Gyunner and N. D. Belykh, *Ukrain. khim. Zhur.*, 1966, **32**, 1270
66Gb A. M. Golub and A. M. Sych, *Zhur. priklad. Khim.*, 1966, **39**, 2658
66Gc A. M. Golub and V. A. Kalibabchuk, *Zhur. neorg. Khim.*, 1966, **11**, 590
66Gd A. M. Golub and V. N. Sergun'kin, *Zhur. neorg. Khim.*, 1966, **11**, 770
66J E. Jósefowicz and H. Ladzińska-Kulińska, *Roczniki Chem.*, 1966, **40**, 1615
66K C. Kiehl, *Z. phys. Chem.* (*Leipzig*), 1966, **232**, 384
66L C. Long and D. A. Skoog, *Inorg. Chem.*, 1966, **5**, 206
66La F. R. Longo, J. D. Kerstetter, T. F. Kumosinski, and E. C. Evers, *J. Phys. Chem.*, 1966, **70**, 431
66M V. E. Mironov, F. Ya. Kul'ba, Yu. I. Rutkovskii, and E. I. Ignatenko, *Zhur. neorg. Khim.*, 1966, **11**, 1786
66N A. V. Novoselova, K. M. Mikheeva, T. I. Pochkaeva, and N. S. Tanm, *Vestnik. Moskov. Univ.*, 1966, Khim, no. 1, 46
66O J. Oleszkiewicz and T. Lipiec, *Roczniki Chem.*, 1966, **40**, 541
66Oa Y. Oka and T. Kato, *J. Chem. Soc. Japan*, 1966, **87**, 580
66S H.-H. Schmidtke, *J. Inorg. Nuclear Chem.*, 1966, **28**, 1735
66Sa A. Swinarski, E. Danilczuk, and R. Gogolin, *Roczniki Chem.*, 1966, **40**, 737
66T V. M. Tarayan and L. G. Mushegyan, *Armyan. khim. Zhur.*, 1966, **19**, 918
66Ta S. Tribalat and M. Grall, *Compt. rend.*, 1966, **263**, *C*, 399
66Tb S. Tribalat and J. M. Calderó, *Bull. Soc. chim. France*, 1966, 774
66V V. N. Vasil'eva and V. P. Vasil'ev, *Izvest. V.U.Z., Khim.*, 1966, **9**, 185
67A R. Alexander, E. C. F. Ko, Y. C. Mac, and A. J. Parker, *J. Amer. Chem. Soc.*, 1967, **89**, 3703
67B R. A. Bailey and S. L. Kozak, *Inorg. Chem.*, 1967, **6**, 419
67Ba K. Burger and B. Pintér, *J. Inorg. Nuclear Chem.*, 1967, **29**, 1717
67Bb W. J. Biermann and R. H. McCorkell, *Canad. J. Chem.*, 1967, **45**, 2846
67Bc B. R. Baker, N. Sutin and T. J. Welch, *Inorg. Chem.*, 1967, **6**, 1948
67Bd R. A. Bailey and S. L. Kozak, *Inorg. Chem.*, 1967, **6**, 2155
67Be A. A. Biryukov and V. I. Shlenskaya, *Zhur. neorg. Khim.*, 1967, **12**, 2579
67Bf T. M. Brown and G. F. Knox, *J. Amer. Chem. Soc.*, 1967, **89**, 5296
67C F. A. Cotton, W. R. Robinson, R. A. Walton, and R. Whyman, *Inorg. Chem.*, 1967, **6**, 929
67Ca F. A. Cotton, W. R. Robinson and R. A. Walton, *Inorg. Chem.*, 1967, **6**, 1257
67Cb T. J. Conocchioli and N. Sutin, *J. Amer. Chem. Soc.*, 1967, **89**, 282
67D G. Dittrich, dissertation, Technische Hochschule, Hannover, 1967
67Da N. V. Duffy and J. E. Earley, *J. Amer. Chem. Soc.*, 1967, **89**, 272
67F D. Forster and W. D. Horrocks, jun., *Inorg. Chem.*, 1967, 6, 339
67G A. M. Golub and A. N. Borshch, *Zhur. neorg. Khim.*, 1967, **12**, 522
67H A. A. Humffray, A. M. Bond, and J. S. Forrest, *J. Electroanalyt. Chem.*, 1967, **15**, 67
67K A. Korczyński, *Roczniki Chem.*, 1967, **41**, 1197
67Ka L. N. Komissarova, V. G. Gulia, T. M. Sas, and N. A. Chernova, *Zhur. neorg. Khim.*, 1967, **12**, 873
67L A. Łodzińska, *Roczniki Chem.*, 1967, **41**, 1007
67La V. V. Lukachina, *Ukrain. khim. Zhur.*, 1967, **33**, 740
67Lb C. H. Langford and W. R. Muir, *J. Amer. Chem. Soc.*, 1967, **89**, 3141
67M W. R. Masontert, E. R. Berger, and R. C. Johnson, *Inorg. Chem.*, 1967, **6**, 248
67N G. H. Nancollas and K. Torrance, *Inorg. Chem.*, 1967, **6**, 1567
67Na B. P. Nikol'skii, A. A. Pendin, and M. S. Zakhar'evskii, *Zhur. neorg. Khim.*, 1967, **12**, 1803
67S H.-H. Schmidtke and D. Garthoff, *Helv. Chim. Acta*, 1967, **50**, 1631
67Sa P. G. Sears, J. A. Caruso and A. I. Popov, *J. Phys. Chem.*, 1967, **71**, 905
67U N. V. Ul'ko and R. A. Savchenko, *Zhur. neorg. Khim.*, 1967, **12**, 328
67W S. Wajda and F. Pruchnik, *Roczniki Chem.*, 1967, **41**, 1169
67Wa S. Wajda and F. Pruchnik, *Roczniki Chem.*, 1967, **41**, 1473
68A S. Ašperger and M. Pribanić, *J. Chem. Soc.* (*A*), 1968, 1503
68D R. C. Das, A. C. Dash, and J. P. Mishra, *J. Inorg. Nuclear Chem.*, 1968, **30**, 2417
68Da M. Della Monica, U. Lamanna, and L. Senatore, *Inorg. Chim. Acta*, 1968, **2**, 363
68G P. Gerding, *Acta Chem. Scand.*, 1968, **22**, 1283
68Ga P. Gerding and I. Jönsson, *Acta Chem. Scand.*, 1968, **22**, 2255

continued overleaf

17 SCN⁻ (contd)

68Gb E. A. Gyunner and N. D. Yakhkind, *Zhur. neorg. Khim.*, 1968, **13**, 245
68Gc E. A. Gyunner and N. D. Yakhkind, *Zhur. neorg. Khim.*, 1968, **13**, 2758
68Gd V. Gutmann and H. Bardy, *Z. anorg. Chem.*, 1968, **361**, 213
68K J. R. Knox and K. Eriks, *Inorg. Chem.*, 1968, **7**, 84
68Ka W. Kruse and D. Thusius, *Inorg. Chem.*, 1968, **7**, 464
68L R. Larsson, *Acta Chem. Scand.*, 1968, **22**, 983
68La R. Larsson and A. Miezis, *Acta Chem. Scand.*, 1968, **22**, 3261
68Lb D. C. Luehrs and K. Abate, *J. Inorg. Nuclear Chem.*, 1968, **30**, 549
68Lc C. H. Langford and F. M. Chung, *J. Amer. Chem. Soc.*, 1968, **90**, 4485
68M A. Miezis and R. Larsson, *Acta Chem. Scand.*, 1968, **22**, 3293
68Ma J. Masłowska, *Roczniki Chem.*, 1968, **42**, 1819
68Mb O. N Malyavinskaya and Ya. I. Tur'yan, *Zhur. fiz. Khim.*, 1968, **42**, 269
68Mc I. N. Marov, Yu. N. Dubrov, A. N. Ermakov, and G. N. Martynova, *Zhur. neorg. Khim.*, 1968, **13**, 3247
68Md E. D. Moorhead and G. M. Frame, jun., *Analyt. Chem.*, 1968, **40**, 280
68Me J. M. Malin and J. H. Swinehart, *Inorg. Chem.*, 1968, **7**, 250
68Mf J. L. Martin, L. C. Thompson, L. J. Radonovich, and M. D. Glick, *J. Amer. Chem. Soc.*, 1968, **90**, 4493
68N T. Nomura, *Bull. Chem. Soc. Japan*, 1968, **41**, 2803
68Na J. F. Neumann, J. R. Paxson, and C. J. Cummiskey, *J. Inorg. Nuclear Chem.*, 1968, **30**, 2243
68O M. Orhanović, H. N. Po, and N. Sutin, *J. Amer. Chem. Soc.*, 1968, **90**, 7224
68P R. Portanova, A. Cassol, L. Magon, and G. Tomat, *Gazzetta*, 1968, **98**, 1290
68Pa B. Prasad, *J. Indian Chem. Soc.*, 1968, **45**, 1037
68Pb N. Payen-Baldy and M. Machtinger, *Bull. Soc. chim. France*, 1968, 2221
68R N. F. Rusin and I. V. Vinarov, *Zhur. neorg. Khim.*, 1968, **13**, 1128
68Ra C. L. Rao, C. J. Shahani, and K. A. Mathew, *Inorg. Nuclear Chem. Letters*, 1968, **4**, 655
68S A. Schlund and H. Wendt, *Ber Bunsengesellschaft Phys. Chem.*, 1968, **72**, 652
68Sa L. G. Sillén and B. Warnqvist, *Acta Chem. Scand.*, 1968, **22**, 3032
68Sb V. M. Samoilenko, *Zhur. neorg. Khim.*, 1968, **13**, 79
68Y N. D. Yakhkind and E. A. Gyunner, *Zhur. neorg. Khim.*, 1968, **13**, 1005

18 Selenocyanate, SeCN⁻

Metal	*Method*	*Temp*	*Medium*	*Log of equilibrium constant, remarks*	*Ref*
general	hyp			K_1 values for various metal ions deduced from electronegativities	67L
H^+	gl	20—33	→0	$K_1 < 1$	66B
Y^{3+}	var		var	ev YL_6^{3-} (pre, con, ir)	66Ba
Ti^{3+}	ir, pre		solid	ev TiL_6^{3-} in solid	67B
Th^{4+}	sp	20	Me_2CO	K_1 3·27, β_4 12·12 in Me_2CO(l), KL var	67G
		20	Me_2NCHO	K_1 3·10, β_2 6·01, β_3 8·85, β_4 11·64, β_5 14·39 in Me_2NCHO(l), NaL var	
				K_1 3·08, β_2 5·92, β_3 8·80, β_4 11·57, β_5 14·36, β_6 17·67 in Me_2NCHO(l), KL var	
V^{3+}	var		var	ev VL_6^{3-} in solid and in $PhNO_2$(l) (con, ir, pre)	67B
VO^{2+}	var		var	ev VOL_4^{2-} in solid and in $PhNO_2$(l) (con, ir, pre)	67B
Cr^{3+}	ir, sp		var	ev $Cr(NCSe)_6^{3-}$ in solid and org	63M
Mo^{3+}	ir, pre		solid	ev MoL_6^{3-}	67S
Mn^{2+}	var		var	ev MnL_4^{2-}, MnL_6^{4-} (pre, ir, sp, mag)	65F
Fe^{2+}	var		var	ev FeL_4^{2-}, FeL_6^{4-} (pre, ir, sp, mag)	65F
	ir		solid	ev FeL_4^{2-}	67S
Fe^{2+}cpx	sp, gl		KL var	$Fe(dimethylglyoxime)_2^{2+}$, β_2 3·43	67Ba
Fe^{3+}	var		var	ev FeL_6^{3-} (pre, con, ir)	66Ba
	ir, pre		solid	ev FeL_6^{3-}	67S

continued

Metal	*Method*	*Temp*	*Medium*	*Log of equilibrium constant, remarks*	*Ref*
Co^{2+}	var		var	ev CoL_4^{2-}, CoL_6^{4-} (pre, ir, sp)	65F
	(con) sp	20	Me_2NCHO	K_1 3·0, β_2 4·91, β_3 5·70, β_4 6·79 in Me_2NCHO(l)	68S
	(con) sp	20	MeCN	K_1 3·93, β_2 6·45, β_3 9·67, β_4 12·81 in MeCN(l)	
Co^{2+}cpx	sp, gl		KL var	Co(dimethylglyoxime)$_2^{2+}$, K_1 5·1, K_2 3·1	67Ba
Ni^{2+}	sp		Me_2NCHO	K_1 2·5 to 2·2, K_2 1·8, K_3 1·5, K_4 1·4, β_4 6·9 in Me_2NCHO(l)	64S
	var		var	ev NiL_4^{2-}, NiL_6^{4-} (pre, ir, sp)	65F
Rh^{3+}	var		var	ev RhL_6^{3-} (pre, con, ir, sp)	66Ba
	var		solid, org	ev RhL_6^{3-} (pre, sp, ir)	66S
	ir, pre		solid	ev RhL_6^{3-}	67S
Pd^{2+}	var		var	ev PdL_4^{2-} (pre, ir)	65F ≈ 66Ba
	var		solid, org	ev PdL_4^{2-} (pre, sp, ir)	66S
	var		solid, org	ev PdL_4^{2-} (pre, con, ir)	67S
Pt^{2+}	var		var	ev PtL_4^{2-} (pre, con, ir, sp)	66Ba
	var		solid, org	ev PtL_4^{2-} (pre, sp, ir)	66S
	ir, pre		solid	ev PtL_4^{2-}	67S
Pt^{4+}	con, ir		solid, org	ev PtL_6^{2-}	67S
Ag^+	sol	20	2 (KNO_3)	K_s [AgL(s) + $2SCN^- \rightleftharpoons AgL(SCN)_2^{2-}$] −1·62, K_s[AgI(s) + $2L^- \rightleftharpoons AgL_2I^{2-}$] −0·4	65G
	Ag	25	2 (KNO_3)	$\beta(Ag^+L^-{}_3I^-)$ 14·5, $\beta(Ag^+L^-{}_3SCN^-)$ 13·4, $\beta(Ag^+L^-I^-{}_3)$ 14·66	
Au^{3+}	var		var	ev AuL_4^- in solid and org (pre, con, ir)	67S
Zn^{2+}	var		var	ev ZnL_4^{2-} (pre, ir)	65F
	pol	30	2 (KNO_3)	K_1 0·76, β_2 1·00	67H
Cd^{2+}	var		var	ev $Cd_2L_6^{4-}$ (pre, con, ir)	66Ba
	pol	30	2 (KNO_3)	K_1 1·30, β_2 2·00, β_3 2·64, β_4 3·00	67H
Hg^{2+}	sol	25	1 ($NaClO_4$)	$\beta(Hg^{2+}I^-{}_2L^-)$ 28·0, $\beta(Hg^{2+}I^-{}_2L^-{}_2)$ 29·42 [$K_s(Hg^{2+}I^-{}_2)$ −27·19]	65C
	sp		1 Na^+	$K(HgI_4^{2-} + L^- \rightleftharpoons HgI_3L^{2-} + I^-)$ 0·44, $\beta(Hg^{2+}I^-{}_3L^-)$ 29·74 [$\beta(Hg^{2+}I^-{}_4)$ 29·3]	
	var		var	ev HgL_4^{2-} in solid and org (pre, con, ir)	67S
Se	*X*		solid	ev $(SeCN)_3^-$ in $K(SeCN)_3(H_2O)_{0.5}$(s)	63F

63F O. Foss and S. Hauge, *Acta Chem. Scand.*, 1963, **17**, 1807
63M K. Michelsen, *Acta Chem. Scand.*, 1963, **17**, 1811
64S V. V. Skopenko and A. I. Brusilovets, *Ukrain. khim. Zhur.*, 1964, **30**, 24
65C D. M. Czakis-Sulikowska, *Roczniki Chem.*, 1965, **39**, 1161
65F D. Forster and D. M. L. Goodgame, *Inorg. Chem.*, 1965, **4**, 1712
65G A. M. Golub, V. V. Skopenko and G. B. Pomerants, *Zhur. neorg. Khim.*, 1965, **10**, 344
66B J. H. Boughton and R. N. Keller, *J. Inorg. Nuclear Chem.*, 1966, **28**, 2851
66Ba J. L. Burmeister and L. E. Williams, *Inorg. Chem.* 1966, **5**, 1113
66S H.-H. Schmidtke, *J. Inorg. Nuclear Chem.*, 1966, **28**, 1735
67B J. L. Burmeister and L. E. Williams, *J. Inorg. Nuclear Chem.*, 1967, **29**, 839
67Ba K. Burger and B. Pintér, *J. Inorg. Nuclear Chem.*, 1967, **29**, 1717
67G A. M. Golub and V. A. Kalibabchuk, *Zhur. neorg. Khim.*, 1967, **12**, 2370
67H A. A. Humffray, A. M. Bond, and J. S. Forrest, *J. Electroanalyt. Chem.*, 1967, **15**, 67
67L A. Łodzińska, *Roczniki Chem.*, 1967, **41**, 1007
67S H.-H. Schmidtke and D. Garthoff, *Helv. Chim. Acta*, 1967, **50**, 1631
68S V. V. Skopenko and A. I. Brusilovets, *Ukrain. khim. Zhur.*, 1968, **34**, 1210

19 Other ligands with CN

Metal	Method	Temp	Medium	Log of equilibrium constant, remarks	Ref
				Tricyanomethyl, $C(CN)_3^-$	
Co^{2+}	spJ		var	ev CoL^+	66K
				Dicyanamide, $N(CN)_2^-$	
Co^{2+}	var		var	ev CoL^+(in H_2O), CoL_3^-, CoL_4^{2-}, in Me_2CO(l) and solids (pre, ir, mag, sp, spJ)	66K

66K H. Köhler and B. Seifert, *Z. anorg. Chem.*, 1966, **344**, 63

20 Carbonyl, carbon monoxide, CO

Metal	Method	Temp	Medium	Log of equilibrium constant, remarks	Ref
general	lit			survey of polynuclear CO cpx	68Cb
Group 6A	var		var	ev $M_2L_{10}X^-$, $MM'L_{10}X^-$(M,M' = Cr, Mo, W; X = Cl, Br, I) in solids and $MeNO_2$(l) (con, pre, ir)	68R
	var		var	ev $ML_4(S_2PF_2)^-$ and $ML_5(SPOF_2^-)$, (M = Cr, Mo, W) in solids and $MeNO_2$(l) (con, ir, pre)	68Ra
Groups 6A, 7A					
	ir, pre		var	ev $L_5MM'L_5^-$ in $(MeOC_2H_4)_2O$(l) *etc.* and solids, M = Cr, Mo, W; M′ = Mn, Re	67A
	pre		var	ev $CrMnL_{10}^-$, $WMnL_{10}^-$	68R
Group 6A, Co	pre, ir		var	ev $CrCoL_9^-$, $WCoL_9^-$	68R
Cr	var		org	ev CrL_5NCS^- in $MeNO_2$(l) (con, ir, pre)	64W
	var		var	ev $L_5CrHCrL_5^-$ (ir, nmr, pre)	65A
	var		var	ev $L_5CrHCrL_5^-$ (*X*, ir, nmr)	66H
	pre, ir		var	ev $CrL_2(NH_3)_4^+$, $CrL_3(NH_3)_2I^-$ in NH_3(l)	67Ba
	var		org	ev CrL_5NCO^- (pre, ir, mag)	67E
	var		var	ev CrL_5CN^-, CrL_5CNH, $CrL_5CNSnMe_3$ (pre, con, ir)	67K
	pre, ir		org	ev $Cr_2L_{10}^{2-}$, CrL_5^{2-}, $Cr_2L_9^{4-}$ in C_4H_8O(l) and $(Me_2N)_3PO$	68K
Mo	var		org	ev MoL_5NCS^- in $MeNO_2$(l) (con, ir, pre)	64W
	var		var	ev $L_5MoHMoL_5^-$ (ir, nmr, pre)	65A
	var		$PhNO_2$	ev $MoL_4Br_3^-$, $MoL_4Br_2I^-$, $MoL_4BrI_2^-$, $MoL_4I_3^-$ in $PhNO_2$(l) (con, ir, mag, pre)	65G
	var		org	ev MoL_5NCO^- (pre, con, ir, mag)	67E
	var		var	ev MoL_5CN^-, MoL_5CNH, $MoL_5CNSnMe_3$ (pre, con, ir)	67K
	var		var	ev $MoL_4Cl_3^-$, $MoL_4Br_3^-$ in solids and Me_2CO(l) (pre, con, ir)	68Ba
W	var		org	ev WL_5NCS^- in $MeNO_2$(l) (con, ir, pre)	64W
	var		var	ev $L_5WHWL_5^-$ (ir, nmr, pre)	65A
	var		$PhNO_2$	ev $WL_4Br_3^-$, $WL_4Br_2I^-$, $WL_4BrI_2^-$ in $PhNO_2$(l) (con, ir, mag, pre)	65G
	var		org	ev WL_5NCO^- (con, ir, fp, pre)	66Ba
	var		var	ev WL_5CN^-, WL_5CNH, $WL_5CNSnMe_3$ (pre, con, ir)	67K
	var		var	ev $WL_4Cl_3^-$, $WL_4Br_3^-$ in solids and Me_2CO(l) (pre, con, ir)	68Ba

continued

Metal	*Method*	*Temp*	*Medium*	*Log of equilibrium constant, remarks*	*Ref*
Mn	pre, ir		$C_2H_4Cl_2$	ev $MnL_4X_2^-$(X = Cl, Br, I, CN) in $C_2H_4Cl_2$(l)	64A
	ir, pre		H_2O, org	ev MnL_4XY^-, $Mn_2L_8X_2^{2-}$ (X, Y = Cl, Br, I)	64Aa
	pre, ir		var	ev $MnL_3(CN)_3^{2-}$ and $MnL_2(CN)_4^{3-}$ in EtOH(l), NH_3(l), K^+ salts	67B
	var		org	ev $Mn_2L_7(N_3)_x(NCO)_{3-x}^-$, (pre, con, ir, mag)	67E
Mn, Fe	ir, pre		var	ev $MnFe_2L_{12}^-$ in solids and EtOH(l)	66Aa
Tc	pre, con		org	ev TcL_6^+ in MeOH, Me_2CO, THF(l)	65H
Tc, Fe	pre, ir		var	ev $TcFe_2L_{12}^-$	68L
Re	var		var	ev $ReL_4I_2^-$, ReL_4IBr^-, $ReL_4Br_2^-$, $ReL_3I_3^{2-}$, $Re_2L_7Cl_3^-$, $Re_2L_7Cl_2Br^-$, $Re_2L_7Br_3^-$, $Re_2L_8I_2^{2-}$ in org, solid (pre, ir, con)	66A
	X		solid	ev $Re_4L_{16}^{2-}$ in $(Bu_4N)_2Re_4(CO)_{16}$(s)	67Bb = 68C
	var		org	ev $ReL_3N_3NCO^-$ (pre, con, ir, mag)	67E
	con, *X*		var	ev $H_2Re_3L_{12}^-$ in MeCN(l) (con) and solid (*X*)	68Cc
	pre, ir		var	ev $ReL_2(CN)_4^{3-}$, $ReL_3(CN)_3^{2-}$	68B
Re, Mn	*X*		solid	ev HRe_2MnL_{14} in solid	67Cb
Re, Fe	pre, ir		var	ev $ReFe_2L_{12}^-$ in solid, Et_2O(l), C_4H_8O(l)	67Ea
Fe	pre, sp		var	ev $HFe_4L_{13}^{2-}$, $Fe_4L_{13}^{2-}$	64S
	X		solid	ev $HFe_3L_{11}^-$ in $Et_3NHHFe_3L_{11}$(s)	65D
	X		solid	ev $Fe_4L_{13}^{2-}$ in Fe $py_6Fe_4L_{13}$(s)	66D
	var		HCl(l)	ev FeL_5H^+, FeL_5Cl^+, FeL_5Br^+, FeL_5NO^+ in HCl(l) and salts (con, pre, sp, ir, mag)	68I
	var		var	ev $FeMnL_9^-$, $FeCoL_8^-$ in solids and $MeNO_2$(l) (con, pre, ir)	68R
Fe, Hg	var		var	ev $FeL_4(HgY)_2X^-$ (X,Y = Cl, Br, I) in solids and $PhNO_2$(l) (con, pre, ir)	68Ca
Co	cix, dis	0	var	ev CoL_4ROH^+, CoL_4^-, R = H, Me, Et *etc.*	64T
	pre, ir		var	ev $Co_6L_{15}^{2-}$ in H_2O, EtOH *etc.*	67Ca
	X		solid	ev CoL_4^- in $Hg(CoL_4)_2$(s)	68S
Co, Zn	*X*		solid	ev $L_4CoZnCoL_4$ units in solid	67L
Co, In	*X*		solid	ev $In_3Br_3Co_4L_{15}$ units	68Cd
Co, Sn	pre, ir		var	ev $(L_4Co)_2SnCl_2$ in solid and $PhNO_2$(l)	65B
	pre, ir		var	ev $(L_4Co)_2SnX_2$ (X = Cl, Br, I) in C_6H_{14}(l)	66B
Ni	pre		C_5H_5N	ev $Ni_5L_9^{2-}$, $Ni_4L_9^{2-}$, $Ni_2L_6^{2-}$ in pyridine(l)	63H
	var		org	ev $Ni_3L_8^{2-}$, $Ni_4L_9^{2-}$ in C_4H_8O(THF)(l) (con, ir, pre)	63Ha
Ru	pre, ir		var	ev RuL_2X_2, $RuL_2X_4^{2-}$, X = Cl, Br, I	67C
	X		solid	ev $Ru_2Br_4L_6$ units in $RuBr_2L_3$(s)	68M
Ru, Sn	var		org	ev $RuX_2L_2(SnX_3)_2^{2-}$(X = Cl or Br) in $PhNO_2$, CH_2Cl_2, EtOH(l) (con, ir, pre, sp)	66K
	pre, con		org	ev $RuCl_2L_2(SnCl_3)_2^{2-}$ in Me_2CO, $MeNO_2$, Me_2NCHO(l)	66S
Rh	ir, pre		org	ev $(RhL_2)_2SO_4$, $(RhL_2NO_3)_2$, $(RhL_2SCN)_2$ in Me_2CO(l) and $CHCl_3$(l)	65L
	var		var	ev $RhL_2X_2^-$(X = Cl, Br, I), $Rh_2L_2X_4^{2-}$ (X = Br, I), $RhLI_4^-$ (pre, mag, con, ir)	65V

continued overleaf

20 CO (contd)

Metal	*Method*	*Temp*	*Medium*	*Log of equilibrium constant, remarks*	*Ref*
Ir	*X*		solid	ev $IrBr_5CO^{2-}$ in $K_2IrBr_5CO(s)$	64B
	var		var	ev IrL_3I_3, $Ir_2L_4I_6$, $Ir_2L_3I_6$, $Ir_2L_4I_6^{2-}$, $IrLI_4^-$, $IrL_2I_4^-$, $IrLI_5^{2-}$, $IrL_2I_2^-$, $IrLI_3^-$, $IrLCl_5^-$, $IrLBr_5^-$	64M

63H W. Hieber, J. Ellermann, and E. Zahn, *Z. Naturforsch*, 1963, **18b**, 589
63Ha W. Hieber and J. Ellermann, *Z. Naturforsch.*, 1963, **18b**, 595
64A R. J. Angelici, *Inorg. Chem.*, 1964, **3**, 1099
64Aa E. W. Abel and I. S. Butler, *J. Chem. Soc.*, 1964, 434
64B M. Bonamico, D. Duranti, A. Vaciago, and L. Zambonelli, *Ricerca sci.*, 1964, *A*, **7**, 613
64M L. Malatesta, L. Naldini and F. Cariati, *J. Chem. Soc.*, 1964, 961
64S E. H. Schubert, dissertation, Technische Hochschule, München, 1964
64T E. R. Tucci and B. H. Gwynn, *J. Amer. Chem. Soc.*, **1964**, **86**, 4838
64W A. Wojcicki and M. F. Farona, *J. Inorg. Nuclear Chem.*, **1964**, **26**, 2291
65A U. Anders and W. A. G. Graham, *Chem. Comm.*, 1965, 499
65B F. Bonati, S. Cenini, D. Morelli and R. Ugo, *Inorg. Nuclear Chem. Letters*, 1965, **1**, 107
65D L. F. Dahl and J. F. Blount, *Inorg. Chem.*, 1965, **4**, 1373
65G M. C. Ganorkar and M. H. B. Stiddard, *J. Chem. Soc.*, 1965, 3494
65H W. Hieber, F. Lux and C. Herget, *Z. Naturforsch.*, 1965, **20b**, 1159
65L D. N. Lawson and G. Wilkinson, *J. Chem. Soc.*, 1965, 1900
65V L. M. Vallarino, *Inorg. Chem.*, 1965, **4**, 161
66A E. W. Abel, I. S. Butler, M. C. Ganorkar, C. R. Jenkins, and M. H. B. Stiddard, *Inorg. Chem.*, 1966, **5**, 25
66Aa U. Anders and W. A. G. Graham, *Chem. Comm.*, 1966, 291
66B F. Bonati, S. Cenini, D. Morelli and R. Ugo, *J. Chem. Soc. (A)*, 1966, 1052
66Ba W. Beck and H. S. Smedal, *Angew. Chem.*, 1966, 78, 267
66D R. J. Doedens and L. F. Dahl, *J. Amer. Chem. Soc.*, 1966, **88**, 4847
66H L. B. Handy, P. H. Treichel, L. F. Dahl and R. G. Hayter, *J. Amer. Chem. Soc.*, 1966, **88**, 366
66K J. V. Kingston and G. Wilkinson, *J. Inorg. Nuclear Chem.*, 1966, **28**, 2709
66S T. A. Stephenson and G. Wilkinson, *J. Inorg. Nuclear Chem.*, 1966, **28**, 945
67A U. Anders and W. A. G. Graham, *J. Amer. Chem. Soc.*, 1967, **89**, 539
67B H. Behrens, E. Ruyter, and E. Lindner, *Z. anorg. Chem.*, 1967, **349**, 251
67Ba H. Behrens and D. Herrmann, *Z. anorg. Chem.*, 1967, **351**, 225
67Bb R. Bau, B. Fontal, H. D. Kaesz, and M. R. Churchill, *J. Amer. Chem. Soc.*, 1967, **89**, 6374
67C R. Colton and R. H. Farthing, *Austral. J. Chem.*, 1967, **20**, 1283
67Ca P. Chini, *Chem. Comm.*, 1967, 29
67Cb M. R. Churchill and R. Bau, *Inorg. Chem.*, 1967, **6**, 2086
67E H. Engelmann, dissertation, Technische Hochschule, München, 1967
67Ea G. O. Evans, J. P. Hargaden and R. K. Sheline, *Chem. Comm.*, 1967, 186
67K R. B. King, *Inorg. Chem.*, 1967, **6**, 25
67L B. Lee, J. M. Burlitch and J. L. Hoard, *J. Amer. Chem. Soc.*, 1967, **89**, 6362
68B H. Behrens, E. Lindner, and P. Pässler, *Z. anorg. Chem.*, 1968, **361**, 125
68Ba J. A. Bowden and R. Colton, *Austral. J. Chem.*, 1968, **21**, 2657
68C M. R. Churchill and R. Bau, *Inorg. Chem.*, 1968, **7**, 2606
68Ca A. A. Chalmers, J. Lewis, and S. B. Wild, *J. Chem. Soc. (A)*, 1968, 1013
68Cb P. Chini, *Inorg. Chim. Acta, Rev.*, 1968, **2**, 31
68Cc M. R. Churchill, P. H. Bird, H. D. Kaesz, R. Bau and B. Fontal, *J. Amer. Chem. Soc.*, 1968, **90**, 7135
68Cd P. D. Cradwick, W. A. G. Graham, D. Hall, and D. J. Patmore, *Chem. Comm.*, 1968, 872
68I Z. Iqbal and T. C. Waddington, *J. Chem. Soc. (A)*, 1968, 2958
68K W. C. Kaska, *J. Amer. Chem. Soc.*, 1968, **90**, 6340
68L M. W. Lindauer, G. O. Evans and R. K. Sheline, *Inorg. Chem.*, 1968, **7**, 1249
68M S. Merlino and G. Montagnoli, *Acta Cryst.*, 1968, **B24**, 424
68R J. K. Ruff, *Inorg. Chem.*, 1968, **7**, 1818, 1821
68Ra J. K. Ruff and M. Lustig, *Inorg. Chem.*, 1968, **7**, 2171
68S G. M. Sheldrick and R. N. F. Simpson, *J. Chem. Soc. (A)*, 1968, 1005

Metal	Method	Temp	Medium	Log of equilibrium constant, remarks	Ref
H^+					

1964 *Errata*: p. 134, *in ref* 43H *for* K_{w12} *read* K_{p12};
p. 135, *in ref* 55Ma *replace the whole entry by*:

Metal	Method	Temp	Medium	Log of equilibrium constant, remarks	Ref
	con, gl	20—65	1 Na^+	$K_{p0}K_1^2$(M atm^{-1}) 2.16(20°), 1·96(35°), 1·81(50°), 1·59(65°)	55Ma
			0·5 Na^+	$K_{p0}K_1^2$ 2·20(20°), 2·04(35°), 1·87(50°), 1·72(65°)	
			0·1 Na^+	$K_{p0}K_1^2$ 2·30(20°), 2·11(35°), 1·97(50°), 1·79(65°)	
	sol	100—350	0	$K_{p12}[CO_2(g)+H_2O \rightleftharpoons H_2CO_3]$ $-1·99(100°)$, $-2·07(150°)$, $-2·05(200°)$, $-1·97(250°)$, $-1·84(300°)$, $-1·57(350°)$, *m* units	63E/67H
	*X*n		solid	ev $H_2L_2^{2-}$ units in $KHCO_3(s)$	64Ha
	E	250—307	MNO_3(l)	$K[CO_3^{2-} \rightleftharpoons CO_2(g, p\ atm)+O^{2-}]$ $-5·41(250°)$, $-4·25(307°)$, $\Delta H = 27·0$, $\Delta S - 27·1$ in $Na_{0.5}K_{0.5}NO_3$(l)	64K
	gl	25	→0	$\Delta V_{12} = 25·5$ cm^3 (measurements at 1000 atm)	65Da
	ΔG	25—218	0 corr	K_1 10·32(25°), 10·17(50°), 10·14(90°), 10·16(100°), 10·36(156°), 10·89(218°)	/67H
	ΔG	25—200	0 corr	K_{12} 6·35(25°), 6·31(50°), 6·45(100°), 6·77(156°), 7·08(200°)	
	ΔG	25—350	0 corr	K_{p12} $-1·465(25°)$, $-1·705(50°)$, $-1·97(100°)$, $-2·07(150°)$, $-2·06(200°)$, $-1·98(250°)$, $-1·83(300°)$, $-1·58(350°)$	
	oth	25	dil	†$K-2·84$, †K_{12} 3·53 (K_{12} 6·37), [*p*-jump method (con)]	68O
K^+	Ram		K_2L var	ev cpx	64L
Be^{2+}					

1964 *Addendum*: p. 136, *for ref* "60G" *read* "60G = 64S"

Metal	Method	Temp	Medium	Log of equilibrium constant, remarks	Ref
	var		var	ev $Be_4OL_6^{6-}$ (fp, gl, pre)	64Fb
Mg^{2+}	gl	≈22	0 corr	$\beta(Mg^{2+}HL^-)$ 0·86 [$K_a(MgHL^+)$ $-8·00$]	41G/63Ha

1964 *Erratum*: p. 136, *in ref* 61Y+/62H, *for* "(nesquehonite, stable)" *read* "(nesquehonite)" *and for* "(s, magnesite)" read "(s) (magnesite, stable)"

Metal	Method	Temp	Medium	Log of equilibrium constant, remarks	Ref
	sol	25	0 corr	†K_{ps0} $-5·20$ (magnesite)	55Yb/63S
	gl	25	0 corr	$\beta(Mg^{2+}HL^-)$ 0·95	63Ha
$Mg^{2+}Ca^{2+}$	sol	25	0 corr	$K[CaMgL_2(s)+2CO_2(g) \rightleftharpoons Mg^{2+}+Ca^{2+}+4HL^-]$ $-13·19$; $K[CaL(s)+MgL(s) \rightleftharpoons CaMgL_2(s)]$ 2·07	55Yb/63S
	ana	natural	dil	$K[2CaL(s)+Mg^{2+} \rightleftharpoons MgCaL_2(s)+Ca^{2+}]$ 0·11	63Hb
		25	0 corr	$K_s(Mg^{2+}Ca^{2+}L^{2-}{}_2)$ $-16·7$, [$K_s(Ca^{2+}L^{2-})$ $-8·29$]	
	ΔG	25	0	$K[CaL(s)+MgL(s) \rightleftharpoons CaMgL_2(s)]$ 1·98, $\Delta H = -2·94$	63S+55G
	ana	10—25	0 corr	$K[2CaL(s)+Mg^{2+} \rightleftharpoons MgCaL_2(s)+Ca^{2+}] \approx 0·54$ $K_s(Mg^{2+}Ca^{2+}L^{2-}{}_2) \approx -17$	64H
	sol	180	Cl^- var	$K[2CaL(s)+Mg^{2+} \rightleftharpoons MgCaL_2(s)+Ca^{2+}]$ 1·28 $K(MgCaL_2(s)+Mg^{2+} \rightleftharpoons 2MgL(s)+Ca^{2+})$ 0·37	64U/
	ana	natural	var	$K[2CaL(s)+Mg^{2+} \rightleftharpoons MgCaL_2(s)+Ca^{2+}] < -0·8$, $\geqslant -1·5$	65D/
	ΔG	25	0	$K[CaL(s)+MgL(s) \rightleftharpoons MgCaL_2(s)]$ 1·26	/65H

continued overleaf

21 CO_3^{2-} (contd)

Metal	*Method*	*Temp*	*Medium*	*Log of equilibrium constant, remarks*	*Ref*
Ca^{2+}	sol	25	0 corr	†K_{ps0} −5·92 (calcite)	55Yb/63S
	X	200—600	*p* bar	K_{s0} (aragonite) = K_{s0} (calcite) if $p = 16t+2400$ bar, (t°C)	56M
	sol	100—250	0 corr	K_{s0} −9·62(100°), −10·06(125°), −10·54(150°), −11·62(200°), −12·96(250°), also data for 0·2, 0·5, and 1·0 m NaCl, [$K(2HL^- \rightleftharpoons H_2L+L^{2-})$ −3·90 at all temp]	63Ea
	lit	0—70	0 corr	K_{s0} $-8{\cdot}260-5{\cdot}10^{-3}t-5{\cdot}10^{-5}t^2$ (t°), −8·416(25°)	/64Z
	sol, gl	28	0 corr	K_{s0}(aragonite) −8·10	66B
Ca^{2+}, Sr^{2+}	ana	90—100	NaCl var	$K[CaL(\text{solid soln})+Sr^{2+} \rightleftharpoons SrL(\text{solid soln})+Ca^{2+}]$ −0·23($Ca_{1-x}Sr_xCO_3$, aragonite, $x \approx 0$) to −0·1 ($x \approx 0{\cdot}33$)	63Hc
	ana	25	0 corr	$K[CaL(\text{solid soln})+Sr^{2+} \rightleftharpoons SrL(\text{solid soln})+Ca^{2+}]$ 0·00 ($Ca_{1-x}Sr_xCO_3$, aragonite), −0·85 ($Ca_{1-x}Sr_xCO_3$, calcite)	64H
	ana	20—90	var	$K[CaL(\text{solid soln}, 1-x)+Sr^{2+} \rightleftharpoons SrL(\text{solid soln}, x)+Ca^{2+}]$ −0·72 (20°, calcite), −0·12(90°, aragonite)	66K
Sr^{2+}	sol	25—40	0 corr	K_{s0} −9·28(25°), −9·29(40°)	37T/39H
	con	25	dil	K_{s0} −8·57	46K
Ba^{2+}	sol	25—40	0 corr	K_{s0} −8·69(25°), −8·56(40°)	37T/39H
	sol	25	0 corr	K_{s0} −9·40 [$K_1(H^+)K_w$ −3·75]	68Ba
Sc^{3+}	fp, gl		var	ev mainly ScL_4^{5-} in solution	64Fa
	aix	20	K_2L var	ev only ScL_4^{5-} in solution	67F
Y^{3+}	sol, gl	25	0 corr	$K_{s0}(Y^{3+}{}_2L^{2-}{}_3)$ −30·6(H^+: K_1 10·25, K_{12} 6·34)	66J
	aix	20	K_2L var	ev only YL_4^{5-} in solution	67Fa
	sol		K_2L var	ev YL_3^{3-}, YL_4^{5-}	67Fc
	sp		var	ev $YL_nF^{(2n-2)-}$	
Group $4f^{3+}$	aix		K_2L var	ev ML_4^{5-}(Ce?, Pr, Nd, Pm, Sm, Eu)	63Sa
	sol, gl	25	0 corr	$K_{s0}(M^{3+}{}_2L^{2-}{}_3)$ −33·4(La), −33·0(Nd), −32·5(Sm), −32·2(Gd), −31·5(Dy), −31·1(Yb), (H^+: K_1 10·25, K_{12} 6·34)	66J
	aix	20	K_2L var	ev only ML_4^{5-} in solution (Sm, Eu, Gd, Tb, Dy, Ho, Er, Tm, Yb, Lu)	67Fa
	sol	20	4·2 or 6 Na^+	$K_s[Na_3ML_3(s)+L^{2-} \rightleftharpoons 3Na^+ + ML_4^{5-}]$ −0·73(M = Lu), −1·07(Yb), −0·82(Tm)	67Fb/
Pr^{3+}	sol	25	K_2L var	K_{s0} −27·0, β_4 11·20, solid = $Pr_2L_3(H_2O)_3$	64F
	sp	25	7 (KCl)	ev PrL_3F^{4-}, $PrL_3F_2^{5-}$, $PrLF_3^{2-}$	
Nd^{3+}	sol	25	K_2L var	K_{s0} −26·75, β_4 11·17, solid = $Nd_2L_3(H_2O)_3$	64F
	sp	25	7 (KCl)	$K(NdL_4^{5-}+F^- \rightleftharpoons NdL_3F^{4-}+L^{2-})$ −0·36 $K(NdL_4^{5-}+2F^- \rightleftharpoons NdLF_2^- + 3L^{2-})$ −0·60	
	aix	rt	K_2L var	K_3 1·89, ev NdL_4OH^{6-}	64Sa
			KHL var	HL^-: K_3 2·71, K_4 1·80, K_5K_6 2·68	
Eu^{3+}	aix	rt	K_2L var	K_3 1·94, ev EuL_4OH^{6-}	64Sa
			KHL var	K_3K_4 4·55, K_5 1·24, K_6K_7 2·00	
Er^{3+}	sol	25	K_2L var	K_{s0} −22·9, β_3 10·60, solid = $Er_2L_3(H_2O)_{7.5}$	64F
	sp	25	7 (KCl)	$K(ErL_3^{3-}+2F^- \rightleftharpoons ErL_2F_2^{3-}+L^{2-})$ 0·89	
	fp		KNO_3 sat	ev ErL_4^{5-}	65Fa
Tm^{3+}	fp		KNO_3 sat	ev TmL_4^{5-}	65Fa
Yb^{3+}	fp		KNO_3 sat	ev YbL_4^{5-}	65Fa

contin

Metal	Method	Temp	Medium	Log of equilibrium constant, remarks	Ref
Zr^{4+}	var	20—25	var	ev ani cpx, 2L/Zr, probably $ZrOL_2^{2-}$ (sol, con, gl)	63L
	var	18	var	ev $ZrOL_2^{2-}$ (sol, gl, con, aix, pre)	65L
Th^{4+}	ir		solid	ev ThL_4^{4-} and L^{2-} in $Na_6ThL_5(H_2O)_{12}(s)$	62K
Cr^{2+}	sp		KHL var	ev yellow dinuclear cpx, $Cr_2L_2^0$?	67C
U^{4+}	sol	60	var	ev $UL_4(OH)_2^{6-}$	63B
NpO_2^+	var		var	ev $NpO_2L_3^{5-}$ in solids (pre, ir), ev cpx ($NpO_2L_2^{3-}$?) in soln (sol, sp)	66Ga
NpO_2^{2+}	var		var	ev $NpO_2L_3^{4-}$ in solid (pre, ir), ev cpx in solution (sol, sp)	66G
PuO_2^+	sol, tp		L^{2-} var	ev PuO_2L^-, PuO_2LOH^{2-}, no ev cpx with > 1L/Pu	68Z
Am^{3+}, Cm^{3+}	aix		Na_2L var	ev ML_2^-, ML_3^{3-}, ML_4^{5-}, ML_4OH^{6-}, (M = Am or Cm)	67Fe
Co^{2+}	sp, pre		var	ev $Co_2(CN)_{10}CO_2^{6-}$	66S
	sol, gl	25	0 corr	K_{s0} −9·98 (H:K_1 10·33, K_{12} 6·35, K_{p12} −1·46)	67B/68B
$Co(NH_3)_5^{3+}$	*X*		solid	ev $Co(NH_3)_5L^+$ in $Co(NH_3)_5CO_3BrH_2O(s)$	65F
	kin	25	1 ($NaClO_4$)	$M^{3+} = Co(NH_3)_5^{3+}$, $\beta(H^+ML^+)$ 8·23	67Fd
	sp	25	1 $(NaClO)_4$	$K(MOH^{2+} + HL^- \rightleftharpoons MHL^{2+} + OH^-)$ −3·53 [$K_1(H^+)$ 9·56, K_w −13·77]	
	kin, gl	25	0?	$K_a[(NH_3)_5CoLH^{2+}]$ −8·22	67J
	X		solid	ev $(NH_3)_5CoOCO_2^+ = Q^+$ in $QIH_2O(s)$	67N
	gl?	15—25	0·5 ($NaClO_4$)	$\beta(H^+ML^+)$ 6·41(kin), 6·7?(gl)	68D
Co^{3+}cpx	gl, kin	20	dil	K_a(*trans*-$en_2CoOH_2L^+$) −7·2	65S
				$K(en_2CoOH_2L^+$, *trans* ⇌ *cis*) 1·23	
				$K(en_2CoOHL$, *trans* ⇌ *cis*) −0·32	
				K_a(*cis*-$en_2CoOH_2LH^{2+}$) −5·32	
				K_a(*cis*-$en_2CoOH_2L^+$) −8·75	
		20	dil	(*cis*-en_2Co = Q), $K_a(QOH_2LH^{2+} \rightleftharpoons QOHLH^+ + H^+)$ −5·32, $K_a(QOHLH^+ \rightleftharpoons QOHL + H^+)$ −8·75	65S/67J
Ni^{2+}	aix, sol	20	KHL var	ev ani cpx	63T
Cu^{2+}	sol	25	0 corr	†K_{ps0}(malachite) −6·75, K_1 6·73, β_2 9·83	57Sc/68S
	X		solid	ev $CuL(NH_3)_2$ units in $Cu(NH_3)_2CO_3(s)$	63Hd
	var	25	*I*($NaClO_4$)	†$K_{ps0}[\frac{1}{2}Cu_2(OH)_2L$ (s, malachite) $+ 2H^+ \rightleftharpoons Cu^{2+} + \frac{1}{2}CO_2(g) + 1\frac{1}{2}H_2O]$ −6·74(*I* = 0·2), −6·49 (0 corr), $K_s(Cu^{2+}{}_2OH^-{}_2L^{2-})$ −33·16(0 corr); †$K_{ps0}[\frac{1}{3}Cu_3(OH)_2L_2$ (s, azurite) $+ 2H^+ \rightleftharpoons Cu^{2+} + \frac{2}{3}CO_2(g) + \frac{4}{3}H_2O]$ −6·72(*I* = 0·2), −6·47(0 corr), $K_s(Cu^{2+}{}_3OH^-{}_2L^{2-}{}_2)$ −44·88(0 corr) (sol, gl, pCO_2)	68S
Zn^{2+}	sol	20	KHL, K_2L var	ev ani cpx	63T
Cd^{2+}	var	25	3 (Na)ClO_4	*$K_{ps0}[CdCO_3(s) + 2H^+ \rightleftharpoons Cd^{2+} + H_2O + CO_2(g)]$ 6·47, K_{s0} −11·18 (sol, pCO_2, gl)	65G+
			0 corr	*K_{ps0} 6·14, K_{s0} −12·00	
				$K[CdCO_3(s) + H_2O \rightleftharpoons Cd(OH)_2(s) + CO_2(g)]$ −7·50	
Pb^{2+}	sol	20	KHL, K_2L var	ev ani cpx	63T
	pol	20	NaHL var	$\beta(Pb^{2+}HL^-{}_2)$ 4·77, $\beta(Pb^{2+}HL^-{}_3)$ 5·19	65B
	pol	18	1 ($NaNO_3$)	$\beta[Pb^{2+}(HL^-)_3]$ 5·19, $\beta[Pb(HL)_2{}^0HL^-]$ 0·42	67Ba
	sol	250—300	0 corr	K_1 11·89 (250°), 12·21(300°)	68Bb+65B
	sol	25	1 ($NaClO_4$)	$\beta_2(Pb^{2+}L^{2-}{}_2)$ 9·09(K_{s0} −13·13)	68Bb

references overleaf

21 CO_3^{2-} (contd)

39H E. Hogge and H. L. Johnston, *J. Amer. Chem. Soc.*, 1939, **61**, 2154
46K A. F. Kapustinsky and I. P. Dezideryeva, *Trans. Faraday Soc.*, 1946, **42**, 69
55G D. L. Graf and J. R. Goldsmith, *Geochim. Cosmochim. Acta*, 1955, **7**, 109
56M G. J. F. MacDonald, *Amer. Min.*, 1956, **41**, 744
62K A. V. Karyakin and M. P. Volynets, *Zhur. strukt Khim.*, 1962, **3**, 714

1964 *Addendum*
62S add "Theory and structure of complex compounds" (symposium papers, Wrocław, 1962), Pergamon, Warszawa, 1964, p. 339.

63B M. Branica, V. Pravdić, and Z. Pučar, *Croat. Chem. Acta*, 1963, **35**, 281
63E A. J. Ellis and R. M. Golding, *Amer. J. Sci.*, 1963, **261**, 47
63Ea E. J. Ellis, *Amer. J. Sci.*, 1963, **261**, 259
63Ha P. B. Hostetler, *J. Phys. Chem.*, 1963, **67**, 720
63Hb K. J. Hsu, *J. Hydrology*, 1963, **1**, 288
63Hc H. D. Holland, M. Borcsik, J. Munoz, and U. M. Oxburgh, *Geochim. Cosmochim. Acta*, **1963**, **27**, 957
63Hd F. Hanic, *Chem. Zvesti*, 1963, **17**, 365
63L T. F. Limar' and K. P. Shatskaya, *Zhur. neorg. Khim.*, 1963, **8**, 2483
63S J. W. Stout and R. A. Robie, *J. Phys. Chem.*, 1963, **67**, 2248
63Sa H. S. Sherry and J. A. Marinsky, *Inorg. Chem.*, 1963, **2**, 957
63T T. Taketatsu, *Bull. Chem. Soc. Japan*, 1963, **36**, 549
64F Ya. D. Fridman and N. V. Dolgashova, *Zhur. neorg. Khim.*, 1964, **9**, 623
64Fa F. Fromage and J. Faucherre, *Compt. rend.*, 1964, **259**, 4043
64Fb J. Faucherre and F. Fromage, *Bull. Soc. chim. France*, 1964, 1244
64H H. D. Holland, T. V. Kirsipu, J. S. Huebner, and U. M. Oxburgh, *J. Geol.*, 1964, **62**, 36
64Ha P. Herpin and P. Meriel, *J. Physique*, 1964, **25**, 484
64K R. N. Kust, *Inorg. Chem.*, 1964, **3**, 1035
64L H. Lee and J. K. Wilmshurst, *Austral. J. Chem.*, 1964, **17**, 943
64S A. K. Sengupta, *J. Inorg. Nuclear Chem.*, 1964, **26**, 1823
64Sa H. S. Sherry and J. A. Marinsky, *Inorg. Chem.*, 1964, **3**, 330
64U H.-E. Usdowski, *Naturwiss.*, 1964, **51**, 357
64Z S. S. Zavodnov, *Zhur. Vsesoyuz. khim. Obshch. im. D.I. Mendeleeva*, 1964, **9**, 472
65B N. N. Baranova and V. L. Barsukov, *Geokhimiya*, 1965, 1093
65D K. S. Deffeyes, F. J. Lucia, and P. K. Weyl, "Dolomitization and limestone diagenesis", a symposium (Soc. Econom. Paleontologists Mineralogists, spec. publ. no. 13), 1965, p. 71
65Da A Distèche and S. Distèche, *J. Electrochem. Soc.*, 1965, **112**, 350
65F H. C. Freeman and G. Robinson, *J. Chem. Soc.*, 1965, 3194
65Fa F. Fromage and J. Faucherre, *Compt. rend.*, 1965, **260**, 572
65G H. Gamsjäger, H. U. Stuber and P. Schindler, *Helv. Chim. Acta*, 1965, **48**, 723
65H F. Halla, *J. Phys. Chem.*, 1965, **69**, 1065
65L T. F. Limar' and K. P. Shatskaya, *Zhur. neorg. Khim.*, 1965, **10**, 115
65S H. Scheidegger and G. Schwarzenbach, *Chimia*, 1965, **19**, 166
66B W. S. Broecker and T. Takahashi, *J. Geophys. Res.*, 1966, **71**, 1575
66G D. S. Gorbenko-Germanov and V. D. Klimov, *Zhur. neorg. Khim.*, 1966, **11**, 516
66Ga D. S. Gorbenko-Germanov and R. A. Zenkova, *Zhur. neorg. Khim.*, 1966, **11**, 520
66J N. Jordanov and I. Havezov, *Z. anorg. Chem.*, 1966, **347**, 101
66K D. Klötzer and H. W. Levi, *Radiochim. Acta*, 1966, **6**, 81
66S T. Suzuki and T. Kwan, *J. Chem. Soc. Japan*, 1966, **87**, 342
67B O. Butkevitsch, *Suomen Kem.*, 1967, **40**, *B*, 148
67Ba N. N. Baranova, *Zhur. neorg. Khim.*, 1967, **12**, 1438
67C R. D. Cannon and J. L. Tsay, *Nature*, 1967, **216**, 681
67F F. Fromage, *Compt. rend.*, 1967, **264**, *C*, 2052
67Fa F. Fromage, *Compt. rend.*, 1967, **265**, *C*, 30
67Fb F. Fromage and A. Morgant, *Bull. Soc. chim. France*, 1967, 2611
67Fc Ya. D. Fridman and S. D. Gorokhov, *Zhur. neorg. Khim.*, 1967, **12**, 1796
67Fd D. J. Francis and R. B. Jordan, *J. Amer. Chem. Soc.*, 1967, **89**, 5591
67Fe D. Fang, dissertation Technische Hochschule, Karlsruhe, 1967
67H H. C. Helgeson, *J. Phys. Chem.*, 1967, **71**, 3121
67J R. B. Jordan and D. J. Francis, *Inorg. Chem.* 1967, **6**, 1605
67N H. Nakai, Y. Kushi, and H. Kuroya, *J. Chem. Soc. Japan*, 1967, **88**, 1126
68B O. Butkevitsch, personal communication (factor 2 left out in eqn (14), ref 67B)
68Ba D. P. Baccanari, B. A. Buckman, M. M. Yevitz, and H. A. Swain, jun., *Talanta*, **1968**, **15**, 416
68Bb N. N. Baranova, *Geokhimiya*, 1968, 17
68D T. P. Dasgupta and G. M. Harris, *J. Amer. Chem. Soc.*, 1968, **90**, 6360
68O J. Osugi, M. Sato, and T. Fujii, *J. Chem. Soc. Japan*, 1968, **89**, 562
68S P. Schindler, M. Reinert, and H. Gamsjäger, *Helv. Chim. Acta*, 1968, **51**, 1845
68Z V. P. Zaitseva, D. P. Alekseeva, and A. D. Gel'man, *Radiokhimiya*, 1968, **10**, 669

Metal	Method	Temp	Medium	Log of equilibrium constant, remarks	Ref
				Boranocarbonate, $H_3BCO_2^{2-}$	
H^+	gl	25	dil	K_1 10·2, $K_{12} \approx 8$	67M
				1,12-$B_{12}H_{10}(COO)_2^{4-}$	
H^+	gl, cal	25	0 corr	K_1 10·24, K_{12} 9·07, $\Delta H_1 = -2{\cdot}30$, $\Delta H_{\beta 12} = -2{\cdot}15$, $\Delta S_1 = 39{\cdot}1$, $\Delta S_{\beta 12} = 34{\cdot}4$	66H
				Carbamide $CO(NH_2)_2$	
H^+	var	25	0 corr	$K(2L \rightleftharpoons L_2)$ 0·26 or better $K(2LH_2O \rightleftharpoons L_2 + 2H_2O)$ 0·51 (act, viscosity), *x* units	67E+
	cal, act	2—40	0 corr	$K(2L \rightleftharpoons L_2)$ −1·19(2°), −1·22(10°), −1·26(20°), −1·28(25°), −1·29(30°), −1·32(40°), M units	67S+
Cu^{2+}	red	25 ?	$I(NaClO_4)$	K_1 −0·49(I = 2), −0·55(I = 0·5), −0·72(I = 0·2)	65S
				Dithiocarbamate, $H_2NCS_2^-$	
Ni^{2+}	*X*		solid	ev NiL_2 in $Ni(S_2CNH_2)_2$(s)	67G
				Thiourea, $CS(NH_2)_2$	
H^+	var	25	0 corr	$K(2L \rightleftharpoons L_2)$ 0·92	67E+
Fe^{3+}	sp	20	1·2 ($NaClO_4$)	β_2 8·4, $\beta(Fe^{3+}L^0SO_4^{2-})$ 6·6 for 0·2M-H^+	68M
	sp	20	1·2 ($NaClO_4$)	$\beta(Fe^{3+}SCN^-L)$ 5·55	68Ma
Co^{2+}	pre, sp		var	ev 5-co-ordinated CoL_3X_2, X = Cl, Br, I	66D
	sp	22—25	95% EtOH	β_2 1·7, β_6 4·5 in 95% (vol) EtOH, I = 0·1($NaClO_4$)	66S
	red	25	90% Me_2CO	K_1 1·05, K_2 0·7 in 90% Me_2CO, I = 2($NaClO_4$)	66Sb
Ni^{2+}	sp	18—22	95% EtOH	β_2 0·7, β_6 1·36 in 95% (vol) EtOH, I = 0·3($NaClO_4$)	66S
		20—23		K_1 −1·4, β_2 0·8 in 95% (vol) EtOH, Cl^- var	
	red	25	90% Me_2CO	K_1 1·05, K_2 0·45 in 90% Me_2CO, I = 2($NaClO_4$)	66Sb
Pd^{2+}	sp		1 or 0·2	$K(PdX_2L_2 + L \rightleftharpoons PdXL_3^+ + X^-)$ 4·86(X = Cl), 4·65(Br), 2·95(SCN); $K(PdXL_3^+ + L \rightleftharpoons PdL_4^{2+} + X^-)$ 4·24(Cl), 4·18(Br), 2·52(SCN)	66Sa
Ag^+	Ag	25	*w*% ROH	R = Me, β_3 14·10(*w* = 25), 14·52(*w* = 50), 15·18(*w* = 75), 15·61(*w* = 85); R = Et, β_3 14·22(*w* = 25), 14·62(*w* = 50), 15·37(*w* = 75), 15·66(*w* = 85); R = Pr, β_3 14·13(*w* = 25), 14·62(*w* = 50), 15·17(*w* = 75), 15·57(*w* = 85); in *w*% ROH, 0·4M($LiNO_3$)	67N
		25	0·4 ($LiNO_3$)	β_3 13·48	
Zn^{2+}	red	25	2 ($NaClO_4$)	K_1 0·5, K_2 0·25, K_3 0·15, also values for Me_2CO–H_2O mixt.	66Sb
Cd^{2+}	red	25	2 ($NaClO_4$)	K_1 1·6, K_2 1·1, K_3 1·0, K_4 0·6, also values for Me_2CO–H_2O mixt.	66Sb

continued overleaf

22 Other ligands with C (contd)

Metal	Method	Temp	Medium	Log of equilibrium, constant remarks	Ref
$Hg(CN)_2$	cal	25	0 corr	K_1 1·97, K_2 0·58, $\Delta H_1 = -1{\cdot}5, \Delta H_2 = -7{\cdot}9, \Delta S_1 = 4{\cdot}0, \Delta S_2 = -23{\cdot}8$; also data for H_2O–EtOH mixt., *m* units	68I
Bi^{3+}	sp		0·98 $HClO_4$	β_6 9·3	66S
	sp	rt	3 (Na, $HClO_4$)	K_1 2·40, K_2 1·15, K_3 0·32	67V
			2 (Na, $HClO_4$)	K_1 2·35, K_2 1·05, K_3 0·30	
			1 ($HClO_4$)	K_1 2·28, K_2 1·04, K_3 0·32	
			0 corr	K_1 2·24, K_2 0·96, K_3 0·30	
	sol	25	0 corr	K_1 2·1, $\beta(Bi^{3+}Cl^-L_n)$ 5·5($n = 1$), 7·6($n = 2$), 7·8($n = 3$), also values for $I = 1$ and 2 ($HClO_4$)	68V
Te^{2+}	X		solid	ev TeL_4^{2+} in $TeL_4Cl_2(H_2O)_2$(s)	65F
	X		solid	ev $Te_2L_6^{4+}$ in $TeL_3(HF_2)_2$(s)	65Fa
	sp		2HCl	β_2 1·7	65T
	X		solid	ev TeL_2Cl_2 and TeL_2Br_2 units in crystals	66F
				Carbon disulphide CS_2	
Co^{2+}	var		var	ev $(CN)_5CoSCSCo(CN)_5^{6-}$ (pre, sp, ir, esr)	67Mb
				Trithiocarbonate, CS_3^{2-}	
Ni^{2+}	*X*		solid	ev $Ni(S_2CS)_2^{2-}$ in $(Ph_4As)_2Ni(CS_3)_2$(s)	67Ma
				Perthiocarbonate (dithiodisulphidocarbonate), CS_4^{2-}	
H^+	gl	0—25	→0	K_1 7·87(0°), 7·54(10°), 7·41(15°), 7·32(20°), 7·24(25°), $\Delta H_1 = -10, \Delta S_1 = 0{\cdot}6(20°)$	66G
	con	2	→0	K_{12} 3·54	
				Triselenocarbonate CSe_3^{2-}	
H^+	con, gl	0—25	0 corr	K_1 7·77(0°), 7·50(10°), 7·41(15°), 7·28(20°), 7·13(25°, gl), K_{12} 1·16(0°, con)	67Ga

65F K. Fosheim, O. Foss, A. Scheie, and S. Solheimsnes, *Acta Chem. Scand.*, 1965, **19**, 2336
65Fa O. Foss and S. Hauge, *Acta Chem. Scand.*, 1965, **19**, 2395
65S V. M. Shul'man and T. V. Kramareva, *Zhur. neorg. Khim.*, 1965, **10**, 1632
65T V. M. Tarayan and A. A. Sarkisyan, *Zhur. neorg. Khim.*, 1965, **10**, 2684
66D K. C. Dash and D. V. R. Rao, *Z. anorg. Chem.*, 1966, **345**, 217
66F O. Foss, K. Johnsen, K. Maartmann-Moe, and K. Marøy, *Acta Chem. Scand.*, 1966, **20**, 113
66G G. Gattow and J. Wortmann, *Z. anorg. Chem.*, 1966, **345**, 172
66H L. D. Hansen, J. A. Partridge, R. M. Izatt, and J. J. Christensen, *Inorg. Chem.*, 1966, **5**, 569
66S K. Swaminathan and H. M. N. H. Irving, *J. Inorg. Nuclear Chem.*, 1966, **28**, 171
66Sa V. I. Shlenskaya, A. A. Biryukov, and E. M. Moskovkina, *Zhur. neorg. Khim.*, 1966, **11**, 600
66Sb V. M. Shul'man, S. V. Larionov, T. V. Kramareva, E. I. Arykova, and V. V. Yudina, *Zhur. neorg. Khim.*, 1966, **11**, 1076
67E H. D. Ellerton and P. J. Dunlop, *Austral. J. Chem.*, 1967, **20**, 2263
67G G. F. Gasparri, M. Nardelli, and A. Villa, *Acta Cryst.*, 1967, **23**, 384
67Ga G. Gattow and M. Dräger, *Z. anorg. Chem.*, 1967, **349**, 202
67M L. J. Malone and R. W. Parry, *Inorg. Chem.*, 1967, **6**, 817
67Ma J. S. McKechnie, S. L. Miesel, and I. C. Paul, *Chem. Comm.*, 1967, 152
67Mb T. Mizuta, T. Suzuki, and T. Kwan, *J. Chem. Soc. Japan*, 1967, **88**, 573
67N L. V. Nazarova and V. I. Prizhilevskaya, *Zhur. neorg. Khim.*, 1967, **12**, 3051
67S R. H. Stokes, *Austral. J. Chem.*, 1967, **20**, 2087
67V V. P. Vasil'ev and N. K. Grechina, *Zhur. neorg. Khim.*, 1967, **12**, 1565
68I R. M. Izatt, D. Eatough, and J. J. Christensen, *J. Phys. Chem.*, 1968, **72**, 2720
68M J. Masłowska, *Roczniki Chem.*, 1968, **42**, 1191
68Ma J. Masłowska, *Roczniki Chem.*, 1968, **42**, 1819
68V V. P. Vasil'ev and N. K. Grechina, *Izvest. V.U.Z., Khim.*, 1968, **11**, 1319

Silicate, $L^{2} = SiO_2(OH)_2^{2-}$ = "SiO_3^{2-}"

Metal	*Method*	*Temp*	*Medium*	*Log of equilibrium constant, remarks*	*Ref*
H^+	H	25	3 Na(ClO_4)?	$K_{12}K_w$ −4·72, K_1K_w −1·23 $*\beta_{44}K_w^{-4}$ 23·75 + series of larger cpx	59La/65A
	X		solid	ev $Si_4O_8(OH)_4^{4-}$ (= $H_4L_4^{4-}$) in "$KHSiO_3$(s)"	64H
	oth		Na, L var	ev species with 1,2,3,4 and more Si (reaction with Me_3SiCl)	64L
	cfu	25	*I* NaCl	ev series of species with up to at least 20Si (I = 0·5—2)	65A
	act	1200— 1500	melt	$K[SiO_4^{4-} + Si_{n-1}O_{3n-2}^{2n-} \rightleftharpoons O^{2-} + Si_nO_{3n+1}^{(2n+2)-}]$ 0·15(M = Fe, 1200−1400°), −0·12(M = Mn, 1300—1500°), ≈ −2·5(M = Ca, only n = 2 important), in MO–SiO_2 melts around composition M_2SiO_4	/65M
	kin	30	H_2SO_4 var	ev cation, $K_{13}(H^+ + H_2L \rightleftharpoons H_3L^+) \approx -3$	65T
	X		solid	ev $SiO_2(OH)_2^{2-}$ in $Na_2H_2SiO_4(H_2O)_8$(s)	66J
	X		solid	ev $SiO_2(OH)_2^{2-}$ in $Na_2H_2SiO_4(H_2O)_4$(s)	66Ja
	X		solid	ev $Si_8O_{20}^{8-}$ units in ekanite, KTh(Na, Ca)$_2$ Si_8O_{20}(s)	66M
	H	25	0·5 Na(ClO_4)	K_{12} 9·46	67B
	con	25—250	0 corr	K_1 11·81(25°), 10·11(50°), 9·24(100°), 9·10(150°), 9·48(200°), 9·81(250°); K_{12} 8·20(100°, in table 9·20), 8·46(150°), 8·59(200°), 8·69(250°) [K_w 14·00(25°), 13·26(50°), 12·25(100°), 11·72(150°), 11·50(200°), 11·51(250°)]	67R
	ir		var	ev mononuclear cpx in 1M-Na_4SiO_4, ev polyn cpx in 1M-$NaHSiO_3$	68I
	sol	25	1 $NaClO_4$	K_{s12} −2·89 (amorphous SiO_2)	68J
	X		solid	ev linear $Si_3O_{10}^{8-}$ in $Na_4Cd_2Si_3O_{10}$(s) and $Na_2Cd_3Si_3O_{10}$(s)	68S
Li^+	fp	≈ 1700	SiO_2(l)	ev units with 6 Li in SiO_2(l)	/66E
Na^+	fp	≈ 1700	SiO_2(l)	ev units with 4 Na in SiO_2(l)	/66E
Rb^+	fp	≈ 1700	SiO_2(l)	ev units with 2 Rb in SiO_2(l)	/66E
Mg^{2+}	cal, ΔG	25—627	0	K[2Mg_2SiO_4(s, forsterite) + 3H_2O(l) ⇌ $Mg(OH)_2$ (s, brucite) + $Mg_3Si_2O_5$(chrysotile)] 9·5(25°), 6·2(100°), −1·5(227°), −7·7(427°), −10·9(627°) at 1 bar	67K
Ca^{2+}	sol, gl	25	0 corr	$K_s(Ca^{2+}L^{2-})$ −7·0 to −7·3, no certain ev for $Ca(HL)_2$(s) (H^+: K_1 11·77, K_{12} 9·7)	65G
Ba^{2+}	fp	≈ 1700	SiO_2(l)	ev units with 4 Ba in SiO_2(l)	/66E
Fe^{3+}	sp	25	0·1 (ClO_4^-)	$K(Fe^{3+} + H_2L \rightleftharpoons H^+ + FeHL^{2+})$ −0·24, $\beta(Fe^{3+}HL^-)$ 9·26 [$K_{12}(H^+)$ 9·5]	65W
Al^{3+}	gl	400	dil	K[1½$NaAlSi_3O_8$ (s, Na-feldspar, albite) + H^+ ⇌ ½$NaAl_3Si_3O_{10}(OH)_2$ (s, Na-mica, paragonite) + 3SiO_2(s) + Na^+] 3·0 K(paragonite + H^+ + 3SiO_2(s) ⇌ 1½$Al_2Si_4O_{10}(OH)_2$ (pyrophyllite, s) + Na^+) 2·25 At 300°C and lower, paragonite is replaced by Na-montmorillonite, $Na_{0.33}Al_{2.33}Si_{3.67}O_{10}(OH)_2$? and pyrophyllite by kaolinite, $Al_2Si_2O_5(OH)_4$	61H
		25		K[3 Na-montmorillonite + H^+ + 3½H_2O(l) ⇌ 3½ kaolinite + 4SiO_2(s) + Na^+] 7·0	61H/65H
Ge^{4+}	ir		solid	ev $SiGeO_7^{2-}$ and $Si_2O_7^{2-}$ in solid solutions $Sc_2(Si, Ge)_2O_7$(s)	68L

references overleaf

23 $L^{2-} = SiO_2(OH)_2^{2-}$ = "SiO_3^{2-}" (contd)

61H J. J. Hemley, C. Meyer, and D. H. Richter, *U.S. Geological Survey, Prof. Paper*, 1961 **424D**, 338
64H W. Hilmer, *Acta Cryst.*, 1964, **17**, 1063; correction, *ibid.* 1965, **18**, 574; *Naturwiss.*, 1963, **50**, 662
64L C. W. Lentz, *Inorg. Chem.*, 1964, **3**, 574
65A J. Aveston, *J. Chem. Soc.*, 1965, 4444
65G S. A. Greenberg and T. N. Chang, *J. Phys. Chem.*, 1965, **69**, 182
65H H. D. Holland, *Proc. Nat. Acad. Sci. U.S.A.*, 1965, **53**, 1173
65M C. R. Mason, *Proc. Roy. Soc.*, 1965, **287**, *A*, 201
65T A.-P. Tai, Y.-S. Chen, and T. Chu, *Scientia Sinica*, 1965, **14**, 1425
65W W. J. Weber, jun., and W. Stumm, *J. Inorg. Nuclear Chem.* 1965, **27**, 237
66E D. A. Everest and E. Napier, *J. Inorg. Nuclear Chem.*, 1966, **28**, 1813
66J P. B. Jamieson and L. S. D. Glasser, *Acta Cryst.*, 1966, **20**, 688
66Ja K.-H. Jost and W. Hilmer, *Acta Cryst.*, 1966, **21**, 583
66M V. I. Mokeeva and N. I. Golovastikov, *Doklady Akad. Nauk S.S.S.R.*, 1966, **167**, 1131
67B H. Bilinski and N. Ingri, *Acta Chem. Scand.*, 1967, **21**, 2503
67K E. G. King, R. Barany, W. W. Weller, and L. B. Pankratz, *U.S. Bureau Mines, Rept. Invest. no.* 6962, 1967
67R B. N. Ryzhenko, *Geokhimiya*, 1967, 161
68I I. Iwasaki and K. Ichikawa (reverse order in English text), *J. Chem. Soc. Japan*, 1968, **89**, 1217
68J S. Storgaard Jørgensen, *Acta Chem. Scand.*, 1968, **22**, 335
68L A. N. Lazarev, T. F. Tenisheva, A. N. Sokolov, A. P. Mirgorodskii, and N. A. Toropov, *Doklady Akad. Nauk S.S.S.R.*, 1968, **183**, 352
68S M. A. Simonov, Yu. K. Egorov-Tismenko, and N. V. Belov, *Doklady Akad. Nauk S.S.S.R.*, 1968, **179**, 1329, and **181**, 92

24 Germanate, $L^{2-} = GeO_2(OH)_2^{2-}$ = "GeO_3^{2-}"

Metal	*Method*	*Temp*	*Medium*	*Log of equilibrium constant, remarks*	*Ref*
H^+	*X*		solid	ev $GeO_2(OH)_2^{2-}$ in $SrH_2GeO_4(s)$	52N
	sol	11—100	dil	$\Delta H_{s2}[GeO_2(s, hex)+2H_2O \rightleftharpoons Ge(OH)_4] = 2{\cdot}54$	63E+31S/65S
	sol, gl	25	0 corr	$K_s[GeO_2(s)+OH^- \rightleftharpoons HGeO_3^-]$ 0·64 K_{s12} −1·80, K_{12} 11·56 (?)	64G
	sol	0—100	dil	$\Delta H_{s2}[GeO_2(s, hex)+2H_2O \rightleftharpoons Ge(OH)_4] = 3{\cdot}6$	64V
	dis	25	8 (Li)Cl	K_{13} 0·4, K_{14} −0·26, K_{15} −0·51, K_{16} ($GeOH^{3+}+H^+ \rightleftharpoons Ge^{4+}$) −0·88 (?, "$Ge^{4+}$" is probably $GeCl_4$, $GeCl_6^{2-}$, *etc.*)	66A
	X		solid	ev $GeO_2(OH)_2^{2-}$ in $Na_2H_2GeO_4(H_2O)_5(s)$	67J
	oth		var	ev cpx with ≈ 8, ≈ 5 and 1 Ge (dialysis method)	68B
	sp	25	1 (K)NO_3	K_{13} ($H_2L+H^+ \rightleftharpoons H_3L^+$) 0·00, K_{14} −0·20, K_{15} −0·45, K_{16} −0·82	68N
			0·1 (K) NO_3	K_{13} 0·70, K_{14} 0·47, K_{15} 0·19, K_{16} −0·18, also values for I = 0·3 and 0·5	
			→ 0	K_{13} 0·72, K_{14} 0·52, K_{15} 0·26, K_{16} −0·11	

52N H. Nowotny and G. Szekely, *Monatsh.*, 1952, **83**, 568
63E D. Ya. Evdokimov and E. A. Kogan, *Ukrain. khim. Zhur.*, 1963, **29**, 1020
64G K. H. Gayer and O. T. Zajicek, *J. Inorg. Nuclear Chem.*, 1964, **26**, 951
64V V. A. Vekhov, B. S. Vitukhnovskaya, and R. F. Doronkina, *Izvest V.U.Z., Khim. i khim. Tekhnol.* 1964, **7**, 1018
65S M. Ya. Shpirt, *Izvest. V.U.Z., Khim. i khim. Tekhnol.*, 1965, **8**, 1036
66A A. M. Andrianov and V. A. Nazarenko, *Zhur. neorg. Khim.*, 1966, **11**, 1527
67J P. B. Jamieson and L. S. D. Glasser, *Acta Cryst*, 1967, **22**, 507
68B A. K. Babko and G. I. Gridchina, *Ukrain. khim. Zhur.*, 1968, **34**, 948
68N V. A. Nazarenko and G. V. Flyantikova, *Zhur. neorg. Khim.*, 1968, **13**, 1855

25 Stannate, $L^{2-} = Sn(OH)_6^{2-}$ = "SnO_3^{2-}"

No data compiled since 1964 volume

Other ligands with 4B elements 26

Metal	Method	Temp	Medium	Log of equilibrium constant, remarks	Ref
				Plumbate	
H^+	ir		solid	ev $Pb(OH)_6^{2-}$ in $K_2Pb(OH)_6$(s)	67M

67M M. Maltese and W. J. Orville-Thomas, *J. Inorg. Nuclear Chem.*, 1967, 29, 2533

Ammonia, ammine, NH_3 27

Metal	Method	Temp	Medium	Log of equilibrium constant, remarks	Ref
H^+	con	−17	ice	K_b −10·4 in ice, M units	63L
	gl	25	1 ($NaClO_4$)	K_1 9·47	64W
	gl	25	MeCN	K_1 16·46 in MeCN(l)	65C
	cal, gl	25	0·1 KCl	K_1 9·29, $\Delta H_1 = -12{\cdot}43$, $\Delta S_1 = 0{\cdot}8$	65P
	oth	25—100	gas	ev L_nH^+, n = 1—6, diagram K_4, $K_5(T)$, $\Delta H_4 = -17{\cdot}8$, $\Delta H_5 = -15{\cdot}9$, $\Delta H_6 \approx -9$, $\Delta S_4 = -38$, $\Delta S_5 = -40{\cdot}5$ in gas phase at 25° (mass spectrometry)	66H
	H	25	w% MeOH	K_1 9·146(10%), 9·044(20%), 8·893(33·4%), 8·687(50%), 8·571(70%) in w% MeOH	66Pa
	ΔG	25—306	0	K_b −4·75(25°), −4·75(50°), −4·86(100°), −5·15(156°), −5·66(218°), −6·87(306°)	/67Ha
	gl	25	0 corr	K_1 9·26	67K
	gl	25	10 NH_4NO_3	K_1 10·45	67L
	H?	25?	var	K_1 9·87(3·55M-KCl), 9·60(1·86M-KCl), 10·00(4·2M-KBr), 9·57(1·85M-KBr), 9·84(4·53M-LiCl), 9·63(2·82M-LiCl), 10·30(5·3M-NaCl), 10·20(6·1M-NaCl)	67La
	gl	25	org	K_1 16·5 in MeCN(l), 10·5 in Me_2SO(l)	68K
	con	0—700	0 corr	K_b −4·80(25°), −4·45(100°), −4·35(150°), −4·30(200°), −4·30(250°), −4·25(300°), at density 1·00 g cm⁻³, also values for other densities and T $K[NH_3(H_2O)_{x+y-n+1} + nH_2O \rightleftharpoons NH_4(H_2O)_x{}^+ + OH(H_2O)_y{}^-]$ −34·0 (400°), −33·0 (500°), −32·9(600°), −33·1(700°), n = 16·5	68Q
	oth	25	gas	K_2 13·8, K_3 6·6 (misprint in table), K_4 4·7, K_5 2·8, K_6 0·15 $\Delta H_2 = -27$, $\Delta H_3 = -17$, $\Delta H_4 = -16{\cdot}5$, $\Delta H_5 = -14{\cdot}5$, $\Delta H_6 = -7{\cdot}5$, $\Delta S_2 = -32$, $\Delta S_3 = -26{\cdot}8$, $\Delta S_4 = -34$, $\Delta S_5 = -36$, $\Delta S_6 = -25$ in gas phase (from mass spectrometric data at −50° to 500°)	68S
	cal	25	I($NaNO_3$)	$\Delta H_1 + \Delta H_w(NH_3 + H_2O \rightleftharpoons NH_4{}^+ + OH^-)$ = 0·11(I = 3), 0·70(I = 1), 0·85(I = 0·5), also values for $\Delta H_1 + \Delta H_w$ and $\Delta S_1 + \Delta S_w$ for these I and 18—40°	68V+
		18—40	0 corr	$\Delta H_1 + \Delta H_w$ = 1·24(18°), 0·88(25°) 0·60(30°), 0·33(35°), 0·08(40°)	
D^+	gl	25	D_2O	K_1 9·873 in D_2O (9·264 in H_2O), also values for mixtures H_2O–D_2O	64S
Be^{2+}	X		solid	ev $Be(NH_3)_4{}^{2+}$ in $Be(NH_3)_4Cl_2$ (s)	65Sa

continued overleaf

27 NH_3 (contd)

Metal	Method	Temp	Medium	Log of equilibrium constant, remarks	Ref
Mg^{2+}	nmr	-78	NH_3(l)	ev MgL_5^{2+} in NH_3(l)	67Sa
Cr^{3+}	red, sp	25?	4·5 NH_4Cl	$\beta[Cr^{3+}{}_2L_6(OH^-)_3]$ 46·42, $\beta_6 \leqslant$ 13·3 $[\beta_4(Cr^{2+}L_4)$ 5, K_w $-14]$	64Sa
Co^{2+}	gl, dis	20	1 NH_4NO_3	K_1 2·18, β_2 3·98, β_3 5·08, β_4 5·98, also β for many mixed cpx Co^{2+}–NH_3–py	66F
	Ag	30	2 (NH_4NO_3)	K_1 2·08, K_2 1·52, K_3 1·17, K_4 0·74, K_5 0·25, K_6 -0·59	66L
	X		solid	ev CoL_5Cl^{2+} and CoL_5CN^{2+} in CoL_5ClSiF_6(s) and CoL_5CNSiF_6(s)	67S
Ni^{2+}	gl, dis	20	1 NH_4NO_3	K_1 3·0, β_2 5·18, β_3 6·82, β_4 7·98, also β for many mixed cpx Ni^{2+}–NH_3–py	66F
	Ag	30	2 (NH_4NO_3)	K_1 2·72, K_2 2·20, K_3 1·71, K_4 1·24, K_5 0·80, K_6 0·14	66L
Ru^{3+}	pre, mag		var	ev RuL_6^{3+}, RuL_5Cl^{2+}	36G
Rh^{3+}	var		var	ev $RhHL_5^{2+}$(hydride H) (pre, ir, nmr, sp)	68T
	gl	25	var	$\beta(RhHL_4^{2+}L)$ 3·7, $\beta(RhEtL_4^{2+}L)$ 9·4, $\beta[Rh(C_2F_4H)L_4^{2+}L]$ 9·7	
Pd^{2+}	Pd	25	0·5 L	β_4 29·6 $\{K[Pd^{2+}+2e^- \rightleftharpoons Pd(s)]$ 33·4$\}$, also β_4 for 10—60°, assuming same *K* and same RTF^{-1} as for 25°!	65F
	gl	25	1 $(NaClO_4)$	K_1 9·6, K_2 8·9, K_3 7·5, K_4 6·8	68R
Pd^{2+}				See also table 67, Cl^-, ref 67R	
$Os^{+,0}$	pre		NH_3(l)	ev OsL_6, OsL_6Br in NH_3(l)	63W
Ir^{3+}	var		var	ev $IrL_5H_2O^{3+}$, IrL_5X^{2+} (X = Cl, Br, I, SCN, NO_2 *etc*) (pre, ir, sp, con, pH_2O)	66S
Cu^+	pol	30	0·5 (KNO_3)	β_2 10·4, $\beta(Cu^+OH^-L)$ 10·9	67F
Cu^{2+}	gl	30	2 NH_4NO_3	K_1 4·13, K_2 3·53, K_3 2·87, K_4 2·15	41B/59Sb
	gl	30	2 NH_4NO_3	$K_1 <$ 2·9, β_2 7·16, K_3 2·75, K_4 2·18	41B/63K
	X		solid	ev CuL_4^{2+}, no ev CuL_5^{2+} in $Cu(NH_3)_5NH_4(ClO_4)_3$(s)	60B
	sp, gl	25	1 (NH_4ClO_4)	K_1 4·16, K_2 3·31, K_3 3·38, K_4 2·20, β_4 13·05	65M
	var		solid	ev CuL_6^{2+} in solids (esr, sp, X)	66E
	gl, dis	20	1 NH_4NO_3	K_1 4·15, β_2 7·65, β_3 10·54, β_4 12·67, also β for many mixed cpx Cu^{2+}–NH_3–py	66F
	pol	30	2 NH_4NO_3	β_2 11·66, β_4 14·38	66G
	X		solid	ev CuL_6^{2+} in $Cu(NH_3)_6X_2$(s) (X = Cl, Br)	67D
	pol	30	0·5 (KNO_3)	β_4 12·3; $\beta(Cu^{2+}OH^-{}_nL_{(4-n)})$ 14·9 (n = 1), 15·7(n = 2), 16·3(n = 3)	67F
	dis, gl	22	2 (Na, NH_4)NO_3	K_1 4·09, β_2 7·54, β_3 10·55, β_4 12·63 $[K_1(H^+)$ 9·614$]$	67H
Ag^+	cal	25	0?	$\Delta H_1 = -4$·9, $\Delta H_2 = -8$·6	66J
	Ag, gl	20—50	var	$K_s(AgL_2{}^+Br^-)$ -6·4(20°), -5·2(50°) β_2 6·5(20°), 5·6(50°), $\Delta H_{\beta 2} = -12$·4	66M
Zn^{2+}	Ag	30	2 (NH_4NO_3)	K_1 2·32, K_2 2·29, K_3 2·36, K_4 2·39	66L
	sol, gl	25	1 $(NaClO_4)$	$\beta_4$10·84, $\beta[Zn^{2+}L(OH^-)_3]$ 16·94, $\beta[Zn^{2+}L_2(OH^-)_2]$ 15·53(K_w -14·00)	68G
Hg^{2+}	gl, Hg	25	1 $(NaClO_4)$	β_2 17·8	64W
$MeHg^+$	gl	20	0·1 (KNO_3)	K_1 7·60 $[K_1(H^+)$ 9·42$]$	63S = 65S
B^{3+}	gl	23	var	$K_a(H_3NBF_3) \approx -12$	65R
$Tl(OH)_2{}^+$	gl, red	25	10 NH_4NO_3	K_1 4·6, K_2 4·7, K_3 2·3, K_4 1·5, β_4 13·0	67L

references on facing page

36G K. Gleu and K. Rehm, *Z. anorg. Chem.*, 1936, **227**, 237
59Sb J. C. Sullivan, J. Rydberg, and W. F. Miller, *Acta Chem. Scand.*, 1959, **13**, 2023
60B M. Bukovska and M. A. Porai-Koshits, *Kristallografiya*, 1960, **5**, 140
63K N. P. Komar', *Uchenye Zapiski Khar'kov Univ.*, 1963, **133** (*Trudy Khim. Fak. Kh.G.U.*, **19**), 66
63L L. Levi, O. Milman, and E. Suraski, *Trans. Faraday Soc.*, 1963, **59**, 2064
63S M. Schellenberg, dissertation, Eidgenössische Technische Hochschule, Zürich, 1963
63W G. W. Watt, E. M. Potrafke, and D. S. Klett, *Inorg. Chem.*, 1963, **2**, 868
64S P. Salomaa, L. L. Schaleger, and F. A. Long, *J. Phys. Chem.*, 1964, **68**, 410
64Sa C. E. Schäffer and P. Andersen in "Theory and Structure of complex compounds" (symposium papers, Wrocław 1962), Pergamon, Warszawa, 1964, p. 571
64W T. H. Wirth and N. Davidson, *J. Amer. Chem. Soc.*, 1964, **86**, 4325
65C J. F. Coetzee and G. R. Padmanabhan, *J. Amer. Chem. Soc.*, 1965, **87**, 5005
65F A. B. Fasman, G. G. Kutyukov, and D. V. Sokol'skii, *Zhur. neorg. Khim.*, 1965, **10**, 1338
65M R. C. Mercier, M. Bonnet, and M. R. Pâris, *Bull. Soc. chim. France*, 1965, 2926
65P P. Paoletti, J. H. Stern, and A. Vacca, *J. Phys. Chem.*, 1965, **69**, 3759
65R I. G. Ryss and N. G. Parkhomenko, *Ukrain. khim. Zhur.*, 1965, **31**, 237
65S G. Schwarzenbach and M. Schellenberg, *Helv. Chim. Acta*, 1965, **48**, 28
65Sa K. N. Semenenko, *Vestnik. Moskov. Univ., Khim.*, 1965, no. 3, 39
66E H. Elliott and B. J. Hathaway, *Inorg. Chem.*, 1966, **5**, 885
66F Ya. D. Fridman, M. G. Levina, and R. I. Sorochan, *Zhur. neorg. Khim.*, 1966, **11**, 1641
66G S. L. Gupta and M. K. Chatterjee, *Indian. J. Chem.*, 1966, **4**, 22
66H A. M. Hogg, R. M. Haynes, and P. Kebarle, *J. Amer. Chem. Soc.*, 1966, **88**, 28
66J S. Johansson, personal communication, quoted by ref 66P
66L C. Luca, V. Magearu, and G. Popa, *J. Electroanalyt. Chem.*, 1966, **12**, 45
66M R. Matejec, *Ber. Bunsengesellschaft Phys. Chem.*, 1966, **70**, 703
66P P. Paoletti, A. Vacca, and D. Arenare, *J. Phys. Chem.*, 1966, **70**, 193
66Pa M. Paabo, R. G. Bates, and R. A. Robinson, *J. Phys. Chem.*, 1966, **70**, 247
66S H.-H. Schmidtke, *Inorg. Chem.*, 1966, **5**, 1682
67D T. Distler and P. A. Vaughan, *Inorg. Chem.*, 1967, **6**, 126
67F J. F. Fisher and J. L. Hall, *Analyt. Chem.*, 1967, **39**, 1550
67H W. J. Haffenden and G. J. Lawson, *J. Inorg. Nuclear Chem.*, 1967, **29**, 1499
67Ha H. C. Helgeson, *J. Phys. Chem.*, 1967, **71**, 3121
67K N. P. Komar' and N. T. Zung, *Zhur. neorg. Khim.*, 1967, **12**, 1265
67L V. I. Lobov, F. Ya. Kul'ba and V. E. Mironov, *Zhur. neorg. Khim.*, 1967, **12**, 334
67La M. Lucas, *Bull. Soc. chim. France*, 1967, 3842
67S J. A. Stanko and I. C. Paul, *Inorg. Chem.*, 1967, **6**, 486
67Sa T. J. Swift and H. H. Lo, *J. Amer. Chem. Soc.*, 1967, **89**, 3988
68G A. O. Gubeli and J. Ste-Marie, *Canad. J. Chem.*, 1968, **46**, 1707
68K I. M. Kolthoff, M. K. Chantooni, jun., and S. Bhowmik, *J. Amer. Chem. Soc.*, 1968, **90**, 23
68Q A. S. Quist and W. L. Marshall, *J. Phys. Chem.*, 1968, **72**, 3122
68R L. Rasmussen and C. K. Jørgensen, *Acta Chem. Scand.*, 1968, **22**, 2313
68S S. K. Searles and P. Kebarle, *J. Phys. Chem.*, 1968, **72**, 742
68T K. Thomas, J. A. Osborn, A. R. Powell, and G. Wilkinson, *J. Chem. Soc. (A).*, 1968, 1801
68V V. P. Vasil'ev and L. A. Kochergina, *Zhur. fiz. Khim.*, 1968, **42**, 373

Other ligands with nitrogen (−III) 28

Metal	*Method*	*Temp*	*Medium*	*Log of equilibrium constant, remarks*	*Ref*
				Amide, NH_2^-	
H^+	H	−60	NH_3	K_1 32 in NH_3(l), 0·1 KI	68H
Co^{3+}	*X*		solid	ev $(NH_3)_5CoNH_2Co(NH_3)_5{}^{5+}$ = Cp^{5+} in $Cp(NO_3)_5$(s) (≈ ref 63V, table 27, NH_3)	68S
$PhHg^+$	pre, ir		var	ev $(PhHg)_2NH_2{}^+$ in solids and MeOH(l)	68D ≈ 09P
Sn^{4+}	*X*		solid	ev $Sn(NH_2)_6{}^{2-}$ in $K_2Sn(NH_2)_6$(s)	64S
Pb^{2+}	pre, con		NH_3	ev $Pb(NH_2)_3{}^-$ in NH_3(l)	67S

continued overleaf

28 N^{3-} (contd)

Metal	Method	Temp	Medium	Log of equilibrium constant, remarks	Ref
				Nitride, N^{3-}	
Re^{5+}	pre, ir		var	ev $ReN(CN)_4OH_2^{2-}$ (misprint in synopsis)	64L
Re^{7+}	ir		solid	ev ReO_3N^{2-} in K_2ReO_3N(s)	68K+
Ru^{4+}	var		var	ev $(H_2OX_4Ru)_2N^{3-}$, X = Cl, Br (pre, ir, Ram, mag)	68C
Se^{6+}	ir, pre		solid, NH_3	ev $H_2NSeO_3^-$ in salts and in NH_3(l), ev $NSeO_3^{3-}$ in Ag^+ salts	66P

09P L. Pesci, *Gazzetta*, 1909, **39**, 147
64L C. J. L. Lock and G. Wilkinson, *J. Chem. Soc.*, 1964, 2281
64S O. Schmitz-DuMont, G. Müller, and W. Schaal, *Z. anorg. Chem.*, 1964, **332**, 263
66P R. Paetzold, K. Dostál, and A. Růžička, *Z. anorg. Chem.*, 1966, **347**, 13; **348**, 1
67S O. Schmitz-DuMont and W. Jansen, *Z. anorg. Chem.*, 1967, **349**, 189
68C M. J. Cleare and W. P. Griffith, *Chem. Comm.*, 1968, 1302
68D G. B. Deacon and J. H. S. Green, *J. Chem. Soc. (A)*, 1968, 1182
68H M. Herlem, *Bull. Soc. chim. France*, 1968, 1687
68K B. Krebs and A. Müller, *J. Inorg. Nuclear Chem.*, 1968, **30**, 463
68S W. P. Schaefer, A. W. Cordes, and R. E. Marsh, *Acta Cryst.*, 1968, **B24**, 283

29 Hydrazine, N_2H_4

Metal	Method	Temp	Medium	Log of equilibrium constant, remarks	Ref
H^+	H	25	w% N_2H_4	K(solvent $\rightleftharpoons LH^+ + OH^-$) −5·4($w$ = 5), −5·0(w = 10), −4·6(w = 25), −5·4(w = 50), −5·9(w = 64), −8·1(w = 85), −11·0(w = 93·5) in w% N_2H_4–H_2O mixt, 1M-KCl	65B
	gl	25	MeCN	K_1 16·61 in MeCN(l)	65C
	X		solid	ev $N_2H_6^{2+}$ in $N_2H_6(H_2PO_4)_2$(s)	66L
	gl	30	1 ($NaClO_4$)	K_1 7·87, K_{12} 3·85 (?, see Fig. 1)	67Ba
	gl	15—35	→0	K_1 8·26(15°), 8·14(20°), 7·965(25°), 7·75(30°); ΔH_1 = −15·1(15°), −10·6(25°), −6·0(35°); ΔS_1 = −14·6(15°), 0·7(25°), 16·1(35°); ΔC_{p1} = 442(15°), 458(25°), 473(35°)	67S
Cr^{2+}	kin	25	1 ($NaClO_4$)	$\beta(CrCl^+L)$ 0·16	68W
Mn^{2+}	gl	30	1 ($NaClO_4$)	K_1 4·76	67Ba
Co^{2+}	sp, gl	23	0 corr	$\beta[H^+Co(CN)_5L^{3-}]$ 4·32(gl), 4·35(sp)	67B
	gl	30	1 ($NaClO_4$)	K_1 1·78, β_2 3·34	67Ba
Ni^{2+}	pre, con	20	dil	ev polyn cpx such as $Ni_2(OH)_2L_n^{2+}$, n = 1 and 5	66K
	gl	30	1 ($NaClO_4$)	K_1 3·18	67Ba
Cu^{2+}	gl	30	1 ($NaClO_4$)	K_1 6·67	67Ba
Zn^{2+}	gl	30	1 ($NaClO_4$)	K_1 3·69, β_2 6·69	67Ba

65B D. Bauer, *Bull. Soc. chim. France*, 1965, 3302
65C J. F. Coetzee and G. R. Padmanabhan, *J. Amer. Chem. Soc.*, 1965, **87**, 5005
66K E. I. Krylov, V. A. Sharov, and V. F. Degtyarev, *Doklady Akad. Nauk S.S.S.R.*, 1966, **166**, 876
66L R. Liminga, *Acta Chem. Scand.*, 1966, **20**, 2483
67B R. Barca, J. Ellis, M.-S. Tsao, and W. K. Wilmarth, *Inorg. Chem.*, 1967, **6**, 243
67Ba D. Banerjea and I. P. Singh, *Z. anorg. Chem.*, 1967, **349**, 213
67S K. Sallavo and P. Lumme, *Suomen Kem.*, 1967, **40**, *B*, 155
68W C. F. Wells and M. A. Salam, *J. Chem. Soc. (A.)*, 1968, 1568

Hydroxylamine, H_2NOH 30

Metal	Method	Temp	Medium	Log of equilibrium constant, remarks	Ref
H^+	con	18	→ 0	K_1 6·1	05R/65L
	gl	20	0·5 ($NaNO_3$)	K_1 6·12	63S
	gl?	4—50	0·5 ($NaClO_4$)	K_1 1517 T^{-1} − 0·0038 T + 1·893	65F
	gl	15—35	2·24 ($NaClO_4$)	K_1 6·44(15°), 6·33(20°), 6·21(25°), 5·99(35°)	65L
			1·00 ($NaClO_4$)	K_1 6·31(15°), 6·18(20°), 6·06(25°), 5·84(35°)	
			0·25 ($NaClO_4$)	K_1 6·21(15°), 6·09(20°), 5·98(25°), 5·76(35°)	
			→ 0	K_1 6·19(15°), 6·06(20°), 5·95(25°), 5·73(35°)	
		25	→ 0	$\Delta H_1 = -9·3$, $\Delta S_1 = -3·8$, $\Delta C_{p1} = 23$	
	gl	25	0·1 KCl	K_1 5·7	66F
Cr^{2+}	kin	25	1 ($NaClO_4$)	$\beta(CrCl^+L)$ 0·18	68W
Mo^{6+}	gl	20	dil	$K_a(H_2Mo_{11}O_{36}L^{4-})$ 3·84	68J
Mn^{2+}	gl	20	0·5 ($NaNO_3$)	K_1 0·5	63S
Tc^{2+}	pre, tp		var	ev cation, probably $TcL_2(NH_3)_3\,H_2O^{2+}$	63E
Co^{2+}	gl	20	0·5 ($NaNO_3$)	K_1 0·9	63S
Ni^{2+}	gl	20	0·5 ($NaNO_3$)	K_1 1·5	63S
	gl, sp	25	var	β_2 9·72(gl), 9·84(sp), β_4 12·53(gl), 12·58(sp), β_6 18·55(gl), 18·58(sp)	66F
Pt^{2+}	gl	25	0·3 $NaNO_3$	$K_s(PtL_4^{2+}H^+_{-2})$ 8·52, $K_a(PtL_4^{2-})$ −6·88	66G
			0 corr	$K_a(PtL_4^{2+})$ −6·45, $K_s(PtL_4^{2+}OH^-_2)$ −20·05	
	gl	25	0·1	$K_a(PtL_3NH_3^{2+})$ −7·22, −9·7, −10·4	
			0 corr	$K_a(PtL_3NH_3^{2+})$ −6·90, $K_a[PtL_2(NH_3)_2^{2+}]$ −8·80 (*trans*), $K_a[Pt(LNH_3)_2^{2+}]$ −7·22(*cis*), $K_a(PtLNH_3\,py_2^{2+})$ −9·41	
Cu^{2+}	gl	20	0·5 ($NaNO_3$)	K_1 2·4, $K_2 \approx 1·7$	63S
Ag^+	gl	20	0·5 ($NaNO_3$)	K_1 1·9, $K_2 < 3$	63S
Zn^{2+}	gl	20	0·5 ($NaNO_3$)	K_1 0·5	63S
B^{3+}	gl	10	0·04 (KCl)	$\beta(H^+F_3BONH_2^-)$ 7·52	65R
Pb^{2+}	pol	25?	1	K_1 0·23, K_2 −0·10, K_3 0·60 with $I = 1$ (NH_4Cl)	68S
				K_1 0·78, K_2 1·40 with $I = 1$ (KCl or KNO_3) (accuracy?)	

05R W. H. Ross, *Proc. Trans. Nova Scotian Inst. Sci.*, 1905, **11**, 95
63E J. D. Eakins, D. G. Humphreys, and C. E. Mellish, *J. Chem. Soc.*, 1963, 6012
63S I. Szilárd, *Acta Chem. Scand.*, 1963, **17**, 2674
65F R. T. M. Fraser, *J. Chem. Soc.*, 1965, 1747
65L P. Lumme, P. Lahermo, and J. Tummavuori, *Acta Chem. Scand.*, 1965, **19**, 2175
65R I. G. Ryss and S. L. Idel's, *Zhur. neorg. Khim.*, 1965, **10**, 714
66F M. T. Falqui, G. Ponticelli, and F. Sotgiu, *Ann. Chim. (Italy)*, 1966, **56**, 464
66G A. A. Grinberg, A. I. Stetsenko, and G. P. Gur'yanova, *Zhur. neorg. Khim.*, 1966, **11**, 1887
68J W. Jakób and M. Dyrek, *Roczniki Chem.*, 1968, **42**, 1393
68S P. Stantscheff, *Z. analyt. Chem.*, 1968, **234**, 109
68W C. F. Wells and M. A. Salam, *J. Chem. Soc. (A.)*, 1968, 1568

Azide, N_3^- 31

Metal	Method	Temp	Medium	Log of equilibrium constant, remarks	Ref
H^+	sp	25	MeOH	K_1 9·20 in MeOH(l), 0 corr	57P
	gl	20—33	→ 0	K_1 4·70(20°), 4·64(26°), 4·58(33°), $\Delta H_1 = -3·8$, $\Delta S_1 = 8·6$	66B
	sp	21	org	K_1 8·9 in MeOH(l), 8·5 in Me_2NCHO(l)	66C
	gl	25	I($NaClO_4$)	K_1 4·78(I = 3), 4·44(I = 1)	67M

continued overleaf

31 N_3^- (contd)

Metal	Method	Temp	Medium	Log of equilibrium constant, remarks	Ref
D^+	gl	25	D_2O	K_1 5·155 in D_2O (→ 0) (4·694 in H_2O), also values for mixtures H_2O–D_2O	64S
Na^+	sol	0—25	org	K_{s0} −0·9 in MeOH(0°), −1·9 in Me_2NCHO(25°), −0·64 in Me_2SO(25°)	67A
Group $4f^{3+}$	Tl, con		Me_3PO_4	ev ML_3 and ML_4^- (M = Pr, Nd) in Me_3PO_4(l)	66Gb
Ti^{3+}	var		org	ev TiL_2^+, TiL_3, TiL_4^-, TiL_6^{3-} in org (sp, con, Tl)	67Ga
V^{3+}	var		org	ev VL^{2+}, VL_3, VL_4^-, VL_6^{3-} in org (sp, con, Tl)	67Ga
VO^{2+}	var		org	ev VOL^+, VOL_2, VOL_3^-, VOL_4^{2-}, VOL_5^{3-} in org (con, sp, Tl)	67Gb
	pre, con		var	ev VOL_4^{2-} in solid, Me_2CO and MeOH(l)	68P
Cr^{2+}	kin	0—25	1 ($NaClO_4$)	$\beta(CrSO_4{}^0L^-)$ 1·06(0°), 1·01(5°), 0·90(10·5°), 0·61(25°), $\Delta H = -7{\cdot}5$, $\Delta S = -23$	68W
	kin	0—25	1 ($NaClO_4$)	$\beta(CrF^+L^-)$ 1·04(0°), 0·98(5°), 0·89(10·5°), 0·86(25°), $\Delta H = -3$, $\Delta S = -7$	
	kin	25	1 ($NaClO_4$)	$\beta(CrCl^+L^-)$ 0·18	
		0—25	1 ($NaClO_4$)	$\beta(CrCl^+HL)$ 0·98(0°), 0·70(5°), 0·34(18°), 0·16(25°), $\Delta H = -12$, $\Delta S = -40$	
		0—25	4 ($NaClO_4$)	$\beta(CrBr^+HL)$ 0·26(0°), 0·01(5°), −0·41(18°), −0·64(25°), $\Delta H = -13$, $\Delta S = -47$	
Cr^{3+}	var		org	ev CrL^{2+}, CrL_2^+, CrL_3, CrL_6^{3-} in org (sp, con, Tl)	67Ga
	kin	40	2 ($LiClO_4$)	$*K_1 < -1{\cdot}0$	68D
	var		var	ev CrL_6^{3-} in solid and Me_2CO(l) (pre, con, mag)	68P
	kin	10	H_2SO_4 var	$\beta[H^+Cr(NH_3)_5L^{2+}]$ −3·26	68S
UO_2^{2+}	pre, con		var	ev $UO_2L_4^{2-}$ in solid and Me_2CO(l) (pre, con, mag)	68P
Mn^{2+}	var		var	ev MnL_4^{2-} in solid and Me_2CO(l) (pre, con, mag, ir)	66Fa
	var		org	ev MnL_2, MnL_4^{2-} in MeCN, $(MeO)_3PO$, propane-diol-1,2-carbonate (= PDC; l) (con, E, sp)	67G
Mn^{3+}, cpx	sp	25	0·25 ($NaClO_4$)	$\beta(MnY^-L^-)$ 1·51, H_4Y = EDTA	67S
Fe^{3+}	pre		var	ev FeL_6^{3-} in solids	64K
	sp, gl	28	I($NaClO_4$)	K_1 0·80(I = 0·25), 1·00(I = 0·15) 1·73(I = 0·05), [$K_1(H^+)$ 4·71, I = 0]	65M
	sp, kin	25	1 ($LiClO_4$)	$*K_1$ −0·29(sp), −0·30(kin), ev FeL_2^+	67C
	var		org	ev FeL^{2+}, FeL_2^+, FeL_3, and FeL_{3+n}^{n-} in MeCN, $(MeO)_3PO$ and PDC(l) (con, E, sp)	67G
Co^{2+}	sp		org	ev CoL^+, CoL_2, CoL_3^- and CoL_4^{2-} in Me_3PO_4, Et_3PO_4, Me_2SO, MeCN(l)	64G = 66G
	pre, ir		var	ev CoL_4^{2-}	66F
Co^{3+}, cpx	sp	25	0·5 (H)ClO_4	$Co(NH_3)_4OH_2L^{2+}$, $K(cis \rightleftharpoons trans)$ −0·66	64H
	sol	25—35	0 corr	$Co(NH_3)_5Cl^{2+}$; K_1 1·07(25°), 0·96(35°)	65A
	kin	25	H_2SO_4 var	$\beta[H^+Co(NH_3)_5L^{2+}]$ −2·78	68S
Ni^{2+}	var		org	ev NiL_2, NiL_4^{2-} in Me_2SO, Me_3PO_4, MeCN(l), NiL_6^{4-} in Me_3PO_4(l) (sp, con, Tl)	66Ga
	gl	25	3 ($NaClO_4$)	K_1 1·04	67Ma
	var		Me_2NCOMe	ev NiL^+, NiL_2, NiL_3^- in Me_2NCOMe(l) (con, sp, Tl)	68G
Rh^{3+}	pre, con		var	ev RhL_6^{3-} in solid and Me_2CO(l)	68P
	kin	60	H_2SO_4 var	$\beta[H^+Rh(NH_3)_5L^{2+}]$ −1·95	68S

contin

Metal	Method	Temp	Medium	Log of equilibrium constant, remarks	Ref
Pd^{2+}	sp		EtOH	ev PdL_5^{3-} and PdL_6^{4-} in EtOH(l)	65Ca
	var		var	ev $Pd_2L_6^{2-}$ and PdL_4^{2-} in solids and Me_2CO(l), $CHCl_3$(l) *etc.* [pre, ir, molar mass (fp, bp?)]	66Fa
	pre		var	ev PdL_4^{2-} in solid	68P
Cu^+	pre		var	ev CuL_2^- in solid, CH_2Cl_2 and Me_2CO(l)	68P
Cu^{2+}	sp		org	ev CuL^+, CuL_2, CuL_3^- and CuL_4^{2-} in Me_3PO_4, MeCN, Me_2SO(l)	64G = 66Ga
	gl	25	$I(NaClO_4)$	K_1 2·75(I = 3), 2·04(I = 1)	67M
Ag^+	Ag	25	org	K_{s0} −11·2 in MeOH, −7·7 in $HCONH_2$, −11·0 in Me_2NCHO, −10·8 in Me_2NCOMe, −6·5 in Me_2SO, −9·6 in MeCN, −8·5 in $(Me_2N)_3PO$; β_2 11·9 in Me_2NCHO, 12·2 in Me_2NCOMe, 7·0 in Me_2SO, 11·4 in $(Me_2N)_3PO$	67A
Au^+	sp		var	ev AuL_2^-	65C
Zn^{2+}	var		var	ev ZnL_4^{2-} (pre, ir, Ram)	66F
Cd^{2+}	cal	25	3 ($NaClO_4$)	$\Delta H_1 = -1·15$, $\Delta H_2 = -1·41$, $\Delta H_3 = -1·7$, $\Delta H_4 = -1·3$; $\Delta S_1 = 3·5$, $\Delta S_2 = 0·2$, $\Delta S_3 = 3·1$, $\Delta S_4 = -1·7$	66G+43L
Hg^{2+}	sp, gl	28	$I(NaClO_4)$	K_1 7·15, K_2 6·97(I = 0·25); K_1 7·30, K_2 7·08(I = 0·15); K_1 7·48, K_2 7·26(I = 0·05); K_1 7·75, K_2 7·49($I \to 0$)	65M
	pre, con		var	ev HgL_3^- and HgL_4^{2-} in solids, Me_2CO(l), $PhNO_2$(l)	68P
Tl^{3+}	red	12—50	1 H^+, 2 ($NaClO_4$)	K_1 3·08(12·7°), 3·0(20°), 2·90(35°), 2·81(50°); β_2 5·60(12·7°), 5·38(20°), 5·18(35°), 4·98(50°); β_3 7·04(12·7°), 6·90(20°), 6·40(35°), 5·70(50°)	65V
	col J		var	ev TlL_6^{3-}, $TlL_4Cl_2^{3-}$, $TlL_2Cl_4^{3-}$	66S
Sn^{4+}	var		var	ev SnL_6^{2-} (pre, ir, Ram)	66F
Pb^{4+}	pre, con		var	ev PbL_6^{2-} in solid, CH_2Cl_2(l) and Me_2CO(l)	68P

I_2

1964 *Addendum*: p. 162, *for ref* "62M" *read* "62M = 63M".

57P R. G. Pearson, P. M. Henry, and F. Basolo, *J. Amer. Chem. Soc.*, 1957, **79**, 5379
63M D. Meyerstein and A. Treinin, *Trans. Faraday Soc.*, 1963, **59**, 1114
64G V. Gutmann, O. Leitmann, and G. Hampel, Proceedings of the Symposium on Co-ordination Chemistry, Tihany, 1964 (Akad Kiadó, Budapest, 1965), p. 199
64H A. Haim, *J. Amer. Chem. Soc.*, 1964, **86**, 2352
64K F. Kröhnke and B. Sander, *Z. anorg. Chem.*, 1964, **334**, 66
64S P. Salomaa, L. L. Schaleger, and F. A. Long, *J. Phys. Chem.*, 1964, **68**, 410
65A D. W. Archer, D. A. East, and C. B. Monk, *J. Chem. Soc.*, 1965, 720
65C R. G. Clem and E. H. Huffman, *Analyt. Chem.*, 1965, **37**, 1155
65Ca R. G. Clem and E. H. Huffman, *J. Inorg. Nuclear Chem.*, 1965, **27**, 365
65M T. R. Musgrave and R. N. Keller, *Inorg. Chem.*, 1965, **4**, 1793
65V G. Vogt, *Ber. Bunsengesellschaft Phys. Chem.*, 1965, **69**, 648
66B J. H. Boughton and R. N. Keller, *J. Inorg. Nuclear Chem.*, 1966, **28**, 2851
66C B. W. Clare, D. Cook, E. C. F. Ko, Y. C. Mac, and A. J. Parker, *J. Amer. Chem. Soc.*, 1966, **88**, 1911
66F D. Forster and W. D. Horrocks, jun., *Inorg. Chem.*, 1966, **5**, 1510
66Fa K. Feldl, dissertation, Technische Hochschule, München, 1966
66G P. Gerding, *Acta Chem. Scand.*, 1966, **20**, 2771
66Ga V. Gutmann and O. Leitmann, *Monatsh*, 1966, **97**, 926
66Gb V. Gutmann, O. Leitmann, and R. Schütz, *Inorg. Nuclear Chem. Letters*, 1966, **2**, 133
66S T. N. Srivastava and N. Singh, *Z. anorg. Chem.*, 1966, **343**, 211
67A R. Alexander, E. C. F. Ko, Y. C. Mac, and A. J. Parker, *J. Amer. Chem. Soc.*, 1967, **89**, 3703

continued overleaf

31 N_3^- (contd)

67C D. W. Carlyle and J. H. Espenson, *Inorg. Chem.*, 1967, **6**, 1370
67G V. Gutmann and W. K. Lux, *J. Inorg. Nuclear Chem.*, 1967, **29**, 2391
67Ga V. Gutmann, O. Leitmann, A. Scherhaufer, and H. Czuba, *Monatsh.*, 1967, 98, 188
67Gb V. Gutmann and H. Laussegger, *Monatsh*, 1967, **98**, 439
67M F. Maggio, V. Romano, and L. Pellerito, *Ann. Chim. (Italy)*, 1967, **57**, 191
67Ma F. Maggio, V. Romano, and L. Pellerito, *J. Electroanalyt. Chem.*, 1967, **15**, 227
67S M. A. Suwyn and R. E. Hamm, *Inorg. Chem.*, 1967, **6**, 2150
68D J. Doyle, A. G. Sykes, and A. Adin, *J. Chem. Soc. (A)*, 1968, 1314
68G V. Gutmann and H. Bardy, *Z. anorg. Chem.*, 1968, **361**, 213
68P P. Pöllmann, dissertation, Technische Hochschule, München, 1968
68S P. J. Staples, *J. Chem. Soc. (A)*, 1968, 2731
68W C. F. Wells and M. A. Salam, *J. Chem. Soc. (A)*, 1968, 1568

32 Nitrosyl, nitrogen oxide, NO

Metal	*Method*	*Temp*	*Medium*	*Log of equilibrium constant, remarks*	*Ref*
V	*X*		solid	ev $VNO(CN)_5^{3-}$ in $K_3VNO(CN)_5(H_2O)_2(s)$ [not $VNO(CN)_5^{5-}$]	68J
Cr	*X*		solid	ev $Cr(CN)_5NO^{3-}$ in $K_3Cr(CN)_5NO(s)$	66V
Mo	var		var	ev $MoL_2Cl_4^{2-}$ in solid and in MeCN(l) (ir, pre, sp)	64C
	var		var	ev $Mo(CN)_5NO^{4-}$, $Mo(CN)_5NO^{3-}$, $MoO(CN)_4NO^{3-}$? (*E*, mag, pre, sp)	64J
	red, mag		var	ev $Mo(CN)_5NO^{3-}$, $MoO(CN)_4NO^{3-}$, probably $Mo(CN)_5NO^{2-}$	68H
	X		solid	ev $Mo(CN)_5L^{4-}$ in $K_4Mo(CN)_6NO(s)$	68S
W	var		var	ev $WL_2Cl_4^{2-}$ in solid and in MeCN(l) (ir, pre, sp)	64C
Mn	*X*		solid	ev $Mn(CN)_5NO^{3-}$ in $K_3Mn(CN)_5NO(H_2O)_2(s)$	67T
Fe^{2+}	kin	25	0·5 (H_2SO_4)	K_1 2·65(kin), 2·68(equil. ref 24M+)	66K
Fe	*X*		solid	ev $FeNO(CN)_5^{2-}$ in $Na_2FeNO(CN)_5(H_2O)_2(s)$	63Ma
	var		Me_2NCHO	ev $Fe(CN)_5NO^{3-}$ in Me_2NCHO(l) (esr, sp, pre)	64H
	esr	25	var	ev FeL_2A_2(paramagnetic), $Fe_2L_4A_2$(diamagnetic) ($A = OH^-$, HPO_4^{2-} *etc.*); $K(Fe_2L_4A_2 + 2A \rightleftharpoons 2FeL_2A_2)$ 0·15 for $A = HPO_4^{2-}$	65M
	var		var	ev $FeL_2Br_2^-$, $FeL_2I_2^-$ (con, ir, mag, pre)	67S
Co	cix		$Ca(NO_3)_2$ var	ev $[(NH_3)_5CoL]_2^{4+}$	64A
	X		solid	ev $CoNO(NH_3)_5^{2+}$ in $Co(NO)(NH_3)_5Cl_2(s)$ see also Table 35, hyponitrite	65D = 65H
	ir, sp		solid	ev $Co(NH_3)_5NO^{2+}$ (black)	66R
	X		solid	ev $(ON)_2CoCl_2Co(NO)_2$ units in $Co(NO)_2Cl(s)$	67J
Ni	pre		var	ev $NiNO^+$	64F
Ir	pre, ir		var	ev $IrLBr_5^-$	63A

63A M. Angoletta, *Ann. Chim. (Italy)*, 1963, **53**, 1208
63Ma P. T. Manoharan and W. C. Hamilton, *Inorg. Chem.*, 1963, **2**, 1043
64A M. Ardon, *Israel J. Chem.*, 1964, **2**, 181
64C F. A. Cotton and B. F. G. Johnson, *Inorg. Chem.*, 1964, **3**, 1609
64F R. F. Feltham and J. T. Carriel, *Inorg. Chem.*, 1964, **3**, 121
64H E. F. Hockings and I. Bernal, *J. Chem. Soc.*, 1964, 5029
64J W. Jakób, E. Hejmo, and A. Kanas, 'Theory and structure of complex compounds (symposium papers, Wrocław 1962),' Pergamon, Warszawa, 1964, p. 333
65D D. Dale and D. C. Hodgkin, *J. Chem. Soc.*, 1965, 1364
65H D. Hall and A. A. Taggart, *J. Chem. Soc.*, 1965, 1359

continued

65M C. C. McDonald, W. D. Phillips, and H. F. Mower, *J. Amer. Chem. Soc.*, 1965, **87**, 3319
66K K. Kustin, I. A. Taub, and E. Weinstock, *Inorg. Chem.*, 1966, **5**, 1079
66R J. B. Raynor, *J. Chem. Soc. (A)*, 1966, 997
66V N.-G. Vannerberg, *Acta Chem. Scand.*, 1966, **20**, 1571
67J S. Jagner and N.-G. Vannerberg, *Acta Chem. Scand.*, 1967, **21**, 1183
67S W. Silverthorn and R. D. Feltham, *Inorg. Chem.*, 1967, **6**, 1662
67T A. Tullberg and N.-G. Vannerberg, *Acta Chem. Scand.*, 1967, **21**, 1462
68H E. Hejmo, W. Jakób, and A. Kanas, *Roczniki. Chem.*, 1968, **42**, 965
68J S. Jagner and N.-G. Vannerberg, *Acta Chem. Scand.*, 1968, **22**, 3330
68S D. H. Svedung and N.-G. Vannerberg, *Acta Chem. Scand.*, 1968, **22**, 1551

Nitrite, NO_2^- 33

Metal	*Method*	*Temp*	*Medium*	*Log of equilibrium constant, remarks*	*Ref*
M^{2+}	*X*		solid	ev ML_6^{4-} ($M = Fe^{2+}, Co^{2+}, Ni^{2+}, Cu^{2+}, Cd^{2+}, Hg^{2+}$) in salts	/51Fa
M^{3+}	*X*		solid	ev ML_6^{3-} ($M^{3+} = Co^{3+}, Rh^{3+}, Ir^{3+}, Bi^{3+}$) in salts	/51Fa
H^+	sp	4—20	0 corr	†$K_{12}(HNO_2+H^+ \rightleftharpoons NO^+ + H_2O)$ $-4{\cdot}58(4{\cdot}5°)$, $-4{\cdot}77(19{\cdot}5°)$ from data for H_2SO_4 aq, $-6{\cdot}06(3{\cdot}5°)$, $-6{\cdot}12(15°)$ from data for $HClO_4$ aq	63B
	gl	25	$I(NaClO_4)$	K_1 3·24(I = 2·00), 3·00(I = 1·00), 2·94(I = 0·50), 2·95(I = 0·25), 2·95(I = 0·09), 3·00(I = 0·04), 3·148($I \to 0$)	65L
	gl	15—35	0·25 ($NaClO_4$)	K_1 3·02(15°), 2·99(20°), 2·95(25°), 2·89(35°)	65La
			0 corr	K_1 3·23(15°), 3·20(20°), 3·15(25°), 3·11(35°)	
		25	0 corr	$\Delta H_1 = -2{\cdot}5$, $\Delta S_1 = 6{\cdot}1$, $\Delta C_p = 117$	
	E	25	0 corr	†$K_{12}(HL+H^+ \rightleftharpoons NO^+ + H_2O)$ $-8{\cdot}10$	66Sb
	kin	20	0 corr	$K(2HL \rightleftharpoons H_2O + N_2O_3)$ $-0{\cdot}80$	67S+60T
	kin	25	0 corr	†$K_{12}[HL+H^+ \rightleftharpoons NO^+ + H_2O$ (or H_2ONO^+)] $-8{\cdot}5$	67Sa
	lit	20—25	0 corr	†K_{12} $-8{\cdot}10$	/67Sa
	gl	15—35	4 Na(NO_3)	K_1 3·09(15°), 3·06(20°), 3·01(25°), 2·95(35°)	68T
			2 Na(NO_3)	K_1 2·94(15°), 2·91(20°), 2·88(25°), 2·81(35°)	
			1 Na(NO_3)	K_1 2·94(15°), 2·91(20°), 2·87(25°), 2·80(35°), also K_1 for I = 0·50, 0·25, 0·09	
			1·87 Na(NO_2)	K_1 3·26(15°), 3·22(20°), 3·19(25°), 3·12(35°), also K_1 for I = 0·94, 0·47, 0·24, 0·09(NaL) [assuming $K_a(CH_3CO_2H)$ to be the same in $NaNO_3$ and $NaNO_2$ of same I]	
			0 corr	K_1 3·22(15°), 3·18(20°), 3·14(25°), 3·10(35°)	
		25	0 corr	$\Delta H_1 = -2{\cdot}2$, $\Delta S_1 = 7{\cdot}1$, $\Delta C_p = 165$	
Group 1A$^+$	con	25	0 corr	K_1 $-0{\cdot}04(Li^+)$, $-0{\cdot}42(Na^+)$, $-0{\cdot}11(K^+)$, $-0{\cdot}53(Rb^+)$, $-0{\cdot}36(Cs^+)$	64P
Th^{4+}	pol, (sp)	23	MeOH	K_1 2·23, β_2 4·17, β_3 5·80 or 5·89 ? in MeOH(l), 0·5M-$NaClO_4$; cpx are probably $Th(OCH_3)_2L_n^{2-n}$	68G
		23	Me_2NCHO	K_1 2·11, β_2 4·03, β_3 5·80, β_4 7·65 or 7·39 ? in Me_2NCHO(l)	
Cr^{3+}	sp, ir	25	2·5 ($NaClO_4$)	K_1 1·80, K_2 0·98, K_3 0·54	67Ga
Mn^{2+}	var		var	ev MnL_4^{2-} in solids and MeCN(l) (pre, ir, sp, mag)	67G
Fe^{2+}	*X*, var		solid	ev FeL_6^{4-} in K_2BaFeL_6(s) (esr, mag, sp, ir)	66E
$Fe(CN)_5^{3-}$	sp	10—35	1 (NaCl), 0·2 OH^-	K_1 3·96(10°), 3·52(25°), 3·24(35°); $\Delta H_1 = -12{\cdot}3$, $\Delta S_1 = -25{\cdot}2$	66S

continued overleaf

33 NO_2^- (contd)

Metal	Method	Temp	Medium	Log of equilibrium constant, remarks	Ref
Co^{2+}	X		solid	ev CoL_6^{4-} in K_2BaCoL_6(s)	66B
	X, var		solid	ev CoL_6^{4-} in K_2BaCoL_6(s) (esr, mag, sp, ir)	66E
	var		var	ev CoL_4^{2-} in solids and MeCN(l) (pre, ir, sp, mag)	67G
Co^{3+}	X		solid	ev CoL_6^{3-} in K_3CoL_6(s) *etc.*	36D
Co^{3+} cpx	kin	25	2·5 ($NaClO_4$)	$K_a[Co(NH_3)_4(LH)_2Cl^{2+}]$ 0·22 $K_a[Co(NH_3)_4LLHCl^+]$ −0·40	66M
	cal, kin	92—111	solid	$K[Co(NH_3)_5ONO^{2+} \rightleftharpoons Co(NH_3)_5NO_2^{2+}]$ in Cl^- salt 1·14(92°), 1·06(102°), 1·03(111°)	68D
Ni^{2+}	X		solid	ev NiL_6^{4-} in K_2CaNiL_6(s)	36D
	sp		10 NaL	ev $NiL_2(H_2O)_4$	63J/66E
	X, var		solid	ev NiL_6^{4-} in K_2BaNiL_6(s) (esr, ir, mag, sp)	66E
Ni^{2+} cpx	sp	1—51	$CHCl_3$	Q = $Ni(EtNHC_2H_4NHEt)_2^{2+}$; $K[Q(ONO)_2 \rightleftharpoons Q(NO_2)_2]$ 0·2(1°), −0·05(51°) in $CHCl_3$	66G
Pd^{2+}	Pd	25	KL var	β_4 21{$K[Pd^{2+}+2e^- \rightleftharpoons Pd(s)]$ 33·4}, also β_4 values for 10—60, assuming same K and same RTF^{-1} as for 25°!	65F
Os^{6+}	X		solid	ev $L_2O_2OsO_2OsO_2L_2^{4-}$ in $K_2OsO_3L_2(H_2O)_3$(s)	67A
Pt^{2+}	Pt	25	KL var	β_4 19·6	67Gb
Cu^{2+}	X		solid	ev CuL_6^{4-} in K_2PbCuL_6(s)	36D
	X, var		solid	ev CuL_6^{4-} in K_2BaCuL_6(s) (esr, ir, mag, sp)	66E
	var		var	ev CuL_4^{2-} in solids and MeCN(l) (pre, ir, sp, mag)	67G
Zn^{2+}	pre, ir		var	ev ZnL_4^{2-} in solids and MeCN(l)	67G
Cd^{2+}	pol	30	1 (KNO_3)	K_1 1·7, β_2 1·85, β_3 3·1	65J
	pol	25	2 ($NaClO_4$)	K_1 1·78, β_2 2·85, β_3 3·53, β_4 2·70	65S
	cal	25	3 ($NaClO_4$)	$\Delta H_1 = -2{\cdot}09, \Delta H_2 = -2{\cdot}11, \Delta H_3 = -1{\cdot}58$, $\Delta S_1 = 1{\cdot}1, \Delta S_2 = -1{\cdot}0, \Delta S_3 = -2{\cdot}4$	66Ga+
	pol	25	2 ($NaClO_4$)	$\beta(Cd^{2+}L^-Cl^-)$ 2·81, $\beta(Cd^{2+}L^-{}_2Cl^-)$ 3·48, $\beta(Cd^{2+}L^-Cl^-{}_2)$ 2·78	66Sa
Pb^{2+}	MHg	20	$I(NaClO_4)$ w% EtOH	K_1 2·13, β_2 2·96, β_3 3·15(I = 3·8), β_4 2·96(I > 5) β_3 4·68 in 50% EtOH, 5·11 in 80% EtOH	64G
	MHg	20	50% MeOH	K_1 2·40, β_2 3·57, β_3 3·80 in 50% MeOH, 1·6 M Li(ClO_4)	65G
			70% MeOH	β_2 3·70, β_3 4·12, in 70% MeOH, 1·6 M Li(ClO_4)	
			80% MeOH	K_1 3·44, β_2 4·05, β_3 5·02 in 80% MeOH, 1·6 M Li(ClO_4)	
			MeOH	K_1 3·6, β_2 5·0, β_3 5·5, β_4 5·6 in MeOH (l), 1·6 M Li(ClO_4)	
	sp		MeOH	K_1 2·41, β_2 3·26, β_4 5·24 in MeOH (l), I var	
	sol	20	KL var	$K_s[PbX_2(s)+L^- \rightleftharpoons PbX_2L^-]$ −0·7(X = SCN), −0·5(X = Cl), −0·8(X = Br), $\beta(Pb^{2+}X^-{}_2L^-)$ values given (main reaction?)	66Gb
	pol	30	1 ($NaClO_4$)	K_1 1·93, β_2 2·36, β_3 2·13	67J

36D M. van Driel and H. J. Verweel, *Z. Krist.*, 1936, *A*, **95**, 308
51Fa A. Ferrari, L. Cavalca, and M. Nardelli, *Gazzetta*, 1951, **81**, 982
63B N. S. Bayliss and D. W. Watts, *Austral. J. Chem.*, 1963, **16**, 927
63J C. K. Jørgensen, 'Inorganic Complexes', Academic Press, New York, 1963, p. 84
64G A. M. Golub and R. Akmyradov, *Ukrain. khim. Zhur.*, 1964, **30**, 1016
64P P. I. Protsenko, O. N. Shokina, and N. P. Chekhunova, *Zhur. fiz. Khim.*, 1964, **38**, 1857
65F A. B. Fasman, G. G. Kutyukov, and D. V. Sokol'skii, *Zhur. neorg. Khim.*, 1965, **10**, 1338
65G A. M. Golub and R. Akmyradov, *Izvest. V.U.Z., Khim. i khim. Tekhnol.*, 1965, **8**, 186

contin

65J D. S. Jain and J. N. Gaur, *Bull. Acad. polon. Sci. Sér. Sci. chim.*, 1965, **13**, 615
65L P. Lumme and J. Tummavuori, *Acta Chem. Scand.*, 1965, **19**, 617
65La P. Lumme, P. Lahermo, and J. Tummavuori, *Acta Chem. Scand.*, 1965, **19**, 2175
65S A. Swinarski and A. Grodzicki, *Roczniki Chem.*, 1965, **39**, 1155
66B J. A. Bertrand and D. A. Carpenter, *Inorg. Chem.*, 1966, **5**, 514
66E H. Elliott, B. J. Hathaway, and R. C. Slade, *Inorg. Chem.*, 1966, **5**, 669
66G D. M. L. Goodgame and M. A. Hitchman, *Inorg. Chem.*, 1966, **5**, 1303
66Ga P. Gerding, *Acta Chem. Scand.*, 1966, **20**, 2771
66Gb A. M. Golub and R. Akmyradov, *Zhur. neorg. Khim.*, 1966, **11**, 2347
66M A. McAuley and U. D. Gomwalk, *J. Chem. Soc.* (*A*), 1966, 1694
66S J. H. Swinehart and P. A. Rock, *Inorg. Chem.*, 1966, **5**, 573
66Sa A. Swinarski and A. Grodzicki, *Roczniki Chem.*, 1966, **40**, 373
66Sb G. Schmid and U. Neumann, *Ber. Bunsengesellschaft Phys. Chem.*, 1966, **70**, 1165
67A L. O. Atovmyan and O. A. D'yachenko, *Zhur. strukt. Khim.*, 1967, **8**, 169
67G D. M. L. Goodgame and M. A. Hitchman, *J. Chem. Soc.* (*A*), 1967, 612
67Ga A. Garnier, *Compt. rend.*, 1967, **265**, *B*, 198
67Gb A. A. Grinberg, M. I. Gel'fman, and D. B. Smolenskaya, *Zhur. neorg. Khim.*, 1967, **12**, 1588
67J D. S. Jain and J. N. Gaur, *Acta Chim. Acad. Sci. Hung.*, 1967, **51**, 165
67S H. Schmid and P. Krenmayr, *Monatsh.*, 1967, **98**, 417
67Sa H. Schmid and P. Krenmayr, *Monatsh.*, 1967, **98**, 423
68D V. Doron, *Inorg. Nuclear Chem. Letters*, 1968, **4**, 601
68G A. M. Golub, V. A. Kalibabchuk, and K. M. Boiko, *Zhur. neorg. Khim.*, 1968, **13**, 2111
68T J. Tummavuori and P. Lumme, *Acta Chem. Scand.*, 1968, **22**, 2003

Nitrate, NO_3^- 34

Metal	*Method*	*Temp*	*Medium*	*Log of equilibrium constant, remarks*	*Ref*
H^+	dis	20	0 corr	$K_D[H^+ + L^- + TH_2O(org) \rightleftharpoons HLT\,H_2O(org)]$ 0·0	59Db
				$K_D[H^+ + L^- + TH_2O(org) \rightleftharpoons HLT(org) + H_2O]$ −1·35	
				$K_D[HLT(org) + H^+ + L^- + H_2O \rightleftharpoons (HL)_2TH_2O(org)]$ −3·22, T = org = TBP	
	O	250—300	MNO_3(l)	$K[(NO_3^-) \rightleftharpoons NO_2^+ + O^{2-}]$ −25·57(250°), −23·24 (300°) in $Na_{0.5}K_{0.5}NO_3$(l), *m* units	63Kb
	gl, sp	25	6 ($NaClO_4$)	K_1 −0·47	63Pb
	sp	24	6 ($NaClO_4$)	K_1 −0·6 (higher [H^+] and [L^-] than with gl)	
	dis	25	HL var	$K_D[H^+ + L^- + T\text{ (in }CCl_4) \rightleftharpoons HLT\text{ (in }CCl_4)]$ −1·2[T = $(BuO)_3PO$], −0·82[T = $(BuO)_2BuPO$], −0·35(T = $BuOBu_2PO$), 0·48(T = Bu_3PO)	59Vb+60Pa/63Pc
	*p*HL	25	0 corr	K_1 −2·19 to −2·27	64Da
	con	25	$MeCO_2H$	K_1 5·1 in $MeCO_2H$(l)	64Fc
	dis	25	0·6 $NaNO_3$	$K_D[H^+ + L^- + T(org) \rightleftharpoons HLT(org)]$ 0·95	64H
				$\Delta H = -3{\cdot}9$, $\Delta S = -8{\cdot}8$ [$T = (C_8H_{17})_3PO$, org = cyclohexane], from data at 10—55°	
	dis	20	→ 0	$K_D[H^+ + L^- + T(org) \rightleftharpoons HLT(org)$, org = T = $(BuO)_nBu_{3-n}PO]$ 0·30(*n* = 3), 0·65(*n* = 2), 1·00(*n* = 1), data for *n* = 0	64Z
	*p*HL, *p*H_2O	25−75	0 corr	K_1 −1·53(25°), −1·31(50°), −1·16(75°), *m* units $\Delta H_1 = 3{\cdot}51$, $\Delta S_1 = 4{\cdot}81$	65Ha+60H
	con	20	HF	ev NO_2^+, $\beta(NO_2^+BF_4^-) \approx 0{\cdot}1$ in HF(l)	65Kb
	H	25	$C_2H_4(NH_2)_2$	K_1 3·20 in $C_2H_4(NH_2)_2$(l)	65Mb
	dis	25	0 corr	$K_D[H^+ + L^- + T(org) \rightleftharpoons HLT(org)]$, org = C_6H_6, −0·55(T = TBP), −0·27 [T = DAMFK = $(isoC_5H_{11}O)_2CH_3PO$], −1·90[T = $(C_6H_5O)_3PO$], also *K* for eleven other T	65Sa

continued overleaf

34 NO_3^- (contd)

Metal	*Method*	*Temp*	*Medium*	*Log of equilibrium constant, remarks*	*Ref*
H^+ (*contd*)	kin	260	MNO_3(l)	ev NO_2^+, $K(NO_3^- + Cr_2O_7^{2-} \rightleftharpoons NO_2^+ + 2CrO_4^{2-})$ −13·07 in (Na, K)NO_3(l, eut)	65Sd
	var		var	ev HL_2^- in solids and org solvents (dis, ir, con, pre, nmr)	66F
	var		var	ev HL_2^- in solids, *e.g.* Co $py_4Cl_2(HL_2)$ and in CH_2Cl_2(l), MeCN(l) (pre, con, ir)	66G
	dis	25	0 corr	$K[H^+ + L^- + T(org) \rightleftharpoons HLT(org)]$ 1·24(org = C_6H_6), 1·30(org = C_6H_5Me), 1·04(org = C_6H_{14}) $K_D[H^+ + L^- + H_2O + HLT(org) \rightleftharpoons H_2L_2TH_2O(org)]$ −2·54(org = C_6H_6), T = Bu_3PO	66Gd
	nmr	26—28	→0	K_1 −1·65 ? also graph of K_1 for dioxan–H_2O mixt	66Ra
	W	450	MCl(l)	ev $N_2O_7^{4-}$(pyronitrate) in (Li, K) Cl(l, eut)	66S
	var	275—350	MNO_3(l)	ev NO_2 ?, no ev NO_2^+ in $Na_{0.5}K_{0.5}NO_3$(l) (i*E*, chronopotentiometry)	66Tb
	dis		0 corr	$K_D[H^+ + L^- + H_2O + T(l) \rightleftharpoons HLTH_2O(\text{in } T)]$ −0·39, T = DAMFK = $(C_5H_{11}O)_2MePO$	66Ua
	X		solid	ev HL_2^- in Rh $py_4Cl_2^+H(NO_3)_2^-$(s)	67D
	nmr	32	C_6H_6	ev cpx HL–H_2O in C_6H_6(l), nmr and dis results compatible	67E
	dis	25	C_6H_6	$K_D[H^+ + L^- \rightleftharpoons HL(org)]$ −4·91, $K_D[H^+ + L^- + H_2O \rightleftharpoons HLH_2O(org)]$ −4·01 $K_D[2H^+ + 2L^- + H_2O \rightleftharpoons (HL)_2H_2O(org)]$ −7·77, org = C_6H_6(l)	
	dis	25	0 corr	$K_D[H^+ + L^- \rightleftharpoons HL(org)]$ −0·66(org = Bu^{i}COMe), −0·92(C_6H_{13}COMe), −0·89(PhCOMe)	67F
	X		solid	ev HL_2^- in Ph_4AsHL_2(s)	67Fa
	ΔG	25	0	K_1 −1·43, ΔH_1 = 4·1, ΔS_1 = 7·2, *m* units	/67H
	O, *X*	350—450	salt(l)	ev $N_2O_7^{4-}$ in (Li, K)Cl(l, eut), KNO_3(l) (*X* powder)	67Sa
	ir		solid	ev HL_4^{3-} ? in $Cr(NH_3)_6HL_4$(s) *etc.*	67U
	kin	25	HF var	$K(HF + H^+ + NO_3^- \rightleftharpoons F^- + H_2O + NO_2^+)$ −1·96, M units	67Vc
	sp, act	25 ?	var	$K_{12}(HL + L^- \rightleftharpoons HL_2^-)$ 0·6	68Db+
	sp	25	C_5H_5N	K_1 4·06 in C_5H_5N(l)	68M
	dis	25 ?	0 corr	$K_D[H^+ + L^- + T(org) \rightleftharpoons HLT(org)]$ 0·69 $K_D[H^+ + L^- + HLT(org) \rightleftharpoons H_2L_2T(org)]$ −0·74 T = MHDPO = $[(C_6H_{13})_2PO]_2CH_2$, org = CCl_4	68O
	dis	25	HL var	$K_D[H^+ + L^- + T(org) \rightleftharpoons HLT(org)]$ −0·72 T = $(C_6H_{13}O)_2PhPO$, org = $C_{10}H_{22}$	68R
	nmr	0—65	0 corr	$K(H_2O + HL \rightleftharpoons H_3O^+ + L^-)$ 1·55(0°), 1·3(25°), 1·0(65°), M units	/68Ra
D^+	nmr	0—65	D_2O, 0 corr	$K(D_2O + HL \rightleftharpoons HD_2O^+ + L^-)$ 1·68(0°), 1·4(25°), 1·0(65°) in D_2O(l), M units	/68Ra
			D_2O, 0 corr	$K(D_2O + DL \rightleftharpoons D_3O^+ + L^-)$ 1·5(0°), 1·25(25°), 0·85(65°), in D_2O(l), M units	
Group 1A$^+$	Ram		ML var	ev cpx, M^+ = Li^+, Na^+, K^+	64Lb
	act	25 ?	0 corr	K_1 −0·2(Na^+), 0·1(K^+)	/66M
Li^+	cix	23	90% Pr^{i}OH	K_1 −0·54 in 90% Pr^{i}OH, 0·5M-HL	66W
	con	25	org	K_1 given for mixtures of dioxan with MeOH Me_2CO, and Me_2SO	67P
	H	25	C_5H_5N	$K_1(Li^+)/K_1(H^+)$ 0·91 in C_5H_5N(l)	68M
Na^+	Ram		NaL conc	no ev inner cpx	64Hb

contin

Metal	*Method*	*Temp*	*Medium*	*Log of equilibrium constant, remarks*	*Ref*
K^+	Ram		KL conc	no ev inner cpx	64Hb
	E	25—39	0·1 (Me_4NCl)	K_1 −0·4(25°), −0·5(39°), (K^+ glass electrode)	66Cb
			0 corr	K_1 −0·15(25°), −0·25(39°), $\Delta H_1 = -3$, $\Delta S_1 = -12$	
Group $2A^{2+}$	act	25 ?	0 corr	K_1 0·6(Ca^{2+}), 0·7(Sr^{2+}), 1.1(Ba^{2+})	/66M
Mg^{2+}	cix	23	90% Pr^iOH	K_1 0·20, β_2 −0·07 in 90% Pr^iOH, 0·5M-HL	66W
Ca^{2+}	con	18—25	0 corr	K_1 0·41(18°), 0·31(25°), $\Delta H_1 = -5{\cdot}7$, $\Delta S_1 = -18$	63V
	Ram		CaL_2 conc	ev cpx	64Hb
	Ram	26—94	10—40 (NH_4ClO_4)	K_1 −0·78 (26°), −0·82(48°, 70°), −0·85(94°)	64Hc
	cix	23	90% Pr^iOH	K_1 0·57, β_2 0·85 in 90% Pr^iOH, 0·5M-HL	66W
Sr^{2+}	con	18—25	0 corr	K_1 0·57(18°), 0·54(25°), $\Delta H_1 = -1{\cdot}7$, $\Delta S_1 = -3$	63V
	cal	25	*c* $LiNO_3$	$\Delta H_1 = 0$, $\Delta S_1 = 2{\cdot}6$($c = 0{\cdot}5$), also data for $c = 0{\cdot}1$—2(K_1 from ref 63V)	64V
			c KNO_3	$\Delta H_1 = -2{\cdot}4$, $\Delta S_1 = -5{\cdot}7$($c = 0{\cdot}5$), also data for $c = 0{\cdot}1$—2, K_1 from ref 63V, difference K–Li ascribed to H_2O cpx	
Ba^{2+}	con	18—25	0 corr	K_1 0·98(18°), 0·94(25°), $\Delta H_1 = -2{\cdot}3$, $\Delta S_1 = -3$	63V
	cal	25	*c* $LiNO_3$	$\Delta H_1 = -1{\cdot}9$, $\Delta S_1 = -2{\cdot}1$($c = 0{\cdot}5$), also data for $c = 0{\cdot}1$—2 (K_1 from ref 63V)	64V
			c KNO_3	$\Delta H_1 = -3{\cdot}2$, $\Delta S_1 = -6{\cdot}5$($c = 0{\cdot}5$), also data for $c = 0{\cdot}1$—2, (K_1 from ref 63V), difference K–Li ascribed to H_2O cpx	
Sc^{3+}	cix		0·5 H, 2 ($NaClO_4$)	K_1 0·55, β_2 0·08	64Sb
	aix		HL var	ev ani cpx in > 3·8M-HL	
	dis	?	5 (NH_4L)	$K_D[Sc^{3+} + 3L^- + 3T(org) \rightleftharpoons ScL_3T_3(org)]$ 2.6, T = TBP, org = ?, units = ?	65S
	cix	25	4 H(ClO_4)	K_1 0·28, β_2 −0·28	66Sa
Group $4f^{3+}$	var		$LiNO_3$ var	ev outer (dis, cix) and inner (nmr, sp) cpx	67A
	dis	25	1 ($NaClO_4$)	K_1 0·26(Pm^{3+}), 0·31(Sm^{3+}), −0·03(Tb^{3+}), —0·28(Dy^{3+}), −0·22(Ho^{3+}), −0·33(Er^{3+})	67Kb

1964 *Erratum*: p. 169, *for ref* "62K" *read* "62K = 63Sc", *and for* "HL var" *read* "3·26 H(ClO_4)".

Metal	*Method*	*Temp*	*Medium*	*Log of equilibrium constant, remarks*	*Ref*
La^{3+}	aix	25, 78	LiL var	ev LaL_7^{4-}	64Ma
Ce^{3+}	*X*		solid	ev CeL_6^{3-} in $Mg_3(CeL_6)_2(H_2O)_{24}$(s)	63Z
	Ram		CeL_3 conc	ev cpx	64Hb
	dis	25	1 H(ClO_4)	K_1 0·21	65C
	dis	25	HNO_3 var	$K_D[Ce^{3+} + 3L^- + 3T(org) \rightleftharpoons CeL_3T_3(org)]$ given for T = 15 compounds $(RO)_3PO$ or $RP(O)(OR)_2$, org = C_6H_6	66O
Ce^{4+}	pre, tp		var	ev CeL_6^{2-} in salts and EtOH(l)	01M
	sp	?	6H(ClO_4)	K_1 −0·24, β_2 −0·48, β_3 −0·72, β_4 −0·96, β_5 −1·22, β_6 −1·46 (all K_n assumed equal)	63L
	X		$(NH_4)_2CeL_6$ conc	ev CeL_6^{2-}	64La
	dis	?	2 H(ClO_4)	$K_D[Ce^{4+} + 5L^- + H^+ + T(org) \rightleftharpoons HCeL_5T(org)]$ 1·81, T = TBP, org = ?; K_1 0·78, β_2 1·20, β_3 1·34, β_4 1·24, β_5 1·07	64Sa
	sp	23	3·56 H(ClO_4)	K_1 0·33 [$K_1(H^+)$ −1·19 at $I = 0$]	65P
	Ram, ir		L^- var	ev cpx	67Mc
	X		solid	ev $Ce(O_2NO)_6^{2-}$ in $(NH_4)_2CeL_6$(s)	68B

continued overleaf

34 NO_3^- (contd)

Metal	Method	Temp	Medium	Log of equilibrium constant, remarks	Ref
Pr^{3+} Nd^{3+}	var		HL var	ev ani cpx in > 3M-HNO_3 (tp, aix, cix)	63Sc

1964 *Addendum*: p. 169, *for ref* "62S" *read* "62S = 63Sc".

Metal	Method	Temp	Medium	Log of equilibrium constant, remarks	Ref
	sp	20	4·2 ($NaClO_4$)	K_1 −0·11	66C
Pm^{3+}	dis	25	1 H(ClO_4)	K_1 0·39	65C
Eu^{3+}	cix	26	1 H(ClO_4)	K_1 0·15, $\beta_2 \approx -0{\cdot}4$	64B
			1 Na(ClO_4)	K_1 0·20	
	aix	25, 78	LiL var	ev EuL_5^{2-}	64Ma
	dis	25	*I* H(ClO_4)	K_1 0·31(I = 1·0), 0·43(I = 0·2), 1·23 ($I \to 0$), also K_1 for intermediate *I* values	65C
		0—55	1 H(ClO_4)	K_1 0·32(0°), 0·30(25°), 0·26(40°), 0·25(55°), $\Delta H_1 = -0{\cdot}57$, $\Delta S_1 = -0{\cdot}54$	
	cix	25	4 H(ClO_4)	K_1 0·17, β_2 −0·72	67S
		25	4 Na(ClO_4)	K_1 0·12, β_2 −0·52	
Tb^{3+}	dis	25	*I* H(ClO_4)	K_1 0·05(I = 1·0), 0·13(I = 0·3), 0·88($I \to 0$), also K_1 for intermediate *I*	65C
Tm^{3+}	dis	25	1 H(ClO_4)	K_1 −0·25	65C
Yb^{3+}					

1964 *Errata*: p. 169, *for ref* "62K" *read* "62K = 63Sc", *in which for* "HL var" *read* "3·26 H(ClO_4)", *in second line, add method* "hyp".

Metal	Method	Temp	Medium	Log of equilibrium constant, remarks	Ref
	aix	25	LiL var	ev YbL_5^{2-}	64Ma
Ac^{3+}	dis	27	1 H(ClO_4)	K_1 0·12, β_2 0·01	68S
Group $4A^{4+}$, 5f					
	lit			empirical equations, including "extraction constants" and "equivalent surface density in first co-ordination sphere" for distribution of ZrL_4, ThL_4, UO_2L_2, NpO_2L_2, PuL_3, PuL_4, PuO_2L_2 between organic and aqueous phase	/64S
Ti^{4+}	*X*		solid	ev TiL_4 units in TiL_4(s)	66Ga
Zr^{4+}	var		solid	ev ZrL_6^{2-} in $(Me_4N)_2ZrL_6$(s) (ir, pre, X)	64Ba
	dis	20	2 ($LiClO_4$)	$K_D[ZrO^{2+} + 2H^+ + 4L^- + nT(org) \rightleftharpoons ZrL_4T_n(org) + H_2O]$ 2·79(n = 1), 3·51(n = 2), T = "DAMFK" = (iso-$C_5H_{11}O)_2$MePO, org = xylene	65Ua
	dis	25	2 H(ClO_4)	no ev cpx up to 2M-NO_3^-	67Sb
Hf^{4+}	var		solid	ev HfL_6^{2-} in $(Me_4N)_2HfL_6$(s) (ir, pre, X)	64Ba
	dis	27	2 H(ClO_4)	K_1 −0·1	65D
	cix	20	2 H(ClO_4)	K_1 −0·15, β_2 −0·96	67Ea
Th^{4+}	dis	25	1 HNO_3	$K_D[Th^{4+} + 4L^- \rightleftharpoons ThL_4(org)]$ −2·0, org = hexol(MeBu^{t}CHOH)	64D
	Ram		ThL_4 conc	ev cpx	64Hb
	cix, tp		4 H(ClO_4)	K_1 0·55, β_2 0·32, β_3 −0·30, β_4 −0·72(cix), ev ani cpx (tp)	64N
	X		solid	ev ThL_6^{2-} in $Mg(H_2O)_6ThL_6(H_2O)_2$(s)	65Sc
	dis	20?	2 ($LiClO_4$)	$K_D[Th^{4+} + 4L^- + nT(org) \rightleftharpoons ThL_4T_n(org)]$ 3·76 (n = 2), 4·49(n = 3), T = "DAMFK" = (iso-$C_5H_{11}O)_2$MePO, org = *p*-xylene	65U

continued

Metal	Method	Temp	Medium	Log of equilibrium constant, remarks	Ref
Th^{4+}	*X*, *X*n		solid	ev $ThL_4(H_2O)_3$, in $Th(NO_3)_4(H_2O)_5$(s)	66T = 66U
(*contd*)	*X*		solid	ev dinuclear $[(H_2O)_3ThL_3]_2(OH)_2$ in $ThOHL_3(H_2O)_4$(s)	68J
	X		2 $Th(NO_3)$	ev cpx in solution	
	cix	25	2 $H(ClO_4)$	K_1 1·22, β_2 1·53, β_3 1·1	68T
VO_2^+	sp	20	1 ($NaClO_4$)	K_1 −0·5	66I
Nb^{5+}	pre, ir		solid	ev $NbOL_4^-$ in Me_4NNbOL_4(s)	66Ba
Ta^{5+}	pre, ir		solid	ev $TaOL_4^-$ in Me_4NTaOL_4(s)	66Ba
Pa^{5+}	cix, aix	20	4 $H(ClO_4)$	K_1 −0·20, β_2 −0·68	63N
			2 $H(ClO_4)$	K_1 −0·10, β_2 −0·13	
			1 $H(ClO_4)$	K_1 −0·17, β_2 0·48, β_4 1·08	
	dis		3 $H(ClO_4$?) var	ev PaL_n^{5-n} in aq, PaL_{n+1}^{4-n} in org, PaL_4^+ and PaL_5 (org) suggested	63Sd
	dis		5 $H(ClO_4)$	K_1 1·23, β_2 2·10, β_3 2·73, β_4 3·14, [Pa^V probably = $Pa(OH)_2^{3+}$]; $K_D[Pa(OH)_2^{3+} + 2H^+ + 5L^- + 2T(\text{in org}) \rightleftharpoons PaL_5T_2(\text{org}) + 2H_2O]$ 3·73, T = TBP, org = C_6H_6	64Sc = 66Ka
	pre, ir		var	ev PaL_6^- in $CsPaL_6$(s) *etc.*, Pa_2OL_8 in MeCN(l)	66Ba
	dis	25	1 $HClO_4$	K_1 0·16, K_2 −1·15	67K
Cr^{3+}	ana	10—35	1 ($HClO_4$), 0·4 H^+	K_1 −2·17(10°), −2·01(25°), −1·91(35°) ΔH_1 = 4·5, ΔS_1 = 5·9	67Aa
MoO_2^{2+}	sp	20	*I* $H(ClO_4)$	K_1 0·14(*I* = 2), 0·10(*I* = 1), 0·14(*I* = 0·5) β_2 −0·17(*I* = 2), −0·26(*I* = 1), −0·08(*I* = 0·5) corrected for assumed mononuclear hydrolysis (MoO_4 table, ref 67Va)	67Va
	cix	20	1 $H(ClO_4)$	K_1 0·15, β_2 −0·15	
	dis	20	HL var	$K_D[MoO_2^{2+} + 2L^- + 2T(\text{org}) \rightleftharpoons MoO_2L_2T_2(\text{org})]$ −0·95, T = TBP, org = kerosene	67Vb
U^{4+}	sp	26·5	0 corr	K_1 1·55	62E/66Sb
	dis	25	*I*(NO_3^-)	$K_D[U^{4+} + 4L^- + 2T(\text{org}) \rightleftharpoons UL_4T_2(\text{org})]$ 0·65(*I* = 1·75) to 1·13(*I* = 2·75), T = TBP, org = kerosene	63Sb
	sp	20	*I*($LiClO_4$)	K_1 0·04(*I* = 1), 0·06(*I* = 2), 0·20(*I* = 3), 0·18(*I* = 4), β_2 −0·3(*I* = 1), 0·0(*I* = 2), 0·3(*I* = 3), 0·8(*I* = 4), β_3 < 1(*I* = 1—4), ev H_2UL_6 in conc HNO_3	64Mc
	sp	20	TBP	$K_D[UL_4(\text{org}) + H_2O \rightleftharpoons UOHL_3(\text{org}) + HL(\text{org})]$ −1·46, org = TBP, ev UL_6^{2-} in TBP	65W
	var		HL var	ev UL_6^{2-} (dis, ir, pre, sp)	66K
UO_2^{2+}	con	25	org	K_2 5·70 in Me_2CO(l), 6·02 in MeCOEt(l), 6·39 in Bu^iCOMe(l) (→ 0), $6H_2O/UO_2$	61J/66J
	dis	25	0 corr	$K_D[UO_2^{2+} + 2L^- + 2T(\text{org}) \rightleftharpoons UO_2L_2T_2(\text{org})]$ 1·68, T = org = TBP, M units	62A
		25	0 corr	K_D 2·37, *x* units in org	62A/68Ma
				$K_D[UO_2^{2+} + 2L^- + 2T(\text{in } CCl_4) \rightleftharpoons UO_2L_2T_2(\text{in } CCl_4)]$ 1·77[T = $(BuO)_3PO$], 3·25[T = $(BuO)_2BuPO$], 4·84 (T = $BuOBu_2PO$), 7·13(T = Bu_3PO);	58Ba/63Pc
	dis	25	0 corr	$K_D[UO_2^{2+} + 2L^- + 2T(\text{in } CCl_4) \rightleftharpoons UO_2L_2T_2(\text{in } CCl_4)]$ 1·74[T = $(BuO)_3PO$], 3·46[T = $(BuO)_2BuPO$], 5·14(T = $BuOBu_2PO$), 7·23(T = Bu_3PO);	59Vb + 60Pa/63Pc

continued overleaf

34 NO_3^- (contd)

Metal	Method	Temp	Medium	Log of equilibrium constant, remarks	Ref
UO_2^{2+} (*contd*)				$K_D[UO_2^{2+}+2L^-+2T(\text{in } CCl_4) \rightleftharpoons UO_2L_2T_2$ (in CCl_4)] 1·90[T = $(BuO)_3PO$], 3·62[T = $(BuO)_2BuPO$], 6·50(T = Bu_3PO);	60Pb+60Ta/63Pc
				$K_D[UO_2^{2+}+2L^-+2T(\text{in } C_6H_6) \rightleftharpoons UO_2L_2T_2$ (in C_6H_6)] 2·27[T = $(BuO)_3PO$], 3·89[T = $(BuO)_2BuPO$], 5·03[T = $(BuO)Bu_2PO$], 5·92(T = Bu_3PO)	59H/63Pc
	dis, *p* org	25	0 corr	$K_D[UO_2^{2+}+2L^-+2T(org) \rightleftharpoons UO_2L_2T_2(org)]$, T = TBP, 0·15(org = $CHCl_3$), 1·1($C_2H_4Cl_2$) 1·75(CCl_4); org = CCl_4: 1·75[T = TBP = $(BuO)_3PO$], 3·41[T = $(BuO)_2BuPO$], 5·08(T = $BuOBu_2PO$), 7·0(T = Bu_3PO); H_2O–T cpx neglected (included in [T]), M units in org; activity coefficients of T and $UO_2L_2T_2$ vary but approximately cancel each other out; also values for K_D with *x* units	63Pc
	dis	25	0 corr	$K_D[UO_2^{2+}+2L^-+2T(org) \rightleftharpoons UO_2L_2T_2(org)]$ 1·68, T = TBP, org = CCl_4	64Aa
	dis	25	0·6 $NaNO_3$	$K[UO_2^{2+}+2L^-+2T(org) \rightleftharpoons UO_2L_2T_2(org)]$ 8·4, $\Delta H = -5{\cdot}8$, $\Delta S = 19$[T = $(C_8H_{17})_3PO$, org = cyclohexane, from data at 10—55°]	64H
	con	25	org	K_2 3·67(in MeOH), 4·47(in EtOH), 5·59(in PrOH) (anhydrous)	64J
				K_2 3·21(in MeOH), 4·16(in EtOH) [soln of $UO_2L_2(H_2O)_6$]	
	dis	25?	0 corr	$K_D[UO_2^{2+}+2L^-+2T(org) \rightleftharpoons UO_2L_2T_2(org)]$ 2·10 (org = C_6H_6), 1·68(CCl_4), −0·20($CHCl_3$), T = TBP, M units; also values for some C_5H_5NO derivatives	64Na
	dis	20	→ 0	$K_D[UO_2^{2+}+2L^-+2T(org) \rightleftharpoons UO_2L_2T_2(org)]$, org = T = $(BuO)_nBu_{3-n}PO$, 1·30(*n* = 3), 2·27(*n* = 2), 3·10(*n* = 1)	64Z
	*X*n		solid	ev $UO_2(NO_3)_3^-$ in $RbUO_2(NO_3)_3(s)$	65B
	*X*n		solid	ev $UO_2L_3^-$ in $RbUO_2L_3(s)$	65Ba
	con	25	ROH	K_2 3·67 (no H_2O), 3·45($2H_2O$), 3·21($6H_2O/UO_2$) in MeOH(l) → 0	66E
				K_2 4·47 (no H_2O), 4·25($2H_2O$), 4·16($6H_2O/UO_2$) in EtOH(l), → 0	
	con	25	org	K_2 6·36 in Me_2CO(l), 6·52 in MeCOEt(l), 6·93 in Bu^iCOMe(l), 6·64 in heptan-2-one(l) (→ 0), water-free	66J
	dis	25	0 corr	$K_D[UO_2^{2+}+2L^-+2T(org) \rightleftharpoons UO_2L_2T_2(org)]$ 2·4—2·5, T = $(C_6H_{13}O)_2PhPO$, org = $C_{10}H_{22}$	68R
	dis	25	0 corr	$K_D[UO_2^{2+}+2L^-+2S(org) \rightleftharpoons UO_2L_2S_2(org)]$ 1·08(org = C_6H_6), −0·08(org = CCl_4), S = $(C_8H_{17})_2SO$	68Ta
Np^{4+}	sp	25	2 ($NaClO_4$)	K_1 0·11	62Sb
	sp, kin	25	2 $H(ClO_4)$	K_1 0·34, β_2 0·18	66R
	dis	20	*I*$H(ClO_4)$	K_1 0·30(*I* = 2), 0·34(*I* = 1), 0·45(*I* = 0·5), β_2 0·34(*I* = 2), 0·08(*I* = 1), β_3 −0·26(*I* = 1)	66Sb
			0 corr	K_1 1·68	

continued

Metal	Method	Temp	Medium	Log of equilibrium constant, remarks	Ref
NpO_2^+	cix	25	2 H(ClO_4)	K_1 −0·25	64G
NpO_2^{2+}	dis	20	→ 0	$K_D[NpO_2^{2+} + 2L^- + 2T(org) \rightleftharpoons NpO_2L_2T_2(org)$, org = T = $(BuO)_nBu_{3-n}PO]$ −0·07(n = 3), 0·60(n = 2), 1·40(n = 1)	64Z
	sp	25	2 H(ClO_4)	K_1 −0·4	66R
Pu^{4+}	sp	25	0 corr	K_1 1·80	49Ha/66Sb
	dis	25	0 corr	$K_D[Pu^{4+} + 4L^- + 2T(in\ CCl_4) \rightleftharpoons PuL_4T_2(in\ CCl_4)]$ 3·06[T = $(BuO)_3PO$], 4·61[T = $(BuO)_2BuPO$], 5·75(T = $BuOBu_2PO$);	58Ba/63Pc
				$K_D[Pu^{4+} + 4L^- + 2T(in\ CCl_4) \rightleftharpoons PuL_4T_2$ (in CCl_4)] 3·55[T = $(BuO)_3PO$], 4·75[T = $(BuO)_2BuPO$], 7·40(T = Bu_3PO)	60Pb + 60Ta/63Pc
	dis	?	4·7 ($NaClO_4$), 0·7 H^+	K_1 0·7, β_2 1·1, β_3 1·1, β_4 0·3 to 0·9	64L
			1·9 ($NaClO_4$), 0·6 H^+	K_1 0·61, β_2 0·85, β_3 0·64, β_4 0·11	
			1·02 H(ClO_4)	K_1 0·72, β_2 0·97, β_3 0·63; criticism in ref. 68F no ev PuL_3^+ or PuL_4, low precision	64L/66D
	dis	20	→ 0	$K_D[Pu^{4+} + 4L^- + 2T(org) \rightleftharpoons PuL_4T_2(org)$, org = T = $(BuO)_nBu_{3-n}PO]$ 3·30(n = 3), 4·22(n = 2), 5·10(n = 1), data for n = 0	64Z
	dis	25	I(HClO_4)	K_1 1·00, β_2 1·36, β_3 0 ?(I = 6) K_1 0·97, β_2 1·43, β_3 −0·4 ?(I = 4)	66D
PuO_2^{2+}	dis	25	0 corr	$K_D[PuO_2^{2+} + 2L^- + 2T(in\ CCl_4) \rightleftharpoons PuO_2L_2(in\ CCl_4)]$ 0·97[T = $(BuO)_3PO$], 2·37[T = $(BuO)_2BuPO$], 3·57(T = $BuOBu_2PO$)	58Ba/63Pc
	dis	20	→ 0	$K_D[PuO_2^{2+} + 2L^- + 2T(org) \rightleftharpoons PuO_2L_2T_2(org)$, org = T = $(BuO)_nBu_{3-n}PO]$ 0·26(n = 3), 1·15(n = 2), 3·05(n = 1)	64Z
	dis	?	4·1 (ClO_4)	K_1 −0·03, β_2 −0·7 ?	65Ma
Am^{3+}	cix	26	1 H(ClO_4)	K_1 0·15, β_2 −0·4	64B
			1 Na(ClO_4)	K_1 0·20	
	dis		1 Na(ClO_4)	K_1 −0·26, β_2 0·18	64K
	aix	25	LiL var	ev AmL_5^{2-} ?	64Ma
	dis	25	1 H(ClO_4)	K_1 0·26	65C
	sp		1 $HClO_4$	K_1 −1·3	66Gc
Mn^{2+}	cix	23	90% Pr^{i}OH	K_1 0·46, β_2 0·36 in 90% Pr^{i}OH, 0·5M-HL	64Fb
Fe^{3+}	con	25	var, p bar	K_1 0·76(1 bar), 0·66(4800 bar)	64Ha
	var		org	ev FeL_4^- in solids, in MeCN(l) and $MeNO_2$(l) (con, ir, pre, mag)	65Aa
Co^{2+}	var		var	ev CoL_3^- in MeCN(l) and $NO^+CoL_3^-$(s) (con, ir, pre, sp)	64A
	X		solid	ev CoL_4^{2-} in $(Ph_4As)_2Co(NO_3)_4$(s)	64C = 66B
Co^{3+}	X		solid	ev $Co(O_2NO)_3$ in $Co(NO_3)_3$(s)	68H
Co^{3+} cpx	sp	25	1·0	$Co(NH_3)_5^{3+}$; K_1 −0·19	63Hb
	sol	25	0 corr	$Co(NH_3)_5Cl^{2+}$; K_1 1·05	65A
	pH$_2$O	37	→ 0	Co $en_2NH_3NO_2^{2+}$; K_1 2·3(*cis*), 2·3(*trans*)	67M
	pH$_2$O	37	→ 0	Co en_2NSCCl$^+$; K_1 0·3	67Ma
	con	25	→ 0	$Co(NH_3)_6^{3+}$; K_1 1·63	68K

continued overleaf

34 NO_3^- (contd)

Metal	Method	Temp	Medium	Log of equilibrium constant, remarks	Ref
Ni^{2+}	sp		Me_2CO	K_3K_4 4·81 in Me_2CO(l)	63Ta
	MHg, sp	125	Me_2SO_2	β_2 3·43, K_3 0·78 in Me_2SO_2(l), 2M-$LiClO_4$	68L
	sp	125	MNO_3(l)	ev mainly NiL_3^- in (Li, Na, K)NO_3(l, eut)	
$RuNO^{3+}$	var		var	ev mixed cpx $RuNO^{3+}-L^-$ with $(NH_2)_2CS$, SO_4^{2-}, Cl^-, not F^-; partial equilibria described, true formulae unknown (dis, sp, paper chromatography)	64Va
	dis	0	0 corr	$K_D[RuNOL_3+4T(org) \rightleftharpoons RuNOL_3T_4(org)+4H_2O]$, 4·3, T = TBP, org = kerosene, also data for $RuNOL_4^-$	65F
	var	20	HL var	ev $RuNOL_n^{3-n}[n = 0, 1, 2$ (*cis*, *trans*), 3, 4], $K_3K_2^{-1}$ −0·8, $K_4K_3^{-1}$ −0·8, $K(RuNOL_2^+$ *trans* $\rightleftharpoons$ *cis*) 0·15(cix, dis, paper chromatography, kin)	65Sb
Cu^{2+}	sp		Me_2CO	K_3K_4 4·23 in Me_2CO(l)	63T
	Ram		CuL_2 conc	ev cpx	64Hb
	X		solid	ev $CuL_2(H_2O)_2^{2+}$ in $CuL_2(H_2O)_{2·5}$(s)	68G
Ag^+	Ag	23	*w*% Pr^iOH	K_1 −0·57(*w* = 0), −0·28(*w* = 40), 0·20(*w* = 70), 0·53(*w* = 80), 0·73(*w* = 90) in *w*% Pr^iOH, ≈ 0·5M-$HClO_4$, also data for 90% Pr^iOH, various $H(ClO_4)$ conc	64F
	cix, aix	23	*w*% Pr^iOH	K_1 0·7(*w* = 90) (cix), ev ani cpx in resin (aix)	
	con	25	org	K_1 3·34 in $C_2H_4(NH_2)_2$(l), 4·09 in $C_3H_6(NH_2)_2$(l) (→ 0)	64Fa
	Ram		AgL conc	ev cpx	64Hb
	Ram		AgL var	ev cpx	64Lb
	Ag, aix	22?	org–H_2O	K_1 1·06(*I* = 0·05) 0·39(*I* = 0·80) in *I*M-$H(ClO_4)$, 80% Pr^iOH, K_1 also for intermediate *I*	65M
	oth	25	AgL var	ev cpx (diffusion method)	66Ja
	act	25?	0 corr	K_1 0·2	/66M
	ir		MeCN	ev AgL in MeCN(l)	67J
Group $2B^{2+}$	Ram, ir		MeCN	ev outer cpx for Zn^{2+}, Cd^{2+}, inner cpx for Hg^{2+} in MeCN(l)	68A
Zn^{2+}	Ram		ZnL_2 conc	ev cpx	64Hb
	ir, Ram		var	ev contact ion pair $Zn^{2+}L^-$	66H
Cd^{2+}	dis	30	3 ($NaClO_4$)	K_1 0·1, β_2 0·1	65H
	pol	125	Me_2SO_2	K_1 2·12, β_2 3·72, β_3 5·2, β_4 6·8 in Me_2SO_2(l)	66Ab
	act	25?	0 corr	K_1 0·2	/66M
	Ram, ir	44	var	K_1 −0·42	68Da
Hg^{2+}	Ram		HgL_2 conc	ev cpx	64Hb
	Ram, ana		Me_2CO	ev HgLOC(Me)CH_2 in Me_2CO(l)	64Mb
	pre, ir		solid	ev HgL_4^{2-} in $(Me_4N)_2HgL_4$(s)	65Ba
	cix	23	90% Pr^iOH	K_1 0·83, β_2 1·56, in 90% Pr^iOH, 0·5M-HL	66W
	Ram, ir	25	var	data agree with K_1 0·11, K_2 0·0 (ref 46I)	68D
$MeHg^+$	Ram		var	ev cpx, K_1 ≈ −1·4(est)	66Ca
B^{3+}	var		MeCN	ev BL_4^- in MeCN(l) (con, fp, ir, pre)	66Gb
Al^{3+}	Ram		AlL_3 conc	ev cpx	64Hb
	var		var	ev AlL_4^- in solids, MeCN(l), $MeNO_2$(l) (pre, ir, con, sp)	66Aa
	nmr		AlL_3 var	ev only $Al(H_2O)_6^{3+}$, hence only outer cpx, no inner cpx	68Mb

continued

Metal	*Method*	*Temp*	*Medium*	*Log of equilibrium constant, remarks*	*Ref*
In^{3+}	Ram		InL_3 conc	ev cpx	64Hb
	pre, ir		solid	ev InL_2^+, InL_4^-	66Ta
	cix	20	0·69 $H(ClO_4)$	K_1 0·18, K_2 −0·49	68Fa
Tl^+	MHg	25	3 $(LiClO_4)$	K_1 −0·48, $\Delta H_1 = -6{\cdot}2$, $\Delta S_1 = -23$	65Ka
	act	25?	0 corr	K_1 0·5	/66M
	con	25	w% dioxan	K_1 0·51(w = 0), 0·60(w = 7·8), 0·69(w = 16),	68Dc
				1·04(w = 36·2), 1·54(w = 52·5), 1·81(w = 59·2),	
				2·61(w = 70·0) 3·27(w = 76·4) in w% dioxan, → 0	
Tl^{3+}	red, gl	25	3 $(LiClO_4)$	K_1 0·90, β_2 0·12, β_3 1·10, $\beta(Tl^{3+}L^-H^+_{-2})$ −2·10,	65K
				$\beta(Tl^{3+}L^-{}_2H^+_{-1})$ −0·32, $\beta(Tl^{3+}L^-{}_3H^+_{-1})$ −0·40	
	ir, sol	25	$PhNO_2$	ev $Tl(dip)_2L_2^+$ in $PhNO_2$(l)	66Fa
	cal, red	25	3 $(LiClO_4)$, 0·5 H^+	K_1 0·90, $\Delta H_1 = 0$, $\Delta S_1 = 4{\cdot}1$	67Mb
Sn^{4+}	pre, ir		solid	ev $SnCl_4L^-$ in $NO^+SnCl_4NO_3^-$(s)	66A
	X		solid	ev $Sn(O_2NO)_4$ in SnL_4(s)	67G
Pb^{2+}	MHg	25	*I* $Na(ClO_4)$	K_1 0·5, β_2 0·0, β_3 −0·3, β_4 −0·2(I = 3)	63Mc
				K_1 0·34, β_2 0·56(I = 1) also data for I = 4, 2, 0·75	
	pol	27	w% Pr^{i}OH	K_1 0·32, β_2 0·02(w = 25, I = 1·5),	64M
				K_1 0·81, β_2 1·15(w = 80, I = 1·5),	
				K_1 1·00, β_3 2·19(w = 90, I = 1·0)	
				in w% Pr^{i}OH *I*M-$H(ClO_4)$	
	MHg	25	2 $(NaClO_4)$	K_1 0·15, β_2 0·39	65Hb
	pol	25	2 $(NaClO_4)$	K_1 0·3, β_2 0·4, $\beta_3 < -2{\cdot}3$	
	pol, aix	?	org	K_1 0·6, β_2 1·0—0·9 in 0·2—0·6M-$H(ClO_4)$, 80% Pr^{i}OH	65M
	pol	125	Me_2SO_2	K_1 2·46, β_2 3·60, β_3 4·6 in Me_2SO_2(l)	66Ab
	act	25?	0 corr	K_1 1·3	/66M
	MHg	2—65	3 $Li(ClO_4)$	K_1 0·56(2°), 0·51(25°), 0·42(43·5°), 0·36(65°);	67Fb
				β_2 0·53(2°), 0·32(25°), 0·30(43·5°), 0·23(65°);	
				β_3 0·30(2°), 0·32(25°), 0·18(43·5°), 0·15(65°),	
				β_4 −0·2(2°)	
		25		$\Delta H_1 = -1{\cdot}4$, $\Delta H_2 = -1{\cdot}3$, $\Delta H_3 = 1{\cdot}3$	
				$\Delta S_1 = -2{\cdot}4$, $\Delta S_2 = -5{\cdot}2$, $\Delta S_3 = 3{\cdot}5$	
As^{5+}	pre, ir		solid	ev AsL_4^+, $AsLOF_3^-$ in solids	65Wa
Sb^{5+}	pre, ir		solid	ev $SbLOF_3^-$ in solid	65Wa
Bi^{3+}	*X*		solid	ev $Bi(O_2NO)_3(H_2O)_3$ in $BiL_3(H_2O)_5$(s)	65Hc
	cix		1 $(NaClO_4)$ }	K_1 0·96, K_2 0·62, K_3 0·35, K_4 0·07, K_5 −0·18,	67Ka
	aix		HNO_3 var }	K_6 −0·56	
	sol	25	H^+ var	†$K_{so}[BiOL(s) + 2H^+ \rightleftharpoons Bi^{3+} + L^- + H_2O]$ −1·92	67V
				K_1 1·38, K_2 0·04	
		25	0 corr	†K_{so} −3·20, K_1 2·32, K_2 0·67	
	Ram, ir	rt	var	ev several species $Bi(O_2NO)_n^{(3-n)+}$, n = 1, 2, 3, 4?	68Oa
Br^+	var		var	ev BrL_2^- in solid and MeCN(l) (con, pre, ir, sp)	66L
I^+	var		var	ev IL_2^- in solid and MeCN(l) (con, pre, ir, sp)	66L
I^{3+}	var		var	ev IL_4^- in solid and MeCN(l) (pre, ir, sp)	66L

01M R. J. Meyer and R. Jacoby, *Z. anorg. Chem.*, 1901, **27**, 359

58Ba L. L. Burger, *J. Phys. Chem.*, 1958, **62**, 590

1964 *Addendum*: p. 177, *in ref* 59C, *to* "H.-C. Chiang" *add* "(H. G. Tsiang)"

59Db L. Damiani and V. Fattore, *Energia nucleare*, 1959, **6**, 793

59H T. V. Healy and J. Kennedy, *J. Inorg. Nuclear Chem.*, 1959, **10**, 128

60Pb K. A. Petrov *et al.*, *Zhur. neorg. Khim.*, 1960, **5**, 498

continued overleaf

34 NO_3^- (contd)

60Ta V. G. Timoshev *et al.*, *Radiokhimiya*, 1960, **2**, 419
62A J. J. van Aartsen, dissertation, Technische Hogeschool, Delft, 1962
62Sb K. W. Sykes and B. L. Taylor, Proc. Seventh Internat. Conference on Co-ordination Chemistry, 1962, p. 31

1964 *Addendum*: p. 178, *to ref* 62M *add* "D. I. Ryabchikov *et al.*, *J. Inorg. Nuclear Chem.*, 1964, **26, 965**"

63Hb A. Haim and H. Taube, *Inorg. Chem.*, 1963, **2**, 1199
63Kb R. N. Kust and F. R. Duke, *J. Amer. Chem. Soc.*, 1963, **85**, 3338
63L V. I. Levin, G. V. Korpusov *et al.*, *Atomnaya Energiya*, 1963, **15**, 138
63Mc V. E. Mironov, F. Ya. Kul'ba, V. A. Fedorov, and O. B. Tikhomirov, *Zhur. neorg. Khim.*, 1963, **8**, 2536
63N J. Nowikow and G. Pfrepper, *Z. Naturforsch.*, 1963, **18b**, 993
63Pb D. Peschanski and J.-M. Fruchart, *Compt. rend.*, 1963, **257**, 1853
63Pc M. F. Pushlenkov and O. N. Shuvalov, *Radiokhimiya*, 1963, **5**, 536, 543, 551
63Sb A. P. Smirnov-Averin, G. S. Kovalenko and N. N. Krot, *Zhur. neorg. khim.*, 1963, **8**, 2400
63Sc Z. A. Sheka, E. E. Kriss, and E. I. Sinyavskaya, in "Redkozemel'nye elementy", Izd. Nauka, Moskva, 1963, p. 240
63Sd I. E. Starik and L. I. Il'menkova, *Radiokhimiya*, 1963, **5**, 679
63T V. N. Tolmachev and E. G. Chmykhalo, *Uchenye Zapiski Khar'kov Univ.*, 1963, **133** (*Trudy Khim. Fak. Khar'kov Gosud. Univ.*, **19**), 140
63Ta V. N. Tolmachev and O. F. Boberov, *Uchenye Zapiski Khar'kov Univ.*, 1963, **133** (*Trudy Khim. Fak. Khar'kov Gosud. Univ.*, **19**), 147
63V V. P. Vasil'ev, V. N. Vasil'eva, N. A. Klindukhova, and A. N. Parfenova, *Izvest. V.U.Z.*, *Khim. i khim. Tekhnol.*, 1963, **6**, 339
63Z A. Zalkin, J. D. Forrester, and D. H. Templeton, *J. Chem. Phys.*, 1963, **39**, 2881
64A C. C. Addison and D. Sutton, *J. Chem. Soc.*, 1964, 5553
64Aa J. J. v. Aartsen and A. E. Korvezee, *Trans. Faraday Soc.*, 1964, **60**, 510
64B B. M. L. Bansal, S. K. Patil, and H. D. Sharma, *J. Inorg. Nuclear Chem.*, 1964, **26**, 993
64Ba K. W. Bagnall, D. Brown, and J. G. H. du Preez, *J. Chem. Soc.*, 1964, 5523
64C F. A. Cotton and J. G. Bergman, *J. Amer. Chem. Soc.*, 1964, **86**, 2941
64D D. Dyrssen and D. H. Liem, *Acta Chem. Scand.*, 1964, **18**, 224
64Da W. Davis, jun., and H. J. de Bruin, *J. Inorg. Nuclear Chem.*, 1964, **26**, 1069
64F J. S. Fritz and H. Waki, *J. Inorg. Nuclear Chem.*, 1964, **26**, 865
64Fa G. W. A. Fowles and W. R. McGregor, *J. Phys. Chem.*, 1964, **68**, 1342
64Fb J. S. Fritz and R. Greene 1964, unpublished, quoted by ref 66W
64Fc Yu. Ya. Fialkov and Yu. Ya. Borovikov, *Ukrain. khim. Zhur.*, 1964, **30**, 119
64G I. Gainar and K. W. Sykes, *J. Chem. Soc.*, 1964, 4452
64H A. H. A. Heyn and Y. D. Soman, *J. Inorg. Nuclear Chem.*, 1964, **26**, 287
64Ha R. A. Horne, B. R. Myers and G. R. Frysinger, *Inorg. Chem.*, 1964, **3**, 452
64Hb R. E. Hester and R. A. Plane, *Inorg. Chem.*, 1964, **3**, 769; results also given in ref 64Hc
64Hc R. E. Hester and R. A. Plane, *J. Chem. Phys.*, 1964, **40**, 411
64J B. Jeżowska-Trzebiatowska and S. Ernst, *J. Inorg. Nuclear Chem.*, 1964, **26**, 837
64K P. K. Khopkar and H. D. Sharma, unpublished, quoted by ref 64B
64L T. S. Laxminarayanan, S. K. Patil, and H. D. Sharma, *J. Inorg. Nuclear Chem.*, 1964, **26**, 1001
64La R. D. Larsen and G. H. Brown, *J. Phys. Chem.*, 1964, **68**, 3060
64Lb H. Lee and J. K. Wilmshurst, *Austral. J. Chem.*, 1964, **17**, 943
64M L. W. Marple, *J. Inorg. Nuclear Chem.*, 1964, **26**, 643
64Ma Y. Marcus and M. Givon, *J. Phys. Chem.*, 1964, **68**, 2230
64Mb R. R. Miano and R. A. Plane, *Inorg. Chem.*, 1964, **3**, 987
64Mc H. A. C. McKay and J. L. Woodhead, *J. Chem. Soc.*, 1964, 717
64N B. I. Nabivanets and L. N. Kudritskaya, *Ukrain. khim. Zhur.*, 1964, **30**, 1007
64Na A. V. Nikolaev, V. G. Torgov, V. A. Mikhailov, and I. L. Kotlyarevskii, *Doklady Akad. Nauk S.S.S.R.*, 1964, **156**, 616
64S A. S. Solovkin, *Zhur. neorg. Khim.*, 1964, **9**, 746
64Sa V. S. Smelov and Yu. I. Vereshchagin, *Zhur. neorg. Khim.*, 1964, **9**, 2775
64Sb A. P. Samodelov, *Radiokhimiya*, 1964, **6**, 568
64Sc V. I. Spitsyn, R. A. D'yachkova, and V. P. Khlebnikov, *Doklady Akad. Nauk S.S.S.R.*, 1964, **157**, 135
64V V. P. Vasil'ev, *Zhur. neorg. khim.*, 1964, **9**, 647
64Va V. M. Vdovenko, L. N. Lazarev, and Ya. S. Khvorostin, *Radiokhimiya*, 1964, **6**, 724
64Z V. I. Zemlyanukhin, G. P. Savoskina, and M. F. Pushlenkov, *Radiokhimiya*, 1964, **6**, 714
65A D. W. Archer, D. A. East, and C. B. Monk, *J. Chem. Soc.*, 1965, 720
65Aa C. C. Addison, P. M. Boorman, and N. Logan, *J. Chem. Soc.*, 1965, 4978, 5146
65B G. A. Barclay, T. M. Sabine, and J. C. Taylor, *Acta Cryst.*, 1965, **19**, 205
65Ba J. I. Bullock and D. G. Tuck, *J. Chem. Soc.*, 1965, 1877
65C G. R. Choppin and W. F. Strazik, *Inorg. Chem.*, 1965, **4**, 1250
65D R. G. Deshpande, P. K. Khopkar, C. L. Rao, and H. D. Sharma, *J. Inorg. Nuclear Chem.*, 1965, **27**, 2171
65F J. M. Fletcher, C. E. Lyon, and A. G. Wain, *J. Inorg. Nuclear Chem.*, 1965, **27**, 1841
65H H. E. Hellwege and G. K. Schweitzer, *J. Inorg. Nuclear Chem.*, 1965, **27**, 99
65Ha R. Haase, K.-H. Dücker, and H. A. Küppers, *Ber. Bunsengesellschaft Phys. Chem.*, 1965, **69**, 97
65Hb R. Hugel, *Bull. Soc. chim. France*, 1965, 971, 2017
65Hc P. Herpin and K. Sudarsanan, *Bull. Soc. Franc. Minéral. Crist.*, 1965, **88**, 590

continued

65K F. Ya. Kul'ba, Yu. B. Yakovlev, and V. E. Mironov, *Zhur. neorg. Khim.*, 1965, **10**, 1624
65Ka F. Ya. Kul'ba, V. E. Mironov, I. F. Mavrin, and Yu. B. Yakovlev, *Zhur. neorg. Khim.*, 1965, **10**, 2053
65Kb M. L. Kilpatrick, M. Kilpatrick, and J. G. Jones, *J. Amer. Chem. Soc.*, 1965, **87**, 2806
65M L. W. Marple, *J. Inorg. Nuclear Chem.*, 1965, **27**, 1693
65Ma A. S. G. Mazumdar and C. K. Sivaramakrishnan, *J. Inorg. Nuclear Chem.*, 1965, **27**, 2423
65Mb L. M. Mukherjee, S. Bruckenstein, and F. A. K. Badawi, *J. Phys. Chem.*, 1965, **69**, 2537
65P D. Peschanski and J.-M. Fruchart, *Compt. rend.*, 1965, **260**, 3073
65S A. P. Samodelov, *Zhur. neorg. Khim.*, 1965, **10**, 2180
65Sa V. P. Shvedov and Yu. F. Orlov, *Zhur. neorg. Khim.*, 1965, **10**, 2774
65Sb D. Scargill, C. E. Lyon, N. R. Large, and J. M. Fletcher, *J. Inorg. Nuclear Chem.*, 1965, **27**, 161
65Sc S. Šćavničar and B. Prodić, *Acta Cryst.*, 1965, **18**, 698
65Sd J. Schlegel, *J. Phys. Chem.*, 1965, **69**, 3638
65U E. V. Ukraintsev, *Radiokhimiya*, 1965, **7**, 641
65Ua E. V. Ukraintsev, *Radiokhimiya*, 1965, **7**, 648
65W J. L. Woodhead and H. A. C. McKay, *J. Inorg. Nuclear Chem.*, 1965, **27**, 2247
65Wa J. Weidlein and K. Dehnicke, *Chem. Ber.*, 1965, **98**, 3053
66A C. C. Addison and W. B. Simpson, *J. Chem. Soc.* (*A*), 1966, 775
66Aa C. C. Addison, P. M. Boorman, and N. Logan, *J. Chem. Soc.* (*A*), 1966, 1434
66Ab C. Auerbach and D. K. McGuire, *J. Inorg. Nuclear Chem.*, 1966, **28**, 2659
66B J. G. Bergman, jun. and F. A. Cotton, *Inorg. Chem.*, 1966, **5**, 1208
66Ba D. Brown and P. J. Jones, *J. Chem. Soc.* (*A*), 1966, 733
66C N. A. Coward and R. W. Kiser, *J. Phys. Chem.*, 1966, **70**, 213
66Ca J. H. R. Clarke and L. A. Woodward, *Trans. Faraday Soc.*, 1966, **62**, 3022
66Cb R. W. Chlebek and M. W. Lister, *Canad. J. Chem.*, 1966, **44**, 437
66D P. R. Danesi, F. Orlandini, and G. Scibona, *J. Inorg. Nuclear Chem.*, 1966, **28**, 1047
66E S. Ernst and B. Jeżowska-Trzebiatowska, *J. Inorg. Nuclear Chem.*, 1966, **28**, 2885
66F B. D. Faithful, R. D. Gillard, D. G. Tuck, and R. Ugo, *J. Chem. Soc.* (*A*), 1966, 1185
66Fa O. Farver and G. Nord, *Acta Chem. Scand.*, 1966, **20**, 1429
66G R. D. Gillard and R. Ugo, *J. Chem. Soc.* (*A*), 1966, 549
66Ga C. D. Garner and S. C. Wallwork, *J. Chem. Soc.* (*A*), 1966, 1496
66Gb C. R. Guibert and M. D. Marshall, *J. Amer. Chem. Soc.*, 1966, **88**, 189
66Gc M. Givon, *Israel J. Chem.*, 1966, **4**, 3p
66Gd J. Goffart and G. Duyckaerts, *Analyt. Chim. Acta*, 1966, **36**, 499
66H R. E. Hester and R. A. Plane, *J. Chem. Phys.*, 1966, **45**, 4588
66I A. A. Ivakin, *Zhur. priklad. Khim.*, 1966, **39**, 2406
66J B. Jeżowska-Trzebiatowska and S. Ernst, *J. Inorg. Nuclear Chem.*, 1966, **28**, 1435
66Ja G. J. Janz, G. R. Lakshminarayanan, M. P. Klotzkin, and G. E. Mayer, *J. Phys. Chem.*, 1966, **70**, 536
66K G. Koch and E. Schwind, *J. Inorg. Nuclear Chem.*, 1966, **28**, 571
66Ka V. P. Khlebnikov, R. A. D'yachkova, and V. I. Spitsyn, *Radiokhimiya*, 1966, **8**, 125
66L M. Lustig and J. K. Ruff, *Inorg. Chem.*, 1966, **5**, 2124
66M W. L. Masterton and L. H. Berka, *J. Phys. Chem.*, 1966, **70**, 1924
66O Yu. F. Orlov and V. P. Shvedov, *Radiokhimiya*, 1966, **8**, 139
66R A. G. Rykov and G. N. Yakovlev, *Radiokhimiya*, 1966, **8**, 27
66Ra R. Radeglia, *Z. phys. Chem.* (*Leipzig*), 1966, **231**, 339
66S A. M. Shams el Din and A. A. El Hosary, *J. Inorg. Nuclear Chem.*, 1966, **28**, 3043
66Sa T. Sekine and Y. Hasegawa, *Bull. Chem. Soc. Japan*, 1966, **39**, 240
66Sb I. V. Shilin and V. K. Nazarov, *Radiokhimiya*, 1966, **8**, 514
66T J. C. Taylor, M. H. Moeller, and R. L. Hitterman, *Acta Cryst.*, 1966, **20**, 842
66Ta D. G. Tuck, E. J. Woodhouse and P. Carty, *J. Chem. Soc.* (*A*), 1966, 1077
66Tb L. E. Topol, R. A. Osteryoung, and J. H. Christie, *J. Phys. Chem.*, 1966, **70**, 2857
66U T. Ueki, A. Zalkin, and D. H. Templeton, *Acta Cryst.*, 1966, **20**, 836
66Ua E. V. Ukraintsev and N. F. Kashcheev, *Zhur. priklad. Khim.*, 1966, **39**, 513
66W H. Waki and J. S. Fritz, *J. Inorg. Nuclear Chem.*, 1966, **28**, 577
67A I. Abrahamer and Y. Marcus, *Inorg. Chem.*, 1967, **6**, 2103
67Aa M. Ardon and N. Sutin, *Inorg. Chem.*, 1967, **6**, 2268
67D G. C. Dobinson, R. Mason, and D. R. Russell, *Chem. Comm.*, 1967, 62
67E J. C. Eriksson, L. Ödberg, and E. Högfeldt, *Acta Chem. Scand.*, 1967, **21**, 1925
67Ea A. N. Ermakov, I. N. Marov, and G. A. Evtikova, *Zhur. neorg. Khim.*, 1967, **12**, 3372
67F V. V. Fomin and V. S. Solomatin, *Radiokhimiya*, 1967, **9**, 310
67Fa B. D. Faithful and S. C. Wallwork, *Chem. Comm.*, 1967, 1211
67Fb V. A. Fedorov, A. M. Robov, and V. E. Mironov, *Zhur. neorg. Khim.*, 1967, **12**, 3307
67H H. C. Helgeson, *J. Phys. Chem.*, 1967, **71**, 3121
67G C. D. Garner, D. Sutton, and S. C. Wallwork, *J. Chem. Soc.* (*A*), 1967, 1949
67J G. J. Janz, M. J. Tait, and J. Meier, *J. Phys. Chem.*, 1967, **71**, 963
67K R. T. Kolarich, V. A. Ryan, and R. P. Schuman, *J. Inorg. Nuclear Chem.*, 1967, **29**, 783
67Ka E. E. Kapantsyan and B. I. Nabivanets, *Ukrain. khim. Zhur.*, 1967, **33**, 961
67Kb Z. Kolařík, *Coll. Czech. Chem. Comm.*, 1967, **32**, 435

continued overleaf

34 NO_3^- (contd)

67M W. L. Masterton, T. I. Munnelly, and L. H. Berka, *J. Phys. Chem.*, 1967, **71**, 942
67Ma W. L. Masterton, *J. Phys. Chem.*, 1967, **71**, 2885
67Mb I. F. Mavrin, F. Ya. Kul'ba, and V. E. Mironov, *Zhur. neorg. Khim.*, 1967, **12**, 324
67Mc J. T. Miller and D. E. Irish, *Canad. J. Chem.*, 1967, **45**, 147
67P G. Pistoia, A. M. Polcaro, and S. Schiavo, *Ricerca sci.*, 1967, **37**, 300
67S T. Sekine, I. Sakamoto, T. Sato, T. Taira, and Y. Hasegawa, *Bull. Chem. Soc. Japan*, 1967, **40**, 251
67Sa A. M. Shams el Din and A. A. El Hosary, *Electrochim. Acta*, 1967, **12**, 1665
67Sb A. S. Solovkin, Z. N. Tsvetkova, and A. I. Ivantsov, *Zhur. neorg. Khim.*, 1967, **12**, 626
67U R. Ugo and R. D. Gillard, *Inorg. Chim. Acta*, 1967, **1**, 311
67V V. P. Vasil'ev and N. K. Grechina, *Zhur. neorg. Khim.*, 1967, **12**, 1372
67Va S. P. Vorob'ev, I. P. Davydov, and I. V. Shilin, *Zhur. neorg. Khim.*, 1967, **12**, 2142
67Vb S. P. Vorob'ev, I. P. Davydov, and I. V. Shilin, *Zhur. neorg. Khim.*, 1967, **6**, 2665
67Vc V. M. Vdovenko, V. A. Shormanov, A. N. Trofimova, and A. S. Krivokhatskii, *Zhur. neorg. Khim.*, 1967, **12**, 2756
68A C. C. Addison, D. W. Amos, and D. Sutton, *J. Chem. Soc.* (*A*), 1968, 2285
68B T. A. Beineke and J. Delgaudio, *Inorg. Chem.*, 1968, **7**, 715
68D A. R. Davis and D. E. Irish, *Inorg. Chem.*, 1968, **7**, 1699
68Da A. R. Davis and R. A. Plane, *Inorg. Chem.*, 1968, **7**, 2565
68Db J. G. Dawber, *J. Chem. Soc.* (*A*), 1968, 1532
68Dc A. D'Aprano and R. M. Fuoss, *J. Phys. Chem.*, 1968, **72**, 4710
68F V. V. Fomin and E. P. Maiorova, *Zhur. neorg. Khim.*, 1968, **13**, 3137
68Fa R. C. Ferguson, P. Dobud, and D. G. Tuck. *J. Chem. Soc.* (*A*), 1968, 1058
68G J. Garaj, *Acta Chem. Scand.*, 1968, **22**, 1710
68H J. Hilton and S. C. Wallwork, *Chem. Comm.*, 1968, 871
68J G. Johansson, *Acta Chem. Scand.*, 1968, **22**, 389, 399
68K S. Katayama and R. Tamamushi, *Bull. Chem. Soc. Japan*, 1968, **41**, 606
68L C. H. Liu, J. Hasson, and G. P. Smith, *Inorg. Chem.*, 1968, **7**, 2244
68M L. M. Mukherjee, J. J. Kelly, W. Baranetzky, and J. Sica, *J. Phys. Chem.*, 1968, **72**, 3410
68Ma V. A. Mikhailov, L. I. Ponomareva, and A. G. Nazin, *Zhur. fiz. Khim.*, 1968, **42**, 318
68Mb N. A. Matwiyoff, P. E. Darley, and W. G. Movius, *Inorg. Chem.*, 1968, **7**, 2173
68O J. W. O'Laughlin, D. F. Jensen, J. W. Ferguson, J. J. Richard, and C. V. Banks, *Analyt. Chem.*, 1968, **40**, 1931
68Oa R. P. Oertel and R. A. Plane, *Inorg. Chem.*, 1968, **7**, 1192
68R V. V. Revyakin, V. V. Chubukov, and N. A. Korableva, *Zhur. neorg. Khim.*, 1968, **13**, 3090
68Ra O. Redlich, R. W. Duerst, and A. Merbach, *J. Chem. Phys.*, 1968, **49**, 2986
68S C. J. Shahani, K. A. Mathew, C. L. Rao, and M. V. Ramaniah, *Radiochim. Acta*, 1968, **10**, 165
68T P. H. Tedesco, V. B. de Rumi, and J. A. González Quintana, *J. Inorg. Nuclear Chem.*, 1968, **30**, 987
68Ta V. G. Torgov, A. V. Nikolaev, V. A. Mikhailov, L. G. Stadnikova, and I. L. Kotlyarevskii, *Zhur. neorg. Khim.*, 1968 **13**, 228

35 Other ligands containing N

Metal	*Method*	*Temp*	*Medium*	*Log of equilibrium constant, remarks*	*Ref*
				Hyponitrite, $N_2O_2^{2-}$	
Co^{3+}	ir, con		var	ev $(NH_3)_5CoLCo(NH_3)_5^{4+}$ (red)	64F
				α-Oxyhyponitrite, $N_2O_3^{2-}$ (*nitrohydroxylaminate*)	
Co^{2+}	pol		0·05 NH_4Cl	ev $Co(NH_3)_4L^0$, $Co(H_2O)_4L^0$	65C
Ni^{2+}	pol		0·1 NH_4Cl	ev $NiL_2NH_3H_2O^{2-}$	67C
				Peroxonitrite, $ONOO^-$	
H^+	kin	2	var	$K_1 \leqslant 8{\cdot}3$	68Ha
				Dinitrogen, N_2	
Fe	pre, ir		var	ev $FeH_2N_2Z_3$ in EtOH(l), C_6H_6(l) *etc.*, Z = Ph_2RP, R = Et, Bu	68S

continue

Other ligands containing N (contd) 35

Metal	Method	Temp	Medium	Log of equilibrium constant, remarks	Ref
Co	pre		EtOH	ev equilibrium $CoH_3Z_3 + N_2 \rightleftharpoons CoHN_2Z_3 + H_2$ in EtOH(l), no *K* given, Z = PPh_3	67S = 68Sa
	pre, ir		org	ev $(Ph_3P)_3CoN_2^0$	67Y
Ru^{2+}	var		var	ev $N_2Ru(NH_3)_5^{2+}$ (con, ir, nmr, pre)	65A = 67Aa
	X		solid	ev $N_2Ru(NH_3)_5^{2+}$ in $Ru(NH_3)_5N_2Cl_2$(s)	68B
	sp, pre		var	$Q^{2+} = Ru(NH_3)_5^{2+}$; ev QN_2Q^{4+} at equil. with QH_2O^{2+} and QN_2^{2+}	68H ≈ 68I
Ru	pre, ir		org	ev equilibrium $H_2Ru(PPh_3)_4 + N_2 \rightleftharpoons$ cpx in C_6H_6(l), no *K* given	68Y
Os^{2+}	var		var	ev $Os(NH_3)_5N_2^{2+}$ in H_2O, Me_2SO(l), *etc.*, and solids (pre, con, ir)	67A
Ir	pre, ir		org	ev $IrClN_2(PPh_3)_2$	68C

64F R. D. Feltham, *Inorg. Chem.*, 1964, **3**, 1038
65A A. D. Allen and C. V. Senoff, *Chem. Comm.*, 1965, 621
65C A. Căluşaru and J. Kůta, *Nature*. 1965, **207**, 751
67A A. D. Allen and J. R. Stevens, *Chem. Comm.*, 1967, 1147
67Aa A. D. Allen, F. Bottomley, R. O. Harris, V. P. Reinsalu, and C. V. Senoff, *J. Amer. Chem. Soc.*, 1967, **89**, 5595
67C A. Căluşaru and J. Kůta, *Coll. Czech. Chem. Comm.*, 1967, **32**, 1331
67S A. Sacco and M. Rossi, *Chem. Comm.*, 1967, 316
67Y A. Yamamoto, S. Kitazume, L. S. Pu, and S. Ikeda, *Chem. Comm.*, 1967, 79
68B F. Bottomley and S. C. Nyburg, *Acta Cryst.*, 1968, **B24**, 1289, *Chem. Comm.*, 1966, 897
68C J. P. Collman, M. Kubota, F. D. Vastine, J. Y. Sun, and J. W. Kang, *J. Amer. Chem. Soc.*, 1968, **90**, 5430
68H D. F. Harrison, E. Weissberger, and H. Taube, *Science*, 1968, **159**, 320
68Ha M. N. Hughes and H. G. Nicklin, *J. Chem. Soc.* (*A*), 1968, 450
68I I. J. Itzkovitch and J. A. Page, *Canad. J. Chem.*, 1968, **46**, 2743
68S A. Sacco and M. Aresta, *Chem. Comm.*, 1968, 1223
68Sa A. Sacco and M. Rossi, *Inorg. Chim. Acta*, 1968, **2**, 127
68Y A. Yamamoto, S. Kitazume, and S. Ikeda, *J. Amer. Chem. Soc.*, 1968, **90**, 1089

Hypophosphite, $PO_2H_2^-$ 36

Metal	Method	Temp	Medium	Log of equilibrium constant, remarks	Ref
H^+	H	20	2 (NaCl)	K_1 1·12	37N
	gl	22	*w*% EtOH	K_1 2·70(*w* = 75), 2·94(*w* = 95), in *w*% EtOH ("pH" on H_2O scale)	65P
	nmr		H_2SO_4	ev H_2L^+ [$H_2P(OH)_2^+$] in H_2SO_4 var	67S
	nmr	25?	1	K_1 1·40	68H
			var	K_{12} −1·7	
Ce^{4+}	kin	25	*I*($NaClO_4$)	$^*K_1 \gg 2$, $^*K_2 \gg 2$, $^*K_3 \approx 2$	66C
	kin	30—35	0·5—1 H_2SO_4	$K_1 \approx$ 1·1—1·3	67M
Eu^{3+}	sp	25	0 corr	K_1 2·27	64B
Cr^{3+}	sp, kin	45—65	1 $HClO_4$	$^*K_1$ 1·32(45°), 1·38(55°), 1·40(65°)	66E
	sp, cix		0·24	β_2 4·14	68L
Cr^{6+}	kin	25	1 $HClO_4$	$K(HCrO_4^- + H_3PO_2 \rightleftharpoons O_3CrOPH_2O^-? + H_2O)$ 1·04	68Ha
Fe^{3+}	red	25	0·13 ($HClO_4$), 0·1 H^+	K_1 3·62, β_2 6·40 (no correction for HL)	64N
	sp	20	0·2	K_1 4·01, K_2 2·78, K_3 2·17	67Ma
Zn^{2+}	nmr	25	4 ($NaNO_3$)	K_1 0·3, β_2 0·0	68H
			4 ($NaClO_4$)	K_1 0·54, β_2 0·2	
Al^{3+}	nmr		var	ev $AlL_n^{(3-n)+}$ and Al_2L^{5+}	68G

references overleaf

36 $PO_2H_2^-$ (contd)

37N P. Nylén, *Z. anorg. Chem.*, 1937, **230**, 385
64B J. C. Barnes, *J. Chem. Soc.*, 1964, 3880
64N R. I. Novoselov, Z. A. Muzykantova, and B. V. Ptitsyn, *Zhur. neorg. Khim.*, 1964, **9**, 799
65P D. F. Peppard, G. W. Mason, and C. M. Andrejasich, *J. Inorg. Nuclear Chem.*, 1965, **27**, 697
66C R. L. Carroll and L. B. Thomas, *J. Amer. Chem. Soc.*, 1966, **88**, 1376
66E J. H. Espenson and D. E. Binau, *Inorg. Chem.*, 1966, **5**, 1365
67M S. K. Mishra and Y. K. Gupta, *J. Inorg. Nuclear Chem.*, 1967, **29**, 1643
67Ma V. M. Masalovich, P. K. Agasyan, and E. R. Nikolaeva, *Zhur. neorg. Khim.*, 1967, **12**, 2041
67S G. M. Sheldrick, *Trans. Faraday Soc.*, 1967, **63**, 1077
68G H. D. Gillman and T. E. Haas, *Chem. Comm.*, 1968, 777
68H T. E. Haas and H. D. Gillman, *Inorg. Chem.*, 1968, **7**, 2051
68Ha G. P. Haight, jun., M. Rose, and J. Preer, *J. Amer. Chem. Soc.*, 1968, **90**, 4809
68L A. M. Lunyatskas and P. K. Norkus *Zhur. neorg. Khim.*, 1968, **13**, 665

37 Phosphite, HPO_3^{2-}

Metal	*Method*	*Temp*	*Medium*	*Log of equilibrium constant, remarks*	*Ref*
H^+	qh, H	20	0 corr	K_1 6·77, K_{12} 1·41	30N
	qh	0	1 (KCl)	K_1 6·18, K_{12} 0·97	37N
	kin	20—30	I var	K_1 6·10(20°, I = 1·15), 6·23(30°, I = 0·4)	40G
	qh?	18—45	0·6	K_{12} 1·01(18°), 1·17(45°)	
	gl	25	0 corr	K_1 6·78—6·90, K_{12} 1·51—1·68	41T
	con	25—45	var	ev HL^-, $H_3L_2^-$, $H_4L_3^{2-}$, $H_5L_3^-$	64E = 65E
	H	20	0	K_1 6·70, K_{12} 1·20	64F
	sp		H_2SO_4 var	$K_{13}[= \beta(H_2L^0H^+)]$ −5·0	64F
	con	25	MeOH	K_{12} 4·48 in MeOH(l)	64P
	gl	22	w% EtOH	K_{12} 3·15(w = 75), 4·01(w = 95) in w% EtOH ("pH" on H_2O scale)	65P
	gl	25	ROH	K_{12} 5·86(R = Me), 7·13(Et), 7·49(Pr) 7·44(Pr^i), 7·75(Bu) in ROH(l), → 0	66K
	nmr		H_2SO_4	ev $H_3L^+[HP(OH)_3^+]$ in H_2SO_4 var	67S
	var		org	ev H_2L, $H_3L_2^-$, HL^-, and L^{2-} in Me_2NCHO(l) and Me_2CO(l), (con, gl, nmr)	68F
	nmr	25?	1	K_1 6·22, K_{12} 1·70	68H
			var	K_{13} ($H_2L + H^+ \rightleftharpoons H_3L^+$) −1·2	
	gl	25	I(KCl)	K_1 6·25(I = 3·5), 6·09(I = 1·0), 6·36(I = 0·1)	68M
		25	I($NaClO_4$)	K_1 6·01(I = 1·0), 6·34(I = 0·1)	
			→ 0	K_1 6·79	
Na^+	con	20	0 corr	K_1 1·05, $\beta(Na^+HL^-)$ 0·96	64F
K^+	con	20	0 corr	K_1 0·80, $\beta(K^+HL^-)$ 0·74	64F
V^{2+}	gl	25	0 corr ?	K_1 4·61, $\beta(V^{2+}HL^-)$ 2·50, $\beta(VHL^+HL^-)$ 1·67, $\beta(VL^0HL^-)$ 2·01	66P
VO^{2+}	gl	20	dil	K_1 3·80, $\beta(VO^{2+}HL^-)$ 1·80, $\beta(VOHL^+HL^-)$ 1·22	67P
	sol	25	var	$\beta(VOL^0H_2L)$ 2·79, ev ani cpx	
Cr^{6+}	sp	25	var	$\beta(HCrO_4^-H_2L)$ 1·42, $\beta(HCrO_4^-HL^-)$ 0·85	65Pa
	sp	25	1 $HClO_4$	$K(HCrO_4^- + H_3PO_3 \rightleftharpoons O_3CrOPHOOH^- + H_2O)$ 1·2, ev $O_3CrOPHO_2^{2-}$	68Ha
Fe^{3+}	con, spJ		var	ev cpx with 0·5, 1, 1·5, and 2 L per Fe^{3+}	64Pa
	sp	21	MeOH	K_1 5·77, K_2 3·34, K_3 3·04 in MeOH(l)	64Pb
	sp, red	24	0 corr	$\beta(Fe^{3+}HL^-)$ 5·0(red), 4·77 to 4·66(sp) $\beta(FeHL^{2+}HL^-)$ 2·17(red), 2·92(sp)	66M

contin[ued]

Metal	*Method*	*Temp*	*Medium*	*Log of equilibrium constant, remarks*	*Ref*
Cu^{2+}	sol	25	3·5 ($NaNO_3$)	K_{s0} −6·72, K_{s2} −2·15, β_2 4·57	64N
Zn^{2+}	var	20	var	K_1 4·76, $\beta(Zn^{2+}HL^-)$ 2·28, $\beta(ZnHL^+HL^-)$ 1·26, $K_a(ZnHL^+)$ −4·07 [$K_1(H^+)$ 6·55] (H, Zn, con)	65F
$MeHg^+$	gl	20	0·1 (KNO_3)	K_1 4·67	63S = 65S

30N P. Nylén, dissertation, Uppsala Univ., 1930
37N P. Nylén, *Z. anorg. Chem.*, 1937, **230**, 385
40G R. O. Griffith and A. McKeown, *Trans. Faraday Soc.*, 1940, **36**, 766
41T K. Takahashi and N. Yui, *Bull. Inst. Phys. Chem. Res.* (*Tokyo*), 1941, **20**, 521
63S M. Schellenberg, dissertation, Eidgenössische Technische Hochschule, Zürich, 1963
64E M. Ebert and F. Škvára, *Coll. Czech. Chem. Comm.*, 1964, **29**, 1945
64F V. Frei, J. Podlahová, and J. Podlaha, *Coll. Czech. Chem. Comm.*, 1964, **29**, 2587
64N J. Nassler, *Coll. Czech. Chem. Comm.*, 1964, **29**, 174
64P J. Podlaha, *Chem. Zvesti*, 1964, **18**, 9
64Pa J. Podlaha, *Coll. Czech. Chem. Comm.*, 1964, **29**, 1128
64Pb J. Podlaha, *Coll. Czech. Chem. Comm.*, 1964, **29**, 3057
65E M. Ebert and J. Čipera, *Chem. Zvesti*, 1965, **19**, 679; *Coll. Czech. Chem. Comm.*, 1966, **31**, 1
65F V. Frai, Ya. Podlagova and Ya. Podlaga (Frei, Podlahová, and Podlaha), *Zhur. neorg. Khim.*, 1965, **10**, 1690
65P D. F. Peppard, G. W. Mason, and C. M. Andrejasich, *J. Inorg. Nuclear Chem.*, 1965, **27**, 697
65Pa J. Preer and G. P. Haight, jun., *J. Amer. Chem. Soc.*, 1965, **87**, 5256
65S G. Schwarzenbach and M. Schellenberg *Helv. Chim. Acta.* 1965, **48**, 28
66K A. P. Kreshkov, V. A. Drozdov, and N. A. Kolchina, *Zhur. fiz. Khim.*, 1966, **40**, 2150
66M V. M. Masalovich, P. K. Agasyan and E. R. Nikolaeva, *Zhur. neorg. Khim.*, 1966, **11**, 272
66P J. Podlaha, *Coll. Czech. Chem. Comm.*, 1966, **31**, 7
67P J. Podlaha and J. Šilha, *Coll. Czech. Chem. Comm.*, 1967, **32**, 3760
67S G. M. Sheldrick, *Trans. Faraday Soc.*, 1967, **63**, 1077
68F A. Francina, A. Lamotte, and J.-C. Merlin, *Compt. rend.*, 1968, **267**, *C*, 763
68H T. E. Haas and H. D. Gillman, *Inorg. Chem.*, 1968, **7**, 2051
68Ha G. P. Haight, jun., M. Rose, and J. Preer, *J. Amer. Chem. Soc.*, 1968, **90**, 4809
68M O. Mäkitie and M.-L. Savolainen, *Suomen Kem.*, 1968, **41**, *B*, 242

Phosphate, PO_4^{3-} 38

Metal	*Method*	*Temp*	*Medium*	*Log of equilibrium constant, remarks*	*Ref*
H^+	H, Ag	0—55	1·25 NaCl	K_{12} 6·334(5°), 6·295(15°), 6·266(25°), 6·244(35°), 6·231(45°), 6·223(55°)	44A
			0·725 NaCl	K_{12} 6·440(5°), 6·396(15°), 6·363(25°), 6·341(35°), 6·328(45°), 6·324(55°)	
			0·36 NaCl	K_{12} 6·580(5°), 6·531(15°), 6·496(25°), 6·472(35°), 6·459(45°), 6·454(55°); also values for 0·07 NaCl, and → 0, *m* units	
	gl	10—37	→ 0	K_{12} 7·21(10°), 7·18(25°), 7·15(37°), $\Delta H_{12} = -0{\cdot}76$, $\Delta S_{12} = 30{\cdot}4$	63P
	gl	20	0·1 (KNO_3)	K_{12} 6·68 (misprint 6·79, p. 62?)	63Sc
	Cu, O	350	KNO_3(l)	$K(2HL^{2-}+O^{2-} \rightleftharpoons 2L^{3-}+H_2O)$ 5·45, $K(2H_2L^-+O^{2-} \rightleftharpoons 2HL^{2-}+H_2O)$ 13·59, $K(P_2O_7^{4-}+O^{2-} \rightleftharpoons 2PO_4^{3-})$ 5·82, $K(2PO_3^-+O^{2-} \rightleftharpoons P_2O_7^{4-})$ 15·49 in KNO_3(l), M units	63Sd = 64Sc
	con, *E*	25	$MeCO_2H$	K_{13} 4·4 in $MeCO_2H$(l)	64F
	H, fp	25—80	H_3L	$K(2H_3L \rightleftharpoons H_4L^+ + H_2L^-)$ −0·85, ev $H_4L_2^{2-}$ (compiler cannot follow derivation of equilibria with H_3O^+ and $H_4L_2^{2-}$)	64M
	ΔG	38—100	H_3L conc	$K(2H_3L \rightleftharpoons H_6L_2+H_2O)$ −3·07(38°), −3·06(40°), −3·02(45°), −2·8(80°), −2·7(100°), *x* units	64M+/65M

continued overleaf

38 PO_4^{3-} (contd)

Metal	Method	Temp	Medium	Log of equilibrium constant, remarks	Ref
H^+ (*contd*)	gl	25	H_2O—D_2O	K_{13} 2·145(x_D = 0), 2·233(x_D = 0·494), 2·350(x_D = 1) in H_2O–D_2O mixtures, also values for other x_D	64Sa
	O	350	KNO_3(l)	$K(PO_3^- + PO_4^{3-} \rightleftharpoons P_2O_7^{4-})$ 4·4—4·7 in KNO_3(l), M units	64Sb/
	con		H_2O_2 var	plot of $[K_{13}/K_{13}(H_2O)]$ *vs*. mole fraction H_2O_2	65Da
	gl	25	→ 0	ΔV_{12} = 24·0 cm^3 (measurements at 1 and 1000 atm)	65Db
	var		H_3L var	$K(H_3L + H_2L^- \rightleftharpoons H_5L_2^-)$ 0·10 $K(H^+ + H_5L_2^- \rightleftharpoons H_6L_2)$ 0·52 (con, H, pH_2O)	65E
	gl	20	dil	K_{12} 7·15, K_{13} 2·75 in H_2O, K_{12} 8·15, K_{13} 3·20 in 44% Me_2CO, K_{12} 8·62, K_{13} 3·85 in 51% dioxan, K_{12} 8·92, K_{13} 4·08 in 49% Me_2NCHO	65H
	cal	25	0 corr	$K_1K_w(L^{3-} + H_2O \rightleftharpoons HL^{2-} + OH^-)$ −1·61, ΔH = 9·0, ΔS = 23	65Ha
	gl	22	75% EtOH	K_{13} 4·17 in 75% EtOH ("pH" on H_2O scale)	65P
	H, Ag	10—40	50% MeOH	K_{12} 8·480(10°), 8·469(15°), 8·452(20°), 8·443(25°), 8·441(30°), 8·442(35°), 8·443(40°) in 50 *w*% MeOH, *m* units	65Pa
	cal	25	*I*(NaCl)	$K_1K_w(L^{3-} + H_2O \rightleftharpoons HL^{2-} + OH^-)$ −2·76(I = 1·1), −2·54(I = 0·6), −2·11(I = 0·1) −1·61($I \to 0$), ΔH = 8·9 (I = 1·1, 0·6, 0·1. Sign misprinted in Ref. 65Pb)	65Pc
	gl	25	→ 0	K_1 12·0, K_{12} 7·18, ΔH_1 = −3·5, ΔH_{12} = −0·8, ΔS_1 = 43, ΔS_{12} = 30 (from data at 10—40°)	65Pb
	cal	25	0 corr	K_1K_w −1·61, $\Delta H(L^{3-} + H_2O \rightleftharpoons HL^{2-} + OH^-)$ = 9·1	66C
	lit	25	0	ΔH_1 = −2·58, ΔH_{12} = −1·0—2	/66I
	cal	25	*I*(Me_4NCl)	ΔH_1 = −5·2(I = 0·20), −2·7(→ 0) ΔH_{12} = −1·2(I = 0·1), −0·6(→ 0) ΔH_{13} = 1·97(I = 2·0), 2·07(I = 1·0, 0·5)	66I
		25	0	ΔH_1 = −2·6, ΔH_{12} = −0·6, ΔH_{13} = 1·9 ΔS_1 = 48, ΔS_{12} = 31, ΔS_{13} = 16	66I+

1964 *Erratum*: p. 181, *in ref* 59Ga, *for* "M units" *read* "*m* units".

Metal	Method	Temp	Medium	Log of equilibrium constant, remarks	Ref
	gl	25	ROH	K_{13} 7·60(R = Me), 8·75(Pr), 8·33(Pr^i), 9·06(Bu) in ROH(l), → 0	66K
	var	≈ 650	MBr(l)	$K(P_2O_7^{4-} + O^{2-} \rightleftharpoons 2PO_4^{3-})$ 1·9 in MBr(l) (fp, *E*, ir), M = Na or K	66Ma
	ΔG	25	0	K_{13} 2·15, ΔH_{13} = 1·84, ΔS_{13} = 16·0	/67H
	gl	25	0·1 ($NaClO_4$)	K_{12} 6·70	67S
		25	10% dioxan	K_{12} 7·00 in 10% dioxan, I = 0·1 ($NaClO_4$)	
	cal	25	0	$\Delta H(H_2P_2O_7^{2-} + H_2O \rightleftharpoons 2H_2PO_4^-)$ = −4·5	67Wa
			dil	$\Delta H(HP_2O_7^{3-} + H_2O \rightleftharpoons HPO_4^{2-} + H_2PO_4^-)$ = −3·7	
D^+	D, Ag	5—50	D_2O, → 0	K_{12} 7·885(5°), 7·850(10°), 7·823(15°), 7·799(20°), 7·780(25°), 7·767(30°), 7·755(35°), 7·748(40°), 7·743(45°), 7·744(50°), in D_2O	64G
Mg^{2+}	sol, gl	25	0 corr	$K_s[Mg^{2+}NH_4^+L^{3-}(H_2O)_6]$ −13·15, $K_s[Mg^{2+}K^+L^{3-}(H_2O)_6]$ −10·62, $K_s[Mg^{2+}HL^{2-}(H_2O)_3]$ −5·82, $K_s[Mg^{2+}{}_3L^{3-}{}_2(H_2O)_n]$ −25·20(n = 8), −23·10(n = 22); $\beta(Mg^{2+}HL^{2-})$ 2·91	63Tb

contir

Metal	Method	Temp	Medium	Log of equilibrium constant, remarks	Ref
Ca^{2+}	qh, H	18	dil	$\beta(Ca^{2+}HPO_4^{2-})$ 2·6, [$K_a(CaHPO_4)$ −8·5]	58B/63Ga
	sol	40	→ 0	$K_s(Ca^{2+}{}_2HL^{2-}OH^-{}_2)$ −28 for "surface complex" on surface of hydroxylapatite, $Ca_5L_3OH(s)$	64D

1964 *Erratum*: p.182, *in ref* 36B, *for values*, "x", *of* K_s *read* "−x" *for the first six values given.*

Metal	Method	Temp	Medium	Log of equilibrium constant, remarks	Ref
	sol, gl	37·5	0 corr	$K_s[Ca^{2+}HL^{2-}(H_2O)_2]$ −6·66, $\beta(Ca^{2+}H_2L^-)$ 0·87, $\beta(Ca^{2+}HL^{2-})$ 2·77	66M
	?	37	0	$\beta(Ca^{2+}H_2L^-)$ 0·89, $\beta(Ca^{2+}HL^{2-})$ 2·77	67M
	?	37	0	$K_1(Ca^{2+}L^{3-})$ 6·3	67N
	gl	25—37	0 corr	$\beta(Ca^{2+}H_2L^-)$ 1·41(25°), 1·50(37°), $\Delta H = 3{\cdot}4$, $\Delta S = 18$, $\beta(Ca^{2+}HL^{2-})$ 2·74(25°), 2·83(37°), $\Delta H = 3{\cdot}3$, $\Delta S = 23$, $K_1 = \beta(Ca^{2+}L^{3-})$ 6·46(25°), 6·54(37°), $\Delta H_1 = 3{\cdot}1$, $\Delta S_1 = 40$ (H^+ equil from refs. 43B, 51B, 61V, 29B)	68C
Sr^{2+}	sol	25	0 corr	$K_s(Sr^{2+}HL^{2-})$ −6·38 (H^+ : K_1 12·32, K_{12} 7·10, K_{13} 2·15)	62Fc
	sol	20	0 corr	$K_s(Sr^{2+}HL^{2-})$ −6·92, almost ideal solid solution (Sr, Ba)HL(s)	66S ≈ 67Ha
Ba^{2+}	sol	20	0 corr	$K_s(Ba^+HL^{2-})$ −7·42	66S ≈ 67Ha
Y^{3+}	cix	25	0·2 (NH_4ClO_4)	$\beta(Y^{3+}H_2L^-)$ 1·84, 2·65(0 corr)	66B
Ce^{3+}	cix	25	0·2 (NH_4ClO_4)	$\beta(Ce^{3+}H_2L^-)$ 1·52, 2·33(0 corr)	66B
Pm^{3+}	cix	25	0·2 (NH_4ClO_4)	$\beta(Pm^{3+}H_2L^-)$ 1·69, 2·51(0 corr)	66B
Gd^{3+}	sol	25	0·5 ($NaClO_4$)	K_{s0} −22·26	67T
Th^{4+}	sol	25?	0·35 H(ClO_4)	$K_s[Th(HL)_2(s)+(4-2n)H^+ \rightleftharpoons Th(HL)_n{}^{(4-2n)+}+(2-n)H_3L]$ −7·64(n = 1), −4·06(n = 2), −3·96(n = 3); $\beta[Th^{4+}(HL^{2-})_n]$ 10·8(n = 1), 22·8(n = 2), 31·3(n = 3); $K_s[Th^{4+}(HL^{2-})_2]$ −26·89 (H^+ : K_{12} 6·60, K_{13} 1·82), solid = $Th(HL)_2(H_2O)_4$	67Mb
VO^{2+}	aix, cix	20	var	ev VOHL, $VOH_3L_2^-$, $VOH_2L_2^{2-}$	66Sa
Pa^{5+}	oth		6 HNO_3	$\beta(Pa^VH_3L)$ 1·54, $\beta[Pa^V(H_3L)_2]$ 2·20 (sorption on silica gel)	66D
Cr^{3+}	pol	25	0·1 $NaClO_4$	$K_a(CrH_xPO_4{}^{x-})$ −4·0	64W
	sp, gl	25	0 corr	$\beta(Cr^{3+}HL^{2-})$ 9·45(gl), 9·41(sp), $\Delta H \approx 0$	66L
Cr^{6+}	sp	25	3 ($NaClO_4$)	$K(HCrO_4{}^- + H_2L^- \rightleftharpoons HLCrO_3{}^{2-} + H_2O)$ 0·8	68F
Mo^{5+}	esr	40	var	†$K[MoOX_4{}^- + nH_3L \rightleftharpoons MoOX_{4-n}(H_2L)_n{}^- + nH^+ + nX^-]$ 0·28(n = 1), 0·04(n = 2), −0·5(n = 3), −1·7(n = 4) for $X^- = Cl^-$	68M
		30	var	†K 0·52(n = 1), 1·11(n = 2), 1·62(n = 3), 1·8(n = 4) for $X^- = Br^-$	
		40	H_3L var	ev dimer, $K\{2MoO(H_2L)_5{}^{2-} + 3H_2O \rightleftharpoons [MoO(H_2L)_3H_2O]_2OH^- + H^+ + 4H_2L^-\} \approx 4(?)$	
U^{4+}	sol	20	HCl var	$K_s[U(HL)_2(s)+(4-2n)H^+ \rightleftharpoons U(HL)_n{}^{(4-2n)+}+(2-n)H_3L]$ −9·96(n = 0), −7·34(n = 1), −5·74(n = 2), −5·60(n = 3), −5·96(n = 4); $K_s[U^{4+}(HL^{2-})_2]$ −26·80,	64Mc/67Mb

continued overleaf

38 PO_4^{3-} (contd)

Metal	*Method*	*Temp*	*Medium*	*Log of equilibrium constant, remarks*	*Ref*
U^{4+} (*contd*)				$\beta[U^{4+}(HL^{2-})_n]$ 12·0(n = 1), 22·0(n = 2), 30·6(n = 3), 38·6(n = 4), (H^+; K_{12} 6·60, K_{13} 1·82) solid = $U(HL)_2(H_2O)_4$	
UO_2^{2+}	sol, gl	25	dil	$K_s(M^+UO_2^{2+}L^{3-})$ −25·6(M = Li), −28·2(Na), −23·1(K), −27·0(Rb)	64Ma
	sol, gl	20	0·32 ($NaNO_3$)	$K_s(UO_2^{2+}HL^{2-})$ −12·17, $K_s(UO_2^{2+}{}_3L^{3-}{}_2)$ −49·7; $K_s(M^+UO_2^{2+}L^{3-})$ −24·21(M = Na), −25·50(M = K), −25·72(M = Rb), −25·41(M = Cs), −26·23(M = NH_4) (complex formation constants from ref 53Ba)	65V
	sol	25?	0·5 ($NaClO_4$)	$K_s[UO_2HL(H_2O)_4(s) \rightleftharpoons UO_2HL^0]$ −2·82(−3·5?) $K_s[UO_2HL(H_2O)_4(s) + 2H^+ \rightleftharpoons UO_2^{2+} + H_3L + 4H_2O]$ −1·96	67Ma
		25—70	→ 0	$K_s[UO_2^{2+}HL^{2-}(H_2O)_4]$ −11·46(25°), −11·05(50°) −11·0(70°) $K_s[UO_2HL(H_2O)_4(s) + 2H^+ \rightleftharpoons UO_2^{2+} + H_3L + 4H_2O]$ −2·12(25°), −1·72(50°), −1·68(70°)	
	est	25	0	$K_s[Na^+UO_2^{2+}L^{3-}(H_2O)_x]$ −19·8	
	sol	25	I HNO_3	$\beta(UO_2^{2+}HL^{2-})$ 7·18(I = 0·5), 8·43(I = 0) $\beta[UO_2^{2+}(HL^{2-})_2]$ 17·30(I = 0·5), 18·57(I = 0)	
Np^{4+}	est	25?		$K_s[Np^{4+}(HL^{2-})_2]$ −28, $\beta[Np^{4+}(HL^{2-})_n]$ 12·4(n = 1), 23·1(n = 2), 32·0(n = 3), 41·0(n = 4)	67Mb
NpO_2^+	cix	20	0·2 (NH_4ClO_4)	$\beta(NpO_2^+HL^{2-})$ 2·85, $\beta(NpO_2^+H_2L^-)$ 0·81 (H^+: K_1 12·66, K_{12} 7·21, K_{13} 2·12)	64Mb
			→ 0	$\beta(NpO_2^+HL^{2-})$ 3·38	
PuO_2^{2+}	sol	?	var	$K_s[NH_4^+PuO_2^{2+}L^{3-}(H_2O)_3]$ −26·6 $K_s(PuO_2^{2+}HL^{2-})$ −12·55 (?) $\beta(NH_4^+PuO_2^{2+}L^{3-})$ 21·43 (?) $\beta(PuO_2^{2+}HL^{2-})$ 8·17 (?) (H^+: K_1 12·66, K_{12} 7·21)	65D
	sol, tp	25	H_3L var	$\beta(PuO_2^{2+}H_2L^-)$ 3·93, ev $PuO_2(H_2L)_2{}^0$; $K_s[PuO_2HL(H_2O)_4(s) \rightleftharpoons 4H_2O + PuO_2HL^0]$ −4·34	67D
Am^{3+}	cix	25	0·2 (NH_4ClO_4)	$\beta(Am^{3+}H_2L^-)$ 1·69, 2·51(0 corr)	66B
Mn^{2+}	sol, gl	25	0·01 H^+	$K_s(Mn^{2+}HL^{2-})$ −12·86(H^+: K_{12} 7·20, K_{13} 2·10), solid = grey, stable $MnHPO_4$(s)	66G
Fe^{2+}	red	20—30	0 corr	$\beta(Fe^{2+}HL^{2-})$ 7·03(20°), 7·34(30°) (Fe^{3+} cpx formation neglected) (?)	64L
Fe^{3+}	sp, cix		var	$\beta(Fe^{3+}{}_2HL^{2-})$ 11·14	66F
Co^{2+}	gl	25	0·1 ($NaClO_4$)	$\beta(Co^{2+}HL^{2-})$ 2·18	67S
		25	10% dioxan	$\beta(Co^{2+}HL^{2-})$ 2·26, $\beta(Co\ bipy^{2+}HL^{2-})$ 2·26 in 10% dioxan, I = 0·1($NaClO_4$)	
Co^{3+}cpx	gl	5	1 $NaClO_4$	$\beta(Co\ en_2L^0H^+)$ 4·25	68L
		5—30	1 $NaClO_4$	$\beta(Co\ en_2OHL^-H^+)$ 9·75(5°), 9·40(23°), 9·25(30°); $\beta(Co\ en_2OHHL^0H^+)$ 7·25(5°), 6·75(23°), 6·67(30°), 6·69(→ 20°, 0·1M-$NaClO_4$); $\beta(Co\ en_2OH_2HL^+H^+)$ 3·30(5°), 3·10(23°), 3·05(30°)	
		5	1 $NaClO_4$	$\beta[(NH_3)_4CoOHL^-H^+]$ 9·2; $\beta[(NH_3)_4CoOHHL^0H^+]$ 6·70, 6·25(→ 20°, 0·1M-$NaClO_4$); $\beta[(NH_3)_4CoOH_2HL^+H^+]$ 3·2	

contin

Metal	Method	Temp	Medium	Log of equilibrium constant, remarks	Ref
Co^{3+}cpx	sp		$\rightarrow 0$	$\beta[(NH_3)_4CoOH_2H_2L^{2+}H^+]$ −0·34	
(contd)	sp		$\rightarrow 0$	$\beta[(NH_3)_5CoH_2L^{2+}H^+]$ −0·23	
	sp		10 $HClO_4$	ev $(NH_3)_5CoH_4L^{4+}$, $(NH_3)_4CoOH_2H_4L^{4+}$	
	kin	22·5	1 $NaClO_4$	$K(Co\,en_2LH^+ + H_2O \rightleftharpoons Co\,en_2OH_2LH^+)$ 1·42	68La
				$K(Co\,en_2L^0 + H_2O \rightleftharpoons Co\,en_2OHHL^0)$ −1·5	
				$\beta(Co\,en_2LH^+H^+)$ 0·0	
Ni^{2+}	gl	25	0·1 ($NaClO_4$)	$\beta(Ni^{2+}HL^{2-})$ 2·08	67S
		25	10% dioxan	$\beta(Ni^{2+}HL^{2-})$ 2·22 in 10% dioxan $I = 0{\cdot}1(NaClO_4)$	
Cu^{2+}	aix		H_3L dil	ev ani cpx	63Ta
	gl	25	0·1 ($NaClO_4$)	$\beta(Cu^{2+}HL^{2-})$ 3·2	67S
		25	10% dioxan	$\beta(Cu^{2+}HL^{2-})$ 3·4, $\beta(Cu\,bipy^{2+}HL^{2-})$ 3·8 in 10% dioxan, $I = 0{\cdot}1$ ($NaClO_4$)	
	MHg, sp	20—36	3 (Na_2SO_4)	$\beta[Cu^{2+}(H_2L^-)_2]$ 2·70(20°, sp), 2·64(36°, MHg)	68B
Zn^{2+}	gl	25	0·1 ($NaClO_4$)	$\beta(Zn^{2+}HL^{2-})$ 2·4	67S
		25	10% dioxan	$\beta(Zn^{2+}HL^{2-})$ 2·4, $\beta(Zn\,bipy^{2+}HL^{2-})$ 2·4 in 10% dioxan, $I = 0{\cdot}1(NaClO_4)$	
$MeHg^+$	gl	20	0·1(KNO_3)	$\beta(MeHg^+HL^{2-})$ 5·03	63Sc = 65S
Al^{3+}	oth	25	var	ev polyn cpx (viscosity, con)	64S
Ga^{3+}	sol	25	1 ($NaClO_4$)	K_{s0} −21·0 (H^+: K_1 12·36, K_{12} 7·21, K_{13} 2·12)	64T
In^{3+}	sol, gl	25	1 ($NaClO_4$)	K_{s0} −21·63	68D
Ge^{4+}	tp		var	ev cat cpx	64Fa
	pre, ir		var	ev $GeOH(HL)_2^-$	65A
Sn^{2+}	sol	25	0·2 ($NaClO_4$)	apparent values: $K_{s0}(Sn^{2+}H_xL)$ −12·81, K_1 2·95, β_3 5·45, at pH = 2·5, solid = $SnHPO_4(H_2O)_{0.5}$ est: $\beta(Sn^{2+}H^+L^{3-})$ 7·83, $\beta(Sn^{2+}H^+{}_3L^{3-}{}_3)$ 10·04	68Ca

44A M. B Alpert, thesis, Yale Univ., 1944
62Fc M. H. Frere, *Proc. Soil. Sci. Soc. Amer.*, 1962, **26**, 48
63Ga I. Greenwald, *J. Phys. Chem.*, 1963, **67**, 2853
63P R. C. Phillips, P. George, and R. J. Rutman, *Biochemistry*, 1963, **2**, 501
63Sc M. Schellenberg, dissertation, Eidgenössische Technische Hochschule, Zürich, 1963
63Sd A. M. Shams el Din and A. A. A. Gerges, *J. Inorg. Nuclear Chem.*, 1963, **25**, 1537
63Ta I. K. Tsitovich and T. A. Lapina, *Izvest. V.U.Z., Khim. i Khim. Tekhnol.*, 1963, **6**, 370
63Tb A. W. Taylor, A. W. Frazier, and E. L. Gurney, *Trans. Faraday Soc.*, 1963, **59**, 1580; A. W. Taylor, A. W. Frazier E. L. Gurney, and J. P. Smith, *ibid.*, p. 1585
64D V. R. Deitz, H. M. Rootare, and F. G. Carpenter, *J. Colloid Sci.*, 1964, **19**, 87
64F Yu. Ya. Fialkov and Yu. Ya. Borovikov, *Ukrain. khim. Zhur.*, 1964, **30**, 119
64Fa N. S. Fortunatov and L. E. Slobtsov, *Ukrain. khim. Zhur.*, 1964, **30**, 1279
64G R. Gary, R. G. Bates, and R. A. Robinson, *J. Phys. Chem.*, 1964, **68**, 3806
64L S. C. Lahiri and S. Aditya, *J. Indian Chem. Soc.*, 1964, **41**, 517
64M R. A. Munson, *J. Phys. Chem.*, 1964, **68**, 3374
64Ma I. A. Murav'eva, K. B. Zaborenko, O. G. Nemkova, and H. D. Pin, *Radiokhimiya*, 1964, **6**, 124
64Mb A. I. Moskvin and V. F. Peretrukhin, *Radiokhimiya*, 1964, **6**, 206
64Mc I. V. Moiseev, in V. K. Markov, *et al.*, 'Uran, methody ego opredeleniya' (2nd edn.), Atomizdat, Moskva, 1964, p. 92
64S V. N. Sveshnikova and S. N. Zaitseva, *Zhur. neorg. Khim.*, 1964, **9**, 1232
64Sa P. Salomaa, L. L. Schaleger and F. A. Long, *J. Amer. Chem. Soc.*, 1964, **86**, 1
64Sb A. M. Shams el Din and A. A. A. Gerges, *Electrochim. Acta*, 1964, **9**, 123
64Sc A. M. Shams el Din and A. A. A. Gerges, *Electrochim. Acta*, 1964, **9**, 613
64T I. V. Tananaev and N. N. Chudinova, *Zhur. neorg. Khim.*, 1964, **9**, 244
64W J. H. Walsh and J. E. Earley, *Inorg. Chem.*, 1964, **3**, 343
65A K. A. Avduevskaya and I. V. Tananaev, *Zhur. neorg. Khim.*, 1965, **10**, 372
65D R. G. Denotkina, V. B. Shevchenko, and A. I. Moskvin, *Zhur. neorg. Khim.*, 1965, **10**, 2449
65Da A. J. Dedman, P. Flood, T. J. Lewis, D. H. Richards, and D. A. Salter, *J. Chem. Soc.*, 1965, 3505

continued overleaf

38 PO_4^{3-} (contd)

65Db A. Distèche and S. Distèche, *J. Electrochem. Soc.*, 1965, **112**, 350
65E K. L. Elmore, J. D. Hatfield, R. L. Dunn, and A. D. Jones, *J. Phys. Chem.*, 1965, **69**, 3520
65H M. K. Hargreaves, E. A. Stevinson, and J. Evans, *J. Chem. Soc.*, 1965, 4582
65Ha L. D. Hansen, J. J. Christensen, and R. M. Izatt, *Chem. Comm.*, 1965, 36
65M R. A. Munson, *J. Phys. Chem.*, 1965, **69**, 1761
65P D. F. Peppard, G. W. Mason, and C. M. Andrejasich, *J. Inorg. Nuclear Chem.*, 1965, **27**, 697
65Pa M. Paabo, R. A. Robinson, and R. G. Bates *J. Amer. Chem. Soc.*, 1965, **87**, 415
65Pb P. Papoff, G. Torsi and P. G. Zambonin, *Gazzetta*, 1965, **95**, 1031
65Pc R. Phillips, P. Eisenberg, P. George, and R. J. Rutman, *J. Biol. Chem.*, 1965, **240**, 4393
65S G. Schwarzenbach and M. Schellenberg *Helv. Chim. Acta*, 1965, **48**, 28
65V V. Veselý, V. Pekárek, and M. Abbrent, *J. Inorg. Nuclear Chem.*, 1965, **27**, 1159
66B M. S. Borisov, A. A. Elesin, I. A. Lebedev, V. T. Filimonov, and G. N. Yakovlev, *Radiokhimiya*, 1966, **8**, 42
66C J. J. Christensen, R. M. Izatt, L. D. Hansen, and J. A. Partridge, *J. Phys. Chem.*, 1966, **70**, 2003
66D A. V. Davydov, I. N. Marov, and P. N. Palei, *Zhur. neorg. Khim.*, 1966, **11**, 1316
66F L. N. Filatova and M. L. Chepelevetskii, *Doklady Akad. Nauk S.S.S.R.*, 1966, **166**, 140, *Zhur. neorg. Khim.*, 1966, **11** 1662
66G M. V. Goloshchapov, B. V. Martynenko, and T. N. Filatova, *Zhur. neorg. Khim.*, 1966, **11**, 935
66I A. R. Irani and T. A. Taulli, *J. Inorg. Nuclear Chem.*, 1966, **28**, 1011
66K A. P. Kreshkov, V. A. Drozdov, and N. A. Kolchina, *Zhur. fiz. Khim.*, 1966, **40**, 2150
66L S. C. Lahiri and S. Aditya, *J. Indian Chem. Soc.*, 1966, **43**, 513
66M E. C. Moreno, T. M. Gregory and W. E. Brown, *J. Res. Nat. Bur. Stand, Sect. A*, 1966, **70**, 545
66Ma A. Mitchell, *Trans. Faraday Soc.*, 1966, **62**, 3470
66S V. I. Spitsyn, N. B. Mikheev, and A. Khermann, *Doklady Akad. Nauk. S.S.S.R.*, 1966, **166**, 658
66Sa J. E. Salmon and D. Whyman, *J. Chem. Soc. (A)*, 1966, 980
67D R. G. Denotkina and V. B. Shevchenko, *Zhur. neorg. Khim.*, 1967, **12**, 2345
67H H. C. Helgeson, *J. Phys. Chem.*, 1967, **71**, 3121
67Ha A. Hermann, N. B. Micheev, and V. I. Spitzin, *Z. anorg. Chem.*, 1967, **350**, 92
67M E. C. Moreno and W. E. Brown, personal communication quoted by ref 67W
67Ma A. I. Moskvin, A. M. Shelyakina, and P. S. Perminov, *Zhur. neorg. Khim.*, 1967, **12**, 3319
67Mb A. I. Moskvin, L. N. Essen, and T. N. Bukhtiyarova, *Zhur. neorg. Khim.*, 1967, **12**, 3390
67N H. G. Nancollas, personal communication, quoted by ref 67W
67S H. Sigel, K. Becker, and D. B. McCormick, *Biochim. Biophys. Acta*, 1967, **148**, 655
67T I. V. Tananaev and S. M. Petushkova, *Zhur. neorg. Khim.*, 1967, **12**, 81
67W A. G. Walton, W. J. Bodin, H. Furedi, and A. Schwartz, *Canad. J. Chem.*, 1967, **45**, 2695
67Wa C.-H. Wu, R. J. Witonsky, P. George, and R. J. Rutman, *J. Amer. Chem. Soc.*, 1967, **89**, 1987
68B R. Bury, *J. Chim. phys.*, 1968, **65**, 1494
68C A. Chughtai, R. Marshall, and G. H. Nancollas, *J. Phys. Chem.*, 1968, **72**, 208
68Ca W. A. Cilley, *Inorg. Chem.* 1968, **7**, 612.
68D E. N. Deichman, I. V. Tananaev, Zh. A. Ezhova, and T. N. Kuz'mina, *Zhur. neorg. Khim.*, 1968, **13**, 47
68F S.-Å. Frennesson, J. K. Beattie, and G. P. Haight, jun., *J. Amer. Chem. Soc.*, 1968, **90**, 6018
68L S. F. Lincoln and D. R. Stranks, *Austral. J. Chem.*, 1968, **21**, 37
68La S. F. Lincoln and D. R. Stranks, *Austral. J. Chem.*, 1968, **21**, 57
68M I. N. Marov, Yu. N. Dubrov, V. K. Belyaeva, and A. N. Ermakov, *Zhur. neorg. Khim.*, 1968, **13**, 2140, 2445

39 Diphosphate (pyrophosphate) $P_2O_7^{4-}$

Metal	*Method*	*Temp*	*Medium*	*Log of equilibrium constant, remarks*	*Ref*
H^+	gl	25	0·1 (Me_4NBr)	$\Delta H_1 = 0{\cdot}5, \Delta H_{12} = 0{\cdot}5, \Delta H_{13}+\Delta H_{14} = 2$	61Ia/66I

1964 *Addendum*: p. 190, *for ref* "62S" *read ref* "62S = 64W".

	gl	25	0·1 $NaNO_3$	K_1 8·3, K_{12} 6·0, K_{13} 2·7, K_{14} 2·5	63J
	gl	25	0·1 (Me_4NCl)	K_1 9·00, K_{12} 6·23	64H
	gl	25	$\to 0$	K_1 9·42, K_{12} 6·54, $\Delta H_1 = -2{\cdot}8, \Delta H_{12} = -0{\cdot}3$, $\Delta S_1 = 34, \Delta S_{12} = 29$, (from data at 10—40°)	65P
	cal	25	I(Me_4NCl)	$\Delta H_1 = -0{\cdot}40 (I = 0{\cdot}22), \Delta H_{12} = -0{\cdot}11 (I = 0{\cdot}13)$, $\Delta H_{13}+\Delta H_{14} = 6 (I > 1)$, K_{14} $-0{\cdot}44 (I = ?$, sign?)	66I
		25	0	$\Delta H_1 = -0{\cdot}4, \Delta H_{12} = -0{\cdot}1, \Delta H_{13} = \Delta H_{14} = 3$, $\Delta S_1 = 42, \Delta S_2 = 31, \Delta S_3 = 22, \Delta S_4 = 8$	66I+

contin

Metal	*Method*	*Temp*	*Medium*	*Log of equilibrium constant, remarks*	*Ref*
H^+	gl	25	1·0 (Me_4NBr)	K_1 8·71, K_{12} 5·76, K_{13} 1·72, K_{14} 0·79	66Ma
(*contd*)			0·05 (Me_4NBr)	K_1 9·10, K_{12} 6·28, K_{13} 2·00, K_{14} 0·88	
			$\rightarrow 0$	K_1 9·32, K_{12} 6·70, K_{13} 2·10, K_{14} 0·91	
		50	1·14 (Me_4NBr)	K_1 8·64, K_{12} 5·74, K_{13} 1·64, K_{14} 0·83	
			0·05 (Me_4NBr)	K_1 9·18, K_{12} 6·37, K_{13} 2·04, K_{14} 1·04	
		70	1·14 (Me_4NBr)	K_1 8·58, K_{12} 5·72, K_{13} 1·50, K_{14} 0·97	
			0·05 (Me_4NBr)	K_1 9·16, K_{12} 6·17, K_{13} 1·91, K_{14} 1·00, also values for $I = 0{\cdot}10$ and 0·42, and for 60 and 65°	
		25	0	$\Delta H_1 = 10{\cdot}0$, $\Delta H_{12} = 10{\cdot}8$, $\Delta H_{13} = 8{\cdot}3$, $\Delta H_{14} = 6{\cdot}7$ $\Delta S_1 = 76{\cdot}1$, $\Delta S_{12} = 66{\cdot}9$, $\Delta S_{13} = 37{\cdot}6$, $\Delta S_{14} = 26{\cdot}6$	
	H	25	1 Na(ClO_4)	K_1 7·36, K_1K_{12} 12·72, K_{13} 1·40, $K_{13}K_{14}$ 2·15	68B
	gl	25	1 ($NaNO_3$)	K_1 7·45	68C
		25	1 ($NaClO_4$)	K_1 6·96	
	gl	25	0·1 (Me_4NCl)	K_1 8·93, K_{12} 6·12	68Ma
Mg^{2+}	gl	25	0·1 Na(NO_3)	K_1 4·7, $\beta(H^+MgL^{2-})$ 6·0	63J

1964 *Erratum*: p. 190, *in ref* 61Ia *for* "β_{12}" *read* "K_{12}". (*misprint in ref.*)

Metal	*Method*	*Temp*	*Medium*	*Log of equilibrium constant, remarks*	*Ref*
Ca^{2+}	tyn	25	I(Me_4NBr)	see table 43	64M

1964 *Erratum*: p. 191, *in ref* 57L *for* "HL^-" *read* "HL^{3-}".

Metal	*Method*	*Temp*	*Medium*	*Log of equilibrium constant, remarks*	*Ref*
Group $4f^{3+}$	kin	25	0 corr	β_{12} 19·53(La), 19·98(Nd), 20·16(Sm), 20·27(Eu), 20·45(Gd), 20·50(Tb), 20·64(Dy), 20·88(Ho), 21·29(Er), 21·88(Yb), 22·23(Lu) [$\beta_{12}(Fe^{3+})$ 23·51]	67Sa
La^{3+}	sol	25	var	ev ani cpx, such as LaL_2^{5-}	64T
	sol	25	0 corr	K_1 16·72, β_2 18·57, β_{12} 19·64; $K_s(La^{3+}LaL^-{}_3)$ −17·36 (dil), $\Delta H_s = 15{\cdot}8$(H^+: K_1 9·25, K_{12} 6·60, K_{13} 2·36, K_{14} 4·5)	66S
	gl	25	0·1 (Me_4NCl)	K_1 4·66, $\beta(H^+LaL^-)$ 5·56, $\beta(La^{3+}HL^{3-})$ 0·85 [?, compare K_1 (H^+)]	68Ma
Ce^{4+}	sol	25	0·1 (Na)ClO_4	$K_{s1}[CeL(s) \rightleftharpoons CeL^0]$ −5·0, K_{s0} −23·03, K_1 18·04(H^+: β_{13} 17·07, β_{14} 17·51); or K_{s0} −23·11, K_1 18·10(H^+: β_{13} 17·09, β_{14} 17·93); or K_{s0} −23·36, K_1 18·41(H^+: β_{13} 17·05, β_{14} 18·35), solid = $CeP_2O_7(H_2O)_9$	67Ma
Nd^{3+}	sp	25	0 corr	K_1 15·0	67Sb
Yb^{3+}	sol	25	0 corr	K_1 17·5, β_2 19·4; $K_{s0}(Yb^{3+}{}_4L^{4-}{}_3)$ −75	66S
Cr^{3+}	pol	25	0·1 $NaClO_4$	$K_a[CrH_xP_2O_7^{(x-1)+}]$ −2·7, −4·8	64Wa
	spJ	25	var	$\beta(Cr^{3+}{}_3L^{4-}{}_2)$ 13·4 (?)	68M
Mo^{3+}	pol		1 $HClO_4$	$\beta(Mo^{III}H_xL)$ 3·3	64S
UO_2^{2+}	var		var	ev UO_2L^{2-}, $UO_2L_2^{6-}$ (sol, gl)	65D
			var	ev$(UO_2)_nL^{(2n-4)+}$, $n = 3, 4, 5$ (sol, tp, sp)	
Mn^{3+}	dis		org	ev $(RNH_3^+)_5MnL_2^{5-}$ in $CHCl_3$(l)	68S
Fe^{3+}	var	20		$\beta[Fe^{3+}(H_2L^{2-})_2]$ 12·07(red), 12·38(sp), 12·74(dis), 12·3(average)	65S
	sp	20	0 corr	$\beta(Fe^{3+}H_2L^{2-})$ 6·62, $\beta(Fe^{3+}H_3L^-)$ 6·05, $\beta(Fe^{3+}H_3L^-{}_2)$ 11·25	66A

continued overleaf

39 $P_2O_7^{4-}$ (contd)

Metal	Method	Temp	Medium	Log of equilibrium constant, remarks	Ref
Fe^{3+} (*contd*)	sp		$I(KNO_3)$	$K(FeSCN^{2+}+H_4L \rightleftharpoons FeH_2L^{+}+SCN^{-}+2H^{+})$ 1·17($I = 2$), 1·26($I = 1·5$), 1·30($I = 1$); $\beta(Fe^{3+}H_2L^{2-})$ 5·58($I = 2$), 5·71($I = 1·5$), 5·81($I = 1$), 8·08(0 corr)	66V
	kin	25	0 corr	$\beta(Fe^{3+}H_3L^{-})$ 6·43, $\beta(Fe^{3+}H_2L^{2-})$ 6·97, $\beta_{12}(Fe^{3+}{}_2L^{4-})$ 23·3 or 23·5	67S
Co^{2+}	gl	25	0·1 $Na(NO_3)$	K_1 6·1, $\beta(H^{+}CoL^{2-})$ 5·7	63J
	gl	25	0·1 (Me_4NCl)	K_1 7·2, $\beta(Co^{2+}HL^{3-})$ 4·05	64H
	sp, con	25	var	ev $Co_2L_4(OH)_3^{15-}$	64J
Ni^{2+}	gl	25	0·1 (Me_4NCl)	K_1 6·98, $\beta(Ni^{2+}HL^{3-})$ 3·83	64H
Cu^{2+}	gl	25	0·1 $Na(NO_3)$	K_1 7·3, $\beta(H^{+}CuL^{2-})$ 5·4	63J
	MHg, gl	25	1 $Na(ClO_4)$	K_1 7·6, β_2 12·45; $\beta(Cu^{2+}L^{4-}H^{+}{}_n)$ 11·81($n = 1$), 14·71($n = 2$); $\beta(Cu^{2+}L^{4-}{}_2H^{+}{}_n)$ 17·3($n = 1$), 22·0($n = 2$), 25·7($n = 3$), 28·4($n = 4$), 30·1($n = 5$)	68B
Zn^{2+}	MHg	25	$I(NaClO_4)$	K_1 10·52($I = 1$), 11·66($I = 0·1$), 12·80($I = 0$), also K_1 for several intermediate I values	67M
	MHg, pol	25	1 $(NaNO_3)$	K_1 5·1, β_2 7·19	68C
Cd^{2+}	dis	30	1 $(NaClO_4)$	K_1 4·0, β_2 6·3	65H
1964 *Addendum*: p. 192, *for* "62S" *read* "62S = 64W".					
In^{3+}	sol, gl	20	var	$K_{s0}(In^{3+}{}_4L^{4-}{}_3)$ −62·5, $K_s(In^{3+}HL^{3-})$ −12·44 (H^{+}: K_1 8·44, K_{12} 6·54, K_{13} 1·96, K_{14} 0·85)	64G
	ana	60	0·18 H_2SO_4	$K[In^{3+}+Fe^{3+}(\text{in solid}) \rightleftharpoons Fe^{3+}+In^{3+}(\text{in solid})]$ −0·7, solid = $Fe_4L_3(s)$	66G
Ge^{4+}	aix	19	var	ev cpx	66L
Sn^{2+}	sol, gl	25	1 $(NaClO_4)$	$K_s[2SnO(s)+H^{+}+H_xL^{x-4} \rightleftharpoons Sn_2L(OH)_{3-x}{}^{(3-x)-}]$ 5.8 $K_s[2SnO(s)+H^{+}+2H_xL^{x-4} \rightleftharpoons Sn_2L_2(OH)_{3-2x}{}^{(7-2x)-}]$ 8·3, $K_s[2SnO(s)+H_xL^{x-4} \rightleftharpoons Sn_2L(OH)_{4-x}{}^{(4-x)-}]$ −0·5, x = 0 or 1	66M
	Sn	?	1 K_4L	β_2 16·42	68P
Pb^{2+}	MHg, pol	25	1 $(NaNO_3)$	K_1 6·4, β_2 9·40	68C
			1 $(NaClO_4)$	K_1 7·3, β_2 10·15	

63J A. Johansson and E. Wänninen, *Talanta*, 1963, **10**, 769
64G R. S. Guzairov, V. A. Leitsin, and S. D. Grekov, *Zhur. neorg. Khim.*, 1964, **9**, 20
64H G. G. Hammes and M. L. Morrell, *J. Amer. Chem. Soc.*, 1964, **86**, 1497
64J H. B. Jonassen and W. J. Connick, jun., *J. Inorg. Nuclear Chem.*, 1964, **26**, 1411
64M M. Miura and Y. Moriguchi, *Bull. Chem. Soc. Japan*, 1964, **37**, 1522
64S P. S. Shetty and P. R. Subbaraman, *Indian J. Chem.*, 1964, **2**, 428
64T I. V. Tananaev and V. P. Vasil'eva, *Zhur. neorg. Khim.*, 1964, **9**, 2111
64W J. I. Watters and R. A. Simonaitis, *Talanta*, 1964, **11**, 247
64Wa J. H. Walsh and J. E. Earley, *Inorg. Chem.*, 1964, **3**, 343
65D C. Drăgulescu, I. Julean, and N. Vîlceanu, *Rev. Roumaine Chim.*, 1965, **10**, 809
65H H. E. Hellwege and G. K. Schweitzer, *J. Inorg. Nuclear Chem.*, 1965, **27**, 99
65P R. Phillips, P. Eisenberg, P. George, and R. J. Rutman, *J. Biol. Chem.*, 1965, **240**, 4393
65S V. B. Spivakovskii and G. B. Makovskaya, *Zhur. neorg. Khim.*, 1965, **10**, 1062
66A L. P. Andrusenko, Z. A. Sheka, and I. A. Sheka, *Zhur. neorg. Khim.*, 1966, **11**, 1266
66G S. D. Grekov, V. A. Leitsin, and R. S. Guzairov, *Zhur. neorg. Khim.*, 1966, **11**, 835
66I R. R. Irani and T. A. Taulli, *J. Inorg. Nuclear Chem.*, 1966, **28**, 1011
66L V. A. Leitsin, *Zhur. neorg. Khim.*, 1966, **11**, 1753
66M R. E. Mesmer and R. R. Irani, *J. Inorg. Nuclear Chem.*, 1966, **28**, 493
66Ma R. P. Mitra, H. C. Malhotra, and D. V. S. Jain, *Trans. Faraday Soc.*, 1966, **62**, 167
66S Z. A. Sheka and E. I Sinyavskaya, *Zhur. neorg. Khim.*, 1966, **11**, 1029
66V V. N. Vasil'eva and V. P. Vasil'ev, *Izvest. V.U.Z., Khim. i khim. Tekhnol.*, 1966, **9**, 185

contini

67M M. G. Mushkina, M. S. Novakovskii, and N. Ch. Uen, *Zhur. neorg. Khim.*, 1967, **12**, 2351
67Ma S. A. Merkusheva, N. A. Skorik, V. N. Kumok, and V. V. Serebrennikov, *Zhur. neorg. Khim.*, 1967, **12**, 3388
67S Z. A. Sheka, L. P. Andrusenko, and I. A. Sheka, *Zhur. neorg. Khim.*, 1967, **12**, 74
67Sa Z. A. Sheka and E. I. Sinyavskaya, *Zhur. neorg. Khim.*, 1967, **12**, 377
67Sb Z. A. Sheka and E. I. Sinyavskaya, *Zhur. neorg. Khim.*, 1967, **12**, 650
68B E. Bottari and L. Ciavatta, *Inorg. Chim. Acta*, 1968, **2**, 74
68C B. D. Costley and J. P. G. Farr, *Chem. and Ind.*, 1968, 1435
68M W. U. Malik and C. L. Sharma, *J. Indian Chem. Soc.*, 1968, **45**, 29
68Ma W. M. McNabb, J. F. Hazel, and R. A. Baxter, *J. Inorg. Nuclear Chem.*, 1968, **30**, 1585
68P B. A. Purin and I. A. Vītiņa, *Izvest. Akad. Nauk Latv. S.S.R., Ser. khim.*, 1968, 277, 372
68S I. A. Shevchuk and T. Ṅ. Ṣimonova, *Zhur. analit. Khim.*, 1968, **23**, 1386

Trimetaphosphate, $P_3O_9^{3-}$ 40

Metal	*Method*	*Temp*	*Medium*	*Log of equilibrium constant, remarks*	*Ref*
H^+	kin	25	0 corr	K_1 1·74	65I
	gl	25	0·2 (NH_4ClO_4)	K_1 2·07, K_{12} 1·64	67E
Pm^{3+}	cix	25	0·2 (NH_4ClO_4)	K_1 3·80, 5·74(0 corr)	67E
Am^{3+}	cix	25	0·2 (NH_4ClO_4)	K_1 3·48, 6·06(0 corr), ev $AmHL^+$	67E
Cm^{3+}	cix	25	0·2 (NH_4ClO_4)	K_1 3·64, 5·92(0 corr)	67E
Fe^{2+}	kin	0—25	1 ($NaClO_4$)	K_1 1·72(0°), 1·43(10°), 1·15(25°)	68W
		35	1·8 ($NaClO_4$)	K_1 0·76	
		0—45	4 ($NaClO_4$)	$\beta(Fe^{2+}HL^{2-})$ 1·79(0°), 1·02(25°), 0·81(35°), 0·49(45°)	

65I A. Indelli and G. Mantovani, *Trans. Faraday Soc.*, 1965, **61**, 909
67E A. A. Elesin, I. A. Lebedev, E. M. Piskunov, and G. N. Yakovlev, *Radiokhimiya*, 1967, **9**, 161
68W C. F. Wells and M. A. Salam, *J. Chem. Soc. (A)*, 1968, 308

Triphosphate, $P_3O_{10}^{5-}$ 41

1964 *Addendum*: p. 195, *penultimate line, for ref* "62Sa" *read* "62Sa ≈ 64W".

Metal	*Method*	*Temp*	*Medium*	*Log of equilibrium constant, remarks*	*Ref*
H^+	gl	25	0·1 (Me_4NBr)	$\Delta H_1 = \Delta H_{12} = 0·5$, $\Delta H_{13}+\Delta H_{14}+\Delta H_{15} = 3$	61Ia/66I
	gl	25	0·1 $Na(NO_3)$	K_1 7·9, K_{12} 5·6, K_{13} 2·6, K_{14} 2·2, K_{15} small	63J
	gl	25	0·1 (KCl)	K_1 8·06, K_{12} 5·43	64E
	gl	25	0·1 (Me_4NCl)	K_1 8·74, K_{12} 6·02	64H
	gl, cal	20	0·1 (Me_4NNO_3)	K_1 8·82, $\Delta H_1 = -0·1$, $\Delta S_1 = 40$; K_{12} 5·93, $K_{13} \approx 2·2$	65A
	gl	25	0 corr	K_1 9·26, K_{12} 6·61, $\Delta H_1 = -3·0$, $\Delta H_{12} = -0·4$, $\Delta S_1 = 34$, $\Delta S_{12} = 29$ (from data at 10—40°)	65P
	cal	25	$I(Me_4NCl)$	$\Delta H_1 = -0·1(I = 0·65)$, $\Delta H_{12} = 1·5(I = 0·4)$, $\Delta H_{13}+\Delta H_{14}+\Delta H_{15} = 7·5(I > 1)$, K_{15} −0·51(I = ?, sign?)	66I
		25	0	$\Delta H_1 = -0·1$, $\Delta H_{12} = 1·5$, $\Delta H_{13} = \Delta H_{14} = \Delta H_{15} = 2·5$; $\Delta S_1 = 42$, $\Delta S_{12} = 31$, $\Delta S_{13} = 19$, $\Delta S_{14} = 14$, $\Delta S_{15} = 6$	66I+
	gl	25	0·1 (Me_4NCl)	K_1 8·81, K_{12} 5·79	68M
Be^{2+}	gl, cal	20	0·1 (Me_4NNO_3)	$K_a(BeHL^{2-})$ −5·35, $\Delta H_1 = 4·7$	65A

continued overleaf

41 $P_3O_{10}^{5-}$ (contd)

Metal	*Method*	*Temp*	*Medium*	*Log of equilibrium constant, remarks*	*Ref*
Mg^{2+}	gl	25	0·1 $Na(NO_3)$	K_1 5·7, $\beta(H^+MgL^{3-})$ 5·8	63J
	gl	25	0·1 (KCl)	K_1 5·65, $\beta(Mg^{2+}HL^{4-})$ 3·27, $\beta(H^+MgL^{3-})$ 5·68	64E
	gl, cal	20	0·1 (Me_4NNO_3)	K_1 7·05, $\Delta H_1 = 4{\cdot}34$, $\Delta S_1 = 47$, $\beta(Mg^{2+}HL^{4-})$ 4·45, $\beta(H^+MgL^{3-})$ 6·22	65A

1964 *Erratum*: p. 196, *in ref* 61Ia *for* "β_{12}" *read* "K_{12}" *(misprint in ref.)*.

Metal	*Method*	*Temp*	*Medium*	*Log of equilibrium constant, remarks*	*Ref*
Ca^{2+}	gl	25	0·1 (KCl)	K_1 5·20, $\beta(Ca^{2+}HL^{4-})$ 3·04, $\beta(H^+CaL^{3-})$ 5·90,	64E
	tyn	25	$I(Me_4NBr)$	see table 43	64M
	gl, cal	20	0·1 (Me_4NNO_3)	K_1 6·31, $\Delta H_1 = 3{\cdot}3$, $\Delta S_1 = 40$, $\beta(Ca^{2+}HL^{4-})$ 4·02, $\beta(H^+CaL^{3-})$ 6·54	65A
Sr^{2+}	gl	25	0·1 (KCl)	K_1 4·35, $\beta(Sr^{2+}HL^{4-})$ 2·53, $\beta(H^+SrL^{3-})$ 6.24	64E
	gl, cal	20	0·1 (Me_4NNO_3)	K_1 5·46, $\Delta H_1 = 3{\cdot}16$, $\Delta S_1 = 35{\cdot}7$, $\beta(Sr^{2+}HL^{4-})$ 3·56, $\beta(H^+SrL^{3-})$ 6·92	65A
Ba^{2+}	gl	25	0·1 (KCl)	$K_1 \approx 3{\cdot}0$, $\beta(Ba^{2+}HL^{4-})$ 1·8 ?, $\beta(H^+BaL^{3-})$ 6·3 ?	64E
Group $4f^{3+}$	gl	30	0·3 $(NaClO_4)$	$\beta(M^{3+}HL^{4-})$ 6·61(La), 6·98(Pr), 7·15(Nd), 7·42(Sm), 7·46(Eu) at pH = 3—4 [$K_{12}(H^+)$?]	65K ≈ 63K
La^{3+}	gl	25	0·1 (Me_4NCl)	K_1 6·56, $\beta(La^{3+}HL^{4-})$ 2·90, $\beta(H^+LaL^{2-})$ 5·17	68M
Mo^{3+}	pol		1 $HClO_4$	$\beta(Mo^{III}H_xL)$ 3·96	64Sa
Mn^{2+}	gl	25	0·1 (KCl)	K_1 7·21, $\beta(Mn^{2+}HL^{4-})$ 3·77, $\beta(H^+MnL^{3-})$ 4·62	64E
	gl, cal	20	0·1 (Me_4NNO_3)	K_1 8·04, $\Delta H_1 = 2{\cdot}8$, $\Delta S_1 = 46{\cdot}4$, $\beta(Mn^{2+}HL^{4-})$ 5·08, $\beta(H^+MnL^{3-})$ 5·86	65A
Fe^{2+}	kin	25	1 $(NaClO_4)$	K_1 2·54, $\beta(Fe^{2+}H_2L^{3-})$ 2·38, $\beta(Fe^{2+}H_3L^{2-})$ 2·12	68W
Fe^{3+}	sp	20	$I(NaClO_4$?)	$\beta(Fe^{3+}H_2L^{3-})$ 5·03(I = 0·1), 7·03(0 corr); $\beta[Fe^{3+}(HL^{4-})_2]$ 18·85(I = 0·1), 20·63(0 corr), (H^+: K_{12} 6·30, K_{13} 2·30)	68A
	sp	20	$I(NaClO_4)$	$\beta(Fe^{3+}H_3L^{2-})$ 5·04(I = 0·1), 6·37(0 corr); $\beta(Fe^{3+}H_2L^{3-})$ 5·10 (I = 0·1), 7·15(0 corr), (H^+: K_{13} 2·3)	68S
Co^{2+}	gl	25	0·1 $Na(NO_3)$	K_1 6·6, $\beta(H^+CoL^{3-})$ 5·4	63J
	gl	25	0·1 (KCl)	K_1 6·89, $\beta(Co^{2+}HL^{4-})$ 3·81, $\beta(H^+CoL^{3-})$ 4·98	64E
	gl	25	0·1 (Me_4NCl)	K_1 8·16, $\beta(Co^{2+}HL^{4-})$ 5·17	64H
	cix, sp	30	1 (KNO_3)	$\beta(Co^{2+}HL^{4-})$ 4·03	64S
	gl, cal	20	0·1 (Me_4NNO_3)	K_1 7·95, $\Delta H_1 = 4{\cdot}51$, $\Delta S_1 = 51{\cdot}7$, $\beta(Co^{2+}HL^{4-})$ 4·93, $\beta(H^+CoL^{3-})$ 5·8	65A
Ni^{2+}	gl	25	0·1 (KCl)	K_1 6·72, $\beta(Ni^{2+}HL^{4-})$ 3·65, $\beta(H^+NiL^{3-})$ 4·99	64E
	gl	25	0·1 (Me_4NCl)	K_1 7·90, $\beta(Ni^{2+}HL^{4-})$ 5·01	64H
	cix, sp	30	1 (KNO_3)	$\beta(Ni^{2+}HL^{4-})$ 4·18	64S
	gl, cal	20	0·1 (Me_4NNO_3)	K_1 7·8, $\Delta H_1 = 5{\cdot}0$, $\Delta S_1 = 52{\cdot}7$, $\beta(Ni^{2+}HL^{4-})$ 4·9, $\beta(H^+NiL^{3-})$ 5·9	65A
Cu^{2+}	gl	25	0·1 $Na(NO_3)$	K_1 7·3, $\beta(H^+CuL^{3-})$ 5·2	63J
	gl	25	0·1 (KCl)	K_1 8·73, $\beta(Cu^{2+}HL^{4-})$ 4·34, $\beta(H^+CuL^{3-})$ 3·67	64E
	gl, cal	20	0·1 (Me_4NNO_3)	K_1 9·3, $\Delta H_1 = 4{\cdot}9$, $\Delta S_1 = 59{\cdot}2$, $\beta(Cu^{2+}HL^{4-})$ 6·1, $\beta(H^+CuL^{3-})$ 5·6	65A
Zn^{2+}	gl	25	0·1 $Na(NO_3)$	K_1 6·9, $\beta(H^+ZnL^{3-})$ 5·3	63J
	gl	25	0·1 (KCl)	K_1 7·62, $\beta(Zn^{2+}HL^{4-})$ 3·92, $\beta(H^+ZnL^{3-})$ 4·36	64E
	gl, cal	20	0·1 (Me_4NNO_3)	K_1 8·35, $\Delta H_1 = 6{\cdot}32$, $\Delta S_1 = 59{\cdot}8$, $\beta(Zn^{2+}HL^{4-})$ 5·13, $\beta(H^+ZnL^{3-})$ 5·6	65A

continued

Metal	Method	Temp	Medium	Log of equilibrium constant, remarks	Ref
Cd^{2+}	gl	25	0·1 (KCl)	K_1 6·60, $\beta(Cd^{2+}HL^{4-})$ 3·60, $\beta(H^+CdL^{3-})$ 5·06	64E
	gl, cal	20	0·1 (Me_4NNO_3)	K_1 8·1, $\Delta H_1 = 2{\cdot}7$, $\Delta S_1 = 46{\cdot}2$, $\beta(Cd^{2+}HL^{4-})$ 4·97, $\beta(H^+CdL^{3-})$ 5·8	65A
Hg_2^{2+}					

1964 *Addendum*: p. 197, *for ref* "62Sa" *read* "62Sa ≈ 64W".

Metal	Method	Temp	Medium	Log of equilibrium constant, remarks	Ref
In^{3+}	sp	20	$I(NaClO_4)$	$\beta[In^{3+}(H_2L^{3-})_2]$ 12·18($I = 0{\cdot}1$), 14·16(0 corr), (H^+: K_1 7·9, K_{12} 5·6, K_{13} 2·7, K_{14} 2·6)	67A
Sn^{2+}	sol, gl	25	1 ($NaClO_4$)	$K_s[2SnO(s)+H^++2HL^{4-} \rightleftharpoons Sn_2L_2OH^{7-}]$ 7·26, $K_s[4SnO(s)+2H^++2HL^{4-} \rightleftharpoons Sn_4L_2(OH)_4^{6-}]$ 11·68, $K_s[2SnO(s)+H^++2H_2L^{3-} \rightleftharpoons Sn_2HL_2^{5-}]$ 5·0, $K_s[4SnO(s)+2H^++2H_2L^{3-} \rightleftharpoons Sn_4L_2(OH)_2^{4-}]$ 6·8 (second and third K exchanged in text)	66M
Pb^{2+}	cix, sp	30	1 ($NaClO_4$)	$\beta(Pb^{2+}HL^{4-})$ 6·32	64S

63J A. Johansson and E. Wänninen, *Talanta*, 1963, **10**, 769
63K S. K. Kundra, *Indian J. Chem.*, 1963, **1**, 362
64E H. Ellison and A. E. Martell, *J. Inorg. Nuclear Chem.*, 1964, **26**, 1555
64H G. G. Hammes and M. L. Morrell, *J. Amer. Chem. Soc.*, 1964, **86**, 1497
64M M. Miura and Y. Moriguchi, *Bull. Chem. Soc. Japan*, 1964, **37**, 1522
64S P. S. Shetty, P. R. Subbaraman, and J. Gupta, *Indian J. Chem.*, 1964, **2**, 8
64Sa P. S. Shetty and P. R. Subbaraman, *Indian J. Chem.*, 1964, **2**, 428
64W J. I. Watters and R. A. Simonaitis, *Talanta*, 1964, **11**, 247
65A G. Anderegg, *Helv. Chim. Acta*, 1965, **48**, 1712
65K S. K. Kundra, P. R. Subbaraman, and J. Gupta, *Indian J. Chem.*, 1965, **3**, 60
65P R. Phillips, P. Eisenberg, P. George, and R. J. Rutman, *J. Biol. Chem.*, 1965, **240**, 4393
66I R. R. Irani and T. A. Taulli, *J. Inorg. Nuclear Chem.*, 1966, **28**, 1011
66M R. E. Mesmer and R. R. Irani, *J. Inorg. Nuclear Chem.*, 1966, **28**, 493
67A L. P. Andrusenko and I. A. Sheka, *Zhur. neorg. Khim.*, 1967, **12**, 638
68A L. P. Andrusenko and I. A. Sheka, *Zhur. neorg. Khim.*, 1968, **13**, 2645
68M W. M. McNabb, J. F. Hazel, and R. A. Baxter, *J. Inorg. Nuclear Chem.*, 1968, **30**, 1585
68S I. A. Sheka and L. P. Andrusenko, *Zhur. neorg. Khim.*, 1968, **13**, 347
68W C. F. Wells and M. A. Salam, *J. Chem. Soc.* (*A*), 1968, 308

42 Tetrametaphosphate, $P_4O_{12}^{4-}$

Metal	Method	Temp	Medium	Log of equilibrium constant, remarks	Ref
H^+	kin	25	0 corr	K_1 2·5	65I

65I A. Indelli and G. Mantovani, *Trans. Faraday Soc.*, 1965, **61**, 909

43 Polyphosphates, $(P_nO_{3n+1})^{(n+2)-}$

Metal	Method	Temp	Medium	Log of equilibrium constant, remarks	Ref
				n = 4, *Tetraphosphate* $P_4O_{13}^{6-}$	
H^+	gl	25	1 (Me_4NCl)	K_1 8·34, K_{12} 6·63	64Wb
	gl	25	0·1 (Me_4NCl)	K_1 8·88, K_{12} 5·91	68M
1964 *Addendum*: p. 200, *for ref* "62S" *read* "62S ≈ 64W".					
Li^+	gl	25	1 Me_4NCl	K_1 2·64, $\beta(Li^+HL^{5-})$ 1·59	67W
Na^+	gl	25	1 Me_4NCl	K_1 1·79, $\beta(Na^+HL^{5-})$ 1·10	67W
K^+	gl	25	1 Me_4NCl	K_1 1·71, $\beta(K^+HL^{5-})$ 1·11	67W
Mg^{2+}	kin	60	var	K_1 1·75	67Wa
	gl	25	1 (Me_4NCl)	K_1 6·04, $\beta(Mg^{2+}HL^{5-})$ 3·74, $K_{12}(Mg^{2+}MgL^{4-})$ 2·19	68W
Ca^{2+}	gl	25	1 (Me_4NCl)	K_1 5·46, $\beta(Ca^{2+}HL^{5-})$ 3·54, $K_{12}(Ca^{2+}CaL^{4-})$ 3·07	68W
Sr^{2+}	gl	25	1 (Me_4NCl)	K_1 4·82, $\beta(Sr^{2+}HL^{5-})$ 3·49, $K_{12}(Sr^{2+}SrL^{4-})$ 3·42	68W
La^{3+}	gl	25	0·1 (Me_4NCl)	K_1 6·59, $\beta(H^+LaL^{3-})$ 5·05, $\beta(La^{3+}HL^{5-})$ 3·29 [?, compare $K_1(H^+)$]	68M
Cu^{2+}	MHg, gl	25	1 (Me_4NNO_3)	K_1 9·44, K_2 1·16, $\beta(CuL^{4-}OH^-)$ 3·86, (equil with H^+ from ref 63W); $K_a(CuH_2L^{2-})$ −3·45, −5·56; $K_a(CuH_4L_2^{6-})$ −3·55, −4·52, −7·28, −8·40	66W
Hg_2^{2+}					
1964 *Addendum*: p. 201, *for ref* "62S" *read* "62S ≈ 64W".					
$C(NH_2)_3^+$ (guanidinium)					
	gl	25	1 (Me_4NCl)	K_1 1·84, $\beta_{12}(2M^+ + L^{6-} \rightleftharpoons M_2L^{4-})$ 2·76, $\beta(M^+HL^{5-})$ 1·16	64Wb
				Polyphosphates with n > 4	
H^+	var		var	ev $P_6O_{19}^{8-}$ in $Ca_4P_6O_{19}$(s, troemelite) (X, paper chromatography, pre)	64Wa
	O	350	KNO_3(l)	ev $P_5O_{17}^{9-}$ in KNO_3(l)	68S
Ca^{2+}	tyn	25	0·1 Me_4NBr	$\beta(Ca^{2+}$, "site" out of θ independent sites per L): 5·42($n = 2, \theta = 2$), 5·87($n = 3$), 5·93($n = 4$), 5·85($n = 5$), 6·00($n = 6, n = 7$), 6·01($n = 8$), $\theta = 2$ in all cases	64M
		25	1 Me_4NBr	$\beta(Ca^{2+}$, "site") 4·49($n = 2, \theta = 2$), 5·15($n = 3, \theta = 3$), 5·7($n = 4, \theta = 4$), 5·65($n = 5$), 5·77($n = 6$), 6·0($n = 7, n = 8$), $\theta = 4$ for $n \geqslant 4$	
Fe^{3+}	var		var	ev cpx with 6P/Fe (con, sp, gl)	65M
Ag^+	sol	25—35	var	K_{s0} −7·07(25°), −6·89(35°), K_1 7·56(25°), 7·34(35°), $L^- = PO_3^-$-unit, n = 60—150	66B

64M M. Miura and Y. Moriguchi, *Bull. Chem. Soc. Japan*, 1964, **37**, 1522
64W J. I. Watters and R. A. Simonaitis, *Talanta*, 1964, **11**, 247
64Wa W. Wieker, A.-R. Grimmer, and E. Thilo, *Z. anorg. Chem.*, 1964, **330**, 78
64Wb J. I. Watters and S. Matsumoto, *J. Amer. Chem. Soc.*, 1964, **86**, 3961
65M M. K. McElroy, J. F. Hazel, and W. M. McNabb, *J. Inorg. Nuclear Chem.*, 1965, **27**, 859
66B R. M. Bhatnagar and A. C. Chatterji, *Z. phys. Chem. (Leipzig)*, 1966, **232**, 391
66W J. I. Watters and S. Matsumoto, *Inorg. Chem.*, 1966, **5**, 361
67W J. I. Watters and S. Matsumoto, *J. Inorg. Nuclear Chem.*, 1967, **29**, 2955
67Wa W. Wieker, *Z. anorg. Chem.*, 1967, **355**, 20
68M W. M. McNabb, J. F. Hazel, and R. A. Baxter, *J. Inorg. Nuclear Chem.*, 1968, **30**, 1585
68S A. M. Shams el Din, H. D. Taki el Din, and A. A. El Hosary, *Electrochim. Acta*, 1968, **13**, 407
68W J. I. Watters and R. Machen, *J. Inorg. Nuclear Chem.*, 1968, **30**, 2163

Amidophosphates 44

Metal	Method	Temp	Medium	Log of equilibrium constant, remarks	Ref
				Monoamidophosphate, $H_2NPO_3^{2-}$	
H^+	*X*		solid	ev $H_3NPO_3^-$ in $NaPO_3NH_3(s)$	64C
	gl	25	0 corr	K_1 8·63, $\Delta H_1 = -5{\cdot}2$ (ev $H_3NPO_3^-$, not $H_2NPO_3H^-$), K_{12} 3·08, $\Delta H_1 \approx 0$, also graphs of K_1 and K_{12} for other *t* and *I*	68L
				Diamidophosphate, $(H_2N)_2PO_2^-$	
Ag^+	pre, ir		solid	ev $Ag_xH_{4-x}N_2PO_2^-$ in K^+ and Ag^+ salts, $x = 1, 2, 3, 4$	66F
				Amidotriphosphate, $P_3O_9NH_2^{4-}$	
H^+	gl	20	0 var	K_1 5·8	64F
	X		solid	ev $O_3POPO_2OPO_2NH_2^{4-}$ in $K_4P_3O_9NH_2(H_2O)_4(s)$	65H
				Others	
	X		solid	ev cyclic $(\cdot NH \cdot PO_2 \cdot)_3^{3-}$ in $Na_3(NHPO_2)_3(H_2O)_4(s)$	65O
	X		solid	ev cyclic $(\cdot NH \cdot PO_2 \cdot)_4^{4-}$ in $H_4(NHPO_2)_4(H_2O)_2(s)$	65M

64C D. W. J. Cruickshank, *Acta Cryst.*, 1964, **17**, 671
64F W. Feldmann and E. Thilo, *Z. anorg. Chem.*, 1964, **328**, 113
65H W. Hilmer, *Acta Cryst.*, 1965, **19**, 362
65M T. Migchelsen, R. Olthof, and A. Vos, *Acta Cryst.*, 1965, **19**, 603
65O R. Olthof, T. Migchelsen, and A. Vos, *Acta Cryst.*, 1965, **19**, 596
66F J. Frei and E. Steger, *Z. anorg. Chem.*, 1966, **342**, 195
68L D. Levine and I. B. Wilson, *Inorg. Chem.*, 1968, **7**, 818

Other ligands with P 45

Metal	Method	Temp	Medium	Log of equilibrium constant, remarks	Ref
				"$P^4P^3P^4$-*Phosphate*", $P_3O_8^{5-}$	
H^+	gl, sp		var	$K_1 \approx 10$, $K_{12} \approx 7$	59W+63Y/
				"P^4–P^4-*phosphate*" ("*hypophosphate*"), $O_3PPO_3^{4-}$	
Na^+	con	25	0 corr	K_1 2·31, $\beta(Na^+HL^{3-})$ 1·32	67N
				Isohypophosphate, P^3–O–P^5 *phosphate, diphosphate* (III, V), $O_3POPHO_2^{3-}$	
H^+	gl	25	$I(Me_4NCl)$	K_1 6·33($I = 1{\cdot}0$), 6·25($I = 0{\cdot}5, 0{\cdot}2, 0{\cdot}1$), 6·26($\rightarrow 0$); K_{12} 1·71($I = 1{\cdot}0$), 1·50($I = 0{\cdot}5$), 1·67($I = 0{\cdot}2, 0{\cdot}1, \rightarrow 0$); K_{13} 0·6($I = 0$, est)	67C
	kin	25—40	0·5 (NaCl)	K_{13} 0·46(25°), 0·3(40°)	
Li^+	gl	25	0·5 (Me_4NCl)	K_1 0·82	67C
Na^+	gl	25	0·5 (Me_4NCl)	K_1 0·50	67C
K^+	gl	25	0·5 (Me_4NCl)	K_1 0·36	67C

continued overleaf

45 Other ligands with P (contd)

Metal	Method	Temp	Medium	Log of equilibrium constant, remarks	Ref
Mg^{2+}	gl	25	0·5 (Me_4NCl)	K_1 2·65	67C
Ca^{2+}	gl	25	0·5 (Me_4NCl)	K_1 2·27	67C
				Pyrophosphite, $O_2HPOPHO_2^{2-}$	
M^{2+}	pol		1·8 KNO_3 *etc.*	ev weak cpx for Cd^{2+}($K_1 \approx 1·8$) and Cu^{2+}, stronger for UO_2^{2+}	64G
				Metaphosphates, $P_nO_{3n}^{n-}$ (cyclic)	
H^+	var		var	ev $P_6O_{18}^{6-}$ (cyclic) (pre, nmr, paper chromatography)	65G
	X		solid	ev $P_6O_{18}^{6-}$ (cyclic) in $Na_6P_6O_{18}(H_2O)_6$	65J
	var		var	ev cyclic $P_6O_{18}^{6-}$, no ev $HP_6O_{18}^{5-}$ (pre, gl, paper chromatography)	65T
	var		var	ev cyclic $P_5O_{15}^{5-}$, no ev $HP_5O_{15}^{4-}$ (pre, gl, paper chromatography)	65T
Ca^{2+}	var		var	ev $P_8O_{24}^{8-}$ (pre, paper chromatography), ev cpx with Ca^{2+} (sol)	68S
				Peroxomonophosphate, PO_5^{3-}	
H^+	sp, gl	25	*I*(Na, $LiClO_4$)	K_1 12·8, K_{12} 5·5 ($I = 0·15$), K_{13} 1·1($I = 0·2$)	65B
				Thiophosphate PO_3S^{3-}	
H^+	gl	?	var	K_1 10·61, K_{12} 5·91	65F
	gl	23	var	K_1 10·4, K_{12} 5·75, $K_{13} < 2$	65N
				μ-Disulphidohexaoxodiphosphate, $O_3PS_2PO_3^{4-}$	
H^+	gl	23	var	$K_1 \approx K_{12}$ 6·65, $K_{13} \approx K_{14}$ 2·5	65N
				Fluorotriphosphate, $P_3O_9F^{4-}$	
H^+	pH	20	var	K_1 6·2	65Fa
				Hexafluorophosphate, PF_6^-	
Cr^{3+} cpx	nmr		var	$\beta(Cr\,en_3^{3+}PF_6^-) \approx 0$, no ev cpx $Cr(C_2O_4)_3^{3-}$–PF_6^-	65S

59W B. Blaser and K.-H. Worms, *Z. anorg. Chem.*, 1959, **300**, 250
63Y N. Yoza and S. Ohashi, *Bull. Chem. Soc. Japan*, 1963, **36**, 1485
64G D. Grant and D. S. Payne, *J. Inorg. Nuclear Chem.*, 1964, **26**, 1985
65B C. J. Battaglia and J. O. Edwards, *Inorg. Chem.*, 1965, **4**, 552
65F F. Fehér and F. Vial, *Z. anorg. Chem.*, 1965, **335**, 113
65Fa W. Feldmann, *Z. anorg. Chem.*, 1965, **338**, 235
65G E. J. Griffith and R. L. Buxton, *Inorg. Chem.*, 1965, **4**, 549
65J K.-H. Jost, *Acta Cryst.*, 1965, **19**, 555
65N H. Neumann, I. Z. Steinberg, and E. Katchalski, *J. Amer. Chem. Soc.*, 1965, **87**, 3841
65S T. R. Stengle and C. H. Langford, *J. Phys. Chem.*, 1965, **69**, 3299
65T E. Thilo and U. Schülke, *Z. anorg. Chem.*, 1965, **341**, 293
67C R. L. Carroll and R. E. Mesmer, *Inorg. Chem.*, 1967, **6**, 1137
67N J. Nassler and A. Sedlák, *Coll. Czech. Chem. Comm.*, 1967, **32**, 2405
68S U. Schülke, *Z. anorg. Chem.*, 1968, **360**, 231

Arsenite, $As(OH)_4^-$ or AsO_2^- 46

Metal	*Method*	*Temp*	*Medium*	*Log of equilibrium constant, remarks*	*Ref*
H^+	gl	25	0·6 Na(Cl)	K_1 9·02, ev polyn anion	63N/62S
	gl	20	dil	K_1 9·38(in H_2O), 10·35(in 44% Me_2CO), 10·68(in 51% dioxan), 10·31(in 49% Me_2NCHO)	65H
	cal	25	0·1—0·3	$\Delta H_1 = -6{\cdot}58$, $\Delta S_1 = 20{\cdot}2$(K_1 9·23)	64S
	gl, con		var	ev AsO_3^{3-}, $As_2O_5^{4-}$, AsO_2^-, and $HAsO_2$ (?, data indicate only OH^-, L^-, and HL)	66B
	Ram	25—80	OH^- var	ev ⩾ 3 species, probably $As(OH)_3$, $AsO(OH)_2^-$, AsO_2OH^{2-}, and AsO_3^{3-}, no ev polyn cpx with charge −1/As	68L

62S L. G. Sillén, Proc. Robert A. Welch Foundation, Nov., 1962, Houston, Texas
63N G. Neumann, unpublished, quoted by ref. 62S
64S P. Sellers, S. Sunner, and I. Wadsö, *Acta Chem. Scand.*, 1964, **18**, 202
65H M. K. Hargreaves, E. A. Stevinson, and J. Evans, *J. Chem. Soc.*, 1965, 4582
66B C. S. Bhatnagar and K. Govind, *Z. Naturforsch.*, 1966, **21b**, 190
68L T. M. Loehr and R. A. Plane, *Inorg. Chem.*, 1968, **7**, 1708

Arsenate, AsO_4^{3-} 47

Metal	*Method*	*Temp*	*Medium*	*Log of equilibrium constant, remarks*	*Ref*
H^+	Cu, O	350	KNO_3(l)	$K(2HL^{2-} + O^{2-} \rightleftharpoons 2L^{3-} + H_2O)$ 2·42 in KNO_3(l), M units	63S = 64Sb
	cal	25	0·1—0·3	$\Delta H_1 = -4{\cdot}35$, $\Delta S_1 = 38{\cdot}5$(K_1 11·53) $\Delta H_{12} = -0{\cdot}77$, $\Delta S_1 = 28{\cdot}4$(K_{12} 6·77) $\Delta H_{13} = 1{\cdot}69$, $\Delta S_{13} = 16{\cdot}0$($K_{13}$ 2·25)	64S
	gl	25	H_2O–D_2O	K_{13} 2·301($x_D = 0$), 2·421($x_D = 0{\cdot}494$), 2·596($x_D = 1$) in $H_{2-2x}D_{2x}O$(l), also K_{13} for other x_D values	64Sa
Pa^{5+}	oth		6 HNO_3	$\beta(Pa^VH_3L)$ 1·65, $\beta[Pa^V(H_3L)_2]$ 2·75 (sorption on silica gel)	66D
Mo^{5+}	esr	0	var	ev $MoOCl_{4-n}(H_2L)_n^-$, $n = 1, 2, 3, 4$, ev dimer	68M
Fe^{2+}	kin, sp	25	1 H_2SO_4	ev cpx, apparent K_1 0·28 in 1M-H_2SO_4	66W

63S A. M. Shams el Din and A. A. A. Gerges, *J. Inorg. Nuclear Chem.*, 1963, **25**, 1537
64S P. Sellers, S. Sunner, and I. Wadsö, *Acta Chem. Scand.*, 1964, **18**, 202
64Sa P. Salomaa, L. L. Schaleger, and F. A. Long, *J. Amer. Chem. Soc.*, 1964, **86**, 1
64Sb A. M. Shams el Din and A. A. A. Gerges, *Electrochim. Acta*, 1964, **9**, 613
66D A. V. Davydov, I. N. Marov, and P. N. Palei, *Zhur. neorg. Khim.*, 1966, **11**, 1316
66W R. Woods, *J. Phys. Chem.*, 1966, **70**, 1446
68M I. N. Marov, Yu. N. Dubrov, V. K. Belyaeva, and A. N. Ermakov, *Zhur. neorg. Khim.*, 1968, **13**, 2140, 2445

Other ligands with As 48

Metal	*Method*	*Temp*	*Medium*	*Log of equilibrium constant, remarks*	*Ref*
				Dialkylarsinate, $R_2AsO_2^-$	
H^+	gl	?	?	K_1 6·51(R = Et), 6·62(n-C_3H_7), 7·13(n-C_4H_9) 7·25(cyclo-C_6H_{11})	66M

66M A. Merijanian and R. A. Zingaro, *Inorg. Chem.*, 1966, **5**, 187

49 Ligands with Sb or Bi

Metal	Method	Temp	Medium	Log of equilibrium, constant remarks	Ref
H^+	X		solid	ev Sb_3 and Sb_5 groups in $Sb_2O_5(H_2O)_4(s)$	68B

1964 *Erratum*: p. 207, *in ref* /57P *for* "K_{s0}" *read* "K_s".

68B L. H. Baetsle and D. Huys, *J. Inorg. Nuclear Chem.*, 1968, **30**, 639

50 Water, H_2O

Metal	Method	Temp	Medium	Log of equilibrium constant, remarks	Ref
general	nmr	33—96	$Ca(NO_3)_2(H_2O)_4(l)$	$\bar{n}(Mg^{2+}) > \bar{n}(Ca^{2+}) > \bar{n}(K^+, Me_4N^+)$ in $Ca(NO_3)_2(H_2O)_4(l)$	67M
	con, sol		var	K_1 and K_{s0} for many L′–M pairs (L′ $\neq H_2O$) is proportional to $[H_2O]^n$, M units. Plots of log K (log $[H_2O]$) and tables of n (apparent hydration number) given.	/68Q
H^+	ir		solid	ev H_3O^+, $H_5O_2^+$ in various solids	64Ga
	lit	25	H_2SO_4 var	ev HL_n^+, $n = 1, 2, 3 \ldots 10$ tentative equil constants given (lit = sp, Ram, pH_2O)	/64R
	oth	15—600	gas	plots log $K_n (n = 2, 3 \ldots 8)$ *versus* T^{-1}, gas phase	67K
		27	gas	$\Delta H_1 = -165$(lit), $\Delta H_2 = -36$, $\Delta H_3 = -22{\cdot}3$, $\Delta H_4 = -17$, $\Delta H_5 = -15{\cdot}3$, $\Delta H_6 = -13$, $\Delta H_7 = -11{\cdot}7$, $\Delta H_8 = -10{\cdot}3$ $\Delta S_2 = -33{\cdot}3$, $\Delta S_3 = -29$, $\Delta S_4 = -28{\cdot}3$, $\Delta S_5 = -32{\cdot}6$, $\Delta S_6 = -30{\cdot}3$, $\Delta S_7 = -29{\cdot}6$, $\Delta S_8 = -27$ in gas phase (mass spectrometry)	
	X		solid	ev $H_5O_2^+$ in $HCl(H_2O)_2(s)$ and $HCl(H_2O)_3(s)$	67L
	sp		MeCN	K_1 2·15, β_2 3·43, β_3 4·52, β_4 4·08 in MeCN(l)	68K
	X		solid	ev $H_7O_3^+$ and $H_9O_4^+$ in $HBr(H_2O)_4(s)$	68La
	X		solid	ev $H_5O_2^+$ in $HClO_4(H_2O)_2(s)$	68O
	X		solid	ev H_3O^+ in $H_2SO_4H_2O(s)$	68T
Group 1A$^+$	sol	25	MeCN	K_1 0·5(Li^+), 0·3(Na^+), 0·0(K^+), −0·3(Cs^+); β_2 0·7(Li^+), 0·5(Na^+) in MeCN(l)	67Ca
Be^{2+}	nmr		var	ev BeL_4^{2+}	63C
Mg^{2+}	X		solid	ev MgL_6^{2+} in $MgSO_4(H_2O)_6(s)$	64Z
Nd^{3+}	sp		MeOH	$K_3K_4[Nd(MeOH)_4L_2^{3+} + 2L \rightleftharpoons Nd(MeOH)_2L_4^{3+} + 2MeOH]$ 1·52, K_5K_6 1·15 in MeOH(l)	67R
V^{2+}	X		solid	ev VL_6^{2+} in $V(NH_4)_2(SO_4)_2(H_2O)_6(s)$	67C
Cr^{3+}	cix	30—60	MeOH	$K_3 \sim 1{\cdot}7$, $K_4 \sim 1{\cdot}4$, $K_5 \sim 0{\cdot}5$, $K_6 \sim 0{\cdot}2$ in MeOH(l), activities (x scale) for MeOH and H_2O	64J
UO_2^{2+}	X		solid	ev $UO_2L_2(NO_3)_2$, not $UO_2L_6^{2+}$ in $UO_2(NO_3)_2(H_2O)_6(s)$	/67V
Mn^{2+}	X		solid	ev MnL_6^{2+} in $Mn(NH_4)_2(SO_4)_2(H_2O)_6(s)$	66Mb
Fe^{2+}	X		solid	ev FeL_6^{2+} in $FeSO_4(H_2O)_7(s)$	64B

continu

Metal	Method	Temp	Medium	Log of equilibrium constant, remarks	Ref
Co^{2+}	*X*		solid	ev CoL_6^{2+} in $CoSO_4(H_2O)_6$(s)	62Z
	kin	32—43	0·4 (K)NO_3	ev CoL_6^{2+} and CoL_4^{2+} (*T*-jump with sp)	64S
	sp	25	Me_2CO	K_1 0·56, K_2 −1·82 in Me_2CO(l) *x* units	65P
		25	EtOH	K_1 0·46, K_2 −2·22 in EtOH(l) *x* units	
	cal	25	C_4H_9OH	K_1 1·76, β_2 3·08, β_3 4·10, β_4 4·64, β_5 5·6?, $\Delta H_1 = -2{\cdot}85$, $\Delta H_{\beta2} = -4{\cdot}70$, $\Delta H_{\beta3} = -5{\cdot}35$, $\Delta H_{\beta4} = -5{\cdot}6$, $\Delta H_{\beta5} = -5{\cdot}8$? $\Delta S_1 = -1{\cdot}5$, $\Delta S_{\beta2} = -1{\cdot}7$, $\Delta S_{\beta3} = 0{\cdot}8$, $\Delta S_{\beta4} = 2{\cdot}4$, $\Delta S_{\beta5} = 6$? in C_4H_9OH(l), $Co(ClO_4)_2$ var; ClO_4^- bound in outer sphere (con)	68H
	nmr	−64 to −38	var	ev CoL_6^{2+}	68M
Ni^{2+}	*X*		solid	ev NiL_6^{2+} in $(NH_4)_2Ni(SO_4)_2(H_2O)_6$(s)	63G
	sp	25	Me_2CO	K_1 0·79, K_2 −1·68 in Me_2CO(l) *x* units	65P
		25	EtOH	K_1 −0·24, K_2 −2·31 in EtOH(l) *x* units	
	cal	25	C_4H_9OH	K_1 1·80, β_2 3·23, β_3 4·36, β_4 5·20, β_5 5·8?; $\Delta H_1 = -3{\cdot}05$, $\Delta H_{\beta2} = -4{\cdot}80$, $\Delta H_{\beta3} = -5{\cdot}25$, $\Delta H_{\beta4} = -5{\cdot}45$, $\Delta H_{\beta5} = -5{\cdot}7$?; $\Delta S_1 = -2{\cdot}0$, $\Delta S_{\beta2} = -1{\cdot}3$, $\Delta S_{\beta3} = 2{\cdot}3$, $\Delta S_{\beta4} = 5{\cdot}5$, $\Delta S_{\beta5} = 8$? in C_4H_9OH(l), $Ni(ClO_4)_2$ var; ClO_4^- bound in outer sphere (con)	68H
Rh^{3+}	kin			ev RhL_6^{3+}	64P
$(NH_3)_2Pt^{2+}$	nmr	5—85	var	ev $Pt(NH_3)_2L_2^{2+}$	68Gb
Me_3Pt^+	nmr	5	var	ev *cis*-$Me_3PtL_3^+$	68Gb ≈ 67Ga
Cu^+	pH_2O	40—80	Cl^- var	$CuCl_2^-$ less hydrated than Cl^-	68A
Cu^{2+}	pol		Me_2CO	K_1 1·75, K_2 1·50, K_3 1·00, K_4 0·75 in Me_2CO(l), 0·1M-$LiClO_4$; K_1 1·75, K_2 1·25, K_3 0·80, K_4 0·65 in Me_2CO(l), 0·1M-Et_4NClO_4.	64N
	X		solid	ev distorted CuL_6^{2+} in $Cu(NH_4)_2(SO_4)_2(H_2O)_6$(s)	66M
	sol	25	org	$K_s[CuSO_4(s)+nL \rightleftharpoons CuL_nSO_4]$ −7·9(n = 4), −10·85(n = 6) in Me_2CO(l); −7·75(n = 4), −10·65(n = 6) in dioxan(l)	67G
	sol	25	dioxan	$K_s[CuCl_2(s)+nL \rightleftharpoons CuL_nCl_2]$ 1·72(n = 1), 2·89(n = 2) in dioxan(l)	68G
Zn^{2+}	*X*		solid	ev ZnL_6^{2+} in $Zn(NH_4)_2(SO_4)_2(H_2O)_6$(s)	64M
	kin	12—34	0·1 KNO_3	ev ZnL_6^{2+} and ZnL_4^{2+} (*T*-jump with sp)	64S
	var	25	MeOH	K_1 1·29, K_2 0·72, K_3 −0·40, K_4 −1·43(pol); K_1 1·07, K_2 0·64, K_3 −0·46, K_4 −1·42 (chronopotentiometry), in MeOH(l)	68L
Cd^{2+}	MHg	20	MeCN	β_5 2·26 in MeCN(l), ev CdL_6^{2+}(65G)	64G ≈ 65G
			Me_2NCHO	β_6 −4·1 in Me_2NCHO(l)	
	X		solid	ev CdL_6^{2+} in $Cd(NH_4)_2(SO_4)_2(H_2O)_6$(s)	66Ma
	sol	25	org	$K_s[CdSO_4(s)+nL \rightleftharpoons CdL_nSO_4]$ −7·45(n = 4), −10·4(n = 6) in Me_2CO(l); −7·35(n = 4), −10·2(n = 6) in dioxan(l)	67G
	sol	25	dioxan	$K_s[CdCl_2(s)+2H_2O \rightleftharpoons CdCl_2(H_2O)_2]$ −4·36 in dioxan(l)	68Ga
Hg_2^{2+}	*X*		solid soln	ev linear $(H_2OHgHgOH_2)^{2+}$ in solid $Hg_2(ClO_4)_2(H_2O)_4$ and in 0·6—2·8M-$Hg_2(ClO_4)_2$	66J

continued overleaf

50 H_2O (contd)

Metal	*Method*	*Temp*	*Medium*	*Log of equilibrium constant, remarks*	*Ref*
Al^{3+}	*X*		solid	ev $AlL_6{}^{3+}$ in three series of alums, $MAl(SO_4)_2(H_2O)_{12}(s)$	35L
	nmr		var	ev $AlL_6{}^{3+}$	63C
	X		solid	ev $AlL_6{}^{3+}$ in $AlCl_3(H_2O)_6(s)$	68B
Ga^{3+}	nmr		var	ev $GaL_6{}^{3+}$	66F
Sb^{5+}	cal	25	$C_2H_4Cl_2$	$\Delta H(SbCl_5 + H_2O \rightarrow SbCl_5H_2O) = -24{\cdot}3$ in $C_2H_4Cl_2(l)$	67O
H_2O	pH_2O	25	org	$K(2L \rightleftharpoons L_2)$ 1·01 in $PhNO_2$, 0·41 in $C_6H_4(CO_2Bu)_2$, −1·22 in $(BuO)_3PO(l)$; $K(3L \rightleftharpoons L_3)$ 2·45 in Bu_3N, 0·63 in $PhCH_2OH$, 1·75 in PhCOMe(l) *etc.*, *x* units	67J
NO_3^-	sol	25	MeCN	K_1 0·3, β_2 0·0 in MeCN(l)	67Ca
Group 7B	sol	25	MeCN	K_1 0·95(Cl^-), −0·15(ClO_4^-), 0·0(IO_4^-); β_2 1·3(Cl^-), in MeCN(l)	67Ca

35L H. Lipson, *Proc. Roy. Soc.*, 1935, *A*, **151**, 347
62Z A. Zalkin, H. Ruben, and D. H. Templeton, *Acta Cryst.*, 1962, **15**, 1219
63C R. E. Connick and D. N. Fiat, *J. Chem. Phys.*, 1963, **39**, 1349
63G N. W. Grimes, H. F. Kay, and M. W. Webb, *Acta Cryst.*, 1963, **16**, 823
64B W. H. Baur, *Acta Cryst.*, 1964, **17**, 1167
64G A. M. Golub and V. M. Samoilenko, *Zhur. neorg. Khim.*, 1964, **9**, 70
64Ga R. D. Gillard and G. Wilkinson, *J. Chem. Soc.*, 1964, 1640
64J J. C. Jayne and E. L. King, *J. Amer. Chem. Soc.*, 1964, **86**, 3989
64M H. Montgomery and E. C. Lingafelter, *Acta Cryst.*, 1964, **17**, 1295
64N I. N. Nelson and R. T. Iwamoto, *Inorg. Chem.*, 1964, **3**, 661
64P W. Plumb and G. M. Harris, *Inorg. Chem.*, 1964, **3**, 542
64R E. B. Robertson and H. B. Dunford, *J. Amer. Chem. Soc.*, 1964, **86**, 5080
64S T. J. Swift, *Inorg. Chem.*, 1964, **3**, 526
64Z A. Zalkin, H. Ruben, and D. H. Templeton, *Acta Cryst.*, 1964, **17**, 235
65G A. M. Golub and V. M. Samoilenko, *Zhur. neorg. Khim.*, 1965, **10**, 328
65P R. F. Pasternack and R. A. Plane, *Inorg. Chem.*, 1965, **4**, 1171
66F D. Fiat and R. E. Connick, *J. Amer. Chem. Soc.*, 1966, **88**, 4754
66J G. Johansson, *Acta Chem. Scand.*, 1966, **30**, 553
66M H. Montgomery and E. C. Lingafelter, *Acta Cryst.*, 1966, **20**, 659
66Ma H. Montgomery and E. C. Lingafelter, *Acta Cryst.*, 1966, **20**, 728
66Mb H. Montgomery, R. V. Chastain, and E. C. Lingafelter, *Acta Cryst.*, 1966, **20**, 731
67C R. V. Chastain, J. J.j Natt, A. M. Witkowska, and E. C. Lingafelter, *Acta Cryst.* 1967, **22**, 775
67Ca M. K. Chantooni, un., and I. M. Kolthoff, *J. Amer. Chem. Soc.*, 1967, **89**, 1582
67G A. M. Golub and V. I. Golovorushkin, *Izvest. V.U.Z., Khim. i khim. Tekhnol.*, 1967, **10**, 754
67Ga G. E. Glass and R. S. Tobias, *J. Amer. Chem. Soc.*, 1967, **89**, 6371
67J J. R. Johnson, S. D. Christian, and H. E. Affsprung, *J. Chem. Soc.* (*A*), 1967, 1924
67K P. Kebarle, S. K. Searles, A. Zolla, J. Scarborough, and M. Arshadi, *J. Amer. Chem. Soc.* 1967, **89**, 6393
67L J.-O. Lundgren and I. Olovsson, *Acta Cryst.*, 1967, **23**, 966, 971
67M C. T. Moynihan and A. Fratiello, *J. Amer. Chem. Soc.*, 1967, **89**, 5546
67O G. Olofsson, *Acta Chem. Scand.*, 1967, **21**, 1887
67R E. D. Romanenko, N. A. Kostromina, and T. V. Ternovaya, *Zhur. neorg. Khim.*, 1967, **12**, 700
67V V. M. Vdovenko, E. V. Stroganov, and A. P. Sokolov, *Radiokhimiya*, 1967, **9**, 127
68A A. S. Tarkhanyan, *Armyan. khim. Zhur.*, 1968, **21**, 825
68B D. R. Buchanan and P. M. Harris, *Acta Cryst.*, 1968, **B24**, 954
68G A. M. Golub and V. I. Golovorushkin, *Zhur. fiz. Khim.*, 1968, **42**, 1902
68Ga A. M. Golub and V. I. Golovorushkin, *Zhur. neorg. Khim.*, 1968, **13**, 3194
68Gb G. E. Glass, W. B. Schwabacher and R. S. Tobias, *Inorg. Chem.*, 1968, **7**, 2471
68H P. C. Harris and T. E. Moore, *Inorg. Chem.*, 1968, **7**, 656
68K I. M. Kolthoff and M. K. Chantooni, jun., *J. Amer. Chem. Soc.*, 1968, **90**, 3320
68L S. L. Levine and W. E. Ohnesorge, *Analyt. Chim. Acta*, 1968, **41**, 293
68La J.-O. Lundgren and I. Olovsson, *J. Chem. Phys.*, 1968, **49**, 1068
68M N. A. Matwiyoff and P. E. Darley, *J. Phys. Chem.*, 1968, **72**, 2659
68O I. Olovsson, *J. Chem. Phys.*, 1968, **49**, 1063
68Q A. S. Quist and W. L. Marshall, *J. Phys. Chem.*, 1968, **72**, 1536
68T I. Taesler and I. Olovsson, *Acta Cryst.*, 1968, **B24**, 299

Peroxide, O_2^{2-} 51

Metal	*Method*	*Temp*	*Medium*	*Log of equilibrium constant, remarks*	*Ref*
H^+	var		org	$K_{13} = \beta(H^+H_2L)$ − 11(est), ev $H_3O_2^+$ in sulpholane(l) (pre)	64A

1964 *Erratum*: p. 210, *in ref* 60C *for* "K_1" *read* "K_{12}"

Metal	*Method*	*Temp*	*Medium*	*Log of equilibrium constant, remarks*	*Ref*
	gl	5	1—6 H_2L	K_{12} 11·67, $K(4H_2L \rightleftharpoons H_8L_4)$ − 2·11, $K_a(H_8L_4)$ − 9·5, $K_a(H_7L_4^-)$ − 10·98 (K_w − 14·62)	64C
			3 K(Cl)	K_{12} 12·13, $K(2H_2L \rightleftharpoons H_4L_2)$ − 0·96, $K_a(H_4L_2)$ − 10·89, $K_a(H_3L_2^-)$ − 13·86 (K_w − 14·93)	
Ti^{4+}	var		var	ev $TiLF_5^{3-}$(fp, ir, Ram), $TiL(SO_4)_2^{2-}$(ir, Ram)	64G
	kin	?	var	ev $TiOHL^+$, $\beta(Ti^{IV}, H_2L)$ 1·82 at pH = 3.3 $K_a(TiOH_2L^{2+})$ − 1·82(?, derivation not clear) $\beta(TiO^{2+}HL^-)$ 14 [$K_{12}(H^+)$ from 29K, Ti data also from 51B]	65B
	sp, gl		var	$K_a(TiOH_2L^{2+})$ − 2, $\beta(TiO^{2+}HL^-)$ 13·85 [$K_1(H^+)$ 11·80, $\beta(TiO^{2+}H_2L)$ 4·05]	66B
	pre, *X*		solid	ev $TiLCl_4H_2O^{2-}$ in $M_2TiLCl_4H_2O$(s) (pre, *X* powder), M = Rb, Cs	67W
	sp		HCl var	ev cpx	
	sp	25	$I(NaClO_4)$, 1 H^+	$K_1 = \beta(TiO^{2+}H_2L)$ 4·17(I = 5·46), 4·01(I = 3·23), 3·94(I = 2·11), 3·86(I = 1·00)	68V
		25	I $HClO_4$	K_1 4·46(I = 5), 4·14(I = 3), 3·87(I = 1), 3·81(I = 0·5) $K_a(TiOH_2L^{2+})$ − 1·81(I = 4), − 1·80(I = 3), − 1·82(I = 1), − 1·46(0 corr)	
			I HCl	K_1 4·07(I = 5), 3·95(I = 3), 3·85(I = 1), 3·78(I = 0·5)	
		5—40	1 $HClO_4$	K_1 4·42(5°), 4·15(15°), 3·87(25°), 3·55(40°)	
		5—40	0 corr	K_1 4·29(5°), 4·01(15°), 3·74(25°), 3·44(40°) ΔH_1 = − 10·0, ΔS_1 = − 15, also K_1 values for various conc of HCl, $HClO_4$, and H_2SO_4 at 5—40°C	
Ti^{4+} cpx	sp		var	$\beta(TiOC_2O_4^0, H_2L)$ 6·15	64J
V^{5+}	var	22—25	H_2L var	$H_3X = H_3VL_4$ or, *e.g.*, H_3VO_3L; $K(H_4X^+ \rightleftharpoons 2H^+ + H_2X^-)$ − 8·3 to − 7·3 $K_a(H_2X^-)$ − 6·8 to − 6·2 in H_2L var, *x* units (sp, gl, kin)	63Fa
	kin?	45—65	var	$K(VO_3^- + H_2L \rightleftharpoons VO_2L^- + H_2O)$ 5·88(45°), 6·74(55°), 7·18(65°)	65Ba
	sp, gl		var	$K_a(VO_2H_2L^+)$ − 0·4, $\beta(VO_2^+HL^-)$ 15·92 [$K_1(H^+)$ 11·80, $\beta(VO_2^+H_2L)$ 4·52]	66B
	pre		solid	ev$(V_2O_{11}^{4-})_n = (V_2O_3L_4^{4-})_n$	67F
	sp, gl	23	1 (Na)ClO_4	$K(VOL^+ + H_2L \rightleftharpoons VOL_2^- + 2H^+)$ 2·15 ([H^+] = 0·1), 0·08([H^+] = 1) (is reaction correct?)	67O
Nb^{5+}	var		var	ev $NbL_2(OH)_2^-$ (fp, ir, Ram)	64G
	var		var	ev ani, unch, and cat cpx, $K_a[Nb(OH)_4H_2L^+]$ − 2·7 (sol, sp, tp, cix)	66Ba
Ta^{5+}	fp		0·5 HCl	ev cation, $TaO_2H_2L^+$ and uncharged polyn cpx	63N
	pol	25	0·34 H_2SO_4	$\beta(HTaO_3{}^0H_2L^0)$ 2·0	64B
	sp		H_2SO_4	$\beta(Ta^VH_2L)$ 3·43 in H_2SO_4(l)	68Va
Cr^{3+}	pre, sp		$HClO_4$ var	ev Cr_2L^{4+}, $Cr_3L_2^{5+}$	68Aa
Cr^{4+}	*X*		solid	ev $CrL_2(NH_3)_3$ units in $CrO_4(NH_3)_3$(s)	64Sb
	X, ir		solid	ev $CrL_2(CN)_3^{3-}$ in $K_3CrO_4(CN)_3$(s)	65S
	var		var	ev Cr en$(NH_3)L_2$, Cr pn$(H_2O)L_2$ (pre, ir, sp)	67H

continued overleaf

51 O_2^{2-} (contd)

Metal	*Method*	*Temp*	*Medium*	*Log of equilibrium constant, remarks*	*Ref*
Cr^{5+}	*X*		solid	ev $CrO_8^{3-} = CrL_4^{3-}$ in K_3CrO_8 (s, red)	41W =
Cr^{6+}	*X*, ir		solid	ev $CrOL_2$ py units in $CrO_5C_5H_5N$ (s, blue)	64Sa
	var		var	ev $CrOL_2Cl^-$ in solid, $MeNO_2$(l) (con, ir, pre)	65T
	X		solid	ev $CrOL_2$ bipy units in solid	68Sa
Mo^{6+}					
1964 *Addendum*: p. 212, *ref* 63D, *add to remarks* "(sp, kin, gl)"					
	var		var	ev $Mo_2O_3L_4^{2-}$, not $HMoO_2L_2^-$ (fp, ir, Ram)	63G
	X		solid	ev $MoOLF_4^{2-}$ in $K_2MoOLF_4H_2O$(s)	65G = 67
	X		solid	ev$(H_2OOL_2Mo)_2O^{2-}$ and $(OL_2Mo)_2(HL)_2^{2-}$ not $MoO_2L_2^{2-}$, $MoOL_3^{2-}$, or $Mo_2O_3L_4^{2-}$ in pyH^+ salts	68Mb
	X		solid	ev $(H_2OMoOL_2)_2O^{2-}$ in $K_2Mo_2O_{11}(H_2O)_4$(s)	68S
W^{6+}	var	22—25	H_2L var	$K(WL_4^{2-}+H^+ \rightleftharpoons HWL_4^-)$ 4·9 to 6·9; $K(WL_4^{2-}+H_2L \rightleftharpoons HWL_4^-+HL^-)$ −6·9 to −8·1 in H_2L var, *x* units (sp, gl, kin)	63Da
	var		var	ev $W_2O_3L_4^{2-}$, not $HWO_2L_2^-$ (fp, ir, Ram)	63G
	X		solid	ev $W_2O_{11}(H_2O)_2^{2-} = (WL_2OOH_2)_2O^{2-}$ in $K_2W_2O_{11}(H_2O)_4$(s)	64E
	var		solid	ev $WLOCl_4^{2-}$ in K_2WLOCl_4(s) *etc.* (pre, *X* powder, ir)	68Y
UO_2^{2+}	gl(sp, dis)	?	1 LiCl	$\beta(H^+UO_2L_3^{4-})$ 11·06 $K[2HUO_2L_3^{3-}+4H_2O \rightleftharpoons (UO_2)_2L_3^{2-}+3H_2L+4OH^-]$ −18·4	65M
	pre		var	ev $(UO_2)_2L_2O^{2-}$	
	sp	?	1 KCl	$K[UO_2(CO_3)_3^{4-}+H_2L+2OH^- \rightleftharpoons UO_2L(CO_3)_2^{4-}+CO_3^{2-}+2H_2O]$ −11·1	65Sa
	X		solid	ev $UO_2L_3^{4-}$ in $Na_4UO_8(H_2O)_9$(s)	66A = 68
	sp, gl		var	$K[(UO_2)_2Y^0+H_2L \rightleftharpoons 2H^++(UO_2)_2YL^{2-}]$ −3·77, $K[2UO_2Y^{2-}+H_2L \rightleftharpoons 2H^++(UO_2)_2Y_2L^{6-}]$ −1·15 $K[(UO_2)_2Y_2L^{6-}+H_2L \rightleftharpoons 2H^++(UO_2)_2YL_2^{4-}+Y^{4-}]$ −16, $K_a[(UO_2)_2YL^{2-}]$ −8, H_4Y = EDTA	68G
	sp, gl		var	equil constants for mixed UO_2^{2+}–L^{2-}–citrate cpx, $K(UO_2^{2+}+3H_2L \rightleftharpoons 6H^++UO_2L_3^{4-})$ −38·4, β_3 71·7(H^+; K_1 25·0, K_{12} 11·74)	68Ga
	sp, gl		var	equil constants given for reactions involving $(UO_2)_2LF_5^{3-}$, $(UO_2)_2L_2F_5^{5-}$, and $(UO_2)_2L_3F_2^{4-}$, β_3 72·95	68Gb
	sol	?	var	†$K_{sn}[UO_2L(H_2O)_4(s)+(2-2n)H^+ \rightleftharpoons UO_2L_n^{(2-2n)+}+(1-n)H_2L]$ −2·0($n = 0$), −6·7($n = 1$), −15·36($n = 2$) K_{s0} −38·74 to −38·48, K_1 32·04, β_2 60·15(H^+: K_1 25·0, K_{12} 11·74)	/68Ma
Np^{4+}	sol		var	equil constant given for reactions involving $NpL_{2.3}H_2L$(s) and aqueous $NpL_{2.3}H_2L$, $NpL_{1.15}H_2L^{2+}$ and $Np(H_2L)_2^{4+}$ (charges not balanced)	/68Ma
Fe^{3+}	kin	70·1	EtOH	$\beta[Fe^{3+}(H_2L)_2]$ 4·86 in EtOH (7% H_2O)	67Ha
Co en_2^{3+}	kin, sp	10—25	0·25 ($NaClO_4$)	$\beta[H^+Co\ en_2(\mu\text{-}L, \mu\text{-}NH_2)Co\ en_2^{3+}]$ (brown B $\rightleftharpoons$ orange) 0·99(kin), 1·05(sp)(10°), 0·98(kin, sp)(15°), 0·91(kin), 0·90(sp)(20°), 0·84(kin), 0·85(sp)(25°);	67Mb

contin

Metal	Method	Temp	Medium	Log of equilibrium constant, remarks	Ref
$Coen_2^{3+}$ (*contd*)			0 corr	β 0·15(10°), 0·07(15°), −0·02(20°), −0·05(25°), $\Delta H = -5{\cdot}5$, $\Delta S = -18$	
	kin	10—25	0·25 ($NaClO_4$)	K[Co en_2(μ-LH, μ-NH_2)Co en_2^{4+}, orange ⇌ red] 0·24(10°), 0·16(15°), 0·10(20°), 0·05(25°)	
	kin	5—10	0·1 ($NaNO_3$)	β[H^+Co en_2(μ-L, μ-NH_2)Co en_2^{3+}] (brown A ⇌ red) 11·0(5°), 10·8(10°)	
	X		solid	ev en_2Co(μ-LH, μ-NH_2)Co en_2^{4+} (red) in NO_3^- salt	67T
Co^{3+} cpx	sp, gl	15—30	0·14	K(2 Co $dien^{2+}$ + O_2 ⇌ $(Co\ dien)_2O_2^{4+}$ 7·50(15°), 7·15(20°), 6·80(25°), 6·40(30°)	63B
	pre, sp	0	KOH var	ev$[(CN)_5Co^{III}]_2L^{6-}$	63Bb
	X		solid	ev$(NH_3)_5CoLCo(NH_3)_5^{4+}$ in $Co_2L(NH_3)_{10}(SCN)_4(s)$	65V
	pre, X		var	ev$[(NH_3)_5Co]_2LH^{5+}$ (red, pre, X powder)	67Ma
	X		solid	ev $M^{4+} = (NH_3)_5CoO_2Co(NH_3)_5^{4+}$ in $M(SO_4)_2(H_2O)_4$ (s)	68Sb
$Co^{3+,4+}$	esr		1 KOH	ev $(CN)_5Co^{III}LCo^{IV}(CN)_5^{5-}$	63Ba = 63Bb
	X		solid	ev $(NH_3)_5CoO_2Co(NH_3)_5^{5+}$ in solid sulphate	66S
	esr, pre		var	ev $(CN)_5CoLCo(CN)_5^{5-}$, $(NH_3)_4CoL(NH_2)Co(NH_3)_4^{4+}$, $(CN)_4CoL(NH_2)Co(CN)_4^{4-}$ in solution and solids	67M
	X		solid	ev en_2Co(μ-L, μ-NH_2)Co en_2^{4+} (green) in NO_3^- salt (probably Co^{3+} and O_2^-, not L^{2-})	67T
	X		solid	ev $(NH_3)_5CoLCo(NH_3)_5^{5+} = M^{5+}$ in $M(NO_3)_5(s)$	68M
Ni^{2+} cpx	sp, gl		var	β(Ni $bipy_2^{2+}$$HL^-$) 4·5	66Z
N^{3+}	kin	23		$\beta(H^+NOL^-) < 6$	64Y
Se^{6+}	pre		var	ev H_2SeO_3L, SeO_3L^{2-}	64S

63B O. Bekâroğlu and S. Fallab, *Helv. Chim. Acta*, 1963 **46**, 2120
63Ba J. H. Bayston F. D. Looney, and M. E. Winfield, *Austral. J. Chem.*, 1963, **16**, 557
63Bb J. H. Bayston, R. N. Beale, N. K. King, and M. E. Winfield, *Austral. J. Chem.*, 1963, **16**, 954
63Da A. J. Dedman, T. J. Lewis, and D. H. Richards, *J. Chem. Soc.*, 1963, 5020
63Fa P. Flood, T. J. Lewis, and D. H. Richards, *J. Chem. Soc.*, 1963, 5024
63G W. P. Griffith, *J. Chem. Soc.*, 1963, 5345
63N B. I. Nabivanets, *Zhur. neorg. Khim.*, 1963, **8**, 2302
63S R. Stomberg, *Acta Chem. Scand.*, 1963, **17**, 1563
64A R. W. Alder and M. C. Whiting, *J. Chem. Soc.*, 1964, 4707
64B L. I. Budarin, T. A. Rumyantseva, and T. T. Sherina, *Izvest. V.U.Z., Khim. i khim. Tekhnol.*, 1964, **7**, 715
64C G. Carpéni, J. Haladjian, and S. Poize, *J. Chim. phys.*, 1964, **61**, 733
64E F. W. B. Einstein and B. R. Penfold, *Acta Cryst.*, 1964, **17**, 1127
64G W. P. Griffith, *J. Chem. Soc.*, 1964, 5248
64J G. V. Jere and C. C. Patel, *Indian J. Chem.*, 1964, **2**, 383
64S M. Schmidt and P. Bornmann, *Z. anorg. Chem.* 1964, **330**, 329
64Sa R. Stomberg, *Arkiv Kemi*, 1964, **22**, 29
64Sb R. Stomberg, *Arkiv Kemi*, 1964, **22**, 49
64Y G. Yagil and M. Anbar, *J. Inorg. Nuclear Chem.*, 1964, **26**, 453
65B A. K. Babko and V. A. Litvinenko, *Zhur. neorg. Khim.*, 1965, **10**, 2075
65Ba G. A. Bogdanov, G. K. Yurchenko, and L. A. Kuzenko, *Zhur. fiz. Khim.*, 1965, **38**, 2359
65G D. Grandjean and R. Weiss, *Compt. rend.*, 1965, **261**, 448
65M H. Martin-Frère, *Bull. Soc. chim. France*, 1965, 2860, 2868
65S R. Stomberg, *Arkiv Kemi*, 1965, **23**, 401, *Nature*, 1965, **205**, 71
65Sa P. Souchay and H. Martin-Frère, *Bull. Soc. chim. France*, 1965, 2874
65T D. G. Tuck and B. D. Faithful, *J. Chem. Soc.*, 1965, 5753
65V N.-G. Vannerberg, *Acta Cryst.*, 1965, **18**, 449
66A N. W. Alcock, *Chem. Comm.*, 1966, 536
66B A. K. Babko, A. I. Volkova, and S. L. Lisichenok, *Zhur. neorg. Khim.*, 1966, **11**, 478
66Ba A. K. Babko, B. I. Nabivanets, and I. G. Lukianets, *Zhur. neorg. Khim.*, 1966, **11**, 1257
66S W. P. Schaefer and R. E. Marsh, *J. Amer. Chem. Soc.*, 1966, **88**, 178
66Z R. Zell and H. Sigel, *Helv. Chim. Acta*, 1966, **49**, 870
67F J. Fuchs, K. F. Jahr, and R. Palm, *Z. Naturforsch.*, 1967, **22b**, 1222

continued overleaf

51 O_2^{2-} (contd)

67G D. Grandjean and R. Weiss, *Bull. Soc. chim. France*, 1967, 3044
67H D. A. House, R. G. Hughes, and C. S. Garner, *Inorg. Chem.*, 1967, **6**, 1077
67Ha C. Heitler, D. B. Scaife, and B. W. Thompson, *J. Chem. Soc. (A)*, 1967, 1409
67M M. Mori, J. A. Weil, and J. K. Kinnaird, *J. Phys. Chem.*, 1967, **71**, 103
67Ma D. P. Mellor and N. C. Stephenson, *Inorg. Nuclear Chem. Letters*, 1967, **3**, 431
67Mb M. Mori and J. A. Weil, *J. Amer. Chem. Soc.*, 1967, **89**, 3733
67O M. Orhanović and R. G. Wilkins, *J. Amer. Chem. Soc.*, 1967, **89**, 278
67T U. Thewalt and R. Marsh, *J. Amer. Chem. Soc.*, 1967, **89**, 6364
67W E. Wendling and J. de Lavillandre, *Bull. Soc. chim. France*, 1967, 2142
68A N. W. Alcock, *J. Chem. Soc. (A)*, 1968, 1588
68Aa A. C. Adams, J. R. Crook, F. Bockhoff, and E. L. King, *J. Amer. Chem. Soc.*, 1968, **90**, 5761
68G A. M. Gurevich, L. P. Polozhenskaya, and L. F. Solntseva, *Radiokhimiya*, 1968, **10**, 195
68Ga A. M. Gurevich and N. P. Osicheva, *Radiokhimiya*, 1968, **10**, 202
68Gb A. M. Gurevich and N. A. Susorova, *Radiokhimiya*, 1968, **10**, 211
68M R. E. Marsh and W. P. Schaefer, *Acta Cryst.*, 1968, **B24**, 246
68Ma A. I. Moskvin, *Radiokhimiya*, 1968, **10**, 13
68Mb A. Mitschler, J. M. Le Carpentier, and R. Weiss, *Chem. Comm.*, 1968, 1260
68S R. Stomberg, *Acta Chem. Scand.*, 1968, **22**, 1076
68Sa R. Stomberg, *Acta Chem. Scand.*, 1968, **22**, 1439
68Sb W. P. Schaefer, *Inorg. Chem.*, 1968, **7**, 725
68V V. P. Vasil'ev and P. N. Vorob'ev, *Izvest. V.U.Z., Khim. i khim. Tekhnol.*, 1968, **11**, 971
68Va V. P. Vasil'ev and G. A. Zaitseva, *Zhur. neorg. Khim.*, 1968, **13**, 84
68Y M.-T. Youinou and J. E. Guerchais, *Bull. Soc. chim. France*, 1968, 40

52 Other ligands with O

No data compiled since 1964 volume

53 Sulphide, S^{2-}

Metal	*Method*	*Temp*	*Medium*	*Log of equilibrium constant, remarks*	*Ref*
H^+	sol	160—330	0	$K_{p12}[H_2S(g) \rightleftharpoons H_2S]$ −1·52(160°), −1·53(202°), −1·52(262°), −1·45(301°), −1·37(330°), *m* units	64K/67H
	Hg	20	1 K(Cl)	K_1 14·15(Hg), K_{12} 6·88 (method?) (K_w −13·97)	64W
	Ag, gl?	25	1 ($NaClO_4$)	K_1 13·48, K_{12} 6·90	64S
	lit	0—200	0	K_1 13·1(0°), 12·6(25°), 12·3(50°), 11·9(100°), 11·7(150°), 11·8(200°) K_{12} 7·48(0°), 6·98(25°), 6·75(50°), 6·88(100°), 7·53(150°), 8·62(200°)	/65K
	sp	21	org	K_{12} 11·9 in MeOH(l), 10·5 in Me_2NCHO(l)	66Cb
	pre	−78°	solid	ev $HS(H_2S)_n^-$, $n = 1$ or 2 ?	66M
	sol	25	(Na, H)ClO_4, var	K_{p12} −0·9918 −0·0590[Na^+]+0·0081[H^+] —0·0001[H^+]^4	67Ga
	ΔG	25–330	0	K_{p12} −0·99(25°), −1·20(50°), −1·43(100°), −1·53(160°), −1·54(202°), −1·51(262°), −1·45(301°), −1·37(330°)	/67H
	sp	−77	NH_3(l)	$K[H_2S(+2NH_3) \rightleftharpoons SHNH_3^- + NH_4^+]$ −2·0 in NH_3(l)	67N
V^{5+}	*X*		solid	ev VS_4^{3-} in K_3VS_4(s)	64B
	X		solid	ev VS_4^{3-} in $(NH_4)_3VS_4$(s)	65Sa
	kin(sp)	20—50	NaOH var	VS_4^{3-} decomposed *via* two intermediates ($VO_2S_2^{3-}$, VO_3S^{3-}?)	65Y

continued

Metal	*Method*	*Temp*	*Medium*	*Log of equilibrium constant, remarks*	*Ref*
Mo^{6+}	sp		var	ev MoS_4^{2-} in aq soln (sp like solid)	66C
	sol	25—60	dil	$K_s(Tl^+{}_2MoS_4^{2-})$ −12·42(25°), −11·87(40°), −11·69(50°), −11·36(60°)	68S
	con, gl		var	ev $Mo_4S_{15}^{6-}$, $Mo_2S_7^{2-}$, $Mo_4S_{13}^{2-}$ (?)	68Sa
W^{6+}	*X*		solid	ev WS_4^{2-} in $(NH_4)_2WS_4(s)$	63Sc
	sp		var	ev WS_4^{2-} in aq soln (sp like solid)	66C
	est		var	$\beta(H^+WS_4^{2-})$ 2, $\beta(H^+HWS_4^-)$ −3	67G
Mn^{2+}	sol	20	0·02	†K_{s0} 8·0	24M/67Gb
	sol, gl	25	3 (Na)ClO₄	†$K_{s0}[\alpha\text{-}MnS(s)+2H^+ \rightleftharpoons Mn^{2+}+H_2S(g)]$ 7·27	67Gb
Re^{7+}	*X*, ir		solid	ev ReO_3S^- in $TlReO_3S(s)$	66Ma
Pt^{4+}	*X*		solid	ev $Pt(S_5)_3^{2-}$ in $(NH_4)_2PtS_{15}(H_2O)_2(s)$	67J
Cu^+	fp, con		var	ev polyn cpx $(CuS_4^-)_n$, $(Cu_2S_7C^{2-})_n$	64F
	var		var	ev S_4^{2-}, no ev CuS_4^- in NH_4CuS_4 (s, and solution) (*X*, sp, ir)	64G
Cu^{2+}	sol	25	0 corr	$K_s[2CuS(\text{covellite}, s)+S_3^{2-}+3S_4^{2-} \rightleftharpoons 2Cu(S_4)_2^{3-}+S^{2-}]$ −1·30; $K_s[2CuS(s)+3S_4^{2-}+S_5^{2-} \rightleftharpoons 2CuS_4S_5^{3-}+S^{2-}]$ −6·44	35H+63C/63Ca
Ag^+	sol	25	0 corr	$K(Ag^++H_2S \rightleftharpoons AgSH+H^+)$ 9·23; $K[Ag^++2H_2S \rightleftharpoons Ag(SH)_2^-+2H^+]$ 4·0	49T/63Ca

1964 *Errata*: p. 219, *for ref* "/49T" *read* "49T", *and for* "49T" *read* "/49T".

Metal	*Method*	*Temp*	*Medium*	*Log of equilibrium constant, remarks*	*Ref*
	sol	30?	0 corr	$K_s[Ag_2S(\alpha = \text{acanthite}, s)+4S_4^{2-} \rightleftharpoons S^{2-}+2Ag(S_4)_2^{3-}]$ −7·67; $K_s[Ag_2S(\alpha, s)+2S_5^{2-}+2S_4^{2-} \rightleftharpoons 2AgS_5S_4^{3-}+S^{2-}]$ −8·76; $K_s[Ag_2S(\alpha, s)+2S_4^{2-}+H^++HS^- \rightleftharpoons 2Ag(SH)S_4^{2-}]$ 4·47	63Ca+63C
	sol	20	1 (NaClO₄)	$K_s[\frac{1}{2}Ag_2S(s)+\frac{1}{2}H_2S \rightleftharpoons AgSH]$ −7·89; $K_s[\frac{1}{2}Ag_2S(s)+\frac{1}{2}H_2S+HS^- \rightleftharpoons Ag(SH)_2^-]$ −4·02; $K_s[Ag_2S(s)+2HS^- \rightleftharpoons Ag_2S_3H_2^{2-}]$ −4·82; $K_{s0}[Ag_2S(s) \rightleftharpoons 2Ag^++S^{2-}]$ −49·7 (H^+: K_1 14·0, K_{12} 6·68)	66S
	Ag	20	NH_3	*$K_{s0}[Ag_2S(s)(+H^+) \rightleftharpoons 2Ag^++HS^-]$ −24·4 in $NH_4NO_3(NH_3)_{1\cdot3}(l)$	66T
Au^{3+}	sol, pH	30	var	$K_s[Au_2S_3(s)+HS^-+OH^- \rightleftharpoons 2AuS_2^-+H_2O]$ −2·81; $K_s[Au_2S_3(s)+S^{2-} \rightleftharpoons 2AuS_2^-]$ −1·89 [$K_1(H^+)$ 14·92]	65D
Zn^{2+}	sol	25—195	1·6 (NaOH)	$K_s[ZnS(s)+HS^- \rightleftharpoons ZnHS_2^-]$ −3·0 at 25—195°C	60B
	sol, gl	25	1 (NaClO₄)	$K_s[ZnS(s)+H_2O \rightleftharpoons ZnHSOH]$ −5·87; K_{s0} −24·37(H^+: K_1 13·48, K_{12} 6·90); $\beta(Zn^{2+}OH^-HS^-)$ 19·02	67Gc
Cd^{2+}	ΔG	25	*I*(Na)ClO₄	†$K_{s0}[CdS(\alpha, s)+2H^+ \rightleftharpoons Cd^{2+}+H_2S(g)]$ −5·8(I = 3), −5·8(I = 1), −6·1(0 corr)	64A+/66Ka
	sol	25	1 (NaClO₄)	K_{s0} −25·76, [$K_1(H^+)$ 13·48] SH^-: K_1 7·55, K_2 7·06, K_3 1·88, K_4 2·36 [authors deduce $\beta(Cd^{2+}OH^-)$ 17·76 (?)]	64S
	sol	25	1 (Na)ClO₄	†K_{s0} −4·3	64S/66Ka

continued overleaf

53 S^{2-} (contd)

Metal	Method	Temp	Medium	Log of equilibrium constant, remarks	Ref
Hg^{2+}	Hg, gl	20	1 (KCl)	K_{s0} (black HgS) −50·96 (β_n for Cl^-–Hg^{2+} from ref 47L, Cl^- table)	63Sa
	sol, gl	20	1 (KCl)	$K[HgS(s)+H_2S \rightleftharpoons Hg(SH)_2]$ −5·97, $K_a[Hg(SH)_2]$ −6·19, $K_a(HgSSH^-)$ −8·30, $K_{s2}[HgS(s)+S^{2-} \rightleftharpoons HgS_2^{2-}]$ 0·57, (H^+: K_1 14·15, K_{12} 6·88)	
	Ram		2 Na_2S	ev linear HgS_2^{2-}	66Ca

1964 *Erratum*: *formula* "$[HgS(s)+S^{2-} \rightleftharpoons HgS_2^{2-}]$ *should be shifted from* $\dagger K_{s0}$ (*ref.* 42M/53R) to K_{s2} (*ref.* 06K/53R)

Metal	Method	Temp	Medium	Log of equilibrium constant, remarks	Ref
$MeHg^+$	gl	20	0·1 (KNO_3)	$^*K_1(M^+ + HS^- \rightleftharpoons MS^- + H^+)$ 7·02, $K_{12}(M^+ + MS^- \rightleftharpoons M_2S)$ 16·34, $K_{13}(M^+ + M_2S \rightleftharpoons M_3S^+)$ 7	63Sb = 65S
B^{3+}	*X*		solid	ev $B_3S_6^{3-}$ in $KBS_2(s)$	65C
Tl^+	sol, gl	25	3 (Na)ClO_4	$\dagger K_{s0}[\frac{1}{2}Tl_2S(s)+H^+ \rightleftharpoons Tl^+ + \frac{1}{2}H_2S(g)]$ 1·36	66G ≈ 65G
		25	0 corr	$\dagger K_{s0}$ 0·395	
C^{4+}	ir		solid	ev S_2CSS^{2-} in salts	66Mc
Sn^{2+}	sol	20	0 corr	K_{s0} −26·6 (H^+: K_1 12·9, K_{12} 6·9)	64Ga
P^{3+}	var	25	1 (KCl)	ev PHO_2S^{2-} (ir, Ram, pre), $K_a(H_2PHO_2S)$ −1·8, $K_a(HPHO_2S^-)$ −4·8 (H)	67M
$P^{3+,\,5+}$	oth		var	ev $P_2S_6^{4-}$ (pre, paper chromatography)	67F
P^{5+}	pre		var	ev $O_3PSSPO_3^{4-}$ and cyclic cpx like $P_3O_6S_3^{3-}$	65L
	X		solid	ev PS_4^{3-} in $(NH_4)_3PS_4(s)$	65Sb
As^{5+}	*X*		solid	ev AsS_4^{3-} in $(NH_4)_3AsS_4(s)$	63S
	ir, Ram		solid	ev AsO_3S^{3-} in $Na_3AsO_3S(H_2O)_{12}(s)$	66Mb
Sb^{3+}	sol	25	Na_2S var	$K_s[2Sb_2S_3(s)+SH^- + OH^- \rightleftharpoons Sb_4S_7^{2-} + H_2O]$ 0·7	66A
S	var	25	0 corr	$K(4S_4^{2-} \rightleftharpoons S^{2-} + 3S_5^{2-})$ −3·23 $K(5S_5^{2-} \rightleftharpoons S^{2-} + 4S_6^{2-})$ −5·33, S_2^{2-}, S_3^{2-} negligible, no ev HS_n^-, $n > 1$ (see also Table 1 e^-) (red, Ag, gl) (H^+: K_1 13·50, K_{12} 6·99)	63C
	kin	100	2·3 (Na)	$K(S^{2-} + S_2O_3^{2-} \rightleftharpoons S_2^{2-} + SO_3^{2-})$ −3·12	64C
	X		solid	ev S_6^{2-} in $Cs_2S_6(s)$	68H
	sp, esr	rt	Me_2NCHO	$K[H_2OS_4^{2-} + 8D \rightleftharpoons 2D_4S_2^-$ (blue)$+H_2O]$ −3·02, D = Me_2NCHO, ev two more S radical species in Me_2NCHO–H_2O mixt	68G
Se	pre, ir		var	ev $SSeO_3^{2-}$	66K
Te^{4+}	sol, gl		var	$K_s[TeS_2(s)+S^{2-} \rightleftharpoons TeS_3^{2-}]$ 7·61 $K_s[TeS_2(s)+2OH^- \rightleftharpoons TeS_2O^{2-})$ 4·26, $\beta(H^+TeS_2O^{2-})$ 8·7, $\beta(H^+HTeS_2O^-)$ 10·5 [$K_1(H^+)$ 14·92]	63D

60B H. L. Barnes, *Bull. Geol. Soc. Amer.*, 1960, **71**, 1821
63C P. L. Cloke, *Geochim. Cosmochim. Acta*, 1963, **27**, 1264
63Ca P. L. Cloke, *Geochim. Cosmochim. Acta*, 1963, **27**, 1299
63D K. P. Dubey and S. Ghosh, *J. Indian Chem. Soc.*, 1963, **40**, 479
63S H. Schäfer, G. Schäfer, and A. Weiss, *Z. Naturforsch.*, 1963, **18b**, 665
63Sa G. Schwarzenbach and M. Widmer, *Helv. Chim. Acta*, 1963, **46**, 2613
63Sb M. Schellenberg, dissertation, Eidgenössische Technische Hochschule, Zürich 1963
63Sc K. Sasvári, *Acta Cryst.*, 1963, **16**, 719

continued

64A L. H. Adami and E. G. King, *Bur. Mines Rept. Invest.*, 1964, no. 6495
64B J. M. van den Berg and R. de Vries, *Proc. k. ned. Akad. Wetenschap.*, 1964, **67**, *B*, 178
64C E. Ciuffarin and W. A. Pryor, *J. Amer. Chem. Soc.*, 1964, **86**, 3621
64F D. de Filippo, G. Peyronel, C. Preti, and G. Marcotrigiano, *Ricerca sci.*, 1964, *A*, **6**, 429
64G G. Gattow and O. Rosenberg, *Z. anorg. Chem.*, 1964, 332, 269
64Ga R. Geyer and H. Mücke, *Z. analyt. Chem.*, 1964, **200**, 210
64K T. N. Kozintseva, *Geokhimiya*, 1964, 758
64S J. Ste-Marie, A. E. Torma, and A. O. Gübeli, *Canad. J. Chem.*, 1964, **42**, 662
64W M. Widmer and G. Schwarzenbach, *Helv. Chim. Acta*, 1964, **47**, 266
65C F. Chopin and A. Hardy, *Compt. rend.*, 1965, **261**, *C*, 142
65D K. P. Dubey, *Z. anorg. Chem.*, 1965, 337, 309
65G H. Gamsjäger, *Monatsh* 1965, **96**, 1224
65K I. L. Khodakovskii, V. V. Zhogina, and B. N. Ryzhenko, *Geokhimiya*, 1965, 827
65L G. Ladwig and E. Thilo, *Z. anorg. Chem.*, 1965, **335**, 126
65S G. Schwarzenbach and M. Schellenberg, *Helv. Chim. Acta*, 1965, **48**, 28
65Sa H. Schäfer, P. Moritz, and A. Weiss, *Z. Naturforsch.*, 1965, **20b**, 603
65Sb H. Schäfer, G. Schäfer, and A. Weiss, *Z. Naturforsch.*, 1965, **20b**, 811
65Y K. B. Yatsimirskii and L. A. Zakharova, *Zhur. neorg. Khim.*, 1965, **10**, 2065
66A R. H. Arntson, F. W. Dickson, and G. Tunell, *Science*, 1966, **153**, 1673
66C G. M. Clark and W. P. Doyle, *J. Inorg. Nuclear Chem.*, 1966, **28**, 381
66Ca R. P. J. Cooney and J. R. Hall, *Austral. J. Chem.*, 1966, **19**, 2179
66Cb B. W. Clare, D. Cook, E. C. F. Ko, Y. C. Mac, and A. J. Parker, *J. Amer. Chem., Soc.*, 1966, **88**, 1911
66G H. Gamsjäger, W. Kraft, and W. Rainer, *Monatsh.*, 1966, **97**, 833
66K A. Kozhakova, E. A. Buketov, M. I. Bakeev, and A. K. Shokanov, *Zhur. neorg. Khim.*, 1966, **11**, 1782
66Ka W. Kraft, H. Gamsjäger, and E. Schwarz-Bergkampf, *Monatsh.*, 1966, **97**, 1134
66M D. H. McDaniel and W. G. Evans, *Inorg. Chem.*, 1966, **5**, 2180
66Ma A. Müller and B. Krebs, *Z. anorg. Chem.*, 1966, **342**, 182
66Mb K. Martin and E. Steger, *Z. anorg. Chem.*, 1966, **345**, 306
66Mc A. Müller and B. Krebs, *Z. anorg. Chem.*, 1966, **347**, 261
66S G. Schwarzenbach and M. Widmer, *Helv. Chim. Acta*, 1966, **49**, 111
66T A. Thiébault, M. Herlem, and J. Badoz-Lambling, *Rev. Chim. minérale*, 1966, **3**, 1005
67F H. Falius, *Z. anorg. Chem.*, 1967, **356**, 189
67G G. Gattow and A. Franke, *Z. anorg. Chem.*, 1967, **352**, 246
67Ga H. Gamsjäger, W. Rainer, and P. Schindler, *Monatsh*, 1967, **98**, 1793
67Gb H. Gamsjäger, *Monatsh*, 1967, **98**, 1803
67Gc A. O. Gubeli and J. Ste-Marie, *Canad. J. Chem.*, 1967, **45**, 2101
67H H. C. Helgeson, *J. Phys. Chem.*, 1967, **71**, 3121
67J P. E. Jones and L. Katz, *Chem. Comm.*, 1967, 842
67M J. D. Murray, G. Nickless, and F. H. Pollard, *J. Chem. Soc.* (*A*), 1967, 1726
67N J. T. Nelson and J. J. Lagowski, *Inorg. Chem.*, 1967, **6**, 862
68G W. Giggenbach, *J. Inorg. Nuclear Chem.*, 1968, **30**, 3189
68H A. Hordvik and E. Sletten, *Acta Chem. Scand.*, 1968, **22**, 3029
68S R. S. Saxena and M. C. Jain, *Indian J. Chem.*, 1968, **6**, 752
68Sa R. S. Saxena, M. C. Jain ,and M. L. Mittal, *Austral. J. Chem.*, 1968, **21**, 91

Thiosulphate, $S_2O_3^{2-}$ 54

Metal	*Method*	*Temp*	*Medium*	*Log of equilibrium constant, remarks*	*Ref*
Na^+	act	25 ?	0 corr	K_1 0·8	/66M
Ca^{2+}	*E*	25	dil	K_1 1·90, β_2 3·98 (aix membrane electrode)	68G
La^{3+}	sp	25	0 corr	K_1 2·99	64B
Eu^{3+}	dis		0 corr ?	K_1 2·82	61M
Cr^{6+}	sp	20	0·11 ($NaClO_4$)	$\beta(HCrO_4^-HL^-)$ 4·09 [$K_1(H^+)$ 1·22]	68B
UO_2^{2+}	sp	15—25	var	K_1 2·47(15°), 2·29(20°), 2·04(25°)	63M
		20—40	60% EtOH	K_1 4·95(20°), 4·8(25°), 4·6(30°), 4·2(40°) in 60% EtOH, I = 0·0064; also some values for 30 and 90 volume % EtOH	

continued overleaf

54 $S_2O_3^{2-}$ (contd)

Metal	*Method*	*Temp*	*Medium*	*Log of equilibrium constant, remarks*	*Ref*
Fe^{3+}	sp	?	0·05 $HClO_4$	K_1 11·98 (?)	64T
Co en_3^{3+}	oth	20	2 ($NaClO_4$)	K_1 2·18, β_2 3·88, β_3 5·18, β_4 5·18 (polarimetry)	64L
	sp		2·88 $Na(ClO_4)$	K_1 0·11	67O
Pt $en_3H_{-1}^{3+}$	var	20	var	ev two cpx (sp, circular dichroism)	67L
Cu en_2^{2+}	sp	25	0 corr	K_1 2·2—2·5	67M
	sp	25	$\rightarrow 0$	K_1 2·34	68H
	oth	25	0 corr	K^{out} ($Maq^{2+} + Laq^{2-} \rightleftharpoons MH_2OL$) 2·22(hyp), $K(MH_2OL \rightleftharpoons ML)$ −0·6, K_1 2·32 (ultrasonic absorption)	
Cu^+	sol	20	3 (Na_2SO_4)	$K_s(CuSCN(s) + L^{2-} \rightleftharpoons CuLSCN^{2-})$ −0·42, $\beta(Cu^+L^{2-}SCN^-)$ 13·90 [$K_s(Cu^+SCN^-)$ −14·32]; $K_s[CuI(s) + L^{2-} \rightleftharpoons CuLI^{2-}]$ 0·42, $\beta(Cu^+L^{2-}I^-)$ 12·38, [$K_s(Cu^+I^-)$ −11·96]	68Ga
Ag^+	Ag	25	1	β_2 12·63, β_3 12·76; $\beta(Ag^+L^{2-}{}_2X^-)$ 10·84(Cl^-), 12·02(Br^-), 13·25(I^-), 15·68(CN^-); $\beta(Ag^+L^{2-}X^-{}_2)$ 11·30(Br^-), 13·62(I^-); $\beta(Ag^+L^{2-}X^-{}_3)$ 9·99(Br^-), 13·52(I^-), 21·28(CN^-); $\beta[Ag^+L^{2-}{}_2(CN^-)_2]$ 18·15	67B
Zn^{2+}	MHg, sp	25	0·78 ($NaClO_4$)	K_1 2·08, β_2 4·60, also β for mixed L^{2-}–$C_2O_4^{2-}$ cpx	68J+
Cd^{2+}	dis	30	0·1 ($NaClO_4$)	K_1 3·2, β_2 5·0	65H
	MHg, sp	25	0·78 ($NaClO_4$)	β_2 5·20, β_3 6·19, also β for mixed L^{2-}–$C_2O_4^{2-}$ cpx	68J+
Hg^{2+}	sol	25	1 ($NaClO_4$)	$K[HgI_2(s) + L^{2-} \rightleftharpoons HgI_2L^{2-}]$ 0·66, $\beta(Hg^{2+}I^-{}_2L^{2-})$ 28·86 [$K_{s0}(Hg^{2+}I_2^-)$ −28·19]	64C
	sp	25	3 ($NaClO_4$)	ev several cpx, $K(HgI_4^{2-} + L^{2-} \rightleftharpoons HgI_3L^{3-} + I^-)$ 0·36, $\beta(Hg^{2+}I^-{}_3L^{2-})$ 29·66 [$\beta(Hg^{2+}I^-{}_4)$ 29·3]	66C
$MeHg^+$	gl, cal	20	0·1 (KNO_3)	K_1 10·90, $\Delta H_1 = -11·7$, $\Delta S_1 = 9·9$	63S = 65S

61M A. R. Manning, thesis, University of Wales 1961, quoted by ref 64B
63M S. L. Melton and E. S. Amis, *Analyt. Chem.*, 1963, **35**, 1626
63S M. Schellenberg, dissertation, Eidgenössische Technische Hochschule, Zürich, 1963
64B J. C. Barnes and C. B. Monk, *Trans. Faraday Soc.*, 1964, **60**, 578
64C D. M. Czakis-Sulikowska, *Roczniki Chem.*, 1964, **38**, 533
64L R. Larsson and L. Johansson, *Proc. Symp. Co-ord. Chem., Tihany*, 1964 (Akad. Kiadó, Budapest 1965) p. 31
64T I. A. Tserkovnitskaya and T. T. Vykhovtseva, *Zhur. analit. Khim.*, 1964, **19**, 922
65H H. E. Hellwege and G. K. Schweitzer, *J. Inorg. Nuclear Chem.*, 1965, **27**, 99
65S G. Schwarzenbach and M. Schellenberg, *Helv. Chim. Acta*, 1965, **48**, 28
66C D. M. Czakis-Sulikowska, *Roczniki Chem.*, 1966, **40**, 1123
66M W. L. Masterton and L. H. Berka, *J. Phys. Chem.*, 1966, **70**, 1924
67B G. A. Boos and A. A. Popel', *Zhur. neorg. Khim.*, 1967, **12**, 2086
67L R. Larsson, *Acta Chem. Scand.*, 1967, **21**, 1081
67M R. A. Matheson, *J. Phys. Chem.*, 1967, **71**, 1302
67O I. Olsen and J. Bjerrum, *Acta Chem. Scand.*, 1967, **21**, 1112
68B I. Baldea and G. Niac, *Inorg. Chem.*, 1968, **7**, 1232
68G A. V. Gordievskii, E. L. Filippov, V. S. Shterman, and A. S. Krivoshein, *Zhur. fiz. Khim.*, 1968, **42**, 1998
68Ga E. A. Gyunner and N. D. Yakhkind, *Zhur. neorg. Khim.*, 1968, **13**, 2758
68H P. Hemmes and S. Petrucci, *J. Phys. Chem.*, 1968, 72, 3986
68J Z. Jabłoński, J. Górnicki, and A. Łodzińska, *Roczniki Chem.*, 1968, **42**, 1809

Metal	*Method*	*Temp*	*Medium*	*Log of equilibrium constant, remarks*	*Ref*
H^+	gl	20	0·1 (KNO_3)	K_1 6·79	63S
	gl	10—50	0 corr	K_1 7·17(10°), 7·30(25°), 7·37(35°), 7·45(50°), ΔH_1 = 2·9(25°)	64A
	sp	25	0 corr	K_{12} 1·845, ΔH_{12} = 3·9	64D
	sp, gl	10—35	0 corr	$K(SO_2 + H_2O \rightleftharpoons H_2SO_3\dagger)$ −0·33(10°), −0·48(25°), −0·57(35°) $H_2SO_3\dagger$ = true H_2SO_3; apparent K_{12} 1·92(10°), 2·00(25°), 2·10(35°), true K_{12} 1·41(10, 35°), 1·40(25°)	65F
	sp	25	w% MeOH	K_{12} 1·94(w = 10), 2·00(w = 20), 2·30(w = 45), 2·72(w = 70) in w% MeOH–H_2O mixt, 0 corr	66D
	sp	25	SO_2 var, p kbar	K_{12} 1·99(p = 0·2 kb), 1·84(0·5 kb), 1·67(1 kb), 1·35(2 kb), 1·11(3 kb), 0·86(4 kb), 0·74(5 kb), 0·62(6 kb), m units	68L
	dis	25	I(Na)ClO_4	$K_D[H_2L \rightleftharpoons H_2L(\text{in } CCl_4)]$ −0·23(I = 3), −0·25(I = 0·1) K_{12} 2·03(I = 3·0), 1·62(I = 0·1)	68S
Cr^{6+}	kin	25	0·5(Na, CH_3CO_2)	$\beta(HCrO_4^- HSO_3^-)$ 1·56	65H
UO_2^{2+}	gl	23	var	K_1 5·85 [$K_1(H^+)$ 6·55]	67Z
	sp		0·1 (NH_4ClO_4)	K_1 6·01	
	sol	23	1 (NaCl)	K_1 5·26 ?, K_2 3·32, β_2 9·17	67Z + 59K
	sol	?	2	β_3(K_3 ?) 1·01	67Za
NpO_2^+	cix ?		0 corr	K_1 2·15, β_2 3·00 [$K_1(H^+)$ 7·19]	65Ma
$Fe(CN)_5NO^{2-}$ = X^{2-}	sp		var	$K_1(X^{2-} + L^{2-} \rightleftharpoons XL^{4-})$ −1·96; $\beta(M^+X^{2-}L^{2-})$ −0·89, (M = Na), −0·41(M = K); $\beta(M^+XL^{4-})$ 0·93(1·07 ?) (M = Na), 1·56(M = K)	65M
	sp		var	$K_1 < -1·77$	66F
Co^{3+}	sp, gl	25 ?	$I(ClO_4)$	$K_a[HCo(CN)_5L^{3-}]$ −1·80(I = 1·0), −2·30(I = 0·18), −4·00(I = 0·0004); $K_a[HCo(CN)_4L^{2-}]$ −1·00(I = 1·0), −2·26(I = 0·006); $\beta[Co(CN)_4L^{3-}OH^-]$ 2·9	66C
	kin	25 ?	1 ($LiClO_4$)	$K_a[HCo(CN)_4L_2^{4-}]$ −2·04, $K_a[H_2Co(CN)_4L_2^{3-}]$ −1·35	
	sp	25	1 ($NaClO_4$)	$K[Co(CN)_4LOH^{4-} + NH_3 \rightleftharpoons Co(CN)_4LNH_3^{3-} + OH^-]$ 1·99	67T
Pd^{2+}	X		solid	ev $PdSO_3(NH_3)_3$ units in solid	67S
Cu^+	X		solid	ev $CuSO_3^-$ tetrahedron in NH_4CuSO_3(s)	67N = 68N
Hg^{2+}	sol	21	0·5 ($NaClO_4$)	$K[Hg(SCN)_2(s) + L^{2-} \rightleftharpoons Hg(SCN)_2L^{2-}]$ 3·95 (H^+: K_1 7·22, K_{12} 1·77)	64C
	sol	25	var	$K_s[HgI_2(s) + L^{2-} \rightleftharpoons HgI_2L^{2-}]$ 0·12, $\beta(Hg^{2+}I^-{}_2L^{2-})$ 28·31 [$K_s(Hg^{2+}I^-{}_2)$ −28·19, $K_1(H^+)$ 7·22]	66Ca
	sp	25	3 ($NaClO_4$)	ev HgL_2^{2-}, $K(HgI_4^{2-} + L^{2-} \rightleftharpoons HgI_3L^{3-} + I^-)$ 0·47, $\beta(Hg^{2+}I^-{}_3L^{2-})$ 29·77 [$\beta(Hg^{2+}I^-{}_4)$ 29·3, $K_1(H^+)$ 7·22]	66Cb
$MeHg^+$	gl	20	0·1 (KNO_3)	K_1 8·11	63S = 65S

63S M. Schellenberg, dissertation, Eidgenössische Technische Hochschule, Zürich, 1963

64A G. P. Arkhipova, I. E. Flis, and K. P. Mishchenko, *Zhur. priklad. Khim.* 1964, 37, 2306

64C M. Czakis-Sulikowska, in "Theory and structure of complex compounds" (symposium papers, Wrocław, 1962), Pergamon, Warszawa 1964, p. 565

continued overleaf

55 SO_3^{2-} (contd)

64D D. Devèze and P. Rumpf, *Compt. rend.*, 1964, **258**, *C*, 6135
65F I. E. Flis, G. P. Arkhipova, and K. P. Mishchenko, *Zhur. priklad. Khim.*, 1965, **38**, 1494
65H G. P. Haight, jun., E. Perchonok, F. Emmenegger, and G. Gordon, *J. Amer. Chem. Soc.*, 1965, **87**, 3835
65M W. Moser, R. A. Chalmers, and A. G. Fogg, *J. Inorg. Nuclear Chem.*, 1965, **27**, 831
65Ma A. I. Moskvin and M. P. Mefod'eva, *Radiokhimiya*, 1965, **7**, 410 (see table 4, p. 417)
65S G. Schwarzenbach and M. Schellenberg, *Helv. Chim. Acta*, 1965, **48**, 28
66C H. H. Chen, M.-S. Tsao, R. W. Gaver, P. H. Tewari, and W. K. Wilmarth, *Inorg. Chem.*, 1966, **5**, 1913
66Ca D. M. Czakis-Sulikowska, *Roczniki Chem.*, 1966, **40**, 521
66Cb D. M. Czakis-Sulikowska, *Roczniki Chem.*, 1966, **40**, 1393
66D D. Devèze, *Compt. rend.*, 1966, **263**, *C*, 392
66F A. G. Fogg, A. D. Jones, and W. Moser, *J. Inorg. Nuclear Chem.*, 1966, **28**, 2427
67N B. Nyberg and P. Kierkegaard, *Acta Chem. Scand.*, 1967, **21**, 823
67S M. A. Spinnler and L. N. Becka, *J. Chem. Soc.* (*A*), 1967, 1194
67T P. H. Tewari, R. W. Gaver, H. K. Wilcox, and W. K. Wilmarth, *Inorg. Chem.*, 1967, **6**, 611
67Z F. A. Zakharova and M. M. Orlova, *Zhur. neorg. Khim.*, 1967, **12**, 3016
67Za F. A. Zakharova and M. M. Orlova, *Zhur. neorg. Khim.*, 1967, **12**, 3211
68L H.-D. Lüdemann and E. U. Franck, *Ber. Bunsengesellschaft Phys. Chem.*, 1968, **72**, 523
68N B. Nyberg and P. Kierkegaard, *Acta Chem. Scand.*, 1968, **22**, 581
68S T. Sekine, H. Iwaki, M. Sakairi, F. Shimada, and M. Inarida, *Bull. Chem. Soc. Japan*, 1968, **41**, 1

56 Sulphate, SO_4^{2-}

Metal	*Method*	*Temp*	*Medium*	*Log of equilibrium constant, remarks*	*Ref*
H^+	cal	25?	0·4 (K^+,Na^+)	$\Delta H(S_2O_7^{2-}+H_2O \rightleftharpoons 2HSO_4^-) = -21{\cdot}0$	63Ta
	con	25	HSO_3F	$K_{13}[H_2L(+HSO_3F) \rightleftharpoons H_3L^+ + SO_3F^-]$ −4, *m* units, in HSO_3F(l)	64Bc
	lit	25	→0	K_1 1·987, *m* units	/64D
	gl	10—75	1 (NaBr)	K_1 $-2{\cdot}925+0{\cdot}01324\,T-1{\cdot}02[H^+]$, $\Delta H_1 = 5{\cdot}39$	64F

1964 *Erratum*: p. 234, *in ref*/60H *for* "9·55" *read* "−9·55"

Metal	*Method*	*Temp*	*Medium*	*Log of equilibrium constant, remarks*	*Ref*
	con	25	1·55 H_2L	$K_1 \approx -1{\cdot}0$(1 bar) to $\approx -0{\cdot}6$(1600 bar)	64Hc
	con	100—218	→0	K_1 3·04(100°), 3·90(156°), 4·87(218°)	64Rb
	H, Hg	25	→0	K_1 1·97	65C
	sp	27	→0	K_1 1·99	65E
	gl	25—45	3 ($LiClO_4$)	K_1 0·92(25°), 0·97(35°), 1·06(45°)	65K
	fp	≈ 10	dioxan	$\beta(H_2O, H_2L)$ or $\beta(H_3O^+HL^-)$ 1·34 in dioxan(l) {$\beta[(H_2O)_n]$ −0·15(n = 2), −0·21(n = 3), −0·39(n = 4)}	65Kb
	con	100—300	0 corr	K_1 2·60(100°), 2·83(150°), 3·13(200°), 3·35(250°), 3·58(300°) at $\rho = 1{\cdot}0$ g cm^{-3}, 4·09(300°, ρ = 0·80), also values for intermediate ρ,	65Q
	con	400—800	0 corr	K_{12} 3·90(400°, ρ = 0·4), 2·24(400°, ρ = 0·6), 0·37(400°, ρ = 0·80), 5·24(600°, ρ = 0·4), 2·94(600°, ρ = 0·6), 1·99(600°, ρ = 0·80), 5·62(800°, ρ = 0·4), 3·66(800°, ρ = 0·6), also K_{12} for intermediate temp and ρ values	
	cal	25	0 corr	K_1 1·91, $\Delta H_1 = 5{\cdot}6$	66C
	pH	32	Na_2SO_4 satd	K_1 [L^{2-}] 0·54	66J
	sol	0—370	→0	K_1 1·78(→ 0°), 1·99(25°), 2·25(50°), 2·54(75°), 2·855(100°), 3·53(150°), 4·25(200°), 4·97(250°), 5·70(300°), 6·42(350°), 6·71(→370°) also ΔH, ΔS, ΔC_p,	66Mb

continu

Metal	Method	Temp	Medium	Log of equilibrium constant, remarks	Ref
H^+ (contd)	con	11—69	H_2SO_4(l)	$\beta(H^+S_2O_7^{2-})$ 1·66(11°), 1·60(25°), 1·54(69°), in H_2SO_4(l), *m* units	66Mc
	con	100	0 corr	K_1 2·94(ρ = 1·0 g cm^{-3}), 3·20 (sat. vapour pressure), 3·18 (sat., v.p., *m* units)	66Q
		200	0 corr	K_1 3·43(ρ = 1·0), 3·91(ρ = 0·95), 4·29(ρ = 0·90), 4·67 (at sat. v.p.), 4·60 (sat v.p., *m* units)	
		300	0 corr	K_1 3·94(ρ = 1·0), 4·32(ρ = 0·95), 4·87(ρ = 0·90), 5·34(ρ = 0·85), 5·81(ρ = 0·80), 6·5(ρ = 0·75)	
	ir		H_2SO_4–H_2O	ev HSO_4^-, H_2SO_4; equil constants in ref 60B give more HL^- than observed	66S
	ir		H_2SO_4–SO_3	no ev $H_3SO_4^+$, $HS_2O_7^-$(?), ev H_2SO_4, $H_2S_2O_7$, $H_2S_3O_{10}$, SO_3, S_3O_9 in H_2SO_4–SO_3(l)	66Sc
	qh	20	2 ($NaClO_4$)	K_1 0·92(0·1M-H^+) to 0·61(0·25M-H^+)	66V
	oth	25—50	0 corr	K_1 1·88(25°), 2·03(35°), 2·21(50°), (membrane equil H^+–Na^+)	66Wa
	sp	25	→ 0	K_1 2·00 (2·01 with NaCl, 1·98 with $BaCl_2$ solutions)	67Ca
	fp	≈ 30	oleum	ev $HS_3O_{10}^-$	67D
	fp	≈ 35	$H_2S_2O_7$	ev $H_2S_3O_{10}$, $H_2S_4O_{13}$; $K[(2H_2S_2O_7) \rightleftharpoons H_2SO_4 + H_2S_3O_{10}]$ −0·6; $K[(2H_2S_2O_7) \rightleftharpoons H_3SO_4^+ + HS_3O_{10}^-]$ −2·44 in $H_2S_2O_7$(l), *m* units	67G
	ΔG	25	0	K_1 1·99, ΔH_1 = 3·85, ΔS_1 = 22·0	/67H
	var	25	H_2L	$K(H_2L + H_2O \rightleftharpoons HL^- + H_3O^+)$ 1·70 $K(H_3O^+ + HL^- \rightleftharpoons H_3O^+HL^-)$ −0·35 in H_2L(l), *x* units (Ram, sp, nmr)	/67Kb
	H?	20—35	2 ($NaClO_4$)	K_1 0·92(20°), 0·95(25°), 1·08(35°)	67V
			1 ($NaClO_4$)	K_1 1·08(20°), 1·12(25°), 1·24(35°)	
			0·5 ($NaClO_4$)	K_1 1·22(20°), 1·26(25°), 1·39(35°)	
			→ 0·25	K_1 1·36(20°), 1·40(25°), 1·50(35°)	
	qh	20—35	2 ($NaClO_4$)	K_1 0·92(20°), 0·95(25°), 1·08(35°)	67Va
			1 ($NaClO_4$)	K_1 1·08(20°), 1·12(25°), 1·24(35°)	
			0·5 ($NaClO_4$)	K_1 1·22(20°), 1·26(25°), 1·39(35°) (not clear whether data are independent of ref. 67V or not)	
			0 corr	K_1 1·92(20°), 1·99(25°), 2·12(35°)	
	oth	750	MCl(l)	$K[SO_3(g) + O^{2-} \rightleftharpoons SO_4^{2-}]$ > 2 in $Na_{0.5}K_{0.5}Cl$(l) (chronopotentiometric method)	67Wa
	gl	20	1 ($NaClO_4$)	K_1 1·11	68A
	gl	25	MeCN	K_1 25·9, K_{12} 7·8, K_2 ($HL^- + L^{2-} \rightleftharpoons HL_2^{3-}$) 1·5, $K(H_2L + HL^- \rightleftharpoons H_3L_2^-$?) 3·6 in MeCN(l)	68K
	gl	25	Me_2SO	K_1 14·5 in Me_2SO(l)	
	dis	25	0 corr	$K_D[H^+ + HL^- + 2T(org) \rightleftharpoons 0{\cdot}6(H_2LT_2)_{1.67}(org)]$ −1·3, T = TBP, org = CCl_4	68N
D^+	sol	25—225	0 corr	K_1 2·53(25°), 2·67(50°), 2·98(75°), 3·29(100°), 3·59(125°), 3·91(150°), 4·21(175°), 4·52(200°), 4·84(225°)	63La
	con	25—69	D_2SO_4(l)	$\beta(D^+S_2O_7^{2-})$ 2·22(25°), 2·13(69°), in D_2SO_4(l), *m* units	66Mc
Group 1A$^+$	act	25?	0 corr	K_1 0·7(Li^+), 0·9(Na^+), 1·0(K^+), 0·8(Rb^+), 0·8(Cs^+), 1·1(NH_4^+)	/66M
Li^+	Ram		2 Li_2L	ev cpx	64Lb
Na^+	Ram		Na_2L conc	no ev (inner) cpx	64Ha
K^+	con	100—300	→ 0	K_1 1·3(100°), 1·96(200°), 1·35(300°, density 1·0), 1·5(300°, density 0·8)	63Q
	E	25—39	0·1 (Me_4NCl)	K_1 0·4(25°), 0·5(39°), (K^+ glass electrode)	66Ca
			0 corr	K_1 0·85(25°), 0·95(39°), ΔH_1 = 3, ΔS_1 = 14	

continued overleaf

56 SO_4^{2-} (contd)

Metal	Method	Temp	Medium	Log of equilibrium constant, remarks	Ref
K^+ (*contd*)	con	400—700	0 corr	$\beta(K^+HL^-)$ 1·16 ($\rho = 0{\cdot}70$ g cm^{-3}, 400°), 1·95($\rho = 0{\cdot}70$, 700°), 3·25($\rho = 0{\cdot}40$, 400°), 4·3($\rho = 0{\cdot}40$, 700°), also β for many other ρ and temp	66Q
Be^{2+}	kin	25	0	K_1^{in}/K_1^{out} −0·22, K_1^{out} 1·70 (K_1 1·95) (from ref 62Bb) (con p-jump)	66K = 65[illegible]
	dis	25	1 Na(ClO$_4$)	β_2 1·78, β_3 2·08	67S
	Hg	35?	→ 0	K_1 2·17	68P
Mg^{2+}	oth	20	MgL var	K_1 1·7 to 2·2, K^{out}/K^{in} 0·7 to 1·0 and 0 (two steps) (sound absorption)	62Ea
	Ram		MgL conc	no ev (inner) cpx	64Ha
	oth	25	0 corr	$K[Mg(aq)^{2+} + Laq^{2-} \rightleftharpoons Mg(aq)^{2+}L(aq)^{2-}]$ 1·4, $\Delta V = 0$ $K[Mg(aq)^{2+}Laq^{2-} \rightleftharpoons MgH_2OL]$ 0, $\Delta V = 18$ ml $K(MgH_2OL \rightleftharpoons MgL)$ −0·95, $\Delta V = 3$ ml (sound absorption)	65Fa + 62E
	con	25	org	K_1 0·98 in H_2NCHO(l), 4·50 in 50 w% Me_2CO in H_2NCHO, values for other mixtures given	65J
	con	25	H_2NCHO	no ev cpx in H_2NCHO(l)	65Ja
	con	25	MeNHCHO	no ev cpx in MeNHCHO(l)	65Jb
	con	25	org	K_1 1·95(20%), 2·33(25%), 2·65(30%), 3·09(35%), 3·58(40%), 4·39(50%), 5·38(60%), 6·42(70%) in w weight % dioxan + H_2NCHO	65T
	oth	25	0 corr	$K[Mg(aq)^{2+} + Laq^{2-} \rightleftharpoons Mg(aq)^{2+}L(aq)^{2-}]$ 1·70 $K[Mg(aq)^{2+}L(aq)^{2-} \rightleftharpoons MgH_2OL]$ 0·29 $K(MgH_2OL \rightleftharpoons MgL)$ −0·76 K_1 2·22 (ultrasonic method)	66Ab
	ΔG	25	0	K_1 2·25, $\Delta H_1 = 4{\cdot}87$, $\Delta S_1 = 26{\cdot}7$, $\Delta C_p = 19$	/67H
	sol	30—200	→ 0	K_1 2·13(→ 0°), 2·40(→ 25°), 2·63(50°), 2·85(75°), 3·06(100°), 3·27(125°), 3·50(150°), 3·74(175°), 4·00(200°), 4·58(→ 250°), 6·27(→ 370°); values for ΔH_1, ΔS_1, ΔC_p; $\Delta C_p \approx 25$	67M
	var		0 corr	K_1 varies with interionic distance assumed (lit: con, fp, sp)	/68M
	Hg	35?	0 corr	K_1 1·97	68P
	con	23—350	MgL dil	K_1(diagram) for $\rho = 0{\cdot}7$ to 1·1 g cm^{-3}, 23—350°	68R
Ca^{2+}	sol	20	0 corr	K_1 2·0	65L
	sol	0—110	→ 0	K_{s0} −4·46(0·5°), −4·37(25°), −4·37(30°, max), −4·41(50°), −4·55(80°), −4·71(110°), solid = $CaSO_4(H_2O)_2$, m units; also K_{s0} for I(NaCl) (I up to 5 or 6), and for intermediate temp	66Ma + 64M
	sol	125	→ 0	K_{s0} −5·02(125°), $CaSO_4(H_2O)_{0.5}$(s)	/66Mb
		150—350	→ 0	K_{s0} −6·00(150°), −6·47(175°), −6·94(200°), −7·48(225°), −8·04(250°), −8·64(275°), −9·22(300°), −9·78(325°), −10·37(350°), $CaSO_4$ (s, anhydrite)	
	ΔG	25	0	K_1 2·31, $\Delta H_1 = 1{\cdot}94$, $\Delta S_1 = 17{\cdot}1$, $\Delta C_p = 46$ (or $\Delta H_1 = 1{\cdot}76$, $\Delta S_1 = 16{\cdot}5$)	/67H
	E, sol	25	→ 0	K_{s0} −4·60, K_1 2·27 (Ca^{2+} membrane electrode)	67Na
	E	25	dil	K_1 2·57 (table captions confused.) (cix membrane electrode)	68G

continu

Metal	Method	Temp	Medium	Log of equilibrium constant, remarks	Ref
Ca^{2+} (*contd*)	sol	30—200	0 corr	K_{s0} 390·9619 − 152·6246 log T − 12545·62 T^{-1} + 0·0818493 T[$CaSO_4(H_2O)_2$, gypsum]; K_{s0} −215·509 + 85·685 log T + 6075·2 T^{-1} − 0·070707 T ($CaSO_4$, anhydrite); K_{s0} 154·527 − 54·958 log T − 6640·0 T^{-1} [$CaSO_4(H_2O)_{0.5}$, hemihydrate] also data for sea water of various salinities	68Me+
Sr^{2+}	sol	20—30	1 H(Cl, ClO_4)	ΔH_{s0} = 4·9 [main reaction is probably SrL(s) + $H^+ \rightleftharpoons Sr^{2+}$ + HL^-! ΔG^0 and ΔS^0 seem dubious]	65Ka
	sol	20	0 corr	K_1 2·1	65L
	cix	25	0·5 NH_4ClO_4	K_1 1·14	68Ca
		25	0 corr	K_1 2·55	
Ba^{2+}	pol	25	0 corr	K_{s0} −9·77	53Sa
	sol	20	0 corr	K_1 2·3	65L
	dis	25	1 Na(ClO_4)	K_1 0·66, β_2 1·42	66Sb
Sc^{3+}	cix		0·5 H, 2 ($NaClO_4$)	K_1 0·62, β_2 1·50	64Sa
	cix	25	0·5 ($NaClO_4$)	K_1 1·66, β_2 3·04, β_3 4·0	65Ta
	cix	20	1 H(ClO_4)	*K_1 1·49, K_1 2·57 [K_1(H^+) 1·08]	66B
	cix	25	0·5 ($NaClO_4$)	K_1 2·59, β_2 3·96	67Ka
Y^{3+}	Ag	25	2 ($NaClO_4$)	K_1 1·24, β_2 1·68	67C
	cal	25	2 ($NaClO_4$)	ΔH_1 = 4·04, ΔH_2 = 1·5, ΔS_1 = 19·2, ΔS_2 = 7·1	
La^{3+}	dis	25	1 Na(ClO_4)	K_1 1·45, β_2 2·46	64S
	dis	25	1 Na(ClO_4)	K_1 1·45, β_2 2·46	65Sa
	con	25	0 corr	K_1 3·70	66E
	sol	120—175	→ 0	K_{s0}($La^{3+}{}_2L^{2-}{}_3$) −16·3(135°), −17·4(150°), −18·2(160°), m units	66L/
	Ag	25	2 ($NaClO_4$)	K_1 1·29	67C
	cal	25	2 ($NaClO_4$)	ΔH_1 = 3·72, ΔS_1 = 18·4	
	con	25	0 corr	K_1 3·62(1 atm), 3·43(500 atm), 3·27(1000 atm), 3·17(1500 atm), 3·07(2000 atm)	67Fa
Ce^{3+}	ir		0·1 Ce_2L_3	K_1^{in}/K_1 −0·9	64La
	con	25	0 corr	K_1 3·72	66E
	Ag	25	2 ($NaClO_4$)	K_1 1·24	67C
	dis	0—55	2 ($NaClO_4$)	K_1 1·11(0°), 1·30(25°), 1·52(40°), 1·62(55°), β_2 1·53(0°), 1·99(25°), 2·13(40°), 2·24(55°), ΔH_1 = 4·2, ΔS_1 = 20	
	cal	25	2 ($NaClO_4$)	ΔH_1 = 4·27, ΔS_1 = 20·0	
	cix	25	0·5—0·6 ($NaClO_4$)	K_1 1·95, β_2 2·84	67La
	dis	24—25	0·5—0·6 ($NaClO_4$)	K_1 1·94, β_2 2·88	
Ce^{4+}	kin	25	var	*K_2(CeL^{2+} + HL^- = CeL_2 + H^+) 2·07	60Mb
	sp	20	H_2L var	K(CeL_2 + $HL^- \rightleftharpoons HCeL_3^-$) −0·2, K($HCeL_3^-$ + $H_2L \rightleftharpoons H_3CeL_4^-$) 0·3	63B
	sp	25	5·9 ($HClO_4$)	K_1 4·78, K_2 3·56, K_3 1·88, ΔH_1 = −6·3 ΔH_2 = 6·8, ΔH_3 = 8·3 [K_1(H^+) 1·15(0°)]	63Bb
	sp, tp	19	var	K_1 2·6, β_2 2·2, ev ani cpx	65B
	kin	35	0·1 H_2L	K_3 1·17 [K_1(H^+) 1·60]	67Mc
Pr^{3+}	sp	25	0 corr	K_1 3·37	64Ba
	con	25	0 corr	K_1 3·85	66E
	Ag	25	2 ($NaClO_4$)	K_1 1·27, β_2 1·88	67C
	cal	25	2 ($NaClO_4$)	ΔH_1 = 3·94, ΔS_1 = 19·0	

continued overleaf

56 SO_4^{2-} (contd)

Metal	Method	Temp	Medium	Log of equilibrium constant, remarks	Ref
Nd^{3+}	Ag	25	2 ($NaClO_4$)	K_1 1·26, β_2 1·79	67C
	cal	25	2 ($NaClO_4$)	$\Delta H_1 = 4{\cdot}18$, $\Delta S_1 = 19{\cdot}8$	
Pm^{3+}	dis	0—55	2 ($NaClO_4$)	K_1 1·08(0°), 1·34(25°), 1·49(40°), 1·60(55°); β_2 1·62(0°), 1·88(25°), 2·00(40°), 2·28(55°) $\Delta H_1 = 3{\cdot}9$, $\Delta S_1 = 19$	67C
Sm^{3+}	sp	25	0 corr	K_1 3·54	64Ba
	Ag	25	2 ($NaClO_4$)	K_1 1·30, β_2 1·91	67C
	cal	25	2 ($NaClO_4$)	$\Delta H_1 = 4{\cdot}27$, $\Delta S_1 = 19{\cdot}7$	
Eu^{2+}	red	24—69	1 KCl	K_{s0} −6·6(24°), −6·3(45°), −6·2(54°)	64K
		25—69	dil, 0 corr	K_{s0} −8·8(25°), −8·3(48°), −8·0(69°)	
	sol	20	0 corr	K_{s0} −8·33	65S/
Eu^{3+}	dis	25	I($NaClO_4$)	(add) K_1 2·23(I = 0·1), 2·53(I = 0·05), 3·56(0 corr)	62Mb
	cix	26	1 H(ClO_4)	K_1 1·23, β_2 1·7 [$K_1(H^+)$ 0·52]	64B
			1 Na(ClO_4)	K_1 1·57, β_2 2·40	
	sp	25	0 corr	K_1 3·35	64Bb
	dis	25	1 Na(ClO_4)	K_1 1·54, β_2 2·69	64S
	dis	25	1 Na(ClO_4)	K_1 1·54, β_2 2·69	65Sa
	Ag	25	2 ($NaClO_4$)	K_1 1·37, β_2 1·96	67C
	dis	0—55	2 ($NaClO_4$)	K_1 1·11(0°), 1·38(25°), 1·56(40°), 1·69(55°); β_2 1·91(0°), 1·98(25°), 2·11(40°), 2·30(55°); $\Delta H_1 = 4{\cdot}3$, $\Delta S_1 = 21$	
	cal	25	2 ($NaClO_4$)	$\Delta H_1 = 3{\cdot}88$, $\Delta H_2 = 2{\cdot}4$, $\Delta S_1 = 19{\cdot}3$, $\Delta S_2 = 11{\cdot}3$	
	cix, dis	24—25	0·5 ($NaClO_4$)	K_1 1·87(cix), 1·88(dis), β_2 2·73(cix), 2·79(dis)	68Aa
Gd^{3+}	Ag	25	2 ($NaClO_4$)	K_1 1·33, β_2 1·75	67C
	cal	25	2 ($NaClO_4$)	$\Delta H_1 = 3{\cdot}95$, $\Delta H_2 = 1{\cdot}7$, $\Delta S_1 = 19{\cdot}3$, $\Delta S_2 = 7{\cdot}6$	
	cix	25	0·5—0·6 ($NaClO_4$)	K_1 1·90, β_2 2·84	67La
	dis	24—25	0·5—0·6 ($NaClO_4$)	K_1 1·90, β_2 2·99	
Tb^{3+}	Ag	25	2 ($NaClO_4$)	K_1 1·27, β_2 1·89	67C
	cal	25	2 ($NaClO_4$)	$\Delta H_1 = 4{\cdot}24$, $\Delta H_2 = 1{\cdot}6$, $\Delta S_1 = 20{\cdot}1$, $\Delta S_2 = 8{\cdot}2$	
Dy^{3+}	Ag	25	2 ($NaClO_4$)	K_1 1·23, β_2 1·72	67C
	cal	25	2 ($NaClO_4$)	$\Delta H_1 = 4{\cdot}41$, $\Delta H_2 = 1{\cdot}3$, $\Delta S_1 = 20{\cdot}4$, $\Delta S_2 = 6{\cdot}6$	
Ho^{3+}	Ag	25	2 ($NaClO_4$)	K_1 1·24, β_2 1·76	67C
	cal	25	2 ($NaClO_4$)	$\Delta H_1 = 4{\cdot}22$, $\Delta H_2 = 1{\cdot}7$, $\Delta S_1 = 19{\cdot}8$, $\Delta S_2 = 8{\cdot}1$	
Er^{3+}	Ag	25	2 ($NaClO_4$)	K_1 1·23, β_2 1·71	67C
	cal	25	2 ($NaClO_4$)	$\Delta H_1 = 4{\cdot}21$, $\Delta H_2 = 1{\cdot}5$, $\Delta S_1 = 19{\cdot}7$, $\Delta S_2 = 7{\cdot}2$	
Tm^{3+}	Ag	25	2 ($NaClO_4$)	K_1 1·15, β_2 1·59	67C
	cal	25	2 ($NaClO_4$)	$\Delta H_1 = 4{\cdot}22$, $\Delta H_2 = 1{\cdot}0$, $\Delta S_1 = 19{\cdot}4$, $\Delta S_2 = 5{\cdot}4$	
Yb^{3+}	cix	25	0 corr	K_1 3·58	66A
	var	25	0 corr	K_1 2·56 (M, Ag, gl, Hg)	66Ab
	Ag	25	2 ($NaClO_4$)	K_1 1·15, β_2 1·59	67C
	cal	25	2 ($NaClO_4$)	ΔH_1 4·14, $\Delta H_2 = 1{\cdot}2$, $\Delta S_1 = 19{\cdot}2$, $\Delta S_2 = 6{\cdot}1$	
Lu^{3+}	dis	25	1 Na(ClO_4)	K_1 1·29, $\beta_2 < 2{\cdot}5$, β_3 3·36	64S
	dis	25	1 Na(ClO_4)	K_1 1·29, $\beta_2 \approx 1{\cdot}9$	65Sa
	Ag	25	2 ($NaClO_4$)	K_1 1·09, β_2 1·61	67C
	cal	25	2 ($NaClO_4$)	$\Delta H_1 = 4{\cdot}25$, $\Delta H_2 = 1{\cdot}5$, $\Delta S_1 = 19{\cdot}2$, $\Delta S_2 = 7{\cdot}4$	
Ac^{3+}	dis	27	1 H(ClO_4)	K_1 1·20, β_2 1·85	68S
Ti^{4+}	pol		2 H_2L	ev two polynuclear cpx	65D

contin

Metal	Method	Temp	Medium	Log of equilibrium constant, remarks	Ref
Ti^{4+}, M	sp	19—26	1·3 H^+, 1·5 NH_4^+, 2·8 HL^-	ev mixed cpx, $\beta(Ti^{IV}Fe^{III})$ 1·4(19°), $\beta(Ti^{IV}Cr^{III})$ 1·5(26°)	66G
	sp	16	80% H_2L	$\beta(Ti^{IV}Fe^{II})$ −0·43 in 80% H_2SO_4	66Ga
	sp, pre		var	ev mixed polyn cpx, ≈ 1 Fe^{2+}/Ti	67J
Zr^{4+}	cix		2·3 $H(ClO_4)$	$*K_1$ 2·67, $*\beta_2$ 3·54, $*\beta_3$ 5·59 (misprint 6·59) [$K_1(H^+)$ 1·08]	62R/64R
Hf^{4+}	cix		2·3 $H(ClO_4)$	$*K_1$ 2·11, $*\beta_2$ 3·32, $*\beta_3$ 6·48 (misprints ?) [$K_1(H^+)$ 1·08]	62R/64R
	dis	27	2 $H(ClO_4)$	$*K_1$ 2·04, $*\beta_2$ 3·7	65Da
	cix	20	2 $H(ClO_4)$	K_1 3·10, β_2 5·42, $\beta(Hf^{4+}L^{2-}Cl^-)$ 3·04 $\beta(Hf^{4+}L^{2-}NO_3^-)$ 3·23, ($\beta(Hf^{4+}L^{2-}{}_2NO_3^-)$ 5·7, [$K_1(H^+)$ 1·08 ?]	67Ea
Th^{4+}	dis	25	Na_2L var	K_3 1·08	64Ra
VO^{2+}	var		oleum, H_2L	ev unch cpx such as $V(HL)_4$ in oleum (con, esr, mag, sp)	63M
	kin	25	var	K_1^{in}/K_1^{out} 0·0, K_1^{out} 2·10 (con p-jump)	66K ≈ 65Kc
	con	25	0 corr	K_1 2·40	
VO_2^+	con, fp	25	H_2SO_4	$\beta[H^+VO(HL)_4^-]$ −2·1, ev polyn cpx in H_2SO_4(l), ([H^+] [HL^-] −3·62), m units	66Gd
	sp	19	1·01 $(NaClO_4)$	K_1 0·97, $*K_1$ −0·14	66Ia
$Nb^{3+,5+}$	sp	22	≈ 10 H_2SO_4	$\beta(Nb^{III}{}_4Nb^{V}{}_2)$ 4·3 ?	66Gb
Nb^{5+}	dis		6 NH_4, 2 HL, 2 L	$K_D[R_3NH^+HL^-(org) + NbO(OH)_2L^- \rightleftharpoons R_3NH^+NbO(OH)_2L^-(org) + HL^-]$ 1·44, R = C_8H_{17}, org = $CHCl_3$	66Na
	sp, tp		1 $HClO_4$	ev ani cpx, apparent K_1 0·48	68Pa
Ta^{5+}	sp, tp		1 $HClO_4$	ev ani cpx, apparent K_1 0·78, β_2 0·25	68Pa
Pa^{4+}	dis		3 $H(ClO_4)$	$*K_1$ 1·62, $*\beta_2$ 2·18	65G
	dis	25	3 $H(ClO_4)$	$*K_1$ 1·62, $*\beta_2$ 2·18	66Ge
	dis	10	0·5	$K[Pa(OH)_2^{2+} + HL^- \rightleftharpoons PaLOH^+]$ 2·51	68Ma
Pa^{5+}	cix	20	2 $H(ClO_4)$	HSO_4^-: (apparent) K_1 0·06, β_2 1·17	63Na
			1 $H(ClO_4)$	HSO_4^-: (apparent) K_1 −0·03, β_2 0·87	
	dis		3 $(LiClO_4)$	$\beta[Pa(OH)_3^{2+}HL^-]$ 1·3, $\beta[Pa(OH)_3^{2+}HL^-{}_2H^+{}_{-1}]$ 2·5, [$K_1(H^+)$ 1·08]	65G
	dis	25	1 $HClO_4$	$*K_1$ 2·08, $*K_2$ 0·23	67K
	dis	25	3 $H(ClO_4)$	$K(PaOOH^{2+} + HL^- \rightleftharpoons PaOL^+ + H_2O)$ 1·29, $K(PaOOH^{2+} + 2HL^- \rightleftharpoons PaOL_2^- + H^+ + H_2O)$ 2·51, [$K_1(H^+)$ 1·06]	66Ge
Cr^{3+}	var		oleum, H_2L	ev cpx such as $Cr(H_3L_2)_3$ in oleum (esr, mag, sp)	63M
	ir, pre		solid	ev $Cr(H_2O)_5L^+$(unidentate) in $Cr(H_2O)_5SO_4Cl$-$(H_2O)_{0.5}$(s)	65F
	pol	25	0·1 $(NaClO_4)$	K_1 1·6	66T
	var		var	ev $(NH_3)_4Cr(L)(OH)Cr(NH_3)_4^{3+}$ (pre, cix, ir)	68F
$Cr(NH_3)_6^{3+}$	pol	25	0·1 $(NaClO_4)$	K_1 1·76	66T
	pol	25	0·1 $(NaClO_4)$	K_1 1·79	67Tb
Cr en_3^{3+}	pol	25	0·1 $(NaClO_4)$	K_1 1·76	67Tb
Cr^{3+} cpx	kin	25	dil	$\beta[Cr(NH_3)_5Cl^{2+}L^{2-}]$ 2·53	66W
Cr^{5+}	var		oleum, H_2L	ev unch cpx such as $OCr(H_3L_2)_3$ in oleum (con, esr, mag, sp)	63M

continued overleaf

56 SO_4^{2-} (contd)

Metal	Method	Temp	Medium	Log of equilibrium constant, remarks	Ref
Cr^{6+}	sp	15—35	3 (Na, HL)	$K(CrO_4^- + HL^- \rightleftharpoons CrO_3L^{2-})$ 0·61, $\Delta H = 0{\cdot}0$, $\Delta S = 2{\cdot}7$	64Hb
Mo^{5+}	esr		12 H_2L	$K[MoOCl_x^{3-x} + nHL^- \rightleftharpoons MoOCl_{x-n}(HL)_n^{3-x} + nCl^-]$ $-2(n = 1)$, $-4(n = 2)$, $-7(n = 3)$ and $-10(n = 4)$ or $-13(n = 5)$	67Md
U^{4+}	nmr	20	2 $HClO_4$	K_1 1·7	63V
	kin	25	1 $(NaClO_4)$	$*K_1$ 2·20 $[K_1(H^+)$ 1·07]	66Sa
UO_2^{2+}	sol	17	8 NH_4NO_3	$K_s[(NH_4)_2UO_2L_2(s) + L^{2-} \rightleftharpoons (2NH_4^+) + UO_2L_3^{4-}]$ $-0{\cdot}97$	63Ka
	oth	25—50	$\rightarrow 0$	K_1 3·14(25°), 3·26(35°), 3·43(50°) $\Delta H_1 = 5{\cdot}1$; β_2 4·21(25°), 4·36(35°), 4·60(50°), $\Delta H_{\beta 2} = 7{\cdot}0$ (membrane equil Na^+–UO_2^{2+})	67W
Np^{4+}	pol	25	3 $(NaClO_4)$ 1, H^+	$*K_1$ 2·49, $*\beta_2$ 3·58	62Mc
	sp	25	2 $(NaClO_4)$	K_1 3·51	62Sb
	cix	20	4 $H(ClO_4)$	$*K_1$ 2·70, $*K_2$ 1·56	66Aa
NpO_2^{2+}	sp	25	2 $(NaClO_4)$	K_1 1·64	62Sb
	dis	21	1 $(NaClO_4)$	$*K_1$ 0·79, $*\beta_2$ 0·56 $[K_1(H^+)$ 1·11], no ev cpx NpO_2^{2+}–HL^-	68A
Pu^{3+}	cix	28	1 $H(ClO_4)$	K_1 1·26, $\beta(Pu^{3+}HL^-{}_2)$ 1·00 $[K_1(H^+)$ 0·52]	67N
Pu^{4+}	dis	?	2 (Na, $HClO_4$)	$*K_1$ 3·2 to 3·3, $*\beta_2$ 4·65 to 4·8	64L
	dis	?	2·2	$*K_1$ 2·74, $*K_2$ 1·63	
	dis	23	1 $(Na)NO_3$, 0·5 H^+	$*K_2(PuL^{2+} + HL^- \rightleftharpoons PuL_2 + H^+)$ 0·58	64Lc
Am^{3+}	dis		1 to 1·3 $Na(ClO_4)$	K_1 1·48, β_2 2·59	64Ka
	cix	26	1 $H(ClO_4)$	K_1 1·18, β_2 1·38 $[K_1(H^+)$ 0·52]	64N/64B
			1 to 1·3 $Na(ClO_4)$	K_1 1·49, β_2 2·48	
	dis	25	1 $Na(ClO_4)$	K_1 1·57, β_2 2·66	64S
	dis	25	1 $Na(ClO_4)$	K_1 1·57, β_2 2·66	65Sa
	dis	0—55	2 $(NaClO_4)$	K_1 1·11(0°) 1·43(25°), 1·58(40°), 1·65(55°), β_2 1·73(0°), 1·85(25°), 2·03(40°), 2·38(55°), $\Delta H_1 = 4{\cdot}4$, $\Delta S_1 = 21$	67C
	cix, dis	24—25	0·5 $(NaClO_4)$	K_1 1·86(cix), 1·85(dis); β_2 2·80(cix), 2·83(dis)	68Aa
	cix	27	1 $H(ClO_4)$	K_1 1·22, $\beta(Am^{3+}HL^-{}_2)$ 0·54; no ev ani cpx (aix, tp)	68Na
	cix	27	1 $Na(ClO_4)$	K_1 1·49, β_2 2·36	
AmO_2^{2+}	aix		var	ev ani cpx	64H
Cm^{3+}	dis	0—55	2 $(NaClO_4)$	K_1 1·08(0°), 1·34(25°), 1·49(40°), 1·61(55°); β_2 1·66(0°), 1·86(25°), 2·05(40°), 2·30(55°); $\Delta H_1 = 4{\cdot}1$, $\Delta S_1 = 20$	67C
	cix, dis	24—25	0·5 $(NaClO_4)$	K_1 1·86(cix), 1·83 to 1·87(dis); β_2 2·37(cix), 2·63 to 2·76(dis)	68Aa
Cf^{3+}	dis	0—55	2 $(NaClO_4)$	K_1 1·04(0°), 1·36(25°), 1·49(40°), 1·62(55°); β_2 1·75(0°), 2·07(25°), 2·09(40°), 2·34(55°); $\Delta H_1 = 4{\cdot}5$, $\Delta S_1 = 21$	67C
Mn^{2+}	con	25	w% Me_2CO	K_1 2·44(w = 9·9), 2·88(w = 19·8), 3·24(w = 29·9), 3·75(w = 40·2) in w (weight) % Me_2CO–H_2O	64A
	oth	25	0 corr	K_1 (over-all) 2·14, (ultrasonic absorption), see also ref 67A	65A

contin

Metal	*Method*	*Temp*	*Medium*	*Log of equilibrium constant, remarks*	*Ref*
Mn^{2+} (*contd*)	con	25	dil	K_1 2·36(1 atm), 2·32(500 atm), 2·23(1000 atm), 2·18(1500 atm), 2·14(2000 atm), ΔV_1 = 7·4 ml(1 atm), 5·3 ml(2000 atm); this is for 0·0005M-MnL, values given also for other conc up to 0·02M	65Fb
	oth	25	0	K_1 2·03, $K[Mn(H_2O)_2L \rightleftharpoons MnH_2OL]$ −0·15 (complex diel. constant)	65P
	con		w% MC	K_1 2·12(w = 0), 2·72(w = 20·14), 3·14(w = 30·13), 3·43(w = 39·9), 3·90(w = 49·95), 4·11(w = 54·93) in w% MC = methyl cellosolve = $CH_3OC_2H_4OH$ (→ 0)	66Ad
	oth	20—30	0 corr	$K[Mn(aq)^{2+} + L(aq)^{2-} \rightleftharpoons Mn(aq)^{2+}L(aq)^{2-}]$ 1·69(20°), 1·72(25°), 1·73(30°), ΔH = 1·5, ΔS = 13; $K[Mn(aq)^{2+}L(aq)^{2-} \rightleftharpoons MnH_2OL]$ −0·40(20°), −0·44(25°), −0·48(30°), ΔH = −3·2, ΔS = −13; $K(MnH_2OL \rightleftharpoons MnL)$ 0·53(20°), 0·55(25°), 0·58(30°), ΔH = 1·6, ΔS = 8 (ultrasonic absorption)	67A ≈ 65A
	nmr	30	var	ev $MnSO_4(H_2O)_4$, bidentate SO_4^{2-}	67F
	ΔG	25	0	K_1 2·26, ΔH_1 = 3·62, ΔS_1 = 22·4, ΔC_p = 26	/67H
	con	25	x $C_2H_4(OH)_2$	K_1 2·13(x = 0), 2·42(x = 0·1), 2·93(x = 0·3), 3·63(x = 0·5) in $[C_2H_4(OH)_2]_x(H_2O)_{1-x}$(l), → 0	67P
	con	25	w% MeOH	K_1 3·59(50%), 3·98(60%), 4·47(70%), 4·95(80%) in w% MeOH, → 0	67T
	Hg	35?	0 corr	K_1 2·27	68P
Mn^{4+}	var		oleum	ev unch cpx such as $Mn(H_3L_2)_4$ in oleum (con, esr, mag, sp)	63M
Fe^{2+}	kin	0—35	1 ($NaClO_4$)	K_1 1·74(0°), 1·0(25°), 0·76(35°)	68W
		45	4 ($NaClO_4$)	K_1 0·40	
		0—25	4 ($NaClO_4$)	$\beta(Fe^{2+}HL^-)$ 0·66(0°), 0·57(5°), 0·48(10°), 0·29(25°)	
Fe^{3+}	sp	25	1 ($NaClO_4$)	K_1 2·24, K_2 0·94, ΔH_1 = −5·15, ΔH_2 = 14·4	63Ba
	sp	20	1·2 ($NaClO_4$)	K_1 2·06, $\beta(Fe^{3+}L^{2-}Br^-)$ 2·50 for 0·2M-H^+ ("$[L^{2-}]$" = $[L^{2-}]+[HL^-]$)	67Ma
Co^{2+}	con	25	H_2NCHO	no ev cpx in H_2NCHO(l)	65Ja
	cix	25	0 corr	K_1 2·41 in H_2O, 2·6 in 20% EtOH	65Sb
	fp	≈ 260?	$LiNO_3$(l)	K_1 −0·4 in $LiNO_3$(l), m units	66I
	sol	160—300	0 corr	$K_{s0}(Co^{2+}L^{2-}H_2O)$ −3·84(160°), −4·15(180°), −4·47(200°), −4·78(220°), −5·11(240°), −5·51(260°), −5·98(280°), −6·58(300°), ΔH_{s0} = −13(→ 25°), ΔS_{s0} = −48(→ 25°)	67Ga
	ΔG	25	0	K_1 2·36, ΔH_1 = 1·80, ΔS_1 = 16·8, ΔC_p = 9	/67H
	Hg	35?	→ 0	K_1 1·93	68P
$Co(NH_3)_6^{3+}$	con	25—40	0 corr	K_1 3·71(25°), 3·74(30°), 3·76(40°) at 1 atm; 3·65(25°), 3·66(30°), 3·60(40°) at 600 atm; also K_1 values for 200 and 400 atm, ΔS and ΔV values	66O
	pol	25	0·1 ($NaClO_4$)	K_1 1·84, no ev cpx with HL^-	66T
	sp	25	0·07 ($NaClO_4$)	K_1 2·06	67Ta
	pol	25	1 ($NaClO_4$)	K_1 1·97	67Tb
$Co(NH_3)_5Cl^{2+}$					
	sol	25—35	0 corr	K_1 2·46(25°), 2·52(35°)	65Aa
	kin	35	0 corr	K_1 2·46	65Lb

continued overleaf

56 SO_4^{2-} (contd)

Metal	*Method*	*Temp*	*Medium*	*Log of equilibrium constant, remarks*	*Ref*
Co en_3^{3+}	pol	25	1 ($NaClO_4$)	K_1 2·00	67Tb
	sp	25	0·07 ($NaClO_4$)	K_1 2·01	67Ta
Co^{3+} cpx	kin	25	var	β(*cis*-$CoCl_2\,en_2^+L^{2-}$) 1·4	68Mb
$Co^{3+,4+}$ cpx	var	var		ev ani cpx SO_4^{2-}–$[Co_2O_2(NH_3)_{10}]^{5+}$ *etc.* (tp, dialysis)	35Ba
	var		var	no ev ani cpx SO_4^{2-}–$[Co_2O_2(NH_3)_{10}]^{5+}$ (con, sol, tp, sp)	67E
Ni^{2+}	pol	25	1 ($NaClO_4$)	K_1 0·57, β_2 1·42, $\beta(Ni^{2+}A^-L^{2-})$ 0·0, $\beta(Ni^{2+}A^-L^{2-}{}_2)$ 0·5 [$\beta(Ni^{2+}A^-)$ 0·28], $A^- = CH_3CO_2^-$	63Tb
	ir		1 $NiSO_4$	K_1^{in}/K_1 −1·1	64La
	oth	25	0	K_1 1·94, $K[M(H_2O)_2L \rightleftharpoons MH_2OL]$ −0·3 (complex diel constant)	65P
	sol	160—300	0 corr	$K_{s0}(Ni^{2+}L^{2-}H_2O)$ −3·69(160°), −4·01(180°), −4·35(200°), −4·70(220°), −5·10(240°), −5·57(260°), −6·12(280°), −6·80(300°), $\Delta H_{s0} = -12(\rightarrow 25°)$, $\Delta S_{s0} = -42(\rightarrow 25°)$	67Ga
	ΔG	25	0	K_1 2·32, $\Delta H_1 = 3·19$, $\Delta S_1 = 21·3$, $\Delta C_p = -6$	/67H
	Hg	35?	→0	K_1 2·05	68P
Ru^{2+}	cix	20	2 ($NaClO_4$)	K_1 1·35	66V
	cix	20—35	I($NaClO_4$)	K_1 1·30($I = 2$), 1·70($I = 1$), 1·88($I = 0·5$), 2·02($I = 0·25$), 2·72(→ 0), $\Delta H = 0$, $\Delta S = 12·4(\rightarrow 0)$	67V
Ru^{3+}	pol	25	2 ($NaClO_4$?)	K_1 2·04, β_2 3·57	68L
$Ru^{3+,4+}$	ana		H_2SO_4 var	ev cpx with average oxidation number +3·5 for Ru	65V
	var		var	ev ani polyn cpx (green), average oxidation number +3·5, 1·5—1·7L^{2-}/Ru (pre, pol, ir, sp)	66Gc
RuO^{2+}	cix		4 ($NaClO_4$)	*$K_1(RuO^{2+} + HL^- \rightleftharpoons H^+ + RuOL)$ 0·82	65V
			2 ($NaClO_4$)	*K_1 1·10	
	cix	20—35	2 ($NaClO_4$)	K_1 1·07(20°), 1·16(25°), 1·36(35°)	67V
			1 ($NaClO_4$)	K_1 1·25(20°), 1·31(25°), 1·58(35°)	
			0·5 ($NaClO_4$)	K_1 1·37(20°), 1·47(25°), 1·69(35°)	
			0·25 ($NaClO_4$)	K_1 1·53(20°), 1·58(25°), 1·78(35°)	
			→0	K_1 2·19(20°), 2·24(25°), 2·45(35°) $\Delta H_1 = 7·7$, $\Delta S_1 = -16(25°)$	
$Rh(NH_3)_5^{3+}$	sp	25—65	4 ($NaClO_4$)	K_1 0·0(25°), −0·15(65°)	68Md
Pt en_3^{4+}	var	20	var	ev three cpx (sp, circular dichroism)	67Lb
Cu^{2+}	Ram		CuL conc	no ev inner cpx	64Ha
	sp	25	I($NaClO_4$)	K_1 1·38($I = 0·091$), 1·64($I = 0·041$) 2·23 to 2·40 (0 corr)	65M
	oth	25	0	K_1 1·93, $K[M(H_2O)_2L \rightleftharpoons MH_2OL]$ −0·18 (complex diel constant)	65P
	pol	25	1 ($NaClO_4$)	$K_1 < 0·5$, β_2 1·5; $\beta(Cu^{2+}A^-L^{2-})$ 1·6, $\beta(Cu^{2+}A^-L^{2-}{}_2)$ 1·85, [$\beta(Cu^{2+}A^-)$ 1·30, $\beta(Cu^{2+}A^-{}_2)$ 2·04], $A^- = CH_3CO_2^-$	65Tb
	con	25	w% dioxan	K_1 3·17($w = 25$), 3·04($w = 20$), 2·75($w = 15$), 2·62($w = 10$), 2·39($w = 5$), 2·25($w \rightarrow 0$) in w% dioxan	65Y
	fp	~ 260	$LiNO_3$(l)	K_1 0·43, ev CuL_2^{2-} in $LiNO_3$(l), m units	66I
	sp	25	→0	K_1 2·35	68H

continued

Metal	Method	Temp	Medium	Log of equilibrium constant, remarks	Ref
Cu^{2+} (contd)	oth	25	0 corr	$K[Cu(aq)^{2+}+L(aq)^{2-} \rightleftharpoons CuH_2OL]$ 2·22 (hyp), $K(CuH_2OL \rightleftharpoons CuL+H_2O)$ −0·6, K_1 2·32, (ultrasonic absorption)	
	var		0 corr	K_1 varies with interionic distance assumed (lit: con, fp, sp)	68M
	sp	25	3 Li(ClO$_4$)	K_1^{in} −0·5, K_1^{out} 0·67, K_1 0·70	68Mc
	Hg	35?	→ 0	K_1 2·17	68P
$Cu\,en_2^{2+}$	con	25	H_2NCHO	no ev cpx in H_2NCHO(l)	65Ja
Ag^+	cal	25	0 corr	$\Delta H_{s0} = 4{\cdot}12$, $\Delta H_1 = 1{\cdot}5$	65Ha
	ΔG	25	0 corr	$\Delta S_{s0} = -8{\cdot}8$, $\Delta S_1 = 11$	65Ha+
	E	25	2 (NaClO$_4$)	K_1 0·31, β_2 0·19, β_3 0·40	67C
	sol	25—100	→ 0	K_{s0} −4·81(25°), −4·67(50°), −4·58(75°), −4·54(100°), (→ 0 from La_2L_3 var)	67L
Ag^{2+}	sp, kin	25	H_2L var	ev two cpx	64Rc
Zn^{2+}	Ram		ZnL conc	no ev inner cpx	64Ha
	ΔG	25	0	K_1 2·38, $\Delta H_1 = 3{\cdot}80$, $\Delta S_1 = 23{\cdot}7$, $\Delta C_p = -56$, or $\Delta H_1 = 4{\cdot}09$, $\Delta S_1 = 24{\cdot}5$	/67H
	oth	25	$C_2H_4(OH)_2$	K_1 2·34(x = 0), 2·55(x = 0·1), 3·05(x = 0·3), 3·25(x = 0·5), $K[Zn(aq)^{2+}L^{2-} \rightleftharpoons ZnL]$ 0·52 in glycol(x)+water(1−x)(l), (ultrasonic absorption + N. Bjerrum's equation for K_1^{out})	67Fb
	MHg	35?	→ 0	K_1 2·12	68P
Cd^{2+}	Ram		CdL conc	no ev inner cpx	64Ha
	dis	30	3 (NaClO$_4$)	K_1 0·1, β_2 1·0, β_3 1·7	65H
	oth	25	0	K_1 2·01, $K[Cd(H_2O)_2L \rightleftharpoons CdH_2OL]$ −0·20 (complex diel constant)	65P
	fp	~ 260	LiNO$_3$(l)	K_1 0·72, ev CdL_2^{2-} in LiNO$_3$(l), *m* units	66I
	MHg	35?	→ 0	K_1 2·11	68P
$MeHg^+$	Ram	25	var	$K_1(MeHg^+) \approx K_1(H^+)$	68C
B^{3+}	fp	≈ 30	oleum	ev $B(HS_2O_7)_4^-$ in oleum	67D
Al^{3+}	kin	25	0·1	$K(Al^{3+}aq+L^{2-}aq \rightleftharpoons AlH_2OL^+)$ 1·28 $K(AlH_2OL^+ \rightleftharpoons AlL^+)$ −1 to −2 (p-jump method)	62Ba
	Ram		Al_2L_3 conc	no ev inner cpx	64Ha
	con, gl	25	→ 0	K_1 3·73	65N
Ga^{3+}	Ram		Ga_2L_3 conc	no ev inner cpx	64Ha
In^{3+}	Ram		In_2L_3 conc	ev inner cpx	64Ha
	ir		0·1 In_2L_3	K_1^{in}/K_1 −0·3	64La
	sol	25	2 NaNO$_3$	K_1 1·78	66D
	dis	25	1 (NaClO$_4$)	K_1 1·79, β_2 2·51	68Ab
Tl^+	Ram		Tl_2L conc	no ev inner cpx	64Ha
	MHg	25	3 (LiClO$_4$)	K_1 −0·48 (−0·52 in summary), β_2 −0·89	65K
Tl^{3+}	sol	25	0 corr	$K_1 \approx 1$, ev higher cpx	60Ha
	red	25	3 (LiClO$_4$)	K_1 1·95, β_2 3·74, $\beta(Tl^{3+}HL^-)$ 1·23, $\beta(Tl^{3+}HL^-{}_2)$ 2·12, $\beta(Tl^{3+}HL^-L^{2-})$ 3·00	65K
	cal, red	25	3 (LiClO$_4$) 0·5 H$^+$	K_1 2·27, $\Delta H_1 = -2{\cdot}7$, $\Delta S_1 = 1{\cdot}1$, [$K_1(H^+)$ 1·0]	67Mb
Sn^{4+}	con, fp		H_2SO_4	ev $Sn(HL)_6^{2-}$, $HSn(HL)_6^-$, $H_2Sn(HL)_6$ in H_2SO_4(l)	66Gd

continued overleaf

56 SO_4^{2-} (contd)

Metal	Method	Temp	Medium	Log of equilibrium constant, remarks	Ref
Pb^{2+}	sol	20	0 corr	K_{s1} −5·38, K_1 2·4, no ev PbL_2^{2-}	65La
	fp	≈ 260	$LiNO_3$(l)	K_1 −0·03 in $LiNO_3$(l), *m* units	66I
	sol	25	4 H(HL)	$^*K_{s0}(Pb^{2+}HL^-H^+_{-1})$ −4·90	/66N
Pb^{4+}	con, fp	25	H_2SO_4	$\beta(H^+Pb(HL)_6^{2-})$ 2·74, $\beta[H^+HPb(HL)_6^-]$ 1·92 in H_2SO_4(l), *m* units ($[H^+][HL^-]$ −3·62)	66Gd
NH_4^+	Ram		$(NH_4)_2L$ var	ev cpx	64Lb
Et_4N^+	con	25	MeCN	$\beta(Et_4N^+HL^-)$ 1·5, $\beta(Et_4N^+L^{2-})$ 2·3 in MeCN(l)	68K
I^{3+}	fp, con		H_2L conc	ev IOHL, $I_2O(HL)_4$, $I(HL)_3$ and polymers in H_2L(l) and $H_2S_2O_7$(l)	64G
I^{5+}				see table 77, IO_3^-	

35Ba H. Brintzinger and H. Osswald, *Z. anorg. Chem.*, 1935, **224,** 283
53Sa N. M. Selivanova and A. F. Kapustinskii, *Zhur. fiz. Khim.*, 1953, **27,** 565
60Ha C. S. Hennings, *Anales Fac. Quím. Farm. Univ. Chile*, 1960, **12,** 150
60Mb W. H. McCurdy, jun., and G. G. Guilbault, *J. Phys. Chem.*, 1960, **64,** 1825
62Ea M. Eigen and K. Tamm, *Z. Elektrochem.*, 1962, **66,** 107
62Mc M. Cl. Musikas, *Radiochim. Acta*, 1962, **1,** 92
62Sb K. W. Sykes and B. L. Taylor, Proc. Seventh Internat. Conference on Co-ordination Chemistry, 1962, p. 31
63B L. T. Bugalenko and G.-L. Khuan, *Zhur. neorg. Khim.*, 1963, **8,** 2479
63Ba K. Bächmann and K. H. Lieser, *Ber. Bunsengesellschaft Phys. Chem.*, 1963, **67,** 802
63Bb K. Bächmann and K. H. Lieser, *Ber. Bunsengesellschaft Phys. Chem.*, 1963, **67,** 810
63Ka I. A. Kuzin, I. A. Galitskaya, and V. P. Taushkanov, *Radiokhimiya*, 1963, **5,** 89
63La M. H. Lietzke and R. W. Stoughton, *J. Phys. Chem.*, 1963, **67,** 652
63M H. C. Mishra and M. C. R. Symons, *J. Chem. Soc.*, 1963, 4490
63Na J. Nowikow and G. Pfrepper, *Z. Naturforsch.*, 1963, **18b,** 993
63Q A. S. Quist, E. U. Franck, H. R. Jolley, and W. L. Marshall, *J. Phys. Chem.*, 1963, **67,** 2453
63Ta E. Thilo and F. von Lampe, *Z. anorg. Chem.*, 1963, **319,** 387
63Tb N. Tanaka, Y. Saito, and H. Ogino, *Bull. Chem. Soc. Japan*, 1963, **36,** 794
63V V. M. Vdovenko, G. A. Romanov, and V. A. Shcherbakov, *Radiokhimiya*, 1963, **5,** 664
64A G. Atkinson and S. Petrucci, *J. Amer. Chem. Soc.*, 1964, **86,** 7
64B B. M. L. Bansal, S. K. Patil, and H. D. Sharma, *J. Inorg. Nuclear Chem.*, 1964, **26,** 993
64Ba J. C. Barnes and C. B. Monk, *Trans. Faraday Soc.*, 1964, **60,** 578
64Bb J. C. Barnes, *J. Chem. Soc.*, 1964, 3880
64Bc J. Barr, R. J. Gillespie, and R. C. Thompson, *Inorg. Chem.*, 1964, **3,** 1149
64D H. S. Dunsmore and G. H. Nancollas, *J. Phys. Chem.*, 1964, **68,** 1579
64F A. N. Fletcher, *J. Inorg. Nuclear Chem.*, 1964, **26,** 955
64G R. J. Gillespie and J. B. Senior, *Inorg. Chem.*, 1964, **3,** 972
64H E. K. Hulet, *J. Inorg. Nuclear Chem.*, 1964, **26,** 1721
64Ha R. E. Hester and R. A. Plane, *Inorg. Chem.*, 1964, **3,** 769
64Hb G. P. Haight, jun., D. C. Richardson, and N. H. Coburn, *Inorg. Chem.*, 1964, **3,** 1777
64Hc R. A. Horne, R. A. Courant, and G. R. Frysinger, *J. Chem. Soc.*, 1964, 1515
64K Yu. A. Koz'min, L. P. Shul'gin, and V. D. Ponomarev, *Zhur. neorg. Khim.*, 1964, **9,** 2532
64Ka P. K. Khopkar and H. D. Sharma, unpublished, quoted by ref 64B
64L T. S. Laxminarayanan, S. K. Patil, and H. D. Sharma, *J. Inorg. Nuclear Chem.*, 1964, **26,** 1001
64La R. Larsson, *Acta Chem. Scand.*, 1964, **18,** 1923
64Lb H. Lee and J. K. Wilmshurst, *Austral. J. Chem.*, 1964, **17,** 943
64Lc M. Lucas, *Radiochim. Acta*, 1964, **3,** 126
64M W. L. Marshall, R. Slusher, and E. V. Jones, *J. Chem. Eng. and Data*, 1964, **9,** 187
64N G. M. Nair and G. A. Welch, unpublished, quoted by ref. 64B
64R D. I. Ryabchikov, I. N. Marov, A. N. Ermakov, and V. K. Belyaeva, *J. Inorg. Nuclear Chem.*, 1964, **26,** 965
64Ra D. Richter and H. Grosse-Ruyken, in "Theory and structure of complex compounds" (symposium papers, Wrocław 1962) Pergamon, Warszawa 1964, p. 457
64Rb B. N. Ryzhenko, *Geokhimiya*, 1964, 23
64Rc G. A. Rechnitz and S. B. Zamochnik, *Talanta*, 1964, **11,** 1645
64S T. Sekine, *J. Inorg. Nuclear Chem.*, 1964, **26,** 1463
64Sa A. P. Samodelov, *Radiokhimiya*, 1964, **6,** 568
65A G. Atkinson and S. K. Kor, *J. Phys. Chem.*, 1965, **69,** 128, discussion: L. G. Jackopin and E. Yeager; G. Atkinson and S. K. Kor, *J. Phys. Chem.*, 1966, **70,** 313, 314
65Aa D. W. Archer, D. A. East, and C. B. Monk, *J. Chem. Soc.*, 1965, 720
65B T. N. Bondareva, V. F. Barkovskii, and T. V. Velikanova, *Zhur. neorg. Khim.*, 1965, **10,** 127
65C A. K. Covington, J. V. Dobson, and Lord Wynne-Jones, *Trans. Faraday Soc.*, 1965, **61,** 2057

continued

65D Yu. D. Dolmatov and Yu. Ya. Bobyrenko, *Zhur. Vsesoyuz. Khim. obshch. im. D.I. Mendeleeva*, 1965, **10**, 352
65Da R. G. Deshpande, P. K. Khopkar, C. L. Rao, and H. D. Sharma, *J. Inorg. Nuclear Chem.*, 1965, **27**, 2171
65E J. T. Edward and I. C. Wang, *Canad. J. Chem.*, 1965, **43**, 2867
65F J. E. Finholt, R. W. Anderson, J. A. Fyfe, and K. G. Caulton, *Inorg. Chem.*, 1965, **4**, 43
65Fa F. H. Fisher, *J. Phys. Chem.*, 1965, **69**, 695
65Fb F. H. Fisher and D. F. Davis, *J. Phys. Chem.*, 1965, **69**, 2595
65G R. Guillaumont, *Compt. rend.*, 1965, **260**, *C*, 4739
65H H. E. Hellwege and G. K. Schweitzer, *J. Inorg. Nuclear Chem.*, 1965, **27**, 99
65Ha H. P. Hopkins, jun. and C. A. Wulff, *J. Phys. Chem.*, 1965, **69**, 9
65J G. P. Johari and P. H. Tewari, *J. Amer. Chem. Soc.*, 1965, **87**, 4691
65Ja G. P. Johari and P. H. Tewari, *J. Phys. Chem.*, 1965, **69**, 696
65Jb G. P. Johari and P. H. Tewari, *J. Phys. Chem.*, 1965, **69**, 3167
65K F. Ya. Kul'ba, Yu. B. Yakovlev, and V. E. Mironov, *Zhur. neorg. Khim.*, 1965, **10**, 2044
65Ka I. V. Kolosov, *Zhur. neorg. Khim.*, 1965, **10**, 2200
65Kb J. Koskikallio and S. Syrjäpalo, *Acta Chem. Scand.*, 1965, **19**, 429
65Kc G. Köhler, dissertation, Univ. Göttingen, 1965
65L K. H. Lieser, *Z. anorg. Chem.*, 1965, **335**, 225
65La K. H. Lieser, G. Beyer, and E. Lakatos, *Z. anorg. Chem.*, 1965, **339**, 208
65Lb S. H. Laurie and C. B. Monk, *J. Chem. Soc.*, 1965, 724
65M R. A. Matheson, *J. Phys. Chem.*, 1965, **69**, 1537
65N T. Nishide and R. Tsuchiya, *Bull. Chem. Soc. Japan*, 1965, **38**, 1398
65P R. Pottel, *Ber. Bunsengesellschaft Phys. Chem.*, 1965, **69**, 363
65Q A. S. Quist, W. L. Marshall, and H. R. Jolley, *J. Phys. Chem.*, 1965, **69**, 2726
65S Yu. S. Sklyarenko and N. S. Stroganova, *Zhur. neorg. Khim.*, 1965, **10**, 647
65Sa T. Sekine, *Acta Chem. Scand.*, 1965, **19**, 1469
65Sb R. G. Seys and C. B. Monk, *J. Chem. Soc.*, 1965, 2452
65T P. H. Tewari and G. P. Johari, *J. Phys. Chem.*, 1965, **69**, 2857
65Ta A. Tateda, *Bull. Chem. Soc. Japan*, 1965, **38**, 165
65Tb N. Tanaka, Y. Saito, and H. Ogino, *Bull. Chem. Soc. Japan*, 1965, **38**, 984
65V V. M. Vdovenko, L. N. Lazarev, and Ya. S. Khvorostin, *Radiokhimiya*, 1965, **7**, 232
65Y M. Yokoi and E. Kubota, *J. Chem. Soc. Japan*, 1965, **86**, 894
66A D. W. Archer and C. B. Monk, *J. Chem. Soc.* (*A*), 1966, 1374
66Aa S. Ahrland and L. Brandt, *Acta Chem. Scand.*, 1966, **20**, 328
66Ab G. Atkinson and S. Petrucci, *J. Phys. Chem.*, 1966, **70**, 3122
66Ac D. W. Archer and C. B. Monk, *Trans. Faraday Soc.*, 1966, **62**, 1583
66Ad G. Atkinson and H. Tsubota, *J. Amer. Chem. Soc.*, 1966, **88**, 3901
66B T. A. Belyavskaya, I. P. Alimarin, and G. D. Brykina, *Vestnik Moskov. Univ.*, 1966, *Ser Khim.*, no. 1, 84
66C J. J. Christensen, R. M. Izatt, L. D. Hansen, and J. A. Partridge, *J. Phys. Chem.*, 1966, **70**, 2003
66Ca R. W. Chlebek and M. W. Lister, *Canad. J. Chem.*, 1966, **44**, 437
66D E. N. Deichman, G. V. Rodicheva, and L. S. Krysina, *Zhur. neorg. Khim.*, 1966, **11**, 2237
66E V. I. Ermolenko, *Dopovidi Akad. Nauk Ukrain. R.S.R., Ser. B*, 1966, 85
66G Ya. G. Goroshchenko and I. A. Sidorenko, *Zhur. neorg. Khim.*, 1966, **11**, 1302
66Ga Ya. G. Goroshchenko and E. K. Sikorskaya, *Zhur. neorg. Khim.*, 1966, **11**, 1591
66Gb Ya. G. Goroshchenko and M. I. Andreeva, *Zhur. neorg. Khim.*, 1966, **11**, 2233
66Gc S. I. Ginzburg, M. I. Yuz'ko, and T. A. Fomina, *Zhur. neorg. Khim.*, 1966, **11**, 2554
66Gd R. J. Gillespie, R. Kapoor, and E. A. Robinson, *Canad. J. Chem.*, 1966, **44**, 1197, 1203
66Ge R. Guillaumont, *Rev. Chim. minérale*, 1966, **3**, 339
66I R. E. Isbell, E. W. Wilson, jun., and D. F. Smith, *J. Phys. Chem.*, 1966, **70**, 2493
66Ia A. A. Ivakin, *Zhur. priklad. Khim.*, 1966, **39**, 277
66J D. V. S. Jain and S. K. Dogra, *J. Chem. Soc.* (*A*), 1966, 284
66K G. Köhler and H. Wendt, *Ber. Bunsengesellschaft Phys. Chem.*, 1966, **70**, 674
66L M. H. Lietzke and R. W. Stoughton, *J. Inorg. Nuclear Chem.*, 1966, **28**, 1063
66M W. L. Masterton and L. H. Berka, *J. Phys. Chem.*, 1966, **70**, 1924
66Ma W. L. Marshall and R. Slusher, *J. Phys. Chem.*, 1966, **70**, 4015
66Mb W. L. Marshall and E. V. Jones, *J. Phys. Chem.*, 1966, **70**, 4028
66Mc G. A. Mountford and P. A. H. Wyatt, *Trans. Faraday Soc.*, 1966, **62**, 3201
66N L. Nilsson and G. P. Haight, jun., *Acta Chem. Scand.*, 1966, **20**, 486
66Na B. I. Nabivanets and E. A. Mazurenko, *Radiokhimiya*, 1966, **32**, 739
66O J. Osugi, K. Shimizu, and H. Takisawa, *J. Chem. Soc. Japan*, 1966, **87**, 1166
66Q A. S. Quist and W. L. Marshall, *J. Phys. Chem.*, 1966, **70**, 3714
66S K. Stopperka, *Z. anorg. Chem.*, 1966, **344**, 263
66Sa J. Sobkowski, *Roczniki Chem.*, 1966, **40**, 271
66Sb T. Sekine, M. Sakairi, and Y. Hasegawa, *Bull. Chem. Soc. Japan*, 1966, **39**, 2141
66Sc K. Stopperka, *Z. anorg. Chem.*, 1966, **345**, 264, 277
66T N. Tanaka, K. Ogino-Ebata, and G. Satô, *Bull. Chem. Soc. Japan*, 1966, **39**, 366
66V V. M. Vdovenko, L. N. Lazarev, and Ya. S. Khvorostin, *Radiokhimiya*, 1966, **30**, 673
66W J. B. Walker and C. B. Monk, *J. Chem. Soc.* (*A*), 1966, 1372

continued overleaf

56 SO_4^{2-} (contd)

66Wa R. M. Wallace, *J. Phys. Chem.*, 1966, **70**, 3922
67A G. Atkinson and S. K. Kor, *J. Phys. Chem.*, 1967, **71**, 673
67C R. G. de Carvalho and G. R. Choppin, *J. Inorg. Nuclear Chem.*, 1967, **29**, 725, 737
67Ca S. Chakraborti, *J. Indian Chem. Soc.*, 1967, **44**, 554
67D B. Dacre, *Trans. Faraday Soc.*, 1967, **63**, 604
67E E. A. V. Ebsworth and R. G. Hughes, *J. Inorg. Nuclear Chem.*, 1967, **29**, 1799
67Ea A. N. Ermakov, I. N. Marov, and G. A. Evtikova, *Zhur. neorg. Khim.*, 1967, **12**, 3372
67F L. S. Frankel, T. R. Stengle, and C. H. Langford, *J. Inorg. Nuclear Chem.*, 1967, **29**, 243
67Fa F. H. Fisher and D. F. Davis, *J. Phys. Chem.*, 1967, **71**, 819
67Fb F. Fittipaldi and S. Petrucci, *J. Phys. Chem.*, 1967, **71**, 3414
67G R. J. Gillespie and K. C. Malhotra, *J. Chem. Soc.* (*A*), 1967, 1994
67Ga H. Gesell and D. Neuschütz, *Z. anorg. Chem.*, 1967, **354**, 172
67H H. C. Helgeson, *J. Phys. Chem.*, 1967, **71**, 3121
67J Z. Jerman, *Coll. Czech. Chem. Comm.*, 1967, **32**, 260
67K R. T. Kolarich, V. A. Ryan, and R. P. Schuman, *J. Inorg. Nuclear Chem.*, 1967, **29**, 783
67Ka I. V. Kolosov, B. N. Ivanov-Emin, L. G. Korotaeva, and Kh. Tetsu, *Radiokhimiya*, 1967, **9**, 473
67Kb O. I. Kachurin, *Zhur. fiz. Khim.*, 1967, **41**, 2245
67L M. H. Lietzke and J. O. Hall, *J. Inorg. Nuclear Chem.*, 1967, **29**, 1249
67La S. J. Lyle and S. J. Naqvi, *J. Inorg. Nuclear Chem.*, 1967, **29**, 2441
67Lb R. Larsson, *Acta Chem. Scand.*, 1967, **21**, 1081
67M W. L. Marshall, *J. Phys. Chem.*, 1967, **71**, 3584
67Ma J. Masłowska, *Roczniki Chem.*, 1967, **41**, 1857
67Mb I. F. Mavrin, F. Ya. Kul'ba, and V. E. Mironov, *Zhur. neorg. Khim.*, 1967, **12**, 324
67Mc S. K. Mishra and Y. K. Gupta, *J. Inorg. Nuclear Chem.*, 1967, **29**, 1643
67Md I. N. Marov, Yu. N. Dubrov, V. K. Belyaeva, A. N. Ermakov, and A. A. Khromov, *Zhur. neorg. Khim.*, 1967, **12**, 1529
67N G. M. Nair, C. L. Rao, and G. A. Welch, *Radiochim. Acta*, 1967, **7**, 77
67Na F. S. Nakayama and B. A. Rasnick, *Analyt. Chem.*, 1967, **39**, 1022
67P S. Petrucci, P. Hemmes, and M. Battistini, *J. Amer. Chem. Soc.*, 1967, **89**, 5552
67S T. Sekine and M. Sakairi, *Bull. Chem. Soc. Japan*, 1967, **40**, 261
67T H. Tsubota and G. Atkinson, *J. Phys. Chem.*, 1967, **71**, 1131
67Ta N. Tanaka, Y. Kobayashi, and M. Kamada, *Bull. Chem. Soc. Japan*, 1967, **40**, 2839
67Tb N. Tanaka and A. Yamada, *Z. analyt. Chem.*, 1967, **224**, 117
67V V. M. Vdovenko, L. N. Lazarev, and Ya. S. Khvorostin, *Radiokhimiya*, 1967, **9**, 460, 464
67Va V. M. Vdovenko, L. N. Lazarev, and Ya. S. Khvorostin, *Zhur. neorg Khim.*, 1967, **12**, 1152
67W R. M. Wallace, *J. Phys. Chem.*, 1967, **71**, 1271
67Wa D. M. Wrench and D. Inman, *Electrochim. Acta*, 1967, **12**, 1601
68A S. Ahrland and L. Brandt, *Acta Chem. Scand.*, 1968, **22**, 1579
68Aa A. Aziz, S. J. Lyle, and S. J. Naqvi, *J. Inorg. Nuclear Chem.*, 1968, **30**, 1013
68Ab A. Aziz and S. J. Lyle, *J. Inorg. Nuclear Chem.*, 1968, **30**, 3223
68C J. H. R. Clarke and L. A. Woodward, *Trans. Faraday Soc.*, 1968, **64**, 1041
68Ca R. Christova and C. Stefanova, *Z. anorg. Chem.*, 1968, **361**, 209
68F J. E. Finholt, K. Caulton, K. Kimball, and E. Uhlenhopp, *Inorg. Chem.*, 1968, **7**, 610
68G A. V. Gordievskii, E. L. Filippov, V. S. Shterman, and A. S. Krivoshein, *Zhur. fiz. Khim.*, 1968, **42**, 1998
68H P. Hemmes and S. Petrucci, *J. Phys. Chem.*, 1968, **72**, 3986
68K I. M. Kolthoff and M. K. Chantooni, jun., *J. Amer. Chem. Soc.*, 1968, **90**, 5961
68L L. N. Lazarev and Ya. S. Khvorostin, *Zhur. neorg. Khim.*, 1968, **13**, 598
68M R. A. Matheson, *J. Phys. Chem.*, 1968, **72**, 3330
68Ma T. Mitsuji, *Bull. Chem. Soc. Japan*, 1968, **41**, 115
68Mb V. I. Muscat and J. B. Walker, *J. Inorg. Nuclear Chem.*, 1968, **30**, 2765
68Mc V. E. Mironov, Yu. A. Makashev, I. Ya. Mavrina, and D. M. Markhaeva, *Zhur. fiz. Khim.*, 1968, **42**, 2987
68Md F. Monacelli, *Inorg. Chim. Acta*, 1968, **2**, 263
68Me W. L. Marshall and R. Slusher, *J. Chem. and Eng. Data*, 1968, **13**, 83
68N S. Nishimura, R. Mitamura, Y. Kondo, and N. C. Li, *J. Inorg. Nuclear Chem.*, 1968, **30**, 3033
68Na G. M. Nair, *Radiochim. Acta*, 1968, **10**, 116
68P B. Prasad, *J. Indian Chem. Soc.*, 1968, **45**, 1037
68Pa Ts. V. Pevzner and I. A. Sheka, *Zhur. neorg. Khim.*, 1968, **13**, 2681
68R G. Ritzert and E. U. Franck, *Ber. Bunsengesellschaft Phys. Chem.*, 1968, **72**, 798
68S C. J. Shahani, K. A. Mathew, C. L. Rao, and M. V. Ramaniah, *Radiochim. Acta*, 1968, **10**, 165
68W C. F. Wells and M. A. Salam, *J. Chem. Soc.* (*A*), 1968, 308

Amidosulphate ("sulphamate") $H_2NSO_3^-$ 57

Metal	Method	Temp	Medium	Log of equilibrium constant, remarks	Ref
H^+	cal	25	0 corr	$\Delta H_1 = -0{\cdot}25$, $\Delta S_1 = 3{\cdot}8$	65H+52K
	gl	20	dil	K_1 1·19 in H_2O, 1·62 in 44% Me_2CO, 1·83 in 51% dioxan, 2·56 in 49% Me_2NCHO	65Ha
	sol	25	0 corr	$K_{s0}(H^+L^-)$ −0·92, $\beta(HL^0L^-)$ −0·52 [$\beta(H^+L^-)$ 1·0]	68C
V^{5+}	var	25	var	no ev cpx (gl, pol, sp)	67T
Mo^{6+}, W^{6+}	var	25	var	no ev cpx (gl, pol, sp)	67T
UO_2^{2+}	var	25	var	no ev cpx (gl, pol, sp)	67T
Pu^{3+}	sp	23	2 $HClO_4$	no ev cpx	68C
Pu^{4+}	sp	23	2·1 ($LiClO_4$)	K_1 0·11, ev cpx PuL^{3+}–HL	68C
PuO_2^{2+}	sp	23	2 $HClO_4$	no ev cpx	68C
Co^{3+}	gl	0—25	0·004	$K_a[Co(NH_3)_5L^{2+}]$ −6·40(0°), −6·03(13°), −5·83(25°)	68P
	gl	25	1 $NaClO_4$	$K_a[Co(NH_3)_5L^{2+}]$ −5·70	

65H H. P. Hopkins, jun., C.-H. Wu, and L. G. Hepler, *J. Phys. Chem.*, 1965, **69**, 2244
65Ha M. K. Hargreaves, E. A. Stevinson, and J. Evans, *J. Chem. Soc.*, 1965, 4582
67T O. Tubertini, M. G. Bettoli, and G. Bertocchi, *Ann. Chim. (Italy)*, 1967, **57**, 555
68C J. M. Cleveland, *Inorg. Chem.*, 1968, **7**, 874
68P L. L. Po and R. B. Jordan, *Inorg. Chem.*, 1968, **7**, 526

Other ligands with S 58

Metal	Method	Temp	Medium	Log of equilibrium constant, remarks	Ref
				Dithionite, $S_2O_4^{2-}$	
	esr	25	0·1 NaOH	$K(2SO_2^- \rightleftharpoons S_2O_4^{2-})$ 9·2	64L
	esr	25—60	var	$K(2SO_2^- \rightleftharpoons S_2O_4^{2-})$ 9·29(25°), 9·04(30°), 8·52(40°), 8·08(50°), 7·66(60°), $\Delta H = -10{\cdot}3$	67B
Mn^{2+}	esr		var	K_1^{out} 1·37, $K(Mn^{2+}L^{2-} \rightleftharpoons MnL)$ 0·72	66B
				Sulphur dioxide, SO_2	
Co^{2+}	var		var	ev $SO_2[Co(CN)_5]_2^{6-}$ (pre, ir, sp)	66V
Ru^{2+}	*X*, ir		solid	ev $Ru(NH_3)_4SO_2Cl^+$, $Ru(NH_3)_5SO_2^{2+}$	65V
				Hexathionate, $S_6O_6^{2-}$	
	X		solid	ev $O_3SSSSSSO_3^{2-}$ in $K_2Ba(S_6O_6)_2$(s)	65F
				Sulphamide, $SO_2(NH_2)_2$	
Cu^{2+}	pre, ir		var, solid	ev very weak cpx	65O
				Imidodisulphuryl fluoride, imidobis (fluorosulphate), $N(SO_2F)_2^-$	
H^+	gl	25	var	K_1 1·28	65R
				Hydroxylamine-O-sulphonate, hydroxylamidosulphate, $H_2NOSO_3^-$	
H^+	gl	45	1 $NaClO_4$	K_1 1·48	65C

continued overleaf

58 Other ligands with S (contd)

Metal	*Method*	*Temp*	*Medium*	*Log of equilibrium constant, remarks*	*Ref*
				Peroxomonosulphate, SO_5^{2-}	
H^+	gl	19	→0	K_1 9·86	65K
D^+	gl	19	D_2O	K_1 10·40 or 10·42 in D_2O, →0	65K
				Peroxodisulphate, $S_2O_8^{2-}$	
K^+	*E*	25—39	0·1 (Me_4NCl)	K_1 0·52(25°), 0·56(39°), K^+ glass electrode	66C
			0 corr	K_1 0·92(25°), 0·97(39°), $\Delta H_1 = 2$, $\Delta S_1 = 10$	
Cr^{3+}	kin	60	0·5H^+, 1 ($NaClO_4$)	ev cpx	66F
				Fluorosulphate, SO_3F^-	
Br^{3+}	con		HSO_3F	ev BrL_3, BrL_2^+ in HSO_3F(l)	68Ga
I^{3+}	con	−78	HSO_3F	$K_3(IL_2^+ + L^- \rightleftharpoons IL_3)$ 5·68 in HSO_3F(l), *m* units	66G
		25	HSO_3F	K_3 4·54, K_4 1·0 in HSO_3F(l), *m* units	
	con	−86·4	HSO_3F	K_4 1·7 in HSO_3F(l), *m* units	68G

64L S. Lynn, R. G. Rinker, and W. H. Corcoran, *J. Phys. Chem.*, 1964, **68,** 2363
65C J. P. Candlin and R. G. Wilkins, *J. Amer. Chem. Soc.*, 1965, **87,** 1490
65F O. Foss and K. Johnsen, *Acta Chem. Scand.*, 1965, **19,** 2207
65K J. R. Kyrki, *Suomen Kem.*, 1965, **38,** *B*, 203
65O A. Ouchi and T. Moeller, *J. Inorg. Nuclear Chem.*, 1965, **27,** 1635
65R J. K. Ruff, *Inorg. Chem.*, 1965, **4,** 1446
65V L. H. Vogt, jun., J. L. Katz, and S. E. Wiberley, *Inorg. Chem.*, 1965, **4** 1157
66B L. Burlamacchi, E. Ferroni, and E. Tiezzi, *Ricerca sci.*, 1966, **36,** 841
66C R. W. Chlebek and M. W. Lister, *Canad. J. Chem.*, 1966, **44,** 437
66F S.-Å. Frennesson and S. Fronaeus, *Acta Chem. Scand.*, 1966, **20,** 2811
66G R. J. Gillespie and J. B. Milne, *Inorg. Chem.*, 1966, **5,** 1236
66V A. A. Vlček and F. Basolo, *Inorg. Chem.*, 1966, **5,** 156
67B L. Burlamacchi, G. Casini, O. Fagioli, and E. Tiezzi, *Ricerca sci.*, 1967, **37,** 97
68G R. J. Gillespie, J. B. Milne, and M. J. Morton, *Inorg. Chem.*, 1968, **7,** 2221
68Ga R. J. Gillespie and M. J. Morton, *Chem. Comm.*, 1968, 1565

59 Selenide, Se^{2-}

Metal	*Method*	*Temp*	*Medium*	*Log of equilibrium constant, remarks*	*Ref*
M	ΔG, est	25	0	K_{s0} −26(Fe^{2+}), −49(Cu^{2+}), −31(Zn^{2+}), −35(Tl^+), −38(Pb^{2+})	52L
	ΔG, est	25	0	K_{s0} −11·5(Mn^{2+}), −26·0(Fe^{2+}), −31·2(Co^{2+}), −32·7(Ni^{2+}), −73·4(Pd^{2+}), −81·4(Pt^{2+}), −60·8(Cu^+), −48·1(Cu^{2+}), −63·7(Ag^+), −29·4(Zn^{2+}), −35·2(Cd^{2+}), −64·5(Hg^{2+}), −33·1(Tl^+), −38·4(Sn^{2+}), −42·1(Pb^{2+})	64B
W^{6+}	*X*		solid	ev WSe_4^{2-} in $(NH_4)_2WSe_4$(s)	68M

52L W. M. Latimer, "Oxidation Potentials," 2nd edn., Prentice-Hall, New York, 1952
64B E. A. Buketov, M. Z. Ugorets, and A. S. Pashinkin, *Zhur. neorg. Khim.*, 1964, **9,** 526
68M A. Müller, B. Krebs, and H. Beyer, *Z. Naturforsch.*, 1968, **23b,** 1537

Metal	Method	Temp	Medium	Log of equilibrium constant, remarks	Ref
H^+	con, tp		$MeNO_2$	ev H_3L^+ in $MeNO_2$(l)	39A
	tp		HCl var	no ev cat cpx in $\leqslant$ 6 M HCl	64G
	Ram		solid	ev $H_3L^+ = Se(OH)_3{}^+$ in $Se(OH)_3ClO_4$(s)	65P
	gl	25	3 K(Cl)	K_1 8·24, K_{12} 2·70, $\beta(H^+HL_2{}^{3-})$ 7·94, $\beta(H^+H_2L_2{}^{2-})$ 3·51, $\beta(H^+H_3L_2{}^-)$ 1·93, $K(2H_2L \rightleftharpoons H_4L_2)$ 0·00, (K_w − 14·17)	66S
		25	3 Na(Cl)	K_1 8·00, K_{12} 2·58, ev polyn cpx (K_w − 14·07)	
	gl	25	0 corr	K_1 8·50, K_{12} 2·75	67K
			ROH	K_1 9·19, K_{12} 2·96 in 23·1 mole% MeOH; K_1 9·95, K_{12} 3·27 in 51·3 mole% MeOH; K_1 9·06, K_{12} 2·98 in 16·9 mole% EtOH; K_1 9·78, K_{12} 3·35 in 41·7 mole% EtOH; K_1 9·17, K_{12} 3·19 in 19·0 mole% Pr^{i}OH; K_1 9·66, K_{12} 3·59 in 35·3 mole% Pr^{i}OH; (→ 0) also K_1 and K_{12} for intermediate and lower concentrations of ROH	
	dis	25	I(Na)ClO_4	$K_D[H_2L \rightleftharpoons H_2L$ (in Bu^{i}CHOHMe)] 0·56 (I = 3), 0·21(I = 0·1) K_{12} 2·24(I = 3), 2·33(I = 0·1)	68S
Group $2A^{2+}$	con	18	dil	K_{s0} − 8·0(Be^{2+}), − 5·74(Mg^{2+}, Ca^{2+}), − 5·10(Sr^{2+}), − 5·21(Ba^{2+})	68R
Mg^{2+}	sol, gl	20	0 corr	K_{s0} − 5·36, $MgSeO_3(H_2O)_6$(s)	56C/66Lb
Sr^{2+}	sol	25	dil	K_{s0} − 5·36 (dil, sat)	65L
Ba^{2+}	sol	25	0 corr	K_{s0} − 6·57(0 corr), − 6·37(dil, sat)	65L
Mn^{2+}	sol, gl	20	0 corr	K_{s0} − 7·27	57C/66La
Fe^{3+}	sp	20—40	1 (Na, $HClO_4$)	$K(Fe^{3+} + H_2L \rightleftharpoons FeHL^{2+} + H^+)$ 0·51(20°), 0·54(25°), 0·56(30°), 0·61(40°), ΔH = 4·0	65H
Co^{2+}	sol, gl	20	0 corr	K_{s0} − 7·08	57C/64S
	sol, gl	?	$I(KNO_3)$	K_{s0} − 7·93, β_2 3·25 (I = 0·3); K_{s0} − 7·94(I = 0·01); solid = $CoSeO_3(H_2O)_2$	66P
	con	18	dil	K_{s0} − 6·93	68R
Co $en_3{}^{3+}$	oth	20	2 ($NaClO_4$)	K_1 2·11, β_2 3·92, β_3 4·46, β_4 5·15 (circular dichroism)	66L
	sp		2·88 Na(ClO_4)	K_1 0·08	67O
Ni^{2+}	con	18	dil	K_{s0} − 5·29	68R
Cu^{2+}	sol, gl	20	0 corr	K_{s0} − 7·78	56C/65La
	con	18	dil	K_{s0} − 7·49	68R
Zn^{2+}	con	18	dil	K_{s0} − 7·72	68R
Cd^{2+}	con	18	dil	K_{s0} − 8·30	68R

39A E. J. Arlman, *Rec. Trav. chim.*, 1939, 58, 871
64G D. S. Gaibakyan and M. V. Darbinyan, *Izvest. Akad. Nauk Armyan. S.S.R., khim. Nauk*, 1964, **17**, 501
64S N. M. Selivanova, Z. L. Leshchinskaya, A. I. Maier, and E. Yu. Muzalev, *Izvest. V.U.Z., Khim. i khim. Tekhnol.*, 1964, **7**, 209
65H S. Hamada, Y. Ishikawa, and T. Shirai, *J. Chem. Soc. Japan.*, 1965, **86**, 1042
65L Z. L. Leshchinskaya and N. M. Selivanova, *Zhur. fiz. Khim.*, 1965, **39**, 2430
65La Z. L. Leshchinskaya, M. A. Averbukh, and N. M. Selivanova, *Zhur. fiz. Khim.*, 1965, **39**, 2036
65P R. Paetzold and M. Garsoffke, *Z. anorg. Chem.*, 1965, 336, 52
66L R. Larsson, S. F. Mason, and B. J. Norman, *J. Chem. Soc.*, 1966, 301
66La Z. L. Leshchinskaya and N. M. Selivanova, *Izvest. V.U.Z. Khim i khim. Tekhnol.*, 1966, **9**, 523

continued overleaf

60 SeO_3^{2-} (contd)

66Lb Z. L. Leshchinskaya and N. M. Selivanova, *Zhur. neorg. Khim.*, 1966, **11**, 260
66P I. V. Pyatnitskii and M. Durdyev, *Ukrain. khim. Zhur.*, 1966, **32**, 77
66S R. Sabbah and G. Carpéni, *J. Chim. phys.*, 1966, **63**, 1549
67K C. Th. Kawassiades, G. E. Manoussakis, and J. A. Tossidis, *J. Inorg. Nuclear Chem.*, 1967, **29**, 401
67O I. Olsen and J. Bjerrum, *Acta Chem. Scand.*, 1967, **21**, 1112
68R R. Ripan and G. Vericeanu, *Studia Univ. Babeş-Bolyai, Ser. Chem.*, 1968, **13**, 31
68S T. Sekine, H. Iwaki, M. Sakairi, F. Shimada, and M. Inarida, *Bull. Chem. Soc. Japan*, 1968, **41**, 1

61 Selenate, SeO_4^{2-}

Metal	*Method*	*Temp*	*Medium*	*Log of equilibrium constant, remarks*	*Ref*
H^+	qh	25	0 corr	K_1 1·92?	50P/65C
	Ram		H_2SeO_4	ev H_2SeO_4, $HSeO_4^-$, SeO_4^{2-} in H_2SeO_4–H_2O mixt, equil similar to H_2SO_4	63W
	var		solid	ev $Se_3O_{10}^{2-}$ in $(NO_2^+)_2Se_3O_{10}^{2-}$(s) (pre, ir, *X* powder)	64K
	H, Ag	0—45	0 corr	K_1 1·36(0°), 1.46(10°), 1.52(15°), 1·59(20°), 1·66(25°), 1·73(30°), 1.82(35°), 1·89(40°), 1·96(45°), $\Delta H_1 = 5{\cdot}5$, $\Delta S_1 = 26{\cdot}3$(25°)	64N
		25	0 corr	K_1 1·70 to 1·78	64N/65C
	ir, Ram		solid	ev $Se_2O_7^{2-}$ in solids	65P
	con, Ram	25	H_2L conc	ev H_3O^+, $H_3SeO_4^+$, $(SeO_3)_4$, $H_2Se_3O_{10}$, $H_2Se_2O_7$, H_2SeO_4, $HSe_2O_7^-$, in H_2SeO_4(l)	66P
Sr^{2+}	sol	18	0 corr	K_{s0} −4·37	63S
Ba^{2+}	pol	25	0 corr	K_{s0} −7·30	53S
Sc^{3+}	cix	25—45	0·5 ($NaClO_4$)	K_1 1·78(25°), 1·75(35°), 1·68(45°); β_2 2·64(25°), 3·15(35°), 3·33(45°); $\Delta H_1 = -2{\cdot}3$, $\Delta H_{\beta 2} = 39{\cdot}2$, $\Delta S_1 = 0{\cdot}5$, $\Delta S_{\beta 2} = 126{\cdot}7$	67K
Fe^{2+}	kin	0—35	4 ($NaClO_4$)	K_1 1·08(25°), 0·71(35°); $\beta(Fe^{2+}HL^-)$ 0·68(0°), 0·64(5°), 0·54(10°), 0·23(25°)	68W
		0—10	1 ($NaClO_4$)	K_1 1·73(0°), 1·40(10°)	
Hg^{2+}	dis	20	H_2L var	ev HgL_2^{2-}	66B

53S N. M. Selivanova and A. F. Kapustinskii, *Zhur. fiz. Khim.*, 1953, **27**, 565
63S E. Schumann, dissertation, Technische Hochschule, Karlsruhe, 1963
63W G. E. Walrafen, *J. Chem. Phys.*, 1963, **39**, 1479
64K G. Kempe, M. Lorenz, and H. Thieme-Schneider, *Wiss. Z. Tech. Hochschule Chem. Leuna-Merseburg*, 1964, **6**, 272
64N V. S. K. Nair, *J. Inorg. Nuclear Chem.*, 1964, **26**, 1911
65C A. K. Covington and J. V. Dobson, *J. Inorg. Nuclear Chem.*, 1965, **27**, 1435
65P R. Paetzold, H. Amoulong, and A. Růžička, *Z. anorg. Chem.*, 1965, **336**, 278
66B J. I. Bullock and D. G. Tuck, *J. Inorg. Nuclear Chem.*, 1966, **28**, 1103
66P R. Paetzold and H. Amoulong, *Z. anorg. Chem.*, 1966, **343**, 70
67K I. V. Kolosov, B. N. Ivanov-Emin, L. G. Korotaeva, and Kh. Tetsu, *Radiokhimiya*, 1967, **9**, 473
68W C. F. Wells and M. A. Salam, *J. Chem. Soc. (A)*, 1968, 308

Telluride, Te^{2-} 62

Metal	Method	Temp	Medium	Log of equilibrium constant, remarks	Ref
H^+	pol	25	var	K_1 12·16 (misprint 12·3 in summary)	63P
M	ΔG, est	25	0	K_{s0} −42(Cd^{2+}), −33·5(Tl^+), −48(Pb^{2+})	52L
	ΔG, est	25	0	K_{s0} −15·9(Mn^{2+}), −30·0(Fe^{2+}), −37·4(Co^{2+}), −38.1(Ni^{2+}), −79(Pd^{2+}), −87(Pt^{2+}), −62.3(Cu^+), −53·8(Cu^{2+}), −71.7(Ag^+), −33·3(Zn^{2+}), −41·5(Cd^{2+}), −69·6(Hg^{2+}), −39·2(Tl^+), −44·7(Sn^{2+}), −46·3(Pb^{2+})	64B

52L W. M. Latimer, "Oxidation Potentials," 2nd edn., Prentice-Hall, New York, 1952
63P A. J. Panson, *J. Phys. Chem.*, 1963, **67**, 2177
64B E. A. Buketov, M. Z. Ugorets, and A. S. Pashinkin, *Zhur. neorg. Khim.*, 1964, **9**, 526

Tellurite, TeO_3^{2-} 63

Metal	Method	Temp	Medium	Log of equilibrium constant, remarks	Ref
H^+	tp		HCl var	ev cat cpx (TeO^{2+}) in > 2M-HCl	64G
	cal	25?	dil	$\Delta H[H_2TeO_3H_2O(s) + 2OH^- \rightleftharpoons L^{2-}] = -7·08$	64Ga
	dis	25	3·1	ev $Te(OH)_3^+$ = $TeOOH^+$ for $[H^+]$ = 0·1—2M	65M
	ir, *X*		solid	ev $Te_2O_3^{2+}$ in $Te_2O_3SO_4(s)$	66L
	sol, ix	18	0·5 ($NaClO_4$)	$K_{s12}[H_2TeO_3(s) \rightleftharpoons H_2L]$ −3·15 $K_{13}(H_2L + H^+ \rightleftharpoons H_3L^+)$ 1·68	68N
	dis	25	3(Na)ClO_4	$K_D[H_2L \rightleftharpoons H_2L(\text{in org})]$ 0·39, K_{12} 3·07, $K_{13}(H_2L + H^+ \rightleftharpoons H_3L^+)$ 4·17,	68S
		25	0·1 (Na)ClO_4	K_D −0·08, K_{12} 2·46, K_{13} 3·16, org = 0·2M-$(C_8H_{17})_3PO$ in hexane	
Ag^+	gl, sol	25	0 corr	$K_{s0}(Ag^+{}_2L^{2-})$ −2·43 or $K_s(Ag^+{}_2H^+{}_2L^{2-}{}_2)$ −7·95; $K_s(Ag^+{}_7OH^-{}_5L^{2-})$ −5·6 (?) (difficult to follow calculation)	65G

64G D. S. Gaibakyan and M. V. Darbinyan, *Izvest. Akad. Nauk Armyan. S.S.R., khim. Nauk*, 1964, **17**, 501
64Ga E. Sh. Ganelina, B. P. Troitskii, and V. S. Yakovleva, *Zhur. priklad. Khim.*, 1964, **37**, 677
65G E. Sh. Ganelina and T. N. Pozhidaeva, *Zhur. priklad. Khim.*, 1965, **38**, 2210
65M H. Mabuchi and I. Okada, *Bull. Chem. Soc. Japan*, 1965, **38**, 1478
66L J. Loub, *Z. Chem.*, 1966, **6**, 425
68N B. I. Nabivanets and E. E. Kapantsyan, *Zhur. neorg. Khim.*, 1968, **13**, 1817
68S T. Sekine, H. Iwaki, M. Sakairi, F. Shimada, and M. Inarida, *Bull. Chem. Soc. Japan*, 1968, **41**, 1

Tellurate, TeO_4^{2-} or $TeO_2(OH)_4^{2-}$ 64

Metal	Method	Temp	Medium	Log of equilibrium constant, remarks	Ref
H^+	*X*		solid	ev TeO_6^{6-} in $Hg_3TeO_6(s)$	63F
	X		solid	ev $TeO(OH)_5^-$ in $KTeO(OH)_5H_2O(s)$	64R
	X		solid	ev $Te_2O_6(OH)_4^{4-}$ in "$K_4H_4Te_2O_{10}(H_2O)_7(s)$"	65L = 66L
	gl	25	1 Na(Cl)	$K_{12}(H^+ + HL^- \rightleftharpoons H_2L)$ 7·32, $K(H^+ + H_3L_2^- \rightleftharpoons 2H_2L)$ 6·25, $K(2H^+ + H_2L_2^{2-} \rightleftharpoons 2H_2L)$ 13·23	66B

continued overleaf

64 Tellurate, TeO_4^{2-} or $TeO_2(OH)_4^{2-}$ (contd)

Metal	Method	Temp	Medium	Log of equilibrium constant, remarks	Ref
Mg^{2+}	sol	20	dil	$K_s(Mg^{2+}{}_3TeO_6^{6-})$ −16·6(?, not corrected for reactions with H^+?)	66K

1964 *Erratum*: p. 255, ref. 61L, *for* "MnL_2^{4-}" *read* "$MnL_2(OH)_4^{4-}$", *and for* "MnL_3OH^{7-}" *read* "$MnL_3(OH)_5^{7-}$"

Metal	Method	Temp	Medium	Log of equilibrium constant, remarks	Ref
Ni^{3+}	pre		solid	ev $Na_6NiL_3(OH)_3$(s)	65La
Pd^{4+}	pre		solid	ev $Na_5PdL_2(OH)_5$(s)	65La
Pt^{4+}	pre		solid	ev $Na_5PtL_2(OH)_5$(s)	65La

63F M. T. Falqui, *Ricerca sci.*, 1963, **3**, *A*, 627
64R S. Raman, *Inorg. Chem.*, 1964, **3**, 634
65L O. Lindqvist and F. Wengelin, *Acta Chem. Scand.*, 1965, **19**, 1510
65La M. W. Lister and P. McLeod, *Canad. J. Chem.*, 1965, **43**, 1720
66B F. Brito, *Anales de Quím.*, 1966, **62**, *B*, 197
66K R. N. Knyazeva, G. N. Chernova, and G. B. Zhukovskaya, *Izvest. V.U.Z., Khim. i khim. Tekhnol.*, 1966, **9**, 869
66L O. Lindqvist and G. Lundgren, *Acta Chem. Scand.*, 1966, **20**, 2138

65 Other ligands with 6B elements

No new data compiled

66 Fluoride, F^-

Metal	Method	Temp	Medium	Log of equilibrium constant, remarks	Ref
H^+	*X*		solid	ev HF_2^- in $NaHF_2$(s)	26A
	lit	25	0·5 ($NaClO_4$)	$\Delta H_1 = 2{\cdot}93$, $\Delta S_1 = 23{\cdot}1$	56C+/67Aa
	con	25—200	0 corr	K_1 3·18(25°), 3·40(50°), 3·64(75°), 3·85(100°), 4·09(125°), 4·34(150°), 4·58(175°), 4·89(200°); K_2 0·53(25°), 0·60(50°), 0·67(75°), 0·68(100°), 0·69(125°), 0·76(150°, 175°), 0·90(200°)	63E
	con	25	→ 0	K_1 3·21	63Eb
	con	25	→ 0	K_1 3·21, ev other cpx	63Ec
	con	−17	ice	K_1 8·1, in ice, M units	63La
	gl	20	0·1 (KNO_3)	K_1 2·85	63Sa
	dis	20	HF var	$K_D[HF + T(org) \rightleftharpoons HFT(org)]$ −0·5, T = TBP, org = kerosene	64B
	qh	25	3 Na(ClO_4)	K_1 3·29, K_2 0·86	64F
			1 Na(ClO_4)	K_1 2·95, K_2 0·58	
	dis	25	→ 0	$K_D[HF \rightleftharpoons HF(\text{in BuOH})]$ 0·26	64Kd
	qh	25	1 ($NaClO_4$)	K_1 2·96, K_2 0·59	65C
	dis	20	HF var	$K_D[nHF + T(\text{in org}) \rightleftharpoons (HF)_nT(\text{in org})]$ −0·40(n = 1), −0·66(n = 2), −2·4(n = 3), −3(n = 4), T = TBP, org = *o*-xylene	66Ka
	lit	25	0 corr	$\Delta H_1 = 3{\cdot}18$, $\Delta S_1 = 25{\cdot}2$	67Aa = 53Ha
	nmr		→ 0	K_2 0·47 (criticized in ref 68Sg)	67H
	E	25	→ 0	K_1 3·19 (LaF_3 membrane electrode)	68Va

continued

Metal	Method	Temp	Medium	Log of equilibrium constant, remarks	Ref
H^+ (contd)	ana	500—700	MF(l)	Values for $K[2F^- + H_2O(g) \rightleftharpoons O^{2-} + 2HF(g)]$ and $K[F^- + H_2O(g) \rightleftharpoons OH^- + HF(g)]$ in $Li_{1-x}Be_xF_{1+x}(l)$	68M
Group $1A^+$	dis	25	$\rightarrow 0$	$K_D[M^+ + F^- \rightleftharpoons M^+ \text{ (in BuOH)} + F^- \text{ (in BuOH)}]$ 1·81(Li^+), 3·84(Na^+), 3·86(K^+) 3·53(Cs^+), 3·69(NH_4^+)	64Kd
	E	25	1 ($NaNO_3$)	K_1 2·90, K_2 0·77 (F^- membrane electrode)	68Sc
Li^+	cal, sol	25	0 corr	K_{s0} −2·77, ΔH_{s0} = 1·07	/64S
Be^{2+}	red	25	0·5 ($NaClO_4$)	ΔH_1 = −0·4, ΔS_1 = 22	55Y/67Aa
		25	0 corr	ΔH_1 = −0·2, ΔS_1 = 27	
	red, qh	25	0·5 (NH_4ClO_4)	K_1 4·71, K_2 3·61, K_3 2·80, K_4 2·27, [$K_1(H^+)$ 2·91]	65B
	con, gl		var	ev $HBeF_4^-$, BeF_4^{2-}	65G
	Be	680—860	MCl(l)	ev BeF_3^-, BeF_4^{2-} in (Na, K)Cl(l), β_4 and β_3 given	65S
	cix		0·16 ($NaNO_3$)	K_1 3·64, β_2 5·90	66Pc
	aix		0·16 ($NaNO_3$)	β_2 5·93, β_3 7·76, β_4 9·12	
	est		0	K_1^{out} 0·18	68Bb
	nmr	2—50	var	K_4 1·1(2—50°), $\Delta H_4 \approx 0$, ΔS_4 = 5	68F
Sc^{3+}	red	25	0·5 ($NaClO_4$)	ΔH_1 = 0·6, ΔS_1 = 30	55P/67Aa
	con, aix		var	ev ScF_6^{3-}	66Ia
Y^{3+}	red	25	0·5 ($NaClO_4$)	ΔH_1 = 2·2, ΔS_1 = 25	61P/67Aa
		25	0 corr	ΔH_1 = 2·3, ΔS_1 = 30	
	X		solid	ev YF_4^- in $LiYF_4$(s)	61Td = 65K
	red, cal	25	1 ($NaClO_4$)	K_1 3·60, ΔH_1 = 8·32, ΔS_1 = 44·5	67W
Group $4f^{3+}$	X		solid	ev MF_4^- in $LiMF_4$(s), M = Eu, Gd . . . Lu,	65K
	sol	25	NH_4F var	ev ani cpx	67Ia
	red	25	1 ($NaClO_4$)	K_1 2·67(La), 2·81(Ce), 3·01(Pr), 3·09(Nd), 3·12(Sm), 3·19(Eu), 3·31(Gd), 3·42(Tb), 3·46(Dy), 3·52(Ho), 3·54(Er), 3·56(Tm), 3·58(Yb), 3·61 (Lu)	67W
	dis	25	1 ($NaClO_4$)	K_1 2·72(Ce^{3+}), 3·16(Pm^{3+}), 3·20(Eu^{3+}), 3·56(Er^{3+}), 3·57(Tm^{3+}), 3·60(Yb^{3+})	
	cal	25	1 ($NaClO_4$)	ΔH_1 = 4·00(La), 4·82(Ce), 5·74(Pr), 6·83(Nd), 9·39(Sm), 9·22(Eu), 8·90(Gd), 7·51(Tb), 7·03(Dy), 7·26(Ho), 7·43(Er), 8·66(Tm), 9·56(Yb), 9·53(Lu); ΔS_1 = 25·6(La), 29·0(Ce), 33·1(Pr), 36·9(Nd), 45·8(Sm), 45·4(Eu), 45·1(Gd), 40·8(Tb), 39·4(Dy), 40·4(Ho), 41·2(Er), 45·4(Tm) 48·5(Yb), 48·5(Lu)	
		25	0	$\Delta H_1 = -2·38 + 0·246\,\Delta S_1$	67W/68Ma
Ce^{3+}	dis	24—25	0·5 ($NaClO_4$)	K_1 3·15, β_2 5·96	67La
	gl		dil	$K[CeF_n^{3-n} + H_2O \rightleftharpoons CeF_nOH^{2-n} + H^+]$ −6·5(n = 1), −6·4(n = 2)	67Sa
Eu^{3+}	dis, gl	24—25	0·5 ($NaClO_4$)	K_1 3·39, β_2 6·48 [$K_1(H^+)$ 2·91]	66L
Gd^{3+}	dis	24—25	0·5 ($NaClO_4$)	K_1 3·43, β_2 6·71	67La
	nmr	20	0·1 ($NaClO_4$)	K_1 4·04	67Vb
Dy^{3+}	red	25	0·5 ($NaClO_4$)	*K_1 0·64, K_1 3·58 [$K_1(H^+)$?]	68I
Er^{3+}	red	25	0·5 ($NaClO_4$)	*K_1 0·61, K_1 3·54 [$K_1(H^+)$?]	68I
Yb^{3+}	red	25	0·5 ($NaClO_4$)	*K_1 0·66, K_1 3·61 [$K_1(H^+)$?]	68I
Lu^{3+}	red	25	0·5 ($NaClO_4$)	*K_1 0·56, K_1 3·47 [$K_1(H^+)$?]	68I

continued overleaf

66 F⁻ (contd)

Metal	Method	Temp	Medium	Log of equilibrium constant, remarks	Ref
Ti^{3+}	sp		solid	ev TiF_6^{3-} in NaK_2TiF_6(s) *etc*	64Bb
	X		solid	ev TiF_6^{3-} in NaK_2TiF_6(s)	64Kb
Ti^{4+}	con, nmr		ROH	ev TiF_5ROH^- in ROH(l), R = Me, Et, Pr^i, Bu	63R
	nmr		HF var	ev TiF_6^{2-}, $TiF_5H_2O^-$, $Ti_2F_4(OH)_4(H_2O)_2{}^0$	67Bc
	var		var	ev TiF_6^{2-} in conc$(NH_4)_2TiF_6$ soln (ir, Ram, nmr)	67Da
TiO^{2+}	cix, aix		0·2 HCl	K_1 6·65, K_2 5·09, K_3 4·58, K_4 4·06 [$K_1(H^+)$ 3·27]	66N
Zr^{4+}	*X*		solid	ev ZrF_8^{4-} in $Li^+{}_6BeF_4{}^{2-}ZrF_8{}^{4-}$(s)	64Sb
	X		solid	ev $Zr_2F_{13}{}^{5-}$ in $K_5Zr_2F_{13}$(s)	65Hb
	qh	25	1 ($NaClO_4$)	K_4 2·8, K_5 1·9, K_6 1·35	66Bb
	var		var	ev ZrF_6^{2-} in conc$(NH_4)_2ZrF_6$ soln (ir, Ram, nmr)	67Da
	X		solid	ev $Zr_2F_{14}{}^{6-}$ in $Cu_3Zr_2F_{14}(H_2O)_{16}$(s)	67F
	X		solid	ev ZrF_8^{4-} in $Cu_2ZrF_8(H_2O)_{12}$(s)	67Fa
	red	20	4 $HClO_4$	*K_2 4·42, *K_3 3·00, *K_4 2·28, *K_5 1·53, *K_6 0·3	67Nb
	X		solid	ev $Zr_2F_{12}{}^{4-}$ in $K_2CuZr_2F_{12}(H_2O)_6$(s)	68Fa
Hf^{4+}	dis	25	3 $HClO_4$	*K_1 4·16, *β_2 7·79, *β_3 10·10, *β_4 12·04, *β_6 12·62, HfF_5^- negligible	64V
	dis	27	2 $H(ClO_4)$	*K_1 4·62	65Da
	var		var	ev HfF_6^{2-} predominant in conc$(NH_4)_2HfF_6$ and $(NH_4)_3HfF_7$ soln (ir, Ram, nmr)	67Da
	dis, red	20	4 $HClO_4$	*K_1 5·52, *K_2 4·04, *K_3 3·04, *K_4 2·20, *K_5 1·7, *K_6 0·5	67N
	cix	20	4 $HClO_4$	*K_1 5·51, *K_2 3·7?	67Na
	aix		0·5 H(Cl)	K_6 3·83	67P
Th^{4+}	sp, sol	25	var	K_1 6·0	59Ta
VO^{2+}	aix		var	ev VOF_3^- in anion exchanger	66Pa
V^{5+}	ir, Ram		HF	ev solvated species such as HVF_6 (not VF_6^-) in HF(l)	67S
VO_2^+	sp	20	1 ($NaClO_4$)	K_1 3·67, β_2 6·32 (H^+: K_1 2·94, β_2 3·52)	67I
Nb^{5+}	nmr		F^- var	ev NbF_6^-, no ev NbF_7^{2-} in soln	63P
	X		solid	ev Nb_4F_{20} units in NbF_5(s)	64E
	sol	18?	3 HNO_3	$K_s[Nb(OH)_3F_2(s) \rightleftharpoons Nb(OH)_3F_2{}^0]$ −4·82, $K[Nb(OH)_3F_2 + HF \rightleftharpoons Nb(OH)_2F_3 + H_2O]$ 4·2	67Bd
			0·5 (KNO_3)	$K_s[Nb(OH)_4F(s) \rightleftharpoons Nb(OH)_4F^0]$ −5·22, $K[Nb(OH)_4F + HF \rightleftharpoons Nb(OH)_4F_2{}^- + H^+]$ 3·64, $\beta[Nb(OH)_4F^0F^-]$ 6·8 [$K_1(H^+)$ 3·17]	
	X		solid	ev $Nb_2F_{11}^-$ in $SeNb_2F_{14}$(s)	68Ea
	X		solid	ev $(NbCl_4)_4F_4$ rings (F bridges) in $NbCl_4F$(s)	68Pc
Ta^{5+}	tp		0·5 HCl	ev uncharged cpx and TaF_7^{2-}	63N
	X		solid	ev Ta_4F_{20} units in TaF_5(s)	64E
	dis	25	3 $HClO_4$	K_4 5·86, K_4K_5 10·77, $\beta_6\beta_3^{-1}$ 15·67, $\beta_7\beta_3^{-1}$ 20·15	65V
	qh	25	1 ($NaClO_4$)	K_6 3·75, K_7 3·10, K_8 0·66, no ev TaF_9^{4-}	66Bb
	Ram		HF var, NH_4F var	ev TaF_6^-, TaF_7^{2-}, $TaOF_6^{3-}$ or $TaOF_5^{2-}$ and one more species, no ev TaF_8^{3-}	66K
			solid	ev TaF_6^-, TaF_7^{2-}, TaF_8^{3-}, $TaOF_6^{3-}$	
	X		solid	ev $Ta_4F_4Cl_{16}$ units (F bridges) in $TaCl_4F$(s)	66P
Pa^{4+}	dis		3 $H(ClO_4)$	*K_1 4·73, *β_2 8·26 [$K_1(H^+)$3·14]	65Ga
	dis	25	3 $H(ClO_4)$	*K_1 4·73, *β_2 8·26	66Gg

contin

Metal	*Method*	*Temp*	*Medium*	*Log of equilibrium constant, remarks*	*Ref*
Pa^{5+}	dis		3 ($LiClO_4$)	$\beta[Pa(OH)_3{}^{2+}(HF)_nH^+{}_{1-n}]$ 3·56(n = 1), 7·65(n = 2), 10·90(n = 3)	65Ga
	pre, *X*		solid	ev $PaF_6{}^-$ in $KPaF_6$(s) *etc.*	66A = 64Aa
	X		solid	ev $PaF_8{}^{3-}$ in K_3PaF_8(s) *etc.* [isostructural with K_3UF_8(s)]	66Ba
	qh	25	1 ($NaClO_4$)	K_1 5·4, K_2 5·0, K_3 4·9, K_4 4·8, K_5 4·5, K_6 4·4, K_7 3·7, K_8 1·7, also ev $PaF_8{}^{3-}$ (con)	66Bb
	dis	25	3 H(ClO_4)	$K(PaOOH^{2+} + HF \rightleftharpoons PaOF^{2+} + H_2O)$ 3·56 $\beta[PaOOH^{2+}(HF)_2H^+{}_{-1}]$ 7·65 $\beta[PaOOH^{2+}(HF)_3H^+{}_{-2}]$ 10·91	66Gg
	dis	25	1 $HClO_4$	$^*K_1$ 3·95, $^*K_2$ 3·48, $^*K_3$ 3·04	67K
Cr^{3+}	cix	77—95	1 ($LiClO_4$)	$^*K_1$ 1·32(77°), 1·35(86°), 1·36(95°), Δ^*H_1 = 1·3, ΔS_1 = 6	65Sa
	aix		var	ev ani cpx charge 1 −	66Pb
Cr^{3+} cpx	nmr		var	$\beta(Cr\,en_3{}^{3+}F^-)$ > 1, $\beta(Cr\,en_2Cl_2{}^+F^-)$ 0·5	65Sc
Mo^{5+}	*X*		solid	ev Mo_4F_{20} units in MoF_5(s)	62E
	X		solid	ev $MoOF_5{}^{2-}$ in $K_2MoOF_5H_2O$(s)	66Gf = 67Ga
Mo^{6+}	pre, ir		solid	ev $MoF_7{}^-$ in $NOMoF_7$(s), NO_2MoF_7(s)	63G
	var		var	ev MoO_3F^-, $MoO_2F_4{}^{2-}$, $Mo_4O_{13}F^{3-}$ (pre, con, fp, sp)	64K
	Ram, ir		5 HF	ev *cis*-$MoO_2F_3H_2O^-$ or *cis*-$MoO_2F_2(H_2O)_2$ predominant	67G
	X		solid	ev $MoO_2F_4{}^{2-}$ in $K_2MoO_2F_4H_2O$(s)	67Ga
	sp, gl		var	$K(H_2MoO_4 + F^- \rightleftharpoons MoO_3F^- + H_2O)$ 4·48 $K[H_2MoO_4 + 4F^- \rightleftharpoons MoO_2F_4{}^{2-}(+2H^+)]$ 10·58 at pH ≈ 3 ([H^+] not counted in *K* values)	67Kd
	*X*e		solid	ev $MoO_2F_4{}^{2-}$ in $K_2MoO_2F_4$(s)	68P
W^{6+}	pre, ir		solid	ev $WF_7{}^-$ in $NOWF_7$(s), NO_2WF_7(s)	63G
	Ram, ir		5 HF	ev *cis*-$WO_2F_3H_2O^-$ or *cis*-$WO_2F_2(H_2O)_2$ predominant	67G
	kin	25	0·2 (KCl)	ev cpx, apparent K_1 3·1(0·2 M-H^+), 2·9(0·05M-H^+)	64Y

1964 *Erratum*: p. 260, ref 60Sa, whole entry to read:

Metal	*Method*	*Temp*	*Medium*	*Log of equilibrium constant, remarks*	*Ref*
U^{4+}	sol	25	0·12 ($HClO_4$)	K_{s2} − 12·46, K_{s3} − 8·23, K_{s4} − 3·96, K_{s5} − 2·38, K_{s6} − 0·08; K_3 4·23, K_4 4·27, K_5 1·59, K_6 2·30 (H^+: K_1 2·97, K_2 0·59)	60Sa
U^{4+}	H	25	UF_4 dil	$K[U(OH)_{5-n}F_{n-1}{}^0 + HF \rightleftharpoons U(OH)_{4-n}F_n{}^0 + H_2O]$ 16·37(n = 1), 10·38(n = 2), 6·21(n = 3), 2·48(n = 4) see also OH^- table, ref 61Na	61N/63Ta
	nmr		2 $HClO_4$	K_1 7·15 > $K_1(Al^{3+})$ (6·13) > K_2 5·26 > $K_2(Al^{3+})$ (5·02)	63Va+
	nmr, hyp	20	2 $HClO_4$	K_1 7·15, β_2 12·4, β_3 17·7(nmr), β_4 23·1(hyp)	63Vb
	sol	25	NH_4F var	$K_s[(NH_4{}^+)_4UF_7{}^{3-}F^-]$? 2·82, *m* units	64Pa
	sp	25	7 NH_4F	ev dinuclear U–F cpx, $K(2U^{IV} \rightleftharpoons U^{IV}{}_2)$ > 0·53	
	sp, pre		1 M(ClO_4, HSO_4)	ev UF^{3+}, $UF_2{}^{2+}$, $UF_3{}^+$	66Sa
	sp, nmr	20	$NaClO_4$ var	$K_1/K_1(Al^{3+})$ 2·64(sp), $K_2/K_1(Al^{3+})$ − 0·26(sp), − 0·55(nmr) K_1 8·78, K_2 5·85 or 5·6 [$K_1(Al^{3+})$ 6·13]	66V
	X		solid	ev $UF_8{}^{4-}$ in Li_4UF_8(s)	67B
	fp, con	9	BrF_3	ev $UF_5{}^-$, $U_2F_9{}^-$, $U_5F_{22}{}^{2-}$ in BrF_3(l)	67Ma
	sol, sp	20	var	ev mainly $UF_2{}^{2+}$	67V

continued overleaf

66 F⁻ (contd)

Metal	Method	Temp	Medium	Log of equilibrium constant, remarks	Ref
U^{5+}	pre, ir		10—27 HF	ev UF_6^-	64A
	X, sp		solid	ev UF_6^- in $CsUF_6$(s)	64P
	X		solid	ev UF_6^- in KUF_6(s) etc	65Cc
U^{6+}	pre, ir		solid	ev UF_7^- in $NOUF_7$(s), NO_2UF_7(s)	63G
	X, pre		solid	ev UF_8^{2-} in Na_2UF_8(s)	66M
UO_2^{2+}	dis	25	2 ($NaClO_4$)	$\Delta H_1 \approx -2, \Delta S_1 \approx 13$	54D/67A
	red	690—830	MCl(l)	β_4 $-3{\cdot}93-4380T^{-1}$ in (Na, K)Cl(l), x units	67Kc
	cix	25	I $HClO_4$	$*K_1$ 1·36(I = 2·10), 1·52(I = 1·04), 1·52(I = 0·51)	68K
			0·2 $HClO_4$	$*K_1$ 1·57, $*\beta_2$ 1·64, $*\beta_3$ 1·68	
	ir, nmr		org	ev ion clusters of R_4N^+ with $(UO_2F_3^-)_n$ ($n \geqslant 20$) and $UO_2F_5^{3-}$ in C_6H_6(l)	68V
Np^{4+}	cix	20	4 H(ClO_4)	$*K_1$ 4·82, $*K_2$ 2·75	66Aa
	red	20	4 H(ClO_4)	$*K_2$ 2·69, $*K_3$ 2·34, $*K_4$ 1·3	
Np^{5+}	pre, X		solid	ev NpF_6^- in $CsNpF_6$(s) and NpF_7^{2-} in Rb_2NpF_7(s)	66Ab
NpO_2^{2+}	dis	21	1 $HClO_4$	$*K_1$ 0·93, $*\beta_2$ 1·11	68A
	cix	25	I $HClO_4$	$*K_1$ 2·11(I = 2·11), 2·20(I = 1·04), K_1 5·29(I = 2·11), 5·37(I = 1·04) [K_1 (H^+) 3·17]	68K
Pu^{5+}	X		solid	ev PuF_7^{2-} in Rb_2PuF_7(s)	65Pa
PuO_2^{2+}	cix	25	2 $HClO_4$	$*K_1$ 2·00, $*\beta_2$ 3·82, $*\beta_3$ 5·52, $*\beta_4$ 6·68	68K
			1 $HClO_4$	$*K_1$ 2·11, $*\beta_2$ 4·15, $*\beta_3$ 6·08, $*\beta_4$ 6·30	
Mn^{2+}	H, qh	25	1 ($NaClO_4$)	K_1 0·79	65C
Mn^{3+}	sp	23	I ($HClO_4$)	$*K_1$ 2·4(I = 5·3), 2·7(I = 6·1)	64Fa
Tc^{5+}	X		solid	ev TcF_6^- in $NaTcF_6$(s), $KTcF_6$(s)	63Ea
	var		solid	ev TcF_6^- in $MTcF_6$(s) (M = Na, K, Rb, Cs) (mag, pre, sp, X)	66Ha
Tc^{6+}	X		solid	ev $(TcOF_4)_3$ units in $TcOF_4$(s)	68Eb
Re^{6+}	X		solid	ev ReF_8^{2-} in K_2ReF_8(s)	64Kc
Fe^{3+}	cal	25	0·5 ($NaClO_4$)	$\Delta H_1 = 2{\cdot}35, \Delta S_1 = 31{\cdot}5$	59Sa/67A
	oth		var	K_1 5·76, K_2 4·44 (refractometry)	64Ba
	red	25	0 corr	$\Delta H_1 = 3{\cdot}4, \Delta S_1 = 39$	56C/67A
	dis		org	ev FeF_6^{3-} (paired with RNH_3^+) in $CHCl_3$(l)	67Sb
	red	25	1 ($NaClO_4$)	K_1 2·23, K_2 0·94, K_3 0·08	67W
	esr	rt	2 NH_4F	ev FeF_6^{3-}	68L
	sp		EtOH	K_1 5·46, K_2 4·75, K_3 3·78, K_4 3·30, K_5 2·83, K_6 2·78 (2·68 ?) in EtOH(l)	68S
	E	25	1 ($NaClO_4$)	K_1 5·06 (F membrane electrode)	68Sd
Co^{3+}	sp	27—49	dil	K(*cis*-Co $en_2H_2OF^{2+} \rightleftharpoons$ *trans*) $-0{\cdot}79$	66Ca
Ru^{5+}	X		solid	ev Ru_4F_{20} units in RuF_5(s)	64H
Os^{7+}	X, pre		solid	ev $OsOF_5$ units in $OsOF_5$(s)	68B
Pt^{5+}	var		solid	ev PtF_6^- in $KPtF_6$(s) (pre, mag, X)	64Bc
Cu^{2+}	pol	25	1 Na(ClO_4)	K_1 0·83	63Ma
Ag^+	E	0	HF	K_1 1·06 in HF(l) (→ 0)	66Cb
Ag^{3+}	var		solid	ev AgF_6^{3-} in Cs_2KAgF_6(s) (pre, mag, X)	66Hd
Zn^{2+}	pol	25	2 Na(ClO_4)	K_1 0·85	63Ma
	pol	18	1 ($NaNO_3$)	K_1 1·53	65Sd
	pol	25	0·3 ($NaNO_3$)	K_1 0·92	67Va

continu

Metal	*Method*	*Temp*	*Medium*	*Log of equilibrium constant, remarks*	*Ref*
Cd^{2+}	pol	25	2 $Na(ClO_4)$	K_1 0·81	63Ma
	dis	30	1 $(NaClO_4)$	K_1 0·3, β_2 0·5, β_3 1·2	65Ha
	cal	25	3 $(NaClO_4)$	$\Delta H_1 = 1{\cdot}02$, $\Delta S_1 = 6{\cdot}0$ (K_1 0·57)	66Ge
		25	1 $(NaClO_4)$	$\Delta H_1 = 1{\cdot}23$, $\Delta H_2 = -0{\cdot}7$, $\Delta S_1 = 6{\cdot}2$, $\Delta S_2 = -2{\cdot}0$ (K_1 0·46, β_2 0·53)	
	fp	≈260	$LiNO_3$(l)	no ev cpx in $LiNO_3$(l)	66I
	cal	25	4 $Li(ClO_4)$	$\Delta H_1 = -0{\cdot}60$, $\Delta H_2 = -0{\cdot}70$, $\Delta H_3 = -1{\cdot}2$, $\Delta H_4 = -0{\cdot}10$; K_1 1·77, K_2 0·80, K_3 0·64, K_4 −0·19; $\Delta S_1 = 10$, $\Delta S_2 = 6$, $\Delta S_3 = 7$, $\Delta S_4 = -0{\cdot}5$	67M
$MeHg^+$	gl	20	0·1 (KNO_3)	K_1 1·50	63Sa = 65Sb
B^{3+}	var	−61	CH_2Cl_2	$K(BF_4^- + BF_3 \rightleftharpoons B_2F_7^-)$ 2·68, $\Delta H = 5{\cdot}6$, $\Delta S = -14$; $K(B_2F_7^- + BF_3 \rightleftharpoons B_3F_{10}^-)$ 0·32, $\Delta H = -2{\cdot}8$, $\Delta S = -12$, in CH_2Cl_2(l), Ag^+ counter ions. Also *K* values for some other temperatures and Bu_4N^+ (pBF_3, nmr, ir)	65Bb
	var		solid	ev $B_2F_7^-$ in $Bu_4NB_2F_7$(s) *etc.* (pre, ir, nmr)	66H
	X		solid	ev BF_4^- in $NaBF_4$(s)	66Hc
	pre		CH_2Cl_2	ev $B_2F_7^-$ in CH_2Cl_2(l)	67Bb
Al^{3+}	oth	25?	var	K_1 6·08, K_2 5·02 (refractometry)	64Ba
	X		solid	ev $F_4Al(OH)_2AlF_4^{4-}$ units in $Sr_2Al_2F_8(OH)_2(H_2O)_2$ (s, tikhonenkovite)	67Pa
	dis		org	ev $R_3NH^+{}_3AlF_6^{3-}$ in $CHCl_3$(l), R = $C_{12}H_{25}$	68Sa
	oth	900—1050	gas	ev AlF_3, $NaAlF_4$, $Na_2Al_2F_8$ in gas phase, equil constants given (mass spectrometry)	68Sb
Ga^{3+}	red	25	0·5 $(NaClO_4)$	$\Delta H_1 = 1{\cdot}8$, $\Delta S_1 = 26$	55Y/67Aa
		25	0 corr	$\Delta H_1 = 1{\cdot}9$, $\Delta S_1 = 34$	
	X		solid	ev GaF_6^{3-} in $(NH_4)_3GaF_6$(s)	64Sa
In^{3+}	red	25	0 corr	$\Delta H_1 = 3{\cdot}5$	54H/55P
	sol, con	25	var	ev InF_4^-	65D
	dis	25	1 $(NaClO_4)$	K_1 3·67, β_2 6·26, β_3 8·61	68Aa
Tl^+	*E*	0	HF	K_1 −0·02 in HF(l)	66Cb
Tl^{3+}	cix	8—32	0·5 $H(ClO_4)$	$K_n < 4 (n = 1, 2)$	58He
Si^{4+}	ir, Ram		var	ev SiF_6^{2-} in solids and satd solutions of $ZnSiF_6$, $CdSiF_6$, Li_2SiF_6, $(NH_4)_2SiF_6$	66B
	X		solid	ev SiF_5^- in Ph_4AsSiF_5(s)	67Be
	con, ir	25	$MeNO_2$	ev SiF_5^- in $MeNO_2$(l)	67Ca
	pre, *X*		solid	ev SiF_5^- in $PtCl(CO)(PEt_3)_2SiF_5$(s)	67Cd
	var		var	ev SiF_6^{2-} in conc$(NH_4)_2SiF_6$ soln *etc.* (ir, Ram, nmr)	67Da
	sol		4 $LiClO_4$	$K_s[2SiO_2(s) + 4SiF_6^{2-} + 8H^+ \rightleftharpoons 6SiF_4 + 4H_2O]$ −7·15, $K_s[SiO_2(s) + 5SiF_6^{2-} + 4H^+ \rightleftharpoons 6SiF_5^- + 2H_2O]$ −1·05	68Ka
	aix		0·5 H(Cl)	K_6 3·96	68Pa
Ge^{4+}	pre, ir		var	ev GeF_5OH^{2-}, $Ge_2F_{10}O^{4-}$	63Ka
	Ram		H_2GeF_6 conc	ev GeF_6^{2-}	64G
	ana	0—50	0·4 (NaCl)	*$K_6(GeF_5H_2O^- + HF \rightleftharpoons GeF_6^{2-} + H_3O^+)$ 0·66(0°), 0·62(10°), 0·58(20°), 0·50(30°), 0·42(40°), 0·34(50°); $\Delta H = 2{\cdot}59$, $\Delta S = -6{\cdot}3$(25°)	64R

continued overleaf

66 F^- (contd)

Metal	*Method*	*Temp*	*Medium*	*Log of equilibrium constant, remarks*	*Ref*
Ge^{4+} (*contd*)	ana, qh	0—25	dil	$K(GeF_4H_2OOH^- + HF + F^- \rightleftharpoons GeF_6^{2-} + 2H_2O)$ 5·9(0°), 5·3(25°), [$K_1(H^+)$ 2·97(0°), 3·17(25°)]	64Ra
	PbHg, qh	25	var	$K[K_2GeF_6(s) + 2H_2O \rightleftharpoons 2K^+ + 4H^+ + 6F^- + GeO_2(s, hex)]$ −30·9, $K[GeF_6^{2-} + 2H_2O \rightleftharpoons 4H^+ + 6F^- + GeO_2(s, hex)]$ −25·8, [$K_s(K^+{}_2GeF_6^{2-})$ −5·07, 0 corr]	65R
	var		var	ev GeF_6^{2-} in conc$(NH_4)_2GeF_6$ soln(ir, Ram, nmr)	67Da
	aix		0·5 H(Cl)	K_6 3·86	68Pa
Sn^{2+}	Sn, pre	25	3 $HClO_4$	ev SnF_3^-, $Sn_2F_5^-$	64D
	X		solid	ev $Sn_2F_5^-$ in $NaSn_2F_5(s)$	64M
	oth	725—895	gas	ev SnF_2, $NaSnF_3$, Na_2SnF_4, $NaSn_2F_5$, NaF in gas phase, equil constants given (mass spectrometry), similar for K^+	68H
	E	25	0·85 $(NaClO_4)$	K_1 6·26, β_2 8·76, β_3 9·25, $\Delta H_{\beta 3} = 10{\cdot}35$, $\Delta S_{\beta 3} = 73{\cdot}2$	68Ha
		45		K_1 5·78, β_2 8·70, β_3 9·82	
		60		K_1 6·21, β_2 9·06, β_3 10·31 (MHg and F^- membrane electrode)	
Sn^{4+}	var		var	ev $SnF_4(OH)_2^{2-}$, $SnF_2(OH)_4^{2-}$, $Sn_2F_{10}O^{4-}$ (pre, fp, ir, *X*)	63Ka
	var		var	ev SnF_6^{2-} in conc$(NH_4)_2SnF_6$ soln (ir, Ram, nmr)	67Da
$MeSn^{3+}$	red	25	0·5 $(NaClO_4)$	K_1 5·10, K_2 4·75, K_3 4·12, K_4 3·09, $K_5 \approx 2{\cdot}2$	67Cb
	aix	25	KF var	ev $MeSnF_5^{2-}$	
Me_2Sn^{2+}	var		var	ev ani cpx such as $Me_2SnF_3^-$, $Me_2SnF_4^{2-}$ (ir, pre, sol)	65W
	qh	25	1 $(NaClO_4)$	K_1 3·70, K_2 2·87, K_3 1·47, β_3 8·04	67Cb
	aix	25	KF var	β_3 8·07, K_4 0·09	
Me_3Sn^+	qh, sol	25	1 $(NaClO_4)$	K_1 2·27(qh), 2·29(sol), K_2 0·61(qh), 0·60(sol)	67Cb
	aix	25	KF var	K_1 2·34, K_2 0·83	
Pb^{2+}	MHg, qh	25	1 $(NaClO_4)$	K_1 1·48, no ev $PbF_2(aq)$ (H^+: K_1 2·95, β_2 3·61)	65Ba
N^+	var		var	ev N_2F^+ in $N_2FAsF_6(s)$ and in HF(l) (pre, ir, *X*, nmr)	65Mb
	pre, ir		solid	ev N_2F^+ in $N_2F^+BF_4^-(s)$ and $N_2F^+SbF_6^-(s)$	68Pb
N^{2+}	var		solid	ev $N_2F_3^+$ in $N_2F_4(AsF_5)_{1\cdot33}(s)$ (nmr, pre, ir)	67Y
N^{3+}	var		var	ev NF_2^- (pH, pol, con)	65Ma
N^{5+}	var		solid	ev NF_4^+ in NF_4AsF_6 (s) (pre, ir, Ram)	66C
	pre, *X*		solid	ev NF_4^+ in $NF_4^+AsF_6^-(s)$	66Gb
	pre, nmr		solid	ev NF_4^+ in $NF_4SbF_6(s)$ and $NF_4AsF_6(s)$	66T
P^{3+}	ana	30	1 KF	$K(H_2P_2O_5^{2-} + F^- \rightleftharpoons HPO_3^{2-} + HPFO_2^-) \approx 0{\cdot}3$	68Ba
P^{5+}	*X*		solid	ev $PO_2F_2^-$ in $KPO_2F_2(s)$	66Hb
	nmr	80	var	$\beta(P_3O_9^{3-}F^-)$ 0·15 (unit?) at pH = 7·5	66Mb
	var		var	ev $O_3POPO_2F^{3-}$ (pre, paper chromatography, nmr)	68Sf
As^{5+}	*X*		solid	ev$(F_4AsO_2AsF_4)^{2-}$, not $As_3O_3F_{12}^{3-}$ in $K_2As_2O_2F_8(s)$	63D
	X		solid	ev AsF_5OH^- in $KAsF_5OH(s)$ and $AsF_4(OH)_2^-$ in $KAsF_4(OH)_2(s)$	65H = 67I
	var		var	ev $As_2O_2F_8^{2-}$ (pre, ir, pH)	65Ka ≈ 65K
	gl, pre		var	ev $As_2F_{10}O^{2-}$, no ev $HAs_2F_{10}O^-$	67Kb
Sb^{5+}	var		var	ev $Sb_2F_{10}O^{2-}$, $HSb_2F_{10}O^-$, $Sb_2F_8O_2^-$ (pre, fp, pH)	64Ka
	var		var	ev $Sb_2OF_{10}^{2-}$, $HSb_2OF_{10}^-$, $Sb_2O_2F_8^{2-}$, (pre, ir, pH)	65Kb

contin

Metal	Method	Temp	Medium	Log of equilibrium constant, remarks	Ref
	var		HSO_3F	ev $SbF_{5-n}X_{1+n}^-$, $n = 0, 1, 2, 3$, and polymers, $X^- = SO_3F^-$, in HSO_3F(l) (con, fp, nmr)	65T
	con	25	HSO_3F	$K[HSbF_5X(+HX) \rightleftharpoons SbF_5X^- + H_2X^+]$ −2·43, $K(2HSbF_5X \rightleftharpoons Sb_2F_{10}X^- + H_2X^+)$ −2·15, $K(Sb_2F_4X_7^- + 2HX \rightleftharpoons 2SbF_2X_4^- + H_2X^+)$ −2·40 in HSO_3F(l) = HX	
	con, nmr		HF(l)	ev $Sb_nF_{5n+1}^-$, (n = 1, 2 and higher)	66Gc
	kin	25	var	ev SbF_5OH^-, $SbF_4(OH)_2^-$, and polyn. anion with ≈ 2F/Sb	66Ma
	gl		var	ev $Sb_2F_{10}O^{2-}$, no ev $HSb_2F_{10}O^-$ (effect of impurities in ref 65Kb)	67Kb
	X		solid	ev $Sb_3F_{16}^-$ in $Br_2Sb_3F_{16}$(s)	68Ec
Bi^{3+}	cix	25	1·89 $H(ClO_4)$	$*K_1$ 1·71, $*\beta_2$ 2·31, K_1 4·7, β_2 8·3 [$K_1(H^+)$ 3·0]	67L
S^{4+}	pre, ir		solid, org	ev SF_5^- in Me_4NSF_5(s) and in Me_2NCHO(l)	63T
	Ram		$SSbF_9$(l)	ev SF_3^+ in $SSbF_9$(l)	68E
S^{6+}	fp	−89	HSO_3F	$K(HSO_3F \rightleftharpoons HF + SO_3)$ < −6·5 in HSO_3F(l)	66G
	con, fp	−78 to 25	HSO_3F	$K(H_3O^+ + SO_3F^- \rightleftharpoons HF + H_2SO_4)$ −1·51 (−78°), −1·38(0°), −0·92(25°) in HSO_3F(l)	66Ga
	cal, ΔG	25	HSO_3F	$K(HSO_3F \rightleftharpoons HF + SO_3)$ −8·0 in HSO_3F(l)	67R+
Se^{4+}	Ram		$SeSbF_9$(l)	ev SeF_3^+ in $SeSbF_9$(l)	68E
	X		solid	ev SeF_3^+ in $SeNb_2F_{14}$(s)	68Ea
Se^{6+}	pre, ir		solid	ev SeO_3F^- in $KSeO_3F$(s) *etc.*	67E
Te^{4+}	var		var	ev TeF_5^- in solids, pyridine(l) and MeOH(l) (pre, ir, Ram, *p*MeOH, con)	66Gd
	Ram		$TeSbF_9$(l)	ev TeF_3^+ or $TeSbF_9$ in $TeSbF_9$(l)	68E
Te^{6+}	var		var	ev $TeF_n(OH)_{6-n}$, n = 1,2,3,4,5, and ions like $TeO(OH)_3F_2^-$, $TeO_2F_4^{2-}$ *etc.* (pre, paper chromatography, ir, gl)	67Ka
Cl^+	var		var	ev ClF_2^- in $NO^+ClF_2^-$(s) and NOF(l) (ir, con, *X*)	65Ca
	X, ir		solid	ev ClF_2^- in $CsClF_2$(s) *etc.*	
Cl^{3+}	pre, con	0	ClF_3	ev ClF_2^+ in ClF_2SbF_6(s) and ClF_3(l)	59Sb
	pre		var	ev ClF_4^- in $MClF_4$(s), M = K, Rb, Cs	64W
	var		var	ev ClF_2^+ in ClF_3(l) and IF_5(l) (con, ir, fp)	65Cb
	pre, ir		solid	ev ClF_2^+ in $ClF_2^+AsF_6^-$(s) and $ClF_2^+BF_4^-$(s)	67C
Cl^{5+}	pre		HF conc	ev $ClO_2F_2^-$	65M
Br^{3+}	pre, con	25	BrF_3	ev BrF_2^+ and BrF_4^- in BrF_3(l)	49W
Br^{5+}	pre		var	ev BrF_6^- in $MBrF_6$(s), M = K, Rb, Cs	64W
	pre		HF conc	ev $BrO_2F_2^-$	65M
I_2	sp	22	var	K_1 0·2—0·4 [$\beta(I^-I_2^0)$ 2·90]	63Mb = 62M
I^{5+}	ir		solid	ev $IO_2F_2^-$ in KIO_2F_2(s)	65P
I^{7+}	ir, Ram		solid	ev IF_6^+ in $IF_6^+AsF_6^-$(s)	67Cc
Xe^{2+}	con, kin		dil	no ev other Xe^{II} species than XeF_2	67A
	X		solid	ev $Xe_2F_3^+$ in $Xe_2F_3AsF_6$(s)	68Se
Xe^{4+}	*X*		solid	ev XeF_5^+ in $XeF_5^+PtF_6^-$(s)	67Ba
Xe^{6+}	var		solid	ev XeF_8^{2-} and NO^+ in $(NO)_2XeF_8$(s) (pre, ir, Ram)	66Mc
	con, pre		$XeOF_4$	ev cpx($XeOF_5^-$?) in $XeOF_4$(l)	66S

references overleaf

66 F⁻ (contd)

26A C. C. Andersen and O. Hassel, *Z. phys. Chem.*, 1926, **123**, 151
49W A. A. Woolf and H. J. Emeléus, *J. Chem. Soc.*, 1949, 2865
58He R. A. Horne, *J. Inorg. Nuclear Chem.*, 1958, **6**, 338
59Sb F. Seel and O. Detmer, *Z. anorg. Chem.*, 1959, **301**, 113
59Ta I. V. Tananaev, *Acta Chim. Sinica*, 1959, **25**, 391
61N N. S. Nikolaev and Yu. A. Luk'yanychev, *Atomnaya Energiya*, 1961, **11**, 67
61Td R. E. Thoma, C. F. Weaver, H. A. Friedmann, H. Insley, L. A. Harris, and H. A. Yakel, jun., *J. Phys. Chem.*, 1961, **65**, 1096
62E A. J. Edwards, R. D. Peacock, and R. W. H. Small, *J. Chem. Soc.*, 1962, 4486
63D H. Dunken and W. Haase, *Z. Chem.*, 1963, **3**, 433
63E A. J. Ellis, *J. Chem. Soc.*, 1963, 4300
63Ea A. J. Edwards, D. Hugill, and R. D. Peacock, *Nature*, 1963, **200**, 672
63Eb T. Erdey-Grúz, L. Majthényi, and E. Kugler, *Acta Chim. Acad. Sci. Hung.*, 1963, **67**, 393
63Ec T. Erdey-Grúz, L. Majthényi, and E. Kugler, *Magyar Kém. Folyóirat*, 1963, **69**, 68
63G J. R. Geichman, E. A. Smith, and P. R. Ogle, *Inorg. Chem.*, 1963, **2**, 1012
63Ka L. Kolditz and H. Preiss, *Z. anorg. Chem.*, 1963, **325**, 252, 263

1964 *Addendum*:
63L add *idem*, *Atomnaya Energiya*, 1963, **15**, 423

63La L. Levi, O. Milman, and E. Suraski, *Trans. Faraday Soc.*, 1963, **59**, 2064
63Ma Š. S. Mesarič and D. N. Hume, *Inorg. Chem.*, 1963, **2**, 1063
63Mb D. Meyerstein and A. Treinin, *Trans. Faraday Soc.*, 1963, **59**, 1114
63N B. I. Nabivanets, *Zhur. neorg. Khim.*, 1963, **8**, 2302
63P K. J. Packer and E. L. Muetterties, *J. Amer. Chem. Soc.*, 1963, **85**, 3035
63R R. O. Ragsdale and B. B. Stewart, *Inorg. Chem.*, 1963, **2**, 1002
63Sa M. Schellenberg, dissertation, Eidgenössische Technische Hochschule, Zürich, 1963
63T R. Tunder and B. Siegel, *J. Inorg. Nuclear Chem.*, 1963, **25**, 1097
63Ta I. V. Tananaev, N. S. Nikolaev, Yu. A. Luk'yanychev, and I. F. Alenchikova, *Khimiya ftoristykh soedinenii aktinidov*, *Izd. Akad. Nauk S.S.S.R.*, 1963
63Va V. M. Vdovenko and G. A. Romanov, *Atomnaya Energiya*, 1963, **15**, 168
63Vb V. M. Vdovenko, G. A. Romanov, and V. A. Shcherbakov, *Radiokhimiya*, 1963, **5**, 581
64A L. B. Asprey and R. A. Penneman, *Inorg. Chem.*, 1964, **3**, 727
64Aa L. B. Asprey and R. A. Penneman, *Science*, 1964, **145**, 924
64B I. I. Baram and B. N. Laskorin, *Zhur. neorg. Khim.*, 1964, **9**, 738
64Ba W. Brandel and A. Swinarski in 'Theory and structure of complex compounds' (symposium papers, Wrocław 1962), Pergamon, Warszawa, 1964, p. 497
64Bb H. D. Bedon, S. M. Horner, and S. Y. Tyree, jun., *Inorg. Chem.*, 1964, **3**, 647
64Bc N. Bartlett and D. H. Lohmann, *J. Chem. Soc.*, 1964, 619
64D J. D. Donaldson and J. D. O'Donoghue, *J. Chem. Soc.*, 1964, 271
64E A. J. Edwards, *J. Chem. Soc.*, 1964, 3714
64F H. N. Farrer and F. J. C. Rossotti, *J. Inorg. Nuclear Chem.*, 1964, **26**, 1959
64Fa J. P. Fackler, jun., and I. D. Chawla, *Inorg. Chem.*, 1964, **3**, 1130
64G J. E. Griffiths and D. E. Irish, *Inorg. Chem.*, 1964, **3**, 1134
64H J. H. Holloway, R. D. Peacock, and R. W. H. Small, *J. Chem. Soc.*, 1964, 644
64K W. Krüger, dissertation, Technical Univ. Berlin, 1964
64Ka L. Kolditz, 'Proc. Symposium Co-ord. Chem.' Tihany 1964, Akad. Kiadó, Budapest, 1965, p. 281
64Kb K. Knox, private communication, quoted by 64Bb
64Kc P. A. Koz'min, *Zhur. strukt. Khim.*, 1964, **5**, 70
64Kd Y. Kurokawa and N. Yui, *J. Chem. Soc. Japan*, 1964, **85**, 397
64M R. R. McDonald, A. C. Larson, and D. T. Cromer, *Acta Cryst.*, 1964, **17**, 1104
64P R. A. Penneman, G. D. Sturgeon, and L. B. Asprey, *Inorg. Chem.*, 1964, **3**, 126
64Pa R. A. Penneman, F. H. Kruse, R. S. George, and J. S. Coleman, *Inorg. Chem.*, 1964, 3, 309
64R I. G. Ryss and N. F. Kulish, *Zhur. neorg. Khim.*, 1964, **9**, 1103
64Ra I. G. Ryss and N. F. Kulish, *Zhur. neorg. Khim.*, 1964, **9**, 1382
64S C. C. Stephenson, H. P. Hopkins, and C. A. Wulff, *J. Phys. Chem.*, 1964, **68**, 1427
64Sa S. Schwarzmann, *Z. Krist*, 1964, **120**, 286
64Sb D. R. Sears and J. H. Burns, *J. Chem. Phys.*, 1964, **41**, 3478
64V L. P. Varga and D. N. Hume, *Inorg. Chem.*, 1964, **3**, 77
64W E. D. Whitney, R. O. MacLaren, C. E. Fogle, and T. J. Hurley, *J. Amer. Chem. Soc.*, 1964, **86**, 2583
64Y K. B. Yatsimirskii and K. E. Prik, *Zhur. neorg. Khim.*, 1964, **9**, 178
65B Yu. A. Buslaev and M. P. Gustyakova, *Zhur. neorg. Khim.*, 1965, **10**, 1524
65Ba E. Bottari and L. Ciavatta, *J. Inorg. Nuclear Chem.*, 1965, **27**, 133
65Bb S. Brownstein and J. Paasivirta, *Canad. J. Chem.*, 1965, **43**, 1645
65C L. Ciavatta and M. Grimaldi, *J. Inorg. Nuclear Chem.*, 1965, **27**, 2019
65Ca K. O. Christe and J. P. Guertin, *Inorg. Chem.*, 1965, **4**, 905, 1785
65Cb K. O. Christe and A. E. Pavlath, *Z. anorg. Chem.*, 1965, **335**, 210
65Cc P. Charpin, *Compt. rend.*, 1965, **260**, 1914

continu

65D E. N. Deichman and L. S. Krysina, *Zhur. neorg. Khim.*, 1965, **10**, 476
65Da R. G. Deshpande, P. K. Khopkar, C. L. Rao, and H. D. Sharma, *J. Inorg. Nuclear Chem.*, 1965, **27**, 2171
65G T. K. Ghose, T. N. Chakrabarty, and N. N. Ray, *J. Indian Chem. Soc.*, 1965, **42**, 847
65Ga R. Guillaumont, *Compt. rend.*, 1965, **260**, 4739
65H W. Haase, *Chem. Zvesti*, 1965, **19**, 167
65Ha H. E. Hellwege and G. K. Schweitzer, *J. Inorg. Nuclear Chem.*, 1965, **27**, 99
65Hb R. M. Herak, S. S. Malčič, and Lj. M. Manojlovič, *Acta Cryst.*, 1965, **18**, 520
65K C. Keller and H. Schmutz, *J. Inorg. Nuclear Chem.*, 1965, **27**, 900
65Ka L. Kolditz, B. Nussbücker, and M. Schönherr, *Z. anorg. Chem.*, 1965, **335**, 189
65Kb L. Kolditz and B. Nussbücker, *Z. anorg. Chem.*, 1965, **337**, 191
65M G. Mitra, *Z. anorg. Chem.*, 1965, **340**, 110
65Ma K. J. Martin, *J. Amer. Chem. Soc.*, 1965, **87**, 394
65Mb D. Moy and A. R. Young, jun., *J. Amer. Chem. Soc.*, 1965, **87**, 1889
65P J. J. Pitts, S. Kongpricha, and A. W. Jache, *Inorg. Chem.*, 1965, **4**, 257
65Pa R. A. Penneman, G. D. Sturgeon, L. B. Asprey, and F. H. Kruse, *J. Amer. Chem. Soc.*, 1965, **87**, 5803
65R I. G. Ryss and N. F. Kulish, *Zhur. neorg. Khim.*, 1965, **10**, 1827
65S M. V. Smirnov, V. E. Komarov, and N. Ya. Chukreev, *Zhur. neorg. Khim.*, 1965, **10**, 200
65Sa T. W. Swaddle and E. L. King, *Inorg. Chem.*, 1965, **4**, 532
65Sb G. Schwarzenbach and M. Schellenberg, *Helv. Chim. Acta*, 1965, **48**, 28
65Sc T. R. Stengle and C. H. Langford, *J. Phys. Chem.*, 1965, **69**, 3299
65Sd A. Swinarski and A. Grodzicki, *Roczniki Chem.*, 1965, **39**, 3
65T R. C. Thompson, J. Barr, R. J. Gillespie, J. B. Milne, and R. A. Rothenburg, *Inorg. Chem.* 1965, **4**, 1641
65V L. P. Varga, W. D. Wakley, L. S. Nicolson, M. L. Madden, and J. Patterson, *Analyt. Chem.*, 1965, **37**, 1003
65W C. J. Wilkins and H. M. Haendler, *J. Chem. Soc.*, 1965, 3174
66A L. B. Asprey, F. H. Kruse, A. Rosenzweig, and R. A. Penneman, *Inorg. Chem.*, 1966, **5**, 659
66Aa S. Ahrland and L. Brandt, *Acta Chem. Scand.*, 1966, **20**, 328
66Ab L. B. Asprey, T. K. Keenan, R. A. Penneman, and G. D. Sturgeon, *Inorg. Nuclear Chem. Letters*, 1966, **2**, 19
66B R. B. Badachhape, G. Hunter, L. D. McCory, and J. L. Margrave, *Inorg. Chem.*, 1966, **5**, 929
66Ba D. Brown and J. F. Easey, *J. Chem. Soc.* (*A*), 1966, 254
66Bb M. N. Bukhsh, J. Flegenheimer, F. M. Hall, A. G. Maddock, and C. Ferreira de Miranda, *J. Inorg. Nuclear Chem.* 1966, **28**, 421
66C K. O. Christe, J. P. Guertin, and A. E. Pavlath, *Inorg. Nuclear Chem. Letters*, 1966, **2**, 83
66Ca S. C. Chan and C. K. Poon, *J. Chem. Soc.* (*A*), 1966, 146
66Cb A. F. Clifford, W. D. Pardieck, and M. W. Wadley, *J. Phys. Chem.*, 1966, **70**, 3241
66G R J. Gillespie, J. B. Milne, and R. C. Thompson, *Inorg. Chem.*, 1966, **5**, 468
66Ga R. J. Gillespie, J. B. Milne, and J. B. Senior, *Inorg. Chem.*, 1966, **5**, 1233
66Gb J. P. Guertin, K. O. Christe, and A. E. Pavlath, *Inorg. Chem.*, 1966, **5**, 1921
66Gc R. J. Gillespie and K. C. Moss, *J. Chem. Soc.* (*A*), 1966, 1171
66Gd N. N. Greenwood, A. C. Sarma, and B. P. Straughan, *J. Chem. Soc.* (*A*), 1966, 1446
66Ge P. Gerding, *Acta Chem. Scand.*, 1966, **20**, 79
66Gf D. Grandjean and R. Weiss, *Compt. rend.*, 1966, **263**, *C*, 58
66Gg R. Guillaumont, *Rev. Chim. minérale*, 1966, **3**, 339
66H J. J. Harris, *Inorg. Chem.*, 1966, **5**, 1627
66Ha D. Hugill and R. D. Peacock, *J. Chem. Soc.* (*A*), 1966, 1339
66Hb R. W. Harrison, R. C. Thompson, and J. Trotter, *J. Chem. Soc.* (*A*), 1966, 1775
66Hc E. Höhne and L. Kutschabsky, *Z. anorg. Chem.*, 1966, **344**, 279
66Hd R. Hoppe and R. Homann, *Naturwiss*, 1966, **53**, 501
66I R. E. Isbell, E. W. Wilson, jun., and D. F. Smith, *J. Phys. Chem.*, 1966, **70**, 2493
66Ia B. N. Ivanov-Emin, T. N. Susanina, and L. A. Ogorodnikova *Zhur. neorg. Khim.*, 1966, **11**, 504
66K O. L. Keller, jun. and A. Chetham-Strode, jun., *Inorg. Chem.*, 1966, **5**, 367
66Ka S. S. Korovin, Yu. I. Kol'tsov, A. M. Reznik, and I. A. Apraksin, *Zhur. neorg. Khim.*, 1966, **11**, 180
66L S. J. Lyle and S. J. Naqvi, *J. Inorg. Nuclear Chem.*, 1966, **28**, 2993
66M J. G. Malm, H. Selig, and S. Siegel, *Inorg. Chem.*, 1966, **5**, 130
66Ma W. A. Mazeika and H. M. Neumann, *Inorg. Chem.*, 1966, **5**, 309
66Mb R. E. Mesmer, *J. Inorg. Nuclear Chem.*, 1966, **28**, 691
66Mc G. J. Moody and H. Selig, *Inorg. Nuclear Chem. Letters*, 1966, **2**, 319
66N B. I. Nabivanets, *Ukrain. khim. Zhur.*, 1966, **32**, 886
66P H. Preiss, *Z. anorg. Chem.*, 1966, **346**, 272
66Pa V. S. Pakholkov, *Izvest. V.U.Z. Khim. i khim. Tekhnol.*, 1966, **9**, 242
66Pb V. S. Pakholkov and V. E. Panikarovskikh, *Izvest. V.U.Z., Khim. i khim. Tekhnol.*, 1966, **9**, 401
66Pc O. E. Presnyakova and R. S. Prishchepo, *Zhur. neorg. Khim.*, 1966, **11**, 1436
66S H. Selig, *Inorg. Chem.*, 1966, **5**, 183
66Sa L. Stein, C. W. Williams, I. M. Fox, and E. Gebert, *Inorg. Chem.*, 1966, **5**, 662
66T W. E. Tolberg, R. T. Rewick, R. S. Stringham, and M. E. Hill, *Inorg. Nuclear Chem. Letters*, 1966, **2**, 79
66V V. M. Vdovenko, G. A. Romanov, and V. A. Shcherbakov, *Zhur. neorg. Khim.*, 1966, **11**, 252
67A E. H. Appelman, *Inorg. Chem.*, 1967, **6**, 1268
67Aa S. Ahrland, *Helv. Chim. Acta*, 1967, **50**, 306

continued overleaf

66 F⁻ (contd)

67B G. Brunton, *J. Inorg. Nuclear Chem.*, 1967, **29**, 1631
67Ba N. Bartlett, F. Einstein, D. F. Stewart, and J. Trotter, *J. Chem. Soc.* (*A*), 1967, 1190; *Chem. Comm.*, 1966, 550 (prel)
67Bb S. Brownstein, *Canad. J. Chem.*, 1967, **45**, 2403
67Bc Yu. A. Buslaev, D. S. Dyer, and R. O. Ragsdale, *Inorg. Chem.*, 1967, **6**, 2208
67Bd A. K. Babko, B. I. Nabivanets, and V. V. Lukachina, *Zhur. neorg. Khim.*, 1967, **12**, 2965
67Be K. Behrends and G. Kiel, *Naturwiss*, 1967, **54**, 537
67C K. O. Christe and W. Sawodny, *Inorg. Chem.*, 1967, **6**, 313
67Ca H. C. Clark and K. R. Dixon, *Chem. Comm.*, 1967, 717
67Cb A. Cassol, L. Magon, and R. Barbieri, *Inorg. Nuclear Chem. Letters*, 1967, **3**, 25
67Cc K. O. Christe and W. Sawodny, *Inorg. Chem.*, 1967, **6**, 1783
67Cd H. C. Clark, P. W. R. Corfield, K. R. Dixon, and J. A. Ibers, *J. Amer. Chem. Soc.*, 1967, **89**, 3360
67D H. Dunken, W. Haase, and F. Schönherr, *Z. Chem.*, 1967, **7**, 399, 442
67Da P. A. W. Dean and D. F. Evans, *J. Chem. Soc.* (*A*), 1967, 698
67E A. J. Edwards, M. A. Mouty, and R. D. Peacock, *J. Chem. Soc.* (*A*), 1967, 557
67F J. Fischer and R. Weiss, *Chem. Comm.*, 1967, 328
67Fa J. Fischer, R. Elchinger, and R. Weiss, *Chem. Comm.*, 1967, 329
67G W. P. Griffith and T. D. Wickins, *J. Chem. Soc.* (*A*), 1967, 675
67Ga D. Grandjean and R. Weiss, *Bull. Soc. chim. France*, 1967, 3049, 3054
67H R. Haque and L. W. Reeves, *J. Amer. Chem. Soc.*, 1967, **89**, 250
67I A. A. Ivakin, *Zhur. neorg. Khim.*, 1967, **12**, 1787
67Ia B. N. Ivanov-Emin and V. A. Zaitseva, *Zhur. neorg. Khim.*, 1967, **12**, 2247
67K R. T. Kolarich, V. A. Ryan, and R. P. Schuman, *J. Inorg. Nuclear Chem.*, 1967, **29**, 783
67Ka L. Kolditz and I. Fitz, *Z. anorg. Chem.*, 1967, **349**, 175
67Kb L. Kolditz and M. Gitter, *Z. anorg. Chem.*, 1967, **354**, 15
67Kc A. P. Koryushin, M. V. Smirnov, and V. E. Komarov, *Zhur. neorg. Khim.*, 1967, **12**, 2511
67Kd Yu. V. Karyakin and E. N. Kryachko, *Zhur. neorg. Khim.*, 1967, **12**, 2567
67L H. Loman and E. van Dalen, *J. Inorg. Nuclear Chem.*, 1967, **29**, 699
67La S. J. Lyle and S. J. Naqvi, *J. Inorg. Nuclear Chem.*, 1967, **29**, 2441
67M V. E. Mironov and A. V. Fokina, *Zhur. neorg. Khim.*, 1967, **12**, 2571
67Ma D. Martin, *Rev. Chim. minérale*, 1967, **4**, 367
67N B. Norén, *Acta Chem. Scand.*, 1967, **21**, 2435
67Na B. Norén, *Acta Chem. Scand.*, 1967, **21**, 2449
67Nb B. Norén, *Acta Chem. Scand.*, 1967, **21**, 2457
67P N. A. Parpiev, I. A. Maslennikov, and Yu. A. Buslaev, *Uzbek, khim. Zhur.*, 1967, No. 3, 9
67Pa Z. V. Pudovkina and Yu. A. Pyatenko, *Doklady Akad. Nauk S.S.S.R.*, 1967, **174**, 193
67R G. W. Richards and A. A. Woolf, *J. Chem. Soc.* (*A*), 1967, 1118
67S H. Selig and B. Frlec, *J. Inorg. Nuclear Chem.*, 1967, **29**, 1887
67Sa P. L. Sarma and M. S. Davis, *J. Inorg. Nuclear Chem.*, 1967, **29**, 2607
67Sb I. A. Shevchuk, N. A. Skripnik, and V. I. Martsokha, *Zhur. analit. Khim.*, 1967, **22**, 891
67V V. M. Vdovenko and G. A. Romanov, *Radiokhimiya*, 1967, **9**, 384
67Va S. Vařvička and J. Koryta, *Coll. Czech. Chem. Comm.*, 1967, **32**, 2346
67Vb V. M. Vdovenko, O. B. Stebunov, and V. A. Shcherbakov, *Zhur. neorg. Khim.*, 1967, **12**, 1706
67W J. B. Walker and G. R. Choppin, *Adv. Chem. Series*, 1967, **71**, 127
67Y A. R. Young, jun., and D. Moy, *Inorg. Chem.*, 1967, **6**, 178
68A S. Ahrland and L. Brandt, *Acta Chem. Scand.*, 1968, **22**, 106
68Aa A. Aziz and S. J. Lyle, *J. Inorg. Nuclear Chem.*, 1968, **30**, 3223
68B N. Bartlett and J. Trotter, *J. Chem. Soc.* (*A*), 1968, 543
68Ba B. Blaser and K.-H. Worms, *Z. anorg. Chem.*, 1968, **360**, 117
68Bb W. G. Baldwin and D. R. Stranks, *Austral. J. Chem.*, 1968, **21**, 2161
68E J. A. Evans and D. A. Long, *J. Chem. Soc.* (*A*), 1968, 1688
68Ea A. J. Edwards and G. R. Jones, *Chem. Comm.*, 1968, 346
68Eb A. J. Edwards, G. R. Jones, and R. J. C. Sills, *Chem. Comm.*, 1968, 1177
68Ec A. J. Edwards, G. R. Jones, and R. J. C. Sills, *Chem. Comm.*, 1968, 1527
68F J. Feeney, R. Haque, L. W. Reeves, and C. P. Yue, *Canad. J. Chem.*, 1968, **46**, 1389
68Fa J. Fischer and R. Weiss, *Chem. Comm.*, 1968, 1137
68H J. W. Hastie, K. F. Zmbov, and J. L. Margrave, *J. Inorg. Nuclear Chem.*, 1968, **30**, 729
68Ha F. M. Hall and S. J. Slater, *Austral. J. Chem.*, 1968, **21**, 2663
68I B. N. Ivanov-Emin, V. A. Zaitseva, and A. M. Egorov, *Zhur. neorg. Khim.*, 1968, **13**, 2655
68K V. N. Krylov, E. V. Komarov, and M. F. Pushlenkov, *Radiokhimiya*, 1968, **10**, 717, 719, 723
68Ka K. Kleboth, *Monatsh.*, 1968, **99**, 1177
68L H. Levanon, G. Stein, and Z. Luz, *J. Amer. Chem. Soc.*, 1968, **90**, 5292
68M A. L. Mathews and C. F. Baes, jun., *Inorg. Chem.*, 1968, **7**, 373
68Ma R. E. Mesmer and C. F. Baes, jun., *J. Phys. Chem.*, 1968, **72**, 4720
68P G. Z. Pinsker and V. G. Kuznetsov, *Kristallografiya*, 1968, **13**, 74
68Pa N. A. Parpiev and I. A. Maslennikov, *Uzbek. khim. Zhur.*, 1968, No. 2, 6
68Pb A. V. Pankratov and N. I. Savenkova, *Zhur. neorg. Khim.*, 1968, **13**, 2610
68Pc H. Preiss, *Z. anorg. Chem.*, 1968, **362**, 13

contin

68S A. Swinarski and R. Gogolin, *Roczniki Chem.*, 1968, **42**, 543
68Sa J. A. Shevchuk and N. A. Skripnik, *Ukrain. khim. Zhur.*, 1968, **34**, 1085
68Sb L. N. Sidorov and E. N. Kolosov, *Zhur. fiz. Khim.*, 1968, **42**, 2617
68Sc K. Srinivasan and G. A. Rechnitz, *Analyt. Chem.*, 1968, **40**, 509
68Sd K. Srinivasan and G. A. Rechnitz, *Analyt. Chem.*, 1968, **40**, 1818
68Se F. O. Sladky, P. A. Bulliner, N. Bartlett, B. G. De Boer, and A. Zalkin, *Chem. Comm.*, 1968, 1048
68Sf U. Schülke, *Z. anorg. Chem.*, 1968, **361**, 225
68Sg K. Schaumburg and C. Deverell, *J. Amer. Chem. Soc.*, 1968, **90**, 2495
68V V. M. Vdovenko, A. I. Skoblo, D. N. Suglobov, L. L. Shcherbakova, and V. A. Shcherbakov, *Radiokhimiya*, 1968, **10**, 527
68Va N. E. Vanderborgh, *Talanta*, 1968, **15**, 1009

Chloride, Cl^- 67

Metal	*Method*	*Temp*	*Medium*	*Log of equilibrium constant, remarks*	*Ref*
H^+	con	25	Me_2NCHO	K_1 3·75 in $Me_2NCHO(l)$, 0 corr	56Se/68Cd
	dis	25	$(ClC_2H_4)_2O$	K_1 7·0 in $(ClC_2H_4)_2O(l)$	58Dc
	dis	25	$(ClC_2H_4)_2O$	K_1 7·4 to 7·5 in $(ClC_2H_4)_2O(l)$	59Md
	con	25	C_6H_5CN	K_1 3·6 in $C_6H_5CN(l)$, → 0	64Ja
	sol	25	0 corr	$K_1 \approx -8{\cdot}4$	65Be
	fp	≈ 10	dioxan	$\beta[(H_2O)_2(HCl)_2]$ or $\beta[H_3O^+{}_2Cl^-{}_2]$ 0·37 in dioxan (l) $\{\beta[(H_2O)_n]$ −0·15(n = 2), −0·21(n = 3), −0·39 (n = 4)$\}$	65K
	var	25	$C_2H_4(NH_2)_2$	K_1 4·01 in $C_2H_4(NH_2)_2(l)$ (sp, fp, con)	65Mi
	ir, Ram		solid	ev HCl_2^- and $ClHBr^-$ in salts with R_4N^+	66Eb
	dis	25	org	$K_D[H^+ + Cl^- \rightleftharpoons HCl(org)]$ −6·46(org = $PhNO_2$), −6·6(p-xylene), −6·63(C_6H_6, PhMe, m-xylene), −6·8(PhCl), −6·7(o-xylene), −7·1(CCl_4)	66Hb
	gl	25	ROH	K_1 4·26(R = Me), 4·21(Et), 4·30(Pr), 4·43(Pr^i), 4·39(Bu) in ROH(l), → 0	66Kc
	dis		0 corr	$K_D[H^+ + Cl^- + nT(org) \rightleftharpoons HClT_n(org)]$ −2.70(n = 1), −2·85(n = 2), −3·01(n = 3), T = TBP, org = C_6H_6	/66Lf
	dis, nmr	25	C_6H_6	ev only HCl, no H_2O-HCl cpx in $C_6H_6(l)$	67Eb
	H, Ag	25	org	K_1 5·51 in 25% EtOH + 75% C_6H_6	68A
	dis	25	0 corr	$K_D[H^+ + Cl^- + T(org) \rightleftharpoons HClT(org)]$ −4·65, T = TBP, org = CCl_4	68Md
	dis	25	$PhNO_2$	K_1 7·53 in $PhNO_2$(l, sat with H_2O)	68Me
	dis		HCl var	$K_D[H^+ + Cl^- + T(org) \rightleftharpoons HClT(org)]$ −1·27, T = TBP, org = C_6H_6	68S
	X		solid	ev HCl_2^- in $Cs_3Cl_3H_3OHCl_2(s)$	68Sk = 66Sm
	con	25	HCO_2H	K_1 1·96 in $HCO_2H(l)$ (→ 0)	68W
Group 1A$^+$	act	25?	0 corr	K_1 −0·3(Rb^+), −0·1(Cs^+)	/66Mc
	con	25	HCO_2H	no ev cpx in $HCO_2H(l)$ (Li^+, Na^+, K^+)	68W
Li^+	con	20	TBP	K_1 5·3 in $(BuO)_3PO$ = TBP	63Mn
	kin	0	DMF	K_1 0·74 in DMF = $Me_2NCHO(l)$	64Wb
	fp	18	Me_2SO	K_1 ~ 0·7 in $Me_2SO(l)$, 0 corr	65Dd
	lit	25	org	K_1 0·5 in MeOH, 1·43 in EtOH, 2·2 in Pr^{i}OH, 2·3 in Bu^{i}OH, 3 in Me_2CO	/67Mc

continued overleaf

67 Cl^- (contd)

Metal	Method	Temp	Medium	Log of equilibrium constant, remarks	Ref
Na^+	sol	25	0 corr	K_{s0} (Na^+Cl^-, halite) 1·553	34A/67Ld
	sol	25	org	K_{s0} −1·5 in MeOH, −4·0 in $(Me_2N)_3PO$	67Ac
	con	400—800	0 corr	Values for K_1 at densities 0·30—0·75 g cm^{-3} $K[Na(H_2O)_x^+ + Cl(H_2O)_y^- \rightleftharpoons NaCl(H_2O)_{x+y-10} + 10H_2O]$ 17·14(400°), 17·32(500°), 17·54(600°), 17·70(700°), 17·83(800°), M units except H_2O	68Q
K^+	sol	25	0 corr	$K_{s0}(K^+Cl^-)$ 0·932	34A/
	con	25	w% EtOH	K_1 0·48 (w = 60·3), 1·06(w = 79·3), 1·37(w = 87·9) in w% EtOH, → 0	65Hd
	sol	25	org	K_{s0} −2·5 in MeOH, −5·4 in Me_2NCHO(l)	67Ac
	ΔG	25	0 corr	$K_{s0}(K^+Cl^-$, sylvite) 0·898	/67Ld
	con	25	w% C_4H_8O	K_1 1·0(w = 50), 1·59(w = 70), 2·52(w = 80), 2·97(w = 85), 3·87(w = 90) in w% C_4H_8O (THF)	68Bd
	con	300—750	KCl var	K_1^{-1} values for 300—750 °C, ρ = 0·5—1·0 g cm^{-3}	68R
Cs^+	con	25	MeOH	$K_1 \approx$ 1·2 in MeOH(l)	25F/65K
	con	25	w% EtOH	K_1 0·48(w = 40·4), 1·26(w = 73·9), 1·83(w = 91·3), 2·20(w = 100) in w% EtOH, → 0, also K_1 for some more w values	65Hd
	con	25	MeOH	K_1 0·95 in MeOH(l)	65Ka
	sol	25	org	K_{s0} −0·53 in H_2NCHO(l), −4·9 in Me_2NCHO(l)	67Ac
Be^{2+}	dis	20	0·691 $H(ClO_4)$	K_1 −0·36	65Me
	var		solid	ev $BeCl_4^{2-}$ in K_2BeCl_4(s) and Na_2BeCl_4(s) (X, ir, Ram)	65S
Mg^{2+}	sol	25	0 corr	$K_s[K^+Mg^{2+}Cl^-_3(H_2O)_3$, carnallite] 4·00, $K_s[Mg^{2+}Cl^-_2(H_2O)_6$, bischofite] 4·445	53Z/67Ld
	tp	25	EtOH var	ev $MgCl^+$ in EtOH–H_2O mixts	66Ga
	Ram		$SOCl_2$	ev $MgCl_4^{2-}$ in $SOCl_2$(l)	68Dc
Ba^{2+}	Ram	700	$KBaCl_3$(l)	ev $BaCl_3^-$ in $KBaCl_3$(l)	64V
	con	255—520	$BaCl_2$ var	K_1 1.4(255°), 1·7(298°), 2·5(420°), 3·0(520°), K_2 1·76(420°), 2·0(520°), at ρ = 1·0 g cm^{-3}, also values at ρ = 0·5 to 0·9	68R
Sc^{3+}	cix	20	0·691 $H(ClO_4)$	K_1 0·04, K_2 −0·15	64R
	cix		2 $(NaClO_4)$, 0·5 H	K_1 0·14, β_2 0·54	64Sh
	aix		10 $(HClO_4?)$	β_4 −1·88	
	dis		6 $H(ClO_4)$	K_1 1·45, β_2 0·70, β_3 −0·22, β_4 −0·37, $K_D[H^+ + Sc^{3+} + 4Cl^- + 2T(org) \rightleftharpoons HScCl_4T_2(org)]$ 0·08, org = T = TBP = $(BuO)_3PO$, [T] = 0·74M	
	cix	20	1 $H(ClO_4)$	K_1 −0·80, β_3 0·16, no ev $ScCl_2^+$	65Ag
	aix		HCl var	no ev ani cpx	66Bc
	cix	25	4 $H(ClO_4)$	K_1 −0·12, β_2 −0·84, β_3 −1·40	66Sf
Y^{3+}	qh	25	2 $(NaClO_4)$	K_1 −0·3, $\beta(Y^{3+}Cl^-CH_3CO_2^-)$ 1·18	63F
Group $4f^{3+}$	ir	400	MCl(l)	ev cpx in $Li_{0.49}Na_{0.17}K_{0.34}Cl$(l) for Pr, (Nd), Sm, Eu, Tb, Ho, not for Dy, Er, Tu, Lu	65Mj
	aix, dis		LiCl conc	ev MCl_2^+ in LiCl conc, MCl_4^- in aix (La, Lu)	66Ma
	pre, sp		var	ev MCl_6^{3-} in $C_2H_4(CN)_2$(l), MeCN(l), $(Ph_3PH)_3MCl_6$(s) and $(pyH)MCl_6$(s), M = Ce, Pr, Nd, Sm, Gd, Dy, Ho, Er, Tm, Yb	66R

contin

Metal	Method	Temp	Medium	Log of equilibrium constant, remarks	Ref
La^{3+}	sol	25	$0{\cdot}5\ H(ClO_4)$	$K_1\ -0{\cdot}46$	62Sd
	qh	25	$2\ (NaClO_4)$	$K_1\ -0{\cdot}39$, $\beta(La^{3+}Cl^-CH_3CO_2^-)\ 0{\cdot}79$	63F
	cal	25	$\rightarrow 0$	$\Delta H_1 = 1{\cdot}2$	64La
	dis	25	$4\ Na(ClO_4)$	$K_1\ -0{\cdot}22$, $\beta_2\ -0{\cdot}64$	64Sb = 65Sd
	Ag	25	$1\ (NaClO_4)$	$K_1\ 0{\cdot}20$	65Ga
Ce^{3+}	cal		var	$\Delta S_1 = 18$	62Mo/67A
Ce^{4+}	*X*		solid	ev $CeCl_6^{2-}$ in $Cs_2CeCl_6(s)$	66Ka
Pr^{3+}	var	25	0 corr	$K_s(Pr^{3+}OH^-{}_{2.5}Cl^-{}_{0.5})\ -19{\cdot}26$ (sol, gl, Ag)	63Ad
	Ag	25	$1\ (NaClO_4)$	$K_1\ 0{\cdot}20$	65Ga
Nd^{3+}	sp	25	HCl var	$K_1\ -2{\cdot}62$ (0 corr)	64Mc
	Ag	25	$1\ (NaClO_4)$	$K_1\ 0{\cdot}21$	65Ga
	sp		80% MeOH	$K_1\ 1{\cdot}39$ in 80% MeOH	67Rc
Sm^{3+}	sol	25	$0{\cdot}5\ H(ClO_4)$	$K_1\ -0{\cdot}46$	62Sd
	Ag	25	$1\ (NaClO_4)$	$K_1\ 0{\cdot}21$	65Ga
Eu^{3+}	cix	26	$1\ H(ClO_4)$	$K_1\ -0{\cdot}10$, $\beta_2 \approx 0{\cdot}82$	64Bb
			$1\ Na(ClO_4)$	$K_1\ 0{\cdot}13$	
	cix, dis	20	$1\ H(ClO_4)$	$K_1\ -0{\cdot}03$ (cix), $-0{\cdot}01$ (dis) (or $K_1 \approx -0{\cdot}2$, $\beta_2 \approx -0{\cdot}6$)	64I
	dis	25	$4\ Na(ClO_4)$	$K_1\ -0{\cdot}15$, $\beta_2\ -0{\cdot}72$	64Sb = 65Sd
	cix	25	$4\ H(ClO_4)$	$K_1\ -0{\cdot}27$, $\beta_2\ -0{\cdot}90$, $\beta_3 \approx -2{\cdot}0$	67Sd
			$4\ Na(ClO_4)$	$K_1\ -0{\cdot}06$, $\beta_2\ -0{\cdot}48$, $\beta_3 \approx -1{\cdot}7$	
Ho^{3+}	sp	22	0 corr	$K_1\ -2{\cdot}03$	66Mg
Er^{3+}	*p*	800—1200	gas	ev KCl(g), $ErCl_3(g)$, $KErCl_4(g)$, equil constants given	65Nb
	sp	22	0 corr	$K_1\ -1{\cdot}93$	66Mg
Yb^{3+}	sol	25	$0{\cdot}5\ H(ClO_4)$	$K_1\ -0{\cdot}56$	62Sd
Lu^{3+}	sol	25	$0{\cdot}5\ H(ClO_4)$	$K_1\ -0{\cdot}48$	62Sd
	dis	25	$4\ Na(ClO_4)$	$K_1\ -0{\cdot}35$, $\beta_2\ -0{\cdot}57$	64Sb = 65Sd
Ac^{3+}	dis	27	$1\ H(ClO_4)$	$K_1\ -0{\cdot}10$, $\beta_2\ -0{\cdot}62$	68Sc
Ti^{3+}	*X*		solid	ev $Ti_2Cl_9^{3-}$ in $Cs_3Ti_2Cl_9(s)$	57Wa
	pre, sp		solid	ev $TiCl_6^{3-}$ in $(C_5H_6N)_3TiCl_6(s)$	67Fa
	sp	25	0 corr	$K_1\ -1{\cdot}15$ to $-0{\cdot}7$, two uncertain sets of K_2—K_6 given	67G
Ti^{4+}	Ram, ir		HCl var	ev $TiCl_6^{2-}$ in > 12 M-HCl	67Gb
	Ram		HCl conc	ev $TiO_2Cl_4^{4-}$ or $TiOCl_5^{3-}$, not $TiCl_6^{2-}$	68De
	con, ir		var	ev $TiOCl_5^{3-}$ and $TiOCl_4^{2-}$ in solids and in MeCN(l)	68Fc
	X		solid	ev pyramidal $TiOCl_4^{2-}$ in $(Et_4N)_2TiOCl_4(s)$	68H
	dis, ir		HCl var	$K_D[Ti^{4+}+4Cl^-+2T(org) \rightleftharpoons TiCl_4T_2(org)]$ $-1{\cdot}0$, T = TBP, org = C_6H_6, "$TiCl_4T_2$" = $(H_3OT^+)_2Ti(OH)_2Cl_4^{2-}$ (does not H^+ enter equilibrium?)	68S
TiO^{2+}	var		(H, Li)Cl var	ev mono- and polyn cpx (sp, dialysis)	67Na
	cix		$3\ (NaClO_4)$	$K_1\ 0{\cdot}55$, $K_2\ -0{\cdot}40$	67Nb
	cix		LiCl var, $0{\cdot}2H^+$	$K_3\ -1{\cdot}03$, $K_4\ -1{\cdot}1$	
Zr^{4+}					

1964 *Erratum*: p. 275, *in ref* 62 Mb *for* "K_2" *read* "K_1"

Metal	Method	Temp	Medium	Log of equilibrium constant, remarks	Ref
	Ram, ir		HCl var	ev $ZrCl_6^{2-}$ in > 12M-HCl	67Gb
	Ram		HCl conc	ev $ZrCl_6^{2-}$	68De

continued overleaf

67 Cl^- (contd)

Metal	Method	Temp	Medium	Log of equilibrium constant, remarks	Ref
Zr^{4+}	pre, ir		solid	ev $Zr_2OCl_{10}^{4-}$ in $(Et_4N)_4Zr_2OCl_{10}$(s)	68Fd
(*contd*)	sp		MeCN	ev $ZrCl_4$, $ZrCl_5^-$, $ZrCl_6^{2-}$ in MeCN(l)	68O
Hf^{4+}	dis	27	2 $H(ClO_4)$	K_1 −0·15, β_2 −0·32	65Da
	cix	20	2 $H(ClO_4)$	K_1 0·07, β_2 −0·48, β_3 −0·40	67Ee
	dis, cix	20	3 $H(ClO_4)$	K_1 0·34, (dis), 0·18(cix), β_2 −0·02(dis)	67H
	Ram		HCl conc	ev $HfCl_6^{2-}$	68De
Th^{4+}					

1964 *Erratum*: p. 275, *in ref* 52W *for* "K_7" *read* "K_2".

Metal	Method	Temp	Medium	Log of equilibrium constant, remarks	Ref
	cix, tp		4 $H(ClO_4)$	K_1 0·20, β_2 −0·80, β_3 −0·85, β_4 −1·46, β_5 −2·46	64Nb
V^{3+}	*X*		solid	ev $V_2Cl_9^{3-}$ in $Cs_3V_2Cl_9$(s)	57Wa
	sp		solid	ev $VCl_2(H_2O)_4^+$ (green) in $KVCl_4(H_2O)_6$(s) and $VCl_3(H_2O)_{6,4}$(s)	64Hd
	var		solid	ev VCl_4^- in $Ph_4As^+VCl_4^-$(s), ev VCl_3Br^- in $Et_4N^+VCl_3Br^-$(s) (pre, sp, ir, *X*)	66Cc
	pre, *X*		solid	ev $VCl_5H_2O^{2-}$ in $K_2VCl_5H_2O$(s) *etc.*	66Sa
	pre, sp		solid	ev VCl_6^{3-} in $(C_5H_6N)_3VCl_6$(s)	67Fa
V^{4+}	var		MeCN	ev VCl_6^{2-} in MeCN(l), (sp, con, mag)	65Fa
	var		solid	ev VCl_6^{2-} in solids (ir, mag, pre, sp)	65Kc
VO^{2+}	var		solid	ev $VOCl_4^{2-}$ and $VOCl_3^-$ in solids (mag, ir, pre, sp)	66K
	X		solid	ev pyramidal $VOCl_4^{2-}$ in $(Et_4N)_2VOCl_4$(s)	68H
V^{5+}	*X*		solid	ev $VOCl_5^{2-}$ in M_2VOCl_5(s) (M = Rb, Cs)	67W
VO_2^+	sp	19	1·01($NaClO_4$)	K_1 −0·38	66Ia
$Nb^{2+,3+}$	var		var	ev $Nb_6Cl_{12}^{2+}$ (con, in EtOH), ev $Nb_6Cl_{12}(OH)_4^{2-}$ (gl) (also sp, ir)	65Ae
	pre		var	ev $Nb_6Cl_{12}^{2+}$ and $Nb_6Cl_{12}^{4+}$ (see also Table 1, e^-)	65Mh
	X		solid	ev $Nb_6Cl_{12}^{2+}$ in Nb_6Cl_{14}(s)	65Sc
	var		org	ev $Nb_6Cl_{18}^{4-}$, $Nb_6Cl_{18}^{3-}$, $Nb_6Cl_{18}^{2-}$ in Me_2SO, $MeNO_2$(l) (con, pol, ir, mag, esr, pre)	67M
	pre		var	ev $Nb_6Cl_{12}^{3+}$	67Sc
	X		solid	ev $Nb_6Cl_{18}^{4-}$ in $K_4Nb_6Cl_{18}$(s)	68Sl
Nb^{3+}	sp		HCl var	ev ani cpx	55Ca
Nb^{4+}	sp, pol		HCl var	ev ani cpx	55Ca
Nb^{5+}	*X*		solid	ev Nb_2Cl_{10} units in $NbCl_5$(s)	58Zc
	var		solid	ev $NbOCl_5^{2-}$ in Cs_2NbOCl_5(s) (ir, mag, *X*)	64Bl
	var		org	ev Nb_2Cl_{10}, $NbCl_5$ no ev $NbCl_4^+$, $NbCl_6^-$ in CCl_4(l), $MeNO_2$(l) (bp, con, ir)	65Kb
	X		solid	ev $NbOCl_5^{2-}$ in M_2NbOCl_5(s) (M = Cs, Rb, NH_4)	65L
	ir		solid	ev $NbOCl_5^{2-}$ in Cs_2NbOCl_5(s)	66S
	Ram, ir		HCl var	ev $NbOCl_5^{2-}$ in > 12M-HCl	67Gb
	dis		HCl var	$K_D[Nb^V + 4Cl^- + 3T(org) \rightleftharpoons NbOHCl_4T_3(H_2O)_x(org)]$ −1·8, T = TBP, org = C_6H_6 (does not H^+ take part in equilibrium?)	68Sb
$Ta^{2+,3+}$	pre		var	ev $Ta_6Cl_{12}^{2+}$ and $Ta_6Cl_{12}^{4+}$	65Mh
	pre, sp		var	ev $Ta_6Cl_{12}^{3+}$ (also $Ta_6Cl_{12}^{2+ \text{ and } 4+}$)	67Sc
Ta^{3+}	fp	500—800	MCl(l)	ev $TaCl_5^{2-}$ in KCl(l), RbCl(l) and CsCl(l)	64S
Ta^{5+}	gl	25	MeOH	$K(TaCl_5 + MeOH \rightleftharpoons TaOMeCl_4 + H^+ + Cl^-)$ −3·55 $K[TaCl_5 + 2MeOH \rightleftharpoons Ta(OMe)_2Cl_3 + 2H^+ + 2Cl^-]$ −7·83 in MeOH, ev other cpx	65G
	Ram, ir		HCl var	ev $TaCl_6^-$ in > 12M-HCl	67Gb

continued

Metal	Method	Temp	Medium	Log of equilibrium constant, remarks	Ref
Pa^{4+}	dis		3 H(ClO₄)	$*K_1 < 0$, $*\beta_2$ 0 (HCl–H^+ as reagent?)	65Gb
	dis	25	3 H(ClO₄)	β_2 0, $\beta_3 < 0$	66Ge
	var		var	ev $PaCl_6^{2-}$ in solids and soln (pre, ir, *X*)	67Bf
Pa^{5+}	var		var	ev $PaCl_6^-$, $PaCl_8^{3-}$ in solids, $SOCl_2$(l), $MeNO_2$(l) (ir, pre, *X*)	64Bk
	dis		3 H, 3 Na(ClO₄)	ev $HPaOCl_4(TBP)_2$ in kerosene(l)	64Bm
			var	ev $PaCl_5H_2O$ in Bu^{i}COMe(l)	
	dis	25	1 HClO₄	K_1 0·21, K_2 −0·89	67K
$Pa(OH)_3^{2+}$	dis		3 (LiClO₄)	K_1 0·08	65Gb
	dis	25	3 H(ClO₄)	K_1 0·08	66Ge
	dis	20?	0 corr	[$M^{2+} = Pa(OH)_3^{2+}$ or PaO_2H^{2+}], *K* or $\beta(M^{2+}H^+)$ 0·0; $\beta(M^{2+}H^+{}_2Cl^-{}_6)$ −11·26, $\beta(M^{2+}H^+{}_3Cl^-{}_6) = K[Pa(OH)_3^{2+} + 3H^+ + 6Cl^- \rightleftharpoons PaCl_6^-]$ −13·70; $K_D[M^{2+} + (2n-1)Cl^- + (2n-3)H^+ + HB_2{}^+Cl^-(org) \rightleftharpoons HB_2{}^+MH_{2n-3}Cl_{2n}{}^-(org)]$ 0·64(n = 1), −1·64(n = 2), −8·49(n = 3), org = C_6H_6, B = Bu$^i_2$CHOH	66Sk
Cr^{3+}	*X*		solid	ev $Cr_2Cl_9^{3-}$ in $Cs_3Cr_2Cl_9$(s)	57Wa
	sp	→ 0	4·4 (HClO₄)	K_2 −1·82	58Ga/67E
	cal	25	5·1 (NaClO₄)	$\Delta S_1 = 19$	58Sc/67A
	cix	30—60	0·418 (NaClO₄)	K_1 −0·96(30°), −0·74(44°), −0·49(60°),	64Bi
		30—60	70·6% MeOH	K_1 1·22(30°), 1·47(44°), 1·72(60°) in 70·6% MeOH, I = 0·418M, values also for other H_2O–MeOH mixtures	
	mag		solid	ev $CrCl_4^-$? in $PCrCl_8$(s)	64Mh
	X		solid	ev *trans*-$CrCl_2(OH_2)_4^+$ in $CrCl_2(OH_2)_4Cl(H_2O)_2$(s)	65Dc
	kin	40	2 (LiClO₄)	K_1 −0·65	66A
	kin, sp	0	1 (H, LiClO₄)	K_2 −2·44	67E
	pre, sp		solid	ev $CrCl_6^{3-}$ in $(C_5H_6N)_3CrCl_6$(s)	67Fa
	sp	40—80	10 H(ClO₄)	$K_1\{H_2O\}$ 0·98(60°), 1·11(80°), $\Delta H = 3·1$, $\Delta S = 14$	67Hc
			6·7 H(ClO₄)	$K_1\{H_2O\}$ 0·34(40°), 0·48(60°), 0·63(80°), $\Delta H = 3·6$, $\Delta S = 13$	
			4 H(ClO₄)	$K_1\{H_2O\}$ −0·34(40°), −0·10(60°), 0·11(80°), $\Delta H = 5·6$, $\Delta S = 16$	
			0·98 H(ClO₄)	$K_1\{H_2O\}$ −0·80(40°), −0·49(60°), −0·29(80°), $\Delta H = 6·0$, $\Delta S = 16$	
	sp	25	9 HClO₄	K_3 −0·13	67N
	ana	25	2·5 (LiClO₄)	K_1 −1·24	68Ea
	dis (sp)	10—50	1 H(ClO₄)	K_1^{out} −0·04(10°), −0·05(20°), −0·06(30°), −0·08(50°); $\Delta H_1^{out} = -0·43$, $\Delta S_1^{out} = -1·7$(25°)	68Mb
		40	1 H(ClO₄)	K_1 0·03, K_1^{in} −0·66	
$Cr(NH_3)_5^{3+}$	sp, kin	30—60	1·71 (NaClO₄)	K_1 −0·7(30°), −0·55(45°), −0·45(60°), $\Delta H_1 = 6$, $\Delta S_1 = 16$	67Db
			0·106 (NaClO₄)	K_1 −0·2(30°), 0·0(45°), 0·1(60°), $\Delta H_1 = 6$, $\Delta S_1 = 19$	
		23	0·16 (NaClO₄)	K_1 0·5	

continued overleaf

67 Cl^- (contd)

Metal	Method	Temp	Medium	Log of equilibrium, constant remarks	Ref
Cr^{3+} cpx	sp	76	org	$CrCl_2en_2^+$: $K(cis \rightleftharpoons trans)$ −0·32, K_1^{out} 3·20(*cis*), 2·3(*trans*) in Me_2NCHO(l)	67P
	con	25—35	Me_2SO	β(*cis*-Cr $en_2Cl_2^+Cl^-$) 2·48(25°), 2·45(30°), 2·43(35°), 2·28(→ 70°) in Me_2SO(l), (→ 0)	68Pe
Cr^{5+}	var		var	ev $CrOCl_5^{2-}$ and $CrOCl_4^-$ in $MeCO_2H$(l) and salts (pre, mag, sp)	63Kh
	var		solid	ev $CrOCl_5^{2-}$ in Cs_2CrOCl_5(s) (ir, mag, *X*)	64Bl
Cr^{6+}	sp	25	*I* H(Cl)	$K(H_2CrO_4+Cl^- \rightleftharpoons CrO_3Cl^-+H_2O)$ 2·0; $K(HCrO_4^-+Cl^-+H^+ \rightleftharpoons CrO_3Cl^-+H_2O)$ 1·2 ($I = 0{\cdot}5$ or 1)	64He
	sp	15—35	1 ($LiClO_4$)	$K(HCrO_4^-+Cl^-+H^+ \rightleftharpoons CrO_3Cl^-+H_2O)$ 1·04(15°), 1·05(25°), 1·09(35°), $\Delta H = 1{\cdot}1$, $\Delta S = 8{\cdot}6$	66T
Mo^{2+}	sp, *X*		HCl var, solid	ev $Mo_3Cl_{12}^{6-}$, $Mo_3Cl_{13}^{7-}$ at equilibrium	67Bd
$Mo^{2+,3+}$	var		var	ev $Mo_3Cl_{12}^{6-}$, $Mo_3Cl_{13}^{7-}$, $Mo_4Cl_{16}^{6-}$, in solids and HCl var (mag, pre, sp)	65Ad
Mo^{4+}	var		var	ev $Mo_3Cl_9^{3+}$ in solid and MeCN(l) (con, mag, sp, tp)	66Sl
Mo^{5+}	*X*		solid	ev Mo_2Cl_{10} units in $MoCl_5$(s)	59Sb
	var		var	ev $MoOCl_4^-$ in solid and SO_2(l) (mag, pre, sp)	63Ae
	var		solid	ev $MoOCl_5^{2-}$ in Cs_2MoOCl_5(s) (ir, mag, *X*)	64Bl
	var		MeCN	ev $MoCl_6^-$ in MeCN(l) (pre, con, sp, mag)	65B
	ir		solid	ev $MoOCl_5^{2-}$ in Cs_2MoOCl_5(s)	66S
	mag	20—40	HCl var	$K[2MoOCl_5^{2-}+4H_2O \rightleftharpoons Mo_2O_2(OH)_4Cl_4^{2-}+4H^+ +6Cl^-]$ 8·03(20°), 7·64(30°), 7·31(40°)	67Ja
	var		var	ev $(Cl_4OMo)_2O^{4-}$ in $(Et_4N)_4Mo_2O_3Cl_8$(s) and solution (pre, ir, mag)	68Cc
Mo^{6+}	Ram, ir		HCl var	ev *cis*-$MoO_2Cl_2(H_2O)_2$ in 6 M-HCl, *cis*-$MoO_2Cl_4^{2-}$ in > 12 M-HCl	67Gb
	X		solid	ev $MoO_2Cl_4^{2-}$ in $M_2MoO_2Cl_4$(s) (M = NH_4, Rb, Cs)	67W
	var		PhCOMe	ev $H_9O_4^+MoO_2Cl_3H_2O^-$ in PhCOMe(l) (dis, con, sp, ir)	68Za
W^{2+}	ir, pre		solid	ev $W_6Cl_{14}^{2-}$ in $(Et_4N)_2W_6Cl_{14}$(s)	67C
W^{3+}	*X*		solid	ev W_6Cl_{18} groups in WCl_3(s)	67Sg
W^{4+}	var		var	ev $(Cl_5W)_2O^{4-}$ (not $WOHCl_5^{2-}$) in $K_4W_2OCl_{10}$(s) (pre, ir, mag)	68Cc
W^{5+}	pre		var	ev $WOCl_4^-$, $WOCl_5^{2-}$	18C
	var		solid	ev $WOCl_5^{2-}$ in Cs_2WOCl_5(s)	64Bl
	var		var	ev $W_3Cl_{12}^{3+}$ in WCl_5(s) and MeCN(l) (mag, con, Ag)	66Cd
	ir		solid	ev $WOCl_5^{2-}$ in Cs_2WOCl_5(s)	66S
	X, ir		solid	ev W_2Cl_{10}, not $W_3Cl_{12}^{3+}$ in WCl_5(s)	67Bi
	mag	1—40	HCl var	$K[2WOCl_5^{2-}+6H_2O \rightleftharpoons W_2O_2(OH)_6Cl_2^{2-}+8Cl^-+ 6H^+]$ 13·23(1°), 12·94(20°), 12·58(40°)	67Jb
W^{6+}	var		var	ev $WOCl_5^-$ in MeCN(l), CH_2Cl_2(l), solid (con, pre, ir, sp, *X* powder)	66Fa
	Ram, ir		HCl var	ev *cis*-$WO_2Cl_4^{2-}$ in > 12 M-HCl	67Gb
	var		solid	ev $WO_2Cl_4^{2-}$ in $Cs_2WO_2Cl_4$(s) *etc.* (pre, ir, *X* powder)	68Pg
Group $5f^{3+}$	aix, dis		LiCl conc	ev MCl_2^+ in LiCl conc, MCl_4^- in aix (Am, Cf)	66Ma+

continued

Metal	Method	Temp	Medium	Log of equilibrium constant, remarks	Ref
U^{3+}	sp	25	LiCl var	K_1 −2·85	62Se
	sp	25	0 corr	K_1 −2·89	65Sh
U^{5+}	con		$SOCl_2$	ev UCl_6^-, UCl_7^{2-}, UCl_8^{3-} in SO_2Cl_2(l)	64Bj
	X		solid	ev U_2Cl_{10} in UCl_5(s)	67Sa
Np^{3+}	sp		LiCl var	K_1 −2·42, β_2 −4·96	64Sd ≈ 66Sd
Np^{4+}	sp	25	2 ($NaClO_4$)	K_1 −0·28	62Sf
	X		solid	ev $NpCl_6^{2-}$ in Cs_2NpCl_6(s)	66Ba
	dis	20	2 H(ClO_4)	K_1 0·04, β_2 −0·15	66Sj
			1 H(ClO_4)	K_1 −0·04, β_2 −0·24, β_3 −0·48	
			0·5 H(ClO_4)	K_1 0·15	
Np^{5+}	ir		solid	ev $NpOCl_5^{2-}$ in Cs_2NpOCl_5(s)	66Ba
NpO_2^+	cix	25	2 H(ClO_4)	K_1 −0·29	64Ga
NpO_2^{2+}	sp	25	2 ($NaClO_4$)	K_1 −0·21	62Sf
Pu^{3+}	sp		LiCl var	K_1 −2·43, β_2 −5·00	64Sd ≈ 66Sd
Pu^{4+}	dis	25	4 ($HClO_4$)	K_1 0·30, β_2 −0·80	66D
PuO_2^{2+}	dis	?	4.1 (ClO_4)	K_1 0·02, β_2 −0·8 ?	65Md
Am^{3+}	cix	26	1 H(ClO_4)	K_1 −0·05	64Bb
			1 Na(ClO_4)	K_1 0·15	
	dis		1 Na(ClO_4)	K_1 −0·60, β_2 −0·42	64K
	dis	25	4 Na(ClO_4)	K_1 −0·15, β_2 −0·69	64Sb
	sp		LiCl var	K_1 −2·21, β_2 −4·70	64Sd
	dis	25	4 Na(ClO_4)	K_1 −0·15, β_2 −0·69	65Sd
	nmr		var	K_1 0·03, K_2 −1·00	66Vb
	X, pre		solid	ev $AmCl_6^{3-}$ in Cs_3AmCl_6(s)	68Bb
AmO_2^+	pre		solid	ev trimeric cpx in "$Cs_8(AmO_2)_3Cl_{11}$(s)", withdrawn in ref 68Bb	67Bj
	X, pre		solid	ev $AmO_2Cl_4^{3-}$ in $Cs_3AmO_2Cl_4$(s)	68Bb
AmO_2^{2+}	pre, X		solid	ev $AmO_2Cl_4^{2-}$ in $Cs_2AmO_2Cl_4$(s)	67Bj
Mn^{2+}	X		solid	ev $MnCl_4(H_2O)_2^{2-}$ in $M_2MnCl_4(H_2O)_2$(s) (M = Cs or Rb)	64J
	X		solid	ev $MnCl_2(H_2O)_4$ in α-$MnCl_2(H_2O)_4$(s)	64Z
	act	25 ?	0 corr	K_1 0·1	/66Mc
	X		solid	ev $Mn_2Cl_6(H_2O)_4^{2-}$ units in $KMnCl_3(H_2O)_2$(s) and β-$RbMnCl_3(H_2O)_2$(s)	68J = 67J
	X		solid	ev $MnCl_4(H_2O)_2^{2-}$ units in $K_2MnCl_4(H_2O)_2$(s)	68Ja
Mn^{4+}	X, sp		solid	ev $MnCl_6^{2-}$ in K_2MnCl_6(s)	66M
$Tc^{2+,\,3+}$	pre, tp		var	ev $Tc_2Cl_8^{3-}$ in conc HCl and solids	63Eb
	X		solid	ev $Tc_2Cl_8^{3-}$ in $(NH_4)_3Tc_2Cl_8(H_2O)_2$(s)	65Cc
Tc^{4+}	Ag	15	3 $HClO_4$	K_6 4·66	65Sj
	X		solid	ev $TcOHCl_5^{2-}$ in $K_2TcOHCl_5$(s)	67Ec
Re^{2+}	kin	25	0·32 H_2SO_4	K_1 2·0	65P
$Re^{2+,\,3+}$	pol	25	MeCN	ev $Re_2Cl_8^{3-}$, $Re_2Cl_8^{4-}$ in MeCN(l)	67Ca
Re^{3+}	pre, sp		var	no ev $ReCl_6^{3-}$, earlier preparations contain $ReCl_6^{2-}$	64Ca
	X		solid	ev Re_3Cl_9 in $ReCl_3$(s),	64Cb
	sp, fp		var	ev Re_3Cl_9 in fresh solutions in H_2O, CH_3CO_2H Me_2CO, sulpholane	
	X		solid	ev Re_3Cl_9 groups in $ReCl_3$(s)	64Cc

continued overleaf

67 Cl⁻ (contd)

Metal	Method	Temp	Medium	Log of equilibrium constant, remarks	Ref
Re^{3+} (contd)	X		solid	ev $Re_3Cl_{11}^{2-}$ in $(Ph_4As)_2Re_3Cl_{11}(s)$	64Fa = 66P
	var		org	ev $Re_3Cl_{11}^{2-}$, $Re_3Cl_{12}^{3-}$ in org, solids (con, pre, pMe₂CO, sp)	64Rb
	var		var	ev $Re_2Cl_8^{2-}$ in HCl conc, MeOH, MeCN(l) *etc.* (pre, sp, con, mag)	65Ca
	X		solid	ev $Re_2Cl_8^{2-}$ in $K_2Re_2Cl_8(H_2O)_2(s)$	65Cb
$Re^{3+,4+}$	var		MeCN	ev $Re_2Cl_9^-$ and $Re_2Cl_9^{2-}$ (con, pre, sp) in MeCN(l) *etc.*	67B
Re^{4+}	var		solid, 6HCl	ev $ReCl_6^{2-}$ in $K_2ReCl_6(s)$ and 6M-HCl (mag, sp, X)	58Db
	mag	5—25	HCl var	$K(Re_2OCl_{10}^{4-}+H_2O \rightleftharpoons 2ReOHCl_5^{2-})$ −2·72(5°), −2·04(15°), −1·41(25°), ΔH = 24, ΔS = 80	65Ja
	Ag	15	3 $HClO_4$	K_6 6·34	65Sj
	cal	25	var	$\Delta H[ReCl_6^{2-}+4\,OH^- \rightleftharpoons ReO_2(H_2O)_2(s)+6Cl^-]$ −74·0 (fresh solid), −79·1 (aged solid, est)	66Be+
Re^{5+}	var		var	ev $Re(OH)_2Cl_4^-$ in 12 M-HCl and Cs^+ salt (pre, mag, sp)	65Ba
	var		var	ev $ReOCl_4^-$ in solid and org (pre, con, ir, sp)	66Ca
	X		solid	ev Re_2Cl_{10} units in $ReCl_5(s)$	68Mc
Re^{6+}	var		var	ev $ReOCl_6^{2-}$ in solids and 12M-HCl (mag, pre, sp)	65Cg
	var		var	ev $ReOCl_5^-$ (pre, con, X, ir) and $Re_2O_3Cl_8^{2-}$ (con, pre, ir) in $MeNO_2$(l) and solids	68Ba
Fe^{2+}	kin, sp		1($HClO_4$)	$K_1 < -0·3$	68Pc
Fe^{3+}	dis	25	0 corr	K_2 −0·57, K_3 −2·96	59Md
	sp	23	5($NaClO_4$), hH⁺	K_1 1·68(h = 5), 1·62(h = 4), 1·54(h = 3), 1·45(h = 2), 1·30(h = 1)	63Hd
			I($HClO_4$)	K_1 1·68(I = 5), 1·30(I = 4), 0·89(I = 3), 0·70(I = 2), 0·59(I = 1); also K_1 values for some other (Na, H)ClO_4 media	
	X, sp		(Fe, HCl, ClO_4) var	ev $FeCl_4(H_2O)_2^-$, $FeCl_4^-$, polyn cpx	64Bf
	con	25	var, p	K_1 1·32 (1 bar), ≈ −0·4 (2100 bar)	64Hc
	oth	350—530	gas	Equil data for $NaCl(s)+FeCl_3(g) \rightleftharpoons NaFeCl_4(g)$	64Ra
	tp		0 corr	K_3K_4 −4·54	64Va
	sp, ir		$MeCONMe_2$	ev $FeCl^{2+}$, $FeCl_2^+$, $FeCl_4^-$, $\beta_4/\beta_2 > 10$ in $MeCONMe_2$(l)	65Db
	X		solid	ev $FeCl_4^-$ in $NaFeCl_4(s)$	65Ra
	var	25	org	K_4/K_3 −1·55 (in pyridine), 1·42 (in Me_2NCHO), ⩾ 3(in MeCN, $MeNO_2$) (sp, emr), ΔH = 11·8, ΔS = 32 in pyridine(l)	65Sg
	var	25	0·5(Na)Cl	ev $Fe(OH)_{2.70}Cl_{0.30}(s)$ (= "β-FeOOH") $K_s(Fe^{3+}H^+_{-2.7})$ 3·04 (red, gl, X) $K_s(Fe^{3+}H^+_{-2.7}Cl^-_{0.3})$ 2·95 (red, gl, X)	66B /66B
	kin	25	var	K_1^{in}/K_1^{out} −0·37 (con, p-jump)	66Kb ≈ 65Kg
	sp	25	Me_2SO	K_1 3·72, β_2 5·76 in Me_2SO(l), I = 0·1($NaClO_4$), 0·02M-H⁺	66W
	var		$FeCl_3$ var	ev polyn uncharged Fe–Cl–OH species (gl, ir, sp, tp)	67Bb
	ir		solid	ev $Cl_3FeCl_3FeCl_3^{3-}$ in α-$Cs_3Fe_2Cl_9(s)$	67Cc
	sp		0 corr	K_4 −2·54 to −2·24 (K_1, K_2, K_3 from ref 42R)	67Da
	X		var	ev *trans*-$FeCl_2(H_2O)_4^+$ in $FeCl_3(H_2O)_6(s)$ and soln	67Lf

continued

Metal	*Method*	*Temp*	*Medium*	*Log of equilibrium constant, remarks*	*Ref*
Fe^{3+} (*contd*)	cal	25	$I(HClO_4)$	$\Delta H_1 = -0{\cdot}77(I = 8), 0{\cdot}40(I = 7), 2{\cdot}14(I = 5), 2{\cdot}89(I = 4), 3{\cdot}33(I = 3), 3{\cdot}98(I = 2)$ Also ΔH_1 for these I and 10°—55°, ΔC_p and ΔS values	67V
		10—55	0 corr	$\Delta H_1 = 4{\cdot}64(10°), 5{\cdot}14(18°), 5{\cdot}60(25°), 6{\cdot}16(35°), 6{\cdot}68(45°), 7{\cdot}20(55°)$, $\Delta S_1 = 22{\cdot}1(10°), 25{\cdot}1(25°), 30{\cdot}2(55°)$	
	lit	25	$I(HClO_4)$	K_1 3·13$(I = 8)$, 2·46$(I = 7)$, 1·50$(I = 5)$, 1·10$(I = 4)$, 0·76$(I = 3)$, 0·53$(I = 2)$, 1·38(0 corr)	/67V
	cal	10—55	4 $(HClO_4)$	$\Delta S_1 = 13{\cdot}0(10°), 17{\cdot}5(55°)$, linear with T, $\Delta C_{p1} = 32{\cdot}1(10°), 27{\cdot}6(55°)$, linear with T, $\Delta S_1 = +\Delta C_{p1} \approx 45{\cdot}1$, independent of T	67V/68Va
			0 corr	$\Delta S_1 = 22{\cdot}1(10°), 30{\cdot}2(55°)$, linear with T $\Delta C_{p1} = 61{\cdot}1(10°), 52{\cdot}8(55°)$, linear with T $\Delta S_1 + \Delta C_{p1} = 83{\cdot}1$, independent of T	
	sp		org	ev $FeCl_3$, no ev Fe_2Cl_6 in Et_2O, C_6H_6, C_5H_5N *etc.*	68Fa+
	dis, red	25	org	ev uncharged clusters of $(C_{12}H_{25})_3NH^+$, Cl^-, $FeCl_4^-$, $FeCl_5^{2-}$ and $FeCl_6^{3-}$ in *o*-xylene (equil constants given)	68K
	kin, sp	0·9	2$(LiClO_4)$	K_1 0·23	68N
	sp	25	Me_2SO	K_1 3·62, K_2 2·10 in Me_2SO(l) $I = 0{\cdot}10$ $(NaClO_4?)$	68Wa
$Fe(C_5H_5)_2^+$	?	25	$2H(ClO_4)$?	K_1 0·48	63Pe
	red	25	0 corr	K_1 0·45	67Nc
Fe cpx	red	25	0 corr	$\beta(Q^{2+}Cl^-)$ 1·0 $[Q^{2+} = C_5H_5FeC_5H_4CH_2N(CH_3)_3^{2+}$ = ferricenylmethyl]	67Le
Co^{2+}	oth	160	MNO_3(l)	K_1 1·11, β_2 2·02, β_3 2·40 in (Li, K)NO_3 (1, eut) [adsorption equil with Al_2O_3(s)]	63La
	sp		0 corr	$K[CoCl_4(H_2O)_2^{2-}$ (octahedral) $\rightleftharpoons CoCl_4^{2-}$ (tetrahedral)$+2H_2O]$ −0·36	63Va
	tp		0 corr	K_2K_3 −6·72	64Va
	fp	≈ 260	$LiNO_3$(l)	$K_1 \approx -0{\cdot}9$ in $LiNO_3$(l)	66I
	cal	25—40	2 $(NaClO_4)$	$\Delta H_1 = 0{\cdot}52(25°), 0{\cdot}50(40°)$, K_1 −0·14(25°), −0·12(40°), $\Delta S_1 = 1{\cdot}11(25°)$	66Kd+60La
	sp	25—500	$CoCl_2$ dil	ev tetrahedral $CoCl_2(H_2O)_2$ at > 300 °C	67La
	sp	25	3 $NaClO_4$	K_1 −0·28, β_4 −2·15	67Mf
			6 $NaClO_4$	K_1 −0·43, β_4 −2·82	
			7·3 $NaClO_4$	K_1 −1, β_4 < −3	
	sp	25—63	H_2O, MeOH	$\Delta H[CoCl_2T_4 + Cl^- \rightleftharpoons CoCl_3T^-(+3T)] = 11{\cdot}7$ $(T = H_2O)$, 13·6(T = MeOH) in T(l)	67S
	sp	25—500	Cl^- var	ev tetrahedral $CoCl_3H_2O^-$ at > 200 °C in 1—4 M NaCl	68L
	con, sp	25	MeCN	$K_3^2\beta_2^{-1}$ $(3CoCl_2T_2 + 2T \rightleftharpoons CoT_6^{2+} + 2CoCl_3T^-)$ −1·81 in T = MeCN(l) (→ 0)	68La
	sp	40—80	MeCN	$K(CoCl_2T_2 + HgCl_2 \rightleftharpoons CoClT_5^+ + HgCl_3^-)$ −0·7(40°), −0·3(50°), 0·0(60°), 0·3(70°), 0·6(80°); $\Delta H = 17$, $\Delta S = 50$, in T = MeCN(l)	68Mi
	nmr	27	HCl var	K_1 −0·8, $\Delta H_1 = 2{\cdot}9$, $\Delta S_1 = 6$; K_2 −2·8, $\Delta H_2 = 2{\cdot}1$, $\Delta S_2 = 6$; K_3 −2·5, $\Delta H_3 = 11{\cdot}3$, $\Delta S_3 = 26$; K_4 −2·06, $\Delta H_3 = 0{\cdot}8$, $\Delta S_4 = -6{\cdot}7$ (0 corr for Cl^-, not for Co species); Co^{2+} and $CoCl^+$ octahedral, others tetrahedral	68Z

continued overleaf

67 Cl^- (contd)

Metal	*Method*	*Temp*	*Medium*	*Log of equilibrium constant, remarks*	*Ref*
Co^{2+}, Al^{3+}	sp	300	(K, Al)Cl(l)	$K[Co(Al_2Cl_7)_2 + AlCl_4^- \rightleftharpoons Co(Al_2Cl_7)AlCl_4 + Al_2Cl_7^-]$ −1·45 in $K_xAl_{1-x}Cl_{3-2x}$(l), x = 0·42—0·50	65O
Co^{2+}, Sn^{2+}	var		var	ev $SnCl_2[Co(CN)_5]_2^{6-}$ and cpx with 1 Sn/Co (pre, ir, sp)	66V
Co^{2+} cpx	sp, gl		MCl var	Co (dimethylglyoxime)$_2^{2+}$; β_2 0·9(M = Li), K_1 0·2, K_2 0·2 (M = Na), β_2 0·12 (M = Rb)	67Ba
Co^{3+}	kin	25	3 ($NaClO_4$)	K_1 1·42	66C
$Co(NH_3)_6^{3+}$	con	25	dil	$K_1^{out}(p\ atm)/K_1^{out}(1\ atm)$ 0·30(100 atm), 0·34(200 atm), 0·42(300 atm), 0·44(400 atm), 0·43(500 atm) 0·41(600 atm) (read from graph) for 10^{-4}–10^{-3} M-MCl_3	63Sc
	con	25	0 corr	K_1 1·5	67T
	sp	25	0·07 ($NaClO_4$)	K_1 0·34	67Ta
Co $en_2ClH_2O^{2+}$					
	sp	0—40	dil	$K(cis \rightleftharpoons trans)$ −0·43	61Bf
	sp	20—40	dil	$K(cis \rightleftharpoons trans)$ −0·43 (20—40°)	63Sb
	sp	25	0·01 HNO_3	$K(cis \rightleftharpoons trans)$ −0·43 (in H_2O), −0·40 (in D_2O)	65Cf
	sp	60	org	K_1 4·22, K_2 1·85 in Me_2NCHO(l)	66Lc
	sp	30	org	K_1^{out} 4·18 (ΔH_1 = 0·93, ΔS_1 = 22·1); β_2^{out} 6·08 ($\Delta H_{\beta 2}$ = −3·8, $\Delta S_{\beta 2}$ = 15·4) in Me_2NCHO(l) (also K_1, β_2 for 1—21°)	66Ld
Co $en_2Cl_2^+$	sp	25	MeOH	K_1 2·13(*cis*), no ev cpx(*trans*) in MeOH(l), $I \approx$ 0·02	57Pe
	sp	60	org	$K(cis \rightleftharpoons trans)$ 0·85, K_1^{out} (Co $en_2Cl_2^+ + Cl^- \rightleftharpoons$ Co $en_2Cl_2^+Cl^-$) 3·26(*cis*), 1·48(*trans*) in Me_2NCHO(l); also data for $MeCONMe_2$(l)	62Ta
	sp	60	Me_2SO	$K(cis \rightleftharpoons trans)$ 0·60, K_1^{out} 2·60(*cis*), 1·43(*trans*) in Me_2SO(l), equil, constants for Co $en_2Me_2SOCl^{2+}$ and for MeOH(l) given	64Te
	kin	35	MeOH	K_1 2·40 in MeOH(l)	65Bf
	sp	60	org	$K(cis \rightleftharpoons trans)$ 1·04, K_1 3·34(*cis*) in Me_2NCHO(l) [$\beta(Et_4N^+Cl^-)$ 2·26]	66Lc
	sp	20—30	org, 0 corr	*cis*: K_1 2·50(20°), 2·44(25°), 2·42(30°) in Me_2SO(l), 3·7(30°) in Me_2NCHO(l), 4·31 (30°) in Me_2NCOMe(l); *trans*: K_1 2·25(30°) in Me_2NCHO(l)	66Md
	sp	30	org, 0 corr	*cis*: K_1 3·72(30°) in Me_2NCHO(l)	66Me
	sp	30—50	sulpholan	$K_1^{out}(cis)$ 4·62(30°), 4·46(40°), 4·32(50°) 4·04(→ 70°), ΔH = −7·1, ΔS = −2·5, in sulpholan	67Fe
	sp	70	sulpholan	K_1^{out} 3·68(*cis*), 0·60 ?(*trans*), in sulpholan(l), I = 0·015; K_1^{out} 3·95, 0 corr; $K(cis \rightleftharpoons trans)$ 1·56 in sulpholane, ΔH = −2·5, ΔS = 0	
	con	25	org	K_1 2·60(*cis*), 2·0(*trans*) in Me_2SO(l) → 0, 3·91(*cis*) in Me_2NCHO(l) → 0, 4·31(*cis*) in Me_2NCOMe(l), → 0	67Mi
	sol	25	org	K_{s0} −7·02(*cis*), −6·74(*trans*) in Me_2NCOMe, −5·92(*cis*), −5·74(*trans*), in Me_2NCHO, −4·10(*cis*), −4·0(*trans*) in Me_2SO, −5·40(*cis*), −3·11(*trans*) in MeOH, −2·05(*cis*), −0·42(*trans*) in H_2O, −6·89(*cis*), −6·02(*trans*) in $C_4H_8SO_2$(30°)	68Fe
	sp	60	org	$K(cis \rightleftharpoons trans)$ −0·05 in Me_2NCOMe, 0·0 in Me_2NCHO, −0·3 in Me_2SO	

continu

Metal	*Method*	*Temp*	*Medium*	*Log of equilibrium constant, remarks*	*Ref*
Co^{3+} cpx	sol	25—35	0 corr	$Co(NH_3)_5Cl^{2+}$; K_1 1·0(25°, 35°)	65Ac
	sp	25—86	1 Na(ClO$_4$)	$Co(NH_3)_3^{3+}$ = M^{3+}; K_1^{out} −0·05(25°), 0·08(35°), 0·15(47°), 0·20(57°), 0·18(66°), 0·11(77°, 86°); $K(M^{3+}Cl^- \rightleftharpoons MCl^{2+})$ 0·18(25—57°), 0·23(66°) 0·36(77°, 86°), $\Delta H \approx$ 0·4, $\Delta S \approx$ 1·6	67Lb
	sp	45	1 (NaClO$_4$)	$Co(NH_3)_5^{3+}$; K_1^{out} 0·5	67Lg
	pH$_2$O	37	→ 0	Co en$_2$NH$_3$NO$_2^{2+}$; K_1 1·15(*cis*), 1·12(*trans*)	67Ma
	pH$_2$O	37	→ 0	Co en$_2$NCSCl$^+$; K_1 0·26	67Mb
	con	25	Me$_2$SO	CoBrCl en$_2^+$; K_1 2·49(*cis*) in Me$_2$SO(l) → 0	67Mi
	con	25	Me$_2$SO	CoCl$_2$ trien$^+$; K_1 2·53(*cis*α), 2·71(*cis*β) in Me$_2$SO(l), → 0	67Mi
	con	25	0 corr	$Co(NH_3)_5NO_2^{2+}$; K_1 1·3	67T
	sp	25	0·07(NaClO$_4$)	Co en$_3^{3+}$; K_1 0·46	67Ta
	kin	25	2 (LiClO$_4$)	$\beta[A_4Co(\mu\text{-}NH_2, \mu\text{-}O_2)CoA_4^{4+}Cl^-]$ −0·03, A = NH$_3$	68Dd
Ni^{2+}	pol	25	MeCN	K_2 0, K_3 −1, K_4 0 in MeCN(l), 0·1M-Et$_4$NClO$_4$	63Na
	sp	25—68	org	K_4 2·55(25°), 2·49(68°), ΔH_4 = −0·6, ΔS_4 = 9·7 in MeCONMe$_2$(l); K_4 4·00(25°), 3·62(68°), ΔH_4 = −3·8, ΔS_4 = 5·7 in CH$_3$CN(l)	64Na
	cix	20	0·691 (HClO$_4$)	K_1 0·23, β_2 −0·04	65Mb
	sp	125—320	MgCl$_2$ var	ev octahedral NiCl$_6^{4-}$, tetrahedral NiCl$_4^{2-}$, $\Delta H(NiCl_6^{4-} \rightleftharpoons NiCl_4^{2-} + 2Cl^-)$ = 8·5	66Aa
	sp	30	10 Li(NO$_3$)	K_1 −1·03	66Fb
	cal	25—40	2 (NaClO$_4$)	ΔH_1 = 0·50(25°), 0·44(40°), K_1 −0·17(25°), −0·15(40°) ΔS_1 = 0·87(25°)	66Kd+60La
	sp	40—85	EtOH	$\Delta H[NiCl_2S_4 + Cl^- \rightleftharpoons NiCl_3S^-(+3S)]$ = 14·6 in S = EtOH(l)	67S
	sp		2·5 Me$_4$NCl	ev tetrahedral NiCl$_4^{2-}$	67Sf
	pol	145	MNO$_3$(l)	K_1 0·9, K_2 1·0 in (Li, Na, K)NO$_3$(l, eut), m units	68I
	sp	25—35	Cl$^-$ var	ev tetrahedral cpx (2 or 3Cl/Ni ?) at 250—350° in 4M-NaCl and 10M-LiCl	68L
	sp	0—40 40—70	Me$_2$NCHO	$\Delta H(NiClT_5^+ + 2AlCl_3 \rightleftharpoons NiCl_3T^- + 2AlCl_2^+)$ = 7·8 $\Delta H(NiCl_3T^- + AlCl_2^+ \rightleftharpoons NiCl_2T_2 + AlCl_3)$ = 4·1 in T = Me$_2$NCHO(l)	68Mi
Ru^{2+}	pre, mag		var	ev RuCl$_4^{2-}$ and RuCl$_5$H$_2$O^{3-} in salts and soln	38Ga
	pol, E		var	β_6 −13	66Bg+60Fb
	sp		HCl var	ev several cpx	67Ab
Ru^{2+}, Sn^{2+}	var		var	ev RuCl$_2$(SnCl$_3$)$_2^{2-}$ (con, pre, sp)	64Ya
Ru^{3+}	var		var	ev RuCl$_2^+$(*cis* and *trans*), RuCl$_3$(*cis* and *trans*), RuCl$_4^-$(*cis* and *trans*), RuCl$_5^{2-}$, RuCl$_6^{3-}$ (aix, cix, fp, pre, sp)	60Fb
	sp	25	3 (CF$_3$CO$_2$H)	K_4 −0·08	
			var	β_6 −4	60Fb/66Bg
	X		solid	ev RuCl$_5$H$_2$O^{2-} in K$_2$RuCl$_5$H$_2$O(s)	60Kd

1964 *Addenda*: p. 283, *for ref* "60Ca" *read* "60 Ca = 60Fb", *and for ref* "61Ca" *read* "61Ca = 60Fb"

	sp	25—26	5·5—6·46 HCl	ev RuCl$_6^{3-}$ and RuCl$_4^-$ (*cis*, *trans*) in 6·5 and 5·5M-HCl, ev also RuCl$_5^{2-}$ only in 5·5M-HCl (?) K_5K_6 given	64Me
	X		solid	ev *cis*-RuCl$_4$(H$_2$O)$_2^-$ in Ph$_4$AsRuCl$_4$(H$_2$O)$_2$H$_2$O(s)	66H

continued overleaf

67 Cl^- (contd)

Metal	*Method*	*Temp*	*Medium*	*Log of equilibrium constant, remarks*	*Ref*
Ru^{3+}	X		solid	ev $RuCl_5H_2O^{2-}$ in $Cs_2RuCl_5H_2O$(s)	
(*contd*)	dis, sp		org	ev $Ru_2Cl_9^{3-}$ (with R_4N^+) in org = $C_2H_4Cl_2$(l)	67I
$Ru(NH_3)_5^{3+}$	ana	35—90	0·1 H^+	K_1 2·18(35°), 2·21(45°), 2·27(64°), 2·32(80°), 2·37(90°)	64Be
	sp	25	0·11 ($NaClO_4$)	K_1 1·85 (withdraws value in ref 62Ea)	65E
Ru^{4+}	X		solid	ev $RuCl_6^{2-}$ in K_2RuCl_6(s)	52A
	dis, sp		org	ev $Ru_2OCl_9H_2O^{3-}$ (with R_4N^+) in $C_2H_4Cl_2$(l)	67I
$RuNO^{3+}$	var		HCl var	ev $RuNOCl_x(H_2O)_{5-x}{}^{3-x}$, x = 1, 2, 3, 4, 5; ev $(RuNO)_2Cl_3(H_2O)_5O^+$, $(RuNO)_2Cl_6(H_2O)_2O^{2-}$ $(RuNOCl_5{}^{2-})_n$ (aix, cix, pre, sp)	62Ca
	var		var	ev $RuNOCl_n{}^{(3-n)+}$, n = 0, 1, 2, 3, 4, 5 (aix, cix, pre, sp)	64Mf
	gl	5	dil	$K_a[RuNOCl_4H_2O^-]$ −6·02 $K_a[RuNOCl_3(H_2O)_2]$ −4·95, −7·5	
Rh^+, Sn^{2+}	var		var	ev $Rh_2Cl_2(SnCl_3)_4{}^{4-}$ (con, pre, sp)	64Ya
Rh^{3+}	oth	40	(H, Na)Cl var	K_3 1·69, K_4 0·47, K_5 −0·51 (electrophoresis), ev $RhCl_6{}^{3-}$ in 7M-HCl (sp)	65Bb
	red	25	0·1 $HClO_4$	K_4 1·39, K_5 0·55, K_6 −0·23	65Bg/
	kin	15—35	4 $H(ClO_4)$	K_6 −0·72(15°), −0·85(20°), −0·93(25°), −1·14(→ 35°)	65R
	Ag, gl	55	0·1 (H, $NaClO_4$)	$K_nK_s(Ag^+Cl^-)$ $[RhCl_{n-1}{}^{4-n} + AgCl(s) \rightleftharpoons RhCl_n{}^{3-n} + Ag^+]$ −3·12(n = 3), −5·42(n = 4), −5·72(n = 5), −5·84(n = 6) $K_a[RhCl_n(H_2O)_{6-n}{}^{3-n}]$ −4·8(n = 3), −6·0(n = 4), −7·3(n = 5)	66Bj
	sp, ana	25—85	0·1 $HClO_4$	K_2K_3 4·25(25°), 4·11(40°), 4·06(55°), 3·83(70°), 3·65(85°), $\Delta H_2 + \Delta H_3$ = −4·8	66Bk
	kin	75—85	2·5 ($NaClO_4$)	K_1 −0·05(75°), −0·10(80°), −0·15(85°); $\beta(RhOH^{2+}Cl^-)$ −0·22(75°), −0·30(80°), −0·40(85°)	66Sn
	kin	30—45	4 ($HClO_4$)	K_5 0·90(30°), 0·83(35°), 0·80(40°), 0·77(45°)	67Rb/
Rh^{3+}, Ag^+	Ag	25—55	0·1 KNO_3	$\beta(Ag^+RhCl_6{}^{3-})$ 5·69(25°), 5·18(35°), 4·68(55°)	66Bi
Rh^{3+},Hg^{2+}	kin	25	0·1 $HClO_4$	$\beta(Hg^{2+}RhCl_6{}^{3-})$ 7·3	65Bg
$Rh(NH_3)_5^{3+}$	ana	84	→ 0	K_1 4·95	39La
	kin	77—97		ΔH_1 = 3·5, ΔS_1 = 27 (?, tables do not agree and must contain misprints)	68Ld
	kin	25—65	4 ($NaClO_4$)	K_1 −0·80(25°), −0·74(65°)	68Mk
		65	var	K_1 −0·60[4M-Li(ClO_4)], −0·66[5M-Na(ClO_4)]	
Rh^{4+}	X		solid	ev $RhCl_6{}^{2-}$ in Cs_2RhCl_6(s)	47D
Pd^{2+}	Pd	18—20	var	β_4 12·2 $\{K[Pd^{2+} + 2e^- \rightleftharpoons Pd(s)]$ 33·4$\}$	63Ge
	sp	25	→ 0	K_1 6·0, K_2 4·6, K_3 2·5, K_4 2·0, β_4 15·1; also β_4 for I = 0·25 to 1·01 ($NaClO_4$)	64B
	var	25	1 $HClO_4$	K_1 3·88(sol), K_2 3·06, K_3 2·14, K_4 1·34, β_4 10·42 (Pd, Ag, sp)	64Bc
	sp	25	0·5 ($NaClO_4$)	K_4 1·35	64Si
	Pd	25	1 KCl	β_4 11·8$\{K[Pd^{2+} + 2e^- \rightleftharpoons Pd(s)]$ 33·4$\}$ (also β_4 values for 10—60°, assuming same K and RTF^{-1} as for 25°!)	65F
	sp	25	1·0 ($NaClO_4$) 0·8H^+	K_1 4·00, K_2 3·45	66Si
		10—45		K_4 1·50(10°), 1·42(25°), 1·28(45°) ΔH_4 = −2·8, ΔS_4 = −2·9	
		25		K_1 4·00, β_2 7·49, β_3 9·73, β_4 11·11	66Si+
	cal	25	0·1 NaCl	$\Delta H_{\beta 4}$ = −5·5	67Ib

continu

Metal	*Method*	*Temp*	*Medium*	*Log of equilibrium constant, remarks*	*Ref*
Pd^{2+} (*contd*)	gl	19—50	$I(NaClO_4)$	$K(PdCl_3OH^{2-}+Cl^- \rightleftharpoons PdCl_4^{2-}+OH^-)$ −5·7 (I = 0·1—1·0, 19—50°)	67Kc
	gl	25	$I\,NH_4(NO_3)$	(A = NH_3) $\beta(PdA_2Cl^+,Cl^-)$ 2·33, $\beta(PdA_3^{2+},Cl^-)$ 3·0 †K_1 −4·21, †K_2 −4·81, †K_3 −5·72 [$K_1(H^+)$ 9·37, $\beta(PdA_2^{2+}Cl^-)$ 2·92]; also values for 30 °C(sp) †K_n; $PdA_{5-n}Cl_{n-1}^{(3-n)}+Cl^- \rightleftharpoons PdA_{4-n}Cl_n^{(2-n)}+A$	67R
	Pd	25	w% dioxan	β_4 12·5 to 12·7(w = 0), 17·7(w = 72), also β_4 for several other w	68Gc
		40	w% dioxan	β_4 12·0 to 12·1(w = 0), 16·6(w = 72), also β_4 for several other w	
	Pd	25	3·4 $(HClO_4)$	β_4 11·4	68Lb
	sp		$I(HClO_4)$	K_4 1·77(I = 3·4), 1·44(I = 1·07)	
	sp	25	$I(LiClO_4)$	K_4 2·00(I = 4), 1·77(I = 3), 1·59(I = 2), 1·43(I = 1)	68Le
		15—40	2 $(LiClO_4)$	K_4 1·68(15°), 1·59(25°), 1·51(40°)	
$Pd(NH_3)_2^{2+}$	con	25	dil	$\beta[Pd(NH_3)_2Cl^+Cl^-]$ 2·55 ?	67Cf
Pd^{2+}, Sn^{2+}	pre, con		var	ev $ClPd(SnCl_3)_3^{2-}$ and $Pd(SnCl_3)_4^{2-}$ in $PhNO_2$(l) and in salts with Ph_4As^+	68Be
Os^{4+}	kin, sp		var	ev $OsCl_5H_2O^-$, $(Cl_5Os)_2O^{4-}$, not $OsOHCl_5^{2-}$	65Mf
	ir		solid	ev $Os_2OCl_{10}^{4-}$, not $OsCl_5OH^{2-}$ in $(NH_4)_4Os_2OCl_{10}$(s)	66Ha
	var		var	ev $Os_2OCl_{10}^{4-}$ in solid, $OsOHCl_5^{2-}$ and $Os(OH)_2Cl_4^{2-}$ in soln (ir, mag, sp, con, fp)	66J
Ir^{3+}	oth	40	HCl var	ev $IrCl_n^{(3-n)+}$, n = 2, 3, 4, 5, $IrOHCl_3^-$ (electrophoresis)	65Bb
	kin, sp	40—50	3·7 (Na, $HClO_4$)	K_5 0·67, ev $IrCl_3$	65C
		50	2·2 (Na, $HClO_4$)	K_5 0·55	
	gl	25	var	$K_a[IrCl_4(H_2O)_2^-]$ −8·5, −10·1	
Ir^{3+}, Sn^{2+}	var		var	ev $Ir_2Cl_6(SnCl_3)_4^{4-}$ (con, pre, sp)	64Ya
	sp, pre		var	ev three or more mixed cpx, one is $IrCl_2(SnCl_3)_4^{3-}$	66Fe
Ir^{4+}	sp	25	3·7 (Na, $HClO_4$)	ev $IrCl_3^+$, $IrCl_4$	65C
Pt^{2+}	*X*		solid	ev Pt_6Cl_{12} units in $PtCl_2$(s)	64Tb = 65Bc
	gl	25—55	0·1 KCl	K_4 5·98(25°), 5·44(35°), 4·92(45°), 4·58(55°), ΔH_4 = −5·35; $K_a(PtCl_3H_2O^-)$ −7·44(25°), −7·25(35°), −7·15(45°), −7·0(55°);	65N
			dil	K_3 5·52(35°), 4·06(45°), 3·13(55°), ΔH_3 = −31·2; $K_a[PtCl_2(H_2O)_2]$ −6·5(35°), −6·25(45°), −6·1(55°); $K_a(PtCl_2OHH_2O^-)$ −8·1(35—55°)	
	sp	25	0·5 $HClO_4$	K_3 2·96, K_4 1·87	66E
	aix	25	0·5 $HClO_4$	K_3 3·0	
	kin	25	0·5 $HClO_4$	K_4 1·89	66Ea
	kin	15—35	0·5 $HClO_4$	K_4 2·00(15°), 1·89(25°), 1·77(35°), ΔH_4 = −4·4, ΔS_4 = −6·0	67D
	ana	60	0·5 $HClO_4$	K_1 1·51	67Ed
Pt^{2+} cpx	con	25	dil	$Pt(NH_3)_2NO_2^+$: K_1 3·77	29Cb
	ana	17—18	var	$Pt(MeNH_2)_2^{2+}$; K_2 2·4(*cis*, I = 0·1), 2·8(*cis*, dil), 3·7(*trans*, I = 0·1); $Pt(EtNH_2)_2^{2+}$; K_2 2·4(*cis*, I = 0·1), 2·8(*cis*, dil) 3·5(*trans*, I = 0·1)	63Gd
	con	20 ?	dil	$Pt(MeNH_2)_2NO_2^+$: K_1 3·85	64Cd

continued overleaf

67 Cl^- (contd)

Metal	Method	Temp	Medium	Log of equilibrium constant, remarks	Ref
Pt^{2+}cpx (contd)	sp	20	0·318 (Na_2SO_4)	$PtNH_3^{2+}$; K_3 1·89(*trans*), 2·96(*cis*)	64Td
		25	0·318 (Na_2SO_4)	K_3 1·88(*trans*), 2·88(*cis*)	
	Ag	25	0·2 $HClO_4$	$\beta(RPtCl_2^+Cl^-)$ 2·60 for R = C_2H_4, also β values for 4 other R	65Ab
	sp	25	0·2 $HClO_4$	β[*trans*-Pt(piperidine)$_2$(PEt$_3$)$H_2O^{2+}Cl^-$] 3·1	66Ed
	Ag	18	1 (KNO_3)	$\beta[Pt(NH_3)_2Cl^+Cl^-]$ 2·72(*cis*), 3·29(*trans*); $\beta[Pt(NH_3)_3^{2+}Cl^-]$ 3·5	66Gc
	Ag	18	0·1 ($NaNO_3$)	$Pt(NH_2OCH_3)_2^{2+}$; K_2 4·20(*cis*), 3·05(*trans*)	67Gc
	Ag	18	0·1 ($NaNO_3$)	$Pt(NH_2OH)_2^{2+}$; K_2 3·44(*cis*), 2·92(*trans*)	
	sp	25	EtOH	$PtC_2H_4^{2+}$; K_3 4·3 in EtOH(l)	68Mg
	var	0—35	dil	*cis*-$Pt(NH_3)_2^{2+}$; K_2 2·39(0°), 2·42(18°, 25°, 30°), $\Delta H_2 \approx 0$ (con, ana, sp)	68P
	kin	30	0 corr	$\beta[Pt(NH_3)_2Cl^+Cl^-]$ 3·9, $\Delta H = -1{\cdot}2$, $\Delta S = 14$	68Pb
Pt^{2+}, Ge^{2+}	pre, ir		var	ev $HPt(GeCl_3)_5^{2-}$, $PtCl_2(GeCl_3)_2^{2-}$ in HCl conc *etc.*	68Wb
Pt^{2+}, Sn^{2+}	var		var	ev $PtCl_2(SnCl_3)_2^{2-}$ (*cis*, *trans*) (con, pre, sp)	64Ya
	X		solid	ev $Pt(SnCl_3)_5^{3-}$ in $(Ph_3MeP)_3PtSn_5Cl_{15}$(s)	65Cd
$Pt^{2+,4+}$				See Table 1, Pt ref 67G	
$Pt^{2+,4+}$, Sn^{2+}	var		var	ev $Pt_3Sn_3Cl_{20}^{4-}$ in org and solid (pre, ir, con)	66L
Pt^{4+}	ana, gl	50	NaCl var	K_6 1·49, $K_a(PtCl_5H_2O^-)$ −3·8	65De
	gl	25—55	KCl var	K_6 2·25, $K_a(PtCl_5H_2O^-)$ −5, K_5 3·7, $K_a[PtCl_4(H_2O)_2]$ −4·2 (25—45°); $K_a(PtCl_4OHH_2O^-)$ −6·2(25—35°)	65Na
	ana, gl	20—50	0·1 (KNO_3)	$\beta[Pt(NH_3)_2Cl_3^+Cl^-]$ 2·24(25°, *cis*), 2·17(50°, *cis*), 2·40(20°, *trans*), 2·29(50°, *trans*); $\beta[Pt(NH_3)_2Cl_2^{2+}Cl^-]$ 3·3(25°, *cis*), 3·7(20°, *trans*)	66Gd
	Cl	25	$HClO_4$ var	K_5K_6 5·60	66Sh
	red	60	3 H(ClO_4)	K_6 1·54	68G
Pt en_3^{4+}	sp	10—40	0 corr	K_1^{out} 1·17(10°), 1·24(25°), 1·29(40°); $\Delta H_1 = 1{\cdot}6$, $\Delta S_1 = 11{\cdot}0$	63Gc
	var	20	var	ev at least one cpx (sp, circular dichroism)	67L
Cu^+	Ram		var	ev linear $CuCl_2^-$ in Et_2O, other cpx in 6—12M-HCl	63Cd
	pol, Cu?	25?	MeCN	K_1 4·9, K_2 5·9 in MeCN(l), 0·1 M-Et_4NClO_4 [$\beta(Et_4N^+Cl^-)$ 1·54, $\beta(Et_4N^+ClO_4^-)$ 1·05]	65Mg
	X		solid	ev $Cu_5Cl_{16}^{11-}$ in $[Co(NH_3)_6]_4Cu_5Cl_{17}$(s)	66Mh
	pol	25	org	β_2 9·3 in MeOH, 12·3 in EtOH, 13·4 in Pr^{i}OH, 14·1 in Bu^{i}OH, 20·9 in Me_2CO, 0·1 M-Li(ClO_4) β_2 9·6 in MeOH, 12·5 in EtOH, 13·4 in Bu^{i}OH, 19·2 in Me_2CO, 1M-Li(ClO_4); correcting for LiCl and $LiClO_4$ pairs; β_2 9·1 in MeOH, 11·6 in EtOH, 11·9 in Pr^{i}OH, 12·2 in Bu^{i}OH, 18·9 in Me_2CO, 0·1M-Li(ClO_4); neglecting LiCl and $LiClO_4$ pairs	67Mc
	Cu, Ag	25	6·5 $NH_4(NO_3)$	β_2 6·04, β_3 5·98, β_4 5·60, β_{32} 12·3, β_{42} 12·2, β_{52} 12·0, β_{62} 11·6, β_{43} 18·6, β_{53} 18·6, β_{63} 18·4, β_{73} 18·3, β_{54} 25·3, β_{64} 25·3, β_{74} 24·7, β_{84} 24·7, β_{65} 32·0, β_{75} 31·7, β_{85} 31·5, β_{95} 31·3 (could not data be explained by fewer cpx?)	68Se
Cu^{2+}	cix	30	1 ($NaClO_4$)	K_1 1·18, β_2 0·87, β_3 0·79, β_4 0·88	62D
	X		solid	ev $Cu_2Cl_6^{2-}$ in $Li_2Cu_2Cl_6(H_2O)_4$(s)	63V = 58V
	X		solid	ev $CuCl_4^{2-}$ in $(NH_4)_2CuCl_4$(s)	64Wc
	spJ	24	org	ev $CuCl_3^-$ (predominant) in γ-butyrolactone	64Y

contin

Metal	*Method*	*Temp*	*Medium*	*Log of equilibrium constant, remarks*	*Ref*
Cu^{2+}	sp		HCl var	ev $CuCl_4(H_2O)_2^{2-}$, $CuCl_6^{4-}$, no ev $CuCl_4^{2-}$ (tetr)	65A
(*contd*)	var	25?	MeCN	K_1 9·7, K_2 7·9, K_3 7·1, K_4 3·7 in MeCN(l), 0·1M-Et_4NClO_4 [$\beta(Et_4N^+Cl^-)$ 1·54, $\beta(Et_4N^+ClO_4^-)$ 1·05] (sp, *iE*, red)	65Mg
	cal	25—40	2 ($NaClO_4$)	$\Delta H_1 = 1{\cdot}58(25°)$, 1·98(40°), K_1 0·09(25°), 0·15(40°), $\Delta S_1 = 5{\cdot}67(25°)$	66Kd+60La
	act	25?	0 corr	K_1 0·3	/66Mc
	nmr		var	K_1 −0·11, K_2 −0·70	66Vb
	sp		$MeCO_2H$	ev $CuCl_4^{2-}$ (planar), $CuCl_4^{2-}$ (distorted tetrahedron), $CuCl_2$ in $MeCO_2H$(l); equil consts given, vary with [LiCl] and [H_2O]	67Ea
	pol	25	org	K_1 4·2, β_2 6·5 in MeOH; $K_1 < 6{\cdot}6$, β_2 10·1, $\beta_3 < 11{\cdot}3$ in EtOH; β_2 12·3, β_3 14·9 in Pr^{i}OH; β_2 13·5, β_3 16·3 in Bu^{i}OH; β_3 25·1 in Me_2CO, 0·1M-Li(ClO_4), correcting for LiCl and $LiClO_4$ pairs;	67Mc
				K_1 4·1, β_2 6·3 in MeOH, β_2 9·4 in EtOH, β_2 10·8, β_3 12·7 in Pr^{i}OH, β_2 11·4, β_3 13·3 in Bu^{i}OH, β_3 22·3 in Me_2CO, 0·1M-Li(ClO_4), neglecting LiCl and $LiClO_4$ pairs	
				K_1 3·3, β_2 5·4 in MeOH, β_2 11·0, β_3 15·1 in Bu^{i}OH, β_2 19·9, β_3 23·5 in Me_2CO, 1M-Li(ClO_4) correcting for ion pairs	
	sp		HCl var	ev $CuCl_n(H_2O)_{6-n}^{2-n}$ (n = 0, 1, 2, 3, 4) and tetrahedral $CuCl_4^{2-}$	68Aa
	sp	25	3 Li(ClO_4)	K_1^{in} −0·35, K_1^{out} −0·38, K_1 −0·06	68Mf
	X		solid	ev trigonal $CuCl_5^{3-}$ in $Cr(NH_3)_6CuCl_5$(s)	68Rb
	sol	25	0 corr	$K_s(Cu^{2+}OH^-{}_{1{\cdot}4}Cl^-{}_{0{\cdot}6})$ −16·10 (fresh), $K_s(Cu^{2+}OH^-{}_{1{\cdot}5}Cl^-{}_{0{\cdot}5})$ −17·52 (aged)	68Sa
	tp	23	LiCl var	$K_2 \approx -1$, $K_3 \approx -1$	68Sj
Ag^+					
1964 *Addendum*: p. 288, *for ref* "60Gc" *read* "60Gc ≈ 64G"					
	sol	20	4 ($NaClO_4$), 3H^+	K_{s0} −10·40, K_1 3·45, β_2 5·67, β_3 6·00, β_4 6·04	64A
	Ag	240	MNO_3(l)	K_1 2·93 in $Li_{0{\cdot}3}K_{0{\cdot}7}NO_3$(l), *x* units	64Bg
	cal	320	MNO_3(l)	ΔH_{s0} = 18·9 in (Na, K)NO_3(eut, l)	64Ha
	con	25	dil	$K_1 < 2$	64Ia
	sol	18	MeNHCHO	β_2 8·92, K_{s0} −10·39 in MeNHCHO(l)	64P
			NH_2CHO	K_{s0} −8·28 in NH_2CHO(l)	
	sol	600K	$NaNO_3$(l)	K_{s0} −6·34, K_{s1} −3·89, K_{s2} −1·80, K_{s12} [AgCl(s)+$Ag^+ \rightleftharpoons Ag_2Cl^+$] −1·96 in $NaNO_3$(l), *x* units	64Sc
	Ag	713K	$CsNO_3$(l)	K_1 2·70 in $CsNO_3$(l), *x* units	64Ta
		658K	MNO_3(l)	K_1 2·72 in $Cs_{0{\cdot}33}K_{0{\cdot}67}NO_3$(l), *x* units	
		664K	MNO_3(l)	K_1 2·58 in $Li_{0{\cdot}33}K_{0{\cdot}67}NO_3$(l), *x* units	
	Ag	225	MNO_3(l)	K_1 1·98, β_2 1·2 in (Li, K)NO_3(l, eut), M units	64Tc
	sol	25	*I* $HClO_4$	K_{s0} −9·62 (*I* = 0·5M), −10·05 (*I* = 3M)	64W
	Ag	350	MNO_3(l)	K_{s0} −3·8 in $NaNO_3$(l), KNO_3(l) and (Na, K)NO_3(l, eut), *m* units	65Af
	Ag	150	MNO_3(l)	K_{s0} −7·4 in $Li_{0{\cdot}43}K_{0{\cdot}57}NO_3$(l, eut), *m* units	65Bi
	dis		org	ev $HAgCl_2(TBP)_3$ in TBP = $(BuO)_3PO$	65Ke+65La
			0 corr	$K_D[H^+ + Ag^+ + 2Cl^- + 3T(org) \rightleftharpoons HAgCl_2T_3(org)]$ 5·37, org = C_6H_6, T = TBP	

continued overleaf

67 Cl⁻ (contd)

Metal	Method	Temp	Medium	Log of equilibrium constant, remarks	Ref
Ag^+ (contd)	Ag	25	$C_2H_4(NH_2)_2$	K_{s0} −4·04 in $C_2H_4(NH_2)_2$(l)	65Mi
	Ag	320—350	MNO_3(l)	K_1 2·71(320°), 2·62(350°) in $K_{0.8}Ca_{0.2}(NO_3)_{1.2}$(l) K_1 2·53(350°), in $K_{0.8}Sr_{0.2}(NO_3)_{1.2}$(l)	66Bd
	fp	210	$AgNO_3$(l)	$\beta_{12}(Ag^+{}_2Cl^-)$ 1·08 in $AgNO_3$(l), *x* units	66Ja
	Ag, *iE*	23	Me_2SO	K_{s0} −10·4, β_2 11·9 to 12·1 in Me_2SO(l), $I = 0{\cdot}1$ (Et_4NClO_4)	66La
		23	MeCN	K_{s0} −12·4, β_2 12·6 to 13·6 in MeCN(l), $I = 0{\cdot}1$ (Et_4NClO_4)	
		23	MeOH	K_{s0} −13·0, β_2 8·0 to 7·7 in MeOH(l), $I = 1{\cdot}0$ ($LiClO_4$)	
		23	Me_2CO	K_{s0} −16·4, β_2 16·7 in Me_2CO(l), $I = 0{\cdot}1$($LiClO_4$)	
		23	$EtNO_2$	K_{s0} −21·1, β_2 22·2 to 22·5 in $EtNO_2$(l), $I = 0{\cdot}1$ (Et_4NClO_4)	
	dis		0 corr	$K_D[Ag^+ + 2Cl^- + H^+ + 3T(org) \rightleftharpoons HAgCl_2T_3(org)]$ 5·39, T = $(BuO)_3PO$, org = C_6H_6	/66Lf
	Ag	350—390	MNO_3(l)	K_1 2·58—2·61(350°), 2·50—2·53(370°), 2·45—2·48(390°) in (K, Ba)NO_3(l, eut), *x* units	66Sg
	sol	25	*w*% EtOH	K_{s0} −9·76($w = 0$), −10·04($w = 10$), −10·22($w = 20$), −10·77($w = 40$), −11·11($w = 50$); K_1 3·32($w = 0$), 3·66($w = 10$), 3·79($w = 20$), 4·27($w = 40$), 4·64($w = 50$); β_2 5·26($w = 0$), 5·58($w = 10$), 5·82($w = 20$), 6·82($w = 40$), 7·36($w = 50$); in *w*% EtOH, 0 corr	67Aa
	Ag	25	org	K_{s0} −13·1 in MeOH; K_{s0} −14·5, β_2 16·3 in Me_2NCHO; K_{s0} −14·3, β_2 17·2 in Me_2NCOMe; K_{s0} −10·4, β_2 11.7 in Me_2SO; K_{s0} −12·9, β_2 13·4 in MeCN; K_{s0} −11·9, β_2 16·1 in $(Me_2N)_3PO$	67Ac
	Ag	438	$NaNO_3$(l)	K_1 4·72 in $NaNO_3$(l), *x* units	67Be
	Ag	25	$C_3H_6CO_3$	K_1 15·15, β_2 20·87, β_3 23·39, K_{s0} −19·87 in propene carbonate(l), 0·1M-Et_4NClO_4; K_1 15·5, β_2 21·18, β_3 23·4, K_{s0} −20·18 in propene carbonate(l), 0 corr (neglecting ion pairs)	67Bg
	Ag	25	1	β_4 5·16	67Bk
	Ag	220—264	$MClO_4$(l)	K_{s0} −6·42(493·8K), −5·70(537·1K)	67F
		250	$MClO_4$(l)	K_{s0} −5·94, ΔH_{s0} = 19·3 in $Li_{0.75}Na_{0.25}ClO_4$(l), *m* units	
	Ag	25	*w*% MeOH	K_{s0} −9·963 in 10% MeOH, −10·780 in 43·12% MeOH → 0, *m* units	67Fb
	Ag	25	20% dioxan	K_{s0} −10·216 in 20% dioxan	67Fc
	Ag	25—75	Me_2SO	K_{s0} −9·66(25°), −9·25(45°), −8·92(55°), −8·49(75°); β_2 10·59(25°), 10·23(45°), 9·92(55°), 9·42(75°) in Me_2SO(l), 0·1M-NH_4NO_3	67Rd
	sol	275	MNO_3(l)	β_2 3·21, β_3 2·65, β_{12} 2·79, β_{13} 2·54 (K_{s0} −4·92, K_{s1} −3·00)	67Sb
		300		β_2 3·04, β_3 2·38, β_{12} 2·71, β_{13} 2·08 (K_{s0} −4·60, K_{s1} −3·24) in $Na_{0.5}K_{0.5}NO_3$(l), *m* units	
	Ag	25	Me_2NCHO	K_1 12·11, β_2 16·295, β_3 < 17·7, K_{s0} −14·49, (0·1M-Et_4NClO_4), K_1 12·80, β_2 16·99, K_{s0} −15·18, K_{s1} −2·38, K_{s2} 1·80 (→ 0) in Me_2NCHO(l)	68B
	sol	25—45	D_2O	K_{s0} −9·88(25·1°), −9·17(45°), ΔH_{s0} = 15·5 K_1 3(25°), 2·5(45°), in D_2O, 0 corr, *m* units	68Bc
	sol	15—25	H_2NCHO	K_{s0} −8·60(15°), −8·28(25°), K_1 3·79(15°), 3·81(25°), ΔH_{s0} = 12·5	68Bf

contin

Metal	Method	Temp	Medium	Log of equilibrium constant, remarks	Ref
Ag^+ (contd)	cal			$\Delta H_{s0} = 11{\cdot}1(15°, 25°)$ in H_2NCHO(l), *m* units	
	Ag	25	$C_3H_6CO_3$	K_{s0} $-20{\cdot}0+3{\cdot}25[H_2O]^{\frac{1}{2}}$, K_{s1} $-4{\cdot}7$, K_1 $15{\cdot}3$—$3{\cdot}25$ $[H_2O]^{\frac{1}{2}}$, β_2 $21{\cdot}2-4{\cdot}87$ $[H_2O]^{\frac{1}{2}}$, β_3 $23{\cdot}75$—$7{\cdot}55$ $[H_2O]^{\frac{1}{2}}$ (for $[H_2O] < 0{\cdot}5$M) in propene carbonate = $C_3H_6CO_3$(l), (0·1M-Et_4NClO_4)	68Bg
	Ag	350	MNO_3(l)	K_{s0} $-4{\cdot}10$, $\Delta H_{S0} = 19{\cdot}15$ in $K_{0.89}Ba_{0.11}(NO_3)_{1.11}$ (l, eut), *m* units	68Gb
	dis	150	MNO_3(l)	K_1 2·49, K_2 2·10, in $Li_{0.42}K_{0.58}NO_3$(l), *m* units	68Gd
	Ag	350	MNO_3(l)	K_1 2·60, K_2 2·16, K_{12} 2·1	68Ge
		370		K_1 2·54, K_2 2·10, K_{12} 2·0	
		390		K_1 2·46, K_2 2·0, K_{12} 2·0 in $K_{0.89}Ba_{0.11}(NO_3)_{1.11}$(l, eut), *x* units	
	sol	55·4	$NH_4NO_3(H_2O)_2$	K_{s0} $-8{\cdot}00$, K_1 2·40, K_2 2·15, K_3 0·69	68Gf
		70·1		K_{s0} $-7{\cdot}61$, K_1 2·33, K_2 2·03, K_3 0·58	
		85·0		K_{s0} $-7{\cdot}31$, K_1 2·31, K_2 1·95, K_3 0·47 in $NH_4NO_3(H_2O)_2$(l, ≈ 28 m NH_4NO_3), *m* units; values for *x* units and quasi-lattice parameters are also given	
	sol	480	KNO_3(l)	$K_{s0}[AgCl(l) \rightleftharpoons Ag^+ + Cl^-]$ $-2{\cdot}85$ in KNO_3(l),	68Ka
	sol	24	Me_2SO	β_{12} 7·73, β_{13} 7·32 in Me_2SO(l), $I = 0{\cdot}5$ (Et_4NClO_4)	68Lc
	sol	150—200	MNO_3(l)	K_{s0} $-7{\cdot}36(150°)$, $-6{\cdot}99(161°)$, $-6{\cdot}64(176°)$, $-6{\cdot}33(190°)$, $-6{\cdot}17(200°)$, $\Delta H_{s0} = 22{\cdot}2$; K_1 2·48(150°), 2·38(161°), 2·25(176°), 2·13(190°), 2·06(200°); K_2 2·08(150°), 1·97(161°), 1·85(176°), 1·72(190°), 1·65(200°); $\Delta H_1 = \Delta H_2 = -1{\cdot}69$, in $Li_{0.43}K_{0.57}NO_3$(l, eut) *m* units (values in *x* units also given)	68Mh
	Ag	25	*w*% Me_2CO	K_{s0} $-10{\cdot}26(w = 20)$, $-10{\cdot}94(w = 40)$, $-11{\cdot}90(w = 60)$ $-13{\cdot}61(w = 80)$ in *w*% Me_2CO, → 0	68Mj
		25	*w*% Me_2SO	K_{s0} $-9{\cdot}81(w = 20)$, $-9{\cdot}86(w = 40)$, $-9{\cdot}82(w = 60)$ $-9{\cdot}82(w = 80)$, in *w*% Me_2SO, → 0	
Au^{3+}	*X*		solid	ev Au_2Cl_6 units in $AuCl_3$(s)	58Cc
	gl, Ag	25	0 corr	†$K_n[Au(OH)_{5-n}Cl_{n-1}{}^- + H^+ + Cl^- \rightleftharpoons Au(OH)_{4-n}Cl_n{}^- + H_2O]$ †K_1 8·7, †K_2 8·0, †K_3 7·15, †K_4 6·15	66Cf
	kin	26	var	†K_4 6·36	66F
	gl	25	3 ($NaClO_4$)	†K_3 7·04, †K_4 6·22, $K_a(AuCl_3H_2O)$ $-2{\cdot}72$	67Cb
	gl, kin			$K_a(AuCl_3OH_2) > -3$	67Ra
	cix		var	ev cation, $AuCl_2{}^+$?	
	Cl	25 ?	0 corr	K_4 4·34	68D
				$K_a(AuCl_3H_2O)$ $-1{\cdot}7$	68D+48B
	dis		dil	$K_D[TFG^+ + AuCl_4{}^- \rightleftharpoons TFGAuCl_4$ (in $C_2H_4Cl_2$)] 5·7, $TFG^+ = (PhHN)_3C^+$	68Ra
Zn^{2+}	*X*		solid	ev $ZnCl_3H_2O^-$ in $KZnCl_3H_2O$(s)	63Bb
	MHg	25	4 (Li)Cl	$\beta(K^+ZnCl_3{}^-)$ $-0{\cdot}28$, $\beta(K^+ZnCl_4{}^{2-})$ 0·0	63Mp
	MHg	25	2·5 $Ca(ClO_4)$	β_3 1·30	64Ba
	MHg	25	4 ($LiClO_4$)	K_1 0·3, β_2 0, β_3 1·0, $\beta_4 \approx -1$; $\beta(Na^+ZnCl_3{}^-)$ $-0{\cdot}7$, $\beta(K^+ZnCl_3{}^-)$ $-0{\cdot}2$ $\beta(K^+{}_2ZnCl_3{}^-)$ -1; $\beta(Rb^+ZnCl_3{}^-)$ $-0{\cdot}1$, $\beta(Rb^+{}_2ZnCl_3{}^-) < -0{\cdot}5$; $\beta(Cs^+ZnCl_3{}^-)$ 0·15, $\beta(Cs^+{}_2ZnCl_3{}^-)$ $-0{\cdot}5$, ev $Cs_3ZnCl_3{}^{2+}$	64Ma

continued overleaf

67 Cl⁻ (contd)

Metal	Method	Temp	Medium	Log of equilibrium constant, remarks	Ref
Zn^{2+}	aix	26	EtOH	ev $H^+ZnCl_3^-$ and $H^+{}_2ZnCl_4^{2-}$ in EtOH(l)	65Pa
(contd)	MHg	25	4 (Li)Cl	$\beta(K^+ZnCl_3^-)$ −0·40, $\beta(K^+ZnCl_4^{2-})$ 0·2	65Rb
	dis	25	var	ev $ZnCl_4^{2-}$ [together with $(C_{12}H_{25})_3NH^+$] in o-xylene; β_n values of ref 56Sc, 61Sa, 63Me and 57Ka may be too high	66Da
	Ram		2·9 $ZnCl_2$, 9HCl	ev tetrahedral $ZnCl_4^{2-}$	66Q
	dis	25	0 corr	K_1 −0·19, β_2 0·18, β_3 −1·4, β_4 −1·52	66Sc
	Ram		(Li, Zn)Cl var	β_2 −1·50, K_3K_4 −1·4	67Ga
	oth	25	var	K_3 0·8 (from molar volumes)	67Jc
	con, sp	25	MeCN	$K_3{}^2\beta_2{}^{-1}(3ZnCl_2T_2+2T \rightleftharpoons ZnT_6{}^{2+}+2ZnCl_3T^-)$ −6·01 (?, −5·01 in diagram) in T = MeCN(l)	68La
	E	25	0 corr	K_s $[Zn(OH^-)_{2-x}Cl^-{}_x(H_2O)_y]$ −13·63(x = 0·5, y = 0·7, after one hour), −14·50(x = 0·4, y = 0·5, after 50 days), −15·28(x = 0·3, y = 0·5, after 4 months) (gl, MHg, Ag)	68Sf
	E	25	0 corr	$K_s[Zn(OH^-)_{1.5}Cl^-{}_{0.5}(H_2O)_{0.66}]$ −13·43 (fresh after 30 min), valid also in EtOH–H_2O mixt (gl, MHg, Ag)	68Sg
	tp	23	LiCl var	$K_2 \approx 0$, $K_3 \approx 0$	68Sj
Cd^{2+}	pol	25	3($NaClO_4$)	K_1 1·46, β_2 2·24, β_3 2·31, β_4 1·65	53Ea/65M
	MHg	39·9	NH_4NO_3 var	K_1 1·49(n = 2), 1·31(n = 4), 1·24(n = 6) in $NH_4NO_3(H_2O)_n$(l), M units	62Bf
	cix	160	MNO_3(l)	K_1 2·92, β_2 3·36, β_3 5·08, β_4 5·78 in (Li, K)NO_3(l, eut)	63L
	MHg	25	2·5 Ca(ClO_4)	β_3 3·38	64Ba
	Ag	160—200	MNO_3(l)	K_1 3·27(160°), 3·18(180°), 3·08(200°) K_2 2·95(160°), 2·85(180°), 2·73(200°) in $Li_{0.5}K_{0.5}NO_3$(l), x units	64Bd
	MHg, Ag	39·9	$NH_4NO_3(H_2O)_2$(l)	K_1 2·50(MHg), 2·53(Ag), K_2 1·88, K_{12} < 1·0(Ag) conc unit mol/mol NH_4NO_3	64H
	oth	15—35	HCl	K_4 −0·3(15°), −0·2(35°), ΔV_4 = −38 ml, ΔV_3 = 4 ml (ultrasonic absorption)	64Vc
	dis	30	1 ($NaClO_4$)	K_1 1·2, β_2 1·8	65H
	MHg	25—35	→ 0	K_1 1·15(25—35°)	65Hc
		25—35	EtOH var	K_1 1·4(30% EtOH), 2·3(60% EtOH), 3·15(90% EtOH), 3·2(100% EtOH), → 0 for 25—35°	
	Pd	254	MNO_3(l)	K_1 1·90, β_2 3·28, β_3 4·26 in $Na_{0.5}K_{0.5}NO_3$(l)	65I
	pol (aix)	?	org–H_2O	K_1 1·53, β_2 1·32, for I = 0·4 M-H(ClO_4), in 80% Pr^{i}OH, β_n values also for I = 0·6 to 2·0	65Mc
	aix	26	EtOH	ev $H^+CdCl_3^-$ and $H^+{}_2CdCl_4^{2-}$? in EtOH(l)	65Pa
	pol	25	I($LiNO_3$)	β_3 1·5 for I = 1 to 8, m units	65Sf
	pol	40	NH_4NO_3	K_1 2·50 in $NH_4NO_3(H_2O)_2$(l)	66Bf
	MHg	425	MNO_3(l)	K_1 3·25, β_2 6·12, β_{12} 5·8 in $Li_{0.43}K_{0.57}NO_3$(l, eut), x units	66Bh
	cal	25	3 ($NaClO_4$)	ΔH_1 = −0·10, ΔH_2 = 0·02, ΔH_3 = 1·85, ΔS_1 = 6·9, ΔS_2 = 3·0, ΔS_3 = 7·0(K_1 1·59, K_2 2·23, K_3 2·42)	66Gb
	pol	27	4·16 ($NaClO_4$)	K_1 1·60, β_2 2·49, β_3 2·91	66Mb
			3·03 ($NaClO_4$)	K_1 1·49, β_2 2·13, β_3 2·42	
			0·76 ($NaClO_4$)	K_1 1·46, β_2 1·83, β_3 1·96 also values for I = 1·52, 2·27, and 3·79	
	pol	25	2 ($NaClO_4$)	K_1 1·36, β_2 1·64, β_3 1·76	66Se
	sol	260—290	MNO_3(l)	K_1 2·92(250°), 2·85(270°), 2·81(290°) in $Li_{0.5}K_{0.5}NO_3$(l) K_1 2·83(260°), 2·82(270°), 2·76(290°) in $LiNO_3$(l), also K_1 for intermediate Li:K ratios, x units	67Fd

contin

Metal	*Method*	*Temp*	*Medium*	*Log of equilibrium constant, remarks*	*Ref*
Cd^{2+} (*contd*)	ΔG	25	0	K_1 2·00, $\Delta H_1 = 0·68$, $\Delta S_1 = 11·4$, $\Delta C_{p1} = 19$ K_2 0·70, $\Delta H_2 = 0·56$, $\Delta S_2 = 5·1$, $\Delta C_{p2} = 9$	67Hd
	cal	25	2 ($NaClO_4$)	$\Delta H_1 = 0·00$, $\Delta H_2 = 0·34$, $\Delta H_3 = 1·93$; $\Delta S_1 = 6·5$, $\Delta S_2 = 3·6$, $\Delta S_3 = 5·7$	68Ga+
			1 ($NaClO_4$)	$\Delta H_1 = 0·13$, $\Delta H_2 = 0·49$, $\Delta H_3 = 1·8$; $\Delta S_1 = 6·6$, $\Delta S_2 = 3·6$, $\Delta S_3 = 4·3$	
			0·5 ($NaClO_4$)	$\Delta H_1 = 0·25$, $\Delta H_2 = 0·67$; $\Delta S_1 = 7·1$, $\Delta S_2 = 4·2$	
			0·25 ($NaClO_4$)	$\Delta H_1 = 0·21$; $\Delta S_1 = 7·3$	
	pol	145	MNO_3(l)	K_1 2·19, β_2 3·7, $\beta_3 \approx 4·3$ in (Li, Na, K)NO_3(l, eut), *m* units	68I
	MHg	15—45	→ 0	K_1 1·96(15°), 1·97(25°), 2·00(35°), 2·06(45°)	68Pa
	tp	23	LiCl var	$K_2 \approx 1$, $K_3 \approx 0$	68Sj
Hg_2^{2+}	Hg	7—40	0·5 (Na)ClO_4, 0·1 H^+	K_{s0} −18·19(7°), −16·16(40°)	63Hc
	Hg	150—210	MNO_3(l)	K_{s0} 4·22 − 7495T^{-1}(*T*K), −12·43(450K) in $Li_{0.43}K_{0.57}NO_3$(l, eut), *m* units	66Fc
Hg^{2+}	aix	23	*I*($NaClO_4$)	K_3 0·70, K_4 0·60 (*I* = 3), K_3 1·1, K_4 1·2 (*I* = 0·3), (K_3 and K_4 for *I* = 0·5 from ref 57Mb)	63Ea
	red	7—40	0·5 (Na)ClO_4, 0·1 H^+	K_1 7·23(7°), 6·62(25°), 6·58(40°); K_2 6·72(7°), 6·36(25°), 6·18(40°)	63Hc
	X		solid	ev tetrahedral $HgCl_4^{2-}$ in (perloline H^+)$_2$-$HgCl_4(H_2O)_2$(s)	63Jb
	Ram		MCl(l)	ev $HgCl_2$, $HgCl_3^-$ and tetrahedral $HgCl_4^{2-}$ in M_xHgCl_{2+x}(l), M = K or NH_4, *x* = 0 to 2·3	63Jc
	X, *X*n		solid	ev $(ClHg)_3O^+$ in Hg_3OCl_4(s)	64Ab
	cal	7—40	0·5 (Na)ClO_4, 0·1 H^+	$\Delta H_1 = -6·75(8°), -5·5(25°), -5·6(40°)$; $\Delta H_{\beta 2} = -14·05(8°), -12·75(25°), -12·50(40°)$;	64C
				$\Delta S_1 = 9·0(8°), 11·9(25°), 12·2(40°)$; $\Delta S_{\beta 2} = 13·7(8°), 16·7(25°), 18·4(40°)$	64C+63Hb
	Ram		TBP	ev $HgCl_2$, $HgCl_3^-$, and $HgCl_4^{2-}$ in $(C_4H_9O)_3PO$(l)	64Sa
	var	37—56	Me_2CO	no ev Hg_2Cl_4 in Me_2CO(l) (bp, *p*Me_2CO)	64Se
	dis	150	MNO_3(l)	K_D[$HgCl_2$(melt) ⇌ $HgCl_2$(org)] 0·38, $\Delta H_D = -3·7$; K_3 0·8, K_4 0·8 in melt = $Li_{0.43}K_{0.57}NO_3$(l, eut), org = polyphenyl eutectic, *m* units	64Za
	red, Hg	25	3 (Na)ClO_4	K_1 7·07, K_2 6·91, K_3 0·75, K_3K_4 2·13	65Aa
	cal	25	3 (Na)ClO_4	$K_3 \approx 1·1$, K_3K_4 2·17; $\Delta H_1 = -5·8$, $\Delta H_2 = -6·5$, $\Delta H_3 = -1·0$, $\Delta H_3 + \Delta H_4 = -2·5$ $\Delta S_1 = 12·9$, $\Delta S_2 = 9·9$, $\Delta S_3 = 0$, $\Delta S_3 + \Delta S_4 = 1·3$	
	red	25	0·1 $HClO_4$	β_2 15·5	65Bg
	aix	26	EtOH	ev $H^+HgCl_3^-$ and $H^+{}_2HgCl_4^{2-}$ in EtOH(l)	65Pa
	gl, cal	25	0 corr	*K*[$Hg(OH)_2 + Cl^-$ ⇌ $HgClOH + OH^-$] −4·09, $\Delta H = 1·2$, $\Delta S = -14·6$; *K*($HgOHCl + Cl^-$ ⇌ $HgCl_2 + OH^-$) −3·77, $\Delta H = 1·21$, $\Delta S = -13·2$	65Pb
	pol	25	MeCN	β_4 39·3 in MeCN(l)	66Fd
			MeCN	β_2 31·1, β_3 37·1 in MeCN(l)	66Fd+52E
	var	25	C_6H_6	ev $(HgCl_2)_n$ in C_6H_6(l), $n \geqslant 2$[dielectric constant, *p*(C_6H_6), Kerr constant]	66Le
	red	0—35	0·5 ($NaClO_4$)	K_1/K_2 0·31(0°), 0·28(12·5°), 0·26(25°), 0·24(35°);	66Va
				K_2 6·87(0°), 6·67(12·5°), 6·48(25°), 6·35(35°)	66Va+

continued overleaf

67 Cl^- (contd)

Metal	Method	Temp	Medium	Log of equilibrium constant, remarks	Ref
Hg^{2+} (contd)	red	25	1 ($NaClO_4$), 0·05H^+	K_1 6·72, K_2 6·51, K_3 1·00, K_4 0·97, β_4 15·20, no ev polyn cpx such as Hg_2Cl_4	68C
	red, gl	25	1 ($NaClO_4$)	$K[Hg(OH)_2+2H^++2Cl^- \rightleftharpoons HgCl_2]$ 19·6, $K(HgOHCl+H^++Cl^- \rightleftharpoons HgCl_2)$ 9·56, $K_a(HgClH_2O^+)$ −3·1	68Ca
	Hg, pol	25?	w% C_5H_5N	K_1 2·2, β_2 4·3(w = 20), β_2 7·9, β_3 10(w = 80) in w% C_5H_5N, also values for w = 10, 40, 60	68Pf
	tp	23	LiCl var	$K_2 > 2$, $K_3 > 1$	68Sj
$Hg(CF_3)_2$	fp	0	KCl var	K_1 −0·5	64D
$EtHg^+$	sol	25	1 $Na(ClO_4)$	K_{s1} −2·4, no ev anionic cpx	65Bd
$MeHg^+$	gl, cal	20	0·1 (KNO_3)	K_1 5·25, ΔH_1 = −6·0, ΔS_1 = 3·6	63Sa = 65S
B^{2+}	con, pre	−85?	HCl(l)	ev $B_2Cl_6^{2-}$ in HCl(l)	68Pd
Al^{3+}	con	25	$PhNO_2$	K_3 4·4, K_4 2·0, $K(2AlCl_3 \rightleftharpoons Al_2Cl_6)$ −0·3 in $PhNO_2$(l)	51Va/64W
	X		solid	ev $AlCl_4^-$ in $Co(AlCl_4)_2$(s)	62Ia
	con	25	$PhNO_2$	K_3 4·45, $K_4 \approx \beta(AlCl_3{^\circ}AlCl_2^+)$ 3·04, $\beta(AlCl_2^+AlCl_4^-) \approx \beta(Al_2Cl_5^+Cl^-)$ 2·95 in $PhNO_2$(l)	66Wa
	con	25	MeCN	ev $AlCl_2(MeCN)_4^+$ and $AlCl_4^-$ in MeCN(l)	67Lc
	nmr		Me_2NCHO	ev $AlT_6^{3+}Cl^-_n$ (n = 1, 2, 3) in T = Me_2NCHO(l)	68M
Ga^{3+}	var		$PhNO_2$	ev $Ga_3Cl_{10}^-$ in $PhNO_2$(l) (con, fp, pre)	65Nc
	X		solid	ev Ga_2Cl_6 units in $GaCl_3$(s)	65W
	dis	20	0·69 $H(ClO_4)$	K_1 0·01	67Md
In^{3+}	cix	?	0·5 ($HClO_4$)	K_1 2·47, β_2 3·11, β_3 3·94	64Vb
			20% EtOH	K_1 2·59, β_2 3·75, β_3 4·53 in 20% EtOH, I = 0·5($HClO_4$)	
			40% EtOH	K_1 2·68, β_2 4·18, β_3 4·84 in 40% EtOH, I = 0·5($HClO_4$)	
	var		var	ev $InCl_5^{2-}$ in MeCN(l), $MeNO_2$(l) *etc.* and solid (pre, con, ir, fp)	68Wc
Tl^+	pol	25	LiCl var	β_2 −0·80, β_3 −1·68, β_4 −2·64	63Kg
	sol	25	0 corr	K_1 0·52 to 0·64 (depending on expression used for activity factors)	64Mi
	sol	25	0 corr	$K_1 \approx$ 0·62, K_{s0} −3·74	66Mf
	sol	30	0 corr	K_1 0·60, K_{s0} −3·62	67Ka
	sol	25	w% MeOH	K_1 0·52(0%), 0·70(16·7%), 0·90(30%), 1·34(60%); K_{s0} −3·74(0%), −4·15(16·7%), −4·49(30%), −5·36(60%), in w% MeOH, 0 corr	67Kb
Tl^{3+}	X		solid	ev $TlCl_6^{3-}$ in $K_3TlCl_6(H_2O)_2$(s)	35Ha
	X		solid	ev $Tl_2Cl_9^{3-}$ in $Cs_3Tl_2Cl_9$(s)	35P = 35H
	red	20	4 ($NaClO_4$), 3 H^+	K_1 7·54, K_2 5·84, K_3 3·41, K_4 2·79, β_4 19·58, $K_5 < -1·2$, no ev $TlCl_5^{2-}$, $TlCl_6^{3-}$	63Ac
		25	4 ($NaClO_4$), 3 H^+	K_1 7·48, β_2 13·26, β_3 16·65, β_4 19·45	63Ac+64L/64
	red	25	3	K_1 7·78, K_2 5·09, K_3 3·29, K_4 2·16, β_4 18·32	63Kf
			0·5	K_1 7·05, K_2 4·97, K_3 2·41, K_4 1·89, β_4 16·32	
	sol	20	4 ($NaClO_4$), 3 H^+	ev $TlCl_4^-$, no ev cpx with $n > 4$	64A
	cal	25	4 ($NaClO_4$), 3 H^+	ΔH_1 = −6·04, ΔH_2 = −4·05, ΔH_3 = −1·08, ΔH_4 = −0·17, ΔS_1 = 13·9, ΔS_2 = 12·9, ΔS_3 = 11·9, ΔS_4 = 12·2	64L+63Ac
	dis	0—35	→ 0	K_3 3·03(25°); K_4 1·51(0°), 1·47(25°), 1·32(35°) ΔH_4 = −5·1, ΔS_4 = −10, no ev cpx with $n > 4$	64N
		25	0·5 ($NaClO_4$)	K_4 1·38	
		25	D_2O, → 0	K_4 1·31 in 96% D_2O, → 0	

continued

Metal	Method	Temp	Medium	Log of equilibrium constant, remarks	Ref
Tl^{3+}	aix	20	$0{\cdot}5\ H^+$, 1 $Na(NO_3)$	K_4 0·6	64Pa
(contd)	dis	25	I(Na)Cl	$\beta(H^+TlCl_4^-)$ 1	64Pb
	var	25	3 $H(ClO_4)$	K_1 7·16, K_2 5·44, K_3 3·55, K_4 2·17, β_4 18·33	64W
				$\Delta H_1 = -5{\cdot}45$, $\Delta H_2 = -4{\cdot}40$, $\Delta H_3 = -1{\cdot}1$, $\Delta H_4 = -0{\cdot}3$	
				$\Delta S_1 = 14{\cdot}5$, $\Delta S_2 = 10{\cdot}1$, $\Delta S_3 = 12{\cdot}6$, $\Delta S_4 = 8{\cdot}9$	
			0·5 $H(ClO_4)$	K_1 6·72, K_2 5·10, K_3 2·68, K_4 1·81, β_4 16·31	
			0 corr	K_1 7·72, K_2 5·76, K_3 3·0, K_4 1·81, no ev $TlCl_5^{2-}$ or $TlCl_6^{3-}$, (red, sol, cal)	
	X, con		var	ev $TlCl_4^-$ in solid and MeCN(l)	65Cc
	red, Ag	25	3 ($LiClO_4$), 1H^+	K_1 7·10 β_2 12·50, β_3 16·00, β_4 18·50	65Kf
	cal	25	3 ($LiClO_4$), 1H^+	$\Delta H_1 = -7{\cdot}8$, $\Delta H_2 = -3{\cdot}7$, $\Delta H_3 = \Delta H_4 = 0$	
				$\Delta S_1 = 7$, $\Delta S_2 = 12$, $\Delta S_3 = 16$, $\Delta S_4 = 6$	
	Ram		var	β_6/β_4 −0·7, ev $TlCl_n^{(3-n)+}$, n = 1, 2, 3 4, no ev polyn cpx	65Sb
	Ram		10 LiCl	ev $TlCl_6^{3-}$	68Db
	Ram, ir		org	ev $TlCl_4^-$ in Et_2O(l), TBP(l) *etc.*	
	Ram	26	$MeNO_2$	K_5 0·66 in $MeNO_2$(l), also ev $TlCl_5^{2-}$ from pre, ir, con, fp	68Wc
$C_6H_5Tl^{2+}$	pre, pMeOH		MeOH	ev $C_6H_5TlCl_3^-$ and $C_6H_5TlCl_4^{2-}$ in MeOH(l)	66Ff
Ge^{2+}	X		solid	ev $GeCl_3^-$ in pilocarpine $H^+GeCl_3^-(H_2O)_{0{\cdot}5}$(s)	68F
Ge^{4+}					

1964 *Addendum*: p. 296, *to remarks for ref* 56Bb *add*: equil[$GeO_2(s) + 4H^+ + 4Cl^- \rightleftharpoons GeCl_4(l) + 2H_2O$] at 8·15 M HCl

Metal	Method	Temp	Medium	Log of equilibrium constant, remarks	Ref
Sn^{4+}	var		var	ev $SnCl_6^{2-}$, $SnCl_5H_2O^-$, $SnCl_4(H_2O)_2$ in HCl var, in $C_8H_{17}OH$(l), Bu_2O(l) *etc.* (sp, Mössbauer effect, dis)	64Ic
	spJ		MeCN	K_5 4·30 in MeCN(l)	65Mk
	ir, Ram		solid	ev $SnCl_5^-$ (bipyramid) in Et_4NSnCl_5(s) and in Me_2CO(l)	68Cb
$MeSn^{3+}$	nmr		HCl var	ev $MeSnCl_n^{3-n}$, n = 1, 3, 5	64Bh
	aix	25	0 corr	K_2 0·35, K_3 −0·25, K_4 −1·79 (from LiCl var)	66Ce = 67Ce
	Ag	25	3 ($NaClO_4$)	K_1 1·69, β_2 2·51	68Ma
Me_2Sn^{2+}	Ram		9 HCl	ev Me_2SnCl^+	64Mg
	gl	25	0·1	K_1 1·45	64Tf+
	Ag	25	3 ($NaClO_4$)	K_1 0·38, β_2 −0·14 (also Ram, nmr)	65Fb
	ir, pre		solid	ev $Me_2SnCl_3^-$, $Me_2SnCl_4^{2-}$ in $Me_4NMe_2SnCl_3$(s) and $Cs_2Me_2SnCl_4$(s)	66Cb
	aix	25	0 corr	K_1 0·37, K_2 −0·23, K_3 −1·45 (from LiCl var)	66Ce = 67Ce
	X		solid	ev $Me_2SnCl_3^-$ in solid	68E
Me_3Sn^+	aix	25	0 corr	K_1 −0·17, K_2 −1·57 (from LiCl var)	66Ce = 67Ce
Ph_3Sn^+	dis	30	0·1 ($NaNO_3$)	$K_D[Ph_3SnOH(org) + Cl^- \rightleftharpoons Ph_3SnCl(org) + OH^-]$ −7·1(org = C_6H_6), −6·9(org = Bu^{i}COMe), K_1 3·0	65Sa
Pb^{2+}	pol	25	1 ($NaClO_4$)	β_4/β_2 0·15?	55Kd/65Hb
	MHg	25	3 $Li(NO_3)$	K_1 0·32, β_2 0·10, β_3 −0·28,	63Ml
			3 $Na(NO_3)$	K_1 0·31, β_2 0·34, β_3 −0·2	
			3 $K(NO_3)$	K_1 0·46, β_2 0·58, β_3 −0·1, β_4 0·0	
			3 $NH_4(NO_3)$	K_1 0·37, β_2 0·48, β_3 < −0·3, β_4 0·3, also data for Li–Rb and Li–Cs mixtures	

continued overleaf

67 Cl⁻ (contd)

Metal	*Method*	*Temp*	*Medium*	*Log of equilibrium constant, remarks*	*Ref*
Pb^{2+} (*contd*)	MHg	25	I $Na(ClO_4)$	K_1 1·16, β_2 1·7, β_3 1·97, β_4 0·7 (I = 3); K_1 0·90, β_2 1·36, β_3 1·45, (I = 1), also data for I = 4, 2, 0·75, 0·5, 0·25	63Mm
			I $Na(NO_3)$	K_1 0·43, β_2 0·5, β_3 −0·3, β_4 −0·4 (I = 4); K_1 0·48, β_2 0·15, β_3 0·49 (I = 1), also data for I = 3, 2, 0·75. See also NO_3^-–Pb^{2+}	
	MHg	25	0 corr	K_{s0} ($Pb^{2+}Cl^-_2$) −4·8	64Aa
	MHg	25	2·5 $Ca(ClO_4)$	β_3 2·81	64Ba
	Ag	160—200	MNO_3(l)	K_1 2·40(160°), 2·36(180°), 2·31(200°); K_2 2·04(160°), 1·97(180°), 1·95(200°), in $Li_{0.5}K_{0.5}NO_3$(l), x units	64Bd
	MHg	25	3 (Li, Na)ClO_4	$K(Li_2PbCl_4 + Na^+ \rightleftharpoons LiNaPbCl_4 + Li^+)$ −0·33 $K(Li_2PbCl_4 + 2Na^+ \rightleftharpoons Na_2PbCl_4 + 2Li^+)$ −1·05(or −0·7?)	64M
			3 (Li)Cl	only small changes in apparent β_n on replacing Li^+ by Mg^{2+} (Ca^{2+}, Sr^{2+}, or Ba^{2+})	
	MHg	25	3 ($LiClO_4$)	K_1 1·16, β_2 1·81, β_3 1·91, $\beta_4 \approx$ 1·2 ΔH_1 = 0·86, $\Delta H_{\beta 2}$ = 1·9, $\Delta H_{\beta 3}$ = 2·6, $\Delta H_{\beta 4} \approx$ 0 ΔS_1 = 8·2, $\Delta S_{\beta 2}$ = 15, $\Delta S_{\beta 3}$ = 18, $\Delta S_{\beta 4} \approx$ 16, also β_n values for temp from 5 to 55°	64Mb
	MHg	25	3 ($LiClO_4$)	K_{s0} −5·0	64Md
	sol	25	3 ($LiClO_4$)	K_1 1·23, β_2 1·87, β_3 1·98, β_4 1·72; $\beta(Na^+PbCl_4^{2-})$ −0·28	
	sol, sp	25	4 $H(ClO_4)$	K_{s2} −3·33 K_1 1·0, β_2 2·47, $K_3 \leqslant$ −0·4, β_4/β_2 0·15, $\beta_6/\beta_4 \leqslant$ −1·7	65Hb
	MHg	25	3 ($LiClO_4$), 1 H^+	K_1 1·19, β_2 1·73, β_3 2·03, β_4 0·85; $\beta(M^+PbCl_4^{2-})$ −0·85(Na^+), −0·64(K^+), 0·0(Cs^+), −0·05(NH_4^+), 0·85(H^+)	65M
	MHg	25	3 ($LiClO_4$) 3 (LiCl)	K_1 1·18, β_2 1·72, β_3 2·00, β_4 1·04 $\beta(M^+PbCl_4^{2-})$ −0·14(Na^+), 0·32(K^+), 0·42(Rb^+), 0·50(Cs^+); $\beta(M^+PbCl_3^-)$ −1·0(Na^+), −0·3(K^+), −0·2(Rb^+), −0·1(Cs^+);	65Ma
	MHg	25	4 ($NaClO_4$) 3 ($NaClO_4$)	K_1 1·24, β_2 1·73, β_3 2·14, β_4 1·39 K_1 1·05, β_2 1·51, β_3 1·83	66Vc
	pol	145	MNO_3(l)	K_1 1·32, K_2 1·04 or $K_2 \approx K_3$ 0·6 in (Li, Na, K)NO_3(l, eut), m units	68I
	tp	23	LiCl var	$K_2 \approx$ 0, $K_3 \approx$ 0	68Sj
	sol	35	dil	$K_s(PbOH^+Cl^-)$ −6·12	68Y
Pb^{4+}					

1964 *Erratum*: p. 299, *in ref* 60Sf *for* "HH^+" *read* "$4H^+$".

Metal	*Method*	*Temp*	*Medium*	*Log of equilibrium constant, remarks*	*Ref*
	sp		var	ev $PbCl_4$ in org, ev $PbCl_5^-$, $PbCl_6^{2-}$, not $PbCl_4$ in HCl var	64Sf
Ph_3Pb^+	dis	30	0·1 ($NaNO_3$)	K_1 4·8, $K_D[Ph_3PbOH(org) + Cl^- \rightleftharpoons Ph_3PbCl(org) + OH^-]$ −3·1 (org = $CHCl_3$), −3·0(org = Bu^{i}COMe) sign misprinted in Table 4?	65Sa
NH_4^+	Ram		5 NH_4Cl	ev cpx	64Lb
Me_4N^+	fp	0	→ 0	agrees with K_1 0·14 (ref 66Ec/68Fb)	34L/68Fb
	con	0—25	→ 0	K_1 0·14(→ 0°), 0·10(10°), 0·04(25°)	66Ec/68Fb

continued

Metal	Method	Temp	Medium	Log of equilibrium constant, remarks	Ref
Bu_4N^+	fp	0	MCl var	agrees with K_1 0·50 (ref 66Ec/68Fb)	34L/68Fb
	con	0—25	→ 0	K_1 0·50(→ 0°), 0·46(10°), 0·40(25°)	66Ec/68Fb
NO^+	sp	25	0 corr	$K(H^+ + Cl^- + HNO_2 \rightleftharpoons NOCl + H_2O)$ −2·94	41S
	sp	0—25	0 corr	$K(H^+ + Cl^- + HNO_2 \rightleftharpoons NOCl + H_2O)$ −3·26(0°), −3·19(5°), −3·12(10°), −3·06(15°) −2·94(25°)	54Sf+/56Sd
N^{5+}	pre, ir		solid	ev $ClHNO_3^-$ in $Me_4NHClNO_3$(s)	66Sb
P^{5+}	*X*		solid	ev PCl_4^+ and PCl_6^- in PCl_5(s)	42C
	pre, ir		solid	ev $PO_2Cl_2^-$ in $Be(PO_2Cl_2)_2$(s)	67Mg
As^{3+}	fp	−19	$AsCl_3$	†$K[(AsOCl)_n + AsOCl \rightleftharpoons (AsOCl)_{n+1}]$ −0·74 in $AsCl_3$(l), *x* units	64T
	bp	130	$AsCl_3$	†K −0·55 in $AsCl_3$(l), *x* units	
	Ram, ir		Bu_2O	ev $AsCl_4^-$ in Bu_2O(l)	68Da
As^{5+}	Ram		solid	ev $AsCl_6^-$ in Et_4NAsCl_6(s)	67Bh
Sb^{3+}	dis	15	0·5 H, 6·3Li(NO_3)	K_4 1·3 to 1·5, K_5K_6 −1·1 to −0·8 $K_D[H^+ + SbCl_4^- \rightleftharpoons HSbCl_4$ (in org)] ≈ 0·6 (org = $C_6H_{13}OH$ or $C_8H_{17}OH$)	64Ib
	sol	25	4 H(HSO_4)	$K_s[(Me_4N^+)_3(SbCl_4^-)_2Cl^-]$ −4·74, K_4 1·0, K_5K_6 −0·77	65Ha
Sb^{5+}	dis	?	0 corr	†K_4 −3·22, †K_5 −3·48, †K_6 −4·06 (from HCl var) (criticised by ref 64F)	63I
	dis		0 corr	†$K_n[SbCl_{n-1}(OH)_{7-n}^- + H^+ + Cl^- \rightleftharpoons SbCl_n(OH)_{6-n}^- + H_2O]$ −3·07(n = 4), −3·46(n = 5), −4·28(n = 6)	65D+
Bi^{3+}	*X*		solid	ev $BiCl_6^{3-}$ in $Co(NH_3)_6BiCl_6$(s)	52Aa

1964 *Erratum*: p. 300, *delete ref* 56N; *for ref* "57Nd" *read* "57Nd ≈ 56N"

Metal	Method	Temp	Medium	Log of equilibrium constant, remarks	Ref
	MHg	20	2 ($NaClO_4$), 1H⁺	β_6/β_4 0·28, β_4/β_2 2·40	57Aa/64Hb
	sp			data compatible with $BiCl_6^{3-}$ or $BiCl_5^{2-}$	57Nd/64Hb
		25	1 $HClO_4$	β_2 4·5 (β_5 6·72 from ref 57Aa)	57Nd/66Lb

1964 *Addendum*: p. 300, *for ref* "62Hb" *read* "62Hb = 63Hb"

Metal	Method	Temp	Medium	Log of equilibrium constant, remarks	Ref
	MHg	25	3 ($LiClO_4$)	$\Delta H_1 \approx 0$, $\Delta S_1 \approx 10$	63Mh/67A
	MHg	25—65	3 (Li)Cl, 1H⁺	$\beta(Na^+BiCl_6^{3-})$ −0·26(25°), −0·3(45°), ≈ −1(65°) $\Delta H \approx -0{\cdot}9$, $\Delta S \approx -5$; $\beta(Na^+{}_2BiCl_6^{3-})$ −0·7(25°), 0(45°), ≈ −1(65°); $\beta(K^+BiCl_6^{3-})$ 0·18(25°), 0·11(45°), 0·08(65°), $\Delta H = -1{\cdot}2$, $\Delta S = -3{\cdot}2$; $\beta(K^+{}_2BiCl_6^{3-})$ −1·0(25°), −1·3(45°), 0·36(65°); $\beta(K^+{}_3BiCl_6^{3-})$ −0·4(25°), −0·7(45°); $\beta(Rb^+BiCl_6^{3-})$ 0·30(25°), 0·23(45°), 0·18(65°), $\Delta H = -1{\cdot}3$, $\Delta S = -3{\cdot}0$; $\beta(Rb^+{}_2BiCl_6^{3-})$ 0·32(25°), 0·20(45°), 0·11(65°); $\beta(Cs^+BiCl_6^{3-})$ 0·40(25°), 0·34(45°), 0·28(65°), $\Delta H = -1{\cdot}4$, $\Delta S = -2{\cdot}9$; $\beta(Cs^+{}_2BiCl_6^{3-})$ 0(25°), 0·2(45°), 0·3(65°); $\beta(Cs^+{}_3BiCl_6^{3-})$ 0·3(25°), 0·2(45°)	63Mk
	sol	25	4 H(HSO_4)	$K_s[(Me_4N^+)_3(BiCl_4^-)Cl_2^-]$ −7·64; β_6/β_4 0·8, no ev $BiCl_5^{2-}$, β_4/β_2 2·58	64Hb
	sp		4 H(HSO_4)	β_6/β_4 0·5, β_4/β_2 2·08	

continued overleaf

67 Cl⁻ (contd)

Metal	Method	Temp	Medium	Log of equilibrium constant, remarks	Ref
Bi^{3+} (contd)	sol	25	2 ($NaClO_4$), 1 H^+	$*K_{s0}[BiOCl(s)+2H^+ \rightleftharpoons Bi^{3+}+Cl^-+H_2O]$ −7·08 β_2 4·30, β_3 5·91, β_4 6·76	65J
	cix	rt	1·89 $H(ClO_4)$	K_1 2·34, β_2 3·89, β_3 5·23	66Lb
	pol	25	1 H^+	β_5 5·25(1 M-HNO_3), 5·36(1 "N" H_2SO_4)	67Cg
	Ram		var	ev $BiCl_4^-$, $BiCl_5^{2-}$, $BiCl_6^{3-}$, and species with 2 and 3 Cl/Bi	67O
	sol	25	$HClO_4$ var	$*K_{s0}$ −6·61	67Va
		25	0 corr	$*K_{s0}$ −7·87	
	cal	−7 to 40	4 ($HClO_4$)	ΔH_1 = −0·63(−7°), −0·27(0°), −0·14(5°), 0·12(10°), 0·34(18°), 0·52(25°), 1·03(40°); ΔS_1 = 10·3(−7°), 11·6(0°), 12·3(5°), 12·9(10°), 13·8(18°), 14·6(25°), 15·1(30°), 15·6(35°), 16·0(40°)	67Vb
		25	I ($HClO_4$)	ΔH_1 = −1·36(I = 6), −0·26(I = 5), 0·52(I = 4), 4·0 (5·0?, I = 0); ΔS_1 = 9·7(I = 6), 12·6(I = 5), 14·6(I = 4), 32·4(I = 0), also ΔC_p values [K_1 3·12(I = 6), 2·94(I = 5), 2·78(I = 4), 3·42(I = 0)]	
	cal	−8 to 40	4 ($HClO_4$)	ΔS_1 = 10·3(−8°), 16·0(40°), linear with T ΔC_{p1} = 37·2(−8°), 31·5(40°), linear with T $\Delta S_1+\Delta C_{p1} \approx$ 47·6, independent of T	67Vb/68V
	sol	25	H^+ var	$*K_{s0}$ −6·72(H^+ var), −7·95(0 corr); $\beta[Bi^{3+}Cl^-(NO_3^-)_2]$ 3·23(H^+ var), 5·04(0 corr)	67Vc
	sol	15—50	3 ($NaClO_4$)	$*K_{s0}[BiOCl(s)+2H^+ \rightleftharpoons Bi^{3+}+Cl^-+H_2O]$ −6·82(15°), −6·75(25°), −6·61(50°)	68V
			2 ($NaClO_4$)	$*K_{s0}$ −6·63(15°), −6·54(25°), −6·52(50°)	
			1 ($NaClO_4$)	$*K_{s0}$ −6·52(15°), −6·47(25°), −6·41(50°)	
			0 corr	$*K_{s0}$ −7·82(15°), −7·80(25°), −7·76(50°)	
S^{2+}	con, pre		Me_2CO	ev SCl^+ in Me_2CO(l)	66N
S^{4+}	Ram		solid	ev SCl_3^+ in $SCl_3^+AsF_6^-$(s)	67Se
S^{6+}	cal, ΔG	25	HSO_3Cl	$K(HSO_3Cl \rightleftharpoons HCl+SO_3)$ −5·3, $K(2HSO_3Cl \rightleftharpoons H_2SO_4+SO_2Cl_2)$ −2·8 in HSO_3Cl(l)	67Re+
Se^{4+}	Ram		var	ev $SeCl_3^+$ in $SeCl_4$(s and l)	54Gb
	con	25	org	ev $SeCl_3^+$ in Me_2NCHO(l), $PhNO_2$(l) *etc.*	67Cd
	X		solid	ev units $Cl_2SeOSbCl_5$ in $SeOCl_2SbCl_5$(s)	67Ha
			solid	ev solvated $Cl(SeOCl_2)_5^-$ in $Me_4NCl(SeOCl_2)_5$(s)	67Hb
	con, Ram		var	ev $SeCl_3^+$ in HSO_3Cl(l), HSO_3F(l), ev $SeCl_4$ in $SeCl_4$(s)	68Rc
Te^{4+}	Ram		var	ev $TeCl_3^+$ in $TeCl_4$ (s and l)	54Gb
	dis	22	var	$K_D[TeO_2H^++3H^++4Cl^-+3T(org) \rightleftharpoons TeCl_4T_3(org)]$ −2·80, T = TBP, org = n-hexane, M units	65Bh
	sp	20	org	K_5 0·77 in Me_2SO(l), 0·7 in Me_2NCHO(l), 1·62 in MeCN(l), 2·05 in $MeNO_2$(l)	65Kd
	ir		solid	ev $TeCl_3^+$ in $TeCl_4$(s)	66G
	sp		I $H(ClO_4)$	K_5K_6 1·55 (I = 8), 0·25(I = 6)	66Ra
	con	25	org	ev $TeCl_3^+$ in Me_2NCHO(l), $PhNO_2$(l) *etc.*	67Cd
	var		org	ev $(TeCl_4)_n$, $n \approx 3$ in C_6H_6(l), C_6H_5Me(l) ev $TeT_2Cl_3^+$ and Cl^- in T = MeCN(l), EtOH(l) *etc.* (fp, con, ir)	68Gg
	sol	18	0·5 ($NaClO_4$)	$\beta(TeOOH^+Cl^-)$ 0·5	68Nb
	sp		var	$\beta(TeOOHCl^0H^+Cl^-)$ 0·23, $\beta(TeOCl_2{}^0Cl^-)$ −0·44, $\beta(TeOCl_3^-Cl^-)$ −0·77, $\beta(TeOCl_4^{2-}H^+{}_2Cl^-{}_2)$ −1·7	

contin

Metal	Method	Temp	Medium	Log of equilibrium constant, remarks	Ref
Te^{4+} (*contd*)	con, Ram		var	ev $TeCl_3^+$ in $HSO_3Cl(l)$, $HSO_3F(l)$, ev $TeCl_4$ in $TeCl_4(s)$	68Rc
	dis	25	7 $H(ClO_4)$	ev $HTeCl_5C_6H_{13}OH$ in $C_6H_6(l)$; K_1 3·24, β_2 6·0, β_3 8·34, β_4 10·18, β_5 12·76, β_6 15·30	68Sd
	sp		0 corr	$\beta(TeOOHCl_3{}^{2-}H^+Cl^-)$ −3·32, $\beta(TeOCl_4{}^{2-}H^+{}_2)$ −2·20, K_5 −1·83, K_6 −2·19 (from data with HCl var)	68Sh
	dis	18	0 corr	$K_D[TeCl_4(H_2O)_2+2T(org) \rightleftharpoons 2H_2O+TeCl_4T_2(org)]$ 0·01, T = TBP, org = iso-C_8H_{18}	68Si
Po^{4+}	cix	22—25	1 $H(ClO_4)$	K_1 2·34, β_2 4·42, β_3 6·34, β_4 8·53, β_5 10·08, β_6 11·57; also values for 10% Me_2CO and 20% Me_2CO	64Sg
	dis		4—6 HCl	$K_D[Po^{4+}+H^++5Cl^-+2T(org) \rightleftharpoons HPoCl_5T_2(org)]$ 1·78, T = TBP, org = C_6H_6; K_1 2·56, β_2 4·80, β_3 6·88, β_4 8·85, β_5 10·60, β_6 11·92	65Si
	dis		0 corr	K_6 2·3, also equil const for Po^{II} and Po^{IV} distribution HCl–C_6H_6+Pr_2CO or $C_6H_{13}OH$	67Ia
Cl_2	sol	30—90	HCl var	K_{p0} −1·31(30°), −1·46(40°), −1·57(50°), −1·75(70°), −1·88(90°); K_{p1} −2·18(30°), −2·29(40°), −2·44(50°), −2·66(70°), −2·86(90°)	68Ha
Br_2	var	25	var	K_1 0·06, $K(Br_2+2Cl^- \rightleftharpoons BrCl_2^-+Br^-)$ −2·14 [red, gl, (sp)]	66Bb
	cal	25	3 $Li(ClO_4)$	$\Delta H_1 = -2{\cdot}2$, $\Delta S_1 = -7$	67Me+
I_2					

1964 *Addendum*: p. 301, *for ref* "62Mm" *read* "62Mm = 63Mo"

Metal	Method	Temp	Medium	Log of equilibrium constant, remarks	Ref
	dis	15—45	3 $Li(ClO_4)$	K_1 0·26(15°), 0·30(25°), 0·34(35°), 0·40(45°), $\Delta H_1 = 1{\cdot}7$, $\Delta S_1 = 7$	67Me
I^+	con	140	$I_2(l)$	K_1 4·00 in $I_2(l)$	67Bc
ICl	pre, ir		solid	ev $I_2Cl_3^-$ in $pyHI_2Cl_3(s)$ *etc.*	65Y = 67Y
IBr	pre, ir		solid	ev $I_2Cl_2Br^-$	67Y
At^+ ?	cix		0·5 HNO_3	K_1 2·85, β_2 5·40	68Na

18C O. Collenberg, *Z. anorg. Chem.*, 1918, **102**, 247
25F J. E. Frazer and H. Hartley, *Proc. Roy. Soc.*, 1925, *A*, **109**, 351
29Cb I. I. Chernyaev and S. I. Khouzlenkov, *Izvest. Inst. po Izucheniya Platiny*, 1929, **7**, 98
34A G. Åkerlöf, *J. Amer. Chem. Soc.*, 1934, **56**, 1439
34L J. Lange, *Z. phys. Chem.*, 1934, *A*, **168**, 147
35H J. L. Hoard and L. Goldstein, *J. Chem. Phys.*, 1935, **3**, 199
35Ha J. L. Hoard and L. Goldstein, *J. Chem. Phys.*, 1935, **3**, 645
35P H. M. Powell and A. F. Wells, *J. Chem. Soc.*, 1935, 1008
38Ga L. W. N. Godward and W. Wardlaw, *J. Chem. Soc.*, 1938, 1422
39La A. B. Lamb, *J. Amer. Chem. Soc.*, 1939, **61**, 699
41S H. Schmid and A. Maschka. *Z. phys. Chem.*, 1941, *B*, **49**, 171
42C D. Clark, H. M. Powell, and A. F. Wells, *J. Chem. Soc.*, 1942, 642
47D F. P. Dwyer, R. S. Nyholm, and L. E. Rogers, *J. Proc. Roy. Soc. New South Wales*, 1947, **81**, 267
51Va R. E. Van Dyke and H. E. Crawford, *J. Amer. Chem. Soc.*, 1951, **73**, 2018
52A C. S. Adams and D. P. Mellor, *Austral. J. Sci. Res.*, 1952, *A*, **5**, 577
52Aa M. Atoji and T. Watanabe, *J. Chem. Phys.*, 1952, **20**, 1045
53Z A. B. Zdanovskii, E. I. Lyakhovskaya, and R. E. Shleimovich, 'Spravochnik eksperimental'nykh dannykh po rastvorimosti mnogo-komponentnykh vodno-solevykh sistem I,' Goskhimizdat, 1953, p. 347–363
54Gb H. Gerding and H. Houtgraaf, *Rec. Trav. chim.*, 1954, **73**, 737
54Sf H. Schmid, *Monatsh.*, 1954, **85**, 424

continued overleaf

67 Cl^- (contd)

55Ca D. Cozzi and S. Vivarelli, *Z. anorg. Chem.*, 1955, **279**, 165
56Sd H. Schmid and E. Hallaba, *Monatsh.*, 1956, **87**, 560
56Se P. G. Sears, R. K. Wolford, and L. R. Dawson, *J. Electrochem. Soc.*, 1956, **103**, 633
57Pe R. G. Pearson, P. M. Henry, and F. Basolo, *J. Amer. Chem. Soc.*, 1957, **79**, 5382
57Wa G. J. Wessel and D. J. W. Ijdo, *Acta Cryst.*, 1957, **10**, 466
58Cc E. S. Clark, D. Templeton, and C. H. MacGillavry, *Acta Cryst.*, 1958, **11**, 284
58Dc R. J. Dietz, jun., thesis, Massachusetts Inst. of Tech. 1958
58Zc A. Zalkin and D. E. Sands, *Acta Cryst.*, 1958, **11**, 615
59Md J. S. Mendez-Schalchi, thesis, Massachusetts Inst. of Tech. 1959
59Sb D. E. Sands and A. Zalkin, *Acta Cryst.*, 1959, **12**, 723
60Fb D. A. Fine, thesis, Univ. of California, Berkeley, 1960, UCRL-9059
60Kd T. S. Khodashova, *Zhur. strukt. Khim.*, 1960, **1**, 333
61Bf M. E. Baldwin, S. C. Chan, and M. L. Tobe, *J. Chem. Soc.*, 1961, 4637
62Bf J. Braunstein and J. M. C. Hess, TID-15741, 1962
62Ca W. M. Campbell, jun., and R. M. Wallace, TID-15360, 1962
62D G. D.'Amore, G. Calabro, and P. Currò, *Atti Soc. Peloritana Sci. fis. mat. nat.*, 1962, **8**, **265**
62Ia J. A. Ibers, *Acta Cryst.*, 1962, **15**, 967
1964 *Addendum*
62Mb add: D. I. Ryabchikov *et al.*, *J. Inorg. Nuclear Chem.*, 1964, **26**, 965
62Sd D .N. Solokov, *Trudy po Khim i khim. Tekh.* 1962, no. 1, 55
62Se M. Shiloh and Y. Marcus, IA-781 (Israel AEC), 1962
62Sf K. W. Sykes and B. L. Taylor, Proc. Seventh International Conference on Co-ordination Chemistry, **1962, p. 31**
62Ta M. L. Tobe and D. W. Watts, *J. Chem. Soc.*, 1962, 4614
63Ac S. Ahrland, I. Grenthe, L. Johansson, and B. Norén, *Acta Chem. Scand.*, 1963, **17**, 1567
63Ad N. V. Aksel'rud in 'Redkozemel'nye elementy', Izd. Nauka, Moskva, 1963, p. 75
63Ae E. A. Allen, B. J. Brisdon, D. A. Edwards, G. W. A. Fowles, and R. G. Williams, *J. Chem. Soc.*, **1963, 4649**
1964 *Addendum*
63Ba add: *idem*, *Inorg. Chem.*, 1963, **2**, 1166
63Bb B. Brehler and P. Süsse, *Naturwiss.*, 1963, **50**, 517
63Cd J. A. Creighton and E. R. Lippincott, *J. Chem. Soc.*, 1963, 5134
63Ea I. Eliezer and Y. Marcus, *J. Inorg. Nuclear Chem.*, 1963, **25**, 1465
63Eb J. D. Eakins, D. G. Humphreys, and C. E. Mellish, *J. Chem. Soc.*, 1963, 6012
63F Ya. D. Fridman, R. K. Drachevskaya, and V. A. Shestakova in 'Redkozemel'nye elementy', Izd. Nauka, Moskva, 1963, p. 166
63Gc D. C. Giedt and C. J. Nyman, *J. Phys. Chem.*, 1963, **67**, 2491
63Gd A. A. Grinberg and E. S. Postnikova, *Doklady Akad. Nauk S.S.S.R.*, 1963, **153**, 340
63Ge A. A. Grinberg, N. V. Kiseleva, and M. I. Gel'fman, *Doklady Akad. Nauk S.S.S.R.*, 1963, **153**, 1327
63Hb A. Hershaft and J. D. Corbett, *Inorg. Chem.*, 1963, **2**, 979
63Hc L. D. Hansen, R. M. Izatt, and J. J. Christensen, *Inorg. Chem.*, 1963, **2**, 1243
63Hd R. N. Heistand and A. Clearfield, *J. Amer. Chem. Soc.*, 1963, **85**, 2566
63I B. Z. Iofa and G. M. Dakar, *Radiokhimiya*, 1963, **5**, 490
63Jb J. A. D. Jeffreys, G. A. Sim, R. H. Burnell *et al.*, *Proc. Chem. Soc* , 1963, 171
63Jc G J. Janz and D. W. James, *J. Chem. Phys.*, 1963, **38**, 905
63Kf E. L. King, private communication quoted by ref 63Ac
63Kg F. Ya. Kul'ba and V. E. Mironov, "Khimiya talliya", Goskhimizdat, Leningrad, 1963, p. 46 (Tl^+)
63Kh H.-L. Krauss, M. Leder, and G. Münster, *Chem. Ber.*, 1963, **96**, 3008
63L J. O. Liljenzin, H. Reinhardt, H. Wirries, and R. Lindner, *Z. Naturforsch.*, 1963, **18a**, 840
63La J. O. Liljenzin, H. Reinhardt, H. Wirries, and R. Lindner, *Radiochim. Acta*, 1963, **1**, 161
63Mk V. E. Mironov, F. Ya. Kul'ba, V. A. Fedorov, and T. F. Nikitenko, *Zhur. neorg. Khim.*, 1963, **8**, 2318
63Ml V. E. Mironov and V. A. Fedorov, *Zhur. neorg. Khim.*, 1963, **8**, 2529
63Mm V. E. Mironov, F. Ya. Kul'ba, V. A. Fedorov, and O. B. Tikhomirov, *Zhur. neorg. Khim.*, 1963, **8**, 2536
63Mn D. F. C. Morris and E. L. Short, *J. Inorg. Nuclear Chem.*, 1963, **25**, 291
63Mo D. Meyerstein and A. Treinin, *Trans. Faraday Soc.*, 1963, **59**, 1114
63Mp V. E. Mironov, dissertation, Leningrad Technological Institute, 1963, quoted by ref 67Mh
63Na I. V. Nelson and R. T. Iwamoto, *J. Electroanalyt. Chem.*, 1963, **6**, 234
63Pe N. C. Petersen and F. R. Duke, *J. Phys. Chem.*, 1963, **67**, 531
63Sa M. Schellenberg, dissertation, Eidgenössische Technische Hochschule, Zürich, 1963
63Sb A. M. Sargeson, *Austral. J. Chem.*, 1963, **16**, 352
63Sc K. Shimizu, H. Takesawa, and J. Osugi, *J. Chem. Soc. Japan*, 1963, **84**, 707
63V P. H. Vossos, D. R. Fitzwater, and R. E. Rundle, *Acta Cryst.*, 1963, **16**, 1037
63Va V. P. Vasil'ev, V. N. Vasil'eva, F. I. Ivanova, and O. N. Morozova, *Izvest. V.U.Z., Khim. i khim. Tekhnol.*, 1963, **6**, 712
64A S. Ahrland and L. Johansson, *Acta Chem. Scand.*, 1964, **18**, 2125
64Aa V. I. Altynov and B. V. Ptitsyn, *Zhur. neorg. Khim.*, 1964, **9**, 2407
64Ab K. Aurivillius, *Arkiv Kemi*, 1964, **22**, 517, 537
64B A. A. Biryukov and V. I. Shlenskaya, *Zhur. neorg. Khim.*, 1964, **9**, 813
64Ba E. Ya. Ben'yash and T. G. Maslakova, *Zhur. neorg. Khim.*, 1964, **9**, 2731
64Bb B. M. L. Bansal, S. K. Patil, and H. D. Sharma, *J. Inorg. Nuclear Chem.*, 1964, **26**, 993

contin

64Bc K. Burger, *Acta Chim. Acad. Sci. Hung.*, 1964, **40**, 261
64Bd J. Braunstein and A. S. Minano, *Inorg. Chem.*, 1964, **3**, 218
64Be J. A. Broomhead, F. Basolo, and R. G. Pearson, *Inorg. Chem.*, 1964, **3**, 826
64Bf G. W. Brady, M. B. Robin, and J. Varimbi, *Inorg. Chem.*, 1964, **3**, 1168
64Bg J. Braunstein, H. Braunstein, and A. S. Minano, *Inorg. Chem.*, 1964, **3**, 1334
64Bh E. V. van den Berghe and G. P. van der Kelen, *Ber. Bunsengesellschaft Phys. Chem.*, 1964, **68**, 652
64Bi R. J. Baltisberger and E. L. King, *J. Amer. Chem. Soc.*, 1964, **86**, 795
64Bj K. W. Bagnall, D. Brown, and J. G. H. du Preez, *J. Chem. Soc.*, 1964, 2603
64Bk K. W. Bagnall and D. Brown, *J. Chem. Soc.*, 1964, 3021
64Bl D. Brown, *J. Chem. Soc.*, 1964, 4944
64Bm I. D. Bulanova and A. M. Borob'ev, *Radiokhimiya*, 1964, **6**, 621, 623
64C J. J. Christensen, R. M. Izatt, L. D. Hansen, and J. D. Hale, *Inorg. Chem.*, 1964, **3**, 130
64Ca F. A. Cotton and B. F. G. Johnson, *Inorg. Chem.*, 1964, **3**, 780
64Cb F. A. Cotton and J. T. Mague, *Inorg. Chem.*, 1964, **3**, 1402
64Cc F. A. Cotton and J. T. Mague, *Proc. Chem. Soc.*, 1964, 233
64Cd I. I. Chernyaev, N. N. Zheligovskaya, and T. N. Leonova, *Zhur. neorg. Khim.*, 1964, **9**, 347
64D A. J. Downs, *J. Inorg. Nuclear Chem.*, 1964, **26**, 41
64F V. V. Fomin, *Radiokhimiya*, 1964, **6**, 378
64Fa J. E. Fergusson, B. R. Penfold, and W. T. Robinson, *Nature*, 1964, **201**, 181
64G M. L. Gavrish and I. S. Galinker, *Zhur. neorg. Khim.*, 1964, **9**, 1289
64Ga I. Gainar and K. W. Sykes, *J. Chem. Soc.*, 1964, 4452
64H J. M. C. Hess, J. Braunstein, and H. Braunstein, *J. Inorg. Nuclear Chem.*, 1964, **26**, 811
64Ha T. Hu, H. C. Ko, and L. G. Hepler, *J. Phys. Chem.*, 1964, **68**, 387
64Hb G. P. Haight, jun., C. H. Springer, and O. J. Heilmann, *Inorg. Chem.*, 1964, **3**, 195
64Hc R. A. Horne, B. R. Myers, and G. R. Frysinger, *Inorg. Chem.*, 1964, **3**, 452
64Hd S. M. Horner and S. Y. Tyree, *Inorg. Chem.*, 1964, **3**, 1173
64He G. P. Haight, jun., D. C. Richardson, and N. H. Coburn, *Inorg. Chem.*, 1964, **3**, 1777
64I H. M. N. H. Irving and P. K. Khopkar, *J. Inorg. Nuclear Chem.*, 1964, **26**, 1561
64Ia M. J. Insley, G. D. Parfitt, and A. L. Smith, *J. Phys. Chem.*, 1964, **68**, 2372
64Ib B. Z. Iofa and G. M. Dakar, *Radiokhimiya*, 1964, **6**, 411
64Ic B. Z. Iofa, K. P. Mitrofanov, M. V. Plotnikova, and S. Kopach, *Radiokhimiya*, 1964, **6**, 419
64J S. J. Jensen, *Acta Chem. Scand.*, 1964, **18**, 2085
64Ja G. J. Janz, I. Ahmad, and H. V. Venkattasetty, *J. Phys. Chem.*, 1964, **68**, 889
64K P. K. Khopkar and H. D. Sharma, unpublished, quoted by ref 64Bb
64L I. Leden and T. Ryhl, *Acta Chem. Scand.*, 1964, **18**, 1196
64La V. A. Latysheva, 'Khimiya redkikh elementov,' Izd. Leningrad Univ., 1964, p. 133
64Lb H. Lee and J. K. Wilmshurst, *Austral. J. Chem.*, 1964, **17**, 943
64M V. E. Mironov, F. Ya. Kul'ba, and V. A. Fedorov, *Zhur. neorg. Khim.*, 1964, **9**, 1487
64Ma V. E. Mironov, F. Ya. Kul'ba, and Yu. E. Ivanov, *Zhur. neorg. Khim.*, 1964, **9**, 1633
64Mb V. E. Mironov, F. Ya. Kul'ba, and V. A. Fedorov, *Zhur. neorg. Khim.*, 1964, **9**, 1641
64Mc T. V. Mal'kova, G. A. Shutova, and K. B. Yatsimirskii, *Zhur. neorg. Khim.*, 1964, **9**, 1833
64Md V. E. Mironov, F. Ya. Kul'ba, V. A. Fedorov, and A. V. Fedorova, *Zhur. neorg. Khim.*, 1964, **9**, 2138
64Me B. E. Marques and M. de L. S. Simões, *Rev. Port. Quim.*, 1964, **6**, 29
64Mf E. E. Mercer, W. M. Campbell, jun., and R. M. Wallace, *Inorg. Chem.*, 1964, **3**, 1018
64Mg M. M. McGrady and R. S. Tobias, *Inorg. Chem.*, 1964, **3**, 1157
64Mh D. J. Machin, D. F. C. Morris, and E. L. Short, *J. Chem. Soc.*, 1964, 4658
64Mi J. B. Macaskill and M. H. Panckhurst, *Austral. J. Chem.*, 1964, **17**, 522
64N G. Nord (Waind) and J. Ulstrup, *Acta Chem. Scand.*, 1964, **18**, 307
64Na C. P. Nash and M. S. Jenkins, *J. Phys. Chem.*, 1964, **68**, 356
64Nb B. I. Nabivanets and L. N. Kudritskaya, *Ukrain. khim. Zhur.*, 1964, **30**, 1007
64P Yu. M. Povarov, V. E. Kazarinov, Yu. M. Kessler, and A. I. Gorbanev, *Zhur. neorg. Khim.*, 1964, **9**, 1008
64Pa V. I. Paramonova, A. N. Mosevich, and Yu. N. Ignat'ev, *Radiokhimiya*, 1964, **6**, 527
64Pb I. Popescu, S. Fişel, A. Crăciun-Ciobanu, and P. Gospodaru, *Rev. Roumaine Chim.*, 1964, **9**, 619
64R G. L. Reed, K. J. Sutton, and D. F. C. Morris, *J. Inorg. Nuclear Chem.*, 1964, **26**, 1227
64Ra R. R. Richards and N. W. Gregory, *J. Phys. Chem.*, 1964, **68**, 3089
64Rb B. H. Robinson and J. E. Ferguson, *J. Chem. Soc.*, 1964, 5683
64S V. V. Safonov, B. G. Korshunov, Z. N. Shevtsova, and S. I. Bakum, *Zhur. neorg. Khim.*, 1964, **9**, 1687
64Sa E. L. Short, D. N. Waters, and D. F. C. Morris, *J. Inorg. Nuclear Chem.*, 1964, **26**, 902
64Sb T. Sekine, *J. Inorg. Nuclear Chem.*, 1964, **26**, 1463
64Sc R. P. Seward and P. E. Field, *J. Phys. Chem.*, 1964, **68**, 1611
64Sd M. Shiloh and Y. Marcus, IA-924 (Israel AEC), 1964
64Se R. S. Satchell, *J. Chem. Soc.*, 1964, 5469
64Sf J. Szychliński, E. Latowska, and W. Moska, *Roczniki Chem.*, 1964, **38**, 1427
64Sg I. E. Starik, N. I. Ampelogova, and B. S. Kuznetsov, *Radiokhimiya*, 1964, **6**, 524
64Sh A. P. Samodelov, *Radiokhimiya*, 1964, **6**, 568
64Si V. I. Shlenskaya and A. A. Bryukov, *Vestnik Moskov. Univ.*, 1964, Khim., No. 3, 65
64T E. Thilo and P. Flögel, *Z. anorg. Chem.*, 1964, **329**, 244

continued overleaf

67 Cl⁻ (contd)

64Ta C. Thomas and J. Braunstein, *J. Phys. Chem.*, 1964, **68**, 957
64Tb G. Thiele, dissertation, Technische Hochschule Aachen, 1964
64Tc H. T. Tien and G. W. Harrington, *Inorg. Chem.*, 1964, **3**, 215, correction *ibid.* p. 1333
64Td M. A. Tucker, C. B. Colvin, and D. S. Martin, jun., *Inorg. Chem.*, 1964 **3**, 1373
64Te M. L. Tobe and D. W. Watts *J. Chem. Soc.*, 1964, 2991
64Tf R. S. Tobias and M. Yasuda, *Canad. J. Chem.*, 1964, **42**, 781
64V J. Vallier and R. Lira, *Compt. rend.*, 1964, **259**, 4579
64Va K. S. Venkateswarlu, C. Gopinathan, and P. C. Das, *Indian J. Chem.*, 1964, **2**, 54
64Vb S. Varva and N. P. Rudenko, *Vestnik Moskov. Univ.*, 1964, Khim., No. 6, 14
64Vc J. P. Valleau and S. J. Turner, *Canad. J. Chem.*, 1964, **42**, 1186
64W M. J. M. Woods, P. K. Gallagher, Z. Z. Hugus, jun., and E. L. King, *Inorg. Chem.*, 1964, **3**, 1313
64Wa H. Wendt, *Ber. Bunsengesellschaft Phys. Chem.*, 1964, **68**, 29
64Wb W. M. Weaver and J. D. Hutchison, *J. Amer. Chem. Soc.*, 1964, **86**, 261
64Wc R. D. Willett, *J. Chem. Phys.*, 1964, **41**, 2243
64Y J. T. Yoke, tert., and G. L. McPherson, *J. Inorg. Nuclear Chem.*, 1964, **26**, 655
64Ya J. F. Young, R. D. Gillard, and G. Wilkinson, *J. Chem. Soc.*, 1964, 5176
64Z A. Zalkin, J. D. Forrester, and D. H. Templeton, *Inorg. Chem.*, 1964, **3**, 529
64Za M. Zangen and Y. Marcus, *Israel J. Chem.*, 1964, **2**, 49
65A S. N. Andreev and O. V. Sapozhnikova, *Zhur. neorg. Khim.*, 1965, **10**, 2538
65Aa R. Arnek, *Arkiv Kemi*, 1965, **24**, 531
65Ab A. D. Allen and T. Theophanides, *Canad. J. Chem.*, 1965, **43**, 290
65Ac D. W. Archer, D. A. East, and C. B. Monk, *J. Chem. Soc.*, 1965, 720
65Ad I. R. Anderson and J. C. Sheldon, *Austral. J. Chem.*. 1965, **18**, 271
65Ae R. J. Allen and J. C. Sheldon, *Austral. J. Chem.*, 1965, **18**, 277
65Af H. J. Arnikar, D. K. Sharma, and R. Tripathi, *Indian J. Chem.*, 1965, **3**, 7
65Ag I. P. Alimarin, T. A. Belyavskaya, and G. D. Brykina, *Vestnik Moskov. Univ.*, 1965, Khim., No. 5, 69
65B B. J. Brisdon and R. A. Walton, *J. Inorg. Nuclear Chem.*, 1965, **27**, 1101
65Ba J. H. Beard, J. Casey, and R. K. Murmann, *Inorg. Chem.*, 1965, **4**, 797
65Bb E. Blasius and W. Preetz, *Z. anorg. Chem.*, 1965, **335**, 1
65Bc K. Brodersen, G. Thiele, and H. G. Schnering, *Z. anorg. Chem.*, 1965, **337**, 120
65Bd R. Barbieri and J. Bjerrum, *Acta Chem. Scand.*, 1965, **19**, 469
65Be R. H. Boyd and C.-H. Wang, *J. Phys. Chem.*, 1965, **69**, 3906
65Bf B. Bosnich, C. Ingold, and M. L. Tobe, *J. Chem. Soc.*, 1965, 4074
65Bg A. V. Belyaev and B. V. Ptitsyn, *Zhur. obshchei Khim.*, 1965, **35**, 1887
65Bh A. V. Belyaev and B. V. Ptitsyn, *Izvest. sibirsk. Otdel. Akad. Nauk, S.S.S.R.*, 1965, No. 11 (= Khim, No. 3), 144
65Bi G. G. Bombi, M. Fiorani, and G.-A. Mazzocchin, *J. Electroanalyt. Chem.*, 1965, **9**, 457
65C J. C. Chang and C. S. Garner, *Inorg. Chem.*, 1965, **4**, 209
65Ca F. A. Cotton, N. F. Curtis, B. F. G. Johnson, and W. R. Robinson, *Inorg. Chem.*, 1965, **4**, 326
65Cb F. A. Cotton and C. B. Harris, *Inorg. Chem.*, 1965, **4**, 330
65Cc F. A. Cotton, B. F. G. Johnson, and R. M. Wing, *Inorg. Chem.*, 1965, **4**, 502
65Cd R. D. Cramer, R. V. Lindsey, jun., C. T. Prewitt, and U. G. Stolberg, *J. Amer. Chem. Soc.*, 1965, **87**, 658
65Ce F. A. Cotton and W. K. Bratton *J. Amer. Chem. Soc.*, 1965, **87**, 921
65Cf S. C. Chan, *J. Chem. Soc.*, 1965, 418
65Cg R. Colton, *Austral. J. Chem.*, 1965, **18**, 435
65D G. M. Dakar and B. Z. Iofa, *Radiokhimiya*, 1965, **7**, 25
65Da R. G. Deshpande, P. K. Khopkar, C. L. Rao, and H. D. Sharma, *J. Inorg. Nuclear Chem.*, 1965, **27**, 2171
65Db R. S. Drago, R. L. Carlson, and K. F. Purcell, *Inorg. Chem.*, 1965, **4**, 15
65Dc I. G. Dance and H. C. Freeman, *Inorg. Chem.*, 1965, **4**, 1555
65Dd J. S. Dunnett and R. P. H. Gasser, *Trans. Faraday Soc.*, 1965, **61**, 922
65De C. M. Davidson and R. F. Jameson, *Trans. Faraday Soc.*, 1965, **61**, 2462
65E J. F. Endicott and H. Taube, *Inorg. Chem.*, 1965, **4**, 437
65F A. B. Fasman, G. G. Kutyukov, and D. V. Sokol'skii, *Zhur. neorg. Khim.*, 1965, **10**, 1338
65Fa G. W. A. Fowles and R. A. Walton, *J. Inorg. Nuclear Chem.*, 1965, **27**, 735
65Fb H. N. Farrer, M. M. McGrady, and R. S. Tobias, *J. Amer. Chem. Soc.*, 1965, **87**, 5019
65G A. M. Golub and A. M. Sych, *Zhur. neorg. Khim.*, 1965, **10**, 889
65Ga T. Goto and M. Smutz, *J. Inorg. Nuclear Chem.*, 1965, **27**, 663
65Gb R. Guillaumont, *Compt. rend.*, 1965, **260**, 4739
65H H. E. Hellwege and G. K. Schweitzer, *J. Inorg. Nuclear Chem.*, 1965, **27**, 99
65Ha G. P. Haight, jun. and B. Y. Ellis, *Inorg. Chem.*, 1965, **4**, 249
65Hb G. P. Haight, jun. and J. R. Peterson, *Inorg. Chem.*, 1965, **4**, 1073
65Hc J. D. Hefley and E. S. Amis, *J. Phys. Chem.*, 1965, **69**, 2082
65Hd J. L. Hawes and R. L. Kay, *J. Phys. Chem.*, 1965, **69**, 2420
65I D. Inman, *Electrochim. Acta*, 1965, **10**, 11
65J E. Jósefowicz and H. Ladzińska-Kulińska, *Roczniki Chem.*, 1965, **39**, 1175
65Ja B. Jeżowska-Trzebiatowska, W. Wojciechowski, and J. Mroziński, *Roczniki Chem.*, 1965, **39**, 1187
65K J. Koskikallio and S. Syrjäpalo, *Acta Chem. Scand.*, 1965, **19**, 429
65Ka R. L. Kay and J. L. Hawes, *J. Phys. Chem.*, 1965, **69**, 2787

contin

65Kb D. L. Kepert and R. S. Nyholm, *J. Chem. Soc.*, 1965, 2871
65Kc P. A. Kilty and D. Nicholls, *J. Chem. Soc.*, 1965, 4915
65Kd R. Korewa and H. Smagowski, *Roczniki Chem.*, 1965, **39**, 1561
65Ke M. D. Kozlova and V. I. Levin, *Radiokhimiya*, 1965, **7**, 534
65Kf F. Ya. Kul'ba, V. E. Mironov, and I. F. Mavrin, *Zhur. fiz. Khim.*, 1965, **39**, 2595
65Kg G. Köhler, dissertation, Univ. Göttingen, 1965
65L N. P. Lipatova and I. S. Morozov, *Zhur. neorg. Khim.*, 1965, **10**, 2817
65La V. I. Levin and M. D. Kozlova, *Radiokhimiya*, 1965, **7**, 437
65M V. E. Mironov, F. Ya. Kul'ba, and V. A. Fedorov, *Zhur. neorg. Khim.*, 1965, **10**, 914
65Ma V. E. Mironov, F. Ya. Kul'ba, and V. A. Fedorov, *Zhur. neorg. Khim.*, 1965, **10**, 1388
65Mb D. F. C. Morris, G. L. Reed, E. L. Short, D. N. Slater, and D. N. Waters, *J. Inorg. Nuclear Chem.*, 1965, **27**, 377
65Mc L. W. Marple, *J. Inorg. Nuclear Chem.*, 1965, **27**, 1693
65Md A. S. G. Mazumdar and C. K. Sivaramakrishnan, *J. Inorg. Nuclear Chem.*, 1965, **27**, 2423
65Me D. F. C. Morris and M. W. Jones, *J. Inorg. Nuclear Chem.*, 1965, **27**, 2454
65Mf R. R. Miano and C. S. Garner, *Inorg. Chem.*, 1965, **4**, 337
65Mg S. E. Manahan and R. T. Iwamoto, *Inorg. Chem.*, 1965, **4**, 1409
65Mh R. E. McCarley, B. G. Hughes, F. A. Cotton, and R. Zimmerman, *Inorg. Chem.*, 1965, **4**, 1491
65Mi L. M. Mukherjee, S. Bruckenstein, and F. A. K. Badawi, *J. Phys. Chem.*, 1965, **69**, 2537
65Mj M. Mamiya, *Bull. Chem. Soc. Japan*, 1965, **38**, 178
65Mk J. Masaguer and V. Coto, *Anales de Quim.*, 1965, **61**, *B*, 905
65N N. M. Nikolaeva, B. V. Ptitsyn, and I. I. Gorbacheva, *Zhur. neorg. Khim.*, 1965, **10**, 1051
65Na N. M. Nikolaeva, B. V. Ptitsyn, and E. D. Pastukhova, *Zhur. neorg. Khim.*, 1965, **10**, 1058
65Nb G. I. Novikov and F. G. Gavryuchenkov, *Zhur. neorg. Khim.*, 1965, **10**, 1668
65Nc R. S. Nyholm and K. Ulm, *J. Chem. Soc.*, 1965, 4199
65O H. A. Øye and D. M. Gruen, *Inorg. Chem.*, 1965, **4**, 1173
65P K. V. Pavlova and K. B. Yatsimirskii, *Zhur. neorg. Khim.*, 1965, **10**, 1027
65Pa J. Penciner, I. Eliezer, and Y. Marcus, *J. Phys. Chem.*, 1965, **69**, 2955
65Pb J. A. Partridge, R. M. Izatt, and J. J. Christensen, *J. Chem. Soc.*, 1965, 4231
65R W. Robb and G. M. Harris, *J. Amer. Chem. Soc.*, 1965, **87**, 4472
65Ra R. R. Richards and N. W. Gregory, *J. Phys. Chem.*, 1965, **69**, 239
65Rb Yu. I. Rutkovskii, dissertation, Leningrad Technological Institute, 1965, quoted by ref 67Mh
65S K. N. Semenenko and A. I. Grigor'ev, *Zhur. neorg. Khim.*, 1965, **10**, 2591
65Sa G. K. Schweitzer and S. W. McCarty, *J. Inorg. Nuclear Chem.*, 1965, **27**, 191
65Sb T. G. Spiro, *Inorg. Chem.*, 1965, **4**, 731, 1290
65Sc A. Simon, H. G. Schnering, H. Wöhrle, and H. Schäfer, *Z. anorg. Chem.*, 1965, **339**, 155
65Sd T. Sekine, *Acta Chem. Scand.*, 1965, **19**, 1435
65Se G. Schwarzenbach and M. Schellenberg, *Helv. Chim. Acta*, 1965, **48**, 28
65Sf D. E. Sellers and N. E. Vanderborgh, *J. Amer. Chem. Soc.*, 1965, **87**, 1206
65Sg T. B. Swanson and V. W. Laurie, *J. Phys. Chem.*, 1965, **69**, 244
65Sh M. Shiloh and Y. Marcus, *Israel J. Chem.*, 1965, **3**, 123
65Si I. E. Starik and N. I. Ampelogova, *Radiokhimiya*, 1965, **7**, 658
65Sj K. Schwochau, *Z. Naturforsch.*, 1965, **20a**, 1286
65W S. C. Wallwork and I. J. Worrall, *J. Chem. Soc.*, 1965, 1816
65Y Y. Yagi and A. I. Popov, *Inorg. Nuclear Chem. Letters*, 1965, **1**, 21
66A A. Adin and A. G. Sykes, *J. Chem. Soc. (A)*, 1966, 1518
66Aa C. A. Angell and D. M. Gruen, *J. Amer. Chem. Soc.*, 1966, **88**, 5192
66B G. Biedermann and J. T. Chow, *Acta Chem. Scand.*, 1966, **20**, 1376
66Ba K. W. Bagnall and J. B. Laidler, *J. Chem. Soc. (A)*, 1966, 516
66Bb R. P. Bell and M. Pring, *J. Chem. Soc. (A)*, 1966, 1607
66Bc G. D. Brykina, I. P. Alimarin, and T. A. Belyavskaya, *Radiokhimiya*, 1966, **8**, 110
66Bd J. Braunstein and J. D. Brill, *J. Phys. Chem.*, 1966, **70**, 1261
66Be R. H. Busey, K. H. Gayer, R. A. Gilbert, and R. B. Bevan, jun., *J. Phys. Chem.*, 1966, **70**, 2609
66Bf J. Braunstein, A. R. Alvarez-Funes, and H. Braunstein, *J. Phys. Chem.*, 1966, **70**, 2734
66Bg R. R. Buckley and E. E. Mercer, *J. Phys. Chem.*, 1966, **70**, 3103
66Bh G. G. Bombi, G.-A. Mazzocchin, and M. Fiorani, *Ricerca sci.*, 1966, **36**, 573
66Bi A. V. Belyaev and B. V. Ptitsyn, *Izvest. sibirsk. Otdel. Akad. Nauk S.S.S.R.*, 1966, No. 3 (= Khim. No. 1), 136
66Bj A. V. Belyaev and B. V. Ptitsyn, *Zhur. neorg. Khim.*, 1966, **11**, 1345
66Bk A. V. Belyaev and B. V. Ptitsyn, *Zhur. neorg. Khim.*, 1966, **11**, 1565
66C T. J. Conocchioli, G. H. Nancollas, and N. Sutin, *Inorg. Chem.*, 1966, **5**, 1
66Ca F. A. Cotton and S. J. Lippard, *Inorg. Chem.*, 1966, **5**, 9
66Cb J. P. Clark and C. J. Wilkins, *J. Chem. Soc. (A)*, 1966, 871
66Cc R. J. H. Clark, R. S. Nyholm, and D. E. Scaife, *J. Chem. Soc. (A)*, 1966, 1296
66Cd R. Colton and I. B. Tomkins, *Austral. J. Chem.*, 1966, **19**, 759
66Ce A. Cassol, R. Portanova, and L. Magon, *Ricerca sci.*, 1966, **36**, 1180
66Cf H. Chateau, M.-C. Gadet, and J. Pouradier, *J. Chim. phys.*, 1966, **63**, 269
66D P. R. Danesi, F. Orlandini, and G. Scibona, *J. Inorg. Nuclear Chem.*, 1966, **28**, 1047
66Da D. Dyrssen and M. de J. Tavares, *Acta Chem. Scand.*, 1966, **20**, 2050

continued overleaf

67 Cl^- (contd)

66E L. I. Elding and I. Leden, *Acta Chem. Scand.*, 1966, **20**, 706
66Ea L. I. Elding, *Acta Chem. Scand.*, 1966, **20**, 2559
66Eb J. C. Evans and G. Y.-S. Lo, *J. Phys. Chem.*, 1966, **70**, 11, 20
66Ec D. F. Evans and R. L. Kay, *J. Phys. Chem.*, 1966, **70**, 366
66Ed R. Ettorre, and M. Martelli, *Inorg. Nuclear Chem. Letters*, 1966, **2**, 289
66F F. H. Fry, G. A. Hamilton, and J. Turkevich, *Inorg. Chem.*, 1966, **5**, 1943
66Fa G. W. A. Fowles and J. L. Frost, *J. Chem. Soc.* (*A*), 1966, 1631
66Fb T. M. Florence, *Austral. J. Chem.*, 1966, **19**, 1343
66Fc M. Fiorani, G. G. Bombi, and G.-A. Mazzocchin, *Ricerca sci.*, 1966, **36**, 580
66Fd C. Furlani and L. Sestili, *Ricerca sci.*, 1966, **36**, 819
66Fe C. Furlani, E. Zinato, and G. Culot, *Atti Accad. naz. Lincei, Rend. Classe sci. fis., mat. nat.*, 1966, **41**, 69
66Ff G. Faraglia, L. R. Fiorani, B. C. L. Pepe, and R. Barbieri, *Inorg. Nuclear Chem. Letters*, 1966, **2**, 277
66G N. N. Greenwood, B. P. Straughan, and A. E. Wilson, *J. Chem. Soc.* (*A*), 1966, 1479
66Ga R. G. Griffin, E. S. Amis, and J. O. Wear, *J. Inorg. Nuclear Chem.*, 1966, **28**, 543
66Gb P. Gerding, *Acta Chem. Scand.*, 1966, **20**, 79
66Gc A. A. Grinberg and M. I. Gel'fman, *Zhur. neorg. Khim.*, 1966, **11**, 1323
66Gd A. A. Grinberg and A. A. Korableva, *Zhur. neorg. Khim.*, 1966, **11**, 1605
66Ge R. Guillaumont, *Rev. Chim. minérale*, 1966, **3**, 339
66H T. E. Hopkins, A. Zalkin, D. H. Templeton, and M. G. Adamson, *Inorg. Chem.*, 1966, **5**, 1427, 1431
66Ha D. J. Hewkin and W. P. Griffith, *J. Chem. Soc.* (*A*), 1966, 472
66Hb E. Högfeldt and K. Rasmusson, *Svensk kem. Tidskr.*, 1966, **78**, 490
66I R. E. Isbell, E. W. Wilson, jun., and D. F. Smith, *J. Phys. Chem.*, 1966, **70**, 2493
66Ia A. A. Ivakin, *Zhur. priklad. Khim.*, 1966, **39**, 277
66J B. Jeżowska-Trzebiatowska, J. Hanuza, and W. Wojciechowski, *J. Inorg. Nuclear Chem.*, 1966, **28**, 2701
66Ja R. Jacoud, V. C. Reinsborough, and F. E. W. Wetmore, *Austral. J. Chem.*, 1966, **19**, 1597
66K P. A. Kilty and D. Nicholls, *J. Chem. Soc.* (*A*), 1966, 1175
66Ka T. Kaatz and M. Marcovich, *Acta Cryst.*, 1966, **21**, 1011
66Kb G. Köhler and H. Wendt, *Ber. Bunsengesellschaft Phys. Chem.*, 1966, **70**, 674
66Kc A. P. Kreshkov, V. A. Drozdov, and N. A. Kolchina, *Zhur. fiz. Khim.*, 1966, **40**, 2150
66Kd M. B. Kennedy and M. W. Lister, *Canad. J. Chem.*, 1966, **44**, 1709
66L R. V. Lindsey, jun., G. W. Parshall, and U. G. Stolberg, *Inorg. Chem.*, 1966, **5**, 109
66La D. C. Luehrs, R. T. Iwamoto, and J. Kleinberg, *Inorg. Chem.*, 1966, **5**, 201
66Lb H. Loman and E. van Dalen, *J. Inorg. Nuclear Chem.*, 1966, **28**, 2037
66Lc I. R. Lantzke and D. W. Watts, *Austral. J. Chem.*, 1966, **19**, 949
66Ld I. R. Lantzke and D. W. Watts, *Austral. J. Chem.*, 1966, **19**, 969
66Le R. J. W. Le Fèvre, D. V. Radford, G. L. D. Ritchie, and J. D. Saxby, *Austral. J. Chem.*, 1966, **19**, 1615
66Lf V. I. Levin and M. D. Kozlova, *Radiokhimiya*, 1966, **8**, 533
66M P. C. Moews, jun., *Inorg. Chem.*, 1966, **5**, 5
66Ma Y. Marcus, *J. Inorg. Nuclear Chem.*, 1966, **28**, 209
66Mb L. W. Marple, *J. Inorg. Nuclear Chem.*, 1966, **28**, 1319
66Mc W. L. Masterton and L. H. Berka, *J. Phys. Chem.*, 1966, **70**, 1924
66Md W. A. Millen and D. W. Watts, *Austral. J. Chem.*, 1966, **19**, 43
66Me W. A. Millen and D. W. Watts, *Austral. J. Chem.*, 1966, **19**, 51
66Mf J. B. Macaskill and M. H. Panckhurst, *Austral. J. Chem.*, 1966, **19**, 915
66Mg T. V. Mal'kova, G. A. Shutova, and K. B. Yatsimirskii, *Zhur. neorg. Khim.*, 1966, **11**, 1556
66Mh P. Murray-Rust, P. Day, and C. K. Prout, *Chem. Comm.*, 1966, 277
66N S. N. Nabi and M. S. Amin, *J. Chem. Soc.* (*A*), 1966, 1018
66P B. R. Penfold and W. T. Robinson, *Inorg. Chem.*, 1966, **5**, 1758
66Q C. O. Quicksall and T. G. Spiro, *Inorg. Chem.*, 1966, **5**, 2232
66R J. L. Ryan and C. K. Jørgensen, *J. Phys. Chem.*, 1966, **70**, 2845
66Ra R. Ripan and M. Marc, *Rev. Roumaine Chim.*, 1966, **11**, 1063
66S A. Sabatini and I. Bertini, *Inorg. Chem.*, 1966, **5**, 204
66Sa H. J. Seifert and H. W. Loh, *Inorg. Chem.*, 1966, **5**, 1822
66Sb J. A. Salthouse and T. C. Waddington, *J. Chem. Soc.* (*A*), 1966, 28
66Sc G. Scibona, F. Orlandini, and P. R. Danesi, *J. Inorg. Nuclear Chem.*, 1966, **28**, 1313
66Sd M. Shiloh and Y. Marcus, *J. Inorg. Nuclear Chem.*, 1966, **28**, 2725
66Se A. Swinarski and A. Grodzicki, *Roczniki Chem.*, 1966, **40**, 373
66Sf T. Sekine and Y. Hasegawa, *Bull. Chem. Soc. Japan*, 1966, **39**, 240
66Sg R. S. Sethi, *Indian J. Chem.*, 1966, **4**, 413
66Sh V. M. Shul'man and V. I. Dubinskii, *Izvest. sibirsk. Otdel. Akad. Nauk S.S.S.R.*, 1966, No. 3 (=Khim, No. 1) 25
66Si V. I. Shlenskaya and A. A. Biryukov, *Zhur. neorg. Khim.*, 1966, **11**, 54
66Sj I. V. Shilin and V. K. Nazarov, *Radiokhimiya*, 1966, **8**, 514
66Sk H.-L. Scherff and G. Herrmann, *Radiochim. Acta*, 1966, **6**, 53
66Sl D. F. Stewart and T. A. O'Donnell, *Nature*, 1966, **210**, 836
66Sm L. W. Schroeder and J. A. Ibers, *J. Amer. Chem. Soc.*, 1966, **88**, 2601
66Sn K. Swaminathan and G. M. Harris, *J. Amer. Chem. Soc.*, 1966, **88**, 4411
66T J. Y. Tong and R. L. Johnson, *Inorg. Chem.*, 1966, **5**, 1902

contin

66V A. A. Vlček and F. Basolo, *Inorg. Chem.*, 1966, **5**, 156
66Va I. M. Vasil'kevich and E. A. Shilov, *Ukrain. khim. Zhur.*, 1966, **32**, 947
66Vb V. M. Vdovenko, V. B. Kolokol'tsov, and O. B. Stebunov, *Radiokhimiya*, 1966, **8**, 286
66Vc F. Vierling, G. Schorsch, and J. Byé, *Rev. Chim. minérale*, 1966, **3**, 875
66W G. Wada and W. Reynolds, *Inorg. Chem.*, 1966, **5**, 1354
66Wa M. Wiebeck, *Electrochim. Acta*, 1966, **11**, 1353
67A S. Ahrland, *Helv. Chim. Acta*, 1967, **50**, 306
67Aa K. P. Anderson, E. A. Butler, D. R. Anderson, and E. M. Woolley, *J. Phys. Chem.*, 1967, **71**, 3566
67Ab M. G. Adamson, *Austral. J. Chem.*, 1967, **20**, 2517
67Ac R. Alexander, E. C. F. Ko, Y. C. Mac, and A. J. Parker, *J. Amer. Chem. Soc.*, 1967, **89**, 3703
67B F. Bonati and F. A. Cotton, *Inorg. Chem.*, 1967, **6**, 1353
67Ba K. Burger and B. Pintér, *J. Inorg. Nuclear Chem.*, 1967, **29**, 1717
67Bb S. Balt, *J. Inorg. Nuclear Chem.*, 1967, **29**, 2307
67Bc D. J. Bearcroft and N. H. Nachtrieb, *J. Phys. Chem.*, 1967, **71**, 4400
67Bd W. van Bronswyk and J. C. Sheldon, *Austral. J. Chem.*, 1967, **20**, 2323
67Be J. Braunstein and R. Lindgren, *Electrochim. Acta*, 1967, **12**, 299
67Bf D. Brown and P. J. Jones, *J. Chem. Soc. (A)*, 1967, 243
67Bg J. N. Butler, *Analyt. Chem.*, 1967, **39**, 1799
67Bh I. R. Beattie, T. Gilson, K. Livingston, V. Fawcett, and G. A. Ozin, *J. Chem. Soc. (A)*, 1967, 712
67Bi P. M. Boorman, N. N. Greenwood, M. A. Hildon, and H. J. Whitfield, *J. Chem. Soc. (A)*, 1967, 2017
67Bj K. W. Bagnall, J. B. Laidler, and M. A. A. Stewart, *Chem. Comm.*, 1967, 24
67Bk G. A. Boos and A. A. Popel', *Zhur. neorg. Khim.*, 1967, **12**, 2086
67C F. A. Cotton, R. M. Wing, and R. A. Zimmerman, *Inorg. Chem.*, 1967, **6**, 11
67Ca F. A. Cotton, W. R. Robinson, and R. A. Walton, *Inorg. Chem.*, 1967, **6**, 1257
67Cb L. Carlsson and G. Lundgren, *Acta Chem. Scand.*, 1967, **21**, 819
67Cc R. J. H. Clark and F. B. Taylor, *J. Chem. Soc. (A)*, 1967, 693
67Cd D. A. Couch, P. S. Elmes, J. E. Fergusson, M. L. Greenfield, and C. J. Wilkins, *J. Chem. Soc. (A)*, 1967, 1813
67Ce A. Cassol, L. Magon, and R. Barbieri, *Inorg. Nuclear Chem. Letters*, 1967, **3**, 25
67Cf J. S. Coe and A. A. Malik, *Inorg. Nuclear Chem. Letters*, 1967, **3**, 99
67Cg E. G. Chikryzova and I. I. Vataman, *Zhur. neorg. Khim.*, 1967, **12**, 2946
67D L. Drougge, L. I. Elding, and L. Gustafson, *Acta Chem. Scand.*, 1967, **21**, 1647
67Da C. Drăgulescu and R. Pomoje, *Rev. Roumaine Chim.*, 1967, **12**, 37
67Db N. V. Duffy and J. E. Earley, *J. Amer. Chem. Soc.*, 1967, **89**, 272
67E J. H. Espenson and S. G. Slocum, *Inorg. Chem.*, 1967, **6**, 906
67Ea R. P. Eswein, E. S. Howald, R. A. Howald, and D. P. Keeton, *J. Inorg. Nuclear Chem.*, 1967, **29**, 437
67Eb J. C. Eriksson, L. Ödberg, and E. Högfeldt, *Acta Chem. Scand.*, 1967, **21**, 1925
67Ec M. Elder, J. E. Fergusson, G. J. Gainsford, J. H. Hickford, and B. R. Penfold, *J. Chem. Soc.*, *(A)*, 1967 1423
67Ed L. I. Elding, to be published, quoted by ref 67D
67Ee A. N. Ermakov, I. N. Marov, and G. A. Evtikova, *Zhur. neorg. Khim.*, 1967, **12**, 3372
67F M. Fiorani, G. G. Bombi, and G.-A. Mazzocchin, *J. Electroanalyt. Chem.*, 1967, **13**, 167
67Fa G. W. A. Fowles and B. J. Russ, *J. Chem. Soc. (A)*, 1967, 517
67Fb D. Feakins, K. G. Lawrence, and R. P. T. Tomkins, *J. Chem. Soc. (A)*, 1967, 753
67Fc D. Feakins, K. G. Lawrence, and R. P. T. Tomkins, *J. Chem. Soc. (A)*, 1967, 1909
67Fd T. P. Flaherty and J. Braunstein, *Inorg. Chim. Acta*, 1967, **1**, 335
67Fe W. R. Fitzgerald and D. W. Watts, *J. Amer. Chem. Soc.*, 1967, **89**, 821
67G H. J. Gardner, *Austral. J. Chem.*, 1967, **20**, 2357
67Ga B. Gilbert, *Bull. Soc. chim. belges*, 1967, **76**, 493
67Gb W. P. Griffith and T. D. Wickins, *J. Chem. Soc. (A)*, 1967, 675
67Gc A. A. Grinberg, M. I. Gel'fman, A. I. Stetsenko, and N. D. Mitkinova, *Zhur. neorg. Khim.*, 1967, **12**, 3097
67H J. Hála and D. Pohanková, *J. Inorg. Nuclear Chem.*, 1967, **29**, 2983
67Ha Y. Hermodsson, *Acta Chem. Scand.*, 1967, **21**, 1313
67Hb Y. Hermodsson, *Acta Chem. Scand.*, 1967, **21**, 1328
67Hc C. F. Hale and E. L. King, *J. Phys. Chem.*, 1967, **71**, 1779
67Hd H. C. Helgeson, *J. Phys. Chem.*, 1967, **71**, 3121
67I S. N. Ivanova, L. M. Gindin, and L. Ya. Mironova, *Izvest. sibirsk. Otdel. Akad. Nauk S.S.S.R.*, 1967, No. 2 (= Khim. No. 1) 97
67Ia B. Z. Iofa and A. S. Yushchenko, *Vestnik Moskov. Univ.*, 1967, **22**, Khim., No. 6, 20
67Ib R. M. Izatt, G. D. Watt, D. Eatough, and J. J. Christensen, *J. Chem. Soc. (A)*, 1967, 1304
67J S. J. Jensen, *Acta Chem. Scand.*, 1967, **21**, 889
67Ja B. Jeżowska-Trzebiatowska and M. Rudolf, *Roczniki Chem.*, 1967, **41**, 453
67Jb B. Jeżowska-Trzebiatowska and M. Rudolf, *Roczniki Chem.*, 1967, **41**, 1879
67Jc V. Jedináková and J. Čeleda, *Coll. Czech. Chem. Comm.*, 1967, **32**, 1679
67K R. T. Kolarich, V. A. Ryan, and R. P. Schuman, *J. Inorg. Nuclear Chem.*, 1967, **29**, 783
67Ka K. H. Khoo, *Austral. J. Chem.*, 1967, **20**, 1287
67Kb K. H. Khoo and M. H. Panckhurst, *Austral. J. Chem.*, 1967, **20**, 2633
67Kc V. I. Kazakova and B. V. Ptitsyn, *Zhur. neorg. Khim.*, 1967, **12**, 620
67L R. Larsson, *Acta Chem. Scand.*, 1967, **21**, 1081

continued overleaf

67 Cl⁻ (contd)

67La H.-D. Lüdemann and E. U. Franck, *Ber Bunsengesellschaft Phys. Chem.*, 1967, **71**, 455
67Lb C. H. Langford and W. R. Muir, *J. Phys. Chem.*, 1967, **71**, 2602
67Lc W. Libuś and D. Puchalska, *J. Phys. Chem.*, 1967, **71**, 3549
67Ld A. Lerman, *Geochim. Cosmochim. Acta*, 1967, **31**, 2309
67Le T. I. L'vova, A. A. Pendin, K. D. Shirko, and B. P. Nikol'skii, *Vestnik Leningrad Univ.*, 1967, No. 16 (*Fiz. Khim.* No. 3) 116
67Lf M. D. Lind, *J. Chem. Phys.*, 1967, **46**, 2010 (prel); *ibid*, **47**, 990
67Lg C. H. Langford and W. R. Muir, *J. Amer. Chem. Soc.*, 1967, **89**, 3141
67M R. A. MacKay and R. F. Schneider, *Inorg. Chem.*, 1967, **6**, 549
67Ma W. L. Masterton, T. I. Munnelly, and L. H. Berka, *J. Phys. Chem.*, 1967, **71**, 942
67Mb W. L. Masterton, *J. Phys. Chem.*, 1967, **71**, 2885
67Mc S. E. Manahan and R. T. Iwamoto, *J. Electroanalyt. Chem.*, 1967, **13**, 411
67Md D. F. C. Morris and B. D. Andrews, *Electrochim. Acta*, 1967, **12**, 41
67Me V. E. Mironov and N. P. Lastovkina, *Zhur. fiz. Khim.*, 1967, **41**, 1850
67Mf K. Mizutani and K. Sone, *Z. anorg. Chem.*, 1967, **350**, 216
67Mg H. Müller and K. Dehnicke, *Z. anorg. Chem.*, 1967, **350**, 231
67Mh V. E. Mironov, A. V. Fokina, and Yu. I. Rutkovskii, *Zhur. neorg. Khim.*, 1967, **12**, 2056
67Mi W. A. Millen and D. W. Watts, *J. Amer. Chem. Soc.*, 1967, **89**, 6858
67N Y. Narusawa, M. Kanazawa, S. Takahashi, K. Morinaga, and K. Nakano, *J. Inorg. Nuclear Chem.*, 1967, **29**, 123
67Na B. I. Nabivanets and L. N. Kudritskaya, *Zhur. neorg. Khim.*, 1967, **12**, 1165
67Nb B. I. Nabivanets and L. N. Kudritskaya, *Zhur. neorg. Khim.*, 1967, **12**, 1500
67Nc B. P. Nikol'skii, A. A. Pendin, and M. S. Zakhar'evskii, *Zhur. neorg. Khim.*, 1967, **12**, 1803
67O R. P. Oertel and R. A. Plane, *Inorg. Chem.*, 1967, **6**, 1960
67P D. A. Palmer and D. W. Watts, *Austral. J. Chem.*, 1967, **20**, 53
67R R. A. Reinhardt, N. L. Brenner, and R. K. Sparkes, *Inorg. Chem.*, 1967, **6**, 254
67Ra W. Robb, *Inorg. Chem.*, 1967, **6**, 382
67Rb W. Robb and M. M. deV. Steyn, *Inorg. Chem.*, 1967, **6**, 616
67Rc E. D. Romanenko and N. A. Kostromina, *Zhur. neorg. Khim.*, 1967, **12**, 516
67Rd N. A. Rumbaut and H. L. Peeters, *Bull. Soc. chim. belges*, 1967, **76**, 33
67Re G. W. Richards and A. A. Woolf, *J. Chem. Soc. (A)*, 1967, 1118
67S D. E. Scaife and K. P. Wood, *Inorg. Chem.*, 1967, **6**, 358
67Sa G. S. Smith, Q. Johnson, and R. E. Elson, *Acta Cryst.*, 1967, **22**, 300
67Sb C. Sinistri and E. Pezzati, *Gazzetta*, 1967, **97**, 1116
67Sc B. Spreckelmeyer and H. Schäfer, *J. Less-Common Metals*, 1967, **13**, 122
67Sd T. Sekine, I. Sakamoto, T. Sato, T. Taira, and Y. Hasegawa, *Bull. Chem. Soc. Japan*, 1967, **40**, 251
67Se W. Sawodny and K. Dehnicke, *Z. anorg. Chem.*, 1967, **349**, 169
67Sf R. K. Scarrow and T. R. Griffiths, *Chem. Comm.*, 1967, 425
67Sg R. Siepmann, H.-G. von Schnering and H. Schäfer, *Angew Chem.*, 1967, **79**, 650
67T R. Tamamushi, T. Isono, and S. Katayama, *Bull. Chem. Soc. Japan*, 1967, **40**, 334
67Ta N. Tanaka, Y. Kobayashi, and M. Kamada, *Bull. Chem. Soc. Japan*, 1967, **40**, 2839
67V V. P. Vasil'ev and G. A. Lobanov, *Zhur. fiz. Khim.*, 1967, **41**, 1969
67Va V. P. Vasil'ev and N. K. Grechina, *Zhur. neorg. Khim.*, 1967, **12**, 605
67Vb V. P. Vasil'ev and G. A. Lobanov, *Zhur. neorg. Khim.*, 1967, **12**, 878
67Vc V. P. Vasil'ev and N. K. Grechina, *Zhur. neorg. Khim.*, 1967, **12**, 1372
67W E. Wendling, *Bull. Soc. chim. France*, 1967, 5
67Y Y. Yagi and A. I. Popov, *J. Inorg. Nuclear Chem.*, 1967, **29**, 2223
68A V. V. Aleksandrov, D. K. Osipenko, and T. A. Berezhnaya, *Elektrokhimiya*, 1968, **4**, 1008
68Aa S. N. Andreev and O. V. Sapozhnikova, *Zhur. neorg. Khim.*, 1968, **13**, 1548
68B J. N. Butler, *J. Phys. Chem.*, 1968, **72**, 3288
68Ba B. J. Brisdon and D. A. Edwards, *Inorg. Chem.*, 1968, **7**, 1898
68Bb K. W. Bagnall, J. B. Laidler, and M. A. A. Stewart, *J. Chem. Soc. (A)*, 1968, 133
68Bc R. W. C. Broadbank, S. Dhabanandana, and K. W. Morcom, *J. Chem. Soc. (A)*, 1968, 213
68Bd R. Bury and C. Treiner, *J. Chim. phys.*, 1968, **65**, 1410
68Be G. E. Batley and J. C. Bailar, jun., *Inorg. Nuclear Chem. Letters*, 1968, **4**, 577
68Bf R. W. C. Broadbank, S. Dhabanandana, K. W. Morcom, and B. L. Muju, *Trans. Faraday Soc.*, 1968, **64**, 3311
68Bg J. N. Butler, D. R. Cogley, and W. Zurosky, *J. Electrochem. Soc.*, 1968, **115**, 445
68C L. Ciavatta and M. Grimaldi, *J. Inorg. Nuclear Chem.*, 1968, **30**, 197
68Ca L. Ciavatta and M. Grimaldi, *J. Inorg. Nuclear Chem.*, 1968, **30**, 563
68Cb J. A. Creighton and J. H. S. Green, *J. Chem. Soc. (A)*, 1968, 808
68Cc R. Colton and G. G. Rose, *Austral. J. Chem.*, 1968, **21**, 883
68Cd S. C. Chan and J. P. Valleau, *Canad. J. Chem.*, 1968, **46**, 853
68D V. I. Dubinskii, V. M. Shul'man, and B. I. Peshchevitskii, *Zhur. neorg. Khim.*, 1968, **13**, 54
68Da J. E. D. Davies and D. A. Long, *J. Chem. Soc. (A)*, 1968, 1761
68Db J. E. D. Davies and D. A. Long, *J. Chem. Soc. (A)*, 1968, 2050
68Dc J. E. D. Davies and D. A. Long, *J. Chem. Soc. (A)*, 1968, 2054
68Dd R. Davies and A. G. Sykes, *J. Chem. Soc. (A)*, 1968, 2237
68De J. E. D. Davies and D. A. Long, *J. Chem. Soc. (A)*, 1968, 2560

contin

68E F. W. B. Einstein and B. R. Penfold, *J. Chem. Soc. (A)*, 1968, 3019
68Ea J. H. Espenson and O. J. Parker, *J. Amer. Chem. Soc.*, 1968, **90**, 3689
68F S. Fregerslev and S. E. Rasmussen, *Acta Chem. Scand.*, 1968, **22**, 2541
68Fa J. Fajer and H. Linschitz, *J. Inorg. Nuclear Chem.*, 1968, **30**, 2259
68Fb R. Fernandez-Prini, *Trans. Faraday Soc.*, 1968, **64**, 2146
68Fc G. W. A. Fowles, D. F. Lewis, and R. A. Walton, *J. Chem. Soc. (A)*, 1968, 1468
68Fd A. Feltz, *Z. anorg. Chem.*, 1968, **358**, 21
68Fe W. R. Fitzgerald, A. J. Parker, and D. W. Watts, *J. Amer. Chem. Soc.*, 1968, **90**, 5744
68G O. Ginstrup and I. Leden, *Acta Chem. Scand.*, 1968, **22**, 1163
68Ga P. Gerding and I. Jönsson, *Acta Chem. Scand.*, 1968, **22**, 2247
68Gb H. C. Gaur and R. S. Sethi, *Electrochim. Acta*, 1968, **13**, 1737
68Gc V. A. Golodov, A. B. Fasman, V. F. Vozdvizhenskii, Yu. A. Kushnikov, and V. V. Roganov, *Zhur. neorg. Khim.*, 1968, **13**, 3306
68Gd I. J. Gal, J. Méndez, and J. W. Irvine, jun., *Inorg. Chem.*, 1968, **7**, 985
68Ge H. C. Gaur and R. S. Sethi, *Trans. Faraday Soc.*, 1968, **64**, 445
68Gf I. J. Gal, *Inorg. Chem.*, 1968, **7**, 1611
68Gg N. N. Greenwood, B. P. Straughan, and A. E. Wilson, *J. Chem. Soc. (A)*, 1968, 2209
68H W. Haase and H. Hoppe, *Acta Cryst.*, 1968, **B24**, 282
68Ha F. Hine and S. Inuta, *Bull. Chem. Soc. Japan*, 1968, **41**, 71
68I D. Inman, D. G. Lovering, and R. Narayan, *Trans. Faraday Soc.*, 1968, **64**, 2476
68J S. J. Jensen, *Acta Chem. Scand.*, 1968, **22**, 641
68Ja S. J. Jensen, *Acta Chem. Scand.*, 1968, **22**, 647
68K L. Kuča and E. Högfeldt, *Acta Chem. Scand.*, 1968, **22**, 183
68Ka W. Kuhrt, dissertation, Technical University, Hannover, 1968
68L H.-D. Lüdemann and E. U. Franck, *Ber Bunsengesellschaft Phys. Chem.*, 1968, **72**, 514
68La W. Libuś, D. Puchalska, and T. Szuchnicka, *J. Phys. Chem.*, 1968, **72**, 2075
68Lb O. G. Levanda, I. I. Moiseev, and M. N. Vargaftik, *Izvest. Akad. Nauk. S.S.S.R., Ser. khim.*, 1968, 2368
68Lc D. C. Luehrs and K. Abate, *J. Inorg. Nuclear Chem.*, 1968, **30**, 549
68Ld G. C. Lalor and G. W. Bushnell, *J. Chem. Soc. (A)*, 1968, 2520
68Le O. G. Levanda, *Zhur. neorg. Khim.*, 1968, **13**, 3311
68M W. G. Movius and N. A. Matwiyoff, *J. Phys. Chem.*, 1968, **72**, 3063
68Ma L. Magon, R. Portanova, A. Cassol, and G. Rizzardi, *Ricerca sci.*, 1968, **38**, 782
68Mb D. F. C. Morris and S. D. Hammond, *Electrochim. Acta*, 1968, **13**, 545
68Mc K. Mucker, G. S. Smith, and Q. Johnson, *Acta Cryst.*, 1968, **B24**, 874
68Md R. Mitamura, I. Tokura, S. Nishimura, Y. Kondo, and N. C. Li, *J. Inorg. Nuclear Chem.*, 1968, **30**, 1019
68Me R. H. McCorkell, M. M. Sein, and J. W. Irvine, jun., *J. Inorg. Nuclear Chem.*, 1968, **30**, 1155
68Mf V. E. Mironov, Yu. A. Makashev, I. Ya. Mavrina, and D. M. Markhaeva, *Zhur. fiz. Khim.*, 1968, **42**, 2987
68Mg D. G. McMane and D. S. Martin, jun., *Inorg. Chem.*, 1968, **7**, 1169
68Mh J. Méndez, I. J. Gal, and J. W. Irvine, jun., *Inorg. Chem.*, 1968, **7**, 1329
68Mi B. S. Magor and T. D. Smith, *J. Chem. Soc. (A)*, 1968, 1753
68Mj J.-P. Morel, *Bull. Soc. chim. France*, 1968, 896
68Mk F. Monacelli, *Inorg. Chim. Acta*, 1968, **2**, 263
68N T. W. Newton, G. E. McCrary, and W. G. Clark, *J. Phys. Chem.*, 1968, **72**, 4333
68Na Yu. V. Norseyev (Norseev) and V. A. Khalkin, *J. Inorg. Nuclear Chem.*, 1968, **30**, 3239
68Nb B. I. Nabivanets and E. E. Kapantsyan, *Zhur. neorg. Khim.*, 1968, **13**, 1817
68O J. V. Olver and R. R. Bessette, *J. Inorg. Nuclear Chem.*, 1968, **30**, 1791
68P J. R. Perumareddi and A. W. Adamson, *J. Phys. Chem.*, 1968, **72**, 414
68Pa B. Prasad, *J. Indian Chem. Soc.*, 1968, **45**, 1037
68Pb V. D. Panasyuk and N. F. Malashok, *Zhur. neorg. Khim.*, 1968, **13**, 2727
68Pc H. N. Po and N. Sutin, *Inorg. Chem.*, 1968, **7**, 621
68Pd M. E. Peach and T. C. Waddington, *J. Chem. Soc. (A)*, 1968, 180
68Pe D. A. Palmer and D. W. Watts, *Austral. J. Chem.*, 1968, **21**, 2895
68Pf N. Payen-Baldy and M. Machtinger, *Bull. Soc. chim. France*, 1968, 2221
68Pg F. Petillon, M.-T. Youinou, and J. E. Guerchais, *Bull. Soc. chim. France*, 1968, 2375
68Q A. S. Quist and W. L. Marshall, *J. Phys. Chem.*, 1968, **72**, 684
68R G. Ritzert and E. U. Franck, *Ber. Bunsengesellschaft Phys. Chem.*, 1968, **72**, 798
68Ra E. E. Rakovskii and B. L. Serebryanyi, *Radiokhimiya*, 1968, **10**, 75
68Rb K. N. Raymond, D. W. Meek, and J. A. Ibers, *Inorg. Chem.*, 1968, **7**, 1111
68Rc E. A. Robinson and J. A. Ciruna, *Canad. J. Chem.*, 1968, **46**, 3197
68S V. N. Startsev, Yu. I. Sannikov, G. N. Ben'yash, and E. I. Krylov, *Zhur. neorg. Khim.*, 1968, **13**, 1122
68Sa V. B. Spivakovskii and G. V. Makovskaya, *Zhur. neorg. Khim.*, 1968, **13**, 1555
68Sb V. N. Startsev, Yu. I. Sannikov, S. S. Stroganov, and E. I. Krylov, *Zhur. neorg. Khim.*, 1968, **13**, 1608
68Sc C. J. Shahani, K. A. Mathew, C. L. Rao, and M. V. Ramaniah, *Radiochim. Acta*, 1968, **10**, 165
68Sd G. G. Shitareva and V. A. Nazarenko, *Zhur. neorg. Khim.*, 1968, **13**, 1808
68Se T. G. Sukhova, O. N. Temkin, R. M. Flid, and T. K. Kaliya, *Zhur. neorg. Khim.*, 1968, **13**, 2073
68Sf V. B. Spivakovskii and L. P. Moisa, *Zhur. neorg. Khim.*, 1968, **13**, 2395
68Sg V. B. Spivakovskii and G. V. Makovskaya, *Zhur. neorg. Khim.*, 1968, **13**, 2764

continued overleaf

67 Cl^- (contd)

68Sh L. V. Shikheeva, *Zhur. neorg. Khim.*, 1968, **13**, 2967
68Si L. V. Shikheeva, *Zhur. neorg. Khim.*, 1968, **13**, 3323
68Sj F. Šmirous and J. Čeleda, *Coll. Czech. Chem. Comm.*, 1968, **33**, 1017
68Sk L. W. Schroeder and J. A. Ibers, *Inorg. Chem.*, 1968, 7, 594
68Sl A. Simon, H.-G. von Schnering, and H. Schäfer, *Z. anorg. Chem.*, 1968, **361**, 235
68V V. P. Vasil'ev and N. K. Grechina, *Izvest. V.U.Z., Khim. i khim. Tekhnol.*, 1968, **11**, 142
68Va V. P. Vasil'ev, *Zhur. fiz. Khim.*, 1968, **42**, 1830
68W T. C. Wehman and A. I. Popov, *J. Phys. Chem.*, 1968, **72**, 4031
68Wa G. Wada, *Bull. Chem. Soc. Japan*, 1968, **41**, 882
68Wb J. K. Wittle and G. Urry, *Inorg. Chem.*, 1968, **7**, 560
68Wc I. Wharf and D. F. Shriver, *Chem. Comm.*, 1968, 526
68Y K. L. Yadava, U. S. Pandey, and K. M. Lal, *J. Inorg. Nuclear Chem.*, 1968, **30**, 2915
68Z A. H. Zeltmann, N. A. Matwiyoff, and L. O. Morgan, *J. Phys. Chem.*, 1968, **72**, 121
68Za A. N. Zelikman, I. G. Kalinina, and R. A. Smol'nikova, *Zhur. neorg. Khim.*, 1968, **13**, 2778

68 Hypochlorite, ClO^-

Metal	*Method*	*Temp*	*Medium*	*Log of equilibrium constant, remarks*	*Ref*
H^+	sp	0—35	0 corr	K_1 7·825(0°), 7·75(5°), 7·69(10°), 7·63(15°), 7·58(20°), 7·537(25°), 7·50(30°), 7·46(35°); [$K_a(H_2PO_4^-)$ from table 38(PO_4^{3-}), ref 43B+]	66M

66M J. C. Morris, *J. Phys. Chem.*, 1966, 70, 3798

69 Chlorite, ClO_2^-

Metal	*Method*	*Temp*	*Medium*	*Log of equilibrium constant, remarks*	*Ref*
H^+	sp	25	0 corr	K_1 1·94	65L
	gl?	25	I	K_1 1·61(I = 1), 1·66(I = 0·5), 1·96(I = 0)	66H
UO_2^{2+}	kin, sp	?	?	ev cpx, $K_1 \geqslant -1$	64K
	sp	25	1·0 ($NaClO_4$)	$K_1 > -1{\cdot}7$ [$K_1(H^+)$ 1·96]	64G
Cl^{4+}	sp	25	dil	$K(ClO_2 + ClO_2^- \rightleftharpoons Cl_2O_4^-)$ 0·20	66G

64G G. Gordon and D. M. H. Kern, *Inorg. Chem.*, 1964, 3, 1055
64K D. M. H. Kern and G. Gordon, in "Theory and Structure of Complex Compounds" (symposium papers, Wrocław 1962), Pergamon, Warszawa, 1964, p. 655
65L D. Leonesi and G. Piantoni, *Ann. Chim. (Italy)*, 1965, **55**, 668
66G G. Gordon and F. Emmenegger, *Inorg. Nuclear Chem. Letters*, 1966, **2**, 395
66H C. C. Hong, thesis, Univ. Toronto, 1966, quoted by ref 68H (data read from graph)
68H C. C. Hong and W. H. Rapson, *Canad. J. Chem.*, 1968, **46**, 2053

70 Chlorate, ClO_3^-

Metal	*Method*	*Temp*	*Medium*	*Log of equilibrium constant, remarks*	*Ref*
H^+	kin	260	(Na, K)NO_3(l)	ev ClO_2^+, $K(ClO_3^- + Cr_2O_7^{2-} \rightleftharpoons ClO_2^+ + 2CrO_4^{2-})$ −9·80 in (Na, K)NO_3 (l, eut)	65S
	var		HSO_3F	ev ClO_2^+ in HSO_3F (con, nmr, sp, ir, tp)	68C

contd

ClO_3^- (contd) 70

Metal	Method	Temp	Medium	Log of equilibrium constant, remarks	Ref
Group 1A$^+$	act	25?	0 corr	K_1 −0·4(Na^+), 0·0(K^+)	/66M
Li^+	con	25	w% dioxan	K_1 1·42(w = 64·5), 6·42(w = 90) in w% dioxan	66C
	con	25	w% dioxan	K_1 1·2(w = 60), 1·53(w = 70), 2·62(w = 80), 3·94(w = 85) in w% $C_4H_4O_2$+water. Also K_1 values for dioxan–MeOH and dioxan–MeCN mixt	67A
Na^+	con	25—35	w% dioxan	K_1 1·22(w = 64·5, 25°), 1·23(w = 64·5, 35°), 6·78(w = 90, 25°) in w% dioxan	66C
Re^{2+}	kin	25	var	$\beta(Re^{2+}L^-I^-)$ 2·4(0·16M-HCl), 3·4(0·33M-H_2SO_4) (β^{-1} in text?)	65P
Cu^{2+}	gl, sol	25	→ 0	$K_s(Cu^{2+}OH^-_{1.5}L^-_{0.5})$ −15·89	63L
			1 Na(ClO_4)	$K_1 \leqslant -0{\cdot}15$, $\beta_2 \leqslant -1{\cdot}12$, K_s −15·69	

63L P. Lumme and H. Lumme, *Suomen Kem.*, 1963, 36, *B*, 176
65P K. V. Pavlova and K. B. Yatsimirskii, *Zhur. neorg. Khim.*, 1965, **10**, 1027
65S J. Schlegel, *J. Phys. Chem.*, 1965, **69**, 3638
66C A. N. Campbell, E. M. Kartzmark, and B. G. Oliver, *Canad. J. Chem.*, 1966, **44**, 925
66M W. L. Masterton and L. H. Berka, *J. Phys. Chem.*, 1966, **70**, 1924
67A F. Accascina, A. D'Aprano, and R. Triolo, *J. Phys. Chem.*, 1967, **71**, 3469; A. D'Aprano and R. Triolo, *J. Phys. Chem.*, 1967, **71**, 3474
68C H. A. Carter, A. M. Qureshi, and F. Aubke, *Chem. Comm.*, 1968, 1461

Perchlorate, ClO_4^- 71

Metal	Method	Temp	Medium	Log of equilibrium constant, remarks	Ref
M	Ram		M, L conc	no ev inner cpx for M = H^+, Li^+, Na^+, Be^{2+}, Mg^{2+}, Ca^{2+}, La^{3+}, Ce^{3+}, Th^{4+}, Cu^{2+}, Ag^+, Hg^{2+}, Al^{3+}, In^{3+}, Pb^{2+}	64H
H^+	con	25	org	K_1 2·08 (in MeCN), 2·49 (in EtCN), 2·80 (in PhCN), 2·70 (in Pr^{i}CN), 2·80 (in $PhCH_2CN$), 2·64 (in Me_2CO) (→ 0)	63C
	p(H_2O)	10—40	0 corr	K_1 −1·61(10°), −1·54(25°), −1·47(40°), m units	64D+60H
	nmr		HL var	less than 9% HL in 12M-$HClO_4$; ^{1}H nmr, ^{35}Cl nmr, and Ram differ	65Aa+
	Ram		HL var	no ev cpx up to 10M-$HClO_4$, $K_1 < -3$	65C ≈ 65Ha
	pH_2O	10—40	0 corr	K_1 −1·61(10°), −1·54(25°), −1·47(40°), ΔH_1 = 1·99, ΔS_1 = −0·36, m units	65H+60H
	fp	≈ 10	dioxan	ev $(H_2O)_2$HL in dioxan(l)	65K
	H	25	$C_2H_4(NH_2)_2$	K_1 3·10 in $C_2H_4(NH_2)_2$(l)	65M
	con	105·7	$MeCO_2H$	$\beta(H_3O^+L^-)$ 4·49 in $MeCO_2H$(l), 0 corr	65Ma
	con	105·7	$MeCO_2H$	K_1 4·8 in $MeCO_2H$(l), 0 corr	65Mb
	fp		$MeCO_2H$	$K(2HL \rightleftharpoons H_2L_2)$ 0·68, $K(3HL \rightleftharpoons H_3L_3)$ 1·45 in $MeCO_2H$(l)	66Ba
	var	var	$HClO_4$(l)	$K(3HL \rightleftharpoons Cl_2O_7 + HLH_2O)$ −4·1(−10°, from viscosity), −3·89(20°, pH_2O, pHL), −3·71(90°, kin), M units	66R+
	nmr	26—28	→ 0	$K_1 \approx -1{\cdot}8$, also graph of K_1 for dioxan–H_2O mixt	66Ra
	E	25	$MeCO_2H$	K_1 5·24 in $MeCO_2H$(l)	67A
	con, oth	25	HL	$K(3HL \rightleftharpoons Cl_2O_7 + H_3O^+ + L^-)$ −6·17(con), −6·14 (from molar volumes) in $HClO_4$(l)	67Ba

continued overleaf

71 ClO_4^- (contd)

Metal	*Method*	*Temp*	*Medium*	*Log of equilibrium constant, remarks*	*Ref*
H^+ (*contd*)	fp		H_3OL	ev mainly H_3O^+ and L^-, $K(2H_3O^+ + L^- \rightleftharpoons H_5O_2^+ + \frac{1}{2}H_2L_2)$ −1·07 in H_3OClO_4(l), M units	67M
	H	25	$(Me_2N)_2CNH$	K_1 3·11 in 1,1,3,3-tetramethylguanidine(l)	68C
	nmr	65	0 corr	$K(H_2O + HL \rightleftharpoons H_3O^+ + L^-) \approx 1{\cdot}0$($x$ units), $\approx 2{\cdot}7$ (M units)	68D
		0—65	HL	$K[H_2O(+HL) \rightleftharpoons H_3O^+ + L^-]$ 1·2(0°), 1·0(25°), 0·1(65°), x units	
	dis	25?	0 corr	$K_D[H^+ + L^- + H_2O + 2T(org) \rightleftharpoons T_2H_3O^+L^-(org)]$ 3·3 T = MHDPO = $[(C_6H_{13})_2PO]_2CH_2$, org = CCl_4, M units, ev polyn cpx in org phase	68O
H^+, D^+	nmr		HL var	ev solvates such as $H_3O(HL)_n^+$	68R
Group $1A^+$	con	25	MeCN	K_1 1·83(Li^+), 1·85(Na^+), 1·99(K^+), 2·02(Rb^+), 2·16(Cs^+) in MeCN(l)	62Ma
	con	25	MeCN	K_1 1·54(Li^+), 1·56(Na^+), 1·75(K^+), 1·71(Rb^+), 1·83(Cs^+) in MeCN(l), → 0; also values for 20 and 30°	66Ma
	con	25	org	no ev cpx in *N*-methyl-2-pyrrolidone C_5H_9NO(l) (Na^+, K^+)	67D
	con	25	MeCN	K_1 1·0(Na^+), 1·13(K^+), 1·28(Rb^+), 1·33(Cs^+) in MeCN(l), → 0	67K
Li^+	con	25	dioxan	no ev cpx in dioxan–H_2O mixt	64A
	con	25	org	K_1 given for mixt dioxan–MeOH, MeCN, Me_2CO	66A
	con	25	org	K_1 given for mixtures Me_2SO–dioxan	67P
	H	20?	C_4H_8O	K_1 0 in C_4H_8O(l), 0·1M-Bu_4NClO_4	67Pb
	H	25	C_5H_5N	$K_1(Li^+)/K_1(H^+)$ 1·12 in C_5H_5N(l)	68M
Na^+	fp		$MeCO_2H$	$K(2NaL \rightleftharpoons Na_2L_2)$ 0·40, $K(3NaL \rightleftharpoons Na_3L_3) \approx 1{\cdot}2$, $\beta(HL^0NaL^0)$ 1·67, $\beta(HL^0{}_2NaL)$ 2·68 in $MeCO_2H$(l)	66Ba
	con	25	org	K_1 given for mixt dioxan–water	67G
	con	25	org	K_1 given for mixtures of dioxan with MeCN, MeOH, and Me_2CO	67P
	H	20?	C_4H_8O	K_1 0 in C_4H_8O(l), 0·1M-Bu_4NClO_4	67Pb
	con	25	org	no ev cpx in $NC(CH_2)_4CN$(l)	67S
K^+	sol	25	org	K_{s0} −4·5 in MeOH, —0·1 in Me_2NCHO	67Aa
	con	25	org	K_1 given for mixt dioxan–water	67G
	con	25	w% C_4H_8O	K_1 1·66(w = 84·5), no ev cpx for $w < 80$, in w% C_4H_8O (THF)	68Ba
	con	25	org	K_1 1·04 in MeOH(l), 1·23 in MeCN(l), also K_1 values for mixtures	68Cb
Rb^+	con	25	org	K_1 given for mixtures dioxan–water	67Pa
Cs^+	con	25	org	K_1 1·52 in MeOH(l), 1·36 in MeCN(l), also K_1 values for mixtures	68Cb
	dis	25	→ 0	$K_D[Cs^+ + L^- \rightleftharpoons Cs^+ (\text{in } MeNO_2) + L^- (\text{in } MeNO_2)]$ −1·74	68H
Group $2A^{2+}$	dis		HL var	$K_D[M^{2+} + 2L^- \rightleftharpoons ML_2(\text{in TBP})]$ 2·3 (M = Ca), 1·85(Sr), 1·62(Ba)	68L
Sc^{3+}	dis		10 H(Cl)	K_1 0·78, β_2 0·61, β_3 0·10, β_4 −0·96, $K_D[H^+ + Sc^{3+} + 4L^- + 4T(org) \rightleftharpoons HScL_4T_4(org)]$ 0·70 $K_D[Sc^{3+} + 3L^- + 3T(org) \rightleftharpoons ScL_3T_3(org)]$ −9·19 (whence?), T = org = TBP = $(BuO)_3PO$, $[T]_{org}$ = 0·74 M	64S

contin

Metal	*Method*	*Temp*	*Medium*	*Log of equilibrium constant, remarks*	*Ref*
Sc^{3+} (*contd*)	dis	25	4 $H(ClO_4)$	$K_D[Sc^{3+}+3L^- \rightleftharpoons ScL_3(org)]$ −2·4, ev mixed cpx Sc^{3+}–L^-–Cl^- and NO_3^- in org = 50% TBP+ $CHCl_3$	66S
Group $4f^{3+}$	dis		(Na)ClO_4 var	$K_D[M^{3+}+3L^- \rightleftharpoons ML_3$(in TBP)] 0·42 (M = La), 0·09(Ce), 0·01(Pr), 0·39(Nd); $K_D[M^{3+}+H^++4L^- \rightleftharpoons HML_4$(in TBP)] 1·57 (M = La), 1·36(Ce), 1·37(Pr), 1·72(Nd)	64G
TiO^{2+}	var		var	ev $TiOL^+$ (fp, cix, sp, pre)	64K
Cr^{3+}	sp, kin	10—20	10·6 $HClO_4$	K_1 −1·68(9·8°), −1·48(20°)	65J
U^{4+}	nmr	20	2 $HClO_4$	K_1 −0·85	63V
UO_2^{2+}	ir	20	C_6H_6	$K[R_4NL+R_4NUO_2L_3 \rightleftharpoons (R_4N)_2UO_2L_4]$ 1·53 in C_6H_6(l), R = $C_{10}H_{21}$	64V
	E	25	$MeCO_2H$	K_2 5·80 in $MeCO_2H$(l)	67A
Mn^{2+}	var		var	ev MnL_2 py_4 *etc.* (ir, mag, sp)	66B
Fe cpx	red	25	0 corr	$\beta(Q^{2+}L^-)$ 1·77, Q^{2+} = $C_5H_5FeC_5H_4CH_2NMe_3^{2+}$, ferricenylmethyl	67L
	red	25	0 corr	$Fe(C_5H_5)_2^+$; K_1 1·2	67N
Co^{2+}	var		var	ev CoL_2 py_4 *etc.* (ir, mag, sp)	66B
Co^{3+}	pre, ir		var	no ev $CoL(NH_3)_5^{2+}$	67J
$Co(NH_3)_6^{3+}$	con	25	→0	K_1 1·40	68K
$Co(NH_3)_5Cl^{2+}$	sol	25—35	0 corr	K_1 1·15(25°), 1·05(35°)	65A
Co $en_2Cl_2^+$	sol	25	MeOH 0 corr	K_{s0} −5·76(*cis*), −5·59(*trans*) in MeOH(l) K_{s0} −2·75(*cis*), −3·88(*trans*)	68F
Ni^{2+}	ir		solid	ev $Ni(MeCN)_4L_2$, $Ni(MeCN)_2L_2$, L mono- and bi-dentate	65W
	var		var	ev NiL_2 py_4 *etc.* (ir, mag, sp)	66B
Cu^{2+}	var		var	ev CuL_2 py_4 *etc.* (ir, mag, sp)	66B
Tl^+	act	25?	0 corr	K_1 0·2	/66M
	sp	23—80	0 corr	K_1 −0·20(23°), −0·26(40°), −0·34(60°), −0·40(80°)	67Z
Tl^{3+}	con, ir		$PhNO_2$	ev Tl $(bipy)_3L^{2+}$, Tl $(bipy)_2L^{2+}$ in $PhNO_2$(l)	66F
Et_4N^+	con	25—50	C_4H_9CN	K_1 2·29(25°), 2·31(30°), 2·34(40°), 2·37(50°), in C_4H_9CN(l)	68B
	con	25	org	K_1 1·61 in MeOH(l), also K_1 for mixts MeOH–CCl_4 and MeOH–C_5H_5N	68Ca
R_4N^+	con	25	*w*% dioxan	R = Et: K_1 0·81(*w* = 50), 1·69(*w* = 70), 2·75(*w* = 78); R = Bu: K_1 1·50(*w* = 50), 2·28(*w* = 70), 3·18(*w* = 78) in *w*% dioxan; also K_1 for other *w* values	67B
Ph_4P^+	sol	5—50	dil	K_{s0} −8·66(5°), −8·56(10°), −8·47(15°), −8·34(20°), −8·25(25°), −8·12(30°), −7·85(40°), −7·60(50°), ΔH_{s0} = 8·7(20°)	68Oa
Ph_4As^+	sol	5—50	dil	K_{s0} −9·09(5°), −8·90(10°), −8·75(15°), −8·58(20°), −8·47(25°), −8·30(30°), −8·00(40°), −7·71(50°), ΔH_{s0} = 11·6(20°)	68Oa
Po^{4+}	dis		$HClO_4$ var	$K_D[Po^{4+}+4L^-+T(org) \rightleftharpoons PoL_4T(org)]$ −0·12, T = TBP, org = C_6H_6; K_1 −0·89, β_2 −1·48, β_3 −2·05, β_4 −2·80	65S

references overleaf

71 ClO_4^- (contd)

62Ma S. Minc and L. Werblan, *Electrochim. Acta*, 1962, 7, 257
63C J. F. Coetzee and D. K. McGuire, *J. Phys. Chem.*, 1963, **67,** 1810
63V V. M. Vdovenko, G. A. Romanov, and V. A. Shcherbakov, *Radiokhimiya*, 1963, 5, 664
64A F. Accascina, G. Pistoia, and S. Schiavo, *Ricerca sci.*, 1964, *A*, **6,** 141
64D K. H. Dücker, dissertation, Technische Hochschule Aachen, 1964
64G A. M. Golub, M. I. Olevinskii, and E. F. Lutsenko, *Ukrain. khim. Zhur.*, 1964, **30,** 1274
64H R. E. Hester and R. A. Plane, *Inorg. Chem.*, 1964, **3,** 769
64K V. Krishnan and C. C. Patel, *Indian J. Chem.*, 1964, **2,** 425
64S A. P. Samodelov, *Radiokhimiya*, 1964, **6,** 568
64V V. M. Vdovenko, A. I. Skoblo, and D. N. Suglobov, *Radiokhimiya*, 1964, **6,** 677
65A D. W. Archer, D. A. East, and C. B. Monk, *J. Chem. Soc.*, 1965, 720
65Aa J. W. Akitt, A. K. Covington, J. G. Freeman, and T. H. Lilley, *Chem. Comm.*, 1965, 349
65C A. K. Covington, M. J. Tait, and W. F. K. Wynne-Jones, *Proc. Roy. Soc.*, 1965, *A*, **286,** 235
65H R. Haase, K.-H. Dücker, and H. A. Küppers, *Ber Bunsengesellschaft Phys. Chem.*, 1965, **69,** 97
65Ha K. Heinzinger and R. E. Weston jun., *J. Chem. Phys.*, 1965, **42,** 272
65J K. M. Jones and J. Bjerrum, *Acta Chem. Scand.*, 1965, **19,** 974
65K J. Koskikallio and S. Syrjäpalo, *Acta Chem. Scand.*, 1965, **19,** 429
65M L. M. Mukherjee, S. Bruckenstein, and F. A. K. Badawi, *J. Phys. Chem.*, 1965, **69,** 2537
65Ma R. J. L. Martin, *Austral. J. Chem.*, 1965, **18,** 321
65Mb R. J. L. Martin, *Austral. J. Chem.*, 1965, **18,** 427
65S I. E. Starik and N. I. Ampelogova, *Radiokhimiya*, 1965, **7,** 658
65W A. E. Wickenden and R. A. Krause, *Inorg. Chem.*, 1965, **4,** 404
66A F. Accascina, G. Pistoia, and S. Schiavo, *Ricerca sci.*, 1966, **36,** 560
66B D. H. Brown, R. H. Nuttall, J. McAvoy, and D. W. A. Sharp, *J. Chem. Soc. (A)*, 1966, 892
66Ba S. Bruckenstein and L. D. Pettit, *J. Amer. Chem. Soc.*, 1966, **88,** 4790
66F O. Farver and G. Nord, *Acta Chem. Scand.*, 1966, **20,** 1429
66M W. L. Masterton and L. H. Berka, *J. Phys. Chem.*, 1966, **70,** 1924
66Ma S. Minc and L. Werblan, *Roczniki Chem.*, 1966, **40,** 1537, 1753
66R V. Ya. Rosolovskii, *Zhur. neorg. Khim.*, 1966, **11,** 2161
66Ra R. Radeglia, *Z. phys. Chem. (Leipzig)*, 1966, **231,** 339
66S T. Sekine, T. Hamada, and M. Sakairi, *Bull. Chem. Soc. Japan*, 1966, **39,** 244
67A M. Alei, jun., Q. C. Johnson, H. D. Cowan, and J. F. Lemons, *J. Inorg. Nuclear Chem.*, 1967, **29,** 2327
67Aa R. Alexander, E. C. F. Ko, Y. C. Mac, and A. J. Parker, *J. Amer. Chem. Soc.*, 1967, **89,** 3703
67B R. Bury and J.-C. Justice, *J. Chim. phys.*, 1967, **64,** 1491
67Ba N. Bout and J. Potier, *Rev. Chim. minérale*, 1967, **4,** 621
67D M. D. Dyke, P. G. Sears, and A. I. Popov, *J. Phys. Chem.*, 1967, **71,** 4140
67G M. Goffredi and R. Triolo, *Ricerca sci.*, 1967, **37,** 1137
67J W. E. Jones and T. W. Swaddle, *Canad. J. Chem.*, 1967, **45,** 2647
67K R. L. Kay, B. J. Hales, and G. P. Cunningham, *J. Phys. Chem.*, 1967, **71,** 3925
67L T. I. L'vova, A. A. Pendin, K. D. Shirko, and B. P. Nikol'skii, *Vestnik Leningrad Univ.*, 1967, No. 16 (*Fiz. Khim.* No. 3) 116
67M G. Mascherpa, S. Gibert, D. Mascherpa, and A. Potier, *Compt. rend.*, 1967, **265,** *C*, 719
67N B. P. Nikol'skii, A. A. Pendin, and M. S. Zakhar'evskii, *Zhur. neorg. Khim.*, 1967, **12,** 1803
67P G. Pistoia, A. M. Polcaro, and S. Schiavo, *Ricerca sci.*, 1967, **37,** 300, 309
67Pa G. Pistoia, *Ricerca sci.*, 1967, **37,** 731
67Pb J. Perichon and R. Buvet, *Bull. Soc. chim. France*, 1967, 3697
67S P. G. Sears, J. A. Caruso, and A. I. Popov, *J. Phys. Chem.*, 1967, **71,** 905
67Z P. A. Zagorets and G. P. Bulgakova, *Zhur. neorg. Khim.*, 1967, **12,** 347
68B J. J. Banewicz, J. A. Maguire, and P. S. Shih, *J. Phys. Chem.*, 1968, **72,** 1960
68Ba R. Bury and C. Treiner, *J. Chim. phys.*, 1968, **65,** 1410
68C J. A. Caruso and A. I. Popov, *J. Phys. Chem.*, 1968, **72,** 918
68Ca F. Conti, P. Delogu, and G. Pistoia, *J. Phys. Chem.*, 1968, **72,** 1396
68Cb F. Conti and G. Pistoia, *J. Phys. Chem.*, 1968, **72,** 2245
68D R. W. Duerst, *J. Chem. Phys.*, 1968, **48,** 2275
68F W. R. Fitzgerald, A. J. Parker, and D. W. Watts, *J. Amer. Chem. Soc.*, 1968, **90,** 5744
68H G. R. Haugen and H. L. Friedman, *J. Phys. Chem.*, 1968, **72,** 4549
68K S. Katayama and R. Tamamushi, *Bull. Chem. Soc. Japan*, 1968, **41,** 606
68L E. V. Lapitskaya and F. P. Gorbenko, *Radiokhimiya*, 1968, **10,** 90
68M L. M. Mukherjee, J. J. Kelly, W. Baranetzky, and J. Sica, *J. Phys. Chem.*, 1968, **72,** 3410
68O J. W. O'Laughlin, D. F. Jensen, J. W. Ferguson, J. J. Richard, and C. V. Banks, *Analyt. Chem.*, 1968, **40,** 1931
68Oa T. Ôkubo, F. Aoki, and R. Teraoka, *J. Chem. Soc. Japan*, 1968, **89,** 432
68R O. Redlich, R. W. Duerst, and A. Merbach, *J. Chem. Phys.*, 1968, **49,** 2986

Metal	Method	Temp	Medium	Log of equilibrium constant, remarks	Ref
H^+	dis	25	$(ClC_2H_4)_2O$	K_1 5·1 in $(ClC_2H_4)_2O(l)$	58Db
	dis	25	$(ClC_2H_4)_2O$	K_1 5·3 in $(ClC_2H_4)_2O(l)$	59Mc
	con	25	PhCN	K_1 4·66 in PhCN(l), → 0	64J
	H, fp	25	$C_2H_4(NH_2)_2$	K_1 3·64(H), 3·73(fp) in $C_2H_4(NH_2)_2(l)$	65Me
	dis		0 corr	$K_D[H^+ + Br^- + nT(org) \rightleftharpoons HBrT_n(org)]$ −2·39 (n = 1, 2, 3), or −1·15(n = 1)	66Le
	sp	25	C_5H_5N	K_1 4·38 in $C_5H_5N(l)$	68Mb
	dis	25	$PhNO_2$	K_1 5·4 in $PhNO_2$ (l, sat with H_2O)	68Md
	con	400—800	0 corr	K_1 given for densities 0·35 to 0·80 g cm^{-3}; $K[H(H_2O)_x^+ + Br(H_2O)_y^- \rightleftharpoons nH_2O + HBr(H_2O)_{x+y-n}]$ 20·4(400°), 23·3(500°), 25·8(600°), 27·4(700°), 28·6(800°), M units; n = 12·2(400°), 13·7(500°), 15·0(600°), 15·7(700°), 16·2(800°)	68Q
	X		solid	ev HBr_2^- in $Cs_3Br_3H_3OHBr_2(s)$	68Sb
Group $1A^+$	con	25	$C_2H_4(NH_2)_2$	K_1 3·18(Li^+), 3·19(Na^+), 3·38(K^+), 3·36(Cs^+) in $C_2H_4(NH_2)_2(l)$, → 0	65Bb
Li^+	bp	35	Et_2O	$K(2Li_2Br_2 \rightleftharpoons Li_4Br_4)$ 1·3 in $Et_2O(l)$	64Ta/65Cd
	kin	0	DMF	K_1 0·41 in DMF = $Me_2NCHO(l)$	64Wa
	pEt_2O	25	Et_2O	ev Li_4Br_4 in $Et_2O(l)$	65Cd
	con	25	MeOH	K_1 1·19 in MeOH(l), also K_1 values for mixt MeOH–H_2O	66Sd
	con	25	org	K_1 given for mixtures of dioxan with H_2O, MeOH, Me_2CO, and Me_2SO	67Pa
	con	25—40	Me_2CO	K_1 2·80(25°), 2·77(30°), 2·85(35°), 2·90(40°), in $Me_2CO(l)$ also K_1 values for mixt Me_2CO–H_2O	67Sd
Na^+	con	400—800	0 corr	K_1 given for densities 0·35 to 0·75 g cm^{-3}; $K[Na(H_2O)_x^+ + Br(H_2O)_y^- \rightleftharpoons nH_2O + NaBr(H_2O)_{x+y-n}]$ 16·37(400°), 16·61(500°), 16·86(600°), 17·09(700°), 17·28(800°), n = 9·85, M units	68Qa
K^+	sol	25	org	K_{s0} −2·4 in Me_2NCHO, −4·0 in $(Me_2N)_3PO$ (l)	67Aa
Cs^+	sol	25	org	K_{s0} −0·29 in $HCONH_2$, −3·3 in $Me_2NCHO(l)$	67Aa
	con	25	→ 0	K_1 0·03	68Hc
Be^{2+}	dis	20	0·691 $H(ClO_4)$	K_1 −0·42	65Mc
Mg^{2+}	*X*		solid	ev Mg_4OBr_6 units in $Mg_4OBr_6(Et_2O)_4(s)$	64Sd
Sc^{3+}	cix	20	0·691 $H(ClO_4)$	K_1 −0·07, K_2 −0·33	64Ma
Group $4f^{3+}$	pre, sp		var	ev MBr_6^{3-} in $C_2H_4(CN)_2(l)$, MeCN(l), $(Ph_3PH)_4MBr_7(s)$, $(pyH)_3MBr_6(s)$, M = La, Ce, Pr, Nd, Sm, Eu, Tb, Tm, Yb	66R
Nd^{3+}	sp	22	LiBr var	K_1 −0·81, K_2 −3·27	65Ma
Ho^{3+}	sp	22	0 corr	K_1 −0·67, K_2 −2·43	66Md
Er^{3+}	sp	22	LiBr var	K_1 −0·58, K_2 −2·39	65Ma
Ac^{3+}	dis	27	1 $H(ClO_4)$	K_1 −0·25, β_2 −0·52	68Sa
Ti^{3+}	pre, sp		solid	ev $TiBr_6^{3-}$ in $(C_5H_6N)_3TiBr_6(s)$	67Fb
Ti^{4+}	pre, *X*		solid	ev $TiBr_6^{2-}$ in $K_2TiBr_6(s)$ *etc.*	64Gb
Hf^{4+}	dis	20	3 $H(ClO_4)$	K_1 −0·1	67H
Th^{4+}	var		solid	ev $ThBr_6^{2-}$ in $(Me_4N)_2ThBr_6(s)$ (pre, ir, *X*)	66Ba
Group $5A^{5+}$	pre, ir		var	ev $NbBr_6^-$, $TaBr_6^-$, $PaBr_6^-$ in salts and org (Me_2CO, MeCN, $MeNO_2$)	67Bg

continued overleaf

72 Br^- (contd)

Metal	*Method*	*Temp*	*Medium*	*Log of equilibrium constant, remarks*	*Ref*
V^{3+}	pre, *X*		solid	ev VBr_4^- in $Et_4N^+VBr_4^-(s)$	66Cb
	pre, sp		solid	ev VBr_6^{3-} in $(C_5H_6N)_3VBr_6(s)$	67Fb
$Nb^{2+,\,3+}$	ir		var	ev $Nb_6Br_{12}^{2+}$ in solid and pyridine(l)	66Bb
	pre, sp		var	ev $Nb_6Br_{12}^{2+}$	67F
Nb^{5+}	ir		solid	ev $NbOBr_5^{2-}$ in $Cs_2NbOBr_5(s)$	66S
$Ta^{2+,\,3+}$	pre		var	ev $Ta_6Br_{12}^{2+}$ and $Ta_6Br_{12}^{4+}$	65Md
	var		var	ev mixed cpx $Nb_xTa_{6-x}Br_{12}^{2+}$ (pre, *X*, sp)	66Sb
	var		solid	ev $Ta_6Br_{12}^{3+}$ in $Ta_6Br_{15}(H_2O)_8(s)$ and $Ta_6Br_{12}^{4+}$ in $Ta_6Br_{16}(H_2O)_7(s)$ (pre, ir, mag, *X*, sp)	68Sd
Pa^{4+}	var		var	ev $PaBr_6^{2-}$ in solids and soln (pre, ir, *X*)	67Bf
Pa^{5+}	*X*		solid	ev Pa_2Br_{10} units in $PaBr_5(s)$	68B
Cr^{3+}					

1964 *Erratum*: p. 320, *in ref* 60E *for* "$\Delta H_1 = 6{\cdot}1, \Delta S_1 = 17{\cdot}2(25°)$" *read* "$\Delta H_1 = 5{\cdot}1, \Delta S_1 = 4{\cdot}9(25°)$"

Metal	*Method*	*Temp*	*Medium*	*Log of equilibrium constant, remarks*	*Ref*
	X, pre		solid	ev $Cr_2Br_9^{3-}$ in $Cs_3Cr_2Br_9(s)$	68Sc
Mo^{5+}	pre, fp		var	ev $MoOBr_5^{2-}$ and $MoOBr_4^-$ in solids and EtOH(l), partly hydrolysed in H_2O	29A
	sp	25	HBr var	ev $Mo_2O_2Br_8^{2-}$ and three other dinuclear species, no ev $MoOBr_5^{2-}$	64Ab
	ir		solid	ev $MoOBr_5^{2-}$ in $Cs_2MoOBr_5(s)$	66S
	sp		HBr var	ev cpx with 2, 4, and 6 Br/Mo	66W
	X		solid	ev $MoOBr_4H_2O^-$ in $Ph_4AsMoOBr_4H_2O(s)$	67Sa
	mag	1—40	HBr var	$K[Mo_4O_4(OH)_4Br_{12}^{4-}(\text{paramagnetic})+6H_2O \rightleftharpoons Mo_4O_6(OH)_8Br_4^{4-}(\text{diamagnetic})+8H^++8Br^-]$ 13·65(1°), 13·36(20°), 13·08(40°)	68J
Mo^{6+}	sp	25	HBr var	ev mainly dinuclear species	64Ab
$W^{2+,\,3+}$	pre, sp		var	ev $W_6Br_8^{6+}$ in solid and in MeCN(s)	66Sc
W^{3+}	var		var	ev $W_2Br_9^{3-}$ in soln and $K_3W_2Br_9(s)$ (pre, sp, *X* powder)	68Hd
W^{5+}	pre		var	ev $WOBr_4^-$, $WOBr_5^{2-}$	31P
	var		var	ev WBr_6^- in MeCN(l), solids (con, ir, mag, pre, sp)	65Ba
	var		var	ev $W_3Br_{12}^{3+}$ in $WBr_5(s)$ and MeCN(l) (Ag, con, mag)	66Cd
	ir		solid	ev $WOBr_5^{2-}$ in $Cs_2WOBr_5(s)$	66S
W^{6+}	var		var	ev $WOBr_5^-$ in MeCN(l), $CH_2Cl_2(l)$ (con, ir, sp)	66Fa
U^{3+}	sp		LiBr var	K_1 −3·96	62Sb
	sp	25	0 corr	K_1 −3·95	65Sb
U^{4+}	var		solid	ev UBr_6^{2-} in $(Me_4N)_2UBr_6(s)$ (*X*, ir, pre)	66Ba
	var		org	ev UBr_6^{2-} and $UCl_4Br_2^{2-}$ in $MeNO_2(l)$, MeCN(l), solid (con, ir, mag, sp)	66D
UO_2^{2+}	var		org	ev $UO_2Br_3^-$ and $UO_2Br_4^{2-}$, ion pairs with R_4N^+, in $C_6H_6(l)$ (dis, ir, sp)	67Bd
Np^{3+}	sp		LiBr var	K_1 −3·39, β_2 −6·48	64Sc ≈ 66S
Pu^{3+}	sp		LiBr var	K_1 −3·45, β_2 −6·54	64Sc ≈ 66S
Pu^{4+}	dis	25	4 (HCl)	K_1 1·00, β_2 0·64	66Da
Am^{3+}	sp		LiBr var	K_1 −3·28	64Sc
Mn^{2+}	cix	20	0·69 $H(ClO_4)$	K_1 0·27, K_2 −0·26	68F
Tc^{4+}	Ag	15	3 $HClO_4$	K_6 3·58	65Sc

contin

Metal	*Method*	*Temp*	*Medium*	*Log of equilibrium constant, remarks*	*Ref*
Re^{3+}	*X*, sp		var	ev Re_3Br_9 groups in various cpx, *e.g.* $Re_3Br_{11}^{2-}$	64Cc
	var		var	ev $Re_3Br_{10}^-$, $Re_3Br_{11}^{2-}$, $Re_3Br_{12}^{3-}$ in org, solids (con, *p*Me_2CO, pre, sp)	64R
	pre, sp		var	ev $Re_2Br_8^{2-}$ in Me_2CO(l) and Bu_4N^+ salt	65Ca
	X		solid	ev $Re_3Br_{11}^{2-}$ in $Cs_2Re_3Br_{11}$(s)	65Ea
	X		solid	ev $Re_3Br_{11}^{2-}$ in $Cs_2Re_3Br_{11}$(s)	66E
$Re^{3+,4+}$	var		MeCN	ev $Re_2Br_9^-$, $Re_2Br_9^{2-}$ in MeCN(l) *etc.*, (con, pre, sp)	67B
Re^{4+}	*X*		solid	ev $ReBr_6^{2-}$ and Re_3Br_9 units in $M_2Re_4Br_{15}$(s) ($M^+ = Et_4N^+$ *etc.*)	65C
	Ag	15	3 $HClO_4$	K_6 5·26	65Sc
Re^{5+}	pre		var	ev $ReBr_4O^-$, not $ReBr_4(H_2O)_2^-$	65Cc
	X		solid	ev $ReBr_4OH_2O^-$ in $Et_4NReBr_4OH_2O$(s)	
	var		var	ev $ReBr_4O^-$ in solid, org (pre, con, ir, sp)	66C
Fe^{3+}	sp	20	1·2 ($NaClO_4$)	K_1 −0·15	67Mb
Co^{2+}	sp		Et_2O	K_3 > 5·7, K_4 4 in Et_2O(l)	63Cd
	cix	20	0·69 H(ClO_4)	K_1 −0·13, β_2 −0·42	65Fc
	cal	25—40	2 ($NaClO_4$)	ΔH_1 = 0·14(25°), 0·15(40°), K_1 −0·11(25°, 40°), ΔS_1 = −0·07(25°)	66K
	var		org	ev $CoBr^+$, $CoBr_2$, $CoBr_3^-$, $CoBr_4^{2-}$ tetrahedral cpx in MeCN and octahedral cpx in Me_3PO_4 (Ag, con, sp)	67G
	sp	25—500	$CoBr_2$ dil	ev tetrahedral $CoBr_2(H_2O)_2$ at > 300°	67La
	sp	25—500	Br^- var	ev tetrahedral $CoBr_3H_2O^-$ at 200° in 1—4M-NaBr	68L
Co^{2+} cpx	sp, gl		MBr var	Co(dimethylglyoxime)$_2{}^{2+}$: K_1 0·60, K_2 0·60 (M = Li), β_2 0·60(M = K); β_2 0·36(M = Rb)	67Ba
Co^{3+}	kin, sp	14·5	5 ($NaClO_4$)	β[($Co^{III}{}_2$)Br_2] 2·46	68W
Co $en_2Cl_2{}^+$	sp	25	MeOH	K_1 1·54(*cis*), no ev cpx(*trans*) in MeOH(l), $I \approx 0{\cdot}02$	57Pa
	sp	30	org	K_1^{out}(*cis*) 3·00 in Me_2NCHO(l), 0 corr	66M = 66Ma
	con	25	org	K_1 2·26(*cis*), 1·34(*trans*) in Me_2SO(l), → 0 2·92 in Me_2NCHO(l) → 0, 3·23 in Me_2NCOMe(l), → 0	67Mg
	sp	55	org	*K*(*cis* ⇌ *trans*) 0·13, K_1^{out} 2·44(*cis*), 1·53(*trans*) in Me_2NCOMe(l)	66Fb
	kin	30—45	org	K_1^{out} 1·70(31°, *trans*), 2·54(45°, *cis*) in Me_2SO(l)	66Fc
Co $en_2Br_2{}^+$	con	25	Me_2SO	K_1 1·86(*cis*) in Me_2SO(l), → 0	67Mg
	sp	30—40	sulpholan	K_1(*cis*) 3·71(30°), 3·70(40°)	68Fb
	kin	60		*K*(*cis* ⇌ *trans*) 1·08, K_1 3·3(*cis*), 0·6(*trans*) in sulpholan = $C_4H_8SO_2$(l)	
Co^{3+} cpx	sp	0—35	dil	*K*(Co $en_2BrH_2O^{2+}$, *cis* ⇌ *trans*) −0·52(0°) −0·50(25°, 35°)	63Cc
	sp	25	0·5	$Co(NH_3)_5{}^{3+}$; K_1 −0·41	62Y/63H
	sp	30	org	Co en_2Cl^{2+}; K_1^{out} 3·96, ΔH_1 = 7·2, ΔS_1 = 42; Co en_2ClBr^+; K_1^{out} 2·84, ΔH_1^{out} = 1·27, ΔS_1^{out} = 8·8, also K_1 values for 1—21°	66Ld
	pH_2O	37	→ 0	Co $en_2NH_3NO_2{}^{2+}$; K_1 1·36(*cis*), 1·26(*trans*)	67M
	pH_2O	37	→ 0	Co en_2NCSCl^+; K_1 0·44	67Ma
	sp	25	0·07 ($NaClO_4$)	$Co(NH_3)_6{}^{3+}$; K_1 0·34	67Tb
	con	25	→ 0	$Co(NH_3)_6{}^{3+}$; K_1 1·65	68K
	con	25	Me_2SO	Co en_2ClBr^+; K_1 2·10(*cis*) in Me_2SO(l), → 0	67Mg

continued overleaf

72 Br^- (contd)

Metal	Method	Temp	Medium	Log of equilibrium constant, remarks	Ref
Ni^{2+}	sp	25	Me_2CO	K_4 2·0 in Me_2CO(l)	65Fa
	sp	25	org	$K_1 \approx$ 2·0 to 2·4 in $MeOCH_2CH_2OH$(l)	66F
	cal	25—40	2 ($NaClO_4$)	ΔH_1 = 0·08(25°), 0·07(40°), K_1 −0·12(25°), −0·11(40°), ΔS_1 = −0·27(25°)	66K
	var		org	ev $NiBr^+$, $NiBr_3^-$, $NiBr_4^{2-}$ tetrahedral cpx in MeCN and propanediol carbonate, octahedral cpx in Me_3PO_4 (Ag, con, sp)	67G
	sp	25—70 *etc.*	ROH	$\Delta H[NiBr_2T_4 + Br^- \rightleftharpoons NiBr_3T^-(+3T)]$ = 12·1 (T = MeOH), 17·1(T = EtOH), 14·8(T = BuOH); $\Delta H[NiBr_2T_4 \rightleftharpoons NiBr_2T_2(+2T)]$ = 17·0(T = BuOH), in T(l)	67S
$Ru(NH_3)_5{}^{3+}$	sp	25	var	$K_1 \approx$ 1·4	65E
Ru^{4+}	ir		solid	ev $Ru_2OBr_{10}{}^{4-}$ in $K_4Ru_2OBr_{10}$(s)	66H
$Rh(NH_3)_5H_2O^{2+}$	sp	35	1·5	K_1 −0·7, see also Table 80	66Sg
$Rh(NH_3)_5{}^{3+}$	ana	84	→ 0	K_1 3·20	39La
	kin	25—65	4 ($NaClO_4$)	K_1 −1·0(25°), −1·1(65°)	68Mg
Pd^{2+}	Pd	18—20	var	β_4 16·1 {$K[Pd^{2+} + 2e^- \rightleftharpoons Pd(s)]$ 33·4}	63Gb
	sol	20	0·60	K_{s0} −12·54, β_3 11·28, β_4 13·42 (?)	64P
			0·43	K_{s0} −12·96, β_3 11·60, β_4 13·40 (?)	
			0·60	$K_{s2} \approx$ −4·4, K_{s3} −1·26, K_3 3·1, K_4 2·1	64P/
			0·43	$K_{s2} \approx$ −4·5, K_{s3} −1·36, K_3 3·1, K_4 1·8	
	sp, hyp	20	0·8 (ClO_4), 0·6 H^+	K_1 4·37(sp), K_2 4·08(hyp), K_3 3·79(hyp), K_4 3·50(sp)	64Sa
	sp	25	0·5 ($NaClO_4$)	K_4 2·20	64Se
	Pd	25	KBr var	β_4 14{$K[Pd^{2+} + 2e^- \rightleftharpoons Pd(s)]$ 33·4} Also values for β_4 at 10°—60° assuming same K and RTF^{-1} as for 25°!	65F
	dis, sp		C_6H_6	ev $(H_3O)_3PdBr_5A_9$, $(H_3O)_2PdBr_4A_6$ in C_6H_6(l), A = cyclohexanone	65G
	sp	10—45	1 ($NaClO_4$), 0·8 H^+	K_4 2·50(10°), 2·30(25°), 2·16(45°) ΔH_4 = −4·3, ΔS_4 = −3·5	66Se
	cal	25	0·1 NaBr	$\Delta H_{\beta 4}$ = −13·11	67I
	gl	19—50	I($NaClO_4$)	$K(PdBr_3OH^{2-} + Br^- \rightleftharpoons PdBr_4^{2-} + OH^-)$ −4·23 (19—50°, I = 0·1—1·0)	67K
	Pd	25	w% dioxan	β_4 16·1 to 16·3(w = 0), 19·0(w = 71), also β_4 for several other w	68Gb
		40	w% dioxan	β_4 15·2 to 15·5(w = 0), 18·1(w = 71), also β_4 for several other w	
Os^{4+}	ir		solid	ev $Os_2OBr_{10}{}^{4-}$, not $OsBr_5OH^{2-}$, in $(Me_4N)_4Os_2OBr_{10}$(s)	66H
Pt^{2+}	sol	25	var	$K_s[Pt(NH_3)_2Br_2]$ −2·96(*cis*), −3·66(*trans*) $\beta[Pt(NH_3)_2BrH_2O^+ Br^-] \approx$ 3·2	64Ga
	X		solid	ev Pt_6Br_{12} in $PtBr_2$(s)	64T
	gl	25	0·1 KBr	†$K_n[Pt(OH)_{5-n}Br_{n-1}{}^{2-} + H^+ + Br^- \rightleftharpoons Pt(OH)_{4-n}Br_n{}^{2-} + H_2O]$ 11·15(n = 1), 10·7(n = 2), 10·0(n = 3), 8·15(n = 4), (K_w −14·15)	67Na
	ana, gl	15—40	0·318 ($NaNO_3$)	K_4 2·76(15°), 2·58(25°), 2·40(40°), $K(2PtBr_3^- \rightleftharpoons Pt_2Br_6{}^{2-}) \approx$ 1·0, $K_a(PtBr_3H_2O^-)$ −7·9	67T
$Pt(NH_3)_2{}^{2+}$	ana	25—35	0·05	K_1 4·05(25°), 3·82(35°), K_2 3·02(25°), 2·92(35°) ΔH_2 = −4, ΔS_2 = 0·4	68Ge

contin

Metal	*Method*	*Temp*	*Medium*	*Log of equilibrium constant, remarks*	*Ref*
Pt $dien^{2+}$	ana	25—35	0·32 $(NaNO_3)$	$K(MCl^+ + Br^- \rightleftharpoons MBr^+ + Cl^-)$ 0·58(25°), 0·54(35°)	67Me
		25—35	dil	K_1 4·02(25°), 4·07(35°)	
	ana	25?	0·318 $(NaNO_3)$	K_1 4·3	67Ta
Pt^{2+} cpx	con	20?	dil	$Pt(MeNH_2)_2NO_2^+$: K_1 4·07	64Ce
Pt^{4+}	ana, gl	50	NaBr var	K_6 2·85, $K_a(PtBr_5H_2O^-)$ −4·4	65D
	ana, gl	25—55	var	$K_6 \approx 2·4$(25—55°), $K_a(PtBr_5H_2O^-)$ −5·7(25°), $K_6K_a^{-1}$ 8·2(25°), 8·0(35°)	67N
	25	0.1 KBr		$K[Pt(OH)_{7-n}Br_{n-1}{}^{2-} + Br^- \rightleftharpoons Pt(OH)_{6-n}Br_n{}^{2-} + OH^-]$ −4·23(n = 1), −4·3(n = 2), −4·5(n = 3), −4·8(n = 4), −4·9(n = 5), −5·3(n = 6)	
Pt $en_3{}^{4+}$	sp	10—40	0 corr	K_1^{out} 1·14(10°), 1·18(25°), 1·25(40°), $\Delta H_1 = 1·3$, $\Delta S_1 = 9·7$	63Ga
Cu^+	Ram		var	ev linear $CuBr_2^-$ in Et_2O(l), other cpx in 5—9M-HBr	63Cb
	sol	25	2 $(NaNO_3)$	K_{s2} −2·42, K_3 1·01, ev $CuBr_4{}^{3-}$	67P
	Cu	25	2 $Na(NO_3)$	no ev polyn cpx	
Cu^{2+}	sp, mag	20	MeCN	ev $CuBr^+$ and $CuBr_3^-$ in MeCN(l)	63Sa
	X		solid	ev $Cu_2Br_6{}^{2-}$ in $K_2Cu_2Br_6$(s)	63Wa
	oth			K_1 0·55 (refractometry)	64B
	cal	25—40	2 $(NaClO_4)$	$\Delta H_1 = 0·90$(25°), 1·00(40°), K_1 −0·07(25°), −0·04(40°), $\Delta S_1 = 2·75$(25°)	66K
	sp		HBr var	ev $Cu(H_2O)_{6-n}Br_n{}^{2-n}$, n = 0, 1, 2, 3, 4, $CuBr_4{}^{2-}$ (tetrahedral), $CuBr_6{}^{4-}$?	67Ab
	sp		$MeCO_2H$	ev $CuBr_3^-$, $CuBr_4{}^{2-}$ in $MeCO_2H$(l)	67E
	sp	25	3 $Li(ClO_4)$	K_1^{in} −1·5, K_1^{out} −0·60, K_1 −0·55	68Me
Ag^+	sol	20	0 corr	(K_1 4·68) K_{12} 2·38, $\beta(AgBr^0Ag^+{}_2)$ 3·45	60Lb/64La
	sol	20	4 $(NaClO_4)$, 3 H^+	K_{s2} −4·90, K_{s3} −3·67, K_{s4} −3·65	64Aa
	Ag	250	MNO_3(l)	K_{s0} −9·07, $\Delta H_{s0} = 22·4$ in $Na_{0.5}K_{0.5}NO_3$(l), *x* units	64Bb
	Ag	280	MNO_3(l)	K_1 2·43, β_2 4·38, β_3 4·77, β_{62} 10·08 ? K_{s0} −6·52 in $Na_{0.5}K_{0.5}NO_3$(l), *m* units	64C
	sol	280	MNO_3(l)	K_1 2·43, β_2 4·36, β_3 4·83, β_{62} 10·34, (K_{s0} −6·52) K_{s0} −6·57 ?, in $Na_{0.5}K_{0.5}NO_3$(l), *m* units	64Ca
	sol	300	NaBr var	K_{s2} −0·95	64G
	Ag	150	MNO_3(l)	K_{s0} −9·72 in $Li_{0.43}K_{0.57}NO_3$(l, eut), *m* units	65Bc
	Ag	25	$C_2H_4(NH_2)_2$	K_{s0} −4·26 in $C_2H_4(NH_2)_2$(l)	65Me
	Ag, *iE*	23	Me_2SO	K_{s0} −10·6, β_2 11·7 to 12·3 in Me_2SO(l), $I = 0·1(Et_4NClO_4)$	66L
		23	MeCN	K_{s0} −13·2, β_2 13·4 to 14·1 in MeCN(l), $I = 0·1(Et_4NClO_4)$	
		23	MeOH	K_{s0} −15·2, β_2 10·9 to 10·2 in MeOH(l), $I = 1·0(LiClO_4)$	
		23	Me_2CO	K_{s0} −18·7, β_2 19·7 to 20·2 in Me_2CO(l), $I = 0·1(LiClO_4)$	
		23	$EtNO_2$	K_{s0} −21·8, β_2 22·5 to 22·4 in $EtNO_2$(l), $I = 0·1(Et_4NClO_4)$	
	dis		0 corr	$K_D[H^+ + Ag^+ + 2Br^- + 3L(org) \rightleftharpoons HAgBr_2L_3(org)]$ 9·79, T = TBP, org = C_6H_6	66Le+
	Ag	25	org	K_{s0} −15·2 in MeOH; K_{s0} −11·4 in $HCONH_2$; K_{s0} −15·0, β_2 16·6 in Me_2NCHO; K_{s0} −14·5, β_2 16·9 in Me_2NCOMe; K_{s0} −10·6, β_2 11·4 in Me_2SO; K_{s0} −12·9, β_2 13·7 in MeCN; K_{s0} −12·3, β_2 16·5 in $(Me_2N)_3PO$(l)	67Aa

continued overleaf

72 Br^- (contd)

Metal	*Method*	*Temp*	*Medium*	*Log of equilibrium constant, remarks*	*Ref*
Ag^+ (*contd*)	Ag	25	1	β_4 8·33	67Bh
	Ag	217—257	$MClO_4$(l)	K_{s0} −9·43(490·0 K), −8·46(530·5 K)	67Fa
		250	$MClO_4$(l)	K_{s0} −8·65, ΔH_{s0} = 26·3 in $Li_{0.75}Na_{0.25}ClO_4$(l), *m* units	
	Ag	412—680	M_2SO_4(l)	K_1 2·20, K_2 1·9, K_{12} 1·9 in $Li_{1.43}K_{0.57}SO_4$(l, eut), *x* units	67Ga
	Ag	25—85	Me_2SO	K_{s0} −10·04(25°), −9·74(45°), −9·35(65°), −9·21(85°) β_2 10·59(25°), 10·43(45°), 10·18(65°), 10·15(85°), in Me_2SO(l), 0·1M-NH_4NO_3	67R
	sol	275	MNO_3(l)	β_2 4·39, β_3 4·89, β_{12} 3·93, β_{13} 4·46 (K_{s0} −6·57, K_{s1} −3·85)	67Sc
		300	MNO_3(l)	β_2 4·24, β_3 4·50, β_{12} 3·75, β_{13} 4·16, (K_{s0} −6·19, K_{s1} −4·3), in $Na_{0.5}K_{0.5}NO_3$(l), *m* units	
	Ag	30	sulpholan	K_{s0} −18·18, β_2 19·32 in sulpholan(l), I = 0·1(Et_4NClO_4)	68Db
	Ag	350	MNO_3(l)	K_{s0} −5·40, ΔH_{s0} = 24·05 in $K_{0.89}Ba_{0.11}(NO_3)_{1.11}$ (l, eut), *m* units	68G
	sol	55·7	$NH_4NO_3(H_2O)_2$	K_{s0} −10·31, K_1 3·74, K_2 2·84?, K_3 1·32	68Gc
		70·0		K_{s0} −10·11, K_1 3·72, K_2 2·88, K_3 1·18	
		85·0		K_{s0} −9·67, K_1 3·65, K_2 2·71, K_3 0·87 in $NH_4NO_3(H_2O)_2$(l) ≈ 28 molal NH_4NO_3, *m* units; values with *x* units, and quasi-lattice parameters are also given	
	sol	24	Me_2SO	β_{12} 8·23 in Me_2SO(l), I = 0·5(Et_4NClO_4)	68La
	sol	150—190	MNO_3(l)	K_{s0} −9·28(150°), −8·83(169°), −8·36(190°); K_1 3·31(150°), 3·20(169°), 3·06(190°); K_2 2·91(150°), 2·82(169°), 2·66(190°); ΔH_{s0} = 22·0, $\Delta H_1 = \Delta H_2$ = −1·25 in $Li_{0.43}K_{0.57}NO_3$ (l, eut), *m* units; values with *x* units also given	68Mf
Au^{3+}	gl, Ag	25	0 corr	†K_3[$Au(OH)_2Br_2^- + H^+ + Br^- \rightleftharpoons AuOHBr_3^- + H_2O$] 9·0 †$K_4$($AuOHBr_3^- + H^+ + Br^- \rightleftharpoons AuBr_4^- + H_2O$) 8·5	66Ce
	Br	25?	0 corr	K_4 5·47	68D
Zn^{2+}	oth		Et_2O	K_3 5·5, K_4 3 in Et_2O(l), LiBr var (from dielectric constants, "Job's method")	63Bd = 64Cc
	MHg	25	4 ($LiClO_4$) 4 (LiBr)	K_1 0·06, β_2 −0·82, β_3 0·28, β_4 −0·74 $\beta(M^+ZnBr_4^{2-})$ −0·34(Na), 0·09(K), 0·17(Rb), 0·39(Cs), 0·15(NH_4); $\beta(M^+ZnBr_3^-)$ −1·0(Na), −0·5(K), −0·4(Rb), −0·3(Cs), −0·5(NH_4)	65Mb
		25	3 ($LiClO_4$) 3 (LiBr)	K_1 −0·19, β_2 −1·15, β_3 −0·55, β_4 −1·5 $\beta(M^+ZnBr_4^{2-})$ −0·41(Na), 0·15(K), 0·20(Rb), 0·21(Cs), 0·08(NH_4); $\beta(M^+ZnBr_3^-)$ −0·85(Na), −0·54(K), −0·37(Rb), −0·07(Cs), −0·37(NH_4)	
	X		$ZnBr_2$ var	ev $ZnBr_2(H_2O)_2$	65W
	X		solid	ev tetrahedral $ZnBr_3H_2O^-$ in α-$KZnBr_3H_2O$(s)	66Be
Cd^{2+}	Ag	171—240	MNO_3(l)	K_1 3·90(171°), 3·51(240°), K_2 3·52(171°), 3·11(240°) in $K_{0.5}Li_{0.5}NO_3$(l); also values for some other K : Li ratios at 240°, *x* units	64Ba
	MHg	25—35	4 ($LiClO_4$)	K_1 1·95(25°, 35°), β_2 3·00(25°, 35°) β_3 4·30(25°), 4·40(35°), β_4 5·2(25°), 5·1(35°); $\beta(K^+CdBr_4^{2-})$ −0·8(25°), −1·0(35°);	64M

continued

Metal	Method	Temp	Medium	Log of equilibrium constant, remarks	Ref
Cd^{2+} (contd)				$\beta(Rb^+CdBr_4^{2-})$ −0·7(25°), −0·8(35°); $\beta(Cs^+CdBr_4^{2-})$ −0·5(25°), −0·6(35°)	
	var	25	0 corr	$K_s(Cd^{2+}OH^-Br^-)$ −10·50(fresh), −10·60(aged) $K_s(Cd^{2+}OH^-_{1.1}Br^-_{0.9})$ −11·25(fresh), −11·40(aged) (MHg, gl, Ag, sol) (K_w −14·00)	64S
	dis	30	1 ($NaClO_4$)	K_1 1·4, β_2 1·9, β_3 2·2	65H
	Pd	254	MNO_3(l)	K_1 2·27, β_2 4·00, β_3 5·33, β_4 6·16 in $Na_{0.5}K_{0.5}NO_3$(l), *m* units	65I
	Ag	240	MNO_3(l)	K_1 3·43 in $Li_{0.5}Na_{0.5}NO_3$(l), *x* units	66B
	E	258—358	MNO_3(l)	K_1 2·81 in KNO_3(l, 358°), 2·80 in $NaNO_3$(l, 331°), 3·18 in $Na_{0.5}K_{0.5}NO_3$(l, 258°), *x* units	66Bc
	MHg, pol	50	$Ca(NO_3)_2$	K_1 3·59, K_2 3·18, no ev Cd_2Br^{3+} in $Ca(NO_3)_2(H_2O)_4$(l)	66Bd
	cal	25	3 ($NaClO_4$)	$\Delta H_1 = -0·98$, $\Delta H_2 = -0·57$, $\Delta H_3 = 1·72$, $\Delta H_4 = 0·30$; $\Delta S_1 = 4·7$, $\Delta S_2 = 0·8$, $\Delta S_3 = 10·2$, $\Delta S_4 = 2·7$, (K_1 1·76, β_2 2·34, β_3 3·32, β_4 3·70)	66Ga
	Ag	119	MNO_3(l)	K_1 4·28(y = 0), 3·96(y = 0·26) to 3·45(y = 1·26) in $Li_{0.5}K_{0.5}NO_3(H_2O)_y$(l)	67Lb
		168		K_1 3·91(y = 0), 3·76(y = 0·1) in $Li_{0.5}K_{0.5}NO_3(H_2O)_y$(l), *x* units; described by quasi-lattice model in ref 67Bb	
	cal	25	4 Li(ClO_4)	$\Delta H_1 = 0·30$, $\Delta H_2 = -1·2$, $\Delta H_3 = 4·6$, $\Delta H_4 = 2·0$; K_1 1·99, K_2 1·10, K_3 1·16, K_4 0·60; $\Delta S_1 = 8·0$, $\Delta S_2 = 9$, $\Delta S_3 = -10$, $\Delta S_4 = -4$	67Mf
	pol	25	2 ($NaClO_4$)	K_1 1·60, β_2 2·26, β_3 2·68, β_4 3·03	67Sb
	pol	145	MNO_3(l)	K_1 2·81, K_2 2·15, K_3 1·90, K_4 1·18 in (Li, Na, K)NO_3 (l, eut), *m* units	68I
	MHg	25	2 ($LiNO_3$)	K_1 1·5, β_2 1·9, β_3 2·2, β_4 2·7	68Ka
			MeOH	K_1 4·0, β_2 6·0, β_3 6·9, β_4 8·2 in MeOH(l), I = 2($LiNO_3$)	
			96% EtOH	K_1 6·5, β_2 7·6, β_3 7·9, β_4 9·5 in 96% EtOH(l), I = 2($LiNO_3$) also β_n values for 25, 50, 75% MeOH and EtOH	
Hg_2^{2+}	Hg	7—40	0·5 (Na)ClO_4, 0·1 H^+	K_{s0} −22·85(7°), −20·17(40°)	63Ha
	Hg	150—210	MNO_3(l)	K_{s0} 5·28 −9785T^{-1} (T K), −16·46(450 K) in $Li_{0.43}K_{0.57}NO_3$(l, eut), *m* units	66Fd
Hg^{2+}	aix	23	I($NaClO_4$)	K_3 1·6, K_4 1·0(I = 3), K_3 2·5, K_4 2·1(I = 0·3) (K_3 and K_4 for I = 0·5 from ref 57Ma)	63E
	cal	25	var	K_3 2·26, K_4 1·38, $\Delta H_3 = -2·08$, $\Delta H_4 = -3·46$	63Bc = 64Lb
	red	7—40	0·5 (Na)ClO_4, 0·1 H^+	K_1 9·53(7°), 8·70(40°); K_2 8·85(7°), 7·76(40°)	63Ha
	cal	8—40	0·5 (Na)ClO_4, 0·1 H^+	$\Delta H_1 = -11·1$(7°), −10·2(25°), −10·0(40°); $\Delta H_{\beta 2} = -22·6$(7°), −21·23(25°), −21·4(40°);	64Cb
				$\Delta S_1 = 4·0$(8°), 7·1(25°), 7·9(40°); $\Delta S_{\beta 2} = 3·4$(8°), 8·2(25°), 7·0(40°)	64Cb+63Ha
	Ram		TBP	ev $HgBr_2$, $HgBr_3^-$, and $HgBr_4^{2-}$ in $(C_4H_9O)_3PO$(l)	64Sb
	dis	150	MNO_3(l)	K_D[$HgBr_2$(salt melt) ⇌ $HgBr_2$("polyphenyl eutectic")] 0·96; K_3 0·90, K_4 0·95 in ($Li_{0.43}K_{0.57}$) NO_3(l, eut)	64Z = 65Z
	red, Hg	25	3 (Na)ClO_4	K_1 9·40, K_2 8·58, K_3 2·76, K_4 1·49	65A
	cal	25	3 (Na)ClO_4	$K_3 \approx 2·3$, $K_3K_4 \approx 4·3$;	

continued overleaf

72 Br^- (contd)

Metal	*Method*	*Temp*	*Medium*	*Log of equilibrium constant, remarks*	*Ref*
Hg^{2+} (*contd*)				$\Delta H_1 = -9{\cdot}6, \Delta H_2 = -9{\cdot}6, \Delta H_3 = -2{\cdot}6, \Delta H_4 = -4{\cdot}4$; $\Delta S_1 = 10{\cdot}9, \Delta S_2 = 7{\cdot}1, \Delta S_3 = 4{\cdot}0, \Delta S_4 = -8{\cdot}1$	
	X		solid	ev $Hg_2Br_6^{2-}$ in $[(Ph_3AsO)_2H]_2Hg_2Br_6$(s)	67Ha
	oth	25	var	K_4 1·2 (from molar volumes)	67J
	sol	20	2·2 [$Ca(NO_3)_2$]	$K_s[HgBr_2(s)+Hg^{2+} \rightleftharpoons Hg_2Br_2^{2+}]$ −0·7 ($[Hg^{II}]$ up to 0·5M)	68Ga
	Hg, pol	25?	w% C_5H_5N	β_2 8·1, β_3 9·5(w = 20), β_2 12·1, β_3 14·5(w = 80) in w% C_5H_5N, also values for w = 10, 40, 60	68P
Hg^{2+}cpx	fp	0	KBr var	$Hg(CF_3)_2$; K_1 0·2	64D
	sp		0·10 ($KClO_4$)	HgY^{2-}; K_1 5·5, H_4Y = EDTA	68N
RHg^+	pol	25	org	K_1 5·28, K_2 0·74(R = C_6H_5); K_1 5·42, K_2 1·10(R = $C_6H_5CH_2$); K_1 7·95, K_2 1·23(R = $C_6H_5CH_2CH_2$); K_1 5·18, K_2 2·75(R = Me_3CCOCH_2); K_1 4·43, K_2 2·93(R = $PhCHCO_2Et$) in 60% Me_2NCHO + 40% H_2O, I = 0·1 (Na, Li, ClO_4)	67Be
$MeHg^+$	gl, cal	20	0·1 (KNO_3)	K_1 6·62, $\Delta H_1 = -9{\cdot}9, \Delta S_1 = -3{\cdot}6$	63S = 6
$EtHg^+$	sol	25	1 Na(ClO_4)	K_{s1} −3·1, no ev ani cpx	65B
	sol	25	4 Na(ClO_4)	ev weak ani cpx	
	aix	25	0 corr	K_2 0·3, K_3 1·1	68S
Al^{3+}	con	25	$PhNO_2$	K_3 4·9, K_4 3·3, $K(2AlBr_3 \rightleftharpoons Al_2Br_6)$ 0·0 in $PhNO_2$(l)	49J + 49V/64V
	nmr		Me_2NCHO	ev $AlT_6^{3+}Br^-_n$ (n = 1, 2, 3) in T = Me_2NCHO(l)	68M
Ga^{3+}	dis	20	0·69 H(ClO_4)	K_1 −0·10 (or K_1 −0·24, β_2 −0·5)	67Mc
In^{3+}	dis	25	0 corr	K_3 −1·22, K_4 −1·92	58Db
Tl^+	sol	25	4 Na(ClO_4)	K_{s0} −4·82	58M
			dil	K_{s0} −5·38	
Tl^{3+}	*X*		solid	ev $TlBr_6^{3-}$ in $Rb_3TlBr_6(H_2O)_{1\cdot14}$(s)	35H
	red	20	4 ($NaClO_4$), 3 H^+	K_1 9·62, K_2 7·44, K_3 5·53, K_4 4·14, β_4 26·73, K_5 < −0·4, no ev $TlBr_5^{2-}$, $TlBr_6^{3-}$	63A
	sol	20	4 ($NaClO_4$), 3 H^+	ev $TlBr_4^-$, no ev cpx with n > 4	64Aa
	cal	25	4 ($NaClO_4$), 3 H^+	$\Delta H_1 = -8{\cdot}96, \Delta H_2 = -6{\cdot}09, \Delta H_3 = -4{\cdot}58, \Delta H_4 = -2{\cdot}14$, $\Delta S_1 = 13{\cdot}4, \Delta S_2 = 13{\cdot}3, \Delta S_3 = 9{\cdot}7, \Delta S_4 = 11{\cdot}7$	64L + 6
		25	4 ($NaClO_4$), 3 H^+	K_1 9·51, β_2 16·88, β_3 22·30, β_4 26·43	64L + 63A/6
	X, con		var	ev $TlBr_4^-$ in solid and MeCN(l)	65Cb
	red	25	3 ($LiClO_4$)	K_1 9·28, K_2 7·42, K_3 5·4, K_4 3·6, β_4 25·7, K_5 1·5, $K(TlBr^{2+} + H_2O \rightleftharpoons TlOHBr^+ + H^+)$ −1·84	67Y
	Ram, ir		org	ev $TlBr_3$ and $TlBr_4^-$ in EtOH(l) and TBP(l)	68Da
Ge^{4+}	sol	25	HBr var	add; equil $[GeO_2(s) + 4H^+ + 4Br^- \rightleftharpoons GeBr_4(s) + 2H_2O]$ at 7·36 M HBr	56B
Sn^{2+}	sol	25	4 H(HSO_4)	K_1 0·90, K_2 0·83, K_3 0·40, K_4 −0·47, K_5 0·32	62H
Sn^{4+}	ir, Ram		var	ev $SnBr_5^-$(bipyramid) in Et_4NSnBr_5(s)	68C
$MeSn^{3+}$	Ag	25	3 ($NaClO_4$)	K_1 0·6	68Mc
Me_2Sn^{2+}	Ag	25	3 ($NaClO_4$)	K_1 < −0·5	65Fb
	ir, pre		solid	ev $Me_2SnBr_3^-$ and $Me_2SnBr_4^{2-}$ in $Me_4NMe_2SnBr_3$(s) and $Cs_2Me_2SnBr_4$(s)	66Ca

cont

Metal	*Method*	*Temp*	*Medium*	*Log of equilibrium constant, remarks*	*Ref*
Ph_3Sn^+	dis	30	0·1 ($NaNO_3$)	K_1 3·3, $K_D[Ph_3SnOH(org)+Br^- \rightleftharpoons Ph_3SnBr(org)+OH^-]$ −6·9(org = C_6H_6), −6·6(org = Bu^iCOMe)	65S
Pb^{2+}					
1964 *Erratum*: p. 328 *delete all of ref* 57Ka					
	MHg	25	*I* $Na(ClO_4)$	K_1 1·70, β_2 3·28, β_3 3·90, β_4 4·65(I = 6); K_1 1·30, β_2 1·90, β_3 2·88, β_4 2·81, β_5 2·3(I = 3); K_1 1·04, β_2 1·45, β_3 2·23, β_4 1·54(I = 1); K_1 1·26, β_2 1·60, β_3 2·58(I = 0·25); also values for I = 4, 2, 0·75, and 0·5	63Mf
			4 $Li(ClO_4)$	K_1 1·54, β_2 2·65, β_3 3·30, β_4 3·76, β_5 2·48, β_6 2·20	
			I $Na(NO_3)$	K_1 0·72, β_2 0·85, β_3 1·0, β_4 0·93, β_5 0·1, β_6 −0·3 (I = 4); K_1 0·60, β_2 1·00, β_3 1·26(I = 1); also values for I = 3, 2, 0·75	
			I Na(Cl)	K_1 0·00, β_2 −0·19, β_3 −1·7, β_4 −1·15(I = 4); K_1 0·04, β_2 0·34(I = 1); also values for I = 3, 2, 0·75, 0·5	
	Ag	160—200	MNO_3(l)	K_1 3·00(160°), 2·86(200°), K_2 2·60(160°), 2·48(200°) in $Li_{0.5}K_{0.5}NO_3$(l), *x* units	64Ba
	sol	25	3 ($NaClO_4$)	ev cpx $PbBr_n(Br_2)_x^{(2-n)}$, x = 1	64Bc
	red, sol	25	3 ($NaClO_4$)	K_1 1·30, (β_2 1·90), β_3 2·5, (β_4 2·81, K_{s0} −5·28)	66Lc
	sol	25	4 $H(HSO_4)$	K_1 1·07, β_2 2·20, β_4 3·43, β_6 2·87	66N
	oth	700	melt	ev $KPbBr_3$(g), K_3 0·7 in KBr–$PbBr_2$(l) ($pPbBr_2$, mass spectrometry), *x* units	68Ba
	pol	145	MNO_3(l)	K_1 1·87, K_2 1·38 in (Li, Na, K)NO_3 (l, eut), *m* units	68I
	MHg	5	3 $Li(ClO_4)$	K_1 1·34, β_2 2·33, β_3 2·92, β_4 3·19	68Fa
		25	3 $Li(ClO_4)$	K_1 1·29, β_2 2·21, β_3 2·78, β_4 2·93; ΔH_1 = −0·52, $\Delta H_{\beta 2}$ = −1·38, $\Delta H_{\beta 3}$ = −1·47, $\Delta H_{\beta 4}$ = −3·7; ΔS_1 = 4·1, $\Delta S_{\beta 2}$ = 5·5, $\Delta S_{\beta 3}$ = 7·8, $\Delta S_{\beta 4}$ = 1·0	
		45	3 $Li(ClO_4)$	K_1 1·28, β_2 2·19, β_3 2·77, β_4 2·78	
		65	3 $Li(ClO_4)$	K_1 1·31, β_2 2·27, β_3 2·88, β_4 2·73; also β_n values for 15°, 35°, 55°, $\Delta H_{\beta n}$, $\Delta S_{\beta n}$ for all temp	
$Pb^{2+,4+}$	red, sol	25	3 ($NaClO_4$)	$K(Pb^{2+}+2Br^-+Br_2 \rightleftharpoons PbBr_4)$ 3·58; $K(Pb^{2+}+4Br^-+Br_2 \rightleftharpoons PbBr_6^{2-})$ 4·23	66Lc
Pb^{4+}	var	25	1 KBr	$\beta_4 \approx$ 25, ev $PbBr_5^-$, $PbBr_6^{2-}$, [$K_{s0}(Pb^{2+}Br^-_2)$ −5·37, $K(Pb^{4+}+2e^- \rightleftharpoons Pb^{2+})$ 57 (1·7 V)] (sol, red, MHg)	64K
Ph_3Pb^+	dis	30	0·1 ($NaNO_3$)	K_1 5·7, $K_D[Ph_3PbOH(org)+Br^- \rightleftharpoons Ph_3PbBr(org)+OH^-]$ −2·2(org = $CHCl_3$), −1·5(org = Bu^iCOMe) (sign misprinted in Table 4)	65S
R_4N^+	oth	25	0 corr	K_1 0·09(R = Me), 0·38(R = Et), 0·49(R = C_3H_7)	67W
		25	var	$K[2MBr \rightleftharpoons (MBr)_2]$ 0·00(R = Et), 0·49(R = C_3H_7) (from molar volumes)	
	con	15—35	EtCOMe	R = C_4H_9; K_1 2·89(15°), ΔH_1 = 2·5, ΔS_1 = 22; R = C_7H_{15}; K_1 2·74(15°), 2·89(35°), ΔH_1 = 3·1, ΔS_1 = 23 in EtCOMe(l), → 0	68Hb
Me_4N^+	pH_2O, con	25	→ 0	K_1 0·15 to 0·17(pH_2O), 0·14—0·22(con)	65L

continued overleaf

72 Br^- (contd)

Metal	*Method*	*Temp*	*Medium*	*Log of equilibrium constant, remarks*	*Ref*
NO^+	sp	0—35	0 corr	$K(HNO_2+H^++Br^- \rightleftharpoons NOBr+H_2O)$ −1·65(0°), −1·57(5°), −1·43(15°), −1·29(25°), −1·16(35°)	54Sc
P^{5+}	X		solid	ev PBr_4^+ in PBr_5(s)	43D
$Sb^{3+,5+}$	X		solid	ev $SbBr_6^{3-}$ and $SbBr_6^-$ in $(NH_4)_4Sb_2Br_{12}$(s) = $(NH_4)_2SbBr_6$(s)	66La
Bi^{3+}	sol	25	2 ($NaClO_4$), 1 H^+	$*K_{s0}[BiOBr(s)+2H^+ \rightleftharpoons Bi^{3+}+Br^-+H_2O]$ −6·24; β_2 4·29, β_3 6·19	65J
	sol, sp	25	4 $H(HSO_4)$	$K_s[(Me_4N^+)_3(BiBr_4^-)_2(Br^-)]$ −14·11; K_1 3·18, β_2 4·98, (4·94?), β_4 8·79, β_6 10·96, β_8 ?9·98	66P
	cix	25	1·89 $H(ClO_4)$	K_1 2·36, β_2 4·42, β_3 6·26, β_4 7·7	67L
	Ram		var	ev $BiBr_6^{3-}$	67O
	cal	10—50	4 ($HClO_4$)	ΔH_1 = −0·56(10°), −0·36(18°), −0·02(25°), 0·10(30°), 0·25(35°), 0·75(50°); ΔS_1 = 10·0(10°), 11·0(18°), 11·8(25°), 12·4(30°), 12·9(35°), 13·5(40°), 14·4(50°)	67V
		25	$I(HClO_4)$	ΔH_1 = −1·52(I = 6), −0·74(I = 5), −0·02(I = 4), 3·30(I = 0); ΔS_1 = 8·5(I = 6), 10·3(I = 5), 11·8(I = 4), 26·1(I = 0); [K_1 3·0(I = 6), 2·80(I = 5), 2·64(I = 4), 3·28(I = 0)]; also values for ΔC_p given	
	cal	10—50	4 ($HClO_4$)	ΔS_1 = 10·0(10°), 14·4(50°), linear with T; ΔC_{p1} = 35·4(10°), 31·0(50°), linear with T; $\Delta S_1+\Delta C_{p1} \approx$ 45·4, independent of T	67V/68
	X		solid	ev $BiBr_6^{3-}$ in $(Me_2NH_2)_3BiBr_6$(s)	68Ma
Se^{4+}	con	25	org	ev $SeBr_3^+$ in Me_2NCHO(l), $PhNO_2$(l), *etc.*; conclusions doubted by ref 68Kb	67C
	Ram		HBr var	ev $SeBr_5^-$ and $SeBr_6^{2-}$	68Ha
Te^{4+}	X		solid	ev $TeBr_6^{2-}$ in $Mg(H_2O)_6TeBr_6$(s)	64A
	X		solid	ev $TeBr_6^{2-}$ in K_2TeBr_2(s)	64Bd
	sp		org	K_2 2·74 in MeCN(l), 1·20 in Me_2NCHO(l), 0·85 in Me_2SO(l)	65K
	ir		solid	ev $TeBr_3^+$ in $TeBr_4$(s)	66G
	sp		$I\,H(ClO_4)$	K_5K_6 3·55(I = 6), 1·55(I = 4)	66Ra
	sp, tp	22	4 $HClO_4$	ev $TeBr_3^+$ (in presence of Li^+ or Na^+), $TeBr_6^{2-}$, β_3(?) and β_6(?) values given	66Sf
	con	25	org	ev $TeBr_3^+$ in Me_2NCHO(l), $PhNO_2$(l) *etc.*; conclusions doubted by ref 68Kb	67C
	dis	25	3 $H(ClO_4)$	K_1 0·97, β_2 1·58, β_3 1·96, β_4 2·15, β_5 2·21, β_6 2·13	67Se
	con, fp		org	ev $(TeBr_4)_n$, $n \approx 3$ in C_6H_6(l) and C_6H_5Me(l); ev $TeBr_3T_2^+$ and Br^- in T = MeCN(l), EtOH(l), *etc.*	68Gd
Br_2	dis	25	CCl_4	$K_D[Br_2(aq, m) \rightleftharpoons Br_2(in\ CCl_4, x)]$ 0·43	17L
			HBr var	K_1 1·21	
		15—45	CCl_4	K_D 0·39(15°), 0·43(25°), 0·46(35°), 0·47(45°), to CCl_4 (m, x units)	56Kc
	sol	20—25	var	K_1 1·3—1·4(20°, I var), 1·1—1·2 [25°, I = 3($NaClO_4$)]	64Bc
	aix	25	0·5 K(Br)	ev Br_3^- and Br_5^-, data agree with K_1 and β_{12} from ref 58S	64I
	dis	25	3 ($LiClO_4$)	K_1 1·03, β_{12} 1·58, $K_D[Br_2 \rightleftharpoons Br_2(in\ CCl_4)]$ 1·52	65M
		25	3 (LiBr)	$\beta(M^+Br_3^-)$ −1·7(Na^+), −0·5(K^+), −0·1(Cs^+)	
	sp	25	MeOH	K_1 2·26 in MeOH(l)	66Lb
	red, sol	25	3 ($NaClO_4$)	$\beta_{12}[Br^-(Br_2)_2]$ 1·40—1·44, $\beta_{22}[Br^-_2(Br_2)_2]$ 2·58	66Lc
	red	0—50	0 corr	K_1 1·39(0°), 1·34(10°), 1·28(20°), 1·22(30°), 1·17(40°), 1·11(50°)	66Mb

contir

Metal	Method	Temp	Medium	Log of equilibrium constant, remarks	Ref
Br_2 (contd)	var	25	3 $Li(ClO_4)$	K_1 1·05(dis, cal), 1·10(sp), ev higher cpx	66Mc
	cal			$\Delta H_1 = -1{\cdot}64$, $\Delta S_1 \approx 0$	
	dis, red		var	K_1 values for various I with ClO_4^- and Li^+, Na^+, Mg^{2+}, Ca^{2+}, Sr^{2+}, and Ba^{2+}	
				ev ion pairs of M^+ and M^{2+} with Br^- and Br_3^-	
			0 corr	K_1 1·50(Li data), 1·55(Na), 1·76(Mg), 1·78(Ca), 1·80(Sr), 1·82(Ba)	
	sp	25	CF_3CO_2H	$K_1 < -0{\cdot}05$ in CF_3CO_2H(l), LiBr var	67A
	cal	25	3 $Li(ClO_4)$	$\Delta H_1 = -1{\cdot}7$, $\Delta S_1 = -0{\cdot}7$	67Md+
	cal	25	CCl_4	K_D 0·44 to CCl_4, (m, x units), $\Delta H_D = 0{\cdot}91$	68H
IBr	sp	20	var	K_1 2·68, $\Delta H_1 = -10{\cdot}8$	64E
I_2					

1964 *Addendum*: p. 330, *for ref* "62Mb" *read* "62Mb = 63Mg"

Metal	Method	Temp	Medium	Log of equilibrium constant, remarks	Ref
	sp	20	var	K_1 1·02, $\Delta H_1 = -1{\cdot}37[K(I_2+Br^- \rightleftharpoons IBr+I^-) - 5{\cdot}66]$	64E
	X		solid	ev $IIBr^-$ in CsI_2Br(s)	66Cc
	sp	25	3 $Li(ClO_4)$	K_1 1·05	67Md
	dis	15—45	3 $Li(ClO_4)$	K_1 1·16(15°), 1·12(25°), 1·10(35°), 1·03(45°); $\Delta H_1 = -1{\cdot}8$, $\Delta S_1 = -1{\cdot}0$(25°)	
	cal	25	3 $Li(ClO_4)$	$\Delta H_1 = -2{\cdot}2$, $\Delta S_1 = -2{\cdot}2$	
I^+	con	140	I_2(l)	$K_1 \approx 5{\cdot}96$ in I_2(l)	67Bc

17L G. N. Lewis and H. Storch, *J. Amer. Chem. Soc.*, 1917, **39**, 2544
29A F. G. Angell, R. G. James, and W. Wardlaw, *J. Chem. Soc.*, 1929, 2578
31P H. Paulssen-v. Beck, *Z. anorg. Chem.*, 1931, **196**, 85
35H J. L. Hoard and L. Goldstein, *J. Chem. Phys.*, 1935, **3**, 645
39La A. B. Lamb, *J. Amer. Chem. Soc.*, 1939, **61**, 699
43D M. van Driel and C. H. MacGillavry, *Rec. Trav. chim.*, 1943, **62**, 167
49J W. J. Jacober and C. A. Kraus, *J. Amer. Chem. Soc.*, 1949, **71**, 2405
49V R. E. Van Dyke and C. A. Kraus, *J. Amer. Chem. Soc.*, 1949, **71**, 2694
54Sc H. Schmid, K. Pinz, and G. Ruess, quoted by H. Schmid, *Monatsh.*, 1954, **85**, 424
56Kc C. M. Kelley and H. V. Tartar, *J. Amer. Chem. Soc.*, 1956, **78**, 5752
57Pa R. G. Pearson, P. M. Henry, and F. Basolo, *J. Amer. Chem. Soc.*, 1957, **79**, 5382
58Db R. J. Dietz, jun., thesis, Massachusetts Institute of Technology, 1958
58M V. E. Mironov, dissertation (kand), 1958, Leningrad Technological Institute, quoted by 63Kb
59Mc J. Mendez-Schalchli, thesis, Massachusetts Institute of Technology, 1959
62H G. P. Haight, jun., Proc. seventh Internat. Conf. Co-ordination Chemistry (1962), p. 318
62Sb M. Shiloh and Y. Marcus, IA-781 (Israel AEC), 1962
63A S. Ahrland, I. Grenthe, L. Johansson, and B. Norén, *Acta Chem. Scand.*, 1963, **17**, 1567
63Bc F. Becker, J. Barthel, N. G. Schmahl, and H. M. Lüschow, *Z. phys. Chem. (Frankfurt)*, 1963, **37**, 52
63Bd P. Bonnet and M. Chabanel, *Compt. rend.*, 1963, **257**, 1280
63Cb J. A. Creighton and E. R. Lippincott, *J. Chem. Soc.*, 1963, 5134
63Cc S. C. Chan and M. L. Tobe, *J. Chem. Soc.*, 1963, 5700
63Cd M. Chabanel, *Bull. Soc. chim. France*, 1963, 673
63E I. Eliezer and Y. Marcus, *J. Inorg. Nuclear Chem.*, 1963, **25**, 1465
63Ga D. C. Giedt and C. J. Nyman, *J. Phys. Chem.*, 1963, **67**, 2491
63Gb A. A. Grinberg, N. V. Kiseleva, and M. I. Gel'fman, *Doklady Akad. Nauk. S.S.S.R.*, 1963, **153**, 1327
63H A. Haim and H. Taube, *Inorg. Chem.*, 1963, **2**, 1199
63Ha L. D. Hansen, R. M. Izatt, and J. J. Christensen, *Inorg. Chem.*, 1963, **2**, 1243
63Kb F. Ya. Kul'ba and V. E. Mironov ,'Khimiya talliya,' Goskhimizdat, Leningrad, 1963, p. 67
63Mf V. E. Mironov, F. Ya. Kul'ba, V. A. Fedorov, and O. B. Tikhomirov, *Zhur. neorg. Khim.*, 1963, **8**, 2524
63Mg D. Meyerstein and A. Treinin, *Trans. Faraday Soc.*, 1963, **59**, 1114
63S M. Schellenberg, dissertation, Eidgenössische Technische Hochschule Zürich, 1963
63Sa W. Schneider and A. von Zelewsky, *Helv. Chim. Acta*, 1963, **46**, 1848
63Wa R. D. Willett, C. Dwiggins, jun., R. F. Kruh, and R. E. Rundle, *J. Chem. Phys.*, 1963, **38**, 2429
64A A. Angoso, H. Onken, and H. Hahn, *Z. anorg. Chem.*, 1964, **328**, 223
64Aa S. Ahrland and L. Johansson, *Acta Chem. Scand.*, 1964, **18**, 2125
64Ab J. F. Allen and H. M. Neumann, *Inorg. Chem.*, 1964, **3**, 1612

continued overleaf

72 Br^- (contd)

64B W. Brandel and A. Swinarski, in 'Theory and structure of complex compounds' (symposium papers, Wrocław 1962), Pergamon, Warszawa, 1964, p. 497
64Ba J. Braunstein and A. S. Minano, *Inorg. Chem.*, 1964, **3**, 218
64Bb M. Blander and E. B. Luchsinger, *J. Amer. Chem. Soc.*, 1964, **86**, 319
64Bc A. Basiński and B. Lenarcik, *Roczniki Chem.*, 1964, **38**, 1035, 1725
64Bd I. D. Brown, *Canad. J. Chem.*, 1964, **42**, 2758
64C R. Cigén and N. G. Mannerstrand, *Acta Chem. Scand.*, 1964, **18**, 1755
64Ca R. Cigén and N. G. Mannerstrand, *Acta Chem. Scand.*, 1964, **18**, 2203
64Cb J. J. Christenson, R. M. Izatt, L. D. Hansen, and J. D. Hale, *Inorg. Chem.*, 1964, **3**, 130
64Cc F. A. Cotton and S. J. Lippard, *J. Amer. Chem. Soc.*, 1964, **86**, 4497
64Cd M. Chabanel and P. Bonnet, *J. Chim. phys.*, 1964, **61**, 1179
64Ce I. I. Chernyaev, N. N. Zheligovskaya, and T. N. Leonova, *Zhur. neorg. Khim.*, 1964, **9**, 347
64D A. J. Downs, *J. Inorg. Nuclear Chem.*, 1964, **26**, 41
64E E. Eyal and A. Treinin, *J. Amer. Chem. Soc.*, 1964, **86**, 4287
64G M. L. Gavrish and I. S. Galinker, *Zhur. neorg. Khim.*, 1964, **9**, 1289
64Ga A. A. Grinberg and A. I. Dobroborskaya, *Zhur. neorg. Khim.*, 1964, 9, 2026
64Gb K. F. Guenther, *Inorg. Chem.*, 1964, **3**, 1788
64I H. Irving and P. D. Wilson, *J. Inorg. Nuclear Chem.*, 1964, **26**, 2235
64J G. J. Janz and I. Ahmad, *Electrochim. Acta*, 1964, **9**, 1539
64K F. Ya. Kul'ba, V. E. Mironov, and G. N. Kolyushenkova, *Zhur. neorg. Khim.*, 1964, **9**, 1638
64L I. Leden and T. Ryhl, *Acta Chem. Scand.*, 1964, **18**, 1196
64La K. H. Lieser, *J. Inorg. Nuclear Chem.*, 1964, **26**, 1571
64Lb H.-M. Lüschow, dissertation, University of Saarbrücken, 1964
64M V. E. Mironov, F. Ya. Kul'ba, A. V. Fokina, V. S. Golubeva, and V. A. Nazarov, *Zhur. neorg. Khim.*, 1964, **9**, 2133
64Ma D. F. C. Morris, G. L. Reed, and K. J. Sutton, *J. Inorg. Nuclear Chem.*, 1964, **26**, 1461
64P N. K. Popovicheva, A. A. Biryukov, and V. I. Shlenskaya, *Zhur. neorg. Khim.*, 1964, **9**, 1482
64R B. H. Robinson and J. E. Fergusson, *J. Chem. Soc.*, 1964, 5683; J. E. Fergusson and B. H. Robinson, *Proc. Chem. Soc.*, 1964, 189.
64S V. B. Spivakovskii and L. P. Moisa, *Zhur. neorg. Khim.*, 1964, 9, 2287
64Sa S. A. Shchukarev, O. A. Lobaneva, M. A. Ivanova, and M. A. Kononova, *Zhur. neorg. Khim.*, 1964, **9**, 2791
64Sb E. L. Short, D. N. Waters, and D. F. C. Morris, *J. Inorg. Nuclear Chem.*, 1964, **26**, 902
64Sc M. Shiloh and Y. Marcus, IA-924 (Israel AEC), 1964
64Sd G. Stucky and R. E. Rundle, *J. Amer. Chem. Soc.*, 1964, **86**, 4821
64Se V. I. Shlenskaya and A. A. Bryukov, *Vestnik Moskov. Univ.*, 1964, Ser. Khim. No. 3, 65
64T G. Thiele, dissertation, Technische Hochschule Aachen, 1964
64Ta T. V. Talalaeva, A. N. Rodionov, and K. A. Kocheshkov, *Doklady Akad. Nauk S.S.S.R.*, 1964, **154**, 174
64W H. Wendt, *Ber. Bunsengesellschaft Phys. Chem.*, 1964, **68**, 29
64Wa W. M. Weaver and J. D. Hutchison, *J. Amer. Chem. Soc.*, 1964, **86**, 261
64Z M. Zangen and Y. Marcus, *Israel J. Chem.*, 1964, **2**, 49
65A R. Arnek, *Arkiv Kemi*, 1965, **24**, 531
65B R. Barbieri and J. Bjerrum, *Acta Chem. Scand.*, 1965, **19**, 469
65Ba B. J. Brisdon and R. A. Walton, *J. Chem. Soc.*, 1965, 2274
65Bb I. R. Bellobono and G. Favini, *Ann. Chim. (Italy)*, 1965, **55**, 32
65Bc G. G. Bombi, M. Fiorani, and G. A. Mazzocchin, *J. Electroanalyt. Chem.*, 1965, **9**, 457
65C F. A. Cotton and S. J. Lippard, *Inorg. Chem.*, 1965, **4**, 59; *Chem. Comm.*, 1965, 245.
65Ca F. A. Cotton, N. F. Curtis, B. F. G. Johnson, and W. R. Robinson, *Inorg. Chem.*, 1965, **4**, 326
65Cb F. A. Cotton, B. F. G. Johnson, and R. M. Wing, *Inorg. Chem.*, 1965, **4**, 502
65Cc F. A. Cotton and S. J. Lippard, *Inorg. Chem.*, 1965, **4**, 1621
65Cd M. Chabanel, *J. Chim. phys.*, 1965, **62**, 678
65D C. M. Davidson and R. F. Jameson, *Trans. Faraday Soc.*, 1965, **61**, 2462
65E J. F. Endicott and H. Taube, *Inorg. Chem.*, 1965, **4**, 437
65Ea M. Elder and B. R. Penfold, *Nature*, 1965, **205**, 276
65F A. B. Fasman, G. G. Kutyukov, and D. V. Sokol'skii, *Zhur. neorg. Khim.*, 1965, **10**, 1338
65Fa D. A. Fine, *Inorg. Chem.*, 1965, **4**, 345
65Fb H. N. Farrer, M. M. McGrady, and R. S. Tobias, *J. Amer. Chem. Soc.*, 1965, **87**, 5019
65Fc J. R. Fryer and D. F. C. Morris, *Electrochim. Acta*, 1965, **10**, 473
65G A. M. Golub and G. V. Pomerants, *Ukrain. khim. Zhur.*, 1965, **31**, 104
65H H. E. Hellwege and G. K. Schweitzer, *J. Inorg. Nuclear Chem.*, 1965, **27**, 99
65I D. Inman, *Electrochim. Acta*, 1965, **10**, 11
65J E. Jósefowicz and H. Ladzińska-Kulińska, *Roczniki Chem.*, 1965, **39**, 1175
65K R. Korewa and Z. Szponar, *Roczniki Chem.*, 1965, **39**, 349
65L B. J. Levien, *Austral. J. Chem.*, 1965, **18**, 1161
65M V. E. Mironov and N. P. Lastovkina, *Zhur. neorg. Khim.*, 1965, **10**, 1082
65Ma T. V. Mal'kova, G. A. Shutova, and K. B. Yatsimirskii, *Zhur. neorg. Khim.*, 1965, **10**, 2611
65Mb V. E. Mironov, Yu. I. Rutkovskii, and E. I. Ignatenko, *Zhur. neorg. Khim.*, 1965, **10**, 2639
65Mc D. F. C. Morris and M. W. Jones, *J. Inorg. Nuclear Chem.*, 1965, **27**, 2454
65Md R. E. McCarley, B. G. Hughes, F. A. Cotton, and R. Zimmerman, *Inorg. Chem.*, 1965, **4**, 1491

continu

65Me L. M. Mukherjee, S. Bruckenstein, and F. A. K. Badawi, *J. Phys. Chem.*, 1965, **69**, 2537
65S G. K. Schweitzer and S. W. McCarty, *J. Inorg. Nuclear Chem.*, 1965, **27**, 191
65Sa G. Schwarzenbach and M. Schellenberg, *Helv. Chim. Acta*, 1965, **48**, 28
65Sb M. Shiloh and Y. Marcus, *Israel J. Chem.*, 1965, **3**, 123
65Sc K. Schwochau, *Z. Naturforsch.*, 1965, **20a**, 1286
65W D. L. Wertz, R. M. Lawrence, and R. F. Kruh, *J. Chem. Phys.*, 1965, **43**, 2163
65Z M. Zangen, *J. Phys. Chem.*, 1965, **69**, 1835
66B J. Braunstein and A. S. Minano, *Inorg. Chem.*, 1966, **5**, 942
66Ba D. Brown, *J. Chem. Soc.* (*A*), 1966, 766
66Bb P. M. Boorman and B. P. Straughan, *J. Chem. Soc.* (*A*), 1966, 1514
66Bc H. Braunstein, J. Braunstein, and D. Inman, *J. Phys. Chem.*, 1966, **70**, 2726
66Bd J. Braunstein, A. R. Alvarez-Funes, and H. Braunstein, *J. Phys. Chem.*, 1966, **70**, 2734
66Be B. Brehler and H. Follner, *Naturwiss.*, 1966, **53**, 177
66C F. A. Cotton and S. J. Lippard, *Inorg. Chem.*, 1966, **5**, 9
66Ca J. P. Clark and C. J. Wilkins, *J. Chem. Soc.* (*A*), 1966, 871
66Cb R. J. H. Clark, R. S. Nyholm, and D. E. Scaife, *J. Chem. Soc.* (*A*), 1966, 1296
66Cc G. B. Carpenter, *Acta Cryst.*, 1966, **20**, 330
66Cd R. Colton and I. B. Tomkins, *Austral. J. Chem.*, 1966, **19**, 759
66Ce H. Chateau, M.-C. Gadet, and J. Pouradier, *J. Chim. phys.*, 1966, **63**, 269
66D J. P. Day and L. M. Venanzi, *J. Chem. Soc.* (*A*), 1966, 197
66Da P. R. Danesi, F. Orlandini, and G. Scibona, *J. Inorg. Nuclear Chem.*, 1966, **28**, 1047
66E M. Elder and B. R. Penfold, *Inorg. Chem.*, 1966, **5**, 1763
66F D. A. Fine, *Inorg. Chem.*, 1966, **5**, 197
66Fa G. W. A. Fowles and J. L. Frost, *J. Chem. Soc.* (*A*), 1966, 1631
66Fb W. R. Fitzgerald and D. W. Watts, *Austral. J. Chem.*, 1966, **19**, 935
66Fc W. R. Fitzgerald and D. W. Watts, *Austral. J. Chem.*, 1966, **19**, 1411
66Fd M. Fiorani, G. G. Bombi, and G.-A. Mazzocchin, *Ricerca sci.*, 1966, **36**, 580
66G N. N. Greenwood, B. P. Straughan, and A. E. Wilson, *J. Chem. Soc.* (*A*), 1966, 1479
66Ga P. Gerding, *Acta Chem. Scand.*, 1966, **20**, 79
66H D. J. Hewkin and W. P. Griffith, *J. Chem. Soc.* (*A*), 1966, 472
66K M. B. Kennedy and M. W. Lister, *Canad. J. Chem.*, 1966, **44**, 1709
66L D. C. Luehrs, R. T. Iwamoto, and J. Kleinberg, *Inorg. Chem.*, 1966, **5**, 201
66La S. L. Lawton and R. A. Jacobson, *Inorg. Chem.*, 1966, **5**, 743; *J. Amer. Chem. Soc.*, 1966, **88**, 616
66Lb S. Lormeau and M.-H. Mannebach, *Bull. Soc. chim. France*, 1966, 2576
66Lc B. Lenarcik and A. Basiński, *Roczniki Chem.*, 1966, **40**, 165
66Ld I. R. Lantzke and D. W. Watts, *Austral. J. Chem.*, 1966, **19**, 969
66Le V. I. Levin and M. D. Kozlova, *Radiokhimiya*, 1966, **8**, 533
66M W. A. Millen and D. W. Watts, *Austral. J. Chem.*, 1966, **19**, 43
66Ma W. A. Millen and D. W. Watts, *Austral. J. Chem.*, 1966, **19**, 51
66Mb T. Mussini and G. Fàita, *Ricerca sci.*, 1966, **36**, 175
66Mc V. E. Mironov and N. P. Lastovkina, *Zhur. neorg. Khim.*, 1966, **11**, 580
66Md T. V. Mal'kova, G. A. Shutova, and K. B. Yatsimirskii, *Zhur. neorg. Khim.*, 1966, **11**, 1556
66N L. Nilsson and G. P. Haight, jun., *Acta Chem. Scand.*, 1966, **20**, 486
66P J. R. Preer and G. P. Haight, jun., *Inorg. Chem.*, 1966, **5**, 656
66R J. L. Ryan and C. K. Jørgensen, *J. Phys. Chem.*, 1966, **70**, 2845
66Ra R. Ripan and M. Marc, *Rev. Roumaine Chim.*, 1966, **11**, 1063
66S A. Sabatini and I. Bertini, *Inorg. Chem.*, 1966, **5**, 204
66Sa M. Shiloh and Y. Marcus, *J. Inorg. Nuclear Chem.*, 1966, **28**, 2725
66Sb H. Schäfer and B. Spreckelmeyer, *J. Less-Common Metals*, 1966, **11**, 73
66Sc H. Schäfer and R. Siepmann, *J. Less-Common Metals*, 1966, **11**, 76
66Sd D. Singh and A. Mishra, *Indian J. Chem.*, 1966, **4**, 308
66Se V. I. Shlenskaya and A. A. Biryukov, *Zhur. neorg. Khim.*, 1966, **11**, 54
66Sf B. D. Stepin and G. M. Serebrennikova, *Zhur. neorg. Khim.*, 1966, **11**, 1807
66Sg K. Shaw, Ph.D. thesis, London University, quoted by H. L. Bott, A J. Poë, and K. Shaw, *Chem. Comm.*, 1968, 793.
66W E. Wendling, *Bull. Soc. chim. France*, 1966, 413
67A P. Alcais, F. Rothenberg, and J.-É. Dubois, *J. Chim. phys.*, 1967, **64**, 1818
67Aa R. Alexander, E. C. F. Ko, Y. C. Mac, and A. J. Parker, *J. Amer. Chem., Soc.*, 1967, **89**, 3703
67Ab S. N. Andreev, V. G. Khaldin, M. V. Andreeva, and M. F. Smirnova, *Zhur. neorg. Khim.*, 1967, **12**, 1791
67B F. Bonati and F. A. Cotton, *Inorg. Chem.*, 1967, **6**, 1353
67Ba K. Burger and B. Pintér, *J. Inorg. Nuclear Chem.*, 1967, **29**, 1717
67Bb J. Braunstein, *J. Phys. Chem.*, 1967, **71**, 3402
67Bc D. J. Bearcroft and N. H. Nachtrieb, *J. Phys. Chem.*, 1967, **71**, 4400
67Bd Yu. I. Belyaev, V. M. Vdovenko, A. I. Skoblo, and D. N. Suglobov, *Radiokhimiya*, 1967, **9**, 720
67Be K. P. Butin, I. P. Beletskaya, A. N. Ryabtsev, and O. A. Reutov, *Elektrokhimiya*, 1967, **3**, 1318
67Bf D. Brown and P. J. Jones, *J. Chem. Soc.* (*A*), 1967, 243
67Bg D. Brown and P. J. Jones, *J. Chem. Soc.* (*A*), 1967, 247
67Bh G. A. Boos and A. A. Popel', *Zhur. neorg. Khim.*, 1967, **12**, 2086

continued overleaf

72 Br⁻ (contd)

67C D. A. Couch, P. S. Elmes, J. E. Fergusson, M. L. Greenfield, and C. J. Wilkins, *J. Chem. Soc. (A)*, 1967, 1813
67E R. P. Eswein, E. S. Howald, R. A. Howald, and D. P. Keeton, *J. Inorg. Nuclear Chem.*, 1967, **29**, 437
67F P. B. Fleming, L. A. Mueller, and R. E. McCarley, *Inorg. Chem.*, 1967, **6**, 1
67Fa M. Fiorani, G. G. Bombi, and G.-A. Mazzocchin, *J. Electroanalyt. Chem.*, 1967, **13**, 167
67Fb G. W. A. Fowles and B. J. Russ, *J. Chem. Soc. (A)*, 1967, 517
67G V. Gutmann and K. Fenkart, *Monatsh.*, 1967, **98**, 1
67Ga J. Guion, *Inorg. Chem.*, 1967, **6**, 1882
67H J. Hála and D. Pohanková, *J. Inorg. Nuclear Chem.*, 1967, **29**, 2983
67Ha G. S. Harris, F. Inglis, J. McKechnie, K. K. Cheung, and G. Ferguson, *Chem. Comm.*, 1967, 442
67I R. M. Izatt, G. D. Watt, D. Eatough, and J. J. Christensen, *J, Chem. Soc. (A)*, 1967, 1304
67J V. Jedináková and J. Čeleda, *Coll. Czech. Chem. Comm.*, 1967, **32**, 1679
67K V. I. Kazakova and B. V. Ptitsyn, *Zhur. neorg. Khim.*, 1967, **12**, 620
67L H. Loman and E. van Dalen, *J. Inorg. Nuclear Chem.*, 1967, **29**, 699
67La H.-D. Lüdemann and E. U. Franck, *Ber. Bunsengesellschaft Phys. Chem.*, 1967, **71**, 455
67Lb P. C. Lammers and J. Braunstein, *J. Phys. Chem.*, 1967, **71**, 2626
67M W. L. Masterton, T. I. Munnelly, and L. H. Berka, *J. Phys. Chem.*, 1967, **71**, 942
67Ma W. L. Masterton, *J. Phys. Chem.*, 1967, **71**, 2885
67Mb J. Masłowska, *Roczniki Chem.*, 1967, **41**, 1857
67Mc D. F. C. Morris and B. D. Andrews, *Electrochim. Acta*, 1967, **12**, 41
67Md V. E. Mironov and N. P. Lastovkina, *Zhur. fiz. Khim.*, 1967, **41**, 1850
67Me D. S. Martin, jun., and E. L. Bahn, *Inorg. Chem.*, 1967, **6**, 1653
67Mf V. E. Mironov and A. V. Fokina, *Zhur. neorg. Khim.*, 1967, **12**, 2571
67Mg W. A. Millen and D. W. Watts, *J. Amer. Chem. Soc.*, 1967, **89**, 6858
67N N. M. Nikolaeva and E. D. Pastukhova, *Zhur. neorg. Khim.*, 1967, **12**, 1227
67Na N. M. Nikolaeva and E. D. Pastukhova, *Zhur. neorg. Khim.*, 1967, **12**, 1514
67O R. P. Oertel and R. A. Plane, *Inorg. Chem.*, 1967, **6**, 1960
67P D. G. Peters and R. L. Caldwell, *Inorg. Chem.*, 1967, **6**, 1478
67Pa G. Pistoia, A. M. Polcaro, and S. Schiavo, *Ricerca sci.*, 1967, **37**, 227
67R N. A. Rumbaut and H. L. Peeters, *Bull. Soc. chim. belges*, 1967, **76**, 33
67S D. E. Scaife and K. P. Wood, *Inorg. Chem.*, 1967, **6**, 358
67Sa J. G. Scane, *Acta Cryst.*, 1967, **23**, 85
67Sb A. Swinarski and A. Grodzicki, *Roczniki Chem.*, 1967, **41**, 1205
67Sc C. Sinistri and E. Pezzati, *Gazzetta*, 1967, **97**, 1116
67Sd D. Singh and A. Mishra, *Bull. Chem. Soc. Japan*, 1967, **40**, 2801
67Se G. G. Shitareva and V. A. Nazarenko, *Zhur. neorg. Khim.*, 1967, **12**, 999
67T J. E. Teggins, D. R. Gano, M. A. Tucker, and D. S. Martin, jun., *Inorg. Chem.*, 1967, **6**, 69
67Ta J. E. Teggins and D. S. Martin, jun., *Inorg. Chem.*, 1967, **6**, 1003
67Tb N. Tanaka, Y. Kobayashi, and M. Kamada, *Bull. Chem. Soc. Japan*, 1967, **40**, 2839
67W H. E. Wirth, *J. Phys. Chem.*, 1967, **71**, 2922
67V V. P. Vasil'ev and G. A. Lobanov, *Zhur. neorg. Khim.*, 1967, **12**, 878
67Y Yu. B. Yakovlev, F. Ya. Kul'ba, and V. E. Mironov, *Zhur. neorg. Khim.*, 1967, **12**, 3283
68B D. Brown, T. J. Petcher, and A. J. Smith, *Nature*, 1968, **217**, 737
68Ba H. Bloom and J. W. Hastie, *Austral. J. Chem.*, 1968, **21**, 583
68C J. A. Creighton and J. H. S. Green, *J. Chem. Soc. (A)*, 1968, 808
68D V. I. Dubinskii, V. M. Shul'man, and B. I. Peshchevitskii, *Zhur. neorg. Khim.*, 1968, **13**, 54
68Da J. E. D. Davies and D. A. Long, *J. Chem. Soc. (A)*, 1968, 2050
68Db M. Della Monica, U. Lamanna, and L. Senatore, *Inorg. Chim. Acta*, 1968, **2**, 363
68F J. R. Fryer and D. F. C. Morris, *Talanta*, 1968, **15**, 1309
68Fa V. A. Fedorov, N. P. Samsonova, and V. E. Mironov, *Zhur. neorg. Khim.*, 1968, **13**, 382
68Fb W. R. Fitzgerald and D. W. Watts, *Austral. J. Chem.*, 1968, **21**, 595
68G H. C. Gaur and R. S. Sethi, *Electrochim. Acta*, 1968, **13**, 1737
68Ga E. A. Gyunner and N. D. Yakhkind, *Zhur. neorg. Khim.*, 1968, **13**, 245
68Gb V. A. Golodov, A. B. Fasman, V. F. Vozdvizhenskii, Yu. A. Kushnikov, and V. V. Roganov, *Zhur. neorg. Khim.*, 1968, **13**, 3306
68Gc I. J. Gal, *Inorg. Chem.*, 1968, **7**, 1611
68Gd N. N. Greenwood, B. P. Straughan, and A. E. Wilson, *J. Chem. Soc. (A)*, 1968, 2209
68Ge D. R. Gano, G. F. Vandegrift, and D. S. Martin, jun., *Inorg. Chim. Acta*, 1968, **2**, 219
68H J. O. Hill, I. G. Worsley, and L. G. Hepler, *J. Phys. Chem.*, 1968, **72**, 3695
68Ha P. J. Hendra and Z. Jović, *J. Chem. Soc. (A)*, 1968, 600
68Hb S. R. C. Hughes and D. H. Price, *J. Chem. Soc. (A)*, 1968, 1404
68Hc K.-L. Hsia and R. M. Fuoss, *J. Amer. Chem. Soc.*, 1968, **90**, 3055
68Hd J. L. Hayden and R. A. D. Wentworth, *J. Amer. Chem. Soc.*, 1968, **90**, 5291
68I D. Inman, D. G. Lovering, and R. Narayan, *Trans. Faraday Soc.*, 1968, **64**, 2476
68J B. Jeżowska-Trzebiatowska and M. Rudolf, *Roczniki Chem.*, 1968, **42**, 1221
68K S. Katayama and R. Tamamushi, *Bull. Chem. Soc. Japan*, 1968, **41**, 606
68Ka O. I. Khotsyanovskii and V. Sh. Telyakova, *Ukrain. khim. Zhur.*, 1968, **34**, 1126
68Kb N. Katsaros and J. W. George, *Chem. Comm.*, 1968, 662

contin

68L H.-D. Lüdemann and E. U. Franck, *Ber. Bunsengesellschaft Phys. Chem.*, 1968, **72**, 514
68La D. C. Luehrs and K. Abate, *J. Inorg. Nuclear Chem.*, 1968, **30**, 549
68M W. G. Movius and N. A. Matwiyoff, *J. Phys. Chem.*, 1968, **72**, 3063
68Ma W. G. McPherson and E. A. Meyers, *J. Phys. Chem.*, 1968, **72**, 3117
68Mb L. M. Mukherjee, J. J. Kelly, W. Baranetzky, and J. Sica, *J. Phys. Chem.*, 1968, **72**, 3410
68Mc L. Magon, R. Portanova, A. Cassol, and G. Rizzardi, *Ricerca sci.*, 1968, **38**, 782
68Md R. H. McCorkell, M. M. Sein, and J. W. Irvine, jun., *J. Inorg. Nuclear Chem.*, 1968, **30**, 1155
68Me V. E. Mironov, Yu. A. Makashev, I. Ya. Mavrina, and D. M. Markhaeva, *Zhur. fiz. Khim.*, 1968, **42**, 2987
68Mf J. Méndez, I. J. Gal, and J. W. Irvine, jun., *Inorg. Chem.*, 1968, **7**, 1329
68Mg F. Monacelli, *Inorg. Chim. Acta*, 1968, **2**, 263
68N T. Nomura, *Bull. Chem. Soc. Japan*, 1968, **41**, 2803
68P N. Payen-Baldy and M. Machtinger, *Bull. Soc. chim., France*, 1968, 2221
68Q A. S. Quist and W. L. Marshall, *J. Phys. Chem.*, 1968, **72**, 1545
68Qa A. S. Quist and W. L. Marshall, *J. Phys. Chem.*, 1968, **72**, 2100
68S G. C. Stocco, E. Rivarola, R. Romeo, and R. Barbieri, *J. Inorg. Nuclear Chem.*, 1968, **30**, 2409
68Sa C. J. Shahani, K. A. Mathew, C. L. Rao, and M. V. Ramaniah, *Radiochim. Acta*, 1968, **10**, 165
68Sb L. W. Schroeder and J. A. Ibers, *Inorg. Chem.*, 1968, **7**, 594
68Sc R. Saillant and R. A. D. Wentworth, *Inorg. Chem.*, 1968, **7**, 1606
68Sd B. Spreckelmeyer, *Z. anorg. Chem.*, 1968, **358**, 147
68V V. P. Vasil'ev, *Zhur. fiz. Khim.*, 1968, **42**, 1830
68W C. F. Wells and D. Mays, *J. Chem. Soc. (A)*, 1968, 2740

Hyprobromite, BrO^- 73

Metal	*Method*	*Temp*	*Medium*	*Log of equilibrium constant, remarks*	*Ref*
H^+	sp	10—50	0 corr	K_1 8·91(10°), 8·66(25°), 8·49(35°), 8·23(50°)	64F

64F I. E. Flis, K. P. Mishchenko, and G. I. Pusenok, *Izvest. V.U.Z., Khim. i khim. Tekhnol.*, 1964, **7**, 764

Bromate, BrO_3^- 74

Metal	*Method*	*Temp*	*Medium*	*Log of equilibrium constant, remarks*	*Ref*
H^+	kin	230—260	MNO_3(l)	$K(BrO_3^- + Cr_2O_7^{2-} \rightleftharpoons BrO_2^+ + 2CrO_4^{2-})$ − 8·0(230°), − 7·3(260°) in $K_{0.5}Na_{0.5}NO_3$(l), *m* units	63D
	kin	260	MNO_3(l)	ev BrO_2^+, $K(BrO_3^- + Cr_2O_7^{2-} \rightleftharpoons BrO_2^+ + 2CrO_4^{2-})$ − 7·46 in (Na, K)NO_3(l, eut), M units	65S
Group 1A$^+$	act	25?	0 corr	K_1 − 0·1(Na^+), 0·1(K^+)	/66M
Cu^{2+}	gl, sol	25	→ 0	$K_s(Cu^{2+}OH^-{}_{1.5}L^-{}_{0.5})$ − 16·53	63L
			1 Na(ClO_4)	K_s − 16·13, $\beta_2 \leqslant$ 0·31	
	spJ		Me_2CO	ev Cu_2L^{3+} in Me_2CO(l) (?, see composition scale of Fig. 1)	66Ma
Ag^+	cal	25	dil	$\Delta H_{s0}[AgL(s) \rightleftharpoons Ag^+ + L^-]$ = 19·3	67S
Hg_2^{2+}	*X*		solid	ev$(O_2BrOHg-)_2$ units in $Hg_2(BrO_3)_2$(s)	67D
Tl^+	sol	30—45	→ 0	K_{s0} − 3·78(30°), − 3·62(35°), —3·47(40°), − 3·34(45°)	68K
				K_1 0·3(30—45°)	

63D F. R. Duke and J. Schlegel, *J. Phys. Chem.*, 1963, **67**, 2487
63L P. Lumme and H. Lumme, *Suomen Kem.*, 1963, **36**, *B*, 192
65S J. Schlegel, *J. Phys. Chem.*, 1965, **69**, 3638
66M W. L. Masterton and L. H. Berka, *J. Phys. Chem.*, 1966, **70**, 1924
66Ma L. Macášková and J. Gažo, *Z. Chem.*, 1966, **6**, 430
67D E. Dorm, *Acta Chem. Scand.*, 1967, **21**, 2834
67S A. A. Shidlovskii, A. A. Voskresenskii, and E. S. Shitikov, *Zhur. fiz. Khim.*, 1967, **41**, 731
68K K. H. Khoo, *J. Inorg. Nuclear Chem.*, 1968, **30**, 2425

75 Iodide, I^-

Metal	Method	Temp	Medium	Log of equilibrium constant, remarks	Ref
H^+	H	25	$C_2H_4(NH_2)_2$	K_1 2·97 in $C_2H_4(NH_2)_2$(l)	62Ba/65[illegible]
	var		solid	ev HI_2^- in Bu_4NHI_2(s) (pre, *p*HI, ir)	63M[illegible]
	*p*T	37	C_5H_5N	K_1 3·11 in T = C_5H_5N(l)	68M[illegible]
Group $1A^+$	con	25	$C_2H_4(NH_2)_2$	K_1 2·97(Li^+), 3·00(Na^+), 2·98(K^+), 3·24(Cs^+) in $C_2H_4(NH_2)_2$(l), → 0	65Bb
	con	25	org	no ev cpx in *N*-methyl-2-pyrrolidone C_5H_9NO(l) (Na^+, K^+)	67D
	dis	25?	0 corr	$K_D[M^+ + I^- \rightleftharpoons M^+(org) + I^-(org)]$, org = TBP = $(BuO)_3PO$: −0·54(Li^+), −1·32(Na^+), −1·64(K^+), −1·70(Cs^+); org = DAMF = $(Am^iO)_2MePO$: −1·03(Li^+), −1·52(Na^+), −1·57(K^+), −1·48(Cs^+); also values for 0·5M-TBP, DAMF, or TOPO = $(C_8H_{17})_3PO$ in C_6H_6 and estimated solvation numbers	67Ra
	con	25	org	no ev cpx in $NC(CH_2)_4CN$(l) (Na^+, K^+)	67S
Li^+	kin	0	Me_2NCHO	K_1 −0·26 in DMF = Me_2NCHO(l)	64W
	con	25	*w*% dioxan	K_1 −0·82(*w* = 40), 0·40(*w* = 60), 1·38(*w* = 70), 2·28(*w* = 80), 2·82(*w* = 87), 3·85(*w* = 91), 4·50(*w* = 95) in *w*% dioxan	66A
Na^+	con	25	var	no ev cpx in H_2O, Me_2SO, Me_2NCOMe, Me_2NCHO	/67J
		25	org	K_1 0·91 in MeCN, 0·97 in $HOC_2H_4NH_2$, 1·88 in PhCN, 1·90 in Me_2CO, 2·03 in PrOH, 2·57 in EtCOMe, 2·08 in PhCOMe, 2·75 in $C_2H_4(NH_2)_2$, 3·31 in C_5H_5N	
	con	25	MeOH	K_1 1·28 in MeOH(l), → 0; also K_1 values for mixt dioxan–MeOH	68S
	con	25?	EtOH	K_1 2·39 in EtOH(l), → 0; also K_1 values for mixt hexane–EtOH	68Sa
K^+	con	25	org	K_1 3·32 in $C_2H_4(NH_2)_2$(l), 3·82 in $C_3H_6(NH_2)_2$(l), → 0	64Fa
	con	0—35	MeCN	K_1 −2·02(0°), −1·05(25°), −1·29(35°), in MeCN(l), → 0	64J
	sol	25	org	K_{s0} 0·5 in Me_2NCHO	67A
	con	140	I_2(l)	K_1 2·07, $K_{12}(K^+ + KI \rightleftharpoons K_2I^+)$ 4·4 in I_2(l)	67Ba
	con	25	var	no ev cpx in H_2O, Me_2SO, Me_2NCOMe, Me_2NCHO	/67J
		25	org	K_1 1·00 in $HOC_2H_4NH_2$, 1·09 in MeCN, 2·02 in PhCN, 1·98 in Me_2CO, 2·41 in PrOH, 2·62 in EtCOMe, 1·83 in PhCOMe, 3·05 in $C_2H_4(NH_2)_2$, 3·56 in C_5H_5N	
	con	5—55	→ 0	K_1 −0·20(5°), −0·19(15, 25°), −0·15(35°), −0·14(45, 55°)	68Ab
	X	23	H_2NCHO	ev ion pairs K^+I^- in H_2NCHO(l)	68D
Rb^+	con	25	→ 0	K_1 0·04 (also K_1 for dioxan–water mixt)	64F
	fp	≈ 18	Me_2SO	K_1 0·64 in Me_2SO(l), *m* units	67C
Cs^+	sol	25	org	K_{s0} −0·23 in $HCONH_2$, −1·7 in Me_2NCHO	67A
	con	25	→ 0	K_1 −0·03	68Ha
Ca^{2+}	dis		NaI var	$K_D[Ca^{2+} + 2I^- \rightleftharpoons CaI_2(\text{in TBP})]$ 0·39	68La
Sr^{2+}	dis		NaI var	$K_D[Sr^{2+} + 2I^- \rightleftharpoons SrI_2(\text{in TBP})]$ 0·12	68La
Ba^{2+}	dis		NaI var	$K_D[Ba^{2+} + 2I^- \rightleftharpoons BaI_2(\text{in TBP})]$ −0·08	68La
Hf^{4+}	dis	20	3 $H(ClO_4)$	K_1 −0·46	67H

contin[illegible]

Metal	*Method*	*Temp*	*Medium*	*Log of equilibrium constant, remarks*	*Ref*
Th^{4+}	var		solid	ev ThI_6^{2-} in solids (mag, pre)	65Ba
$Nb^{+,2+}$	*X*		solid	ev $Nb_6I_8^{3+}$ in Nb_6I_{11}(s)	66B
$Ta^{2+,3+}$	oth		solid	ev $Ta_6I_{12}^{2+}$ in Ta_6I_{14}(s) (pre, analogy)	64Sa
	pre, sp		var	ev $Ta_6I_{12}^{2+}$ in aq solution	65M+65K
Pa^{4+}	var		var	ev PaI_6^{2-} in solids and soln (pre, ir, *X*)	67Bc
Cr^{3+}	pre, sp		var	ev $CrI(H_2O)_5^{2+}$ in solids and soln	65Ma
	sp	15—45	4·2 (KI), 0·26 H^+	K_1 −4·35(15°), −4·16(25°), −3·98(35°), −3·80(45°), ΔH_1 = 7·6, ΔS_1 = 6·6(25°)	68Sd
		25	*I*(KI)	K_1 −4·00(*I* = 5·6), −4·16(*I* = 4·2), −4·42(*I* = 2·6), −4·60(*I* = 2·0), −5·0(*I* → 1·0)	
	X, pre		solid	ev $Cr_2I_9^{3-}$ in $Cs_3Cr_2I_9$(s)	68Sc
U^{4+}	var		solid	ev UI_6^{2-} in solids (mag, pre)	65Ba
Re^{3+}	*X*		solid	ev Re_3I_9 units in ReI_3(s)	68B
Re^{4+}	mag, pre		var	ev ReI_5OH^{2-}, not ReI_5^-	65Ca
Re^{5+}	pre		var	ev ReI_4O^- in solid and org	66C
Fe^{3+}	red	25	0·1 ($HClO_4$)	K_1 2·85, β_2 1·57	65N
Co^{2+} cpx	sp, gl		M′I var	Co(dimethylglyoxime)$_2^{2+}$: β_2 3·8(M′ = Li or Na), 4·0(M′ = K), 3·7(M′ = Rb); K_1 2·04, K_2 2·04(M′ = Cs); β_2 4·0(M′ = Mg)	67B
$Co(NH_3)_6^{3+}$	sp	25	0·07 ($NaClO_4$)	K_1 −0·15	67T
	con	25	→ 0	K_1 1·38	68K
Co^{3+} cpx	sp	30	org	*cis*-Co $en_2Cl_2^+$: K_1 2·93 in Me_2NCHO(l), 0 corr	66M
	pH_2O	37	→ 0	Co $en_2NH_3NO_2^{2+}$: K_1 1·27(*cis*), 1·15(*trans*)	67M
	pH_2O	37	→ 0	Co en_2NCSCl^+: K_1 0·54	67Ma
$Ru(NH_3)_5^{3+}$	sp	25	var	K_1 ~ 1·6	65E
Pd^{2+}	Pd	18—20	var	β_4 24·9{$K[Pd^{2+}+2e^- \rightleftharpoons Pd(s)]$ 33·4}	63Gb
	Pd	25	1 KI	β_4 24{$K[Pd^{2+}+2e^- \rightleftharpoons Pd(s)]$ 33·4} Also β_4 for 10—60°, assuming same *K* and same RTF^{-1} as for 25°!	65F
	dis		C_6H_6	ev mixed cpx, H_2PdI_4+2BuOH, or $2C_5H_{11}OH$, or 6 cyclohexanone, in C_6H_6(l)	65Ga
	sp, hyp	20	0·8 (0·6 H^+, ClO_4^-)	K_1 4·95(sp), K_2 4·27(hyp), K_3 3·60(hyp), K_4 2·92(sp), β_4 15·74(sp+hyp)	65Sb
Pt^{2+}	sp	25	1 ($NaClO_4$)	K_4 1·70	67Ca
		25	0·001	K_3 3·5	
	gl	25	var	†$K_n[Pt(OH)_{5-n}I_{n-1}^{2-}+H^++I^- \rightleftharpoons Pt(OH)_{4-n}I_n^{2-}+H_2O]$ 12(*n* = 1), 11·7(*n* = 2), 11(*n* = 3), 10(*n* = 4) (K_w −14·17)	67Nb
Pt^{4+}	var	25	dil	K_4 4·8, K_5 4·4, K_6 3·4, $K_a(PtI_5H_2O^-)$ −8·6; $K(PtI_6^{2-} \rightleftharpoons PtI_4^{2-}+I_2)$ 8·1, $K(PtI_4+I^- \rightleftharpoons PtI_3^-+I_2)$ 0·8 (sp, Ag, gl, kin)	67Cb
	ana, gl	25	0·1 KI	$K[PtOH_{7-n}I_{n-1}^{2-}+I^- \rightleftharpoons Pt(OH)_{6-n}I_n^{2-}+OH^-]$ −1·57(*n* = 1), −1·82(*n* = 2), −1·87(*n* = 3), −2·00(*n* = 4), −2·38(*n* = 5), −3·38(*n* = 6)	67Na
Pt en_3^{4+}	sp	10—40	0 corr	K_1^{out} 1·11(10°), 1·15(25°), 1·20(40°), ΔH_1 = 1·3, ΔS_1 = 9·6	63Ga

continued overleaf

75 I⁻ (contd)

Metal	Method	Temp	Medium	Log of equilibrium constant, remarks	Ref
Cu^+	sol	200–320	KI var	K_{s2} −1·17(200°), −0·68(260°), −0·32(300°), −0·13(320°)	64G
	sol	20	3·9 ($NaNO_3$)	K_{s2} −2·28, K_{s4} −2·52, β_2 8·68, (9·68 ?) β_4 8·44 (9·44 ?) (K_{s0} −11·96)	68Gd
Ag^+	sol	20	0 corr	(K_1 6·58), K_{12} 4·42, $\beta(AgI^0Ag^+{}_2)$ 7·04	60L/64L
	Ag	350	KNO_3(l)	K_1 4·08(350°) in KNO_3(l)	63B
	Ag	350—375	MNO_3(l)	K_1 4·20(350°), 4·11(375°) in $Na_{0.5}K_{0.5}NO_3$(l)	
	est	350	$NaNO_3$(l)	K_1 4·69 in $NaNO_3$(l), *x* units (mol/mol NO_3^-)	
	Ag	25	Me_2NCHO	K_{s0} −16·44, β_2 18·12, $\beta_{43}(Ag^+{}_3I^-{}_4)$ 51·55 in Me_2NCHO(l)	63Ca
	Ag	349	$NaNO_3$(l)	K_{s0} −10·36, ΔH_{s0} = 29·5 in $NaNO_3$(l), *x* units	64Bb
		360	KNO_3(l)	K_{s0} −9·22(359°), −9·24(361°) in KNO_3(l), *x* units	
		250	MNO_3(l)	K_{s0} −11·85, ΔH_{s0} = 27·9 in $Na_{0.5}K_{0.5}NO_3$(l), *x* units	
	con	25	org	K_1 4·95 in $C_2H_4(NH_2)_2$(l) (→ 0)	64Fa
	sol	300	NaI var	K_{s2} −0·70	64G
	Ag	150	MNO_3(l)	K_{s0} −13·0 in $Li_{0.43}K_{0.57}NO_3$(l, eut), *m* units	65Bc
	Ag	25	$C_2H_4(NH_2)_2$	K_{s0} −5·03 in $C_2H_4(NH_2)_2$(l)	65Mb
	Ag, *iE*	23	Me_2SO	K_{s0} −12·0, β_2 13·1 to 13·2 in Me_2SO(l), I = 0·1(Et_4NClO_4)	66L
		23	MeCN	K_{s0} −14·2, β_2 14·6 to 15·7 in MeCN(l), I = 0·1(Et_4NClO_4)	
		23	MeOH	K_{s0} −18·2, β_2 14·8 in MeOH(l), I = 1·0($LiClO_4$)	
		23	Me_2CO	K_{s0} −20·9, β_2 22·2 to 22·3 in Me_2CO(l), I = 0·1($LiClO_4$)	
		23	$EtNO_2$	K_{s0} −22·6, β_2 23·5 to 23·6 in $EtNO_2$(l), I = 0·1(Et_4NClO_4)	
	Ag	20	NH_3	K_{s0} −5·6 in $NH_4NO_3(NH_3)_{1.3}$(l)	66Tb
	Ag	25	org	K_{s0} −18·3 in MeOH, K_{s0} −14·5 in $HCONH_2$, K_{s0} −15·8, β_2 17·8 in Me_2NCHO, K_{s0} −14·7, β_2 17·3 in Me_2NCOMe K_{s0} −11·4, β_2 12·5 in Me_2SO(l)	67A
	con	140	I_2(l)	K_1 3·49 in I_2(l)	67Ba
	Ag	25	1	β_4 13·37	67Be
	Ag	222—239	$MClO_4$(l)	K_{s0} −13·21(495·9 K ?), −12·62(512·4 K)	67F
		250	$MClO_4$(l)	K_{s0} −12·3, ΔH_{s0} = 40 in $Li_{0.75}Na_{0.25}ClO_4$(l), *m* units	
	Ag	564–718	M_2SO_4(l)	K_1 3·23, K_2 2·93 in $Li_{1.43}K_{0.57}SO_4$(l, eut), *x* units	67Ga
	Ag	25—85	Me_2SO	K_{s0} −11·47(25°), $\beta_{32}(Ag^+{}_2I^-{}_3)$ 23·95(25°), 23·04(45°), 21·83(65°), 20·83(85°) in Me_2SO(l), 0·1M-NH_4NO_3	67R
	Ag	30	sulpholan	K_{s0} −18·48, $\beta_{43}(Ag^+{}_3I^-{}_4)$ 56·52 in sulpholan(l), I = 0·1(Et_4NClO_4)	68Da
	Ag	350	MNO_3(l)	K_{s0} −7·71, ΔH_{s0} = 32·83 in $K_{0.89}Ba_{0.11}(NO_3)_{1.11}$(l, eut)	68Gc
	sol, sp	24	org	β_{12} 10·04, β_{13} 10·61 in Me_2SO(l), I = 0·5(Et_4NClO_4); β_{12} 9·11, β_{13} 10·00 in MeCN(l), I = 1($LiClO_4$); β_{13} 18·96 in Me_2CO(l), I = 0·5($LiClO_4$);	68Lb
	Ag	428—476	MCl(l)	K_1 0·73, β_2 0·82(428°), K_1 0·63, β_2 0·75(450°), K_1 0·58, β_2 0·8(476°) in $Li_{0.59}K_{0.41}Cl$(l), *m* units	68W
Au^+	con		var	ev AuI_6^{5-} (?)	66T
Zn^{2+}	MHg	25	4 Li(ClO_4) 4 (Li)I	K_1 −0·47, K_2 −1·53, K_3 1·26, K_4 −0·51 $\beta(K^+ZnI_4^{2-})$ −0·52, $\beta(K^+ZnI_3^-)$ −1·1	67Me

contin

Metal	*Method*	*Temp*	*Medium*	*Log of equilibrium constant, remarks*	*Ref*
Cd^{2+}	cal	25	2 ($NaClO_4$), 0·1 H^+	K_1 2·11, β_2 2·72, β_3 4·67, β_4 5·08, $\Delta H_1 = -2{\cdot}32$, $\Delta H_2 = 3{\cdot}45$, $\Delta H_3 = -7{\cdot}65$, $\Delta H_4 = -1{\cdot}43$, $\Delta H_{\beta 4} = -7{\cdot}95$; $\Delta S_1 = 1{\cdot}9$, $\Delta S_2 = 14{\cdot}3$, $\Delta S_3 = -16{\cdot}7$, $\Delta S_4 = -3{\cdot}0$	64La = 64Bc
	MHg	25	$\to 0$	K_1 2·17, β_2 3·67, β_3 4·34, β_4 5·35, β_5 5·15; Also β_n values for $I = 0{\cdot}25$ to 4·5 with $LiNO_3$, $NaNO_3$, KNO_3, $Mg(NO_3)_2$, $NaClO_4$, and empirical equations log $\beta(I)$	64V
	oth	25	CdI_2 var	$K_1 \approx 2{\cdot}2$, $\beta_2 \approx 3{\cdot}5$ (diffusion method)	65G
	dis	30	1 ($NaClO_4$)	K_1 1·4, β_2 2·7, β_3 4·2	65H
	cal	25	3 ($NaClO_4$)	$\Delta H_1 = -2{\cdot}26$, $\Delta H_2 + \Delta H_3 = -0{\cdot}93$, $\Delta H_2 \approx -0{\cdot}2$ $\Delta H_4 = -3{\cdot}81$, $\Delta S_1 = 2{\cdot}0$, $\Delta S_2 = 2{\cdot}5$, $\Delta S_3 = 7{\cdot}3$ $\Delta S_4 = -5{\cdot}5$ (K_1 2·08, β_2 2·78, β_3 4·91, β_4 6·52)	66Ga
	pol	25	w% PrOH	$w = 25$; K_1 2·25, β_2 3·7, β_3 5·4, β_4 6·8; $w = 50$; K_1 2·9, β_2 4·8, β_3 6·7, β_4 8·3; $w = 75$; β_4 9·5 in w% PrOH, 2M-$Li(NO_3)$	67Kb
	MHg	25	4 $Li(ClO_4)$	K_1 2·40, K_2 1·72, K_3 1·90, K_4 2·18	67Me
			4 (Li)I	no ev cpx K^+–CdI_4^{2-}	
	cal	25	4 $Li(ClO_4)$	$\Delta H_1 = 2{\cdot}3$, $\Delta H_2 = -2{\cdot}2$, $\Delta H_3 = 7{\cdot}9$, $\Delta H_4 = 2{\cdot}8$; K_1 2·36, K_2 1·70, K_3 1·97, K_4 1·70; $\Delta S_1 = 3{\cdot}0$, $\Delta S_2 = 18$, $\Delta S_3 = -18$, $\Delta S_4 = -1{\cdot}6$	67Mf
	Ag	25	2 ($LiClO_4$)	K_1 1·74, β_2 2·94, β_3 4·30, β_4 5·00	67Pa
			w% MeOH	K_1 1·88, β_2 3·50, β_3 5·09, β_4 6·25($w = 10$); K_1 3·40, β_2 6·15, β_3 9·48, β_4 11·19($w = 90$); values also for w = 20, 40, 50, 60, 70, 80% MeOH, $I = 2(LiClO_4)$	
	pol	25	2 ($NaClO_4$)	K_1 1·90, β_2 3·00, β_3 4·51, β_4 5·92	67Sa
	cal	25	$\to 0$	$\Delta H_1 = -2{\cdot}34$	67V
			3	$\Delta H_1 = -2{\cdot}39(LiNO_3)$, $-2{\cdot}02[Mg(NO_3)_2]$, $-2{\cdot}11$ ($NaClO_4$), also for other I values	
	cal	25	3 ($NaClO_4$)	K_1 2·06, β_2 3·74, β_3 5·18, β_4 6·7	68G
	MHg	25	2 ($NaClO_4$)	K_1 1·97, β_2 2·6, β_3 4·71, β_4 6·04	
			1 ($NaClO_4$)	K_1 1·88, β_2 2·65, β_3 4·34, β_4 5·62	
			0·5 ($NaClO_4$)	K_1 1·89, β_2 2·63, β_3 4·28, β_4 5·49	
			0·25 ($NaClO_4$)	K_1 1·94, β_2 2·64, β_3 4·32, β_4 5·51	
	cal	25	3 ($NaClO_4$)	$\Delta H_1 = -2{\cdot}26$, $\Delta H_2 = -0{\cdot}20$, $\Delta H_3 = -0{\cdot}73$, $\Delta H_4 = -3{\cdot}81$; $\Delta S_1 = 1{\cdot}9$, $\Delta S_2 = 2{\cdot}5$, $\Delta S_3 = 7{\cdot}3$, $\Delta S_4 = -5{\cdot}4$	68Ga+68G
			2 ($NaClO_4$)	$\Delta H_1 = -2{\cdot}32$, $\Delta H_2 = -0{\cdot}3$, $\Delta H_3 = -1{\cdot}2$, $\Delta H_4 = 3{\cdot}80$; $\Delta S_1 = 1{\cdot}2$, $\Delta S_2 = 1{\cdot}9$, $\Delta S_3 = 5{\cdot}6$, $\Delta S_4 = -6{\cdot}6$	
			1 ($NaClO_4$)	$\Delta H_1 = -2{\cdot}45$, $\Delta H_2 = -0{\cdot}5$, $\Delta H_3 = -1{\cdot}4$, $\Delta H_4 = -4{\cdot}0$; $\Delta S_1 = 0{\cdot}4$, $\Delta S_2 = 1{\cdot}9$, $\Delta S_3 = 3{\cdot}1$, $\Delta S_4 = -7{\cdot}5$	
			0·5 ($NaClO_4$)	$\Delta H_1 = -2{\cdot}43$, $\Delta H_2 = -0{\cdot}9$, $\Delta H_3 = -0{\cdot}75$, $\Delta H_4 = -4{\cdot}2$; $\Delta S_1 = 0{\cdot}5$, $\Delta S_2 = 0{\cdot}3$, $\Delta S_3 = 5{\cdot}0$, $\Delta S_4 = -8{\cdot}7$	
			0·25 ($NaClO_4$)	$\Delta H_1 = -2{\cdot}32$, $\Delta H_2 = -1{\cdot}7$, $\Delta S_1 = 1{\cdot}1$, $\Delta S_2 = -2{\cdot}5$	
Hg_2^{2+}	Hg	7—40	0·5 (Na)ClO_4, 0·1 H^+	K_{s0} −29·34(7°), −26·46(40°)	63H
	Hg	177	MNO_3(l)	K_{s0} −22? in $Li_{0.43}K_{0.57}NO_3$(l, eut), m units	66Fa

continued overleaf

75 I^- (contd)

Metal	Method	Temp	Medium	Log of equilibrium constant, remarks	Ref
Hg_2^{2+}, Hg^{2+}	sol	25		$K[Hg_2I_2(s) \rightleftharpoons Hg(l) + HgI_2(s)] \approx -1{\cdot}75$	67Ka,
Hg^{2+}	sol	25	ROH	K_{s2} $-1{\cdot}16$ in MeOH(l), $-1{\cdot}40$ in EtOH(l)	08Ha
	aix	23	$I(NaClO_4)$	K_3 3·0, K_4 1·4(I = 3), K_3 3·8, K_4 2·6(I = 0·3), (K_3 and K_4 for I = 0·5 from ref 57Ma)	63E
	red, sol	7—40	0·5 (Na)ClO_4, 0·1 H^+	K_1 13·58(7°), 12·40(40°), K_2 11·63(7°), 10·53(40°), K_{s2} $-4{\cdot}43$(8°), $-4{\cdot}09$(25°), $-3{\cdot}87$(40°) K_{s1} $-16{\cdot}20$(7°), $-14{\cdot}42$(40°)	63H
	sol			ΔH_{s2} = 7·6(16°), 7·0(25°), 6·4(32°)	63H/64C
	cal	8—40	0·5 (Na)ClO_4, 0·1 H^+	ΔH_1 = $-18{\cdot}9$(8°), $-18{\cdot}0$(25°), $-17{\cdot}3$(40°); $\Delta H_{\beta 2}$ = $-34{\cdot}0$(8°), $-34{\cdot}15$(25°), $-34{\cdot}5$(40°); ΔH_{s0} = 41·5(8°), 41·1(25°), 41·0(40°); ΔH_{s1} = 22·6(8°), 23·15(25°), 23·65(40°); ΔH_{s2} = 7·2(8°), 6·9(25°)	64C
				ΔS_1 = $-5{\cdot}3$(8°), $-1{\cdot}5$(25°), 1·5(40°); $\Delta S_{\beta 2}$ = $-6{\cdot}7$(8°), $-5{\cdot}7$(25°), $-4{\cdot}9$(40°); ΔS_{s0} = 11·8(8°), 10·1(25°), 9·3(40°) ΔS_{s1} = 6·5(8°), 8·6(25°), 10·8(40°);	64C+63
	sol	25	1 ($NaClO_4$)	β_2 26·0, β_3 27·65, β_4 29·3 (K_{s0} $-28{\cdot}19$)	64Ca
	Ram		TBP	ev HgI_2, HgI_3^-, and HgI_4^{2-} in $(C_4H_9O)_3PO$(l)	64S
	dis	150	MNO_3(l)	$K_D[HgI_2(melt) \rightleftharpoons HgI_2(org)]$ 1·58, ΔH_D = $-2{\cdot}7$; K_3 1·18, K_4 1·11 in melt = $Li_{0.43}K_{0.57}NO_3$(l, eut), org = polyphenyl eutectic, m units	64Z
	Ram		var	ev HgI^+ and (for ⩾ 2M Hg^{II}) Hg_2I^{3+}	65Cb
	X		solid	ev planar HgI_3^- in Me_3SHgI_3(s) and ev tetrahedral HgI_4^{2-} in $(Me_3S)_2HgI_4$(s)	66F
	var	25	Me_2SO	ev HgI^+, HgI_2, HgI_3^-, HgI_4^{2-} (sp, con, Ram), $\Delta H_3 + \Delta H_4 \approx -10{\cdot}5$ in Me_2SO(l)	67Bd
	var	25	org	ev HgI_2, HgI_3^-, and heteropoly cpx like $Bi(HgI_3)_3$ in Me_2SO and Me_2NCHO(l) (con, sp, sol, pre)	67G
	sol	25	org	K_{s2} $-3{\cdot}99$ in iso-octane(l) †$K_{s2}[Hg_2I_2(s) \rightleftharpoons Hg(l) + HgI_2(org)]$ $-5{\cdot}85$ in iso-octane(l), $-2{\cdot}84$ in MeOH(l)	67Ka
	dis	25	0·5 $NaClO_4$	$K_D[HgI_2 \rightleftharpoons HgI_2(in\ CCl_4)]$ 1·06 to 1·21	67Md
	sp	25	96% MeOH	K_3 4·87 in 96% MeOH	68A
	X		Me_2SO	ev HgI_2, HgI_3^-, and HgI_4^{2-} in Me_2SO(l)	68Gb
	Hg, pol	25?	w% C_5H_5N	β_2 14, β_3 17·5, β_4 19·3(w = 20), β_2 16·1, β_3 20·1, β_4 20·9(w = 90) in w% C_5H_5N, also values for w = 10, 40, 60	68Pa
Hg^{2+}, Ag^+	sol	20	5·2[$Ca(NO_3)_2$]	$K_s[AgI(s) + Hg^{2+} \rightleftharpoons AgIHg^{2+}]$ $-1{\cdot}14$	68Y
$MeHg^+$	gl	20	0·1 (KNO_3)	K_1 8·60, K_2 0·26? $K_{s0}(MeHg^+I^-)$ $-11{\cdot}46$	63S = 65S.
$EtHg^+$	sol	25	1 Na(ClO_4)	K_{s1} $-4{\cdot}11$, K_2 $-0{\cdot}67$, K_3 0·75	65B
$Hg(CF_3)_2$	fp	0	KI var	K_1 0·98	64D
Al^{3+}	nmr		Me_2NCHO	ev $AlT_6^{3+}I^-_n$ (n = 1, 2, 3) in T = Me_2NCHO(l)	68M
Ga^{3+}	con	140	I_2(l)	$K(GaI_3 + I_2 \rightleftharpoons I^+ + GaI_4^-)$ $-1{\cdot}89$ in I_2(l)	67Bb
	dis	20	0·69 H(ClO_4)	K_1 $-0{\cdot}24$ (or K_1 $-0{\cdot}41$, β_2 $-0{\cdot}6$)	67Mb
In^{3+}	X		solid	ev In_2I_6 in InI_3(s)	64Fb
	con	140	I_2(l)	$K(InI_3 + I_2 \rightleftharpoons I^+ + InI_4^-)$ $-1{\cdot}89$ in I_2(l)	67Bb

contir

Metal	Method	Temp	Medium	Log of equilibrium constant, remarks	Ref
Tl^+	sol	25	KI var	$\Delta H_1 = -4$, $\Delta H_{\beta 2} = -7{\cdot}2$, $\Delta S_1 = -7$, $\Delta S_{\beta 2} = -16$	58Kb/58M
	E	25	4 Na(ClO₄)	K_{s0} −6·73	58M
	sol	20−70	dil	K_{s0} −7·49(20°), −7·24(25°), −6·69(40°), −6·31(50°), −5·96(60°), −5·63(70°)	
	sol	25	ZnI_2 var	β_4 −0·92	
	red	25	$3H^+$, 4 (NaClO₄)	$K_s(Tl^+I_3^-)$ −7·74	66J
				$K_{s0}(Tl^+I^-)$ −6·77, $K_s[Tl^+I^-{}_{0.83}(I_3^-)_{0.17}]$ −6·16	66J+57Ka
	con	140	I_2(l)	K_1 2·91, K_{12} 3·10 in I_2(l)	67Ba
Tl^{3+}	X		solid	ev square TlI_4^- in $CsTlI_4$(s) *etc*	48W = 67K
	con, pre		var	ev TlI_4^- in solid and MeCN(l)	65C
	red	25	$3H^+$, 4 (NaClO₄)	β_4 35·66, lower cpx decompose to Tl^+	66J
Ge^{4+}	sol	25	HI var	equil[$GeO_2(s)+4H^++4I^- \rightleftharpoons GeI_4(s)+2H_2O$] at 4·95M-HI	56Ba
Sn^{2+}	sol	25	4 (NaClO₄)	$K_{s0}(Sn^{2+}I^-{}_2)$ −5·08, $K_s(Me_4N^+{}_2Sn^{2+}I^-{}_4)$ −12·60; K_1 0·70, β_2 1·13, β_3 2·13, β_4 2·30, β_6 2·59, β_8 2·08	68H
Ph_3Sn^+	dis	30	0·1 (NaNO₃)	K_1 3·7, $K_D[Ph_3SnOH(org)+I^- \rightleftharpoons Ph_3SnI(org)+OH^-]$ −6·1 (org = C_6H_6), −6·2(org = Bu^{i}COMe)	65S
R_3SnBr^0	kin	11—20	Me_2CO	K_1 1·94(11°), 1·96(20°), R = Me; 2·23(20°), Et; 1·88(11°), 1·85(20°), Pr^i; 1·95(11°), 2·09(20°), Bu; in Me_2CO(l)	63Gd
Ph_3Pb^+	dis	30	0·1 (NaNO₃)	K_1 7·3, $K_D[Ph_3PbOH(org)+I^- \rightleftharpoons Ph_3PbI(org)+OH^-]$ −0·6(org = $CHCl_3$), −0·1(org = Bu^{i}COMe) (sign misprinted in Table 4)	65S
R_4N^+	fp	0	MI var	agrees with K_1 values (ref 66E/68F) (R = Me, Pr, Bu)	34L/68F
	con	0—25	→ 0	R = Me; K_1 0·39(→ 0°), 0·36(10°), 0·31(25°); R = Pr; K_1 0·74(→ 0°), 0·71(10°), 0·67(25°); R = Bu; K_1 0·82(→ 0°), 0·80(10°), 0·78(25°)	66E/68F
	con	27—55	Me_2CO	R = Me; K_1 2·55(27°, 1 bar), 2·14(27°, 1103 bar); R = Et; K_1 2·22(27°, 1 bar), 1·95(27°, 1103 bar); R = Pr; K_1 2·19(27°, 1 bar), 1·97(27°, 1103 bar); in Me_2CO(l) → 0, also K_1 for intermediate pressures and for 35, 45, and 55°C.	68Aa
	con	15−35	EtCOMe	R = Et; K_1 2·54(15°), 2·61(25°), 2·64(35°), $\Delta H_1 = 2{\cdot}1$, $\Delta S_1 = 19$; R = C_7H_{15}; K_1 2·44(15°), 2·56(35°), $\Delta H_1 = 2{\cdot}6$, $\Delta S_1 = 20$; in EtCOMe(l), → 0, also data for 5 other R_4N^+ ions	68Hb
Me_4N^+	var	25	→ 0	K_1 0·33 to 0·35(pH₂O), 0·25 to 0·31(con)	65L
$MeEt_3N^+$	con	−95—0	CH_2Cl_2	K_1 4·56(−94·7°), 4·77(−47·7°), 5·04 (0°) in CH_2Cl_2(l) (→ 0)	64Ba
Et_4N^+	var	25	0 corr	K_1 0·46(con), 0·42(pH₂O)	63Ba/68F
Sb^{3+}	con	140	I_2(l)	$K(SbI_2^+ + I^- \rightleftharpoons SbI_3\,?)$ 6·27 in I_2(l)	67Ba
Bi^{3+}	sol, MHg	20	2 (NaClO₄), 1 H⁺	K_5K_6 3·85, no ev BiI_5^{2-}	57A/67E
	sp	25	Me_2NCHO	ev mixed cpx, $2HgI_2/BiI_3$, in Me_2NCHO(l)	64B
	sp	24	1 HClO₄	K_1 2·68	64E
	sp	24	2 (NaClO₄) 1 H⁺	K_4 2·42, K_5K_6 2·43, K_7 −0·85, no ev BiI_5^{2-}; K_{s3} −2·45	67E
	sp			ev cpx with > 6 I per Bi, values for K_7 or K_7K_8 cannot be deduced from data	67E/68Hc

continued overleaf

75 I^- (contd)

Metal	Method	Temp	Medium	Log of equilibrium constant, remarks	Ref
Bi^{3+}	cix	25	1·89 $H(ClO_4)$	K_1 2·90	67L
(contd)	X		solid	ev $Bi_2I_9^{3-}$ in $Cs_3Bi_2I_9$(s)	67N = 6
	sol, sp	25	4 $(NaClO_4)$	ev cpx with > 6 I per Bi, K_7 −1·3 or K_7K_8 −2·05	68Hc
O_2	sp	24	var	ev cpx O_2–I^- (also O_2–Br^-, Cl^-, SCN^-)	64N
R_3S^+	con	10—25	0 corr	K_1 0·56(R = Bu) no ev cpx for R = Me, Et, Pr	68E
	con	10—25	MeCN	K_1 1·50(10°), 1·55(25°), (R = Me); 1·25 (10°), 1·30(25°), (R = Et); 1·13(10°), 1·27(25°) (R = Pr); in MeCN(l)	
	con	10—25	MeOH	K_1 1·37(R = Me), 1·38(R = Et), 1·38(R = Pr), 1·49(R = Bu) in MeOH(l)	
Te^{4+}	X		solid	ev TeI_6^{2-} in $Mg(H_2O)_6TeI_6$(s)	64A
	ir		solid	ev TeI_3^+ in TeI_4(s)	66G
	sp		1 HCl	apparent β_6 7·4	66Ma
	con, fp		org	ev ions, $TeI_3T_2^+$? (by analogy with Cl^-) in T = MeCN(l), EtOH(l), *etc.*	68Ge
I_2	dis	0—40	var	K_1 3·25(0°), 3·03(10°), 2·92(20°), 2·81(30°), 2·71(40°), ΔH_1 = −4·34, ΔS_1 = −1·47; $K_D[I_2 \rightleftharpoons I_2(\text{in } CCl_4)]$ 1·86(0°), 1·91(10°), 1·93(20°), 1·94(30°), 1·94(40°)	62Ra
	sp	10—20	?	K_1 and ΔH_1 values given for 14 medium cations (concentration ?)	63Gc
	dis	25	org	$K_{12}(I_3^- + I_2 \rightleftharpoons I_5^-)$ 1·51 in $MeNO_2$(l, Cs^+); 2·57(Cs^+), 2·30(NH_4^+) in $PhNO_2$(l)	64Va
	con	25	org	$\beta(Cs^+I_{3\text{ or }5}^-)$ 6·1 in Et_2O(l), 11 in Bu_2O(l), 5·5 in $C_5H_{11}OCOMe$(l), no ev ion pair in $PhNO_2$(l)	
	iE, red	20	Me_2NCHO	K_1 7·35 in Me_2NCHO(l) 0·1M-$HClO_4$	65Bd =
	sol	5—50	D_2O	K_{s0} −3·21(5°), −3·09(15°), −2·96(25°), −2·84(35°), −2·71(45°), −2·64(50°), ΔH_{s0} = 5·2, ΔS_{s0} = 3·7, ΔC_p = 42 in D_2O	65R
	sol	2—50	H_2O	$K_{s0}[I_2(s) \rightleftharpoons I_2]$ −3·17(2°), −3·07(11°), −2·95(20°), −2·88(25°), −2·81(30°), −2·75(35°), −2·67(40°), −2·61(45°), −2·55(49°), ΔH_{s0} = 5·4, ΔS_{s0} = 5·0, ΔC_p = 57	
	sol	25	0 corr	K_{12} 0·95(K_1 from 60D), 0·954 + 0·0308(t − 25) (t° C)	
	dis, sol	25	3 H^+, 4 $(NaClO_4)$	K_1 2·97, $K_D[I_2(aq) \rightleftharpoons I_2(CCl_4)]$ 2·12, K_{s0} −3·05	66J
	sp	25	MeOH	$K_1 \geqslant$ 4·6 in MeOH(l)	66La
	dis		var	$K_D[I_2 \rightleftharpoons I_2(org)]$ 2·30, $K_D[I^- + I_2(org) \rightleftharpoons I_3^-]$ 0·54; org = $PhNO_2$ $K_D[nI_2(org) + M^+ + I^- \rightleftharpoons M^+(org) + I_{2n+1}^-(org)]$ n = 1; −1·27(Na^+), 2·00(Cs^+); n = 2; 1·30(Na^+), 4·70(Cs^+), 1·34(H^+); $K_1 \approx 8$, $K_{12} \approx 2·7$ in $PhNO_2$(l)	66Ta
	dis	15—45	3 $Li(ClO_4)$	K_1 2·61(15°), 2·59(25°), 2·49(35°), 2·46(45°), ΔH_1 = −3·1, ΔS_1 = 1·4(25°)	67Mc
	cal	25	3 $Li(ClO_4)$	ΔH_1 = −3·3, ΔS_1 = 0·7	
	red	25	Me_2NCHO	K_1 4·9 in Me_2NCHO(l)	67P
	red	20?	org	K_1 7·9, K_{12} 1·9 in $C_3H_6O_2CO$(l), I = 0·1 (Et_4NClO_4)	68C
	sol	10—40	KI var	K_{s0} −3·18(10°), −3·01(20°), −2·95(25°), −2·87(30°), −2·76(40°), ΔH_{s0} = 5·80; K_{s1} −0·05(10°), 0·02(20°), 0·06(25°), 0·11(30°), 0·17(40°), ΔH_{s1} = 3·00;	68Sb

cont

Metal	Method	Temp	Medium	Log of equilibrium constant, remarks	Ref
				K_1 3·13(10°), 3·04(20°), 3·00(25°), 2·98(30°), 2·93(40°), $\Delta H_1 = -2{\cdot}80$	
I^+	con	140	I_2(l)	K_1 6·98 in I_2(l)	67Ba
ICl	*iE*	25	1 ($HClO_4$)	K_1 8·48	68P+

08Ha W. Herz and F. Kuhn, *Z. anorg. Chem.*, 1908, **58,** 159
34L J. Lange, *Z. phys. Chem.* 1934, *A*, **168,** 147
48W T. Watanabe, Y. Saito, R. Shiono, and M. Atoji, Abstracts of papers, First Congress of the International Union of Crystallography, 1948, 30, *Structure Reports*, 1947–8, **11,** 394
56Ba G. Brauer and H. Müller, *Z. anorg. Chem.*, 1956, **287,** 71
58M V. E. Mironov, dissertation (kand) 1958, Leningrad Technological Institute, quoted by ref. 63Ka, pp. 67–68, 78
62Ba S. Bruckenstein and L. M. Mukherjee, *J. Phys. Chem.*, 1962, **66,** 2228
62Ra E. N. Rengevich and E. A. Shilov, *Ukrain, khim. Zhur.*, 1962, **28,** 1080
63B J. Braunstein and R. E. Hagman, *J. Phys. Chem.*, 1963, **67,** 2881
63Ba V. E. Bower and R. A. Robinson, *Trans. Faraday Soc.*, 1963, **59,** 1717
63Ca H. Chateau and M. C. Moncet, *J. Chim. phys.*, 1963, **60,** 1060
63E I. Eliezer and Y. Marcus, *J. Inorg. Nuclear Chem.*, 1963, **25,** 1465
63Ga D. C. Giedt and C. J. Nyman, *J. Phys. Chem.*, 1963, **67,** 2491
63Gb A. A. Grinberg, N. V. Kiseleva, and M. I. Gel'fman, *Doklady Akad. Nauk S.S.S.R.*, 1963, **153,** 1327
63Gc I. P. Gorelov and V. V. Serebrennikov, *Zhur. fiz. Khim.*, 1963, **37,** 2322
63Gd M. Gielen, J. Nasielski, and R. Yernaux, *Bull. Soc. chim. belges*, 1963, **72,** 594
63H L. D. Hansen, R. M. Izatt, and J. J. Christensen, *Inorg. Chem.*, 1963, **2,** 1243
63Ka F. Ya. Kul'ba and V. E. Mironov, 'Khimiya talliya', Goskhimizdat, Leningrad, 1963
63Ma D. H. McDaniel and R. E. Vallée, *Inorg. Chem.*, 1963, **2,** 996
63S M. Schellenberg, dissertation, Eidgenössische Technische Hochschule, Zürich, 1963
64A A. Angoso, H. Onken, and H. Hahn, *Z. anorg. Chem.*, 1964, **328,** 223
64B M. T. Beck and F. Gaizer, *J. Inorg. Nuclear Chem.*, 1964, **26,** 1616
64Ba J. H. Beard and P. H. Plesch, *J. Chem. Soc.*, 1964, 4879
64Bb M. Blander and E. B. Luchsinger, *J. Amer. Chem. Soc.*, 1964, **86,** 319
64Bc F. Becker and H. M. Lüschow, Proceedings of the 8th International Conference on Co-ordination Chemistry, Vienna 1964, p. 334
64C J. J. Christensen, R. M. Izatt, L. D. Hansen, and J. D. Dale, *Inorg. Chem.*, 1964, **3,** 130
64Ca D. M. Czakis-Sulikowska, *Roczniki Chem.*, 1964, **38,** 533
64D A. J. Downs, *J. Inorg. Nuclear Chem.*, 1964, **26,** 41
64E A. J. Eve and D. N. Hume, *Inorg. Chem.*, 1964, **3,** 276
64F T. L. Fabry and R. M. Fuoss, *J. Phys. Chem.*, 1964, **68,** 974
64Fa G. W. A. Fowles and W. R. McGregor, *J. Phys. Chem.*, 1964, **68,** 1342
64Fb J. D. Forrester, A. Zalkin, and D. H. Templeton, *Inorg. Chem.*, 1964, **3,** 63
64G M. L. Gavrish and I. S. Galinker, *Zhur. neorg. Khim.*, 1964, **9,** 1289
64J G. J. Janz, A. E. Marcinkowsky, and I. Ahmad, *Electrochim. Acta*, 1964, **9,** 1687
64L K. H. Lieser, *J. Inorg. Nuclear Chem.*, 1964, **26,** 1571
64La H.-M. Lüschow, dissertation, University Saarbrücken, 1964
64N G. Navon, *J. Phys. Chem.*, 1964, **68,** 969
64S E. L. Short, D. N. Waters, and D. F. C. Norris, *J. Inorg. Nuclear Chem.*, 1964, **26,** 902
64Sa H. Schäfer, D. Bauer, *et al.*, *Naturwiss.*, 1964, **51,** 241
64V V. P. Vasil'ev and N. K. Grechina, *Zhur. neorg. Khim.*, 1964, **9,** 647
64Va V. M. Vdovenko and L. S. Bulyanitsa, *Radiokhimiya*, 1964, **6,** 385, 399
64W W. M. Weaver and J. D. Hutchison, *J. Amer. Chem., Soc.*, 1964, **86,** 261
64Z M. Zangen and Y. Marcus, *Israel J. Chem.*, 1964, **2,** 49
65B R. Barbieri and J. Bjerrum, *Acta Chem. Scand.*, 1965, **19,** 469
65Ba K. W. Bagnall, D. Brown, P. J. Jones, and J. G. H. du Preez, *J. Chem. Soc.*, 1965, 350
65Bb I. R. Bellobono and G. Favini, *Ann. Chim. (Italy)*, 1965, **55,** 32
65Bc G. G. Bombi, M. Fiorani, and G.-A. Mazzocchin, *J. Electroanalyt. Chem.*, 1965, **9,** 457
65Bd M. Bréant and C. Sinicki, *Compt. rend.*, 1965, **260,** 5016
65C F. A. Cotton, B. F. G. Johnson, and R. M. Wing, *Inorg. Chem.*, 1965, **4,** 502
65Ca R. Colton, *Austral. J. Chem.*, 1965, **18,** 435
65Cb J. H. R. Clarke and L. A. Woodward, *Trans. Faraday Soc.*, 1965, **61,** 207
65E J. F. Endicott and H. Taube, *Inorg. Chem.*, 1965, **4,** 437
65F A. B. Fasman, G. G. Kutyukov, and D. V. Sokol'skii, *Zhur. neorg. Khim.*, 1965, **10,** 1338
65G C. W. Garland, S. Tong, and W. H. Stockmayer, *J. Phys. Chem.*, 1965, **69,** 1718
65Ga A. M. Golub and G. V. Pomerants, *Ukrain. khim. Zhur.*, 1965, **31,** 104
65H H. E. Hellwege and G. K. Schweitzer, *J. Inorg. Nuclear Chem.*, 1965, **27,** 99
65K P. J. Kuhn and R. E. McCarley, *Inorg. Chem.*, 1965, **4,** 1482
65L B. J. Levien, *Austral. J. Chem.*, 1965, **18,** 1161

continued overleaf

75 I^- (contd)

65M R. E. McCarley and J. C. Boatman, *Inorg. Chem.*, 1965, **4**, 1486
65Ma P. Moore and F. Basolo, *Inorg. Chem.*, 1965, **4**, 1670
65Mb L. M. Mukherjee, S. Bruckenstein, and F. A. K. Badawi, *J. Phys. Chem.*, 1965, **69**, 2537
65N R. I. Novoselov and B. V. Ptitsyn, *Zhur. neorg. Khim.*, 1965, **10**, 2282
65R R. W. Ramette and R. W. Sandford, jun., *J. Amer. Chem. Soc.*, **1965**, **87**, 5001
65S G. K. Schweitzer and S. W. McCarty, *J. Inorg. Nuclear Chem.*, 1965, **27**, 191
65Sa G. Schwarzenbach and M. Schellenberg, *Helv. Chim. Acta*, 1965, **48**, 28
65Sb S. A. Shchukarev, O. A. Lobaneva, and M. A. Kononova, *Vestnik Leningrad Univ.*, 1965, No. 4, (= Fiz. Khim. No. 1), 149
66A G. Atkinson and Y. Mori, *J. Chem. Phys.*, 1966, **45**, 4716
66B L. R. Bateman, J. F. Blount, and L. F. Dahl, *J. Amer. Chem. Soc.*, 1966, **88**, 1082
66C F. A. Cotton and S. J. Lippard, *Inorg. Chem.*, 1966, **5**, 9
66E D. F. Evans and R. L. Kay, *J. Phys. Chem.*, 1966, **70**, 366
66F R. H. Fenn, *Acta Cryst.*, 1966, **20**, 20, 24
66Fa M. Fiorani, G. G. Bombi, and G. A. Mazzocchin, *Ricerca sci.*, 1966, **36**, 580
66G N. N. Greenwood, B. P. Straughan, and A. E. Wilson, *J. Chem. Soc.* (*A*), 1966, 1479
66Ga P. Gerding, *Acta Chem. Scand.*, 1966, **20**, 79
66J L. Johansson, *Acta Chem. Scand.*, 1966, **20**, 2156
66L D. C. Luehrs, R. T. Iwamoto, and J. Kleinberg, *Inorg. Chem.*, 1966, **5**, 201
66La S. Lormeau and M.-H. Mannebach, *Bull. Soc. chim. France*, 1966, 2576
66M W. A. Millen and D. W. Watts, *Austral. J. Chem.*, 1966, **19**, 51
66Ma V. I. Murashova, *Zhur. analit. Khim.*, 1966, **21**, 345
66S C. Sinicki, *Bull. Soc. chim. France*, 1966, 194
66T V. M. Tarayan, G. N. Shaposhnikova, and E. N. Ovsepyan, *Armyan. khim. Zhur.*, 1966, **19**, 22
66Ta S. Tribalat and M. Grall, *Compt. rend.*, 1966, **263**, *C*, 399
66Tb A. Thiébault, M. Herlem, and J. Badoz-Lambling, *Rev. Chim. minérale*, 1966, **3**, 1005
67A R. Alexander, E. C. F. Ko, Y. C. Mac, and A. J. Parker, *J. Amer. Chem. Soc.*, 1967, **89**, 3703
67B K. Burger and B. Pintér, *J. Inorg. Nuclear Chem.*, 1967, **29**, 1717
67Ba D. J. Bearcroft and N. H. Nachtrieb, *J. Phys. Chem.*, 1967, **71**, 316
67Bb D. J. Bearcroft and N. H. Nachtrieb, *J. Phys. Chem.*, 1967, **71**, 4400
67Bc D. Brown and P. J. Jones, *J. Chem. Soc.* (*A*), 1967, 243
67Bd A. Buckingham and R. P. H. Gasser, *J. Chem. Soc.* (*A*), 1967, 1964
67Be G. A. Boos and A. A. Popel', *Zhur. neorg. Khim.*, 1967, **12**, 2086.
67C J. M. Crawford and R. P. H. Gasser, *Trans. Faraday Soc.*, 1967, **63**, 2758
67Ca B. Corain and A. J. Poë, *J. Chem. Soc.* (*A*), 1967, 1318
67Cb B. Corain and A. J. Poë, *J. Chem. Soc.* (*A*), 1967, 1633
67D M. D. Dyke, P. G. Sears, and A. I. Popov, *J. Phys. Chem.*, 1967, **71**, 4140
67E A. J. Eve and D. N. Hume, *Inorg. Chem.*, 1967, **6**, 331
67F M. Fiorani, G. G. Bombi, and G.-A. Mazzocchin, *J. Electroanalyt. Chem.*, 1967, **13**, 167
67G F. Gaizer and M. T. Beck, *J. Inorg. Nuclear Chem.*, 1967, **29**, 31
67Ga J. Guion, *Inorg. Chem.*, 1967, **6**, 1882
67H J. Hála and D. Pohanková, *J. Inorg. Nuclear Chem.*, 1967, **29**, 2983
67J G. J. Janz and M. J. Tait, *Canad. J. Chem.*, 1967, **45**, 1101
67K H. Kashiwagi, D. Nakamura, and M. Kubo, *J. Phys. Chem.*, 1967, **71**, 4443
67Ka M. M. Kreevoy and E. M. Wewerka, *J. Phys. Chem.*, 1967, **71**, 4534
67Kb O. I. Khotsyanovskii, *Zhur. neorg. Khim.*, 1967, **12**, 1179
67L H. Loman and E. van Dalen, *J. Inorg. Nuclear Chem.*, 1967, **29**, 699
67M W. L. Masterton, T. I. Munnelly, and L. H. Berka, *J. Phys. Chem.*, 1967, **21**, 942
67Ma W. L. Masterton, *J. Phys. Chem.*, 1967, **71**, 2885
67Mb D. F. C. Morris and B. D. Andrews, *Electrochim. Acta*, 1967, **12**, 41
67Mc V. E. Mironov and N. P. Lastovkina, *Zhur. fiz. Khim.*, 1967, **41**, 1850
67Md M. D. Morris and L. R. Whitlock, *Analyt. Chem.*, 1967, **39**, 1180
67Me V. E. Mironov, A. V. Fokina, and Yu. I. Rutkovskii, *Zhur. neorg. Khim.*, 1967, **12**, 2056
67Mf V. E. Mironov and A. V. Fokina, *Zhur. neorg. Khim.*, 1967, **12**, 2571
67N A. Nyström and O. Lindqvist, *Acta Chem. Scand.*, 1967, **21**, 2570
67Na N. M. Nikolaeva and E. D. Pastukhova, *Zhur. neorg. Khim.*, 1967, **12**, 1227
67Nb N. M. Nikolaeva and E. D. Pastukhova, *Zhur. neorg. Khim.*, 1967, **12**, 1510
67P Yu. M. Povarov and I. E. Barbasheva, *Elektrokhimiya*, 1967, **3**, 745
67Pa G. Popa, E. Iosif, and C. Luca, *Rev. Roumaine Chim.*, 1967, **12**, 169
67R N. A. Rumbaut and H. L. Peeters, *Bull. Soc. chim. belges*, 1967, **76**, 33
67Ra A. M. Rozen and A. I. Mikhailichenko, *Zhur. neorg. Khim.*, 1967, **12**, 741
67S P. G. Sears, J. A. Caruso, and A. I. Popov, *J. Phys. Chem.*, 1967, **21**, 905
67Sa A. Swinarski and A. Grodzicki, *Roczniki Chem.*, 1967, **41**, 1205
67T N. Tanaka, Y. Kobayashi, and M. Kamada, *Bull. Chem. Soc. Japan*, 1967, **40**, 2839
67V V. P. Vasil'ev and P. S. Mukhina, *Izvest. V.U.Z., Khim. i khim. Tekhnol.*, 1967, **10**, 263
68A M. H. Abraham, C. F. Johnston, and T. R. Spalding, *J. Inorg. Nuclear Chem.*, 1968, **30**, 2167
68Aa W. A. Adams and K. J. Laidler, *Canad. J. Chem.*, 1968, **46**, 1977, 2005

contin

I^- (contd) 75

68Ab E. Andalaft, R. P. T. Tomkins, and G. J. Janz, *Canad. J. Chem.*, 1968, **46**, 2959
68B M. J. Bennett, F. A. Cotton, and B. M. Foxman, *Inorg. Chem.*, 1968, **7**, 1563
68C J. Courtot-Coupez and M. L'Her, *Compt. rend.*, 1968, **266**, *C*, 1286
68D R. J. De Sando and G. H. Brown, *J. Phys. Chem.*, 1968, **72**, 1088
68Da M. Della Monica, U. Lamanna, and L. Senatore, *Inorg. Chim. Acta*, 1968, **2**, 363
68E D. F. Evans and T. L. Broadwater, *J. Phys. Chem.*, 1968, **72**, 1037
68F R. Fernandez-Prini, *Trans. Faraday Soc.*, 1968, **64**, 2146
68G P. Gerding, *Acta Chem. Scand.*, 1968, **22**, 1283
68Ga P. Gerding and I. Jönsson, *Acta Chem. Scand.*, 1968, **22**, 2247
68Gb F. Gaizer and G. Johansson, *Acta Chem. Scand.*, 1968, **22**, 3013
68Gc H. C. Gaur and R. S. Sethi, *Electrochim. Acta*, 1968, **13**, 1737
68Gd E. A. Gyunner and N. D. Yakhkind, *Zhur. neorg. Khim.*, 1968, **13**, 2758
68Ge N. N. Greenwood, B. P. Straughan, and A. E. Wilson, *J. Chem. Soc. (A)*, 1968, 2209
68H G. P. Haight, jun., and L. Johansson, *Acta Chem. Scand.*, 1968, **22**, 961
68Ha K.-L. Hsia and R. M. Fuoss, *J. Amer. Chem. Soc.*, 1968, **90**, 3055
68Hb S. R. C. Hughes and D. H. Price, *J. Chem. Soc. (A)*, 1968, 1464
68Hc G. P. Haight, jun., and L. Johansson, *Inorg. Chem.*, 1968, **7**, 1255
68K S. Katayama and R. Tamamushi, *Bull. Chem. Soc. Japan*, 1968, **41**, 606
68L O. Lindqvist, *Acta Chem. Scand.*, 1968, **22**, 2943
68La E. V. Lapitskaya, E. D. Kuchkina, and F. P. Gorbenko, *Zhur. neorg. Khim.*, 1968, **13**, 2774
68Lb D. C. Luehrs and K. Abate, *J. Inorg. Nuclear Chem.*, 1968, **30**, 549
68M W. G. Movius and N. A. Matwiyoff, *J. Phys. Chem.*, 1968, **72**, 3063
68Ma L. M. Mukherjee, J. J. Kelly, W. Baranetzky, and J. Sica, *J. Phys. Chem.*, 1968, **72**, 3410
68P G. Piccardi and R. Guidelli, *J. Phys. Chem.*, 1968, **72**, 2782
68Pa N. Payen-Baldy and M. Machtinger, *Bull. Soc. chim. France*, 1968, 2221
68S A. M. Shkodin and N. K. Levitskaya, *Ukrain. khim. Zhur.*, 1968, **34**, 330
68Sa A. M. Shkodin, N. K. Levitskaya, and T. F. Lisachenko, *Ukrain. khim. Zhur.*, 1968, **34**, 1113
68Sb M. S. Sytilin, *Zhur. fiz. Khim.*, 1968, **42**, 1138
68Sc R. Saillant and R. A. D. Wentworth, *Inorg. Chem.*, 1968, **7**, 1606
68Sd T. W. Swaddle and G. Guastalla, *Inorg. Chem.*, 1968, **7**, 1915
68W S. H. White, D. Inman, and B. Jones, *Trans. Faraday Soc.*, 1968, **64**, 2841
68Y N. D. Yakhkind and E. A. Gyunner, *Zhur. neorg. Khim.*, 1968, **13**, 1005

Hypoiodite, IO^- 76

No new entries

Iodate, IO_3^- 77

Metal	*Method*	*Temp*	*Medium*	*Log of equilibrium constant, remarks*	*Ref*
H^+	fp, con		H_2SO_4	ev $(IO_2HSO_4)_n$, $(IO_2)_n(HSO_4)_{n-1}^+$, $n = 1, 2, 3$; $H_2IO_3^+$ gives best (but not satisfactory) agreement among models with single mononuclear reaction	64G
	var	0—50	var	no ev polyn cpx in 0—0·2M-HL, or 0—0·2M-KHL_2 (sp, gl, con, aix)	64H
	$p$$H_2O$, Ram	32	0 corr	K_1 0·74	65D
	ir		solid	ev IO_2^+ in $IO_2^+AsF_6^-$(s)	65P
	X		solid	ev HIO_3 and I_2O_5 in HI_3O_8(s)	66Fa
	var	25	0 corr	K_1 0·80, $K_{12}[=\beta(HL^0L^-)]$ 0·6 (con, gl, kin)	67P
Sr^{2+}	sol	20	1 $HClO_4$	K_{s0} −5·55 (HL neglected)	65K
Y^{3+}	sol	25	0 corr	K_{s0} −9·96	66F

continued overleaf

77 IO_3^- (contd)

Metal	*Method*	*Temp*	*Medium*	*Log of equilibrium constant, remarks*	*Ref*
Group $4f^{3+}$	sol	25	0 corr	K_{s0} −10·92(La), −10·77(Pr), −10·92(Nd), −11·19(Sm), −11·32(Eu), −11·13(Gd), −11·11(Tb), −10·92(Dy), −10·70(Ho), −10·41(Er), −10·36(Tm), −10·21(Yb)	66F
Zr^{4+}	var		var	ev ani cpx, ZrL_5^- ? (pre, sol, aix)	68D
Hf^{4+}	var		var	ev ani cpx, HfL_5^- ? (pre, sol, aix)	68D
Pa^{5+}	dis	25	1 $HClO_4$	$*K_1$ 2·11, $*K_2$ 1·54	67K
Cr^{6+}	*X*		solid	ev $O_3CrOIO_2^-$ in $KCrIO_6$(s)	67L

64G R. J. Gillespie and J. B. Senior, *Inorg. Chem.*, 1964, **3**, 440
64H J. F. Harvey, J. P. Redfern, and J. E. Salmon, *J. Inorg. Nuclear Chem.*, 1964, **26**, 1326
65D J. R. Durig, O. D. Bonner, and W. H. Breazeale, *J. Phys. Chem.*, 1965, **69**, 3886
65K I. V. Kolosov, *Zhur. neorg. Khim.*, 1965, **10**, 2200
65P J. J. Pitts, S. Kongpricha, and A. W. Jache, *Inorg. Chem.*, 1965, **4**, 257
66F F. H. Firsching and T. R. Paul, *J. Inorg. Nuclear Chem.*, 1966, **28**, 2414
66Fa Y. D. Feikema and A. Vos, *Acta Cryst.*, 1966, **20**, 769
67K R. T. Kolarich, V. A. Ryan, and R. P. Schuman, *J. Inorg. Nuclear Chem.*, 1967, **29**, 783
67L P. Löfgren, *Acta Chem. Scand.*, 1967, **21**, 2781
67P A. D. Pethybridge and J. E. Prue, *Trans. Faraday Soc.*, 1967, **63**, 2019
68D J. Deabriges and R. Rohmer, *Bull. Soc. chim. France*, 1968, 521

78 Periodate, IO_4^- = HL^-

Metal	*Method*	*Temp*	*Medium*	*Log of equilibrium constant, remarks*	*Ref*
H^+	sol	5—40	0 corr	$\dagger K$ 0·78(5·3°), 1·63(24·9°), 2·04(40·3°) (?)	64L
			0 corr	$\dagger K_1$ 6·62(5·3°), 6·5(24·9°), 6·6(40·3°)	
			0 corr	$\dagger K_{12}$ 3·16(5·3°), 3·0(24·9°, 40·3°)	
	sol	5—75	0 corr	$K_s[(C_6H_5)_4As^+IO_4^-]$ −8·97(5·3°), −8·41(24·9°), −8·05(40·3°), −7·61(60·3°), −7·29(75·8°)	
	sp	1—25	0·1 (KCl)	$K(2L^{2-} \rightleftharpoons L_2^{4-})$ 3·38(1°), 2·82(25°), no ev $H_2L_2^{2-}$	65B
	gl	0—45	0 corr	K_1 8·34(0°), 8·33(25°), 8·43(45°)	
	X		solid	ev $I_2O_8(OH)_2^{4-}$ in $K_4I_2O_8(OH)_2(H_2O)_8$(s)	65F
	X		solid	ev $IO(OH)_5$ in H_5IO_6(s)	66F
	gl	25	→ 0	K_1 8·25, K_{12} 1·61	66S
	gl	25	D_2O, → 0	K_1 8·91, K_{12} 1·69 in D_2O, also values for H_2O–D_2O mixt	
	sp	5—45	H_2O	$\dagger K$ 0·95(5°), 1·18(15°), 1·43(25°), 1·76(35°), 2·21(45°), $\Delta\dagger H$ = 11·4, $\Delta\dagger S$ = 44·9	
	sp	25	H_2O–D_2O	$\dagger K$ 1·43(x = 0), 1·46(x = 0·2), 1·49(x = 0·4), 1·76(x = 0·8) in $H_{2-2x}D_{2x}O$	
	X		solid	ev $I_2O_9^{4-}$ in $K_4I_2O_9$(s)	68B
	gl	25	dil	K_1 8·20, K_{12} 1·64, $K_a(L^{2-})$ −12·10, $\beta(2L^{2-} \rightleftharpoons I_2O_9^{4-})$ 2·09	68H
	sp, con	25	dil	K_{12} 1·61(sp), 1·73(con)	
	nmr	5—40	0 corr	$\dagger K(H_4IO_6^- \rightleftharpoons IO_4^- + 2H_2O)$ 0·67(5°), 1·20(18°), 1·46(→ 25°), 1·97(→ 40°), $\Delta\dagger H$ = 14·8, $\Delta\dagger S$ = 56	68K
	gl, sp	25	0·13 ($NaClO_4$)	K_{12} 2·99(gl), 3·01(sp)	68M
Cs^+	sol	5—40	0 corr	$K_{s0}(Cs^+IO_4^-)$ −3·40(5°), −2·65(25°), −2·19(40°), ΔH_{s0} = 13·1, ΔS_{s0} = 37·6	68K
Mn^{4+}	*X*		solid	ev $Mn(IO_6)_3^{11-}$ group in $Na_7H_4Mn(IO_6)_3(H_2O)_{17}$(s)	63L

contin

IO_4^- (contd) 78

Metal	Method	Temp	Medium	Log of equilibrium constant, remarks	Ref
Pd^{4+}	pre, ir		var	ev $Pd(IO_6)_2^{6-}$	67S
Pt^{4+}	pre, ir		var	ev $Pt(IO_6)_2^{6-}$	67S
Ag^{3+}	var		var	ev $Ag(IO_5OH)_2H_2O^{5-}$ (*X*, pre, mag, sp)	64C

63L A. Línek, *Czech. J. Phys.*, 1963, *B*, **13**, 398 (prel)
64C G. H. Cohen and G. Atkinson, *Inorg. Chem.*, 1964, **3**, 1741
64L S. H. Laurie, J. M. Williams, and C. J. Nyman, *J. Phys. Chem.*, 1964, **68**, 1311
65B G. J. Buist and J. D. Lewis, *Chem. Comm.*, 1965, 66
65F A. Ferrari, A. Braibanti, and A. Tiripicchio, *Acta Cryst.*, 1965, **19**, 629; *Ricerca sci*,, 1964, **A7**, 709 (prel.)
66F Y. D. Feikema, *Acta Cryst.*, 1966, **20**, 765
66S P. Salomaa and A. Vesala, *Acta Chem. Scand.*, 1966, **20**, 1414
67S H. Siebert and W. Mader, *Z. anorg. Chem.*, 1967, **351**, 146
68B B. Brehler, H. Jacobi, and H. Siebert, *Z. anorg. Chem.*, 1968, **362**, 301
68H J. Haladjian, R. Sabbah, and P. Bianco, *J. Chim. phys.*, 1968, **65**, 1751
68K R. M. Kren, H. W. Dodgen, and C. J. Nyman, *Inorg. Chem.*, 1968, **7**, 446
68M E. E. Mercer and D. T. Farrar, *Canad. J. Chem.*, 1968, **46**, 2679

Other ligand with 7B elements 79

Trioximidochlorate(VII), $NClO_3^{2-}$

Metal	Method	Temp	Medium	Log of equilibrium constant, remarks	Ref
H^+	gl	?	0 corr	K_1 11·95, K_{12} 5·54	68R

68R V. Ya. Rosolovskii and I. V. Kolesnikov, *Zhur. neorg. Khim.*, 1968, **13**, 1801

Halogenides: mixed and comparative 80

Metal	Method	Temp	Medium	Log of equilibrium constant, remarks	Ref
H^+	ir, pre		solid	ev $ClHBr^-$, $ClHI^-$ in Bu_4N^+ salts	64S
	ir		var	ev $ClHF^-$, $BrHF^-$, IHF^- in solids, in MeCN(l), and CH_2Cl_2(l)	66E
	H	25	C_5H_5N	$\beta(H^+X^-)$ 1·60 + $x(Cl^-)$, 0·30 + $x(Br^-)$, −0·64 + $x(I^-)$, $x(NO_3^-)$, 0·78 + $x(ClO_4^-)$ in C_5H_5N(l)	67M
Group 1A$^+$	dis	25 ?	0 corr	$K_D[M^+ + L^+ = M^+(org) + L^+(org)]$, org = $Me_2CHC_2H_4OH$; −4·84(LiCl), −4·51(LiBr), −3·92(LiI), −5·54(NaCl), −4·41(NaI), —5·45(KCl), −4·39(KI), −5·23(CsCl), −4·22(CsI)	67R
	con	25 ?	$Me_2CHC_2H_4OH$	$K_1(Li^+L^-)$ 3·56(Cl^-), 3·49(Br^-), 3·41(I^-) in $Me_2CHC_2H_4OH$(l)	
Li^+	con	25	Me_2CO	K_1 5·48(Cl^-), 3·66(Br^-), 2·16(I^-) in Me_2CO(l) → 0	66Sa
Ti^{4+}	var		solid	ev $TiCl_4Br_2^{2-}$, $TiCl_2Br_4^{2-}$, $TiCl_4I_2^{2-}$, all *cis* (ir, Ram, pre)	68C

continued overleaf

80 Halogenides: mixed and comparative (contd)

Metal	*Method*	*Temp*	*Medium*	*Log of equilibrium constant, remarks*	*Ref*
VO^{2+} cpx	sp		org	$M^+ = VO(acac)_2^+$; $K_1(L)$ in MeCN(l) 3·1(N_3^-), 1·96(SCN^-),0·43(Cl^-), < −0·1(Br^-, I^-); $K_1(L)$ in CH_2Cl_2(l) 2·64(N_3^-), 1·74(SCN^-), 0·38(Cl^-), < −0·1(Br^-, I^-)	68G
Cr^{2+}	kin	25	var	$\beta(Cr^{2+}Cl^-)/\beta(Cr^{2+}Br^-)$ 0·4	63M
Mo^{5+}	esr	20	≈ 8 H^+	ev $MoOX_{5-n}L_n^{2-}$, n = 0, 1, 2, 3, and 4 or 5; $X^- = Cl^-$, $L^- = HSO_4^-$, F^-, Br^-, I^-; $X^- = Br^-$, $L^- = HSO_4^-$, F^-, I^-; †$\beta_n(MoOCl_5^{2-} + nL^- \rightleftharpoons MoOCl_{5-n}L_n^{2-} + nCl^-)$; $L^- = Br^-$; †β_1 −0·5, †β_2 −0·7, †β_3 −1·7, †β_4 −3·2, or †β_5 −4·3; $L^- = HSO_4^-$; †β_1 −2, †β_2 −4, †β_3 −7, †β_4 −10, or †β_5 −13	66M
U^{4+}	nmr	20	2 $HClO_4$	K_1 0·8(Cl^-), 0·3(Br^-), 0·2(I^-)	63V
Mn^{2+}	esr	22	0 corr	K_1^{out} 0·18(F^-), −1·05(Cl^-), −1·35(Br^-), < −2(I^-)	68M
Co^{2+}	pre, sp		solid	ev $CoCl_xBr_{4-x}^{2-}$, x = 0, 1, 2, 3, 4,	67Ga
Co^{3+} cpx	pH_2O	37	0 corr	$Co(NH_3)_5MeCO_2^{2+}$: K_1 1·15(Cl^-), 1·12(Br^-), 1·01(I^-), 1·10(NO_3^-); $Co(NH_3)_5EtCO_2^{2+}$: K_1 1·10(Cl^-), 1·14(Br^-), 1·15(I^-), 1·19(NO_3^-); $Co(NH_3)_5Me_2CHCO_2^{2+}$: K_1 1·12(Cl^-), 1·19(Br^-), 1·15(I^-), 1·27(NO_3^-)	66Bb
	sp	40	1 Na(ClO_4)	$Co(CN)_5^{2-}$; K_1 −0·6(Cl^-), −0·06(Br^-), 1·56(I^-)	67G
	kin	30	MeOH	*cis*-Co $en_2H_2OCl^{2+}$; $K_2^{out}(Cl^-)$ 1·90, $K_2^{out}(Br^-)$ 1·83, $K_1^{out}(NO_2^-)$ 2·51 in MeOH(l)	68A
	kin	25	3 ($LiClO_4$)	$en_2Co(\mu NH_2, \mu\text{-}OH)Co\ en_2^{4+}$: K_1 0·0(Cl^-), −0·3(Br^-)	68Da
	sp		1 ($NaClO_4$)	$M^- = Co(dimethylglyoximate)_2SO_3H_2O^-$: $\beta(M^-SCN^-)$ 2·36, $\beta(M^-I^-)$ 0·40, $\beta(M^-Br^-) \leqslant 0$	68T
	gl		1 ($NaClO_4$)	$\beta(M^-OH^-)$ 3·46	
Ni^{2+}	MHg	125	Me_2SO_2	$\beta_4(Cl^-)$ 11·08, $\beta_4(Br^-)$ 11·97, $\beta_4(I^-)$ 9·66 in Me_2SO_2(l), 2M-$LiClO_4$	68L
Rh^{3+} cpx	sp	85	1·5 K(Cl, Br)	Q^{3+} = *trans*-Rh en_2^{3+}: $K(QCl_2^+ + Br^- \rightleftharpoons QBrCl^+ + Cl^-)$ 0·29; $K(QBrCl^+ + Br^- \rightleftharpoons QBr_2^+ + Cl^-)$ −0·10	65B
	sp	85—96	2·0 K^+	Q^{3+} = *trans*-Rh en_2^{3+}; $K(QCl_2^+ + I^- \rightleftharpoons QClI^+ + Cl^-)$ 0·78 ?(85°), 0·85(90°), 0·80 ?(96°), $\Delta H \approx 0$	66B
		45—90	0·5—0·6 K^+	$K(QClI^+ + I^- \rightleftharpoons QI_2^+ + Cl^-)$ 0·82(45°), 0·76(50°), 0·69(55°), 0·34(→ 90°), $\Delta H \approx -6$	
		70—90	1·5 K^+	$K(QBr_2^+ + I^- \rightleftharpoons QBrI^+ + Br^-)$ 0·93(70°), 0·84(90°), $\Delta H = -2·6$	
		45—90	0·5 K^+	$K(QBrI^+ + I^- \rightleftharpoons QI_2^+ + Br^-)$ 0·55(45°), 0·51(50°), 0·48(55°), 0·28(→ 90°), $\Delta H = -3·4$	
	kin	50	var	Q^{3+} = *trans*-Rh en_2^{3+} ("QX^{2+}" = QXH_2O^{2+}); $\beta(QCl^{2+}Cl^-)$ 2·7, $\beta(QCl^{2+}Br^-)$ 3·2, $\beta(QBr^{2+}Cl^-)$ 3·2, $\beta(QBr^{2+}Br^-)$ 3·0, $\beta(QI^{2+}Cl^-)$ 3·0, $\beta(QI^{2+}Br^-)$ 3·2, $\beta(QI^{2+}I^-)$ 3·7; ΔH and ΔS given	67Ba
	kin	50	0·2 ($NaClO_4$)	$Rh(NH_3)_5^{3+}$: K_1 2·25(Cl^-), 2·16(Br^-), 2·68(I^-); ΔH_1 = −1·3(Cl^-), −0·7(Br^-), 1·8(I^-), ΔS_1 = −14·3(Cl^-), −11·8(Br^-), −6·4(I^-)	67Pa

continued

Metal	*Method*	*Temp*	*Medium*	*Log of equilibrium constant, remarks*	*Ref*
Rh^{3+} cpx (*contd*)	kin	35—65	1·5 ($NaClO_4$)	$Rh(NH_3)_5OH_2^{3+}$: K_1^{out} − 1·02(Cl^-, 35°), − 0·81(Cl^-, 65°), − 1·27(Br^-, 35°)	68Ba
		45—75	0·2 ($NaClO_4$)	K_1^{out} 0·42(F^-, 45°), 1·06(F^-, 75°)	
		45	1·5 ($NaClO_4$)	*trans*-Rh $en_2H_2OBr^{2+}$: K_1^{out} − 0·19(Cl^-), − 0·1(NO_3^-)	
		45	0·2 ($NaClO_4$)	K_1^{out} 1·04(F^-, 75°)	
		70	1·5 ($NaClO_4$)	*trans*-Rh $en_2(OH_2)_2^{3+}$; K_1^{out}(Br^-) − 0·29	
Pd^{2+}	sp	25	1 Na^+, 0·1 H^+	$K(PdCl_4^{2-} + 2Br^- \rightleftharpoons PdCl_2Br_2^{2-} + 2Cl^-)$ 1·99 $K(PdCl_2Br_2^{2-} + 2Br^- \rightleftharpoons PdBr_4^{2-} + 2Cl^-)$ − 0·06 $\beta(Pd^{2+}Cl^-{}_2Br^-{}_2)$ 13·11 (misprint 13·01) [$\beta(Pd^{2+}Cl^-{}_4)$ 11·12]	66Bc
	sp	25	4·5 ($LiClO_4$)	$PdCl_4^{2-}$–(Br^-–Cl^-); †K_1 1·55, †K_2 1·09, †K_3 0·95, †K_4 0·55; †K_n: $PdCl_{5-n}Br_{n-1}^{2-} + Br^- \rightleftharpoons PdCl_{4-n}Br_n^{2-} + Cl^-$	66S
	sol	20	0·1 ($HClO_4$)	K_1 5·1(Cl^-), 6·8(Br^-), 10·0(I^-)	67Gc
			0·1 ($NaClO_4$)	K_1 10·5(CN^-) [$K_s(Ag^+L^-)$ − 10·0(Cl^-), − 12·74(Br^-), − 16·4(I^-), − 15·22(CN^-)]	
	sp	25 ?	4·5 ($LiClO_4$)	$PdBr_4^{2-}$–(I^-–Br^-): †K_1 2·75, †K_2 3·00, †K_3 1·70, †K_4 0·80, †K_n: $PdBr_{5-n}I_{n-1}^{2-} + I^- \rightleftharpoons PdBr_{4-n}I_n^{2-} + Br^-$	67S
Pd^{2+} cpx	sp	27	0·5	Q^{2+} = Pd $dien^{2+}$, $K(X, Y) = K(QX^+ + Y^- \rightleftharpoons QY^+ + X^-)$: K(Cl, I) 1·95, $\Delta H = -3·7$, $\Delta S = -3·4$; K(Br, I) 1·48, $\Delta H = -2·5$, $\Delta S = -1·6$; K(Br, SCN) 2·23, $\Delta H = -4·7$, $\Delta S = -5·4$;	67H
	sp	27		$Q^{2+} = Pd(Et_2NC_2H_4NHC_2H_4NEt_2)^{2+}$;	
			1·0	K(Cl, Br) 0·11, $\Delta H = 0·6$, $\Delta S = 3·0$	
			0·5	K(Br, NCS) 2·02, $\Delta H = -4·4$, $\Delta S = 4·4$	
	kin	27	0·5?	K(Cl, I) 1, $\Delta H \approx -7$; K(Br, I) 1, $\Delta H \approx -8$	
	kin	24	?	$\beta(Q^{2+}X^-)$ 1·74(Cl), 1·72(Br), 1·70(I)	
Os^{4+}	oth	100	H^+ conc	ev $OsI_nBr_{6-n}^{2-}$ (n = 0, 1, 2, 3), $OsCl_nBr_{6-n}^{2-}$ (n = 0, 1 . . . 6), (electrophoresis, sp)	65Ba
Ir^{4+}	var	100	H(Cl, Br) conc	†K_1 0·90, †K_2 0·74, †K_3 0·52, †K_4 0·29, †K_5 − 0·01, †K_6 − 0·21, $\beta_6(Br^-)/\beta_6(Cl^-)$ 2·23; †K_1: $IrCl_6^{2-} + Br^- \rightleftharpoons IrBrCl_5^{2-} + Cl^-$ *etc.* (electrophoresis, sp, *X*)	65Ba
Pt^{2+} cpx	sp	23	MeOH	Pt $diars_2^{2+}$: K_1 2·52(L = Cl^-), 3·83(Br^-), 5·68(I^-), 3·68(SCN^-), 1·60(N_3^-), 1·30[$(NH_2)_2CS$] in MeOH(l) (→ 0)	66Da
	sp	23 (to 56)	MeOH	ΔH_1 = 0(Cl^-), − 1·02(Br^-), − 3·8(I^-), − 2·25(SCN^-), 0(N_3^-), − 4·6[$(NH_2)_2CS$]; ΔS_1 = 11·5(Cl^-), 14(Br^-), 13·2(I^-), 9·3(SCN^-), 7·3(N_3^-), − 9·5[$(NH_2)_2CS$] in MeOH(l) (→ 0)	
	sol	25	dil	$Q^{2+} = Pt(NH_3)_2^{2+}$: $K_s[QX_2(s) \rightleftharpoons QX_2]$ − 2·08(Cl, *cis*), − 2·92(Cl, *trans*), − 2·60(Br, *cis*), − 3·48(Br, *trans*), − 3·02(I, *cis*), − 4·00(I, *trans*); also values for 10 and 50°	67Gb
	sp		1 ($NaClO_4$)	$Q^{2+} = Pt(H_2NOH)_2^{2+}$; $K(QCl_2 + Br^- \rightleftharpoons QClBr + Cl^-)$ 1·29, $K(QClBr + Br^- \rightleftharpoons QBr_2 + Cl^-)$ 0·75	67Sb

continued overleaf

80 Halogenides: mixed and comparative (contd)

Metal	*Method*	*Temp*	*Medium*	*Log of equilibrium constant, remarks*	*Ref*
Pt^{2+} cpx (*contd*)	sp	rt	0·1 ($HClO_4$)	Q^{2+} = *trans* $Pt(H_2NOH)_2{}^{2+}$ $\dagger K_1 = K(QX_2+Y^- \rightleftharpoons QXY+X^-)$, $\dagger K_2 = K(QXY+Y^- \rightleftharpoons QY_2+X^-)$; X = Br, Y = I: $\dagger K_1$ 2·22, $\dagger K_2$ 1·63; X = Cl, Y = I: $\dagger K_1$ 3·37 (misprint in Table 2), $\dagger K_2$ 2·78 X = Cl, Y = Br: $\dagger K_1$ 1·29, $\dagger K_2$ 0·75;	68P 68P = 67Sb
	kin	30—50	0·0385 ($NaNO_3$)	Pt $diars_2{}^{2+}$: $K_1(Cl^-)$ 1·43(20°), 1·20(30°), 1·37(50°); $K_1(Br^-)$ 2·62(20°), 2·55 to 2·58(30°), 2·60(50°)	68Pa
Pt^{4+}	var		solid	ev $PtF_3Cl_3{}^{2-}$ (pre, ir, sp, *X*)	66Ba
	nmr	28	2 H_2PtX_6	$\dagger K_n$; $PtCl_{7-n}Br_{n-1}{}^{2-} + Br^- \rightleftharpoons PtCl_{6-n}Br_n{}^{2-} + Cl^-$; $\dagger K_1/\dagger K_2$ 0·39, $\dagger K_2/\dagger K_3$ 0·25, $\dagger K_3/\dagger K_4$ 0·20, $\dagger K_4/\dagger K_5$ 0·26, $\dagger K_5/\dagger K_6$ 0·30	68Z
	oth	100	H^+ conc	ev various mixed cpx Pt^{IV}–Cl^-–Br^-–OH^- (electrophoresis)	65Ba
Pt^{4+} cpx	sp, kin	25	0·2 Na(Cl, Br)	$Pt(NH_3)_4{}^{4+} = M^{4+}$: $K(MCl_2{}^{2+} + Br^- \rightleftharpoons MClBr^{2+} + Cl^-)$ 1·20, $K(MClBr^{2+} + Br^- \rightleftharpoons MBr_2{}^{2+} + Cl^-)$ 0·64	65R
	sp	25	$I(NaClO_4)$	Q = $Pt(CN)_4$; $K(QI_2{}^{2-} + X^- \rightleftharpoons Q^{2-} + I_2X^-)$ −4·22(Cl^-, I = 5), −3·55(Br^-, I = 5), −1·63(I^-, I = 0·5)	68Pb
Cu^{2+}	sp		Me_2CO	ev $CuCl_3Br^{2-}$ and $CuCl_2Br^-$ in Me_2CO(l)	62S
	pre, sp		Me_2CO	ev $CuCl_{4-x}Br_x{}^{2-}$ (x = 0, 1, 2, 3) in Me_2CO(l)	63I
	pre, sp		solid	ev $CuCl_xBr_{4-x}{}^{2-}$, x = 0, 1, 2, 3, 4 and $CuCl_xBr_{4-x}(H_2O)_2{}^{2-}$ with x = 1, 2, 3	67Ga
Ag^+	Ag	438	$NaNO_3$(l)	$K(AgCl_2{}^- + AgBr_2{}^- \rightleftharpoons 2AgClBr^-)$ 0·71 in $NaNO_3$(l)	67B
	Ag	20	$MeNO_2$	$K_{s0}(Ag^+X^-)$ −19·2(Cl^-), −19·7(Br^-), −20·5(I^-), −16·9(SCN^-), < −24(CN^-); β_2 19·5(Cl^-), 19·7(Br^-), 22·0(I^-), 16·4(SCN^-) > 34(CN^-), in $MeNO_2$(l), I = 0·1(Et_4NClO_4)	68B
Au^{3+}	Ag	25	var	$\dagger K_n = K(AuCl_{5-n}Br_{n-1}{}^- + Br^- \rightleftharpoons AuCl_{4-n}Br_n{}^- + Cl^-)$: $\dagger K_1$ 2·57, $\dagger K_2$ 1·80, $\dagger K_3$ 1·80, $\dagger K_4$ 1·13	66P
	sp	17—21	0·1	$\dagger K_1$ 2·53, $\dagger K_2$ 2·04, $\dagger K_3$ 1·70, $\dagger K_4$ 1·5	67P
	hyp			$\Delta\dagger H_1 \approx \Delta\dagger H_2 \approx \Delta\dagger H_3 \approx \Delta\dagger H_4 \approx -3{\cdot}4$	
Zn^{2+}	Ram		org	ev ZnClBr, ZnClI, ZnBrI, $ZnCl_nI_{4-n}{}^{2-}$, $ZnBr_nI_{4-n}{}^{2-}$	55Da
	X, ir		solid	ev $ZnBr_nCl_{4-n}{}^{2-}$, $ZnBr_nI_{4-n}{}^{2-}$ (n = 1, 2, 3) $ZnCl_nI_{4-n}{}^{2-}$ (n = 1, 2, no ev 3)	66D
	ir		solid	ev $ZnCl_nBr_{4-n}{}^{2-}$, $ZnBr_nI_{4-n}{}^{2-}$ (n = 0, 1, 2, 3, 4) ev. $ZnCl_nI_{4-n}{}^{2-}$, (n = 1, 2)	67D
	dis	150—200	MNO_3(l)	$K(ZnCl_2 + ZnBr_2 \rightleftharpoons 2ZnClBr)$ −1·08(150°), 0·38(180°), 1·28(200°); $K(ZnCl_2 + ZnI_2 \rightleftharpoons 2ZnClI)$ −1·22(150°), 0·38(180°), 1·11(200°); $K(ZnBr_2 + ZnI_2 \rightleftharpoons 2ZnBrI)$ −1·35(150°), 0·18(180°), 0·96(200°); $\beta_2(Br^-)/\beta_2(Cl^-)$ −0·1(150°—200°); $\beta_2(I^-)/\beta_2(Cl^-)$ −0·2(150°—200°); $\beta(ZnCl_2{}^0Br^-)$ 0·8(150°), 0·7(180, 200°); $\beta(ZnCl_2{}^0Cl^-{}_{-1}Br^-{}_2)$ 0·8(150°), 0·7(180°, 200°); $\beta(ZnCl_2{}^0Cl^-Br^-)$ 1·45(150°), 1·30(180°), 1·18(200°); $\beta(ZnCl_2{}^0Br^-{}_2)$ 1·61(150°), 1·54(180°), 1·52(200°); $\beta(ZnCl_2{}^0Cl^-{}_{-1}Br^-{}_3)$ small(150°), 0·60(180°), 0·57(200°); $\beta(ZnCl_2{}^0I^-)$ 0·45(150°), 0·38(180°), 0·36(200°);	68Z+68Z

continued

Metal	*Method*	*Temp*	*Medium*	*Log of equilibrium constant, remarks*	*Ref*
Zn^{2+} (*contd*)				$\beta(ZnCl_2{}^0Cl^-{}_{-1}I^-{}_2)$ 0·15(150°), 0·08(180°), 0·04(200°); $\beta(ZnCl_2{}^0Cl^-I^-)$ 1·11(150°), 0·90(180°), 0·85(200°); $\beta(ZnCl_2{}^0I^-{}_2)$ small(150°), 0·51(180°), 0·56(200°); $\beta(ZnBr_2{}^0I^-)$ small(150°), 0·11(180°), 0·15(200°); no ev $ZnBrI_2{}^-$, $ZnBr_3I^{2-}$, $ZnBr_2I_2{}^{2-}$, $ZnBrI_3{}^{2-}$ in $Li_{0.43}K_{0.57}NO_3$(l, eut)	
	dis	150—200	MNO_3(l)	$K_D = K_D[ZnX_2(\text{salt melt}) \rightleftharpoons ZnX_2(\text{org})]$, "salt melt" = $Li_{0.43}K_{0.57}NO_3$(l, eut), org = "polyphenyl eutectic"; X = Cl: K_1 7·5(150°), 5·2(200°), K_2 7·3(150°), 4·9(200°), K_3 1·04(150°), 0·85(200°), K_4 0·9(150°), 0·7(200°), K_D 1·11(150°), 0·34(200°); X = Br; K_1 7·5(150°), 5·1(200°), K_2 7·25(150°), 4·85(200°), K_3 0·4(150°), 0·25(200°), K_D 1·38(150°), 1·08(200°); X = I, K_1 7·4(150°), 5·0(200°), K_2 7·2(150°), 4·8(200°) no ev ani cpx, K_D 1·59(150°), 1·45(200°); $\Delta H_1 = \Delta H_2$ = 11(Cl, Br, I), $\Delta H_3 = \Delta H_4$ = 1·1(Cl), ΔH_3 = 0·6(Br), ΔH_D = 3·6(Cl), 1·4(Br), 0·75(I); $K_D[ZnXY(\text{salt melt}) \rightleftharpoons ZnXY(\text{org})]$; ZnClBr: 2·24(150°), 0·53(200°), ZnClI; 2·31(150°), 0·69(200°), ZnBrI: 2·41(150°), 1·14(200°), $\Delta H_D \approx -30$ †K(ZnClBr) −1·04(150°), 1·30(200°), †K(ZnClI) −1·18(150°), 1·15(200°), †K(ZnBrI) −1·28(150°), 0·98(200°), Δ†$H \approx 43$	68Za
			org	†K(ZnClBr) 0·93(150°, 200°), †K(ZnClI) 0·75(150°, 200°). †K(ZnBrI) 0·57(150°, 200°), Δ†$H \approx 0$; †K(ZnXY): $ZnX_2 + ZnY_2 \rightleftharpoons 2ZnXY$; also data for 165 and 180°, *m* units	
Cd^{2+}					

1964 *Erratum*: p. 355, *in ref* 55Da, *for* "ev $CdBr_nI_{4-n}{}^{2-}$, n = 0, 1, 2, 3, 4" *read* "ev $CdCl_nI_{4-n}{}^{2-}$ and $CdBr_nI_{4-n}{}^{2-}$, n = 0, 1, 2, 3, 4"; *for* "sp" *read* "Ram".

Metal	*Method*	*Temp*	*Medium*	*Log of equilibrium constant, remarks*	*Ref*
	pol	25	2 ($NaClO_4$)	$\beta(Cd^{2+}Br^-I^-)$ 3·32, $\beta(Cd^{2+}Br^-I^-{}_2)$ 4·51, $\beta(Cd^{2+}Br^-I^-{}_3)$ 5·83, $\beta(Cd^{2+}Br^-{}_2I^-)$ 3·75, $\beta(Cd^{2+}Br^-{}_2I^-{}_2)$ 5·33, $\beta(Cd^{2+}Br^-{}_3I^-)$ 4·18	67Sa
Hg^{2+}					

1964 *Addendum*: p. 355, *for ref* "55D" *read* "55D = 55Da"

Metal	*Method*	*Temp*	*Medium*	*Log of equilibrium constant, remarks*	*Ref*
Hg^{2+}	dis	25	C_6H_6	†K(HgClBr) 1·16, †K(HgClI) 1·50, †K(HgBrI) 0·76 in C_6H_6(l)	57M/64E
	dis	25	0·5 $NaClO_4$	†K(HgClBr) 1·20	57M/64E
	Ram, sol		MeOH	$K[(F_3CS)_2Hg + HgX_2 \rightleftharpoons 2F_3CSHgX]$ 1·3(X = Cl), 0·36(Br), 0·22(I), 0·13(SCN), low for X = CN, ev $(F_3CS)_2HgX^-$, in MeOH(l)	61D+61Da
	sp	5—60	0·001 $HClO_4$	†ΔH(HgClBr) = −1·0, †ΔH(HgClI) = −1·4, †ΔH(HgBrI) = −0·8	64E
	dis	150	MNO_3(l)	$K_D[HgXY(\text{melt}) \rightleftharpoons HgXY(\text{org})]$ 0·59(HgClBr), 1·95(HgClI), 2·57(HgBrI); †K(HgClBr) 1·85(melt), 1·69(org); †K(HgClI) −0·35(melt), 1·65(org);	64Z

continued overleaf

80 Halogenides: mixed and comparative (contd)

Metal	*Method*	*Temp*	*Medium*	*Log of equilibrium constant, remarks*	*Ref*
Hg^{2+} (*contd*)				†K(HgBrI) −1·0(melt), 1·58(org), m units; melt = $Li_{0.43}K_{0.57}NO_3$(l, eut), org = polyphenyl eutectic; †K(HgXY); $HgX_2+HgY_2 \rightleftharpoons 2HgXY$	
	dis	150	MNO_3(l)	$\beta(HgCl_2, mCl^-, nBr^-)$ 0·78(m = −1, n = 1), −0·32(−2, 2), 0·78(1, 0), 1·86(0, 1), 1·43(−1, 2), 0·58(−2, 3), 1·56(2, 0), 2·79(1, 1), 2·61(0, 2), 2·36(−1, 3), 1·54(−2, 4); $\beta(HgCl_2, mCl^-, nI^-)$ −0·54(m = −1, n = 1), −0·73(−2, 2), 0·88(0, 1), 0·28(−1, 2), 0·45(−2, 3), 1·95(1, 1), 1·77(0, 2), 1·98(−1, 3), 1·56(−2, 4); $\beta(HgBr_2, mBr^-, nI^-)$ −0·94(m = −1, n = 1), −0·92(−2, 2), 0·90(1, 0), 0·76(0, 1), 0·16(−1, 2), 0·26(−2, 3), 1·86(2, 0), 2·04(1, 1), 1·52(0, 2), 1·62(−1, 3), 1·38(−2, 4) in $Li_{0.43}K_{0.57}NO_3$(l)	64Za
	var		MeOH	$K[(F_3CSe)_2Hg+HgX_2 \rightleftharpoons 2F_3CSeHgX]$ 1·3(X = Cl), 1·18(Br), 0·31(I), 0·37(SCN), low for X = CN, ev $(F_3CSe)_2HgX_2^{2-}$, $[(F_3CSe)_2Hg]_2X^-$ in MeOH(l), (Ram, sol, con, ir)	65C
	sol	25	org	K_{s2} values for $HgCl_2$, $HgBr_2$ and HgI_2 in dioxan and 11 hydrocarbons	65E
	pol	20?	Me_2NCHO	$\beta_4(Cl^-)$ 32·9, $\beta_4(Br^-)$ 35·2, $\beta_4(I^-)$ 36·6, $\beta_4(SCN^-)$ 22·5, $\beta_4[SC(NH_2)_2]$ 26·5 in Me_2NCHO(l), I = 0·1($LiClO_4$)	66Bd
	sol	25	1 $M(NO_3)$	$\beta(Hg^{2+}Br^-I^-{}_2)$ 25·7(M = Li^+), 26·0(Na^+, K^+), 26·3(NH_4^+), 26·5(Rb^+); $\beta(Hg^{2+}Br^-{}_2I^-{}_2)$ 27·35(Li^+), 27·37(Na^+), 27·43(K^+), 27·45(NH_4^+), 27·48(Rb^+), [$K_s(Hg^{2+}I^-{}_2)$ −28·19]	66C
RHg^+	gl	25	0·1 (KNO_3)	$K_1(Cl^-)$ 4·90(Me), 4·78(Et), 4·65(Pr), 4·55(Bu); $K_1(Br^-)$, 5·98(Me), 5·90(Et), 5·80(Pr), 5·74(Bu); $K_1(I^-)$, 7·70(Me), 7·85(Et), 8·20(Pr);	68Zc
		20	0·1 (KNO_3)	$K_1(I^-)$ 8·70 (R = Me)	
Tl^{3+}	var		var	ev $TlCl_nBr_{4-n}^-$, $TlCl_nI_{4-n}^-$, $TlBr_nI_{4-n}^-$, n = 1, 2, 3 in solids and MeCN(l) (pre, sp, con, X)	68Ma
Sn^{4+}	var		solid	ev $SnCl_4Br_2^{2-}$, $SnCl_2Br_4^{2-}$, $SnCl_4I_2^{2-}$, $SnCl_2I_4^{2-}$, $SnBr_4I_2^{2-}$, $SnBr_2I_4^{2-}$, all *cis* (ir, Ram, pre)	68C
	nmr	22	org	$K(2SnF_5X^{2-} \rightleftharpoons cis\text{-}SnF_4X_2^{2-}+SnF_6^{2-})$ −0·52 (X = CNO^-, in MeOH), −0·72(Cl^-), −0·42(Br^-), 0·04(I^-), in $CHCl_3$(l); $K(2SnF_5X^{2-} \rightleftharpoons trans\text{-}SnF_4X_2^{2-}+SnF_6^{2-})$ −1·03 (X = CNO^-, in MeOH), −1·77(Cl^-), −1·92(Br^-), < −1·96(I^-), in $CHCl_3$(l)	68D
	nmr	22	var	$K(SnF_5X^{2-}+F^- \rightleftharpoons SnF_6^{2-}+X^-)$ 3·26(X^- = Cl^-) 4·24(Br^-), 5·64(H_2O); $K(SnF_4Cl_2^{2-}+F^- \rightleftharpoons SnF_5Cl^{2-}+Cl^-)$ 3·89, $K_6(F^-)$ 5·64	
	qh	22?	var	$K(SnF_5OH^{2-}+F^- \rightleftharpoons SnF_6^{2-}+OH^-)$ −6·89	
Bi^{3+}	sp	25	dioxan	$K(BiICl_2+BiI_3 \rightleftharpoons 2BiI_2Cl)$ −0·8 $K(BiCl_3+BiI_2Cl \rightleftharpoons 2BiICl_2)$ 0·56 in dioxan(l)	66G

contin

Metal	*Method*	*Temp*	*Medium*	*Log of equilibrium constant, remarks*	*Ref*
Group 7B$_2$	*iE*	25	org	$\beta(X_2{}^oX^-) > 13$(Cl), 7·3(Br), 6·7(I) in $MeNO_2$(l); > 12(Cl), 9·3(Br), 8·3(I) in Me_2CO(l); 10(Cl), 7(Br), 6·6(I) in MeCN(l); in all solvents $I = 0{\cdot}1(Et_4NClO_4)$	64N
	iE	22	sulpholan	$\beta(X_2{}^oX^-)$ 3·1(Cl), 6·8(Br), 7·4(I) in sulpholan, 0·1M-$LiClO_4$	68Bb
		22	MeOH	$\beta(X_2{}^oX^-)$ 2·3(Br), 4·2(I) in MeOH(l) 0·1M-$LiClO_4$	
	red	22	sulpholan	$\beta(Cl_2{}^oCl^-)$ 3·1, $\beta(Br_2{}^oCl^-)$ 4·8, $\beta(Br_2{}^oBr^-)$ 6·4, $\beta(I_2{}^oCl^-)$ 4·8, $\beta(I_2{}^oBr^-)$ 6·5 $\beta(I_2{}^oI^-)$ 7·5 in sulpholan, 0·1M-$LiClO_4$	68Bc

61D A. J. Downs, E. A. V. Ebsworth, and H. J. Emeléus, *J. Chem. Soc.*, 1961, 3187
61Da A. J. Downs, thesis, Cambridge University (U.K.) 1961, quoted by ref 65C
62S M. Serátor, *Sbornik Prac. chem. Fak. S.V.Š.T. (Bratislava)*, 1962, 31
63I L. A. Il'yukevich and G. A. Shagisultanova, *Zhur. neorg. Khim.*, 1963, **8**, 2308
63M P. V. Manning and R. C. Jarnagin, *J. Phys. Chem.*, 1963, **67**, 2884
63V V. M. Vdovenko, G. A. Romanov, and V. A. Shcherbakov, *Radiokhimiya*, 1963, **5**, 664
64E I. Eliezer, *J. Phys. Chem.*, 1964, **68**, 2722
64N I. V. Nelson and R. T. Iwamoto, *J. Electroanalyt. Chem.*, 1964, **7**, 218
64S J. A. Salthouse and T. C. Waddington, *J. Chem. Soc.*, 1964, 4664
64Z M. Zangen, *Israel J. Chem.*, 1964, **2**, 91
64Za M. Zangen and Y. Marcus, *Israel J. Chem.*, 1964, **2**, 155
65B H. L. Bott and A. J. Poë, *J. Chem. Soc.*, 1965, 5931
65Ba E. Blasius and W. Preetz, *Z. anorg. Chem.*, 1965, **335**, 16
65C H. J. Clase and E. A. V. Ebsworth, *J. Chem. Soc.*, 1965, 940
65E I. Eliezer, *J. Chem. Phys.*, 1965, **42**, 3625
65R R. R. Rettew and R. C. Johnson, *Inorg. Chem.*, 1965, **4**, 1565
66B E. J. Bounsall and A. J. Poë, *J. Chem. Soc.*, 1966, 286
66Ba D. H. Brown, K. R. Dixon, and D. W. A. Sharp, *J. Chem. Soc. (A)*, 1966, 1244
66Bb L. H. Berka and W. L. Masterton, *J. Phys. Chem.*, 1966, **70**, 1641
66Bc A. A. Biryukov, V. I. Shlenskaya, and I. P. Alimarin, *Izvest. Akad. Nauk S.S.S.R., Ser. khim.*, 1966, 3
66Bd M. Bréant and N. Van Kiet, *Compt. rend.*, 1966, **262**, 955
66C D. M. Czakis-Sulikowska, *Roczniki Chem.*, 1966, **40**, 521
66D G. B. Deacon and F. B. Taylor, *J. Chem. Soc.*, 1966, 463
66Da G. Dolcetti, A. Peloso and L. Sindellari, *Gazzetta*, 1966, **96**, 1648
66E J. C. Evans and G. Y.-S. Lo, *J. Phys. Chem.*, 1966, **70**, 543
66G F. Gaizer and M. T. Beck, *J. Inorg. Nuclear Chem.*, 1966, **28**, 503
66M I. N. Marov, Yu. N. Dubrov, V. K. Belyaeva, A. N. Ermakov, and D. I. Ryabchikov, *Zhur. neorg. Khim.*, **1966, 11, 2443**
66P J. Pouradier and M. Coquard, *J. Chim. phys.*, 1966, **63**, 1072
66S S. C. Srivastava and L. Newman, *Inorg. Chem.*, 1966, **5**, 1506
66Sa L. G. Savedoff, *J. Amer. Chem. Soc.*, 1966, **88**, 664
67B J. Braunstein and R. Lindgren, *Electrochim. Acta*, 1967, **12**, 299
67Ba H. L. Bott and A. J. Poë, *J. Chem. Soc. (A)*, 1967, 205
67D G. B. Deacon, J. H. S. Green, and F. B. Taylor, *Austral. J. Chem.*, 1967, **20**, 2069
67G R. Grassi, A. Haim, and W. K. Wilmarth, *Inorg. Chem.*, 1967, **6**, 237
67Ga P. S. Gentile, T. A. Shankoff, and J. Carlotto, *J. Inorg. Nuclear Chem.*, 1967, **29**, 1393
67Gb A. A. Grinberg and A. I. Dobroborskaya, *Zhur. neorg. Khim.*, 1967, **12**, 276
67Gc A. A. Grinberg, M. I. Gel'fman, and N. V. Kiseleva, *Zhur. neorg. Khim.*, 1967, **12**, 1171
67H D. J. Hewkin and A. J. Poë, *J. Chem. Soc. (A)*, 1967, 1884
67M L. M. Mukherjee and J. J. Kelley, *J. Phys. Chem.*, 1967, **71**, 2348
67P B. I. Peshchevitskii and V. I. Belevantsev, *Zhur. neorg. Khim.*, 1967, **12**, 312
67Pa A. J. Poë, K. Shaw and M. J. Wendt, *Inorg. Chim. Acta*, 1967, **1**, 371
67R A. M. Rozen and A. I. Mikhailichenko, *Zhur. neorg. Khim.*, 1967, **12**, 729
67S S. C. Srivastava and L. Newman, *Inorg. Chem.*, 1967, **6**, 762
67Sa A. Swinarski and A. Grodzicki, *Roczniki Chem.*, 1967, **41**, 1205
67Sb V. M. Shul'man, R. L. Shchekochikhina, and B. I. Peshchevitskii, *Zhur. neorg. Khim.*, 1967, **12**, 281
68A S. Ašperger and M. Flögel, *J. Chem. Soc. (A)*, 1968, 769
68B J. Badoz-Lambling and J.-C. Bardin, *Compt. rend.*, 1968, **266**, *C*, 95
68Ba H. L. Bott, A. J. Poë, and K. Shaw, *Chem. Comm.*, 1968, 793
68Bb R. L. Benoit, M. Guay, and J. Desbarres, *Canad. J. Chem.*, 1968, **46**, 1261
68Bc R. L. Benoit and M. Guay, *Inorg. Nuclear Chem. Letters*, 1968, **4**, 215
68C R. J. H. Clark, L. Maresca, and R. J. Puddephatt, *Inorg. Chem.*, 1968, **7**, 1603
68D P. A. W. Dean and D. F. Evans, *J. Chem. Soc. (A)*, 1968, 1154

continued overleaf

80 Halogenides: mixed and comparative (contd)

68Da R. Davies and A. G. Sykes, *J. Chem. Soc. (A)*, 1968, 2840
68G V. Gutmann and U. Mayer, *Monatsh.*, 1968, **99**, 1383
68L C. H. Liu, L. Newman, and J. Hasson, *Inorg. Chem.*, 1968, **7**, 1868
68M D. C. McCain and R. J. Myers, *J. Phys. Chem.*, 1968, **72**, 4115
68Ma R. W. Matthews and R. A. Walton, *J. Chem. Soc. (A)*, 1968, 1639
68P B. I. Peshchevitskii, R. L. Shchekochikhina, and V. M. Shul'man, *Zhur. neorg. Khim.*, 1968, **13**, 2194
68Pa A. Peloso and R. Ettorre, *J. Chem. Soc. (A)*, 1968, 2253
68Pb A. J. Poë and D. H. Vaughan, *Inorg. Chim. Acta*, 1968, **2**, 159
68T H. G. Tsiang and W. K. Wilmarth, *Inorg. Chem.*, 1968, **7**, 2535
68Z A. v. Zelewsky, *Helv. Chim. Acta*, 1968, **51**, 803
68Za M. Zangen, *Inorg. Chem.*, 1968, **7**, 133, 138
68Zb M. Zangen, *Inorg. Chem.*, 1968, **7**, 1202
68Zc P. Zanella, G. Plazzogna, and G. Tagliavini, *Inorg. Chim. Acta*, 1968, **2**, 340

81 Ligands with 8B (or 0) group elements ("inert gases")

Metal	*Method*	*Temp*	*Medium*	*Log of equilibrium constant, remarks*	*Ref*
				L = Zerovalent "inert gas"	
H^+	gl	25?	dil	$K_1(H^+ + L \rightleftharpoons HL^+) \approx 2$($L = He^0, Ar^0$, also N_2), less for Ne^0	67H
				Xenon(VI) *species*	
H^+	gl, sp	25	0·5 Na(ClO_4)	$K(XeO_3 + OH^- \rightleftharpoons HXeO_4^-)$ 3·17	64A
	Ram		2 XeO_3	ev trigonal XeO_3 and other species	64C
				Xenon(VIII) *species*	
H^+	gl, sp	24	0·1 Na(ClO_4)	$K(H_2XeO_6^{2-} + OH^- \rightleftharpoons HXeO_6^{3-} + H_2O)$ 3·62	64A
	gl	24	var	$K_a(H_4XeO_6)$ −2, −6, −10·5	
	X		solid	ev XeO_6^{4-} in $K_4XeO_6(H_2O)_9$(s)	64Z
Li^+	est			$K_s(Li^+{}_4XeO_6^{4-}) < -5·9$	64K

64A E. H. Appelman and J. G. Malm, *J. Amer. Chem. Soc.*, 1964, **86**, 2141
64C H. H. Claassen and G. Knapp, *J. Amer. Chem. Soc.*, 1964, **86**, 2341
64K C. W. Koch and S. M. Williamson, *J. Amer. Chem. Soc.*, 1964, **86**, 5439
64Z A. Zalkin, J. D. Forrester, D. H. Templeton, S. M. Williamson, and C. W. Koch, *J. Amer. Chem. Soc.*, 1964, **86**, 3569
67H E. M. Holleran, J. T. Hennessy, and F. R. La Pietra, *J. Phys. Chem.*, 1967, **71**, 3081

PART II—ORGANIC LIGANDS

TABLES

CH_2O_2 Methanoic acid (formic acid) HL 101

Metal	*Method*	*Temp*	*Medium*	*Log of equilibrium constant, remarks*	*Ref*
H^+	gl	25	0·1 ($NaClO_4$) 50% dioxan	K_1 4·75±0·01	68G
Al^{3+}	ix	25	1	K_1 1·78±0·03	62T
Ba^{2+}	gl	25	~0	K_1 1·38	48S
	gl	25, 35	0 corr	K_1 (25°) 1·38±0·01, (35°) 1·34±0·01 ΔH_1 −1·89, ΔS_1 −0·3	56N
Ca^{2+}	gl	25	~0	K_1 1·43	48S
	gl	25, 35	0 corr	K_1 (25°) 1·43±0·01, (35°) 1·46±0·01 ΔH_1 0·98, ΔS_1 10	56N
Cd^{2+}	pol	30	1 (KNO_3)	K_1 0·65, β_2 0·40, β_3 1·32	65G
Ce^{3+}	ix	25	1	K_1 1·65±0·05	62T
Cu^{2+}	sp	28	0·5	K_1 2·0	65D
	gl	25	0·1 ($NaClO_4$) 50% dioxan	K_1 2·80, $K(CuA^{2+}+L^- \rightleftharpoons CuAL^+)$ 2·84 where A = 2,2′-bipyridyl	68G
Fe^{3+}	ix	25	1	K_1 1·85±0·03, β_2 3·61±0·05, β_3 3·95±0·15, β_4 5·4±0·2	62T
K^+	con	30	acetic acid	K_1 6·94	54J
Li^+	con	30	acetic acid	K_1 7·06	54J
Mg^{2+}	gl	25	~0	K_1 1·43	48S
	gl	25, 35	0 corr	K_1 (25°) 1·43±0·01, (35°) 1·39±0·01 ΔH_1 −1·77, ΔS_1 0·6	56N
Mn^{2+}	ix	25	1	K_1 0·80±0·05	62T
Na^+	con	30	acetic acid	K_1 7·19	54J
Pb^{2+}	pol	30	1 (KNO_3)	K_1 0·85, β_2 0·98, β_3 1·15	66J
Sr^{2+}	gl	25	~0	K_1 1·39	48S
	gl	25, 35	0 corr	K_1 (25°) 1·39±0·01, (35°) 1·40±0·01 ΔH_1 0·59, ΔS_1 8·3	56N
UO_2^{2+}	pol	?	1 ($NaNO_3$)	K_1 2·4(?), β_2 3·0	62H
	sp	20	1 ($NaClO_4$)	K_1 1·89±0·01, K_2 1·08±0·03	67M
Zn^{2+}	gl	25	0·1 ($NaClO_4$) 50% dioxan	K_1 1·97, $K(ZnA^{2+}+L^- \rightleftharpoons ZnAL^+)$ 1·83 where A = 2,2′-bipyridyl	68G

48S J. Schubert, *J. Phys. & Colloid Chem.*, 1948, **52**, 340
54J M. M. Jones and E. Griswold, *J. Amer. Chem. Soc.*, 1954, **76**, 3247
56N G. N. Nancollas, *J. Chem. Soc.*, 1956, 744
62H J. Hala and A. Okac, *Coll. Czech. Chem. Comm.*, 1962, **27**, 1697
62T H. Tsubota, *Bull. Chem. Soc. Japan*, 1962, **35**, 640
65D V. P. Devendran and M. Santappa, *Current Sci.*, 1965, **34**, 145
65G J. N. Gaur and D. S. Jain, *Austral. J. Chem.*, 1965, **18**, 1687
66J D. S. Jain and J. N. Gaur, *J. Indian Chem. Soc.*, 1966, **43**, 425
67M C. Miyake and H. W. Nurnberg, *J. Inorg. Nuclear Chem.*, 1967, **29**, 2411
68G R. Griesser, B. Prijs, and H. Sigel, *Inorg. Nuclear Chem. Letters*, 1968, **4**, 443

101.1 CH_2N_4 Tetrazole L

Metal	Method	Temp	Medium	Log of equilibrium constant, remarks	Ref
Ni^{2+}	sp	24	0·179 $Ni(ClO_4)_2$ dimethylformamide	K_1 0·4±0·5, β_2 1·12±0·03 (?)	66H

66H R. D. Holm and P. L. Donnelly, *J. Inorg. Nuclear Chem.*, 1966, **28**, 1887

103.1 CH_4O Methanol HL

Metal	Method	Temp	Medium	Log of equilibrium constant, remarks	Ref
H^+	H	20	1·0 $(CH_3)_4NCl$ methanol	K_1 16·60±0·05 (Reciprocal of solvent ion product constant)	64G
			1·0 LiCl methanol	K_1 16·05±0·05	
Al^{3+}	H	20	1 $(CH_3)_4NCl$ methanol	$K(2Al^{3+}+3L^- \rightleftharpoons Al_2L_3^{3+})$ 42·0±0·2 $K(Al_2L_3^{3+}+L^- \rightleftharpoons 2AlL_2^+)$ 11·1±0·2 $K_3 \sim$ 10·5, $K_4 \sim$ 5·5	64G
B^{3+}	H	20	1 $(CH_3)_4NCl$ methanol	K_4 5·62 $K(BL_4^-+H^+ \rightleftharpoons BL_3)$ 10·98	64G
Ga^{3+}	H	20	1 $(CH_3)_4NCl$ methanol	$K_1 \sim$ 10·2	64G
Ge^{4+}	H	20	1 $(CH_3)_4NCl$ methanol	K_4 13·65 $K(GeL_4+H^+ \rightleftharpoons GeL_3^+)$ 2·95	64G
Nb^{5+}	H	20	1 $(CH_3)_4NCl$ methanol	K_5 10·45, K_6 5·45 $K(NbL_6^-+H^+ \rightleftharpoons NbL_5)$ 6·15 $K(NbL_7^{2-}+H^+ \rightleftharpoons NbL_6^-)$ 11·15	64G
	H	20	1 $(CH_3)_4NCl$ methanol	$K(NbAL_4+2L^- \rightleftharpoons NbL_6^-+A^-)$ 5·1±0·2 $K(NbL_4^++A^- \rightleftharpoons NbAL_4)$ 10·84 $K(NbAL_3^++L^- \rightleftharpoons NbAL_4)$ 12·4±0·1 $K(NbL_5+HA \rightleftharpoons NbAL_4)$ 5·18 where HA = acetylacetone $K(NbBL_3+H_2B+L^- \rightleftharpoons NbB_2L_2^-)$ 13·9±0·1 $K(NbB_2L_2^-+H_2B+L^- \rightleftharpoons NbB_3L^{2-})$ 7·0±0·1 $K(NbBL_3+L^- \rightleftharpoons NbBL_4^-)$ 7·89 $K(NbBL_4^-+NbBL_3 \rightleftharpoons Nb_2B_2L_7^-)$ 2·5±0·2 where H_2B = catechol	65G
	H	20	1 $(CH_3)_4NCl$ methanol	$K(NbOL_2^++L^- \rightleftharpoons NbOL_3)$ 10·51 $K(NbOL_4^-+H^+ \rightleftharpoons NbOL_3)$ 6·03	64G
Ta^{5+}	H	20	1 $(CH_3)_4NCl$ methanol	K_5 11·47, K_6 6·67 $K(TaL_6^-+H^+ \rightleftharpoons TaL_5)$ 5·13 $K(TaL_7^{2-}+H^+ \rightleftharpoons TaL_6^-)$ 9·93	64G
	H	20	1 $(CH_3)_4NCl$ methanol	$K(TaAL_4+2L^- \rightleftharpoons TaL_6^-+A^-)$ 7·1±0·2 $K(TaL_4^++A^- \rightleftharpoons TaAL_4)$ 11·04 $K(TaAL_3^++L^- \rightleftharpoons TaAL_4)$ 12·95±0·1 $K(TaL_5+HA \rightleftharpoons TaAL_4)$ 4·36 where HA = acetylacetone $K(TaBL_3+H_2B+L^- \rightleftharpoons TaB_2L_2^-)$ 14·2±0·1 $K(TaB_2L_2^-+H_2B+L^- \rightleftharpoons TaB_3L^{2-})$ 7·85±0·1 $K(TaBL_3+L^- \rightleftharpoons TaBL_4^-)$ 9·04 $K(TaBL_4^-+TaBL_3 \rightleftharpoons Ta_2B_2L_7^-)$ 2·5±0·2 where H_2B = catechol	65G

cont

CH_4O (contd) 103.1

Metal	Method	Temp	Medium	Log of equilibrium constant, remarks	Ref
Th^{4+}	H	20	1 $(CH_3)_4NCl$ methanol	K_1 12·35 $K(ThL_2^{2+}+H^+ \rightleftharpoons ThL^{3+})$ 4·25	64G
Ti^{4+}	H	20	1 $(CH_3)_4NCl$ methanol	$K(Ti_2L_8+L^- \rightleftharpoons Ti_2L_9^-)$ 11·3	64G
W^{6+}	H	20	1 $(CH_3)_4NCl$ methanol	$K(WOL_3^++L^- \rightleftharpoons WOL_4)$ 12·51 $K(WOL_5^-+H^+ \rightleftharpoons WOL_4)$ 4·09	64G
Zn^{2+}	H	20	1 $(CH_3)_4NCl$ methanol	$K(2Zn^{2+}+3L^- \rightleftharpoons Zn_2L_3^+)$ 12·1±0·1	64G

64G R. Gut, *Helv. Chim. Acta*, 1964, **47**, 2262
65G R. Gut, H. Buser, and E. Schmid, *Helv. Chim. Acta*, 1965, **48**, 878

CH_5N Methylamine L 104

Metal	Method	Temp	Medium	Log of equilibrium constant, remarks	Ref
H^+	gl	25	0 corr	K_1 10·59±0·01	66P
	cal	25	0 corr	$\Delta H_1°$ −13·25±0·01	66P
Ag^+	Ag	25	0·02 0—90% methanol	β_2 (0%) 6·72, (30%) 7·10, (50%) 7·13, (60%) 7·15, (80%) 7·43, (90%) 7·51	63P
	sol	25	Cl^- or Br^-	β_2 (Cl^-) 8·86, (Br^-) 6·67	64S
Hg^{2+}	gl, cal	25	0 corr	$K(HgCl_2+L \rightleftharpoons HgClL^++Cl^-)$ (gl) 2·40±0·07 $\Delta H°$ (cal) −6·8±0·2, $\Delta S°$ −12 $K(HgClL^++L \rightleftharpoons HgL_2^{2+}+Cl^-)$ (gl) 2·21±0·05 $\Delta H°$ (cal) −1·0±0·3, $\Delta S°$ 7	66P
Pt^{2+}	Pt	18	1 (KNO_3)	β_4 40·1	61G

61G A. A. Grinberg and M. L. Gelfman, *Proc. Acad. Sci. (U.S.S.R.)*, 1961, **137**, 257 (87)
63P G. Popa, C. Luca, and V. Magearu, *J. Chim. phys.*, 1963, **60**, 355
64S R. L. Seth and A. K. Dey, *Indian J. Chem.*, 1964, **2**, 291
66P J. A. Portridge, J. J. Christiansen, and R. M. Izatt, *J. Amer. Chem. Soc.*, 1966, **88**, 1649

CH_5N_3 Diaminomethylimine (guanidine) L 104.1

Metal	Method	Temp	Medium	Log of equilibrium constant, remarks	Ref
H^+	sp	27	1 $(NaClO_4)$	K_1 13·54	64W
Hg^{2+}	gl	27	1 $(NaClO_4)$	β_2 24·96	64W

64W T. H. Wirth and N. Davidson, *J. Amer. Chem. Soc.*, 1964, **86**, 4325

104.2 CH_3ON Methanoic acid amide (formamide) L

Metal	*Method*	*Temp*	*Medium*	*Log of equilibrium constant, remarks*	*Ref*
Cd^{2+}	pol	?	0·05 $NaClO_4$ 40—100% methanol	K_1 (40%) 0·36±0·02, (77%) 0·79±0·05, (100%) 2·0±0·2 β_2 (77%) 0·42±0·06, (85%) 1·43±0·06 (92%) 1·57±0·05, (100%) 2·5±0·2	62M
			0·05 $NaClO_4$ 71—100% ethanol	K_1 (71%) 0·5±0·1, (82%) 0·81±0·05, (90%) 0·90±0·05, (96%) 1·3±0·1, (100%) 1·96±0·03, β_2 (90%) 1·48±0·08, (96%) 1·7±0·2, (100%) 2·8±0·1, β_3 (96%) 1·9±0·1, (100%) 4·1±0·1	
Pb^{2+}	pol	?	0·05 $NaClO_4$ 40—100% methanol	K_1 (40%) −0·90±0·05, (77%) 1·00±0·03, β_2 (40%) 0·80±0·04, (77%) 1·11±0·03, (92%) 2·35±0·06, (100%) 3·4±0·1, β_3 (85%) 2·11±0·06	62M
			0·05 $NaClO_4$ 13—100% ethanol	K_1 (13%) 0·67±0·03, (40%) 1·11±0·06, (71%) 1·85±0·05, (82%) 1·85±0·05, (90%) 1·7±0·1 β_2 (71%) 1·63±0·05, (82%) 1·83±0·04, (90%) 2·75±0·10, (96%) 4·20±0·05, β_3 (96%) 4·49±0·06, (100%) 4·6±0·1, β_4 (100%) 4·9±0·2	
Zn^{2+}	pol	?	0·05 $NaClO_4$ 40—100% methanol	β_2 (40%) 0·62±0·05, (77%) 0·79±0·03, (85%) 0·93±0·08, (92%) 1·9±0·1, (100%) 3·7±0·2	62M
			0·05 $NaClO_4$ 90—100% ethanol	K_1 (90%) 1·64±0·05, (96%) 2·1±0·2, β_2 (90%) 1·96±0·05, (96%) 3·5±0·1, β_3 (96%) 4·23±0·08, β_4 (100%) 9·0±0·4	

62M P. K. Migal and N. Kh. Grinberg, *Russ. J. Inorg. Chem.*, 1962, **7**, 268 (527), 675 (1309)

104.3 CH_4ON_2 Carbamide (urea) L

Metal	*Method*	*Temp*	*Medium*	*Log of equilibrium constant, remarks*	*Ref*
Cu^{2+}	pH	?	0·2—2	K_1 (0·2) −0·7±0·1, (0·5) −0·55±0·07, (2) −0·49±0·06	65S

65S V. M. Shulman and T. V. Kramareva, *Russ. J. Inorg. Chem.*, 1965, **10**, 890 (1632)

105 CH_4N_2S Thiocarbamide (thiourea) L

Metal	*Method*	*Temp*	*Medium*	*Log of equilibrium constant, remarks*	*Ref*
Ag^+	Ag	25	0·052 $NaC_2H_3O_2$ 40% ethanol	β_3 13·2±0·1	61T
	Ag	28	2 NH_4NO_3 80% dioxan	β_3 13·6	64K
	Ag	25	0·4 ($LiNO_3$) 0—85% methanol	β_3 (0%) 13·48, (25%) 14·10, (50%) 14·52, (75%) 15·18, (85%) 15·61	67N
			0·4 ($LiNO_3$) 25—85% ethanol	β_3 (25%) 14·22, (50%) 14·62, (75%) 15·37, (85%) 15·66	

cont

Metal	Method	Temp	Medium	Log of equilibrium constant, remarks	Ref
			0·4 ($LiNO_3$) 25—85% propanol	β_3 (25%) 14·13, (50%) 14·62, (75%) 15·17, (85%) 15·57	
Bi^{3+}	sp	18—22	0·9 $HClO_4$	β_6 9·3±0·1	66Sa
	sp	?	1 $HClO_4$ 0—2 $NaClO_4$	K_1 (0) 2·28±0·04, (1) 2·35±0·04, (2) 2·40±0·05 K_2 (0) 1·04±0·08, (1) 1·05±0·08, (2) 1·15±0·08, K_3 (0) 0·32±0·06, (1) 0·30±0·06, (2) 0·32±0·07	67V
			→ 0	K_1 2·24±0·05, K_2 0·96±0·08, K_3 0·30±0·06	
Cd^{2+}	pol	25	0·01 NH_4Cl 0—100% ethanol	K_1 (0%) 1·30, (5%) 1·32, (25%) 1·30, β_2 (0%) 1·89, (5%) 1·86, (25%) 2·43, (55%) 2·60, (77%) 3·92, (91%) 5·00, β_3 (0%) 2·20, (5%) 2·00, (25%) 3·20, (55%) 3·78, (77%) 4·78, (91%) 6·6, β_4 (0%) 3·20, (5%) 3·30, (25%) 4·04, (55%) 5·38, (77%) 6·18, (91%) 8·20, (100%) 10·62	63M
	pol	25	0·01 NH_4NO_3 0—100% methanol	K_1 (0%) 1·32, (20%) 0·60 (?), (40%) 1·60, (60%) 1·70 (?), (80%) 3·00, β_2 (0%) 2·04, (20%) 2·51, (40%) 2·60, (60%) 3·30, (80%) 4·08, (100%) 5·88, β_3 (0%) 2·20, (20%) 3·38, (40%) 3·68, (60%) 4·26, (80%) 5·11, (100%) 7·00, β_4 (0%) 3·04, (40%) 4·48, (60%) 5·04, (80%) 5·74, (100%) 8·21, β_5 (20%) 4·23, (40%) 4·92, (60%) 5·66, (80%) 6·76, (100%) 10·28	64M
	red	25	2 0—90% acetone	K_1 (0%) 1·6, (50%) 2·1, (80%) 2·65, (90%) 3·05, K_2 (0%) 1·1, (50%) 1·65, (80%) 2·15, (90%) 2·65, K_3 (0%) 1·0, (50%) 1·40, (80%) 1·9, (90%) 2·35, K_4 (0%) 0·6, (50%) 1·1, (80%) 1·55, (90%) 2·0, K_5 (80%) ~ 0·45, (90%) ~ 0·7, K_6 (90%) ~ 0·6	66Sb
Co^{2+}	sp	18—22	0·1 ($NaClO_4$) 95% ethanol	β_2 1·7, β_6 4·5	66Sa
	red	25	2 90% acetone	K_1 1·05, K_2 0·7	66Sb
Hg^{2+}	cal	25	~ 0 0—92·3% ethanol	K [$Hg(CN)_2 + L \rightleftharpoons Hg(CN)_2L$] (0·0%) 1·97±0·06, (20·0%) 2·03±0·08, (40·0%) 1·94±0·10, (60·0%) 1·86±0·05, (80·0%) 2·11±0·06, (92·3%) 2·28±0·04 $\Delta H°$ (0·0%) −1·5±0·1, (20·0%) −2·4±0·1, (40·0%) −4·51±0·09, (60·0%) −5·93±0·04, (80·0%) −5·96±0·04, (92·3%) −6·16±0·03 $\Delta S°$ (0·0%) 4·0±0·3, (20·0%) 1·3±0·3, (40·0%) −6·3±0·3, (60·0%) −11·1±0·2, (80·0%) −10·3±0·2, (92·3%) −10·3±0·1 K [$Hg(CN)_2L + L \rightleftharpoons Hg(CN)_2L_2$] (0·0%) 0·58±0·04, (20·0%) 0·77±0·08, (40·0%) 1·04±0·05, (60·0%) 1·26±0·04, (80·8%) 1·30±0·08, (92·3%) 1·34±0·08 $\Delta H°$ (0·0%) −7·9±0·2, (20·0%) −8·6±0·3, (40·0%) −8·2±0·2, (60·0%) −5·21±0·07, (80·0%) −5·0±0·2, (92·3%) −4·8±0·4 $\Delta S°$ (0·0%) −23·8±0·6, (20·0%) −25·5±0·9, (40·0%) −22·7±0·6, (60·0%) −11·5±0·2, (80·0%) −11·1±0·6, (92·3%) −10·1±0·2	68I

continued overleaf

105 CH_4N_2S (contd)

Metal	*Method*	*Temp*	*Medium*	*Log of equilibrium constant, remarks*	*Ref*
Ni^{2+}	red	25	2 90% acetone	K_1 1·05, K_2 0·45	66Sb
	sp	18—22	~ 0 ($NiCl_2$) 95% ethanol	K_1 −1·4 (?), β_2 0·84 (?)	66Sa
			0·3 ($NaClO_4$) 95% ethanol	β_2 0·7, β_6 1·36 (?)	
Pd^{2+}	sp	?	1 NaCl	K_3 4·86, K_4 4·24	66S
			1 NaBr	K_3 4·65, K_4 4·18	
			0·2 NaSCN	K_3 2·95, K_4 2·52	
Te^{2+}	sp	?	2 HCl	β_2 1·7	65T
Zn^{2+}	red	25	2 0—90% acetone	K_1 (0%) 0·5, (50%) 1·0, (80%) 1·75, (90%) 2·05, K_2 (0%) 0·25, (50%) 0·7, (80%) 1·45, (90%) 1·8, K_3 (0%) 0·15, (50%) 0·5, (80%) 1·20, (90%) 1·65, K_4 (80%) 0·4, (90%) 1·45, K_5 (90%) ~ 0·75	66Sb

61T V. F. Toropova, Yu. P. Kitaev, and G. K. Budnikov, *Russ. J. Inorg. Chem.*, 1961, **6**, 330 (647)
63M P. K. Migal and V. A. Tsiplyakova, *Russ. J. Inorg. Chem.*, 1963, **8**, 319 (629)
64K S. N. Khodaskar and D. D. Khanolkar, *Current Sci.*, 1964, **33**, 399
64M P. K. Migal and V. A. Tsiplyakova, *Russ. J. Inorg. Chem.*, 1964, **9**, 333 (601)
65T V. M. Tarayan and A. A. Sarkisyan, *Russ. J. Inorg. Chem.*, 1965, **10**, 1457 (2684)
66S V. I. Shlenskaya, A. A. Biryukov, and E. M. Moskovkina, *Russ. J. Inorg. Chem.*, 1966, **11**, 325 (600)
66Sa K. Swaminathan and H. M. N. H. Irving, *J. Inorg. Nuclear Chem.*, 1966, **28**, 171
66Sb V. M. Shulman, S. V. Larionov, T. V. Kramareva, E. I. Arykova, and V. V. Yudina, *Russ. J. Inorg. Chem.*, 1966, **11**, 580 (1076)
67N I. V. Natarova and V. I. Prizlielevskaya, *Russ. J. Inorg. Chem.*, 1967, **12**, 1614 (3051)
67V V. P. Vasilev and N. K. Grechina, *Russ. J. Inorg. Chem.*, 1967, **12**, 823 (1865)
68I R. M. Izatt, D. Eaught, and J. J. Christensen, *J. Phys. Chem.*, 1968, **72**, 2720

105.1 CH_5ON_3 Semicarbazide L

Metal	*Method*	*Temp*	*Medium*	*Log of equilibrium constant, remarks*	*Ref*
Cd^{2+}	pol	25	1 $NaNO_3$	β_2 3·3±0·2	60T

60T V. F. Toropova and K. V. Naimushina, *Russ. J. Inorg. Chem.*, 1960, **5**, 421 (874)

105.2 CH_5O_4P Methylphosphoric acid (Methyl phosphate) H_2L

Metal	*Method*	*Temp*	*Medium*	*Log of equilibrium constant, remarks*	*Ref*
Ca^{2+}	sp	20, 65	0·1	K_1 (20°) 1·49, (65°) 1·74	65B
Co^{2+}	sp	20, 65	0·1	K_1 (20°) 2·00, (65°) 2·28	65B
Mg^{2+}	sp	20, 65	0·1	K_1 (20°) 1·57, (65°) 2·09	65B
Mn^{2+}	sp	20, 65	0·1	K_1 (20°) 2·19, (65°) 2·55	65B
Ni^{2+}	sp	20, 65	0·1	K_1 (20°) 1·91, (65°) 2·28	65B
Zn^{2+}	sp	20, 65	0·1	K_1 (20°) 2·16, (65°) 2·67	65B

65B H. Brintzinger, *Helv. Chim. Acta*, 1965, **48**, 47

CH_5N_3S Thiosemicarbazide L 106

Metal	Method	Temp	Medium	Log of equilibrium constant, remarks	Ref
Ag^+	Hg	20—50	0·8 $NaNO_3$	β_4 (20°) 13·35±0·06, (25°) 13·10±0·05, (30°) 12·85±0·06, (40°) 12·42±0·07, (50°) 12·03±0·08	60T
	Ag	25	0·05 $NaC_2H_3O_2$ 40% ethanol	β_3 13·3±0·1	61T
Cd^{2+}	pol	25	1 $NaNO_3$	β_2 5·5±0·2	60Ta
	Hg(Cd)	25	1 $NaNO_3$	K_1 2·57±0·02, β_2 4·70±0·01, β_3 5·86±0·02	63C
Hg^{2+}	Hg	20—50	0·8 $NaNO_3$	β_4 (20°) 26·77±0·08, (25°) 26·25±0·07, (30°) 25·75±0·08, (40°) 24·80±0·08, (50°) 23·93±0·07	60T
	Hg	25	0·05 $NaC_2H_3O_2$ 40% ethanol	β_4 27·3±0·1	61T

60T V. F. Toropova and L. S. Kirillova, *Russ. J. Inorg. Chem.*, 1960, **5**, 276 (575)
60Ta V. F. Toropova and K. V. Naimushina, *Russ. J. Inorg. Chem.*, 1960, **5**, 421 (874)
61T V. F. Toropova, Yu. P. Kitaev, and G. K. Budnikov, *Russ. J. Inorg. Chem.*, 1961, **6**, 330 (647)
63C A. N. Christensen and S. E. Rasmussen, *Acta Chem. Scand.*, 1963, **17**, 1315

CH_6ON_4 Carbohydrazide L 106.1

Metal	Method	Temp	Medium	Log of equilibrium constant, remarks	Ref
H^+	gl	20	0·1 ($NaClO_4$)	K_1 4·14, K_1K_2 6·24	64C
Cd^{2+}	gl	20	0·1 ($NaClO_4$)	K_1 2·37	64C
	pol	25	1 ($NaClO_4$)	K_1 2·70, β_2 3·65, β_3 5·61, β_4 5·26	66K
Co^{2+}	gl	20	0·1 ($NaClO_4$)	K_1 2·83, β_2 5·38	64C
Cu^{2+}	gl	20	0·1 ($NaClO_4$)	K_1 4·92, β_2 8·97	64C
Ni^{2+}	gl	20	0·1 ($NaClO_4$)	K_1 3·44, β_2 6·62, β_3 8·64	64C
Zn^{2+}	gl	20	0·1 ($NaClO_4$)	K_1 2·77	64C

64C E. Campi, G. Ostacoli, A. Vanni, and E. Casorati, *Ricerca sci.*, 1964, **34** (II-A, 6), 341
66K A. F. Krivis, G. R. Supp, and R. L. Doerr, *Analyt. Chem.*, 1966, **38**, 936

$CH_6O_6P_2$ Methylenediphosphonic acid H_4L 106.2

Metal	Method	Temp	Medium	Log of equilibrium constant, remarks	Ref
H^+	gl	25	0—1 $(CH_3)_4NBr$	K_1 (→ 0) 10·69±0·07, (0·1) 10·42±0·07, (1) 10·32±0·07, K_2 (→ 0) 7·45±0·07, (0·1) 7·33±0·07, (1) 7·28±0·07, K_3 (→ 0) 2·87±0·07, (0·1) 2·75±0·07, (1) 2·67±0·07, K_4 (→ 0) 2·2±0·2, (0·1) 1·7±0·2, (1) 1·7±0·2	62I
		50	0—1 $(CH_3)_4NBr$	K_1 (→ 0) 10·47±0·07, (0·1) 10·25±0·07, (1) 10·07±0·07, K_2 (→ 0) 7·57±0·07, (0·1) 7·40±0·07, (1) 7·38±0·07, K_3 (→ 0) 3·08±0·07, (0·1) 2·88±0·07, (1) 2·85±0·07, K_4 (→ 0) 1·5±0·2, (0·1) 1·5±0·2, (1) 1·7±0·2	
	gl	25	0·1	K_1 10·10, K_2 6·80, K_3 1·55, K_4 >2(?)	63K

continued overleaf

106.2 $CH_6O_6P_2$ (contd)

Metal	*Method*	*Temp*	*Medium*	*Log of equilibrium constant, remarks*	*Ref*
	gl	25	0·5 $(CH_3)_4NCl$	K_1 10·54±0·02, K_2 6·87±0·02, K_3 2·49±0·04	67C
	gl	25	0·1 KCl	K_1 10·42±0·04, K_2 7·33±0·03, K_3 2·75±0·06, K_4 1·7±0·2	67K
Al^{3+}	gl	25	0·1 KCl	K_1 14·08, β_2 23·01 $K(Al^{3+}+HL^{3-} \rightleftharpoons AlHL)$ 9·05 $K(Al^{3+}+2HL^{3-} \rightleftharpoons Al(HL)_2^{3-})$ 13·66	67K
Be^{2+}	gl	25	0·1 KCl	$K(Be^{2+}+HL^{3-} \rightleftharpoons BeHL^-)$ 8·82 $K(2Be^{2+}+L^{4-} \rightleftharpoons Be_2L)$ 19·15	67K
Ca^{2+}	oth	25	0—1 $(CH_3)_4NBr$	K_1 (→ 0) 6·5±0·1, (0·1) 6·02±0·05, (1) 5·01±0·06, nephelometric	62I
		25	→ 0	$\Delta H_1°$ 0±2, $\Delta S_1°$ 22±7	
		50	0·1—1 $(CH_3)_4NBr$	K_1 (0·1) 5·93, (1) 4·77, nephelometric	
	gl	25	0·1	K_1 5·51, $K(Ca^{2+}+HL^{3-} \rightleftharpoons CaHL^-)$ 2·76 $K(Ca^{2+}+CaL^{2-} \rightleftharpoons Ca_2L)$ 3·31	63K
	gl	25	0·1 KCl	K_1 6·03, $K(Ca^{2+}+HL^{3-} \rightleftharpoons CaHL^-)$ 3·88	67K
Co^{2+}	gl	25	0·1 KCl	K_1 12·03, β_2 18·99 $K(Co^{2+}+HL^{3-} \rightleftharpoons CoHL^-)$ 6·11 $K[Co^{2+}+2HL^{3-} \rightleftharpoons Co(HL)_2^{4-}]$ 10·67 $K(2Co^{2+}+L^{4-} \rightleftharpoons Co_2L)$ 14·98 $K(2Co^{2+}+HL^{3-} \rightleftharpoons Co_2HL^+)$ 8·65	67K
Cs^+	gl	25	0·5 $(CH_3)_4NCl$	K_1 0·84±0·05, $K(Cs^++HL^{3-} \rightleftharpoons CoHL^{2-})$ 0·04±0·02	67C
Cu^{2+}	gl	25	0·1 KCl	K_1 13·29, β_2 23·98 $K(Cu^{2+}+HL^{3-} \rightleftharpoons CuHL^-)$ 6·78 $K[Cu^{2+}+2HL^{3-} \rightleftharpoons Cu(HL)_2^{4-}]$ 12·88 $K(2Cu^{2+}+L^{4-} \rightleftharpoons Cu_2L)$ 18·54 $K(2Cu^{2+}+HL^{3-} \rightleftharpoons Cu_2HL^+)$ 11·57	67K
Fe^{2+}	gl	25	0·1 KCl	K_1 12·6, β_2 18·8 $K(Fe^{2+}+HL^{3-} \rightleftharpoons FeHL^-)$ 6·6 $K[Fe^{2+}+2HL^{3-} \rightleftharpoons Fe(HL)_2^{4-}]$ 11·9 $K(2Fe^{2+}+L^{4-} \rightleftharpoons Fe_2L)$ 15·4 $K(2Fe^{2+}+HL^{3-} \rightleftharpoons Fe_2HL^+)$ 9·1	67K
Fe^{3+}	gl	25	0·1 KCl	K_1 19·9, β_2 26·6	67K
K^+	gl	25	0·5 $(CH_3)_4NCl$	K_1 1·02±0·04 $K(K^++HL^{3-} \rightleftharpoons KHL^{2-})$ 0·20±0·02	67C
Li^+	gl	25	0·5 $(CH_3)_4NCl$	K_1 2·48±0·05, $K(Li^++HL^{3-} \rightleftharpoons LiHL^{2-})$ 0·82±0·03	67C
Mg^{2+}	gl	25, 50	1 $(CH_3)_4NBr$	K_1 (25°) 4·82±0·03, (50°) 5·07 $K(Mg^{2+}+HL^{3-} \rightleftharpoons MgHL^-)$ (25°) 2·97±0·05, (50°) 3·33	62I
		25	0 corr	K_1 6·3, $\Delta H_1°$ 4·4, $\Delta S_1°$ 34	
	gl	25	0·1	K_1 5·51, $K(Mg^{2+}+HL^{3-} \rightleftharpoons MgHL^-)$ 2·76, $K(Mg^{2+}+MgL^{2-} \rightleftharpoons Mg_2L)$ 2·60	63K
	gl	25	0·1 KCl	K_1 6·38, $K(Mg^{2+}+HL^{3-} \rightleftharpoons MgHL^-)$ 4·02	67K
Mn^{2+}	gl	25	0·1 KCl	K_1 12·95, β_2 18·77 $K(Mn^{2+}+HL^{3-} \rightleftharpoons MnHL^-)$ 7·20 $K[Mn^{2+}+2HL^{3-} \rightleftharpoons Mn(HL)_2^{4-}]$ 13·26 $K(2Mn^{2+}+L^{4-} \rightleftharpoons Mn_2L)$ 15·85 $K(2Mn^{2+}+HL^{3-} \rightleftharpoons Mn_2HL^+)$ 9·60	67K
Na^+	gl	25	0·5 $(CH_3)_4NCl$	K_1 1·13±0·08, $K(Na^++HL^{3-} \rightleftharpoons NaHL^{2-})$ 0·39±0·03	67C

contin

$CH_6O_6P_2$ (contd) 106.2

Metal	Method	Temp	Medium	Log of equilibrium constant, remarks	Ref
Ni^{2+}	gl	25	0·1 KCl	K_1 8·16, β_2 15·04	67K
				$K(Ni^{2+}+HL^{3-} \rightleftharpoons NiHL^{-})$ 4·87	
				$K[Ni^{2+}+2HL^{3-} \rightleftharpoons Ni(HL)_2{}^{4-}]$ 10·01	
				$K(2Ni^{2+}+L^{4-} \rightleftharpoons Ni_2L)$ 12·70	
				$K(2Ni^{2+}+HL^{3-} \rightleftharpoons Ni_2HL^{+})$ 8·07	
Sr^{2+}	gl	25	0·1	K_1 4·48,	63K
				$K(Sr^{2+}+HL^{3-} \rightleftharpoons SrHL^{-})$ 1·77,	
				$K(Sr^{2+}+SrL^{2-} \rightleftharpoons Sr_2L)$ 3·70	
	gl	25	0·1 KCl	K_1 5·87,	67K
				$K(Sr^{2+}+HL^{3-} \rightleftharpoons SrHL^{-})$ 3·63	
Th^{4+}	gl	25	0·1 KCl	K_1 23·9, β_2 36·7	67K
Zn^{2+}	gl	25	0·1 KCl	K_1 13·99, β_2 20·55	67K
				$K(Zn^{2+}+HL^{3-} \rightleftharpoons ZnHL^{-})$ 7·50	
				$K[Zn^{2+}+2HL^{3-} \rightleftharpoons Zn(HL)_2{}^{4-}]$ 13·54	
				$K(2Zn^{2+}+L^{4-} \rightleftharpoons Zn_2L)$ 18·16	
				$K(2Zn^{2+}+HL^{3-} \rightleftharpoons Zn_2HL^{+})$ 11·23	
Zr^{4+}	gl	25	0·1 KCl	$K(ZrO^{2+}+L^{4-} \rightleftharpoons ZrOL^{2-})$ 13·13	67K
				$K(ZrO^{2+}+2L^{4-} \rightleftharpoons ZrOL_2{}^{6-})$ 19·45	
				$K(ZrO^{2+}+HL^{3-} \rightleftharpoons ZrOHL^{-})$ 9·01	
				$K[ZrO^{2+}+2HL^{3-} \rightleftharpoons ZrO(HL)_2{}^{4-})$ 12·18	

62I R. R. Irani and K. Moedritzer, *J. Phys. Chem.*, 1962, **66**, 1349

63K H. Kroll, V. Elkind, and R. Davis, US-AEC Report TID-19989, Dec. 9, 1963

67C R. L. Carroll and R. R. Irani, *Inorg. Chem.*, 1967, **6**, 1994

67K M. I. Kabachnik, R. P. Laslovskii, T. Ja. Medved, V. V. Medyntseo, I. D. Kolpakovo, and N. M. Dyatlova, *Proc. Acad. Sci.* (*U.S.S.R.*) 1967, **177**, 1060 (582)

$CH_7O_{10}P_3$ Methyltriphosphoric acid H_4L 106.3

Metal	Method	Temp	Medium	Log of equilibrium constant, remarks	Ref
H^+	gl	20	0·1 ($NaClO_4$)	K_1 6·45±0·02	64S
Cu^{2+}	gl	20	0·1 ($NaClO_4$)	K_1 6·17±0·05	64S
				$K(CuL^{2-}+H^{+} \rightleftharpoons CuLH^{-})$ 3·93	

64S P. W. Schneider, H. Brintzinger, and H. Erlenmeyer, *Helv. Chim. Acta.*, 1964, **47**, 992

C_2H_2 Ethyne (acetylene) L 106.4

Metal	Method	Temp	Medium	Log of equilibrium constant, remarks	Ref
Ag^+	sol	25	→0	K_1 1·63	65T
				$K_s(Ag^{+}+L+OH^{-})$ −19·29	
Hg^{2+}	sol	25	→0	$K_s(Hg^{2+}+2L+2OH^{-})$ −37·10	65T
Na^+	con	−40	Liquid NH_3	K_1 3·42	63B

63B G. Bombara and M. Troyli, *J. Electroanalyt. Chem.*, 1963, **5**, 379

65T Ya. I. Tur'yan, *Russ. J. Inorg. Chem.*, 1965, **10**, 297 (549)

107 C_2H_4 Ethene (ethylene) L

Metal	Method	Temp	Medium	Log of equilibrium constant, remarks	Ref
Ag^+	kin	25	1 (CF_3COONa)	K_1 1·97	59B
				$K(AgA^+ + L \rightleftharpoons AgAL^+)$ 1·76	
				$K(AgA_2^+ + L \rightleftharpoons AgAL^+ + A)$ 0·42	
				where A = triethanolamine	
Pd^{2+}	sol	13	2 $HClO_4$	$K(PdCl_4^{2-} + L \rightleftharpoons PdCl_3L^- + Cl^-)$ 1·19±0·01	66P
				$K[PdCl_4^{2-} + L \rightleftharpoons PdCl_2(H_2O)L + 2Cl^-]$ −0·7± 0·7	
				$K[PdCl_3L^- \rightleftharpoons PdCl_2(H_2O)L + Cl^-]$ −1·5	
			3 $HClO_4$	$K(PdCl_4^{2-} + L \rightleftharpoons PdCl_3L^- + Cl^-)$ 1·20±0·03	
				$K[PdCl_4^{2-} + L \rightleftharpoons PdCl_2(H_2O)L + 2Cl^-]$ 0·4±0·3	
				$K[PdCl_3L^- \rightleftharpoons PdCl_2(H_2O)L + Cl^-]$ −0·7	
			4·5 $LiClO_4$+ $HClO_4$	$K(PdCl_4^{2-} + L \rightleftharpoons PdCl_3L^- + Cl^-)$ 1·21±0·02	
				$K[PdCl_4^{2-} + L \rightleftharpoons PdCl_2(H_2O)L + 2Cl^-]$ 0·81±0·07	
				$K[PdCl_3L^- \rightleftharpoons PdCl_2(H_2O)L + Cl^-]$ −0·4	

59B P. Brandt, *Acta Chem. Scand.*, 1959, **13**, 1639
66P S. V. Pestrikov, I. I. Moiseev, and L. M. Sverzh, *Russ. J. Inorg. Chem.*, 1966, **11**, 1113 (2081)

110 $C_2H_2O_3$ Oxoacetic acid (glyoxylic acid) HL

Metal	Method	Temp	Medium	Log of equilibrium constant, remarks	Ref
H^+	gl	20	0·10 ($NaClO_4$)	K_1 3·178±0·002	64P
	gl	25	0·5 KCl	K_1 2·98	66L
Ce^{3+}	gl	20	0·10 ($NaClO_4$)	K_1 2·39, K_2 1·78, K_3 0·9	64P
Dy^{3+}	gl	20	0·10 ($NaClO_4$)	K_1 2·56, K_2 1·90, K_3 1·7	64P
Er^{3+}	gl	20	0·10 ($NaClO_4$)	K_1 2·60, K_2 2·00, K_3 1·6	64P
Eu^{3+}	gl	20	0·10 ($NaClO_4$)	K_1 2·50, K_2 2·08, K_3 1·5	64P
Gd^{3+}	gl	20	0·10 ($NaClO_4$)	K_1 2·49, K_2 2·04, K_3 1·5	64P
Ho^{3+}	gl	20	0·10 ($NaClO_4$)	K_1 2·58, K_2 1·90, K_3 1·7	64P
La^{3+}	gl	20	0·10 ($NaClO_4$)	K_1 2·36, K_2 1·60, K_3 0·8	64P
Lu^{3+}	gl	20	0·10 ($NaClO_4$)	K_1 2·68, K_2 2·15, K_3 1·7	64P
Nd^{3+}	gl	20	0·10 ($NaClO_4$)	K_1 2·48, K_2 2·00, K_3 1·3	64P
Ni^{2+}	gl	25	0·5 KCl	K_1 0·94	66L
				See glycine (130), alanine (171) and 2-aminoiso-butanoic acid (249.1) for mixed ligands	
Pr^{3+}	gl	20	0·10 ($NaClO_4$)	K_1 2·44, K_2 1·90, K_3 1·0	64P
Sm^{3+}	gl	20	0·10 ($NaClO_4$)	K_1 2·55, K_2 2·04, K_3 1·5	64P
Tb^{3+}	gl	20	0·10 ($NaClO_4$)	K_1 2·52, K_2 1·90, K_3 1·7	64P
Tm^{3+}	gl	20	0·10 ($NaClO_4$)	K_1 2·61, K_2 2·00, K_3 1·7	64P
Y^{3+}	gl	20	0·10 ($NaClO_4$)	K_1 2·56, K_2 1·85, K_3 1·5	64P
Yb^{3+}	gl	20	0·10 ($NaClO_4$)	K_1 2·65, K_2 2·08, K_3 1·7	64P
Zn^{2+}	gl	25	0·5 KCl	K_1 0·64,	66L
				See glycine (130), alanine (171) and 2-aminoiso-butanoic acid (249·1) for mixed ligands	

64P J. E. Powell and Y. Suzuki, *Inorg. Chem.*, 1964, **3**, 690
66L D. L. Leussing and E. M. Hanna, *J. Amer. Chem. Soc.*, 1966, **88**, 696

$C_2H_2O_4$ Ethanedioic acid (oxalic acid) H_2L 111

Metal	Method	Temp	Medium	Log of equilibrium constant, remarks	Ref
H^+	gl	25	0·1 ($NaClO_4$)	K_1 3·81, K_2 1·37	60M
	H	0—45	→ 0	K_2 (0°) 1·244, (15°) 1·252, (25°) 1·252, (35°) 1·286, (45°) 1·295	61Mb
	?	?	?	K_1 4·22	61G
	?	?	?	K_1 4·19, K_2 1·23	61Z, 62Y
	gl	25	0·15 ($NaClO_4$)	K_1 2·59, K_2 1·33	62B
	?	?	?	K_1 4·19, K_2 1·21	63P
	?	?	?	K_1 4·14, K_2 1·25	63S
	gl	25	1 $NaClO_4$	K_1 3·54, K_2 ~ 1·0	64S, 65S
	gl	25	1 $NaClO_4$	K_1 3·57±0·02, K_2 1·07±0·05	65B
	gl	25	0·1 ($NaClO_4$)	K_1 3·85, K_2 1·13	65N
	gl	25	0·5 ($NaClO_4$)	K_1 3·51	66L
	gl	25	1—3 ($NaClO_4$)	K_1 (1·0) 3·554±0·002, (3·0) 3·801±0·003, K_2 (1·0) 1·08±0·01, (3·0) 1·26±0·03	66Ma
	sp	5—25	0·5	K_1 (5°) 3·63, (15°) 3·65, (25°) 3·67, K_2 (5°) 1·32, (15°) 1·23, (25°) 1·20	65Bb
	gl	25	1·00 (KNO_3)	K_1 3·62±0·01, K_2 1·1	67R
	pH	25	0·5 $LiClO_4$	K_1 3·50±0·03, K_2 1·00±0·05	68D
Ag^+	sol	25	2 (KNO_3)	$K(Ag^+ + L^{2-} + A \rightleftharpoons AgLA^-)$ 7·2±0·1 where A = ethylenediamine	63F
	dis	20	0·1 ($KClO_4$)	K_1 < 2·0	63S
Al^{3+}	dis	20	0·1 ($KClO_4$)	β_3 15·60±0·05	63S
Am^{3+}	ix	20—25	0·2	K_1 5·99, β_2 10·15	60L
	dis	25	1 $NaClO_4$	K_1 4·63±0·08, β_2 8·35±0·09, β_3 11·15±0·07	64S
	oth	25	0·1	K_1 6·15, β_2 10·54, electromigration	65Sa
	ix	25	0·500 ($NaClO_4$)	K_1 4·82±0·02, β_2 8·60±0·02	68A
Ba^{2+}	dis	20	0·1 ($KClO_4$)	K_{so} −6·0	63S
	dis	25	1 $NaClO_4$	K_1 0·58, β_2 2·20 (?)	66S
Be^{2+}	gl	25	0·15 ($NaClO_4$)	K_1 4·08, β_2 5·91	62B
	dis	20	0·1 ($KClO_4$)	K_1 4·12±0·06	63S
	dis	25	1 $NaClO_4$	K_1 3·55, β_2 5·40	67Sa
Bi^{3+}	dis	20	0·1 ($KClO_4$)	K_{so} −35·4	63S
Ca^{2+}	dis	20	0·1 ($KClO_4$)	K_{so} −7·9	63S
	dis	25	1 $NaClO_4$	K_1 1·66, β_2 2·69	67H
Cd^{2+}	pol	25	1 $NaNO_3$	K_1 2·61, β_2 4·11, β_3 5·06	62Ma
			1 $NaNO_3$ heavy water	K_1 2·66, β_2 4·20, β_3 5·17	
	sol	25	2 (KNO_3)	β_2 5·37 $K(Cd^{2+} + L^{2-} + A \rightleftharpoons CuLA)$ 8·29 where A = ethylenediamine	63F
	dis	20	0·1 ($KClO_4$)	K_1 3·71±0·03	63S
	dis	30	1·0 ($NaClO_4$)	K_1 3·0±0·2, β_2 4·7±0·2	65H
	gl	25	1 (KNO_3)	K_1 3·20, K_2 1·37, K_3 0·96, $K(Cd^{2+} + L^{2-} + A \rightleftharpoons CdLA)$ 7·73, $K(Cd^{2+} + 2L^{2-} + A \rightleftharpoons CdL_2A^{2-})$ 9·49, $K(Cd^{2+} + L^{2-} + 2A \rightleftharpoons CdLA_2)$ 11·24, $K(CdLA_2 + A \rightleftharpoons CdA_3^{2+} + L^{2-})$ 0·99, $K(CdA^{2+} + L^{2-} \rightleftharpoons CdLA)$ 2·12, $K(CdLA + A \rightleftharpoons CdA_2^{2+} + L^{2-})$ 2·45, where A = ethylenediamine	67Kb
Cm^{3+}	ix	20—25	0·2	K_1 5·96, β_2 10·15	60L

continued overleaf

111 $C_2H_2O_4$ (contd)

Metal	Method	Temp	Medium	Log of equilibrium constant, remarks	Ref
	ix	25	0·500 ($NaClO_4$)	K_1 4·80±0·01, β_2 8·61±0·01	68A
Co^{2+}	H	0—45	→ 0	$\Delta H_1°$ 0·59±0·07, $\Delta S_1°$ 23·9±0·3 K_1 6·810−(1·500×10^{-2} T)+(2·760×10^{-5} T^2) (±3%)	61Mb

Errata. 1964, p. 361 *for* 60Ma *read* 61M

Metal	Method	Temp	Medium	Log of equilibrium constant, remarks	Ref
	dis	25	0·00—0·20 (NaCl)	K_1 (0 corr) 4·64±0·02, (0·02) 4·174±0·003, (0·04) 4·027±0·005, (0·08) 3·858±0·007, (0·10) 3·809±0·003, (0·16) 3·688±0·004, (0·20) 3·630 ±0·002, K_2 (0·16) 2·14	61Ma
	dis	20	0·1 ($KClO_4$)	β_2 6·79±0·07	63S
	gl	25	0 corr	K_1 4·69±0·01, K_2 2·46	65M
	ix	25	0 corr	K_1 4·75±0·07, K_2 2·16±0·04	65S
	ix	20	0·1 ($NaClO_4$)	K_1 3·87	67W
Co^{3+}	sp	25	I ($KClO_4$)	$K[Co(NH_3)_4^{3+}+L^{2-} \rightleftharpoons Co(NH_3)_4L^+]$ 62·24 I −25·98$\sqrt{I}$+9·69	61M
	sp	25	I ($KClO_4$)	$K[Co(NH_3)_5^{3+}+HL^- \rightleftharpoons Co(NH_3)_5L^{2+}]$ 3·790 I −3·954$\sqrt{I}$+3·428	62T
Cr^{3+}	gl	25	0·1 ($NaClO_4$)	K_1 5·34, K_2 5·17, K_3 4·93	65N
	kin	50	0·5—2·7	K_3 0·28±0·02 (?) $K(CrL_2^- + HL^- \rightleftharpoons CrHL_3^{2-})$ −0·07	67Ka
			0—1	$K(CrL_2^- + HL^- \rightleftharpoons CrL_3^{3-} + H^+)$ (0·00) −0·73, (0·10) 0·26, (0·20) 0·59, (0·50) 0·92, (1·00) 1·15	
Cu^{2+}	gl+pol	25	0—0·1	K_1 (→ 0) 6·23, (0·1) 4·85, K_2 (→ 0) 4·04, (0·1) 4·36 $K(Cu^{2+}+HL^- \rightleftharpoons CuHL^+)$ (→ 0) 3·18, (0·1) 2·49	60M
	pol	25	1 $NaNO_3$	β_2 9·27	62Ma
			1 $NaNO_3$ heavy water	β_2 9·51	
	sol	25	2 (KNO_3)	β_2 9·70, $K(Cu^{2+}+L^{2-}+A \rightleftharpoons CuLA)$ 15·44 where A = ethylenediamine	63F
	dis	20	0·1 ($KClO_4$)	β_2 10·46±0·07	63S
	Hg(Cu)	25	1 $NaClO_4$	K_1 5·53±0·05, β_2 9·54±0·05	65C
Er^{3+}	sol	?	0·1 ($KClO_4$)	K_1 4·82, β_2 8·21, β_3 10·03	62Aa
Eu^{3+}	ix	25	0·5	K_1 4·81, β_2 8·57	63K
	dis	25	1 $NaClO_4$	K_1 4·77±0·10, β_2 8·72±0·09, β_3 11·4±0·1	64S
	dis	24—25	0·5 ($NaClO_4$)	K_1 4·86±0·03, β_2 8·67±0·02	66L
	ix	25	0·5 ($NaClO_4$)	K_1 4·86±0·03, β_2 8·65±0·02	66L
	dis	25	→ 0	K_1 6·52	66M
Fe^{2+}	gl	25	1 $NaClO_4$	K_1 3·05±0·05, β_2 5·15±0·1	65B
Fe^{3+}	ix	?	0·5	K_1 7·54, β_2 14·59, β_3 20·0	63P
	dis	20	0·1 ($KClO_4$)	β_3 20·46	63S
	Ag, AgC_2O_4–H	25	→ 0	K_3 4·27	65P
	sp	25	1,3 ($NaClO_4$)	K_1 (1·0) 7·59±0·01, (3·0) 7·74±0·02, $K(FeOH^{2+}+HL^- \rightleftharpoons FeL^+)$ (1·0) 6·83	66Ma
	sp	5—25	0·5	K_1 (5°) 7·75, (15°) 7·70, (25°) 7·39 $K(Fe^{3+}+HL^- \rightleftharpoons FeHL^{2+})$ (5°) 4·10, (15°) 4·34, (25°) 4·35	65Bb
	Pt	25	0·5 $LiClO_4$	K_1 7·53, β_2 13·64, β_3 18·49	68D
Ga^{3+}	dis	20	0·1 ($KClO_4$)	β_3 17·98±0·08	63S
Gd^{3+}	kin	25	~ 0	K_1 7·01	62Yb
	ix	25	0·5	K_1 4·78±0·04, β_2 8·68±0·02	67L
	dis	24—25	0·5	K_1 4·76±0·03, β_2 8·65±0·02	67L

contin

Metal	*Method*	*Temp*	*Medium*	*Log of equilibrium constant, remarks*	*Ref*
Ge^{IV}	ix	25	?	K_3 3·5	64K
Hf^{4+}	ix	?	2—4 $HClO_4$	$K(Hf^{4+}+H_2L \rightleftharpoons HfL^{2+}+2H^+)$ (2·0) 5·2±0·1, (4·0) 5·2±0·1 $K(Hf^{4+}+2H_2L \rightleftharpoons HfL_2+4H^+)$ (2·0) 9·7±0·1	62M
Hg^{2+}	dis	20	0·1 ($KClO_4$)	K_1 < 4	63S
In^{3+}	dis	20	0·1 ($KClO_4$)	β_3 14·7±0·1	63S
	dis	25	1 $NaClO_4$	K_1 5·30, β_2 10·52	66H
La^{3+}	dis	20	0·1 ($KClO_4$)	K_{so} −25·0	63S
	dis	25	1 $NaClO_4$	K_1 4·3±0·1, β_2 7·9±0·1, β_3 10·3±0·2	64S
Lu^{3+}	dis	25	1 $NaClO_4$	K_1 5·11±0·08, β_2 9·2±0·1, β_3 12·79±0·08	64S
Mg^{2+}	Ag-$Ag_2C_2O_4$	25	0·03—0·09	β_2 (0·03–0·05) 4·54, (0·09) 4·24	59T
	oth	~ 23	0·2 KCl 0·1 $(HOCH_2)_3CNH_2$	K_1 2·61, interferometer	62A
	dis	20	0·1 ($KClO_4$)	K_1 2·39±0·05	63S
Mn^{2+}	H	0—35	→ 0	$\Delta H_1°$ 1·4±0·2, $\Delta S_1°$ 22·9±0·7 K_1 8·141−(3·146×10⁻²T)+(5·857×10⁻⁵T²) (±3%)	61Mb
	dis	20	0·1 ($KClO_4$)	K_1 3·75±0·05	63S
Mo^{VI}	pol	25	0·1—0·3H^+	$K(H_2MoO_4+H_2L \rightleftharpoons MoO_2L)$ (0·112) 3·91±0·03, (0·179) 3·80±0·04, (0·345) 3·49±0·04	61Y
	dis	20	0·1 ($KClO_4$)	$K(H_2MoO_4+2HL^- \rightleftharpoons MoO_2L_2^{2-})$ 7·37±0·05	63S
Nb^V	dis	20	4·5 (NaCl or $NaNO_3$+2·5 HCl)	$K[Nb(OH)_4^+ + H_2L \rightleftharpoons Nb(OH)_4HL+H^+]$ 3·55 $K[Nb(OH)_4^+ + 2H_2L \rightleftharpoons Nb(OH)_2L_2^- + 2H^+]$ 5·13	67K
	pH	25	0·5	$K[Nb(OH)_4^+ + 2HL^- \rightleftharpoons Nb(OH)_2L_2^-]$ 12·11 $K[Nb(OH)_4^+ + 2HL^- + L^{2-} \rightleftharpoons Nb(OH)_2L_3^{3-}]$ 17·15	67N
Ni^{2+}	H	0—45	→ 0	$\Delta H_1°$ 0·15±0·10, $\Delta S_1°$ 24·2±0·4 K_1 9·065−(2·655×10⁻²T)+(4·512×10⁻⁵T²) (±3%)	61Mb
	sol	25	2 (KNO_3)	β_2 7·64, $K(Ni^{2+}+L^{2-}+A \rightleftharpoons NiLA)$ 11·20 where A = ethylenediamine	63F
	dis	20	0·1 ($KClO_4$)	β_2 7·88±0·04	63S
	spJ	17—21	0·01 ($NaNO_3$)	$K[2NiA^{2+}+L^{2-} \rightleftharpoons Ni_2A_2L^{2+}]$ ~ 5 where A = triethylenetetramine $K(NiA^{2+}+L^{2-} \rightleftharpoons NiAL)$ 1·00 where A = 4,6,6-trimethyl-3,7-diazanon-3-ene-1, 9-diamine See nitrilotriacetic acid (449) and diethylenetriamine (231) for mixed ligands	63C
Np^{4+}	sol	20	?	K_{so} −22·07	58M
	sol	24—28	1 $HClO_4$	K_1 9·63±0·01, β_2 16·88±0·01, β_3 23·69±0·02	64B
	sol	23	?	K_1 8·64, β_2 16·8, β_3 23·2, β_4 27·0	67M
NpO_2^+	ix	18—22	0·05 NH_4ClO_4	K_1 4·04, K_2 7·36, $K(NpO_2^+ + HL^- \rightleftharpoons NpO_2HL)$ 2·70	61Z
Pa^{4+}	hyp	25	→ 0	K_1 10·7, β_2 20·3, β_3 26·5, β_4 29·2	67Ma
Pa^V	ix	25	0·25	$K[Pa_2O(OH)_3^{5+} + HL^- \rightleftharpoons Pa_2O(OH)_3HL^{4+}]$ (?) 2·60, $K[Pa_2O(OH)_2^{6+} + 2HL^- \rightleftharpoons Pa_2O(OH)_2(HL)_2^{4+}]$ (?) 3·95	66G
Pb^{2+}	dis	20	0·1 ($KClO_4$)	β_2 6·56±0·04	63S
	pol	30	1·5 (KNO_3)	K_1 3·32, β_2 5·03	68J
Pu^{4+}	sol	20	0·5—1·0	K_{so} −21·3	58M
PuO_2^{2+}	sol	20	?	K_1 6·66, β_2 11·4, K_{so} −9·85	58G
Sc^{3+}	dis	20	0·1 ($KClO_4$)	β_3 16·28±0·05	63S
	sp	18—20	0·01 (HNO_3)	K_1 8·3±0·1	66Ka

continued overleaf

111 $C_2H_2O_4$ (contd)

Metal	*Method*	*Temp*	*Medium*	*Log of equilibrium constant, remarks*	*Ref*
Sr^{2+}	dis	20	0·1 ($KClO_4$)	K_{so} −6·4	63S
	dis	25	1 $NaClO_4$	K_1 1·25, β_2 1·90	67H
Ta^V	sol	18—20	?	$K[Ta(OH)_2^{3+}+L^{2-} \rightleftharpoons Ta(OH)_2L^+]$ 11·10,	65Ba
				$K[Ta(OH)_2^{3+}+2L^{2-} \rightleftharpoons Ta(OH)_2L_2^-]$ 18·52,	
				$K[Ta(OH)_2L^+ + OH^- \rightleftharpoons Ta(OH)_3L]$ 13·33	
Th^{4+}	kin	25	~ 0	K_1 7·16+0·07	62Yb
				$K(2ThOH^{3+}+HL^- \rightleftharpoons Th_2LOH^{5+})$ 22·9±0·6	
	sol	25	1 (NH_4ClO_4)	K_1 8·23, β_2 16·8, β_3 22·8, β_4 27·2	67Ma
				K_{so} −21·38,	
			→ 0	K_1 10·6, β_2 20·2, β_3 26·4, β_4 29·6	
				K_{so} −24·96	
Ti^{4+}	sp	?	0·02 $HClO_4$	$K(TiO^{2+}+L^{2-} \rightleftharpoons TiOL)$ 6·6±0·1	59B
				$K(TiOL+L^{2-} \rightleftharpoons TiOL_2^{2-})$ 3·3	
	gl	25	0·03	$K(TiO^{2+}+L^{2-} \rightleftharpoons TiOL)$ ~ 9·7	60G
				$K(TiOL+L^{2-} \rightleftharpoons TiOL_2^{2-})$ 5·11	
	HgC_2O_4	25	0·03	$K(TiOL+L^{2-} \rightleftharpoons TiOL_2^{2-})$ 4·4	60G
	dis	20	0·1 ($KClO_4$)	$K(TiO^{2+}+2L^{2-} \rightleftharpoons TiOL_2^{2-})$ 10·7±0·1	63S
Tl^{3+}	dis	20	0·1 ($KClO_4$)	β_3 ~ 16·9	63S
UO_2^{2+}	sol	20	1 $HClO_4$	K_1 6·72±0·02, β_2 11·92±0·07,	59M
				K_{so} −8·66	
			1 HNO_3	K_1 6·85, β_2 12·10, K_{so} −8·52	
	Ag, sol	25	→ 0	K_1 4·44, K_2 6·00	59P
	Ag-$Ag_2C_2O_4$	25	0·07	$K[(UO_2)_2L_3^{2-}+2L^{2-} \rightleftharpoons (UO_2)_2L_5^{6-}]$ 4·42	59T
			0·02	$K[(UO_2)_2L_3^{2-}+L^{2-} \rightleftharpoons (UO_2)_2L_4^{4-}]$ 1·32	
	dis	20	0·1 ($KClO_4$)	β_2 11·08±0·03	60S
	sp	20	1 ($NaClO_4$)	K_1 4·63±0·02, K_2 4·05±0·06, K_3 3·31±0·09	67Mb
	gl	25	1·00 (KNO_3)	β_2 9·1±0·2	67R
	oth	25	0·1 (KCl)	K_1 6·7±0·1, β_2 11·8±0·3, electromigration	67S
VO^{2+}	Hg	25	saturated $K_2C_2O_4$	β_2 12·3±0·2	59Z
	gl	25	0·1 ($NaClO_4$)	$K[VOL(OH)^- + H^+ \rightleftharpoons VOL]$ 5·7	66Kb
VO_2^+	dis	20	0·1 ($KClO_4$)	K_1 ~ 6·4, β_2 ~ 9·0	63S
W^{VI}	pol	25	0·08—0·24 H_2SO_4	$K(H_2WO_4+H_2L \rightleftharpoons H_2WO_3L)$ 4·85±0·09	62Y
				$K(H_2WO_3L+H_2L \rightleftharpoons H_2WO_2L_2)$ 7·5±0·1	
	kin	25	?	$K(H_2WO_4+H_2L \rightleftharpoons H_2WO_3L)$ 5·13±0·05	62Ya
Zn^{2+}	dis	20	0·1 ($KClO_4$)	β_2 7·59±0·05	63S
	dis	25	0 corr	K_1 4·85±0·05, K_2 2·7±0·2	66R
	sp	43	1(?) $Zn(NO_3)_2$	K_3 0·57±0·08	67G
	gl	25	1 (KNO_3)	K_1 3·44, K_2 3·04, K_3 0·76,	67Kb
				$K(Zn^{2+}+2L^{2-}+A \rightleftharpoons ZnL_2A^{2-})$ 10·76,	
				$K(Zn^{2+}+L^{2-}+A \rightleftharpoons ZnLA)$ 9·21,	
				$K(Zn^{2+}+L^{2-}+2A \rightleftharpoons ZnLA_2)$ 12·31,	
				$K(ZnLA+A \rightleftharpoons ZnA_2^{2+}+L^{2-})$ 1·64,	
				$K(ZnLA_2+A \rightleftharpoons ZnA_3^{2+}+L^{2-})$ 0·38,	
				$K(ZnA^{2+}+L^{2-} \rightleftharpoons ZnLA)$ 3·49,	
				where A = ethylenediamine	
Zr^{4+}	gl	25	?	K_3(?) 4·0±0·1	61G
	ix	?	2—4 $HClO_4$	$K(Zr^{4+}+H_2L \rightleftharpoons ZrL^{2+}+2H^+)$ (2·0) 5·50±0·04,	62M
				(4·0) 5·60±0·07	
				$K(Zr^{4+}+2H_2L \rightleftharpoons ZrL_2+4H^+)$ (2·0) 9·7±0·2	
	ix	~ 20	1—5 HNO_3	K_1 (1) 11·1±0·2, (5) 11·3±0·2, K_2 (1) 20·3±0·3	64C
	sp	18—20	1 (HCl)	K_1 10·26±0·05	66K

continued

58G A. D. Gelman, L. E. Drabkina, and A. I. Moskvin, *J. Inorg. Chem. USSR*, 1958, **3,** No. 7, 96 (1546); No. 8, 290 (1934)
58M A. I. Moskvin and A. D. Gelman, *J. Inorg. Chem. USSR*, 1958, **3,** No. 4, 188 (956)
59B A. K. Babko and L. I. Dubovenko, *Russ. J. Inorg. Chem.*, 1959, **4,** 165 (372)
59M A. I. Moskvin and F. A. Zakharova, *Russ. J. Inorg. Chem.*, 1959, **4,** 975 (2151)
59P B. V. Ptitsyn and E. N. Tekster, *Russ. J. Inorg. Chem.*, 1959, **4,** 1024 (2248)
59T E. N. Takster, L. I. Vinogradova, and B. V. Ptitsyn, *Russ. J. Inorg. Chem.*, 1959, **4,** 347 (764)
59Z V. L. Zolotavin, *Russ. J. Inorg. Chem.*, 1959, **4,** 1254 (2713)
60G A. A. Grinberg and L. V. Shikheeva, *Russ. J. Inorg. Chem.*, 1960, **5,** 287 (599)
60L I. A. Lebediv, S. V. Pirozhkov, and G. N. Yakovlev, *Radiokhimiya*, 1960, **2,** 559
60M A. McAuley and G. H. Nancollas, *Trans. Faraday Soc.*, 1960, **56,** 1165
60S J. Stary, *Coll. Czech. Chem. Comm.*, 1960, **25,** 2630
61G A. A. Grinberg and V. I. Astapovich, *Russ. J. Inorg. Chem.*, 1961, **6,** 164 (321)
61M M. Mori, R. Tsuchya, and E. Matsuda, *Bull. Chem. Soc. Japan*, 1961, **34,** 1761
61Ma P. G. Manning and C. B. Monk, *Trans. Faraday Soc.*, 1961, **57,** 1996
61Mb A. McAuley and G. H. Nancollas, *J. Chem. Soc.*, 1961, 2215
61Y K. B. Yatsimirskii and L. I. Budarin, *Russ. J. Inorg. Chem.*, 1961, **6,** 944 (1840)
61Z Yu. A. Zolotov, I. N. Marov, and A. I. Moskvin, *Russ. J. Inorg. Chem.*, 1961, **6,** 539 (1055)
62A H. Asai and M. Morales, *Arch. Biochem. Biophys.*, 1962, **99,** 383
62Aa Z. F. Andreeva and I. V. Kolosor, *Izu. Timirazeusk Selskokhoz. Acad.*, 1962, 212; (*Chem. Abs.* 1963, **58,** 64a)
62B H. J. de Bruin, D. Kaitis, and R. B. Temple, *Austral. J. Chem.*, 1962, **15,** 457
62M I. N. Marov and D. I. Ryabchikov, *Russ. J. Inorg. Chem.*, 1962, **7,** 533 (1036)
62Ma D. L. McMasters, J. C. Di Raimondo, L. H. Jones, R. P. Lindley, and E. W. Zeltmann, *J. Phys. Chem.*, 1962, **66,** 249 (688)
62T R. Tsuchita, *Bull. Chem. Soc. Japan*, 1962, **35,** 666
62Y K. B. Yatsimirskii and L. I. Budarin, *Russ. J. Inorg. Chem.*, 1962, **7,** 942 (1824)
62Ya K. B. Yatsimirskii and K. E. Prik, *Russ. J. Inorg. Chem.*, 1962, **7,** 821 (1589)
62Yb K. B. Yatsimirskii and Yu. A. Zhukov, *Russ. J. Inorg. Chem.*, 1962, **7,** 818 (1583), 1463 (2807)
63C N. F. Curtis, *J. Chem. Soc.*, 1963, 4115
63F Ya. D. Fridman, R. A. Veresova, N. V. Dolgashova, and R. I. Sorochan, *Russ. J. Inorg. Chem.*, 1963, **8,** 344 (676)
63K A. S. Kerechuk and V. I. Paramonova, *Radiokhimiya*, 1963, **5,** 464
63P V. I. Paramonova and S. A. Bartenev, *Russ. J. Inorg. Chem.*, 1963, **8,** 157 (311)
63S J. Stary, *Analyt. Chim. Acta*, 1963, **28,** 132
64B B. M. L. Bansal and H. D. Sharma, *J. Inorg. Nuclear Chem.*, 1964, **26,** 799
64C R. Caletka, M. Kyrs, and J. Rais, *J. Inorg. Nuclear Chem.*, 1964, **26,** 1443
64K G. I. Kurnevich and G. A. Shagisultanova, *Russ. J. Inorg. Chem.*, 1964, **9,** 1383 (2559)
64S T. Sekine, *J. Inorg. Nuclear Chem.*, 1964, **26,** 1463; *Acta Chem. Scand.*, 1965, **19,** 1476
65B E. Bottari and L. Ciavatta, *Gazzetta*, 1965, **95,** 908
65Ba A. K. Babko, V. V. Lukachina, and B. I. Nabivanets, *Russ. J. Inorg. Chem.*, 1965, **10,** 467 (865)
65Bb R. F. Bauer and W. M. Smith, *Canad. J. Chem.*, 1965, **43,** 2755
65C L. Ciavatta and M. Villafiorita, *Gazzetta*, 1965, **95,** 1247
65H H. E. Hellwege and G. K. Schweitzer, *J. Inorg. Nuclear Chem.*, 1965, **27,** 99
65M C. B. Monk, *J. Chem. Soc.*, 1965, 2456
65N K. Nagata, A. Umayahara, and R. Tsuchiya, *Bull. Chem. Soc. Japan*, 1965, **38,** 1059
65P B. V. Ptitsyn, L. I. Vinogradova, and E. A. Maksimyuk, *Russ. J. Inorg. Chem.*, 1965, **10,** 1050 (1929)
65S R. G. Seys and C. B. Monk, *J. Chem. Soc.*, 1965, 2452
65Sa A. V. Stepanov and T. P. Makarova, *Radiokhimiya*, 1965, **7,** 670
66G I. Galateanu, *Canad. J. Chem.*, 1966, **44,** 647
66H Y. Hasegawa and T. Sekine, *Bull. Chem. Soc. Japan*, 1966, **39,** 2776
66K I. M. Korenman, F. R. Sheyanova, and Z. M. Gureva, *Russ. J. Inorg. Chem.*, 1966, **11,** 1485 (2761)
66Ka I. M. Korenman and N. V. Zaglyadimova, *Russ. J. Inorg. Chem.*, 1966, **11,** 1491 (2774)
66Kb Th. Kaden and S. Fallab, *Chimia* (*Switz.*), 1966, **20,** 51
66L S. J. Lyte and S. J. Naqui, *J. Inorg. Nuclear Chem.*, 1966, **28,** 2993
66M P. G. Manning, *Canad. J. Chem.*, 1966, **44,** 3057
66Ma E. G. Moorhead and N. Sutin, *Inorg. Chem.*, 1966, **5,** 1866
66R D. L. G. Rowlands and C. B. Monk, *Trans. Faraday Soc.*, 1966, **62,** 945
66S T. Sekine, M. Sakairi, and Y. Hasegawa, *Bull. Chem. Soc. Japan*, 1966, **39,** 2141
67G E. C. Gruen and R. A. Plane, *Inorg. Chem.*, 1967, **6,** 1123
67H Y. Hasegawa, K. Maki and T. Sekine, *Bull. Chem. Soc. Japan*, 1967, **40,** 1845
67K C. Konecny, *Z. phys. Chem.* (*Leipzig*), 1967, **235,** 39
67Ka H. Kelm and G. M. Harris, *Inorg. Chem.*, 1967, **6,** 706, 1743
67Kb Y. Kanemura and J. I. Watters, *J. Inorg. Nuclear Chem.*, 1967, **29,** 1701
67L S. J. Lyle and S. J. Naqvi, *J. Inorg. Nuclear Chem.*, 1967, **29,** 2441
67M A. I. Moskvin, calculated from data of J. A. Porter, *Ind. and Eng. Chem.*, (*Process Design and Development*), 1964, **3,** 289
67Ma A. I. Moskvin and L. N. Essen, *Russ. J. Inorg. Chem.*, 1967, **12,** 359 (688)
67Mb C. Miyake and H. W. Nurnberg, *J. Inorg. Nuclear Chem.*, 1967, **29,** 2411
67N A. N. Nevzorov and O. A. Songina, *Russ. J. Inorg. Chem.*, 1967, **12,** 1259 (2388)
67R K. S. Rajan and A. E. Martell, *J. Inorg. Nuclear Chem.*, 1967, **29,** 523
67S A. V. Stepanov and T. P. Makarova, *Russ. J. Inorg. Chem.*, 1967, **12,** 1262 (2395)
67Sa T. Sekine and M. Sakairi, *Bull. Chem. Soc. Japan*, 1967, **40,** 261

continued overleaf

111 $C_2H_2O_4$ (contd)

67W H. Waki, 'Selected Topics on Stability Constants of Metal Complexes', Proceedings of the Informal Meeting, Tenth International Conference on Co-ordination Chemistry, ed. S. Misumi, Sept. 1967, Faculty of Science, Tohoku Univ., Sendai, Japan. p 15

68A A. Aziz, S. J. Lyle, and S. J. Naqvi, *J. Inorg. Nuclear. Chem.*, 1968, **30**, 1013

68D M. Deneux, R. Meilleur, and R. L. Benoit, *Canad. J. Chem.*, 1968, **46**, 1383

68J D. S. Jain, A. Kumar, and J. N. Gaur, *J. Electroanalyt. Chem.*, 1968, **17**, 201

113 C_2H_3N Cyanomethane (acetonitrile) L

Metal	*Method*	*Temp*	*Medium*	*Log of equilibrium constant, remarks*	*Ref*
Ag^+	kin	40	0·01	K_1 0·89±0·04	64Y
	sol	25	~0	K_1 0·75±0·05	64Y
	Ag	25	0·01(?) $LiClO_4$	K_1 0·7, β_2 0·8	67M
			methanol	K_1 1·1, β_2 1·2, β_3 1·2	
			butan-2-ol	K_1 1·0, β_2 1·4, β_3 1·3	
			acetone	K_1 1·0, β_2 1·5, β_3 1·6	
			0·01(?) Et_4NClO_4 nitroethane	K_1 1·1, β_2 2·8, β_3 3·7, β_4 4·2	
Cu^+	pol	25	0·1 ($NaClO_4$)	β_2 4·35	63H
	pH	25	0·1 (?) $LiClO_4$	β_2 3·9, β_3 4·1	67M
			methanol	K_1 2·5, β_2 3·9, β_3 4·5, β_4 4·2	
			ethanol	K_1 3·7, β_2 5·4, β_3 5·9, β_4 5·9	
			propan-2-ol-	K_1 3·1, β_2 5·3, β_3 5·9, β_4 6·1	
			butan-2-ol	K_1 3·1, β_2 4·6, β_3 5·7, β_4 6·4	
			acetone	K_1 4·4, β_2 6·3, β_3 6·7, β_4 7·2	
			0·1 (?) Et_4NClO_4 nitroethane	K_1 2·3, β_2 4·5, β_3 5·5, β_4 5·9	

63H P. Hemmerich and Ch. Sigwart, *Experimentia*, 1963, **19**, 488

64Y K. B. Yatsimirskii and V. D. Korableva, *Russ. J. Inorg. Chem.*, 1964, **9**, 195 (357)

67M S. E. Manahan and R. T. Iwamoto, *J. Electroanalyt. Chem.*, 1967, **14**, 213

114 $C_2H_4O_2$ Ethanoic acid (acetic acid) HL

Metal	*Method*	*Temp*	*Medium*	*Log of equilibrium constant, remarks*	*Ref*
H^+	gl	20	0·1 ($NaClO_4$)	K_1 4·55	62K
	gl	25	0·2 (KNO_3)	K_1 4·64	63F
	gl	30	1 $NaClO_4$	K_1 4·30	64B
	pH	25	3 ($LiClO_4$)	K_1 4·52	65Ka
	H, Ag-AgCl	10—40	0 corr 50% methanol	K_1 (10°) 5·673±0·002, (15°) 5·665±0·002, (20°) 5·662±0·001, (25°) 5·660±0·001, (30°) 5·662±0·001, (35°) 5·665±0·001, (40°) 5·677±0·002 $\Delta H_1°$ (25°) 0·052, $\Delta S_1°$ (25°) 26·1	65P
	gl	25	0·500 ($NaClO_4$) 95% methanol	K_1 6·97±0·01, K ($2L^- + H^+ \rightleftharpoons HL_2^-$) 7·05±0·03 K($2H^+ + 2L^- \rightleftharpoons H_2L_2$) 13·40±0·06	67G
	pH	25	0·5 $LiClO_4$	K_1 4·48±0·03	68D

contir

$C_2H_4O_2$ (contd) 114

Metal	*Method*	*Temp*	*Medium*	*Log of equilibrium constant, remarks*	*Ref*
	gl	25	0·1 $NaClO_4$ 50% dioxan	K_1 6·01 ± 0·01	68G, 68E
Ag^+	Ag	25	→ 0 0—30% ethanol	K_1 (0%) 0·73, (10%) 0·89, (20%) 1·10, (30%) 1·31, β_2 (0%) 0·64, (10%) 1·0, (20%) 1·25, (30%) 1·72	52M
			→ 0 10—30% acetone	K_1 (10%) 0·88, (20%) 1·10, (30%) 1·26, β_2 (10%) 0·92, (20%) 1·24, (30%) 1·75	
Am^{3+}	ix	20	0·5 $NaClO_4$	K_1 1·99 ± 0·02, β_2 3·28 ± 0·05, β_3 3·9	62G
Ba^{2+}	gl	25, 35	0 corr	K_1 (25°) 1·15 ± 0·01, (35°) 1·10 ± 0·01 ΔH_1 −2·32, ΔS_1 −2·5	56N
	gl	25	0 corr	K_1 0·979	64A
Ca^{2+}	gl	25, 35	0 corr	K_1 (25°) 1·24 ± 0·01, (35°) 1·26 ± 0·01 ΔH_1 0·91, ΔS_1 8·7	56N
	gl	25	0 corr	K_1 1·12	64A
Cd^{2+}	gl	20	0·1 ($NaClO_4$)	K_1 1·61, K_2 1·07	62K
	gl	25	0 corr	K_1 1·928, K_2 1·22	64A
	dis	30	1·0 ($NaClO_4$)	K_1 0·7 ± 0·2, β_2 1·4 ± 0·2	65H
	Hg(Cd)	25	0·25—2 ($NaClO_4$)	K_1 (0·25) 1·26 ± 0·02, (0·50) 1·19 ± 0·01, (1·00) 1·17 ± 0·02, (2·00) 1·23 ± 0·02, β_2 (0·25) 2·00 ± 0·08, (0·50) 1·90 ± 0·03, (1·00) 1·82 ± 0·03, (2·00) 1·98 ± 0·03, β_3 (0·25) 2·70 ± 0·15, (0·50) 2·17 ± 0·06, (1·00) 2·04 ± 0·04, (2·00) 2·13 ± 0·03	68Ga
	cal	25	3·00 ($NaClO_4$)	K_1 1·32 ± 0·04, β_2 2·32 ± 0·06	68Ga
Ce^{3+}	gl	20	0·1 ($NaClO_4$)	K_1 2·09, K_2 1·44	62K
Ce^{4+}	sp	25	0·6—1·0 ($NaClO_4$)	$K[CeOH^{3+} + HL \rightleftharpoons CeOH(HL)^{3+}]$ −0·41 $K(2CeOH^{3+} + HL \rightleftharpoons CeOCeL^{5+} + H^+)$ −1·43 $K[3CeOH^{3+} + HL \rightleftharpoons Ce_3O_3(HL)^{6+} + 3H^+]$ 5·21	68W
Cm^{3+}	dis	20	0·5 $NaClO_4$	K_1 2·06 ± 0·04, β_2 3·09 ± 0·09	63Ga
Co^{2+}	gl	25	0 corr	K_1 1·46	64A
	ix	25	0 corr	K_1 1·29 ± 0·02	65S
Co^{3+}	sol	25, 35	0 corr	$K[CoCl(NH_3)_5^{2+} + L^- \rightleftharpoons CoCl(NH_3)_5L^-]$ (25°) 0·7, (35°) 0·46	65Aa
Cu^{2+}	gl	20	0·1 ($NaClO_4$)	K_1 1·89, K_2 1·20	62K
	gl	25	4 ($NaNO_3$)	K_1 2·52, β_2 3·33	63S
	gl	25	0 corr	K_1 2·23, K_2 1·40	64A
	ix	20	?	K_1 > 1·65, K_2 1·1 ± 0·2, K_3 0·4 ± 0·3	64L
	pol	25	1 ($NaClO_4$)	K_1 1·30, β_2 2·04 $K(Cu^{2+} + L^- + SO_4^{2-} \rightleftharpoons CuLSO_4^-)$ 1·6 $K[Cu^{2+} + L^- + 2SO_4^{2-} \rightleftharpoons CuL(SO_4)_2^{3-}]$ ~ 1·8	65T
	qh	25	3 ($NaClO_4$)	K_1 1·87 ± 0·01, β_2 3·12 ± 0·01, β_3 3·58 ± 0·01, β_4 3·33 ± 0·04	66G
			1 ($NaClO_4$)	K_1 1·71 ± 0·02, β_2 2·71 ± 0·04	
	sp	35	0—1·65	K_1 (→ 0) 1·92, (0·05) 1·51, (1·65) 1·28	67A
	gl	25	0·1 $NaClO_4$ 50% dioxan	K_1 3·36	68E
	gl	25	0·1 ($NaClO_4$) 50% dioxan	$K(CuA^{2+} + L^- \rightleftharpoons CuAL^+)$ 3·51 where A = 2,2′-bipyridyl	68G
Dy^{3+}	qh	20	2 ($NaClO_4$)	K_1 1·67 ± 0·02, β_2 2·97 ± 0·05, β_3 3·79 ± 0·08, β_4 ~ 3·9	58S
	gl	20	0·1 ($NaClO_4$)	K_1 2·03, K_2 1·61	62K
	gl	25	0·500 ($NaClO_4$) 95% methanol	K_1 5·03 ± 0·01, β_2 9·01 ± 0·01, β_3 11·79 ± 0·01, β_4 13·18 ± 0·03	67G
	cal	25	0·500 ($NaClO_4$) 95% methanol	$\Delta H_1°$ 4·09 ± 0·07, $\Delta S_1°$ 36·7, $\Delta H_2°$ 2·72 ± 0·07, $\Delta S_2°$ 27·4, $\Delta H_3°$ 1·00 ± 0·06, $\Delta S_3°$ 16·0, $\Delta H_4°$ 0·3 ± 0·1, $\Delta S_4°$ 7	67G

continued overleaf

114 $C_2H_4O_2$ (contd)

Metal	*Method*	*Temp*	*Medium*	*Log of equilibrium constant, remarks*	*Ref*
Er^{3+}	qh	20	2 ($NaClO_4$)	K_1 1·60±0·02, β_2 2·83±0·04, β_3 3·65±0·07, β_4 ~ 3·6	58S
	sp	20	2 ($NaClO_4$)	K_1 1·65±0·05, β_2 2·88±0·10, β_3 ~ 3·7	58S
	gl	20	0·1 ($NaClO_4$)	K_1 2·01, K_2 1·59	62K
	gl	25	0·500 ($NaClO_4$) 95% methanol	K_1 5·15±0·01, β_2 9·11±0·01, β_3 11·88±0·01, β_4 13·06±0·04	67G
	cal	25	0·500 ($NaClO_4$) 95% methanol	$\Delta H_1°$ 4·58±0·07, $\Delta S_1°$ 38·9, $\Delta H_2°$ 2·92±0·08, $\Delta S_2°$ 27·9, $\Delta H_3°$ 1·9±0·1, $\Delta S_3°$ 19·0, $\Delta H_4°$ −4±1, $\Delta S_4°$ −6	67G
Eu^{3+}	qh	20	0·5 $NaClO_4$	K_1 1·94±0·01, β_2 3·19±0·02, β_3 3·79±0·02	62G
	gl	20	0·1 ($NaClO_4$)	K_1 2·31, K_2 1·60	62K
	dis	25	0·07 ($NaClO_4$)	K_1 1·91±0·02	62Ma
	pol	25	1 ($NaClO_4$)	K_1 2·51, K_2 1·31, acetate buffer	65M
	sp	18—20	0·1	K_1 1·95, β_2 3·84, β_3 5·62	66Ga
Fe^{2+}	red	25	~ 0·5 HCl	K_1 1·82	61N
Fe^{3+}	red	25	~ 0·5 HCl	β_3 10·32	61N
	gl	18—22	0·1	K_1 3·38±0·04, β_2 6·1, β_3 8·7	61S
	sp	18—22	0·1	K_1 3·2, β_2 6·5, β_3 8·3	61S
			0·01	K_2 2·8	
	sp	25	0·4 $LiClO_4$	K_1 2·63	68D
Gd^{3+}	ix	20	1·0 ($NaClO_4$)	K_1 1·82±0·02, β_2 3·13±0·09, β_3 4·0±0·2	59S
	gl	20	0·1 ($NaClO_4$)	K_1 2·16, K_2 1·60	62K
	gl	25	0·500 ($NaClO_4$) 95% methanol	K_1 4·98±0·01, β_2 8·98±0·01, β_3 11·66±0·01, β_4 12·83±0·03	67G
	cal	25	0·500 ($NaClO_4$) 95% methanol	$\Delta H_1°$ 3·38±0·04, $\Delta S_1°$ 34·1, $\Delta H_2°$ 1·71±0·04, $\Delta S_2°$ 24·0, $\Delta H_3°$ 1·59±0·06, $\Delta S_3°$ 17·7, $\Delta H_4°$ −0·3±0·3, $\Delta S_4°$ 4	67G
Hg^{2+}	gl	30	1 $NaClO_4$	K_1 5·55, β_2 9·30, β_3 13·28, β_4 17·06	64B
Ho^{3+}	qh	20	2 ($NaClO_4$)	K_1 1·63±0·02, β_2 2·86±0·04, β_3 3·75±0·07, β_4 ~ 3·6	58S
	gl	20	0·1 ($NaClO_4$)	K_1 2·01, K_2 1·59	62K
La^{3+}	gl	20	0·1 ($NaClO_4$)	K_1 2·02, K_2 1·24	62K
	gl	25	0 corr	K_1 2·545, K_2 1·57	64A
	gl	25	0·500 ($NaClO_4$) 95% methanol	K_1 4·75±0·01, β_2 8·32±0·01, β_3 10·90±0·01, β_4 11·76±0·05	67G
	cal	25	0·500 ($NaClO_4$) 95% methanol	$\Delta H_1°$ 3·28±0·04, $\Delta S_1°$ 32·7, $\Delta H_2°$ 2·60±0·04, $\Delta S_2°$ 25·0, $\Delta H_3°$ 1·08±00·4, $\Delta S_3°$ 15·5, $\Delta H_4°$ 0·4±0·1, $\Delta S_4°$ 5	67G
	qh	25	2 $NaClO_4$ 0—85% ethanol	K_1 (0%) 1·54±0·02, (20%) 1·94±0·005, (40%) 2·22±0·02, (60%) 2·59±0·01, β_2 (0%) 2·46±0·06, (20%) 3·06±0·02, (40%) 3·74±0·04, (60%) 4·65±0·01, (80%) 6·79±0·09, (85%) 6·85±0·07, β_3 (0%) 3·16±0·05, (20%) 3·88±0·02, (40%) 5·06±0·02, (60%) 6·36±0·02, (80%) 8·24±0·06, (85%) 9·23±0·08, β_4 (20%) 4·09±0·01, (40%) 5·35±0·05, (60%) 7·36±0·06, (80%) 10·61±0·01, (85%) 11·48±0·005, β_6 (60%) 7·95±0·07, (80%) 12·08±0·08, (85%) 12·55±0·02, β_7 (85%) 13·26±0·03 (no β_5 observed in any solution)	67M
	gl	25	2 ($NaClO_4$)	K_1 1·45±0·05, K_2 0·8±0·1, K_3 ~ 0·3	62B
Li^+	gl	25	0 corr	K_1 0·26	64A
	con	18	0·1—0·7 LiAc	K_1 (0·1) −0·53, (0·2) −0·54, (0·5) −0·52, (0·7) −0·49	64S
Lu^{3+}	gl	20	0·1 ($NaClO_4$)	K_1 2·05, K_2 1·64	62K

contin

Metal	Method	Temp	Medium	Log of equilibrium constant, remarks	Ref
Mg^{2+}	gl	25, 35	0 corr	K_1 (25°) 1·25±0·01, (35°) 1·21±0·01 ΔH_1 −1·52, ΔS_1 0·6	56N
	gl	25	0 corr	K_1 1·28	64A
Mn^{2+}	gl	25	0 corr	K_1 1·40	64A
Na^+	gl	25	0 corr	K_1 −0·18	64A
Nd^{3+}	gl	20	0·1 ($NaClO_4$)	K_1 2·22, K_2 1·54	62K
	sp	18—20	0·1—0·25	K_1 1·95, β_2 3·59, β_3 5·02, See hexamethylenediamine-*NNN′N′*-tetra-acetic acid (1004) for mixed ligands	63Gb
	gl	25	0 corr	K_1 2·668, K_2 1·870	64A
	gl	25	0·500 ($NaClO_4$) 95% methanol	K_1 5·23±0·01, β_2 9·18±0·01, β_3 11·81±0·01, β_4 13·12±0·02	67G
	cal	25	0·500 ($NaClO_4$) 95% methanol	$\Delta H_1°$ 2·20±0·06, $\Delta S_1°$ 31·3, $\Delta H_2°$ 2·21±0·08, $\Delta S_2°$ 25·6, $\Delta H_3°$ 1·9±0·1, $\Delta S_3°$ 18·4, $\Delta H_4°$ −1·7±0·5, $\Delta S_4°$ 0·3	67G
Ni^{2+}	pol	25	1 ($NaClO_4$)	K_1 0·28 $K[Ni^{2+} + L^- + SO_4^{2-} \rightleftharpoons NiL(SO_4)^-]$ 0·0 $K[Ni^{2+} + L^- + 2SO_4^{2-} \rightleftharpoons NiL(SO_4)_2^{3-}]$ ~ 0·5	63T
	gl	25	0 corr	K_1 1·43	64A
Pb^{2+}	gl	20	0·1 ($NaClO_4$)	K_1 2·20, K_2 1·39	62K
	Hg (Pb)	25	3 ($NaClO_4$)	K_1 2·33±0·00, β_2 3·60±0·00, β_3 3·59±0·02, β_4 2·87±0·08	63G
	gl	25	0 corr	K_1 2·68, K_2 1·40	64A
	gl	30	1 $NaClO_4$	K_1 2·02, β_2 2·98	64B
Pr^{3+}	qh	20	2 ($NaClO_4$)	K_1 1·81±0·02, β_2 2·81±0·04, β_3 3·28±0·08, β_4 ~ 3·3	58S
	gl	20	0·1 ($NaClO_4$)	K_1 2·18, K_2 1·45	62K
	gl	25	0·500 ($NaClO_4$) 95% methanol	K_1 5·11±0·01, β_2 8·93±0·01, β_3 11·54±0·01, β_4 12·49±0·08	67G
	cal	25	0·500 ($NaClO_4$) 95% methanol	$\Delta H_1°$ 2·89±0·05, $\Delta S_1°$ 33·1, $\Delta H_2°$ 2·64±0·07, $\Delta S_2°$ 26·3, $\Delta H_3°$ 1·6±0·1, $\Delta S_3°$ 17·2, $\Delta H_4°$ 0·7±0·6, $\Delta S_4°$ 7	67G
Pu^{3+}	Au	25	0·1 ($HClO_4$)	β_5 16·70	62S
Pu^{4+}	Au	25	0·1 ($HClO_4$)	β_1 ~ 5·3, β_2 ~ 9·0, β_3 ~ 13·9, β_4 ~ 18·3, β_5 22·60	62S
	sp	25	0·5	β_1 ~ 4·9, β_2 ~ 9·8, β_3 ~ 14·6, β_4 ~ 19·4, β_5 22·9 ±0·4	63N
Sm^{3+}	qh	20	2 ($NaClO_4$)	K_1 2·01±0·01, β_2 3·26±0·05, β_3 3·85±0·08, β_4 ~ 3·8	58S
	gl	20	0·1 ($NaClO_4$)	K_1 2·30, K_2 1·58	62K
	gl	25	0 corr	K_1 2·839, K_2 1·96	64A
	gl	25	0·500 ($NaClO_4$) 95% methanol	K_1 5·14±0·01, β_2 9·20±0·01, β_3 11·89±0·01, β_4 13·00±0·05	67G
	sp	18—20	0·1	K_1 1·95, β_2 3·74, β_3 5·28 See hexamethylenediamine-*NNN′N′*-tetra-acetic acid (1004) for mixed ligands	66Ga
	cal	25	0·500 ($NaClO_4$) 95% methanol	$\Delta H_1°$ 2·77±0·05, $\Delta S_1°$ 32·8, $\Delta H_2°$ 1·54±0·05, $\Delta S_2°$ 23·7, $\Delta H_3°$ 1·93±0·06, $\Delta S_3°$ 18·8, $\Delta H_4°$ −0·4±0·2, $\Delta S_4°$ 4	67G
Sr^{2+}	gl	25, 35	0 corr	K_1 (25°) 1·19±0·01, (35°) 1·18±0·01 ΔH_1 0·74, ΔS_1 7·8	56N
	gl	25	0 corr	K_1 1·08	64A
Tb^{3+}	gl	20	0·1 ($NaClO_4$)	K_1 2·07, K_2 1·59	62K

continued overleaf

114 $C_2H_4O_2$ (contd)

Metal	Method	Temp	Medium	Log of equilibrium constant, remarks	Ref
Tl^{3+}	pH	25	3 ($LiClO_4$)	K_1 6·17±0·05, β_2 11·28±0·05, β_3 15·10±0·08, β_4 18·3±0·1, $K(Tl^{3+}+HL+L^- \rightleftharpoons TlHL_2^{2+})$ 7·97 ±0·07	65Ka
				$K(Tl^{3+}+OH^-+L^- \rightleftharpoons TlOHL^+)$ 18·41±0·07	
				$K(Tl^{3+}+OH^-+2L^- \rightleftharpoons TlOHL_2)$ 22·9±0·1	
				$K[Tl^{3+}+2OH^-+L^- \rightleftharpoons Tl(OH)_2L]$ 30·1±0·1	
Tm^{3+}	gl	20	0·1 ($NaClO_4$)	K_1 2·02, K_2 1·59	62K
UO_2^{2+}	dis	18—22	0·1 $NaClO_4$	K_1 2·61±0·04, β_2 4·9±0·1, β_3 6·3±0·1	60S
	gl	25	0·2 (KNO_3)	K_1 2·70±0·05	63F
	gl	30	1 $NaClO_4$	K_1 1·48 (?), β_2 4·82, β_3 6·00, β_4 7·54	64B
	sp	20	1 ($NaClO_4$)	K_1 2·40±0·02, K_2 2·03±0·04, K_3 1·95±0·06	67Ma
Y^{3+}	qh	20	2 $NaClO_4$	K_1 1·53, β_2 2·66, β_3 3·4, β_4 ~ 3·3	60Sa
	gl	20	0·1 ($NaClO_4$)	K_1 1·97, K_2 1·63	62K
Yb^{3+}	qh	20	2 ($NaClO_4$)	K_1 1·64±0·02, β_2 2·83±0·04, β_3 3·54±0·08, β_4 ~ 3·6	58S
	gl	20	0·1 ($NaClO_4$)	K_1 2·03, K_2 1·64	62K
	gl	25	0 corr	K_1 2·56, K_2 1·82	64A
	ix	?	0 corr	K_1 2·46±0·05, K_2 1·3±0·1	66A
	H	25	0 corr	K_1 2·51±0·02, K_2 1·48±0·02	66Aa
	gl	25	0 corr	K_1 2·57±0·02, K_2 1·77±0·04	66Aa
	gl	25	0·500 ($NaClO_4$) 95% methanol	K_1 5·53±0·01, β_2 10·00±0·01, β_3 13·37±0·01, β_4 15·51±0·01	67G
	cal	25	0·500 ($NaClO_4$) 95% methanol	$\Delta H_1°$ 4·84±0·05, $\Delta S_1°$ 41·5, $\Delta H_2°$ 3·50±0·09, $\Delta S_2°$ 32·2, $\Delta H_3°$ 4·2±0·2, $\Delta S_3°$ 29·4, $\Delta H_4°$ −0·1±0·1, $\Delta S_4°$ 9	67G
Zn^{2+}	gl	20	0·1 ($NaClO_4$)	K_1 1·28, K_2 0·81	62K
	qh	25	1 $NaClO_4$	K_1 1·03	63L
	pol	25	4 ($NaNO_3$)	K_1 0·96	63M
	gl	25	0 corr	K_1 1·57	64A
	gl	25	0·1 ($NaClO_4$) 50% dioxan	K_1 2·32 $K(ZnA^{2+}+L^- \rightleftharpoons ZnAL^+)$ 2·21 where A = 2,2′-bipyridyl	68G
H^+	oth	25	acetic acid	K_1 14·45, chloranil electrode (Reciprocal of solvent ion product constant)	56B
	gl	25	→0 95—100+% acetic acid	K_1 (95%) 8·42, (97·5%) 9·73, (99·0%) 10·88, (99·4%) 11·40, (100%) 12·68 (95%+5·0% anhydride) 13·17 (90%+10% anhydride) 13·02	65K
Ba^{2+}	sp	25	acetic acid	β_2 9·20	61P
	gl	25	acetic acid	K_2 6·48±0·02	64K
Ca^{2+}	gl	25	acetic acid	K_2 6·77±0·02	64K
Cd^{2+}	gl	25	acetic acid	K_2 7·54±0·03	64K
	pol	?	acetic acid 0, 90% acetic anhydride	$K[Cd(ClO_4)_2+2NaL \rightleftharpoons CdL_2+NaClO_4]$ (0%) 5·9±0·1, (90%) 12·6±0·2 $K(CdL_2+2NaL \rightleftharpoons Na_2CdL_4)$ (0%) 1·5±0·2	65A
Co^{2+}	gl	25	acetic acid	K_2 7·56±0·02	64K
Cr^{3+}	gl	25	acetic acid	$K(CrL^{2+}+2L^- \rightleftharpoons CrL_3)$ 5·03±0·02	64K
Cs^+	sp	25	acetic acid	K_1 6·78	61P
Cu^{2+}	gl	25	acetic acid	K_2 7·90±0·03	64K
Fe^{3+}	gl	25	acetic acid	K_3 6·17±0·03	64K
Hg_2^{2+}	pol	?	acetic acid	$K[Hg_2(ClO_4)_2+2HL \rightleftharpoons Hg_2L_2+2HClO_4]$ 7·3±0·2	63D
K^+	con	30	acetic acid	K_1 6·44	54J
	oth	25	acetic acid	K_1 6·15, chloranil electrode	56B
	sp	25	acetic acid	K_1 6·93	61P
	gl	25	acetic acid	K_1 6·11±0·02	64K

contin

Metal	*Method*	*Temp*	*Medium*	*Log of equilibrium constant, remarks*	*Ref*
Li^+	con	30	acetic acid	K_1 6·82	54J
	oth	25	acetic acid	K_1 6·79, chloranil electrode	56B
	sp	25	acetic acid	K_1 6·20	61P
	gl	25	acetic acid	K_1 6·78±0·03	64K
Mg^{2+}	sp	25	acetic acid	β_2 9·92	61P
	gl	25	acetic acid	K_2 7·22±0·02	64K
Mn^{2+}	sp	25	acetic acid	β_2 10·27	61P
	gl	25	acetic acid	K_2 7·53±0·02	64K
Na^+	con	30	acetic acid	K_1 6·68	54J
	oth	25	acetic acid	K_1 6·58±0·02, chloranil electrode	56B, 64K
	con	106	acetic acid	K_1 5·88	62M
	gl	25	→ 0 97—100+% acetic acid	K_1 (97·5%) 3·90, (99·0%) 4·35, (99·4%) 4·56, (100%) 4·77, (95%+5·0% anhydride) 4·63 (90%+10% anhydride) 4·49	65K
NH_4^+	sp	25	acetic acid	K_1 7·00	61P
Ni^{2+}	gl	25	acetic acid	K_2 7·63±0·03	64K
Pb^{2+}	sp	25	acetic acid	β_2 8·03	61P
	gl	25	acetic acid	K_2 7·55±0·02	64K
Rb^+	sp	25	acetic acid	K_1 6·89	61P
	gl	25	acetic acid	K_1 6·04±0·02	64K
Sr^{2+}	sp	25	acetic acid	β_2 9·48	61P
	gl	25	acetic acid	K_2 6·65±0·02	64K

Errata. 1964, p. 367. *For* 43L *read* 34L

52M F. H. MacDougall and L. E. Topol, *J. Phys. Chem.*, 1952, **56**, 1090
54J M. M. Jones and E. Griswold, *J. Amer. Chem. Soc.*, 1954, **76**, 3247
56B S. Bruckenstein and I. M. Kolthoff, *J. Amer. Chem. Soc.*, 1956, **78**, 2974
56N G. N. Nancollas, *J. Chem. Soc.*, 1956, 744
58S A. Sonesson, *Acta Chem. Scand.*, 1958, **12**, 1937
59S A. Sonesson, *Acta Chem. Scand.*, 1959, **13**, 1437
60S J. Stary, *Coll. Czech. Chem. Comm.*, 1960, **25**, 2630
60Sa A. Sonesson, *Acta Chem. Scand.*, 1969, **14**, 1495
61N B. P. Nikolskii, V. V. Palchevskii, and R. G. Gorbunova, *Russ. J. Inorg. Chem.*, 1961, **6**, 309 (606)
61P P. J. Proll and L. H. Sutcliffe, *Trans. Faraday Soc.*, 1961, **57**, 1078
61S L. Sommer and K. Pliska, *Coll. Czech. Chem. Comm.*, 1961, **26**, 2754
62B J. L. Bear, G. R. Choppin, and J. V. Quagliano, *J. Inorg. Nuclear Chem.*, 1962, **24**, 1601
62G I. Grenthe, *Acta Chem. Scand.*, 1962, **16**, 1695
62K R. S. Kolat and J. E. Powell, *Inorg. Chem.*, 1962, **1**, 293
62M R. J. L. Martin, *Austral. J. Chem.*, 1962, **15**, 409
62Ma P. G. Manning and C. B. Monk, *Trans. Faraday Soc.*, 1962, **58**, 938
62S K. Schwabe and D. Nebel, *Z. phys. Chem. (Leipzig)*, 1962, **220**, 339
63D G. Durand and B. Tremillon, *Bull. Soc. chim. France*, 1963, 2855
63F I. Feldman and L. Koval, *Inorg. Chem.*, 1963, **2**, 145
63G S. Gobom, *Acta Chem. Scand.*, 1963, **17**, 2181; *Nature*, 1963, **197**, 283
63Ga I. Grenthe, *Acta Chem. Scand.*, 1963, **17**, 1814
63Gb Yn. P. Galakfionov and K. V. Astakhov, *Russ. J. Inorg. Chem.*, 1963, **8**, 1309 (2498)
63L A. Liberti, P. Curro, and G. Calabro, *Ricerca sci.*, 1963, **33** (II-A), 36
63M H. Matsuda, Y. Ayabe, and K. Adachi, *Z. Elektrochem.*, 1963, **67**, 503
63N E. Nebel and K. Schwabe, *Z. phys. Chem. (Leipzig)*, 1963, **224**, 29
63S A. Swinarski and J. Wojtczakove, *Z. phys. Chem. (Leipzig)*, 1963, **223**, 345
63T N. Tanaka, Y. Saito, and H. Ogino, *Bull. Chem. Soc. Japan*, 1963, **36**, 794
64A D. W. Archer and C. B. Monk, *J. Chem. Soc.*, 1964, 3117
64B D. Banerjea and I. P. Singh, *Z. anorg. Chem.*, 1964, **331**, 225
64K O. W. Kolling and J. L. Lambert, *Inorg. Chem.*, 1964, **3**, 202
64L I. Lundquist, *Acta Chem. Scand.*, 1964, **18**, 858
64S M. Suryanarayana, *Current Sci.* 1964, **33**, 520
65A B. Adli-Bloch and B. Tremillon, *Bull. Soc. chim. France*, 1965, 1683
65Aa D. W. Archer, D. A. East, and C. B. Monk, *J. Chem. Soc.*, 1965, 720

continued overleaf

114 $C_2H_4O_2$ (contd)

65H H. E. Hellwege and G. K. Schweitzer, *J. Inorg. Nuclear Chem.*, 1965, **27**, 99
65K S. Kilpi and E. Lindell, *Acta Chem. Scand.*, 1965, **19**, 1420
65Ka F. Ya. Kulba, Yu. B. Yakovlev, and V. E. Mironov, *Russ. J. Inorg. Chem.*, 1965, **10**, 886 (1624)
65M D. J. Macero, H. B. Herman, and A. J. Dukat, *Analyt. Chem.*, 1965, **37**, 675
65P M. Paabo, R. A. Robinson, and R. G. Bates, *J. Amer. Chem. Soc.*, 1965, **87**, 415
65S R. G. Seys and C. B. Monk, *J. Chem. Soc.*, 1965, 2452
65T N. Tanaka, Y. Saito, and H. Ogino, *Bull. Chem. Soc. Japan*, 1965, **38**, 984
66A D. W. Archer and C. B. Monk, *J. Chem. Soc.* (*A*), 1966, 1374
66Aa D. W. Archer and C. B. Monk, *Trans. Faraday Soc.*, 1966, **62**, 1583
66G P. Gerding, *Acta Chem. Scand.*, 1966, **20**, 2624
66Ga Yu. P. Galaktionov and K. V. Astakhov, *Russ. J. Inorg. Chem.*, 1966, **11**, 969 (1813)
67A S. Aditga, *J. Inorg. Nuclear Chem.*, 1967, **29**, 1901
67G I. Grenthe and D. R. Williams, *Acta Chem. Scand.*, 1967, **21**, 341, 347
67M P. K. Migal' and N. G. Chebotar', *Russ. J. Inorg. Chem.*, 1967, **12**, 630 (1190)
67Ma C. Miyake and H. W. Nürnberg, *J. Inorg. Nuclear Chem.*, 1967, **29**, 2411
68D M. Deneux, R. Meilleur, and R. L. Benoit, *Canad. J. Chem.*, 1968, **46**, 1383
68E H. Erlenmeyer, R. Griesser, B. Prijs, and H. Sigel, *Helv. Chim. Acta*, 1968, **51**, 339
68G R. Griesser, B. Prijs, and H. Sigel, *Inorg. Nuclear Chem. Letters*, 1968, **4**, 443
68Ga P. Gerding, *Acta Chem. Scand.*, 1968, **22**, 1283
68W K. B. Wiberg and P. C. Ford, *Inorg. Chem.*, 1968, **7**, 369

115 $C_2H_4O_3$ Hydroxyacetic acid (glycollic acid) HL

Metal	*Method*	*Temp*	*Medium*	*Log of equilibrium constant, remarks*	*Ref*
H^+	gl	30	0·1 (KCl)	K_1 3·70	62Ca
	?	?	?	K_1 3·86	62S
	gl	25	1 $NaClO_4$	K_1 3·62	63M
	gl	25	0·1 ($NaClO_4$) 50% dioxan	K_1 4·88±0·03	68G
Am^{3+}	ix	20	0·5 $NaClO_4$	K_1 2·82±0·01, β_2 4·86±0·07, β_3 6·3±0·3	62G
Be^{2+}	ix	18	0·1 ($NaClO_4$)	K_1 1·49±0·01	65B
Cd^{2+}	pol	30	1 KNO_3	K_1 1·26, β_2 2·15	66J
Ce^{3+}	qh	20	2 ($NaClO_4$)	K_1 2·35, β_2 4·02, β_3 5·15, β_4 5·5, β_5 ~ 5·3	59S
	ix	20	0·2 ($NaClO_4$)	K_1 2·43±0·05, β_2 4·11±0·10, β_3 5·3±0·3	60Sa
	gl	30	0·1 (KCl)	K_1 2·84, K_2 2·45	62Ca
	gl	20	0·1 ($NaClO_4$)	K_1 2·695, β_2 4·55, β_3 5·4	64P
Cm^{3+}	dis	20	0·5 $NaClO_4$	K_1 2·85±0·04, β_2 4·75±0·05	63G
Co^{2+}	ix	25	0 corr	K_1 1·96±0·02, K_2 1·05±0·02	65S
Co^{3+}	sol	25	0 corr	$K[CoCl(NH_3)_5^{2+} + L^- \rightleftharpoons CoCl(NH_3)_5L^-]$ 1·5	65A
Cu^{2+}	gl	25	1 $NaClO_4$	K_1 2·36, β_2 3·70 $K(Cu^{2+} + L^- + A^- \rightleftharpoons CuLA)$ 10·2 where HA = glycine	63M
	ix	20	?	K_2 1·4±0·2, K_3 0·3±0·3	64L
	gl	25	0·1 ($NaClO_4$) 50% dioxan	K_1 3·96 $K(CuA^{2+} + L^- \rightleftharpoons CuAL^+)$ 3·86 where A = 2,2′-bipyridyl	68G
Dy^{3+}	qh	20	2 ($NaClO_4$)	K_1 2·52, β_2 4·48, β_3 5·9, β_4 6·5, β_5 ~ 6·3	59S
	gl	20	0·1 ($NaClO_4$)	K_1 2·924, β_2 4·97, β_3 6·55	64P
Er^{3+}	qh	20	2 ($NaClO_4$)	K_1 2·60, β_2 4·58, β_3 6·0, β_4 6·5, β_5 ~ 6·5	59S
	gl	20	0·1 ($NaClO_4$)	K_1 3·004, β_2 5·19, β_3 6·85	64P
Eu^{3+}	qh	20	0·5 $NaClO_4$	K_1 2·57±0·01, β_2 4·61±0·03, β_3 5·91±0·04, β_4 6·4	62G
	dis	25	0·08 ($NaClO_4$)	K_1 2·69	62M
	gl	20	0·1 ($NaClO_4$)	K_1 2·935, β_2 5·07, β_3 6·52	64P

conti

Metal	Method	Temp	Medium	Log of equilibrium constant, remarks	Ref
Gd^{3+}	qh	20	2·0 ($NaClO_4$)	K_1 2·48±0·03, β_2 4·43±0·07, β_3 5·78±0·06, β_4 6·42±0·10, β_5 ~ 6·0	59S
	ix	20	1·0 ($NaClO_4$)	K_1 2·54±0·02, β_2 4·48±0·06, β_3 5·8±0·1	59S
	qh	20	3 ($NaClO_4$)	K_1 2·59, β_2 4·60, β_3 6·1, β_4 6·8, β_5 6·8, β_6 6·6	61S
			0·2 ($NaClO_4$)	K_1 2·73, β_2 4·79, β_3 6·3, β_4 7·1, β_5 7·0	
	gl	20	0·1 ($NaClO_4$)	K_1 2·792, β_2 4·86, β_3 6·00	64P
Ho^{3+}	qh	20	2 ($NaClO_4$)	K_1 2·54, β_2 4·48, β_3 5·9, β_4 6·4, β_5 ~ 6·5	59S
	gl	25	2 ($NaClO_4$)	K_1 2·46, K_2 1·95, K_3 1·25	62B
	gl	20	0·1 ($NaClO_4$)	K_1 2·991, β_2 5·04, β_3 6·57	64P
La^{3+}	qh	20	2 ($NaClO_4$)	K_1 2·19, β_2 3·76, β_3 4·81, β_4 5·1, β_5 ~ 4·8	59S
	gl	30	0·1 (KCl)	K_1 2·68, K_2 2·42	62Ca
	gl	20	0·1 ($NaClO_4$)	K_1 2·55, β_2 4·24, β_3 5·0	64P
Lu^{3+}	gl	20	0·1 ($NaClO_4$)	K_1 3·146, β_2 5·48, β_3 7·28	64P
Nd^{3+}	qh	20	2 ($NaClO_4$)	K_1 2·51, β_2 4·34, β_3 5·6, β_4 6·0, β_5 ~ 5·7	59S
	sp	20	2 ($NaClO_4$)	K_1 2·54, β_2 4·4, β_3 ~ 5·3	59S
	gl	30	0·1 (KCl)	K_1 3·07, K_2 2·81	62Ca
	gl	20	0·1 ($NaClO_4$)	K_1 2·89, β_2 4·86, β_3 6·1	64P
Pb^{2+}	pH	25	3 ($NaClO_4$)	K_1 2·23±0·02, β_2 3·24±0·05, β_3 3·26±0·07	65Ba
	pol	30	1 KNO_3	K_1 1·90, β_2 3·16	66J
Pr^{3+}	qh	20	2 ($NaClO_4$)	K_1 2·43, β_2 4·19, β_3 5·4, β_4 5·9, β_5 ~ 5·7	59S
	gl	30	0·1 (KCl)	K_1 2·98, K_2 2·69	62Ca
	gl	20	0·1 ($NaClO_4$)	K_1 2·78, β_2 4·68, β_3 5·9	64P
Sm^{3+}	qh	20	2 ($NaClO_4$)	K_1 2·56, β_2 4·53, β_3 5·9, β_4 6·4, β_5 ~ 6·0	59S
	gl	20	0·1 ($NaClO_4$)	K_1 2·914, β_2 5·01, β_3 6·57	64P
Tm^{3+}	gl	20	0·1 ($NaClO_4$)	K_1 3·055, β_2 5·33, β_3 7·00	64P
UO_2^{2+}	gl	30	0·1 KCl	K_1 2·97, K_2 2·40	62C
	dis	20	1 $NaClO_4$	K_1 2·71±0·07, β_2 4·08±0·07, β_3 5·5±0·3	62S
Y^{3+}	qh	20	2 ($NaClO_4$)	K_1 2·47, β_2 4·40, β_3 5·7, β_4 6·3, β_5 ~ 6·3	60S
	ix	20	0·2 ($NaClO_4$)	K_1 2·78±0·03, β_2 4·70±0·06, β_3 6·0±0·1	60Sa
	gl	20	0·1 ($NaClO_4$)	K_1 2·785, β_2 4·88, β_3 5·78	64P
Yb^{3+}	qh	20	2 ($NaClO_4$)	K_1 2·72, β_2 4·82, β_3 6·3, β_4 6·8, β_5 ~ 7·0	59S
	gl	20	0·1 ($NaClO_4$)	K_1 3·130, β_2 5·37, β_3 7·11	64P
Zn^{2+}	qh	25	1 $NaClO_4$	K_1 1·92, β_2 2·93, β_3 3·00, β_4 4·04	63L
	gl	25	0·1 ($NaClO_4$)	K_1 3·26	68G
			50% dioxan	$K(ZnA^{2+} + L^- \rightleftharpoons ZnAL^+)$ 3·02 where A = 2,2′-bipyridyl	

Errata. 1964, p. 368. Under Ce^{3+}, Gd^{3+}, Eu^{3+} and Ho^{3+} *for* ref. 60C *read* 61C. Insert ref. 61C, G. R. Choppin and J. A. Chopoorian, *J. Inorg. Nuclear Chem.*, 1961, **22**, 97

59S A. Sonesson, *Acta Chem. Scand.*, 1959, **13**, 998, 1437
60S A. Sonesson, *Acta Chem. Scand.*, 1960, **14**, 1495
60S V. G. Spitsyn and O. Voitekh, *Proc. Acad. Sci. (U.S.S.R.)*, 1960, **133**, 859 (613)
61S A. Sonesson, *Acta Chem. Scand.*, 1961, **15**, 1
62B J. L. Bear, G. R. Choppin, and J. V. Quagliano, *J. Inorg. Nuclear Chem.* 1962, **24**, 1601
62C M. Cefola, R. C. Taylor, P. S. Gentile, and A. V. Celiano, *J. Phys. Chem.*, 1962, **66**, 790
62Ca M. Cefola, A. S. Tompa, A. V. Celiano, and P. S. Gentile, *Inorg. Chem.*, 1962, **1**, 290
62G I. Grenthe, *Acta Chem. Scand.*, 1962, **16**, 1965
62M P. G. Manning and C. B. Monk, *Trans. Faraday Soc.*, 1962, **58**, 938
62S J. Stary and V. Balek, *Coll. Czech. Chem. Comm.*, 1962, **27**, 809
63G I. Grenthe, *Acta Chem. Scand.*, 1963, **17**, 1814
63L A. Liberti, P. Curro, and G. Calabro, *Ricerca sci.*, 1963, **33**, (II-A), 36
63M R.-P. Martin and R. A. Paris, *Bull. Soc. chim. France.*, 1963, 1600
64L I. Lundquist, *Acta Chem. Scand.*, 1964, **18**, 858
64P J. E. Powell, R. H. Karraker, R. S. Kolat, and J. L. Farrell, 'Rare Earth Research II', ed. K. S. Vorres; Gordon and Breach, New York, 1964, p. 512
65A D. W. Archer, D. A. East, and C. B. Monk, *J. Chem. Soc.*, 1965, 720

continued overleaf

115 $C_2H_4O_3$ (contd)

65B T. A. Belyarskaya and I. F. Kolosova, *Russ. J. Inorg. Chem.*, 1965, **10**, 236 (441)
65Ba A. Basinski and Z. Warnke, *Roczniki Chem.*, 1965, **39**, 1776
65S R. G. Seys and C. B. Monk, *J. Chem. Soc.*, 1965, 2452
66J D. S. Jain and J. N. Gaur, *J. Polarog. Soc.*, 1966, **12**, 59
68G R. Griesser, B. Prijs, and H. Sigel, *Inorg. Nuclear Chem. Letters*, 1968, **4**, 443

115.1 $C_2H_4N_4$ 1-Methyltetrazole L

Metal	*Method*	*Temp*	*Medium*	*Log of equilibrium constant, remarks*	*Ref*
Co^{2+}	sp	25	tetrahydrofuran	K_1 2·13±0·01, β_2 3·52±0·02	63G
Ni^{2+}	sp	25	ethanol	K_1 0·6±0·6, β_2 2·05±0·04	63G

63G G. L. Gilbert and C. H. Brubaker jun., *Inorg. Chem.*, 1963, **2**, 1216.

115.2 C_2H_6O Ethanol L

Metal	*Method*	*Temp*	*Medium*	*Log of equilibrium constant, remarks*	*Ref*
Cu^{2+}	sp	27	?	K_1 0·26±0·06, K_2 −4·41±0·04	63F

63F N. J. Friedman and R. A. Plane, *Inorg. Chem.*, 1963, **2**, 11

115.3 $C_2H_6O_2$ Ethane-1,2-diol (ethylene glycol) L

Metal	*Method*	*Temp*	*Medium*	*Log of equilibrium constant, remarks*	*Ref*
As^{III}	gl	25	0·1	$K[As(OH)_4^- + L \rightleftharpoons As(OH)_2(H_{-2}L)^-]$ −1·15	57R
B^{III}	gl	25	0·1	$K[B(OH)_4^- + L \rightleftharpoons B(OH)_2(H_{-2}L)^-]$ 0·27	57R
				$K[B(OH)_4^- + 2L \rightleftharpoons B(H_{-2}L)_2^-]$ −1·0	
	gl	0—35	0·025 borax	$K[B(OH)_4^- + L \rightleftharpoons B(OH)_2(H_{-2}L)^-]$ (0°) 0·52, (13°) 0·46, (25°) 0·33, (35°) 0·27	67C
				$\Delta H°$ −2·7±0·3, $\Delta S°$ −8±2	
				$K[B(OH)_4^- + 2L \rightleftharpoons B(H_{-2}L)_2^-]$ (0°) 0·14, (13°) 0·08, (25°) 0·06, (35°) −0·05	
				$\Delta H°$ −2·0±0·2, $\Delta S°$ −6±2	
	gl	25	I KCl	$K[B(OH)_4^- + 2L \rightleftharpoons B(H_{-2}L)_2^-]$ $-0{\cdot}007+1{\cdot}334\sqrt{I}$	67N
Ge^{IV}	qh	25	0·1 KCl	$K[HGeO_3^- + L \rightleftharpoons HGeO_2(H_{-2}L)^-]$ 0·17	59A
				$K[HGeO_3^- + 2L \rightleftharpoons HGeO(H_{-2}L)_2^-]$ −0·37	
Te^{VI}	gl	25	0·1	$K[H_5TeO_6^- + L \rightleftharpoons H_3TeO_4(H_{-2}L)^-]$ 1·21	57R

57R G. L. Roy, A. L. Laferriere, and J. O. Edwards, *J. Inorg. Nuclear Chem.*, 1957, **4**, 106
59A P. J. Antikainen, *Acta Chem. Scand.*, 1959, **13**, 312
67C J. M. Conner and V. C. Bulgrin, *J. Inorg. Nuclear Chem.*, 1967, **29**, 1953
67N V. A. Nazarenko and L. D. Ermak, *Russ. J. Inorg. Chem.*, 1967, **12**, 335 (643), 1079 (2051)

C_2H_7N Dimethylamine L 117

Metal	Method	Temp	Medium	Log of equilibrium constant, remarks	Ref
Ag^+	Ag	25	0·02 0—90% methanol	β_2 (0%) 5·22, (30%) 5·32, (50%) 5·54, (60%) 5·55, (80%) 5·85, (90%) 5·84	63P

63P G. Popa, C. Luca, and V. Magearu, *J. Chim. phys.*, 1963, **60**, 355

C_2H_7N Ethylamine L 118

Metal	Method	Temp	Medium	Log of equilibrium constant, remarks	Ref
H^+	gl	30	0·5 (KNO_3)	K_1 10·70	67F
	gl	20	0·5 KNO_3	K_1 10·98±0·01	68A
	H	20	0 corr	K_1 10·79±0·02	68A
Ag^+	sol	25	Cl^- or Br^-	β_2 (Cl^-) 7·39, (Br^-) 7·48	64S
	gl	20	0·500 KNO_3	K_1 3·46±0·06, β_2 7·57±0·02	68A
	Ag-AgCl	20	0·500 KNO_3	β_2 7·51±0·01	68A
Cu^+	pol	30	0·5 (KNO_3)	β_2 10·1, $K(Cu^+ + L + OH^- \rightleftharpoons CuLOH)$ 10·8	67F
Cu^{2+}	pol	30	0·5 (KNO_3)	β_4 11·5, $K[Cu^{2+} + 2L + 2OH^- \rightleftharpoons CuL_2(OH)_2]$ 15·9	67F
Pt^{2+}	Pt	18	1 (KNO_3)	β_4 37·0	61G
Zn^{2+}				See dimethyldithiocarbamic acid (176) for mixed ligands	

61G A. A. Grinberg and M. I. Gelfman, *Proc. Acad. Sci. (U.S.S.R.)*, 1961, **137**, 257 (87)
64S R. L. Seth and A. K. Dey, *Indian J. Chem.*, 1964, **2**, 291
67F J. F. Fisher and J. L. Hall, *Analyt. Chem.*, 1967, **39**, 1550
68A D. J. Alner, R. C. Lansbury, and A. G. Smeeth, *J. Chem. Soc. (A)*, 1968, 417

$C_2H_7N_5$ Biguanide L 119

Metal	Method	Temp	Medium	Log of equilibrium constant, remarks	Ref
H^+	gl	32	0·01	K_1 11·52, K_2 2·93	59R
Cu^{2+}	sp	28—32	0·5 KCl	β_2 18·31	59R

59R M. M. Ray and P. Ray, *J. Indian Chem. Soc.* 1959, **36**, 849

$C_2H_8N_2$ 1,2-Diaminoethane (ethylenediamine) (en) L 120

Metal	Method	Temp	Medium	Log of equilibrium constant, remarks	Ref
H^+	cal	25	0·1 KCl	ΔH_1 −11·9±0·1, ΔH_2 −10·9±0·1	54D
	gl	25	0·3 (K_2SO_4)	K_1 9·83±0·03, K_2 7·30±0·05	61J
	sp	25	1·2	K_1 10, K_2 7·3	63C
	gl	25	I ($NaClO_4$)	K_1 9·93+0·308 I K_2 6·84+1·018 $\sqrt{I}/(1+1·381\sqrt{I})$+0·209 I	63N, 65N, 67N

continued overleaf

120 $C_2H_8N_2$ (contd)

Metal	Method	Temp	Medium	Log of equilibrium constant, remarks	Ref
H^+ (contd)	gl	20	~ 0	K_1 9·27, K_2 6·17	66Pa
	gl	25	2 L·2HNO$_3$	K_1 16·64 (?), K_2 9·49 (?)	67L
	gl	37	0·15 KNO_3	K_1 9·696, K_2 6·928	67P, 68
	gl	25	0·5 $NaNO_3$	K_1 10·18, K_2 7·39	67S
Ag^+	Ag, sol	25	< 0·01	β_2 7·73, β_3 9·75	61A
Al^{3+}	fp	?	benzene	$K(4Al_2A_6L \rightleftharpoons Al_8A_{24}L_4)$ 6·2 where HA = isopropyl alcohol	62B
Cd^{2+}	cal	25	0·1 KCl	$\Delta H_{\beta 2}$ −13·3, $\Delta S_{\beta 2}$ −1·7 $\Delta H_{\beta 3}$ −19·7, $\Delta S_{\beta 3}$ −15·3	54D
	gl	0—49	0·15	$\Delta H_1°$ (25°) −5·3±0·4, $\Delta S_1°$ 7±1 $\Delta H_2°$ (25°) −4·3±0·5, $\Delta S_2°$ 6±2	55C
	gl	25	0·5 $NaNO_3$	K_1 5·69, K_2 4·67, K_3 2·44 $K(Cd^{2+}+L+A \rightleftharpoons CdLA^{2+})$ 12·54±0·04 where A = 2,2′-diaminodiethylamine See oxalic acid (111) for mixed ligands	67S
Co^{2+}	pol	20	2·7	K_1 6·26, K_2 5·07, K_3 3·57	63K
	pol	0·5	1 L·2HCl	K_3 3·51	68F
			1 L·2HClO$_4$	K_3 4·17 See EDTA (836) for mixed ligands	
Cr^{3+}	sp	25	0·1	K_1 16·5, K_2 < 14	64O
Cu^+	Pt	25	0·3 (K_2SO_4)	β_2 11·4	61J
Cu^{2+}	gl	0—25	0·5 KNO_3	ΔH_1 (0°) −8·6, ΔS_1 21, ΔH_2 (0°) −8·6, ΔS_2 14	52B
	cal	0	0·50 KNO_3	$\Delta H_{\beta 2}°$ −24·6, $\Delta S_{\beta 2}°$ 7	54B
	cal	25	0·1 KCl	$\Delta H_{\beta 2}$ −25·2, $\Delta S_{\beta 2}$ 7·1	54D
	gl	0—49	0·15	$\Delta H_1°$ (25°) −11·9±0·6, $\Delta S_1°$ 9±3, $\Delta H_2°$ (25°) −11·3±0·3, $\Delta S_2°$ 4±1	55C
	cal	25	1 KNO_3	ΔH_1 −13·0±0·1, ΔS_1 5·4, $\Delta H_{\beta 2}$ −25·4, $\Delta S_{\beta 2}$ 6·3	55P
	pol	25	0·17	β_2 8·48 (?), phosphate buffer	61K
	gl	25	I ($NaClO_4$)	K_1 10·48+0·646 I −0·254 $I^{1·5}$+0·052 I^2 K_2 9·07+0·626 I+0·122 $I^{1·5}$−0·2020 I^2	63N
	gl	20	~ 0	K_1 10·66, K_2 9·33	66Pa
	gl	25	→ 0	$K(Cu^{2+}+L+A \rightleftharpoons CuLA^{2+})$ 19·75 where A = 1,2-propanediamine 17·87 where A = *NN′*-diethylethylenediamine 18·58 where A = 1,3-propanediamine $K(CuL_2^{2+}+CuA_2^{2+} \rightleftharpoons 2CuLA)$ 0·31 where A = 1,2-propanediamine 1·83 where A = *NN′*-diethylethylenediamine 0·98 where A = 1,3-propanediamine	67N
	gl	37	0·15 KNO_3	K_1 10·175, K_2 8·765 $K(CuA^{2+}+L \rightleftharpoons CuAL^{2+})$ 8·15 where A = histamine $K(CuA^++L \rightleftharpoons CuAL^+)$ 9·31 where HA = serine $K(CuA+L \rightleftharpoons CuAL)$ 9·16 where H_2A = salicylic acid See 1,2-diaminopropane (158), pentane-2,4-dione (283) and EDTA (836) for mixed ligands	67P
Hg^{2+}	pol	10—40	0·1 KNO_3	β_2 (10°) 24·36±0·04, (25°) 23·18±0·02, (40°) 21·94 ±0·01, β_3 (10°) 24·1±0·1, (25°) 23·09±0·01, (40°) 21·74±0·01 $\Delta H_{\beta 2}°$ −32·9±0·6, $\Delta S_{\beta 2}°$ −5±2	61R
	gl	25	0 corr	$K(HgCl_2+L \rightleftharpoons HgClL^++Cl^-)$ 5·54±0·01 $K(HgClL^++L \rightleftharpoons HgL_2^{2+}+Cl^-)$ 4·19±0·04	66P

contd

Metal	Method	Temp	Medium	Log of equilibrium constant, remarks	Ref
Hg^{2+} (contd)	cal	25	0 corr	$\Delta H°(HgCl_2+L \rightleftharpoons HgClL^+ + Cl^-)$ −8·7±0·2, $\Delta S°$ −4 $\Delta H°(HgClL^+ + L \rightleftharpoons HgL_2^{2+} + Cl^-)$ −9·0±0·1, $\Delta S°$ −11	66P
Ni^{2+}	gl	0—25	0·5 KNO_3	ΔH_1 (0°) −4·8, ΔS_1 19, ΔH_2 (0°) −4·3, ΔS_2 15, ΔH_3 (0°) −4·9, ΔS_3 7	52B
	cal	0	0·50 KNO_3	$\Delta H_{\beta 3}°$ −25·1, $\Delta S_{\beta 3}°$ 1	54B
	cal	25	0·1 KCl	$\Delta H_{\beta 2}$ −17·3, $\Delta S_{\beta 2}$ 3·4, $\Delta H_{\beta 3}$ −28·0, $\Delta S_{\beta 3}$ −14·0	54D
	gl	0—49	0·15	$\Delta H_1°$ (25°) −7·9±0·3, $\Delta S_1°$ 10±1, $\Delta H_2°$ (25°) −7·8±0·5, $\Delta S_2°$ 2±2, $\Delta H_3°$ (25°) −7±1, $\Delta S_3°$ −5±5	55C
	cal	25	1 KNO_3	ΔH_1 −9·0±0·1, ΔS_1 4·1, $\Delta H_{\beta 2}$ −18·2±0·1, $\Delta S_{\beta 2}$ 2·3, $\Delta H_{\beta 3}$ −27·9±0·1, $\Delta S_{\beta 3}$ −10·1	55P
	sp	25	1·2	K_1 7·55, K_2 6·20, K_3 4·77	63C
	gl	25	I ($NaClO_4$)	K_1 7·32+0·290 I, K_2 6·18+0·343 I, K_3 4·11+0·409 I	65N
	gl	37	0·15 KNO_3	K_1 6·982, K_2 5·807, K_3 3·662 $K(NiA^{2+} + L \rightleftharpoons NiAL^{2+})$ 5·78 $K(NiA_2^{2+} + L \rightleftharpoons NiA_2L^{2+})$ 3·93 $K(NiAL^{2+} + L \rightleftharpoons NiAL_2^{2+})$ 3·29 where A = histamine $K(NiA^+ + L \rightleftharpoons NiAL^+)$ 6·57 $K(NiA_2 + L \rightleftharpoons NiA_2L)$ 4·99 $K(NiAL^+ + L \rightleftharpoons NiAL_2^+)$ 4·30 where HA = serine See nitrilotriacetic acid (449) and EDTA (836) for mixed ligands	68P
Pt^{2+}	Pt	18	1 (KNO_3)	β_2 36·5	61G
Ti^{4+}	fp	?	benzene cyclohexane	$K(TiA_4 + L \rightleftharpoons TiA_4L)$ 2·8 $K(TiA_4 + L \rightleftharpoons TiA_4L)$ 3·7 $K(TiA_4L + TiA_4 \rightleftharpoons Ti_2A_8L)$ −5·3 where HA = isopropyl alcohol	62B
Tl^{3+}	gl	25	2 L·2HNO_3	$K(Tl(OH)_2^+ + L \rightleftharpoons Tl(OH)_2L^+)$ 13·0 $K(Tl^{+3} + 2OH^- + L \rightleftharpoons Tl(OH)_2L^+)$ 41·64	67L
VO^{2+}	con	20	0·017 $VOSO_4$	β_2 5·7	61B
Zn^{2+}	cal	25	0·1 KCl	$\Delta H_{\beta 2}$ −11·5, $\Delta S_{\beta 2}$ 7·6, $\Delta H_{\beta 3}$ −18·5, $\Delta S_{\beta 3}$ −9·8	54D
	pol	30	0·5 KCl	β_2 11·2, β_3 12·3 See oxalic acid (111) for mixed ligands	67Sa

52B F. Basolo and R. K. Murmann, *J. Amer. Chem. Soc.*, 1952, **74**, 5243
54B F. Basolo and R. K. Murmann, *J. Amer. Chem. Soc.*, 1954, **76**, 211
54D T. Davies, S. S. Singer, and L. A. K. Staveley, *J. Chem. Soc.*, 1954, 2304
55C F. A. Cotton and F. E. Harris, *J. Phys. Chem.*, 1955, **59**, 1203
55P I. Poulsen and J. Bjerrum, *Acta Chem. Scand.* 1955, **9**, 1407
61A V. Armeanu and C. Luca, *Z. Phys. Chem. (Leipzig)*, 1961, **217**, 389
61B P. K. Bhattacharya, M. C. Saxena, and S. N. Banerji, *J. Indian Chem. Soc.*, 1961, **38**, 801
61G A. A. Grinberg and M. I. Gelfman, *Proc. Acad. Sci. (U.S.S.R.)*, 1961, **137**, 257 (87)
61J B. R. James and R. J. P. Williams, *J. Chem. Soc.*, 1961, 2007
61K E. C. Knoblock and W. C. Purdy, *J. Electroanalyt. Chem.*, 1961, **2**, 493
61R D. K. Roe, D. B. Masson, and C. J. Nyman, *Analyt. Chem.*, 1961, **33**, 1464
62B M. S. Bains and D. C. Bradley, *Canad. J. Chem.*, 1962, **40**, 2218
63C C. Caullet, *Bull. Soc. chim. France*, 1963, 688
63K D. Konrad and A. A. Vleck, *Coll. Czech. Chem. Comm.*, 1963, **28**, 595
63N R. Nasanen and P. Merilainen, *Suomen Kem.*, 1963, **B36**, 97 (*Acta Chem. Scand.* 1964, **18**, 1337)
64O M. Ohta, H. Matsukawa, and R. Tsuchiya, *Bull. Chem. Soc. Japan*, 1964, **37**, 692.
65N R. Nasanen, M. Koskinen, and K. Kajander, *Suomen Kem.*, 1965, **B38**, 103
66P J. A. Partridge, J. J. Christensen, and R. M. Izatt, *J. Amer. Chem. Soc.*, 1966, **88**, 1649
66Pa D. D. Perrin and V. S. Sharma, *J. Inorg. Nuclear Chem.*, 1966, **28**, 1271
67L B. I. Lobov, F. Ya. Kul'ba and V. E. Mironov, *Russ. J. Inorg. Chem.*, 1967, **12**, 176 (341

continued overleaf

120 $C_2H_8N_2$ (contd)

67N R. Nasanen and M. Koskinen, *Suomen Kem.*, 1967, **B40**, 108, 23
67P D. D. Perrin, I. G. Sayce, and V. S. Sharma, *J. Chem. Soc. (A)*, 1967, 1755
67S J. P. Scharf and M. R. Paris, *Compt. Rend.*, 1967, **265C**, 488
67Sa R. Sundaresan, S. C. Saraiya, and A. K. Sundaram, *Proc. Indian Acad. Sci.*, 1967, **66A**, 120
68F E. Fischerova, O. Dracka, and M. Meloun, *Coll. Czech. Chem. Comm.*, 1968, **33**, 473
68P D. D. Perrin and V. S. Sharma, *J. Chem. Soc. (A)*, 1968, 446

121.1 $C_2HO_2F_3$ Trifluoroacetic acid HL

Metal	*Method*	*Temp*	*Medium*	*Log of equilibrium constant, remarks*	*Ref*
Pb^{2+}	pol	25	2 ($NaClO_4$)	K_1 0·02, β_2 −0·14	64C
	Pb	25	2 ($NaClO_4$)	K_1 −0·03, β_2 −0·48	64C

64C S. A. Carrano, K. A. Chem, and R. F. O'Malley, *Inorg. Chem.*, 1964, **3**, 1047

122.1 $C_2H_3O_2N_3$ 1,2,4-Triazolidin-3,5-dione (urazole) HL

Metal	*Method*	*Temp*	*Medium*	*Log of equilibrium constant, remarks*	*Ref*
H^+	gl	20	0·1 ($NaClO_4$)	K_1 5·69±0·01	63C
Co^{2+}	gl	20	0·1 ($NaClO_4$)	K_1 2·07±0·02	63C
Ni^{2+}	gl	20	0·1 ($NaClO_4$)	K_1 2·45±0·02	63C
Zn^{2+}	gl	20	0·1 ($NaClO_4$)	K_1 1·87±0·01	63C

63C E. Campi, G. Ostacoli, and A. Vanni, *Ricerca sci.*, 1963, 33 (II-A), 1073

123 $C_2H_3O_2Cl$ Chloroacetic acid HL

Metal	*Method*	*Temp*	*Medium*	*Log of equilibrium constant, remarks*	*Ref*
H^+	E	20	1 KCl	K_1 2·60	60J
	qh	20	1 ($NaClO_4$)	K_1 2·66	62H
	gl	25	0·1 $NaClO_4$ 50% dioxan	K_1 4·08±0·02	68E
Cu^{2+}	gl	25	0·1 $NaClO_4$ 50% dioxan	K_1 2·57	68E
UO_2^{2+}	pol	?	1 ($NaClO_4$)	K_1 1·6±0·1, β_2 2·3±0·1	62H
V^{3+}	gl	22	1 KCl	$K[V^{3+}+2VOH^{2+}+6L^- \rightleftharpoons V(VOH)_2L_6^+]$ 18·52 ±0·05	60J
Zn^{2+}	qh	25	1 $NaClO_4$	K_1 0·96, β_2 0·0 (?), β_3 0·7 (?), β_4 1·5 (?)	63L

60J B. Jezowska-Trzebiatowska and L. Pajdowski, *Roczniki Chem.*, 1960, **34**, 787
63L A. Liberti, P. Curro, and G. Calabro, *Ricerca sci.*, 1963, **33** (II-A), 36
62H J. Hala and A. Okac, *Coll. Czech. Chem. Comm.*, 1962, **27**, 1697.
68E H. Erlenmeyer, R. Griesser, B. Prijs, and H. Sigel, *Helv. Chim. Acta.*, 1968, **51**, 339

$C_2H_4O_2N_4$ 4-Amino-1,2,4-triazolidin-3,5-dione (urazine) HL 125.1

Metal	Method	Temp	Medium	Log of equilibrium constant, remarks	Ref
H^+	gl	20	0·1 ($NaClO_4$)	K_1 5·49±0·09	63C
Co^{2+}	gl	20	0·1 ($NaClO_4$)	K_1 2·34±0·02	63C
Ni^{2+}	gl	20	0·1 ($NaClO_4$)	K_1 2·65±0·01, β_2 4·80±0·03	63C
Zn^{2+}	gl	20	0·1 ($NaClO_4$)	K_1 2·17±0·02	63C

63C E. Campi, G. Ostacoli, and A. Vanni, *Ricerca sci.*, 1963, 33 (II-A), 1073

$C_2H_4O_2S$ Mercaptoacetic acid (thioglycolic acid) H_2L 126

Metal	Method	Temp	Medium	Log of equilibrium constant, remarks	Ref
H^+	gl	30	0·1 (KCl)	K_2 3·55	62Ca
	gl	20	0·1 ($NaClO_4$)	K_2 3·41	64P
	gl	20	0·1 ($NaClO_4$)	*K_1 10·203±0·006, K_2 3·516±0·008	67P
Am^{3+}	ix	20	0·5 $NaClO_4$	*K_1 1·55±0·07, β_2 2·6±0·1	62G
Ce^{3+}	gl	25	2 ($NaClO_4$)	*K_1 1·43±0·05, K_2 0·7±0·1	62B
	gl	30	0·1 (KCl)	*K_1 2·28, K_2 2·36	62Ca
	gl	20	0·1 ($NaClO_4$)	*K_1 1·99, K_2 1·04	64P
Dy^{3+}	gl	20	0·1 ($NaClO_4$)	*K_1 1·94, K_2 1·32	64P
Er^{3+}	gl	25	2 ($NaClO_4$)	*K_1 1·28±0·05, K_2 0·9±0·1	62B
	gl	20	0·1 ($NaClO_4$)	*K_1 1·94, K_2 1·32	64P
Eu^{3+}	gl	25	2 ($NaClO_4$)	*K_1 1·75±0·05, K_2 0·8±0·1	62B
	gl	20	0·5 ($NaClO_4$)	*K_1 1·55, β_2 2·27	62G
	gl	20	0·1 ($NaClO_4$)	*K_1 2·07, K_2 1·34	64P
Gd^{3+}	gl	25	2 ($NaClO_4$)	*K_1 1·64±0·05, K_2 0·8±0·1	62B
	gl	20	0·1 ($NaClO_4$)	*K_1 2·01, K_2 1·30	64P
Ho^{3+}	gl	25	2 ($NaClO_4$)	*K_1 1·32±0·05, K_2 0·8±0·1	62B
	gl	20	0·1 ($NaClO_4$)	*K_1 1·92, K_2 1·32	64P
La^{3+}	gl	25	2 ($NaClO_4$)	*K_1 1·42±0·05, K_2 0·7±0·1	62B
	gl	30	0·1 (KCl)	*K_1 2·27, K_2 2·35	62Ca
	gl	20	0·1 ($NaClO_4$)	*K_1 1·98, K_2 1·00	64P
Lu^{3+}	gl	20	0·1 ($NaClO_4$)	*K_1 2·01, K_2 1·30	64P
Nd^{3+}	gl	25	2 ($NaClO_4$)	*K_1 1·49±0·05, K_2 0·8±0·1	62B
	gl	30	0·1 (KCl)	*K_1 2·48, K_2 2·52	62Ca
	gl	20	0·1 ($NaClO_4$)	*K_1 2·07, K_2 1·20	64P
Ni^{2+}	gl	0—40	0·10 KCl	$\Delta H_{\beta 2}$ (25°) −3·5±1, $\Delta S_{\beta 2}$ (25°) 48 ΔH ($4Ni^{2+} + 6L^- \rightleftharpoons Ni_4L_6^{4-}$) (25°) −31±2, ΔS (25°) 124	60L
	gl	20	0·1 ($NaClO_4$)	$K_1 \leqslant$ 6·2±0·6, β_2 13·01±0·02, β_3 14·99±0·08 K($2Ni^{2+} + 3L^{2-} \rightleftharpoons Ni_2L_3^{2-}$) 22·7±0·2 K($3Ni^{2+} + 4L^{2-} \rightleftharpoons Ni_3L_4^{2-}$) 33·27±0·05 K($4Ni^{2+} + 6L^{2-} \rightleftharpoons Ni_4L_6^{4-}$) 49·85±0·06	67P
Pr^{3+}	gl	30	0·1 (KCl)	*K_1 2·40, K_2 2·44	62Ca
	gl	20	0·1 ($NaClO_4$)	*K_1 2·03, K_2 1·04	64P
Sm^{3+}	gl	25	2 ($NaClO_4$)	*K_1 1·81±0·05, K_2 0·9±0·1	62B
	gl	20	0·1 ($NaClO_4$)	*K_1 2·11, K_2 1·36	64P
Tb^{3+}	gl	25	2 ($NaClO_4$)	*K_1 1·63±0·05, K_2 0·8±0·1	62B
	gl	20	0·1 ($NaClO_4$)	*K_1 1·96, K_2 1·26	64P
Tm^{3+}	gl	20	0·1 ($NaClO_4$)	*K_1 1·98, K_2 1·11	64P
UO_2^{2+}	gl	30	0·1 KCl	*K_1 2·88±0·02, K_2 2·40±0·05	62C

continued overleaf

126 $C_2H_4O_2S$ (contd)

Metal	Method	Temp	Medium	Log of equilibrium constant, remarks	Ref
Y^{3+}	gl	25	2 ($NaClO_4$)	$*K_1$ 1·49±0·05, K_2 0·7±0·1	62B
	gl	20	0·1 ($NaClO_4$)	$*K_1$ 1·91, K_2 1·28	64P
Yb^{3+}	gl	25	2 ($NaClO_4$)	$*K_1$ 1·32±0·05, K_2 0·9±0·1	62B
	gl	20	0·1 ($NaClO_4$)	$*K_1$ 1·98, K_2 1·32	64P
Zn^{2+}	Hg(Zn)	20	0·1 ($NaClO_4$)	K_1 7·80±0·02, β_2 14·96±0·02, β_3 17·80±0·05 $K(2Zn^{2+}+3L^{2-} \rightleftharpoons Zn_2L_3^{2-})$ 25·2±0·3 $K(3Zn^{2+}+4L^{2-} \rightleftharpoons Zn_3L_4^{2-})$ 36·47±0·09	67P

60L D. L. Leussing, R. E. Laramy, and G. S. Alberts, *J. Amer. Chem. Soc.*, 1960, **82**, 4826
62B J. L. Bear, G. R. Choppin, and J. V. Quagliano, *J. Inorg. Nuclear Chem.*, 1962, **24**, 1601
62C M. Cefola, R. C. Taylor, P. S. Gentile, and A. V. Celiano, *J. Phys. Chem.*, 1962, **66**, 790
62Ca M. Cefola, A. S. Tompa, A. V. Celiano, and P. S. Gentile, *Inorg. Chem.*, 1962, **1**, 290
62G I. Grenthe, *Acta Chem. Scand.*, 1962, **16**, 1695
64P J. E. Powell, R. S. Kolat, and G. S. Paul, *Inorg. Chem.*, 1964, **3**, 518
67P D. D. Perrin and I. G. Sayce, *J. Chem. Soc.* (*A*), 1967, 82

Special Note. For this Table *only* and for these entries $*K$, refers to $M^3+HL^- \rightleftharpoons MHL^2$; $*K_2$ refers to $MHL^{2+}+HL^- \rightleftharpoons M(HL)_2{}^+$; β_2 correspon[ds to] $M^{3+}+2HL^- \rightleftharpoons M(HL)_2{}^+$. For $U0_2{}^{2+}$, replace M^{3+} by $U0_2{}^{2+}$.

130 $C_2H_5O_2N$ Aminoacetic acid (glycine) HL

Metal	Method	Temp	Medium	Log of equilibrium constant, remarks	Ref
H^+	gl	22	~0	K_1 9·73	52P
	H, Ag-AgCl	5—45	→0	K_1 (5°) 10·340, (15°) 10·048, (25°) 9·778, (35°) 9·528, (37°) 9·480, (45°) 9·296 (±0·002) $\Delta H_1°$ (25°) −44·22 kJ mole^{-1}, $\Delta S_1°$ (25°) 38·9 J mole^{-1} deg^{-1}.	58D
	gl	5—55	0·01	K_1 (5°) 10·26, (15°) 9·97, (25°) 9·66, (35°) 9·44, (45°) 9·21, (55°) 8·98	59D
	gl	25	0·3 (K_2SO_4)	K_1 9·53±0·03, K_2 < 3·0	61J
	gl	25	→0	K_1 9·78	62A
	gl	30	0·1 (KCl)	K_1 9·53	62Ca
	gl	25	0·15—1 $NaClO_4$	K_1 (0·15) 9·62, (1) 9·65, K_2 (0·15) 2·44, (1) 2·45	63M
	gl	20, 30	→0	K_1 (20°) 9·73, (30°) 9·65, K_2 (20°) 2·34, (30°) 2·33 $\Delta H_1°$ −10·6±0·1, $\Delta S_1°$ 9, $\Delta H_2°$ −1·4±0·1, $\Delta S_2°$ 6	64I
	gl	10, 25	0·65 (KCl)	K_1 (10°) 10·10, (25°) 9·70, K_2 (10°) 2·76, (25°) 2·46	64L, 66L, [cut off]
	pol	30	1·0 KNO_3	K_1 9·64±0·02	64R
	gl	25	1 NH_4ClO_4	K_1 9·80, K_2 2·50	65B
	sp	25	1 $NaClO_4$	K_1 9·75, K_2 2·45	65M
	gl	25	0·1 $NaClO_4$	K_1 9·62, K_2 2·43	65Ma
	gl	15—40	0·2 (KCl)	K_1 (15°) 9·92, (25°) 9·68, (40°) 9·34, K_2 (15°) 2·46, (25°) 2·47, (40°) 2·40	65S
	gl	10—40	0 corr	K_1 (10°) 10·20, (25°) 9·77, (40°) 9·46, K_2 (10°) 2·41, (25°) 2·39, (40°) 2·33 $\Delta H_1°$ −10·0, $\Delta S_1°$ 11·2	66A, 67A
	cal	10—40	0 corr	$\Delta H_1°$ (10°) −11·6±0·1, (25°) −10·76±0·05, (40°) −10·22±0·08, $\Delta S_1°$ (10°) 5·8, (25°) 8·6, (40°) 10·7	66A, 67A
	Hg	25	0·6	K_1 9·96±0·05	67Aa
	gl	20—30	0·1 (KNO_3)	K_1 (20°) 9·84, (25°) 9·70, (30°) 9·60, K_2 (20°) 2·52, (25°) 2·51, (40°) 2·50	67G
	gl	25	0·1 ($NaClO_4$)	K_1 9·68±0·01, K_2 2·33±0·01	67Sa
	gl	20	0·5 KNO_3	K_1 9·86±0·01	68A

contin[ued]

Metal	Method	Temp	Medium	Log of equilibrium constant, remarks	Ref
H^+ (contd)	gl	2—40	2·0 ($NaClO_4$)	K_1 (2°) 9·64±0·03, (25°) 9·56±0·08, (40°) 9·6±0·1 K_2 (2°) 2·79±0·02, (25°) 2·76±0·05, (40°) 2·72±0·05	68T
	gl	25	0·5 KNO_3	K_1 9·61	68Ta
Ag^+	gl, sol	25	→ 0	K_1 3·51, K_2 3·38	51M
	gl	22	0·02 $AgNO_3$	K_1 3·7	52P
	gl	5—55	0·01	K_1 −7686·3/T+57·454−0·094725T, K_2 1506·7/T −0·763−0·002879T ΔH_1 (5°) 7±2, (15°) −3·4, (25°) −14·0±0·8, (35°) −25·0, (45°) −36·3, (55°) −48±3, ΔH_2 (25°) −34±3 kJ mole^{-1} ΔS_1 (5°) 91±8, (15°) 55, (25°) 19±3, (35°) 18, (45°) 54, (55°) 90±8, ΔS_2 (25°) 47±8 J mole^{-1}deg^{-1}	59D
	Ag	20	0 corr	K_1 4·00, K_2 3·3±0·1, β_2 7·19±0·02	62A
	sol	20	0 corr	β_2 7·26±0·01	62A
	Ag	25	0·6	K_1 3·54±0·03, β_2 6·82±0·03	67Aa
	gl	20	0·500 KNO_3	K_1 3·22±0·03, β_2 6·75±0·01	68A
	Ag-AgCl	20	0·500 KNO_3	β_2 6·85±0·02	68A
	gl	25	0·5 KNO_3	K_1 5·15, K_2 3·38	68Ta
Am^{3+}	dis	0—40	2·0 ($NaClO_4$)	$K(Am^{3+}+HL \rightleftharpoons AmHL^{3+})$ (0°) 0·48±0·05, (11°) 0·57±0·02, (25°) 0·69±0·02, (40°) 0·78±0·02 ΔH (25°) 2·9±0·4, ΔS 13±2	68T
Be^{2+}	gl	22	0·01 $BeSO_4$	β_2 13·3	52P
Cd^{2+}	gl	22	0·01 $CdSO_4$	β_2 7·9	52P
	pol	30	1	β_2 9·80	62R
	oth	20	0·1 (KNO_3)	K_1 6·0, K_2 3·9, K_3 2·6, paper electrophoresis	64J
	pol	30	1·0 KNO_3	β_2 8·08, β_3 9·78 $K(Cd^{2+}+2L^-+OH^- \rightleftharpoons CdL_2OH^-)$9·27 $K(Cd^{2+}+2L^-+CO_3^{2-} \rightleftharpoons CdL_2CO_3^{2-})$8·89 $K[Cd^{2+}+2L^-+4NH_3 \rightleftharpoons CdL_2(NH_3)_4]$ 9·38 See Solochrom Violet R (1050·1) for mixed ligands	64R
Ce^{3+}	gl	30	0·1 (KCl)	K_1 3·40, K_2 3·00	62Ca
	dis	0—55	2·0 ($NaClO_4$)	$K(Ce^{3+}+HL \rightleftharpoons CeHL^{3+})$ (0°) 0·34±0·09, (25°) 0·53±0·05, (40°) 0·70±0·05, (55°) 0·76±0·04 ΔH (25°) 3·3±0·4, ΔS 14±2	68T
Ce^{4+}	pol	20	2 $NaClO_4$	$K(Ce^{4+}+H_2L^+ \rightleftharpoons CeL^{3+}+2H^+)$ 0·95 $K(CeL^{3+}+H_2L^+ \rightleftharpoons CeL_2^{2+}+2H^+)$ −0·46 $K(CeL_2^{2+}+H_2L^+ \rightleftharpoons CeL_3^++2H^+)$ −0·92 $K(CeL_3^++H_2L^+ \rightleftharpoons CeL_4+2H^+)$ 1·19 (?)	64T
Cm^{3+}	dis	0—40	2·0 ($NaClO_4$)	$K(Cm^{3+}+HL \rightleftharpoons CmHL^{3+})$ (0°) 0·62±0·05, (11°) 0·66±0·06, (25°) 0·80±0·02, (40°) 0·95±0·02 ΔH (25°) 3·3±0·4, ΔS 15±2	68T
Co^{2+}	gl	22	0·01 $CoCl_2$	β_2 8·8	52P
	H	0—45	→ 0	K_1 (0°) 5·276±0·010, (15°) 5·143±0·006, (25°) 5·072±0·008, (35°) 5·009±0·005, (45°) 4·953±0·005 K_2 (0°) 4·23±0·02, (15°) 4·07±0·02, (25°) 4·97±0·02, (35°) 4·90±0·02, (45°) 4·82±0·02 ΔH_1 −2·8±0·1, ΔS_1 13·7±0·5 ΔH_2 −3·6±0·2, ΔS_2 6·3±0·6	64B
	oth	20	0·1 (KNO_3)	K_1 5·5, K_2 3·5, K_3 2·3, paper electrophoresis	64J
	gl	15, 40	0·2 (KCl)	K_1 (15°) 4·76, (40°) 4·64, K_2 (15°) 3·56, (40°) 3·34 ΔH_1 −2·0±0·5, ΔS_1 15±2, ΔH_2 −3·6±0·5, ΔS_2 4±2	65S

continued overleaf

130 $C_2H_5O_2N$ (contd)

Metal	Method	Temp	Medium	Log of equilibrium constant, remarks	Ref
Co^{2+} (contd)	cal	25	0·1 (KCl)	ΔH_1 $-2{\cdot}48\pm0{\cdot}05$, ΔS_1 $12{\cdot}8\pm0{\cdot}3$, ΔH_2 $-2{\cdot}55\pm0{\cdot}02$, ΔS_2 $9{\cdot}6\pm0{\cdot}2$	67B
	cal	20	0·1 KNO_3	$\Delta H_{\beta 2}$ $-6{\cdot}6$, $\Delta S_{\beta 2}$ 16·1	67S
				See nitrilotriacetic acid (449) for mixed ligands	
Cr^{3+}	gl	25	~ 0·5	K_1 8·4, K_2 6·4, K_3 5·7	63K
	gl	25	0·1 ($NaClO_4$)	K_1 8·62, K_2 7·65	65M
Cu^+	Pt	25	0·3 (K_2SO_4)	β_2 ~ 10·0	61J
Cu^{2+}	Pt	25	0·3 (K_2SO_4)	β_2 15·2, K_3 0·47	61J
	pol	25	0·25 0—50% dioxan	β_2 (0%) 14·6, (20%) 15·6, (35%) 16·0, (50%) 16·3	63G
	gl	25	0·15—1 $NaClO_4$	K_1 (0·15) 8·18, (1) 8·33, β_2 (0·15) 15·02, (1) 15·20	63M
	gl	20, 30	→ 0	K_1 (20°) 8·59, (30°) 8·47, K_2 (20°) 7·24, (30°) 7·04	64I
				$\Delta H_1°$ $-6{\cdot}0\pm0{\cdot}1$, $\Delta S_1°$ 19, $\Delta H_2°$ $-6{\cdot}4\pm0{\cdot}1$, $\Delta S_2°$ 11	
	oth	20	0·1 (KNO_3)	K_1 8·6, K_2 7·2, K_3 0·15, paper electrophoresis	64J
	gl	25	1 NH_4ClO_4	K_1 8·29, β_2 15·30,	65B
				$K(Cu^{2+}+NH_3+L^- \rightleftharpoons CuNH_3L^+)$ 12·50,	
				$K[Cu^{2+}+2NH_3+L^- \rightleftharpoons Cu(NH_3)_2L^+]$ 14·85	
	sp	25	1 $NaClO_4$	K_1 8·33, β_2 15·20	65M
	gl	15—40	0·2 (KCl)	K_1 (15°) 8·54, (25°) 8·46, (40°) 8·25, K_2 (15°) 7·00, (25°) 6·83, (40°) 6·64	65S
				ΔH_1 $-4{\cdot}8\pm0{\cdot}5$, ΔS_1 22 ± 2, ΔH_2 $-6{\cdot}8\pm0{\cdot}5$, ΔS_2 11 ± 2	
	gl	10—40	0 corr	K_1 (10°) $8{\cdot}85\pm0{\cdot}03$, (25°) $8{\cdot}58\pm0{\cdot}03$, (40°) $8{\cdot}42\pm0{\cdot}03$, K_2 (10°) $7{\cdot}36\pm0{\cdot}03$, (25°) $7{\cdot}09\pm0{\cdot}03$, (40°) $6{\cdot}85\pm0{\cdot}03$	66A
				$\Delta H_1°$ $-5{\cdot}83$, $\Delta S_1°$ 19·7, $\Delta H_2°$ $-6{\cdot}83$, $\Delta S_2°$ 9·5	
	cal	10—40	0 corr	$\Delta H_1°$ (10°) $-7{\cdot}28\pm0{\cdot}04$, (25°) $-6{\cdot}22\pm0{\cdot}08$, (40°) $-5{\cdot}75\pm0{\cdot}06$, $\Delta S_1°$ (10°) 14·8, (25°) 18·4, (40°) 20·2, $\Delta H_2°$ (10°) $-6{\cdot}92\pm0{\cdot}08$, (25°) $-7{\cdot}0\pm0{\cdot}2$, (40°) $-7{\cdot}33\pm0{\cdot}07$, $\Delta S_2°$ (10°) 9·2, (25°) 9·1, (40°) 7·9	66A
	gl	25	0·5 KCl	K_1 8·11, β_2 14·43	66L
	cal	25	0·1 ($NaClO_4$)	ΔH_1 $-6{\cdot}76\pm0{\cdot}04$, ΔS_1 $16{\cdot}6\pm0{\cdot}3$, ΔH_2 $-6{\cdot}89\pm0{\cdot}09$, ΔS_2 $8{\cdot}7\pm0{\cdot}3$	67B
	gl	20—30	0·1 (KNO_3)	K_1 (20°) 8·313, (25°) 8·23, (30°) 8·17, β_2 (20°) 15·363, (25°) 15·19, (30°) 15·06	67G
				ΔH_1 $-5{\cdot}9$, ΔS_1 18, ΔH_2 $-6{\cdot}5$, ΔS_2 11	
	cal	20	0·1 KNO_3	$\Delta H_{\beta 2}$ $-12{\cdot}8$, $\Delta S_{\beta 2}$ 25·2	67S
	gl	25	0·1 ($NaClO_4$)	K_1 $8{\cdot}27\pm0{\cdot}01$, K_2 $6{\cdot}92\pm0{\cdot}05$	67Sa
				$K(CuA^{2+}+L^- \rightleftharpoons CuAL^+)$ $7{\cdot}88\pm0{\cdot}01$ where A=2,2′-bipyridyl	
	gl	25	0·5 KCl	K_1 8·12, β_2 14·87, β_3 15·3	68L
				See nitrilotriacetic acid (449), Solochrom Violet R (1050·1), glycollic acid (115), salicylaldehyde (501) and 5-sulphosalicylic acid (552) for mixed ligands	
Eu^{3+}	dis	0—55	2·0 ($NaClO_4$)	$K(Eu^{3+}+HL \rightleftharpoons EuHL^{3+})$ (0°) $0{\cdot}61\pm0{\cdot}03$, (25°) $0{\cdot}70\pm0{\cdot}03$, (40°) $0{\cdot}78\pm0{\cdot}03$, (55°) $0{\cdot}90\pm0{\cdot}02$	68T
				ΔH (25°) $2{\cdot}3\pm0{\cdot}4$, ΔS 11 ± 2	
Hg^{2+}	gl	22	0·01 $Hg(NO_3)_2$	β_2 18·2	52P
	gl, cal	25	0 corr	$K(HgCl_2+L^- \rightleftharpoons HgClL+Cl^-)$ (gl) $3{\cdot}42\pm0{\cdot}01$	66P
				$\Delta H°$ (cal) $-6{\cdot}10\pm0{\cdot}07$, $\Delta S°$ -5	
				$K(HgClL+L^- \rightleftharpoons HgL_2+Cl^-)$ (gl) $2{\cdot}61\pm0{\cdot}02$	
				$\Delta H°$ (cal) $-2{\cdot}94\pm0{\cdot}06$, $\Delta S°$ 2	
	pol	25	0·6 KNO_3	β_2 $18{\cdot}36\pm0{\cdot}04$	66T

continued

Metal	Method	Temp	Medium	Log of equilibrium constant, remarks	Ref
La^{3+}	gl	30	0·1 (KCl)	K_1 3·23, K_2 2·92	62Ca
Mn^{2+}	H	0—45	→ 0	K_1 (0°) 3·199±0·009, (15°) 3·179±0·009, (25°) 3·167±0·006, (35°) 3·161±0·006, (45°) 3·161±0·006 ΔH_1 −0·29±0·08, ΔS_1 13·5±0·3	64B
	oth	20	0·1 (KNO_3)	K_1 3·9, K_2 1·7, paper electrophoresis	64J
	gl	10, 25	0·65 (KCl)	K_1 (10°) 2·66, (25°) 2·60, β_2 (10°) 4·71, (25°) 4·58, β_3 (10°) 6·0, (25°) 5·7 ΔH_1 (25°) −1·4, $T\Delta S_1$ 2·2 kcal mole^{-1} $\Delta H_{\beta 2}$ (25°) −3·3, $T\Delta S_\beta$ 2·9 kcal mole^{-1} $K(Mn^{2+}+A^-+L^- \rightleftharpoons MnAL)$ (10°) 5·51±0·02, (25°) 5·36±0·02 ΔH (25°) −4·0, $T\Delta S$ 3·3 kcal mole^{-1} $K(Mn^{2+}+A^-+2L^- \rightleftharpoons MnAL_2^-)$ (10°) 7·7±0·6, (25°) 6·9±0·03 $K(Mn^{2+}+2A^-+2L^- \rightleftharpoons MnA_2L_2^{2-})$ (10°) 10·25 ±0·03, (25°) 9·79±0·03 ΔH (25°) −11·8, $T\Delta S$ 1·6 kcal mole^{-1} where HA = pyruvic acid	64L
	gl	25	0·5 KCl	K_1 2·65, β_2 4·7 See nitrilotriacetic acid (449) and salicylaldehyde (501) for mixed ligands	68L
Nd^{3+}	gl	30	0·1 (KCl)	K_1 3·71, K_2 3·30	62Ca
Ni^{2+}	gl	22	0·01 $NiCl_2$	β_2 11·0	52P
	H	0—45	→ 0	K_1 (0°) 6·465±0·003, (15°) 6·286±0·003, (25°) 6·179±0·003, (35°) 6·083±0·004, (45°) 6·000±0·009 K_2 (0°) 5·286±0·010, (15°) 5·076±0·008, (25°) 5·951±0·005, (35°) 5·836±0·007, (45°) 5·75±0·02 ΔH_1 −4·09±0·03, ΔS_1 14·5±0·1 ΔH_2 −4·7±0·3, ΔS_2 6·9±0·8	64B
	oth	20	0·1 (KNO_3)	K_1 6·4, K_2 4·4, K_3 3·0, paper electrophoresis	64J
	gl	10, 25	0·65 (KCl)	K_1 (10°) 5·73, (25°) 5·66, β_2 (10°) 10·80, (25°) 10·51, β_3 (10°) 14·4, (25°) 14·0 ΔH_1 (25°) −1·9, $T\Delta S$ 5·9 kcal mole^{-1} $\Delta H_{\beta 2}$ (25°) −7·5, $T\Delta S$ 6·8 kcal mole^{-1} $K(Ni^{2+}+A^-+L^- \rightleftharpoons NiAL)$ (10°) 8·19±0·07, (25°) 8·09±0·02, ΔH (25°) −2·6, $T\Delta S$ 8·5 kcal mole^{-1} $K(Ni^{2+}+A^-+2L^- \rightleftharpoons NiAL_2^-)$ (10°) 13·57±0·09, (25°) 13·00±0·04 $K(Ni^{2+}+2A^-+2L^- \rightleftharpoons NiA_2L_2^{2-})$ (10°) 15·76±0·05, (25°) 15·29±0·02, ΔH (25°) −11·9, $T\Delta S$ 9·0 kcal mole^{-1} where HA = pyruvic acid	64L
	gl	25	1 $NaClO_4$	K_1 5·70, β_2 10·50, β_3 13·96 $K(Ni^{2+}+L^-+A^- \rightleftharpoons NiLA)$ 10·85 $K(Ni^{2+}+2L^-+A^- \rightleftharpoons NiL_2A^-)$ 15·50 $K(Ni^{2+}+L^-+2A^- \rightleftharpoons NiLA_2^-)$ 14·40 where HA = alanine	64M
	gl	15—40	0·2 (KCl)	K_1 (15°) 6·04, (25°) 5·94, (40°) 5·78, K_2 (15°) 4·98, (25°) 4·84, (40°) 4·71	65S

continued overleaf

130 $C_2H_5O_2N$ (contd)

Metal	*Method*	*Temp*	*Medium*	*Log of equilibrium constant, remarks*	*Ref*
Ni^{2+} (*contd*)				ΔH_1 $-4{\cdot}3\pm0{\cdot}5$, ΔS_1 13 ± 2, ΔH_2 $-4{\cdot}4\pm0{\cdot}5$, ΔS_2 7 ± 2	
	gl	25	0·5 KCl	K_1 5·65, β_2 10·51, β_3 13·95 $K(Ni^{2+}+A^-+L^- \rightleftharpoons NiAL)$ $8{\cdot}07\pm0{\cdot}01$ $K(Ni^{2+}+A^-+2L^- \rightleftharpoons NiAL_2^-)$ $12{\cdot}95\pm0{\cdot}04$ $K(Ni^{2+}+2A^-+2L^- \rightleftharpoons NiA_2L_2^{2-})$ $15{\cdot}283\pm0{\cdot}002$ where HA = pyruvic acid $K(Ni^{2+}+A^-+L^- \rightleftharpoons NiAL)$ $8{\cdot}08\pm0{\cdot}01$ $K(Ni^{2+}+A^-+2L^- \rightleftharpoons NiAL_2^-)$ $12{\cdot}92\pm0{\cdot}09$ $K(Ni^{2+}+2A^-+2L^- \rightleftharpoons NiA_2L_2^{2-})$ $14{\cdot}69\pm0{\cdot}01$ where HA = glyoxylic acid	66L
	gl	10—40	0 corr	K_1 (10°) 6·36, (25°) 6·18, (40°) 6·09, K_2 (10°) 5·29, (25°) 5·07, (40°) 4·92 $\Delta H_1°$ -4 ± 1, $\Delta S_1°$ 16·1, $\Delta H_2°$ $-5{\cdot}1\pm0{\cdot}8$, $\Delta S_2°$ 6·1	67A
	cal	10—40	0 corr	$\Delta H_1°$ (10°) $-5{\cdot}2\pm0{\cdot}1$, (25°) $-4{\cdot}9\pm0{\cdot}1$, (40°) $-4{\cdot}3\pm0{\cdot}2$, $\Delta S_1°$ (10°) 10·7, (25°) 11·9, (40°) 14·2, $\Delta H_2°$ (10°) $-5{\cdot}8\pm0{\cdot}1$, (25°) $-4{\cdot}7\pm0{\cdot}2$, (40°) $-4{\cdot}7\pm0{\cdot}3$, $\Delta S_2°$ (10°) 3·7, (25°) 7·6, (40°) 7·4	67A
	cal	25	0·1 (KCl)	ΔH_1 $-4{\cdot}1\pm0{\cdot}1$, ΔS_1 $14{\cdot}4\pm0{\cdot}4$	67B
	gl	20—30	0·1 (KNO_3)	K_1 (20°) 5·80, (25°) 5·73, (30°) 5·70, β_2 (20°) 10·70, (25°) 10·56, (30°) 10·47 ΔH_1 $-4{\cdot}1$, ΔS_1 13, ΔH_2 $-5{\cdot}3$, ΔS_2 4	67G
	cal	20	0·1 KNO_3	$\Delta H_{\beta 2}$ $-8{\cdot}8$, $\Delta S_{\beta 2}$ 18·0	67S
	gl	25	0·5 KCl	K_1 5·63, β_2 10·48, β_3 14·0 See nitrilotriacetic acid (449), Solochrom Violet R (1050·1), diethylenetriamine (231), salicylaldehyde (501), pyridoxal (667·3), and 5-sulphosalicylic acid (552) for mixed ligands	68L
Pb^{2+}	gl	22	0·01 $Pb(NO_3)_2$	β_2 9·3	52P
	pol	30	1·0 KNO_3	K_1 5·11, β_2 7·08 $K(Pb^{2+}+2L^-+CO_3^{2-} \rightleftharpoons PbL_2CO_3^{2-})$ 8·61 See nitrilotriacetic acid (449) for mixed ligands	64R
Pm^{3+}	dis	0—40	2·0 ($NaClO_4$)	$K(Pm^{3+}+HL \rightleftharpoons PmHL^{3+})$ (0°) $0{\cdot}45\pm0{\cdot}05$, (11°) $0{\cdot}52\pm0{\cdot}08$, (25°) $0{\cdot}67\pm0{\cdot}03$, (40°) $0{\cdot}79\pm0{\cdot}03$ ΔH (25°) $3{\cdot}5\pm0{\cdot}4$, ΔS 15 ± 2	68T
Pr^{3+}	gl	30	0·1 (KCl)	K_1 3·64, K_2 3·32	62Ca
UO_2^{2+}	gl	30	0·1 KCl	K_1 7·53, K_2 7·15	62C
Zn^{2+}	gl	22	0·01 $ZnSO_4$	β_2 9·2	52P
	oth	20	0·1 (KNO_3)	K_1 5·9, K_2 4·2, K_3 3·1, paper electrophoresis	64J
	gl	10, 25	0·65 (KCl)	K_1 (10°) 4·96, (25°) 4·88, β_2 (10°) 9·24, (25°) 9·01, β_3 (10°) 11·9, (25°) 11·0 ΔH_1 (25°) $-2{\cdot}1$, $T\Delta S_1$ 4·6 kcal mole^{-1} $\Delta H_{\beta 2}$ (25°) $-5{\cdot}9$, $T\Delta S_{\beta 2}$ 6·4 kcal mole^{-1} $K(Zn^{2+}+A^-+L^- \rightleftharpoons ZnAL)$ (10°) $7{\cdot}61\pm0{\cdot}02$, (25°) $7{\cdot}53\pm0{\cdot}02$, ΔH (25°) $-1{\cdot}9$, $T\Delta S$ 8·4 kcal mole^{-1} $K(Zn^{2+}+A^-+2L^- \rightleftharpoons ZnAL_2^-)$ (10°) $11{\cdot}5\pm0{\cdot}3$, (25°) $12{\cdot}0\pm0{\cdot}1$ $K(Zn^{2+}+2A^-+2L^- \rightleftharpoons ZnA_2L_2^{2-})$ (10°) $14{\cdot}54\pm0{\cdot}02$, (25°) $14{\cdot}25\pm0{\cdot}02$, ΔH (25°) $-7{\cdot}5$, $T\Delta S$ 12·0 kcal mole^{-1} where HA = pyruvic acid	64L

contin

Metal	Method	Temp	Medium	Log of equilibrium constant, remarks	Ref
Zn^{2+} (*contd*)	gl	15—40	0·2 (KCl)	K_1 (15°) 5·27, (25°) 5·19, (40°) 5·07, K_2 (15°) 4·31, (25°) 4·21, (40°) 4·07	65S
				ΔH_1 −3·3±0·5, ΔS_1 13±2, ΔH_2 −4·0±0·5, ΔS_2 6±2	
	gl	25	0·5 KCl	K_1 4·88, β_2 9·01, β_3 11·02	66L
				$K(Zn^{2+}+A^-+L^- \rightleftharpoons ZnAL)$ 7·60±0·02	
				$K(Zn^{2+}+A^-+2L^- \rightleftharpoons ZnAL_2^-)$ 11·9±0·1	
				$K(Zn^{2+}+2A^-+2L^- \rightleftharpoons ZnA_2L_2^{2-})$ 14·35±0·02 where HA = pyruvic acid	
				$K(Zn^{2+}+A^-+L^- \rightleftharpoons ZnAL)$ 6·75±0·01	
				$K(Zn^{2+}+A^-+2L^- \rightleftharpoons ZnAL_2^-)$ 10·70±0·04	
				$K(Zn^{2+}+2A^-+2L^- \rightleftharpoons ZnA_2L_2^{2-})$ 12·23±0·03 where HA = glyoxylic acid	
	cal	25	0·1 ($NaClO_4$)	ΔH_1 −3·4±0·2, ΔS_1 13·8±0·7	67B
	cal	20	0·1 KNO_3	$\Delta H_{\beta 2}$ −6·3, $\Delta S_{\beta 2}$ 22·1	67S
	gl	25	0·5 KCl	K_1 4·88, β_2 9·11, β_3 11·56	68L
				See nitritotriacetic acid (449), salicylaldehyde (501), pyridoxal (667·3), and Solochrom Violet R (1050·1) for mixed ligands	

Errata. 1964, p. 381. Under Cd^{2+}, Cu^{2+}, Pb^{2+}, and Zn^{2+} *for* ref. 55M *read* 61M. Insert ref. 61M H. A. McKenzie and D. P. Mellor, *Austral. J. Chem.*, 1961, **14,** 562.

51M C. M. Monk, *Trans. Faraday Soc.*, 1951, 285, 292, 297
52P D. J. Perkins, *Biochem. J.*, 1952, **51,** 487
58D S. P. Datta and A. K. Grzybowski, *Trans. Faraday Soc.*, 1958, **54,** 1179
59D S. P. Datta and A. K. Grzybowski, *J. Chem. Soc.*, 1959, 1091
61J B. R. James and R. J. P. Williams, *J. Chem. Soc.*, 1961, 2007
62A D. J. Alner, *J. Chem. Soc.*, 1962, 3282
62C M. Cefola, R. C. Taylor, P. S. Gentile, and A. V. Celiano, *J. Phys. Chem.*, 1962, **66,** 790
62Ca M. Cefola, A. S. Tompa, A. V. Celiano, and P. S. Gentile, *Inorg. Chem.*, 1962, **1,** 290
62R G. N. Rao and R. S. Subrahmanya, *Current Sci.* 1962, **31,** 55
63G R. R. Gutierrez-Flores and B. Tremillon, *Bull. Soc. chim., France*, 1963, 2878
63K A. A. Khan and W. U. Malik, *J. Indian Chem. Soc.*, 1963, **40,** 564
63M R.-P. Martin and R. A. Paris, *Bull. Soc. chim. France*, 1963, 570
64B J. R. Brannan, H. S. Dunsmore, and G. H. Nancollas, *J. Chem. Soc.*, 1964, 304
64I R. M. Izatt, J. J. Christensen, and V. Kothari, *Inorg. Chem.*, 1964, **3,** 1565
64J V. Jokl, *J. Chromatog.*, 1964, **14,** 71
64L D. L. Leussing and D. C. Shultz, *J. Amer. Chem. Soc.*, 1964, **86,** 4846
64M R.-P. Martin and R. A. Paris, *Bull. Soc. chim. France*, 1964, 3170
64R G. N. Rao and R. S. Subrahmanya, *Proc. Indian Acad. Sci.*, 1964, **60,** 165, 185
64T L. Treindl, *Coll. Czech. Chem. Comm.*, 1964, **29,** 2927
65B M. Bonnet, R.-P. Martin, and R. A. Paris, *Bull. Soc. chim. France*, 1965, 176
65M R. C. Mercier, M. Bonnet, and M. R. Paris, *Bull. Soc. chim. France*, 1965, 2926
65Ma M. Matsukawa, M. Ohta, S. Takata, and R. Tsuchiya, *Bull. Chem. Soc. Japan.*, 1965, **38,** 1235
65S V. S. Sharma, H. B. Mathur, and P. S. Kulkarni, *Indian J. Chem.*, 1965, **3,** 146, 475
66A K. P. Anderson, W. O. Greenhalgh, and R. M. Izatt, *Inorg. Chem.*, 1966, **5,** 2106
66L D. L. Leussing and E. M. Hanna, *J. Amer. Chem. Soc.*, 1966, **88,** 693, 696
66P J. A. Partridge, J. J. Christensen, and R. M. Izatt, *J. Amer. Chem. Soc.*, 1966, **88,** 1649
66T V. F. Toropova and Yu. M. Azizov, *Russ. J. Inorg. Chem.*, 1966, **11,** 288 (531)
67A K. P. Anderson, W. O. Greenhalgh, and E. A. Butler, *Inorg. Chem.*, 1967, **6,** 1056
67Aa Yu. M. Azizov, A. Kh. Miftakhova, and V. F. Toropova, *Russ. J. Inorg. Chem.*, 1967, **12,** 345 (661)
67B S. Boyd, J. R. Brannan, H. S. Dunsmore, and G. H. Nancollas, *J. Chem. and Eng. Data*, 1967, **12,** 601
67G A. Gregely, I. Nagypal, and J. Mojzes, *Acta Chim. Acad. Sci. Hung.*, 1967, **51,** 381
67S W. F. Stack and H. A. Skinner, *Trans. Faraday Soc.*, 1967, **63,** 1136
67Sa H. Sigel and R. Griesser, *Helv. Chim. Acta*, 1967, 50, 1842
68A D. J. Alner, R. C. Lansbury, and A. G. Smeeth, *J. Chem. Soc.* (*A*), 1968, 417
68L D. L. Leussing and K. S. Bai, *Analyt. Chem.*, 1968, **40,** 575
68T S. P. Tanner and G. R. Choppin, *Inorg. Chem.*, 1968, **7,** 2046
68Ta G. F. Thiers, L. C. van Poucke, and M. A. Herman, *J. Inorg. Nuclear Chem.*, 1968, **30,** 1543

131.1 C_2H_5NS Thioacetic acid amide (thioacetamide) HL

Metal	Method	Temp	Medium	Log of equilibrium constant, remarks	Ref
Ag^+	pH	25	0·3—0·9 HNO_3	K_1 6·1, β_2 13·7, β_3 14·7, β_4 13·4	66P
			0·2—0·4 H_2SO_4	K_1 4·7, β_2 13·7	

66P S. Petri and T. Lipiec, *Roczniki Chem.*, 1966, **40**, 1795

131.2 $C_2H_6ON_2$ Acetic acid hydrazide L

Metal	Method	Temp	Medium	Log of equilibrium constant, remarks	Ref
Cd^{2+}	pol	25	1 $NaClO_4$	K_1 1·93, β_2 3·64, β_3 4·39	65K
Zn^{2+}	pol	25	1 $NaClO_4$	K_1 2·48, β_2 3·65, β_3 4·68	68S

65K A. F. Krivis, G. R. Supp, and R. L. Doerr, *Analyt Chem.*, 1965, **37**, 52
68S G. R. Supp, *Analyt. Chem.*, 1968, **40**, 981

132 $C_2H_6ON_2$ Aminoacetic acid amide L

Metal	Method	Temp	Medium	Log of equilibrium constant, remarks	Ref
H^+	gl	25	0·16 (KNO_3)	K_1 8·05	60M
	gl	25	0·1 ($NaClO_4$)	K_1 8·04±0·02	68S
Cu^{2+}	gl	25	0·1 ($NaClO_4$)	K_1 5·40±0·04	68S
				$K[Cu(H_{-1}L)^+ + H^+ \rightleftharpoons CuL^{2+}]$ 7·01±0·04	
				$K[Cu(H_{-1}L)OH + H^+ \rightleftharpoons Cu(H_{-1}L)^+]$ 8·07±0·06	
				$K(CuA^{2+} + L \rightleftharpoons CuAL^{2+})$ 5·01±0·02	
				$K[CuA(H_{-1}L)^+ + H^+ \rightleftharpoons CuAL^{2+}]$ 7·71±0·06	
				$K[CuA_2^{2+} + CuL_2^{2+} \rightleftharpoons 2CuAL^{2+})$ 2·8	
				where A = 2,2′-bipyridyl	
Ni^{2+}	gl	25	0·16 (KNO_3)	$K[Ni(H_{-1}L) + H^+ \rightleftharpoons NiL^+]$ 9·8	60M
				$K[Ni(H_{-1}L)OH^- + H^+ \rightleftharpoons Ni(H_{-1}L)]$ 10·1	

Errata. 1964, p. 381. Table 132 line 2, delete 60M–58L.
Under Ni^{2+}, *for* 60C *read* 60M

60M R. B. Martin, M. Chamberlin, and J. T. Edsall, *J. Amer. Chem. Soc.*, 1960, **82** 495
68S H. Sigel, *Angew. Chem.*, 1968, **80**, 124; *Angew. Chem. Internat. Edn.*, 1968, **7**, 137

133.1 C_2H_6OS 2-Mercaptoethanol HL

Metal	Method	Temp	Medium	Log of equilibrium constant, remarks	Ref
H^+	gl	?	→0	K_1 9·77	61A
	gl	25	→0	K_1 9·44	63A
Cd^{2+}	gl	?	→0	β_4 20·56	61A
Co^{2+}	gl	?	→0	β_3 13·08	61A
Cu^{2+}	pol	25	0·17	β_3 20·13, phosphate buffer	61K

contin

C_2H_6OS (contd) 133.1

Metal	Method	Temp	Medium	Log of equilibrium constant, remarks	Ref
Fe^{2+}	gl	?	$\to 0$	β_2 6·6	61A
Ge^{IV}	gl	25	0·1 KCl	$K[H_2GeO_3 + 2HL \rightleftharpoons GeOH(H_{-1}L)_2^- + H^+]$ −4·22	63A
Mn^{2+}	gl	?	$\to 0$	β_2 5·41	61A
Ni^{2+}	gl	?	$\to 0$	β_2 11·19	61A
Pb^{2+}	gl	?	$\to 0$	β_2 14·53	61A
Zn^{2+}	gl	?	$\to 0$	β_3 17·31	61A

61A S. P. Armanet and J. C. Merlin, *Bull. Soc. chim. France*, 1961, 440

61K E. C. Knoblock and W. C. Purdy, *J. Electroanalyt. Chem.*, 1961, **2**, 493

63A P. J. Antikainen and K. Tevanen, *Suomen Kem.*, 1963, **36B**, 199; **35B**, 224

C_2H_7ON 2-Aminoethanol (ethanolamine) L 136

Metal	Method	Temp	Medium	Log of equilibrium constant, Remarks	Ref
H^+	gl	25	0·1 (KNO_3)	K_1 9·50	62C
	gl	30	0·5 (KNO_3)	K_1 9·55	62F, 63S
	gl	25	~ 0·1	K_1 9·74	65D
	gl	25	0·4 (KNO_3)	K_1 9·62	66S
	gl	30	0·5 (KNO_3)	K_1 9·54	67F
	gl	20	0·5 KNO_3	K_1 9·80±0·01	68A
	H	20	$\to 0$	K_1 9·56±0·03	68A
Ag^+	Ag	25	< 0·01	β_2 6·78	61A
	ix	25	0·1 (KNO_3)	β_2 6·56	62C
	Ag	25	0·4 $LiNO_3$ 0—100% ethanol	K_1 (0%) 4·18±0·06, (20%) 4·60±0·04, (40%) 5·15±0·03, (60%) 5·5±0·2, (80%) 5·8±0·2, (90%) 5·9±0·2, (95%) 6·0±0·2, (100%) 6·0±0·1, β_2 (0%) 6·45±0·05, (20%) 6·50±0·04, (40%) 6·85±0·03, (60%) 7·0±0·2, (80%) 7·23±0·05, (90%) 7·3±0·2, (95%) 7·4±0·1, (100%) 7·48±0·03, β_3 (0%) 6·30±0·04, (20%) 6·84±0·04, (40%) 7·11±0·04, (60%) 7·5±0·2, (80%) 7·8±0·2, (90%) 7·9±0·2, (95%) 8·0±0·1, (100%) 8·36±0·02	65Ma
	gl	20	0·500 KNO_3	K_1 3·22±0·06, β_2 6·79±0·03	68A
	Ag-AgCl	20	0·500 KNO_3	β_2 6·83±0·02	68A
Cd^{2+}	pol	25	0·1 KNO_3	K_1 2·77, β_2 4·09, β_3 5·46	60M
	pol	25	0·1 $NaClO_4$ 0—100% ethanol	β_2 (0%) 4·78, (20%) 5·45, (40%) 5·78, (50%) 6·30, β_3 (0%) 5·30, (20%) 5·70, (40%) 6·08, (50%) 6·42, (80%) 8·18, (94%) 9·70, (100%) 12·30, β_4 (0%) 6·26, (20%) 6·28, (40%) 6·97, (50%) 7·28, (80%) 8·74, (94%) 9·85, (100%) 13·95, β_5 (94%) 10·56, (100%) 14·73	62M
	pol	25	0·01 $NaClO_4$ 40—100% methanol	β_2 (40%) 4·54, (80%) 6·00, (94%) 6·38, β_3 (40%) 5·66, (80%) 6·93, (94%) 6·78, (100%) 8·30, β_4 (40%) 7·40, (80%) 7·36, (94%) 7·40, (100%) 8·60, β_5 (100%) 9·99	65M
Co^{2+}	gl	25	0·43 ($L \cdot HNO_3$)	K_1 2·42, K_2 1·68, K_3 1·27	66S

continued overleaf

136 C_2H_7ON (contd)

Metal	*Method*	*Temp*	*Medium*	*Log of equilibrium constant, remarks*	*Ref*
Cu^{2+}	pol	25	0·1 KNO_3	β_4 15·44±0·08	59M
	pol	30	0·5 KNO_3	$K[Cu^{2+}+2L+2OH^- \rightleftharpoons CuL_2(OH)_2]$ 19·9	62F
	pol	30	0·5 KNO_3	$K(Cu^{2+}+2L+OH^- \rightleftharpoons CuL_2OH^+)$ 16·67 $K[Cu^{2+}+2L+2OH^- \rightleftharpoons CuL_2(OH)_2]$ 19·60	63S
	gl	25	~ 0·1	K_1 5·7, K_2 4·1, K_3 3·2, K_4 2·0	65D
	gl	25	0·43 ($L{\cdot}HNO_3$)	K_1 4·73, K_2 3·79, K_3 2·87	66S
	pol	30	0·5 (KNO_3)	$K[Cu^{2+}+L+2OH^- \rightleftharpoons CuL(OH)_2]$ 17·4 $K[Cu^{2+}+2L+2OH^- \rightleftharpoons CuL_2(OH)_2]$ 19·6	67F
Fe^{2+}	hyp	25	0·43 ($L{\cdot}HNO_3$)	K_1 1·90, K_2 1·15	66S
Mn^{2+}	hyp	25	0·43 ($L{\cdot}HNO_3$)	K_1 0·87, K_2 0·23	66S
Ni^{2+}	gl, ix	25	0·1 (KNO_3)	K_1 2·98±0·02, K_2 2·35, K_3 2·00	62C
	gl	25	0·4	K_1 3·06±0·04, β_2 5·52±0·04, β_3 6·95±0·04	62S
	sp	?	dil	K_1 4·18±0·08, K_2 3·20±0·09, K_3 2·88±0·09	66U
Pb^{2+}	pol	25	0·1 KNO_3	β_2 7·56±0·05	59M
	pol	25	0·01 $NaClO_4$ 0—60% ethanol	K_1 (0%) 6·70, (20%) 8·08, β_2 (0%) 7·58, (20%) 8·48, (40%) 9·03, (60%) 9·71	62M
Zn^{2+}	pol	25	0·1 KNO_3	β_4 9·2±0·2	59M
	pol	25	0·01 $NaClO_4$ 0—100% ethanol	β_2 (0%) 8·00, (20%) 8·95, (40%) 9·73, (60%) 8·70, (80%) 9·70, (88%) 10·18, (94%) 11·85, (97%) 12·00, (100%) 12·30, β_3 (0%) 9·48, (20%) 9·78, (40%) 10·08, (60%) 10·42, (80%) 10·60, (88%) 11·49, (94%) 12·23, (97%) 12·64, (100%) 13·30, β_4 (94%) 13·00, (100%) 13·72	62M
	gl	25	~ 0·1	K_1 3·7, K_2 2·3, K_3 1·9, K_4 1·5	65D

59M P. K. Migal and A. N. Pushnyak, *Russ. J. Inorg. Chem.*, 1959, **4**, 601 (1336)
60M P. K. Migal and A. N. Pushnyak, *Russ. J. Inorg. Chem.*, 1960, **5**, 293 (610)
61A V. Armeann and C. Luca, *Z. Phys. Chem.*, (*Leipzig*), 1961, **217**, 389
62C L. Cockerell and H. F. Walton, *J. Phys. Chem.* 1962, **66**, 75
62F J. F. Fisher and J. L. Hall, *Analyt. Chem.*, 1962, **34**, 1094
62M P. K. Migal and G. F. Serova, *Russ. J. Inorg. Chem.*, 1962, **7**, 827 (1601)
62S A. Ya. Sychev and A. P. Gerbeleu, *Russ. J. Inorg. Chem.*, 1962, **7**, 138 (269)
63S P. E. Sturrock, *Analyt. Chem.*, 1963, **35**, 1092
65D G. Douheret, *Bull. Soc. chim. France*, 1965, 2915
65M P. K. Migal and G. F. Serova, *Russ. J. Inorg. Chem.*, 1965, **10**, 1366 (2513)
65Ma P. K. Migal and K. I. Ploae, *Russ. J. Inorg. Chem.*, 1965, **10**, 1368 (2517)
66S E. V. Sklenskaya and M. Kh. Karapet'yants, *Russ. J. Inorg. Chem.*, 1966, **11**, 1478 (2749)
66U V. V. Udovenko and Yu. S. Duchinskii, *Russ. J. Inorg. Chem.*, 1966, **11**, 1481 (2755)
67F J. F. Fisher and J. L. Hall, *Analyt. Chem.*, 1967, **39**, 1550
68A D. J. Alner, R. C. Lansbury, and A. G. Smeeth, *J. Chem. Soc.* (*A*), 1968, 417

137 C_2H_7NS 2-Aminoethanthiol HL

Metal	*Method*	*Temp*	*Medium*	*Log of equilibrium constant, remarks*	*Ref*
H^+	gl	25	0·1 (KNO_3)	K_1 10·73, K_2 8·19	63T
Ba^{2+}	gl	25	0·1 (KNO_3)	K_1 1·37	63T
Ca^{2+}	gl	25	0·1 (KNO_3)	K_1 2·21	63T
Cd^{2+}	pol	25	0·264	β_2 9·02, phosphate buffer	61K
Co^{2+}	pol	25	0·264	β_4 12·89, phosphate buffer	61K

contin

C_2H_7NS (contd) 137

Metal	Method	Temp	Medium	Log of equilibrium constant, remarks	Ref
Cu^{2+}	pol	25	0·264	β_2 16·74, phosphate buffer	61K
	pol	25	0·17	β_2 16·24, phosphate buffer	61Ka
Mg^{2+}	gl	25	0·1 (KNO_3)	K_1 2·30	63T
Sr^{2+}	gl	25	0·1 (KNO_3)	K_1 1·55	63T
Zn^{2+}	pol	25	0·264	β_2 6·17, phosphate buffer	61K

61K E. C. Knoblock and W. C. Purdy, *Radiation Res.*, 1961, **15**, 94
61Ka E. C. Knoblock and W. C. Purdy, *J. Electroanalyt Chem.*, 1961, **2**, 493
63T Y. Tsuchitani, T. Ando, and K. Ueno, *Bull. Chem. Soc. Japan*, 1963, **36**, 1534

$C_2H_8O_6P_2$ Ethane-1,2-diphosphonic acid H_4L 137.1

Metal	Method	Temp	Medium	Log of equilibrium constant, remarks	Ref
H^+	gl	25	1 $(CH_3)_4NBr$	K_1 8·96±0·07, K_2 7·42±0·07, K_3 2·74±0·07, K_4 1·5±0·2	62I
Ca^{2+}	gl	25	1 $(CH_3)_4NBr$	K_1 2·80, K($Ca^{2+}+HL^{3-} \rightleftharpoons CaHL^-$) 2·60	62I
Mg^{2+}	gl	25	1 $(CH_3)_4NBr$	K_1 2·85, K($Mg^{2+}+HL^{3-} \rightleftharpoons MgHL^-$) 2·67	62I

62I R. R. Irani and K. Moedritzer, *J. Phys. Chem.*, 1962, **66**, 1349

$C_2H_8O_6P_2$ Ethane-1,1-diphosphonic acid H_4L 137.2

Metal	Method	Temp	Medium	Log of equilibrium constant, remarks	Ref
H^+	gl	25	0·5 $(CH_3)_4NCl$	K_1 11·54±0·04, K_2 7·18±0·03, K_3 2·66±0·03	67C
Cs^+	gl	25	0·5 $(CH_3)_4NCl$	K_1 1·02±0·09 K($HL^{3-}+Cs^+ \rightleftharpoons CsHL^{2-}$) 0·09±0·02	67C
K^+	gl	25	0·5 $(CH_3)_4NCl$	K_1 1·2±0·1 K($HL^{3-}+K^+ \rightleftharpoons KHL^{2-}$) 0·28±0·03	67C
Li^+	gl	25	0·5 $(CH_3)_4NCl$	K_1 3·1±0·1 K($HL^{3-}+Li^+ \rightleftharpoons LiHL^{2-}$) 0·99±0·04	67C
Na^+	gl	25	0·5 $(CH_3)_4NCl$	K_1 1·51±0·09 K($HL^{3-}+Na^+ \rightleftharpoons NaHL^{2-}$) 0·50±0·03	67C

67C R. L. Carrol and R. R. Irani, *Inorg. Chem.*, 1967, **6**, 1994

$C_2H_8O_7P_2$ 1-Hydroxyethane-1,1-diphosphonic acid H_4L 137.3

Metal	Method	Temp	Medium	Log of equilibrium constant, remarks	Ref
H^+	gl	25	0·5 $(CH_3)_4NCl$	K_1 11·41±0·05, K_2 6·97±0·05, K_3 2·54±0·05	67C
	gl	25	0·1 (KCl)	K_1 10·29±0·07, K_2 7·28±0·06, K_3 2·47±0·04, K_4 1·7±0·2 K[$(H_{-1}L)^{5-}+H^+ \rightleftharpoons L^{4-}$] 11·13±0·09	67K

continued overleaf

137.3 $C_2H_8O_7P_2$ (contd)

Metal	Method	Temp	Medium	Log of equilibrium constant, remarks	Ref
Al^{3+}	gl	25	0·1 (KCl)	$K[Al^{3+}+(H_{-1}L)^{5-} \rightleftharpoons Al(H_{-1}L)^{2-}]$ 21·37 $K[Al^{3+}+2(H_{-1}L)^{5-} \rightleftharpoons Al(H_{-1}L)_2^{7-}]$ 25·87 K_1 15·29, β_2 22·26 $K[2Al^{3+}+(H_{-1}L)^{5-} \rightleftharpoons Al_2(H_{-1}L)^{+}]$ 27·25 $K(2Al^{3+}+L^{4-} \rightleftharpoons Al_2L^{2+})$ 19·33	67K
Be^{2+}	gl	25	0·1 (KCl)	$K[Be^{2+}+(H_{-1}L)^{5-} \rightleftharpoons Be(H_{-1}L)^{3-}]$ 16·55 K_1 13·40, $K(Be^{2+}+HL^{3-} \rightleftharpoons BeHL^{-})$ 7·00 $K[2Be^{2+}+(H_{-1}L)^{5-} \rightleftharpoons Be_2(H_{-1}L)^{-}]$ 25·74 $K(2Be^{2+}+L^{4-} \rightleftharpoons Be_2L)$ 18·01	67K
Ca^{2+}	gl	25	0·1 (KCl)	K_1 6·04 $K[2Ca^{2+}+(H_{-1}L)^{5-} \rightleftharpoons Ca_2(H_{-1}L)^{-}]$ 15·59 $K(2Ca^{2+}+L^{4-} \rightleftharpoons Ca_2L)$ 9·67	67K
Co^{2+}	gl	25	0·1 (KCl)	K_1 9·36, $K(Co^{2+}+HL^{3-} \rightleftharpoons CoHL^{-})$ 5·29 $K[2Co^{2+}+(H_{-1}L)^{5-} \rightleftharpoons Co_2(H_{-1}L)^{-}]$ 19·65 $K(2Co^{2+}+L^{4-} \rightleftharpoons Co_2L)$ 12·77 $K(2Co^{2+}+HL^{3-} \rightleftharpoons Co_2HL^{+})$ 7·51	67K
Cs^{+}	gl	25	0·5 $(CH_3)_4NCl$	K_1 1·6±0·1 $K(Cs^{+}+HL^{3-} \rightleftharpoons CsHL^{2-})$ 0·24±0·04	67C
Cu^{2+}	gl	25	0·1 (KCl)	K_1 12·48, $K(Cu^{2+}+HL^{3-} \rightleftharpoons CuHL^{-})$ 6·26 $K[2Cu^{2+}+(H_{-1}L)^{5-} \rightleftharpoons Cu_2(H_{-1}L)^{-}]$ 25·03 $K(2Cu^{2+}+L^{4-} \rightleftharpoons Cu_2L)$ 16·86 $K(2Cu^{2+}+HL^{3-} \rightleftharpoons Cu_2HL^{+})$ 9·55	67K
Fe^{2+}	gl	25	0·1 (KCl)	K_1 9·05, $K(Fe^{2+}+HL^{3-} \rightleftharpoons FeHL^{-})$ 5·31 $K[2Fe^{2+}+(H_{-1}L)^{5-} \rightleftharpoons Fe_2(H_{-1}L)^{-}]$ 19·59 $K(2Fe^{2+}+L^{4-} \rightleftharpoons Fe_2L)$ 13·89 $K(2Fe^{2+}+HL^{3-} \rightleftharpoons Fe_2HL^{+})$ 7·99	67K
Fe^{3+}	gl	25	0·1 (KCl)	K_1 16·21, β_2 25·25 $K[Fe^{3+}+(H_{-1}L)^{5-} \rightleftharpoons Fe(H_{-1}L)^{2-}]$ 21·60 $K[2Fe^{3+}+(H_{-1}L)^{5-} \rightleftharpoons Fe_2(H_{-1}L)^{+}]$ 29·1	67K
K^{+}	gl	25	0·5 $(CH_3)_4NCl$	K_1 1·79±0·04 $K(K^{+}+HL^{3-} \rightleftharpoons KHL^{2-})$ 0·36±0·03	67C
Li^{+}	gl	25	0·5 $(CH_3)_4NCl$	K_1 3·35±0·06 $K(Li^{+}+HL^{3-} \rightleftharpoons LiHL^{2-})$ 1·08±0·03	67C
Mg^{2+}	gl	25	0·1 (KCl)	K_1 6·55 $K[2Mg^{2+}+(H_{-1}L)^{5-} \rightleftharpoons Mg_2(H_{-1}L)]$ 14·95 $K(2Mg^{2+}+L^{4-} \rightleftharpoons Mg_2L)$ 10·50	67K
Mn^{2+}	gl	25	0·1 (KCl)	K_1 9·16, $K(Mn^{2+}+HL^{3-} \rightleftharpoons MnHL^{-})$ 5·26 $K[2Mn^{2+}+(H_{-1}L)^{5-} \rightleftharpoons Mn_2(H_{-1}L)^{-}]$ 19·64 $K(2Mn^{2+}+L^{4-} \rightleftharpoons Mn_2L)$ 13·23 $K(2Mn^{2+}+HL^{3-} \rightleftharpoons Mn_2HL^{+})$ 8·06	67K
Na^{+}	gl	25	0·5 $(CH_3)_4NCl$	K_1 2·07±0·05 $K(Na^{+}+HL^{3-} \rightleftharpoons NaHL^{2-})$ 0·54±0·03	67C
Ni^{2+}	gl	25	0·1 (KCl)	K_1 9·24, $K(Ni^{2+}+HL^{3-} \rightleftharpoons NiHL^{-})$ 5·14 $K[2Ni^{2+}+(H_{-1}L)^{5-} \rightleftharpoons Ni_2(H_{-1}L)^{-}]$ 18·53 $K(2Ni^{2+}+L^{4-} \rightleftharpoons Ni_2L)$ 12·18 $K(2Ni^{2+}+HL^{3-} \rightleftharpoons Ni_2HL^{+})$ 7·70	67K

contin

$C_2H_8O_7P_2$ (contd) 137.3

Metal	*Method*	*Temp*	*Medium*	*Log of equilibrium constant, remarks*	*Ref*
Sr^{2+}	gl	25	0·1 (KCl)	K_1 5·52	67K
				$K[2Sr^{2+}+(H_{-1}L)^{5-} \rightleftharpoons Sr_2(H_{-1}L)^-]$ 14·37	
				$K(2Sr^{2+}+L^{4-} \rightleftharpoons Sr_2L)$ 9·11	
Th^{4+}	gl	25	0·1 (KCl)	$K[Th^{4+}+(H_{-1}L)^{5-} \rightleftharpoons Th(H_{-1}L)^-]$ 27·8	67K
				$K(Th^{4+}+2(H_{-1}L)^{5-} \rightleftharpoons Th(H_{-1}L)_2{}^{6-}]$ 39·9	
Zn^{2+}	gl	25	0·1 (KCl)	K_1 10·73	67K
				$K(Zn^{2+}+HL^{3-} \rightleftharpoons ZnHL^-)$ 5·66	
				$K[2Zn^{2+}+(H_{-1}L)^{5-} \rightleftharpoons Zn_2(H_{-1}L)^-]$ 22·36	
				$K(2Zn^{2+}+L^{4-} \rightleftharpoons Zn_2L)$ 15·03	
				$K(2Zn^{2+}+HL^{3-} \rightleftharpoons Zn_2HL^+)$ 8·13	
Zr^{4+}	gl	25	0·1 (KCl)	$K(ZrO^{2+}+L^{4-} \rightleftharpoons ZrOL^{2-})$ 15·18	67K
				$K[2ZrO^{2+}+(H_{-1}L)^{5-} \rightleftharpoons (ZrO)_2(H_{-1}L)^-]$ 26·04	
				$K[2ZrO^{2+}+L^{4-} \rightleftharpoons (ZrO)_2L]$ 20·40	
				$K(ZrO^{2+}+2(H_{-1}L)^{5-} \rightleftharpoons ZrO(H_{-1}L)_2{}^{8-}]$ 21·92	
				$K(ZrO^{2+}+2L^{4-} \rightleftharpoons ZrOL_2{}^{6-})$ 18·63	

67C R. L. Carroll and R. R. Irani, *Inorg. Chem.*, 1967, **6**, 1994

67K M. I. Kabachnik, R. P. Lastovskii, T. Ya. Medved, V. V. Medynstev, I. D. Kolpakova, and N. M. Dyatlova, *Proc. Acad. Sci. (U.S.S.R.)*, 1967, **177**, 1060 (582)

$C_2H_8O_4NP$ 2-Aminoethanephosphoric acid (*O*-phosphorylethanolamine) H_2L 140

Metal	*Method*	*Temp*	*Medium*	*Log of equilibrium constant, remarks*	*Ref*
H^+	gl	25	0·15 KCl	K_1 10·10, K_2 5·57	60O
	gl	25	0·15 KCl	K_1 10·13, K_2 5·57	62O
	gl	20	0·1 $(C_3H_7)_4NI$	K_1 10·26±0·05, K_2 5·77±0·05	65H
Ca^{2+}	gl	25	0·15 KCl	K_1 1·57	62O
				$K(Ca^{2+}+HL^- \rightleftharpoons CaHL^+)$ 1·11	
	gl	20	0·1 $(C_3H_7)_4NI$	K_1 2·0±0·1	65H
				$K(Ca^{2+}+HL^- \rightleftharpoons CaHL^+)$ 1·4±0·2	
Cu^{2+}	gl	25	0·15 KCl	K_1 6·39, β_2 12·39	60O
				$K(Cu^{2+}+HL^- \rightleftharpoons CuHL^+)$ 1·94	
				$K(CuHL^++L^{2-} \rightleftharpoons CuHL_2{}^-)$ 6·32	
Mg^{2+}	gl	25	0·15 KCl	K_1 1·70	62O
				$K(Mg^{2+}+HL^- \rightleftharpoons MgHL^+)$ 1·23	
	gl	20	0·1 $(C_3H_7)_4NI$	K_1 2·2±0·1	65H
				$K(Mg^{2+}+HL^- \rightleftharpoons MgHL^+)$ 1·5±0·1	
Mn^{2+}	gl	25	0·15 KCl	K_1 2·55	62O
				$K(Mn^{2+}+HL^- \rightleftharpoons MnHL^-)$ 1·72	
Ni^{2+}	gl	20	0·1 $(C_3H_7)_4NI$	K_1 4·6±0·1	65H
				$K(Ni^{2+}+HL^- \rightleftharpoons NiHL^+)$ 1·8±0·2	

Errata. 1964, p. 384, Table 140. *For* $C_2H_7O_4NP$ *read* $C_2H_8O_4NP$.

60O R. Osterberg, *Acta Chem. Scand.*, 1960, **14**, 471

62O R. Osterberg, *Acta Chem. Scand.*, 1962, **16**, 2434

65H H. S. Henderson and J. G. Fullington, *Biochemistry*, 1965, **4**, 1599

141.1 $C_3H_2O_5$ Oxopropanedioic acid (mesoxalic acid) H_2L

Metal	Method	Temp	Medium	Log of equilibrium constant, remarks	Ref
Hf^{4+}	ix	?	2 $HClO_4$	$K_1 \sim 4$	60R
Zr^{4+}	ix	?	2 $HClO_4$	$K_1 \sim 4$	60R

60R D. I. Ryabchikov, A. N. Ermakov, V. K. Belyaeva, and I. N. Marov, *Russ. J. Inorg. Chem.*, 1960, **5**, 505 (1051)

141.2 C_3H_3N Cyanoethene (acrylonitrile) L

Metal	Method	Temp	Medium	Log of equilibrium constant, remarks	Ref
Ag^+	oth	61	4 $LiNO_3$	K_1 −0·15, gas chromatography	68S

68S H. Schnecko, *Analyt. Chem.*, 1968, **40**, 1391

142 $C_3H_4O_3$ 2-Oxopropanoic acid (pyruvic acid) HL

Metal	Method	Temp	Medium	Log of equilibrium constant, remarks	Ref
H^+	gl	10, 25	0·65 (KCl)	K_1 (10°) 2·39, (25°) 2·39	64L
	gl	25	0·5 KCl	K_1 2·35	66L
Mn^{2+}	gl	10, 25	0·65 (KCl)	K_1 (10°) 1·20, (25°) 1·26 See glycine (130) for mixed ligands	64L
Ni^{2+}	gl	10, 25	0·65 (KCl)	K_1 (10°) 1·40, (25°) 1·15	64L
	sol	25	0·5 KCl	K_1 1·12, β_2 0·46	66L
				See glycine (130), sarcosine (174), β-alanine (172), isoleucine (470.2), and alanine (171) for mixed ligands	
Zn^{2+}	gl	10, 25	0·65 (KCl)	K_1 (10°) 0·90, (25°) 1·28	64L
	sol	25	0·5 KCl	K_1 1·26, β_2 1·98	66L
				See glycine (130), β-alanine (172), isoleucine (470.2), and alanine (171) for mixed ligands	

64L D. L. Leussing and D. C. Shultz, *J. Amer. Chem. Soc.*, 1964, **86**, 4846
66L D. L. Leussing and E. M. Hanna, *J. Amer. Chem. Soc.*, 1966, **88**, 693

$C_3H_4O_4$ Propanedioic acid (malonic acid) H_2L 143

Metal	Method	Temp	Medium	Log of equilibrium constant, remarks	Ref
H^+	H, Ag-AgCl	0—60	→ 0	K_1 (0°) 5·6696±0·0003, (5°) 5·6646±0·0005, (10°) 5·6671±0·0004, (15°) 5·6729±0·0005, (20°) 5·6828±0·0003, (25°) 5·6960±0·0001, (30°) 5·7104±0·0003, (35°) 5·7297±0·0003, (40°) 5·7525±0·0004, (45°) 5·7772±0·0001, (50°) 5·8026±0·0005, (55°) 5·8331±0·0002, (60°) 5·8658±0·0006 $\Delta H_1°$ (25°) 1·161, $\Delta S_1°$ 30·0	40H, 61N, 65N
	gl	0—45	→ 0	K_2 (0°) 1·29±0·01, (15°) 1·38±0·01, (25°) 1·43±0·01, (35°) 1·43±0·01, (45°) 1·41±0·01	61N, 65N
	gl	25	0·15 ($NaClO_4$)	K_1 5·34, K_2 2·85	62B
	gl	20	0·1 ($NaClO_4$)	K_1 5·32, K_2 2·66	63C
	kin	7, 12	0·1 ($NaClO_4$)	K_1 (7°) 5·23, (12°) 5·23, K_2 (7°) 2·66, (12°) 2·65	65C
	?	?	?	K_1 5·29, K_2 2·76	66M
	gl	30	0·2 ($NaClO_4$)	K_1 5·10, K_2 2·60	67A
	gl	25	1·00 (KNO_3)	K_1 5·11±0·01, K_2 2·74±0·04	67R
	pH	25	0·5 $LiClO_4$	K_1 5·01±0·03, K_2 2·53±0·03	68D
	gl	25	0·15 ($NaClO_4$)	K_2 2·77	68K
	gl	25	0·10 (KNO_3)	K_1 5·27, K_2 2·61	68P
Al^{3+}	gl	30	0·2 ($NaClO_4$)	K_1 5·24, K_2 4·16	67A
Ba^{2+}	gl	20	0·1 ($NaClO_4$)	K_1 1·34, $K(Ba^{2+}+HL^- \rightleftharpoons BaHL^+)$ 0·61	63C
Be^{2+}	gl	25	0·15 ($NaClO_4$)	K_1 5·73, β_2 9·28	62B
	gl	30	0·2 ($NaClO_4$)	K_1 5·15, K_2 3·33	67A
Ca^{2+}	gl	20	0·1 ($NaClO_4$)	K_1 1·85, $K(Ca^{2+}+HL^- \rightleftharpoons CaHL^+)$ 0·80	63C
Ce^{3+}	gl	25	0·10 (KNO_3)	K_1 3·83, β_2 6·17	68P
Cd^{2+}	gl	20	0·1 ($NaClO_4$)	K_1 2·51, $K(Cd^{2+}+HL^- \rightleftharpoons CdHL^+)$ 1·05	63C
Co^{2+}	dis	25	0·00—0·16 (NaCl)	K_1 (→ 0) 3·60±0·01, (0·02) 3·135±0·006, (0·04) 2·981±0·003, (0·08) 2·820±0·003, (0·16) 2·658±0·009	61M
	gl	20	0·1 ($NaClO_4$)	K_1 2·98, $K(Co^{2+}+HL^- \rightleftharpoons CoHL^+)$ 2·21	63C
	cal	25	0 corr	$\Delta H°$ 2·90±0·08, $\Delta S°$ 26·9	63M
	kin	7, 12	0·1 ($NaClO_4$)	K_1 (7°) 2·78, (12°) 2·80	65C
	gl	25	0 corr	K_1 3·74±0·02, K_2 1·4	65M
	ix	25	0 corr	K_1 3·77±0·05, K_2 1·35±0·04	65S
Co^{3+}	gl	25, 35	0 corr	$K[CoCl(NH_3)_5^{2+}+L^{2-} \rightleftharpoons CoCl(NH_3)_5L]$ (25°) 2·27, (35°) 2·31	65A
	sol	25, 35	0 corr	$K[CoCl(NH_3)_5^{2+}+L^{2-} \rightleftharpoons CoCl(NH_3)_5L]$ (25°) 2·32, (35°) 2·30	65A
Cr^{3+}	gl	25	0·1 ($NaClO_4$)	K_1 7·06, K_2 5·79, K_3 3·30	66M
Cu^{2+}	gl	20	0·1 ($NaClO_4$)	K_1 5·55 $K(Cu^{2+}+HL^- \rightleftharpoons CuHL^+)$ 2·76	63C
	gl	30	0·2 ($NaClO_4$)	K_1 4·42, K_2 2·78	67A
Dy^{3+}	ix	25	0·15 ($NaClO_4$)	$K(Dy^{3+}+HL^- \rightleftharpoons DyHL^{2+})$ 2·0±0·1 $K[DyHL^{2+}+HL^- \rightleftharpoons Dy(HL)_2^+]$ 1·1±0·1	68K
	gl	25	0·10 (KNO_3)	K_1 4·47, β_2 7·17	68P
Er^{3+}	ix	25	0·15 ($NaClO_4$)	$K(Er^{3+}+HL^- \rightleftharpoons ErHL^{2+})$ 2·1±0·1 $K[ErHL^{2+}+HL^- \rightleftharpoons Er(HL)_2^+]$ 0·90±0·09	68K
	gl	25	0·10 (KNO_3)	K_1 4·42, β_2 7·04	68P
Eu^{3+}	ix	25	0·15 ($NaClO^4$)	$K(Eu^{3+}+HL^- \rightleftharpoons EuHL^{2+})$ 1·9±0·1 $K[EuHL^{2+}+HL^- \rightleftharpoons Eu(HL)^{2+}]$ 1·1±0·1	68K
	gl	25	0·10 (KNO_3)	K_1 4·30, β_2 6.99	68P
Fe^{3+}	pH, sp	25	0·5 $LiClO_4$	K_1 7·46	68D

continued overleaf

143 $C_3H_4O_4$ (contd)

Metal	*Method*	*Temp*	*Medium*	*Log of equilibrium constant, remarks*	*Ref*
Gd^{3+}	ix	25	0·15 ($NaClO_4$)	$K(Gd^{3+}+HL^- \rightleftharpoons GdHL^{2+})$ 1·65±0·08	68K
				$K[GdHL^{2+}+HL^- \rightleftharpoons Gd(HL)_2{}^+]$ 1·2±0·1	
	gl	25	0·10 (KNO_3)	K_1 4·32, β_2 6·97	68P
Ho^{3+}	gl	25	0·10 (KNO_3)	K_1 4·39, β_2 6·97	68P
La^{3+}	gl	25	0·10 (KNO_3)	K_1 3·69, β_2 5·90	68P
	ix	25	0·15 ($NaClO_4$)	$K(La^{3+}+HL^- \rightleftharpoons LaHL^{2+})$ 1·34±0·07	68K
				$K[LaHL^{2+}+HL^- \rightleftharpoons La(HL)_2{}^+]$ 0·85±0·08	
Lu^{3+}	ix	25	0·15 ($NaClO_4$)	$K(Lu^{3+}+HL^- \rightleftharpoons LuHL^{2+})$ 2·2±0·1	68K
				$K[LuHL^{2+}+HL^- \rightleftharpoons Lu(HL)_2{}^+]$ 1·1±0·1	
	gl	25	0·10 (KNO_3)	K_1 4·45, β_2 7·13	68P
Mg^{2+}	gl	20	0·1 ($NaClO_4$)	K_1 1·95, $K(Mg^{2+}+HL^- \rightleftharpoons MgHL^+)$ 0·83	63C
Mn^{2+}	cal	25	0 corr	$\Delta H°$ 3·7±0·1, $\Delta S°$ 27·4	63M
Na^+	gl	25	0 corr	K_1 0·74	65A
Nb^V	dis	20	4·5 (NaCl or $NaNO_3$+2·5 HCl)	$K[Nb(OH)_4{}^+ + H_2L \rightleftharpoons Nb(OH)_4HL+H^+]$ 1·72	67K
Nd^{3+}	gl	25	0·10 (KNO_3)	K_1 3·95, β_2 6·41	68P
Ni^{2+}	gl	20	0·1 ($NaClO_4$)	K_1 3·30, $K(Ni^{2+}+HL^- \rightleftharpoons NiHL^+)$ 1·41	63C
	gl	25	0–0·2 ($NaClO_4$)	K_1 (0 corr) 4·102±0·010, (→0) 4·097, (0·03) 3·516±0·009, (0·05) 3·393±0·007, (0·10) 3·196±0·008, (0·15) 3·090±0·004, (0·20) 3·02±0·02	62Ba
Pr^{3+}	ix	25	0·15 ($NaClO_4$)	$K(Pr^{3+}+HL^- \rightleftharpoons PrHL^{2+})$ 1·48±0·07	68K
				$K[PrHL^{2+}+HL^- \rightleftharpoons Pr(HL)_2{}^+]$ 1·04±0·10	
	gl	25	0·10 (KNO_3)	K_1 3·91, β_2 6·30	68P
Sc^{3+}	sp	18—20	0·01 (HNO_3)	K_1 4·19±0·06	66K
Sm^{2+}	ix	25	0·15 ($NaClO_4$)	$K(Sm^{3+}+HL^- \rightleftharpoons SmHL^{2+})$ 1·81±0·09	68K
				$K[SmHL^{2+}+HL^- \rightleftharpoons Sm(HL)_2{}^+]$ 1·1±0·1	
	gl	25	0·10 (KNO_3)	K_1 4·19, β_2 6·84	68P
Tb^{3+}	ix	25	0·15 ($NaClO_4$)	$K(Tb^{3+}+HL^- \rightleftharpoons TbHL^{2+})$ 1·95±0·10	68K
				$K[TbHL^{2+}+HL^- \rightleftharpoons Tb(HL)_2{}^+]$ 1·04±0·10	
	gl	25	0·10 (KNO_3)	K_1 4·44, β_2 7·15	68P
Th^{4+}	kin	25	~0	K_1 7·25±0·01	63Y
				$K(2Th^{4+}+L^{2-}+OH^- \rightleftharpoons Th_2LOH^{5+})$ 22·46±0·02	
Tm^{3+}	gl	25	0·10 (KNO_3)	K_1 4·42, β_2 7·01	68P
$UO_2{}^{2+}$	gl	25	1·00 (KNO_3)	K_1 5·66±0·02, K_2 4·00±0·02	67R
	gl	30	0·2 ($NaClO_4$)	K_1 4·88, K_2 3·75	67A
W^{VI}	kin	25	0·05	$K(H_2WO_4+H_2L \rightleftharpoons H_2WO_3L)$ 3·09±0·05	62Y
Y^{3+}	gl	25	0·10 (KNO_3)	K_1 4·40, β_2 7·04	68P
Yb^{3+}	ix	?	→0	K_1 5·44	66A
	H, gl	25	0 corr	K_1 5·70±0·03, K_2 2·9±0·1	66Aa
	ix	25	0·15 ($NaClO_4$)	$K(Yb^{3+}+HL^- \rightleftharpoons YbHL^{2+})$ 2·1±0·1	68K
				$K[YbHL^{2+}+HL^- \rightleftharpoons Yb(HL)_2{}^+]$1·2±0·1	
	gl	25	0·10 (KNO_3)	K_1 4·53, β_2 7·27	68P
Zn^{2+}	gl	20	0·1 ($NaClO_4$)	K_1 2·97, $K(Zn^{2+}+HL^- \rightleftharpoons ZnHL^+)$ 1·24	63C
	H, Ag-AgCl	0—45	0 corr	K_1 (25°) 3·82±0·02	65N
				K_1 $8{\cdot}89-4{\cdot}15\times10^{-2}T+9{\cdot}18\times10^{-5}\,T^2$	
				$\Delta H_1°$ 3·0±0·1, $\Delta S_1°$ 27·4±0·4	
	dis	25	0 corr	K_1 3·85±0·07	66R
				K_2 2·1±0·08	
Zr^{4+}	kin	25°	var	$K[Zr(OH)_3{}^+ + H_2L \rightleftharpoons Zr(OH)_3L^- + 2H^+]$ 1·46±0·06	61Y

Erratum. 1964, p. 386, Table 143. Under Mg^{2+} fourth line. *For* log δ K_1 *read* δ log K_1.

references on facing p

40H W. J. Hamer, J. O. Burton, and S. F. Acree, *J. Res. Nat. Bur. Stand.*, 1940, **24**, 269
61M P. G. Manning and C. B. Monk, *Trans. Faraday Soc.*, 1961, **57**, 1996
61N V. S. K. Nair and G. H. Nancollas, *J. Chem. Soc.*, 1961, 4367
61Y K. B. Yatsimirskii and L. P. Raizman, *Russ. J. Inorg. Chem.*, 1961, **6**, 1263 (2496)
62B H. J. deBruin, D. Kaitis, and R. B. Temple, *Austral. J. Chem.*, 1962, **15**, 457
62Ba J. R. Brannan and G. H. Nancollas, *Trans. Faraday Soc.*, 1962, **58**, 354
62Y K. B. Yatsimirskii and K. E. Prik, *Russ. J. Inorg. Chem.*, 1962, **1**, 821 (1589)
63C E. Campi, *Ann. Chim. (Italy)*, 1963, **53**, 96
63M A. McAuley and G. H. Nancollas, *J. Chem. Soc.*, 1963, 989
63Y K. B. Yatsimirskii and Yu. A. Zhukov, *Russ. J. Inorg. Chem.*, 1963, **8**, 149 (295).
65A D. W. Archer, D. A. East, and C. B. Monk, *J. Chem. Soc.*, 1965, 720
65C F. P. Cavasino, *Ricerca sci.*, 1965, **35** (II-A), 1120
65M C. B. Monk, *J. Chem. Soc.*, 1965, 2456
65N V. S. K. Nair, *J. Chem. Soc.*, 1965, 1450
65S R. G. Seys and C. B. Monk, *J. Chem. Soc.*, 1965, 2452
66A D. W. Archer and C. B. Monk, *J. Chem. Soc. (A)*, 1966, 1374
66Aa D. W. Archer and C. B. Monk, *Trans. Faraday Soc.*, 1966, **62**, 1583
66K I. M. Korenman and N. V. Zaglyadimova, *Russ. J. Inorg. Chem.*, 1966, **11**, 1491
66M H. Muro and R. Tsuchiya, *Bull. Chem. Soc. Japan*, 1966, **39**, 1589
66R D. L. G. Rowlands and C. B. Monk, *Trans. Faraday Soc.*, 1966, **62**, 945
67A V. T. Athavole, N. Mahadevan, P. K. Mathur, and R. M. Sathe, *J. Inorg. Nuclear Chem.*, 1967, **29**, 1947
67K C. Konecny, *Z. phys. Chem. (Leipzig)*, 1967, **235**, 39
67R K. S. Rajan and A. E. Martell, *J. Inorg. Nuclear Chem.*, 1967, **29**, 523
68D M. Deneux, R. Meilleur, and R. L. Benoit, *Canad. J. Chem.*, 1968, **46**, 1383
68K C. M. Ke, P. C. Kong, M. S. Cheng, and N. C. Li, *J. Inorg. Nuclear Chem.*, 1968, **30**, 961
68P J. E. Powell, J. L. Farrell, W. F. S. Neillie, and R. Russell, *J. Inorg. Nuclear Chem.*, 1968, **30**, 2223

$C_3H_4O_5$ Hydroxypropanedioic acid (tartronic acid) H_2L 143.1

Metal	*Method*	*Temp*	*Medium*	*Log of equilibrium constant, remarks*	*Ref*
H^+	gl	20	0·1 ($NaClO_4$)	K_1 4·24, K_2 2·02	63C
	gl	20	?	K_1 4·28 ± 0·06, K_2 1·89 ± 0·05	64Z
	?	25	0·05 ($NaClO_4$)	K_1 4·35	67M
Ba^{2+}	gl	20	0·1 ($NaClO_4$)	K_1 1·80, $K(Ba^{2+}+HL^- \rightleftharpoons BaHL^+)$ 0·87	63C
Ca^{2+}	gl	20	0·1 ($NaClO_4$)	K_1 2·27, $K(Ca^{2+}+HL^- \rightleftharpoons CaHL^+)$ 1·30	63C
Cd^{2+}	gl	20	0·1 ($NaClO_4$)	K_1 2·85, $K(Cd^{2+}+HL^- \rightleftharpoons CdHL^+)$1·61	63C
Co^{2+}	gl	20	0·1 ($NaClO_4$)	K_1 3·25, $K(Co^{2+}+HL^- \rightleftharpoons CoHL^+)$ 1·91	63C
Cu^{2+}	gl	20	0·1 ($NaClO_4$)	K_1 5·34, $K(Cu^{2-}+HL^- \rightleftharpoons CuHL^+)$ 3·62 $K[Cu(H_{-1}L)^- + H^+ \rightleftharpoons CuL]$ 4·03	63C
Eu^{3+}	dis	25	0·1 ($NaClO_4$)	K_1 4·85, K_2 3·77	67M
Mg^{2+}	gl	20	0·1 ($NaClO_4$)	K_1 2·17, $K(Mg^{2+}+HL^- \rightleftharpoons MgHL^+)$ 1·23	63C
Nd^{3+}	gl	20	?	K_1 6·7 ± 0·1	64Z
Ni^{2+}	gl	20	0·1 ($NaClO_4$)	K_1 3·45, $K(Ni^{2+}+HL^- \rightleftharpoons NiHL^+)$ 2·10	63C
Pr^{3+}	gl	20	?	K_1 6·06 ± 0·07	64Z
Zn^{2+}	gl	20	0·1 ($NaClO_4$)	K_1 3·22, $K(Zn^{2+}+HL^- \rightleftharpoons ZnHL^+)$ 1·91	63C

63C E. Campi, *Ann. Chim. (Italy)*, 1963, **53**, 96
64Z O. E. Zuyagintsen and V. P. Tikhonov, *Russ. J. Inorg. Chem.*, 1964, **9**, 865 (1597)
67M P. G. Manning, *Canad. J. Chem.*, 1967, **45**, 1643

143.2 $C_3H_4N_2$ 1,2-Diazole (pyrazole) L

Metal	*Method*	*Temp*	*Medium*	*Log of equilibrium constant, remarks*	*Ref*
Cd^{2+}	pol	0—45	0·1 KNO_3	K_1 (0°) 1·76, (25°) 1·50, (45°) 1·28, β_2 (0°) 2·73, (25°) 2·18, (45°) 1·80, β_3 (0°) 3·21, (25°) 2·32, (45°) 1·83 ΔH_1 −3·94, $\Delta H_{\beta 2}$ −8·10, $\Delta H_{\beta 3}$ −12·14	63A
	pol	25	0·1 KNO_3 ?% methanol	K_1 1·18, β_2 1·48, β_3 2·22	65C
	pol	25	0·1 KNO_3	K_1 1·11, β_2 1·60, β_3 1·83, β_4 1.54	65C
Co^{2+}	pol	25	0·1 $NaNO_3$	K_1 1·50, β_2 1·78, β_3 2·23, β_4 1·78	68S
Fe^{2+}	pol	25	0·1 $NaNO_3$	K_1 0·84, β_2 1·08, β_3 1·28, β_4 1·52	68C
Mn^{2+}	pol	25	0·1 $NaNO_3$	K_1 0·25, β_2 0·34	68C
Pb^{2+}	pol	25	0·1 KNO_3	K_1 −0·40, β_2 −0·47	66C

63A A. C. Andrews and J. K. Romary, *Inorg. Chem.*, 1963, **2**, 1060
65C D. R. Crow, *J. Polarog. Soc.*, 1965, **11**, 22, 67
66C D. R. Crow, *J. Polarog. Soc.*, 1966, **12**, 101
68C D. R. Crow and J. V. Westwood, *J. Inorg. Nuclear Chem.*, 1968, **30**, 179

144 $C_3H_4N_2$ 1,3-Diazole (imidazole) L

Metal	*Method*	*Temp*	*Medium*	*Log of equilibrium constant, remarks*	*Ref*
H^+	gl	25	0·3 (K_2SO_4)	K_1 7·01±0·01	61J
	gl	20	0·15 ($NaClO_4$)	K_1 7·20±0·03	62H
	gl	25	0·2 KNO_3	K_1 7·12	63C
	gl	25	0·1 (KCl)	K_1 7·09	64B
	sp	25	0 corr	K ($H_{-1}L^- + H^+ \rightleftharpoons L$) 14·44±0·04 $\Delta H°$ −17·6±0·2, $\Delta S°$ 7±5	64G, 64H, 6
	gl	0—50	0·1 (KNO_3)	K_1 (0°) 7·53, (10°) 7·32, (20°) 7·11, (30°) 6·90, (40°) 6·70, (50°) 6·49 ΔH_1 (25°) −35±2 kJ mole^{-1}, ΔS_1 16±6 J mole^{-1} deg^{-1}	66D
	H, Ag-AgCl	0—50	0 corr	K_1 (0°) 7·580±0·001, (5°) 7·455±0·002, (10°) 7·334±0·001, (15°) 7·216±0·001, (20°) 7·1028±0·0004, (25°) 6·9929±0·0004, (30°) 6·887±0·001, (35°) 6·784±0·001, (40°) 6·685±0·002, (45°) 6·589±0·002, (50°) 6·497±0·001 $\Delta H_1°$ (25°) −36·8±0·1 kJ mole^{-1}, $\Delta S_1°$ 10·5±0·4 J mole^{-1} deg^{-1}	66D
Ag^+	gl	25	1 KNO_3	K_1 3·11, K_2 3·73	60G
	gl	25	1 (KNO_3)	K_1 3·05, K_2 3·83	64B
	gl	0—50	0·1 (KNO_3)	K_1 (0°) 3·52, (10°) 3·38, (20°) 3·24, (30°) 3·10, (40°) 2·96, (50°) 2·81, K_2 (0°) 4·36, (10°) 4·15, (20°) 3·89, (30°) 3·64, (40°) 3·41, (50°) 3·24 ΔH_1 (0°) −19, (10°) −21, (20°) −23, (30°) −25, (40°) −28, (50°) −30 kJ mole^{-1} ΔS_1 (0°) −2, (10°) −9, (20°) −17, (30°) −24, (40°) −32, (50°) −39 J mole^{-1} deg^{-1}, ΔH_2 (0°) −25, (10°) −37, (20°) −43, (30°) −43, (40°) −37, (50°) −25 kJ mole^{-1} ΔS_2 (0°) −9, (10°) −51, (20°) −73, (30°) −74, (40°) −54, (50°) −14 J mole^{-1} deg^{-1}	66D

contin

Metal	Method	Temp	Medium	Log of equilibrium constant, remarks	Ref
Co^{2+}	sp	10·7—29·8	0·042	$K(CoA^+ + L \rightleftharpoons CoAL^+)$ (10·7°) 4·88, (18·5°) 4·75, (25·0°) 4·59, (29·8°) 4·53	64H
		25	0 corr	$\Delta H_1°$ −7·2±0·6, $\Delta S_1°$ −3±2 $K[CoA(H_{-1}L) + H^+ \rightleftharpoons CoAL^+]$ 10·25±0·02 $\Delta H°$ −11±1, $\Delta S°$ 12±3 where CoA^+ = aquocobalamin	
	sp	15—34·2	0·11	$K(CoA^+ + L \rightleftharpoons CoAL^+)$ (15°) 4·25±0·08, (25°) 4·09±0·08, (28°) 4·04±0·08, (34·2°) 3·95±0·08	66H
		25	0 corr	$\Delta H°$ −6·3±0·5, $\Delta S°$ −2±2 $K[CoA(H_{-1}L) + H^+ \rightleftharpoons CoAL^+]$ 4·49±0·02, $\Delta H°$ −4·7±0·9, $\Delta S°$ 5±3 $K[CoA(H_{-2}L) + H^+ \rightleftharpoons CoA(H_{-1}L)]$ 11·00±0·02, $\Delta H°$ −12±1, $\Delta S°$ 11±3 where CoA^+ = cobalamin Factor B	
	gl	25	0·16 KNO_3	K_1 2·47±0·02, K_2 1·93±0·04, K_3 1·45±0·06, K_4 1·00±0·10 K_5 0·5, K_6 0	66S
		10—50	0·16 KNO_3	ΔH_1 −4·2±0·3, ΔS_1 −2·6, ΔH_2 −3·9±0·5, ΔS_2 −4·7, ΔH_3 −3·5±0·7, ΔS_3 −5·1, ΔH_4 −4±1, ΔS_4 −8, ΔH_5 −3, ΔS_5 −7, ΔH_6 −4, ΔS_6 −12	
Cu^+	Pt	25	0·3 (K_2SO_4)	β_2 10·44	61J
	Pt, Au	20	0·15 (Na_2SO_4)	K_1 5·78, β_2 10·98	62H
Cu^{2+}	Pt	25	0·3 (K_2SO_4)	K_1 4·33, K_2 3·27, K_3 2·7, K_4 1·9	61J
	gl	20	0·15 ($NaClO_4$)	K_1 4·26, β_2 7·87, β_3 10·73, β_4 12·98	62H
	gl	25	0·2 KNO_3	K_1 4·15, K_2 3·52	63C
	gl	25	0·16 KNO_3	K_1 4·31±0·02, K_2 3·53±0·04, K_3 2·92±0·06, K_4 2·14±0·10	66S
		10—50	0·16 KNO_3	ΔH_1 −7·2±0·3, ΔS_1 −4·2, ΔH_2 −5·4±0·5, ΔS_2 −1·8, ΔH_3 −4·6±0·7, ΔS_3 −2·1, ΔH_4 −3±1, ΔS_4 −0·5 See glycylglycine (245) for mixed ligands	
Fe^{2+}	hyp	30	0·16 KNO_3	K_1 1·81, K_2 1·23	66S
Fe^{3+}	sp	25	0 corr	$K(FeA^+ + L \rightleftharpoons FeAL^+)$ 2·20±0·03 $\Delta H_1°$ −4·1±0·7, $\Delta S_1°$ −4±3 $K[FeA(H_{-1}L) + H^+ \rightleftharpoons FeAL^+]$ 10·34±0·03 $\Delta H°$ −11±1, $\Delta S°$ 9±4 where FeA^+ = ferrimyoglobin	64G
Mn^{2+}	hyp	30	0·16 KNO_3	K_1 1·25, K_2 0·70	66S
Ni^{2+}	gl	25	0·2 KNO_3	K_1 3·01	63C
	gl	25	0·16 KNO_3	K_1 3·09±0·02, K_2 2·47±0·04, K_3 2·00±0·06, K_4 1·54±0·10, K_5 1·1, K_6 0·5	66S
		10—50	0·16 KNO_3	ΔH_1 −5·2±0·3, ΔS_1 −2·9, ΔH_2 −4·6±0·5, ΔS_2 −4·1, ΔH_3 −4·3±0·7, ΔS_3 −5·3, ΔH_4 −4±1, ΔS_4 −5, ΔH_5 −3, ΔS_5 −6, ΔH_6 −3, ΔS_6 −7	
Zn^{2+}	gl	25	0·2 KNO_3	K_1 2·13	63C
	nmr	36	$(CH_3)_2SO$	β_4 6·2±0·3 $K(ZnAL^{2+} + 3L \rightleftharpoons ZnL_4^{2+} + A)$ 4·5 where A = cytosine $K(ZnL^{2+} + A \rightleftharpoons ZnAL^{2+})$ 1·23 where A = purine	66W

60G D. H. Gold and H. P. Gregor, *J. Phys. Chem.*, 1960, **64**, 1461
61J B. R. James and R. J. P. Williams, *J. Chem. Soc.*, 1961, 2007
62H C. J. Hawkins and D. D. Perrin, *J. Chem. Soc.*, 1962, 1351
63C A. Chakravorty and F. A. Cotton, *J. Phys. Chem.*, 1963, **67**, 2878

continued overleaf

144 $C_3H_4N_2$ (contd)

64B J. E. Bauman, jun. and J. C. Wang, *Inorg. Chem.*, 1964, **3**, 368
64G P. George, G. I. H. Hanania, D. H. Irvine, and I. Abu-Issa, *J. Chem. Soc.*, 1964, 5689
64H G. I. H. Hanania and D. H. Irvine, *J. Chem. Soc.*, 1964, 5694
66D S. P. Datta and A. K. Grzbowski, *J. Chem. Soc. (A)*, 1966, 1059, *(B)* 1966, 136
66H G. I. H. Hanania, D. H. Irvine, and M. V. Irvine, *J. Chem. Soc. (A)*, 1966, 296
66S E. V. Sklenskaya and M. Kh. Karapet'yants, *Russ. J. Inorg. Chem.*, 1966, **11**, 1102, (2061)
66W S. M. Wang and N. C. Li, *J. Amer. Chem. Soc.*, 1966, **88**, 4592

144.1 C_3H_5Cl 3-Chloropropene (allyl chloride) L

Metal	*Method*	*Temp*	*Medium*	*Log of equilibrium constant, remarks*	*Ref*
Cu^+	Cu	5—25	?	$K(CuCl_3^{2-} + L \rightleftharpoons CuCl_3L^{2-})$ (5°) 0·89, (15°) 0·43, (25°) 0·00 $\Delta H°$ (25°) −17·1, $\Delta S°$ −57 $K(CuCl_3^{2-} + L \rightleftharpoons CuCl_2L^- + Cl^-)$ (5°) 0·48, (15°) 0·18, (25°) −0·05 $\Delta H°$ (25°) −7·1, $\Delta S°$ −24	65T

65T Yu. A. Treger, R. M. Flid, L. V. Antonova, and S. S. Spektor, *Russ. J. Phys. Chem.* 1965, **39**, 1515 (2831)

145 C_3H_6O Propan-2-one (acetone) L

Metal	*Method*	*Temp*	*Medium*	*Log of equilibrium constant, remarks*	*Ref*
Cd^{2+}	pol	18	0·05 $NaClO_4$ 90% Acetone	K_1 1·08±0·08, β_2 1·70±0·05, β_3 2·11±0·03, β_4 2·78±0·08, β_5 2·9±0·1, β_6 2·9±0·2	62M
			0·05 $NaClO_4$ 90% Acetone 10% Methanol	K_1 0·95±0·5, β_2 0·74±0·4, β_3 −0·78±0·8, β_4 0·93±0·5	
Cu^{2+}	sp	27	?	K_1 0·57, K_2 −2·4±0·1	63F
Pb^{2+}	pol	18	0·05 $NaClO_4$ 90% Acetone	K_1 1·2±0·1, β_2 1·5±0·1, β_3 1·70±0·05, β_4 1·7±0·1, β_5 1·1±0·1	62M
			0·05 $NaClO_4$ 90% Acetone 10% Methanol	K_1 0·40±0·04, β_2 0·43±0·03	
			0·05 $NaClO_4$ 90% Acetone 10% Ethanol	K_1 0·0±0·1, β_2 −0·7±0·1, β_3 0·45±0·03	

62M P. K. Migal and N. Kh. Grinberg, *Russ. J. Inorg. Chem.*, 1962, **7**, 270 (531)
63F N. J. Friedman and R. A. Plane, *Inorg. Chem.*, 1963, **2**, 11

C_3H_6O Propen-3-ol (allyl alcohol) L 146

Metal	Method	Temp	Medium	Log of equilibrium constant, remarks	Ref
Ag^+	Ag	25	2 $(NaClO_4)$	K_1 1·36 0±0·002, K_2 −0·24±0·05	67H
Cu^+	pol	25	0·1 $NaClO_4$	K_1 4·7±0·2	66M
	Cu	25—85	?	$K(CuCl_3^{2-}+L \rightleftharpoons CuCl_2L^-+Cl^-)$ (25°) 1·32, (50°) 1·00, (70°) 0·80, (85°) 0·65 $\Delta H°$ (25°) −6·0, $\Delta S°$ −14 $K(CuCl_2^-+L \rightleftharpoons CuCl_2L^-)$ (25°) 1·83, (50°) 1·72, (70°) 1·65, (85°) 1·60 $\Delta H°$ (25°) −2·0, $\Delta S°$ 2	65T
Pt^{2+}	sp	30—60	2 (NaCl)	$K(PtCl_4^{2-}+L \rightleftharpoons PtCl_3L^-+Cl^-)$ (30°) 4·11, (44·5°) 3·86, (60°) 3·59 ΔH −8·1±0·5, ΔS −7·6±1·5	67H

65T Yu. A. Treger, R. M. Flid, L. V. Antonova, and S. S Spektor, *Russ J. Phys. Chem.*, 1965, **39**, 1515 (2831)
66M S. E. Manahan, *Inorg. Chem.*, 1966, **5**, 482
67H F. R. Hartley and L. M. Venanzi, *J. Chem. Soc. (A)*, 1967, 330, 333

$C_3H_6O_2$ Propanoic acid HL 147

Metal	Method	Temp	Medium	Log of equilibrium constant, remarks	Ref
H^+	con	25	→0	K_1 4·87	62H
	H	30	→0	K_1 4·89	63M
	gl	20	0·1 $(NaClO_4)$	K_1 4·66	64P
	gl	25	0·1 $(NaClO_4)$ 50% dioxan	K_1 6·29±0·01	68G
Cd^{2+}	Hg	35	→0	K_1 2·89, K_2 1·28	66A
Ce^{3+}	gl	20	0·1 $(NaClO_4)$	K_1 2·05, K_2 1·23	64P
	gl	25	2 $(NaClO_4)$	K_1 1·67±0·03, K_2 1·00±0·05	65C
Cr^{3+}	gl	30	0·15 (KCl)	K_1 2·74	63M
Cu^{2+}	sp	30	0·1	K_1 2·3	65D
	Hg	30	→0	K_1 2·2, K_2 1·31	66A
	sp	35	0—1·65	K_1 (→ 0) 1·98, (0·05) 1·66, (1·65) 1·43	67A
	gl	25	0·1 $(NaClO_4)$ 50% dioxan	K_1 3·45±0·01 $K(CuA^{2+}+L^- \rightleftharpoons CuAL^+)$ 3·60 where A = 2,2′-bipyridyl	68G
Dy^{3+}	g	20	0·1 $(NaClO_4)$	K_1 1·93, K_2 1·52	64P
	gl	25	2 $(NaClO_4)$	K_1 1·63±0·03, K_2 1·34±0·05	65C
Er^{3+}	gl	20	0·1 $(NaClO_4)$	K_1 1·94, K_2 1·54	64P
	gl	25	2 $(NaClO_4)$	K_1 1·60±0·03, K_2 1·12±0·05	65C
Eu^{3+}	gl	20	0·1 $(NaClO_4)$	K_1 2·23, K_2 1·52	64P
	gl	25	2 $(NaClO_4)$	K_1 1·98±0·03, K_2 1·30±0·05	65C
Gd^{3+}	gl	20	0·1 $(NaClO_4)$	K_1 2·09, K_2 1·48	64P
	gl	25	2 $(NaClO_4)$	K_1 1·84±0·03, K_2 1·33±0·05	65C
Ho^{3+}	gl	20	0·1 $(NaClO_4)$	K_1 1·96, K_2 1·50	64P
	gl	25	2 $(NaClO_4)$	K_1 1·62±0·03, K_2 1·23±0·05	65C
La^{3+}	gl	20	0·1 $(NaClO_4)$	K_1 1·89, K_2 1·18	64P
	gl	25	2 $(NaClO_4)$	K_1 1·53±0·03, K_2 0·89±0·05	65C
Lu^{3+}	gl	20	0·1 $(NaClO_4)$	K_1 2·00, K_2 1·53	64P
	gl	25	2 $(NaClO_4)$	K_1 1·66±0·03, K_2 1·12±0·05	65C

continued overleaf

147 $C_3H_6O_2$ (contd)

Metal	Method	Temp	Medium	Log of equilibrium constant, remarks	Ref
Nd^{3+}	gl	20	0·1 ($NaClO_4$)	K_1 2·20, K_2 1·32	64P
	gl	25	2 ($NaClO_4$)	K_1 1·93±0·03, K_2 1·15±0·05	65C
Pb^{2+}	Hg	35	→0	K_1 2·64, K_2 1·41	66A
Pr^{3+}	gl	20	0·1 ($NaClO_4$)	K_1 2·12, K_2 1·34	64P
	gl	25	2 ($NaClO_4$)	K_1 1·78±0·03, K_2 1·08±0·05	65C
Sm^{3+}	gl	20	0·1 ($NaClO_4$)	K_1 2·21, K_2 1·49	64P
	gl	25	2 ($NaClO_4$)	K_1 2·02±0·03, K_2 1·22±0·05	65C
Tb^{3+}	gl	20	0·1 ($NaClO_4$)	K_1 2·00, K_2 1·59	64P
	gl	25	2 ($NaClO_4$)	K_1 1·73±0·03, K_2 1·37±0·05	65C
Tm^{3+}	gl	20	0·1 ($NaClO_4$)	K_1 1·91, K_2 1·52	64P
	gl	25	2 ($NaClO_4$)	K_1 1·61±0·03, K_2 1·05±0·05	65C
UO_2^{2+}	pol	?	1 ($NaNO_3$)	$K(UO_2L + UO_2^{2+} \rightleftharpoons UO_2L^+ + UO_2^+)$ 4·7±0·1 $K(UO_2L_2^- + UO_2^{2+} \rightleftharpoons UO_2L_2 + UO_2^+)$ 5·2±0·1	62H
	sp	20	1 ($NaClO_4$)	K_1 2·53±0·01, K_2 2·15±0·04, K_3 1·81±0·05, K_4 1·76±0·06	67M
Y^{3+}	gl	20	0·1 ($NaClO_4$)	K_1 1·88, K_2 1·18	64P
	gl	25	2 ($NaClO_4$)	K_1 1·61±0·03, K_2 1·20±0·05	65C
Yb^{3+}	gl	20	0·1 ($NaClO_4$)	K_1 1·93, K_2 1·45	64P
	gl	25	2 ($NaClO_4$)	K_1 1·63±0·03, K_2 1·07±0·05	65C
Zn^{2+}	gl	25	0·1 ($NaClO_4$) 50% dioxan	K_1 2·41 $K(ZnA^{2+} + L^- \rightleftharpoons ZnAL^-)$ 2·38 where A = 2,2′-bipyridyl	68G

62H J. Hala and A. Okac, *Coll. Czech. Chem. Comm.*, 1962, **27**, 1697
63M M. Muzaffaruddin, Salahuddin, and W. U. Malik, *J. Indian Chem. Soc.*, 1963, **40**, 467
64P J. E. Powell, R. S. Kolat, and C. S. Paul, *Inorg. Chem.*, 1964, 3, 518
65C G. R. Choppin and A. J. Graffeo, *Inorg. Chem.*, 1965, **4**, 1254
65D V. P. Devendran and M. Santappa, *Current Sci.*, 1965, **34**, 145
66A S. Aditya, S. Aditya, and S. K. Mukherjee, *J. Electrochem. Soc. Japan*, 1966, **34**, 203
67A S. Aditya, *J. Inorg. Nuclear Chem.*, 1967, **29**, 1901
67M C. Miyake and H. W. Nurnberg, *J. Inorg. Nuclear Chem.*, 1967, **29**, 2411
68G R. Griesser, B. Prijs, and H. Sigel, *Inorg. Nuclear Chem. Letters*, 1968, **4**, 443

148 $C_3H_6O_3$ 3-Hydroxypropanoic acid HL

Metal	Method	Temp	Medium	Log of equilibrium constant, remarks	Ref
H^+	gl	30	0·1 ($NaClO_4$)	K_1 4·52±0·03	62C
	gl	30	0·1 (KCl)	K_1 4·49	62Cb
Ce^{3+}	gl	30	0·1 (KCl)	K_1 2·61, K_2 2·60	62Cb
Cu^{2+}	sp	30	0·1	K_1 2·05	65D
La^{3+}	gl	30	0·1 (KCl)	K_1 2·58, K_2 2·49	62Cb
Nd^{3+}	gl	30	0·1 (KCl)	K_1 2·80, K_2 2·72	62Cb
Pr^{3+}	gl	30	0·1 (KCl)	K_1 2·65, K_2 2·58	62Cb
UO_2^{2+}	gl	30	0·12 ($NaClO_4$)	K_1 2·74±0·05, K_2 2·2±0·3, K_3 2±1	62C
	gl	30	0·1 KCl	K_1 3·25, K_2 2·88	62Ca

62C C. A. Crutchfield Jun., W. M. McNabb, and J. F. Hazel, *J. Inorg. Nuclear Chem.*, 1962, **24**, 291
62Ca M. Cefola, R. C. Taylor, P. S. Gentile, and A. V. Celiano, *J. Phys. Chem.*, 1962, 790
62Cb M. Cefola, A. S. Tompa, A. V. Celiano, and P. S. Gentile, *Inorg. Chem.*, 1962, **1**, 290
65D V. P. Devendran and M. Santappa, *Current Sci.*, 1965, **34**, 145

$C_3H_6O_3$ 2-Hydroxypropanoic acid (lactic acid) HL 149

Metal	*Method*	*Temp*	*Medium*	*Log of equilibrium constant, remarks*	*Ref*
H^+	?	?	?	K_1 3·86	57V, 62S
	gl	31	0·1 ($NaClO_4$)	K_1 3·83±0·03	62C
	?	?	?	K_1 3·85	64R
	gl	20	1 $NaClO_4$	K_1 3·626±0·002	65J
	qh	25	1 ($NaClO_4$)	K_1 3·64	67T
	?	?	?	K_1 3·73	67V
	oth	25	0·1 KCl	K_1 3·81, circular dichroism	68B
Ba^{2+}	gl	25	1	K_1 0·34, β_2 0·42	65V
Be^{2+}	ix	18	0·1 ($NaClO_4$)	K_1 1·53±0·01	65B
Ca^{2+}	qh	25	1	K_1 0·90, β_2 1·24	65V
Cd^{2+}	qh	25	1 ($NaClO_4$)	K_1 1·21, K_2 0·87, K_3 ~ 0·2	67T
Ce^{3+}	ix	20	0·2 ($NaClO_4$)	K_1 2·43±0·04, β_2 4·11±0·09, β_3 5·3±0·3	60S
	gl	25	0·2 $NaClO_4$	K_1 2·49, K_2 1·57	64D
	gl	20	0·1 ($NaClO_4$)	K_1 2·756, β_2 4·72, β_3 5·95	64P
Co^{2+}	qh	25	1 ($NaClO_4$)	K_1 1·37, K_2 0·95, K_3 ~ 0·2,	67T
	oth	25	0·05—0·5	β_2 (0·05) 3·33, (0·10) 2·61, (0·20) 2·15, (0·50) 1·63, circular dichroism	68B
Cu^{2+}	sp	31	0·1	K_1 2·55	65D
	gl	25	1 ($NaClO_4$)	K_1 2·49, K_2 1·49, K_3 ~ 0·3	67T
	oth	25	0·05—0·07	K_1 (0·05) 3·40, (0·07) 3·01, circular dichroism	68B
Dy^{3+}	gl	25	0·2 $NaClO_4$	K_1 2·59, K_2 2·23, K_3 1·09, K_4 0·63	64D
	gl	20	0·1 ($NaClO_4$)	K_1 3·009, β_2 5·35, β_3 6·67	64P
	gl	25	0·1 ($NaClO_4$)	K_1 3·09±0·01, β_2 5·38±0·04, β_3 5·95±0·10	66G
Er^{3+}	gl	25	0·2 $NaClO_4$	K_1 2·86, K_2 2·33, K_3 1·33, K_4 0·66	64D
	gl	20	0·1 ($NaClO_4$)	K_1 3·164, β_2 5·62, β_3 7·20	64P
	gl	25	0·1 ($NaClO_4$)	K_1 3·21±0·01, β_2 5·57±0·05, β_3 7·2±0·1	66G
Eu^{3+}	gl	25	0·2 $NaClO_4$	K_1 2·55, K_2 2·12, K_3 0·88, K_4 0·51	64D
	gl	20	0·1 ($NaClO_4$)	K_1 2·949, β_2 5·18, β_3 6·43	64P
Gd^{3+}	gl	25	0·2 $NaClO_4$	K_1 2·57, K_2 2·15, K_3 1·02, K_4 0·47	64D
	gl	20	0·1 ($NaClO_4$)	K_1 2·892, β_2 5·04, β_3 6·24	64P
	gl	25	0·1 ($NaClO_4$)	K_1 2·96±0·01, β_2 5·09±0·05	66G
Ge^{IV}	con	18	1 NaCl	$K(H_2GeO_3+2HL \rightleftharpoons GeOL_2)$ (?) 0·6±0·1	57V
	gl	18	→ 0	$K(H_2GeO_3+2HL \rightleftharpoons GeOL_2)$ (?) 1·9	57V
Hf^{4+}	ix	?	2 $HClO_4$	$K(Hf^{4+}+HL \rightleftharpoons HfL^{3+}+H^+)$ 1·73	60R
	ix	?	2·0 $HClO_4$	$K(Hf^{4+}+HL \rightleftharpoons HfL^{3+}+H^+)$ 2·03±0·03, $K(Hf^{4+}+2HL \rightleftharpoons HfL_2^{2+}+2H^+)$ 2·0±0·5	64R
Ho^{3+}	gl	25	0·2 $NaClO_4$	K_1 2·73, K_2 2·30, K_3 1·27, K_4 0·67	64D
	gl	20	0·1 ($NaClO_4$)	K_1 3·021, β_2 5·42, β_3 6·83	64P
La^{3+}	gl	25	0·2 $NaClO_4$	K_1 2·44, K_2 1·49	64D
	gl	20	0·1 ($NaClO_4$)	K_1 2·60, β_2 4·34, β_3 5·6	64P
	gl	25	0·1 ($NaClO_4$)	K_1 2·44±0·01, β_2 4·32±0·04	66G
Lu^{3+}	gl	25	0·2 $NaClO_4$	K_1 3·05, K_2 2·51, K_3 1·41, K_4 0·86	64D
	gl	20	0·1 ($NaClO_4$)	K_1 3·273, β_2 5·88, β_3 7·78	64P
	gl	25	0·1 ($NaClO_4$)	K_1 3·40±0·01, β_2 5·82±0·05, β_3 7·9±0·1	66G
Mg^{2+}	qh	25	1	K_1 0·73, β_2 1·30	65V
Mn^{2+}	qh	25	1 ($NaClO_4$)	K_1 0·92, K_2 0·54, K_3 ~ 0·1	67T
Nd^{3+}	gl	25	0·2 $NaClO_4$	K_1 2·65, K_2 1·79, K_3 0·93, K_4 0·08	64D
	gl	20	0·1 ($NaClO_4$)	K_1 2·87, β_2 4·97, β_3 6·4	64P
Ni^{2+}	qh	25	1 ($NaClO_4$)	K_1 1·59, K_2 1·08, K_3 ~ 0·3	67T
	oth	25	0·20	β_2 1·89, circular dichroism	68B
Pa^V	ix	?	0·25	$K[PaO(OH)^{2+}+L^- \rightleftharpoons PaO(OH)L^+]$ (?) 2·24	62G
Pb^{2+}	sol	25	3 ($NaClO_4$)	K_1 1·71±0·01, β_2 2·38±0·01	65Ba
	pH	25	3 ($NaClO_4$)	K_1 2·26±0·03, β_2 3·30±0·04, β_3 3·33±0·06	65Bb
	qh	25	1 ($NaClO_4$)	K_1 1·98, K_2 1·00	67T

continued overleaf

149 $C_3H_6O_3$ (contd)

Metal	*Method*	*Temp*	*Medium*	*Log of equilibrium constant, remarks*	*Ref*
Pr^{3+}	gl	25	0·2 $NaClO_4$	K_1 2·58, K_2 1·70, K_3 0·85	64D
	gl	20	0·1 ($NaClO_4$)	K_1 2·85, β_2 4·90, β_3 6·1	64P
	gl	25	0·1 ($NaClO_4$)	K_1 2·69±0·01, β_2 4·96±0·05	66G
Sm^{3+}	gl	25	0·2 $NaClO_4$	K_1 2·50, K_2 2·04, K_3 0·90, K_4 0·41	64D
	gl	20	0·1 ($NaClO_4$)	K_1 2·88, β_2 5·09, β_3 6·35	64P
Sr^{2+}	qh	25	1	K_1 0·53, β_2 0·69	65V
Tb^{3+}	gl	25	0·2 $NaClO_4$	K_1 2·65, K_2 2·17, K_3 1·07, K_4 0·44	64D
	gl	20	0·1 ($NaClO_4$)	K_1 2·90, β_2 5·20, β_3 6·35	64P
Tm^{3+}	gl	25	0·2 $NaClO_4$	K_1 2·92, K_2 2·34, K_3 1·34, K_4 0·65	64D
	gl	20	0·1 ($NaClO_4$)	K_1 3·190, β_2 5·71, β_3 7·43	64P
UO_2^{2+}	gl	31	0·12 ($NaClO_4$)	K_1 3·36±0·05, K_2 2·2±0·3, K_3 2±1	62C
	dis	20	1 $NaClO_4$	K_1 2·81±0·06, β_2 4·56±0·10, β_3 5·46±0·04	62S
	sp	20	1 ($NaClO_4$)	K_1 2·43±0·02, K_2 2·06±0·06, K_3 1·86±0·08	67M
	gl	25	1 ($NaClO_4$)	K_1 2·77, K_2 1·68, K_3 1·33	67T
	gl	25	0·1 KNO_3	K_1 2·48±0·02 $K[UO_2(OH)L+H^+ \rightleftharpoons UO_2L^+]$ 4·0±0·2 $K\{2UO_2L \rightleftharpoons [UO_2(OH)L]_2+2H^+\}$ −5·13	67V
VO^{2+}	sp, oth	20	1 $NaClO_4$	K_1 2·68, β_2 4·83, circular dichroism	65J
Y^{3+}	ix	20	0·2 ($NaClO_4$)	K_1 2·83±0·03, β_2 4·92±0·06, β_3 6·8±0·1	60S
	gl	25	0·2 $NaClO_4$	K_1 2·80, K_2 2·14, K_3 1·22, K_4 0·55	64D
	gl	20	0·1 ($NaClO_4$)	K_1 3·017, β_2 5·33, β_3 6·95	64P
Yb^{3+}, Yb^{2+}	pol	?	0·1 $(CH_3)_4NI$	$K(YbL_6^{3-}+e^- \rightleftharpoons YbL_4^{2-}+2L^-)$ −7	58K
Yb^{3+}	gl	25	0·2 $NaClO_4$	K_1 3·03, K_2 2·42, K_3 1·34, K_4 0·69	64D
	gl	20	0·1 ($NaClO_4$)	K_1 3·230, β_2 5·82, β_3 7·58	64P
Zn^{2+}	qh	25	1 ($NaClO_4$)	K_1 1·61, K_2 1·24, K_3 ~ 0·3	67T
Zr^{4+}	ix	?	2 $HClO_4$	$K(Zr^{4+}+HL \rightleftharpoons ZrL^{3+}+H^+)$ 1·98	60R
	ix	?	2·0 $HClO_4$	$K(Zr^{4+}+HL \rightleftharpoons ZrL^{3+}+H^+)$ 2·28±0·03, $K(Zr^{4+}+2HL \rightleftharpoons ZrL_2^{2+}+2H^+)$ 2·5±0·2	64R

57V O. Vartapetian, *Ann. Chim.*, 1957, **2,** 916
58K N. A. Kostromina and S. I. Yakubson, *J. Inorg. Chem. (U.S.S.R.)*, 1958, **3,** No. 11, 104 (2506)
60R D. I. Ryabchikov, A. N. Ermakov, V. K. Belyaeva, and I. N. Marov, *Russ. J. Inorg. Chem.*, 1960, **5,** 505 (1051)
60S V. T. Spitsyn and O. Voitekh, *Proc. Acad. Sci. (U.S.S.R.)*, 1960, **133,** 859 (613)
62C C. A. Crutchfield jun, W. M. McNabb, and J. F. Hazel, *J. Inorg. Nuclear Chem.*, 1962, **24,** 291
62G I. Geletseanu and A. V. Lapitskii, *Proc. Acad. Sci. (U.S.S.R.)*, (Chem. Sect.) 1962, **147,** 983 (372)
62S J. Stary and V. Balek, *Coll. Czech. Chem. Comm.*, 1962, **27,** 809
64D H. Deelstra and F. Verbeck, *Analyt. Chim. Acta*, 1964, **31,** 251
64P J. E. Powell, R. H. Karraker, R. S. Kolat, and J. L. Farrell, 'Rare Earth Research II', ed. K. S. Vorres, Gordon and Breach, New York, 1964, p. 513
64R D. I. Ryabchikov, I. N. Marov, A. N. Ermakov, and V. K. Belyaeva, *J. Inorg. Nuclear Chem.*, 1964, **26,** 965
65B T. A. Belyavskaya and I. F. Kolosova, *Russ. J. Inorg. Chem.*, 1965, **10,** 236
65Ba A. Basinski and Z. Warnke, *Trans. Faraday Soc.*, 1965, **61,** 129
65Bb A. Basinski and Z. Warnke, *Roczniki Chem.*, 1965, **39,** 1776
65D V. P. Devendran and M. Santappa, *Current Sci.*, 1965, **34,** 145
65J K. M. Jones and E. Larsen, *Acta Chem. Scand.*, 1965, **19,** 1205
65V F. Verbeck and H. Thun, *Analyt. Chim. Acta*, 1965, **33,** 378
66G M. A. Gouveia and R. Guedes De Carvalho, *J. Inorg. Nuclear Chem.*, 1966, **28,** 1683
67M C. Miyake and H. W. Nurnberg, *J. Inorg. Nuclear Chem.*, 1967, **29,** 2411
67T H. Thun, W. Guns, and F. Verbeck, *Analyt. Chim. Acta*, 1967, **37,** 332
67V S. K. Verma and R. P. Agarwal, *J. Less-Common Metals*, 1967, **12,** 221
68B A. Bonniol and P. Vieles, *J. Chim. phys.*, 1968, **65,** 414

$C_3H_6O_3$ Methoxyacetic acid HL 150

Metal	Method	Temp	Medium	Log of equilibrium constant, remarks	Ref
H^+	qh	20	1 $NaClO_4$	K_1 3·37±0·01	61S
	gl	20	0·1 ($NaClO_4$)	K_1 3·31	64P
Ce^{3+}	gl	20	0·1 ($NaClO_4$)	K_1 2·06, K_2 1·00	64P
Cu^{2+}	Cu(Hg)	20	1·00 $NaClO_4$	K_1 1·82±0·02, β_2 2·81±0·04, β_3 3·1±0·1, β_4 2·8±0·3	61S
Dy^{3+}	gl	20	0·1 ($NaClO_4$)	K_1 2·05, K_2 1·08	64P
Er^{3+}	gl	20	0·1 ($NaClO_4$)	K_1 2·08, K_2 1·15	64P
Eu^{3+}	gl	20	0·1 ($NaClO_4$)	K_1 2·12, K_2 1·30	64P
Gd^{3+}	gl	20	0·1 ($NaClO_4$)	K_1 2·06, K_2 0·95	64P
Ho^{3+}	gl	20	0·1 ($NaClO_4$)	K_1 2·07, K_2 1·15	64P
La^{3+}	gl	20	0·1 ($NaClO_4$)	K_1 2·03, K_2 0·85	64P
Lu^{3+}	gl	20	0·1 ($NaClO_4$)	K_1 2·09, K_2 1·28	64P
Nd^{3+}	gl	20	0·1 ($NaClO_4$)	K_1 2·11, K_2 1·23	64P
Pr^{3+}	gl	20	0·1 ($NaClO_4$)	K_1 2·07, K_2 1·18	64P
Sm^{3+}	gl	20	0·1 ($NaClO_4$)	K_1 2·13, K_2 1·26	64P
Tb^{3+}	gl	20	0·1 ($NaClO_4$)	K_1 2·05, K_2 0·95	64P
Tm^{3+}	gl	20	0·1 ($NaClO_4$)	K_1 2·08, K_2 1·15	64P
Y^{3+}	gl	20	0·1 ($NaClO_4$)	K_1 2·00, K_2 1·11	64P
Yb^{3+}	gl	20	0·1 ($NaClO_4$)	K_1 2·08, K_2 1·28	64P

61S A. Sandell, *Acta Chem. Scand.*, 1961, **15**, 190
64P J. E. Powell, R. S. Kolat, and G. S. Paul, *Inorg. Chem.*, 1964, 3, 518

C_3H_7N 3-Aminopropene (allylamine) L 152

Metal	Method	Temp	Medium	Log of equilibrium constant, remarks	Ref
Ag^+	Ag	25	2 ($NaClO_4$)	K_1 0·114±0·003	67H
Pt^{2+}	sp	30—59	2 (NaCl)	$K(PtCl_4^{2-}+HL^+ \rightleftharpoons PtCl_3HL+Cl^-)$ (30·2°) 3·45, (44·0°)3·24, (59°) 3·01	67D
				ΔH −7·1±0·2, ΔS −7·6±0·7	
	sp	24·5	2 (KBr)	$K(PtBr_4^{2-}+HL^+ \rightleftharpoons PtBr_3HL+Br^-)$ 2·49	67Da

67D R. G. Denning, F. R. Hartley, and L. M. Venanzi, *J. Chem. Soc. (A)*, 1967, 324
67Da R. G. Denning and L. M. Venanzi, *J. Chem. Soc. (A)*, 1967, 336
67H F. R. Hartley and L. M. Venanzi, *J. Chem. Soc. (A)*, 1967, 333

C_3H_8O Propan-2-ol (isopropyl alcohol) L 152.1

Metal	Method	Temp	Medium	Log of equilibrium constant, remarks	Ref
Al^{3+}				See ethylenediamine (120) for mixed ligands	
Ti^{4+}				See ethylenediamine (120) for mixed ligands	

152.2 $C_3H_8O_2$ Propan-1,2-diol (propylene glycol) L

Metal	Method	Temp	Medium	Log of equilibrium constant, remarks	Ref
As^{III}	gl	25	0·1	$K[As(OH)_4^- + L \rightleftharpoons As(OH)_2(H_{-2}L)^-]$ −1·00	57R
B^{III}	gl	25	0·1	$K[B(OH)_4^- + L \rightleftharpoons B(OH)_2(H_{-2}L)^-]$ 0·49	57R
				$K[B(OH)_4^- + 2L \rightleftharpoons B(H_{-2}L)_2^-]$ 0·21	
	gl	0—35	0·025 borax	$K[B(OH)_4^- + L \rightleftharpoons B(OH)_2(H_{-2}L)^-]$ (0°) 0·80 (13°) 0·64, (25°) 0·61, (35°) 0·53	67C
				ΔH° −3·0±0·3, ΔS° −7±2	
				$K[B(OH)_4^- + 2L \rightleftharpoons B(H_{-2}L)_2^-]$ (0°) 0·92, (13°) 0·78, (25°) 0·59, (35°) 0·37	
				ΔH° −7·2±0·7, ΔS° −22±5	
Ge^{IV}	qh	25	0·1 KCl	$K[HGeO_3^- + L \rightleftharpoons HGeO_2(H_{-2}L)^-]$ 0·28	59A
				$K[HGeO_3^- + 2L \rightleftharpoons HGeO(H_{-2}L)_2^-]$ 0·06	
Te^{VI}	gl	25	0·1	$K[H_5TeO_6^- + L \rightleftharpoons H_3TeO_4(H_{-2}L)^-]$ 1·47	57R

57R G. L. Roy, A. L. Laferriere, and J. O. Edwards, *J. Inorg. Nuclear Chem.*, 1957, **4**, 106
59A P. J. Antikainen, *Acta Chem. Scand.*, 1959, **13**, 312
67C J. M. Connor and V. C. Bulgrin, *J. Inorg. Nuclear Chem.*, 1967, **29**, 1953

152.3 $C_3H_8O_2$ Propan-1,3-diol L

Metal	Method	Temp	Medium	Log of equilibrium constant, remarks	Ref
B^{III}	gl	0—35	0·025 borax	$K[B(OH)_4^- + L \rightleftharpoons B(OH)_2(H_{-2}L)^-]$ (0°) 0·45, (13°) 0·25, (25°) 0·10, (35°) −0·02	67C
				ΔH° −4·6±0·5, ΔS° −15±7	
				$K[B(OH)_4^- + 2L \rightleftharpoons B(H_{-2}L)_2^-]$ (0°) −0·70, (13°) −0·92, (25°) −0·96, (35°) −1·25	
				ΔH° −6·2±0·6, ΔS° −16±8	

67C J. M. Conner and V. C. Bulgrin, *J. Inorg. Nuclear Chem.*, 1967, **29**, 1953

153 $C_3H_8O_3$ Propan-1,2,3-triol (glycerol) L

Metal	Method	Temp	Medium	Log of equilibrium constant, remarks	Ref
As^{III}	gl	25	0·1	$K[As(OH)_4^- + L \rightleftharpoons As(OH)_2(H_{-2}L)^-]$ 0·06	57R
B^{III}	qh	25	0·1 KCl	$K[B(OH)_4^- + L \rightleftharpoons B(OH)_2(H_{-2}L)^-]$ 1·56	56A
				$K[B(OH)_4^- + 2L \rightleftharpoons B(H_{-2}L)_2^-]$ 1·91	
	gl	25	0·1	$K[B(OH)_4^- + L \rightleftharpoons B(OH)_2(H_{-2}L)^-]$ 1·21	57R
				$K[B(OH)_4^- + 2L \rightleftharpoons B(H_{-2}L)_2^-]$ 1·62	
	gl	0—35	0·025 borax	$K[B(OH)_4^- + L \rightleftharpoons B(OH)_2(H_{-2}L)^-]$ (0°) 1·36, (13°) 1·24, (25°) 1·15, (35°) 1·10	67C
				ΔH° −2·9±0·3, ΔS° −4±1	
				$K[B(OH)_4^- + 2L \rightleftharpoons B(H_{-2}L)_2^-]$ (0°) 1·99, (13°) 1·84, (25°) 1·76, (35°) 1·62	
				ΔH° −3·9±0·4, ΔS° −5±1	
	gl	25	*I* KCl	$K[B(OH)_4^- + 2L \rightleftharpoons B(H_{-2}L)_2^-]$ 1·584+0·730 $\sqrt{I}$	67N

conti

$C_3H_8O_3$ (contd) 153

Metal	Method	Temp	Medium	Log of equilibrium constant, remarks	Ref
Cr^{VI}	sp	25—35	0·1	$K[HCrO_4^- + L \rightleftharpoons HCrO_3(H_{-2}L)^-]$ (25°) 14·0, (35°) 11·6	67R
Ge^{IV}	gl	25	0·1 KCl	$K[HGeO_3^- + L \rightleftharpoons HGeO_2(H_{-2}L)^-]$ 1·21	57A
				$K[HGeO_3^- + 2L \rightleftharpoons HGeO(H_{-2}L)_2^-]$ 1·94	
	gl	25	*I* KCl	$K[HGeO_3^- + 2L \rightleftharpoons HGeO(H_{-2}L)_2^-]$ $1{\cdot}105 - 1{\cdot}700\sqrt{I}$	63N
Pb^{2+}	Hg(Pb)	25	1 $NaClO_4$	K_1 1·15 ± 0·15	67V
Te^{VI}	qh	25	0·1 KCl	$K[H_5TeO_6^- + L \rightleftharpoons H_3TeO_4(H_{-2}L)^-]$ 1·77	56A
	gl	25	0·1	$K[H_5TeO_6^- + L \rightleftharpoons H_3TeO_4(H_{-2}L)^-]$ 1·86	57R

56A P. J. Antikainen, *Suomen Kem.*, 1956, **B29**, 135, 179
57A P. J. Antikainen, *Suomen Kem.*, 1957, **B30**, 147
57R R. L. Roy, A. L. Laferriere, and J. O. Edwards, *J. Inorg. Nuclear Chem.*, 1957, **4**, 106
63N J. A. Nazarenko and C. V. Flgantikova, *Russ. J. Inorg. Chem.*, 1963, **8**, 1189 (2271)
67C J. M. Conner and V. C. Bulgrin, *J. Inorg. Nuclear Chem.*, 1967, **29**, 1953
67N V. A. Nazarenko and L. D. Ermak, *Russ. J. Inorg. Chem.*, 1967, **12**, 335 (643), 1079 (2051)
67R K. K. Rohatgi and P. K. Bhattacharya, *Indian J. Chem.*, 1967, **5**, 195
67V M. Vicedomini and A. Liberti, *Gazzetta*, 1967, **97**, 1627

C_3H_9N 1-Propylamine L 154

Metal	Method	Temp	Medium	Log of equilibrium constant, remarks	Ref
H^+	gl	20	0·5 KNO_3	K_1 10·92 ± 0·01	68A
	H	20	0 corr	K_1 10·74 ± 0·03	68A
Ag^+	qh	20	0·500 KNO_3	K_1 3·47 ± 0·03, β_2 7·54 ± 0·01	68A
	Ag-AgCl	20	0·500 KNO_3	β_2 7·48 ± 0·02	68A

68A D. J. Alner, R. C. Lansbury, and A. G. Smeeth, *J. Chem. Soc. (A)*, 1968, 417

C_3H_9N 1-Methylethylamine (isopropylamine) L 154.1

Metal	Method	Temp	Medium	Log of equilibrium constant, remarks	Ref
Ag^+	oth	?	0·1 $NaClO_4$ acetone	β_2 10·5 ± 0·1, coulometric titration	65M

65M K. K. Mead, D. L. Maricle, and C. A. Streuli, *Analyt. Chem.*, 1965, 37, 237

$C_3H_9N_5$ Methylbiguanide L 156

Metal	Method	Temp	Medium	Log of equilibrium constant, remarks	Ref
H^+	gl	32	0·01	K_1 11·44, K_2 3·00	59R
Cu^{2+}	sp	28—32	0·25 KCl	β_2 17·12	59R

59R M. M. Ray and P. Ray, *J. Indian Chem. Soc.*, 1959, **36**, 849

157 $C_3H_{10}N_2$ *N*-Methyl-1,2-diaminoethane (*N*-methylethylenediamine) L

Metal	Method	Temp	Medium	Log of equilibrium constant, remarks	Ref
H^+	gl	0—25	0·50 KNO_3	ΔH_1 (0°) −6·3, ΔH_2 −6·6	52B
Cu^{2+}	gl	0—25	0·5 KNO_3	ΔH_1 (0°) −8·5, ΔS_1 20, ΔH_2 (0°) −7·0, ΔS_2 16	52B
Ni^{2+}	gl	0—25	0·5 KNO_3	ΔH_1 (0°) −8·8, ΔS_1 4·0, ΔH_2 (0°) −6·1, ΔS_2 5·9, ΔH_3 (0°) −6·0, ΔS_3 −11·0	52B

52B F. Basolo and R. K. Murmann, *J. Amer. Chem. Soc.*, 1952, **74**, 2373, 5243

158 $C_3H_{10}N_2$ 1,2-Diaminopropane L

Metal	Method	Temp	Medium	Log of equilibrium constant, remarks	Ref
H^+	gl	25	I ($NaClO_4$)	K_1 9·82−(0·509√I)/(1+2·2√I)+(0·509√I)/(1+√I)+0·214 I, K_2 6·61+(1·018√I)/(1+1·42√I) +0·180 I	61N, 6 65N, 67
	gl	10—55	→ 0	K_1 4193·2/T−9·673+0·01823 T, K_2 3492·3/T−8·692+0·01204T	62Nb
				$\Delta H_1°$ (25°) −11·8±0·2, $\Delta S_1°$ 5·4±0·7, $\Delta H_2°$ (25°) −11·1±0·1, $\Delta S_2°$ −6·9±0·3	
Co^{2+}				See EDTA (836) for mixed ligands	
Cu^{2+}	gl	25	I ($NaClO_4$)	K_1 10·56+0·89 I−0·52 $I^{3/2}$+0·13I^2, β_2 19·64+2·11 I−1·24 $I^{3/2}$+0·28 I^2	62Nc
	gl	25	I ($NaClO_4$)	K_2 9·08+1·22 I−0·72 $I^{3/2}$+0·15 I^2	65N
	gl	10—55	→ 0	K_1 4519·8/T−10·181+0·01878T	62Nb
				K_2 4238·8/T−9·697+0·01530T	
				$\Delta H_1°$ (25°) −13·0±0·3, $\Delta S_1°$ 5±1	
				$\Delta H_2°$ (25°) −13·2±0·2, $\Delta S_2°$ −2·7±0·7	
	gl	25	→ 0	$K(Cu^{2+}+L+A \rightleftharpoons CuLA^{2+})$	67N
				19·75 where A = ethylenediamine	
				17·98 where A = N,N′-diethylethylenediamine	
				18·72 where A = 1,3-propanediamine	
				$K(CuL_2^{2+}+CuA_2^{2+} \rightleftharpoons 2CuLA^{2+})$	
				0·31 where A = ethylenediamine	
				1·92 where A = N,N′-diethylethylenediamine	
				1·15 where A = 1,3-propanediamine	
				See 5-sulphosalicylic acid (552) and EDTA (836) for mixed ligands	
Hg^{2+}	pol	10—40	0·1 KNO_3	β_2 (10°) 24·75±0·02, (25°) 23·51±0·02, (40°) 22·23±0·02, β_3 (10°) 24·57±0·06, (25°) 23·30±0·02, (40°) 21·85±0·07	61R
				$\Delta H_{\beta 2}°$ −33·8±0·8, $\Delta S_{\beta 2}°$ −6±2	
Ni^{2+}	gl	25	I ($NaClO_4$)	K_1 7·29±1·32I−1·72 $I^{3/2}$+0·69I^2, β_2 13·43+1·99I −2·18$I^{3/2}$+0·84I^2, β_3 17·61+2·99 I−3·19$I^{3/2}$+1·21 I^2	62Na
				See EDTA (836) for mixed ligands	
Zn^{2+}	gl	25	I ($NaClO_4$)	K_1 5·64+1·31 I−1·11 $I^{3/2}$+0·37 I^2	61N
				β_2 10·58+2·44 I−2·18 $I^{3/2}$+0·70 I^2	
			I (KCl)	K_1 5·69+0·458 I−0·245 $I^{3/2}$−0·0286 I^2	62N
			I ($NaClO_4$)	K_1 5·69+0·908 I−0·851 $I^{3/2}$+0·285 I^2	
				K_2 4·94+1·31 I−1·07 $I^{3/2}$+0·325 I^2	

Erratum. 1964, p. 393, Table 158, under Co^{3+}. Insert parenthesis before *dextro*-.

references on facing p

61N R. Nasanen and P. Merilainen, *Suomen Kem.*, 1961, **B34**, 127
61R D. K. Roe, D. B. Masson, and C. J. Nyman, *Analyt. Chem.*, 1961, 33, 1464
62N R. Nasanen, P. Merilainen, and O. Butkewitsch, *Suomen Kem.*, 1962, **B35**, 219
62Na R. Nasanen, R. Merilainen, and E. Heinanen, *Suomen Kem.*, 1962, **B35**, 15
62Nb R. Nasanen, R. Merilainen, and M. Koskinen, *Suomen Kem.*, 1962, **B35**, 59
62Nc R. Nasanen, P. Merilainen, and S. Lukkari, *Acta Chem. Scand.*, 1962, **17**, 2384
65N R. Nasanen, M. Koskinen, R. Salonen, and A. Kiiski, *Suomen Kem.*, 1965, **B38**, 81
67N R. Nasanen and M. Koskinen, *Suomen Kem.*, 1967, **B40**, 108, 23

$C_3H_{10}N_2$ 1,3-Diaminopropane L 159

Metal	*Method*	*Temp*	*Medium*	*Log of equilibrium constant, remarks*	*Ref*
H^+	gl	25	I ($NaClO_4$)	K_1 10·51+0·301I	65N, 65Na, 67N
				K_2 8·48+1·018$\sqrt{I}/(1+1·338\sqrt{I})$+0·247 I	
Cd^{2+}	gl	0—49	0·15	$\Delta H_1°$ (25°) −5·3±0·5, $\Delta S_1°$ 3±2, $\Delta H_2°$ (25°) −4·4±0·5, $\Delta S_2°$ −1±2	55C
Cu^{2+}	gl	0—49	0·15	$\Delta H_1°$ (25°) −12·5±0·5, $\Delta S_1°$ 2±2, $\Delta H_2°$ (25°) −12·3±0·5, $\Delta S_2°$ −9±2	55C
	cal	25	1 KNO_3	$\Delta H_{\beta 2}$ −22·8, $\Delta S_{\beta 2}$ 2·0	55P
	gl	25	I ($NaClO_4$)	K_1 9·63+0·402 I+0·076 $I^{3/2}$−0·085 I^2, K_2 7·02+0·530 I−0·018 $I^{3/2}$−0·068 I^2	65Na
			→ 0	$K(CuOHL^+ + H^+ \rightleftharpoons CuL^{2+})$ 7·42	
			→ 0	$K[Cu_2(OH)_2L_2^{2+} + 2H^+ \rightleftharpoons 2CuL^{2+}]$ 12·67	
			→ 0	$K[2CuOHL^+ \rightleftharpoons Cu_2(OH)_2L_2^{2+}]$ 2·17	
			→ 0	$K[Cu(OH)_2L + H^+ \rightleftharpoons CuOHL^+]$ 11·70	
	gl	25	→ 0	$K(Cu^{2+} + L + A \rightleftharpoons CuLA^{2+})$, 18·58 where A = ethylenediamine; 16·26 where A = *NN′*-diethylethylenediamine; 18·72 where A = 1,2-propanediamine	67N
				$K(CuL_2^{2+} + CuA_2^{2+} \rightleftharpoons 2CuLA^{2+})$, 0·98 where A = ethylenediamine; 1·49 where A = N,N′-diethylethylenediamine; 1·15 where A = 1,2-propanediamine	
Ni^{2+}	gl	0—49	0·15	$\Delta H_1°$ (25°) −8·8±0·3, $\Delta S_1°$ 0±1, $\Delta H_2°$ (25°) −8·2±0·5, $\Delta S_2°$ −7±2, $\Delta H_3°$ (25°) −6±2, $\Delta S_2°$ −13±6	55C
	cal	25	1 KNO_3	ΔH_1 −7·77, ΔS_1 3·0, $\Delta H_{\beta 2}$ −15·0±0·1, $\Delta S_{\beta 2}$ −1·1, $\Delta H_{\beta 3}$ −21·3±0·4, $\Delta S_{\beta 3}$ −16·6	55P
	gl	25	I ($NaClO_4$)	K_1 6·28+0·272 I, K_2 4·20+0·321 I	65N

Errata. 1964, p. 394, Table 159. Under Cd^{2+} *for* (4·91°) *read* (49·1°)

55C F. A. Cotton and F. E. Harris, *J. Phys. Chem.*, 1955, **59**, 1203
55P I. Poulsen and J. Bjerrum, *Acta Chem. Scand.*, 1955, **9**, 1407
65N R. Nasanen, M. Koskinen, and K. Kajander, *Suomen Kem.*, 1965, **B38**, 103
65Na R. Nasanen, M. Koskinen, R. Salonen, and A. Kiiski, *Suomen Kem.*, 1965, **B38**, 81
67N R. Nasanen and M. Koskinen, *Suomen Kem.*, 1967, **B40**, 108, 23

160.1 $C_3HO_2F_5$ Pentafluoropropanoic acid HL

Metal	Method	Temp	Medium	Log of equilibrium constant, remarks	Ref
Pb^{2+}	pol	25	2 ($NaClO_4$)	K_1 −0·03, β_2 −0·01	64C
	Pb	25	2 ($NaClO_4$)	K_1 −0·21, β_2 −0·34	

64C S. A. Carrano, K. A. Chem, and R. F. O'Malley, *Inorg. Chem.*, 1964, **3**, 1947

160.2 C_3H_2NCl 1-Cyano-2-chloropropene (2-chloroacrylonitrile) L

Metal	Method	Temp	Medium	Log of equilibrium constant, remarks	Ref
Ag^{+}	oth	61	4 $LiNO_3$	K_1 −0·55, gas chromatography	68S

68S H. Schnecko, *Analyt. Chem.*, 1968, **40**, 1391

164.1 $C_3H_4O_2N_2$ 1,3-Diazolidin-2,4-dione (hydantoin) HL

Metal	Method	Temp	Medium	Log of equilibrium constant, remarks	Ref
H^{+}	gl	25	0·1 (KNO_3)	K_1 8·93	65C
Ag^{+}	gl, Ag	25	0·1 (KNO_3)	K_1 4·29, β_2 9·20	65C

65C C. E. Campi, G. Ostacoli, and A. Vanni, *Gazzetta*, 1965, **95**, 796

167.1 $C_3H_5O_6P$ 2-Phosphonopropanoic acid (*O*-phospho-enolpyruvic acid) H_3L

Metal	Method	Temp	Medium	Log of equilibrium constant, remarks	Ref
H^{+}	gl	25	0·1—0·4 [$(C_3H_7)_4NI$]	K_1 (0·1) 6·35, (0·4) 6·4, K_2 (0·1) 3·4, (0·4) 3·5	57W
Cd^{2+}	gl	25	0·1—0·4 $(C_3H_7)_4NI$	K_1 2·96	57W
Co^{2+}	gl	25	0·1—0·4 $(C_3H_7)_4NI$	K_1 2·54	57W
K^{+}	gl	25	0·1—0·4 $(C_3H_7)_4NI$	K_1 1·08	57W
Mg^{2+}	gl	25	0·1—0·4 $(C_3H_7)_4NI$	K_1 2·26	57W
Mn^{2+}	gl	25	0·1—0·4 $(C_3H_7)_4NI$	K_1 2·75	57W
Ni^{2+}	gl	25	0·1—0·4 $(C_3H_7)_4NI$	K_1 2·34	57W
Zn^{2+}	gl	25	0·1—0·4 $(C_3H_7)_4NI$	K_1 2·96	57W

57W F Wold and C. E. Ballou, *J. Biol. Chem.*, 1957, **227**, 301

C_3H_6OS Thiopropanoic acid HL 167.2

Metal	Method	Temp	Medium	Log of equilibrium constant, remarks	Ref
Pb^{2+}	gl	30	0·007 ClO_4^-	K_1 6·74	67M

67M M. B. Mishra, S. C. Sinha, and H. L. Nigam, *Indian J. Chem.*, 1967, **5**, 649

$C_3H_6OS_2$ (Ethoxy)dithiomethanoic acid (ethylxanthic acid) HL 167.3

Metal	Method	Temp	Medium	Log of equilibrium constant, remarks	Ref
Cd^{2+}	pol	25	1 (KNO_3)	β_4 11·05	67K
			2 (KNO_3) 50% DMF	β_3 14·40	
Zn^{2+}	pol	25	1 (LiCl)	β_3 4·63	67K

67K H. Kodama and K. Hayashi, *J. Electroanalyt. Chem.*, 1967, **14**, 209

$C_3H_6O_2S$ β-Mercaptopropionic acid H_2L 168

Erratum. 1964, p. 397, Table 168. This table should be numbered **170.2** to place it in the correct ligand sequence.

$C_3H_6O_2N_2$ D-4-Amino-1,2-oxazolidin-3-one (D-cycloserine) L 169

Metal	Method	Temp	Medium	Log of equilibrium constant, remarks	Ref
H^+	Hg	25	0·4 KNO_3	K_1 4·57, $K[(H_{-1}L)^- + H^+ \rightleftharpoons L]$ 7·40	63L
Hg^{2+}	Hg	25	0·4 KNO_3	β_2 14·7 $K[Hg(H_{-1}L)_2 + H^+ \rightleftharpoons Hg(H_{-1}L)L^+]$ 5·93 $K[Hg(H_{-1}L)L^+ + H^+ \rightleftharpoons HgL_2^{2+}]$ 5·44	63L

63L N. G. Lordi, *J. Pharm. Sci.*, 1963, **52**, 397

$C_3H_6O_2S$ 2-Mercaptopropanoic acid H_2L 170.1

Metal	Method	Temp	Medium	Log of equilibrium constant, remarks	Ref
Er^{3+}	gl	31	2 ($NaClO_4$)	$K(Er^{3+} + HL^- \rightleftharpoons ErHL^{2+})$ 1·38 ± 0·05, $K[ErHL^{2+} + HL^- \rightleftharpoons Er(HL)_2^+]$ 1·0 ± 0·1	63B
Eu^{3+}	gl	31	2 ($NaClO_4$)	$K(Eu^{3+} + HL^- \rightleftharpoons EuHL^{2+})$ 1·81 ± 0·05, $K[EuHL^{2+} + HL^- \rightleftharpoons Eu(HL)_2^+]$ 0·8 ± 0·1	63B

continued overleaf

170.1 $C_3H_6O_2S$ (contd)

Metal	*Method*	*Temp*	*Medium*	*Log of equilibrium constant, remarks*	*Ref*
Gd^{3+}	gl	31	2 ($NaClO_4$)	$K(Gd^{3+}+HL^- \rightleftharpoons GdHL^{2+})$ 1·62±0·05, $K[GdHL^{2+}+HL^- \rightleftharpoons Gd(HL)_2{}^+]$ 0·8±0·1	63B
Ho^{3+}	gl	31	2 ($NaClO_4$)	$K(Ho^{3+}+HL^- \rightleftharpoons HoHL^{2+})$ 1·45±0·05, $K[HoHL^{2+}+HL^- \rightleftharpoons Ho(HL)_2{}^+]$ 1·0±0·2	63B
La^{3+}	gl	31	2 ($NaClO_4$)	$K(La^{3+}+HL^- \rightleftharpoons LaHL^{2+})$ 1·49±0·05, $K[LaHL^{2+}+HL^- \rightleftharpoons La(HL)_2{}^+]$ 0·7±0·1	63B
Nd^{3+}	gl	31	2 ($NaClO_4$)	$K(Nd^{3+}+HL^- \rightleftharpoons NdHL^{2+})$ 1·56±0·05, $K[NdHL^{2+}+HL^- \rightleftharpoons Nd(HL)_2{}^+]$ 0·8±0·1	63B
Sm^{3+}	gl	31	2 ($NaClO_4$)	$K(Sm^{3+}+HL^- \rightleftharpoons SmHL^{2+})$ 1·83±0·05, $K[SmHL^{2+}+HL^- \rightleftharpoons Sm(HL)_2{}^+]$ 0·9±0·1	63B
Tb^{3+}	gl	31	2 ($NaClO_4$)	$K(Tb^{3+}+HL^- \rightleftharpoons TbHL^{2+})$ 1·60±0·05, $K(TbHL^{2+}+HL^- \rightleftharpoons Tb(HL)_2{}^+]$ 0·9±0·1	63B
Y^{3+}	gl	31	2 ($NaClO_4$)	$K(Y^{3+}+HL^- \rightleftharpoons YHL^{2+})$ 1·38±0·05, $K[YHL^{2+}+HL^- \rightleftharpoons Y(HL)_2{}^+]$ 0·7±0·1	63B
Yb^{3+}	gl	31	2 ($NaClO_4$)	$K(Yb^{3+}+HL^- \rightleftharpoons YbHL^{2+})$ 1·43±0·05, $K[YbHL^{2+}+HL^- \rightleftharpoons Yb(HL)_2{}^+]$ 1·0±0·1	63B

63B J. L. Bear, G. R. Choppin, and J. V. Quagliano, *J. Inorg. Nuclear Chem.*, 1963, **25**, 513

170.2 $C_3H_6O_2S$ 3-Mercaptopropanoic acid H_2L

Metal	*Method*	*Temp*	*Medium*	*Log of equilibrium constant, remarks*	*Ref*
H^+	gl	30	0·1 (KCl)	K_2 4·27	62C
	?	?	?	K_1 10·54, K_2 8·35	63S
Ce^{3+}	gl	30	0·1 (KCl)	$K(Ce^{3+}+HL^- \rightleftharpoons CeHL^{2+})$ 2·41 $K[CeHL^{2+}+HL^- \rightleftharpoons Ce(HL)_2{}^+]$ 2·48	62C
Er^{3+}	gl	31	2 ($NaClO_4$)	$K(Er^{3+}+HL^- \rightleftharpoons ErHL^{2+})$ 1·70±0·05, $K[ErHL^{2+}+HL^- \rightleftharpoons Er(HL)_2{}^+]$ 1·4±0·1	63B
Eu^{3+}	gl	31	2 ($NaClO_4$)	$K(Eu^{3+}+HL^- \rightleftharpoons EuHL^{2+})$ 2·15±0·05, $K[EuHL^{2+}+HL^- \rightleftharpoons Eu(HL)_2{}^+]$ 1·3±0·1	63B
Gd^{3+}	gl	31	2 ($NaClO_4$)	$K(Gd^{3+}+HL^- \rightleftharpoons GdHL^{2+})$ 2·03±0·05, $K[GdHL^{2+}+HL^- \rightleftharpoons Gd(HL)_2{}^+]$ 1·3±0·1	63B
Ho^{3+}	gl	31	2 ($NaClO_4$)	$K(Ho^{3+}+HL^- \rightleftharpoons HoHL^{2+})$ 1·68±0·05, $K[HoHL^{2+}+HL^- \rightleftharpoons Ho(HL)_2{}^+]$ 1·3±0·1	63B
La^{3+}	gl	30	0·1 (KCl)	$K(La^{3+}+HL^- \rightleftharpoons LaHL^{2+})$ 2·30 $K[LaHL^{2+}+HL^- \rightleftharpoons La(HL)_2{}^+]$ 2·40	62C
	gl	31	2 ($NaClO_4$)	$K(La^{3+}+HL^- \rightleftharpoons LaHL^{2+})$ 1·82±0·05, $K[LaHL^{2+}+HL^- \rightleftharpoons La(HL)_2{}^+]$ 1·3±0·2	63B
Mo^{VI}	spJ	25	?	$K(MoO_4{}^{2-}+3HL^- \rightleftharpoons MoOL_3{}^{2-}+3OH^-)$ (?), 23, acetate buffer	63S
Nd^{3+}	gl	30	0·1 (KCl)	$K(Nd^{3+}+HL^- \rightleftharpoons NdHL^{2+})$ 2·58 $K[NdHL^{2+}+HL^- \rightleftharpoons Nd(HL)_2{}^+]$ 2·49	62C
	gl	31	2 ($NaClO_4$)	$K(Nd^{3+}+HL^- \rightleftharpoons NdHL^{2+})$ 1·94±0·05, $K[NdHL^{2+}+HL^- \rightleftharpoons Nd(HL)_2{}^+]$ 1·3±0·1	63B
Pr^{3+}	gl	30	0·1 (KCl)	$K(Pr^{3+}+HL^- \rightleftharpoons PrHL^{2+})$ 2·56 $K[PrHL^{2+}+HL^- \rightleftharpoons Pr(HL)_2{}^+]$ 2·49	62C
Sm^{3+}	gl	31	2 ($NaClO_4$)	$K(Sm^{3+}+HL^- \rightleftharpoons SmHL^{2+})$ 2·19±0·05, $K[SmHL^{2+}+HL^- \rightleftharpoons Sm(HL)_2{}^+]$ 1·3±0·1	63B

contin.

$C_3H_6O_2S$ (contd) 170.2

Metal	Method	Temp	Medium	Log of equilibrium constant, remarks	Ref
Tb^{3+}	gl	31	2 ($NaClO_4$)	$K(Tb^{3+}+HL^- \rightleftharpoons TbHL^{2+})$ 1·79±0·05, $K[TbHL^{2+}+HL^- \rightleftharpoons Tb(HL)_2{}^+]$ 1·3±0·1	63B
Yb^{3+}	gl	31	2 ($NaClO_4$)	$K(Yb^{3+}+HL^- \rightleftharpoons YbHL^{2+})$ 1·75±0·05, $K[YbHL^{2+}+HL^- \rightleftharpoons Yb(HL)_2{}^+]$ 1·3±0·1	63B

62C M. Cefola, A. S. Tompa, A. V. Celiano, and P. S. Gentile, *Inorg. Chem.*, 1962, **1**, 290
63B J. L. Bear, G. R. Choppin, and J. V. Quagliano, *J. Inorg. Nuclear Chem.*, 1963, **25**, 513
63S J. T. Spence and H. H. Y. Chang, *Inorg. Chem.*, 1963, **25**, 319

$C_3H_6O_3S$ Prop-2-enesulphonic acid (allylsulphonic acid) HL 170.3

Metal	Method	Temp	Medium	Log of equilibrium constant, remarks	Ref
H^+	sp	25	2 (NaCl)	K_1 −0·66	68M
Pt^{2+}	sp	25—56	2 (NaCl)	$K(PtCl_4{}^{2-}+L^- \rightleftharpoons PtCl_3L^{2-}+Cl^-)$ (→25·0°) 3·61±0·01, (35·0°) 3·46±0·01, (45·0°) 3·33±0·01, (55·6°) 3·19±0·01 ΔH_1 −6·1±0·2, ΔS_1 −4·1±0·7	68M

68M R. M. Milburn and L. M. Venanzi, *Inorg. Chim. Acta*, 1968, **2**, 97

$C_3H_6N_2S$ 1,3-Diazolidin-2-thione (ethylene thiourea) HL 170.4

Metal	Method	Temp	Medium	Log of equilibrium constant, remarks	Ref
H^+				H^+ constants not given.	
Ag^+	Ag	28	$2NH_4NO_3$ 80% dioxan	β_3 11·5	64K
Cd^{2+}	Cd (Hg)	25	1 $NaNO_3$	K_1 1·31±0·02, β_2 2·14±0·09, β_3 2·7±0·2, β_4 3·4±0·1	63C

63C A. N. Christensen and S. E. Rasmussen, *Acta Chem. Scand.*, 1963, **17**, 1315
64K S. N. Khodaskar and D. D. Khanolkar, *Current Sci.*, 1964, **33**, 399

$C_3H_7O_2N$ 2-Aminopropanoic acid (α-alanine) HL 171

Metal	Method	Temp	Medium	Log of equilibrium constant, remarks	Ref
H^+	gl	21	~ 0	K_1 9·90	52P, 53P
	gl	0—40	0 corr	K_1 (0°) 10·57, (10°) 10·30, (20°) 10·04, (30°) 9·78, (40°) 9·48, K_2 (0°) 2·43, (10°) 2·39, (20°) 2·37, (30°) 2·36, (40°) 2·36 $\Delta H_1°$ −10·6, $\Delta S_1°$ 9·8, $\Delta H_2°$ −0·7, $\Delta S_2°$ 8·6	61I
	gl	25	0·3 (K_2SO_4)	K_1 9·62±0·03, K_2 < 3·0	61J
	gl	20	0·1 (KCl)	K_1 9·93, K_2 2·30	63I

continued overleaf

171 $C_3H_7O_2N$ (contd)

Metal	Method	Temp	Medium	Log of equilibrium constant, remarks	Ref
H^+ (contd)	gl	25	0·1 KNO_3	K_1 9·86	64R
	pol	30	1·0 KNO_3	K_1 9·70±0·02	64Ra
	gl	25	0·1 ($NaClO_4$)	K_1 9·59, K_2 2·53	65M
	gl	15—40	0·2 (KCl)	K_1 (15°) 10·01, (25°) 9·77, (40°) 9·41, K_2 (15°) 2·47, (25°) 2·40, (40°) 2·44	65S
	gl	10—40	→ 0	K_1 (10°) 10·29, (25°) 9·89, (40°) 9·51, K_2 (10°) 2·40, (25°) 2·37, (40°) 2·35 $\Delta H_1°$ −10·4, $\Delta S_1°$ 10·3, $\Delta H_2°$ −0·7	66A, 67
	cal	10—40	~ 0·01	$\Delta H_1°$ (10°) −11·7±0·1, (25°) −11·3±0·1, (40°) −11·26±0·06 $\Delta S_1°$ (10°) 5·8, (25°) 7·4, (40°) 10·8	66A, 67
	gl	20	0·1 (KCl)	K_1 9·84, K_2 2·20	66G
	gl	25	0·5 KCl	K_1 9·83, K_2 2·44	66L
	pol	25	0·6	K_1 9·88±0·03	67Aa
	gl	25	0·1	K_1 9·97	67E
	pol	25	0·5 $NaClO_4$	K_1 9·84, K_2 2·39	67R
	gl	25	0·01	K_1 9·87, K_2 2·34	67S
	gl	25	0·1	K_1 9·75, K_2 2·40	68B
Ag^+	Ag	25	0·6	K_1 3·60±0·05, β_2 7·06±0·02	67Aa
Be^{2+}	gl	21	0·005—0·01 $BeSO_4$	β_2 13·1	52P, 53
Cd^{2+}	gl	21	0·005—0·01 $CdSO_4$	β_2 7·6	52P, 53
	pol	30	1	β_3 9·15	62R
	oth	20	0·1 (KNO_3)	K_1 5·9, K_2 3·5, K_3 2·4, paper electrophoresis	64J
	pol	30	0·1 KNO_3	K_1 4·49, β_2 8·00, β_3 9·49	64R
	pol	30	1·0 KNO_3	β_2 7·56, β_3 9·15 $K(Cd^{2+}+2L^-+OH^- \rightleftharpoons CdL_2OH^-)$ 8·42 $K[Cd^{2+}+2L^-+4NH_3 \rightleftharpoons CdL_2(NH_3)_4]$ 8·91	64Ra
	oth	25	0·5 $NaClO_4$	K_1 3·96, β_2 7·57, optical rotation	67R
Co^{2+}	gl	20	0·1 (KCl)	K_1 4·32, K_2 3·6	63I
	oth	20	0·1 (KNO_3)	K_1 5·0, K_2 3·2, K_3 2·4, paper electrophoresis	64J
	gl	15—40	0·2 (KCl)	K_1 (15°) 4·41, (25°) 4·36, (40°) 4·25, K_2 (15°) 3·27, (25°) 3·20, (40°) 3·08 ΔH_1 −2·6±0·5, ΔS_1 11±2, ΔH_2 −3·1±0·5, ΔS_2 4±2	65S
	cal	22	0·1 KNO_3	$\Delta H_{\beta 2}$ −5·9, $\Delta S_{\beta 2}$ 20·0	67Sa
Cr^{3+}	gl	25	~ 0·5	K_1 8·6, K_2 6·6, K_3 5·6	63K
	gl	25	0·1 ($NaClO_4$)	K_1 8·53, K_2 7·44	65M
Cu^+	Pt	25	0·3 (K_2SO_4)	β_2 ~ 9·6	61J
Cu^{2+}	gl	0—40	0 corr	K_1 (0°) 8·95, (10°) 8·76, (20°) 8·66, (30°) 8·56, (40°) 8·34 K_2 (0°) 7·33, (20°) 7·13, (30°) 7·02, (30°) 6·90, (40°) 6·66 $\Delta H_1°$ −5·6, $\Delta S_1°$ 20·3, $\Delta H_2°$ −6·2, $\Delta S_2°$ 10·9	61I
	Pt	25	0·3 (K_2SO_4)	β_2 15·0, K_3 ~ 0·05	61J
	gl	25	3·0 (K_2SO_4)	$K(CuL^+ + H^+ \rightleftharpoons CuHL^{2+})$ 0·57	59C
			0·375 (K_2SO_4)	$K(CuL^+ + H^+ \rightleftharpoons CuHL^{2+})$ 0·72	
	gl	20	0·1 (KCl)	K_1 8·15, K_2 6·78	63I
	oth	20	0·1 (KNO_3)	K_1 8·5, K_2 6·7, paper electrophoresis	64J
	gl	15—40	0·2 (KCl)	K_1 (15°) 8·40, (25°) 8·29, (40°) 8·10, K_2 (15°) 6·86, (25°) 6·72, (40°) 6·51 ΔH_1 −4·9±0·5, ΔS_1 21±2, ΔH_2 −5·8±0·5, ΔS_2 11±2	65S
	gl	10—40	→ 0	K_1 (10°) 8·70, (25°) 8·54, (40°) 8·32, K_2 (10°) 7·26, (25°) 6·98, (40°) 6·76 $\Delta H_1°$ −5·1, $\Delta S_1°$ 21·8, $\Delta H_2°$ −6·8, $\Delta S_2°$ 9·0	66A

conti

Metal	Method	Temp	Medium	Log of equilibrium constant, remarks	Ref
Cu^{2+} (*contd*)	cal	10—40	0·02	$\Delta H_1°$ (10°) $-5{\cdot}4\pm0{\cdot}2$, (25°) $-4{\cdot}5\pm0{\cdot}1$, (40°) $-3{\cdot}99\pm0{\cdot}09$, $\Delta S_1°$ (10°) 20·8, (25°) 23·9, (40°) 25·3, $\Delta H_2°$ (10°) $-5{\cdot}5\pm0{\cdot}3$, (25°) $-5{\cdot}2\pm0{\cdot}1$, (40°) $-5{\cdot}65\pm0{\cdot}09$, $\Delta S_2°$ (10°) 13·9, (25°) 14·3, (40°) 12·9	66A
	gl	20	0·1 (KCl)	K_1 8·22, K_2 6·85	66G
	oth	25	0·5 $NaClO_4$	K_1 8·21, β_2 15·00, optical rotation	67R
	cal	22	0·1 KNO_3	$\Delta H_{\beta2}$ $-11{\cdot}9$, $\Delta S_{\beta2}$ 28·4	67Sa
	oth	25	0·1 (ClO_4^-)	K_1 7·59, β_2 14·76, circular dichroism. See nitrilotriacetic acid (449) and salicylic acid (506) for mixed ligands	68B
Dy^{3+}	gl	25	0·1 (KNO_3)	K_1 $4{\cdot}7\pm0{\cdot}1$	67E
Er^{3+}	gl	25	0·1 (KNO_3)	K_1 $4{\cdot}7\pm0{\cdot}1$	67E
Eu^{3+}	gl	25	0·1 (KNO_3)	K_1 $4{\cdot}7\pm0{\cdot}1$	67E
Fe^{3+}	pol	30	1 KCl	K_1 10·98	67K
Gd^{3+}	gl	15	0·1 (KNO_3)	K_1 $4{\cdot}6\pm0{\cdot}1$	67E
Hg^{2+}	gl	21	0·01 Hg $(NO_3)_2$	β_2 18·4	52P
	gl	21	0·005 Hg $(NO_3)_2$	β_2 18·8	53P
	pol	25	0·6 KNO_3	β_2 $18{\cdot}25\pm0{\cdot}03$	66T
Ho^{3+}	gl	25	0·1 (KNO_3)	K_1 $4{\cdot}6\pm0{\cdot}1$	67E
La^{3+}	gl	25	0·1 (KNO_3)	K_1 $4{\cdot}5\pm0{\cdot}1$	67E
Lu^{3+}	gl	25	0·1 (KNO_3)	K_1 $4{\cdot}8\pm0{\cdot}1$	67E
Mn^{2+}	oth	20	0·1 (KNO_3)	K_1 3·4, K_2 1·9, paper electrophoresis	64J
Nd^{3+}	gl	25	0·1 (KNO_3)	K_1 $4{\cdot}8\pm0{\cdot}2$	67E
Ni^{2+}	gl	20	0·1 (KCl)	K_1 5·40, K_2 4·50	63I
	oth	20	0·1 (KNO_3)	K_1 6·0, K_2 4·3, K_3 2·9, paper electrophoresis	64J
	gl	25	1 $NaClO_4$	K_1 5·40, β_2 9·91, β_3 13·02	64M
	gl	15—40	0·2 (KCl)	K_1 (15°) 5·65, (25°) 5·53, (40°) 5·38, K_2 (15°) 4·57, (25°) 4·45, (40°) 4·28 ΔH_1 $-4{\cdot}4\pm0{\cdot}5$, ΔS_1 10 ± 2, ΔH_2 $-4{\cdot}8\pm0{\cdot}5$, ΔS_2 4 ± 2	65S
	gl	25	0·5 KCl	K_1 5·31, β_2 9·73, β_3 12·73 $K(Ni^{2+}+A^-+L^- \rightleftharpoons NiAL)$ $7{\cdot}14\pm0{\cdot}05$ $K(Ni^{2+}+A^-+2L^- \rightleftharpoons NiAL_2^-)$ $11{\cdot}6\pm0{\cdot}1$ $K(Ni^{2+}+2A^-+2L^- \rightleftharpoons NiA_2L_2^{2-})$ $14{\cdot}03\pm0{\cdot}06$ where HA = pyruvic acid $K(Ni^{2+}+A^-+L^- \rightleftharpoons NiAL)$ $7{\cdot}56\pm0{\cdot}01$ $K(Ni^{2+}+A^-+2L^- \rightleftharpoons NiAL_2^-)$ $12{\cdot}30\pm0{\cdot}01$ $K(Ni^{2+}+2A^-+2L^- \rightleftharpoons NiA_2L_2^{2-})$ $13{\cdot}73\pm0{\cdot}01$ where HA = glyoxalic acid	66L
	gl	10—40	$\rightarrow 0$	K_1 (10°) 5·93, (25°) 5·81, (40°) 5·69, K_2 (10°) 4·87, (25°) 4·73, (40°) 4·50 $\Delta H_1°$ $-3{\cdot}2$, $\Delta S_1°$ 15·9, $\Delta H_2°$ $-5{\cdot}1$, $\Delta S_2°$ 4·7	67A
	cal	10—40	$\rightarrow 0$	$\Delta H_1°$ (10°) $-3{\cdot}4\pm0{\cdot}2$, (25°) $-3{\cdot}3\pm0{\cdot}2$, (40°) $-3{\cdot}6\pm0{\cdot}2$, $\Delta S_1°$ (10°) 15·1, (25°) 15·5, (40°) 15·4, $\Delta H_2°$ (10°) $-4{\cdot}6\pm0{\cdot}2$, (25°) $-3{\cdot}9\pm0{\cdot}3$, (40°) $-2{\cdot}6\pm0{\cdot}2$, $\Delta S_2°$ (10°) 5·9, (25°) 8·7, (40°) 11·4	67A
	oth	25	0·5 $NaClO_4$	K_1 5·40, K_2 4·15, K_3 3·17, optical rotation	67R
	cal	22	0·1 KNO_3	$\Delta H_{\beta2}$ $-7{\cdot}0$, $\Delta S_{\beta2}$ 22·0 See glycine (130) and Solochrom Violet R (1050·1) for mixed ligands	67Sa
Pb^{2+}	pol	30	1·0 KNO_3	K_1 4·18, β_2 6·83 $K(Pb^{2+}+2L^-+OH^- \rightleftharpoons PbL_2OH^-)$ 9·85	64Ra

continued overleaf

171 $C_3H_7O_2N$ (contd)

Metal	*Method*	*Temp*	*Medium*	*Log of equilibrium constant, remarks*	*Ref*
Pr^{3+}	gl	25	0·1 (KNO_3)	K_1 4·7±0·2	67E
Sc^{3+}	gl	25	0·1 (KNO_3)	K_1 7·7±0·2	67E
Sm^{3+}	gl	25	0·1 (KNO_3)	K_1 4·7±0·1	67E
Tb^{3+}	gl	25	0·1 (KNO_3)	K_1 4·7±0·2	67E
Tm^{3+}	gl	25	0·1 (KNO_3)	K_1 4·7±0·1	67E
Y^{3+}	gl	25	0·1 (KNO_3)	K_1 5·0±0·3	67E
Yb^{3+}	gl	25	0·1 (KNO_3)	K_1 4·9±0·1	67E
Zn^{2+}	gl	21	0·005—0·01 $ZnSO_4$	β_2 9·1	52P, 53P
	gl	20	0·1 (KCl)	K_1 4·55, K_2 4·1	63I
	oth	20	0·1 (KNO_3)	K_1 5·7, K_2 3·9, K_3 2·3 paper electrophoresis	64J
	gl	15—40	0·2 (KCl)	K_1 (15°) 4·98, (25°) 4·88, (40°) 4·80, K_2 (15°) 4·12, (25°) 3·96, (40°) 3·86	65S
				ΔH_1 −3·0±5, ΔS_1 12±2, ΔH_2 −4·3±0·5, ΔS_2 4±2	
	gl	25	0·5 KCl	K_1 4·56, β_2 8·52, β_3 10·51	66L
				$K(Zn^{2+}+A^-+L^- \rightleftharpoons ZnAL)$ 6·79±0·02	
				$K(Zn^{2+}+A^-+2L^- \rightleftharpoons ZnAL_2^-)$ 9·4±0·1	
				$K(Zn^{2+}+2A^-+2L^- \rightleftharpoons ZnA_2L_2^{2-})$ 13·01±0·03	
				where HA = pyruvic acid	
				$K(Zn^{2+}+A^-+L^- \rightleftharpoons ZnAL)$ 6·32±0·02	
				$K(Zn^{2+}+A^-+2L^- \rightleftharpoons ZnAL_2^-)$ 10·17±0·05	
				$K(Zn^{2+}+2A^-+2L^- \rightleftharpoons ZnA_2L_2^{2-})$ 12·03±0·02	
				where HA = glyoxylic acid	
	oth	25	0·5 $NaClO_4$	K_1 4·67, β_2 8·95, optical rotation	67R
	pol	30	0·5 KCl	β_2 8·85	67S

52P D. J. Perkins, *Biochem. J.*, 1952, **51**, 487
53P D. J. Perkins, *Biochem. J.*, 1953, **55**, 649
59C E. J. Corey and J. C. Bailar, *J. Amer. Chem. Soc.*, 1959, **81**, 2620
61I R. M. Izatt, J. W. Wrathall, and K. P. Anderson, *J. Phys. Chem.*, 1961, **65**, 1914
61J B. R. James and R. J. P. Williams, *J. Chem. Soc.*, 1961, 2007
62R G. N. Rao and R. S. Subrahmanya, *Current Sci.*, 1962, **31**, 55
63I H. Irving and L. D. Pettit, *J. Chem. Soc.*, 1963, 1546
63K A. A. Khan and W. U. Malik, *J. Indian Chem. Soc.*, 1963, **40**, 564
64J V. Jokl, *J. Chromatog.*, 1964, **14**, 71
64M R. P. Martin and R. A. Paris, *Bull. Soc. chim. France*, 1964, 3170
64R T. P. Radhakrishnan, S. C. Saraiya, and A. K. Sundaram, *J. Indian Chem. Soc.*, 1964, **41**, 521
64Ra G. N. Rao and R. S. Subrahmanya, *Proc. Indian Acad. Sci*, 1964, **60**, 165, 185
65M M. Matsukawa, M. Ohta, S. Takata, and R. Tsuchiya, *Bull. Chem. Soc. Japan*, 1965, **38**, 1235
65S V. S. Sharma, H. B. Mathur, and P. S. Kulkarni, *Indian J. Chem.*, 1965, **3**, 146, 475
66A K. P. Anderson, D. A. Newell, and R. M. Izatt, *Inorg. Chem.* 1966, **5**, 62
66G R. D. Gillard, H. M. Irving, R. M. Parkins, N. C. Payne, and L. D. Pettit, *J. Chem. Soc. (A)*, 1966, 1159
66L D. L. Leussing and E. M. Hanna, *J. Amer. Chem. Soc.*, 1966, **88**, 693, 696
66T V. F. Toropova and Yu. M. Azizov, *Russ. J. Inorg. Chem.*, 1966, **11**, 288
67A K. P. Anderson, W. O. Greenhalgh, and E. A. Butler, *Inorg. Chem.*, 1967, **6**, 1056
67Aa Yu. M. Azizov, A. Kh. Miftakhova, and V. F. Toropova, *Russ. J. Inorg. Chem.*, 1967, **12**, 345
67E A. E. Elkhilyali, L. I. Martynenko, and V. I. Spitsyn, *Proc. Acad. Sci. (U.S.S.R.)* 1967, **176**, 886, (855)
67K R. C. Kapoor and K. C. K. Mathur, *J. Polarog. Soc.*, 1967, **13**, 86
67R M. M. Ramel and M. R. Paris, *Bull. Soc. chim. France*, 1967, 1359
67S R. Sundaresan, S. C. Saraiya, and A. K. Sundaram, *Proc. Indian Acad. Sci.*, 1967, **66A**, 184
67Sa W. F. Stack and H. A. Skinner, *Trans. Faraday Soc.*, 1967, **63**, 1136
68B A. Bonniol and P. Vieles, *J. Chim. phys.*, 1968, **65**, 414

$C_3H_7O_2N$ 3-Aminopropanoic acid (β-alanine) HL 172

Metal	Method	Temp	Medium	Log of equilibrium constant, remarks	Ref
H^+	gl	30	0·1 (KCl)	K_1 10·55	62Ca
	pol	30	1·0 KNO_3	K_1 10·09	64R
	gl	15—40	0·2 (KCl)	K_1 (15°) 10·44, (25°) 10·16, (40°) 9·80, K_2 (15°) 3·68, (25°) 3·66, (40°) 3·65	65S
	gl	25	0·5 KCl	K_1 10·21, K_2 3·62	66L
	gl	20	0·01	K_1 10·36, K_2 3·60	67A
	H	0—45	→ 0	K_1 (0°) 11·05, (15°) 10·58, (25°) 10·29, (35°) 10·03, (45°) 9·78, K_2 (0°) 3·65, (15°) 3·58, (25°) 3·55, (35°) 3·53, (45°) 3·52	67B
	gl	20	0·1 KCl	K_1 10·26, K_2 3·52	67S
	gl	20	0·5 KNO_3	K_1 10·39±0·01	68A
	gl	20	0·5 KNO_3	K_1 10·17	68T
Ag^+	Ag	25	0·6	K_1 3·76±0·03, β_2 7·21±0·04	67A
	gl	20	0·500 KNO_3	K_1 3·44±0·05, β_2 7·25±0·03	68A
	Ag-AgCl	20	0·500 KNO_3	β_2 7·32±0·02	68A
	gl	25	0·5 KNO_3	K_1 3·33, K_2 3·79	68T
Cd^{2+}	pol	30	1	β_3 6·80	62R
	pol	30	1·0 KNO_3	β_2 5·70, β_3 6·78 $K(Cd^{2+}+3L^-+OH^- \rightleftharpoons CdL_3OH^{2-})$ 7·20 $K(Cd^{2+}+2L^-+CO_3^{2-} \rightleftharpoons CdL_2CO_3^{2-})$ 6·60 $K[Cd^{2+}+2L^-+4NH_3 \rightleftharpoons CdL_2(NH_3)_4]$ 7·98	64R
Ce^{3+}	gl	30	0·1 (KCl)	K_1 2·63	62Ca
Co^{2+}	gl	15—40	0·2 (KCl)	K_1 (15°) 3·69, (25°) 3·58, (40°) 3·53, K_2 (15°) 2·59, (25°) 2·56, (40°) 2·45 ΔH_1 −2·6±0·5, ΔS_1 8±2, ΔH_2 −2·3±0·5, ΔS_2 4±2	65S
	ix	40	0·2 (KCl)	K_1 3·56	65S
	H	0—45	→ 0	K_1 (0°) 4·47±0·01, (15°) 4·31±0·01, (25°) 4·21±0·01, (35°) 4·13±0·01, (45°) 4·06±0·01 ΔH_1 −3·60	67B
	cal	25	0·1 (KCl)	ΔH_1 −3·3±0·2, ΔS_1 8·1±0·8	67B
Cu^{2+}	gl	15—40	0·2 (KCl)	K_1 (15°) 7·16, (25°) 7·10, (40°) 6·93, K_2 (15°) 5·59, (25°) 5·40, (40°) 5·22 ΔH_1 −4·0±0·5, ΔS_1 20±2, ΔH_2 −6·0±0·5, ΔS_2 4±2	65S
	cal	22	0·1 KNO_3	$\Delta H_{\beta 2}$ −10·9, $\Delta S_{\beta 2}$ 20·3 See nitrilotriacetic acid (449) for mixed ligands	67Sa
La^{3+}	gl	30	0·1 (KCl)	K_1 2·43	62Ca
Nd^{3+}	gl	30	0·1 (KCl)	K_1 3·04	62Ca
Ni^{2+}	gl	15—40	0·2 (KCl)	K_1 (15°) 4·80, (25°) 4·71, (40°) 4·65, K_2 (15°) 3·54, (25°) 3·41, (40°) 3·24 ΔH_1 −4·0±0·5, ΔS_1 8±2, ΔH_2 −5·0±0·5, ΔS_2 −1±2	65S
	gl	25	0·5 KCl	K_1 4·46, β_2 7·84, β_3 9·55 $K(Ni^{2+}+A^-+L^- \rightleftharpoons NiAL)$ 8·34±0·01 $K(Ni^{2+}+A^-+2L^- \rightleftharpoons NiAL_2^-)$ 11·95±0·01 $K(Ni^{2+}+2A^-+2L^- \rightleftharpoons NiA_2L_2^{2-})$ 15·17±0·02 where HA = pyruvic acid	66L
	H	0—45	→ 0	K_1 (0°) 5·22±0·01, (15°) 5·08±0·01, (25°) 4·99±0·01, (35°) 4·92±0·01, (45°) 4·86±0·01 ΔH_1 −3·46	67B
	cal	25	0·1 (KCl)	ΔH_1 −3·8±0·1, ΔS_1 10·2±0·5 See Solochrom Violet R (1050·1) for mixed ligands	67B
	cal	22	0·1 KNO_3	$\Delta H_{\beta 2}$ −6·1, $\Delta S_{\beta 2}$ 15·9	67Sa

continued overleaf

172 $C_3H_7O_2N$ (contd)

Metal	Method	Temp	Medium	Log of equilibrium constant, remarks	Ref
Pb^{2+}	pol	30	1·0 KNO_3	$K(Pb^{2+}+2L^-+2OH^- \rightleftharpoons PbL_2(OH)_2{}^{2-})$ 12·11	64R
Pr^{3+}	gl	30	0·1 (KCl)	K_1 2·92	62Ca
$UO_2{}^{2+}$	gl	30	0·1 KCl	K_1 7·78, K_2 7·53	62C
Zn^{2+}	gl	15—40	0·2 (KCl)	K_1 (15°) 4·10, (40°) 3·82	65S
	gl	25	0·5 KCl	K_1 3·9	66L
				$K(Zn^{2+}+A^-+L^- \rightleftharpoons ZnAL)$ 7·08	
				$K(Zn^{2+}+2A^-+2L^- \rightleftharpoons ZnA_2L_2{}^{2-})$ 12·1	
				where HA = pyruvic acid	
	pol	30	0·5 KCl	K_1 4·2, β_2 7·0	67S
	cal	22	0·1 KNO_3	$\Delta H_{\beta 2} < -4{\cdot}1$	67Sa

62C M. Cefola, R. C. Taylor, P. S. Gentile, and A. V. Celiano, *J. Phys. Chem.*, 1962, **66**, 790
62Ca M. Cefola, A. S. Tompa, A. V. Celiano, and P. S. Gentile, *Inorg. Chem.*, 1962, **1**, 290
62R G. N. Rao and R. S. Subrahmanya, *Current Sci.*, 1962, **31**, 55
64R G. N. Rao and R. S. Subrahmanya, *Proc. Indian Acad. Sci.*, 1964, **60**, 165, 185
65S V. S. Sharma, H. B. Mathur, and P. S. Kilkarni, *Indian J. Chem.*, 1965, **3**, 146, 475
66L D. L. Leussing and E. M. Hanna, *J. Amer. Chem. Soc.*, 1966, **88**, 693
67A Yu. M. Azizov, A. Kh. Miftakhova, and V. F. Toropova, *Russ. J. Inorg. Chem.*, 1967, **12**, 345
67B S. Boyd, J. R. Brannan, H. S. Dunsmore, and G. H. Nancollas, *J. Chem. and Eng. Data*, 1967, **12**, 601
67S R. Sundaresan, S. C. Saraiya, and A. K. Sundaram, *Proc. Indian Acad. Sci.*, 1967, **66A**, 184
67Sa W. F. Stack and H. A. Skinner, *Trans. Faraday Soc.*, 1967, **63**, 1136
68A D. J. Alner, R. C. Lansbury, and A. G. Smeeth, *J. Chem. Soc.* (*A*), 1968, 417
68T G. F. Thiers, L. C. van Poncke, and M. A. Herman, *J. Inorg. Nuclear Chem.*, 1968, **30**, 1543

173 $C_3H_7O_2N$ Aminoacetic acid methyl ester (methyl glycinate) L

Metal	Method	Temp	Medium	Log of equilibrium constant, remarks	Ref
Cu^{2+}				See nitrilotriacetic acid (449) for mixed ligands	

174 $C_3H_7O_2N$ N-Methylaminoacetic acid (sarcosine) HL

Metal	Method	Temp	Medium	Log of equilibrium constant, remarks	Ref
H^+	gl	20	~ 0	K_1 10·22	52P
	H, Ag-AgCl	5—45	→ 0	K_1 (5°) 10·712, (15°) 10·447, (25°) 10·200, (35°) 9·970, (37°) 9·926, (45°) 9·755 (±0·004)	58D
				$\Delta H_1°$ (25°) −40·51 kJ mole^{-1}, $\Delta S°$ (25°) 59·4 J mole^{-1} deg^{-1}	
	gl	25	0·015, 1·0	K(0·015) 10·31, (1·0) 10·28, K_2 (0·015) 2·63, (1·0) 2·64	60K
	gl	25	0·3 (K_2SO_4)	K_1 9·73±0·03, K_2 < 3·0	61J
	gl	20, 30	→ 0	K_1 (20°) 10·19, (30°) 10·08, K_2 (20°) 2·12, (30°) 2·09	64I
	cal	25	→ 0	$\Delta H_1°$ −9·8±0·1, $\Delta S_1°$ 13, $\Delta H_2°$ −1·7±0·1, $\Delta S_2°$ 4	64I
	gl	25	0·5 KCl	K_1 10·14, K_2 2·30	66L

contin

$C_3H_7O_2N$ (contd) 174

Metal	Method	Temp	Medium	Log of equilibrium constant, remarks	Ref
Be^{2+}	gl	20	0·01 $BeSO_4$	β_2 13·9	52P
Cd^{2+}	gl	20	0·01 $CdSO_4$	β_2 7·5	52P
Cu^{+}	Pt	25	0·3 (K_2SO_4)	β_2 ~ 9·2	61J
Cu^{2+}	gl	25	0·015, 1·0	K_1 (0·015) 8·08, (1·0) 7·84, K_2 (0·015) 6·70, (1·0) 6·50	60K
	gl	20, 30	→ 0	K_1 (20°) 8·16, (30°) 8·12, K_2 (20°) 6·89, (30°) 6·76	64I
	cal	25	→ 0	$\Delta H_1°$ −4·6±0·1, $\Delta S_1°$ 22, $\Delta H_2°$ −5·4±0·1, $\Delta S_2°$ 13	64I
Hg^{2+}	gl	20	0·01 $Hg(NO_3)_2$	β_2 18·7	52P
Ni^{2+}	gl	25	0·5 KCl	K_1 5·24, β_2 9·54, β_3 12·4	66L
				$K(Ni^{2+}+A^-+L^- \rightleftharpoons NiAL)$ 5·97±0·06	
				$K(Ni^{2+}+A^-+2L^- \rightleftharpoons NiAL_2^-)$ 7·8±0·1	
				$K(Ni^{2+}+2A^-+2L^- \rightleftharpoons NiA_2L_2^{2-})$ 8·7±0·1	
				where HA = pyruvic acid	
Zn^{2+}	gl	20	0·01 $ZnSO_4$	β_2 8·8	52P
	gl	25	0·5 KCl	K_1 4·31, β_2 8·3	66L
				$K(Zn^{2+}+A^-+L^- \rightleftharpoons ZnAL)$ 4·6±0·4	
				$K(Zn^{2+}+A^-+2L^- \rightleftharpoons ZnAL_2^-)$ 9·85±0·06	
				$K(Zn^{2+}+2A^-+2L^- \rightleftharpoons ZnA_2L_2^{2-})$ 10·1 (pptn)	
				where HA = pyruvic acid	

52P D. J. Perkins, *Biochem. J.*, 1952, **51**, 487
58D S. P. Datta and A. K. Grzybowski, *Trans. Faraday Soc.*, 1958, **54**, 1179
60K W. L. Koltun, M. Fried, and F. R. N. Gurd, *J. Amer. Chem. Soc.*, 1960, **82**, 233
61J B. R. James and R. J. P. Williams, *J. Chem. Soc.*, 1961, 2007
64I R. M. Izatt, J. J. Christensen, and V. Kothari, *Inorg. Chem.*, 1964, **3**, 1565
66L D. L. Leussing and E. M. Hanna, *J. Amer. Chem. Soc.*, 1966, **88**, 693

$C_3H_7O_3N$ 2-Amino-3-hydroxypropanoic acid (serine) HL 175

Metal	Method	Temp	Medium	Log of equilibrium constant, remarks	Ref
H^{+}	gl	20	~ 0	K_1 9·34	53P
	gl	25	0·1	K_1 9·83	65P
	Hg	25	0·6	K_1 9·24±0·03	67A
	gl	37	0·15 KNO_3	K_1 8·841, K_2 2·180	67P, 68P
	gl	15—40	0·2 KNO_3	K_1 (15°) 9·34, (25°) 9·12, (40°) 8·78, K_2 (15°) 2·30, (25°) 2·29, (40°) 2·27	68R
Ag^{+}	Ag	25	0·6	K_1 3·40, β_2 6·67±0·02	67A
Be^{2+}	gl	20	0·005 $BeSO_4$	β_2 12·1	53P
Cd^{2+}	gl	20	0·005 $CdSO_4$	β_2 7·4	53P
Co^{2+}	gl	25	0—0·05	K_1 (0) 4·90, (0·01) 4·84, (0·02) 4·74, (0·05) 4·47, K_2 (0) 4·20, (0·01) 4·41, (0·02) 4·11, (0·05) 3·78	64S
	gl	15—40	0·2 KNO_3	K_1 (15°) 4·37, (25°) 4·33, (40°) 4·25, K_2 (15°) 3·38, (25°) 3·33, (40°) 3·26	68R
				$\Delta H_{\beta 2}$ −4·0±0·5, $\Delta S_{\beta 2}$ 22±2	
Cr^{3+}	gl	25	~ 0·5	K_1 8·0, K_2 6·2, K_3 5·2	63K

continued overleaf

175 $C_3H_7O_3N$ (contd)

Metal	Method	Temp	Medium	Log of equilibrium constant, remarks	Ref
Cu^{2+}	gl	25	0—0·1	K_1 (0) 8·40, (0·01) 8·20, (0·02) 8·0, (0·05) 7·65, (0·1) 7·57, K_2 (0) 6·10, (0·01) 6·06, (0·02) 6·02, (0·05) 5·85, (0·1) 5·75	64S
	gl	37	0·15 KNO_3	K_1 7·565, β_2 14·012 $K(CuA+L^- \rightleftharpoons CuAL^-)$ 6·41 $K(CuB^{2+}+L^- \rightleftharpoons CuBL^+)$ 6·70 $K(CuC^{2+}+L^- \rightleftharpoons CuCL^+)$ 6·94 where H_2A = salicylic acid, B = ethylenediamine, C = histamine	67P
	cal	22	0·1 KNO_3	$\Delta H_{\beta 2}$ −14·1, $\Delta S_{\beta 2}$ 19·0	67S
	gl	15—50	0·2 KNO_3	K_1 (15°) 8·02, (25°) 7·89, (40°) 7·73, K_2 (15°) 6·62, (25°) 6·51, (40°) 6·33 $\Delta H_{\beta 2}$ −9·6±0·5, $\Delta S_{\beta 2}$ 34±2	68R
Er^{3+}	gl	25	0·1	K_1 3·89	65P
Fe^{2+}	hyp	25	→ 0	β_2 7·7	64S
	gl	15—40	0·2 KNO_3	K_1 (15°) 3·67, (40°) 3·62, K_2 (15°) 2·78, (40°) 2·74 $\Delta H_{\beta 2}$ −1·5±0·5, $\Delta S_{\beta 2}$ 24±2	68R
Gd^{3+}	gl	25	0·1	K_1 3·59	65P
Hg^{2+}	gl	20	0·005 $Hg(NO_3)_2$	β_2 17·5	53P
	pol	25	0·6 KNO_3	β_2 17·34±0·04	66T
Ho^{3+}	gl	25	0·1	K_1 4·00	65P
Lu^{3+}	gl	25	0·1	K_1 3·98	65P
Mn^{2+}	hyp	25	→ 0	K_1 3·4, β_2 6·7	64S
	gl	15—40	0·2 KNO_3	K_1 (15°) 2·51, (40°) 2·48, K_2 (15°) 1·49, (40°) 1·47 $\Delta H_{\beta 2}$ −0·8±0·5, $\Delta S_{\beta 2}$ 15±2	68R
Ni^{2+}	hyp	25	→ 0	K_1 6·0, β_2 10·6	64S
	cal	22	0·1 KNO_3	$\Delta H_{\beta 2}$ −8·0, $\Delta S_{\beta 2}$ 19·0	67S
	gl	37	0·15 (KNO_3)	K_1 5·211, β_2 9·590, β_3 12·491 See 4(2-aminoethyl)imidazole (289) for mixed ligands	68P
	gl	15—40	0·2 KNO_3	K_1 (15°) 5·50, (25°) 5·42, (40°) 5·28, K_2 (15°) 4·44, (25°) 4·34, (40°) 4·19 $\Delta H_{\beta 2}$ −7·8±0·5, $\Delta S_{\beta 2}$ 19±2	68R
Tb^{3+}	gl	25	0·1	K_1 3·77	65P
Y^{3+}	gl	25	0·1	K_1 3·50	65P
Yb^{3+}	gl	25	0·1	K_1 3·98	65P
Zn^{2+}	gl	20	0·005 $ZnSO_4$	β_2 8·6	53P
	gl	25	0—0·1	K_1 (0) 5·30, (0·01) 5·22, (0·02) 5·19, (0·05) 5·08, (0·1) 4·94, K_2 (0) 4·45, (0·01) 4·46, (0·02) 4·45, (0·05) 4·40, (0·1) 4·28	64S
	cal	22	0·1 KNO_3	$\Delta H_{\beta 2}$ < 6·7	67S
	gl	15—40	0·2 KNO_3	K_1 (15°) 4·71, (25°) 4·66, (40°) 4·58, K_2 (15°) 3·77, (25°) 3·72, (40°) 3·64 $\Delta H_{\beta 2}$ −4·3±0·5, $\Delta S_{\beta 2}$ 24±2	68R

53P D. J. Perkins, *Biochem. J.*, 1953, **55**, 649
63K A. A. Khan and W. U. Malik, *J. Indian Chem. Soc.*, 1963, **40**, 564
64S A. Ya. Sychev, *Russ. J. Inorg. Chem.*, 1964, **9**, 1270 (2343)
65P V. D. Panasynk and V. A. Golub, *Russ. J. Inorg. Chem.*, 1965, **10**, 1482 (2732)
66T V. F. Toropova and Yu. M. Azizov. *Russ. J. Inorg. Chem.*, 1966, **11**, 288
67A Yu. M. Azizov, A. Kh. Miftakhova, and V. F. Toropova, *Russ. J. Inorg. Chem.*, 1967, **12**, 345 (661)
67P D. D. Perrin, I. G. Sayce, and V. S. Sharma, *J. Chem. Soc. (A)*, 1967, 1755
67S W. F. Stack and H. A. Skinner, *Trans. Faraday Soc.*, 1967, **63**, 1136
68P D. D. Perrin and V. S. Sharma, *J. Chem. Soc. (A)*, 1968, 446
68R E. V. Raju and M. B. Mathur, *J. Inorg. Nuclear chem.*, 1968, **30**, 2181

$C_3H_7O_7P$ D-2, 3-dihydroxypropanoic acid 2-phospate (D-2-phosphoglyceric acid) H_3L 175.1

Metal	*Method*	*Temp*	*Medium*	*Log of equilibrium constant, remarks*	*Ref*
H^+	gl	25	0·1—0·4 $[(C_3H_7)_4NI]$	K_1 (0·1) 7·0, (0·4) 7·1, K_2 (0·1) 3·55, (0·4) 3·6	57W
Cd^{2+}	gl	25	0·1—0·4 $(C_3H_7)_4NI$	K_1 3·40	57W
Co^{2+}	gl	25	0·1—0·4 $(C_3H_7)_4NI$	K_1 2·97	57W
K^+	gl	25	0·1—0·4 $(C_3H_7)_4NI$	K_1 1·18	57W
Mg^{2+}	gl	25	0·1—0·4 $(C_3H_7)_4NI$	K_1 2·45	57W
Mn^{2+}	gl	25	0·1—0·4 $(C_3H_7)_4NI$	K_1 3·09	57W
Ni^{2+}	gl	25	0·1—0·4 $(C_3H_7)_4NI$	K_1 2·88	57W
Zn^{2+}	gl	25	0·1—0·4 $(C_3H_7)_4NI$	K_1 3·40	57W

57W F. Wold and C. E. Ballou, *J. Biol. Chem.*, 1957, **227**, 301

$C_3H_7NS_2$ *NN*-Dimethyldithiocarbamic acid HL 176

Metal	*Method*	*Temp*	*Medium*	*Log of equilibrium constant, remarks*	*Ref*
H^+				H^+ constants not given.	
Cu^{2+}	sp	23—27	0·01 NaOH 75% ethanol	K_1 14·4±0·2, K_2 13·4±0·2	56J
	sp	18—22	0·01 NaOH 0–89% ethanol	K_1 (→ 0%) 11·4±0·2, (51·7%) 13·2±0·2, (75%) 14·4±0·2, (89%) 15·4±0·2, K_2 (→ 0%) 10·3±0·2, (51·7%) 12·5±0·2, (75%) 13·3±0·2, (89%) 14·1±0·2	57J
Zn^{2+}	sol	25	toluene	$K(ZnL^+ + A \rightleftharpoons ZnLA^+)$ 3·41 where A = ethylamine, 2·58, ΔH −9·8, ΔS −21 where A = t-butylamine, 3·16, ΔH −7·5, ΔS −11 where A = n-butylamine, 3·4 where A = pyrrolidine, 0·04 where A = aniline, 1·91, ΔH −7·6, ΔS −17 where A = pyridine, 2·34, ΔH −9·6, ΔS −21 where A = 4-methylpyridine, 1·05, ΔH ~ −7, ΔS ~ −17 where A = 2-methylpyridine, −0·28, ΔH ~ −7, ΔS ~ −26 where A = 2,4,6-trimethylpyridine, 2·47, ΔH −8·9, ΔS −19 where A = di-n-butylamine, 3·5 where A = piperidine, 3·11 where A = morpholine	65C

56J M. J. Janssen, *Rec. Trav. chim.*, 1956, **75**, 1411
57J M. J. Janssen, *Rec. Trav. chim.*, 1957, **76**, 827
65C E. Coates, B. Rigg, B. Saville, and D. Skelton, *J. Chem. Soc.*, 1965, 5613

179 $C_3H_8OS_2$ 2,3-Dimercaptopropanol H_2L

Metal	Method	Temp	Medium	Log of equilibrium constant, remarks	Ref
H^+	gl	30	0·1 KCl	K_1 10·72, K_2 8·69	62L
Fe^{2+}	gl	30	0·1 KCl	β_2 15·78,	62L
				$K(2Fe^{2+}+3L^{2-} \rightleftharpoons Fe_2L_3{}^{2-}) \sim 28$	

62L D. L. Leussing and J. Jayne, *J. Phys. Chem.*, 1962, **66,** 426

180 $C_3H_8O_2N_2$ 2,3-Diaminopropanoic acid HL

Metal	Method	Temp	Medium	Log of equilibrium constant, remarks	Ref
H^+	gl	25	0·1 (KCl)	K_1 9·95, K_2 6·69, K_3 1·23	68H
Cu^{2+}	gl	25	0·1 (KCl)	K_1 11·46±0·01, β_2 19·95±0·01	68H
				$K(Cu^{2+}+HL \rightleftharpoons CuHL^{2+})$ 6·31±0·01	
				$K(Cu^{2+}+HL+L^- \rightleftharpoons CuHL_2{}^+)$ 15·74±0·02	
	gl	20	0·025	K_1 12·02±0·04, β_2 20·34±0·04	68H
				$K(Cu^{2+}+HL \rightleftharpoons CuHL^{2+})$ 6·64±0·04	
				$K(Cu^{2+}+HL+L^- \rightleftharpoons CuHL_2{}^+)$ 16·14±0·05	
				(Calculated from data of A. Albert, *Biochem. J.*, 1952, **50**, 690)	
	gl	25	0·1	K_1 10·6, K_2 8·4	68Ha
				$K(Cu^{2+}+L^-+A \rightleftharpoons CuLA^+)$ 23·91±0·02	
				where A = 2,3-diaminopropanoic acid methyl ester	

68H R. W. Hay, P. J. Morris, and D. D. Perrin, *Austral. J. Chem.*, 1968, **21,** 1073
68Ha R. W. Hay and P. J. Morris, *Chem. Comm.*, 1968, 732

180.1 $C_3H_8O_2S$ 3-Mercaptopropan-1,2-diol HL

Metal	Method	Temp	Medium	Log of equilibrium constant, remarks	Ref
H^+	gl	25	→0	K_1 9·46	64A
B^{III}	gl	25	0·1 KCl	$K[H_3BO_3+HL \rightleftharpoons B(OH)_2(H_{-1}L)^-+H^+]$ −7·79	64A
				$K[H_3BO_3+2HL \rightleftharpoons B(H_{-1}L)_2{}^-+H^+]$ −6·12	

64A P. J. Antikainen and K. Tevanen, *Suomen Kem.*, 1964, **B37**, 6: 1962, **B35**, 224

180.2 C_3H_9ON 1-Aminopropan-2-ol L

Metal	Method	Temp	Medium	Log of equilibrium constant, remarks	Ref
H^+	H	20	0 corr	K_1 9·57±0·01	64A
Ag^+	gl	20	0 corr	K_1 3·23±0·08, K_2 3·55±0·08	64A
	sol	20	0 corr	β_2 6·78±0·01	64A

64A D. J. Alner and M. A. A. Kahn, *J. Chem. Soc.*, 1964, 5265

C_3H_9ON 3-Aminopropanol L 180.3

Metal	Method	Temp	Medium	Log of equilibrium constant, remarks	Ref
H^+	gl	20	0·5 KNO_3	K_1 10·37±0·02	68A
	H	20	0 corr	K_1 10·16±0·03	
Ag^+	gl	20	0·500 KNO_3	K_1 3·25±0·04, β_2 7·04±0·01	68A
	Ag-AgCl	20	0·500 KNO_3	β_2 7·04±0·02	

68A D. J. Alner, R. C. Lansbury, and A. G. Smeeth, *J. Chem. Soc.* (*A*), 1968, 417

C_3H_9ON *N*-Methyl-2-aminoethanol L 180.4

Metal	Method	Temp	Medium	Log of equilibrium constant, remarks	Ref
H^+	gl	25	~ 0·1	K_1 10·10	65D
Cu^{2+}	gl	25	~ 0·1	K_1 5·0, K_2 4·0, K_3 3·2, K_4 2·5	65D
Zn^{2+}	gl	25	~ 0·1	K_1 3·9, K_2 2·9, K_3 2·4, K_4 2·2	65D

65D G. Douheret, *Bull. Soc. chim. France*, 1965, 2915

C_3H_9NS 3-Aminopropanthiol HL 182.1

Metal	Method	Temp	Medium	Log of equilibrium constant, remarks	Ref
H^+				H^+ constants not given	
Cu^{2+}	pol	25	0·17	β_2 16·28, phosphate buffer	61K

61K E. C. Knoblock and W. C. Purdy, *J. Electroanalyt. Chem.*, 1961, **2**, 493

C_3H_9NS 2-(Methylthio)ethylamine L 183

Metal	Method	Temp	Medium	Log of equilibrium constant, remarks	Ref
H^+	gl	20	0·15 ($NaClO_4$)	K_1 9·56±0·02	62H
Cu^+	Pt, Au	20	0·15 (Na_2SO_4)	K_1 5·65, β_2 10·98	62H
Cu^{2+}	gl	20	0·15 ($NaClO_4$)	K_1 5·30, β_2 9·68	62H

Erratum. 1964, p. 403, Table 183. Heading should read **2-(Methylthio)ethylamine (2-aminoethyl methyl sulphide)**

62H C. J. Hawkins and D. D. Perrin, *J. Chem. Soc.*, 1962, 1351

183.1 $C_3H_9N_3S$ Aminoethylisothiourea L

Metal	Method	Temp	Medium	Log of equilibrium constant, remarks	Ref
H^+				H^+ constants not given	
Cu^{2+}	pol	25	0·17	K_1 14·04, phosphate buffer	61K

61K E. C. Knoblock and W. C. Purdy, *J. Electroanalyt. Chem.*, 1961, **2**, 493

184 $C_3H_{10}ON_2$ 1,3-Diaminopropan-2-ol L

Metal	Method	Temp	Medium	Log of equilibrium constant, remarks	Ref
H^+	sp	30	0·16 (NaCl)	$K[(H_{-1}L)^- + H^+ \rightleftharpoons L]$ 14·06	65M
Cu^{2+}	gl	30	0·16 (NaCl)	$K[Cu^{2+} + (H_{-1}L)^- \rightleftharpoons Cu(H_{-1}L)^+]$ 18·40	65M
Ni^{2+}	gl	30	0·16 (NaCl)	K_1 5·47, K_2 4·14	65M

65M E. Mario and S. M. Bolton, *Analyt. Chem.*, 1965, 37, 165

184.1 $C_3H_{10}O_6P_2$ Propane-1,2-diphosphonic acid H_4L

Metal	Method	Temp	Medium	Log of equilibrium constant, remarks	Ref
H^+	gl	20	0·1 KCl	K_1 9·27, K_2 7·00, K_3 2·6, K_4 < 2	51S
Ba^{2+}	gl	20	0·1 KCl	K_1 2·20, $K(Ba^{2+} + HL^{3-} \rightleftharpoons BaHL^-)$ 1·3	51S
Ca^{2+}	gl	20	0·1 KCl	K_1 2·65 $K(Ca^{2+} + HL^{3-} \rightleftharpoons CaHL^-)$ 1·7	51S
Mg^{2+}	gl	20	0·1 KCl	K_1 3·04, $K(Mg^{2+} + HL^{3-} \rightleftharpoons MgHL^-)$ 2·08	51S

51S G. Schwarzenbach, P. Ruckstuhl, and J. Zurc, *Helv. Chim. Acta*, 1951, **34**, 455

184.2 $C_3H_{10}O_6P_2$ Propane-1,3-diphosphonic acid H_4L

Metal	Method	Temp	Medium	Log of equilibrium constant, remarks	Ref
H^+	gl	20	0·1 KCl	K_1 8·35, K_2 7·34, K_3 2·6, K_4 < 2	51S
	gl	25	0·1 $(CH_3)_4NBr$	K_1 8·43±0·07, K_2 7·50±0·07, K_3 2·81±0·07, K_4 1·6±0·2	62I
Ba^{2+}	gl	20	0·1 KCl	K_1 2·34, $K(Ba^{2+} + HL^{3-} \rightleftharpoons BaHL^-)$ 1·6	51S
Ca^{2+}	gl	20	0·1 KCl	K_1 2·58, $K(Ca^{2+} + HL^{3-} \rightleftharpoons CaHL^-)$ 1·8	51S
	gl	25	0·1	K_1 2·6	62I
Mg^{2+}	gl	20	0·1 KCl	K_1 2·84, $K(Mg^{2+} + HL^{3-} \rightleftharpoons MgHL^-)$ 2·08	51S
	gl	25	0·1	K_1 2·8	62I

51S G. Schwarzenbach, P. Ruckstuhl, and J. Zurc, *Helv. Chim. Acta*, 1951, **34**, 455
62I R. R. Irani and K. Moedritzer, *J. Phys. Chem.*, 1962, **66**, 1349

$C_3H_{10}O_6P_2$ Propane-2,2-diphosphonic acid H_4L 184.3

Metal	Method	Temp	Medium	Log of equilibrium constant, remarks	Ref
H^+	gl	25	0·5 $(CH_3)_4NCl$	K_1 12·4±0·2, K_2 7·75±0·05, K_3 2·94±0·02	67C
Cs^+	gl	25	0·5 $(CH_3)_4HCl$	K_1 1·40±0·05,	67C
				$K(Cs^+ + HL^{3-} \rightleftharpoons CsHL^{2-})$ 0·35±0·03	
K^+	gl	25	0·5 $(CH_3)_4NCl$	K_1 1·60±0·07	67C
				$K(K^+ + HL^{3-} \rightleftharpoons KHL^{2-})$ 0·35±0·03	
Li^+	gl	25	0·5 $(CH_3)_4NCl$	K_1 3·8±0·1	67C
				$K(Li^+ + HL^{3-} \rightleftharpoons LiHL^{2-})$ 1·38±0·04	
Na^+	gl	25	0·5 $(CH_3)_4NCl$	K_1 2·08±0·04	67C
				$K(Na^+ + HL^{3-} \rightleftharpoons NaHL^{2-})$ 0·57±0·05	

67C R. L. Carroll and R. R. Irani, *Inorg. Chem.*, 1967, 6, 1994

$C_3H_3ONS_2$ 2-Thioxo-1,3-thiazolidin-4-one (rhodanine) HL 184.4

Metal	Method	Temp	Medium	Log of equilibrium constant, remarks	Ref
H^+	dis	20	0·1 $(NaClO_4)$	K_1 5·18	65N
Ag^+	dis	20	0·1 $(NaClO_4)$	K_1 5·47, K_2 4·21	65N

65N O. Navratil and J. Kotas, *Coll. Czech. Chem. Comm.*, 1965, **30**, 2736

$C_3H_5O_2NS_2$ N-(Dithiocarboxy)aminoacetic acid H_2L 184.5

Metal	Method	Temp	Medium	Log of equilibrium constant, remarks	Ref
H^+	Hg	25	0·1 (KNO_3)	K_1 6·67, K_2 3·75	67B
Pb^{2+}	Hg	25	0·1 (KNO_3)	K_1 7·32, β_2 13·03	67B

67B O. Budevsky and E. Platikanova, *Talanta*, 1967, **14**, 901

185 $C_3H_7O_2NS$ 2-Amino-3-mercaptopropanoic acid (cysteine) H_2L

Metal	Method	Temp	Medium	Log of equilibrium constant, remarks	Ref
H^+	gl	20	~0	K_1 10·53, K_2 8·53	53P
	gl	20	0·15 ($NaClO_4$)	K_1 8·80, K_2 8·03	63H
	sp	23	?	K_1 10·36, K_2 8·53, K_3 2·00, phosphate-borate buffer	63S
	gl	25	0·10 (KNO_3)	K_1 10·11, K_2 8·13	64L
	gl	20	0·1 $NaClO_4$	K_1 10·50±0·01, K_2 8·33±0·01	68P
Cd^{2+}	pol	25	0·2	β_2 9·89, phosphate buffer	66S
Cu^{2+}	pol	25	0·17	β_2 16·0, phosphate buffer	61K
	gl	20	0·15 ($NaClO_4$)	$K(Cu^{2+}+HL^- \rightleftharpoons CuHL^+)$ 7·00 $K(Cu^{2+}+2HL^- \rightleftharpoons CuH_2L_2)$ 13·72 $K(2Cu^{2+}+L^{2-} \rightleftharpoons Cu_2L^{2+})$ 14·00 $K(2Cu^{2+}+L^{2-}+HL^- \rightleftharpoons Cu_2HL_2{}^+)$ 21·33 $K[2Cu^{2+}+2L^{2-} \rightleftharpoons (CuL)_2]$ 28·05 $K(2Cu^{2+}+2HL^-+L^{2-} \rightleftharpoons Cu_2H_2L_3)$ 28·05	63H
Hg^{2+}	gl	20	0·0025 $Hg(NO_3)_2$	β_2 20.5	53P
	gl	25	0·10 (KNO_3)	K_1 14·21	64L
Mn^{2+}	gl	25	0·10 (KNO_3)	K_1 4·56	64L
Mo^{V}	spJ	25	?	K(?) 6·0±0·1, acetate buffer	63S
Mo^{VI}	spJ	25	?	$K(MoO_4{}^{2-}+3HL^- \rightleftharpoons MoOL_3{}^{2-}+3OH^-)$ (?) 18±1, acetate buffer	63S
Ni^{2+}	gl	25	0·10 (KNO_3)	K_1 9·64, K_2 9·40	64L
	gl	20	0·1 $NaClO_4$	K_1 9·0±0·3, K_2 20·16±0·01 $K(Ni^{2+}+HL^- \rightleftharpoons NiHL^+)$ 15·43±0·05 $K(2Ni^{2+}+3L^{2-} \rightleftharpoons Ni_2L_3{}^{2-})$ 33·01±0·07 $K(3Ni^{2+}+4L^{2-} \rightleftharpoons Ni_3L_4{}^{2-})$ 45·72±0·09	68P
Pb^{2+}	gl	25	0·10 (KNO_3)	K_1 11·39	64L
Zn^{2+}	gl	20	0·0025 $Hg(NO_3)_2$	β_2 17·1	53P
	gl	25	0·10 (KNO_3)	K_1 9·04, K_2 8·50	64L
	gl	20	0·1 $NaClO_4$	β_2 18·21±0·02 $K(Zn^{2+}+HL^-+L^{2-} \rightleftharpoons ZnHL_2{}^-)$ 24·79±0·05 $K(Zn^{2+}+2HL^- \rightleftharpoons ZnH_2L_2)$ 30·6±0·1 $K(3Zn^{2+}+4L^{2-} \rightleftharpoons Zn_3L_4{}^{2-})$ 43·5±0·1 $K(3Zn^{2+}+3L^{2-}+HL^- \rightleftharpoons Zn_3HL_4{}^-)$ 49·5±0·2	68P

53P D. J. Perkins, *Biochem. J.*, 1953, **55**, 649
61K E. C. Knoblock and W. C. Purdy, *J. Electroanalyt. Chem.*, 1961, **2**, 493
63H C. J. Hawkins and D. D. Perrin, *Inorg. Chem.*, 1963, **2**, 843
63S J. T. Spence and H. H. Y. Chang, *Inorg. Chem.*, 1963, **2**, 319
64L G. R. Lenz and A. E. Martell, *Biochemistry*, 1964, **3**, 745
66S I. H. Suffet and W. C. Purdy, *J. Electroanalyt. Chem.*, 1966, **11**, 302
68P D. D. Perrin and I. G. Sayce, *J. Chem. Soc.* (*A*), 1968, 53

186 $C_3H_8O_6NP$ 2-Amino-3-hydroxypropanoic acid 3-phosphate (*O*-phosphoserine) H_3L

Metal	Method	Temp	Medium	Log of equilibrium constant, remarks	Ref
H^+	gl	25	0·15 (KCl)	K_1 9·71±0·05, K_2 5·65±0·02, K_3 2·07±0·08	59O
	gl	20	0·1 $(C_3H_7)_4NI$	K_1 9·90±0·05, K_2 5·80±0·05	65H
Ca^{2+}	gl	25	0·15 (KCl)	K_1 2·2 $K(Ca^{2+}+HL^{2-} \rightleftharpoons CaHL)$ 1·43	59O
	gl	20	0·1 $(C_3H_7)_4NI$	K_1 3·2±0·1 $K(Ca^{2+}+HL^{2-} \rightleftharpoons CaHL)$ 2·5±0·1	65H

conti

$C_3H_8O_6NP$ (contd) 186

Metal	Method	Temp	Medium	Log of equilibrium constant, remarks	Ref
Cu^{2+}	gl	25	0·15 (KCl)	K_1 9·64, K_2 5·88 $K(Cu^{2+}+HL^{2-} \rightleftharpoons CuHL)$ 4·81	59O
Fe^{3+}	gl	25	0·16 (KCl)	K_1 14·0, K_2 7·4 $K(Fe^{3+}+HL^{2-} \rightleftharpoons FeHL^{+})$ 6·7	59O
Mg^{2+}	gl	25	0·15 (KCl)	K_1 2·4 $K(Mg^{2+}+HL^{2-} \rightleftharpoons MgHL)$ 1·60	59O
	gl	20	0·1 $(C_3H_7)_4NI$	K_1 3·3±0·1 $K(Mg^{2+}+HL^{2-} \rightleftharpoons MgHL)$ 2·5±0·1	65H
Mn^{2+}	gl	25	0·15 (KCl)	K_1 3·9 $K(Mn^{2+}+HL^{2-} \rightleftharpoons MnHL)$ 1·91	59O
Ni^{2+}	gl	20	0·1 $(C_3H_7)_4NI$	K_1 6·7±0·1 $K(Ni^{2+}+HL^{2-} \rightleftharpoons NiHL)$ 2·6±0·1	65H

59O R. Osterberg, *Arkiv Kemi*, 1959, **13**, 393 (*Nature*, 1957, **179**, 476)
65H H. S. Hendrickson and J. G. Fullington, *Biochemistry*, 1965, **4**, 1599

$C_3H_{12}O_9NP_3$ Nitrilotri(methylenephosphonic acid) H_6L 186.1

Metal	Method	Temp	Medium	Log of equilibrium constant, remarks	Ref
H^+	gl	25	1 KNO_3	K_1 12·3±0·1, K_2 6·66±0·04, K_3 5·46±0·05, K_4 4·30±0·05, K_5 < 2, K_6 < 2	67C
Ca^{2+}	gl	25	1 KNO_3	K_1 6·68±0·10 $K(Ca^{2+}+HL^{5-} \rightleftharpoons CaHL^{3-})$ 2·85±0·10 $K(Ca^{2+}+H_2L^{4-} \rightleftharpoons CaH_2L^{2-})$ 2·3±0·3 $K(Ca^{2+}+H_3L^{3-} \rightleftharpoons CaH_3L^{-})$ 1·8±0·2	67C
Mg^{2+}	gl	25	1 KNO_3	K_1 6·49±0·10 $K(Mg^{2+}+HL^{5-} \rightleftharpoons MgHL^{3-})$ 3·24±0·09 $K(Mg^{2+}+H_2L^{4-} \rightleftharpoons MgH_2L^{2-})$ 2·7±0·3 $K(Mg^{2+}+H_3L^{3-} \rightleftharpoons MgH_3L^{-})$ 1·9±0·2	67C

67C R. P. Carter, R. L. Carroll, and R. R. Irani, *Inorg. Chem.*, 1967, **6**, 939

$C_3H_{12}O_{10}NP_3$ Nitrilotri(methylenephosphonic acid) *N*-oxide H_6L 186.2

Metal	Method	Temp	Medium	Log of equilibrium constant, remarks	Ref
H^+	gl	25	1 KNO_3	K_1 12·1±0·2, K_2 6·95±0·07, K_3 5·26±0·07, K_4 3·28±0·07, K_5 < 2, K_6 < 2	67C
Ca^{2+}	gl	25	1 KNO_3	K_1 5·7±0·3 $K(Ca^{2+}+HL^{5-} \rightleftharpoons CaHL^{3-})$ 2·9±0·1 $K(Ca^{2+}+H_2L^{4-} \rightleftharpoons CaH_2L^{2-})$ 1·7±0·3 $K(Ca^{2+}+H_3L^{3-} \rightleftharpoons CaH_3L^{-}) \ll 1$	67C
Mg^{2+}	gl	25	1 KNO_3	K_1 8·3±0·2 $K(Mg^{2+}+HL^{5-} \rightleftharpoons MgHL^{3-})$ 3·6±0·1 $K(Mg^{2+}+H_2L^{4-} \rightleftharpoons MgH_2L^{2-})$ 2·1±0·3 $K(Mg^{2+}+H_3L^{3-} \rightleftharpoons MgH_3L^{-})$ 1·05±0·02	67C

67C R. P. Carter, M. M. Crutchfield, and R. R. Irani, *Inorg. Chem.*, 1967, **6**, 943

187 C_4H_8 Butene L

Metal	Method	Temp	Medium	Log of equilibrium constant, remarks	Ref
Pd^{2+}	sol	15	2—5 ($LiClO_4$+ $HClO_4$)	$K(PdCl_4^{2-}+L \rightleftharpoons PdCl_3L^-+Cl^-)$ (2) 1·14±0·05, (3) 1·13±0·02, (4) 1·13±0·02, (5) 1·05±0·03, $K[PdCl_4^{2-}+L \rightleftharpoons PdCl_2(H_2O)L+2Cl^-]$ (2) −0·5 ±0·9, (3) 0·1±0·3, (4) 0·65±0·09, (5) 0·95±0·07	66P

66P S. V. Pestrikov, I. I. Moiseev, and L. M. Sverzh, *Russ. J. Inorg. Chem.*, 1966, **11**, 1113 (2081)

191 $C_4H_4O_4$ cis-Butenedioic acid (maleic acid) H_2L

Metal	Method	Temp	Medium	Log of equilibrium constant, remarks	Ref
H^+	gl	25	0·15 ($NaClO_4$)	K_1 5·91, K_2 1·92	62B
	gl	25	0·2 $NaClO_4$	K_1 5·81, K_2 1·98	67N
	gl	25	1·00 (KNO_3)	K_1 5·61±0·01, K_2 1·65±0·01	67R
	gl	25	0·1 ($NaClO_4$)	K_1 5·79, K_2 1·92	68T
Be^{2+}	gl	25	0·15 ($NaClO_4$)	K_1 4·33, β_2 6·46	62B
Cd^{2+}	pol	25	0·2, 0·4 $NaClO_4$	K_1 (0·2) 2·2, (0·4) 2·0, β_2 (0·2) 3·6, (0·4) 3·2, β_3 (0·2) 3·8	67N
Co^{3+}	gl	25, 35	0 corr	$K[CoCl(NH_3)_5^{2+}+L^{2-} \rightleftharpoons CoCl(NH_3)_5L]$ (25°) 2·52, (35°) 2·52	65A
	sol	25, 35	0 corr	$K[CoCl(NH_3)_5^{2+}+L^{2-} \rightleftharpoons CoCl(NH_3)_5L]$ (25°) 2·47, (35°) 2·51	65A
Cr^{3+}	gl	25	0·1 ($NaClO_4$)	K_1 5·4, K_2 3·0, K_3 1·9	68T
Cu^{2+}	pol	25	0·2 $NaClO_4$	K_1 3·4, β_2 4·9, β_3 6·2	67N
In^{3+}	pol	25	0·2 $NaClO_4$	K_1 5·0, β_2 7·1, β_3 6·2	67N
Na^+	gl	25	0 corr	K_1 0·7	65A
Pb^{2+}	pol	25	0·1, 0·2 $NaClO_4$	K_1 (0·1) 3·2, (0·2) 3·0, β_2 (0·2) 4·5, β_3 (0·2) 5·4	67N
UO_2^{2+}	gl	25	1·00 (KNO_3)	K_1 4·46±0·05	67R

62B H. J. deBruin, D. Kaitis, and R. B. Temple, *Austral. J. Chem.*, 1962, **15**, 457
65A D. W. Archer, D. A. East, and C. B. Monk, *J. Chem. Soc.*, 1965, 720
67N T. Nozaki, T. Mise, and K. Higaki, *Nippon Kagaku Zasshi*, 1967, **88**, 1168
67R K. S. Rajan and A. E. Martell, *J. Inorg. Nuclear Chem.*, 1967, **29**, 523
68T T. Tomita, E. Kyuno, and R. Tsuchiya, *Bull. Chem. Soc. Japan*, 1968, **41**, 1130

193 $C_4H_4O_5$ Oxobutanedioic acid (oxaloacetic acid) H_3L

Metal	Method	Temp	Medium	Log of equilibrium constant, remarks	Ref
H^+	gl	25	0·1 KCl	K_1 13·06, K_2 3·89, K_3 2·22	64T
	sp	25	0·1 KCl	K_1 (H_nL_K) 13·0±0·1, K_2 (H_nL_K) 3·90±0·01, K_3 (H_nL_K) 2·24±0·02 K_1 (H_nL_E) 12·2±0·1, K_2 (H_nL_E) 3·7±0·2, K_3 (H_nL_E) 1·9±0·1 $K(HL_E^{2-} \rightleftharpoons HL_K^{2-})$ 0·79 $K(H_2L_E^- \rightleftharpoons H_2L_K^-)$ 0·91±0·07 $K(H_3L_E \rightleftharpoons H_3L_K)$ 1·28 where H_nL_E and H_nL_K represent enol and keto forms of the ligand, respectively	64T

cont

$C_4H_4O_5$ (contd) 193

Metal	Method	Temp	Medium	Log of equilibrium constant, remarks	Ref
Mg^{2+}	gl	25	0·1 KCl	K_1 6·27±0·05, K_2 4·82±0·04	64T
				$K(Mg^{2+}+HL^{2-} \rightleftharpoons MgHL)$ 1·96±0·09	
	sp	25	0·1 KCl	$K(Mg^{2+}+HL_K^{2-} \rightleftharpoons MgHL_K)$ 1·91±0·02	64T
				$K(Mg^{2+}+HL_E^{2-} \rightleftharpoons MgHL_E)$ 2·20±0·04	
				$K(MgHL_E \rightleftharpoons MgHL_K)$ 0·49±0·06	
				where H_nL_E and H_nL_K represent enol and keto forms of the ligand, respectively	

64T S. S. Tate, A. K. Grzybowski, and S. P. Datta, *J. Chem. Soc.*, 1964, 1381

$C_4H_4N_2$ 1,4-Diazine (pyrazine) L 193.1

Metal	Method	Temp	Medium	Log of equilibrium constant, remarks	Ref
Ag^+	sol	30	~ 1% Pyrazine	K_1 1·50±0·05, K_2 0·62±0·05	62S

62S J. G. Schmidt and R. F. Trimble, *J. Phys. Chem.*, 1962, **66,** 1063

C_4H_5N 2-Cyanopropene (methacrylonitrile) L 194.1

Metal	Method	Temp	Medium	Log of equilibrium constant, remarks	Ref
Ag^+	oth	61	4 $LiNO_3$	K_1 −0·11, gas chromatography	68S

68S H. Schnecko, *Analyt. Chem.*, 1968, **40,** 1391

C_4H_5N 3-Cyanopropene (crotononitrile) L 194.2

Metal	Method	Temp	Medium	Log of equilibrium constant, remarks	Ref
Ag^+	oth	61	4 $LiNO_3$	K_1 −0·03, gas chromatography	68S

68S H. Schnecko, *Analyt. Chem.*, 1968, **40,** 1391

195.1 C_4H_6O But-1-en-3-one (methylvinylketone) L

Metal	Method	Temp	Medium	Log of equilibrium constant, remarks	Ref
Ag^+	dis	25	1 (KNO_3)	K_1 −0·4	68F

68F T. Fueno, O. Kajimoto, and J. Furukawa, *Bull. Chem. Soc. Japan*, 1968, **41**, 782

197.1 $C_4H_6O_2$ 2-Propenoic acid methyl ester (methyl acrylate) L

Metal	Method	Temp	Medium	Log of equilibrium constant, remarks	Ref
Ag^+	dis	25	1 (KNO_3)	K_1 −0·4, K_2 −0·6	68F

68F T. Fueno, O. Kajimoto, and J. Furukawa, *Bull. Chem. Soc. Japan*, 1968, **41**, 782

197.2 $C_4H_6O_2$ Acetic acid vinyl ester (vinyl acetate) L

Metal	Method	Temp	Medium	Log of equilibrium constant, remarks	Ref
Ag^+	dis	25	1 (KNO_3)	K_1 0·01, K_2 −0·4	68F

68F T. Fueno, O. Kajimoto, and J. Furukawa, *Bull. Chem. Soc. Japan*, 1968, **41**, 782

198 $C_4H_6O_4$ Butanedioic acid (succinic acid) H_2L

Metal	Method	Temp	Medium	Log of equilibrium constant, remarks	Ref
H^+	gl	25	0·15 ($NaClO_4$)	K_1 4·88, K_2 4·05	62B
	gl	20	0·10 ($NaClO_4$)	K_1 5·28, K_2 4·00	63C
	gl	25	0·2 (KNO_3)	K_2 4·07	63F
	?	?	?	K_1 5·21, K_2 4·00	66M
	gl	25	1·00 (KNO_3)	K_1 5·20±0·04, K_2 3·98±0·04	67R
	pH	25	0·5 $LiClO_4$	K_1 5·05, K_2 3·92±0·03	68D
	gl	25	0·15 ($NaClO_4$)	K_2 4·20	68K
Be^{2+}	gl	25	0·15 ($NaClO_4$)	K_1 4·69, β_2 6·43	62B
Ca^{2+}	gl	20	0·1 ($NaClO_4$)	K_1 1·20, $K(Ca^{2+}+HL^- \rightleftharpoons CaHL^+)$ 0·54	63C
Co^{2+}	dis	25	0·00—0·16	K_1 (→ 0) 2·80±0·02, (0·04) 2·207±0·004, (0·06) 2·111±0·007, (0·08) 2·034±0·009, (0·16) 1·916±0·004	61M
	gl	20	0·1 ($NaClO_4$)	K_1 1·70, $K(Co^{2+}+HL^- \rightleftharpoons CoHL^+)$ 0·99	63C
	gl	25	0 corr	K_1 2·37±0·03	65M
	ix	25	0 corr	K_1 2·41±0·07	65S
	cal	25	0·1 KCl	ΔH_1 3·2±0·2, ΔS_1 21	67M

contin

Metal	Method	Temp	Medium	Log of equilibrium constant, remarks	Ref
Co^{3+}	gl	25, 35	0 corr	$K[CoCl(NH_3)_5^{2+} + L^{2-} \rightleftharpoons CoCl(NH_3)_5L]$ (25°) 2·00, (35°) 2·05	65A
	sol	25, 35	0 corr	$K[CoCl(NH_3)_5^{2+} + L^{2-} \rightleftharpoons CoCl(NH_3)_5L]$ (25°) 1·94, (35°) 1·92	65A
Cr^{3+}	gl	25	0·1 ($NaClO_4$)	K_1 6·42, K_2 4·57, K_3 2·86	66M
Cu^{2+}	gl	20	0·1 ($NaClO_4$)	K_1 2·93, $K(Cu^{2+} + HL^- \rightleftharpoons CuHL^+)$ 1·70	63C
	Ag-AgCl	25	0·0—0·2 ($NaClO_4$)	K_1 (→ 0) 3·22±0·01, (0·2) 2·26±0·01	67M
	cal	25	0·1 KCl	ΔH_1 4·56±0·07, ΔS_1 30·1	67M
Dy^{3+}	ix	25	0·15 ($NaClO_4$)	$K(Dy^{3+} + HL^- \rightleftharpoons DyHL^{2+})$ 1·72±0·09 $K[DyHL^{2+} + HL^- \rightleftharpoons Dy(HL)_2^+]$ 1·4±0·1	68K
Er^{3+}	ix	25	0·15 ($NaClO_4$)	$K(Er^{3+} + HL^- \rightleftharpoons ErHL^{2+})$ 1·71±0·09 $K[ErHL^{2+} + HL^- \rightleftharpoons Er(HL)_2^+]$ 1·3±0·1	68K
Eu^{3+}	ix	25	0·15 ($NaClO_4$)	$K(Eu^{3+} + HL^- \rightleftharpoons EuHL^{2+})$ 1·99±0·10 $K[EuHL^{2+} + HL^- \rightleftharpoons Eu(HL)_2^+]$ 1·3±0·1	68K
Fe^{3+}	sp	25	0·5 $LiClO_4$	K_1 6·88	68D
Gd^{3+}	ix	25	0·15 ($NaClO_4$)	$K(Gd^{3+} + HL^- \rightleftharpoons GdHL^{2+})$ 1·83±0·09 $K[GdHL^{2+} + HL^- \rightleftharpoons Gd(HL)_2^+]$ 1·3±0·1	68K
La^{3+}	ix	25	0·15 ($NaClO_4$)	$K(La^{3+} + HL^- \rightleftharpoons LaHL^{2+})$ 1·48±0·08 $K[LaHL^{2+} + HL^- \rightleftharpoons La(HL)_2^+]$ 1·2±0·1	68K
Lu^{3+}	ix	25	0·15 ($NaClO_4$)	$K(Lu^{3+} + HL^- \rightleftharpoons LuHL^{2+})$ 1·76±0·09 $K[LuHL^{2+} + HL^- \rightleftharpoons Lu(HL)_2^+]$ 1·3±0·1	68K
Mn^{2+}	cal	25	0·1 KCl	ΔH_1 3·0±0·2, ΔS_1 20·5	67M
Na^+	gl	25	0 corr	K_1 0·3	65A
Nb^V	dis	20	4·5 (NaCl or $NaNO_3$ + 2·5 HCl)	$K[Nb(OH)_4^+ + H_2L \rightleftharpoons Nb(OH)_4HL + H^+]$ 1·53	67K
Ni^{2+}	cal	25	0·1 KCl	ΔH_1 2·5±0·2, ΔS_1 19	67M
Pr^{3+}	ix	25	0·15 ($NaClO_4$)	$K(Pr^{3+} + HL^- \rightleftharpoons PrHL^{2+})$ 1·72±0·09 $K[PrHL^{2+} + HL^- \rightleftharpoons Pr(HL)_2^+]$ 1·3±0·1	68K
Sc^{3+}	sp	18—20	0·01 (HNO_3)	K_1 4·48±0·04	66K
Sm^{3+}	ix	25	0·15 ($NaClO_4$)	$K(Sm^{3+} + HL^- \rightleftharpoons SmHL^{2+})$ 2·00±0·10 $K[SmHL^{2+} + HL^- \rightleftharpoons Sm(HL)_2^+)]$ 1·4±0·2	68K
Tb^{3+}	ix	25	0·15 ($NaClO_4$)	$K(Tb^{3+} + HL^- \rightleftharpoons TbHL^{2+})$ 1·66±0·08 $K[TbHL^{2+} + HL^- \rightleftharpoons Tb(HL)_2^+]$ 1·5±0·2	68K
Th^{4+}	kin	25	~ 0	$K(2Th^{4+} + L^{2-} \rightleftharpoons Th_2L^{6+})$ 11·78±0·03	63Y
UO_2^{2+}	gl	25	0·2 (KNO_3)	$K(UO_2^{2+} + HL^- \rightleftharpoons UO_2HL^+)$ 2·62±0·03	63F
	sp	20	1·0 ($NaClO_4$)	$K(UO_2^{2+} + HL^- \rightleftharpoons UO_2HL^+)$ 2·53±0·03	67Ma
	gl	25	1·00 (KNO_3)	K_1 3·68±0·03, separation of solid phase	67R
Yb^{3+}	ix	25	0·15 ($NaClO_4$)	$K(Yb^{3+} + HL^- \rightleftharpoons YbHL^{2+})$ 1·72±0·09 $K[YbHL^{2+} + HL^- \rightleftharpoons Yb(HL)_2^+]$ 1·2±0·1	68K
Zn^{2+}	gl	20	0·1 ($NaClO_4$)	K_1 1·76, $K(Zn^{2+} + HL^- \rightleftharpoons ZnHL^+)$ 0·96	63C
	dis	25	0 corr	K_1 3·22±0·08	66R
	Pt, Ag-AgCl	25	→ 0	K_1 2·47±0·02, $K(Zn^{2+} + HL^- \rightleftharpoons ZnHL^+)$ 1·51	67M
	cal	25	0·1 KCl	ΔH_1 4·4±0·1, ΔS_1 26	67M
Zr^{4+}	kin	25	var.	$K[Zr(OH)_3^+ + H_2L \rightleftharpoons ZrOHL^+)]$ 1·46±0·08	61Y

61M P. G. Manning and C. B. Monk, *Trans. Faraday Soc.*, 1961, **57**, 1996

61Y K. B. Yatsimirskii and L. P. Raizman, *Russ. J. Inorg. Chem.*, 1961, **6**, 1263 (2496)

62B H. J. deBruin, D. Kaitis, and R. B. Temple, *Austral. J. Chem.*, 1962, **15**, 457

63C E. Campi, *Ann. Chim. (Italy)*, 1963, **53**, 96

63F I. Feldman and L. Koval, *Inorg. Chem.*, 1963, **2**, 145

63Y K. B. Yatsimirskii and Yu. A. Khukov, *Russ. J. Inorg. Chem.*, 1963, **8**, 149 (295)

continued overleaf

198 $C_4H_6O_4$ (contd)

65A D. W. Archer, D. A. East, and C. B. Monk, *J. Chem. Soc.*, 1965, 720
65M C. B. Monk, *J. Chem. Soc.*, 1965, 2456
65S R. G. Seys and C. B. Monk, *J. Chem. Soc.*, 1965, 2452
66K I. M. Korenman and N. V. Zaglyadimova, *Russ. J. Inorg. Chem.*, 1966, **11**, 1491
66M H. Muro and R. Tsuchiya, *Bull. Chem. Soc., Japan*, 1966, **39**, 1589
66R D. L. G. Rowlands and C. B. Monk, *Trans. Faraday Soc.*, 1966, **62**, 945
67K C. Konecny, *Z. phys. Chem. (Leipzig)*, 1967, **235**, 39
67M A. McAuley, G. H. Nancollas, and K. Torrance, *Inorg. Chem.*, 1967, **6**, 136
67Ma C. Miyake and H. W. Nurnberg, *J. Inorg. Nuclear Chem.*, 1967, **29**, 2411
67R K. S. Rajan and A. E. Martell, *J. Inorg. Nuclear Chem.*, 1967, **29**, 523
68D M. Deneux, R. Meilleur, and R. L. Benoit, *Canad. J. Chem.*, 1968, **46**, 1383
68K C. M. Ke, P. C. Kong, M. S. Cheng, and N. C. Li, *J. Inorg. Nuclear Chem.*, 1968, **30**, 961

201 $C_4H_6O_5$ Oxydiacetic acid (diglycolic acid) H_2L

Metal	*Method*	*Temp*	*Medium*	*Log of equilibrium constant, remarks*	*Ref*
H^+	gl	20	0·1	K_1 4·11, K_2 3·06	61C
Ce^{3+}	qh	20	1·00 $NaClO_4$	K_1 5·16±0·01, β_2 8·92±0·01, β_3 11·23±0·03	63G
Cu^{2+}	gl	20	0·1	K_1 3·70, $K(Cu^{2+}+HL^- \rightleftharpoons CuHL^+)$ 2·67	61C
Dy^{3+}	qh	20	1·00 $NaClO_4$	K_1 5·31±0·01, β_2 9·98±0·01, β_3 13·36±0·01	63G
Er^{3+}	qh	20	1·00 $NaClO_4$	K_1 5·34±0·01, β_2 10·02±0·01, β_3 13·23±0·02	63G
Eu^{3+}	qh	20	1·00 $NaClO_4$	K_1 5·53±0·02, β_2 10·04±0·02, β_3 13·20±0·02	63G
Gd^{3+}	qh	20	1·00 $NaClO_4$	K_1 5·40±0·01, β_2 9·93±0·01, β_3 13·04±0·02	63G
Ho^{3+}	qh	20	1·00 $NaClO_4$	K_1 5·28±0·02, β_2 9·95±0·01, β_3 13·31±0·01	63G
La^{3+}	qh	20	1·00 $NaClO_4$	K_1 4·93±0·01, β_2 8·41±0·01, β_3 10·25±0·02	63G
Lu^{3+}	qh	20	1·00 $NaClO_4$	K_1 5·64±0·01, β_2 10·55±0·01, β_3 13·16±0·02	63G
Nd^{3+}	qh	20	1·00 $NaClO_4$	K_1 5·45±0·01, β_2 9·50±0·01, β_3 12·16±0·01	63G
Pb^{2+}	pol	25	0—0·2 ($NaClO_4$)	K_1 (→ 0) 4·95, (0·04) 4·92, (0·2) 4·53, K_2 (→ 0) 2·45, (0·04) 2·38, (0·2) 2·29	64K
Pr^{3+}	qh	20	1·00 $NaClO_4$	K_1 5·33±0·01, β_2 9·23±0·02, β_3 11·63±0·03	63G
Sm^{3+}	qh	20	1·00 $NaClO_4$	K_1 5·55±0·01, β_2 9·89±0·01, β_3 12·79±0·02	63G
Tb^{3+}	qh	20	1·00 $NaClO_4$	K_1 5·23±0·02, β_2 9·98±0·01, β_3 13·25±0·01	63G
Tm^{3+}	qh	20	1·00 $NaClO_4$	K_1 5·49±0·01, β_2 10·22±0·01, β_3 13·29±0·01	63G
Y^{3+}	qh	20	1·00 $NaClO_4$	K_1 5·24±0·01, β_2 9·76±0·02, β_3 13·03±0·02	63G
Yb^{3+}	qh	20	1·00 $NaClO_4$	K_1 5·55±0·01, β_2 10·36±0·01, β_3 13·17±0·01	63G

61C E. Campi, G. Ostacoli, N. Cibrario, and G. Saini, *Gazzetta*, 1961, **91**, 361
63G I. Grenthe and I. Tobiasson, *Acta Chem. Scand.*, 1963, **17**, 2101
64K D. H. Klein and G. J. Kersels, *J. Inorg. Nuclear Chem.*, 1964 **26**, 1325

202 $C_4H_6O_5$ Hydroxybutanedioic acid (malic acid) H_2L

Metal	*Method*	*Temp*	*Medium*	*Log of equilibrium constant, remarks*	*Ref*
H^+	?	?	?	K_2 3·4	57V
	gl	30	0·1 (KCl)	K_1 4·78, K_2 3·22	62Ca
	gl	?	?	K_1 5·11, K_2 3·41	64R
	gl	20	0·1 ($NaClO_4$)	K_1 4·71, K_2 3·22	63C
	gl	?	0·28 [$(CH_3)_4NBr$]	K_1 4·86	63E
	gl	20	0·1 ($NaClO_4$)	K_1 4·72±0·02, K_2 3·28±0·02	64T

contin

Metal	Method	Temp	Medium	Log of equilibrium constant, remarks	Ref
Al^{3+}	sp	27—30	1 ($NaClO_4$)	K_1 3·32	65M
Ba^{2+}	gl	20	0·1 ($NaClO_4$)	$K(Ba^{2+}+H_2L \rightleftharpoons BaH_2L^{2+})$ 0·67	63C
				$K(Ba^{2+}+HL^- \rightleftharpoons BaHL^+)$ 1·38	
Ca^{2+}	gl	20	0·1 ($NaClO_4$)	$K(Ca^{2+}+H_2L \rightleftharpoons CaH_2L^{2+})$ 1·06	63C
				$K(Ca^{2+}+HL^- \rightleftharpoons CaHL^+)$ 1·96	
Cd^{2+}	gl	20	0·1 ($NaClO_4$)	$K(Cd^{2+}+H_2L \rightleftharpoons CdH_2L^{2+})$ 1·34	63C
				$K(Cd^{2+}+HL^- \rightleftharpoons CdHL^+)$ 2·36	
Ce^{3+}	gl	30	0·1 (KCl)	K_1 5·00, K_2 3·28, K_3 2·75	62Ca
	pH	25	~ 0·2 (KCl)	K_1 4·11 ± 0·07	64D
Co^{2+}	dis	25	0·00—0·16	K_1 (→ 0) 3·00 ± 0·02, (0·04) 2·373 ± 0·002, (0·06) 2·281 ± 0·004, (0·08) 2·198 ± 0·001, (0·16) 2·012 ± 0·002	61M
	gl	20	0·1 ($NaClO_4$)	$K(Co^{2+}+H_2L \rightleftharpoons CoH_2L^{2+})$ 1·64	63C
				$K(Co^{2+}+HL^- \rightleftharpoons CoHL^+)$ 2·86	
Co^{3+}	gl	25	0 corr	$K[CoCl(NH_3)_5{}^{2+}+L^{2-} \rightleftharpoons CoCl(NH_3)_5L]$ 2·01	65A
	sol	25, 35	0 corr	$K[CoCl(NH_3)_5{}^{2+}+L^{2-} \rightleftharpoons CoCl(NH_3)_5L]$ (25°) 1·99, (35°) 1·98	65A
Cs^+	gl	?	0·28 [$(CH_3)_4NBr$]	K_1 −0·1 ± 0·1	63E
	oth	25	0·1	K_1 −0·15 ± 0·06, cation-sensitive glass electrode	64Ra
Cu^{2+}	gl	20	0·1 ($NaClO_4$)	$K(Cu^{2+}+H_2L \rightleftharpoons CuH_2L^{2+})$ 2·00	63C
				$K(Cu^{2+}+HL^- \rightleftharpoons CuHL^+)$ 3·42	
				$K(CuL+H^+ \rightleftharpoons CuHL^+)$ 4·54	
	sp	27—30	1 ($NaClO_4$)	K_1 3·43,	65M
	gl	25	1·0 (KNO_3)	$K(2Cu^{2+}+2L^{2-} \rightleftharpoons Cu_2L_2)$ 8·0 ± 0·1	67R
				$K[Cu_2(H_{-1}L)_2{}^{2-}+2H^+ \rightleftharpoons Cu_2L_2]$ 7·8	
	sp	30	0·1 ($NaClO_4$)	K_1 3·97, $K(2Cu^{2+}+L^{2-} \rightleftharpoons Cu_2L^{2+})$ 7·71	68R
Eu^{3+}	pH	25	~ 0·2 (KCl)	K_1 4·34 ± 0·04	64D
Fe^{2+}	gl	20	0·1 ($NaClO_4$)	K_1 2·5 ± 0·1	64T
	ix	20	0·1 ($NaClO_4$)	K_1 2·68 ± 0·08	64T
Fe^{3+}	gl	20	0·1 ($NaClO_4$)	K_1 7·1 ± 0·2	64T
				$K[2Fe^{3+}+2L^{2-} \rightleftharpoons Fe_2(H_{-1}L)_2+2H^+]$ 12·85	
				$K[2Fe^{3+}+3L^{2-} \rightleftharpoons Fe_2L(H_{-1}L)_2{}^{2-}+2H^+]$ 17·85	
				$K[3Fe^{3+}+5L^{2-} \rightleftharpoons Fe_3L(H_{-1}L)_4{}^{5-}+4H^+]$ 25·97	
	sp	20	0·1 ($NaClO_4$)	K_1 7·09	64T
Gd^{3+}	pH	25	~ 0·2 (KCl)	K_1 4·28 ± 0·08	64D
Ge^{IV}	con	18	1 NaCl	$K[H_2GeO_3+2H_2L \rightleftharpoons GeO(HL)_2]$ 0·68 ± 0·09	57V
	gl	18	→ 0	$K[H_2GeO_3+2H_2L \rightleftharpoons GeO(HL)_2]$ 2·92	57V
Hf^{4+}	ix	?	1—2 $HClO_4$	$K(Hf^{4+}+H_2L \rightleftharpoons HfHL^{3+}+H^+)$ (1) 2·16 ± 0·03 (2) 1·53 ± 0·03	60R
	ix	?	2·0 $HClO_4$	$K(Hf^{4+}+H_2L \rightleftharpoons HfHL^{3+}+H^+)$ 1·83 ± 0·03	64R
K^+	gl	?	0·28 [$(CH_3)_4NBr$]	K_1 0·23 ± 0·05	63E
	oth	25	0·1	K_1 0·18 ± 0·03, cation-sensitive glass electrode	64Ra
La^{3+}	gl	30	0·1 (KCl)	K_1 4·73, K_2 3·22, K_3 2·68	62Ca
	gl	21—22	0·12	K_1 4·38 ± 0·04	62D
	pH	25	~ 0·2 (KCl)	K_1 4·04 ± 0·04	64D
Li^+	gl	?	0·28 [$(CH_3)_4NBr$]	K_1 0·45 ± 0·05	63E
	oth	25	0·1	K_1 0·38 ± 0·02, cation-sensitive glass electrode	64Ra
Mg^{2+}	gl	20	0·1 ($NaClO_4$)	$K(Mg^{2+}+H_2L \rightleftharpoons MgH_2L^{2+})$ 0·90	63C
				$K(Mg^{2+}+HL^- \rightleftharpoons MgHL^+)$ 1·70	
Na^+	gl	?	0·28[$(CH_3)_4NBr$]	K_1 0·30 ± 0·07	63E
	oth	25	0·1	K_1 0·28 ± 0·02, cation-sensitive glass electrode	64Ra
Nb^V	dis	20	4·5 (NaCl or $NaNO_3$ + 2·5 HCl)	$K[Nb(OH)_4{}^++H_2L \rightleftharpoons Nb(OH)_4HL+H^+]$ 2·01	67K

continued overleaf

202 $C_4H_6O_5$ (contd)

Metal	*Method*	*Temp*	*Medium*	*Log of equilibrium constant, remarks*	*Ref*
Nd^{3+}	gl	30	0·1 (KCl)	K_1 5·12, K_2 3·64, K_3 2·92	62Ca
	pH	25	~ 0·2 (KCl)	K_1 4·45 ± 0·05	64D
Ni^{2+}	gl	20	0·1 ($NaClO_4$)	$K(Ni^{2+} + H_2L \rightleftharpoons NiH_2L^{2+})$ 1·83	63C
				$K(Ni^{2+} + HL^- \rightleftharpoons NiHL^+)$ 3·17	
Pa^V	ix	25	0·25	$K[Pa_2O(OH)_{1\cdot5}{}^{6\cdot5+} + HL^- \rightleftharpoons Pa_2O(OH)_{1\cdot5}HL^{5\cdot5+}]$ (?) 2·42	66G
				$K[Pa_2O(OH)_{1\cdot5}{}^{6\cdot5+} + 2HL^- \rightleftharpoons Pa_2O(OH)_{1\cdot5}(HL)_2{}^{4\cdot5+}]$ (?) 4·80	
Pr^{3+}	gl	30	0·1 (KCl)	K_1 5·04, K_2 3·40, K_3 2·80	62Ca
	pH	25	~ 0·2 (KCl)	K_1 4·28 ± 0·05	64D
Rb^+	gl	?	0·28 [$(CH_3)_4NBr$]	K_1 0·18 ± 0·06	63E
	oth	25	0·1	K_1 0·04 ± 0·04, cation-sensitive glass electrode	64Ra
Sm^{3+}	pH	25	~ 0·2 (KCl)	K_1 4·41 ± 0·03	64D
Sr^{2+}	ix	?	0·1 NH_4Cl	K_1 2·4 ± 0·3	62B
Th^{4+}	ix	?	0·3	K_1 5·15, β_2 6·70	62G
	kin	25	~ 0	$K(2Th^{4+} + L^{2-} \rightleftharpoons Th_2L^{6+})$ 13·34 ± 0·08	63Y
$UO_2{}^{2+}$	gl	30	0·1 KCl	K_1 5·50, K_2 3·63	62C
	gl	25	1·0 (KNO_3)	$K[UO_2{}^{2+} + L^{2-} \rightleftharpoons UO_2(H_{-1}L)^- + H^+]$ 1·66	64Rb
				$K[UO_2{}^{2+} + H_2L \rightleftharpoons UO_2(H_{-1}L)^- + 3H^+]$ −5·55	
				$K[2UO_2{}^{2+} + 2H_2L \rightleftharpoons (UO_2)_2(H_{-1}L)_2{}^{2-} + 6H^+]$ −7·75	
				$K[(UO_2)_2(H_{-1}L)_2(OH)^{3-} + H^+ \rightleftharpoons (UO_2)_2(H_{-1}L)_2{}^{2-}]$ 6·1 ± 0·2	
				$K[(2UO_2)_3(H_{-1}L)_3(OH)_2{}^{5-} + 4H^+ \rightleftharpoons 3(UO_2)_2(H_{-1}L)_2{}^{2-}]$ 19·35 ± 0·05	
Yb^{3+}	gl	21—22	0·12	K_1 4·92 ± 0·04	62D
Zn^{2+}	gl	20	0·1 ($NaClO_4$)	$K(Zn^{2+} + H_2L \rightleftharpoons ZnH_2L^{2+})$ 1·66	63C
				$K(Zn^{2+} + HL^- \rightleftharpoons ZnHL^+)$ 2·93	
Zr^{4+}	ix	?	2 $HClO_4$	$K(Zr^{4+} + H_2L \rightleftharpoons ZrHL^{3+} + H^+)$ 1·94 ± 0·04	60R
	ix	?	2·0 $HClO_4$	$K(Zr^{4+} + H_2L \rightleftharpoons ZrHL^{3+} + H^+)$ 2·24 ± 0·05	64R

57V O. Vartapetian, *Ann. Chim. (France)*, 1957, **2**, 916
60R E. I. Ryabchikov, A. N. Ermakov, V. K. Belyaeva, and I. N. Marov, *Russ. J. Inorg. Chem.*, 1960, **5**, 505 (1051)
61M P. G. Manning and C. B. Monk, *Trans. Faraday Soc.*, 1961, **57**, 1996
62B I. E. Bukolov, K. V. Astakhov, V. I. Zimin, and V. S. Tairov, *Russ. J. Inorg. Chem.*, 1962, **7**, 816 (1577)
62C M. Cefola, R. C. Taylor, P. S. Gentile, and A. V. Celiano, *J. Phys. Chem.*, 1962, **66**, 790
62Ca M. Cefola, A. S. Tompa, A. V. Celiano, and P. S. Gentile, *Inorg. Chem.*, 1962, **1**, 290
62D N. K. Davidenko, *Russ. J. Inorg. Chem.*, 1962, **7**, 1412 (2709)
62G I. Geletseanu and A. V. Lapitskii, *Proc. Acad. Sci. (U.S.S.R.)*, 1962, **144**, 460 (573)
63C E. Campi, *Ann. Chim. (Italy)*, 1963, **53**, 96
63E L. E. Erikson and J. A. Dembo, *J. Phys. Chem.*, 1963, **67**, 707
63Y K. B. Yatsimirskii and Yu. A. Zhukov, *Russ. J. Inorg. Chem.*, 1963, **8**, 149 (295)
64D N. K. Davidenko, *Russ. J. Inorg. Chem.*, 1964, **9**, 859 (1584)
64R D. I. Ryabchikov, I. N. Marov, A. N. Ermakov, and V. K. Belyaeva, *J. Inorg. Nuclear Chem.*, 1964, **26**, 965
64Ra G. A. Rechnitz and S. B. Zamochnick, *Talanta*, 1964, **11**, 1061
64Rb K. S. Rajan and A. E. Martell, *J. Inorg. Nuclear Chem.*, 1964, **26**, 1927
64T C. F. Timberlake, *J. Chem. Soc.*, 1964, 5078
65A D. W. Archer, D. A. East, and C. B. Monk, *J. Chem. Soc.*, 1965, **720**
65M M. K. Misra and R. K. Nanda, *J. Indian Chem. Soc.*, 1965, **42**, 267

contin

66G I. Galateanu, *Canad. J. Chem.*, 1966, **44**, 647
67K C. Kovecny, *Z. phys. Chem. (Leipzig)*, 1967, **235**, 39
67R K. S. Rajan and A. E. Martell, *J. Inorg. Nuclear Chem.*, 1967, **29**, 463
68R S. Ramamoorthy and M. Sautappa, *J. Inorg. Nuclear Chem.*, 1968, **30**, 1855

$C_4H_6O_6$ 2,3-Dihydroxybutanedioic acid (tartaric acid) H_2L 203

Metal	*Method*	*Temp*	*Medium*	*Log of equilibrium constant, remarks*	*Ref*
H^+	gl	18	1 NaCl	K_1 3·69, K_2 2·48	57V
	?	?	?	K_1 4·34, K_2 2·98	61Z
	?	?	?	$K(H_{-1}L^{3-}+H^+ \rightleftharpoons L^{2-})$ 13·26	63F
	H	25	0·2 (KCl)	K_1 3·94, K_2 2·88	63S, 66D
	gl	25	1 (KNO_3)	K_1 3·77, K_2 2·60	64R
	?	?	?	K_1 4·81, K_2 3·22	64Ra
	gl	20	0·1 ($NaClO_4$)	K_1 3·96, K_2 2·80	64T
	gl	20	$\rightarrow 0$	K_1 4·52 ± 0·02, K_2 2·89 ± 0·02	65F
	gl	20	?	K_1 4·62, K_2 3·19	65Fa
	H, AgCl	25	0·1	K_1 3·03, K_2 1·36	65Sc
	H	25	var	K_1 4·09, K_2 2·96	66P
	?	?	?	$K(H_{-1}L^{3}+H^+ \rightleftharpoons L^{2-})$ 15·79	66P
	oth	20	$\rightarrow 0$	K_1 4·5, K_2 3·2, optical rotation	67F
	?	?	?	K_1 3·75, K_2 3·05	67T
Ag^+	dis	20	0·1 ($KClO_4$)	K_1 < 2·0	63S
Al^{3+}	dis	20	0·1 ($KClO_4$)	β_2 9·56 ± 0·04	63S
	oth	20	$\rightarrow 0$	$K(Al^{3+}+HL^- \rightleftharpoons AlHL^{2+})$ 3·43	67F
				K_1 6·35	
				$K(AlOH^{2+}+L^{2-} \rightleftharpoons AlH_{-1}L)$ 9·05	
				$K[Al(OH)_2^+ + L^{2-} \rightleftharpoons Al(H_{-2}L)^-]$ 10·92	
				$K[Al(OH)_3 + L^{2-} \rightleftharpoons Al(H_{-2}L)OH^{2-}]$ 8·37	
				$K[Al(OH)_4^- + L^{2-} \rightleftharpoons Al(H_{-2}L)(OH)_2^{3-}]$ 8·89	
				$K(Al^{3+}+AlL^+ \rightleftharpoons Al_2L^{4+})$ ~ 2	
				$K[Al^{3+}+Al(H_{-1}L) \rightleftharpoons Al_2(H_{-1}L)^{3+}]$ ~ 3	
				$K[Al^{3+}+Al(H_{-2}L)^- \rightleftharpoons Al_2(H_{-2}L)^{2+}]$ ~ 4	
				$K[AlOH^{2+}+Al(H_{-2}L)^- \rightleftharpoons Al_2(H_{-2}L)OH^+]$ ~ 3	
				$K[Al(OH)_2^+ + Al(H_{-2}L)^- \rightleftharpoons Al_2(H_{-2}L)(OH)_2]$ ~ 2	
				$K[Al(OH)_3 + Al(H_{-2}L)^- \rightleftharpoons Al_2(H_{-2}L)(OH)_3^-]$ 0	
				$K[Al(OH)_3 + Al(H_{-2}L)OH^{2-} \rightleftharpoons Al_2(H_{-2}L)(OH)_4^{2-}]$ 0	
				$K[Al(OH)_3 + Al(H_{-2}L)(OH)_2^{3-} \rightleftharpoons Al_2(H_{-2}L)(OH)_5^{3-}]$ 0	
				$K[Al(OH)_4^- + Al(H_{-2}L)(OH)_2^{3-} \rightleftharpoons Al_2(H_{-2}L)(OH)_6^{4-}]$ 0	
				$K(AlL^+ + H_2L \rightleftharpoons AlH_2L_2^+)$ 0·66	
				$K[Al(H_{-1}L)+H_2L \rightleftharpoons AlHL_2]$ 1·12	
				$K[Al(H_{-2}L)^- + H_2L \rightleftharpoons AlL_2^-]$ 2·11	
				$K[Al(H_{-2}L)^- + HL^- \rightleftharpoons Al(H_{-1}L)^{2-}]$ 1·92	
				$K[Al(H_{-2}L)^- + L^{2-} \rightleftharpoons Al(H_{-1}L)_2^{3-}]$ 1·54	
				$K[Al(H_{-2}L)OH^{2-} + L^{2-} \rightleftharpoons AP(H_{-1}L)(H_{-2}L)^{4-}]$ ~ 0·3	
				$K[Al(H_{-2}L)(OH)_2^{3-} + L^{2-} \rightleftharpoons Al(H_{-2}L)_2^{5-}]$ ~ 0·0	
				optical rotation	
Am^{3+}	oth	?	0·1 (NaCl)	β_2 7·80, paper electrophoresis	65M
B^{III}	gl	20	?	$K[B(OH)_3 + H_2L \rightleftharpoons B(H_{-1}L)]$ 0·77 ± 0·03	65Fa
				$K[B(OH)_3 + HL^- \rightleftharpoons B(H_{-2}L)^-]$ 1·60 ± 0·05	
				$K[B(OH)_3 + L^{2-} \rightleftharpoons B(H_{-2}L)OH^{2-}]$ 0·61 ± 0·09	
				$K[B(OH)_4^- + L^{2-} \rightleftharpoons B(H_{-2}L)(OH)_2^{3-}]$ 0·77 ± 0·03	

continued overleaf

203 $C_4H_6O_6$ (contd)

Metal	Method	Temp	Medium	Log of equilibrium constant, remarks	Ref
B^{III} (contd)	sol	22	?	$K[B(OH)_4^- + HL^- \rightleftharpoons BOL^- + OH^-]$ 6·97	67S
				$K[2B(OH)_4^- + HL^- \rightleftharpoons B_2O_3L^{2-} + OH^-]$ 14·07	
Ba^{2+}	dis	20	0·1 ($KClO_4$)	K_1 < 2·0	63S
	gl	37	0·2 $NaClO_4$	K_1 2·19	67Ta
Be^{2+}	dis	20	0·1 ($KClO_4$)	K_1 2·89±0·03	63S
	pH	18	0·1	K_1 2·57±0·01	65K
Bi^{3+}	dis	20	0·1 ($KClO_4$)	β_2 11·3±0·1	63S
Ca^{2+}	dis	20	0·1 ($KClO_4$)	K_1 < 2·0	63S
	gl	37	~ 0·09 $NaClO_4$	K_1 2·17	65T
Cd^{2+}	dis	20	0·1 ($KClO_4$)	K_1 (?) 3·71±0·03	63S
	dis	30	1·0 ($NaClO_4$)	K_1 (?) 1·1±0·2, β_2 (?) 2·2±0·2, β_3 (?) 2·4±0·2	65H
Ce^{3+}	gl (?)	25	0·3	$K(Ce^{3+} + HL^- \rightleftharpoons CeHL^{2+})$ 2·54±0·03	65Ba
	gl	25	0·1 ($NaNO_3$)	K_1 5·5±0·2, K_2 2·9±0·2	65Sc
				$K(2Ce^{3+} + 2L^{2-} \rightleftharpoons Ce_2L_2^{2+})$ 10·8	
				$K(2Ce^{3+} + 3L^{2-} \rightleftharpoons Ce_2L_3)$ 14·7	
	gl	23—25	0·2 KCl	K_1 3·09	66D
	H	25	var	K_1 3·84, β_2 6·72	66P
				$K(2Ce^{3+} + L^{2-} \rightleftharpoons Ce_2L^{4+})$ 5·80	
				$K[Ce^{3+} + (H_{-1}L)^{3-} \rightleftharpoons Ce(H_{-1}L)]$ 11·42	
Cm^{3+}	oth	?	0·1 (NaCl)	β_2 7·40, paper electrophoresis	65Ma
Co^{2+}	dis	25	0·000—0·155	K_1 (→ 0) 3·08±0·02 (0·035) 2·50±0·01, (0·055) 2·377±0·005, (0·075) 2·288±0·002, (0·155) 2·098±0·007	61M
	dis	20	0·1 ($KClO_4$)	K_1 2·8±0·1	63S
	gl	25	0 corr	K_1 3·08±0·02, K_2 0·7	65M
	ix	25	0 corr	K_1 3·02±0·02, K_2 1·19±0·09	65Sb
Co^{3+}	gl	25, 35	0 corr	$K[CoCl(NH_3)_5^{2+} + L^{2-} \rightleftharpoons CoCl(NH_3)_5L]$ (25°) 2·00, (35°) 1·98	65A
	sol	25, 35	0 corr	$K[CoCl(NH_3)_5^{2+} + L^{2-} \rightleftharpoons CoCl(NH_3)_5L]$ (25°) 2·12, (35°) 2·09	65A
Cu^{2+}	pol	25	0·3—2·0	$K[Cu^{2+} + 2OH^- + 2L^{2-} \rightleftharpoons Cu(OH)_2L_2^{4-}]$ 9·85	49M
	ix	20	var	K_1 3·1±0·1, K_2 1·8±0·3, K_3 0·8±0·3 (+ acid)	64L
	gl	25	1·0 (KNO_3)	K_1 2·6	67R
				$K(2Cu^{2+} + 2L^{2-} \rightleftharpoons Cu_2L_2)$ 8·2	
Dy^{3+}	gl	23—25	0·2 KCl	K_1 3·28	66D
Er^{3+}	gl	23—25	0·2 KCl	K_1 3·33	66D
Eu^{3+}	dis	25	0·054 ($NaClO_4$)	K_1 3·92, β_2 6·70	62M
	oth	?	0·1 (NaCl)	β_2 6·20, paper electrophoresis	65Ma
	gl	23—25	0·2 KCl	K_1 3·40	66D
Fe^{2+}	ix	20	0·1 ($NaClO_4$)	K_1 2·24	64T
Fe^{3+}	dis	20	0·1 ($KClO_4$)	β_2 11·86±0·05	63S
	gl	20	0·1 ($NaClO_4$)	K_1 6·49	64T
				$K[2Fe^{3+} + 2L^{2-} \rightleftharpoons Fe_2(H_{-1}L)_2 + 2H^+]$ 11·87	
				$K[2Fe^{3+} + 2L^{2-} \rightleftharpoons Fe_2(H_{-1}L)(H_{-2}L)^- + 3H^+]$ 9·05	
				$K[3Fe^{3+} + 3L^{2-} \rightleftharpoons Fe_3(H_{-2}L)_3^{3-} + 6H^+]$ 9·48	
	sp	25	1	$K(Fe^{3+} + H_2L \rightleftharpoons FeL^+ + 2H^+)$ (?) −0·66±0·04	67N
Ga^{3+}	dis	20	0·1 ($KClO_4$)	β_2 9·76±0·05	63S
Gd^{3+}	gl	23—25	0·2 KCl	K_1 3·32	66D
Ge^{IV}	gl	18	1 NaCl	$K[H_2GeO_3 + HL^- \rightleftharpoons GeO(H_{-1}L)^-]$ 5·2±0·2	57V
Hf^{4+}	dis	?	2 $HClO_4$	$K(Hf^{4+} + H_2L \rightleftharpoons HfHL^{3+} + H^+)$ 1·69±0·01	59R
	ix	?	1 $HClO_4$	$K(Hf^{4+} + H_2L \rightleftharpoons HfHL^{3+} + H^+)$ 2·45±0·03	60R
	ix	?	2·0 $HClO_4$	$K(Hf^{4+} + H_2L \rightleftharpoons HfHL^{3+} + H^+)$ 2·99±0·01	64Ra
Hg^{2+}	dis	20	0·1 ($KClO_4$)	K_1 < 4	63S

contin[ued]

Metal	*Method*	*Temp*	*Medium*	*Log of equilibrium constant, remarks*	*Ref*
Ho^{3+}	gl	23—25	0·2 KCl	K_1 3·38	66D
In^{3+}	dis	20	0·1 $(KClO_4)$	K_1 4·48±0·04	63S
La^{3+}	dis	20	0·1 $(KClO_4)$	β_2 6·72±0·05	63S
	con (?)	25	?	$K(La^{3+}+HL^- \rightleftharpoons LaHL^{2+})$ 2·5±0·1	65B
	gl	23—25	0·2 KCl	K_1 2·89	66D
	H	25	var	K_1 3·68, β_2 6·37	66P
				$K(2La^{3+}+L^{2-} \rightleftharpoons La_2L^{4+})$ 5·32	
				$K[La^{3+}+(H_{-1}L)^{3-} \rightleftharpoons La(H_{-1}L)]$ 10·73	
Lu^{3+}	gl	23—25	0·2 KCl	K_1 3·76	66D
Mg^{2+}	dis	20	0·1 $(KClO_4)$	$K_1 < 2$	63S
	gl	37	0·2 $NaClO_4$	K_1 1·91	67Ta
Mn^{2+}	dis	20	0·1 $(KClO_4)$	K_1 (?) 2·92±0·05	63S
	gl	32	0·1 $NaClO_4$	K_1 1·44	67T
				$K[Mn(H_{-1}L)^- + H^+ \rightleftharpoons MnL]$ 7·62	
				$K[Mn(H_{-2}L)^{2-} + H^+ \rightleftharpoons Mn(H_{-1}L)^-]$ 10·14	
Mo^{III}	pol	25	0·2 Na_2SO_4	K(?) 3·17	62Z
			0·1 H_2SO_4		
			0·04 KNO_3		
Mo^{IV}	pol	25	1 $HClO_4$	K (?) 2·06	62Z
Mo^{VI}	dis	20	0·1 $(KClO_4)$	$K(H_2MoO_4+2L^{2-} \rightleftharpoons MoO_2(H_{-1}L)_2^{4-}$(?) 7·66±0·03	63S
Na^+	gl	20	→ 0	K_1 1·98±0·04	65F
				$K(Na^+ + HL^- \rightleftharpoons NaHL)$ 1·47±0·02	
Nb^V	dis	20	4·5 (NaCl or $NaNO_3$+2·5 HCl)	$K[Nb(OH)_4^+ + H_2L \rightleftharpoons Nb(OH)_4HL + H^+]$ 2·34	67K
Nd^{3+}	gl	23—25	0·2 KCl	K_1 3·45	66D
Ni^{2+}	dis	20	0·1 $(KClO_4)$	β_2 5·42±0·06	63S
NpO_2^+	ix	18—22	0·5 NH_4ClO_4	K_1 2·32, β_2 4·30, β_3 6·18	61Ma
				$K(NpO_2^+ + HL^- \rightleftharpoons NpO_2HL)$ 2·36	
Pa^V	ix	25	0·25	$K[Pa_2O(OH)_{3\cdot5}^{4\cdot5+} + HL^- \rightleftharpoons Pa_2O(OH)_{3\cdot5}HL^{3\cdot5+}]$ (?) 2·34,	66G
				$K[Pa_2O(OH)_{4\cdot5}^{3\cdot5+} + 2HL^- \rightleftharpoons Pa_2O(OH)_{4\cdot5}(HL)_2^{1\cdot5+}]$ (?) 4·96	
Pb^{2+}	dis	20	0·1 $(KClO_4)$	K_1 2·92±0·04	63S
Pm^{3+}	oth	?	0·1 (NaCl)	β_2 5·81, paper electrophoresis	65Ma
Pr^{3+}	gl	23—25	0·2 KCl	K_1 3·25	66D
Sc^{3+}	dis	20	0·1 $(KClO_4)$	β_2 12·5±0·1	63S
	sp	18—20	0·01 (HNO_3)	K_1 6·20±0·06	66Ka
Sm^{3+}	gl	23—25	0·2 KCl	K_1 3·50	66D
Sn^{2+}	gl	20	0·1 KCl (?)	K_1 5·2, β_2 9·91	65S
Sr^{2+}	ix	?	0·1 NH_4Cl	K_1 1·8±0·3	62B
	dis	20	0·1 $(KClO_4)$	$K_1 < 2{\cdot}0$	63S
	gl	37	0·2 $NaClO_4$	K_1 2·25	67Ta
Tb^{3+}	gl	23—25	0·2 KCl	K_1 3·33	66D
Te^{VI}	oth	22	?	* $K[H_5TeO_6^- + L^{2-} \rightleftharpoons TeO(OH)_3(H_{-2}L)^{3-}]$ 1·64, 1·79, optical rotation	63L
Th^{4+}	ix	?	0·3	K_1 4·64	62G
	kin	25	~ 0	$K[2ThOH^{3+} + L^{2-} \rightleftharpoons Th_2(H_{-2}L)^{4+}]$ 13·2±0·2	63Y
Ti^{4+}	dis	20	0·1 $(KClO_4)$	$K(TiO^{2+} + 2L^{2-} \rightleftharpoons TiOL_2^{2-})$ 9·7±0·1	63S
Tl^+	con	28	?	K_1 1·39	65Sa
Tl^{3+}	pH	20	1 $(NaClO_4)$	K_1 11·57, β_2 12·81, β_3 13·34	62Ba

continued overleaf

203 $C_4H_6O_6$ (contd)

Metal	Method	Temp	Medium	Log of equilibrium constant, remarks	Ref
UO_2^{2+}	dis	20	0·1 ($KClO_4$)	β_2 9·73±0·05	63S
	gl	25	1·0 (KNO_3)	$K[UO_2^{2+}+L^{2-} \rightleftharpoons UO_2(H_{-1}L)^- + H^+]$ 0·75	64R
				$K[UO_2^{2+}+H_2L \rightleftharpoons UO_2(H_{-1}L)^- + 3H^+]$ −5·62	
				$K[UO_2^{2+}+2H_2L \rightleftharpoons (UO_2)_2(H_{-1}L)_2^{2-} + 6H^+]$ −8·00	
				$K[(UO_2)_2(H_{-1}L)_2(OH)^{3-} + H^+ \rightleftharpoons (UO_2)_2(H_{-1}L)_2^{2-}]$ 5·26±0·10	
				$K[(UO_2)_3(H_{-1}L)_3(OH)_2^{5-} + 4H^+ \rightleftharpoons 3(UO_2)_2(H_{-1}L)_2^{2-}]$ 17·91±0·06	
V^V	sp	25	?	$K[2HVO_3+H_2L \rightleftharpoons (HVO_2)_2(H_{-2}L)]$ (?) 4·9±0·2	63G
W^{VI}	kin	25	0·1	$K(H_2WO_4+H_2L \rightleftharpoons H_2WO_3L)$ 3·93±0·06	62Y
Y^{3+}	H	25	var	K_1 4·07, β_2 6·89	66P
				$K(2Y^{3+}+L^{2-} \rightleftharpoons Y_2L^{4+})$ 5·97	
				$K[Y^{3+}+(H_{-1}L)^{3-} \rightleftharpoons Y(H_{-1}L)]$ 12·87	
				$K(Y^{3+}+HL^- \rightleftharpoons YHL^{2+})$ 2·82	
Yb^{3+}	gl	23—25	0·2 KCl	K_1 3·48	66D
	pol	?	0·1 $(CH_3)_4NI$	$K(YbL_6^{9-}+e^- \rightleftharpoons YbL_4^{6-}+2L^{2-})$ −7	58K
Zn^{2+}	Zn	20	0·01—0·3	$K[Zn^{2+}+(H_{-1}L)^{3-} \rightleftharpoons Zn(H_{-1}L)^-]$ 6·33±0·02	63F
				K_1 3·09±0·05, K_2 1·89±0·07	
				$K(Zn^{2+}+HL^- \rightleftharpoons ZnHL^+)$ 1·22±0·08	
				K_{so} −6·54	
	dis	20	0·1 ($KClO_4$)	K_1 2·69±0·08	63S
	dis	25	0 corr	K_1 3·31±0·08, K_2 1·85±0·06	66R
Zr^{4+}	dis	?	2 $HClO_4$	$K(Zr^{4+}+H_2L \rightleftharpoons ZrHL^{3+}+H^+)$ 2·19±0·03	59R
	ix	?	1 $HClO_4$	$K(Zr^{4+}+H_2L \rightleftharpoons ZrHL^{3+}+H^+)$ 3·15±0·02	60R
	kin	25	var	$K[Zr(OH)_3^+ + H_2L \rightleftharpoons ZrOHL^+]$ 5·51±0·06	61Y
				$K[Zr(OH)L+H_2L \rightleftharpoons ZrL_2+H^+]$ 0·95±0·07	
	ix	?	2·0 $HClO_4$	$K(Zr^{4+}+H_2L \rightleftharpoons ZrHL^{3+}+H^+)$ 2·49±0·03	64Ra
	sp	18—20	1 (HCl)	$K(Zr^{4+}+HL^- \rightleftharpoons ZrHL^{3+})$ 6·86±0·07	66K

* Apparent constants varying with pH.

49M L. Meites, *J. Amer. Chem. Soc.*, 1949, **71**, 3269
57V O. Vartapetian, *Ann. Chim. (France)*, 1957, **2**, 916
58K N. A. Kostromina and S. I. Yakubson, *J. Inorg. Chem. (U.S.S.R.)*, 1958, **3**, No. 11, 104 (2506)
59R D. I. Ryabchikov, A. N. Ermakov, V. K. Belyaeva, and I. N. Marov, *Russ. J. Inorg. Chem.*, 1959, **4**, 818 (1814)
60R D. I. Ryabchikov, A. N. Ermakov, V. K. Belyaeva, and I. N. Marov, *Russ. J. Inorg. Chem.*, 1960, **5**, 505 (1051)
61M P. G. Manning and C. B. Monk, *Trans. Faraday Soc.*, 1961, **57**, 1996
61Ma A. I. Moskvin, I. N. Marov, and Yu. A. Zolotov, *Russ. J. Inorg. Chem.*, 1961, **6**, 926, (1813)
61Y K. B. Yatsimirskii and L. P. Raizman, *Russ. J. Inorg. Chem.*, 1961, **6**, 1263 (2496)
62B I. E. Bukolov, K. V. Astakhov, V. I. Zimin, and U. S. Tairov, *Russ. J. Inorg. Chem.*, 1962, **7**, 816 (1577)
62Ba A. I. Busev, V. G. Tiptsova, and L. M. Sonokina, *Russ. J. Inorg. Chem.*, 1962, **7**, 1098 (2122)
62G I. Geletseanu and A. V. Lapitskii, *Proc. Acad. Sci. (U.S.S.R.)*, 1962, **144**, 460 (573)
62M P. G. Manning and C. B. Monk, *Trans. Faraday Soc.*, 1962, **58**, 938
62Y K. B. Yatsimirskii and K. E. Prik, *Russ. J. Inorg. Chem.*, 1962, **7**, 821 (1589)
62Z E. W. Zahnow and R. J. Robinson, *J. Electroanalyt. Chem.*, 1962, **3**, 263
63F V. Frei and J. Loub, *Z. phys. Chem. (Leipzig)*, 1963, **222**, 249
63G B. N. Ghosh, S. P. Moulik, K. K. Sengupta, and P. K. Pal, *J. Indian Chem. Soc.*, 1963, **40**, 509
63L J. G. Lanese and B. Jaselskis, *Analyt. Chem.*, 1963, **35**, 1878
63S J. Stary, *Analyt. Chim. Acta*, 1963, **28**, 132
63Y K. B. Yatsimirskii and Yu. A. Zhukov, *Russ. J. Inorg. Chem.*, 1963, **8**, 149 (295)
64L I. Lundquist, *Acta Chem. Scand.*, 1964, **18**, 858
64R K. S. Rajan and A. E. Martell, *J. Inorg. Nuclear Chem.*, 1964, **26**, 1927
64Ra D. I. Ryabchikov, I. N. Marov, A. N. Ermakov, and V. K. Belyaeva, *J. Inorg. Nuclear Chem.* 1964, **26**, 965
64T C. F. Timberlake, *J. Chem. Soc.*, 1964, 1229
65A D. W. Archer, D. A. East, and C. B. Monk, *J. Chem. Soc.*, 1965, 720
65B F. Brezina and J. Rosicky, *Monatsh.*, 1965, **96**, 1025

contin

65Ba F. Brezina, J. Rosicky, and R. Pastorek, *Monatsh.*, 1965, **96**, 533
65F V. Frei, *Coll. Czech. Chem. Comm.*, 1965, **30**, 1402
65Fa V. Frei and A. Solcova, *Coll. Czech. Chem. Comm.*, 1965, **30**, 961
65H H. E. Hellwege and G. K. Schweitzer, *J. Inorg. Nuclear Chem.*, 1965, **27**, 99
65K I. F. Kolosova and T. A. Belgarskaya, *Russ. J. Inorg. Chem.*, 1965, **10**, 411 (764)
65M C. B. Monk, *J. Chem. Soc.*, 1965, 2456
65Ma G. Marcu and K. Samochocka, *Studia Univ. Babes-Bolyai*, 1965, **10**, 71
65S T. D. Smith, *J. Chem. Soc.*, 1965, 2145
65Sa K. N. Sahn and A. K. Bhattacharya, *J. Indian Chem. Soc.*, 1965, **42**, 247
65Sb R. G. Seys and C. B. Monk, *J. Chem. Soc.*, 1965, 2452
65Sc A. V. Stepanov, V. P. Shvedov, and A. P. Rozhnov, *Russ. J. Inorg. Chem.*, 1965, **10**, 750 (1379)
65T N. Tripathy and R. K. Patnaik, *J. Indian Chem. Soc.*, 1965, **42**, 712
66D N. K. Davidenko and V. F. Deribon, *Russ. J. Inorg. Chem.*, 1966, **11**, 53
66G I. Galateanu, *Canad. J. Chem.*, 1966, **44**, 647
66K I. M. Korenman, E. R. Sheyanova, and Z. M. Gur'eva, *Russ. J. Inorg. Chem.*, 1966, **11**, 1485 (2761)
66Ka I. M. Korenman and N. V. Zaglyadimova, *Russ. J. Inorg. Chem.*, 1966, **11**, 1491 (2774)
66P R. Pastorek and F. Brezina, *Monatsh.*, 1966, **97**, 1095
66R D. L. G. Rowlands and C. B. Monk, *Trans. Faraday Soc.*, 1966, **62**, 945
67F V. Frei, *Coll. Czech. Chem. Comm.*, 1967, **32**, 1815
67K C. Kovecny, *Z. phys. Chem. (Leipzig)*, 1967, **235**, 39
67N G. Niac and Z. Andrei, *Rev. Roumaine Chim.*, 1967, **12**, 801
67R K. S. Rajan and A. E. Martell, *J. Inorg. Nuclear Chem.*, 1967, **29**, 463
67S M. B. Shchigol and N. B. Burchinskaya, *Russ. J. Inorg. Chem.*, 1967, **12**, 626 (1183)
67T K. K. Tripathy and R. K. Patnaik, *Indian J. Chem.*, 1967, **5**, 511
67Ta N. Tripathy, K. K. Tripathy, and R. K. Patnaik, *J. Indian Chem. Soc.*, 1967, **44**, 329

$C_4H_6N_2$ *N*-Methyl-1,3-diazole (*N*-methylimidazole) L 204

Metal	*Method*	*Temp*	*Medium*	*Log of equilibrium constant, remarks*	*Ref*
Ag^+	gl	25—27	1·0 (KNO_3)	K_1 (→ 25°) 3·00, (25·6°) 2·98, (27°) 2·94 K_2 (→ 25°) 3·89, (25·6°) 3·87, (27°) 3·82	64B

64B J. E. Bauman jun. and J. C. Wang, *Inorg. Chem.*, 1964, **3**, 368

C_4H_8O *trans*-But-2-en-1-ol L 207.1

Metal	*Method*	*Temp*	*Medium*	*Log of equilibrium constant, remarks*	*Ref*
Ag^+	Ag	25	2 ($NaClO_4$)	K_1 0·896±0·001, K_2 −0·24±0·06	67H
Pt^{2+}	sp	60	2 (NaCl)	$K(PtCl_4^{2-}+L \rightleftharpoons PtCl_3L^-+Cl^-)$ 3·48	67H

67H F. R. Hartley and L. M. Venanzi, *J. Chem. Soc. (A)*, 1967, 330, 333

C_4H_8O 2-Methoxypropene L 207.2

Metal	*Method*	*Temp*	*Medium*	*Log of equilibrium constant, remarks*	*Ref*
Ag^+	dis	10, 20	ethylene glycol	K_1 (10°) 0·98, (20°) 0·87, gas chromatography ΔH_1 −4·7±0·2, ΔS_1 −12·0±0·6	68F

68F T. Fueno, O. Kajimoto, T. Okuyama, and J. Furukawa, *Bull. Chem. Soc. Japan*, 1968, **41**, 785

207.3 C_4H_8O Ethoxyethene (ethylvinyl ether) L

Metal	Method	Temp	Medium	Log of equilibrium constant, remarks	Ref
Ag^+	dis	10—30	ethylene glycol	K_1 (10°) 0·96, (20°) 0·83, (30°) 0·74 gas chromatography ΔH_1 −4·4±0·2, ΔS_1 −11·0±0·6	68Fa
	dis	25	1 (KNO_3)	K_1 1·11, K_2 −0·2	68F

68F T. Fueno, O. Kajimoto, and J. Furukawa, *Bull. Chem. Soc. Japan*, 1968, **41,** 782
68Fa T. Fueno, O. Kajimoto, T. Okuyama, and J. Furukawa, *Bull. Chem. Soc. Japan*, 1968, **41,** 785

211 $C_4H_8O_2$ 2-Methylpropanoic acid (isobutyric acid) HL

Metal	Method	Temp	Medium	Log of equilibrium constant, remarks	Ref
H^+	gl	25	0·5 ($NaClO_4$)	K_1 4·64±0·01	64S
Ce^{3+}	gl	25	0·5 ($NaClO_4$)	K_1 1·79, β_2 2·32	64S
	gl	25	2·00 ($NaClO_4$)	K_1 1·62±0·03, K_2 1·10±0·05	65C
	cal	25	2·00 ($NaClO_4$)	ΔH_1 3·33±0·07, ΔS_1 18·6±1, ΔH_2 2·6±0·2, ΔS_2 13·6±1	65C
Cu^{2+}	sp	30	0·1	K_1 2·44	65D
Dy^{3+}	gl	25	0·5 ($NaClO_4$)	K_1 1·74, β_2 2·57	64S
	gl	25	2·00 ($NaClO_4$)	K_1 1·65±0·03, K_2 1·32±0·05	65C
	cal	25	2·00 ($NaClO_4$)	ΔH_1 5·0±0·1, ΔS_1 24·5±1, ΔH_2 1·8±0·1, ΔS_2 12·0±1	65C
Er^{3+}	gl	25	0·5 ($NaClO_4$)	K_1 1·69, β_2 2·59	64S
	gl	25	2·00 ($NaClO_4$)	K_1 1·61±0·03, K_2 1·11±0·05	65C
	cal	25	2·00 ($NaClO_4$)	ΔH_1 5·5±0·1, ΔS_1 25·8±1, ΔH_2 3·4±0·2, ΔS_2 16·6±1	65C
Eu^{3+}	gl	25	0·5 ($NaClO_4$)	K_1 1·98, β_2 3·10	64S
	gl	25	2·00 ($NaClO_4$)	K_1 1·98±0·03, K_2 1·31±0·05	65C
	cal	25	2·00 ($NaClO_4$)	ΔH_1 2·91±0·09, ΔS_1 18·8±1, ΔH_2 1·9±0·1 ΔS_2 12·2±1	65C
Gd^{3+}	gl	25	0·5 ($NaClO_4$)	K_1 1·87, β_2 3·28	64S
	gl	25	2·00 ($NaClO_4$)	K_1 1·86±0·03, K_2 1·34±0·05	65C
	cal	25	2·00 ($NaClO_4$)	ΔH_1 3·45±0·07, ΔS_1 20·1±1, ΔH_2 1·7±0·1, ΔS_2 11·7±1	65C
Ho^{3+}	gl	25	0·5 ($NaClO_4$)	K_1 1·70, β_2 2·92	64S
	gl	25	2·00 ($NaClO_4$)	K_1 1·63±0·03, K_2 1·21±0·05	65C
	cal	25	2·00 ($NaClO_4$)	ΔH_1 5·3±0·1, ΔS_1 25·3±1, ΔH_2 2·6±0·2, ΔS_2 14·1±1	65C
La^{3+}	gl	25	0·5 ($NaClO_4$)	K_1 1·64, β_2 2·16	64S
	gl	25	2·00 ($NaClO_4$)	K_1 1·57±0·03, K_2 0·90±0·05	65C
	cal	25	2·00 ($NaClO_4$)	ΔH_1 3·47±0·07, ΔS_1 18·8±1, ΔH_2 2·5±0·2, ΔS_2 12·5±1	65C
Lu^{3+}	gl	25	0·5 ($NaClO_4$)	K_1 1·81, β_2 2·32	64S
	gl	25	2·00 ($NaClO_4$)	K_1 1·65±0·03, K_2 1·12±0·05	65C
	cal	25	2·00 ($NaClO_4$)	ΔH_1 5·4±0·1, ΔS_1 25·5±1, ΔH_2 3·7±0·3, ΔS_2 17·4±1	65C

conti

$C_4H_8O_2$ (contd) 211

Metal	Method	Temp	Medium	Log of equilibrium constant, remarks	Ref
Nd^{3+}	gl	25	0·5 ($NaClO_4$)	K_1 1·98, β_2 3·10	64S
	gl	25	2·00 ($NaClO_4$)	K_1 1·91±0·03, K_2 1·18±0·05	65C
	cal	25	2·00 ($NaClO_4$)	ΔH_1 2·84±0·08, ΔS_1 18·3±1, ΔH_2 2·4±0·2, ΔS_2 13·3±1	65C
Pr^{3+}	gl	25	0·5 ($NaClO_4$)	K_1 1·92, β_2 3·18	64S
	gl	25	2·00 ($NaClO_4$)	K_1 1·80±0·03, K_2 1·11±0·05	65C
	cal	25	2·00 ($NaClO_4$)	ΔH_1 3·02±0·09, ΔS_1 18·4±1, ΔH_2 2·5±0·2, ΔS_2 13·5±1	65C
Sm^{3+}	gl	25	0·5 ($NaClO_4$)	K_1 2·05, β_2 3·30	64S
	gl	25	2·00 ($NaClO_4$)	K_1 2·00±0·03, K_2 1·25±0·05	65C
	cal	25	2·00 ($NaClO_4$)	ΔH_1 2·66±0·08, ΔS_1 18·1±1, ΔH_2 2·1±0·2, ΔS_2 12·7±1	65C
Tb^{3+}	gl	25	0·5 ($NaClO_4$)	K_1 1·82, β_2 2·84	64S
	gl	25	2·00 ($NaClO_4$)	K_1 1·73±0·03, K_2 1·39±0·05	65C
	cal	25	2·00 ($NaClO_4$)	ΔH_1 4·4±0·1, ΔS_1 22·6±1, ΔH_2 1·5±0·1, ΔS_2 11·3±1	65C
Tm^{3+}	gl	25	0·5 ($NaClO_4$)	K_1 1·69, β_2 2·28	64S
	gl	25	2·00 ($NaClO_4$)	K_1 1·61±0·03, K_2 1·06±0·05	65C
	cal	25	2·00 ($NaClO_4$)	ΔH_1 5·4±0·1, ΔS_1 25·5±1, ΔH_2 4·1±0·3, ΔS_2 18·6±1	65C
Y^{3+}	gl	25	0·5 ($NaClO_4$)	K_1 1·60, β_2 2·71	64S
	gl	25	2·00 ($NaClO_4$)	K_1 1·64±0·03, K_2 1·15±0·05	65C
	cal	25	2·00 ($NaClO_4$)	ΔH_1 5·4±0·1, ΔS_1 25·5±1, ΔH_2 3·2±0·2, ΔS_2 16·0±1	65C
Yb^{3+}	gl	25	0·5 ($NaClO_4$)	K_1 1·78, β_2 3·10	64S
	gl	25	2·00 ($NaClO_4$)	K_1 1·62±0·03, K_2 1·05±0·05	65C
	cal	25	2·00 ($NaClO_4$)	ΔH_1 5·4±0·1, ΔS_1 25·4±1, ΔH_2 4·0±0·2, ΔS_2 18·1±1	65C

64S W. R. Stagg and J. E. Powell, *Inorg. Chem.*, 1964, **3**, 242

65C G. R. Choppin and A. J. Graffeo, *Inorg. Chem.*, 1965, **4**, 1254

65D V. P. Devendran and M. Santappa, *Current Sci.*, 1965, **34**, 145

$C_4H_8O_3$ Ethoxyacetic acid HL 212

Metal	Method	Temp	Medium	Log of equilibrium constant, remarks	Ref
H^+	qh	20	1·00 $NaClO_4$	K_1 3·46±0·01	61S
Cu^{2+}	qh	20	1·00 $NaClO_4$	K_1 1·79±0·02, β_2 2·87±0·03, β_3 3·20±0·08, β_4 2·8±0·3	61S

61S A. Sandell, *Acta Chem. Scand.*, 1961, **15**, 190

213 $C_4H_8O_3$ 2-Hydroxybutanoic acid HL

Metal	*Method*	*Temp*	*Medium*	*Log of equilibrium constant, remarks*	*Ref*
H^+	gl	31	0·1 ($NaClO_4$)	K_1 3·81±0·03	62C
	?	?	?	K_1 3·97	62S
Pb^{2+}	pH	25	3 ($NaClO_4$)	K_1 2·04±0·02, β_2 2·88±0·05, β_3 2·7±0·1	66W
UO_2^{2+}	gl	31	0·1 ($NaClO_4$)	K_1 3·29±0·05, K_2 1·7±0·3	62C
	dis	20	1 $NaClO_4$	K_1 3·58±0·01, β_2 5·3±0·2, β_3 7·01±0·02	62S

62C C. A. Crutchfield, jun., W. M. McNabb, and J. F. Hazel, *J. Inorg. Nuclear Chem.*, 1962, **24**, 291

62S J. Stary and V. Balek, *Coll. Czech. Chem. Comm.*, 1962, 27, 809

66W Z. Warnke and A. Basinski, *Roczniki Chem.*, 1966, **40**, 1141

214 $C_4H_8O_3$ 3-Hydroxybutanoic acid HL

Metal	*Method*	*Temp*	*Medium*	*Log of equilibrium constant, remarks*	*Ref*
H^+	gl	31	0·1 ($NaClO_4$)	K_1 4·40±0·01	62C
UO_2^{2+}	gl	31	0·1 ($NaClO_4$)	K_1 2·70±0·05, K_2 1·4±0·3	62C

62C C. A. Crutchfield, jun., W. M. McNabb, and J. F. Hazel, *J. Inorg. Nuclear Chem.*, 1962, **24**, 291

215 $C_4H_8O_3$ 2-Hydroxy-2-methylpropanoic acid HL

Metal	*Method*	*Temp*	*Medium*	*Log of equilibrium constant, remarks*	*Ref*
H^+	gl	25	0·5 ($NaClO_4$)	K_1 3·75±0·01	64S
	gl	25	0·5 ($NaClO_4$)	K_1 3·61	66L
	qh	25	1 ($NaClO_4$)	K_1 3·76	67T
Ba^{2+}	qh	25	1 ($NaClO_4$)	K_1 0·36±0·01, β_2 0·51±0·02	65V
Ca^{2+}	qh	25	1 ($NaClO_4$)	K_1 0·92±0·02, β_2 1·42±0·03	65V
Cd^{2+}	qh	25	1 ($NaClO_4$)	K_1 1·24, K_2 0·92, K_3 ~ 0·3	67T
Ce^{3+}	ix	20	0·2 ($NaClO_4$)	K_1 2·43±0·03, β_2 4·34±0·07, β_3 5·3±0·2	60S
	gl	25	0·2 $NaClO_4$	K_1 2·55±0·04, K_2 1·53±0·09, K_3 1·4±0·2	64D
	gl	20	0·1 ($NaClO_4$)	K_1 2·80, β_2 4·74, β_3 5·95	64P
	gl	25	0·5 ($NaClO_4$)	K_1 2·37, β_2 4·01	64S
	ix	25	1·0 ($NaClO_4$)	K_1 2·36, β_2 3·96, β_3 4·7, β_4 ~ 5·6	67L
	dis	24—25	1·0 ($NaClO_4$)	K_1 2·37, β_2 3·93	67L
Co^{2+}	qh	25	1 ($NaClO_4$)	K_1 1·45, K_2 0·98, K_3 ~ 0·3	67T
Cu^{2+}	qh	25	1 ($NaClO_4$)	K_1 2·74, K_2 1·60, K_3 ~ 0·4	67T
Dy^{3+}	gl	25	0·2 $NaClO_4$	K_1 2·94±0·07, K_2 2·51±0·09, K_3 1·84±0·10, K_4 1·2±0·1	64D
	qh	25	1·0 ($NaClO_4$)	K_1 2·83, K_2 2·50, K_3 1·72, K_4 1·28	64E
	gl	20	0·1 ($NaClO_4$)	K_1 2·272, β_2 5·90, β_3 7·87	64P
	gl	25	0·5 ($NaClO_4$)	K_1 2·95, β_2 5·32, β_3 7·16	64S
Er^{3+}	gl	25	0·2 $NaClO_4$	K_1 3·01±0·06, K_2 2·69±0·08, K_3 1·88±0·07, K_4 1·45±0·06	64D

cont

Metal	*Method*	*Temp*	*Medium*	*Log of equilibrium constant, remarks*	*Ref*
Er^{3+}	gl	20	0·1 ($NaClO_4$)	K_1 3·350, β_2 6·04, β_3 8·13	64P
(*contd*)	gl	25	0·5 ($NaClO_4$)	K_1 3·03, β_2 5·54, β_3 7·56	64S
Eu^{3+}	gl	25	0·2 $NaClO_4$	K_1 2·79±0·04, K_2 2·07±0·07, K_3 1·5±0·2, K_4 1·3±0·2	64D
	gl	20	0·1 ($NaClO_4$)	K_1 3·090, β_2 5·54, β_3 7·32	64P
	gl	25	0·5 ($NaClO_4$)	K_1 2·71, β_2 4·92, β_3 5·91	64S
	dis	24—25	0·5 ($NaClO_4$)	K_1 2·72, β_2 5·08±0·04, β_3 6·40±0·08	66L
	ix	25	0·5 ($NaClO_4$)	K_1 2·71±0·01, β_2 4·97±0·01	66L
Gd^{3+}	gl	25	0·2 $NaClO_4$	K_1 2·79±0·06, K_2 2·2±0·1, K_3 1·5±0·2, K_4 1·2±0·2	64D
	gl	20	0·1 ($NaClO_4$)	K_1 3·076, β_2 5·51, β_3 7·19	64P
	gl	25	0·5 ($NaClO_4$)	K_1 2·71, β_2 4·97, β_3 6·01	64S
Ho^{3+}	gl	25	0·2 $NaClO_4$	K_1 2·98±0·08, K_2 2·56±0·09, K_3 1·9±0·1, K_4 1·3 ±0·1	64D
	gl	20	0·1 ($NaClO_4$)	K_1 3·314, β_2 5·98, β_3 7·96	64P
	gl	25	0·5 ($NaClO_4$)	K_1 2·98, β_2 5·42, β_3 7·41	64S
La^{3+}	gl	25	0·2 $NaClO_4$	K_1 2·30±0·02, K_2 1·74±0·06	64D
	qh	25	1·0 ($NaClO_4$)	K_1 2·16, K_2 1·47, K_3 0·75, K_4 0·46	64E
	gl	20	0·1 ($NaClO_4$)	K_1 2·62, β_2 4·42, β_3 5·53	64P
	gl	25	0·5 ($NaClO_4$)	K_1 2·22, β_2 3·67	64S
Lu^{3+}	gl	25	0·2 $NaClO_4$	K_1 3·18±0·06, K_2 2·87±0·06, K_3 2·03±0·06, K_4 1·92±0·05	64D
	gl	20	0·1 ($NaClO_4$)	K_1 3·665, β_2 6·47, β_3 8·82	64P
	gl	25	0·5 ($NaClO_4$)	K_1 3·21, β_2 5·85, β_3 8·21	64S
Mg^{2+}	qh	25	1 ($NaClO_4$)	K_1 0·81±0·02, β_2 1·47±0·04	65V
Mn^{2+}	qh	25	1 ($NaClO_4$)	K_1 0·90, K_2 0·58, K_3 ~ 0·2	67T
Nd^{3+}	gl	25	0·2 $NaClO_4$	K_1 2·74±0·04, K_2 1·68±0·08, K_3 1·56±0·10, K_4 0·6±0·2	64D
	gl	20	0·1 ($NaClO_4$)	K_1 2·88, β_2 5·02, β_3 6·30	64P
	gl	25	0·5 ($NaClO_4$)	K_1 2·54, β_2 4·32	64S
Ni^{2+}	qh	25	1 ($NaClO_4$)	K_1 1·67, K_2 1·13, K_3 ~ 0·4	67T
Pa^{V}	ix	?	0·25	K (?) 3·47, K (?) 7·00	62G
Pb^{2+}	pH	25	3 ($NaClO_4$)	K_1 2·23±0·01, β_2 3·23±0·01, β_3 3·29±0·03	66W
	qh	25	1 ($NaClO_4$)	K_1 2·03, K_2 1·17, K_3 ~ 0·2	67T
Pr^{3+}	gl	25	0·2 $NaClO_4$	K_1 2·59±0·02, K_2 1·78±0·03, K_3 1·23±0·05, K_4 0·78±0·07	64D
	gl	20	0·1 ($NaClO_4$)	K_1 2·84, β_2 4·91, β_3 6·21	64P
	gl	25	0·5 ($NaClO_4$)	K_1 2·48, β_2 4·12	64S
Sm^{3+}	gl	25	0·2 $NaClO_4$	K_1 2·75±0·02, K_2 2·02±0·03, K_3 1·40±0·07, K_4 1·21±0·10	64D
	qh	25	1·0 ($NaClO_4$)	K_1 2·63, K_2 2·13, K_3 1·33, K_4 1·15	64E
	gl	20	0·1 ($NaClO_4$)	K_1 2·993, β_2 5·39, β_3 6·77	64P
	gl	25	0·5 ($NaClO_4$)	K_1 2·63, β_2 4·60	64S
Sr^{2+}	qh	25	1 ($NaClO_4$)	K_1 0·55±0·01, β_2 0·73±0·03	65V
Tb^{3+}	gl	25	0·2 $NaClO_4$	K_1 2·92±0·03, K_2 2·32±0·02, K_3 1·62±0·03, K_4 1·23±0·03	64D
	gl	20	0·1 ($NaClO_4$)	K_1 3·107, β_2 5·63, β_3 7·43	64P
	gl	25	0·5 ($NaClO_4$)	K_1 2·87, β_2 5·21, β_3 6·19	64S
Th^{4+}	ix	?	0·2	K_1 3·56, β_2 5·53, β_3 7·08	62G
Tm^{3+}	gl	25	0·2 $NaClO_4$	K_1 3·10±0·08, K_2 2·69±0·10, K_3 1·92±0·08, K_4 1·62±0·09	64D
	gl	20	0·1 ($NaClO_4$)	K_1 3·515, β_2 6·22, β_3 8·39	64P
	gl	25	0·5 ($NaClO_4$)	K_1 3·13, β_2 5·62, β_3 7·84	64S
UO_2^{2+}	qh	25	1 ($NaClO_4$)	K_1 3·02, K_2 1·83, K_3 1·54	67T

continued overleaf

215 $C_4H_8O_3$ (contd)

Metal	*Method*	*Temp*	*Medium*	*Log of equilibrium constant, remarks*	*Ref*
Y^{3+}	ix	20	0·2 ($NaClO_4$)	β_1 3·11±0·03, β_2 5·60±0·06, β_3 7·3±0·1	60S
	gl	25	0·2 $NaClO_4$	K_1 2·9±0·1, K_2 2·7±0·1, K_3 1·7±0·2, K_4 1·4±0·2	64D
			0—2 $NaClO_4$	K_1 (0) 3·22, (0·05) 3·13, (0·1) 3·12, (0·2) 3·13, (0·5) 3·05, (1·0) 3·08, (2·0) 3·11, K_2 (0) 2·60, (0·05) 2·55, (0·1) 2·49, (0·2) 2·44, (0·5) 2·45, (1·0) 2·44, (2·0) 2·43, K_3 (0) 1·98, (0·05) 1·97, (0·1) 1·86, (0·2) 1·81, (0·5) 1·83, (1·0) 1·81, (2·0) 1·74, K_4 (0) 1·37, (0·05) 1·39, (0·1) 1·22, (0·2) 1·20, (0·5) 1·23, (1·0) 1·17, (2·0) 1·06	
	gl	20	0·1 ($NaClO_4$)	K_1 3·204, β_2 5·79, β_3 7·51	64P
	gl	25	0·5 ($NaClO_4$)	K_1 2·88, β_2 5·32, β_3 6·75	64S
Yb^{3+}	gl	25	0·2 $NaClO_4$	K_1 3·13±0·10, K_2 2·7±0·1, K_3 2·1±0·1, K_4 1·78±0·09	64D
			0—0·5 $NaClO_4$	K_1 (0) 3·40, (0·05) 3·32, (0·1) 3·35, (0·2) 3·32, (0·5) 3·29, K_2 (0) 2·83, (0·05) 2·79, (0·1) 2·75, (0·2) 2·73, (0·5) 2·71, K_3 (0) 2·26, (0·05) 2·25, (0·1) 2·14, (0·2) 2·15, (0·5) 2·13, K_4 (0) 1·69, (0·05) 1·72, (0·1) 1·54, (0·2) 1·55, (0·5) 1·56	
	qh	25	1·0 ($NaClO_4$)	K_1 3·00, K_2 2·79, K_3 2·09, K_4 1·65	64E
	gl	20	0·1 ($NaClO_4$)	K_1 3·643, β_2 6·42, β_3 8·69	64P
	gl	25	0·5 ($NaClO_4$)	K_1 3·18, β_2 5·76, β_3 8·02	64S
Zn^{2+}	qh	25	1 ($NaClO_4$)	K_1 1·70, K_2 1·29, K_3 ~ 0·4	67T

Errata. 1964, p. 417, Table 215. Insert **HL** in the heading.
Under Ce^{3+}, Er^{3+}, Eu^{3+}, Gd^{3+} Ho^{3+}, Nd^{3+}, Sm^{3+}, Tb^{3+} Y^{3+}, and Yb^{3+} *for* 60C *read* 61C.
Ref. 56O should read *J. Chem. Educ.*, 1959, **36**, 462.
Ref. 60C should read 61C G. R. Choppin and J. A. Chopoorian, *J. Inorg. Nuclear Chem.*, 1961, **22**, **97**.

60S V. T. Spitsyn and O. Voitekh, *Proc. Acad. Sci. (U.S.S.R.)*, 1960, **133**, 859 (613)
62G I. Geletseanu and A. V. Lapitskii, *Proc. Acad. Sci. (U.S.S.R.)*, 1962, **144**, 460 (573); **147**, 983 (372)
64D H. Deelstra and F. Verbeek, *Analyt. Chim. Acta*, 1964, **31**, 251
64E L. Eeckhaut, F. Verbeek, H. Deelstra, and J. Hoste, *Analyt. Chim. Acta*, 1964, **30**, 369
64P J. E. Powell, R. H. Karraker, R. S. Kolat, and J. L. Farrell, 'Rare Earth Research II', ed. K. S. Vorres, Gordon and Breach, New York, 1964, p. 514
64S W. R. Stagg and J. E. Powell, *Inorg. Chem.*, 1964, **3**, 242
65V F. Verbeek and H. Thun, *Analyt. Chim. Acta*, 1965, **33**, 378
66L S. J. Lyle and S. J. Naqvi, *J. Inorg. Nuclear Chem.*, 1966, **28**, 2993
66W Z. Warnke and A. Basinski, *Roczniki Chem.*, 1966, **40**, 1141
67L S. J. Lyle and S. J. Naqvi, *J. Inorg. Nuclear Chem.*, 1967, **29**, 2441
67T H. Thun, W. Guns, and F. Verbeek, *Analyt. Chim. Acta*, 1967, **37**, 332

215.1 $C_4H_8O_5$ 2,3-Dihydroxy-2-hydroxymethylpropanoic acid HL

Metal	*Method*	*Temp*	*Medium*	*Log of equilibrium constant, remarks*	*Ref*
H^+	gl	25	0·5 ($NaClO_4$)	K_1 3·29±0·01	64S
Ce^{3+}	gl	25	0·5 ($NaClO_4$)	K_1 2·61, β_2 4·45, β_3 5·98	64S
Dy^{3+}	gl	25	0·5 ($NaClO_4$)	K_1 2·66, β_2 4·87, β_3 6·37	64S
Er^{3+}	gl	25	0·5 ($NaClO_4$)	K_1 2·79, β_2 4·83, β_3 6·57	64S
Eu^{3+}	gl	25	0·5 ($NaClO_4$)	K_1 2·80, β_2 5·00, β_3 6·45	64S
Gd^{3+}	gl	25	0·5 ($NaClO_4$)	K_1 2·69, β_2 4·99, β_3 6·41	64S

conti

$C_4H_8O_5$ (contd) 215.1

Metal	Method	Temp	Medium	Log of equilibrium constant, remarks	Ref
Ho^{3+}	gl	25	0·5 ($NaClO_4$)	K_1 2·71, β_2 4·89, β_3 6·22	64S
La^{3+}	gl	25	0·5 ($NaClO_4$)	K_1 2·40, β_2 3·88, β_3 4·90	64S
Lu^{3+}	gl	25	0·5 ($NaClO_4$)	K_1 2·94, β_2 5·19, β_3 6·90	64S
Nd^{3+}	gl	25	0·5 ($NaClO_4$)	K_1 2·81, β_2 4·62, β_3 6·36	64S
Pr^{3+}	gl	25	0·5 ($NaClO_4$)	K_1 2·75, β_2 4·69, β_3 6·15	64S
Sm^{3+}	gl	25	0·5 ($NaClO_4$)	K_1 2·86, β_2 5·07, β_3 6·51	64S
Tb^{3+}	gl	25	0·5 ($NaClO_4$)	K_1 2·71, β_2 4·88, β_3 6·58	64S
Tm^{3+}	gl	25	0·5 ($NaClO_4$)	K_1 2·85, β_2 4·97, β_3 6·51	64S
Y^{3+}	gl	25	0·5 ($NaClO_4$)	K_1 2·65, β_2 4·67, β_3 5·26	64S
Yb^{3+}	gl	25	0·5 ($NaClO_4$)	K_1 2·90, β_2 5·07, β_3 6·50	64S

64S W. R. Stagg and J. E. Powell, *Inorg. Chem.*, 1964, 3, 242

$C_4H_8N_3$ 4-Aminomethyl-1,3-diazole (4-aminomethylimidazole) L 215.2

Metal	Method	Temp	Medium	Log of equilibrium constant, remarks	Ref
H^+	gl	25	0·015	K_1 9·37, K_2 4·71	60H
Co^{2+}	gl	25	0·01—0·02	$K_1 \sim 4{\cdot}8$	60H
Cu^{2+}	gl	25	0·01—0·02	K_1 9·05, β_2 16·8	60H
Ni^{2+}	gl	25	0·01—0·02	K_1 6·0, β_2 11·0, β_3 14·3	60H

60H F. Holmes and F. Jones, *J. Chem. Soc.*, 1960, 2398

C_4H_9N Azolidine (pyrrolidine) L 216

Metal	Method	Temp	Medium	Log of equilibrium constant, remarks	Ref
H^+	gl	25	0·20 (KNO_3)	K_1 11·4	61B
Cu^{2+}	gl	25	0·20 (KNO_3)	K_1 6·4, K_2 6·0, K_3 5·4, K_4 5·2	61B
Zn^{2+}				See dimethyldithiocarbamic acid (176) for mixed ligands	

61B J. A. Broomhead, H. A. McKenzie, and D. P. Mellor, *Austral. J. Chem.*, 1961, **14**, 649

C_4H_9N 3-Aminobut-1-ene L 216.1

Metal	Method	Temp	Medium	Log of equilibrium constant, remarks	Ref
Pt^{2+}	sp	30—60	2 (NaCl)	$K(PtCl_4^{2-} + HL^+ \rightleftharpoons PtCl_3HL + Cl^-)$ (30°) 3·34, (45·3°) 3·11, (60°) 2·91 ΔH −6·7±0·3, ΔS −6·9±1·0	67D

67D R. G. Denning, F. R. Hartley, and L. M. Venanzi, *J. Chem. Soc.* (*A*), 1967, 328

216.2 C_4H_9N 4-Aminobut-1-ene L

Metal	Method	Temp	Medium	Log of equilibrium constant, remarks	Ref
Pt^{2+}	sp	30—60	2 (NaCl)	$K(PtCl_4^{2-} + HL^+ \rightleftharpoons PtCl_3HL + Cl^-)$ (30°) 3·64, (44·5°) 3·48, (60°) 3·31 ΔH $-5{\cdot}1 \pm 0{\cdot}4$, ΔS $-0{\cdot}2 \pm 1{\cdot}2$	67D
	sp	24·5	2 (KBr)	$K(PtBr_4^{2-} + HL^+ \rightleftharpoons PtBr_3HL + Br^-)$ 3·08	67Da

67D R. G. Denning, F. R. Hartley, and L. M. Venanzi, *J. Chem. Soc. (A)*, 1967, 324
67Da R. G. Denning and L. M. Venanzi, *J. Chem. Soc. (A)*, 1967, 336

216.3 C_4H_9N *trans*-4-Aminobut-2-ene L

Metal	Method	Temp	Medium	Log of equilibrium constant, remarks	Ref
Ag^+	Ag	25	2 ($NaClO_4$)	K_1 $0{\cdot}107 \pm 0{\cdot}007$	67H
Pt^{2+}	sp	30—60	2 (NaCl)	$K(PtCl_4^{2-} + HL^+ \rightleftharpoons PtCl_3HL + Cl^-)$ (30°) 2·65, (44·5°) 2·48, (60·2°) 2·32 ΔH $-5{\cdot}1 \pm 0{\cdot}3$, ΔS $-4{\cdot}6 \pm 0{\cdot}7$	67D

67D R. G. Denning, F. R. Hartley, and L. M. Venanzi, *J. Chem. Soc. (A)*, 1967, 328
67H F. R. Hartley and L. M. Venanzi, *J. Chem. Soc. (A)*, 1967, 333

216.4 C_4H_9N 3-Amino-2-methylprop-1-ene L

Metal	Method	Temp	Medium	Log of equilibrium constant, remarks	Ref
Pt^{2+}	sp	60	2 (NaCl)	$K(PtCl_4^{2-} + HL^+ \rightleftharpoons PtCl_3HL + Cl^-)$ 0·51	67D

67D R. G. Denning, F. R. Hartley, and L. M. Venanzi, *J. Chem. Soc. (A)*, 1967, 328

216.5 $C_4H_{10}O$ Butanol (n-butanol) L

Metal	Method	Temp	Medium	Log of equilibrium constant, remarks	Ref
Ce^{4+}	sp	?	1·7 $HClO_4$	K_1 1·20	65O

65O H. G. Offner and D. A. Skoog, *Analyt. Chem.*, 1965, 37, 1018

216.6 $C_4H_{10}O$ Butan-2-ol (s-butanol) L

Metal	Method	Temp	Medium	Log of equilibrium constant, remarks	Ref
Ce^{4+}	sp	?	1·7 $HClO_4$	K_1 1·04	65O

65O H. G. Offner and D. A. Skoog, *Analyt. Chem.*, 1965, 37, 1018

$C_4H_{10}O$ 1,1-Dimethylethanol (t-butanol) L 216.7

Metal	Method	Temp	Medium	Log of equilibrium constant, remarks	Ref
Ce^{4+}	sp	?	1·6 $HClO_4$	K_1 1·12	65O

65O H. G. Offner and D. A. Skoog, *Analyt. Chem.*, 1965, **37**, 1018

$C_4H_{10}O$ Ethoxyethane (diethyl ether) L 216.8

Metal	Method	Temp	Medium	Log of equilibrium constant, remarks	Ref
Al^{3+}	oth	20	0·07 AlI_3	K_1 1·34±0·06, mass spectrograph	65P

65P A. Polaczek and E. Baranowska, *J. Inorg. Nuclear Chem.*, 1965, **27**, 1649

$C_4H_{10}O_2$ Butanediol (butylene glycol*) L 216.9

Metal	Method	Temp	Medium	Log of equilibrium constant, remarks	Ref
Ge^{IV}	qh	25	0·1 KCl	$K[HGeO_3^- + L \rightleftharpoons HGeO_2(H_{-2}L)^-]$ 0·64 $K[HGeO_3^- + 2L \rightleftharpoons HGeO(H_{-2}L)_2^-]$ 0·04	59A

* 1,2- or 2,3- not stated.

59A P. J. Antikainen, *Acta Chem. Scand.*, 1959, **13**, 312

$C_4H_{10}O_2$ *meso*-Butan-2,3-diol L 216.10

Metal	Method	Temp	Medium	Log of equilibrium constant, remarks	Ref
As^{III}	gl	25	0·1	$*K[As(OH)_4^- + L \rightleftharpoons As(OH)_2(H_{-2}L)^-]$ −0·89	57R
B^{III}	gl	25	0·1	$*K[B(OH)_4^- + L \rightleftharpoons B(OH)_2(H_{-2}L)^-]$ 0·54 $*K[B(OH)_4^- + 2L \rightleftharpoons B(H_{-2}L)_2^-]$ 0·69	57R
	gl	0—35	0·025 borax	$K[B(OH)_4^- + L \rightleftharpoons B(OH)_2(H_{-2}L)^-]$ (0°) 0·71, (13°) 0·51, (25°) 0·43, (35°) 0·36 ΔH° −3·4±0·3, ΔS° −10±2 $K[B(OH)_4^- + 2L \rightleftharpoons B(H_{-2}L)_2^-]$ (0°) 1·11, (13°) 0·88, (25°) 0·66, (35°) 0·43 ΔH° −7·2±0·7, ΔS° −21±5	67C
Te^{VI}	gl	25	0·1	$*K[H_5TeO_6^- + L \rightleftharpoons H_3TeO_4(H_{-2}L)^-]$ 1·16	57R

* DL- or *meso*- not stated.

57R G. L. Roy, A. L. Laferriere, and J. O. Edwards, *J. Inorg. Nuclear Chem.*, 1957, **4**, 106
67C J. M. Connor and V. C. Bulgrin, *J. Inorg. Nuclear Chem.*, 1967, **29**, 1953

216.11 $C_4H_{10}O_2$ 1-Butane-2,3-diol L

Metal	*Method*	*Temp*	*Medium*	*Log of equilibrium constant, remarks*	*Ref*
B^{III}	gl	0—35	0·025 borax	$K[B(OH)_4^- + L \rightleftharpoons B(OH)_2(H_{-2}L)^-]$ (0°) 1·79, (13°) 1·63, (25°) 1·57, (35°) 1·40 $\Delta H°$ $-4·3 \pm 0·4$, $\Delta S°$ -7 ± 2 $K[B(OH)_4^- + 2L \rightleftharpoons B(H_{-2}L)_2^-]$ (0°) 2·60, (13°) 2·45, (25°) 2·21, (35°) 2·10 $\Delta H°$ $-5·5 \pm 0·5$, $\Delta S°$ -8 ± 2	67C

67C J. M. Conner and V. C. Bulgrin, *J. Inorg. Nuclear Chem.*, 1967, **29**, 1953

217.1 $C_4H_{10}O_3$ 3-Methoxypropan-1,2-diol L

Metal	*Method*	*Temp*	*Medium*	*Log of equilibrium constant, remarks*	*Ref*
As^{III}	gl	25	0·1	$K[As(OH)_4^- + L \rightleftharpoons As(OH)_2(H_{-2}L)^-]$ $-0·18$	57R
B^{III}	gl	25	0·1	$K[B(OH)_4^- + L \rightleftharpoons B(OH)_2(H_{-2}L)^-]$ 1·28 $K[B(OH)_4^- + 2L \rightleftharpoons B(H_{-2}L)_2^-]$ 1·13	57R
Ge^{IV}	qh	25	0·1 KCl	$K[HGeO_3^- + L \rightleftharpoons HGeO_2(H_{-2}L)^-]$ 0·84 $K[HGeO_3^- + 2L \rightleftharpoons HGeO(H_{-2}L)_2^-]$ 0·58	59A
Te^{VI}	gl	25	0·1	$K[H_5TeO_6^- + L \rightleftharpoons H_3TeO_4(H_{-2}L)^-]$ 1·40	57R

57R G. L. Roy, A. L. Laferriere, and J. O. Edwards, *J. Inorg. Nuclear Chem.*, 1957, **4**, 106
59A P. J. Antikainen, *Acta Chem. Scand.*, 1959, **13**, 312

218 $C_4H_{11}N$ Diethylamine L

Metal	*Method*	*Temp*	*Medium*	*Log of equilibrium constant, remarks*	*Ref*
Ag^+	Ag	25	0·02 0—80% methanol	β_2 (0%) 6·27, (20%) 6·38, (30%) 6·51, (40%) 6·59, (50%) 6·63, (60%) 6·64, (70%) 6·72, (80%) 7·81	63P
	Ag	25	$2NH_4NO_3$	K_1 4·93, β_2 7·08	68P
Cd^{2+}	Ag	25	2 NH_4NO_3	K_1 2·62, K_2 2·11, K_3 1·52, K_4 1·06	68P
Co^{2+}	Ag-$Ag(Et_2NH)_2$	25	2 (NH_4NO_3)	K_1 2·10, K_2 1·42, K_3 1·25, K_4 1·07	67P
Ni^{2+}	Ag-$Ag(Et_2NH)_2$	25	2 (NH_4NO_3)	K_1 2·78, K_2 2·19, K_3 1·75, K_4 1·21, K_5 0·94	67P
Zn^{2+}	Ag	25	2 NH_4NO_3	K_1 2·51, K_2 2·45, K_3 2·53, K_4 2·33	68P

63P G. Popa, C. Luca, and V. Magearu, *J. Chim. phys.*, 1963, **60**, 355
67P G. Popa and V. Magearu, *Rev. Roumaine Chim.*, 1967, **12**, 1107
68P G. Popa, V. Magearu, and C. Luca, *J. Electroanalyt. Chem.*, 1968, **17**, 335

$C_4H_{11}N$ 1-Aminobutane (n-butylamine) L 219

Metal	Method	Temp	Medium	Log of equilibrium constant, remarks	Ref
H^+	gl	20	0·5 KNO_3	K_1 10·93 ± 0·02	68A
	H	20	→ 0	K_1 10·77 ± 0·02	68A
Ag^+	oth	?	0·1 $NaClO_4$ acetone	β_2 10·3 ± 0·1, coulometric titration	65M
	gl	20	0·500 KNO_3	K_1 3·50 ± 0·10, β_2 7·60 ± 0·05	68A
	Ag-AgCl	20	0·500 KNO_3	β_2 7·59 ± 0·04	68A
Zn^{2+}				See dimethyldithiocarbamic acid (176) for mixed ligands	

65M K. K. Mead, D. L. Maricle and C. A. Strenli, *Analyt. Chem.*, 1965, 37, 237
68A D. J. Alner, R. C. Lansbury and A. G. Smeeth, *J. Chem. Soc.* (*A*), 1968, 417

$C_4H_{11}N$ 2-Aminobutane (s-butylamine) L 219.1

Metal	Method	Temp	Medium	Log of equilibrium constant, remarks	Ref
Ag^+	oth	?	0·1 $NaClO_4$ acetone	β_2 10·76 ± 0·08, coulometric titration	65M

65M K. K. Mead, D. L. Maricle and C. A. Strenli, *Analyt. Chem.*, 1965, 37, 237

$C_4H_{11}N$ 1-Amino-2-methylpropane (isobutylamine) L 220

Metal	Method	Temp	Medium	Log of equilibrium constant, remarks	Ref
Ag^+	oth	?	0·1 $NaClO_4$ acetone	β_2 9·68 ± 0·04, coulometric titration	65M

64M K. K. Mead, D. L. Maricle and C. A. Strenli, *Analyt. Chem.*, 1965, 37, 237

$C_4H_{11}N$ 2-Amino-2-methylpropane (t-butylamine) L 221

Metal	Method	Temp	Medium	Log of equilibrium constant, remarks	Ref
Zn^{2+}				See dimethyldithiocarbamic acid (176) for mixed ligands	

223.1 $C_4H_{12}O_3$ 2,2′-Oxydiethanol (diethyleneglycol) L (*see Corrigenda*)

Metal	*Method*	*Temp*	*Medium*	*Log of equilibrium constant, remarks*	*Ref*
Ce^{4+}	sp	?	1·7 $HClO_4$	K_1 1·60	65O

65O H. G. Offner and D. A. Skoog, *Analyt. Chem.*, 1965, **37**, 1018

224.1 $C_4H_{12}N_2$ 1,3-Diaminobutane L

Metal	*Method*	*Temp*	*Medium*	*Log of equilibrium constant, remarks*	*Ref*
H^+	gl	25	I ($NaClO_4$)	K_1 10·50+0·257 I K_2 8·42+1·018$\sqrt{I}/(1+1{\cdot}440\sqrt{I})$+0·222 I	65N
	gl	25	I ($NaClO_4$)	K_1K_2 18·93+1·018$\sqrt{I}/(1+0{\cdot}79\sqrt{I})$+0·411 I K_2 8·43+1·018$\sqrt{I}/(1+1{\cdot}24\sqrt{I})$+0·211 I	68N
Cu^{2+}	gl	25	I ($NaClO_4$)	K_1 9·67+0·905 I−0·822 $I^{3/2}$+0·307 I^2 K_2 7·25+1·169 I−1·116 $I^{3/2}$+0·402 I^2	65N
Ni^{2+}	gl	25	I ($NaClO_4$)	K ($Ni^{2+}+H_2L^{2+} \rightleftharpoons NiL^{2+}+2H^+$) −12·69−1·018$\sqrt{I}/(1+1{\cdot}48\sqrt{I})$−0·250 I K($NiL^{2+}+H_2L^{2+} \rightleftharpoons NiL_2^{2+}+2H^+$) −14·76−1·018$\sqrt{I}/(1+1{\cdot}36\sqrt{I})$−0·192 I	68N
			→ 0	K_1 6·25, K_2 4·18	

65N R. Nasanen, M. Koskinen, R. Salonen, and A. Kiiski, *Suomen Kem.*, 1965, **B38**, 81
68N R. Nasanen, P. Tilus, and T. Uro, *Suomen Kem.*, 1968, **B41**, 314

225 $C_4H_{12}N_2$ *meso*-2,3-Diaminobutane L (*see Corrigenda*)

Metal	*Method*	*Temp*	*Medium*	*Log of equilibrium constant, remarks*	*Ref*
H^+	?	20	0·1	K_1 11·1, K_2 6·26, K_3 2·31, K_4 1·77	62S
Hg^{2+}	pol	20	0·1 KNO_3	K_1 20·33	62S

62S S. Stankoviansky and J. Konigstein, *Coll. Czech. Chem. Comm.*, 1962, **27**, 1997

226 $C_4H_{12}N_2$ D,L-2,3-Diaminobutane L (*see Corrigenda*)

Metal	*Method*	*Temp*	*Medium*	*Log of equilibrium constant, remarks*	*Ref*
H^+	?	20	0·1	K_1 11·5, K_2 6·13, K_3 3·66, K_4 2·65	62S
Hg^{2+}	pol	20	0·1 KNO_3	K_1 21·51	62S

62S S. Stankoviansky and J. Konigstein, *Coll. Czech. Chem. Comm.*, 1962, **27**, 1997

$C_4H_{12}N_2$ 1,2-Diamino-2-methylpropane L 226.1

Metal	Method	Temp	Medium	Log of equilibrium constant, remarks	Ref
H^+	gl	25	I ($NaClO_4$)	K_1 9·82+0·297 I+0·0287 $I^{1·5}$−0·046 I^2	64N
				K_2 6·25+1·018$\sqrt{I}$/(1+1·383$\sqrt{I}$)+0·169 I	
Cu^{2+}	gl	25	I ($NaClO_4$)	K_1 10·40+0·404 I+0·234 $I^{1·5}$−0·190 I^2	64N
				K_2 9·07+1·261 I−1·050 $I^{1·5}$+0·324 I^2	

64N R. Nasanen and M. Koskinen, *Acta Chem. Scand.*, 1964, **18**, 1337

$C_4H_{12}N_2$ *NN*-Dimethyl-1,2-diaminoethane L 227

Metal	Method	Temp	Medium	Log of equilibrium constant, remarks	Ref
H^+	gl	25	→0	K_1 9·54	67N
			I ($NaClO_4$)	K_2 6·178+(1·018$\sqrt{I}$)/(1+0·823$\sqrt{I}$)+0·258 I	
				K_1K_2 15·713+(1·018$\sqrt{I}$)/(1+0·641$\sqrt{I}$)+0·612 I	
Cu^{2+}	gl	25	→0	K_1 9·08, K_2 6·83	67N
			I ($NaClO_4$)	$K(Cu^{2+}+H_2L^{2+} \rightleftharpoons CuL^{2+}+2H^+)$ −6·635 −(1·018$\sqrt{I}$)/(1+1·294$\sqrt{I}$)−0·267 I	
				$K(CuL^{2+}+H_2L^{2+} \rightleftharpoons CuL_2{}^{2+}+2H^+)$ −8·888 −(1·108$\sqrt{I}$)/(1+1·447$\sqrt{I}$)−0·301 I	

Erattum. 1964, p. 420, Table 227, Add **L** to heading.

67N R. Nasanen, M. Koskinen, R. Jarvinen, and R. Penttonen, *Suomen Kem.*, 1967, **B40**, 25

$C_4H_{12}N_2$ *NN'*-Dimethyl-1,2-diaminoethane L 228

Metal	Method	Temp	Medium	Log of equilibrium constant, remarks	Ref
H^+	gl	25	I ($NaClO_4$)	K_1 10·032+1·018$\sqrt{I}$/(H 0·589$\sqrt{I}$)−1·018$\sqrt{I}$/(1+0·0897$\sqrt{I}$)+0·292 I	66N
				K_2 6·795+1·018$\sqrt{I}$/(1+0·897$\sqrt{I}$)+0·236 I	
Cu^{2+}	gl	25	→0	K_1 9·96, K_2 6·94	66N
			I ($NaClO_4$)	$K(Cu^{2+}+H_2L^{2+} \rightleftharpoons CuL^{2+}+2H^+)$ −6·869−1·018$\sqrt{I}$/(1+1·023$\sqrt{I}$)−0·219 I	
				$K(CuL^{2+}+H_2L^{2+} \rightleftharpoons CuL_2{}^{2+}+2H^+)$ −9·884 −1·018$\sqrt{I}$/(1+1·604$\sqrt{I}$)−0·216 I	
				$K(CuOHL^++H^+ \rightleftharpoons CuL^{2+})$ 7·89±1·018$\sqrt{I}$/(1+$\sqrt{I}$)	
				$K[Cu_2(OH)_2L^{2+}+2H^+ \rightleftharpoons 2CuL^{2+}]$ 12·09+1·018$\sqrt{I}$/(1+$\sqrt{I}$)	
			→0	$K[2CuOHL^+ \rightleftharpoons Cu_2(OH)_2L_2{}^{2+}]$ 3·69	
				$K[Cu(OH)_2L+H^+ \rightleftharpoons CuOHL^+]$ 10·81	
Ni^{2+}	gl	25	→0	K_1 6·84, K_2 3·85	66N
			I ($NaClO_4$)	$K(Ni^{2+}+H_2L^{2+} \rightleftharpoons NiL^{2+}+2H^+)$ −9·986−1·018$\sqrt{I}$/(1+0·983$\sqrt{I}$)−0·315 I	
				$K(NiL^{2+}+H_2L^{2+} \rightleftharpoons NiL_2{}^{2+}+2H^+)$ −12·977−1·018$\sqrt{I}$/(1+1·897$\sqrt{I}$)−0·314 I	

66N R. Nasanen, M. Koskinen, L. Anttila, and M. L. Korvola, *Suomen Kem.*, 1966, **B39**, 122

229 $C_4H_{12}N_2$ *N*-Ethyl-1,2-diaminoethane L

Metal	Method	Temp	Medium	Log of equilibrium constant, remarks	Ref
H^+	gl	0—25	0·50 KNO_3	ΔH_1 (0°) −8·2, ΔH_2 −7·9	52B
Cu^{2+}	gl	0—25	0·5 KNO_3	ΔH_1 (0°) −5·4, ΔS_1 29, ΔH_2 (0°) −6·4, ΔS_2 17	52B
Ni^{2+}	gl	0—25	0·5 KNO_3	ΔH_1 (0°) −6·1, ΔS_1 10·6, ΔH_2 (0°) −7·2, ΔS_2 0·4, ΔH_3 (0°) −7·6, ΔS_3 −16·5	52B

52B F. Basolo and R. K. Murmann, *J. Amer. Chem. Soc.*, 1952, **74**, 2373, 5243

230 $C_4H_{13}N_3$ 1,3-Diamino-2-aminomethylpropane [tris(aminomethyl)methane] L

Metal	Method	Temp	Medium	Log of equilibrium constant, remarks	Ref
H^+	gl	22	1·0 KNO_3	K_1 10·51, K_2 8·86, K_3 6·90	61S
Ni^{2+}	gl	22	1·0 KNO_3	K_1 9·23, $K(Ni^{2+}+HL^+ \rightleftharpoons NiHL^{3+})$ 5·90	61S

Erratum. 1964, p. 421, Table 230. Under Zn^{2+} insert reference 62A.

61S T. G. Spiro and C. J. Ballhausen, *Acta Chem. Scand.*, 1961, **15**, 1707

231 $C_4H_{13}N_3$ 2,2′-Iminobis(ethylamine) (diethylenetriamine) L

Metal	Method	Temp	Medium	Log of equilibrium constant, remarks	Ref
H^+	gl	30, 40	1 (KNO_3)	ΔH_1 −11·7, ΔH_2 −12·6, ΔH_3 −8·2	50J
	gl	20, 26	0·1 (KCl)	K_1 (20°) 9·94, (26°) 9·79, K_2 (20°) 9·13, (26°) 8·95, K_3 (20°) 4·34, (26°) 4·22	65P
	gl	25	0·5 $NaNO_3$	K_1 10·03, K_2 9·24, K_3 4·59	67S
Cd^{2+}	gl	25	0·5 $NaNO_3$	K_1 8·05, K_2 5·79 $K(Cd^{2+}+L+A \rightleftharpoons CdLA^{2+})$ 12·54±0·04 where A = ethylenediamine	67S
Co^{2+}	gl	30, 40	1	$\Delta H_1°$ (35°) −9, $\Delta H_2°$ −10	52J
Cr^{2+}	gl	20, 26	0·1 (KCl)	K_1 (20°) 6·71, (26°) 6·78, K_2 (20°) 2·69, (26°) 2·6	65P
Cu^{2+}	gl	30, 40	1	$\Delta H_1°$ (35°) −20	52J
Fe^{2+}	gl	30, 40	1·3 KCl	$\Delta H_1°$ (35°) −9, $\Delta H_2°$ −8	52J
Hg^{2+}	pol	25, 40	0·1 KNO_3	β_2 (25°) 25·02±0·02, (40°) 23·76±0·01, β_3 (25°) 24·5±0·2 (40°) 23·43±0·03 $\Delta H_{\beta 2}°$ −36±1, $\Delta S_{\beta 2}°$ −7±4	61R
Mn^{2+}	gl	30, 40	1	$\Delta H_1°$ (35°) −4, $\Delta H_2°$ −5	52J
Ni^{2+}	gl	30, 40	1	$\Delta H_1°$ (35°) −12, $\Delta H_2°$ −13	52J
	sp	25	0·5 (NaCl)	$K(NiL^{2+}+A^- \rightleftharpoons NiLA^+)$ 5·13±0·09 where HA = glycine	67J
			1 (KCl)	$K(NiL^{2+}+A^{2-} \rightleftharpoons NiLA)$ 4·59±0·05 where H_2A = oxalic acid	
Zn^{2+}	gl	30, 40	1	$\Delta H_1°$ (35°) −8	52J

50J H. B. Jonassen, R. B. Le Blanc, A. W. Meibohm, and R. M. Rogan, *J. Amer. Chem. Soc.*, 1950, **72**, 2430
52J H. B. Jonassen, G. G. Hurst, R. B. Le Blanc, and A. W. Meibohm, *J. Phys. Chem.*, 1952, **56**, 16
61R D. K. Roe, D. B. Masson, and C. J. Nyman, *Analyt. Chem.*, 1961, **33**, 1464
65P R. L. Pecsok, R. A. Garber, and L. D. Shields, *Inorg. Chem.*, 1965, **4**, 447
67J N. E. Jacobs and D. W. Margerum, *Inorg. Chem.*, 1967, **6**, 2038
67S J. P. Scharf and M. R. Paris, *Compt rend.*, 1967, **265C**, 488

$C_4HO_2F_7$ Heptafluorobutanoic acid HL 231.1

Metal	Method	Temp	Medium	Log of equilibrium constant, remarks	Ref
Pb^{2+}	pol	25	2 ($NaClO_4$)	K_1 −0·21, β_2 0·00	64C
	Pb	25	2 ($NaClO_4$)	K_1 −0·36, β_2 −0·35	64C

64C S. A. Carrano, K. A. Chem, and R. F. O'Malley, *Inorg. Chem.*, 1964, **3**, 1047

$C_4H_5ON_3$ 4-Amino-1,3-diazin-2(1*H*)-one (cytosine) L 232.1

Metal	Method	Temp	Medium	Log of equilibrium constant, remarks	Ref
H^+	?	27	0·1 $NaClO_4$	K_1 4·5	61F
	gl	25	0·16	K_1 4·69	61M
Cu^{2+}	gl	25	0·16	K_1 2·0	61M
Hg^{2+}	Hg	27	0·1 $NaClO_4$	β_2 10·90	61F
Zn^{2+}	nmr	36	$(CH_3)_2SO$	K_1 1·39±0·06	66W
				See imidazole (144) for mixed ligands	

61F R. Ferreira, E. Ben-Zvi, T. Yamane, J. Vasileuskis, and N. Davidson, 'Advances in Chemistry of Co-ordination Compounds', ed. S. Kirschner, MacMillan, New York, 1961, p. 457
61M R. B. Martin, *Fed. Proc.*, 1961, **20**, No. 3, Suppl. 10, 54
66W W. M. Wang and N. C. Li, *J. Amer. Chem. Soc.*, 1966, **88**, 4592

$C_4H_5O_2N$ Pyrrolidine-2,5-dione (succinimide) HL 232.2

Metal	Method	Temp	Medium	Log of equilibrium constant, remarks	Ref
H^+	gl	25	0·1 (KNO_3)	K_1 9·38	65C
	gl	30	?	K_1 9·29	65J
Ag^+	gl, Ag	25	0·1 (KNO_3)	K_1 4·45, β_2 9·54	65C
Cu^{2+}	gl	30	?	K_1 3·5, K_2 4·7	65J

65C E. Campi, G. Ostacoli, and A. Vanni, *Gazetta*, 1965, **95**, 796
65J M. C. Jain, A. A. Khan, and W. U. Malik, *J. Indian Chem. Soc.*, 1965, **42**, 597

$C_4H_6O_2N_2$ 1-Methylimidazolidine-2,4-dione (1-methylhydantoin) HL 232.3

Metal	Method	Temp	Medium	Log of equilibrium constant, remarks	Ref
H^+	gl	25	0·1 (KNO_3)	K_1 9·09	65C
Ag^+	gl, Ag	25	0·1 (KNO_3)	K_1 4·37, β_2 9·34	65C

65C E. Campi, G. Ostacoli, and A. Vanni, *Gazzetta*, 1965, **95**, 796

232.4 $C_4H_6O_2N_2$ 5-Methylimidazolidine-2,4-dione (5-methylhydantoin) HL

Metal	Method	Temp	Medium	Log of equilibrium constant, remarks	Ref
H^+	gl	25	0·1 (KNO_3)	K_1 9·07	65C
Ag^+	gl, Ag	25	0·1 (KNO_3)	K_1 4·34, β_2 9·27	65C

65C E. Campi, G. Ostacoli, and A. Vanni, *Gazzetta*, 1965, **95**, 796

233 $C_4H_6O_4S$ Thiodiacetic acid (thiodiglycollic acid) H_2L

Metal	Method	Temp	Medium	Log of equilibrium constant, remarks	Ref
H^+	gl	20	0·1	K_1 4·35, K_2 3·30	61C
Cu^{2+}	gl	20	0·1	K_1 4·57, $K(Cu^{2+}+HL^- \rightleftharpoons CuHL^+)$ 3·18	61C
	gl	20	1·00 $NaClO_4$	K_1 4·18±0·02, β_2 7·08±0·08, β_3 8·6±0·3, β_4 11·9±0·2 (?)	61S

61C E. Campi, G. Ostacoli, N. Cibrario, and G. Saini, *Gazzetta*, 1961, **91**, 361
61S A. Sandell, *Acta Chem. Scand.*, 1961, **15**, 190

234 $C_4H_6O_4S$ Mercaptobutanedioic acid (thiomalic acid) H_3L

Metal	Method	Temp	Medium	Log of equilibrium constant, remarks	Ref
H^+	gl	30	0·1 (KCl)	K_2 4·68, K_3 3·15	62Ca
	gl	25	0·1—1 (KNO_3)	K_1 (0·10) 10·37, (1·0) 9·73, K_2 (0·10) 4·64, (1·0) 4·44, K_3 (0·10) 3·64, (1·0) 2·53	65L
	gl	?	~ 0·005	K_1 10·55, K_2 4·94, K_3 3·29	65S
	gl	31, 45	0·1 ($NaClO_4$)	K_2 (31°) 4·51, (45°) 4·45, K_3 (31°) 3·01, (45°) 2·95	68R
	gl	?	0·1 KNO_3	K_1 10·64, K_2 4·94, K_3 3·30	68S, 68S
Ag^+	gl	25—35	0·1 (KNO_3)	K_1 (25°) 6·89, (30°) 6·95, (35°) 7·17 ΔH_1 (30°) −7·5(?), ΔS_1 (30°) 7	68Sb
Ce^{3+}	gl	30	0·1 (KCl)	$K(Ce^{3+}+HL^{2-} \rightleftharpoons CeHL^+)$ 3·22 $K[CeHL^+ + HL^{2-} \rightleftharpoons Ce(HL)_2^-]$ 2·88 $K[Ce(HL)_2^- + HL^{2-} \rightleftharpoons Ce(HL)_3^{3-}]$ 2·53	62Ca
Co^{3+}	sp	30	?	β_2 (?) 7·7, ammonia buffer	65N
La^{3+}	gl	30	0·1 (KCl)	$K(La^{3+}+HL^{2-} \rightleftharpoons LaHL^+)$ 3·01 $K[LaHL^+ + HL^{2-} \rightleftharpoons La(HL)_2^-]$ 2·57 $K[La(HL)_2^- + HL^{2-} \rightleftharpoons La(HL)_3^{3-}]$ 2·27	62Ca
Nd^{3+}	gl	30	0·1 (KCl)	$K(Nd^{3+}+HL^{2-} \rightleftharpoons NdHL^+)$ 3·38 $K[NdHL^+ + HL^{2-} \rightleftharpoons Nd(HL)_2^-]$ 2·99 $K[Nd(HL)_2^- + HL^{2-} \rightleftharpoons Nd(HL)_3^{3-}]$ 2·57	62Ca
Ni^{2+}	gl	25	0·10 (KNO_3)	$K(NiLOH^{2-}+H^+ \rightleftharpoons NiL^-)$ 9·37	65L
	sp	30	0·4	β_3 (?) 9·6, ammonia buffer	65N
	pol	30	0·4	β_3 (?) 10·2, ammonia buffer	65N
	gl	20—30	0·1 KNO_3	K_1 (20°) 7·86, (25°) 7·87, (30°) 7·96, K_2 (20°) 6·24, (25°) 6·31, (30°) 6·39 $\Delta H_{\beta 2}$ −8·8(?), $\Delta S_{\beta 2}$ 35(?)	68S

contin

Metal	Method	Temp	Medium	Log of equilibrium constant, remarks	Ref
Pb^{2+}	gl	30	0·007 ClO_4^-	K_1 10·80	67M
Pr^{3+}	gl	30	0·1 (KCl)	$K(Pr^{3+}+HL^{2-} \rightleftharpoons PrHL^+)$ 3·31	62Ca
				$K[PrHL^+ + HL^{2-} \rightleftharpoons Pr(HL)_2^-]$ 2·86	
				$K[Pr(HL)_2^- + HL^{2-} \rightleftharpoons Pr(HL)_3^{3-}]$ 2·59	
Rh^{3+}	sp	30	?	K_1 21·6,	66S
	gl	30	?	K_2 8·4	66S
UO_2^{2+}	gl	30	0·1 KCl	$K(UO_2^{2+}+HL^{2-} \rightleftharpoons UO_2HL)$ 3·56	62C
				$K[UO_2HL+HL^{2-} \rightleftharpoons UO_2(HL)_2^{2-}]$ 3·42	
	sp	25	'fixed' (?)	$K(UO_2^{2+}+HL^{2-} \rightleftharpoons UO_2HL)$ 3·0	63M
	gl	31, 45	0·1 ($NaClO_4$)	$K(UO_2^{2+}+HL^{2-} \rightleftharpoons UO_2HL)$ (31°) 3·82 ± 0·08, (45°) 3·91 ± 0·08	68R
Zn^{2+}	gl	25	0·10, 1·0 (KNO_3)	K_1 (0·10) 8·24, (1·0) 7·52, K_2 (0·10) 6·32	65L
				$K(ZnLOH^{2-}+H^+ \rightleftharpoons ZnL^-)$ (0·10) 8·36, (1·0) 8·01	
				$K(ZnLOH^{2-}+H^+ \rightleftharpoons ZnL^-)$ 8·01	
	gl	25—35	0·1 KNO_3	K_1 (25°) 8·75, (30°) 8·85, (35°) 8·86, K_2 (25°) 6·82, (30°) 6·75, (35°) 6·93	68Sa
				$\Delta H_{\beta 2}$ −5·3, $\Delta S_{\beta 2}$ 54(?)	
Zr^{4+}	gl	?	~ 0·005	$K(ZrO^{2+}+L^{3-} \rightleftharpoons ZrOL^-)$ 9·6	65S

Errata. 1964, p. 424, Table 234. *For* $Ag^{(I)}$ *read* Ag^+.
For $Hg^{(II)}$ *read* Hg^{2+}.
For ref. 64L *read* ref. 65L, for details see below.

62C M. Cefola, R. C. Taylor, P. S. Gentile, and A. V. Celiano, *J. Phys. Chem.*, 1962, **66**, 790
62Ca M. Cefola, A. S. Tompa, A. V. Celiano, and P. S. Gentile, *Inorg. Chem.*, 1962, **1**, 290
63M V. K. Mathur, H. L. Nigam, and S. C. Srivastava, *Bull. Chem. Soc. Japan*, 1963, **36**, 1658
65L G. R. Lenz and A. E. Martell, *Inorg. Chem.*, 1965, **4**, 378
65N H. L. Nigam, R. C. Kapoor, U. Kapoor, and S. C. Srivastava, *Indian J. Chem.*, 1965, **3**, 443, 527
65S S. C. Sinha and H. L. Nigam, *Vijnana Parishad Anusandhan Patrika*, 1965, **8**, 83
66S S. C. Sinha, H. L. Nigam, and S. P. Sangal, *Chim. analyt.*, 1966, **48**, 515
67M M. B. Mishra, S. C. Sinha, and H. L. Nigam, *Indian J. Chem.*, 1967, **5**, 649
68R S. Ramamoorthy and M. Santappa, *Bull. Chem. Soc. Japan*, 1968, **41**, 1330
68S R. S. Saxena, K. C. Gupta, and M. L. Mittal, *Austral. J. Chem.*, 1968, **21**, 641
68Sa R. S. Saxena, K. C. Gupta, and M. L. Mittal, *Canad. J. Chem.*, 1968, **46**, 311
68Sb R. S. Saxena, K. C. Gupta, and M. L. Mittal, *J. Inorg. Nuclear Chem.*, 1968, **30**, 189

$C_4H_6O_4S_2$ ***meso*-2,3-Dimercaptobutanedioic acid (*meso*-dithiotartaric acid) H_4L 235**

Metal	Method	Temp	Medium	Log of equilibrium constant, remarks	Ref
Ni^{2+}	gl	25	0·10 (KNO_3)	$K(NiL^{2-}+H^+ \rightleftharpoons NiHL^-)$ 8·7	65L
				$K(NiHL^-+H^+ \rightleftharpoons NiH_2L)$ 3·3	
Zn^{2+}	gl	25	0·10 (KNO_3)	$K(ZnL^{2-}+Zn^{2+} \rightleftharpoons Zn_2L)$ 4·07	65L
				$K(ZnL^{2-}+H^+ \rightleftharpoons ZnHL^-)$ 5·7	
				$K(ZnHL^-+H^+ \rightleftharpoons ZnH_2L)$ 3·4	

Errata. 1964, p. 424, Table 235. *For* ref. 64L *read* 65L, see below; *under* Ni^{2+} *for* 10·4 *read* 2·64.

65L G. R. Lenz and A. E. Martell, *Inorg. Chem.*, 1965, **4**, 378

235.1 $C_4H_6O_4S_2$ Dithiodiacetic acid H_2L

Metal	Method	Temp	Medium	Log of equilibrium constant, remarks	Ref
H^+	gl	25	0·1 ($NaClO_4$)	K_1 3·81, K_2 2·88	68S
Cd^{2+}	gl	25	0·1 ($NaClO_4$)	K_1 1·9	68S
Co^{2+}	gl	25	0·1 ($NaClO_4$)	K_1 1·5	68S
Mn^{2+}	gl	25	0·1 ($NaClO_4$)	K_1 1·7	68S
Ni^{2+}	gl	25	0·1 ($NaClO_4$)	K_1 1·8	68S
Pb^{2+}	gl	25	0·1 ($NaClO_4$)	K_1 2·4	68S
Zn^{2+}	gl	25	0·1 ($NaClO_4$)	K_1 1·6	68S

68S K. Suzuki, C. Karaki, S. Mori, and K. Yamasaki, *J. Inorg. Nuclear Chem.*, 1968, **30**, 167

235.2 $C_4H_6O_4Se$ Selenodiacetic acid H_2L

Metal	Method	Temp	Medium	Log of equilibrium constant, remarks	Ref
H^+	gl	25	0·1 ($NaClO_4$)	K_1 4·31, K_2 3·17	66S
Ca^{2+}	gl	25	0·1 ($NaClO_4$)	K_1 2·3	66S
Co^{2+}	gl	25	0·1 ($NaClO_4$)	K_1 2·3	66S
Cu^{2+}	gl	25	0·1 ($NaClO_4$)	K_1 3·6	66S
Mn^{2+}	gl	25	0·1 ($NaClO_4$)	K_1 1·6	66S
Ni^{2+}	gl	25	0·1 ($NaClO_4$)	K_1 2·9	66S
Pb^{2+}	gl	25	0·1 ($NaClO_4$)	K_1 3·2	66S
Zn^{2+}	gl	25	0·1 ($NaClO_4$)	K_1 2·1	66S

66S K. Suzuki and K. Yamasaki, *J. Inorg. Nuclear Chem.*, 1966, **28**, 473

239.1 $C_4H_7O_3N$ Acetylglycine HL

Metal	Method	Temp	Medium	Log of equilibrium constant, remarks	Ref
H^+	gl	20	0·015, 1·0	K_1 (0·015) 3·84, (1·0) 3·68	60K
Cu^{2+}	gl	20	0·015, 1·0	K_1 (0·015) 2·14, (1·0) 1·71	60K

60K W. L. Koltun, M. Fried, and F. R. N. Gurd, *J. Amer. Chem. Soc.*, 1960, **82**, 233

$C_4H_7O_4N$ Aminobutanedioic acid (aspartic acid) H_2L 240

Metal	Method	Temp	Medium	Log of equilibrium constant, remarks	Ref
H^+	gl	15	~ 0	K_1 10·06	53P
	gl	25	0·1 (KCl)	K_1 9·87, K_2 3·87	61B
	gl	30	0·1 (KCl)	K_1 9·63, K_2 3·79	62Ca
	gl	25	0·2 (KNO_3)	K_2 3·69, K_3 1·92	63F
	pol	30	1·0 KNO_3	K_1 9·59	64R
	gl	25	0·1 (KNO_3)	K_1 9·63, K_2 3·71, K_3 1·94	65R
Be^{2+}	gl	15	0·005 $BeSO_4$	β_2 13·4	53P
Cd^{2+}	gl	15	0·005 $CdSO_4$	β_2 8·8	53P
	pol	30	1	β_3 10·30	62R
	pol	30	1·0 KNO_3	β_2 8·89, β_3 10·31	64R
				$K(Cd^{2+}+2L^{2-}+OH^- \rightleftharpoons CdL_2OH^{3-})$ 9·80	
				$K(Cd^{2+}+2L^{2-}+CO_3^{2-} \rightleftharpoons CdL_2CO_3^{4-})$ 8·14	
				$K[Cd^{2+}+2L^{2-}+2NH_3 \rightleftharpoons CdL_2(NH_3)_2^{2-}]$ 9·81	
Ce^{3+}	gl	25	0·1 (KCl)	K_1 5·2, K_2 4·6	61B
	gl	30	0·1 (KCl)	K_1 5·13, K_2 3·65, K_3 2·75	62Ca
Co^{2+}	gl	25	0·1 (KNO_3)	K_1 5·96±0·04, K_2 4·27±0·04	65R
La^{3+}	gl	25	0·1 (KCl)	K_1 5·0, K_2 4·2	61B
	gl	30	0·1 (KCl)	K_1 4·84, K_2 3·42	62Ca
Nd^{3+}	gl	25	0·1 (KCl)	K_1 5·5, K_2 4·9	61B
	gl	30	0·1 (KCl)	K_1 5·40, K_2 4·08, K_3 3·06	62Ca
Ni^{2+}	gl	25	0·1 (KNO_3)	K_1 7·14±0·04, K_2 5·20±0·04	65R
Pb^{2+}	pol	30	1·0 KNO_3	K_1 5·88, β_2 7·38	64R
				$K(Pb^{2+}+2L^{2-}+CO_3^{2-} \rightleftharpoons PbL_2CO_3^{4-})$ 8·88	
Pr^{3+}	gl	25	0·1 (KCl)	K_1 5·4, K_2 4·8	61B
	gl	30	0·1 (KCl)	K_1 5·23, K_2 3·84, K_3 2·72	62Ca
UO_2^{2+}	gl	30	0·1 KCl	K_1 8·00	62C
	gl	25	0·2 (KNO_3)	$K(UO_2^{2+}+HL^- \rightleftharpoons UO_2HL^+)$ 2·61±0·06	63F
Zn^{2+}	gl	15	0·005 $ZnSO_4$	β_2 10·4	53P

53P D. J. Perkins, *Biochem. J.*, 1953, **55**, 649
61B I. M. Batgaeu, S. V. Larionov, and V. M. Shul'man, *Russ. J. Inorg. Chem.*, 1961, **6**, 75
62C M. Cefola, R. C. Taylor, P. S. Gentile, and A. V. Celiano, *J. Phys. Chem.*, 1962, **66**, 790
62Ca M. Cefola, A. S. Tompa, A. V. Celiano, and P. S. Gentile, *Inorg. Chem.*, 1962, **1**, 290
62R G. N. Rao and R. S. Subrahmanya, *Current Sci.*, 1962, **31**, 55
63F I. Feldman and L. Koval, *Inorg. Chem.*, 1963, **2**, 145
64R G. N. Rao and R. S. Subrahmanya, *Proc. Indian Acad. Sci.*, 1964, **60**, 165, 185
65R J. H. Ritsma, G. A. Wiegers, and F. Jellinek, *Rec. Trav. chim.*, 1965, **84**, 1577

$C_4H_7O_4N$ Iminodiacetic acid (IDA) H_2L 241

Metal	Method	Temp	Medium	Log of equilibrium constant, remarks	Ref
H^+	gl	25	0·1 (KNO_3)	K_1 9·33, K_2 2·58	62T, 67S
	gl	20	0·1 (KNO_3)	K_1 9·45	64A
	cal	20	0·1 (KNO_3)	$\Delta H_1°$ −8·15, $\Delta S_1°$ 15·4	64A
	gl	25	0·1—1 (KNO_3)	K_1 (0·1) 9·40±0·01, (1·0) 9·38±0·01, K_2 (0·1) 2·50±0·01, (1·0) 2·55±0·01	64R
	gl	25	0·3 (KCl)	K_1 9·60, K_2 3·01	66M
	sp	25	0·1 (NaCl)	K_2 2·39, K_3 1·82	67S
	gl	0·4	0·10 (KNO_3)	K_1 10·70, K_2 2·84	67T

continued overleaf

241 $C_4H_7O_4N$ (contd)

Metal	*Method*	*Temp*	*Medium*	*Log of equilibrium constant, remarks*	*Ref*
Ba^{2+}	gl	20	0·1 (KNO_3)	K_1 1·67	64A
	cal	20	0·1 (KNO_3)	$\Delta H_1°$ 0·1, $\Delta S_1°$ 8·0	64A
Ca^{2+}	gl	20	0·1 (KNO_3)	K_1 2·59	64A
	cal	20	0·1 (KNO_3)	$\Delta H_1°$ 0·3, $\Delta S_1°$ 12·7	64A
Cd^{2+}	gl	20	0·1 (KNO_3)	K_1 5·73, β_2 10·19	64A
	cal	20	0·1 (KNO_3)	$\Delta H_1°$ −1·46, $\Delta S_1°$ 21·3, $\Delta H_{\beta 2}°$ −5·48, $\Delta S_{\beta 2}°$ 27·9	64A
Ce^{3+}	gl	25	0·1 (KNO_3)	K_1 6·18±0·03, K_2 4·53±0·02	62T
				See *N*-(2-hydroxyethyl)ethylenediamine-*NN′N′*-triacetic acid (838) for mixed ligands	
Co^{2+}	gl	20	0·1 (KNO_3)	K_1 6·97, β_2 12·31	64A
	cal	20	0·1 (KNO_3)	$\Delta H_1°$ −2·14, $\Delta S_1°$ 24·6, $\Delta H_{\beta 2}°$ −6·0, $\Delta S_{\beta 2}°$ 35·8	64A
Cu^{2+}	gl	20	0·1 (KNO_3)	K_1 10·63, β_2 16·68	64A
	cal	20	0·1 (KNO_3)	$\Delta H_1°$ −4·5, $\Delta S_1°$ 33·3, $\Delta H_{\beta 2}°$ −10·9, $\Delta S_{\beta 2}°$ 39·2	64A
	gl	0·4	0·10 (KNO_3)	K_1 11·70	67T
Dy^{3+}	gl	25	0·1 (KNO_3)	K_1 6·88±0·02, K_2 5·43±0·03	62T
				See *N*-(2-hydroxyethyl)ethylenediamine-*NN′N′*-triacetic acid (838) for mixed ligands	
Er^{3+}	gl	25	0·1 (KNO_3)	K_1 7·09±0·02, K_2 5·59±0·01	62T
				See *N*-(2-hydroxyethyl)ethylenediamine-*NN′N′*-triacetic acid (838) for mixed ligands	
Eu^{3+}	gl	25	0·1 (KNO_3)	K_1 6·73±0·03, K_2 5·38±0·03	62T
	gl	25	0·3 (KCl)	K_1 6·22, K_2 4·72	66M
				See *N*-(2-hydroxyethyl)ethylenediamine-*NN′N′*-triacetic acid (838) for mixed ligands	
Fe^{2+}	gl	20	0·1 (KNO_3)	K_1 5·8, β_2 10·1	64A
Gd^{3+}	gl	25	0·1 (KNO_3)	K_1 6·68±0·01, K_2 5·39±0·01	62T
				See *N*-(2-hydroxyethyl)ethylenediamine-*NN′N′*-triacetic acid (838) for mixed ligands	
Hg_2^{2+}	sol	25	0·1 ($NaClO_4$)	K_1 10·81	67S
Hg^{2+}	sol	25	0·1 ($NaClO_4$)	K_1 11·76	67S
Ho^{3+}	gl	25	0·1 (KNO_3)	K_1 6·97±0·03, K_2 5·50±0·03	62T
				See *N*-(2-hydroxyethyl)ethylenediamine-*NN′N′*-triacetic acid (838) for mixed ligands	
In^{3+}	gl	25	0·3 (KCl)	K_1 9·54, K_2 8·87	66M
La^{3+}	gl	25	0·1 (KNO_3)	K_1 5·88±0·01, K_2 4·09±0·02	62T
	gl	20	0·1 (KNO_3)	K_1 5·70, β_2 9·67	64A
	cal	20	0·1 (KNO_3)	$\Delta H_1°$ 0·17, $\Delta S_1°$ 25·6, $\Delta H_{\beta 2}°$ −0·16, $\Delta S_{\beta 2}°$ 43·6	64A
				See *N*-(2-hydroxyethyl)ethylenediamine-*NN′N′*-triacetic acid (838) for mixed ligands	
Lu^{3+}	gl	25	0·1 (KNO_3)	K_1 7·61±0·03, K_2 6·12±0·02	62T
				See *N*-(2-hydroxyethyl)ethylenediamine-*NN′N′*-triacetic acid (838) for mixed ligands	
Mg^{2+}	gl	20	0·1 (KNO_3)	K_1 2·94	64A
	cal	20	0·1 (KNO_3)	$\Delta H_1°$ 2·94, $\Delta S_1°$ 23·5	64A
Nd^{3+}	gl	25	0·1 (KNO_3)	K_1 6·50±0·01, K_2 4·89±0·01	62T
				See *N*-(2-hydroxyethyl)ethylenediamine-*NN′N′*-triacetic acid (838) for mixed ligands	
Ni^{2+}	gl	20	0·1 (KNO_3)	K_1 8·19, β_2 14·3	64A
	cal	20	0·1 (KNO_3)	$\Delta H_1°$ −5·05, $\Delta S_1°$ 20·0, $\Delta H_{\beta 2}°$ −9·5, $\Delta S_{\beta 2}°$ 33·1	64A
Pb^{2+}	gl	20	0·1 (KNO_3)	K_1 7·45	64A
	cal	20	0·1 (KNO_3)	$\Delta H_1°$ −3·34, $\Delta S_1°$ 22·7	64A
Pr^{3+}	gl	25	0·1 (KNO_3)	K_1 6·44±0·03, K_2 4·78±0·03	62T
				See *N*-(2-hydroxyethyl)ethylenediamine-*NN′N′*-triacetic acid (838) for mixed ligands	

cont

Metal	Method	Temp	Medium	Log of equilibrium constant, remarks	Ref
Sm^{3+}	gl	25	0·1 (KNO_3)	K_1 6·64±0·02, K_2 5·24±0·02	62T
				See *N*-(2-hydroxyethyl)ethylenediamine-*NN'N'*-triacetic acid (838) for mixed ligands	
Sr^{2+}	gl	20	0·1 (KNO_3)	K_1 2·23	64A
	cal	20	0·1 (KNO_3)	$\Delta H_1°$ 0·1, $\Delta S_1°$ 10·5	64A
Tb^{3+}	gl	25	0·1 (KNO_3)	K_1 6·78±0·03, K_2 5·46±0·04	62T
				See *N*-(2-hydroxyethyl)ethylenediamine-*NN'N'*-triacetic acid (838) for mixed ligands	
Th^{4+}				See EDTA (836) and CDTA (997) for mixed ligands	
Tm^{3+}	gl	25	0·1 (KNO_3)	K_1 7·22±0·02, K_2 5·68±0·01	62T
				See *N*-(2-hydroxyethyl)ethylenediamine-*NN'N'*-triacetic acid (838) for mixed ligands	
U^{4+}				See EDTA (836) for mixed ligands	
UO_2^{2+}	gl	25	0·1—1 (KNO_3)	K_1 (0·1) 8·93±0·05, (1·0) 8·73±0·03	64R
	pol	30	0·15 $NaClO_4$	$K[UO_2^{2+}+2HL^- \rightleftharpoons UO_2(HL)_2]$ 3·92±0·01	67L
VO^{2+}	gl	25	0·1 ($NaClO_4$)	K ($VOLOH^- + H^+ \rightleftharpoons VOL$) 5·50	66K
Y^{3+}	gl	25	0·1 (KNO_3)	K_1 6·78±0·03, K_2 5·25±0·04	62T
				See *N*-(2-hydroxyethyl)ethylenediamine-*NN'N'*-triacetic acid (838) for mixed ligands	
Yb^{3+}	gl	25	0·1 (KNO_3)	K_1 7·42±0·04, K_2 5·85±0·03	62T
				See *N*-(2-hydroxyethyl)ethylenediamine-*NN'N'*-triacetic acid (838) for mixed ligands	
Zn^{2+}	gl	20	0·1 (KNO_3)	K_1 7·27, β_2 12·60	64A
	cal	20	0·1 (KNO_3)	$\Delta H_1°$ −2·2, $\Delta S_1°$ 25·7, $\Delta H_{\beta 2}°$ −5·9, $\Delta S_{\beta 2}°$ 37·5	64A
Zr^{4+}	ix	?	2	K (?) 3·45	64P

62T L. C. Thompson, *Inorg. Chem.*, 1962, **1**, 490
64A G. Anderegg, *Helv. Chim. Acta*, 1964, **47**, 1801
64P L. N. Pankratova, L. G. Vlasov and A. V. Lapitskii, *Russ. J. Inorg. Chem.*, 1964, **9**, 954 (1763)
64R K. S. Rajan and A. E. Martell, *J. Inorg. Nuclear Chem.*, 1964, **26**, 789
66K Th. Kaden and S. Fallab, *Chimia (Switz.)*, 1966, **20**, 51
66M S. Misumi and M. Aihara, *Bull. Chem. Soc. Japan*, 1966, **39**, 2677
67L T. T. Lai and T. Y. Chen, *J. Inorg. Nuclear Chem.*, 1967, **29**, 2975
67S N. A. Skorik and V. N. Kumok, *J. Gen. Chem. (U.S.S.R.)*, 1967, **37**, 1461 (1722)
67T M. M. Taqui Khan and A. E. Martell, *J. Amer. Chem. Soc.*, 1967, **89**, 7104

$C_4H_8O_2N_2$ Butane-2,3-dione dioxime (dimethylglyoxime) HL 242

Metal	Method	Temp	Medium	Log of equilibrium constant, remarks	Ref
H^+	gl	25	0·1 75% dioxan	K_1 13·53	63B
Co^{2+}	gl	25	0·1 0—75% dioxan	K_1 (→ 0%) 8·35, (50%) 11·01, (75%) 12·20, β_2 (→ 0%) 16·98, (50%) 20·68, (75%) 22·44	63B
	sp	?	1—6 $NaClO_4$	$K(CoHL_2^+ + I^- \rightleftharpoons CoHL_2I)$, (1) 1·38, (2) 1·54, (3) 1·80, (4) 2·40, (5) 3·04, (6) 3·86	68B
			1—5 $NaNO_3$	(1) 1·56, (2) 1·62, (3) 2·20, (4) 2·60, (5) 3·20	
			1—6 $LiNO_3$	(1) 1·58, (2) 2·0, (3) 2·50, (4) 3·24, (5) 3·46, (6) 4·04	

continued overleaf

242 $C_4H_8O_2N_2$ (contd)

Metal	*Method*	*Temp*	*Medium*	*Log of equilibrium constant, remarks*	*Ref*
Co^{2+} (*contd*)			1—6 $NaClO_4$	$K(CoHL_2^+ + 2I^- \rightleftharpoons CoHL_2I_2^-)$, (1) 3·4±0·1, (2) 3·7±0·1, (3) 4·2±0·1, (4) 4·8±0·1, (5) 5·5±0·1, (6) 6·3±0·1	
			1—5 $NaNO_3$	(1) 3·7±0·1, (2) 3·8±0·1, (3) 4·4±0·1, (4) 4·6±0·1, (5) 5·0±0·1	
			1—6 $LiNO_3$	(1) 3·8±0·1, (2) 4·1±0·1, (3) 4·40±0·05, (4) 5·08±0·04, (5) 5·52±0·05, (6) 6·08±0·05	
Cu^{2+}	gl	25	0·1 0—75% dioxan	K_1 (→ 0%) 9·05, (75%) 12·23; β_2 (→ 0%) 18·50, (75%) 24·34	63B
Ni^{2+}	gl	20	0·1 (KCl) 50% methanol	β_2 21·8±0·3	61V
	gl	25	0·1 0—75% dioxan	K_1 (75%) 11·0±0·3; β_2 (→ 0%) 17·0, (75%) 23·1±0·1; K_{so} −23·66	63B
	sol	20	0·1 ($NaClO_4$)	K_1 8·08±0·05, K_2 8·57±0·05	64A
	dis	25	0·1 ($NaClO_4$)	K_1 9·0±0·2, K_2 8·62±0·02	64S
	ix	?	0·2 (KCl)	$K(NiL_2 + L^{2-} \rightleftharpoons NiL_3^{2-})$ 3·02	67B
Pd^{2+}	dis	25	1 $NaClO_4$	β_2 34·1, $K(PdL_2 + OH^- \rightleftharpoons PdL_2OH^-)$ 5·50	63Ba

Errata. 1964, p. 427, Table 242. Formula, *for* $C_4H_8O_2N$ *read* $C_4H_8O_2N_2$.
p. 428. Under Ni^{2+}, *for* ref. 60V *read* ref. 61V.
p. 429. Reference 58J, *for* R. J. P. Williamson *read* R. J. P. Williams.
pp. 427—429. *For* ref. 60D *read* ref. 61D. D. Dyrssen and M. Hennichs, *Acta Chem. Scand.*, 1961, **15,** 47.

61V F. Vlacil, *Coll. Czech. Chem. Comm.*, 1961, **26**, 658
63B C. V. Banks and S. Anderson, *Inorg. Chem.*, 1963, **2**, 112
63Ba K. Burger and D. Dyrssen, *Acta Chem. Scand.*, 1963, **17**, 1489
64A E. K. Astakhova, V. M. Savostina, and V. M. Peshkova, *Russ. J. Inorg. Chem.*, 1964, **9**, 452 (817)
64S V. M. Savostina, E. K. Astakhova, and V. M. Peshkova, *Russ. J. Inorg. Chem.*, 1964, **9**, 42 (80)
67B M. K. Boreiko, E. I. Kazantsev, and I. I. Kalinichenko, *Russ. J. Inorg. Chem.*, 1967, **12**, 137 (269)
68B K. Burger, B. Pinter, E. Papp-Molnar, and S. Nemes-kosa, *Acta Chim. Acad. Sci. Hung.*, 1968, **57**, 363

243 $C_4H_8O_2S$ (Ethylthio)acetic acid HL

Metal	*Method*	*Temp*	*Medium*	*Log of equilibrium constant, remarks*	*Ref*
H^+	gl	20	1·00 $NaClO_4$	K_1 3·61±0·01	61S
	gl	31	2·0 ($NaClO_4$)	K_1 3·62	63B
Cu^{2+}	gl	20	1·00 $NaClO_4$	K_1 2·56±0·01, β_2 4·76±0·02, β_3 4·85±0·04	61S
Er^{3+}	gl	31	2·0 ($NaClO_4$)	K_1 1·42±0·05, K_2 1·0±0·1	63B
Eu^{3+}	gl	31	2·0 ($NaClO_4$)	K_1 1·79±0·05, K_2 0·9±0·1	63B
Gd^{3+}	gl	31	2·0 ($NaClO_4$)	K_1 1·70±0·05, K_2 0·9±0·1	63B
Ho^{3+}	gl	31	2·0 ($NaClO_4$)	K_1 1·43±0·05, K_2 1·0±0·2	63B
La^{3+}	gl	31	2·0 ($NaClO_4$)	K_1 1·70±0·05, K_2 0·8±0·1	63B
Nd^{3+}	gl	31	2·0 ($NaClO_4$)	K_1 1·72±0·05, K_2 0·8±0·1	63B
Sm^{3+}	gl	31	2·0 ($NaClO_4$)	K_1 1·85±0·05, K_2 0·9±0·1	63B
Tb^{3+}	gl	31	2·0 ($NaClO_4$)	K_1 1·53±0·05, K_2 1·0±0·2	63B
Y^{3+}	gl	31	2·0 ($NaClO_4$)	K_1 1·42±0·05, K_2 0·7±0·1	63B
Yb^{3+}	gl	31	2·0 ($NaClO_4$)	K_1 1·40±0·05, K_2 1·0±0·1	63B

61S A. Sandell, *Acta Chem. Scand.*, 1961, **15**, 190
63B J. L. Bear, G. R. Choppin, and J. V. Quagliano, *J. Inorg. Nuclear Chem.*, 1963, **25**, 513

$C_4H_8O_3N_2$ 2-Aminobutanedioic acid 4-amide (asparagine) HL 244

Metal	Method	Temp	Medium	Log of equilibrium constant, remarks	Ref
H^+	gl	15	~ 0	K_1 9·13	53P
	pol	30	1·0 KNO_3	K_1 8·88	64R
	gl	25	0·1 (KNO_3)	K_1 8·72, K_2 2·14	65R
	Hg	25	0·6	K_1 9·04±0·04	67A
Ag^+	Ag	25	0·6	K_1 3·30±0·03, β_2 6·45±0·04	67A
Be^{2+}	gl	15	0·005 $BeSO_4$	β_2 11·7	53P
Cd^{2+}	gl	15	0·005 $CdSO_4$	β_2 7·1	53P
	pol	30	1	β_3 8·60	62R
	pol	30	1·0 KNO_3	β_2 6·90, β_3 8·58 $K(Cd^{2+}+3L^-+OH^- \rightleftharpoons CdL_3OH^{2-})$ 9·22 $K[Cd^{2+}+2L^-+2NH_3 \rightleftharpoons CdL_2(NH_3)_2]$ 9·08	64R
Co^{2+}	gl	25	0·1 (KNO_3)	K_1 4·51±0·04, K_2 3·50±0·04	65R
Cr^{3+}	gl	25	~ 0·5	K_1 7·7, K_2 5·9, K_3 4·9	63K
Cu^{2+}	gl	25	0·1 (KNO_3)	K_1 7·86±0·04, K_2 6·56±0·04	65R
Ni^{2+}	gl	25	0·1 (KNO_3)	K_1 5·68±0·04, K_2 4·55±0·04	65R
Pb^{2+}	pol	30	1·0 KNO_3	K_1 4·36, β_2 6·23 $K(Pb^{2+}+2L^-+OH^- \rightleftharpoons PbL_2OH^-)$ 10·02	64R
Zn^{2+}	gl	15	0·005 $ZnSO_4$	β_2 8·5	53P

53P D. J. Perkins, *Biochem. J.*, 1953, **55**, 649
62R G. N. Rao and R. S. Subrahmanya, *Current Sci.*, 1962, **31**, 55
63K A. A. Khan and W. U. Malik, *J. Indian Chem. Soc.*, 1963, **40**, 564
64R G. N. Rao and R. S. Subrahmanya, *Proc. Indian Acad. Sci.*, 1964, **60**, 165, 185
65R J. H. Ritsma, G. A. Wiegers, and F. Jellinek, *Rec. Trav. chim.*, 1965, **84**, 1577
67A Yu. M. Azizov, A. Kh. Miftakhova, and V. F. Toropova, *Russ. J. Inorg. Chem.*, 1967, **12**, 345 (661)

$C_4H_8O_3N_2$ Glycylglycine HL 245

Metal	Method	Temp	Medium	Log of equilibrium constant, remarks	Ref
H^+	gl	21	~ 0	K_1 8·21	52P
	gl	25	0·100 (NaCl)	K_1 8·17±0·02, K_2 3·22±0·02	59B
	gl	25	0·16 KCl	K_1 8·13, K_2 3·19	60K
	gl	25	0·1 (KCl)	K_1 8·10±0·01, K_2 3·15±0·02	64D
	gl	25	0·16 KCl	K_1 8·16, K_2 3·23	65B
	gl	10—40	0·06	K_1 (10°) 8·64±0·02, (25°) 8·22±0·02, (30°) 8·11±0·02, (35°) 7·97±0·02, (40°) 7·86±0·02, K_2 (10°) 3·21±0·02, (25°) 3·20±0·02, (30°) 3·20±0·02, (35°) 3·19±0·02, (40°) 3·19±0·02 ΔH_1 −10·5±0·2, ΔS_1 2·2±0·6 ΔH_2 −0·26±0·05, ΔS_2 13·8±0·2	66V
	gl	25	39·1, 70% methanol	K_1 (39·1%) 8·03±0·02, (70%) 7·93±0·02 K_2 (39·1%) 3·69±0·02, (70%) 4·28±0·02	66V
			45, 60% dioxan	K_1 (45%) 8·36±0·02, (60%) 8·73±0·02, K_2 (45%) 4·06±0·02, (60%) 4·74±0·02	
	gl	25	0·1 KNO_3	K_1 8·13, K_2 3·14	67M
	gl	25	0·1 ($NaClO_4$)	K_1 8·00±0·01, K_2 3·08±0·01	68B
	cal	25	0·1 ($NaClO_4$)	ΔH_1 −10·6±0·1, ΔS_1 1·5±0·4, ΔH_2 −0·03±0·02, ΔS_2 13·9±0·6	68B

continued overleaf

245 $C_4H_8O_3N_2$ (contd)

Metal	*Method*	*Temp*	*Medium*	*Log of equilibrium constant, remarks*	*Ref*
Ag^+	gl	21	0·02 $AgNO_3$	K_1 3·1	52P
Be^{2+}	gl	21	0·01 $BeSO_4$	β_2 9·8	52P
Cd^{2+}	gl	21	0·01 $CdSO_4$	β_2 5·4	52P
	gl	10—40	0·06	K_1 (10°) 3·16±0·05, (25°) 3·08±0·04, (30°) 3·04±0·04, (40°) 2·99±0·04 β_2 (10°) 5·84±0·04, (25°) 5·65±0·03, (30°) 5·57±0·04, (40°) 5·49±0·04 ΔH_1 −2·5±0·1, ΔS_1 5·7±0·6, $\Delta H_{\beta 2}$ −4·7±0·2, $\Delta S_{\beta 2}$ 10±1	66V
	gl	25	39·1, 70% methanol 45, 60% dioxan	K_1 (39·1%) 3·57±0·02, (70%) 4·26±0·02 β_2 (39·1%) 6·59±0·02, (70%) 7·62±0·02 K_1 (45%) 4·14±0·02, (60%) 4·93±0·08 β_2 (45%) 7·47±0·01, (60%) 8·90±0·03	66V
Co^{2+}	gl	21	0·01 $CoCl_2$	β_2 5·8	52P
	gl	25	0·100 (NaCl)	K_1 2·94±0·01, β_2 5·42±0·02	59B
Cu^{2+}	gl	25	0·100 (NaCl)	K_1 5·43±0·03, β_2 8·64±0·08 $K[Cu(H_{-1}L)+H^+ \rightleftharpoons CuL^+]$ 4·17 $K[Cu(H_{-1}L)_2^{2-}+H^+ \rightleftharpoons CuL(H_{-1}L)^-]$ 9·67	59B
	gl	25	0·16 KCl	K_1 4·96, $K[Cu(H_{-1}L)+H^+ \rightleftharpoons CuL^+]$ 3·90 $K[Cu(H_{-1}L)+L^- \rightleftharpoons CuL(H_{-1}L)^-]$ 3·07 $K[Cu(H_{-1}L)OH^- +H^+ \rightleftharpoons Cu(H_{-1}L)]$ 9·37 $K[Cu(H_{-1}L)(OH)_2^{2-}+H^+ \rightleftharpoons Cu(H_{-1}L)OH^-]$ 12·2 $K[Cu(H_{-1}L)OH+Cu(H_{-1}L) \rightleftharpoons (CuH_{-1}L)_2OH^-]$ 2·30 $K[Cu(H_{-1}L)+A \rightleftharpoons Cu(H_{-1}L)A]$ 3·85 where A = imidazole	60K
	gl	25	0·1 (KCl)	K_1 6·52±0·05 $K[Cu(H_{-1}L)+H^+ \rightleftharpoons CuL^+]$ 4·79±0·05 $K[Cu^{2+}+L^- \rightleftharpoons Cu(H_{-1}L)+H^+]$ 1·73±0·05	64D
	gl	25	1 (KCl)	K_1 5·42±0·02 $K[Cu(H_{-1}L)+H^+ \rightleftharpoons CuL^+]$ 4·38±0·02 $K[Cu(H_{-1}L)OH^- +H^+ \rightleftharpoons Cu(H_{-1}L)]$ 9·52±0·02 $K[Cu(H_{-1}L)(OH)_2^{2-}+H^+ \rightleftharpoons Cu(H_{-1}L)OH^-]$ 12·8±0·1 $K[Cu(H_{-1}L)+Cu(H_{-1}L)OH^- \rightleftharpoons Cu_2(H_{-1}L)_2OH^-]$ −2·07±0·02 $K[Cu(H_{-1}L)+L^- \rightleftharpoons CuL(H_{-1}L)^-]$ 2·92±0·02	64K
	gl	25	0·16 KCl	K_1 5·44, $K[Cu(H_{-1}L)+H^+ \rightleftharpoons CuL^+]$ 4·20	65B
	gl	25	0·1 KNO_3	K_1 5·56 $K[Cu(H_{-1}L)+H^+ \rightleftharpoons CuL^+]$ 4·12 $K[Cu(H_{-1}L)OH^- +H^+ \rightleftharpoons Cu(H_{-1}L)]$ 9·38 $K[Cu(H_{-1}L)+L^- \rightleftharpoons CuL(H_{-1}L)^-]$ 3·17 $K[Cu(H_{-1}L)+Cu(H_{-2}L)^- \rightleftharpoons Cu_2(H_{-3}L_2)^-]$ 2·20	67M
	cal	22	0·1 KNO_3	$\Delta H_{\beta 2}$ −12·5	67S
	gl	25	0·1 ($NaClO_4$)	K_1 5·56±0·01 $K[Cu(H_{-1}L)+H^+ \rightleftharpoons CuL^+]$ 4·06±0·01 $K[Cu(H_{-1}L)OH^- +H^+ \rightleftharpoons Cu(H_{-1}L)]$ 9·29±0·02 $K[Cu(H_{-1}L)+Cu(H_{-1}L)OH^- \rightleftharpoons Cu_2(H_{-1}L)_2OH^-]$ 2·12±0·03	68B
	cal	25	0·1 ($NaClO_4$)	ΔH_1 −6·1±0·2, ΔS_1 5·0±0·7 $\Delta H[Cu(H_{-1}L)+H^+ \rightleftharpoons CuL^+]$ −6·9±0·2, ΔS −4·5±0·7	68B
Hg^{2+}	gl	21	0·01 $Hg(NO_3)_2$	β_2 12·4	52P

contin

$C_4H_8O_3N_2$ (contd) 245

Metal	Method	Temp	Medium	Log of equilibrium constant, remarks	Ref
Ni^{2+}	gl	21	0·01 $NiCl_2$	β_2 7·6	52P
	gl	25	0·16 (KNO_3)	$K[Ni(H_{-1}L)+H^+ \rightleftharpoons NiL^+]$ 9·35	60M
				$K[Ni(H_{-1}L)OH^- + H^+ \rightleftharpoons Ni(H_{-1}L)]$ 9·95	
	gl	25	0·1 (KNO_3)	K_1 3·34, K_2 4·07, K_3 2·5	67K
	cal	22	0·1 KNO_3	$\Delta H_{\beta 2}$ −6	67S
Pb^{2+}	gl	21	0·01 $Pb(NO_3)_2$	β_2 5·8	52P
Zn^{2+}	gl	21	0·01 $ZnSO_4$	β_2 6·4	52P

Erratum. 1964, p. 431, Table 245 (cont.). Under Cu^{2+} line 18, *for* HL^- *read* L^-.

52P D. J. Perkins, *Biochem. J.*, 1952, **51**, 487
59B J. L. Biester and P. M. Ruoff, *J. Amer. Chem. Soc.*, 1959, **81**, 6517
60K W. L. Koltun, M. Fried, and F. R. N. Gurd, *J. Amer. Chem. Soc.*, 1960, **82**, 233
60M R. B. Martin, M. Chamberlin, and J. T. Edsall, *J. Amer. Chem. Soc.*, 1960, **82**, 495
64D M. A. Doran, S. Chaberek, and A. E. Martell, *J. Amer. Chem. Soc.*, 1964, **86**, 2129
64K M. K. Kim and A. E. Martell, *Biochemistry*, 1964, **3**, 1169
65B G. F. Bryce, J. M. H. Pinkerton, L. K. Steinrauf, and F. R. N. Gurd, *J. Biol. Chem.*, 1965, **240**, 3829
66V J. Vaissermann and M. Quintin, *J. Chim. phys.*, 1966, **63**, 731
67K M. K. Kim and A. E. Martell, *J. Amer. Chem. Soc.*, 1967, **89**, 5138
67M R. P. Martin, *Bull. Soc. chim. France*, 1967, 2217
67S W. F. Stack and H. A. Skinner, *Trans. Faraday Soc.*, 1967, **63**, 1136
68B A. P. Brunetti, M. C. Lim, and G. H. Nancollas, *J. Amer. Chem. Soc.*, 1968, **90**, 5120

$C_4H_8N_2S$ 4-Methylimidazolidine-2-thione (propylene thiourea) HL 247.1

Metal	Method	Temp	Medium	Log of equilibrium constant, remarks	Ref
Ag^+	Ag	28	2NH_4NO_3 80% dioxan	$K[Ag^+ + 3HL \rightleftharpoons Ag(HL)_3^+]$ 9·5	64K

64K S. N. Khodaskar and D. D. Khodanolkar, *Current Sci.*, 1964, **33**, 399

C_4H_9ON Perhydro-1,4-oxazine (morpholine) L 248

Metal	Method	Temp	Medium	Log of equilibrium constant, remarks	Ref
Ag^+	oth	?	0·1 $NaClO_4$ acetone	β_2 9·17±0·05, coulometric titration	65M
Zn^{2+}				See dimethyldithiocarbamic acid (176) for mixed ligands	

64M K. K. Mead, D. L. Maricle, and C. A. Streali, *Analyt. Chem.*, 1965, **37**, 237

249 $C_4H_9O_2N$ 2-Aminobutanoic acid HL

Metal	*Method*	*Temp*	*Medium*	*Log of equilibrium constant, remarks*	*Ref*
H^+	gl	15—20	~ 0	K_1 9·62	52P, 53P
	gl	15—40	0·2 (KCl)	K_1 (15°) 9·97, (25°) 9·73, (40°) 9·33, K_2 (15°) 2·39, (25°) 2·42, (40°) 2·46	65S
Be^{2+}	gl	15—20	0·005—0·01 $BeSO_4$	β_2 12·9	52P, 53P
Cd^{2+}	gl	15—20	0·005—0·01 $CdSO_4$	β_2 6·8	52P, 53P
Co^{2+}	gl	15—40	0·2 (KCl)	K_1 (15°) 4·31, (40°) 4·16, K_2 (15°) 3·19, (40°) 3·01	65S
				ΔH_1 −2·5±0·5, ΔS_1 11±2, ΔH_2 −3·0±0·5, ΔS_2 4±2	
Cu^{2+}	gl	15—40	0·2 (KCl)	K_1 (15°) 8·34, (25°) 8·21, (40°) 8·01, K_2 (15°) 6·85, (25°) 6·72, (40°) 6·53	65S
				ΔH_1 −5·4±0·5, ΔS_1 19±2, ΔH_2 −5·3±0·5, ΔS_2 13±2	
Hg^{2+}	gl	15—20	0·005—0·01 $Hg(NO_3)_2$	β_2 18·5	52P, 53P
Ni^{2+}	gl	15—40	0·2 (KCl)	K_1 (15°) 5·58, (25°) 5·46, (40°) 5·29, K_2 (15°) 4·48, (25°) 4·36, (40°) 4·19	65S
				ΔH_1 −4·8±0·5, ΔS_1 9±2, ΔH_2 −4·8±0·5, ΔS_2 4±2	
Zn^{2+}	gl	15	0·005 $ZnSO_4$	β_2 7·2	53P
	gl	20	0·01 $ZnSO_4$	β_2 8·3	52P
	gl	15—40	0·2 (KCl)	K_1 (15°) 4·87, (25°) 4·78, (40°) 4·68, K_2 (15°) 3·96, (25°) 3·90, (40°) 3·76	65S
				ΔH_1 −3·1±0·5, ΔS_1 11±2, ΔH_2 −3·3±0·5 ΔS_2 7±2	

52P D. J. Perkins, *Biochem. J.*, 1952, **51**, 487

53P D. J. Perkins, *Biochem. J.*, 1953, **55**, 649

65S V. S. Sharma, H. B. Mathur, and P. S. Kulkarni, *Indian J. Chem.*, 1965, **3**, 146, 475

249.1 $C_4H_9O_2N$ 2-Amino-2-methylpropanoic acid HL (*see Corrigenda*)

Metal	*Method*	*Temp*	*Medium*	*Log of equilibrium constant, remarks*	*Ref*
H^+	gl	19	~ 0	K_1 10·24	52P
	gl	20	0·1 (KCl)	K_1 10·25, K_2 2·35	63I
	gl	20, 30	→ 0	K_1 (20°) 10·10, (30°) 10·08, K_2 (20°) 2·38, (30°) 2·35	64I
	cal	25	→ 0	$\Delta H_1°$ −11·6±0·1, $\Delta S_1°$ 7, $\Delta H_2°$ −1·4±0·1, $\Delta S_2°$ 6	64I
	gl	25	0·5 KCl	K_1 10·22, K_2 2·48	66L
Be^{2+}	gl	19	0·01 $BeSO_4$	β_2 12·4	52P
Cd^{2+}	gl	19	0·01 $CdSO_4$	β_2 7·2	52P
Co^{2+}	gl	20	0·1 (KCl)	K_1 4·11, K_2 3·4	63I
Cu^{2+}	gl	20	0·1 (KCl)	K_1 8·26, K_2 6·84	63I
				$K(CuL^+ + H^+ \rightleftharpoons CuHL^{2+})$ 1·2	
	gl	20, 30	→ 0	K_1 (20°) 8·55, (30°) 8·53, K_2 (20°) 7·05, (30°) 7·04	64I
	cal	25	→ 0	$\Delta H_1°$ −5·4±0·1, $\Delta S_1°$ 21, $\Delta H_2°$ −5·7±0·1, $\Delta S_2°$ 13	64I
Hg^{2+}	gl	19	0·01 $Hg(NO_3)_2$	β_2 18·3	52P

contin

Metal	*Method*	*Temp*	*Medium*	*Log of equilibrium constant, remarks*	*Ref*
Ni^{2+}	gl	20	0·1 (KCl)	K_1 5·16, K_2 4·23	63I
	gl	25	0·5 KCl	K_1 5·12, β_2 9·32, β_3 11·91 $K(Ni^{2+}+A^-+L^- \rightleftharpoons NiAL)$ 7·36±0·02 $K(Ni^{2+}+A^-+2L^- \rightleftharpoons NiAL_2^-)$ 12·35±0·03 $K(Ni^{2+}+2A^-+2L^- \rightleftharpoons NiA_2L_2^{2-})$ 14·17±0·01 where HA = glyoxylic acid	66L
Zn^{2+}	gl	19	0·01 $ZnSO_4$	β_2 8·8	52P
	gl	20	0·1 (KCl)	K_1 4·55, K_2 4·0	63I
	gl	25	0·5 KCl	K_1 3·85, β_2 8·88 ? $K(Zn^{2+}+A^-+L^- \rightleftharpoons ZnAL)$ 6·28 $K(Zn^{2+}+2A^-+2L^- \rightleftharpoons ZnA_2L_2^{2-})$ 12·6 (pptn.) where HA = glyoxylic acid	66L

52P D. J. Perkins, *Biochem. J.*, 1952, **51**, 487
63I H. Irving and L. D. Pettit, *J. Chem. Soc.*, 1963, 1546
64I R. M. Izatt, J. J. Christensen, and V. Kothari, *Inorg. Chem.*, 1964, **3**, 1565
66L D. L. Leussing and E. M. Hanna, *J. Amer. Chem. Soc.*, 1966, **88**, 696

$C_4H_9O_2N$ 3-Aminobutanoic acid HL 249.2

Metal	*Method*	*Temp*	*Medium*	*Log of equilibrium constant, remarks*	*Ref*
H^+	gl	15—40	0·2 (KCl)	K_1 (15°) 10·44, (25°) 10·19, (40°) 9·80, K_2 (15°) 3·50, (25°) 3·55, (40°) 3·55	65S
Co^{2+}	gl	15—40	0·2 (KCl)	K_1 (15°) 3·54, (25°) 3·52, (40°) 3·44, K_2 (15°) 2·46, (25°) 2·36, (40°) 2·26 ΔH_1 −1·7±0·5, ΔS_1 11±2, ΔH_2 −3·3±0·5, ΔS_2 0±2	65S
	ix	40	0·2 (KCl)	K_1 3·44	65S
Cu^{2+}	gl	15—40	0·2 (KCl)	K_1 (15°) 7·30, (25°) 7·18, (40°) 7·00, K_2 (15°) 5·66, (25°) 5·59, (40°) 5·38 ΔH_1 −5·0±0·5, ΔS_1 16±2, ΔH_2 −4·6±0·5, ΔS_2 10±2	65S
Ni^{2+}	gl	15—40	0·2 (KCl)	K_1 (15°) 4·72, (25°) 4·60, (40°) 4·49, K_2 (15°) 3·44, (25°) 3·32, (40°) 3·15 ΔH_1 −3·8±0·5, ΔS_1 8±2, ΔH_2 −4·8±0·5, ΔS_2 −1±2	65S

65S V. S. Sharma, H. B. Mathur, and P. S. Kulkarni, *Indian J. Chem.*, 1965, **3**, 146, 475

249.3 $C_4H_9O_2N$ 4-Aminobutanoic acid HL

Metal	Method	Temp	Medium	Log of equilibrium constant, remarks	Ref
H^+	gl	20	0·5 KNO_3	K_1 10·66±0·02	68A
	gl	25	0·5 KNO_3	K_1 10·46	68T
Ag^+	gl	20	0·500 KNO_3	K_1 3·47±0·05, β_2 7·24±0·03	68A
	Ag-AgCl	20	0·500 KNO_3	β_2 7·32±0·02	68A
	gl	25	0·5 KNO_3	K_1 3·46, K_2 3·75	68T

68A D. J. Alner, R. C. Lansbury, and A. G. Smeeth, *J. Chem. Soc.* (*A*), 1968, 417
68T G. F. Thiers, L. C. van Poucke, and M. A. Herman, *J. Inorg. Nuclear Chem.*, 1968, **30**, 1543

250 $C_4H_9O_2N$ *NN*-Dimethylglycine HL

Metal	Method	Temp	Medium	Log of equilibrium constant, remarks	Ref
H^+	H, Ag-AgCl	5—45	→ 0	K_1 (5°) 10·336, (15°) 10·133, (25°) 9·940, (35°) 9·757, (37°) 9·721, (45°) 9·581 (±0·003)	58D
				$\Delta H_1°$ (25°) −32·03 kJ mole^{-1}, $\Delta S_1°$ (25°) 82·9 J mole^{-1} deg^{-1}	
Ni^{2+}				See Solochrom Violet R (1050.1) for mixed ligands	

58D S. P. Datta and A. K. Grzybowski, *Trans. Faraday Soc.*, 1958, **54**, 1179, 1188

251 $C_4H_9O_2N$ α-Amino-α-methylpropionic acid HL

Erratum. 1964, p. 433, Table 251. Ref 63I, for *J. Chem. Soc.*, 1963, 1946, read *J. Chem. Soc.*, 1963, 1546.

253 $C_4H_9O_2N$ Glycine ethyl ester (ethyl glycinate) L

Metal	Method	Temp	Medium	Log of equilibrium constant, remarks	Ref
H^+	gl	20—45	0·1 $NaNO_3$	K_1 (20°) 7·84±0·05, (25°) 7·75±0·05, (30°) 7·65±0·05, (35°) 7·56±0·05, (40°) 7·48±0·05, (45°) 7·38±0·05	65C
				$\Delta H_1°$ −7·6, $\Delta S_1°$ 10	
	gl	30	1·00 $NaClO_4$	K_1 7·66±0·01	66H
Co^{2+}	gl	30	1·0 $NaClO_4$	K_1 1·43±0·02, K_2 1·2±0·2	66H
Cu^{2+}	gl	20—45	0·1 $NaClO_4$	K_1 (20°) 4·14, (25°) 4·04, (30°) 3·99, (35°) 3·85, (40°) 4·15, (45°) 4·78, K_2 (20°) 3·29, (25°) 3·89, (30°) 4·97, (35°) 4·31, (40°) 4·18, (45°) 3·53, K_3 (20°) 4·38, (25°) 4·80, (30°) 4·24, (35°) 4·21, (40°) 4·01, (45°) 4·36	65C
				See nitrilotriacetic acid (449) for mixed ligands	
Ni^{2+}	gl	30	1·0 $NaClO_4$	K_1 2·30±0·01, K_2 1·92±0·02	66H
Zn^{2+}	gl	30	1·0 $NaClO_4$	K_1 1·79±0·01, K_2 1·90±0·04	66H

65C W. A. Connor, M. M. Jones, and D. L. Tuleen, *Inorg. Chem.*, 1965, **4**, 1129
66H J. E. Hix, jun., and M. M. Jones, *Inorg. Chem.*, 1966, **5**, 1863

$C_4H_9O_3N$ 2-Amino-3-hydroxybutanoic acid (threonine) HL 255

Metal	Method	Temp	Medium	Log of equilibrium constant, remarks	Ref
H^+	gl	20	~ 0	K_1 9·12	53P
	gl	20, 30	→ 0	K_1 (20°) 9·26, (30°) 9·01, K_2 (20°) 2·21, (30°) 2·14	64I
	cal	25	→ 0	$\Delta H_1°$ −10·0±0·1, $\Delta S_1°$ 8, $\Delta H_2°$ −1·4±0·1, $\Delta S_2°$ 5	64I
	gl	15—40	0·2 KNO_3	K_1 (15°) 9·26, (25°) 9·03, (40°) 8·71, K_2 (15°) 2·32, (25°) 2·29, (40°) 2·27	68R
Be^{2+}	gl	20	0·005 $BeSO_4$	β_2 11·9	53P
Cd^{2+}	gl	20	0·005 $CdSO_4$	β_2 7·2	53P
Co^{2+}	gl	15—40	0·2 KNO_3	K_1 (15°) 4·50, (25°) 4·43, (40°) 4·37, K_2 (15°) 3·48, (25°) 3·41, (40°) 3·35 $\Delta H_{\beta 2}$ −4·3±0·5, $\Delta S_{\beta 2}$ 22±2	68R
Cu^{2+}	gl	20, 30	→ 0	K_1 (20°) 8·44, (30°) 8·41, K_2 (20°) 6·96, (30°) 6·91	64I
	cal	25	→ 0	$\Delta H_1°$ −5·3±0·1, $\Delta S_1°$ 21, $\Delta H_2°$ −6·1±0·1, $\Delta S_2°$ 11	64I
	gl	15—40	0·2 KNO_3	K_1 (15°) 8·20, (25°) 8·06, (40°) 8·78, K_2 (15°) 6·74, (25°) 6·63, (40°) 6·47, $\Delta H_{\beta 2}$ −9·9±0·5, $\Delta S_{\beta 2}$ 34±2	68R
Fe^{2+}	gl	15—40	0·2 KNO_3	K_1 (15°) 3·76, (40°) 3·69, K_2 (15°) 2·86, (40°) 2·81 $\Delta H_{\beta 2}$ −2·0±0·5, $\Delta S_{\beta 2}$ 23±2	68R
Hg^{2+}	gl	20	0·005 $Hg(NO_3)_2$	β_2 17·5	53P
Mn^{2+}	gl	15—40	0·2 KNO_3	K_1 (15°) 2·59, (40°) 2·56, K_2 (15°) 1·39, (40°) 1·37 $\Delta H_{\beta 2}$ −0·8±0·5, $\Delta S_{\beta 2}$ 15±2	68R
Ni^{2+}	gl	15—40	0·2 KNO_3	K_1 (15°) 5·50, (25°) 5·42, (40°) 5·28, K_2 (15°) 4·44, (25°) 4·34, (40°) 4·19 $\Delta H_{\beta 2}$ −8·1±0·5, $\Delta S_{\beta 2}$ 19±2	68R
Zn^{2+}	gl	20	0·005 $ZnSO_4$	β_2 8·6	53P
	gl	15—40	0·2 KNO_3	K_1 (15°) 4·79, (25°) 4·74, (40°) 4·76, K_2 (15°) 3·85, (25°) 3·77, (40°) 3·68 $\Delta H_{\beta 2}$ −4·5±0·5, $\Delta S_{\beta 2}$ 24±2	68R

53P D. J. Perkins, *Biochem. J.*, 1953, **55**, 649

64I R. M. Izatt, J. J. Christensen, and V. Kothari, *Inorg. Chem.*, 1964, **3**, 1565

68R E. V. Raju and M. B. Mathur, *J. Inorg. Nuclear Chem.*, 1968, **30**, 2181

$C_4H_9N_3S$ 5-Methyl-2-thioxo-1,3,4-triazahex-4-ene (acetone thiosemicarbazone) L 257.1

Metal	Method	Temp	Medium	Log of equilibrium constant, remarks	Ref
Ag^+	Ag	25	0·052 $NaC_2H_3O_2$ 40% ethanol	β_3 13·2±0·1	61T
Hg^{2+}	Hg	25	0·052 $NaC_2H_3O_2$ 40% ethanol	β_4 27·3±0·1	61T

61T V. F. Toropova, Yu. P. Kitaev, and G. K. Budnikov, *Russ. J. Inorg. Chem.*, 1961, **6**, 330 (647)

260 $C_4H_{10}O_2N_2$ 2,4-Diaminobutanoic acid HL

Metal	*Method*	*Temp*	*Medium*	*Log of equilibrium constant, remarks*	*Ref*
H^+	gl	20	0·025	K_1 10·44, K_2 8·21, K_3 1·85	68H
Cu^{2+}	gl	20	0·025	K_1 ~ 10·4, β_2 19·48±0·03	68H
				$K(Cu^{2+}+HL \rightleftharpoons CuHL^{2+})$ 7·15±0·02	
				$K(Cu^{2+}+2HL \rightleftharpoons CuH_2L_2^{2+})$ 13·00±0·03	
				$K(Cu^{2+}+HL+L^- \rightleftharpoons CuHL_2^+)$ 17·22±0·04	

68H R. W. Hay, P. J. Morris, and D. D. Perrin, *Austral. J. Chem.*, 1968, **21**, 1073 (calculated from data of A. Albert, *Biochem. J.*, 1952, **50**, 690)

260.1 $C_4H_{10}O_2N_2$ 2,3-Diaminopropanoic acid methyl ester L

Metal	*Method*	*Temp*	*Medium*	*Log of equilibrium constant, remarks*	*Ref*
H^+	gl	25	0·1	K_1 8·250, K_2 4·412	68H
Cu^{2+}	gl	25	0·1	K_1 8·99, K_2 7·76	68H
				$K(CuLOH^++H^+ \rightleftharpoons CuL^{2+})$ 6·83	
				See 2,3-diaminopropionic acid (180) for mixed ligands	
Hg^{2+}	gl	25	0·1	K_1 6·38, K_2 5·10	68H
				$K(HgLOH^++H^+ \rightleftharpoons HgL^{2+})$ 7·81	

68H R. W. Hay and P. J. Morris, *Chem. Comm.*, 1968, 732

260.2 $C_4H_{10}O_2N_4$ Butanedioic acid dihydrazide L

Metal	*Method*	*Temp*	*Medium*	*Log of equilibrium constant, remarks*	*Ref*
Cd^{2+}	pol	25	1 ($NaClO_4$)	K_1 1·90, β_2 3·65, β_3 4·99, β_4 4·95	66K
Zn^{2+}	pol	25	1 ($NaClO_4$)	K_1 2·26, β_2 4·34, β_3 5·10, β_4 4·63	68S

66K A. F. Krivis, G. R. Supp, and R. L. Doerr, *Analyt. Chem.*, 1966, **38**, 936
68S G. R. Supp, *Analyt. Chem.*, 1968, **40**, 981

260.3 $C_4H_{11}ON$ 4-Aminobutanol L

Metal	*Method*	*Temp*	*Medium*	*Log of equilibrium constant, remarks*	*Ref*
H^+	gl	20	0·5 KNO_3	K_1 10·61±0·02	68A
	H	20	0 corr	K_1 10·38±0·02	68A
Ag^+	gl	20	0·500 KNO_3	K_1 3·39±0·05, β_2 7·27±0·02	68A
	Ag-AgCl	20	0·500 KNO_3	β_2 7·32±0·02	68A

68A D. J. Alner, R. C. Lansbury, and A. G. Smeeth, *J. Chem. Soc. (A)*, 1968, 417

$C_4H_{11}ON$ 2-(Ethylamino)ethanol L 261

Metal	Method	Temp	Medium	Log of equilibrium constant, remarks	Ref
H^+	gl	25	~ 0·1	K_1 10·21	65D
	gl	30	0·5 (KNO_3)	K_1 10·03	67F
Cu^{2+}	gl	25	~ 0·1	K_1 5·0, K_2 4·1, K_3 3·5, K_4 2·8	65D
	pol	30	0·5 (KNO_3)	$K[Cu^{2+}+L+2OH^- \rightleftharpoons CuL(OH)_2]$ 17·2 $K[Cu^{2+}+2L+2OH^- \rightleftharpoons CuL_2(OH)_2]$ 18·9	67F

65D G. Douheret, *Bull. Soc. chim. France*, 1967, 2915
67F J. F. Fisher and J. L. Hall, *Analyt. Chem.*, 1967, 39, 1550

$C_4H_{11}ON$ *NN*-Dimethyl-2-aminoethanol L 261.1

Metal	Method	Temp	Medium	Log of equilibrium constant, remarks	Ref
H^+	H	20	0 corr	$K(L \rightleftharpoons HL^+ + OH^-)$ 4·86±0·01	61A
	gl	25	~ 0·1	K_1 9·49	65D
Ag^+	sol	20	0 corr	β_2 3·83±0·05	61A
	gl	20	0 corr	β_2 3·80±0·05	61A
Cu^{2+}	gl	25	~ 0·1	K_1 4·7, K_2 4·0, K_3 3·3, K_4 2·9	65D

61A D. J. Alner and R. C. Lansbury, *J. Chem. Soc.*, 1961, 3169
65D G. Douheret, *Bull. Soc. chim. France*, 1965, 2915

$C_4H_{11}O_2N$ 2,2′-Iminodiethanol (diethanolamine) L 264

Metal	Method	Temp	Medium	Log of equilibrium constant, remarks	Ref
H^+	H, Ag-AgCl	0—50	→ 0	K_1 (0°) 9·550±0·001, (5°) 9·404±0·002, (10°) 9·268±0·001, (15°) 9·133±0·001, (20°) 9·005±0·001, (25°) 8·883±0·001, (30°) 8·759±0·001, (35°) 8·632±0·002, (40°) 8·518±0·001, (45°) 8·406±0·001, (50°) 8·297±0·001 $\Delta H_1°$ (25°) −42·4 kJ mole^{-1}, $\Delta S_1°$ (25°) 27·8 J mole^{-1} deg^{-1}	62B
	gl	30	0·5 KNO_3	K_1 8·96	62F, 63S
	gl	25	~ 0·1	K_1 9·12	65D
	gl	25	0·4 (KNO_3)	K_1 9·09	66S
	gl	30	0·5 (KNO_3)	K_1 8·95	67F

continued overleaf

264 $C_4H_{11}O_2N$ (contd)

Metal	Method	Temp	Medium	Log of equilibrium constant, remarks	Ref
Ag^+	Ag	25	< 0·01	β_2 5·44	61A
	Ag	25	0·4 $LiNO_3$ 0—100% ethanol	K_1 (0%) 3·48±0·08, (20%) 4·00±0·05, (40%) 4·30±0·09, (60%) 4·70±0·09, (80%) 4·90±0·10, (90%) 5·04±0·09, (95%) 5·20±0·05, (100%) 5·40±0·10, β_2 (0%) 5·60±0·08, (20%) 6·00±0·05, (40%) 6·20±0·09, (60%) 6·46±0·03, (80%) 6·68±0·10, (90%) 6·74±0·10, (95%) 6·81±0·05, (100%) 6·98±0·10, β_3 (0%) 6·2±0·2, (20%) 6·34±0·04, (40%) 6·53±0·08, (60%) 6·92±0·09, (80%) 7·08±0·08, (90%) 7·28±0·08, (95%) 7·51±0·06, (100%) 7·68±0·10	65M
Cd^{2+}	pol	25	0·1 KNO_3	K_1 2·40, β_2 4·52	60M
	pol	25	0·01 $NaClO_4$ 0—100% ethanol	β_2 (0%) 4·30, (20%) 5·00, (40%) 5·08, (60%) 5·30, (80%) 6·36, (94%) 7·93, (100%) 10·84, β_3 (0%) 5·08, (20%) 5·61, (40%) 5·83, (60%) 6·30, (80%) 6·83, (94%) 8·51, (100%) 11·30, β_4 (100%) 12·72	64M
Co^{2+}	gl	25	0·43 (CH_2OHCH_2· NH_3NO_3)	K_1 2·72, K_2 1·75	66S
Cu^{2+}	pol	30	0·5 KNO_3	$K[Cu^{2+}+2L+2OH^- \rightleftharpoons CuL_2(OH)_2]$ 19·4	62F
	pol	30	0·5 KNO_3	$K[Cu^{2+}+L+3OH^- \rightleftharpoons CuL(OH)_3^-]$ 18·97 $K(Cu^{2+}+2L+OH^- \rightleftharpoons CuL_2OH^+)$ 14·10 $K[Cu^{2+}+2L+2OH^- \rightleftharpoons CuL_2(OH)_2]$ 19·28	63S
	gl	25	~ 0·1	K_1 5·4, K_2 4·2, K_3 3·2, K_4 1·8	65D
	gl	25	0·43 (CH_2OHCH_2· NH_3NO_3)	K_1 4·75, K_2 3·67, K_3 2·75	66S
	pol	30	0·5 (KNO_3)	$K[Cu^{2+}+L+2OH^- \rightleftharpoons CuL(OH)_2]$ 18·2 $K[Cu^{2+}+2L+2OH^- \rightleftharpoons CuL_2(OH)_2]$ 19·8	67F
Fe^{2+}	hyp	25	0·43 (CH_2OHCH_2· NH_3NO_3)	K_1 2·37, K_2 1·30	66S
Mn^{2+}	hyp	25	0·43 (CH_2OHCH_2. NH_3NO_3)	K_1 1·55, K_2 0·45	66S
Ni^{2+}	gl	25	0·43 (CH_2OHCH_2 NH_3NO_3)	K_1 3·31, K_2 2·13, K_3 1·42	66S
Pb^{2+}	pol	25	0·01 $NaClO_4$ 0—100% ethanol	β_2 (0%) 8·70, (20%) 9·04, (40%) 9·50, (60%) 9·52, (80%) 10·00, (100%) 12·40, β_3 (0%) 9·00, (20%) 9·82, (94%) 11·52, (100%) 13·56, β_4 (0%) 0·91	64M
Zn^{2+}	pol	25	0·01 $NaClO_4$ 0—100% ethanol	β_2 (0%) 6·60, (20%) 7·48, (40%) 8·15, (60%) 8·18, (80%) 9·18, (94%) 11·60, (100%) 14·85, β_3 (0%) 8·08, (20%) 8·43, (40%) 9·46, (60%) 9·43, (80%) 10·51, (94%) 12·48, (100%) 15·30, β_4 (0%) 9·11, (20%) 8·88, (100%) 16·52	64M

60M P. K. Migal' and A. N. Pushnyak, *Russ. J. Inorg. Chem.*, 1960, **5**, 293 (610)
61A V. Armeanu and C. Luca, *Z. phys. Chem. (Leipzig)*, 1961, **217**, 389
62B V. E. Bower, R. A. Robinson, and R. G. Bates, *J. Res. Nat. Bur. Stand. Sect. A*, 1962, **66**, 71
62F J. F. Fisher and J. L. Hall, *Analyt. Chem.*, 1962, **34**, 1094
63S P. E. Sturrock, *Analyt. Chem.*, 1963, **35**, 1092
64M P. K. Migal and G. F. Serova, *Russ. J. Inorg. Chem.*, 1964, **9**, 978 (1806)
65D G. Douheret, *Bull. Soc. chim. France*, 1965, 2915
65M P. K. Migal' and K. I. Ploae, *Russ. J. Inorg. Chem.*, 1965, **10**, 1368 (2517)
66S E. V. Skenskaya and M. Kh. Karapet'yants, *Russ. J. Inorg. Chem.*, 1966, **11**, 1478 (2749)
67F J. F. Fisher and J. L. Hall, *Analyt. Chem.*, 1967, **39**, 1550

$C_4H_{11}O_3N$ 2-Amino-2-(hydroxymethyl)propan-1,3-diol [tris(hydroxymethyl)methylamine] L 265

Metal	Method	Temp	Medium	Log of equilibrium constant, remarks	Ref
H^+	H, Ag-AgCl	0—50	→ 0	K_1 (0°) 8·8500±0·0008, (5°) 8·6774±0·0008, (10°) 8·5164±0·0008, (15°) 8·3616±0·0008, (20°) 8·2138±0·0009, (25°) 8·0745±0·0006, (30°) 7·9344±0·0010, (35°) 7·8031±0·0010, (40°) 9·677±0·001, (45°) 7·554±0·001, (50°) 7·437±0·001	61B
				$\Delta H_1°$ (25°) −47·6 kJ mole⁻¹, $\Delta S_1°$ (25°) −5·1 J mole⁻¹ deg⁻¹	
	H, Ag-AgCl	0—60	0 corr	K_1 (0°) 8·8553, (5°) 8·6792, (10°) 8·5158, (15°) 8·3602, (20°) 8·2124, (25°) 8·0686, (30°) 7·9336, (35°) 7·8006, (40°) 7·6756, (45°) 7·5515, (50°) 7·4380, (55°) 7·3204, (60°) 7·2098	63D
				$\Delta H_1°$ (25°) −47·4 kJ mole⁻¹, $\Delta S_1°$ (25°) −4·5 J mole⁻¹ deg⁻¹	
	H, Ag-AgCl	10—40	→ 0 50% methanol	K_1 (10°) 8·273±0·002, (15°) 8·113±0·001, (20°) 7·962±0·001, (25°) 7·818±0·001, (30°) 7·681±0·001, (35°) 7·550±0·001, (40°) 7·426±0·001	65W
				$\Delta H_1°$ (25°) −47·86 kJ mole⁻¹, $\Delta S_1°$ (25°) −10·9 J mole⁻¹ deg⁻¹	
	gl	0—60	0·05 (KNO_3)	K_1 (0°) 8·88, (10°) 8·54, (20°) 8·24, (30°) 7·96, (40°) 7·70, (50°) 7·46, (60°) 7·25	66D
Ag^+	gl	0—60	0·05 (KNO_3)	K_1 (0°) 3·50, (10°) 3·39, (20°) 3·22, (30°) 3·05, (40°) 2·90, (50°) 2·90, (60°) 2·79, K_2 (0°) 4·04, (10°) 3·77, (20°) 3·53, (30°) 3·33, (40°) 3·16, (50°) 3·02, (60°) 2·90	66D
				ΔH_1 (0°) −10, (10°) −22, (20°) −29, (30°) −29, (40°) −24, (50°) −12, (60°) 7 kJ mole⁻¹, ΔS_1 (0°) 32, (10°) −12, (20°) −36, (30°) −39, (40°) −21, (50°) 17, (60°) 75 J mole⁻¹ deg⁻¹	
				ΔH_2 (0°) −42, (10°) −39, (20°) −36, (30°) −33, (40°) −29, (50°) −25, (60°) −22 kJ mole⁻¹, ΔS_2 (0°) −78, (10°) −67, (20°) −55, (30°) −44, (40°) −32, (50°) −21, (60°) −10 J mole⁻¹ deg⁻¹	
Cu^{2+}	gl	?	0·1 (KNO_3)	K_1 3·98, K_2 3·49, K_3 3·2	62H
Ni^{2+}	gl	?	0·1 (KNO_3)	K_1 2·86	62H

61B R. G. Bates and H. B. Hetzer, *J. Phys. chem.*, 1961, 65, 667
62H J. L. Hall, J. A. Swisher, D. G. Brannon, and T. M. Ledem, *Inorg. Chem.*, 1962, 1, 409
63D S. P. Datta, A. K. Grzybowski, and B. A. Weston, *J. Chem. Soc.*, 1963, 792
65W M. Woodhead, M. Paabo, R. A. Robinson, and R. G. Bates, *J. Res. Nat. Bur. Stand. Sect. A*, 1965, 69, 263
66D S. P. Datta and A. K. Grzybowski, *J. Chem. Soc. (A)*, 1966, 1059

$C_4H_{11}S_2P$ Diethylphosphinodithioic acid HL 265.1

Metal	Method	Temp	Medium	Log of equilibrium constant, remarks	Ref
H^+	gl	25	0·8 KNO_3 40% ethanol	K_1 2·62±0·04	67T
Hg^{2+}	Hg	25	0·8 KNO_3 40% ethanol	β_2 33·6, β_3 36·0, β_4 38·2	67T

67T V. F. Toropova, M. K. Saikina, and R. Sh. Aleshov, *J. Gen. Chem. (U.S.S.R.)*, 1967, 37, 679 (725)

267 $C_4H_{12}ON_2$ *N*-(2′-Hydroxyethyl)-1,2-diaminoethane L

Metal	Method	Temp	Medium	Log of equilibrium constant, remarks	Ref
H^+	gl	25	$I(NaClO_4)$	K_1 $9{\cdot}56+(1{\cdot}018\sqrt{I})/(1+0{\cdot}904\sqrt{I})-(1{\cdot}018\sqrt{I})/(1+1{\cdot}227\sqrt{I})+0{\cdot}256\,I$, K_2 $6{\cdot}34+(1{\cdot}018\sqrt{I})/(1+1{\cdot}227\sqrt{I})+0{\cdot}233\,I$	66N
Cu^{2+}	gl	25	$I(NaClO_4)$	K_1 $10{\cdot}02+(1{\cdot}018\sqrt{I})/(1+0{\cdot}904\sqrt{I})-(1{\cdot}018\sqrt{I})/(1+2{\cdot}36\sqrt{I})+0{\cdot}282\,I$, K_2 $7{\cdot}43+(1{\cdot}018\sqrt{I})/(1+9{\cdot}04\sqrt{I})-(1{\cdot}018\sqrt{I})/(1+3{\cdot}00\sqrt{I})+0{\cdot}234\,I$. $K(CuOHL^+ + H^+ \rightleftharpoons CuL^{2+})$ $7{\cdot}09+1{\cdot}018\sqrt{I}/(1+\sqrt{I})$ $K[Cu_2(OH)_2L_2 + 2H^+ \rightleftharpoons 2CuL^{2+}]$ $12{\cdot}31+1{\cdot}018\sqrt{I}/(1+\sqrt{I})$	66N
			$\rightarrow 0$	$K[2CuOHL \rightleftharpoons Cu_2(OH)_2L_2]$ 1·87 $K[Cu(OH)_2L_2^{2+} + 2H^+ \rightleftharpoons CuL^{2+}]$ 9·84	
Ni^{2+}	gl	25	$I(NaClO_4)$	K_1 $6{\cdot}76+(1{\cdot}018\sqrt{I})/(1+0{\cdot}904\sqrt{I})-(1{\cdot}018\sqrt{I})/(1+1{\cdot}56\sqrt{I})+0{\cdot}231\,I$, K_2 $5{\cdot}52+(1{\cdot}018\sqrt{I})/(1+0{\cdot}904\sqrt{I})-(1{\cdot}018\sqrt{I})/(1+2{\cdot}52\sqrt{I})+0{\cdot}262\,I$	66N

66N R. Nasanen, P. Tilus, and A. M. Rinne, *Suomen Kem.*, 1966, **B39**, 45

267.1 $C_4H_{12}O_6P_2$ Butane-1,4-diphosphonic acid H_4L

Metal	Method	Temp	Medium	Log of equilibrium constant, remarks	Ref
H^+	gl	20	0·1 KCl	K_1 8·38, K_2 7·54, K_3 2·7, K_4 < 2	51S
	gl	25	0·1 $(CH_3)_4NBr$	K_1 8·38±0·07, K_2 7·58±0·07, K_3 2·85±0·07, K_4 1·7±0·2	62I
Ba^{2+}	gl	20	0·1 KCl	K_1 2·28, $K(Ba^{2+} + HL^{3-} \rightleftharpoons BaHL^-)$ 1·5	51S
Ca^{2+}	gl	20	0·1 KCl	K_1 2·54, $K(Ca^{2+} + HL^{3-} \rightleftharpoons CaHL^-)$ 1·7	51S
	gl	25	0·1	K_1 2·5	62I
Mg^{2+}	gl	20	0·1 KCl	K_1 2·77, $K(Mg^{2+} + HL^{3-} \rightleftharpoons MgHL^-)$ 2·05	51S
	gl	25	0·1	K_1 2·7	62I

51S G. Schwarzenbach, P. Ruckstuhl, and J. Zurc, *Helv. Chim. Acta*, 1951, **34**, 455
62I R. R. Irani and K. Moedritzer, *J. Phys. Chem.*, 1962, **66**, 1349

268 $C_4H_{12}N_2S$ 2,2′-Thiobis(ethylamine) 2,2′-diaminodiethyl sulphide L

Erratum. 1964, p. 438, Table 268. Under Zn^{2+}, *for* 1 KNO_6 *read* 1 KNO_3.

$C_4H_{12}N_2S_2$ 2,2′-Dithiobis(ethylamine) [bis-(2-aminoethyl)disulphide] L 268.1

Metal	*Method*	*Temp*	*Medium*	*Log of equilibrium constant, remarks*	*Ref*
H^+	gl	20	0·15 ($NaClO_4$)	K_1 9·16, K_2 8·82	63H
Cu^{2+}	gl	20	0·15 ($NaClO_4$)	K_1 6·70±0·01 $K(Cu^{2+} + HL^+ \rightleftharpoons CuHL^{3+})$ 3·79	63H

63H C. J. Hawkins and D. D. Perrin, *Inorg. Chem.*, 1963, **2**, 843

$C_4H_9O_2NS$ L-2-Amino-3-(methylthio)propanoic acid (*S*-methyl-L-cysteine) HL 270.1

Metal	*Method*	*Temp*	*Medium*	*Log of equilibrium constant, remarks*	*Ref*
H^+	gl	25	0·10 (KNO_3)	K_1 8·73	64L
Ag^+	gl	25	0·10 (KNO_3)	K_1 5·25	64L
Cd^{2+}	gl	25	0·10 (KNO_3)	K_1 3·77, K_2 3·32	64L
Co^{2+}	gl	25	0·10 (KNO_3)	K_1 4·12, K_2 3·49	64L
Cu^{2+}	gl	25	0·10 (KNO_3)	K_1 7·88, K_2 6·84	64L
Hg^{2+}	gl	25	0·10 (KNO_3)	K_1 7·20, K_2 5·81	64L
Mn^{2+}	gl	25	0·10 (KNO_3)	K_1 2·52, K_2 1·75	64L
Ni^{2+}	gl	25	0·10 (KNO_3)	K_1 5·26, K_2 4·56	64L
Pb^{2+}	gl	25	0·10 (KNO_3)	K_1 4·43, K_2 3·54	64L
Zn^{2+}	gl	25	0·10 (KNO_3)	K_1 4·46, K_2 4·06	64L

64L G. R. Lenz and A. E. Martell, *Biochemistry*, 1964, **3**, 745

$C_4H_{10}O_5N_3P$ Creatinephosphoric acid (phosphocreatine) H_3L 270.2

Metal	*Method*	*Temp*	*Medium*	*Log of equilibrium constant, remarks*	*Ref*
H^+				H^+ constants not given.	
Mg^{2+}	sp	30	0·1 buffer	K_1 1·6±0·2 (buffer = *N*-ethylmorpholine)	64O

64O W. J. O'Sullivan and D. D. Perrin, *Biochemistry*, 1964, **3**, 18

$C_4H_{11}OSP$ Diethylphosphinothioic acid HL 271.1

Metal	*Method*	*Temp*	*Medium*	*Log of equilibrium constant, remarks*	*Ref*
H^+	gl	25	0·8 KNO_3 40% ethanol	K_1 3·60±0·05	67T
Hg^{2+}	Hg	25	0·8 KNO_3 40% ethanol	β_2 26·2, β_3 29·8, β_4 33·5	67T

67T V. F. Toropova, M. K. Saikina, and R. Sh. Aleshov, *J. Gen. Chem. (U.S.S.R.)*, 1967, **37**, 679 (725)

271.2 $C_4H_{11}O_2S_2P$ Phosphorodithioic acid *OO*-diethyl ester HL

Metal	*Method*	*Temp*	*Medium*	*Log of equilibrium constant, remarks*	*Ref*
H^+	gl	25	0·8 KNO_3 40% ethanol	K_1 2·11 ± 0·01	67T
Cd^{2+}	red	25	1 ($NaClO_4$) 60—80% acetone	K_1 (60%) 3·45, (70%) 3·80, (80%) 4·65, β_2 (60%) 5·85, (70%) 6·7, (80%) 8·65	67L
	pol	25	0·12 ($LiNO_3$) 50—90% ethanol	β_2 (50%) 6·48, (60%) 6·93, (70%) 7·40, (80%) 8·18, (90%) 9·15	67S
Hg^{2+}	Hg	25	0·8 KNO_3 40% ethanol	β_2 29·1, β_3 30·9, β_4 33·7	67T
Ni^{2+}	sp	20	80—95% ethanol	β_2 (80%) 4·0, (85%) 4·8, (90%) 6·3, (95%) 7·8	67L
			75—95% acetone	β_2 (75%) 3·9, (80%) 4·8, (85%) 5·9, (90%) 6·7, (95%) 9·0	
Pb^{2+}	pol	25	0·12 ($LiNO_3$) 50—90% ethanol	β_2 (50%) 7·98, (60%) 8·56, (70%) 9·04, (80%) 9·81, (90%) 10·53	67S

67L S. V. Larionov, V. M. Shul'man, and L. A. Podol'skaya, *Russ. J. Inorg. Chem.*, 1967, **12**, 1295 (2456)
67S P. S. Shetty and Q. Fernando, *J. Inorg. Nuclear Chem.*, 1967, **29**, 1921
67T V. F. Toropova, M. K. Saikina, and R. Sh. Aleshov, *J. Gen. Chem.* (*U.S.S.R.*), 1967, **37**, 679 (725)

271.3 $C_4H_{11}O_3SP$ Phosphorothioic acid *OO*-diethyl ester HL

Metal	*Method*	*Temp*	*Medium*	*Log of equilibrium constant, remarks*	*Ref*
H^+	gl	25	0·8 KNO_3 40% ethanol	K_1 1·83 ± 0·07	67T
Ag^+	Ag	25	1	β_3 10·81 ± 0·05	61T
Hg^{2+}	pol	25	1	β_2 23·5 ± 0·1, β_3 22·6 ± 0·1, β_4 20·1 ± 0·1	61T
	Hg	25	0·8 KNO_3 40% ethanol	β_2 24·0, β_3 26·4, β_4 27·7	67T

61T V. F. Toropova, M. K. Saikina, and N. K. Lutskaga, *Russ. J. Inorg. Chem.*, 1961, **6**, 1066 (2086)
67T V. F. Toropova, M. K. Saikina, and R. Sh. Aleshov, *J. Gen. Chem.* (*U.S.S.R.*), 1967, **37**, 679 (725)

271.4 $C_4H_{11}O_3PSe$ Phosphoroselenoic acid *OO*-diethyl ester HL

Metal	*Method*	*Temp*	*Medium*	*Log of equilibrium constant, remarks*	*Ref*
H^+	pH	25	?	K_1 1·77	66T
Hg^{2+}	Hg	25	1 KNO_3	β_2 28·6, β_3 30·3, β_4 32·7	66T

66T V. F. Toropova, M. K. Saikina, and R. Sh. Aleshov, *Russ. J. Inorg. Chem.*, 1966, **11**, 605 (1130)

$C_4H_{11}O_8NP_2$ *N*-(Carboxymethyl)iminobis(methylphosphonic acid) H_5L 271.5

Metal	*Method*	*Temp*	*Medium*	*Log of equilibrium constant, remarks*	*Ref*
H^+	gl	25	0·10 (KNO_3)	K_1 10·80, K_2 6·37, K_3 5·01, K_4 2·00, K_5 1·73	65W
Ca^{2+}	gl	25	0·10 (KNO_3)	K_1 6·17	65W
Cu^{2+}	gl	25	0·10 (KNO_3)	K_1 12·53 ± 0·05	65W
				$K(CuHL^{2-} + H^+ \rightleftharpoons CuH_2L^-)$ 4·14 ± 0·05	
				$K(CuL^{3-} + H^+ \rightleftharpoons CuHL^{2-})$ 5·89 ± 0·02	
Fe^{3+}	gl	25	0·10 (KNO_3)	K_1 14·65 ± 0·05	65W
				$K(Fe^{3+} + HL^{4-} \rightleftharpoons FeHL^-)$ 8·65 ± 0·03	
				$K(FeLOH^{3-} + H^+ \rightleftharpoons FeL^{2-})$ 7·20 ± 0·05	
Mn^{2+}	gl	25	0·10 (KNO_3)	K_1 7·0 ± 0·5	65W

65W S. Westerback, K. S. Rajan, and A. E. Martell, *J. Amer. Chem. Soc.*, 1965, **87**, 2567

$C_4H_{13}O_6NP_2$ *N*-Ethyliminobis(methylphosphonic acid) H_4L 271.6

Metal	*Method*	*Temp*	*Medium*	*Log of equilibrium constant, remarks*	*Ref*
H^+	gl	25	1 KNO_3	K_1 12·42 ± 0·10, K_2 5·92 ± 0·04, K_3 4·70 ± 0·04, K_4 < 2	67C
Ca^{2+}	gl	25	1 KNO_3	K_1 3·36 ± 0·05	67C
				$K(Ca^{2+} + HL^{3-} \rightleftharpoons CaHL^-)$ 1·29 ± 0·08	
				$K(Ca^{2+} + H_2L^{2-} \rightleftharpoons CaH_2L) \ll 1$	
Mg^{2+}	gl	25	1 KNO_3	K_1 4·42 ± 0·09	67C
				$K(Mg^{2+} + HL^{3-} \rightleftharpoons MgHL^-)$ 2·33 ± 0·07	
				$K(Mg^{2+} + H_2L^{2-} \rightleftharpoons MgH_2L)$ 1·9 ± 0·3	

67C R. P. Carter, R. L. Carrol, and R. R. Irani, *Inorg. Chem.*, 1967, **6**, 939

$C_4H_{14}O_6N_2P_2$ Ethylenebis(iminomethylphosphonic acid) H_4L 271.7

Metal	*Method*	*Temp*	*Medium*	*Log of equilibrium constant, remarks*	*Ref*
H^+	gl	25	0·1 KCl	K_1 10·47, K_2 8·02, K_3 5·72, K_4 4·61	65D
Be^{2+}	gl	25	0·1 KCl	K_1 > 7	65D
Ca^{2+}	gl	25	0·1 KCl	K_1 < 2	65D
Co^{2+}	gl	25	0·1 KCl	K_1 10·80,	65D
				$K(Co^{2+} + HL^{3-} \rightleftharpoons CoHL^-)$ 3·84	
Cu^{2+}	gl	25	0·1 KCl	K_1 18·58,	65D
				$K(Cu^{2+} + HL^{3-} \rightleftharpoons CuHL^-)$ 8·72	
Dy^{3+}	gl	25	0·1 KCl	$K(Dy^{3+} + HL^{3-} \rightleftharpoons DyHL)$ 8·79	65D
Er^{3+}	gl	25	0·1 KCl	$K(Er^{3+} + HL^{3-} \rightleftharpoons ErHL)$ > 8·79	65D
Eu^{3+}	gl	25	0·1 KCl	$K(Eu^{3+} + HL^{3-} \rightleftharpoons EuHL)$ 8·54	65D
Fe^{3+}	gl	25	0·1 KCl	K_1 > 10	65D
Gd^{3+}	gl	25	0·1 KCl	$K(Gd^{3+} + HL^{3-} \rightleftharpoons GdHL)$ 8·31	65D
La^{3+}	gl	25	0·1 KCl	$K(La^{3+} + HL^{3-} \rightleftharpoons LaHL)$ 7·49	65D
Lu^{3+}	gl	25	0·1 KCl	$K(Lu^{3+} + HL^{3-} \rightleftharpoons LuHL)$ > 8·79	65D

continued overleaf

271.7 $C_4H_{14}O_6N_2P_2$ (contd)

Metal	Method	Temp	Medium	Log of equilibrium constant, remarks	Ref
Mg^{2+}	gl	25	0·1 KCl	K_1 < 2	65D
Mn^{2+}	gl	25	0·1 KCl	K_1 7·55, $K(Mn^{2+}+HL^{3-} \rightleftharpoons MnHL^-)$ 3·63	65D
Nd^{3+}	gl	25	0·1 KCl	$K(Nd^{3+}+HL^{3-} \rightleftharpoons NdHL)$ 8·31	65D
Ni^{2+}	gl	25	0·1 KCl	K_1 12·02, $K(Ni^{2+}+HL^{3-} \rightleftharpoons NiHL^-)$ 4·71	65D
Pr^{3+}	gl	25	0·1 KCl	$K(Pr^{3+}+HL^{3-} \rightleftharpoons PrHL)$ 7·98	65D
Sm^{3+}	gl	25	0·1 KCl	$K(Sm^{3+}+HL^{3-} \rightleftharpoons SmHL)$ 8·43	65D
Sr^{2+}	gl	25	0·1 KCl	K_1 < 1	65D
Y^{3+}	gl	25	0·1 KCl	$K(Y^{3+}+HL^{3-} \rightleftharpoons YHL)$ 8·79	65D

65D N. M. Dyatlova, M. I. Kabachnik, T. Ja. Nedved, M. V. Rudomino, and Yu. F. Belugin, *Proc. Acad. Sci. (U.S.S.R.)*, 1965, **161**, 307 (607)

271.8 C_5H_8 1-Methylcyclobutene L

Metal	Method	Temp	Medium	Log of equilibrium constant, remarks	Ref
Ag^+	oth	30	1·77 $AgNO_3$ ethylene glycol	K_1 −0·27, gas chromatography	62G

62G E. Gil-Av and J. Herling, *J. Phys. Chem.*, 1962, **66**, 1208

271.9 C_5H_8 Methylenecyclobutane L

Metal	Method	Temp	Medium	Log of equilibrium constant, remarks	Ref
Ag^+	oth	30	1·77 $AgNO_3$ ethylene glycol	K_1 0·91, gas chromatography	62G

62G E. Gil-Av and J. Herling, *J. Phys. Chem.*, 1962, **66**, 1208

272.1 $C_5H_2O_5$ 4,5-Dihydroxycyclopent-4-ene-1,2,3-trione (croconic acid) H_2L

Metal	Method	Temp	Medium	Log of equilibrium constant, remarks	Ref
H^+	gl	25	2 NaCl	K_1 1·5, K_2 0·4	62C
			0·1 NaCl	K_1 1·8, K_2 0·6	
	sp	25	2 NaCl	K_1 1·51, K_2 0·32	62C
Ba^{2+}	sol	25	0·3 KCl	K_1 1·55±0·03, K_{so} −8·28	65C
Ca^{2+}	sol	25	0·3 KCl	K_1 1·29±0·03, K_{so} −4·05	65C
Sr^{2+}	sol	25	0·3 KCl	K_1 1·21±0·03, K_{so} −5·08	65C

62C B. Carlquist and D. Dyrssen, *Acta Chem. Scand.*, 1962, **16**, 94
65C B. Carlquist and D. Dyrssen, *Acta Chem. Scand.*, 1965, **19**, 1293

$C_5H_4O_3$ Furan-2-carboxylic acid (2-furoic acid) HL 273

Metal	Method	Temp	Medium	Log of equilibrium constant, remarks	Ref
H^+	gl	25	0·1 ($NaClO_4$) 50% dioxan	K_1 4·56±0·01	68E
Cu^{2+}	gl	25	0·1 ($NaClO_4$) 50% dioxan	K_1 2·79	68E

68E H. Erlenmeyer, R. Griesser, B. Prijs, and H. Sigel, *Helv. Chim. Acta*, 1968, **51**, 339

$C_5H_4O_3$ 3-Hydroxy-4*H*-pyran-4-one (pyromeconic acid) HL 273.1

Metal	Method	Temp	Medium	Log of equilibrium constant, remarks	Ref
H^+	gl, sp	25	0·5	K_1 7·69±0·03	67C
Ge^{IV}	sp	25	0·5 (NaCl)	$K[Ge(OH)_4+2HL \rightleftharpoons Ge(OH)_2L_2]$ 2·86±0·05	67C
Zn^{2+}	gl	25	0·5 ($NaClO_4$)	K_1 5·03±0·06, β_2 9·18±0·10	67C

67C G. Choux and R. L. Benoit, *Bull. Soc. chim. France*, 1967, 2920

$C_5H_4N_4$ 7,9-Diazolo[4,5-*d*]-1,3-diazine (7-imidazo[4,5-*d*]pyrimidinepurine) L 274

Metal	Method	Temp	Medium	Log of equilibrium constant, remarks	Ref
Zn^{2+}	nmr	36	$(CH_3)_2SO$	K_1 0·72±0·05, See imidazole (144) for mixed ligands.	66W

66W S. M. Wang and N. C. Li, *J. Amer. Chem. Soc.*, 1966, **88**, 4592

C_5H_5N Azine (pyridine) L 275

Metal	Method	Temp	Medium	Log of equilibrium constant, remarks	Ref
H^+	?	?	?	K_1 5·17	61D
	gl	25	0·3 (K_2SO_4)	K_1 5·52±0·03	61J
	gl	20	0·15 ($NaClO_4$)	K_1 5·32±0·01	62H
	gl	25	1 $NaClO_4$	K_1 5·15	63A
	gl	25	0—1 $C_5H_5N{\cdot}HNO_3$	K_1 (→ 0) 5·247, (0·5) 5·215±0·003, (1) 5·183	64B
	gl	25	0·1 ($NaClO_4$)	K_1 5·33±0·01	64K
	gl	35, 45	?	K_1 (35°) 5·50, (45°) 5·00	67R
	gl	25	0·61 (KNO_3)	K_1 5·44	67S
	?	?	?	K_1 5·55	67T
Ag^+	Ag-AgBr	25	0·1 (KNO_3)	K_1 1·90, β_2 4·25	61C
	Ag	?	0·1 ($NaClO_4$)	K_1 1·87, K_2 2·22	62N
	cal	25	?	K_1 2·25, β_2 4·19 $\Delta H_1°$ −4·77, $\Delta S_1°$ −5·7, $\Delta H_2°$ −6·76, $\Delta S_2°$ −13·8	63B

continued overleaf

275 C_5H_5N (contd)

Metal	*Method*	*Temp*	*Medium*	*Log of equilibrium constant, remarks*	*Ref*
Ag^+ (*contd*)	pH	25	0·2 ($LiNO_3$) 0—75% ethanol	K_1 (0%) 1·96, β_2 (0%) 4·19, (25%) 4·06, (50%) 3·86, (75%) 3·88	65N
	Ag	25	0—90% acetone	β_2 (0%) 4·21, (10%) 4·20, (20%) 4·18, (30%) 4·16, (40%) 4·18, (50%) 4·23, (60%) 4·35, (70%) 4·47, (80%) 4·72, (90%) 5·12	65P
			10—90% dioxan	β_2 (10%) 4·14, (20%) 4·09, (30%) 4·05, (40%) 4·04, (50%) 4·06, (60%) 4·15, (70%) 4·30, (80%) 4·49, (90%) 4·80	
			10—90% ethanol	β_2 (10%) 4·26, (20%) 4·25, (30%) 4·19, (40%) 4·12, (50%) 4·02, (60%) 3·97, (70%) 3·99, (80%) 4·06, (90%) 4·16	
	cal	25	0·5 KNO_3	K_1 2·00±0·04, K_2 2·11±0·08 ΔH_1 −4·83±0·05, ΔS_1 −7·0±0·3 ΔH_2 −6·51±0·06, ΔS_2 −12·2±0·5	66P
	gl	25	0·61 (KNO_3)	K_1 2·12±0·02, β_2 4·25±0·01	67S
	cal	25	~ 0	K_1 2·05±0·05, β_2 4·10±0·07 $\Delta H_1°$ −4·6±0·2, $\Delta S_1°$ −6·0±0·8, $\Delta H_{\beta 2}°$ −11·2±0·1, $\Delta S_{\beta 2}°$ −18·9±0·4	68I
Be^{2+}	gl	35, 45	?	K_1 (35°) 2·40, (45°) 2·30 ΔH_1 −4·5, ΔS_1 −3	67R
Cd^{2+}	qh, Hg	30	0·1 $NaClO_4$	K_1 1·26, β_2 1·95, β_3 2·29	61D
	pH	25	0·2 ($LiNO_3$) 0—75% ethanol	β_4 (0%) 2·64, (25%) 2·37, (50%) 2·10, (75%) 2·09	65N
	pol	30	0·1 (KNO_3)	K_1 1·36, β_2 1·86, β_3 1·90	65S
	gl, dis	20	1·0 (NH_4NO_3)	K_1 1·51, β_2 2·46, β_4 2·50 $K(Cd^{2+}+NH_3+L \rightleftharpoons CdNH_3L^{2+})$ 3·25 $K[Cd^{2+}+2NH_3+L \rightleftharpoons Cd(NH_3)_2L^{2+}]$ 5·60 $K(Cd^{2+}+NH_3+2L \rightleftharpoons CdNH_3L_2^{2+})$ 4·04 $K[Cd^{2+}+3NH_3+L \rightleftharpoons Cd(NH_3)_3L^{2+}]$ 6·69 $K[Cd^{2+}+2NH_3+2L \rightleftharpoons Cd(NH_3)_2L_2^{2+}]$ 5·90 $K(Cd^{2+}+NH_3+3L \rightleftharpoons CdNH_3L_3^{2+})$ 4·08	67F
	Cd(Hg)	25	0·3 (KNO_3)	K_1 1·04	67N
	pol	30	0·5 KNO_3	K_1 1·10, β_2 1·48, β_3 1·91, β_4 1·95, β_5 1·48	67Sa
Co^{2+}	sp	?	0—100% acetone	K_1 (0%) 1·26, (50%) 1·34, (85%) 1·47, (100%) 3·49	59A
			100% methanol	K_1 1·49	
			85—100% ethanol	K_1 (85%) 1·43, (100%) 2·36	
			50—100% n-propanol	K_1 (50%) 1·28, (85%) 1·36, (100%) 2·75	
	sp, cal	20	$CHCl_3$	$K(CoL_2Cl_2+2L \rightleftharpoons CoL_4Cl_2)$ 1·10 ΔH −15·2±0·2, ΔS −46·8±0·7 $K(CoL_2Br_2+2L \rightleftharpoons CoL_4Br_2)$ 0·956 ΔH −15·6±0·2, ΔS −48·8±0·8 $K(CoL_2I_2+2L \rightleftharpoons CoL_4L_2)$ 0·365 ΔH −16·6±0·2, ΔS −54·9±0·9 $K[CoL_2(NCO)_2+2L \rightleftharpoons CoL_4(NCO)_2]$ 1·37 ΔH −13·7±0·2, ΔS −40·5±0·8 $K[CoL_2(NCS)_2+2L \rightleftharpoons CoL_4(NCS)_2]$ 4·92 ΔH −16·6±0·3, ΔS −34·1±1·0 $K[CoL_2(NCSe)_2+2L \rightleftharpoons CoL_4(NCSe)_2] \gg 5$	63K
			CH_3NO_2	$K(CoL_2Cl_2+2L \rightleftharpoons CoL_4Cl_2)$ −0·40 $K[CoL_2(NCS)_2+2L \rightleftharpoons CoL_4(NCS)_2]$ 3·55 $K[CoL_2(NCSe)_2+2L \rightleftharpoons CoL_4(NCSe)_2]$ 4·44	

contin

Metal	*Method*	*Temp*	*Medium*	*Log of equilibrium constant, remarks*	*Ref*
Co^{2+}	sp	20	$CHCl_3$	$K(CoL_2I_2+2L \rightleftharpoons CoL_4I_2)$ 0·37	65Na
(contd)	cal	20	$CHCl_3$	ΔH −16·6±0·2, ΔS −55±1	65Na
	dis	20	1 NH_4NO_3	K_1 1·35, β_2 1·95, β_3 2·25, β_4 2·35	66F
				$K[Co^{2+}+XNH_3+YL \rightleftharpoons Co(NH_3)_XL_Y{}^{2+}]$ (X = 1) (Y = 1) 3·22, (Y = 2) 3·50, (Y = 3) 3·85; (X = 2) (Y = 1) 4·2, (Y = 2) 4·50, (Y = 3) 5·40, (Y = 4) 5·35; (X = 3) (Y = 1) 5·3, (Y = 2) 6·56, (Y = 3) 6·5	
				$K[Co(NH_3)_XL_Y{}^{2+} \rightleftharpoons Co(NH_3)_X{}^{2+}+CoL_Y{}^{2+}]$ (X = 1) (Y = 1) 0·255, (Y = 2) 0·31, (Y = 3) 0·59, (Y = 4) 0·8; (X = 2) (Y = 1) 0·06, (Y = 2) 0·335, (Y = 3) 1·5, (Y = 4) 2·0; (X = 3) (Y = 1) 0·228, (Y = 2) 1·9, (Y = 3) 2·6	
				See EDTA (836) for mixed ligands	
Cu^+	Pt	25	0·3 (K_2SO_4)	K_1 3·9±0·1, K_2 2·7±0·1, K_3 1·3±0·2, K_4 0·8±0·2, β_3 8·29	61J
	Pt, Au	20	0·15 (Na_2SO_4)	K_1 3·17, β_2 6·64	62H
	pol	30	2·0 NH_4NO_3 0·1 $C_5H_5N{\cdot}HCl$	β_2 8·14	66G
Cu^{2+}	sp	?	?	K_1 2·36; K_1 2·39, K_2 1·96, K_3 1·35	61A
			methanol	K_1 2·92	
			ethanol	K_1 3·37	
			acetone	K_1 4·42	
	gl	20	0·15 ($NaClO_4$)	K_1 2·65, β_2 4·86, β_3 6·90, β_4 8·45	62H
	gl	25	1 $NaClO_4$	K_1 2·46, K_2 1·95, K_3 1·27, K_4 0·84	63A
	cal	25	1 $NaClO_4$	$\Delta H_{i(ave)}$ −3·0, ΔS_1 1, ΔS_2 −1, ΔS_3 −4, ΔS_4 −6	63A
	gl	25	0·5 $C_5H_5N{\cdot}HNO_3$	K_1 2·408, K_2 1·880, K_3 1·137, K_4 0·605	64B
	gl	25	0·1 ($NaClO_4$)	K_1 2·54	64K
	dis	20	1 NH_4NO_3	K_1 2·4, β_2 4·3, β_3 5·68, β_4 6·58	66F
				$[Cu^{2+}+XNH_3+YL \rightleftharpoons Cu(NH_3)_XL_Y{}^{2+}]$ (X = 1) (Y = 1) 6·6, (Y = 2) 8·1, (Y = 3) 9·09; (X = 2) (Y = 1) 9·86, (Y = 2) 10·83; (X = 3) (Y = 1) 11·62	
				$[Cu(NH_3)_XL_Y{}^{2+} \rightleftharpoons Cu(NH_3)_X{}^{2+}+CuL_Y{}^{2+}]$ (X = 1) (Y = 1) 0·63, (Y = 2) 0·80, (Y = 3) 1·0; (X = 2) (Y = 1) 0·94, (Y = 2) 1·2; (X = 3) (Y = 1) 0·47	
	gl	25	0·61 (KNO_3)	K_1 2·60±0·01, β_2 4·54±0·03, β_3 5·8±0·1, β_4 6·7±0·2	67S
	cal	25	~ 0	K_1 2·50±0·02, β_2 4·30±0·05, β_3 5·16±0·06, β_4 6·04±0·1	68I
				$\Delta H_1{}^\circ$ −4·02±0·08, $\Delta S_1{}^\circ$ −2·0±0·3, $\Delta H_{\beta 2}{}^\circ$ −8·9±0·1, $\Delta S_{\beta 2}{}^\circ$ −10·0±0·7, $\Delta H_{\beta 3}{}^\circ$ −16·1±0·6, $\Delta S_{\beta 3}{}^\circ$ −31±2, $\Delta H_{\beta 4}{}^\circ$ −22±2, $\Delta S_{\beta 4}{}^\circ$ −41±5	
				See EDTA (836) for mixed ligands	
Fe^{2+}				See acetylacetone (283) for mixed ligands	
Ge^{4+}	cal	25	n-hexane	$\Delta H[GeF_4(l)+2L(l) \rightarrow GeF_4L_2(c)]$ −48·4	67M
				$\Delta H[GeF_4(g)+2L(l) \rightarrow GeF_4L_2(c)]$ −53·6	
				$\Delta H[GeCl_4(g)+2L(l) \rightarrow GeCl_4L_2(c)]$ −49·6	
				$\Delta H[GeCl_4(l)+2L(l) \rightarrow GeCl_4L_2(c)]$ −41·4	
				$\Delta H[GeBr_4(l)+2L(l) \rightarrow GeBr_4L_2(c)]$ −33·8	
				$\Delta H[GeBr_4(g)+2L(l) \rightarrow GeBr_4L_2(c)]$ −44·4	
Mn^{2+}	gl	25	1 $NaClO_4$	K_1 1·86, K_2 1·59, K_3 0·90, K_4 0·60	63A
	cal	25	1 $NaClO_4$	$\Delta H_{i(ave)}$ −2·4, ΔS_1 1, ΔS_2 −2, ΔS_3 −4, ΔS_4 −5	63A
Ni^{2+}	sp	?	0—100% acetone	K_1 (0%) 2·04, (50%) 2·19, (85%) 2·28, (100%) 3·49	59A

continued overleaf

275 C_5H_5N (contd)

Metal	Method	Temp	Medium	Log of equilibrium constant, remarks	Ref
Ni^{2+} (contd)			methanol	K_1 2·35	
			85—100% ethanol	K_1 (85%) 2·21, (100%) 2·96	
			n-butanol	K_1 3·19	
	gl	25	1 $NaClO_4$	K_1 2·13, K_2 1·66, K_3 1·12, K_4 0·64	63A
	cal	25	1 $NaClO_4$	$\Delta H_{1(ave)}$ −2·7, ΔS_1 1, ΔS_2 −1, ΔS_3 −4, ΔS_4 −6	63A
	gl	25	0·1 ($NaClO_4$)	K_1 1·85	64K
	sp	20	$CHCl_3$	$K(NiL_2I_2 + 2L \rightleftharpoons NiL_4I_2)$ 4·49	65N
	cal	20	$CHCl_3$	ΔH −23·8±0·4, ΔS −61±2	65N
	dis	20	1 NH_4NO_3	K_1 1·98, β_2 3·02, β_3 3·42, β_4 3·44	66F
				$[Ni^{2+} + XNH_3 + YL \rightleftharpoons Ni(NH_3)_XL_Y^{2+}]$ (X = 1) (Y = 1) 4·54, (Y = 2) 5·4, (Y = 3) 5·14; (X = 2) (Y = 1) 6·65, (Y = 2) 6·3; (X = 3) (Y = 1) 7·10, (Y = 2) 7·0	
				$[Ni(NH_3)_XL_Y^{2+} \rightleftharpoons Ni(NH_3)_X^{2+} + NiL_Y^{2+}]$ (X = 1) (Y = 1) 0·44, (Y = 2) 0·85, (Y = 3) 0·565; (X = 2) (Y = 1) 0·96, (Y = 2) 0·59; (X = 3) (Y = 1) 0·47, (Y = 2) 0·7	
	gl	25	0·61 (KNO_3)	K_1 1·91±0·02, β_2 3·19±0·06, β_3 3·7±0·3	67S
	gl	20	1 Cl^- or Br^-	K_1 2·10, K_2 1·29, K_3 −1·3	67T
			1 SO_4^{2-}	K_1 1·42, K_2 0·74, K_3 −1·4	
				See EDTA (836) and 3-butylacetylacetone (714) for mixed ligands	
Si^{4+}	cal	25	n-hexane	$\Delta H[SiF_4(l) + 2L(l) \rightarrow SiF_4L_2(c)]$ −29·0	67M
				$\Delta H[SiF_4(g) + 2L(l) \rightarrow SiF_4L_2(c)]$ −33·1	
				$\Delta H[SiCl_4(l) + 2L(l) \rightarrow SiCl_4L_2(c)]$ −51·7	
				$\Delta H[SiCl_4(g) + 2L(l) \rightarrow SiCl_4L_2(c)]$ −58·9	
				$\Delta H[SiBr_4(l) + 2L(l) \rightarrow SiBr_4L_2(c)]$ −46·1	
				$\Delta H[SiBr_4(g) + 2L(l) \rightarrow SiBr_4L_2(c)]$ −54·2	
Sn^{4+}	cal	25	n-hexane	$\Delta H[SnCl_4(l) + 2L(l) \rightarrow SnCl_4L_2(c)]$ −52·9	67M
				$\Delta H[SnCl_4(g) + 2L(l) \rightarrow SnCl_4L_2(c)]$ −60·7	
Ti^{3+}	pol	127	Fused ethyl-pyridinium bromide	K_1 2·75±0·1	67L
Tl^{3+}	gl	25	4 $C_5H_5N{\cdot}HNO_3$	$K[Tl(OH)_2^+ + L \rightleftharpoons Tl(OH)_2L^+]$ 0·7	66L
				$K[Tl^{3+} + 2OH^- + L \rightleftharpoons Tl(OH)_2L^+]$ 29·1	
				$K[Tl(OH)_2^+ + 2L \rightleftharpoons Tl(OH)_2L_2^+]$ 2·4	
				$K[Tl^{3+} + 2OH^- + 2L \rightleftharpoons Tl(OH)_2L_2^+]$ 30·8	
				$K[Tl(OH)_2^+ + 4L \rightleftharpoons Tl(OH)_2L_4^+]$ 2·5	
				$K[Tl^{3+} + 2OH^- + 4L \rightleftharpoons Tl(OH)_2L_4^+]$ 31·0	
Zn^{2+}	gl	25	1 $NaClO_4$	K_1 2·08, K_2 1·69, K_3 1·03, K_4 0·64	63A
	cal	25	1 $NaClO_4$	$\Delta H_{i(ave)}$ −2·6, ΔS_1 1, ΔS_2 −1, ΔS_3 −4, ΔS_4 −6	63A
	gl	25	0·1 $NaClO_4$	K_1 1·07	64K
	pol	30	0·1 (KNO_3)	K_1 0·90, β_2 1·53	65S
	Hg(Zn)	30	0·1 $NaClO_4$	K_1 1·10, β_2 1·71, β_3 1·92	66D
	gl, dis	20	1·0 (NH_4NO_3)	K_1 1·45, β_2 2·01, β_3 1·8, β_4 2·12	67F
				$K(Zn^{2+} + NH_3 + L \rightleftharpoons ZnNH_3L^{2+})$ 4·07	
				$K[Zn^{2+} + 2NH_3 + L \rightleftharpoons Zn(NH_3)_2L^{2+}]$ 6·05	
				$K(Zn^{2+} + NH_3 + 2L \rightleftharpoons ZnNH_3L_2^{2+})$ 4·07	
				$K[Zn^{2+} + 3NH_3 + L \rightleftharpoons Zn(NH_3)_3L^{2+}]$ 7·69	
				$K[Zn^{2+} + 2NH_3 + 2L \rightleftharpoons Zn(NH_3)_2L_2^{2+}]$ 6·27	
				$K(Zn^{2+} + NH_3 + 3L \rightleftharpoons ZnNH_3L_3^{2+})$ 4·14	
				See dimethyldithiocarbamic acid (176), 8-hydroxyquinoline (720), 2-methyl-8-hydroxyquinoline (818), and 4-methyl-8-hydroxyquinoline (819) for mixed ligands	

references on facing p

59A A. V. Ablov and L. V. Nazarova, *Russ. J. Inorg. Chem.*, 1959, **4**, 1140 (2480)
61A A. V. Ablov and L. V. Nazarova, *Russ. J. Inorg. Chem.*, 1961, **6**, 1044 (2043)
61C G. Curthoys and D. A. J. Swinkels, *Analyt. Chim. Acta*, 1961, **24**, 589
61D A. G. Desai and M. B. Kabadi, *J. Indian Chem. Soc.*, 1961, **38**, 805
61J B. R. James and R. J. P. Williams, *J. Chem. Soc.*, 1961, 2007
62H C. J. Hawkins and D. D. Perrin, *J. Chem. Soc.*, 1962, 1351
62N L. V. Nazarova and A. V. Ablov, *Russ. J. Inorg. Chem.*, 1962, **7**, 673 (1305)
63A G. Atkinson and J. E. Bauman, jun., *Inorg. Chem.*, 1963, **2**, 64
63B F. Becker, J. Barthel, N. G. Schmahl, and H. M. Luschow, *Z. phys. Chem. (Frankfurt)*, 1963, **37**, 52
63K H. C. A. King, E. Körös, and S. M. Nelson, *J. Chem. Soc.*, 1963, 5449
64B J. Bjerrum, *Acta Chem. Scand.*, 1964, **18**, 843
64K K. Kahmann, H. Sigel, and H. Erlenmeyer, *Helv. Chim. Acta*, 1964, **47**, 1754
65N L. V. Nazarova, *Russ. J. Inorg. Chem.*, 1965, **10**, 1364 (2509)
65Na S. M. Nelson and T. M. Shepherd, *J. Chem. Soc.*, 1965, 3284
65P G. Popa, C. Luca, and V. Magearu, *J. Chim. phys.*, 1965, **62**, 449, 853
65S V. K. Sharma and J. N. Gaur, *J. Electroanalyt. Chem.*, 1965, **9**, 321
66D A. G. Desai and M. B. Kabadi, *J. Inorg. Nuclear Chem.*, 1966, **28**, 1279
66F Ya. D. Fridman, M. G. Levina, and R. I. Sorochan, *Russ. J. Inorg. Chem.*, 1966, **11**, 877 (1641)
66G S. L. Gupta and M. K. Chatterjee, *Indian J. Chem.*, 1966, **4**, 22
66L B. I. Lobov, F. Ya. Kul'ba, and V. E. Mironov, *Russ. J. Phys. Chem.*, **40**, 1353 (2527)
66P P. Paoletti, A. Vacca, and D. Arenare, *J. Phys. Chem.*, 1966, **70**, 193
67F Ya. D. Fridman and M. G. Levine, *Russ. J. Inorg. Chem.*, 1967, **12**, 1425 (2704)
67L M. Lapidus and B. Tremillon, *J. Electroanalyt. Chem.*, 1967, **15**, 359
67M J. M. Miller and M. Onyszchuk, *J. Chem. Soc. (A)*, 1967, 1132
67N L. V. Nazarova, *Russ. J. Inorg. Chem.*, 1967, **12**, 1620 (3062)
67R P. S. Relan and P. K. Bhattacharya, *J. Indian Chem. Soc.*, 1967, **44**, 536
67S M. S. Sun and D. G. Brewer, *Canad. J. Chem.*, 1967, **45**, 2729
67Sa R. Sundaresan, S. C. Saraiya, and A. K. Sundaram, *Proc. Indian Acad. Sci.*, 1967, **66A**, 246
67T E. G. Timofeeva, A. A. Knyazeva, and I. I. Kalinichenko, *Russ. J. Inorg. Chem.*, 1967, **12**, 1117 (2119)
68I R. M. Izatt, D. Eatough, R. L. Snow, and J. J. Christensen, *J. Phys. Chem.*, 1968, **72**, 1208

$C_5H_6O_4$ Methylenebutanedioic acid (itaconic acid) H_2L 278

Metal	*Method*	*Temp*	*Medium*	*Log of equilibrium constant, remarks*	*Ref*
H^+	gl	28	0·1 ($NaClO_4$)	K_1 5·08, K_2 3·01	68R
Cu^{2+}	sp	30	0·1 ($NaClO_4$)	K_1 2·96, $K(2Cu^{2+} + L^{2-} \rightleftharpoons Cu_2L^{2+})$ 3·82	68Ra
Pb^{2+}	pol	30	0·3 KNO_3	β_2 4·08	67L
UO_2^{2+}	gl	28	0·1 ($NaClO_4$)	K_1 4·7 $\pm$ 0·1	68R

67L T.-T. Lai, S.-N. Chen, and E. Lin, *Talanta*, 1967, **14**, 251
68R S. Ramamoorthy and M. Santappa, *Bull. Chem. Soc. Japan*, 1968, **41**, 1330
68Ra S. Ramamoorthy and M. Santappa, *J. Inorg. Nuclear Chem.*, 1968, **30**, 1855

280 $C_5H_6N_2$ 2-Aminopyridine L

Metal	Method	Temp	Medium	Log of equilibrium constant, remarks	Ref
H^+	gl	35, 45	?	K_1 (35°) 6·86, (45°) 6·75	67R
	gl	25	0·61 (KNO_3)	K_1 6·96	67S
Ag^+	gl	25	0·61 (KNO_3)	K_1 2·38±0·01, β_2 4·79±0·01	67S
Be^{2+}	gl	35, 45	?	K_1 (35°) 4·37, (45°) 4·24 ΔH_1 −6·8, ΔS_1 −4(?)	67R
Cu^{2+}	gl	25	0·61 (KNO_3)	K_1 1·71±0·03, β_2 3·25±0·07	67S

67R P. S. Relan and P. K. Bhattacharya, *J. Indian Chem. Soc.*, 1967, **44**, 536
67S M. S. Sun and D. G. Brewer, *Canad. J. Chem.*, 1967, **45**, 2729

281 $C_5H_6N_2$ 3-Aminopyridine L

Metal	Method	Temp	Medium	Log of equilibrium constant, remarks	Ref
H^+	gl	30	0·1	K_1 5·98	63D
	gl	25	0·61 (KNO_3)	K_1 6·26	67S
Ag^+	gl	25	0·61 (KNO_3)	K_1 2·21±0·02, β_2 4·41±0·01	67S
Cd^{2+}	gl	30	0·1	K_1 1·52, K_2 0·67, K_3 0·69	63D
Cu^{2+}	gl	25	0·61 (KNO_3)	K_1 2·80±0·01, β_2 4·84±0·03, β_3 6·48±0·05, β_4 7·5±0·1	67S
Ni^{2+}	gl	25	0·61 (KNO_3)	K_1 1·97±0·02, β_2 3·23±0·06, β_3 4·1±0·2	67S
Zn^{2+}	Hg(Zn)	30	0·1 $NaClO_4$	K_1 1·34, β_2 2·16, β_3 2·78	66D

63D A. G. Desai and M. B. Kabadi, *Current Sci.*, 1963, **32**, 15
66D A. G. Desai and M. B. Kabadi, *J. Inorg. Nuclear Chem.*, 1966, **28**, 1279
67S M. S. Sun and D. G. Brewer, *Canad. J. Chem.*, 1967, **45**, 2729

282 $C_5H_6N_2$ 4-Aminopyridine L

Metal	Method	Temp	Medium	Log of equilibrium constant, remarks	Ref
H^+	H, Ag-AgCl	0—50	→ 0	K_1 (0°) 9·8731±0·0005, (5°) 9·7043±0·0006, (10°) 9·5486±0·0006, (15°) 9·3979±0·0006, (20°) 9·2524±0·0006, (25°) 9·1141±0·0007, (30°) 8·978±0·001, (35°) 8·8455±0·0010, (40°) 8·717±0·001, (45°) 8·594±0·001, (50°) 8·477±0·001, $\Delta H_1°$ (25°) −47·1±0·1 kJ mole⁻¹, $\Delta S_1°$ (25°) 16·5±0·5 J mole⁻¹ deg⁻¹	60B

60B R. G. Bates and H. B. Hetzer, *J. Res. Nat. Bur. Stand. Sect. A*, 1960, **64**, 427

$C_5H_8O_2$ Pentane-2,4-dione (acetylacetone) HL 283

Metal	*Method*	*Temp*	*Medium*	*Log of equilibrium constant, remarks*	*Ref*
H^+	gl	30	0·025 $NaClO_4$ ethanol	K_1 11·5	60D
	pH	25	1·0	K_1 8·76 ± 0·01	61P
	dis	25	< 0·02, 1 $NaClO_4$	K_1 (< 0·02) 9·03, (1) 8·71	63G
	gl	25	0·05—0·25 mole fraction dioxan	(0·0526) (1/D = 0·0167) K_1 9·25, (0·1285) (1/D = 0·0238) K_1 9·85, (0·1800) (1/D = 0·0302) K_1 10·31, (0·2467) (1/D = 0·0408) K_1 10·92	63Ga, 63G
			0·06—0·27 mole fraction propan-2-ol	(0·0603) (1/D = 0·0152) K_1 9·21, (0·1442) (1/D = 0·0190) K_1 9·46, (0·2002) (1/D = 0·0221) K_1 9·71, (0·2712) (1/D = 0·0264) K_1 10·01 D = dielectric constant	
	?	?	?	K_1 8·95	63P
	H	20	1 $(CH_3)_4NCl$ methanol	K_1 11·81	65G
	gl	25	0·1 $(C_4H_9)_4NClO_4$ methanol	K_1 11·78 ± 0·05	65L
			0·1 $(C_4H_9)_4NClO_4$ ethanol	K_1 11·81 ± 0·05	
	gl	25	1·00 (KCl)	K_1 8·878 ± 0·005	65S, 65Sa
	gl	25	0·0172 NaCl 0—0·610 mole fraction methanol	K_1 (0) 8·98, (0·099) 9·14, (0·222) 9·32, (0·295) 9·45, (0·383) 9·59, (0·485) 9·77, (0·610) 10·03	68G
			0·0172 NaCl 0·070—0·517 mole fraction ethanol	K_1 (0·070) 9·12, (0·163) 9·35, (0·223) 9·53, (0·270) 9·68, (0·393) 9·94, (0·517) 10·31	
			0·0172 NaCl 0·056—0·473 mole fraction propan-1-ol	K_1 (0·056) 9·16, (0·134) 9·43, (0·188) 9·64, (0·256) 9·91, (0·346) 10·26, (0·473) 10·66	
	gl	25	0·1 $(NaClO_4)$	K_1 8·82	68Ga
	cal	25	0·1 $(NaClO_4)$	ΔH_1 −3·21, ΔS_1 29·6	68Ga
Be^{2+}	dis	25	< 0·02—1 $NaClO_4$	K_1 (< 0·02) 7·96, (1) 7·55, β_2 (< 0·02) 14·67, (1) 14·35, $K(BeLOH + H^+ \rightleftharpoons BeL^+)$ (< 0·02) 6·4, $K[BeL(OH)_2^- + H^+ \rightleftharpoons BeLOH]$ (< 0·02) 9·8	63G
Cd^{2+}	gl	30	0·025 $NaClO_4$ ethanol	K_1 2·7	60D
	dis	?	0·1	K_1 4·0, K_2 3·8	60S
	pol	30	0·7 (KNO_3)	β_2 6·12	62S
Co^{2+}	gl	25	0·0172 NaCl 0—0·610 mole fraction methanol	(0) K_1 5·51, K_2 4·23, (0·099) K_1 5·74, K_2 4·45, (0·222) K_1 6·04, K_2 4·60, (0·295) K_1 6·29, K_2 4·72, (0·383) K_1 6·58, K_2 5·00, (0·485) K_1 6·84, K_2 5·20, (0·610) K_1 7·27, K_2 5·42	68G
			0·0172 NaCl 0·056—0·473 mole fraction propan-1-ol	(0·056) K_1 5·76, K_2 4·43, (0·134) K_1 6·15, K_2 4·72, (0·188) K_1 6·39, K_2 4·95, (0·256) K_1 6·84, K_2 5·32, (0·346) K_1 7·33, K_2 5·75, (0·473) K_1 7·85, K_2 6·22	
Cr^{2+}	gl	25	1·00 (KCl)	K_1 5·96 ± 0·02, K_2 11·70 ± 0·03	65Sa
Cu^{2+}	dis	20	0·18 (NaOAc)	K_1 8·70, β_2 15·24	65I
	sp	?	20—100% acetone	$K(CuL_2 + CuA_2^{2+} \rightleftharpoons 2CuLA^+)$ (20%) 1·03, (30%) 0·82, (40%) 0·93, (50%) 0·74, (60%) 0·60, (70%) 0·44, (80%) 0·42, (90%) 0·27 where A = ethylenediamine	67F

continued overleaf

283 $C_5H_8O_2$ (contd)

Metal	Method	Temp	Medium	Log of equilibrium constant, remarks	Ref
Cu^{2+} *(contd)*				$K(CuL_2+CuA_2 \rightleftharpoons 2CuLA)$ (20%) 2·12, (40%) 1·99, (60%) 1·90, (80%) 1·82, (100%) 1·70 where HA = 8-hydroxyquinoline	
			CCl_4	$K(CuL_2+CuA_2 \rightleftharpoons 2CuLA)$ 2·02	
			3-methylbutanol	$K(CuL_2+CuA_2 \rightleftharpoons 2CuLA)$ 1·97	
			benzene	$K(CuL_2+CuA_2 \rightleftharpoons 2CuLA)$ 1·15	
			$CHCl_3$	$K(CuL_2+CuA_2 \rightleftharpoons 2CuLA)$ 0·94 where HA = 8-hydroxyquinoline	
	gl	25	50% methanol	K_1 3·59±0·02 (?), K_2 2·55±0·02 (?)	67M
	cal	25	50% methanol	ΔH_1 −0·9±0·2 (?), ΔS_1 13±1 (?), ΔH_2 −1·9±0·1 (?), ΔS_2 5±4 (?)	67M
	gl	25	0·1 ($NaClO_4$)	K_1 8·16, β_2 14·76	68Ga
	cal	25	0·1 ($NaClO_4$)	ΔH_1 −4·8, ΔS_1 21, $\Delta H_{\beta 2}$ −10·1, $\Delta S_{\beta 2}$ 34	68Ga
Dy^{3+}	gl	25	2·0 ($NaClO_4$)	K_1 5·74±0·05, K_2 4·48±0·10	64Y
Er^{3+}	gl	25	2·0 ($NaClO_4$)	K_1 5·70±0·05, K_2 4·38±0·10	64Y
Eu^{3+}	gl	25	2·0 ($NaClO_4$)	K_1 5·41±0·05, K_2 4·30±0·10	64Y
Fe^{2+}	sp	25	benzene	$K(FeL_2+A \rightleftharpoons FeL_2A)$ > 4 $K(FeL_2A+A \rightleftharpoons FeL_2A_2)$ ~ 2·25 $K(FeL_2A_2+A \rightleftharpoons FeL_2A_3)$ ~ 1·40 where A = pyridine	67B
Fe^{3+}	pol	25	0·1 ($NaClO_4$)	K_2 8·4, K_3 6·5	63P
Gd^{3+}	gl	25	2·0 ($NaClO_4$)	K_1 5·42±0·05, K_2 4·39±0·10	64Y
Hf^{4+}	dis	25	1·0	K_1 7·40, K_2 7·28, K_3 6·74, K_4 6·68	61P
Ho^{3+}	ix	30	0·1 $NaClO_4$	K_1 5·65, K_2 4·76	64P
In^{3+}	dis	?	0·1	K_1 8·08, β_2 14·3, β_3 18·6	60S
	pol	25	0·5 $NaClO_4$	K_1 8·8, K_2 7·4, K_3 6·0	66C
K^+	gl	25	0·1 KI methanol	K_1 0·9± > 0·2	65L
			0·1 KSCN ethanol	K_1 2·1±0·2	
La^{3+}	gl	30	0·1 ($NaClO_4$) 67% acetone	K_1 6·12, K_2 4·55, K_3 3·83	64D
	gl	25	2·0 ($NaClO_4$)	K_1 4·71±0·05, K_2 3·45±0·10	64Y
Li^+	gl	25	0·1 $LiClO_4$ methanol	K_1 2·8±0·2	65L
			0·1 $LiClO_4$ ethanol	K_1 4·6±0·2	
Lu^{3+}	gl	25	2·0 ($NaClO_4$)	K_1 5·98±0·05, K_2 4·78±0·10	64Y
Mn^{2+}	gl	25	0·1 ($NaClO_4$)	K_1 4·07	68Ga
	cal	25	0·1 ($NaClO_4$)	ΔH_1 −1·5, ΔS_1 14	68Ga
Na^+	gl	25	0·1 $NaClO_4$ methanol	K_1 1·6±0·2	65L
			0·1 $NaClO_4$ ethanol	K_1 2·8±0·2	
Nb^{5+}				See methanol (103·1) for mixed ligands	
Nd^{3+}	gl	30	0·1 ($NaClO_4$) 67% acetone	K_1 6·84, K_2 5·62, K_3 4·64	64D
Ni^{2+}	gl	30	0·025 $NaClO_4$ ethanol	K_1 8·0, K_2 5·1	60D
	gl	25	0·0172 NaCl 0—0·610 mole fraction methanol	(0) K_1 6·05, K_2 4·61; (0·099) K_1 6·37, K_2 4·85; (0·222) K_1 6·64, K_2 5·06; (0·295) K_1 6·85, K_2 5·21; (0·383) K_1 7·07, K_2 5·30; (0·485) K_1 7·38, K_2 5·50; (0·610) K_1 7·92, K_2 5·78,	68G

continued

Metal	Method	Temp	Medium	Log of equilibrium constant, remarks	Ref
Ni^{2+} (contd)			0·0172 NaCl 0·070—0·517 mole fraction ethanol	(0·070) K_1 6·44, K_2 4·88, (0·163) K_1 6·73, K_2 5·14, (0·223) K_1 7·03, K_2 5·47, (0·270) K_1 7·34, K_2 5·72, (0·393) K_1 7·82, K_2 6·12, (0·517) K_1 8·30, K_2 6·55	
			0·0172 NaCl 0·056—0·473 mole fraction 1-propanol	(0·056) K_1 6·19, K_2 4·75, (0·134) K_1 6·52, K_2 5·00, (0·188) K_1 6·89, K_2 5·29, (0·256) K_1 7·32, K_2 5·70, (0·346) K_1 7·79, K_2 6·09, (0·473) K_1 8·40, K_2 6·72	
			0·06—0·3 mole fraction 2-propanol	(0·060) ($1/D$ = 0·0149) K_1 6·23, K_2 4·85, (0·14) ($1/D$ = 0·0184) K_1 6·48, K_2 5·10, (0·18) ($1/D$ = 0·0208) K_1 6·74, K_2 5·25, (0·24) ($1/D$ = 0·0245) K_1 7·16, K_2 5·58, (0·31) ($1/D$ = 0·0300) K_1 7·65, K_2 6·01	68G
			0·05—0·4 mole fraction dioxan	(0·053) ($1/D$ = 0·0168) K_1 6·48, K_2 4·98, (0·128) ($1/D$ = 0·0236) K_1 7·01, K_2 5·40, (0·180) ($1/D$ = 0·0292) K_1 7·47, K_2 5·77, (0·23) ($1/D$ = 0·0380) K_1 8·20, K_2 6·26, (~ 0·3) ($1/D$ = 0·0520) K_1 8·88, K_2 7·08, (~ 0·4) ($1/D$ = 0·0915) K_1 10·56, K_2 8·70 D = dielectric constant	
	gl	25	0·1 ($NaClO_4$)	K_1 5·72, β_2 9·66	68Ga
	cal	25	0·1 ($NaClO_4$)	ΔH_1 −3·4, ΔS_1 15, $\Delta H_{\beta 2}$ −7·6, $\Delta S_{\beta 2}$ 19	68Ga
Pb^{2+}	pol	30	0·7 (KNO_3)	β_2 6·32	62S
Pr^{3+}	gl	30	0·1 ($NaClO_4$) 67% acetone	K_1 6·85, K_2 5·59, K_3 4·61	64D
	gl	25	2·0 ($NaClO_4$)	K_1 5·01±0·05, K_2 3·83±0·10	64Y
Sc^{3+}	sp	18—20	0·01 (HNO_3)	K_1 8·4±0·1	66Ka
	dis	25	0·1 $NaClO_4$	K_1 8·3	66S
Sm^{3+}	gl	25	2·0 ($NaClO_4$)	K_1 5·32±0·05, K_2 4·40±0·10	64Y
$Sn(CH_3)_2^{2+}$	Ag-AgCl	25	0·1 (KNO_3)	K_1 6·6±0·4	63Y
Ta^{5+}				See methanol (103·1) for mixed ligands	
Ti^{3+}	sp	25	1	K_1 10·43, β_2 18·82, β_3 24·9	67V
Tm^{3+}	gl	25	2·0 ($NaClO_4$)	K_1 6·03±0·05, K_2 4·72±0·10	64Y
V^{2+}	gl	25	1·00 (KCl)	K_1 5·38±0·01, β_2 10·19±0·03, β_3 14·70±0·01	65S
Y^{3+}	dis	?	0·1	K_1 6·4, β_2 11·1, β_3 13·9	60S
	gl	30	0·1 ($NaClO_4$) 67% acetone	K_1 7·70, K_2 5·92, K_3 4·91	64D
	gl	25	2·0 ($NaClO_4$)	K_1 5·57±0·05, K_2 4·59±0·10	64Y
Yb^{3+}	ix	30	0·1 $NaClO_4$	K_1 5·7, K_2 4·45	64P
Zn^{2+}	gl	30	0·025 $NaClO_4$ ethanol	K_1 5·7, K_2 4·8	60D
	gl	25	50% methanol	K_1 5·36±0·02, K_2 4·49±0·02	67M
	cal	25	50% methanol	ΔH_1 −1·3±0·1, ΔS_1 20±1, ΔH_2 −1·4±0·8, ΔS_2 16±3	67M
	gl	25	0·1 ($NaClO_4$)	K_1 4·68, β_2 7·92	68Ga
	cal	25	0·1 ($NaClO_4$)	ΔH_1 −1·5, ΔS_1 16, $\Delta H_{\beta 2}$ −3·4, $\Delta S_{\beta 2}$ 24	68Ga
Zr^{4+}	sp	18—20	1 (HCl)	K_1 11·25±0·10	66K

60D M. C. Day and G. M. Rouayheb, *J. Chem. and Eng. Data*, 1960, **5**, 508
60S J. Stary, *Coll. Czech. Chem. Comm.*, 1960, **25**, 86
61P V. M. Peshkova and P. Ang., *Russ. J. Inorg. Chem.*, 1961, **6**, 1064 (2082)
62S S. C. Saraiya, V. S. Srinivasan, and A. K. Sundaram, *Current Sci.*, 1962, **31**, 187
63G R. W. Green and P. W. Alexander, *J. Phys. Chem.*, 1963, **67**, 905
63Ga P. S. Gentile, M. Cefola, and A. V. Celiano, *J. Phys. Chem.*, 1963, **67**, 1083

continued overleaf

283 $C_5H_8O_2$ (contd)

63P M. Petek and M. Branica, *J. Polarog. Soc.*, 1963, **9**, 1
63Y M. Yasuda and R. S. Tobias, *Inorg. Chem.*, 1963, **2**, 207
64D N. K. Dutt and P. Bandyopadhyay, *J. Inorg. Nuclear Chem.*, 1964, **26**, 729
64P J. Prasilova, *J. Inorg. Nuclear Chem.*, 1964, **26**, 661
64Y H. Yoneda, G. R. Choppin, J. L. Bear, and J. V. Quagliano, *Inorg. Chem.*, 1964, **3**, 1642
65G R. Gut, H. Buser, and E. Schmid, *Helv. Chim. Acta.*, 1965, **48**, 878
65I H. M. N. H. Irving and N. S. Al-Niami, *J. Inorg. Nuclear Chem.*, 1965, **27**, 419
65L D. C. Luehrs, R. T. Iwamoto, and J. Kleinberg, *Inorg. Chem.*, 1965, **4**, 1739
65S W. P. Schaefer, *Inorg. Chem.*, 1965, **4**, 642
65Sa W. P. Schaefer and M. E. Mathison, *Inorg. Chem.*, 1965, **4**, 431
66C B. Cosovic and M. Branica, *J. Polarog. Soc.*, 1966, **12**, 5
66K I. M. Korenman, F. R. Sheyanova, and Z. M. Gureva, *Russ. J. Inorg. Chem.*, 1966, **11**, 1485 (2761)
66Ka I. M. Korenman and N. V. Zaglyadimova, *Russ. J. Inorg. Chem.*, 1966, **11**, 1491 (2774)
66S T. Sekine, A. Koizumi, and M. Sakairi, *Bull. Chem. Soc. Japan*, 1966, **39**, 2681
67B D. A. Buckingham, R. C. Gorges, and J. T. Henry, *Austral. J. Chem.*, 1967, **20**, 497
67F Ya. D. Fridman and N. V. Dolgashova, *Russ. J. Inorg. Chem.*, 1967, **12**, 639 (1206)
67M G. Matsubayashi, Y. Kawasaki, and T. Tanaka, *Nippon Kagaku Zasshi*, 1967, **88**, 1251
67V W. E. Van der Linder and G. Den Boef, *Analyt. Chim. Acta*, 1967, **37**, 179
68G P. S. Gentile and A. Dadgar, *J. Chem. and Eng. Data*, 1968, **13**, 236
68Ga G. Gutnikov and H. Freiser, *Analyt. Chem.*, 1968, **40**, 39

283.1 $C_5H_8O_3$ Oxolane-2-carboxylic acid (tetrahydrofuran-2-carboxylic acid) HL

Metal	*Method*	*Temp*	*Medium*	*Log of equilibrium constant, remarks*	*Ref*
H^+	gl	25	0·1 ($NaClO_4$) 50% dioxan	K_1 4·95±0·01	68G, 6
Cu^{2+}	gl	25	0·1 ($NaClO_4$) 50% dioxan	K_1 3·72	68E
	gl	25	0·1 ($NaClO_4$) 50% dioxan	$K(CuA^{2+}+L^- \rightleftharpoons CuAL^+)$ 3·52 where A = 2,2′-bipyridyl	68G
Zn^{2+}	gl	25	0·1 ($NaClO_4$) 50% dioxan	K_1 3·07; $K(ZnA^{2+}+L^- \rightleftharpoons ZnAL^+)$ 2·79 where A = 2,2′-bipyridyl	68G

68E H. Erlenmeyer, R. Griesser, B. Prijs, and H. Sigel, *Helv. Chim. Acta*, 1968, **51**, 339
68G R. Griesser, B. Prijs, and H. Sigel, *Inorg. Nuclear Chem. Letters*, 1968, **4**, 443

286 $C_5H_8O_4$ Pentanedioic acid (glutaric acid) H_2L

Metal	*Method*	*Temp*	*Medium*	*Log of equilibrium constant, remarks*	*Ref*
H^+	pH	25	0·5 $LiClO_4$	K_1 4·84±0·03, K_2 4·09±0·03	68D
Co^{2+}	gl	25	0 corr	K_1 2·21±0·02	65M
	ix	25	0 corr	K_1 2·35±0·09	65S
Fe^{3+}	sp	25	0·5 $LiClO_4$	K_1 6·78	68D
Sc^{3+}	sp	18—20	0·01 (HNO_3)	K_1 4·1±0·2	66K
Zn^{2+}	dis	25	0 corr	K_1 2·85±0·07	66R

65M C. B. Monk, *J. Chem. Soc.*, 1965, 2456
65S R. G. Seys and C. B. Monk, *J. Chem. Soc.*, 1965, 2452
66K I. M. Korenman and N. V. Zaglyadimova, *Russ. J. Inorg. Chem.*, 1966, **11**, 1491 (2774)
66R D. L. G. Rowlands and C. B. Monk, *Trans. Faraday Soc.*, 1966, **62**, 945
68D M. Deneux, R. Meilleur, and R. L. Benoit, *Canad. J. Chem.*, 1968, **46**, 1383

$C_5H_8O_7$ 2,3,4-Trihydroxypentanedioic acid (trihydroxyglutaric acid) H_2L 287

Metal	Method	Temp	Medium	Log of equilibrium constant, remarks	Ref
H^+	?	?	?	K_1 4·21, K_2 3·08	61Z, 64R
	gl	23—25	0·2 KCl	K_1 4·93, K_2 4·05	66D
Ce^{3+}	gl	23—25	0·2 KCl	K_1 3·41	66D
Cu^{2+}	pol	25	0·5 K_2SO_4	K_1 2·81 ± 0·02, β_2 4·10 ± 0·04	63C
	pol	25	0·8 (Na_2SO_4)	$K[Cu^{2+} + L^{2-} \rightleftharpoons Cu(H_{-2}L)^{2-} + 2H^+]$ 10·5 (?)	67C
Dy^{3+}	gl	23—25	0·2 KCl	K_1 3·70	66D
Er^{3+}	gl	23—25	0·2 KCl	K_1 3·85	66D
Eu^{3+}	gl	23—25	0·2	K_1 3·69	66D
Gd^{3+}	gl	23—25	0·2 KCl	K_1 3·64	66D
Hf^{4+}	ix	?	2 $HClO_4$	$K(Hf^{4+} + H_2L \rightleftharpoons HfHL^{3+} + H^+)$ 2·42, $K[Hf^{4+} + 2H_2L \rightleftharpoons Hf(HL)_2^{2+} + 2H^+]$ 4·9	60R
	ix	?	2·0 $HClO_4$	$K(Hf^{4+} + H_2L \rightleftharpoons HfHL^{3+} + H^+)$ 2·72 ± 0·03, $K[Hf^{4+} + 2H_2L \rightleftharpoons Hf(HL)_2^{2+} + 2H^+]$ 5·2 ± 0·1	64R
Ho^{3+}	gl	23—25	0·2 KCl	K_1 3·82	66D
La^{3+}	gl	23—25	0·2 KCl	K_1 3·20	66D
Lu^{3+}	gl	23—25	0·2 KCl	K_1 4·04	66D
Nd^{3+}	gl	23—25	0·2 KCl	K_1 3·71	66D
Pa^V	ix	25	0·25	$K[Pa_2O(OH)_3^{5+} + HL^- \rightleftharpoons Pa_2O(OH)_3HL^{4+}]$ (?) 2·80	66G
Pr^{3+}	gl	23—25	0·2 KCl	K_1 3·53	66D
Sm^{3+}	gl	23—25	0·2 KCl	K_1 3·81	66D
Sr^{2+}	ix	?	0·1 NH_4Cl	K_1 2·5 ± 0·3	62B
Tb^{3+}	gl	23—25	0·2 KCl	K_1 3·68	66D
Th^{4+}	gl	20	~ 0·06	K_1 3·70	61Z
	ix	?	0·3	K_1 4·52	62G
VO^{2+}	sp	?	0·01 $HClO_4$	K_1 3·82	62Ga
Yb^{3+}	gl	23—25	0·2 KCl	K_1 3·85	66D
Zr^{4+}	ix	?	2 $HClO_4$	$K(Zr^{4+} + H_2L \rightleftharpoons ZrHL^{3+} + H^+)$ 3·09, $K[Zr^{4+} + 2H_2L \rightleftharpoons Zr(HL)_2^{2+} + 2H^+]$ 4·14	60R
	ix	?	2·0 $HClO_4$	$K(Zr^{4+} + H_2L \rightleftharpoons ZrHL^{3+} + H^+)$ 3·41 ± 0·01, $K[Zr^{4+} + 2H_2L \rightleftharpoons Zr(HL)_2^{2+} + 2H^+]$ 5·40 ± 0·07	64R

60R D. I. Ryabchikov, A. N. Ermakov, V. K. Belyaeva, and I. N. Marov, *Russ. J. Inorg. Chem.*, 1960, **5**, 505 (1051)
61Z O. E. Zvyagintsev and L. G. Khromenkov, *Russ. J. Inorg. Chem.*, 1961, **6**, 548 (1074)
62B I. E. Bukolov, K. V. Astakhov, V. I. Zimin, and V. S. Tairov, *Russ. J. Inorg. Chem.*, 1962, **7**, 816 (1577)
62G I. Geletseanu and A. V. Lapitskii, *Proc. Acad. Sci.* (*U.S.S.R.*), 1962, **144**, 460 (573)
62Ga V. V. Grigoreva and I. M. Maister, *Russ. J. Inorg. Chem.*, 1962, **7**, 1107 (2140)
63C E. G. Chikryzova, *Russ. J. Inorg. Chem.*, 1963, **8**, 41 (83)
64R D. I. Ryabchikov, I. N. Morov, A. N. Ermakov, and V. K. Betyaeva, *J. Inorg. Nuclear Chem.*, 1964, **26**, 965
66D N. K. Davidenko and V. F. Deribon, *Russ. J. Inorg. Chem.*, 1966, **11**, 53 (99)
66G I. Galateanu, *Canad. J. Chem.*, 1966, **44**, 647
67C E. G. Chikryzova, B. A. Orgiyan, and L. G. Kiriyak, *Russ. J. Inorg. Chem.*, 1967, **12**, 1448 (2747)

$C_5H_8N_2$ 3,5-Dimethylpyrazole L 287.1

Metal	Method	Temp	Medium	Log of equilibrium constant, remarks	Ref
Cd^{2+}	pol	25	0·1 KNO_3 ?% methanol	K_1 1·15, β_2 0·6 (?), β_3 1·88	65C
Co^{2+}	pol	25	0·1 KNO_3 ?% methanol	K_1 −0·22, β_2 0·43	66C
Ni^{2+}	pol	25	0·1 KNO_3 ?% methanol	K_1 0·40, β_2 0·8, β_3 0·81, β_4 0·88	66C

65C D. R. Crow, *J. Polarog. Soc.*, 1965, **11**, 22
66C D. R. Crow, *J. Polarog. Soc.*, 1966, **12**, 101

288.1 $C_5H_9N_3$ 3-(2′-Aminoethyl)pyrazole (betazole) L

Metal	Method	Temp	Medium	Log of equilibrium constant, remarks	Ref
H^+	gl	25	0·015	K_1 9·61	60H
	sp	20	0·015	K_2 2·02	60H
Cu^{2+}	gl	25	0·01—0·02	K_1 ~ 7·5	60H
Ni^{2+}	gl	25	0·01—0·02	K_1 ~ 5·7	60H

60H F. Holmes and F. Jones, *J. Chem. Soc.*, 1960, 2398

289 $C_5H_9N_3$ 4-(2′-Aminoethyl)imidazole (histamine) L

Metal	Method	Temp	Medium	Log of equilibrium constant, remarks	Ref
H^+	gl	20	~ 0	K_1 9·89, K_2 6·08	53P
	gl	0	0·015	K_1 10·42, K_2 6·37	62H
	gl	25	0·015	K_1 9·80, K_2 5·94	60H, 62H
	gl	0—45	0·1 (KNO_3)	K_1 (0°) 10·71, (25°) 9·88, (45°) 9·10, K_2 (0°) 6·59, (25°) 6·13, (45°) 5·63	64A
	gl	25	0·1 (KCl)	K_1 9·70±0·01, K_2 6·02±0·01	64D
	gl	25	0·1 ($NaClO_4$) 0·19 CH_3CN	K_1 9·84, K_2 6·08	66K
	gl	20	~ 0	K_1 10·03, K_2 6·08	66P
	gl	22	→ 0	K_1 9·84, K_2 5·86	66Z
			1·5	K_1 10·11, K_2 6·67	
	gl	37	0·15 KNO_3	K_1 9·569, K_2 6·012	67P, 68P
Cd^{2+}	gl	20	0·005 $CdSO_4$	β_2 8·0	53P
	pol	0—45	0·10 (KNO_3)	β_2 (0°) 9·60, (25°) 8·57, (45°) 8·30 $\Delta H_{\beta 2}°$ −11·8, $\Delta S_{\beta 2}°$ 1	64A
Co^{2+}	gl	25	0·01—0·02	K_1 5·2	60H
Cu^+	gl	25	0·1 ($NaClO_4$) 0·19 CH_3CN	$K[Cu^+ + 2HL^+ \rightleftharpoons Cu(HL)_2^{3+}]$ 10·32 K_1 8·87	66K
Cu^{2+}	gl	0—25	0·01—0·02	K_1 (0°) 10·10, (25°) 9·55, β_2 (0°) 17·00, (25°) 16·04 $\Delta H_1°$ (25°) −8, $\Delta S_1°$ 16, $\Delta H_{\beta 2}°$ −14, $\Delta S_{\beta 2}°$ 25	62H
	gl	20	~ 0	K_1 9·76, K_2 6·71, $K[Cu_2(OH)_2L_2^{2+} + 2H^+ \rightleftharpoons 2CuL^{2+}]$ 11·99	66P
	sp	22	1·5 (K_2SO_4)	K_1 9·83, K_2 16·6	66Z
	Hg(Cu)	36	1·5 (K_2SO_4)	K_1 9·45, K_2 15·96 $K(CuLOH^+ + H^+ \rightleftharpoons CuL^{2+})$ 7·1	66Z
	gl	37	0·15 KNO_3	K_1 9·278, β_2 15·577 $K(CuA^{2+} + L \rightleftharpoons CuAL^{2+})$ 7·86 where A = ethylenediamine $K(CuA^+ + L \rightleftharpoons CuAL^+)$ 8·71 where HA = L-serine $K(CuA + L \rightleftharpoons CuAL)$ 8·41 where H_2A = salicylic acid	67P
Ni^{2+}	gl	25	0·01—0·02	K_1 6·84, β_2 11·92, β_3 14·98	62H
	gl	37	0·15 (KNO_3)	K_1 6·601, β_2 11·440, β_3 14·370 $K(Ni^{2+} + L + A^- \rightleftharpoons NiLA^+)$ 10·99±0·05 $K(Ni^{2+} + 2L + A^- \rightleftharpoons NiL_2A^+)$ 14·19±0·09 $K(Ni^{2+} + L + 2A^- \rightleftharpoons NiLA_2)$ 13·09±0·09 where HA = serine See ethylenediamine (120) for mixed ligands	68P
Zn^{2+}	gl	20	0·005 $ZnSO_4$	β_2 9·6	53P

references on opposite

53P D. J. Perkins, *Biochem. J.*, 1953, **55**, 649
60H F. Holmes and F. Jones, *J. Chem. Soc.*, 1960, 2398
62H F. Holmes and F. Jones, *J. Chem. Soc.*, 1962, 2818
64A A. C. Andrews and J. Romary, *J. Chem. Soc.*, 1964, 405
64D M. A. Doran, S. Chaberek, and A. E. Martell, *J. Amer. Chem. Soc.*, 1964, **86**, 2129
66K Th. Kaden and A. Zuberbühler, *Helv. Chim. Acta*, 1966, **49**, 2189
66P D. D. Perrin and V. S. Sharma, *J. Inorg. Nuclear Chem.*, 1966, **28**, 1271
66Z J. Zarembowitch, *J. Chim. phys.*, 1966, **63**, 420
67P D. D. Perrin, I. G. Sayce, and V. S. Sharma, *J. Chem. Soc. (A)*, 1967, 1755
68P D. D. Perrin and V. S. Sharma, *J. Chem. Soc. (A)*, 1968, 446

$C_5H_9N_3$ 4(5)-Aminomethyl-2-methylimidazole L 289.1

Metal	*Method*	*Temp*	*Medium*	*Log of equilibrium constant, remarks*	*Ref*
H^+	gl	0—25	0·015	K_1 (0°) 10·04, (25°) 9·32, K_2 (0°) 6·02, (25°) 5·59	62H
Cu^{2+}	gl	0—25	0·01—0·02	K_1 (0°) 9·09, (25°) 8·56, β_2 (0°) 16·57, (25°) 15·60 $\Delta H_1°$ (25°) −8, $\Delta S_1°$ 13, $\Delta H_{\beta 2}°$ (25°) −15, $\Delta S_{\beta 2}°$ 23	62H
Ni^{2+}	gl	25	0·01—0·02	K_1 5·99, β_2 11·10, β_3 15·16	62H

62H F. Holmes and F. Jones, *J. Chem. Soc.*, 1962, 2818

$C_5H_{10}O$ Pent-4-en-1-ol HL 290.1

Metal	*Method*	*Temp*	*Medium*	*Log of equilibrium constant, remarks*	*Ref*
Pt^{2+}	sp	60	2 (NaCl)	$K(PtCl_4^{2-} + L \rightleftharpoons PtCl_3L^- + Cl^-)$ 3·40	67H

67H F. R. Hartley and L. M. Venanzi, *J. Chem. Soc. (A)*, 1967, 330

$C_5H_{10}O$ *cis*-1-Ethoxypropene L 290.2

Metal	*Method*	*Temp*	*Medium*	*Log of equilibrium constant, remarks*	*Ref*
Ag^+	dis	10—30	ethylene glycol	K_1 (10°) 0·60, (20°) 0·50, (30°) 0·41, gas chromatography ΔH_1 −3·9±0·2, ΔS_1 −11·0±0·6	68F

68F T. Fueno, O. Kajimoto, T. Okuyama, and J. Furukawa, *Bull. Chem. Soc. Japan*, 1968, **41**, 785

$C_5H_{10}O$ *trans*-1-Ethoxypropene L 290.3

Metal	*Method*	*Temp*	*Medium*	*Log of equilibrium constant, remarks*	*Ref*
Ag^+	dis	10—30	ethylene glycol	K_1 (10°) −0·17, (20°) −0·27, (30°) −0·33, gas chromatography ΔH_1 −3·2±0·2, ΔS_1 −12·1±0·6	68F

68F T. Fueno, O. Kajimoto, T. Okuyama, and J. Furukawa, *Bull. Chem. Soc. Japan*, 1968, **41**, 785

290.4 $C_5H_{10}O_2$ *cis*-Cyclopentane-1,2-diol L

Metal	Method	Temp	Medium	Log of equilibrium constant, remarks	Ref
B^{III}	gl	0—35	0·025 borax	$K[B(OH)_4^- + L \rightleftharpoons B(OH)_2(H_{-2}L)^-]$ (0°) 1·65, (13°) 1·49, (25°) 1·42, (35°) 1·32 $\Delta H°$ $-3{\cdot}5 \pm 0{\cdot}4$, $\Delta S°$ -5 ± 1 $K[B(OH)_4^- + 2L \rightleftharpoons B(H_{-2}L)_2^-]$ (0°) 2·56, (13°) 2·36, (25°) 2·15, (35°) 2·01 $\Delta H°$ $-6{\cdot}1 \pm 0{\cdot}6$, $\Delta S°$ -11 ± 3	67C

67C J. M. Conner and V. C. Bulgrin, *J. Inorg. Nuclear Chem.*, 1967, **29**, 1953

290.5 $C_5H_{10}O_2$ Methylbutanoic acid* HL

Metal	Method	Temp	Medium	Log of equilibrium constant, remarks	Ref
Na^+	con	−40	liquid NH_3	K_1 4·64 (?)	63B

* 2- or 3-Methyl not stated.

63B G. Bombara and M. Troyli, *J. Electroanalyt. Chem.*, 1963, **5**, 379

294.1 $C_5H_{10}O_3$ 2-Hydroxypentanoic acid HL

Metal	Method	Temp	Medium	Log of equilibrium constant, remarks	Ref
H^+	gl	25	1·0 ($NaClO_4$)	*K_1 3·59	68G
Dy^{3+}	gl	25	1·0 ($NaClO_4$)	K_1 2·63 ± 0·02	68G
Er^{3+}	gl	25	1·0 ($NaClO_4$)	K_1 2·68 ± 0·02	68G
Eu^{3+}	gl	25	1·0 ($NaClO_4$)	K_1 2·43 ± 0·02	68G
Gd^{3+}	gl	25	1·0 ($NaClO_4$)	K_1 2·45 ± 0·02	68G
Ho^{3+}	gl	25	1·0 ($NaClO_4$)	K_1 2·64 ± 0·02	68G
La^{3+}	gl	25	1·0 ($NaClO_4$)	K_1 1·98 ± 0·02	68G
Lu^{3+}	gl	25	1·0 ($NaClO_4$)	K_1 2·76 ± 0·02	68G
Nd^{3+}	gl	25	1·0 ($NaClO_4$)	K_1 2·31 ± 0·02	68G
Pr^{3+}	gl	25	1·0 ($NaClO_4$)	K_1 2·24 ± 0·02	68G
Sm^{3+}	gl	25	1·0 ($NaClO_4$)	K_1 2·40 ± 0·02	68G
Tb^{3+}	gl	25	1·0 ($NaClO_4$)	K_1 2·53 ± 0·02	68G
Y^{3+}	gl	25	1·0 ($NaClO_4$)	K_1 2·46 ± 0·02	68G
Yb^{3+}	gl	25	1·0 ($NaClO_4$)	K_1 2·76 ± 0·02	68G

* K_1 extrapolated to zero ligand concentration.

68G M. A. Gouveia and R. G. Carvahlo, *J. Inorg. Nuclear Chem.*, 1968, **30**, 2219

$C_5H_{10}O_3$ 2-Hydroxy-3-methylbutanoic acid HL 294.2

Metal	Method	Temp	Medium	Log of equilibrium constant, remarks	Ref
Ce^{3+}	ix	20	0·2 ($NaClO_4$)	K_1 2·23±0·05, β_2 3·50±0·08, β_3 4·78±0·08	60S
Y^{3+}	ix	20	0·2 ($NaClO_4$)	K_1 2·60±0·03, β_2 4·95±0·05, β_3 > 6·0	60S

60S V. J. Spitsyn and O. Voitekh, *Proc. Acad. Aci.* (*U.S.S.R.*), 1960, **133**, 859 (613)

$C_5H_{10}O_5$ L-Arabinose L 295

Metal	Method	Temp	Medium	Log of equilibrium constant, remarks	Ref
As^{III}	gl	25	0·1 KCl	$K[As(OH)_4^- + 2L \rightleftharpoons As(H_{-2}L)_2^-]$ 1·24	60A
B^{III}	gl	25	0·1 KCl	$K[B(OH)_4^- + 2L \rightleftharpoons B(H_{-2}L)_2^-]$ 3·55	59A
Ge^{IV}	gl	25	0·1 KCl	$K[HGeO_3^- + 2L \rightleftharpoons HGeO(H_{-2}L)_2^-]$ 3·63	59A

Note: Constants should be identical to those listed for D-Arabinose.

59A P. J. Antikainen and K. Tevanen, *Suomen Kem.*, 1959, **B32**, 214
60A P. J. Antikainen and K. Tevanen, *Suomen Kem.*, 1960, **B33**, 7

$C_5H_{10}O_5$ D-Arabinose L 295.1

Metal	Method	Temp	Medium	Log of equilibrium constant, remarks	Ref
As^{III}	gl	25	0·1 KCl	$K[As(OH)_4^- + 2L \rightleftharpoons As(H_{-2}L)_2^-]$ 1·28	60A
B^{III}	gl	25	0·1 KCl	$K[B(OH)_4^- + 2L \rightleftharpoons B(H_{-2}L)_2^-]$ 3·28	59A
Ge^{IV}	gl	25	0·1 KCl	$K[HGeO_3^- + 2L \rightleftharpoons HGeO(H_{-2}L)_2^-]$ 3·52	59A

Note: Constants should be identical to those listed for L- Arabinose.

59A P. J. Antikainen and K. Tevanen, *Suomen Kem.*, 1959, **B32**, 214
60A P. J. Antikainen and K. Tevanen, *Suomen Kem.*, 1960, **B33**, 7

$C_5H_{10}O_5$ D-Xylose L 295.2

Metal	Method	Temp	Medium	Log of equilibrium constant, remarks	Ref
As^{III}	gl	25	0·1 KCl	$K[As(OH)_4^- + 2L \rightleftharpoons As(H_{-2}L)_2^-]$ 0·74	60A
B^{III}	gl	25	0·1 KCl	$K[B(OH)_4^- + 2L \rightleftharpoons B(H_{-2}L)_2^-]$ 4·01	59A
Ge^{IV}	gl	25	0·1 KCl	$K[HGeO_3^- + 2L \rightleftharpoons HGeO(H_{-2}L)_2^-]$ 3·38	59A

59A P. J. Antikainen and K. Tevanen, *Suomen Kem.*, 1959, **B32**, 214

296 $C_5H_{11}N$ Piperidine (perhydropyridine) L

Metal	*Method*	*Temp*	*Medium*	*Log of equilibrium constant, remarks*	*Ref*
Ag^+	Ag	25	0·02 0—90% methanol	β_2 (0%) 6·68, (20%) 6·70, (40%) 6·88, (50%) 6·92, (60%) 6·92, (80%) 7·04, (90%) 7·10	63P
	Ag, qh	?	0·1 $NaClO_4$ acetone	β_2 10·97±0·05, coulometric titration	65M
Zn^{2+}				See dimethyldithiocarbamic acid (176) for mixed ligands.	

63P G. Popa, C. Luca, and V. Magearu, *J. Chim. phys.*, 1963, **60**, 355
65M K. K. Mead, D. L. Maricle, and C. A. Streuli, *Analyt. Chem.*, 1965, **37**, 237

296.1 $C_5H_{11}N$ 1-Amino-3-methylbut-2-ene L

Metal	*Method*	*Temp*	*Medium*	*Log of equilibrium constant, remarks*	*Ref*
Pt^{2+}	sp	60	2 (NaCl)	$K(PtCl_4^{2-}+HL^+ \rightleftharpoons PtCl_3HL+Cl^-)$ 0·41	67D

67D R. G. Denning, F. R. Hartley, and L. M. Venanzi, *J. Chem. Soc.* (*A*), 1967, 328

296.2 $C_5H_{11}N$ 5-Aminopent-1-ene L

Metal	*Method*	*Temp*	*Medium*	*Log of equilibrium constant, remarks*	*Ref*
Pt^{2+}	sp	60	2 (NaCl)	$K(PtCl_4^{2-}+HL^+ \rightleftharpoons PtCl_3HL+Cl^-)$ 3·04	67H

67H F. R. Hartley and L. M. Venanzi, *J. Chem. Soc.* (*A*), 1967, 330

296.3 $C_5H_{11}N$ *N*-Ethyl-3-aminoprop-1-ene (*N*-allylethylamine) L

Metal	*Method*	*Temp*	*Medium*	*Log of equilibrium constant, remarks*	*Ref*
Pt^{2+}	sp	24—59	2 (NaCl)	$K(PtCl_4^{2-}+HL^+ \rightleftharpoons PtCl_3HL+Cl^-)$ (24°) 3·37, (44°) 3·09, (59°) 2·91 ΔH −5·9±0·2, ΔS −4·4±0·7	67D
	sp	0—35	2 (KBr)	$K(PtBr_4^{2-}+HL^+ \rightleftharpoons PtBr_3HL+Br^-)$ (0°) 2·70, (25°) 2·38, (35°) 2·26 ΔH° −4·8±0·2, ΔS −5·3±0·7	67Da

67D R. G. Denning, F. R. Hartley, and L. M. Venanzi, *J. Chem. Soc.* (*A*), 1967, 324
67Da R. G. Denning and L. M. Venanzi, *J. Chem. Soc.* (*A*), 1967, 336

$C_5H_{12}O_2$ 2-Methylbutane-2,3-diol L 296.4

Metal	Method	Temp	Medium	Log of equilibrium constant, remarks	Ref
B^{III}	gl	0—35	0·025 borax	$K[B(OH)_4^- + L \rightleftharpoons B(OH)_2(H_{-2}L)^-]$ (0°) 1·59, (13°) 1·38, (25°) 1·26, (35°) 1·11 ΔH° −4·9±0·5, ΔS° −11±3, $K[B(OH)_4^- + 2L \rightleftharpoons B(H_{-2}L)_2^-]$ (0°) 2·76, (13°) 2·53, (25°) 2·32, (35°) 2·09 ΔH° −8·0±0·8, ΔS° −16±4	67C

67C J. M. Conner and V. C. Bulgrin, *J. Inorg. Nuclear Chem.*, 1967, **29**, 1953

$C_5H_{12}O_3$ 2-Hydroxy-2-methylbutanoic acid (methylethylglycolic acid) HL 296.5

Metal	Method	Temp	Medium	Log of equilibrium constant, remarks	Ref
Dy^{3+}	qh	25	1 $NaClO_4$	K_1 3·03, K_2 2·48, K_3 1·66, K_4 1·37	64E
La^{3+}	qh	25	1 $NaClO_4$	K_1 2·01, K_2 1·26, K_3 0·62	64E
Sm^{3+}	qh	25	1 $NaClO_4$	K_1 2·60, K_2 2·14, K_3 1·26, K_4 1·14	64E
Yb^{3+}	qh	25	1 $NaClO_4$	K_1 3·20, K_2 2·67, K_3 1·98, K_4 1·42	64E

64E L. Eeckhaut, F. Verbeek, H. Deelstra, and J. Hoste, *Analyt. Chim. Acta.*, 1964, **30**, 369

$C_5H_{12}O_4$ Pentaerythritol L 297

Metal	Method	Temp	Medium	Log of equilibrium constant, remarks	Ref
As^{III}	gl	25	0·1	$K[As(OH)_4^- + L \rightleftharpoons As(OH)_2(H_{-2}L)^-]$ 0·00	57R
	gl	25	0·1 KCl	$K[As(OH)_4^- + L \rightleftharpoons As(OH)_2(H_{-2}L)^-]$ 0·94	60A
B^{III}	gl	25	0·1 KCl	$K[B(OH)_4^- + L \rightleftharpoons B(OH)_2(H_{-2}L)^-]$ 2·699 $K[B(OH)_4^- + 2L \rightleftharpoons B(H_{-2}L)_2^-]$ 3·651	60A
Te^{VI}	gl	25	0·1 KCl	$K[H_5TeO_6^- + L \rightleftharpoons H_3TeO_4(H_{-2}L)^-]$ 0·58	60A

57R G. L. Roy, A. L. Laferriere, and J. O. Edwards, *J. Inorg. Nuclear Chem.*, 1957, **4**, 106
60A P. J. Antikainen and V. M. K. Rossi, *Suomen Kem.*, 1960, **B33**, 94

$C_5H_{13}N$ 1-Pentylamine L 297.1

Metal	Method	Temp	Medium	Log of equilibrium constant, remarks	Ref
H^+	gl	20	0·5 KNO_3	K_1 10·93±0·06	68A
	H	20	→0	K_1 10·79±0·03	68A
Ag^+	gl	20	0·500 KNO_3	K_1 3·55±0·09, β_2 7·70±0·03	68A
	Ag-AgCl	20	0·500 KNO_3	β_2 7·67±0·03	68A

68A D. J. Alner, R. C. Lansbury, and A. G. Smeeth, *J. Chem. Soc.* (*A*), 1968, 417

300 $C_5H_{14}N_2$ *N*-Propyl-1,2-diaminoethane (*N*-n-propylethylenediamine) L

Metal	Method	Temp	Medium	Log of equilibrium constant, remarks	Ref
H^+	gl	0—25	0·50 KNO_3	ΔH_1 (0°) −10·4, ΔH_2 −10·4	52B
Cu^{2+}	gl	0—25	0·5 KNO_3	ΔH_1 (0°) −7·6, ΔS_1 20, ΔH_2 (0°) −8·0, ΔS_2 11	52Ba
Ni^{2+}	gl	0—25	0·5 KNO_3	ΔH_1 (0°) −7·5, ΔS_1 5·1, ΔH_2 (0°) −9·9, ΔS_2 −9·5, ΔH_3 (0°) −3·9, ΔS_3 −4·0	52Ba

52B F. Basolo and R. K. Murmann, *J. Amer. Chem. Soc.*, 1952, **74**, 2373

52Ba F. Basolo and R. K. Murmann, *J. Amer. Chem. Soc.*, 1952, **74**, 5243

301 $C_5H_{14}N_2$ *N*-Isopropyl-1,2-diaminoethane (*N*-isopropylethylenediamine) L

Metal	Method	Temp	Medium	Log of equilibrium constant, remarks	Ref
H^+	gl	0—25	0·50 KNO_3	ΔH_1 (0°) −7·9, ΔH_2 −9·0	52B
Cu^{2+}	gl	0—25	0·5 KNO_3	ΔH_1 (0°) −5·8, ΔS_1 22, ΔH_2 (0°) −8·2, ΔS_2 7	52Ba
Ni^{2+}	gl	0—25	0·5 KNO_3	ΔH_1 (0°) −6·7, ΔS_1 1·1, ΔH_2 (0°) −5·5, ΔS_2 −2·6	52Ba

52B F. Basolo and R. K. Murmann, *J. Amer. Chem. Soc.*, 1952, **74**, 2373

52Ba F. Basolo and R. K. Murmann, *J. Amer. Chem. Soc.*, 1952, **74**, 5243

301.1 $C_5H_{16}N_4$ Tetrakis(aminomethyl)methane L

Metal	Method	Temp	Medium	Log of equilibrium constant, remarks	Ref
H^+	gl	25	0·1 (KNO_3)	K_1 9·89±0·03, K_2 8·17±0·02, K_3 5·67±0·01, K_4 3·03±0·01	66Z, 68Z
Cd^{2+}	gl	25	0·1 (KNO_3)	K_1 5·7 $K(Cd^{2+}+HL^+ \rightleftharpoons CdHL^{3+})$ 3·7	68Z
Co^{2+}	gl	25	0·1 (KNO_3)	K_1 7·6 $K(CoL^{2+}+H^+ \rightleftharpoons CoHL^{3+})$ 7·8 $K(CoHL^{3+}+H^+ \rightleftharpoons CoH_2L^{4+})$ 5·5	68Z
Cu^{2+}	gl	25	0·1 (KNO_3)	K_1 11·0 β_2 19·43±0·03 $K(Cu^{2+}+HL^+ = CuHL^{3+})$ 7·6(?) $K(Cu^{2+}+H_2L^{2+} = CuH_2L^{4+})$ 5·4 $K(2Cu^{2+}+L = Cu_2L^{4+})$ 17·59±0·06 $K(CuL^{2+}+HL^+ = CuHL_2^{3+})$ 6·9 $K[CuL^{2+}+H_2L^{2+} = Cu(HL)_2^{4+}]$ 6·3	66Z
	gl	25	0·1 (KNO_3)	$K(Cu^{2+}+HL^+ \rightleftharpoons CuHL^{3+})$ 8·6	68Z
Ni^{2+}	gl	25	0·1 (KNO_3)	K_1 10·8, $K(Ni^{2+}+HL^+ = NiHL^{3+})$ 8·5	66Z
	gl	25	0·1 (KNO_3)	K_1 10·7, K_2 8·1 $K(NiL^{2+}+H^+ \rightleftharpoons NiHL^{3+})$ 7·6 $K[NiL_2^{2+}+H^+ \rightleftharpoons NiL(HL)^{3+}]$ 8·0 $K[NiL(HL)^{3+}+H^+ \rightleftharpoons Ni(HL)_2^{4+}]$ 7·6	68Z
Zn^{2+}	gl	25	0·1 (KNO_3)	$K(Zn^{2+}+HL^+ \rightleftharpoons ZnHL^{3+})$ 5·0 $K(Zn^{2+}+H_2L^{2+} \rightleftharpoons ZnH_2L^{4+})$ 3·2	68Z

66Z L. J. Zompa and R. F. Bogucki, *J. Amer. Chem. Soc.*, 1966, **88**, 5186

68Z L. J. Zompa and R. F. Bogucki, *J. Amer. Chem. Soc.*, 1968, **90**, 4569

$C_5H_2O_2F_6$ 1,1,1,5,5,5-Hexafluoropentane-2,4-dione (hexafluoracetylacetone) HL 302

Erratum. 1964, p. 453, Table 302. In columns 1 and 2 *for* Al^{3+} → Al^{3+} *read* in column 1 Al^{3+}.

$C_5H_3O_6N_3$ 5-Nitro-1,3*H*-pyrimidine-2,4-dione-6-carboxylic acid (5-nitro-orotic acid) H_2L 302.1

Metal	*Method*	*Temp*	*Medium*	*Log of equilibrium constant, remarks*	*Ref*
H^+	gl	25	0·1 [$(CH_3)_4NBr$]	K_1 4·94, K_2 < 1·5	61T
Cd^{2+}	gl	25	0·1 KCl	K_1 1·91	61T
Co^{2+}	gl	25	0·1 KCl	K_1 2·42	61T
	ix	25	0·1 ($KClO_4$)	K_1 2·44	66D
Mn^{2+}	gl	25	0·1 KCl	K_1 1·79	61T
	ix	25	0·1 ($KClO_4$)	K_1 1·74	66D
Ni^{2+}	gl	25	0·1 KCl	K_1 3·04	61T
	sp	25	0·1 (KNO_3)	K_1 3·1 ± 0·1	64T
Zn^{2+}	gl	25	0·1 KCl	K_1 2·51	61T
	ix	25	0·1 KCl	K_1 2·54	61T
	sp	25	0·1 (KNO_3)	K_1 2·58	64T

61T E. R. Tucci, E. Doody, and N. C. Li, *J. Phys. Chem.*, 1961, **65**, 1570
64T E. R. Tucci, F. Tskahashi, V. A. Tucci, and N. C. Li, *J. Inorg. Nuclear Chem.*, 1964, **26**, 1263
66D E. Doody, E. R. Tucci, R. Scruggs, and N. C. Li, *J. Inorg. Nuclear Chem.*, 1966, **28**, 833

$C_5H_4O_2S$ Thiole-2-carboxylic acid (thiophene-2-carboxylic acid) HL 305

Metal	*Method*	*Temp*	*Medium*	*Log of equilibrium constant, remarks*	*Ref*
H^+	gl	25	0·1 ($NaClO_4$) 50% dioxan	K_1 5·03 ± 0·02	68E
Co^{2+}	gl	25	0·1 ($NaClO_4$) 50% dioxan	K_1 1·82	68E
Cu^{2+}	gl	25	0·1 ($NaClO_4$) 50% dioxan	K_1 2·91	68E
Mn^{2+}	gl	25	0·1 ($NaClO_4$) 50% dioxan	K_1 1·65	68E
Ni^{2+}	gl	25	0·1 ($NaClO_4$) 50% dioxan	K_1 1·85	68E
Zn^{2+}	gl	25	0·1 ($NaClO_4$) 50% dioxan	K_1 2·05	68E

68E H. Erlenmeyer, R. Griesser, B. Prijs, and H. Sigel, *Helv. Chim. Acta*, 1968, **51**, 339

305.1 $C_5H_4O_4N_2$ 1,3*H*-pyrimidine-2,4-dione-6-carboxylic acid (orotic acid) H_2L

Metal	*Method*	*Temp*	*Medium*	*Log of equilibrium constant, remarks*	*Ref*
H^+	gl	25	0·1 KCl	K_1 9·45, K_2 2·07	61T
Cd^{2+}	gl	25	0·1 $[(CH_3)_4NBr]$	K_1 5·87	67T
Co^{2+}	gl	25	0·1 $[(CH_3)_4NBr]$	K_1 6·39	67T
Ni^{2+}	gl	25	0·1 $[(CH_3)_4NBr]$	K_1 6·82	67T
Zn^{2+}	gl	25	0·1$[(CH_3)_4NBr]$	K_1 6·42	67T

61T E. R. Tucci, E. Doody, and N. C. Li, *J. Phys. Chem.*, 1961, **65**, 1570
67T E. R. Tucci, C. H. Ke, and N. C. Li, *J. Inorg. Nuclear Chem.*, 1967, **29**, 1657

305.2 $C_5H_4O_4N_2$ 1,3*H*-pyrimidine-2,4-dione-5-carboxylic acid (iso-orotic acid) H_2L

Metal	*Method*	*Temp*	*Medium*	*Log of equilibrium constant, remarks*	*Ref*
H^+	gl	25	0·1 KCl	K_1 8·89, K_2 4·16	61T
Cd^{2+}	gl	25	0·1 KCl	$K(Cd^{2+}+HL^- \rightleftharpoons CdHL^+)$ 2·02	61T
Co^{2+}	ix	25	0·1 $(KClO_4)$	$K(Co^{2+}+HL^- \rightleftharpoons CoHL^+)$ 2·48	66D
Cu^{2+}	gl	25	0·1 KCl	$K(Cu^{2+}+HL^- \rightleftharpoons CuHL^+)$ 4·12	61T
Mn^{2+}	gl	25	0·1 KCl	$K(Mn^{2+}+HL^- \rightleftharpoons MnHL^+)$ 2·19	61T
	ix	25	0·1 $(KClO_4)$	$K(Mn^{2+}+HL^- \rightleftharpoons MnHL^+)$ 2·16	66D
Ni^{2+}	gl	25	0·1 KCl	$K(Ni^{2+}+HL^- \rightleftharpoons NiHL^+)$ 2·95	61T
Zn^{2+}	gl	25	0·1 KCl	$K(Zn^{2+}+HL^- \rightleftharpoons ZnHL^+)$ 2·69	61T
	ix	25	0·1 KCl	$K(Zn^{2+}+HL^- \rightleftharpoons ZnHL^+)$ 2·65	61T

61T E. R. Tucci, E. Doody, and N. C. Li, *J. Phys. Chem.*, 1961, **65**, 1570
66D E. Doody, E. R. Tucci, R. Scruggs, and N. C. Li, *J. Inorg. Nuclear Chem.*, 1966, **28**, 833

305.3 C_5H_4NBr 3-Bromopyridine L

Metal	*Method*	*Temp*	*Medium*	*Log of equilibrium constant, remarks*	*Ref*
H^+	gl	25	0·3 (KNO_3)	K_1 3·48	67N
Cd^{2+}	Cd(Hg)	25	0·3 (KNO_3)	K_1 0·64	67N

67N L. V. Nazarova, *Russ. J. Inorg. Chem.*, 1967, **12**, 1620 (3062)

306.1 C_5H_5ON Pyridine-1-oxide L

Metal	*Method*	*Temp*	*Medium*	*Log of equilibrium constant, remarks*	*Ref*
Cu^{2+}	sp	25	0·5 $(NaClO_4)$ 50% dioxan	K_1 $-0·69 \pm 0·06$	63S

63S H. Sigel and H. Brintzinger, *Helv. Chim. Acta*, 1963, **46**, 701

$C_5H_5O_2N$ 3-Furancarboxaldehyde oxime (3-furfuraldoxime) HL 308.1

Metal	*Method*	*Temp*	*Medium*	*Log of equilibrium constant, remarks*	*Ref*
H^+	gl	15—35	0·01—0·1 (NaClO₄) 75% dioxan	K_1 (15°) (0·01) 13·40, (0·1) 12·55, (25°) (0·01) 13·50(?), (0·1) 12·10, (35°) (0·01) 12·50, (0·1) 11·80	63A
Co^{3+}	gl	15—35	0·104 (NaClO₄) 75% dioxan	K_1 (15°) 7·33, (25°) 7·32, (35°) 7·62(?)	63A
Cr^{3+}	gl	15	0—0·1 (NaClO₄) 75% dioxan	K_1 (→ 0) 10·78, (0·01) 10·54, (0·1) 9·41, K_2 (→ 0) 10·14, (0·01) 9·73, (0·1) 8·98, K_3 (→ 0) 9·62, (0·01) 9·43, (0·1) 8·55	63A
		25	0—0·1 (NaClO₄) 75% dioxan	K_1 (→ 0) 10·64, (0·01) 10·96(?), (0·1) 9·26, K_2 (→ 0) 10·04, (0·01) 10·23(?), (0·1) 8·43, K_3 (→ 0) 9·42, (0·01) 9·51(?), (0·1) 7·61	
		35	0—0·1 (NaClO₄) 75% dioxan	K_1 (→ 0) 10·50, (0·01) 10·46, (0·1) 9·65(?), K_2 (→ 0) 9·94, (0·01) 9·84, (0·1) 9·04(?), K_3 (→ 0) 9·32, (0·01) 8·80, (0·1) 8·47	
			→ 0 75% dioxan	$\Delta H_1°$ −5·35, $\Delta S_1°$ 30·7, $\Delta H_2°$ −4·07, $\Delta S_2°$ 32·3, $\Delta H_3°$ −5·81, $\Delta S_3°$ 23·7	
Cu^{2+}	gl	15	0—0·104 (NaClO₄) 75% dioxan	K_1 (→ 0) 10·39, (0·01) 10·09, (0·104) 8·99, K_2 (→ 0) 10·18, (0·01) 9·57	63A
		25	0—0·104 (NaClO₄) 75% dioxan	K_1 (→ 0) 10·28, (0·01) 10·34(?), (0·104) 8·35(?), K_2 (→ 0) 9·88, (0·01) 9·94(?)	
		35	0—0·104 (NaClO₄) 75% dioxan	K_1 (→ 0) 9·90, (0·01) 9·64, (0·104) 8·71, K_2 (→ 0) 9·45, (0·01) 9·24	
			→ 0 75% dioxan	$\Delta H_1°$ −9·86, $\Delta S_1°$ 13·6, $\Delta H_2°$ −13·98, $\Delta S_2°$ −1·9	
Fe^{3+}	gl	15	0—0·1 (NaClO₄) 75% dioxan	K_1 (→ 0) 13·26, (0·01) 12·75, (0·104) 10·91, K_2 (→ 0) 12·52, (0·01) 11·58, (0·104) 8·89, K_3 (→ 0) 11·60, (0·01) 10·41, (0·104) 6·88	63A
		25	0—0·1 (NaClO₄) 75% dioxan	K_1 (→ 0)12·64, (0·01) 12·06, (0·104) 9·64, K_2 (→ 0) 10·80 (0·01) 10·10, (0·104) 8·48, K_3 (→ 0) 8·52, (0·01) 8·14, (0·104) 7·32	
		35	0—0·1 (NaClO₄) 75% dioxan	K_1 (0·01) 12·83, (0·104) 9·03, K_2 (→ 0) 9·94, (0·01) 9·51, (0·104) 8·83, K_3 (0·01) 6·19, (0·104) 8·62(?)	
			→ 0 75% dioxan	$\Delta H_1°$ −22·9, $\Delta S_1°$ −18·8, $\Delta H_2°$ −51·7, $\Delta S_2°$ highly negative, $\Delta H_3°$ −114·3, $\Delta S_3°$ highly negative	
Ni^{2+}	gl	15	0—0·104 (NaClO₄) 75% dioxan	K_1 (→ 0) 7·70, (0·01) 6·39(?), (0·104) 6·76, K_2 (→ 0) 7·00, (0·01) 5·99, (0·104) 6·30	63A
		25	0—0·104 (NaClO₄) 75% dioxan	K_1 (→ 0) 7·60, (0·01) 7·97(?), (0·104) 6·02(?), K_2 (→ 0) 7·52, (0·01) 7·45, (0·104) 5·54	
		35	0—0·104 (NaClO₄) 75% dioxan	K_1 (→ 0) 6·66, (0·01) 6·56, (0·104) 6·57, K_2 (→ 0) 6·40, (0·01) 6·30, (0·104) 6·14	
			→ 0 75% dioxan	$\Delta H_1°$ −20·4, $\Delta S_1°$ 35, $\Delta H_2°$ −11·0, $\Delta S_2°$ 5	
Pb^{2+}	gl	15—35	0·104 (NaClO₄) 75% dioxan	K_1 (15°) 7·53, (25°) 7·97(?), (35°) 6·59, K_2 (15°) 7·88, (25°) 9·96	63A
Zn^{2+}	gl	15—35	0·104 (NaClO₄) 75% dioxan	K_1 (15°) 6·79, (25°) 6·71, (35°) 6·74(?), K_2 (15°) 6·38, (25°) 6·02	63A

63A K. M. J. Al-Komser and B. Sen., *Inorg. Chem.*, 1963, 2, 1219

308.2 $C_5H_5O_2N$ Pyrrole-2-carboxylic acid HL

Metal	Method	Temp	Medium	Log of equilibrium constant, remarks	Ref
H^+	gl	25	0·1 ($NaClO_4$) 50% dioxan	K_1 5·98±0·01	68E
Cu^{2+}	gl	25	0·1 ($NaClO_4$) 50% dioxan	K_1 3·37	68E

68E H. Erlenmeyer, R. Griesser, B. Prijs, and H. Sigel, *Helv. Chim. Acta.*, 1968, **51**, 339

308.3 $C_5H_5O_4N_3$ 5-Amino-1,3*H*-pyrimidine-2,4-dione-6-carboxylic acid (5-amino-orotic acid) H_2L

Metal	Method	Temp	Medium	Log of equilibrium constant, remarks	Ref
H^+				H^+ constants not given.	
Cd^{2+}	gl	25	0·1 [$(CH_3)_4NBr$]	K_1 4·48	67T
Co^{2+}	gl	25	0·1 [$(CH_3)_4NBr$]	K_1 5·23	67T
Ni^{2+}	gl	25	0·1 [$(CH_3)_4NBr$]	K_1 6·01	67T
Zn^{2+}	gl	25	0·1 [$(CH_3)_4NBr$]	K_1 5·38	67T

67T E. R. Tucci, C. H. Ke, and N. C. Li, *J. Inorg. Nuclear Chem.*, 1962, **29**, 1657

310 $C_5H_6ON_2$ 2-Aminopyridine-1-oxide HL

Metal	Method	Temp	Medium	Log of equilibrium constant, remarks	Re,
H^+	gl	25	0·5 ($NaClO_4$) 50% dioxan	K_2 3·14±0·04	63S
Cu^{2+}	gl	25	0·1 ($NaClO_4$)	$K(CuL^+ + H^+ \rightleftharpoons CuHL^{2+})$ 4·48	63Sa
Mn^{2+}	sp	25	0·5 ($NaClO_4$) 50% dioxan	$K(Mn^{2+} + HL \rightleftharpoons MnHL^{2+})$ 0·75±0·06	63S

Errata, 1964, p. 456, Table 310. Under Ba^{2+}, Ca^{2+}, Co^{2+}, Mg^{2+}, Ni^{2+} (first line), and Zn^{2+} (first line), *for* 0·5 ($NaClO_4$) *read* 0·5 ($NaClO_4$) 50% dioxan.
Under Cu^{2+}, first line, *for* 0·5 ($NaClO_4$) *read* 0·2 ($NaClO_4$).
Under Cu^{2+}, second line, *for* 0·2 ($NaClO_4$) *read* 0·5 ($NaClO_4$).
Under Cu^{2+}, fourth line, *for* gl *read* sp; *also for* var 50% dioxan *read* 0·5 ($NaClO_4$) 50% dioxan.

63S H. Sigel and H. Brintzinger, *Helv. Chim. Acta*, 1963, **46**, 701
63Sa H. Sigel, H. Brintzinger, and H. Erlenmeyer, *Helv. Chim. Acta*, 1963, **46**, 712

310.1 $C_5H_6O_2N_2$ 4(or 5)-Imidazolylacetic acid [4(or 5)-Carboxymethyl imidazole] HL

Metal	Method	Temp	Medium	Log of equilibrium constant, remarks	Ref
H^+	gl	0—40	0·25 (KCl)	K_1 (0°) 7·86, (15°) 7·58, (25°) 7·40, (40°) 7·15, K_2 (0°) 3·13, (15°) 3·15, (25°) 3·24, (40°) 3·25	65A
Co^{2+}	gl	0—40	0·25 (KCl)	K_1 (0°) 3·94, (15°) 4·00, (25°) 3·83, (40°) 3·68, K_2 (0°) 3·04, (15°) 3·03, (25°) 2·98, (40°) 2·63	65A
				$\Delta H_1°$ (15°) −2·3, $T\Delta S_1°$ (15°) 3·0 kcal mole^{-1}, $\Delta H_2°$ (15°) −3·4, $T\Delta S_2°$ (15°) 0·6 kcal mole^{-1}	

cont

$C_5H_6O_2N_2$ (contd) 310.1

Metal	Method	Temp	Medium	Log of equilibrium constant, remarks	Ref
Cu^{2+}	gl	0—40	0·25 (KCl)	K_1 (0°) 7·02, (15°) 7·34, (25°) 7·00, (40°) 6·72, K_2 (0°) 5·71, (15°) 5·81, (25°) 5·69, (40°) 5·40	65A
				$\Delta H_1°$ (15°) −3·4, $T\Delta S_1°$ (15°) 6·3 kcal mole^{-1}, $\Delta H_2°$ (15°) −2·3, $T\Delta S_2°$ (15°) 5·4 kcal mole^{-1}	
Ni^{2+}	gl	0—40	0·25 (KCl)	K_1 (0°) 4·65, (15°) 4·83, (25°) 4·70, (40°) 4·34, K_2 (0°) 3·84, (15°) 3·71, (25°) 3·55, (40°) 3·44, K_3 (0°) 2·28	65A
				$\Delta H_1°$ (15°) −2·3, $T\Delta S_1°$ (15°) 4·1 kcal mole^{-1}, $\Delta H_2°$ (15°) −3·4, $T\Delta S_2°$ (15°) 1·5 kcal mole^{-1}	
Zn^{2+}	gl	0—40	0·25 (KCl)	K_1 (0°) 3·86, (15°) 3·83, (25°) 3·86, (40°) 3·59, K_2 (0°) 3·43, (15°) 3·32, (25°) 3·24, (40°) 3·33, K_3 (0°) 2·80, (15°) 2·63, (25°) 2·70	65A
				$\Delta H_1°$ (15°) −2·3, $T\Delta S_1°$ (15°) 2·7 kcal mole^{-1}, $\Delta H_2°$ (15°) 0·0, $T\Delta S_2°$ (15°) 4·4 kcal mole^{-1}, $\Delta H_3°$ (15°) −3·4, $T\Delta S_3°$ (15°) 1·5 kcal mole^{-1}	

65A A. C. Andrews and D. M. Zebolsky, *J. Chem. Soc.*, 1965, 742

$C_5H_6O_2N_2$ 5-Methyl-1,3*H*-pyrimidine-2,4-dione (thymine) HL 310.2

Metal	Method	Temp	Medium	Log of equilibrium constant, remarks	Ref
H^+	?	27	0·1 $NaClO_4$	K_1 9·8	61F
Hg^{2+}	Hg	27	0·1 $NaClO_4$	β_2 21·2	61F

61F R. Ferreira, E. Ben-Zvi, T. Yamane, J. Vasileuskis, and N. Davidson, 'Advances in Chemistry of Co-ordination Compounds', ed. S. Kirschner, Macmillan, New York, 1961, 457

$C_5H_8O_2S$ Tetrahydrothiophen-2-carboxylic acid HL 312.1

Metal	Method	Temp	Medium	Log of equilibrium constant, remarks	Ref
H^+	gl	25	0·1 $NaClO_4$ 50% dioxan	K_1 5·58±0·01	68E
Cu^{2+}	gl	25	0·1 $NaClO_4$ 50% dioxan	K_1 4·31	68E

68E H. Erlenmeyer, R. Griesser, B. Prijs, and H. Sigel, *Helv. Chim. Acta.*, 1968, **51**, 339

312.2 $C_5H_8O_4S$ 3-(Carboxymethylthio)propanoic acid H_2L

Metal	Method	Temp	Medium	Log of equilibrium constant, remarks	Ref
H^+	gl	20	0·1	K_1 4·70, K_2 3·64	61C
Cu^{2+}	gl	20	0·1	K_1 3·52, $K(Cu^{2+}+HL^- \rightleftharpoons CuHL^+)$ 2·40	61C

61C E. Campi, G. Ostacoli, N. Cibrario, and G. Saini, *Gazzetta*, 1961, **91**, 361

313 $C_5H_8O_4S_2$ Dithiomethylene-*S*,*S*′-diacetic acid H_2L

Metal	Method	Temp	Medium	Log of equilibrium constant, remarks	Ref
H^+	gl	20	0·1	K_1 4·19, K_2 3·37	61O
Cu^{2+}	gl	20	0·1	K_1 3·17, $K(Cu^{2+}+HL^- \rightleftharpoons CuHL^+)$ 1·98	61O

61O G. Ostacoli, E. Campi, N. Cibrario, and G. Saini, *Gazzetta*, 1961, **91**, 349

313.1 $C_5H_8N_2S$ 4-(2′-Aminoethyl)-1,3-thiazole L

Metal	Method	Temp	Medium	Log of equilibrium constant, remarks	Ref
H^+	gl	25	0·015	K_1 9·53	60H
	sp	20	0·015	K_2 1·68	60H
Co^{2+}	gl	25	0·01—0·02	K_1 ~ 4·1	60H
Cu^{2+}	gl	25	0·01—0·02	K_1 7·4, β_2 12·7	60H
Ni^{2+}	gl	25	0·01—0·02	K_1 5·6, β_2 9·6	60H

60H F. Holmes and F. Jones, *J. Chem. Soc.*, 1960, 2398

314 $C_5H_9O_2N$ Pyrrolidine-2-carboxylic acid (proline) HL

Metal	Method	Temp	Medium	Log of equilibrium constant, remarks	Ref
H^+	gl	17	~ 0	K_1 10·46	52P
	gl	20	0·1 (KCl)	K_1 10·63, K_2 1·80	66G
Be^{2+}	gl	17	0·01 $BeSO_4$	β_2 14·2	52P
Cd^{2+}	gl	17	0·01 $CdSO_4$	β_2 8·0	52P
Cu^{2+}	gl	20	0·1 (KCl)	K_1 8·92, K_2 7·66	66G
Hg^{2+}	gl	17	0·01 $Hg(NO_3)_2$	β_2 20·5	52P
	pol	25	0·6 KNO_3	β_2 19·25 ± 0·05	66T
Zn^{2+}	gl	17	0·01 $ZnSO_4$	β_2 9·9	52P

52P D. J. Perkins, *Biochem. J.*, 1952, **51**, 487
66G R. D. Gillard, H. M. Irving, R. M. Parkins, N. C. Payne, and L. D. Pettit, *J. Chem. Soc.* (*A*), 1966, 1159
66T V. F. Toropova and Yu. M. Azizov, *Russ. J. Inorg. Chem.*, 1966, **11**, 288 (531)

$C_5H_9O_2N$ *N*-Methyl-2,2′-iminodiethanol L (*see Corrigenda*) 314.1

Metal	*Method*	*Temp*	*Medium*	*Log of equilibrium constant, remarks*	*Ref*
H^+	gl	25	~ 0·1	K_1 8·78	65D
Cu^{2+}	gl	25	~ 0·1	K_1 4·9, K_2 4·2, K_3 3·4, K_4 2·0	65D
Zn^{2+}	gl	25	~ 0·1	K_1 4·3, K_2 2·9, K_3 1·9, K_4 1·0	65D

65D G. Douheret, *Bull. Soc. chim. France*, 1965, 2915

$C_5H_9O_3N$ 4-Hydroxypyrrolidine-2-carboxylic acid (hydroxyproline) HL 315

Metal	*Method*	*Temp*	*Medium*	*Log of equilibrium constant, remarks*	*Ref*
H^+	gl	17	~ 0	K_1 9·70	52P
Be^{2+}	gl	17	0·01 $BeSO_4$	β_2 12·7	52P
Cd^{2+}	gl	17	0·01 $CdSO_4$	β_2 8·2	52P
Hg^{2+}	gl	17	0·01 $Hg(NO_3)_2$	β_2 17·7	52P
Zn^{2+}	gl	17	0·01 $ZnSO_4$	β_2 9·6	52P

ıtum. 1964, p. 457, *Table* 315. *In first column, move* Ca^{2+} *from third line to fourth line.*

52P D. J. Perkins, *Biochem. J.*, 1952, **51**, 487

$C_5H_9O_4N$ 2-Aminopentanedioic acid (glutamic acid) H_2L 317

Metal	*Method*	*Temp*	*Medium*	*Log of equilibrium constant, remarks*	*Ref*
H^+	gl	15	~ 0	K_1 10·06	53P
	gl	30	0·1	K_1 9·46, K_2 4·23	59N
	?	?	?	K_1 9·60	61J
	gl	25	0·2 (KNO_3)	K_2 4·26, K_3 2·06	63F
	pol	30	1·0 KNO_3	K_1 9·44	64R
	gl	25	0·1 $NaClO_4$	K_1 9·41, K_2 4·07	65N
	gl	25	0·1 (KNO_3)	K_1 9·64, K_2 4·18, K_3 2·18	65R
Be^{2+}	gl	15	0·005 $BeSO_4$	β_2 13·0	53P
Cd^{2+}	gl	15	0·005 $CdSO_4$	β_2 7·9	53P
	pol	30	1	β_3 9·75	62R
	oth	20	0·1 (KNO_3)	K_1 5·3, K_2 2·9, paper electrophoresis	64J
	pol	30	1·0 KNO_3	β_2 7·10, β_3 8·28 $K(Cd^{2+}+2L^-+OH^- \rightleftharpoons CdL_2OH^-)$ 7·97 $K[Cd^{2+}+2L^-+2NH_3 \rightleftharpoons CdL_2(NH_3)_2]$ 9·01	64R
Co^{2+}	gl	30	0·1	K_1 4·49	59N
	oth	20	0·1 (KNO_3)	K_1 4·9, K_2 3·1, paper electrophoresis	64J
	gl	20	0·1 $NaNO_3$	K_1 4·6, K_2 2·8	65P
	gl	25	0·1 (KNO_3)	K_1 4·56±0·04, K_2 3·29±0·04	65R
Cu^{2+}	gl	30	0·1	K_1 7·74	59N
	oth	20	0·1 (KNO_3)	K_1 10·1, K_2 6·3, paper electrophoresis	64J
	gl	25	0·1 $NaClO_4$	K_1 7·87, K_2 6·29	65N
Mn^{2+}	oth	20	0·1 (KNO_3)	K_1 3·4, paper electrophoresis	64J
Ni^{2+}	gl	30	0·1	K_1 5·62	59N
	oth	20	0·1 (KNO_3)	K_1 5·8, K_2 3·4, paper electrophoresis	64J
	gl	25	0·1 (KNO_3)	K_1 5·58±0·04, K_2 4·12±0·04	65R

continued overleaf

317 $C_5H_9O_4N$ (contd)

Metal	Method	Temp	Medium	Log of equilibrium constant, remarks	Ref
Pb^{2+}	pol	30	1·0 KNO_3	K_1 4·60, β_2 6·22	64R
UO_2^{2+}	gl	25	0·2 (KNO_3)	$K(UO_2^{2+}+HL^- \rightleftharpoons UO_2HL^+)$ 2·66±0·05	63F
Zn^{2+}	gl	15	0·005 $ZnSO_4$	β_2 8·8	53P
	pol	20	0·2 $NaClO_4$	K_1 5·73, β_2 9·6	61J
	gl	25	0·02	K_1 5·45	61J
	oth	20	0·1 (KNO_3)	K_1 5·6, K_2 3·2, paper electrophoresis	64J

53P D. J. Perkins, *Biochem. J.*, 1953, **55**, 649
59N M. H. T. Nyberg, M. Cefola, and D. Sabine, *Arch. Biochem. Biophys.*, 1959, **85**, 82
61J C. Je-Sia, J. Dolezal, and J. Zyka, *Coll. Czech. Chem. Comm.*, 1961, **26**, 1768
62R G. N. Rao and R. S. Subrahmanya, *Current Sci.* 1962, **31**, 55
63F I. Feldman and L. Koval, *Inorg. Chem.*, 1963, **2**, 145
64J V. Jokl, *J. Chromatog.*, 1964, **14**, 71
65N M. H. T. Nyberg and M. Cefola, *Arch. Biochem. Biophys.*, 1965, **111**, 321
64R G. N. Rao and R. S. Subrahmanya, *Proc. Indian Acad. Sci.*, 1964, **60**, 165, 185
65P V. D. Danosyuk and L. C. Reiter, *Russ. J. Inorg. Chem.*, 1965, **10**, 730 (1344)
65R J. H. Ritsma, G. A. Wiegers, and F. Jellinek, *Rec. Trav. chim.*, 1965, **84**, 1577

318 $C_5H_9O_4N$ *N*-Methyliminodiacetic acid H_2L

Metal	Method	Temp	Medium	Log of equilibrium constant, remarks	Ref
H^+	cal	20	0·1 (KNO_3)	ΔH_1 −6·94, ΔS_1 20·5	65A
	gl	25	?	K_1 9·65±0·05	66K
	nmr	33—37	?	K_1 9·6±0·2	66K
	gl	25	0·1 (KCl)	K_1 9·57	68N
Ba^{2+}	cal	20	0·1 (KNO_3)	ΔH_1 −0·79, ΔS_1 9·2	65A
	gl	25	0·1 (KCl)	K_1 2·61, K_2 2·33	68N
	cal	25	0·1 (KCl)	ΔH_1 −1·06±0·10, ΔS_1 8·4±0·4	68N
Ca^{2+}	cal	20	0·1 (KNO_3)	ΔH_1 −1·64, ΔS_1 11·6	65A
	gl	25	0·1 (KCl)	K_1 3·85, K_2 2·72	68N
	cal	25	0·1 (KCl)	ΔH_1 −1·2±0·1, ΔS_1 13·6±0·4, ΔH_2 −0·7±0·3, ΔS_2 10±2	68N
Cd^{2+}	cal	20	0·1 (KNO_3)	ΔH_1 −1·89, ΔS_1 24·5, $\Delta H_{\beta 2}$ −7·27, $\Delta S_{\beta 2}$ 32·5	65A
Co^{2+}	cal	20	0·1 (KNO_3)	ΔH_1 −1·85, ΔS_1 28·6, $\Delta H_{\beta 2}$ −5·48, $\Delta S_{\beta 2}$ 45·0	65A
Cu^{2+}	cal	20	0·1 (KNO_3)	ΔH_1 −3·84, ΔS_1 37·7, $\Delta H_{\beta 2}$ −12·11, $\Delta S_{\beta 2}$ 40·8	65A
Mg^{2+}	cal	20	0·1 (KNO_3)	ΔH_1 3·12, ΔS_1 26·4	65A
	gl	25	0·1 (KCl)	K_1 3·48, K_2 2·35	68N
	cal	25	0·1 (KCl)	ΔH_1 2·85±0·10, ΔS_1 26·5±0·5, ΔH_2 −0·7±0·3, ΔS_2 8±2	68N
Mn^{2+}	cal	20	0·1 (KNO_3)	ΔH_1 0·56, ΔS_1 26·6, $\Delta H_{\beta 2}$ 0·23, $\Delta S_{\beta 2}$ 44·6	65A
Mo^{IV}	nmr	33—37	?	$K(MoO_4^{2-}+HL^-+H^+ \rightleftharpoons MoO_3L^{2-})$ 8·5±0·1 $K(MoO_3+L^{2-} \rightleftharpoons MoO_3L^{2-})$ 10·4 $K(Mo_7O_{24}^{6-}+7HL^- \rightleftharpoons 7MoO_3L^{2-}+H^+)$ 8·9 $K(MoO_3L^{2-}+H^+ \rightleftharpoons HMoO_3L^-)$ 2·8±0·2 $K(2MoO_3L^{2-}+2H^+ \rightleftharpoons Mo_2O_5L_2^{2-})$ 7±0·5	66K
Ni^{2+}	cal	20	0·1 (KNO_3)	ΔH_1 −4·7, ΔS_1 23·9, $\Delta H_{\beta 2}$ −7·65, $\Delta S_{\beta 2}$ 46·9	65A
Pb^{2+}	cal	20	0·1 (KNO_3)	ΔH_1 −3·56, ΔS_1 24·6	65A
Sr^{2+}	cal	20	0·1 (KNO_3)	ΔH_1 −1·23, ΔS_1 8·9	65A
	gl	25	0·1 (KCl)	K_1 2·96, K_2 1·80	68N
	cal	25	0·1 (KCl)	ΔH_1 −0·83±0·06, ΔS_1 10·7±0·2	68N
Zn^{2+}	cal	20	0·1 (KNO_3)	ΔH_1 −2·17, ΔS_1 27·7, $\Delta H_{\beta 2}$ −5·83, $\Delta S_{\beta 2}$ 44·4	65A

references on opposite

65A G. Anderegg, *Helv. Chim. Acta.*, 1965, **48**, 1718
66K R. J. Kula, *Analyt. Chem.*, 1966, **38**, 1382
68N G. H. Nancollas and A. C. Park, *Inorg. Chem.*, 1968, **7**, 58

$C_5H_9O_5N_3$ Semicarbazide-1,1-diacetic acid H_2L 318.1

Metal	*Method*	*Temp*	*Medium*	*Log of equilibrium constant, remarks*	*Ref*
H^+	gl	30	0·1 (KCl)	K_1 4·04, K_2 2·96	67G
	cal	30	0·1 (KNO_3)	ΔH_1 0·6, ΔS_1 20, ΔH_2 0·2, ΔS_2 14	67Ga
Cd^{2+}	gl	30	0·1 (KCl)	K_1 5·7 $K(Cd^{2+}+HL^- \rightleftharpoons CdHL^+)$ 4·1	67G
Co^{2+}	gl	30	0·1 (KCl)	K_1 5·9 $K(Co^{2+}+HL^- \rightleftharpoons CoHL^+)$ 4·6	67G
	cal	30	0·1 (KNO_3)	ΔH_1 0·7, ΔS_1 29	67Ga
Cu^{2+}	sp	30	0·1 (KNO_3)	K_1 8·4	67G
	cal	30	0·1 (KNO_3)	ΔH_1 −0·5, ΔS_1 37	67Ga
Mg^{2+}	gl	30	0·1 (KCl)	K_1 1·4 $K(Mg^{2+}+HL^- \rightleftharpoons MgHL^+)$ 0	67G
	cal	30	0·1 (KNO_3)	ΔH_1 9·8, ΔS_1 39	67Ga
Mn^{2+}	gl	30	0·1 (KCl)	K_1 2·6 $K(Mn^{2+}+HL^- \rightleftharpoons MnHL^+)$ 1·6	67G
	cal	30	0·1 (KNO_3)	ΔH_1 3·2, ΔS_1 22	67Ga
Ni^{2+}	gl	30	0·1 (KCl)	K_1 6·4 $K(Ni^{2+}+HL^- \rightleftharpoons NiHL^+)$ 5·1	67G
	cal	30	0·1 (KNO_3)	ΔH_1 −1·1, ΔS_1 26	67Ga
Sr^{2+}	gl	30	0·1 (KCl)	K_1 2·3 $K(Sr^{2+}+HL^- \rightleftharpoons SrHL^+)$ 0·7	67G
	cal	30	0·1 (KNO_3)	ΔH_1 −0·7, ΔS_1 9	67Ga
Zn^{2+}	gl	30	0·1 (KCl)	K_1 6·6 $K(Zn^{2+}+HL^- \rightleftharpoons ZnHL^+)$ 5·1	67G
	cal	30	0·1 (KNO_3)	ΔH_1 −0·4, ΔS_1 29	67Ga

67G D. R. Goddard and S. I. Nwankwo, *J. Chem. Soc. (A)*, 1967, 1371
67Ga D. R. Goddard, S. I. Nwankwo, and L. A. K. Staveley, *J. Chem. Soc. (A)*, 1967, 1376

$_5H_9NS_2$ Pyrrolidine-*N*-carboxydithioic acid (*NN*-cyclotetramethylenedithiocarbamic acid) HL 318.2

Metal	*Method*	*Temp*	*Medium*	*Log of equilibrium constant, remarks*	*Ref*
Cu^{2+}	sp	23—27	0·01 NaOH 75% ethanol	K_1 13·9±0·2, K_2 12·9±0·2	56J
	sp	18—22	0·01 NaOH 0—89% ethanol	K_1 (→ 0) 10·9±0·2, (51·7%) 12·6±0·2, (75%) 13·9±0·2, (89%) 14·8±0·2, K_2 (→ 0) 9·9±0·2, (51·7%) 12·1±0·2, (75%) 12·9±0·2, (89%) 13·6±0·2	57J

56J M. J. Janssen, *Rec. Trav. chim.*, 1956, **75**, 1411
57J M. J. Janssen, *Rec. Trav. chim.*, 1957, **76**, 827

320 $C_5H_{10}O_2N_2$ Pentane-2,3-dione dioxime (ethylmethylglyoxime) HL

Metal	Method	Temp	Medium	Log of equilibrium constant, remarks	Ref
H^+	gl	25	0·1 75% dioxan	K_1 13·55	63B
Ni^{2+}	gl	25	0·1 0—75% dioxan	K_1 (75%) 10·4±0·5, β_2 (→ 0) 17·26, (75%) 23·97±0·03, K_{so} −23·27	63B

63B C. V. Banks and S. Anderson, *Inorg. Chem.*, 1963, **2**, 112

323 $C_5H_{10}O_3N_2$ L-Alanylglycine HL

Metal	Method	Temp	Medium	Log of equilibrium constant, remarks	Ref
H^+	gl	25	0·16 KCl	K_1 8·15, K_2 3·23	65B
Cu^{2+}	gl	25	0·16 KCl	K_1 5·44, $K[Cu(H_{-1}L)+H^+ \rightleftharpoons CuL^+]$ 4·16	65B

Erratum, 1964, p. 461, Table 323. Under Cd^{2+}, *for* K_1 3·0 *read* K_1 2·6.

65B G. F. Bryce, J. M. H. Pinkerton, L. K. Steinrauf, and F. R. N. Gurd, *J. Biol. Chem.*, 1965, **240**, 3829

324 $C_5H_{10}O_3N_2$ Glycyl-DL-alanine HL

Metal	Method	Temp	Medium	Log of equilibrium constant, remarks	Ref
H^+	pH	25	0·1 ($NaClO_4$)	K_1 8·21, K_2 3·21	67S
Cu^{2+}	pH	25	0·1 ($NaClO_4$)	K_1 5·92, $K[Cu(H_{-1}L)+H^+ \rightleftharpoons CuL^+]$ 4·33 $K[Cu(H_{-1}L)+L^- \rightleftharpoons CuL(H_{-1}L)^-]$ 3·62 $K[Cu(H_{-1}L)_2^{2-}+H^+ \rightleftharpoons CuL(H_{-1}L)^-]$ 9·28	67S
Ni^{2+}	pH	25	0·1 ($NaClO_4$)	K_1 4·08, K_2 3·79	67S

67S A. Ya. Sychev and N. S. Mitsul, *Russ. J. Inorg. Chem.*, 1967, **12**, 1120 (2127)

324.1 $C_5H_{10}O_3N_2$ Glycyl-L-alanine HL

Metal	Method	Temp	Medium	Log of equilibrium constant, remarks	Ref
H^+	gl	25	0·100 (NaCl)	K_1 8·19±0·04, K_2 3·34±0·02	59B
	gl	25	0·16 KCl	K_1 8·24, K_2 3·22	65B
	gl	25	0·12 (NaCl)	K_1 8·28±0·02, K_2 3·31±0·02	67S
Co^{2+}	gl	25	0·100 (NaCl)	K_1 3·10±0·02, β_2 5·68±0·03	59B
Cu^{2+}	gl	25	0·100 (NaCl)	K_1 5·81±0·03 $K[Cu(H_{-1}L)+L^- \rightleftharpoons CuL(H_{-1}L)^-]$ 9·0±0·1 $K[Cu(H_{-1}L)+H^+ \rightleftharpoons CuL^+]$ 4·35 $K[Cu(H_{-1}L)_2^{2-}+H^+ \rightleftharpoons CuL(H_{-1}L)^-]$ 9·28	59B
	gl	25	0·16 KCl	K_1 5·67, $K[Cu(H_{-1}L)+H^+ \rightleftharpoons CuL^+]$ 4·03	65B

conti

$C_5H_{10}O_3N_2$ (contd) 324.1

Metal	Method	Temp	Medium	Log of equilibrium constant, remarks	Ref
Cu^{2+} (contd)	gl	25	0·12 (NaCl)	K_1 5·61, $K[Cu(H_{-1}L)+H^+ \rightleftharpoons CuL^+]$ 4·81	67S
				$K[Cu(H_{-1}L)+L^- \rightleftharpoons CuL(H_{-1}L)^-]$ 3·84	
				$K[Cu(H_{-1}L)_2{}^{2-}+H^+ \rightleftharpoons CuL(H_{-1}L)^-]$ 9·81	
				$K[Cu(H_{-1}L)OH^-+H^+ \rightleftharpoons Cu(H_{-1}L)]$ 10·42	

Errata. 1964, p. 461, Table 324. Under Cd^{2+}, *for* K_1 2·6 *read* K_1 3·6. Reference column, *for* 53P *read* 54P.

59B J. L. Biester and P. M. Ruoff, *J. Amer. Chem. Soc.*, 1959, **81**, 6517

65B G. F. Bryce, J. M. H. Pinkerton, L. K. Steinrauf, and F. R. N. Gurd, *J. Biol. Chem.*, 1965, **240**, 3829

67S U. I. Salakhutdinov, A. P. Borisova, Yu. V. Granovskii, I. A. Savich, and V. I. Spitsyn, *Proc. Acad. Sci.* (*U.S.S.R.*), 1967, **177**, 1039 (365)

$C_5H_{10}O_3N_2$ Glycyl-β-alanine HL 324.2

Metal	Method	Temp	Medium	Log of equilibrium constant, remarks	Ref
H^+	gl	25	0·12 (NaCl)	K_1 8·27±0·02, K_2 4·07±0·03	67S
Cu^{2+}	gl	25	0·12 (NaCl)	K_1 6·11	67S
				$K[Cu(H_{-1}L)+H^+ \rightleftharpoons CuL^+]$ 4·69	
				$K[Cu(H_{-1}L)+L^- \rightleftharpoons CuL(H_{-1}L)^-]$ 3·66	
				$K[Cu(H_{-1}L)_2{}^{2-}+H^+ \rightleftharpoons CuL(H_{-1}L)^-]$ 9·53	
				$K[Cu(H_{-1}L)OH^-+H^+ \rightleftharpoons Cu(H_{-1}L)]$ 9·79	

67S U. I. Salakhutdinov, A. P. Borisova, Yu. V. Granovskii, I. A. Savich, and V. I. Spitsyn, *Proc. Acad. Sci.* (*U.S.S.R.*), 1967, **177**, 1039 (365)

$C_5H_{10}O_3N_2$ 2-Aminopentanedioic acid 5-amide (glutamine) HL 324.3

Metal	Method	Temp	Medium	Log of equilibrium constant, remarks	Ref
H^+	gl	15	~0	K_1 9·34	53P
	gl	25	0·1 (KNO_3)	K_1 9·01, K_2 2·17	65R
Be^{2+}	gl	15	0·005 $BeSO_4$	β_2 12·4	53P
Cd^{2+}	gl	15	0·005 $CdSO_4$	β_2 7·4	53P
Co^{2+}	gl	25	0·1 (KNO_3)	K_1 4·05±0·04, K_2 3·30±0·04	65R
Cu^{2+}	gl	25	0·1 (KNO_3)	K_1 7·74±0·04, K_2 6·46±0·04	65R
Ni^{2+}	gl	25	0·1 (KNO_3)	K_1 5·17±0·04, K_2 4·28±0·02	65R
Zn^{2+}	gl	15	0·005 $ZnSO_4$	β_2 8·4	53P

53P D. J. Perkins, *Biochem, J.*, 1953, **55**, 649

65R J. H. Ritsma, G. A. Wiegers, and F. Jellinck, *Rec. Trav. chim.*, 1965, **84**, 1577

325 $C_5H_{10}O_3N_2$ Glycylsarcosine HL

Metal	Method	Temp	Medium	Log of equilibrium constant, remarks	Ref
H^+	gl	25	0·16 KCl	K_1 8·57, K_2 2·98	60K
Cu^{2+}	gl	25	0·16 KCl	K_1 6·13, K_2 4·62	60K
Ni^{2+}	gl	25	0·16 (KNO_3)	$K(NiLOH+H^+ \rightleftharpoons NiL^+)$ 10·7	60M

60K W. L. Koltun, M. Fried, and F. R. N. Gurd, *J. Amer. Chem. Soc.*, 1960, **82**, 233
60M R. B. Martin, M. Chamberlin, and J. T. Edsall, *J. Amer. Chem. Soc.*, 1960, **82**, 495

326 $C_5H_{10}O_3N_2$ Sarcosylglycine HL

Metal	Method	Temp	Medium	Log of equilibrium constant, remarks	Ref
H^+	gl	25	0·16 KCl	K_1 8·56, K_2 3·15	60K
Cu^{2+}	gl	25	0·16 KCl	K_1 4·39	60K
				$K[Cu(H_{-1}L)+H^+ \rightleftharpoons CuL^+]$ 3·45	
				$K[Cu(H_{-1}L)+L^- \rightleftharpoons CuL(H_{-1}L)^-]$ 3·42	
				$K[Cu(H_{-1}L)OH^- +H^+ \rightleftharpoons Cu(H_{-1}L)]$ 9·19	
				$K[Cu(H_{-1}L)(OH)_2^{2-} +H^+ \rightleftharpoons Cu(H_{-1}L)OH^-]$ 11·9	
				$K[Cu(H_{-1}L)OH^- + Cu(H_{-1}L) \rightleftharpoons (CuH_{-1}L)_2OH^-]$ 1·48	

60K W. L. Koltun, M. Fried, and F. R. N. Gurd, *J. Amer. Chem. Soc.*, 1960, **82**, 233

327 $C_5H_{10}O_4N_2$ Glycylserine HL

Errata. 1964, p. 462, Table 327. Under Cd^{2+}, *for* K_1 1·5 *read* K_1 3·3.
For ref. 53P *read* ref. 54P D. J. Perkins, *Biochem. J.*, 1954, **57**, 702.

328 $C_5H_{11}O_2N$ 2-Aminopentanoic acid (norvaline) HL

Metal	Method	Temp	Medium	Log of equilibrium constant, remarks	Ref
H^+	gl	19—20	~0	K_1 9·87	52P, 53P
Be^{2+}	gl	19—20	0·0005 $BeSO_4$	β_2 12·6	52P, 53P
Cd^{2+}	gl	19—20	0·0005 $CdSO_4$	β_2 6·6	52P, 53P
Hg^{2+}	gl	19	0·005 $Hg(NO_3)_2$	β_2 17·7	53P
	gl	20	0·0005 $Hg(NO_3)_2$	β_2 17·6	52P
Zn^{2+}	gl	19—20	0·0005 $ZnSO_4$	β_2 8·1	52P, 53P

52P D. J. Perkins, *Biochem. J.*, 1952, **51**, 487
53P D. J. Perkins, *Biochem. J.*, 1953, **55**, 649

$C_5H_{11}O_2N$ 5-Aminopentanoic acid HL 328.1

Metal	Method	Temp	Medium	Log of equilibrium constant, remarks	Ref
H^+	gl	20	0·5 KNO_3	K_1 10·91 ± 0·02	68A
	gl	25	0·5 KNO_3	K_1 10·66	68T
Ag^+	gl	20	0·500 KNO_3	K_1 3·5 ± 0·1, β_2 7·41 ± 0·05	68A
	Ag-AgCl	20	0·500 KNO_3	β_2 7·51 ± 0·01	68A
	gl	25	0·5 KNO_3	K_1 3·56, K_2 3·79	68T

68A D. J. Alner, R. C. Lansbury, and A. G. Smeeth, *J. Chem. Soc.* (*A*), 1968, 417
68T G. F. Thiers, L. C. van Poucke, and M. A. Herman, *J. Inorg. Nuclear Chem.*, 1968, **30**, 1543

$C_5H_{11}O_2N$ 2-Amino-3-methylbutanoic acid (valine) HL 329

Metal	Method	Temp	Medium	Log of equilibrium constant, remarks	Ref
H^+	gl	20	~ 0	K_1 9·59	52P
	pol	30	1·0 KNO_3	K_1 9·44	64R
	gl	20	0·1 (KCl)	K_1 9·65, K_2 2·24	66G
	pol	25	0·5 $NaClO_4$	K_1 9·62, K_2 2·35	67R
Be^{2+}	gl	20	0·01 $BeSO_4$	β_2 12·4	52P
Cd^{2+}	gl	20	0·01 $CdSO_4$	β_2 6·7	52P
	pol	30	1	β_3 8·60	62R
	pol	30	1·0 KNO_3	β_2 6·66, β_3 8·48	64R
				$K(Cd^{2+} + 2L^- + OH^- \rightleftharpoons CdL_2OH^-)$ 8·35	
				$K(Cd^{2+} + 2L^- + CO_3^{2-} \rightleftharpoons CdL_2CO_3^{2-})$ 7·45	
				$K[Cd^{2+} + 2L^- + 4NH_3 \rightleftharpoons CdL_2(NH_3)_4]$ 8·75	
	oth	25	0·5	K_1 3·46, β_2 6·46, optical rotation	67R
Cr^{3+}	gl	25	~ 0·5	K_1 8·3, K_2 6·4, K_3 5·4	63K
Cu^{2+}	gl	20	0·1 (KCl)	K_1 8·19, β_2 15·18	66G
	oth	25	0·5	K_1 7·98, β_2 14·71, optical rotation	67R
	oth	25	0·5 $NaClO_4$	$K(Cu^{2+} + L^- + A^- \rightleftharpoons CuAL)$	68R
				15·75 ± 0·04	
				where HA = glycine	
				16·86 ± 0·09, optical rotation	
				where HA = L-proline	
				See salicylic acid (506) and nitrilotriacetic acid (449) for mixed ligands	
Ni^{2+}	pol	25	0·5 $NaClO_4$	K_1 5·26, K_2 4·16, K_3 2·53	67R
Pb^{2+}	pol	30	1·0 KNO_3	K_1 4·02, β_2 5·89	64R
				$K(Pb^{2+} + 2L^- + OH^- \rightleftharpoons PbL_2OH^-)$ 9·41	
Zn^{2+}	gl	20	0·01 $ZnSO_4$	β_2 8·2	52P
	oth	25	0·5	K_1 4·40, β_2 8·17, optical rotation	67R

Erratum. 1964, p. 463, *Table* 329, *under* Cd^{2+}, *third line, insert* Co^{2+} *in first column; same line, for* K_1 8·60 *read* β_2 8·60.

52P D. J. Perkins, *Biochem. J.*, 1952, **51**, 487
62R G. N. Rao and R. S. Subrahmanya, *Current Sci.*, 1962, **31**, 55
63K A. A. Kahn and W. U. Malik, *J. Indian Chem. Soc.*, 1963, **40**, 564
64R G. N. Rao and R. S. Subrahmanya, *Proc. Indian Acad. Sci.*, 1964, **60**, 165, 185
66G R. D. Gillard, H. M. Irving, R. M. Parkins, N. C. Payne, and L. D. Pettit, *J. Chem. Soc.* (*A*), 1966, 1159
67R M. M. Ramel and M. R. Paris, *Bull. Soc. chim. France*, 1967, 1359
68R M. M. Ramel-Petit and M. R. Paris, *Bull. Soc. chim. France*, 1968, 2791

329.1 $C_5H_{11}O_2N$ 2-Aminopropanoic acid ethyl ester (ethyl alanate) L

Metal	Method	Temp	Medium	Log of equilibrium constant, remarks	Ref
Cu^{2+}				See nitrilotriacetic acid (449) for mixed ligands	

329.2 $C_5H_{11}O_2N$ 3-Aminopropanoic acid ethyl ester (ethyl β-alanate) L

Metal	Method	Temp	Medium	Log of equilibrium constant, remarks	Ref
Cu^{2+}				See nitrilotriacetic acid (449) for mixed ligands	

331.1 $C_5H_{11}O_3N$ *N*-(2′-Hydroxyethyl)-3-aminopropanoic acid (*N*-hydroxyethyl-β-alanine) HL

Metal	Method	Temp	Medium	Log of equilibrium constant, remarks	Ref
H^+	gl	20	0·1 KCl	K_1 9·72	64U
Co^{2+}	gl	20	0·1 KCl	K_1 3·80, K_2 2·00	64U
Cu^{2+}	gl	20	0·1 KCl	K_1 7·40, K_2 4·85 $K[Cu(H_{-1}L)+H^+ \rightleftharpoons CuL^+]$ 7·15	64U
Ni^{2+}	gl	20	0·1 KCl	K_1 4·75, K_2 3·40	64U

64U E. Uhlig and D. Linke, *Z. anorg. Chem.*, 1964, 331, 112

332 $C_5H_{11}NS_2$ Diethyldithiocarbamic acid HL

Metal	Method	Temp	Medium	Log of equilibrium constant, remarks	Ref
Ag^+	Ag	25	0·01 (KNO_3) 75% dioxan	K_1 8·3	68B
	sp	?	CCl_4	$K(AgHA+HL \rightleftharpoons AgL+H_2A)$ 2·58±0·03 where H_2A = dithizone	68S
As^{3+}	sp	?	CCl_4	$K(AsAL+2HL \rightleftharpoons AsL_3+H_2A)$ 7·93±0·06 where H_2A = dithizone	68S
Bi^{3+}	sp	?	CCl_4	$K[BiL_3+2H_2A \rightleftharpoons Bi(HA)_3+3HL]$ 5·72±0·08 where H_2A = dithizone	68S
Cd^{2+}	sp	?	CCl_4	$K[Cd(HA)_2+2HL \rightleftharpoons CdL_2+2H_2A]$ 2·53±0·05 where H_2A = dithizone	68S
	sp	23—27	0·01 NaOH 75% ethanol	K_1 14·9±0·2, K_2 13·9±0·2	56J
Cu^{2+}	sp	?	CCl_4	$K[Cu(HA)_2+HL \rightleftharpoons CuHAL+H_2A]$ 2·5±0·1 $K[CuHAL+HL \rightleftharpoons CuL_2+H_2A]$ 2·1±0·1 $K[Cu(HA)_2+2HL \rightleftharpoons CuL_2+2H_2A]$ 4·6±0·2 $K[Cu(HA)_2+CuL_2 \rightleftharpoons 2CuHAL]$ 0·35±0·05 where H_2A = dithizone	68S
Mg^{2+}	sp	?	CCl_4	$K[Hg(HA)_2+2HL \rightleftharpoons HgL_2+2H_2A]$ 2·31±0·05 where H_2A = dithizone	68S
Lu^{3+}	sp	?	CCl_4	$K[Lu(HA)_3+3HL \rightleftharpoons LuL_3+3H_2A]$ 4·75 where H_2A = dithizone	68S

cont

$C_5H_{11}NS_2$ (contd) 332

Metal	Method	Temp	Medium	Log of equilibrium constant, remarks	Ref
Pb^{2+}	sp	?	CCl_4	$K[Pb(HA)_2+2HL \rightleftharpoons PbL_2+2H_2A]$ 4·98±0·05 where H_2A = dithizone	68S
Pd^{2+}	sp	?	CCl_4	$K[Pd(HA)_2+2HL \rightleftharpoons PdL_2+2H_2A]$ 1·6±0·2 where H_2A = dithizone	68S
Sb^{3+}	sp	?	CCl_4	$K(SbAL+2HL \rightleftharpoons SbL_3+H_2A)$ 2·47±0·07 where H_2A = dithizone	68S
Se^{4+}	sp	?	CCl_4	$K[Se(HA)_4+4HL \rightleftharpoons SeL_4+4H_2A]$ ~ 6·5 where H_2A = dithizone	68S
Te^{4+}	sp	?	CCl_4	$K[Te(HA)_4+4HL \rightleftharpoons TeL_4+4H_2A]$ ~ 5·5 where H_2A = dithizone	68S
Tl^{+}	sp	?	CCl_4	$K(TlHA+HL \rightleftharpoons TlL+H_2A)$ 3·53±0·05 where H_2A = dithizone	68S
Zn^{2+}	sp	?	CCl_4	$K[Zn(HA)_2+2HL \rightleftharpoons ZnL_2+2H_2A]$ 0·24±0·03 where H_2A = dithizone	68S

56J M. J. Janssen, *Rec. Trav. Chim.*, 1956, **75**, 1411
68B I. M. Bhatt and P. K. Soni, *Indian J. Chem.*, 1968, **6**, 115
68S J. Stary and J. Ruzicha, *Talanta*, 1968, **15**, 505

$C_5H_{12}O_2N_2$ 2,5-Diaminopentanoic acid (ornithine) HL 335

Metal	Method	Temp	Medium	Log of equilibrium constant, remarks	Ref
H^{+}	gl	20	~0	K_1 10·73, K_2 8·75	53P
Be^{2+}	gl	20	0·005 $BeSO_4$	β_2 11·7	53P
Cd^{2+}	gl	20	0·005 $CdSO_4$	β_2 6·1	53P
Zn^{2+}	gl	20	0·005 $ZnSO_4$	β_2 7·6	53P

53P D. J. Perkins, *Biochem. J.*, 1953, 55, 649

$C_5H_{13}ON$ 5-Aminopentan-1-ol L 335.1

Metal	Method	Temp	Medium	Log of equilibrium constant, remarks	Ref
H^{+}	gl	20	0·5 KNO_3	K_1 10·91±0·01	68A
	H	20	0 corr	K_1 10·61±0·02	68A
Ag^{+}	gl	20	0·500 KNO_3	K_1 3·42±0·06, β_2 7·55±0·02	68A
	Ag-AgCl	20	0·500 KNO_3	β_2 7·55±0·02	68A

68A D. J. Alner, R. C. Lansbury, and A. G. Smeeth, *J. Chem. Soc. (A)*, 1968, 417

341.1 $C_5H_{14}ON_2$ *N*-(2′-Hydroxyethyl)-1,3-diaminopropane L

Metal	Method	Temp	Medium	Log of equilibrium constant, remarks	Ref
H^+	gl	25	I ($NaClO_4$)	K_1 10·19+0·327 I	63N, 64N
				K_2 7·95+1·018$\sqrt{I}/(1+1{\cdot}241\sqrt{I})$+0·280 I	
Cu^{2+}	gl	25	I ($NaClO_4$)	K_1 10·42+0·690 I−0·252 $I^{3/2}$+0·055 I^2	63N
	gl	25	I ($NaClO_4$)	$K(CuOHL^+ + H^+ \rightleftharpoons CuL^{2+})$ 7·07+1·018$\sqrt{I}/(1+\sqrt{I})$	64N
				$K[Cu_2(OH)_2L_2^{2+} + 2H^+ \rightleftharpoons 2CuL^{2+}]$ 11·55+1·018$\sqrt{I}/(1+\sqrt{I})$	
			→ 0	$K[2CuOHL^+ \rightleftharpoons Cu_2(OH)_2L_2^{2+}]$ 2·59	
				$K[Cu(OH)_2L + H^+ \rightleftharpoons CuL^{2+}]$ 10·17	
				$K(Cu^{2+} + OH^- + L \rightleftharpoons CuOHL^+)$ 17·35	

63N R. Nasanen, P. Merilainen, and S. Lukkari, *Suomen Kem.*, 1963, **B36**, 135
64N R. Nasanen, P. Merilainen, and S. Lukkari, *Suomen Kem.*, 1964, **B37**, 1

342.1 $C_5H_3O_4N_2Br$ 5-Bromo-1,3*H*-pyrimidine-2,4-dione-6-carboxylic acid (5-bromo-orotic acid) H_2L

Metal	Method	Temp	Medium	Log of equilibrium constant, remarks	Ref
H^+	gl	25	0·1 [$(CH_3)_4NBr$]	K_1 7·33, K_2 2·38	64T
Cd^{2+}	gl	25	0·1 [$(CH_3)_4NBr$]	K_1 2·43	64T
Co^{2+}	gl	25	0·1 [$(CH_3)_4NBr$]	K_1 3·27	64T
Cu^{2+}	gl	25	0·1 [$(CH_3)_4NBr$]	K_1 5·58	64T
Mn^{2+}	gl	25	0·1 [$(CH_3)_4NBr$]	K_1 1·88	64T
Ni^{2+}	gl	25	0·1 [$(CH_3)_4NBr$]	K_1 4·19	64T
Zn^{2+}	gl	25	0·1 [$(CH_3)_4NBr$]	K_1 3·26	64T

64T E. R. Tucci, F. Takahashi, V. A. Tucci, and N. C. Li, *J. Inorg. Nuclear Chem.*, 1964, **26**, 1263

342.2 $C_5H_3O_4N_2I$ 5-Iodo-1,3*H*-pyrimidine-2,4-dione-6-carboxylic acid (5-iodo-orotic acid) H_2L

Metal	Method	Temp	Medium	Log of equilibrium constant, remarks	Ref
H^+	gl	25	0·1 [$(CH_3)_4NBr$]	K_1 7·63, K_2 1·88	64T
Cd^{2+}	gl	25	0·1 [$(CH_3)_4NBr$]	K_1 2·90	64T
Co^{2+}	gl	25	0·1 [$(CH_3)_4NBr$]	K_1 3·78	64T
Cu^{2+}	gl	25	0·1 [$(CH_3)_4NBr$]	K_1 6·59	64T
Mn^{2+}	gl	25	0·1 [$(CH_3)_4NBr$]	K_1 2·25	64T
Ni^{2+}	gl	25	0·1 [$(CH_3)_4NBr$]	K_1 4·65	64T
Zn^{2+}	gl	25	0·1 [$(CH_3)_4NBr$]	K_1 3·77	64T

64T E. R. Tucci, F. Takahashi, V. A. Tucci, and N. C. Li, *J. Inorg. Nuclear Chem.*, 1964, **26**, 1263

342.3 $C_5H_4O_3N_2S$ 2-Thioxo-1,3*H*-pyrimidine-4-one-5-carboxylic acid (2-thio-iso-orotic acid) H_2L

Metal	Method	Temp	Medium	Log of equilibrium constant, remarks	Ref
H^+				H^+ constants not given.	
Zn^{2+}	ix	25	0·1 ($KClO_4$)	K_1 3·94	66D

66D B. E. Doody, E. R. Tucci, R. Scruggs, and N. C. Li, *J. Inorg. Nuclear Chem.*, 1966, **28**, 833

$C_5H_7O_4NS_2$ *NN*-Bis(carboxymethyl)dithiocarbamic acid H_3L 342.4

Metal	*Method*	*Temp*	*Medium*	*Log of equilibrium constant, remarks*	*Ref*
H^+				H^+ constants not given.	
Cd^{2+}	dis	20	0·1 KNO_3	β_2 11·2±0·2	67H
Cu^{2+}	dis	20	0·1 KNO_3	β_2 21·5±0·3	67H
Ni^{2+}	dis	20	0·1 KNO_3	β_2 7·9±0·1	67H
Pb^{2+}	dis	20	0·1 KNO_3	β_2 15·5±0·1	67H
Zn^{2+}	dis	20	0·1 KNO_3	β_2 < 4·5	67H

67H A. Hulanicki and M. Minczewska, *Talanta*, 1967, **14**, 677

$C_5H_9O_2NS$ L-1,4-Thiazine-3-carboxylic acid HL 342.5

Metal	*Method*	*Temp*	*Medium*	*Log of equilibrium constant, remarks*	*Ref*
H^+	gl	20	0·1 KNO_3	K_1 8·49	68H
Cu^{2+}	gl	20	0·1 KNO_3	K_1 6·66, K_2 5·88	68H

68H P. Herman and K. Lemke, *Z. physiol. Chem.*, 1968, **349**, 390

$C_5H_9O_4N_3S$ Thiosemicarbazide-1,1-diacetic acid H_2L 342.6

Metal	*Method*	*Temp*	*Medium*	*Log of equilibrium constant, remarks*	*Ref*
H^+	gl	30	0·1 (KCl)	K_1 4·07, K_2 2·94	67G
	cal	30	0·1 (KNO_3)	ΔH_1 0·2, ΔS_1 19, ΔH_2 0·04, ΔS_2 14	67Ga
Cd^{2+}	gl	30	0·1 (KCl)	K_1 7·4, $K(Cd^{2+}+HL^- \rightleftharpoons CdHL^+)$ 6·1	67G
Co^{2+}	gl	30	0·1 (KCl)	K_1 5·4, $K(Co^{2+}+HL^- \rightleftharpoons CoHL^+)$ 4·1	67G
	cal	30	0·1 (KNO_3)	ΔH_1 2·6, ΔS_1 33	67Ga
Cu^{2+}	sp	30	0·1 (KNO_3)	K_1 8·1	67G
	cal	30	0·1 (KNO_3)	ΔH_1 −2·6, ΔS_1 28	67Ga
Mg^{2+}	gl	30	0·1 (KCl)	K_1 0·7, $K(Mg^{2+}+HL^- \rightleftharpoons MgHL^+)$ 0	67G
	cal	30	0·1 (KNO_3)	ΔH_1 −1·4(?), ΔS_1 −1(?)	67Ga
Mn^{2+}	gl	30	0·1 (KCl)	K_1 2·0, $K(Mn^{2+}+HL^- \rightleftharpoons MnHL^+)$ 1·5	67G
	cal	30	0·1 (KNO_3)	ΔH_1 7·2, ΔS_1 33	67Ga
Ni^{2+}	gl	30	0·1 (KCl)	K_1 5·9, $K(Ni^{2+}+HL^- \rightleftharpoons NiHL^+)$ 4·6	67G
	cal	30	0·1 (KNO_3)	ΔH_1 −0·8, ΔS_1 24	67Ga
Sr^{2+}	gl	30	0·1 (KCl)	K_1 2·2	67G
	cal	30	0·1 (KNO_3)	ΔH_1 −0·1, ΔS_1 10	67Ga
Zn^{2+}	gl	30	0·1 (KCl)	K_1 5·8, $K(Zn^{2+}+HL^- \rightleftharpoons ZnHL^+)$ 4·4	67G
	cal	30	0·1 (KNO_3)	ΔH_1 3·1, ΔS_1 37	67Ga

67G D. R. Goddard and S. I. Nwankwo, *J. Chem. Soc. (A)*, 1967, 1371
67Ga D. R. Goddard, S. I. Nwankwo, and L. A. K. Staveley, *J. Chem. Soc. (A)*, 1967, 1376

344 $C_5H_{11}O_2NS$ 2-Amino-4-(methylthio)butanoic acid (methionine) HL

Metal	Method	Temp	Medium	Log of equilibrium constant, remarks	Ref
H^+	gl	18	~ 0	K_1 9·31	53P
	gl	20	0·15 ($NaClO_4$)	K_1 9·20, K_2 2·17	63H
	gl	25	0·10 (KNO_3)	K_1 9·04	64L
	pol	25	0·6	K_1 9·15 ± 0·04	67A
Ag^+	gl	25	0·10 (KNO_3)	K_1 3·17	64L
	Ag	25	0·6	K_1 (?) 6·45 ± 0·02	67A
Be^{2+}	gl	18	0·005 $BeSO_4$	β_2 12·0	53P
Cd^{2+}	gl	18	0·005 $CdSO_4$	β_2 7·1	53P
	oth	20	0·1 (KNO_3)	K_1 5·4, K_2 3·3, K_3 2·1, paper electrophoresis	64J
	gl	25	0·10 (KNO_3)	K_1 3·67, K_2 3·36	64L
Co^{2+}	oth	20	0·1 (KNO_3)	K_1 4·5, K_2 3·1, K_3 1·9, paper electrophoresis	64J
	gl	25	0·10 (KNO_3)	K_1 4·12, K_2 3·44	64L
Cr^{3+}	gl	25	~ 0·5	K_1 8·3, K_2 6·2, K_3 5·3	63K
Cu^{2+}	gl	20	0·15 ($NaClO_4$)	K_1 8·00, β_2 15·23	63H
	oth	20	0·1 (KNO_3)	K_1 8·1, K_2 6·7, paper electrophoresis	64J
	gl	25	0·10 (KNO_3)	K_1 7·87, K_2 6·85	64L
Hg^{2+}	gl	25	0·10 (KNO_3)	K_1 6·52, K_2 4·93	64L
	pol	25	0·6 KNO_3	β_2 (?) 17·62 ± 0·05	66T
Mn^{2+}	oth	20	0·1 (KNO_3)	K_1 3·2, K_2 ~ 1·5, paper electrophoresis	64J
	gl	25	0·10 (KNO_3)	K_1 2·77, K_2 1·80	64L
Ni^{2+}	oth	20	0·1 (KNO_3)	K_1 5·7, K_2 3·7, K_3 2·3, paper electrophoresis	64J
	gl	25	0·10 (KNO_3)	K_1 5·19, K_2 4·65	64L
Pb^{2+}	gl	25	0·10 (KNO_3)	K_1 4·38, K_2 4·24	64L
Zn^{2+}	gl	25	0·10 (KNO_3)	K_1 4·37, K_2 3·96	64L
	gl	18	0·005 $ZnSO_4$	β_2 8·3	53P
	oth	20	0·1 (KNO_3)	K_1 4·9, K_2 3·6, K_3 ~ 3·2, paper electrophoresis	64J

Erratum. 1964, p. 468, Table 344. In the heading *for* **L** *read* **HL**.

53P D. J. Perkins, *Biochem. J.*, 1953, **55**, 649
63H C. J. Hawkins and D. D. Perrin, *Inorg. Chem.*, 1963, **2**, 843
63K A. A. Khan and W. U. Malik, *J. Indian Chem. Soc.*, 1963, **40**, 564
64J V. Jokl, *J. Chromatog.*, 1964, **14**, 71
64L G. R. Lenz and A. E. Martell, *Biochemistry*, 1964, **3**, 745
66T V. F. Toropova and Yu. M. Azizov, *Russ. J. Inorg. Chem.*, 1966, **11**, 288 (531)
67A Yu. M. Azizov, A. Kh. Miftakhova, and V. F. Toropova, *Russ. J. Inorg. Chem.*, 1967, **12**, 345 (661)

344.1 $C_5H_{11}O_2NS$ 2-Amino-3-mercapto-3-methylbutanoic acid (penicillamine) H_2L

Metal	Method	Temp	Medium	Log of equilibrium constant, remarks	Ref
H^+	gl	25	0·15 KNO_3	K_1 10·46, K_2 7·97, K_3 2·44	62K
	gl	25	0·10 (KNO_3)	K_1 10·43, K_2 7·88	64L
	gl	20	0·1 $NaClO_4$	K_1 10·68 ± 0·01, K_2 8·03 ± 0·01	68P
Ag^+	gl	25	0·10 (KNO_3)	$K(Ag^+ + HL^- \rightleftharpoons AgHL)$ 12·42	64L
Cd^{2+}	gl	25	0·15 KNO_3	K_1 11·4, K_2 7·1	62K
	gl	25	0·10 (KNO_3)	K_1 10·88	64L
	pol	25	0·2	β_3 13·08, phosphate buffer	66S
Cu^+	pol	25	0·1 (NH_4Cl)	K_1 19·53	68V
Cu^{2+}	gl	25	0·15 KNO_3	K_1 16·5, K_2 5·2	62K
	pol	25	0·17	β_2 15·13, phosphate buffer	61K

conti

$C_5H_{11}O_2NS$ (contd) 344.1

Metal	Method	Temp	Medium	Log of equilibrium constant, remarks	Ref
Hg^{2+}	gl	25	0·15 KNO_3	K_1 17·5, K_2 6·0	62K
	gl	25	0·10 (KNO_3)	K_1 16·15	64L
Ni^{2+}	gl	25	0·15 KNO_3	K_1 11·4, K_2 10·9	62K
	gl	25	0·10 (KNO_3)	K_1 11·11, K_2 10·68	64L
	gl	20	0·1 $NaClO_4$	K_1 10·75±0·05, β_2 22·89±0·01	68P
Pb^{2+}	gl	25	0·15 KNO_3	K_1 13·0, K_2 4·3	62K
	gl	25	0·10 (KNO_3)	K_1 12·37	64L
Zn^{2+}	gl	25	0·15 KNO_3	K_1 10·0, K_2 8·9	62L
	gl	25	0·10 (KNO_3)	K_1 9·51, K_2 9·00	64L
	pol	25	0·2	β_3 16·11, phosphate buffer	66S
	gl	20	0·1 $NaClO_4$	K_1 9·59±0·02, β_2 19·56±0·02 $K(Zn^{2+}+HL^-+L^{2-} \rightleftharpoons ZnHL_2^-)$ 25·55±0·03 $K(Zn^{2+}+2HL^- \rightleftharpoons ZnH_2L_2)$ 31·17±0·06	68P

61K E. C. Knoblock and W. C. Purdy, *J. Electroanalyt. Chem.*, 1961, **2**, 493
62K E. J. Kuchinkos and Y. Rosen, *Arch. Biochem. Biophys*, 1962, **97**, 370
64L G. R. Lenz and A. E. Martell, *Biochemistry*, 1964, **3**, 745
66S I. H. Suffet and W. C. Purdy, *J. Electroanalyt. Chem.*, 1966, **11**, 302
68P D. D. Perrin and I. G. Sayce, *J. Chem. Soc. (A)*, 1968, 53
68V J. J. Vallon and A. Badinand, *Analyt. Chim. Acta*, 1968, **42**, 445

$C_5H_{11}O_2NS$ Cysteine ethyl ester HL 344.2

Metal	Method	Temp	Medium	Log of equilibrium constant, remarks	Ref
H^+	sp	23	?	K_1 8·41, K_2 7·45, phosphate–borate buffer	63S
Mo^{VI}	spJ	25	?	$K(MoO_4^{2-}+3HL \rightleftharpoons MoOL_3^++3OH^-)$ (?) 23, acetate buffer	63S

63S J. T. Spence and H. H. Chang, *Inorg. Chem.*, 1963, **25**, 319

$C_5H_{11}O_3NS$ L-2-Amino-3-(2′-hydroxyethylthio)propanoic acid HL 345.1 [*S*-(2′-hydroxyethyl)-L-cysteine]

Metal	Method	Temp	Medium	Log of equilibrium constant, remarks	Ref
H^+	gl	20	0·1 KNO_3	K_1 8·64	68H
Cu^{2+}	gl	20	0·1 KNO_3	K_1 7·64, K_2 6·51	68H

68H P. Herman and K. Lemke, *Z. physiol. Chem.*, 1968, **349**, 390

345.2 $C_5H_{11}O_7N_2P$ Glycyl-*O*-phosphoryl-D,L-serine (glycyl-D,L-phosphoserine) H_3L

Metal	*Method*	*Temp*	*Medium*	*Log of equilibrium constant, remarks*	*Ref*
H^+	gl	25	0·15 (KCl)	K_1 8·42, K_2 6·04, K_3 2·90	62O
Ca^{2+}	gl	25	0·15 (KCl)	K_1 1·77	62O
				$K(Ca^{2+}+HL^{2-} \rightleftharpoons CaHL)$ 1·48	
Mg^{2+}	gl	25	0·15 (KCl)	K_1 1·86	62O
				$K(Mg^{2+}+HL^{2-} \rightleftharpoons MgHL)$ 1·64	
				$K(Mg^{2+}+MgL^{-} \rightleftharpoons Mg_2L^{+})$ 1·4	

62O R. Österberg, *Acta Chem. Scand.*, 1962, **16**, 2434

345.3 $C_5H_{11}O_7N_2P$ *O*-Phosphorylserylglycine H_3L

Metal	*Method*	*Temp*	*Medium*	*Log of equilibrium constant, remarks*	*Ref*
H^+	gl	25	0·15 (KCl)	K_1 8·01, K_2 5·41, K_3 3·13	62O
	gl	25	0·15 (KNO_3)	K_1 8·02, K_2 5·44, K_3 3·20	66O
Ca^{2+}	gl	25	0·15 (KCl)	K_1 1·85	62O
				$K(Ca^{2+}+HL^{2-} \rightleftharpoons CaHL)$ 1·30	
Cu^{2+}	gl, Hg(Cu)	25	0·15 (KNO_3)	K_1 7·89 ± 0·01, β_2 11·6	66O
				$K(Cu^{2+}+H^{+}+L^{3-} \rightleftharpoons CuHL)$ 11·58 ± 0·05	
				$K(Cu^{2+}+HL^{2-}+L^{3-} \rightleftharpoons CuHL_2{}^{3-})$ 18·0 ± 0·2	
				$K[Cu(H_{-1}L)^{2-}+H^{+} \rightleftharpoons CuL^{-}]$ 5·79 ± 0·01	
Mg^{2+}	gl	25	0·15 (KCl)	K_1 1·94	62O
				$K(Mg^{2+}+HL^{2-} \rightleftharpoons MgHL)$ 1·40	
				$K(Mg^{2+}+MgL^{-} \rightleftharpoons Mg_2L^{+})$ 1·25	
Mn^{2+}	gl	25	0·15 (KCl)	K_1 2·63	62O
				$K(Mn^{2+}+HL^{2-} \rightleftharpoons MnHL)$ 1·89	
				$K(Mn^{2+}+MnL^{-} \rightleftharpoons Mn_2L^{+})$ 1·54	

62O R. Österberg, *Acta Chem. Scand.*, 1962, **16**, 2434
66O R. Österberg, *Arkiv Kemi*, 1966, **25**, 177

345.4 $C_5H_{12}O_2N_2S$ 2-Amino-3-(2′-aminoethylthio)propanoic acid [*S*-(2′-aminoethyl)-cysteine] HL

Metal	*Method*	*Temp*	*Medium*	*Log of equilibrium constant, remarks*	*Ref*
H^+	gl	20	0·1 KNO_3	K_1 9·52, K_2 8·22	68H
Cu^{2+}	gl	20	0·1 KNO_3	K_1 7·11, K_2 6·25	68H

68H P. Herman and K. Lemke, *Z. physiol. Chem.*, 1968, **349**, 390

$C_5H_{12}O_2N_2S_2$ L-2-Amino-3-(2′-aminoethyldithio)propanoic acid HL 345.5

Metal	Method	Temp	Medium	Log of equilibrium constant, remarks	Ref
H^+	gl	20	0·15 ($NaClO_4$)	K_1 9·30, K_2 8·28	63H
Cu^{2+}	gl	20	0·15 ($NaClO_4$)	K_1 7·08, β_2 13·80	63H

63H C. J. Hawkins and D. D. Perrin, *Inorg. Chem.*, 1963, **2,** 843

$C_5H_{12}O_3N_2S$ 2-Amino-3-(2′-aminoethylsulphinyl)propanoic acid HL 345.6

Metal	Method	Temp	Medium	Log of equilibrium constant, remarks	Ref
H^+	gl	20	0·1 KNO_3	K_1 8·37, K_2 7·18	68H
Cu^{2+}	gl	20	0·1 KNO_3	K_1 6·56, K_2 5·44	68H

68H P. Herman and K. Lemke, *Z. physiol. Chem.*, 1968, **349,** 390

$C_5H_{12}O_4N_2S$ 2-Amino-3-(2′-aminoethylsulphonyl)propanoic acid HL 345.7

Metal	Method	Temp	Medium	Log of equilibrium constant, remarks	Ref
H^+	gl	20	0·1 KNO_3	K_1 8·02, K_2 6·86	68H
Cu^{2+}	gl	20	0·1 KNO_3	K_1 6·30, K_2 5·32	68H

68H P. Herman and K. Lemke, *Z. physiol. Chem.*, 1968, **349,** 390

$C_5H_{13}OSP$ Diethylphosphinothioic acid *O*-methyl ester (methyl diethylthiophosphinate) L 345.8

Metal	Method	Temp	Medium	Log of equilibrium constant, remarks	Ref
Hg^{2+}	Hg	25	0·8 KNO_3 40% ethanol	β_2 13·7, β_3 16·0, β_4 17·4	67T

67T V. F. Toropova, M. K. Saikina, and R. Sh. Aleshou, *J. Gen. Chem.* (*U.S.S.R.*), 1967, 37, 679 (725)

$C_5H_{14}O_3NP$ *NN*-Diethylaminomethylphosphonic acid H_2L 345.9

Metal	Method	Temp	Medium	Log of equilibrium constant, remarks	Ref
H^+	gl	25	1 KNO_3	K_1 12·3±0·2, K_2 5·8±0·1	67C
Ca^{2+}	gl	25	1 KNO_3	K_1 1·3±0·2, $K(Ca^{2+}+HL^- \rightleftharpoons CaHL^+) \ll 1$	67C
Mg^{2+}	gl	25	1 KNO_3	$K_1 \sim 2$, $K(Mg^{2+}+HL^- \rightleftharpoons MgHL^+)$ 1·3±0·2	67C

67C R. P. Carter, R. L. Carroll, and R. R. Irani, *Inorg. Chem.*, 1962, **6,** 939

346 C_6H_6 Benzene L

Metal	Method	Temp	Medium	Log of equilibrium constant, remarks	Ref
Ag^+	sp	1·4, 20	benzene, ClO_4^-	K_1 (1·4°) 1·04, (20°) 1·66 ΔH 4·0	64T
Hg^{2+}	sp	19—25	1·0—1·2 ($NaClO_4$)	$K[Hg^{2+}+L \rightleftharpoons Hg(H_{-1}L)^+ + H^+]$ 2·5±0·4	64W

64T B. G. Torre-Mori, D. Janjic, and B. P. Susz, *Helv. Chim. Acta*, 1964, **47**, 1172
64W T. H. Wirth and N. Davidson, *J. Amer. Chem. Soc.*, 1964, **86**, 4322

347.1 C_6H_{10} 1-Methylcyclopentene L

Metal	Method	Temp	Medium	Log of equilibrium constant, remarks	Ref
Ag^+	oth	30	1·77 $AgNO_3$ ethylene glycol	K_1 0·46, gas chromatography	62G

62G E. Gil-Av and J. Herling, *J. Phys. Chem.*, 1962, **66**, 1208

347.2 C_6H_{10} 3-Methylcyclopentene L

Metal	Method	Temp	Medium	Log of equilibrium constant, remarks	Ref
Ag^+	oth	30	1·77 $AgNO_3$ ethylene glycol	K_1 1·08, gas chromatography	62G

62G E. Gil-Av and J. Herling, *J. Phys. Chem.*, 1962, **66**, 1208

347.3 C_6H_{10} 4-Methylcyclopentene L

Metal	Method	Temp	Medium	Log of equilibrium constant, remarks	Ref
Ag^+	oth	30	1·77 $AgNO_3$ ethylene glycol	K_1 0·74, gas chromatography	62G

62G E. Gil-Av and J. Herling, *J. Phys. Chem.*, 1962, **66**, 1208

347.4 C_6H_{10} Methylenecyclopentane L

Metal	Method	Temp	Medium	Log of equilibrium constant, remarks	Ref
Ag^+	oth	30	1·77 $AgNO_3$ ethylene glycol	K_1 0·78, gas chromatography	62G

62G E. Gil-Av and J. Herling, *J. Phys. Chem.*, 1962, **66**, 1208

$C_6H_4O_4$ 2,5-Dihydroxy-1,4-benzoquinone H_2L 347.5

Metal	Method	Temp	Medium	Log of equilibrium constant, remarks	Ref
H^+	red	25	?	K_1 5·18, K_2 2·71	66B
	sp	25	0·5 NaCl	K_1 5·22±0·03, K_2 2·81±0·05	67B
Ge^{IV}	sp	25	0·5 (NaCl)	$K[Ge(OH)_4 + 2HL^- \rightleftharpoons Ge(OH)_2L_2^{2-}]$ 9·1	66B
	sp	25	0·5 NaCl	$K[Ge(OH)_4 + 2HL^- \rightleftharpoons Ge(OH)_2L_2^{2-}]$ 8·09	67B
				$K[Ge(OH)_2L_2^{2-} + H^+ \rightleftharpoons Ge(OH)_2L_2H^-]$ 1·8	

66B A. Beauchamp and R. L. Benoit, *Canad. J. Chem.*, 1966, **44**, 1615
67B A. Beauchamp and R. Benoit, *Bull. Soc. chim. France*, 1967, 672

$C_6H_4O_5$ 3-Hydroxypyran-4-one-6-carboxylic acid (comenic acid) H_2L 347.6

Metal	Method	Temp	Medium	Log of equilibrium constant, remarks	Ref
H^+	gl, sp	25	0·5	K_1 7·29±0·03	67C
Ge^{IV}	sp	25	0·5 (NaCl)	$K[Ge(OH)_4 + 2HL^- \rightleftharpoons Ge(OH)_2L_2^{2-}]$ 2·25±0·05	67C
Zn^{2+}	gl	25	0·5 $(NaClO_4)$	K_1 4·86±0·06, β_2 8·76±0·10	67C

67C G. Choux and R. L. Benoit, *Bull. Soc. chim. France*, 1967, 2920

$C_6H_4N_2$ 3-Cyanopyridine (*see Corrigenda*) L 348

Metal	Method	Temp	Medium	Log of equilibrium constant, remarks	Ref
	sp	24	1 $(NaNO_3)$	K_1 1·45	64M

64M P. Moore and R. G. Wilkins, *J. Chem. Soc.*, 1964, 3455

$C_6H_4N_2$ 4-Cyanopyridine (*see Corrigenda*) L 349

Metal	Method	Temp	Medium	Log of equilibrium constant, remarks	Ref
	est	24	1 $(NaNO_3)$	K_1 1·7	64M

64M P. Moore and R. G. Wilkins, *J. Chem. Soc.*, 1964, 3455

$C_6H_4I_2$ 1,3-Di-iodobenzene L 353

Erratum. 1964, p. 471, Table 353. *For* formula $C_6H_7I_2$ *read* $C_6H_4I_2$.

359 C_6H_6O Hydroxybenzene (phenol) HL

Metal	*Method*	*Temp*	*Medium*	*Log of equilibrium constant, remarks*	*Ref*
H^+	gl	20	0·1 (KNO_3)	K_1 9·62	65B, 66
	sp	25	0·10 ($NaClO_4$)	K_1 9·78	66P
	gl	25	0·5 KNO_3	K_1 9·79	68B
			0·027 $NaClO_4$	K_1 9·85	
Fe^{3+}	sp	~ 23	0·5 KNO_3	K_1 7·81,	68B
			0·027 $NaClO_4$	K_1 8·11	
La^{3+}	sp	25	0·10 ($NaClO_4$)	K_1 1·51 ± 0·02	66P
UO_2^{2+}	gl	20	0·1 (KNO_3)	K_1 5·8	65B, 66
Y^{3+}	sp	25	0·10 ($NaClO_4$)	K_1 2·40 ± 0·02	66P

65B M. Bartusek and L. Sommer, *J. Inorg. Nuclear Chem.*, 1965, **27**, 2397

66B M. Bartusek and J. Ruzickova, *Coll. Czech. Chem. Comm.*, 1966, **31**, 207

66P C. Postmus jun., L. B. Magnusson, and C. A. Craig, *Inorg. Chem.*, 1966, **5**, 1154

68B W. A. E. McBryde, *Canad. J. Chem.*, 1968, **46**, 2385

360 $C_6H_6O_2$ 1,2-Dihydroxybenzene (catechol, pyrocatechol) H_2L

Metal	*Method*	*Temp*	*Medium*	*Log of equilibrium constant, remarks*	*Ref*
H^+	gl	25	→ 0	K_1 12·8, K_2 9·45	62A
	gl	20	0·10—0·15	K_1 ~ 13, K_2 9·43	62Sa
	gl	30	0·1 (KNO_3)	K_1 11·59, K_2 9·13	63M
	gl	18—22	0·1	K_1 13·4, K_2 9·17	63S
	gl	25	→ 0	K_1 12·92 (?), K_2 9·45	64D, 64
	Ag	20	1 $N(CH_3)_4Cl$ methanol	K_1 15·5, K_2 13·15	64G
	gl	20	0·1 (KNO_3)	K_1 13·7, K_2 9·37	65S, 65B,
	gl	25	0·1 (KCl)	K_1 12·8 ± 0·2, K_2 9·37 ± 0·01	65J
	gl	30	0·1 ($NaClO_4$)	K_1 ~ 12·6, K_2 9·34	66A
	gl	25	0·10 (KNO_3)	K_1 11·93 ± 0·01, K_2 9·20 ± 0·01	66L, 67
	gl	25	0·1 $NaClO_4$	K_1 12·62, K_2 9·33	67O, 68
	sp	25	0·10 KCl	K_1 13·1, K_2 9·39	67P
	gl	25	1·0 KNO_3	K_1 13·05, K_2 9·229	68T
			0·1 KNO_3 50% methanol	K_1 14·1, K_2 10·012	
Al^{3+}	gl	?	0·2 KNO_3	K_1 16·56, K_2 15·64, K_3 13·65	64D
As^{III}	gl	25	0·1 KCl	$K[As(OH)_4^- + H_2L \rightleftharpoons As(OH)_2L^-]$ 2·24 $K[As(OH)_4^- + 2H_2L \rightleftharpoons AsL_2^-]$ 2·71	59Aa
B^{III}	gl	0—45	0·1 KCl	$K[B(OH)_4^- + H_2L \rightleftharpoons B(OH)_2L^-]$ 4·36 − (1·45 × 10^{-2})t t = Centigrade temperature $K[B(OH)_4^- + 2H_2L \rightleftharpoons BL_2^-]$ 4·61 − (1·40 × 10^{-2})t	59Ab
	gl	0—35	0·025 borax	$K[B(OH)_4^- + H_2L \rightleftharpoons B(OH)_2L^-]$ (0°) 3·86, (13°) 3·76, (25°) 3·70, (35°) 3·62 ΔH° −2·7 ± 0·3, ΔS° 8 ± 2	67C
	gl	0—45	0·1 KCl	$K[B(OH)_4^- + H_2L \rightleftharpoons B(OH)_2L^-]$ (0°) 4·361, (25°) 3·972, (30°) 3·945, (35°) 3·843, (40°) 3·773, (45°) 3·748 ΔH° −5·6, ΔS° −0·4 $K[B(OH)_4^- + 2H_2L \rightleftharpoons BL_2^-]$ (0°) 4·637, (25°) 4·263, (30°) 4·194, (35°) 4·077, (40°) 4·040, (45°) 3·996 ΔH° −5·8, ΔS° 0·0	68A
	gl	20	0·1 (KNO_3)	$K[H_3BO_3 + H_2L \rightleftharpoons B(OH)_2L^- + H^+]$ −5·17	68H

conti

Metal	Method	Temp	Medium	Log of equilibrium constant, remarks	Ref
Be^{2+}	gl	?	0·2 KNO_3	K_1 13·70, K_2 12·02	64Da
	gl	20	0·1 (KNO_3)	K_1 13·52, K_2 9·83,	67B
				$K(Be^{2+}+HL^- \rightleftharpoons BeHL^+)$ 5·0,	
				$K(BeL+HL^- \rightleftharpoons BeHL_2^-)$ 2·8	
Cd^{2+}	gl	30	0·1 ($NaClO_4$)	K_1 7·70	66A
Co^{2+}	gl	25	0·1 (KCl)	K_1 8·40±0·03, K_2 5·8±0·1	66J
	gl	25	1·0 (KNO_3)	$K(Co^{2+}+H_2L \rightleftharpoons CoL+2H^+)$ −13·959	68T
				$K(CoL+H_2L \rightleftharpoons CoL_2^{2-}+2H^+)$ −15·856	
Cu^{2+}	gl	30	0·1 (KNO_3)	K_1 12·52, K_2 9·66	63M
	gl	25	0·1 (KCl)	K_1 13·76±0·02, K_2 10·75±0·05	65J
	gl	30	0·1 ($NaClO_4$)	K_1 13·58, K_2 10·49	66A
	gl	?	0·2 KNO_3	K_1 14·27, K_2 13·36	66D
	gl	25	0·10 (KNO_3)	K_1 12·74±0·03	66L
				$K(CuA^{2+}+L^{2-} \rightleftharpoons CuAL)$ 13·10±0·02	
				where A = 2,2′-bipyridyl	
	sp	25	0·1 $NaClO_4$	K_1 13·88, K_2 10·44	67O
	gl	25	1·0 (KNO_3)	$K(Cu^{2+}+H_2L \rightleftharpoons CuL+2H^+)$ −8·679	68T
				$K(CuL+H_2L \rightleftharpoons CuL_2^{2-}+2H^+)$ −10·955	
			0·1 (KNO_3)	$K(Cu^{2+}+H_2L \rightleftharpoons CuL+2H^+)$ −7·85±0·01	
			50% methanol	$K(CuL+H_2L \rightleftharpoons CuL_2^{2-}+2H^+)$ −9·02±0·01	
Fe^{2+}	gl	25	1·0 (KNO_3)	$K(Fe^{2+}+H_2L \rightleftharpoons FeHL^++H^+)$ −5·71	68T
				$K(Fe^{2+}+H_2L \rightleftharpoons FeL+2H^+)$ −14·332	
				$K(FeL+H_2L \rightleftharpoons FeL_2^{2-}+2H^+)$ −16·740	
Ge^{IV}	gl	25	0·1 KCl	$K(H_2GeO_3+3H_2L \rightleftharpoons GeL_3^{2-}+2H^+)$ −0·77±0·02	59A
	gl	25	→ 0	$K(HGeO_3^-+3H_2L \rightleftharpoons HGeL_3^-)$ 8·67±0·03	63A
Hf^{4+}	ix	?	1·0 $HClO_4$	K_1 22·58	67E
Mg^{2+}	gl	30	0·1 ($NaClO_4$)	K_1 5·24	66A
Mn^{2+}	gl	25	0·1 (KCl)	K_1 7·52±0·03, K_2 5·7±0·1	66J
	gl	25	1·0 (KNO_3)	$K(Mn^{2+}+H_2L \rightleftharpoons MnHL^++H^+)$ −6·41	68T
				$K(Mn^{2+}+H_2L \rightleftharpoons MnL+2H^+)$ −14·807	
				$K(MnL+H_2L \rightleftharpoons MnL_2^{2-}+2H^+)$ −16·996	
			0·1 (KNO_3)	$K[Mn^{2+}+H_2L \rightleftharpoons Mn(OH)(HL)+2H^+]$ −14·66±0·02	
			50% methanol	$K[Mn^{2+}+2H_2L \rightleftharpoons Mn(HL)_2+2H^+]$ −11·46±0·01	
Nb^{5+}				See methanol (103.1) for mixed ligands	
Ni^{2+}	gl	30	0·1 ($NaClO_4$)	K_1 8·36, K_2 5·15	66A
	gl	25	0·1 (KCl)	K_1 8·74±0·03, K_2 5·1±0·1	66J
	sp	25	0·1	K_1 9·34	68O
	gl	25	1·0 (KNO_3)	$K(Ni^{2+}+H_2L \rightleftharpoons NiL+2H^+)$ −13·511	68T
Ta^{5+}				See methanol (103.1) for mixed ligands	
Th^{4+}	gl	~ 30 (?)	0·1 KNO_3	K_1 17·72	62A
				See EDTA (836) and CDTA (997) for mixed ligands	
Ti^{4+}	sp	22	0·5	$K(TiO^{2+}+L^{2-} \rightleftharpoons TiOL)$ 22·5±0·1, acetate buffer	61S
				$K(TiOL+L^{2-} \rightleftharpoons TiOL_2^{2-})$ 15·9±0·3 acetate buffer	
			0·05	$K(TiO^{2+}+L^{2-} \rightleftharpoons TiOL)$ 18·8, acetate buffer	
				$K(TiOL+L^{2-} \rightleftharpoons TiOL_2^{2-})$ 17·7 acetate buffer	
	sp	18—22	0·1 ($NaClO_4$)	$K[TiO^{2+}+2H_2L \rightleftharpoons TiO(HL)_2+2H^+]$ −1·9	63S
				$K[TiO(HL)_2+H_2L \rightleftharpoons TiL_3^{2-}+2H^+]$ −4·7	
				$K(TiO^{2+}+2H^++3L^{2-} \rightleftharpoons TiL_3^{2-})$ 61·6	
	sp	18—22	0—100% methanol	$K(TiO^{2+}+HL^- \rightleftharpoons TiOHL^+)$ (0%) 4·2 (25%) 4·7, (50%) 5·1, (75%) 5·5, (100%) 6·1	66S
			25—100% ethanol	(25%) 4·56, (50%) 4·98, (75%) 5·17, (100%) 6·15	

continued overleaf

360 $C_6H_6O_2$ (contd)

Metal	*Method*	*Temp*	*Medium*	*Log of equilibrium constant, remarks*	*Ref*
Ti^{4+} (*contd*)			25—100% propanol	(25%) 4·36, (50%) 4·64, (75%) 5·07, (100%) 6·39	
			25—100% 2-propanol	(25%) 4·28, (50%) 4·73, (75%) 4·9, (100%) 6·75	
Tl^{+}	gl	?	0·1 KCl	K_1 7·05	66T
U^{4+}				See EDTA (836) for mixed ligands	
UO_2^{2+}	sp	?	~ 0	$K(UO_2L+H_2L \rightleftharpoons UO_2L_2^{2-}+2H^+)$ −10·5 $K(UO_2L+H^+ \rightleftharpoons UO_2HL^+)$ 3·76	62Sa
	gl	20	0·1 (KNO_3)	K_1 15·9 $K(UO_2^{2+}+HL^- \rightleftharpoons UO_2HL^+)$ 6·2 $K(UO_2L+HL^- \rightleftharpoons UO_2HL_2^-)$ 4·9 $K(UO_2HL_2^-+HL^- \rightleftharpoons UO_2H_2L_3^{2-})$ 3·7	65B, 66J
	sp	18—22	0·1 ($NaClO_4$)	K_1 15·9 $K(UO_2^{2+}+HL^- \rightleftharpoons UO_2HL^+)$ 6·3 $K(UO_2L+HL^- \rightleftharpoons UO_2HL_2^-)$ 4·9	65B, 65S
VO^{2+}	gl	30	0·1	K_1 15·28, K_2 13·02 $K[VO(OH)L^-+H^+ \rightleftharpoons VOL]$ 5·90 $K(VOA^{2+}+L^{2-} \rightleftharpoons VOAL$ 16·69) where A = 1,10-phenanthroline	67L
Zn^{2+}	gl	30	0·1 (KNO_3)	K_1 8·46, K_2 6·78	63M
	gl	30	0·1 ($NaClO_4$)	K_1 9·08, K_2 7·24	66A
	gl	25	0·1 (KCl)	K_1 9·50 ± 0·03, K_2 7·7 ± 0·1	66J
	gl	25	1·0 (KNO_3)	$K(Zn^{2+}+H_2L \rightleftharpoons ZnL+2H^+)$ −12·744 $K(ZnL+H_2L \rightleftharpoons ZnL_2^{2-}+2H^+)$ −14·315	68T
Zr^{4+}	ix	?	1·0 $HClO_4$	K_1 22·63	67E

Errata. 1964, p. 472, Table 360. References, *for* HJ *read* H; *for* 47R *read* 57R.

59A P. J. Antikainen and P. J. Malkonen, *Suomen Kem.*, 1959, **B32**, 179
59Aa P. J. Antikainen and V. M. K. Rossi, *Suomen Kem.*, 1959, **B32**, 185
59Ab P. J. Antikainen and A. Kauppila, *Suomen Kem.*, 1959, **B32**, 141
61S S. Ya. Shnaiderman and I. E. Kalinichenko, *Russ. J. Inorg. Chem.*, 1961, **6**, 941 (1843)
62A R. P. Agarwal and R. C. Mehrotra, *J. Inorg. Nuclear Chem.*, 1962, **24**, 821
63A A. M. Andrianov and V. A. Nazarenko, *Russ. J. Inorg. Chem.*, 1963, **8**, 1194 (2281)
63M Y. Murakami, K. Nakamura, and M. Tokunda, *Bull. Chem. Soc. Japan*, 1963, **36**, 669
63S L. Sommer, *Coll. Czech. Chem. Comm.*, 1963, **28**, 2102
63Sa S. Ya. Shnaiderman and E. V. Galinker, *Russ. J. Inorg. Chem.*, 1963, **7**, 142 (279)
64D S. N. Dubey and R. C. Mehrotra, *J. Inorg. Nuclear Chem.*, 1964, **26**, 1543
64Da S. N. Dubey and R. C. Mehrotra, *J. Less-Common Metals*, 1964, **7**, 169
64G R. Gut, *Helv. Chim. Acta*, 1964, **47**, 2262
65B M. Bartusek and L. Sommer, *J. Inorg. Nuclear Chem.*, 1965, **27**, 2397
65J R. F. Jameson and W. F. S. Neillie, *J. Inorg. Nuclear Chem.*, 1965, **27**, 2623
65S L. Sommer, T. Sepel, and L. Kurilova, *Coll. Czech. Chem. Comm.*, 1965, **30**, 3834
66A V. T. Athavale, L. H. Prebhu, and D. G. Vartak, *J. Inorg. Nuclear Chem.*, 1966, **28**, 1237
66B M. Bartusek and J. Ruzickova, *Coll. Czech. Chem. Comm.*, 1966, **31**, 207
66D S. N. Dubey and R. C. Mehrotra, *J. Indian Chem. Soc.*, 1966, **43**, 73
66J R. F. Jameson and W. F. S. Neillie, *J. Inorg. Nuclear Chem.*, 1966, **28**, 2667
66L G. A. L'Heuveux and A. E. Martell, *J. Inorg. Nuclear Chem.*, 1966, **28**, 481
66S S. Ya. Shnaiderman and N. V. Chernaya, *Russ. J. Inorg. Chem.*, 1966, **11**, 72 (134)
66T O. S. Tomar and P. K. Bhattacharya, *J. Indian Chem. Soc.*, 1966, **43**, 250
67B M. Bartusek and J. Zelinka, *Coll. Czech. Chem. Comm.*, 1967, **32**, 992
67C J. M. Conner and V. C. Bulgrin, *J. Inorg. Nuclear Chem.*, 1967, **29**, 1953
67E A. N. Ermakov, N. B. Kalinichenko, and I. N. Marov, *Russ. J. Inorg. Chem.*, 1967, **12**, 812 (1545)
67L K. Lal and R. P. Agarwal, *Bull. Chem. Soc. Japan*, 1967, **40**, 1148
67O Y. Oka and H. Harada, *Nippon Kagaku Zasshi*, 1967, **88**, 441
67P P. Pichet and R. L. Benoit, *Inorg. Chem.*, 1967, **6**, 1505
68A P. J. Antikainen and I. P. Pitkanen, *Suomen Kem.*, 1968, **41**, 65
68O Y. Oka, M. Okamura, and H. Harada, *Nippon Kagaku Zasshi*, 1968, **89**, 171
68T C. A. Tyson and A. E. Martell, *J. Amer. Chem. Soc.* 1968, **90**, 3379
68H L. Havelkova and M. Bartusek, *Coll. Czech. Chem. Comm.*, 1968, **33**, 4188

$C_6H_6O_2$ 1,3-Dihydroxybenzene (resorcinol) H_2L 360.1

Metal	Method	Temp	Medium	Log of equilibrium constant, remarks	Ref
H^+	gl	20	0·1 (KNO_3)	K_1 11·06, K_2 9·30	65B
UO_2^{2+}	gl	20	0·1 (KNO_3)	$K(UO_2^{2+}+HL^- \rightleftharpoons UO_2HL^+)$ 5·9	65B
	gl	20	0·1 (KNO_3)	K_1 16·9	66B

65B M. Bartusek and L. Sommer, *J. Inorg. Nuclear Chem.*, 1965, **27**, 2397

66B M. Bartusek and J. Ruzickova, *Coll. Czech. Chem. Comm.*, 1966, **31**, 207

$C_6H_6O_2$ 1,4-Dihydroxybenzene (hydroquinone) H_2L 360.2

Metal	Method	Temp	Medium	Log of equilibrium constant, remarks	Ref
H^+				H^+ constants not given.	
UO_2^{2+}	gl	25	?	K_1 10·32	67R

67R R. R. Rao and P. K. Bhattacharya, *Current Sci.* 1967, **36**, 71

$C_6H_6O_3$ 1,2,3-Trihydroxybenzene (pyrogallol) H_3L 361

Metal	Method	Temp	Medium	Log of equilibrium constant, remarks	Ref
H^+	gl	20	0·1 (KNO_3)	$K_1 \sim 14$, K_2 11·19, K_3 9·05	67B
	gl	?	0·2 KNO_3	K_2 11·09, K_3 8·81	67D
Al^{3+}	gl	?	0·2 KNO_3	$K(Al^{3+}+HL^{2-} \rightleftharpoons AlHL^+)$ 14·3, $K[AlHL^+ + HL^{2-} \rightleftharpoons Al(HL)_2^-]$ 13·5, $K[Al(HL)_2^- + HL^{2-} \rightleftharpoons Al(HL)_3^{3-}]$ 11·9	67D
As^{III}	gl	25	0·1 KCl	$K[As(OH)_4^- + H_3L \rightleftharpoons As(OH)_2HL^-]$ 2·81, $K[As(OH)_4^- + 2H_3L \rightleftharpoons As(HL)_2^-]$ 3·09	59Aa
B^{III}	gl	20	0·1 (KNO_3)	$K[H_3BO_3 + H_3L \rightleftharpoons B(OH)_2HL^- + H^+]$ −4·98	68H
Be^{2+}	gl	20	0·1 (KNO_3)	$K(Be^{2+}+HL^{2-} \rightleftharpoons BeHL) \sim 13·5$, $K(Be^{2+}+H_2L^- \rightleftharpoons BeH_2L^+)$ 4·6	67B
	gl	?	0·2 KNO_3	$K(Be^{2+}+HL^{2-} \rightleftharpoons BeHL)$ 11·4, $K[BeHL+HL^- \rightleftharpoons Be(HL)_2^-]$ 10·0	67D
Cu^{2+}	gl	?	0·2 KNO_3	$K(Cu^{2+}+HL^{2-} \rightleftharpoons CuHL)$ 12·4, $K[CuHL+HL^{2-} \rightleftharpoons Cu(HL)_2^{2-}]$ 11·8	66D
Ge^{IV}	gl	25	0·1 KCl	$K[H_2GeO_3 + 3H_3L \rightleftharpoons Ge(HL)_3^{2-} + 2H^+]$ −0·22±0·02	59A
	gl	25	→0	$K[HGeO_3^- + 3H_3L \rightleftharpoons HGe(HL)_3^-]$ 9·05±0·05	63A
Hf^{4+}	sp	18—20	0—0·1 KCl	$K[Hf(OH)_3^+ + H_3L \rightleftharpoons Hf(OH)_2HL + H^+]$ (→ 0) 4·44, (0·1) 4·33	66P
Th^{4+}	gl	30—34	0·1 KNO_3	$K[Th^{4+} + H_3L \rightleftharpoons ThL^+ + 3H^+]$ −6·32, $K[Th^{4+} + 2H_3L \rightleftharpoons ThH_2L_2 + 4H^+]$ −7·30	65A
Ti^{4+}	sp	18—22	0—100% methanol	$K[TiO^{2+} + H_2L^- \rightleftharpoons TiOH_2L^+]$ (0%) 4·5, (25%) 5·1, (50%) 5·4, (75%) 5·8, (100%) 6·2	66S
			25—100% ethanol	(25%) 4·8, (50%) 5·7, (75%) 6·0, (100%) 6·4	
			25—100% propanol	(25%) 4·6, (50%) 4·8, (75%) 5·6, (100%) 6·6	

continued overleaf

361 $C_6H_6O_3$ (contd)

Metal	Method	Temp	Medium	Log of equilibrium constant, remarks	Ref
Ti^{4+} (contd)			25—100% 2-propanol	(25%) 4·7, (50%) 4·7, (75%) 5·6, (100%) 6·8	
UO_2^{2+}	gl	25	0·1 (KNO_3)	$K[2UO_2^{2+}+H_3L \rightleftharpoons (UO_2)_2L^+ + 3H^+]$ −6·84 $K(UO_2L^- + H_3L \rightleftharpoons UO_2H_2L_2^{2-} + H^+)$ −4·69	65B
Zr^{4+}	sp	18—20	0—0·1 KCl	$K[Zr(OH)_3^+ + H_3L \rightleftharpoons Zr(OH)_2HL + H^+]$, (→ 0) 4·17, (0·1) 4·06	66P

59A P. J. Antikainen and P. J. Malkonen, *Suomen Kem.*, 1959, **B32**, 179
59Aa P. J. Antikainen and V. M. K. Rossi, *Suomen Kem.*, 1959, **B32**, 185
63A A. M. Andrianov and V. A. Nazarenko, *Russ. J. Inorg. Chem.*, 1963, **8**, 1194 (2281)
65A R. P. Agarwal and R. C. Mehrotra, *J. Indian Chem. Soc.*, 1965, **42**, 61
65B M. Bartusek, *Coll. Czech. Chem. Comm.*, 1965, **30**, 2746
66D S. N. Dubey and R. C. Mehrotra, *J. Indian Chem. Soc.*, 1966, **43**, 73
66P A. D. Pakhomova, Ya. L. Rudyakova, and I. A. Sheka, *Russ. J. Inorg. Chem.*, 1966, **11**, 621 (1161)
66S S. Ya. Shnaiderman and N. V. Chernaya, *Russ. J. Inorg. Chem.*, 1966, **11**, 72 (134)
67B M. Bartusek and J. Zelinka, *Coll. Czech. Chem. Comm.*, 1967, **32**, 992
67D S. N. Dubey and R. C. Mehrotra, *Indian J. Chem.*, 1967, **5**, 327
68H L. Havelkova and M. Bartusek, *Coll. Czech. Chem. Comm.*, 1968, **33**, 4188

362 $C_6H_6O_3$ 3-Hydroxy-2-methyl-4-pyrone (maltol) HL

Metal	Method	Temp	Medium	Log of equilibrium constant, remarks	Ref
H^+	gl	25	0·5 (NaCl)	K_1 8·36±0·03	66B, 67
	sp	25	0·5 (NaCl)	K_1 8·37±0·03, K_2 −0·71±0·10	66B
	sp	25	0·1 $NaClO_4$	K_1 8·61±0·01	68C, 68
	gl	20	0·1 KNO_3	K_1 8·62	68C
Fe^{3+}	sp	18—22	0·1 ($NaClO_4$)	K_1 11·5, K_2 9·9, K_3 8·3	68S
Ge^{IV}	sp	25	0·5 (NaCl)	$K[Ge(OH)_4 + 2HL \rightleftharpoons Ge(OH)_2L_2]$ 3·90±0·04 $K[Ge(OH)_4 + 3HL + H^+ \rightleftharpoons GeL_3^+]$ 8·05±0·07	66B
UO_2^{2+}	sp	25	0·1 $NaClO_4$	K_1 8·3, K_2 6·7, K_3 3·26	68C
Zn^{2+}	gl	25	0·5 ($NaClO_4$)	K_1 5·53±0·06, β_2 10·20±0·10	67C

66B A. Beauchamp and R. L. Benoit, *Canad. J. Chem.*, 1966, **44**, 1607, 1615
67C G. Choux and R. L. Benoit, *Bull. Soc. chim. France*, 1967, 2920
68C E. Chiacchierini, J. Havel, and L. Sommer, *Coll. Czech. Chem. Comm.*, 1968, **33**, 4215
68S A. Stefanovic, J. Havel, and L. Sommer, *Coll. Czech. Chem. Comm.*, 1968, **33**, 4198

362.1 $C_6H_6O_3$ 3-Hydroxy-6-methyl-4-pyrone (allomaltol) HL

Metal	Method	Temp	Medium	Log of equilibrium constant, remarks	Ref
H^+	gl, sp	25	0·5	K_1 7·98±0·03	67C
Ge^{IV}	sp	25	0·5 (NaCl)	$K[Ge(OH)_4 + 2HL \rightleftharpoons Ge(OH)_2L_2]$ 3·43±0·05	67C
Zn^{2+}	gl	25	0·5 ($NaClO_4$)	K_1 5·28±0·06, β_2 9·57±0·10	67C

67C G. Choux and R. L. Benoit, *Bull. Soc. chim. France*, 1967, 2920

$C_6H_6O_4$ 3-Hydroxy-5-(hydroxymethyl)-4-pyrone (kojic acid) HL 363

Metal	*Method*	*Temp*	*Medium*	*Log of equilibrium constant, remarks*	*Ref*
H^+	gl	25	0·5	K_1 7·73 ± 0·005	61M
			→ 0	K_1 7·88	
	gl	20	0·1	K_1 7·75	61S, 66S
	gl	25	0·1 (KNO_3)	K_1 7·68	62M
	gl, sp	25	0·5	K_1 7·66 ± 0·03	67C
Dy^{3+}	gl	25	2·0 ($NaClO_4$)	K_1 5·74 ± 0·05, K_2 5·39 ± 0·10	64Y
Er^{3+}	gl	25	2·0 ($NaClO_4$)	K_1 5·72 ± 0·05, K_2 5·60 ± 0·10	64Y
Eu^{3+}	gl	25	2·0 ($NaClO_4$)	K_1 5·35 ± 0·05, K_2 5·10 ± 0·10	64Y
Fe^{3+}	sp	25	0·5	K_1 10·16, K_2 8·29, K_3 6·90	61M
	sp	20	0—0·2	K_1 (0·1) 9·2, β_2 (→ 0) 17·5, (0·2) 16·8, β_3 (→ 0) 24·6	61S
	sp	15	0·1 HNO_3	K_1 10·2 ± 0·1	62Ma
	gl	25	0·1 KNO_3	K_2 8·78, K_3 7·53,	62Ma
				$K(FeOHL^+ + H^+ \rightleftharpoons FeL^{2+})$ 3·16,	
				$K[(FeOHL)_2^{2+} + 2H^+ \rightleftharpoons 2FeL^{2+}]$ 3·92,	
				$K[Fe(OH)_2L + 2H^+ \rightleftharpoons FeL^{2+}]$ 7·40,	
				$K\{[Fe(OH)_2L]_2 + 4H^+ \rightleftharpoons 2FeL^{2+}\}$ 9·92	
Gd^{3+}	gl	25	2·0 ($NaClO_4$)	K_1 5·49 ± 0·05, K_2 5·17 ± 0·10	64Y
Ge^{IV}	sp	25	0·5 (NaCl)	$K[Ge(OH)_4 + 2HL \rightleftharpoons Ge(OH)_2L_2]$ 2·81 ± 0·05	67C
La^{3+}	gl	25	2·0 ($NaClO_4$)	K_1 5·14 ± 0·05, K_2 4·13 ± 0·10	64Y
Lu^{3+}	gl	25	2·0 ($NaClO_4$)	K_1 6·24 ± 0·05, K_2 5·60 ± 0·10	64Y
Mg^{2+}	gl	25	0·1 (KNO_3)	K_1 2·92, K_2 2·19	62M
Mn^{2+}	gl	25	0·1 (KNO_3)	K_1 3·95, K_2 2·83	62M
Pr^{3+}	gl	25	2·0 ($NaClO_4$)	K_1 5·18 ± 0·05, K_2 4·58 ± 0·10	64Y
Sm^{3+}	gl	25	2·0 ($NaClO_4$)	K_1 5·38 ± 0·05, K_2 5·20 ± 0·10	64Y
Tm^{3+}	gl	25	2·0 ($NaClO_4$)	K_1 6·00 ± 0·05, K_2 5·55 ± 0·10	64Y
U^{4+}	pol	18—22	1 $NaNO_3$	K_1 4·9 ± 0·5(?)	67H
UO_2^{2+}	gl	20	0·1 (KNO_3)	K_1 7·2, K_2 5·45, K_3 3·4	65B
	sp	~ 20	0·1 ($NaClO_4$)	K_1 7·05, K_2 5·4, K_3 3·5	65B
	gl	20	0·1 (KNO_3)	$K(UO_2^{2+} + HL \rightleftharpoons UO_2L^+ + H^+)$ −0·5	66B
				$K(UO_2L^+ + HL \rightleftharpoons UO_2L_2 + H^+)$ −2·10	
				$K(UO_2L_2 + HL \rightleftharpoons UO_2L_3^- + H^+)$ −4·33	
	sp	18—22	0·1 ($NaClO_4$)	K_1 7·05, K_2 5·61, K_3 3·53	66S
	pol	18—22	1 $NaNO_3$	K_1 7·27	67H
				$K(UO_2L^+ + A^- \rightleftharpoons UO_2LA)$ 1·57	
				where HA = formic acid	
Y^{3+}	gl	25	2 ($NaClO_4$)	K_1 5·43, K_2 5·38	64Y
Zn^{2+}	gl	25	0·5 ($NaClO_4$)	K_1 4·98 ± 0·06, β_2 8·95 ± 0·10	67C

61M W. A. E. McBryde and G. F. Atkinson, *Canad. J. Chem.*, 1961, **39**, 510
61S L. Sommer and A. Losmanova, *Coll. Czech. Chem. Comm.*, 1961, **26**, 2781
62M Y. Murakami, *Bull. Chem. Soc. Japan*, 1962, **35**, 52
62Ma Y. Murakami, *J. Inorg. Nuclear Chem.*, 1962, **24**, 679
64Y H. Yoneda, G. R. Choppin, J. L. Bear, and J. V. Quagliano, *Inorg. Chem.*, 1964, **3**, 1642
65B M. Bartusek and L. Sommer, *J. Inorg. Nuclear Chem.*, 1965, **27**, 2397
66B M. Bartusek and J. Ruzickova, *Coll. Czech. Chem. Comm.*, 1966, **31**, 207
66S L. Sommer, L. Kurilova-Navratilova, and T. Sepel, *Coll. Czech. Chem. Comm.*, 1966, **31**, 1288
67C G. Choux and R. L. Benoit, *Bull. Soc. chim. France*, 1967, 2920
67H J. Hala, *Coll. Czech. Chem. Comm.*, 1967, 32, 2565

365.1 $C_6H_6O_6$ Propene-1,2,3-tricarboxylic acid (aconitic acid) H_3L

Metal	*Method*	*Temp*	*Medium*	*Log of equilibrium constant, remarks*	*Ref*
Pa^{V}	ix	25	0·25	K (?) 2·39	66G

66G I. Galateanu, *Canad. J. Chem.*, 1966, **44**, 647

367 C_6H_7N Aminobenzene (aniline) L

Metal	*Method*	*Temp*	*Medium*	*Log of equilibrium constant, remarks*	*Ref*
H^+	pH	25	0·3 (KNO_3)	K_1 4·55	64N
	gl	27	1·0 ($NaClO_4$)	K_1 4·78±0·02	64W
Ag^+	Ag	25	ethanol	β_2 3·07, β_3 3·53	60A
Cd^{2+}	Hg (Cd)	25	0·3 (KNO_3)	K_1 0·10, β_2 −0·35	64N
Hg_2^{2+}	gl	27	1·0 ($NaClO_4$)	K_1 3·71	64W
Hg^{2+}	gl, sp	27	1·0 ($NaClO_4$)	K_1 4·61, β_2 9·21	64W
Zn^{2+}				See dimethyldithiocarbamic acid (176) for mixed ligands.	

60A V. Armeanu and C. Luca, *Z. phys. Chem. (Leipzig)*, 1960, **214**, 81
64N L. V. Nazarova, A. V. Ablov, and V. A. Dagaev, *Russ. J. Inorg. Chem.*, 1964, **9**, 1150 (2129)
64W T. H. Wirth and N. Davidson, *J. Amer. Chem. Soc.*, 1964, **86**, 4314, 4318

368 C_6H_7N 2-Methylpyridine (α-picoline) L

Metal	*Method*	*Temp*	*Medium*	*Log of equilibrium constant, remarks*	*Ref*
H^+	gl	25	0·1 ($NaClO_4$)	K_1 6·06±0·04	64K
	gl	35, 45	?	K_1 (35°) 5·98, (45°) 5·93	67R
	gl	25	0·61 (KNO_3)	K_1 6·21	67S
Ag^+	Ag	25	0—90% acetone	β_2 (0%) 4·61, (10%) 4·58, (20%) 4·55, (30%) 4·54, (40%) 4·55, (50%) 4·60, (60%) 4·67, (70%) 4·78, (80%) 5·00, (90%) 5·48	65P
			10—90% dioxan	β_2 (10%) 4·53, (20%) 4·48, (30%) 4·44, (40%) 4·43, (50%) 4·46, (60%) 4·52, (70%) 4·64, (80%) 4·82, (90%) 5·09	
			10—90% ethanol	β_2 (10%) 4·69, (20%) 4·65, (30%) 4·56, (40%) 4·44, (50%) 4·36, (60%) 4·31, (70%) 4·32, (80%) 4·38, (90%) 4·49	
	gl	25	0·61 (KNO_3)	K_1 2·36±0·01, β_2 4·71±0·01	67S
Be^{2+}	gl	35, 45	?	K_1 (35°) 3·53, (45°) 3·42 ΔH_1 −5·0, ΔS_1 0	67R
Cu^+	pol	25	0·1	K_1 5·40, β_2 7·65, β_3 8·5	64P
Cu^{2+}	gl	25	0·1 ($NaClO_4$)	K_1 1·3	64K
	gl	25	1·3 ($NaNO_3$+ picoline HNO_3)	K_1 1·75, K_2 0·9	64P
	gl	25	0·61 (KNO_3)	K_1 1·69±0·03, β_2 2·8±0·1	67S
Ni^{2+}	gl	25	0·1 $NaClO_4$	K_1 ⩽ 1	64K

conti

Metal	Method	Temp	Medium	Log of equilibrium constant, remarks	Ref
Zn^{2+}	gl	25	0·1 ($NaClO_4$)	$K_1 \leqslant 1$	64K
				See dimethyldithiocarbamic acid (176), 8-hydroxyquinoline (720), 2-methyl-8-hydroxyquinoline (818) and 4-methyl-8-hydroxyquinoline (819) for mixed ligands	

64K K. Kahmann, H. Sigel, and H. Erlenmeyer, *Helv. Chim. Acta*, 1964, **47**, 1754
64P F. Pantani, *Ricerca sci.*, 1964, **34** (II-A-6), 417
65P G. Popa, C. Luca, and V. Magearu, *J. Chim. phys.*, 1965, **62**, 449, 853
67R P. S. Relan and P. K. Bhattacharya, *J. Indian Chem. Soc.*, 1967, **44**, 536
67S M. S. Sun and D. G. Brewer, *Canad. J. Chem.*, 1967, **45**, 2729

C_6H_7N 3-Methylpyridine (β-picoline) L 369

Metal	Method	Temp	Medium	Log of equilibrium constant, remarks	Ref
H^+	?	?	?	K_1 5·68	61D
	gl	25	0·1 ($NaClO_4$)	K_1 6·00±0·02	64K
	gl	35, 45	?	K_1 (35°) 5·66, (45°) 5·65	67R
	gl	25	0·61 (KNO_3)	K_1 5·88	67S
Ag^+	Ag	25	0—90% ethanol	β_2 (0%) 4·36, (10%) 4·41, (20%) 4·40, (30%) 4·35, (40%) 4·27, (50%) 4·19, (60%) 4·14, (70%) 4·17, (80%) 4·24, (90%) 4·34	65P
	gl	25	0·61 (KNO_3)	K_1 2·15±0·02, β_2 4·44±0·01	67S
Be^{2+}	gl	34, 45	?	K_1 (35°) 2·89, (45°) 2·80	67R
				ΔH_1 −4·1, ΔS_1 0	
Cd^{2+}	qh, Hg	30	0·1 $NaClO_4$	K_1 1·41, β_2 2·16, β_3 2·54	61D
	Hg(Cd)	25	0·3 (KNO_3)	K_1 1·28	67N
	pol	30	0·1 KNO_3	K_1 1·27±0·05, β_2 2·35±0·07	68G
Co^{2+}	sp	20	$CHCl_3$	$K(CoL_2Cl_2+2L \rightleftharpoons CoL_4Cl_2)$ 0·35	66C
				ΔH −13·4±0·3, ΔS −44±1	
				$K[CoL_2(NCO)_2+2L \rightleftharpoons CoL_4(NCO)_2]$ 0·79	
				ΔH −10·4±0·4, ΔS −32±2	
				$K[CoL_2(NCS)_2+2L \rightleftharpoons CoL_4(NCS)_2]$ 4·20	
				ΔH −15·2±0·3, ΔS −33±1	
Cu^+	pol	25	0·1	K_1 5·60, β_2 7·78, β_3 8·6, β_4 9·0	64P
Cu^{2+}	gl	25	0·1 ($NaClO_4$)	K_1 2·77	64K
	gl	25	1·3 ($NaNO_3$+ picoline HNO_3)	K_1 2·76, K_2 1·93, K_3 1·44, K_4 0·90	64P
	gl	25	0·61 (KNO_3)	K_1 2·70±0·01, β_2 4·72±0·03, β_3 6·12±0·08, β_4 6·9±0·2	67S
Ni^{2+}	gl	25	0·1 ($NaClO_4$)	K_1 1·85	64K
	sp, cal	20	$CHCl_3$	$K(NiL_2I_2+2L \rightleftharpoons NiL_4I_2)$ (sp) 3·98	65N
				ΔH (cal) −21·7±0·4, ΔS −56±2	65N
	con	25	?	β_4 4·03	66G
	gl	25	0·61	K_1 1·97±0·02, β_2 3·21±0·06, β_3 3·9±0·2	67S
Zn^{2+}	gl	25	0·1 ($NaClO_4$)	$K_1 \leqslant 1$	64K
	Hg(Zn)	30	0·1 $NaClO_4$	K_1 1·23, β_2 1·91, β_3 2·18	66D

references overleaf

369 C_6H_7N (contd)

61D A. G. Desai and M. B. Kabadi, *J. Indian Chem. Soc.*, 1961, **38**, 805
64K K. Kahmann, H. Sigel, and H. Erlenmeyer, *Helv. Chim. Acta*, 1964, **47**, 1754
64P F. Pantani, *Ricerca sci.*, 1964, **34** (II-A-6), 417
65N S. M. Nelson and T. M. Shepherd, *J. Chem. Soc.*, 1965, 3284
65P G. Popa, V. Magearu, and C. Luca, *J. Chim. phys.*, 1965, **62**, 853
66C J. de O. Cabral, H. C. A. King, S. M. Nelson, T. M. Shepherd, and E. Koros, *J. Chem. Soc. (A)*, 1966, 1348
66D A. C. Desai and M. B. Kabadi, *J. Inorg. Nuclear Chem.*, 1966, **28**, 1279
66G G. Gawalek, W. Jugel, and H. -G. Konnecke, *J. prakt. Chem.*, 1966, **34**, 41
67N L. N. Nazarone, *Russ. J. Inorg. Chem.*, 1967, **12**, 1620 (3062)
67R P. S. Relan and P. K. Bhattacharya, *J. Indian Chem. Soc.*, 1967, **44**, 536
67S M. S. Sun and D. G. Brewer, *Canad. J. Chem.*, 1967, **45**, 2729
68G J. N. Gaur and V. K. Sharma, *Acta Chim. Acad. Sci. Hung.*, 1968, **55**, 249

370 C_6H_7N 4-Methylpyridine (γ-picoline) L

Metal	*Method*	*Temp*	*Medium*	*Log of equilibrium constant, remarks*	*Ref*
H^+	?	?	?	K_1 6·02	61D
	gl	20	0·15 ($NaClO_4$)	K_1 6·19±0·02	62H
	gl	25	0·1 ($NaClO_4$)	K_1 6·18±0·02	64K
	gl	35, 45	?	K_1 (35°), 6·28, (45°) 6·20	67R
	gl	25	0·61 (KNO_3)	K_1 6·24	67S
	gl	25	0·1 ($NaClO_4$) 50% dioxan	K_1 5·11±0·03	67Sa
Ag^+	Ag	25	0—90% ethanol	β_2 (0%) 4·49, (10%) 4·57, (20%) 4·53, (30%) 4·46, (40%) 4·37, (50%) 4·29, (60%) 4·24, (70%) 4·25, (80%) 4·31, (90%) 4·41	65P
	gl	25	0·61 (KNO_3)	K_1 2·21±0·04, β_2 4·70±0·02	67S
Be^{2+}	gl	35, 45	?	K_1 (35°) 3·54, (45°) 3·43 ΔH_1 −5·0, ΔS_1 0	67R
Cd^{2+}	qh, Hg	30	0·1 ($NaClO_4$)	K_1 1·50, β_2 2·17, β_3 2·97	61D
	pol	30	0·1 KNO_3	K_1 1·52±0·04, β_2 2·47±0·01, β_3 2·82±0·01	68Ga
Co^{2+}	sp	20	$CHCl_3$	$K(CoL_2Cl_2+2L \rightleftharpoons CoL_4Cl_2)$ 1·05 ΔH −15·7±0·2, ΔS −48·8±0·8 $K[CoL_2(NCO)_2+2L \rightleftharpoons CoL_4(NCO)_2]$ 1·18 ΔH −13·3±0·4, ΔS −40±2 $K[CoL_2(NCS)_2+2L \rightleftharpoons CoL_4(NCS)_2]$ 4·89 ΔH −16·7±0·3, ΔS −34±1	66C
	gl	25	0·1 ($NaClO_4$) 50% dioxan	K_1 1·53±0·05 $K[CoA^{2+}+L \rightleftharpoons CoAL^{2+})$ 1·3±0·1 where A = 2,2′-bipyridyl	67Sa
Cu^+	Pt, Au	20	0·15 (Na_2SO_4)	K_1 4·30, β_2 7·65	62H
	pol	25	0·1	K_1 5·65, β_2 8·20, β_3 8·8, β_4 9·2	64P
Cu^{2+}	gl	20	0·15 ($NaClO_4$)	K_1 2·56, β_2 5·39, β_3 7·66, β_4 9·54	62H
	gl	25	0·1 ($NaClO_4$)	K_1 2·88	64K
	gl	25	1·3 ($NaNO_3$+ picoline HNO_3)	K_1 2·99, K_2 2·20, K_3 1·63, K_4 1·05, K_5 0·4	64P
	gl	25	0·61 (KNO_3)	K_1 2·93±0·01, β_2 5·16±0·02, β_3 6·77±0·06, β_4 8·08±0·07	67S
	gl	25	0·1 ($NaClO_4$) 50% dioxan	K_1 2·70±0·05 $K[CuA^{2+}+L \rightleftharpoons CuAL^{2+})$ 2·09±0·05 where A = 2,2′-bipyridyl	67Sa

contin

Metal	Method	Temp	Medium	Log of equilibrium constant, remarks	Ref
Cu^{2+} (contd)	gl	22	$CHCl_3$	$K(CuA_2+L \rightleftharpoons CuA_2L)$	68G
				1·46 where HA = salicylaldehyde	
				0·53 where HA = 2-hydroxyacetophenone	
				0·34 where HA = 2-hydroxy-5-methylacetophenone	
				−0·22 where HA = 2-hydroxypropiophenone	
				0·66 where HA = 2-hydroxybenzophenone	
				0·59 where HA = 5-methyl-2-hydroxybenzophenone	
				1·57 where HA = 5-chloro-2-hydroxybenzophenone	
				−0·22 where HA = methyl salicylate	
				−0·30 where HA = ethyl salicylate	
				−1·35 where HA = 2-hydroxy-1-naphthaldehyde	
				1·13 where HA = 1-acetyl-2-naphthol	
				0·96 where HA = 1-propionyl-2-naphthol	
				0·82 where HA = 1-benzoyl-2-naphthol	
				0·78 where HA = 2-acetyl-1-naphthol	
				0·57 where HA = 2-propionyl-1-naphthol	
				1·01 where HA = methyl-1-hydroxy-2-naphthoate	
Ni^{2+}	gl	25	0·1 ($NaClO_4$)	K_1 2·11	64K
	con	25	?	β_4 4·03	66G
	gl	25	0·61 (KNO_3)	K_1 2·09±0·02, β_2 3·51±0·05, β_3 4·3±0·2	67S
	gl	25	0·1 ($NaClO_4$) 50% dioxan	K_1 2·03±0·05; $K(NiA^{2+}+L \rightleftharpoons NiAL^{2+})$ 1·83±0·05 where A = 2,2′-bipyridyl	67Sa
	sp	21	toluene	$K(NiA_2+L \rightleftharpoons NiA_2L)$	68G
				1·67 where HA = 2-hydroxy-5-methylbenzophenone	
				0·98 where HA = 2-hydroxypropiophenone	
				0·81 where HA = ethyl-2-hydroxybenzoate	
				$K(NiA_2L+L \rightleftharpoons NiA_2L_2)$	
				−0·50 where HA = 2-hydroxy-5-methylacetophenone	
				0·08 where HA = 2-hydroxy-5-methylbenzophenone	
				−0·38 where HA = 2-hydroxypropiophenone	
				0·62 where HA = methyl-2-hydroxybenzoate	
				−0·85 where HA = ethyl-2-hydroxybenzoate	
Zn^{2+}	gl	25	0·1 ($NaClO_4$)	K_1 1·40	64K
	Hg(Zn)	30	0·1 $NaClO_4$	K_1 1·30, β_2 2·11, β_3 2·85	66D
	gl	25	0·1 ($NaClO_4$) 50% dioxan	K_1 1·42±0·10; $K(ZnA^{2+}+L \rightleftharpoons ZnAL^{2+})$ 1·2±0·1 where A = 2,2′-bipyridyl	67Sa

See dimethyldithiocarbamic acid (176), 8-hydroxyquinoline (720), 2-methyl-8-hydroxyquinoline (818), and 4-methyl-8-hydroxyquinoline (819) for mixed ligands.

61D A. G. Desai and M. B. Kabaki, *J. Indian Chem. Soc.*, 1961, **38**, 805
62H C. J. Hawkins and D. D. Perrin, *J. Chem. Soc.*, 1962, 1351
64K K. Kahmann, H. Sigel, and H. Erlenmeyer, *Helv. Chim. Acta*, 1964, **47**, 1754.
64P F. Pantani, *Ricerca sci.*, 1964, **34** (II-A-6), 417
65P G. Popa, V. Magearu, and C. Luca, *J. Chim. phys.*, 1965, **62**, 853
66C J. de O. Cabral, H. C. A. King, S. M. Nelson, T. M. Shepherd, and E. Koros, *J. Chem. Soc.* (*A*), 1966, 1348
66D A. C. Desai and M. B. Kabaki, *J. Inorg. Nuclear Chem.*, 1966, **28**, 1279
66G G. Gawalek, W. Jugel, and H. G. Konuecke, *J. prakt. Chem.*, 1966, **34**, 41
67R P. S. Relan and P. K. Bhattacharya, *J. Indian Chem. Soc.*, 1967, **44**, 536
67S M. S. Sun and D. G. Brewer, *Canad. J. Chem.*, 1967, **45**, 2729
67Sa H. Sigel, *Chimia* (*Switz.*), 1967, **21**, 489
68G D. P. Graddon and G. M. Mockler, *Austral. J. Chem.*, 1968, **21**, 617, 907
68Ga J. N. Gaur and V. K. Sharma, *Acta Chim. Acad. Sci. Hung.*, 1968, **55**, 249

370.1 $C_6H_8O_4$ Cyclobutane-1,1-dicarboxylic acid H_2L

Metal	*Method*	*Temp*	*Medium*	*Log of equilibrium constant, remarks*	*Ref*
H^+	gl	25	0·1 ($NaClO_4$)	K_1 5·55, K_2 3·01	66O
Ba^{2+}	gl	25	0·1 ($NaClO_4$)	K_1 1·46	66O
Ca^{2+}	gl	25	0·1 ($NaClO_4$)	K_1 1·54	66O
Cd^{2+}	gl	25	0·1 ($NaClO_4$)	K_1 2·68,	66O
				$K(Cd^{2+}+Hl^- \rightleftharpoons CdHL^+)$ 1·30	
Co^{2+}	gl	25	0·1 ($NaClO_4$)	K_1 2·23	66O
Cu^{2+}	gl	25	0·1 ($NaClO_4$)	K_1 5·02, K_2 3·47	66O
				$K(Cu^{2+}+HL^- \rightleftharpoons CuHL^+)$ 1·37	
Mg^{2+}	gl	25	0·1 ($NaClO_4$)	K_1 2·11,	66O
				$K(Mg^{2+}+HL^- \rightleftharpoons MgHL^+)$ 0·95	
Ni^{2+}	gl	25	0·1 ($NaClO_4$)	K_1 2·20	66O
Zn^{2+}	gl	25	0·1 ($NaClO_4$)	K_1 2·48	66O

66O G. Ostacoli, E. Campi, A. Vanni, and E. Roletto, *Ricerca sci.*, 1966, **36**, 427

372 $C_6H_8O_6$ Ascorbic acid H_2L

Metal	*Method*	*Temp*	*Medium*	*Log of equilibrium constant, remarks*	*Ref*
H^+	?	?	?	K_1 11·57, K_2 4·14	63S
Ag^+	Ag	25	0·1 KNO_3	K_1 3·66±0·04	64N
Ti^{4+}	sp	18—22	0·1 ($NaClO_4$)	$K[TiO^{2+}+2HL^- \rightleftharpoons TiO(HL)_2]$ 24·8	63S
				$K(TiO^{2+}+H_2L \rightleftharpoons TiOH_2L^{2+})$ 3·1	
				$K[TiO^{2+}+2H_2L \rightleftharpoons TiO(H_2L)_2{}^{2+}]$ 6·25	
				$K[TiO^{2+}+2H^++3HL^- \rightleftharpoons Ti(HL)_3{}^+]$ 39·3	

Errata. 1964, p. 477, Table 372. In heading *for* HL *read* H_2L.
Under Sr^{2+}, *for* 52L *read* 52S

63S L. Sommer, *Coll. Czech. Chem. Comm.*, 1963, **28**, 449
64N R. I. Novoselov, Z. A. Muzykantova, and B. V. Ptitsyn, *Russ. J. Inorg. Chem.*, 1964, **9**, 1399 (2590)

373 $C_6H_8O_6$ Propane-1,2,3-tricarboxylic acid (tricarballylic acid) H_3L

Metal	*Method*	*Temp*	*Medium*	*Log of equilibrium constant, remarks*	*Ref*
H^+	gl	20	0·1 ($NaClO_4$)	K_1 5·89, K_2 4·54, K_3 3·47	64C
Ba^{2+}	gl	20	0·10 ($NaClO_4$)	K_1 1·95,	64C
				$K(Ba^{2+}+HL^{2-} \rightleftharpoons BaHL)$ 1·15	
				$K(Ba^{2+}+H_2L^- \rightleftharpoons BaH_2L^+)$ 0·73	
Ca^{2+}	gl	20	0·10 ($NaClO_4$)	K_1 2·17,	64C
				$K(Ca^{2+}+HL^{2-} \rightleftharpoons CaHL)$ 1·46	
				$K(Ca^{2+}+H_2L^- \rightleftharpoons CaH_2L^+)$ 0·88	
Co^{2+}	gl	20	0·10 ($NaClO_4$)	K_1 2·44,	64C
				$K(Co^{2+}+HL^{2-} \rightleftharpoons CoHL)$ 1·60	
				$K(Co^{2+}+H_2L^- \rightleftharpoons CoH_2L^+)$ 0·95	
Cu^{2+}	gl	20	0·10 ($NaClO_4$)	K_1 3·70,	64C
				$K(Cu^{2+}+HL^{2-} \rightleftharpoons CuHL)$ 2·57	

conti

Metal	Method	Temp	Medium	Log of equilibrium constant, remarks	Ref
Cu^{2+} (contd)				$K(Cu^{2+}+H_2L^- \rightleftharpoons CuH_2L^+)$ 1·40	
				$K(Cu^{2+}+CuL^- \rightleftharpoons Cu_2L^+)$ 1·60	
Mg^{2+}	gl	20	0·10 ($NaClO_4$)	K_1 2·06,	64C
				$K(Mg^{2+}+HL^{2-} \rightleftharpoons MgHL)$ 1·20	
				$K(Mg^{2+}+H_2L^- \rightleftharpoons MgH_2L^+)$ 0·77	
Ni^{2+}	gl	20	0·10 ($NaClO_4$)	K_1 2·65,	64C
				$K(Ni^{2+}+HL^{2-} \rightleftharpoons NiHL)$ 1·66	
				$K(Ni^{2+}+H_2L^- \rightleftharpoons NiH_2L^+)$ 1·07	
Zn^{2+}	gl	20	0·10 ($NaClO_4$)	K_1 2·43,	64C
				$K(Zn^{2+}+HL^{2-} \rightleftharpoons ZnHL)$ 1·61	
				$K(Zn^{2+}+H_2L^- \rightleftharpoons ZnH_2L^+)$ 0·94	

64C E. Campi, G. Ostacoli, M. Meirone, and G. Saini, *J. Inorg. Nuclear Chem.*, 1964, **26**, 553

$C_6H_8O_7$ 2-Hydroxypropane-1,2,3-tricarboxylic acid (citric acid) H_3L 374

Metal	Method	Temp	Medium	Log of equilibrium constant, remarks	Ref
H^+	?	?	?	K_2 4·31, K_3 2·90	62R
	?	?	?	K_2 4·74	62S
	gl	25	3·0 ($NaClO_4$)	K_1 4·88, K_2 3·83, K_3 2·40	63D
	gl	20	0·1 ($NaClO_4$)	K_1 5·68, K_2 4·35, K_3 2·87	64C
	?	?	?	K_2 4·74, K_3 3·08	64R
	gl	20	0·1 ($NaClO_4$)	K_1 5·68±0·02, K_2 4·38±0·02, K_3 2·96±0·03	64T
	gl	25	0·1, 1 (KNO_3)	K_1 (0·1) 5·65, (1) 5·34, K_2 (0·1) 4·30, (1) 4·11, K_3 (0·1) 2·79, (1) 2·63	65R
	gl	25	0·1 [$(CH_3)_4NCl$]	K_1 5·84, K_2 4·36, K_3 2·88	65T
	gl	20	0·1 KCl	K_1 5·49, K_2 4·39, K_3 3·08	66T
Ag^+	sol	25	→ 0	K_{so} −12·2	58P
Al^{3+}	gl	33	0·25 $NaClO_4$	$K(Al^{3+}+H_3L \rightleftharpoons AlL+3H^+)$ −4·7	61P
				$K[Al(H_{-1}L)^-+H^+ \rightleftharpoons AlL]$ 3·5	
				$K[Al(OH)(H_{-1}L)^{2-}+H^+ \rightleftharpoons Al(H_{-1}L)^-]$ 6·8	
Ba^{2+}	gl	20	0·1 ($NaClO_4$)	K_1 2·89,	64C
				$K(Ba^{2+}+HL^{2-} \rightleftharpoons BaHL)$ 1·75	
				$K(Ba^{2+}+H_2L^- \rightleftharpoons BaH_2L^+)$ 0·79	
	gl	32	~ 0·1	K_1 3·6	65P
Be^{2+}	gl	32	0·25 $KClO_4$	$K(Be^{2+}+H_3L \rightleftharpoons BeHL+2H^+)$ −3·3	61P
				$K(BeL^-+H^+ \rightleftharpoons BeHL)$ 3·6	
				$K[Be(H_{-1}L)^{2-}+H^+ \rightleftharpoons BeL^-]$ 5·3	
	pH	18	0·1	$K(Be^{2+}+HL^{2-} \rightleftharpoons BeHL)$ 2·56±0·01 (?)	65K
				$K[Be^{2+}+2HL^{2-} \rightleftharpoons Be(HL)_2^{2-}]$ 3·95±0·01 (?)	
				$K[Be^{2+}+3HL^{2-} \rightleftharpoons Be(HL)_3^{4-}]$ 6·97±0·01 (?)	
Ca^{2+}	gl	33	0·25 $KClO_4$	K_1 3·4	61P
	gl	20	0·1 ($NaClO_4$)	K_1 3·55,	64C
				$K(Ca^{2+}+HL^{2-} \rightleftharpoons CaHL)$ 2·10	
				$K(Ca^{2+}+H_2L^- \rightleftharpoons CaH_2L^+)$ 1·05	
Cd^{2+}	pol	25	var	K_1 4·2	51M
	gl	33	0·25 KNO_3	$K(Cd^{2+}+H_3L \rightleftharpoons CdHL+2H^+)$ −6·2	57P
				$K(CdL^-+H^+ \rightleftharpoons CdHL)$ 4·8	
				$K[Cd(H_{-1}L)^{2-}+H^+ \rightleftharpoons CdL^-]$ 8·3	

continued overleaf

374 $C_6H_8O_7$ (contd)

Metal	Method	Temp	Medium	Log of equilibrium constant, remarks	Ref
Cd^{2+} (*contd*)	gl	20	0·1 ($NaClO_4$)	K_1 3·75, $K(Cd^{2+}+HL^{2-} \rightleftharpoons CdHL)$ 2·20 $K(Cd^{2+}+H_2L^- \rightleftharpoons CdH_2L^+)$ 0·97	64C
	dis	30	1·0 ($NaClO_4$)	K_1 (?) 2·6±0·2, β_2 (?) 3·6±0·2	65H
Ce^{3+}	gl	25	0·5	K_1 8·82, β_2 12·23	66Na
	sol	25	0·1 $NaClO_4$	K_1 7·38±0·06, K_{so} −10·78±0·01	66S
Ce^{4+}	Au	25	0·5	K_1 11·84, β_2 22·32	66Na
Co^{2+}	gl	33	0·25 $KClO_4$	$K(Co^{2+}+H_3L \rightleftharpoons CoHL+2H^+)$ −4·1 $K(CoL^-+H^+ \rightleftharpoons CoHL)$ 4·2 $K[Co(H_{-1}L)^{2-}+H^+ \rightleftharpoons CoL^-]$ 8·0	61P
	gl	20	0·1 ($NaClO_4$)	K_1 5·00, $K(Co^{2+}+HL^{2-} \rightleftharpoons CoHL)$ 3·02 $K(Co^{2+}+H_2L^- \rightleftharpoons CoH_2L^+)$ 1·25	64C
Cr^{3+}	pH	22—25	0·1 $NaClO_4$	$K(Cr^{3+}+H_3L \rightleftharpoons CrL+3H^+)$ −5·55 $K[Cr(H_{-1}L)^-+H^+ \rightleftharpoons CrL]$ 5·3 $K[CrOH(H_{-1}L)^{2-}+H^+ \rightleftharpoons Cr(H_{-1}L)^-]$ 6·5	66T
Cs^+	oth	25	0·1	K_1 0·32±0·02, cation-sensitive glass electrode	64Ra
Cu^{2+}	pol	25	?	$K(Cu^{2+}+H_2L^- \rightleftharpoons CuL^-+2H^+)$ −3·08 $K(Cu^{2+}+HL^{2-} \rightleftharpoons CuL^-+H^+)$ 2·62	50M
	gl	32	0·25 KNO_3	$K(Cu^{2+}+H_3L \rightleftharpoons CuHL+2H^+)$ −3·3 $K(CuL^-+H^+ \rightleftharpoons CuHL)$ 3·4 $K[Cu(H_{-1}L)^{2-}+H^+ \rightleftharpoons CuL^-]$ 4·5	60D
	gl	20	0·1 ($NaClO_4$)	K_1 5·90, $K(Cu^{2+}+HL^{2-} \rightleftharpoons CuHL)$ 3·42 $K(Cu^{2+}+H_2L^- \rightleftharpoons CuH_2L^+)$ 2·26 $K[Cu(H_{-1}L)^{2-}+H^+ \rightleftharpoons CuL^-]$ 4·34 $K(2Cu^{2+}+L^{3-} \rightleftharpoons Cu_2L^+)$ 8·10 $K[CuL^-+Cu^{2+} \rightleftharpoons Cu_2(H_{-1}L)+H^+]$ −0·87	64C
	gl	25	1·0 KNO_3	$K(2Cu^{2+}+2L^{3-} \rightleftharpoons Cu_2L_2^{2-})$ 13·2±0·1 $K[Cu_2(H_{-1}L)_2^{4-}+2H^+ \rightleftharpoons Cu_2L_2^{2-}]$ 8·03±0·05	67R
Dy^{3+}	sol	25	0·1 $NaClO_4$	K_1 7·58±0·07, K_{so} −11·49±0·01	66S
Eu^{3+}	sol	25	0·1 $NaClO_4$	K_1 8·4±0·2, K_{so} −12·01±0·03	66S
Fe^{2+}	gl	20	0·1 ($NaClO_4$)	K_1 4·4, $K(Fe^{2+}+HL^{2-} \rightleftharpoons FeHL)$ 2·65	64T
Fe^{3+}	gl	20	0·1 ($NaClO_4$)	K_1 11·40±0·02 $K[2Fe^{3+}+2L^{3-} \rightleftharpoons Fe_2(H_{-1}L)_2^{2-}+2H^+]$ 21·2±0·1	64T
	sp	20	0·1 ($NaClO_4$)	K_1 11·56	64T
Gd^{3+}	sol	25	0·1 $NaClO_4$	K_1 8·4±0·1, K_{so} −11·88±0·01	66S
Hf^{4+}	ix	?	1,2 $HClO_4$	$K(Hf^{4+}+H_3L \rightleftharpoons HfH_2L^{3+}+H^+)$ (1) 3·05±0·01 (?), (2) 2·24±0·04 (?)	60R
	ix	?	2·0 $HClO_4$	$K(Hf^{4+}+H_3L \rightleftharpoons HfH_2L^{3+}+H^+)$ 2·54±0·05 (?)	64R
Hg^{2+}	Hg	25	0·1 ($NaClO_4$)	K_1 10·9±0·2	67S
Ho^{3+}	sol	25	0·1 $NaClO_4$	K_1 7·9±0·6, K_{so} −11·52±0·03	66S
In^{3+}	ix	?	0·5 ($NaClO_4$)	K_1 6·18	62R
K^+	oth	25	0·1	K_1 0·59±0·01, cation-sensitive glass electrode	64Ra
La^{3+}	sol	25	→ 0	K_1 8·37, β_2 11·05, K_{so} −12·06	65Sa
			0·1	K_1 6·9±0·2, β_2 10·7, K_{so} −10·71±0·01	
	ix	20	0·6 NH_4Cl	β_2 8·4±0·2, pH = 6, static, β_2 8·7±0·2, pH = 4·3, chromatographic	66Sa
Li^+	oth	25	0·1	K_1 0·83±0·01, cation-sensitive glass electrode	64Ra
Mg^{2+}	ix	25	0·1 NH_4(Cl)	K_1 3·16	63Ta
	gl	20	0·1 ($NaClO_4$)	K_1 3·40, $K(Mg^{2+}+HL^{2-} \rightleftharpoons MgHL)$ 1·84	64C

contin

Metal	*Method*	*Temp*	*Medium*	*Log of equilibrium constant, remarks*	*Ref*
Mg^{2+}				$K(Mg^{2+}+H_2L^- \rightleftharpoons MgH_2L^+)$ 0·84	
(*contd*)	ix	25	→ 0	K_1 3·96	64Ta
	gl	32	~ 0·1	K_1 3·6	65P
	gl	25	0·1 [$(CH_3)_4NCl$]	K_1 3·73, $K(Mg^{2+}+HL^{2-} \rightleftharpoons MgHL)$ 1·85	65T
Mn^{2+}	ix	25	0·15 NaCl	K_1 3·72	58W
	gl	33	0·25 $NaClO_4$	$K(Mn^{2+}+H_3L \rightleftharpoons MnHL+2H^+)$ −4·9 $K(MnL^-+H^+ \rightleftharpoons MnHL)$ 4·7 $K[Mn(H_{-1}L)^{2-}+H^+ \rightleftharpoons MnL^-]$ 8·5	61P
	conJ	28	?	K_1 2·84	62K
Mo^{III}	pol	25	0·2 Na_2SO_4 0·1 H_2SO_4 0·04 KNO_3	K (?) 3·47	62Z
Mo^{IV}	pol	25	1 $HClO_4$	K (?) 2·42	62Z
Na^+	oth	25	0·1	K_1 0·70±0·01, cation-sensitive glass electrode	64Ra
Nb^V	dis	20	4·5 (NaCl or $NaNO_3$+2·5HCl)	$K[Nb(OH)_4^+ + H_3L \rightleftharpoons Nb(OH)_4H_2L+H^+]$ 2·94 (?)	67K
Nd^{3+}	sol	25	→ 0	K_1 8·87, β_2 12·92, K_{so} −12·24	65Sa
			0·1	K_1 7·59±0·07, β_2 11·6, K_{so} −10·89±0·03	
Ni^{2+}	gl	32	0·25 KNO_3	$K(Ni^{2+}+H_3L \rightleftharpoons NiHL+2H^+)$ −4·1 $K(NiL^-+H^+ \rightleftharpoons NiHL)$ 3·7 $K[Ni(H_{-1}L)^{2-}+H^+ \rightleftharpoons NiL^-]$ 7·9	57P
	gl	20	0·1 ($NaClO_4$)	K_1 5·40, $K(Ni^{2+}+HL^{2-} \rightleftharpoons NiHL)$ 3·30 $K(Ni^{2+}+H_2L^- \rightleftharpoons NiH_2L^+)$ 1·75	64C
	sol	35	?	$K[Ni(H_{-1}L)^{2-}+H^+ \rightleftharpoons NiL^-]$ 7·9	65P
NpO_2^+	ix	18—22	0·05 NH_4ClO_4	K_1 3·67, $K(NpO_2^+ + HL^{2-} \rightleftharpoons NpO_2HL^-)$ 2·69	61M
Pa^V	ix	25	0·25	K (?) 3·43, K (?) 5·92	66G
Pb^{2+}	gl	?	0·3 KNO_3	$K[Pb^{2+}+L^{3-} \rightleftharpoons Pb(H_{-1}L)^{2-}+H^+]$ −1·1	57P
	sol	35	?	$K[Pb(H_{-1}L)^{2-}+H^+ \rightleftharpoons PbL^-]$ 7·1 $K[Pb(H_{-1}L)OH^{3-}+H^+ \rightleftharpoons Pb(H_{-1}L)^{2-}]$ 9·5	
	Hg	25	3·0 ($NaClO_4$)	K_1 4·34, β_2 6·08, β_3 6·97	63D
Pr^{3+}	sol	25	→ 0	K_1 8·72, β_2 12·60, K_{so} −12·34	65Sa
			0·1	K_1 7·4±0·4, β_2 11·2, K_{so} −10·98±0·05	
Pu^{4+}	sp	25	0·5 ($NaClO_4$)	K_1 15·2, β_2 30·1	66N
	gl	25	0·5 ($NaClO_4$)	K_1 15·7, β_2 29·5	66N
Rb^+	oth	25	0·1	K_1 0·49±0·02, cation-sensitive glass electrode	64Ra
Ru^{3+}				See 1-Nitroso-2-naphthol (808) and 2-Nitroso-2-naphthol (809) for mixed ligands.	
Sc^{3+}	sp	18—20	0·01 (HNO_3)	K_1 7·00	66Ka
Sr^{2+}	gl	32	0·25 $NaClO_4$	K_1 2·7	61P
	ix	22	1 NH_4^+	K_1 2·24±0·01	62T
Tb^{3+}	sol	25	0·1 $NaClO_4$	K_1 8·1±0·5, K_{so} −11·82±0·02	66S
Th^{4+}	gl	25	0·5	K_1 13·0, β_2 20·97	66Na
Tl^+	con	28	?	K_1 2·82	65S
U^{4+}	sp	20	0·1 $NaClO_4$	$K[U(OH)_2^{2+}+L^{3-} \rightleftharpoons U(OH)_2L^-]$ 13·5	60A
	gl	25	0·5	K_1 11·53, β_2 19·46	66Na
UO_2^{2+}	dis	20	1 $NaClO_4$	β_2 ~ 11·2	62S
	gl	25	0·1, 1 (KNO_3)	K_1 (0·1) 7·4±0·2, (1) 6·9±0·1 $K[2UO_2^{2+}+2L^{3-} \rightleftharpoons (UO_2)_2L_2^{2-}]$, (0·1) 18·87±0·06, (1) 17·70±0·04	65R

continued overleaf

374 $C_6H_8O_7$ (contd)

Metal	*Method*	*Temp*	*Medium*	*Log of equilibrium constant, remarks*	*Ref*
UO_2^{2+} (*contd*)				$K[2UO_2L^- \rightleftharpoons (UO_2)_2L_2^{2-}]$ (0·1) 4·1±0·4, (1) 4·0±0·2	
				$K\{3(UO_2)_2L_2^{2-} \rightleftharpoons (UO_2)_2L_2[(OH)_5(UO_2)_2L_2]_2^{16-} + 10H^+\}$ −47·9	
VO^{2+}	sp	18—22	1 NaCl	K_1 8·83±0·02	67N
				$K[2VO^{2+} + L^{3-} \rightleftharpoons (VO)_2L^+]$ 11·50±0·08	
Y^{3+}	sol	25	0·1 $NaClO_4$	K_1 7·81±0·06, K_{so} −11·03±0·02	66S
Yb^{3+}	ix	20	0·6 NH_4Cl	β_2 9·2±0·2, pH = 6, static, β_2 9·6±0·2, pH = 4·3, chromatographic	66Sa
Zn^{2+}	pol	25	0·3 $Na_3C_6H_5O_7$	$K[ZnL^- + OH^- \rightleftharpoons Zn(H_{-1}L)^{2-}]$ 5·5	51M
	Zn	23	?	$K[Zn^{2+} + OH^- + L^{3-} \rightleftharpoons Zn(H_{-1}L)^{2-}]$ 10·92	59K
	gl	33	0·25 $NaClO_4$	$K(Zn^{2+} + H_3L \rightleftharpoons ZnHL + 2H^+)$ −4·2	61P
				$K(ZnL^- + H^+ \rightleftharpoons ZnHL)$ 4·3	
				$K[Zn(H_{-1}L)^{2-} + H^+ \rightleftharpoons ZnL^-]$ 7·7	
	gl	20	0·1 ($NaClO_4$)	K_1 4·98,	64C
				$K(Zn^{2+} + HL^{2-} \rightleftharpoons ZnHL)$ 2·98	
				$K(Zn^{2+} + H_2L^- \rightleftharpoons ZnH_2L^+)$ 1·25	
Zr^{4+}	ix	?	1, 2 $HClO_4$	$K(Zr^{4+} + H_3L \rightleftharpoons ZrH_2L^{3+} + H^+)$ (1) 3·87±0·03, (2) 3·12±0·06	60R
	ix	?	2·0 $HClO_4$	$K(Zr^{4+} + H_3L \rightleftharpoons ZrH_2L^{3+} + H^+)$ 3·41±0·08	64R
	sp	18—20	1 (HCl)	$K(Zr^{4+} + HL^{2-} \rightleftharpoons ZrHL^{2+})$ 10·78±0·04	66K

Errata. 1964, p. 478, Table 374. Under Ba^{2+}, 2nd line, *for* HL_3^- *read* HL^{3-}.
p. 480. Under UO_2^{2+}, *for* ref. 64R *read* 65R.
p. 481. *Delete* ref. 64R, *Substitute* ref. 65R K. S. Rajan and A. E. Martell, *Inorg. Chem.*, 1965, **4,** 462.

50M L. Meites, *J. Amer. Chem. Soc.*, 1950, **72,** 180
51M L. Meites, *J. Amer. Chem. Soc.*, 1951, **73,** 3727
57P R. K. Patnaik and S. Pani, *J. Indian Chem. Soc.*, 1957, **34,** 19, 619, 673
58P B. V. Ptitsyn, L. N. Sheronov, and V. P. Komlev, *J. Inorg. Chem.* (*U.S.S.R.*), 1958, **3,** No. 11, 25 (2450)
58W J. S. Wiberg, *Arch. Biochem. Biophys.*, 1958, **73,** 337
59K K. N. Kovalenko and L. I. Vistyak, *Russ. J. Inorg. Chem.*, 1959, **4,** 364 (801)
60A A. Adams and T. D. Smith, *J. Chem. Soc.*, 1960, 4846
60D R. Das, R. K. Patnaik, and S. Panit, *J. Indian Chem. Soc.*, 1960, **37,** 59
60R D. I. Ryabchikov, A. N. Ermakov, V. K. Belyaeva, and I. N. Marov, *Russ. J. Inorg. Chem.*, 1960, **5,** 505 (1051)
61M A. I. Moskvin, I. N. Marov, and Yu. A. Zolotov, *Russ. J. Inorg. Chem.*, 1961, **6,** 926 (1813)
61P R. K. Patnaik and S. Pani, *J. Indian Chem. Soc.*, 1961, **38,** 229, 233, 364, 379, 709, 896
62K M. S. Kachhawaha and A. K. Bhattacharya, *Z. Anorg. Chem.*, 1962, **315,** 104
62R D. I. Ryabchikov, I. N. Marov, and Y. K'o-min, *Russ. J. Inorg. Chem.*, 1962, **7,** 1415 (2716)
62S J. Stary and V. Balek, *Coll. Czech. Chem. Comm.*, 1962, **27,** 809
62T L. I. Takhonova, *Russ. J. Inorg. Chem.*, 1962, **1,** 424 (822)
62Z E. W. Zahnow and R. J. Robinson, *J. Electroanalyt. Chem.*, 1962, **3,** 263
63D A. F. Donda and A. M. Giuliani, *Ricerca sci.*, 1963, **33** (II-A), 819
63T S. K. Tobia and N. E. Milad, *J. Chem. Soc.*, 1963, 734
64C E. Campi, G. Ostacoli, M. Meirone, and G. Saini, *J. Inorg. Nuclear Chem.*, 1964, **26,** 553
64R R. I. Ryabchikov, I. N. Marov, A. N. Ermakov, and V. K. Belyaeva, *J. Inorg. Nuclear Chem.*, 1964, **26, 965**
64Ra G. A. Rechnitz and S. B. Zamochnick, *Talanta*, 1964, **11,** 1061
64T C. F. Timberlake, *J. Chem. Soc.*, 1964, 5078
64Ta S. K. Tobia and N. E. Milad, *J. Chem. Soc.*, 1964, 1915
65H H. E. Hellwege and G. K. Schweitzer, *J. Inorg. Nuclear Chem.*, 1965, **27,** 99
65K I. F. Kolosova, T. A. Belgarskaya, *Russ. J. Inorg. Chem.*, 1965, **10,** 411 (764)
65P R. K. Patnaik and S. Pani, *J. Indian Chem. Soc.*, 1965, **42,** 527, 793
65R K. S. Rajan and A. E. Martell, *Inorg. Chem.*, 1965, **4,** 462
65S K. N. Sahu and A. K. Bhattacharya, *J. Indian Chem. Soc.*, 1965, **42,** 247
65Sa N. A. Skorik, V. N. Kumok, E. I. Peror, K. P. Augustan, and V. V. Sevebrennikov, *Russ. J. Inorg. Chem.*, 1965, **10,** 351 (653)
65T S. S. Tate, A. K. Grzybowski, and S. P. Datta, *J. Chem. Soc.*, 1965, 3905
66G I. Galateanu, *Canad. J. Chem.*, 1966, **44,** 647
66K I. M. Korenman, F. R. Sheyanova, and F. M. Gureva, *Russ. J. Inorg. Chem.*, 1966, **11,** 1485 (2761)
66Ka I. M. Korenman and N. V. Zaglyadimova, *Russ. J. Inorg. Chem.*, 1966, **11,** 1491 (2774)

contin

66N D. Nebel, *Z. phys. Chem. (Leipzig)*, 1966, **232**, 161, 368
66Na D. Nebel and G. Urban, *Z. phys. Chem. (Leipzig)*, 1966, **233**, 73
66S N. A. Skorik and V. V. Serebrennikov, *Russ. J. Inorg. Chem.*, 1966, **11**, 416 (764)
66Sa A. M. Sorochan and M. M. Senyavin, *Russ. J. Inorg. Chem.*, 1966, **11**, 753 (1410)
66T K. K. Tripathy and R. K. Patnaik, *J. Indian Chem. Soc.*, 1966, **43**, 772
67K C. Konecny, *Z. phys. Chem. (Leipzig)*, 1967, **235**, 39
67N B. M. Nikolova and G. St. Nikolov, *J. Inorg. Nuclear Chem.*, 1967, **29**, 1013
67R K. S. Rajan and A. E. Martell, *J. Inorg. Nuclear Chem.*, 1967, **29**, 463
67S N. A. Skorik, V. N. Kumok, and V. V. Serebrennikov, *Russ. J. Inorg. Chem.*, 1967, **12**, 1429 (2711)

$C_6H_8N_2$ 2-Aminomethylpyridine L 376

Metal	*Method*	*Temp*	*Medium*	*Log of equilibrium constant, remarks*	*Ref*
H^+	gl	25	0·015	K_1 8·62	60H
	sp	20	0·015	K_2 1·85	60H
	gl	25	0·1 (KNO_3)	K_1 8·57, K_2 2·14	64L
Cd^{2+}	gl	25	0·1 (KNO_3)	K_1 4·5	64L
	pol	25	0·1 KNO_3 50% dioxan	β_3 9·68	66W
	cal	25	0·1 KNO_3 50% dioxan	$\Delta H_{\beta2}°(?)$ −12·6, $\Delta S_{\beta2}°(?)$ 2·1	66W
Co^{2+}	gl	25	0·01—0·02	K_1 ~ 5·8	60H
	gl	25	0·1 (KNO_3)	K_1 5·3	64L
	pol	25	0·1 KNO_3 50% dioxan	β_2 10·39	66W
	cal	25	0·1 KNO_3 50% dioxan	$\Delta H_{\beta2}°$ −16·7, $\Delta S_{\beta2}°$ −8·4	66W
Cu^+	pol	25	0·1 KNO_3 50% dioxan	β_2 10·66	66W
Cu^{2+}	gl	25	0·01—0·02	K_1 ~ 9·3, β_2 ~ 17·2	60H
	gl	25	0·1 (KNO_3)	K_1 9·5	64L
	pol, gl	25	0·1 KNO_3 50% dioxan	β_2 (pol) 15·47, (gl) 15·68	66W
	cal	25	0·1 KNO_3 50% dioxan	$\Delta H_{\beta2}°$ −21·6, $\Delta S_{\beta2}°$ −0·6	66W
Ni^{2+}	gl	25	0·01—0·02	K_1 7·3, β_2 13·6, β_3 19·4	60H
	gl	25	0·1 (KNO_3)	K_1 7·1	64L
	pol	25	0·1 KNO_3 50% dioxan	β_2 13·15	66W
	cal	25	0·1 KNO_3 50% dioxan	$\Delta H_{\beta2}°$ −20·5, $\Delta S_{\beta2}°$ −8·6	66W
Zn^{2+}	gl	25	0·1 (KNO_3)	K_1 5·2	64L
	pol	25	0·1 KNO_3 50% dioxan	β_2 9·84	66W
	cal	25	0·1 KNO_3 50% dioxan	$\Delta H_{\beta2}°$ −12·4, $\Delta S_{\beta2}°$ 3·3	66W

Errata. 1964, p. 482, Table 376. *For* references 60I and 60L *read* 64L.
Under Medium—fourth column, *for* 0·1 *read* 0·1 (KNO_3).
Under references, *delete* 60L, *substitute* 64L R. G. Lacoste and A. E. Martell, *Inorg. Chem.*, 1964, **3**, 881.

60H F. Holmes and F. Jones, *J. Chem. Soc.*, 1960, 2398
64L R. G. Lacoste and A. E. Martell, *Inorg. Chem.*, 1964, **3**, 881
66W J. L. Walter and S. M. Rosalie, *J. Inorg. Nuclear Chem.*, 1966, **28**, 2969

376.1 $C_6H_8N_2$ 2-Amino-3-methylpyridine (2-amino-3-picoline) L

Metal	Method	Temp	Medium	Log of equilibrium constant, remarks	Ref
H^+	gl	25	0·61 (KNO_3)	K_1 7·23	67S
Ag^+	gl	25	0·61 (KNO_3)	K_1 2·42±0·01, β_2 4·85±0·02	67S
Cu^{2+}	gl	25	0·61 (KNO_3)	K_1 1·91±0·04	67S

67S M. S. Sun and D. G. Brewer, *Canad. J. Chem.*, 1967 **45**, 2729

377 $C_6H_{10}O_2$ Hexane-2,4-dione (propionylacetone) HL

Metal	Method	Temp	Medium	Log of equilibrium constant, remarks	Ref
H^+				H^+ constants not given.	
La^{3+}	gl	30	0·1 ($NaClO_4$) 67% acetone	K_1 6·22, K_2 4·87, K_3 4·16	64D
Nd^{3+}	gl	30	0·1 ($NaClO_4$) 67% acetone	K_1 6·94, K_2 6·24, K_3 5·01	64D
Pr^{3+}	gl	30	0·1 ($NaClO_4$) 67% acetone	K_1 6·72, K_2 5·84, K_3 5·38	64D
Y^{3+}	gl	30	0·1 ($NaClO_4$) 67% acetone	K_1 7·14, K_2 5·80, K_3 5·62	64D

64D N. K. Dutt and P. Bandyopadhyay, *J. Inorg. Nuclear Chem.*, 1964, 26, 729

379.1 $C_6H_{10}O_3$ 1-Hydroxycyclopentanecarboxylic acid HL

Metal	Method	Temp	Medium	Log of equilibrium constant, remarks	Ref
H^+	gl	25	0·1 ($NaClO_4$)	K_1 3·97±0·02	66P
	gl	25	0·100 ($NaClO_4$)	K_1 3·95	67P
Cd^{2+}	gl	25	0·100 ($NaClO_4$)	K_1 1·45±0·02, K_2 0·93±0·05	67P
Ce^{3+}	gl	25	0·1 ($NaClO_4$)	K_1 2·511±0·002, K_2 1·827±0·004, K_3 1·08±0·04	66P
Co^{2+}	gl	25	0·100 ($NaClO_4$)	K_1 1·57±0·01, K_2 1·0±0·1	67P
Cu^{2+}	gl	25	0·100 ($NaClO_4$)	K_1 2·799±0·004, K_2 1·78±0·02	67P
Dy^{3+}	gl	25	0·1 ($NaClO_4$)	K_1 2·981±0·002, K_2 2·467±0·006, K_3 1·77±0·02, K_4 1·67±0·02	66P
Er^{3+}	gl	25	0·1 ($NaClO_4$)	K_1 3·066±0·004, K_2 2·581±0·007, K_3 1·82±0·02, K_4 1·72±0·02	66P
Eu^{3+}	gl	25	0·1 ($NaClO_4$)	K_1 2·803±0·004, K_2 2·196±0·006, K_3 1·57±0·03	66P
Gd^{3+}	gl	25	0·1 ($NaClO_4$)	K_1 2·844±0·005, K_2 2·22±0·01, K_3 1·65±0·03, K_4 1·49±0·06	66P
Ho^{3+}	gl	25	0·1 ($NaClO_4$)	K_1 3·021±0·002, K_2 2·480±0·006, K_3 1·87±0·01, K_4 1·49±0·03	66P
La^{3+}	gl	25	0·1 ($NaClO_4$)	K_1 2·377±0·004, K_2 1·68±0·01, K_3 0·95±0·10	66P
Lu^{3+}	gl	25	0·1 ($NaClO_4$)	K_1 3·221±0·003, K_2 2·723±0·006, K_3 2·15±0·02, K_4 1·54±0·02	66P
Nd^{3+}	gl	25	0·1 ($NaClO_4$)	K_1 2·666±0·003, K_2 1·960±0·009, K_3 1·08±0·08	66P
Ni^{2+}	gl	25	0·100 ($NaClO_4$)	K_1 1·82±0·01, K_2 1·30±0·07	67P

cont

$C_6H_{10}O_3$ (contd) 379.1

Metal	Method	Temp	Medium	Log of equilibrium constant, remarks	Ref
Pm^{3+}	hyp	25	0·1 ($NaClO_4$)	K_1 2·73, K_2 2·04, K_3 1·40	66P
Pr^{3+}	gl	25	0·1 ($NaClO_4$)	K_1 2·604±0·001, K_2 1·921±0·003, K_3 1·18±0·02	66P
Sm^{3+}	gl	25	0·1 ($NaClO_4$)	K_1 2·789±0·002, K_2 2·146±0·003, K_3 1·50±0·01, K_4 1·08±0·04	66P
Tb^{3+}	gl	25	0·1 ($NaClO_4$)	K_1 2·898±0·004, K_2 2·36±0·01, K_3 1·73±0·03, K_4 1·23±0·07	66P
Tm^{3+}	gl	25	0·1 ($NaClO_4$)	K_1 3·110±0·002, K_2 2·612±0·005, K_3 1·99±0·01, K_4 1·30±0·02	66P
Y^{3+}	gl	25	0·1 ($NaClO_4$)	K_1 2·998±0·005, K_2 2·436±0·007, K_3 1·84±0·03, K_4 1·69±0·04	66P
Yb^{3+}	gl	25	0·1 ($NaClO_4$)	K_1 3·175±0·003, K_2 2·676±0·006, K_3 2·05±0·01, K_4 1·32±0·03	66P
Zn^{2+}	gl	25	0·100 ($NaClO_4$)	K_1 1·89±0·01, K_2 1·30±0·07	67P

66P J. E. Powell and D. L. G. Rowlands, *Inorg. Chem.*, 1966, **5**, 819
67P J. E. Powell and D. L. G. Rowlands, *J. Inorg. Nuclear Chem.*, 1967, **29**, 1729

$C_6H_{10}O_4$ Hexanedioic acid (adipic acid) H_2L 380

Metal	Method	Temp	Medium	Log of equilibrium constant, remarks	Ref
Co^{2+}	dis	25	0—0·08 (NaCl)	K_1 (→ 0) 2·40±0·02, (0·04) 1·78±0·01, (0·08) 1·60±0·01	61M
	gl	25	0 corr	K_1 2·15±0·02	65M
	ix	25	0 corr	K_1 2·23±0·05	65S
	pol	30, 38	0·5 ($NaClO_4$)	β_2 (30·5°) 4·07, (38°) 4·15; $\Delta H_{\beta 2}$ 2·03	68G
Sc^{3+}	sp	18—20	0·01 (HNO_3)	K_1 4·4±0·2	66K
Zn^{2+}	dis	25	0 corr	K_1 2·67±0·07	66R

61M P. G. Manning and C. B. Monk, *Trans. Faraday Soc.*, 1961, **57**, 1996
65M C. B. Monk, *J. Chem. Soc.*, 1965, 2456
65S R. G. Seys and C. B. Monk, *J. Chem. Soc.*, 1965, 2452
66K I. M. Korenman and N. V. Zadlyadimova, *Russ. J. Inorg. Chem.*, 1966, **11**, 1491 (2774)
66R D. L. G. Rowlands and C. B. Monk, *Trans. Faraday Soc.*, 1966, **62**, 945
68G J. K. Gupta and C. M. Gupta, *Indian J. Chem.*, 1968, **6**, 50

$C_6H_{10}O_6$ Ethylenedioxydiacetic acid H_2L 382.1

Metal	Method	Temp	Medium	Log of equilibrium constant, remarks	Ref
H^+	gl	25	0·1	K_1 3·73, K_2 2·34	61K
	gl	20	0·1	K_1 3·97, K_2 3·15	61O
Ca^{2+}	gl	25	0·1	K_1 4·12	61K
Cu^{2+}	gl	20	0·1	K_1 3·15, $K(Cu^{2+}+HL^- \rightleftharpoons CuHL^+)$ 2·61	61O
Mg^{2+}	gl	25	0·1	K_1 2·78	61K
Sr^{2+}	gl	25	0·1	K_1 3·18	61K

61K H. Kroll, V. Elkind, and R. Davis, *US-AEC Report TID-14373*, 1961
61O G. Ostacoli, E. Campi, N. Cibrario, and G. Saini, *Gazzetta*, 1961, **91**, 349

382.2 $C_6H_{10}O_8$ Mucic acid H_2L

Metal	Method	Temp	Medium	Log of equilibrium constant, remarks	Ref
H^+	gl	25	1 ($NaNO_3$)	K_1 3·63, K_2 3·08	68B
Cu^{2+}	gl	25	1 ($NaNO_3$)	$K[Cu^{2+}+L^{2-} \rightleftharpoons Cu(H_{-1}L)^- + H^+]$ −9·36 $K[Cu^{2+}+L^{2-} \rightleftharpoons Cu(H_{-2}L)^{2-} + 2H^+]$ −18·11	68B
Ni^{2+}	gl	25	1 ($NaNO_3$)	$K[Ni^{2+}+L^{2-} \rightleftharpoons Ni(H_{-1}L)^- + H^+]$ −9·34 $K[Ni^{2+}+L^{2-} \rightleftharpoons Ni(H_{-2}L)^{2-} + 2H^+]$ −18·08	68B

68B E. Bottari, *Monatsh.*, 1968, **99**, 176

382.3 $C_6H_{10}O_8$ D-Saccharic acid H_2L

Metal	Method	Temp	Medium	Log of equilibrium constant, remarks	Ref
Hf^{4+}	ix	20	2 $HClO_4$	$K(Hf^{4+}+H_2L \rightleftharpoons HfL^{2+}+2H^+)$ 2·29±0·02	63R
Zr^{4+}	ix	20	2 $HClO_4$	$K(Zr^{4+}+H_2L \rightleftharpoons ZrL^{2+}+2H^+)$ 2·43±0·02	63R

63R H. Roth, L. Czegledi, and C. G. Macarovici, *Acad. Rep. Populare Romine, Studii Cercetari Chimie (Cluj)*, 1963, **14**, 69

384.1 $C_6H_{11}N$ Di-allylamine L

Metal	Method	Temp	Medium	Log of equilibrium constant, remarks	Ref
Ag^+	oth	?	0·1 $NaClO_4$ acetone	β_2 8·96±0·05, coulometric titration	65M

65M K. K. Mead, D. L. Maricle, and C. A. Streuli, *Analyt. Chem.*, 1965, **37**, 237

385.1 $C_6H_{12}O$ *cis*-1-Ethoxybutene L

Metal	Method	Temp	Medium	Log of equilibrium constant, remarks	Ref
Ag^+	dis	10—30	ethylene glycol	K_1 (10°) 0·75, (20°) 0·63, (30°) 0·49, gas chromatography ΔH_1 −4·5±0·2, ΔS_1 −12·0±0·6	68F

68F T. Fueno, O. Kajimoto, T. Okuyama, and J. Furukawa, *Bull. Chem. Soc. Japan*, 1968, **41**, 785

385.2 $C_6H_{12}O$ *trans*-1-Ethoxybutene L

Metal	Method	Temp	Medium	Log of equilibrium constant, remarks	Ref
Ag^+	dis	10—30	ethylene glycol	K (10°) −0·02, (20°) −0·14, (30°) −0·20, gas chromatography ΔH_1 −3·3±0·2, ΔS_1 −12·0±0·6	68F

68F T. Fueno, O. Kajimoto, T. Okuyama, and J. Furukawa, *Bull. Chem. Soc. Japan*, 1968, **41**, 785

$C_6H_{12}O$ 1-Ethoxy-2-methylprop-1-ene L 385.3

Metal	Method	Temp	Medium	Log of equilibrium constant, remarks	Ref
Ag^+	dis	10—30	ethylene glycol	K_1 (10°) −0·57, (20°) −0·68, (30°) −0·77, gas chromatography ΔH_1 −3·5±0·2, ΔS_1 −12	68F

68F T. Fueno, O. Kajimoto, T. Okuyama, and J. Furukawa, *Bull. Chem. Soc. Japan*, 1968, **41**, 785

$C_6H_{12}O$ Butoxyethene (butyl vinyl ether) L 385.4

Metal	Method	Temp	Medium	Log of equilibrium constant, remarks	Ref
Ag^+	dis	10—30	ethylene glycol	K_1 (10°) 0·94, (20°) 0·89, (30°) 0·75, gas chromatography ΔH_1 −3·7±0·2, ΔS_1 −8·8±0·6	68F

68F T. Fueno, O. Kajimoto, T. Okuyama, and J. Furukawa, *Bull. Chem. Soc. Japan*, 1968, **41**, 785

$C_6H_{12}O$ (2-Methylpropoxy)ethene (isobutyl vinyl ether) L 583.5

Metal	Method	Temp	Medium	Log of equilibrium constant, remarks	Ref
Ag^+	dis	10—30	ethylene glycol	K_1 (10°) 0·94, (20°) 0·83, (30°) 0·75, gas chromatography ΔH_1 −4·0±0·2, ΔS_1 9·8±0·6	68F

68F T. Fueno, O. Kajimoto, T. Okuyama, and J. Furukawa, *Bull. Chem. Soc. Japan*, 1968, **41**, 785

$C_6H_{12}O_2$ *cis*-Cyclohexan-1,2-diol L 386.1

Metal	Method	Temp	Medium	Log of equilibrium constant, remarks	Ref
B^{III}	gl	0—35	0·025 borax	$K[B(OH)_4^- + L \rightleftharpoons B(OH)_2(H_{-2}L)^-]$ (0°) 0·3, (13°) 0·0, (25°) 0·0, (35°) 0·0 $\Delta H°$ −4±1, $\Delta S°$ −10±6, $K[B(OH)_4^- + 2L \rightleftharpoons B(H_{-2}L)_2^-]$ (0°) 0·3, (13°) 0·3, (25°) −0·2, (35°) −0·5 $\Delta H°$ −10±3, $\Delta S°$ −40±20	67C

67C J. M. Conner and V. C. Bulgrin, *J. Inorg. Nuclear Chem.*, 1967, **29**, 1953

386.2 $C_6H_{12}O_3$ 2-Ethyl-2-hydroxybutanoic acid (diethylglycollic acid) HL

Metal	*Method*	*Temp*	*Medium*	*Log of equilibrium constant, remarks*	*Ref*
H^+	qh	25	1·0 ($NaClO_4$)	K_1 3·64	65T
Er^{3+}	qh	25	1·0 ($NaClO_4$)	K_1 3·11±0·02, K_2 2·16±0·03, K_3 1·33±0·04, K_4 0·95±0·05	65T
Gd^{3+}	qh	25	1·0 ($NaClO_4$)	K_1 2·71±0·01, K_2 1·94±0·02, K_3 0·98±0·02, K_4 0·86±0·02	65T
Ho^{3+}	qh	25	1·0 ($NaClO_4$)	K_1 3·11±0·03, K_2 2·14±0·03, K_3 1·46±0·09, K_4 0·7±0·2	65T
La^{3+}	qh	25	1·0 ($NaClO_4$)	K_1 1·85±0·01, K_2 1·26±0·02, K_3 0·56±0·04, K_4 0·42±0·05	65T
Nd^{3+}	qh	25	1·0 ($NaClO_4$)	K_1 2·28±0·02, K_2 1·61±0·04, K_3 1·21±0·07, K_4 0·94±0·08	65T
Pr^{3+}	qh	25	1·0 ($NaClO_4$)	K_1 2·31±0·03, K_2 1·49±0·04, K_3 1·02±0·07, K_4 0·57±0·08	65T
Tb^{3+}	qh	25	1·0 ($NaClO_4$	K_1 3·01±0·02, K_2 2·07±0·02, K_3 1·37±0·02, K_4 0·53±0·03	65T
Yb^{3+}	qh	25	1·0 ($NaClO_4$)	K_1 3·10±0·04, K_2 2·26±0·04, K_3 1·31±0·05, K_4 1·09±0·05	65T

65T H. Thun, F. Verbeek and W. Vanderleen, *J. Inorg. Nuclear Chem.*, 1965, 27, 1813

386.3 $C_6H_{12}O_3$ 2-Hydroxy-2,3-dimethylbutanoic acid (isopropylmethylglycollic acid) HL

Metal	*Method*	*Temp*	*Medium*	*Log of equilibrium constant, remarks*	*Ref*
H^+	qh	25	1·0 ($NaClO_4$)	K_1 3·80	65T
Eu^{3+}	qh	25	1·0 ($NaClO_4$)	K_1 2·68±0·02, K_2 1·97±0·03, K_3 1·41±0·04, K_4 0·99±0·04	65T
Gd^{3+}	qh	25	1·0 ($NaClO_4$)	K_1 2·65±0·01, K_2 2·05±0·01, K_3 1·34±0·02, K_4 0·94±0·02	65T
Ho^{3+}	qh	25	1·0 ($NaClO_4$)	K_1 3·02±0·03, K_2 2·35±0·04, K_3 1·53±0·04, K_4 1·34±0·03	65T
Nd^{3+}	qh	25	1·0 ($NaClO_4$)	K_1 2·57±0·06, K_2 1·71±0·08, K_3 1·2±0·2, K_4 1·1±0·1	65T
Pr^{3+}	qh	25	1·0 ($NaClO_4$)	K_1 2·25±0·02, K_2 1·62±0·02, K_3 1·00±0·04, K_4 0·82±0·05	65T
Yb^{3+}	qh	25	1·0 ($NaClO_4$)	K_1 3·12±0·03, K_2 2·44±0·04, K_3 1·65±0·05, K_4 1·39±0·04	65T

65T H. Thun, F. Verbeek and W. Vanderleen, *J. Inorg. Nuclear Chem.*, 1965, 27, 1813

$C_6H_{12}O_5$ L-Rhamnose L 386.4

Metal	Method	Temp	Medium	Log of equilibrium constant, remarks	Ref
As^{III}	gl	25	0·1 KCl	$K[As(OH)_4^- + 2L \rightleftharpoons As(H_{-2}L)_2^-]$ 0·68	60A
B^{III}	gl	25	0·1 KCl	$K[B(OH)_4^- + 2L \rightleftharpoons B(H_{-2}L)_2^-]$ 2·61	59A
Ge^{IV}	gl	25	0·1 KCl	$K[HGeO_3^- + 2L \rightleftharpoons HGeO(H_{-2}L)_2^-]$ 3·24	59A

59A P. J. Antikainen and K. Tevanen, *Suomen Kem.*, 1959, **B32**, 214
60A P. J. Antikainen and K. Tevanen, *Suomen Kem.*, 1960, **B33**, 7

$C_6H_{12}O_6$ D-Fructose L 387

Metal	Method	Temp	Medium	Log of equilibrium constant, remarks	Ref
As^{III}	gl	25	0·1 KCl	$K[As(OH)_4^- + 2L \rightleftharpoons As(H_{-2}L)_2^-]$ 1·08	60A
	gl	15—45	0·1 KCl	$K[As(OH)_4^- + L \rightleftharpoons As(OH)_2(H_{-2}L)^-]$ (15°) 0·779, (25°) 0·739, (35°) 0·724, (45°) 0·703 ΔH° −1·02, ΔS° 0·0	68A
B^{III}	gl	25	0·1 KCl	$K[B(OH)_4^- + 2L \rightleftharpoons B(H_{-2}L)_2^-]$ 5·04	58A
	pH	25	I KCl	$K[B(OH)_4^- + 2L \rightleftharpoons B(H_{-2}L)_2^-]$ $4·723 + 0·470\sqrt{I}$	67N
	gl	0—45	0·1 KCl	$K[B(OH)_4^- + L \rightleftharpoons B(OH)_2(H_{-2}L)^-]$ (0°) 4·142, (15°) 3·642, (25°) 3·416, (35°) 3·178, (45°) 2·976 ΔH° −9·4, ΔS° −15·9 $K[B(OH)_4^- + 2L \rightleftharpoons B(H_{-2}L)_2^-]$ (0°) 5·109, (15°) 5·062, (25°) 4·917, (35°) 4·772, (45°) 4·643 ΔH° −5·9, ΔS° 2·8	68A
Ge^{IV}	gl	25	0·1 KCl	$K[HGeO_3^- + 2L \rightleftharpoons HGeO(H_{-2}L)_2^-]$ 5·48	58A
	gl	25	I KCl	$K[HGeO_3^- + 2L \rightleftharpoons HGeO(H_{-2}L)_2^-]$ $4·273 + 1·155\sqrt{I}$	63N
Te^{VI}	gl	25	0·1 KCl	$K[H_5TeO_6^- + L \rightleftharpoons H_3TeO_4(H_{-2}L)^-]$ 1·92	57A

57A P. J. Antikainen, *Suomen Kem.*, 1957, **B30**, 45
58A P. J. Antikainen, *Suomen Kem.*, 1958, **B31**, 255
60A P. J. Antikainen and K. Tevanen, *Suomen Kem.*, 1960, **B33**, 7
63N V. A. Nazarenko and G. V. Flyantikova, *Russ. J. Inorg. Chem.*, 1963, **8**, 1189 (2271)
67N V. A. Nazarenko and L. D. Ermak, *Russ. J. Inorg. Chem.*, 1967, **12**, 1304 (2472)
68A P. J. Antikainen and I. P. Pitkanen, *Suomen Kem.*, 1968, **B41**, 65, 108

$C_6H_{12}O_6$ D-Galactose L 388

Metal	Method	Temp	Medium	Log of equilibrium constant, remarks	Ref
B^{III}	gl	25	0·1 KCl	$K[B(OH)_4^- + 2L \rightleftharpoons B(H_{-2}L)_2^-]$ 2·39	58A
Ge^{IV}	gl	25	0·1 KCl	$K[HGeO_3^- + 2L \rightleftharpoons HGeO(H_{-2}L)_2^-]$ 3·29	58A
	gl	25	I KCl	$K[HGeO_3^- + 2L \rightleftharpoons HGeO(H_{-2}L)_2^-]$ $2·117 + 1·297\sqrt{I}$	63N

58A P. J. Antikainen, *Suomen Kem.*, 1958, **B31**, 255
63N V. A. Nazarenko and G. V. Flyantikova, *Russ. J. Inorg. Chem.*, 1963, **8**, 1189 (2271)

389 $C_6H_{12}O_6$ D-Glucose L

Metal	*Method*	*Temp*	*Medium*	*Log of equilibrium constant, remarks*	*Ref*
B^{III}	gl	25	0·1 KCl	$K[B(OH)_4^- + 2L \rightleftharpoons B(H_{-2}L)_2^-]$ 2·86	58A
	gl	0—35	0·025 borax	$K[B(OH)_4^- + L \rightleftharpoons B(OH)_2H_{-2}L^-]$ (0°) 2·33, (13°) 2·24, (25°) 2·13, (35°) 2·10 ΔH° −3·5±0·4, ΔS° −2·0±0·5, $K[B(OH)_4^- + 2L \rightleftharpoons B(H_{-2}L)_2^-]$ (0°) 2·95, (13°) 2·95, (25°) 2·94, (35°) 2·95 ΔH° −0·15±0·02, ΔS° 13±3	67C
	pH	25	I KCl	$K[B(OH)_4^- + 2L \rightleftharpoons B(H_{-2}L)_2^-]$ 2·376+1·073$\sqrt{I}$	67N
	gl	0—45	0·1 KCl	$K[B(OH)_4^- + L \rightleftharpoons B(OH)_2H_{-2}L^-]$ (0°) 2·305, (15°) 2·071, (25°) 2·022, (35°) 1·985, (45°) 1·978 ΔH° −1·3, ΔS° 4·8 $K[B(OH)_4^- + 2L \rightleftharpoons B(H_{-2}L)_2^-]$ (0°) 2·894, (15°) 2·750, (25°) 2·633, (35°) 2·560, (45°) 2·407 ΔH° −4·6, ΔS° −3·4	68A
Ge^{IV}	gl	25	0·1 KCl	$K[HGeO_3^- + 2L \rightleftharpoons HGeO(H_{-2}L)_2^-]$ 3·46	58A
	gl	25	I KCl	$K[HGeO_3^- + 2L \rightleftharpoons HGeO(H_{-2}L)_2^-]$ 1·451+1·178$\sqrt{I}$	63N

58A P. J. Antikainen, *Suomen Kem.*, 1958, **B32**, 255
63N V. A. Nazarenko and G. V. Flyantikova, *Russ. J. Inorg. Chem.*, 1963, **8**, 1189 (2271); 712 (1370)
67C J. M. Conner and V. C. Bulgrin, *J. Inorg. Nuclear Chem.*, 1967, **29**, 1953
67N V. A. Nazarenko and L. D. Ermak, *Russ. J. Inorg. Chem.*, 1967, **12**, 1304 (2472)
68A P. J. Antikainen and I. P. Pitkanen, *Suomen Kem.*, 1968, **B41**, 65

389.1 $C_6H_{12}O_6$ Inositol L

Metal	*Method*	*Temp*	*Medium*	*Log of equilibrium constant, remarks*	*Ref*
B^{III}	gl	25	0·1 KCl	$K[B(OH)_4^- + L \rightleftharpoons B(OH)_2(H_{-1}L)^-]$ 1·637	67F
Ge^{IV}	gl	25	0·1 KCl	$K[HGeO_3^- + 2L \rightleftharpoons HGeO(H_{-2}L)_2^-]$ 2·140	67F
Te^{VI}	gl	25	0·1 KCl	$K[H_5TeO_6^- + L \rightleftharpoons H_3TeO_4(H_{-2}L)^-]$ 1·773 $K[H_5TeO_6^- + 2L \rightleftharpoons HTeO_2(H_{-2}L)_2^-]$ 1·853	67F

67F R. Frostell and P. J. Antikainen, *Suomen Kem.*, 1967, **B40**, 86

390 $C_6H_{12}O_6$ D-Mannose L

Metal	*Method*	*Temp*	*Medium*	*Log of equilibrium constant, remarks*	*Ref*
As^{III}	gl	25	0·1 KCl	$K[As(OH)_4^- + L \rightleftharpoons As(OH)_2(H_{-2}L)^-]$ 2·22 $K[As(OH)_4^- + 2L \rightleftharpoons As(H_{-2}L)_2^-]$ 2·97	59A
B^{III}	gl	25	0·1 KCl	$K[B(OH)_4^- + 2L \rightleftharpoons B(H_{-2}L)_2^-]$ 4·52	58A
Ge^{IV}	gl	25	0·1 KCl	$K[HGeO_3^- + 2L \rightleftharpoons HGeO(H_{-2}L)_2^-]$ 4·13	58A

58A P. J. Antikainen, *Suomen Kem.*, 1958, **B31**, 255
59A P. J. Antikainen and V. M. K. Rossi, *Suomen Kem.*, 1959, **B32**, 185

$C_6H_{12}O_6$ L-Sorbose L 390.1

Metal	Method	Temp	Medium	Log of equilibrium constant, remarks	Ref
As^{III}	gl	25	0·1 KCl	$K[As(OH)_4^- + 2L \rightleftharpoons As(H_{-2}L)_2^-]$ 1·08	60A
B^{III}	gl	25	0·1 KCl	$K[B(OH)_4^- + 2L \rightleftharpoons B(H_{-2}L)_2^-]$ 5·80	59A
Ge^{IV}	gl	25	0·1 KCl	$K[HGeO_3^- + 2L \rightleftharpoons HGeO(H_{-2}L)_2^-]$ 5·35	59A

59A P. J. Antikainen and K. Tevanen, *Suomen Kem.*, 1959, **B32**, 214
60A P. J. Antikainen and K. Tevanen, *Suomen Kem.*, 1960, **B33**, 7

$C_6H_{12}O_7$ Gluconic acid HL 391

Metal	Method	Temp	Medium	Log of equilibrium constant, remarks	Ref
H^+	H	20	0·2 (KCl)	K_1 3·56	62K, 63K, 68L
	pH	25	50—95% methanol	K_1 (50%) 5·22, (80%) 5·8, (90%) 6·8, (95%) 7·5	66K
Cd^{2+}	pol	25	1 $NaNO_3$	β_2 2·09	63Z
Ce^{3+}	oth	20	0·1	K_1 3·64, paper electrophoresis	67M
Dy^{3+}	gl	25	0·2 KCl	K_1 2·30, β_2 4·57 ± 0·03	63K
	pH	25	50, 80% methanol	K_1 (50%) 4·67 ± 0·04, (80%) 5·49 ± 0·10	66K
Er^{3+}	gl	25	0·2 KCl	K_1 2·36, β_2 4·53 ± 0·04	63K
	pH	25	50, 80% methanol	K_1 (50%) 4·68 ± 0·03, (80%) 5·56 ± 0·07	66K
	sp	20	0, 80% methanol	K_1 (0%) 2·9 ± 0·1, (80%) 4·9 ± 0·2	67R
Eu^{3+}	gl	25	0·2 KCl	K_1 2·69, β_2 4·97 ± 0·07	63K
	pH	25	50, 80% methanol	K_1 (50%) 4·77 ± 0·08, (80%) 5·47 ± 0·09	66K
Gd^{3+}	gl	25	0·2 KCl	K_1 2·56, β_2 4·76 ± 0·07	63K
	pH	25	50, 80% methanol	K_1 (50%) 4·64 ± 0·06, (80%) 5·37 ± 0·10	66K
Hf^{4+}	ix	20	2 $HClO_4$	$K(Hf^{4+} + HL \rightleftharpoons HfL^{3+} + H^+)$ 1·49 ± 0·03	63R
Ho^{3+}	gl	25	0·2 KCl	K_1 2·36, β_2 4·51 ± 0·08	63K
	pH	25	50, 80% methanol	K_1 (50%) 4·62 ± 0·02, (80%) 5·53 ± 0·08	66K
La^{3+}	ix	?	0·2	K_1 2·38, K_2 1·90, K_3 1·42	62K
	oth	?	0·2	K_2 1·98, K_3 1·43, K_4 0·84, K_5 0·17, K_6 −0·4, electrostatic (?)	62K
	gl	25	0·2 KCl	K_1 2·43, β_2 4·25 ± 0·04	63K
	pH	25	50—90% methanol	K_1 (50%) 4·48 ± 0·05, (80%) 5·17 ± 0·05, (90%) 6·11	66K
	oth	20	0·1	K_1 2·60, K_2 2·15, K_3 1·54, paper electrophoresis	67M
Lu^{3+}	gl	25	0·2 KCl	K_1 2·72, β_2 4·78 ± 0·09	63K
	pH	25	50, 80% methanol	K_1 (50%) 5·09 ± 0·08, (80%) 6·05 ± 0·07	66K
Nb^V	gl	25	0·1	$K[Nb(OH)_n^{5-n} + L^- \rightleftharpoons Nb(OH)_nL^{4-n}]$ 2·78 $K[Nb(OH)_n(H_{-1}L)^{3-n} + H^+ \rightleftharpoons Nb(OH)_nL^{4-n}]$ 7·82	68L
Nd^{3+}	gl	25	0·2 KCl	K_1 2·71, β_2 4·70 ± 0·03	63K
	pH	25	50—95% methanol	K_1 (50%) 4·76 ± 0·05, (80%) 5·51 ± 0·03, (90%) 6·6, (95%) 7·0	66K
	sp	20	0, 80% methanol	K_1 (0%) 3·10 ± 0·09, (80%) 5·26 ± 0·04	67R
	pH	20	80% methanol	K_1 5·2 ± 0·1	67R

continued overleaf

391 $C_6H_{12}O_7$ (contd)

Metal	Method	Temp	Medium	Log of equilibrium constant, remarks	Ref
Ni^{2+}	gl	25	0·12 (KNO_3)	$K[2Ni^{2+}+4OH^-+L^- \rightleftharpoons Ni_2(OH)_4L^-]$ 29·4, pH > 9 (?)	65J
	sp	25	0·12 (KNO_3)	K_1 1·82, pH < 7 (?) $K_s[Ni_2(OH)_3L]$ −27, pH 7—9	65J
Pr^{3+}	gl	25	0·2 KCl	K_1 2·55, β_2 4·55±0·02	63K
	pH	25	50—95% methanol	K_1 (50%) 4·71±0·03, (80%) 5·45±0·04, (90%) 6·49, (95%) 7·2	66K
Sm^{3+}	gl	25	0·2 KCl	K_1 2·65, β_2 4·88±0·06	63K
	pH	25	80% methanol	K_1 5·51±0·07	66K
Tb^{3+}	gl	25	0·2 KCl	K_1 2·34, β_2 4·67±0·05	63K
	pH	25	50% methanol	K_1 4·68±0·04	66K
Y^{3+}	gl	25	0·2 KCl	K_1 2·38, β_2 4·52±0·03	63K
Yb^{3+}	gl	25	0·2 KCl	K_1 2·72, β_2 4·68±0·02	62K
	pH	25	50, 80% methanol	K_1 (50%) 5·10±0·09, (80%) 5·97±0·09	66K
	pol	?	0·1 $(CH_4)_4NI$	$K(YbL_6^{3-}+e^- \rightleftharpoons YbL_4^{2-}+2L^-)$ −5·8	58K
Zr^{4+}	ix	20	2 $HClO_4$	$K(Zr^{4+}+HL \rightleftharpoons ZrL^{3+}+H^+)$ 1·73±0·04	63R

58K N. A. Kostromina and S. I. Yakubson, *J. Inorg. Chem. (U.S.S.R.)*, 1958, **3**, No. 11, 104 (2506)
62K N. A. Kostromina, *Russ. J. Inorg. Chem.*, 1962, **7**, 806 (1559)
63K N. A. Kostromina, *Russ. J. Inorg. Chem.*, 1963, **8**, 988 (1900)
63R H. Roth, L. Czegledi, and C. G. Macarovici, *Acad. Rep. Populare Romine, Studii Cercetari Chimie (Cluj)*, 1963, **14**, 69
63Z V. K. Zolotukhin and Z. G. Galanets, *Visn. L'vivsk. Derzh. Univ. Ser. Khim.*, 1963, 19 (*Chem. Abs.*, 1965, **62**, 4909a)
65J L. G. Joyce and W. F. Pickering, *Austral. J. Chem.*, 1965, **18**, 783
66K N. A. Kostromina and E. D. Romanenko, *Russ. J. Inorg. Chem.*, 1966, **11**, 598 (1116)
67M Gh. Marcu and Gh. Murgu, *Rev. Roumaine Chim.*, 1967, **12**, 957
67R E. D. Romanenko and N. A. Kostromino, *Russ. J. Inorg. Chem.*, 1967, **12**, 266 (516)
68L J. E. Land and C. V. Osborne, *J. Less-Common Metals*, 1968, **14**, 349

394 $C_6H_{12}N_4$ Hexamethylenetetramine L

Metal	Method	Temp	Medium	Log of equilibrium constant, remarks	Ref
Ag^+	sol	20	?	K_2 3·24, K_3 2·89	63G
	sol	20	0·3 ClO_3^-	$K(AgLClO_3+L \rightleftharpoons AgL_2^++ClO_3^-)$ 3·65 $K(AgLClO_3+2L \rightleftharpoons AgL_3^++ClO_3^-)$ 3·10	67M

63G E. A. Gyunner, *Russ. J. Inorg. Chem.*, 1963, **8**, 218
67M L. M. Mel'nichenko and E. A. Gyunner, *Russ. J. Inorg. Chem.*, 1967, **12**, 801 (1524)

394.1 $C_6H_{14}O_2$ 2,3-Dimethylbutan-2,3-diol L

Metal	Method	Temp	Medium	Log of equilibrium constant, remarks	Ref
B^{III}	gl	0—35	0·025 borax	$K[B(OH)_4^-+L \rightleftharpoons B(OH)_2(H_{-2}L)^-]$ (0°) 1·38, (13°) 1·28, (25°) 1·04, (35°) 0·70 ΔH° −7·1±0·7, ΔS° −19±5, $K[B(OH)_4^-+2L \rightleftharpoons B(H_{-2}L)_2^-]$ (0°) 2·79, (13°) 2·60, (25°) 2·33, (35°) 1·95 ΔH° −9·0±0·9, ΔS° −20±5	67C

67C J. M. Conner and V. C. Bulgrin, *J. Inorg. Nuclear Chem.*, 1967, **29**, 1953

$C_6H_{14}O_3$ 2-Hydroxy-2-methylpentanoic acid (methylpropylglycollic acid) HL 394.2

Metal	Method	Temp	Medium	Log of equilibrium constant, remarks	Ref
Ce^{3+}	qh	25	1 $NaClO_4$	K_1 2·21, K_2 1·66, K_3 1·03, K_4 0·81	64E
Dy^{3+}	qh	25	1 $NaClO_4$	K_1 3·00, K_2 2·47, K_3 1·86, K_4 1·12	64E
La^{3+}	qh	25	1 $NaClO_4$	K_1 2·20, K_2 1·25, K_3 0·77	64E
Nd^{3+}	qh	25	1 $NaClO_4$	K_1 2·38, K_2 1·85, K_3 1·17, K_4 1·06	64E
Sm^{3+}	qh	25	1 $NaClO_4$	K_1 2·59, K_2 2·12, K_3 1·50	64E
Yb^{3+}	qh	25	1 $NaClO_4$	K_1 3·29, K_2 2·55, K_3 2·21, K_4 1·12	64E

64E L. Eeckhaut, F. Verbeek, H. Deelstra, and J. Hoste, *Analyt. Chim. Acta*, 1964, 30, 369

$C_6H_{14}O_6$ D-Galactitol (D-dulcitol) L 394.3

Metal	Method	Temp	Medium	Log of equilibrium constant, remarks	Ref
B^{III}	gl	25	0·1 KCl	$K[B(OH)_4^- + 2L \rightleftharpoons B(H_{-2}L)_2^-]$ 5·23	59A
Ge^{IV}	gl	25	0·1 KCl	$K[HGeO_3^- + 2L \rightleftharpoons HGeO(H_{-2}L)_2^-]$ 4·71	59A

59A P. J. Antikainen and V. M. K. Rossi, *Suomen Kem.*, 1959, B32, 182

$C_6H_{14}O_6$ D-Mannitol L 394.4

Metal	Method	Temp	Medium	Log of equilibrium constant, remarks	Ref
As^{III}	gl	25	0·1 KCl	$K[As(OH)_4^- + L \rightleftharpoons As(OH)_2(H_{-2}L)^-]$ 0·85	59A
B^{III}	gl	25	0·1 (KCl)	$K[H_3BO_3 + 2L \rightleftharpoons B(H_{-2}L)_2^- + H^+]$ −4·00	49R
	gl	25	0—2 KCl	$K[H_3BO_3 + L \rightleftharpoons B(OH)_2(H_{-2}L)^- + H^+]$ (→ 0) −5·22±0·05, (0·1) −5·10, (0·4) −5·02, (2) −5·13 $K[H_3BO_3 + 2L \rightleftharpoons B(H_{-2}L)_2^- + H^+]$ (→ 0) −4·36±0·02, (0·1) −4·18, (0·4) −4·15, (2) −4·29	55A
	gl	25	0·1 KCl	$K[B(OH)_4^- + 2L \rightleftharpoons B(H_{-2}L)_2^-]$ 4·92	59A
	gl	0—35	0·025 borax	$K[B(OH)_4^- + L \rightleftharpoons B(OH)_2(H_{-2}L)^-]$ (0°) 3·62, (13°) 3·36, (25°) 3·04, (35°) 2·90 $\Delta H°$ −8·1±0·8, $\Delta S°$ −13±3, $K[B(OH)_4^- + 2L \rightleftharpoons B(H_{-2}L)_2^-]$ (0°) 5·43, (13°) 5·31, (25°) 5·14, (35°) 5·05 $\Delta H°$ −4·5±0·4, $\Delta S°$ 9±2	67C
	gl	25	I KCl	$K[B(OH)_4^- + 2L \rightleftharpoons B(H_{-2}L)_2^-]$ 4·225+0·554$\sqrt{I}$	67N
	gl	0—45	0·1 KCl	$K[B(OH)_4^- + L \rightleftharpoons B(OH)_2(H_{-2}L)^-]$ (0°) 4·205, (25°) 4·000, (40°) 3·617, (45°) 3·398 $\Delta H°$ (0°) 0·8, (15°) −4·0, (25°) −7·7, (35°) −11·9, (45°) −16·7, $\Delta S°$ (0°) 22·1, (15°) 5·1, (25°) −7·6, (35°) −21·6, (45°) −36·6 $K[B(OH)_4^- + 2L \rightleftharpoons B(H_{-2}L)_2^-]$ (0°) 5·408, (25°) 4·888, (40°) 4·610, (45°) 4·551 $\Delta H°$ (0°) −8·6, (15°) −8·1, (25°) −7·6, (35°) −7·0, (45°) −6·2, $\Delta S°$ (0°) −6·9, (15°) −4·9, (25°) −3·1, (35°) −1·1, (45°) 1·4	68A

continued overleaf

394.4 $C_6H_{14}O_6$ (contd)

Metal	Method	Temp	Medium	Log of equilibrium constant, remarks	Ref
Ge^{IV}	gl	25	0·1 KCl	$K[HGeO_3^- + 2L \rightleftharpoons HGeO(H_{-2}L)_2^-]$ 4·53	59A
	gl	25	I KCl	$K[HGeO_3^- + 2L \rightleftharpoons HGeO(H_{-2}L)_2^-]$ 3·394+1·055$\sqrt{I}$	63N
Te^{VI}	qh	25	0·1 KCl	$K[H_5TeO_6^- + L \rightleftharpoons H_3TeO_4(H_{-2}L)^-]$ 3·19	56A

49R S. D. Ross and A. J. Catotti, *J. Amer. Chem. Soc.*, 1949, **71**, 3563
55A P. J. Antikainen, *Acta Chem. Scand.*, 1955, **9**, 1008
56A P. J. Antikainen, *Suomen Kem.*, 1956, **B29**, 14
59A P. J. Antikainen and V. M. K. Rossi, *Suomen Kem.*, 1959, **B32**, 182, 185
63N J. A. Nazarenko and G. V. Flyantikova, *Russ. J. Inorg. Chem.*, 1963, **8**, 1189 (2271)
67C J. M. Conner and V. C. Bulgrin, *J. Inorg. Nuclear Chem.*, 1967, **29**, 1953
67N V. A. Nazarenko and L. D. Ermak, *Russ. J. Inorg. Chem.*, 1967, **12**, 335 (643), 1079 (2051)
68A P. J. Antikainen and I. P. Pitkanen, *Suomen Kem.*, 1968, **B41**, 65

394.5 $C_6H_{14}O_6$ D-Sorbitol L

Metal	Method	Temp	Medium	Log of equilibrium constant, remarks	Ref
B^{III}	gl	25	0·1 KCl	$K[B(OH)_4^- + 2L \rightleftharpoons B(H_{-2}L)_2^-]$ 5·65	59A
	gl	25	I KCl	$K[B(OH)_4^- + 2L \rightleftharpoons B(H_{-2}L)_2^-]$ 4·533+1·341$\sqrt{I}$	67N
Ge^{IV}	gl	25	0·1 KCl	$K[HGeO_3^- + 2L \rightleftharpoons HGeO(H_{-2}L)_2^-]$ 5·09	59A

59A P. J. Antikainen and V. M. K. Rossi, *Suomen Kem.*, 1959, **B32**, 182
67N V. A. Nazarenko and L. D. Ermak, *Russ. J. Inorg. Chem.*, 1967, **12**, 335 (643), 1079 (2051)

394.6 $C_6H_{14}N^+$ *NNN*-Allyltrimethylammonium cation L^+

Metal	Method	Temp	Medium	Log of equilibrium constant, remarks	Ref
Pt^{2+}	sp	30—60	2 (NaCl)	$K(PtCl_4^{2-} + L^+ \rightleftharpoons PtCl_3L + Cl^-)$ (30°) 2·40, (44·5°) 2·24, (60°) 2·07 ΔH −5·3±0·2, ΔS −6·6±0·6	67D

67D R. G. Denning, F. R. Hartley, and L. M. Venanzi, *J. Chem. Soc. (A)*, 1967, 324

398.1 $C_6H_{15}N$ Dipropylamine L

Metal	Method	Temp	Medium	Log of equilibrium constant, remarks	Ref
Ag^+	oth	?	0·1 $NaClO_4$ acetone	β_2 10·05±0·04, coulometric titration	65M

65M K. K. Mead, D. L. Maricle, and C. A. Streuli, *Analyt. Chem.*, 1965, **37**, 237

$C_6H_{16}N_2$ *N*-Butyl-1,2-diaminoethane L 401

Metal	Method	Temp	Medium	Log of equilibrium constant, remarks	Ref
H^+	gl	0—25	0·50 KNO_3	ΔH_1 (0°) −9·4, ΔH_2 −8·7	52B
Cu^{2+}	gl	0—25	0·5 KNO_3	ΔH_1 (0°) −8·0, ΔS_1 19, ΔH_2 (0°) −8·2, ΔS_2 12	52B
Ni^{2+}	gl	0—25	0·5 KNO_3	ΔH_1 (0°) −7·8, ΔS_1 4·8, ΔH_2 (0°) −6·1, ΔS_2 5·1, ΔH_3 (0°) −8·8, ΔS_3 −19·4	52B

52B F. Basolo and R. K. Murmann, *J. Amer. Chem. Soc.*, 1952, **74**, 2373, 5243

$C_6H_{16}N_2$ *NN*-Diethyl-1,2-diaminoethane L 402

Metal	Method	Temp	Medium	Log of equilibrium constant, remarks	Ref
H^+	gl	25	*I* ($NaClO_4$)	K_1 9·93+0·439 *I*, K_2 6·51+1·018$\sqrt{I}/(1+0·977\sqrt{I})$+ 0·263 *I*	63N, 64N
Cu^{2+}	gl	25	*I* ($NaClO_4$)	K_1 8·05+0·576 *I*−0·117 $I^{3/2}$+0·021 I^2, K_2 5·47+ 0·537 *I*+0·192 $I^{3/2}$−0·129 I^2	63N
	gl	25	*I* ($NaClO_4$)	$K(CuOHL^+ + H^+ \rightleftharpoons CuL^{2+})$ 7·30 +1·018$\sqrt{I}/(1+\sqrt{I})$	64N
				$K[Cu_2(OH)_2L_2^{2+} + 2H^+ \rightleftharpoons 2CuL^{2+}]$ 11·42+1·018$\sqrt{I}/(1+\sqrt{I})$	
			→ 0	$K[2CuOHL^+ \rightleftharpoons Cu_2(OH)_2L_2^{2+}]$ 3·18	
				$K(Cu^{2+} + OH^- + L \rightleftharpoons CuOHL^+)$ 14·74	

63N R. Nasanen, P. Merilainen, and M. Koskinen, *Suomen Kem.*, 1963, **B36**, 9, 110
64N R. Nasanen, P. Merilainen, and M. Koskinen, *Suomen Kem.*, 1964, **B37**, 41

$C_6H_{16}N_2$ *NN'*-Diethyl-1,2-diaminoethane L 403

Metal	Method	Temp	Medium	Log of equilibrium constant, remarks	Ref
H^+	gl	25	*I* ($NaClO_4$)	K_1 10·16+0·775 *I*−0·581 $I^{3/2}$+0·203 I^2, K_2 6·96+ 1·018$\sqrt{I}/(1+0·868\sqrt{I})$+0·226 *I*	63N, 64N, 65N
Cu^{2+}	cal	0	0·50 KNO_3	$\Delta H_{\beta2}$° −17·5, $\Delta S_{\beta2}$° 21	54B
	gl	25	*I* ($NaClO_4$)	K_1 8·79+0·580 *I*−0·064 $I^{3/2}$, −0·024 I^2	63N
			→ 0	K_2 5·57	
	gl	25	*I* ($NaClO_4$)	K_2 5·41+0·626 *I*+0·122 $I^{3/2}$, −0·202 I^2	64N
				$K(CuOHL^+ + H^+ \rightleftharpoons CuL^{2+})$ 7·77+1·018$\sqrt{I}/(1+\sqrt{I})$	
				$K[Cu_2(OH)_2L_2^{2+} + 2H^+ \rightleftharpoons 2CuL^{2+}]$ 11·96+1·018$\sqrt{I}/(1+\sqrt{I})$	
			→ 0	$K[2CuOHL \rightleftharpoons Cu_2(OH)_2L^+]$ 3·58	
				$K(Cu^{2+} + OH^- + L \rightleftharpoons CuOHL^+)$ 15·01	
				See ethylenediamine (120), 1,2-propanediamine (158), and 1,3-propanediamine (159) for mixed ligands	
Ni^{2+}	cal	0	0·50 KNO_3	$\Delta H_{\beta2}$° −7·8, $\Delta S_{\beta2}$° 27	54B
	gl	25	*I* ($NaClO_4$)	K_1 5·50+1·112 *I*−1·078 $I^{3/2}$+0·380 I^2, K_2 2·36+ 1·168 *I* −1·318 $I^{3/2}$+0·536 I^2	65N

54B F. Basolo and R. K. Murmann, *J. Amer. Chem. Soc.*, 1954, **76**, 211
63N R. Nasanen and P. Merilainen, *Suomen Kem.*, 1963, **B36**, 205
64N R. Nasanen and P. Merilainen, *Suomen Kem.*, 1964, **B37**, 1, 54
65N R. Nasanen, M. Koskinen, and P. Tilus, *Suomen Kem.*, 1965, **B38**, 125

404 $C_6H_{16}N_2$ *NNN′N′*-Tetramethyl-1,2-diaminoethane L

Metal	Method	Temp	Medium	Log of equilibrium constant, remarks	Ref
H^+	gl	25	$\to 0$	K_1 9·15	67N
			I ($NaClO_4$)	K_2 5·583+(1·018$\sqrt{I}$)/(1+0·550$\sqrt{I}$)+0·300 I	
				K_1K_2 14·732+(1·018$\sqrt{I}$)/(1+0·194$\sqrt{I}$)+0·612 I	
Cu^{2+}	gl	25	$\to 0$	K_1 7·19	67N
			I ($NaClO_4$)	$K(Cu^{2+}+H_2L^{2+} \rightleftharpoons CuL^{2+}+2H^+)$ −7·541+ (1·018$\sqrt{I}$)/(1+0·751$\sqrt{I}$)−0·316 I	

Errata. 1964, p. 491, Table 404. Under H^+ *delete* first line (gl . . . 54B).
Under Cu^{2+} *delete* first line (gl . . . 54B).
Under references *delete* ref. 54B.

67N R. Nasanen, M. Koskinen, M. L. Alatalo, L. Adler, and S. Koskela, *Suomen Kem.*, 1967, **B40**, 124

404.1 $C_6H_{16}N_2$ 2,3-Dimethyl-2,3-diaminobutane L

Metal	Method	Temp	Medium	Log of equilibrium constant, remarks	Ref
H^+	gl	0, 25	0·50 KNO_3	K_1 (0°) 10·73, (25°) 10·13, K_2 (0°) 7·18, (25°) 6·56 ΔH_1 (0°) −8·9, ΔH_2 −9·2	53B
	gl	0—40	0·1—1	K_1 (0°) (0·1) 10·40, (0·5) 10·60, (1) 10·80, (10°) (0·1) 10·18, (0·5) 10·35, (1) 10·55, (20°) (0·1) 9·88, (0·5) 10·05, (1) 10·18, (30°) (0·1) 9·70, (0·5) 9·80, (1) 9·90, (40°) (0·1), 9·54, (0·5) 9·65, (1) 9·75, K_2 (0°) (0·1) 6·90, (0·5) 7·10, (1) 7·30 (10°) (0·1) 6·70, (0·5) 6·90, (1) 7·10, (20°) (0·1) 6·40, (0·5) 6·55, (1) 6·70, (30°) (0·1) 6·25, (0·5) 6·35, (1) 6·50, (40°) (0·1) 6·10, (0·5) 6·20, (1) 6·30	68P
Cd^{2+}	gl	0—40	0·1—1	β_2 (0°) (0·1) 12·80, (0·5) 13·32, (1) 13·80, (10°) (0·1) 11·94, (0·5) 12·50, (1) 12·86, (20°) (0·1) 11·24, (0·5) 11·86, (1) 12·25, (30°) (0·1) 10·44, (0·5) 10·90, (1) 11·40, (40°) (0·1) 9·60, (0·5) 10·10, (1) 10·66 $\Delta H_{\beta 2}°$ 1·26 (?), $\Delta S_{\beta 2}°$ 0·4 (?)	68P
Cu^{2+}	cal	0	0·50 KNO_3	$\Delta H_{\beta 2}°$ −24·4, $\Delta S_{\beta 2}°$ 19	54B
	gl	0, 25	0·50 KNO_3	K_1 (0°) 12·22, (25°) 11·63, β_2 (0°) 23·10, (25°) 21·87	54B
Ni^{2+}	cal	0	0·50 KNO_3	$\Delta H_{\beta 2}°$ −14·1, $\Delta S_{\beta 2}°$ 20	54B
	sp	15—39	0·50 KNO_3	β_2 (15°) 14·87, (25°) 14·68, (39°) 14·32	54B
Zn^{2+}	gl	0—40	0·1—1	β_2 (0°) (0·1) 14·08, (0·5) 14·80, (1) 15·20, (10°) (0·1) 13·30, (0·5) 14·02, (1) 14·30, (20°) (0·1) 12·24, (0·5) 12·95, (1) 13·20, (30°) (0·1) 11·60, (0·5) 12·00, (1) 12·30, (40°) (0·1) 10·76, (0·5) 11·20, (1) 11·60 $\Delta H_{\beta 2}°$ 1·45 (?), $\Delta S_{\beta 2}°$ 0 (?)	68P

53B F. Basolo, R. K. Murmann, and Y. T. Chen, *J. Amer. Chem. Soc.*, 1953, **75**, 1478
54B F. Basolo, Y. T. Chen, and R. K. Murmann, *J. Amer. Chem. Soc.*, 1954, **76**, 956
68P N. G. Podder, *Current Sci.*, 1968, **37**, 48

$C_6H_{17}N_3$ 3,3′-Iminobis(propylamine) L 406

Metal	Method	Temp	Medium	Log of equilibrium constant, remarks	Ref
H^+	gl	25	0·1 KCl	K_1 10·65±0·02, K_2 9·57±0·03, K_3 7·72±0·03	66V
	cal	25	0·1 KCl	$\Delta H_1°$ −12·29±0·01, $\Delta S_1°$ 7·5±0·1, $\Delta H_2°$ −12·99±0·02, $\Delta S_2°$ 0·2±0·1, $\Delta H_3°$ −10·47±0·02, $\Delta S_3°$ 0·2±0·1	66P
Co^{2+}	gl	25	0·1 KCl	K_1 6·92±0·01	66V
	cal	25	0·1 KCl	$\Delta H_1°$ −7·8±0·1, $\Delta S_1°$ 5·6±0·4	66P
Cu^{2+}	gl	25	0·1 KCl	K_1 14·20±0·06	66V
	cal	25	0·1 KCl	$\Delta H_1°$ −16·09±0·08, $\Delta S_1°$ 11·0±0·5	66P
	gl	25	0·1 KCl	$K(CuL^{2+}+OH^- \rightleftharpoons CuLOH^+)$ 4·1±0·2	66V
	cal	25	0·1 KCl	$\Delta H°$ −2·3±0·3, $\Delta S°$ 11±1	66P
Ni^{2+}	gl	25	0·1 KCl	K_1 9·19±0·02, K_2 3·55±0·08	66V
	cal	25	0·1 KCl	$\Delta H_1°$ −10·56±0·06, $\Delta S_1°$ 6·6±0·3, $\Delta H_2°$ −7·1±0·1, $\Delta S_2°$ −7·5±0·7	66P
Zn^{2+}	gl	25	0·1 KCl	K_1 7·92±0·03	66V
	cal	25	0·1 KCl	$\Delta H_1°$ −5·4±0·2, $\Delta S_1°$ 18·0±0·7	66P
	gl	25	0·1 KCl	$K(ZnL^{2+}+OH^- \rightleftharpoons ZnLOH^+)$ 5·2±0·1	66V
	cal	25	0·1 KCl	$\Delta H°$ −3·8±0·2, $\Delta S°$ 11·2±0·9	66P

66P P. Paoletti, F. Nuzzi, and A. Vacca, *J. Chem. Soc. (A)*, 1966, 1385
66V A. Vacca, D. Arenare, and P. Paoletti, *Inorg. Chem.*, 1966, 5, 1384

$C_6H_{18}N_4$ Ethylenebis(2′-aminoethylamine) (triethylenetetramine) (trien) L 407

Metal	Method	Temp	Medium	Log of equilibrium constant, remarks	Ref
H^+	gl	30, 40	1 (KNO_3)	ΔH_1 −10·0, ΔH_2 −9·6, ΔH_3 −9·6, ΔH_4 −5·6	50J
	gl	25	0·1	K_1 9·80, K_2 9·08, K_3 6·55, K_4 3·25	63M
	gl	20, 26	0·1 (KCl)	K_1 (20°) 9·80, (26°) 9·69, K_2 (20°) 9·26, (26°) 9·09, K_3 (20°) 6·76, (26°) 6·62, K_4 (20°) 3·55, (26°) 3·37	65P
Cd^{2+}	gl	30, 40	1 (KNO_3+KCl)	$\Delta H_1°$ (35°) −4 $K(3Cd^{2+}+2L \rightleftharpoons Cd_3L_2^{6+})$ (30°) 3·19, (40°) 3·07 $\Delta H°$ −4	52J
	cal	25	0·1 KNO_3	ΔH_1 −9·2, ΔS_1 19	65W
Co^{2+}	gl	30, 40	1 (KNO_3+KCl)	$\Delta H_1°$ (35°) −9 $K(3Co^{2+}+2L \rightleftharpoons Co_3L_2^{6+})$ (30°) 3·19, (40°) 3·07 $\Delta H°$ −4	52J
Cr^{3+}	gl	20, 26	0·1 (KCl)	K_1 (20°) 7·71, (26°) 8·0	65P
Cu^{2+}	gl	30, 40	1 (KNO_3+KCl)	$\Delta H_1°$ (35°) −22	52J
	cal	25	0·1 KNO_3	ΔH_1 −21·4, ΔS_1 21	65W
Fe^{2+}	gl	30, 40	1·3 KCl	$\Delta H_1°$ (35°) −9 $K(3Fe^{2+}+2L \rightleftharpoons Fe_3L_2^{6+})$ (30°) 3·92, (40°) 3·70 $\Delta H°$ −9	52J

continued overleaf

407 $C_6H_{18}N_4$ (contd)

Metal	Method	Temp	Medium	Log of equilibrium constant, remarks	Ref
Mn^{2+}	gl	30, 40	1 (KNO_3+KCl)	$\Delta H_1°$ (35°) −4 $K(3Mn^{2+}+2L \rightleftharpoons Mn_3L_2^{6+})$ (30°) 2·84, (40°) 2·72 $\Delta H°$ −4	52J
Ni^{2+}	gl	30, 40	1 (KNO_3+KCl)	$\Delta H_1°$ (35°) −13 $K(3Ni^{2+}+2L \rightleftharpoons Ni_3L_2^{6+})$ (30°) 5·63, (40°) 5·41 $\Delta H°$ −9	52J
	kin	25	0·1 ($NaClO_4$)	K_1 13·82, $K(Ni^{2+}+HL^+ \rightleftharpoons NiHL^{3+})$ 8·7, $K(NiL^{2+}+H^+ \rightleftharpoons NiHL^{3+})$ 4·7	63M
	cal	25	0·1 KNO_3	ΔH_1 −13·9, ΔS_1 18 See oxalic acid (111) for mixed ligands	65W
Pb^{2+}	pol	25	1 KNO_3	K_1 9·9	68L
Zn^{2+}	gl	30, 40	1 (KNO_3+KCl)	$\Delta H_1°$ (35°) −4	52J
	cal	25	0·1 KNO_3	ΔH_1 −8·3, ΔS_1 27	65W

50J H. B. Jonassen, R. B. LeBlanc, A. W. Meibohm, and R. M. Rogan, *J. Amer. Chem. Soc.*, 1950, **72**, 2430
52J H. B. Jonassen, G. G. Hurst, R. B. LeBlanc, and A. W. Meibohm, *J. Phys. Chem.*, 1952, **56**, 16
63M D. W. Margerum, D. B. Rarabacher, and J. F. G. Clarke jun., *Inorg. Chem.*, 1963, **2**, 667
65P R. L. Pecsok, R. A. Garber, and L. D. Shields, *Inorg. Chem.*, 1965, **4**, 447
65W D. L. Wright, J. H. Holloway, and C. N. Reilley, *Analyt. Chem.*, 1965, **37**, 884
68L T.-T. Lai and J.-Y. Chen, *J. Electroanalyt. Chem.*, 1968, **16**, 413

408 $C_6H_{18}N_4$ 2,2′2″-Triaminotriethylamine L

Errata. 1964, p. 493, Table 408. Under Fe^{2+} *for* ΔS_1 −18·5 *read* ΔS_1 18·5.
Under Mn^{2+} *for* ΔS_1 −16·5 *read* ΔS_1 16·5.
Under Ag^+ *for* $K(2Ag^+ + L \rightleftharpoons Ag_2L^{2+})$ *read* $K(AgL^+ + Ag^+ \rightleftharpoons Ag_2L^{2+})$

408.1 $C_6HO_3Cl_3$ Trichlorohydroxy-1,4-benzoquinone HL

Metal	Method	Temp	Medium	Log of equilibrium constant, remarks	Ref
H^+	gl	25	0·5 (NaCl)	K_1 1·05±0·04	66B
	sp	25	var H_2SO_4	K_2 −8·8±1·0	66B
Ge^{IV}	sp	25	0·5 (NaCl)	$K[Ge(OH)_4+2HL \rightleftharpoons Ge(OH)_2L_2] \leqslant 1·4$	66B

66B A. Beauchamp and R. L. Benoit, *Canad. J. Chem.*, 1966, **44**, 1607, 1615

$C_6H_2O_4Cl_2$ 2,5-Dichloro-3,6-dihydroxy-1,4-benzoquinone (chloranilic acid) H_2L 409

Metal	*Method*	*Temp*	*Medium*	*Log of equilibrium constant, remarks*	*Ref*
H^+	sp	25	?	K_1 2·97, K_2 0·49	62N
	gl	25	0·5 HCl	K_1 2·37, K_2 0·66	64B
	sp	25	0·37 ($NaClO_4$)	K_1 2·75	64L
	sp	25	0·15	K_1 2·72±0·01, K_2 0·81±0·05	67C
Fe^{3+}	sp	15—35	0·15	K_1 (15°) 5·91, (25°) 5·81, (35°) 5·75, β_2 (15°) 9·95, (25°) 9·84, (35°) 9·83, $\Delta H_1°$ −3·36, $\Delta S_1°$ −15·4(?), $\Delta H_{\beta 2}°$ −2·60, $\Delta S_{\beta 2}$ −3·62(?)	67C
	sp	15—35	0·15	K_1 (15°) 5·82, (25°) 5·51, (35°) 5·16, β_2 (15°) 9·74, (25°) 9·47, (35°) 9·24 $\Delta H_1°$ −14·7, $\Delta S_1°$ 24·4(?), $\Delta H_{\beta 2}°$ −2·60, $\Delta S_{\beta 2}°$ −7·4(?)	67C
Ge^{IV}	sp	25	2·5	$K[Ge(OH)_4+HL^- \rightleftharpoons Ge(OH)_3L^-]$ 7·64±0·06, $K[Ge(OH)_3L^-+HL^- \rightleftharpoons Ge(OH)_2L_2^{2-}]$ 6·30±0·05, $K[Ge(OH)_2L_2^{2-}+HL^- \rightleftharpoons Ge(OH)L_3^{3-}]$ 5·65±0·04	62N
	gl	25	0·5 NaCl	$K[Ge(OH)_4+2HL^- \rightleftharpoons Ge(OH)_2L_2^{2-}]$ 6·57	64B
	sp	25	0·5 HCl	$K[Ge(OH)_2L_2^{2-}+H^+ \rightleftharpoons Ge(OH)_2L_2H^-]$ 0·8±0·1	64B
Hf^{4+}	sp	25	3 $HClO_4$	$K(Hf^{4+}+H_2L \rightleftharpoons HfL^{2+}+2H^+)$ 3·73±0·02, $K(Hf^{4+}+3H_2L \rightleftharpoons HfL_3^{2-}+6H^+)$ 11·63±0·02	67V
Mo^{VI}	sp	25	0·37 ($NaClO_4$)	$K[Mo_3O_{11}^{4-}+3HL^- \rightleftharpoons Mo_3O_{11}(HL)_3^{7-}]$ 2·30	64L
Ni^{2+}	sp	15—35	0·15	K_1 (15°) 4·08, (25°) 4·02, (35°) 3·95 $\Delta H_1°$ −1·66, $\Delta S_1°$ −12·8(?)	67C

62N V. A. Nazarenko and G. V. Flyantikova, *Russ. J. Inorg. Chem.*, 1962, **1**, 1210
64B A. Beauchamp and R. L. Benoit, *Canad. J. Chem.*, 1964, **42**, 2161
64L W. F. Lee, N. K. Shastri, and E. S. Amis, *Talanta*, 1964, **11**, 685
67C D. K. Cabbiness and E. S. Amis, *Bull. Chem. Soc. Japan*, 1967, **40**, 435
67V L. P. Varga and F. C. Veatch, *Analyt. Chem.*, 1967, **39**, 1101

$C_6H_2O_8N_2$ 3,6-Dinitro-2,5-dihydroxy-1,4-benzoquinone (nitroanilic acid) H_2L 409.1

Metal	*Method*	*Temp*	*Medium*	*Log of equilibrium constant, remarks*	*Ref*
H^+	sp	25	3 $LiClO_4$	K_1 −0·5, K_2 −3·0 (?)	67B
Ge^{IV}	sp	25	3 $LiClO_4$	$K[Ge(OH)_4+2HL^- \rightleftharpoons Ge(OH)_2L_2^{2-}]$ 4·9±0·2	67B

67B A. Beauchamp and R. Benoit, *Bull. Soc. chim. France*, 1967, 672

$C_6H_4O_5N_2$ 2,4-Dinitrophenol HL 413

Metal	*Method*	*Temp*	*Medium*	*Log of equilibrium constant, remarks*	*Ref*
H^+	sp	25	0·1	K_1 3·93	66P
La^{3+}	sp	25	0·1	K_1 1·01±0·02	66P
Ni^{2+}	pol	20	~ 0·3 25% ethanol	K_1 0·6, acetate buffer	67C

66P C. Postmus jun., L. B. Magnusson, and C. A. Craig, *Inorg. Chem.*, 1966, **5**, 1154
67C E. Casassas and L. Eek, *J. Chim. phys.*, 1967, **64**, 971

413.1 $C_6H_4O_{10}S_2$ 3,6-Disulpho-2,5-dihydroxy-1,4-benzoquinone (euthiochronic acid) H_4L

Metal	Method	Temp	Medium	Log of equilibrium constant, remarks	Ref
H^+	sp	25	0·5 NaCl	K_1 3·81±0·05, K_2 0·75±0·15	67B
Ge^{IV}	sp	25	0·5 NaCl	$K[Ge(OH)_4+2HL^{3-} \rightleftharpoons Ge(OH)_2L_2^{6-}]$ 6·35±0·10	67B

67B A. Beauchamp and R. Benoit, *Bull. Soc. chim. France*, 1967, 672

413.2 C_6H_5OCl 2-Chlorophenol HL

Metal	Method	Temp	Medium	Log of equilibrium constant, remarks	Ref
H^+	sp	25	0 corr	K_1 8·53	65E
	gl	25	0·10 ($NaClO_4$)	K_1 8·33	66J
Fe^{3+}	sp	25	0 corr	K_1 7·32±0·02	65E
	gl	25	0·10 ($NaClO_4$)	K_1 7·26	66J

65E Z. L. Ernst and F. G. Herring, *Trans. Faraday Soc.*, 1965, **61**, 454
66J K. E. Jabalpurwala and R. M. Milburn, *J. Amer. Chem. Soc.*, 1966, **88**, 3224

413.3 C_6H_5OCl 3-Chlorophenol HL

Metal	Method	Temp	Medium	Log of equilibrium constant, remarks	Ref
H^+	sp	25	0 corr	K_1 9·13	65E
	gl	25	0·10 ($NaClO_4$)	K_1 8·76	66J
Fe^{3+}	sp	25	0 corr	K_1 7·89±0·03	65E
	gl	25	0·10 ($NaClO_4$)	K_1 7·52	66J

65E Z. L. Ernst and F. G. Herring, *Trans. Faraday Soc.*, 1965, **61**, 454
66J K. E. Jabalpurwala and R. M. Milburn, *J. Amer. Chem. Soc.*, 1966, **88**, 3224

413.4 C_6H_5OCl 4-Chlorophenol HL

Metal	Method	Temp	Medium	Log of equilibrium constant, remarks	Ref
H^+	sp	25	0 corr	K_1 9·42	65E
	gl	25	0·10 ($NaClO_4$)	K_1 9·10	66J
Fe^{3+}	sp	25	0 corr	K_1 7·92±0·04	65E
	gl	25	0·10 ($NaClO_4$)	K_1 7·95	66J

65E Z. L. Ernst and F. G. Herring, *Trans. Faraday Soc.*, 1965, **61**, 454
66J K. E. Jabalpurwala and R. M. Milburn, *J. Amer. Chem. Soc.*, 1966, **88**, 3224

C_6H_5OBr 2-Bromophenol HL 413.5

Metal	Method	Temp	Medium	Log of equilibrium constant, remarks	Ref
H^+	sp	25	0 corr	K_1 8·44	65E
	gl	25	0·10 ($NaClO_4$)	K_1 8·22	66J
Fe^{3+}	sp	25	0 corr	K_1 7·19±0·03	65E
	gl	25	0·10 ($NaClO_4$)	K_1 6·98	66J

65E Z. L. Ernst and G. F. Herring, *Trans. Faraday Soc.*, 1965, **61**, 454
66J K. E. Jabalpurwala and R. M. Milburn, *J. Amer. Chem. Soc.*, 1966, **88**, 3224

C_6H_5OBr 3-Bromophenol HL 413.6

Metal	Method	Temp	Medium	Log of equilibrium constant, remarks	Ref
H^+	gl	25	0·10 ($NaClO_4$)	K_1 8·75	66J
Fe^{3+}	gl	25	0·10 ($NaClO_4$)	K_1 7·65	66J

66J K. E. Jabalpurwala and R. M. Milburn, *J. Amer. Chem. Soc.*, 1966, 88, 3224

C_6H_5OBr 4-Bromophenol HL 414

Metal	Method	Temp	Medium	Log of equilibrium constant, remarks	Ref
H^+	gl	25	0·10 ($NaClO_4$)	K_1 9·06	66J
	gl	25	0·5 (KNO_3)	K_1 9·17	68M
Fe^{3+}	gl	25	0·10 ($NaClO_4$)	K_1 8·00	66J
	sp	~ 23	0·5 KNO_3	K_1 7·72	68M
			0·02 $NaClO_4$	K_1 7·89	

66J K. E. Jabalpurwala and R. M. Milburn, *J. Amer. Chem. Soc.*, 1966, **88**, 3224
68M W. A. E. McBryde, *Canad. J. Chem.*, 1968, **46**, 2385

C_6H_5OI 2-Iodophenol HL 414.1

Metal	Method	Temp	Medium	Log of equilibrium constant, remarks	Ref
H^+	sp	25	0 corr	K_1 8·51	65E
Fe^{3+}	sp	25	0 corr	K_1 7·43±0·05	65E

65E Z. L. Ernst and F. G. Herring, *Trans. Faraday Soc.*, 1965, **61**, 454

414.2 C_6H_5OI 3-Iodophenol HL

Metal	*Method*	*Temp*	*Medium*	*Log of equilibrium constant, remarks*	*Ref*
H^+	gl	25	0·10 ($NaClO_4$)	K_1 8·74	66J
Fe^{3+}	gl	25	0·10 ($NaClO_4$)	K_1 7·57	66J

66J K. E. Jabalpurwala and R. M. Milburn, *J. Amer. Chem. Soc.*, 1966, **88**, 3224

414.3 C_6H_5OI 4-Iodophenol HL

Metal	*Method*	*Temp*	*Medium*	*Log of equilibrium constant, remarks*	*Ref*
H^+	sp	25	0 corr	K_1 9·31	65E
Fe^{3+}	sp	25	0 corr	K_1 8·63±0·05	65E

65E Z. L. Ernst and F. G. Herring, *Trans. Faraday Soc.*, 1965, **61**, 454

414.4 C_6H_5OF 2-Fluorophenol HL

Metal	*Method*	*Temp*	*Medium*	*Log of equilibrium constant, remarks*	*Ref*
H^+	sp	25	0 corr	K_1 8·71	65E
	gl	25	0·10 ($NaClO_4$)	K_1 8·49	66J
Fe^{3+}	sp	25	0 corr	K_1 7·33±0·01	65E
	gl	25	0·10 ($NaClO_4$)	K_1 7·19	66J

65E Z. L. Ernst and F. G. Herring, *Trans. Faraday Soc.*, 1965, **61**, 454
66J K. E. Jabalpurwala and R. M. Milburn, *J. Amer. Chem. Soc.*, 1966, **88**, 3224

414.5 C_6H_5OF 3-Fluorophenol HL

Metal	*Method*	*Temp*	*Medium*	*Log of equilibrium constant, remarks*	*Ref*
H^+	sp	25	0 corr	K_1 9·21	65E
	gl	25	0·10 ($NaClO_4$)	K_1 8·81	66J
Fe^{3+}	sp	25	0 corr	K_1 7·77±0·01	65E
	gl	25	0·10 ($NaClO_4$)	K_1 7·77	66J

65E Z. L. Ernst and F. G. Herring, *Trans. Faraday Soc.*, 1965, **61**, 454
66J K. E. Jabalpurwala and R. M. Milburn, *J. Amer. Chem. Soc.*, 1966, **88**, 3224

C_6H_5OF 4-Fluorophenol HL 414.6

Metal	Method	Temp	Medium	Log of equilibrium constant, remarks	Ref
H^+	sp	25	0 corr	K_1 9·91	65E
	gl	25	0·10 ($NaClO_4$)	K_1 9·46	66J
Fe^{3+}	sp	25	0 corr	K_1 9·38±0·02	65E
	gl	25	0·10 ($NaClO_4$)	K_1 8·29	66J

65E Z. L. Ernst and F. G. Herring, *Trans. Faraday Soc.*, 1965, **61**, 454
66J K. E. Jabalpurwala and R. M. Milburn, *J. Amer. Chem. Soc.*, 1966, **88**, 3224

$C_6H_5O_2N$ Pyridine-2-carboxylic acid (picolinic acid) HL 416

Metal	Method	Temp	Medium	Log of equilibrium constant, remarks	Ref
H^+	?	?	?	K_1 5·52	58B
	gl	25	~ 0·02	K_1 5·43, K_2 1·5	63I
	gl	25	0·5	K_1 5·40, K_2 1·60	64M
	gl	25	0·1 (KNO_3)	K_1 5·20	64T
	gl	25	0·1 ($HClO_4$) 50% dioxan	K_1 5·86, K_2 1·31	66W
	gl	25	3·0 KBr	K_1 5·80, K_2 1·36	68P
Cd^{2+}	pol	25	0·1 KNO_3 50% dioxan	β_4 3·52	66W
Ce^{3+}	gl	25	0·1 (KNO_3)	K_1 3·74±0·02, β_2 6·56±0·04, β_3 9·5±0·3	64T
	gl	25	2 ($NaClO_4$)	K_1 3·63±0·05, K_2 2·10±0·08	65Y
Cr^{2+}	sp	25	0·5 $NaClO_4$	K_1 5·96	66M
Cr^{3+}	sp	25	0·5 $NaClO_4$	β_2 10·22	66M
Cu^+	pol	25	0·1 KNO_3 50% dioxan	β_2 6·15	66W
Cu^{2+}	sp	?	0·02	$K(Cu^{2+} + L^- + A^{3-} \rightleftharpoons CuLA^{2-})$ 15·94 where H_3A = nitrilotriacetic acid	63I
	sp	25	1 $NaClO_4$	K_1 8·73, β_2 15·51	65M
	gl	25	0·1 ($HClO_4$) 50% dioxan	K_2 6·15 See phenanthroline (906) for mixed ligands	66W
Dy^{3+}	gl	25	0·1 (KNO_3)	K_1 4·22±0·02, β_2 7·76±0·04, β_3 10·8±0·3	64T
	gl	25	2 ($NaClO_4$)	K_1 3·83±0·05, K_2 3·09±0·08	65Y
Er^{3+}	gl	25	0·1 (KNO_3)	K_1 4·28±0·02, β_2 7·86±0·04, β_3 10·9±0·3	64T
Eu^{3+}	gl	25	0·1 (KNO_3)	K_1 4·07±0·02, β_2 7·48±0·04, β_3 10·6±0·3	64T
	gl	25	2 ($NaClO_4$)	K_1 3·80±0·05, K_2 2·93±0·08	65Y
	gl	25	0·5 NaCl	K_1 2·86, K_2 2·34, K_3 1·55	66M
Fe^{2+}	sp	25 (?)	0·2 KNO_3	β_3 11·30	58B
	sp	13—17	0·1	K (?) (0·02 KCN) 3·3, K (?) (no KCN) 3·8	63M
Fe^{3+}	gl	20	0·1 ($NaClO_4$)	β_2 12·88±0·03; $K(Fe^{3+} + 2L^- + OH^- \rightleftharpoons FeL_2OH)$ 23·92±0·03; $K[2Fe^{3+} + 4L^- + 2OH^- \rightleftharpoons Fe_2L_4(OH)_2]$ 50·72	64Ta

416 $C_6H_5O_2N$ (contd)

Metal	*Method*	*Temp*	*Medium*	*Log of equilibrium constant, remarks*	*Ref*
Gd^{3+}	gl	25	0·1 (KNO_3)	K_1 4·06±0·02, β_2 7·34±0·04, β_3 10·5±0·3	64T
	gl	25	2 ($NaClO_4$)	K_1 3·77±0·05, K_2 3·07±0·08	65Y
Ho^{3+}	gl	25	0·1 (KNO_3)	K_1 4·22±0·02, β_2 7·72±0·04, β_3 10·7±0·3	64T
La^{3+}	gl	25	0·02, 0·5	K_1 (0·02) 4·19, (0·5) 3·21, β_2 (0·02) 7·20, (0·5) 6·06, β_3 (0·02) 10·47, (0·5) 7·94	64M
	gl	25	0·1 (KNO_3)	K_1 3·54±0·02, β_2 6·28±0·04, β_3 8·9±0·3	64T
	gl	25	2 ($NaClO_4$)	K_1 3·50±0·05, K_2 2·39±0·08	65Y
Lu^{3+}	gl	25	0·1 (KNO_3)	K_1 4·45±0·02, β_2 8·20±0·04, β_3 11·4±0·3	64T
	gl	25	2 ($NaClO_4$)	K_1 4·54±0·05, K_2 3·11±0·08	65Y
Nd^{3+}	gl	25	0·02, 0·5	K_1 (0·02) 4·27, (0·5) 3·69, β_2 (0·02) 7·91, (0·5) 6·80, β_3 (0·02) 10·95, (0·5) 9·33	64M
	gl	25	0·1 (KNO_3)	K_1 3·88±0·02, β_2 6·92±0·04, β_3 10·0±0·3	64T
	gl	25	2 ($NaClO_4$)	K_1 3·79±0·05, K_2 2·86±0·08	65Y
Pr^{3+}	gl	25	0·02, 0·5	K_1 (0·02) 4·38, (0·5) 3·43, β_2 (0·02) 7·90, (0·5) 6·65, β_3 (0·02) 11·00, (0·5) 8·94	64M
	gl	25	0·1 (KNO_3)	K_1 3·85±0·02, β_2 6·96±0·04, β_3 9·9±0·3	64T
	gl	25	2 ($NaClO_4$)	K_1 3·62±0·05, K_2 2·63±0·08	65Y
Sm^{3+}	gl	25	0·1 (KNO_3)	K_1 4·06±0·02, β_2 7·40±0·04, β_3 10·6±0·3	64T
	gl	25	2 ($NaClO_4$)	K_1 3·85±0·05, K_2 2·99±0·08	65Y
$Sn(CH_3)_2^{2+}$	Ag-AgCl	25	0·1 (KNO_3)	K_1 5·1±0·2	63Y
Tb^{3+}	gl	25	0·1 (KNO_3)	K_1 4·15±0·02, β_2 7·62±0·04, β_3 10·6±0·3	64T
Ti^{3+}	gl	25	3·0 KBr	K_1 5·62, K_2 5·48, K_3 5·48	68P
Tm^{3+}	gl	25	0·1 (KNO_3)	K_1 4·36±0·02, β_2 7·98±0·04, β_3 11·2±0·3	64T
	gl	25	2 ($NaClO_4$)	K_1 4·31±0·05, K_2 3·17±0·08	65Y
V^{2+}	sp	25	0·5 KCl	K_1 4·43, β_2 9·00, β_3 12·84	65M
	sp	25	0·5 K_2SO_4	β_3 12·77	65M
	pol	25	0·5 KCl	β_2 8·54, β_3 12·46	65M
V^{3+}	sp	25	0·5 KCl	β_3 15·41	65M
VO^{2+}	gl	25	0·1 ($NaClO_4$)	$K(VOLOH + H^+ \rightleftharpoons VOL^+)$ 5·03 $K(VOL_2OH^- + H^+ \rightleftharpoons VOL_2)$ 6·95	66K
	sp	25	0·1 ($NaClO_4$)	$K(VOL_2OH^- + H^+ \rightleftharpoons VOL_2)$ 6·98	66K
Y^{3+}	gl	25	0·1 (KNO_3)	K_1 4·03±0·02, β_2 7·36±0·04, β_3 10·0±0·3	64T
Yb^{3+}	gl	25	0·1 (KNO_3)	K_1 4·43±0·02, β_2 8·12±0·04, β_3 11·3±0·3	64T
	gl	25	2 ($NaClO_4$)	K_1 4·28±0·05, K_2 3·38±0·08	65Y
Zn^{2+}	gl	25	0·1 ($NClO_4$) 50% dioxan	K_2 6·08	66W

58B B. Banerjee and P. Ray, *J. Indian Chem. Soc.*, 1958, **35**, 493
63I Y. J. Israeli, *Bull. Soc. chim. France*, 1963, 1273
63M I. Morimoto and T. Sato, *Bull. Chem. Soc. Japan*, 1963, **36**, 605
63Y M. Yasuda and R. S. Tobias, *Inorg. Chem.*, 1963, **2**, 207
64M L. Moyne and G. Thomas, *Analyt. Chim. Acta*, 1964, **31**, 583
64T L. C. Thompson, *Inorg. Chem.*, 1964, **3**, 1319
64Ta C. F. Timberlake, *J. Chem. Soc.*, 1964, 1229
65M R. C Mercier, M. Bonnet, and M. R. Paris, *Bull. Soc. chim. France*, 1965, 2926, 3577
65Y H. Yoneda, G. R. Choppin, J. L. Bear, and A. J. Graffeo, *Inorg. Chem.*, 1965, **4**, 244
66K Th. Kadem and S. Fallab, *Chimia* (*Switz.*), 1966, **20**, 51
66M R. C. Mercier and M. R. Paris, *Compt. rend.*, 1966, 349, 598
66W J. L. Walter and S. M. Rosalie, *J. Inorg. Nuclear Chem.*, 1966, **28**, 2969
67S J. P. Scharff and M. R. Paris, *Bull. Soc. chim. France*, 1967, 1782
68P M. R. Paris and C. Gregoire, *Analyt. Chim. Acta*, 1968, **42**, 439

$C_6H_5O_2Cl$ 4-Chloro-1,2-dihydroxybenzene (4-chlorocatechol) H_2L 416.1

Metal	*Method*	*Temp*	*Medium*	*Log of equilibrium constant, remarks*	*Ref*
H^+	gl	30	0·1 (KNO_3)	K_1 11·54, K_2 8·43	64M
	sp	25	0·10 NaCl	K_1 12·55, K_2 8·62	67P
Co^{2+}	gl	30	0·1 (KNO_3)	K_1 7·64, K_2 6·37, K_3 4·23	64M
Cu^{2+}	gl	30	0·1 (KNO_3)	K_1 12·56, K_2 9·83	64M
Ge^{IV}	sp	25	0·10 NaCl	$K[Ge(OH)_4+3H_2L \rightleftharpoons GeL_3^{2-}+2H^+]$ 0·65	67P
Mn^{2+}	gl	30	0·1 (KNO_3)	K_1 6·82, K_2 4·66	64M
Mo^{VI}	sp	25	0·1 (KCl)	$K(MoO_4^{2-}+2H_2L \rightleftharpoons MoO_2L_2^{2-})$ 5·85±0·01	62H
Ni^{2+}	gl	30	0·1 (KNO_3)	K_1 7·90, K_2 5·00, K_3 4·16	64M
W^{VI}	spJ	25	0·1 (KCl)	$K(WO_4^{2-}+2H_2L \rightleftharpoons WO_2L_2^{2-})$ 7·1±0·2	63H
Zn^{2+}	gl	30	0·1 (KNO_3)	K_1 8·63, K_2 6·82	64M

62H J. Halmekoski, *Suomen Kem.*, 1962, **B35**, 171
63H J. Halmekoski, *Suomen Kem.*, 1963, **B36**, 29
64M Y. Murakami and M. Tokunaga, *Bull. Chem. Soc. Japan*, 1964, 37, 1562
67P P. Pichet and R. L. Benoit, *Inorg. Chem.*, 1967, **6**, 1505

$C_6H_5O_3N$ 2-Nitrophenol HL 416.2

Metal	*Method*	*Temp*	*Medium*	*Log of equilibrium constant, remarks*	*Ref*
H^+	gl	25	0·10 ($NaClO_4$)	K_1 7·04	66J
	sp	25	0·10 ($NaClO_4$)	K_1 7·06	66P
Fe^{3+}	gl	25	0·10 ($NaClO_4$)	K_1 5·99	66J
La^{3+}	sp	25	0·10 ($NaClO_4$)	K_1 2·20±0·02	66P

66J K. E. Jabalpurwala and R. M. Milburn, *J. Amer. Chem. Soc.*, 1966, **88**, 3224
66P C. Postmus jun., L. B. Magnusson, and C. A. Craig, *Inorg. Chem.*, 1966, **5**, 1154

$C_6H_5O_3N$ 3-Nitrophenol HL 417

Metal	*Method*	*Temp*	*Medium*	*Log of equilibrium constant, remarks*	*Ref*
H^+	gl	25	0·10 ($NaClO_4$)	K_1 8·04	66J
	gl	25	0·5 KNO_3	K_1 8·18	68B
Fe^{3+}	gl	25	0·10 ($NaClO_4$)	K_1 6·95	66J
	sp	~ 23	0·5 KNO_3	K_1 6·69	68B
			0·027 $NaClO_4$	K_1 7·05	

66J K. E. Jabalpurwala and R. M. Milburn, *J. Amer. Chem. Soc.*, 1966, **88**, 3224
68B W. A. E. McBryde, *Canad. J. Chem.*, 1968, **46**, 7385

418 $C_6H_5O_3N$ 4-Nitrophenol HL

Metal	Method	Temp	Medium	Log of equilibrium constant, remarks	Ref
H^+	gl	25	0·10 ($NaClO_4$)	K_1 7·02	66J
	gl	20	0·1 (KNO_3)	K_1 7·15	67B
	gl	25	0·5 KNO_3	K_1 6·96	68B
Fe^{3+}	gl	25	0·10 ($NaClO_4$)	K_1 5·74	66J
	sp	~ 23	0·5 KNO_3	K_1 5·60	68B
			0·027 $NaClO_4$	K_1 5·86	
UO_2^{2+}	gl	20	0·1 (KNO_3)	K_1 4·40	67B, 67

66J K. E. Jabalpurwala and R. M. Milburn, *J. Amer. Chem. Soc.*, 1966, **88**, 3224
67B M. Bartusek, *Coll. Czech. Chem. Comm.*, 1967, **32**, 757
67Ba M. Bartusek, *J. Inorg. Nuclear Chem.*, 1967, **29**, 1089
68B W. A. E. McBryde, *Canad. J. Chem.*, 1968, **46**, 2385

418.1 $C_6H_5O_3N$ Pyridine-2-carboxylic acid *N*-oxide (α-picolinic acid *N*-oxide) HL

Metal	Method	Temp	Medium	Log of equilibrium constant, remarks	Ref
Ce^{3+}	gl	25	2 ($NaClO_4$)	K_1 2·71±0·05	65Y
Dy^{3+}	gl	25	2 ($NaClO_4$)	K_1 3·00±0·05, K_2 2·55±0·08	65Y
Eu^{3+}	gl	25	2 ($NaClO_4$)	K_1 2·94±0·05, K_2 2·29±0·08	65Y
Gd^{3+}	gl	25	2 ($NaClO_4$)	K_1 2·93±0·05, K_2 2·27±0·08	65Y
La^{3+}	gl	25	2 ($NaClO_4$)	K_1 2·53±0·05, K_2 1·76±0·08	65Y
Lu^{3+}	gl	25	2 ($NaClO_4$)	K_1 3·48±0·05, K_2 2·84±0·08	65Y
Nd^{3+}	gl	25	2 ($NaClO_4$)	K_1 2·91±0·05, K_2 2·15±0·08	65Y
Pr^{3+}	gl	25	2 ($NaClO_4$)	K_1 2·75±0·05, K_2 2·26±0·08	65Y
Tm^{3+}	gl	25	2 ($NaClO_4$)	K_1 3·16±0·05, K_2 2·69±0·08	65Y
Yb^{3+}	gl	25	2 ($NaClO_4$)	K_1 3·15±0·05, K_2 2·60±0·08	65Y

65Y H. Yoneda, G. R. Choppin, J. L. Bear, and A. J. Graffeo, *Inorg. Chem.*, 1965, **4**, 244

419 $C_6H_5O_4Cl$ 3-Chloro-5-hydroxy-2-hydroxymethyl-4-pyrone (chlorokojic acid) HL

Metal	Method	Temp	Medium	Log of equilibrium constant, remarks	Ref
H^+	gl, sp	25	0·5	K_1 7·40±0·03	67C
Ge^{IV}	sp	25	0·5 (NaCl)	$K[Ge(OH)_4+2HL \rightleftharpoons Ge(OH)_2L_2]$ 2·33±0·05	67C
Zn^{2+}	gl	25	0·5 ($NaClO_4$)	K_1 4·88±0·06	67C

67C G. Choux and R. L. Benoit, *Bull. Soc. chim. France*, 1967, 2920

$C_6H_5O_4I$ 3-Iodo-5-hydroxy-2-hydroxymethyl-4-pyrone (iodokojic acid) HL 419.1

Metal	Method	Temp	Medium	Log of equilibrium constant, remarks	Ref
H^+	gl, sp	25	0·5	K_1 7·50±0·03	67C
Ge^{IV}	sp	25	0·5 (NaCl)	$K[Ge(OH)_4+2HL \rightleftharpoons Ge(OH)_2L_2]$ 2·49±0·05	67C
Zn^{2+}	gl	25	0·5 ($NaClO_4$)	K_1 4·92±0·06	67C

67C G. Choux and R. L. Benoit, *Bull. Soc. chim. France*, 1967, 2920

$C_6H_5O_4N$ 4-Nitro-1,2-dihydroxybenzene (4-nitrocatechol) H_2L 419.2

Metal	Method	Temp	Medium	Log of equilibrium constant, remarks	Ref
H^+	gl	30	0·1 (KNO_3)	K_1 10·75, K_2 6·59	64M
	gl	20	0·1 (KNO_3)	K_1 11·1, K_2 6·84	67B
	sp	25	0·10 KCl	K_1 10·90, K_2 6·78	67P
	gl	20	0·1 (KNO_3)	K_2 6·89	68B
B^{III}	gl	20	0·1 (KNO_3)	$K[H_3BO_3+H_2L \rightleftharpoons BL(OH)_2^-+H^+]$ −4·0±0·1	68B
Co^{2+}	gl	30	0·1 (KNO_3)	K_1 7·48, K_2 5·31, K_3 3·14	64M
Cu^{2+}	gl	30	0·1 (KNO_3)	K_1 11·65, K_2 9·28	64M
Ge^{IV}	sp	25	0·10 KCl	$K[Ge(OH)_4+3H_2L \rightleftharpoons GeL_3^{2-}+2H^+]$ 3·90	67P
Mn^{2+}	gl	30	0·1 (KNO_3)	K_1 6·51, K_2 4·74	64M
Ni^{2+}	gl	30	0·1 (KNO_3)	K_1 7·82, K_2 5·27, K_3 3·81	64M
UO_2^{2+}	gl	20	0·1 (KNO_3)	K_1 12·9, K_2 9·8	67B, 67Ba
				$K(UO_2L+H^+ \rightleftharpoons UO_2LH^+)$ 2·7	
				$K(UO_2L_2^{2-}+H^+ \rightleftharpoons UO_2L_2H^-)$ 4·97	
Zn^{2+}	gl	30	0·1 (KNO_3)	K_1 8·20, K_2 6·80	64M

64M Y. Murakami and M. Tokunaga, *Bull. Chem. Soc. Japan*, 1964, **37**, 1562
67B M. Bartusek, *Coll. Czech. Chem. Comm.*, 1967, **32**, 757
67Ba M. Bartusek, *J. Inorg. Nuclear Chem.*, 1967, **29**, 1089
67P P. Pichet and R. L. Benoit, *Inorg. Chem.*, 1967, **6**, 1505
68B M. Bartusek and L. Havelkova, *Coll. Czech. Chem. Comm.*, 1968, **33**, 385

$C_6N_6ON_2$ Pyridine-3-carboxylic acid amide (nicotinamide) L 420

Metal	Method	Temp	Medium	Log of equilibrium constant, remarks	Ref
Ni^{2+}	sp	24	1 ($NaNO_3$)	K_1 3·40	64M

64M P. Moore and R. G. Wilkins, *J. Chem. Soc.*, 1964, 3455

422 $C_6H_6ON_2$ Pyridine-2-carboxaldoxime HL

Metal	Method	Temp	Medium	Log of equilibrium constant, remarks	Ref
H^+	gl	24	0·1 (KNO_3)	K_1 10·02, K_2 3·69	62B
	sp	25	I (NaCl or $NaClO_4$)	K_1 $10{\cdot}22-0{\cdot}5\sqrt{I}/(1+\sqrt{I})\pm0{\cdot}02$, K_2 $3{\cdot}42+0{\cdot}5\sqrt{I}/(1+\sqrt{I})\pm0{\cdot}02$	62H
		15—41·5	→ 0	K_1 (18·5°) 10·29±0·02, (25°) 10·22±0·02, (34°) 10·06±0·02, K_2 (15°) 3·51±0·02, (25°) 3·42±0·02, (34°) 3·34±0·02, (41·5°) 3·29±0·02	
				ΔH_1 (25°) −6·8±0·8, ΔS_1 (25°) 24±3, ΔH_2 (25°) −3·3±0·2, ΔS_2 (25°) 4·5±0·9	
	sp	25	0·3 $NaClO_4$	K_1 10·0, K_2 3·4	66B
Cd^{2+}	gl	24	0·1 (KNO_3)	K_1 5·2, K_2 4·4	62B
Co^{2+}	gl	24	0·1 (KNO_3)	K_1 9·6, K_2 8·7	62B
	gl	25	0·3 $NaClO_4$	K_1 8·8±0·1, K_2 8·8±0·1	66B
	sp	25	0·3 $NaClO_4$	K_1 8·6±0·2, K_2 8·6±0·2	66B
Cu^{2+}	gl	24	0·1 (KNO_3)	K_1 10·8, K_2 6·0	62B
				See nitrilotriacetic acid (449) for mixed ligands	
Fe^{2+}	gl	24	0·1 (KNO_3)	K_1 9·4, K_2 8·0, K_3 5·1	62B
	sp	18—32·5	0·045	β_3 (18·0°) 25·13±0·06, (25·0°) 24·85±0·09, (28·5°) 24·70±0·10, (32·5°) 24·43±0·09	62H
				ΔH −20±4, ΔS 47±14	
		17·5—33·5	→ 0	$K(FeL_3^- + H^+ \rightleftharpoons FeHL_3)$ (17·5°) 7·14±0·02, (25·0°) 7·13±0·02, (33·5°) 7·11±0·01	
				$\Delta H°$ (25°) −0·8±0·4, $\Delta S°$ 30±2	
				$K(FeHL_3 + H^+ \rightleftharpoons FeH_2L_3^+)$ (24·5°) 3·36, (33·0°) 3·33	
				$\Delta H°$ (25°) 1±1, $\Delta S°$ 18±6	
Fe^{3+}	gl	24	0·1 (KNO_3)	K_1 11·4, K_2 10·3, K_3 8·4	62B
	gl	25	~0	$K(FeL_3 + H^+ \rightleftharpoons FeHL_3^+)$ 3·5	62H
Hg^{2+}	gl	24	0·1 (KNO_3)	K_1 6·5, K_2 5·7	62B
Mn^{2+}	gl	25	0·3 $NaClO_4$	K_1 5·2±0·2, K_2 3·9±0·2	66B
Ni^{2+}	gl	24	0·1 (KNO_3)	K_1 8·1, K_2 6·1, K_3 5·0	62B
	gl	25	0·3 $NaClO_4$	K_1 9·4±0·2, K_2 7·1±0·2, K_3 5·5±0·2	66B
				See nitrilotriacetic acid (449) for mixed ligands	
Zn^{2+}	gl	24	0·1 (KNO_3)	K_1 5·5, K_2 5·3	62B
	gl	25	0·3 $NaClO_4$	K_1 5·8±0·1, K_2 5·3±0·1	66B

62B S. Bolton and R. I. Ellin, *J. Pharm. Sci.*, 1962, **51**, 533

62H G. I. H. Hanania and D. H. Irvine, *J. Chem. Soc.*, 1962, 2745

66B K. Burger, I. Egyed, and I. Raff, *J. Inorg. Nuclear Chem.*, 1966, **28**, 139

423.1 $C_6H_6ON_4$ 5-Methyl-7-hydroxy-[1,2,4]-triazolo[1,5-*a*]pyrimidine L

Metal	Method	Temp	Medium	Log of equilibrium constant, remarks	Ref
H^+	gl	20	0·1 (KNO_3)	K_1 6·22	66O
Co^{2+}	gl	20	0·1 (KNO_3)	K_1 2·15	66O
Cu^{2+}	gl	20	0·1 (KNO_3)	K_1 3·19, β_2 5·90	66O
Ni^{2+}	gl	20	0·1 (KNO_3)	K_1 2·53, β_2 5·11	66O

66O G. Ostacoli, E. Campi, and M. C. Gennaro, *Gazzetta*, 1966, **96**, 741

$C_6H_6O_2N_2$ **3-Nitroaniline** L 424

Metal	Method	Temp	Medium	Log of equilibrium constant, remarks	Ref
Ag^+	Ag	25	ethanol	β_2 1·88	60A
Zn^{2+}	sp	25	acetone	$K(ZnCl_2+L \rightleftharpoons ZnCl_2L)$ 1·69	65S
				$K(ZnBr_2+L \rightleftharpoons ZnBr_2L)$ 1·98	
				$K(ZnI_2+L \rightleftharpoons ZnI_2L)$ 1·88	

60A V. Armeanu and C. Luca, *Z. phys. Chem.* (*Leipzig*), 1960, **214,** 81
65S D. P. N. Satchell and R. S. Satchell, *Trans. Faraday Soc.*, 1965, **61,** 1118

$C_6H_6O_2N_2$ **4-Nitroaniline** L 425

Metal	Method	Temp	Medium	Log of equilibrium constant, remarks	Ref
Ag^+	Ag	25	ethanol	β_2 1·55	60A
Zn^{2+}	sp	25	acetone	$K(ZnCl_2+L \rightleftharpoons ZnCl_2L)$ 0·74	65S
				$K(ZnBr_2+L \rightleftharpoons ZnBr_2L)$ 0·95	
				$K(ZnI_2+L \rightleftharpoons ZnI_2L)$ 0·78	

60A V. Armeanu and C. Luca, *Z. phys. Chem.* (*Leipzig*), 1960, **714,** 81
65S D. P. N. Satchell and R. S. Satchell, *Trans. Faraday Soc.*, 1965, **61,** 1118

$C_6H_6O_2N_2$ ***N*-Nitrosophenylhydroxylamine (cupferron)** HL 426

Metal	Method	Temp	Medium	Log of equilibrium constant, remarks	Ref
H^+	?	?	?	K_1 4·28	61K
Nb^V	sp	25	0·1 $[(NH_4)_2SO_4]$ 50% ethanol	$K(NbOL_2^+ + L^- \rightleftharpoons NbOL_3)$ 4·83	67L
UO_2^{2+}	sp	26	~ 0	β_2 11·0±0·2	61K
	sol	25	?	$K_s(NH_4^+ + UO_2L_2 + L^-)$ −9·2±0·2	61K

61K A. E. Klygin and N. S. Kolyada, *Russ. J. Inorg. Chem.*, 1961, **6,** 107 (216)
67L J. E. Land and J. R. Sanchez-Caldas, *J. Less-Common Metals*, 1967, **12,** 41

$C_6H_6O_2N_2$ **Pyridine-3-carbohydroxamic acid (nicotinoyl hydroxylamine)** HL 426.1

Metal	Method	Temp	Medium	Log of equilibrium constant, remarks	Ref
H^+	sp	25	0·1 $(NaClO_4)$	K_1 8·09	64R
Mo^{VI}	sp	25	0·1 $(NaClO_4)$	K(?) 6·3, K(?) 6·7	64R

64R R. Rowland and C. E. Meloan, *Analyt. Chem.*, 1964, **36,** 1997

426.2 $C_6H_6O_2S$ 2-Acetyl-3-hydroxythiophene HL

Metal	*Method*	*Temp*	*Medium*	*Log of equilibrium constant, remarks*	*Ref*
H^+	sp	25	0·1 ($NaClO_4$) 10% dioxan	K_1 6·73±0·07	66P
	gl	25	0·1 ($NaClO_4$) 10% dioxan	K_1 6·77±0·03	66P
	gl	25	0·1 ($NaClO_4$) 50% dioxan	K_1 8·11±0·02	67S
Co^{2+}	sp	25	0·1 ($NaClO_4$) 10% dioxan	K_1 3·98±0·05	66P
	gl	25	0·1 ($NaClO_4$) 10% dioxan	K_1 3·92±0·05	66P
	gl	25	0·1 ($NaClO_4$) 50% dioxan	K_1 5·13±0·05 $K(CoA^{2+}+L^- \rightleftharpoons CoAL^+)$ 5·20±0·05 where A = 2,2′-bipyridyl	67S
Cu^{2+}	sp	25	0·1 ($NaClO_4$) 10% dioxan	K_1 6·30±0·07	66P
	gl	25	0·1 ($NaClO_4$) 10% dioxan	K_1 6·46±0·09	66P
	gl	25	0·1 ($NaClO_4$) 50% dioxan	K_1 7·86±0·05 $K(CuA^{2+}+L^- \rightleftharpoons CuAL^+)$ 8·32±0·05 where A = 2,2′-bipyridyl	67S
Mn^{2+}	sp	25	0·1 ($NaClO_4$) 10% dioxan	K_1 2·9±0·1	66P
	gl	25	0·1 ($NaClO_4$) 10% dioxan	K_1 3·0±0·2	66P
Ni^{2+}	sp	25	0·1 ($NaClO_4$) 10% dioxan	K_1 4·38±0·05	66P
	gl	25	0·1 ($NaClO_4$) 10% dioxan	K_1 4·34±0·05	66P
	gl	25	0·1 ($NaClO_4$) 50% dioxan	K_1 5·44±0·05 $K(NiA^{2+}+L^- \rightleftharpoons NiAL^+)$ 5·39±0·05 where A = 2,2′-bipyridyl	67S
Zn^{2+}	sp	25	0·1 ($NaClO_4$) 10% dioxan	K_1 3·73±0·08	66P
	gl	25	0·1 ($NaClO_4$) 10% dioxan	K_1 3·58±0·07	66P
	gl	25	0·1 ($NaClO_4$) 50% dioxan	K_1 4·86±0·05 $K(ZnA^{2+}+L^- \rightleftharpoons ZnAL^+)$ 4·86±0·05 where A = 2,2′-bipyridyl	67S

66P S. Petri, H. Sigel, and H. Erlenmeyer, *Helv. Chim. Acta*, 1966, **49**, 1612
67S H. Sigel, *Chimia (Switz.)*, 1967, **21**, 489

426.3 $C_6H_6O_2S$ 3-Acetyl-4-hydroxythiophene HL

Metal	*Method*	*Temp*	*Medium*	*Log of equilibrium constant, remarks*	*Ref*
H^+	sp	25	0·1 ($NaClO_4$) 10% dioxan	K_1 8·04±0·07	66P
Cu^{2+}	sp	25	0·1 ($NaClO_4$) 10% dioxan	K_1 5·4±0·3	66P
Ni^{2+}	sp	25	0·1 ($NaClO_4$) 10% dioxan	K_1 3·2±0·2	66P

66P S. Petri, H. Sigel, and H. Erlenmeyer, *Helv. Chim. Acta*, 1966, **49**, 1612

$C_6H_6O_3N_2$ 4-Nitro-2-aminophenol HL 426.4

Metal	Method	Temp	Medium	Log of equilibrium constant, remarks	Ref
H^+	gl	30	0·1 $NaClO_4$ 50% dioxan	K_1 8·00±0·02, K_2 1·92±0·20	66V
Be^{2+}	gl	30	0·1 $NaClO_4$ 50% dioxan	K_1 4·47±0·05, K_2 3·23±0·05	66V
Cd^{2+}	gl	30	0·1 $NaClO_4$ 50% dioxan	K_1 2·37±0·05	66V
Co^{2+}	gl	30	0·1 $NaClO_4$ 50% dioxan	K_1 4·77±0·05, K_2 3·63±0·05	66V
Cu^{2+}	gl	30	0·1 $NaClO_4$ 50% dioxan	K_1 6·15±0·05, K_2 4·93±0·05	66V
Ni^{2+}	gl	30	0·1 $NaClO_4$ 50% dioxan	K_1 4·10±0·05, K_2 3·16±0·05	66V
UO_2^{2+}	gl	30	0·1 $NaClO_4$ 50% dioxan	K_1 7·59±0·05, K_2 7·13±0·05	66V
Zn^{2+}	gl	30	0·1 $NaClO_4$ 50% dioxan	K_1 3·64±0·05, K_2 2·49±0·05	66V

66V D. G. Vartak and N. G. Menon, *J. Inorg. Nuclear Chem.*, 1966, **28**, 2911

$C_6H_6O_4S$ 4-Hydroxybenzenesulphonic acid H_2L 427.1

Metal	Method	Temp	Medium	Log of equilibrium constant, remarks	Ref
H^+	H, Ag-AgCl	0—60	→ 0	K_1 (0°) 9·352±0·001, (5°) 9·284±0·001, (10°) 9·220±0·002, (15°) 9·160±0·002, (20°) 9·105±0·002, (25°) 9·053±0·001, (30°) 9·005±0·002, (35°) 8·961±0·002, (40°) 8·921±0·002, (45°) 8·883±0·002, (50°) 8·849±0·002, (55°) 8·816±0·002, (60°) 8·787±0·002 $\Delta H_1°$ (25°) −4·04, $\Delta S_1°$ (25°) 27·9	43B
	gl	25	0·5 KNO_3	K_1 8·56	68B
Fe^{3+}	sp	~ 23	0·5 KNO_3	K_1 6·72	68B

43B R. G. Bates, G. L. Siegel, and S. F. Acree, *J. Res. Nat. Bur. Stand.*, 1943, **31**, 205
68B W. A. E. McBryde, *Canad. J. Chem.*, 1968, **46**, 2385

$C_6H_6O_5S$ 3,4-Dihydroxybenzenesulphonic acid H_3L 427.2

Metal	Method	Temp	Medium	Log of equilibrium constant, remarks	Ref
H^+	gl	30	0·1 (KNO_3)	K_1 12·16, K_2 8·26	63M, 63Ma
	gl	20	0·1 (KNO_3)	K_1 12·8, K_2 8·50	65B, 65Ba
B^{III}	gl	20	0·1 (KNO_3)	$K[H_3BO_3 + H_2L^- \rightleftharpoons B(OH)_2L^{2-} + H^+]$ −4·60	68H
Ca^{2+}	gl	30	0·1 (KNO_3)	K_1 4·40, K_2 3·59	63M
Co^{2+}	gl	30	0·1 (KNO_3)	K_1 8·54, K_2 5·86, K_3 3·08	63M
Cu^{2+}	gl	30	0·1 (KNO_3)	K_1 13·29, K_2 10·23	63M

continued overleaf

427.2 $C_6H_6O_5S$ (contd)

Metal	Method	Temp	Medium	Log of equilibrium constant, remarks	Ref
Fe^{3+}	gl	30	0·1 (KNO_3)	K_1 > 17, K_2 14·0, K_3 9·1	63Ma
Mg^{2+}	gl	30	0·1 (KNO_3)	K_1 6·27, K_2 4·14	63M
Mn^{2+}	gl	30	0·1 (KNO_3)	K_1 7·87, K_2 4·66	63M
Ni^{2+}	gl	30	0·1 (KNO_3)	K_1 8·85, K_2 5·56, K_3 4·73	63M
Sr^{2+}	gl	30	0·1 (KNO_3)	K_1 3·61	63M
UO_2^{2+}	gl	20	0·1 (KNO_3)	$K(UO_2^{2+} + HL^{2-} \rightleftharpoons UO_2HL)$ 6·4	65Ba
	gl	20	0·1 (KNO_3)	$K[(UO_2)_2L_2OH^{3-} + 3H^+ \rightleftharpoons 2UO_2HL]$ 9·0	65B
Zn^{2+}	gl	30	0·1 (KNO_3)	K_1 9·40, K_2 7·20	63M

63M Y. Murakami, K. Nakamura, and M. Tokunada, *Bull. Chem. Soc. Japan*, 1963, **36**, 669
63Ma Y. Murakami and K. Nakamura, *Bull. Chem. Soc. Japan*, 1963, **36**, 1408
65B M. Bartusek and O. Stankova, *Coll. Czech. Chem. Comm.*, 1965, **30**, 3415
65Ba M. Bartusek and L. Sommer, *J. Inorg. Nuclear Chem.*, 1965, **27**, 2397
68H L. Havelkova and M. Bartusek, *Coll. Czech. Chem. Comm.*, 1968, **33**, 4188

427.3 $C_6H_6O_6S$ 2,3,4-Trihydroxybenzenesulphonic acid H_4L

Metal	Method	Temp	Medium	Log of equilibrium constant, remarks	Ref
H^+				H^+ constants not given	
Er^{3+}	sp	?	1 KOH	K_1 5·86	66T
Ho^{3+}	sp	?	1 KOH	K_1 5·81	66T
Nd^{3+}	sp	?	1 KOH	K_1 5·72	66T

66T K. V. Tserkasevich, N. P. Kfryushina, and N. S. Poluektov, *Russ. J. Inorg. Chem.*, 1966, **11**, 49 (93)

428 $C_6H_6O_8S_2$ 4,5-Dihydroxybenzene-1,3-disulphonic acid (tiron) H_4L

Metal	Method	Temp	Medium	Log of equilibrium constant, remarks	Ref
H^+	gl	20—22	0·1 KCl	K_1 12·6±0·2, K_2 7·66±0·02	51W, 55 65D, 65 66K, 67
	gl	30	0·1 (KNO_3)	K_1 12·26, K_2 7·54	63M
	sp	25	0—1·0	K_1 (→ 0) 12·9, (0·1) 12·51, (0·5) 12·10, (1·0) 11·78, K_2 (→ 0) 7·9, (0·1) 7·67, (0·5) 7·40, (1·0) 7·17	64M
	sp	25	0·1, 0·35 $NaClO_4$	K_1 (0·1) 11·96, (0·35) 12·18	65O, 68O
	gl	25	0·1, 0·35 $NaClO_4$	K_2 (0·1) 7·81, (0·35) 7·52	65O, 68O
	gl	25	0·1 (KNO_3)	K_1 12·48, K_2 7·57±0·01	66L, 67L
	gl	25—35	0·10 (KNO_3)	K_1 (25°) 12·60±0·06, (35°) 12·08±0·05, K_2 (25°) 7·60±0·04, (35°) 7·49±0·02	66M
Al^{3+}	con	?	0·1 KNO_3	K_1 16·81	65Da
	gl	?	0·1 KNO_3	K_1 16·79, K_2 16·58, K_3 14·34	65Da

contin

Metal	*Method*	*Temp*	*Medium*	*Log of equilibrium constant, remarks*	*Ref*
As^{III}	gl	25	I (KCl)	$K(H_3AsO_3+2H_2L^{2-} \rightleftharpoons AsL_2^{5-}+H^+)$ $-8{\cdot}186+9{\cdot}162\sqrt{I}/(1+0{\cdot}553\sqrt{I})-1{\cdot}61\,I$	64A
B^{III}	gl	25	I (KCl)	$K[B(OH)_3+H_2L^{2-} \rightleftharpoons B(OH)_2L^{3-}+H^+]$ $-4{\cdot}34+3{\cdot}05\sqrt{I}/(1+1{\cdot}3\sqrt{I})-0{\cdot}16\,I$	60N
			$\rightarrow 0$	$K[B(OH)_4^-+H_2L^{2-} \rightleftharpoons B(OH)_2L^{3-}]$ 4·90	
	gl	25	0·1 (KNO_3)	$K[H_3BO_3+H_2L^{2-} \rightleftharpoons B(OH)_2L^{3-}+H^+]$ $-3{\cdot}72$	68H
Be^{2+}	gl	20 (?)	0·1 KNO_3	K_1 13·5±0·1, K_2 12·5±0·1	65Db
	gl	20	0·1 (KNO_3)	K_1 12·88, K_2 9·37, $K(Be^{2+}+HL^{3-} \rightleftharpoons BeHL^-)$ 4·2, $K(BeL^{2-}+HL^{3-} \rightleftharpoons BeHL_2^{5-})$ 2·3	67B
Cu^{2+}	gl	30	0·1 (KNO_3)	K_1 13·99, K_2 11·17	63M
	sp	25	0·1 $NaClO_4$	K_1 14·43, $K(Cu^{2+}+HL^{3-} \rightleftharpoons CuHL^-)$ 5·14	65O
			0·35 $NaClO_4$	K_2 10·93	
	gl	25	0·1 (KNO_3)	K_1 14·27±0·02 $K(CuA^{2+}+L^{4-} \rightleftharpoons CuAL^{2-})$ 15·14±0·03 where A = 2,2′-bipyridyl See triethylenetetraminehexa-acetic acid (1085) for mixed ligands	66L
Fe^{3+}	sp	18	0·1	K_1 20·8±0·2, K_2 15·4±0·4, K_3 11·5±0·4	55V
	sp	25	0·1—1·0 (KNO_3)	K_1 (0·1) 20·4, (0·5) 19·3, (1·0) 18·8, K_2 (0·1) 15·1, (0·5) 15·1, (1·0) 14·7, K_3 (0·1) 10·8, (0·5) 11·6, (1·0) 11·6	64M
			0·1—1·0 ($NaClO_4$)	K_1 (0·1) 20·4, (0·5) 19·7, (1·0) 19·1, K_2 (0·1) 14·9, (0·5) 15·0, (1·0) 15·1, K_3 (0·1) 10·1, (0·5) 11·3, (1·0) 11·3 See triethylenetetraminehexa-acetic acid (1085) for mixed ligands	
Ga^{3+}	sp	23	0—0·2	K_1 (→ 0) 5·24, (0·02) 4·97, (0·05) 4·80, (0·1) 4·65, (0·2) 4·51	65D
		17, 40	0·1	K_1 (17°) 4·63, (40°) 4·72, ΔH_1 (25°) 0·39±0·04, ΔS_1 21±2	
Ge^{IV}	gl	25	0·1 KCl	$K[Ge(OH)_4+3H_2L^{2-} \rightleftharpoons GeL_3^{8-}+2H^+]$ $-2{\cdot}74\pm0{\cdot}02$	59A
	gl	25	I (KCl)	$K[Ge(OH)_4+3H_2L^{2-} \rightleftharpoons GeL_3^{8-}+2H^+]$ $-2{\cdot}307+27{\cdot}49\sqrt{I}/(1+2{\cdot}851\sqrt{I})-0{\cdot}370\,I$	66A
	sp	25	0·1—1 NaCl	$K[Ge(OH)_4+2H_2L^{2-} \rightleftharpoons GeL_2^{4-}]$ (1·00) 3·89 $K[Ge(OH)_4+3H_2L^{2-} \rightleftharpoons GeL_3^{8-}+2H^+]$ (0·11) 2·30, (0·26) 3·10, (0·50) 3·50, (1·00) 3·70	67P
Hf^{4+}	dis	?	0·2 $HClO_4$	K_1 24·66, β_3 66·92	66K
	ix	?	0·5—2 $HClO_4$	K_1 (1·0) 23·00 $K(Hf^{4+}+H_2L^{2-} \rightleftharpoons HfL+2H^+)$ (0·5) 3·15±0·07, (1·0) 2·61±0·03, (2·0) 2·28±0·05 $K(Hf^{4+}+2H_2L^{2-} \rightleftharpoons HfL_2^{4-}+4H^+)$ (0·5) 4·4±0·2, (1·0) 4·05±0·04	67E
In^{3+}	sp	29	0—0·2	K_1 (→ 0) 4·45, (0·05) 3·91, (0·1) 3·79, (0·2) 3·71	65N
		20, 45	0·1	K_1 (20°) 3·75, (45°) 3·84 ΔH_1 1·4±0·1, ΔS_1 22·2±0·3	
Ni^{2+}	sp	25	0·1	K_1 9·40	68O
Th^{4+}	gl	25	0·10 KNO_3	$K[Th_2L_3(OH)_2^{6-}+2H^+ \rightleftharpoons Th_2L_3^{4-}]$ 12·8±0·1 $K[Th_2L_3(OH)_2^{6-}+4H^+ \rightleftharpoons 2ThL+H_2L^{2-}]$ 11·9±0·1 See EDTA (836) and CDTA (997) for mixed ligands	66Ma
Ti^{4+}	sp	18—22	0·1 ($NaClO_4$)	$K(TiO^{2+}+2H_2L^{2-} \rightleftharpoons TiO(HL)_2^{4-}+2H^+]$ $-0{\cdot}3$ $K[TiO(HL)_2^{2-}+H_2L^{2-} \rightleftharpoons TiL_3^{8-}+2H^+]$ $-2{\cdot}9$ $K(TiO^{2+}+2H^++3L^{4-} \rightleftharpoons TiL_3^{8-})$ 57·6	63S

continued overleaf

428 $C_6H_6O_8S_2$ (contd)

Metal	*Method*	*Temp*	*Medium*	*Log of equilibrium constant, remarks*	*Ref*
U^{4+}				See EDTA (836) for mixed ligands	
UO_2^{2+}	gl	20	0·1 (KNO_3)	$K(UO_2^{2+}+HL^{3-} \rightleftharpoons UO_2HL^-)$ 6·3	65Ba
				$K[(UO_2)_2L_2OH^{5-}+3H^+ \rightleftharpoons 2UO_2HL^-]$ 8·9	65B
	sp	18—22	0·1 ($NaClO_4$)	$K(UO_2^{2+}+HL^{3-} \rightleftharpoons UO_2HL^-)$ 6·5	65S
VO^{2+}	gl	25—35	0·10 (KNO_3)	K_1 (25°) 16·74±0·03, (35°) 16·05±0·05, K_2 (25°) 14·20±0·04	66M
				$\Delta H_1°$ −2·5±0·5, $\Delta S_1°$ 67±3	
				$K[VO(OH)L^{2-}+H^+ \rightleftharpoons VOL^-]$ 6·3±0·2	
				$K\{2VO(OH)L \rightleftharpoons [VO(OH)L]_2\}$4·3±0·2	
	gl	30	0·1 KCl	K_1 16·61	67L
				$K(VOA^{2+}+L^{2-} \rightleftharpoons VOAL)$ 17·19 where A=1,10-phenanthroline	
Zn^{2+}	gl	30	0·1 (KNO_3)	K_1 10·19, K_2 8·33	63M
Zr^{4+}	ix	?	0·5—2 $HClO_4$	K_1 (1·0) 24·15	67E
				$K(Zr^{4+}+H_2L^{2-} \rightleftharpoons ZrL+2H^+)$ (0·5) 3·95±0·03, (1·0) 3·89±0·08, (2·0) 3·00±0·01	

51W A. Willi and G. Schwarzenbach, *Helv. Chim. Acta*, 1951, **34,** 528
55V L. Vareille, *Bull. Soc. chim. France*, 1955, 1496
59A P. J. Antikainen and P. J. Malkonen, *Suomen Kem.*, 1959, **B32,** 179
60N R. Nasanen, *Suomen Kem.*, 1960, **B33,** 1
63M Y. Marakami, K. Nakamura, and M. Tokunda, *Bull. Chem. Soc. Japan*, 1963, **36,** 669
63S L. Sommer, *Coll. Czech. Chem. Comm.*, 1963, **28,** 2102
64A P. J. Antikainen and K. Tevanen, *Suomen Kem.*, 1964, **B37,** 213
64M W. A. E. McBryde, *Canad. J. Chem.*, 1964, **42,** 1917
65B M. Bartusek and O. Stankova, *Coll. Czech. Chem. Comm.*, 1965, **30,** 3415
65Ba M. Bartusek and L. Sommer, *J. Inorg. Nuclear Chem.*, 1965, **27,** 2397
65D R. C. Das, *Indian J. Chem.*, 1965, **3,** 179
65Da S. N. Dubey and R. C. Mehrotra, *J. Indian Chem. Soc.*, 1965, **42,** 685
65Db S. N. Dubey and R. C. Mehrotra, *J. Less-Common Metals*, 1965, **9,** 123
65N R. N. Nanda, R. C. Das, and R. K. Nanda, *Indian J. Chem.*, 1965, **3,** 278
65O Y. Oka, N. Nakazawa, and H. Harada, *Nippon Kagaku Zasshi*, 1965, **86,** 1158
65S L. Sommer, T. Sepel, and L. Kurilova, *Coll. Czech. Chem. Comm.*, 1965, **30,** 3834
66A P. J. Antikainen and K. Tevanen, *Suomen Kem.*, 1966, **B39,** 2
66K N. B. Kalinichenko, A. N. Ermakov, D. I. Ryabchikov, and I. N. Marov, *Russ. J. Inorg. Chem.*, 1966, **11,** 425 (781)
66L G. A. L'Heureux and A. E. Martell, *J. Inorg. Nuclear Chem.*, 1966, **28,** 481
66M C. E. Mont and A. E. Martell, *J. Amer. Chem. Soc.*, 1966, **88,** 1387
66Ma Y. Murakami and A. E. Martell, *Bull. Chem. Soc. Japan*, 1966, **39,** 1077
67B M. Bartusek and J. Zelinka, *Coll. Czech. Chem. Comm.*, 1967, **32,** 992
67E A. N. Ermakov, N. B. Kalinichenko, and I. N. Marov, *Russ. J. Inorg. Chem.*, 1967, **12,** 812 (1545)
67L K. Lal and R. P. Agarwal, *J. Less-Common Metals*, 1967, **12,** 269
67P P. Pichet and R. L. Benoit, *Inorg. Chem.*, 1967, **6,** 1505
68H L. Havelkova and M. Bartusek, *Coll. Czech. Chem. Comm.*, 1968, **33,** 4188
68O Y. Oka, M. Okamura, and H. Harada, *Nippon Kagaku Zasshi*, 1968, **89,** 171

429 C_6H_6NCl 2-Chloroaniline L

Metal	*Method*	*Temp*	*Medium*	*Log of equilibrium constant, remarks*	*Ref*
Ag^+	Ag	25	ethanol	β_2 1·71	60A

60A V. Armeanu and C. Luca, *Z. phys. Chem. (Leipzig)*, 1960, **214,** 81

C_6H_6NCl 3-Chloroaniline L 430

Metal	Method	Temp	Medium	Log of equilibrium constant, remarks	Ref
Ag^+	Ag	25	ethanol	β_2 2·13	60A

60A V. Armeanu and C. Luca, *Z. phys. Chem.* (*Leipzig*), 1960, **214**, 81

C_6H_6NCl 4-Chloroaniline L 431

Metal	Method	Temp	Medium	Log of equilibrium constant, remarks	Ref
Ag^+	Ag	25	ethanol	β_2 2·65	60A

60A V. Armeanu and C. Luca, *Z. phys. Chem.* (*Leipzig*), 1960, **214**, 81

C_6H_6NBr 4-Bromoaniline L 434

Metal	Method	Temp	Medium	Log of equilibrium constant, remarks	Ref
H^+	gl	25	0·3 (KNO_3)	K_1 3·92	64N
Cd^{2+}	Hg(Cd)	25	0·3 (KNO_3)	K_1 −0·49	64N

64N L. V. Nazarova, A. V. Ablov, and V. A. Dagaer, *Russ. J. Inorg. Chem.*, 1964, **9**, 1150 (2129)

C_6H_6NI 4-Iodoaniline L 434.1

Metal	Method	Temp	Medium	Log of equilibrium constant, remarks	Ref
Ag^+	Ag	25	ethanol	β_2 2·50	60A

60A V. Armeanu and C. Luca, *Z. phys. Chem.* (*Leipzig*), 1960, **214**, 81

C_6H_7ON 2-Hydroxymethylpyridine (2-pyridylmethanol) L 438

Metal	Method	Temp	Medium	Log of equilibrium constant, remarks	Ref
H^+	gl	0—25	50% dioxan	K_1 (0°) 4·18, (15°) 4·19, (25°) 4·16	64L
				$K[(H_{-1}L)^- + H^+ \rightleftharpoons L]$ (0°) 13·68, (15°) 13·14, (25°) 12·90	
	gl	25	0·1 (KNO_3)	K_1 4·89	65M
	gl	25	0·61 (KNO_3)	K_1 5·10	67S
Ag^+	gl	25	0·61 (KNO_3)	K_1 2·14±0·02, β_2 4·37±0·01	67S
Cd^{2+}	gl	25	0·10 (KNO_3)	K_1 < 1	65M
Co^{2+}	gl	25	0·10 (KNO_3)	K_1 2·1	65M

continued overleaf

438 C_6H_7ON (contd)

Metal	Method	Temp	Medium	Log of equilibrium constant, remarks	Ref
Cu^{2+}	gl	0—25	50% dioxan	$K[Cu^{2+}+(H_{-1}L)^- \rightleftharpoons Cu(H_{-1}L)^+]$ (0°) 10·45, (15°) 10·22, (25°) 10·19 $K[Cu^{2+}+2(H_{-1}L)^- \rightleftharpoons Cu(H_{-1}L)_2]$ (0°) 20·02, (15°) 19·36, (25°) 19·11 $\Delta H°$ (25°) −14·6, $\Delta S°$ 39	64L
	cal	25	50% dioxan	$\Delta H°$ −23·3±0·1, $\Delta S°$ 9·4	64L
	gl	25	0·10 (KNO_3)	K_1 3·41, K_2 2·81 $K[Cu(H_{-1}L)^+ + H^+ \rightleftharpoons CuL^{2+}]$ 5·5 $K[CuL(H_{-1}L)^+ + H^+ \rightleftharpoons CuL_2^{2+}]$ 5·55±0·01 $K[Cu(H_{-1}L)_2 + H^+ \rightleftharpoons CuL(H_{-1}L)^+]$ 6·36±0·01	65M
	gl	25	0·61 (KNO_3)	K_1 3·56±0·01, β_2 6·23±0·01, β_3 8·00±0·02, β_4 8·3±0·3	67S
Mn^{2+}	gl	25	0·10 (KNO_3)	K_1 ~ 1	65M
Ni^{2+}	gl	25	0·10 (KNO_3)	K_1 2·79, K_2 2·60	65M
	gl	25	0·61 (KNO_3)	K_1 2·90±0·02, β_2 5·26±0·02, β_3 7·04±0·03, β_4 7·1±0·5	67S
Zn^{2+}	gl	25	0·10 (KNO_3)	K_1 1·9	65M

64L T. J. Lane, P. J. Kandathil, and S. M. Rosalie, *Inorg. Chem.*, 1964, **3**, 487
65M Y. Murakami and M. Takagi, *Bull. Chem. Soc. Japan*, 1965, **38**, 828
67S M. S. Sun and D. G. Brewer, *Canad. J. Chem.*, 1967, **45**, 2729

438.1 C_6H_7ON 3-Hydroxymethylpyridine (3-pyridylmethanol) L

Metal	Method	Temp	Medium	Log of equilibrium constant, remarks	Ref
H^+	gl	25	0·61 (KNO_3)	K_1 5·19	67S
Ag^+	gl	25	0·61 (KNO_3)	K_1 2·01±0·02, β_2 4·09±0·01	67S
Cu^{2+}	gl	25	0·61 (KNO_3)	K_1 2·43±0·01, β_2 4·27±0·03, β_3 5·0±0·3, β_4 6±1	67S
Ni^{2+}	gl	25	0·61 (KNO_3)	K_1 1·85±0·01, β_2 2·99±0·04	67S

67S M. S. Sun and D. G. Brewer, *Canad. J. Chem.*, 1967, **45**, 2729

438.2 C_6H_7ON 4-Hydroxymethylpyridine (4-pyridylmethanol) L

Metal	Method	Temp	Medium	Log of equilibrium constant, remarks	Ref
H^+	gl	25	0·61 (KNO_3)	K_1 5·66	67S
Ag^+	gl	25	0·61 (KNO_3)	K_1 2·15±0·02, β_2 4·23±0·02	67S
Cu^{2+}	gl	25	0·61 (KNO_3)	K_1 2·65±0·01, β_2 4·53±0·04, β_3 5·7±0·3, β_4 6±3	67S
Ni^{2+}	gl	25	0·61 (KNO_3)	K_1 1·97±0·01, β_2 3·02±0·05	67S

67S M. S. Sun and D. G. Brewer, *Canad. J. Chem.*, 1967, **45**, 2729

C_6H_7ON 3-Methoxypyridine L 439

Metal	Method	Temp	Medium	Log of equilibrium constant, remarks	Ref
Zn^{2+}	Hg(Zn)	30	0·1 $NaClO_4$	K_1 0·90, β_2 1·40, β_3 1·54	66D

66D A. G. Desai and M. B. Kabadi, *J. Inorg. Nuclear Chem.*, 1966, **28**, 1279

C_6H_7ON 4-Methoxypyridine L 440

Metal	Method	Temp	Medium	Log of equilibrium constant, remarks	Ref
Zn^{2+}	Hg(Zn)	30	0·1 $NaClO_4$	K_1 1·53, β_2 2·31, β_3 3·08	66D

66D A. G. Desai and M. B. Kabadi, *J. Inorg. Nuclear Chem.*, 1966, **28**, 1279

$C_6H_7ON_3$ Pyridine-4-carboxylic acid hydrazide (isonicotinic hydrazide) L 441

Metal	Method	Temp	Medium	Log of equilibrium constant, remarks	Ref
Cd^{2+}	pol	25	1 $NaClO_4$	K_1 1·20, β_2 2·44, β_3 3·29	65K

65K A. F. Krivis, G. R. Supp, and R. L. Doerr, *Analyt. Chem.*, 1965, 37, 52

$C_6H_7O_3As$ Benzenearsonic acid H_2L 443.1

Metal	Method	Temp	Medium	Log of equilibrium constant, remarks	Ref
Th^{4+}	sol	15—21	~ 0·1	K_1 3·8±0·3	60M
UO_2^{2+}	sol	15—21	~ 0·1	K_1 2·2±0·2	60M
Zr^{4+}	sol	15—21	~ 0·1	K_1 11·5±0·3	60M
	sol	?	2 HNO_3	β_2 30·6±0·3	67K

60M V. A. Michajlov, *J. Analyt. Chem. (U.S.S.R.)*, 1960, 15, 605 (528)
67K M. Kyrs, P. Pistek, and P. Selucky, *Coll. Czech. Chem. Comm.*, 1967, 32, 747

446.1 $C_6H_8O_4Se$ *cis*-Tetrahydroselenophene-2,5-dicarboxylic acid H_2L

Metal	Method	Temp	Medium	Log of equilibrium constant, remarks	Ref
H^+	gl	25	0·1 ($NaClO_4$)	K_1 4·47, K_2 3·24	68S
Cd^{2+}	gl	25	0·1 ($NaClO_4$)	K_1 2·7	68S
Cu^{2+}	gl	25	0·1 ($NaClO_4$)	K_1 3·8, K_2 3·6	68S
Ni^{2+}	gl	25	0·1 ($NaClO_4$)	K_1 3·0, K_2 2·9	68S
Pb^{2+}	gl	25	0·1 ($NaClO_4$)	K_1 3·6, K_2 3·2	68S
Zn^{2+}	gl	25	0·1 ($NaClO_4$)	K_1 2·4	68S

68S K. Suzuki, I. Nakano, and K. Yamasaki, *J. Inorg. Nuclear Chem.*, 1968, **30**, 545

446.2 $C_6H_8O_6S$ (Carboxymethylthio)butanedioic acid (*S*-carboxymethylmercaptosuccinic acid)

Metal	Method	Temp	Medium	Log of equilibrium constant, remarks	Ref
Pt^{4+}	sp	30	~ 0	K (?) 2·65	66N

66N B. C. Nair and H. L. Nigam, *J. Less-Common Metals*, 1966, **10**, 250

447 $C_6H_9O_2N_3$ 2-Amino-3-(4′-imidazolyl)propanoic acid (histidine) HL

Metal	Method	Temp	Medium	Log of equilibrium constant, remarks	Ref
H^+	gl	20	~ 0	K_1 9·40, K_2 6·14	53P
	gl	25	0·3 (K_2SO_4)	K_1 9·05 ± 0·02, K_2 6·16, K_3 < 3·0	61J
	gl	20	1	K_2 6·18	61V
	gl	0—45	0·1 (KNO_3)	K_1 (0°) 9·97, (25°) 9·18, (45°) 8·63, K_2 (0°) 6·65, (25°) 6·08, (45°) 5·66	64A
	gl	25	0·1 (KCl)	K_1 9·08 ± 0·01, K_2 5·98 ± 0·01	64D
	gl	0—40	0·25 (KCl)	K_1 (0°) 9·81, (15°) 9·46, (25°) 9·17, (40°) 8·86, K_2 (0°) 6·62, (15°) 6·24, (25°) 6·12, (40°) 5·88, K_3 (0°) 1·98, (15°) 1·79, (25°) 1·96, (40°) 1·85	65A
	gl	37	0·15 (KNO_3)	K_1 8·92, K_2 5·96, K_3 2·24	67P
Cd^{2+}	gl	20	0·0025 $CdSO_4$	β_2 11·1	53P
	pol	0—45	0·1 (KNO_3)	β_2 (0°) 11·40, (25°) 10·20, (45°) 9·90 $\Delta H_{\beta 2}°$ −14·2, $\Delta S_{\beta 2}°$ 0	64A
Co^{2+}	gl	0—40	0·25 (KCl)	K_1 (0°) 7·30, (15°) 7·10, (25°) 6·77, (40°) 6·56, K_2 (0°) 6·07, (15°) 5·62, (25°) 5·13, (40°) 4·94 $\Delta H_1°$ (15°) −8·0, $T\Delta S_1°$ (15°) 1·4 kcal mole⁻¹ $\Delta H_2°$ (15°) −12·6, $T\Delta S_2°$ (15°) −5·2 kcal mole⁻¹	65A
	gl	15	0·25 (KCl)	$\Delta H°(CoA^{2+} + L^- \rightarrow CoL^- + A)$ +8·0 $T\Delta S°$ 10·8 kcal mole⁻¹ where A = histidine methyl ester	65A
	gl	37	0·15 (KNO_3)	K_1 6·71 ± 0·02, K_2 5·35 ± 0·02	67P

Metal	Method	Temp	Medium	Log of equilibrium constant, remarks	Ref
Co^{3+}	sp	25	0 corr	$K(CoAL^+ + H^+ \rightleftharpoons CoAHL^{2+})$ 4·49±0·02	66H
				$\Delta H°$ 4·7±0·9, $\Delta S°$ −5±3	
				$K[CoA(H_{-1}L) + H^+ \rightleftharpoons CoAL^+]$ 11·00±0·02	
				$\Delta H°$ 11·7±1·0, $\Delta S°$ −11±3	
				where CoA = cobalamin factor B	
Cu^{2+}	gl	25	0·3 (K_2SO_4)	K_1 10·3	61J
	Pt	25	0·3 (K_2SO_4)	K_1 10·5	61J
	gl	20	1	β_2 28·0,	61V
				$K(Cu^{2+} + HL + L^- \rightleftharpoons CuHL_2^+)$ 20·0	
	gl	25	0·1 (KCl)	K_1 10·21±0·05, K_2 8·32±0·05	64D
	gl	0—40	0·25 (KCl)	K_1 (0°) 12·8, (15°) 11·2, (25°) 10·7, (40°) 10·5	65A
				$\Delta H_1°$ (15°) −27·4, $T\Delta S_1°$ (15°) −12·6 kcal mole^{-1}	
	gl	15	0·25 (KCl)	$\Delta H°(CuA^{2+} + L^- \rightarrow CuL^- + A)$ −9·1	65A
				$T\Delta S°$ −6·3 kcal mole^{-1}	
				where A = histidine methyl ester	
	gl	37	0·15 (KNO_3)	K_1 9·79±0·05, K_2 7·62±0·05	67P
	cal	22	0·1 KNO_3	$\Delta H_{\beta 2}$ −21·3, $\Delta S_{\beta 2}$ 13·9	67S
				See nitrilotriacetic acid (449) for mixed ligands	
Hg^{2+}	pol	25	0·6 KNO_3	β_2 20·62±0·06	66T
Mn^{2+}	gl	37	0·15 (KNO_3)	K_1 3·24±0·10, K_2 2·92±0·09	67P
Ni^{2+}	gl	0—40	0·25 (KCl)	K_1 (0°) 9·28, (15°) 8·79, (25°) 8·50, (40°) 8·30,	65A
				K_2 (0°) 7·71, (15°) 7·08, (25°) 6·69, (40°) 6·40	
				$\Delta H_1°$ (15°) −11·4, $T\Delta S_1°$ (15°) 0·2 kcal mole^{-1}	
				$\Delta H_2°$ (15°) −13·7, $T\Delta S_2°$ (15°) −4·4 kcal mole^{-1}	
	gl	15	0·25 (KCl)	$\Delta H°(NiA^{2+} + L^- \rightarrow A + NiL^+)$ 4·6	65A
				$T\Delta S°$ 7·4 kcal mole^{-1}	
				$\Delta H°(NiA_3^{2+} + 2L^- \rightarrow 3A + NiL_2)$ +24·1	
				$T\Delta S°$ 25·8 kcal mole^{-1}	
				where A = histidine methyl ester	
	gl	37	0·15 (KNO_3)	K_1 8·43±0·02, K_2 6·71±0·03	67P
	cal	22	0·1 KNO_3	$\Delta H_{\beta 2}$ −16·6, $\Delta S_{\beta 2}$ 14·9	67S
Pb^{2+}	gl	37	0·15 (KNO_3)	K_1 5·96±0·08, K_2 ~ 3·0±0·3	67P
Zn^{2+}	gl	20	0·0025 $ZnSO_4$	β_2 12·0	53P
	gl	0—40	0·25 (KCl)	K_1 (0°) 7·00, (15°) 6·78, (25°) 6·40, (40°) 6·52,	65A
				K_2 (0°) 5·96, (15°) 5·55, (25°) 5·02, (40°) 5·07	
				$\Delta H_1°$ (15°) −6·9, $T\Delta S_1°$ (15°) 2·0 kcal mole^{-1}	
				$\Delta H_2°$ (15°) −11·4, $T\Delta S_2°$ (15°) −4·1 kcal mole^{-1}	
	gl	15	0·25 (KCl)	$\Delta H°(ZnA^{2+} + L^- \rightarrow A + ZnL^+)$ +2·3	65A
				$T\Delta S°$ 4·8 kcal mole^{-1}	
				where A = histidine methyl ester	
	gl	37	0·15 (KNO_3)	K_1 6·34±0·01, K_2 5·35±0·01	67P
	cal	21	0·1 KNO_3	$\Delta H_{\beta 2}$ −11·7, $\Delta S_{\beta 2}$ 14·6	67S

53P D. J. Perkins, *Biochem. J.*, 1953, **55**, 649
61J B. R. James and R. J. P. Williams, *J. Chem. Soc.*, 1961, 2007
61V S. Valladas-Dubois, *Bull. Soc. chim. France*, 1961, 967
64A A. C. Andrews and J. K. Romary, *J. Chem. Soc.*, 1964, 405
64D M. A. Doran, S. Chaberek, and A. E. Martell, *J. Amer. Chem. Soc.*, 1964, **86**, 2129
65A A. C. Andrews and D. M. Zebolsky, *J. Chem. Soc.*, 1965, 742
66H G. I. H. Hanania, D. H. Irvine, and M. V. Irvine, *J. Chem. Soc. (A)*, 1966, 296
66T V. F. Toropova and Yu. M. Azizov, *Russ. J. Inorg. Chem.*, 1966, **11**, 288 (531)
67P D. D. Perrin and V. S. Sharma, *J. Chem. Soc. (A)*, 1967, 724
67S W. F. Stack and H. A. Skinner, *Trans. Faraday Soc.*, 1967, **63**, 1136

447.1 $C_6H_9O_3N$ 2-Methyl-5-oxoazolidine-2-carboxylic acid (α-methylpyroglutamic acid) H_2L

Metal	*Method*	*Temp*	*Medium*	*Log of equilibrium constant, remarks*	*Ref*
H^+	gl	25	0·1 $NaClO_4$	K_1 11·82, K_2 3·15	65N
Cu^{2+}	gl	25	0·1 $NaClO_4$	K_1 8·04	65N

65N M. H. T. Nyberg and M. Cefola, *Arch. Biochem. Biophys.*, 1965, **111**, 321

449 $C_6H_9O_6N$ Nitrilotriacetic acid (NTA) H_3L

Metal	*Method*	*Temp*	*Medium*	*Log of equilibrium constant, remarks*	*Ref*
H^+	gl	20	0·1 KCl	K_1 9·73, K_2 2·49, K_3 1·89	57N, 63
	gl	25	0·1 50% dioxan	K_1 10·11, K_2 4·18	66B
	gl	25	1 NaCl	K_1 8·70	66C
	gl	20	0·1 (KCl)	K_1 9·71±0·02, K_2 2·47±0·02, K_3 1·75±0·05	66I
	?	?	?	K_2 2·50, K_3 1·88	66K
	gl	20	0·1 [$(CH_3)_4$ NCl]	K_1 9·87	67A
	gl	20	0·1 (KNO_3)	K_1 9·73, K_2 2·5, K_3 1·9	67A
	H	20	1 [$(CH_3)_4$ NCl]	K_1 9·67, K_2 2·4, K_3 1·7	67A
	gl	20	1 ($NaClO_4$)	K_1 8·96, K_2 2·27, K_3 1·99	67A
	sol	20	1 ($NaClO_4$)	K_2 2·14, K_3 1·97, K_4 1·10	67A
	gl	25	0·2 ($NaClO_4$)	K_1 9·45, K_2 2·60, K_3 1·97	67Ba
	gl	0·4	0·10 (KNO_3)	K_1 10·76, K_2 3·00, K_3 2·30	67T
Ag^+	dis	20	0·1 ($KClO_4$)	K_1 5·16±0·05	63S
Al^{3+}	dis	20	0·1 ($KClO_4$)	K_1 9·5±0·1	63S
	sp	25	0·2 ($NaClO_4$)	K_1 11·37	67Ba
	gl	25	0·2 ($NaClO_4$)	$K(AlL+H^+ \rightleftharpoons AlHL^+)$ 1·90 $K(AlLOH^- + H^+ \rightleftharpoons AlL)$ 5·09 $K(AlL(OH)_2^{2-} + H^+ \rightleftharpoons AlLOH^-)$ 8·28	67Ba
Ba^{2+}	gl	20	0·1 (KCl)	K_1 4·83±0·03	66I
Be^{2+}	dis	20	0·1 ($KClO_4$)	K_1 7·11±0·05	63S
Ca^{2+}	cal	20	0·1 (KNO_3)	ΔH_1 −1·36, $\Delta S_1°$ 24·7	64H
	gl	20	0·1 (KCl)	K_1 6·46±0·01	66I
Cd^{2+}	dis	20	0·1 ($KClO_4$)	β_2 15·45±0·04	63S
	oth	20	0·1 (KNO_3)	K_1 10·0, K_2 4·6, paper electrophoresis	64J
	dis	30	1·0 ($NaClO_4$)	K_1 9·2±0·2	65H
Ce^{3+}	ix	18—20	0·1	K_1 10·97±0·05, β_2 20·85±0·07	65V
	pol	20	0·1 (KNO_3)	$K(2Ce^{3+} + 3L^{3-} \rightleftharpoons Ce_2L_3^{3-})$ 36·0±0·2	57N
Co^{2+}	dis	20	0·1 ($KClO_4$)	K_1 10·81±0·03, β_2 14·28±0·05	63S
	cal	20	0·1 (KNO_3)	ΔH_1 −0·15, $\Delta S_1°$ 47·1	64H
	oth	20	0·1 (KNO_3)	K_1 10·0, K_2 3·9, paper electrophoresis	64J
	gl	25	0·07—0·08 (KNO_3)	$K(CoL^- + A \rightleftharpoons CoLA^-)$ 1·88 where A = ethyl valinate, $K(CoL^- + A^- \rightleftharpoons CoLA^{2-})$ 3·65 where HA = glycine	68H
	sp	25	0·2 ($NaClO_4$)	K_1 10·44	67Ba
Cu^{2+}	gl	?	0·5 ($NaNO_3$)	$K(CuL^- + A^{2-} \rightleftharpoons CuLA^{3-})$ 5·32 where H_2A = salicylic acid, $K(CuL^- + A^- \rightleftharpoons CuLA^{2-})$ 5·44 where HA = glycine, 6·20 where HA = pyridyl carbaldoxime	63I

con

Metal	Method	Temp	Medium	Log of equilibrium constant, remarks	Ref
Cu^{2+}	dis	20	0·1 ($KClO_4$)	K_1 13·05 ± 0·01	63S
(contd)	cal	20	0·1 (KNO_3)	ΔH_1 −1·84, $\Delta S_1°$ 53·0	64H
				$\Delta H_{\beta2}$ −8·3, $\Delta S_{\beta2}$ 55	
	oth	20	0·1 (KNO_3)	K_1 11·5, K_2 3·3, paper electrophoresis	64J
	pol	25	0·1 50% dioxan	β_3 17·02	66B
	gl	0·4	0·10 (KNO_3)	K_1 13·11	67T
	gl	25	0·07—0·08	$K[CuL^- + OH^- \rightleftharpoons CuL(OH)^{2-}]$ 4·39	68H
			(KNO_3)	$K(CuL^- + A \rightleftharpoons CuLA^-)$	
				3·06 where A = methylglycinate	
				3·15 where A = ethylglycinate	
				3·33 where A = n-butylglycinate	
				7·91 where A = ethylalaninate	
				2·77 where A = ethylphenylalaninate	
				2·79 where A = ethylleucinate	
				2·88 where A = ethylvalinate	
				3·65 where A = ethyl 3-alaninate	
				3·98 and 4·90 where A = methylhistidinate	
				$K(CuL^- + A^- \rightleftharpoons CuLA^{2-})$	
				5·44 where HA = glycine	
				5·42 where HA = alanine	
				4·99 where HA = phenylalanine	
				5·35 where HA = leucine	
				5·10 where HA = valine	
				4·56 where HA = β-alanine	
				5·73 and 4·16 where HA = histidine	
				See picolinic acid (416) for mixed ligands	
Dy^{3+}	pol	20	0·1 (KNO_3)	$K(2Dy^{3+} + 3L^{3-} \rightleftharpoons Dy_2L_3^{3-})$ 37·43 ± 0·10	57N
Er^{3+}	pol	20	0·1 (KNO_3)	$K(2Er^{3+} + 3L^{3-} \rightleftharpoons Er_2L_3^{3-})$ 37·96 ± 0·09	57N
Eu^{3+}	pol	20	0·1 (KNO_3)	$K(2Eu^{3+} + 3L^{3-} \rightleftharpoons Eu_2L_3^{3-})$ 36·84 ± 0·10	57N
Fe^{3+}	dis	20	0·1 ($KClO_4$)	K_1 15·91 ± 0·03, β_2 24·61 ± 0·05	63S
	red	20	0·1 ($NaClO_4$)	K_1 16·26	67B
Ga^{3+}	dis	20	0·1 ($KClO_4$)	β_2 25·81 ± 0·04	63S
	red	20	0·1 ($NaClO_4$)	K_1 13·6	67B
Gd^{3+}	pol	20	0·1 (KNO_3)	$K(2Gd^{3+} + 3L^{3-} \rightleftharpoons Gd_2L_3^{3-})$ 36·89 ± 0·10	57N
Hf^{4+}	ix	?	1,2 $HClO_4$	$K(Hf^{4+} + H_3L \rightleftharpoons HfL^+ + 3H^+)$ (1) 5·05 ± 0·09, (2) 3·83 ± 0·02	64E
	ix	?	0·23 $HClO_4$	K_1 20·34	66E
Hg^{2+}	Hg	25	0·1 ($NaClO_4$)	K_1 14·6 ± 0·1	67S
Ho^{3+}	pol	20	0·1 (KNO_3)	$K(2Ho^{3+} + 3L^{3-} \rightleftharpoons Ho_2L_3^{3-})$ 37·68 ± 0·09	57N
In^{3+}	ix	?	0·5	K_1 14·88 ± 0·09	63R
	dis	20	0·1 ($KClO_4$)	β_2 24·4	63S
	sp	20—22	?	K_1 15·88	65Z
	red	20	0·1 ($NaClO_4$)	K_1 16·9	67B
La^{3+}	pol	20	0·1 (KNO_3)	$K(2La^{3+} + 3L^{3-} \rightleftharpoons La_2L_3^{3-})$ 35·5 ± 0·2	57N
	dis	20	0·1 ($KClO_4$)	β_2 17·15 ± 0·10	63S
	Hg	25	0·1 ($NaClO_4$)	K_1 10·5 ± 0·2	67S
	sol	25	0·1 ($NaClO_4$)	K_1 10·6 ± 0·2	67S
Li^+	gl	20	0·1 [$(CH_3)_4NNO_3$]	K_1 2·51 ± 0·01	63Ia
Lu^{3+}	pol	20	0·1 (KNO_3)	$K(2Lu^{3+} + 3L^{3-} \rightleftharpoons Lu_2L_3^{3-})$ 38·81 ± 0·08	57N
Mg^{2+}	gl	20	0·1 (KCl)	K_1 5·46 ± 0·02	66I
Mn^{2+}	dis	20	0·1 ($KClO_4$)	K_1 7·36 ± 0·05	63S
	cal	20	0·1 (KNO_3)	ΔH_1 1·14, $\Delta S_1°$ 37·9	64H
	oth	20	0·1 (KNO_3)	K_1 8·6, K_2 3·0, paper electrophoresis	64J

continued overleaf

449 $C_6H_9O_6N$ (contd)

Metal	*Method*	*Temp*	*Medium*	*Log of equilibrium constant, remarks*	*Ref*
Mn^{2+}	gl	25	0·07—0·08	$K(MnL^- + A \rightleftharpoons MnLA^-)$ 2·39	68H
(*contd*)			(KNO_3)	where A = ethylvalinate,	
				$K(MnL^- + A^- \rightleftharpoons MnLA^{2-})$ 2·24	
				where HA = glycine	
Mo^{VI}	nmr	28	1·3	$K(MoO_4^{2-} + WO_3L^{3-} \rightleftharpoons MoO_3L^{3-} + WO_4^{2-})$ 0·15	67M
Na^+	gl	20	0·1	K_1 1·22±0·02	63Ia
			[$(CH_3)_4$ NNO_3]		
Nd^{3+}	pol	20	0·1 (KNO_3)	$K(2Nd^{3+} + 3L^{3-} \rightleftharpoons Nd_2L_3{}^{3-})$ 36·5±0·1	57N
	sp	18—20	0·02	K_1 10·49, β_2 19·47	61A
Ni^{2+}	gl	?	0·5 ($NaNO_3$)	$K(NiL^- + A^{2-} \rightleftharpoons NiLA^{3-})$ 3·02	63I
				where H_2A = salicylic acid,	
				$K(NiL^- + A^- \rightleftharpoons NiLA^{2-})$ 4·41	
				where HA = glycine,	
				$K(NiL^- + A \rightleftharpoons NiLA^-)$ 5·18	
				where A = pyridine aldoxime	
	dis	20	0·1 ($KClO_4$)	K_1 11·54±0·08	63S
	cal	20	0·1 (KNO_3)	ΔH_1 −2·53, $\Delta S_1°$ 44·1	64H
	sp	25	0·5 (KCl)	$K(NiL^- + A \rightleftharpoons NiLA^-)$ 2·54±0·04	67J
				where A = NH_3	
			0·5 ($NaClO_4$)	7·20±0·06 where A = ethylenediamine	
			0·5 (NaCl)	$K(NiL^- + A^- \rightleftharpoons NiLA^{2-})$ 4·89±0·04	
				where HA = glycine	
			0·5 (KNO_3)	$K(NiL^- + A^{2-} \rightleftharpoons NiLA^{3-})$ 2·17±0·04	
				where H_2A = oxalic acid	
	gl	25	0·07—0·08	$K(NiL^- + A \rightleftharpoons NiLA^-)$ 2·03	68H
			(KNO_3)	where A = ethylvalinate,	
				$K(NiL^- + A^- \rightleftharpoons NiLA^{2-})$ 4·95	
				where HA = glycine	
Pb^{2+}	dis	20	0·1 ($KClO_4$)	K_1 11·47±0·04	63S
	nmr	28	0·6	$K(PbL^- + Zn^{2+} \rightleftharpoons Pb^{2+} + ZnL^-)$ 0·86	67M
	gl	25	0·07—0·08	$K(PbL^- + A \rightleftharpoons PbLA^-)$ 1·55	68H
			(KNO_3)	where A = ethylvalinate,	
				$K(PbL^- + A^- \rightleftharpoons PbLA^{2-})$ 1·93	
				where HA = glycine	
Pr^{3+}	pol	20	0·1 (KNO_3)	$K(2Pr^{3+} + 3L^{3-} \rightleftharpoons Pr_2L_3{}^{3-})$ 36·2±0·2	57N
	sp	18—20	0·02	K_1 10·28, β_2 19·25	61A
Sc^{3+}	pol	20	0·1 (KNO_3)	$K(2Sc^{3+} + 3L^{3-} \rightleftharpoons Sc_2L_3{}^{3-})$ 42·20±0·08	57N
	dis	20	0·1 ($KClO_4$)	β_2 ~ 24·1	63S
	Hg	25	0·1 ($NaClO_4$)	K_1 12·7±0·2	67S
Sm^{3+}	pol	20	0·1 (KNO_3)	$K(2Sm^{3+} + 3L^{3-} \rightleftharpoons Sm_2L_3{}^{3-})$ 36·7±0·1	57N
	sp	18—20	0·02	K_1 10·78, β_2 20·54	61A
Sr^{2+}	cal	20	0·1 (KNO_3)	ΔH_1 −0·54, $\Delta S_1°$ 20·9	64H
	gl	20	0·1 (KCl)	K_1 5·01±0·01	66I
Tb^{3+}	pol	20	0·1 (KNO_3)	$K(2Tb^{3+} + 3L^{3-} \rightleftharpoons Tb_2L_3{}^{3-})$ 37·20±0·10	57N
Th^{4+}	red	20	0·1 ($NaClO_4$)	K_1 16·9	67B
	Hg	25	0·1 ($NaClO_4$)	K_1 13·3±0·2	67S
Ti^{4+}	dis	20	0·1 ($KClO_4$)	$K(TiO^{2+} + L^{3-} \rightleftharpoons TiOL^-)$ 12·3±0·1	63S
Tl^+	gl	20	0·1 [$(CH_3)_4NNO_3$]	K_1 4·74±0·01	63Ia
	gl	20	0·1 (KNO_3)	K_1 4·75	67Aa
Tl^{3+}	gl	25	1 (HNO_3)	K_1 18,	65K
				$K[Tl^{3+} + 3H_2L^- \rightleftharpoons Tl(H_2L)_3]$ 17·64	
			1 NaCl	$K(TlLCl_n + H^+ \rightleftharpoons TlHLCl_n)$ 2·5	

contd

Metal	*Method*	*Temp*	*Medium*	*Log of equilibrium constant, remarks*	*Ref*
Tl^{3+}	sp	20	?	$K(Tl^{3+}+H_2L^- \rightleftharpoons TlH_2L^{2+})$ 4·38	66K
(*contd*)				$K(TlL+H_3L \rightleftharpoons TlH_3L_2)$ 5·37	
	red	20	1 ($HClO_4$)	K_1 20·9	67Aa
			1 ($NaClO_4$)	β_2 32·5	
Tm^{3+}	pol	20	0·1 (KNO_3)	$K(2Tm^{3+}+3L^{3-} \rightleftharpoons Tm_2L_3{}^{3-})$ 38·28 ± 0·09	57N
$UO_2{}^{2+}$	dis	20	0·1 ($KClO_4$)	K_1 9·56 ± 0·03	63S
VO^{2+}	gl	25	0·1 ($NaClO_4$)	$K(VOLOH^{2-}+H^+ \rightleftharpoons VOL^-)$ 7·38	66Kb
W^{VI}				See Mo^{VI} above.	
Y^{3+}	pol	20	0·1 (KNO_3)	$K(2Y^{3+}+3L^{3-} \rightleftharpoons Y_2L_3{}^{3-})$ 36·8 ± 0·1	57N
Yb^{3+}	pol	20	0·1 (KNO_3)	$K(2Yb^{3+}+3L^{3-} \rightleftharpoons Yb_2L_3{}^{3-})$ 38·56 ± 0·08	57N
Zn^{2+}	cal	20	0·1 (KNO_3)	ΔH_1 −0·84, $\Delta S_1°$ 46·0	64H
	gl	25	1 NaCl	K_1 9·18 ± 0·06	66C
	gl	20	0·1 (KCl)	K_1 10·44 ± 0·05	66I
	gl	25	0·07—0·08	$K(ZnL^-+OH^- \rightleftharpoons ZnLOH^{2-})$ 3·55	68H
			(KNO_3)	$K(ZnL^-+A \rightleftharpoons ZnLA^-)$ 1·58	
				where A = ethylvalinate,	
				$K(ZnL^-+A^- \rightleftharpoons ZnLA^{2-})$ 3·64	
				where HA = glycine	
				See Pb^{2+} above.	
Zr^{4+}	ix	?	1,2 $HClO_4$	$K(Zr^{4+}+H_3L \rightleftharpoons ZrL^+ +3H^+)$ (1) 5·35 ± 0·05, (2) 4·08 ± 0·04	64E
	ix	?	2	K(?) 6·58	64P
	ix	?	0·23, 1 $HClO_4$	K_1 (0·23) 20·81, (1) 19·51	66E
	sp	18—20	1 (HCl)	K_1 18·93 ± 0·07	66Ka

Erratum. 1964, p. 509, Table 449. Under Fe^{3+} equation should *read* $K\{2Fe(OH)L^- \rightleftharpoons [Fe(OH)L]_2{}^{2}\}$ 4·0.

57N W. Noddak and G. Oertel, *Z. Electrochem.*, 1957, **61**, 1216
61A K. V. Astakhov, V. B. Verenikin, V. I. Zinin, and A. D. Zverkova, *Russ. J. Inorg. Chem.*, 1961, **6**, 1057 (2069)
63I Y. J. Israeli, *Canad. J. Chem.*, 1963, **41**, 2710
63Ia H. Irving and J. J. R. F. da Silva, *J. Chem. Soc.*, 1963, 458
63R D. I. Ryabchikov, I. N. Marov, and Y. Ko-min, *Russ. J. Inorg. Chem.*, 1963, **8**, 326 (641)
63S J. Stary, *Analyt. Chim. Acta*, 1963, **28**, 132
64E A. N. Ermakov, I. N. Marov, and G. A. Evitikova, *Russ. J. Inorg. Chem.*, 1964, **9**, 275 (499)
64H J. A. Hull, R. H. Davies, and L. A. K. Staveley, *J. Chem. Soc.*, 1964, 5422
64J V. Jokl, *J. Chromatog.*, 1964, **14**, 71
64P L. N. Pankratova, L. G. Vlasov, and A. V. Lapitskii, *Russ. J. Inorg. Chem.*, 1964, **9**, 954 (1763)
65H H. E. Hellwege and G. K. Schweitzer, *J. Inorg. Nuclear Chem.*, 1965, **27**, 99
65K F. Ya. Kulba and Yu. A. Makashev, *Russ. J. Inorg. Chem.*, 1965, **10**, 634 (1172)
65V V. B. Verenikin and K. V. Astakhov, *Russ. J. Inorg. Chem..* 1965, **10**, 1493 (2753)
65Z N. M. Zhirnova, K. V. Astakhov, and S. A. Barkov, *Russ. J. Phys. Chem.*, 1965, **39**, 647 (1224)
66B H. Berge, *J. prakt. Chem.*, 1966, **34**, 15
66C S. R. Cohen and I. B. Wilson, *Biochemistry*, 1966, **5**, 904
66E A. N. Ermakov, I. N. Marov, and G. A. Evtikova, *Russ. J. Inorg. Chem.*, 1966, **11**, 618 (1155)
66I H. M. N. H. Irving and M. G. Miles, *J. Chem. Soc. (A)*, 1966, 727
66K V. I. Kornev, K. V. Astakhov, and V. I. Rybina, *Russ. J. Phys. Chem.*, 1966, **40**, 594 (1106)
66Ka I. M. Korenman, F. R. Sheyanova, and Z. M. Gureva, *Russ. J. Inorg. Chem.*, 1966, **11**, 1485 (2761)
66Kb Th. Kaden and S. Fallab, *Chimia (Switz.)*, 1966, **20**, 51
67A G. Anderegg, *Helv. Chim. Acta*, 1967, **50**, 2333
67Aa G. Anderegg and E. Bottari, *Helv. Chim. Acta*, 1967, **50**, 2341
67B E. Bottari and G. Anderegg, *Helv. Chim. Acta*, 1967, **50**, 2349
67Ba T. R. Bhat, R. R. Das, and J. Shankar, *Indian J. Chem.*, 1967, **5**, 324
67J N. E. Jacobs and D. W. Margerum, *Inorg. Chem.*, 1967, **6**, 2038
67M A. Merbach, *Helv. Chim. Acta*, 1967, **50**, 1431
67S N. A. Skorik, V. N. Kumok, and V. V. Serebrennikov, *Russ. J. Inorg. Chem.*, 1967, **12**, 1429 (2711), 1788 (3381)
67T M. M. Taqui Khan and A. E. Martell, *J. Amer. Chem. Soc.*, 1967, **89**, 7104
68H D. Hopgood and R. J. Augelici, *J. Amer. Chem. Soc.*, 1968, **90**, 2508

449.1 $C_6H_9O_7N$ Nitrilotriacetic acid *N*-oxide H_3L

Metal	Method	Temp	Medium	Log of equilibrium constant, remarks	Ref
H^+	gl	25	0·1 KNO_3	K_1 7·89±0·01, K_2 2·57±0·01, K_3 < 2	67C
Ca^{2+}	gl	25	0·1 KNO_3	K_1 2·46±0·06	67C
Mg^{2+}	gl	25	0·1 KNO_3	K_1 2·83±0·05	67C

67C R. P. Carter, M. M. Crutchfield, and R. R. Irani, *Inorg. Chem.*, 1967, **6**, 943

450 $C_6H_{10}O_2N_2$ Cyclohexan-1,2-dione dioxime (nioxime) HL

Metal	Method	Temp	Medium	Log of equilibrium constant, remarks	Ref
H^+	gl	25	0·1 75% dioxan	K_1 13·11	63B
Co^{2+}	sp	?	1—6 $NaClO_4$	$K(CoHL_2^+ + I^- \rightleftharpoons CoHL_2 I)$ (1) 2·66, (2) 2·80, (3) 2·92, (4) 3·07, (5) 3·52, (6) 4·00 $K(CoHL_2^+ + 2I^- \rightleftharpoons CoHL_2 I_2^-)$ (1) 4·7±0·1, (2) 5·00±0·05, (3) 5·24±0·05, (4) 5·5±0·1, (5) 5·6±0·1, (6) 6·00±0·05	68B
Ni^{2+}	gl	25	0·1 0, 75% dioxan	K_1 (75%) 11·1±0·2, β_2 (→ 0%) 17·3, (75%) 22·5±0·1 K_{so} −28·39	63B
	dis	25	0·1 ($NaClO_4$)	K_1 11·0±0·1, K_2 10·6±0·1	64S

63B C. V. Banks and S. Anderson, *Inorg. Chem.*, 1963, **2**, 112
64S V. M. Savostina, E. K. Astakhova, and V. M. Peshova, *Russ. J. Inorg. Chem.*, 1964, **9**, 42 (80)
68B K. Burger, B. Pinter, E. Papp-Molnar, and S. Nemes-Kosa, *Acta Chim. Acad. Sci. Hung.*, 1968, **57**, 363

453.1 $C_6H_{10}O_4N_2$ *N*-(Iminomethyl)-2-aminopentanedioic acid [*N*-(formimidoyl)glutamic acid] H_2L

Metal	Method	Temp	Medium	Log of equilibrium constant, remarks	Ref
H^+	gl	25	0·1 $NaClO_4$	K_1 11·0, K_2 4·50	65N
Cu^{2+}	gl	25	0·1 $NaClO_4$	K_1 9·14, K_2 7·42	65N
Ni^{2+}	gl	25	0·1 $NaClO_4$	K_1 5·94, K_2 4·81	65N

65N M. H. T. Nyberg and M. Cefola, *Arch. Biochem. Biophys.*, 1965, **111**, 321

453.2 $C_6H_{10}O_4N_2$ *N*-Acetylglycylglycine HL

Metal	Method	Temp	Medium	Log of equilibrium constant, remarks	Re
H^+	gl	20	0·015, 1·0	K_1 (0·015) 3·66, (1·0) 3·50	60K
Cu^{2+}	gl	20	0·015, 1·0	K_2 (0·015) 2·07, (1·0) 1·41	60K

60K W. L. Koltun, M. Fried, and F. R. N. Gurd, *J. Amer. Chem. Soc.*, 1960, **82**, 233

$C_6H_{10}O_4S$ 3,3′-Thiodipropanoic acid H_2L 454

Metal	Method	Temp	Medium	Log of equilibrium constant, remarks	Ref
H^+	gl	20	0·1	K_1 4·91, K_2 4·09	61C
	gl	25	0·1 ($NaClO_4$)	K_1 4·66, K_2 3·84	68S
Cd^{2+}	gl	25	0·1 ($NaClO_4$)	K_1 2·0	68S
Co^{2+}	gl	25	0·1 ($NaClO_4$)	K_1 1·6	68S
Cu^{2+}	gl	20	0·1	K_1 2·53, $K(Cu^{2+}+HL^- \rightleftharpoons CuHL^+)$ 1·65	61C
	gl	25	0·1 ($NaClO_4$)	K_1 3·0	68S
Mn^{2+}	gl	25	0·1 ($NaClO_4$)	K_1 0·5	68S
Ni^{2+}	gl	25	0·1 ($NaClO_4$)	K_1 1·6	68S
Pb^{2+}	gl	25	0·1 ($NaClO_4$)	K_1 2·7	68S
Zn^{2+}	gl	25	0·1 ($NaClO_4$)	K_1 1·6	68S

61C E. Campi, G. Ostacoli, N. Cibrario, and G. Saini, *Gazzetta*, 1961, **91**, 361
68S K. Suzuki, C. Kavaki, S. Mori, and K. Yamasaki, *J. Inorg. Nuclear Chem.*, 1968, **30**, 167

$C_6H_{10}O_4S_2$ Ethylenedithiodiacetic acid (ethylenebisthioglycollic acid] H_2L 455

Metal	Method	Temp	Medium	Log of equilibrium constant, remarks	Ref
H^+	gl	25	0·3 (K_2SO_4)	K_1 3·81, K_2 3·19	61J
	gl	20	0·1	K_1 4·21, K_2 3·39	61S
Ca^{2+}	gl	20	0·1	K_1 1·74, $K(Ca^{2+}+HL^- \rightleftharpoons CaHL^+)$ 1·16	61S
Cd^{2+}	gl	20	0·1	K_1 2·85, $K(Cd^{2+}+HL^- \rightleftharpoons CdHL^+)$ 1·93	61S
Cu^+	Pt	25	0·3 (K_2SO_4)	β_n 4·37+2·9n $K(CuHL^+ +H^+ \rightleftharpoons CuH_2L^{2+})$ (?) 2·44 $K(CuL+H^+ \rightleftharpoons CuHL^+)$ (?) 2·71	61J
Cu^{2+}	gl	25	0·3 (K_2SO_4)	$K_1 \sim$ 4·8	61J
	gl	20	0·1	K_1 5·66, $K(Cu^{2+}+HL^- \rightleftharpoons CuHL^+)$ 3·98	61S
Zn^{2+}	gl	20	0·1	K_1 2·61, $K(Zn^{2+}+HL^- \rightleftharpoons ZnHL^+)$ 1·74	61S

61J B. R. James and R. J. P. Williams, *J. Chem. Soc.*, 1961, 2007
61S G. Saini, G. Ostacoli, E. Campi, and N. Cibrario, *Gazzetta*, 1961, **91**, 242

$C_6H_{10}O_4S_2$ 3,3′-Dithiodipropanoic acid H_2L 455.1

Metal	Method	Temp	Medium	Log of equilibrium constant, remarks	Ref
H^+	gl	20	0·1	K_1 4·77, K_2 4·02	61O
	gl	20	0·15 ($NaClO_4$)	K_1 4·47, K_2 3·88	63H
Cu^{2+}	gl	20	0·1	K_1 1·61, $K(Cu^{2+}+HL^- \rightleftharpoons CuHL^+)$ 0·88	61O
	gl	20	0·15 ($NaClO_4$)	K_1 3·02±0·06 $K(Cu^{2+}+HL^- \rightleftharpoons CuHL^+)$ 2·54	63H

61O G. Ostacoli, E. Campi, N. Cibrario, and G. Saini, *Gazzetta*, 1961, **91**, 349
63H C. J. Hawkins and D. D. Perrin, *Inorg. Chem.*, 1963, **2**, 843

457 $C_6H_{11}O_2N$ 1-Aminocyclopentanecarboxylic acid HL

Metal	*Method*	*Temp*	*Medium*	*Log of equilibrium constant, remarks*	*Ref*
Cu^{2+}	gl	20	0·1 (KCl)	$K(CuL^+ + H^+ \rightleftharpoons CuHL^{2+})$ 1·9	63I

Erratum. 1964, p. 514, Table 457. Insert **HL** in heading.

63I H. Irving and L. D. Pettit, *J. Chem. Soc.*, 1963, 1546

457.1 $C_6H_{11}O_2N$ Piperidine-2-carboxylic acid HL

Metal	*Method*	*Temp*	*Medium*	*Log of equilibrium constant, remarks*	*Ref*
H^+	gl	20	0·1 KNO_3	K_1 10·38	68H
Cu^{2+}	gl	20	0·1 KNO_3	K_1 7·5, K_2 6·4	68H

68H P. Herman and K. Lemke, *Z. physiol. Chem.*, 1968, **349**, 390

457.2 $C_6H_{11}O_2N$ Glycine butyl ester (n-butyl glycinate) L

Metal	*Method*	*Temp*	*Medium*	*Log of equilibrium constant, remarks*	*Ref*
Cu^{2+}				See nitrilotriacetic acid (449) for mixed ligands	

458.1 $C_6H_{11}O_4N$ 2-Amino-2-methylpentanedioic acid (α-methylglutamic acid) H_2L

Metal	*Method*	*Temp*	*Medium*	*Log of equilibrium constant, remarks*	*Ref*
H^+	gl	25	0·1 $NaClO_4$	K_1 9·71, K_2 4·05	65N
Cu^{2+}	gl	25	0·1 $NaClO_4$	K_1 7·97, K_2 6·34	65N

65N M. H. T. Nyberg and M. Cefola, *Arch. Biochem. Biophys.*, 1965, **111**, 321

$C_6H_{11}O_4N_3$ Glycylglycylglycine (triglycine) HL 459

Metal	Method	Temp	Medium	Log of equilibrium constant, remarks	Ref
H^+	gl	25	~0	K_1 8·02	54P
	gl	20	~0	K_1 8·20, K_2 3·24	55D
	gl	25	0·1 (KNO_3)	K_1 7·90, K_2 3·27	66K
	gl	25	0·1 ($NaClO_4$)	K_1 7·87±0·01, K_2 3·18±0·01	68B
	cal	25	0·1 ($NaClO_4$)	ΔH_1 −10·0±0·1, ΔS_1 2·1±0·4, ΔH_2 −0·2±0·1, ΔS_2 13·9±0·6	68B
	gl	25	3·0 ($NaClO_4$)	K_1 8·547±0·005, K_2 3·712±0·006	68O
Cd^{2+}	gl	25	0·01 $CdSO_4$	K_1 2·0	54P
Cu^{2+}	gl	20	~0	$K[Cu(H_{-1}L)+L^- \rightleftharpoons CuL(H_{-1}L)^-]$ 3·6	55D
				$K[Cu(H_{-1}L)+H^+ \rightleftharpoons CuL^+]$ 5·2	
				$K[Cu(H_{-2}L)^- +H^+ \rightleftharpoons Cu(H_{-1}L)]$ 7·0	
				$K[Cu(H_{-1}L)_2^{2-} +H^+ \rightleftharpoons CuL(H_{-1}L)^-]$ 8·6	
	gl	25	0·1 (KNO_3)	K_1 5·5±0·1	66K
				$K[Cu(H_{-1}L)+H^+ \rightleftharpoons CuL^+)$ 5·4±0·1	
				$K[Cu(H_{-2}L)^- +H^+ \rightleftharpoons Cu(H_{-1}L)]$ 6·63±0·02	
				$K[Cu(H_{-2}L)OH^{2-} +H^+ \rightleftharpoons Cu(H_{-2}L)^-]$ 10·9±0·1	
	gl	25	0·1 ($NaClO_4$)	K_1 5·04±0·01	68B
				$K[Cu(H_{-1}L)+H^+ \rightleftharpoons CuL^+]$ 5·06±0·01	
				$K[Cu(H_{-2}L)^- +H^+ \rightleftharpoons Cu(H_{-1}L)]$ 6·78±0·02	
	cal	25	0·1 ($NaClO_4$)	ΔH_1 −6·3±0·2, ΔS_1 1·9±1·0	68B
				$\Delta H[Cu(H_{-1}L)+H^+ \rightleftharpoons CuL^+)]$ −7·5±0·2	
				ΔS −2·0±0·7	
				$\Delta H[Cu(H_{-2}L)^- +H^+ \rightleftharpoons Cu(H_{-1}L)]$ −7·4±0·2	
				ΔS 6·2±0·8	
	gl, Hg(Cu)	25	3·0 ($NaClO_4$)	K_1 5·66±0·03, β_2 10·17±0·04	68O
				$K(Cu^{2+}+H^+ +L^- \rightleftharpoons CuHL^{2+})$ 10·13±0·03	
				$K(Cu^{2+}+2H^+ +2L^- \rightleftharpoons CuH_2L_2^{2+})$ 19·0	
				$K(2Cu^{2+}+2L^- \rightleftharpoons Cu_2L_2^{2+})$ 13·12±0·01	
				$K(2Cu^{2+}+H^+ +2L^- \rightleftharpoons Cu_2HL_2^{3+})$ 17·3±0·1	
				$K(2Cu^{2+}+2H^+ +2L^- \rightleftharpoons Cu_2H_2L_2^{4+})$ 21·0	
				$K[Cu(H_{-1}L)+H^+ \rightleftharpoons CuL^+)]$ 5·79±0·03	
				$K[Cu(H_{-2}L)^- +H^+ \rightleftharpoons Cu(H_{-1}L)]$ 6·73±0·03	
				$K[Cu(H_{-1}L)L^- +H^+ \rightleftharpoons CuL_2]$ 6·26±0·04	
				$K[Cu(H_{-1}L)_2^{2-} +H^+ \rightleftharpoons Cu(H_{-1}L)L^-]$ 8·72±0·04	
				$K[Cu_2(H_{-1}L)_2+2H^+ \rightleftharpoons Cu_2L_2^{2+}]$ 11·7±0·1	
				$K[Cu_2(H_{-2}L)_2^{2-} +2H^+ \rightleftharpoons Cu_2(H_{-1}L)_2]$ 14·8±0·2	
Ni^{2+}	gl	25	0·16 (KNO_3)	$K[Ni(H_{-1}L)+H^+ \rightleftharpoons NiL^+]$ 8·25	60M
				$K[Ni(H_{-2}L)^- +H^+ \rightleftharpoons Ni(H_{-1}L)]$ 8·45	
	gl	25	0·1 (KNO_3)	K_1 3·76, K_2 3·10	67K
				$K[Ni(H_{-2}L)^- +2H^+ \rightleftharpoons NiL^+]$ 16·9	
				$K[Ni(H_{-2}L)OH^{2-} +H^+ \rightleftharpoons Ni(H_{-2}L)^-]$ 10·5	
Zn^{2+}	gl	25	0·01 $ZnSO_4$	K_1 2·6	54P

Erratum. 1964, p. 515, Table 459. Under Cu^{2+} *delete* K_2 3·6

54P D. J. Perkins, *Biochem. J.*, 1954, **57**, 702

55D H. Dobbie and W. O. Kermack, *Biochem. J.*, 1955, **59**, 257

60M R. B. Martin, M. Chamberlin, and J. T. Edsall, *J. Amer. Chem. Soc.*, 1960, **82**, 495

66K M. K. Kim and A. E. Martell, *J. Amer. Chem. Soc.*, 1966, **88**, 914

67K M. K. Kim and A. E. Martell, *J. Amer. Chem. Soc.*, 1967, **89**, 5138

68B A. P. Brunetti, M. C. Lim, and G. H. Nancollas, *J. Amer. Chem. Soc.*, 1968, **90**, 5120

68O R. Osterberg and B. Sjoberg, *J. Biol. Chem.*, 1968, **243**, 3038

460 $C_6H_{11}O_5N$ *N*-(2′-Hydroxyethyl)iminodiacetic acid (HIMDA) H_2L

Metal	*Method*	*Temp*	*Medium*	*Log of equilibrium constant, remarks*	*Ref*
H^+	gl	25	0·1 (KNO_3)	K_1 8·72, K_2 1·91	63T, 66J
	gl	25	0·10 (KCl)	K_1 8·57, K_2 2·25	65D
	gl	0·4	0·10 (KNO_3)	K_1 9·92, K_2 2·44	67T
Cd^{2+}	oth	20	0·1 (KNO_3)	K_1 8·6, K_2 5·2, electrophoresis	65J
	pol	25	0·1 (KNO_3)	K_1 7·41, β_2 13·75	65V
Ce^{3+}	gl	25	0·1 (KNO_3)	K_1 8·46 ± 0·03, K_2 6·56 ± 0·02	63T
	pol	25	0·10 (KCl)	β_2 14·12	65D
	oth	20	0·1 ($NaNO_3$)	K_1 8·4, K_2 6·6, paper electrophoresis	66J
				See HEDTA (838) for mixed ligands	
Co^{2+}	oth	20	0·1 (KNO_3)	K_1 9·0, K_2 4·4, electrophoresis	65J
Cu^{2+}	oth	20	0·1 (KNO_3)	K_1 11·2, K_2 4·0, electrophoresis	65J
	pol	25	0·1 (KNO_3)	K_1 13·38, β_2 15·62	65V
	gl	0·4	0·10 (KNO_3)	K_1 12·00	67T
Dy^{3+}	Hg(Cu)	25	0·1 (KNO_3)	K_1 9·08 ± 0·02, K_2 8·30 ± 0·02	63T
	gl	25	0·10 (KCl)	K_1 8·88, K_2 7·52	65D
	oth	20	0·1 ($NaNO_3$)	K_1 9·1, K_2 8·1, paper electrophoresis	66J
				See HEDTA (838) for mixed ligands	
Er^{3+}	Hg(Cu)	25	0·1 (KNO_3)	K_1 9·24 ± 0·02, K_2 7·98 ± 0·01	63T
	pol	25	0·10 (KCl)	β_2 16·30	65D
	oth	20	0·1 ($NaNO_3$)	K_1 9·1, K_2 8·0, paper electrophoresis	66J
				See HEDTA (838) for mixed ligands	
Eu^{3+}	Hg(Cu)	25	0·1 (KNO_3)	K_1 9·10 ± 0·02, K_2 7·91 ± 0·02	63T
	gl	25	0·10 (KCl)	K_1 8·99, K_2 7·27	65D
	oth	20	0·1 ($NaNO_3$)	K_1 8·95, K_2 7·85, paper electrophoresis	66J
				See HEDTA (838) for mixed ligands	
Fe^{3+}	pol	25	0·1 (KNO_3)	$K(Fe^{3+} + L^{2-} + OH^- \rightleftharpoons FeLOH)$ 24·94	65V
				$K[Fe^{3+} + L^{2-} + 2OH^- \rightleftharpoons FeL(OH)_2^-]$ 32·33	
Gd^{3+}	Hg(Cu)	25	0·1 (KNO_3)	K_1 9·01 ± 0·03, K_2 8·04 ± 0·03	63T
	pol	25	0·10 (KCl)	β_2 16·08	65D
	oth	20	0·1 ($NaNO_3$)	K_1 9·0, K_2 7·95, paper electrophoresis	66J
				See HEDTA (838) for mixed ligands	
Hf^{4+}	ix	?	0·23 $HClO_4$	K_1 14·63	66E
Ho^{3+}	Hg(Cu)	25	0·1 (KNO_3)	K_1 9·18 ± 0·03, K_2 8·13 ± 0·02	63T
	pol	25	0·10 (KCl)	β_2 16·36	65D
	oth	20	0·1 ($NaNO_3$)	K_1 9·1, K_2 8·05, paper electrophoresis	66J
				See HEDTA (838) for mixed ligands	
In^{3+}	ix	?	0·5	K_1 11·0 ± 0·1	63R
La^{3+}	gl	25	0·1 (KNO_3)	K_1 8·00 ± 0·01, K_2 5·98 ± 0·01	63T
	gl	25	0·10 (KCl)	K_1 7·33, K_2 6·57	65D
	oth	20	0·1 ($NaNO_3$)	K_1 8·15, K_2 5·8, paper electrophoresis	66J
				See HEDTA (838) for mixed ligands	
Lu^{3+}	Hg(Cu)	25	0·1 (KNO_3)	K_1 9·50 ± 0·02, K_2 8·02 ± 0·02	63T
	gl	25	0·10 (KCl)	K_1 9·40, K_2 6·98	65D
	oth	20	0·1 ($NaNO_3$)	K_1 9·1, K_2 7·95, paper electrophoresis	66J
				See HEDTA (838) for mixed ligands	
Mn^{2+}	oth	20	0·1 (KNO_3)	K_1 6·4, K_2 3·3, electrophoresis	65J
Nd^{3+}	gl	25	0·1 (KNO_3)	K_1 8·80 ± 0·02, K_2 7·13 ± 0·01	63T
	gl	25	0·10 (KCl)	K_1 8·12, K_2 6·56	65D
	oth	20	0·1 ($NaNO_3$)	K_1 8·65, K_2 7·2, paper electrophoresis	66J
				See HEDTA (838) for mixed ligands	
Ni^{2+}	oth	20	0·1 (KNO_3)	K_1 10·2, K_2 5·3, electrophoresis	65J
Pb^{2+}	oth	20	0·1 (KNO_3)	K_1 10·2, K_2 3·0, electrophoresis	65J
	pol	25	0·1 (KNO_3)	K_1 9·51	65V

continu

Metal	Method	Temp	Medium	Log of equilibrium constant, remarks	Ref
Pr^{3+}	gl	25	0·1 (KNO_3)	K_1 8·64±0·03, K_2 6·86±0·02	63T
	pol	25	0·10 (KCl)	β_2 14·45	65D
	oth	20	0·1 ($NaNO_3$)	K_1 8·55, K_2 6·9, paper electrophoresis	66J
				See HEDTA (838) for mixed ligands	
Sc^{3+}	oth	20	0·1 ($NaNO_3$)	K_2 8·65, paper electrophoresis	66J
Sm^{3+}	gl	25	0·1 (KNO_3)	K_1 9·10±0·03, K_2 7·77±0·02	63T
	pol	25	0·10 (KCl)	β_2 15·74	65D
	oth	20	0·1 ($NaNO_3$)	K_1 8·8, K_2 7·65, paper electrophoresis	66J
				See HEDTA (838) for mixed ligands	
Tb^{3+}	Hg(Cu)	25	0·1 (KNO_3)	K_1 9·08±0·02, K_2 8·19±0·02	63T
	pol	25	0·10 (KCl)	β_2 16·42	65D
	oth	20	0·1 ($NaNO_3$)	K_1 9·0, K_2 8·05, paper electrophoresis	66J
				See HEDTA (838) for mixed ligands	
Tm^{3+}	Hg(Cu)	25	0·1 (KNO_3)	K_1 9·35±0·03, K_2 7·88±0·01	63T
	gl	25	0·10 (KCl)	K_1 9·03, K_2 7·31	65D
	oth	20	0·1 ($NaNO_3$)	K_1 9·1, K_2 7·8, paper electrophoresis	66J
				See HEDTA (838) for mixed ligands	
U^{4+}	gl	25	0·10 (KCl)	$K[U(HL)_2OH^- + H^+ \rightarrow U(HL)_2]$ 3·67±0·02	68C
Y^{3+}	Hg(Cu)	25	0·1 (KNO_3)	K_1 9·22±0·01, K_2 7·61±0·02	63T
	gl	25	0·10 (KCl)	K_1 8·38, K_2 7·31	65D
	oth	20	0·1 ($NaNO_3$)	K_1 8·6, K_2 7·6, paper electrophoresis	66J
				See HEDTA (838) for mixed ligands	
Yb^{3+}	Hg(Cu)	25	0·1 (KNO_3)	K_1 9·38±0·02, K_2 7·74±0·02	63T
	pol	25	0·1 (KCl)	β_2 16·37	65D
	oth	20	0·1 ($NaNO_3$)	K_1 9·1, K_2 7·8, paper electrophoresis	66J
				See HEDTA (838) for mixed ligands	
Zn^{2+}	oth	20	0·1 (KNO_3)	K_1 9·4, K_2 3·5, electrophoresis	65J

Errata. 1964, p. 516, Table 460. Under UO_2^{2+} third line, equation should *read* $K[2UO_2(OH)L^- \rightleftharpoons UO_2(OH)_2L_2^{2-}]$ 3·50. Ref 64R. *Add* page number 789 at end of line.

63R D. I. Ryabchikov, I. N. Marov, and Y. K'o-min, *Russ. J. Inorg. Chem.*, 1963, **8**, 326 (641)
63T L. C. Thompson and J. A. Loraas, *Inorg. Chem.*, 1963, **2**, 594
65D N. M. Dyatlova, V. Ya. Temkina, Yu. F. Belugin, O. Yu. Lavrova, L. E. Bertina, F. D. Iozefovich, N. N. Kalmykova, and E. P. Zhirov, *Russ. J. Inorg. Chem.*, 1965, **10**, 612 (1131)
65J V. Jokl and J. Majer, *Acta Fac. Pharm. Brun. Bratislav.* 1965, **10**, 55
65V E. Verdier and J. Fournier, *J. Chim. phys.*, 1965, **62**, 1196
66E A. N. Ermakov, I. N. Marov, and G. A. Evtikova, *Russ. J. Inorg. Chem.*, 1966, **11**, 618 (1155)
66J V. Jokl, J. Majer, H. Scharf, and H. Kroll, *Mikrochim. Acta*, 1966, 63
67T M. M. Taqui Khan and A. E. Martell, *J. Amer. Chem. Soc.*, 1967, **89**, 7104
68C G. H. Carey and A. E. Martell, *J. Amer. Chem. Soc.*, 1968, **90**, 32

2.1 $C_6H_{11}NS_2$ Piperidine-*N*-dithiocarboxylic acid (*N,N*-pentamethylenedithiocarbamic acid) HL

Metal	Method	Temp	Medium	Log of equilibrium constant, remarks	Ref
Cu^{2+}	sp	23—27	0·01 NaOH 75% ethanol	K_1 14·7±0·2, K_2 13·7±0·2	56J
	sp	18—22	0·01 NaOH 0—89% ethanol	K_1 (0%) 11·9±0·2, (51·7%) 13·9±0·2, (75%) 15·1 ±0·2, (89%) 16·2±0·2, K_2 (0%) 10·5±0·2, (51·7%) 12·9±0·2, (75%) 13·9±0·2, (89%) 14·8±0·2	57J

56J M. J. Janssen, *Rec. Trav. chim.*, 1956, **75**, 1411
57J M. J. Janssen, *Rec. Trav. chim.*, 1957, **76**, 827

462.2 $C_6H_{12}O_2N_2$ Hexan-3,4-dione dioxime (diethylglyoxime) HL

Metal	*Method*	*Temp*	*Medium*	*Log of equilibrium constant, remarks*	*Ref*
H^+	gl	25	0·1 75% dioxan	K_1 13·93	63B
Ni^{2+}	gl	25	0·1 0—75% dioxan	K_1 (75%) 12·4±0·2, β_2 (→ 0%) 17·2, (75%) 23·7±0·1, K_{so} −24·21	63B

63B C. V. Banks and S. Anderson, *Inorg. Chem.*, 1963, 2, 112

463 $C_6H_{12}O_3N_2$ L-Alanyl-D-alanine HL

Erratum. 1964, p. 517, Table 463. *For* K_1 9·51 *read* K_1 8·51.

466 $C_6H_{12}O_3N_2$ Glycylglycine ethyl ester L

Metal	*Method*	*Temp*	*Medium*	*Log of equilibrium constant, remarks*	*Ref*
Ni^{2+}	gl	25	0·16 (KNO_3)	$K[Ni(H_{-1}L)+H^+ \rightleftharpoons NiL^+]$ ~ 9·2 $K[Ni(H_{-1}L)OH^- + H^+ \rightleftharpoons Ni(H_{-1}L)]$ ~ 9·8	60M

60M R. B. Martin, M. Chamberlin, and J. T. Edsall, *J. Amer. Chem. Soc.*, 1960, **82**, 495

468 $C_6H_{12}O_4N_2$ *NN'*-Ethylenediglycine (ethylenediamine-*NN'*-diacetic acid) H_2L

Metal	*Method*	*Temp*	*Medium*	*Log of equilibrium constant, remarks*	*Ref*
Ce^{3+}, La^{3+}, Nd^{3+}, Pr^{3+}, Sm^{3+}				See HEDTA (838) for mixed ligands	
Ni^{2+}	sp	25	0·5 ($NaClO_4$)	$K(NiL + NH_3 \rightleftharpoons NiLNH_3)$ 2·00±0·06	67J

67J N. E. Jacobs and D. W. Margerum, *Inorg. Chem.*, 1967, **6**, 2038

$C_6H_{12}O_4N_2$ DL-Alanyl-DL-serine HL 468.1

Metal	Method	Temp	Medium	Log of equilibrium constant, remarks	Ref
H^+	pH	25	0·1 ($NaClO_4$)	K_1 8·25, K_2 2·98	67S
Cu^{2+}	pH	25	0·1 ($NaClO_4$)	K_1 6·23,	67S
				$K[Cu(H_{-1}L)+H^+ \rightleftharpoons CuL^+]$ 4·25	
				$K[Cu(H_{-1}L)+L^- \rightleftharpoons CuL(H_{-1}L)^-]$ 2·58	
				$K[Cu(H_{-1}L)_2{}^{2-}+H^+ \rightleftharpoons CuL(H_{-1}L)^-]$ 10·16	
Ni^{2+}	pH	25	0·1 ($NaClO_4$)	K_1 3·41, K_2 3·82	67S

67S A. Ya. Sychev and N. S. Mitsul, *Russ. J. Inorg. Chem.*, 1967, **12**, 1120 (2127)

$C_6H_{13}O_2N$ 2-Aminohexanoic acid (norleucine) HL 470

Metal	Method	Temp	Medium	Log of equilibrium constant, remarks	Ref
H^+	gl	18—20	~ 0	K_1 9·92	52P, 53P
	gl	25	~ 0·1	K_1 9·86, K_2 2·39	66M
	E	25	?	K_1 9·76	67A
Ag^+	Ag	25	0·6	K_1 3·48±0·07, β_2 6·76±0·05	67A
Be^{2+}	gl	18—20	0·0005—0·005 $BeSO_4$	β_2 12·8	52P, 53P
Cd^{2+}	gl	18	0·005 $CdSO_4$	β_2 7·3	53P
	gl	20	0·0005 $CdSO_4$	β_2 6·9	52P
Hg^{2+}	gl	18—20	0·0005—0·005 $Hg(NO_3)_2$	β_2 17·8	52P, 53P
Ni^{2+}	gl	25	~ 0·1	K_1 5·51, β_2 9·67, β_3 13·86	66M
Zn^{2+}	gl	18	0·005 $ZnSO_4$	β_2 8·7	53P
	gl	20	0·0005 $ZnSO_4$	β_2 8·5	52P

52P D. J. Perkins, *Biochem. J.*, 1952, **51**, 487
53P D. J. Perkins, *Biochem. J.*, 1953, **55**, 649
66M S. Morazzani-Pelletier and S. Meriaux, *J. Chim. phys.*, 1966, 63, 278
67A Yu. M. Azizov, A. Kh. Miftakhova, and V. F. Toropova, *Russ. J. Inorg. Chem.*, 1967, **12**, 345 (661)

$C_6H_{13}O_2N$ 6-Aminohexanoic acid HL 470.1

Metal	Method	Temp	Medium	Log of equilibrium constant, remarks	Ref
H^+	gl	20	0·5 KNO_3	K_1 11·04±0·02	68A
	gl	25	0·5 KNO_3	K_1 10·77	68T
Ag^+	gl	20	0·500 KNO_3	K_1 3·6±0·1, β_2 7·54±0·01	68A
	Ag-AgCl	20	0·500 KNO_3	β_2 7·65±0·02	68A
	gl	25	0·5 KNO_3	K_1 3·59, K_2 3·95	68T

68A D. J. Alner, R. C. Lansbury, and A. G. Smeeth, *J. Chem. Soc.* (*A*), 1968, 417
68T G. F. Thiers, L. C. Van Poucke, and M. A. Herman, *J. Inorg. Nuclear Chem.*, 1968, **30**, 1543

470.2 $C_6H_{13}O_2N$ 2-Amino-3-methylpentanoic acid (DL-isoleucine) HL

Metal	Method	Temp	Medium	Log of equilibrium constant, remarks	Ref
H^+	gl	20	~0	K_1 9·86	52P
	gl	25	~0·1	K_1 9·76, K_2 2·38	66M
Be^{2+}	gl	20	0·01 $BeSO_4$	β_2 12·6	52P
Cd^{2+}	gl	20	0·01 $CdSO_4$	β_2 6·6	52P
Hg^{2+}	gl	20	0·01 $Hg(NO_3)_2$	β_2 17·6	52P
Ni^{2+}	gl	25	~0·1	K_1 5·28, β_2 9·56, β_3 12·40	66M
Zn^{2+}	gl	20	0·01 $ZnSO_4$	β_2 8·2	52P

52P D. J. Perkins, *Biochem. J.*, 1952, **51**, 487
66L D. L. Leussing and E. M. Hanna, *J. Amer. Chem. Soc.*, 1966, **88**, 693
66M S. Morazzani-Pelletier and S. Meriaux, *J. Chim. phys.*, 1966, **63**, 278

470.3 $C_6H_{13}O_2N$ L(+)-Isoleucine HL

Metal	Method	Temp	Medium	Log of equilibrium constant, remarks	Ref
H^+	gl	25	0·5 KCl	K_1 9·70, K_2 2·66	66L
	gl	25	~0·1	K_1 9·69, K_2 2·31	66M
Ni^{2+}	gl	25	0·5 (KCl)	K_1 5·22, β_2 9·45,	66L
				$K(NiL^+ + A^- \rightleftharpoons NiAL)$ 0·99	
				$K(Ni^{2+} + A^- + L^- \rightleftharpoons NiAL)$ 6·2±0·1	
				$K(Ni^{2+} + A^- + 2L^- \rightleftharpoons NiAL_2^-)$ 11·1±0·1	
				$K(Ni^{2+} + 2A^- + 2L^- \rightleftharpoons NiA_2L_2^{2-})$ 13·05±0·07	
				where HA = pyruvic acid	
	gl	25	~0·1	K_1 5·52, β_2 9·83, β_3 12·58	66M
Zn^{2+}	gl	25	0·5 KCl	K_1 4·49, β_2 8·49, β_3 10·9	66L
				$K(ZnL^+ + A^- \rightleftharpoons ZnAL)$ 2·27	
				$K(Zn^{2+} + A^- + L^- \rightleftharpoons ZnAL)$ 6·76±0·02	
				$K(Zn^{2+} + A^- + 2L^- \rightleftharpoons ZnAL_2^-)$ 9·58±0·04	
				$K(Zn^{2+} + 2A^- + 2L^- \rightleftharpoons ZnA_2L_2^{2-})$ 12·34±0·05	
				where HA = pyruvic acid	

Note: Constants should be identical to those listed for D(−)-Isoleucine.

66M S. Morazzani-Pelletier and S. Meriaux, *J. Chim. phys.*, 1966, **63**, 278

470.4 $C_6H_{13}O_2N$ D(−)-Isoleucine HL

Metal	Method	Temp	Medium	Log of equilibrium constant, remarks	Ref
H^+	gl	25	~0·1	K_1 9·73, K_2 2·39	66M
Cd^{2+}	pol	25	1	β_2 6·9, β_3 8·8	65V
Cu^{2+}	pol	25	1·1	K_1 8·4, β_2 15·4	65V
Ni^{2+}	gl	25	~0·1	K_1 5·48, β_2 9·69, β_3 13·03	66M
Zn^{2+}	pol	25	1	β_2 11·2	65V

Note: Constants should be identical to those listed for L(−)-Isoleucine.

65V E. Verdier and G. Zalessky, *J. Chim. phys.*, 1965, **62**, 479
66M S. Morazzani-Pelletier and S. Meriaux, *J. Chim. phys.*, 1966, **63**, 278

$C_6H_{13}O_2N$ 2-Amino-4-methylpentanoic acid (DL-leucine) HL 471

Metal	Method	Temp	Medium	Log of equilibrium constant, remarks	Ref
H^+	gl	20	~ 0	K_1 9·92	52P
	gl	25	~ 0·1	K_1 9·71, K_2 2·39	66M
Be^{2+}	gl	20	0·01 $BeSO_4$	β_2 13·2	52P
Cd^{2+}	gl	20	0·01 $CdSO_4$	β_2 7·8	52P
	oth	20	0·1 (KNO_3)	K_1 5·8, K_2 3·6, K_3 2·4, paper electrophoresis	64J
Co^{2+}	oth	20	0·1 (KNO_3)	K_1 5·2, K_2 3·2, K_3 2·3, paper electrophoresis	64J
Cr^{3+}	gl	25	~ 0·5	K_1 8·8, K_2 6·8, K_3 5·9	63K
Cu^{2+}	oth	20	0·1 (KNO_3)	K_1 8·6, K_2 7·0, paper electrophoresis	64J
				See nitrilotriacetic acid (449) for mixed ligands	
Hg^{2+}	gl	20	0·01 $Hg(NO_3)_2$	β_2 17·5	52P
Mn^{2+}	oth	20	0·1 (KNO_3)	K_1 3·9, K_2 1·8, paper electrophoresis	64J
Ni^{2+}	oth	20	0·1 (KNO_3)	K_1 6·3, K_2 4·0, K_3 2·5, paper electrophoresis	64J
	gl	25	~ 0·1	K_1 5·71, β_2 10·26, β_3 14·97	66M
Zn^{2+}	gl	20	0·01 $ZnSO_4$	β_2 9·1	52P
	oth	20	0·1 (KNO_3)	K_1 5·8, K_2 4·2, K_3 3·3, paper electrophoresis	64J

52P D. J. Perkins, *Biochem J.*, 1952, **51**, 487
63K A. A. Khan and W. U. Malik, *J. Indian Chem. Soc.*, 1963, **40**, 564
64J V. Jokl, *J. Chromatog.*, 1964, **14**, 71
66M S. Morazzani-Pelletier and S. Meriaux, *J. Chim. phys.*, 1966, **63**, 278

$C_6H_{13}O_2N$ D(+)-Leucine HL 471.1

Metal	Method	Temp	Medium	Log of equilibrium constant, remarks	Ref
H^+	gl	25	~ 0·1	K_1 9·76, K_2 2·33	66M
Ni^{2+}	gl	25	~ 0·1	K_1 5·68, β_2 10·02, β_3 14·27	66M

Note: Constants should be identical to those listed for L(−)-Leucine.

66M S. Morazzani-Pelletier and S. Meriaux, *J. Chim. phys.*, 1966, **63**, 278

$C_6H_{13}O_2N$ L(−)-Leucine HL 471.2

Metal	Method	Temp	Medium	Log of equilibrium constant, remarks	Ref
H^+	gl	25	~ 0·1	K_1 9·69, K_2 2·36	66M
Ni^{2+}	gl	25	~ 0·1	K_1 5·53, β_2 9·46, β_3 14·38	66M

Note: Constants should be identical to those listed for D(+)-Leucine.

66M S. Morazzani-Pelletier and S. Meriaux, *J. Chim. phys.*, 1966, **63**, 278

473.1 $C_6H_{13}O_2N_3$ *N*-(2-Dimethylaminoethyl)oxamide L

Metal	Method	Temp	Medium	Log of equilibrium constant, remarks	Ref
H^+	gl	23	0·5 (KCl)	K_1 8·65	68K
Cu^{2+}	gl	23	0·5 (KCl)	K_1 4·54	68K
				$K[Cu(H_{-1}L)^+ + H^+ \rightleftharpoons CuL^{2+}]$ 4·92	
				$K[CuOH(H_{-1}L) + H^+ \rightleftharpoons Cu(H_{-1}L)^+]$ 8·44	
				$K[CuOH(H_{-2}L)^- + H^+ \rightleftharpoons CuOH(H_{-1}L)]$ 10·31	
				$K[Cu(H_{-1}L)^+ + CuOH(H_{-1}L) \rightleftharpoons Cu_2OH(H_{-1}L)_2{}^+]$ 2·94	

68K Th. Kaden and A. Zuberbuhler, *Helv. Chim. Acta*, 1968, **51**, 1797

475.1 $C_6H_{13}O_3N_3$ 2-Amino-5-ureidopentanoic acid (L-citrulline) HL

Metal	Method	Temp	Medium	Log of equilibrium constant, remarks	Ref
H^+	gl	20	~ 0	K_1 9·69	53P
Be^{2+}	gl	20	0·005 $BeSO_4$	β_2 13·0	53P
Cd^{2+}	gl	20	0·005 $CdSO_4$	β_2 7·3	53P
Hg^{2+}	gl	20	0·005 $Hg(NO_3)_2$	β_2 18·8	53P
Zn^{2+}	gl	20	0·005 $ZnSO_4$	β_2 8·7	53P

53P D. J. Perkins, *Biochem. J.*, 1953, **55**, 649

476 $C_6H_{13}O_4N$ *NN*-Bis(2′-hydroxyethyl)glycine HL

Metal	Method	Temp	Medium	Log of equilibrium constant, remarks	Ref
H^+	?	20	0·1	K_1 8·11, K_2 2·5	62S
	?	20	0·1	K_1 8·41, K_2 1·99	64J, 65J, 6
	sp	20	0·1 ($NaClO_4$)	K_1 8·14, K_2 1·68	67S
Cd^{2+}	oth	20	0·1 (KNO_3)	K_1 6·3, K_2 4·0, paper electrophoresis	65J
Ce^{3+}	oth	20	0·1 ($NaNO_3$)	K_1 7·5, K_2 5·6, paper electrophoresis	66J
Co^{2+}	oth	20	0·1 (KNO_3)	K_1 6·1, K_2 3·5, paper electrophoresis	64J
	sp	20	0·1 ($NaClO_4$)	K_1 5·5	67S
	oth	20	0·1 ($NaClO_4$)	$K[Co(H_{-2}L)^- + L^- + 2H^+ \rightleftharpoons CoL_2]$ 11·9 paper electrophoresis	67S
Cu^{2+}	oth	20	0·1 (KNO_3)	K_1 8·6, K_2 5·0, paper electrophoresis	64J
	sp	20	0·1 ($NaClO_4$)	K_1 8·37, K_2 5·47	67S
	oth	20	0·1 ($NaClO_4$)	$K[Cu(H_{-2}L)^- + L^- + 2H^+ \rightleftharpoons CuL_2]$ 12·0 paper electrophoresis	67S
Dy^{3+}	oth	20	0·1 ($NaNO_3$)	K_1 7·6, K_2 6·2, paper electrophoresis	66J
Er^{3+}	oth	20	0·1 ($NaNO_3$)	K_1 7·7, K_2 5·9, paper electrophoresis	66J
Eu^{3+}	oth	20	0·1 ($NaNO_3$)	K_1 8·0, K_2 6·3, paper electrophoresis	66J
Fe^{2+}	oth	20	0·1 (KNO_3)	K_1 ~ 5, paper electrophoresis	65J
Fe^{3+}	gl	25	0·01	$K[Fe^{3+} + HL + H_2O \rightleftharpoons Fe(OH)_2L + 3H^+]$ −7·4	64N
Gd^{3+}	oth	20	0·1 ($NaNO_3$)	K_1 7·7, K_2 5·8, paper electrophoresis	66J
Hg^{2+}	pol	20	0·1 KNO_3	K_1 14·17	62S
Ho^{3+}	oth	20	0·1 ($NaNO_3$)	K_1 7·6, K_2 5·8, paper electrophoresis	66J
La^{3+}	oth	20	0·1 ($NaNO_3$)	K_1 7·3, K_2 4·8, paper electrophoresis	66J
Lu^{3+}	oth	20	0·1 ($NaNO_3$)	K_1 7·5, K_2 6·0, paper electrophoresis	66J

contin

$C_6H_{13}O_4N$ (contd) 476

Metal	Method	Temp	Medium	Log of equilibrium constant, remarks	Ref
Mn^{2+}	oth	20	0·1 (KNO_3)	K_1 3·9, K_2 2·1, paper electrophoresis	65J
Nd^{3+}	oth	20	0·1 ($NaNO_3$)	K_1 7·6, K_2 5·7, paper electrophoresis	66J
Ni^{2+}	oth	20	0·1 (KNO_3)	K_1 7·7, K_2 5·0, paper electrophoresis	64J
	sp	20	0·1 ($NaClO_4$)	K_1 6·5	67S
	oth	20	0·1 ($NaClO_4$)	$K[NiL(H_{-1}L)^- + H^+ \rightleftharpoons NiL_2]$ 11·9 paper electrophoresis	67S
Pb^{2+}	oth	20	0·1 (KNO_3)	K_1 7·5, paper electrophoresis	65J
Pr^{3+}	oth	20	0·1 ($NaNO_3$)	K_1 7·7, K_2 5·5, paper electrophoresis	66J
Sc^{3+}	oth	20	0·1 ($NaNO_3$)	K_1 8·0, K_2 7·6, paper electrophoresis	66J
Sm^{3+}	oth	20	0·1 ($NaNO_3$)	K_1 7·8, K_2 6·0, paper electrophoresis	66J
Tb^{3+}	oth	20	0·1 ($NaNO_3$)	K_1 8·0, K_2 6·3, paper electrophoresis	66J
Th^{4+}	oth	20	0·1 ($NaNO_3$)	K_1 7·8, K_2 6·0, paper electrophoresis	66J
Y^{3+}	oth	20	0·1 ($NaNO_3$)	K_1 7·2, K_2 5·6, paper electrophoresis	66J
Yb^{3+}	oth	20	0·1 ($NaNO_3$)	K_1 7·7, K_2 6·0, paper electrophoresis	66J
Zn^{2+}	oth	20	0·1 (KNO_3)	K_1 6·5, K_2 4·2, paper electrophoresis	65J

62S S. Stankoviansky and J. Konigstein, *Coll. Czech. Chem. Comm.*, 1962, **27**, 1997
64J V. Jokl, J. Majer, and M. Mazacova, *Chem. Zvesti*, 1964, **18**, 584
64N E. R. Nightingale and R. F. Benck, *Talanta*, 1964, **11**, 241
65J V. Jokl and J. Majer, *Acta Fac. Pharm. Brun. Bratislav.*, 1965, **10**, 55
66J V. Jokl, J. Majer, H. Scharf, and H. Kroll, *Mikrochim. Acta*, 1966, 63
67S V. Springer, R. Karlicek, and J. Majer, *Coll. Czech. Chem. Comm.*, 1967, **32**, 774

$C_6H_{13}O_9P$ α-D-Glucose-1′-phosphate H_2L 476.1

Metal	Method	Temp	Medium	Log of equilibrium constant, remarks	Ref
Co^{2+}	ix	25	0·1 ($KClO_4$)	K_1 2·18	66D
	gl	25	0·1 (KNO_3)	K_1 2·12	66D
Mn^{2+}	ix	25	0·1 ($KClO_4$)	K_1 2·19	66D
Zn^{2+}	ix	25	0·1 ($KClO_4$)	K_1 2·37	66D
	gl	25	0·1 (KNO_3)	K_1 2·34	66D

66D B. E. Doody, E. R. Tucci, R. Scruggs, and N. C. Li, *J. Inorg. Nuclear Chem.*, 1966, **28**, 833

$C_6H_{14}O_2N_2$ 2,6-Diaminohexanoic acid (lysine) HL 479

Metal	Method	Temp	Medium	Log of equilibrium constant, remarks	Ref
H^+	gl	20	~0	K_2 9·03	53P
	gl	20	0·1 KNO_3	K_1 10·47, K_2 9·13	68H
	gl	25	0·1 $NaClO_4$	K_1 10·54, K_2 9·06	65N
Be^{2+}	gl	20	0·005 $BeSO_4$	β_2 11·4	53P
Cd^{2+}	gl	20	0·005 $CdSO_4$	β_2 5·8	53P
Cr^{3+}	gl	25	~0·5	K_1 8·1, K_2 6·2, K_3 5·3	63K
Cu^{2+}	gl	25	0·1 $NaClO_4$	β_2 13·90	65N
	gl	20	0·1 KNO_3	K_1 7·56, K_2 6·46	68H
Zn^{2+}	gl	20	0·005 $ZnSO_4$	β_2 7·3	53P

53P D. J. Perkins, *Biochem. J.*, 1953, **55**, 649
63K A. A. Khan and W. U. Malik, *J. Indian Chem. Soc.*, 1963, **40**, 564
65N M. H. T. Nyberg and M. Cefola, *Arch. Biochem. Biophys.*, 1965, **111**, 327
68H P. Herman and K. Lemke, *Z. physiol. Chem.*, 1968, **349**, 390

480 $C_6H_{14}O_2N_4$ 2-Amino-5-guanidopentanoic acid (arginine) HL

Metal	Method	Temp	Medium	Log of equilibrium constant, remarks	Ref
H^+	gl	19	~0	K_1 9·11	53P
	gl	5—55	0·024	K_1 $-4349·7/T+12·654-0·023694\,T$	59D
				ΔH_1 (25°) -43 ± 3 kJ mole^{-1}, ΔS_1 (25°) 28 ± 9 J mole^{-1} deg^{-1}	
	H (?)	25	0	ΔH_1 $-44·89\pm0·08$ kJ mole^{-1}, ΔS_1 $21·6\pm0·3$ J mole^{-1} deg^{-1}	59D
Ag^+	gl	5—55	0·024	K_1 $3855·2/T-18·452+0·029208\,T$, K_2 $4898·35/T-23·905+0·037321\,T$	59D
				ΔH_1 (25°) -24 ± 4 kJ mole^{-1}, ΔS_1 (25°) -20 ± 12 J mole^{-1} deg^{-1}, ΔH_2 (25°) -30 ± 4 kJ mole^{-1}, ΔS_2 (25°) 32 ± 12 J mole^{-1} deg^{-1}	
Be^{2+}	gl	19	0·005 $BeSO_4$	β_2 12·4	53P
Cd^{2+}	gl	19	0·005 $CdSO_4$	β_2 6·7	53P
Cr^{3+}	gl	25	~0·5	K_1 8·0, K_2 6·1, K_3 5·2	63K
Hg^{2+}	gl	19	0·005 $Hg(NO_3)_2$	β_2 17·4	53P
Zn^{2+}	gl	19	0·005 $ZnSO_4$	β_2 8·0	53P

53P D. J. Perkins, *Biochem. J.*, 1953, **55**, 649
59D S. P. Dutta and A. K. Grzybowski, *J. Chem. Soc.*, 1959, 1091
63K A. A. Khan and W. U. Malik, *J. Indian Chem. Soc.*, 1963, **40**, 564

481 $C_6H_{14}O_2N_4$ *NN'*-Diglycyl-1,2-diaminoethane (diglycylethylenediamine, dgen)

Metal	Method	Temp	Medium	Log of equilibrium constant, remarks	Ref
H^+	gl	23	0·5 KCl	K_1 8·51, K_2 7·81	67Z
	gl	25	0·10 (KNO_3)	K_1 8·22, K_2 7·48	69B
Co^{2+}	gl	25	0·10 (KNO_3)	K_1 3·30	69B
Cu^{2+}	gl	25	1 KCl	$K[Cu(H_{-1}L)^+ + H^+ \rightleftharpoons CuL^{2+}]$ 6·58	53C
				$K[Cu(H_{-2}L) + H^+ \rightleftharpoons Cu(H_{-1}L)^+]$ 8·43	
	gl	23	0·5 KCl	K_1 8·26	67Z
				$K[Cu^{2+} + HL \rightleftharpoons CuHL^{3+})$ 5·24±0·04	
				$K[CuL^{2+} + H^+ \rightleftharpoons CuHL^{3+})$ 5·49	
				$K[Cu(H_{-1}L)^+ + H^+ \rightleftharpoons CuL^{2+}]$ 6·51	
				$K[Cu(H_{-2}L) + H^+ \rightleftharpoons Cu(H_{-1}L)^+]$ 8·21	
	gl	25	0·10 (KNO_3)	K_1 7·50	69B
				$K[(CuH_{-1}L)_2^{2+} + 2H^+ \rightleftharpoons 2CuL^{2+}]$ 9·2	
				$K[2Cu(H_{-2}L) + 2H^+ \rightleftharpoons Cu_2(H_{-1}L)_2^{2+}]$ 18·40	
				$K[Cu(H_{-2}L) + 2H^+ \rightleftharpoons CuL^{2+}]$ 13·8	
Ni^{2+}	gl	25	1 KCl	$K[Ni(H_{-1}L)^+ + H^+ \rightleftharpoons NiL^{2+}]$ 7·04	53C
				$K[Ni(H_{-2}L) + H^+ \rightleftharpoons Ni(H_{-1}L)^+]$ 8·94	
	gl	25	0·10 (KNO_3)	K_1 5·38, β_2 8·50	69B
				$K[Ni(H_{-2}L) + 2H^+ \rightleftharpoons NiL^{2+}]$ 16·04	
Zn^{2+}	gl	25	0·10 (KNO_3)	K_1 3·95	69B

53C A. K. Chakraburtty, N. N. Ghosh, and P. Ray, *J. Indian Chem. Soc.*, 1953, **30**, 185
67Z A. Zuberbuhler and S. Fallab, *Helv. Chim. Acta*, 1967, **50**, 889
69B K. S. Bai and A. E. Martell, *J. Amer. Chem. Soc.*, 1969, **91**, 4412

$C_6H_{14}O_2N_4$ *NN'*-Bis(2-aminoethyl)oxamide L 481.1

Metal	Method	Temp	Medium	Log of equilibrium constant, remarks	Ref
H^+	gl	25	1·0 (KNO_3)	K_1 9·54, K_2 8·30	65W
	gl	22	0·1 (KNO_3)	K_1 9·31, K_2 8·43	68G
Cu^{2+}	gl	22	0·1 (KNO_3)	K_1 9·41	68G
				$K[Cu(H_{-1}L)^+ + H^+ \rightleftharpoons CuL^{2+}]$ 7·51	
				$K[Cu(H_{-2}L) + H^+ \rightleftharpoons Cu(H_{-1}L)^+]$ 8·10	
				$K[Cu(H_{-2}L) + Cu^{2+} \rightleftharpoons Cu_2(H_{-2}L)^{2+}]$ 7·37	

65W J. C. Wang and J. E. Bauman, *Inorg. Chem.*, 1965, **4**, 1613
68G R. Griesser and S. Fallab, *Chimia* (*Switz.*), 1968, **22**, 90

$C_6H_{14}O_2N_4$ Hexanedioic acid dihydrazide L 481.2

Metal	Method	Temp	Medium	Log of equilibrium constant, remarks	Ref
Cd^{2+}	pol	25	1 ($NaClO_4$)	K_1 2·36, β_2 4·22, β_3 5·20, β_4 5·60	66K

66K A. F. Krivis, G. R. Supp, and R. L. Doerr, *Analyt. Chem.*, 1966, **38**, 936

$C_6H_{14}O_3N_2$ 2,6-Diamino-5-hydroxyhexanoic acid HL 481.3

Metal	Method	Temp	Medium	Log of equilibrium constant, remarks	Ref
H^+	gl	25	0·1 $NaClO_4$	K_1 9·83, K_2 8·85	65N
Co^{2+}	gl	25	0·1 $NaClO_4$	K_1 3·73, K_2 3·21	65N
Cu^{2+}	gl	25	0·1 $NaClO_4$	K_1 7·46, K_2 6·29	65N
Fe^{2+}	gl	25	0·1 $NaClO_4$	K_1 3·1	65N
Mn^{2+}	gl	25	0·1 $NaClO_4$	K_1 2·3	65N
Ni^{2+}	gl	25	0·1 $NaClO_4$	K_1 4·76, K_2 3·98	65N
Zn^{2+}	gl	25	0·1 $NaClO_4$	K_1 4·00, K_2 3·75	65N

65N M. H. T. Nyberg and M. Cefola, *Arch. Biochem. Biophys.*, 1965, **111**, 327

$C_6H_{14}O_{12}P_2$ Fructose-1,6-diphosphoric acid (fructose 1,6-diphosphate) H_4L 481.4

Metal	Method	Temp	Medium	Log of equilibrium constant, remarks	Ref
H^+	gl	25	0·08	K_1 6·76, K_2 5·96	65M
Mg^{2+}	gl	25	0·08	K_1 2·7,	65M
				$K(Mg^{2+} + HL^{3-} \rightleftharpoons MgHL^-)$ 2·12	

65M R. W. McGilvery, *Biochem rv*, 1965, **4**, 1924

482 $C_6H_{15}ON$ *NN*-Diethyl-2-aminoethanol L

Metal	Method	Temp	Medium	Log of equilibrium constant, remarks	Ref
H^+	H	20	0 corr	$K(L \rightleftharpoons HL^+ + OH^-)$ 4·30±0·01	61A
	gl	25	~ 0·1	K_1 9·98	65D
Ag^+	sol	20	0 corr	β_2 4·62±0·03	61A
	gl	20	0 corr	β_2 4·66±0·05	61A
Cu^{2+}	gl	25	~ 0·1	K_1 4·9, K_2 4·1, K_3 3·2, K_4 2·4	65D

61A D. J. Alner and R. C. Lansbury, *J. Chem. Soc.*, 1961, 3169
65D G. Douheret, *Bull. Soc. chim. France*, 1965, 2915

483.1 $C_6H_{15}O_2N$ 3,3′-Iminodipropanol L

Metal	Method	Temp	Medium	Log of equilibrium constant, remarks	Ref
H^+	H	20	0 corr	K_1 8·97±0·01	64A
Ag^+	gl	20	0 corr	K_1 2·95±0·05, K_2 2·76±0·05	64A
	sol	20	0 corr	β_2 5·71±0·01	64A

64A D. J. Alner and M. A. A. Kahn, *J. Chem. Soc.*, 1964, 5265

483.2 $C_6H_{15}O_2N$ *N*-Ethyl-2,2′-iminodiethanol L

Metal	Method	Temp	Medium	Log of equilibrium constant, remarks	Ref
H^+	gl	30	0·5 (KNO_3)	K_1 8·88	67F
Cu^{2+}	pol	30	0·5 (KNO_3)	$K[Cu^{2+} + L + 2OH^- \rightleftharpoons CuL(OH)_2]$ 17·4 $K[Cu^{2+} + 2L + 2OH^- \rightleftharpoons CuL_2(OH)_2]$ 19·0	67F

67F J. F. Fisher and J. L. Hall, *Analyt. Chem.*, 1967, 39, 1550

484 $C_6H_{15}O_3N$ 2,2′,2″-Nitrilotriethanol (triethanolamine) L

Metal	Method	Temp	Medium	Log of equilibrium constant, remarks	Ref
H^+	gl	30	0·5 KNO_3	K_1 7·91	62F, 6[illegible]
	gl	25	~ 0·1	K_1 8·08	65D
	gl	25	0·4 (KNO_3)	K_1 8·02	66S
	gl	30	0·5 (KNO_3)	K_1 7·83	67F
Ag^+	Ag	25	< 0·01	β_2 5·28	61A
	Ag	25	0·4 $LiNO_3$ 0—100% ethanol	K_1 (0%) 2·78±0·06, (20%) 2·90±0·04, (40%) 3·00±0·10, β_2 (0%) 4·54±0·05, (20%) 4·78±0·05, (40%) 4·93±0·10, (60%) 5·2±0·2, (80%) 5·38±0·06, (90%) 5·49±0·03, (95%) 5·6±0·1, (100%) 5·81±0·06, β_3 (0%) 5·08±0·04, (20%) 5·15±0·04, (40%) 5·32±0·09, (60%) 5·53±0·05, (80%) 6·04±0·04, (90%) 6·18±0·03, (95%) 6·4±0·1, (100%) 6·59±0·05 See ethylene (107) for mixed ligands	65M

cont

Metal	*Method*	*Temp*	*Medium*	*Log of equilibrium constant, remarks*	*Ref*
Cd^{2+}	pol	25	0·1 KNO_3	K_1 2·70, β_2 4·60, β_3 5·21	60M
	pol	?	1 $NaClO_4$	K_1 2·3, β_2 5·0	63C
				$K(Cd^{2+}+2L+OH^- \rightleftharpoons CdL_2OH^+)$ 8	
				$K[Cd^{2+}+2L+2OH^- \rightleftharpoons CdL_2(OH)_2]$ 11	
				$K[Cd^{2+}+L+3OH^- \rightleftharpoons CdL(OH)_3^-]$ 11·7	
				$K[Cd^{2+}+2L+3OH^- \rightleftharpoons CdL_2(OH)_3^-]$ 13·1	
				$K[Cd^{2+}+2L+2PO_4^{3-} \rightleftharpoons CdL_2(PO_4)_2^{4-}]$ 9·7	
				$K(Cd^{2+}+L+CO_3^{2-} \rightleftharpoons CdLCO_3)$ 5·2	
				$K(Cd^{2+}+2L+CO_3^{2-} \rightleftharpoons CdL_2CO_3)$ 6·2	
				$K[Cd^{2+}+L+2CO_3^{2-} \rightleftharpoons CdL(CO_3)_2^{2-}]$ 6·5	
				$K[Cd^{2+}+2L+2CO_3^{2-} \rightleftharpoons CdL_2(CO_3)_2^{2-}]$ 7·7	
Co^{2+}	gl	25	0·43 (CH_2OHCH_2· NH_3NO_3)	K_1 2·70, K_2 1·65	66S
Cu^{2+}	pol	30	0·5 KNO_3	$K[Cu^{2+}+L+2OH^- \rightleftharpoons CuL(OH)_2]$ 18·4	62F
				$K[Cu^{2+}+L+3OH^- \rightleftharpoons CuL(OH)_3^-]$ 20·7	
	Cu	?	2 $NaClO_4$	K_1 4·79	63C
				$K[2Cu^{2+}+2L+2OH^- \rightleftharpoons Cu_2L_2(OH)_2^{2+}]$ 27·9	
				$K[2Cu^{2+}+2L+4OH^- \rightleftharpoons Cu_2L_2(OH)_4]$ 40·3	
				$K[4Cu^{2+}+4L+5OH^- \rightleftharpoons Cu_4L_4(OH)_5^{3+}]$ 63·4	
				$K[2Cu^{2+}+L+2OH^- \rightleftharpoons Cu_2L(OH)_2^{2+}]$ 20·4	
	pol	30	0·5 KNO_3	$K(Cu^{2+}+L+OH^- \rightleftharpoons CuLOH^+)$ 11·83	63S
				$K[Cu^{2+}+L+2OH^- \rightleftharpoons CuL(OH)_2]$ 18·02	
				$K[Cu^{2+}+L+3OH^- \rightleftharpoons CuL(OH)_3^-]$ 20·18	
				$K[Cu^{2+}+2L+2OH^- \rightleftharpoons CuL_2(OH)_2]$ 19·00	
	gl	25	~ 0·1	K_1 3·9, K_2 2·1	65D
	gl	25	0·43 (CH_2OHCH_2· NH_3NO_3)	K_1 4·44, K_2 3·14, K_3 2·14	66S
	pol	30	0·5 (KNO_3)	$K(Cu^{2+}+L+OH^- \rightleftharpoons CuLOH^+)$ 11·9	67F
				$K[Cu^{2+}+L+2OH^- \rightleftharpoons CuL(OH)_2]$ 18·3	
				$K[Cu^{2+}+L+3OH^- \rightleftharpoons CuL(OH)_3^-]$ 20·7	
				$K[Cu^{2+}+2L+2OH^- \rightleftharpoons CuL_2(OH)_2]$ 18·6	
Fe^{2+}	hyp	25	0·43 (CH_2OHCH_2· NH_3NO_3)	K_1 2·27, K_2 1·32	66S
Mn^{2+}	hyp	25	0·43 (CH_2OHCH_2· NH_3NO_3)	K_1 1·47, K_2 0·67	66S
Ni^{2+}	Cu	?	2 $NaClO_4$	K_1 2·95	63C
				$K[2Ni^{2+}+2L+2OH^- \rightleftharpoons Ni_2L_2(OH)_2^{2+}]$ 18·2	
				$K[4Ni^{2+}+4L+6OH^- \rightleftharpoons Ni_4L_4(OH)_6^{2+}]$ 47·8	
	sp	?	2 $NaClO_4$	K_1 3·06	63C
				$K[Ni_4L_4(OH)_6^{2+}+2L+2OH^- \rightleftharpoons 2Ni_2L_3(OH)_4]$ −1·7	
				$K[Ni_2L_3(OH)_4+4OH^- \rightleftharpoons 2NiL(OH)_4^{2-}+3L]$ −2·4	
	gl	25	0·5	K_1 2·27, β_2 3·09	63Sa
	gl	25	0·43 (CH_2OHCH_2· NH_3NO_3)	K_1 3·43, K_2 2·20, K_3 1·37	66S

Errata. 1964, p. 525, Table 484. Under H^+ in second column *for E read* H, Ag-AgCl. Under H^+ first entry fourth column *for* — *read* → 0.

60M P. K. Migal' and A. N. Pushnyak, *Russ. J. Inorg. Chem.*, 1960, **5**, 293 (610)
61A V. Armeanu and C. Luca, *J. phys. Chem.* (*Leipzig*), 1961, **217**, 389
62F J. F. Fisher and J. L. Hall, *Analyt. Chem.*, 1962, **34**, 1094
63C M. Cadiot-Smith, *J. Chim. phys.*, 1963, **60**, 957, 976, 991
63S P. E. Sturrock, *Analyt. Chem.*, 1963, **35**, 1092
63Sa A. Ya. Sychev, A. P. Gerbeleu, and P. K. Migal', *Russ. J. Inorg. Chem.*, 1963, **8**, 1081 (2070)

continued overleaf

484 $C_6H_{15}O_3N$ (contd)

65D G. Douheret, *Bull. Soc. chim. France*, 1965, 2915
65M P. K. Migal' and K. I. Ploae, *Russ. J. Inorg. Chem.*, 1965, **10**, 1368 (2517)
66S E. V. Sklenskaya and M. Kh. Karapet'yants, *Russ. J. Inorg. Chem.*, 1966, **11**, 1478 (2749)
67F J. F. Fisher and J. L. Hall, *Analyt. Chem.*, 1967, **39**, 1550

486.1 $C_6H_{16}O_6P_2$ Hexane-1,6-diphosphonic acid H_4L

Metal	*Method*	*Temp*	*Medium*	*Log of equilibrium constant, remarks*	*Ref*
H^+	gl	25	0·1 KCl	K_1 8·37±0·04, K_2 7·65±0·04, K_3 3·07±0·04, K_4 1·8±0·2	67K
Al^{3+}	gl	25	0·1 KCl	K_1 14·66	67K
Be^{2+}	gl	25	0·1 KCl	$K(Be^{2+}+HL^{3-} \rightleftharpoons BeHL^-)$ 8·31	67K
				$K(2Be^{2+}+L^{4-} \rightleftharpoons Be_2L)$ 15·55	
Ca^{2+}	gl	25	0·1 KCl	K_1 < 3	67K
Co^{2+}	gl	25	0·1 KCl	$K(Co^{2+}+HL^{3-} \rightleftharpoons CoHL^-)$ 4·57	67K
				$K(2Co^{2+}+L^{4-} \rightleftharpoons Co_2L)$ 9·86	
				$K(2Co^{2+}+HL^{3-} \rightleftharpoons Co_2HL^+)$ 7·73	
Cu^{2+}	gl	25	0·1 KCl	$K(Cu^{2+}+HL^{3-} \rightleftharpoons CuHL^-)$ 6·05	67K
				$K(2Cu^{2+}+L^{4-} \rightleftharpoons Cu_2L)$ 11·09	
Mg^{2+}	gl	25	0·1 KCl	K_1 < 3	67K
Mn^{2+}	gl	25	0·1 KCl	$K(Mn^{2+}+HL^{3-} \rightleftharpoons MnHL^-)$ 5·82	67K
				$K(2Mn^{2+}+L^{4-} \rightleftharpoons Mn_2L)$ 12·51	
				$K(2Mn^{2+}+HL^{3-} \rightleftharpoons Mn_2HL^+)$ 9·62	
Ni^{2+}	gl	25	0·1 KCl	K_1 3·71	67K
Sr^{2+}	gl	25	0·1 KCl	K_1 < 2	67K
Zn^{2+}	gl	25	0·1 KCl	$K(Zn^{2+}+HL^{3-} \rightleftharpoons ZnHL^-)$ 7·34	67K
				$K(2Zn^{2+}+L^{4-} \rightleftharpoons Zn_2L)$ 14·73	
				$K(2Zn^{2+}+HL^{3-} \rightleftharpoons Zn_2HL^+)$ 10·31	

67K M. I. Kabechuik, R. P. Lastovskii, T. Ja. Medved, U. V. Medyntsev, I. D. Kopekova, and N. M. Dyatlova, *Proc. Acad. Sci. (U.S.S.R.)*, 1967, 177, 1060 (582)

487 $C_6H_{16}N_2S_2$ 3,6-Dithiaoctane-1,8-diamine L

Erratum. 1964, p. 526, Table 487. Insert **L** in heading.

$C_6H_4O_2NCl$ 4-Chloro-2-nitrosophenol HL 487.1

Metal	Method	Temp	Medium	Log of equilibrium constant, remarks	Ref
H^+	gl	25	0·1 KNO_3 50% dioxan	K_1 6·36	61S
Cd^{2+}	gl	25	0·1 KNO_3 50% dioxan	K_1 4·03	61S
Cu^{2+}	gl	25	0·1 KNO_3 50% dioxan	K_1 6·20	61S
Ni^{2+}	gl	25	0·1 KNO_3 50% dioxan	K_1 6·08	61S
Pb^{2+}	gl	25	0·1 KNO_3 50% dioxan	K_1 4·59	61S
Zn^{2+}	gl	25	0·1 KNO_3 50% dioxan	K_1 4·53	61S

61S H. Shimura, *Nippon Kagaku Zasshi*, 1961, **82**, 641

$C_6H_5O_2N_2Cl$ 4-Chloro-3-nitroaniline L 487.2

Metal	Method	Temp	Medium	Log of equilibrium constant, remarks	Ref
Zn^{2+}	sp	25	acetone	$K(ZnCl_2+L \rightleftharpoons ZnCl_2L)$ 1·37 $K(ZnBr_2+L \rightleftharpoons ZnBr_2L)$ 1·52 $K(ZnI_2+L \rightleftharpoons ZnI_2L)$ 1·39	65S

65S D. P. N. Satchell and R. S. Satchell, *Trans. Faraday Soc.*, 1965, **61**, 1118

$C_6H_5O_9NS_2$ 4-Nitroso-5,6-dihydroxybenzene-1,3-disulphonic acid H_4L 487.3

Metal	Method	Temp	Medium	Log of equilibrium constant, remarks	Ref
H^+	gl	?	0·10 (NaCl)	K_2 4·41±0·03	67B
	sp	?	0·10 (NaCl)	K_1 10·6±0·3, K_2 4·45±0·05	67B
Fe^{3+}	sp	?	0·10 (NaCl)	K_1 16·42±0·06, K_2 12·03±0·03, K_3 6·49±0·10	67B

67B M. Beran and S. Havelka, *Coll. Czech. Chem. Comm.*, 1967, 32, 2944

$C_6H_7O_4NS$ 1-Hydroxy-1-(2′-pyridyl)methanesulphonic acid H_2L 490.1

Metal	Method	Temp	Medium	Log of equilibrium constant, remarks	Ref
H^+	gl	25	0·1 ($NaClO_4$)	K_1 9·70, K_2 4·05	64B
Co^{2+}	gl	25	0·1 ($NaClO_4$)	K_1 4·92, K_2 3·53	64B
Cu^{2+}	gl	25	0·1 ($NaClO_4$)	K_1 9·43, K_2 7·57	64B
Ni^{2+}	gl	25	0·1 ($NaClO_4$)	K_1 5·13, K_2 3·55	64B
Zn^{2+}	gl	25	0·1 ($NaClO_4$)	K_1 5·08, K_2 4·21	64B

64B L. Banford and W. J. Geary, *J. Chem. Soc.*, 1964, 378

490.2 $C_6H_7O_4NS$ 1-Hydroxy-1-(3′-pyridyl)methanesulphonic acid H_2L

Metal	*Method*	*Temp*	*Medium*	*Log of equilibrium constant, remarks*	*Ref*
H^+	gl	25	0·1 ($NaClO_4$)	K_1 9·88, K_2 4·83	64B
Co^{2+}	gl	25	0·1 ($NaClO_4$)	K_1 7·54, K_2 6·97	64B
Cu^{2+}	gl	25	0·1 ($NaClO_4$)	K_1 7·81, K_2 6·98	64B
Ni^{2+}	gl	25	0·1 ($NaClO_4$)	K_1 7·67, K_2 6·72	64B
Zn^{2+}	gl	25	0·1 ($NaClO_4$)	K_1 7·58, K_2 6·71	64B

64B L. Banford and W. J. Geary, *J. Chem. Soc.*, 1964, 378

490.3 $C_6H_8O_4NP$ 2-Pyridylmethanephosphoric acid (α-picolyl phosphate) H_2L

Metal	*Method*	*Temp*	*Medium*	*Log of equilibrium constant, remarks*	*Ref*
H^+	gl	25	0·1 (KNO_3)	K_1 6·30, K_2 4·42, K_3 1·8	68M
Co^{2+}	gl	25	0·1 (KNO_3)	K_1 2·27±0·04	68M
Cu^{2+}	gl	25	0·1 (KNO_3)	K_1 4·44±0·01, K_2 2·33±0·03	68M
Mg^{2+}	gl	25	0·1 (KNO_3)	K_1 ~ 1·7	68M
Mn^{2+}	gl	25	0·1 (KNO_3)	K_1 2·44±0·05	68M
Ni^{2+}	gl	25	0·1 (KNO_3)	K_1 2·85±0·02	68M
Zn^{2+}	gl	25	0·1 (KNO_3)	K_1 2·83±0·04	68M

68N Y. Murakami and M. Takagi, *J. Phys. Chem.*, 1968, **72**, 116

493.1 $C_6H_{12}O_2N_2S_2$ *NN′*-Bis(2-hydroxyethyl)dithio-oxamide H_2L

Metal	*Method*	*Temp*	*Medium*	*Log of equilibrium constant, remarks*	*Ref*
H^+	sp	25	0·0—0·5 ($NaClO_4$)	K_1 (0·5) 13·92±0·08, K_2 (→ 0) 11·07, (0·01) 11·02±0·02, (0·1) 10·92±0·02, (0·2) 10·90±0·01, (0·3) 10·84±0·01, (0·4) 10·75±0·01, (0·5) 10·71±0·01	64V
Ag^+	Ag	25	0·1 ($NaClO_4$) 0·01—0·05 H^+	K_1 (0·05H^+) 10·01, (0·01H^+) 10·09, $K(2Ag^+ + 3L^{2-} \rightleftharpoons Ag_2L_3{}^{4-})$ (0·05H^+) 20·53, (0·01H^+) 20·69, $K(3Ag^+ + 4L^{2-} \rightleftharpoons Ag_3L_4{}^{5-})$ (0·05H^+) 31·05, (0·01H^+) 31·29	68V

64V L. C. Van Poucke and M. A. Herman, *Analyt. Chim. Acta*, 1964, **30**, 569
68V L. C. Van Poucke, M. A. Herman, and Z. Eeckhaut, *Analyt. Chim. Acta*, 1968, **40**, 55

$C_6H_{12}O_3N_2S$ Glycyl-DL-methionine HL 494

Metal	Method	Temp	Medium	Log of equilibrium constant, remarks	Ref
H^+	pH	25	0·1 ($NaClO_4$)	K_1 8·51, K_2 3·16	67S
Cu^{2+}	pH	25	0·1 ($NaClO_4$)	K_1 7·17,	67S
				$K[Cu(H_{-1}L)+H^+ \rightleftharpoons CuL^+]$ 4·60	
				$K[Cu(H_{-1}L)+L^- \rightleftharpoons CuL(H_{-1}L)^-]$ 3·73	
				$K[Cu(H_{-1}L)_2{}^{2-}+H^+ \rightleftharpoons CuL(H_{-1}L)^-]$ 10·63	
Ni^{2+}	pH	25	0·1 ($NaClO_4$)	K_1 4·44, K_2 3·88	67S

67S A. Ya. Sychev and N. S. Mitsul, *Russ. J. Inorg. Chem.*, 1967, **12**, 1120 (2127)

$C_6H_{12}O_7NP$ *N*-(2′-Carboxyethyl)iminobis(methylenephosphonic acid) H_4L 495.1

Metal	Method	Temp	Medium	Log of equilibrium constant, remarks	Ref
H^+	gl	25	0·10 (KNO_3)	K_1 10·41, K_2 5·59, K_3 3·48, K_4 2·72	65W
Ca^{2+}	gl	25	0·10 (KNO_3)	K_1 4·88	65W
Cu^{2+}	gl	25	0·10 (KNO_3)	K_1 13·0±0·1	65W
				$K(Cu^{2+}+HL^{3-} \rightleftharpoons CuHL^-)$ 7·2±0·1	
				$K(CuL^{2-}+H^+ \rightleftharpoons CuHL^-)$ 4·71±0·01	
Fe^{3+}	gl	25	0·10 (KNO_3)	K_1 16·3±0·4	65W
Mn^{2+}	gl	25	0·10 (KNO_3)	K_1 7·24	65W

65W S. Westerback, K. S. Rajan, and A. E. Martell, *J. Amer. Chem. Soc.*, 1965, **87**, 2567

$C_6H_{13}O_2NS$ 2-Amino-4-(ethylthio)butanoic acid (ethionine) HL 495.2

Metal	Method	Temp	Medium	Log of equilibrium constant, remarks	Ref
H^+	gl	25	0·10 (KNO_3)	K_1 9·02	64L
Ag^+	gl	25	0·10 (KNO_3)	K_1 5·25	64L
Cd^{2+}	gl	25	0·10 (KNO_3)	K_1 4·68, K_2 4·54	64L
Co^{2+}	gl	25	0·10 (KNO_3)	K_1 5·13, K_2 4·34	64L
Cu^{2+}	gl	25	0·10 (KNO_3)	K_1 8·43	64L
Hg^{2+}	gl	25	0·10 (KNO_3)	K_1 7·25, K_2 5·92	64L
Ni^{2+}	gl	25	0·10 (KNO_3)	K_1 6·15, K_2 5·18	64L
Zn^{2+}	gl	25	0·10 (KNO_3)	K_1 5·22, K_2 5·14	64L

64L G. R. Lenz and A. E. Martell. *Biochemistry*, 1964, **3**, 745

$C_6H_{15}O_2NS$ 2-Aminoethyl-2-hydroxyethyl sulphide L 496

Erratum. 1964, p. 529, Table 496. This Table is included in Table No. **271** on page 439.

496.1 $C_6H_{15}O_5N_4P$ Phosphoarginine* H_2L

Metal	Method	Temp	Medium	Log of equilibrium constant, remarks	Ref
H^+				H^+ constants not given	
Mg^{2+}	sp	30	0·1 buffer	K_1 2·0 (buffer = *N*-ethylmorpholine)	64O

* Structure not given.

64O W. J. O'Sullivan and D. D. Perrin, *Biochemistry*, 1964, 3, 18

498 $C_6H_{20}O_{12}N_2P_4$ Ethane-1,2-bis[iminobis(methylenephosphonic acid)] H_8L

Metal	Method	Temp	Medium	Log of equilibrium constant, remarks	Ref
H^+	gl	25	0·10 (KNO_3)	K_1 10·60, K_2 9·22, K_3 7·43, K_4 6·63, K_5 6·18, K_6 5·05, K_7 2·72, K_8 1·46	65W
	gl	25	0·1 KCl	K_1 12·10, K_2 10·18, K_3 8·08, K_4 6·54, K_5 5·23, K_6 3·0	67K
Ca^{2+}	gl	25	0·10 (KNO_3)	K_1 5·74	65W
	gl	25	0·1 KCl	K_1 9·33, $K[Ca^{2+} + H_nL^{(8-n)-} \rightleftharpoons CaH_nL^{(6-n)-}]$ 7·00 (n = 1), 4·95 (n = 2), 3·87 (n = 3), 2·17 (n = 4)	67K
Cd^{2+}	gl	25	0·1 KCl	K_1 13·88, $K[Cd^{2+} + H_nL^{(8-n)-} \rightleftharpoons CdH_nL^{(6-n)-}]$ 11·18 (n = 1), 8·18 (n = 2), 6·99 (n = 3), 5·45 (n = 4), 2·77 (n = 5)	67K
Co^{2+}	gl	25	0·1 KCl	K_1 15·49, $K[Co^{2+} + H_nL^{(8-n)-} \rightleftharpoons CoH_nL^{(6-n)-}]$ 11·79 (n = 1), 8·51 (n = 2), 6·09 (n = 3), 4·75 (n = 4), 1·92 (n = 5)	67K
Cu^{2+}	gl	25	0·10 (KNO_3)	K_1 16·1 ± 0·4 $K(Cu^{2+} + H_3L^{5-} \rightleftharpoons CuH_3L^{3-})$ 9·1 ± 0·3 $K(H^+ + CuH_2L^{4-} \rightleftharpoons CuH_3L^{3-})$ 5·85 ± 0·06 $K(H^+ + CuHL^{5-} \rightleftharpoons CuH_2L^{4-})$ 6·71 ± 0·02 $K(H^+ + CuL^{6-} \rightleftharpoons CuHL^{5-})$ 7·77 ± 0·05	65W
	gl	25	0·1 KCl	K_1 18·95, $K[Cu^{2+} + H_nL^{(8-n)-} \rightleftharpoons CuH_nL^{(6-n)-}]$ 14·82 (n = 1), 11·14 (n = 2), 8·31 (n = 3), 5·67 (n = 4), 3·28 (n = 5)	67K
Fe^{3+}	gl	25	0·10 (KNO_3)	K_1 19·6 ± 0·5 $K(Fe^{3+} + H_2L^{6-} \rightleftharpoons FeH_2L^{3-})$ 13·2 ± 0·4 $K(H^+ + FeHL^{4-} \rightleftharpoons FeH_2L^{3-})$ 6·32 ± 0·07 $K(H^+ + FeL^{5-} \rightleftharpoons FeHL^{4-})$ 7·14 ± 0·07	65W
Gd^{3+}	gl	25	0·1 KCl	K_1 21·80, $K[Gd^{3+} + H_nL^{(8-n)-} \rightleftharpoons GdH_nL^{(5-n)-}]$ 17·09 (n = 1), 13·12 (n = 2), 10·34 (n = 3), 8·04 (n = 4), 4 (n = 5)	67K
Ho^{3+}	gl	25	0·1 KCl	K_1 21·85, $K[Ho^{3+} + H_nL^{(8-n)-} \rightleftharpoons HoH_nL^{(5-n)-}]$ 17·15 (n = 1), 13·17 (n = 2), 10·38 (n = 3), 8·04 (n = 4), 4 (n = 5)	67K
La^{3+}	gl	25	0·1 KCl	K_1 20·15, $K[La^{3+} + H_nL^{(8-n)-} \rightleftharpoons LaH_nL^{(5-n)-}]$ 15·38 (n = 1), 12·13 (n = 2), 10·00 (n = 3), 8·04 (n = 4), 4 (n = 5)	67K

conti

$C_6H_{20}O_{12}N_2P_4$ (contd) 498

Metal	Method	Temp	Medium	Log of equilibrium constant, remarks	Ref
Mg^{2+}	gl	25	0·1 KCl	K_1 8·63, $K[Mg^{2+}+H_nL^{(8-n)-} \rightleftharpoons MgH_nL^{(6-n)-}]$ 6·58 (n = 1), 5·00 (n = 2), 4·07 (n = 3), 2·45 (n = 4)	67K
Mn^{2+}	gl	25	0·10 (KNO_3)	K_1 9·40	65W
	gl	25	0·1 KCl	K_1 12·70, $K[Mn^{2+}+H_nL^{(8-n)-} \rightleftharpoons MnH_nL^{(6-n)-}]$ 9·66 (n = 1), 6·99 (n = 2), 5·13 (n = 3), 3·19 (n = 4)	67K
Nd^{3+}	gl	25	0·1 KCl	K_1 21·47, $K[Nd^{3+}+H_nL^{(8-n)-} \rightleftharpoons NdH_nL^{(5-n)-}]$ 16·77 (n = 1), 13·05 (n = 2), 10·43 (n = 3), 8·04 (n = 4), 4 (n = 5)	67K
Ni^{2+}	gl	25	0·1 KCl	K_1 15·30, $K[Ni^{2+}+H_nL^{(8-n)-} \rightleftharpoons NiH_nL^{(6-n)-}]$ 12·00 (n = 1), 9·12 (n = 2), 6·77 (n = 3), 4·72 (n = 4), 2·34 (n = 5)	67K
Sm^{3+}	gl	25	0·1 KCl	K_1 22·39, $K[Sm^{3+}+H_nL^{(8-n)-} \rightleftharpoons SmH_nL^{(5-n)-}]$ 17·56 (n = 1), 13·68 (n = 2), 10·72 (n = 3), 8·04 (n = 4), 4 (n = 5)	67K
Tm^{3+}	gl	25	0·1 KCl	K_1 21·41, $K[Tm^{3+}+H_nL^{(8-n)-} \rightleftharpoons TmH_nL^{(5-n)-}]$ 16·17 (n = 1), 12·88 (n = 2), 10·28 (n = 3), 8·04 (n = 4), 4 (n = 5)	67K
Zn^{2+}	gl	25	0·1 KCl	K_1 17·05, $K[Zn^{2+}+H_nL^{(8-n)-} \rightleftharpoons ZnH_nL^{(6-n)-}]$ 13·52 (n = 1), 9·90 (n = 2), 6·99 (n = 3), 4·73 (n =4), 2·18 (n = 5)	67K

65W S. Westerback, K. S. Rajan, and A. E. Martell, *J. Amer. Chem. Soc.*, 1965, **87**, 2567
67K M. I. Kubechnik, I. M. Dyatlova, T. A. Medved, Yu. F. Begulin, and V. V. Gidorenko, *Proc. Acad. Sci.* (*U.S.S.R.*), 1967, **175**, 621 (351)

C_7H_8 Methylbenzene (toluene) L 499

Metal	Method	Temp	Medium	Log of equilibrium constant, remarks	Ref
Ag^+	sp	1·4, 20	toluene ClO_4^-	K_1 (1·4°) 2·67, (20°) 3·97 ΔH_1 3·4	64T

64T B. G. Torre-Mori, D. Janjic, and B. P. Susz, *Helv. Chim. Acta*, 1964, **47**, 1172

C_7H_{12} 1-Methylcyclohexene L 499.1

Metal	Method	Temp	Medium	Log of equilibrium constant, remarks	Ref
Ag^+	oth	30	1·77 $AgNO_3$ ethylene glycol	K_1 0·10, gas chromatography	62G

62G E. Gil-Av and J. Herling, *J. Phys. Chem.*, 1962, **66**, 1208

499.2 C_7H_{12} 3-Methylcyclohexene L

Metal	*Method*	*Temp*	*Medium*	*Log of equilibrium constant, remarks*	*Ref*
Ag^+	oth	30	1·77 $AgNO_3$ ethylene glycol	K_1 0·74, gas chromatography	62G

62G E. Gil-Av and J. Herling, *J. Phys. Chem.*, 1962, **66**, 1208

499.3 C_7H_{12} 4-Methylcyclohexene L

Metal	*Method*	*Temp*	*Medium*	*Log of equilibrium constant, remarks*	*Ref*
Ag^+	oth	30	1·77 $AgNO_3$ ethylene glycol	K_1 0·71, gas chromatography	62G

62G E. Gil-Av and J. Herling, *J. Phys. Chem.*, 1962, **66**, 1208

499.4 C_7H_{12} Methylenecyclohexane L

Metal	*Method*	*Temp*	*Medium*	*Log of equilibrium constant, remarks*	*Ref*
Ag^+	oth	30	1·77 $AgNO_3$ ethylene glycol	K_1 0·98, gas chromatography	62G

62G E. Gil-Av and J. Herling, *J. Phys. Chem.*, 1962, **66**, 1208

499.5 C_7H_{12} 1-Ethylcyclopentene L

Metal	*Method*	*Temp*	*Medium*	*Log of equilibrium constant, remarks*	*Ref*
Ag^+	oth	30	1·77 $AgNO_3$ ethylene glycol	K_1 0·56, gas chromatography	62G

62G E. Gil-Av and J. Herling, *J. Phys. Chem.*, 1962, **66**, 1208

C_7H_{12} 3-Ethylcyclopentene L 499.6

Metal	Method	Temp	Medium	Log of equilibrium constant, remarks	Ref
Ag^+	oth	30	1·77 $AgNO_3$ ethylene glycol	K_1 1·07, gas chromatography	62G

62G E. Gil-Av and J. Herling, *J. Phys. Chem.*, 1962, **66**, 1208

C_7H_{12} 4-Ethylcyclopentene L 499.7

Metal	Method	Temp	Medium	Log of equilibrium constant, remarks	Ref
Ag^+	oth	30	1·77 $AgNO_3$ ethylene glycol	K_1 0·85, gas chromatography	62G

62G E. Gil-Av and J. Herling, *J. Phys. Chem.*, 1962, **66**, 1208

$C_7H_4O_7$ 3-Hydroxy-4-pyrone-2,6-dicarboxylic acid (meconic acid) H_3L 499.8

Metal	Method	Temp	Medium	Log of equilibrium constant, remarks	Ref
H^+	gl	20	→ 0, 30% ethanol	K_1 11·21, K_2 2·11, K_3 1·83	61S, 66S
	sp	20	→ 0, 30% ethanol	K_1 11·30 ± 0·02	61S, 66S
	gl, sp	25	0·5	K_1 9·35 ± 0·03	67C
Fe^{3+}	sp	20	0—0·03	K_1 (→ 0) 15·0, β_2 (→ 0) 25·3, β_3 (→ 0) 31·6, (0·03) 30·9	61S
				$K(Fe^{3+} + H_2L^- \rightleftharpoons FeH_2L^{2+})$ (0·1) 5·1	
			→ 0, 30% ethanol	$K(Fe^{3+} + H_2L^- \rightleftharpoons FeHL^+ + H^+)$ 5·2	
				$K(FeL + HL^{2-} \rightleftharpoons FeL_2^{3-} + H^+)$ 0·4	
				$K(FeL_2 + HL^{2-} \rightleftharpoons FeL_3^{6-} + H^+)$ −4·0	
Ge^{IV}	sp	25	0·5 (NaCl)	$K[Ge(OH)_4 + 2HL^{2-} \rightleftharpoons Ge(OH)_2L_2^{4-}] < 1$	67C
UO_2^{2+}	sp	~ 20	0·1 ($NaClO_4$)	K_1 12·5, K_2 8·5	65B
	sp	18—22	0·1 ($NaClO_4$) 0—30% ethanol	K_1 (0%) 11·8, (30%) 12·4, K_2 (0%) 8·9, (30%) 9·0	66S
				$K(UO_2^{2+} + H_3L \rightleftharpoons UO_2H_2L^+ + H^+)$ (0%) 0·6, (30%) 1·0	
Zn^{2+}	gl	25	0·5 ($NaClO_4$)	K_1 7·25 ± 0·06	67C

61S* L. Sommer and A. Losmanova, *Coll. Czech. Chem. Comm.*, 1961, **26**, 2781
65B M. Bartusek and L. Sommer, *J. Inorg. Nuclear Chem.*, 1965, **27**, 2397
66S L. Sommer, L. Kurilova-Navratilova, and T. Sepel, *Coll. Czech. Chem. Comm.*, 1966, **31**, 1288
67C G. Choux and R. L. Benoit, *Bull. Soc. chim. France*, 1967, 2920

* Calculation error in the original paper.

500 $C_7H_6O_2$ Benzenecarboxylic acid (benzoic acid) HL

Metal	*Method*	*Temp*	*Medium*	*Log of equilibrium constant, remarks*	*Ref*
H^+	gl	25	0·1 ($NaClO_4$) 50% dioxan	K_1 5·79±0·01	68E, 68
	gl	31	0·1 ($NaClO_4$)	K_1 4·01	68R
Ag^+	Ag-AgCl	30	1 ($NaClO_4$)	K_1 3·4, β_2 4·2	67V
Cd^{2+}	pol	30	0·1 ($NaNO_3$)	K_1 1·08, β_2 1·18, β_3 1·64, β_4 1·87	66J
	qh	30	1 ($NaClO_4$)	K_1 0·99, β_2 1·76	65V
Cu^{2+}	gl	25	0·1 ($NaClO_4$) 50% dioxan	K_1 3·30	68E
	gl	25	0·1 ($NaClO_4$) 50% dioxan	$K(CuA^{2+}+L^- \rightleftharpoons CuAL^+)$ 3·58 where A = 2,2′-bipyridyl	68G
Pb^{2+}	pol	30—40	1 ($NaClO_4$)	β_2 (30°) 3·30, (40°) 3·28 $\Delta H_{\beta 2}$ 7·1	64J
UO_2^{2+}	gl	31	0·1 ($NaClO_4$)	K_1 2·59	68R
Zn^{2+}	gl	25	0·1 ($NaClO_4$) 50% dioxan	K_1 2·35 $K(ZnA^{2+}+L^- \rightleftharpoons ZnAL^+)$ 2·40 where A = 2,2′-bipyridyl	68G

64J D. S. Jain and J. N. Gaur, *Indian J. Chem.*, 1964, **2**, 503
65V D. G. Vartak and R. S. Shetiya, *Indian J. Chem.*, 1965, **3**, 533
66J D. S. Jain and J. N. Gaur, *J. Electroanalyt. Chem.*, 1966, **11**, 310
67V D. G. Vartak and R. S. Shetiya, *J. Inorg. Nuclear Chem.*, 1967, **29**, 1261
68E H. Erlenmeyer, R. Griesser, B. Prijs, and H. Sigel, *Helv. Chim. Acta*, 1968, **51**, 339
68G R. Griesser, B. Prijs, and H. Sigel, *Inorg. Nuclear Chem. Letters*, 1968, **4**, 443
68R S. Ramamoorthy and M. Santappa, *Bull. Chem. Soc. Japan*, 1968, **7**, 1330

501 $C_7H_6O_2$ 2-Hydroxybenzaldehyde (salicylaldehyde) HL

Metal	*Method*	*Temp*	*Medium*	*Log of equilibrium constant, remarks*	*Ref*
H^+	pH	20	0·3 $NaClO_4$ 50% dioxan	K_1 9·35±0·02	63C
	gl	30	0·1 $NaClO_4$ 75% dioxan	K_1 10·17	64J
	sp	25	0·10 ($NaClO_4$)	K_1 8·14	66P
	gl	25	0·5 KCl	K_1 8·22±0·01	68L
Ca^{2+}	sp	25	0·10 ($NaClO_4$)	K_1 1·1±0·1	66P
Cu^{2+}	pH	20	0·3 $NaClO_4$ 50% dioxan	K_1 6·6, K_2 5·4	63C
	gl	30	0·1 $NaClO_4$ 75% dioxan	K_1 5·77, K_2 2·01	64J
	sp	25	0·10 ($NaClO_4$)	K_1 5·56±0·02	66P
	gl	25	0·5 KCl	K_1 5·36, β_2 10·11 $K(Cu^{2+}+L^-+A^- \rightleftharpoons CuLA)$ 16·15 where HA = glycine	68L
La^{3+}	sp	25	0·10 ($NaClO_4$)	K_1 3·40±0·02	66P
Mg^{2+}	pH	20	0·3 $NaClO_4$ 50% dioxan	K_1 2	63C
	gl	30	0·1 $NaClO_4$ 75% dioxan	K_1 3·88	64J

cont

Metal	Method	Temp	Medium	Log of equilibrium constant, remarks	Ref
Mn^{2+}	gl	25	0·5 KCl	K_1 2·15, β_2 4·0 $K(Mn^{2+}+L^-+A^- \rightleftharpoons MnLA)$ 7·26 $K(Mn^{2+}+L^-+2A^- \rightleftharpoons MnLA_2^-)$ 9·15 $K(Mn^{2+}+2L^-+2A^- \rightleftharpoons MnL_2A_2^{2-})$ 13·04 where HA = glycine	68L
Ni^{2+}	pH	20	0·3 $NaClO_4$ 50% dioxan	K_1 4·5	63C
	gl	30	0·1 $NaClO_4$ 75% dioxan	K_1 5·36, K_2 3·75	64J
	gl	25	0·5 KCl	K_1 3·58, β_2 6·5 $K(Ni^{2+}+L^-+A^- \rightleftharpoons NiLA)$ 10·75 $K(Ni^{2+}+L^-+2A^- \rightleftharpoons NiLA_2^-)$ 15·62 $K(Ni^{2+}+2L^-+2A^- \rightleftharpoons NiL_2A_2^{2-})$ 18·89 where HA = glycine	68L
Y^{3+}	sp	25	0·10 ($NaClO_4$)	K_1 4·34 ± 0·02	66P
Zn^{2+}	gl	30	0·1 $NaClO_4$ 75% dioxan	K_1 6·26	64J
	gl	25	0·5 KCl	K_1 2·87, β_2 5·00 $K(Zn^{2+}+L^-+A^- \rightleftharpoons ZnLA)$ 9·65 $K(Zn^{2+}+L^-+2A^- \rightleftharpoons ZnLA_2^-)$ 13·42 $K(Zn^{2+}+2L^-+2A^- \rightleftharpoons ZnL_2A_2^{2-})$ 16·73 where HA = glycine	68L

63C K. Clarke, R. A. Cowen, G. W. Gray, and E. H. Osborne, *J. Chem. Soc.*, 1963, 245
64J K. E. Jabalpurwala, K. A. Venkatachalam, and M. B. Kabadi, *J. Inorg. Nuclear Chem.*, 1964, **26**, 1011, 1027
66P C. Postmus jun., L. B. Magnusson, and C. A. Craig, *Inorg. Chem.*, 1966, **5**, 1154
68L D. L. Leussing and K. S. Bai, *Analyt. Chem.*, 1968, **40**, 575

$C_7H_6O_2$ 3-Hydroxybenzaldehyde HL 501.1

Metal	Method	Temp	Medium	Log of equilibrium constant, remarks	Ref
H^+	sp	25	0 corr	K_1 9·02	65E
Fe^{3+}	sp	25	0 corr	K_1 8·11 ± 0·03	65E

65E Z. L. Ernst and F. G. Herring, *Trans. Faraday Soc.*, 1965, **61**, 454

$C_7H_6O_2$ 4-Hydroxybenzaldehyde HL 501.2

Metal	Method	Temp	Medium	Log of equilibrium constant, remarks	Ref
H^+	sp	25	0 corr	K_1 7·62	65E
Fe^{3+}	sp	25	0 corr	K_1 7·56 ± 0·02	65E

65E Z. L. Ernst and F. G. Herring, *Trans. Faraday Soc.*, 1965, **61**, 454

502 $C_7H_6O_2$ 2-Hydroxycyclohepta-2,4,6-trien-1-one (tropolone) HL

Metal	Method	Temp	Medium	Log of equilibrium constant, remarks	Ref
H^+	sp	25	2 $NaClO_4$	K_1 6·42	62O
	gl	25	0·5 (NaCl)	K_1 6·70 ± 0·04	66B
	gl	25	0·5 (NaCl)	K_1 6·64 ± 0·03	66B
	sp	25	0·10	K_1 6·68	68O
	sp	25	var H_2SO_4	K_2 −0·53 ± 0·10	66B
Co^{2+}	sp	25	0·1 $NaClO_4$	K_1 5·59	68Oa
Cu^{2+}	sp	25	0·10	K_1 8·34	68O
Fe^{3+}	sp	25	2 $NaClO_4$	K_1 10·50	62O
Ge^{IV}	sp	25	0·5 (NaCl)	$K[Ge(OH)_4 + 2HL \rightleftharpoons Ge(OH)_2L_2]$ 8·03 ± 0·05 $K[Ge(OH)_4 + 3HL + H^+ \rightleftharpoons GeL_3^+]$ 13·3 ± 0·1	66B
Mn^{2+}	sp	25	0·1 $NaClO_4$	K_1 4·60	68Oa
Ni^{2+}	sp	25	0·1 $NaClO_4$	K_1 5·97	68Oa
Zn^{2+}	sp	25	0·1 $NaClO_4$	K_1 5·84	68Oa

62O Y. Oka, M. Umehara, and T. Nozoe, *Nippon Kagaku Zasshi*, 1962, **83**, 1197
66B A. Beauchamp and R. L. Benoit, *Canad. J. Chem.*, 1966, **44**, 1607, 1615
68O Y. Oka and Y. Hirai, *Nippon Kagaku Zasshi*, 1968, **89**, 589
68Oa Y. Oka, I. Watanabe, and M. Hirai, *Nippon Kagaku Zasshi*, 1968, **89**, 1220

503 $C_7H_6O_3$ 3,4-Dihydroxybenzaldehyde (protocatechualdehyde) H_2L

Metal	Method	Temp	Medium	Log of equilibrium constant, remarks	Ref
H^+	gl	25	0·1 KCl	K_2 7·22	68A
As^{III}	gl	25	0·1 KCl	$K[As(OH)_4^- + H_2L \rightleftharpoons As(OH)_2L^-]$ 2·96	68A
B^{III}	gl	25	0·1 KCl	$K[B(OH)_4^- + H_2L \rightleftharpoons B(OH)_2L^-]$ 5·15	68A
Ge^{IV}	gl	25	0·1 KCl	$K[HGeO_3^- + 3H_2L \rightleftharpoons HGeL_3^-]$ 2·78	68A

68A P. J. Antikainen and H. Oksanen, *Acta Chem. Scand.*, 1968, **22**, 2867

506 $C_7H_6O_3$ 2-Hydroxybenzoic acid (salicylic acid) H_2L

Metal	Method	Temp	Medium	Log of equilibrium constant, remarks	Ref
H^+	gl	22	50% ethanol	K_1 13·13, K_2 3·91	61A
	?	?	?	K_1 13·4, K_1K_2 16·32	62B
	gl	30	0·1 (KCl)	K_2 2·88	62C
	sp	25	0 corr	K_1 13·592 ± 0·008, K_2 2·973 ± 0·002	63E
	gl	?	0·5 ($NaNO_3$)	K_1 11·6	63I
	gl	30	0·1 $NaClO_4$ 75% dioxan	K_1 13·27, K_2 4·73	64J
	?	?	?	K_1 13·4	65Ka
	gl	25	0·10 (KNO_3)	K_2 2·88 ± 0·02	66L
		20	0·1 (KCl)	K_1 13·6	
	gl	25, 35	0·10 (KNO_3)	K_1 (25°) 13·9 ± 0·1, (35°) 13·4 ± 0·1, K_2 (25°) 2·70 ± 0·02, (35°) 2·45 ± 0·02	66M
	gl	25	0·1 ($NaClO_4$)	K_2 2·82 ± 0·005	66P
	?	?	?	K_1 13·59, K_2 2·98	66R
	gl	20	0·1 (KNO_3)	K_1 13·12, K_2 2·92	67B, 67B[illegible]
	gl	37	0·15 KNO_3	K_1 13·0, K_2 2·814	67P

conti[illegible]

Metal	Method	Temp	Medium	Log of equilibrium constant, remarks	Ref
Al^{3+}	sp	26—28	0—0·20 $NaClO_4$	$*K(Al^{3+}+HL^- \rightleftharpoons AlL^+ + H^+)$ (→ 0) 0·26, (0·02) 0·06, (0·05) 0·01, (0·10) −0·13, (0·20) −0·18	59D
	qh	18	1 $NaNO_3$	$K(Al^{3+}+3HL^- \rightleftharpoons AlHL_2+H_2L)$ 4·5	61C
Be^{2+}	gl	22	50% ethanol	K_1 12·45, K_2 8·50	61A
	sp	28—32	0—0·20 $NaClO_4$	$*K(Be^{2+}+HL^- \rightleftharpoons BeL+H^+)$ (→ 0) −0·27, (0·02) −0·38, (0·05) −0·46, (0·10) −0·51, (0·20) −0·53	61D, 64D
	gl	25	0·15 ($NaClO_4$)	K_1 12·61, β_2 22·60	62B
	sp	20—45	0·2 $NaClO_4$	$*K(Be^{2+}+HL^- \rightleftharpoons BeL+H^+)$ (20°) −0·55, (30°) −0·52, (45°) −0·48 $*\Delta H$ 1·2, ΔS 6	64D
	pH	18	0·1	$K(Be^{2+}+HL^- \rightleftharpoons BeHL^+)$ 2·51±0·01 $K[Be^{2+}+2HL^- \rightleftharpoons Be(HL)_2]$ 4·4±0·1 $K[Be^{2+}+3HL^- \rightleftharpoons Be(HL)_3^-]$ 6·6±0·2	65K
	gl	20	0·1 (KNO_3)	K_1 12·37, K_2 9·65	67Ba
Cd^{2+}	pol	25	1 $NaNO_3$	$K(Cd^{2+}+HL^- \rightleftharpoons CdHL^+)$ (?) 0·96	63Z
	pol	30	1 (KNO_3) 0—60% formaldehyde	$K(Cd^{2+}+HL^- \rightleftharpoons CdHL^+)$ (?) (0%) 0·60, (20%) 0·60, (40%) 0·78, (60%) 0·90 $K[Cd^{2+}+2HL^- \rightleftharpoons Cd(HL)_2]$ (?) (0%) 1·20, (20%) 1·19, (40%) 1·27, (60%) 1·48	66G
Ce^{3+}	con	29	?	$K[Ce^{3+}+3HL^- \rightleftharpoons Ce(HL)_3]$ (?) 7·61	60B
	gl	30	0·1 (KCl)	$K(Ce^{3+}+HL^- \rightleftharpoons CeHL^{2+})$2·66	62C
Cu^{2+}	gl	30	0·1 $NaClO_4$ 75% dioxan	K_1 12·03, K_2 8·67	64J
	sp	23—30	0—0·5 $NaClO_4$	$*K(Cu^{2+}+HL^- \rightleftharpoons CuL+H^+)$ (→ 0) −2·11, (0·02) −2·19, (0·08) −2·24, (0·12) −2·30, (0·2) −2·35, (0·5) −2·32	65D
		15—35	0·2 $NaClO_4$	$*K(Cu^{2+}+HL^- \rightleftharpoons CuL+H^+)$ (15°) −2·46, (23°) −2·37, (35°) −2·23 $*\Delta H$ 4·4, ΔS 26	
	oth	20	0·1 ($NaNO_3$)	$K_1 \geqslant$ 10, K_2 7, electrophoresis	65Ka
	gl	25	0·10 (KNO_3)	K_1 10·64±0·03 $K(CuA^{2+}+L^{2-} \rightleftharpoons CuAL)$ 10·91±0·01 where A = 2,2′-bipyridyl	66L
	sp	25	0·5 ($NaClO_4$)	$K(Cu^{2+}+L^{2-}+A^- \rightleftharpoons CuLA^-)$ 19·04 where HA = alanine, 18·22 where HA = serine, 18·68 where HA = valine	67M
	gl	37	0·15 KNO_3	K_1 10·133, β_2 ~ 18·2 $K(CuA^{2+}+L^{2-} \rightleftharpoons CuAL)$ 9·18 where A = ethylenediamine, 9·27 where A = histamine $K(CuA^+ +L^{2-} \rightleftharpoons CuAL^-)$ 8·98 where HA = L-serine See ethylenediamine (120), 1,3-propanediamine (159) and *NN′*-diethylethylenediamine (403) and EDTA (836) for mixed ligands	67P
Fe^{3+}	sp	15	~ 0	K_1 16·35±0·05	52B
	sp	25	0 corr	K_1 17·436±0·001	63E
	sp	20—25	0·05	K_1 15·35, β_2 27·20, β_3 36·27	66O
	gl, Pt	25	0·1 ($NaClO_4$)	$K(Fe^{3+}+HL^- \rightleftharpoons FeL^+ + H^+)$ 2·92±0·04, $K(FeL^+ +HL^- \rightleftharpoons FeL_2^- + H^+)$ 1·2±0·1 $K(FeL_2^- +HL^- \rightleftharpoons FeL_3^{3-} + H^+)$ −4·2±0·2 $K(Fe^{3+}+HL^- \rightleftharpoons FeHL^{2+})$ 4·4	66P

continued overleaf

506 $C_7H_6O_3$ (contd)

Metal	*Method*	*Temp*	*Medium*	*Log of equilibrium constant, remarks*	*Ref*
Ga^{3+}	sp	28—32	0·02—0·2 $NaClO_4$	$^*K(Ga^{3+}+HL^- \rightleftharpoons GaL^+ + H^+)$ (0·02) 1·04, (0·05) 0·74, (0·1) 0·62 (0·2) 0·47	63D
		20—40	0·05 $NaClO_4$	$^*K(Ga^{3+}+HL^- \rightleftharpoons GaL^+ + H^+)$ (20°) 0·62, (29°) 0·69, (40°) 0·80	
				$^*\Delta H$ 3·7, ΔS 9	
La^{3+}	con	25	?	$K(La^{3+}+HL^- \rightleftharpoons LaHL^{2+})$ 2·82	59F
	gl	20—22	2·86 NaHL	$K[LaL(HL)_2^- + H^+ \rightleftharpoons La(HL)_3]$ 7·5	59F
				$K[LaL_2HL^{2-} + H^+ \rightleftharpoons LaL(HL)_2^-]$ 10·0	
				$K(La^{3+}+H^+ + 3L^{2-} \rightleftharpoons LaL_2HL)$ 16·57	
				β_2 16·82	
	gl	30	0·1 (KCl)	$K(La^{3+}+HL^- \rightleftharpoons LaHL^{2+})$ 2·64	62C
	sol	25	?	K_{s3} −10·48	59F
Mg^{2+}	gl	30	0·1 $NaClO_4$ 75% dioxan	K_1 3·30	64J
Nd^{3+}	gl	30	0·1 (KCl)	$K(Nd^{3+}+HL^- \rightleftharpoons NdHL^{2+})$ 2·70	62C
	con	25	0·003	$K[Nd^{3+}+HL^- \rightleftharpoons NdHL^{2+})$ 2·85	65E
				$K[NdHL^{2+}+HL^- \rightleftharpoons Nd(HL)_2^+]$ 2·38	
				$K[Nb(HL)_2^+ + HL^- \rightleftharpoons Nd(HL)_3]$ 1·89	
			0·001 methanol	$K(Nd^{3+}+HL^- \rightleftharpoons NdHL^{2+})$ 4·4±0·2	
			0·001 butan	$K(Nd^{3+}+HL^- \rightleftharpoons NdHL^{2+})$ 5·0±0·2	
	gl	20	0·01	$K(NdL^+ + OH^- \rightleftharpoons NdOHL)$ 7·21±0·01	65E
				$K[NdOHL+OH^- \rightleftharpoons Nd(OH)_2L^-]$ 5·11±0·01	
			3	$K(NdH_2L_3^- + H^+ \rightleftharpoons NdH_3L_3)$ 9·32±0·06	
				$K(NdH_3L_3 + H^+ \rightleftharpoons NdH_4L_3^+)$ (?) 6·5±0·1	
	sol	25	?	K_1 9·7, β_2 17·7, K_{so} −11·0	65E
				$K(Nd^{3+}+2L^{2-}+H_2L \rightleftharpoons NdH_2L_3^-)$ 17·75±0·01	
			1	K_4 −0·85	
Nb^V	sp	25	0·04	$K(NbO^{3+}+2L^{2-} \rightleftharpoons NbOL_2^-)$ 22·60±0·04	62Ba
Ni^{2+}	gl	30	0·1 $NaClO_4$ 75% dioxan	K_1 8·41, K_2 7·04 See nitrilotriacetic acid (449) for mixed ligands	64J
Pr^{3+}	gl	30	0·1 (KCl)	$K(Pr^{3+}+HL^- \rightleftharpoons PrHL^{2+})$ 2·68	62C
Ti^{IV}	sp	20	0·1	$K(TiO^{2+}+L^{2-} \rightleftharpoons TiOL)$ 15·66	62Bb
				$K(TiO^{2+}+2L^{2-} \rightleftharpoons TiOL_2^{2-})$ 24·36	
	sol	20	2—3 (KNO_3)	$K(TiO^{2+}+2L^{2-} \rightleftharpoons TiOL_2^{2-})$ 24·62	62Bb
Tl^+	con	28	?	β_2 5·32	67S
Tl^{3+}	gl	25	0·1 (KNO_3)	K_1 12·96	67A
UO_2^{2+}	gl	28	0·1 KNO_3	K_1 13·12	66R
	gl	20	0·1 (KNO_3)	K_1 12·08, β_2 20·83	67B
VO^{2+}	sp	?	0·05	K_1 15·4	62L
	gl	25, 35	0·10 (KNO_3)	K_1 (25°) 13·38±0·03, (35°) 12·89±0·05	66M
				$\Delta H_1°$ −2·1±0·5, $\Delta S_1°$ 54±3	
Zn^{2+}	gl	30	0·1 $NaClO_4$ 75% dioxan	K_1 9·20	64J

* These constants have been recalculated by the compilers on the basis of the equations indicated

52B C. Bertin-Batsch, *Ann. Chim.* (*France*), 1952, **1**, 481
59D B. Das and S. Aditya, *J. Indian Chem. Soc.*, 1959, **36**, 473
59F Ya. A. Fialkov and V. I. Ermolenko, *Russ. J. Inorg. Chem.*, 1959, **4**, 159 (359), 615 (1369)
60B A. K. Bhattacharya and M. C. Saxena, *Current Sci.*, 1960, **29**, 128
61A R. P. Agarwal and R. C. Mehrotra, *J. Less-Common Metals*, 1961, **3**, 398
61C I. Cadariu and L. Oniciu, *Studii si Cercetari de Chimie*, (*Cluj*). 1961, **12**, 69
61D R. C. Das and S. Aditya, *J. Indian Chem. Soc.*, 1961, **38**, 19
62B H. J. deBruin, D. Kaitis and R. B. Temple, *Austral J. Chem.*, 1962, **15**, 457
62Ba A. K. Babko and A. I. Volkova, *Russ. J. Inorg. Chem.*, 1962, **7**, 1216 (2345)

contd

62Bb A. K. Babko, A. I. Volkova, and T. E. Get'man, *Russ. J. Inorg. Chem.*, 1962, 7, 145 (284), 1121 (2167)
62C M. Cefola, A. S. Tompa, A. V. Celiano, and P. S. Gentile, *Inorg. Chem.*, 1962, **1**, 290
62L L. B. Levashova and V. L. Zolotavin, *Russ. J. Inorg. Chem.*, 1962, **7**, 418 (811)
63D R. C. Das, R. K. Nanda, and S. Aditya, *J. Indian Chem. Soc.*, 1963, **40**, 739
63E Z. L. Ernst and J. Menashi, *Trans. Faraday Soc.*, 1963, **59**, 1794
63I Y. J. Israeli, *Canad. J. Chem.*, 1963, **41**, 2710
63Z V. K. Zolotukhin and Z. G. Galanets, *Visn. L'vivsk. Derzh. Univ., Ser. Khim.*, 1963, 91 (*Chem. Abs.* 1965, **62**, 4909a)
64D R. C. Das and S. Aditya, *J. Indian Chem. Soc.*, 1964, **41**, 765
64E V. I. Ermolenko, *Russ. J. Inorg. Chem.*, 1964, **9**, 25 (48)
64J K. E. Jabalpurwala, K. A. Venkatachalam, and M. B. Kabadi, *J. Inorg. Nuclear Chem.*, 1964, **26**, 1011, 1027
65D R. C. Das and S. Aditya, *J. Indian Chem. Soc.*, 1965, **42**, 15
65E V. I. Ermolenko, *Russ. J. Inorg. Chem.*, 1965, **10**, 1423 (2617)
65K I. F. Kolosova and T. A. Belyavskaya, *Russ. J. Inorg. Chem.*, 1965, **10**, 411 (764)
65Ka J. Kratsmar-Smogrovic and V. Jokl, *Chem. Zvesti*, 1965, **19**, 881
66G J. N. Gaur, N. K. Goswami, and D. S. Jain, *Analyt. Chem.*, 1966, **38**, 626
66L G. A. L'Heureux and A. E. Martell, *J. Inorg. Nuclear Chem.*, 1966, **28**, 481
66M G. E. Mont and A. E. Martell, *J. Amer. Chem. Soc.*, 1966, **88**, 1387
66O K. Ogawa and N. Tobe, *Bull. Chem. Soc. Japan*, 1966, **39**, 227
66P M. V. Park, *J. Chem. Soc.* (*A*), 1966, 816; *Nature*, 1963, **197**, 283
66R J. M. Rao and U. V. Seshaiah, *Bull. Chem. Soc. Japan*, 1966, **39**, 2668
67A R. C. Aggarwal and A. K. Srivastava, *Indian J. Chem.*, 1967, **5**, 114
67B M. Bartusek, *Coll. Czech. Chem. Comm.*, 1967, **32**, 116
67Ba M. Bartusek and J. Zelinka, *Coll. Czech. Chem. Comm.*, 1967, **32**, 992
67M P. K. Migal and A. P. Gerbelen, *Russ. J. Inorg. Chem.*, 1967, **12**, 1421 (2695)
67P D. D. Perrin, I. G. Sayce, and V. S. Sharma, *J. Chem. Soc.* (*A*), 1967, 1755
67S K. N. Sahn and A. K. Bhattacharya, *Current Sci.*, 1967, **36**, 70

$C_7H_6O_4$ 2,4-Dihydroxybenzoic acid (β-resorcylic acid) H_3L 508

Metal	*Method*	*Temp*	*Medium*	*Log of equilibrium constant, remarks*	*Ref*
H^+	gl	25	0·1 $(NaClO_4)$	K_3 3·10	66P
	gl	28	0·1 KNO_3	K_1 10·94, K_2 9·35, K_3 3·65	66R
Cu^{2+}	pol	30	0·1 $NaC_2H_3O_2$	$K(Cu^{2+}+H_2L^- \rightleftharpoons CuHL+H^+)$ 1·83	65G
	spJ	30	0·1 $NaClO_4$	$K(Cu^{2+}+H_2L^- \rightleftharpoons CuHL+H^+)$ 2·70	66G
Fe^{3+}	sp	25	?	$K(Fe^{3+}+H_2L^- \rightleftharpoons FeHL^+ +H^+)$ 4·04	64R
	pol	30	0·1 KCl	K_1 15·05	65G
	gl, Pt	25	0·1 $(NaClO_4)$	$K(Fe^{3+}+H_2L^- \rightleftharpoons FeHL^+ +H^+)$ 3·135, $K[FeHL^+ +H_2L^- \rightleftharpoons Fe(HL)_2^- +H^+]$ 1·75, $K[Fe(HL)_2^- +H_2L^- \rightleftharpoons Fe(HL)_3^{3-} +H^+]$ −3·7 $K(Fe^{3+}+H_2L^- \rightleftharpoons FeH_2L^{2+})$ 4·8	66P
UO_2^{2+}	sp	25	?	$K(UO_2^{2+}+H_2L^- \rightleftharpoons UO_2HL+H^+)$ 3·70	65D
	gl	28	0·1 KNO_3	K_1 11·98	66R

64R M. L. N. Reddy and U. V. Seshaiah, *J. Indian Chem. Soc.*, 1964, **41**, 289
65D M. N. Desai and B. M. Desai, *J. Indian Chem. Soc.*, 1965, **42**, 643
65G S. L. Gupta, J. N. Jaitly, and R. N. Soni, *J. Indian Chem. Soc.*, 1965, **42**, 384
66G S. L. Gupta, R. N. Soni, and J. N. Jaitly, *J. Indian Chem. Soc.*, 1966, **43**, 331
66P M. V. Park, *J. Chem. Soc.* (*A*), 1966, 816; *Nature*, 1963, **197**, 283
66R J. M. Rao and U. V. Seshaiah, *Bull. Chem. Soc. Japan*, 1966, **39**, 2668

509.1 $C_7H_6O_4$ 2,6-Dihydroxybenzoic acid (γ-resorcylic acid) H_3L

Metal	Method	Temp	Medium	Log of equilibrium constant, remarks	Ref
H^+	gl	25	0·1 ($NaClO_4$)	K_3 1·08	66P
Fe^{3+}	gl, Pt	25	0·1 ($NaClO_4$)	$K(Fe^{3+}+H_2L^- \rightleftharpoons FeHL^+ + H^+)$ 2·76, $K[FeHL^+ + H_2L^- \rightleftharpoons Fe(HL)_2^- + H^+]$ 1·2	66P

66P M. V. Park, *J. Chem. Soc. (A)*, 1966, 816; *Nature*, 1963, **197**, 283

510 $C_7H_6O_4$ 3,4-Dihydroxybenzoic acid (protocatechuic acid) H_3L

Metal	Method	Temp	Medium	Log of equilibrium constant, remarks	Ref
H^+	gl	30	0·1 (KNO_3)	K_1 11·94, K_2 8·67, K_3 4·38	63M
	gl	18—22	0·1	K_1 13·00±0·04, K_2 8·32±0·08, K_3 4·47	63S, 6
	gl	30	0·1 ($NaClO_4$)	K_1 12·6, K_2 8·83, K_3 4·40	66A
	gl	25	0·1 KCl	K_2 8·84, K_3 4·35	68A
	gl	30	0·1 $NaNO_3$	K_1 11·66, K_2 9·16, K_3 4·40	68J
B^{III}	gl	25	0·1 KCl	$K[H_3BO_3 + H_2L^- \rightleftharpoons B(OH)_2L^{2-} + H^+]$ 5·01	68A
Cd^{2+}	gl	30	0·1 ($NaClO_4$)	K_1 7·97, K_2 4·75	66A
Co^{2+}	gl	30	0·10 (KNO_3)	K_3 4·06	63M
	gl	30	0·1 ($NaNO_3$)	K_1 7·46, K_2 4·28, K_3 3·27	68J
Cu^{2+}	gl	30	0·1 ($NaClO_4$)	K_1 14·53, K_2 10·76	66A
	gl	30	0·1 ($NaNO_3$)	K_1 12·58, K_2 9·64	68J
Mg^{2+}	gl	30	0·1 ($NaClO_4$)	K_1 6·30	66A
Nb^V	sp	18—22	0·1	$K(NbO_2^+ + H_3L \rightleftharpoons NbO_2LH_2 + H^+)$ 2·3 $K(NbO_2LH_2 + H_3L \rightleftharpoons NbOL_2H_2^- + H^+)$ 1·3 $K(NbO_2^+ + 3HL^{2-} + 4H^+ \rightleftharpoons NbL_3H_3^-)$ 63·1	64S
Ni^{2+}	gl	30	0·10 (KNO_3)	K_3 3·89	63M
	gl	30	0·1 ($NaClO_4$)	K_1 8·96, K_2 5·38	66A
	gl	30	0·1 ($NaNO_3$)	K_1 7·92, K_2 3·65, K_3 3·10	68J
Ti^{IV}	sp	18—22	0·1 ($NaClO_4$)	$K[TiO^{2+} + 2H_3L \rightleftharpoons TiO(H_2L)_2 + 2H^+]$ −1·35 $K[TiO(H_2L)_2 + H_3L \rightleftharpoons TiL_3^{5-} + 5H^+]$ −3·9 $K(TiO^{2+} + 2H^{2+} + 3L^{3-} \rightleftharpoons TiL_3^{5-})$ 58·6	63S
Tl^+	con J	28	?	$K(Tl^+ + HL^{2-} \rightleftharpoons TlHL^-)$ (?) 2·38	64S
Zn^{2+}	gl	30	0·1 ($NaClO_4$)	K_1 9·84, K_2 7·55	66A
	gl	30	0·1 ($NaNO_3$)	K_1 8·13, K_2 4·55	68J

63M Y. Murakami, K. Nakamura, and M. Tokunada, *Bull. Chem. Soc. Japan*, 1963, **36**, 669
63S L. Sommer, *Coll. Czech. Chem. Comm.*, 1963, **28**, 2102
64S K. N. Sahu and A. K. Bhattacharya, *J. Indian Chem. Soc.*, 1964, **41**, 787
64Sa L. Sommer and J. Havel, *Coll. Czech. Chem. Comm.*, 1964, **29**, 690
66A V. T. Athavale, L. H. Prabhu, and D. G. Vastak, *J. Inorg. Nuclear Chem.*, 1966, **28**, 1237
68A P. J. Antikainen and M. Viro, *Suomen Kem.*, 1968, **B41**, 206
68J E. S. Jayadevappa and P. B. Hukkevi, *J. Inorg. Nuclear Chem.*, 1968, **30**, 157

$C_7H_6O_5$ 2,3,4-Trihydroxybenzoic acid H_4L 510.1

Metal	*Method*	*Temp*	*Medium*	*Log of equilibrium constant, remarks*	*Ref*
Be^{2+}	pH	18	0·1	$K(Be^{2+}+H_3L^- \rightleftharpoons BeH_3L^+)$ (?) 2·51 ± 0·01, $K[Be^{2+}+2H_3L^- \rightleftharpoons Be(H_3L)_2]$ (?) 4·07 ± 0·01, $K[Be^{2+}+3H_3L^- \rightleftharpoons Be(H_3L)_3^-]$ (?) 7·15 ± 0·01	65K

65K I. F. Kolosova and T. A. Kelgarskaya, *Russ. J. Inorg. Chem.*, 1965, **10**, 411 (764)

$C_7H_6O_5$ 3,4,5-Trihydroxybenzoic acid (gallic acid) H_4L 511

Metal	*Method*	*Temp*	*Medium*	*Log of equilibrium constant, remarks*	*Ref*
Tl^+	con J	28	?	$K(Tl^+ + H_2L^{2-} \rightleftharpoons TlH_2L^-)$ (?) 2·55	64S

64S K. N. Sahu and A. K. Bhattacharya, *J. Indian Chem. Soc.*, 1964, **41**, 787

$C_7H_6N_2$ Benzimidazole L 511.1

Metal	*Method*	*Temp*	*Medium*	*Log of equilibrium constant, remarks*	*Ref*
H^+	gl	20	0·15 ($NaClO_4$)	K_1 5·68 ± 0·02	62H
Cd^{2+}	pol	25	50% methanol	K_1 1·93, β_2 3·45, β_3 4·19	68C
Cu^+	Pt, Au	20	0·15 (Na_2SO_4)	K_1 4·47, β_2 9·73	62H
Cu^{2+}	gl	20	0·15 ($NaClO_4$)	K_1 3·56, β_2 6·34, β_3 9·00, β_4 10·97	62H
Ni^{2+}	sp	20	acetone	$K(NiL_2Cl_2+2L \rightleftharpoons NiL_4Cl_2)$ 2·34, $K(NiL_2Br_2+2L \rightleftharpoons NiL_4Br_2)$ 2·72, $K(NiL_2I_2+2L \rightleftharpoons NiL_4I_2)$ 3·19	67G
			nitromethane	$K(NiL_2Br_2+2L \rightleftharpoons NiL_4Br_2)$ 4·10, $K(NiL_2I_2+2L \rightleftharpoons NiL_4I_2)$ 5·28	

62H C. J. Hawkins and D. D. Perrin, *J. Chem. Soc.*, 1962, 1351
67G D. M. L. Goodgame, M. Goodgame, and M. J. Weeks, *J. Chem. Soc.* (*A*), 1967, 1676
68C D. R. Crow, *J. Electroanalyt. Chem.*, 1968, **16**, 137

$C_7H_6N_4$ 1-Phenyltetrazole L 511.2

Metal	*Method*	*Temp*	*Medium*	*Log of equilibrium constant, remarks*	*Ref*
Ni^{2+}	sp	25	tetrahydrofuran	K_1 ~ 0·6, β_2 ~ 3·54	63G

63G G. L. Gilbert and C. G. Brubaker jun., *Inorg. Chem.*, 1963, **2**, 1216

513.1 C_7H_8O 3-Methylphenol (*m*-cresol) HL

Metal	Method	Temp	Medium	Log of equilibrium constant, remarks	Ref
H^+	sp	25	0 corr	K_1 10·09	65E
Fe^{3+}	sp	25	0 corr	K_1 8·51 ± 0·03	65E

65E Z. L. Ernst and F. G. Herring, *Trans. Faraday Soc.*, 1965, **61**, 454

514 C_7H_8O 4-Methylphenol (*p*-cresol) HL

Metal	Method	Temp	Medium	Log of equilibrium constant, remarks	Ref
H^+	gl	25	0·5 (KNO_3)	K_1 10·02	68B
Fe^{3+}	sp	~ 23	0·5 KNO_3	K_1 8·33	68B
			0·027 $NaClO_4$	K_1 8·73	

68B W. A. E. McBryde, *Canad. J. Chem.*, 1968, **46**, 2385

514.1 $C_7H_8O_2$ 1,2-Dihydroxy-4-methylbenzene (4-methylcatechol) H_2L

Metal	Method	Temp	Medium	Log of equilibrium constant, remarks	Ref
Mo^{VI}	sp	20	0·014	$K(MoO_4^{2-} + 2H_2L \rightleftharpoons MoO_2L_2^{2-})$ 4·74 ± 0·01	62H
W^{VI}	sp	25	0·1 (KCl)	$K(WO_4^{2-} + 2H_2L \rightleftharpoons WO_2L_2^{2-})$ 6·31 ± 0·07	63H

62H J. Halmekoski, *Suomen Kem.*, 1962, **B35**, 41
63H J. Halmekoski, *Suomen Kem.*, 1963, **B36**, 46

514.2 $C_7H_8O_2$ 2-(Hydroxymethyl)phenol (salicyl alcohol, saligenin) HL

Metal	Method	Temp	Medium	Log of equilibrium constant, remarks	Ref
H^+	gl	30	0·1 $NaClO_4$ 75% dioxan	K_1 11·93	64J
Cu^{2+}	gl	30	0·1 $NaClO_4$ 75% dioxan	K_1 9·19	64J
Mg^{2+}	gl	30	0·1 $NaClO_4$ 75% dioxan	K_1 4·95	64J
Ni^{2+}	gl	30	0·1 $NaClO_4$ 75% dioxan	K_1 6·82	64J
Zn^{2+}	gl	30	0·1 $NaClO_4$ 75% dioxan	K_1 7·63	64J

64J K. E. Jabalpurwala, K. A. Venkatachalam, and M. B. Kabadi, *J. Inorg. Nuclear Chem.*, 1964, **26**, 1011, 1027

$C_7H_8N_4$ 4,4′-(5,5′)-Bisimidazolylmethane L 514.3

Metal	Method	Temp	Medium	Log of equilibrium constant, remarks	Ref
H^+	gl	30	0·16 KNO_3	K_1 7·39±0·03, K_2 5·61±0·02	65D
Cd^{2+}	gl	30	0·16 KNO_3	K_1 5·50±0·05, K_2 4·55±0·04	65D
Co^{2+}	gl	30	0·16 KNO_3	K_1 5·72±0·04, K_2 4·81±0·03	65D
Cu^{2+}	gl	30	0·16 KNO_3	K_1 10·41±0·04, K_2 8·18±0·03	65D
Fe^{2+}	gl	30	0·16 KNO_3	K_1 4·47±0·05, K_2 3·90±0·04, sulphate present, slightly low?	65D
Mn^{2+}	gl	30	0·16 KNO_3	K_1 2·96±0·03, K_2 2·84±0·05	65D
Ni^{2+}	gl	30	0·16 KNO_3	K_1 7·33±0·02, K_2 6·30±0·03	65D
Zn^{2+}	gl	30	0·16 KNO_3	K_1 5·62±0·02, K_2 4·86±0·03	65D

65D C. N. C. Drey and J. S. Fruton, *Biochemistry*, 1965, **4**, 1258

C_7H_9N (Aminomethyl)benzene (benzylamine) L 515

Metal	Method	Temp	Medium	Log of equilibrium constant, remarks	Ref
Ag^+	Ag	25	ethanol	β_2 7·79	60A

60A V. Armeanu and C. Luca, *Z. phys. Chem.* (*Leipzig*), 1960, **214**, 81

C_7H_9N 2,4-Dimethylpyridine (2,4-lutidine) L 516

Metal	Method	Temp	Medium	Log of equilibrium constant, remarks	Ref
Cu^{2+}	pol	25	0·05 L·HNO_3 20% methanol	β_2 8·9	65P

65P F. Pantani and R. Guidelli, *Ricerca sci.*, 1965, **35** (II-A), 713

C_7H_9N 2,5-Dimethylpyridine (2,5-lutidine) L 516.1

Metal	Method	Temp	Medium	Log of equilibrium constant, remarks	Ref
H^+	gl	25	0·61 (KNO_3)	K_1 6·63	67S
Ag^+	gl	25	0·61 (KNO_3)	K_1 2·42±0·02, β_2 4·95±0·01	67S
Cu^{2+}	gl	25	0·61 (KNO_3)	K_1 1·78±0·03, β_2 2·8±0·2	67S

67S M. S. Sun and D. G. Brewer, *Canad. J. Chem.*, 1967, **45**, 2729

516.2 C_7H_9N 2,6-Dimethylpyridine (2,6-lutidine) L

Metal	Method	Temp	Medium	Log of equilibrium constant, remarks	Ref
Ag^+	Ag	25	0—90% acetone	β_2 (0%) 4·95, (10%) 4·90, (20%) 4·85, (30%) 4·81, (40%) 4·80, (50%) 4·83, (60%) 4·89, (70%) 5·01, (80%) 5·28	65P
			10—90% dioxan	β_2 (10%) 4·83, (20%) 4·75, (30%) 4·70, (40%) 4·68, (50%) 4·70, (60%) 4·76, (70%) 4·89, (80%) 5·10, (90%) 5·36	
			10—90% ethanol	β_2 (10%) 5·01, (20%) 4·99, (30%) 4·92, (40%) 4·82, (50%) 4·71, (60%) 4·60, (70%) 4·59, (80%) 4·66, (90%) 4·81	

65P G. Popa, C. Luca and V. Magearu, *J. Chim. phys.*, 1965, **62**, 449, 853

516.3 C_7H_9N 3,4-Dimethylpyridine (3,4-lutidine) L

Metal	Method	Temp	Medium	Log of equilibrium constant, remarks	Ref
H^+	gl	25	0·61 (KNO_3)	K_1 6·64	67S
Co^{2+}	sp, cal	20	$CHCl_3$	$K(CoL_2Cl_2+2L \rightleftharpoons CoL_4Cl_2)$ 0·20 $K[CoL_2(NCS)_2+2L \rightleftharpoons CoL_4(NCS)_2]$ 4·26 ΔH −16·8±0·3, ΔS −38±1	66C
Cu^{2+}	pol	25	0·05 L·HNO_3 20% methanol	β_2 9·1	65P
	gl	25	0·61 (KNO_3)	K_1 3·11±0·01, β_2 5·38±0·03, β_3 7·46±0·05, β_4 8·7±0·1	67S
Ni^{2+}	gl	25	0·61 (KNO_3)	K_1 2·26±0·02, β_2 3·2±0·3, β_3 5·21±0·08	67S

65P F. Pantani and R. Guidelli, *Ricerca sci.*, 1965, **35** (II-A), 713
66C J. de O. Cabral, H. C. A. King, S. M. Nelson, T. M. Shepherd, and E. Koros, *J. Chem. Soc.* (*A*), 1966, 1348
67S M. S. Sun and D. G. Brewer, *Canad. J. Chem.*, 1967, **45**, 2729

516.4 C_7H_9N 3,5-Dimethylpyridine (3,5-lutidine) L

Metal	Method	Temp	Medium	Log of equilibrium constant, remarks	Ref
H^+	gl	25	0·61 (KNO_3)	K_1 6·31	67S
Ag^+	gl	25	0·61 (KNO_3)	K_1 2·37±0·01, β_2 4·65±0·02	67S
Co^{2+}	sp, cal	20	$CHCl_3$	$K(CoL_2Cl_2+2L \rightleftharpoons CoL_4Cl_2)$ −0·46 $K[CoL_2(NCS)_2+2L \rightleftharpoons CoL_4(NCS)_2]$ 3·70 ΔH −14·6±0·3, ΔS −33±1	66C
Cu^{2+}	pol	25	0·05 L·HNO_3 20% methanol	β_2 9·1	65P
	gl	25	0·61 (KNO_3)	K_1 2·94±0·01, β_2 5·10±0·04, β_3 6·8±0·1, β_4 8·1±0·3	67S
Ni^{2+}	gl	25	0·61 (KNO_3)	K_1 2·13±0·01, β_2 3·1±0·1, β_3 4·87±0·07	67S

65P F. Pantani and R. Guidelli, *Ricerca Sci.*, 1965, **35** (II-A), 713
66C J. de O. Cabral, H. C. A. King, S. M. Nelson, T. M. Shepherd, and E. Koros, *J. Chem. Soc.* (*A*), 1966, 1348
67S M. S. Sun and D. G. Brewer, *Canad. J. Chem.*, 1967, **45**, 2729

C_7H_9N 3-Ethylpyridine L 516.5

Metal	Method	Temp	Medium	Log of equilibrium constant, remarks	Ref
Co^{2+}	sp, cal	20	$CHCl_3$	$K(CoL_2Cl_2+2L \rightleftharpoons CoL_4Cl_2)$ 0·36 ΔH −13·2±0·3, ΔS −43±1 $K[CoL_2(NCS)_2+2L \rightleftharpoons CoL_4(NCS)_2]$ 4·34 ΔH −15·3±0·2, ΔS −32·3±0·8	66C

66C J. de O. Cabral, H. C. A. King, S. M. Nelson, T. M. Shepherd, and E. Koros, *J. Chem. Soc.* (*A*), 1966, 1348

C_7H_9N 4-Ethylpyridine L 516.6

Metal	Method	Temp	Medium	Log of equilibrium constant, remarks	Ref
Co^{2+}	sp, cal	20	$CHCl_3$	$K(CoL_2Cl_2+2L \rightleftharpoons CoL_4Cl_2)$ 1·05 ΔH −16·0±0·3, ΔS −50±1 $K[CoL_2(NCS)_2+2L \rightleftharpoons CoL_4(NCS)_2]$ 4·89 ΔH −16·5±0·3, ΔS −34±2	66C

66C J. de O. Cabral, H. C. A. King, S. M. Nelson, T. M. Shepherd, and E. Koros, *J. Chem. Soc.* (*A*), 1966, 1348

C_7H_9N *N*-Methylaniline L 517

Metal	Method	Temp	Medium	Log of equilibrium constant, remarks	Ref
Ag^+	Ag	25	< 0·01 96% ethanol	K_1 1·38, β_2 1·74	61A

61A V. Armeanu and C. Luca, *Z. phys. Chem.* (*Leipzig*), 1961, **217**, 389

C_7H_9N 2-Methylaniline (*o*-toluidine) L 518

Metal	Method	Temp	Medium	Log of equilibrium constant, remarks	Ref
H^+	pH	25	0·3 (KNO_3)	K_1 4·44	64N
Ag^+	Ag	25	ethanol	K_1 2·04, β_2 3·25, β_3 3·48	60A
Cd^{2+}	Hg(Cd)	25	0·3 (KNO_3)	K_1 −0·10	64N

60A V. Armeanu and C. Luca, *Z. phys. Chem.* (*Leipzig*), 1960, **214**, 81
64N L. V. Nazarova, A. V. Ablov, and V. A. Dagaev, *Russ. J. Inorg. Chem.*, 1964, **9**, 1150 (2129)

C_7H_9N 3-Methylaniline (*m*-toluidine) L 519

Metal	Method	Temp	Medium	Log of equilibrium constant, remarks	Ref
Ag^+	Ag	25	ethanol	K_1 2·35, β_2 3·63, β_3 3·98	60A

60A V. Armeanu and C. Luca, *Z. phys. Chem.* (*Leipzig*), 1960, **214**, 81

520 C_7H_9N 4-Methylaniline (*p*-toluidine) L

Metal	*Method*	*Temp*	*Medium*	*Log of equilibrium constant, remarks*	*Ref*
H^+	gl	25	0·3 (KNO_3)	K_1 4·79	64N
Ag^+	Ag	25	ethanol	β_2 3·86, β_3 4·22	60A
Cd^{2+}	Hg(Cd)	25	0·3 (KNO_3)	K_1 0·26, β_2 −0·01	64N

60A V. Armeanu and C. Luca, *Z. phys. Chem.* (*Leipzig*), 1960, **214,** 81
64N L. V. Nazarova, A. V. Ablov, and V. A. Dagaev, *Russ. J. Inorg. Chem.*, 1964, **9,** 1150 (2129)

520.1 $C_7H_{10}O_5$ 3,4,5-Trihydroxycyclohex-1-ene-1-carboxylic acid (dihydroshikimic acid) HL*

Metal	*Method*	*Temp*	*Medium*	*Log of equilibrium constant, remarks*	*Ref*
H^+	gl	25	→ 0	K_1 4·39	59T
	gl	25	0·05	K_1 4·29	59T
Cu^{2+}	gl	25	→ 0	K_1 2·08	59T
	gl	25	0·05	K_1 1·65	59T

* Note. This was Table 523 on page 540 in the 1964 edition. Geometric configuration of hydroxy-groups not given.

59T C. F. Timberlake, *J. Chem. Soc.*, 1959, 2795

521 $C_7H_{10}N_2$ 2-(2′-Aminoethyl)pyridine L

Metal	*Method*	*Temp*	*Medium*	*Log of equilibrium constant, remarks*	*Ref*
H^+	gl	25	0·015	K_1 9·52, K_2 3·80	60H
	gl	25	0·1 (KNO_3)	K_1 9·59, K_2 3·92	64L
Co^{2+}	gl	25	0·01—0·02	K_1 ~ 3·8	60H
Cu^{2+}	gl	25	0·01—0·02	K_1 7·3, β_2 12·7	60H
	gl	25	0·1 (KNO_3)	K_1 7·3, K_2 5·6	64L
Ni^{2+}	gl	25	0·01—0·02	K_1 5·2, β_2 8·5	60H
	gl	25	0·1 (KNO_3)	K_1 5·2	64L

Errata. 1964, p. 539, Table 521. Throughout *for* Medium 0·1 *read* 0·1 (KNO_3).
For ref. 60L *read* 64L R. G. Lacoste and A. E. Martell, *Inorg. chem.*, 1964, **3,** 881.

60H F. Holmes and F. Jones, *J. Chem. Soc.*, 1960, 2398
64L R. C. Lacoste and A. E. Martell, *Inorg. Chem.*, 1964, **3,** 881

$C_7H_{10}N_2$ 6-Methyl-2-(aminomethyl)pyridine L 521.1

Metal	Method	Temp	Medium	Log of equilibrium constant, remarks	Ref
Cd^{2+}	pol	25	0·1 KNO_3 50% dioxan	β_2 7·07	66W
	cal	25	0·1 KNO_3 50% dioxan	$\Delta H_{\beta 2}°$ −6·0, $\Delta S_{\beta 2}°$ 12·3	66W
Co^{2+}	cal	25	0·1 KNO_3 50% dioxan	$\Delta H_{\beta 2}°$ −10·6, $\Delta S_{\beta 2}°$ −3·4	66W
	pol	25	0·1 KNO_3 50% dioxan	β_2 7·00	66W
Cu^{+}	pol	25	0·1 KNO_3 50% dioxan	β_2 12·04	66W
Cu^{2+}	cal	25	0·1 KNO_3 50% dioxan	$\Delta H_{\beta 2}°$ −15·3, $\Delta S_{\beta 2}°$ 6·8	66W
	pol	25	0·1 KNO_3 50% dioxan	β_2 12·69	66W
	gl	25	0·1 KNO_3 50% dioxan	β_2 13·27	66W
Ni^{2+}	cal	25	0·1 KNO_3 50% dioxan	$\Delta H_{\beta 2}°$ −9·0, $\Delta S_{\beta 2}°$ 8·3	66W
	pol	25	0·1 KNO_3 50% dioxan	β_2 8·43	66W
Zn^{2+}	cal	25	0·1 KNO_3 50% dioxan	$\Delta H_{\beta 2}°$ −5·3, $\Delta S_{\beta 2}°$ 17·8	66W
	pol	25	0·1 KNO_3 50% dioxan	β_2 7·80	66W

66W J. L. Walter and S. M. Rosalie, *J. Inorg. Nuclear Chem.*, 1966, **28**, 2969

$C_7H_{10}N_2$ 2-(Methylaminomethyl)pyridine (2-picolylmethylamine) L 522

Metal	Method	Temp	Medium	Log of equilibrium constant, remarks	Ref
H^{+}	gl	25	0·1 KNO_3 50% dioxan	K_1 8·40, K_2 2·99	66W
Cd^{2+}	var	25	0·1 KNO_3 50% dioxan	β_2 (pol) 8·54 $\Delta H_{\beta 2}$ (cal) −10·0, $\Delta S_{\beta 2}$ 5·5	66W
Co^{2+}	cal	25	0·1 KNO_3 50% dioxan	$\Delta H_{\beta 2}$ −16·9	66W
Cu^{+}	pol	25	0·1 KNO_3 50% dioxan	β_2 11·40	66W
Cu^{2+}	var	25	0·1 KNO_3 50% dioxan	β_2 (gl) 14·60, (pol) 14·82 $\Delta H_{\beta 2}$ (cal) −19·2, $\Delta S_{\beta 2}$ 2·3	66W
Ni^{2+}	cal	25	0·1 KNO_3 50% dioxan	$\Delta H_{\beta 2}$ −16·9	66W
Zn^{2+}	cal	25	0·1 KNO_3 50% dioxan	$\Delta H_{\beta 2}$ −11·0	66W

66W J. L. Walter and S. M. Rosalie, *J. Inorg. Nuclear Chem.*, 1966, **28**, 2969

$C_7H_{10}O_5$ Dihydroshikimic acid HL 523

Erratum. 1964, p. 540, Table 523. For correct ligand sequence this Table is renumbered **520.2** *q.v.*

525.1 $C_7H_{12}O_2$ 3-Ethylpentan-2,4-dione (3-ethylacetylacetone) HL

Metal	*Method*	*Temp*	*Medium*	*Log of equilibrium constant, remarks*	*Ref*
H^+				H^+ constants not given.	
Ni^{2+}	sp	30	toluene	$K(Ni_3L_6 \rightleftharpoons 3NiL_2)$ − 1·23	68A

68A A. W. Addison and D. P. Graddon, *Austral. J. Chem.*, 1968, **21**, 2003

525.2 $C_7H_{12}O_3$ 2-Hydroxycyclohexanecarboxylic acid* HL

Metal	*Method*	*Temp*	*Medium*	*Log of equilibrium constant, remarks*	*Ref*
H^+	gl	25	1 ($NaClO_4$)	K_1 5·00	67S
Ce^{3+}	gl	25	1 ($NaClO_4$)	K_1 1·99±0·01, K_2 1·49±0·01	67S
Dy^{3+}	gl	25	1 ($NaClO_4$)	K_1 2·43±0·01, K_2 1·90±0·01	67S
Er^{3+}	gl	25	1 ($NaClO_4$)	K_1 2·48±0·02, K_2 2·35±0·01	67S
Eu^{3+}	gl	25	1 ($NaClO_4$)	K_1 2·21±0·01, K_2 1·86±0·01	67S
Gd^{3+}	gl	25	1 ($NaClO_4$)	K_1 2·29±0·01, K_2 1·88±0·01	67S
Ho^{3+}	gl	25	1 ($NaClO_4$)	K_1 2·38±0·02, K_2 2·29±0·01	67S
La^{3+}	gl	25	1 ($NaClO_4$)	K_1 1·76±0·01, K_2 1·43±0·01	67S
Nd^{3+}	gl	25	1 ($NaClO_4$)	K_1 2·16±0·01, K_2 1·55±0·01	67S
Pr^{3+}	gl	25	1 ($NaClO_4$)	K_1 2·05±0·01, K_2 1·56±0·01	67S
Sm^{3+}	gl	25	1 ($NaClO_4$)	K_1 2·18±0·01, K_2 1·73±0·01	67S
Tb^{3+}	gl	25	1 ($NaClO_4$)	K_1 2·25±0·01, K_2 2·01±0·01	67S
Yb^{3+}	gl	25	1 ($NaClO_4$)	K_1 2·61±0·02, K_2 2·45±0·01	67S

* Geometric configuration not given

67S H. Schurmans, H. Thun, and F. Verbeek, *J. Inorg. Nuclear Chem.*, 1967, **29**, 1759

526 $C_7H_{12}O_4$ Pentane-3,3-dicarboxylic acid (diethylmalonic acid) H_2L

Metal	*Method*	*Temp*	*Medium*	*Log of equilibrium constant, remarks*	*Ref*
H^+	gl	25	0·10 (KNO_3)	*K_1 6·98, K_2 1·96	68P
Ce^{3+}	gl	25	0·10 (KNO_3)	K_1 3·78, β_2 6·32	68P
Dy^{3+}	gl	25	0·10 (KNO_3)	K_1 4·61, β_2 7·29	68P
Er^{3+}	gl	25	0·10 (KNO_3)	K_1 4·66, β_2 7·26	68P
Eu^{3+}	gl	25	0·10 (KNO_3)	K_1 4·46, β_2 7·05	68P
Gd^{3+}	gl	25	0·10 (KNO_3)	K_1 4·49, β_2 7·05	68P
Ho^{3+}	gl	25	0·10 (KNO_3)	K_1 4·63, β_2 7·16	68P
La^{3+}	gl	25	0·10 (KNO_3)	K_1 3·61, β_2 5·95	68P
Lu^{3+}	gl	25	0·10 (KNO_3)	K_1 4·69, β_2 7·40	68P
Nd^{3+}	gl	25	0·10 (KNO_3)	K_1 4·01, β_2 6·63	68P
Pr^{3+}	gl	25	0·10 (KNO_3)	K_1 3·91, β_2 6·49	68P
Sm^{3+}	gl	25	0·10 (KNO_3)	K_1 4·33, β_2 6·92	68P
Tb^{3+}	gl	25	0·10 (KNO_3)	K_1 4·63, β_2 7·24	68P
Tm^{3+}	gl	25	0·10 (KNO_3)	K_1 4·70, β_2 7·31	68P
Y^{3+}	gl	25	0·10 (KNO_3)	K_1 4·60, β_2 7·05	68P
Yb^{3+}	gl	25	0·10 (KNO_3)	K_1 4·76, β_2 7·43	68P

* K_1 extrapolated to zero ligand concentration.

68P J. E. Powell, J. L. Farrell, W. F. S. Neillie, and R. Russell, *J. Inorg. Nuclear Chem.*, 1968, **30**, 2223

$C_7H_{12}O_4$ Pentane-1,1-dicarboxylic acid (n-butylmalonic acid) H_2L 526.1

(see *Corrigenda*)

Metal	Method	Temp	Medium	Log of equilibrium constant, remarks	Ref
Ni^{2+}	gl	25	0 corr	K_1 3·424 ± 0·004	62B
			0—0·2 ($NaClO_4$)	K_1 (→ 0) 3·412, (0·03) 2·829 ± 0·004, (0·05) 2·709 ± 0·005, (0·10) 2·49 ± 0·01, (0·15) 2·414 ± 0·007, (0·20) 2·347 ± 0·010	

62B J. R. Brannan and G. M. Nancollas, *Trans. Faraday Soc.*, 1962, **58,** 354

$C_7H_{12}O_6$ 1,3,4,5-Tetrahydroxycyclohexanecarboxylic acid* (quinic acid) HL 528

Metal	Method	Temp	Medium	Log of equilibrium constant, remarks	Ref
Dy^{3+}	qh	25	1 ($NaClO_4$)	K_1 2·76, K_2 2·09, K_3 1·53, K_4 0·90	67O
Er^{3+}	qh	25	1 ($NaClO_4$)	K_1 2·85, K_2 2·20, K_3 1·63, K_4 0·96	67O
Eu^{3+}	qh	25	1 ($NaClO_4$)	K_1 2·67, K_2 2·02, K_3 1·46, K_4 0·75	67O
Gd^{3+}	qh	25	1 ($NaClO_4$)	K_1 2·64, K_2 1·99, K_3 1·44, K_4 0·73	67O
Ho^{3+}	qh	25	1 ($NaClO_4$)	K_1 2·79, K_2 2·14, K_3 1·58, K_4 0·94	67O
La^{3+}	qh	25	1 ($NaClO_4$)	K_1 2·30, K_2 1·64, K_3 1·15, K_4 0·61	67O
Pr^{3+}	qh	25	1 ($NaClO_4$)	K_1 2·50, K_2 1·83, K_3 1·27, K_4 0·67	67O
Sm^{3+}	qh	25	1 ($NaClO_4$)	K_1 2·66, K_2 1·99, K_3 1·43, K_4 0·72	67O
Tb^{3+}	qh	25	1 ($NaClO_4$)	K_1 2·68, K_2 2·05, K_3 1·50, K_4 0·84	67O
Yb^{3+}	qh	25	1 ($NaClO_4$)	K_1 2·97, K_2 2·33, K_3 1·77, K_4 1·08	67O

* Geometric configuration not given

67O W. Ooghe, H. Thun, and F. Verbeek, *Analyt. Chim. Acta*, 1967, 39, 397

$C_7H_{12}N_4$ 1-Cyclohexyltetrazole L 528.1

Metal	Method	Temp	Medium	Log of equilibrium constant, remarks	Ref
Co^{2+}	sp	25	tetrahydrofuran	K_1 2·16 ± 0·08, β_2 3·50 ± 0·07	63G
Ni^{2+}	sp	25	ethanol	K_1 1·1 ± 0·1, β_2 2·05 ± 0·05	63G

63G G. L. Gilbert and C. H. Brubaker jun., *Inorg. Chem.*, 1963, **2,** 1216

$C_7H_{13}N$ 3 *N*-Allylpyrrolidine L 528.2

Metal	Method	Temp	Medium	Log of equilibrium constant, remarks	Ref
Pt^{2+}	sp	60	2 (NaCl)	$K(PtCl_4^{2-} + HL^+ \rightleftharpoons PtCl_3HL + Cl^-)$ 2·81	67D

67D R. G. Denning, F. R. Hartley, and L. M. Venanzi, *J. Chem. Soc.* (*A*), 1967, 324

528.3 $C_7H_{14}O$ *cis*-1-Ethoxy-3-methylbut-1-ene L

Metal	Method	Temp	Medium	Log of equilibrium constant, remarks	Ref
Ag^+	dis	10—30	ethylene glycol	K_1 (10°) 0·78, (20°) 0·65, (30°) 0·54 gas chromatography ΔH_1 −4·7±0·2, ΔS_1 −12·9±0·6	68F

68F T. Fueno, O. Kajimoto, T. Okuyama, and J. Furukawa, *Bull. Chem. Soc. Japan*, 1968, **41**, 785

528.4 $C_7H_{14}O$ *trans*-1-Ethoxy-3-methylbut-1-ene L

Metal	Method	Temp	Medium	Log of equilibrium constant, remarks	Ref
Ag^+	dis	10—30	ethylene glycol	K_1 (10°) −0·11, (20°) −0·23, (30°) −0·30, gas chromatography ΔH_1 −3·5±0·2, ΔS_1 −12·9±0·6	68F

68F T. Fueno, O. Kajimoto, T. Okuyama, and J. Furukawa, *Bull. Chem. Soc. Japan*, 1968, **41**, 785

528.5 $C_7H_{14}O$ *cis*-1-(2′-Methylpropoxy)propene L

Metal	Method	Temp	Medium	Log of equilibrium constant, remarks	Ref
Ag^+	dis	10—30	ethylene glycol	K_1 (10°) 0·55, (20°) 0·46, (30°) 0·37 gas chromatography ΔH_1 −3·7±0·2, ΔS_1 −10·4±0·6	68F

68F T. Fueno, O. Kajimoto, T. Okuyama, and J. Furukawa, *Bull. Chem. Soc. Japan*, 1968, **41**, 785

528.6 $C_7H_{14}O$ *trans*-1-(2′-Methylpropoxy)propene L

Metal	Method	Temp	Medium	Log of equilibrium constant, remarks	Ref
Ag^+	dis	10—30	ethylene glycol	K_1 (10°) −0·11, (20°) −0·22, (30°) −0·26 gas chromatography ΔH_1 −2·8±0·2, ΔS_1 −10·6±0·6	68F

68F T. Fueno, O. Kajimoto, T. Okuyama, and J. Furukawa, *Bull. Chem. Soc. Japan*, 1968, **41**, 785

$C_7H_{14}O_3$ 2-Hydroxy-2,4-dimethylpentanoic acid (isobutylmethylglycollic acid) HL 528.7

Metal	Method	Temp	Medium	Log of equilibrium constant, remarks	Ref
H^+	qh	25	1 ($NaClO_4$)	K_1 3·80	65T
Ce^{3+}	qh	25	1 ($NaClO_4$)	K_1 2·23±0·02, K_2 1·34±0·04, K_3 1·31±0·07	65T
Dy^{3+}	qh	25	1 ($NaClO_4$)	K_1 3·18±0·03, K_2 2·46±0·05, K_3 1·75±0·05, K_4 1·18±0·05	65T
Er^{3+}	qh	25	1 ($NaClO_4$)	K_1 3·24±0·01, K_2 2·63±0·01, K_3 1·89±0·01, K_4 1·09±0·01	65T
Eu^{3+}	qh	25	1 ($NaClO_4$)	K_1 2·71±0·02, K_2 2·35±0·02, K_3 1·37±0·04, K_4 1·25±0·04	65T
Gd^{3+}	qh	25	1 ($NaClO_4$)	K_1 2·77±0·03, K_2 2·36±0·03, K_3 1·39±0·05, K_4 1·14±0·06	65T
La^{3+}	qh	25	1 ($NaClO_4$)	K_1 2·07±0·02, K_2 1·30±0·04, K_3 1·10±0·07	65T
Nd^{3+}	qh	25	1 ($NaClO_4$)	K_1 2·53±0·03, K_2 1·89±0·05, K_3 1·32±0·07	65T
Pr^{3+}	qh	25	1 ($NaClO_4$)	K_1 2·51±0·03, K_2 1·52±0·04, K_3 1·48±0·04	65T
Sm^{3+}	qh	25	1 ($NaClO_4$)	K_1 2·71±0·02, K_2 2·19±0·03, K_3 1·48±0·04, K_4 1·02±0·04	65T
Yb^{3+}	qh	25	1 ($NaClO_4$)	K_1 3·21±0·02, K_2 2·74±0·02, K_3 1·75±0·03, K_4 1·40±0·03	65T

65T H. Thun, F. Verbeek, and W. Vanderleen, *J. Inorg. Nuclear Chem.*, 1965, 27, 1813

$C_7H_{14}O_6$ Methyl α-D-glucopyranoside L 528.8

Metal	Method	Temp	Medium	Log of equilibrium constant, remarks	Ref
B^{III}	gl	25	var	$K[H_3BO_3+L \rightleftharpoons B(OH)_2(H_{-2}L)^- + H^+]$ −9·2	65L

65L S. Lormeau and M. Ahond, *Bull. Soc. chim. France*, 1965, 505

$C_7H_{14}O_6$ Methyl α-D-mannopyranoside L 528.9

Metal	Method	Temp	Medium	Log of equilibrium constant, remarks	Ref
B^{III}	oth	30	0·105	$K[B(OH)_4^- + L \rightleftharpoons B(OH)_2(H_{-2}L)^-]$ 1·7 $K[B(OH)_4^- + 2L \rightleftharpoons B(H_{-2}L)_2^-]$ 2·78 refractive index and optical rotation	64M
	gl	25	var	$K[H_3BO_3+L \rightleftharpoons B(OH)_2(H_{-2}L)^- + H^+]$ −8 $K[H_3BO_3+2L \rightleftharpoons B(H_{-2}L)_2^- + H^+]$ −7·2	65L

64M E. W. Malcolm, J. W. Green, and H. A. Swenson, *J. Chem. Soc.*, 1964, 4669
65L S. Lormeau and M. Ahond, *Bull. Soc. chim. France*, 1965, 505

528.10 $C_7H_{14}O_6$ Methyl α-D-galactopyranoside L

Metal	*Method*	*Temp*	*Medium*	*Log of equilibrium constant, remarks*	*Ref*
B^{III}	oth	30	0·105	$K[B(OH)_4^- + L \rightleftharpoons B(OH)_2(H_{-2}L)^-]$ 2·00 $K[B(OH)_4^- + 2L \rightleftharpoons B(H_{-2}L)_2^-]$ 2·60 refractive index and optical rotation	64M

64M E. W. Malcolm, J. W. Green, and H. A. Swenson, *J. Chem. Soc.*, 1964, 4669

528.11 $C_7H_{15}N$ *NN*-Diethyl-3-aminopropene (*N*-allyldiethylamine) L

Metal	*Method*	*Temp*	*Medium*	*Log of equilibrium constant, remarks*	*Ref*
Pt^{2+}	sp	30—59	2 (NaCl)	$K(PtCl_4^{2-} + HL^+ \rightleftharpoons PtCl_3HL + Cl^-)$ (30°) 2·93, (45·3°) 2·74, (59·0°) 2·59 ΔH −5·6±0·3, ΔS −5·0±0·6	67D
	sp	24·5	2(KBr)	$K(PtBr_4^{2-} + HL^+ \rightleftharpoons PtBr_3HL + Br^-)$ 2·10	67Da

67D R. G. Denning, F. R. Hartley, and L. M. Venanzi, *J. Chem. Soc.* (*A*), 1967, 324
67Da R. G. Denning and L. M. Venanzi, *J. Chem. Soc.* (*A*), 1967, 336

530 $C_7H_{16}N_2$ 2-2′-(Aminoethyl)pipendine L

Errata. 1964, p. 541, Table 530. Under Cu^{2+}, fourth line should *read* ΔH_1 8, ΔH_2 8, ΔS_1 12, ΔS_2 0. Under Ni^{2+} fourth line should *read* ΔH_1 1, ΔH_2 −4, ΔS_1 21, ΔS_2 5.

531.1 $C_7H_{18}N_2$ *NN*-Diethyl-*N*′-methyl-1,2-diaminoethane L

Metal	*Method*	*Temp*	*Medium*	*Log of equilibrium constant, remarks*	*Ref*
H^+	gl	25	→ 0 I ($NaClO_4$)	K_1 9·92 K_2 $6{\cdot}478+(1{\cdot}018\sqrt{I})/(1+0{\cdot}853\sqrt{I})+0{\cdot}282\,I$ K_1K_2 $16{\cdot}395+(1{\cdot}018\sqrt{I})/(1+0{\cdot}792\sqrt{I})+0{\cdot}667\,I$	67N
Cu^{2+}	gl	25	→ 0 I ($NaClO_4$)	K_1 7·64, K_2 4·84 $K(Cu^{2+} + H_2L^{2+} \rightleftharpoons CuL^{2+} + 2H^+)$ $8{\cdot}751+(1{\cdot}018\sqrt{I})/(1+0{\cdot}905\sqrt{I})+0{\cdot}252\,I$	67N

67N R. Nasanen, M. Koskinen, M. L. Alatalo, L. Adler, and S. Koskela, *Suomen Kem.*, 1967, **B40**, 124

$C_7H_4O_2Cl_2$ 2-Hydroxy-3,5-dichlorobenzaldehyde (3,5-dichlorosalicylaldehyde) HL 532.1

Metal	Method	Temp	Medium	Log of equilibrium constant, remarks	Ref
$^{+}$	pH	20	0·3 $NaClO_4$ 50% dioxan	K_1 6·54	63C
u^{2+}	pH	20	0·3 $NaClO_4$ 50% dioxan	K_1 4·6, K_2 3·0	63C
i^{2+}	pH	20	0·3 $NaClO_4$ 50% dioxan	K_1 3·0	63C

63C K. Clarke, R. A. Cowen, G. W. Gray, and E. H. Osborne, *J. Chem. Soc.*, 1963, 245

$C_7H_4O_2Br_2$ 2-Hydroxy-3,5-dibromobenzaldehyde (3,5-dibromosalicylaldehyde) HL 532.2

Metal	Method	Temp	Medium	Log of equilibrium constant, remarks	Ref
H^{+}	pH	20	0·3 $NaClO_4$ 50% dioxan	K_1 6·54	63C
Cu^{2+}	pH	20	0·3 $NaClO_4$ 50% dioxan	K_1 4·6, K_2 3·6	63C

63C K. Clarke, R. A. Cowen, G. W. Gray, and E. H. Osborne, *J. Chem. Soc.*, 1963, 245

$C_7H_4O_2I_2$ 2-Hydroxy-3,5-di-iodobenzaldehyde (3,5-di-iodosalicylaldehyde) HL 532.3

Metal	Method	Temp	Medium	Log of equilibrium constant, remarks	Ref
H^{+}	pH	20	0·3 $NaClO_4$ 50% dioxan	K_1 6·89	63C
Cu^{2+}	pH	20	0·3 $NaClO_4$ 50% dioxan	K_1 5·0, K_2 3·9	63C

63C K. Clarke, R. A. Cowen, G. W. Gray, and E. H. Osborne, *J. Chem. Soc.*, 1963, 245

$C_7H_4O_6N_2$ 2-Hydroxy-3,5-dinitrobenzaldehyde (3,5-dinitrosalicylaldehyde) HL 533

Metal	Method	Temp	Medium	Log of equilibrium constant, remarks	Ref
H^{+}	pH	20	0·3 $NaClO_4$ 50% dioxan	K_1 2·05	63C
	sp	25	0·10 ($NaClO_4$)	K_1 2·09	66P
Cu^{2+}	pH	20	0·3 $NaClO_4$ 50% dioxan	K_1 2·1, K_2 1·7	63C
	sp	25	0·10 ($NaClO_4$)	K_1 1·68±0·02	66P
La^{3+}	sp	25	0·10 ($NaClO_4$)	K_1 1·84±0·02	66P
Y^{3+}	sp	25	0·10 ($NaClO_4$)	K_1 1·75±0·02	66P
Zn^{2+}	sp	25	0·10 ($NaClO_4$)	K_1 0·75±0·1	66P

63C K. Clarke, R. A. Cowen, G. W. Gray, and E. H. Osborne, *J. Chem. Soc.*, 1963, 245
66P C. Postmus jun., L. B. Magnusson, and C. A. Craig, *Inorg. Chem.*, 1966, 5, 1154

533.1 $C_7H_4O_7N_2$ 2-Hydroxy-3,5-dinitrobenzoic acid (3,5-dinitrosalicylic acid) H_2L

Metal	Method	Temp	Medium	Log of equilibrium constant, remarks	Ref
H^+	gl	20	0·1 (KNO_3)	K_1 7·40, K_2 2·25	67B
UO_2^{2+}	gl	20	0·1 (KNO_3)	K_1 7·55, β_2 13·05	67B

67B M. Bartusek, *Coll. Czech. Chem. Comm.*, 1967, 32, 116

533.2 C_7H_5ON 2-Cyanophenol (salicylnitrile) HL

Metal	Method	Temp	Medium	Log of equilibrium constant, remarks	Ref
H^+	gl	30	0·1 $NaClO_4$ 75% dioxan	K_1 9·13, K_2 2·52	64J
Cu^{2+}	gl	30	0·1 $NaClO_4$ 75% dioxan	K_1 5·94	64J

64J K. E. Jabalpurwala, K. A. Venkatachalam, and M. B. Kabadi, *J. Inorg. Nuclear Chem.*, 1964, **26**, 1011, 1027

536.1 $C_7H_5O_2Cl$ 3-Chlorobenzoic acid HL

Metal	Method	Temp	Medium	Log of equilibrium constant, remarks	Ref
H^+	gl	25	0·1 ($NaClO_4$) 50% dioxan	K_1 5·24 ± 0·02	68E, 68G
Cu^{2+}	gl	25	0·1 ($NaClO_4$) 50% dioxan	K_1 3·03	68E
	gl	25	0·1 ($NaClO_4$) 50% dioxan	$K(CuA^{2+} + L^- \rightleftharpoons CuAL^+)$ 3·20 where A = 2,2′-bipyridyl	68G
Zn^{2+}	gl	25	0·1 ($NaClO_4$) 50% dioxan	K_1 1·99 $K(ZnA^{2+} + L^- \rightleftharpoons ZnAL^+)$ 2·10 where A = 2,2′-bipyridyl	68G

68E H. Erlenmeyer, R. Griesser, B. Prijs, and H. Sigel, *Helv. Chim. Acta*, 1968, **51**, 339
68G R. Griesser, B. Prijs, and H. Sigel, *Inorg. Nuclear Chem. Letters*, 1968, **4**, 443

$C_7H_5O_2Cl$ 2-Hydroxy-3-chlorobenzaldehyde (3-chlorosalicylaldehyde) HL 537

Metal	Method	Temp	Medium	Log of equilibrium constant, remarks	Ref
H^+	pH	20	0·3 NaClO₄ 50% dioxan	K_1 7·72	63C
	sp	25	0·10 (NaClO₄)	K_1 6·61	66P
Ca^{2+}	sp	25	0·10 (NaClO₄)	K_1 0·7±0·1	66P
Cu^{2+}	pH	20	0·3 NaClO₄ 50% dioxan	K_1 5·6, K_2 3·8	63C
	sp	25	0·10 (NaClO₄)	K_1 4·74±0·02	66P
La^{3+}	sp	25	0·10 (NaClO₄)	K_1 3·16±0·02	66P
Ni^{2+}	pH	20	0·3 NaClO₄ 50% dioxan	K_1 3·8	63C
Y^{3+}	sp	25	0·10 (NaClO₄)	K_1 3·77±0·02	66P
Zn^{2+}	sp	25	0·10 (NaClO₄)	K_1 2·39±0·02	66P

63C K. Clarke, R. A. Cowen, G. W. Gray, and E. H. Osborne, *J. Chem. Soc.*, 1963, 245
66P C. Postmus jun., L. B. Magnusson, and C. A. Craig, *Inorg. Chem.*, 1966, **5**, 1154

$C_7H_5O_2Cl$ 2-Hydroxy-4-chlorobenzaldehyde (4-chlorosalicylaldehyde) HL 538

Metal	Method	Temp	Medium	Log of equilibrium constant, remarks	Ref
H^+	pH	20	0·3 NaClO₄ 50% dioxan	K_1 9·42	63C
	sp	25	0·10 (NaClO₄)	K_1 7·18	66P
Ca^{2+}	sp	25	0·10 (NaClO₄)	K_1 1·1±0·1	66P
Cu^{2+}	pH	20	0·3 NaClO₄ 50% dioxan	K_1 6·8, K_2 5·5	63C
	sp	25	0·10 (NaClO₄)	K_1 5·11±0·02	66P
La^{3+}	sp	25	0·10 (NaClO₄)	K_1 3·38±0·02	66P
Mg^{2+}	pH	20	0·3 NaClO₄ 50% dioxan	K_1 3	63C
Y^{3+}	sp	25	0·10 (NaClO₄)	K_1 4·09±0·02	66P

63C K. Clarke, R. A. Cowen, G. W. Gray, and E. H. Osborne, *J. Chem. Soc.*, 1963, 245
66P C. Postmus jun., L. B. Magnusson, and C. A. Craig, *Inorg. Chem.*, 1966, **5**, 1154

$C_7H_5O_2Cl$ 2-Hydroxy-5-chlorobenzaldehyde (5-chlorosalicylaldehyde) HL 539

Metal	Method	Temp	Medium	Log of equilibrium constant, remarks	Ref
H^+	pH	20	0·3 NaClO₄ 50% dioxan	K_1 8·47	63C
	sp	25	0·10 (NaClO₄)	K_1 7·41	66P
Ca^{2+}	sp	25	0·10 (NaClO₄)	K_1 0·9±0·1	66P
Cu^{2+}	pH	20	0·3 NaClO₄ 50% dioxan	K_1 5·7, K_2 4·6	63C
	sp	25	0·10 (NaClO₄)	K_1 4·96±0·02	66P
La^{3+}	sp	25	0·10 (NaClO₄)	K_1 3·20±0·02	66P
Y^{3+}	sp	25	0·10 (NaClO₄)	K_1 3·94±0·02	66P

63C K. Clarke, R. A. Cowen, G. W. Gray, and E. H. Osborne, *J. Chem. Soc.*, 1963, 245
66P C. Postmus jun., L. B. Magnusson, and C. A. Craig, *Inorg. Chem.*, 1966, **5**, 1154

539.1 $C_7H_5O_2Cl$ 2-Hydroxy-6-chlorobenzaldehyde (6-chlorosalicylaldehyde) HL

Metal	*Method*	*Temp*	*Medium*	*Log of equilibrium constant, remarks*	*Ref*
H^+	pH	20	0·3 $NaClO_4$ 50% dioxan	K_1 8·99	63C
	sp	25	0·10 ($NaClO_4$)	K_1 8·26	66P
Ca^{2+}	sp	25	0·10 ($NaClO_4$)	K_1 1·74±0·02	66P
Cu^{2+}	pH	20	0·3 $NaClO_4$ 50% dioxan	K_1 6·3, K_2 5·1	63C
	sp	25	0·10 ($NaClO_4$)	K_1 5·88±0·02	66P
La^{3+}	sp	25	0·10 ($NaClO_4$)	K_1 4·08±0·02	66P
Mg^{2+}	pH	20	0·3 $NaClO_4$ 50% dioxan	K_1 2	63C
Ni^{2+}	pH	20	0·3 $NaClO_4$ 50% dioxan	K_1 4·5	63C
Y^{3+}	sp	25	0·10 ($NaClO_4$)	K_1 4·87±0·02	66P

63C K. Clarke, R. A. Cowen, G. W. Gray, and E. H. Osborne, *J. Chem. Soc.*, 1963, 245
66P C. Postmus jun., L. B. Magnusson, and C. A. Craig, *Inorg. Chem.*, 1966, **5**, 1154

539.2 $C_7H_5O_2Cl$ 5-Chlorotropolone HL

Metal	*Method*	*Temp*	*Medium*	*Log of equilibrium constant, remarks*	*Ref*
H^+	sp	25	2 $NaClO_4$	K_1 5·62	62O
Fe^{3+}	sp	25	2 $NaClO_4$	K_1 9·92	62O

62O Y. Oka, M. Umehara, and T. Nozoe, *Nippon Kagaku Zasshi*, 1962, **83**, 1197

540 $C_7H_5O_2Br$ 2-Hydroxy-5-bromobenzaldehyde (5-bromosalicylaldehyde) HL

Metal	*Method*	*Temp*	*Medium*	*Log of equilibrium constant, remarks*	*Ref*
H^+	pH	20	0·3 $NaClO_4$ 50% dioxan	K_1 8·43	63C
Cu^{2+}	pH	20	0·3 $NaClO_4$ 50% dioxan	K_1 5·8, K_2 5·0	63C
Mg^{2+}	pH	20	0·3 $NaClO_4$ 50% dioxan	K_1 2	63C

63C K. Clarke, R. A. Cowen, G. W. Gray, and E. H. Osborne, *J. Chem. Soc.*, 1963, 245

541 $C_7H_5O_2Br$ 3-Bromotropolone HL

Metal	*Method*	*Temp*	*Medium*	*Log of equilibrium constant, remarks*	*Ref*
H^+	sp	25	2 $NaClO_4$	K_1 5·14	65O
Fe^{3+}	sp	25	2 $NaClO_4$	K_1 9·25	65O

65O Y. Oka and M. Yanai, *Nippon Kagaku Zasshi*, 1965, **86**, 929

$C_7H_5O_2Br$ 4-Bromotropolone HL 541.1

Metal	Method	Temp	Medium	Log of equilibrium constant, remarks	Ref
H^+	sp	25	2 $NaClO_4$	K_1 5·72	64O
Fe^{3+}	sp	25	2 $NaClO_4$	K_1 10·04	64O

64O Y. Oka, M. Yanai, and C. Suzuki, *Nippon Kagaku Zasshi*, 1964, **85**, 873

$C_7H_5O_2Br$ 5-Bromotropolone HL 541.2

Metal	Method	Temp	Medium	Log of equilibrium constant, remarks	Ref
H^+	sp	25	2 $NaClO_4$	K_1 5·54	62O
Fe^{3+}	sp	25	2 $NaClO_4$	K_1 9·74	62O

62O Y. Oka, M. Umehara, and T. Nozoe, *Nippon Kagaku Zasshi*, 1962, **83**, 1197

$C_7H_5O_3Cl$ 2-Hydroxy-5-chlorobenzoic acid (5-chlorosalicylic acid H_2L) 542.1

Metal	Method	Temp	Medium	Log of equilibrium constant, remarks	Ref
H^+	sp	25	0 corr	K_1 12·951 ± 0·004, K_2 2·652 ± 0·004	63E
Fe^{3+}	sp	25	0 corr	K_1 16·842 ± 0·004	63E

63E Z. L. Ernst and J. Menashi, *Trans. Faraday Soc.*, 1963, **59**, 2838

$C_7H_5O_3Br$ 2-Hydroxy-5-bromobenzoic acid (5-bromosalicylic acid) H_2L 542.2

Metal	Method	Temp	Medium	Log of equilibrium constant, remarks	Ref
H^+	sp	25	0 corr	K_1 12·835 ± 0·009, K_2 2·658 ± 0·002	63E
Fe^{3+}	sp	25	0 corr	K_1 16·762 ± 0·002	63E

63E Z. L. Ernst and J. Menashi, *Trans. Faraday Soc.*, 1963, **59**, 2838

$C_7H_5O_4N$ Pyridine-2,3-dicarboxylic acid H_2L 543

Metal	Method	Temp	Medium	Log of equilibrium constant, remarks	Ref
Fe^{2+}	sp	13—17	0·1	K_1 (0·02 KCN) 3·2, β_2 (no KCN) 3·8	63M

63M I. Morimoto and T. Sato, *Bull. Chem. Soc. Japan*, 1963, **36**, 605

544.1 $C_7H_5O_4N$ Pyridine-2,5-dicarboxylic acid H_2L

Metal	Method	Temp	Medium	Log of equilibrium constant, remarks	Ref
Fe^{2+}	sp	13—17	0·1	K_1 (0·02 KCN) 3·4, β_2 (no KCN) 3·8	63M

63M I. Morimoto and T. Sato, *Bull. Chem. Soc. Japan*, 1963, **36**, 605

545 $C_7H_5O_4N$ Pyridine-2,6-dicarboxylic acid (dipicolinic acid) H_2L

Metal	Method	Temp	Medium	Log of equilibrium constant, remarks	Ref
H^+	gl	20	0·1 ($NaNO_3$)	K_1 4·68, K_2 2·10	64T
	gl	25	0·10	K_1 4·67, K_2 2·24	66B
	sp	?	$\sim 5\times10^{-4}$	K_1 4·81, K_2 2·36, K_3 < 0·5	67F
	gl	25	0·5 $NaClO_4$	K_1 4·532 ± 0·004, K_2 2·092 ± 0·006	68N
Al^{3+}	gl	25	0·5 $NaClO_4$	K_1 4·87, β_2 8·32	68N
	sp	25	0·5 $NaClO_4$	K_1 4·85	68N
Ba^{2+}	gl	25	0·10	K_1 3·43	66B
	ix	25	0·10	K_2 0·5	66B
Ca^{2+}	gl	25	0·10	K_1 4·60, K_2 2·6	66B
	ix	25	0·10	K_2 2·98	66B
Cu^{2+}	sp	?	0·2	K_1 8·9	67F
	pol	?	0·07	β_2 17·1	67F
				$K[Cu^{2+} + 2HL^- \rightleftharpoons Cu(HL)_2]$ 9·4	
Fe^{2+}	sp	13—17	0·1	K_1 (?) (0·02 KCN) 3·3, β_2 (?) (no KCN) 3·6	63M
Fe^{3+}	gl	20	0·1 ($NaClO_4$)	β_2 16·74 ± 0·01	64T
Mg^{2+}	gl	25	0·10	K_1 2·32	66B
	ix	25	0·10	K_2 0·7	66B
Sr^{2+}	gl	25	0·10	K_1 3·80, K_2 1·7	66B
	ix	25	0·10	K_2 1·96	66B

63M I. Morimoto and T. Sato, *Bull. Chem. Soc. Japan*, 1963, **36**, 605
64T C. F. Timberlake, *J. Chem. Soc.*, 1964, 1229
66B W. E. Bennett and D. O. Skovlin, *J. Inorg. Nuclear Chem.*, 1966, **28**, 591
67F J. Faucherre, C. Petitfaux, and B. Charlier, *Bull. Soc. chim. France*, 1967, 1091
68N A. Napoli, *Talanta*, 1968, **15**, 189

545.1 $C_7H_5O_4N$ Pyridine-3,4-dicarboxylic acid H_2L

Metal	Method	Temp	Medium	Log of equilibrium constant, remarks	Ref
Fe^{2+}	sp	13—17	0·1	β_2 (0·02 KCN) 5·0, β_3 (no KCN) 4·9	63M

63M I. Morimoto and T. Sato, *Bull. Chem. Soc. Japan*, 1963, **36**, 605

$C_7H_5O_4N$ 2-Nitrobenzoic acid HL 545.2

Metal	Method	Temp	Medium	Log of equilibrium constant, remarks	Ref
H^+	?	25	?	K_1 2·17	67V
Ag^+	Ag-AgCl	30	1 ($NaClO_4$)	K_1 2·0, K_2 1·4	67V

67V D. G. Vartak and R. S. Shetiya, *J. Inorg. Nuclear Chem.*, 1967, **29**, 1261

$C_7H_5O_4N$ 3-Nitrobenzoic acid HL 545.3

Metal	Method	Temp	Medium	Log of equilibrium constant, remarks	Ref
H^+	?	25	?	K_1 3·49	67V
Ag^+	Ag-AgCl	30	1 ($NaClO_4$)	K_1 2·4, K_2 1·5	67V

67V D. G. Vartak and R. S. Shetiya, *J. Inorg. Nuclear Chem.*, 1967, **29**, 1261

$C_7H_5O_4N$ 4-Nitrobenzoic acid HL 545.4

Metal	Method	Temp	Medium	Log of equilibrium constant, remarks	Ref
H^+	?	25	?	K_1 3·43	67V
	gl	25	0·1 ($NaClO_4$) 50% dioxan	K_1 4·64±0·01	68E, 68G
Ag^+	Ag-AgCl	30	1 ($NaClO_4$)	K_1 2·1, K_2 3·0	67V
Cu^{2+}	gl	25	0·1 ($NaClO_4$) 50% dioxan	K_1 2·76	68E
	gl	25	0·1 ($NaClO_4$) 50% dioxan	$K(CuA^{2+}+L^- \rightleftharpoons CuAL^+)$ 2·88 where A = 2,2′-bipyridyl	68G
Zn^{2+}	gl	25	0·1 ($NaClO_4$) 50% dioxan	K_1 1·83; $K(ZnA^{2+}+L^- \rightleftharpoons ZnAL^+)$ 1·86 where A = 2,2′-bipyridyl	68G

67V D. G. Vartak and R. S. Shetiya, *J. Inorg. Nuclear Chem.*, 1967, **29**, 1261
68E H. Erlenmeyer, R. Griesser, B. Prijs, and H. Sigel, *Helv. Chim. Acta*, 1968, **51**, 339
68G R. Griesser, B. Prijs, and H. Sigel, *Inorg. Nuclear Chem. Letters*, 1968, **4**, 443

$C_7H_5O_4N$ 2-Hydroxy-3-nitrobenzaldehyde (3-nitrosalicylaldehyde) HL 546

Metal	Method	Temp	Medium	Log of equilibrium constant, remarks	Ref
H^+	sp	25	0·10 ($NaClO_4$)	K_1 5·21	66P
Ca^{2+}	sp	25	0·10 ($NaClO_4$)	K_1 1·0±0·1	66P
Cu^{2+}	sp	25	0·10 ($NaClO_4$)	K_1 3·61±0·02	66P
La^{3+}	sp	25	0·10 ($NaClO_4$)	K_1 3·03±0·02	66P
Y^{3+}	sp	25	0·10 ($NaClO_4$)	K_1 3·27±0·02	66P

66P C. Postmus jun., L. B. Magnusson, and C. A. Craig, *Inorg. Chem.*, 1966, **5**, 1154

547 $C_7H_5O_4N$ 2-Hydroxy-4-nitrobenzaldehyde (4-nitrosalicylaldehyde) HL

Metal	*Method*	*Temp*	*Medium*	*Log of equilibrium constant, remarks*	*Ref*
H^+	pH	20	0·3 $NaClO_4$ 50% dioxan	K_1 7·33	63C
Cu^{2+}	pH	20	0·3 $NaClO_4$ 50% dioxan	K_1 4·8, K_2 3·7	63C

63C K. Clarke, R. A. Cowen, G. W. Gray, and E. H. Osborne, *J. Chem. Soc.*, 1963, 245

548 $C_7H_5O_4N$ 2-Hydroxy-5-nitrobenzaldehyde (5-nitrosalicylaldehyde) HL

Metal	*Method*	*Temp*	*Medium*	*Log of equilibrium constant, remarks*	*Ref*
H^+	pH	20	0·3 $NaClO_4$ 50% dioxan	K_1 5·96	63C
	sp	25	0·10 ($NaClO_4$)	K_1 5·32	66P
Ca^{2+}	sp	25	0·10 ($NaClO_4$)	K_1 0·8±0·1	66P
Cu^{2+}	pH	20	0·2 $NaClO_4$ 50% dioxan	K_1 4·2, K_2 3·7	63C
	sp	25	0·10 ($NaClO_4$)	K_1 3·83±0·02	66P
La^{3+}	sp	25	0·10 ($NaClO_4$)	K_1 2·73±0·02	66P
Y^{3+}	sp	25	0·10 ($NaClO_4$)	K_1 3·17±0·02	66P
Zn^{2+}	sp	25	0·10 ($NaClO_4$)	K_1 2·01±0·02	66P

63C K. Clarke, R. A. Cowen, G. W. Gray, and E. H. Osborne, *J. Chem. Soc.*, 1963, 245
66P C. Postmus jun., L. B. Magnusson, and C. A. Craig, *Inorg. Chem.*, 1966, **5**, 1154

548.1 $C_7H_5O_4N$ 5-Nitrotropolone HL

Metal	*Method*	*Temp*	*Medium*	*Log of equilibrium constant, remarks*	*Ref*
H^+	sp	25	2 $NaClO_4$	K_1 2·64	62O
	sp	25	0·10	K_1 3·21	68O
Cu^{2+}	sp	25	0·10	K_1 5·53	68O
Fe^{3+}	sp	25	2 $NaClO_4$	K_1 6·57	62O

62O Y. Oka, M. Umehara, and T. Mozoe, *Nippon Kagaku Zasshi*, 1962, **83**, 1197
68O Y. Oka and Y. Hirai, *Nippon Kagaku Zasshi*, 1968, **89**, 589

$C_7H_5O_5N$ 4-Hydroxypyridine-2,6-dicarboxylic acid H_3L 549

Metal	Method	Temp	Medium	Log of equilibrium constant, remarks	Ref
H^+	gl	22	0·1 ($NaClO_4$)	K_1 10·85, K_2 3·18, K_3 1·9	64B
Ba^{2+}	gl	22	0·1 ($NaClO_4$)	K_1 3·98	64B
Ca^{2+}	gl	22	0·1 ($NaClO_4$)	K_1 5·40	64B
Mg^{2+}	gl	22	0·1 ($NaClO_4$)	K_1 3·68	64B
Sr^{2+}	gl	22	0·1 ($NaClO_4$)	K_1 4·34	64B

64B E. Blasius and B. Brozio, *Ber. Bunsengesellschaft Phys. Chem.*, 1964, **68**, 52

$C_7H_5O_5N$ 2-Hydroxy-3-nitrobenzoric acid (3-nitrosalicylic acid) H_2L 549.1

Metal	Method	Temp	Medium	Log of equilibrium constant, remarks	Ref
H^+	sp	25	0 corr	K_1 10·328±0·004	63E
	con	25	0 corr	K_2 1·873±0·010	63E
	?	?	?	K_1 10·33, K_2 1·90	66R
Fe^{3+}	sp	25	0 corr	K_1 14·193±0·003	63E
UO_2^{2+}	gl	28	0·1 KNO_3	K_1 8·57	66R

63E Z. L. Ernst and J. Menashi, *Trans. Faraday Soc.*, 1963, **59**, 2838
66R J. M. Rao and U. V. Seshaiah, *Bull. Chem. Soc. Japan*, 1966, **39**, 2668

$C_7H_5O_5N$ 2-Hydroxy-5-nitrobenzoic acid (5-nitrosalicylic acid H_2L) 549.2

Metal	Method	Temp	Medium	Log of equilibrium constant, remarks	Ref
H^+	sp	25	0 corr	K_1 10·339±0·004	63E
	con	25	0 corr	K_2 2·121±0·003	63E
Fe^{3+}	sp	25	0 corr	K_1 14·339±0·002	63E

63E Z. L. Ernst and J. Menashi, *Trans. Faraday Soc.*, 1963, **59**, 2838

$C_7H_5NS_2$ 2-Mercaptobenzo-1,3-thiazole HL 549.3

Metal	Method	Temp	Medium	Log of equilibrium constant, remarks	Ref
Zn^{2+}	dis	20	0·1 ($NaClO_4$)	K_1 3·25, K_2 2·49	68N

68N O. Navratil and J. Liska, *Coll. Czech. Chem. Comm.*, 1968, 33, 987

550.1 $C_7H_6O_2S$ 2-Mercaptobenzoic acid (thiosalicylic acid) H_2L

Metal	Method	Temp	Medium	Log of equilibrium constant, remarks	Ref
H^+	gl	25	0·1 50% dioxan	K_1 9·96±0·04, K_2 4·92±0·04	64L
	pol	?	?	K_1 9·52	67K
Cd^{2+}	pol	30	0·2 50% ethanol	K_1 7·20, ammonia buffer	67Ka
	gl	30	0·2 50% ethanol	K_1 7·85, K_2 8·10 (?)	67Ka
Co^{2+}	sp	30	40% ethanol	β_3 11·76	66K
	gl	30—50	0·15 45% ethanol	K_1 (30°) 6·03, (40°) 6·20, (50°) 6·35, K_2 (30°) 4·44, (40°) 4·55, (50°) 4·70	68R
				$\Delta H_1°$ (30—40°) 7·3, (40—50°) 7·1, $\Delta S_1°$ (30°) 52, (40°) 51, (50°) 51, $\Delta H_2°$ (30—40°) 5·0, (40—50°) 6·8, $\Delta S_2°$ (30°) 37, (40°) 40, (50°) 42	
Fe^{2+}	gl	30—50	0·15 45% ethanol	K_1 (30°) 5·45, (40°) 5·57, (50°) 5·72, K_2 (30°) 4·41, (40°) 4·48, (50°) 4·62	68R
				$\Delta H_1°$ (30—40°) 5·1, (40—50°) 6·8, $\Delta S_1°$ (30°) 42, (40°) 44, (50°) 47, $\Delta H_2°$ (30—40°) 3·1, (40—50°) 6·5, $\Delta S_2°$ (30°) 31, (40°) 36, (50°) 41	
Mn^{2+}	gl	30—50	0·15 45% ethanol	K_1 (30°) 5·04, (40°) 5·15, (50°) 5·28, K_2 (30°) 4·05, (40°) 4·18, (50°) 4·28	68R
				$\Delta H_1°$ (30—40°) 4·4, (40—50°) 6·1, $\Delta S_1°$ (30°) 38, (40°) 40, (50°) 43, $\Delta H_2°$ (30—40°) 5·7, (40—50°) 4·4, $\Delta S_2°$ (30°) 37, (40°) 35, (50°) 33	
Ni^{2+}	gl	25	0·1 50% dioxan	K_1 8·1, K_2 5·25	64L
	gl	30—50	0·15 45% ethanol	K_1 (30°) 7·08, (40°) 7·34, (50°) 7·64, K_2 (30°) 4·46, (40°) 4·64, (50°) 4·80	68R
				$\Delta H_1°$ (30—40°) 11·5, (40—50°) 8·0, $\Delta S_1°$ (30°) 70, (40°) 76, (50°) 78, $\Delta H_2°$ (30—40°) 8·0, (40—50°) 7·5, $\Delta S_2°$ (30°) 47, (40°) 46, (50°) 45	
Pb^{2+}	pol	30	0·2 KNO_3 50% ethanol	K_1 2·29(?), acetate buffer	67K
Zn^{2+}	gl	30	0·001 $NaClO_4$ 50% ethanol	K_1 9·1, K_2 11·2 (?)	67K
	gl	30—50	0·15 45% ethanol	K_1 (30°) 8·45, (40°) 8·60, (50°) 8·84, K_2 (30°) 5·99, (40°) 6·20, (50°) 6·38	68R
				$\Delta H_1°$ (30—40°) 6·5, (40—50°) 11·1, $\Delta S_1°$ (30°) 60, (40°) 68, (50°) 75, $\Delta H_2°$ (30—40°) 9·1, (40—50°) 8·4, $\Delta S_2°$ (30°) 58, (40°) 56, (50°) 55	

64L S. V. Larionov, V. M. Shul'man, and L. A. Podolskaya, *Russ. J. Inorg. Chem.*, 1964, **9**, 1264 (2333)
66K A. N. Kumar, H. L. Nigam, and M. Katyal, *J. prakt. Chem.*, 1966, 33, 160
67K A. N. Kumar and H. L. Nigam, *Indian J. Chem.*, 1967, 5, 48
67Ka A. N. Kumar and H. L. Nigam, *Proc. Indian Acad Sci.*, 1967, **65A**, 119
68R R. Ramamani and S. Shanmuganathan, *Current Sci.*, 1968, 37, 39

$C_7H_6O_4N_2$ 4-Aminopyridine-2,6-dicarboxylic acid (4-aminodipicolinic acid) H_2L 550.2

Metal	Method	Temp	Medium	Log of equilibrium constant, remarks	Ref
H^+	gl	22	0·1 ($NaClO_4$)	K_1 9·19, K_2 1·8	64B
	gl	20	0·1 (KNO_3)	K_1 9·05, K_2 2·29	65A
Ba^{2+}	gl	22	0·1 ($NaClO_4$)	K_1 3·76	64B
	gl	20	0·1 (KNO_3)	K_1 3·68	65A
Ca^{2+}	gl	22	0·1 ($NaClO_4$)	K_1 5·28	64B
	gl	20	0·1 (KNO_3)	K_1 5·18	65A
Cd^{2+}	gl	20	0·1 (KNO_3)	K_1 7·85, K_2 5·78	65A
Co^{2+}	gl	20	0·1 (KNO_3)	K_1 7·36, K_2 6·97	65A
Cu^{2+}	gl	20	0·1 (KNO_3)	K_1 11·36, K_2 8·99	65A
Fe^{2+}	gl	20	0·1 (KNO_3)	K_1 6·68, K_2 5·85	65A
Fe^{3+}	redox	20	0·1 (KNO_3)	K_1 13·15, K_2 9·74	65A
Hg^{2+}	Hg	20	0·1 (KNO_3)	K_1 15·80, K_2 8·69	65A
La^{3+}	sol, gl	20	0·1 (KNO_3)	K_1 9·6, K_2 6·67, K_3 4·50	65A
Mg^{2+}	gl	22	0·1 ($NaClO_4$)	K_1 2·88	64B
	gl	20	0·1 (KNO_3)	K_1 2·91	65A
Mn^{2+}	gl	20	0·1 (KNO_3)	K_1 5·89, K_2 4·81	65A
Ni^{2+}	gl	20	0·1 (KNO_3)	K_1 8·18, K_2 7·34	65A
Sr^{2+}	gl	22	0·1 ($NaClO_4$)	K_1 4·22	64B
	gl	20	0·1 (KNO_3)	K_1 4·17	65A
Zn^{2+}	gl	20	0·1 (KNO_3)	K_1 8·15, K_2 7·90	65A

64B E. Blasius and B. Brozio, *Ber. Bunsengesellschaft Phys. Chem.* 1964, **68**, 52
65A G. Anderegg and E. Bottari, *Helv. Chim. Acta*, 1965, **48**, 887

$C_7H_6O_4N_2$ 2-Hydroxy-5-nitrobenzaldehyde oxime (5-nitrosalicylaldoxime) HL 550.3

Metal	Method	Temp	Medium	Log of equilibrium constant, remarks	Ref
H^+	gl	20	0·1 ($NaClO_4$) 75% dioxan	K_1 8·72±0·05	65B
Co^{2+}	gl	20	0·1 ($NaClO_4$) 75% dioxan	K_1 6·3±0·2, K_2 6·6±0·2 (?)	65B
Fe^{2+}	gl	20	0·1 ($NaClO_4$) 75% dioxan	K_1 6·9±0·1, K_2 5·6±0·1	65B
Mn^{2+}	gl	20	0·1 ($NaClO_4$) 75% dioxan	K_1 4·42±0·05, K_2 3·90±0·05	65B
Ni^{2+}	gl	20	0·1 ($NaClO_4$) 75% dioxan	K_1 6·5±0·1, K_2 7·3±0·1 (?)	65B
Zn^{2+}	gl	20	0·1 ($NaClO_4$) 75% dioxan	K_1 5·3±0·1, K_2 5·6±0·1 (?)	65B

65B K. Burger and I. Egyed, *J. Inorg. Nuclear Chem.*, 1965, **27**, 2361

551.1 $C_7H_6O_5S$ Tropolone-5-sulphonic acid H_2L

Metal	Method	Temp	Medium	Log of equilibrium constant, remarks	Ref
H^+	sp	25	2 $NaClO_4$	K_1 4·68	62O, 63O
	sp	25	0·10 $NaClO_4$	K_1 4·92	68O
Cu^{2+}	sp	25	0·10 $NaClO_4$	K_1 6·90	68O
Fe^{3+}	sp	25	2 $NaClO_4$	K_1 8·71, K_2 7·43, K_3 5·72	62O
Th^{4+}	sp	25	2 $NaClO_4$	K_1 7·95, K_2 6·14	63O

62O Y. Oka, M. Umehara, and T. Nozoe, *Nippon Kagaku Zasshi*, 1962, **83**, 703
63O Y. Oka and M. Umehara, *Nippon Kagaku Zasshi*, 1963, **84**, 928
68O Y. Oka and Y. Kirai, *Nippon Kagaku Zasshi*, 1968, **89**, 589

552 $C_7H_6O_6S$ 3-Carboxy-4-hydroxybenzenesulphonic acid (5-sulphosalicylic acid) H_3L

Metal	Method	Temp	Medium	Log of equilibrium constant, remarks	Ref
H^+	gl	25	0·1 ($NaClO_4$)	K_1 11·74, K_2 2·49	59B
	gl	22	50% ethanol	K_1 12·88, K_2 3·41, K_3 1·60	61A
	sp	25	0·1—0·2	K_1 (0·1) 11·97, (0·16—0·19) 12·23	61N
	gl	25	I ($NaClO_4$)	K_1 $12{\cdot}53-(3{\cdot}054\sqrt{I})/(1+1{\cdot}31\sqrt{I})+0{\cdot}262\,I$, K_2 $2{\cdot}84-(2{\cdot}04\sqrt{I})/(1+2{\cdot}1\sqrt{I})+0{\cdot}17\,I$	61Na, 62I
	gl	18—22	0·1	K_1 11·95, K_2 2·62	63S
	gl	25	1 $NaClO_4$	K_1 11·47, K_2 2·30	64M
	gl	25	0·1 (KNO_3)	K_1 11·70±0·05, K_2 2·50±0·02	66L
	gl	25—35	0·10 (KNO_3)	K_1 (25°) 11·67±0·03, (35°) 11·59±0·03, K_2 (25°) 2·67±0·04, (35°) 2·36±0·02	66M
	?	?	?	K_1 11·75, K_2 2·50	66R
	gl	20	0·1 ($NaClO_4$)	K_1 11·72±0·04, K_2 2·51±0·01	66T, 68K
	gl	30	0·2 ($NaClO_4$)	K_1 12·00, K_2 2·50	67A
	gl	20	0·1 (KNO_3)	K_1 11·97, K_2 2·48	67B, 67Ba
Al^{3+}	sp	?	0·02—0·2	*$K(Al^{3+}+HL^{2-} \rightleftharpoons AlL+H^+)$ (0·02) 0·36, (0·05) 0·31, (0·2) 0·19	57N
	sp	28—34	0·02—0·20 $NaClO_4$	*$K(Al^{3+}+HL^{2-} \rightleftharpoons AlL+H^+)$ (0·02) 0·79, (0·05) 0·55, (0·10) 0·37, (0·20) 0·23	63N
		15—40	0·05 $NaClO_4$	(15°) 0·49, (28°) 0·64, (40°) 0·81	
			0·10 $NaClO_4$	(15°) 0·22, (28°) 0·41, (40°) 0·53 *ΔH 4·9, ΔS 14	
	gl	30	0·2 ($NaClO_4$)	K_1 12·20, K_2 10·01	67A
Be^{2+}	sp	25	0·1 ($NaClO_4$)	K_1 11·46, K_2 8·62	59B
	gl	25	0·1 ($NaClO_4$)	K_1 11·50, K_2 8·84	59B
	gl	22	50% ethanol	K_1 11·52, K_2 8·90	61A
	sp	28—32	0·02—0·20 $NaClO_4$	*$K(Be^{2+}+HL^{2-} \rightleftharpoons BeL^-+H^+)$ (0·02) −0·14, (0·05) −0·27, (0·10) −0·33, (0·20) −0·39	61D, 64D
	sp	20—45	0·2 $NaClO_4$	*$K(Be^{2+}+HL^{2-} \rightleftharpoons BeL^-+H^+)$ (20°) −0·45, (30°) −0·42, (45°) −0·37 *ΔH 1·4, ΔS 7	64D
	gl	30	0·2 ($NaClO_4$)	K_1 11·30, K_2 9·07	67A
	gl	20	0·1 (KNO_3)	K_1 11·54, K_2 8·89	67Ba
Ce^{3+}	sp, gl	20	0·1 ($NaClO_4$)	K_1 6·83, K_2 5·57, $K(Ce^{3+}+HL^{2-} \rightleftharpoons CeHL^+)$ 1·93	68K
Co^{2+}	gl, Ag-AgCl	25	0·1 (KCl)	K_1 6·47, K_2 4·3	62N

contin…

Metal	Method	Temp	Medium	Log of equilibrium constant, remarks	Ref
Cu^{2+}	gl	25	0·1 ($NaClO_4$)	K_1 9·27, K_2 6·64	59B
	gl, Ag-AgCl	25	0·1 (KCl)	K_1 9·35, K_2 6·92	62N
	sp	25	0·2 $NaClO_4$	K_1 9·7	61N
	gl	25	I ($NaClO_4$)	K_1 $10{\cdot}74-6{\cdot}11\sqrt{I}/(1+2{\cdot}4\sqrt{I})+0{\cdot}033\,I$, K_2 $6{\cdot}43\pm3{\cdot}054\sqrt{I}/(1+3{\cdot}4\sqrt{I})+0{\cdot}004\,I$	62Na
	gl	25	I ($NaClO_4$)	$K(CuA^{2+}+HL^{2-}\rightleftharpoons CuAL^{-}+H^{+})$ $-2{\cdot}70+(3{\cdot}054\sqrt{I})/(1+1{\cdot}25\sqrt{I})-0{\cdot}035\,I$ $K(CuL^{-}+A\rightleftharpoons CuAL^{-})$ $9{\cdot}48+0{\cdot}109\,I$ $K(CuL_2^{4-}+CuA_2^{2+}\rightleftharpoons 2CuAL)$ (I = 0) 3·94, (I = 0·1) 1·83, (I = 1) 0·09, (I = 2) −0·25 $K(Cu^{2+}+L^{3-}+A\rightleftharpoons CuAL^{-})$ (I = 0) 20·39, (I = 0·1) 19·09, (I = 1) 18·49, (I = 2) 18·74 where A = 1,2-diaminopropane	61Na
	gl	25	1 $NaClO_4$	K_1 8·91, β_2 15·86 $K(Cu^{2+}+L^{3-}+A^{-}\rightleftharpoons CuLA^{2-})$ 16·04 where HA = glycine	64M
	sp	28	0·02—0·50 $NaClO_4$	$^{*}K(Cu^{2+}+HL^{2-}\rightleftharpoons CuL^{-}+H^{+})$ (0·02) −1·91, (0·08) −2·00, (0·12) −2·04, (0·20) −2·10, (0·50) −2·10	65D
		20—40	0·2 $NaClO_4$	(20°) −2·03, (34°) −1·97, (40°) −1·93 $^{*}\Delta H$ 2·1 ΔS 16	
	gl	25	1 NH_4ClO_4	K_1 9·04, β_2 16·69 $K(Cu^{2+}+NH_3+L^{3-}\rightleftharpoons CuNH_3L^{-})$ 12·0 $K[Cu^{2+}+2NH_3+L^{3-}\rightleftharpoons Cu(NH_3)_2L^{-}]$ 15·60 $K[CuL_2^{4-}+Cu(NH_3)_4^{2+}\rightleftharpoons 2Cu(NH_3)_2L^{-}]$ 1·56 $K(Cu^{2+}+L^{3-}+A^{-}\rightleftharpoons CuLA^{2-})$ 16·04 $K(CuL^{-}+A^{-}\rightleftharpoons CuLA^{2-})$ 1·02 where HA = glycine	66B
	gl	25	0·1 (KNO_3)	K_1 9·41±0·02 $K(CuA^{2+}+L^{3-}\rightleftharpoons CuAL^{-})$ 9·86±0·03 where A = 2,2′-bipyridyl	66L
	gl	30	0·2 ($NaClO_4$)	K_1 9·75, K_2 7·14	67A
Dy^{3+}	sp, gl	20	0·1 ($NaClO_4$)	K_1 8·29, K_2 6·60 $K(Dy^{3+}+HL^{2-}\rightleftharpoons DyHL^{+})$ 2·42	68K
Er^{3+}	sp, gl	20	0·1 ($NaClO_4$)	K_1 8·15, K_2 6·30 $K(Er^{3+}+HL^{2-}\rightleftharpoons ErHL^{+})$ 2·12	68K
Eu^{3+}	sp, gl	20	0·1 ($NaClO_4$)	K_1 7·87, K_2 6·03, $K(Eu^{3+}+HL^{2-}\rightleftharpoons EuHL^{+})$ 2·26	68K
Fe^{3+}	sp	?	0·02—0·2	$^{*}K(Fe^{3+}+HL^{2-}\rightleftharpoons FeL+H^{+})$ (0·02), 2·54, (0·05) 2·46, (0·2) 2·40	57N
	sp	25—28	0·05	K_1 14·05, β_2 24·33, β_3 33·10	66O
Ga^{3+}	sp	29	0·02—0·20 $NaClO_4$	$^{*}K(Ga^{3+}+HL^{2-}\rightleftharpoons GaL+H^{+})$ (0·02) 1·61 (0·05), 1·48, (0·10) 1·20, (0·20) 0·98	63N
		20—40	0·05 $NaClO_4$	(20°) 1·32, (29°) 1·39, (40°) 1·52 $^{*}\Delta H$ 4·2, ΔS 8	
Gd^{3+}	gl	20	0·1 ($NaClO_4$)	K_1 7·58, K_2 6·07±0·03 $K(GdL+H^{+}\rightleftharpoons GdHL^{+})$ 6·20±0·02	66T
Ho^{3+}	sp, gl	20	0·1 ($NaClO_4$)	K_1 8·40, K_2 6·75 $K(Ho^{3+}+HL^{2-}\rightleftharpoons HoHL^{+})$ 2·23	68K
Lu^{3+}	sp, gl	20	0·1 ($NaClO_4$)	K_1 8·43, K_2 7·03 $K(Lu^{3+}+HL^{2-}\rightleftharpoons LuHL^{+})$ 2·47	68K
Mn^{2+}	gl, Ag-AgCl	25	0·1 (KCl)	K_1 5·25, K_2 3·4	62N

continued overleaf

552 $C_7H_6O_6S$ (contd)

Metal	Method	Temp	Medium	Log of equilibrium constant, remarks	Ref
Nd^{3+}	sp, gl	20	0·1 ($NaClO_4$)	K_1 7·39, K_2 5·62 $K(Nd^{3+}+HL^{2-} \rightleftharpoons NdHL^+)$ 2·09	68K
Ni^{2+}	gl, Ag-AgCl	25	0·1 (KCl)	K_1 6·61, K_2 4·2	62N
Pr^{3+}	sp, gl	20	0·1 ($NaClO_4$)	K_1 7·08, K_2 5·61 $K(Pr^{3+}+HL^{2-} \rightleftharpoons PrHL^+)$ 1·99	68K
Sc^{3+}	sp	18—20	0·01 (HNO_3)	K_1 3·96 $\pm$ 0·08	66K
Sm^{3+}	sp, gl	20	0·1 ($NaClO_4$)	K_1 7·65, K_2 5·93 $K(Sm^{3+}+HL^{2-} \rightleftharpoons SmHL^+)$ 2·23	68K
Tb^{3+}	sp, gl	20	0·1 ($NaClO_4$)	K_1 8·42, K_2 6·19 $K(Tb^{3+}+HL^{2-} \rightleftharpoons TbHL^+)$ 2·47	68K
Th^{4+}	con J	28	0·01	$K(Th^{4+}+HL^{2-} \rightleftharpoons ThL^+ + H^+)$ (?) 2·42 See EDTA (836) and CDTA (997) for mixed ligands	62S
Ti^{IV}	sp	18—22	0·1	$K(TiO^{2+}+2H^+ + 3L^{3-} \rightleftharpoons TiL_3^{5-})$ 42·2 $K(TiO^{2+}+2HL^{2-} \rightleftharpoons TiOL_2H_2^{2-})$ 5·4 $K(TiO^{2+}+HL^{2-} \rightleftharpoons TiOLH)$ 3·1	63S
Tl^+	con J	25	0·01	$K(Tl^+ + HL^{2-} \rightleftharpoons TlL^{2-} + H^+)$ (?) 2·38	62Sa
Tl^{3+}	gl	25	0·1 (KNO_3)	K_1 12·41	67Aa
Tm^{3+}	sp, gl	20	0·1 ($NaClO_4$)	K_1 8·34, K_2 6·61 $K(Tm^{3+}+HL^{2-} \rightleftharpoons TmHL^+)$ 2·27	68K
U^{4+}				See EDTA (836) for mixed ligands	
UO_2^+	pol	18—22	1 ($NaNO_3$)	K_1 5·1 $\pm$ 0·2	64H
UO_2^{2+}	pol	18—22	1 ($NaNO_3$)	K_1 11·7 $\pm$ 0·2, β_2 17·6 $\pm$ 0·2	64H
	gl	28	0·1 KNO_3	K_1 10·70	66R
	gl	30	0·2 ($NaClO_4$)	K_1 10·85, K_2 8·53	67A
	gl	20	0·1 (KNO_3)	K_1 11·25, β_2 18·75	67B
VO^{2+}	gl	25, 35	0·10 (KNO_3)	K_1 (25°) 11·71 $\pm$ 0·07, (35°) 11·29 $\pm$ 0·08 $\Delta H_1°$ $-1·7 \pm 0·5$, $\Delta S_1°$ 45 $\pm$ 3 $K[VO(OH)L^{2-} + H^+ \rightleftharpoons VOL^-]$ (25°) 7·22 $\pm$ 0·05 $K\{2VO(OH)L^{2-} \rightleftharpoons [VO(OH)L]_2^{4-}\}$ (25°) 5·33 $\pm$ 0·05	66M
Yb^{3+}	sp, gl	20	0·1 ($NaClO_4$)	K_1 8·35, K_2 6·81 $K(Yb^{3+}+HL^{2-} \rightleftharpoons YbHL^+)$ 2·30	68K
Zr^{4+}	sp	18—20	1 (HCl)	K (?) 4·79 $\pm$ 0·03	66K

* These constants have been recalculated by the compilers on the basis of the equations indicated

Errata. 1964, p. 548, Table 552. Under Al^{3+}, *delete* lines one, two, and three.
Under Fe^{3+} *delete* third, seventh, and tenth lines.
Under references *delete* 57N.
Reference 64R. *Add* page number 789 at end.

57N R. K. Nanda and S. Aditya, *J. Indian Chem. Soc.*, 1957, **34,** 577
59B C. V. Banks and R. S. Singh, *J. Amer. Chem. Soc.*, 1959, **81,** 6159
61A R. P. Agarwal and R. C. Mehrotra, *J. Less-Common Metals*, 1961, 3, 398
61D R. C. Das and S. Aditya, *J. Indian Chem. Soc.*, 1961, **38,** 19
61N V. S. K. Nair, *Trans. Faraday Soc.*, 1961, **57,** 1988
61Na R. Nasanen and P. Merilainen, *Suomen Kem.*, 1961, **B34,** 75
62N V. S. K. Nair, *Talanta*, 1962, **9,** 27
62Na R. Nasanen and P. Merilainen, *Suomen Kem.*, 1962, **B35,** 79
62S D. D. Sharma and A. K. Bhattacharya, *J. Indian Chem. Soc.*, 1962, **39,** 299
62Sa K. N. Sahu, M. C. Saxena, and A. K. Bhattacharya, *J. Indian Chem. Soc.*, 1962, **39,** 731
63N R. K. Nanda and S. Aditya, *J. Indian Chem. Soc.*, 1963, **40,** 660, 755
63S L. Sommer, *Coll. Czech. Chem. Comm.*, 1963, **28,** 2716
64D R. C. Das and S. Aditya, *J. Indian Chem. Soc.*, 1964, **41,** 765
64H J. Hala, *Coll. Czech. Chem. Comm.*, 1964, **29,** 905
64M R. P. Martin and R. A. Paris, *Bull. Soc. chim. France*, 1964, 80
65D R. C. Das, R. K. Nanda, and S. Aditya, *J. Indian Chem. Soc.*, 1965, **42,** 307

contd

66B M. Bonnet and R. A. Paris, *Bull. Soc. chim. France*, 1966, 747
66K I. M. Korenman, F. R. Sheyanova, and Z. M. Gur'eva, *Russ. J. Inorg. Chem.*, 1966, **11**, 1485 (2761)
66Ka I. M. Korenman and N. V. Zaglyadimova, *Russ. J. Inorg. Chem.*, 1966, **11**, 1491 (2774)
66L G. A. L'Heureux and A. E. Martell, *J. Inorg. Nuclear Chem.*, 1966, **28**, 481
66M G. E. Mont and A. E. Martell, *J. Amer. Chem. Soc.*, 1966, **88**, 1387
66O K. Ogawa and N. Tobe, *Bull. Chem. Soc. Japan*, 1966, **39**, 223
66R J. M. Rao and U. V. Seshaiah, *Bull. Chem. Soc. Japan*, 1966, **39**, 2668
66T N. V. Thakur, S. M. Jogdeo, and C. R. Kanekar, *J. Inorg. Nuclear Chem.*, 1966, **28**, 2297
67A V. T. Athavale, N. Mahadevan, P. K. Mathur, and R. M. Sathe, *J. Inorg. Nuclear Chem.*, 1967, **29**, 1947
67Aa R. C. Agarwal and A. K. Srivastava, *Indian J. Chem.*, 1967, **5**, 114
67B M. Bartusek, *Coll. Czech. Chem. Comm.*, 1967, **32**, 116
67Ba M. Bartusek and J. Zelinka, *Coll. Czech. Chem. Comm.*, 1967, **32**, 992
68K C. R. Kanekar, N. V. Thakar, and S. M. Jogdeo, *Bull. Chem. Soc. Japan*, 968, **41**, 759

$C_7H_7O_2N$ 2-Aminobenzoic acid (anthranilic acid) HL 553

Metal	*Method*	*Temp*	*Medium*	*Log of equilibrium constant, remarks*	*Ref*
H^+	gl	30	0·1 (KCl)	K_1 4·86	62C
	gl	22	0·2 ($NaClO_4$)	K_1 4·76±0·07	67S
Ce^{3+}	gl	30	0·1 (KCl)	K_1 3·18	62C
Dy^{3+}	gl	22	0·2 ($NaClO_4$)	K_1 3·2±0·2	67S
La^{3+}	gl	30	0·1 (KCl)	K_1 3·14	62C
Nd^{3+}	gl	30	0·1 (KCl)	K_1 3·23	62C
Pr^{3+}	gl	30	0·1 (KCl)	K_1 3·22	62C

Erratum. 1964, p. 550, Table 553. Under Cu^{2+} third line *for* 0·44 *read* 0 corr.

62C M. Cefola, A. S. Tompa, A. V. Celiano, and P. S. Gentile, *Inorg. Chem.*, 1962, **1**, 290
67S M. B. Silber and J. M. Swinehart, *J. Phys. Chem.*, 1967, **71**, 4344

$C_7H_7O_2N$ 3-Aminobenzoic acid HL 553.1

Metal	*Method*	*Temp*	*Medium*	*Log of equilibrium constant, remarks*	*Ref*
Co^{2+}	pH	25	1 KNO_3	K_1 3·0, K_2 3·6 (?)	61G

61G A. A. Grinberg and Kh. Kh. Khakimov, *Russ. J. Inorg. Chem.*, 1961, **6**, 71 (144)

$C_7H_7O_2N$ 4-Aminobenzoic acid HL 553.2

Metal	*Method*	*Temp*	*Medium*	*Log of equilibrium constant, remarks*	*Ref*
Co^{2+}	pH	25	1 KNO_3	K_1 3·4, K_2 2·9	61G

61G A. A. Grinberg and Kh. Kh. Khakimov, *Russ. J. Inorg. Chem.*, 1961, **6**, 71 (144)

553.3 $C_7H_7O_2N$ 4-Methyl-2-nitrosophenol HL

Metal	*Method*	*Temp*	*Medium*	*Log of equilibrium constant, remarks*	*Ref*
H^+	gl	25	0·1 KNO_3 50% dioxan	K_1 7·55	61S
Cd^{2+}	gl	25	0·1 KNO_3 50% dioxan	K_1 4·59	61S
Cu^{2+}	gl	25	0·1 KNO_3 50% dioxan	K_1 7·30	61S
Ni^{2+}	gl	25	0·1 KNO_3 50% dioxan	K_1 7·03	61S
Pb^{2+}	gl	25	0·1 KNO_3 50% dioxan	K_1 5·53	61S
Zn^{2+}	gl	25	0·1 KNO_3 50% dioxan	K_1 5·13	61S

61S H. Shimura, *Nippon Kagaku Zasshi*, 1961, **82**, 641

553.4 $C_7H_7O_2N$ Benzohydroxamic acid HL

Metal	*Method*	*Temp*	*Medium*	*Log o equilibrium constant, remarks*	*Ref*
H^+	gl	25	1	K_1 8·97±0·03	65B, 66B
	gl	20	1 (KCl)	K_1 8·68	66M
Fe^{3+}	dis	25	1 $HClO_4$	K_1 12·18	65B
Pu^{4+}	dis	25	7 HNO_3	K_1 12·73±0·01	66B
Th^{4+}	gl	25	0·025 HNO_3	K_1 9·60, β_2 19·81, β_3 28·76	66B
U^{4+}	gl	25	0·01 $HClO_4$	K_1 9·89, β_2 18·00, β_3 26·32, β_4 32·95	66B
UO_2^{2+}	gl	25	0·1 $HClO_4$	K_1 8·72, β_2 16·77	65B
	sp	20	1 ($NaClO_4$)	K_1 7·72	66M
Zr^{4+}	dis	25	1 $HClO_4$	K_1 12·43, β_2 24·08	65B

65B F. Baroncelli and G. Grossi, *J. Inorg. Nuclear Chem.*, 1965, **27**, 1085
66B A. Barocas, F. Baroncelli, G. B. Biondi, and G. Grossi, *J. Inorg. Nuclear Chem.*, 1966, **28**, 2961
66M F. Maggio, V. Romano, and R. Cefalu, *J. Inorg. Nuclear Chem.*, 1966, **28**, 1979

553.5 $C_7H_7O_2N$ 2-Pyridylacetic acid HL

Metal	*Method*	*Temp*	*Medium*	*Log of equilibrium constant, remarks*	*Ref*
H^+	gl	25	0·1 ($HClO_4$) 50% dioxan	K_1 6·24, K_2 3·83	66W
Co^{2+}	gl	15—35	0·1 KNO_3 50% dioxan	K_2 (15°) 6·25, (25°) 5·55, (35°) 3·77	66W
Cu^{2+}	gl	15—35	0·1 KNO_3 50% dioxan	K_1 (35°) 6·05, K_2 (15°) 7·11, (25°) 6·94, (35°) 4·18	66W
Ni^{2+}	gl	15—35	0·1 KNO_3 50% dioxan	K_1 (35°) 4·89, K_2 (15°) 6·47, (25°) 6·06, (35°) 3·95	66W
Zn^{2+}	gl	15—35	0·1 KNO_3 50% dioxan	K_1 (15°) 7·15, (25°) 6·41, (35°) 4·33, K_2 (15°) 6·22, (25°) 5·36, (35°) 3·40	66W

66W J. L. Walter and S. M. Rosalie, *J. Inorg. Nuclear Chem.*, 1966, **28**, 2969

$C_7H_7O_2N$ 6-Methylpyridine-2-carboxylic acid HL 556

Erratum. 1964, p. 551, Table 556. *For* 55Ha *read* 55H.

$C_7H_7O_2N$ 2-Hydroxybenzaldehyde oxime (salicylaldoxime) H_2L 557

Metal	*Method*	*Temp*	*Medium*	*Log of equilibrium constant, remarks*	*Ref*
H^+	gl	30	0·1 $NaClO_4$ 75% dioxan	K_1 12·08, K_2 11·72, K_3 1·72 (?)	64J
	gl	20	0·1 ($NaClO_4$) 75% dioxan	K_2 10·70±0·05	65B
	gl	25	0—0·1 (KCl)	K_1 (→ 0) 12·1, (0·01) 11·90, (0·025) 11·85, (0·05) 11·62, (0·075) 11·50, (0·1) 11·07, K_2 (→ 0) 9·3, (0·01) 9·20, (0·025) 9·05, (0·05) 8·98, (0·075) 8·90, (0·1) 8·85	68M
Co^{2+}	gl	20	0·1 ($NaClO_4$) 75% dioxan	*K_1 6·4±0·1, *K_2 7·1±0·1 (?)	65B
Cu^{2+}	gl	30	0·1 $NaClO_4$ 75% dioxan	K_1 12·64, K_2 11·17	64J
	sp	20	0·1 ($NaClO_4$) 75% dioxan	*β_2 21·5±0·2	65B
Fe^{2+}	gl	20	0·1 ($NaClO_4$) 75% dioxan	*K_1 9·38±0·05, *K_2 7·35±0·05	65B
Fe^{3+}	sp	20	0·26	$K(Fe^{3+}+H_2L \rightleftharpoons FeH_2L^{3+})$ 3·89	67M
Hf^{4+}	gl	25	0—0·1 (KCl)	K_1 (→ 0) 16·7, (0·01) 15·82, (0·025) 14·15, (0·05) 13·00, (0·075) 12·30, (0·1) 11·05	68M
Mn^{2+}	con, sp	26	?	K_1 3·01	63K
	gl	20	0·1 ($NaClO_4$) 75% dioxan	*K_1 5·8±0·2, *K_2 6·1±0·2 (?)	65B
Ni^{2+}	gl	20	0·1 ($NaClO_4$) 75% dioxan	*K_1 6·9±0·2, *K_2 7·4±0·2 (?)	65B
Ti^{4+}	gl	25	0—0·1 (KCl)	K_1 (→ 0) 18·5, (0·01) 18·29, (0·025) 17·74 (0·05) 17·35, (0·075) 16·86, (0·1) 16·30, K_2 (→ 0) 17·2, (0·01) 16·88, (0·025) 16·62, (0·05) 16·07, (0·075) 15·66, (0·1) 14·85	68M
Zn^{2+}	gl	20	0·1 ($NaClO_4$) 75% dioxan	*K_1 6·3±0·2, *K_2 7·2±0·2 (?)	65B
Zr^{4+}	gl	25	0—0·1 (KCl)	K_1 (→ 0) 17·9, (0·01) 17·35, (0·025) 16·45, (0·05) 15·13, (0·075) 13·90, (0·1) 12·43	68M

* For these constants only, $K_1 = K(M^{2+}+HL^- \rightleftharpoons MHL^+)$, $K_2 = K[MHL^+ + HL^- \rightleftharpoons (MHL)_2]$, $\beta_2 = K[M^{2+}+2HL^- \rightleftharpoons M(HL)_2]$

Errata. 1964, p. 551, Table 557. Under Co^{2+} equation should *read* $K[Co^{2+}+2HL^- \rightleftharpoons Co(HL)_2]$ 8·13.
Under Cu^{2+} transfer spJ *from* third column *to* second column.
page 552. Under Ni^{2+} transfer spJ *from* third column *to* second column.
Under Ni^{2+} equation should *read* $K[Ni^{2+}+2HL^- \rightleftharpoons Ni(HL)_2]$ 3·77.
Under Sr^{2+} equation should *read* $K[Sr^{2+}+2HL^- \rightleftharpoons Sr(HL)_2]$ 3·77.

63K M. S. Kachhawaha and A. K. Bhattacharya, *Z. anorg. Chem.*, 1963, **325**, 321
64J K. E. Jabalpurwala, K. A. Venkatachalam, and M. B. Kabadi, *J. Inorg. Nuclear Chem.*, 1964, **26**, 1011, 1027
65B K. Burger and I. Egyed, *J. Inorg. Nuclear Chem.*, 1965, **27**, 2361
67M R. Manolov, *Russ. J. Inorg. Chem.*, 1967, **12**, 1431 (2715)
68M S. Mandel and A. K. Dey, *J. Inorg. Nuclear Chem.*, 1968, **30**, 1221

558 $C_7H_7O_2N$ 2-Hydroxybenzamide (salicylamide) HL

Metal	Method	Temp	Medium	Log of equilibrium constant, remarks	Ref
H^+	gl	30	0·1 $NaClO_4$ 75% dioxan	K_1 10·27	64J
Cu^{2+}	gl	30	0·1 $NaClO_4$ 75% dioxan	K_1 7·80	64J
Mg^{2+}	gl	30	0·1 $NaClO_4$ 75% dioxan	K_1 2·79	64J
Ni^{2+}	gl	30	0·1 $NaClO_4$ 75% dioxan	K_1 5·65	64J
Zn^{2+}	gl	30	0·1 $NaClO_4$ 75% dioxan	K_1 6·17	64J

64J K. E. Jabalpurwala, K. A. Venkatachalam, and M. B. Kabadi, *J. Inorg. Nuclear Chem.*, 1964, **26**, 1011, 1927

558.1 $C_7H_7O_2N$ 4-Aminotropolone HL

Metal	Method	Temp	Medium	Log of equilibrium constant, remarks	Ref
H^+	sp	25	2 $NaClO_4$	K_1 7·47	64O
Fe^{3+}	sp	25	2 $NaClO_4$	K_1 12·58	64O

64O Y. Oka, M. Yanai, and C. Suzuki, *Nippon Kagaku Zasshi*, 1964, **85**, 873

558.2 $C_7H_7O_2N_3$ Pyridine-2,6-dicarboxaldehyde dioxime H_2L

Metal	Method	Temp	Medium	Log of equilibrium constant, remarks	Ref
H^+	gl	25	0·01	K_1 10·7, K_2 9·7	63B
			0·005	K_3 2·34	
	sp	20—35	0·05	K_1 (20·3°) 10·68, (25°) 10·63, (30°) 10·51, (35°) 10·45, K_2 (20·3°) 10·04, (25°) 10·00, (30°) 9·92, (35°) 9·85	65H
		25	0 corr	K_1 10·88 ± 0·03, K_2 10·08 ± 0·03 $\Delta H_1°$ −6·7 ± 1·0, $\Delta S_1°$ 27 ± 3, $\Delta H_2°$ −5·3 ± 0·9, $\Delta S_2°$ 28 ± 3	
Cu^{2+}	sp	25	0·1 ($NaClO_4$)	K_1 9·9 ± 0·1	63B
Fe^{2+}	sp	25	0 corr	$K(FeL_2^{2-} + H^+ \rightleftharpoons FeHL_2^-)$ 7·40 ± 0·03 $\Delta H°$ 0 ± 0·9, $\Delta S°$ 34 ± 3 $K(FeHL_2^- + H^+ \rightleftharpoons FeH_2L_2)$ ~ 5	65H
Mn^{2+}	gl	25	0·1 ($NaClO_4$)	K_1 4·4 ± 0·1, K_2 4·1 ± 0·1	63B
Ni^{2+}	sp	25	0·1 ($NaClO_4$)	K_1 8·4 ± 0·1	63B
	gl	25	0·1 ($NaClO_4$)	K_1 8·6 ± 0·1, K_2 7·1 ± 0·1	63B
Zn^{2+}	gl	25	0·1 ($NaClO_4$)	K_1 5·9 ± 0·1, K_2 5·5 ± 0·1	63B

63B S. P. Bag, Q. Fernando, and H. Freiser, *Analyt. Chem.*, 1963, **35**, 719
65H G. I. H. Hanania, D. H. Irvine, and F. Shurayh, *J. Chem. Soc.*, 1965, 1149

$C_7H_7O_3N$ 4-Amino-2-hydroxybenzoic acid (4-aminosalicylic acid) H_2L 558.3

Metal	Method	Temp	Medium	Log of equilibrium constant, remarks	Ref
H^+	gl	25	1 KNO_3	K_2 4·36	61G
	?	?	?	K_1 13·75, K_2 4·08	66R
Co^{2+}	gl	25	1 KNO_3	K_1 4·2, K_2 3·7	61G
Fe^{3+}	sp	30	0—0·20	$K(Fe^{3+}+HL^- \rightleftharpoons FeL+H^+)$ (0) −3·43, (0·02) −3·51, (0·05) −3·59, (0·10) −3·63, (0·20) −3·66	60D
UO_2^{2+}	gl	28	0·1 KNO_3	K_1 14·41	66R
Zn^{2+}	pol	20	1·0 (KNO_3)	K_1 0·65 (?), K_2 1·02 (?), K_3 1·34 (?), K_4 1·6 (?)	66N

60D R. C. Das and S. Aditya, *J. Indian Chem. Soc.*, 1960, **37**, 557

61G A. A. Grinberg and Kh. Kh. Khakimov, *Russ. J. Inorg. Chem.*, 1961, **6**, 71 (144)

66N M. S. Novakovskii and V. V. Voinova, *Russ. J. Inorg. Chem.*, 1966, **11**, 1409 (2624)

66R J. M. Rao and U. V. Seshaiah, *Bull. Chem. Soc. Japan*, 1966, **39**, 2668

$C_7H_7O_3N$ 2-Hydroxybenzohydroxamic (salicylohydroxamic acid) H_2L 560.1

Metal	Method	Temp	Medium	Log of equilibrium constant, remarks	Ref
H^+	gl	30	0·1 $NaClO_4$ 75% dioxan	K_1 12·06, K_2 9·00, K_3 1·30	64J
Cu^{2+}	gl	30	0·1 $NaClO_4$ 75% dioxan	K_1 9·60, K_2 8·19	64J

64J K. E. Jabalpurwala, K. A. Venkatachalam, and M. B. Kabadi, *J. Inorg. Nuclear Chem.*, 1964, **26**, 1011, 1027

$C_7H_7O_3N$ 2,4-Dihydroxybenzoic acid amide (β-resorcylamide) H_2L 560.2

Metal	Method	Temp	Medium	Log of equilibrium constant, remarks	Ref
H^+				H^+ constants not given.	
Fe^{3+}	sp	25	0·1 ($NaClO_4$)	K_1 3·58	66D

66D M. N. Desai, *Indian J. Chem.*, 1966, **4**, 218

$C_7H_7O_3N$ 6-(Hydroxymethyl)pyridine-2-carboxylic acid HL 560.3

Metal	Method	Temp	Medium	Log of equilibrium constant, remarks	Ref
H^+	?	25	?	K_1 4·61, K_2 1·2	62G
Cd^{2+}	gl	25	?	K_1 4·81, K_2 3·29	62G
Co^{2+}	gl	25	?	K_1 4·28, K_2 4·23	62G
Cu^{2+}	sp	25	?	K_1 6·74, K_2 5·05	62G
Fe^{2+}	gl	25	?	K_1 3·91, K_2 3·22	62G
Ni^{2+}	sp	25	?	K_1 5·21, K_2 4·84	62G
Zn^{2+}	gl	25	?	K_1 4·34, K_2 4·11	62G

62G R. W. Green and G. K. S. Ooi, *Austral. J. Chem.*, 1962, **15**, 786

561 $C_7H_7O_6P$ 2-Carboxyphenyl phosphate (salicyl phosphate) H_3L

Metal	Method	Temp	Medium	Log of equilibrium constant, remarks	Ref
H^+	gl	25, 35	0·10 (KNO_3)	K_1 (25°) 6·61±0·04, (35°) 6·45±0·01, K_2 (25°) 3·69±0·05, (35°) 3·60±0·03	66M
VO^{2+}	gl	25, 35	0·10 (KNO_3)	K_1 (25°) 5·81±0·04, (35°) 5·68±0·02 $\Delta H_1°$ −0·5±0·5, $\Delta S_1°$ 25±3 $K[VO(OH)L^{2-} + H^+ \rightleftharpoons VOL^-]$ (25°) 5·7±0·2 $K\{2VO(OH)L \rightleftharpoons [VO(OH)L]_2\}$ (25°) 2·3±0·3	66M

Erratum. 1964, p. 553, Table 561. Under Cu^{2+} *add* reference 63M.

66M G. E. Mont and A. E. Martell, *J. Amer. Chem. Soc.*, 1966, **88**, 1387

562.1 $C_7H_8ON_2$ Salicylaldehyde-hydrazone (salicylaldazone) HL

Metal	Method	Temp	Medium	Log of equilibrium constant, remarks	Ref
H^+	gl	30	0·1 $NaClO_4$ 75% dioxan	K_1 12·57, K_2 3·53	64J
Cu^{2+}	gl	30	0·1 $NaClO_4$ 75% dioxan	K_1 14·24, K_2 11·10, K_3 10·95, K_4 5·99	64J

64J K. E. Jabalpurwala, K. A. Venkatachalam, and M. B. Kabadi, *J. Inorg. Nuclear Chem.*, 1964, **26**, 1011, 1027

564.1 $C_7H_8O_2N_2$ Salicylamidoxime H_2L

Metal	Method	Temp	Medium	Log of equilibrium constant, remarks	Ref
H^+	gl	30	0·1 $NaClO_4$ 75% dioxan	K_1 12·16, K_2 11·93, K_3 3·46	64J
Cu^{2+}	gl	30	0·1 $NaClO_4$ 75% dioxan	K_1 13·71, K_2 11·49	64J

64J K. E. Jabalpurwala, K. A. Venkatachalam, and M. B. Kabadi, *J. Inorg. Nuclear Chem.*, 1964, **26**, 1011, 1027

564.2 $C_7H_8O_2N_2$ 2-Hydroxybenzohydrazide (salicyloylhydrazine) HL

Metal	Method	Temp	Medium	Log of equilibrium constant, remarks	Ref
H^+	gl	30	0·1 $NaClO_4$ 75% dioxan	K_1 9·52, K_2 2·39	64J
Cu^{2+}	gl	30	0·1 $NaClO_4$ 75% dioxan	K_1 9·27	64J

64J K. E. Jabalpurwala, K. A. Venkatachalam, and M. B. Kabadi, *J. Inorg. Nuclear Chem.*, 1964, **26**, 1011, 1027

$C_7H_8O_2N_2$ 2-Methyl-4-nitroaniline L 564.3

Metal	Method	Temp	Medium	Log of equilibrium constant, remarks	Ref
Zn^{2+}	sp	25	acetone	$K(ZnBr_2+L \rightleftharpoons ZnBr_2L)$ 1·18	65S

65S D. P. N. Satchell and R. S. Satchell, *Trans. Faraday Soc.*, 1965, **61**, 1118

$C_7H_8O_2N_2$ 3-Methyl-4-nitroaniline L 564.4

Metal	Method	Temp	Medium	Log of equilibrium constant, remarks	Ref
Zn^{2+}	sp	25	acetone	$K(ZnBr_2+L \rightleftharpoons ZnBr_2L)$ 1·27	65S

65S D. P. N. Satchell and R. S. Satchell, *Trans. Faraday Soc.*, 1965, **61**, 1118

$C_7H_8O_2N_2$ 4-Methyl-2-nitroaniline L 564.5

Metal	Method	Temp	Medium	Log of equilibrium constant, remarks	Ref
Zn^{2+}	sp	25	acetone	$K(ZnBr_2+L \rightleftharpoons ZnBr_2L)$ 0·40	65S

65S D. P. N. Satchell and R. S. Satchell, *Trans. Faraday Soc.*, 1965, **61**, 1118

$C_7H_8O_2N_2$ 4-Methyl-3-nitroaniline L 564.6

Metal	Method	Temp	Medium	Log of equilibrium constant, remarks	Ref
Zn^{2+}	sp	25	acetone	$K(ZnBr_2+L \rightleftharpoons ZnBr_2L)$ 2·40 $K(ZnCl_2+L \rightleftharpoons ZnCl_2L)$ 1·90 $K(ZnI_2+L \rightleftharpoons ZnI_2L)$ 2·02	65S

65S D. P. N. Satchell and R. S. Satchell, *Trans. Faraday Soc.*, 1965, **61**, 1118

$C_7H_8O_2N_2$ 6-Methyl-3-nitroaniline L 564.7

Metal	Method	Temp	Medium	Log of equilibrium constant, remarks	Ref
Zn^{2+}	sp	25	acetone	$K(ZnCl_2+L \rightleftharpoons ZnCl_2L)$ 1·88 $K(ZnBr_2+L \rightleftharpoons ZnBr_2L)$ 2·12 $K(ZnI_2+L \rightleftharpoons ZnI_2L)$ 1·95	65S

65S D. P. N. Satchell and R. S. Satchell, *Trans. Faraday Soc.*, 1965, **61**, 1118

564.8 $C_7H_8O_3S$ 3-Hydroxythiophene-2-carboxylic acid ethyl ester L

Metal	Method	Temp	Medium	Log of equilibrium constant, remarks	Ref
H^+	sp	25	0·1 $NaClO_4$ 10% dioxan	K_1 7·93	65C
Co^{2+}	sp	25	0·1 $NaClO_4$ 10% dioxan	K_1 4·17	65C
Cu^{2+}	sp	25	0·1 $NaClO_4$ 10% dioxan	K_1 5·84	65C
Ni^{2+}	sp	25	0·1 $NaClO_4$ 10% dioxan	K_1 4·35	65C
Mn^{2+}	sp	25	0·1 $NaClO_4$ 10% dioxan	K_1 3·79	65C
Zn^{2+}	sp	25	0·1 $NaClO_4$ 10% dioxan	K_1 4·12	65C

65C A. Courtin and H. Sigel, *Helv. Chim. Acta*, 1965, **48**, 617

566 $C_7H_8O_5S$ 4-Hydroxy-3-methoxybenzenesulphonic acid H_3L

Erratum. 1964, p. 554, Table 566. Reference 55V should *read Chem. Abs.*, 1958, **52**, 954i.

566.1 C_7H_9ON 2-Methoxyaniline (*o*-anisidine) L

Metal	Method	Temp	Medium	Log of equilibrium constant, remarks	Ref
H^+	pH	25	0·3 (KNO_3)	K_1 4·52	64N
Cd^{2+}	Hg(Cd)	25	0·3 (KNO_3)	K_1 0·05, β_2 −0·35	64N

64N L. V. Nazarova, A. V. Ablov, and V. A. Dagaev, *Russ. J. Inorg. Chem.*, 1964, **9**, 1150 (2129)

566.2 C_7H_9ON 4-Methoxyaniline (*p*-anisidine) L

Metal	Method	Temp	Medium	Log of equilibrium constant, remarks	Ref
H^+	pH	25	0·3 (KNO_3)	K_1 5·01	64N
Cd^{2+}	Hg(Cd)	25	0·3 (KNO_3)	K_1 0·45, β_2 0·32, β_3 0·16	64N

64N L. V. Nazarova, A. V. Ablov, and V. A. Dagaev, *Russ. J. Inorg. Chem.*, 1964, **9**, 1150 (2129)

C_7H_9ON 2-(Aminomethyl)phenol (salicylamine) HL 566.3

Metal	Method	Temp	Medium	Log of equilibrium constant, remarks	Ref
H^+	gl	30	0·1 $NaClO_4$ 75% dioxan	K_1 12·87, K_2 6·27	64J
Cu^{2+}	gl	30	0·1 $NaClO_4$ 75% dioxan	K_1 14·62, K_2 11·60	64J

64J K. E. Jabalpurwala, K. A. Venkatachalam, and M. B. Kabadi, *J. Inorg. Nuclear Chem.*, 1964, 26, 1011, 1027

C_7H_9ON 2-(2′-Hydroxyethyl)pyridine HL 567.1

Metal	Method	Temp	Medium	Log of equilibrium constant, remarks	Ref
H^+	gl	0—25	50% dioxan	K_1 (0°) 13·55, (15°) 13·11, (25°) 12·75, K_2 (0°) 4·73, (15°) 4·65, (25°) 4·52	64L
Cu^{2+}	gl	0—25	50% dioxan	K_1 (0°) 10·12, (15°) 9·63, (25°) 9·52, β_2 (0°) 19·30, (15°) 18·45, (25°) 18·04 $\Delta H_{\beta 2}°$ (25°) $-13·7$, $\Delta S_{\beta 2}°$ 39	64L
	cal	25	50% dioxan	$\Delta H_{\beta 2}°$ $-13·82 \pm 0·03$, $\Delta S_{\beta 2}°$ 38·5	64L

64L T. J. Lane, A. J. Kandathil, and S. M. Rosalie, *Inorg. Chem.*, 1964, 3, 487

C_7H_9ON 6-Methyl-2-(hydroxymethyl)pyridine HL 567.2

Metal	Method	Temp	Medium	Log of equilibrium constant, remarks	Ref
H^+	gl	0—25	50% dioxan	K_1 (0°) 13·77, (15°) 13·14, (25°) 12·83, K_2 (0°) 4·94, (15°) 4·71, (25°) 4·56	64L
Cu^{2+}	gl	0—25	50% dioxan	K_1 (0°) 10·13, (15°) 9·71, (25°) 9·81, β_2 (0°) 19·31, (15°) 18·47, (25°) 18·37 $\Delta H_{\beta 2}°$ (25°) $-14·5$, $\Delta S_{\beta 2}°$ 35	64L
	cal	25	50% dioxan	$\Delta H_{\beta 2}°$ $-13·5 \pm 0·1$, $\Delta S_{\beta 2}°$ 38·6	64L

64L T. J. Lane, A. J. Kandathil, and S. M. Rosalie, *Inorg. Chem.*, 1964, 3, 487

C_7H_9NS 2-(Methylthiomethyl)pyridine (methyl-α-picolyl sulphide) L 567.3

Metal	Method	Temp	Medium	Log of equilibrium constant, remarks	Ref
H^+	gl	25	0·1 ($NaClO_4$)	K_1 4·64	64K
	sp	25	0·1 ($NaClO_4$)	K_1 4·67	64K
	gl	25	0·1 ($NaClO_4$) 50% dioxan	K_1 3·77 ± 0·03	67S
Co^{2+}	gl	25	0·1 ($NaClO_4$) 50% dioxan	K_1 1·1 ± 0·1 $K(CoA^{2+} + L \rightleftharpoons CoAL^{2+})$ 1·1 ± 0·1 where A = 2,2′-bipyridyl	67S

continued overleaf

567.3 C_7H_9NS (contd)

Metal	*Method*	*Temp*	*Medium*	*Log of equilibrium constant, remarks*	*Ref*
Cu^{2+}	gl	25	0·1 ($NaClO_4$)	K_1 3·27	64K
	gl	25	0·1 ($NaClO_4$) 50% dioxan	K_1 2·90±0·05 $K(CuA^{2+}+L \rightleftharpoons CuAL^{2+})$ 1·92±0·05 where A = 2,2′-bipyridyl	67S
Ni^{2+}	gl	25	0·1 ($NaClO_4$)	K_1 2·06	64K
	gl	25	0·1 ($NaClO_4$) 50% dioxan	K_1 1·97±0·05 $K(NiA^{2+}+L \rightleftharpoons NiAL^{2+})$ 1·69±0·05 where A = 2,2′-bipyridyl	67S
Zn^{2+}	gl	25	0·1 ($NaClO_4$)	$K_1 \leqslant 1$	64K
	gl	25	0·1 ($NaClO_4$) 50% dioxan	K_1 1·1±0·1 $K(ZnA^{2+}+L \rightleftharpoons ZnAL^{2+})$ 1·1±0·1 where A = 2,2′-bipyridyl	67S

64K K. Kahmann, H. Sigel, and H. Erlenmeyer, *Helv. Chim. Acta*, 1964, **47**, 1754
67S H. Sigel, *Chimia* (*Switz.*), 1967, **21**, 489

569.1 $C_7H_{10}O_4N_2$ 4-Carboxycyclohexane-1,2-dione dioxime H_3L

Metal	*Method*	*Temp*	*Medium*	*Log of equilibrium constant, remarks*	*Ref*
H^+	gl	?	0 corr	K_1 12·37, K_2 10·45, K_3 4·85	62B
Ni^{2+}	sp	25	9 KOH	K_1 28·7±0·6, K_2 0·8±0·2, K_3 3·7±0·7 (?)	62B

62B C. V. Banks and J. P. Laplante, *Analyt. Chim. Acta*, 1962, **27**, 80

570 $C_7H_{11}O_2N_3$ Histidine methyl ester L

Metal	*Method*	*Temp*	*Medium*	*Log of equilibrium constant, remarks*	*Ref*
H^+	gl	0—45	0·10 (KNO_3)	K_1 (0°) 7·82, (25°) 7·32, (45°) 6·94, K_2 (0°) 5·62, (25°) 5·30, (45°) 5·12	64A
	gl	0—40	0·25 (KCl)	K_1 (0°) 8·02, (15°) 7·58, (25°) 7·34, (40°) 7·06, K_2 (0°) 5·84, (15°) 5·51, (25°) 5·39, (40°) 5·16	65A
	gl	25	0·16 (KNO_3)	K_1 7·30±0·03, K_2 5·35±0·02	65C
	gl	25	0·1	K_1 7·23, K_2 5·01	67H
Cd^{2+}	pol	0—45	0·10 (KNO_3)	β_2 (0°) 8·34, (25°) 7·42, (45°) 7·10 $\Delta H_{\beta 2}°$ (0°) −11·2, $\Delta S_{\beta 2}°$ −3±1	64A
	gl	25	0·16 (KNO_3)	K_1 3·98±0·04, K_2 2·81±0·04, K_3 ~ 1	65C

contd

Metal	*Method*	*Temp*	*Medium*	*Log of equilibrium constant, remarks*	*Ref*
Co^{2+}	gl	0—40	0·25 (KCl)	K_1 (0°) 5·68, (15°) 5·00, (25°) 4·24, (40°) 4·10, K_2 (0°) 4·50, (15°) 3·57, (25°) 3·12, (40°) 2·96, K_3 (0°) 2·67, (15°) 2·18 $\Delta H_1°$ (15°) −16·0, $T\Delta S_1°$ (15°) −9·4 kcal mole^{-1}, $\Delta H_2°$ (15°) −16·0, $T\Delta S_2°$ (15°) −11·3 kcal mole^{-1} See histidine (447) for mixed ligands	65A
Cu^{2+}	gl	0—40	0·25 (KCl)	K_1 (0°) 9·84, (15°) 9·12, (25°) 8·32, (40°) 8·14, K_2 (0°) 7·10, (15°) 6·54, (25°) 5·82, (40°) 5·72, K_3 (0°) 2·42 $\Delta H_1°$ (15°) −18·3, $T\Delta S_1°$ (15°) −6·3 kcal mole^{-1}, $\Delta H_2°$ (15°) −16·0, $T\Delta S_2°$ (15°) −7·4 kcal mole^{-1} See histidine (447) and NTA (449) for mixed ligands	65A
	gl	25	0·16 (KNO_3)	K_1 8·48±0·04, K_2 5·90±0·04, K_3 1·6±0·2	65C
	gl	25	0·1	K_1 8·55, K_2 5·97	67H
Ni^{2+}	gl	0—40	0·25 (KCl)	K_1 (0°) 7·32, (15°) 6·65, (25°) 6·02, (40°) 5·95, K_2 (0°) 5·88, (15°) 5·14, (25°) 4·30, (40°) 4·34, K_3 (0°) 3·59, (15°) 2·76, (25°) 2·50, (40°) 2·11 $\Delta H_1°$ (15°) −16·0, $T\Delta S_1°$ (15°) −7·2 kcal mole^{-1}, $\Delta H_2°$ (15°) −16·0, $T\Delta S_2°$ (15°) −9·2 kcal mole^{-1}, $\Delta H_3°$ (15°) −17·2, $T\Delta S_3°$ (15°) −13·6 kcal mole^{-1}	65A
	gl	25	0·16 (KNO_3)	K_1 6·19±0·04, K_2 4·91±0·04, K_3 2·90±0·06 See histidine (447) for mixed ligands	65C
Zn^{2+}	gl	0—40	0·25 (KCl)	K_1 (0°) 5·29, (15°) 4·82, (25°) 4·40, (40°) 4·36, K_2 (0°) 4·71, (15°) 3·93, (25°) 3·78, (40°) 3·69, K_3 (0°) 1·90 $\Delta H_1°$ (15°) −9·2, $T\Delta S_1°$ (15°) −2·8 kcal mole^{-1}, $\Delta H_2°$ (15°) −9·2, $T\Delta S_2°$ (15°) −4·0 kcal mole^{-1}	65A
	gl	25	0·16 (KNO_3)	K_1 4·46±0·08, K_2 4·20±0·08, K_3 0·0 See histidine (447) for mixed ligands	65C

64A A. C. Andrews and J. K. Romary, *J. Chem. Soc.*, 1964, 405
65A A. C. Andrews and D. M. Bebolsky, *J. Chem. Soc.*, 1965, 742
65C H. L. Conley and R. B. Martin, *J. Phys. Chem.*, 1965, **69**, 2923
67H R. W. Hay and P. J. Morris, *Chem. Comm.*, 1967, 23

$C_7H_{11}O_4N$ ***N*-(Prop-2′-enyl)iminodiacetic acid(*N*-allyliminodiacetic acid) H_2L 571.1**

Metal	*Method*	*Temp*	*Medium*	*Log of equilibrium constant, remarks*	*Ref*
H^+	gl	25	0·1 KCl	K_1 9·28, K_2 2·41	66S
Ca^{2+}	gl	25	0·1 KCl	K_1 3·4, K_2 1·91	66S
Co^{2+}	gl	25	0·1 KCl	K_1 7·20, K_2 5·55	66S
Cu^{2+}	gl	25	0·1 KCl	K_1 10·22, K_2 6·07	66S
Ni^{2+}	gl	25	0·1 KCl	K_1 8·55, K_2 6·75	66S

66S P. Souchey, N. Israily, and P. Gouzerh, *Bull. Soc. chim. France*, 1966, **12**, 3917

572 $C_7H_{11}O_5N$ *N*-(Acetonyl)iminodiacetic acid H_2L

Metal	*Method*	*Temp*	*Medium*	*Log of equilibrium constant, remarks*	*Ref*
H^+	gl	25	0·1 (KNO_3)	K_1 7·71, K_2 2·62	63A, 6[illegible]
Ca^{2+}	gl	25	0·1 (KNO_3)	K_1 4·08	63A
Cd^{2+}	gl	25	0·1 (KNO_3)	K_1 6·84, K_2 4·12	63A
	gl	25	0·1 (KNO_3)	K_1 6·77, K_2 4·13	65A
Co^{2+}	gl	25	0·1 (KNO_3)	K_1 6·40, K_2 4·52	63A
	gl	25	0·1 (KNO_3)	K_1 6·37, K_2 4·50	65A
Cu^{2+}	gl	25	0·1 (KNO_3)	K_1 9·05, K_2 4·94	63A
	gl	25	0·1 (KNO_3)	K_1 9·10, K_2 4·88	65A
Mg^{2+}	gl	25	0·1 (KNO_3)	K_1 2·72	63A
Ni^{2+}	gl	25	0·1 (KNO_3)	K_1 7·32, K_2 5·29	63A
	gl	25	0·1 (KNO_3)	K_1 7·43, K_2 5·27	65A
Pb^{2+}	gl	25	0·1 (KNO_3)	K_1 7·70	63A
Sr^{2+}	gl	25	0·1 (KNO_3)	K_1 3·62	63A
Zn^{2+}	gl	25	0·1 (KNO_3)	K_1 7·01, K_2 3·64	63A
	gl	25	0·1 (KNO_3)	K_1 6·89, K_2 3·83	65A

63A T. Ando, *Bull. Chem. Soc. Japan*, 1963, **36**, 1593
65A T. Ando and K. Ueno, *Inorg. Chem.*, **4**, 375

572.1 $C_7H_{11}O_5N$ *N*-Acetyl-2-aminopentanedioic acid (*N*-acetylglutamic acid) H_2L

Metal	*Method*	*Temp*	*Medium*	*Log of equilibrium constant, remarks*	*Ref*
H^+	gl	20	0·1 NaCl	K_1 4·60, K_2 3·28	65N
Cu^{2+}	gl	25	0·1 $NaClO_4$	K_1 2·2	65N

65N M. H. T. Nyberg and M. Cefola, *Arch. Biochem. Biophys.*, 1965, **111**, 321

572.2 $C_7H_{11}O_6N$ *N*-(1′-Carboxyethyl)iminodiacetic acid H_3L

Metal	*Method*	*Temp*	*Medium*	*Log of equilibrium constant, remarks*	*Ref*
H^+	gl	20	0·1 (KCl)	K_1 10·47±0·02, K_2 2·46±0·02, K_3 1·57±0·05	66I
Ba^{2+}	gl	20	0·1 (KCl)	K_1 4·86±0·03	66I
Ca^{2+}	gl	20	0·1 (KCl)	K_1 6·97±0·02	66I
Mg^{2+}	gl	20	0·1 (KCl)	K_1 5·84±0·02	66I
Sr^{2+}	gl	20	0·1 (KCl)	K_1 5·18±0·02	66I
Zn^{2+}	gl	20	0·1 (KCl)	K_1 10·89±0·03	66I

66I H. M. N. H. Irving and M. G. Miles, *J. Chem. Soc.* (*A*), 1966, 1268

$C_7H_{11}O_6N$ *N*-(2′-Carboxyethyl)iminodiacetic acid H_3L 573

Metal	Method	Temp	Medium	Log of equilibrium constant, remarks	Ref
H^+	gl	25	0·1 (KNO_3)	K_1 9·61, K_2 3·71, K_3 1·90	67U
Cd^{2+}	gl	25	0·1 (KNO_3)	K_1 8·24 $K(Cd^{2+}+HL^{2-} \rightleftharpoons CdHL)$ 2·15	67U
Co^{2+}	gl	25	0·1 (KNO_3)	K_1 10·00	67U
Cu^{2+}	pol	25	0·1 (KNO_3)	K_1 13·22	67U
Mn^{2+}	gl	25	0·1 (KNO_3)	K_1 7·33, $K(Mn^{2+}+HL^{2-} \rightleftharpoons MnHL)$ 1·51	67U
Ni^{2+}	pol	25	0·1 (KNO_3)	K_1 11·37	67U
Zn^{2+}	pol	25	0·1 (KNO_3)	K_1 9·98	67U

67U E. Uhlig and R. Krannich, *J. Inorg. Nuclear Chem.*, 1967, 29, 1164

$C_7H_{12}O_2N_2$ Cycloheptane-1,2-dione dioxime (heptoxime) HL 575

Metal	Method	Temp	Medium	Log of equilibrium constant, remarks	Ref
H^+	gl	25	0·1 75% dioxan	K_1 13·33	63B
Ni^{2+}	gl	25	0·1 0—75% dioxan	K_1 (75%) 12·3±0·2, β_2 (→ 0%) 19·4, (75%) 24·7±0·2, K_{so} −26·64	63B
	sol	20	0·1 ($NaClO_4$)	K_1 10·77±0·05, K_2 10·39±0·02	64A
	dis	25	0·1 ($NaClO_4$)	K_1 11·2±0·2, K_2 10·68±0·10	64S

63B C. V. Banks and S. Anderson, *Inorg. Chem.*, 1963, 2, 112
64A E. K. Astkhova, V. M. Savostina, and V. M. Peshkova, *Russ. J. Inorg. Chem.*, 1964, 9, 452 (817)
64S V. M. Savostina, E. K. Astakhova, and V. M. Peshkova, *Russ. J. Inorg. Chem.*, 1964, 9, 42 (80)

$C_7H_{12}O_2N_2$ 3-Methylcyclohexane-1,2-dione dioxime (3-methylnioxime) HL 576

Metal	Method	Temp	Medium	Log of equilibrium constant, remarks	Ref
H^+	gl	25	0·1 75% dioxan	K_1 13·23	63B
Ni^{2+}	gl	25	0·1 0—75% dioxan	K_1 (75%) 11·3±0·1, β_2 (→ 0%) 18·28, (75%) 23·52±0·09, K_{so} −27·62	63B

63B C. V. Banks and S. Anderson, *Inorg. Chem.*, 1963, 2, 112

577 $C_7H_{12}O_2N_2$ 4-Methylcyclohexane-1,2-dione dioxime HL

Metal	Method	Temp	Medium	Log of equilibrium constant, remarks	Ref
H^+	gl	25	0·1 75% dioxan	K_1 13·07	63B
Ni^{2+}	gl	25	0·1 0—75% dioxan	K_1 (75%) 10·8±0·3, β_2 (→ 0%) 17·9, (75%) 23·0±0·3, K_{so} −28·25	63B

63B C. V. Banks and S. Anderson, *Inorg. Chem.*, 1963, **2**, 112

578 $C_7H_{12}O_3N_2$ Glycyl-L-proline HL

Metal	Method	Temp	Medium	Log of equilibrium constant, remarks	Ref
H^+	gl	25	0·16 KCl	K_1 8·48, K_2 2·97	60K
Cu^{2+}	gl	25	0·16 KCl	K_1 6·43, K_2 5·02	60K
Ni^{2+}	gl	25	0·16 (KNO_3)	$K(NiLOH+H^+ \rightleftharpoons NiL^+)$ 10·7	60M

60K W. L. Koltun, M. Fried, and F. R. N. Gurd, *J. Amer. Chem. Soc.*, 1960, **82**, 233
60M R. B. Martin, M. Chamberlin, and J. T. Edsall, *J. Amer. Chem. Soc.*, 1960, **82**, 495

578.1 $C_7H_{12}O_3N_2$ L-Prolylgylcine HL

Metal	Method	Temp	Medium	Log of equilibrium constant, remarks	Ref
H^+	gl	25	0·16 KCl	K_1 8·97, K_2 3·19	60K
Cu^{2+}	gl	25	0·16 KCl	K_1 6·39, $K[Cu(H_{-1}L)+H^+ \rightleftharpoons CuL^+]$ 3·95 $K[Cu(H_{-1}L)OH^- + H^+ \rightleftharpoons Cu(H_{-1}L)]$ 9·34 $K[Cu(H_{-1}L)+L^- \rightleftharpoons CuL(H_{-1}L)^-]$ 2·66 $K[Cu(H_{-1}L)+Cu(H_{-1}L)OH^- \rightleftharpoons (CuH_{-1}L)_2OH^-]$ 1·88	60K

60K W. L. Koltun, M. Fried, and F. R. N. Gurd, *J. Amer. Chem. Soc.*, 1960, **82**, 233

579 $C_7H_{12}O_4S_2$ 3,7-Dithianonanedioic acid [(trimethylenedithio)diacetic acid H_2L

Metal	Method	Temp	Medium	Log of equilibrium constant, remarks	Ref
H^+	gl	20	0·1	K_1 4·25, K_2 3·44	61O
Cu^{2+}	gl	20	0·1	K_1 4·41, $K(Cu^{2+}+HL^- \rightleftharpoons CuHL^+)$ 3·05	61O

61O G. Ostacoli, E. Campi, N. Cibrario, and G. Saini, *Gazzetta*, 1961, **91**, 349

$C_7H_{12}O_4S_2$ 4,6-Dithianonanedioic acid [3,3′-(methylenedithio)dipropanoic acid] H_2L 579.1

Metal	*Method*	*Temp*	*Medium*	*Log of equilibrium constant, remarks*	*Ref*
H^+	gl	20	0·1	K_1 4·86, K_2 4·08	61O
Cu^{2+}	gl	20	0·1	K_1 2·06, $K(Cu^{2+} + HL^- \rightleftharpoons CuHL^+)$ 1·38	61O

61O G. Ostacoli, E. Campi, N. Cibrario, and G. Saini, *Gazzetta*, 1961, **91**, 349

$C_7H_{13}O_2N$ Piperidine-*N*-acetic acid HL 581.1

Metal	*Method*	*Temp*	*Medium*	*Log of equilibrium constant, remarks*	*Ref*
H^+	H	20	0·1 (KCl)	K_1 10·25, K_2 2·13	63I
Cu^{2+}	H	20	0·1 (KCl)	K_1 6·33, β_2 ~ 12	63I
Ni^{2+}	H	20	0·1 (KCl)	K_1 3·7, β_2 ~ 7	63I

63I H. Irving and L. D. Pettit, *J. Chem. Soc.*, 1963, 3051

$C_7H_{13}O_4N$ *N*-Propyliminodiacetic acid H_2L 581.2

Metal	*Method*	*Temp*	*Medium*	*Log of equilibrium constant, remarks*	*Ref*
H^+	gl	25	0·1 KCl	K_1 10·09, K_2 2·43	66S
Ca^{2+}	gl	25	0·1 KCl	K_1 3·4, K_2 2·0	66S
Co^{2+}	gl	25	0·1 KCl	K_1 7·55, K_2 5·85	66S
Cu^{2+}	gl	25	0·1 KCl	K_1 10·47, K_2 6·13	66S
Ni^{2+}	gl	25	0·1 KCl	K_1 8·80, K_2 6·80	66S

66S P. Souchey, N. Israily, and P. Gouzerh, *Bull. Soc. chim. France*, 1966, **12**, 3917

$C_7H_{14}O_3N_2$ Glycyl-L-valine HL 584

Erratum. 1964, p. 559, Table 584. This Table has been renumbered **587.1** for correct ligand sequence.

$C_7H_{13}O_4N_3$ L-Alanylglycylglycine HL 584.1

Metal	*Method*	*Temp*	*Medium*	*Log of equilibrium constant, remarks*	*Ref*
H^+	gl	25	0·16 KCl	K_1 8·05, K_2 3·36	66B
Cu^{2+}	gl	25	0·16 KCl	K_1 4·81 $K[Cu(H_{-1}L) + H^+ \rightleftharpoons CuL^+]$ 4·98 $K[Cu(H_{-2}L)^- + H^+ \rightleftharpoons Cu(H_{-1}L)]$ 6·84 $K[Cu(H_{-2}L)OH^{2-} + H^+ \rightleftharpoons Cu(H_{-2}L)^-]$ 11·2	66B

66B G. F. Bryce and F. R. N. Gurd, *J. Biol. Chem.*, 1966, **241**, 1439

584.2 $C_7H_{13}O_4N_3$ Glycyl-L-alanylglycine HL

Metal	Method	Temp	Medium	Log of equilibrium constant, remarks	Ref
H^+	gl	25	0·16 KCl	K_1 8·08, K_2 3·46	66B
Cu^{2+}	gl	25	0·16 KCl	K_1 5·18	66B
				$K[Cu(H_{-1}L)+H^+ \rightleftharpoons CuL^+]$ 5·32	
				$K[Cu(H_{-2}L)^- + H^+ \rightleftharpoons Cu(H_{-1}L)]$ 6·62	
				$K[Cu(H_{-2}L)OH^{2-} + H^+ \rightleftharpoons Cu(H_{-2}L)^-]$ 11·4	

66B G. F. Bryce and F. R. N. Gurd, *J. Biol. Chem.*, 1966, **241**, 1439

584.3 $C_7H_{13}O_4N_3$ Glycylglycyl-L-alanine HL

Metal	Method	Temp	Medium	Log of equilibrium constant, remarks	Ref
H^+	gl	25	0·16 KCl	K_1 8·02, K_2 3·30	66B
Cu^{2+}	gl	25	0·16 KCl	K_1 5·08	66B
				$K[Cu(H_{-1}L)+H^+ \rightleftharpoons CuL^+]$ 5·10	
				$K[Cu(H_{-2}L)^- + H^+ \rightleftharpoons Cu(H_{-1}L)]$ 6·89	

66B G. F. Bryce and F. R. N. Gurd, *J. Biol. Chem.*, 1966, **241**, 1439

587 $C_7H_{14}O_3N_2$ L-Valylglycine HL

Metal	Method	Temp	Medium	Log of equilibrium constant, remarks	Ref
H^+	gl	25	0·16 KCl	K_1 8·00, K_2 3·23	60K
Cu^{2+}	gl	25	0·16 KCl	K_1 4·87	60K
				$K[Cu(H_{-1}L)+H^+ \rightleftharpoons CuL^+]$ 3·85	
				$K[Cu(H_{-1}L)OH^- + H^+ \rightleftharpoons Cu(H_{-1}L)]$ 9·13	
				$K[Cu(H_{-1}L)(OH)_2^{2-} + H^+ \rightleftharpoons Cu(H_{-1}L)OH^-]$ 11·8	
				$K[Cu(H_{-1}L)+L^- \rightleftharpoons CuL(H_{-1}L)^-]$ 2·22	
				$K[Cu(H_{-1}L)+Cu(H_{-1}L)OH^- \rightleftharpoons (CuH_{-1}L)_2OH^-]$ 2·24	
Ni^{2+}	gl	25	0·16 (KNO_3)	$K[Ni(H_{-1}L)+H^+ \rightleftharpoons NiL^+]$ 9·0	60M
				$K[Ni(H_{-1}L)OH^- + H^+ \rightleftharpoons Ni(H_{-1}L)]$ 9·6	

60K W. K. Koltun, M. Fried, and F. R. N. Gurd, *J. Amer. Chem. Soc.*, 1960, **82**, 233
60M R. B. Martin, M. Chamberlin, and J. T. Edsall, *J. Amer. Chem. Soc.*, 1960, **82**, 495

$C_7H_{14}O_3N_2$ Glycyl-L-valine HL 587.1*

Metal	Method	Temp	Medium	Log of equilibrium constant, remarks	Ref
H^+	gl	25	0·16 KCl	K_1 8·18, K_2 3·15	60K
	gl	25	0·16 KCl	K_1 8·20, K_2 3·26	65B
Cu^{2+}	gl	25	0·16 KCl	K_1 5·62,	60K
				$K[Cu(H_{-1}L)+H^+ \rightleftharpoons CuL^+]$ 4·75	
				$K[Cu(H_{-1}L)OH^- + H^+ \rightleftharpoons Cu(H_{-1}L)]$ 9·30	
				$K[Cu(H_{-1}L)(OH)_2{}^{2-} + H^+ \rightleftharpoons Cu(H_{-1}L)OH^-]$ 12·0	
				$K[Cu(H_{-1}L)+L^- \rightleftharpoons CuL(H_{-1}L)^-]$ 2·94	
				$K[Cu(H_{-1}L)+Cu(H_{-1}L)OH^- \rightleftharpoons (CuH_{-1}L)_2OH^-]$ 2·18	
	gl	25	0·16 KCl	K_1 5·68,	65B
				$K[Cu(H_{-1}L)+H^+ \rightleftharpoons CuL^+]$ 4·85	
Ni^{2+}	gl	25	0·16 (KNO_3)	$K[Ni(H_{-1}L)+H^+ \rightleftharpoons NiL^+]$ 10·4	60M
				$K[Ni(H_{-1}L)OH^- + H^+ \rightleftharpoons Ni(H_{-1}L)]$ 11·0	

* Note. This Table was numbered 584 on page 559 in the 1964 edition.

60K W. L. Koltun, M. Fried, and F. R. N. Gurd, *J. Amer. Chem. Soc.*, 1960, **82**, 233
60M R. B. Martin, M. Chamberlin, and J. T. Edsall, *J. Amer. Chem. Soc.*, 1960, **82**, 495
65B G. F. Bryce, J. M. H. Pinkerton, L. K. Steinrauf, and F. R. N. Gurd, *J. Biol. Chem.*, 1965, **240**, 3829

$C_7H_{14}O_3N_2$ Glycyl-DL-norvaline HL 587.2

Metal	Method	Temp	Medium	Log of equilibrium constant, remarks	Ref
H^+	gl	25	0·12 (NaCl)	K_1 8·32±0·01, K_2 3·24±0·01	67S
Cu^{2+}	gl	25	0·12 (NaCl)	K_1 5·88,	67S
				$K[Cu(H_{-1}L)+H^+ \rightleftharpoons CuL^+]$ 4·82	
				$K[Cu(H_{-1}L)OH^- + H^+ \rightleftharpoons Cu(H_{-1}L)]$ 9·77	
				$K[Cu(H_{-1}L)+L^- \rightleftharpoons CuL(H_{-1}L)^-]$ 3·59	

67S U. I. Salakhutdinov, A. P. Borisona, Yu. V. Granovskii, I. A. Savich, and V. I. Spitsyn, *Proc. Acad. Sci.* (*U.S.S.R.*), 1967, **177**, 1039 (365)

$C_7H_{15}O_2N$ Valine ethyl ester L 587.3

Metal	Method	Temp	Medium	Log of equilibrium constant, remarks	Ref
Co^{2+}				See nitrilotriacetic acid (449) for mixed ligands	
Cu^{2+}				See nitrilotriacetic acid (449) for mixed ligands	
Mn^{2+}				See nitrilotriacetic acid (449) for mixed ligands	
Ni^{2+}				See nitrilotriacetic acid (449) for mixed ligands	
Pb^{2+}				See nitrilotriacetic acid (449) for mixed ligands	
Zn^{2+}				See nitrilotriacetic acid (449) for mixed ligands	

588 $C_7H_{15}NS_2$ ***NN*-Dipropyldithiocarbamic acid HL**

Metal	Method	Temp	Medium	Log of equilibrium constant, remarks	Ref
Cu^{2+}	sp	23—27	0·01 NaOH 75% ethanol	K_1 15·1±0·2, K_2 14·2±0·2	56J
	sp	18—22	0·01 NaOH 0—89% ethanol	K_1 (→ 0%) 12·5±0·2, (51·7%) 14·1±0·2, (75%) 15·1±0·2, (89%) 16·1±0·2, K_2 (→ 0%) 11·5±0·2, (51·7%) 13·5±0·2, (75%) 14·3±0·2, (89%) 14·9±0·2	57J

Erratum. 1964, p. 560, Table 588. *Insert* **HL** in the heading.

56J M. J. Janssen, *Rec. Trav. chim.*, 1956, **75**, 1411

57J M. J. Janssen, *Rec. Trav. chim.*, 1957, **76**, 827

588.1 $C_7H_{16}O_2N_4$ ***NN'*-Diglycyl-1,3-diaminopropane L**

Metal	Method	Temp	Medium	Log of equilibrium constant, remarks	Ref
H^+	gl	23	0·5 KCl	K_1 8·53, K_2 7·84	67Z
Cu^{2+}	gl	23	0·5 KCl	K_1 8·25 $K(CuL^{2+} + H^+ \rightleftharpoons CuHL^{3+})$ 5·61 $K[Cu(H_{-2}L) + 2H^+ \rightleftharpoons CuL^{2+}]$ 11·33	67Z

67Z A. Zuberbuhler and S. Fallab, *Helv. Chim. Acta*, 1967, **50**, 889

588.2 $C_7H_{16}O_2N_4$ ***NN'*-Bis(2-aminoethyl)malonamide L**

Metal	Method	Temp	Medium	Log of equilibrium constant, remarks	Ref
H^+	gl	22	0·1 (KNO_3)	K_1 9·40, K_2 8·68	68G
Cu^{2+}	gl	22	0·1 (KNO_3)	K_1 7·93 $K[Cu(H_{-1}L)^+ + H^+ \rightleftharpoons CuL^{2+}]$ 6·37	68G

68G R. Griesser and S. Fallab, *Chimia (Switz.)*, 1968, **22**, 90

588.3 $C_7H_{17}ON_5$ **Methoxyethylbiguanide (?) L**

Metal	Method	Temp	Medium	Log of equilibrium constant, remarks	Ref
H^+	gl	28—32	0·2 KCl	K_1 11·47, K_2 3·00	60S
Cu^{2+}	sp	28—32	0·2 KCl	K_1 9·77, K_2 7·50	60S
Ni^{2+}	sp	28—32	0·2 KCl	β_2 11·96	60S

60S N. R. Sengupta and P. Ray, *J. Indian Chem. Soc.*, 1960, **37**, 303

$C_7H_4O_4NCl$ 4-Chloropyridine-2,6-dicarboxylic acid H_2L 589.1

etal	*Method*	*Temp*	*Medium*	*Log of equilibrium constant, remarks*	*Ref*
+	gl	22	0·1 ($NaClO_4$)	K_1 3·75, K_2 1·7	64B
$^{2+}$	gl	22	0·1 ($NaClO_4$)	K_1 3·19	64B
ı$^{2+}$	gl	22	0·1 ($NaClO_4$)	K_1 3·61	64B
g^{2+}	gl	22	0·1 ($NaClO_4$)	K_1 2·38	64B
$^{2+}$	gl	22	0·1 ($NaClO_4$)	K_1 3·29	64B

64B E. Blasius and B. Brozio, *J. Electrochem. Soc.*, 1964, **68**, 52

$C_7H_5O_2NS_2$ 5-(2-Furyl)-2-thioxo-1,3-thiazolidine-4-one [5-(furylidene)rhodanine*] HL 590.1

etal	*Method*	*Temp*	*Medium*	*Log of equilibrium constant, remarks*	*Ref*
+	dis	20	0·1 ($NaClO_4$)	K_1 6·37	65N
g$^+$	dis	20	0·1 ($NaClO_4$)	K_1 7·19	65N

2- or 3-Furylidene not specified.

65N O. Navratil and J. Kotas, *Coll. Czech. Chem. Comm.*, 1965, **30**, 2736

$C_7H_6O_2NCl$ 5-Chloro-2-hydroxybenzaldehyde oxime (5-chlorosalicylaldoxime) HL 592.1

etal	*Method*	*Temp*	*Medium*	*Log of equilibrium constant, remarks*	*Ref*
$^+$	gl	20	0·1 ($NaClO_4$) 75% dioxan	K_1 10·25±0·05	65B
o^{2+}	gl	20	0·1 ($NaClO_4$) 75% dioxan	K_1 6·3±0·2, K_2 7·0±0·1	65B
u^{2+}	sp	20	0·1 ($NaClO_4$) 75% dioxan	β_2 21·4±0·2	65B
e^{2+}	gl	20	0·1 ($NaClO_4$) 75% dioxan	K_1 8·2±0·1, K_2 6·4±0·1	65B
In^{2+}	gl	20	0·1 ($NaClO_4$) 75% dioxan	K_1 4·8±0·1, K_2 5·7±0·1	65B
Ni^{2+}	gl	20	0·1 ($NaClO_4$) 75% dioxan	K_1 6·6±0·1, K_2 7·1±0·1	65B
n^{2+}	gl	20	0·1 ($NaClO_4$) 75% dioxan	K_1 5·8±0·2, K_2 5·8±0·2	65B

65B K. Burger and I. Egyed, *J. Inorg. Nuclear Chem.*, 1965, **27**, 2361

$C_7H_7O_5NS$ 5-Sulpho-2-aminobenzoic acid H_2L 594

Metal	*Method*	*Temp*	*Medium*	*Log of equilibrium constant, remarks*	*Re*
I$^+$	gl	22	0·1 KCl	K_1 4·35	60U
Cu^{2+}	gl	22	0·1 KCl	K_1 2·75, K_2 2·15	60U

60U E. Uhlig, *Chem. Ber.*, 1960, **93**, 2470

595.1 $C_7H_8O_3N_2S$ 2-Ethylthio-1*H*-1,3-diazin-4-one-5-carboxylic acid (2-ethylthio-iso-orotic acid) H

Metal	Method	Temp	Medium	Log of equilibrium constant, remarks	Re
H^+	gl	25	0·1 $[(CH_3)_4NBr]$	K_1 10·52, K_2 6·01	61
Cd^{2+}	gl	25	0·1 KCl	$K(Cd^{2+}+HL^- \rightleftharpoons CdHL^+)$ 1·98	61
Co^{2+}	gl	25	0·1 KCl	$K(Co^{2+}+HL^- \rightleftharpoons CoHL^+)$ 2·47	61
Cu^{2+}	gl	25	0·1 KCl	$K(Cu^{2+}+HL^- \rightleftharpoons CuHL^+)$ 3·14	61
Mn^{2+}	gl	25	0·1 KCl	$K(Mn^{2+}+HL^- \rightleftharpoons MnHL^+)$ 2·07	61
Ni^{2+}	gl	25	0·1 KCl	$K(Ni^{2+}+HL^- \rightleftharpoons NiHL^+)$ 2·70	61
Zn^{2+}	gl	25	0·1 KCl	$K(Zn^{2+}+HL^- \rightleftharpoons ZnHL^+)$ 2·33	61

61T E. R. Tucci, E. Doody, and N. C. Li, *J. Phys. Chem.*, 1961, **65**, 1570

595.2 $C_7H_9O_4NS$ Hydroxy(6-methyl-2-pyridyl)methanesulphonic acid H_2L

Metal	Method	Temp	Medium	Log of equilibrium constant, remarks	Re
H^+	gl	25	0·1 $(NaClO_4)$	K_1 9·88, K_2 4·83	64B
Co^{2+}	gl	25	0·1 $(NaClO_4)$	K_1 4·25, K_2 3·58	64B
Cu^{2+}	gl	25	0·1 $(NaClO_4)$	K_1 7·75, K_2 5·32	64B
Ni^{2+}	gl	25	0·1 $(NaClO_4)$	K_1 4·27, K_2 2·84	64B
Zn^{2+}	gl	25	0·1 $(NaClO_4)$	K_1 4·79, K_2 3·59	64B

64B L. Banford and W. J. Geary, *J. Chem. Soc.*, 1964, 378

598.1 $C_7H_{14}O_3N_2S$ 2-Amino-3-(*N*′-acetyl-2′-aminoethylthio)propanoic acid [*S*-(2-acetylaminoethyl)cysteine] H

Metal	Method	Temp	Medium	Log of equilibrium constant, remarks	Ref
H^+	gl	20	0·1 KNO_3	K_1 8·65	68H
Cu^{2+}	gl	20	0·1 KNO_3	K_1 7·50, K_2 6·47	68H

68H P. Herman and K. Lemke, *Z. physiol. Chem.*, 1968, **349**, 390

598.2 $C_7H_{14}O_4N_2S$ 2-Amino-3-(*N*′-acetyl-2′-aminoethylsulphinyl)propanoic acid HL

Metal	Method	Temp	Medium	Log of equilibrium constant, remarks	Ref
H^+	gl	20	0·1 KNO_3	K_1 7·57	68H
Cu^{2+}	gl	20	0·1 KNO_3	K_1 6·86, K_2 5·82	68H

68H P. Herman and K. Lemke, *Z. physiol. Chem.*, 1968, **349**, 390

$C_7H_{14}O_5N_2S$ 2-Amino-3-(*N*′-acetyl-2′-aminoethylsulphonyl)propanoic acid HL 598.3

etal	Method	Temp	Medium	Log of equilibrium constant, remarks	Ref
$^{+}$	gl	20	$0{\cdot}1$ KNO_3	K_1 7·44	68H
$^{2+}$	gl	20	$0{\cdot}1$ KNO_3	K_1 6·89, K_2 5·76	68H

68H P. Herman and K. Lemke, *Z. physiol. Chem.*, 1968, **349**, 390

$C_7H_{14}O_8N_3P$ Glycyl-*O*-phosphoryl-DL-serylglycine (glycyl-DL-phosphoserylglycine) H_3L 598.4

letal	Method	Temp	Medium	Log of equilibrium constant, remarks	Ref
$^{+}$	gl	25	0·15 (KCl)	K_1 8·22, K_2 5·78, K_3 3·29	62O
a^{2+}	gl	25	0·15 (KCl)	K_1 1·81	62O
				$K(Ca^{2+} + HL^{2-} \rightleftharpoons CaHL)$ 1·45	
lg^{2+}	gl	25	0·15 (KCl)	K_1 1·79	62O
				$K(Mg^{2+} + HL^{2-} \rightleftharpoons MgHL)$ 1·46	
				$K(MgL^{-} + Mg^{2+} \rightleftharpoons Mg_2L^{+})$ 0·9	
In^{2+}	gl	25	0·15 (KCl)	$K(Mn^{2+} + HL^{2-} \rightleftharpoons MnHL)$ 2·08	62O

62O R. Osterberg, *Acta Chem. Scand.*, 1962, **16**, 2434

C_8H_8 Vinylbenzene (styrene) L 599

letal	Method	Temp	Medium	Log of equilibrium constant, remarks	Ref
g^{+}	dis	25	1 (KNO_3)	K_1 1·28, K_2 0·2	68F

68F T. Fueno, O. Kajimoto, and J. Furukawa, *Bull. Chem. Soc. Japan*, 1968, **41**, 782

C_8H_{10} 1,3-Dimethylbenzene (*m*-xylene) L 602

Metal	Method	Temp	Medium	Log of equilibrium constant, remarks	Ref
g^{+}	sp	1·4—20	*m*-xylene ClO_4^-	K_1 (1·4°) 3·62, (20°) 6·07 ΔH 4·5	64T

64T B. G. Torre-Mori, D. Janjic, and B. P. Susz, *Helv. Chim. Acta*, 1964, **47**, 1172

603.1 C_8H_{14} Ethylidenecyclohexane L

Metal	Method	Temp	Medium	Log of equilibrium constant, remarks	Re
Ag^+	oth	30	1·77 $AgNO_3$ ethylene glycol	K_1 0·48, gas chromatography	62

62G E. Gil-Av and J. Herling, *J. Phys. Chem.*, 1962, **66,** 1208

603.2 C_8H_{14} 1-Ethylcyclohexene L

Metal	Method	Temp	Medium	Log of equilibrium constant, remarks	Re
Ag^+	oth	30	1·77 $AgNO_3$ ethylene glycol	K_1 0·11, gas chromatography	62

62G E. Gil-Av and J. Herling, *J. Phys. Chem.*, 1962, **66,** 1208

604 $C_8H_6O_4$ Benzene-1,2-dicarboxylic acid (phthalic acid) H_2L

Metal	Method	Temp	Medium	Log of equilibrium constant, remarks	Ref
H^+	gl	25	0·15 ($NaClO_4$)	K_1 5·13, K_2 2·91	62E
	H, Ag-AgCl	0—60	→ 0	K_1 (0°) 5·432, (5°) 5·418, (10°) 5·410, (15°) 5·405, (20°) 5·405, (25°) 5·408, (30°) 5·416, (35°) 5·427, (40°) 5·442, (45°) 5·462, (50°) 5·485, (55°) 5·512, (60°) 5·541 K_2 (0°) 2·925, (5°) 2·927, (10°) 2·931, (15°) 2·937, (20°) 2·934, (25°) 2·950, (30°) 2·958, (35°) 2·967, (40°) 2·978, (45°) 2·988, (50°) 3·001, (55°) 3·014, (60°) 3·028 $\Delta H_1°$ (25°) 0·50, $\Delta S_1°$ (25°) 26·4, $\Delta H_2°$ (25°) 0·64, $\Delta S_2°$ 15·6	45H, 62D
	gl	30	0·1 ($NaClO_4$)	K_1 5·40, K_2 3·10	66K
	gl	25	1·00 (KNO_3)	K_1 4·73±0·01, K_2 2·63±0·01	67R
Be^{2+}	gl	25	0·15 ($NaClO_4$)	K_1 3·97, β_2 5·69	62B
Ce^{3+}	gl	30	0·1 ($NaClO_4$)	K_1 3·96, K_2 3·16	66K
Co^{2+}	H, Ag-AgCl	0—45	0 corr	K_1 (25°), 2·831±0·003 K_1 $5{\cdot}690-2{\cdot}374\times10^{-2}T+4{\cdot}752\times10^{-5}T^2$ $\Delta H_1°$ 1·87±0·05, $\Delta S_1°$ 19·2±0·2	62D
	ix	25	0 corr	K_1 2·76±0·06, K_2 0·9±0·1 (?)	65S
	gl	25	0 corr	K_1 2·86±0·02	65M

co

$C_8H_6O_4$ (contd) 604

Metal	Method	Temp	Medium	Log of equilibrium constant, remarks	Ref
Co^{3+}	gl	25, 35	0 corr	$K[CoCl(NH_3)_5^{2+} + L^{2-} \rightleftharpoons CoCl(NH_3)_5L]$ (25°) 2·54, (35°) 2·63	65A
	sol	25, 35	0 corr	$K[CoCl(NH_3)_5^{2+} + L^{2-} \rightleftharpoons CoCl(NH_3)_5L]$ (25°) 2·51, (35°) 2·52	65A
Cr^{3+}	gl	25	0·1	K_1 5·52, K_2 4·48, K_3 2·48	67H
Dy^{3+}	gl	30	0·1 ($NaClO_4$)	K_1 4·53, K_2 3·49	66K
Er^{3+}	gl	30	0·1 ($NaClO_4$)	K_1 4·43, K_2 3·36	66K
Eu^{3+}	gl	30	0·1 ($NaClO_4$)	K_1 4·12, K_2 3·15	66K
Gd^{3+}	gl	30	0·1 ($NaClO_4$)	K_1 4·28, K_2 3·30	66K
Hg_2^{2+}	sp	18	0·1	K_1 4·90	66G
Ho^{3+}	gl	30	0·1 ($NaClO_4$)	K_1 3·97, K_2 3·15	66K
La^{3+}	gl	30	0·1 ($NaClO_4$)	K_1 4·10, K_2 3·18	66K
Mn^{2+}	H, Ag-AgCl	0—45	0 corr	K_1 (25°) 2·741 ± 0·004 K_1 6·365 − 2·975 × $10^{-2}T$ + 5·897 × $10^{-5}T^2$ $\Delta H_1°$ 2·20 ± 0·05, $\Delta S_1°$ 19·9 ± 0·3	62D
Na^+	gl	25	0 corr	K_1 0·7	65A
Nd^{3+}	gl	30	0·1 ($NaClO_4$)	K_1 4·22, K_2 3·25	66K
Ni^{2+}	H, Ag-AgCl	0—45	0 corr	K_1 (25°) 2·952 ± 0·003 K_1 7·795 − 3·867 × $10^{-2}T$ + 6·912 × $10^{-5}T^2$ $\Delta H_1°$ 1·77 ± 0·05, $\Delta S_1°$ 19·4 ± 0·3	62D
Pr^{3+}	gl	30	0·1 ($NaClO_4$)	K_1 4·22, K_2 3·21	66K
Sm^{3+}	gl	30	0·1 ($NaClO_4$)	K_1 4·99, K_2 3·38	66K
Tb^{3+}	gl	30	0·1 ($NaClO_4$)	K_1 4·06, K_2 3·09	66K
Th^{4+}				See EDTA (836) and CDTA (997) for mixed ligands.	
Tm^{3+}	gl	30	0·1 ($NaClO_4$)	K_1 4·45, K_2 3·35	66K
U^{4+}				See EDTA (836) for mixed ligands.	
UO_2^{2+}	gl	25	1·00 (KNO_3)	K_1 4·38 ± 0·03	67R
	gl	31	0·1 ($NaClO_4$)	K_1 4·81	67S
Y^{3+}	gl	30	0·1 ($NaClO_4$)	K_1 4·04, K_2 3·08	66K
Zn^{2+}	H, Ag-AgCl	0—45	0 corr	K_1 (25°) 2·893 ± 0·006 K_1 7·635 − 3·956 × $10^{-2}T$ + 7·937 × $10^{-5}T^2$ $\Delta H_1°$ 3·2 ± 0·1, $\Delta S_1°$ 23·8 ± 0·4	65N

45H W. J. Hamer, G. D. Pinching, and S. F. Acree, *J. Res. Nat. Bur. Standards*, 1945, **35**, 381, 539
62B H. J. de Bruin, D. Kaitis, and R. B. Temple, *Austral. J. Chem.*, 1962, **15**, 457
62D I. R. Desai and V. S. K. Nair, *J. Chem. Soc.*, 1962, 2360
65A D. W. Archer, D. A. East, and C. B. Monk, *J. Chem. Soc.*, 1965, 720
65M C. B. Monk, *J. Chem. Soc.*, 1965, 2456
65N V. S. K. Nair, *J. Chem. Soc.*, 1965, 1450
65S R. G. Seys and C. B. Monk, *J. Chem. Soc.*, 1965, 2452
66G Z. Gregorowicz and J. Ciba, *Roczniki Chem.*, 1966, **40**, 1377
66K M. Krishnamurthy and N. S. K. Prasad, *Indian J. Chem.*, 1966, **4**, 316
67H K. Higashi, K. Hori, and R. Tsuchiya, *Bull. Chem. Soc. Japan*, 1967, **40**, 2569
67R K. S. Rajan and A. E. Martell, *J. Inorg. Nuclear Chem.*, 1967, **29**, 523
67S A. J. Singh and N. S. K. Prasad, *Indian J. Chem.*, 1967, **5**, 573

$C_8H_6N_4$ 2,2′-Bis(1,3-diazine) (2,2′-bipyrimidine) L 605.1

Metal	Method	Temp	Medium	Log of equilibrium constant, remarks	Ref
Fe^{2+}	spJ	?	0·2 (NaCl)	β_3 7·53	63B

63B D. D. Bly and M. G. Mellon, *Analyt. Chem.*, 1963, **35**, 1386

605.2 $C_8H_7N_3$ 2-(2′-Pyridyl)imdiazole L

Metal	*Method*	*Temp*	*Medium*	*Log of equilibrium constant, remarks*	*Ref*
H^+	pH	25	0·1 NO_3^-	K_1 5·470±0·002	67E
	sp	25	0·1 NO_3^-	K_2 −0·70	67E
Cd^{2+}	pH	25	0·1 NO_3^-	K_1 4·70±0·01, β_2 8·16±0·01, β_3 10·74±0·01	67E
Co^{2+}	pH	25	0·1 NO_3^-	K_1 5·263±0·005, β_2 10·048±0·005, β_3 13·871±0·005	67E
Cu^{2+}	pH	25	0·1 NO_3^-	K_1 7·94±0·02, β_2 13·64±0·02, β_3 16·92±0·02	67E
Fe^{2+}	pH	25	0·1 NO_3^-	K_1 4·097±0·005, β_2 7·90±0·01, β_3 11·600±0·005	67E
Hg^{2+}	pH	25	0·1 NO_3^-	K_1 10·07, β_2 18·28	67E
Ni^{2+}	pH	25	0·1 NO_3^-	K_1 6·39±0·03, β_2 12·61±0·01, β_3 17·80±0·02	67E
Zn^{2+}	pH	25	0·1 NO_3^-	K_1 4·39±0·01, β_2 8·96±0·01, β_3 12·07±0·02	67E

67E W. J. Eilbeck and F. Holmes, *J. Chem. Soc.* (*A*), 1967, 1777

605.3 $C_8H_7N_3$ 4-(2′-Pyridyl)imdiazole L

Metal	*Method*	*Temp*	*Medium*	*Log of equilibrium constant, remarks*	*Ref*
H^+	gl	25	0·1 (KNO_3)	K_1 5·492±0·003	67E
	sp	25	0·1 (KNO_3)	K_2 1·33	67E
Co^{2+}	gl	25	0·1 (KNO_3)	K_1 5·811±0·007, β_2 11·321±0·005, β_3 15·71±0·02	67E
Cu^{2+}	gl,Cu(Hg)	25	0·1 (KNO_3)	K_1 8·76±0·04, β_2 15·16±0·03, β_3 18·41±0·03	67E
Fe^{2+}	gl	25	0·1 (KNO_3)	K_1 4·93±0·01, β_2 9·02±0·02, β_3 13·76±0·01	67E
Ni^{2+}	gl,Cu(Hg)	25	0·1 (KNO_3)	K_1 7·20±0·04, β_2 13·95±0·03, β_3 19·82±0·03	67E
Zn^{2+}	gl	25	0·1 (KNO_3)	K_1 5·419±0·005, β_2 10·232±0·005, β_3 13·836±0·010	67E

67E W. J. Eilbeck, F. Holmes, G. G. Phillips, and A. E. Underhill, *J. Chem. Soc.* (*A*), 1967, 1161

606.1 C_8H_8O Phenoxyethene (phenyl vinyl ether) L

Metal	*Method*	*Temp*	*Medium*	*Log of equilibrium constant, remarks*	*Ref*
Ag^+	dis	25	1 (KNO_3)	K_1 0·72, K_2 −0·1	68F

68F T. Fueno, O. Kajimoto, and J. Furukawa, *Bull. Chem. Soc. Japan*, 1968, **41**, 782

607.1 $C_8H_8O_2$ 3-Acetylphenol (3-hydroxyacetophenone) HL

Metal	*Method*	*Temp*	*Medium*	*Log of equilibrium constant, remarks*	*Ref*
H^+	sp	25	0 corr	K_1 9·25	65E
Fe^{3+}	sp	25	0 corr	K_1 8·36±0·04	65E

65E Z. L. Ernst and F. G. Herring, *Trans. Faraday Soc.*, 1965, **61**, 454

$C_8H_8O_2$ 4-Acetylphenol (4-hydroxyacetophenone) HL 607.2

Metal	Method	Temp	Medium	Log of equilibrium constant, remarks	Ref
H^+	sp	25	0 corr	K_1 8·05	65E
Fe^{3+}	sp	25	0 corr	K_1 7·20±0·02	65E

65E Z. L. Ernst and F. G. Herring, *Trans. Faraday Soc.*, 1965, **61**, 454

$C_8H_8O_2$ 4-Methylbenzoic acid (*p*-toluic acid) HL 608

Metal	Method	Temp	Medium	Log of equilibrium constant, remarks	Ref
H^+	gl	25	0·1 ($NaClO_4$) 50% dioxan	K_1 6·02±0·01	68G
Cu^{2+}	gl	25	0·1 ($NaClO_4$) 50% dioxan	K_1 3·44 $K(CuA^{2+}+L^- \rightleftharpoons CuAL^+)$ 3·66 where A = 2,2′-bipyridyl	68G
Zn^{2+}	gl	25	0·1 ($NaClO_4$) 50% dioxan	K_1 2·43 $K(ZnA^{2+}+L^- \rightleftharpoons ZnAL^+)$ 2·48 where A = 2,2′-bipyridyl	68G

Erratum. 1964, p. 566, Table 608. *For* $C_8H_{10}O_2$ *read* $C_8H_8O_2$.

68G R. Griesser, B. Prijs, and H. Sigel, *Inorg. Nuclear Chem. Letters*, 1968, **4**, 443

$C_8H_8O_2$ 3-Methyltropolone HL 612

Metal	Method	Temp	Medium	Log of equilibrium constant, remarks	Ref
H^+	gl	?	0·5 (KNO_3) 50% ethanol	K_1 7·46	65D
	gl	?	0·1 $NaClO_4$ 40% dioxan	K_1 7·55	66S
	sp	?	0·2 $NaClO_4$ 3% ethanol	K_1 7·07	67G
Ce^{3+}	sp	?	0·2 $NaClO_4$ 3% ethanol	K_1 6·42	67G
Dy^{3+}	sp	?	0·2 $NaClO_4$ 3% ethanol	K_1 7·36	67G
Er^{3+}	sp	?	0·2 $NaClO_4$ 3% ethanol	K_1 7·57	67G
Fe^{3+}	sp	?	0·1 $NaClO_4$ 40% dioxan	K_1 11·88, K_2 10·94, K_3 9·6	66S
Gd^{3+}	sp	?	0·2 $NaClO_4$ 3% ethanol	K_1 7·10	67G
Ho^{3+}	sp	?	0·2 $NaClO_4$ 3% ethanol	K_1 7·42	67G
La^{3+}	sp	?	0·2 $NaClO_4$ 3% ethanol	K_1 6·12	67G
Lu^{3+}	sp	?	0·2 $NaClO_4$ 3% ethanol	K_1 8·01	67G

continued overleaf

612 $C_8H_8O_2$ (contd)

Metal	*Method*	*Temp*	*Medium*	*Log of equilibrium constant, remarks*	*Ref*
Nd^{3+}	sp	?	0·2 $NaClO_4$ 3% ethanol	K_1 6·78	67G
Pr^{3+}	sp	?	0·2 $NaClO_4$ 3% ethanol	K_1 6·71	67G
Sm^{3+}	sp	?	0·2 $NaClO_4$ 3% ethanol	K_1 7·07	67G
UO_2^{2+}	sp, gl	?	0·5 (KNO_3) 50% ethanol	K_1 (sp) 9·62, K_2 (sp) 6·98, (gl) 6·93	65D
Y^{3+}	sp	?	0·2 $NaClO_4$ 3% ethanol	K_1 7·38	67G
Yb^{3+}	sp	?	0·2 $NaClO_4$ 3% ethanol	K_1 7·74	67G

65D Y. Dutt and R. P. Singh, *J. Indian Chem. Soc.*, 1965, **42**, 767
66S R. P. Singh and Y. Dutt, *Indian J. Chem.*, 1966, **4**, 214
67G B. P. Gupta, Y. Dutt, and R. P. Singh, *Indian J. Chem.*, 1967, **5**, 214

613 $C_8H_8O_2$ Phenylacetic acid HL

Metal	*Method*	*Temp*	*Medium*	*Log of equilibrium constant, remarks*	*Ref*
H^+	gl	31	0·1 ($NaClO_4$)	K_1 4·20	68R
UO_2^{2+}	gl	31	0·1 ($NaClO_4$)	K_1 3·25	68R

68R S. Ramamoorthy and M. Santappa, *Bull. Chem. Soc. Japan*, 1968, **41**, 1330

614 $C_8H_8O_3$ 2-Hydroxy-3-methylbenzoic acid (*o*-cresotic acid) H_2L

Metal	*Method*	*Temp*	*Medium*	*Log of equilibrium constant, remarks*	*Ref*
H^+	sp	25	0 corr	K_1 14·597±0·009, K_2 2·945±0·002	63E
	gl	25	0·1 ($NaClO_4$)	K_2 2·82±0·02	66P
Fe^{3+}	sp	25	0 corr	K_1 18·130±0·003	63E
	gl, Pt	25	0·1 ($NaClO_4$)	$K(Fe^{3+}+HL^- \rightleftharpoons FeL^+ + H^+)$ 2·58±0·01, $K(FeL^+ + HL^- \rightleftharpoons FeL_2^- + H^+)$ 0·5±0·2, $K(FeL_2^- + HL^- \rightleftharpoons FeL_3^{3-} + H^+)$ −3·7 $K(Fe^{3+} + HL^- \rightleftharpoons FeHL^{2+})$ 4·6±0·04	66P
Mn^{2+}	con J, sp J	26	?	$K(Mn^{2+} + HL^- \rightleftharpoons MnL + H^+)$ (?) 2·87	62K

62K M. S. Kachhawaha and A. K. Bhattacharya, *J. Indian Chem. Soc.*, 1962, **39**, 399
63E Z. L. Ernst and J. Menashi, *Trans. Faraday Soc.*, 1963, **59**, 2838
66P M. V. Park, *J. Chem. Soc.* (*A*), 1966, 816; *Nature*, 1963, **197**, 283

$C_8H_8O_3$ 2-Hydroxy-4-methylbenzoic acid (*m*-cresotic acid) H_2L 615

Metal	Method	Temp	Medium	Log of equilibrium constant, remarks	Ref
H^+	?	25	?	K_2 3·17	63K
	gl	25	0·1 ($NaClO_4$)	K_2 2·97±0·005	66P
Fe^{3+}	sp	35	0—0·2	* K(?) (0 corr) 6·01, 5·65, 4·64, (0·2) 5·00, 4·67, 3·68	63K
	gl, Pt	25	0·1 ($NaClO_4$)	$K(Fe^{3+}+HL^- \rightleftharpoons FeL^+ + H^+)$ 2·99±0·01, $K(FeL^+ + HL^- \rightleftharpoons FeL_2^- + H^+)$ 1·3±0·1 $K(Fe^{3+}+HL^- \rightleftharpoons FeHL^{2+})$ 4·7	66P

* Apparent constants varying with pH.

63K S. S. Katiyar and V. B. S. Chauhan, *J. Inorg. Nuclear Chem.*, 1963, **25**, 1375

66P M. V. Park, *J. Chem. Soc.* (*A*), 1966, 816; *Nature*, 1963, **197**, 283

$C_8H_8O_3$ 2-Hydroxy-5-methylbenzoic acid (*p*-cresotic acid) H_2L 615.1

Metal	Method	Temp	Medium	Log of equilibrium constant, remarks	Ref
H^+	gl	25	0·1 ($NaClO_4$)	K_2 2·90±0·005	66P
Fe^{3+}	gl, Pt	25	0·1 ($NaClO_4$)	$K(Fe^{3+}+HL^- \rightleftharpoons FeL^+ + H^+)$ 2·98±0·005, $K(FeL^+ + HL^- \rightleftharpoons FeL_2^- + H^+)$ 1·2±0·03 $K(Fe^{3+}+HL^- \rightleftharpoons FeHL^{2+})$ 4·4±0·2	66P

66P M. V. Park, *J. Chem. Soc.* (*A*), 1966, 816; *Nature*, 1963, **197**, 283

$C_8H_8O_3$ 2-Hydroxy-6-methylbenzoic acid (6-methylsalicylic acid) H_2L 615.2

Metal	Method	Temp	Medium	Log of equilibrium constant, remarks	Ref
H^+	gl	25	0·1 ($NaClO_4$)	K_2 3·16±0·01	66P
Fe^{3+}	gl, Pt	25	0·1 ($NaClO_4$)	$K(Fe^{3+}+HL^- \rightleftharpoons FeL^+ + H^+)$ 2·58±0·01, $K(FeL^+ + HL^- \rightleftharpoons FeL_2^- + H^+)$ 0·6±0·03, $K(FeL_2^- + HL^- \rightleftharpoons FeL_3^{3-} + H^+)$ −3·0	66P

66P M. V. Park, *J. Chem. Soc.* (*A*), 1966, 816; *Nature*, 1963, **197**, 283

$C_8H_8O_3$ 2-Phenyl-2-hydroxyacetic acid (mandelic acid) HL 618

Metal	Method	Temp	Medium	Log of equilibrium constant, remarks	Ref
H^+	?	?	?	K_1 3·37	57V, 67V
	gl	25	0·1—1 (KNO_3)	K_1 (0·1) 3·19, (1) 3·14	67P
Be^{2+}	ix	18	0·1 ($NaClO_4$)	K_1 1·64±0·01	65B
Ce^{3+}	gl	25	1 $NaClO_4$	K_1 2·24±0·01, K_2 1·51±0·01, K_3 1·27±0·01, K_4 0·64±0·01	66T
	gl	25	0·1—1 (KNO_3)	K_1 (0·1) 2·34±0·02, (1) 2·03±0·03, K_2 (0·1) 1·8±0·1, (1) 1·5±0·1	67P

continued overleaf

618 $C_8H_8O_3$ (contd)

Metal	*Method*	*Temp*	*Medium*	*Log of equilibrium constant, remarks*	*Ref*
Eu^{3+}	dis	25	0—0·1 ($NaClO_4$)	K_1 (→ 0) 3·37, (0·05) 2·86, (0·1) 2·70, K_2 (0·05) 2·3, (0·1) 2·20	67M
Fe^{3+}	sp	25	?	K_1 3·71	61B
Ge^{IV}	con	18	1 NaCl	$K(H_2GeO_3+2HL \rightleftharpoons GeOL_2)$ 2·0±0·1	57V
	gl	18	→ 0	$K(H_2GeO_3+2HL \rightleftharpoons GeOL_2)$ 2·92	57V
La^{3+}	gl	25	1 $NaClO_4$	K_1 2·18±0·01, K_2 1·44±0·01, K_2 1·23±0·01, K_4 0·74±0·01	66T
	gl	25	0·1—1 (KNO_3)	K_1 (0·1) 2·24±0·03, (1) 1·94±0·04, K_2 (0·1) 1·7±0·1, (1) 1·4±0·1	67P
Nd^{3+}	gl	25	1 $NaClO_4$	K_1 2·59±0·01, K_2 1·70±0·02, K_3 1·32±0·02, K_4 1·20±0·03	66T
	gl	25	0·1—1 (KNO_3)	K_1 (0·1) 2·49±0·02, (1) 2·12±0·02, K_2 (0·1) 1·90±0·05, (1) 1·6±0·1	67P
Pa^V	ix	?	0·25	K (?) 2·85	62G
Pr^{3+}	gl	25	1 $NaClO_4$	K_1 2·48±0·01, K_2 1·62±0·01, K_3 1·35±0·01, K_4 0·89±0·01	66T
	gl	25	0·1—1 (KNO_3)	K_1 (0·1) 2·43±0·02, (1) 2·11±0·03, K_2 (0·1) 1·84±0·08, (1) 1·6±0·1	67P
Sc^{3+}	sp	18—20	0·01 (HNO_3)	K_1 5·40±0·06	66Ka
Sm^{3+}	gl	25	1 $NaClO_4$	K_1 2·56±0·01, K_2 1·76±0·02, K_3 1·42±0·02, K_4 1·26±0·03	66T
	gl	25	0·1—1 (KNO_3)	K_1 (0·1) 2·56±0·01, (1) 2·23±0·02, K_2 (0·1) 1·67±0·09, (1) 2·00±0·04	67P
Th^{4+}	ix	?	0·2	K_1 2·94, β_2 4·98, β_3 5·91	62G
UO_2^{2+}	gl	25	0·1 KNO_3	K_1 2·47±0·03 $K[UO_2(OH)L+H^+ \rightleftharpoons UO_2L^+]$ 3·9±0·1 $K\{[UO_2(OH)L]_2+2H^+ \rightleftharpoons 2UO_2L^+\}$4·94	67V
Zn^{2+}	oth	20	2 $NaClO_4$	K_1 1·48, β_2 2·41, β_3 3·588 optical rotation	65L
Zr^{4+}	sp	18—20	1 (HCl)	K_1 5·64±0·07	66K

57V O. Vartapetian, *Ann. Chim. (France)*, 1957, **2**, 916
61B S. D. Bhardwaj and G. V. Bakore, *J. Indian Chem. Soc.*, 1961, **38**, 967
62G I. Geletseanu and A. V. Lapitskii, *Proc. Acad. Sci. (U.S.S.R.)*, 1962, **144**, 460 (573); **147**, 983 (372)
65B T. A. Belyauskaya and I. F. Kolosova, *Russ. J. Inorg. Chem.*, 1965, **10**, 236 (441)
65L R. Larsson and B. Folkeson, *Acta Chem. Scand.*, 1965, **19**, 53
66K I. M. Korenman, F. E. Sheyanova, and Z. M. Gur'eva, *Russ. J. Inorg. Chem.*, 1966, **11**, 1485 (2761)
66Ka I. M. Korenman and N. V. Zaglyadimova, *Russ. J. Inorg. Chem.*, 1966, **11**, 1491 (2774)
66T T. H. Thun, F. Verbeek, and W. Vanderleen, *J. Inorg. Nuclear Chem.*, 1966, **28**, 1949
67M P. G. Manning, *Canad. J. Chem.*, 1967, **45**, 1643
67P J. E. Powell and W. F. S. Neillie, *J. Inorg. Nuclear Chem.*, 1967, **29**, 2371
67V S. K. Verma and R. P. Agarwal, *J. Less-Common Metals*, 1967, **12**, 221

622 $C_8H_8O_3$ 2-Hydroxy-4-methoxybenzaldehyde (4-methoxysalicylaldehyde) HL

Metal	*Method*	*Temp*	*Medium*	*Log of equilibrium constant, remarks*	*Ref*
H^+	gl	30	0·1 $NaClO_4$ 75% dioxan	K_1 9·82	67K
Cd^{2+}	gl	30	0·1 $NaClO_4$ 75% dioxan	K_1 6·85, K_2 5·33	67K
Co^{2+}	gl	30	0·1 $NaClO_4$ 75% dioxan	K_1 4·97, K_2 2·58	67K

continued

$C_8H_8O_3$ (contd) 622

Metal	Method	Temp	Medium	Log of equilibrium constant, remarks	Ref
Cu^{2+}	gl	30	0·1 $NaClO_4$ 75% dioxan	K_1 8·28, K_2 3·28	67K
Fe^{2+}	gl	30	0·1 $NaClO_4$ 75% dioxan	K_1 4·30, K_2 3·25	67K
Mg^{2+}	gl	30	0·1 $NaClO_4$ 75% dioxan	K_1 3·29	67K
Mn^{2+}	gl	30	0·1 $NaClO_4$ 75% dioxan	K_1 3·98, K_2 3·10	67K
Ni^{2+}	gl	30	0·1 $NaClO_4$ 75% dioxan	K_1 5·05, K_2 2·65	67K
Zn^{2+}	gl	30	0·1 $NaClO_4$ 75% dioxan	K_1 6·48, K_2 5·90	67K

67K P. V. Kamat, M. R. Bapat, and M. G. Datar, *J. Indian Chem. Soc.*, 1967, **44**, 731

$C_8H_8O_3$ Phenoxyacetic acid HL 625

Metal	Method	Temp	Medium	Log of equilibrium constant, remarks	Ref
H^+	gl	31	0·1 ($NaClO_4$)	K_1 2·96	68R
UO_2^{2+}	gl	31	0·1 ($NaClO_4$)	K_1 2·41	68R

68R S. Ramamoorthy and M. Santappa, *Bull. Chem. Soc. Japan*, 1968, **41**, 1330

$C_8H_8O_3$ 3,4-Dihydroxyacetophenone (4-acetylcatechol) H_2L 625.1

Metal	Method	Temp	Medium	Log of equilibrium constant, remarks	Ref
Mo^{VI}	sp	20	?	$K(MoO_4^{2-}+2H_2L \rightleftharpoons MoO_2L_2^{2-})$ 6·74±0·07	61H

61H J. Halmekoski, *Suomen Kem.*, 1961, **B34**, 169

$C_8H_8O_4$ 3′,4′-Dihydroxyphenylacetic acid (homoprotocatechuic acid) H_3L 625.2

Metal	Method	Temp	Medium	Log of equilibrium constant, remarks	Ref
H^+	gl	30	0·1 ($NaClO_4$)	K_1 12·0, K_2 9·44, K_3 4·25	66A
Cd^{2+}	gl	30	0·1 ($NaClO_4$)	K_1 7·35, K_2 4·28	66A
Cu^{2+}	gl	30	0·1 ($NaClO_4$)	K_1 12·82, K_2 9·50	66A
Mg^{2+}	gl	30	0·1 ($NaClO_4$)	K_1 4·94	66A
Ni^{2+}	gl	30	0·1 ($NaClO_4$)	K_1 8·04, K_2 4·35	66A
Zn^{2+}	gl	30	0·1 ($NaClO_4$)	K_1 8·80, K_2 6·37	66A

66A V. T. Athavale, L. H. Prabhu, and D. G. Vartak, *J. Inorg. Nuclear Chem.*, 1966, **28**, 1237

626.1 $C_8H_9N_3$ 2-(Aminomethyl)benzimidazole L

Metal	*Method*	*Temp*	*Medium*	*Log of equilibrium constant, remarks*	*Ref*
H^+	gl	15—45	50% dioxan	K_1 (15°) 7·34, (25°) 7·14, (35°) 6·86, (45°) 6·68, K_2 (15—45°) 2	63L
Cu^{2+}	gl	15—45	50% dioxan	K_1 (15°) 7·59, (25°) 7·38, (35°) 7·18, (45°) 6·92, K_2 (15°) 7·26, (25°) 6·85, (35°) 6·66, (45°) 6·55 ΔH_1 −19·2	63L

63L T. J. Lane and H. B. Durham, *Inorg. Chem.*, 1963, **2**, 632

628.1 $C_8H_{10}O_2$ 1,4-Dimethoxybenzene (*p*-methoxyanisole) L

Metal	*Method*	*Temp*	*Medium*	*Log of equilibrium constant, remarks*	*Ref*
Hg^{2+}	sp	26	1—2 ($NaClO_4$)	$K[Hg^{2+}+L \rightleftharpoons Hg(H_{-1}L)^+ + H^+]$ 2·7±0·4	64W

64W T. H. Wirth and N. Davidson, *J. Amer. Chem. Soc.*, 1964, **86**, 4322

628.2 $C_8H_{11}N$ 4-Propylpyridine L

Metal	*Method*	*Temp*	*Medium*	*Log of equilibrium constant, remarks*	*Ref*
Co^{2+}	sp	20	$CHCl_3$	$K(CoL_2Cl_2+2L \rightleftharpoons CoL_4Cl_2)$ 1·07 ΔH −15·8±0·3, ΔS −49±1 $K[CoL_2(NCS)_2+2L \rightleftharpoons CoL_4(NCS)_2]$ 4·90 ΔH −16·4±0·3, ΔS −34±1	66C

66C J. de O. Cabral, H. C. A. King, S. M. Nelson, T. M. Shepherd, and E. Koros, *J. Chem. Soc.* (*A*), 1966, 1348

629 $C_8H_{11}N$ 2,4,6-Trimethylpyridine L

Metal	*Method*	*Temp*	*Medium*	*Log of equilibrium constant, remarks*	*Ref*
Zn^{2+}				See dimethyldithiocarbamic acid (176), 8-hydroxyquinoline (720), 2-methyl-8-hydroxyquinoline (818), and 4-methyl-8-hydroxyquinoline (819) for mixed ligands.	

629.1 $C_8H_{11}N$ 2,4-Dimethylaniline (2,4-xylidine) L

Metal	*Method*	*Temp*	*Medium*	*Log of equilibrium constant, remarks*	*Ref*
Ag^+	Ag	25	ethanol	K_1 2·49, β_2 3·71, β_3 3·99	60A

60A V. Arneanu and C. Luca, *Z. phys. Chem.* (*Leipzig*), 1960, **214**, 81

$C_8H_{11}N$ **N-Ethylaniline** L 631.1

Metal	Method	Temp	Medium	Log of equilibrium constant, remarks	Ref
Ag^+	Ag	25	< 0·01 96% ethanol	K_1 1·95, β_2 2·95	61A

61A V. Arneanu and C. Luca, *Z. phys. Chem.* (*Leipzig*), 1961, **217**, 389

$C_8H_{11}N$ **N,N-Dimethylaniline** L 631.2

Metal	Method	Temp	Medium	Log of equilibrium constant, remarks	Ref
Ag^+	Ag	25	< 0·01 96% ethanol	K_1 0·98	61A

61A V. Arneanu and C. Luca, *Z. phys. Chem.* (*Leipzig*), 1961, **217**, 389

$C_8H_{11}N_5$ **1-Phenylbiguanide** L 632

Metal	Method	Temp	Medium	Log of equilibrium constant, remarks	Ref
H^+	gl	32	0·01	K_1 10·72, K_2 2·16	59R
Cu^{2+}	sp	28—32	0·25 KCl	β_2 16·41	59R

59R M. M. Ray and P. Ray, *J. Indian Chem. Soc.*, 1959, **36**, 849

$C_8H_{12}O_2$ **5,5-Dimethylcyclohexane-1,3-dione (dimedone)** HL 633.1

Metal	Method	Temp	Medium	Log of equilibrium constant, remarks	Ref
H^+	gl	25	0·1 $(C_4H_9)_4NClO_4$ methanol	K_1 8·40±0·05	65L
			0·1 $(C_4H_9)_4NClO_4$ ethanol	K_1 8·37±0·05	
K^+	gl	25	0·1 KI methanol	K_1 0·8± > 0·2	65L
			0·1 KSCN ethanol	K_1 1·8±0·2	
Li^+	gl	25	0·1 $LiClO_4$ methanol	K_1 1·5±0·2	65L
			0·1 $LiClO_4$ ethanol	K_1 2·1±0·2	
Na^+	gl	25	0·1 $NaClO_4$ methanol	K_1 1·5±0·2	65L
			0·1 $NaClO_4$ ethanol	K_1 2·2±0·2	

65L D. C. Luehrs, R. I. Iwamoto, and J. Kleinberg, *Inorg. Chem.*, 1965, **4**, 1739

634.1 $C_8H_{12}O_4$ Cyclohexane-1,2-dicarboxylic acid* H_2L

Metal	*Method*	*Temp*	*Medium*	*Log of equilibrium constant, remarks*	*Ref*
Tl^+	con, spJ	28	?	K_1 1·74	66S

* Geometric configuration of ligand not stated.

66S K. N. Sahn and A. K. Bhattacharya, *J. Indian Chem. Soc.*, 1966, **43**, 781

635 $C_8H_{12}N_2$ 6-Methyl-2-(methylaminomethyl)pyridine (6-methyl-2-picolylmethylamine) L

Metal	*Method*	*Temp*	*Medium*	*Log of equilibrium constant, remarks*	*Ref*
H^+	gl	25	0·1 ($HClO_4$) 50% dioxan	K_1 8·40, K_2 2·99	66W
Cd^{2+}	pol	25	0·1 KNO_3 50% dioxan	β_2 6·96	66W
	gl	25	0·1 KNO_3 50% dioxan	K_1 4·26, K_2 3·30	66W
Co^{2+}	gl	15—35	0·1 KNO_3 50% dioxan	K_1 (15°) 4·39, (25°) 4·10, (35°) 3·84, K_2 (15°) 4·09, (25°) 3·87, (35°) 3·43	66W
	cal	25	0·1 KNO_3 50% dioxan	$\Delta H_{\beta 2}°$ −11·4, $\Delta S_{\beta 2}°$ −2·4	66W
Cu^+	pol	25	0·1 KNO_3 50% dioxan	β_2 12·14	66W
Cu^{2+}	pol	25	0·1 KNO_3 50% dioxan	β_2 12·37	66W
	gl	25	0·1 KNO_3 50% dioxan	K_1 7·29, K_2 5·48	66W
	cal	25	0·1 KNO_3 50% dioxan	$\Delta H_{\beta 2}°$ −12·0, $\Delta S_{\beta 2}°$ 15·8	66W
Ni^{2+}	gl	15—35	0·1 KNO_3 50% dioxan	K_1 (15°) 4·76, (25°) 4·66, (35°) 4·45, K_2 (15°) 3·33, (25°) 3·31, (35°) 3·26	66W
	cal	25	0·1 KNO_3 50% dioxan	$\Delta H_{\beta 2}°$ −7·5, $\Delta S_{\beta 2}°$ 9·4	66W
Zn^{2+}	gl	15—35	0·1 KNO_3 50% dioxan	K_1 (15°) 4·67, (25°) 3·93, (35°) 3·92, K_2 (15°) 4·58, (25°) 3·81, (35°) 3·76	66W
	cal	25	0·1 KNO_3 50% dioxan	$\Delta H_{\beta 2}°$ −3·9, $\Delta S_{\beta 2}°$ 22·4	66W

66W J. L. Walter and S. M. Rosalie, *J. Inorg. Nuclear Chem.*, 1966, **28**, 2969

641.1 $C_8H_{15}N$ *N*-Allylpiperidine L

Metal	*Method*	*Temp*	*Medium*	*Log of equilibrium constant, remarks*	*Ref*
Pt^{2+}	sp	60	2 (NaCl)	$K(PtCl_4^{2-} + HL^+ \rightleftharpoons PtCl_3HL + Cl^-)$ 2·64	67D

67D R. G. Denning, F. R. Hartley, and L. M. Venanzi, *J. Chem. Soc. (A)*, 1967, 324

$C_8H_{16}O_3$ 2-Hydroxy-2-propylpentanoic acid (dipropylglycollic acid) HL 641.2

Metal	*Method*	*Temp*	*Medium*	*Log of equilibrium constant, remarks*	*Ref*
H^+	qh	25	1 ($NaClO_4$)	K_1 3·80	65T
Dy^{3+}	qh	25	1 ($NaClO_4$)	K_1 3·08±0·03, K_2 2·49±0·03	65T
Er^{3+}	qh	25	1 ($NaClO_4$)	K_1 3·29±0·03, K_2 2·24±0·07, K_3 2·1±0·1	65T
Eu^{3+}	qh	25	1 ($NaClO_4$)	K_1 2·81±0·03, K_2 2·18±0·03	65T
Gd^{3+}	qh	25	1 ($NaClO_4$)	K_1 2·83±0·04, K_2 2·20±0·04	65T
Nd^{3+}	qh	25	1 ($NaClO_4$)	K_1 2·61±0·05, K_2 1·8±0·2	65T
Pr^{3+}	qh	25	1 ($NaClO_4$)	K_1 2·53±0·05, K_2 1·5±0·3	65T
Sm^{3+}	qh	25	1 ($NaClO_4$)	K_1 2·65±0·02, K_2 2·17±0·03	65T
Yb^{3+}	qh	25	1 ($NaClO_4$)	K_1 3·36±0·03, K_2 2·23±0·08, K_3 2·1±0·1	65T

65T H. Thun, F. Verbeek, and W. Vanderleen, *J. Inorg. Nuclear Chem.*, 1965, 27, 1813

$C_8H_{16}O_6$ Methyl-4-*O*-methyl-α-D-mannopyranoside L 641.3

Metal	*Method*	*Temp*	*Medium*	*Log of equilibrium constant, remarks*	*Ref*
B^{III}	oth	30	0·105	$K[B(OH)_4^- + L \rightleftharpoons B(OH)_2(H_{-2}L)^-]$ 1·5 $K[B(OH)_4^- + 2L \rightleftharpoons B(H_{-2}L)_2]$ 3·5 refractive index, optical rotation	64M

64M E. W. Malcolm, J. W. Green, and H. A. Swenson, *J. Chem. Soc.*, 1964, 4669

$C_8H_{16}O_6$ Methyl-4-*O*-methyl-β-D-mannopyranoside L 641.4

Metal	*Method*	*Temp*	*Medium*	*Log of equilibrium constant, remarks*	*Ref*
B^{III}	oth	30	0·105	$K[B(OH)_4^- + L \rightleftharpoons B(OH)_2(H_{-2}L)^-]$ 0·3 $K[B(OH)_4^- + 2L \rightleftharpoons B(H_{-2}L)_2^-]$ 2·60 refractive index, optical rotation	64M

64M E. W. Malcolm, J. W. Green, and H. A. Swenson, *J. Chem. Soc.*, 1964, 4669

$C_8H_{19}N$ Dibutylamine L 642

Metal	*Method*	*Temp*	*Medium*	*Log of equilibrium constant, remarks*	*Ref*
Ag^+	oth	?	0·1 $NaClO_4$ acetone	β_2 10·18±0·05, coulometric titration	65M
Zn^{2+}				See dimethyldithiocarbamic acid (176) for mixed ligands.	

65M K. K. Mead, D. L. Maricle, and C. A. Streuli, *Analyt. Chem.*, 1965, 37, 237

642.1 $C_8H_{19}N$ Bis(2-methylpropyl)amine (di-isobutylamine) L

Metal	Method	Temp	Medium	Log of equilibrium constant, remarks	Ref
Ag^+	oth	?	0·1 $NaClO_4$ acetone	β_2 8·8±0·1 coulometric titration	65M

65M K. K. Mead, D. L. Maricle, and C. A. Streuli, *Analyt. Chem.*, 1965, 37, 237

646 $C_8H_{23}N_5$ 1,4,7,10,13-Penta-azatridecane (tetraethylenepentamine) (tetren) L

Metal	Method	Temp	Medium	Log of equilibrium constant, remarks	Ref
H^+	gl	25	0·1 KCl	K_1 9·68, K_2 9·10, K_3 8·08, K_4 4·72, K_5 2·98	63P
	cal	25	0·1 KCl	ΔH_1 −10·76, ΔS_1 8·2, ΔH_2 −11·32, ΔS_2 3·7, ΔH_3 −10·71, ΔS_3 1·0, ΔH_4 −7·89, ΔS_4 −4·9, ΔH_5 −6·83, ΔS_5 −9·3	64P
Cd^{2+}	pol	25	0·5	K_1 14·7, phosphate buffer	62J
	cal	25	0·1 KNO_3	ΔH_1 −12·8, ΔS_1 22	65W
Co^{2+}	gl	25	0·1 KCl	K_1 13·30	63P
				$K(Co^{2+}+H_2L^{2+} \rightleftharpoons CoH_2L^{4+})$ 4·9	
	cal	25	0·1 KCl	ΔH_1 −13·85, ΔS_1 14·5	64P
Cu^{2+}	gl	25	0·1 KCl	K_1 22·80	63P
				$K(Cu^{2+}+HL^+ \rightleftharpoons CuHL^{3+})$ 18·30	
				$K(Cu^{2+}+H_2L^{2+} \rightleftharpoons CuH_2L^{4+})$ 13·0	
	cal	25	0·1 KCl	ΔH_1 −24·95, ΔS_1 20·5	64P
	cal	25	0·1 KNO_3	ΔH_1 −24, ΔS_1 25	65W
Fe^{2+}	gl	25	0·1 KCl	K_1 9·85,	63P
				$K(Fe^{2+}+HL^+ \rightleftharpoons FeHL^{3+})$ 4·2	
	cal	25	0·1 KCl	ΔH_1 −8·70, ΔS_1 16·0	64P
Mn^{2+}	gl	25	0·1 KCl	K_1 6·55	63P
	cal	25	0·1 KCl	ΔH_1 −3·70, ΔS_1 17·5	64P
Ni^{2+}	gl	25	0·1 KCl	K_1 17·43	63P
				$K(Ni^{2+}+HL^+ \rightleftharpoons NiHL^{3+})$ 11·78	
				$K(Ni^{2+}+H_2L^{2+} \rightleftharpoons NiH_2L^{4+})$ 6·7	
	cal	25	0·1 KCl	ΔH_1 −18·90, ΔS_1 16·5	64P
	cal	25	0·1 KNO_3	ΔH_1 −18·4, ΔS_1 20	65W
	sp	25	0·5 ($NaClO_4$)	$K(NiL^{2+}+NH_3 \rightleftharpoons NiLNH_3^{2+})$ 0·83±0·05	67J
Pb^{2+}	pol	25	1 NH_3	K_1 10·9	62J
Zn^{2+}	gl	25	0·1 KCl	K_1 15·10	63P
				$K(Zn^{2+}+H_2L^{2+} \rightleftharpoons ZnH_2L^{4+})$ 5·7	
	cal	25	0·1 KCl	ΔH_1 −13·85, ΔS_1 22·5	64P
	cal	25	0·1 KNO_3	ΔH_1 −14·0, ΔS_1 24	65W

62J E. Jacobsen and K. Schroder, *Acta Chem. Scand.*, 1962, **16**, 1393
63P P. Paoletti, A. Vacca, and I. Giusti, *Ricerca Sci.*, 1963, **33** (II-A), 523
64P P. Paoletti and A. Vacca, *J. Chem. Soc.*, 1964, 5051
65W D. L. Wright, J. H. Holloway, and C. N. Reilley, *Analyt. Chem.*, 1965, 37, 884
67J N. E. Jacobs and D. W. Margerum, *Inorg. Chem.*, 1967, **6**, 2038

$C_8H_5O_2N$ 3-Cyanotropolone HL 647.1

Metal	*Method*	*Temp*	*Medium*	*Log of equilibrium constant, remarks*	*Ref*
H^+	sp	25	2 $NaClO_4$	K_1 3·41	65O
Fe^{3+}	sp	25	2 $NaClO_4$	K_1 7·28	65O

65O Y. Oka and M. Yanai, *Nippon Kagaku Zasshi*, 1965, **86**, 929

$C_8H_5O_2N$ 5-Cyanotropolone HL 647.2

Metal	*Method*	*Temp*	*Medium*	*Log of equilibrium constant, remarks*	*Ref*
H^+	sp	25	2 $NaClO_4$	K_1 3·72	64O
Fe^{3+}	sp	25	2 $NaClO_4$	K_1 7·53	64O

64O Y. Oka and K. Yamamoto, *Nippon Kagaku Zasshi*, 1964, **83**, 779

$C_8H_5O_6N$ 3-Nitrobenzene-1,2-dicarboxylic acid (3-nitrophthalic acid) H_2L 652

Metal	*Method*	*Temp*	*Medium*	*Log of equilibrium constant, remarks*	*Ref*
UO_2^{2+}	gl	31	0·1 ($NaClO_4$)	K_1 3·82	67S

67S A. J. Singh and N. S. K. Prasad, *Indian J. Chem.*, 1967, **5**, 573

$C_8H_5O_6N$ 4-Nitrobenzene-1,2-dicarboxylic acid (4-nitrophthalic acid) H_2L 653

Metal	*Method*	*Temp*	*Medium*	*Log of equilibrium constant, remarks*	*Ref*
UO_2^{2+}	gl	31	0·1 ($NaClO_4$)	K_1 4·02	67S

67S A. J. Singh and N. S. K. Prasad, *Indian J. Chem.*, 1967, **5**, 573

$C_8H_5O_6N$ Pyridine-2,4,6-tricarboxylic acid H_3L 653.1

Metal	*Method*	*Temp*	*Medium*	*Log of equilibrium constant, remarks*	*Ref*
H^+				H^+ constants not given	
Fe^{2+}	sp	13—17	0·1	K_1 (0·02 KCN) 2·2, β_2 (no KCN) 3·7	63M

63M I. Morimoto and T. Sato, *Bull. Chem. Soc. Japan*, 1963, **36**, 605

654 $C_8H_5O_6N_5$ *N*-(4′Hydroxy-2′,6′-dioxo-1′,3′-diazin-5′-yl)-5-imino(perhydro -1,3-diazine-2,4,6-trione) (purpuric acid) (murexide = L·NH_3) H_3L

Metal	*Method*	*Temp*	*Medium*	*Log of equilibrium constant, remarks*	*Ref*
Ca^{2+}	sp	25	0·1 $CaCl_2$	$K(Ca^{2+}+H_2L^- \rightleftharpoons CaH_2L^+)$ 2·68 ± 0·01	61N
Cd^{2+}	sp	rt	~ 0·1	$K(Cd^{2+}+H_2L^- \rightleftharpoons CdH_2L^+)$ 4·2	49S
Ce^{3+}	sp	12	0·1 (KNO_3)	$K(Ce^{3+}+H_2L^- \rightleftharpoons CeH_2L^{2+})$ 3·65	65G
Co^{2+}	sp	12	0·1 (KNO_3)	$K(Co^{2+}+H_2L^- \rightleftharpoons CoH_2L^+)$ 2·46	65G
Cu^{2+}	sp	25	0 corr	$K(Cu^{2+}+H_2L^- \rightleftharpoons CuH_2L^+)$ 3·5 (?)	60C
Dy^{3+}	sp	12	0·1 (KNO_3)	$K(Dy^{3+}+H_2L^- \rightleftharpoons DyH_2L^{2+})$ 3·78	65G
Er^{3+}	sp	12	0·1 (KNO_3)	$K(Er^{3+}+H_2L^- \rightleftharpoons ErH_2L^{2+})$ 3·48	65G
Eu^{3+}	sp	12	0·1 (KNO_3)	$K(Eu^{3+}+H_2L^- \rightleftharpoons EuH_2L^{2+})$ 4·17	65G
Gd^{3+}	sp	12	0·1 (KNO_3)	$K(Gd^{3+}+H_2L^- \rightleftharpoons GdH_2L^{2+})$ 4·08	65G
Ho^{3+}	sp	12	0·1 (KNO_3)	$K(Ho^{3+}+H_2L^- \rightleftharpoons HoH_2L^{2+})$ 3·71	65G
In^{3+}	sp	12	0·1 (KNO_3)	$K(In^{3+}+H_2L^- \rightleftharpoons InH_2L^{2+})$ 4·61	65G
La^{3+}	sp	12	0·1 (KNO_3)	$K(La^{3+}+H_2L^- \rightleftharpoons LaH_2L^{2+})$ 3·43	65G
Lu^{3+}	sp	12	0·1 (KNO_3)	$K(Lu^{3+}+H_2L^- \rightleftharpoons LuH_2L^{2+})$ 3·45	65G
Mg^{2+}	sp	15—25 (?)	~ 0·1	$K(Mg^{2+}+H_2L^- \rightleftharpoons MgH_2L^+)$ 1 (?), 2·2 (?)	49S
Nd^{3+}	sp	12	0·1 (KNO_3)	$K(Nd^{3+}+H_2L^- \rightleftharpoons NdH_2L^{2+})$ 4·04	65G
Ni^{2+}	sp	12	0·1 (KNO_3)	$K(Ni^{2+}+H_2L^- \rightleftharpoons NiH_2L^+)$ 3·36	65G
Pr^{3+}	sp	12	0·1 (KNO_3)	$K(Pr^{3+}+H_2L^- \rightleftharpoons PrH_2L^{2+})$ 3·78	65G
Sc^{3+}	sp	12	0·1 (KNO_3)	$K(Sc^{3+}+H_2L^- \rightleftharpoons ScH_2L^{2+})$ 4·50	65G
Sm^{3+}	sp	12	0·1 (KNO_3)	$K(Sm^{3+}+H_2L^- \rightleftharpoons SmH_2L^{2+})$ 4·20	65G
Tb^{3+}	sp	12	0·1 (KNO_3)	$K(Tb^{3+}+H_2L^- \rightleftharpoons TbH_2L^{2+})$ 3·95	65G
Tm^{3+}	sp	12	0·1 (KNO_3)	$K(Tm^{3+}+H_2L^- \rightleftharpoons TmH_2L^{2+})$ 3·36	65G
Y^{3+}	sp	12	0·1 (KNO_3)	$K(Y^{3+}+H_2L^- \rightleftharpoons YH_2L^{2+})$ 3·36	65G
Yb^{3+}	sp	12	0·1 (KNO_3)	$K(Yb^{3+}+H_2L^- \rightleftharpoons YbH_2L^{2+})$ 3·41	65G
Zn^{2+}	sp	15—25 (?)	~ 0·1	$K(Zn^{2+}+H_2L^- \rightleftharpoons ZnH_2L^+)$ 3·1	49S

Erratum. 1964, p. 578, Table 654. Equation should *read* $K[Zn^{2+}+2HL^{2-} \rightleftharpoons Zn(HL)_2^{2-}]$ 9·3.

49S G. Schwarzenbach and H. Gysling, *Helv. Chim. Acta*, 1949, **32**, 1314
60C R. K. Chaturvedi, *Current Sci.*, 1960, **29**, 128
61N L. B. Nanninga, *Biochim. Biophys. Acta*, 1961, **54**, 330
65G G. Geier, *Ber. Bunsengesellschaft Phys. Chem.*, (*Z. Elektrochem*), 1965, **69**, 617

660.1 $C_8H_6N_2S$ 2-(2′-Pyridyl)-1,3-thiazole L

Metal	*Method*	*Temp*	*Medium*	*Log of equilibrium constant, remarks*	*Ref*
H^+	sp	25	0·1 ($NaClO_4$)	K_1 2·30 ± 0·04	65K
Cu^{2+}	sp	25	0·1 ($NaClO_4$)	K_1 5·65 ± 0·05	65K

65K K. Kahmann, H. Sigel, and H. Erlenmeyer, *Helv. Chim. Acta*, 1965, **48**, 295

660.2 $C_8H_6N_2S$ 4-(2′-Pyridyl)-1,3-thiazole L

Metal	*Method*	*Temp*	*Medium*	*Log of equilibrium constant, remarks*	*Ref*
H^+	sp	25	0·1 ($NaClO_4$)	K_1 4·10 ± 0·04	65K
Cu^{2+}	sp	25	0·1 ($NaClO_4$)	K_1 7·20 ± 0·05	65K

65K K. Kahmann, H. Sigel, and H. Erlenmeyer, *Helv. Chim. Acta*, 1965, **48**, 295

$C_8H_6N_2S$ 5-(2′Pyridyl)-1,3-thiazole L 660.3

Metal	Method	Temp	Medium	Log of equilibrium constant, remarks	Ref
H^+	sp	25	0·1 ($NaClO_4$)	K_1 2·79±0·04	65K
Cu^{2+}	sp	25	0·1 ($NaClO_4$)	K_1 1·70±0·05	65K
Ni^{2+}	sp	25	0·1 ($NaClO_4$)	K_1 1·35±0·05	65K
Zn^{2+}	sp	25	0·1 ($NaClO_4$)	K_1 0·3	65K

65K K. Kahmann, H. Sigel, and H. Erlenmeyer, *Helv. Chim. Acta*, 1965, 48 295

H_7O_2Cl 5-Chloro-2-hydroxy-4-methylbenzaldehyde (5-chloro-4-methylsalicylaldehyde) HL 660.4

Metal	Method	Temp	Medium	Log of equilibrium constant, remarks	Ref
H^+	gl	20	0·3 $NaClO_4$ 50% dioxan	K_1 8·90	63C
Ni^{2+}	pH	20	0·3 $NaClO_4$ 50% dioxan	K_1 4·1	63C

63C K. Clarke, R. A. Cowen, G. W. Gray, and E. H. Osborne, *J. Chem. Soc.*, 1963, 245

$C_8H_7O_3Cl$ 4-Chloro-2,3-dihydroxyacetophenone (3-acetyl-6-chlorocatechol) H_2L 660.5

Metal	Method	Temp	Medium	Log of equilibrium constant, remarks	Ref
Mo^{VI}	sp	25	0·1 (KCl)	$K(MoO_4^{2-}+2H_2L \rightleftharpoons MoO_2L_2^{2-})$ 7·03±0·04	63H

63H J. Halmekoski, *Suomen Kem.*, 1963, **B36**, 19

$C_8H_7O_3Br$ 2-(4′-Bromophenyl)-2-hydroxyacetic acid (*p*-bromomandelic acid) HL 660.6

Metal	Method	Temp	Medium	Log of equilibrium constant, remarks	Ref
H^+	dis	25	1 $HClO_4$	K_1 2·87±0·03	61A
Hf^{4+}	dis	25	1 $HClO_4$	K_1 7·00±0·05, K_2 6·15±0·07, K_3 6·61±0·05, K_4 6·26±0·08	61A
Zr^{4+}	dis	25	1 $HClO_4$	K_1 7·15±0·04, K_2 6·28±0·05, K_3 6·65±0·05, K_4 5·52±0·06	61A

61A I. P. Alimarin and S. Hanshi, *Russ. J. Inorg. Chem.*, 1961, **6**, 1054 (2062)

660.7 $C_8H_8ON_2$ 2-(Hydroxymethyl)benzimidazole HL

Metal	Method	Temp	Medium	Log of equilibrium constant, remarks	Ref
H^+	gl	15—45	50% dioxan	K_1 (15°) 13·03, (25°) 12·76, (40°) 12·19, (45°) 12·04, K_2 (15°) 4·89, (25°) 4·82, (40°) 4·68, (45°) 4·58	63L
Cu^{2+}	gl	15—45	50% dioxan	K_1 (15°) 11·56, (25°) 11·00, (40°) 10·20, (45°) 9·82 ΔH_1 −16·7	63L

63L T. J. Lane and H. B. Durham, *Inorg. Chem.*, 1963, **2**, 632

661.1 $C_8H_8O_2N_2$ 1-(2′-Hydroxyphenyl)-4-oxo-2,3-diazabut-1-ene (salicylaldehyde formylhydrazone) H_2L

Metal	Method	Temp	Medium	Log of equilibrium constant, remarks	Ref
H^+	gl	18—20	0·05 ($NaClO_4$) 40% ethanol	K_1 11·95±0·04, K_2 9·75±0·09, K_3 2·8±0·2	66S
Al^{3+}	sp	18—20	0·025 28% ethanol	K (?) 4·7, K (?) ~ 9, acetate buffer	63H
Cu^{2+}	sp	18—20	0·05 ($NaClO_4$) 40% ethanol	$K(Cu^{2+}+HL^- \rightleftharpoons CuHL^+)$ 11·2±0·3 $K(Cu^{2+}+H_2L \rightleftharpoons CuH_2L^{2+})$ 7·4±0·2	66S
Ni^{2+}	sp	18—20	0·05 ($NaClO_4$) 40% ethanol	K_1 11·2±0·4 $K(Ni^{2+}+HL^- \rightleftharpoons NiHL^+)$ 6·4±0·1 $K(Ni^{2+}+H_2L \rightleftharpoons NiH_2L^{2+})$ 3·8±0·1 $K[Ni^{2+}+L^{2-}+OH^- \rightleftharpoons Ni(OH)L^-]$ 14±2	66S
Sc^{3+}	sp	18—20	0·025 28% ethanol	K (?) 4·88, acetate buffer	63H
Zn^{2+}	sp	18—20	0·025 28% ethanol	K (?) 5·8, acetate buffer	63H
	sp	18—20	0·05 ($NaClO_4$) 40% ethanol	$K(Zn^{2+}+HL^- \rightleftharpoons ZnHL^+)$ 6·28±0·02 $K(Zn^{2+}+H_2L \rightleftharpoons ZnH_2L^{2+})$ 4·41±0·01	66S

63H Z. Holzbecher, *Coll. Czech. Chem. Comm.*, 1963, **28**, 716
66S L. Sucha and M. Suchanek, *Coll. Czech. Chem. Comm.*, 1966, **31**, 4539

663 $C_8H_8O_2S$ 1-(3′-Thienyl)butane-1,3-dione (3-thenoylacetone) HL

Metal	Method	Temp	Medium	Log of equilibrium constant, remarks	Ref
H^+	gl	30	→0 75% dioxan	K_1 12·41±0·01	65R
Co^{2+}	gl	30	→0 75% dioxan	K_1 10·29±0·05, K_2 8·10±0·01	65R
Cu^{2+}	gl	30	→0 75% dioxan	K_1 12·43±0·10, K_2 10·1±0·3	65R
Ni^{2+}	gl	30	→0 75% dioxan	K_1 10·73±0·07, K_2 8·5±0·1	65R
UO_2^{2+}	gl	30	→0 75% dioxan	K_1 12·3±0·1, K_2 10·32±0·04	65R
Zn^{2+}	gl	30	→0 75% dioxan	K_1 10·0±0·2, K_2 8·0±0·2	65R

65R J. L. Rosenstreich and D. E. Goldberg, *Inorg. Chem.*, 1965, **4**, 909

$_8O_3N_2$ 1-(2′,4′-Dihydroxyphenyl)-4-oxo-2,3-diazabut-1-ene(resorcylaldehyde formylhydrazone) H_3L 663.1

Metal	Method	Temp	Medium	Log of equilibrium constant, remarks	Ref
H^+				H^+ constants not given	
Sc^{3+}	sp	18—20	0·025 28% ethanol	K(?) 4·72, acetate buffer	63H
Zn^{2+}	sp	18—20	0·025 28% ethanol	K(?) 4·08, K(?) 9·79, acetate buffer	63H

63H Z. Holzbecher, *Coll. Czech. Chem. Comm.*, 1963, **28**, 716

$C_8H_8O_4N_2$ 4-(Methylamino)pyridine-2,6-dicarboxylic acid H_2L 663.2

Metal	Method	Temp	Medium	Log of equilibrium constant, remarks	Ref
H^+	gl	22	0·1 ($NaClO_4$)	K_1 9·68 (?), K_2 1·3	64B
Ba^{2+}	gl	22	0·1 ($NaClO_4$)	K_1 3·81	64B
Ca^{2+}	gl	22	0·1 ($NaClO_4$)	K_1 5·37	64B
Mg^{2+}	gl	22	0·1 ($NaClO_4$)	K_1 3·09	64B
Sr^{2+}	gl	22	0·1 ($NaClO_4$)	K_1 4·32	64B

64B E. Blasius and B. Brazio, *Ber. Bunsengellschaft Phy. Chem.*, 1964, **68**, 52

C_8H_9ON *N*-(Salicylidene)aminomethane (*N*-methyliminesalicylaldehyde) HL 663.3

Metal	Method	Temp	Medium	Log of equilibrium constant, remarks	Ref
H^+	gl	30	0·1 $NaClO_4$ 75% dioxan	K_1 11·65, K_2 6·73	64J
Cu^{2+}	gl	30	0·1 $NaClO_4$ 75% dioxan	K_1 10·81, K_2 6·53, K_3 7·48	64J

64J K. E. Jabalpurwala, K. A. Venkatachalam, and M. B. Kabadi, *J. Inorg. Nuclear Chem.*, 1964, **26**, 1011, 1027

$C_8H_9O_2N$ 2-Hydroxyacetophenone oxime HL 664.1

Metal	Method	Temp	Medium	Log of equilibrium constant, remarks	Ref
H^+				H^+ constants not given	
Cd^{2+}	gl	30	0·1 $NaClO_4$ 75% dioxan	K_1 6·95	58K
Co^{2+}	gl	30	0·1 $NaClO_4$ 75% dioxan	K_1 11·20, K_2 9·48, K_3 7·65	58K
Cu^{2+}	gl	30	0·1 $NaClO_4$ 75% dioxan	K_1 10·15, K_2 11·23 (?)	58K
Mg^{2+}	gl	30	0·1 $NaClO_4$ 75% dioxan	K_1 5·23, K_2 4·97	58K

continued overleaf

664.1 $C_8H_9O_2N$ (contd)

Metal	Method	Temp	Medium	Log of equilibrium constant, remarks	Ref
Mn^{2+}	gl	30	0·1 $NaClO_4$ 75% dioxan	K_1 7·57, K_2 7·35, K_3 7·13	58K
Ni^{2+}	gl	30	0·1 $NaClO_4$ 75% dioxan	K_1 7·55, K_2 7·25	58K
Zn^{2+}	gl	30	0·1 $NaClO_4$ 75% dioxan	K_1 8·85	58K

58K M. B. Kabadi and K. A. Venkatachalam, *Current Sci.*, 1958, **27**, 337

664.2 $C_8H_9O_2N$ 2-Hydroxy-5-methylbenzaldehyde oxime (5-methylsalicylaldoxime) HL

Metal	Method	Temp	Medium	Log of equilibrium constant, remarks	Ref
H^+	gl	20	0·1 ($NaClO_4$)	K_1 11·06±0·05	65B
Co^{2+}	gl	20	0·1 ($NaClO_4$)	K_1 6·8±0·2, K_2 7·5±0·2	65B
Cu^{2+}	sp	20	0·1 ($NaClO_4$)	β_2 22·2±0·2	65B
Fe^{2+}	gl	20	0·1 ($NaClO_4$)	K_1 9·7±0·1, K_2 7·7±0·1	65B
Mn^{2+}	gl	20	0·1 ($NaClO_4$)	K_1 6·1±0·2, K_2 6·1±0·2	65B
Ni^{2+}	gl	20	0·1 ($NaClO_4$)	K_1 7·2±0·2, K_2 7·5±0·2	65B
Zn^{2+}	gl	20	0·1 ($NaClO_4$)	K_1 7·0±0·2, K_2 7·3±0·2	65B

65B K. Burger and I. Egyed, *J. Inorg. Nuclear Chem.*, 1965, **27**, 2361

666 $C_8H_9O_2N$ *N*-Methyl-2-aminobenzoic acid (*N*-methylanthranilic acid) HL

Metal	Method	Temp	Medium	Log of equilibrium constant, remarks	Ref
H^+	gl	22	0·1 KCl	K_1 5·34	60U
Cu^{2+}	gl	22	0·1 KCl	K_1 3·45, K_2 2·85	60U

60U E. Uhlig, *Chem. Ber.*, 1960, **93**, 2470

667.1 $C_8H_9O_2N$ Pyridine-3-carboxylic acid ethyl ester (ethyl nicotinate) L

Metal	Method	Temp	Medium	Log of equilibrium constant, remarks	Ref
H^+	gl	25	0·3 (KNO_3)	K_1 3·33	67N
Cd^{2+}	Cd(Hg)	25	0·3 (KNO_3)	K_1 0·60	67N

67N L. V. Nazarova, *Russ. J. Inorg. Chem.*, 1967, **12**, 1620 (3062)

$C_8H_9O_3N$ 1-(2′,4′-Dihydroxyphenyl)acetophone oxime (resacetophenone oxime) H_2L 667.2

Metal	Method	Temp	Medium	Log of equilibrium constant, remarks	Ref
H^+	pol	29	60% ethoxy—ethanol	K_1 7·7, K_2 not given	67S
Co^{2+}	gl	~ 30 (?)	60% dioxan	β_2 10·50	67S
Cu^{2+}	dis	30	?	$K(Cu^{2+}+HL^- \rightleftharpoons CuL+H^+)$ −3·41, KH phthalate buffer	64B
	gl	~ 30 (?)	60% dioxan	β_2 11·90	67S
Mn^{2+}	gl	~ 30 (?)	60% dioxan	β_2 6·50	67S
Ni^{2+}	dis	30	?	$K(Ni^{2+}+HL^- \rightleftharpoons NiL+H^+)$ −10·74, KH phthalate buffer	64B
	gl	~ 30 (?)	60% dioxan	β_2 8·10	67S

64B K. S. Bhatki, A. T. Rane, and M. B. Kabadi, *J. Chem. and Eng. Data*, 1964, **9**, 175
67S V. Seshagiri and S. B. Rao, *Austral. J. Chem.*, 1967, **20**, 2783; *J. Electroanalyt. Chem.*, 1967, **13**, 330

$C_8H_9O_3N$ 2-Methyl-3-hydroxy-4-formyl-5-hydroxymethylpyridine (pyridoxal) HL 667.3

Metal	Method	Temp	Medium	Log of equilibrium constant, remarks	Ref
H^+	gl	25	0·5 HCl	K_1 8·54, K_2 4·25	66L
Cu^{2+}	gl	25	0·5 KCl	K_1 3·51, β_2 7·0	66L
Ni^{2+}	gl	25	0·5 KCl	K_1 1·85 $K(Ni^{2+}+L^-+A^- \rightleftharpoons NiLA)$ 10·30 $K(Ni^{2+}+2L^-+2A^- \rightleftharpoons NiL_2A_2^{2-})$ 19·84 $K(NiLA+H^+ \rightleftharpoons NiLAH^+)$ 7·18 where HA = glycine	66L
Zn^{2+}	gl	25	0·5 KCl	K_1 2·32 $K(Zn^{2+}+L^-+A^- \rightleftharpoons ZnLA)$ 8·43 $K(Zn^{2+}+2L^-+2A^- \rightleftharpoons ZnL_2A_2^{2-})$ 16·86 $K(ZnLA+H^+ \rightleftharpoons ZnLAH^+)$ 7·34 where HA = glycine	66L

66L D. L. Leussing and N. Hug, *Analyt. Chem.*, 1966, **38**, 1388

$C_8H_9O_7N_3$ *N*-(2′,4′,6′-Trioxopyrimidin-5′-yl)iminodiacetic acid (uramil-*N*,*N*-diacetic acid) H_3L 668

Metal	Method	Temp	Medium	Log of equilibrium constant, remarks	Ref
H^+	gl	20—39	0·1 [$(CH_3)_4NNO_3$]	K_1 (20°) 9·63±0·03, (27°) 9·47, (34°) 9·38, (39°) 9·31	63I
			0·1 (KNO_3)	K_2 (20°) 2·67±0·03, (27°) 2·83, (34°) 2·88, (39°) 2·90, K_3 (20°) 1·7±0·3, (27°) 1·82, (34°) 1·90, (39°) 2·21 ΔH_1 −6·9, ΔS_1 19±2	
		20	0 corr	K_1 10·33, K_2 3·1, K_3 1·9	
Ba^{2+}	gl	20	0·1 (KNO_3)	K_1 6·13±0·02, K_2 3·7±0·1	63I
Be^{2+}	gl	20	0·1 (KNO_3)	K_1 10·36±0·02 $K(Be^{2+}+HL^{2-} \rightleftharpoons BeHL)$ 3·44	63I

continued overleaf

668 $C_8H_9O_7N_3$ (contd)

Metal	*Method*	*Temp*	*Medium*	*Log of equilibrium constant, remarks*	*Ref*
Ca^{2+}	gl	20	0·1 (KNO_3)	K_1 8·31±0·01, K_2 5·27±0·02	63I
K^+	gl	20—39	0·1 (KNO_3)	K_1 (20°) 1·23±0·03, (27°) 1·00, (34°) 0·81, (39°) 0·70	63I
				ΔH_1 −11·8, ΔS_1 −35±2	
		20	0 corr	K_1 1·94	
Li^+	gl	20—39	0·1 [$(CH_3)_4NNO_3$]	K_1 (20°) 4·90±0·02, (27°) 4·70, (34°) 4·57, (39°) 4·60	63I
				ΔH_1 −7·0, ΔS_1 −1±5	
		20	0 corr	K_1 5·61	
Mg^{2+}	gl	20	0·1 (KNO_3)	K_1 8·19±0·02, K_2 3·62±0·05	63I
Na^+	gl	20—34	0·1 ($NaNO_3$)	K_1 (20°) 2·72±0·01, (27°) 2·54, (34°) 2·42	63I
				ΔH_1 −8·7, ΔS_1 −18±2	
		20	0 corr	K_1 3·33	
Pb^{2+}	gl	20	0·1 (KNO_3)	K_1 ~ 12	63I
Sr^{2+}	gl	20	0·1 (KNO_3)	K_1 6·93±0·02, K_2 4·1±0·1	63I
Tl^+	gl	20—39	0·1 (KNO_3)	K_1 (20°) 5·99±0·02, (27°) 5·76, (34°) 5·41, (39°) 5·33	63I
				ΔH_1 −15·4, ΔS_1 −25±4	
		20	0 corr	K_1 6·70	

63I H. Irving and J. J. R. F. da Silva, *J. Chem. Soc.*, 1963, 448, 458

668.1 $C_8H_{10}ON_2$ 2-Hydroxy-2-phenylacetamidine (mandelamidine) HL

Metal	*Method*	*Temp*	*Medium*	*Log of equilibrium constant, remarks*	*Ref*
H^+	gl	25	0·1 (KCl)	K_1 12·52±0·05, K_2 10·82±0·01	63G
Cu^{2+}	gl J	25	0·1 (KCl)	K_1 12·50, K_2 11·30	63G
Ni^{2+}	gl J	25	0·1 (KCl)	K_1 7·38, K_2 7·02	63G

63G R. O. Gould and R. J. Jameson, *J. Chem. Soc.*, 1963, **15**, 5211

668.2 $C_8H_{10}ON_4$ 5-Propyl-7-hydroxy[1,2,4]triazolo[1,5-a][1,3]diazine HL

Metal	*Method*	*Temp*	*Medium*	*Log of equilibrium constant, remarks*	*Ref*
H^+	gl	20	0·1 (KNO_3)	K_1 6·31	66O
Co^{2+}	gl	20	0·1 (KNO_3)	K_1 2·39	66O
Cu^{2+}	gl	20	0·1 (KNO_3)	K_1 3·41, β_2 6·45	66O
Ni^{2+}	gl	20	0·1 (KNO_3)	K_1 2·78, β_2 5·65	66O

66O G. Ostacoli, E. Campi, and M. C. Gennaro, *Gazzetta*, 1966, **96**, 741

$C_8H_{10}ON_4$ 5-Methyl-6-ethyl-7-hydroxy[1,2,4]triazolo[1,5-a][1,3]diazine HL 668.3

Metal	*Method*	*Temp*	*Medium*	*Log of equilibrium constant, remarks*	*Ref*
H^+	gl	20	0·1 (KNO_3)	K_1 6·84	66O
Co^{2+}	gl	20	0·1 (KNO_3)	K_1 2·40	66O
Cu^{2+}	gl	20	0·1 (KNO_3)	K_1 3·46, β_2 6·84	66O
Ni^{2+}	gl	20	0·1 (KNO_3)	K_1 2·73, β_2 5·60	66O

66O G. Ostacoli, E. Campi, and M. C. Gennaro, *Gazzetta*, 1966, **96**, 741

$C_8H_{10}O_2N_2$ 2,6-Dimethyl-4-nitroaniline L 668.4

Metal	*Method*	*Temp*	*Medium*	*Log of equilibrium constant, remarks*	*Ref*
Zn^{2+}	sp	25	acetone	$K(ZnCl_2 + L \rightleftharpoons ZnCl_2L)$ 0·36	65S

65S D. P. N. Satchell and R. S. Satchell, *Trans. Faraday Soc.*, 1965, **61**, 1118

$C_8H_{10}O_2N_2$ *N*-(2′-Pyridylmethyl)glycine HL 669

Errata. 1964, p. 584, Table 669. *For* reference 57L *read* 65L R. G. Lacoste, G. V. Christoffers, and A. E. Martell, *J. Amer. Chem. Soc.*, 1965, **87**, 2385.
Under Medium, *for* 0·1 KNO_3 *read* 0·1 (KNO_3) all through.

$C_8H_{10}O_3S$ 3-Hydroxy-5-methylthiophene-2-carboxylic acid ethyl ester HL 669.1

Metal	*Method*	*Temp*	*Medium*	*Log of equilibrium constant, remarks*	*Ref*
H^+	sp	25	0·1 ($NaClO_4$) 10% dioxan	K_1 7·88	65C
Cu^{2+}	sp	25	0·1 ($NaClO_4$) 10% dioxan	K_1 5·95	65C

65C A. Courtin and H. Sigel, *Helv. Chim. Acta*, 1965, **48**, 617

669.2 $C_8H_{10}O_3S$ 4-Hydroxy-1-methylthiophen-3-carboxylic acid ethyl ester HL

Metal	Method	Temp	Medium	Log of equilibrium constant, remarks	Ref
H^+	sp	25	0·1 ($NaClO_4$) 10% dioxan	K_1 8·88	65C
Cu^{2+}	sp	25	0·6 ($NaClO_4$) 10% dioxan	K_1 4·72	65C

65C A. Courtin and H. Sigel, *Helv. Chim. Acta*, 1965, **48**, 617

671.1 $C_8H_{11}ON$ 4-Ethoxyaniline (*p*-phenetidine) L

Metal	Method	Temp	Method	Log of equilibrium constant, remarks	Ref
H^+	pH	25	0·3 (KNO_3)	K_1 4·86	64N
Cd^{2+}	Hg(Cd)	25	0·3 (KNO_3)	K_1 0·30, β_2 0·40, β_3 0·04	64N

64N L. V. Nazarova, A. V. Ablov, and V. A. Dagaer, *Russ. J Inorg. Chem.*, 1964, **9**, 1150 (2129)

671.2 $C_8H_{11}ON$ 2-Amino-1-phenylethanol HL

Metal	Method	Temp	Medium	Log of equilibrium constant, remarks	Ref
H^+	gl	25	0·1 (KCl)	K_1 11·90±0·05, K_2 8·79±0·01	65J
Cu^{2+}	gl	25	0·1 (KCl)	K_1 9·50±0·05, K_2 6·0±0·1	65J

65J R. F. Jameson and W. F. S. Neillie, *J. Inorg. Nuclear Chem.*, 1965, **27**, 2623

671.3 $C_8H_{11}ON$ 2-(2′-Hydroxyethyl)-6-methylpyridine HL

Metal	Method	Temp	Medium	Log of equilibrium constant, remarks	Ref
H^+	gl	0—25	50% dioxan	K_1 (0°) 13·86, (15°) 13·20, (25°) 12·81, K_2 (0°) 5·58, (15°) 5·32, (25°) 5·14	64L
Cu^{2+}	gl	0—25	50% dioxan	K_1 (0°) 9·44, (15°) 9·36, (25°) 9·34, K_2 (0°) 9·12, (15°) 9·10, (25°) 8·88 $\Delta H_{\beta 2}°$ −12·2, $\Delta S_{\beta 2}°$ 42	64L
	cal	25	50% dioxan	$\Delta H_{\beta 2}°$ −12·4±0·1, $\Delta S_{\beta 2}°$ 41·2	64L

64L T. J. Lane, A. J. Kandithil, and S. M. Rosalie, *Inorg. Chem.*, 1964, **3**, 487

$C_8H_{11}O_2N$ 1-(2′-Aminoethyl)-3,4-dihydroxybenzene (3-hydroxytyramine) H_2L 671.4

Metal	Method	Temp	Medium	Log of equilibrium constant, remarks	Ref
Mo^{VI}	sp	25	0·1 (KCl)	$K(MoO_4^{2-}+2H_2L \rightleftharpoons MoO_2L_2^{2-})$ 5·57	63H

63H J. Halmekoski, *Suomen Kem.*, 1963, **B36**, 55

$C_8H_{11}O_3N$ (−)-2-Amino-1-(3′,4′-dihydroxyphenyl)ethanol (noradrenaline) H_2L 672

Metal	Method	Temp	Medium	Log of equilibrium constant, remarks	Ref
H^+	gl	0, 25	0·06 KCl	K_2 (0°) 10·69, (25°) 9·98, K_3 (0°) 9·34, (25°) 8·82, K(?) (25°) 3·30	62A
	gl	25	0·1 (KCl)	$K_1 \sim 13$, K_2 9·70±0·05, K_3 8·64±0·01 $K(H_{-1}L^{3-}+H^+ \rightleftharpoons L^{2-}) \sim 13$	65J, 66J
Co^{2+}	gl	0, 25	0·06 KCl	K_1 (0°) 5·32, (25°) 4·82, K_2 (0°) 3·62, (25°) 3·50 β_2(0°) 8·64, (25°) 7·36 $\Delta H_{\beta 2}$ −19·0, $\Delta S_{\beta 2}$ −30	62A
	gl	25	0·1 (KCl)	*K_1 9·36±0·03, K_2 7·0±0·1	66J
Cu^{2+}	gl	0, 25	0·06 KCl	K_1 (0°) 10·23, (25°) 9·13, K_2 (0°) 6·50, (25°) 5·87 β_2(0°) 18·36, (25°) 16·32 $\Delta H_{\beta 2}$ −30·3, $\Delta S_{\beta 2}$ −27	62A
	gl	25	0·1 (KCl)	*K_1 14·75±0·03, K_2 12·50±0·05, K_3 10·40±0·05	65J
Mn^{2+}	gl	25	0·1 (KCl)	*K_1 8·58±0·03, K_2 6·2±0·1	66J
Mo^{VI}	sp	25	0·1 (KCl)	$K(MoO_4^{2-}+2H_2L \rightleftharpoons MoO_2L_2^{2-})$ (?) 5·82±0·02	62H
Ni^{2+}	gl	0, 25	0·06 KCl	K_1 (0°) 5·76, (25°) 5·28, K_2 (0°) 3·75, (25°) 3·50 β_2(0°) 9·50, (25°) 8·00 $\Delta H_{\beta 2}$ −22·4, $\Delta S_{\beta 2}$ −38	62A
	gl	25	0·1 (KCl)	*K_1 10·45±0·03, K_2 8·5±0·1	66J
Zn^{2+}	gl	25	0·1 (KCl)	*K_1 10·69±0·03, K_2 9·1±0·1	66J

* Equilibrium constants adjusted to give hypothetical microscopic constant for combination of metal ion with ligand species HL^-, in which both phenolic groups are dissociated.

62A A. C. Andrews, T. D. Lyons, and T. D. O'Brien, *J. Chem. Soc.*, 1962, 1776
62H J. Halmekoski, *Suomen Kem.*, 1962, **B35**, 209
65J R. F. Jameson and W. F. S. Neillie, *J. Inorg. Nuclear Chem.*, 1965, **27**, 2623
66J R. F. Jameson and W. F. S. Neillie, *J. Inorg. Nuclear Chem.*, 1966, **28**, 2667

$C_8H_{12}ON_2$ *N*-(2-Hydroxyethyl)-2-amino methylpyridine [*N*-(2-picolyl)ethandamine] L 675

Errata. 1964, p. 585, Table 675. *For* reference 57L *read* 64L R. G. Lacoste and A. E. Martell, *Inorg. Chem.*, 1964, **3**, 881. Under Medium, *for* 0·1 *read* 0·1 (KNO_3) all through.

676.1 $C_8H_{12}O_3N_4$ Glycyl-L-histidine HL

Metal	Method	Temp	Medium	Log of equilibrium constant, remarks	Ref
H^+	gl	25	0·16 KCl	K_1 8·24, K_2 6·77, K_3 2·66	66B
	gl	25	0·1 ($NaClO_4$) 0·19 CH_3CN	K_1 8·17, K_2 6·74	66K
Cu^+	gl	25	0·1 ($NaClO_4$) 0·19 CH_3CN	K_1 8·61 $K[Cu^+ + 2HL \rightleftharpoons Cu(HL)_2{}^+]$ 11·57 $K(CuL + Cu^+ \rightleftharpoons Cu_2L^+)$ 4·8	66K
Ni^{2+}	gl	25	0·16 KCl	$K[Ni(H_{-1}L) + H^+ \rightleftharpoons NiL^+]$ 6·10 $K[Ni(H_{-1}L)OH^- + H^+ \rightleftharpoons Ni(H_{-1}L)]$ 6·70 $K[Ni(H_{-1}L)(OH)_2{}^{2-} + H^+ \rightleftharpoons Ni(H_{-1}L)OH^-]$ 9·25	66B

66B G. F. Bryce, R. W. Roeske, and F. R. N. Gurd, *J. Biol. Chem.*, 1966, **241**, 1072
66K Th. Kaden and A. Zuberbuhler, *Helv. Chim. Acta*, 1966, **49**, 2189

676.2 $C_8H_{12}O_4N_2$ Perhydro-1,4-diazine-1,4-diacetic acid (Piperazine-*NN*′-diacetic acid) H_2L

Metal	Method	Temp	Medium	Log of equilibrium constant, remarks	Ref
H^+	H	20	0·1 (KCl)	K_1 8·68, K_2 4·40, K_3 1·78	63I
Ba^{2+}	H	20	0·1 (KCl)	K_1 1·6	63I
Ca^{2+}	H	20	0·1 (KCl)	K_1 2·5	63I
Cu^{2+}	H	20	0·1 (KCl)	K_1 7·37, $K(CuL + H^+ \rightleftharpoons CuHL^+)$ 5·41	63I
Mg^{2+}	H	20	0·1 (KCl)	K_1 1·5	63I
Ni^{2+}	H	20	0·1 (KCl)	K_1 3·64, $K(NiL + H^+ \rightleftharpoons NiHL^+)$ 6·35	63I
Sr^{2+}	H	20	0·1 (KCl)	K_1 2·2	63I
Zn^{2+}	H	20	0·1 (KCl)	K_1 3·05	63I

63I H. Irving and L. D. Pettit, *J. Chem. Soc.*, 1963, 3051

677.1 $C_8H_{13}O_6N$ *N*-(2′-Carboxy-2′-propyl)iminodiacetic acid H_3L

Metal	Method	Temp	Medium	Log of equilibrium constant, remarks	Ref
H^+	gl	20	0·1 (KCl)	K_1 11·86±0·03, K_2 2·52±0·02, K_3 1·51±0·05	66I
Ba^{2+}	gl	20	0·1 (KCl)	K_1 5·61±0·03	66I
Ca^{2+}	gl	20	0·1 (KCl)	K_1 8·32±0·01	66I
Mg^{2+}	gl	20	0·1 (KCl)	K_1 6·30±0·03	66I
Sr^{2+}	gl	20	0·1 (KCl)	K_1 6·14±0·03	66I
Zn^{2+}	gl	20	0·1 (KCl)	K_1 12·45±0·03	66I

66I H. M. N. H. Irving and M. G. Miles, *J. Chem. Soc. (A)*, 1966, 1268

$C_8H_{14}O_5N_4$ Tetraglycine HL 680

Metal	*Method*	*Temp*	*Medium*	*Log of equilibrium constant, remarks*	*Ref*
H^+	gl	25	~ 0	K_1 7·91	54P
	gl	25	0·1 (KNO_3)	K_1 7·89, K_2 3·24	66K, 67K
Cd^{2+}	gl	25	0·01 $CdSO_4$	K_1 2·8	54P
Cu^{2+}	gl	25	0·16 (KNO_3)	K_1 5·05	60M
				$K[Cu(H_{-1}L)+H^+ \rightleftharpoons CuL^+] < 6{\cdot}0$	
				$K[Cu(H_{-2}L)^- +H^+ \rightleftharpoons Cu(H_{-1}L)]$ 6·95	
				$K[Cu(H_{-3}L)^{2-} +H^+ \rightleftharpoons Cu(H_{-2}L)^-]$ 9·45	
	gl	25	0·1 (KNO_3)	K_1 $5{\cdot}4 \pm 0{\cdot}1$	66K
				$K[Cu(H_{-1}L)+H^+ \rightleftharpoons CuL^+]$ $5{\cdot}6 \pm 0{\cdot}1$	
				$K[Cu(H_{-2}L)^- +H^+ \rightleftharpoons Cu(H_{-1}L)]$ $6{\cdot}77 \pm 0{\cdot}02$	
				$K[Cu(H_{-3}L)^{2-} +H^+ \rightleftharpoons Cu(H_{-2}L)^-]$ $9{\cdot}0 \pm 0{\cdot}1$	
Ni^{2+}	gl	25	0·16 (KNO_3)	$K[Ni(H_{-1}L)+H^+ \rightleftharpoons NiL^+]$ 8·10	60M
				$K[Ni(H_{-2}L)^- +H^+ \rightleftharpoons Ni(H_{-1}L)]$ 8·20	
				$K[Ni(H_{-3}L)^{2-} +H^+ \rightleftharpoons Ni(H_{-2}L)^-]$ 8·25	
	gl	25	0·1 (KNO_3)	K_1 3·65, K_2 3·3	67K
				$K[Ni(H_{-3}L)^{2-} +3H^+ \rightleftharpoons NiL^+]$ 24·4	
				$K[Ni(OH)(H_{-3}L)^{3-} +H^+ \rightleftharpoons Ni(H_{-3}L)^{2-}]$ 10·0	
Zn^{2+}	gl	25	0·01 $ZnSO_4$	K_1 2·9	54P

54P D. J. Perkins, *Biochem. J.*, 1954, **57**, 702

60M R. B. Martin, M. Chamberlin, and J. T. Edsall, *J. Amer. Chem. Soc.*, 1960, **82**, 495

66K M. K. Kim and A. E. Martell, *J. Amer. Chem. Soc.*, 1966, **88**, 914

67K M. K. Kim and A. E. Martell, *J. Amer. Chem. Soc.*, 1967, **89**, 5138

$C_8H_{15}O_2N$ 1-Aminocycloheptanecarboxylic acid HL 682

Metal	*Method*	*Temp*	*Medium*	*Log of equilibrium constant, remarks*	*Ref*
Cu^{2+}	gl	20	0·1 (KCl)	$K(CuL^+ +H^+ \rightleftharpoons CuHL^{2+})$ 1·1	63I

63I H. Irving and L. D. Pettit, *J. Chem. Soc.*, 1963, 1546

$C_8H_{16}O_2N_2$ Octane-4,5-dione dioxime (di-n-propylglyoxime) HL 682.1

Metal	*Method*	*Temp*	*Medium*	*Log of equilibrium constant, remarks*	*Ref*
H^+	gl	25	0·1 75% dioxan	K_1 14·18	63B
Ni^{2+}	gl	25	0·1 0—75% dioxan	K_1 (75%) $13{\cdot}1 \pm 0{\cdot}4$, β_2 (→ 0%) 17·1, (75%) $23{\cdot}9 \pm 0{\cdot}7$, K_{so} −25·14	63B

63B C. V. Banks and S. Anderson, *Inorg. Chem.*, 1963, **2**, 112

684.1 $C_8H_{16}O_3N_2$ Glycyl-DL-norleucine HL

Metal	*Method*	*Temp*	*Medium*	*Log of equilibrium constant, remarks*	*Ref*
H^+	gl	25	0·12 (NaCl)	K_1 8·37±0·02, K_2 3·42±0·02	67S
Cu^{2+}	gl	25	0·12 (NaCl)	K_1 5·92,	67S
				$K[Cu(H_{-1}L)+H^+ \rightleftharpoons CuL^+]$ 4·73	
				$K[Cu(H_{-1}L)OH^- + H^+ \rightleftharpoons Cu(H_{-1}L)]$ 10·19	
				$K[Cu(H_{-1}L)+L^- \rightleftharpoons CuL(H_{-1}L)^-]$ 3·73	
				$K[Cu(H_{-1}L)_2^{2-} + H^+ \rightleftharpoons CuL(H_{-1}L)^-]$ 9·51	

67S U. I. Salakhutdinov, A. O. Borisova, Yu. V. Granovskii, I. A. Savich, and V. I. Spitsyn, *Proc. Acad. Sci.* (*U.S.S.R.*), 1967, **177**, 1039 (365)

684.2 $C_8H_{16}O_3N_2$ 6-Acetylamino-2-aminohexanoic acid (*N*-ε-acetyl-lysine) HL

Metal	*Method*	*Temp*	*Medium*	*Log of equilibrium constant, remarks*	*Ref*
H^+	gl	20	0·1 KNO_3	K_1 9·46	68H
Cu^{2+}	gl	20	0·1 KNO_3	K_1 7·86, K_2 6·72	68H

68H P. Herman and K. Lemke, *Z. physiol. Chem.*, 1968, **349**, 390

687 $C_8H_{16}O_4N_2$ 2,7-Diamino-octanedioic acid (2,7-diaminosuberic acid) H_2L

Metal	*Method*	*Temp*	*Medium*	*Log of equilibrium constant, remarks*	*Ref*
H^+	gl	20	0·15 ($NaClO_4$)	K_1 9·89, K_2 9·23, K_3 2·62, K_4 1·84	63H
Cu^{2+}	gl	20	0·15 ($NaClO_4$)	$K(Cu^{2+}+HL^- \rightleftharpoons CuHL^+)$ 8·03	63H
				$K(Cu^{2+}+2HL^- \rightleftharpoons CuH_2L_2)$ 14·20	
				$K(Cu^{2+}+L^{2-}+HL^- \rightleftharpoons CuHL_2^-)$ 22·83	
				$K[2Cu^{2+}+2L^{2-} \rightleftharpoons (CuL)_2]$ 29·00	
				$K(2Cu^{2+}+L^{2-} \rightleftharpoons Cu_2L^{2+})$ 16·06	
				$K(2Cu^{2+}+L^{2-}+2HL^- \rightleftharpoons Cu_2H_2L_3)$ 29·00	

63H C. J. Hawkins and D. D. Perrin, *Inorg. Chem.*, 1963, **2**, 839

688.1 $C_8H_{17}O_2N$ Leucine ethyl ester L

Metal	*Method*	*Temp*	*Medium*	*Log of equilibrium constant, remarks*	*Ref*
Cu^{2+}				See nitrilotriacetic acid (449) for mixed ligands.	

$C_8H_{18}O_2N_4$ ***NN'*-Diglycyl-1,4-diaminobutane L 689.1**

Metal	Method	Temp	Medium	Log of equilibrium constant, remarks	Ref
H^+	gl	23	0·5 KCl	K_1 8·56, K_2 7·93	67Z
Cu^{2+}	gl	23	0·5 KCl	K_1 8·64	67Z
				$K(CuL^{2+}+H^+ \rightleftharpoons CuHL^{3+})$ 5·27	
				$K[Cu(H_{-1}L)^+ + H^+ \rightleftharpoons CuL^{2+}]$ 7·77	
				$K[Cu(H_{-2}L)+H^+ \rightleftharpoons Cu(H_{-1}L)^+]$ 7·52	

67Z A. Zuberbuhler and S. Fallab, *Helv. Chim. Acta*, 1967, **50**, 889

$C_8H_{19}ON$ ***NN*-Bis(2'-propyl)-2-aminoethanol L 689.2**
(*NN*-di-isopropyl-2-hydroxyethylamine)

Metal	Method	Temp	Medium	Log of equilibrium constant, remarks	Ref
H^+	H	20	0 corr	$K(L \rightleftharpoons HL^+ + OH^-)$ 4·09±0·02	61A
Ag^+	sol	20	0 corr	β_2 3·84±0·06	61A
	gl	20	0 corr	β_2 4·07±0·04	61A

61A D. J. Alner and R. C. Lansbury, *J. Chem. Soc.*, 1961, 3169

$C_8H_{19}O_2N$ ***N*-Butyl-2,2'-iminodiethanol (butyldiethanolamine) L 689.3**

Metal	Method	Temp	Medium	Log of equilibrium constant, remarks	Ref
H^+	gl	25	0·4 (KNO_3)	K_1 9·03	66S
Co^{2+}	gl	25	0·43 $(CH_2OHCH_2NH_3NO_3)$	K_1 2·50, K_2 1·57, K_3 1·23	66S
Cu^{2+}	gl	25	0·43 $(CH_2OHCH_2NH_3NO_3)$	K_1 4·28, K_2 3·51, K_3 2·65	66S
Fe^{2+}	hyp	25	0·43 $(CH_2OHCH_2NH_3NO_3)$	K_1 2·10, K_2 1·25	66S
Mn^{2+}	hyp	25	0·43 $(CH_2OHCH_2NH_3NO_3)$	K_1 1·35, K_2 0·45	66S
Ni^{2+}	gl	25	0·43 $(CH_2OHCH_2NH_3NO_3)$	K_1 3·17, K_2 2·03, K_3 1·47	66S

66S E. V. Sklenskaya and M. Kh. Karapet'yants, *Russ. J. Inorg. Chem.*, 1966, **11**, 1478 (2749)

689.4 $C_8H_{19}O_4P$ Phosphoric acid *O,O*-dibutyl ester (dibutylphosphoric acid) HL

Metal	*Method*	*Temp*	*Medium*	*Log of equilibrium constant, remarks*	*Ref*
H^+	?	?	1	K_1 −0·5	65S
Ce^{3+}	dis	18—22	?	K_1 1·48±0·06	61S
La^{3+}	sol	?	→ 0	K_1 1·80±0·05, β_2 3·5±0·3, β_3 4·62±0·05	65S
Nd^{3+}	dis	?	1·1 HNO_3	$K(Nd^{3+}+3HL+3L^- \rightleftharpoons NdH_3L_6)$ 15·4±0·2	62S
Pr^{3+}	dis	?	0·5 HNO_3	$K[Pr^{3+}+3HL+3L^- \rightleftharpoons Pr(HL_2)_3]$ 15·0	61S
Y^{3+}	dis	18—22	?	K_1 1·91±0·05	61S
Yb^{3+}	dis	?	0·5—1·2 HNO_3	$K(Yb^{3+}+3HL+3L^- \rightleftharpoons YbH_3L_6)$ 18·6±0·4	62S
	sol	?	?	$K(YbL_3+1{\cdot}5H_2L_2 \rightleftharpoons YbH_3L_6)$ −0·9±0·1	62S

61S Z. A. Sheka and E. E. Kriss, *Russ. J. Inorg. Chem.*, 1961, **6**, 984 (1930)
61Sa V. B. Shevchenko and V. S. Smelov, *Russ. J. Inorg. Chem.*, 1961, **6**, 372 (732)
62S Z. A. Sheka and E. E. Kriss, *Russ. J. Inorg. Chem.*, 1962, **7**, 333 (658)
65S Z. A. Sheka and E. I. Sinyavskaya, *Russ. J. Inorg. Chem.*, 1965, **10**, 212 (394)

691 $C_8H_5O_2SF_3$ 1-(2′-Thienlyl)-4,4,4-trifluorobutane-1,3-dione (2-thenoyltrifluoroacetone) HL

Metal	*Method*	*Temp*	*Medium*	*Log of equilibrium constant, remarks*	*Ref*
H^+	gl	25	1·0	K_1 6·53	62P, 67M
	gl	25	0·1 $(C_4H_9)_4NClO_4$ methanol	K_1 8·59±0·05	65L
			0·1 $(C_4H_9)_4NClO_4$ ethanol	K_1 8·20±0·05	
	gl	30	→ 0 75% dioxan	K_1 8·64±0·02	65R
	dis	25	0·1 $NaClO_4$	K_1 6·2	68S
Be^{2+}	dis	25	0·1, 2-xylene	K_1 5·54, β_2 11·11	62B
Co^{2+}	gl	30	→ 0 75% dioxan	K_1 7·81±0·01, K_2 7·1±0·2	65R
Cu^{2+}	gl	30	→ 0 75% dioxan	K_1 8·2±0·2, K_2 8·1±0·2	65R
Hf^{4+}	dis	25	1·0	K_1 10·60, β_2 21·44, β_3 31·50, β_4 41·52	62P
K^+	gl	25	0·1 KI methanol	K_1 1·6±0·2	65L
			0·1 KSCN ethanol	K_1 3·2±0·2	
Li^+	gl	25	0·1 $LiClO_4$ methanol	K_1 3·2±0·2	65L
			0·1 $LiClO_4$ ethanol	K_1 5·3±0·2	
Lu^{3+}	dis	25	0·1 $NaClO_4$	K_1 6·0, β_2 12·0, β_3 17·6	68S
				$K(Lu^{3+}+L^-+OH^- \rightleftharpoons LuLOH^+)$ 16·8	
				$K[Lu^{3+}+L^-+2OH^- \rightleftharpoons LuL(OH)_2]$ 26·0	
				$K(Lu^{3+}+2L^-+OH^- \rightleftharpoons LuL_2OH)$ 22·3	

cont

Metal	Method	Temp	Medium	Log of equilibrium constant, remarks	Ref
Na^+	gl	25	0·1 $NaClO_4$ methanol	K_1 2·4±0·2	65L
			0·1 $NaClO_4$ ethanol	K_1 4·2±0·2	
Ni^{2+}	gl	30	→ 0 75% dioxan	K_1 7·93±0·07, K_2 7·3±0·2	65R
Pa^V	dis	25	3 $LiClO_4$	$K[Pa(OH)_3^{2+}+L^- \rightleftharpoons Pa(OH)_3L^+]$ 2·26 $K[Pa(OH)_3^{2+}+2L^- \rightleftharpoons Pa(OH)_3L_2]$ 2·2 $K[Pa(OH)_3^{2+}+OH^-+L^- \rightleftharpoons Pa(OH)_4L]$ −0·9 $K[Pa(OH)_3^{2+}+OH^-+L^-+2HL \rightleftharpoons Pa(OH)_4L(HL)_2]$ ~ 6·2 $K[Pa(OH)_3^{2+}+2L^-+2HL \rightleftharpoons Pa(OH)_3L_2(HL)_2]$ ⩽ 9·6	65G
Sc^{3+}	dis	25	0·1 $NaClO_4$	K_1 7·1	66S
UO_2^{2+}	gl	30	→ 0 75% dioxan	K_1 8·7±0·1, K_2 7·92±0·04	65R
Zn^{2+}	gl	30	→ 0 75% dioxan	K_1 7·75±0·03, K_2 6·3±0·1	65R
	dis	25	?	$K(ZnL_2+A \rightleftharpoons ZnL_2A)$ 8·07 $K(Zn^{2+}+2HL+A \rightleftharpoons ZnL_2A+2H^+)$ −0·27 where A = tri-n-octylphosphine oxide	67C
Zr^{4+}	dis	25	1·0	K_1 10·98, K_2 10·90, K_3 10·36, K_4 9·93	67M

62B H. J. deBruin and R. B. Temple, *Austral. J. Chem.*, 1962, **15**, 153
62P V. M. Peshkova and P. Aug, *Russ. J. Inorg. Chem.*, 1962, **7**, 765 (1484)
65G R. Guillaumont, *Bull. Soc. chim. France*, 1965, 135
65L D. C. Luehrs, R. T. Iwamoto, and J. Kleinberg, *Inorg. Chem.*, 1965, **4**, 1739
65R J. L. Rosenstreich and D. E. Goldberg, *Inorg. Chem.*, 1965, **4**, 909
66S T. Sekine, A. Koizumi, and M. Sakairi, *Bull Chem. Soc. Japan*, 1966, **39**, 2681
67C G. C. Curthoys and W. R. Walker, *Austral. J. Chem.*, 1967, **20**, 2541
67M N. V. Mel'chakova, G. P. Ozerova and, V. M. Peshkova, *Russ. J. Inorg. Chem.*, 1967, **12**, 577 (1096)
68S G. K. Schweitzer and M. M. Anderson, *J. Inorg. Nuclear Chem.*, 1968, **30**, 1051

$C_8H_5O_2F_3Se$ 4,4,4-Trifluoro-1-(2′-selenoyl)-butane-1,3-dione (selenoyltrifluoroacetone) HL 691.1

Metal	Method	Temp	Medium	Log of equilibrium constant, remarks	Ref
H^+	gl	25	1·0	K_1 6·32	62P, 63M
Hf^{4+}	dis	20	1·0	K_1 10·46, β_2 20·74, β_3 30·22, β_4 39·70	62P
Zr^{4+}	dis	25	1·0	K_1 11·35, K_2 10·75, K_3 10·15, K_4 9·55	63M

62P V. M. Peshkova and P. Aug., *Russ. J. Inorg. Chem.*, 1962, **7**, 765 (1484)
63M N. V. Mel'chakova and V. M. Peshkova, *Russ. J. Inorg. Chem.*, 1963, **8**, 663 (1280)

691.2 $C_8H_8O_6N_2S$ *N*-(2′-Nitrobenzenesulphonyl)aminoacetic acid HL

Metal	Method	Temp	Medium	Log of equilibrium constant, remarks	Ref
H^+	gl	30	50% ethanol	K_1 4·40	67G
Cd^{2+}	gl	30	50% ethanol	$K(Cd^{2+}+HL \rightleftharpoons CdHL^{2+})$ 2·14	
				$K(CdHL^{2+}+HL \rightleftharpoons CdHL_2^{2+})$ 1·46	
				$K(CdL^++H^+ \rightleftharpoons CdHL^{2+})$ 6·57	
				$K[Cd(H_{-1}L)+H^+ \rightleftharpoons CdL^+]$ 8·57	
Cu^{2+}	gl	30	50% ethanol	$K(Cu^{2+}+HL \rightleftharpoons CuHL^{2+})$ 2·20	67G
				$K[CuHL^{2+}+HL \rightleftharpoons Cu(HL)_2^{2+}]$ 3·20 (?)	
				$K(CuL^++H^+ \rightleftharpoons CuHL^{2+})$ 4·33	
				$K[Cu(H_{-1}L)+H^+ \rightleftharpoons CuL^+]$ 7·00	
Zn^{2+}	gl	30	50% ethanol	$K(Zn^{2+}+HL \rightleftharpoons ZnHL^{2+})$ 1·94	67G
				$K[ZnHL^{2+}+HL \rightleftharpoons Zn(HL)_2^{2+}]$ 1·46	
				$K(ZnL^++H^+ \rightleftharpoons ZnHL^{2+})$ 5·97	
				$K[Zn(H_{-1}L)+H^+ \rightleftharpoons ZnL^+]$ 7·60	

67G N. Ghosh and M. N. Majumder, *J. Indian Chem. Soc.*, 1967, **44**, 559

691.3 $C_8H_8O_6N_2S$ *N*-(3′-Nitrophenylsulphonyl)aminoacetic acid HL

Metal	Method	Temp	Medium	Log of equilibrium constant, remarks	Ref
H^+	gl	30	50% ethanol	K_1 4·36	67G
Cd^{2+}	gl	30	50% ethanol	$K(Cd^{2+}+HL \rightleftharpoons CdHL^{2+})$ 2·25	67G
				$K[CdHL^{2+}+HL \rightleftharpoons Cd(HL)_2^{2+}]$ 1·71	
				$K(CdL^++H^+ \rightleftharpoons CdHL^{2+})$ 7·17	
				$K[Cd(H_{-1}L)+H^+ \rightleftharpoons CdL^+]$ 9·19	
Cu^{2+}	gl	30	50% ethanol	$K(Cu^{2+}+HL \rightleftharpoons CuHL^{2+})$ 2·66	67G
				$K[CuHL^{2+}+HL = Cu(HL)_2^{2+}]$ 3·10 (?)	
				$K(CuL^++H^+ = CuHL^{2+})$ 4·34	
				$K[Cu(H_{-1}L)+H^+ = CuL^+]$ 6·80	
Zn^{2+}	gl	30	50% ethanol	$K(Zn^{2+}+HL = ZnHL^{2+})$ 2·18	67G
				$K[ZnHL^{2+}+HL = Zn(HL)_2^{2+}]$ 1·46	
				$K(ZnL^++H^+ = ZnHL^{2+})$ 5·83	
				$K[Zn(H_{-1}L)+H^+ = ZnL^+]$ 7·21	

67G N. Ghosh and M. N. Majumder, *J. Indian Chem. Soc.*, 1967, **44**, 559

691.4 $C_8H_8O_6N_2S$ *N*-(4′-Nitrophenylsulphonyl)aminoacetic acid HL

Metal	Method	Temp	Medium	Log of equilibrium constant, remarks	Ref
H^+	gl	30	50% ethanol	K_1 4·35	67G
Cd^{2+}	gl	30	50% ethanol	$K(Cd^{2+}+HL \rightleftharpoons CdHL^{2+})$ 2·34,	67G
				$K(CdHL^{2+}+HL \rightleftharpoons CdHL^{2+})$ 1·58,	
Cu^{2+}	gl	30	50% ethanol	$K(Cu^{2+}+HL \rightleftharpoons CuHL^{2+})$ 2·25,	67G
				$K[CuHL^{2+}+HL \rightleftharpoons Cu(HL)_2^{2+}]$ 3·21 (?),	
				$K(CuL^++H^+ \rightleftharpoons CuHL^{2+})$ 4·24,	

conti…

$C_8H_8O_6N_2S$ (contd) 691.4

Metal	Method	Temp	Medium	Log of equilibrium constant, remarks	Ref
Zn^{2+}	gl	30	50% ethanol	$K(Zn^{2+}+HL \rightleftharpoons ZnHL^{2+})$ 2·23, $K[ZnHL^{2+}+HL \rightleftharpoons Zn(HL)_2^{2+}]$ 1·79, $K(ZnL^{+}+H^{+} \rightleftharpoons ZnHL^{2+})$ 5·82, $K[Zn(H_{-1}L)+H^{+} \rightleftharpoons ZnL^{+}]$ 7·06	67G

67G N. Ghosh and M. N. Majumder, *J. Indian Chem. Soc.*, 1967, **44**, 559

$C_8H_9O_4NS$ *N*-(Phenylsulphonyl)aminoacetic acid HL 693.1

Metal	Method	Temp	Medium	Log of equilibrium constant, remarks	Ref
H^+	gl	30	50% ethanol	K_1 4·50	63G, 67G
Cd^{2+}	gl	30	50% ethanol	$K(Cd^{2+}+HL \rightleftharpoons CdHL^{2+})$ 1·23 $K[CdHL^{2+}+HL \rightleftharpoons Cd(HL)_2^{2+}]$ 2·77 (?)	63G
	gl	30	50% ethanol	$K(Cd^{2+}+HL \rightleftharpoons CdHL^{2+})$ 1·71 $K[CdHL^{2+}+HL \rightleftharpoons Cd(HL)_2^{2+}]$ 2·21 $K(CdL^{+}+H^{+} \rightleftharpoons CdHL^{2+})$ 7·28 $K[Cd(H_{-1}L)+H^{+} \rightleftharpoons CdL^{+}]$ 8·89	67G
Cu^{2+}	gl	30	50% ethanol	$K(Cu^{2+}+HL \rightleftharpoons CuHL^{2+})$ 1·86 $K[CuHL^{2+}+HL \rightleftharpoons Cu(HL)_2^{2+}]$ 3·40 (?)	63G
	gl	30	50% ethanol	$K(Cu^{2+}+HL \rightleftharpoons CuHL^{2+})$ 2·40 $K[CuHL^{2+}+HL \rightleftharpoons Cu(HL)_2^{2+}]$ 2·80 (?) $K(CuL^{+}+H^{+} \rightleftharpoons CuHL^{2+})$ 4·19 $K[Cu(H_{-1}L)+H^{+} \rightleftharpoons CuL^{+}]$ 7·70	67G
Zn^{2+}	gl	30	50% ethanol	$K(Zn^{2+}+HL \rightleftharpoons ZnHL^{2+})$ 1·57 $K[ZnHL^{2+}+HL \rightleftharpoons Zn(HL)_2]$ 2·53 (?)	63G
	gl	30	50% ethanol	$K(Zn^{2+}+HL \rightleftharpoons ZnHL^{2+})$ 1·87 $K[ZnHL^{2+}+HL \rightleftharpoons Zn(HL)_2^{2+}]$ 2·10 (?) $K(ZnL^{+}+H^{+} \rightleftharpoons ZnHL^{2+})$ 6·61 $K[Zn(H_{-1}L)+H^{+} \rightleftharpoons ZnL^{+}]$ 7·88	67G

63G N. N. Ghosh and M. N. Majumder, *J. Indian Chem. Soc.*, 1963, **40**, 945
67G N. Ghosh and M. N. Majumder, *J. Indian Chem. Soc.*, 1967, **44**, 559

$C_8H_{15}O_9N_2P$ *O*-Phosphoryl-L-seryl-L-glutamic acid(L-phosphoseryl-L-glutamic acid) H_4L 694.1

Metal	Method	Temp	Medium	Log of equilibrium constant, remarks	Ref
H^+	gl	25	0·15 (KCl)	K_1 8·25, K_2 5·68, K_3 4·39, K_4 3·03	62O
	gl	25	0·15 (KNO_3)	K_1 8·19±0·01, K_2 5·69±0·01, K_3 4·40±0·01, K_4 3·04±0·01	65O
Ca^{2+}	gl	25	0·15 (KCl)	K_1 2·14 $K(Ca^{2+}+HL^{3-} \rightleftharpoons CaHL^{-})$ 1·64±0·02 $K(Ca^{2+}+CaL^{2-} \rightleftharpoons Ca_2L)$ 1·46 $K(Ca^{2+}+CaHL^{-} \rightleftharpoons Ca_2HL^{+})$ 1·0±0·1 $K(Ca_2L+H^{+} \rightleftharpoons Ca_2HL^{+})$ 7·29 $K(Ca^{2+}+H_2L^{2-} \rightleftharpoons CaH_2L)$ 1·08	62O

continued overleaf

694.1 $C_8H_{15}O_9N_2P$ (contd)

Metal	*Method*	*Temp*	*Medium*	*Log of equilibrium constant, remarks*	*Ref*
Cu^{2+}	gl, Cu(Hg)	25	0·15 (KNO_3)	K_1 8·34±0·01, β_2 12·4±0·1	65O
				$K(Cu^{2+}+HL^{3-} \rightleftharpoons CuHL^-)$ 13·04±0·05	
				$K(Cu^{2+}+H_2L^{2-} \rightleftharpoons CuH_2L)$ 16·7±0·1	
				$K(Cu^{2+}+2HL^{3-} \rightleftharpoons CuH_2L_2^{4-})$ 24·3±0·2	
				$K[Cu(H_{-1}L)^{3-}+H^+ \rightleftharpoons CuL^{2-}]$ 6·05±0·02	
Mg^{2+}	gl	25	0·15 (KCl)	K_1 2·09	62O
				$K(Mg^{2+}+HL^{3-} \rightleftharpoons MgHL^-)$ 1·63	
				$K(Mg^{2+}+MgL^{2-} \rightleftharpoons Mg_2L)$ 1·81	
				$K(Mg^{2+}+MgHL^- \rightleftharpoons Mg_2HL^+)$ 1·51	
				$K(Mg_2L+H^+ \rightleftharpoons Mg_2HL^+)$ 7·49	
				$K(Mg^{2+}+H_2L^{2-} \rightleftharpoons MgH_2L)$ 1·00	
Mn^{2+}	gl	25	0·15 (KCl)	K_1 2·984	62O
				$K(Mn^{2+}+HL^{3-} \rightleftharpoons MnHL^-)$ 2·24	
				$K(Mn^{2+}+MnL^{2-} \rightleftharpoons Mn_2L)$ 1·95	
				$K(Mn_2L+H^+ \rightleftharpoons Mn_2HL^+)$ 6·88	
				$K(Mn^{2+}+MnHL^- \rightleftharpoons Mn_2HL^+)$ 1·32	
				$K(Mn^{2+}+H_2L^{2-} \rightleftharpoons MnH_2L)$ 1·42	

62O R. Osterberg, *Acta Chem. Scand.*, 1962, **16**, 2434
65O R. Osterberg, *Acta Chem. Scand.*, 1965, **19**, 1445

694.2 $C_8H_{18}O_{10}N_2P_2$ Ethylenedinitro-*NN'*-bis(methylphosphonic acid)-*NN'*-diacetic acid **H_6L**

Metal	*Method*	*Temp*	*Medium*	*Log of equilibrium constant, remarks*	*Ref*
H^+	gl	25	0·1 KCl	K_1 10·34, K_2 8·36, K_3 6·13, K_4 4·65, K_5 2·30, K_6 1·5	65D
	gl	20	0·1 KCl	K_1 10·77, K_2 8·33, K_3 6·12, K_4 4·66, K_5 2·68, K_6 2·00	65T
Ca^{2+}	gl	25	0·1 KCl	K_1 7·91	65D
	gl	20	0·1 KCl	K_1 8·9±0·1	65T
	ix	20	0·1 NH_4^+	K_1 8·88±0·01	65T
Ce^{3+}	ix	20	0·1 NH_4^+	K_1 18·48±0·01	65T
Cu^{2+}	gl	25	0·1 KCl	K_1 18·5,	65D
				$K(Cu^{2+}+HL^{5-} \rightleftharpoons CuHL^{3-})$ 9·49	
Er^{3+}	gl	25	0·1 KCl	$K(Er^{3+}+HL^{5-} \rightleftharpoons ErHL^{2-})$ 17·7	65D
				$K(Er^{3+}+H_4L^{2-} \rightleftharpoons ErH_4L^+)$ 9·2	
Fe^{3+}	gl	25	0·1 KCl	$K_1 > 12$	65D
La^{3+}	gl	25	0·1 KCl	$K(La^{3+}+HL^{5-} \rightleftharpoons LaHL^{2-})$ 15·6	65D
				$K(La^{3+}+H_2L^{4-} \rightleftharpoons LaH_2L^-)$ 8·43	
Lu^{3+}	gl	25	0·1 KCl	$K(Lu^{3+}+HL^{5-} \rightleftharpoons LuHL^{2-})$ 17·7	65D
				$K(Lu^{3+}+H_4L^{2-} \rightleftharpoons LuH_4L^+)$ 9·2	
Mg^{2+}	gl	25	0·1 KCl	K_1 8·11	65D
Ni^{2+}	gl	25	0·1 KCl	K_1 15·23	65D
				$K(Ni^{2+}+HL^{5-} \rightleftharpoons NiHL^{3-})$ 9·49	
Sm^{3+}	gl	25	0·1 KCl	$K(Sm^{3+}+HL^{5-} \rightleftharpoons SmHL^{2-})$ 17·8	65D
				$K(Sm^{3+}+H_2L^{4-} \rightleftharpoons SmH_2L^-)$ 9·6	
Sr^{2+}	gl	25	0·1 KCl	K_1 6·89	65D
	gl	20	0·1 KCl	K_1 8·15±0·1	65T
	ix	20	0·1 NH_4^+	K_1 8·31±0·03	65T

conti

$C_8H_{18}O_{10}N_2P_2$ (contd) 694.2

Metal	Method	Temp	Medium	Log of equilibrium constant, remarks	Ref
Y^{3+}	gl	25	0·1 KCl	$K(Y^{3+}+HL^{5-} \rightleftharpoons YHL^{2-})$ 17·7	65D
				$K(Y^{3+}+H_2L^{4-} \rightleftharpoons YH_2L^{-})$ 9·2	
	ix	20	0·1 NH_4^+	K_1 24·04±0·02	65T

65D N. M. Dyatlova, M. I. Kabachnik, T. Ya. Medved, M. V. Rudomino, and Yu. F. Belugin, *Proc. Acad. Sci.* (*U.S.S.R.*), 1965, **161**, 307 (607)
65T L. I. Tikhonova, *Russ. J. Inorg. Chem.*, 1965, **10**, 70 (132)

$C_8H_{19}O_2S_2P$ Phosphorodithioic acid *S,S′*-dibutyl ester (dibutyldithiophosphoric acid) HL 694.3

Metal	Method	Temp	Medium	Log of equilibrium constant, remarks	Ref
H^+	dis	25	1 NaCl	K_1 0·22	63H
Zn^{2+}	dis	23—27	1 NaCl	β_2 3·81	63H

63H T. H. Handley, R. H. Zucal, and J. A. Dean, *Analyt. Chem.*, 1963, 35, 1163

$C_8H_{19}O_2S_2P$ Phosphorodithioic acid *S,S′*-bis(2-methylpropyl) ester (bis(2-methylpropyl)dithiophosphoric acid) HL 694.4

Metal	Method	Temp	Medium	Log of equilibrium constant, remarks	Ref
H^+	dis	25	1 NaCl	K_1 0·10	63H
Zn^{2+}	dis	23—27	1 NaCl	β_2 4·00	63H

63H T. H. Handley, R. H. Zucal, and J. A. Dean, *Analyt. Chem.*, 1963, 35, 1163

$C_8H_{22}O_6N_2P_2$ 2,2′-(Ethylenedi-imino)bis(propylphosphonic acid) H_4L 694.5

Metal	Method	Temp	Medium	Log of equilibrium constant, remarks	Ref
H^+	gl	25	0·1 KCl	K_1 11·68, K_2 8·55, K_3 6·00, K_4 4·95	65D
Be^{2+}	gl	25	0·1 KCl	K_1 7	65D
Ca^{2+}	gl	25	0·1 KCl	K_1 < 2	65D
Co^{2+}	gl	25	0·1 KCl	K_1 11·39,	65D
				$K(Co^{2+}+HL^{3-} \rightleftharpoons CoHL^{-})$ 3·84	
Cu^{2+}	gl	25	0·1 KCl	K_1 20·35,	65D
				$K(Cu^{2+}+HL^{3-} \rightleftharpoons CuHL^{-})$ 8·83	
Dy^{3+}	gl	25	0·1 KCl	K_1 12·89,	65D
				$K(Dy^{3+}+HL^{3-} \rightleftharpoons DyHL)$ 6·40	
Er^{3+}	gl	25	0·1 KCl	K_1 13·39,	65D
				$K(Er^{3+}+HL^{3-} \rightleftharpoons ErHL)$ 6·88	
Fe^{3+}	gl	25	0·1 KCl	$K(Fe^{3+}+HL^{3-} \rightleftharpoons FeHL)$ > 10	65D
Gd^{3+}	gl	25	0·1 KCl	K_1 12·27,	65D
				$K(Gd^{3+}+HL^{3-} \rightleftharpoons GdHL)$ 6·16	

continued overleaf

694.5 $C_8H_{22}O_6N_2P_2$ (contd)

Metal	*Method*	*Temp*	*Medium*	*Log of equilibrium constant, remarks*	*Ref*
La^{3+}	gl	25	0·1 KCl	K_1 10·13, $K(La^{3+}+HL^{3-} \rightleftharpoons LaHL)$ 5·37	65D
Lu^{3+}	gl	25	0·1 KCl	K_1 13·37, $K(Lu^{3+}+HL^{3-} \rightleftharpoons LuHL)$ 7·67	65D
Mg^{2+}	gl	25	0·1 KCl	$K_1 < 2$	65D
Mn^{2+}	gl	25	0·1 KCl	K_1 8·00, $K(Mn^{2+}+HL^{3-} \rightleftharpoons MnHL^-)$ 3·57	65D
Nd^{3+}	gl	25	0·1 KCl	K_1 11·60, $K(Nd^{3+}+HL^{3-} \rightleftharpoons NdHL)$ 5·82	65D
Ni^{2+}	gl	25	0·1 KCl	K_1 11·13, $K(Ni^{2+}+HL^{3-} \rightleftharpoons NiHL^-)$ 3·84	65D
Sm^{3+}	gl	25	0·1 KCl	K_1 12·56, $K(Sm^{3+}+HL^{3-} \rightleftharpoons SmHL)$ 6·20	65D
Sr^{2+}	gl	25	0·1 KCl	$K_1 < 1$	65D
Y^{3+}	gl	25	0·1 KCl	K_1 12·87, $K(Y^{3+}+HL^{3-} \rightleftharpoons YHL)$ 6·48	65D
Zn^{2+}	gl	25	0·1 KCl	K_1 13·38, $K(Zn^{2+}+HL^{3-} \rightleftharpoons ZnHL^-)$ 4·81	65D

65D N. M. Dyatlova, M. I. Kabachnik, T. Ya. Medved, M. V. Rudomino, and Yu. F. Belugin, *Proc. Acad. Sci.* (*U.S.S.R.*), 1965, **161**, 307 (607)

699.1 C_9H_{16} 1-(2′-Propyl)cyclohexene L

Metal	*Method*	*Temp*	*Medium*	*Log of equilibrium constant, remarks*	*Ref*
Ag^+	oth	30	1·77 $AgNO_3$ ethylene glycol	K_1 0·02, gas chromatography	62G

62G E. Gil-Av and J. Herling, *J. Phys. Chem.*, 1962, **66**, 1208

699.2 $C_9H_6O_3$ 5-Hydroxybenzo[*b*]-4-pyrone (5-hydroxychromone) HL

Metal	*Method*	*Temp*	*Medium*	*Log of equilibrium constant, remarks*	*Ref*
H^+				H^+ constants not given.	
Fe^{3+}	sp	20	0·1 ($NaClO_4$) 10% methanol	K_1 12·74, K_2 10·51	68M

68M A. Murata, T. Ito, and T. Suzuki, *Nippon Kagaku Zasshi*, 1968, **89**, 54

$C_9H_6O_4$ 6,7-Dihydroxybenzo[*b*]-2-pyrone (esculetin) H_2L 699.3

Metal	*Method*	*Temp*	*Medium*	*Log of equilibrium constant, remarks*	*Ref*
H^+	gl	?	50% ethanol	K_2 7·60, K_1 not given	63J
Mo^{VI}	sp	?	50% ethanol	$K(MoO_4^{2-} + 2H_2L \rightleftharpoons MoO_2L_2^{2-})$ (?) 3·65	63J
Ti^{IV}	sp	20	0·4 ($NaClO_4$) 20% ethanol	K (?) 8·8	64J

63J B. D. Jain and H. B. Singh, *Indian J. Chem.*, 1963, **1**, 369
64J B. D. Jain and H. B. Singh, *J. Indian Chem. Soc.*, 1964, **41**, 29

C_9H_7N Quinoline L 702

Metal	*Method*	*Temp*	*Medium*	*Log of equilibrium constant, remarks*	*Ref*
Cd^{2+}	Cd(Hg)	25	0·3 (KNO_3)	K_1 0·25	67N

67N L. V. Nazarova, *Russ. J. Inorg. Chem.*, 1967, **12**, 1620 (3062)

C_9H_7N Isoquinoline L 703

Metal	*Method*	*Temp*	*Medium*	*Log of equilibrium constant, remarks*	*Ref*
Co^{2+}	sp	20	$CHCl_3$	$K(CoCl_2 + 2L \rightleftharpoons CoL_2Cl_2)$ 0·99	64K
				$K(CoBr_2 + 2L \rightleftharpoons CoL_2Br_2)$ 0·862	
				$K(CoI_2 + 2L \rightleftharpoons CoL_2I_2)$ 0·36	
				$K[Co(NCO)_2 + 2L \rightleftharpoons CoL_2(NCO)_2]$ 1·03	
				$K[Co(NCS)_2 + 2L \rightleftharpoons CoL_2(NCS)_2]$ 4·38	
Ge^{4+}	cal	25	n-hexane	$\Delta H[GeF_4(l) + 2L(l) \rightarrow GeF_4L_2(c)]$ −35·7	67M
				$\Delta H[GeF_4(g) + 2L(l) \rightarrow GeF_4L_2(c)]$ −40·9	
				$\Delta H[GeCl_4(l) + 2L(l) \rightarrow GeCl_4L_2(c)]$ −22·3	
				$\Delta H[GeCl_4(g) + 2L(l) \rightarrow GeCl_4L_2(c)]$ −30·5	
				$\Delta H[GeBr_4(l) + 2L(l) \rightarrow GeBr_4L_2(c)]$ −18·0	
				$\Delta H[GeBr_4(g) + 2L(l) \rightarrow GeBr_4L_2(c)]$ −28·6	
Si^{4+}	cal	25	n-hexane	$\Delta H[SiF_4(l) + 2L(l) \rightarrow SiF_4L_2(c)]$ −27·8	67M
				$\Delta H[SiF_4(g) + 2L(l) \rightarrow SiF_4L_2(c)]$ −31·9	
				$\Delta H[SiCl_4(l) + 2L(l) \rightarrow SiCl_4L_2(c)]$ −17·4	
				$\Delta H[SiCl_4(g) + 2L(l) \rightarrow SiCl_4L_2(c)]$ −24·6	
				$\Delta H[SiBr_4(l) + 2L(l) \rightarrow SiBr_4L_2(c)]$ −12·8	
				$\Delta H[SiBr_4(g) + 2L(l) \rightarrow SiBr_4L_2(c)]$ −21·9	
Sn^{4+}	cal	25	n-hexane	$\Delta H[SnCl_4(l) + 2L(l) \rightarrow SnCl_4L_2(c)]$ −37·4	67M
				$\Delta H[SnCl_4(g) + 2L(l) \rightarrow SnCl_4L_2(c)]$ −45·2	
Zn^{2+}	Hg(Zn)	30	0·1 $NaClO_4$	K_1 1·08, β_2 1·65, β_3 2·01	66D

64K H. C. A. King, E. Koros, and S. M. Nelson, *J. Chem. Soc.*, 1964, 4832
66D A. G. Desai and M. B. Kabadi, *J. Inorg. Nuclear Chem.*, 1966, 28, 1279
67M J. M. Miller and M. Onyszchuk, *J. Chem. Soc (A)*, 1966, 1132

704.1 $C_9H_8O_3$ 4-Acetyltropolone HL

Metal	Method	Temp	Medium	Log of equilibrium constant, remarks	Ref
H^+	sp	25	2 $NaClO_4$	K_1 5·54	64O
Fe^{3+}	sp	25	2 $NaClO_4$	K_1 9·60	64O

64O Y. Oka, M. Yanai, and C. Suzuki, *Nippon Kagaku Zasshi*, 1964, **85**, 873

704.2 $C_9H_{10}O_4$ 3-(3′,4′-Dihydroxyphenyl)propanoic acid (3,4-dihydroxyhydrocinnamic acid) H_3L

(see Corrigenda)

Metal	Method	Temp	Medium	Log of equilibrium constant, remarks	Ref
H^+	gl	30	0·1 ($NaClO_4$)	K_1 11·6, K_2 9·36, K_3 4·56	66A
Cd^{2+}	gl	30	0·1 ($NaClO_4$)	K_1 7·14	66A
Cu^{2+}	gl	30	0·1 ($NaClO_4$)	K_1 12·74, K_2 9·42	66A
Mg^{2+}	gl	30	0·1 ($NaClO_4$)	K_1 4·90	66A
Ni^{2+}	gl	30	0·1 ($NaClO_4$)	K_1 7·45, K_2 4·08	66A
Zn^{2+}	gl	30	0·1 ($NaClO_4$)	K_1 8·64, K_2 6·15	66A

66A V. T. Athavale, L. H. Prabhu, and D. G. Vartak, *J. Inorg. Nuclear Chem.*, 1966, **28**, 1237

704.3 $C_9H_8O_4$ 3-Carboxy-4-methyltropolone H_2L

(see Corrigenda)

Metal	Method	Temp	Medium	Log of equilibrium constant, remarks	Ref
H^+	gl	?	0·5 (KNO_3)	K_1 7·31, K_2 2·68	65D
	gl	25	0·2 ($NaClO_4$)	K_1 7·45, K_2 2·48	66G, 67G
	gl	?	0·1 $NaClO_4$	K_1 7·40, K_2 2·50	66S
	sp	?	0·2 ($NaClO_4$)	K_1 9·93 (?), K_2 7·45 (?)	67G
Ba^{2+}	sp	?	0·2 ($NaClO_4$)	K_1 2·43	67G
Ca^{2+}	sp	?	0·2 ($NaClO_4$)	K_1 3·05	67G
Cd^{2+}	sp	?	0·2 ($NaClO_4$)	K_1 5·23	67G
	gl	?	0·2 ($NaClO_4$)	K_1 5·28, K_2 3·83	67G
Ce^{3+}	gl	25	0·2 ($NaClO_4$)	K_1 7·42, K_2 5·72, K_3 3·34	66G
	sp	?	0·2 ($NaClO_4$)	K_1 7·20, $K(Ce^{3+}+L^{2-}+H^+ \rightleftharpoons CeHL^{2+})$ 9·85	67Ga
Co^{2+}	sp	?	0·2 ($NaClO_4$)	K_1 6·07	67G
	gl	?	0·2 ($NaClO_4$)	K_1 6·30, K_2 4·82, K_3 2·82	67G
Cu^{2+}	sp	?	0·2 ($NaClO_4$)	K_1 9·32	67G
Dy^{3+}	gl	25	0·2 ($NaClO_4$)	K_1 8·28, K_2 6·78, K_3 4·00	66G
Er^{3+}	gl	25	0·2 ($NaClO_4$)	K_1 8·35, K_2 6·85, K_3 4·18	66G
	sp	?	0·2 ($NaClO_4$)	K_1 8·65, $K(Er^{3+}+H^++L^{2-} \rightleftharpoons ErHL^{2+})$ 10·78	67Ga
Fe^{3+}	sp	?	0·1 $NaClO_4$	K_1 11·65, K_2 10·25, K_3 6·90	66S
Gd^{3+}	gl	25	0·2 ($NaClO_4$)	K_1 8·02, K_2 6·38, K_3 3·82	66G
	sp	?	0·2 ($NaClO_4$)	K_1 7·91, $K(Gd^{3+}+H^++L^{2-} \rightleftharpoons GaHL^{2+})$ 10·56	67Ga
Ho^{3+}	gl	25	0·2 ($NaClO_4$)	K_1 8·24, K_2 6·76, K_3 3·98	66G
	sp	?	0·2 ($NaClO_4$)	K_1 8·58, $K(Ho^{3+}+H^++L^{2-} \rightleftharpoons HoHL^{2+})$ 10·62	67Ga
La^{3+}	gl	25	0·2 ($NaClO_4$)	K_1 7·20, K_2 5·56, K_3 3·40	66G
	sp	?	0·2 ($NaClO_4$)	K_1 7·07, $K(La^{3+}+L^{2-}+H^+ \rightleftharpoons LaHL^{2+})$ 9·48	67Ga

contin[ued]

$C_9H_8O_4$ (contd) 704.3

Metal	*Method*	*Temp*	*Medium*	*Log of equilibrium constant, remarks*	*Ref*
Lu^{3+}	gl	25	0·2 ($NaClO_4$)	K_1 8·64, K_2 7·06, K_3 4·60	66G
	sp	?	0·2 ($NaClO_4$)	K_1 8·85, $K(Lu^{3+}+H^++L^{2-} \rightleftharpoons LuHL^{2+})$ 10·97	67Ga
Mg^{2+}	sp	?	0·2 ($NaClO_4$)	K_1 4·14	67G
Mn^{2+}	sp	?	0·2 ($NaClO_4$)	K_1 4·96	67G
Nd^{3+}	gl	25	0·2 ($NaClO_4$)	K_1 7·76, K_2 6·04, K_3 3·70	66G
	sp	?	0·2 ($NaClO_4$)	K_1 7·69, $K(Nd^{3+}+H^++L^{2-} \rightleftharpoons NdHL^{2+})$ 10·14	67Ga
Ni^{2+}	sp	?	0·2 ($NaClO_4$)	K_1 6·73	67G
	gl	?	0·2 ($NaClO_4$)	K_1 6·83, K_2 5·22, K_3 3·11	67G
Pr^{3+}	gl	25	0·2 ($NaClO_4$)	K_1 7·74, K_2 5·96, K_3 3·56	66G
	sp	?	0·2 ($NaClO_4$)	K_1 7·45, $K(Pr^{3+}+H^++L^{2-} \rightleftharpoons PrHL^{2+})$ 10·07	67Ga
Sm^{3+}	gl	25	0·2 ($NaClO_4$)	K_1 7·98, K_2 6·42, K_3 3·78	66G
	sp	?	0·2 ($NaClO_4$)	K_1 7·99, $K(Sm^{3+}+H^++L^{2-} \rightleftharpoons SmHL^{2+})$ 10·36	67Ga
Sr^{2+}	sp	?	0·2 ($NaClO_4$)	K_1 2·61	67G
UO_2^{2+}	sp, gl	?	0·5 (KNO_3)	K_1 (sp) 9·22, K_2 (sp) 6·75, (gl) 6·80	65D
	sp	?	0·2 ($NaClO_4$)	K_1 9·60	67G
	gl	?	0·2 ($NaClO_4$)	K_1 9·72, K_2 6·78	67G
Y^{3+}	gl	25	0·2 ($NaClO_4$)	K_1 8·26, K_2 6·62, K_3 3·96	66G
	sp	?	0·2 ($NaClO_4$)	K_1 8·47, $K(Y^{3+}+H^++L^{2-} \rightleftharpoons YHL^{2+})$ 10·61	67Ga
Yb^{3+}	gl	25	0·2 ($NaClO_4$)	K_1 8·60, K_2 7·00, K_3 4·42	66G
	sp	?	0·2 ($NaClO_4$)	K_1 8·86, $K(Yb^{3+}+H^++L^{2-} \rightleftharpoons YbHL^{2+})$ 10·95	67Ga
Zn^{2+}	sp	?	0·2 ($NaClO_4$)	K_1 6·52	67G
	gl	?	0·2 ($NaClO_4$)	K_1 6·68, K_2 5·44, K_3 3·24	67G

65D Y. Dutt and R. P. Singh, *J. Indian Chem. Soc.*, 1965, **42,** 767
66G B. P. Gupta, Y. Dutt, and R. P. Singh, *J. Indian Chem. Soc.*, 1966, **43,** 610
66S R. P. Singh and Y. Dutt, *Indian J. Chem.*, 1966, **4,** 214
67G B. P. Gupta, Y. Dutt, and R. P. Singh, *Indian J. Chem.*, 1967, **5,** 322
67Ga B. P. Gupta, Y. Dutt, and R. P. Singh, *J. Inorg. Nuclear Chem.*, 1967, **29,** 1806

$C_9H_8N_2$ 8-Aminoquinoline L 707

Metal	*Method*	*Temp*	*Medium*	*Log of equilibrium constant, remarks*	*Ref*
H^+	gl	20	0·1 (KCl)	K_1 4·04	57W
Ca^{2+}	gl	20	0·1 (KCl)	K_1 1·49	57W
Cd^{2+}	gl	20	0·1 (KCl)	K_1 2·37	57W
Co^{2+}	gl	20	0·1 (KCl)	K_1 2·66	57W
Cu^{2+}	gl	20	0·1 (KCl)	K_1 6·06, K_2 4·73, K_3 3·69	57W
	pol	30	0·1 KNO_3 20—60% dioxan	K_1 (20%) 9·53, (30%) 9·20, (40%) 9·00, (50%) 8·79, (60%) 8·62	64S
Fe^{3+}	gl	20	0·1 (KCl)	K_1 ~ 3	57W
Mg^{2+}	gl	20	0·1 (KCl)	K_1 1·43	57W
Ni^{2+}	gl	20	0·1 (KCl)	K_1 4·90, K_2 3·64, K_3 3·29	57W
Sr^{2+}	gl	20	0·1 (KCl)	K_1 1·27	57W
Zn^{2+}	gl	20	0·1 (KCl)	K_1 2·42	57W

57W K. Wallenfels and H. Sund, *Biochem. J.*, 1957, **329,** 41
64S R. P. Singh and Y. Dutt, *J. Indian Chem. Soc.*, 1964, **41,** 267

707.1 $C_9H_9N_5$ 2,4-Diamino-6-phenyl-1,3,5-triazine (benzoquanamine) L

Metal	*Method*	*Temp*	*Medium*	*Log of equilibrium constant, remarks*	*Ref*
H^+	gl	25	0·1	K_1 4·3, K_2 2·9	63M
Cu^{2+}	gl	25	0·1	K_1 4·6	63M

63M C. E. Meloan and J. Butel, *Analyt. Chem.*, 1963, **35**, 768

709.1 $C_9H_{10}O_2$ Acetic acid 4-methylphenyl ester (4-tolyl acetate) L

Metal	*Method*	*Temp*	*Medium*	*Log of equilibrium constant, remarks*	*Ref*
Ti^{4+}	sp	25, 60	1,2-dichloroethane	$K(TiCl_4+L \rightleftharpoons TiCl_4L)$ (25°) 2·60±0·09, (60°) 1·93±0·05 ΔH° −9±2, ΔS° −17±6	66G

66G J. Gohring and B. P. Susz, *Helv. Chim. Acta*, 1966, **49**, 486

709.2 $C_9H_{10}O_3$ Salicylic acid ethyl ester HL

Metal	*Method*	*Temp*	*Medium*	*Log of equilibrium constant, remarks*	*Ref*
H^+	gl	30	0·1 $NaClO_4$ 75% dioxan	K_1 12·56	64J
Cu^{2+}	gl	30	0·1 $NaClO_4$ 75% dioxan	K_1 9·77	64J
Mg^{2+}	gl	30	0·1 $NaClO_4$ 75% dioxan	K_1 5·36	64J
Ni^{2+}	gl	30	0·1 $NaClO_4$ 75% dioxan	K_1 7·57	64J
Zn^{2+}	gl	30	0·1 $NaClO_4$ 75% dioxan	K_1 8·48	64J

64J K. E. Jabalpurwala, K. A. Venkatachalam, and M. B. Kabadi, *J. Inorg. Nuclear Chem.*, 1964, **26**, 1011, 1027

$C_9H_{10}O_3$ 2-Hydroxy-2-phenylpropanoic acid (atrolactic acid) HL 710.1

Metal	Method	Temp	Medium	Log of equilibrium constant, remarks	Ref
Er^{3+}	gl	25	1 $NaClO_4$	K_1 3·03±0·01, K_2 2·48±0·02, K_3 2·01±0·01, K_4 1·90±0·02	66T
Eu^{3+}	gl	25	1 $NaClO_4$	K_1 2·55±0·01, K_2 2·17±0·01	66T
Gd^{3+}	gl	25	1 $NaClO_4$	K_1 2·54±0·01, K_2 2·07±0·01, K_3 1·70±0·02, K_4 1·33±0·01	66T
Ho^{3+}	gl	25	1 $NaClO_4$	K_1 2·97±0·01, K_2 2·38±0·01, K_3 1·92±0·01, K_4 1·76±0·02	66T
Nd^{3+}	gl	25	1 $NaClO_4$	K_1 2·55±0·01, K_2 1·64±0·02, K_3 1·42±0·02, K_4 1·21±0·03	66T
Pr^{3+}	gl	25	1 $NaClO_4$	K_1 2·40±0·01, K_2 1·56±0·02, K_3 1·36±0·01, K_4 0·92±0·01	66T
Sm^{3+}	gl	25	1 $NaClO_4$	K_1 2·57±0·01, K_2 1·89±0·02, K_3 1·54±0·02, K_4 1·31±0·02	66T
Yb^{3+}	gl	25	1 $NaClO_4$	K_1 3·05±0·01, K_2 2·56±0·02, K_3 2·07±0·01, K_4 1·86±0·02	66T

66T H. Thun, E. Verbeek, and W. Vanderleen, *J. Inorg. Nuclear Chem.*, 1966, **28**, 1949

$C_9H_{11}N_3$ 2-(*N′*-Methylaminomethyl)benzimidazole L 710.2

(*See Corrigenda*)

Metal	Method	Temp	Medium	Log of equilibrium constant, remarks	Ref
H^+	gl	15—45	50% dioxan	K_1 (15°) 7·40, (25°) 7·32, (35°) 7·12, (45°) 6·97, K_2 (15—45°) 2	63L
Cu^{2+}	gl	15—45	50% dioxan	K_1 (15°) 7·41, (25°) 7·30, (35°) 6·81, (45°) 5·83, K_2 (15°) 4·45, (25°) 4·38, (35°) 4·15, (45°) 4·00 ΔH_1 −13·7	63L

63L T. L. Lane and H. B. Durham, *Inorg. Chem.*, 1963, **2**, 632

$C_9H_{13}N$ 2-*NN*-Trimethylaniline L 711.1

Metal	Method	Temp	Medium	Log of equilibrium constant, remarks	Ref
Ag^+	Ag	25	< 0·01 96% ethanol	K_1 1·01	61A

61A V. Armeanu and C. Luca, *J. phys. Chem.* (*Leipzig*), 1961, **217**, 389

$C_9H_{15}N_3$ 2-[2′*N*′-(2″-Aminoethyl)aminoethyl]pyridine L 713

Errata. 1964, p. 595, Table 713. *For* reference 57L *read* 64L R. G. Lacoste and A. E. Martell, *Inorg. Chem.*, 1964, **3**, 881. Under Medium, *for* 0·1 *read* 0·1 (KNO_3) all through.

714 $C_9H_{16}O_2$ 3-Butylpentane-2,4-dione HL

Metal	Method	Temp	Medium	Log of equilibrium constant, remarks	Ref
Ni^{2+}	sp	30	toluene	$K(2NiL_2+A \rightleftharpoons Ni_2L_2A)$ 5·1 $K(Ni_2L_2A+3A \rightleftharpoons 2NiL_2A_2)$ 2·9 where A = pyridine	68A

68A A. W. Adisson and D. P. Graddon, *Austral. J. Chem.*, 1968, **21**, 2003

716 $C_9H_{16}O_4$ 2,2-Dimethyl malonic acid H_2L

Erratum. 1964, p. 596, Table 716. *Delete* entire Table.

717 $C_9H_{16}O_4$ Heptane-4,4-dicarboxylic acid (di-n-propylmalonic acid) H_2L

Metal	Method	Temp	Medium	Log of equilibrium constant, remarks	Ref
H^+	gl	25	0·10 (KNO_3)	*K_1 7·15, K_2 1·82	68P
Ce^{3+}	gl	25	0·10 (KNO_3)	K_1 3·96, β_2 6·62	68P
Dy^{3+}	gl	25	0·10 (KNO_3)	K_1 4·78, β_2 7·47	68P
Er^{3+}	gl	25	0·10 (KNO_3)	K_1 4·73, β_2 7·35	68P
Eu^{3+}	gl	25	0·10 (KNO_3)	K_1 4·57, β_2 7·42	68P
Gd^{3+}	gl	25	0·10 (KNO_3)	K_1 4·58, β_2 7·30	68P
Ho^{3+}	gl	25	0·10 (KNO_3)	K_1 4·72, β_2 7·28	68P
La^{3+}	gl	25	0·10 (KNO_3)	K_1 3·66	68P
Lu^{3+}	gl	25	0·10 (KNO_3)	K_1 4·78, β_2 7·53	68P
Nd^{3+}	gl	25	0·10 (KNO_3)	K_1 4·06, β_2 7·05	68P
Pr^{3+}	gl	25	0·10 (KNO_3)	K_1 4·01, β_2 6·93	68P
Sm^{3+}	gl	25	0·10 (KNO_3)	K_1 4·38, β_2 7·34	68P
Tb^{3+}	gl	25	0·10 (KNO_3)	K_1 4·73, β_2 7·48	68P
Tm^{3+}	gl	25	0·10 (KNO_3)	K_1 4·76, β_2 7·39	68P
Y^{3+}	gl	25	0·10 (KNO_3)	K_1 4·74, β_2 7·36	68P
Yb^{3+}	gl	25	0·10 (KNO_3)	K_1 4·81, β_2 7·56	68P

*K_1 extrapolated to zero ligand concentration.

68P J. E. Powell, J. L. Farrell, W. F. S. Neillie, and R. Russell, *J. Inorg. Nuclear Chem.*, 1968, **30**, 2223

718.1 $C_9H_{20}N^+$ *NNN*-Triethylallylammonium cation L^+

Metal	Method	Temp	Medium	Log of equilibrium constant, remarks	Ref
Pt^{2+}	sp	25—59	2 (NaCl)	$K(PtCl_4^{2-}+L^+ \rightleftharpoons PtCl_3L+Cl^-)$ (25°) 2·41, (45°) 2·18, (59°) 2·05 ΔH $-4·9\pm0·3$, ΔS $-4·6\pm1·0$	67D
	sp	25	2 (KBr)	$K(PtBr_4^{2-}+L^+ \rightleftharpoons PtBr_3L+Br^-)$ 1·64	67Da

67D R. G. Denning, F. R. Hartley, and L. M. Venanzi, *J. Chem. Soc.* (*A*), 1967, 324
67Da R. G. Denning and L. M. Venanzi, *J. Chem. Soc.* (*A*), 1967, 336

$C_9H_{20}As^+$ *AsAsAs*-Triethylallylarsinium cation L$^+$ 718.2

Metal	Method	Temp	Medium	Log of equilibrium constant, remarks	Ref
Pt^{2+}	sp	45—58	2 (NaCl)	$K(PtCl_4^{2-}+L^+ \rightleftharpoons PtCl_3L+Cl^-)$ (45°) 3·12, (58°) 2·96 ΔH −5·9±0·6, ΔS −4±2	67D

67D R. G. Denning, F. R. Hartley, and L. M. Venanzi, *J. Chem. Soc.* (*A*), 1967, 324

$C_9H_{20}P^+$ *PPP*-Triethylallylphosphinium cation L$^+$ 718.3

Metal	Method	Temp	Medium	Log of equilibrium constant, remarks	Ref
Pt^{2+}	sp	59	2 (NaCl)	$K(PtCl_4^{2-}+L^+ \rightleftharpoons PtCl_3L+Cl^-)$ 2·70	67D

67D R. G. Denning, F. R. Hartley, and L. M. Venanzi, *J. Chem. Soc.* (*A*), 1967, 324

$C_9H_{24}N_4$ Tris(3-aminopropyl)amine L 719

Metal	Method	Temp	Medium	Log of equilibrium constant, remarks	Ref
H^+	gl	25	0·1 KCl	K_1 10·511±0·004, K_2 9·824±0·007, K_3 9·129±0·006, K_4 5·615±0·009	68D
Co^{2+}	gl	25	0·1 KCl	K_1 6·360±0·004 $K(CoL^{2+}+OH^- \rightleftharpoons CoLOH^+)$ 2·99±0·04	68D
Cu^{2+}	gl	25	0·1 KCl	K_1 13·117±0·009 $K(Cu^{2+}+HL^+ \rightleftharpoons CuHL^{3+})$ 10·757±0·008 $K(CuL^{2+}+OH^- \rightleftharpoons CuLOH^+)$ 3·99±0·02	68D
Ni^{2+}	gl	25	0·1 KCl	K_1 8·702±0·004 $K(Ni^{2+}+HL^+ \rightleftharpoons NiHL^{3+})$ 5·3±0·1	68D
Zn^{2+}	gl	25	0·k KCl	K_1 10·702±0·003	68D

68P A. Dei, P. Paoletti, and A. Vacca, *Inorg. Chem.*, 1968, 7, 865

$C_9H_6O_3N_2$ 5-Nitro-8-hydroxyquinoline HL 719.1

Metal	Method	Temp	Medium	Log of equilibrium constant, remarks	Ref
H^+	gl	25	0·1	K_1K_2 8·79	68C
Zn^{2+}	dis	25	0·1	β_2 12·14	68C

68C F. C. Chou and H. Freiser, *Analyt. Chem.*, 1968, 40, 34

720 C_9H_7ON 8-Hydroxyquinoline (oxine) HL

Metal	*Method*	*Temp*	*Medium*	*Log of equilibrium constant, remarks*	*Ref*
H^+	sol	25	?	K_1 9·75, K_2 5·06	58K
	gl	15—40	50% dioxan	K_1 (15°) 11·61±0·03, (25°) 11·46±0·03, (40°) 11·35±0·03, K_2 (15°) 4·20±0·03, (25°) 4·07±0·03, (40°) 3·98±0·03 ΔH_1 −6·0, ΔH_2 −5·2	59F
	cal	25	50% dioxan	ΔH_1 −6·0, ΔS_1 33, ΔH_2 −5·7, ΔS_2 −0·3	59F
	gl	25	0·1 50% dioxan	K_1 11·20±0·02, K_2 4·16±0·02	67S
	gl	25	0·5 NaCl	K_1 9·81, K_2 4·91	67T
	gl	25	0·1	K_1K_2 14·9	68C
	gl	25	0·1 $NaClO_4$ 50% dioxan	K_1 10·95, K_2 4·13	68G
	cal	25	0·1 $NaClO_4$ 50% dioxan	ΔH_1 −6·41, ΔS_1 28·6, ΔH_2 − 5·83, ΔS_2 −0·7	68G
	dis	25	0·1 $NaClO_4$	K_1 9·7	68S
Ag^+	dis	18—25	0·1 ($NaClO_4$)	K_1 5·20, K_2 4·36	65H
Cd^{2+}	gl	25	0·1 $NaClO_4$ 50% dioxan	K_1 8·22, β_2 15·22	68G
	cal	25	0·1 $NaClO_4$ 50% dioxan	ΔH_1 −5·5, ΔS_1 19, $\Delta H_{\beta 2}$ −11·6, $\Delta S_{\beta 2}$ 31	68G
Co^{2+}	gl	15—40	50% dioxan	β_2 (15°) 19·8±0·2, (25°) 19·50, (40°) 19·17±0·01 $\Delta H_{\beta 2}$ −11·1, $\Delta S_{\beta 2}$ 52	59F
	cal	25	50% dioxan	β_2 19·20±0·05 $\Delta H_{\beta 2}$ −21·6 (−20·4), $\Delta S_{\beta 2}$ 21	59F
	gl	25	0·1 50% dioxan	K_1 9·65, β_2 18·05	67S
	cal	25	0·1 $NaClO_4$ 50% dioxan	ΔH_1 −7·2, ΔS_1 20, $\Delta H_{\beta 2}$ −15·4, $\Delta S_{\beta 2}$ 31	68G
Cu^{2+}	gl	15—40	50% dioxan	β_2 (15°) 25·9±0·2, (25°) 25·35±0·03, (40°) 24·8±0·2 $\Delta H_{\beta 2}$ −19·2, $\Delta S_{\beta 2}$ 53	59F
	cal	25	50% dioxan	$\Delta H_{\beta 2}$ −20·6 (−20·8), $\Delta S_{\beta 2}$ 48	59F
	Pt	25	0·3 (KNO_3) 50% dioxan	β_2 ~ 14·7	61J
	sol	25	0·1	K_1 12·10, K_2 10·90, K_{so} −29·26	64F
	gl	25	0·1 50% dioxan	K_1 13·29, β_2 25·90	67S
	cal	25	0·1 $NaClO_4$ 50% dioxan	ΔH_1 −10·2, ΔS_1 27, $\Delta H_{\beta 2}$ −19·6, $\Delta S_{\beta 2}$ 53	68G
				See propane-2,4-dione (283) for mixed ligands.	
Fe^{2+}	sp	?	5—90% N_2H_4	β_2 (5%) 6·5, (10%) 6·5, (25%) 6·9, (50%) 8·6, (64%) 10·7, (85%) 13·0, (90%) 13·7	66B
Fe^{3+}	sp	25	0·1 ($NaClO_4$)	K_1 13·69, β_2 26·3, β_3 36·9 K_{so} −43·51	68T
Ga^{3+}	dis	20	0·1 ($NaClO_4$)	K_1 14·51, K_2 13·50, K_3 12·49	65S
Ge^{IV}	sp	25	0·5 NaCl	$K[Ge(OH)_4 + 2HL \rightleftharpoons Ge(OH)_2L_2]$ 6·61	67T
Lu^{3+}	dis	25	0·1 $NaClO_4$	K_1 12, β_2 23·9, β_3 35·3	68S
Mg^{2+}	sp	16	0·1 (KNO_3)	K_1 4·35	64H
Mn^{2+}	gl	25	0·1 50% dioxan	K_1 7·30, β_2 13·49	67S
	cal	25	0·1 $NaClO_4$ 50% dioxan	ΔH_1 −3·5, ΔS_1 22, $\Delta H_{\beta 2}$ −10·5, $\Delta S_{\beta 2}$ 27	68G
Ni^{2+}	gl	15—40	50% dioxan	β_2 (15°) 22·0±0·2, (25°) 21·54, (40°) 20·68±0·04 $\Delta H_{\beta 2}$ −19·0, $\Delta S_{\beta 2}$ 35	59F

conti

C_9H_7ON (contd) 720

Metal	Method	Temp	Medium	Log of equilibrium constant, remarks	Ref
Ni^{2+} (*contd*)	cal	25	50% dioxan	$\Delta H_{\beta 2}$ −21·3 (−20·8), $\Delta S_{\beta 2}$ 29	59F
	gl	25	50% dioxan	K_1 10·50, β_2 20·27	67S
	cal	25	0·1 $NaClO_4$ 50% dioxan	ΔH_1 −9·3, ΔS_1 17, $\Delta H_{\beta 2}$ −19·3, $\Delta S_{\beta 2}$ 28	68G
Pb^{2+}	gl	25	0·1 $NaClO_4$ 50% dioxan	K_1 10·03, β_2 17·34	68G
	cal	25	0·1 $NaClO_4$ 50% dioxan	ΔH_1 −6·6, ΔS_1 24, $\Delta H_{\beta 2}$ −15·1, $\Delta S_{\beta 2}$ 29	68G
UO_2^{2+}	sol	25	?	K_s (UO_2HL_3) −28·72	58K
VO^{2+}	gl	25	0·1 ($NaClO_4$)	$K(VOLOH + H^+ \rightleftharpoons VOL^+)$ 5·3	66K
Zn^{2+}	sol	25	0·1	K_1 8·52, K_2 7·32, K_{so} −23·34	64F
	dis	25	0·1	β_2 17·1±0·2	65C
	gl	25	0·1 50% dioxan	K_1 9·45, β_2 18·15	67S
	cal	25	0·1 $NaClO_4$ 50% dioxan	ΔH_1 −5·9, ΔS_1 23, $\Delta H_{\beta 2}$ −9·6, $\Delta S_{\beta 2}$ 51	68G
	dis	25	chloroform	$K(ZnL^+ + A \rightleftharpoons ZnLA^+)$ 3·05 where A = pyridine, 2·10 where A = 2-methylpyridine, 3·40 where A = 4-methylpyridine, 1·50 where A = 2,4,6-trimethylpyridine	68C

58K A. E. Klygin and N. S. Kolyada, *J. Inorg. Chem.* (*U.S.S.R.*), 1958, **3**, No. 12, 223 (2767)
59F D. Fleischer and H. Freiser, *J. Phys. Chem.*, 1959, **63**, 260
61J B. R. James and R. J. P. Williams, *J. Chem. Soc.*, 1961, 2007
64F J. Fresco and H. Freiser, *Analyt. Chem.*, 1964, **36**, 631
65C F. C. Chou, Q. Fernando, and H. Freiser, *Analyt. Chem.*, 1965, **37**, 361
65H H. Hala, *J. Inorg. Nuclear Chem.*, 1965, **27**, 2659
65S A. P. Savostin, *Russ. J. Inorg. Chem.*, 1965, **10**, 1394 (2565)
66B D. Bauer, *Bull. Soc. chim. France*, 1966, 2631
66H D. N. Hange and M. Eigen, *Trans. Faraday Soc.*, 1966, **62**, 1236
66K Th. Kaden and S. Fallab, *Chimia* (*Switz.*), 1966, **20**, 51
67S R. L. Stevenson and H. Freiser, *Analyt. Chem.*, 1967, **39**, 1354
67T J. Tsau, S. Matsouo, P. Clere, and R. Benoit, *Bull. Soc. chim. France*, 1967, 1039
68C F. C. Chou and H. Freiser, *Analyt. Chem.*, 1968, **40**, 34
68G G. Gutnikov and H. Freiser, *Analyt. Chem.*, 1968, **40**, 39
68S G. K. Schweitzer and M. M. Anderson, *J. Inorg. Nuclear Chem.*, 1968, **30**, 1057
68T T. D. Turnquist and E. B. Sandell, *Analyt. Chim. Acta*, 1968, 42, 239

C_9H_7NS 2-(2′-Pyridyl)thiophene L 721.1

Metal	Method	Temp	Medium	Log of equilibrium constant, remarks	Ref
H^+	gl	25	0·1 ($NaClO_4$)	K_1 5·59	64K
	sp	25	0·1 ($NaClO_4$)	K_1 5·58	64K
	sp	25	0·1 ($NaClO_4$)	K_1 5·59	65K
Cu^{2+}	gl	25	0·1 ($NaClO_4$)	K_1 2·58	64K
	sp	25	0·1 ($NaClO_4$)	K_1 2·58	65K
Ni^{2+}	gl	25	0·1 ($NaClO_4$)	K_1 1·91	64K
Zn^{2+}	gl	25	0·1 ($NaClO_4$)	K_1 1·10	64K

64K K. Kahmann, H. Sigel, and H. Erlenmeyer, *Helv. Chim. Acta*, 1964, **47**, 1754
65K K. Kahmann, H. Sigel, and H. Erlenmeyer, *Helv. Chim. Acta*, 1965, **48**, 295

721.2 C_9H_7NS 8-Mercaptoquinoline HL

Metal	*Method*	*Temp*	*Medium*	*Log of equilibrium constant, remarks*	*Ref*
H^+	gl	25	0·1 50% dioxan	K_1 9·20	63C
	sp	27	0·1 50% dioxan	K_2 1·74	63C
	gl	25	0·1 ($NaClO_4$) 50% dioxan	K_1 9·22, K_2 1·79	66K
	cal	25	0·1 ($NaClO_4$) 50% dioxan	ΔH_1 −3·02, ΔS_1 32·1, ΔH_2 −1·73, ΔS_2 2·3	68G
Co^{2+}	gl	25	0·1 ($NaClO_4$) 50% dioxan	K_1 7·9	66K
	cal	25	0·1 ($NaClO_4$) 50% dioxan	ΔH_1 −15·4, ΔS_1 −15	68G
Cu^{2+}	sp	27	⩾0·1 50% dioxan	K_1 12—14	63C
	gl	25	0·1 ($NaClO_4$) 50% dioxan	K_1 12·7	66K
	cal	25	0·1 ($NaClO_4$) 50% dioxan	ΔH_1 −14·7, ΔS_1 9	68G
Mn^{2+}	gl	27	⩾ 0·1 50% dioxan	K_1 6·74	63C
	cal	25	0·1 ($NaClO_4$) 50% dioxan	ΔH_1 −3·5, ΔS_1 19	68G
Ni^{2+}	sp	27	⩾ 0·1 50% dioxan	K_1 10·95	63C
	gl	25	0·1 ($NaClO_4$) 50% dioxan	K_1 11·0	66K
	cal	25	0·1 ($NaClO_4$) 50% dioxan	ΔH_1 −11·4, ΔS_1 12	68G
Pb^{2+}	sp	27	⩾ 0·1 50% dioxan	K_1 11·85	63C
	cal	25	0·1 ($NaClO_4$) 50% dioxan	ΔH_1 −10·1, ΔS_1 20	68G
Zn^{2+}	sp	27	⩾ 0·1 50% dioxan	K_1 11·05	63C
	gl	25	0·1 ($NaClO_4$) 50% dioxan	K_1 11·0	66K
	cal	25	0·1 ($NaClO_4$) 50% dioxan	ΔH_1 −7·2, ΔS_1 26	68G

63C A. Corsini, Q. Fernando, and H. Freiser, *Analyt. Chem.*, 1963, **35**, 1424

66K D. Kealey and H. Freiser, *Analyt. Chem.*, 1966, **38**, 1577

68G G. Gutnikov and H. Freiser, *Analyt. Chem.*, 1968, **40**, 39

C_9H_7NSe 8-Hydroselenylquinoline HL 721.3

Metal	Method	Temp	Medium	Log of equilibrium constant, remarks	Ref
H^+	?	?	?	K_1 7·86, K_2 0·76	65S
Cd^{2+}	sp	25	0·1 ($NaClO_4$) 50% dioxan	K_1 10·5±0·1 $K(CdL^+ + H^+ \rightleftharpoons CdHL^{2+})$ 0·2±0·1	65S
Pb^{2+}	sp	25	0·1 ($NaClO_4$) 50% dioxan	K_1 10·4±0·1 $K(PbL^+ + H^+ \rightleftharpoons PbHL^{2+})$ 0·4±0·1	65S
Zn^{2+}	sp	25	0·1 ($NaClO_4$) 50% dioxan	K_1 10·2±0·1	65S

65S E. Sekido, Q. Fernando, and H. Freiser, *Analyt. Chem.*, 1965, 37, 1556

$C_9H_9O_2Cl$ 5-Chloro-2-hydroxy-4,6-dimethylbenzaldehyde (5-chloro-4,6-dimethylsalicylaldehyde) HL 727.1

Metal	Method	Temp	Medium	Log of equilibrium constant, remarks	Ref
H^+	gl	20	0·3 $NaClO_4$ 50% dioxan	K_1 9·51	63C
Ni^{2+}	pH	20	0·3 $NaClO_4$ 50% dioxan	K_1 4·1	63C

63C K. Clarke, R. A. Cowen, G. W. Gray, and E. H. Osborne, *J. Chem. Soc.*, 1963, 245

$C_9H_9O_4N$ *N*-(2-Hydroxybenzoyl)glycine (salicyluric acid) H_2L 729.1

Metal	Method	Temp	Medium	Log of equilibrium constant, remarks	Ref
H^+	gl	25	0·1 ($NaClO_4$)	K_2 3·41	66P
Fe^{3+}	gl, Pt	25	0·1 ($NaClO_4$)	$K(Fe^{3+} + HL^- \rightleftharpoons FeL^+ + H^+)$ 2·09, $K(FeL^+ + HL^- \rightleftharpoons FeL_2^- + H^+)$ 0·57, $K(FeL_2^- + HL^- \rightleftharpoons FeL_3^{3-} + H^+)$ −4·1 $K(Fe^{3+} + HL^- \rightleftharpoons FeHL^{2+})$ 3·9	66P

66P M. V. Park, *J. Chem. Soc. (A)*, 1966, 816

$C_9H_{10}ON_2$ 2-(2′-Hydroxyethyl)benzimidazole HL 730.1

Metal	Method	Temp	Medium	Log of equilibrium constant, remarks	Ref
H^+	gl	15—45	50% dioxan	K_1 (15°) 13·06, (25°) 12·77, (40°) 12·27, (45°) 12·05, K_2 (15°) 4·88, (25°) 4·82, (40°) 4·66, (45°) 4·59	63L
Cu^{2+}	gl	15—45	50% dioxan	K_1 (15°) 11·80, (25°) 11·43, (40°) 10·80, (45°) 10·50 ΔH_1 −13·5	63L

63L T. J. Lane and H. B. Durham, *Inorg. Chem.*, 1963, 2, 632

731 $C_9H_{10}O_2N_2$ 1-Acetyl-3-(2′-hydroxyphenyl)-1,2-diazaprop-2-ene (salicylaldehyde acetylhydrazone) H_2L

Metal	Method	Temp	Medium	Log of equilibrium constant, remarks	Ref
H^+				H^+ constants not given	
Sc^{3+}	sp	18—20	0·025 28% ethanol	K (?) 4·59, acetate buffer	63H
Zn^{2+}	sp	18—20	0·025 28, 50% ethanol,	K (?) (28%) 4·0, (50%) 4·2, K (?) (28%) < 7, (50%) 6·32 acetate buffer	63H

63H Z. Holzbecher, *Coll. Czech. Chem. Comm.*, 1963, **28,** 716

731.1 $C_9H_{10}O_2N_2$ *N*-(2′-Pyridyl)-3-oxobutanamide HL

Metal	Method	Temp	Medium	Log of equilibrium constant, remarks	Ref
H^+	gl	25	0·10 KNO_3	K_1 9·72±0·02	67H
Be^{2+}	gl	25	0·10 KNO_3	K_1 6·55, K_2 4·68	67H
Co^{2+}	gl	25	0·10 KNO_3	K_1 4·52, K_2 3·75	67H
Cu^{2+}	gl	25	0·10 KNO_3	K_1 6·92	67H
Fe^{2+}	gl	25	0·10 KNO_3	K_1 5·65	67H
Mn^{2+}	gl	25	0·10 KNO_3	K_1 3·38, K_2 2·9	67H
UO_2^{2+}	gl	25	0·10 KNO_3	K_1 7·26, K_2 6·63	67H
Zn^{2+}	gl	25	0·10 KNO_3	K_1 4·30	67H

67H H. J. Harries, *J. Inorg. Nuclear Chem.*, 1967, **29,** 2484

734.1 $C_9H_{10}O_4N_2$ 4-(*N′N′*-Dimethylamino)pyridine-2,6-dicarboxylic acid H_2L

Metal	Method	Temp	Medium	Log of equilibrium constant, remarks	Ref
H^+	gl	22	0·1 ($NaClO_4$)	K_1 9·77, K_2 1·5	64B
Ba^{2+}	gl	22	0·1 ($NaClO_4$)	K_1 3·86	64B
Ca^{2+}	gl	22	0·1 ($NaClO_4$)	K_1 5·42	64B
Mg^{2+}	gl	22	0·1 ($NaClO_4$)	K_1 3·08	64B
Sr^{2+}	gl	22	0·1 ($NaClO_4$)	K_1 4·28	64B

64B E. Blasius and B. Brozio, *Ber. Bunsengesellschaft Phys. Chem.* 1964, **68,** 52

735.1 $C_9H_{11}ON$ *N*-(Salicylidene)aminoethane HL

Metal	Method	Temp	Medium	Log of equilibrium constant, remarks	Ref
H^+	sp	25	?	K_1 11·8	66G
Be^{2+}	dis	25 (?)	?	K_1 10·4, β_2 18·3 $K[BeL(OH)_2^- + 2H^+ \rightleftharpoons BeL^+]$ 18·4	66G

66G R. W. Green and R. J. Sleet, *Austral. J. Chem.*, 1966, **19,** 2101

$C_9H_{11}O_2N$ 3-Phenylalanine HL 736

Metal	Method	Temp	Medium	Log of equilibrium constant, remarks	Ref
H^+	gl	20	~0	K_1 9·33	53P
	gl	0—40	0 corr	K_1 (0°) 9·95, (10°) 9·66, (20°) 9·38, (30°) 9·15, (40°) 8·89, K_2 (0°) 2·28, (10°) 2·21, (20°) 2·20, (30°) 2·23, (40°) 2·20 $\Delta H_1°$ −10·3, $\Delta S_1°$ 7·8, $\Delta H_2°$ −0·6, $\Delta S_2°$ 8·3	61I
	gl	10—40	0 corr	K_1 (10°) 9·75, (25°) 9·31, (40°) 8·96, K_2 (10°) 2·14, (25°) 2·20, (40°) 2·21 $\Delta H_1°$ −10·63, $\Delta S_1°$ 7·3	66A, 67A
	cal	10—40	0 corr	$\Delta H_1°$ (10°) −11·4±0·1, (25°) −10·67±0·09, (40°) −10·5±0·1 $\Delta S_1°$ (10°) 4·3, (25°) 6·7, (40°) 7·4	66A, 67A
	E	25	?	K_1 9·13, K_2 1·83	67Aa
Ag^+	Ag	25	0·6	K_1 5·30±0·03, β_2 7·8±0·1	67Aa
Be^{2+}	gl	20	0·005 $BeSO_4$	β_2 11·9	53P
Cd^{2+}	gl	20	0·005 $CdSO_4$	β_2 7·2	53P
Cu^{2+}	gl	0—40	0 corr	K_1 (0°) 8·64, (10°) 8·42, (20°) 8·31, (30°) 8·30, (40°) 8·06, K_2 (0°) 7·45, (10°) 7·24, (20°) 7·08, (30°) 6·95, (40°) 6·78 $\Delta H_1°$ −5·1, $\Delta S_1°$ 20·7, $\Delta H_2°$ −6·4, $\Delta S_2°$ 10·7	61I
	gl	10—40	0 corr	K_1 (10°) 8·48±0·03, (25°) 8·25±0·03, (40°) 8·13±0·03, K_2 (10°) 7·43±0·03, (25°) 7·43±0·03, (40°) 6·94±0·03 $\Delta H_1°$ −4·48, $\Delta S_1°$ 22·8, $\Delta H_2°$ −6·4, $\Delta S_2°$ 10·9	66A
	cal	10—40	0 corr	$\Delta H_1°$ (10°) −6·0±0·5, (25°) −5·3±0·3, (40°) −4·9±0·1, $\Delta S_1°$ (10°) 17·7, (25°) 19·8, (40°) 21·8, $\Delta H_2°$ (10°) −6±1, (25°) −6·4±0·8, (40°) −5·5±0·2, $\Delta S_2°$ (10°) 10·3, (25°) 13·7, (40°) 14·2 See nitrilotriacetic acid (449) for mixed ligands.	66A
Hg^{2+}	gl	20	0·005 $Hg(NO_3)_2$	β_2 18·7	53P
	pol	25	0·6 KNO_3	β_2 18·06±0·05	66T
Ni^{2+}	gl	10—40	0 corr	K_1 (10°) 5·61, (20°) 5·56, (40°) 5·52, K_2 (10°) 4·95, (20°) 4·66, (40°) 4·39 $\Delta H_1°$ −1·2, $\Delta S_1°$ 21·6, $\Delta H_2°$ −7·6, $\Delta S_2°$ −4·4	67A
	cal	10—40	0 corr	$\Delta H_1°$ (10°) −3·4±0·3, (25°) −3·2±0·2, (40°) −2·6±0·3, $\Delta S_1°$ (10°) 13·8, (25°) 14·8, (40°) 17·0, $\Delta H_2°$ (10°) −4±1, (25°) −3·3±0·4, (40°) −2·9±0·8, $\Delta S_2°$ (10°) 10·0, (25°) 10·3, (40°) 10·7	67A
Zn^{2+}	gl	20	0·005 $ZnSO_4$	β_2 8·4	53P

53P D. J. Perkins, *Biochem. J.*, 1953, **55**, 649
61I R. M. Izatt, J. W. Wrathall, and K. P. Anderson, *J. Phys. Chem.*, 1961, **65**, 1914
66A K. P. Anderson, W. O. Greenhalgh, and R. M. Izatt, *Inorg. Chem.*, 1966, **5**, 2106
66T V. F. Toropova and Yu. M. Azizov, *Russ. J. Inorg. Chem.*, 1966, **11**, 288 (531)
67A K. P. Anderson, W. O. Greenhalgh, and E. A. Butler, *Inorg. Chem.*, 1967, **6**, 1056
67Aa Yu. M. Azizov, A. Kh. Miftakhova, and V. F. Toropova, *Russ. J. Inorg. Chem.*, 1967, **12**, 345 (661)

736.1 $C_9H_{11}O_2N$ 2-(Ethylamino)benzoic acid HL

Metal	*Method*	*Temp*	*Medium*	*Log of equilibrium constant, remarks*	*Ref*
H^+	gl	22	0·1 KCl	K_1 5·75	60U
Cu^{2+}	gl	22	0·1 KCl	K_1 3·45, K_2 2·8	60U
Ni^{2+}	gl	22	0·1 KCl	K_1 1·9	60U

60U E. Uhlig, *Chem. Ber.*, 1960, **93**, 2470

736.2 $C_9H_{11}O_2N$ 3-2′,4′-Dihydroxypropiophenone oxime H_2L

Metal	*Method*	*Temp*	*Medium*	*Log of equilibrium constant, remarks*	*Ref*
H^+				H^+ constants not given	
Fe^{3+}	sp	?	0·1 $NaClO_4$	K(?) 3·20	67G

67G M. H. Gandhi and M. N. Desai, *Analyt. Chem.*, 1967, **39**, 1643

737 $C_9H_{11}O_3N$ 2-Amino-3-(4′-hydroxyphenyl)propanoic acid (tyrosine) H_2L

Metal	*Method*	*Temp*	*Medium*	*Log of equilibrium constant, remarks*	*Ref*
H^+	gl	20	~ 0	K_1 10·43, K_2 9·23	53P
Be^{2+}	gl	20	0·002 $BeSO_4$	β_2 11·1	53P
Cd^{2+}	gl	20	0·002 $CdSO_4$	β_2 6·4	53P
Hg^{2+}	gl	20	0·002 $Hg(NO_3)_2$	β_2 17·1	53P
Ni^{2+}				See Solochrom Violet R (1050.1) for mixed ligands.	
Zn^{2+}	gl	20	0·002 $ZnSO_4$	β_2 7·9	53P

53P D. J. Perkins, *Biochem. J.*, 1953, **55**, 649

737.1 $C_9H_{11}O_3N$ 2-[(2′-Hydroxyethyl)amino]benzoic acid HL

Metal	*Method*	*Temp*	*Medium*	*Log of equilibrium constant, remarks*	*Ref*
H^+	gl	22	0·1 KCl	K_1 4·87	60U
Cu^{2+}	gl	22	0·1 KCl	K_1 3·85, K_2 2·9	60U
Ni^{2+}	gl	22	0·1 KCl	K_1 2·4, K_2 2·55	60U

60U E. Uhlig, *Chem. Ber.*, 1960, **93**, 2470

$C_9H_{11}O_3N$ 3-Phenylserine HL 737.2

Metal	Method	Temp	Medium	Log of equilibrium constant, remarks	Ref
H^+	gl	17	~ 0	K_1 8·86	53P
Be^{2+}	gl	17	0·005 $BeSO_4$	β_2 11·1	53P
Cd^{2+}	gl	17	0·005 $CdSO_4$	β_2 7·0	53P
Hg^{2+}	gl	17	0·005 $Hg(NO_3)_2$	β_2 17·3	53P
Zn^{2+}	gl	17	0·005 $ZnSO_4$	β_2 8·5	53P

53P D. J. Perkins, *Biochem. J.*, 1953, **55**, 649

$C_9H_{11}O_4N$ 3,4-Dihydroxyphenylalanine (DOPA) H_3L 737.3

Metal	Method	Temp	Medium	Log of equilibrium constant, remarks	Ref
H^+	gl	20	~ 0	K_3 9·03	53P
Be^{2+}	gl	20	0·005 $BeSO_4$	β_2 11·6 $K\,[Be^{2+}+2H_2L^- \rightleftharpoons Be(H_2L)_2]$ (?)	53P
Cd^{2+}	gl	20	0·005 $CdSO_4$	β_2 7·9 $K\,[Cd^{2+}+2H_2L^- \rightleftharpoons Cd(H_2L)_2]$ (?)	53P
Zn^{2+}	gl	20	0·005 $ZnSO_4$	β_2 8·7 $K\,[Zn^{2+}+2H_2L^- \rightleftharpoons Zn\,(H_2L)_2]$ (?)	53P

53P D. J. Perkins, *Biochem. J.*, 1953, **55**, 649

$C_9H_{11}O_5N$ *N*-(2′-Furfuryl)iminodiacetic acid H_2L 737.4

Metal	Method	Temp	Medium	Log of equilibrium constant, remarks	Ref
H^+	gl	20	0·1 (KNO_3)	K_1 8·40±0·01, K_2 2·17±0·04	63I
Ag^+	gl	20	0·1 (KNO_3)	K_1 3·92±0·03	63I
Ba^{2+}	gl	20	0·1 (KNO_3)	K_1 2·68±0·02	63I
Ca^{2+}	gl	20	0·1 (KNO_3)	K_1 3·58±0·01	63I
Mg^{2+}	gl	20	0·1 (KNO_3)	K_1 2·78±0·03	63I
Sr^{2+}	gl	20	0·1 (KNO_3)	K_1 2·79±0·02	63I
Tl^+	gl	20	0·1 (KNO_3)	K_1 3·11±0·03	63I

63I H. Irving and J. J. R. F. da Silva, *J. Chem. Soc.*, 1963, 1144

737.5 $C_9H_{11}O_7N_3$ *N*-(1-Methyl-2,4,6-trioxo-perhydropyrimidinyl)iminodiacetic acid (1-methyluramil-*NN*-diacetic acid) H_3L

Metal	*Method*	*Temp*	*Medium*	*Log of equilibrium constant, remarks*	*Ref*
H^+	gl	20	0·1 [$(CH_3)_4NNO_3$]	K_1 9·81±0·01, K_2 2·67±0·02, K_3 1·85±0·05	63I
Ba^{2+}	gl	20	0·1 (KNO_3)	K_1 6·06±0·01, K_2 3·85±0·05	63I
Be^{2+}	gl	20	0·1 (KNO_3)	K_1 10·42±0·02 $K(Be^{2+}+HL^{2-} \rightleftharpoons BeHL)$ 3·32	63I
Ca^{2+}	gl	20	0·1 (KNO_3)	K_1 8·22±0·01, K_2 5·38±0·01	63I
K^+	gl	20	0·1 [$(CH_3)_4NNO_3$]	K_1 1·11	63I
Li^+	gl	20	0·1 [$(CH_3)_4NNO_3$]	K_1 4·86±0·01	63I
Mg^{2+}	gl	20	0·1 (KNO_3)	K_1 8·23±0·01, K_2 3·72±0·06	63I
Na^+	gl	20	0·1 [$(CH_3)_4NNO_3$]	K_1 2·67±0·01	63I
Sr^{2+}	gl	20	0·1 (KNO_3)	K_1 6·83±0·02, K_2 4·19±0·02	63I
Tl^+	gl	20	0·1 [$(CH_3)_4NNO_3$]	K_1 5·79±0·01	63I

63I H. Irving and J. J. R. F. da Silva, *J. Chem. Soc.*, 1963, 458

737.6 $C_9H_{12}ON_2$ 2-Hydroxy-2-phenylpropanoylamidine (atrolactamidine) HL

Metal	*Method*	*Temp*	*Medium*	*Log of equilibrium constant, remarks*	*Ref*
H^+	gl	25	0·1 (KCl)	K_1 12·72±0·05, K_2 10·96±0·01	63G
Cu^{2+}	glJ	25	0·1 (KCl)	K_1 12·73, K_2 11·57	63G
Ni^{2+}	glJ	25	0·1 (KCl)	K_1 7·87, K_2 7·53	63G

63G R. O. Gould and R. F. Jameson, *J. Chem. Soc.*, 1963, 15, 5211

737.7 $C_9H_{12}O_6N_2$ Pyrimidine-2,4-dione 1-riboside (uracil ribonucleoside) HL

Metal	*Method*	*Temp*	*Medium*	*Log of equilibrium constant, remarks*	*Ref*
H^+	gl	20	1 $NaNO_3$	K_1 9·20±0·01	65F
Cu^{2+}	gl	20	1 $NaNO_3$	K_1 4·2±0·2	65F
Pb^{2+}	gl	20	1 $NaNO_3$	K_1 3·4±0·3	65F

65F A. M. Fiskin and M. Beer, *Biochemistry*, 1965, 4, 1289

738 $C_9H_{13}O_3N$ (−)-1-(3′,4′-Dihydroxyphenyl)-2-(methylamino)ethanol [(−)-adrenaline] H_2L

Metal	*Method*	*Temp*	*Medium*	*Log of equilibrium constant, remarks*	*Ref*
H^+	gl	0, 25	0·06 KCl	K_2 (0°) 10·98, (25°) 10·02, K_3 (0°) 9·61, (25°) 8·78, K (?) (0°) 2·75, (25°) 2·58	62A
	gl	25	0·1 (KCl)	K_1 ~ 13, K_2 9·95±0·05, K_3 8·66±0·01 $K[(H_{-1}L)^{3-}+H^+ \rightleftharpoons L^{2-}]$ ~ 13	65J, 66J
	gl	25	0—1 KCl	K_2 (→ 0) 10·05, (0·005) 10·00, (0·09) 9·87, (0·2) 9·79, (0·4) 9·79, (0·9) 9·75, K_3 (→ 0) 8·62, (0·001) 8·60, (0·1) 8·66, (0·2) 8·62, (0·5) 8·67, (0·9) 8·74	66A

contin

$C_9H_{13}O_3N$ (contd) 738

Metal	Method	Temp	Medium	Log of equilibrium constant, remarks	Ref
B^{III}	gl	25	0·1 KCl	$K[H_3BO_3+H_2L \rightleftharpoons B(OH)_2L^-+H^+]$ −4·67 $K[H_3BO_3+2H_2L \rightleftharpoons BL_2^-+H^+]$ −3·70	66A
Co^{2+}	gl	0, 25	0·06 KCl	K_1 (0°) 6·09, (25°) 5·42, K_2 (0°) 4·19, (25°) 3·80, β_2 (0°) 10·30, (25°) 8·94 $\Delta H_{\beta 2}$ −20·2, $\Delta S_{\beta 2}$ −27	62A
	gl	25	0·1 (KCl)	*K_1 9·61±0·03, K_2 7·1±0·1	66J
Cu^{2+}	gl	0, 25	0·06 KCl	K_1 (0°) 10·96, (25°) 10·50, K_2 (0°) 8·60, (25°) 7·90, β_2 (0°) 19·06, (25°) 17·72 $\Delta H_{\beta 2}$ −19·9, $\Delta S_{\beta 2}$ 14	62A
	gl	25	0·1 (KCl)	*K_1 14·95±0·03, K_2 12·50±0·05, K_3 9·65±0·05, K_4 6·0±0·2	65J
Mn^{2+}	gl	25	0·1 (KCl)	*K_1 8·80±0·03, K_2 6·3±0·1	66J
Mo^{VI}	sp	25	0·1 (KCl)	$K[MoO_4^{2-}+2H_3L \rightleftharpoons MoO_2(HL)_2^{2-}]$ (?) 5·76±0·02	62H
Ni^{2+}	gl	0, 25	0·06 KCl	K_1 (0°) 6·17, (25°) 5·65, K_2 (0°) 3·71, (25°) 3·52, β_2 (0°) 8·90, (25°) 8·40 $\Delta H_{\beta 2}$ −7·4, $\Delta S_{\beta 2}$ 14	62A
	gl	25	0·1 (KCl)	*K_1 10·40±0·03	66J
Zn^{2+}	gl	25	0·1 (KCl)	*K_1 10·92±0·03, K_2 9·2±0·1	66J

* Equilibrium constants adjusted to give hypothetical microscopic constant for combination of metal ion with ligand species HL^+, in which both phenolic groups are dissociated.

62A A. C. Andrews, T. D. Lyons, and T. D. O'Brien, *J. Chem. Soc.*, 1962, 1776
62H J. Halmekoski, *Suomen Kem.*, 1962, **B35**, 238
65J R. F. Jameson and W. F. S. Neillie, *J. Inorg. Nuclear Chem.*, 1965, **27**, 2623
66A P. J. Antikainen and K. Tevanen, *Suomen Kem.*, 1966, **B39**, 247, 285
66J R. F. Jameson and W. F. S. Neillie, *J. Inorg. Nuclear Chem.*, 1966, **28**, 2667

$C_9H_{13}O_3N$ (+)-Adrenaline H_2L 738.1

Metal	Method	Temp	Medium	Log of equilibrium constant, remarks	Ref
H^+	gl	0, 25	0·06 KCl	K_2 (0°) 10·57, (25°) 9·96, K_3 (0°) 9·17, (25°) 8·64	62A
Co^{2+}	gl	0, 25	0·06 KCl	K_1 (0°) 5·68, (25°) 5·76, K_2 (0°) 4·07, (25°) 4·29, β_2 (0°) 9·60, (25°) 10·06 $\Delta H_{\beta 2}$ 6·9, $\Delta S_{\beta 2}$ 69	62A
Cu^{2+}	gl	0, 25	0·06 KCl	K_1 (0°) 11·42, (25°) 10·70, K_2 (0°) 8·93, (25°) 6·70, β_2 (0°) 20·30, (25°) 18·44 $\Delta H_{\beta 2}$ −27·7, $\Delta S_{\beta 2}$ −8	62A
Ni^{2+}	gl	0, 25	0·06 KCl	K_1 (0°) 6·17, (25°) 6·22, K_2 (0°) 3·58, (25°) 3·66, β_2 (0°) 9·00, (25°) 9·26 $\Delta H_{\beta 2}$ 3·9, $\Delta S_{\beta 2}$ 56	62A

62A A. C. Andrews, T. D. Lyons, and T. D. O'Brien, *J. Chem. Soc.*, 1962, 1776

$C_9H_{13}O_3N$ 1-(3′,4′-Dihydroxyphenyl)-2-aminopropanol (corbadrine) H_2L 738.2

Metal	Method	Temp	Medium	Log of equilibrium constant, remarks	Ref
Mo^{VI}	sp	25	0·1 (KCl)	$K(MoO_4^{2-}+2H_2L \rightleftharpoons MoO_2L_2^{2-})$ 5·92±0·02	62H

62H J. Halmekoski, *Suomen Kem.*, 1962, **B35**, 241

738.3 $C_9H_{13}O_5N_3$ 4-Amino-1,3-diazin-2-one 1-riboside (cytidine) L

Metal	*Method*	*Temp*	*Medium*	*Log of equilibrium constant, remarks*	*Ref*
H^+	gl	25	0·16	K_1 4·21	61M
	gl	20	1 $NaNO_3$	K_1 4·23 ± 0·005	65F
Cu^{2+}	gl	25	0·16	K_1 1·4	61M
	gl	20	1 $NaNO_3$	K_1 1·58 ± 0·05	65F
Pb^{2+}	gl	20	1 $NaNO_3$	K_1 0·96 ± 0·05	65F

61M R. B. Martin, *Fed. Proc.*, 1961, **20**, No. 3, Suppl., 10, 54
65F A. M. Fiskin and M. Beer, *Biochemistry*, 1965, **4**, 1289

738.4 $C_9H_{13}O_6N$ 2,6-Dicarboxypiperidyl-*N*-acetic acid H_3L

Metal	*Method*	*Temp*	*Medium*	*Log of equilibrium constant, remarks*	*Ref*
H^+	gl	25	0·1 (KNO_3)	K_1 9·33 ± 0·01, K_2 2·71 ± 0·01, K_3 1·3 ± 0·1	68K, 6[illegible]
Ba^{2+}	gl	25	0·1 (KNO_3)	K_1 3·40 ± 0·01	68K
Ca^{2+}	gl	25	0·1 (KNO_3)	K_1 5·41 ± 0·01	68K
Cd^{2+}	gl	25	0·1 (KNO_3)	K_1 8·81 ± 0·01, K_2 2·99 ± 0·01	68K
Ce^{3+}	gl	25	0·1 (KNO_3)	K_1 9·72 ± 0·05, K_2 6·65 ± 0·05	68T
Co^{2+}	gl	25	0·1 (KNO_3)	K_1 9·64 ± 0·01	68K
Cu^{2+}	gl	25	0·1 (KNO_3)	K_1 13·25 ± 0·05	68K
Dy^{3+}	gl	25	0·1 (KNO_3)	K_1 10·98 ± 0·05, K_2 8·13 ± 0·05	68T
Er^{3+}	gl	25	0·1 (KNO_3)	K_1 11·35 ± 0·05, K_2 8·42 ± 0·05	68T
Eu^{3+}	gl	25	0·1 (KNO_3)	K_1 10·63 ± 0·05, K_2 7·91 ± 0·05	68T
Gd^{3+}	gl	25	0·1 (KNO_3)	K_1 10·66 ± 0·05, K_2 7·96 ± 0·05	68T
Ho^{3+}	gl	25	0·1 (KNO_3)	K_1 11·18 ± 0·05, K_2 8·17 ± 0·05	68T
La^{3+}	gl	25	0·1 (KNO_3)	K_1 9·17 ± 0·05, K_2 6·17 ± 0·05	68T
Lu^{3+}	gl	25	0·1 (KNO_3)	K_1 11·74 ± 0·05, K_2 8·92 ± 0·05	68T
Mg^{2+}	gl	25	0·1 (KNO_3)	K_1 5·06 ± 0·01	68K
Mn^{2+}	gl	25	0·1 (KNO_3)	K_1 7·40 ± 0·01	68K
Nd^{3+}	gl	25	0·1 (KNO_3)	K_1 10·18 ± 0·05, K_2 7·32 ± 0·05	68T
Ni^{2+}	gl	25	0·1 (KNO_3)	K_1 10·87 ± 0·05	68K
Pb^{2+}	gl	25	0·1 (KNO_3)	K_1 11·24 ± 0·05	68K
Pr^{3+}	gl	25	0·1 (KNO_3)	K_1 10·01 ± 0·05, K_2 7·04 ± 0·05	68T
Sm^{3+}	gl	25	0·1 (KNO_3)	K_1 10·59 ± 0·05, K_2 7·77 ± 0·05	68T
Sr^{2+}	gl	25	0·1 (KNO_3)	K_1 3·81 ± 0·01	68K
Tb^{3+}	gl	25	0·1 (KNO_3)	K_1 10·85 ± 0·05, K_2 8·04 ± 0·05	68T
Tm^{3+}	gl	25	0·1 (KNO_3)	K_1 11·54 ± 0·05, K_2 8·63 ± 0·05	68T
Y^{3+}	gl	25	0·1 (KNO_3)	K_1 10·83 ± 0·05, K_2 7·75 ± 0·05	68T
Yb^{3+}	gl	25	0·1 (KNO_3)	K_1 11·73 ± 0·05, K_2 8·91 ± 0·05	68T
Zn^{2+}	gl	25	0·1 (KNO_3)	K_1 10·25 ± 0·05	68K

68K S. K. Kundra and L. C. Thompson, *J. Inorg. Nuclear Chem.*, 1968, **30**, 1847
68T L. C. Thompson and S. K. Kundra, *Inorg. Chem.*, 1968, **7**, 338

$C_9H_{14}O_3N_4$ 3-Alanyl-L-histidine (L-carnosine) HL 740

Metal	Method	Temp	Medium	Log of equilibrium constant, remarks	Ref
H^+	gl	25	0·10 (KNO_3)	K_1 9·36±0·01, K_2 6·76±0·01	64L
	gl	25	0·1 $NaClO_4$ 0·19 CH_3CN	K_1 9·32, K_2 6·57	66K
Ca^{2+}	gl	25	0·10 (KNO_3)	K_1 3·22±0·05	64L
Cd^{2+}	gl	25	0·10 (KNO_3)	K_1 3·19±0·02	64L
Co^{2+}	gl	25	0·10 (KNO_3)	$K(Co^{2+}+HL \rightleftharpoons CoHL^{2+})$ 3·69±0·02	64L
Cu^+	gl	25	0·1 ($NaClO_4$) 0·19 CH_3CN	K_1 10·55 $K[Cu^+ + 2HL \rightleftharpoons Cu(HL)_2^+]$ 11·62	66K
Cu^{2+}	gl	25	0·10 (KNO_3)	K_1 9·72±0·06 $K(Cu^{2+}+HL \rightleftharpoons CuHL^{2+})$ 5·01±0·06 $K[Cu(H_{-1}L)+H^+ \rightleftharpoons CuL^+]$ 5·14±0·02	64L
Hg^{2+}	gl	25	0·10 (KNO_3)	K_1 8·08±0·02 $K(Hg^{2+}+HL \rightleftharpoons HgHL^{2+})$ 5·27±0·02	64L
Mg^{2+}	gl	25	0·10 (KNO_3)	K_1 3·10±0·04	64L
Mn^{2+}	gl	25	0·10 (KNO_3)	K_1 4·40±0·02 $K(Mn^{2+}+HL \rightleftharpoons MnHL^{2+})$ 3·14±0·02	64L
Ni^{2+}	gl	25	0·10 (KNO_3)	K_1 5·42±0·05 $K[Ni(H_{-1}L)+H^+ \rightleftharpoons NiL^+]$ 9·14±0·02	64L
Pb^{2+}	gl	25	0·10 (KNO_3)	$K(Pb^{2+}+HL \rightleftharpoons PbHL^{2+})$ 3·40±0·02	64L
Sr^{2+}	gl	25	0·10 (KNO_3)	K_1 3·34±0·02	64L
Zn^{2+}	gl	25	0·10 (KNO_3)	$K(Zn^{2+}+HL \rightleftharpoons ZnHL^{2+})$ 3·39±0·02	64L

64L G. R. Lenz and A. E. Martell, *Biochemistry*, 1964, **3**, 750
66K Th. Kaden and A. Zuberbuhler, *Helv. Chim. Acta*, 1966, **49**, 2189

$C_9H_{16}O_2N_2$ 4-(2′-Propyl)cyclohexane-1,2-dione dioxime (4-isopropylnioxime) HL 744.1

Metal	Method	Temp	Medium	Log of equilibrium constant, remarks	Ref
H^+	gl	25	0·1 75% dioxan	K_1 12·27	63B
Ni^{2+}	gl	25	0·1 0—75% dioxan	K_1 (75%) 10·1±0·3, β_2 (→ 0%) 16·8, (75%) 20·4±0·2 K_{so} −27·84	63B

63B C. V. Banks and S. Anderson, *Inorg. Chem.*, 1963, **2**, 112

$C_9H_{17}O_4N_3$ L-Alanyl-L-alanyl-L-alanine HL 746.1

Metal	Method	Temp	Medium	Log of equilibrium constant, remarks	Ref
H^+	gl	25	0·16 KCl	K_1 8·08, K_2 3·36	66B
Cu^{2+}	gl	25	~ 0	K_1 4·65 $K[Cu(H_{-1}L)+H^+ \rightleftharpoons CuL^+]$ 4·47 $K[Cu(H_{-2}L)^- + H^+ \rightleftharpoons Cu(H_{-1}L)]$ 6·63 $K[Cu(H_{-2}L)OH^{2-} + H^+ \rightleftharpoons Cu(H_{-2}L)^-]$ 12·3	66B

66B G. F. Bryce and F. R. N. Gurd, *J. Biol. Chem.*, 1966, **241**, 1439

750.1 $C_9H_{20}O_2N_4$ **NN′-Diglycyl-1,5-diaminopentane L**

Metal	Method	Temp	Medium	Log of equilibrium constant, remarks	Ref
H^+	gl	23	0·5 KCl	K_1 8·51, K_2 7·88	67Z
Cu^{2+}	gl	23	0·5 KCl	K_1 8·46	67Z
				$K(CuL^{2+}+H^+ \rightleftharpoons CuHL^{3+})$ 5·49	
				$K[Cu(H_{-1}L)^+ + H^+ \rightleftharpoons CuL^{2+}]$ 7·47	
				$K[Cu(H_{-2}L)+H^+ \rightleftharpoons Cu(H_{-1}L)^+]$ 10·30	

67Z A. Zuberbuhler and S. Fallab, *Helv. Chim. Acta*, 1967, **50**, 889

750.2 $C_9H_{21}O_3N$ **3,3′,3″-Nitrilotripropan-2-ol [tris(2-hydroxypropyl)amine] L**

Metal	Method	Temp	Medium	Log of equilibrium constant, remarks	Ref
H^+	H	20	0 corr	K_1 7·97 $(L \rightleftharpoons LH^+ + OH^-)$ 6·20±0·01	64A
Ag^+	gl	20	0 corr	K_1 2·30±0·04, K_2 1·97±0·04	64A
	sol	20	0 corr	β_2 4·27±0·01	64A

64A D. J. Alner and M. A. A. Kahn, *J. Chem. Soc.*, 1964, 5265

750.3 $C_9H_{21}O_{17}P_3$ **1′-Glycerylphosphorylinositol-3,4-diphosphoric acid H_5L**

Metal	Method	Temp	Medium	Log of equilibrium constant, remarks	Ref
H^+	gl	20	0·1 $(C_3H_7)_4NI$	K_1 8·05±0·05, K_2 5·70±0·05	65H
Ca^{2+}	gl	20	0·1 $(C_3H_7)_4NI$	K_1 3·3±0·1	65H
				$K(Ca^{2+}+HL^{4-} \rightleftharpoons CaHL^{2-})$ 2·2±0·1	
Mg^{2+}	gl	20	0·1 $(C_3H_7)_4NI$	K_1 3·5±0·1	65H
				$K(Mg^{2+}+HL^{4-} \rightleftharpoons MgHL^{2-})$ 2·4±0·1	

65H H. S. Henderson and J. G. Fullington, *Biochemistry*, 1965, **4**, 1599

751 $C_9H_5ONCl_2$ **5,7-Dichloro-8-hydroxyquinoline HL**

Metal	Method	Temp	Medium	Log of equilibrium constant, remarks	Ref
H^+	?	25	?	K_1 7·77, K_2 2·79	66R
	sp	25	~ 0·05	K_1 7·62, K_2 2·89	67T
Ge^{IV}	sp	25	0·5 NaCl	$K[Ge(OH)_4+2HL \rightleftharpoons Ge(OH)_2L_2]$ 6·7	67T
Nd^{3+}	dis	25	1 $NaClO_4$	K_1 6·6, β_2 12·8, β_3 18·4	66R

66R T. I. Romantseva, M. I. Gromova, and V. M. Peshkova, *Russ. J. Inorg. Chem.*, 1966, **11**, 935 (1748)
67T J. Tsau, S. Matsouo, P. Clerc, and R. Benoit, *Bull. Soc. chim. France*, 1967, 1039

$C_9H_5ONBr_2$ 5,7-Dibromo-8-hydroxyquinoline HL 752

Metal	Method	Temp	Medium	Log of equilibrium constant, remarks	Ref
H^+				H^+ constants not given.	
Cu^{2+}	dis	18	0·2 ($NaClO_4$)	K_1 14·18, K_2 10·32	65N
Sr^{2+}	dis	18	0·2 ($NaClO_4$)	K_1 7·1, K_2 6·5	65N
Zn^{2+}	dis	18	0·2 ($NaClO_4$)	K_1 7·76, K_2 7·56	65N

Erratum. 1964, p. 607, Table 752. Insert **HL** in heading.

65N O. Navratil and J. Kotas, *Coll. Czech. Chem. Comm.*, 1965, **30**, 1824

C_9H_6ONCl 5-Chloro-8-hydroxyquinoline HL 754

Metal	Method	Temp	Medium	Log of equilibrium constant, remarks	Ref
H^+	gl	25	0·1	K_1K_2 13·00	68C
Zn^{2+}	dis	25	0·1	β_2 15·58	68C

68C F. C. Chou and H. Freiser, *Analyt. Chem.*, 1968, **40**, 34

C_9H_6ONBr 5-Bromo-8-hydroxyquinoline HL 755

Metal	Method	Temp	Medium	Log of equilibrium constant, remarks	Ref
H^+	gl	25	0·1	K_1K_2 12·80	68C
Zn^{2+}	dis	25	0·1	β_2 14·62	68C

68C F. C. Chou and H. Freiser, *Analyt. Chem.*, 1968, **40**, 34

C_9H_6ONI 5-Iodo-8-hydroxyquinoline HL 756

Metal	Method	Temp	Medium	Log of equilibrium constant, remarks	Ref
H^+	gl	25	0·1	K_1K_2 11·90	68C
Zn^{2+}	dis	25	0·1	β_2 14·86	68C

68C F. C. Chou and H. Freiser, *Analyt. Chem.*, 1968, **40**, 34

758 $C_9H_6O_6N_2S$ 7-Nitro-8-hydroxyquinoline-5-sulphonic acid H_2L

Metal	*Method*	*Temp*	*Medium*	*Log of equilibrium constant, remarks*	*Ref*
H^+	gl	25	0·005 $HClO_4$	K_1 5·62	63F
	sp	28	0·005 $HClO_4$	K_3 0·4	63F
	gl	25	0·1 50% dioxan	K_1 6·23, K_2 3·28	66B
Co^{2+}	gl	25	0·005 $HClO_4$	K_2 5·41, K_3 < 3·8	63F
Cu^{2+}	gl	25	0·005 $HClO_4$	K_2 > 6·4, K_3 < 3·8	63F
	pol	25	0·1 50% dioxan	β_2 (?) 16·10	66B
Ni^{2+}	gl	25	0·005 $HClO_4$	K_2 ~ 6·2, K_3 4·74	63F
Zn^{2+}	gl	25	0·005 $HClO_4$	K_1 5·90, K_2 4·90, K_3 < 3·8	63F

63F J. Fresco and H. Freiser, *Inorg. Chem.*, 1963, **2**, 82

66B H. Berge, *J. prakt. Chem.*, 1966, **34**, 15

758.1 $C_9H_7O_2N_3S$ 4-(2′-Thiazolylazo)-1,3-dihydroxybenzone (thiazolylazoresorcinol) (TAR) H_2

Metal	*Method*	*Temp*	*Medium*	*Log of equilibrium constant, remarks*	*Ref*
H^+	gl	25	50% dioxan	K_1 12·80±0·04, K_2 7·37±0·03, K_3 1·65±0·11	66S
	sp	18—22	0·1 ($NaClO_4$)	K_1 9·44, K_2 6·23, K_3 0·96	66H, 67
	gl, sp	25	0·1 $NaClO_4$ 50% methanol	K_1 (gl) 10·76, K_2 (gl) 6·53, K_3 (sp) 0·9	67N
Au^{III}	sp	25	0·1 $NaClO_4$ 50% methanol	*K(?) 12±1	67N
Ba^{2+}	gl	25	0·1 $NaClO_4$ 50% methanol	*$K(Ba^{2+}+HL^- \rightleftharpoons BaHL^+)$ < 3	67N
Bi^{3+}	sp	18—22	0·1 ($NaClO_4$)	*$K(Bi^{3+}+HL^- \rightleftharpoons BiHL^{2+})$ 13·11	66H
Ca^{2+}	gl	25	0·1 $NaClO_4$ 50% methanol	*$K(Ca^{2+}+HL^- \rightleftharpoons CaHL^+)$ 3·5±0·3	67N
Cd^{2+}	sp	18—22	0·1 ($NaClO_4$)	*$K(Cd^{2+}+HL^- \rightleftharpoons CdHL^+)$ 6·96	66H
	gl	25	0·1 $NaClO_4$ 50% methanol	*$K[Cd^{2+}+2HL^- \rightleftharpoons Cd(HL)_2]$ 16·0±0·2	67N
Co^{2+}	gl	25	50% dioxan	*$K(Co^{2+}+HL^- \rightleftharpoons CoHL^+)$ 12·05±0·10 *$K[CoHL^+ +HL^- \rightleftharpoons Co(HL)_2]$ 11·23±0·06	66S
Cr^{3+}	sp	25	0·1 $NaClO_4$ 50% methanol	*$K(Cr^{3+}+HL^- \rightleftharpoons CrHL^{2+})$ 10	67N
Cu^{2+}	sp	18—22	0·1 ($NaClO_4$)	K_1 13·55, $K(CuL+H^+ \rightleftharpoons CuHL^+)$ 4·24 *$K(Cu^{2+}+HL^- \rightleftharpoons CuHL^+)$ 11·56	66H
	gl	25	50% dioxan	*$K(Cu^{2+}+HL^- \rightleftharpoons CuHL^+)$ 14·2±0·3 $K(CuL+H^+ \rightleftharpoons CuHL^+)$ 4·34±0·06 $K(CuOHL^- +H^+ \rightleftharpoons CuL)$ 10·13±0·06	66S
	gl	25	0·1 $NaClO_4$ 50% methanol	*$K(Cu^{2+}+HL^- \rightleftharpoons CuHL^+)$ 12·3±0·2, *$K[CuHL^+ +HL^- \rightleftharpoons Cu(HL)_2]$ 9·9±0·1	67N
Fe^{2+}	gl	25	0·1 $NaClO_4$ 50% methanol	*$K[Fe^{2+}+2HL^- \rightleftharpoons Fe(HL)_2]$ 21·6±0·3	67N
Ga^{3+}	sp	25	0·1 $NaClO_4$ 50% methanol	*$K(Ga^{3+}+HL^- \rightleftharpoons GaHL^{2+})$ 12·0±0·05	67N
In^{3+}	gl	25	0·1 $NaClO_4$ 50% methanol	*$K(In^{3+}+HL^- \rightleftharpoons InHL^{2+})$ 10·8±0·2	67N

cont

Metal	*Method*	*Temp*	*Medium*	*Log of equilibrium constant, remarks*	*Ref*
Mg^{2+}	gl	25	0·1 $NaClO_4$ 50% methanol	*$K(Mg^{2+}+HL^- \rightleftharpoons MgHL^+) < 3$	67N
Mn^{2+}	gl	25	50% dioxan	*$K(Mn^{2+}+HL^- \rightleftharpoons MnHL^+)$ 9·43±0·02	66S
				*$K[MnHL^+ + HL^- \rightleftharpoons Mn(HL)_2]$ 8·6±0·2	
				$K(MnL+H^+ \rightleftharpoons MnHL^+)$ 7·88±0·05	
				$K(MnOHL^- + H^+ \rightleftharpoons MnL)$ 9·4±0·1	
	gl	25	0·1 $NaClO_4$ 50% methanol	*$K[Mn^{2+}+2HL^- \rightleftharpoons Mn(HL)_2]$ 13·1±0·2	67N
Nb^V	sp	25	0·1 $NaClO_4$ 50% methanol	*$K(NbO_3^- + H_2L \rightleftharpoons NbO_2L^-)$ (?) 9·5±1	67N
Ni^{2+}	gl	25	50% dioxan	*$K(Ni^{2+}+HL^- \rightleftharpoons NiHL^+)$ 12·94±0·08	66S
				*$K[NiHL^+ + HL^- \rightleftharpoons Ni(HL)_2]$ 11·82±0·04	
				$K(NiL+H^+ \rightleftharpoons NiHL^+)$ 6·84±0·07	
				$K(NiOHL^- + H^+ \rightleftharpoons NiL)$ 8·55±0·10	
Pb^{2+}	sp	18—22	0·1 ($NaClO_4$)	*$K(Pb^{2+}+HL^- \rightleftharpoons PbHL^+)$ 8·34	66H
	gl	25	0·1 $NaClO_4$ 50% methanol	*$K(Pb^{2+}+HL^- \rightleftharpoons PbHL^+)$ 9·7±0·2	67N
Pt^{2+}	gl	25	0·1 $NaClO_4$ 50% methanol	*K (?) 12±1	67N
Rh^{3+}	sp	25	0·1 $NaClO_4$ 50% methanol	*K (?) 12±1	67N
Sc^{3+}	gl	25	0·1 $NaClO_4$ 50% methanol	*$K(Sc^{3+}+HL^- \rightleftharpoons ScHL^{2+})$ 10·4±0·1,	67N
				*$K[ScHL^{2+}+HL^- \rightleftharpoons Sc(HL)_2^+]$ 9·9±0·1	
Sr^{2+}	gl	25	0·1 $NaClO_4$ 50% methanol	*$K(Sr^{2+}+HL^- \rightleftharpoons SrHL^+) \leqslant 3$	67N
Ti^{IV}	sp	25	0·1 $NaClO_4$ 50% methanol	*$K(TiO^{2+}+HL^{2-} \rightleftharpoons TiOHL)$ 13±1	67N
Tl^+	gl	25	0·1 $NaClO_4$ 50% methanol	*$K(Tl^+ + HL^- \rightleftharpoons TlHL) < 3$	67N
Tl^{3+}	sp	25	0·1 $NaClO_4$ 50% methanol	*$K(Tl^{3+}+HL^- \rightleftharpoons TlHL^{2+})$ 12·0±0·05	67N
UO_2^{2+}	gl	25	0·1 $NaClO_4$ 50% methanol	*$K[UO_2^{2+}+HL^- \rightleftharpoons UO_2(HL)^+]$ 10·7±0·2,	67N
				*$K[UO_2(HL)^+ + HL^- \rightleftharpoons UO_2(HL)_2]$ 9·7±0·1	
	sp	18—22	0·1 ($NaClO_4$)	K_1 11·35	67S
				$K(UO_2L+H^+ \rightleftharpoons UO_2HL^+)$ 4·5	
				*$K(UO_2^{2+}+HL^- \rightleftharpoons UO_2HL^+)$ 9·8	
VO^{2+}	gl	25	0·1 $NaClO_4$ 50% methanol	*$K[VO^{2+}+HL^- \rightleftharpoons VO(HL)^+]$ 11·2±0·1,	67N
				*$K[VO(HL)^+ + HL^- \rightleftharpoons VO(HL)_2]$ 9·8±0·2	
V^V	sp	25	0·1 $NaClO_4$ 50% methanol	*$K(VO_3^- + H_2L \rightleftharpoons VO_2L^-)$ (?) 12·5±0·05	67N
Zn^{2+}	sp	18—22	0·1 ($NaClO_4$)	*$K(Zn^{2+}+HL^- \rightleftharpoons ZnHL^+)$ 7·19	66H
			50% dioxan	*$K(Zn^{2+}+HL^- \rightleftharpoons ZnHL^+)$ 11·08±0·04	66S
				*$K[ZnHL^+ + HL^- \rightleftharpoons Zn(HL)_2]$ 10·11±0·02	
				$K(ZnL+H^+ \rightleftharpoons ZnHL^+)$ 7·12±0·10	
				$K(ZnOHL^- + H^+ \rightleftharpoons ZnL)$ 8·74±0·11	
				$K[Zn(OH)_2L^{2-} + H^+ \rightleftharpoons ZnOHL^-]$ 8·98±0·03	
	gl	25	0·1 $NaClO_4$ 50% methanol	*$K[Zn^{2+}+2HL^- \rightleftharpoons Zn(HL)_2]$ 17·2±0·2	67N
Zr^{IV}	sp	25	0·1 $NaClO_4$ 50% methanol	*$K[ZrO^{2+}+HL^{2-} \rightleftharpoons ZrO(HL)]$ 13±1	67N

* Formation constants K_{MHL} adjusted to give assumed microscopic formation constants for combination of metal ion with ligand HL^- having hydroxyl proton at 1-position

references overleaf

758.1 $C_9H_7O_2N_3S$ (contd)

66H M. Hnilickova and L. Sommer, *Talanta*, 1966, **13**, 667
66S R. W. Stanley and G. E. Cheney, *Talanta*, 1966, **13**, 1619
67N G. Nickless, F. H. Pollard, and T. J. Samuelson, *Analyt. Chim. Acta*, 1967, **39**, 37
67S L. Sommer and V. M. Ivanov, *Talanta*, 1967, **14**, 171

759 $C_9H_7O_4NS$ 8-Hydroxyquinoline-5-sulphonic acid (sulphoxine) H_2L

Metal	*Method*	*Temp*	*Medium*	*Log of equilibrium constant, remarks*	*Ref*
H^+	gl	25	0·005 $HClO_4$	K_1 8·66, K_2 4·09	63F
	sp	28	0·005 $HClO_4$	$K_3 < 0$	63F
	gl	25	0·1 50% dioxan	K_1 10·48, K_2 3·38	66B[illegible]
	gl	25	0·10 (KNO_3)	K_1 8·35, K_2 3·84	66L
	gl	25	0·10 (KNO_3)	K_1 8·47±0·01, K_2 3·98±0·01	66L[illegible]
	gl	25, 35	0·10 (KNO_3)	K_1 (25°) 8·48±0·05, (35°) 8·41±0·02, K_2 (25°) 3·92±0·03, (35°) 3·85±0·04	66M
	gl	25	0·5 $NaNO_3$	K_1 8·38, K_2 3·98	67S
	gl	25	0·5 NaCl	K_1 8·23, K_2 3·84	67T
	gl	25	0·1 $NaClO_4$	K_1 8·43, K_2 3·88	68G
			0·1 $NaClO_4$ 50% dioxan	K_1 9·85, K_2 3·56	
	cal	25	0·1 $NaClO_4$	ΔH_1 −4·00, ΔS_1 25·2, ΔH_2 −4·37, ΔS_2 3·1,	68G
			0·1 $NaClO_4$ 50% dioxan	ΔH_1 −4·95, ΔS_1 28·5, ΔH_2 −4·04, ΔS_2 2·7	
Co^{2+}	gl	25	0·005 $HClO_4$	K_1 8·54, K_2 7·22, K_3 5·39	63F
	gl	25	0·1 $NaClO_4$ 50% dioxan	K_1 7·38, β_2 17·55	68G
	cal	25	0·1 $NaClO_4$ 50% dioxan	ΔH_1 −6·3 (?), ΔS_1 22 (?), $\Delta H_{\beta 2}$ −14·5, $\Delta S_{\beta 2}$ 32	68G
Cr^{3+}	gl	30	0·1 KCl	K_1 10·99, K_2 10·05; $K(CrOHL + H^+ \rightleftharpoons CrL^+)$ 5·14	66L
Cu^{2+}	pol	25	0·1 0—75% methanol	β_2 (0%) 23·6 (20%) 24·0, (50%) 24·5, (75%) 25·3	65B
			0·1 20—60% ethanol	β_2 (20%) 24·1, (50%) 25·0, (60%) 26·1	
			0·1 20—50% n-propanol	β_2 (20%) 24·0, (50%) 25·6	
			0·1 20—50% isopropanol	β_2 (20%) 24·1, (50%) 25·6	
	pol	25	0·1 50% dioxan	β_2 (?) 24·73	66Ba
	gl	25	0·10 (KNO_3)	K_1 11·92±0·05	66La
	gl	25	0·5 $NaNO_3$	K_1 11·57, β_2 21·63	67S
	gl	25	0·1 $NaClO_4$	K_1 12·45, β_2 22·99	68G
	cal	25	0·1 $NaClO_4$	ΔH_1 −8·1, ΔS_1 30, $\Delta H_{\beta 2}$ −17·6, $\Delta S_{\beta 2}$ 46	68G
	gl	25	0·1 $NaClO_4$ 50% dioxan	K_1 13·16, β_2 25·55	68G
	cal	25	0·1 $NaClO_4$ 50% dioxan	ΔH_1 −9·0, ΔS_1 30, $\Delta H_{\beta 2}$ −18·6, $\Delta S_{\beta 2}$ 54	68G

See 1,10-phenanthroline (906) and triethylenetetramine-hexa-acetic acid (1085) for mixed ligands.

conti[illegible]

$C_9H_7O_4NS$ (contd) 759

Metal	Method	Temp	Medium	Log of equilibrium constant, remarks	Ref
Fe^{2+}	sp	?	5—64% N_2H_4	β_2 (5%) 6·2, (10%) 5·7, (25%) 6·2, (50%) 8·0, (64%) 9·3	66B
Fe^{3+}				See triethylenetetraminehexa-acetic acid (1085) for mixed ligands.	
Ge^{IV}	sp	25	0·5 NaCl	$K[Ge(OH)_4 + 2HL \rightleftharpoons Ge(OH)_2L_2]$ 6·55	67T
Mn^{2+}	gl	25	0·1 $NaClO_4$ 50% dioxan	K_1 7·05, β_2 13·18	68G
	cal	25	0·1 $NaClO_4$ 50% dioxan	ΔH_1 −3·2, ΔS_1 22, $\Delta H_{\beta 2}$ −6·7 (?), $\Delta S_{\beta 2}$ 46 (?)	68G
Ni^{2+}	gl	25	0·005 $HClO_4$	K_1 9·57, K_2 8·58, K_3 7·42	63F
	gl	25	0·1 $NaClO_4$	K_1 9·11, β_2 17·34, β_3 23·23	68G
	cal	25	0·1 $NaClO_4$	ΔH_1 −6·4, ΔS_1 20, $\Delta H_{\beta 2}$ −14·8, $\Delta S_{\beta 2}$ 30, $\Delta H_{\beta 3}$ −25·6, $\Delta S_{\beta 3}$ 20	68G
	gl	25	0·1 $NaClO_4$ 50% dioxan	K_1 10·22, β_2 19·25, β_3 25·55	68G
	cal	25	0·1 $NaClO_4$ 50% dioxan	ΔH_1 −7·3, ΔS_1 22, $\Delta H_{\beta 2}$ −15·9, $\Delta S_{\beta 2}$ 35, $\Delta H_{\beta 3}$ −25·4, $\Delta S_{\beta 3}$ 32	68G
Th^{4+}				See EDTA (836) and CDTA (997) for mixed ligands.	
U^{4+}				See EDTA (836) for mixed ligands.	
VO^{2+}	gl	25, 35	0·10 (KNO_3)	K_1 (25°) 11·79±0·03, (35°) 11·32±0·04 $\Delta H_1°$ −1·7±0·5, $\Delta S_1°$ 45±3 $K[VO(OH)L^- + H^+ \rightleftharpoons VOL]$ (25°) 6·45±0·03 $K\{2VO(OH)L^- \rightleftharpoons [VO(OH)L]_2^{2-}\}$ (25°) 4·84±0·04	66M
Zn^{2+}	gl	25	0·5 NaCl	K_1 7·45, K_2 6·5	67T
	gl	25	0·1 $NaClO_4$	K_1 7·95, β_2 14·97	68G
	cal	25	0·1 $NaClO_4$	ΔH_1 −5·1, ΔS_1 19, $\Delta H_{\beta 2}$ −9·6, $\Delta S_{\beta 2}$ 36	68G
	gl	25	0·1 $NaClO_4$ 50% dioxan	K_1 9·23, β_2 17·56	68G
	cal	25	0·1 $NaClO_4$ 50% dioxan	ΔH_1 −5·2, ΔS_1 25, $\Delta H_{\beta 2}$ −10·3, $\Delta S_{\beta 2}$ 46	68G

63F J. Fresco and H. Freiser, *Inorg. Chem.*, 1963, **2**, 82
65B H. Berge and P. Yeroschewski, *Z. phys. Chem. (Leipzig)*, 1965, **228**, 239
66B D. Bauer, *Bull. Soc. chim. France*, 1966, 2631
66Ba H. Berge, *J. prakt. Chem.*, 1966, **34**, 15
66L K. Lal and R. P. Agarwal, *J. Indian Chem. Soc.*, 1966, **43**, 169
66La G. A. L'Heureux and A. E. Martell, *J. Inorg. Nuclear Chem.*, 1966, **28**, 481
66M G. E. Mont and A. E. Martell, *J. Amer. Chem. Soc.*, 1966, **88**, 1387
67S J. P. Scharff and M. A. Paris, *Bull. Soc. chim. France*, 1967, 1782
67T J. Tsau, S. Matsouo, P. Clerc, and R. Benoit, *Bull. Soc. chim. France*, 1967, 1039
68G G. Gutnikov and H. Freiser, *Analyt. Chem.*, 1968, **40**, 39

$C_9H_8O_4N_2S$ 7-Amino-8-hydroxyquinoline-5-sulphonic acid H_2L 759.1

Metal	Method	Temp	Medium	Log of equilibrium constant, remarks	Ref
H^+	gl	25	0·1 50% dioxan	K_1 9·78, K_2 2·03	66B
Cu^{2+}	pol	25	0·1 50% dioxan	β_2 (?) 23·17	66B

66B H. Berge, *J. prakt. Chem.*, 1966, **34**, 15

763 $C_9H_9O_3NI_2$ 2-Amino-3-(4′-hydroxy-3′,5′-di-iodophenyl)propanoic acid (3,5-di-iodotyrosine)

Metal	Method	Temp	Medium	Log of equilibrium constant, remarks	Ref
H^+	gl	20	~0	K_1 9·48, K_2 5·32	53P
Cd^{2+}	gl	20	0·002 $CdSO_4$	β_2 6·9	53P
Zn^{2+}	gl	20	0·002 $ZnSO_4$	β_2 8·0	63P

53P D. J. Perkins, *Biochem. J.*, 1953, **55**, 649

766 $C_9H_{14}O_8N_3P$ Cytidine-5′-monophosphoric acid (CMP) H_2L

Metal	Method	Temp	Medium	Log of equilibrium constant, remarks	Ref
H^+	gl	25	0·1 KCl	K_1 6·35, K_2 4·35	58W
Zn^{2+}	gl	25	0·1 KCl	K_1 2·54	58W

58W G. Weitzel and T. Speer, *Z. physiol. Chem.*, 1958, **313**, 212

767.1 $C_9H_{15}O_{11}N_3P_2$ Cytidine-5′-diphosphoric acid (CDP) H_3L

Metal	Method	Temp	Medium	Log of equilibrium constant, remarks	Ref
Mg^{2+}	sp	?	0·05 $(CH_3)_4NCl$	K (?) 1·5	61H

61H K. Hotta, J. Brahms, and M. Morales, *J. Amer. Chem. Soc.*, 1961, **83** 997

767.2 $C_9H_{15}O_{15}N_2P_3$ Uridine-5′-triphosphoric acid (UTP) H_5L

Metal	Method	Temp	Medium	Log of equilibrium constant, remarks	Ref
H^+	sp	25	natural I	K_1 10·2	68S
	sp, gl	25	0·1 $(NaClO_4)$	K_1 (sp) 9·6, (gl) 9·5	68S
Cu^{2+}	sp	25	natural I	$K(CuL^{3-} + H^+ \rightleftharpoons CuHL^{2-})$ 8·4	68S
	sp, gl	25	0·1 $(NaClO_4)$	$K(CuL^{3-} + H^+ \rightleftharpoons CuHL^{2-})$ (sp) 7·8, (gl) 7·8 $K(CuLOH^{4-} + H^+ \rightleftharpoons CuL^{3-})$ (gl) ~8·4 $K(CuAL^{3-} + H^+ \rightleftharpoons CuAHL^{2-})$ (gl) 9·0 where A = 2,2′-bipyridyl	68S

68S H. Sigel, *European J. Biochem.*, 1968, **3**, 530

768 $C_{10}H_{15}O_{14}N_4P_3$ Inosine-5′-triphosphoric acid

Errata. 1964, p. 613, Table 768. *For* formula $C_9H_{15}O_{14}N_4P_3$ *read* $C_{10}H_{15}O_{14}N_4P_3$.
For correct ligand sequence this Table has been renumbered **859.1** *q.v.*

$C_9H_{16}O_{14}N_3P_3$ Cytidine-5′-triphosphoric acid (CTP) H_4L 770

Metal	Method	Temp	Medium	Log of equilibrium constant, remarks	Ref
Cu^{2+}	gl	25	0·1 ($NaClO_4$)	$K(CuLOH^{3-} + H^+ \rightleftharpoons CuL^{2-})$ ~7·6	68S
Mg^{2+}	sp	?	0·05 $(CH_3)_4NCl$	K(?) 1·95	61H

61H K. Hotta, Y. Brahms, and M. Morales, *J. Amer. Chem. Soc.*, 1961, **83**, 997

68S H. Sigel, *European J. Biochem.*, 1968, **3**, 530

$C_9H_{20}O_7N_3P$ *O*-Phosphoryl-L-seryl-L-lysine H_3L 770.1

Metal	Method	Temp	Medium	Log of equilibrium constant, remarks	Ref
H^+	gl	25	0·15 (KCl)	K_1 not given. K_2 7·59, K_3 5·34, K_4 2·99	62O
Ca^{2+}	gl	25	0·15 (KCl)	K_1 1·53 $K(Ca^{2+} + HL^{2-} \rightleftharpoons CaHL)$	62O
Mg^{2+}	gl	25	0·15 (KCl)	K_1 1·63 $K(Mg^{2+} + HL^{2-} \rightleftharpoons MgHL)$	62O
Mn^{2+}	gl	25	0·15 (KCl)	K_1 2·33 $K(Mn^{2+} + HL^{2-} \rightleftharpoons MnHL)$	62O

62O R. Osterberg, *Acta Chem. Scand.*, 1962, **16**, 2434

$H_{28}O_{15}N_3P_5$ *N*-(Phosphonomethyl)-2,2′-iminobis[ethylnitrilobis(methylphosphonic acid)] [diethylenetriamine-*NNN′N″N″*-penta(methylenephosphonic acid)] $H_{10}L$ 770.2

Metal	Method	Temp	Medium	Log of equilibrium constant, remarks	Ref
H^+	gl	25	0·1 KCl	K_1 12·04, K_2 10·10, K_3 8·15, K_4 7·17, K_5 6·38, K_6 5·50, K_7 4·45, K_8 2·8	67K
Ca^{2+}	gl	25	0·1 KCl	K_1 7·11 $K[Ca^{2+} + H_nL^{(10-n)-} \rightleftharpoons CaH_nL^{(8-n)-}]$, ($n$ = 1) 5·42, (n = 2) 4·49, (n = 3) 4·04, (n = 4) 3·11, (n = 5) 2·29	67K
Cd^{2+}	gl	25	0·1 KCl	K_1 13·37 $K[Cd^{2+} + H_nL^{(10-n)-} \rightleftharpoons CdH_nL^{(8-n)-}]$, ($n$ = 1) 10·76, (n = 2) 7·68, (n = 3) 6·36, (n = 4) 5·33, (n = 5) 4·40, (n = 6) 3·70, (n = 7) 1·99	67K
Co^{2+}	gl	25	0·1 KCl	K_1 15·73 $K[Co^{2+} + H_nL^{(10-n)-} \rightleftharpoons CoH_nL^{(8-n)-}]$, ($n$ = 1) 12·07, (n = 2) 9·17, (n = 3) 7·35, (n = 4) 5·74, (n = 5) 4·30, (n = 6) 3·10	67K
Cu^{2+}	gl	25	0·1 KCl	K_1 19·47 $K[Cu^{2+} + H_nL^{(10-n)-} \rightleftharpoons CuH_nL^{(8-n)-}]$, ($n$ = 1) 15·88, (n = 2) 13·38, (n = 3) 11·60, (n = 4) 9·66, (n = 5) 7·48, (n = 6) 5·22, (n = 7) 3·19	67K
Mg^{2+}	gl	25	0·1 KCl	K_1 6·40 $K[Mg^{2+} + H_nL^{(10-n)-} \rightleftharpoons MgH_nL^{(8-n)-}]$, ($n$ = 1) 5·40, (n = 2) 4·70, (n = 3) 3·94, (n = 4) 3·13, (n = 5) 2·36	67K

continued overleaf

770.2 $C_9H_{28}O_{15}N_3P_5$ (contd)

Metal	Method	Temp	Medium	Log of equilibrium constant, remarks	Ref
Mn^{2+}	gl	25	0·1 KCl	K_1 11·15 $K[Mn^{2+} + H_nL^{(10-n)-} \rightleftharpoons MnH_nL^{(8-n)-}]$, ($n$ = 1) 8·41, (n = 2) 6·31, (n = 3) 5·34, (n = 4) 4·64, (n = 5) 3·94, (n = 6) 2·64	67K
Zn^{2+}	gl	25	0·1 KCl	K_1 16·45 $K[Zn^{2+} + H_nL^{(10-n)-} \rightleftharpoons ZnH_nL^{(8-n)-}]$, ($n$ = 1) 13·36, (n = 2) 10·41, (n = 3) 8·44, (n = 4) 6·77, (n = 5) 5·23, (n = 6) 3·91	67K

67K M. I. Kabachnik, I. M. Dyatlova, T. A. Medved, Yu. F. Belugin, and V. V. Sidorenko, *Proc. Acad. Sci.* (*U.S.S.R.*), 1967, **175**, 621 (351)

771.1 $C_9H_6O_4NClS$ 7-Chloro-8-hydroxyquinoline-5-sulphonic acid H_2L

Metal	Method	Temp	Medium	Log of equilibrium constant, remarks	Ref
H^+	gl	25	0·1 50% dioxan	K_1 9·98, K_2 3·48	66B
Cu^{2+}	pol	25	0·1 50% dioxan	β_2 (?) 24·44	66B

66B H. Berge, *J. prakt. Chem.*, 1966, **34**, 15

771.2 $C_9H_6O_4NBrS$ 7-Bromo-8-hydroxyquinoline-5-sulphonic acid H_2L

Metal	Method	Temp	Medium	Log of equilibrium constant, remarks	Ref
H^+	gl	25	0·1 50% dioxan	K_1 8·67, K_2 2·68	66B
Cu^{2+}	pol	25	0·1 50% dioxan	β_2 (?) 22·41	66B

66B H. Berge, *J. prakt. Chem.*, 1966, **34**, 15

772 $C_9H_6O_4NIS$ 7-Iodo-8-hydroxyquinoline-5-sulphonic acid (Ferron) H_2L

Metal	Method	Temp	Medium	Log of equilibrium constant, remarks	Ref
H^+	gl	25	0·1 KCl	K_1 7·11, K_2 2·50	61L
	gl	25	0·1 50% dioxan	K_1 8·83, K_2 2·18	66B
	gl	28	0·1 KNO_3	K_1 7·05, K_2 2·50	67L
	gl	25	0·5 NaCl	K_1 6·90, K_2 2·22	67T
Al^{3+}	gl	25	0·1 KCl	K_1 7·6, K_2 7·1, K_3 5·6 $K[Al(OH)L_2^{2-} + H^+ \rightleftharpoons AlL_2^-]$ 5·0	61L

conti

$C_9H_6O_4NIS$ (contd) 772

Metal	Method	Temp	Medium	Log of equilibrium constant, remarks	Ref
Co^{2+}	gl	25	0·1 KCl	K_1 7·3, K_2 6·3, K_3 5·0	63S
	gl	28	0·1 KNO_3	K_1 6·70, K_2 4·17	67L
Cu^{2+}	pol	25	0·1 50% dioxan	β_2 (?) 24·15	66B
	gl	28	0·1 KNO_3	K_1 8·33, K_2 8·25	67L
Fe^{3+}	gl	25	0·1 KCl	K_1 8·9, K_2 8·4, K_3 7·9	61S
Ge^{IV}	sp	25	0·5 NaCl	$K[Ge(OH)_4 + 2HL \rightleftharpoons Ge(OH)_2L_2]$ 6·78	67T
Mn^{2+}	gl	25	0·1 KCl	K_1 5·3, K_2 4·3	63S
	gl	28	0·1 KNO_3	K_1 4·95, K_2 3·15	67L
Ni^{2+}	gl	25	0·1 KCl	K_1 8·2, K_2 7·0, K_3 5·6	63S
	gl	28	0·1 KNO_3	K_1 7·70, K_2 6·26	67L
Pd^{2+}	sp	25	~ 0	K (?) 9·05	67M
Zn^{2+}	gl	25	0·1 KCl	K_1 7·1, K_2 6·1	63S
	gl	28	0·1 KNO_3	K_1 7·25, K_2 6·15	67L
	gl	25	0·5 NaCl	K_1 6·83, K_2 5·85	67T
	sp	25	0·5 NaCl	K_1 6·87, K_2 6·22	67T

61L F. J. Langmyhr and A. R. Storm, *Acta Chem. Scand.*, 1961, **15**, 1461
61S A. R. Storm and F. J. Langmyhr, *Acta Chem. Scand.*, 1961, **15**, 1765
63S A. R. Storm, *Acta Chem. Scand.*, 1963, **17**, 667
66B H. Berge, *J. prakt. Chem.*, 1966, **34**, 15
67L P. Lingaiah, J. Mohan Rao, and U. V. Seshaiah, *Current Sci.*, 1967, **36**, 197
67M J. N. Mathur and S. N. Banerji, *J. Indian Chem. Soc.*, 1967, **44**, 513
67T J. Tsau, S. Matsouo, P. Clerc, and R. Benoit, *Bull. Soc. chim. France*, 1967, 1039

$C_{10}H_6O_3$ 2-Hydroxy-1,4-naphthaquinone (Lawsone) HL 779

Metal	Method	Temp	Medium	Log of equilibrium constant, remarks	Ref
H^+	gl	25	0·5 (NaCl)	K_1 4·02±0·04	66B
	gl	25	0·5 (NaCl)	K_1 3·97±0·04,	66B
	sp	25	var H_2SO_4	K_2 −5·6±0·2	66B
Cd^{2+}	pol	25	0·2	β_2 8·51, phosphate buffer	66S
Ge^{IV}	sp	25	0·5 (NaCl)	$K[Ge(OH)_4 + 2HL \rightleftharpoons Ge(OH)_2L_2] \leqslant 3·0$	66B
Zn^{2+}	pol	25	0·2	β_2 8·49, phosphate buffer	66S

66B A. Beauchamp and R. L. Benoit, *Canad. J. Chem.*, 1966, **44**, 1607, 1615
66S I. H. Suffet and W. C. Purdy, *J. Electroanalyt. Chem.*, 1966, **11**, 302

$C_{10}H_6O_3$ 5-Hydroxy-1,4-naphthaquinone HL 780

Erratum. 1964, p. 615, Table 780. Ref. 60K should *read* H. Kido, W. C. Fernelius, and C. G. Haas jun., *Analyt. Chim. Acta*, 1960, **23**, 116.

780.1 $C_{10}H_8O_2$ 2,3-Dihydroxynaphthalene H_2L

Metal	Method	Temp	Medium	Log of equilibrium constant, remarks	Ref
H^+	gl	20	0·1 (KNO_3)	K_1 12·5, K_2 8·68	67B
B^{III}	gl	20	0·1 (KNO_3)	$K[H_3BO_3+H_2L \rightleftharpoons B(OH)_2L^- + H^+]$ −4·13	68H
UO_2^{2+}	gl	20	0·1 (KNO_3)	K_1 15·0, K_2 10·8 $K(UO_2L + + H^+ \rightleftharpoons UO_2HL^+)$ 3·9 $K(UO_2HL^+ + H^+ \rightleftharpoons UO_2H_2L^{2+})$ 6·5	67B

67B M. Bartusek, *Coll. Czech. Chem. Comm.*, 1967, **32**, 757; *J. Inorg. Nuclear chem.*, 1967, **29**, 1089
68H L. Havelkova and M. Bartusek, *Coll. Czech. Chem. Comm.*, 1968, **33**, 4188

782.1 $C_{10}H_8O_4$ 4-Methyl-6,7-dihydroxybenzo[*b*]2-pyrone (4-methyl-esculetin) H_2L

Metal	Method	Temp	Medium	Log of equilibrium constant, remarks	Ref
H^+	gl	?	50% ethanol	K_1 7·90	63J
Mo^{VI}	sp	?	50% ethanol	$K(MoO_4^{2-} + 2H_2L \rightleftharpoons MoO_2L_2^{2-})$ 7·55	63J
Ti^{IV}	sp	20	0·4 ($NaClO_4$) 20% ethanol	β_3 10·7	64J

63J B. D. Jain and H. B. Singh, *Indian J. Chem.*, 1963, **1**, 369
64J B. D. Jain and H. B. Singh, *J. Indian Chem. Soc.*, 1964, **41**, 29

783 $C_{10}H_8N_2$ 2,2′-Bipyridyl L

Metal	Method	Temp	Medium	Log of equilibrium constant, remarks	Ref
H^+	gl	25	0·3 0, 50% dioxan	K_1 (0%) 4·50±0·02, (50%) 3·62±0·02	61J
	?	25	0·1 KNO_3	K_1 4·40	62C, 62C
	gl	25	0·1 KCl	K_1 4·47	62I
	cal	20	0·1 ($NaNO_3$)	ΔH_1 −3·66, ΔS_1 8·2	63A
	gl	30	1 KNO_3	K_1 4·62	65D
	cal	30	1 KNO_3	ΔH_1 −4·02	65D
	gl	20	0·1 (KNO_3)	K_1 4·59	66P
Ag^+	Ag	35	0·1 KNO_3	K_1 3·03, β_2 6·67	67L
Cd^{2+}	Hg(Cd)	25	0·05 KNO_3	K_1 4·26, K_2 3·56, K_3 2·66	58C
	gl	25	0·05 KNO_3	K_1 4·28, K_2 3·51, K_3 2·69	58C
	dis	25	0·1 KCl	K_1 4·12, K_2 3·50, K_3 2·60	62I
	cal	20	0·1 ($NaNO_3$)	ΔH_1 −5·1, ΔS_1 2·1, $\Delta H_{\beta 2}$ −9·4, $\Delta S_{\beta 2}$ 3·7, $\Delta H_{\beta 3}$ −14·0, $\Delta S_{\beta 3}$ 0·3	63A
	Ag	35	0·1 KNO_3	K_1 3·52, β_2 6·86, β_3 9·27	67L
Co^{2+}	dis	25	0·1 KCl	K_1 5·65, K_2 5·60, K_3 4·80	62I
	cal	20	0·1 ($NaNO_3$)	ΔH_1 −8·2, ΔS_1 −0·35, $\Delta H_{\beta 2}$ −15·2, $\Delta S_{\beta 2}$ 0·35, $\Delta H_{\beta 3}$ −21·3, $\Delta S_{\beta 3}$ 1·4	63A
	gl	30	1 KNO_3	K_1 5·72, K_2 5·41, K_3 4·80	65D
	cal	30	1 KNO_3	ΔH_1 −7·20, ΔS_1 2·5, $\Delta H_{\beta 2}$ −14·40, $\Delta S_{\beta 2}$ 3·5, $\Delta H_{\beta 3}$ −19·66, $\Delta S_{\beta 3}$ 8·1	65D
				See 2-acetyl-3-hydroxythiole (426.2), 2-(methylthiomethyl)pyridine (567.3), AMP-5 (858), and ATP (860) for mixed ligands.	

contin

Metal	Method	Temp	Medium	Log of equilibrium constant, remarks	Ref
Cu^+	Pt	25	0·3 (K_2SO_4)	β_2 13·18	61J
	pol	25	0·1 KNO_3 50% methanol	β_2 15·5	67P
				See Cu^{2+} below.	
Cu^{2+}	gl	25	0·3 (K_2SO_4)	K_1 8·15, K_2 5·50, K_3 3·30	61J
	Pt	25	0·3 (K_2SO_4)	K_2 ~ 5·6, K_3 3·20	61J
	Pt	25	0·3 (KNO_3) 50% dioxan	K_2 ~ 5·25, $K(CuL_2^{2+} + Cu^+ \rightleftharpoons CuL_2^+ + Cu^{2+})$ 0·0	61J
	Ag	12—40	0·05 KNO_3 41·5% ethanol	K_1 (12°) 8·44, (20°) 8·2±0·2, (25°) 8·06±0·05, (30°) 7·92, (36°) 7·78, (40°) 7·68, K_2 (12°) 5·38±0·04, (20°) 5·14±0·04, (25°) 5·04±0·04, (30°) 4·89±0·04, (36°) 4·77±0·04, (40°) 4·70±0·04, K_3 (12°) 3·38±0·06, (20°) 3·18±0·06, (25°) 3·13±0·06, (30°) 3·00±0·06, (36°) 2·91±0·06, (40°) 2·88±0·06 $\Delta H_2°$ −10·0±0·3, $\Delta S_2°$ −10±1, $\Delta H_3°$ −7·4±0·5, $\Delta S_3°$ −11±2	62C
		25	0·05 KNO_3 0—30% ethanol	K_1 (0%) 8·47±0·05, (20%) 8·30±0·05, (30%) 8·20±0·05, K_2 (0%) 5·58±0·05, (20%) 5·35±0·05, (30%) 5·25±0·05, K_3 (0%) 3·51±0·05, (20%) 3·34±0·05, (30%) 3·28±0·05	
	Hg(Cd)	25	?	K_2 5·52, K_3 3·40	62C
	dis	25	0·1 KCl	K_1 8·15, K_2 5·50, K_3 3·30	62I
	red	25	0·1 K_2SO_4	K_2 ~ 5·6, K_3 3·20	62I
	cal	20	0·1 ($NaNO_3$)	ΔH_1 −11·9, ΔS_1 −4·1, $\Delta H_{\beta 2}$ −17·3, $\Delta S_{\beta 2}$ 3·1, $\Delta H_{\beta 3}$ −23·8, $\Delta S_{\beta 3}$ −3·1	63A
	gl	30	1 KNO_3	K_1 8·39, K_2 5·63, K_3 3·63	65D
	cal	30	1 KNO_3	ΔH_1 −10·16, ΔS_1 4·9, $\Delta H_{\beta 2}$ −19·02, $\Delta S_{\beta 2}$ 1·5, $\Delta H_{\beta 3}$ −21·62, $\Delta S_{\beta 3}$ 9·5	65D
	gl	20	0·1 (KNO_3)	$K[Cu_2(OH)_2L_2^{2+} + 2H^+ \rightleftharpoons 2CuL^{2+}]$ 11·06	66P
	pol	25	0·1 KNO_3 50% methanol	β_3 18·0	67P
				See formic (101), acetic (114), propionic (147), benzoic (500), 4-nitrobenzoic (545.4), 3-chlorobenzoic (536.1), 4-toluic (608), hydroxyacetic (115), tetrahydrofuran-2-carboxylic (283.1), salicylic (506), 5-sulphosalicylic (552), and Chromotropic (817) acids, and glycine amide (132), catechol (360), Tiron (428), 2-(methylthiomethyl)pyridine (567.3), 2-acetyl-3-hydroxythiole (426.2), UTP (767.2), AMP-5 (858), AMP-5-*N*-oxide (858.1), ITP (859.1), ATP (860), GTP (861), and TTP (862.1) for mixed ligands.	
Fe^{2+}	dis	25	0·1 KCl	K_1 4·20, K_2 3·70, K_3 9·55	62I
	cal	20	0·1 ($NaNO_3$)	$\Delta H_{\beta 3}$ −31·35, $\Delta S_{\beta 3}$ −27·0	63A
	dis	30	1 KNO_3	K_1 4·65, β_3 17·14	65D
	cal	30	1 KNO_3	$\Delta H_{\beta 3}$ −28·0, $\Delta S_{\beta 3}$ −13·8	65D
	sp	?	0—50% N_2H_4	β_3 (0%) 17·9, (5%) 9·0, (10%) 8·2, (25%) 7·9, (50%) 6·8	66B
Mn^{2+}	dis	25	0·1 KCl	K_1 2·62, K_2 2·00, K_3 ~ 1·1	62I
	cal	20	0·1 ($NaNO_3$)	ΔH_1 −3·5, ΔS_1 0	63A
	dis	30	1 KNO_3	K_1 2·54, K_2 1·85, K_3 1·51	65D

continued overleaf

783 $C_{10}H_8N_2$ (contd)

Metal	*Method*	*Temp*	*Medium*	*Log of equilibrium constant, remarks*	*Ref*
Mn^{2+} *(contd)*	cal	30	1 KNO_3	ΔH_1 −5·73, ΔS_1 −7·3, $\Delta H_{\beta2}$ −6·11, $\Delta S_{\beta2}$ 0, $\Delta H_{\beta3}$ −6·23, $\Delta S_{\beta3}$ 6·5	65D
Ni^{2+}	dis	25	0·1 KCl	K_1 7·07, K_2 6·86, K_3 6·20	62I
	sp	20	0·05 KNO_3 41·5% ethanol	K_1 6·9±0·3, β_2 13·7±0·1, β_3 19·6±0·1	62C
	cal	20	0·1 ($NaNO_3$)	ΔH_1 −9·6, ΔS_1 0, $\Delta H_{\beta2}$ −19·0, $\Delta S_{\beta2}$ −0·7, $\Delta H_{\beta3}$ −28·2, $\Delta S_{\beta3}$ −2·1	63A
	dis	30	1 KNO_3	K_1 6·95, K_2 6·83, K_3 6·35	65D
	cal	30	1 KNO_3	ΔH_1 −8·90, ΔS_1 2·5, $\Delta H_{\beta2}$ −17·80, $\Delta S_{\beta2}$ 4·4, $\Delta H_{\beta3}$ −26·70, $\Delta S_{\beta3}$ 4·1	65D
				See 2-acetyl-3-hydroxythiole (426.2), 2-(methylthiomethyl)pyridine (567.3), and ATP (860) for mixed ligands.	
Tl^{3+}	red	25	1	K_1 9·40, β_2 16·10	61K
	dis	25	1 $NaNO_3$	K_2 5·5, K_3 4·89, β_3 20·05	62K
Zn^{2+}	Ag	10—40	0·1 KNO_3	K_1 (10°) 5·39±0·02, (20°) 5·20±0·02, (25°) 5·16±0·02, (32°) 5·02±0·02, (40°) 4·86±0·02, K_2 (10°) 4·74±0·02, (20°) 4·50±0·02, (25°) 4·46±0·02, (32°) 4·32±0·02, (40°) 4·24±0·02, K_3 (10°) 4·01±0·02, (20°) 3·83±0·02, (25°) 3·74±0·02, (32°) 3·71±0·02, (40°) 3·56±0·02	62Ca
				$\Delta H_1°$ −7·0±0·5, $\Delta S_1°$ 0±2, $\Delta H_2°$ −6·7±0·6, $\Delta S_2°$ −2±2, $\Delta H_3°$ −5·8±0·6, $\Delta S_3°$ −2±2	
			0·05 KNO_3 41·5% ethanol	K_1 (10°) 5·08±0·02, (17·5°) 5·00±0·02, (20°) 4·95±0·02, (25°) 4·88±0·02, (32·5°) 4·68±0·02, (40°) 4·54±0·02, K_2 (10°) 4·43±0·02, (17·5°) 4·22±0·02, (20°) 4·17±0·02, (25°) 4·08±0·02, (32·5°) 3·98±0·02, (40°) 3·86±0·02, K_3 (10°) 3·28±0·02, (17·5°) 3·18±0·02, (20°) 3·13±0·02, (25°) 3·05±0·02, (32·5°) 2·96±0·02, (40°) 2·91±0·02	
				$\Delta H_1°$ −7·6±0·5, $\Delta S_1°$ −3±2, $\Delta H_2°$ −7·4±0·4, $\Delta S_2°$ −6±1, $\Delta H_3°$ −5·2±0·4, $\Delta S_3°$ −4±2	
	dis	25	0·1 KCl	K_1 5·04, K_2 4·35, K_3 3·57	62I
	cal	20	0·1 ($NaNO_3$)	ΔH_1 −7·1, ΔS_1 0, $\Delta H_{\beta2}$ −12·5, $\Delta S_{\beta2}$ 2·4, $\Delta H_{\beta3}$ −17·5, $\Delta S_{\beta3}$ 2·7	63A
	gl	30	1 KNO_3	K_1 5·26, K_2 4·55, K_3 3·96	65D
	cal	30	1 KNO_3	ΔH_1 −6·25, ΔS_1 3·5, $\Delta H_{\beta2}$ −11·75, $\Delta S_{\beta2}$ 6·2, $\Delta H_{\beta3}$ −15·92, $\Delta S_{\beta3}$ 10·6	65D
				See formic (101), acetic (114), propionic (147), 4-nitrobenzoic (545.4), 3-chlorobenzoic (536.1), benzoic (500), 4-toluic (608), hydroxyacetic (115), tetrahydrofuran-2-carboxylic (283.1) acids, 2-acetyl-3-hydroxythiole (426.2), 2-(methylthiomethyl)pyridine (567.3), AMP-5 (858), and ATP (860) for mixed ligands.	

Erratum. 1964, p. 616, Table 783. Under Ag^+, *for* 3·70 (20°) *read* 3·70 (25°).

58G S. Cabani and E. Scrocco, *J. Inorg. Nuclear Chem.*, 1958, **8**, 332
61J B. R. James and R. J. P. Williams, *J. Chem. Soc.*, 1961, 2007
61K F. Ya. Kul'ba, Yu. A. Makashev, and V. E. Mironov, *Russ. J. Inorg. Chem.*, 1961, **6**, 321 (630)
62C S. Cabani, G. Moretti, and E. Scrocco, *J. Chem. Soc.*, 1962, 88
62Ca S. Cabani and M. Landucci, *J. Chem. Soc.*, 1962, 278
62K F. Ya. Kul'ba, Yu. A. Makashev, B. D. Guller, and G. V. Kiselev, *Russ. J. Inorg. Chem.*, 1962, **7**, 351 (689)

contd

62I H. Irving and D. H. Mellor, *J. Chem. Soc.*, 1962, 5222
63A G. Anderegg, *Helv. Chim. Acta*, 1963, **46**, 2813
65D R. L. Davies and K. W. Dunning, *J. Chem. Soc.*, 1965, 4168
66B D. Bauer, *Bull. Soc. chim. France*, 1966, 2631
66P D. D. Perrin and V. S. Sharma, *J. Inorg. Nuclear Chem.*, 1966, **28**, 1271
67L C. Luca, *Bull. Soc. chim. France*, 1967, 2556
67P F. Pantani, *Ricerca sci.*, 1967, **37**, 33

$C_{10}H_9N$ 1-Aminonaphthalene L 783.1

Metal	*Method*	*Temp*	*Medium*	*Log of equilibrium constant, remarks*	*Ref*
Ag^+	Ag	25	ethanol	K_1 2·20, β_2 2·76	60A

60A V. Armeanu and C. Luca, *Z. phys. Chem.* (*Leipzig*), 1960, **214**, 81

$C_{10}H_9N_5$ 1-(Pyrimidin-2′-yl)-3-pyridyl-1,2-diazaprop-2-ene HL 784.1 (pyridine-2-aldehyde-2′-pyrimidylhydrazone)

Metal	*Method*	*Temp*	*Medium*	*Log of equilibrium constant, remarks*	*Ref*
H^+	sp	25	?	K_1 13·03±0·01, K_3 1·10±0·06	64G
	gl	25	?	K_2 4·48±0·005	64G
Fe^{2+}	gl	25	?	β_2 30	64G
				$K[FeHL_2^+ + H^+ \rightleftharpoons Fe(HL)_2^{2+}]$ 4·56±0·02	
				$K[FeL_2 + H^+ \rightleftharpoons FeHL_2^+)$ 6·09±0·02	
	sp	25	?	$K(Fe^{2+} + HL \rightleftharpoons FeHL^{2+})$ 6·0±0·3	64G
				$K[Fe^{2+} + 2HL \rightleftharpoons Fe(HL)_2^{2+}]$ 14·00±0·01	

64G R. W. Green, P. S. Hallman, and F. Lions, *Inorg. Chem.*, 1964, **3**, 1541

$C_{10}H_{10}O_2$ 1-Phenylbutane-1,3-dione (benzoylacetone) HL 785

Metal	*Method*	*Temp*	*Medium*	*Log of equilibrium constant, remarks*	*Ref*
H^+	pH	25	1	K_1 8·24±0·01	61P
	gl	25	0·1 $(C_4H_9)_4NClO_4$ 100% methanol	K_1 12·02±0·05	65L
			0·1 $(C_4H_9)_4NClO_4$ 100% ethanol	K_1 12·02±0·05	
	gl	24	0·1 (NaCl) 80% methanol	K_1 9·53±0·01	66Y, 67D
	gl	22	0·1 (NaCl) 100% methanol	K_1 11·95±0·01	67Z

continued overleaf

785 $C_{10}H_{10}O_2$ (contd)

Metal	*Method*	*Temp*	*Medium*	*Log of equilibrium constant, remarks*	*Ref*
Cd^{2+}	dis	?	0·1 $NaClO_4$	K_1 3·96, β_2 4·0	60S
Dy^{3+}	gl	25	0·1 (NaCl) 80% methanol	K_1 8·36±0·02, K_2 6·47±0·01, K_3 4·45±0·01	67D
Er^{3+}	gl	25	0·1 (NaCl) 80% methanol	K_1 8·47±0·02, K_2 6·55±0·01, K_3 4·55±0·01	67D
Eu^{3+}	gl	25	0·1 (NaCl) 80% methanol	K_1 8·17±0·02, K_2 6·30±0·02, K_3 4·36±0·02	67D
	gl	22	0·1 (NaCl) 100% methanol	K_1 11·1±0·2, K_2 8·7±0·1, K_3 4·6±0·1, K_4 2·9±0·1	67Z
Gd^{3+}	gl	25	0·1 (NaCl) 80% methanol	K_1 8·09±0·02, K_2 6·21±0·02, K_3 4·18±0·02	67D
Ho^{3+}	gl	25	0·1 (NaCl) 80% methanol	K_1 8·41±0·02, K_2 6·50±0·01, K_3 4·50±0·01	67D
In^{3+}	dis	?	0·1	K_1 8·4, β_2 15·5, β_3 20·8	60S
K^+	gl	25	0·1 KI 100% methanol	K_1 1·2±0·2	65L
			0·1 KSCN 100% ethanol	K_1 2·4±0·2	
La^{3+}	dis	?	0·1	K_1 5·2, β_2 9·3, β_3 12·1	60S
	gl	30	0·1 ($NaClO_4$) 67% acetone	K_1 6·24, K_2 5·21, K_3 4·76	64D
	gl	25	0·1 (NaCl) 80% methanol	K_1 7·02±0·01, K_2 5·25±0·01, K_3 3·36±0·01	67D
Li^+	gl	25	0·1 $LiClO_4$ 100% methanol	K_1 3·1±0·2	65L
			0·1 $LiClO_4$ 100% ethanol	K_1 3·2±0·2	
Lu^{3+}	gl	25	0·1 (NaCl) 80% methanol	K_1 8·37±0·02, K_2 6·44±0·01, K_3 4·45±0·02	67D
Na^+	gl	25	0·1 $NaClO_4$ 100% methanol	K_1 1·8±0·2	65L
			0·1 $NaClO_4$ 100% ethanol	K_1 3·2±0·2	
Nd^{3+}	gl	30	0·1 ($NaClO_4$) 67% acetone	K_1 6·94, K_2 6·61, K_3 5·58	64D
	gl	25	0·1 (NaCl) 80% methanol	K_1 7·83±0·02, K_2 6·13±0·01, K_3 4·13±0·02	67D
	gl	22	0·1 (NaCl) 100% methanol	K_1 10·8±0·2, K_2 8·5±0·1, K_3 4·4±0·1, K_4 2·6±0·1	67Z
Pr^{3+}	gl	30	0·1 ($NaClO_4$) 67% acetone	K_1 6·84, K_2 6·25, K_3 5·21	64D
	gl	23	0·1 NaCl 80% methanol	K_1 7·76±0·01, K_2 6·06±0·01, K_3 4·06±0·01, K_4 2·15	66Y
	gl	22	0·1 (NaCl) 100% methanol	K_1 10·7±0·2, K_2 8·4±0·1, K_3 4·4±0·1, K_4 2·6±0·1	67Z
Sc^{3+}	sp	18—20	0·01 (HNO_3)	K_1 8·8±0·2	66K
Sm^{3+}	gl	25	0·1 NaCl 80% methanol	K_1 8·03±0·02, K_2 6·28±0·02, K_3 4·30±0·02	67D
Tb^{3+}	gl	25	0·1 NaCl 80% methanol	K_1 8·27±0·02, K_2 6·44±0·02, K_3 4·41±0·02	67D
UO_2^{2+}	dis	20	0·1 $NaClO_4$	K_1 7·2 $K(UO_2^{2+}+L^-+OH^- \rightleftharpoons UO_2LOH)$ 15·9 $K[UO_2^{2+}+L^-+2OH^- \rightleftharpoons UO_2L(OH)_2^-]$ 24·1	60S

conti

$C_{10}H_{10}O_2$ (contd) 785

Metal	Method	Temp	Medium	Log of equilibrium constant, remarks	Ref
$^{3+}$	dis	?	0·1	K_1 6·55, β_2 11·4, β_3 14·4	60S
	gl	30	0·1 ($NaClO_4$) 67% acetone	K_1 8·21, K_2 6·68, K_3 5·68	64D
	gl	25	0·1 NaCl 80% methanol	K_1 8·29±0·02, K_2 6·35±0·01, K_3 4·42±0·01	67D
b^{3+}	gl	25	0·1 NaCl 80% methanol	K_1 8·44±0·01, K_2 6·49±0·01, K_3 4·46±0·02	67D
r^{4+}	dis	25	1	K_1 12·71±0·04, K_2 11·86±0·02, K_3 11·34±0·04, K_4 11·08±0·04	61P

60S J. Stary, *Coll. Czech. Chem. Comm.*, 1960, **25**, 86, 890
61P V. M. Peshkova, N. V. Mel'chakova, and S. G. Zhemchuzin, *Russ. J. Inorg. Chem.*, 1961, **6**, 630 (1233)
64D N. K. Dutt and P. Bandyopadhyay, *J. Inorg. Nuclear Chem.*, 1964, **26**, 729
65L D. C. Luehrs, R. T. Iwamoto, and J. Kleinberg, *Inorg. Chem.*, 1965, **4**, 1739
66K I. M. Korenman and N. V. Zaglyadimova, *Russ. J. Inorg. Chem.*, 1966, **11**, 1491 (2774)
66Y K. B. Yatsimirskii, N. K. Davidenko, and L. N. Lugine, *Proc. Acad. Sci.* (*U.S.S.R.*), 1966, **170**, 954 (864)
67D N. K. Davidenko and A. A. Zholdakov, *Russ. J. Inorg. Chem.*, 1967, **12**, 633 (1195)
67Z A. A. Zholdakov and N. K. Davidenko, *Russ. J. Inorg. Chem.*, 1967, **12**, 1622 (3066)

$C_{10}H_{10}O_6$ Phenyl-1,2-bis(oxyacetic acid) H_2L 787.1

Metal	Method	Temp	Medium	Log of equilibrium constant, remarks	Ref
H^+	gl	25	0·1 ($NaClO_4$)	K_1 3·45, K_2 2·40	68S
Ba^{2+}	gl	25	0·1 ($NaClO_4$)	K_1 2·0	68S
Ca^{2+}	gl	25	0·1 ($NaClO_4$)	K_1 3·1	68S
Cd^{2+}	gl	25	0·1 ($NaClO_4$)	K_1 3·8	68S
Co^{2+}	gl	25	0·1 ($NaClO_4$)	K_1 1·1	68S
Cu^{2+}	gl	25	0·1 ($NaClO_4$)	K_1 3·3	68S
Mg^{2+}	gl	25	0·1 ($NaClO_4$)	$K_1 < 1·5$	68S
Mn^{2+}	gl	25	0·1 ($NaClO_4$)	K_1 2·8	68S
Ni^{2+}	gl	25	0·1 ($NaClO_4$)	K_1 1·6	68S
Sr^{2+}	gl	25	0·1 ($NaClO_4$)	K_1 2·3	68S
Zn^{2+}	gl	25	0·1 ($NaClO_4$)	K_1 2·0	68S

68S K. Suzuki, T. Mattori, and K. Yamasaki, *J. Inorg. Nuclear Chem.*, 1968, **30**, 161

$C_{10}H_{12}O_2$ 3-Isopropyltropolone HL 791

Metal	Method	Temp	Medium	Log of equilibrium constant, remarks	Ref
H^+	gl	?	0·5 (KNO_3) 50% ethanol	K_1 7·42	65D
	gl	?	0·1 $NaClO_4$ 40% dioxan	K_1 7·55	66S
	gl	23	0·1 NaCl 80% methanol	K_1 7·85	66Y
	sp	?	0·2 $NaClO_4$ 3% ethanol	K_1 7·01	67G

continued overleaf

791 $C_{10}H_{12}O_2$ (contd)

Metal	*Method*	*Temp*	*Medium*	*Log of equilibrium constant, remarks*	*Ref*
Ce^{3+}	sp	?	0·2 $NaClO_4$ 3% ethanol	K_1 6·53	67G
Dy^{3+}	sp	?	0·2 $NaClO_4$ 3% ethanol	K_1 7·33	67G
Er^{3+}	sp	?	0·2 $NaClO_4$ 3% ethanol	K_1 7·49	67G
Fe^{3+}	sp	?	0·1 $NaClO_4$ 40% dioxan	K_1 11·80, K_2 10·96, K_3 9·54	66S
Gd^{3+}	sp	?	0·2 $NaClO_4$ 3% ethanol	K_1 7·15	67G
Ho^{3+}	sp	?	0·2 $NaClO_4$ 3% ethanol	K_1 7·40	67G
La^{3+}	sp	?	0·2 $NaClO_4$ 3% ethanol	K_1 6·29	67G
Lu^{3+}	sp	?	0·2 $NaClO_4$ 3% ethanol	K_1 7·95	67G
Nd^{3+}	sp	?	0·2 $NaClO_4$ 3% ethanol	K_1 6·70	67G
Pr^{3+}	sp	?	0·2 $NaClO_4$ 3% ethanol	K_1 6·74	67G
	gl	23	0·1 KCl 80% methanol	K_1 8·31, K_2 6·88, K_3 5·48, K_4 4·20	66Y
Sm^{3+}	sp	?	0·2 $NaClO_4$ 3% ethanol	K_1 7·04	67G
UO_2^{2+}	sp, gl	?	0·5 (KNO_3) 50% ethanol	K_1 (sp) 9·62, K_2 (sp) 6·92, (gl) 6·93	65D
Y^{3+}	sp	?	0·2 $NaClO_4$ 3% ethanol	K_1 7·28	67G
Yb^{3+}	sp	?	0·2 $NaClO_4$ 3% ethanol	K_1 7·62	67G

65D Y. Dutt and R. P. Singh, *J. Indian Chem. Soc.*, 1965, **42**, 767

66S R. P. Singh and Y. Dutt, *Indian J. Chem.*, 1966, **4**, 214

66Y K. B. Yatsimirskii, N. K. Davidenko, and L. N. Lugina, *Proc. Acad. Sci.* (*U.S.S.R.*), 1966, **170**, **954** (**864**)

67G B. P. Gupta, Y. Dutt, and R. P. Singh, *Indian J. Chem.*, 1967, **5**, 214

792 $C_{10}H_{12}O_2$ 4-Isopropyltropolone HL

Metal	*Method*	*Temp*	*Medium*	*Log of equilibrium constant, remarks*	*Ref*
H^+	sp	25	2·00 $NaClO_4$	K_1 6·72	64O
Fe^{3+}	sp	25	2·00 $NaClO_4$	K_1 11·55	64O

64O Y. Oka, M. Yanai, and C. Suzuki, *Nippon Kagaku Zasshi*, 1964, **85**, 873

$C_{10}H_{12}O_2$ 5-Isopropyltropolone HL 792.1

Metal	Method	Temp	Medium	Log of equilibrium constant, remarks	Ref
H^+	sp	25	2·00 $NaClO_4$	K_1 6·86	62O
Fe^{3+}	sp	25	2·00 $NaClO_4$	K_1 10·64	62O

62O Y. Oka, M. Umehara, and T. Nozoe, *Nippon Kagaku Zasshi*, 1962, **83**, 1197

$C_{10}H_{12}O_3$ 2-Hydroxy-3-(2′-propyl) benzoic acid (3-isopropylsalicylic acid) H_2L 793.1

Metal	Method	Temp	Medium	Log of equilibrium constant, remarks	Ref
H^+	gl	25	0·1 ($NaClO_4$)	K_2 2·76±0·02	66P
Fe^{3+}	gl, Pt	25	0·1 ($NaClO_4$)	$K(Fe^{3+} + HL^- \rightleftharpoons FeL^+ + H^+)$ 2·56±0·02	66P

66P M. V. Park, *J. Chem. Soc.* (*A*), 1966, 816; *Nature*, 1963, **197**, 283

$C_{10}H_{12}O_5$ 3,4,5-Trihydroxybenzoic acid propyl ester H_3L 794.1

Metal	Method	Temp	Medium	Log of equilibrium constant, remarks	Ref
H^+	sp	21—23	5% 1-propanol	K_1 7·23	68L
Nb^V	sp	21—23	5% 1-propanol	K (?) 3·48, carbonate buffer	68L

68L J. E. Land and C. M. Stillwell, *J. Less-Common Metals*, 1968, **14**, 231

$C_{10}H_{13}N_3$ 2-(*N*-Ethylaminomethyl)benzimidazole L 794.2

Metal	Method	Temp	Medium	Log of equilibrium constant, remarks	Ref
H^+	gl	15—45	50% dioxan	K_1 (15°) 7·49, (25°) 7·33, (35°) 7·15, (45°) 6·99, K_2 (15—45°) 2	63L
Cu^{2+}	gl	15—45	50% dioxan	K_1 (15°) 7·20, (25°) 6·98, (35°) 6·85, (45°) 6·60, K_2 (15°) 4·73, (25°) 4·58, (35°) 4·38, (45°) 4·27 ΔH_1 −12·8	63L

63L T. J. Lane and H. B. Durham, *Inorg. Chem.*, 1963, **2**, 632

795.1 $C_{10}H_{15}N$ *NN*-Diethylaniline L

Metal	*Method*	*Temp*	*Medium*	*Log of equilibrium constant, remarks*	*Ref*
Ag^+	Ag	25	< 0·01 96% ethanol	K_1 1·28	61A

61A V. Armeanu and C. Luca, *Z. phys Chem. (Leipzig)*, 1961, 217, 389

795.2 $C_{10}H_{16}N_6$ *NN'*-Bis[4'-(5')-imidazolylmethyl]-1,2-diaminoethane L

Metal	*Method*	*Temp*	*Medium*	*Log of equilibrium constant, remarks*	*Ref*
H^+	gl	25	0·1 (KCl)	K_1 9·05, K_2 6·56, K_3 4·26, K_4 3·21	68G
Co^{2+}	gl	25	0·1 (KCl)	K_1 11·43	68G
Cu^{2+}	gl	25	0·1 (KCl)	K_1 16·5	68G
Ni^{2+}	gl	25	0·1 (KCl)	K_1 14·02	68G
Zn^{2+}	gl	25	0·1 (KCl)	K_1 10·39	68G

68G D. W. Gruenwedel, *Inorg. Chem.*, 1968, 7, 495

802.1 $C_{10}H_{22}N^+$ *NNN*-Triethylbut-3-enylammonium cation L^+

Metal	*Method*	*Temp*	*Medium*	*Log of equilibrium constant, remarks*	*Ref*
Pt^{2+}	sp	30—60	2 (NaCl)	$K(PtCl_4^{2-} + L^+ \rightleftharpoons PtCl_3L + Cl^-)$ (30°) 3·89, (44·8°) 3·77, (60°) 3·65 ΔH $-3·8 \pm 0·4$, ΔS 5 ± 1	67D

67D R. G. Denning, F. R. Hartley, and L. M. Venanzi, *J. Chem. Soc. (A)*, 1967, 324

802.2 $C_{10}H_{22}N_2$ 1-(Cyclohexylamino)-2-methyl-2-propylamine L

Metal	*Method*	*Temp*	*Medium*	*Log of equilibrium constant, remarks*	*Ref*
H^+	gl	20	0·1 (KCl)	K_1 10·18, K_2 6·77	65T
Cu^{2+}	gl	20	0·1 (KCl)	K_1 9·01, K_2 7·14	65T
Ni^{2+}	gl	20	0·1 (KCl)	$K_1 \sim 0$, $K_2 \sim 9·6$, $K_3 \sim 3·9$	65T

65T M. L. Tomlinson, M. L. R. Sharp, and H. M. N. H. Irving, *J. Chem. Soc.*, 1965, 603

$C_{10}H_{22}As^+$ *As,As,As*-Triethylbut-3-enylarsinium cation L⁺ 802.3

Metal	Method	Temp	Medium	Log of equilibrium constant, remarks	Ref
Pt^{2+}	sp	30—60	2(NaCl)	$K(PtCl_4^{2-}+L^+ \rightleftharpoons PtCl_3L+Cl^-)$ (30°) 3·95, (44·8°) 3·85, (60°) 3·74 ΔH −3·3±0·3, ΔS 7±1	67D

67D R. G. Denning, F. R. Hartley, and L. M. Venanzi, *J. Chem. Soc. (A)*, 1967, 324

$C_{10}H_{24}N_4$ 1,9-Diamino-4,6,6-trimethyl-3,7-diazanon-3-ene L 803.1

Metal	Method	Temp	Medium	Log of equilibrium constant, remarks	Ref
Ni^{2+}				See oxalic acid (111) for mixed ligands.	

$C_{10}H_{28}N_6$ 2,2′,2″,2‴-(Ethylenedinitrito)tetrakis(ethylamine) L 805

Metal	Method	Temp	Medium	Log of equilibrium constant, remarks	Ref
H^+	cal	25	?	ΔH_1 −11·30, ΔS_1 8·1, ΔH_2 −11·45, ΔS_2 5·3, ΔH_3 −13·15, ΔS_3 −3·1, ΔH_4 −12·00, ΔS_4 −1·8, ΔH_5 −4·50, ΔS_5 −9·0	63P
Co^{2+}	var	25	0·1 KCl	K_1 11·55 (?) (calculated) ΔH_1 −14·75±0·06 (cal), ΔS_1 21·5±0·4 (cal) $K(Co^{2+}+L+H^+ \rightleftharpoons CoLH^{3+})$ 12·40 (calculated) ΔH −14·00±0·09 (cal), ΔS 10·0±0·5 (cal)	64S
Cu^{2+}	var	25	0·1 KCl	K_1 22·15 (calculated) ΔH_1 −24·5±0·1 (cal), ΔS_1 19±1 (cal) $K(Cu^{2+}+L+H^+ \rightleftharpoons CuLH^{3+})$ 20·15 (calculated) ΔH −24·80±0·03 (cal), ΔS 9±1 (cal)	64S
Fe^{2+}	var	25	0·1 KCl	K_1 11·05 (calculated) ΔH_1 −9·65±0·02 (cal), ΔS_1 19±1 (cal)	64S
Mn^{2+}	var	25	0·1 KCl	K_1 9·30 (calculated) ΔH_1 −8·85±0·00 (cal), ΔS_1 12·5±0·2 (cal)	64S
Ni^{2+}	var	25	0·1 KCl	K_1 19·05 (calculated) ΔH_1 −19·7±0·2 (cal), ΔS_1 22±2 (cal) $K(Ni^{2+}+L+H^+ \rightleftharpoons NiLH^{3+})$ 15·65 (calculated) ΔH −18·35±0·10 (cal), ΔS 10±1 (cal)	64S
Zn^{2+}	var	25	0·1 KCl	K_1 16·05 (calculated) ΔH_1 −14·50±0·02 (cal), ΔS 25·0±0·3 (cal) $K(Zn^{2+}+L+H^+ \rightleftharpoons ZnLH^{3+})$ 14·00 (calculated) ΔH −14·65±0·07 (cal), ΔS 15·0±0·5 (cal)	64S

63P P. Paoletti and M. Ciampolini, *Ricerca sci.*, 1963, 33 (II-A), 405
64S L. Sacconi, P. Paoletti, and M. Ciampolini, *J. Chem. Soc.*, 1964, 5046

805.1 $C_{10}H_{28}N_6$ 1,4,7,10,13,16-Hexa-azahexadecane (pentaethylenehexamine) L

Metal	Method	Temp	Medium	Log of equilibrium constant, remarks	Ref
Cd^{2+}	pol	25	0·5	K_1 19, phosphate buffer	62J
Pb^{2+}	pol	25	1 NH_3	K_1 11·0	62J

62J E. Jacobsen and K. Schroder, *Acta Chem. Scand.*, 1962, **16,** 1393

808 $C_{10}H_7O_2N$ 1-Nitroso-2-naphthol HL

Metal	Method	Temp	Medium	Log of equilibrium constant, remarks	Ref
H^+				H^+ constants not given.	
Co^{3+}	sol	20	var	β_3 46·9±0·6	64A
Cu^{2+}	pol	20	~ 0·3 50% dioxan	K_1 10·27, acetate buffer	67C
Ni^{2+}	pol	20	~ 0·35 50% dioxan	K_1 8·56, acetate buffer	67C
Ru^{3+}	sp	?	0·2 30% ethanol	K_1 10·2, β_3 24·2 citric acid buffer	63K
	sp	?	0·2 30% ethanol	$K[Ru(NO)^{3+} + L^- \rightleftharpoons Ru(NO)L^{2+}]$ 11·8 $K[Ru(NO)^{3+} + 2L^- \rightleftharpoons Ru(NO)L_2^{+}]$ 21·2	64K

63K C. Konecny, *Analyt. Chim. Acta*, 1963, **29,** 423
64A E. K. Astakhova, V. M. Savostina, and V. M. Peshkova, *Russ. J. Inorg. Chem.*, 1964, **9,** 452 (817)
64K C. Konecny, *Analyt. Chim. Acta*, 1964, **31,** 352
67C E. Casassas and L. Eek, *J. Chim. phys.*, 1967, **64,** 971

809 $C_{10}H_7O_2N$ 2-Nitroso-1-naphthol HL

Metal	Method	Temp	Medium	Log of equilibrium constant, remarks	Ref
H^+				H^+ constants not given.	
Ru^{3+}	sp	?	0·2 30% ethanol	K_1 10·0, β_3 24·0 citric acid buffer	63K
	sp	?	0·2 30% ethanol	$K[Ru(NO)^{3+} + L^- \rightleftharpoons Ru(NO)L^{2+}]$ 11·8 $K[Ru(NO)^{3+} + 2L^- \rightleftharpoons Ru(NO)L_2^{+}]$ 20·5	64K

63K C. Konecny, *Analyt. Chim. Acta*, 1963, **29,** 423
64K C. Konecny, *Analyt. Chim. Acta*, 1964, **31,** 352

810 $C_{10}H_7O_2N$ Quinoline-2-carboxylic acid HL

Metal	Method	Temp	Medium	Log of equilibrium constant, remarks	Ref
H^+	gl	25	?	K_1 4·96	58B
Fe^{2+}	sp	25	0·2 KNO_3	β_2 5·44	58B

58B B. Banerjee and P. Ray, *J. Indian Chem. Soc.*, 1958, **35,** 297

$C_{10}H_7O_3N$ 4-Hydroxyquinoline-2-carboxylic acid H_2L 812.1

Metal	Method	Temp	Medium	Log of equilibrium constant, remarks	Ref
H^+	sp	25	~ 0·005 50% dioxan	K_1 13·9±0·2	64B
	gl	25	~ 0·005 50% dioxan	K_2 3·65±0·01, K_3 < 2	64B
Co^{2+}	gl	25	~ 0·01 50% dioxan	K_1 3·3, K_2 2·9 $K(CoOHL^- + H^+ \rightleftharpoons CoL)$ 7·3 $K[Co(OH)_2L^{2-} + H^+ \rightleftharpoons CoOHL^-]$ 9·0	64B
Cu^{2+}	gl	25	~ 0·01 50% dioxan	$K(CuOHL^- + H^+ \rightleftharpoons CuL)$ 6·2 $K[Cu(OH)_2L^{2-} + H^+ \rightleftharpoons CuOHL^-]$ 7·3	64B
Ni^{2+}	gl	25	~ 0·01 50% dioxan	K_1 3·5, K_2 2·7 $K(NiOHL^- + H^+ \rightleftharpoons NiL)$ 7·2 $K[Ni(OH)_2L^{2-} + H^+ \rightleftharpoons NiOHL^-]$ 9·5	64B
Zn^{2+}	gl	25	~ 0·01 50% dioxan	K_1 3·4, K_2 3·1 $K(ZnOHL^- + H^+ \rightleftharpoons ZnL)$ 7·0 $K[Zn(OH)_2L^{2-} + H^+ \rightleftharpoons ZnOHL^-]$ 8·1	64B

64B S. P. Bag, Q. Fernando, and H. Freiser, *Inorg. Chem.*, 1964, 3, 93

$C_{10}H_7O_4N$ 4,8-Dihydroxyquinoline-2-carboxylic acid H_3L 812.2

Metal	Method	Temp	Medium	Log of equilibrium constant, remarks	Ref
H^+	sp	25	~ 0·02 50% dioxan	K_1 14·5±0·2	64B
	gl	25	0·005—0·01 50% dioxan	K_2 (0·01) 10·05±0·1, K_3 (0·005) 3·55±0·01, K_4 < 2	64B
Co^{2+}	gl	25	~ 0·01 50% dioxan	K_1 6·7, K_2 5·6 $K(CoOHL^- + H^+ \rightleftharpoons CoL)$ 9·9 $K[Co(OH)_2L^{2-} + H^+ \rightleftharpoons CoOHL^-]$ 11·7	64B
Cu^{2+}	gl	25	~ 0·01 50% dioxan	K_1 9·4, K_2 7·2 $K(CuOHL^- + H^+ \rightleftharpoons CuL)$ 9·5 $K[Cu(OH)_2L^{2-} + H^+ \rightleftharpoons CuOHL^-]$ 11·0	64B
Mn^{2+}	gl	25	~ 0·01 50% dioxan	K_1 5·5, K_2 5·0 $K(MnOHL^- + H^+ \rightleftharpoons MnL)$ 10·7	64B
Ni^{2+}	gl	25	~ 0·01 50% dioxan	K_1 7·1, K_2 5·5 $K(NiOHL^- + H^+ \rightleftharpoons NiL)$ 10·0 $K[Ni(OH)_2L^{2-} + H^+ \rightleftharpoons NiOHL^-]$ 11·8	64B
Zn^{2+}	gl	25	~ 0·01 50% dioxan	K_1 7·7, K_2 6·0 $K(ZnOHL^- + H^+ \rightleftharpoons ZnL)$ 9·7 $K[Zn(OH)_2L^{2-} + H^+ \rightleftharpoons ZnOHL^-]$ 11·3	64B

64B S. P. Bag, Q. Fernando, and H. Freiser, *Inorg. Chem.*, 1964, 3, 93

813 $C_{10}H_8O_2N_2$ 8-Hydroxyquinoline-2-carboxaldehyde oxime HL

Metal	Method	Temp	Medium	Log of equilibrium constant, remarks	Ref
H^+	gl	25	0·1 50% dioxan	K_1 10·72±0·04, K_2 < 3	67S
Co^{2+}	gl	25	0·1 50% dioxan	K_1 7·83, K_2 7·71	67S
Cu^{2+}	gl	25	0·1 50% dioxan	K_1 > 9	67S
Mn^{2+}	gl	25	0·1 50% dioxan	K_1 5·83, K_2 5·8	67S
Ni^{2+}	gl	25	0·1 50% dioxan	K_1 8·47, K_2 7·5	67S
Zn^{2+}	gl	25	0·1 50% dioxan	K_1 8·09, K_2 7·66	67S

67S R. L. Stevenson and H. Freiser, *Analyt. Chem.*, 1967, **39**, 1354

815 $C_{10}H_8N_4O_2$ 1,2-Di(2′-furyl)ethane-1,2-dione dioxime (2-furil dioxine) HL

Metal	Method	Temp	Medium	Log of equilibrium constant, remarks	Ref
Ni^{2+}	dis	25	0·1 ($NaCO_4$)	K_1 8·18±0·03, K_2 6·67±0·01	64S

64S V. M. Savostina, E K. Astalakova, and V. M. Peshkova, *Russ. J. Inorg. Chem.*, 1964, **9**, 42(80)

816 $C_{10}H_8O_5S$ 2,3-Dihydroxynaphthalene-6-sulphonic acid H_3L

Metal	Method	Temp	Medium	Log of equilibrium constant, remarks	Ref
H^+	gl	18—22	0·1	K_1 12·16±0·08, K_2 8·19±0·02	63S, 65
	gl	25	0·1	K_1 11·85, K_2 8·09	65O
B^{III}	gl	25	0·1 (KNO_3)	$K[H_3BO_3+H_2L^- \rightleftharpoons B(OH)_2L^{2-}+H^+]$ −3·98	68H
Cu^{2+}	sp	25	0·1	K_1 13·13, K_2 10·33	65O
Ge^{IV}	sp	25	0·1 KCl	$K(H_2GeO_3+3H_2L^- \rightleftharpoons GeL_3^{5-}+2H^+)$ 2·0	67P
Ti^{IV}	gl	18—22	0·1	$K(TiO^{2+}+2L^{3-} \rightleftharpoons TiOL_2^{4-})$ 38·1 $K(TiO^{2+}+3L^{3-} \rightleftharpoons TiOL_3^{7-})$ 54·7 $K(TiO^{2+}+2H^++3L^{3-} \rightleftharpoons TiL_3^{5-})$ 56·5 $K[TiO^{2+}+2H_2L^- \rightleftharpoons TiO(HL)_2^{2-}+2H^+]$ −0·69	63S
UO_2^{2+}	sp	18—22	0·1 ($NaClO_4$)	K_1 15·6, K_2 10·6, $K(UO_2^{2+}+HL^{2-} \rightleftharpoons UO_2HL)$ 6·2 $K(UO_2L^-+HL^{2-} \rightleftharpoons UO_2HL_2^{3-})$ 4·2	65S, 65B
	gl	20	0·1 (KNO_3)	K_1 15·5, K_2 10·65, $K(UO_2^{2+}+HL^{2-} \rightleftharpoons UO_2HL)$ 5·6 $K(UO_2L^-+HL^{2-} \rightleftharpoons UO_2HL_2^{3-})$ 4·2	65B, 66B

63S L. Sommer, *Coll. Czech. Chem. Comm.*, 1963, **28**, 3057
65B M. Bartusek and L. Sommer, *J. Inorg. Nuclear Chem.*, 1965, **27**, 2397
65O Y. Oka, N. Nakazawa, and H. Harata, *Nippon Kagaku Zasshi*, 1965, **86**, 1162
65S L. Sommer, T. Sepel, and L. Kurilova, *Coll. Czech. Chem. Comm.*, 1965, **30**, 3426
66B M. Bartusek and J. Ruzickova, *Coll. Czech. Chem. Comm.*, 1966, **31**, 207
67P P. Pichet and R. L. Benoit, *Inorg. Chem.*, 1967, **6**, 1505
68H L. Havelkova and M. Bartusek, *Coll. Czech. Chem. Comm.*, 1968, **33**, 4188

$C_{10}H_8O_7S_2$ 2-Hydroxynaphthalene-3,6-disulphonic acid H_3L 816.1

Metal	Method	Temp	Medium	Log of equilibrium constant, remarks	Ref
H^+	gl	25	0·1 ($NaClO_4$)	K_1 9·23	68B
Fe^{3+}	sp	25	?	K (?) 8·8±0·1	62B
UO_2^{2+}	gl	25	0·1 ($NaClO_4$)	K_1 7·42, K_2 5·70	68B

62B S. K. Banerji, *Z. anorg. Chem.*, 1962, **315**, 229

68B A. Banerjee and A. K. Dey, *Analyt. Chim. Acta*, 1968, **42**, 473

$C_{10}H_8O_8S_2$ 1,8-Dihydroxynaphthalene-3,6-disulphonic acid (chromotropic acid) H_4L 817

Metal	Method	Temp	Medium	Log of equilibrium constant, remarks	Ref
H^+	gl	20	0·1	K_1 15·6, K_2 5·36	63S, 64S, 65Da, 65S, 66L, 67A, 67Ba, 67L
	sp	16—23	0·1 ($NaClO_4$)	K_2 5·44	63Sa
	gl	25	0·10 (KNO_3)	K_2 5·38±0·01	66L
	gl	30	0·2 ($NaClO_4$)	K_2 5·00	67A
	gl	20	0·1 (KNO_3)	K_2 5·41	67Ba
	gl	25	0·1 ($NaClO_4$)	K_1 12·99, K_2 5·56	68B
Al^{3+}	con	?	0·1 KNO_3	K_1 17·22	65D
	gl	?	0·1 KNO_3	K_1 17·40, K_2 16·86	65D
	gl	30	0·2 ($NaClO_4$)	K_1 17·16, K_2 13·25	67A
B^{III}	gl	20	0·1 (KNO_3)	$K[H_3BO_3+H_2L^{2-} \rightleftharpoons BL(OH)_2^{3-}+H^+]$ −1·55±0·04 $K(H_3BO_3+2H_2L^{2-} \rightleftharpoons BL_2^{5-}+H^+)$ −2·4±0·2	67B
Be^{2+}	gl	20 (?)	0·1 KNO_3	K_1 16·89±0·09, K_2 15·9±0·1	65Da
	gl	30	0·2 ($NaClO_4$)	K_1 16·69, K_2 12·45	67A
	gl	20	0·1 (KNO_3)	K_1 16·34, K_2 11·85, $K(Be^{2+}+HL^{3-} \rightleftharpoons BeHL^-)$ 2·9	67Ba
Co^{2+}	sp	22	?	β_3 12·97 (?)	66M
Cr^{III}	sp	25	?	K_1 8·21	65B
Cu^+	sp	30	?	$K(2CuCl+H_2L^{2-} \rightleftharpoons Cu_2L^{2-}+2H^++2Cl^-)$ 4·33	63R
Cu^{2+}	gl	25	0·10 (KNO_3)	K_1 13·44±0·03 $K(CuA^{2+}+L^{4-} \rightleftharpoons CuAL^{2-})$ 13·78±0·02 where A = 2,2′-bipyridyl	66L
	gl	30	0·2 ($NaClO_4$)	K_1 13·91, K_2 9·92	67A
Fe^{3+}	pol	?	1	K_1 22·8	60B
Ge^{IV}	sp	25	0·1 KCl	$K(H_3GeO_4+3H_2L^{2-} \rightleftharpoons HGeL_3^{8-}+2H^+)$ 2·30	67P
Nb^V	sp	18—22	0·1 NaCl 3 $NaClO_4$	$K(NbO_2^++3L^{4-}+4H^+ \rightleftharpoons NbL_3^{7-})$ 64·7 $K(NbO_2^++2H^++2L^{4-} \rightleftharpoons NbOL_2^{5-})$ 42·5	64S
Ni^{2+}	sp	22	?	β_3 12·00 (?)	66M
Th^{4+}	sp	16—23	0·1 ($NaClO_4$)	$^*K(Th^{4+}+H_2L \rightleftharpoons ThHL^{3+}+H^+)$ 4·11, 4·70	63Sa
	gl	25	0·1 ($NaClO_4$)	K_1 16·46, K_2 12·68 See EDTA (836) and CDTA (997) for mixed ligands.	68B

continued overleaf

817 $C_{10}H_8O_8S_2$ (contd)

Metal	Method	Temp	Medium	Log of equilibrium constant, remarks	Ref
Ti^{IV}	sp	18—22	0·1 ($NaClO_4$)	$K(TiO^{2+}+2L^{4-} \rightleftharpoons TiOL_2^{6-})$ 40·5 $K(TiO^{2+}+3L^{4-} \rightleftharpoons TiOL_3^{10-})$ 56·4 $K(TiO^{2+}+2H^{+}+3L^{4-} \rightleftharpoons TiL_3^{8-})$ 60·5 $K[TiOL_2^{6-}+2H^{+} \rightleftharpoons TiO(HL)_2^{4-}]$ 4·4	63S
	sp	18—22	0—100% $HCON(CH_3)_2$	$K[TiO^{2+}+HL^{-} \rightleftharpoons TiO(HL)^{-}]$ ($5{\cdot}0\times10^{-3}$ TiO^{2+}) (0%) 4·80, (25%) 5·60, (50%) 6·26, (74%) 6·26, (100%) 7·34, ($2{\cdot}5\times10^{-3}$ TiO^{2+}) (0%) 5·05, (25%) 5·57, (50%) 6·12, (74%) 6·57, (100%) 7·38 $K[TiO^{2+}+2HL^{-} \rightleftharpoons TiO(HL)_2]$ ($5{\cdot}0\times10^{-3}$ TiO^{2+}) (0%) 10·08, (25%) 10·49, (50%) 10·85, (74%) 11·83, (100%) 12·17, ($2{\cdot}5\times10^{-3}$ TiO^{2+}) (0%) 9·79, (25%) 10·49, (50%) 11·04, (74%) 11·86, (100%) 11·11	66C
UO_2^{2+}	gl	20	0·1 (KNO_3)	K_1 16·1, $K(UO_2^{2+}+HL^{3-} \rightleftharpoons UO_2HL^{-})$ 3·9	65Ba, 6
	sp	18—22	0·1 ($NaClO_4$)	K_1 16·6, K_2 11·5, $K(UO_2^{2+}+HL^{3-} \rightleftharpoons UO_2HL^{-})$ 4·0 $K(UO_2L^{2-}+HL^{3-} \rightleftharpoons UO_2HL_2^{5-})$ 1·5	65Ba, 6
	gl	30	0·2 ($NaClO_4$)	K_1 16·60, K_2 11·40	67A
	gl	25	0·1 ($NaClO_4$)	K_1 13·58, K_2 8·54	68B
VO^{2+}	gl	30	0·1 KCl	K_1 17·17 $K(VOA^{2+}+L^{4-} \rightleftharpoons VOAL^{2-})$ 18·09 where A = 1,10-phenanthroline	67L
Zr^{4+}	sp	16—23	0·1 (KCl)	$K[Zr(OH)_2^{2+}+HL^{3-} \rightleftharpoons Zr(OH)L^{-}]$ 18·68	63Sa

* Apparent constant varies with pH.

60B L. Benisek, *Coll. Czech. Chem. Comm.*, 1960, **25**, 2688
63R S. M. F. Rahman and A. U. Malik, *Indian J. Chem* , 1963, **1**, 424
63S L. Sommer, *Coll. Czech. Chem. Comm.*, 1963, **28**, 2393
63Sa M. Sakaguchi, A. Mizote, H. Miyata, and K. Toel, *Bull. Chem. Soc., Japan*, 1963, **36**, 885
64S L. Sommer and J. Havel, *Coll. Czech. Chem. Comm.*, 1964, **29**, 690
65B S. K. Banerji and S. Z. Qureshi, *Bull. Chem. Soc. Japan*, 1965, **38**, 720
65Ba M. Bartusek and L. Sommer, *J. Inorg. Nuclear Chem.*, 1965, **27**, 2397
65D S. N. Dubey and R. C. Mehrotra, *J. Indian Chem. Soc.*, 1965, **42**, 685
65Da S. N. Dubey and R. C. Mehrotra, *J. Less-Common Metals*, 1965, **9**, 123
65S L. Sommer, T. Sepel, and L. Kurilova, *Coll. Czech. Chem. Comm.*, 1965, **30**, 3426
66B M. Bartusek and J. Ruzickova, *Coll. Czech. Chem. Comm.*, 1966, **31**, 207
66C N. V. Chernaya and S. Ya. Shnaiderman, *J. Gen. Chem. (U.S.S.R.)*, 1966, **36**, 1179 (1165)
66L G. A. L'Heureux and A. E. Martell, *J. Inorg. Nuclear Chem.*, 1966, **28**, 481
66M A. K. Majumdar and A. B. Chatterjee, *Talanta*, 1966, **13**, 821
67A V. T. Athavale, N. Mahadevan, P. K. Mathur, and R. M. Sathe, *J. Inorg. Nuclear Chem.*, 1967, **29**, 1947
67B M. Bartusek and L. Havelkova, *Coll. Czech. Chem. Comm.*, 1967, **32**, 3853
67Ba M. Bartusek and J. Zelinka, *Coll. Czech. Chem. Comm.*, 1967, **32**, 992
67L K. Lal and R. P. Agarwal, *J. Less-Common Metals*, 1967, **12**, 269
67P P. Pichet and R. L. Benoit, *Inorg. Chem.*, 1967, **6**, 1505
68B A. Banerjee and A. K. Dey, *J. Inorg. Nuclear Chem.*, 1968, **30**, 995

818 $C_{10}H_9ON$ 2-Methyl-8-hydroxyquinoline HL

Metal	Method	Temp	Medium	Log of equilibrium constant, remarks	Ref
H^{+}	gl	1—40	50% dioxan	K_1 (0·7°) 12·18, (25°) 11·71, (40°) 11·44, K_2 (0·7°) 5·00, (25°) 4·58, (40°) 4·31 ΔH_1 −7·5, ΔS_1 28, ΔH_2 −6·9, ΔS_2 −2	54J

cont

Metal	Method	Temp	Medium	Log of equilibrium constant, remarks	Ref
H^+ (contd)	gl	25	50% dioxan	K_1 11·80±0·03, K_2 4·65±0·03	59F
	cal	25	50% dioxan	ΔH_1 −6·4, ΔS_1 33, ΔH_2 −6·9, ΔS_2 −1·7	59F
	gl	25	0·1 50% dioxan	K_1 11·30±0·05, K_2 4·68±0·02	67S
	gl	25	→ 0	K_1 10·16, K_2 5·61	67T
	gl	25	0·1	K_1K_2 15·9	68C
	gl	25	0·1 $NaClO_4$ 50% dioxan	K_1 11·31, K_2 4·72	68G
	cal	25	0·1 $NaClO_4$ 50% dioxan	ΔH_1 −6·92, ΔS_1 28·5, ΔH_2 −9·96, ΔS_2 −1·7	68G
Al^{3+}	sp	?	ethanol	K_1 8 (?)	62O
Co^{2+}	gl	1—40	50% dioxan	K_1 (0·7°) 9·97, (25°) 9·63, (40°) 9·37, K_2 (0·7°) 9·17, (25°) 8·87, (40°) 8·74 $\Delta H_{\beta 2}$ −10·3, $\Delta S_{\beta 2}$ 50	54J
	cal	25	50% dioxan	ΔH_1 −4·6, $\Delta H_{\beta 2}$ −11·6, $\Delta S_{\beta 2}$ 45	59F
	gl	25	0·1 50% dioxan	K_1 8·59, β_2 17·38	67S
	cal	25	0·1 $NaClO_4$ 50% dioxan	ΔH_1 −4·2, ΔS_1 25, $\Delta H_{\beta 2}$ −13·8, $\Delta S_{\beta 2}$ 33	68G
Cu^{2+}	gl	1—40	50% dioxan	K_1 (0·7°) 12·94, (25°) 12·48, (40°) 12·12, K_2 (0·7°) 11·74, (25°) 11·32, (40°) 10·82 $\Delta H_{\beta 2}$ −17·3, $\Delta S_{\beta 2}$ 50	54J
	cal	25	50% dioxan	ΔH_1 −8·7 (−6·5), $\Delta H_{\beta 2}$ −15·5 (−17·0), $\Delta S_{\beta 2}$ 52	59F
	gl	25	0·1 50% dioxan	K_1 11·92, β_2 22·82	67S
	cal	25	0·1 $NaClO_4$ 50% dioxan	ΔH_1 −8·0, ΔS_1 28, $\Delta H_{\beta 2}$ −15·5, $\Delta S_{\beta 2}$ 53	68G
Ge^{IV}	sp	25	0·5 NaCl	$K[Ge(OH)_4 + 2HL \rightleftharpoons Ge(OH)_2L_2]$ 3·4	67T
In^{3+}	sp	?	ethanol	K_1 12·2±0·4, β_2 23·9±0·9, β_3 35±1	63O
Mn^{2+}	gl	1—40	50% dioxan	K_1 (0·7°) 7·75, (25°) 7·44, (40°) 7·40, K_2 (0·7°) 6·85, (25°) 6·55, (40°) 6·50 $\Delta H_{\beta 2}$ −6·6, $\Delta S_{\beta 2}$ 42	54J
	gl	25	0·1 50% dioxan	K_1 6·81, β_2 13·10	67S
	cal	25	0·1 $NaClO_4$ 50% dioxan	ΔH_1 −3·3, ΔS_1 20, $\Delta H_{\beta 2}$ −6·3, $\Delta S_{\beta 2}$ 39	68G
Ni^{2+}	gl	1—40	50% dioxan	K_1 (0·7°) 9·67, (25°) 9·41, (40°) 9·07, K_2 (0·7°) 8·71, (25°) 8·35, (40°) 8·22 $\Delta H_{\beta 2}$ −10·7, $\Delta S_{\beta 2}$ 46	54J
	cal	25	50% dioxan	ΔH_1 −6·5 (−5·2), $\Delta H_{\beta 2}$ −10·3 (−10·6), $\Delta S_{\beta 2}$ 47	59F
	gl	25	0·1 50% dioxan	K_1 8·96, β_2 16·94	67S
	cal	25	0·1 $NaClO_4$ 50% dioxan	ΔH_1 −5·0, ΔS_1 24, $\Delta H_{\beta 2}$ −12·1, $\Delta S_{\beta 2}$ 37	68G
Pb^{2+}	gl	25	0·1 50% dioxan	K_1 9·97, β_2 17·18	68G
	cal	25	0·1 $NaClO_4$ 50% dioxan	ΔH_1 −6·3, ΔS_1 25, $\Delta H_{\beta 2}$ −13·7, $\Delta S_{\beta 2}$ 33	68G
Zn^{2+}	gl	1—40	50% dioxan	K_1 (0·7°) 10·07, (25°) 9·82, (40°) 9·47, K_2 (0·7°) 9·27 $\Delta H_{\beta 2}$ −10·3, $\Delta S_{\beta 2}$ 51	54J
	sp	?	ethanol	K_1 11·6, β_2 23·2	64C
	dis	25	0·1 50% dioxan	β_2 15·68±0·08, β_2 18·28	65C

continued overleaf

818 $C_{10}H_9ON$ (contd)

Metal	*Method*	*Temp*	*Medium*	*Log of equilibrium constant, remarks*	*Ref*
Zn^{2+} (*contd*)	dis	25	chloroform	$K(ZnL^+ + A \rightleftharpoons ZnLA^+)$ 1·60 where A = pyridine 1·00 where A = 2-methylpyridine 1·75 where A = 4-methylpyridine 0·20 where A = 2,4,6-trimethylpyridine	68C
	gl	25	0·1 50% dioxan	K_1 9·06, β_2 17·90	67S
	cal	25	0·1 $NaClO_4$ 50% dioxan	ΔH_1 −3·8, ΔS_1 29, $\Delta H_{\beta 2}$ −11·1, $\Delta S_{\beta 2}$ 45	68G

54J W. D. Johnson and H. Freiser, *Analyt. Chim. Acta*, 1954, **11**, 201
59F D. Fleischer and H. Freiser, *J. Phys. Chem.*, 1959, **63**, 260
62O W. E. Ohnesorge and A. L. Burlingame, *Analyt. Chem.*, 1962, **34**, 1086
63O W. E. Ohnesorge, *Analyt. Chem.*, 1963, **35**, 1137
64C D. A. Carter and W. E. Ohnesorge, *Analyt. Chem.*, 1964, **36**, 327
65C F. C. Chou, Q. Fernando, and H. Freiser, *Analyt. Chem.*, 1965, **37**, 361
67S R. L. Stevenson and H. Freiser, *Analyt. Chem.*, 1967, **39**, 1354
67T J. Tsau, S. Matsouo, P. Clerc, and R. Benoit, *Bull. Soc. chim. France*, 1967, 1039
68C F. C. Chou and H. Freiser, *Analyt. Chem.*, 1968, **40**, 34
68G G. Gutnikov and H. Freiser, *Analyt. Chem.*, 1968, **40**, 39

819 $C_{10}H_9ON$ 4-Methyl-8-hydroxyquinoline HL

Metal	*Method*	*Temp*	*Medium*	*Log of equilibrium constant, remarks*	*Ref*
H^+	gl	1—40	50% dioxan	K_1 (0·7°) 12·22, (25°) 11·62, (40°) 11·31, K_2 (0·7°) 5·18, (25°) 4·67, (40°) 4·45 ΔH_1 −9·1, ΔS_1 23, ΔH_2 −7·1, ΔS_2 −2	54J
	gl	25	0·1	K_1K_2 15·66	68C
	gl	25	0·1 50% dioxan	K_1 11·10, K_2 4·83	68G
	cal	25	0·1 $NaClO_4$ 50% dioxan	ΔH_1 −6·74, ΔS_1 30·0, ΔH_2 −6·18, ΔS_2 −0·5	68G
Co^{2+}	gl	1—40	50% dioxan	K_1 (0·7°) 11·29, (25°) 10·55, (40°) 10·22, K_2 (0·7°) 10·08, (25°) 9·45 $\Delta H_{\beta 2}$ −20·3, $\Delta S_{\beta 2}$ 23	54J
	cal	25	50% dioxan	$\Delta H_{\beta 2}$ −25·0, $\Delta S_{\beta 2}$ 8	59F
	gl	25	0·1 $NaClO_4$ 50% dioxan	K_1 9·95, β_2 18·92	68G
	cal	25	0·1 $NaClO_4$ 50% dioxan	ΔH_1 −6·8, ΔS_1 23, $\Delta H_{\beta 2}$ −17·8, $\Delta S_{\beta 2}$ 27	68G
Cu^{2+}	gl	1—40	50% dioxan	K_1 values not given, K_2 (0·7°) 14·32, (25°) 13·52, (40°) 12·88 $\Delta H_{\beta 2}$ −34·1, $\Delta S_{\beta 2}$ 12	54J
	cal	25	50% dioxan	$\Delta H_{\beta 2}$ −24·4, $\Delta S_{\beta 2}$ 44	59F
	gl	25	0·1 $NaClO_4$ 50% dioxan	K_1 14·04, β_2 26·96	68G
	cal	25	0·1 $NaClO_4$ 50% dioxan	ΔH_1 −10·8, ΔS_1 28, $\Delta H_{\beta 2}$ −21·6, $\Delta S_{\beta 2}$ 51	68G
Mn^{2+}	gl	1—40	50% dioxan	K_1 (0·7°) 8·63, (25°) 8·31, (40°) 8·12, K_2 (0·7°) 7·60, (25°) 7·24, (40°) 6·94 $\Delta H_{\beta 2}$ −11·6, $\Delta S_{\beta 2}$ 32	54J
	gl	25	0·1 $NaClO_4$ 50% dioxan	K_1 7·74, β_2 14·81	68G

cont

$C_{10}H_9ON$ (contd) 819

Metal	Method	Temp	Medium	Log of equilibrium constant, remarks	Ref
	cal	25	0·1 $NaClO_4$ 50% dioxan	ΔH_1 −4·1, ΔS_1 22, $\Delta H_{\beta2}$ −6·4, $\Delta S_{\beta2}$ 46	68G
Ni^{2+}	gl	1—40	50% dioxan	K_1 (0·7°) 12·36, (25°) 11·57, (40°) 11·15, K_2 (25°) 10·72, (40°) 10·17; $\Delta H_{\beta2}$ −26·0, $\Delta S_{\beta2}$ 15	54J
	cal	25	50% dioxan	$\Delta H_{\beta2}$ −25·8, $\Delta S_{\beta2}$ 15	59F
	gl	25	0·1 $NaClO_4$ 50% dioxan	K_1 10·56, β_2 20·47	68G
	cal	25	0·1 $NaClO_4$ 50% dioxan	ΔH_1 −9·4, ΔS_1 17, $\Delta H_{\beta2}$ −18·3, $\Delta S_{\beta2}$ 32	68G
Pb^{2+}	gl	25	0·1 $NaClO_4$ 50% dioxan	K_1 10·46, β_2 18·55	68G
	cal	25	0·1 $NaClO_4$ 50% dioxan	ΔH_1 −6·8, ΔS_1 25, $\Delta H_{\beta2}$ −15·4, $\Delta S_{\beta2}$ 33	68G
Zn^{2+}	gl	1—40	50% dioxan	K_1 (0·7°) 11·25, (25°) 10·67, (40°) 10·25, K_2 (0·7°) 10·28, (25°) 9·57, (40°) 9·19; $\Delta H_{\beta2}$ −19·6, $\Delta S_{\beta2}$ 26	54J
	dis	25	0·1	β_2 18·1±0·2	65C
	dis	25	chloroform	$K(ZnL^+ + A \rightleftharpoons ZnLA^+)$ 2·47 where A = pyridine; 2·00 where A = 2-methylpyridine; 2·87 where A = 4-methylpyridine; 1·50 where A = 2,4,6-trimethylpyridine	68C
	gl	25	0·1 $NaClO_4$ 50% dioxan	K_1 9·76, β_2 18·96	68G
	cal	25	0·1 $NaClO_4$ 50% dioxan	ΔH_1 −5·6, ΔS_1 26, $\Delta H_{\beta2}$ −12·8, $\Delta S_{\beta2}$ 44	68G

54J W. D. Johnson and H. Freiser, *Analyt. Chim Acta.*, 1954, **11**, 201

59F D. Fleischer and H. Freiser, *J. Phys. Chem.*, 1959, **63**, 260

65C F. C. Chou, Q. Fernando and H. Freiser, *Analyt. Chem.*, 1965, **37**, 361

68C F. C. Chou and H. Freiser, *Analyt. Chem.*, 1968, **40**, 34

68G G. Gutnikov and H. Freiser, *Analyt. Chem.*, 1968, **40**, 39

$C_{10}H_9O_2N$ 2-Hydroxymethyl-8-hydroxyquinoline HL 823.1

Metal	Method	Temp	Medium	Log of equilibrium constant, remarks	Ref
H^+	gl	25	0·1 50% dioxan	K_1 11·27±0·05, K_2 3·43±0·01	67S
Co^{2+}	gl	25	0·1 50% dioxan	K_1 8·68, K_2 8·4	67S
Cu^{2+}	gl	25	0·1 50% dioxan	K_1 >9	67S
Mg^{2+}	gl	25	0·1 50% dioxan	K_1 3·99, K_2 4·09	67S
Mn^{2+}	gl	25	0·1 50% dioxan	K_1 7·5	67S
Ni^{2+}	gl	25	0·1 50% dioxan	K_1 9·7, K_2 9	67S
Zn^{2+}	gl	25	0·1 50% dioxan	K_1 10	67S

67S R. L. Stevenson and H. Freiser, *Analyt. Chem.*, 1967, **39**, 1354

823.2 $C_{10}H_9NS$ 2-Methyl-8-mercaptoquinoline HL

Metal	Method	Temp	Medium	Log of equilibrium constant, remarks	Ref
H^+	gl	25	0·1 ($NaClO_4$) 50% dioxan	K_1 9·76, K_2 1·96	66K
	cal	25	0·1 ($NaClO_4$) 50% dioxan	$\Delta H_1 -3{\cdot}58$, ΔS_1 33, $\Delta H_2 -2{,}3$, ΔS_2 1	68G
Co^{2+}	gl	25	0·1 ($NaClO_4$) 50% dioxan	K_1 9·6	66K
	cal	25	0·1 ($NaClO_4$) 50% dioxan	$\Delta H_1 -5{\cdot}7$, ΔS_1 25	68G
Cu^{2+}	gl	25	0·1 ($NaClO_4$) 50% dioxan	K_1 11·7	66K
Ni^{2+}	gl	25	0·1 ($NaClO_4$) 50% dioxan	K_1 9·2	66K
	cal	25	0·1 ($NaClO_4$) 50% dioxan	$\Delta H_1 -8{\cdot}1$, ΔS_1 15	68G
Zn^{2+}	gl	25	0·1 ($NaClO_4$) 50% dioxan	K_1 11·1	66K
	cal	25	0·1 ($NaClO_4$) 50% dioxan	$\Delta H_1 -8{\cdot}1$, ΔS_1 24	68G

66K D. Kealey and H. Freiser, *Analyt. Chem.*, 1966, **38**, 1577
68G G. Gutnikov and H. Freiser, *Analyt. Chem.*, 1968, **40**, 39

823.3 $C_{10}H_{10}ON_2$ 2-Aminomethyl-8-hydroxyquinoline HL

Metal	Method	Temp	Medium	Log of equilibrium constant, remarks	Ref
H^+	gl	25	0·1 50% dioxan	K_1 11·25±0·01, K_2 8·56±0·01	67S
Co^{2+}	gl	25	0·1 50% dioxan	K_1 11·7, K_2 10·8	67S
Cu^{2+}	sp	25	0·1 50% dioxan	K_1 15·9	67S
	gl	25	0·1 50% dioxan	K_2 8·0	67S
Mn^{2+}	gl	25	0·1 50% dioxan	K_1 8·32, K_2 7·52	67S
Ni^{2+}	gl	25	0·1 50% dioxan	K_1 13·42, K_2 12·7	67S
Zn^{2+}	gl	25	0·1 50% dioxan	K_1 12·31, K_2 10·8	67S

67S R. L. Stevenson and H. Freiser, *Analyt. Chem.*, 1967, **39**, 1354

$C_{10}H_{11}O_4N$ *N*-Phenyliminodiacetic acid H_2L 827

Metal	*Method*	*Temp*	*Medium*	*Log of equilibrium constant, remarks*	*Ref*
H^+	H	20	0·1 (KCl)	K_1 4·96, K_2 2·40	64P
Ag^+	gl	20	0·1 (KNO_3)	K_1 1·0	64P
Hg^{2+}	Hg	20	0·1 (KNO_3)	β_2 12·9	64P

64P L. D. Pettit and H. M. N. H. Irving, *J. Chem. Soc.*, 1964, 5336

$C_{10}H_{11}O_4As$ *As*-Phenylarsinodiacetic acid H_2L 827.1

Metal	*Method*	*Temp*	*Medium*	*Log of equilibrium constant, remarks*	*Ref*
H^+	gl	25	0·3 (K_2SO_4)	K_1 4·93, K_2 3·61	61J
	gl	20	0·1 (KNO_3)	K_1 5·03, K_2 3·60	64P
Ag^+	Ag	20	0·1 (KNO_3)	K_1 5·37	64P
				$K(Ag^+ + HL^- \rightleftharpoons AgHL)$ 3·73	
				$K(AgL^- + H^+ \rightleftharpoons AgHL)$ 3·40	
Cd^{2+}	gl	20	0·1 (KNO_3)	K_1 1·0	64P
Cu^+	Pt	25	0·3 (K_2SO_4)	β_n 3·95+1·7n	61J
				$K(CuHL^+ + H^+ \rightleftharpoons CuH_2L^{2+})$ (?) 2·70	
				$K(CuL + H^+ \rightleftharpoons CuHL^+)$ (?) 3·96	
	gl	20	0·1 (KNO_3)	$K(CuL + H^+ \rightleftharpoons CuHL^+)$ 4·0	64P
Cu^{2+}	gl	20	0·1 (KNO_3)	K_1 2·51	64P
Hg^{2+}	Hg	20	0·1 (KNO_3)	K_1 14·7±0·1, β_2 19·92±0·06	64P
Ni^{2+}	gl	20	0·1 (KNO_3)	K_1 1·5	64P
Zn^{2+}	gl	20	0·1 (KNO_3)	K_1 1·4	64P

61J B. R. James and R. J. P. Williams, *J. Chem. Soc.*, 1961, 2007
64P L. D. Pettit and H. M. N. H. Irving, *J. Chem. Soc.*, 1964, 5336

$C_{10}H_{11}O_5N$ *N*-(2′-Hydroxyphenyl)iminodiacetic acid H_3L 827.2

Metal	*Method*	*Temp*	*Medium*	*Log of equilibrium constant, remarks*	*Ref*
H^+	gl	20	0·1 (KNO_3)	K_1 11·08±0·03, K_2 5·43±0·01, K_3 2·98±0·01	63I
Ba^{2+}	gl	20	0·1 (KNO_3)	K_1 4·27±0·01	63I
				$K(Ba^{2+} + HL^{2-} \rightleftharpoons BaHL)$ 2·50±0·04	
Ca^{2+}	gl	20	0·1 (KNO_3)	K_1 6·27±0·03	63I
				$K(Ca^{2+} + HL^{2-} \rightleftharpoons CaHL)$ 3·21±0·01	
Li^+	gl	20	0·1 (KNO_3)	K_1 2·20±0·02	63I
Mg^{2+}	gl	20	0·1 (KNO_3)	K_1 6·86±0·01	63I
				$K(Mg^{2+} + HL^{2-} \rightleftharpoons MgHL)$ 2·67±0·02	
Na^+	gl	20	0·1 (KNO_3)	K_1 1·0±0·1	63I
Sr^{2+}	gl	20	0·1 (KNO_3)	K_1 4·65±0·01	63I
				$K(Sr^{2+} + HL^{2-} \rightleftharpoons SrHL)$ 2·67±0·02	
Tl^+	gl	20	0·1 (KNO_3)	K_1 4·79±0·03	63I
				$K(Tl^+ + HL^{2-} \rightleftharpoons TlHL^-)$ 2·34±0·02	

63I H. Irving and J. J. R. F. da Silva, *J. Chem. Soc.*, 1963, 3308

827.3 $C_{10}H_{11}O_5N$ 2-*N*[-(2′-Hydroxyethyl)amino]benzene-1,4-dicarboxylic acid H_2L

Metal	*Method*	*Temp*	*Medium*	*Log of equilibrium constant, remarks*	*Ref*
H^+	gl	20	0·1 KCl	K_1 4·73, K_2 3·40	64U
Cu^{2+}	gl	20	0·1 KCl	K_1 4·05, K_2 2·3 $K(Cu^{2+}+HL^- \rightleftharpoons CuHL^+)$ 2·75	64U
Ni^{2+}	gl	20	0·1 KCl	K_1 2·45, K_2 < 2·4 $K(Ni^{2+}+HL^- \rightleftharpoons NiHL^+)$ 1·50	64U

64U E. Uhlig and D. Linke, *Z. anorg. Chem.*, 1964, **331**, 112

827.4 $C_{10}H_{12}ON_2$ 2-(2′-Hydroxypropyl)benzimidazole HL

Metal	*Method*	*Temp*	*Medium*	*Log of equilibrium constant, remarks*	*Ref*
H^+	gl	15—45	50% dioxan	K_1 (15°) 13·07, (25°) 12·77, (40°) 12·26, (45°) 12·06, K_2 (15°) 4·89, (25°) 4·82, (40°) 4·68, (45°) 4·60	63L
Cu^{2+}	gl	15—45	50% dioxan	K_1 (15°) 11·69, (25°) 11·24, (40°) 10·52, (45°) 10·38 ΔH_1 −11·3	63L

63L T. J. Lane and H. B. Durham, *Inorg. Chem.*, 1963, **2**, 632

828.1 $C_{10}H_{12}O_4N_2$ *N*-(2′-Pyridylmethyl)iminodiacetic acid H_2L

Metal	*Method*	*Temp*	*Medium*	*Log of equilibrium constant, remarks*	*Ref*
H^+	gl	20	0·1 (KNO_3)	K_1 8·25±0·01, K_2 2·85±0·02	63I
	gl	25	0·1 (KNO_3)	K_1 8·21, K_2 2·6	64T
	gl	25	0·1 KCl	K_1 8·22, K_2 3·05	66S
Ag^+	gl	20	0·1 (KNO_3)	K_1 6·09±0·01	63I
Ba^{2+}	gl	20	0·1 (KNO_3)	K_1 3·40±0·01	63I
Ca^{2+}	gl	20	0·1 (KNO_3)	K_1 4·92±0·02	63I
	gl	25	0·1 KCl	K_1 4·80, K_2 2·92	66S
Cd^{2+}	gl	20	0·1 (KNO_3)	K_1 9·45±0·02, K_2 5·29±0·01	63I
Ce^{3+}	gl	25	0·1 (KNO_3)	K_1 8·30, K_2 6·44	64T
Co^{2+}	gl	20	0·1 (KNO_3)	K_1 10·16±0·02, K_2 3·18±0·05	63I
	gl	25	0·1 KCl	K_1 10·39, K_2 3·20	66S
Cu^{2+}	gl	20	0·1 (KNO_3)	K_2 ~ 3·5	63I
	gl	25	0·1 KCl	K_1 10·75, K_2 3·35	66S
Dy^{3+}	gl	25	0·1 (KNO_3)	K_1 9·00, K_2 8·03	64T
Er^{3+}	gl	25	0·1 (KNO_3)	K_1 9·25, K_2 7·65	64T
Eu^{3+}	gl	25	0·1 (KNO_3)	K_1 8·92, K_2 8·02	64T
Fe^{2+}	gl	20	0·1 (KNO_3)	K_1 8·94±0·01	63I

cont

$C_{10}H_{12}O_4N_2$ **(contd) 828.1**

Metal	*Method*	*Temp*	*Medium*	*Log of equilibrium constant, remarks*	*Ref*
Gd^{3+}	gl	25	0·1 (KNO_3)	K_1 8·76, K_2 8·01	64T
Ho^{3+}	gl	25	0·1 (KNO_3)	K_1 9·07, K_2 7·80	64T
La^{3+}	gl	25	0·1 (KNO_3)	K_1 7·80, K_2 5·90	64T
Li^{+}	gl	20	0·1 (KNO_3)	K_1 1·71±0·01	63I
Lu^{3+}	gl	25	0·1 (KNO_3)	K_1 9·72, K_2 7·75	64T
Mg^{2+}	gl	20	0·1 (KNO_3)	K_1 3·90±0·01	63I
Mn^{2+}	gl	20	0·1 (KNO_3)	K_1 6·97±0·01, K_2 3·63±0·01	63I
Na^{+}	gl	20	0·1 (KNO_3)	K_1 0·85±0·03	63I
Nd^{3+}	gl	25	0·1 (KNO_3)	K_1 8·64, K_2 7·18	64T
Ni^{2+}	gl	25	0·1 KCl	K_1 11·22, K_2 3·68	66S
Pb^{2+}	gl	20	0·1 (KNO_3)	K_1 10·31±0·04	63I
Pr^{3+}	gl	25	0·1 (KNO_3)	K_1 8·53, K_2 6·95	64T
Sm^{3+}	gl	25	0·1 (KNO_3)	K_1 8·92, K_2 7·96	64T
Sr^{2+}	gl	20	0·1 (KNO_3)	K_1 3·65±0·01	63I
Tb^{3+}	gl	25	0·1 (KNO_3)	K_1 8·87, K_2 8·17	64T
Tl^{+}	gl	20	0·1 (KNO_3)	K_1 3·84±0·01	63I
Tm^{3+}	gl	25	0·1 (KNO_3)	K_1 9·40, K_2 7·61	64T
VO^{2+}	gl	25	0·1 ($NaClO_4$)	$K[VOL(OH)^- + H^+ \rightleftharpoons VOL]$ 6·45	66K
Y^{3+}	gl	25	0·1 (KNO_3)	K_1 8·63, K_2 7·38	64T
Yb^{3+}	gl	25	0·1 (KNO_3)	K_1 9·60, K_2 7·73	64T
Zn^{2+}	gl	20	0·1 (KNO_3)	K_1 10·87±0·03	63I

63I H. Irving and J. J. R. F. da Silva, *J. Chem. Soc.*, 1963, 945

64T L. C. Thompson, *Inorg. Chem.*, 1964, **3**, 1015

66K Th. Kaden and S. Fallab, *Chimia (Switz.)*, 1966, **20**, 51

66S P. Souchey, N. Israily, and P. Gouzerh, *Bull. Soc. chim. France*, 1966, **12**, 3917

$C_{10}H_{12}O_6N_4$ Purin-6-one 9-riboside *N*(1)-oxide HL 829.1 [inosine *N*(1)-oxide]

Metal	*Method*	*Temp*	*Medium*	*Log of equilibrium constant, remarks*	*Ref*
H^{+}	sp	25	0·1 ($NaClO_4$)	K_1 5·40±0·02	65S
Ba^{2+}	sp	25	0·1 ($NaClO_4$)	K_1 1·2±0·2	65S
Ca^{2+}	sp	25	0·1 ($NaClO_4$)	K_1 1·5±0·2	65S
Co^{2+}	sp	25	0·1 ($NaClO_4$)	K_1 3·46±0·05	65S
Cu^{2+}	sp	25	0·1 ($NaClO_4$)	K_1 5·27±0·05	65S
Mg^{2+}	sp	25	0·1 ($NaClO_4$)	K_1 1·7±0·2	65S
Mn^{2+}	sp	25	0·1 ($NaClO_4$)	K_1 2·5±0·2	65S
Ni^{2+}	sp	25	0·1 ($NaClO_4$)	K_1 3·50±0·05	65S
Zn^{2+}	sp	25	0·1 ($NaClO_4$)	K_1 3·60±0·05	65S

65S H. Sigel, *Helv. Chim. Acta*, 1965, **48**, 1513, 1519

830.1 $C_{10}H_{13}O_4N_5$ 6-Aminopurine 9-riboside (adenosine) L

Metal	*Method*	*Temp*	*Medium*	*Log of equilibrium constant, remarks*	*Ref*
H^+	gl	20	0·1 (KCl)	K_1 3·60	57W
	?	27	0·1 $NaClO_4$	K_1 3·5	61F
	gl	20	1 $NaNO_3$	K_1 3·70±0·01	65F
Co^{2+}	sp	20	1—3 $Co(ClO_4)_2$	K_1 −0·30	64S
Cu^{2+}	sp	20	0·1—2 $Cu(ClO_4)_2$	K_1 0·84	64S
	gl	20	1 $NaNO_3$	K_1 0·71±0·05	65F
Hg^{2+}	Hg	27	0·1 $NaClO_4$	β_2 8·50	61F
Mn^{2+}	sp	20	1—3 $Mn(ClO_4)_2$	K_1 −0·82	64S
Ni^{2+}	sp	20	1—3 $Ni(ClO_4)_2$	K_1 −0·17	64S
Pb^{2+}	gl	20	1 $NaNO_3$	K_1 −0·5	65F
Zn^{2+}	gl	20	0·1 (KCl)	K_1 1·51	57W
	sp	20	0·5—3 $Zn(ClO_4)_2$	K_1 −0·28	64S

57W K. Wallenfels and H. Sund, *Biochem. Z.*, 1957, **329**, 41
61F R. Ferreira, E. Ben-Zvi, T. Yamane, J. Vasilevskis, and N. Davidson, 'Advances in Chemistry of Coordination Compounds', ed. S. Kirschner, Macmillan, New York, 1961, 457
64S P. W. Schneider, H. Brintzinger, and H. Erlenmeyer, *Helv. Chim. Acta*, 1964, **67**, 992
65F A. M. Fiskin and M. Beer, *Biochemistry*, 1965, **4**, 1289

830.2 $C_{10}H_{13}O_4N_5$ 2-Aminopurine 9-riboside (deoxyguanosine) L

Metal	*Method*	*Temp*	*Medium*	*Log of equilibrium constant, remarks*	*Ref*
Cu^{2+}	epr	19—25	0·02 $NaClO_4$	K_1 2·91, β_2 5·58	65R

65R C. Ropars and R. Viovy, *J. Chim. phys.*, 1965, **62**, 408

832 $C_{10}H_{13}O_5N_5$ 2-Aminopurin-6-one 9-riboside (guanosine) HL

Metal	*Method*	*Temp*	*Medium*	*Log of equilibrium constant, remarks*	*Ref*
H^+	gl	20	1 $NaNO_3$	K_1 9·24±0·01, K_2 2·20±0·05	65F
Cu^{2+}	gl	20	1 $NaNO_3$	K_1 4·3±0·4 (?)	65F
				$K(Cu^{2+}+HL \rightleftharpoons CuHL^{2+})$ 2·15±0·02	
Pb^{2+}	gl	20	1 $NaNO_3$	K_1 3·5	65F
				$K(Pb^{2+}+HL \rightleftharpoons PbHL^{2+})$ 0·5±0·2	
Th^{4+}	gl	20	1 $NaNO_3$	$K(Th^{4+}+HL \rightleftharpoons ThHL^{4+})$ 0·9±0·1	65F
UO_2^{2+}	gl	20	1 $NaNO_3$	$K(UO_2^{2+}+HL \rightleftharpoons UO_2HL^{2+})$ 0·7±0·3	65F

65F A. M. Fiskin and M. Beer, *Biochemistry*, 1965, **4**, 1289

$C_{10}H_{13}O_7N_3$ *N*-(1′,3′-Dimethyl-2′,4′,6′-trioxoperhydropyrinidin-5-yl)iminodiacetic acid H₃L 832.1 (1,3-dimethyluramil-*NN*-diacetic acid)

Metal	*Method*	*Temp*	*Medium*	*Log of equilibrium constant, remarks*	*Ref*
H^+	gl	20	0·1 [$(CH_3)_4NNO_3$]	K_1 10·12±0·01, K_2 2·67±0·02, K_3 2·05±0·05	63I
Ba^{2+}	gl	20	0·1 (KNO_3)	K_1 6·00±0·01, K_2 3·88±0·10	63I
Be^{2+}	gl	20	0·1 (KNO_3)	K_1 10·54±0·02	63I
				$K(Be^{2+}+HL^{2-} \rightleftharpoons BeHL)$ 3·54	
Ca^{2+}	gl	20	0·1 (KNO_3)	K_1 8·13±0·01, K_2 5·40±0·02	63I
K^+	gl	20	0·1 [$(CH_3)_4NNO_3$]	K_1 0·94±0·01	63I
Li^+	gl	20	0·1 [$(CH_3)_4NNO_3$]	K_1 4·91±0·01	63I
Mg^{2+}	gl	20	0·1 (KNO_3)	K_1 8·29±0·01, K_2 3·78±0·02	63I
Na^+	gl	20	0·1 [$(CH_3)_4NNO_3$]	K_1 2·53±0·01	63I
Sr^{2+}	gl	20	0·1 (KNO_3)	K_1 6·82±0·02, K_2 4·27±0·10	63I
Tl^+	gl	20	0·1 [$(CH_3)_4NNO_3$]	K_1 5·73±0·01	63I

63I H. Irving and J. J. R. F. da Silva, *J. Chem. Soc.*, 1963, 458

$C_{10}H_{14}ON_2$ 2-Hydroxy-2-phenylbutanoylamidine HL 832.2

Metal	*Method*	*Temp*	*Medium*	*Log of equilibrium constant, remarks*	*Ref*
H^+	gl	25	0·1 (KCl)	K_1 12·86±0·05, K_2 11·06±0·01	63G
Cu^{2+}	gl J	25	0·1 (KCl)	K_1 12·86, K_2 11·70	63G
Ni^{2+}	gl J	25	0·1 (KCl)	K_1 8·06, K_2 7·74	63G

63G R. P. Gould and R. J. Jameson, *J. Chem. Soc.*, 1963, 15, 5211

$C_{10}H_{14}O_5N_2$ 5-Methyl-pyramidin-2,4-dione 1-riboside (thymidine) HL 833.1

Metal	*Method*	*Temp*	*Medium*	*Log of equilibrium constant, remarks*	*Ref*
H^+	?	27	0·1 $NaClO_4$	K_1 9·6	61F
	gl	20	1 $NaNO_3$	K_1 9·65±0·01	65F
Cu^{2+}	gl	20	1 $NaNO_3$	K_1 4·7±0·1	65F
Hg^{2+}	Hg	27	0·1 $NaClO_4$	β_2 21·2	61F
Pb^{2+}	gl	20	1 $NaNO_3$	K_1 4·7±0·5	65F

61F R. Ferreira, E. Ben-Zvi, T. Yamane, J. Vasileuskis, and N. Davidson, 'Advances in Chemistry of Coordination Compounds', ed. S. Kirschner, Macmillan, New York, 1961, 457

65F A. M. Fiskin and M. Beer, *Biochemistry*, 1965, 4, 1289

833.2 $C_{10}H_{14}O_8S_4$ 1,1,2,2-Tetrathioethane-*SS′S″S‴*-tetra-acetic acid H_4L

Metal	Method	Temp	Medium	Log of equilibrium constant, remarks	Ref
H^+	gl	20	0·1	K_1 4·93, K_2 3·99, K_3 3·56, K_4 3·24	61S
Cu^{2+}	gl	20	0·1	K_1 5·00, K_2 2·33	61S
				$K(Cu^{2+}+HL^{3-} \rightleftharpoons CuHL^-)$ 4·08	
				$K(Cu^{2+}+H_2L^{2-} \rightleftharpoons CuH_2L)$ 3·24	
				$K(Cu^{2+}+H_3L^- \rightleftharpoons CuH_3L^+)$ 2·64	

61S G. Saini, G. Ostacoli, E. Campi, and N. Cibrario, *Gazzetta*, 1961, **91**, 904

834 $C_{10}H_{15}ON$ 2-Methylamino-1-phenylpropan-1-ol (ephedrine) L

Metal	Method	Temp	Medium	Log of equilibrium constant, remarks	Ref
H^+	gl	0—45	0·1 (KNO_3)	K_1 (0°) 10·47, (25°) 9·56, (45°) 8·94	64A
Cd^{2+}	pol	0—45	0·1 (KNO_3)	β_2 (0°) 6·94, (25°) 6·49, (45°) 5·55	64A
				$\Delta H_{\beta 2}°$ −10·8, $\Delta S_{\beta 2}°$ −8	

64A A. C. Andrews and J. K. Romary, *J. Chem. Soc.*, 1964, 405

834.1 $C_{10}H_{15}O_4N_5$ Glycylglycyl-L-histidine HL

Metal	Method	Temp	Medium	Log of equilibrium constant, remarks	Ref
H^+	gl	25	0·16 KCl	K_1 8·22, K_2 6·87, K_3 2·84	66B
Ni^{2+}	gl	25	0·16 KCl	$K[Ni(H_{-1}L)+H^+ \rightleftharpoons NiL^+]$ 6·20	66B
				$K[Ni(H_{-2}L)^- + H^+ \rightleftharpoons Ni(H_{-1}L)]$ 6·30	
				$K[Ni(H_{-2}L)OH^{2-} + H^+ \rightleftharpoons Ni(H_{-2}L)^-]$ 6·35	

66B G. F. Bryce, R. W. Roeske, and F. R. N. Gurd, *J. Biol. Chem.*, 1966, **241**, 1072

834.2 $C_{10}H_{15}O_6N$ *N*-(1′-Carboxycyclopentyl)iminodiacetic acid H_3L

Metal	Method	Temp	Medium	Log of equilibrium constant, remarks	Ref
H^+	gl	20	0·1 (KCl)	K_1 11·69±0·03, K_2 2·56±0·02, K_3 1·61±0·05	66I
Ba^{2+}	gl	20	0·1 (KCl)	K_1 5·50±0·04	66I
Ca^{2+}	gl	20	0·1 (KCl)	K_1 8·22±0·01	66I
Mg^{2+}	gl	20	0·1 (KCl)	K_1 6·75±0·02	66I
Sr^{2+}	gl	20	0·1 (KCl)	K_1 6·08±0·03	66I
Zn^{2+}	gl	20	0·1 (KCl)	K_1 12·31±0·04	66I

66I H. M. N. H. Irving and M. G. Miles, *J. Chem. Soc. (A)*, 1966, 1268

$C_{10}H_{17}O_5N$ *N*-(2-Hydroxycyclohexyl)iminodiacetic acid H_2L 835

.rrata. **1964**, p. 634, Table 835. *For* formula $C_{10}H_{16}O_5N$ *read* $C_{10}H_{17}O_5N$.
For correct ligand sequence this table has been renumbered **836.2** *q.v.*

$C_{10}H_{16}O_8N_2$ (Ethylenedinitrito) tetra-acetic acid) (EDTA) H_4L 836

Metal	*Method*	*Temp*	*Medium*	*Log of equilibrium constant, remarks*	*Ref*
H^+	cal	25	?	$\Delta H°\ (HL^{3-} + OH^- \rightleftharpoons L^{4-})\ -8{\cdot}5$	54C
	sol	25	?	K_1 10·26, K_2 6·16, K_3 2·77, K_4 2·21, K_5 1·91	59K
	?	20	0·4	K_1 9·92, K_2 5·89, K_3 2·49, K_4 1·91	60B
	H	20	0·1 (KCl)	K_1 10·26, K_2 6·16, K_3 2·67, K_4 1·99	62K, 63S, 63T, 64A, 64P, 65J, 65P, 67K, 68F
	pH	25	0·1	K_5 1·51 ± 0·05	62K, 67T
			1—3	K_6 (1) 0·04 ± 0·02, (2) −0·19 ± 0·05, (3) −0·41 ± 0·05	
	gl	25	3 ($NaClO_4$)	K_1 8·50, K_2 6·40, K_3 2·32, K_4 1·77	63D
	gl	30	0·1 KCl	K_1 10·25, K_2 6·22, K_3 2·86, K_4 2·4	63Ga
	gl	?	0·02—3 $(CH_3)_4NOH$	K_1 11·33, K_2 6·81	63P
	gl	25	0·32 CsCl	K_1 10·46, K_2 6·29	65Bc
	gl	?	0·1 NaCl	K_1 9·60, K_2 6·22, K_3 2·67, K_4 1·99	66B
	gl	25	0·1 50% dioxan	K_1 9·98, K_2 6·33	66Ba
	nmr	33—37	?	K_1 9·8 ± 0·1, K_2 6·2	66K
	gl	20—40	0·1 (KNO_3)	K_1 (20°) 10·31, (30°) 10·12, (40°) 9·91, K_2 (20°) 6·21, (30°) 6·14, (40°) 5·98, K_3 (20°) 2·66, K_4 (20°) 2·02	66M
				$\Delta H_1°$ (25°) −8 ± 2, $\Delta S_1°$ 20 ± 6, $\Delta H_2°$ (25°) −4·8 ± 0·5, $\Delta S_2°$ 12 ± 2	
	gl	20	0·1 [$(CH_3)_4NCl$]	K_1 10·44, K_2 6·16	67Aa
	gl	20	0·1 (KNO_3)	K_1 10·23, K_2 6·16, K_3 2·7, K_4 2·0	67Aa
	H	20	1 [$(CH_3)_4NCl$]	K_1 10·12, K_2 6·07, K_3 2·7, K_4 2·2	67Aa
	H	20	1 (KCl)	K_1 9·95, K_2 6·26	67Aa
	gl	20	1 ($NaClO_4$)	K_1 8·85, K_2 6·28, K_3 2·3	67Aa
	sol	20	1 ($NaClO_4$)	K_4 2·2, K_5 1·4, K_6 −0·12	67Aa
	gl	25	→ 0	K_1 10·1, K_2 6·47, K_3 5·8	67Ba
	oth	15—35	0·01	K_1 (15°) 10·39, (25°) 10·25, (35°) 10·11, K_2 (15°) 6·28, (25°) 6·17, (35°) 6·06 (Calculated)	67R
	?	20	1·2 HCl	K_1 9·37, K_2 6·21, K_3 2·44, K_4 1·80	67T
	gl	25	0·1 (KNO_3)	K_1 10·34, K_2 6·24, K_3 2·75, K_4 2·07	68W
Ag^+	dis	20	0·1 ($KClO_4$)	K_1 7·11 ± 0·06	63S
	gl	20	0·1 KNO_3	K_1 7·72, K_2 4·0	64J
	oth	15—35	0·01	K_1 (15°) 7·37, (25°) 7·28, (35°) 7·15	67R
				$K(Ag^+ + HL^{3-} \rightleftharpoons AgHL^{2-})$ (15°) 3·46, (25°) 3·36, (35°) 3·29, silver-sensitive glass electrode	
	Ag	25	0·1 (KNO_3)	K_1 7·31	68W
				$K(AgL^{3-} + H^+ \rightleftharpoons AgHL^{2-})$ 6·49	
				See oxalic acid (111) for mixed ligands.	

continued overleaf

836 $C_{10}H_{16}O_8N_2$ (contd)

Metal	*Method*	*Temp*	*Medium*	*Log of equilibrium constant, remarks*	*Ref*
Al^{3+}	cal	20	0·1 (KNO_3)	ΔH_1° 12·58, ΔS_1° 116·6	58S
	dis	20	0·1 ($KClO_4$)	$K[Al^{3+}+L^{4-}+OH^- \rightleftharpoons AlL(OH)^{2-}]$ 25·04±0·04	63S
	Hg	20—40	0·1 (KNO_3)	K_1 (20°) 16·7±0·1, (30°) 16·84±0·06, (40°) 17·26±0·06,	66M
				ΔH_1° 12±2, ΔS_1° 117±6	
				$K(AlL+H^+ \rightleftharpoons AlHL^+)$ (20°) 2·77	
	Hg	25	0·1 KNO_3	K_1 16·5	67A
				$K(Al^{3+}+HL^{3-} \rightleftharpoons AlHL)$ 3·4	67A
				$K(AlL^-+OH^- \rightleftharpoons AlOHL^{2-})$ 8·0	
	sp	25	0·2 ($NaClO_4$)	K_1 16·01	67B
	gl	25	0·2 ($NaClO_4$)	$K(AlL+H^+ \rightleftharpoons AlHL^+)$ 2·63	67B
				$K(AlLOH^-+H^+ \rightleftharpoons AlL)$ 5·87	
				$K[AlL(OH)_2^{2-}+H^+ \rightleftharpoons AlLOH^-]$ 10·31	
Ba^{2+}	cal	25	~ 0·05 $BaCl_2$	ΔH_1° −5·1, ΔS_1° 18	54C
	oth	20	0·1 (KNO_3)	K_1 8, electrophoresis	65J
	cal	25	0·1 KNO_3	ΔH_1 −5·3, ΔS_1 18	65W
Be^{2+}	dis	20	0·1 ($KClO_4$)	K_1 9·27±0·03	63S
	sol	20	0·3	K_1 10·2,	63Sa
				$K(BeL^{2-}+OH^- \rightleftharpoons BeOHL^{3-})$ 5·4	
	oth	20	0·1	K_1 10·8,	63Sa
				$K(BeL^{2-}+OH^- \rightleftharpoons BeOHL^{3-})$ 5·2	
				calculated from Davies equation	
	ix	~ 20	0·1 NaCl	K_1 8·4,	66B
				$K(Be^{2+}+HL^{3-} \rightleftharpoons BeHL^-)$ 2·1	
				$K(Be^{2+}+H_2L^{2-} \rightleftharpoons BeH_2L)$ 3·7	
				$K(Be^{2+}+H_3L^- \rightleftharpoons BeH_3L^+)$ 2·7	
Bi^{3+}	sp	?	0·1	K_1 22·8	60Ka
	pol	20	0·1	K_1 27·9	61M
	dis	20	0·1 ($KClO_4$)	$K[Bi^{3+}+L^{4-}+OH^- \rightleftharpoons BiL(OH)^{2-}]$ 32·45±0·10	63S
	Hg	20	0·1 ($NaClO_4$)	K_1 27·4	64E
	pol	25	0·1—1·0 ($NaClO_4$)	K_1 (0·1) 28·8, (1·0) 30·5	64E
	sp	25	1 ($NaClO_4$)	$K(BiO^++L^{4-}+2H^+ \rightleftharpoons BiL^-)$ 26·5	65B
	pol	25	0·5 (KNO_3)	K_1 28·2±0·2	66Bb
	red	20	1 ($NaClO_4$)	K_1 26·7,	67B
				$K(BiL^-+H^+ \rightleftharpoons BiHL)$ 1·7	
				$K(BiL^-+OH^- \rightleftharpoons BiLOH^{2-})$ 2·96	
Ca^{2+}	cal	25	~ 0·05 $Ca(NO_3)_2$	ΔH_1° −5·8, ΔS_1° 31	54C
	ix	?	0·3	K_1 10·45	60M
				$K(2Ca^{2+}+L^{4-} \rightleftharpoons Ca_2L)$ 12·52	
	gl	30	0·1 KCl	K_1 10·59	63Ga
	gl	20	0·1 (KNO_3)	$K(Ca^{2+}+HL^{3-} \rightleftharpoons CaHL^-)$ 3·51	64A
	oth	20	0·1 (KNO_3)	K_1 11, electrophoresis	65J
	nmr	20	0·2 EDTA	$K(CaL^{2-}+Ca^{2+} \rightleftharpoons Ca_2L)$ 1·1±0·1	68La
				$K(Ca_3L^{2+} \rightleftharpoons Ca^{2+}+Ca_2L)$ 0·4±0·1	
Cd^{2+}	cal	25	~ 0·05 $Cd(NO_3)_2$	ΔH_1° −9·1, ΔS_1° 38	54C
	sol	25	2 (KNO_3)	K (?) 10·02	63F
	dis	20	0·1 ($KClO_4$)	K_1 16·9±0·1	63S
	gl	20	0·1 (KNO_3)	K_1 16·46,	64A
				$K(Cd^{2+}+HL^{3-} \rightleftharpoons CdHL^-)$ 9·1	
	oth	20	0·1 (KNO_3)	K_1 17·5, electrophoresis	65J
	pol	25	0·2 (KNO_3)	K_1 15·98	65O
	cal	25	0·1 KNO_3	ΔH_1 −10·1, ΔS_1 42	65W

cont

Metal	*Method*	*Temp*	*Medium*	*Log of equilibrium constant, remarks*	*Ref*
Cd^{2+} (contd)	oth	32—42	?	K_1 (32°) 15·32, (42°) 14·60, ultrasonic	68L
				ΔH_1 25·2 (?), ΔS_1 113·8 (?)	
				See oxalic acid (111) for mixed ligands.	
Ce^{3+}	cal	20	0·1 (KNO_3)	$\Delta H_1°$ −2·43, $\Delta S_1°$ 64·8	58S
	ix	22	0·5 NH_4^+	K_1 15·49±0·04	62T
Ce^{4+}	sp	25	1 ($NaClO_4$)	K_1 24·2, complex unstable	65Bb
				$K(CeL+OH^- \rightleftharpoons CeLOH^-)$ 11·2	
Co^{2+}	cal	25	~ 0·05 $Co(NO_3)_2$	$\Delta H_1°$ −4·1, $\Delta S_1°$ 58	54C
	dis	20	0·1 ($KClO_4$)	K_1 16·55±0·04	63S
	gl	20	0·1 (KNO_3)	K_1 16·31,	64A
				$K(Co^{2+}+HL^{3-} \rightleftharpoons CoHL^-)$ 9·15	
	sp	25	1·0 ($NaClO_4$)	$K(CoL^{2-}+A \rightleftharpoons CoLA^{2-})$	65Ba
				1·40±0·04 where A = hydroxylamine	
				1·56±0·05 where A = hydrazine	
				1·64±0·05 where A = pyridine	
				1·68±0·05 where A = ethylenediamine	
				1·68±0·05 where A = propylenediamine	
	oth	20	0·1 (KNO_3)	K_1 16·5, electrophoresis	65J
	pol	25	0·2 (KNO_3)	K_1 15·71	65O
	sp	25	0·2 ($NaClO_4$)	K_1 16·14	67Ba
Co^{3+}	pol	25	0·2 (KNO_3)	K_1 40·6	65T
Cr^{2+}	gl	20	0·1 (KCl)	K_1 13·61,	64P
				$K(CrL^{2-}+H^+ \rightleftharpoons CrHL^-)$ 3·00	
	pol	?	2·5 NaCl	K_1 13·61	68F
Cr^{3+}	pol	20	0·1 (KCl)	K_1 23·40	64P
Cs^+	gl	25	0·32 CsCl	K_1 0·15	65Bc
Cu^{2+}	cal	25	~ 0·05 $Cu(NO_3)_2$	$\Delta H_1°$ −8·2, $\Delta S_1°$ 55	54C
	cal	20	0·1 (KNO_3)	$\Delta S_1°$ 56·4	56C
	sol	25	2 (KNO_3)	K (?) 20·04	63F
				See oxalic acid (111) for mixed ligands.	
	dis	20	0·1 ($KClO_4$)	K_1 18·92±0·04	63S
	gl	20	0·1 (KNO_3)	K_1 18·80,	64A
				$K(Cu^{2+}+HL^{3-} \rightleftharpoons CuHL^-)$ 11·54	
	sp	25	1·0 ($NaClO_4$)	$K(CuL^{2-}+A \rightleftharpoons CuLA^{2-})$	65Ba
				1·53±0·05 where A = hydroxylamine	
				1·77±0·07 where A = hydrazine	
				1·88+0·08 where A = pyridine	
				3·3±0·1 where A = ethylenediamine	
				3·3±0·1 where A = propylenediamine	
	oth	20	0·1 (KNO_3)	K_1 > 18, electrophoresis	65J
	pol	25	0·2 (KNO_3)	K_1 18·46	65O
	cal	25	0·1 KNO_3	ΔH_1 −8·8, ΔS_1 56	65W
	pol	25	0·1 50% dioxan	K_1 19·13	66Ba
	oth	30	0·5 (KNO_3)	$K(CuA^{2-}+L^{4-} \rightleftharpoons CuL^{2-}+A^{4-})$ −0·98	67C
				where H_4A = propylenedinitrilotetra-acetic acid, polarimetry	
Eu^{3+}	sp	18—20	0·04	K_1 16·43,	63G
				$K(Eu^{3+}+HL^{3-} \rightleftharpoons EuHL)$ 8·18	
				$K(Eu^{3+}+H_3L^- \rightleftharpoons EuH_3L^{2+})$ 3·20	
Fe^{2+}	gl	20	0·1 (KNO_3)	K_1 14·2,	64A
				$K(Fe^{2+}+HL^{3-} \rightleftharpoons FeHL^-)$ 6·86	
	cal	25	0·1 KNO_3	ΔH_1 −4·0, ΔS_1 51	65W

continued overleaf

836 $C_{10}H_{16}O_8N_2$ (contd)

Metal	Method	Temp	Medium	Log of equilibrium constant, remarks	Ref
Fe^{3+}	dis	20	0·1 ($KClO_4$)	$K[Fe^{3+}+L^{4-}+OH^- \rightleftharpoons FeL(OH)^{2-}]$ 34·0±0·1	63S
	sp	25	1 ($NaClO_4$)	$K(Fe^{3+}+HL^{3-} \rightleftharpoons FeHL)$ 15·2 $K(FeL^-+OH^- \rightleftharpoons FeLOH^{2-})$ 7·1	65Bb
	red	20	0·1 ($NaClO_4$)	K_1 25·1, $K(FeL^-+H^+ \rightleftharpoons FeHL)$ 1·2 $K(FeL^-+OH^- \rightleftharpoons FeLOH^{2-})$ 6·50	67B
	red	20	1 ($NaClO_4$)	K_1 25·15	67B
	sp	18—20	0·1	K_1 23·75, $K(Fe^{3+}+HL^{3-} \rightleftharpoons FeHL)$ 14·59 $K(Fe^{3+}+H_6L^{2+} \rightleftharpoons FeHL+5H^+)$ 1·35 $K(Fe^{3+}+H_5L^+ \rightleftharpoons FeL^-+5H^+)$ 1·30	67Z
Ga^{3+}	dis	20	0·1 ($KClO_4$)	$K[Ga^{3+}+L^{4-}+OH^- \rightleftharpoons GaL(OH)^{2-}]$ 34·0±0·1	63S
	Hg	20	0·1 (KNO_3)	K_1 20·5, $K(GaL^-+H^+ \rightleftharpoons GaHL)$ 1·91	66M
	red	20	0·1 ($NaClO_4$)	K_1 21·1, $K(GaL^-+H^+ \rightleftharpoons GaHL)$ 1·83 $K(GaL^-+OH^- \rightleftharpoons GaLOH^{2-})$ 8·36	67B
Ge^{IV}	pol	25	0·1 ($KClO_4$)	$K[Ge(OH)_4+H_4L \rightleftharpoons Ge(OH)_2H_2L]$ (?) 4·80 $K[Ge(OH)_4+H_3L^- \rightleftharpoons Ge(OH)_2HL^-]$ (?) 4·58	67K
Gd^{3+}	cal	20	0·1 (KNO_3)	$\Delta H_1°$ −1·11, $\Delta S_1°$ 75·7	58S
Hf^{4+}	ix	?	0·23 $HClO_4$	K_1 29·5	66E
Hg^{2+}	gl	20	0·1 (KNO_3)	K_1 21·8, $K(Hg^{2+}+HL^{3-} \rightleftharpoons HgHL^-)$ 14·6	64A
	cal	25	0·1 KNO_3	ΔH_1 −19·2, ΔS_1 37	65W
	Hg	20—40	0·1 (KNO_3)	K_1 (20°) 21·7±0·2, (30°) 21·44±0·10, (40°) 21·23±0·08 $K(HgL^{2-}+H^+ \rightleftharpoons HgHL^-)$ (20°) 3·19 $\Delta H_1°$ (25°) −9±2, $\Delta S_1°$ 67±5	66M
In^{3+}	cal	20	0·1 (KNO_3)	$\Delta H_1°$ −7·23, $\Delta S_1°$ 89·5	58S
	ix	?	0·5	K_1 23·06±0·10	63R
	dis	20	0·1 ($KClO_4$)	$K[In^{3+}+L^{4-}+OH^- \rightleftharpoons InL(OH)^{2-}]$ 32·0±0·1	63S
	sp	25	1 ($NaClO_4$)	$K[In^{3+}+HL^{3-} \rightleftharpoons InHL)$ 15·0	65Bb
	sp	20—22	?	K_1 25·62	65Z
	red	20	0·1 ($NaClO_4$)	K_1 25·3 $K(InL^-+H^+ \rightleftharpoons InHL)$ 1·5 $K(InL^-+OH^- \rightleftharpoons InLOH^{2-})$ 5·33	67B
K^+	gl	25	0·32 CsCl	K_1 0·96, $K(K^++HL^{3-} \rightleftharpoons KHL^{2-})$ −0·31	65Bc
La^{3+}	dis	20	0·1 ($KClO_4$)	K_1 15·3±0·2	63S
	gl	20	0·1 (KNO_3)	K_1 15·5	64A
	oth	20	0·1 (KNO_3)	K_1 15·5, electrophoresis	65J
Li^+	cal	25	~ 0·05 LiCl	$\Delta H_1°$ 0·1, $\Delta S_1°$ 13	54C
	gl	25	0·32 CsCl	K_1 2·85, K_2 0·83, $K(Li^++HL^{3-} \rightleftharpoons LiHL^{2-})$ 0·86	65Bc
Mg^{2+}	cal	25	~ 0·05 $Mg(NO_3)_2$	$\Delta H_1°$ 3·1, $\Delta S_1°$ 52	54C
	cal	20	0·1 (KNO_3)	$\Delta H_1°$ 3·14, $\Delta S_1°$ 50·5	56C
	gl	20	0·1 (KNO_3)	K_1 8·69, $K(Mg^{2+}+HL^{3-} \rightleftharpoons MgHL^-)$ 2·28	64A
	oth	20	0·1 (KNO_3)	K_1 ~ 11, electrophoresis	65J
Mn^{2+}	cal	25	~ 0·05 $Mn(NO_3)_2$	$\Delta H_1°$ −5·2, $\Delta S_1°$ 41	54C
	dis	20	0·1 ($KClO_4$)	K_1 12·88±0·05	63S
	gl	20	0·1 (KNO_3)	K_1 14·04, $K(Mn^{2+}+HL^{3-} \rightleftharpoons MnHL^-)$ 6·9	64A
	oth	20	0·1 (KNO_3)	K_1 14·5, electrophoresis	65J
	pol	25	0·2 (KNO_3)	K_1 13·64	65O

conti

Metal	Method	Temp	Medium	Log of equilibrium constant, remarks	Ref
Mn^{3+}	gl	25	?	$K[Mn(OH)L^{2-}+H^+ \rightleftharpoons MnL^-]$ 5·5	62Y
				$K(MnL^- + H^+ \rightleftharpoons MnHL)$ 2·7	
	sp	?	0·1	$K[Mn(OH)L^{2-}+H^+ \rightleftharpoons MnL^-]$ 5·3	
	pol	25	0·2	K_1 24·9	65Ta
	sp	25	0·2 ($NaClO_4$)	K_1 24·8	67H
Mo^{VI}	nmr	33—37	?	$K(MoO_4^{2-}+HL^{3-}+H^+ \rightleftharpoons MoO_3L^{4-})$ 8·8±0·3	66K
				$K(MoO_3L^{4-}+H^+ \rightleftharpoons MoO_3HL^{3-})$ 7·5±0·2	
				$K[2MoO_4^{2-}+L^{4-}+4H^+ \rightleftharpoons (MoO_3)_2L^{4-}]$ 35·1±0·3	
				$K(MoO_3+L^{4-} \rightleftharpoons MoO_3L^{4-})$ 10·7±0·3	
				$K[2MoO_3+L^{4-} \rightleftharpoons (MoO_3)_2L^{4-}]$ 19·5±0·3	
				$K[(MoO_3)_2L^{4-}+H_2L^{2-} \rightleftharpoons 2(MoO_3)HL^{3-}]$ 0·26	
Na^+	cal	25	~ 0·05 NaCl	$\Delta H_1°$ −1·4, $\Delta S_1°$ 3	54C
	oth	?	0·01 NaCl 0·03—5 $(CH_3)_4NOH$	K_1 2·61, $K(Na^+ + HL^{3-} \rightleftharpoons NaHL^{2-})$ −0·03 sodium-sensitive glass electrode	63P
	gl	25	0·32 CsCl	K_1 1·79, K_2 0·68, $K(Na^+ + HL^{3-} \rightleftharpoons NaHL^{2-})$ 0·49	65Bc
Nb^V	pol	20	1	$K[Nb(OH)_2^{3+}+L^{4-} \rightleftharpoons Nb(OH)_2L^-]$ 40·78	67V
Nd^{3+}	cal	20	0·1 (KNO_3)	$\Delta H_1°$ −2·98, $\Delta S_1°$ 65·8	58S
Ni^{2+}	cal	25	~ 0·05 $Ni(NO_3)_2$	$\Delta H_1°$ −7·6, $\Delta S_1°$ 55	54C
	cal	25	→ 0	K_1 20·33	59Y
	sol	25	2 (KNO_3)	K (?) 14·06	63F
				See oxalic acid (111) for mixed ligands.	
	dis	20	0·1 ($KClO_4$)	K_1 18·36±0·06	63S
	gl	20	0·1 (KNO_3)	K_1 18·62, $K(Ni^{2+}+HL^{3-} \rightleftharpoons NiHL^-)$ 11·56	64A
	sp	25	1·0 ($NaClO_4$)	$K(NiL^{2-}+A \rightleftharpoons NiLA^{2-})$	65Ba
				1·5±0·1 where A = hydroxylamine	
				1·66±0·05 where A = hydrazine	
				1·69±0·03 where A = pyridine	
				2·3±0·1 where A = ethylenediamine	
				2·3±0·1 where A = propylenediamine	
	oth	20	0·1 (KNO_3)	K_1 19, electrophoresis	65J
	pol	25	0·2 (KNO_3)	K_1 18·12	65O
	cal	25	0·1 KNO_3	ΔH_1 −8·5, ΔS_1 57	66W
	sp	25	0·5 (KCl)	$K(NiL^{2-}+NH_3 \rightleftharpoons NiLNH_3^{2-})$ 1·35±0·05	67J
			1 ($NaClO_4$)	1·39±0·01	
NpO_2^+	ix	18—22	0·05 NH_4ClO_4	K_1 9·7±0·1	61Z
Pa^V	ix	25	0·25	K (?) 8·19, K (?) 11·96	66G
Pb^{2+}	cal	25	~ 0·05 $Pb(NO_3)_2$	$\Delta H_1°$ −13·1, $\Delta S_1°$ 35	54C
	pol	15—35	0·2	K_s $(Pb^{2+}+PbL^{2-})$ (15°) −5·55, (25°) −5·64, (35°) −5·76	61Ta
	Hg	25	3 ($NaClO_4$)	K_1 15·99, $K(Pb^{2+}+HL^{3-} \rightleftharpoons PbHL^-)$ 12·00	63D
	dis	20	0·1 ($KClO_4$)	K_1 18·32±0·03	63S
	gl	20	0·1 (KNO_3)	K_1 18·04, $K(Pb^{2+}+HL^{3-} \rightleftharpoons PbHL^-)$ 10·61	64A
	oth	20	0·1 (KNO_3)	K_1 > 18, electrophoresis	65J
	pol	25	0·2 (KNO_3)	K_1 17·76	65O
	cal	25	0·1 KNO_3	ΔH_1 −13·1, ΔS_1 35	65W
	oth	25	0·5 (KNO_3)	$K(PbA^{2-}+L^{4-} \rightleftharpoons PbL^{2-}+A^{4-})$ −0·96 where H_4A = propylenedinitrilotetra-acetic acid, polarimetry	67C

continued overleaf

836 $C_{10}H_{16}O_8N_2$ (contd)

Metal	*Method*	*Temp*	*Medium*	*Log of equilibrium constant, remarks*	*Ref*
Pu^{3+}	ix	20	0·1	K_1 25·75	62K
Pu^{4+}	sol	25	?	K_1 26·1	59K
PuO_2^+	ix	20 (?)	0·05	K_1 10·2±0·2	59Q
	gl	20 (?)	0·1 KCl	K_1 12·9±0·1	61K
Rb^+	gl	25	0·32 CsCl	K_1 0·59, $K(Rb^+ + HL^{3-} \rightleftharpoons RbHL^{2-})$ −0·57	65B
Sb^{3+}	sp	25	1 ($NaClO_4$)	$K(SbO^+ + L^{4-} + 2H^+ \rightleftharpoons SbL^-)$ 24·8 $K[SbL^- + OH^- \rightleftharpoons SbL(OH)^{2-}]$ −8·7	65B
Sc^{3+}	dis	20	0·1 ($KClO_4$)	K_1 ~ 23·0 $K[Sc^{3+} + L^{4-} + OH^- \rightleftharpoons ScL(OH)^{2-}]$ 27·43±0·03	63S
	red	20	0·1 ($NaClO_4$)	K_1 25·05, $K(ScL^- + H^+ \rightleftharpoons ScHL)$ 2 $K[ScL^- + OH^- \rightleftharpoons ScL(OH)^{2-}]$ 3·3	67B
Sm^{3+}	con, pH	25	?	K_1 16·43 $K(SmL^- + H^+ \rightleftharpoons SmHL)$ 2·22	61T
	sp	18—20	0·04	K_1 16·34, K_2 6·49, $K(Sm^{3+} + HL^{3-} \rightleftharpoons SmHL)$ 8·18 $K(Sm^{3+} + H_4L \rightleftharpoons SmH_4L^{3+})$ 2·58 $K(Sm^{3+} + H_3L^- \rightleftharpoons SmH_3L^{2+})$ 3·62 $K(SmL^- + HL^{3-} \rightleftharpoons SmHL_2^{4-})$ 2·31	63G
Sn^{2+}	Hg(Sn), gl	20	1 (Na)ClO_4	K_1 18·3±0·1, $K(SnL^{2-} + H^+ \rightleftharpoons SnHL^-)$ 2·5±0·1 $K(SnHL^- + H^+ \rightleftharpoons SnH_2L)$ 1·5±0·2	68B
Sr^{2+}	cal	25	~ 0·05 $Sr(NO_3)_2$	$\Delta H_1°$ −4·2, $\Delta S_1°$ 26	54C
	ix	?	0·3	K_1 8·28, $K(Sr^{2+} + HL^{3-} \rightleftharpoons SrHL^-)$ 1·90 $K(Sr^{2+} + H_2L^{2-} \rightleftharpoons SrH_2L)$ 0·96	60M
	ix	22	0·165 NH_4^+	K_1 8·42±0·01	62T
	gl	30	0·1 KCl	K_1 8·80	63G
	oth	20	0·1 (KNO_3)	K_1 8, electrophoresis	65J
	oth	32, 42	?	K_1 (32°) 7·99, (42°) 7·62, ultrasonic ΔH_1 16·26 (?), ΔS_1 69·2 (?)	68L
	nmr	29	0·2 EDTA	$K(Sr^{2+} + SrL^{2-} \rightleftharpoons Sr_2L)$ 1·0±0·1 $K(Sr_3L^{2+} \rightleftharpoons Sr_2L + Sr^{2+})$ 0·4±0·1	68La
Th^{4+}	gl	25	0·1 KNO_3	$K(ThLA^{4-} + H^+ \rightleftharpoons ThLHA^{3-})$ 4·46±0·05 $K(ThL + HA^{3-} \rightleftharpoons ThLHA^{3-})$ 5·35±0·05 $K(ThL + A^{4-} \rightleftharpoons ThLA^{4-})$ 13·4±0·1 where H_4A = 1,2-dihydroxybenzene-3,5-disulphonic acid 13·66±0·02 where H_4A = 1,8-dihydroxynaphthalene-3,6-disulphonic acid $K(ThL + A^{3-} \rightleftharpoons ThLA^{3-})$ 9·29±0·02 where H_3A = 5-sulphosalicylic acid $K(ThL + A^{2-} \rightleftharpoons ThLA^{2-})$ 12·90±0·05 where H_2A = catechol 6·98±0·02 where H_2A = 8-hydroxyquinoline-5-sulphonic acid 6·70±0·02 where H_2A = iminodiacetic acid 3·09±0·02 where H_2A = 2-phthalic acid	64Ca
	gl	25	0·1 (KNO_3)	$K[ThL(OH) + ThL(OH) \rightleftharpoons Th_2L_2(OH)_2]$ 4·3	66M
	red	20	0·1 ($NaClO_4$)	K_1 25·3, $K(ThL + H^+ \rightleftharpoons ThHL^+)$ 1·98 $K[2ThL + 2OH^- \rightleftharpoons Th_2L_2(OH)_2^{2-}]$ 7·92	67B

con

Metal	*Method*	*Temp*	*Medium*	*Log of equilibrium constant, remarks*	*Ref*
Ti^{4+}	dis	20	0·1 ($KClO_4$)	$K(TiO^{2+}+L^{4-} \rightleftharpoons TiOL^{2-})$ 17·5±0·1	63S
Tl^{+}	gl	20	0·1 [$(CH_3)_4NNO_3$]	K_1 6·55±0·01	63I
				$K(Tl^{+}+HL^{3-} \rightleftharpoons TlHL^{2-})$ 2·06±0·02	
	gl	20	0·1 (KNO_3)	K_1 6·53	67Ab
Tl^{3+}	gl	20	0·4	$K(TlOH^{2-}+H^{+} \rightleftharpoons TlL^{-})$ 7·8	60B
	sp	18—20	dil	K_1 24·0	66Ka
				$K(FeL^{-}+Tl^{3+} \rightleftharpoons TlL^{-}+Fe^{3+})$ 0·086	
	red	20	1 ($HClO_4$)	K_1 38·9	67Ab
			1 ($NaClO_4$)	K_1 37·8	
U^{4+}	sol	25	?	K_1 25·6±0·4	59K
	sp	25	0·1	K_1 25·83	62K
	gl	25	0—1	$K(UL+OH^{-} \rightleftharpoons ULOH^{-})$ (→ 0) 8·95, (0·01) 9·00, (0·1) 9·07, (0·25) 9·08, (0·5) 9·17, (1) 9·13	63E
				$K[2ULOH^{-} \rightleftharpoons (ULOH)_2^{2-}]$ (→ 0) 2·78, (0·01) 2·84, (0·1) 2·75, (0·25) 2·79, (0·5) 2·48, (1) 2·86	
				$K(ULOH^{-}+H^{+} \rightleftharpoons UL)$ (→ 0) 4·94,	
				$K[(ULOH)_2^{2-}+2H^{+} \rightleftharpoons 2UL_2]$ (→ 0) 7·01	
				$K[ULOH^{-}+OH^{-} \rightleftharpoons UL(OH)_2^{2-}]$ (0·01) 5·91, (0·1) 6·29, (0·25) 6·41, (0·5) 6·49 (1) 6·87	
				$K[UL(OH)_2^{2-}+2H^{+} \rightleftharpoons UL]$ (0·01) 12·92, (0·1) 12·01, (0·25) 11·81, (0·5) 11·59, (1) 12·83	
				$K(2UL_2+H_2L^{2-} \rightleftharpoons H_2U_2L_3^{2-})$ (0·04) 3·64±0·06, (0·1) 3·57±0·11, (0·5) 2·85±0·15, (2) 3·3±0·2	
				$K(HU_2L_3^{3-}+H^{+} \rightleftharpoons H_2U_2L_3^{2-})$ (0·04) 3·2±0·1, (0·1) 2·8±0·1, (0·5) 3·0±0·3, (2) 2·2±0·4	
				$K(2UL_2+L^{4-} \rightleftharpoons U_2L_3^{4-})$ (0·1) 12·93	
				$K[(ULOH)_2^{2-}+HL^{3-}+OH^{-} \rightleftharpoons (UOH)_2L_3^{6-}]$ (0·1) 9·00	
				$K[U_2L_3^{4-}+2OH^{-} \rightleftharpoons (UOH)_2L_3^{6-}]$ (0·1) 12·11	
	gl	25	0·10 (KCl)	K_1 25·8±0·2	67Ca
				$K(UL+A^{4-} \rightleftharpoons ULA^{4-})$ 5·61±0·05 where H_4A = 1,2-dihydroxybenzene-3,5-disulphonic acid 16·22±0·01 where H_4A = 1,8-dihydroxynaphthalene-3,6-disulphonic acid	
				$K(UL+A^{3-} \rightleftharpoons ULA^{3-})$ 11·08±0·05 where H_3A = 5-sulphosalicylic acid	
				$K(UL+A^{2-} \rightleftharpoons ULA^{2-})$ 14·2±0·5 where H_2A = catechol	
				8·2±0·1 where H_2A = iminodiacetic acid	
				4·2±0·1 where H_2A = 2-phthalic acid	
				9·72±0·04 where H_2A = 8-hydroxyquinoline-5-sulphonic acid	
				$K[ULA(OH)^{3-}+H^{+} \rightleftharpoons ULA^{2-}]$ 7·14±0·01 where H_2A = 8-hydroxyquinoline-5-sulphonic acid	
	gl	20	0·10 (KCl)	$K(ULOH^{-}+H^{+} \rightleftharpoons UL)$ 4·72±0·01	68C
				$K[(ULOH)_2^{2-}+2H^{+} \rightleftharpoons 2UL]$ 6·53±0·04	
				$K[2ULOH^{-} \rightleftharpoons (ULOH)_2^{2-}]$ 2·9±0·05	
UO_2^{2+}	sol	25	?	$K(UO_2^{2+}+HL^{3-} \rightleftharpoons UL_2HL^{-})$ 4·13±0·02	59K
				K_s (UO_2H_2L) −5·64±0·04	
	sp	24	0·1 (NH_4Cl)	K_1 10·4±0·3	60K
				$K[2UO_2^{2+}+L^{4-} \rightleftharpoons (UO_2)_2L]$ 15·2±0·3	
	dis	?	0·1 $NaClO_4$	$K(UO_2^{2+}+HL^{3-} \rightleftharpoons UO_2HL^{-})$ 7·32±0·02	60S

continued overleaf

836 $C_{10}H_{16}O_8N_2$ (contd)

Metal	*Method*	*Temp*	*Medium*	*Log of equilibrium constant, remarks*	*Ref*
UO_2^{2+} (*contd*)	sp	25	0·15 ($NaClO_4$)	$K(UO_2^{2+}+HL^{3-} \rightleftharpoons UO_2HL^-)$ 7·8±0·1	64B
				$K[2UO_2^{2+}+L^{4-} \rightleftharpoons (UO_2)_2L]$ 17·8±0·1	
				$K[UO_2(OH)LH^{2-}+H^+ \rightleftharpoons UO_2LH^-]$ 5·6±0·1	
				$K[(UO_2OH)_2L^{2-}+2H^+ \rightleftharpoons (UO_2)_2L]$ 11·1±0·2	
				$K[2UO_2LH^-+Ca^{2+} \rightleftharpoons (UO_2)_2L+2H^++CaL^{2-}]$ −8·2±0·1	
VO_2^+	sp	20	0·1 (KCl)	$K(VO_2^++L^{4-} \rightleftharpoons VO_2L^{3-})$ 15·55	65P
				$K(VO_2^++HL^{3-} \rightleftharpoons VO_2HL^{2-})$ 9·60	
				$K(VO_2^++H_2L^{2-} \rightleftharpoons VO_2H_2L^-)$ 6·93	
				$K(VO_2^++H_3L^- \rightleftharpoons VO_2H_3L)$ 5·6	
Y^{3+}	cal	20	0·1 (KNO_3)	$\Delta H_1°$ 0·32, $\Delta S_1°$ 83·8	58S
	ix	22	0·5 NH_4^+	K_1 17·70±0·01	62T
	sol	20	0·15	K_1 18·21	63T
			0·1	K_{so} −25·13	
	sp	19	0·1	K_1 16·9	65V
Zn^{2+}	cal	25	~ 0·05 $Zn(NO_3)_2$	$\Delta H_1°$ −4·5, $\Delta S_1°$ 55	54C
	gl	20	0·1 (KNO_3)	K_1 16·26,	64A
				$K(Zn^{2+}+HL^{3-} \rightleftharpoons ZnHL^-)$ 9·0	
	oth	20	0·1 (KNO_3)	K_1 17·5, electrophoresis	65J
	nmr	34—38	?	$K[Zn(OH)^++L^{4-} \rightleftharpoons Zn(OH)L^{3-}]$ 2·2	65K
				$K[ZnL^{2-}+OH^- \rightleftharpoons Zn(OH)L^{3-}]$ 2·0	
	pol	25	0·2 (KNO_3)	K_1 15·94	65O
	cal	25	0·1 KNO_3	ΔH_1 −5·6, ΔS_1 57	65W
	oth	32, 42	?	K_1 (32°) 15·52, (42°) 14·80, ultrasonic	68L
				ΔH_1 25·14 (?), ΔS_1 115·5 (?)	
Zr^{4+}	ix	20	0·1	K_1 29·5±0·05	63K
			1—5 HNO_3	K_1 (1) 28·5±0·3, (5) 30·6±0·2	
	ix	?	1—5 HNO_3	K_1 (1) 28·46, (2), 30·58, (3) 31·11, (4) 30·92, (5) 30·63	64C
	ix	?	2	K(?) 5·91	64Pa
	ix	?	0·23—1 $HClO_4$	K_1 (0·23) 29·0, (1) 28·0	66E
	red	20	1 ($NaClO_4$)	K_1 27·7,	67B
				$K(ZrL+OH^- \rightleftharpoons ZrLOH^-)$ 7·9	
	oth	20	1·2 HCl	K_1 28·96±0·04, chromatographic	67T

Errata. 1964, Table 836. Under H^+ fourth line, (a) *for* ~ 0 *read* ~ 0 $HClO_4$; (b) *for* K_{13} *read* K_{15}; (c) *for* K_{14} *read* K_{16}; (d) *Add* reference 60B.

p. 635. Under H^+ ninth line, *delete* the entire line.

p. 635. Under H^+ eleventh line, (a) *for* 0·1 (KNO_3), 0·2 ($HClO_4$) *read* 0·2 (KCl); (b) *for* K_{13} 8·62 *read* K_{13} 2·62; (c) *add* in fifth column: calculated from 47S.

p. 635. Under Ba^{2+} first line, *for* ΔH_{14} *read* ΔH_1−4

p. 638. Under In^{3+} first line, *delete* K_1 24·95.

p. 638. Under La^{3+} second line, *for* ΔH_1 0·8 *read* ΔH_1 −0·8.

p. 638. Under Mg^{2+} first line, *for* 54S *read* 47S.

p. 638. Under Mg^{2+} third line, *for* 47S *read* 54C.

p. 638. Under Mo^{5+}, *for* K_1 6·36 *read* K_1 6·36 (?).

p. 638. Under Nd^{3+} second line, *for* ΔH_1 0·80 *read* ΔH_1 −0·80.

p. 638. Under Ni^{2+} eighth line, *for* 8·35 *read* −8·35.

p. 639. Under Sr^{2+} sixth line, *for* 54S *read* 47S.

p. 640. *For* Ti^{3+} *read* Ti^{4+}.

p. 640. Under Ti^{3+} third line, *for* K_1 17·3 *read* $K(TiO^{2+}+L^{4-} \rightleftharpoons TiOL^{2-})$ 17·3.

p. 641. Reference 53G, *for* Dnyckaerts *read* Duyckaerts.

p. 641. Reference 56C, *for* Stavely *read* Staveley.

cont

p. 641. Reference 57J, *for* Jordon *read* Jordan.

p. 641. Reference 59Bb, *delete Acta. chim. Acad. Sci. Hung.* 1960, **22,** 159.

p. 641. Reference 64I, add B. J. Intorre and A. E. Martell, *Inorg. Chem.*, 1964, **3,** 81.

54C R. C. Charles, *J. Amer. Chem. Soc.*, 1954, **76,** 5854
56C R. A. Care and L. A. K. Staveley, *J. Chem. Soc.*, 1956, 4571
58S L. A. K. Staveley and T. Randall, *Diss. Faraday Soc.*, 1958, **26,** 157
59G A. D. Gel'man, P. I. Artyukhin, and A. I. Moskvin, *Russ. J. Inorg. Chem.*, 1959, **4,** 599 (1332)
59K A. E. Klygin, I. D. Smirnova, and N. A. Nikol'skaya, *Russ. J. Inorg. Chem.*, 1959, **4,** 1209 (2623), 1279 (2766)
59Y K. B. Yatsimirskii and G. A. Karacheva, *Russ. J. Inorg. Chem.*, 1959, **4,** 127 (294)
60B A. I. Busev, V. G. Tiptsova, and T. A. Sokolova, *Russ. J. Inorg. Chem.*, 1960, **5,** 1326 (2749)
60K A. G. Kozlov and N. N. Krot, *Russ. J. Inorg. Chem.*, 1960, **5,** 954 (1959)
60Ka S. Kotrly and J. Vrestal, *Coll. Czech. Chem. Comm.*, 1960, **25,** 1148
60M N. N. Matorina and N. D. Safonova, *Russ. J. Inorg. Chem.*, 1960, **5,** 151 (313)
60S J. Stary, *Coll. Czech. Chem. Comm.*, 1960, **25,** 2630
61K O. L. Kabanova, *Russ. J. Inorg. Chem.*, 1961, **6,** 401 (786)
61M I. Miklos and R. Szegedi, *Acta Chim. Acad. Sci. Hung.*, 1961, **26,** 365
61T I. V. Tanangev and G. V. Shevchenko, *Russ. J. Inorg. Chem.*, 1961, **6,** 974 (1909)
61Ta N. Tanaka, M. Kamada, and G. Sato, *Bull. Chem. Soc. Japan*, 1961, **34,** 541
61Z Yu. A. Zolotov, I. N. Marov, and A. I. Moskvin, *Russ. J. Inorg. Chem.*, 1961, **6,** 539 (1055)
62K N. N. Krot, N. P. Ermolaev, and A. D. Gel'man, *Russ. J. Inorg. Chem.*, 1962, 7, 1062 (2054)
62T L. I. Tikhonova, *Russ. J. Inorg. Chem.*, 1962, **7,** 424 (822)
62Y Y. Yoshino, A. Ouchi, Y. Tsunoda, and M. Kojima, *Canad. J. Chem.*, 1962, **40,** 775
63D A. F. Donda and A. M. Giuliani, *Ricerca sci.*, 1963, **33** (II-A) 819
63E N. P. Ermolaev and N. N. Krot, *Russ. J. Inorg. Chem.*, 1963, **8,** 1282 (2447)
63F Ya. D. Fridman, R. A. Veresova, N. V. Dolgashova, and R. I. Sorochan, *Russ. J. Inorg. Chem.*, 1963, **8,** 344 (676)
63G Yu. P. Galaktionov and K. V. Astakhov, *Russ. J. Inorg. Chem.*, 1963, **8,** 460 (896)
63Ga J. H. Grimes, A. J. Huggard, and S. P. Wilford, *J. Inorg. Nuclear Chem.*, 1963, **25,** 1225
63I H. Irving and J. J. R. F. da Silva, *J. Chem. Soc.*, 1963, 458
63K M. Kyrs and R. Caletka, *Talanta*, 1963, **10,** 1115
63P V. Palaty, *Canad. J. Chem.*, 1963, **41,** 18
63R D. I. Ryabchikov, I. N. Marov, and K. K'o-min, *Russ. J. Inorg. Chem.*, 1963, **8,** 326 (641)
63S J. Stary, *Analyt. Chim. Acta*, 1963, **28,** 132
63Sa V. V. Starostin, V. I. Spitsyn, and G. F. Silina, *Russ. J. Inorg. Chem.*, 1963, **8,** 335 (660)
63T I. V. Tananaev and G. S. Tereshin, *Russ. J. Inorg. Chem.*, 1963, **8,** 1182 (2258)
64A G. Anderegg, *Helv. Chim. Acta*, 1964, **47,** 1801
64B T. R. Bhat and M. Krishnamurthy, *J. Inorg. Nuclear Chem.*, 1964, **26,** 587
64C R. Caletka, M. Kyrs, and J. Rais, *J. Inorg. Nuclear Chem.*, 1964, **26,** 1443
64Ca G. H. Carey, R. F. Bogucki, and A. E. Martell, *Inorg. Chem.*, 1964, **3,** 1288
64E N. Elenkova and L. Ilcheya, *Godishnik Klim.-Tekhnol. Inst.*, 1964, **11,** 143; (*Chem. Abs.*, 1967, **66,** 14524y
64J J. Joussot-Dubien and M. Cotrait, *J. Chim. phys.*, 1964, **61,** 1211
64P R. L. Pecsok, L. D. Shields, and W. P. Schaefer, *Inorg. Chem.*, 1964, **3,** 114
64Pa L. N. Pankratova, L. G. Vlasov, and A. V. Lapitskii, *Russ. J. Inorg. Chem.*, 1964, **9,** 954 (1763)
65B T. R. Bhat and R. K. Iyer, *Z. anorg. Chem.*, 1965, **335,** 331
65Ba T. R. Bhat, D. Radhamma, and J. Shankar, *J. Inorg. Nuclear Chem.*, 1965, **27,** 2641
65Bb T. R. Bhat and D. Radhamma, *Indian J. Chem.*, 1965, **3,** 151
65Bc J. Botts, A. Chashin, and H. L. Young, *Biochemistry*, 1965, **4,** 1788
65J V. Jokl and J. Majer, *Chem. Zvesti*, 1965, **19,** 249
65K R. J. Kula, *Analyt. Chem.*, 1965, **37,** 989
65O H. Ogino, *Bull. Chem. Soc. Japan*, 1965, **38,** 771
65P L. Przyborowski, G. Schwarzenbach, and Th. Zimmerman, *Helv. Chim. Acta*, 1965, **48,** 1556
65T N. Tanaka and H. Ogino, *Bull. Chem. Soc. Japan*, 1965, **38,** 1054
65Ta N. Tanaka, T. Shirakashi, and H. Ogino, *Bull. Chem. Soc. Japan*, 1965, **38,** 1515
65V V. B. Verenikin, K. V. Astakhov, and F. G. Malanichev, *Russ. J. Inorg. Chem.*, 1965, **10,** 1344 (2471)
65W D. L. Wright, J. H. Holloway, and C. N. Reilley, *Analyt. Chem.*, 1965, **37,** 884
65Z N. M. Zhirnova, K. V. Astakhov, and S. A. Barkov, *Russ. J. Phys. Chem.*, 1965, **39,** 647 (1224)
66B C. Bamberger and F. Laguna, *J. Inorg. Nuclear Chem.*, 1966, **28,** 1067
66Ba H. Berge, *J. prakt. Chem* , 1966, **34,** 15
66Bb M. T. Beck and A. Gergely, *Acta Chim. Acad. Sci. Hung.*, 1966, **50,** 155
66E A. N. Ermakov, I. N. Marov, and G. A. Evtikova, *Russ. J. Inorg. Chem.*, 1966, **11,** 618 (1155)
66G I. Galateanu, *Canad. J. Chem.*, 1966, **44,** 647
66K R. J. Kula, *Analyt. Chem.*, 1966, **38,** 1581
66Ka V. I. Kornev, K. V. Astakhov, and V. I. Rybina, *Russ. J. Inorg. Chem.*, 1966, **11,** 988 (1851)
66M T. Moeller and S. K. Chu, *J. Inorg. Nuclear Chem.*, 1966, **28,** 153
67A D. A. Aikens and F. J. Bahbah, *Analyt. Chem.*, 1967, **39,** 646
67Aa G. Anderegg, *Helv. Chim. Acta*, 1967, **50,** 2333

continued overleaf

836 $C_{10}H_{16}O_8N_2$ (contd)

67Ab G. Anderegg and E. Bottari, *Helv. Chim. Acta*, 1967, **50**, 2341
67B E. Bottari and G. Anderegg, *Helv. Chim. Acta*, 1967, **50**, 2349
67Ba T. R. Bhat, R. R. Das, and J. Shankar, *Indian J. Chem.*, 1967, **5**, 324
67C J. D. Carr, K. Torrance, C. J. Cruz, and C. N. Reilley, *Analyt. Chem.*, 1967, **39**, 1358
67Ca G. H. Carey and A. E. Martell, *J. Amer. Chem. Soc.*, 1967, **89**, 2859
67H R. E. Hamm and M. A. Suwyn, *Inorg. Chem.*, 1967, **6**, 139
67J N. E. Jacobs and D. W. Margerum, *Inorg. Chem.*, 1967, **6**, 2038
67K N. Konopik, *Z. analyt. Chem.*, 1967, **224**, 107
67R G. A. Rechnitz and Z. F. Lin, *Analyt. Chem.*, 1967, **39**, 1406
67T L. I. Tikhonova, *Russ. J. Inorg. Chem.*, 1967, **12**, 494 (939)
67V G. A. Volkova and V. G. Sochevanov, *Russ. J. Inorg. Chem.*, 1967, **12**, 222 (433)
67Z N. M. Zhirnova, K. V. Astakhov, and S. A. Barkov, *Russ. J. Phys. Chem.*, 1967, **41**, 366 (710)
68B E. Bottari, A. Liberti, and A. Rufulo, *J. Inorg. Nuclear Chem.*, 1968, **30**, 2173
68C G. H. Carey and A. E. Martell, *J. Amer. Chem. Soc.*, 1968, **90**, 32
68F E. Fischerova, O. Dracka, and M. Meloun, *Coll. Czech. Chem. Comm.*, 1968, **33**, 473
68L S. Laemi, S. Prakash, and S. Prakash, *Indian J. Chem.*, 1968, **6**, 31
68La D. E. Leyden and J. F. Whidley, *Analyt. Chim. Acta*, 1968, **42**, 271
68W H. Wikberg and A. Ringbom, *Suomen Kem.*, 1968, **B41**, 177

836.1 $C_{10}H_{17}O_4N$ *N*-(Cyclohexyl)iminodiacetic acid H_2L

Metal	*Method*	*Temp*	*Medium*	*Log of equilibrium constant, remarks*	*Ref*
H^+	gl	25	0·5 ($NaClO_4$)	K_1 10·58±0·01, K_2 2·26±0·03, K_3 1·49±0·07	67F
	gl	20	0·1 (KNO_3)	K_1 10·81±0·01, K_2 2·15±0·02	63I, 64P
Ag^+	gl	20	0·1 (KNO_3)	K_1 4·94±0·02	63I
Ba^{2+}	gl	20	0·1 (KNO_3)	K_1 2·37±0·02	63I
Ca^{2+}	gl	20	0·1 (KNO_3)	K_1 3·34±0·02	63I
Cd^{2+}	gl	20	0·1 (KNO_3)	K_1 6·94	64P
	gl	25	0·5 ($NaClO_4$)	K_1 7·09±0·09	67F
Co^{2+}	gl	25	0·5 ($NaClO_4$)	K_1 7·19±0·02, β_2 12·87±0·02	67F
Cu^{2+}	gl	20	0·1 (KNO_3)	K_1 11·04	64P
	gl	25	0·5 ($NaClO_4$)	K_1 10·92±0·05, β_2 16·35±0·06	67F
Li^+	gl	20	0·1 (KNO_3)	K_1 1·74±0·02	63I
Mg^{2+}	gl	20	0·1 (KNO_3)	K_1 3·46±0·02	63I
Na^+	gl	20	0·1 (KNO_3)	K_1 0·90±0·03	63I
Ni^{2+}	gl	20	0·1 (KNO_3)	K_1 8·08	64P
Sr^{2+}	gl	20	0·1 (KNO_3)	K_1 2·55±0·01	63I
Tl^+	gl	20	0·1 (KNO_3)	K_1 3·40±0·01	63I
Zn^{2+}	gl	20	0·1 (KNO_3)	K_1 7·60	64P
	gl	25	0·5 ($NaClO_4$)	K_1 7·42±0·03, β_2 12·73±0·05	67F

63I H. Irving and J. J. R. F. da Silva, *J. Chem. Soc.*, 1963, 3308
64P L. D. Pettit and H. M. N. H. Irving, *J. Chem. Soc.*, 1964, 5336
67F A. Furlani, M. Maltese, E. Mantovani, and C. Maremmani, *Gazzetta*, 1967, **97**, 1423

$C_{10}H_{17}O_5N$ *N*-(2-Hydroxycyclohexyl)iminodiacetic acid* H_2L 836.2

Metal	Method	Temp	Medium	Log of equilibrium constant, remarks	Ref
H^+	gl	20	0·1 (KNO_3)	K_1 9·57±0·01, K_2 2·32±0·01	63I
Ag^+	gl	20	0·1 (KNO_3)	K_1 3·83±0·01	63I
Ba^{2+}	gl	20	0·1 (KNO_3)	K_1 3·26±0·01	63I
Li^+	gl	20	0·1 (KNO_3)	K_1 2·19±0·01	63I
Na^+	gl	20	0·1 (KNO_3)	K_1 0·76±0·01	63I
Sr^{2+}	gl	20	0·1 (KNO_3)	K_1 3·81±0·01	63I
Tl^+	gl	20	0·1 (KNO_3)	K_1 3·07±0·01	63I

* Steric configuration of ligand not given.

Erratum. 1964 p. 634, Table 835. Table **836.2** was numbered **835** in the 1964 edition.

63I H. Irving and J. J. R. F. da Silva, *J. Chem. Soc.*, 1963, 3308

$C_{10}H_{17}O_5N$ *N*-(Tetrahydropyran-2-ylmethyl)iminodiacetic acid H_2L 836.3

Metal	Method	Temp	Medium	Log of equilibrium constant, remarks	Ref
H^+	gl	20	0·1 (KNO_3)	K_1 9·04±0·02, K_2 1·88±0·02	63I
Ag^+	gl	20	0·1 (KNO_3)	K_1 4·83±0·03	63I
Ba^{2+}	gl	20	0·1 (KNO_3)	K_1 3·61±0·01	63I
Ca^{2+}	gl	20	0·1 (KNO_3)	K_1 4·86±0·01	63I
Cd^{2+}	gl	20	0·1 (KNO_3)	K_1 7·93±0·01, K_2 6·06±0·02	63I
				$K(Cd^{2+}+HL^- \rightleftharpoons CdHL^+)$ 1·47±0·02	
Co^{2+}	gl	20	0·1 (KNO_3)	K_1 8·51±0·01, K_2 4·5±0·9	63I
				$K(Co^{2+}+HL^- \rightleftharpoons CoHL^+)$ 2·60±0·02	
Fe^{2+}	gl	20	0·1 (KNO_3)	K_1 7·40±0·01	63I
Li^+	gl	20	0·1 (KNO_3)	K_1 1·7±0·1	63I
Mg^{2+}	gl	20	0·1 (KNO_3)	K_1 3·70±0·01	63I
Mn^{2+}	gl	20	0·1 (KNO_3)	K_1 5·89±0·03, K_2 4·35±0·02	63I
Na^+	gl	20	0·1 (KNO_3)	K_1 0·85±0·04	63I
Pb^{2+}	gl	20	0·1 (KNO_3)	K_1 10·30±0·02	63I
				$K(Pb^{2+}+HL^- \rightleftharpoons PbHL^+)$ 5·16±0·03	
Sr^{2+}	gl	20	0·1 (KNO_3)	K_1 3·97±0·01	63I
Tl^+	gl	20	0·1 (KNO_3)	K_1 4·06±0·01	63I
Zn^{2+}	gl	20	0·1 (KNO_3)	K_1 9·06±0·01, K_2 4·52±0·05	63I
				$K(Zn^{2+}+HL^- \rightleftharpoons ZnHL^+)$ 2·75±0·02	

63I H. Irving and J. J. R. F. da Silva, *J. Chem. Soc.*, 1963, 1144

838 $C_{10}H_{18}O_7N_2$ Hydroxyethylethylenediaminetriacetic acid (*N*-[*N*′-(2″-hydroxyethyl)-*N*′-(carboxymethyl)-2′-aminoethyl]iminodiacetic acid (HEDTA) H

Metal	*Method*	*Temp*	*Medium*	*Log of equilibrium constant, remarks*	*Ref*
H^+	gl	30	0·1 (KCl)	K_1 9·73, K_2 5·33, K_3 2·64	62S, 65
	gl	25	1 ($NaClO_4$)	K_1 8·65, K_2 5·11, K_3 2·30	66B
	gl	20—40	0·1 (KNO_3)	K_1 (20°) 9·86, (30°) 9·53, (40°) 9·43, K_2 (20°) 5·31, (30°) 5·28, (40°) 5·20, K_3 (20°) 2·51 $\Delta H_1°$ (25°) -9 ± 1, $\Delta S_1°$ 14 ± 2, $\Delta H_2°$ (25°) $-2·4\pm0·4$, $\Delta S_2°$ 16 ± 1	66M
	gl	25	0·2 ($NaClO_4$)	K_1 9·70, K_2 5·35, K_3 2·63	67B
	?	?	?	$K_1K_2K_3$ 17·15	67K
	sp	20	0·5 $NaClO_4$	K_1 9·13, K_2 5·24, K_3 1·83	67N
	gl	25	0·1 (KNO_3)	K_1 9·81, K_2 5·41, K_3 2·72	68W
Ag^+	Ag	25	0·1 (KNO_3)	K_1 6·71	68W
Al^{3+}	Hg	20—40	0·1 (KNO_3)	K_1 (20°) $12·43\pm0·06$, (30°) $12·6\pm0·2$, (40°) $12·9\pm0·1$ $\Delta H_1°$ 9 ± 2, $\Delta S_1°$ 89 ± 7 $K(AlL+H^+\rightleftharpoons AlHL^+)$ (20°) 5·08	66M
	Hg	25	0·1 KNO_3	K_1 14·4, $K(AlL+H^+\rightleftharpoons AlHL^+)$ 2·4 $K(AlL+OH^-\rightleftharpoons AlOHL^-)$ 9·3	67A
	sp	25	0·2 ($NaClO_4$)	K_1 13·96	67B
	gl	25	0·2 ($NaClO_4$)	$K(AlL+H^+\rightleftharpoons AlHL^+)$ 2·14 $K(AlLOH^-+H^+\rightleftharpoons AlL)$ 4·89 $K[AlL(OH)_2^{2-}+H^+\rightleftharpoons AlLOH^-]$ 9·19	67B
Ba^{2+}	cal	25	0·1 KNO_3	ΔH_1 $-5·4$, ΔS_1 10	65W
Bi^{3+}	sp	25	1 ($NaClO_4$)	$K(BiO^++L^{3-}+2H^+\rightleftharpoons BiL)$ 22·3	66B
	sp	20	0·5 $NaClO_4$	K_1 21·8	67N
Ca^{2+}	cal	25	0·1 KNO_3	ΔH_1 $-6·5$, ΔS_1 15	65W
Cd^{2+}	cal	25	0·1 KNO_3	ΔH_1 $-10·3$, ΔS_1 25	65W
	pol	25	0·1 $NaNO_3$	K_1 $13·6\pm0·1$	67K
Ce^{3+}	gl	25	0·1 (KNO_3)	$K(CeL+A^{2-}\rightleftharpoons CeLA^{2-})$ $3·50\pm0·01$ where H_2A = iminodiacetic acid $4·07\pm0·02$ where H_2A = *N*-hydroxyethyliminodiacetic acid $3·60\pm0·04$ where H_2A = ethylenediamine-*NN*′-diacetic acid	63T
Co^{2+}	cal	25	0·1 KNO_3	ΔH_1 $-6·5$, ΔS_1 44	65W
	sp	25	0·2 ($NaClO_4$)	K_1 14·12	67B
Cu^{2+}	pol	25	0·1 NH_3–NH_4Cl	K_1 18·8 $K(Cu^{2+}+HL^{2-}\rightleftharpoons CuHL)$ 12·4	62S
	sp	25	1·25 ($NaClO_4$)	$K(CuL^-+Ni^{2+}\rightleftharpoons NiL^-+Cu^{2+})$ 2·0 $K(CuL^-+H^+\rightleftharpoons CuHL)$ 2·32	63B
	cal	25	0·1 KNO_3	ΔH_1 $-9·4$, ΔS_1 48	65W
Dy^{3+}	gl	25	0·1 (KNO_3)	$K(DyL+A^{2-}\rightleftharpoons DyLA^{2-})$ $5·33\pm0·01$ where H_2A = iminodiacetic acid $4·81\pm0·03$ where H_2A = *N*-hydroxyethyliminodiacetic acid	63T
Er^{3+}	gl	25	0·1 (KNO_3)	$K(ErL+A^{2-}\rightleftharpoons ErLA^{2-})$ $5·30\pm0·01$ where H_2A = iminodiacetic acid $4·62\pm0·03$ where H_2A = *N*-hydroxyethyliminodiacetic acid	63T
Eu^{3+}	gl	25	0·1 (KNO_3)	$K(EuL+A^{2-}\rightleftharpoons EuLA^{2-})$ $4·77\pm0·01$ where H_2A = iminodiacetic acid $4·61\pm0·02$ where H_2A = *N*-hydroxyethyliminodiacetic acid	63T
Fe^{2+}	cal	25	0·1 KNO_3	ΔH_1 $-6·0$, ΔS_1 33	65W
Fe^{3+}	sp	20	?	K_1 19·06	67Ka
Ga^{3+}	Hg	20	0·1 (KNO_3)	K_1 16·9 $K(GaL+H^+\rightleftharpoons GaHL^+)$ 4·17	66M

cont

Metal	Method	Temp	Medium	Log of equilibrium constant, remarks	Ref
Gd^{3+}	gl	25	0·1 (KNO_3)	$K(GdL+A^{2-} \rightleftharpoons GdLA^{2-})$	63T
				4·96±0·01 where H_2A = iminodiacetic acid	
				4·57±0·03 where H_2A = *N*-hydroxyethyliminodiacetic acid	
Hg^{2+}	cal	25	0·1 KNO_3	ΔH_1 −20·0, ΔS_1 25	65W
	Hg	20—40	0·1 (KNO_3)	K_1 (20°) 19·47±0·09, (30°) 19·3±0·1, (40°) 19·1±0·1	66M
				$\Delta H_1°$ (25°) −7±2, $\Delta S_1°$ 65±7	
				$K(HgL^- + H^+ \rightleftharpoons HgHL)$ (20°) 2·57	
Ho^{3+}	gl	25	0·1 (KNO_3)	$K(HoL+A^{2-} \rightleftharpoons HoLA^{2-})$	63T
				5·37±0·01 where H_2A = iminodiacetic acid	
				4·76±0·03 where H_2A = *N*-hydroxyethyliminodiacetic acid	
				$K(HoLOH^- + H^+ \rightleftharpoons HoL)$ 9·06±0·04	
In^{3+}	ix	?	0·5	K_1 17·16±0·03	63R
La^{3+}	gl	25	0·1 (KNO_3)	$K(LaL+A^{2-} \rightleftharpoons LaLA^{2-})$	63T
				3·22±0·03 where H_2A = iminodiacetic acid	
				4·04±0·02 where H_2A = *N*-hydroxyethyliminodiacetic acid	
				3·90±0·05 where H_2A = ethylenediamine-*NN′*-diacetic acid	
Lu^{3+}	gl	25	0·1 (KNO_3)	$K(LuL+A^{2-} \rightleftharpoons LuLA^{2-})$	63T
				4·51±0·02 where H_2A = iminodiacetic acid	
				3·88±0·03 where H_2A = *N*-hydroxyethyliminodiacetic acid	
Mg^{2+}	cal	25	0·1 KNO_3	ΔH_1 3·4, ΔS_1 43	65W
Mn^{2+}	cal	25	0·1 KNO_3	ΔH_1 −5·2, ΔS_1 32	65W
Mn^{3+}	sp	25	0·2 ($NaClO_4$)	K_1 22·7	67H
Nd^{3+}	gl	25	0·1 (KNO_3)	$K(NdL+A^{2-} \rightleftharpoons NdLA^{2-})$	63T
				4·07±0·01 where H_2A = iminodiacetic acid	
				4·23±0·03 where H_2A = *N*-hydroxyethyliminodiacetic acid	
				3·41±0·06 where H_2A = ethylenediamine-*NN′*-diacetic acid	
Ni^{2+}	sp	25	1·25 ($NaClO_4$)	$K(NiL^- + H^+ \rightleftharpoons NiHL)$ 2·54	63B
				also see Cu^{2+} above	
	cal	25	0·1 KNO_3	ΔH_1 −10·3, ΔS_1 45	65W
	sp	25	0·5 (KCl)	$K(NiL^- + NH_3 \rightleftharpoons NiLNH_3^-)$ 2·00±0·07	67J
Pb^{2+}	cal	25	0·1 KNO_3	ΔH_1 −12·6, ΔS_1 29	65W
Pr^{3+}	gl	25	0·1 (KNO_3)	$K(PrL+A^{2-} \rightleftharpoons PrLA^{2-})$	63T
				3·84±0·01 where H_2A = iminodiacetic acid	
				4·20±0·02 where H_2A = *N*-hydroxyethyliminodiacetic acid	
				3·43±0·05 where H_2A = ethylenediamine-*NN′*-diacetic acid	
Sb^{3+}	sp	25	1 ($NaClO_4$)	$K(SbO^+ + L^{3-} + 2H^+ \rightleftharpoons SbL)$ 20·2	66B
				$K[Sb(H_{-1}L)^- + H^+ \rightleftharpoons SbL]$ −3·2	
				$K[Sb(H_{-1}L)^- + OH^- \rightleftharpoons SbLOH^{2-}]$ −8·1	
Sm^{3+}	gl	25	0·1 (KNO_3)	$K(SmL+A^{2-} \rightleftharpoons SmLA^{2-})$	63T
				4·57±0·02 where H_2A = iminodiacetic acid	
				4·49±0·03 where H_2A = *N*-hydroxyethyliminodiacetic acid	
				3·02±0·06 where H_2A = ethylenediamine-*NN′*-diacetic acid	
Sr^{2+}	cal	25	0·1 KNO_3	ΔH_1 −5·2, ΔS_1 14	65W
Tb^{3+}	gl	25	0·1 (KNO_3)	$K(TbL+A^{2-} \rightleftharpoons TbLA^{2-})$	63T
				5·18±0·01 where H_2A = iminodiacetic acid	
				4·73±0·03 where H_2A = *N*-hydroxyethyliminodiacetic acid	
Th^{4+}	ix	20	0·1 NH_4^+	K_1 19·24±0·01	65R
Tl^{3+}	sp	20	?	K_1 19·72	67Ka
				$K(Tl^{3+} + FeL \rightleftharpoons TlL + Fe^{3+})$ 0·66	
Tm^{3+}	gl	25	0·1 (KNO_3)	$K(TmL+A^{2-} \rightleftharpoons TmLA^{2-})$	63T
				4·95±0·02 where H_2A = iminodiacetic acid	
				4·26±0·01 where H_2A = *N*-hydroxyethyliminodiacetic acid	

continued overleaf

838 $C_{10}H_{18}O_7N_2$ (contd)

Metal	*Method*	*Temp*	*Medium*	*Log of equilibrium constant, remarks*	*Ref*
Y^{3+}	gl	25	0·1 (KNO_3)	$K(YL+A^{2-} \rightleftharpoons YLA^{2-})$	63T
				5·10±0·01 where H_2A = iminodiacetic acid	
				4·39±0·01 where H_2A = *N*-hydroxyethyliminodiacetic acid	
Yb^{3+}	gl	25	0·1 (KNO_3)	$K(YbL+A^{2-} \rightleftharpoons YbLA^{2-})$	63T
				4·74±0·04 where H_2A = iminodiacetic acid	
				4·05±0·03 where H_2A = *N*-hydroxyethyliminodiacetic acid	
Zn^{2+}	cal	25	0·1 KNO_3	ΔH_1 −8·4, ΔS_1 38	65W

62S K. H. Schroder, *Acta Chem. Scand.*, 1962, **16**, 1315
63B T. J. Bydalek and D. W. Margerum, *Inorg. Chem.*, 1963, **2**, 678
63R R. I. Ryabchikov, I. N. Marov, and Y. K'o-min, *Russ. J. Inorg. Chem.*, 1963, **8**, 326 (641)
63T L. C. Thompson and J. A. Loraas, *Inorg. Chem.*, 1963, **2**, 89
65R D. I. Ryabchikov and M. P. Volynets, *Russ. J. Inorg. Chem.*, 1965, **10**, 334 (619)
65W D. L. Wright, J. H. Holloway, and C. N. Reilley, *Analyt. Chem.*, 1965, **37**, 884
66B T. R. Bhat, R. K. Iyer, and J. Shankar, *Z. anorg. Chem.*, 1966, **343**, 329
66M T. Moeller and S. Chu, *J. Inorg. Nuclear Chem.*, 1966, **28**, 153
67A D. A. Aikens and F. J. Bahbah, *Analyt. Chem.*, 1967, **39**, 646
67B T. R. Bhat, R. R. Das, and J. Shankar, *Indian J. Chem.*, 1967, **5**, 324
67H R. E. Hamm and M. A. Suwyn, *Inorg. Chem.*, 1967, **6**, 139
67J N. E. Jacobs and D. W. Margerum, *Inorg. Chem.*, 1967, **6**, 2038
67K Y. Koike and H. Hamaguchi, *J. Inorg. Nuclear Chem.*, 1967, **29**, 473
67Ka V. I. Kornev, A. V. Astakhov, and V. I. Rybina, *Russ. J. Phys. Chem.*, 1967, **41**, 730 (1378)
67N T. Nozaki and K. Kosiba, *Nippon Kagaku Zasshi*, 1967, **88**, 1287
68W H. Wikberg and A. Ringbom, *Suomen Kem.*, 1968, **B41**, 177

843.1 $C_{10}H_{22}O_2N_4$ *NN'*-Diglycyl-1,6-diaminohexane L

Metal	*Method*	*Temp*	*Medium*	*Log of equilibrium constant, remarks*	*Ref*
H^+	gl	23	0·5 (KCl)	K_1 8·56, K_2 7·93	67Z
Cu^{2+}	gl	23	0·5 (KCl)	K_1 8·93	67Z
				$K(CuL^{2+}+H^+ \rightleftharpoons CuHL^{3+})$ 5·18	
				$K[Cu(H_{-1}L)+H^+ \rightleftharpoons CuL^{2+})$ 7·85	
				$K[Cu(H_{-2}L)+H^+ \rightleftharpoons Cu(H_{-1}L)^+]$ 10·62	

67Z A. Zuberbuhler and S. Fallab *Helv. Chim. Acta*, 1967, **50**, 889

843.2 $C_{10}H_{22}O_2N_4$ *NN'*-Bis(2-dimethylaminoethyl)oxamide L

Metal	*Method*	*Temp*	*Medium*	*Log of equilibrium constant, remarks*	*Ref*
H^+	gl	25	0·5 (KCl)	K_1 8·93, K_2 8·19	68Z
Cu^{2+}	gl	25	0·5 (KCl)	K_1 8·26	68Z
				$K(CuOHL^+ +H^+ \rightleftharpoons CuL^{2+})$ 7·71	
				$K[Cu(H_{-2}L)+H^+ \rightleftharpoons CuOHL^+)$ 7·99	
				$K[Cu(H_{-2}L)+Cu^{2+} \rightleftharpoons Cu_2(H_{-2}L)^{2+}]$ 8·93	
				$K[Cu_2(H_{-2}L)^{2+}+OH^- \rightleftharpoons Cu_2OH(H_{-2}L)^+]$ 5·21	
				$K[Cu_2OH(H_{-2}L)^+ +OH^- \rightleftharpoons Cu_2(OH)_2(H_{-2}L)]$ 4·64	

68Z A. Zuberbuhler and Th. Kaden, *Helv. Chim. Acta.*, 1968, **51**, 1805

$C_{10}H_{24}O_4N_2$ 2,2′,2″,2‴-(Ethylenedinitrito)tetraethanol L 844

Metal	Method	Temp	Medium	Log of equilibrium constant, remarks	Ref
H^+	gl	25	0·5 (KNO_3)	K_1 8·45, K_2 4·45	64P
	gl	?	0·5 ($NaNO_3$)	K_1 8·50, K_2 4·50	65I
Co^{2+}	gl	?	0·5 ($NaNO_3$)	K_1 5·30±0·05	65I
Cu^{2+}	gl	?	0·5 ($NaNO_3$)	K_1 8·45±0·05	65I
Ni^{2+}	gl	25	0·5 (KNO_3)	K_1 6·50, $K[Ni(H_{-1}L)^+ + H^+ \rightleftharpoons NiL^{2+}]$ 9·07	64P

64P K. H. Pearson and K. H. Gayer, *Inorg. Chem.*, 1964, 3, 476
65I Y. J. Israeli, *J. Inorg. Nuclear Chem.*, 1965, 27, 2271

$C_{10}H_{24}O_6P_2$ Decane-1,10-diphosphonic acid H_4L 844.1

Metal	Method	Temp	Medium	Log of equilibrium constant, remarks	Ref
H^+	gl	25	1 $(CH_3)_4NBr$	K_1 8·73±0·07, K_2 7·68±0·07, K_3 3·06±0·07, K_4 2·0±0·2	62I
Ca^{2+}	gl	25	1 $(CH_3)_4NBr$	K_1 < 1, $K(Ca^{2+} + HL^{3-} \rightleftharpoons CaHL^-)$ < 1	62I
Mg^{2+}	gl	25	1 $(CH_3)_4NBr$	K_1 < 1, $K(Mg^{2+} + HL^{3-} \rightleftharpoons MgHL^-)$ < 1	62I

62I R. R. Irani and K. Moedritzer, *J. Phys. Chem.*, 1962, 66, 1349

$C_{10}H_7ONS_2$ 5-Benzylidene-2-thioxo-1,3-thiazolidin-4-one (5-(benzylidene)rhodanine) HL 844.2

Metal	Method	Temp	Medium	Log of equilibrium constant, remarks	Ref
H^+	dis	20	0·1 ($NaClO_4$)	K_1 7·58	65N
Ag^+	dis	20	0·1 ($NaClO_4$)	K_1 8·35, K_2 7·50	65N

65N O. Navratil and J. Kotas, *Coll. Czech. Chem. Comm.*, 1965, 30, 2736

$C_{10}H_7O_2NS$ 3-Phenyl-1,2-thiazole-5-carboxylic acid HL 844.3

Metal	Method	Temp	Medium	Log of equilibrium constant, remarks	Ref
H^+	gl	25	0·1 ($NaClO_4$) 50% dioxan	K_1 3·66±0·01	68E
	sp	25	0·1 ($NaClO_4$) 50% dioxan	K_1 3·7	68E
Co^{2+}	gl	25	0·1 ($NaClO_4$) 50% dioxan	K_1 1·73	68E
Cu^{2+}	gl	25	0·1 ($NaClO_4$) 50% dioxan	K_1 2·35	68E

continued overleaf

844.3 $C_{10}H_7O_2NS$ (contd)

Metal	*Method*	*Temp*	*Medium*	*Log of equilibrium constant, remarks*	*Ref*
Mn^{2+}	gl	25	0·1 ($NaClO_4$) 50% dioxan	K_1 1·51	68E
Ni^{2+}	gl	25	0·1 ($NaClO_4$) 50% dioxan	K_1 1·8	68E
Zn^{2+}	gl	25	0·1 ($NaClO_4$) 50% dioxan	K_1 1·66	68E

68E H. Erlenmeyer, R. Griesser, B. Prijs, and H. Sigel, *Helv. Chim. Acta*, 1968, **51**, 339

846 $C_{10}H_7O_5NS$ 1-Hydroxy-2-nitrosonaphthalene-4-sulphonic acid H_2L

Metal	*Method*	*Temp*	*Medium*	*Log of equilibrium constant, remarks*	*Ref*
H^+	gl	25	0—0·1 (KCl)	K_1 (→ 0) 6·50, (0·01) 6·32, (0·1) 6·09, K_2 (→ 0) 1·94, (0·01) 2·26, (0·1) 2·63	66M, 67
Cd^{2+}	sp	25	→ 0	K_1 3·12	66M
Ce^{3+}	gl	25	0—0·1 (KCl)	K_1 (→ 0) 3·79, (0·1) 2·70	67M
Cu^{2+}	sp	25	→ 0	K_1 7·8	66M
Dy^{3+}	gl	25	0—0·1 (KCl)	K_1 (→ 0) 4·27, (0·1) 3·18	67M
Er^{3+}	gl	25	0—0·1 (KCl)	K_1 (→ 0) 4·19, (0·1) 3·09	67M
Gd^{3+}	gl	25	0—0·1 (KCl)	K_1 (→ 0) 4·50, (0·1) 3·41	67M
Ho^{3+}	gl	25	0—0·1 (KCl)	K_1 (→ 0) 4·22, (0·1) 3·13	67M
La^{3+}	gl	25	0—0·1 (KCl)	K_1 (→ 0) 4·56, (0·01) 4·06, (0·1) 3·46, β_3 (0·01) 10·8, (0·1) 9·9	67M
Mn^{2+}	sp	25	→ 0	K_1 2·07	66M
Nd^{3+}	gl	25	0—0·1 (KCl)	K_1 (→ 0) 4·57, (0·1) 3·47	67M
Ni^{2+}	sp	25	→ 0	K_1 6·31	66M
Pb^{2+}	sp	25	→ 0	K_1 4·74	66M
			I (KCl)	$K(Pb^{2+}+HL^- \rightleftharpoons PbL+H^+)$ $1·76+2·036\sqrt{I}/(1+0·95\sqrt{I})-0·04\,I$	
Sm^{3+}	gl	25	0—0·1 (KCl)	K_1 (→ 0) 4·57, (0·1) 3·48	67M
Y^{3+}	gl	25	0—0·1 (KCl)	K_1 (→ 0) 3·97, (0·1) 2·87	67M
Yb^{3+}	gl	25	0—0·1 (KCl)	K_1 (→ 0) 4·18, (0·1) 3·09	67M
Zn^{2+}	sp	25	→ 0	K_1 3·86	66M

66M O. Makitie, *Suomen Kem.*, 1966, **B39**, 218
67M O. Makitie, *Suomen Kem.*, 1967, **B40**, 128, 267

847 $C_{10}H_7O_8NS_2$ 1,2-Naphthaquinone-3,6-disulphonic acid 1-oxime (nitroso-*R* acid) H_3L

Metal	*Method*	*Temp*	*Medium*	*Log of equilibrium constant, remarks*	*Ref*
H^+	gl	25	I (KCl)	K_1 $7·51-3·054\sqrt{I}/(1+1·49\sqrt{I})+0·27\,I$	66M, 67 68M
Cd^{2+}	?	25	0—0·1 (KCl)	K_1 (→ 0) 4·7, (0·1) 3·4, β_2 (→ 0) 6·6, (0·1) 6·0	67M
Ce^{3+}	gl	25	0·1 (KCl)	K_1 4·42	68M

cont

$C_{10}H_7O_8NS_2$ (contd) 847

Metal	Method	Temp	Medium	Log of equilibrium constant, remarks	Ref
Co^{2+}	sp	25	?	K (?) 13·3	66Ma
Cu^{2+}	?	25	0—0·1 (KCl)	K_1 (→ 0) 9·9, (0·1) 7·7, β_2 (→ 0) 15·6, (0·1) 15·0	67M
Dy^{3+}	gl	25	0·1 (KCl)	K_1 4·73	68M
Er^{3+}	gl	25	0·1 (KCl)	K_1 4·65	68M
Gd^{3+}	gl	25	0·1 (KCl)	K_1 4·92	68M
Ho^{3+}	gl	25	0·1 (KCl)	K_1 4·70	68M
La^{3+}	gl	25	0—0·1 (KCl)	K_1 (→ 0) 6·19, (0·1) 4·37, β_2 (0·1) 7·83, β_3 (0·1) 11·24	67M
Mn^{2+}	?	25	0—0·1 (KCl)	K_1 (→ 0) 3·7, (0·1) 2·7	67M
Nd^{3+}	gl	25	0·1 (KCl)	K_1 5·01	68M
Ni^{2+}	?	25	0—0·1 (KCl)	K_1 (→ 0) 8·3, (0·1) 6·9, β_2 (→ 0) 13·4, (0·1) 12·5, β_3 (0·1) 17·3	67M
Pb^{2+}	gl, sp	25	0—0·1 (KCl)	K_1 (→ 0) 6·07, (0·1) 4·64, β_2 (→ 0) 8·34, (0·1) 7·37	66M
Pd^{2+}	sp	25	?	K (?) 8·8±0·5	63B
	sp	25	0·1 ($NaClO_4$)	K (?) 8·9	64M
Ru^{3+}	sp	25	0·3	K (?) 9·7±0·2, acetate buffer	65M
Sm^{3+}	gl	25	0·1 (KCl)	K_1 5·15	68M
Y^{3+}	gl	25	0—0·1 (KCl)	K_1 (→ 0) 6·24, (0·1) 4·48, β_2 (0·1) 7·83, β_3 (0·1) 11·29	67M
Yb^{3+}	gl	25	0·1 (KCl)	K_1 4·74	68M
Zn^{2+}	?	25	0—0·1 (KCl)	K_1 (→ 0) 5·7, (0·1) 4·5, β_2 (→ 0) 7·6, (0·1) 7·1	67M

63B S. K. Banerjee and M. Garg, *Z. anorg. Chem.*, 1963, **325**, 315
64M K. N. Munshi, S. P. Sangal, and A. K. Dey, *J. Indian Chem. Soc.*, 1964, **41**, 701
65M D. J. Miller, S. C. Srivastava, and M. L. Good, *Analyt. Chem.*, 1965, **37**, 739
66M O. Makitie, *Suomen Kem.*, 1966, **B39**, 171
66Ma S. P. Mushran, P. Sanyal, and J. D. Pandey, *J. Indian Chem Soc.*, 1966, **43**, 273
67M O. Makitie, *Suomen Kem.*, 1967, **B40**, 27
68M O. Makitie, *Suomen Kem.*, 1968, **B41**, 31

$C_{10}H_8O_6N_2S$ 2-Methyl-7-nitro-8-hydroxyquinoline-5-sulphonic acid H_2L 847.1

Metal	Method	Temp	Medium	Log of equilibrium constant, remarks	Ref
H^+	gl	25	0·005 $HClO_4$	K_1 6·24	63F
	sp	28	0·005 $HClO_4$	K_2 0·7	63F
Co^{2+}	gl	25	0·005 $HClO_4$	K_1 5·50, K_2 4·34, K_3 < 3·5	63F
Cu^{2+}	gl	25	0·005 $HClO_4$	K_2 ~ 6·4, K_3 < 3·5	63F
Ni^{2+}	gl	25	0·005 $HClO_4$	K_1 5·92, K_2 4·85, K_3 < 3·5	63F
Zn^{2+}	gl	25	0·005 $HClO_4$	K_1 5·31, K_2 4·32, K_3 < 3·5	63F

63F J. Fresco and H. Freiser, *Inorg. Chem.*, 1963, **2**, 82

847.2 $C_{10}H_9ON_3S$ 3-(2′-Thiazolylazo)-4-methylphenol HL

Metal	*Method*	*Temp*	*Medium*	*Log of equilibrium constant, remarks*	*Ref*
H^+	gl	25	0·1 $NaClO_4$ 50% methanol	K_1 8·95	67N
Co^{2+}	gl	25	0·1 $NaClO_4$ 50% methanol	β_2 14·5	67N
Cu^{2+}	gl	25	0·1 $NaClO_4$ 50% methanol	K_1 10·5, K_2 5·8	67N
Mn^{2+}	gl	25	0·1 $NaClO_4$ 50% methanol	β_2 7·6	67N
Ni^{2+}	gl	25	0·1 $NaClO_4$ 50% methanol	K_1 8·3, K_2 7·9	67N
Zn^{2+}	gl	25	0·1 $NaClO_4$ 50% methanol	K_1 6·1, K_2 5·4	67N

67N G. Nickless, F. H. Pollard, and T. J. Samuelson, *Analyt. Chim. Acta*, 1967, **39**, 37

848.1 $C_{10}H_9O_2N_3S$ 2-(2′-Thiazolylazo)-4-methoxyphenol HL

Metal	*Method*	*Temp*	*Medium*	*Log of equilibrium constant, remarks*	*Ref*
H^+	sp	20	0·1 0, 30% ethanol	K_1 (0%) 8·12±0·05, (30%) 8·45	67C, 6
UO_2^{2+}	sp	20	0·1 30% ethanol	K_1 8·8	68S

67C V. Chromy and L. Sommer, *Talanta*, 1967, **14**, 393
68S L. Sommer, T. Sepel, and V. M. Ivanov, *Talanta*, 1968, **15**, 949

848.2 $C_{10}H_9O_2N_3S$ 2-(2′-Thiazolylazo)-5-methoxyphenol HL

Metal	*Method*	*Temp*	*Medium*	*Log of equilibrium constant, remarks*	*Ref*
H^+	sp	~ 20	0·1 0, 30% ethanol	K_1 (0%) 7·08±0·04, (30%) 7·63±0·03	68S
UO_2^{2+}	sp	~ 20	0·1 30% ethanol	K_1 8·1	68S

68S L. Sommer, T. Sepel, and V. M. Ivanov, *Talanta*, 1968, **15**, 949

$C_{10}H_9O_4NS$ 2-Methyl-8-hydroxyquinoline-5-sulphonic acid H_2L 848.3

Metal	Method	Temp	Medium	Log of equilibrium constant, remarks	Ref
H^+	gl	25	0·005 $HClO_4$	K_1 8·99, K_2 4·73	63F
	sp	28	0·005 $HClO_4$	$K_3 < 0$	63F
	gl	25	0·5 NaCl	K_1 8·72, K_2 4·63	67T
Co^{2+}	gl	25	0·005 $HClO_4$	K_1 7·54, K_2 6·52, $K_3 < 3·5$	63F
Cu^{2+}	gl	25	0·005 $HClO_4$	K_1 9·86, K_2 8·70, $K_3 < 3·5$	63F
Ge^{IV}	sp	25	0·5 NaCl	$K[Ge(OH)_4+2HL \rightleftharpoons Ge(OH)_2L_2] \sim 2·2$	67T
Ni^{2+}	gl	25	0·005 $HClO_4$	K_1 7·69, K_2 6·45, K_3 4·48	63F
Zn^{2+}	gl	25	0·005 $HClO_4$	K_1 7·50, K_2 7·14, $K_3 < 3·5$	63F
	gl	25	0·5 NaCl	K_1 7·32, K_2 6·69	67T

63F J. Fresco and H. Freiser, *Inorg. Chem.*, 1963, **2**, 82
67T J. Tsau, S. Matsouo, P. Clerc, and R. Benoit, *Bull. Soc. chim. France*, 1967, 1039

$C_{10}H_9O_7NS_2$ 8-Amino-1-hydroxynaphthalene-3,6-disulphonic acid H_3L 849

Metal	Method	Temp	Medium	Log of equilibrium constant, remarks	Ref
Fe^{3+}	sp	25	?	$K(Fe^{3+}+HL^{2-} \rightleftharpoons FeL+H^+)$ (?) 3·67	63R

63R M. L. N. Reddy and U. V. Seshaiah, *Indian J. Chem.*, 1963, **1**, 536

$C_{10}H_9O_7NS_2$ 7-Amino-1-hydroxynaphthalene-3,6-disulphonic acid H_3L 849.1

Metal	Method	Temp	Medium	Log of equilibrium constant, remarks	Ref
H^+	gl	25	0·1 ($NaClO_4$)	K_1 8·54, K_2 2·49	68B
UO_2^{2+}	gl	25	0·1 ($NaClO_4$)	K_1 6·19, K_2 4·87	68B

68B A. Banerjee and A. K. Dey, *Analyt. Chim. Acta*, 1968, **42**, 473

$C_{10}H_{10}O_2NCl$ *N*-2-Chlorophenylacetoacetamide HL 850

Erratum. 1964, p. 647, Table 850. Under Metal column third line, *insert* Be^{2+}.

850.1 $C_{10}H_{11}O_4NS$ *N*-(2′-Mercaptophenyl)iminodiacetic acid H_3L

Metal	*Method*	*Temp*	*Medium*	*Log of equilibrium constant, remarks*	*Ref*
H^+	gl	20	0·1 (KNO_3)	K_1 9·54±0·02, K_2 6·30±0·01, K_3 2·85±0·01	63I
Ca^{2+}	gl	20	0·1 (KNO_3)	K_1 (?) 2·79±0·04	63I
Mg^{2+}	gl	20	0·1 (KNO_3)	K_1 (?) 1·84±0·03	63I

63I H. Irving and J. J. R. F. da Silva, *J. Chem. Soc.*, 1963, 3308

850.2 $C_{10}H_{11}O_5NS$ *N*-(2-Thenoylmethyl)iminodiacetic acid (2-glycylthiophene-*NN*-diacetic acid) H_2L

Metal	*Method*	*Temp*	*Medium*	*Log of equilibrium constant, remarks*	*Ref*
H^+	gl	25	0·1 (KNO_3)	K_1 7·46, K_2 1·95	65A
Ca^{2+}	gl	25	0·1 (KNO_3)	K_1 4·26	65A
Cd^{2+}	gl	25	0·1 (KNO_3)	K_1 7·43, K_2 4·93	65A
Co^{2+}	gl	25	0·1 (KNO_3)	K_1 6·93, K_2 4·97	65A
Cu^{2+}	gl	25	0·1 (KNO_3)	K_1 8·79, K_2 5·31	65A
Mg^{2+}	gl	25	0·1 (KNO_3)	K_1 2·80	65A
Ni^{2+}	gl	25	0·1 (KNO_3)	K_1 8·14, K_2 5·78	65A
Sr^{2+}	gl	25	0·1 (KNO_3)	K_1 3·40	65A
Zn^{2+}	gl	25	0·1 (KNO_3)	K_1 7·35, K_2 4·26	65A

65A T. Ando and K. Ueno, *Inorg. Chem.*, 1965, **4**, 375

856.1 $C_{10}H_{13}O_9N_4P$ Inosine-5′-monophosphoric acid-*N*(1)-oxide H_3L

Metal	*Method*	*Temp*	*Medium*	*Log of equilibrium constant, remarks*	*Ref*
H^+	gl	25	0·1 ($NaClO_4$)	K_1 6·31, K_2 5·43	65S
	sp	25	0·1 ($NaClO_4$)	K_2 5·53±0·04	65S
Ba^{2+}	sp	25	0·1 ($NaClO_4$)	$K(Ba^{2+}+HL^{2-} \rightleftharpoons BaHL)$ 1·6±0·2	65S
Ca^{2+}	sp	25	0·1 ($NaClO_4$)	$K(Ca^{2+}+HL^{2-} \rightleftharpoons CaHL)$ 2·0±0·2	65S
Co^{2+}	sp	25	0·1 ($NaClO_4$)	$K(Co^{2+}+HL^{2-} \rightleftharpoons CoHL)$ 3·73±0·05	65S
Cu^{2+}	sp	25	0·1 ($NaClO_4$)	$K(Cu^{2+}+HL^{2-} \rightleftharpoons CuHL)$ 5·46±0·09	65S
Mg^{2+}	sp	25	0·1 ($NaClO_4$)	$K(Mg^{2+}+HL^{2-} \rightleftharpoons MgHL)$ 2·1±0·2	65S
Mn^{2+}	sp	25	0·1 ($NaClO_4$)	$K(Mn^{2+}+HL^{2-} \rightleftharpoons MnHL)$ 2·85±0·09	65S
Ni^{2+}	sp	25	0·1 ($NaClO_4$)	$K(Ni^{2+}+HL^{2-} \rightleftharpoons NiHL)$ 3·90±0·05	65S
Zn^{2+}	sp	25	0·1 ($NaClO_4$)	$K(Zn^{2+}+HL^{2-} \rightleftharpoons ZnHL)$ 3·83±0·05	65S

65S H. Sigel, *Helv. Chim. Acta*, 1965, 48, 1513, 1519

$C_{10}H_{14}O_7N_5P$ Adenosine-2′-monophosphoric acid (AMP-2) H_2L 856.2

Metal	*Method*	*Temp*	*Medium*	*Log of equilibrium constant, remarks*	*Ref*
H^+	gl	0·4—40	0·10 (KNO_3)	K_1 (0·4°) 6·12±0·02, (12°) 6·07±0·02, (25°) 6·01±0·02, (40°) 5·95±0·02, K_2 (0·4°) 4·03±0·01, (12°) 3·88±0·01, (25°) 3·71±0·01, (40°) 3·54±0·01	67T
				ΔH_1 (25°) −1·6±0·1, ΔS_1 (25°) 22·2+0·5, ΔH_2 (25°) −4·7±0·1, ΔS_2 (25°) 1·2±0·3	
Ba^{2+}	gl	0·4—40	0·10 (KNO_3)	K_1 (0·4°) 1·82±0·02, (12°) 1·77±0·02, (25°) 1·71±0·02, (40°) 1·64±0·02	67T
				ΔH_1 (25°) −2·0±0·2, ΔS_1 (25°) 1·2±0·5	
Ca^{2+}	gl	0·4—40	0·10 (KNO_3)	K_1 (0·4°) 1·87±0·02, (12°) 1·85±0·02, (25°) 1·83±0·02, (40°) 1·81±0·02	67T
				ΔH_1 (25°) −0·6±0·2, ΔS_1 (25°) 6·5±0·5	
Co^{2+}	gl	0·4—40	0·10 (KNO_3)	K_1 (0·4°) 2·15±0·02, (12°) 2·19±0·02, (25°) 2·24±0·02, (40°) 2·28±0·02	67T
				ΔH_1 (25°) −0·7±0·2 (?), ΔS_1 (25°) 8·5±0·5 (?)	
Cu^{2+}	gl	0·4—40	0·10 (KNO_3)	K_1 (0·4°) 3·28±0·02, (12°) 3·23±0·02, (25°) 3·16±0·02, (40°) 3·10±0·02	67T
				ΔH_1 (25°) −1·9±0·1, ΔS_1 (25°) 8·0±0·3	
Mg^{2+}	gl	0·4—40	0·10 (KNO_3)	K_1 (0·4°) 1·71±0·02, (12°) 1·82±0·02, (25°) 1·93±0·02, (40°) 2·05±0·02	67T
				ΔH_1 (25°) 3·5±0·1, ΔS_1 (25°) 20·5±0·3	
Mn^{2+}	gl	0·4—40	0·10 (KNO_3)	K_1 (0·4°) 2·43±0·02, (12°) 2·41±0·02, (25°) 2·38±0·02, (40°) 2·35±0·02	67T
				ΔH_1 (25°) −1·0±0·2, ΔS_1 (25°) 7·5±0·5	
Ni^{2+}	gl	0·4—40	0·10 (KNO_3)	K_1 (0·4°) 2·86±0·02, (12°) 2·84±0·02, (25°) 2·81±0·02, (40°) 2·78±0·02	67T
				ΔH_1 (25°) −1·0±0·2, ΔS_1 (25°) 9·5±0·5	
Sr^{2+}	gl	0·4—40	0·10 (KNO_3)	K_1 (0·4°) 1·85±0·02, (12°) 1·79±0·02, (25°) 1·74±0·02, (40°) 1·71±0·02	67T
				ΔH_1 (25°) −1·0±0·2, ΔS_1 (25°) 4·5±0·5	
Zn^{2+}	gl	0·4—40	0·10 (KNO_3)	K_1 (0·4°) 2·72±0·02, (12°) 2·68±0·02, (25°) 2·64±0·02, (40°) 2·60±0·02	67T
				ΔH_1 (25°) −1·2±0·2, ΔS_1 (25°) 8·0±0·5	

67T M. M. Taqui Khan and A. E. Martell, *J. Amer. Chem. Soc.*, 1967, 89, 5585

$C_{10}H_{14}O_7N_5P$ Adenosine-3′-monophosphoric acid (AMP-3) H_2L 857

Metal	*Method*	*Temp*	*Medium*	*Log of equilibrium constant, remarks*	*Ref*
H^+	gl	25	0·1 KCl	K_1 6·55, K_2 3·93	58W
	gl	0·4—40	0·10 (KNO_3)	K_1 (0·4°) 5·93±0·02, (12°) 5·88±0·02, (25°) 5·83±0·02, (40°) 5·78±0·02, K_2 (0·4°) 3·95±0·01, (12°) 3·80±0·01, (25°) 3·65±0·01, (40°) 3·49±0·01	67T
				ΔH_1 (25°) −1·5±0·1, ΔS_1 (25°) 22·3±0·05, ΔH_2 (25°) −4·6±0·1, ΔS_2 (25°) 1·3±0·3	
Ba^{2+}	gl	0·4—40	0·10 (KNO_3)	K_1 (0·4°) 1·81±0·02, (12°) 1·75±0·02, (25°) 1·69±0·02, (40°) 1·62±0·02	67T
				ΔH_1 (25°) −1·9±0·2, ΔS_1 (25°) 1·2±0·5	

continued overleaf

857 $C_{10}H_{14}O_7N_5P$ (contd)

Metal	Method	Temp	Medium	Log of equilibrium constant, remarks	Ref
Ca^{2+}	gl	0·4—40	0·10 (KNO_3)	K_1 (0·4°) 1·86±0·02, (12°) 1·84±0·02, (25°) 1·80±0·02, (40°) 1·78±0·02 ΔH_1 (25°) −0·6±0·2, ΔS_1 (25°) 6·5±0·5	67T
Co^{2+}	ix	25	0·1 ($KClO_4$)	K_1 2·08	66D
	gl	25	0·1 (KNO_3)	K_1 2·10	66D
			0·1 [$(CH_3)_4NBr$]	K_1 2·19	
	gl	0·4—40	0·10 (KNO_3)	K_1 (0·4°) 2·11±0·02, (12°) 2·15±0·02, (25°) 2·20±0·02, (40°) 2·24±0·02 ΔH_1 (25°) −0·6±0·2 (?), ΔS_1 (25°) 8·5±0·5 (?)	67T
Cu^{2+}	gl	0·4—40	0·10 (KNO_3)	K_1 (0·4°) 3·06±0·02, (12°) 3·00±0·02, (25°) 2·96±0·02, (40°) 2·90±0·02 ΔH_1 (25°) −1·7±0·1, ΔS_1 (25°) 8·0±0·3	67T
Mg^{2+}	gl	25	0·1 KCl	K_1 1·73	58W
	gl	0·4—40	0·10 (KNO_3)	K_1 (0·4°) 1·68±0·02, (12°) 1·78±0·02, (25°) 1·86±0·02, (40°) 2·01±0·02 ΔH_1 (25°) 3·5±0·1, ΔS_1 (25°) 20·5±0·3	67T
Mn^{2+}	ix	25	0·1 ($KClO_4$)	K_1 1·86	66D
	gl	25	0·1 [$(CH_3)_4NBr$]	K_1 1·98	66D
	gl	0·4—40	0·10 (KNO_3)	K_1 (0·4°) 2·34±0·02, (12°) 2·31±0·02, (25°) 2·28±0·02, (40°) 2·25±0·02 ΔH_1 (25°) −0·9±0·2, ΔS_1 (25°) 7·6±0·5	67T
Ni^{2+}	gl	0·4—40	0·10 (KNO_3)	K_1 (0·4°) 2·85±0·02, (12°) 2·82±0·02, (25°) 2·79±0·02, (40°) 2·75±0·02 ΔH_1 (25°) −1·0±0·2, ΔS_1 (25°) 9·6±0·5	67T
Sr^{2+}	gl	0·4—40	0·10 (KNO_3)	K_1 (0·4°) 1·81±0·02, (12°) 1·75±0·02, (25°) 1·71±0·02, (40°) 1·68±0·02 ΔH_1 (25°) −0·9±0·2, ΔS_1 (25°) 4·5±0·5	67T
Zn^{2+}	gl	25	0·1 KCl	K_1 2·69	58W
	ix	25	0·1 ($KClO_4$)	K_1 2·48	66D
	gl	0·4—40	0·10 (KNO_3)	K_1 (0·4°) 2·65±0·02, (12°) 2·62±0·02, (25°) 2·60±0·02, (40°) 2·56±0·02 ΔH_1 (25°) −1·1±0·2, ΔS_1 (25°) 8·2±0·5	67T

58W G. Weitzel and T. Speer, *Z. physiol. Chem.*, 1958, **313**, 212
66D E. Doody, E. R. Tucci, R. Scruggs, and N. C. Li, *J. Inorg. Nuclear Chem.*, 1966, **28**, 833
67T M. M. Taqui Khan and A. E. Martell, *J. Amer. Chem. Soc.*, 1967, **89**, 5585

858 $C_{10}H_{14}O_7N_5P$ Adenosine-5′-monophosphoric acid (AMP-5) H_2L

Metal	Method	Temp	Medium	Log of equilibrium constant, remarks	Ref
H^+	gl	25	0·1 $(CH_3)_4NBr$	K_1 6·40	61T
			0·1 KCl	K_1 6·30	
	gl	25	0·1 ($NaClO_4$)	K_1 6·14±0·03	64S
	gl	25	0·1 ($NaClO_4$) 10% dioxan	K_1 6·43±0·04	67S

contd

Metal	*Method*	*Temp*	*Medium*	*Log of equilibrium constant, remarks*	*Ref*
H^+ (contd)	gl	0·4—40	0·10 (KNO_3)	K_1 (0·4°) 6·38±0·02, (12°) 6·31±0·02, (25°) 6·23±0·02, (40°) 6·16±0·02, K_2 (0·4°) 4·15±0·01, (12°) 3·98±0·02, (25°) 3·80±0·01, (40°) 3·62±0·01	67T
				ΔH_1 (25°) −1·9±0·1, ΔS_1 (25°) 22·0±0·5, ΔH_2 (25°) −4·9±0·1, ΔS_2 (25°) 1·0±0·3	
Ba^{2+}	gl	25	0·1 ($NaClO_4$)	K_1 1·14	64S
	gl	0·4—40	0·10 (KNO_3)	K_1 (0·4°) 1·85±0·02, (12°) 1·80±0·02, (25°) 1·73±0·02, (40°) 1·66±0·02	67T
				ΔH_1 (25°) −2·0±0·2, ΔS_1 (25°) 1·2±0·5	
Ca^{2+}	gl	25	0·1 ($NaClO_4$)	K_1 1·39	64S
	gl	0·4—40	0·10 (KNO_3)	K_1 (0·4°) 1·88±0·02, (12°) 1·87±0·02, (25°) 1·85±0·02, (40°) 1·83±0·02	67T
				ΔH_1 (25°) −0·6±0·2, ΔS_1 (25°) 6·4±0·5	
Co^{2+}	gl	25	0·1 ($NaClO_4$)	K_1 2·19	64S
	gl	25	0·1 (KNO_3)	K_1 2·57	66D
	gl	25	0·1 ($NaClO_4$)	K_1 2·34	67S
			10% dioxan	$K(CoA^{2+}+L^{2-} \rightleftharpoons CoAL)$ 2·37 where A = 2,2′-bipyridyl	
	gl	0·4—40	0·10 (KNO_3)	K_1 (0·4°) 2·44±0·02, (12°) 2·49±0·02, (25°) 2·53±0·02, (40°) 2·57±0·02	67T
				ΔH_1 (25°) −1·1±0·2 (?), ΔS_1 (25°) 8·4±0·5 (?)	
Cu^{2+}	gl	25	0·1 ($NaClO_4$)	K_1 3·04	64S
	gl	25	0·1 ($NaClO_4$)	K_1 3·22	67S
			10% dioxan	$K(CuA^{2+}+L^{2-} \rightleftharpoons CuAL)$ 3·72 where A = 2,2′-bipyridyl	
	gl	0·4—40	0·10 (KNO_3)	K_1 (0·4°) 3·30±0·02, (12°) 3·24±0·02, (25°) 3·18±0·02, (40°) 3·12±0·02	67T
				ΔH_1 (25°) −2·0±0·1, ΔS_1 (25°) 8·0±0·3	
Mg^{2+}	gl	25	0·1 KCl	K_1 2·14	58W
	gl	25	0·1 ($NaClO_4$)	K_1 1·63	64S
	gl	0·4—40	0·10 (KNO_3)	K_1 (0·4°) 1·75±0·02, (12°) 1·85±0·02, (25°) 1·97±0·02, (40°) 2·09±0·02	67T
				ΔH_1 (25°) 3·4±0·2, ΔS_1 (25°) 20·4±0·5	
Mn^{2+}	ix	25	0·1	K_1 2·19, Veronal buffer	61T
	gl	25	0·1 ($NaClO_4$)	K_1 2·14	64S
	gl	25	0·1 (KNO_3)	K_1 2·35	66D
	gl	0·4—40	0·10 (KNO_3)	K_1 (0·4°) 2·46±0·02, (12°) 2·43±0·02, (25°) 2·40±0·02, (40°) 2·37±0·02	67T
				ΔH_1 (25°) −1·0±0·2, ΔS_1 (25°) 7·6±0·5	
Ni^{2+}	ix	25 (?)	0·25 $(CH_3)_4NBr$	K_1 2·8±0·1	61T
	gl	25	0·1 ($NaClO_4$)	K_1 2·62	64S
	gl	25	0·1 (KNO_3)	K_1 2·67	66D
	gl	25	0·1 ($NaClO_4$) 10% dioxan	K_1 2·88	67S
	gl	0·4—40	0·10 (KNO_3)	K_1 (0·4°) 2·90±0·02, (12°) 2·87±0·02, (25°) 2·84±0·02, (40°) 2·84±0·02	67T
				ΔH_1 (25°) −1·0±0·2, ΔS_1 (25°) 9·6±0·5	
Sr^{2+}	ix	~ 25	0·15 NaCl	K_1 1·5	60O
	gl	0·4—40	0·10 (KNO_3)	K_1 (0·4°) 1·88±0·02, (12°) 1·83±0·02, (25°) 1·79±0·02, (40°) 1·74±0·02	67T
				ΔH_1 (25°) −1·4±0·2, ΔS_1 (25°) 4·4±5	

continued overleaf

858 $C_{10}H_{14}O_7N_5P$ (contd)

Metal	*Method*	*Temp*	*Medium*	*Log of equilibrium constant, remarks*	*Ref*
Y^{3+}	ix	~ 25	0·15 NaCl	K_1 5·7	60O
Zn^{2+}	gl	25	0·1 ($NaClO_4$)	K_1 2·23	64S
	gl	25	0·1 ($NaClO_4$) 10% dioxan	K_1 2·40 $K(ZnA^{2+}+L^{2-} \rightleftharpoons ZnAL)$ 2·40 where A = 2,2′-bipyridyl	67S
	gl	0·4—40	0·10 (KNO_3)	K_1 (0·4°) 2·80±0·02, (12°) 2·76±0·02, (25°) 2·72±0·02, (40°) 2·68±0·02 ΔH_1 (25°) −1·2±0·2, ΔS_1 (25°) 8·2±0·5	67T

58W G. Weitzel and T. Speer, *Z. physiol. Chem.*, 1958, **313**, 212
60O J. Olivard, *Arch. Biochem., Biophys.*, 1960, **88**, 382
61T E. R. Tucci, E. Doody, and N. C. Li, *J. Phys. Chem.*, 1961, **65**, 1570
64S H. Sigel and H. Brintzinger, *Helv. Chim. Acta*, 1964, **47**, 1701
66D B. E. Doody, E. R. Tucci, R. Scruggs, and N. C. Li, *J. Inorg. Nuclear Chem.*, 1966, **28**, 833
67S H. Sigel, K. Becker, and D. B. McCormick, *Biochem. Biophys. Acta*, 1967, **148**, 655
67T M. M. Taqui Khan and A. E. Martell, *J. Amer. Chem. Soc.*, 1967, **89**, 5585

858.1 $C_{10}H_{14}O_8N_5P$ Adenosine-5′-monophosphoric acid *N*(1)-oxide H_2L

Metal	*Method*	*Temp*	*Medium*	*Log of equilibrium constant, remarks*	*Ref*
H^+	sp	25	0·1 ($NaClO_4$)	K_1 12·49±0·04, K_3 2·58±0·02	64S, 67S
	gl	25	0·1 ($NaClO_4$)	K_2 6·13±0·02	64S, 67S
Ba^{2+}	gl	25	0·1 ($NaClO_4$)	$K(Ba^{2+}+HL^- \rightleftharpoons BaHL^+)$ 1·15 $K(BaL+H^+ \rightleftharpoons BaHL^+)$ > 10·79	64S
	sp	25	0·1 ($NaClO_4$)	K_1 < 2·86	
Ca^{2+}	gl	25	0·1 ($NaClO_4$)	$K(Ca^{2+}+HL^- \rightleftharpoons CaHL^+)$ 1·43 $K(CaL+H^+ \rightleftharpoons CaHL^+)$ > 10·93	64S
	sp	25	0·1 ($NaClO_4$)	K_1 < 3·00	64S
Co^{2+}	gl	25	0·1 ($NaClO_4$)	$K(Co^{2+}+HL^- \rightleftharpoons CoHL^+)$ 2·11 $K(CoL+H^+ \rightleftharpoons CoHL^+)$ 7·77	64S
	sp	25	0·1 ($NaClO_4$)	K_1 6·8±0·1	64S
Cu^{2+}	gl	25	0·1 ($NaClO_4$)	K_1 11·87 $K(CuL+H^+ \rightleftharpoons CuHL^+)$ 5·44±0·04	64S
	gl	25	0·1 ($NaClO_4$)	K_1 11·77 $K(CuL+H^+ \rightleftharpoons CuHL^+)$ 5·55±0·08 $K(CuLOH^-+H^+ \rightleftharpoons CuL)$ 8·2 $K(CuA^{2+}+L^{2-} \rightleftharpoons CuAL)$ 11·43 $K(CuA^{2+}+HL^- \rightleftharpoons CuAHL^+)$ 11·19 $K(CuAL+H^+ \rightleftharpoons CuAHL^+)$ 5·9±0·1 where A = 2,2′-bipyridyl	67S
Mg^{2+}	gl	25	0·1 ($NaClO_4$)	$K(Mg^{2+}+HL^- \rightleftharpoons MgHL^+)$ 1·62 $K(MgL+H^+ \rightleftharpoons MgHL^+)$ > 10·39	64S
	sp	25	0·1 ($NaClO_4$)	K_1 <3·72	
Mn^{2+}	gl	25	0·1 ($NaClO_4$)	$K(Mn^{2+}+HL^- \rightleftharpoons MnHL^+)$ 2·14 $K(MnL+H^+ \rightleftharpoons MnHL^+)$ 8·93	64S
	sp	25	0·1 ($NaClO_4$)	K_1 5·71±0·09	64S

contd

$C_{10}H_{14}O_8N_5P$ (contd) 858.1

Metal	Method	Temp	Medium	Log of equilibrium constant, remarks	Ref
Ni^{2+}	gl	25	0·1 ($NaClO_4$)	$K(Ni^{2+}+HL^- \rightleftharpoons NiHL^+)$ 2·66	64S
				$K(NiL+H^+ \rightleftharpoons NiHL^+)$ 7·70	
	sp	25	0·1 ($NaClO_4$)	K_1 7·45±0·07	64S
Zn^{2+}	gl	25	0·1 ($NaClO_4$)	$K(Zn^{2+}+HL^- \rightleftharpoons ZnHL^+)$ 2·20	64S
				$K(ZnL+H^+ \rightleftharpoons ZnHL^+)$ 6·90	
	sp	25	0·1 ($NaClO_4$)	K_1 7·79±0·07	64S

64S H. Sigel and H. Brintzinger, *Helv. Chim. Acta*, 1964, **47**, 1701
67S H. Sigel and B. Prijs, *Helv. Chim. Acta*, 1967, **50**, 2357

$C_{10}H_{14}O_{11}N_4P_2$ Inosine-5′-diphosphoric acid (IDP) H_4L 858.2

Metal	Method	Temp	Medium	Log of equilibrium constant, remarks	Ref
Mg^{2+}	sp	?	0·05 $(CH_3)_4NCl$	$K(Mg^{2+}+HL^{3-} \rightleftharpoons MgHL^-)$ (?) 2·38	61H
	gl	?	0·05 $(CH_3)_4NCl$	K_1 3·76	61H

61H K. Hotta, J. Brahms, and M. Morales, *J. Amer. Chem. Soc.*, 1961, **83**, 997

$C_{10}H_{15}O_{10}N_5P_2$ Adenosine-5′-diphosphoric acid (ADP) H_3L 859

Metal	Method	Temp	Medium	Log of equilibrium constant, remarks	Ref
H^+	gl	25	0·1 KCl	K_1 6·61, K_2 4·21	58W
	gl	10—37	$I(C_4H_9)_4NBr$	K_1 (10°) $7{\cdot}15-2{\cdot}48\sqrt{I}+3{\cdot}85\,I$, (25°) $7{\cdot}20-2{\cdot}54\sqrt{I}+3{\cdot}84\,I$, (37°) $7{\cdot}24-2{\cdot}60\sqrt{I}+3{\cdot}34\,I$ (±0·04)	66P
			→ 0	$\Delta H_1°$ 1·4±0·3, $\Delta S_1°$ 38±1	
	gl	0·4—40	0·10 (KNO_3)	K_1 (0·4°) 6·51±0·02, (12°) 6·48±0·02, (25°) 6·44±0·02, (40°) 6·41±0·02, K_2 (0·4°) 4·20±0·01, (12°) 4·09±0·01, (25°) 3·93±0·01, (40°) 3·73±0·01	67T
				$\Delta H_1°$ (25°) −1·2±0·1, $\Delta S_1°$ (25°) 25·4±0·5, $\Delta H_2°$ (25°) −4·8±0·1, $\Delta S_2°$ (25°) 1·9±0·5	

continued overleaf

859 $C_{10}H_{15}O_{10}N_5P_2$ (contd)

Metal	*Method*	*Temp*	*Medium*	*Log of equilibrium constant, remarks*	*Ref*
Ba^{2+}	gl	0·4—40	0·1 (KNO_3)	K_1 (0·4°) 2·53±0·02, (12°) 2·45±0·02, (25°) 2·36±0·02, (40°) 2·25±0·02	67T
				$\Delta H_1°$ (25°) −2·9±0·4, $\Delta S_1°$ (25°) 1±1	
				$K(Ba^{2+}+HL^{2-} \rightleftharpoons BaHL)$ (0·4°) 1·55±0·02, (12°) 1·50±0·02, (25°) 1·44±0·02, (40°) 1·37±0·02	
				$\Delta H°$ (25°) −1·8±0·3, $\Delta S°$ (25°) 1±1	
Ca^{2+}	ix	37	0·1	K_1 3·74	53D
	sp	25	0·1 $(C_4H_9)_3(C_2H_5)NBr$	K_1 2·86	59B
	oth	~ 23	0·06 $(CH_3)_4NCl$ 0·05 $(HOCH_2)_3\cdot CNH_2$	K_1 2·93 interferometer	62A
			0·06 $(CH_3)_4NCl$ 0·05 CH_3COO^-	$K(Ca^{2+}+HL^{2-} \rightleftharpoons CaHL)$ 1·52 interferometer	
	gl	0·4—40	0·10 (KNO_3)	K_1 (0·4°) 2·91±0·02, (12°) 2·88±0·02, (25°) 2·86±0·02, (40°) 2·80±0·02	67T
				$\Delta H_1°$ (25°) −1·2±0·4, $\Delta S_1°$ (25°) 9±1	
				$K(Ca^{2+}+HL^{2-} \rightleftharpoons CaHL)$ (0·4°) 1·61±0·02, (12°) 1·60±0·02, (25°) 1·58±0·02, (40°) 1·54±0·02	
				$\Delta H°$ (25°) −0·6±0·4, $\Delta S°$ (25°) 5±1	
Co^{2+}	gl	0·4—40	0·10 (KNO_3)	K_1 (0·4°) 4·63±0·02 (?), (12°) 4·27±0·02, (25°) 4·20±0·02, (40°) 4·12±0·02	67T
				$\Delta H_1°$ (25°) −2·0±0·4, $\Delta S_1°$ (25°) 13±1	
				$K(Co^{2+}+HL^{2-} \rightleftharpoons CoHL)$ (0·4°) 2·12±0·02, (12°) 2·07±0·02, (25°) 2·01±0·02, (40°) 1·93±0·02	
				$\Delta H°$ (25°) −1·9±0·3, $\Delta S°$ (25°) 3±1	
Cu^{2+}	gl	0·4—40	0·10 (KNO_3)	K_1 (0·4°) 6·16±0·02, (12°) 6·04±0·02, (25°) 5·90±0·02, (40°) 5·75±0·02	67T
				$\Delta H_1°$ (25°) −4·1±0·3, $\Delta S_1°$ (25°) 13±1	
				$K(Cu^{2+}+HL^{2-} \rightleftharpoons CuHL)$ (0·4°) 2·80±0·02, (12°) 2·72±0·02, (25°) 2·62±0·02, (40°) 2·52±0·02	
				$\Delta H°$ (25°) −2·7±0·3, $\Delta S°$ (25°) 3±1	
				$K(CuOHL^{2-}+H^+ \rightleftharpoons CuL^-)$ (0·4°) 7·65±0·05, (12°) 7·38±0·05, (25°) 7·08±0·05, (40°) 6·76±0·05	
				$\Delta H°$ (25°) −9·0±0·5, $\Delta S°$ (25°) 2±1	
				$K[(CuOHL)_2^{4-}+2H^+ \rightleftharpoons 2CuL^-]$ (0·4°) 3·85±0·06, (12°) 3·65±0·06, (25°) 3·42±0·06, (40°) 3·19±0·06	
				$\Delta H°$ (25°) −11±1, $\Delta S°$ (25°) 11±3	
				$K[2CuOHL^{2-} \rightleftharpoons (CuOHL)_2^{4-}]$ (0·4°) 11·45±0·07, (12°) 11·11±0·07, (25°) 10·73±0·07 (40°) 10·33±0·07	
				$\Delta H°$ (25°) −6·2±0·6, $\Delta S°$ (25°) 5±2	
K^+	Ag–AgBr	25	0·2 (Br^-)	K_1 0·67±0·01	54M
Mg^{2+}	gl	25	0·1 KCl	K_1 3·23, $K(Mg^{2+}+HL^{2-} \rightleftharpoons MgHL)$ 1·58	58W
	sp	?	0·05 $(CH_3)_4NCl$	$K(Mg^{2+}+HL^{2-} \rightleftharpoons MgHL)$ (?) 1·5	61H

conti

Metal	Method	Temp	Medium	Log of equilibrium constant, remarks	Ref
Mg^{2+}	gl	?	0·05 $(CH_3)_4NCl$	K_1 3·34	61H
(contd)	ix	5—65	0·07—0·17 $(C_4H_9)_4NBr$	K_1 (5°) (0·07) 3·48, (0·1) 3·24, (0·17) 3·14, (25°) (0·07) 3·65, (0·1) 3·44, (0·17) 3·33, (45°) (0·07) 3·83, (0·1) 3·60, (0·17) 3·46, (65°) (0·07) 4·00, (0·1) 3·76, (0·17) 3·64	66P
		25	I $(C_4H_9)_4NBr$ $\rightarrow 0$	K_1 $4{\cdot}27-4{\cdot}06\sqrt{I}+6{\cdot}36\sqrt{I}-2{\cdot}04\sqrt{I}/(1+6{\cdot}02\sqrt{I})\pm0{\cdot}10$ $\Delta H°$ 4·3±0·3, $\Delta S_1°$ 34±2	
			I $(C_4H_9)_4NBr$	$K(Mg^{2+}+HL^{2-}\rightleftharpoons MgHL)$ $2{\cdot}45-2{\cdot}03\sqrt{I}+3{\cdot}34\,I-2{\cdot}04\sqrt{I}/(1+6{\cdot}02\sqrt{I})\pm0{\cdot}20$	
			0·1 $(C_4H_9)_4NBr$ $\rightarrow 0$	$K(Mg^{2+}+HL^{2-}\rightleftharpoons MgHL)$ 2·0±0·2 $\Delta H°$ 1±1, $\Delta S°$ 14±5	
			I $MgCl_2$ $\rightarrow 0$	$K(MgL^{-}+H^{+}\rightleftharpoons MgHL)$ $5{\cdot}38-0{\cdot}51\sqrt{I}+0{\cdot}82\,I\pm0{\cdot}06$ $\Delta H°$ −2·0±0·8, $\Delta S°$ 18±3	
	gl	0·4—40	0·10 (KNO_3)	K_1 (0·4°) 2·94±0·02, (12°) 3·05±0·02, (25°) 3·17±0·02, (40°) 3·30±0·02 $\Delta H_1°$ (25°) 3·6±0·2, $\Delta S_1°$ (25°) 27±1 $K(Mg^{2+}+HL^{2-}\rightleftharpoons MgHL)$ (0·4°) 1·39±0·02, (12°) 1·51±0·02, (25°) 1·64±0·02, (40°) 1·78±0·02 $\Delta H°$ (25°) 3·9±0·2, $\Delta S°$ (25°) 21±1	67T
Mn^{2+}	gl	0·4—40	0·10 (KNO_3)	K_1 (0·4°) 4·47±0·02, (12°) 4·24±0·02, (25°) 4·16±0·02, (40°) 4·06±0·02 $\Delta H_1°$ (25°) −2·4±0·4, $\Delta S_1°$ (25°) 11±1 $K(Mn^{2+}+HL^{2-}\rightleftharpoons MnHL)$ (0·4°) 2·00±0·02, (12°) 1·95±0·02, (25°) 1·89±0·02, (40°) 1·81±0·02 $\Delta H°$ (25°) −1·9±0·3, $\Delta S°$ (25°) 2±1	67T
Na^{+}	Ag–AgBr	25	0·2 (Br^{-})	K_1 0·65±0·07	54M
Ni^{2+}	gl	0·4—40	0·10 (KNO_3)	K_1 (0·4°) 4·62±0·02, (12°) 4·57±0·02, (25°) 4·50±0·02, (40°) 4·42±0·02 $\Delta H_1°$ (25°) −1·9±0·3, $\Delta S_1°$ (25°) 14±1 $K(Ni^{2+}+HL^{2-}\rightleftharpoons NiHL)$ (0·4°) 2·43±0·02, (12°) 2·37±0·02, (25°) 2·30±0·02, (40°) 2·22±0·02 $\Delta H°$ (25°) −2·1±0·2, $\Delta S°$ (25°) 3±1	67T
Sr^{2+}	gl	0·4—40	0·10 (KNO_3)	K_1 (0·4°) 2·70±0·02, (12°) 2·63±0·02, (25°) 2·54±0·02, (40°) 2·43±0·02 ΔH_1 (25°) −2·7±0·5 (?), $\Delta S_1°$ (25°) 6±1 (?) $K(Sr^{2+}+HL^{2-}\rightleftharpoons SrHL)$ (0·40°) 1·60±0·02, (12°) 1·57±0·02, (25°) 1·53±0·02, (40°) 1·48±0·02 $\Delta H°$ (25°) −1·2±0·3, $\Delta S°$ (25°) 3±1	67T
Zn^{2+}	gl	25	0·1 KCl	K_1 4·13, $K(Zn^{2+}+HL^{2-}\rightleftharpoons ZnHL)$ 2·34	58W
	gl	0·4—40	0·10 (KNO_3)	K_1 (0·4°) 4·40±0·02, (12°) 4·35±0·02, (25°) 4·28±0·02, (40°) 4·20±0·02 $\Delta H_1°$ (25°) −2·0±0·4, $\Delta S_1°$ (25°) 13±1 $K(Zn^{2+}+HL^{2-}\rightleftharpoons ZnHL)$ (0·4°) 2·15±0·02, (12°) 2·11±0·02, (25°) 2·04±0·02, (40°) 1·96±0·02 $\Delta H°$ (25°) −1·9±0·3, $\Delta S°$ (25°) 3±1 $K(ZnOHL^{2-}+H^{+}\rightleftharpoons ZnL^{-})$ (0·4°) 9·14±0·05, (12°) 8·83±0·05, (25°) 8·51±0·05, (40°) 8·18±0·05 $\Delta H°$ (25°) −9·9±0·3, $\Delta S°$ (25°) 6±1	67T

continued overleaf

859 $C_{10}H_{15}O_{10}N_5P_2$ (contd)

Metal	*Method*	*Temp*	*Medium*	*Log of equilibrium constant, remarks*	*Ref*
Zn^{2+} (*contd*)				$K[2ZnOHL^{2-} \rightleftharpoons (ZnOHL)_2^{4-}]$ (0·4°) 14·62±0·07, (12°) 14·16±0·07, (25°) 13·68±0·07, (40°) 13·20±0·07	
				$\Delta H°$ (25°) −5·1±0·2, $\Delta S°$ (25°) −2±1	
				$K[(ZnOHL)_2^{4-} + 2H^+ \rightleftharpoons 2ZnL^-]$ (0·4°) 3·66±0·06, (12°) 3·50±0·06, (25°) 3·34±0·06, (40°) 3·16±0·06	
				$\Delta H°$ (25°) −13±1, $\Delta S°$ (25°) − 17±3	

53D V. Di Stefano and W. F. Neuman, *J. Biol. Chem.*, 1953, **200**, 759
54M N. C. Melchior, *J. Biol. Chem.*, 1954, **208**, 615
58W G. Weitzel and T. Speer, *Z. physiol. Chem.*, 1958, **313**, 212
59B K. Burton, *Biochem. J.*, 1959, **71**, 388
61H K. Hotta, J. Brahms, and M. Morales, *J. Amer. Chem. Soc.*, 1961, **83**, 997
62A H. Asai and M. Morales, *Arch. Biochem. Biophys.*, 1962, **99**, 383
66P R. C. Phillips, P. George, and R. J. Rutman, *J. Amer. Chem. Soc.*, 1966, **88**, 2631
67T M. M. Taqui Khan and A. E. Martell, *J. Amer. Chem. Soc.*, 1967, **89**, 5585

859.1 $C_{10}H_{15}O_{14}N_4P_3$ Inosine-5-triphosphoric acid (ITP) H_5L

Metal	*Method*	*Temp*	*Medium*	*Log of equilibrium constant, remarks*	*Ref*
H^+	sp	25	natural I	K_1 9·5	68S
	sp, gl	25	0·1 ($NaClO_4$)	K_1 (sp) 9·0, (gl) 9·2	68S
Cu^{2+}	sp	25	natural I	$K(CuL^{3-} + H^+ \rightleftharpoons CuHL^{2-})$ 7·7	68S
	sp, gl	25	0·1 ($NaClO_4$)	$K(CuL^{3-} + H^+ \rightleftharpoons CuHL^{2-})$ (sp) 7·2, (gl) 7·5	68S
				$K(CuOHL^{4-} + H^+ \rightleftharpoons CuL^{3-})$ (gl) ~ 9·2	
				$K(CuAL^{3-} + H^+ \rightleftharpoons CuAHL^{2-})$ (gl) 8·7 where A = 2,2′-bipyridyl	
Mg^{2+}	sp	?	0·05 $(CH_3)_4NCl$	$K(Mg^{2+} + HL^{4-} \rightleftharpoons MgHL^{2-})$ (?) 2·42	61H
	gl	?	0·05 $(CH_3)_4NCl$	K_1 4·08	61H

Erratum. 1964, p. 613, Table 768. This Table was numbered **768** in the 1964 edition and has been renumbered to correct the ligand sequence.

61H K. Hotta, J. Brahms, and M. Morales, *J. Amer. Chem. Soc.*, 1961, **83**, 997
68S H. Sigel, *European J. Biochem.*, 1968, **3**, 530

860 $C_{10}H_{16}O_{13}N_5P_3$ Adenosine-5′-triphosphoric acid (ATP) H_4L

Metal	*Method*	*Temp*	*Medium*	*Log of equilibrium constant, remarks*	*Ref*
H^+	Ag–AgBr	25	0·2 (Br^-)	K_1 6·90	54M
	gl	25	0·1 KCl	K_1 6·73, K_2 4·26	58W
	gl	30	0·1 $(C_2H_5)_4NBr$	K_1 6·97±0·02, K_2 3·93±0·02	61O, 64O
	gl	20	0·1 KCl	K_1 6·50, K_2 3·95	62H
	gl	20	0·1 ($NaClO_4$)	K_1 6·47±0·02, K_2 4·10±0·03	64S

contd

Metal	Method	Temp	Medium	Log of equilibrium constant, remarks	Ref
H^+ (contd)	gl	25	0·32 CsCl	K_1 6·75, K_2 4·17	65B
	gl	10—37	I $(C_4H_9)_4NBr$	K_1 (10°) $7{\cdot}62-3{\cdot}48\sqrt{I}+4{\cdot}95\,I$, (25°) $7{\cdot}68-3{\cdot}56\sqrt{I}+4{\cdot}90\,I$, (37°) $7{\cdot}73-3{\cdot}64\sqrt{I}+4{\cdot}25\,I$ ($\pm$0·04)	66P
			→ 0	$\Delta H_1°$ 1·7±0·3, $\Delta S_1°$ 41±1	
	gl	30	0·1 $[(CH_3)_4NBr]$	K_1 6·81±0·03, K_2 3·83±0·03	66Pa
	gl	0·4—40	0·1 (KNO_3)	K_1 (0·4°) 6·56±0·01, (12°) 6·54±0·01, (25°) 6·53±0·01, (40°) 6·52±0·01	66T
				$\Delta H_1°$ (25°) −0·5±0·1, $\Delta S_1°$ (25°) 27·8±0·5	
				K_2 (0·4°) 4·29±0·01, (12°) 4·14±0·01, (25°) 4·06±0·01, (40°) 3·87±0·01	
				$\Delta H_2°$ (25°) −4·1±0·1, $\Delta S_2°$ (25°) 4·5±0·5	
	gl	25	0·1 $(NaClO_4)$	K_1 6·42±0·05	67S
Ba^{2+}	gl	25	0·1 $(C_2H_5)_4NBr$	K_1 3·73±0·06	61N
	gl	0·4—40	0·1 (KNO_3)	K_1 (0·4°) 3·58±0·02, (12°) 3·42±0·02, (25°) 3·29±0·02, (40°) 3·12±0·02,	66T
				$\Delta H_1°$ (25°) −3·9±0·2, $\Delta S_1°$ (25°) 2±1	
				$K(Ba^{2+}+HL^{3-} \rightleftharpoons BaHL^-)$ (0·4°) 2·02±0·02, (12°) 1·92±0·02, (25°) 1·85±0·02, (40°) 1·75±0·02	
				$\Delta H°$ (25°) −2·1±0·2, $\Delta S°$ (25°) 2±1	
Ca^{2+}	ix	37	0·1	K_1 4·06	53D
	sp	25	0·1 $(C_4H_9)_3(C_2H_5)NBr$	K_1 3·45	59B
	sp	25	0·1 NaCl	K_1 3·90±0·08	61N
	gl	25	0·1 $(C_2H_5)_4NBr$	K_1 3·92±0·03	61N
	ix	25	0·1 $(C_2H_5)_4NBr$	K_1 3·97	61N
	oth	~ 23	0·06 $(CH_3)_4NCl$ 0·05 $(HOCH_2)_3.CNH_2$	K_1 3·88, interferometer	62A
			0·06 $(CH_3)_4NCl$ 0·05 CH_3COO^-	$K(Ca^{2+}+HL^{3-} \rightleftharpoons CaHL^-)$ 1·45, interferometer	
	gl	30	0·1 $(C_2H_5)_4NBr$	K_1 4·51	64O
	sp	30	0·1 buffer	K_1 4·49 (buffer = *N*-ethylmorpholine)	64O
	gl	0·4—40	0·1 (KNO_3)	K_1 (0·4°) 4·10±0·02, (12°) 3·99±0·02, (25°) 3·97±0·02, (40°) 3·94±0·02	66T
				$\Delta H_1°$ (25°) −0·9±0·3 (?), $\Delta S_1°$ (25°) 12±1 (?)	
				$K(Ca^{2+}+HL^{3-} \rightleftharpoons CaHL^-)$ (0·4°) 2·34±0·02, (12°) 2·21±0·02, (25°) 2·13±0·02, (40°) 2·13±0·02	
				$\Delta H°$ (25°) −0·3±0·2, $\Delta S°$ (25°) 9±1	
Co^{2+}	gl	22	0·1 KCl	K_1 4·71	61B
				$K[Co(OH)L^{3-}+H^+ \rightleftharpoons CoL^{2-}]$ 9·4	
	gl	30	0·1 $[(CH_3)_4NBr]$	K_1 5·21±0·02	66Pa
				$K(Co^{2+}+HL^{3-} \rightleftharpoons CoHL^-)$ 2·65±0·05	
	gl	0·4—40	0·1 (KNO_3)	K_1 (0·4°) 4·80±0·02, (12°) 4·69±0·02, (25°) 4·66±0·02, (40°) 4·55±0·02,	66T
				$\Delta H_1°$ (25°) −2·2±0·3, $\Delta S_1°$ (25°) 14±1	
				$K(Co^{2+}+HL^{3-} \rightleftharpoons CoHL^-)$ (0·4°) 2·45±0·2, (12°) 2·39±0·02, (25°) 2·32±0·02, (40°) 2·24±0·02	
				$\Delta H°$ (25°) −2·1±0·3, $\Delta S°$ (25°) 4±1	
	gl	25	0·1 $(NaClO_4)$	K_1 4·86±0·07	67S
				$K(CoA^{2+}+L^{4-} \rightleftharpoons CoAL^{2-})$ 4·79±0·05 where A = 2,2′-bipyridyl	
Cs^+	gl	25	0·32 CsCl	K_1 0·9, K_2 0	65B
				$K(Cs^++HL^{3-} \rightleftharpoons CsHL^{2-})$ < −0·3	

continued overleaf

860 $C_{10}H_{16}O_{13}N_5P_3$ (contd)

Metal	*Method*	*Temp*	*Medium*	*Log of equilibrium constant, remarks*	*Ref*
Cu^{2+}	gl	22	0·1 KCl	K_1 5·77 $K[Cu(OH)L^{3-} + H^+ \rightleftharpoons CuL^{2-}]$ 7·7	61B
	gl	20	0·1 KCl	K_1 5·82, K_2 1·88 $K(Cu^{2+} + HL^{3-} \rightleftharpoons CuHL^-)$ 3·25 $K(Cu^{2+} + H_2L^{2-} \rightleftharpoons CuH_2L)$ 1·94	62H
	gl	20	0·1 ($NaClO_4$)	K_1 6·30 ± 0·05 $K(CuL^{2-} + H^+ \rightleftharpoons CuHL^-)$ 3·93	64S
	gl	30	0·1 [$(CH_3)_4NBr$]	K_1 6·83 ± 0·02 $K(Cu^{2+} + HL^{3-} \rightleftharpoons CuHL^-)$ 3·97 ± 0·05	66Pa
	gl	0·4—40	0·1 (KNO_3)	K_1 (0·4°) 6·42 ± 0·02, (12°) 6·20 ± 0·02, (25°) 6·13 ± 0·02, (40°) 5·97 ± 0·02 $\Delta H_1°$ (25°) −4·3 ± 0·2, $\Delta S_1°$ (25°) 14 ± 1 $K(Cu^{2+} + HL^{3-} \rightleftharpoons CuHL^-)$ (0·4°) 3·32 ± 0·02, (12°) 3·20 ± 0·02, (25°) 3·12 ± 0·02, (40°) 3·01 ± 0·02 $\Delta H°$ (25°) −3·0 ± 0·2, $\Delta S°$ (25°) 4 ± 1 $K[Cu(OH)L^{3-} + H^+ \rightleftharpoons CuL^{2-}]$ (0·4°) 7·05 ± 0·04, (12°) 6·74 ± 0·04, (25°) 6·47 ± 0·04, (40°) 6·19 ± 0·04 $\Delta H°$ (25°) −8·0 ± 0·3, $\Delta S°$ (25°) 3 ± 1 $K[Cu(OH)_2L^{4-} + 2H^+ \rightleftharpoons CuL^{2-}]$ (0·4°) 14·37 ± 0·07, (12°) 13·92 ± 0·07, (25°) 13·50 ± 0·07, (40°) 13·10 ± 0·07 $\Delta H°$ (25°) −12 ± 1, $\Delta S°$ (25°) 22 ± 3 $K\{[Cu(OH)L]_2^{6-} + 2H^+ \rightleftharpoons 2CuL^{2-}\}$ (0·4°) 11·00 ± 0·05, (12°) 10·58 ± 0·05, (25°) 10·35 ± 0·05, (40°) 10·00 ± 0·05 $\Delta H°$ (25°) −9·9 ± 0·3, $\Delta S°$ (25°) 14 ± 1 $K\{2Cu(OH)L^{3-} \rightleftharpoons [Cu(OH)L]_2^{6-}\}$ (0·4°) 3·10 ± 0·04, (12°) 2·80 ± 0·04, (25°) 2·59 ± 0·04, (40°) 2·38 ± 0·04 $\Delta H°$ (25°) −6·4 ± 0·5, $\Delta S°$ (25°) −10 ± 2	66T
	gl	25	0·1 ($NaClO_4$)	K_1 6·38 ± 0·09 $K(CuA^{2+} + L^{4-} \rightleftharpoons CuAL^{2-})$ 6·91 ± 0·01 where A = 2,2′-bipyridyl	67S
	gl	25	0·1 ($NaClO_4$)	$K(CuOHL^{3-} + H^+ \rightleftharpoons CuL^{2-})$ ~ 7·9	68S
K^+	Ag–AgBr	25	0·2 (Br^-)	K_1 0·99 ± 0·03	54M
	gl	25	0·32 CsCl	K_1 0·9, K_2 −0·2 $K(K^+ + HL^{3-} \rightleftharpoons KHL^{2-})$ ~ −0·3	65B
Li^+	gl	25	0·32 CsCl	K_1 1·7, K_2 0·53 $K(Li^+ + HL^{3-} \rightleftharpoons LiHL^{2-})$ 0·8	65B
Mg^{2+}	gl	25	0·1 KCl	K_1 4·04 $K(Mg^{2+} + HL^{3-} \rightleftharpoons MgHL^-)$ 2·16	58W
	sp	25—64	0·1 $(C_4H_9)_3(C_2H_5)NBr$	K_1 (25°) 4·58, (→ 37°) 4·74, (64°) 4·99	59B
		25	0·2 $(C_4H_9)_3(C_2H_5)NBr$	K_1 4·35 ± 0·05	
			0·1 KCl	K_1 4·25 ± 0·04	
	sp	?	0·5 $(CH_3)_4NCl$	$K(Mg^{2+} + HL^{3-} \rightleftharpoons MgHL^-)$ (?) 2·23	61H
	gl	?	0·5 $(CH_3)_4NCl$	K_1 3·90	61H
	gl	25	0·1 $(C_2H_5)_4NBr$	K_1 4·43 ± 0·04	61N
	ix	25	0·1 $(C_2H_5)_4NBr$	K_1 4·37 ± 0·04	61N
	gl	30	0·1 $(C_2H_5)_4NBr$	K_1 5·02 ± 0·06 $K(Mg^{2+} + HL^{3-} \rightleftharpoons MgHL^-)$ 2·90	61O
	sp	30	0·1 buffer	K_1 4·30 [buffer = $(HOCH_2)_3CNH_2$] K_1 4·89 (buffer = triethanolamine) K_1 4·93 (buffer = *N*-ethylmorpholine)	61O

contd

Metal	Method	Temp	Medium	Log of equilibrium constant, remarks	Ref
Mg^{2+} (contd)	oth	~ 23	0·1 $(HOCH_2)_3CNH_2$	K_1 4·9, interferometer	62A
	gl	20	0·1 KCl	K_1 3·84, $K(Mg^{2+}+HL^{3-} \rightleftharpoons MgHL^-)$ 2·09 $K(Mg^{2+}+H_2L^{2-} \rightleftharpoons MgH_2L)$ 1·58	62H
	gl	30	0·1 $(C_2H_5)_4NBr$	K_1 4·88, $K(Mg^{2+}+HL^{3-} \rightleftharpoons MgHL^-)$ 2·7	64O
	sp	30	0·1 buffer	K_1 4·90 (buffer = *N*-ethylmorpholine)	64O
	ix	5—65	0·07—0·17 $(C_4H_9)_4NBr$	K_1 (5°) (0·07) 4·46, (0·1) 4·45, (0·17) 4·38, (25°) (0·07) 4·60, (0·1) 4·63, (0·17) 4·54, (45°) (0·07) 4·78, (0·1) 4·81, (0·17) 4·70, (65°) (0·07) 4·96, (0·1) 4·92, (0·17) 4·85	66P
		25	*I* $(C_4H_9)_4NBr$	K_1 $5{\cdot}83-6{\cdot}10\sqrt{I}+8{\cdot}74\,I-(2{\cdot}04\sqrt{I})/(1+6{\cdot}02\sqrt{I})\pm0{\cdot}10$ $\Delta H_1°$ 5·1 ± 0·3, $\Delta S_1°$ 44 ± 2 $K(Mg^{2+}+HL^{3-} \rightleftharpoons MgHL^-)$ $3{\cdot}59-4{\cdot}06\sqrt{I}+6{\cdot}36\,I-2{\cdot}04\sqrt{I}/(1+6{\cdot}02\sqrt{I})\pm0{\cdot}12$	
			0·1 $(C_4H_9)_4NBr$	$K(Mg^{2+}+HL^{3-} \rightleftharpoons MgHL^-)$ 2·9 ± 0·2 $\Delta H°$ 2 ± 1, $\Delta S°$ 24 ± 4	
			I $MgCl_2$	$K(MgL^{2-}+H^+ \rightleftharpoons MgHL^-)$ $5{\cdot}44-1{\cdot}52\sqrt{I}+2{\cdot}52\,I\pm0{\cdot}03$	
			→ 0	$\Delta H°$ −1·2 ± 0·6, $\Delta S°$ 21 ± 2	
	gl	0·4—40	0·10 (KNO_3)	K_1 (0·4°) 3·97 ± 0·02, (12°) 4·10 ± 0·02, (25°) 4·22 ± 0·02, (40°) 4·28 ± 0·02 $\Delta H_1°$ (25°) 2·6 ± 0·1, $\Delta S_1°$ (25°) 27·5 ± 0·5 $K(Mg^{2+}+HL^{3-} \rightleftharpoons MgHL^-)$ (0·4°) 1·95 ± 0·02, (12°) 2·16 ± 0·02, (25°) 2·24 ± 0·02, (40°) 2·29 ± 0·02 $\Delta H°$ (25°) 3·4 ± 0·1, $\Delta S°$ (25°) 21·5 ± 0·5	66T
Mn^{2+}	gl	22	0·1 KCl	K_1 4·78 $K[Mn(OH)L^{3-}+H^+ \rightleftharpoons MnL^{2-}]$ 10·4	61B
	gl	20	0·1 KCl	K_1 4·52, K_2 1·37 $K(Mn^{2+}+HL^{3-} \rightleftharpoons MnHL^-)$ 2·61 $K(Mn^{2+}+H_2L^{2-} \rightleftharpoons MnH_2L)$ 2·03	62H
	gl	30	0·1 $[(CH_3)_4NBr]$	K_1 5·19 ± 0·02 $K(Mn^{2+}+HL^{3-} \rightleftharpoons MnHL^-)$ 2·62 ± 0·05	66Pa
	gl	0·4—40	0·1 (KNO_3)	K_1 (0·4°) 4·97 ± 0·02, (12°) 4·82 ± 0·02, (25°) 4·78 ± 0·02, (40°) 4·63 ± 0·02 $\Delta H_1°$ (25°) −3·0 ± 0·3, $\Delta S_1°$ (25°) 12 ± 1 $K(Mn^{2+}+HL^{3-} \rightleftharpoons MnHL^-)$ (0·4°) 2·55 ± 0·02, (12°) 2·48 ± 0·02, (25°) 2·39 ± 0·02, (40°) 2·30 ± 0·02 $\Delta H°$ (25°) −2·3 ± 0·3, $\Delta S°$ (25°) 3 ± 1	66T
Na^+	Ag–AgBr	25	0·2 (Br^-)	K_1 0·98 ± 0·05	54M
	gl	25	0·32 CsCl	K_1 1·2, K_2 0·93 $K(Na^++HL^{3-} \rightleftharpoons NaHL^{2-})$ 0·7	65B
Ni^{2+}	gl	22	0·1 KCl	K_1 4·54 $K[Ni(OH)L^{3-}+H^+ \rightleftharpoons NiL^{2-}]$ 9·3	61B
	gl	30	0·1 $[(CH_3)_4NBr]$	K_1 5·32 ± 0·02 $K(Ni^{2+}+HL^{3-} \rightleftharpoons NiHL^-)$ 2·98 ± 0·05	66Pa
	gl	0·4—40	0·1 (KNO_3)	K_1 (0·4°) 5·18 ± 0·02, (12°) 5·05 ± 0·02, (25°) 5·02 ± 0·02, (40°) 4·90 ± 0·02 $\Delta H_1°$ (25°) −2·5 ± 0·2, $\Delta S_1°$ (25°) 15 ± 1 $K(Ni^{2+}+HL^{3-} \rightleftharpoons NiHL^-)$ (0·4°) 2·88 ± 0·02, (12°) 2·80 ± 0·02, (25°) 2·72 ± 0·02, (40°) 2·59 ± 0·02 $\Delta H°$ (25°) −2·4 ± 0·2, $\Delta S°$ (25°) 4 ± 1	66T

continued overleaf

860 $C_{10}H_{16}O_{13}N_5P_3$ (contd)

Metal	*Method*	*Temp*	*Medium*	*Log of equilibrium constant, remarks*	*Ref*
Ni^{2+} (*contd*)	gl	25	0·1 ($NaClO_4$)	K_1 4·85±0·04 $K(NiA^{2+}+L^{4-} \rightleftharpoons NiAL^{2-})$ 4·45±0·05 where A = 2,2′-bipyridyl	67S
Rb^+	gl	25	0·32 CsCl	K_1 0·9, K_2 0 $K(Rb^{+}+HL^{3-} \rightleftharpoons RbHL^{2-})$ < −0·3	65B
Sr^{2+}	ix	~ 25	0·15 NaCl	K_1 3·2	60O
	gl	25	0·1 $(C_2H_5)_4NBr$	K_1 3·60±0·04	61N
	gl	0·4—40	0·1 (KNO_3)	K_1 (0·4°) 3·80±0·02, (12°) 3·66±0·02, (25°) 3·54±0·02, (40°) 3·45±0·02 ΔH_1 (25°) −3·0±0·3, ΔS_1 (25°) 6±1 $K(Sr^{2+}+HL^{3-} \rightleftharpoons SrHL^{-})$ (0·4°) 2·17±0·02, (12°) 2·11±0·02, (25°) 2·05±0·02, (40°) 2·00±0·02 ΔH° (25°) −1·6±0·3, ΔS° (25°) 4±1	66T
Y^{3+}	ix	~ 25	0·15 NaCl	K_1 11·1	60O
Zn^{2+}	gl	25	0·1 KCl	K_1 4·76, $K(Zn^{2+}+HL^{3-} \rightleftharpoons ZnHL^{-})$ 2·75	58W
	gl	22	0·1 KCl	K_1 4·80 $K[Zn(OH)L^{3-}+H^{+} \rightleftharpoons ZnL^{2-}]$ 8·5	61B
	gl	20	0·1 KCl	K_1 4·75, K_2 1·41 $K(Zn^{2+}+HL^{3-} \rightleftharpoons ZnHL^{-})$ 2·78 $K(Zn^{2+}+H_2L^{2-} \rightleftharpoons ZnH_2L)$ 2·09	62H
	gl	30	0·1 [$(CH_3)_4NBr$]	K_1 5·52±0·02 $K(Zn^{2+}+HL^{3-} \rightleftharpoons ZnHL^{-})$ 2·91±0·05	66Pa
	gl	0·4—40	0·1 (KNO_3)	K_1 (0·4°) 5·00±0·02, (12°) 4·88±0·02, (25°) 4·85±0·02, (40°) 4·71±0·02 ΔH_1° (25°) −2·7±0·3, ΔS_1° (25°) 13±1 $K(Zn^{2+}+HL^{3-} \rightleftharpoons ZnHL^{-})$ (0·4°) 2·81±0·02, (12°) 2·73±0·02, (25°) 2·67±0·02, (40°) 2·58±0·02 ΔH° (25°) −2·4±0·3, ΔS° (25°) 4±1	66T
	gl	25	0·1 ($NaClO_4$)	K_1 5·21±0·06 $K(ZnA^{2+}+L^{4-} \rightleftharpoons ZnAL^{2-})$ 5·26±0·03 where A = 2,2′-bipyridyl	67S

Errata. 1964, p. 651, Table 860. Under Ca^{2+}, *delete* sixth and seventh lines.
p. 652. Under Mg^{2+}, *delete* sixth line.
p. 652. Under Mn^{2+}, *delete* fourth to twelfth lines inclusive.

53D V. Di Stefano and W. F. Neuman, *J. Biol. Chem.*, 1953, **200**, 759
54M N. C. Melchior, *J. Biol. Chem.*, 1954, **208**, 615
58W G. Weitzel and T. Speer, *Z. physiol. Chem.*, 1958, **313**, 212
59B K. Burton, *Biochem. J.*, 1959, **71**, 388
60O J. Olivard, *Arch. Biochem. Biophys.*, 1960, **88**, 382
61B H. Brintzinger, *Helv. Chim. Acta*, 1961, **44**, 935, 1199
61H K. Hotta, J. Brahms, and M. Morales, *J. Amer. Chem. Soc.*, 1961, **83**, 997
61N L. B. Nanninga, *Biochim. Biophys. Acta*, 1961, **54**, 330
61O W. J. O'Sullivan and D. D. Perrin, *Biochem. Biophys. Acta*, 1961, **52**, 612
62A H. Asai and M. Morales, *Arch. Biochem. Biophys.*, 1962, **99**, 383
62H Y. Handschin and H. Brintzinger, *Helv. Chim. Acta*, 1962, **45**, 1037
64O W. J. O'Sullivan and D. D. Perrin, *Biochemistry*, 1964, **3**, 18
64S P. W. Schneider, H. Brintzinger, and H. Erlenmeyer, *Helv. Chim. Acta*, 1964, **47**, 992
65B J. Botts, A. Chashin, and H. L. Young, *Biochemistry*, 1965, **4**, 1788
66P R. C. Phillips, P. George, and R. J Rutman, *J. Amer. Chem. Soc.*, 1966, **88**, 2631
66Pa D. D. Perrin and V. S. Sharma, *Biochem. Biophys. Acta*, 1966, **127**, 35
66T M. M. Taqui Khan and A. E. Martell, *J. Amer. Chem. Soc.*, 1966, **88**, 668
67S H. Sigel, K. Becker, and D. B. McCormick, *Biochim. Biophys. Acta*, 1967, **148**, 655
68S H. Sigel, *European J. Biochem.*, 1968, **3**, 530

$C_{10}H_{16}O_{14}N_5P_3$ Guanosine-5′-triphosphoric acid (GTP) H_5L 861

Metal	Method	Temp	Medium	Log of equilibrium constant, remarks	Ref
H^+	sp	25	natural *I*	K_1 10·1	68S
	sp, gl	25	0·1 ($NaClO_4$)	K_1 (sp) 9·5, (gl) 9·6	68S
Cu^{2+}	sp	25	natural *I*	$K(CuL^{3-} + H^+ \rightleftharpoons CuHL^{2-})$ 8·1	68S
	sp, gl	25	0·1 ($NaClO_4$)	$K(CuL^{3-} + H^+ \rightleftharpoons CuHL^{2-})$ (sp) 7·5, (gl) 7·7	68S
				$K(CuOHL^{4-} + H^+ \rightleftharpoons CuL^{3-})$ (gl) ~ 9·3	
				$K(CuAL^{3-} + H^+ \rightleftharpoons CuAHL^{2-})$ (gl) 9·0	
				where A = 2,2′-bipyridyl	

Erratum. 1964, p. 652, Table 861. *For* Guanosine trophosphoric acid *read* Guanosine triphosphoric acid.

68S H. Sigel, *European J. Biochem.*, 1968, **3**, 530

$C_{10}H_{17}O_6N_3S$ L-Glutamyl-L-cysteinylglycine (glutathione) H_3L 862

Metal	Method	Temp	Medium	Log of equilibrium constant, remarks	Ref
H^+	gl	25	0·16 (KNO_3)	K_1 9·62, K_2 8·74	58M
Co^{2+}	gl	25	0·16 (KNO_3)	K_1 4·2	59M
Cu^{2+}	pol	25	0·17	β_2 4·66 (?), phosphate buffer	61K
Mn^{2+}	gl	25	0·16 (KNO_3)	K_1 2·7	59M
Ni^{2+}	gl	25	0·16 (KNO_3)	K_1 5·0	59M
Zn^{2+}	gl	25	0·16 (KNO_3)	K_1 5·1	59M

58M R. B. Martin and J. T. Edsall, *Bull. Soc. Chim. biol.*, 1958, **40**, 1763
59M R. B. Martin and J. T. Edsall, *J. Amer. Chem. Soc.*, 1959, **81**, 4044
61K E. C. Knoblock and W. C. Purdy, *J. Electroanalyt. Chem.*, 1961, **2**, 493

$C_{10}H_{17}O_{14}N_2P_3$ Thymidine-5′-triphosphoric acid (TTP) H_5L 862.1

Metal	Method	Temp	Medium	Log of equilibrium constant, remarks	Ref
H^+	sp	25	natural *I*	K_1 10·7	68S
	sp, gl	25	0·1 ($NaClO_4$)	K_1 (sp) 10·1, (gl) 9·8	68S
Cu^{2+}	sp	25	natural *I*	$K(CuL^{3-} + H^+ \rightleftharpoons CuHL^{2-})$ 8·5	68S
	sp, gl	25	0·1 ($NaClO_4$)	$K(CuL^{3-} + H^+ \rightleftharpoons CuHL^{2-})$ (sp) 7·9, (gl) 7·7	68S
				$K(CuOHL^{4-} + H^+ \rightleftharpoons CuL^{3-})$ (gl) ~ 8·2	
				$K(CuAL^{3-} + H^+ \rightleftharpoons CuAHL^{2-})$ (gl) 9·1	
				where A = 2,2′-bipyridyl	

68S H. Sigel, *European J. Biochem.*, 1968, **3**, 530

864.1 $C_{10}H_8O_4NBrS$ 7-Bromo-8-hydroxy-2-methylquinoline-5-sulphonic acid H_2L

Metal	Method	Temp	Medium	Log of equilibrium constant, remarks	Ref
H^+	gl	25	0·005 $HClO_4$	K_1 7·96, K_2 3·32	63F
	sp	28	0·005 $HClO_4$	K_3 < 2·0	63F
Co^{2+}	gl	25	0·005 $HClO_4$	K_1 6·56, K_2 5·48, K_3 < 3·5	63F
Cu^{2+}	gl	25	0·005 $HClO_4$	K_2 8·53, K_3 < 3·5	63F
Ni^{2+}	gl	25	0·005 $HClO_4$	K_1 6·80, K_2 5·50, K_3 < 3·5	63F
Zn^{2+}	gl	25	0·005 $HClO_4$	K_1 6·61, K_2 5·68, K_3 < 3·5	63F

63F J. Fresco and H. Freiser, *Inorg. Chem.*, 1963, **2**, 82

870 $C_{11}H_8O_3$ 2-Hydroxy-3-methyl-1,4-naphthaquinone (phthiocol) HL

Metal	Method	Temp	Medium	Log of equilibrium constant, remarks	Ref
H^+				H^+ constants not given.	
Cd^{2+}	pol	25	0·2	β_3 11·7, phosphate buffer	66S
Zn^{2+}	pol	25	0·2	β_3 12·15, phosphate buffer	66S

66S I. H. Suffet and W. C. Purdy, *J. Electroanalyt. Chem.*, 1966, **11**, 302

870.1 $C_{11}H_8O_3$ 1-Hydroxynaphthalene-2-carboxylic acid H_2L

Metal	Method	Temp	Medium	Log of equilibrium constant, remarks	Ref
UO_2^{2+}	sp J	30	?	$K(UO_2^{2+}+HL^- \rightleftharpoons UO_2L+H^+)$ (?) 1·62	59T

59T S. C. Tripathi and S. Prakash, *J. Indian Chem. Soc.*, 1959, **36**, 19

870.2 $C_{11}H_8O_3$ 3-Hydroxynaphthalene-2-carboxylic acid H_2L

Metal	Method	Temp	Medium	Log of equilibrium constant, remarks	Ref
H^+	gl	22	50% ethanol	K_1 12·87, K_2 3·65	61A
	sp	25	I (KCl)	K_1 $12{\cdot}84-2{\cdot}036\sqrt{I}/(1+2{\cdot}15\sqrt{I})+0{\cdot}20\,I$, K_2 $2{\cdot}79-1{\cdot}018\sqrt{I}/(1+0{\cdot}55\sqrt{I})+0{\cdot}28\,I$	66M
			→0	K_1 13·38	
Al^{3+}	sp	25	I (KCl)	$K(Al^{3+}+HL^- \rightleftharpoons AlL^++H^+)$ (?) $-0{\cdot}54+4{\cdot}072\sqrt{I}/(1+2{\cdot}22\sqrt{I})-0{\cdot}03\,I$	66M
	sp	20—60	0·02	$K(Al^{3+}+HL^- \rightleftharpoons AlL^++H^+)$ (20°) 4·55±0·02, (30°) 4·77±0·05, (40°) 4·90±0·09, (50°) 5·18±0·02, (60°) 5·42±0·05 ΔH −9·5±0·2(?), ΔS 52±2(?)	67G
		20	0·06—0·21	$K(Al^{3+}+HL^- \rightleftharpoons AlL^++H^+)$ (?) (0·06) 4·38±0·02, (0·09) 4·26±0·01, (0·11) 4·17±0·01, (0·16) 4·07±0·01, (0·21) 3·97±0·09	

con

$C_{11}H_8O_3$ (contd) 870.2

Metal	Method	Temp	Medium	Log of equilibrium constant, remarks	Ref
Be^{2+}	gl	22	50% ethanol	K_1 11·98, K_2 7·92	61A
	sp	25	→ 0	K_1 12·51	66M
			I (KCl)	$K(Be^{2+}+HL^- \rightleftharpoons BeL+H^+)$ $0{\cdot}33+2{\cdot}026\sqrt{I}/(1+1{\cdot}75\sqrt{I})-0{\cdot}05\,I$	
Cu^{2+}	gl	26	→ 0	K_1 10·28, β_2 19·8	66M
			I (KCl)	$K(Cu^{2+}+HL^- \rightleftharpoons CuL+H^+)$ $2{\cdot}56+2{\cdot}036\sqrt{I}/(1+1{\cdot}23\sqrt{I})-0{\cdot}32\,I$	
	sp	25	0·026 (KCl)	$K(Cu^{2+}+HL^- \rightleftharpoons CuL+H^+)$ 2·86	66M
Fe^{3+}	sp	30	?	$K(Fe^{3+}+HL^- \rightleftharpoons FeL^++H^+)$ (?) 5·01	66G
UO_2^{2+}	sp	25	?	$K(UO_2^{2+}+HL^- \rightleftharpoons UO_2L+H^+)$ 3·45	65D
Zn^{2+}	gl	25	?	$K(Zn^{2+}+HL^- \rightleftharpoons ZnL+H^+)$ 4·35	66M
				$K(Zn^{2+}+2HL^- \rightleftharpoons ZnL_2+2H^+)$ 4·60	

61A R. P. Agarwal and R. C. Mehrotra, *J. Less-Common Metals*, 1961, **3**, 398
65D M. N. Desai, *Current Sci.*, 1965, **34**, 312
66G S. L. Gupta and R. N. Soni, *J. Indian Chem. Soc.*, 1966, **43**, 473
66M O. Makitie, *Suomen Kem.*, 1966, **B39**, 26, 175
67G S. L. Gupta and R. N. Soni, *J. Indian Chem. Soc.*, 1967, **44**, 195

$C_{11}H_9N$ 2-Phenylpyridine L 871.1

Metal	Method	Temp	Medium	Log of equilibrium constant, remarks	Ref
H^+	gl	25	0·1 ($NaClO_4$)	K_1 4·77±0·02	64K
Cu^{2+}	gl	25	0·1 ($NaClO_4$)	K_1 1·3	64K
Ni^{2+}	gl	25	0·1 ($NaClO_4$)	$K_1 \leqslant 1$	64K
Zn^{2+}	gl	25	0·1 ($NaClO_4$)	$K_1 \leqslant 1$	64K

64K K. Kahmann, H. Sigel, and H. Erlenmeyer, *Helv. Chim. Acta*, 1964, **47**, 1754

$C_{11}H_{10}N_4$ 1,3-Bis(2-pyridyl)-1,2-diazaprop-2-ene HL 873.1 (pyridine-2-aldehyde 2-pyridylhydrazone)

Metal	Method	Temp	Medium	Log of equilibrium constant, remarks	Ref
H^+	gl	25	0 corr	K_1 ~ 14·5, K_2 5·71±0·005, K_3 2·87±0·01	64G
	gl	5—50	→ 0	K_2 (5°) 5·93, (15°) 5·77, (25°) 5·62, (30°) 5·53, (40°) 5·34, (50°) 5·18, (60°) 5·11	68G
				$\Delta H_2°$ −6·9±0·8, $\Delta S_2°$ 3±3	
				K_3 (5°) 3·20, (15°) 3·07, (25°) 2·91, (30°) 2·83, (40°) 2·73, (50°) 2·61, (60°) 2·53	
				$\Delta H_3°$ −5·3±0·7, $\Delta S_3°$ −4±2	

continued overleaf

873.1 $C_{11}H_{10}N_4$ (contd)

Metal	*Method*	*Temp*	*Medium*	*Log of equilibrium constant, remarks*	*Ref*
Cd^{2+}	gl	25	0 corr	β_2 20, $K(Cd^{2+}+HL \rightleftharpoons CdHL^{2+})$ 4·8 $K[Cd^{2+}+2HL \rightleftharpoons Cd(HL)_2{}^{2+}]$ 10·1 $K[CdHL_2{}^{+}+H^+ \rightleftharpoons Cd(HL)_2{}^{2+}]$ 8·93 $K(CdL_2+H^+ \rightleftharpoons CdHL_2{}^{+})$ 10·22	64G
Cu^{2+}	sp	25	0 corr	K_1 20, $K(Cu^{2+}+HL \rightleftharpoons CuHL^{2+})$ 11·6 $K[CuHL_2{}^{+}+H^+ \rightleftharpoons Cu(HL)_2{}^{2+}]$ 5·96 $K(CuL_2+H^+ \rightleftharpoons CuHL_2{}^{+})$ 8·74	64G
Fe^{2+}	sp	25	0 corr	β_2 33, $K[Fe^{2+}+2HL \rightleftharpoons Fe(HL)_2{}^{2+}]$ 16·7 $K[FeHL_2{}^{+}+H^+ \rightleftharpoons Fe(HL)_2{}^{2+}]$ 5·68 $K(FeL_2+H^+ \rightleftharpoons FeHL_2{}^{+})$ 6·57	64G
	gl, sp	5—60	→ 0	$K(Fe^{2+}+2HL \rightleftharpoons Fe(HL)_2{}^{2+})$ (5°) 17·70, (15°) 17·12, (25°) 16·57, (30°) 16·18, (40°) 15·59, (50°) 15·05, (60°) 14·39 $\Delta H°$ −25·3±0·9, $\Delta S°$ −9±3 $K[FeHL_2{}^{+}+H^+ \rightleftharpoons Fe(HL)_2{}^{2+}]$ (5°) 6·36, (15°) 6·19, (25°) 6·08, (30°) 6·00, (40°) 5·87, (50°) 5·73, (60°) 5·60 $\Delta H°$ −5·7±0·2, $\Delta S°$ 9±1 $K(FeL_2+H^+ \rightleftharpoons FeHL_2{}^{+})$ (5°) 7·71, (15°) 7·53, (25°) 7·39, (30°) 7·33, (40°) 7·19, (50°) 7·05, (60°) 6·88 $\Delta H°$ −6·2±0·6, $\Delta S°$ 13±2	68G
Mn^{2+}	gl	25	0 corr	$K(Mn^{2+}+HL \rightleftharpoons MnHL^{2+})$ 3·3 $K[Mn^{2+}+2HL \rightleftharpoons Mn(HL)_2{}^{2+}]$ 6·9	64G
Ni^{2+}	gl (?)	25	0 corr	β_2 32, $K(Ni^{2+}+2HL \rightleftharpoons NiHL^{2+})$ 8·3 $K[Ni^{2+}+2HL \rightleftharpoons Ni(HL)_2{}^{2+}]$ 18·5 $K[NiHL_2{}^{+}+H^+ \rightleftharpoons Ni(HL)_2{}^{2+}]$ 7·37 $K(NiL_2+H^+ \rightleftharpoons NiHL_2{}^{+})$ 8·50	64G
	sp	25	0 corr	$K[NiHL_2{}^{+}+H^+ \rightleftharpoons Ni(HL)_2{}^{2+}]$ 7·42 $K(NiL_2+H^+ \rightleftharpoons NiHL_2{}^{+})$ 8·61	64G
Zn^{2+}	gl	25	0 corr	β_2 23, $K(Zn^{2+}+HL \rightleftharpoons ZnHL^{2+})$ 5·7 $K[Zn^{2+}+2HL \rightleftharpoons Zn(HL)_2{}^{2+}]$ 11·2 $K[ZnHL_2{}^{+}+H^+ \rightleftharpoons Zn(HL)_2{}^{2+}]$ 7·94 $K(ZnL_2+H^+ \rightleftharpoons ZnHL_2{}^{+})$ 8·85	64G
	gl	5—60	→ 0	$K(Zn^{2+}+HL \rightleftharpoons ZnHL^{2+})$ (5°) 6·40, (15°) 6·02, (25°) 5·82, (30°) 5·71, (40°) 5·48, (50°) 5·15, (60°) 4·93 $\Delta H°$ −11·0±0·2, $\Delta S°$ −10±1 $K[Zn^{2+}+2HL \rightleftharpoons Zn(HL)_2{}^{2+}]$ (5°) 11·77, (15°) 11·37, (25°) 11·08, (30°) 10·87, (40°) 10·63, (50°) 10·23, (60°) 9·94 $\Delta H°$ −13·9±0·2, $\Delta S°$ 4±1	68G

64G R. W. Green, P. S. Hallman, and F. Lions, *Inorg. Chem.*, 1964, **3**, 376
68G R. W. Green and W. G. Goodwin, *Austral. J. Chem.*, 1968, **21**, 1165

$C_{11}H_{11}N_5$ 1-(3′-Methyl-2-pyridyl)-3-2″-pyridyl)-1,2-diazaprop-2-ene HL 873.2
(pyridine-2-aldehyde 3′-methyl-2′-pyrazinylhydrazone)

Metal	*Method*	*Temp*	*Medium*	*Log of equilibrium constant, remarks*	*Ref*
H^+	sp	25	~0	K_1 12·90±0·01, K_3 1·40±0·05	64G
	gl	25	~0	K_2 4·52±0·01	64G
Fe^{2+}	gl	25	~0	β_2 32	64G
				$K[FeHL_2^+ + H^+ \rightleftharpoons Fe(HL)_2^{2+}]$ 4·12±0·003	
				$K(FeL_2 + H^+ \rightleftharpoons FeHL_2^+)$ 5·61±0·003	
	sp	25	~0	$K(Fe^{2+} + HL \rightleftharpoons FeHL^{2+})$ 7·90±0·06	64G
				$K[Fe^{2+} + 2HL \rightleftharpoons Fe(HL)_2^{2+}]$ 15·60±0·05	

64G R. W. Green, P. S. Hallman, and F. Lions, *Inorg. Chem.*, 1964, 3, 1541

$C_{11}H_{15}N_3$ 2-(*N*′-Propylaminomethyl)benzimidazole L 874.1

Metal	*Method*	*Temp*	*Medium*	*Log of equilibrium constant, remarks*	*Ref*
H^+	gl	15—45	50% dioxan	K_1 (15°) 7·38, (25°) 7·22, (35°) 7·00, (45°) 6·86, K_2 (15—45°) 2	63L
Cu^{2+}	gl	15—45	50% dioxan	K_1 (15°) 7·50, (25°) 7·33, (35°) 7·14, (45°) 6·83, K_2 (15°) 5·00, (25°) 4·82, (35°) 4·68, (45°) 4·55	63L
				ΔH_1 −11·4	

63L T. J. Lane and H. B. Durham, *Inorg. Chem.*, 1963, 2, 632

$C_{11}H_{20}O_4$ Nonane-5,5-dicarboxylic acid (di-n-butylmalonic acid) H_2L 874.2

Metal	*Method*	*Temp*	*Medium*	*Log of equilibrium constant, remarks*	*Ref*
H^+	gl	25	0·100 (KNO_3)	K_1 7·19, K_2 1·89, K_1 extrapolated to zero ligand concentration.	68P
Dy^{3+}	gl	25	0·100 (KNO_3)	K_1 4·75, β_2 7·43±0·02	68P
Er^{3+}	gl	25	0·100 (KNO_3)	K_1 6·74, β_2 7·53±0·04	68P
Eu^{3+}	gl	25	0·100 (KNO_3)	K_1 4·53, β_2 7·28±0·05	68P
Gd^{3+}	gl	25	0·100 (KNO_3)	K_1 4·54, β_2 7·32±0·04	68P
Ho^{3+}	gl	25	0·100 (KNO_3)	K_1 4·69, β_2 7·36±0·02	68P
Lu^{3+}	gl	25	0·100 (KNO_3)	K_1 4·78, β_2 7·76±0·02	68P
Sm^{3+}	gl	25	0·100 (KNO_3)	K_1 4·43	68P
Tb^{3+}	gl	25	0·100 (KNO_3)	K_1 4·70, β_2 7·45±0·03	68P
Tm^{3+}	gl	25	0·100 (KNO_3)	K_1 4·73, β_2 7·34±0·04	68P
Y^{3+}	gl	25	0·100 (KNO_3)	K_1 4·67, β_2 7·23±0·05	68P
Yb^{3+}	gl	25	0·100 (KNO_3)	K_1 4·80, β_2 7·61±0·03	68P

68P J. E. Powell, J. L. Farrell, W. F. S. Neillie, and R. Russell, *J. Inorg. Nuclear Chem.*, 1968, 30, 2223

875 $C_{11}H_8O_2S_2$ 1,3-Di(2-thienyl)propane-1,3-dione (di-2-thenoylmethane) HL

Metal	Method	Temp	Medium	Log of equilibrium constant, remarks	Ref
H^+	gl, sp	25	1 $NaClO_4$ 30% tetrahydrofuran	K_1 9·5	65C
Co^{2+}	sp	25	1 $NaClO_4$ 30% tetrahydrofuran	β_2 11·1	65C
Ni^{2+}	sp	25	1 $NaClO_4$ 30% tetrahydrofuran	β_2 11·4	65C

65C C. Caullet, *Bull. Soc. chim. France*, 1965, 3459

877.1 $C_{11}H_9ON_3$ 2-(2′-Pyridylazo)phenol HL

Metal	Method	Temp	Medium	Log of equilibrium constant, remarks	Ref
H^+	gl	25	0·1 ($NaClO_4$) 50% methanol	K_1 9·42	67A
	sp	18—22	0·1 ($NaClO_4$) 50% methanol	K_2 1·85	67A
Ag^+	gl	25	0·1 ($NaClO_4$) 50% methanol	K_1 5·4	67A
Cd^{2+}	gl	25	0·1 ($NaClO_4$) 50% methanol	K_1 7·8, K_2 6·6	67A
Co^{2+}	gl	25	0·1 ($NaClO_4$) 50% methanol	K_1 8·9, K_2 9·3	67A
Cu^{2+}	sp	18—22	0·1 ($NaClO_4$) 50% methanol	K_1 13·8	67A
	gl	25	0·1 ($NaClO_4$) 50% methanol	K_2 7·7	67A
Fe^{2+}	sp	18—22	0·1 ($NaClO_4$) 50% methanol	β_2 26·3	67A
Mn^{2+}	gl	25	0·1 ($NaClO_4$) 50% methanol	K_1 5·6, K_2 7·0	67A
Ni^{2+}	sp	18—22	0·1 ($NaClO_4$) 50% methanol	β_2 22·8	67A
Pb^{2+}	gl	25	0·1 ($NaClO_4$) 50% methanol	K_1 9·4, K_2 4·8	67A
Pd^{2+}	sp	18—22	0·1 ($NaClO_4$) 50% methanol	K_1 17·1	67A
UO_2^{2+}	sp	18—22	0·1 ($NaClO_4$) 50% methanol	K_1 10·7	67A
Zn^{2+}	gl	25	0·1 ($NaClO_4$) 50% methanol	K_1 8·8, K_2 8·1	67A

67A R. G. Anderson and G. Nickless, *Analyt. Chim. Acta*, 1967, 39, 469

$C_{11}H_9ON_3$ 4-(2′-Pyridylazo)phenol HL 877.2

Metal	Method	Temp	Medium	Log of equilibrium constant, remarks	Ref
H^+	gl	25	0·1 ($NaClO_4$) 50% methanol	K_1 8·29	67A
	sp	18—22	0·1 ($NaClO_4$) 50% methanol	K_2 2·47	67A
Co^{2+}	gl	25	0·1 ($NaClO_4$) 50% methanol	K_1 3·5, K_2 3·8	67A
Cu^{2+}	gl	25	0·1 ($NaClO_4$) 50% methanol	K_1 5·8, K_2 5·2	67A
Fe^{2+}	gl	25	0·1 ($NaClO_4$) 50% methanol	K_1 5·6, K_2 4·8	67A
Ni^{2+}	gl	25	0·1 ($NaClO_4$) 50% methanol	K_1 ~ 5·0, K_2 ~ 4·5	67A
Zn^{2+}	gl	25	0·1 ($NaClO_4$) 50% methanol	K_1 < 3, K_2 < 3	67A

67A R. G. Anderson and G. Nickless, *Analyt. Chim. Acta*, 1967, **39**, 469

$C_{11}H_9O_2N_3$ 1, 3-Dihydroxy-4-(2′-pyridylazo)benzene(pyridylazoresorcinol) (PAR) H_2L 879

Metal	Method	Temp	Medium	Log of equilibrium constant, remarks	Ref
H^+	sp	25	0·1 KNO_3	K_1 12·5	61I
	gl	25	0·1 KNO_3	K_2 5·83	61I
	gl	25	~ 0·005 50% dioxan	K_1 12·4, K_2 6·9, K_3 2·3	62C, 63C
	sp	25	0·1 0, 50% dioxan	K_1 (0%) 12·31 ± 0·03, (50%) 13·00 ± 0·05	62G, 68T
	gl	25	0·1 0, 50% dioxan	K_2 (0%) 5·50, (50%) 7·15, K_3 (0%) 2·69, (50%) 2·41	62G, 68T
	sp	18—22	0·1	K_1 11·9, K_2 5·6, K_3 3·1	62S, 63H, 66B, 66H, 67S, 67Sa
	gl	25	< 0·01 50% dioxan	K_1 13·42 ± 0·01, K_2 6·87 ± 0·01, K_3 2·31 ± 0·02	66S
Al^{3+}	sp	18—22	0·1 ($NaClO_4$)	K_1 11·5	67Sa
Bi^{3+}	sp	18—22	0·1 ($NaClO_4$)	$*K(Bi^{3+} + HL^- \rightleftharpoons BiHL^{2+})$ 17·2	66H
Cd^{2+}	sp	18—22	0·1 ($NaClO_4$)	$*K(Cd^{2+} + HL^- \rightleftharpoons CdHL^+)$ 10·5	66H
Co^{2+}	gl	25	~ 0·005 50% dioxan	$*K(Co^{2+} + HL^- \rightleftharpoons CoHL^+)$ > 12 $K(CoL + H^+ \rightleftharpoons CoHL^+)$ 4·7 $K(CoOHL^- + H^+ \rightleftharpoons CoL)$ 6·0	62C, 63C
	gl	25	0·1 0, 50% dioxan	K_1 (0%) 10·0, (50%) 14·8, K_2 (0%) 7·1, (50%) 8·2	62G

continued overleaf

879 $C_{11}H_9O_2N_3$ (contd)

Metal	*Method*	*Temp*	*Medium*	*Log of equilibrium constant, remarks*	*Ref*
Cu^{2+}	sp	25	0·1 KNO_3	K_1 11·7 $K(CuL+H^+ \rightleftharpoons CuHL^+)$ 5·3	61I
	gl	25	~ 0·005 50% dioxan	$K(CuL+H^+ \rightleftharpoons CuHL^+)$ 5·5	63C
	gl	25	0·1 0, 50% dioxan	K_1 (0%) 14·8, (50%) 16·4, K_2 (0%) 9·1, (50%) 8·9	62G
	sp	18—22	0·1 ($NaClO_4$)	*$K(Cu^{2+}+HL^- \rightleftharpoons CuHL^+)$ 16·5	66H
	gl	25	< 0·01 50% dioxan	*$K(Cu^{2+}+HL^- \rightleftharpoons CuHL^+)$ 15·4±0·2 $K(CuL+H^+ \rightleftharpoons CuHL^+)$ 5·56±0·01 $K(CuOHL^-+H^+ \rightleftharpoons CuL)$ 10·09±0·06	66S
Dy^{3+}	sp	18—22	0·1 ($NaClO_4$)	K_1 10·6, *$K(Dy^{3+}+HL^- \rightleftharpoons DyHL^{2+})$ 11·2	67Sa
Er^{3+}	sp	18—22	0·1 ($NaClO_4$)	K_1 10·1, *$K(Er^{3+}+HL^- \rightleftharpoons ErHL^{2+})$ 11·0	67Sa
Ga^{3+}	sp	18—22	0·2 ($NaClO_4$)	β_2 30·3, *$K(Ga^{3+}+HL^- \rightleftharpoons GaHL^{2+})$ 14·6	63H
	sp	25	?	K(?) 10·3±0·2	66D
In^{3+}	sp	25	?	K(?) 9·3±0·05	66D
La^{3+}	sp	18—22	0·1 ($NaClO_4$)	K_1 9·2	67Sa
Mn^{2+}	gl	25	~ 0·005 50% dioxan	*$K(Mn^{2+}+HL^- \rightleftharpoons MnHL^+)$ 9·7 *$K[MnHL^++HL^- \rightleftharpoons Mn(HL)_2]$ 9·2 $K(MnL+H^+ \rightleftharpoons MnHL^+)$ 8·8 $K(MnOHL^-+H^+ \rightleftharpoons MnL)$ 10·3	62C, 63C
			< 0·01 50% dioxan	*$K(Mn^{2+}+HL^- \rightleftharpoons MnHL^+)$ 9·79 *$K[MnHL^++HL^- \rightleftharpoons Mn(HL)_2]$ 9·13	66S
Nb^V	sp	25	?	K(?) 4·3	67A
Nd^{3+}	sp	18—22	0·1 ($NaClO_4$)	K_1 9·8, *$K(Nd^{3+}+HL^- \rightleftharpoons NdHL^{2+})$ 11·1	67Sa
Ni^{2+}	gl	25	~ 0·005 50% dioxan	*$K(Ni^{2+}+HL^- \rightleftharpoons NiHL^+)$ 13·2 *$K[NiHL^++HL^- \rightleftharpoons Ni(HL)_2]$ 12·8 $K(NiL+H^+ \rightleftharpoons NiHL^+)$ 7·7 $K(NiOHL^-+H^+ \rightleftharpoons NiL)$ 9·2	62C, 63C
Pb^{2+}	sp	?	0·01	K(?) 6·48	59K
	gl	25	0·1	K_1 8·6, 11·2, K_2 7·1 (?)	62G
	sp	18—22	0·1 ($NaClO_4$)	*$K(Pb^{2+}+HL^- \rightleftharpoons PbHL^+)$ 11·9	66H
Pr^{3+}	sp	18—22	0·1 ($NaClO_4$)	K_1 9·3, *$K(Pr^{3+}+HL^- \rightleftharpoons PrHL^{2+})$ 10·5	67Sa
Sc^{3+}	sp	18—22	?	*$K(Sc^{3+}+HL^- \rightleftharpoons ScHL^{2+})$ 12·8	62S
Sm^{3+}	sp	18—22	0·1 ($NaClO_4$)	K_1 10·1, *$K(Sm^{3+}+HL^- \rightleftharpoons SmHL^{2+})$ 11·4	67Sa
Ta^V	sp	25	?	K (?) 4·5	67A
Tl^{3+}	sp	25	?	K(?) 9·9±0·05	66D
UO_2^{2+}	gl	25	0·1 0, 50% dioxan	K_1 (0%) 12·5, (50%) 16·2, K_2 (0%) 8·4, (50%) 9·6	62G
	sp	18—22	0·1	K_1 11·9, *$K(UO_2^{2+}+HL^- \rightleftharpoons UO_2HL^+)$ 12·9	67S
VO_2^+	sp	15	0·01	K_1 16·49±0·03	66B
	sp	25	?	K(?) 4·2	67A
Y^{3+}	sp	18—22	0·1 ($NaClO_4$)	K_1 9·1, *$K(Y^{3+}+HL^- \rightleftharpoons YHL^{2+})$ 10·2	67Sa
Yb^{3+}	sp	18—22	0·1 ($NaClO_4$)	K_1 10·2, *$K(Yb^{3+}+HL^- \rightleftharpoons YbHL^{2+})$ 11·1	67Sa

contin

Metal	Method	Temp	Medium	Log of equilibrium constant, remarks	Ref
Zn^{2+}	gl	25	~ 0·005 50% dioxan	$^*K(Zn^{2+}+HL^- \rightleftharpoons ZnHL^+)$ 12·4	62C, 63C
				$^*K(ZnHL^+ + HL^- \rightleftharpoons Zn(HL)_2]$ 11·1	
				$K(ZnL+H^+ \rightleftharpoons ZnHL^+)$ 7·7	
				$K(ZnOHL^- + H^+ \rightleftharpoons ZnL)$ 9·3	
	gl	25	0·1 0, 50% dioxan	K_1 (0%) 10·5, (50%) 11·2, K_2 (0%) 6·6, (50%) 7·8	62G
	sp	18—22	0·1 ($NaClO_4$)	$^*K(Zn^{2+}+HL^- \rightleftharpoons ZnHL^+)$ 11·6	66H
	sp	25	0·1 ($NaClO_4$)	K_1 11·9±0·1, K_2 10·3±0·2	68T
				$K(ZnL+H^+ \rightleftharpoons ZnHL^+)$ 5·90±0·05	
				$K(ZnL_2^{2-}+H^+ \rightleftharpoons ZnHL_2^-)$ 7·55±0·05	
				$K(ZnHL_2^- + H^+ \rightleftharpoons ZnH_2L_2)$ 6·45±0·05	

* Formation constants K_{MHL} adjusted to give microscopic constants assumed for ligand HL^- with hydroxyl proton at 1-position.

59K H. Kristiansen and F. J. Langmyhr, *Acta Chem. Scand.*, 1959, **13**, 1473
61I T. Iwamoto, *Bull. Chem. Soc. Japan*, 1961, **34**, 605
62C A. Corsini, I. M. L. Yih, Q. Fernando, and H. Freiser, *Analyt. Chem.*, 1962, **34**, 1090
62G W. J. Geary, G. Nickless, and F. H. Pollard, *Analyt. Chim. Acta*, 1962, **26**, 575; **27**, 71
62S L. Sommer and M. Hnilickova, *Analyt. Chim. Acta*, 1962, **27**, 241
63C A. Corsini, Q. Fernando, and H. Freiser, *Inorg. Chem.*, 1963, **2**, 224
63H H. Hnilickova and L. Sommer, *Z. analyt. Chem.*, 1963, **193**, 171
66B A. K. Babko, A. I. Volkova, and T. E. Get'man, *Russ. J. Inorg. Chem.*, 1966, **11**, 203 **(374)**
66D C. D. Dwivedi, K. N. Munshi, and A. K. Dey, *J. Inorg. Nuclear Chem.*, 1966, **28**, 245
66H M. Hnilickova and L. Sommer, *Talanta*, 1966, **13**, 667
66S R. W. Stanley and G. E. Cheney, *Talanta*, 1966, **13**, 1619
67A B. V. Agarwalla and A. K. Dey, *Current Sci.*, 1967, **36**, 544
67S L. Sommer, V. M. Ivanov, and H. Novotna, *Talanta*, 1967, **14**, 329
67Sa L. Sommer and H. Novotna, *Talanta*, 1967, **14**, 457
68T M. Tanaka, S. Funabashi, and K. Shirai, *Inorg. Chem.*, 1968, **7**, 573

$C_{11}H_{10}ON_4$ 3-(2′-Hydroxyphenyl)1-(pyrimidin-2″-yl)-1,2-diazaprop-2-ene HL 879.1 (salicylaldehyde 2-pyrimidylhydrazone)

Metal	Method	Temp	Medium	Log of equilibrium constant, remarks	Ref
H^+	gl	25	0·1 ($NaClO_4$) 50% methanol	K_1 10·20, K_2 2·92	67A
Co^{2+}	gl	25	0·1 ($NaClO_4$) 50% methanol	K_1 10·4, K_2 8·6	67A
Cu^{2+}	sp	18—22	0·1 ($NaClO_4$) 50% methanol	K_1 16·0	67A
	gl	25	0·1 ($NaClO_4$) 50% methanol	K_2 7·0	67A
Fe^{2+}	gl	25	0·1 ($NaClO_4$) 50% methanol	K_1 9·9, K_2 8·6	67A
Ni^{2+}	gl	25	0·1 ($NaClO_4$) 50% methanol	K_1 10·7, K_2 8·1	67A
Zn^{2+}	gl	25	0·1 ($NaClO_4$) 50% methanol	K_1 9·2, K_2 7·1	67A

67A R. G. Anderson and G. Nickless, *Talanta*, 1967, **14**, 1221

880 $C_{11}H_{11}ON$ 8-Methoxy-2-methylquinoline L

Errata. 1964, p. 657, Table 880. *For* formula $C_{11}H_{21}ON$ *read* $C_{11}H_{11}ON$.
Insert **L** in heading.

880.1 $C_{11}H_{11}ON$ 2-Ethyl-8-hydroxyquinoline HL

Metal	*Method*	*Temp*	*Medium*	*Log of equilibrium constant, remarks*	*Ref*
H^+	gl	25	~ 0	K_1 11·83, K_2 4·63	66K
Cd^{2+}	gl	25	~ 0	K_1 9·18, K_2 8·21	66K
Cu^{2+}	gl	25	~ 0	K_1 12·42, K_2 11·46	66K
Ni^{2+}	gl	25	~ 0	K_1 9·35, K_2 8·91	66K
Zn^{2+}	gl	25	~ 0	K_1 9·89, K_2 9·22	66K

66K H. Kaneko and K. Ueno, *Bull. Chem. Soc. Japan*, 1966, **39**, 1910

880.2 $C_{11}H_{11}O_3N$ 5-Oxo-2-phenylpyrrolidine-2-carboxylic acid (α-phenylpyroglutamic acid) H_2L

Metal	*Method*	*Temp*	*Medium*	*Log of equilibrium constant, remarks*	*Ref*
H^+	gl	25	0·1 $NaClO_4$	K_1 11·52, K_2 2·72	65N
Cu^{2+}	gl	25	0·1 $NaClO_4$	K_1 7·83	65N

65N M. H. T. Nyberg and M. Cefola, *Arch. Biochem. Biophys.*, 1965, **111**, 321

881 $C_{11}H_{11}O_6N$ *N*-(2′-Carboxyphenyl)iminodiacetic acid (anthranilic acid *NN*-diacetic acid) H_3L

Metal	*Method*	*Temp*	*Medium*	*Log of equilibrium constant, remarks*	*Ref*
H^+	gl	22	0·1 KCl	K_1 7·75, K_2 3·01, K_3 2·20	61U
	gl	20	0·1 (KNO_3)	K_1 7·75 ± 0·02, K_2 2·98 ± 0·01, K_3 2·33 ± 0·01	63I
	gl	25	0·1 (KNO_3)	K_1 7·77, K_2 3·00, K_3 2·20	67U
Ag^+	gl	20	0·1 [$(CH_3)_4NNO_3$]	K_1 3·54 ± 0·01	63I
Cd^{2+}	sp	20	0·1 $NaNO_3$	K(?) 5·12	61D
	gl	25	0·1 (KNO_3)	K_1 7·44, $K(Cd^{2+} + HL^{2-} \rightleftharpoons CdHL)$ 2·37	67U
	pol	25	0·1 (KNO_3)	K_1 7·41	67U
Co^{2+}	sp	20	0·1 $NaNO_3$	K (?) 5·45	61D
	gl	22	0·1 KCl	K_1 8·17	61U
	gl	25	0·1 (KNO_3)	K_1 8·42	67U
Cu^{2+}	sp	20	0·1 $NaNO_3$	K(?) 7·52	61D
	gl	22	0·1 KCl	K_1 > 10, $K(Cu^{2+} + HL^{2-} \rightleftharpoons CuHL)$ > 4	61U
	pol	25	0·1 (KNO_3)	K_1 10·93	67U

conti

$C_{11}H_{11}O_6N$ (contd) 881

Metal	Method	Temp	Medium	Log of equilibrium constant, remarks	Ref
Fe^{3+}	sp	20	0·1 $NaNO_3$	$K(?)$ 9·62	60D
La^{3+}	sp	20	0·1 $NaNO_3$	$K(?)$ 5·41	61D
Li^{+}	gl	20	0·1 $[(CH_3)_4NNO_3]$	K_1 2·05 ± 0·01	63I
Mn^{2+}	sp	20	0·1 $NaNO_3$	$K(?)$ 5·37	61D
	gl	25	0·1 (KNO_3)	K_1 5·85, $K(Mn^{2+} + HL^{2-} \rightleftharpoons MnHL) < 1$	67U
Na^{+}	gl	20	0·1 $[(CH_3)_4NNO_3]$	K_1 0·89 ± 0·01	63I
Ni^{2+}	gl	22	0·1 KCl	K_1 9·6, $K(Ni^{2+} + HL^{2-} \rightleftharpoons NiHL)$ 3·9	61U
	pol	25	0·1 (KNO_3)	K_1 9·48	67U
Pb^{2+}	sp	20	0·1 $NaNO_3$	$K(?)$ 6·14	61D
Tl^{+}	gl	20	0·1 $[(CH_3)_4NNO_3]$	K_1 2·93 ± 0·01	63I
Zn^{2+}	sp	20	0·1 $NaNO_3$	$K(?)$ 5·61	61D
	gl	25	0·1 (KNO_3)	K_1 8·42	67U

60D C. Dragulescu, T. Simonescu, I. Menessy, and R. Anton, *Studii si cercet-Stiinte Chimice Timisoara*, 1960, **7**, 9

61D C. Dragulescu, T. Simonescu, I. Menessy, and R. Anton, *Studii si cercet-Stiinte Chimice Timisoara*, 1961, **8**, 10

61U E. Uhlig, *Z. anorg. Chem.*, 1961, **311**, 249

63I H. Irving and J. J. R. F. da Silva, *J. Chem. Soc.*, 1963, 3308

67U E. Uhlig and R. Krannich, *J. Inorg. Nuclear Chem.*, 1967, **29**, 1164

$C_{11}H_{12}O_2N_2$ 2-Amino-3-(indol-3′-yl)propanoic acid (tryptophan) HL 884

Metal	Method	Temp	Medium	Log of equilibrium constant, remarks	Ref
H^{+}	gl	20	~ 0	K_1 9·57	53P
Be^{2+}	gl	20	0·005 $BeSO_4$	β_2 11·6	53P
Cd^{2+}	gl	20	0·005 $CdSO_4$	β_2 7·0	53P
Zn^{2+}	gl	20	0·005 $ZnSO_4$	β_2 8·2	53P

53P D. J. Perkins, *Biochem. J.*, 1953, **55**, 649

$C_{11}H_{12}O_6N_2$ *N*-(2-Nitrobenz)iminodiacetic acid H_2L 885.1

Metal	Method	Temp	Medium	Log of equilibrium constant, remarks	Ref
H^{+}	gl	25	0·1 KNO_3	K_1 8·29, K_2 1·99	62A
Ca^{2+}	gl	25	0·1 KNO_3	K_1 2·93	62A
Mg^{2+}	gl	25	0·1 KNO_3	K_1 2·65	62A

62A T. Ando, *Bull. Chem. Soc. Japan*, 1962, **35**, 1395

885.2 $C_{11}H_{12}O_6N_2$ *N*-(4′-Nitrobenzyl)iminodiacetic acid H_2L

Metal	*Method*	*Temp*	*Medium*	*Log of equilibrium constant, remarks*	*Ref*
H^+	gl	25	0·1 KNO_3	K_1 7·64, K_2 1·80	62A
Ca^{2+}	gl	25	0·1 KNO_3	K_1 2·53	62A
Mg^{2+}	gl	25	0·1 KNO_3	K_1 1·6	62A

62A T. Ando, *Bull. Chem. Soc. Japan*, 1962, **35**, 1395

892.1 $C_{11}H_{13}O_4N$ *N*-Benzylaminobutanedioic acid (*N*-benzylaspartic acid) H_2L

Metal	*Method*	*Temp*	*Medium*	*Log of equilibrium constant, remarks*	*Ref*
H^+	gl	30	0·1 KCl	K_1 9·07±0·03, K_2 3·34±0·02, K_3 2·06	66S
Mg^{2+}	gl	30	0·1 KCl	K_1 1·74±0·07	66S
Ni^{2+}	gl	30	0·1 KCl	K_1 6·84±0·03, K_2 5·46±0·04	66S

66S G. Shtacher, *J. Inorg. Nuclear Chem.*, 1966, **28**, 845

892.2 $C_{11}H_{13}O_4N$ *N*-Benzyliminodiacetic acid H_2L

Metal	*Method*	*Temp*	*Medium*	*Log of equilibrium constant, remarks*	*Ref*
H^+	gl	25	0·1 KNO_3	K_1 8·90, K_2 2·24	62A
	gl	30	0·1 KCl	K_1 8·96±0·02, K_2 2·09, K_3 1·49	66S
	gl	25	0·1 KCl	K_1 8·91, K_2 2·30	66Sa
	gl	10—40	0·1 KCl	K_1 (10°) 9·07, (25°) 8·88, (40°) 8·76, K_2 (10°) 2·67, (25°) 2·32, (40°) 1·90	68E
Ca^{2+}	gl	25	0·1 KNO_3	K_1 3·13	62A
	gl	30	0·1 KCl	K_1 3·26±0·07	66S
	gl	25	0·1 KCl	K_1 3·17	66Sa
Co^{2+}	gl	25	0·1 KCl	K_1 6·78, K_2 5·35	66Sa
	gl	10—40	0·1 KCl	K_1 (10°) 7·01, (25°) 6·87, (40°) 6·65, K_2 (10°) 5·75, (25°) 5·46, (40°) 5·27	68E
Cu^{2+}	gl	25	0·1 KCl	K_1 9·88, K_2 4·85	66Sa
	gl	10—40	0·1 KCl	K_1 (25°) 10·61, K_2 (10°) 5·22, (25°) 5·03, (40°) 4·77	68E
Mg^{2+}	gl	25	0·1 KNO_3	K_1 2·63	62A
	gl	30	0·1 KCl	K_1 3·02±0·06	66S
Ni^{2+}	gl	30	0·1 KCl	K_1 7·62±0·02, K_2 6·34±0·02	66S
	gl	25	0·1 KCl	K_1 7·85, K_2 6·10	66Sa
	gl	10—40	0·1 KCl	K_1 (10°) 8·10, (25°) 7·92, (40°) 7·66, K_2 (10°) 6·56, (25°) 6·32, (40°) 6·10	68E
Zn^{2+}	gl	10—40	0·1 KCl	K_1 (10°) 7·09, (25°) 7·00, (40°) 6·96, K_2 (10°) 5·65, (25°) 5·45, (40°) 5·29	68E

62A T. Ando, *Bull. Chem. Soc. Japan*, 1962, **35**, 1395
66S G. Shtacher, *J. Inorg. Nuclear Chem.*, 1966, **28**, 845
66Sa P. Souchey, N. Israily, and P. Gouzerk, *Bull. Soc. chim. France*, 1966, **12**, 3917
68E C. Eger, W. M. Auspach, and J. A. Marinsky, *J. Inorg. Nuclear Chem.*, 1968, **30**, 1899, 1911

$C_{11}H_{13}O_5N$ *N*-(2′-Methoxyphenyl)iminodiacetic acid H_2L 892.3

letal	Method	Temp	Medium	Log of equilibrium constant, remarks	Ref
$($^{+}$	gl	20	0·1 (KNO_3)	K_1 5·58 ± 0·01, K_2 2·69 ± 0·01	63I
g^{+}	gl	20	0·1 $[(CH_3)_4NNO_3]$	K_1 2·75 ± 0·02	63I
a^{2+}	gl	20	0·1 (KNO_3)	K_1 2·08 ± 0·02	63I
a^{2+}	gl	20	0·1 (KNO_3)	K_1 2·75 ± 0·01	63I
r^{2+}	gl	20	0·1 (KNO_3)	K_1 2·13 ± 0·02	63I
l^{+}	gl	20	0·1 $[(CH_3)_4NNO_3]$	K_1 2·46 ± 0·01	63I

63I H. Irving and J. J. R. F. da Silva, *J. Chem. Soc.*, 1963, 3308

$C_{11}H_{13}O_5N$ *N*-(Carboxymethyl)-*N*-(2′-hydroxyethyl)-2-aminobenzoic acid H_2L 893.1

Metal	Method	Temp	Medium	Log of equilibrium constant, remarks	Ref
I^{+}	gl	22	0·1 KCl	K_1 7·73, K_2 2·41	63U
o^{2+}	gl	22	0·1 KCl	K_1 6·40	63U
u^{2+}	gl	22	0·1 KCl	K_1 9·80	63U
i^{2+}	gl	22	0·1 KCl	K_1 7·80	63U

63U E. Uhlig, *Z. anorg. Chem.*, 1963, **320**, 283

$C_{11}H_{14}ON_2$ 2-(2′-Hydroxybutyl)benzimidazole HL 893.2

Metal	Method	Temp	Medium	Log of equilibrium constant, remarks	Ref
I^{+}	gl	15—45	50% dioxan	K_1 (15°) 13·07, (25°) 12·77, (40°) 12·30, (45°) 12·06, K_2 (15°) 4·88, (25°) 4·80, (40°) 4·68, (45°) 4·58	63L
u^{2+}	gl	15—45	50% dioxan	K_1 (15°) 11·47, (25°) 11·05, (40°) 10·47, (45°) 10·28 ΔH_1 −11·1	63L

63L T. J. Lane and H. B. Durham, *Inorg. Chem.*, 1963, 2, 632

$C_{13}H_{18}O_4N_2$ Pyridoxylidenevaline H_2L 894

Errata. 1964, p. 661, Table 894. *For* formula $C_{11}H_{14}O_3N$ *read* $C_{13}H_{18}O_4N_2$.
In the heading, *for* H_3L *read* H_2L
Renumber this table to **962.1** to correct the ligand sequence.

895 $C_{11}H_{14}O_3N_2$ Glycyl-DL-phenylalanine HL

Metal	*Method*	*Temp*	*Medium*	*Log of equilibrium constant, remarks*	*Ref*
H^+	gl	25	0·12 (NaCl)	K_1 8·2±0·2, K_2 3·20±0·02	67S
Cu^{2+}	gl	25	0·12 (NaCl)	K_1 6·06,	67S
				$K[Cu(H_{-1}L)+H^+ \rightleftharpoons CuL^+]$ 4·60	
				$K[Cu(H_{-1}L)OH^- + H^+ \rightleftharpoons Cu(H_{-1}L)]$ 9·94	
				$K[Cu(H_{-1}L)+L^- \rightleftharpoons CuL(H_{-1}L)^-]$ 3·67	

67S U. I. Salaklutdinov, A. P. Bonsova, Yu. V. Granovskii, I. A. Savich, and V. I. Spitsyn, *Proc. Acad. Sci.* (*U.S.S.R.*), 196 **177,** 1039, (365)

895.1 $C_{11}H_{14}O_3N_2$ Glycyl-L-phenylalanine HL

Metal	*Method*	*Temp*	*Medium*	*Log of equilibrium constant, remarks*	*Ref*
H^+	gl	25	0·100 (NaCl)	K_1 8·16±0·02, K_2 3·12±0·02	59B
Co^{2+}	gl	25	0·100 (NaCl)	K_1 2·96±0·01, β_2 5·27±0·02	59B
Cu^{2+}	gl	25	0·100 (NaCl)	K_1 5·45±0·01, β_2 8·63±0·06	59B
				$K[Cu(H_{-1}L)+H^+ \rightleftharpoons CuL^+]$ 3·86	
				$K[Cu(H_{-1}L)_2^{2-} + H^+ \rightleftharpoons CuL(H_{-1}L)^-]$ 9·78	

59B J. L. Biester and P. M. Ruoff, *J. Amer. Chem. Soc.*, 1959, **81,** 6517

895.2 $C_{11}H_{14}O_3N_2$ Glycyl-D-phenylalanine HL

Metal	*Method*	*Temp*	*Medium*	*Log of equilibrium constant, remarks*	*Ref*
H^+	gl	25	0·100 (NaCl)	K_1 8·18±0·01, K_2 3·11±0·02	59B
Co^{2+}	gl	25	0·100 (NaCl)	K_1 2·91±0·02, β_2 5·35±0·02	59B
Cu^{2+}	gl	25	0·100 (NaCl)	K_1 5·26±0·01, β_2 8·39±0·08	59B
				$K[Cu(H_{-1}L)+H^+ \rightleftharpoons CuL^+]$ 3·70	
				$K[Cu(H_{-1}L)_2^{2-} + H^+ \rightleftharpoons CuL(H_{-1}L)^-]$ 9·77	

59B J. L. Biester and P. M. Ruoff, *J. Amer. Chem. Soc.*, 1959, **81,** 6517

895.3 $C_{11}H_{14}O_3N_2$ L-Phenylalanylglycine HL

Metal	*Method*	*Temp*	*Medium*	*Log of equilibrium constant, remarks*	*Ref*
H^+	gl	25	0·100 (NaCl)	K_1 7·62±0·02, K_2 3·13±0·04	59B
Co^{2+}	gl	25	0·100 (NaCl)	K_1 2·12±0·02, β_2 4·14±0·01	59B
Cu^{2+}	gl	25	0·100 (NaCl)	K_1 4·66±0·02	59B
				$K[Cu(H_{-1}L)+H^+ \rightleftharpoons CuL^+]$ 3·50	

59B J. L. Biester and P. M. Ruoff, *J. Amer. Chem. Soc.*, 1959, **81,** 6517

$C_{11}H_{14}O_4N_2$ *N*-(6′-Methyl-2′-pyridylmethyl)iminodiacetic acid H_2L 896.1

Metal	Method	Temp	Medium	Log of equilibrium constant, remarks	Ref
H^+	gl	25	0·1 (KNO_3)	K_1 8·30, K_2 3·46	64T
Ce^{3+}	gl	25	0·1 (KNO_3)	K_1 6·00, K_2 4·07	64T
Dy^{3+}	gl	25	0·1 (KNO_3)	K_1 7·23, K_2 4·92	64T
Er^{3+}	gl	25	0·1 (KNO_3)	K_1 7·42, K_2 5·22	64T
Eu^{3+}	gl	25	0·1 (KNO_3)	K_1 6·76, K_2 4·50	64T
Gd^{3+}	gl	25	0·1 (KNO_3)	K_1 6·71, K_2 4·61	64T
Ho^{3+}	gl	25	0·1 (KNO_3)	K_1 7·30, K_2 5·03	64T
La^{3+}	gl	25	0·1 (KNO_3)	K_1 5·72, K_2 3·85	64T
Lu^{3+}	gl	25	0·1 (KNO_3)	K_1 7·60, K_2 5·39	64T
Nd^{3+}	gl	25	0·1 (KNO_3)	K_1 6·28, K_2 4·26	64T
Pr^{3+}	gl	25	0·1 (KNO_3)	K_1 6·18, K_2 4·24	64T
Sm^{3+}	gl	25	0·1 (KNO_3)	K_1 6·57, K_2 4·48	64T
Tb^{3+}	gl	25	0·1 (KNO_3)	K_1 7·16, K_2 4·94	64T
Tm^{3+}	gl	25	0·1 (KNO_3)	K_1 7·54, K_2 5·27	64T
Y^{3+}	gl	25	0·1 (KNO_3)	K_1 6·84, K_2 4·74	64T
Yb^{3+}	gl	25	0·1 (KNO_3)	K_1 7·65, K_2 5·33	64T

64T L. C. Thompson, *Inorg. Chem.*, 1964, **3**, 1015

$C_{11}H_{15}ON$ (Salicylideneaminobutane (*N*-n-butylsalicylideneimine) HL 896.2

Metal	Method	Temp	Medium	Log of equilibrium constant, remarks	Ref
H^+	sp	25	~ 0	K_1 12·0	65G
Be^{2+}	dis	25 (?)	~ 0	K_1 11·11, β_2 20·44	65G

65G R. W. Green and P. W. Alexander, *Austral. J. Chem.*, 1965, **18**, 329, 651

$C_{11}H_{15}O_2N$ 3-Phenylalanine ethyl ester L 896.3

Metal	Method	Temp	Medium	Log of equilibrium constant, remarks	Ref
Cu^{2+}				See nitrilotriacetic acid (449) for mixed ligands.	

$C_{11}H_{17}O_3N$ 3,4-Dihydroxy-1-[1′-hydroxy-2′-(propylamino)ethyl]benzene (isoprenaline) H_2L 897.1

Metal	Method	Temp	Medium	Log of equilibrium constant, remarks	Ref
Mo^{VI}	sp	25	0·1 (KCl)	$K(MoO_4^{2-}+2H_2L \rightleftharpoons MoO_2L_2^{2-})$ 5·87±0·01	63H

63H J. Halmekoski, *Suomen Kem.*, 1963, **B36**, 40

897.2 $C_{11}H_{17}O_6N$ *N*-(2′-Carboxycyclohexyl)iminodiacetic acid H_3L

Metal	*Method*	*Temp*	*Medium*	*Log of equilibrium constant, remarks*	*Ref*
H^+	gl	20	0·1 (KCl)	K_1 11·24±0·02, K_2 2·59±0·02, K_3 1·63±0·05	66I
Ba^{2+}	gl	20	0·1 (KCl)	K_1 5·07±0·04	66I
Ca^{2+}	gl	20	0·1 (KCl)	K_1 7·66±0·01	66I
Mg^{2+}	gl	20	0·1 (KCl)	K_1 5·3±0·3	66I
Sr^{2+}	gl	20	0·1 (KCl)	K_1 5·56±0·03	66I
Zn^{2+}	gl	20	0·1 (KCl)	K_1 11·70±0·03	66I

66I H. M. N. H. Irving and M. G. Miles, *J. Chem. Soc.* (*A*), 1966, 1268

898 $C_{11}H_{18}O_8N_2$ (Trimethylenedinitro)tetra-acetic acid) H_4L

Metal	*Method*	*Temp*	*Medium*	*Log of equilibrium constant, remarks*	*Ref*
H^+	gl	20	0·1 (KNO_3)	K_1 10·46, K_2 8·02, K_3 2·57, K_4 1·88	64A, 6
	cal	20	0·1 (KNO_3)	$\Delta H_1°$ −5·16, $\Delta S_1°$ 30·2 $\Delta H_2°$ −4·43, $\Delta S_2°$ 21·6	64A
	gl	20	0·1 ($NaNO_3$)	K_1 10·27, K_2 7·90, K_3 2·67, K_4 2·0	65O, 6
	gl	20	0·1 [$(CH_3)_4NCl$]	K_1 10·56, K_2 8·00	67A
	H	20	1 [$(CH_3)_4NCl$]	K_1 10·23, K_2 7·80, K_3 2·55, K_4 2·3	67A
Al^{3+}	gl	20	0·1 (KNO_3)	K_1 16·33	64L
	pol	20	0·1 (KNO_3)	K_1 16·31	64L
Ba^{2+}	gl	20	0·1 (KNO_3)	K_1 3·95, $K(Ba^{2+}+HL^{3-} \rightleftharpoons BaHL^-)$ 2·21	64L
Ca^{2+}	cal	20	0·1 (KNO_3)	$\Delta H_1°$ −1·74, $\Delta S_1°$ 27·4	64A
	gl	20	0·1 (KNO_3)	K_1 7·28, $K(Ca^{2+}+HL^{3-} \rightleftharpoons CaHL^-)$ 3·16	64L
Cd^{2+}	gl	20	0·1 (KNO_3)	$K(Cd^{2+}+HL^{3-} \rightleftharpoons CdHL^-)$ 6·5	64A
	cal	20	0·1 (KNO_3)	$\Delta H_1°$ −5·44, $\Delta S_1°$ 45·0	64A
	gl	20	0·1 (KNO_3)	K_1 13·90, $K(CdL^{2-}+H^+ \rightleftharpoons CdHL^-)$ 3·06	64L
	gl	20	0·1 (KNO_3)	K_1 14·09	64L
	pol	25	0·2 (KNO_3)	K_1 12·69	65O
Ce^{3+}	gl	20	0·1 (KNO_3)	K_1 11·75, $K(CeL^-+H^+ \rightleftharpoons CeHL)$ 4·55	64L
	gl	20	0·1 (KNO_3)	K_1 11·66	64L
Co^{2+}	gl	20	0·1 (KNO_3)	$K(Co^{2+}+HL^{3-} \rightleftharpoons CoHL^-)$ 7·4	64A
	cal	20	0·1 (KNO_3)	$\Delta H_1°$ −2·6, $\Delta S_1°$ 62·2	64A
	gl	20	0·1 (KNO_3)	K_1 15·54, $K(CoL^{2-}+H^+ \rightleftharpoons CoHL^-)$ 2·4	64L
	pol	20	0·1 (KNO_3)	K_1 15·56	64L
	pol	25	0·2 (KNO_3)	K_1 14·48	65O
Co^{3+}	pol	25	0·2 (KNO_3)	K_1 40·7	65T
Cu^{2+}	gl	20	0·1 (KNO_3)	$K(Cu^{2+}+HL^{3-} \rightleftharpoons CuHL^-)$ 10·7	64A
	cal	20	0·1 (KNO_3)	$\Delta H_1°$ −7·74, $\Delta S_1°$ 60·1	64A
	gl	20	0·1 (KNO_3)	K_1 18·92, $K(CuL^{2-}+H^+ \rightleftharpoons CuHL^-)$ 2·2	64L
	pol	25	0·2 (KNO_3)	K_1 18·04	65O

con

Metal	Method	Temp	Medium	Log of equilibrium constant, remarks	Ref
Dy^{3+}	gl	20	0·1 (KNO_3)	K_1 14·72	64L
	pol	20	0·1 (KNO_3)	K_1 14·60	64L
	gl	20	0·1 (KNO_3)	K_1 14·67	64L
Er^{3+}	pol	20	0·1 (KNO_3)	K_1 15·10	64L
	gl	20	0·1 (KNO_3)	K_1 15·15	64L
Eu^{3+}	gl	20	0·1 (KNO_3)	K_1 13·58	64L
	gl	20	0·1 (KNO_3)	K_1 13·49	64L
Fe^{2+}	gl	20	0·1 (KNO_3)	K_1 13·42, $K(Fe^{2+}+HL^{3-} \rightleftharpoons FeHL^-)$ 6·30	64L
Fe^{3+}	gl	20	0·1 (KNO_3)	$K(FeL^- + H^+ \rightleftharpoons FeHL)$ 2·45	64L
	red	20	0·1 (KNO_3)	K_1 21·61	64L
	red	20	0·1 ($NaClO_4$)	K_1 21·4, $K(FeL^- + H^+ \rightleftharpoons FeHL)$ 2·4	67B
Ga^{3+}	red	20	0·1 ($NaClO_4$)	K_1 20·8, $K[GaL^- + 2\,OH^- \rightleftharpoons GaL(OH)_2{}^{3-}]$ 10·7	67B
Gd^{3+}	gl	20	0·1 (KNO_3)	K_1 13·74	64L
	pol	20	0·1 (KNO_3)	K_1 13·80	64L
	gl	20	0·1 (KNO_3)	K_1 13·73	64L
Hg^{2+}	gl	20	0·1 (KNO_3)	$K(Hg^{2+}+HL^{3-} \rightleftharpoons HgHL^-)$ 13·46	64A
	cal	20	0·1 (KNO_3)	$\Delta H_1°$ −18·9, $\Delta S_1°$ 26·6	64A
	gl	20	0·1 (KNO_3)	$K(HgL^{2-} + H^+ \rightleftharpoons HgHL^-)$ 4·00	64L
	Hg	20	0·1 (KNO_3)	K_1 19·92	64L
In^{3+}	red	20	0·1 ($NaClO_4$)	K_1 21·15, $K(InL^- + H^+ \rightleftharpoons InHL)$ 1·64, $K(InL^- + OH^- \rightleftharpoons InLOH^{2-})$ 5·60	67B
La^{3+}	gl	20	0·1 (KNO_3)	$K(La^{3+}+HL^{3-} \rightleftharpoons LaHL)$ 5·44	64A
	cal	20	0·1 (KNO_3)	$\Delta H_1°$ 3·76, $\Delta S_1°$ 64·2	64A
	gl	20	0·1 (KNO_3)	K_1 11·23, $K(LaL^- + H^+ \rightleftharpoons LaHL)$ 4·67	64L
Mg^{2+}	cal	20	0·1 (KNO_3)	$\Delta H_1°$ 9·09, $\Delta S_1°$ 59·0	64A
	gl	20	0·1 (KNO_3)	K_1 6·21, $K(Mg^{2+}+HL^{3-} \rightleftharpoons MgHL^-)$ 3·05	64L
Mn^{2+}	cal	20	0·1 (KNO_3)	$\Delta H_1°$ −0·72 (?), $\Delta S_1°$ 52·9 (?)	64A
	gl	20	0·1 (KNO_3)	K_1 9·99, $K(Mn^{2+}+HL^{3-} \rightleftharpoons MnHL^-)$ 4·82	64L
	pol	25	0·2 (KNO_3)	K_1 < 10·8	65O
Nd^{3+}	gl	20	0·1 (KNO_3)	K_1 12·36, $K(NdL^- + H^+ \rightleftharpoons NdHL)$ 4·03	64L
	pol	20	0·1 (KNO_3)	K_1 12·34	64L
	gl	20	0·1 (KNO_3)	K_1 12·32	64L
Ni^{2+}	gl	20	0·1 (KNO_3)	$K(Ni^{2+}+HL^{3-} \rightleftharpoons NiHL^-)$ 9·9	64A
	cal	20	0·1 (KNO_3)	$\Delta H_1°$ −6·66, $\Delta S_1°$ 60·3	64A
	pol	20	0·1 (KNO_3)	K_1 18·15	64L
	gl	20	0·1 (KNO_3)	$K(NiL^{2-} + H^+ \rightleftharpoons NiHL^-)$ 2·2	64L
	pol	25	0·2 (KNO_3)	K_1 17·29	65O
Pb^{2+}	gl	20	0·1 (KNO_3)	$K(Pb^{2+}+HL^{3-} \rightleftharpoons PbHL^-)$ 7·18	64A
	cal	20	0·1 (KNO_3)	$\Delta H_1°$ −6·4, $\Delta S_1°$ 40·8	64A
	gl	20	0·1 (KNO_3)	K_1 13·78, $K(PbL^{2-} + H^+ \rightleftharpoons PbHL^-)$ 3·86	64L
	Hg	20	0·1 (KNO_3)	K_1 13·69	64L
	gl	20	0·1 (KNO_3)	K_1 13·64	64L
	pol	25	0·2 (KNO_3)	K_1 13·04	65O

continued overleaf

898 $C_{11}H_{18}O_8N_2$ (contd)

Metal	Method	Temp	Medium	Log of equilibrium constant, remarks	Ref
Sm^{3+}	gl	20	0·1 (KNO_3)	K_1 13·21	64L
	pol	20	0·1 (KNO_3)	K_1 13·12	64L
	gl	20	0·1 (KNO_3)	K_1 13·08	64L
Sr^{2+}	gl	20	0·1 (KNO_3)	K_1 5·28, $K(Sr^{2+}+HL^{3-} \rightleftharpoons SrHL^-)$ 2·58	64L
Tl^+	gl	20	0·1 (KNO_3)	K_1 3·90, $K(Tl^+ + HL^{3-} \rightleftharpoons TlHL^{2-})$ 2·7	67A
	red	20	1 ($NaClO_4$)	K_1 30·9	67A
Y^{3+}	gl	20	0·1 (KNO_3)	K_1 14·40	64L
	gl	20	0·1 (KNO_3)	K_1 14·26	64L
Yb^{3+}	pol	20	0·1 (KNO_3)	K_1 15·88	64L
	gl	20	0·1 (KNO_3)	K_1 15·94	64L
Zn^{2+}	gl	20	0·1 (KNO_3)	$K(Zn^{2+}+HL^{3-} \rightleftharpoons ZnHL^-)$ 7·3	64A
	cal	20	0·1 (KNO_3)	$\Delta H_1°$ −2·27, $\Delta S_1°$ 61·8	64A
	gl	20	0·1 (KNO_3)	K_1 15·26, $K(ZnL^{2-}+H^+ \rightleftharpoons ZnHL^-)$ 2·5	64L
	pol	20	0·1 (KNO_3)	K_1 15·22	64L
	Hg	20	0·1 (KNO_3)	K_1 15·25	64L
	gl	20	0·1 (KNO_3)	K_1 15·29, K_1 (with Ca^{2+}) 15·06	64L
	pol	25	0·2 (KNO_3)	K_1 14·26	65O

64A G. Anderegg, *Helv. Chim. Acta*, 1964, **47**, 1801
64L F. L'Eplattenier and G. Anderegg, *Helv. Chim. Acta*, 1964, **47**, 1792
65O H. Ogino, *Bull. Chem. Soc. Japan*, 1965, **38**, 771
65T N. Tanaka and H. Ogino, *Bull. Chem. Soc. Japan*, 1965, **38**, 1054
67A G. Anderegg, *Helv. Chim. Acta*, 1967, **50**, 2333
67Aa G. Anderegg and E. Bottari, *Helv. Chim. Acta*, 1967, **50**, 2341
67B E. Bottari and G. Anderegg, *Helv. Chim. Acta*, 1967, **50**, 2349

899 $C_{11}H_{18}O_8N_2$ (1,2-Propylenedinitro)tetra-acetic acid) H_4L

Metal	Method	Temp	Medium	Log of equilibrium constant, remarks	Ref
H^+	Hg	20	0·1 (KNO_3)	K_1 10·86, K_2 6·25, K_3 2·79, K_4 1·83	64I
	gl	30	0·1 (KCl)	K_1 10·84, K_2 6·20, K_3 3·03, K_4 2·60	65O, 65[illegible]
Cd^{2+}	Hg	20	0·1 (KNO_3)	K_1 18·83	64I
	pol	25	0·2 (KNO_3)	K_1 17·43 See Zn^{2+} below.	65O
Ce^{3+}	pol	20	0·1 (KNO_3)	K_1 16·79	64I
Co^{2+}	pol	25	0·2 (KNO_3)	K_1 17·07	65O
Co^{3+}	pol	25	0·2 (KNO_3)	K_1 42·1 See Zn^{2+} below.	65T
Cu^{2+}	pol	20	0·1 (KNO_3)	K_1 19·94	64I
	pol	25	0·2 (KNO_3)	K_1 19·64 See EDTA (836) for ligand exchange. See Zn^{2+} below.	65O
Dy^{3+}	pol	20	0·1 (KNO_3)	K_1 19·05	64I
Er^{3+}	pol	20	0·1 (KNO_3)	K_1 19·61	64I
Eu^{3+}	pol	20	0·1 (KNO_3)	K_1 18·26	64I
Gd^{3+}	pol	20	0·1 (KNO_3)	K_1 18·21	64I

con

$C_{11}H_{18}O_8N_2$ (contd) 899

Metal	Method	Temp	Medium	Log of equilibrium constant, remarks	Ref
Hg^{2+}	Hg	20	0·1 (KNO_3)	K_1 22·81, $K(Hg^{2+}+HL^{3-} \rightleftharpoons HgHL^-)$ 3·12	64I
Ho^{3+}	pol	20	0·1 (KNO_3)	K_1 19·30	64I
K^+	nmr	100	0·5 K_4L	K_1 0·1±0·1	68S
La^{3+}	pol	20	0·1 (KNO_3)	K_1 16·42	64I
Lu^{3+}	pol	20	0·1 (KNO_3)	K_1 20·56	64I
Mn^{2+}	pol	25	0·2 (KNO_3)	K_1 14·85	65O
Nd^{3+}	pol	20	0·1 (KNO_3)	K_1 17·54	64I
Ni^{2+}	pol	25	0·2 (KNO_3)	K_1 19·42	65O
				See Zn^{2+} below.	
Pb^{2+}	pol	20	0·1 (KNO_3)	K_1 18·97	64I
	pol	25	0·2 (KNO_3)	K_1 18·69	65O
				See EDTA (836) for ligand exchange.	
				See Zn^{2+} below.	
Pr^{3+}	pol	20	0·1 (KNO_3)	K_1 17·17	64I
Rb^+	nmr	100	0·5 Rb_4L	K_1 −0·8±0·01	68S
Sm^{3+}	pol	20	0·1 (KNO_3)	K_1 17·97	64I
Tb^{3+}	pol	20	0·1 (KNO_3)	K_1 18·64	64I
Tm^{3+}	pol	20	0·1 (KNO_3)	K_1 20·08	64I
Y^{3+}	pol	20	0·1 (KNO_3)	K_1 18·78	64I
Yb^{3+}	pol	20	0·1 (KNO_3)	K_1 20·25	64I
Zn^{2+}	pol	25	0·2 (KNO_3)	K_1 17·14	65O
	pol	25	0·2 KNO_3 0·01% gelatin	$K(Co^{2+}+ZnL^{2-} \rightleftharpoons CoL^{2-}+Zn^{2+})$ −0·07 $K(Ni^{2+}+ZnL^{2-} \rightleftharpoons NiL^{2-}+Zn^{2+})$ 2·28 $K(Cu^{2+}+ZnL^{2-} \rightleftharpoons CuL^{2-}+Zn^{2+})$ 2·50 $K(Cd^{2+}+ZnL^{2-} \rightleftharpoons CdL^{2-}+Zn^{2+})$ 0·29 $K(Pb^{2+}+ZnL^{2-} \rightleftharpoons PbL^{2-}+Zn^{2+})$ 1·56	65T

64I H. M. N. H. Irving and J. P. Conesa, *J. Inorg. Nuclear Chem.*, 1964, **26**, 1945
65O H. Ogino, *Bull. Chem. Soc. Japan*, 1965, **38**, 771
65T N. Tanaka and H. Ogino, *Bull. Chem. Soc. Japan*, 1965, **38**, 439, 1054
68S J. L. Sudmeier and A. J. Sengel, *Analyt. Chem.*, 1968, **40**, 1693

$C_{11}H_{18}O_9N_2$ (2-Hydroxytrimethylenedinitrilo)tetra-acetic acid H_4L 900

Metal	Method	Temp	Medium	Log of equilibrium constant, remarks	Ref
H^+	gl	20	0·1 KCl	K_1 9·43, K_2 6·90	59K
	?	20	0·1	K_1 9·37, K_2 6·87, K_3 2·46, K_4 1·71	62S
	gl	20	0·1 (KCl)	K_1 9·60, K_2 7·00, K_3 2·6, K_4 2·3	64D, 67T
	gl	20	0·1 (KCl)	K_1 9·44, K_2 6·93, K_3 2·47, K_4 1·78	65J
	H	20	0·1 KCl	K_1 9·70, K_2 7·10, K_3 2·90, K_4 1·85	66P
	gl	25	0·1 (KNO_3)	K_1 9·49±0·01, K_2 6·96±0·01, K_3 2·60±0·01, K_4 ~1·6	66T
Ba^{2+}	gl	20	0·1 KCl	K_1 3·99	59K
	gl	20	0·1 (KCl)	K_1 4·92	64D
	pol	20	0·1 (KCl)	K_1 5·45	64D
	oth	20	0·1 (KNO_3)	K_1 5, electrophoresis	65J
	gl	25	0·1 (KNO_3)	K_1 4·91±0·02 $K(BaL^{2-}+H^+ \rightleftharpoons BaHL^-)$ 7·34±0·03	66T

continued overleaf

900 $C_{11}H_{18}O_9N_2$ (contd)

Metal	*Method*	*Temp*	*Medium*	*Log of equilibrium constant, remarks*	*Ref*
Ca^{2+}	gl	20	0·1 (KCl)	K_1 6·90	64D
	pol	20	0·1 (KCl)	K_1 6·98	64D
	oth	20	0·1 (KNO_3)	K_1 5·5, electrophoresis	65J
	gl	25	0·1 (KNO_3)	K_1 6·69 ± 0·02	66T
				$K(CaL^{2-} + H^+ \rightleftharpoons CaHL^-)$ 6·54 ± 0·04	
Cd^{2+}	pol	20	0·1 (KCl)	K_1 11·73	64D
	oth	20	0·1 (KNO_3)	K_1 12·5, electrophoresis	65J
	H	20	0·1 KCl	K_1 12·60	66P
	gl	25	0·1 (KNO_3)	K_1 12·10 ± 0·03	66T
				$K(CdL^{2-} + H^+ \rightleftharpoons CdHL^-)$ 4·12 ± 0·04	
Co^{2+}	oth	20	0·1 (KNO_3)	K_1 14·5, electrophoresis	65J
	H	20	0·1 KCl	K_1 12·95	66P
	gl	25	0·1 (KNO_3)	K_1 13·92 ± 0·03	66T
				$K(CoL^{2-} + H^+ \rightleftharpoons CoHL^-)$ 3·33 ± 0·02	
Cu^{2+}	pol	20	0·1 (KCl)	K_1 14·82	64D
	oth	20	0·1 (KNO_3)	K_1 > 18, electrophoresis	65J
	H	20	0·1 KCl	K_1 13·10	66P
	gl	25	0·1 (KNO_3)	K_1 17·21 ± 0·05	66T
				$K(CuL^{2-} + H^+ \rightleftharpoons CuHL^-)$ 3·15 ± 0·03	
Fe^{3+}	oth	20	0·1 (KNO_3)	K_1 > 18, electrophoresis	65J
Hg^{2+}	pol	20	0·1 KNO_3	K_1 20·59	62S
La^{3+}	oth	20	0·1 (KNO_3)	K_1 12·5, electrophoresis	65J
Mg^{2+}	gl	20	0·1 (KCl)	K_1 4·93	64D
	pol	20	0·1 (KCl)	K_1 5·25	64D
	oth	20	0·1 (KNO_3)	K_1 4·5, electrophoresis	65J
Mn^{2+}	pol	20	0·1 (KCl)	K_1 8·20	64D
	oth	20	0·1 (KNO_3)	K_1 9, electrophoresis	65J
	H	20	0·1 KCl	K_1 8·90	66P
	gl	25	0·1 (KNO_3)	K_2 9·06 ± 0·04	66T
				$K(MnL^{2-} + H^+ \rightleftharpoons MnHL^-)$ 5·1 ± 0·1	
Ni^{2+}	oth	20	0·1 (KNO_3)	K_1 16, electrophoresis	65J
	H	20	0·1 KCl	K_1 13·48	66P
	gl	25	0·1 (KNO_3)	K_1 16·63 ± 0·08	66T
				$K(NiL^{2-} + H^+ \rightleftharpoons NiHL^-)$ 2·44 ± 0·06	
Pb^{2+}	oth	20	0·1 (KNO_3)	K_1 17, electrophoresis	65J
Sr^{2+}	gl	20	0·1 (KCl)	K_1 5·84	64D
	pol	20	0·1 (KCl)	K_1 6·12	64D
	oth	20	0·1 (KNO_3)	K_1 5·5, electrophoresis	65J
	gl	25	0·1 (KNO_3)	K_1 5·33 ± 0·03	66T
				$K(SrL^{2-} + H^+ \rightleftharpoons SrHL^-)$ 7·16 ± 0·06	
Zn^{2+}	pol	20	0·1 (KCl)	K_1 12·95	64D
	oth	20	0·1 (KNO_3)	K_1 13·5, electrophoresis	65J
	H	20	0·1 KCl	K_1 11·51	66P
	gl	25	0·1 (KNO_3)	K_1 13·70 ± 0·04	66T
				$K(ZnL^{2-} + H^+ \rightleftharpoons ZnHL^-)$ 3·58 ± 0·02	
Zr^{4+}	oth	20	0·1 (HCl)	K_1 23·58 ± 0·04, chromatographic	67T

59K H. Kroll, A.E.C. Contract (30-1) 2096, Annual Report, 1959

62S S. Stankoviansky and J. Konigstein, *Coll. Czech. Chem. Comm.*, 1962, **27**, 1997

64D N. M. Dyatlova, I. A. Seliverstova, V. G. Yashunskii, and O. I. Samoilova, *J. Gen. Chem. (U.S.S.R.)*, 1964, **34**, 4061 (400

65J V. Jokl and M. Majer, *Chem. Zvesti*, 1965, **19**, 249

66P I. V. Podgornaya, A. A. Ivakin, and K. N. Klyachina, *J. Gen. Chem. (U.S.S.R.)*, 1966, **36**, 2044 (2052)

66T L. C. Thompson and S. K. Kundra, *J. Inorg. Nuclear Chem.*, 1966, **28**, 2945

67T L. I. Tikhonova, *Russ. J. Inorg. Chem.*, 1967, **12**, 494 (939)

$C_{11}H_{22}O_3N_2$ DL-Valyl-DL-leucine HL 900.1

Metal	Method	Temp	Medium	Log of equilibrium constant, remarks	Ref
H^+	gl	25	0·12 (NaCl)	K_1 8·16±0·01, K_2 3·3±0·2	67S
Cu^{2+}	gl	25	0·12 (NaCl)	K_1 5·72, $K[Cu(H_{-1}L)+H^+ \rightleftharpoons CuL^+]$ 6·41 $K[Cu(H_{-1}L)OH^- + H^+ \rightleftharpoons Cu(H_{-1}L)]$ 9·69 $K[Cu(H_{-1}L)+L^- \rightleftharpoons CuL(H_{-1}L)^-]$ 3·58	67S

67S U. I. Salakhutdinov, A. P. Borisova, Yu. V. Granovskii, I. A. Savich, and V. I. Spitsyn, *Proc. Acad. Sci.* (*U.S.S.R.*), 1967, 177, 1039 (365)

$C_{11}H_9O_2NS_2$ 5-(4′Methoxybenzylidene)-2-thioxo-1,3-thiazolidin-4-one HL 900.2
5-(4′-methoxybenzylidene)rhodanine

Metal	Method	Temp	Medium	Log of equilibrium constant, remarks	Ref
H^+	dis	20	0·1 ($NaClO_4$)	K_1 7·76	65N
Ag^+	dis	20	0·1 ($NaClO_4$)	K_1 8·80	65N

65N O. Navratil and J. Kotas, *Coll. Czech. Chem. Comm.*, 1965, 30, 2736

$C_{11}H_{12}ONCl$ [*N*-(4′-Chlorophenyl)imino]pentan-2-one L 901

Erratum. 1964, p. 663, Table 901. *Insert* L into heading.

$C_{11}H_{12}ON_2Cl_2$ 2-(2′,6′-Dichlorophenylhydroxymethyl)-1,4,5,6 tetrahydropyrimidine HL (?) 901.1
(2′,6′-Dichlorophenyl-1″,4″,5″,6″-tetrahydropyrimidin-2″-yl-methanol)

Metal	Method	Temp	Medium	Log of equilibrium constant, remarks	Ref
H^+	gl	25	0·1 KCl	K_1 11·87, K_2 10·81	68J
Cu^{2+}	gl	25	0·1 KCl	K_1 12·16, K_2 9·81	68J

68J R. J. Jameson and I. A. Khan, *J. Chem. Soc.* (*A*), 1968, 921

$C_{11}H_{13}ON_2Cl$ 2-(2′-Chlorophenylhydroxymethyl)-1,4,5,6 tetrahydropyrimidine HL (?) 901.2
(2′-chlorophenyl-1″,4″,5″, 6″-tetrahydropyrimidin-2″-yl-methanol)

Metal	Method	Temp	Medium	Log of equilibrium constant, remarks	Ref
H^+	gl	25	0·1 KCl	K_1 12·04, K_2 11·08	68J
Cu^{2+}	gl	25	0·1 KCl	K_1 12·62, K_2 10·44	68J

68J R. F. Jameson and I. A. Khan, *J. Chem. Soc.* (*A*), 1968, 921

906 $C_{12}H_8N_2$ 1,10-Phenanthroline L

Metal	*Method*	*Temp*	*Medium*	*Log of equilibrium constant, remarks*	*Ref*
H^+	gl	?	0·1 0, 50% ethanol	K_1 (0%) 5·07, (50%) 4·27	61H
	gl	25	0·3 0, 50% dioxan	K_1 (0%) 4·97 ± 0·02, (50%) 4·63 ± 0·03	61J
	gl	25	0·1 KCl	K_1 4·98	62I
	cal	20	0·1 ($NaNO_3$)	ΔH_1 −3·95, ΔS_1 9·2	63A
	sp	25—45	→ 0	K_1 (25°) 4·96, (33°) 4·89, (45°) 4·77 $\Delta H_1°$ −4·1 ± 0·3, $\Delta S_1°$ 10 ± 2	64L
	gl	20	0·1 (KNO_3)	K_1 4·96	66P
	gl	30	0·1 KCl	K_1 5·00	67L
	gl	25	0·5 ($NaNO_3$)	K_1 5·03	67S
Ag^+	Ag	25	0·5 ($NaNO_3$)	$K_S(AgL_2NO_3)$ −18·95	67S
Cd^{2+}	dis	25	0·1 KCl	K_1 5·17, K_2 4·83, K_3 4·26	62I
	cal	20	0·1 ($NaNO_3$)	ΔH_1 −6·3, ΔS_1 4·8, $\Delta H_{\beta 2}$ −13·1, $\Delta S_{\beta 2}$ 4·8, $\Delta H_{\beta 3}$ −16·1, $\Delta S_{\beta 3}$ 13·3	63A
Co^{2+}	dis	25	0·1 KCl	K_1 7·02, K_2 6·70, K_3 6·38	62I
	cal	20	0·1 ($NaNO_3$)	ΔH_1 − 9·1, ΔS_1 2·1, $\Delta H_{\beta 2}$ −15·8, $\Delta S_{\beta 2}$ 9·9, $\Delta H_{\beta 3}$ −23·8, $\Delta S_{\beta 3}$ 9·9	63A
Cu^+	Pt	25	0·3 (K_2SO_4)	β_2 15·82 See Cu^{2+} below.	61J
Cu^{2+}	sp	25	0·3 (K_2SO_4)	K_2 6·64, K_3 4·90	61J
	gl	25	0·3 (K_2SO_4)	K_1 8·82, K_2 6·57	61J
	Pt	25	0·3 (K_2SO_4) 0·3 (KNO_3) 50% dioxan	K_3 5·02 K_3 4·50, $K(CuL_2^{2+} + Cu^+ \rightleftharpoons CuL_2^+ + Cu^{2+})$ 0·77	61J
	dis	25	0·1 KCl	K_1 8·82, K_2 6·57, K_3 5·02	62I
	cal	20	0·1 ($NaNO_3$)	ΔH_1 −11·7, ΔS_1 2·4, $\Delta H_{\beta 2}$ −18·2, $\Delta S_{\beta 2}$ 11·2, $\Delta H_{\beta 3}$ −26·4, $\Delta S_{\beta 3}$ 7·5	63A
	gl	20	~ 0	$K[Cu_2(OH)_2L_2^{2+} + 2H^+ \rightleftharpoons 2CuL^{2+}]$ 10·61	66P
	Hg, sol	25	0·5 ($NaNO_3$)	K_1 9·16, K_2 6·96 $K(Cu^{2+} + L + A^{2-} \rightleftharpoons CuAL)$ 20·73 where H_2A = 8-hydroxyquinoline-5-sulphonic acid $K(Cu^{2+} + L + A^- \rightleftharpoons CuLA^+)$ 17·37 where HA = pyridine-2-carboxylic acid	67S
Fe^{2+}	sp	?	0·1 0, 50% ethanol	β_3 (0%) 18·5, (50%) 16·2, acetate buffer	61H
	dis	25	0·1 KCl	K_1 5·86, K_2 5·25, K_3 10·03,	62I
	cal	20	0·1 ($NaNO_3$)	$\Delta H_{\beta 3}$ −33·0, $\Delta S_{\beta 3}$ −15·4	63A
	sp	25—45	→ 0	β_3 (25°) 20·22, (33°) 19·58, (45°) 18·77 $\Delta H_{\beta 3}°$ −31·3 ± 0·7, $\Delta S_{\beta 3}°$ −11 ± 4	64L
	sp	?	0—64% N_2H_4	β_3 (0%) 20·2, (5%) 11·3, (10%) 10·5, (25%) 10·1, (50%) 9·15, (64%) 8·8	66B
Fe^{3+}	sp	25—45	→ 0	$\Delta H_{\beta 3}°$ −9·9 ± 0·8, $\Delta S_{\beta 3}°$ 29 ± 4, $\Delta G°_{\beta 3}$ −18·8 ± 0·1	64L
Hg_2^{2+}	Hg	25	0·5 ($NaNO_3$)	$K_S[Hg_2L_2(NO_3)_2]$ −23·20	67S
Mn^{2+}	red	25	0·1 K_2SO_4	K_1 3·5, K_2 3·25, K_3 3·0	62I
	dis	25	0·1 KCl	K_1 4·50, K_2 4·15, K_3 4·05	62I
	cal	20	0·1 ($NaNO_3$)	ΔH_1 −3·5, ΔS_1 6·8, $\Delta H_{\beta 2}$ −7·0, $\Delta S_{\beta 2}$ 10·9, $\Delta H_{\beta 3}$ −9·0, $\Delta S_{\beta 3}$ 10·4	63A
Ni^{2+}	sp	25	0·1 KCl	β_3 23·9	55I
	dis	25	0·1 KCl	K_1 8·0, K_2 8·0, K_3 7·9	62I
	cal	20	0·1 ($NaNO_3$)	ΔH_1 −11·2, ΔS_1 2·1, $\Delta H_{\beta 2}$ −20·5, $\Delta S_{\beta 2}$ 8·2, $\Delta H_{\beta 3}$ −30·0, $\Delta S_{\beta 3}$ 11·2	63A
	Hg	25	0·5 ($NaNO_3$)	K_1 8·65, K_2 8·43, K_3 7·83	67S

con

$C_{12}H_8N_2$ (contd) 906

Metal	Method	Temp	Medium	Log of equilibrium constant, remarks	Ref
$Sn(CH_3)_2^+$	Ag–AgCl	25	0·1 (KNO_3)	K_1 4·2±0·2	63Y
Tl^{3+}	red	25	1	K_1 11·57, β_2 18·30	61K
	dis	25	1 $NaNO_3$	K_2 7·4, K_3 5·82, β_3 24·3	62K
VO^{2+}	gl	30	0·1 KCl	K_1 5·88 $K[VO(OH)L^+ + H^+ \rightleftharpoons VOL^{2+}]$ 3·04 See catechol (360), tiron (428), and chromotropic acid (817) for mixed ligands.	67L
Zn^{2+}	dis	25	0·1 KCl	K_1 6·30, K_2 5·65, K_3 5·10	62I
	cal	20	0·1 ($NaNO_3$)	ΔH_1 −7·5, ΔS_1 4·4, $\Delta H_{\beta 2}$ −15·0, $\Delta S_{\beta 2}$ 5·5, $\Delta H_{\beta 3}$ −19·3, $\Delta S_{\beta 3}$ 14·3	63A
	Hg	25	0·5 ($NaNO_3$)	K_1 6·73	67S

55I H. Irving and D. H. Miller, *J. Chem. Soc.*, 1955, 3457

61H C. J. Hawkins, H. Duewell, and W. F. Pickering, *Analyt. Chim. Acta*, 1961, **25**, 257

61J B. R. James and R. J. P. Williams, *J. Chem. Soc.*, 1961, 2007

61K F. Ya. Kul'ba, Yu. A. Makshev, and V. E. Moronov, *Russ. J. Inorg. Chem.*, 1961, **6**, 321 (630)

62I H. Irving and D. H. Miller, *J. Chem. Soc.*, 1962, 5222

62K F. Ya. Kul'ba, Yu. A. Makashev, B. D. Culler, and C. V. Kiselev, *Russ. J. Inorg. Chem.*, 1962, **7**, 351 (689)

63A G. Anderegg, *Helv. Chim. Acta*, 1963, **46**, 2813

63Y M. Yasuda and R. S. Tobias, *Inorg. Chem.*, 1963, **2**, 207

64L S. C. Lahiri and S. Aditya, *Z. phys. Chem. (Frankfurt)*, 1964, **41**, 173

66B D. Bauer, *Bull. Soc. chim. France*, 1966, 2631

66P D. D. Perrin and V. S. Sharma, *J. Inorg. Nuclear Chem.*, 1966, **28**, 1271

67L K. Lal and R. P. Agarwal, *J. Less-Common Metals*, 1967, **12**, 269

67S J. P. Scharff and M. R. Paris, *Bull. Soc. chim. France*. 1967, 1782

$C_{12}H_9N_3$ 5-Amino-1,10-phenanthroline L 906.1

Metal	Method	Temp	Medium	Log of equilibrium constant, remarks	Ref
H^+	gl	25	0·3 (KNO_3) 50% dioxan	K_1 5·23±0·03	61J
Cu^+				See Cu^{2+} below.	
Cu^{2+}	Pt	25	0·3 (KNO_3) 50% dioxan	K_3 5·02, $K(CuL_2^{2+} + Cu^+ \rightleftharpoons CuL_2^+ + Cu^{2+})$ −0·03	61J

61J B. R. James and R. J. P. Williams, *J. Chem. Soc.*, 1961, 2007

$C_{12}H_{12}N_2$ 4,4′-Dimethyl-2,2′-bipyridyl L 912

Metal	Method	Temp	Medium	Log of equilibrium constant, remarks	Ref
H^+	gl	25	0·3 0, 50% dioxan	K_1 (0%) 5·45±0·02, (50%) 4·40±0·03	61J
Cu^+				See Cu^{2+} below.	
Cu^{2+}	Pt	25	0·3 (K_2SO_4)	K_3 3·72	61J
	Pt	25	0·3 (KNO_3) 50% dioxan	K_3 3·25, $K(CuL_2^{2+} + Cu^+ \rightleftharpoons CuL_2^+ + Cu^{2+})$ −0·77	61J

61J B. R. James and R. J. P. Williams, *J. Chem. Soc.*, 1961, 2007

912.1 $C_{12}H_{12}N_2$ 5,5′-Dimethyl-2,2′-bipyridyl L

Metal	*Method*	*Temp*	*Medium*	*Log of equilibrium constant, remarks*	*Ref*
H^+	gl	25	0·3 (KNO_3) 50% dioxan	K_1 3·97±0·03	61J
Cu^+				See Cu^{2+} below.	
Cu^{2+}	Pt	25	0·3 (KNO_3) 50% dioxan	K_2 5·48, $K(CuL_2^{2+}+Cu^+ \rightleftharpoons CuL_2^++Cu^{2+})$ 0·10	61J

61J B. R. James and R. J. P. Williams, *J. Chem. Soc.*, 1961, 2007

912.2 $C_{12}H_{12}N_2$ 6,6′-Dimethyl-2,2′-bipyridyl L

Metal	*Method*	*Temp*	*Medium*	*Log of equilibrium constant, remarks*	*Ref*
H^+	gl	25	0·3 (KNO_3) 50% dioxan	K_1 4·23±0·02	61J
Cu^+	Pt	25	0·3 (KNO_3) 50% dioxan	β_2 ~ 15·8	61J
Cu^{2+}	gl	25	0·3 (KNO_3) 50% dioxan	K_1 4·88	61J
	Pt	25	0·3 (KNO_3) 50% dioxan	K_2 ~ 2·6	61J

61J B. R. James and R. J. P. Williams, *J. Chem. Soc.*, 1961, 2007

912.3 $C_{12}H_{12}N_4$ 1-(2′-Pyridyl)-3-(6″-methyl-2″-pyridyl)-1,2-diazaprop-2-ene (6-methylpyridine-2-aldehyde 2′-pyridylhydrazone) HL

Metal	*Method*	*Temp*	*Medium*	*Log of equilibrium constant, remarks*	*Ref*
H^+	sp	25	0 corr	K_1 > 14	64G
	gl	25	0 corr	K_2 6·06±0·005, K_3 3·65±0·01	64G
Fe^{2+}	gl	25	0 corr	β_2 26 $K[FeHL_2^++H^+ \rightleftharpoons Fe(HL)_2^{2+}]$ 6·28±0·02 $K(FeL_2+H^+ \rightleftharpoons FeHL_2^+)$ 7·95±0·02 $K(Fe^{2+}+HL \rightleftharpoons FeHL^{2+})$ 6·30±0·06 $K[Fe^{2+}+2HL \rightleftharpoons Fe(HL)_2^{2+}]$ 12·60±0·05	64G

64G R. W. Green, P. S. Hallman, and F. Lions, *Inorg. Chem.*, 1964, 13, 1541

$C_{12}H_{13}N_3$ Bis(2′-pyridylmethyl)amine (di-2-picolylamine) L 912.4

Metal	Method	Temp	Medium	Log of equilibrium constant, remarks	Ref
H^+	gl	25	0·1 (KCl)	K_1 7·27, K_2 2·41, K_3 1·75	68G
	gl	25	0·10 KNO_3	K_1 7·30, K_2 2·60, K_3 1·12	68R
Ag^+	gl	25	0·10 KNO_3	K_1 5·1, K_2 3·1	68R
Cd^{2+}	gl	25	0·10 KNO_3	K_1 6·44, K_2 5·30	68R
Co^{2+}	gl	25	0·1 (KCl)	K_1 5·2	68G
	gl	25	0·10 KNO_3	K_1 7·74, K_2 5·31	68R
Cu^{2+}	gl	25	0·10 KNO_3	K_1 9·31, K_2 4·54	68R
Mn^{2+}	gl	25	0·10 KNO_3	K_1 4·16, K_2 2·91	68R
Ni^{2+}	gl	25	0·1 (KCl)	K_1 8·5	68G
	gl	25	0·10 KNO_3	K_1 8·70, K_2 7·90	68R
Zn^{2+}	gl	25	0·1 (KCl)	K_1 6·8	68G
	gl	25	0·10 KNO_3	K_1 7·57, K_2 4·36	68R

68G D. V. Gruenwedel, *Inorg. Chem.*, 1968, 7, 495
68R J. K. Romary, J. D. Barger, and J. E. Bunds, *Inorg. Chem.*, 1968, 7, 1142

$C_{12}H_{14}O_2$ 3-Benzylpentane-2,4-dione (3-benzylacetylacetone) HL 912.5

Metal	Method	Temp	Medium	Log of equilibrium constant, remarks	Ref
Ni^{2+}	sp	30	toluene	$K(Ni_3L_6 \rightleftharpoons 3NiL_2)$ 2·09	68A

68A A. W. Addison and D. P. Graddon, *Austral. J. Chem.*, 1968, 21, 2003

$C_{12}H_{16}O_6$ Phenyl β-D-glucopyranoside L 912.6

Metal	Method	Temp	Medium	Log of equilibrium constant, remarks	Ref
B^{III}	gl	25	var	$K(H_3BO_3 + L \rightleftharpoons H_2BO_3L^- + H^+)$ −8	65L

65L S. Lormeau and M. Ahond, *Bull. Soc. chim. France*, 1965, 505

$C_{12}H_{17}N_3$ 2-(Butylaminomethyl)benzimidazole L 912.7

Metal	Method	Temp	Medium	Log of equilibrium constant, remarks	Ref
H^+	gl	15—45	50% dioxan	K_1 (15°) 7·42, (25°) 7·16, (35°) 6·98, (45°) 6·84, K_2 (15—45°) 2	63L
Cu^{2+}	gl	15—45	50% dioxan	K_1 (15°) 7·25, (25°) 6·85, (35°) 6·75, (45°) 6·58, K_2 (15°) 4·74, (25°) 4·60, (35°) 4·43, (45°) 4·40 ΔH_1 −11·6	63L

63L T. J. Lane and H. B. Durham, *Inorg. Chem.*, 1963, 2, 632

913.1 $C_{12}H_{24}N_2$ 1,1′-Diaminobicyclohexyl* L

Metal	*Method*	*Temp*	*Medium*	*Log of equilibrium constant, remarks*	*Ref*
H^+	gl	20	0·1 (KCl)	K_1 10·41, K_2 5·62	65T
			0·1 (K_2SO_4)	K_1 10·41	
Co^{2+}	gl	20	0·1 (KCl)	K_1 5·3±0·2, K_2 4·8±0·4, K_3 5·3±0·4	65T
Cu^{2+}	gl	20	0·1 (KCl)	K_1 12·20, K_2 10·95	65T
Ni^{2+}	gl	20	0·1 (K_2SO_4)	β_2 14·9	65T
Zn^{2+}	gl	20	0·1 (KCl)	K_1 6·35, K_2 6·95	65T

* Geometric isomer not stated.

65T M. L. Tomlinson, M. L. R. Sharp, and H. M. N. H. Irving, *J. Chem. Soc.*, 1965, 603

913.2 $C_{12}H_{26}N^+$ *NNN*-Tripropylallylammonium cation L^+

Metal	*Method*	*Temp*	*Medium*	*Log of equilibrium constant, remarks*	*Ref*
Pt^{2+}	sp	60	2 (NaCl)	$K(PtCl_4^{2-} + L^+ \rightleftharpoons PtCl_3L + Cl^-)$ 2·12	67D

67D R. G. Denning, F. R. Hartley, and L. M. Venanzi, *J. Chem. Soc.* (*A*), 1967, 324

914.1 $C_{12}H_{29}N_3$ 3,6,9,-Triazaundecane (*NN*″-diethyldiethylenetriamine) L

Metal	*Method*	*Temp*	*Medium*	*Log of equilibrium constant, remarks*	*Ref*
H^+	gl	25	0·1 (KCl)	K_1 9·66±0·06, K_2 8·62±0·05, K_3 3·55±0·02	68M
Cu^{2+}	sp, gl	25	0·1 ($NaClO_4$)	K_1 10·11±0·06	68M

68M D. W. Margerum, B. L. Powell, and J. A. Luthy, *Inorg. Chem.*, 1968, 7, 800

915.1 $C_{12}H_6N_2Cl_2$ 4,7-Dichloro-1,10-phenanthroline L

Metal	*Method*	*Temp*	*Medium*	*Log of equilibrium constant, remarks*	*Ref*
H^+	gl	25	~ 0	K_1 3·03	61H
	gl	25	0·3 (KNO_3) 50% dioxan	K_1 2·65±0·10	61J
Cu^+	Pt	25	0·3 (KNO_3) 50% dioxan	β_2 ~ 12·5 See Cu^{2+} below.	61J
Cu^{2+}	gl	25	0·3 (KNO_3) 50% dioxan	β_2 ~ 10·2	61J
	Pt	25	0·3 (KNO_3) 50% dioxan	K_3 3·25, $K(CuL_2^{2+} + Cu^+ \rightleftharpoons CuL_2^+ + Cu^{2+})$ 2·28	61J
Fe^{2+}	sp	?	0·1 ethanol	β_3 12·1, acetate buffer H^+ constant not given for ethanol solution.	61H

61H C. J. Hawkins, H. Duewell, and W. F. Pickering, *Analyt. Chim. Acta*, 1961, **25**, 257
61J B. R. James and R. J. P. Williams, *J. Chem. Soc.*, 1961, 2007

$C_{12}H_7O_2N_3$ 5-Nitro-1,10-phenanthroline L 916

Metal	Method	Temp	Medium	Log of equilibrium constant, remarks	Ref
H^+	gl	25	0·3 0, 50% dioxan	K_1 (0%) 3·25±0·10, (50%) 2·80±0·05	61J
	sp	25—45	→ 0	K_1 (25°) 3·23±0·01, (30°) 3·21±0·01, (35°) 3·18±0·02, (45°) 3·12±0·01 $\Delta H_1°$ (25°) −2·3±0·1, $\Delta S_1°$ 7±1	64L
Cu^+				See Cu^{2+} below.	
Cu^{2+}	Pt	25	0·3 (KNO_3) 50% dioxan	K_3 3·25, $K(CuL_2^{2+} + Cu^+ \rightleftharpoons CuL_2^+ + Cu^{2+})$ 2·15	61J
Fe^{2+}	sp	25—45	→ 0	K_1 (35°) 4·57, β_3 (25°) 15·64±0·02, (35°) 14·99±0·02, (45°) 14·47±0·01 $\Delta H_{\beta 3}°$ (25°) −25·2±0·6, $\Delta S_{\beta 3}°$ 13±4	64L
Fe^{3+}	sp	25	→ 0	β_3 7·46	64L
Ni^{2+}	sp	35	0·1	β_3 16·48, acetate buffer	67L

61J B. R. James and R. J. P. Williams, *J. Chem. Soc.*, 1961, 2007
64L S. C. Lahiri and S. Aditya, *Z. phys. Chem. (Frankfurt)*, 1964, **43**, 282
67L S. C. Lahiri and S. Aditya, *J. Indian Chem. Soc.*, 1967, **44**, 9

$C_{12}H_7N_2Cl$ 2-Chloro-1,10-phenanthroline L 916.1

Metal	Method	Temp	Medium	Log of equilibrium constant, remarks	Ref
H^+	gl	25	0·3 (K_2SO_4)	K_1 4·20±0·02	61J
Cu^+	Pt	25	0·3 (K_2SO_4)	β_2 14·6	61J
Cu^{2+}	Pt	25	0·3 (K_2SO_4)	K_1 5·60, K_2 4·85	61J

61J R. B. James and R. J. P. Williams, *J. Chem. Soc.*, 1961, 2007

$C_{12}H_7N_2Cl$ 4-Chloro-1,10-phenanthroline L 916.2

Metal	Method	Temp	Medium	Log of equilibrium constant, remarks	Ref
H^+	sp	?	0·1	K_1 4·30	61H
Fe^{2+}	sp	?	0·1	β_3 14·8, acetate buffer	61H

61H C. J. Hawkins, H. Duewell, and W. F. Pickering, *Analyt. Chim. Acta*, 1961, **25**, 257

917 $C_{12}H_7N_2Cl$ 5-Chloro-1,10-phenanthroline L

Metal	*Method*	*Temp*	*Medium*	*Log of equilibrium constant, remarks*	*Ref*
H^+	gl	25	0·3 (KNO_3) 50% dioxan	K_1 3·43±0·03	61J
Cu^+	Pt	25	0·3 (KNO_3) 50% dioxan	β_2 ~ 12·2 See Cu^{2+} below.	61J
Cu^{2+}	Pt	25	0·3 (KNO_3) 50% dioxan	K_3 3·72, $K(CuL_2^{2+}+Cu^+ \rightleftharpoons CuL_2^+ + Cu^{2+})$ 1·18	61J
	gl	25	0·3 (KNO_3) 50% dioxan	β_2 ~ 11·0	61J

61J B. R. James and R. J. P. Williams, *J. Chem. Soc.*, 1961, 2007

918 $C_{12}H_7N_2Br$ 5-Bromo-1,10-phenanthroline L

Erratum, 1964, p. 668 Table 918. *Insert* L into heading.

918.1 $C_{12}H_8ON_2$ 5-Hydroxy-1,10-phenanthroline HL

Metal	*Method*	*Temp*	*Medium*	*Log of equilibrium constant, remarks*	*Ref*
H^+	gl	?	0·1	K_2 2·17	61H
Fe^{2+}	gl	?	0·1	$K[Fe^{2+}+3HL \rightleftharpoons Fe(HL)_3^{2+}]$(?) 11·7, acetate buffer	61H

61H C. J. Hawkins, H. Duewell, and W. F. Pickering, *Analyt. Chim. Acta*, 1961, **25**, 257

918.2 $C_{12}H_{10}ON_2$ 2-[*N*2′-Hydroxyphenyl)aminomethyl]pyridine HL

Metal	*Method*	*Temp*	*Medium*	*Log of equilibrium constant, remarks*	*Ref*
H^+	gl	25	0·1 50% dioxan	K_1 11·80, K_2 4·5, K_3 3·05	62G
Co^{2+}	gl	25	0·1 50% dioxan	K_1 12·2, K_2 6·8	62G
Cu^{2+}	gl	25	0·1 50% dioxan	K_1 13·8, K_2 6·4	62G
Ni^{2+}	gl	25	0·1 50% dioxan	K_1 12·9, K_2 8·3	62G
Pb^{2+}	gl	25	0·1 50% dioxan	K_1 10·9	62G
UO_2^{2+}	gl	25	0·1 50% dioxan	K_1 12·4, K_2 ~ 9·1	62G
Zn^{2+}	gl	25	0·1 50% dioxan	K_1 10·8, K_2 8·0	62G

62G W. J. Geary, G. Nickless, and F. H. Pollard, *Analyt. Chim. Acta*, 1962, **27**, 71

$C_{12}H_{10}ON_2$ 2-Salicylideneaminopyridine HL 918.

Metal	Method	Temp	Medium	Log of equilibrium contrast, remarks	Ref
H^+	gl	25	0·1 50% dioxan	K_1 9·77, K_2 6·37	62G
Co^{2+}	gl	25	0·1 50% dioxan	K_1 5·3, K_2 5·0	62G
Cu^{2+}	gl	25	0·1 50% dioxan	K_1 7·3, K_2 5·1	62G
Ni^{2+}	gl	25	0·1 50% dioxan	K_1 4·9, K_2 4·1	62G
UO_2^{2+}	gl	25	0·1 50% dioxan	K_1 8·1	62G
Zn^{2+}	gl	25	0·1 50% dioxan	K_1 5·8, K_2 4·3	62G

62G W. J. Geary, G. Nickless, and F. H. Pollard, *Analyt. Chim. Acta*, 1962, 27, 71

$C_{12}H_{10}ON_2$ 2-Benzoylpyridine oxime HL 918.4

Metal	Method	Temp	Medium	Log of equilibrium constant, remarks	Ref
H^+	gl	20—40	0—0·1 ($NaClO_4$) 40% acetone	K_1 (0) (20°) 12·35, (30°) 12·14, (40°) 11·87, (0·050) (20°) 12·23, (30°) 11·96, (40°) 11·69, (0·075) (20°) 12·21, (30°) 11·92, (40°) 11·65, (0·1) (20°) 12·19, (30°) 11·89, (40°) 11·61, K_2 (0) (20°) 2·85, (30°) 3·15, (40°) 3·28, (0·025) (20°) 2·85, (30°) 3·03, (40°) 3·13, (0·050) (20°) 2·85, (30°) 3·00, (40°) 3·06, (0·075) (20°) 2·84, (30°) 2·97, (40°) 3·03, (0·1) (20°) 2·84, (30°) 2·95, (40°) 3·01	65S
Cd^{2+}	gl	20—40	0—0·1 ($NaClO_4$) 40% acetone	K_1 (0) (20°) 7·24, (30°) 6·95, (40°) 6·65, (0·025) (20°) 7·20, (30°) 6·92, (40°) 6·62, (0·050) (20°) 7·14, (30°) 6·84, (40°) 6·54, (0·075) (20°) 7·11, (30°) 6·82, (40°) 6·53, (0·1) (20°) 7·08, (30°) 6·80, (40°) 6·50, K_2 (0) (20°) 6·65, (30°) 6·44, (40°) 6·45, (0·025) (20°) 6·64, (30°) 6·42, (40°) 6·40, (0·050) (20°) 6·60, (30°) 6·38, (40°) 6·33, (0·075) (20°) 6·56, (30°) 6·35, (40°) 6·31, (0·1) (20°) 6·55, (30°) 6·34, (40°) 6·30	65S
			→ 0	$\Delta H_1°$ −12·4, $\Delta S_1°$ −9·0, $\Delta H_2°$ −4·2, $\Delta S_2°$ 16	
Hg^{2+}	gl	20—40	0—0·1 ($NaClO_4$) 40% acetone	K_1 (0) (20°) 9·80, (30°) 10·12, (40°) 10·44, (0·025) (20°) 9·74, (30°) 10·06, (40°) 10·38, (0·050) (20°) 9·52, (30°) 9·84, (40°) 10·16, (0·075) (20°) 9·44, (30°) 9·79, (40°) 10·10, 0·1) (20°) 9·40, (30°) 9·74, (40°) 10·00, K_2 (0) (20°) 7·12, (30°) 7·40, (40°) 8·04, (0·025) (20°) 7·15, (30°) 7·36, (40°) 8·00, (0·050) (20°) 6·94, (30°) 7·20, (40°) 7·82, (0·075) (20°) 6·90, (30°) 7·20, (40°) 7·78, (0·1) (20°) 6·88, (30°) 7·12, (40°) 7·72	65S
			→ 0	$\Delta H°$ −high, $\Delta S°$ +high	

continued overleaf

918.4 $C_{12}H_{10}ON_2$ (contd)

Metal	Method	Temp	Medium	Log of equilibrium constant, remarks	Ref
Ni^{2+}	gl	30—40	0—0·1 ($NaClO_4$) 40% acetone	K_1 (0) (30°) 9·21, (40°) 9·10, (0·025) (30°) 9·15, (40°) 9·00, (0·050) (30°) 8·90, (40°) 8·70, (0·075) (30°) 8·81, (40°) 8·62, (0·1) (30°) 8·75, (40°) 8·52, K_2 (0) (30°) 7·05, (40°) 7·03, (0·025) (30°) 7·00, (40°) 6·95, (0·050) (30°) 6·70, (40°) 6·66, (0·075) (30°) 6·61, (40°) 6·54, (0·1) (30°) 6·52, (40°) 6·45	65S
			→ 0	$\Delta H_1°$ −2·31, $\Delta S_1°$ 34, $\Delta H_2°$ −0·42, $\Delta S_2°$ 30	
Pb^{2+}	gl	20—40	0—0·1 ($NaClO_4$) 40% acetone	K_1 (0) (20°) 7·74, (30°) 7·52, (40°) 7·41, (0·025) (20°) 7·72, (30°) 7·50, (40°) 7·39, (0·050) (20°) 7·68, (30°) 7·47, (40°) 7·33, (0·075) (20°) 7·66, (30°) 7·45, (40°) 7·30, (0·1) (20°) 7·65, (30°) 7·43, (40°) 7·29, K_2 (0) (20°) 7·20, (30°) 7·11, (40°) 7·01, (0·025) (20°) 7·18, (30°) 7·08, (40°) 6·97, (0·050) (20°) 7·10, (30°) 7·01, (40°) 6·90, (0·075) (20°) 7·07, (30°) 6·98, (40°) 6·88, (0·1) (20°) 7·04, (30°) 6·96, (40°) 6·85	65S
			→ 0	$\Delta H_1°$ −6·93, $\Delta S_1°$ 11·5, $\Delta H_2°$ −3·99, $\Delta S_2°$ 19·1	
Zn^{2+}	gl	20—40	0—0·1 ($NaClO_4$) 40% acetone	K_1 (0) (20°) 8·44, (30°) 8·40, (40°) 8·14, (0·025) (20°) 8·40, (30°) 8·34, (40°) 8·08, (0·050) (20°) 8·20, (30°) 8·15, (40°) 7·91, (0·075) (20°) 8·15, (30°) 8·08, (40°) 7·85, (0·1) (20°) 8·11, (30°) 8·04, (40°) 7·78, K_2 (0) (20°) 6·90, (30°) 6·95, (40°) 6·92, (0·025) (20°) 6·86, (30°) 6·89, (40°) 6·85, (0·050) (20°) 6·75, (30°) 6·70, (40°) 6·70, (0·075) (20°) 6·70, (30°) 6·66, (40°) 6·63, (0·1) (20°) 6·66, (30°) 6·62, (40°) 6·58	65S
			→ 0	$\Delta H_1°$ −6·30, $\Delta S_1°$ 17·0, $\Delta H_2°$ −0·63, $\Delta S_2°$ 29·2	

65S D. C. Shuman and B. Sen, *Analyt. Chim. Acta*, 1965, **33**, 487

919 $C_{12}H_{10}O_2N_2$ 2,2′-Dihydroxyazobenzene H_2L

Metal	Method	Temp	Medium	Log of equilibrium constant, remarks	Ref
H^+	sp	25	0·1 (KCl)	K_1 9·3	62K
	sp	?	0·100 (KCl)	K_1 11·5, K_2 7·8	60D, 67C
Al^{3+}	sp	25	0·1 (KCl)	$K(Al^{3+}+H_2L \rightleftharpoons AlL^+ + 2H^+)$ 3·1	62K
				$K(AlL^+ + H_2L \rightleftharpoons AlL_2^- + 2H^+)$ (?) 7·4	
Ga^{3+}	sp	25	0·1 (KCl)	$K(Ga^{3+}+H_2L \rightleftharpoons GaL^+ + 2H^+)$ 0·61,	62K
				$K(GaL^+ + H_2L \rightleftharpoons GaL_2^- + 2H^+)$ (?) 5·2	
In^{3+}	sp	25	0·1 (KCl)	$K(In^{3+}+H_2L \rightleftharpoons InL^+ + 2H^+)$ 5·2,	62K
				$K(InL^+ + H_2L \rightleftharpoons InL_2^- + 2H^+)$ (?) 8·0	
Mg^{2+}	sp	?	0·100 (KCl)	K_1 4·85	60D
Ni^{2+}	pol	20°	~ 0·3 50% ethanol	K_1 11·23, acetate buffer	67C
	sp	20	~ 0·3 50% ethanol	K_1 11·64	67C

60D H. Diehl and J. Ellingboe, *Analyt. Chem.*, 1960, **32**, 1120
62K J. R. Kirby, R. M. Milburn, and J. H. Saylor, *Analyt. Chim. Acta*, 1962, **26**, 458
67C E. Casassas and L. Eek, *J. Chim. phys.*, 1967, **64**, 978

$C_{12}H_{10}O_2N_2$ 2,4-Dihydroxyazobenzene H_2L 919.1

Metal	*Method*	*Temp*	*Medium*	*Log of equilibrium constant, remarks*	*Ref*
H^+	sp	25	~ 0 50% dioxan	K_1 13·60±0·10	62G
	gl	25	~ 0 50% dioxan	K_2 8·33	62G
Cu^{2+}	gl	25	~ 0 50% dioxan	K_1 (?) 13·9	62G
Ni^{2+}	gl	25	~ 0 50% dioxan	K_1 (?) 8·2, K_2 9·1 (?)	62G
Zn^{2+}	gl	25	~ 0 50% dioxan	K_1 (?) 10·8, K_2 11·9 (?)	62G

62G W. J. Geary, G. Nickless, and F. H. Pollard, *Analyt. Chim. Acta*, 1962, **27**, 71

$C_{12}H_{10}O_3N_2$ 2, 2′, 4′-Trihydroxyazobenzene H_3L 919.2

Metal	*Method*	*Temp*	*Medium*	*Log of equilibrium constant, remarks*	*Ref*
H^+	sp	?	0·100 (KCl)	K_1 12·2, K_2 8·7, K_3 6·6	60D
Ca^{2+}	sp	?	0·100 (KCl)	K_1 1·23	60D
Mg^{2+}	sp	?	0·100 (KCl)	K_1 3·50	60D

60D H. Diehl and J. Ellingboe, *Analyt. Chem.*, 1960, 32, 1120

$C_{12}H_{11}ON_3$ 1-Pyridyl-3-(2′-hydroxyphenyl)-1,2-diazaprop-2-ene HL 920.1 (salicylaldehyde 2-pyridylhydrazone)

Metal	*Method*	*Temp*	*Medium*	*Log of equilibrium constant, remarks*	*Ref*
H^+	sp	18—22	0·1 ($NaClO_4$) 50% methanol	K_1 10·87	67A
	gl	25	0·1 ($NaClO_4$) 50% methanol	K_2 5·08	67A
Co^{2+}	gl	25	0·1 ($NaClO_4$) 50% methanol	K_1 12·1	67A
Cu^{2+}	sp	18—22	0·1 ($NaClO_4$) 50% methanol	K_1 17·9	67A
	gl	25	0·1 ($NaClO_4$) 50% methanol	K_2 5·9	67A
Ni^{2+}	gl	25	0·1 ($NaClO_4$) 50% methanol	K_1 13·8, K_2 11·3	67A
Zn^{2+}	gl	25	0·1 ($NaClO_4$) 50% methanol	K_1 11·1	67A

67A R. G. Anderson and G. Nickless, *Talanta*, 1967, **14**, 1221

920.2 $C_{12}H_{11}ON_3$ 3-Hydroxy-1,3-diphenyltriazene HL

Metal	*Method*	*Temp*	*Medium*	*Log of equilibrium constant, remarks*	*Ref*
H^+	sp	?	?	K_1 11·41	65P
Cu^{2+}	gl	25	0·1 KCl 70% dioxan	K_1 11·27, K_2 11·04	65P

65P D. N. Purohit and N. C. Sogani, *Z. Naturforsch.*, 1965, **20b**, 206

920.3 $C_{12}H_{11}ON_3$ 4-Methyl-2-(2′-pyridylazo)phenol HL

Metal	*Method*	*Temp*	*Medium*	*Log of equilibrium constant, remarks*	*Ref*
H^+	sp	25	2·0 0·4% dioxan	K_1 8·90, K_2 2·55	68W
In^{3+}	sp	25	2·0 0·4% dioxan	K_1 11·8 $K(InL^{2+}+A^- \rightleftharpoons InAL^+)$ 3·0 $K(InL_2^+ + A^- \rightleftharpoons InAL_2)$ 1·9 $K(InL_3 + A^- \rightleftharpoons InAL_3^-)$ 1·3 where HA = acetic acid	68W

68W H. Wada and M. Kakagawa, *Nippon Kagaku Zasshi*, 1968, **89**, 499

921.1 $C_{12}H_{11}O_9N$ *N*-(2′,5′-Dicarboxy-4′-hydroxyphenyl)iminodiacetic acid {2-[bis(carboxymethyl)amino]-5-hydroxyterephthalic acid} H_5L

Metal	*Method*	*Temp*	*Medium*	*Log of equilibrium constant, remarks*	*Ref*
H^+	gl	22	0·1 KCl	K_2 8·56, K_3 3·15, K_4 2·55, K_5 1·6	61U
	gl	25	0·1 (KNO_3)	K_2 8·61, K_3 3·18, K_4 2·20, K_5 1·60	67U
Ba^{2+}	gl	25	0·1 (KNO_3)	$K(Ba^{2+}+HL^{4-} \rightleftharpoons BaHL^{2-})$ 3·90	67U
Ca^{2+}	gl	25	0·1 (KNO_3)	$K(Ca^{2+}+HL^{4-} \rightleftharpoons CaHL^{2-})$ 5·45	67U
Cd^{2+}	gl	25	0·1 (KNO_3)	$K(Cd^{2+}+HL^{4-} \rightleftharpoons CdHL^{2-})$ 7·97 $K(Cd^{2+}+H_2L^{3-} \rightleftharpoons CdH_2L^-)$ 2·61	67U
	pol	25	0·1 (KNO_3)	$K(Cd^{2+}+HL^{4-} \rightleftharpoons CdHL^{2-})$ 7·92	67U
Co^{2+}	gl	22	0·1 KCl	$K(Co^{2+}+HL^{4-} \rightleftharpoons CoHL^{2-})$ 9·0 $K(Co^{2+}+H_2L^{3-} \rightleftharpoons CoH_2L^-)$ 3·2	61U
	gl	25	0·1 (KNO_3)	$K(Co^{2+}+HL^{4-} \rightleftharpoons CoHL^{2-})$ 9·18 $K(Co^{2+}+H_2L^{3-} \rightleftharpoons CoH_2L^-)$ 3·24	67U
Cu^{2+}	gl	22	0·1 KCl	$K(Cu^{2+}+HL^{4-} \rightleftharpoons CuHL^{2-})$ 11·1 $K(Cu^{2+}+H_2L^{3-} \rightleftharpoons CuH_2L^-)$ 5·0	61U
	pol	25	0·1 (KNO_3)	$K(Cu^{2+}+HL^{4-} \rightleftharpoons CuHL^{2-})$ 11·62	67U

contin

$C_{12}H_{11}O_9N$ (contd) 921.1

Metal	Method	Temp	Medium	Log of equilibrium constant, remarks	Ref
Mg^{2+}	gl	25	0·1 (KNO_3)	$K(Mg^{2+}+HL^{4-} \rightleftharpoons MgHL^{2-})$ 4·59	67U
Mn^{2+}	gl	25	0·1 (KNO_3)	$K(Mn^{2+}+HL^{4-} \rightleftharpoons MnHL^{2-})$ 6·49	67U
				$K(Mn^{2+}+H_2L^{3-} \rightleftharpoons MnH_2L^{-})$ 1·41	
Ni^{2+}	gl	22	0·1 KCl	$K(Ni^{2+}+HL^{4-} \rightleftharpoons NiHL^{2-})$ 10·6	61U
				$K(Ni^{2+}+H_2L^{3-} \rightleftharpoons NiH_2L^{-})$ 4·7	
	pol	25	0·1 (KNO_3)	$K(Ni^{2+}+HL^{4-} \rightleftharpoons NiHL^{2-})$ 10·38	67U
Sr^{2+}	gl	25	0·1 (KNO_3)	$K(Sr^{2+}+HL^{4-} \rightleftharpoons SrHL^{2-})$ 4·31	67U
Zn^{2+}	gl	25	0·1 (KNO_3)	$K(Zn^{2+}+HL^{4-} \rightleftharpoons ZnHL^{2-})$ 9·19	67U
				$K(Zn^{2+}+H_2L^{3-} \rightleftharpoons ZnH_2L^{-})$ 3·20	

61U E. Uhlig, *Z. anorg. Chem.*, 1961, **311**, 249
67U E. Uhlig and R. Krannich, *J. Inorg. Nuclear Chem.*, 1967, **29**, 1164

$C_{12}H_{12}O_2N_2$ *NN'*-Di(furfurylidene)ethylenediamine L 921.2

Metal	Method	Temp	Medium	Log of equilibrium constant, remarks	Ref
Ag^{+}	Ag	20	0·1 $NaNO_3$ methanol	K_1 6·70, β_2 8·95	66H
Cu^{2+}	pol	20	0·1 KNO_3 methanol	K_1 7·60, β_2 9·95	66H

66H E. Hoyer and V. V. Skopenko, *Russ. J. Inorg. Chem.*, 1966, **11**, 436 (803)

$C_{12}H_{12}N_2S_2$ *NN'*-Di(thienylidene)ethylenediamine L 921.3

Metal	Method	Temp	Medium	Log of equilibrium constant, remarks	Ref
Ag^{+}	Ag	20	0·1 $NaNO_3$ methanol	K_1 7·54, β_2 9·42	66H
Cu^{2+}	pol	20	0·1 KNO_3	K_1 8·20, β_2 10·08	66H

66H E. Hoyer and V. V. Skopenko, *Russ. J. Inorg. Chem.*, 1966, **11**, 436 (803)

921.4 $C_{12}H_{13}ON$ 8-Hydroxy-2-propyl-quinoline HL

Metal	*Method*	*Temp*	*Medium*	*Log of equilibrium constant, remarks*	*Ref*
H^+	gl	25	~ 0	K_1 11·95, K_2 4·44	66K
Cd^{2+}	gl	25	~ 0	K_1 8·84, K_2 8·70	66K
Cu^{2+}	gl	25	~ 0	K_1 12·06, K_2 11·81	66K
Ni^{2+}	gl	25	~ 0	K_1 9·01, K_2 9·46	66K
Zn^{2+}	gl	25	~ 0	K_1 8·96, K_2 10·49 (?)	66K

66K H. Kaneko and K. Ueno, *Bull. Chem. Soc. Japan*, 1966, **39**, 1910

922.1 $C_{12}H_{13}O_5N$ *N*-(2′-Acetylphenyl)iminodiacetic acid (2-aminoacetophenone-*NN*-diacetic acid) H_2L

Metal	*Method*	*Temp*	*Medium*	*Log of equilibrium constant, remarks*	*Ref*
H^+	gl	25	0·1 (KNO_3)	K_1 7·89, K_2 1·9	65A
Ca^{2+}	gl	25	0·1 (KNO_3)	K_1 4·11	65A
Cd^{2+}	gl	25	0·1 (KNO_3)	K_1 7·37, K_2 4·97	65A
Co^{2+}	gl	25	0·1 (KNO_3)	K_1 6·99, K_2 4·85	65A
Cu^{2+}	gl	25	0·1 (KNO_3)	K_1 9·22, K_2 5·03	65A
Mg^{2+}	gl	25	0·1 (KNO_3)	K_1 3·06	65A
Ni^{2+}	gl	25	0·1 (KNO_3)	K_1 8·12, K_2 5·72	65A
Sr^{2+}	gl	25	0·1 (KNO_3)	K_1 3·24	65A
Zn^{2+}	gl	25	0·1 (KNO_3)	K_1 7·13, K_2 3·97	65A

65A T. Ando and K. Ueno, *Inorg. Chem.*, 1965, **4**, 375

922.2 $C_{12}H_{13}O_5N$ *N*-(Benzoylmethyl)iminodiacetic acid H_2L

Metal	*Method*	*Temp*	*Medium*	*Log of equilibrium constant, remarks*	*Ref*
H^+	gl	30	0·1 KCl	K_1 8·02±0·07, K_2 1·85, K_3 1·25	66S
Ca^{2+}	gl	30	0·1 KCl	K_1 4·70±0·02	66S
Mg^{2+}	gl	30	0·1 KCl	K_1 3·11±0·05	66S
Ni^{2+}	gl	30	0·1 KCl	K_1 6·87±0·10, K_2 5·90±0·04	66S

66S G. Shtacher, *J. Inorg. Nuclear Chem.*, 1966, **28**, 845

$C_{12}H_{13}O_6N$ α(Carboxybenzyl)iminodiacetic acid H_3L 922.3

Metal	Method	Temp	Medium	Log of equilibrium constant, remarks	Ref
H^+	gl	20	0·1 (KCl)	K_1 9·26±0·02, K_2 2·39±0·02, K_3 1·45±0·05	66I
Ba^{2+}	gl	20	0·1 (KCl)	K_1 4·28±0·01	66I
Ca^{2+}	gl	20	0·1 (KCl)	K_1 6·17±0·02	66I
Mg^{2+}	gl	20	0·1 (KCl)	K_1 4·64±0·02	66I
Sr^{2+}	gl	20	0·1 (KCl)	K_1 4·44±0·02	66I
Zn^{2+}	gl	20	0·1 (KCl)	K_1 9·78±0·02	66I

66I H. M. N. H. Irving and M. G. Miles, *J. Chem. Soc.* (*A*), 1966, 727

$C_{12}H_{15}ON$ 4-[*N*-(4′-Methylphenyl)amino]pentan-2-one L 924

Errata. 1964, p. 670, Table 924. *Insert* **L** into heading.

This table has been renumbered **922·4** to correct the ligand sequence
For $C_{12}H_{15}O_2N$ *read* $C_{12}H_5ON$.
Under Be^{2+}, *for* K_1 10·9, K_2 10·63, *read* K_1 10·43, K_2 10·61

$C_{12}H_{15}O_5N$ *N*-(2′-Phenoxyethyl)iminodiacetic acid H_2L 924.1

Metal	Method	Temp	Medium	Log of equilibrium constant, remarks	Ref
H^+	gl	30	0·1 KCl	K_1 8·49±0·02	66S
Ca^{2+}	gl	30	0·1 KCl	K_1 3·31±0·05	66S
Mg^{2+}	gl	30	0·1 KCl	K_1 3·03±0·01	66S
Ni^{2+}	gl	30	0·1 KCl	K_1 7·62±0·05, K_2 6·36±0·03	66S

66S G. Shtacher, *J. Inorg. Nuclear Chem.*, 1966, **28**, 845

$C_{12}H_{16}ON_2$ 1,4,5,6-Tetrahydro-2-(1′-phenyl-1′-hydroxyethyl)-pyrimidine HL 924.2 [1-phenyl-1-(1′,4′,5′,6′-tetrahydropyrimidin-2′-yl)ethanol]

Metal	Method	Temp	Medium	Log of equilibrium constant, remarks	Ref
H^+	gl	25	0·1 KCl	K_1 12·31±0·06, K_2 11·49±0·02	68J
Cu^{2+}	gl	25	0·1 KCl	K_1 13·18, K_2 11·13	68J

68J R. F. Jameson and I. A. Khan, *J. Chem. Soc.* (*A*), 1968, 921

924.3 $C_{12}H_{16}ON_2$ 2-(2′-Hydroxypentyl)benzimidazole HL

Metal	*Method*	*Temp*	*Medium*	*Log of equilibrium constant, remarks*	*Ref*
H^+	gl	15—45	50% dioxan	K_1 (15°) 13·07, (25°) 12·77, (40°) 12·30, (45°) 12·06, K_2 (15°) 4·88, (25°) 4·80, (40°) 4·66, (45°) 4·58	63L
Cu^{2+}	gl	15—45	50% dioxan	K_1 (15°) 12·07, (25°) 11·45, (40°) 10·75, (45°) 10·58 ΔH_1 −10·1	63L

63L T. J. Lane and H. B. Durham, *Inorg. Chem.*, 1963, **2**, 632

925.1 $C_{12}H_{16}O_3N_2$ *N*-Benzylalanylglycine HL

Metal	*Method*	*Temp*	*Medium*	*Log of equilibrium constant, remarks*	*Ref*
H^+	gl	25	0·12 (NaCl)	K_1 7·05±0·03, K_2 3·28±0·03	67S
Cu^{2+}	gl	25	0·12 (NaCl)	K_1 4·51, $K[Cu(H_{-1}L)+H^+ \rightleftharpoons CuL^+]$ 4·85 $K[Cu(H_{-1}L)OH^- + H^+ \rightleftharpoons Cu(H_{-1}L)]$ 7·84 $K[Cu(H_{-1}L)+L^- \rightleftharpoons CuL(H_{-1}L)^-]$ 3·38	67S

67S U. I. Salakhutdinov, A. P. Borisona, Yu. V. Granovskii, I. A. Savich, and V. I. Spitsyn, *Proc. Acad. Sci.* (*U.S.S.R.*), 1967, **177**, 1039, (365)

926 $C_{12}H_{16}O_3N_6$ Histidylhistidine HL

Metal	*Method*	*Temp*	*Medium*	*Log of equilibrium constant, remarks*	*Ref*
H^+	gl	25	0·1 (KCl)	K_1 7·92±0·01, K_2 6·70±0·01, K_3 5·36±0·01, K_4 2·16±0·05	64D
	gl	25	0·1 ($NaClO_4$) 1·14 CH_3CN	K_1 7·94, K_2 6·82, K_3 5·60	66K
Cu^+	gl	25	0·1 ($NaClO_4$) 1·14 CH_3CN	$K[Cu^+ + 2H_2L^+ \rightleftharpoons Cu(H_2L)_2{}^{3+}]$ 12·47 $K(CuHL^+ + Cu^+ \rightleftharpoons Cu_2HL^{2+})$ 6·18 $K(Cu_2L^+ + H^+ \rightleftharpoons Cu_2HL^{2+})$ 6·47	66K
Cu^{2+}	gl	25	0·1 (KCl)	K_1 12·0±0·2 $K[Cu(H_{-1}L)+H^+ \rightleftharpoons CuL^+]$ 6·21±0·1	64D

64D M. A. Doran, S. Chaberek, and A. E. Martell, *J. Amer. Chem. Soc.*, 1964, **86**, 2129
66K Th. Kaden and A. Zuberbuhler, *Helv. Chim. Acta*, 1966, **49**, 2189

$C_{12}H_{16}O_6N_2$ 2,5-Bis[*N*-(2′-hydroxyethyl)amino]benzene-1,4-dicarboxylic acid H_2L 926.1

Metal	Method	Temp	Medium	Log of equilibrium constant, remarks	Ref
H^+	gl	22	0·1 KCl	K_1 6·79, K_2 3·71	61U
Ni^{2+}	gl	22	0·1 KCl	K_1 4·40,	61U
				$K(Ni^{2+}+HL^- \rightleftharpoons NiHL^+)$ 1·50	
	gl	20	0·1 KCl	K_2 2·90	64U

61U E. Uhlig, *Z. anorg. Chem.*, 1961, **312**, 332
64U E. Uhlig and D. Linke, *Z. anorg. Chem.*, 1964, **331**, 112

$C_{12}H_{17}ON$ *erythro*-2-(Hydroxybenzyl)piperidine HL 927.1
(*erythro*-phenyl-2′-piperidylmethanol)

Metal	Method	Temp	Medium	Log of equilibrium constant, remarks	Ref
H^+	gl	25	→ 0	K_1 16·6, K_2 9·57	65T
Cu^{2+}	pol	25	0·1 KNO_3	β_2 25·6	65T

65T G. Tissier, *Bull. Soc. chim. France*, 1965, 124

$C_{12}H_{17}ON$ *threo*-2-(Hydroxybenzyl)piperidine HL 927.2
(*threo*-phenyl-2′-piperidylmethanol)

Metal	Method	Temp	Medium	Log of equilibrium constant, remarks	Ref
H^+	gl	25	→ 0	K_1 17·6, K_2 9·77	65T
Cu^{2+}	pol	25	0·1 KNO_3	β_2 28·4	65T

65T G. Tissier, *Bull. Soc. chim. France*, 1965, 124

$C_{12}H_{17}O_5N_5$ *N*-Acetylglycylglycyl-L-histidine HL 927.3

Metal	Method	Temp	Medium	Log of equilibrium constant, remarks	Ref
H^+	gl	25	0·16 KCl	K_1 7·18, K_2 3·08	66B
Ni^{2+}	gl	25	0·16 KCl	K_1 2·94, K_2 2·30	66B
				$K[Ni(H_{-1}L)+H^+ \rightleftharpoons NiL^+]$ 8·65	
				$K[Ni(H_{-2}L)^- + H^+ \rightleftharpoons Ni(H_{-1}L)]$ 8·95	
				$K[Ni(H_{-3}L)^{2-} + H^+ \rightleftharpoons Ni(H_{-2}L)^-]$ 9·20	

66B G. F. Bryce, R. W. Roeske, and F. R. N. Gurd, *J. Biol. Chem.*, 1966, **241**, 1072

927.4 $C_{12}H_{18}O_5N_6$ Glycylglycylglycyl-L-histidine HL

Metal	*Method*	*Temp*	*Medium*	*Log of equilibrium constant, remarks*	*Ref*
H^+	gl	25	0·16 KCl	K_1 8·11, K_2 6·85, K_3 3·02	66B
Ni^{2+}	gl	25	0·16 KCl	$K[Ni(H_{-1}L)+H^+ \rightleftharpoons NiL^+]$ 8·50	66B
				$K[Ni(H_{-2}L)^- + H^+ \rightleftharpoons Ni(H_{-1}L)]$ 8·75	
				$K[Ni(H_{-3}L)^{2-} + H^+ \rightleftharpoons Ni(H_{-2}L)^-]$ 9·10	

66B G. F. Bryce, R. W. Roeske, and F. R. N. Gurd, *J. Biol. Chem.*, 1966, **241**, 1072

928.1 $C_{12}H_{19}O_6N$ *N*-(2′-Carboxycycloheptyl)iminodiacetic acid* H_3L

Metal	*Method*	*Temp*	*Medium*	*Log of equilibrium constant, remarks*	*Ref*
H^+	gl	20	0·1 (KCl)	K_1 12·06±0·04, K_2 2·61±0·02, K_3 1·60±0·05	66I
Ba^{2+}	gl	20	0·1 (KCl)	K_1 5·54±0·04	66I
Ca^{2+}	gl	20	0·1 (KCl)	K_1 8·24±0·01	66I
Mg^{2+}	gl	20	0·1 (KCl)	K_1 6·15±0·03	66I
Sr^{2+}	gl	20	0·1 (KCl)	K_1 6·04±0·03	66I
Zn^{2+}	gl	20	0·1 (KCl)	K_1 12·28±0·02	66I

* Geometric isomerism of ligand not specified.

66I H. M. N. H. Irving and M. G. Miles, *J. Chem. Soc.* (*A*), 1966, 1268

932 $C_{12}H_{20}O_8N_2$ (Tetramethylenedinitrilo)tetra-acetic acid H_4L

Metal	*Method*	*Temp*	*Medium*	*Log of equilibrium constant, remarks*	*Ref*
H^+	gl	20	0·1 (KNO_3)	K_1 10·66, K_2 9·05, K_3 2·45, K_4 1·90	64A, 64
	cal	20	0·1 (KNO_3)	$\Delta H_1°$ −6·68, $\Delta S_1°$ 26·0, $\Delta H_2°$ −5·81, $\Delta S_2°$ 21·6	64A
	gl	20	1 (KNO_3)	K_1 10·35, K_2 8·96, K_3 2·4, K_4 2·4	67A
Ba^{2+}	gl	20	0·1 (KNO_3)	K_1 3·77	64L
				$K(Ba^{2+} + HL^{3-} \rightleftharpoons BaHL^-)$ 2·58	
Ca^{2+}	gl	20	0·1 (KNO_3)	$K(Ca^{2+} + CaL^{2-} \rightleftharpoons Ca_2L)$ 1·42	64A
	cal	20	0·1 (KNO_3)	$\Delta H_1°$ −0·9 (?), $\Delta S_1°$ 29·7 (?)	64A
	gl	20	0·1 (KNO_3)	K_1 5·66,	64L
				$K(Ca^{2+} + HL^{3-} \rightleftharpoons CaHL^-)$ 3·65	
Cd^{2+}	gl	20	0·1 (KNO_3)	$K(Cd^{2+} + CdL^{2-} \rightleftharpoons Cd_2L)$ 2·20	64A
	cal	20	0·1 (KNO_3)	$\Delta H_1°$ −2·88, $\Delta S_1°$ 45·2	64A
	gl	20	0·1 (KNO_3)	K_1 12·02,	64L
				$K(Cd^{2+} + HL^{3-} \rightleftharpoons CdHL^-)$ 6·79	
Co^{2+}	cal	20	0·1 (KNO_3)	$\Delta H_1°$ −1·6, $\Delta S_1°$ 66·3	64A
	gl	20	0·1 (KNO_3)	K_1 15·69	64L
Cu^{2+}	gl	20	0·1 (KNO_3)	$K(Cu^{2+} + HL^{3-} \rightleftharpoons CuHL^-)$ 10·86	64A
	cal	20	0·1 (KNO_3)	$\Delta H_1°$ −6·52, $\Delta S_1°$ 57·0	64A
	gl	20	0·1 (KNO_3)	K_1 17·33	64L
Fe^{2+}	gl	20	0·1 (KNO_3)	K_1 13·27	64L
Hg^{2+}	cal	20	0·1 (KNO_3)	$\Delta H_1°$ −19·1, $\Delta S_1°$ 30·8	64A
	Hg	20	0·1 (KNO_3)	K_1 20·99	64L
La^{3+}	gl	20	0·1 (KNO_3)	$K(La^{3+} + HL^{3-} \rightleftharpoons LaHL)$ 6·17	64A
	cal	20	0·1 (KNO_3)	$\Delta H_1°$ 1·88, $\Delta S_1°$ 48·2	64A
	gl	20	0·1 (KNO_3)	K_1 9·13	64L

cont

$C_{12}H_{20}O_8N_2$ (contd) 932

Metal	Method	Temp	Medium	Log of equilibrium constant, remarks	Ref
Mg^{2+}	gl	20	0·1 (KNO_3)	$K(Mg^{2+}+HL^{3-} \rightleftharpoons MgHL^-)$ 3·50	64A
	cal	20	0·1 (KNO_3)	$\Delta H_1°$ 8·5, $\Delta S_1°$ 54·0	64A
	gl	20	0·1 (KNO_3)	K_1 6·23	64L
Mn^{2+}	gl	20	0·1 (KNO_3)	$K(Mn^{2+}+MnL^{2-} \rightleftharpoons Mn_2L)$ 1·82	64A
	cal	20	0·1 (KNO_3)	$\Delta H_1°$ 3·41, $\Delta S_1°$ 55·2	64A
	gl	20	0·1 (KNO_3)	K_1 9·53, $K(Mn^{2+}+HL^{3-} \rightleftharpoons MnHL^-)$ 5·44	64L
Ni^{2+}	cal	20	0·1 (KNO_3)	$\Delta H_1°$ −6·95, $\Delta S_1°$ 55·5	64A
	gl	20	0·1 (KNO_3)	K_1 17·36, $K(Ni^{2+}+HL^{3-} \rightleftharpoons NiHL^-)$ 9·05	64L
Pb^{2+}	gl	20	0·1 (KNO_3)	$K(Pb^{2+}+PbL^{2-} \rightleftharpoons Pb_2L)$ 5·41	64A
	cal	20	0·1 (KNO_3)	$\Delta H_1°$ −4·85, $\Delta S_1°$ 31·6	64A
	gl	20	0·1 (KNO_3)	K_1 10·53, $K(Pb^{2+}+HL^{3-} \rightleftharpoons PbHL^-)$ 7·50	64L
Sr^{2+}	gl	20	0·1 (KNO_3)	K_1 4·42, $K(Sr^{2+}+HL^{3-} \rightleftharpoons SrHL^-)$ 2·82	64L
Zn^{2+}	cal	20	0·1 (KNO_3)	$\Delta H_1°$ −3·48, $\Delta S_1°$ 56·8	64A
	gl	20	0·1 (KNO_3)	K_1 15·04, $K(Zn^{2+}+HL^{3-} \rightleftharpoons ZnHL^-)$ 7·42	64L

64A G. Anderegg, *Helv. Chim. Acta*, 1964, **47**, 1801
64L F. L'Eplattenier and G. Anderegg, *Helv. Chim. Acta*, 1964, **47**, 1792
67A G. Anderegg, *Helv. Chim. Acta*, 1967, **50**, 2333

$C_{12}H_{20}O_8N_2$ (DL-2,3-Butylenedinitrilo)tetra-acetic acid H_4L 932.1

Metal	Method	Temp	Medium	Log of equilibrium constant, remarks	Ref
H^+	?	20	0·1	K_1 11·5, K_2 6·13, K_3 3·66, K_4 2·65	62S
	gl	20	0·1 KCl	K_1 11·74, K_2 6·12, K_3 3·54, K_4 2·41	63M, 64M, 65J
	gl	20	0·1 (KCl)	K_1 11·61, K_2 6·05, K_3 3·44, K_4 2·38	66I
Al^{3+}	oth	20	0·1 (KNO_3)	K_1 18·5, electrophoresis	65J
Ba^{2+}	gl	20	0·1 KCl	K_1 8·53	63M
	oth	20	0·1 (KNO_3)	K_1 ~ 11, electrophoresis	65J
	gl	20	0·1 (KCl)	K_1 8·49	66I
Ca^{2+}	gl	20	0·1 KCl	K_1 12·34	63M
	gl	20	0·1 (KCl)	K_1 11·49	66I
Cd^{2+}	gl	20	0·10 (KNO_3)	K_1 18·71±0·06	64M
	oth	20	0·1 (KNO_3)	K_1 18·51, electrophoresis	65J
Ce^{3+}	oth	20	0·1 (KNO_3)	K_1 18·5, electrophoresis	65J
Co^{2+}	pol	20	0·10 (KNO_3)	K_1 18·89±0·09	64M
	oth	20	0·1 (KNO_3)	K_1 19, electrophoresis	65J
Cu^{2+}	pol	20	0·10 (KNO_3)	K_1 21·4±0·2	64M
	oth	20	0·1 (KNO_3)	K_1 22, electrophoresis	65J
Hg^{2+}	pol	20	0·1 KNO_3	K_1 21·51	62S
La^{3+}	oth	20	0·1 (KNO_3)	K_1 18·5, electrophoresis	65J
Mg^{2+}	gl	20	0·1 KCl	K_1 11·44	63M
	gl	20	0·1 (KCl)	K_1 11·33	66I
Mn^{2+}	pol	20	0·10 (KNO_3)	K_1 16·3±0·1	64M
	oth	20	0·1 (KNO_3)	K_1 17·5, electrophoresis	65J
Ni^{2+}	oth	20	0·1 (KNO_3)	K_1 22·5, electrophoresis	65J
Pb^{2+}	pol	20	0·10 (KNO_3)	K_1 19·4±0·1	64M
	oth	20	0·1 (KNO_3)	K_1 19·5, electrophoresis	65J

continued overleaf

932.1 $C_{12}H_{20}O_8N_2$ (contd)

Metal	Method	Temp	Medium	Log of equilibrium constant, remarks	Ref
Sr^{2+}	gl	20	0·1 KCl	K_1 10·20	63M
	oth	20	0·1 (KNO_3)	K_1 11, electrophoresis	65J
	gl	20	0·1 (KCl)	K_1 10·10	66I
Zn^{2+}	pol	20	0·10 (KNO_3)	K_1 18·85±0·09	64M
	oth	20	0·1 (KNO_3)	K_1 18·5, electrophoresis	65J

63S S. Stankoviansky and J. Konigstein, *Coll. Czech. Chem. Comm.*, 1962, **27**, 1997
63M J. Majer and E. Dvorakova, *Chem. Zvesti*, 1963, **17**, 402
64M J. Majer, V. Novak, and M. Svicekova, *Chem. Zvesti*, 1964, **18**, 481
65J V. Jokl and J. Majer, *Chem. Zvesti*, 1965, **19**, 281
66I H. M. N H. Irving and R. Parkins, *J. Inorg. Nuclear Chem.*, 1966, **28**, 1629

932.2 $C_{12}H_{20}O_8N_2$ (*meso*-2,3-Butylenedinitrilo)tetra-acetic acid H_4L

Metal	Method	Temp	Medium	Log of equilibrium constant, remarks	Ref
H^+	?	20	0·1	K_1 11·1, K_2 6·26, K_3 2·31, K_4 1·77	62S
	gl	20	0·1 KCl	K_1 11·25, K_2 6·27, K_3 2·53, K_4 1·80	63M
	gl	20	0·1 (KCl)	K_1 11·22, K_2 6·29, K_3 2·69, K_4 1·76	66I
Al^{3+}	oth	20	0·1 (KNO_3)	K_1 16·5, electrophoresis	65J
Ba^{2+}	gl	20	0·1 KCl	K_1 6·53, $K(Ba^{2+}+HL^{3-} \rightleftharpoons BaHL^-)$ 1·83	63M
	oth	20	0·1 (KNO_3)	K_1 7, electrophoresis	65J
	gl	20	0·1 (KCl)	K_1 6·45	66I
Ca^{2+}	gl	20	0·1 KCl	K_1 9·67, $K(Ca^{2+}+HL^{3-} \rightleftharpoons CaHL^-)$ 2·76	63M
	oth	20	0·1 (KNO_3)	K_1 11, electrophoresis	65J
	gl	20	0·1 (KCl)	K_1 9·60	66I
Cd^{2+}	gl	20	0·10 (KNO_3)	K_1 16·77±0·04	64M
	oth	20	0·1 (KNO_3)	K_1 18, electrophoresis	65J
Ce^{3+}	oth	20	0·1 (KNO_3)	K_1 16, electrophoresis	65J
Co^{2+}	pol	20	0·10 (KNO_3)	K_1 17·09±0·09	64M
	oth	20	0·1 (KNO_3)	K_1 17·5, electrophoresis	65J
Cu^{2+}	pol	20	0·10 (KNO_3)	K_1 19·8±0·2	64M
	oth	20	0·1 (KNO_3)	K_1 20, electrophoresis	65J
Hg^{2+}	pol	20	0·1 KNO_3	K_1 20·33	62S
La^{3+}	oth	20	0·1 (KNO_3)	K_1 15·5, electrophoresis	65J
Mg^{2+}	gl	20	0·1 KCl	K_1 8·85, $K(Mg^{2+}+HL^{3-} \rightleftharpoons MgHL^-)$ 2·07	
	oth	20	0·1 (KNO_3)	K_1 10·5, electrophoresis	65J
	gl	20	0·1 (KCl)	K_1 8·84	66I
Mn^{2+}	pol	20	0·10 (KNO_3)	K_1 14·2±0·1	64M
	oth	20	0·1 (KNO_3)	K_1 15, electrophoresis	65J
Ni^{2+}	oth	20	0·1 (KNO_3)	K_1 21, electrophoresis	65J
Pb^{2+}	pol	20	0·10 (KNO_3)	K_1 16·83±0·07	64M
	oth	20	0·1 (KNO_3)	K_1 17·5, electrophoresis	65J
Sr^{2+}	gl	20	0·1 KCl	K_1 7·65, $K(Sr^{2+}+HL^{3-} \rightleftharpoons SrHL^-)$ 1·81	63M
	oth	20	0·1 (KNO_3)	K_1 7, electrophoresis	65J
	gl	20	0·1 (KCl)	K_1 7·62	66I
Zn^{2+}	pol	20	0·10 (KNO_3)	K_1 17·35±0·10	64M
	oth	20	0·1 (KNO_3)	K_1 18, electrophoresis	65J

63S S. Stankoviansky and J. Konigstein, *Coll. Czech. Chem. Comm.*, 1962, **27**, 1997
63M J. Majer and E. Dvorakova, *Chem. Zvesti*, 1963, **17**, 402
64M J. Majer, V. Novak, and M. Svicekova, *Chem. Zvesti*, 1964, **18**, 481
65J V. Jokl and J. Majer, *Chem. Zvesti*, 1965, **19**, 281
66I H. M. N. H. Irving and R. Parkins, *J. Inorg. Nuclear Chem.*, 1966, **28**, 1629

$C_{12}H_{20}O_9N_2$ [(2,2′-Oxydiethylene)dinitrilo]tetra-acetic acid) (EEDTA) H_4L 933

Metal	Method	Temp	Medium	Log of equilibrium constant, remarks	Ref
H^+	gl	20	0·1 (KNO_3)	K_1 9·47, K_2 8·84, K_3 2·76, K_4 1·8	64A
	cal	20	0·1 (KNO_3)	$\Delta H_1°$ −6·23, $\Delta S_1°$ 22·1, $\Delta H_2°$ −7·25, $\Delta S_2°$ 15·7	64A
	gl	20	0·1 [$(CH_3)_4NCl$]	K_1 9·75, K_2 8·98	67A
	H	20	1 [$(CH_3)_4NCl$]	K_1 9·49, K_2 8·77, K_3 2·7, K_4 2·3	67A
	gl	20	1 (KNO_3)	K_1 9·16, K_2 8·67, K_3 2·5, K_4 2·4	67A
	gl	20	1 ($NaClO_4$)	K_1 8·58, K_2 8·44, K_3 2·44, K_4 2·4	67A
	H	20	0·1	K_1 9·49, K_2 8·82, K_3 2·67, K_4 1·9	67T
Ba^{2+}	cal	25	0·1 KNO_3	ΔH_1 −6·5, ΔS_1 15	65W
Ca^{2+}	gl	20	0·1 (KNO_3)	K_1 10·0, $K(Ca^{2+}+HL^{3-} \rightleftharpoons CaHL^-)$ 4·9	64A
	cal	20	0·1 (KNO_3)	$\Delta H_1°$ −6·85, $\Delta S_1°$ 22·4	64A
	cal	25	0·1 KNO_3	ΔH_1 −6·4, ΔS_1 24	65W
Cd^{2+}	gl	20	0·1 (KNO_3)	K_1 16·2, $K(Cd^{2+}+HL^{3-} \rightleftharpoons CdHL^-)$ 9·9	64A
	cal	20	0·1 (KNO_3)	$\Delta H_1°$ −9·42, $\Delta S_1°$ 42·0	64A
	cal	25	0·1 KNO_3	ΔH_1 −9·7, ΔS_1 41	65W
Ce^{3+}	ix	22	0·5 NH_4^+	K_1 17·87±0·02	62T
Co^{2+}	gl	20	0·1 (KNO_3)	K_1 15·27, $K(Co^{2+}+HL^{3-} \rightleftharpoons CoHL^-)$ 8·55	64A
	cal	20	0·1 (KNO_3)	$\Delta H_1°$ −6·35, $\Delta S_1°$ 48·2	64A
	cal	25	0·1 KNO_3	ΔH_1 −6·6, ΔS_1 45	65W
Cu^{2+}	gl	20	0·1 (KNO_3)	K_1 18·1, $K(Cu^{2+}+HL^{3-} \rightleftharpoons CuHL^-)$ 12·85	64A
	cal	20	0·1 (KNO_3)	$\Delta H_1°$ −9·82, $\Delta S_1°$ 49·0	64A
	cal	25	0·1 KNO_3	ΔH_1 −9·7, ΔS_1 48	65W
Fe^{2+}	gl	20	0·1 (KNO_3)	K_1 14·3, $K(Fe^{2+}+HL^{3-} \rightleftharpoons FeHL^-)$ 8·2	64A
	cal	25	0·1 KNO_3	ΔH_1 −6·4, ΔS_1 46	65W
Fe^{3+}	sp	18—20	?	K_1 23·03	65Z
	red	20	0·1 ($NaClO_4$)	K_1 24·7	67B
Ga^{3+}	red	20	0·1 ($NaClO_4$)	K_1 21·0, $K(GaL^-+H^+ \rightleftharpoons GaHL)$ 1·54, $K[GaL^-+2OH^- \rightleftharpoons GaL(OH)_2^{3-}]$ 11·9	67B
Hg^{2+}	gl	20	0·1 (KNO_3)	K_1 23·09, $K(Hg^{2+}+HL^{3-} \rightleftharpoons HgHL^-)$ 16·1	64A
	cal	20	0·1 (KNO_3)	$\Delta H_1°$ −20·5, $\Delta S_1°$ 35·7	64A
	cal	25	0·1 KNO_3	ΔH_1 −19·5, ΔS_1 40	65W
In^{3+}	sp	18—20	?	K_1 22·67, $K(FeL^-+In^{3+} \rightleftharpoons InL^-+Fe^{3+})$ 0·37	65Z
	red	20	0·1 ($NaClO_4$)	K_1 25·5, $K(InL^-+H^+ \rightleftharpoons InHL)$ 2·1, $K(InL^-+OH^- \rightleftharpoons InLOH^{2-})$ 3·90	67B
La^{3+}	gl	20	0·1 (KNO_3)	K_1 16·6	64A
	cal	20	0·1 (KNO_3)	$\Delta H_1°$ −3·35, $\Delta S_1°$ 64·5	64A
Mg^{2+}	gl	20	0·1 (KNO_3)	K_1 8·32, $K(Mg^{2+}+HL^{3-} \rightleftharpoons MgHL^-)$ 3·8	64A
	cal	20	0·1 (KNO_3)	$\Delta H_1°$ 3·51, $\Delta S_1°$ 50·0	64A
	cal	25	0·1 KNO_3	ΔH_1 3·6, ΔS_1 50	65W
Mn^{2+}	gl	20	0·1 (KNO_3)	K_1 13·76	64A
	cal	20	0·1 (KNO_3)	$\Delta H_1°$ −5·9, $\Delta S_1°$ 46	64A
	cal	25	0·1 KNO_3	ΔH_1 −5·6, ΔS_1 41	65W
Nd^{3+}	sp	18—20	?	K_1 18·33, $K(Nd^{3+}+HL^{3-} \rightleftharpoons NdHL)$ 10·77, $K(Nd^{3+}+H_2L^{2-} \rightleftharpoons NdH_2L^+)$ 3·21	65Z

continued overleaf

933 $C_{12}H_{20}O_9N_2$ (contd)

Metal	*Method*	*Temp*	*Medium*	*Log of equilibrium constant, remarks*	*Ref*
Ni^{2+}	gl	20	0·1 (KNO_3)	K_1 15·07, $K(Ni^{2+}+HL^{3-} \rightleftharpoons NiHL^-)$ 8·9	64A
	cal	20	0·1 (KNO_3)	$\Delta H_1°$ −4·75, $\Delta S_1°$ 52·8	64A
	cal	25	0·1 KNO_3	ΔH_1 −5·1, ΔS_1 50	65W
Pb^{2+}	gl	20	0·1 (KNO_3)	K_1 15·03, $K(Pb^{2+}+HL^{3-} \rightleftharpoons PbHL^-)$ 9·4	64A
	cal	20	0·1 (KNO_3)	$\Delta H_1°$ −13·15, $\Delta S_1°$ 23·9	64A
	cal	25	0·1 KNO_3	ΔH_1 −12·2, ΔS_1 25	65W
Sr^{2+}	cal	25	0·1 KNO_3	ΔH_1 −8·1, ΔS_1 12	65W
Th^{4+}	red	20	0·1 ($NaClO_4$)	K_1 24·9, $K(ThL+H^+ \rightleftharpoons ThHL^+)$ 2·09, $K(ThL+OH^- \rightleftharpoons ThLOH^-)$ 7·44	67B
Tl^+	gl	20	0·1 (KNO_3)	K_1 4·47, $K(Tl^++HL^{3-} \rightleftharpoons TlHL^{2-})$ 4·0	67Aa
Tl^{3+}	sp	18—22	?	K_1 23·08, $K(FeL^-+Tl^{3+} \rightleftharpoons TlY^-+Fe^{3+})$ 0·51	67K
	red	20	1 ($HClO_4$)	K_1 33·4	67Aa
			1 ($NaClO_4$)	K_1 32·8	
Y^{3+}	ix	20	0·1 NH_4^+	K_1 17·92±0·04, pH 3·0; K_1 17·77±0·03, pH 3·5	62S
			0·5 NH_4^+	K_1 17·66±0·02, pH 2·6	
Zn^{2+}	gl	20	0·1 (KNO_3)	K_1 15·3, $K(Zn^{2+}+HL^{3-} \rightleftharpoons ZnHL^-)$ 8·5	64A
	cal	20	0·1 (KNO_3)	$\Delta H_1°$ −5·99, $\Delta S_1°$ 49·6	64A
	cal	25	0·1 KNO_3	ΔH_1 −7·6, ΔS_1 44	65W
Zr^{4+}	oth	20	0·1	K_1 24·72±0·02, chromatography	67T

62S M. M. Senyavin and L. I. Tikhonova, *Russ. J. Inorg. Chem.*, 1962, **7**, 562 (1095)
62T L. I. Tikhonova, *Russ. J. Inorg. Chem.*, 1962, **7**, 424 (822)
64A G. Anderegg, *Helv. Chim. Acta*, 1964, **47**, 1801
65W D. L. Wright, J. H. Holloway, and C. N. Reilley, *Analyt. Chem.*, 1965, **37**, 884
65Z N. M. Zhirnova, K. V. Astakhov, and S. A. Barkov, *Russ. J. Phys. Chem.*, 1965, **39**, 1489 (2791)
67A G. Anderegg, *Helv. Chim. Acta*, 1967, **50**, 2333
67Aa G. Anderegg and E. Bottari, *Helv. Chim. Acta*, 1967, **50**, 2341
67B E. Bottari and G. Anderegg, *Helv. Chim. Acta*, 1967, **50**, 2349
67K V. I. Kornev, K. V. Astakhov, and V. I. Rybina, *Russ. J. Inorg. Chem.*, 1967, **12**, 73 (148)
67T L. I. Tikhonova, *Russ. J. Inorg. Chem.*, 1967, **12**, 494 (939)

933.1 $C_{12}H_{20}O_{10}N_2$ (2,3-Dihydroxytetramethylenedinitrilo)tetra-acetic acid H_6L

Metal	*Method*	*Temp*	*Medium*	*Log of equilibrium constant, remarks*	*Ref*
H^+	gl	~ 20	0·1 KNO_3	K_1 11·79, K_2 11·24, K_3 9·26, K_4 7·52, K_5 3·18, K_6 2·2	67D
Ba^{2+}	gl	~ 20	0·1 KNO_3	K_1 3·61, $KBa^{2+}+HL^{5-} \rightleftharpoons BaHL^{3-})$ 2·94, $K(BaL^{4-}+Ba^{2+} \rightleftharpoons Ba_2L^{2-})$ 2·26	67D
Ca^{2+}	gl	~ 20	0·1 KNO_3	K_1 5·73, $K(Ca^{2+}+HL^{5-} \rightleftharpoons CaHL^{3-})$ 4·15, $K(CaL^{4-}+Ca^{2+} \rightleftharpoons Ca_2L^{2-})$ 3·69	67D

conti.

$C_{12}H_{20}O_{10}N_2$ (contd) 933.1

Metal	Method	Temp	Medium	Log of equilibrium constant, remarks	Ref
Mg^{2+}	gl	~ 20	0·1 KNO_3	K_1 4·11, $K(Mg^{2+}+HL^{5-} \rightleftharpoons MgHL^{3-})$ 3·3 $K(MgL^{4-}+Mg^{2+} \rightleftharpoons Mg_2L^{2-})$ 2·95	67D
Sr^{2+}	gl	~ 20	0·1 KNO_3	K_1 3·93, $K(Sr^{2+}+HL^{5-} \rightleftharpoons SrHL^{3-})$ 3·18 $K(SrL^{4-}+Sr^{2+} \rightleftharpoons Sr_2L^{2-})$ 2·85	67D

67D N. M. Dyatlova, I. A. Siliverstova, O. I. Samoilova, and V. G. Yashunskii, *Proc. Acad. Sci.* (*U.S.S.R.*), 1967, **172**, 4, (94)

$C_{12}H_{24}O_2N_2$ *NN′*-Bis(2′-hydroxypent-3′-enyl)-1,2-diaminoethane L 934.1

Metal	Method	Temp	Medium	Log of equilibrium constant, remarks	Ref
Nd^{3+}	sp	18—20	0·05	K_1 1·50	61A

61A K. V. Astakhov, V. B. Verenikin, and V. I. Zimin, *Russ. J. Inorg. Chem.*, 1961, **6**, 1062 (2077)

$C_{12}H_{26}O_2N_4$ *NN′*-Diglycyl-1,8-diamino-octane L 935.1

Metal	Method	Temp	Medium	Log of equilibrium constant, remarks	Ref
H^+	gl	23	0·5 (KCl)	K_1 8·56, K_2 7·90	67Z
Cu^{2+}	gl	23	0·5 (KCl)	K_1 8·97 $K(CuL^{2+}+H^+ \rightleftharpoons CuHL^{3+})$ 5·15	67Z

67Z A. Zuberbuhler and S. Fallab, *Helv. Chim. Acta*, 1967, **50**, 889

$C_{12}H_{27}O_4P$ Tributyl phosphate L 935.2

Metal	Method	Temp	Medium	Log of equilibrium constant, remarks	Ref
Hf^{3+}	dis	20	?	K_1 −0·12	62P
La^{3+}	dis	?	1·3 HNO_3	β_3 ~ 0	61S
Nd^{3+}	dis	?	1·2 HNO_3	β_3 ~ 0	61S
Y^{3+}	dis	?	3 HNO_3	β_3 ~ 0	61S
Yb^{3+}	dis	?	1 HNO_3	β_3 ~ 0	61S

61S Z. A. Sheka and E. E. Kriss, *Russ. J. Inorg. Chem.*, 1961, **6**, 984 (1930)
62P N. P. Prokhorova and N. E. Brezhneva, *Russ. J. Inorg. Chem.*, 1962, **7**, 953 (1846)

935.3 $C_{12}H_8O_3N_2S$ 1,10-Phenanthroline-5-sulphonic acid HL

Metal	*Method*	*Temp*	*Medium*	*Log of equilibrium constant, remarks*	*Ref*
Fe^{2+}	sp	?	5—64% N_2H_4	β_3 (5%) 11·7, (10%) 11·0, (25%) 9·9, (50%) 9·1, (64%) 8·4	66B

66B D. Bauer, *Bull. Soc. chim. France*, 1966, 2631

935.4 $C_{12}H_9ONS_2$ 5-(3′-Phenylallylidene)-2-thioxo-1,3-thiazolidin-4-one (5-(cinnamylidene)rhodanine) HL

Metal	*Method*	*Temp*	*Medium*	*Log of equilibrium constant, remarks*	*Ref*
H^+	dis	20	0·1 ($NaClO_4$)	K_1 7·68	65N
Ag^+	dis	20	0·1 ($NaClO_4$)	K_1 9·08	65N

65N O. Navratil and J. Kotas, *Coll. Czech. Chem. Comm.*, 1965, 30, 2736

937.1 $C_{12}H_{10}ON_3Cl$ 1-(2′-chlorophenyl)-3-hydroxy-3-phenyltriazine HL

Metal	*Method*	*Temp*	*Medium*	*Log of equilibrium constant, remarks*	*Ref*
H^+	sp	?	?	K_1 10·52	64P
Cu^{2+}	gl	25	0·1 KCl 70% dioxan	K_1 10·35, K_2 9·95	64P
Mn^{2+}	gl	25	0·1 KCl 70% dioxan	K_1 5·36, K_2 4·42	64P
Ni^{2+}	gl	25	0·1 KCl 70% dioxan	K_1 7·70, K_2 6·41	64P
Pd^{2+}	gl	25	0·1 KCl 70% dioxan	K_1 10·43, K_2 10·00	64P
Zn^{2+}	gl	25	0·1 KCl 70% dioxan	K_1 7·26, K_2 5·91	64P

64P D. N. Purohit and N. C. Sogani, *Z. anorg. Chem.*, 1964, 331, 220

$C_{12}H_{10}ON_3Cl$ 1-(4′-Chlorophenyl)-3-hydroxy-3-phenyltriazene HL 937.2

Metal	Method	Temp	Medium	Log of equilibrium constant, remarks	Ref
H^+	sp	?	?	K_1 10·72	64P
Cu^{2+}	gl	25	0·1 KCl 70% dioxan	K_1 10·67, K_2 9·98	64P
Mn^{2+}	gl	25	0·1 KCl 70% dioxan	K_1 5·83, K_2 4·67	64P
Ni^{2+}	gl	25	0·1 KCl 70% dioxan	K_1 8·25, K_2 6·33	64P
Pd^{2+}	gl	25	0·1 KCl 70% dioxan	K_1 10·70, K_2 10·25	64P
Zn^{2+}	gl	25	0·1 KCl 70% dioxan	K_1 7·41, K_2 6·12	64P

64P D. N. Purohit and N. C. Sogani, *Bull. Chem. Soc. Japan*, 1964, 37, 476

$C_{12}H_{10}ON_3Cl$ 3-(4′-Chlorophenyl)-3-hydroxy-1-phenyltriazene HL 937.3

Metal	Method	Temp	Medium	Log of equilibrium constant, remarks	Ref
H^+	sp	?	?	K_1 10·65	65P
Cu^{2+}	gl	25	0·1 KCl 70% dioxan	K_1 10·55, K_2 9·72	65P

65P D. N. Purohit and N. C. Sogani, *Z. Naturforsch.*, 1965, 20b, 206

$C_{12}H_{10}ON_3Br$ 1-(4′-Bromophenyl)-3-hydroxy-3-phenyltriazene HL 937.4

Metal	Method	Temp	Medium	Log of equilibrium constant, remarks	Ref
H^+	sp	?	?	K_1 10·86	65P
Cu^{2+}	gl	25	0·1 KCl 70% dioxan	K_1 10·85, β_2 21·04	65P
Mn^{2+}	gl	25	0·1 KCl 70% dioxan	K_1 5·90, β_2 10·85	65P
Ni^{2+}	gl	25	0·1 KCl 70% dioxan	K_1 8·69, β_2 15·44	65P
Pd^{2+}	gl	25	0·1 KCl 70% dioxan	K_1 10·86, β_2 21·30	65P
Zn^{2+}	gl	25	0·1 KCl 70% dioxan	K_1 7·64, β_2 13·94	65P

65P D. N. Purohit and N. C. Sogani, *J. & Proc. Inst. Chemists (India)*, 1965, 37, 212, *Z. Naturforsch.*, 1965, 20b, 206

937.5 $C_{12}H_{11}ONS$ 2-Mercapto-*N*-(2′-naphthyl)acetamide (thionalide) HL

Metal	Method	Temp	Medium	Log of equilibrium constant, remarks	Ref
H^+	gl	20	0·1 ($NaClO_4$) 75% dioxan	K_1 10·20±0·05	68B
Co^{2+}	gl	20	0·1 ($NaClO_4$) 75% dioxan	K_1 7·3±0·1, β_2 14·1±0·1, β_3 20·1±0·3	68B
Cu^{2+}	gl	20	0·1 ($NaClO_4$) 75% dioxan	K_1 10·3±0·2, β_2 20·0±0·1, β_3 28·4±0·3	68B
Fe^{2+}	gl	20	0·1 ($NaClO_4$) 75% dioxan	K_1 7·0±0·2, β_2 13·7±0·1, β_3 19·2±0·3	68B
Mn^{2+}	gl	20	0·1 ($NaClO_4$) 75% dioxan	K_1 4·4±0·3, β_2 8·8±0·1	68B
Ni^{2+}	gl	20	0·1 ($NaClO_4$) 75% dioxan	K_1 6·9±0·3, β_2 13·4±0·1	68B
Zn^{2+}	gl	20	0·1 ($NaClO_4$) 75% dioxan	K_1 7·8±0·1, β_2 14·9±0·1	68B

68B K. Burger, L. Korecz, and A. Toth, *Acta Chim. Acad. Sci. Hung.*, 1968, **55**, 1

937.6 $C_{12}H_{11}O_4N_3S$ 3-Hydroxy-3-phenyl-1(4′-sulphophenyl)triazine H_2L

Metal	Method	Temp	Medium	Log of equilibrium constant, remarks	Ref
H^+	sp	?	?	K_1 9·99	64P
Co^{2+}	sp	25	?	K (?) 12·59	63D
Cu^{2+}	gl	25	0·1 KCl 70% dioxan	K_1 9·55, K_2 9·22	64P
	sp	25	~ 0·003	K (?) 8·74, acetate buffer	59D
Fe^{3+}	sp	20	?	K (?) 11·62	60M
Mn^{2+}	gl	25	0·1 KCl 70% dioxan	K_1 4·83, K_2 3·38	64P
Mo^{VI}	sp	25	?	K (?) 12·87	58D
Ni^{2+}	sp	25	?	K (?) 10·45	63D
	gl	25	0·1 KCl 70% dioxan	K_1 7·96, K_2 5·98	64P
Pd^{2+}	sp	25	?	K (?) 11·52, acetate buffer	58D
	gl	25	0·1 KCl 70% dioxan	K_1 9·71, K_2 9·32	64P
Zn^{2+}	gl	25	0·1 KCl 70% dioxan	K_1 6·52, K_2 5·33	64P

58D U. C. Durgapal and N. C. Sogani, *J. Indian Chem. Soc.*, 1958, **35**, 542, 842
59D U. C. Durgapal and N. C. Sogani, *J. Indian Chem. Soc.*, 1959, **36**, 263
60M R. N. Mathur and N. C. Sogani, *J. Indian Chem. Soc.*, 1960, **37**, 117
63D S. M. Dugar, D. N. Purohit, and N. C. Sogani, *J. Indian Chem. Soc.*, 1963, **40**, 213, 667
64P D. N. Purohit and N. C. Sogani, *Z. analyt. Chem.*, 1964, **203**, 97

$C_{12}H_{12}ON_2S_2$ [5-4′(Dimethylamino)benzylidene]-2-thioxo-1,3-thiazolidin-4-one [5-(4′-dimethylaminobenzylidene)rhodanine] HL 937.7

Metal	Method	Temp	Medium	Log of equilibrium constant, remarks	Ref
H^+	dis	20	0·1 ($NaClO_4$)	K_1 8·20	65N
Ag^+	dis	20	0·1 ($NaClO_4$)	K_1 9·15	65N
Cu^{2+}	dis	20	0·1 ($NaClO_4$)	K_1 6·08	65N

65N O. Navratil and J. Kotas, *Coll. Czech. Chem. Comm.*, 1965, **30**, 2736

$C_{12}H_{12}O_6NCl$ (α-Carboxy-4′chlorobenzyl)iminodiacetic acid H_3L 937.8

Metal	Method	Temp	Medium	Log of equilibrium constant, remarks	Ref
H^+	gl	20	0·1 (KCl)	K_1 8·88±0·02, K_2 2·31±0·02, K_3 1·55±0·05	66I
Ba^{2+}	gl	20	0·1 (KCl)	K_1 4·21±0·01	66I
Ca^{2+}	gl	20	0·1 (KCl)	K_1 6·05±0·02	66I
Mg^{2+}	gl	20	0·1 (KCl)	K_1 4·45±0·02	66I
Sr^{2+}	gl	20	0·1 (KCl)	K_1 4·35±0·02	66I
Zn^{2+}	gl	20	0·1 (KCl)	K_1 9·51±0·02	66I

66I H. M. N. H. Irving and M. G. Miles, *J. Chem. Soc.* (*A*), 1966, 727

$C_{12}H_{13}O_5N_2Br$ [2′-(4″-Bromoanilino)-2′-oxoethyl]iminodiacetic acid H_2L 937.9

Metal	Method	Temp	Medium	Log of equilibrium constant, remarks	Ref
H^+	gl	30	0·1 KCl	K_1 5·90±0·03, K_2 2·20, K_3 1·60	66S
Ca^{2+}	gl	30	0·1 KCl	K_1 3·30±0·04	66S
Mg^{2+}	gl	30	0·1 KCl	K_1 2·06±0·05	66S
Ni^{2+}	gl	30	0·1 KCl	K_1 6·04±0·04, K_2 5·07±0·04	66S

66S G. Shtacher, *J. Inorg. Nuclear Chem.*, 1966, **28**, 845

$C_{12}H_{14}O_4ClAs$ *As*-(4′-Chlorophenyl)-3,3′-arsinodipropanoic acid H_2L 937.10

Metal	Method	Temp	Medium	Log of equilibrium constant, remarks	Ref
H^+	gl	20	0·1 (KNO_3)	K_1 5·08, K_2 4·17	64P
Ag^+	Ag	20	0·1 (KNO_3)	K_1 5·00, $K(Ag^+ + HL^- \rightleftharpoons AgHL)$ 3·98	64P
Cu^{2+}	gl	20	0·1 (KNO_3)	K_1 1·5	64P

64P L. D. Pettit and H. M. N. H. Irving, *J. Chem. Soc.*, 1964, 5336

939 $C_{12}H_{20}O_8N_2S$ [(2,2′-Thiodiethylene)dinitrilo]tetra-acetic acid H_4L

Metal	Method	Temp	Medium	Log of equilibrium constant, remarks	Ref
H^+	gl	20	0·1 (KNO_3)	K_1 9·42, K_2 8·47, K_3 2·52, K_4 1·8	64A
	cal	20	0·1 (KNO_3)	$\Delta H_1°$ −6·69, $\Delta S_1°$ 20·3 $\Delta H_2°$ −6·59, $\Delta S_2°$ 16·3	64A
	gl	20	0·1 [$(CH_3)_4NCl$]	K_1 9·61, K_2 8·52	67A
	H	20	1 [$(CH_3)_4NCl$]	K_1 9·29, K_2 8·29, K_3 2·5, K_4 2·1	67A
	gl	20	1 ($NaClO_4$)	K_1 9·05, K_2 8·24, K_3 2·35, K_4 2·35	67A
	H	20	0·1	K_1 9·42, K_2 8·38, K_3 2·52, K_4 1·8	67T
Ca^{2+}	gl	20	0·1 (KNO_3)	K_1 6·21, $K(Ca^{2+}+HL^{3-} \rightleftharpoons CaHL^-)$ 3·5	64A
	cal	20	0·1 (KNO_3)	$\Delta H_1°$ −2·5, $\Delta S_1°$ 23	64A
Cd^{2+}	gl	20	0·1 (KNO_3)	K_1 14·38, $K(Cd^{2+}+HL^{3-} \rightleftharpoons CdHL^-)$ 8·28	64A
	cal	20	0·1 (KNO_3)	$\Delta H_1°$ −8·2, $\Delta S_1°$ 37·8	64A
Co^{2+}	gl	20	0·1 (KNO_3)	K_1 13·99, $K(Co^{2+}+HL^{3-} \rightleftharpoons CoHL^-)$ 8·37	64A
	cal	20	0·1 (KNO_3)	$\Delta H_1°$ −4·63, $\Delta S_1°$ 48·2	64A
Cu^{2+}	gl	20	0·1 (KNO_3)	K_1 16·57, $K(Cu^{2+}+HL^{3-} \rightleftharpoons CuHL^-)$ 12·09	64A
	cal	20	0·1 (KNO_3)	$\Delta H_1°$ −9·13, $\Delta S_1°$ 44·7	64A
Fe^{2+}	gl	20	0·1 (KNO_3)	K_1 11·57, $K(Fe^{2+}+HL^{3-} \rightleftharpoons FeHL^-)$ 6·91	64A
Fe^{3+}	sp	18—20	dil	K_1 20·67, $K(Fe^{3+}+HL^{3-} \rightleftharpoons FeHL)$ 13·19	66Z
	red	20	0·1 ($NaClO_4$)	K_1 20·41	67B
Ga^{3+}	red	20	0·1 ($NaClO_4$)	K_1 17·3, $K(GaL^-+H^+ \rightleftharpoons GaHL)$ 3·2 $K(GaL^-+OH^- \rightleftharpoons GaLOH^{2-})$ 7·14	67B
Hg^{2+}	gl	20	0·1 (KNO_3)	K_1 23·9, $K(Hg^{2+}+HL^{3-} \rightleftharpoons HgHL^-)$ 17·5	64A
	cal	20	0·1 (KNO_3)	$\Delta H_1°$ −22·8, $\Delta S_1°$ 31·6	64A
In^{3+}	sp	18—20	~0	K_1 24·1 $K(FeL^-+In^{3+} \rightleftharpoons InL^-+Fe^{3+})$ 0·76	66Z
	red	20	0·1 ($NaClO_4$)	K_1 20·26, $K(InL^-+H^+ \rightleftharpoons InHL)$ 1·88 $K(InL^-+OH^- \rightleftharpoons InLOH^{2-})$ 4·2	67B
La^{3+}	gl	20	0·1 (KNO_3)	K_1 12·8	64A
	cal	20	0·1 (KNO_3)	$\Delta H_1°$ −0·2, $\Delta S_1°$ 58	64A
Mg^{2+}	gl	20	0·1 (KNO_3)	K_1 4·61, $K(Mg^{2+}+HL^{3-} \rightleftharpoons MgHL^-)$ 3·2	64A
	cal	20	0·1 (KNO_3)	$\Delta H_1°$ 4·13, $\Delta S_1°$ 35·2	64A
Mn^{2+}	gl	20	0·1 (KNO_3)	K_1 10·07, $K(Mn^{2+}+HL^{3-} \rightleftharpoons MnHL^-)$ 5·53	64A
	cal	20	0·1 (KNO_3)	$\Delta H_1°$ −1·53, $\Delta S_1°$ 41·9	64A
Nd^{3+}	sp	18—20	dil	K_1 14·7, $K(Nd^{3+}+H_2L^{2-} \rightleftharpoons NdH_2L^+)$ 2·1	66Z
Ni^{2+}	gl	20	0·1 (KNO_3)	K_1 15·7, $K(Ni^{2+}+HL^{3-} \rightleftharpoons NiHL^-)$ 9·4	64A
	cal	20	0·1 (KNO_3)	$\Delta H_1°$ −7·7, $\Delta S_1°$ 45·5	64A
Pb^{2+}	gl	20	0·1 (KNO_3)	K_1 13·86, $K(Pb^{2+}+HL^{3-} \rightleftharpoons PbHL^-)$ 8·39	64A
	cal	20	0·1 (KNO_3)	$\Delta H_1°$ −13·0, $\Delta S_1°$ 19·1	64A
Th^{4+}	red	20	0·1 ($NaClO_4$)	K_1 19·8, $K(ThL+H^+ \rightleftharpoons ThHL^+)$ 2·43 $K(ThL+OH^- \rightleftharpoons ThLOH^-)$ 7·24	67B

cont.

$C_{12}H_{20}O_8N_2S$ (contd) 939

Metal	Method	Temp	Medium	Log of equilibrium constant, remarks	Ref
$^{+}$	gl	20	0·1 (KNO_3)	K_1 4·47, $K(Tl^+ + HL^{3-} \rightleftharpoons TlHL^{2-})$ 3·85	67Aa
$^{3+}$	red	20	1 ($HClO_4$)	K_1 32·3	67Aa
			1 ($NaClO_4$)	K_1 31·8	
$^{2+}$	gl	20	0·1 (KNO_3)	K_1 13·44, $K(Zn^{2+} + HL^{3-} \rightleftharpoons ZnHL^-)$ 8·05	64A
	cal	20	0·1 (KNO_3)	$\Delta H_1°$ −3·7, $\Delta S_1°$ 49	64A
$^{4+}$	oth	20	0·1	K_1 23·17±0·03, chromatography	67T

64A G. Anderegg, *Helv. Chim. Acta*, 1964, **47**, 1801
66Z N. M. Zhirnova, K. V. Astakhov, and S. A. Barkov, *Russ. J. Inorg. Chem.*, 1966, **11**, 1417 (2638)
67A G. Anderegg, *Helv. Chim. Acta*, 1967, **50**, 2333
67Aa G. Anderegg and E. Bottari, *Helv. Chim. Acta*, 1957, **50**, 2341
67B E. Bottari and G. Anderegg, *Helv. Chim. Acta*, 1967, **50**, 2349
67T L. I. Tikhonova, *Russ. J. Inorg. Chem.*, 1967, **12**, 494 (939)

$C_{12}H_{20}O_8N_2Se$ [(2,2′-Selenodiethylene)dinitrilo]tetra-acetic acid) H_4L 939.1

Metal	Method	Temp	Medium	Log of equilibrium constant, remarks	Ref
$^{+}$	gl	25	0·1	K_1 9·37, K_2 8·78, K_3 2·65, K_4 2·02	66K
$^{2+}$	gl	25	0·1	K_1 5·12, $K(Ca^{2+} + HL^{3-} \rightleftharpoons CaHL^-)$ 3·42	66K
g^{2+}	gl	25	0·1	K_1 6·15, $K(Mg^{2+} + HL^{3-} \rightleftharpoons MgHL^-)$ 3·17	66K
$^{2+}$	gl	25	0·1	K_1 4·94, $K(Sr^{2+} + HL^{3-} \rightleftharpoons SrHL^-)$ 2·85	66K

66K H. Kroll, M. A. Lipson, and E. F. Bolton, *U.S. A.E.C. Report TID-22717*, March 11, 1966

$C_{12}H_{17}ON_4SCl$ 3-(4′-Amino-2′-methyl-pyrimidin-5′-yl-methyl)-5-(2″-hydroxyethyl)-4-methyl 1,3-thiazolium chloride (thiamine) L 940.1

Metal	Method	Temp	Medium	Log of equilibrium constant, remarks	Ref
	gl	25	1 KNO_3	K_1 5·4±0·2, K_2 1·98±0·05	61G
$^{2+}$	gl	25	1 KNO_3	K_1 2·71	61G

61G A. A. Grinberg and Kh. Kh. Khakimov, *Russ. J. Inorg. Chem.*, 1961, **6**, 71 (144)

941.1 $C_{13}H_7N_3$ 4-Cyano-1,10-phenanthroline L

Metal	*Method*	*Temp*	*Medium*	*Log of equilibrium constant, remarks*	*Ref*
H^+	sp	?	0·1 ethanol	K_1 3·60	61H
Fe^{2+}	sp	?	0·1 ethanol	β_3 15·1, acetate buffer	61H

61H C. J. Hawkins, H. Duewell, and W. F. Pickering, *Analyt. Chim. Acta*, 1961, **25**, 257

944.1 $C_{13}H_{10}O_3$ 3,4-Dihydroxybenzophenone H_2L

Metal	*Method*	*Temp*	*Medium*	*Log of equilibrium constant, remarks*	*Ref*
Mo^{VI}	sp	25	0·01	$K(MoO_4^{2-} + 2H_2L \rightleftharpoons MoO_2L_2^{2-})$ 6·75 ± 0·02	62H

62H J. Halmekoski, *Suomen Kem.*, 1962, **B35**, 108

945 $C_{13}H_{10}N_2$ 2-Methyl-1,10-phenanthroline L

Metal	*Method*	*Temp*	*Medium*	*Log of equilibrium constant, remarks*	*Ref*
H^+	gl	25	0·3 (K_2SO_4)	K_1 5·30 ± 0·05	61J
	gl	25	0·1 KCl	K_1 5·42	62I
Co^{2+}	dis	25	0·1 KCl	K_1 5·1, K_2 4·9, K_3 3·9	62I
Cd^{2+}	dis	25	0·1 KCl	K_1 5·15, K_2 4·50, K_3 3·65	62I
Cu^+	Pt	25	0·3 (K_2SO_4)	β_2 16·95	61J
Cu^{2+}	Pt	25	0·3 (K_2SO_4)	K_1 7·4, K_2 6·4	61J
	sp	25	0·2 (K_2SO_4)	K_2 6·21	61J
	dis	25	0·1 KCl	K_1 7·40, K_2 6·45	62I
Mn^{2+}	dis	25	0·1 KCl	K_1 3·0, K_2 2·5, K_3 2·4	62I
Ni^{2+}	dis	25	0·1 KCl	K_1 5·95, K_2 6·85, K_3 4·9	62I
Zn^{2+}	dis	25	0·1 KCl	K_1 4·96, K_2 4·40, K_3 3·35	62I

61J B. R. James and R. J. P. Williams, *J. Chem. Soc.*, 1961, 2007
62I H. Irving and D. H. Mellor, *J. Chem. Soc.*, 1962, 5237

$C_{13}H_{10}N_2$ 5-Methyl-1,10-phenanthroline L 946

Metal	Method	Temp	Medium	Log of equilibrium constant, remarks	Ref
H^+	gl	25	0·3 (KNO_3) 50% dioxan	K_1 4·65±0·05	61J
	sp	25	0·1 KCl	K_1 5·28	62M
	sp	15—35	→ 0	K_1 (15°) 5·22±0·02, (25°) 5·11±0·01, (35°) 4·99±0·02 $\Delta H_1°$ −4·4±0·3, $\Delta S_1°$ 9±2	67L
Cd^{2+}	dis	25	0·1 KNO_3	K_1 6·13, K_2 4·90, K_3 5·00	62M
Co^{2+}	dis	25	0·1 KCl	K_1 7·14, K_2 6·86, K_3 6·60	62M
Cu^+				See Cu^{2+} below.	
Cu^{2+}	Pt	25	0·3 (KNO_3) 50% dioxan	K_1 ~ 6·9, K_3 4·60, $K(CuL_2^{2+} + Cu^+ \rightleftharpoons CuL_2^+ + Cu^{2+})$ 0·81	61J
	dis	25	0·1 KCl	K_1 8·55, K_2 6·47, K_3 5·10	62M
Fe^{2+}	dis	25	0·1 KCl	K_1 6·46, β_2 ~ 13·5, β_3 21·94	62M
	sp	15—45	→ 0	K_1 (22°) 6·11, β_3 (15°) 21·87±0·04, (22°) 21·33, (35°) 20·25±0·03, (45°) 19·45 $\Delta H_{\beta 3}°$ −33·8±0·7, $\Delta S_{\beta 3}°$ −17±4	67L
Fe^{3+}	sp	25	→ 0	β_3 16·77	67L
Mn^{2+}	dis	25	0·1 KCl	K_1 4·28, K_2 3·3, K_3 3·60	62M
Ni^{2+}	dis	25	0·1 KCl	K_1 8·30, K_2 8·65, K_3 7·73	62M
Zn^{2+}	dis	25	0·1 KCl	K_1 6·62, K_2 5·96, K_3 5·67	62M

61J B R. James and R. J. P. Williams, *J. Chem. Soc.*, 1961, 2007
62M W. A. E. McBryde, D. A. Brisbin, and H. Irving, *J. Chem. Soc.*, 1962, 5245
67L S. C. Lahiri and S. Aditya, *Z. phys. Chem. (Frankfurt)*, 1967, 55, 6

$C_{13}H_{13}N_3$ Diphenylguanidine L 947.1

Metal	Method	Temp	Medium	Log of equilibrium constant, remarks	Ref
Ag^+	oth	?	0·05 Bu_4NClO_4 acetone	β_2 10·3±0·1, coulometric titration	65M

65M K. K. Mead, D. L. Maricle, and C. A. Streuli, *Analyt. Chem.*, 1965, 37, 237

$C_{13}H_{14}N_4$ 2-[4′-(Dimethylamino)phenylazo]pyridine L 949

Metal	Method	Temp	Medium	Log of equilibrium constant, remarks	Ref
Cd^{2+}	kin	16	0·1 KNO_3	K_1 2·7	64W
Co^{2+}	kin	16	0·1 KNO_3	K_1 3·8	64W
Cu^{2+}	kin	16	0·1 KNO_3	K_1 5·00	64W
Zn^{2+}	kin	16	0·1 KNO_3	K_1 2·62	64W

64W R. G. Wilkins, *Inorg. Chem.*, 1964, 3, 520

949.1 $C_{13}H_{16}O_2$ 1-(2′,4′,6′-Trimethylphenyl)butane-1,3-dione (mesitoylacetone) HL

Metal	*Method*	*Temp*	*Medium*	*Log of equilibrium constant, remarks*	*Ref*
H^+	gl	30	75% dioxan	K_1 11·95	65U
Be^{2+}	gl	30	75% dioxan	K_1 11·02, K_2 10·05	65U
Cu^{2+}	gl	30	75% dioxan	K_1 10·74, K_2 9·86	65U
Ni^{2+}	gl	30	75% dioxan	K_1 9·19, K_2 7·86	65U
Zn^{2+}	gl	30	75% dioxan	K_1 8·89, K_2 7·73	65U

65U E. Uhlemann and E. Frank, *Z. anorg. Chem.*, 1965, **340**, 319

950.1 $C_{13}H_9N_2Cl$ 4-Chloro-2-methyl-1,10-phenanthroline L

Metal	*Method*	*Temp*	*Medium*	*Log of equilibrium constant, remarks*	*Ref*
H^+	sp	?	0·1 ethanol	K_1 4·57	61H
Fe^{2+}	sp	?	0·1 ethanol	β_2 4·9, acetate buffer	61H

61H C. J. Hawkins, H. Duewell, and W. F. Pickering, *Analyt. Chim. Acta*, 1961, **25**, 257

952.1 $C_{13}H_{10}O_4N_2$ 2-(3′,4′-Dihydroxyphenylazo)benzoic acid (2-Carboxy-3′,4′-dihydroxyazobenzene) H_3L

Metal	*Method*	*Temp*	*Medium*	*Log of equilibrium constant, remarks*	*Ref*
H^+	?	18—20	1 (HCl)	K_1 3·12	66K
Zr^{4+}	sp	18—20	1 (HCl)	$K(Zr^{4+}+2H_3L \rightleftharpoons ZrH_4L_2^{2+}+2H^+)$ 14·34	66K

66K I. M. Korenman, F. R. Sheyanova, and Z. M. Gur'eva, *Russ. J. Inorg. Chem.*, 1966, **11**, 1485 (2761)

$C_{13}H_{10}O_4N_2$ 4-(Phenylamino)pyridine-2,6-dicarboxylic acid H_2L 952.2

Metal	Method	Temp	Medium	Log of equilibrium constant, remarks	Ref
H^+	gl	22	0·1 ($NaClO_4$)	K_1 8·65, K_2 1·4	64B
Ba^{2+}	gl	22	0·1 ($NaClO_4$)	K_1 3·75	64B
Mg^{2+}	gl	22	0·1 ($NaClO_4$)	K_1 2·85	64B
Sr^{2+}	gl	22	0·1 ($NaClO_4$)	K_1 4·18	64B

64B E. Blasius and B. Brozio, *Ber. Bunsengesellschaft Phys. Chem.*, 1964, **68**, 52

$C_{13}H_{10}N_4S$ 1-(2′-Benzothiazolyl)-3-(2″-pyridyl)-1,2-diazaprop-2-ene HL 952.3
(2-picolinylaldehyde 2-benzothiazolylhydrazone)

Metal	Method	Temp	Medium	Log of equilibrium constant, remarks	Ref
H^+	gl	25	50% dioxan	K_1 10·81, K_2 3·00	65H
Cd^{2+}	gl	25	50% dioxan	K_1 8·76, K_2 8·11	65H
Cu^{2+}	gl	25	50% dioxan	K_1 10·48, K_2 8·59	65H
Ni^{2+}	gl	25	50% dioxan	K_1 9·86, K_2 9·37	65H
Pd^{2+}	gl	25	50% dioxan	K_1 10·33	65H
Zn^{2+}	gl	25	50% dioxan	K_1 9·33, K_2 9·05	65H

65H M. L. Heit and D. E. Ryan, *Analyt. Chim. Acta*, 1965, **32**, 448

$C_{13}H_{11}ON$ *N*-(Salicylidene)aniline (salicylanil) HL 953

Metal	Method	Temp	Medium	Log of equilibrium constant, remarks	Ref
H^+	gl	30	0·1 $NaClO_4$ 75% dioxan	K_1 10·44, K_2 3·19	64J
Cu^{2+}	gl	20	0·1 $NaClO_4$ 75% dioxan	K_1 8·82, K_2 5·77	64J

64J K. E. Jabalpurwala, K. A. Venkatachalam, and M. B. Kabadi, *J. Inorg. Nuclear Chem.*, 1964, **26**, 1011, 1027

$C_{13}H_{11}ON_3$ 2-(2′-Hydroxy-5′-methylphenyl)benzotriazole HL 953.1

Metal	Method	Temp	Medium	Log of equilibrium constant, remarks	Ref
Cu^{2+}	dis	20	0·1 $NaClO_4$	K_1 9·22, K_2 8·01	68N

68N O. Navratil and J. Liska, *Coll. Czech. Chem. Comm.*, 1968, **33**, 991

955 $C_{13}H_{11}O_2N$ *N*-Phenylbenzohydroxamic acid HL

Metal	*Method*	*Temp*	*Medium*	*Log of equilibrium constant, remarks*	*Ref*
H^+	gl	25	0·1 $NaClO_4$	K_1 8·2	68S
Ga^{3+}	dis	18—22	1	K_1 10·22, K_2 9·99, K_3 9·83 H^+ constants not given.	65A
In^{3+}	dis	25	0·1 $NaClO_4$	K_1 9·2, β_2 18·4, β_3 26·3	68S

65A I. P. Alimarin, S. A. Khamid, and I. V. Puzdrenkova, *Russ. J. Inorg. Chem.*, 1965, **10**, 209 (389)
68S G. K. Schweitzer and M. M. Anderson, *J. Inorg. Nuclear Chem.*, 1968, **30**, 1051

956.1 $C_{13}H_{11}O_2N$ *N*-Phenyl-2-hydroxybenzamide (salicylanilide) HL

Metal	*Method*	*Temp*	*Medium*	*Log of equilibrium constant, remarks*	*Ref*
H^+	gl	30	0·1 $NaClO_4$ 75% dioxan	K_1 9·38	64J
Cu^{2+}	gl	30	0·1 $NaClO_4$ 75% dioxan	K_1 6·59	64J
Ni^{2+}	gl	30	0·1 $NaClO_4$ 75% dioxan	K_1 4·21	64J
Zn^{2+}	gl	30	0·1 $NaClO_4$ 75% dioxan	K_1 5·06	64J

64J K. E. Jabalpurwala, K. A. Venkatachalam, and M. B. Kabadi, *J. Inorg. Nuclear Chem.*, 1964, **26**, 1011, 1027

956.2 $C_{13}H_{11}O_3N_3$ 1-(4′-Carboxyphenyl)-3-hydroxy-3-phenyltriazene H_2L

Metal	*Method*	*Temp*	*Medium*	*Log of equilibrium constant, remarks*	*Ref*
H^+	sp	?	?	K_1 10·97	65P
Cu^{2+}	gl	25	0·1 KCl 70% dioxan	K_1 11·28, K_2 10·60	65P

65P D. N. Purohit and N. C. Sogani, *Z. Naturforsch.*, 1965, **20b**, 206

$C_{13}H_{12}ON_4$ 1,5-Diphenyl-1,2,4,5-tetraazapent-1-en-3-one (1,5-diphenylcarbazone) L 958.1

Metal	Method	Temp	Medium	Log of equilibrium constant, remarks	Ref
H^+	sp	25	0·1 50% dioxan	K_1 9·26	64M
	dis	?	?	K_1 8·54	66G
Cu^{2+}	dis	28	~ 0	K_1 9·8, β_2 19·5, β_3 29	66G
Ni^{2+}	sp	25	0·1 50% dioxan	K_1 6·02	64M
Zn^{2+}	sp	25	0·1 50% dioxan	K_1 5·76	64M

64M K. S. Math, Q. Fernando, and H. Freiser, *Analyt. Chem.*, 1964, **36**, 1762

66G H. R. Geering and J. F. Hodgson, *Analyt. Chim. Acta*, 1966, **36**, 537

$C_{13}H_{12}N_4S$ 1,5-Diphenyl-1,2,4,5-tetraazapent-1-en-3-thione HL 959 (diphenylthiocarbazone) (dithizone)

Metal	Method	Temp	Medium	Log of equilibrium constant, remarks	Ref
H^+	sp	25	0·1 50% dioxan	K_1 5·80	64M
Ag^+				See diethyldithiocarbamic acid (332) for mixed ligands.	
As^{3+}				See diethyldithiocarbamic acid (332) for mixed ligands.	
Bi^{3+}				See diethyldithiocarbamic acid (332) for mixed ligands.	
Cd^{2+}				See diethyldithiocarbamic acid (332) for mixed ligands.	
Cu^{2+}				See diethyldithiocarbamic acid (332) for mixed ligands.	
Lu^{3+}				See diethyldithiocarbamic acid (332) for mixed ligands.	
Mg^{2+}				See diethyldithiocarbamic acid (332) for mixed ligands.	
Ni^{2+}	sp	25	0·1 50% dioxan	K_1 5·83	64M
Pb^{2+}	dis	?	0·1	β_2 (?) 15·85, K_{so} −23·7	64Ma
				See diethyldithiocarbamic acid (332) for mixed ligands.	
Pd^{2+}				See diethyldithiocarbamic acid (332) for mixed ligands.	
Sb^{3+}				See diethyldithiocarbamic acid (332) for mixed ligands.	
Se^{4+}				See diethyldithiocarbamic acid (332) for mixed ligands.	
Te^{4+}				See diethyldithiocarbamic acid (332) for mixed ligands.	
Tl^+				See diethyldithiocarbamic acid (332) for mixed ligands.	
Zn^{2+}	dis	25	1 ($NaClO_4$)	K_1 5·05	62H
				See diethyldithiocarbamic acid (332) for mixed ligands.	
	sp	25	0·1 50% dioxan	K_1 6·18	64M

62H C. B. Honaker and H. Freiser, *J. Phys. Chem.*, 1962, **66**, 127

64M K. S. Math, Q. Fernando, and H. Freiser, *Analyt. Chem.*, 1964, **36**, 1762

64Ma O. B. Mathre and E. B. Sandell, *Talanta*, 1964, **11**, 295

960.1 $C_{13}H_{13}ON_3$ 3-Hydroxy-1-(2′-methylphenyl)-3-phenyltriazene HL

Metal	*Method*	*Temp*	*Medium*	*Log of equilibrium constant, remarks*	*Ref*
H^+	sp	?	?	K_1 11·55	64P
Cu^{2+}	gl	25	0·1 KCl	K_1 11·35, β_2 22·47	64P
Mn^{2+}	gl	25	0·1 KCl	K_1 6·68, β_2 12·07	64P
Ni^{2+}	gl	25	0·1 KCl	K_1 10·17, β_2 17·76	64P
Pd^{2+}	gl	25	0·1 KCl	K_1 11·70, β_2 22·97	64P
Zn^{2+}	gl	25	0·1 KCl	K_1 8·38, β_2 15·46	64P

64P D. N. Purohit and N. C. Sogani, *Bull. Chem. Soc. Japan*, 1964, **37**, 1727

960.2 $C_{13}H_{13}ON_3$ 3-Hydroxy-1-(4′-methylphenyl)-3-phenyltriazene HL

Metal	*Method*	*Temp*	*Medium*	*Log of equilibrium constant, remarks*	*Ref*
H^+	sp	?	?	K_1 11·78	64P
Cu^{2+}	gl	25	0·1 KCl	K_1 11·69, β_2 23·09	64P
Mn^{2+}	gl	25	0·1 KCl	K_1 7·02, β_2 12·74	64P
Ni^{2+}	gl	25	0·1 KCl	K_1 10·46, β_2 21·23	64P
Pd^{2+}	gl	25	0·1 KCl	K_1 11·89, β_2 23·35	64P
Zn^{2+}	gl	25	0·1 KCl	K_1 8·62, β_2 15·99	64P

64P D. N. Purohit and N. C. Sogani, *Bull. Chem. Soc. Japan*, 1964, **37**, 1727

960.3 $C_{13}H_{13}ON_3$ 3-Hydroxy-3-(4′-methylphenyl)-1-phenyltriazene HL

Metal	*Method*	*Temp*	*Medium*	*Log of equilibrium constant, remarks*	*Ref*
H^+	sp	?	?	K_1 11·86	65P
Cu^{2+}	gl	25	0·1 KCl 70% dioxan	K_1 11·77, K_2 11·53	65P

65P D. N. Purohit and N. C. Sogani, *Z. Naturforsch.*, 1965, **20b**, 206

$C_{13}H_{13}O_2N_3$ 3-Hydroxy-1-(4′-methoxyphenyl)-3-phenyltriazene HL 960.4

Metal	Method	Temp	Medium	Log of equilibrium constant, remarks	Ref
H^+	sp	?	?	K_1 11·95	65P
Cu^{2+}	gl	25	0·1 KCl 70% dioxan	K_1 11·91, K_2 11·67	65P
Mn^{2+}	gl	25	0·1 KCl 70% dioxan	K_1 7·24, K_2 6·10	65P
Ni^{2+}	gl	25	0·1 KCl 70% dioxan	K_1 10·74, K_2 8·04	65P
Pd^{2+}	gl	25	0·1 KCl 70% dioxan	K_1 12·06, K_2 11·68	65P
Zn^{2+}	gl	25	0·1 KCl 70% dioxan	K_1 8·84, K_2 7·79	65P

65P D. N. Purohit and N. C. Sogani, *Indian J. Chem.*, 1965, **3**, 58

$C_{13}H_{14}O_7N_2$ 2 Methyl-*N*-4′-nitrobenzoyl-glutamic acid H_2L 960.5

Metal	Method	Temp	Medium	Log of equilibrium constant, remarks	Ref
H^+	gl	25	0·1 $NaClO_4$	K_1 4·57, K_2 3·23	65N
Cu^{2+}	gl	25	0·1 $NaClO_4$	K_1 2·1	65N

65N M. H. T. Nyberg and M. Cefola, *Arch. Biochem. Biophys.*, 1965, **111**, 321

$C_{13}H_{15}ON$ 2-Butyl-8-hydroxyquinoline HL 960.6

Metal	Method	Temp	Medium	Log of equilibrium constant, remarks	Ref
H^+	gl	25	~ 0	K_1 12·07, K_2 4·41	66K
Cd^{2+}	gl	25	~ 0	K_1 9·28, K_2 8·98	66K
Cu^{2+}	gl	25	~ 0	K_1 12·05, K_2 12·23	66K
Ni^{2+}	gl	25	~ 0	K_1 9·18, K_2 9·89	66K
Zn^{2+}	gl	25	~ 0	K_1 9·75, K_2 10·24	66K

66K H. Kaneko and K. Ueno, *Bull. Chem. Soc. Japan*, 1966, **39**, 1910

961.1 $C_{13}H_{15}O_5N$ **N-Benzoyl-2-methyglutamic acid** H_2L

Metal	Method	Temp	Medium	Log of equilibrium constant, remarks	Ref
H^+	gl	25	0·1 $NaClO_4$	K_1 4·63, K_2 3·51	65N
Cu^{2+}	gl	25	0·1 $NaClO_4$	K_1 2·2	65N

65N M. H. T. Nyberg and M. Cefola, *Arch. Biochem. Biophys.*, 1965, **111**, 321

961.2 $C_{13}H_{15}O_6N$ **N-(α-Carboxy-4′-methylbenzyl)iminodiacetic acid** H_3L

Metal	Method	Temp	Medium	Log of equilibrium constant, remarks	Ref
H^+	gl	20	0·1 (KCl)	K_1 9·45±0·02, K_2 2·40±0·02, K_3 1·45±0·05	66I
Ba^{2+}	gl	20	0·1 (KCl)	K_1 4·31±0·03	66I
Ca^{2+}	gl	20	0·1 (KCl)	K_1 6·22±0·01	66I
Mg^{2+}	gl	20	0·1 (KCl)	K_1 4·74±0·02	66I
Sr^{2+}	gl	20	0·1 (KCl)	K_1 4·48±0·02	66I
Zn^{2+}	gl	20	0·1 (KCl)	K_1 9·90±0·03	66I

66I H. M. N. H. Irving and M. G. Miles, *J. Chem. Soc. (A)*, 1966, 727

961.3 $C_{13}H_{15}O_6N$ **N-(1′-Carboxy-1′-phenylethyl)iminodiacetic acid** H_3L

Metal	Method	Temp	Medium	Log of equilibrium constant, remarks	Ref
H^+	gl	20	0·1 (KCl)	K_1 11·07±0·02, K_2 2·66±0·02, K_3 1·46±0·05	66I
Ba^{2+}	gl	20	0·1 (KCl)	K_1 4·93±0·02	66I
Ca^{2+}	gl	20	0·1 (KCl)	K_1 7·46±0·01	66I
Mg^{2+}	gl	20	0·1 (KCl)	K_1 5·17±0·02	66I
Sr^{2+}	gl	20	0·1 (KCl)	K_1 5·38±0·02	66I
Zn^{2+}	gl	20	0·1 (KCl)	K_1 11·45±0·06	66I

66I H. M. N. H. Irving and M. G. Miles, *J. Chem. Soc. (A)*, 1966, 1268

961.4 $C_{13}H_{15}O_7N$ **N-(α-Carboxy-4′-methoxybenzyl)iminodiacetic acid** H_3L

Metal	Method	Temp	Medium	Log of equilibrium constant, remarks	Ref
H^+	gl	20	0·1 (KCl)	K_1 9·51±0·02, K_2 2·44±0·22, K_3 1·47±0·05	66I
Ba^{2+}	gl	20	0·1 (KCl)	K_1 4·32±0·01	66I
Ca^{2+}	gl	20	0·1 (KCl)	K_1 6·24±0·01	66I
Mg^{2+}	gl	20	0·1 (KCl)	K_1 4·75±0·01	66I
Sr^{2+}	gl	20	0·1 (KCl)	K_1 4·49±0·01	66I
Zn^{2+}	gl	20	0·1 (KCl)	K_1 9·95±0·02	66I

66I H. M. N. H. Irving and M. G. Miles, *J. Chem. Soc. (A)*, 1966, 727

$C_{13}H_{18}O_4N_2$ Pyridoxylidenevaline H_2L 962.1

Metal	Method	Temp	Medium	Log of equilibrium constant, remarks	Ref
H^+	sp	27	methanol	K_1 3·6	67M
Zn^{2+}	sp	27	methanol	β_2 11·5	67M

Erratum. 1964, p. 661, Table 894. This Table was numbered **894** in the 1964 edition but has been renumbered to correct the ligand sequence.

67M Y. Matsushima and A. E. Martell, *J. Amer. Chem. Soc.*, 1967, 89, 1322

$C_{13}H_{20}O_2N_2$ 2-(Diethylamino)ethyl 4-aminobenzoate (procaine) L 962.2

Metal	Method	Temp	Medium	Log of equilibrium constant, remarks	Ref
H^+	gl	25	1 KNO_3	K_1 8·8±0·2, K_2 2·3±0·1	61G
Co^{2+}	gl	25	1 KNO_3	K_1 7·21, K_2 6·18, K_3 5·90, K_4 5·80	61G

61G A. A. Grinberg and Kh. Kh. Khakimov, *Russ. J. Inorg. Chem.*, 1961, 6, 71 (144)

$C_{13}H_{20}O_8N_2$ (*trans*-Cyclopentylenedinitrilo)tetra-acetic acid H_4L 963

Metal	Method	Temp	Medium	Log of equilibrium constant, remarks	Ref
H^+	gl	20	0·1	K_1 10·31, K_2 7·42, K_3 2·92, K_4 2·41	60K
Ba^{2+}	gl	20	0·1	K_1 7·75	60K
Ca^{2+}	gl	20	0·1	K_1 11·08	60K
Mg^{2+}	gl	20	0·1	K_1 9·05	60K
Sr^{2+}	gl	20	0·1	K_1 9·45	60K
	ix	20	0·1 NH_4^+	K_1 10·12±0·03	62S
			0·165 NH_4^+	K_1 9·71±0·01	

60K H. Kroll and M. Gordon, *Ann. New York Acad. Sci.*, 1960, 88, 341
62S M. M. Senyavin and L. I. Tikhonova, *Russ. J. Inorg. Chem.*, 1962, 7, 562 (1095)

$C_{13}H_{22}O_8N_2$ (Pentamethylenedinitrilo)tetra-acetic acid H_4L 964

Metal	Method	Temp	Medium	Log of equilibrium constant, remarks	Ref
H^+	gl	20	0·1 (KNO_3)	K_1 10·70, K_2 9·52, K_3 2·7, K_4 2·3	64A
	cal	20	0·1 (KNO_3)	$\Delta H_1°$ −7·5, $\Delta S_1°$ 23·3, $\Delta H_2°$ −6·3, $\Delta S_2°$ 22·1	64A
	gl	20	0·1 [$(CH_3)_4NCl$]	K_1 10·75, K_2 9·58	67A
	H	20	1 [$(CH_3)_4NCl$]	K_1 10·49, K_2 9·42, K_3 2·6, K_4 2·2	67A
	gl	20	1 (KNO_3)	K_1 10·50, K_2 9·46, K_3 2·6, K_4 2·4	67A
Ca^{2+}	gl	20	0·1 (KNO_3)	K_1 5·2, $K(Ca^{2+}+HL^{3-} \rightleftharpoons CaHL^-)$ 3·6	64A

continued overleaf

964 $C_{13}H_{22}O_8N_2$ (contd)

Metal	*Method*	*Temp*	*Medium*	*Log of equilibrium constant, remarks*	*Ref*
Cd^{2+}	gl	20	0·1 (KNO_3)	K_1 11·6, $K(Cd^{2+}+HL^{3-} \rightleftharpoons CdHL^-)$ 6·9	64A
	cal	20	0·1 (KNO_3)	$\Delta H_1°$ −4·46, $\Delta S_1°$ 37·5	64A
Co^{2+}	gl	20	0·1 (KNO_3)	K_1 13·38, $K(Co^{2+}+HL^{3-} \rightleftharpoons CoHL^-)$ 7·94	64A
	cal	20	0·1 (KNO_3)	$\Delta H_1°$ −3·1, $\Delta S_1°$ 50·7	64A
Cu^{2+}	gl	20	0·1 (KNO_3)	K_1 16·24, $K(Cu^{2+}+HL^{3-} \rightleftharpoons CuHL^-)$ 11·35	64A
	cal	20	0·1 (KNO_3)	$\Delta H_1°$ −10·9, $\Delta S_1°$ 37·2	64A
Fe^{2+}	gl	20	0·1 (KNO_3)	$K_1 \sim$ 10·8, $K(Fe^{2+}+HL^{3-} \rightleftharpoons FeHL^-) \sim$ 6·4	64A
Mg^{2+}	gl	20	0·1 (KNO_3)	K_1 5·2, $K(Mg^{2+}+HL^{3-} \rightleftharpoons MgHL^-)$ 3·6	64A
Mn^{2+}	gl	20	0·1 (KNO_3)	K_1 8·7, $K(Mn^{2+}+HL^{3-} \rightleftharpoons MnHL^-)$ 5·6	64A
	cal	20	0·1 (KNO_3)	$\Delta H_1°$ 0·9, $\Delta S_1°$ 43·0	
Ni^{2+}	gl	20	0·1 (KNO_3)	K_1 13·9, $K(Ni^{2+}+HL^{3-} \rightleftharpoons NiHL^-)$ 9·0	64A
	cal	20	0·1 (KNO_3)	$\Delta H_1°$ −6·7, $\Delta S_1°$ 40·8	64A
Pb^{2+}	gl	20	0·1 (KNO_3)	$K(Pb^{2+}+HL^{3-} \rightleftharpoons PbHL^-)$ 7·83	64A
Tl^+	gl	20	0·1 (KNO_3)	K_1 3·73, $K(Tl^++HL^{3-} \rightleftharpoons TlHL^{2-})$ 2·88	67Aa
Tl^{3+}	red	20	1 ($NaClO_4$)	K_1 31·3	67Aa
Zn^{2+}	gl	20	0·1 (KNO_3)	K_1 12·67, $K(Zn^{2+}+HL^{3-} \rightleftharpoons ZnHL^-)$ 7·8	64A
	cal	20	0·1 (KNO_3)	$\Delta H_1°$ −2·7, $\Delta S_1°$ 48·8	64A

64A G. Anderegg, *Helv. Chim. Acta*, 1964, **47,** 1801
67A G. Anderegg, *Helv. Chim. Acta*, 1967, **50,** 2333
67Aa G. Anderegg and E. Bottari, *Helv. Chim. Acta*, 1967, **50,** 2341

968.1 $C_{13}H_9ON_3S$ 1-(1′,3′-Thiazol-2′-ylazo)-2-naphthol HL

Metal	*Method*	*Temp*	*Medium*	*Log of equilibrium constant, remarks*	*Ref*
H^+	sp	20	0·05 ($NaClO_4$)	K_1 9·10, K_2 0·88	64N
Ag^+	dis	18—22	0·05	K_1 8·67	66N
Cd^{2+}	sp	18—22	0·05	K_1 9·18, β_2 17·88	67N
Co^{2+}	sp	18—22	0·05	K_1 9·50, β_2 19·00	67N
Cu^{2+}	sp	20	0·05 ($NaClO_4$)	K_1 10·92, β_2 22·52	64N
Eu^{3+}	dis	18—22	0·05	K_1 9·56, β_2 18·76, β_3 27·60, β_4 36·08	66N
Ho^{3+}	dis	18—22	0·05	K_1 12·76, β_2 24·36, β_3 34·80, β_4 44·08	66N
Yb^{3+}	dis	18—22	0·05	K_1 9·81, β_2 19·32, β_3 28·53, β_4 37·44	66N
Zn^{2+}	sp	20	0·05 ($NaClO_4$)	K_1 9·87, β_2 19·74	64N

64N O. Navratil, *Coll. Czech. Chem. Comm.*, 1964, **29,** 2490
66N O. Navratil, *Coll. Czech. Chem. Comm.*, 1966, **31,** 2492
67N O. Navratil, *Coll. Czech. Chem. Comm.*, 1967, **32,** 2004

$C_{13}H_9O_4N_3S_2$ 1-(1′,3′-Thiazol-2′-ylazo)-2-naphthol-6-sulphonic acid H_2L 968.2

Metal	Method	Temp	Medium	Log of equilibrium constant, remarks	Ref
H^+	gl	25	0·1 $NaClO_4$ 0—50% methanol	K_1 (0%) 8·38, (50%) 8·44	67N
Co^{2+}	gl	25	0·1 $NaClO_4$ 0—50% methanol	K_1 (0%) 7·7, K_2 (0%) 6·6, β_2 (50%) 15·4	67N
Cu^{2+}	gl	25	0·1 $NaClO_4$ 0—50% methanol	K_1 (0%) 11·1, (50%) 11·3, K_2 (50%) 6·3	67N
Fe^{2+}	gl	25	0·1 $NaClO_4$ 0—50% methanol	β_2 (0%) 16·7, (50%) 17·3	67N
Mn^{2+}	gl	25	0·1 $NaClO_4$ 0—50% methanol	K_1 (0%) 4·3, (50%) 4·9, K_2 (0%) 3·3, (50%) 4·0	67N
Ni^{2+}	gl	25	0·1 $NaClO_4$ 0—50% methanol	K_1 (0%) 8·5, K_2 (0%) 8·3, β_2 (50%) 17·1	67N
Pd^{2+}	gl	25	0·1 $NaClO_4$ 0—50% methanol	K_1 (0%) 13, (50%) 13, K_2 (0%) 5·7, (50%) 6·4	67N
UO_2^{2+}	gl	25	0·1 $NaClO_4$ 0—50% methanol	K_1 (0%) 8·2, (50%) 8·7, K_2 (0%) 5·5, (50%) 7·2	67N
Zn^{2+}	gl	25	0·1 $NaClO_4$ 0—50% methanol	K_1 (0%) 6·3, (50%) 7·0, K_2 (0%) 5·7, (50%) 6·3	67N

67N G. Nickless, F. H. Pollard, and T. J. Samuelson, *Analyt. Chim. Acta*, 1967, **39**, 37

$C_{13}H_{10}O_2NCl$ *N*-(4′-Chlorophenyl)benzohydroxamic acid L 968.3

Metal	Method	Temp	Medium	Log of equilibrium constant, remarks	Ref
V^V	sp	28	?	$K[VO_3^- + L \rightleftharpoons VO_2(H_{-2}L)^-]$ (?) 7·05	65M

65M A. K. Majumdar and G. Das, *J. Indian Chem. Soc.*, 1965, **42**, 189

968.4 $C_{13}H_{10}O_6N_2S$ 5-(4′-Sulphophenylazo)salicylic acid (3-Carboxy-4-hydroxy-4′-sulphoazobenzene) **H_3L**

Metal	*Method*	*Temp*	*Medium*	*Log of equilibrium constant, remarks*	*Ref*
H^+	gl	25	0·1 (KNO_3)	K_1 11·04, K_2 2·38	64M
Ca^{2+}	gl	25	0·1 (KNO_3)	K_1 3·10	64M
Co^{2+}	gl	25	0·1 (KNO_3)	K_1 5·84, K_2 3·93	64M
Fe^{3+}	gl	25	0·1 (KNO_3)	K_2 10·70	64M
Mg^{2+}	gl	25	0·1 (KNO_3)	K_1 4·45, K_2 3·04	64M
Mn^{2+}	gl	25	0·1 (KNO_3)	K_1 4·94, K_2 3·5	64M
Ni^{2+}	gl	25	0·1 (KNO_3)	K_1 6·17, K_2 4·05	64M

64M Y. Murakami and M. Takagi, *Bull. Chem. Soc. Japan*, 1964, **37**, 268

970.1 $C_{13}H_{12}O_5N_2S$ 4-(2′,4′-Dihydroxyphenylazomethyl)benzenesulphonic acid (tropaeolin O) **H_3L**

Metal	*Method*	*Temp*	*Medium*	*Log of equilibrium constant, remarks*	*Ref*
H^+				H^+ constants not given.	
Pd^{2+}	sp	25	?	$K[Pd^{2+} + 2HL^{2-} \rightleftharpoons Pd(HL)_2^{2-}]$ (?) 9·4	63S

63S R. L. Seth and A. K. Dey, *J. Indian Chem. Soc.*, 1963, **40**, 794

970.2 $C_{13}H_{14}O_2N_2S$ 2-(4′-Methylphenylsulphonamido)aniline HL

Metal	*Method*	*Temp*	*Medium*	*Log of equilibrium constant, remarks*	*Ref*
H^+	gl	?	0·01 (KNO_3) 50% dioxan	K_1 10·16±0·02, K_2 2·24±0·04	68B
	dis	?	0·1 (KNO_3)	K_1 10·49±0·08, K_2 1·72±0·1	68B
	sp	?	0·1 (KNO_3)	K_1 9·20±0·06, K_2 2·01±0·04	68B
Co^{2+}	gl	?	0·01 (KNO_3) 50% dioxan	K_1 9·57, β_2 18·73±0·03	68B
Cu^{2+}	gl	?	0·01 (KNO_3) 50% dioxan	K_1 9·50, β_2 19·11±0·06	68B
	dis	?	0·1 (KNO_3)	K_1 8·03, β_2 16·06±0·05	68B
Fe^{2+}	gl	?	0·01 (KNO_3) 50% dioxan	K_1 9·31, β_2 17·45±0·04	68B
Ni^{2+}	gl	?	0·01 (KNO_3) 50% dioxan	K_1 9·67, β_2 18·94±0·08	68B
Zn^{2+}	gl	?	0·01 (KNO_3) 50% dioxan	K_1 9·33, β_2 18·24±0·07	68B

68B D. Betteridge and R. Rangaswamy, *Analyt. Chim. Acta*, 1968, **42**, 293

$C_{14}H_8O_6$ 1,2,5,8-Tetrahydroxyanthraquinone (quinalizarin) H_4L 979.1

Metal	Method	Temp	Medium	Log of equilibrium constant, remarks	Ref
H^+				H^+ constants not given.	
Pb^{2+}	sp	30	50% ethanol	K (?) $4{\cdot}1 \pm 0{\cdot}1$	67S

67S K. C. Srivastava and S. K. Banerjee, *Austral. J. Chem.*, 1967, 20, 1385

$C_{14}H_{12}N_2$ 2,9-Dimethyl-1,10-phenanthroline L 981

Metal	Method	Temp	Medium	Log of equilibrium constant, remarks	Ref
H^+	gl	25	0·3 (K_2SO_4)	K_1 $5{\cdot}88 \pm 0{\cdot}02$	61J
	gl	25	0·1 KCl	K_1 5·85	62I
Cd^{2+}	dis	25	0·1 KCl	K_1 4·1, K_2 3·3, K_3 ~ 3·0	62I
Co^{2+}	dis	25	0·1 KCl	K_1 4·2, K_2 2·8	62I
Cu^+	Pt	25	0·3 (K_2SO_4)	β_2 19·1	61J
Cu^{2+}	Pt	25	0·3 (K_2SO_4)	K_1 ~ 6·1, K_2 ~ 5·6	61J
	dis	25	0·1 KCl	K_1 5·2, K_2 5·8	62I
Fe^{2+}	dis	25	0·1 KCl	$K_1 \ll 4$	62I
Mn^{2+}	dis	25	0·1 KCl	$K_1 \ll 3$	62I
Ni^{2+}	dis	25	0·1 KCl	K_1 5·0, K_2 3·5	62I
Zn^{2+}	dis	25	0·1 KCl	K_1 4·1, K_2 3·6	62I

61J R. B. James and R. J. P. Williams, *J. Chem. Soc.*, 1961, 2007
62I H. Irving and D. H. Mellor, *J. Chem. Soc.*, 1962, 5237

$C_{14}H_{12}N_2$ 4,7-Dimethyl-1,10-phenanthroline L 982

Metal	Method	Temp	Medium	Log of equilibrium constant, remarks	Ref
H^+	gl	25	0·3 (KNO_3) 50% dioxan	K_1 $5{\cdot}40 \pm 0{\cdot}02$	61J
	sp	25	0·1	K_1 5·95	63B
Co^{2+}	dis	25	0·1	K_1 8·08, K_2 8·00, K_3 8·43	63B
Cu^+				See Cu^{2+} below.	
Cu^{2+}	Pt	25	0·3 (KNO_3) 50% dioxan	K_1 ~ 7·20, K_3 4·86; $K(CuL_2^{2+} + Cu^+ \rightleftharpoons CuL_2^+ + Cu^{2+})$ −0·50	61J
	dis	25	0·1	K_1 8·76, K_2 7·26, K_3 5·97	63B
Fe^{2+}	sp	25	0·1	K_1 5·60	63B
Ni^{2+}	dis	25	0·1	K_1 8·44, K_2 8·20, K_3 8·40	63B
Zn^{2+}	dis	25	0·1	K_1 6·90, K_2 6·18, K_3 6·04	63B

61J B. R. James and R. J. P. Williams, *J. Chem. Soc.*, 1961, 2007
63B D. A. Brisbin and W. A. E. McBryde, *Canad. J. Chem.*, 1963, 41, 1135

982.1 $C_{14}H_{12}N_2$ 5,6-Dimethyl-1,10-phenanthroline L

Metal	Method	Temp	Medium	Log of equilibrium constant, remarks	Ref
H^+	gl	25	0·3 (KNO_3) 50% dioxan	K_1 5·00±0·03	61J
	sp	25	0·1	K_1 5·60	63B
Co^{2+}	dis	25	0·1	K_1 7·47, K_2 8·00, K_3 8·14	63B
Cu^+				See Cu^{2+} below.	
Cu^{2+}	Pt	25	0·3 (KNO_3) 50% dioxan	K_1 7·15, K_3 4·93; $K(CuL_2^{2+}+Cu^+ \rightleftharpoons CuL_2^+ + Cu^{2+})$ 0·0	61J
	dis	25	0·1	K_1 8·71, K_2 6·99, K_3 5·41	63B
Fe^{2+}	sp	25	0·1	K_1 6·37	63B
Ni^{2+}	dis	25	0·1	K_1 8·25, K_2 8·30, K_3 8·21	63B
Zn^{2+}	dis	25	0·1	K_1 6·87, K_2 6·02, K_3 5·71	63B

61J B. R. James and R. J. P. Williams, *J. Chem. Soc.*, 1961, 2007
63B D. A. Brisbin and W. A. E. McBryde, *Canad. J. Chem.*, 1963, **41**, 1135

982.2 $C_{14}H_{14}N_4$ *NN′*-Bis(2′-picolinylidene)-1,2-diaminoethane L

Metal	Method	Temp	Medium	Log of equilibrium constant, remarks	Ref
Ag^+	Ag	20	0·1 $NaNO_3$ methanol	K_1 9·78, β_2 11·00	66H

66H E. Hoyer and V. V. Skopenko, *Russ. J. Inorg. Chem.*, 1966, **11**, 436 (803)

982.3 $C_{14}H_{15}N$ Dibenzylamine L

Metal	Method	Temp	Medium	Log of equilibrium constant, remarks	Ref
H^+	gl	25	50 mole % C_2H_5OH	K_1 8·09	55A
Ag^+	gl	25	50 mole % C_2H_5OH	K_1 2·99, K_2 3·71	55A

Note. This Table was No. **993**, p. 689, in the 1964 edition and has been renumbered to correct the ligand sequence.

55A C. T. Anderson, doctoral dissertation, Ohio State Univ., 1955

984.1 $C_{14}H_{16}N_2$ 4,4′-Diethyl-2,2′-bipyridyl L

Metal	Method	Temp	Medium	Log of equilibrium constant, remarks	Ref
H^+	gl	25	0·3 (KNO_3) 50% dioxan	K_1 4·38±0·02	61J
Cu^+				See Cu^{2+} below.	
Cu^{2+}	Pt	25	0·3 (KNO_3) 50% dioxan	K_2 ~ 6·9, K_3 ~ 2·9; $K(CuL_2^{2+}+Cu^+ \rightleftharpoons CuL_2^+ + Cu^{2+})$ −0·58	61J

61J B. R. James and R. J. P. Williams, *J. Chem. Soc.*, 1961, 2007

$C_{14}H_{18}N_4$ *NN*′-Bis(2′-pyridylmethyl)-1,2-diaminoethane L 985

Metal	*Method*	*Temp*	*Medium*	*Log of equilibrium constant, remarks*	*Ref*
H^+	gl	25	0·1 (KCl)	K_1 8·28, K_2 5·47, K_3 2·0, K_4 1·8	68G
Co^{2+}	gl	25	0·1 (KCl)	K_1 12·0	68G
Ni^{2+}	gl	25	0·1 (KCl)	K_1 14·4	68G
Zn^{2+}	gl	25	0·1 (KCl)	K_1 11·4	68G

Errata. 1964, p. 687, Table 985. *For* reference L *read* ref. 64L R. G. Lacoste and A. E. Martell, *Inorg. Chem.*, 1964, **3,** 881.
Under Medium *for* 0·1 *read* 0·1 (KNO_3).

68G D. W. Gruenwedel, *Inorg. Chem.*, 1968, **7**, 495

$C_{14}H_{22}O_2$ 1,2-Dihydroxy-3,5-bis(1′,1′-dimethylethyl)benzene H_2L 985.1

Metal	*Method*	*Temp*	*Medium*	*Log of equilibrium constant, remarks*	*Ref*
H^+	gl	25	0·1 KNO_3 50% methanol	K_1 14·7, K_2 10·354	68T
Cu^{2+}	gl	25	0·1 KNO_3 50% methanol	$K(Cu^{2+}+H_2L \rightleftharpoons CuHL^+ + H^+)$ −0·94±0·01	68T
Mn^{2+}	gl	25	0·1 KNO_3 50% methanol	$K[Mn^{2+}+H_2L \rightleftharpoons Mn(OH)HL+2H^+]$ −14·68±0·01 $K[Mn^{2+}+2H_2L \rightleftharpoons Mn(HL)_2+2H^+]$ −12·23±0·01	68T

68T C. A. Tyson and A. E. Martell, *J. Amer. Chem. Soc.*, 1968, **90**, 3379

$C_{14}H_8O_7S$ 1,2-Dihydroxyanthraquinone-3-sulphonic acid (Alizarin Red S) H_3L 986

Metal	*Method*	*Temp*	*Medium*	*Log of equilibrium constant, remarks*	*Ref*
H^+	gl	20	0·1 (KNO_3)	K_1 11·1, K_2 6·07	67B, 68B
	sp	25	0·5	K_1 10·85±0·03, K_2 5·49±0·01	67Z
B^{III}	gl	20	0·1 (KNO_3)	$K[H_3BO_3+H_3L \rightleftharpoons B(OH)_2HL^- + H^+]$ −3·4±0·1	68B
Be^{2+}	gl	20	0·1 (KNO_3)	K_1 10·96	67B
Cr^{VI}	sp	25	?	K (?) 4·72	61Ba
	sp	25	?	K (?) 4·6±0·2	64S
Gd^{3+}	sp	25	?	K (?) 8·6±0·4	67S
Hf^{4+}	sp	25	?	K (?) 10·4±0·3	62B
Lu^{3+}	sp	25	?	K (?) 9·2±0·4	67S
Mo^{VI}	sp	25	0·1 ($NaClO_4$)	$K(MoO_4^{2-}+H_2L^- \rightleftharpoons MoO_2L_2^{3-})$ (?) 9·2	63S

continued overleaf

986 $C_{14}H_8O_7S$ (contd)

Metal	*Method*	*Temp*	*Medium*	*Log of equilibrium constant, remarks*	*Ref*
Tb^{3+}	sp	25	?	K (?) 9·1±0·4	67S
Th^{4+}	sp	30	0·1 (Na_4NO_3)	K (?) 8·2	63Sa
UO_2^{2+}	sp	30	0·15 ($NaClO_4$)	K (?) 4·5	63Sa
V^{IV}	sp	25	?	$K(VO_3^{2-}+H_2L^- \rightleftharpoons VO_2L^{3-})$ (?) 8·4±0·2	66S
V^{V}	sp	25	?	$K(VO_3^-+H_2L^- \rightleftharpoons VO_2L^{2-})$ (?) 8·6±0·3	61B
	sp	25	0·1 ($NaClO_4$)	$K(VO_3^-+H_2L^- \rightleftharpoons VO_2L^{2-})$ (?) 8·5	62S
Yb^{3+}	sp	25	?	K (?) 8·7±0·5	67S
Zr^{4+}	sp	25	1·6	$K[Zr^{4+}+2OH^-+L^{2-} \rightleftharpoons Zr(OH)_2L]$ 49·0	67Z

61B S. K. Banerji and A. K. Dey, *Z. anorg. Chem.*, 1961, **309**, 226
61Ba S. K. Banerji and A K. Dey, *J. Indian Chem. Soc.*, 1961, **38**, 121
62B S. K. Banerji and A. K. Dey, *Bull. Chem. Soc. Japan*, 1962, **35**, 2051
62S R. L. Seth and A. K. Dey, *J. Indian Chem. Soc.*, 1962, 39, 724
63S R. L. Seth and A. K. Dey, *Z. anorg. Chem.*, 1963, **321**, 278
63Sa S. C. Srivastava and A. K. Dey, *Indian J. Chem.*, 1963, **1**, 200, 242
64S S. P. Sangal, *Chim. Analyt.*, 1964, **46**, 492
66S P. Sanyal and S. P. Mushran, *Analyt. Chim. Acta*, 1966, **35**, 400
67B M. Bartusek and J. Zelinka, *Coll. Czech. Chem. Comm.*, 1967, **32**, 992
67S S. P. Sangal, *J. prakt. Chem.*, 1967, **36**, 126
67Z H. E. Zittel and T. M. Florence, *Analyt. Chem.*, 1967, **39**, 320
68B M. Bartusek and L. Havelkova, *Coll. Czech. Chem. Comm.*, 1968, 33, 385

986.1 $C_{14}H_8O_7S$ 1,4-Dihydroxyanthraquinone-2-sulphonic acid (quinizarin-2-sulphonic acid) H_3L

Metal	*Method*	*Temp*	*Medium*	*Log of equilibrium constant, remarks*	*Ref*
H^+				H^+ constants not given.	
Fe^{3+}	sp	29	0·1 (NaCl)	$K(Fe^{3+}+H_2L^- \rightleftharpoons FeHL^++H^+)$ (?) 3·73	64J

64J D. P. Joshi and D. V. Jain, *J. Indian Chem. Soc.*, 1964, **41**, 33

986.2 $C_{14}H_9O_2N$ 2-(2′-Pyridyl)indane-1,3-dione HL

Metal	*Method*	*Temp*	*Medium*	*Log of equilibrium constant, remarks*	*Ref*
H^+	gl	30	75% dioxan	K_1 13·56	64C
Be^{2+}	gl	30	75% dioxan	K_1 10·96±0·02, K_2 10·75±0·02	64C
Co^{2+}	gl	30	75% dioxan	K_1 9·8±0·2, K_2 9·6±0·2	64C
Cu^{2+}	gl	30	75% dioxan	K_1 12·89±0·02	64C
Mg^{2+}	gl	30	75% dioxan	K_1 6·36±0·02, K_2 5·27±0·02	64C
Mn^{2+}	gl	30	75% dioxan	K_1 8·06±0·02, K_2 7·68±0·02	64C
Ni^{2+}	gl	30	75% dioxan	K_1 10·60±0·02, K_2 10·09±0·02	64C
UO_2^{2+}	gl	30	75% dioxan	K_1 11·76±0·02, K_2 10·61±0·02	64C
Zn^{2+}	gl	30	75% dioxan	K_1 9·64±0·02, K_2 9·34±0·02	64C

64C J. R. Cook and D. F. Martin, *J. Inorg. Nuclear Chem.*, 1964, **26**, 571

$C_{14}H_{10}O_4N_2$ 4,5-Diamino-1,8-dihydroxyanthraquinone (diaminochrysazin) H_2L 987.1

Metal	Method	Temp	Medium	Log of equilibrium constant, remarks	Ref
B^{III}	sp	25	?	$K[B(OH)_3+H_2L \rightleftharpoons B(OH)_2HL]$ 3·54±0·05	64B

64B R. S. Brown, *Canad. J. Chem.*, 1964, **42**, 2635

$C_{14}H_{11}O_2N$ 1-Phenyl-3-(2′-pyridyl)propane-1,3-dione (benzoyl-2-picolinoylmethane) HL 987.2

Metal	Method	Temp	Medium	Log of equilibrium constant, remarks	Ref
H^+				H^+ constants not given.	
Cu^{2+}	gl	25	0·002—0·01 75% dioxan	K_1 (0·002) 13·09, (0·01) 12·67, K_2 (0·002) 13·10, (0·01) 11·94	67W
Ni^{2+}	gl	25	0·002—0·01 75% dioxan	K_1 (0·002) 12·13, (0·01) 11·85, K_2 (0·002) 11·85, (0·01) 11·04	67W
Zn^{2+}	gl	25	0·002—0·01 75% dioxan	K_1 (0·002) 11·15, (0·01) 10·86, K_2 (0·002) 10·85, (0·01) 10·04	67W

67W L. Wolf, H. Hennig, and I. P. Sereda, *Russ. J. Inorg. Chem.*, 1967, **12**, 231 (450)

$C_{14}H_{11}O_2N$ 1-Phenyl-3-(3′-pyridyl)propane-1,3-dione (benzoylnicotinoylmethane) HL 987.3

Metal	Method	Temp	Medium	Log of equilibrium constant, remarks	Ref
H^+				H^+ constants not given.	
Cu^{2+}	gl	25	0·002—0·01 75% dioxan	K_1 (0·002) 12·41, (0·01) 11·83, K_2 (0·002) 12·33, (0·01) 11·44	67W
Ni^{2+}	gl	25	0·002—0·01 75% dioxan	K_1 (0·002) 10·77, (0·01) 10·97, K_2 (0·002) 10·68, (0·01) 9·64	67W
Zn^{2+}	gl	25	0·002—0·01 75% dioxan	K_1 (0·002) 10·39, (0·01) 10·15, K_2 (0·002) 9·86, (0·01) 9·32	67W

67W L. Wolf, H. Hennig, and I. P. Sereda, *Russ. J. Inorg. Chem.*, 1967, **12**, 231 (450)

$C_{14}H_{11}O_2N$ 1-Phenyl-3-(4′-pyridyl)propane-1,3-dione (benzoylisonicotinoylmethane) HL 987.4

Metal	Method	Temp	Medium	Log of equilibrium constant, remarks	Ref
H^+				H^+ constants not given.	
Cu^{2+}	gl	25	0·002—0·01 75% dioxan	K_1 (0·002) 11·75, (0·01) 11·37, K_2 (0·002) 11·69, (0·01) 10·88	67W
Ni^{2+}	gl	25	0·002 75% dioxan	K_1 10·88, K_2 10·55	67W
Zn^{2+}	gl	25	0·002—0·01 75% dioxan	K_1 (0·002) 9·92, (0·01) 9·62, K_2 (0·002) 9·50, (0·01) 8·81	67W

67W L. Wolf, H. Hennig, and I. P. Sereda, *Russ. J. Inorg. Chem.*, 1967, **12**, 231 (450)

989.1 $C_{14}H_{12}O_2N_2$ 4-Phenylazo-2-acetylphenol (3-acetyl-4-hydroxyazobenzene) HL

Metal	Method	Temp	Medium	Log of equilibrium constant, remarks	Ref
H^+	gl	30	75% dioxan	K_1 12·30	67U
Be^{2+}	gl	30	75% dioxan	K_1 9·95, K_2 8·56	67U
Cu^{2+}	gl	30	75% dioxan	K_1 9·64, K_2 9·05	67U
Ni^{2+}	gl	30	75% dioxan	K_1 7·47, K_2 6·71	67U

67U E. Uhlemann and F. Dietze, *Z. anorg. Chem.*, 1967, **353**, 26

990 $C_{14}H_{12}O_3N_2$ 2-(2′-Hydroxy-5′-methylphenylazo)benzoic acid (2-hydroxy-5-methyl-2-carboxyazobenzene) H_2L

Metal	Method	Temp	Medium	Log of equilibrium constant, remarks	Ref
H^+	sp	?	0·100 (KCl)	K_1 11·4	60D
Ca^{2+}	sp	?	0·100 (KCl)	K_1 1·39	60D
Mg^{2+}	sp	?	0·100 (KCl)	K_1 3·68	60D

60D H. Diehl and J. Ellingboe, *Analyt. Chem.*, 1960, **32**, 1120

992.1 $C_{14}H_{13}O_2N$ *N*-2′(Tolyl)benzohydroxamic acid HL

Metal	Method	Temp	Medium	Log of equilibrium constant, remarks	Ref
H^+				H^+ constants not given.	
V^V	sp	28	?	$K[VO_3^- + 2HL \rightleftharpoons VOL_2^-]$ (?) 8·1	64M

64M A. K. Majumdar and G. Das, *Analyt. Chim. Acta*, 1964, **31**, 147

992.2 $C_{14}H_{13}O_2N_3$ 1-(4′-acetylphenyl)-3-hydroxy-3-phenyltriazene HL

Metal	Method	Temp	Medium	Log of equilibrium constant, remarks	Ref
H^+	sp	?	?	K_1 10·96	64P
Cu^{2+}	gl	25	0·1 (KCl) 70% dioxan	K_1 10·91, K_2 10·25	64P
Mn^{2+}	gl	25	0·1 (KCl) 70% dioxan	K_1 6·09, K_2 4·80	64P
Ni^{2+}	gl	25	0·1 (KCl) 70% dioxan	K_1 8·79, K_2 7·15	64P
Pd^{2+}	gl	25	0·1 (KCl) 70% dioxan	K_1 10·97, K_2 10·54	64P
Zn^{2+}	gl	25	0·1 (KCl) 70% dioxan	K_1 7·79, K_2 6·45	64P

64P D. N. Purohit and N. C. Sogani, *J. Indian Chem. Soc.*, 1964, **41**, 20

$C_{14}H_{14}O_2N_4$ 1-[4′-(Acetylamino)phenyl]-3-hydroxy-3-phenyltriazene HL 992.3

Metal	Method	Temp	Medium	Log of equilibrium constant, remarks	Ref
H^+	sp	?	?	K_1 11·66	65P
Cu^{2+}	gl	25	0·1 KCl 70% dioxan	K_1 11·34, K_2 11·15	65P

65P D. N. Purohit and N. C. Sogani, *Z. Naturforsch.*, 1965, **20B**, 206

$C_{14}H_{14}O_2N_4$ *NN*′-Bis(2-pyridylmethyl)oxamide L 992.4

Metal	Method	Temp	Medium	Log of equilibrium constant, remarks	Ref
Cu^{2+}	gl	22	0·1 (KNO_3)	$K[Cu(H_{-2}L)+Cu^{2+} \rightleftharpoons Cu_2(H_{-2}L)^{2+}]$ 6·54	68G

68G R. Griesser and S. Fallab, *Chimia (Switz.)*, 1968 **22**, 90

$C_{14}H_{15}N$ Dibenzylamine L 993

Erratum. 1964, p. 689, Table 993. This Table has been renumbered **982.3** to correct the ligand sequence.

$_4H_{20}O_6N_2$ Ethyl hydrogen-2,5-bis-[*N*-(2′-hydroxyethyl)amino]benzene-1,4-dicarboxylate HL 995.1

Metal	Method	Temp	Medium	Log of equilibrium constant, remarks	Ref
H^+	gl	22	0·1 KCl	K_1 5·95	60U
Co^{2+}	gl	22	0·1 KCl	K_1 2·3, K_2 3·05	60U
Cu^{2+}	gl	22	0·1 KCl	K_1 4·85, K_2 3·9	60U
Ni^{2+}	gl	22	0·1 KCl	K_1 3·3, K_2 4·65	60U

60U E. Uhlig, *Z. anorg. Chem.*, 1960, **306**, 71

995.2 $C_{14}H_{20}O_6N_6$ *N*-Acetylglycylglycyl-L-histidylglycine HL

Metal	Method	Temp	Medium	Log of equilibrium constant, remarks	Ref
H^+	gl	25	0·16 KCl	K_1 6·90, K_2 3·29	66B
Cu^{2+}	gl	25	0·16 KCl	K_1 3·95, K_2 3·80	66B
				$K[Cu(H_{-1}L)+H^+ \rightleftharpoons CuL^+]$ 5·95	
				$K[Cu(H_{-2}L)^- + H^+ \rightleftharpoons Cu(H_-L)]$ 6·45	
				$K[Cu(H_{-3}L)^{2-} + H^+ \rightleftharpoons Cu(H_{-2}L)^-]$ 9·00	
Ni^{2+}	gl	25	0·16 KCl	K_1 2·93	66B
				$K[Ni(H_{-1}L)+H^+ \rightleftharpoons NiL^+]$ 8·40	
				$K[Ni(H_{-2}L)^- + H^+ \rightleftharpoons Ni(H_{-1}L)]$ 8·50	
				$K[Ni(H_{-3}L)^{2-} + H^+ \rightleftharpoons Ni(H_{-2}L)^-]$ 8·60	

66B G. F. Bryce, R. W. Roeske, and F. R. N. Gurd, *J. Biol. Chem.*, 1966, **241**, 1072

995.3 $C_{14}H_{20}O_7N_2$ 2-[*N*-(2′-Hydroxyethyl)amino]-5-[*NN*-bis(2″-hydroxyethyl)amino]benzene-1,4-dicarboxylic acid H_2L

Metal	Method	Temp	Medium	Log of equilibrium constant, remarks	Ref
H^+	gl	22	0·1 KCl	K_1 8·29, K_2 3·65	61U
Cu^{2+}	gl	22	0·1 KCl	K_1 5·95,	61U
				$K(Cu^{2+} + HL^- \rightleftharpoons CuHL^+)$ 2·15	
Ni^{2+}	gl	22	0·1 KCl	K_1 4·10,	61U
				$K(Ni^{2+} + HL^- \rightleftharpoons NiHL^+)$ 1·45	

61U E. Uhlig, *Z. anorg. Chem.*, 1961, **312**, 332

997 $C_{14}H_{22}O_8N_2$ (*trans*-1,2-Cyclohexylenedinitrilo)tetra-acetic acid (CDTA) H_4L

Metal	Method	Temp	Medium	Log of equilibrium constant, remarks	Ref
H^+	cal, sp	20	0·1 (KNO_3)	K_1 12·35±0·2	63A
				$\Delta H_1°$ −6·65, $\Delta S_1°$ 33·9, $\Delta H_2°$ −2·06, $\Delta S_2°$ 21·1	
	sol	20	→0	K_3 1·92, K_4 0·96	63R
	?	?	?	K_1 11·70, K_2 6·12, K_3 3·52, K_4 2·43	63S
	gl	20—40	0·1 (KNO_3)	K_1 (20°) 11·70, (30°) 11·52, (40°) 11·34, K_2 (20°) 6·14, (30°) 6·10, (40°) 6·07	66M

conti

Metal	Method	Temp	Medium	Log of equilibrium constant, remarks	Ref
H^+ (contd.)	gl	25	0·5 (KNO_3) 10% methanol	K_1 11·40, K_2 5·93, K_3 3·51, K_4 2·83	66P
	gl	20	0·1 [$(CH_3)_4NCl$]	K_2 6·15	67Aa
	gl	20	0·1 (KNO_3)	K_2 6·12	67Aa
	H	20	1 [$(CH_3)_4NCl$]	K_2 6·15, K_3 3·58, K_4 2·45	67Aa
	gl	20	1 ($NaClO_4$)	K_1 9·30, K_2 5·87, K_3 3·52	67Aa
	sol	20	1 ($NaClO_4$)	K_4 2·41, K_5 1·72	67Aa
	gl	25	0·2 ($NaClO_4$)	K_1 11·30, K_2 6·05, K_3 3·45, K_4 2·50	67B
	oth	25	0·1	K_1 12·27, K_2 6·09, calculated from 63A	67M
	sp	20	0·5 $NaClO_4$	K_1 9·26, K_2 5·75, K_3 3·20, K_4 1·34	67N
	gl	25	0·1 (KNO_3)	K_1 11·78, K_2 6·20, K_3 3·60, K_4 2·51	68W
Ag^+	dis	20	0·1 ($KClO_4$)	K_1 8·15 ± 0·03	63S
	Ag	25	0·1 (KNO_3)	K_1 8·41	68W
Al^{3+}	dis	20	0·1 ($KClO_4$)	$K[Al^{3+} + L^{4-} + OH^- \rightleftharpoons AlL(OH)^{2-}]$ 26·61 ± 0·08	63S
	Hg	20—40	0·1 (KNO_3)	K_1 (20°) 18·63 ± 0·08, (30°) 18·8 ± 0·3, (40°) 19·15 ± 0·06 $\Delta H_1°$ 11 ± 2, $\Delta S_1°$ 122 ± 7 $K(AlL + H^+ \rightleftharpoons AlHL^+)$ (20°) 2·59	66M
	Hg	25	0·1 KNO_3	K_1 18·9, $K(Al^{3+} + HL^{3-} \rightleftharpoons AlHL)$ 3·4 $K(AlL^- + OH^- \rightleftharpoons AlOHL^{2-})$ 6·3	67A
	sp	25	0·2 ($NaClO_4$)	K_1 18·50	67B
	gl	25	0·2 ($NaClO_4$)	$K(AlL^- + H^+ \rightleftharpoons AlHL)$ 2·29 $K(AlLOH^{2-} + H^+ \rightleftharpoons AlL^-)$ 7·82	67B
Am^{3+}	ix	25	0·1 NH_4ClO_4	K_1 18·79	66Ba
Ba^{2+}	cal	20	0·1 (KNO_3)	$\Delta H_1°$ 0·33, $\Delta S_1°$ 40·9	63A
	cal	25	0·1 KNO_3	ΔH_1 −2·2, ΔS_1 29	65W
Be^{2+}	dis	20	0·1 ($KClO_4$)	K_1 10·81 ± 0·06	63S
Bi^{3+}	pol	20	0·1	K_1 24·1 ± 0·4	61S
	dis	20	0·1 ($KClO_4$)	$K[Bi^{3+} + L^{4-} + OH^- \rightleftharpoons BiL(OH)^{2-}]$ 34·6 ± 0·1	63S
	pol	25	0·5 (KNO_3)	K_1 31·2 ± 0·2	66B
	red	20	1 ($NaClO_4$)	$K(BiL^- + H^+ \rightleftharpoons BiHL)$ 1·25 $K(BiL^- + OH^- \rightleftharpoons BiLOH^{2-})$ 3·0	67B
	sp	20	0·5 $NaClO_4$	K_1 23·8, $K(Bi^{3+} + HL^{3-} \rightleftharpoons BiHL)$ 15·7	67N
Bk^{3+}	ix	25	0·1 NH_4ClO_4	K_1 19·16	66Ba
Ca^{2+}	cal	20	0·1 (KNO_3)	$\Delta H_1°$ −3·70, $\Delta S_1°$ 47·5	63A
	cal	25	0·1 KNO_3	ΔH_1 −6·2, ΔS 35	65W
Cd^{2+}	cal	20	0·1 (KNO_3)	$\Delta H_1°$ −7·40, $\Delta S_1°$ 65·7	63A
	dis	20	0·1 ($KClO_4$)	K_1 ~ 19·0	63S
	pol	30	0·1—1·0 KNO_3	K_1 (0·1) 19·12, (1·0) 18·87	65J
	cal	25	0·1 KNO_3	ΔH_1 −11·2, ΔS_1 50	65W
Cf^{3+}	ix	25	0·1 NH_4ClO_4	K_1 19·42	66Ba
Cm^{3+}	ix	25	0·1 NH_4ClO_4	K_1 18·81	66Ba
Co^{2+}	sp	20	0·08	K_1 21·9 ± 0·2	61J
	cal	20	0·1 (KNO_3)	$\Delta H_1°$ −2·80, $\Delta S_1°$ 80·0	63A
	dis	20	0·1 ($KClO_4$)	K_1 18·92 ± 0·04	63S
	cal	25	0·1 KNO_3	ΔH_1 −5·4, ΔS_1 68	65W
	sp	25	0·2 ($NaClO_4$)	K_1 18·78	67Ba
	pol	25	0·1—1 KNO_3	K_1 (0·1) 18·6, K (1·0) 18·3	67J
Cu^{2+}	cal	20	0·1 (KNO_3)	$\Delta H_1°$ −6·07, $\Delta S_1°$ 79·7	63A
	gl	25	1·25 ($NaClO_4$)	$K(CuL^{2-} + H^+ \rightleftharpoons CuHL^-)$ 2·68 $K(CuHL^- + H^+ \rightleftharpoons CuH_2L)$ 1·72	63M
	dis	20	0·1 ($KClO_4$)	K_1 21·59 ± 0·02	63S
	cal	25	0·1 KNO_3	ΔH_1 −8·8, ΔS_1 67	65W
	pol	25	0·1 KNO_3	K_1 21·1	67J

continued overleaf

997 $C_{14}H_{22}O_8N_2$ (contd)

Metal	Method	Temp	Medium	Log of equilibrium constant, remarks	Ref
Eu^{3+}	ix	25	0·1 NH_4ClO_4	K_1 18·87	66Ba
Fe^{2+}	pol	30	1 $NaClO_4$	K_1 16·27	63R
	cal	25	1 KNO_3	ΔH_1 −6·6, ΔS_1 61	65W
Fe^{3+}	sp	30	1·0	K_1 26·93	63R
	dis	20	0·1 ($KClO_4$)	$K[Fe^{3+}+L^{4-}+OH^- \rightleftharpoons FeL(OH)^{2-}]$ 36·6±0·1	63S
	red	20	0·1 ($NaClO_4$)	K_1 28·05, $K(FeL^- + OH^- \rightleftharpoons FeLOH^{2-})$ 9·70	67B
Ga^{3+}	dis	20	0·1 ($KClO_4$)	$K(Ga^{3+}+L^{4-}+OH^- \rightleftharpoons GaL(OH)^{2-}]$ 35·6±0·1	63S
	Hg	20	0·1 (KNO_3)	K_1 22·5, $K(GaL+H^+ \rightleftharpoons GaHL^+)$ 2·43	66M
	red	20	0·1 ($NaClO_4$)	K_1 23·10, $K(GaL^- + OH^- \rightleftharpoons GaLOH^{2-})$ 6·46	67B
Hg^{2+}	cal	20	0·1 (KNO_3)	$\Delta H_1°$ −16·60, $\Delta S_1°$ 59·0	63A
	Hg	10—30	0·10 (KNO_3)	K_1 (10°) 24·16, (20°) 23·77, (30°) 23·47 $\Delta H_1°$ (25°) −13·7, $\Delta S_1°$ 62	65H
	cal	25	0·1 KNO_3	ΔH_1 −18·9, ΔS_1 48	65W
	sp	25	0·10 ($NaClO_4$)	$K(HgL^{2-}+A^- \rightleftharpoons HgLA^{3-})$ 3·20±0·05 where A = OH^-; 2·16±0·05 where A = Cl^-; 3·20±0·04 where A = Br^-; 5·3±0·1 where A = I^-; 4·29±0·05 where A = SCN^-	66J
In^{3+}	ix	?	0·5	K_1 25·05±0·09	63Ra
	dis	20	0·1 ($KClO_4$)	$K[In^{3+}+L^{4-}+OH^- \rightleftharpoons InL(OH)^{2-}]$ 33·46±0·05	63S
	red	20	0·1 ($NaClO_4$)	K_1 28·74, $K(InL^- + OH^- \rightleftharpoons InLOH^{2-})$ 5·00	67B
La^{3+}	cal	20	0·1 (KNO_3)	$\Delta H_1°$ 1·60, $\Delta S_1°$ 82·8	63A
	dis	20	0·1 ($KClO_4$)	K_1 16·75±0·05	63S
Li^+	pol	30	0·1 KNO_3	K_1 4·13	67S
Mg^{2+}	cal	20	0·1 (KNO_3)	$\Delta H_1°$ 3·80, $\Delta S_1°$ 63·1	63A
	cal	25	0·1 KNO_3	ΔH_1 1·6, ΔS_1 52	65W
Mn^{2+}	cal	20	0·1 (KNO_3)	$\Delta H_1°$ −4·14, $\Delta S_1°$ 65·6	63A
	dis	20	0·1 ($KClO_4$)	K_1 14·70±0·06	63S
	cal	25	0·1 KNO_3	ΔH_1 −7·1, ΔS_1 52	65W
Mn^{3+}	sp	25	0·2 ($NaClO_4$)	K_1 28·9	67H
Na^+	pol	30	0·1 KNO_3	K_1 2·70	67S
Ni^{2+}	cal	20	0·1 (KNO_3)	$\Delta H_1°$ −5·37	63A
	kin	25	1·25 ($NaClO_4$)	K_1 19·9±0·1	63M
	gl	25	1·25 ($NaClO_4$)	$K(NiL^{2-}+H^+ \rightleftharpoons NiHL^-)$ 2·74 $K(NiHL^- + H^+ \rightleftharpoons NiH_2L)$ 1·80	
	dis	20	0·1 ($KClO_4$)	K_1 19·68±0·04	63S
	cal	25	0·1 KNO_3	ΔH_1 −7·5, ΔS_1 63	65W
	kin	25	0·10 ($NaClO_4$)	K_1 21·35±0·03	67M
Pb^{2+}	cal	20	0·1 (KNO_3)	$\Delta H_1°$ −11·36, $\Delta S_1°$ 54·2	63A
	dis	20	0·1 ($KClO_4$)	K_1 19·5±0·1	63S
	pol	30	0·1—1·0	K_1 (0·1) 19·60, (1·0) 19·16	65J
		40	0·1	K_1 19·32	
	cal	25	0·1 KNO_3	ΔH_1 −12·4, ΔS_1 47	65W
Sc^{3+}	dis	20	0·1 ($KClO_4$)	K_1 ~ 25·4 $K[Sc^{3+}+L^{4-}+OH^- \rightleftharpoons ScL(OH)^{2-}]$ ~ 28·0	63S
Sr^{2+}	gl	20	0·1 KCl	K_1 10·69	59K
	cal	20	0·1 (KNO_3)	$\Delta H_1°$ −0·74, $\Delta S_1°$ 45·7	63A
	cal	25	0·1 KNO_3	ΔH_1 −3·6, ΔS_1 34	65W

conti

Metal	Method	Temp	Medium	Log of equilibrium constant, remarks	Ref
Th^{4+}	gl	25	0·1 KNO_3	$K(ThL+A^{4-} \rightleftharpoons ThLA^{4-})$	
				12·67±0·7, where H_4A = 1,2-dihydroxybenzene-3,5-disulphonic acid	
				13·13±0·02, where H_4A = 1·8-dihydroxynaphthalene-3,6-disulphonic acid	
				$K(ThL+A^{3-} \rightleftharpoons ThLA^{3-})$	
				8·87±0·08, where H_3A = 5-sulphosalicylic acid	
				$K(ThL+A^{2-} \rightleftharpoons ThLA^{2-})$	
				12·26±0·05, where H_2A = catechol	
				6·69±0·02, where H_2A = 8-hydroxyquinoline-5-sulphonic acid	
				6·11±0·08, where H_2A = iminodiacetic acid	
				2·63±0·03, where H_2A = 2-phthalic acid	
	Hg	10—30	0·10 (KNO_3)	K_1 (10°) 23·79, (20°) 23·78, (30°) 23·77	65H
				$\Delta H_1°$ −0·5, $\Delta S_1°$ 107	
	red	20	0·1 ($NaClO_4$)	K_1 29·25	67B
				$K(ThL+H^+ \rightleftharpoons ThHL^+)$ 2·50	
				$K[2ThL+2OH^- \rightleftharpoons Th_2L_2(OH)_2{}^{2-}]$ 5·70	
Ti^{4+}	dis	20	0·1 ($KClO_4$)	$K(TiO^{2+}+L^{4-} \rightleftharpoons TiOL^{2-})$ 19·9±0·1	63S
Tl^+	pol	25	0·5 (KNO_3)	K_1 5·33	66P
	gl	25	0·5 (KNO_3) 10% methanol	K_1 5·58	66P
	gl	20	0·1 (KNO_3)	K_1 6·7	67Ab
	pol	30	0·1 KNO_3	K_1 5·84	67S
Tl^{3+}	red	20	1 ($NaClO_4$)	K_1 38·3	67Ab
U^{4+}	gl	20	0·10 (KCl)	K_1 26·9±0·2,	68C
				$K(ULOH^- + H^+ \rightleftharpoons UL)$ 4·85±0·01	
				$K[(ULOH)_2{}^{2-}+2H^+ \rightleftharpoons 2UL]$ 6·24±0·04	
				$K[2ULOH^- \rightleftharpoons (ULOH)_2{}^{2-}]$ 3·5±0·05	
Zn^{2+}	cal	20	0·1 (KNO_3)	$\Delta H_1°$ −1·94, $\Delta S_1°$ 81·8	63A
	cal	25	0·1 KNO_3	ΔH_1 −7·7, ΔS_1 59	65W
	pol	25	0·1—1·0 (KNO_3)	K_1 (0·1) 18·5, (1·0) 18·0	67J
Zr^{4+}	ix	?	2·1	K_2 4·36	64P
	Hg	10—30	0·10 (KNO_3)	K_1 (10°) 20·85, (20°) 20·74, (30°) 20·64	65H
				$\Delta H_1°$ −4·2, $\Delta S_1°$ 81	

59K H. Kroll, *US-AEC Report, AECU-4322*, 1959
61J E. Jacobsen and A. R. Selmer-Olsen, *Analyt. Chim. Acta*, 1961, **25**, 476
61S A. R. Selmer-Olsen, *Acta Chem. Scand.*, 1961, **15**, 2052
63A G. Anderegg, *Helv. Chim. Acta*, 1963, **46**, 1833
63M D. W. Margerum and T. J. Bydalek, *Inorg. Chem.*, 1963, **2**, 683
63R T. P. Radhakrishran, S. C. Saraiya, and A. K. Sundaram, *Current Sci.*, 1963, **32**, 450
63Ra D. I. Ryabchikov, I. N. Marov, and Y. K'o-min, *Russ. J. Inorg. Chem.*, 1963, **8**, 326 (641)
63S J. Stary, *Analyt. Chim. Acta*, 1963, **28**, 132
64C G. H. Carey, R. F. Bogucki, and A. E. Martell, *Inorg. Chem.*, 1964, **3**, 1288
64P L. N. Pankratova, L. G. Vlasov, and A. V. Lapitskii, *Russ. J. Inorg. Chem.*, 1964, **9**, 742 (1363)
65H T. M. Hseu, S. F. Wu, and T. J. Chuang, *J. Inorg. Nuclear Chem.*, 1965, **27**, 1655
65J D. S. Jain and J. N. Gaur, *J. Indian Chem. Soc.*, 1965, **42**, 753, 759
65W D. L. Wright, J. H. Holloway, and C. N. Reilley, *Anal. Chem.*, 1965, **37**, 884
66B M. T. Beck and A. Gergely, *Acta Chim. Hung.*, 1966, **50**, 155
66Ba R. D. Baybarz, *J. Inorg. Nuclear Chem.*, 1966, **28**, 1055
66J D. L. Janes and D. W. Margerum, *Inorg. Chem.*, 1966, **5**, 1135
66M T. Moeller and S. Chu, *J. Inorg. Nuclear Chem.*, 1966, **28**, 153
66P F. Pantani, *Ricerca sci.*, 1966, **36**, 702
67A D. A. Aikens and F. J. Bahbah, *Analyt. Chem.*, 1967, **39**, 646
67Aa G. Anderegg, *Helv. Chim. Acta*, 1967, **50**, 2333
67Ab G. Anderegg and E. Bottari, *Helv. Chim. Acta*, 1967, **50**, 2341
67B E. Bottari and G. Anderegg, *Helv. Chim. Acta*, 1967, **50**, 2349

continued overleaf

997 $C_{14}H_{22}O_8N_2$ (contd)

67Ba T. R. Bhat, R. R. Das, and J. Shankar, *Indian J. Chem.*, 1967, **5**, 324
67H R. E. Hamm and M. A. Suwyn, *Inorg. Chem.*, 1967, **6**, 139
67J D. S. Jain and J. N. Gaur, *J. Indian Chem. Soc.*, 1967, **44**, 436
67M D. W. Margerum, P. J. Menardi, and D. L. Janes, *Inorg. Chem.*, 1967, **6**, 283
67N T. Nozaki and K. Koshiba, *Nippon Kagaku Zasshi*, 1967, **88**, 1287
67S R. Sundaresan, S. C. Saraiya, and A. K. Sundaram, *Current Sci.*, 1967, **36**, 255
68C G. H. Carey and A. E. Martell, *J. Amer. Chem. Soc.*, 1968, **90**, 32
68W H. Wikberg and A. Ringbom, *Suomen Kem.*, 1968, **B41**, 177

1000 $C_{14}H_{23}O_{10}N_3$ [(*N*-Carboxymethyl-2,2′-iminodiethylene)dinitrilo]tetra-acetic acid (diethylenetriaminepenta-acetic acid) (DTPA) H

Metal	*Method*	*Temp*	*Medium*	*Log of equilibrium constant, remarks*	*Ref*
H^+	?	20	0·1	K_1 10·53, K_2 8·6, K_3 4·26, K_4 2·38, K_5 2·14	62S
	cal	20	0·1 (KNO_3)	$\Delta H_1°$ −7·96, $\Delta S_1°$ 21·2, $\Delta H_2°$ −4·32 (?), $\Delta S_2°$ 17·7 (?), $\Delta H_3°$ −1·73, $\Delta S_3°$ 13·6	65A
	gl	20—40	0·1 (KNO_3)	K_1 (20°) 10·58, (30°) 10·34, (40°) 10·23, K_2 (20°) 8·60, (30°) 8·46, (40°) 8·37, K_3 (20°) 4·33, (30°) 4·30, (40°) 4·27, K_4 (20°) 2·55, K_5 (20°) 1·80	65B, 66M
	gl	25	0·1	K_1 10·42, K_2 8·76, K_3 4·42, K_4 2·56, K_5 1·79	65R, 67T
	gl	20	0·1 [$(CH_3)_4NCl$]	K_1 10·81, K_2 8·64	67A
	gl	20	0·1 (KNO_3)	K_1 10·58, K_2 8·60, K_3 4·27, K_4 2·6, K_5 1·5	67A
	H	20	1 [$(CH_3)_4NCl$]	K_1 10·46, K_2 8·41, K_3 4·14, K_4 2·7, K_5 2·2	67A
	gl	20	1 ($NaClO_4$)	K_1 9·48, K_2 8·26, K_3 4·19, K_4 2·5, K_5 2·5	67A
	sp	20	0·5 $NaClO_4$	K_1 9·49, K_2 8·18, K_3 3·39, K_4 2·35, K_5 1·13	67N
	cal	27	0·1	ΔH_1 −8·4, ΔH_2 −2·4	68Ca
	?	?	0·4	K_1 10·02, K_2 8·45, K_3 4·20, K_4 2·43, K_5 1·75	68K
	gl	25	0·1 (KNO_3)	K_1 10·56, K_2 8·69, K_3 4·37, K_4 2·87, K_5 1·94	68W
Ag^+	Ag	25	0·1 (KNO_3)	K_1 8·70	68W
Al^{3+}	Hg	20—40	0·1 (KNO_3)	K_1 (20°) 18·4±0·1, (25°) 18·51±0·02, (30°) 18·62±0·06, (40°) 18·80±0·08 $\Delta H_1°$ (25°) 8±1, $\Delta S_1°$ 113±3 $K(AlL + H^+ \rightleftharpoons AlHL^+)$ (20°) 4·63	66M
	Hg	25	0·1 KNO_3	K_1 18·7, $K(Al^{3+} + HL^{4-} \rightleftharpoons AlHL^-)$ 4·3 $K[AlL^{2-} + OH^- \rightleftharpoons Al(OH)L^{3-}]$ 6·6	67Ab
Am^{3+}	ix	25	0·1 NH_4ClO_4	K_1 22·92	65B
Ba^{2+}	cal	25	0·1 KNO_3	ΔH_1 −7·3, ΔS_1 14	65W
	cal	27	0·1	ΔH_1 −6·9, ΔS_1 16	68Ca
Bi^{3+}	red	20	1 ($NaClO_4$)	K_1 35·6, $K(BiL^{2-} + H^+ \rightleftharpoons BiHL^-)$ 2·6 $K(BiL^{2-} + OH^- \rightleftharpoons BiLOH^{3-})$ 2·7	67B
	sp	20	0·5 $NaClO_4$	K_1 29·7, $K(Bi^{3+} + HL^{4-} \rightleftharpoons BiHL^-)$ 22·5	67N
Bk^{3+}	ix	25	0·1 NH_4ClO_4	K_1 22·79	65B

conti

Metal	Method	Temp	Medium	Log of equilibrium constant, remarks	Ref
Ca^{2+}	cal	20	0·1 (KNO_3)	ΔH_1 −5·95, ΔS_1 29·5	65A
	cal	25	0·1 KNO_3	ΔH_1 −6·1, ΔS_1 29	65W
	cal	27	0·1	ΔH_1 −5·6, ΔS_1 30	68Ca
	gl	25	0·1 (KNO_3)	K_1 10·6	68W
Cd^{2+}	cal	20	0·1 (KNO_3)	ΔH_1 −12·35, ΔS_1 46·4	65A
	cal	25	0·1 KNO_3	ΔH_1 −12·4, ΔS_1 45	65W
	sp	18—22	~0	K_1 19·1±0·06	68Ka
Ce^{3+}	cal	27	0·1 (KNO_3)	ΔH_1 −5·8±0·3	68Ca
Cf^{3+}	ix	25	0·1 NH_4ClO_4	K_1 22·57	65B
Cm^{3+}	ix	25	0·1 NH_4ClO_4	K_1 22·99	65B
Co^{2+}	cal	20	0·1 (KNO_3)	ΔH_1 −9·41, ΔS_1 56·2	65A
	cal	25	0·1 KNO_3	ΔH_1 −9·5, ΔS_1 56	65W
	sp	18—22	~0	K_1 19·72, $K(Co^{2+}+HL^{4-} \rightleftharpoons CoHL^{2-})$ 11·89	68Ka
Cu^{2+}	cal	20	0·1 (KNO_3)	ΔH_1 −13·6, ΔS_1 52·2	65A
	cal	25	0·1 KNO_3	ΔH_1 −13·4, ΔS_1 53	65W
Dy^{3+}	cal	27	0·1 (KNO_3)	ΔH_1 −7·9±0·3, ΔS_1 78·0	68Ca
Er^{3+}	cal	27	0·1 (KNO_3)	ΔH_1 −7·4±0·3, ΔS_1 79·3	68Ca
Es^{3+}	ix	25	0·1 NH_4ClO_4	K_1 22·62±0·07	65B
Eu^{3+}	sp	18—20	0·1	K_1 23·17, $K(2Eu^{3+}+L^{5-} \rightleftharpoons Eu_2L^+)$ 26·23	63G
	ix	25	0·1 NH_4ClO_4	K_1 22·40	65B
	cal	27	0·1 (KNO_3)	ΔH_1 −7·9, ΔS_1 76·0	68Ca
Fe^{2+}	cal	25	0·1 KNO_3	ΔH_1 −7·7, ΔS_1 49	65W
Fe^{3+}	red	20	0·1 ($NaClO_4$)	K_1 27·3, $K(FeL^{2-}+H^+ \rightleftharpoons FeHL^-)$ 3·58, $K(FeL^{2-}+OH^- \rightleftharpoons FeLOH^{3-})$ 3·9	67B
Fm^{3+}	ix	25	0·1 NH_4ClO_4	K_1 22·7±0·1	65B
Ga^{3+}	Hg	20	0·1 (KNO_3)	K_1 23·0, $K(GaL^{2-}+H^+ \rightleftharpoons GaHL^-)$ 3·92	66M
	red	20	0·1 ($NaClO_4$)	K_1 25·54, $K(GaL^{2-}+H^+ \rightleftharpoons GaHL^-)$ 4·35, $K(GaL^{2-}+OH^- \rightleftharpoons GaLOH^{3-})$ 6·52	67B
Gd^{3+}	cal	27	0·1 (KNO_3)	ΔH_1 −7·8±0·3, ΔS_1 76·7	68Ca
Hf^{4+}	ix	?	1—2 $HClO_4$	$K(Hf^{4+}+H_5L \rightleftharpoons HfL^-+5H^+)$ (1) 4·86±0·04, (2) 3·13±0·04	64E
	x	?	0·23 $HClO_4$	K_1 35·40	66E
Hg^{2+}	pol	20	0·1 KNO_3	K_1 25·4	62S
	cal	20	0·1 (KNO_3)	ΔH_1 −23·7, ΔS_1 41·3	65A
	cal	25	0·1 KNO_3	ΔH_1 −23·6, ΔS_1 44	65W
	sp	18—22	< 0·1	K_1 28·4±0·4	67K
Ho^{3+}	cal	27	0·1 (KNO_3)	ΔH_1 −7·5±0·3, ΔS_1 79·0	68Ca
In^{3+}	ix	?	0·5	K_1 27·65±0·04	63R
	sp	18—20	?	K_1 28·42, $K(In^{3+}+FeL^{2-} \rightleftharpoons InL^{2-}+Fe^{3+})$ 0·91	66Z
	red	20	0·1 ($NaClO_4$)	K_1 29·0, $K(InL^{2-}+OH^- \rightleftharpoons InLOH^{3-})$ 2·06	67B
La^{3+}	cal	20	0·1 (KNO_3)	ΔH_1 −4·7, ΔS_1 73·5	65A
	cal	27	0·1 (KNO_3)	ΔH_1 −5·2±0·3, ΔS_1 71·8	68Ca
Lu^{3+}	cal	27	0·1 (KNO_3)	ΔH_1 −5·1±0·3, ΔS_1 85·7	68Ca
Mg^{2+}	cal	20	0·1 (KNO_3)	ΔH_1 3·0, ΔS_1 52·4	65A
	cal	25	0·1 KNO_3	ΔH_1 3·6, ΔS_1 54	65W
	cal	27	0·1	ΔH_1 2·6, ΔS_1 50	68Ca
	gl	25	0·1 (KNO_3)	K_1 9·3	68W

continued overleaf

1000 $C_{14}H_{23}O_{10}N_3$ (contd)

Metal	Method	Temp	Medium	Log of equilibrium constant, remarks	Ref
Mn^{2+}	cal	20	0·1 (KNO_3)	ΔH_1 −7·18, ΔS_1 47·0	65A
	cal	25	0·1 KNO_3	ΔH_1 −7·5, ΔS_1 46	65W
Nd^{3+}	sp	18—20	0·1	K_1 21·96	63G
				$K(2Nd^{3+}+H_5L \rightleftharpoons Nd_2L^{+}+5H^{+})$ 26·25	
	cal	27	0·1 (KNO_3)	ΔH_1 −7·1±0·3, ΔS_1 75·0	68Ca
Ni^{2+}	cal	20	0·1 (KNO_3)	ΔH_1 −11·7, ΔS_1 52·7	65A
	cal	25	0·1 KNO_3	ΔH_1 −11·2, ΔS_1 54	65W
Pb^{2+}	cal	20	0·1 (KNO_3)	ΔH_1 −18·8, ΔS_1 21·8	65A
	cal	25	0·1 KNO_3	ΔH_1 −18·8, ΔS_1 22	65W
Pr^{3+}	cal	27	0·1 (KNO_3)	ΔH_1 −6·5±0·3, ΔS_1 74·8	68Ca
Sm^{3+}	sp	18—20	0·1	K_1 22·36,	63G
				$K(Sm^{3+}+HL^{4-} \rightleftharpoons SmHL^{-})$ 13·35	
				$K(2Sm^{3+}+L^{5-} \rightleftharpoons Sm_2L^{+})$ 25·47	
	cal	27	0·1 (KNO_3)	ΔH_1 −7·9±0·3, ΔS_1 75·7	68Ca
Sr^{2+}	ix	22	0·165 NH_4^+	K_1 9·57±0·07	62T
	cal	25	0·1 KNO_3	ΔH_1 −7·5, ΔS_1 19	65W
	cal	27	0·1	ΔH_1 −6·7, ΔS_1 21	68Ca
Tb^{3+}	cal	27	0·1 (KNO_3)	ΔH_1 −7·7±0·3, ΔS_1 78·0	68Ca
Th^{4+}	ix	20	0·1 NH_4^+	K_1 30·34±0·03	65R
	red	20	0·1 ($NaClO_4$)	K_1 28·78,	67B
				$K(ThL^{-}+H^{+} \rightleftharpoons ThHL)$ 2·16	
				$K(ThL^{-}+OH^{-} \rightleftharpoons ThLOH^{2-})$ 4·9	
Tl^{+}	gl	20	0·1 (KNO_3)	K_1 5·97,	67Aa
				$K(Tl^{+}+HL^{4-} \rightleftharpoons TlHL^{3-})$ 4·2	
	pol	25 (?)	0·4 ($NaClO_4$)	K_1 (d.c.) 5·45, (a.c.) 5·53	68K
				$K(TlL^{4-}+H^{+} \rightleftharpoons TlHL^{3-})$ (d.c.) 8·81, (a.c.) 8·78	
Tl^{3+}	red	20	1 ($HClO_4$)	K_1 48·0	67Aa
			1 ($NaClO_4$)	K_1 46·0	
Tm^{3+}	cal	27	0·1 (KNO_3)	ΔH_1 −6·6±0·3, ΔS_1 82·0	68Ca
U^{4+}	gl	25	0·10 (KCl)	$K(ULOH^{2-}+H^{+} \rightleftharpoons UL^{-})$ 7·69±0·02	68C
Y^{3+}	ix	20	0·1 NH_4^+	K_1 22·28±0·01	62Sa
Yb^{3+}	cal	27	0·1 (KNO_3)	ΔH_1 −6·2±0·3, ΔS_1 82·7	68Ca
Zn^{2+}	cal	20	0·1 (KNO_3)	ΔH_1 −8·8, ΔS_1 55·0	65A
	cal	25	0·1 KNO_3	ΔH_1 −10·6, ΔS_1 50	65W
	sp	18—22	~0	K_1 18·73±0·02	68Ka
Zr^{4+}	ix	?	1—2 $HClO_4$	$K(Zr^{4+}+H_5L \rightleftharpoons ZrL^{-}+5H^{+})$ (1) 5·67±0·08, (2) 3·63±0·06	64E
	ix	?	2·1	K_2 4·89	64P
	ix	?	0·23—1 $HClO_4$	K_1 (0·23) 35·81, (1) 35·40	66E
	red	20	1 ($NaClO_4$)	K_1 36·9,	67B
				$K(ZrL^{-}+OH^{-} \rightleftharpoons ZrLOH^{2-})$ 8·1	
	oth	20	0·39	K_1 33·96±0·02, chromatography	67T

Errata. 1964, p. 693—695, Table 1000. *For* 56H *read* 59H;
Under H^+ second line, *after* 56F, 59C, 59V, *add* 59H;
Under Hg^{2+} second line, *for* 20 *read* 25;
Under Nd^{3+}, *for* 22·4 *read* 22·24;
Under Sr^{2+} *delete* second line;
Under Y^{3+}, *for* 23·01 *read* 22·40;
Under Yb^{3+}, *for* 22·40 *read* 23·01.

62S S. Stankoviansky and J. Konigstein, *Coll. Czech. Chem. Comm.*, 1962, **27**, 1997
62Sa M. M. Senyavin and L. I. Tikhonova, *Russ. J. Inorg. Chem.*, 1962, 7, 562 (1095)
62T L. I. Tikhonova, *Russ. J. Inorg. Chem.*, 1962, 7, 424 (822)

contd

64E A. N. Ermakov, I. N. Morov, and C. A. Evtikova, *Russ. J. Inorg. Chem.*, 1964, **9**, 277 (502)
63G Yu. P. Galaktionov and K. V. Astakhov, *Russ. J. Inorg. Chem.*, 1963, 8, 724 (1395); 1306 (1963)
63R D. I. Ryabchikov, I. N. Marov, and Y. K'o-min, *Russ. J. Inorg. Chem.*, 1963, **8**, 326 (641)
64P L. N. Pankratova, L. G. Vlasov, and A. V. Lapitskii, *Russ. J. Inorg. Chem.*, 1964, **9**, 742 (1363)
65A G. Anderegg, *Helv. Chim. Acta*, 1965, **48**, 1722
65B R. D. Baybarz, *J. Inorg. Nuclear Chem.*, 1965, **27**, 1831
65R D. I. Ryabchikov and M. P. Volynets, *Russ. J. Inorg. Chem.*, 1965, **10**, 334 (619)
65W D. L. Wright, J. H. Holloway and C. N. Reilley, *Analyt. Chem.*, 1965, **37**, 884
66E A. N. Ermakov, I. N. Marov, and G. A. Evtikova, *Russ. J. Inorg. Chem.*, 1966, **11**, 618 (1155)
66M T. Moeller and S. Chu, *J. Inorg. Nuclear Chem.*, 1966, **28**, 153
66Z N. M. Zhirnova, K. V. Astakhov, and S. A. Barkov, *Russ. J. Phys. Chem.*, 1966, **40**, 222 (417)
67A G. Anderegg, *Helv. Chim. Acta*, 1967, **50**, 2333
67Aa G. Anderegg and E. Bottari, *Helv. Chim. Acta*, 1967, **50**, 2341
67Ab D. A. Aikens and F. J. Bahbah, *Analyt. Chem.*, 1967, **39**, 646
67B E. Bottari and G. Anderegg, *Helv. Chim. Acta*, 1967, **50**, 2349
67K V. T. Krumina, K. V. Astakhov, S. A. Barkov, and V. I. Kornev, *Russ. J. Inorg. Chem.*, 1967, **12**, 1780 (3356)
67N T. Nozaki and K. Koshiba, *Nippon Kagaku Zasshi*, 1967, **88**, 1287
67T L. I. Tikhonova, *Russ. J. Inorg. Chem.*, 1967, **12**, 494 (939)
68C G. H. Carey and A. E. Martell, *J. Amer. Chem. Soc.*, 1968, **90**, 32
68Ca A. S. Carson, P. G. Laye, and P. N. Smith, *J. Chem. Soc. (A)*, 1968, 141, 1384
68K M. Kodama, T. Noda, and M. Murata, *Bull. Chem. Soc. Japan*, 1968, **41**, 354
68Ka V. T. Krumina, K. V. Astakhov, S. A. Barkov, and V. I. Kornev, *Russ. J. Phys. Chem.*, 1968, **42**, 1334 (2524)
68W H. Wikberg and A. Ringbom, *Suomen Kem.*, 1968, **B41**, 177

$C_{14}H_{24}O_8N_2$ (Hexamethylenedinitrilo)tetra-acetic acid (HDTA) H_4L 1004

Metal	*Method*	*Temp*	*Medium*	*Log of equilibrium constant, remarks*	*Ref*
H^+	gl	20	0·1 (KNO_3)	K_1 10·81, K_2 9·79, K_3 2·7, K_4 2·2	64A
	cal	20	0·1 (KNO_3)	$\Delta H_1°$ −7·91, $\Delta S_1°$ 22·5 $\Delta H_2°$ −6·24, $\Delta S_2°$ 23·5	64A
	gl	20	1 (KNO_3)	K_1 10·56, K_2 9·69, K_3 2·35, K_4 2·45 (?)	67A
Ca^{2+}	gl	20	0·1 (KNO_3)	K_1 4·6, $K(Ca^{2+}+HL^{3-} \rightleftharpoons CaHL^-)$ 3·7	64A
Cd^{2+}	gl	20	0·1 (KNO_3)	K_1 11·9, $K(Cd^{2+}+HL^{3-} \rightleftharpoons CdHL^-)$ 6·99 $K(CdL^{2-}+Cd^{2+} \rightleftharpoons Cd_2L)$ 2·2	64A
	cal	20	0·1 (KNO_3)	$\Delta H_1°$ −4·26, $\Delta S_1°$ 39·9	64A
Co^{2+}	gl	20	0·1 (KNO_3)	K_1 13·05, $K(Co^{2+}+HL^{3-} \rightleftharpoons CoHL^-)$ 7·92 $K(CoL^{2-}+Co^{2+} \rightleftharpoons Co_2L)$ 2·9	64A
	cal	20	0·1 (KNO_3)	$\Delta H_1°$ −4·56, $\Delta S_1°$ 44·1	64A
Fe^{2+}	gl	20	0·1 (KNO_3)	K_1 ~ 11·0, $K(Fe^{2+}+HL^{3-} \rightleftharpoons FeHL^-)$ ~ 6·6	64A
Fe^{3+}	sp	18—20	?	$K(Fe^{3+}+HL^{3-} \rightleftharpoons FeHL)$ 10·3	65Z
Hg^{2+}	gl	20	0·1 (KNO_3)	K_1 21·58	64A
	cal	20	0·1 (KNO_3)	$\Delta H_1°$ −20·97, $\Delta S_1°$ 27·3	64A
In^{3+}	sp	18—20	?	$K(In^{3+}+HL^{3-} \rightleftharpoons InHL)$ 9·03	65Z
Mg^{2+}	gl	20	0·1 (KNO_3)	K_1 4·8, $K(Mg^{2+}+HL^{3-} \rightleftharpoons MgHL^-)$ 3·66	64A
Mn^{2+}	gl	20	0·1 (KNO_3)	K_1 9·03, $K(Mn^{2+}+HL^{3-} \rightleftharpoons MnHL^-)$ 5·69	64A
	cal	20	0·1 (KNO_3)	$\Delta H_1°$ 0·87, $\Delta S_1°$ 44·2	64A
Nd^{3+}	sp	18—20	0·1—0·25	$K(Nd^{3+}+H_2L^{2-} \rightleftharpoons NdH_2L^+)$ 2·54 $K(Nd^{3+}+HL^{3-} \rightleftharpoons NdHL)$ 9·43 $K(Nd^{3+}+2HL^{3-} \rightleftharpoons NdH_2L_2^{3-})$ 14·07 $K(NdHL+A^- \rightleftharpoons NdHLA^-)$ 1·22 $K(Nd^{3+}+HL^{3-}+A^- \rightleftharpoons NdHLA^-)$ 10·65 $K(Nd^{3+}+2HL^{3-}+A^- \rightleftharpoons NdH_2L_2A^{4-})$ 15·36 where HA = acetic acid	63G

continued overleaf

1004 $C_{14}H_{24}O_8N_2$ (contd)

Metal	*Method*	*Temp*	*Medium*	*Log of equilibrium constant, remarks*	*Ref*
Ni^{2+}	gl	20	0·1 (KNO_3)	K_1 13·82, $K(Ni^{2+}+HL^{3-} \rightleftharpoons NiHL^-)$ 9·16 $K(NiL^{2-}+Ni^{2+} \rightleftharpoons Ni_2L)$ 4·5	64A
	cal	20	0·1 (KNO_3)	$\Delta H_1°$ −8·5, $\Delta S_1°$ 34·2	64A
Pb^{2+}	gl	20	0·1 (KNO_3)	$K(Pb^{2+}+HL^{3-} \rightleftharpoons PbHL^-)$ 8·24	64A
	cal	20	0·1 (KNO_3)	$\Delta H[Pb^{2+}+L^{4-}+H_2O \rightleftharpoons PbOH(HL)^{2-}]$ −7·53	64A
Sm^{3+}	sp	18—20	0·1	$K(Sm^{3+}+H_3L^- \rightleftharpoons SmH_3L^{2+})$ 1·52 $K(2Sm^{3+}+HL^{3-}+H_2L^{2-} \rightleftharpoons Sm_2H_3L_2{}^+)$ 11·32 $K(Sm_2H_3L_2{}^++HL^{3-} \rightleftharpoons Sm_2H_4L_3{}^{2-})$ 4·61 $K(Sm_2H_3L_2{}^++2A^- \rightleftharpoons Sm_2H_3L_2A_2{}^-)$ 2·00 $K(Sm_2H_3L_2A_3{}^{2-}+HL^{3-} \rightleftharpoons Sm_2H_4L_3A_3{}^{5+})$ 4·06 $K(Sm_2H_2L_2A_3{}^{3-}+HL^{3-} \rightleftharpoons Sm_2H_3L_3A_3{}^{6-})$ 7·15 $K(Sm_2H_3L_2{}^++3A^- \rightleftharpoons Sm_2H_3L_2A_3{}^{2-})$ 2·78 where HA = acetic acid	66G
Tl^{3+}	sp	18—20	0·0006	$K(Tl^{3+}+HL^{3-} \rightleftharpoons TlHL)$ 9·72 $K(Tl^{3+}+H_2L^{2-} \rightleftharpoons TlH_2L^+)$ 2·52 $K(Tl^{3+}+H_3L^- \rightleftharpoons TlH_3L^{2+})$ 2·28	67K
Zn^{2+}	gl	20	0·1 (KNO_3)	K_1 12·68 $K(Zn^{2+}+HL^{3-} \rightleftharpoons ZnHL^-)$ 8·15 $K(ZnL^{2-}+Zn^{2+} \rightleftharpoons Zn_2L)$ 3·7	64A
	cal	20	0·1 (KNO_3)	$\Delta H_1°$ −4·0, $\Delta S_1°$ 44·4	64A

63G Yu. P. Galaktionov and K. V. Astakhov, *Russ. J. Inorg. Chem.*, 1963, **8**, 1309 (2498)
64A G. Anderegg, *Helv. Chim. Acta*, 1964, **47**, 1801
65Z N. M. Zhirnova, K. V. Astakhov, and S. A. Barkov, *Russ. J. Phys. Chem.*, 1965, **39**, 952
66G Yu. P. Galaktionov and K. V. Astakhov, *Russ. J. Inorg. Chem.*, 1966, **11**, 1216 (2269)
67A G. Anderegg, *Helv. Chim. Acta*, 1967, **50**, 2333
67K V. J. Kornev, K. V. Astakhov, and V. I. Rybina, *Russ. J. Inorg. Chem.*, 1967, **12**, 76 (152)

1005 $C_{14}H_{24}O_9N_2$ [(2,2′-Oxypropylene)dinitrilo]tetra-acetic acid H_4L

Metal	*Method*	*Temp*	*Medium*	*Log of equilibrium constant, remarks*	*Ref*
H^+	gl	20	0·1 (KCl)	K_1 10·14, K_2 9·64, K_3 2·74, K_4 2·0	61I
Ba^{2+}	gl	20	0·1 (KCl)	K_1 3·77, $K(Ba^{2+}+HL^{3-} \rightleftharpoons BaHL^-)$ 2·69	61I
Ca^{2+}	gl	20	0·1 (KCl)	K_1 5·38, $K(Ca^{2+}+HL^{3-} \rightleftharpoons CaHL^-)$ 4·04	61I
Cd^{2+}	gl	20	0·1 (KCl)	K_1 14·22, $K(Cd^{2+}+HL^{3-} \rightleftharpoons CdHL^-)$ 8·20	61I
Hg^{2+}	gl	20	0·1 (KCl)	K_1 21·67, $K(Hg^{2+}+HL^{3-} \rightleftharpoons HgHL^-)$ 15·0	61I
Mg^{2+}	gl	20	0·1 (KCl)	K_1 4·8, $K(Mg^{2+}+HL^{3-} \rightleftharpoons MgHL^-)$ 3·92	61I
Sr^{2+}	gl	20	0·1 (KCl)	K_1 4·15, $K(Sr^{2+}+HL^{3-} \rightleftharpoons SrHL^-)$ 3·06	61I
Zn^{2+}	gl	20	0·1 (KCl)	K_1 13·53, $K(Zn^{2+}+HL^{3-} \rightleftharpoons ZnHL^-)$ 9·19	61I

61I H. Irving and M. H. Stacey, *J. Chem. Soc.*, 1961, 2019

$C_{14}H_{24}O_{10}N_2$ [(Ethylenedioxy)diethylenedinitrilo]tetra-acetic acid (EGTA) H_4L 1006

Metal	*Method*	*Temp*	*Medium*	*Log of equilibrium constant, remarks*	*Ref*
H^+	gl	20	0·1 (KNO_3)	K_1 9·46, K_2 8·78, K_3 2·65, K_4 < 2	63D
	gl	20	0·1 (KNO_3)	K_1 9·46, K_2 8·85, K_3 2·65, K_4 2·0	64A
	cal	20	0·1 (KNO_3)	$\Delta H_1°$ −5·84, $\Delta S_1°$ 23·3, $\Delta H_2°$ −5·76, $\Delta S_2°$ 20·8	64A
	gl	5—35	0·1 (KCl)	K_1 (5°) 9·95, (15°) 9·69, (25°) 9·53, (35°) 9·38	65B
				K_2 (5°) 9·20, (15°) 9·01, (25°) 8·88, (35°) 8·77	
				ΔH_1 (25°) −6·33, ΔS_1 (25°) 22·4, ΔH_2 (25°) −4·88, ΔS_2 (25°) 24·2	
	gl	25	0·5 (KNO_3)	K_1 9·38, K_2 8·55, K_3 3·2, K_4 1·9	66P
	gl	20	1 (KNO_3)	K_1 9·22, K_2 8·67, K_3 2·5, K_4 2·4	67A
	gl	25	0·2 ($NaClO_4$)	K_1 9·42, K_2 8·65, K_3 2·87, K_4 1·91	67B
	sp	20	0·5 $NaClO_4$	K_1 8·94, K_2 8·40, K_3 2·40, K_4 1·15	67N
	?	25	1·5 (KCl)	K_1 9·28±0·05, K_2 8·78±0·05	68T
	gl	25	0·1 (KNO_3)	K_1 9·54, K_2 8·93, K_3 2·73, K_4 2·08	68W
Ag^+	gl	20	0·1 (KNO_3)	K_1 6·88,	63D
				$K(Ag^+ + HL^{3-} \rightleftharpoons AgHL^{2-})$ 4·93	
	Ag	25	0·1 (KNO_3)	K_1 7·06	68W
Al^{3+}	sp	25	0·2 ($NaClO_4$)	K_1 13·90	67B
	gl	25	0·2 ($NaClO_4$)	$K(AlL + H^+ \rightleftharpoons AlHL^+)$ 3·97	67B
				$K(AlLOH^- + H^+ \rightleftharpoons AlL)$ 5·20	
				$K[AlL(OH)_2^{2-} + H^+ \rightleftharpoons AlLOH^-]$ 8·42	
Ba^{2+}	cal	25	0·1 KCl	ΔH_1 −9·0±0·2, ΔS_1 7·8	65B
	cal	25	0·1 KNO_3	ΔH_1 −8·8, ΔS_1 7·1	65W
Bi^{3+}	sp	20	0·5 $NaClO_4$	K_1 23·8	67N
				$K(Bi^{3+} + HL^{3-} \rightleftharpoons BiHL)$ 16·0	
Ca^{2+}	gl	20	0·1 (KNO_3)	K_1 10·97,	64A
				$K(Ca^{2+} + HL^{3-} \rightleftharpoons CaHL^-)$ 5·3	
	cal	20	0·1 (KNO_3)	$\Delta H_1°$ −8·38, $\Delta S_1°$ 21·6	64A
	cal	25	0·1 KCl	ΔH_1 −7·94±0·07, ΔS_1 23·2	65B
	cal	25	0·1 KNO_3	ΔH_1 −8·0, ΔS_1 23	65W
	gl	25	0·1 (KNO_3)	K_1 11·0	68W
Cd^{2+}	gl	20	0·1 (KNO_3)	K_1 16·1,	64A
				$K(Cd^{2+} + HL^{3-} \rightleftharpoons CdHL^-)$ 10·14	
	cal	20	0·1 (KNO_3)	$\Delta H_1°$ −14·8, $\Delta S_1°$ 23·2	64A
	cal	25	0·1 KCl	ΔH_1 −14·9±0·2, ΔS_1 25·7	65B
	cal	25	0·1 KNO_3	ΔH_1 −14·1, ΔS_1 29	65W
Co^{2+}	gl	20	0·1 (KNO_3)	K_1 12·50,	63D
				$K(Co^{2+} + HL^{3-} \rightleftharpoons CoHL^-)$ 7·99	
	gl	20	0·1 (KNO_3)	K_1 12·28	64A
				$K(Co^{2+} + HL^{3-} \rightleftharpoons CoHL^-)$ 7·98	
				$K(Co^{2+} + CoL^{2-} \rightleftharpoons Co_2L)$ 3·3	
	cal	20	0·1 (KNO_3)	$\Delta H_1°$ −2·83, $\Delta S_1°$ 46·5	64A
	cal	25	0·1 KNO_3	ΔH_1 −3·4, ΔS_1 45	65W
	sp	25	0·2 ($NaClO_4$)	K_1 15·6,	67B
				$K(Co^{2+} + HL^{3-} \rightleftharpoons CoHL^-)$ 8·64	
Cr^{3+}	sp	25	0·5 $NaClO_4$	K_1 2·54	66C
				$K(2Cr^{3+} + L^{4-} \rightleftharpoons Cr_2L^{2+})$ 3·51	
Cu^{2+}	gl	20	0·1 (KNO_3)	K_1 17·71,	64A
				$K(Cu^{2+} + HL^{3-} \rightleftharpoons CuHL^-)$ 12·61	
	cal	20	0·1 (KNO_3)	$\Delta H_1°$ −11·0, $\Delta S_1°$ 43·5	64A
	gl	25	0·1 ($NaClO_4$)	K_2 4·31,	65S
				$K(CuL^{2-} + H^+ \rightleftharpoons CuHL^-)$ 4·28	
				$K(Cu_2L + OH^- \rightleftharpoons Cu_2LOH^-)$ 6·9	
				$K[Cu_2LOH^- + OH^- \rightleftharpoons Cu_2L(OH)_2^{2-}]$ 5·8	
				$K(CuL^{2-} + OH^- \rightleftharpoons CuLOH^-)$ negligible	

continued overleaf

1006 $C_{14}H_{24}O_{10}N_2$ (contd)

Metal	*Method*	*Temp*	*Medium*	*Log of equilibrium constant, remarks*	*Ref*
Cu^{2+} (*contd*)	cal	25	0·1 KNO_3	ΔH_1 −10·5, ΔS_1 46	65W
Fe^{2+}	gl	20	0·1 (KNO_3)	K_1 11·92, $K(Fe^{2+}+HL^{3-} \rightleftharpoons FeHL^-)$ 6·93	63D
	gl	20	0·1 (KNO_3)	K_1 11·81, $K(Fe^{2+}+HL^{3-} \rightleftharpoons FeHL^-)$ 6·4	64A
	cal	25	0·1 KNO_3	ΔH_1 −5·2, ΔS_1 37	65W
Fe^{3+}	sp	25	0·1 ($NaClO_4$)	K_1 20·5	63S
Hg^{2+}	gl	20	0·1 (KNO_3)	K_1 23·2, $K(Hg^{2+}+HL^{3-} \rightleftharpoons HgHL^-)$ 16·8	64A
	cal	20	0·1 (KNO_3)	$\Delta H_1°$ −23·7, $\Delta S_1°$ 25·2	64A
	cal	25	0·1 KNO_3	ΔH_1 −23·3, ΔS_1 31	65W
La^{3+}	gl	20	0·1 (KNO_3)	K_1 15·79	64A
	cal	20	0·1 (KNO_3)	$\Delta H_1°$ −5·46, $\Delta S_1°$ 53·6	64A
Li^+	kin	25	1·5 (KCl)	K_1 1·17	68T
Mg^{2+}	gl	20	0·1 (KNO_3)	K_1 5·2, $K(Mg^{2+}+HL^{3-} \rightleftharpoons MgHL^-)$ 3·4	64A
	cal	20	0·1 (KNO_3)	$\Delta H_1°$ 5·18, $\Delta S_1°$ 41·5	64A
	cal	25	0·1 KCl	ΔH_1 5·5±0·1, ΔS_1 42·6	65B
	cal	25	0·1 KNO_3	ΔH_1 4·4, ΔS_1 40	65W
	gl	25	0·1 (KNO_3)	K_1 5·2	68W
Mn^{2+}	gl	20	0·1 (KNO_3)	K_1 12·11, $K(Mn^{2+}+HL^{3-} \rightleftharpoons MnHL^-)$ 6·59	63D
	gl	20	0·1 (KNO_3)	K_1 12·28, $K(Mn^{2+}+HL^{3-} \rightleftharpoons MgHL^-)$ 7·02	64A
	cal	20	0·1 (KNO_3)	$\Delta H_1°$ −8·16, $\Delta S_1°$ 21·5	64A
	cal	25	0·1 KNO_3	ΔH_1 −8·8, ΔS_1 27	65W
Na^+	kin	25	1·5 (KCl)	K_1 1·38	68T
Ni^{2+}	gl	20	0·1 (KNO_3)	K_1 13·55, $K(Ni^{2+}+HL^{3-} \rightleftharpoons NiHL^-)$ 9·19	63D
	gl	20	0·1 (KNO_3)	K_1 11·82, $K(Ni^{2+}+HL^{3-} \rightleftharpoons NiHL^-)$ 8·3, $K(Ni^{2+}+NiL^{2-} \rightleftharpoons Ni_2L)$ 4·9	64A
	cal	20	0·1 (KNO_3)	$\Delta H_1°$ −3·83, $\Delta S_1°$ 41·0	64A
	cal	25	0·1 KNO_3	ΔH_1 −5·0, ΔS_1 45	65W
Pb^{2+}	gl	20	0·1 (KNO_3)	K_1 14·71, $K(Pb^{2+}+HL^{3-} \rightleftharpoons PbHL^-)$ 10·28	63D
	gl	20	0·1 (KNO_3)	K_1 11·8 $K(Pb^{2+}+HL^{3-} \rightleftharpoons PbHL^-)$ 7·5 $K(Pb^{2+}+PbL^{2-} \rightleftharpoons Pb_2L)$ 4·6	64A
	cal	20	0·1 (KNO_3)	$\Delta H_1°$ −13·2, $\Delta S_1°$ 9·1	64A
	cal	25	0·1 KNO_3	ΔH_1 −12·5, ΔS_1 25	65W
Sr^{2+}	cal	25	0·1 KCl	ΔH_1 −5·7±0·2, ΔS_1 19·3	65B
	cal	25	0·1 KNO_3	ΔH_1 −6·4, ΔS_1 15	65W
Tl^+	gl	20	0·1 (KNO_3)	K_1 4·38, $K(Tl^++HL^{3-} \rightleftharpoons TlHL^{2-})$ 3·85	63D
	pol	25	0·5 (KNO_3)	K_1 5·37	66P
	gl	25	0·5 (KNO_3)	K_1 5·63, $K(Tl^++HL^{3-} \rightleftharpoons TlHL^{2-})$ 3·38	66P
Zn^{2+}	gl	20	0·1 (KNO_3)	K_1 12·49, $K(Zn^{2+}+HL^{3-} \rightleftharpoons ZnHL^-)$ 7·97	63D
	gl	20	0·1 (KNO_3)	K_1 12·91	64A

cont

$C_{14}H_{24}O_{10}N_2$ **(contd) 1006**

Metal	Method	Temp	Medium	Log of equilibrium constant, remarks	Ref
Zn^{2+}				$K(Zn^{2+}+HL^{3-} \rightleftharpoons ZnHL^{-})$ 8·42	
(contd)				$K(Zn^{2+}+ZnL^{2-} \rightleftharpoons Zn_2L)$ 3·3	
	cal	20	0·1 (KNO_3)	$\Delta H_1°$ −4·28, $\Delta S_1°$ 44·4	64A
	cal	25	0·1 KCl	ΔH_1 −5·0±0·1, ΔS_1 42·1	65B
	cal	25	0·1 KNO_3	ΔH_1 −3·8, ΔS_1 53	65W

63D J. J. R. F. da Silva and J. G. Calado, *Rev. Port. Quim.*, 1963, 5, 121
63S K. H. Schroder, *Acta Chem. Scand.*, 1963, 17, 1509
64A G. Anderegg, *Helv. Chim. Acta*, 1964, 47, 1801
65B S. Boyd, A. Bryson, G. H. Nancollas, and K. Torrance, *J. Chem. Soc.*, 1965, 7353
65S K. H. Schroder, *Acta Chem. Scand.*, 1965, 19, 1347
65W D. L. Wright, J. H. Holloway, and C. N. Reilley, *Analyt. Chem.*, 1965, 37, 884
66C E. Chiacchierini, *Ricerca sci.*, 1966, 36, 1016
66P F. Pantani, *Ricerca sci.*, 1966, 36, 702
67A G. Anderegg, *Helv. Chim. Acta*, 1967, 50, 2333
67B T. R. Bhat, R. R. Das, and J. Shankar, *Indian J. Chem.*, 1967, 5, 324
67N T. Nozaki and K. Koshiba, *Nippon Kagaku Zasshi*, 1967, 88, 1287
68T M. Tanaka, S. Funabashi, and K. Shirai, *Inorg. Chem.*, 1968, 7, 573
68W H. Wikberg and A. Ringbom, *Suomen Kem.*, 1968, B41, 177

$C_{15}H_{10}O_4$ 6,7-Dihydroxy-4-phenyl-1-benzopyran-2-one (4-phenylesculetin) H_2L 1011.1

Metal	Method	Temp	Medium	Log of equilibrium constant, remarks	Ref
H^+	gl	?	50% ethanol	K_1 7·95	63J
Mo^{VI}	sp	?	50% ethanol	$K(MoO_4^{2-}+2H_2L \rightleftharpoons MoO_2L_2^{2-})$ (?) 8·20	63J

63J B. D. Jain and H. B. Singh, *Indian J. Chem.*, 1963, 1, 369

$C_{15}H_{10}O_4$ 7,8-Dihydroxy-3-phenyl-1-benzopyran-2-one (3-phenyldaphnetin) H_2L 1011.2

Metal	Method	Temp	Medium	Log of equilibrium constant, remarks	Ref
Mo^{VI}	sp	30—34	0·2 KCl 50% ethanol	$K(MoO_4^{2-}+H_2L \rightleftharpoons MoO_3L^{2-})$ (?) 4·5	66J

66J B. D. Jain and R. Kumar, *Current Sci.*, 1966, 35, 557

$C_{15}H_{10}O_5$ 3,5,7-Trihydroxy-2-phenyl-1-benzopyran-4-one (galangin) H_3L 1011.3

Metal	Method	Temp	Medium	Log of equilibrium constant, remarks	Ref
Mo^{VI}	sp	28	?	$K(MoO_4^{2-}+H_3L \rightleftharpoons MoO_3HL^{2-})$ (?) 4·58	64K

64K M. Katyal and R. P. Singh, *Indian J. Chem.*, 1964, 2, 454

1011.4 $C_{15}H_{10}O_7$ 3,7,8-Trihydroxy-2-(3′,4′-dihydroxyphenyl)-1-benzopyran-4-one (melanoxetin)

Metal	Method	Temp	Medium	Log of equilibrium constant, remarks	Ref
Th^{4+}	sp	20	40% ethanol	$K[ThO^{2+}+H_5L \rightleftharpoons ThO(H_4L)^{+}+H^{+}]$ (?) 3·68	66K
Zr^{4+}	sp	20	1 HCl 50% ethanol	$K[ZrO^{2+}+H_5L \rightleftharpoons ZrO(H_4L)^{+}+H^{+}]$ 5·56	65K

65K M. Katyal, B. P. Gupta, and R. P. Singh, *Current Sci.*, 1965, **34**, 456
66K M. Katyal, S. Prakash, R. P. Singh, and K. K. Malik, *Current Sci.*, 1966, **35**, 388

1011.5 $C_{15}H_{10}O_8$ 3,5,7-Trihydroxy-2-(3′,4′,5′-trihydroxyphenyl)-1-benzopyran-4-one (myricetin)

Metal	Method	Temp	Medium	Log of equilibrium constant, remarks	Ref
Mo^{IV}	sp	20	?	$K(MoO_4^{2-}+H_6L \rightleftharpoons MoO_3H_4L^{2-})$ (?) 4·62	65G
Th^{4+}	sp	20	?	$K[ThO^{2+}+H_6L \rightleftharpoons ThO(H_5L)^{+}+H^{+}]$ (?) 4·55	65G

65G B. P. Gupta, M. Katyal, and R. P. Singh, *J. Indian Chem. Soc.*, 1965, **42**, 811

1012 $C_{15}H_{11}N_3$ 2,2′: 6′,2″-Terpyridine L

Metal	Method	Temp	Medium	Log of equilibrium constant, remarks	Ref
H^{+}	gl	25	0·3 (K_2SO_4)	K_1 4·69, K_2 3·99	61J
Cd^{2+}	kin	25	var	K_1 5·1	66H
Co^{2+}	kin	25	var	K_1 8·4, β_2 18·3	66H
Cu^{+}	Pt	25	0·3 (K_2SO_4)	K_1 ~ 9·3	61J
Cu^{2+}	sp	25	0·3 (K_2SO_4)	K_1 ~ 13·0	61J
Fe^{2+}	kin	25	var	K_1 7·1, β_2 20·9	66H
Mn^{2+}	kin	25	var	K_1 4·4	66H
Ni^{2+}	kin	25	var	K_1 10·7, β_2 21·8	66H
Zn^{2+}	kin	25	var	K_1 6·0	66H

61J B. R. James and R. J. P. Williams, *J. Chem. Soc.*, 1961, 2007
66H R. H. Holyer, C. D. Hubbard, S. F. A. Kettle, and R. G. Wilkins, *Inorg. Chem.*, 1966, **5**, 622

$C_{15}H_{12}O_2$ 1,3-Diphenylpropane-1,3-dione (dibenzoylmethane) HL 1013

Metal	Method	Temp	Medium	Log of equilibrium constant, remarks	Ref
H^+	gl	25	0·1 $(C_4H_9)_4NClO_4$ methanol	K_1 13·03±0·05	65L
			0·1$(C_4H_9)_4NClO_4$ ethanol	K_1 13·4±0·05	
K^+	gl	25	0·1 KI methanol	K_1 1·6±0·2	65L
Li^+	gl	25	0·1 $LiClO_4$ methanol	K_1 4·1±0·2	65L
Na^+	gl	25	0·1 $NaClO_4$ methanol	K_1 2·4±0·05	65L

65L D. C. Luehrs, R. T. Iwamoto, and J. Kleinberg, *Inorg. Chem.*, 1965, **4**, 1739

$C_{15}H_{12}N_4$ 3-(2′-Pyridyl)-1-(2′-quinolyl)-1,2-diazaprop-2-ene (2-picolinealdehyde 2′-quinolylhydrazone) HL 1014.1

Metal	Method	Temp	Medium	Log of equilibrium constant, remarks	Ref
H^+	gl	25	50% dioxan	K_1 12·91, K_2 5·26	65H
Cd^{2+}	gl	25	50% dioxan	K_1 9·52, K_2 8·49	65H
Cu^{2+}	gl	25	50% dioxan	K_1 11·60, K_2 8·65	65H
Fe^{2+}	gl	25	50% dioxan	K_1 10·44, K_2 10·18	65H
Ni^{2+}	gl	25	50% dioxan	K_1 10·46, K_2 9·39	65H
Pd^{2+}	gl	25	50% dioxan	K_1 10·57	65H
Zn^{2+}	gl	25	50% dioxan	K_1 10·24, K_2 9·26	65H

65H M. L. Heit and D. E. Ryan, *Analyt. Chim. Acta*, 1965, 32, 448

$C_{15}H_{32}N^+$ *NNN*-Tributylallylammonium cation L^+ 1016.1

Metal	Method	Temp	Medium	Log of equilibrium constant, remarks	Ref
Pt^{2+}	sp	60	2 (NaCl)	$K(PtCl_4^{2-}+L^+ \rightleftharpoons PtCl_3L+Cl^-)$ 2·49	67D

67D R. G. Denning, F. R. Hartley, and L. M. Venanzi, *J. Chem. Soc. (A)*, 1967, 324

1016.2 $C_{15}H_{10}O_6S$ 3-Hydroxy-2-(2′-sulphophenyl)-1-benzopyran-4-one H_2L (flavonol-2′-sulphonic acid)

Metal	Method	Temp	Medium	Log of equilibrium constant, remarks	Ref
H^+	sp	25	0·5 $NaClO_4$	K_1 8·51	64O, 68Y
	sp	25	0·1 $NaClO_4$	K_1 8·92	67Y
Bi^{3+}	sp	25	0·5 $NaClO_4$	K_1 12·3, K_2 8·1	68Y
Th^{4+}	sp	25	0·5 $NaClO_4$	K_1 10·28, K_2 7·78	64O
Tl^{3+}	sp	25	0·1 $NaClO_4$	K_1 9·2, K_2 7·2	67Y

64O Y. Oka, K. Yamamoto, and T. Aoki, *Nippon Kagaku Zasshi*, 1964, **85,** 430
67Y K. Yamamoto and K. Takamizawa, *Nippon Kagaku Zasshi*, 1967, **88,** 345
68Y K. Yamamoto and T. Nishio, *Nippon Kagaku Zasshi*, 1968, **89,** 1214

1017.1 $C_{15}H_{11}ON$ 8-Hydroxy-2-phenylquinoline HL

Metal	Method	Temp	Medium	Log of equilibrium constant, remarks	Ref
H^+	gl	25	50% dioxan	K_1 11·87, K_2 2·07	54J
Co^{2+}	gl	25	50% dioxan	K_1 7·75	54J
Cu^{2+}	gl	25	50% dioxan	K_1 11·40, K_2 10·89	54J
Mn^{2+}	gl	25	50% dioxan	K_1 6·22	54J
Ni^{2+}	gl	25	50% dioxan	K_1 7·57, K_2 6·93	54J
Zn^{2+}	gl	25	50% dioxan	K_1 9·00 K_2 8·52	54J

54J W. D. Johnston and H. Freiser, *Analyt. Chim. Acta*, 1954, **11,** 201

1018 $C_{15}H_{11}ON_3$ 1-(2′-Pyridylazo)-2-naphthol (PAN) HL

Metal	Method	Temp	Medium	Log of equilibrium constant, remarks	Ref
H^+	sp	?	0·08 20% dioxan	K_1 12·2, K_2 2·90	59P, 64N
	gl	25	50% dioxan	K_1 12·3, K_2 < 2	62C
	dis	29—33	0·1 ($NaClO_4$)	K_1 11·2, K_2 2·9	63B
	gl	25	20% dioxan	K_1 12·2, K_2 2·3	67P
Co^{2+}	gl	25	50% dioxan	K_1 >12	62C
	sp	18—22	0·05	K_1 12·15, β_2 24·16	67N
Cu^{2+}	sp	?	0·08 20% dioxan	K_1 16	59P
	dis	29—33	0·1 ($NaClO_4$)	K_1 12·6, $K(CuLOH+H^+ \rightleftharpoons CuL^+)$ 6·9	63B
Eu^{3+}	dis	18—22	0·05	K_1 12·39, β_2 23·80, β_3 34·23, β_4 43·68	67N
Ho^{3+}	dis	18—22	0·05	K_1 12·76, β_2 24·36, β_3 34·80, β_4 44·08	66N
Mn^{2+}	gl	25	50% dioxan	K_1 8·5, K_2 7·9	62C
	dis	29—33	0·1 ($NaClO_4$)	β_2 15·3	63B

contin

$C_{15}H_{11}ON_3$ (contd) 1018

etal	Method	Temp	Medium	Log of equilibrium constant, remarks	Ref
$^{2+}$	gl	25	50% dioxan	K_1 12·7, K_2 12·6	62C
$(C_2H_5)_2{}^{2+}$	sp	25	0·1 (ClO_4^-) 20% dioxan	K_1 12·08	67P
$(CH_3)_2{}^{2+}$	sp	25	0·1 (ClO_4^-) 20% dioxan	K_1 12·55	67P
$(C_2H_5)_2{}^{2+}$	sp	25	0·1 (ClO_4^-) 20% dioxan	K_1 13·73	67P
$(C_4H_9)_2{}^{2+}$	sp	25	0·1 (ClO_4^-) 20% dioxan	K_1 14·37	67P
$(C_6H_5)_2{}^{2+}$	sp	25	0·1 (ClO_4^-) 20% dioxan	K_1 14·68	67P
$^{2+}$	gl	25	50% dioxan	K_1 11·2, K_2 10·5	62C
	dis	29—33	0·1 $(NaClO_4)$	β_2 21·8	63B
	sp	20	0·05 $(NaClO_4)$	K_1 12·72, β_2 24·54	64N

59P B. F. Pease and M. B. Williams, *Analyt. Chem.*, 1959, **31**, 1044
62C A. Corsini, I. M. L. Yih, Q. Fernando, and H. Freiser, *Analyt. Chem.*, 1962, **34**, 1090
63B D. Betteridge, Q. Fernando, and H. Freiser, *Analyt. Chem.*, 1963, **35**, 294
64N O. Navratil, *Coll.Czech. Chem. Comm.*, 1964, **29**, 2490
66N O. Navratil, *Coll. Czech. Chem. Comm.*, 1966, **31**, 2492
67N O. Navratil, *Coll. Czech. Chem. Comm.*, 1967, **32**, 2004
67P G. Pilloni, *Analyt. Chim. Acta*, 1967, **37**, 497

$C_{15}H_{11}ON_3$ 1-(2′-Pyridylazo)-4-naphthol (*p*-PAN) HL 1018.1

etal	Method	Temp	Medium	Log of equilibrium constant, remarks	Ref
$^{+}$	gl	30—36	50% dioxan	K_1 10·74, K_2 2·54	63B
	sp	30—36	0·1	K_1 9·1, K_2 3·0	63B
	dis	30—36	?	K_1 9·5, K_2 3·1, CCl_4	63B
$^{2+}$	sp	30—36	0·1 50% dioxan	β_2 20	63B
$^{2+}$	sp	30—36	0·1 50% dioxan	β_2 23	63B
$^{2+}$	dis	30—36	?	β_2 19·4	63B
	sp	30—36	0·1 50% dioxan	β_2 19	63B

63B D. Betteridge, P. K. Todd, Q. Fernando, and H. Freiser, *Analyt. Chem.*, 1963, 35, 729

$C_{15}H_{11}ON_3$ 8-Hydroxy-5-(phenylazo)quinoline HL 1018.2

etal	Method	Temp	Medium	Log of equilibrium constant, remarks	Ref
$^{+}$	gl	25	0·1 $NaClO_4$ 50% dioxan	K_1 8·84, K_2 3·07	65T
$^{2+}$	gl	25	0·1 $NaClO_4$ 50% dioxan	K_1 7·5±0·1, β_2 14·35±0·05	65T

continued overleaf

1018.2 $C_{15}H_{11}ON_3$ (contd)

Metal	*Method*	*Temp*	*Medium*	*Log of equilibrium constant, remarks*	*Ref*
Co^{2+}	gl	25	0·1 $NaClO_4$ 50% dioxan	K_1 8·8±0·1, β_2 16·74±0·05	65T
Mn^{2+}	gl	25	0·1 $NaClO_4$ 50% dioxan	K_1 6·2±0·1, β_2 12·57±0·05	65T
Ni^{2+}	gl	25	0·1 $NaClO_4$ 50% dioxan	K_1 9·7±0·1, β_2 18·33±0·05	65T
Pb^{2+}	gl	25	0·1 $NaClO_4$ 50% dioxan	K_1 8·6±0·1, β_2 15·09±0·05	65T
Zn^{2+}	gl	25	0·1 $NaClO_4$ 50% dioxan	K_1 8·6±0·1, β_2 16·44±0·05	65T

65T S. Takamoto, Q. Fernando, and H. Freiser, *Analyt. Chem.*, 1965, 37, 1249

1018.3 $C_{15}H_{11}O_2N$ 2-(6′-Methyl-2′-pyridyl)indane-1,3-dione HL

Metal	*Method*	*Temp*	*Medium*	*Log of equilibrium constant, remarks*	*Ref*
H^+	gl	30	75% dioxan	K_1 14·41	64C
Be^{2+}	gl	30	75% dioxan	K_1 11·89±0·02, K_2 11·61±0·02	64C
Cu^{2+}	gl	30	75% dioxan	K_1 10·48±0·02, K_2 12·25±0·02	64C
Mg^{2+}	gl	30	75% dioxan	K_1 6·86±0·02, K_2 6·44±0·02	64C
Mn^{2+}	gl	30	75% dioxan	K_1 8·72±0·02	64C
UO_2^{2+}	gl	30	75% dioxan	K_1 12·54±0·02, K_2 11·58±0·02	64C
Zn^{2+}	gl	30	75% dioxan	K_1 8·88±0·02, K_2 11·56±0·02	64C

64C J. R. Cook and D. F. Martin, *J. Inorg. Nuclear Chem.*, 1964, 26, 571

1018.4 $C_{15}H_{11}O_2N_3$ 8-Hydroxy-5-(2′-hydroxyphenylazo)quinoline H_2L

Metal	*Method*	*Temp*	*Medium*	*Log of equilibrium constant, remarks*	*Ref*
H^+	gl	25	0·1 $NaClO_4$ 50% dioxan	K_1 12·0, K_2 8·51, K_3 2·91	65T
Cd^{2+}	gl	25	0·1 $NaClO_4$ 50% dioxan	K_1 7·1±0·1, β_2 13·81±0·05	65T
Co^{2+}	gl	25	0·1 $NaClO_4$ 50% dioxan	K_1 8·3±0·1, β_2 16·12±0·05	65T
Mn^{2+}	gl	25	0·1 $NaClO_4$ 50% dioxan	K_1 7·1±0·1, β_2 13·01±0·05	65T
Ni^{2+}	gl	25	0·1 $NaClO_4$ 50% dioxan	K_1 9·7±0·1, β_2 18·58±0·05	65T
Pb^{2+}	gl	25	0·1 $NaClO_4$ 50% dioxan	K_1 8·5±0·1, β_2 15·04±0·05	65T
Zn^{2+}	gl	25	0·1 $NaClO_4$ 50% dioxan	K_1 8·2±0·1, β_2 15·93±0·05	65T

65T S. Takamoto, Q. Fernando, and H. Freiser, *Analyt. Chem.*, 1965, 37, 1249

$C_{15}H_{11}O_2N_3$ 8-Hydroxy-5-(3′-hydroxyphenylazo)quinoline H_2L 1018.5

Metal	Method	Temp	Medium	Log of equilibrium constant, remarks	Ref
H^+	gl	25	0·1 $NaClO_4$ 50% dioxan	K_1 11·2, K_2 8·84, K_3 3·11	65T
Cd^{2+}	gl	25	0·1 $NaClO_4$ 50% dioxan	K_1 7·7±0·1, β_2 14·30±0·05	65T
Co^{2+}	gl	25	0·1 $NaClO_4$ 50% dioxan	K_1 8·8±0·1, β_2 16·98±0·05	65T
Mn^{2+}	gl	25	0·1 $NaClO_4$ 50% dioxan	K_1 6·6±0·1, β_2 12·52±0·05	65T
Ni^{2+}	gl	25	0·1 $NaClO_4$ 50% dioxan	K_1 9·7±0·1, β_2 18·26±0·05	65T
Pb^{2+}	gl	25	0·1 $NaClO_4$ 50% dioxan	K_1 8·8±0·1, β_2 15·05±0·05	65T
Zn^{2+}	gl	25	0·1 $NaClO_4$ 50% dioxan	K_1 8·9±0·1, β_2 16·60±0·05	65T

65T S. Takamoto, Q. Fernando, and H. Freiser, *Analyt. Chem.*, 1965, 37, 1249

$C_{15}H_{11}O_2N_3$ 8-Hydroxy-5-(4′-hydroxyphenylazo)quinoline H_2L 1018.6

Metal	Method	Temp	Medium	Log of equilibrium constant, remarks	Ref
H^+	gl	25	0·1 $NaClO_4$ 50% dioxan	K_1 10·6, K_2 9·15, K_3 3·31	65T
Cd^{2+}	gl	25	0·1 $NaClO_4$ 50% dioxan	K_1 7·9, β_2 14·46±0·05	65T
Co^{2+}	gl	25	0·1 $NaClO_4$ 50% dioxan	K_1 9·1±0·1, β_2 17·19±0·05	65T
Mn^{2+}	gl	25	0·1 $NaClO_4$ 50% dioxan	K_1 6·6±0·1, β_2 12·66±0·05	65T
Ni^{2+}	gl	25	0·1 $NaClO_4$ 50% dioxan	K_1 9·9±0·1, β_2 18·78±0·05	65T
Pb^{2+}	gl	25	0·1 $NaClO_4$ 50% dioxan	K_1 9·2±0·1, β_2 16·00±0·05	65T
Zn^{2+}	gl	25	0·1 $NaClO_4$ 50% dioxan	K_1 9·2±0·1, β_2 17·12±0·05	65T

65T S. Takamoto, Q. Fernando, and H. Freiser, *Analyt. Chem.*, 1965, 37, 1249

1018.7 $C_{15}H_{12}OS$ 3-Mercapto-1,3-diphenylprop-2-en-1-one (monothiodibenzoylmethane) HL

Metal	*Method*	*Temp*	*Medium*	*Log of equilibrium constant, remarks*	*Ref*
H^+	gl	30	X mole fraction dioxan (0·18—0·44)	K_1 7·0+10·9X	66
	gl	30	75% dioxan	K_1 11·35	66U
Be^{2+}	gl	30	75% dioxan	K_1 9·00, K_2 8·86	66U
Cd^{2+}	gl	30	75% dioxan	K_1 10·57, K_2 8·96	66U
Cu^{2+}	gl	30	var 75% dioxan	β_2 22·2	66C
	gl	30	75% dioxan	K_2 >12·0	66U
Mn^{2+}	gl	30	75% dioxan	K_1 7·67, K_2 6·53	66U
Ni^{2+}	gl	30	var 75% dioxan	β_2 21·7	66C
	gl	30	75% dioxan	K_1 >11·0	66U
Pb^{2+}	gl	30	75% dioxan	K_1 10·20, K_2 8·95	66U
UO_2^{2+}	gl	30	75% dioxan	K_1 10·34, K_2 9·47	66U
Zn^{2+}	gl	30	75% dioxan	K_1 11·94, K_2 10·70	66U

66C S. H. Chaston and S. E. Livingstone, *Austral. J. Chem.*, 1966, **19**, 2035
66U E. Uhlemann and W. W. Suchan, *Z. anorg. Chem.*, 1966, **342**, 41

1020.1 $C_{15}H_{12}O_3N_2$ Ethyl 4-hydroxy-1,10-phenanthroline-3-carboxylate HL

Metal	*Method*	*Temp*	*Medium*	*Log of equilibrium constant, remarks*	*Ref*
H^+				H^+ constants not given	
Fe^{2+}	sp	?	0·1 Ethanol	$K[Fe^{2+}+3HL \rightleftharpoons Fe(HL)_3^{2+}]$(?) 13·0, acetate buffer	61H

61H C. J. Hawkins, H. Duewell, and W. F. Pickering, *Analyt. Chim. Acta.*, 1961, **25**, 257

1020.2 $C_{15}H_{13}O_2N$ *N*,3-Diphenylpropenohydroxamic acid (*N*-cinnamoyl-*N*-phenylhydroxylamine) HL

Metal	*Method*	*Temp*	*Medium*	*Log of equilibrium constant, remarks*	*Ref*
H^+				H^+ constants not given.	
Th^{4+}	dis	20	0·1 ($NaClO_4$)	K_1 12·76, β_2 24·70, β_3 35·72	67Z

67Z F. G. Zharovskii, R. I. Sukhomlin, and M. S. Ostrovskaya, *Russ. J. Inorg. Chem.*, 1967, **12**, 1306 (2476)

1022.1 $C_{15}H_{15}O_2P$ 3-(Diphenylphosphino)propanoic acid HL

Metal	*Method*	*Temp*	*Medium*	*Log of equilibrium constant, remarks*	*Ref*
H^+	gl	20	0·10 (KNO_3) 20% dioxan	K_1 5·03	64P
Ag^+	gl	20	0·10 (KNO_3) 20% dioxan	K_1 3·80; $K(Ag^+ + HL \rightleftharpoons AgHL^+)$ 2·7	64P

64P L. D. Pettit and H. M. N. H. Irving, *J. Chem. Soc.*, 1964, 5336

$C_{15}H_{15}O_2As$ 3-(Diphenylarsino)propanoic acid HL 1022.2

Metal	Method	Temp	Medium	Log of equilibrium constant, remarks	Ref
H^+	gl	20	0·10 (KNO_3) 20% dioxan	K_1 5·12	64P
Ag^+	gl	20	0·10 (KNO_3) 20% dioxan	K_1 3·87 $K(Ag^+ + HL \rightleftharpoons AgHL^+)$ 2·3	64P

64P L. D. Pettit and H. M. N. H. Irving, *J. Chem. Soc.*, 1964, 5336

$C_{15}H_{16}N_4S$ 1,5-Bis(2′-tolyl)-1,2,4,5-tetra-azapent-1-ene-3-thione HL 1022.3 (di-2-tolylthiocarbazone)

Metal	Method	Temp	Medium	Log of equilibrium constant, remarks	Ref
H^+	sp	25	0·1 50% dioxan	K_1 6·23	64M
Ni^{2+}	sp	25	0·1 50% dioxan	K_1 5·90	64M
Zn^{2+}	sp	25	0·1 50% dioxan	K_1 4·50	64M

64M K. S. Math, Q. Fernando, and H. Freiser, *Analyt. Chem.*, 1964, **36**, 1762

$C_{15}H_{16}N_4S$ 1,5-Bis(4′-tolyl)-1,2,4,5-tetra-azapent-1-ene-3-thione HL 1022.4 (di-4-tolylthiocarbazone)

Metal	Method	Temp	Medium	Log of equilibrium constant, remarks	Ref
H^+	sp	25	0·1 50% dioxan	K_1 6·40	64M
Ni^{2+}	sp	25	0·1 50% dioxan	K_1 6·60	64M
Zn^{2+}	sp	25	0·1 50% dioxan	K_1 6·45	64M

64M K. S. Math, Q. Fernando, and H. Freiser, *Analyt. Chem.*, 1964, **36**, 1762

1025.1 $C_{15}H_{23}O_{12}N_3$ 1,2,3-Tris[*NN*-bis(carboxymethyl)amino]propane H_6L

Metal	*Method*	*Temp*	*Medium*	*Log of equilibrium constant, remarks*	*Ref*
H^+	gl	25	0·1 (KNO_3)	K_1 9·88, K_2 8·26, K_3 4·30, K_4 3·52, K_5 2·43, K_6 ~ 2	68M
Ba^{2+}	gl	25	0·1 (KNO_3)	K_1 7·41±0·09	68M
				$K(Ba^{2+}+HL^{5-} \rightleftharpoons BaHL^{3-})$ 5·42±0·09	
				$K(Ba^{2+}+H_2L^{4-} \rightleftharpoons BaH_2L^{2-})$ 1·4	
				$K(2Ba^{2+}+L^{6-} \rightleftharpoons Ba_2L^{2-})$ 1·6	
Ca^{2+}	gl	25	0·1 (KNO_3)	K_1 10·50±0·04	68M
				$K(Ca^{2+}+HL^{5-} \rightleftharpoons CaHL^{3-})$ 8·03±0·05	
				$K(Ca^{2+}+H_2L^{4-} \rightleftharpoons CaH_2L^{2-})$ 2·9	
				$K(2Ca^{2+}+L^{6-} \rightleftharpoons Ca_2L^{2-})$ 2·3	
Cd^{2+}	gl	25	0·1 (KNO_3)	K_1 16·18±0·09	68M
				$K(Cd^{2+}+HL^{5-} \rightleftharpoons CdHL^{3-})$ 13·1±0·1	
				$K(CdL^{4-}+Ca^{2+} \rightleftharpoons CdCaL^{2-})$ 2·19±0·01	
Co^{2+}	gl	25	0·1 (KNO_3)	K_1 13·8±0·1	68M
				$K(Co^{2+}+HL^{5-} \rightleftharpoons CoHL^{3-})$ 10·5±0·1	
				$K(CoL^{4-}+Ca^{2+} \rightleftharpoons CoCaL^{2-})$ 2·05±0·01	
Cu^{2+}	gl	25	0·1 (KNO_3)	K_1 18·37±0·02	68M
				$K(Cu^{2+}+HL^{5-} \rightleftharpoons CuHL^{3-})$ 14·93±0·03	
				$K(CuL^{4-}+Ca^{2+} \rightleftharpoons CuCaL^{2-})$ 2·08±0·03	
Mg^{2+}	gl	25	0·1 (KNO_3)	K_1 9·21±0·08	68M
				$K(Mg^{2+}+HL^{5-} \rightleftharpoons MgHL^{3-})$ 6·46±0·07	
				$K(Mg^{2+}+H_2L^{4-} \rightleftharpoons MgH_2L^{2-})$ 2·8	
Ni^{2+}	gl	25	0·1 (KNO_3)	K_1 18·00±0·09	68M
				$K(Ni^{2+}+HL^{5-} \rightleftharpoons NiHL^{3-})$ 14·6±0·1	
				$K(NiL^{4-}+Ca^{2+} \rightleftharpoons NiCaL^{2-})$ 2·08±0·01	
Sr^{2+}	gl	25	0·1 (KNO_3)	K_1 8·56±0·05	68M
				$K(Sr^{2+}+HL^{5-} \rightleftharpoons SrHL^{3-})$ 6·32±0·05	
				$K(Sr^{2+}+H_2L^{4-} \rightleftharpoons SrH_2L^{2-})$ 1·7	
				$K(2Sr^{2+}+L^{6-} \rightleftharpoons Sr_2L^{2-})$ 1·4	
Th^{4+}	gl	25	0·1 (KNO_3)	$K(ThL^{2-}+H^+ \rightleftharpoons ThHL^-)$ 5·99±0·09	68M
Zn^{2+}	gl	25	0·1 (KNO_3)	K_1 15·80±0·08	68M
				$K(Zn^{2+}+HL^{5-} \rightleftharpoons ZnHL^{3-})$ 12·7±0·1	
				$K(ZnL^{4-}+Ca^{2+} \rightleftharpoons ZnCaL^{2-})$ 2·62±0·01	

68M Y. Moriguchi, M. Miyazaki, and K. Ueno, *Bull. Chem. Soc. Japan*, 1968, **41**, 1344

1025.2 $C_{15}H_{12}O_3N_2S$ 2-(3′-Benzoylthioureido)benzoic acid HL

Metal	*Method*	*Temp*	*Medium*	*Log of equilibrium constant, remarks*	*Ref*
H^+				H^+ constants not given.	
Os^{VIII}	sp J	~ 30	95% ethanol	K(?) 4·38	66M

66M A. K. Majumdar and S. K. Bhowal, *Analyt. Chim. Acta*, 1966, **35**, 479

$C_{16}H_{12}O_4$ 3-Benzyl-4,5-dihydroxy-1-benzopyran-2-one H_2L 1030.1

Metal	Method	Temp	Medium	Log of equilibrium constant, remarks	Ref
H^+				H^+ constants not given.	
UO_2^{2+}	sp	20—22	0·4 ($NaClO_4$) 40% ethanol	K(?) 5·0	66J

66J B. D. Jain and R. Kumar, *Current Sci.*, 1966, 35, 173

$C_{16}H_{14}O_5$ 5,7-Dihydroxy-8-methoxy-2-phenyl-1-benzopyran-4-one (3-methylgalangin) H_2L 1030.2

Metal	Method	Temp	Medium	Log of equilibrium constant, remarks	Ref
H^+				H^+ constants not given.	
Th^{4+}	sp	20	~ 0·001 50% ethanol	K(?) 4·54	65K

65K M. Katyal and R. P. Singh, *Indian J. Chem.*, 1965, 3, 281

$C_{16}H_{16}N_2$ 3,5,6,8-Tetramethyl-1,10-phenanthroline L 1031.1

Metal	Method	Temp	Medium	Log of equilibrium constant, remarks	Ref
H^+	gl	25	0·3 0, 50% dioxan	K_1 (0%) 5·80±0·05, (50%) 5·30±0·02	61J
Cu^+				See Cu^{2+} below.	
Cu^{2+}	Pt	25	0·3 (KNO_3) 50% dioxan	K_1 7·30, K_3 4·75; $K(CuL_2^{2+} + Cu^+ \rightleftharpoons CuL_2^+ + Cu^{2+})$ 0·30	61J

61J B. R. James and R. J. P. Williams, *J. Chem. Soc.*, 1961, 2007

$C_{16}H_{16}N_2$ 1,2-Bis(benzylideneamino)ethane L 1031.2

Metal	Method	Temp	Medium	Log of equilibrium constant, remarks	Ref
H^+				H^+ constants not given.	
Ag^+	Ag	20	0·1 $NaNO_3$ methanol	K_1 6·48, β_2 8·36	66H
Cu^{2+}	pol	20	0·1 KNO_3 methanol	K_1 6·48, β_2 9·00	66H

66H E. Hoyer and V. V. Skopenko, *Russ. J. Inorg. Chem.*, 1966, 11, 436 (803)

1033.1 $C_{16}H_{20}N_2$ 5,5′-Diethyl-4,4′-dimethyl-2,2′-bipyridyl L

Metal	*Method*	*Temp*	*Medium*	*Log of equilibrium constant, remarks*	*Ref*
H^+	gl	25	0·3 (KNO_3) 50% dioxan	K_1 4·59±0·03	61J
Cu^+				See Cu^{2+} below.	
Cu^{2+}	Pt	25	0·3 (KNO_3) 50% dioxan	K_3 ~ 3·25 $K(CuL_2^{2+} + Cu^+ \rightleftharpoons CuL_2^+ + Cu^{2+})$ −0·67	61J

61J B. R. James and R. J. P. Williams, *J. Chem. Soc.*, 1961, 2007

1034.1 $C_{16}H_{22}N_4$ 1,2-Bis(2′-aminobenzylamino)ethane L

Metal	*Method*	*Temp*	*Medium*	*Log of equilibrium constant, remarks*	*Ref*
H^+	gl	25	0·1 (KCl)	K_1 9·00, K_2 5·90, K_3 2·32, K_4 2·00	68G
Co^{2+}	gl	25	0·1 (KCl)	K_1 7·0	68G
Ni^{2+}	gl	25	0·1 (KCl)	K_1 10·0	68G
Zn^{2+}	gl	25	0·1 (KCl)	K_1 7·17	68G

68G D. W. Gruenwedel, *Inorg. Chem.*, 1968, **7**, 495

1036 $C_{16}H_{12}O_2N_2$ 1-(2′-Hydroxyphenylazo)-2-naphthol H_2L

Metal	*Method*	*Temp*	*Medium*	*Log of equilibrium constant, remarks*	*Ref*
H^+	sp	?	0·100 (KCl)	K_1 12·4, K_2 7·7	60D
Ca^{2+}	sp	?	0·100 (KCl)	K_1 2·26	60D
Mg^{2+}	sp	?	0·100 (KCl)	K_1 4·59	60D

60D H. Diehl and J. Ellingboe, *Analyt. Chem.*, 1960, **32**, 1120

1036.1 $C_{16}H_{13}ON_3$ 3-(2′-Hydroxyphenyl)-1-quinolyl-1,2-diazaprop-2-ene HL (salicylaldehyde 8-quinolylhydrazone)

Metal	*Method*	*Temp*	*Medium*	*Log of equilibrium constant, remarks*	*Ref*
H^+	sp	18—22	0·1 ($NaClO_4$) 50% methanol	K_1 10·67	67A
	gl	25	0·1 ($NaClO_4$) 50% methanol	K_2 2·27	67A
Co^{2+}	gl	25	0·1 $NaClO_4$ 50% methanol	K_1 6·3, K_2 5·4	67A
Ni^{2+}	gl	25	0·1 $NaClO_4$ 50% methanol	K_1 ~ 7	67A

67A R. G. Anderson and G. Nickless, *Talanta*, 1967, **14**, 1221

$C_{16}H_{13}O_3N_5$ 5-Methyl-4-(2′-nitrophenylazo)-1-phenyl-pyrazol-3(2*H*)-one HL 1039.1

Metal	*Method*	*Temp*	*Medium*	*Log of equilibrium constant, remarks*	*Ref*
H^+	gl	30	75% dioxan	K_1 9·73	67S
Cd^{2+}	gl	30	75% dioxan	K_1 4·96±0·09, β_2 10·26	67S
Co^{2+}	gl	30	75% dioxan	K_1 6·1±0·1, β_2 12·57	67S
Cu^{2+}	gl	30	75% dioxan	K_1 8·8±0·1, β_2 17·93	67S
Ni^{2+}	gl	30	75% dioxan	K_1 6·62±0·08, β_2 13·02	67S
Zn^{2+}	gl	30	75% dioxan	K_1 5·8±0·2, β_2 11·61	67S

67S F. A. Snavely, D. A. Sweigart, C. H. Yoder, and A. Terzis, *Inorg. Chem.*, 1967, **6**, 1831

$C_{16}H_{13}O_3N_5$ 5-Methyl-4-(3′-nitrophenylazo)-1-phenyl-pyrazol-3(2*H*)-one HL 1039.2

Metal	*Method*	*Temp*	*Medium*	*Log of equilibrium constant, remarks*	*Ref*
H^+	gl	30	75% dioxan	K_1 9·77	67S
Cd^{2+}	gl	30	75% dioxan	K_1 5·5±0·2, β_2 12·31	67S
Co^{2+}	gl	30	75% dioxan	K_1 7·0±0·3, β_2 14·08	67S
Cu^{2+}	gl	30	75% dioxan	K_1 8·98±0·06, β_2 17·56	67S
Ni^{2+}	gl	30	75% dioxan	K_1 7·0±0·2, β_2 14·65	67S
Zn^{2+}	gl	30	75% dioxan	K_1 6·2±0·2, β_2 12·91	67S

67S F. A. Snavely, D. A. Sweigart, C. H. Yoder, and A. Terzis, *Inorg. Chem.*, 1967, **6**, 1831

$C_{16}H_{13}O_3N_5$ 5-Methyl-4-(4′-nitrophenylazo)-1-phenyl-pyrazol-3(2*H*)-one HL 1039.3

Metal	*Method*	*Temp*	*Medium*	*Log of equilibrium constant, remarks*	*Ref*
H^+	gl	30	75% dioxan	K_1 9·45	67S
Cd^{2+}	gl	30	75% dioxan	K_1 5·6±0·2, β_2 11·63	67S
Co^{2+}	gl	30	75% dioxan	K_1 6·5±0·1, β_2 13·35	67S
Cu^{2+}	gl	30	75% dioxan	K_1 8·4±0·2, β_2 16·67	67S
Ni^{2+}	gl	30	75% dioxan	K_1 6·8±0·2, β_2 13·66	67S
Zn^{2+}	gl	30	75% dioxan	K_1 5·8±0·2, β_2 11·87	67S

67S F. A. Snavely, D. A. Sweigart, C. H. Yoder, and A. Terzis, *Inorg. Chem.*, 1967, **6**, 1831

$C_{16}H_{14}ON_4$ 5-Methyl-4-(phenylazo)-1-phenyl-pyrazol-3(2*H*)-one HL 1040.1

Metal	*Method*	*Temp*	*Medium*	*Log of equilibrium constant, remarks*	*Ref*
H^+	gl	30	75% dioxan	K_1 10·38	67S
Cd^{2+}	gl	30	75% dioxan	K_1 6·1±0·2, β_2 12·47	67S
Co^{2+}	gl	30	75% dioxan	K_1 6·90±0·05, β_2 14·00	67S
Cu^{2+}	gl	30	75% dioxan	K_1 9·7±0·1, β_2 18·62	67S
Ni^{2+}	gl	30	75% dioxan	K_1 7·2±0·2, β_2 14·62	67S
Zn^{2+}	gl	30	75% dioxan	K_1 6·7±0·1, β_2 13·67	67S

67S F. A. Snavely, D. A. Sweigart, C. H. Yoder, and A. Terzis, *Inorg. Chem.*, 1967, **6**, 1831

1041.1 $C_{16}H_{14}N_4S$ 3-Methyl-1-phenyl-(phenylazo)-pyrazol-3(2*H*)-one HL

Metal	*Method*	*Temp*	*Medium*	*Log of equilibrium constant, remarks*	*Ref*
H^+	gl	30	75% dioxan	K_1 10·74	64S
Cd^{2+}	gl	30	75% dioxan	K_1 7·60±0·07, K_2 9·06±0·04	64S
Co^{2+}	gl	30	75% dioxan	K_1 7·09±0·06, K_2 9·3±0·1	64S
Cu^{2+}	gl	30	75% dioxan	K_1 > 13	64S
Ni^{2+}	gl	30	75% dioxan	K_1 10·8±0·2, K_2 9·8±0·2	64S
Zn^{2+}	gl	30	75% dioxan	K_1 6·6±0·1, K_2 9·5±0·2	64S

64S F. A. Snavely, W. S. Trahanovsky, and F. H. Suydam, *Inorg. Chem.*, 1964, **3**, 123

1042.1 $C_{16}H_{15}O_7N$ N-(3-Carboxy-2-hydroxynaphthy-1-ylmethyl)iminodiacetic acid H_4L

Metal	*Method*	*Temp*	*Medium*	*Log of equilibrium constant, remarks*	*Ref*
H^+	sp	17—23	0·1 ($NaClO_4$)	K_1 13·8, K_2 9·7, K_3 8·5, K_4 2·7, K_5 0·7	68B
Be^{2+}	sp	17—23	0·1 ($NaClO_4$)	$K(Be^{2+}+2H^++L^{4-} \rightleftharpoons BeH_2L)$ 33·1 $K(Be^{2+}+H^++L^{4-} \rightleftharpoons BeHL^-)$ 27·2	68B
La^{3+}	sp	17—23	0·1 ($NaClO_4$)	K_1 12·8	68B

68B B. Budisinsky and T. S. West, *Analyt. Chim. Acta*, 1968, **42**, 455

1042.2 $C_{16}H_{16}O_2N_2$ Bis(salicylidene)-1,2-diaminoethane H_2L

Metal	*Method*	*Temp*	*Medium*	*Log of equilibrium constant, remarks*	*Ref*
H^+	sp	20	0·3	K_1 11·4, K_2 8·1, K_3 4·8, K_4 4·4, Theorell-Steinhagen buffer	66S
Ni^{2+}	dis	20	0·3 (acetate)	K (?) 5·25	66S
UO_2^{2+}	dis	20	0·3 (acetate)	K_1 24·35	66S

66S T. Stronski, A. Zielinski, A. Samotus, Z. Stasick, and B. Budesinsky, *Z. analyt. Chem.*, 1966, **222**, 14

1043.1 $C_{16}H_{16}O_4N_2$ 2,2′-Bipyridyl-4,4′-bis(carboxylic acid ethyl ester) L

Metal	*Method*	*Temp*	*Medium*	*Log of equilibrium constant, remarks*	*Ref*
H^+	gl	25	0·3 (KNO_3) 50% dioxan	K_1 2·45±0·10	61J
Cu^+	Pt	25	0·3 (KNO_3) 50% dioxan	$\beta_2 \sim 11{\cdot}4$ See Cu^{2+} below.	61J
Cu^{2+}	Pt	25	0·3 (KNO_3) 50% dioxan	$\beta_2 \sim 9{\cdot}6$ $K(CuL_2^{2+}+Cu^+ \rightleftharpoons CuL_2^++Cu^{2+})$ 1·80	61J

61J B. R. James and R. J. P. Williams, *J. Chem. Soc.*, 1961, 2007

$C_{16}H_{16}O_4N_2$ Diethyl 2,2′-bipyridyl-5,5′-dicarboxylate H_2L 1043.2

Metal	*Method*	*Temp*	*Medium*	*Log of equilibrium constant, remarks*	*Ref*
H^+	sp	25	0·3 (KNO_3) 50% dioxan	K_1 0·85	61J
Cu^+, Cu^{2+}	Pt	25	0·3 (KNO_3) 50% dioxan	$K(CuL_2^{2+} + Cu^+ \rightleftharpoons CuL_2^+ + Cu^{2+})$ 2·17	61J
Fe^{2+}	sp	?	0·1 H_2SO_4 75% dioxan	K_1 2·5±0·3, β_2 4·5±0·3, β_3 6·5±0·3	61Ja

61J B. R. James and R. J. P. Williams, *J. Chem. Soc.*, 1961, 2007
61Ja B. R. James, M. Parris, and R. J. P. Williams, *J. Chem. Soc.*, 1961, 4630

$C_{16}H_{17}O_2N_3$ 6-(4′-Dimethylaminophenylazo)-2-acetylphenol HL 1043.3

Metal	*Method*	*Temp*	*Medium*	*Log of equilibrium constant, remarks*	*Ref*
H^+	gl	30	75% dioxan	K_1 14·55	67U
Be^{2+}	gl	30	75% dioxan	K_1 12·09, K_2 10·64	67U
Cu^{2+}	gl	30	75% dioxan	K_1 11·80, K_2 11·17	67U
Ni^{2+}	gl	30	75% dioxan	K_1 9·16, K_2 8·82	67U

67U E. Uhlemann and F. Dietze, *Z. anorg. Chem.*, 1967, 353, 26

$C_{16}H_{17}O_2N_3$ 5-(4′-Dimethylaminophenylazo)-2-acetylphenol HL 1043.4

Metal	*Method*	*Temp*	*Medium*	*Log of equilibrium constant, remarks*	*Ref*
H^+	gl	30	75% dioxan	K_1 14·42	67U
Be^{2+}	gl	30	75% dioxan	K_1 11·84, K_2 10·49	67U
Cu^{2+}	gl	30	75% dioxan	K_1 11·56, K_2 10·15	67U
Ni^{2+}	gl	30	75% dioxan	K_1 9·01, K_2 8·43	67U

67U E. Uhlemann and F. Dietze, *Z. anorg. Chem.*, 1967, 353, 26

$C_{16}H_{20}O_2N_2$ 1,2-Bis(2″-hydroxybenzlamino)ethane H_2L 1044.1

Metal	*Method*	*Temp*	*Medium*	*Log of equilibrium constant, remarks*	*Ref*
H^+	gl	25	0·1 (KCl)	K_1 10·50, K_2 9·80, K_3 8·37, K_4 6·17	68G
Co^{2+}	gl	25	0·1 (KCl)	K_1 12·78	68G
Cu^{2+}	gl	25	0·1 (KCl)	K_1 20·5	68G
Ni^{2+}	gl	25	0·1 (KCl)	$K_1 > 13$	68G
Zn^{2+}	gl	25	0·1 (KCl)	K_1 11·97	68G

68G D. W. Gruenwedel, *Inorg. Chem.*, 1968, 7, 495

1044.2 $C_{16}H_{20}O_8N_2$ (1-Phenylethylenedinitrilo)tetra-acetic acid H_4L

Metal	*Method*	*Temp*	*Medium*	*Log of equilibrium constant, remarks*	*Ref*
H^+	gl	25	0·1 (KCl)	K_1 9·60±0·02, K_2 5·42±0·02, K_3 3·21±0·01, K_4 1·87±0·05	67O
Ba^{2+}	gl	25	0·1 (KCl)	K_1 8·06±0·01	67O
Ca^{2+}	gl	25	0·1 (KCl)	K_1 10·90±0·03	67O
Co^{2+}	gl	25	0·1 (KCl)	K_1 15·6±0·2	67O
Cu^{2+}	gl	25	0·1 (KCl)	K_1 18·7±0·1	67O
Mg^{2+}	gl	25	0·1 (KCl)	K_1 9·14±0·02	67O
Ni^{2+}	gl	25	0·1 (KCl)	K_1 18·5±0·2	67O
Sr^{2+}	gl	25	0·1 (KCl)	K_1 8·98±0·02	67O
Zn^{2+}	gl	25	0·1 (KCl)	K_1 16·46±0·02	67O

67O N. Okaku, K. Toyoda, Y. Moriguchi, and K. Ueno, *Bull. Chem. Soc. Japan*, 1967, **40**, 2326

1044.3 $C_{16}H_{20}O_{10}N_2$ 2,5-Bis[*N*-carboxymethyl-*N*-(2-hydroxyethyl)amino]benzene-1, 4-dicarboxylic acid H_4L

Metal	*Method*	*Temp*	*Medium*	*Log of equilibrium constant, remarks*	*Ref*
H^+	gl	22	0·1 KCl	K_1 9·06, K_2 5·25, K_3 2·68, K_4 1·95	63U
Co^{2+}	gl	22	0·1 KCl	K_1 7·85 $K(2Co^{2+}+L^{4-} \rightleftharpoons Co_2L)$ 13·00 $K(Co^{2+}+HL^{3-} \rightleftharpoons CoHL^-)$ 5·20 $K(Co^{2+}+H_2L^{2-} \rightleftharpoons CoH_2L)$ 2·30	63U
Ni^{2+}	gl	22	0·1 KCl	K_1 9·20 $K(2Ni^{2+}+L^{4-} \rightleftharpoons Ni_2L)$ 15·70 $K(Ni^{2+}+HL^{3-} \rightleftharpoons NiHL^-)$ 6·50 $K(Ni^{2+}+H_2L^{2-} \rightleftharpoons NiH_2L)$ 3·25	63U

63U E. Uhlig, *Z. anorg. Chem.*, 1963, **320**, 283

1045.1 $C_{16}H_{24}O_8N_2$ 2,5-Bis-[*NN*-bis(2′-hydroxyethyl)amino]benzene-1,4-dicarboxylic acid H_2L

Metal	*Method*	*Temp*	*Medium*	*Log of equilibrium constant, remarks*	*Ref*
H^+	gl	22	0·1 KCl	K_1 7·98, K_2 5·04	61U
Cu^{2+}	gl	22	0·1 KCl	K_1 6·50, $K(Cu^{2+}+HL^- \rightleftharpoons CuHL^+)$ 3·75	61U
Ni^{2+}	gl	22	0·1 KCl	K_1 4·25, $K(Ni^{2+}+HL^- \rightleftharpoons NiHL^+)$ 2·30	61U

61U E. Uhlig, *Z. anorg. Chem.*, 1961, **312**, 332

$C_{16}H_{28}O_8N_2$ (Octamethylenedinitrilo)tetra-acetic acid H_4L 1046

etal	Method	Temp	Medium	Log of equilibrium constant, remarks	Ref
+	gl	20	0·1 (KNO_3)	K_1 10·77, K_2 9·91, K_3 2·75, K_4 2·0	64A
	cal	20	0·1 (KNO_3)	$\Delta H_1°$ −8·09, $\Delta S_1°$ 22·0, $\Delta H_2°$ −5·79, $\Delta S_2°$ 25·5	64A
a^{2+}	gl	20	0·1 (KNO_3)	K_1 4·6, $K(Ca^{2+}+HL^{3-} \rightleftharpoons CaHL^-)$ 3·7	64A
d^{2+}	gl	20	0·1 (KNO_3)	K_1 11·99 $K(Cd^{2+}+HL^{3-} \rightleftharpoons CdHL^-)$ 7·02 $K(Cd^{2+}+CdL^{2-} \rightleftharpoons Cd_2L)$ 2·4	64A
	cal	20	0·1 (KNO_3)	$\Delta H_1°$ −4·59, $\Delta S_1°$ 39·2	64A
o^{2+}	gl	20	0·1 (KNO_3)	K_1 12·91, $K(Co^{2+}+HL^{3-} \rightleftharpoons CoHL^-)$ 7·99 $K(Co^{2+}+CoL^{2-} \rightleftharpoons Co_2L)$ 3·4	64A
	cal	20	0·1 (KNO_3)	$\Delta H_1°$ −4·76, $\Delta S_1°$ 42·7	64A
u^{2+}	gl	20	0·1 (KNO_3)	K_1 15·8, $K(Cu^{2+}+HL^{3-} \rightleftharpoons CuHL^-)$ 11·51	64A
	cal	20	0·1 (KNO_3)	$\Delta H_1°$ −10·28, $\Delta S_1°$ 37·2	64A
e^{2+}	gl	20	0·1 (KNO_3)	K_1 10·96, $K(Fe^{2+}+HL^{3-} \rightleftharpoons FeHL^-)$ 6·71	64A
g^{2+}	gl	20	0·1 (KNO_3)	K_1 21·83	64A
	cal	20	0·1 (KNO_3)	$\Delta H_1°$ −19·74, $\Delta S_1°$ 32·5	64A
Ig^{2+}	gl	20	0·1 (KNO_3)	K_1 4·8, $K(Mg^{2+}+HL^{3-} \rightleftharpoons MgHL^-)$ 3·66	64A
In^{2+}	gl	20	0·1 (KNO_3)	K_1 9·0, $K(Mn^{2+}+HL^{3-} \rightleftharpoons MnHL^-)$ 5·7	64A
	cal	20	0·1 (KNO_3)	$\Delta H_1°$ 0·5, $\Delta S_1°$ 43	
i^{2+}	gl	20	0·1 (KNO_3)	K_1 13·62 $K(Ni^{2+}+HL^{3-} \rightleftharpoons NiHL^-)$ 9·16 $K(Ni^{2+}+NiL^{2-} \rightleftharpoons Ni_2L)$ 4·9	64A
	cal	20	0·1 (KNO_3)	$\Delta H_1°$ −8·5, $\Delta S_1°$ 33·2	64A
b^{2+}	gl	20	0·1 (KNO_3)	$K(Pb^{2+}+HL^{3-} \rightleftharpoons PbHL^-)$ 8·26	64A
	cal	20	0·1 (KNO_3)	$\Delta H_1°$ −8·18	64A
n^{2+}	gl	20	0·1 (KNO_3)	K_1 12·66 $K(Zn^{2+}+HL^{3-} \rightleftharpoons ZnHL^-)$ 8·28 $K(Zn^{2+}+ZnL^{2-} \rightleftharpoons Zn_2L)$ 4·1	64A
	cal	20	0·1 (KNO_3)	$\Delta H_1°$ −4·36, $\Delta S_1°$ 43·1	64A

64A G. Anderegg, *Helv. Chim. Acta*, 1964, **47**, 1801

$C_{16}H_{11}O_3N_3S$ Perhydro-1,3-diphenyl-2-thipyrinidine-2,5,6-trione 5-oxime HL 1049.1 (diphenylthiovioluric acid)

Aetal	Method	Temp	Medium	Log of equilibrium constant, remarks	Ref
I^+	gl	?	50% dioxan	K_1 5·00	60S
e^{2+}	sp	?	0·2 KNO_3 33% dioxan	β_2 5·20, buffer	60S

60S R. P. Singh and N. R. Banerjee, *J. Indian Chem. Soc.*, 1960, **37**, 713

1049.2 $C_{16}H_{11}O_{10}N_3S_2$ 2-(2′-Nitrophenylazo)chromotropic acid H_4L

Metal	*Method*	*Temp*	*Medium*	*Log of equilibrium constant, remarks*	*Ref*
H^+	sp	25	0·1	K_1 8·96	67T
Th^{4+}	sp	25	0·1	$K(Th^{4+} + H_2L^{2-} \rightleftharpoons ThHL^+ + H^+)$ 3·8	67T

67T K. Toel, H. Miyata, and H. Kimara, *Bull. Chem. Soc. Japan*, 1967, **40**, 2085

1049.3 $C_{16}H_{11}O_{10}N_3S_2$ 2-(3′-Nitrophenylazo)chromotropic acid H_4L

Metal	*Method*	*Temp*	*Medium*	*Log of equilibrium constant, remarks*	*Ref*
H^+	sp	25	0·1	K_1 8·60	67T
Th^{4+}	sp	25	0·1	$K(Th^{4+} + H_2L^{2-} \rightleftharpoons ThHL^+ + H^+)$ 4·1	67T

67T K. Toel, H. Miyata, and H. Kimura, *Bull. Chem. Soc. Japan*, 1967, **40**, 2085

1049.4 $C_{16}H_{11}O_{10}N_3S_2$ 2-(4′-Nitrophenylazo)chromotropic acid (chromotrope 2B) H_4L

Metal	*Method*	*Temp*	*Medium*	*Log of equilibrium constant, remarks*	*Ref*
H^+	sp	25	0·1	K_1 8·77	67T
Cu^{2+}	sp	25	0·1 ($NaClO_4$)	*K (?) 9·7±0·1	63S
Eu^{3+}	sp	25	?	*K (?) 4·7±0·3	67S
La^{3+}	sp	25	?	*K (?) 4·9±0·2	63Sa
Lu^{3+}	sp	25	?	*K (?) 5·2±0·2	67S
Sc^{3+}	sp	25	?	*K (?) 4·9±0·1	64M
Th^{4+}	sp	25	?	*K (?) 10·1±0·2	61B
	sp	25	0·1	$K(Th^{4+} + H_2L^{2-} \rightleftharpoons ThHL^+ + H^+)$ 4·0	67T
Tm^{3+}	sp	25	?	*K (?) 5·0±0·2	67S
Y^{3+}	sp	25	?	*K (?) 4·7±0·2	64M
Yb^{3+}	sp	25	?	*K (?) 4·3±0·2	67S

* H constants not given.

61B S. K. Banerjee and A. K. Dey, *J. Indian Chem. Soc.*, 1961, 38, 139
63S S. P. Sangal and A. K. Dey, *Indian J. Chem.*, 1963, **1**, 270
63Sa S. P. Sangal, S. C. Srivastava, and A. K. Dey, *J. Indian Chem. Soc.*, 1963, **40**, 275
64M K. N. Munshi and A. K. Dey, *J. Inorg. Nuclear Chem.*, 1964, **26**, 1603
67S S. P. Sangal, *J. prakt. Chem.*, 1967, **36**, 126
67T K. Toel, H. Miyata, and H. Kimura, *Bull. Chem. Soc. Japan*, 1967, **40**, 2085

$C_{16}H_{12}O_4N_2S$ 1-(4′-Sulphophenylazo)-2-naphthol (tropoeolin OOO) H_2L 1050

Metal	Method	Temp	Medium	Log of equilibrium constant, remarks	Ref
H^+				H^+ constants not given.	
Pd^{2+}	sp	25	?	K (?) 9·8	68S

68S K. K. Saxena and A. K. Dey, *Analyt. Chem.*, 1968, **40**, 1280

$C_{16}H_{12}O_5N_2S$ 1-(2′-Hydroxy-5′-sulphophenylazo)-2-naphthol (Solochrome Violet R) H_3L 1050.1

Metal	Method	Temp	Medium	Log of equilibrium constant, remarks	Ref
H^+	sp	20—60	→0	K_1 (20°) 13·04±0·3, (30°) 12·78±0·03, (40°) 12·56±0·02, (50°) 12·36±0·02, (60°) 12·19±0·02, K_2 (20°) 7·03±0·01, (25°) 7·25, (30°) 6·94±0·01, (40°) 6·88±0·01, (50°) 6·83±0·01, (60°) 6·79±0·01, $K[2H_2L^- \rightleftharpoons (H_2L)_2^{2-}]$ (25°) 3·29	62C, 63C
Al^{3+}	sp, gl	25	→0	K_1 18·4±0·1, K_2 13·2±0·1	62C
Ca^{2+}	sp, gl	25	→0	K_1 6·6±0·1, K_2 ~ 3	62C
Cd^{2+}	gl	25	→0	$K(CdL^- + A^- \rightleftharpoons CdLA^{2-})$ 3·32±0·01, where HA = glycine	63C
Cr^{3+}	sp, pol	25—100	→0	β_2 (75°) 17·25±0·06, (85°) 17·05±0·06, (95°) 17·05±0·06, (100°) 16·93±0·06 $K(CrLOH^- + H^+ \rightleftharpoons CrL)$ (25°) 6·88, (40°) 6·58 ΔH −7·5, ΔS 7 $K[CrL(OH)_2^{2-} + H^+ \rightleftharpoons CrL(OH)^-]$ (25°) 9·82, (40°) 9·41 ΔH −9·7, ΔS 13 $K[CrL(OH)_3^{3-} + H^+ \rightleftharpoons CrL(OH)_2^{2-}]$ (25°) 12·12	62C
Cu^{2+}	sp, gl	25	→0	K_1 21·8±0·1	62C
	gl	25	→0	$K(CuL^- + A^- \rightleftharpoons CuLA^{2-})$ 3·12±0·03, where HA = glycine	63C
Mg^{2+}	sp, gl	25	→0	K_1 8·6±0·1, K_2 5·0±0·1	62C
Ni^{2+}	sp, gl	25	→0	K_1 15·9±0·1, K_2 10·45±0·1	62C
	gl	25	→0	$K(NiL^- + A^- \rightleftharpoons NiLA^{2-})$ 4·02±0·005, where HA = glycine 3·98±0·006, where HA = tyrosine 3·79±0·01, where HA = *NN′*-dimethylglycine 3·62±0·008, where HA = alanine 3·38±0·02, where HA = β-alanine	63C
		15—42	→0	$K(NiL^- + A^- \rightleftharpoons NiLA^{2-})$ (15°) 4·22, (25°) 4·02, (35°) 3·85, (42°) 3·71, where HA = glycine	
Pb^{2+}	sp, gl	25	→0	K_1 12·5±0·1, K_2 5·3	62C
Zn^{2+}	sp, gl	25	→0	K_1 13·5±0·1, K_2 7·4±0·1	62C
	gl	25	→0	$K(ZnL^- + A^- \rightleftharpoons ZnLA^{2-})$ 3·44±0·03, where HA = glycine	63C

62C E. Coates and B. Rigg, *Trans. Faraday Soc.*, 1962, **58**, 88, 2058
63C E. Coates, J. R. Evans, and B. Rigg, *Trans. Faraday Soc.*, 1963, **59**, 2369

1053 $C_{16}H_{12}O_8N_2S_2$ 2-(Phenylazo)chromotropic acid (chromotrope 2R) H_4L

Metal	Method	Temp	Medium	Log of equilibrium constant, remarks	Ref
H^+	sp	25	0·1 ($NaClO_4$)	K_2 9·30	63M
Co^{2+}	sp	25	?	K (?) 9·2	67P
Fe^{3+}	sp	25	?	K (?) 4·4	67P
Ni^{2+}	sp	25	?	K (?) 9·2	67P
Th^{4+}	sp	25	0·1 ($NaClO_4$)	$K(Th^{4+}+H_2L^{2-} \rightleftharpoons ThHL^{+}+H^{+})$ 3·41, 4·42 K varies with pH.	63M

63M H. Miyata, *Bull. Chem. Soc. Japan*, 1963, **36**, 382
67P O. Prakash and S. P. Mushran, *Chim. analyt.*, 1967, **49**, 473

1053.1 $C_{16}H_{12}O_9N_2S_2$ 2-(2′-Hydroxyphenylazo)chromotropic acid H_5L

Metal	Method	Temp	Medium	Log of equilibrium constant, remarks	Ref
H^+	gl	25	0·1	K_2 10·60, K_3 7·60	67T
Th^{4+}	sp	25	0·1	$K(Th^{4+}+H_3L^{2-} \rightleftharpoons ThHL+2H^{+})$ 4·2, 6·0 K varies with pH.	67T

67T K. Toel, H. Miyata, S. Nakashima, and S. Kiguchi, *Bull. Chem. Soc. Japan*, 1967, **40**, 1145

1053.2 $C_{16}H_{12}O_{11}N_2S_3$ 2-(2′-Sulphophenylazo)chromotropic acid H_5L

Metal	Method	Temp	Medium	Log of equilibrium constant, remarks	Ref
H^+	sp	25	0·1 ($NaClO_4$)	K_2 9·76	63M
Th^{4+}	sp	25	0·1 ($NaClO_4$)	$K(Th^{4+}+H_2L^{3-} \rightleftharpoons ThHL+H^{+})$ 3·30, 3·18 K varies with pH.	63M

63M H. Miyata, *Bull. Chem. Soc. Japan*, 1963, **36**, 386

1053.3 $C_{16}H_{12}O_{12}N_2S_3$ 2-(2′-Hydroxy-5′-sulphophenylazo)chromotropic acid H_6L

Metal	Method	Temp	Medium	Log of equilibrium constant, remarks	Ref
H^+	sp	18—22	0·2 $NaNO_3$	K_1 14·40, K_2 10·80, K_3 6·30	66B
Mg^{2+}	sp	18—22	0·2 $NaNO_3$	$K(Mg^{2+}+3H^{+}+2L^{6-} \rightleftharpoons MgH_3L_2^{7-})$ 46·6	66B

66B A. Bezdekova and B. Budesinsky, *Coll. Czech. Chem. Comm.*, 1966, **31**, 199

$C_{16}H_{13}ON_4Cl$ 4-(2′-Chlorophenylazo)-1-phenyl-5-methylpyrazol-3(2*H*)-one HL 1055.1

Metal	*Method*	*Temp*	*Medium*	*Log of equilibrium constant, remarks*	*Ref*
H^+	gl	30	75% dioxan	K_1 10·36	67S
Cd^{2+}	gl	30	75% dioxan	K_1 5·0±0·2, β_2 10·60	67S
Co^{2+}	gl	30	75% dioxan	K_1 5·9±0·2, β_2 12·07	67S
Cu^{2+}	gl	30	75% dioxan	K_1 9·48±0·08, β_2 18·72	67S
Ni^{2+}	gl	30	75% dioxan	K_1 6·5±0·1, β_2 13·04	67S
Zn^{2+}	gl	30	75% dioxan	K_1 5·3±0·1, β_2 11·59	67S

67S F. A. Snavely, D. A. Sweigart, C. H. Yoder, and A. Terzis, *Inorg. Chem.*, 1967, 6, 1831

$C_{16}H_{13}ON_4Cl$ 4-(3′-Chlorophenylazo)-1-phenyl-5-methylpyrazol-3(2*H*)-one HL 1055.2

Metal	*Method*	*Temp*	*Medium*	*Log of equilibrium constant, remarks*	*Ref*
H^+	gl	30	75% dioxan	K_1 10·12	67S
Cu^{2+}	gl	30	75% dioxan	K_1 9·31±0·04, β_2 18·22	67S

67S F. A. Snavely, D. A. Sweigart, C. H. Yoder, and A. Terzis, *Inorg. Chem.*, 1967, 6, 1831

$C_{16}H_{13}ON_4Cl$ 4-(4′-Chlorophenylazo)-1-phenyl-5-methylpyrazol-3(2*H*)-one HL 1055.3

Metal	*Method*	*Temp*	*Medium*	*Log of equilibrium constant, remarks*	*Ref*
H^+	gl	30	75% dioxan	K_1 10·12	67S
Cu^{2+}	gl	30	75% dioxan	K_1 9·4±0·1, β_2 18·33	67S

67S F. A. Snavely, D. A. Sweigart, C. H. Yoder, and A. Terzis, *Inorg. Chem.*, 1967. 6, 1831

$C_{16}H_{13}ON_4Br$ 4-(2′-Bromophenylazo)-1-phenyl-5-methylpyrazole-3(2*H*)-one HL 1057.1

Metal	*Method*	*Temp*	*Medium*	*Log of equilibrium constant, remarks*	*Ref*
H^+	gl	30	75% dioxan	K_1 10·40	67S
Cd^{2+}	gl	30	75% dioxan	K_1 4·8±0·3, β_2 10·32	67S
Co^{2+}	gl	30	75% dioxan	K_1 5·6±0·1, β_2 11·45	67S
Cu^{2+}	gl	30	75% dioxan	K_1 9·3±0·1, β_2 18·70	67S
Ni^{2+}	gl	30	75% dioxan	K_1 6·20±0·04, β_2 12·53	67S
Zn^{2+}	gl	30	75% dioxan	β_2 11·25	67S

67S F. A. Snavely, D. A. Sweigart, C. H. Yoder, and A. Terzis, *Inorg. Chem.*, 1967, 6, 1831

1057.2 $C_{16}H_{13}ON_4Br$ 4-(3′-Bromophenylazo)-1-phenyl-5-methylpyrazol-3(2*H*)-one HL

Metal	*Method*	*Temp*	*Medium*	*Log of equilibrium constant, remarks*	*Ref*
H^+	gl	30	75% dioxan	K_1 10·05	67S
Cu^{2+}	gl	30	75% dioxan	K_1 9·3±0·1, β_2 18·15	67S

67S F. A. Snavely, D. A. Sweigart, C. H. Yoder, and A. Terzis, *Inorg. Chem.*, 1967, **6,** 1831

1057.3 $C_{16}H_{13}ON_4Br$ 4-(4′-Bromophenylazo)-1-phenyl-5-methylpyrazol-3(2*H*)-one HL

Metal	*Method*	*Temp*	*Medium*	*Log of equilibrium constant, remarks*	*Ref*
H^+	gl	30	75% dioxan	K_1 10·12	67S
Cu^{2+}	gl	30	75% dioxan	K_1 9·28±0·04, β_2 18·23	67S

67S F. A. Snavely, D. A. Sweigart, C. H. Yoder, and A. Terzis, *Inorg. Chem.*, 1967, **6,** 1831

1058.1 $C_{16}H_{13}ON_4I$ 4-(2′-Iodophenylazo)-1-phenyl-5-methylpyrazol-3(2*H*)-one HL

Metal	*Method*	*Temp*	*Medium*	*Log of equilibrium constant, remarks*	*Ref*
H^+	gl	30	75% dioxan	K_1 10·22	67S
Cd^{2+}	gl	30	75% dioxan	K_1 4·2±0·2, β_2 9·44	67S
Co^{2+}	gl	30	75% dioxan	K_1 5·2±0·3, β_2 10·87	67S
Cu^{2+}	gl	30	75% dioxan	K_1 8·8±0·2, β_2 18·03	67S
Ni^{2+}	gl	30	75% dioxan	K_1 5·7±0·1, β_2 11·39	67S
Zn^{2+}	gl	30	75% dioxan	β_2 1068	67S

67S F. A. Snavely, D. A. Sweigart, C. H. Yoder, and A. Terzis, *Inorg. Chem.*, 1967, **6,** 1831

1058.2 $C_{16}H_{13}ON_4I$ 4-(4′-Iodophenylazo)-1-phenyl-5-methylpyrazol-3(2*H*)-one HL

Metal	*Method*	*Temp*	*Medium*	*Log of equilibrium constant, remarks*	*Ref*
H^+	gl	30	75% dioxan	K_1 10·09	67S
Cu^{2+}	gl	30	75% dioxan	K_1 9·40±0·01, β_2 18·20	67S

67S F. A. Snavely, D. A. Sweigart, C. H. Yoder, and A. Terzis, *Inorg. Chem.*, 1967, **6,** 1831

$C_{16}H_{13}ON_4F$ 4-(2′-Fluorophenylazo)-1-phenyl-5-methylpyrazol-3(2*H*)-one HL 1058.3

Metal	Method	Temp	Medium	Log of equilibrium constant, remarks	Ref
H^+	gl	30	75% dioxan	K_1 10·02	67S
Cd^{2+}	gl	30	75% dioxan	K_1 5·4±0·1, β_2 10·84	67S
Co^{2+}	gl	30	75% dioxan	K_1 6·1±0·3, β_2 13·09	67S
Cu^{2+}	gl	30	75% dioxan	K_1 9·31±0·04, β_2 18·48	67S
Ni^{2+}	gl	30	75% dioxan	K_1 7·0±0·1, β_2 13·76	67S
Zn^{2+}	gl	30	75% dioxan	K_1 6·2±0·1, β_2 12·70	67S

67S F. A. Snavely, D. A. Sweigart, C. H. Yoder, and A. Terzis, *Inorg. Chem.*, 1967, **6**, 1831

$C_{16}H_{13}ON_4F$ 4-(3′-Fluorophenylazo)-1-phenyl-5-methylpyrazol-3(2*H*)-one HL 1058.4

Metal	Method	Temp	Medium	Log of equilibrium constant, remarks	Ref
H^+	gl	30	75% dioxan	K_1 10·10	67S
Cu^{2+}	gl	30	75% dioxan	K_1 9·35±0·05, β_2 17·97	67S

67S F. A. Snavely, D. A. Sweigart, C. H. Yoder, and A. Terzis, *Inorg. Chem.*, 1967, **6**, 1831

$C_{16}H_{13}ON_4F$ 4-(4′-Fluorphenylazo)-1-phenyl-5-methylpyrazol-3(2*H*)-one HL 1058.5

Metal	Method	Temp	Medium	Log of equilibrium constant, remarks	Ref
H^+	gl	30	75% dioxan	K_1 10·30	67S
Cu^{2+}	gl	30	75% dioxan	K_1 9·57±0·06, β_2 18·52	67S

67S F. A. Snavely, D. A. Sweigart, C. H. Yoder, and A. Terzis, *Inorg. Chem.*, 1967, **6**, 1831

$C_{16}H_{13}O_8N_3S_2$ 8-amino-1-hydroxy-2-(2′-Hydroxyphenylazo)-naphthalene-3,6-disulphonic acid 1058.6 H_4L

Metal	Method	Temp	Medium	Log of equilibrium constant, remarks	Ref
H^+	sp	?	0·100 (KCl)	K_1 11·9, K_2 8·4	60D
Ca^{2+}	sp	?	0·100 (KCl)	K_1 2·50	60D
Mg^{2+}	sp	?	0·100 (KCl)	K_1 3·81	60D

60D H. Diehl and J. Ellingboe, *Analyt. Chem.*, 1960, **32**, 1120

1058.7 $C_{16}H_{13}O_8N_3S_2$ 8-Amino-1-hydroxy-2-(2′-hydroxyphenylazo)-naphthalene-5,7-disulphonic a

Metal	Method	Temp	Medium	Log of equilibrium constant, remarks	Ref
H^+	sp	?	0·100 (KCl)	K_1 11·6, K_2 7·4	60D
Ca^{2+}	sp	?	0·100 (KCl)	K_1 3·08	60D
Mg^{2+}	sp	?	0·100 (KCl)	K_1 4·50	60D

60D H. Diehl and J. Ellingboe, *Analyt. Chem.*, 1960, **32,** 1120

1062 $C_{16}H_{13}O_{10}N_2S_2As$ 1-(2′-Arsonophenylazo)-2-hydroxy-naphthalene-3,6-disulphonic acid H_5L (thoron)

Metal	Method	Temp	Medium	Log of equilibrium constant, remarks	Ref
H^+				H^+ constants not given by any of the references listed below.	
Al^{3+}	sp	25	?	K (?) 10·5	68G
Eu^{3+}	sp	25	?	K (?) 8·2±0·2	67S
Ga^{3+}	sp	25	?	K (?) 9·9	68G
Hf^{4+}	sp	25	?	K (?) 8·2±0·1	66Sa
In^{3+}	sp	25	?	K (?) 9·9	68G
La^{3+}	sp	25	?	K (?) 10·1±0·3	63Sa, 6
	sp	25	?	K (?) 4·4±0·1	64Sa
Lu^{3+}	sp	25	?	K (?) 9·7±0·2	67S
Pd^{2+}	sp	25	?	K (?) 4·4±0·2	64S
Sc^{3+}	sp	25	?	K (?) 9	66Sb
Tb^{3+}	sp	25	?	K (?) 8·9±0·3	67S
Th^{4+}	sp	25	?	K (?) 9·8±0·3	63S, 6
UO_2^{2+}	sp	25	?	K (?) 4·3±0·3	66S
Yb^{3+}	sp	25	?	K (?) 9·6±0·2	67S

Erratum. 1964, p. 714, Table 1062. *For* 4-(p-Arsono *etc. read* 4-(*o*-Arsono *etc.*

63S S. P. Sangal and A. K. Dey, *J. Indian Chem. Soc.*, 1963, **40,** 279
63Sa S. P. Sangal and A. K. Dey, *Z. anorg. Chem.*, 1963, **322,** 107
64S S. P. Sangal and A. K. Dey, *J. Indian Chem. Soc.*, 1964, **41,** 306
64Sa S. P. Sangal and A. K. Dey, *Chim. analyt.*, 1964, **46,** 223
66S S. P. Sangal, *J. prakt. Chem.*, 1966, **31,** 68
66Sa S. P. Sangal, *Chim. analyt.*, 1966, **48,** 566
66Sb S. P. Sangal and A. K. Dey, *J. Indian Chem. Soc.*, 1966, **43,** 440
67S S. P. Sangal, *J. prakt. Chem.*, 1967, **36,** 126
68G V. C. Garg, S. C. Shrivastawa, and A. K. Dey, *Current Sci.*, 1968, **37,** 47

$C_{16}H_{13}O_{11}N_2S_2As$ 2-(2′-Arsonophenylazo)chromotropic acid (arsenazo) H_6L 1062.1

Metal	Method	Temp	Medium	Log of equilibrium constant, remarks	Ref
H^+	?	?	?	K_4 2·72, K_5 2·16, K_6 0·07	61K
La^{3+}	sp	~ 20	?	$K(La^{3+}+L^{6-}+2H^+ \rightleftharpoons LaH_2L^-)$ 28·8 H^+ constants not given.	65B
Pu^{4+}	sp	20	?	$K(Pu^{4+}+H_3L^{3-} \rightleftharpoons PuH_3L^+)$ 7·7±0·2 $K[Pu(OH)^{3+}+H_4L^{2-} \rightleftharpoons Pu(OH)H_4L^+]$ 6·6±0·2	61K
Th^{4+}	sp	20	?	$K(Th^{4+}+H_4R^{2-} \rightleftharpoons ThH_4R^{2+})$ 6·8	60K

60K A. E. Klygin and V. K. Pavlova, *Russ. J. Inorg. Chem.*, 1960, **5**, 734 (1516)
61K A. E. Klygin and V. K. Pavlova, *Russ. J. Inorg. Chem.*, 1961, **6**, 536 (1050)
65B B. Budesinksy, *Z. analyt. Chem.*, 1965, **207**, 105

$C_{16}H_{14}O_6N_3SAs$ 4-Amino-2-(2′-arsonophenylazo)-naphthalene-1-sulphonic acid H_3L 1062.2

Metal	Method	Temp	Medium	Log of equilibrium constant, remarks	Ref
H^+	sp	~ 20	?	K_1 9·30, K_2 6·35, K_3 3·90	65B
Cu^{2+}	sp	~ 20	?	$K(Cu^{2+}+H^++L^{3-} \rightleftharpoons CuHL)$ 14·8	65B

65B B. Budesinsky, *Z. analyt. Chem.*, 1965, **207**, 241

$C_{17}H_{16}O_6$ 2-Hydroxy-2′,4′,4-trimethoxydibenzoyl HL 1064.1

Metal	Method	Temp	Medium	Log of equilibrium constant, remarks	Ref
H^+	gl	?	0·1 ($NaClO_4$)	K_1 10·15	63D
Be^{2+}	gl	?	0·1 ($NaClO_4$)	K_1 7·45, K_2 6·55	63D
Cd^{2+}	gl	?	0·1 ($NaClO_4$)	β_2 8·80	63D
Co^{2+}	gl	?	0·1 ($NaClO_4$)	K_1 5·15, K_2 4·85	63D
Cu^{2+}	gl	?	0·1 ($NaClO_4$)	K_1 7·55, K_2 6·68	63D
Mg^{2+}	gl	?	0·1 ($NaClO_4$)	K_1 3·61, K_2 3·23	63D
Mn^{2+}	gl	?	0·1 ($NaClO_4$)	K_1 4·35, K_2 4·04	63D
Ni^{2+}	gl	?	0·1 ($NaClO_4$)	K_1 5·40, K_2 4·96	63D
UO_2^{2+}	gl	?	0·1 ($NaClO_4$)	K_1 8·43, K_2 7·02	63D
Zn^{2+}	gl	?	0·1 ($NaClO_4$)	β_2 11·25	63D

63D Y. Dutt and R. P. Singh, *Indian J. Chem.*, 1963, **1**, 402

1065 $C_{17}H_{19}N_3$ 2-[*N*-(Benzyl)-*N*-phenylaminomethyl]-1,4,5*H*-1,3-diazole (antistine) L

Metal	*Method*	*Temp*	*Medium*	*Log of equilibrium constant, remarks*	*Ref*
H^+	gl	0—45	0·1 (KNO_3)	K_1 (0°) 10·81, (25°) 10·10, (45°) 9·43, K_2 (0°) 2·30, (25°) 2·37, (45°) 3·15	64A
Cd^{2+}	pol	0—45	0·1 (KNO_3)	β_2 (0°) 10·08, (25°) 8·73, (45°) 8·10 $\Delta H_{\beta 2}°$ −17·7, $\Delta S_{\beta 2}°$ −19	64A

64A A. C. Andrews and J. K. Romary, *J Chem. Soc.*, 1964, 405

1066 $C_{17}H_{12}O_3N_2$ 2-(2′-Hydroxy-1′-naphthylazo)benzoic acid H_2L

Metal	*Method*	*Temp*	*Medium*	*Log of equilibrium constant, remarks*	*Ref*
H^+	sp	?	0·100 (KCl)	K_1 12·0	60D
Mg^{2+}	sp	?	0·100 (KCl)	K_1 2·10	60D

60D H. Diehl and J. Ellingboe, *Analyt. Chem.*, 1960, **32**, 1120

1069.1 $C_{17}H_{16}ON_4$ 4-(2′-Tolylazo)-1-phenyl-5-methylpyrazol-3-(2*H*)-one HL

Metal	*Method*	*Temp*	*Medium*	*Log of equilibrium constant, remarks*	*Ref*
H^+	gl	30	75% dioxan	K_1 10·54	67S
Cd^{2+}	gl	30	75% dioxan	K_1 5·1±0·1, β_2 10·77	67S
Co^{2+}	gl	30	75% dioxan	K_1 6·15±0·07, β_2 12·34 (?)	67S
Cu^{2+}	gl	30	75% dioxan	K_1 9·50±0·09, β_2 18·31	67S
Ni^{2+}	gl	30	75% dioxan	K_1 6·56±0·05, β_2 12·60	67S
Zn^{2+}	gl	30	75% dioxan	K_1 5·6±0·4, β_2 11·74	67S

67S F. A. Snavely, D. A. Sweigart, C. H. Yoder, and A. Terzis, *Inorg. Chem.*, 1967, **6**, 1831

1069.2 $C_{17}H_{16}ON_4$ 4-(3′-Tolylazo)-1-phenyl-5-methylpyrazol-3(2*H*)-one HL

Metal	*Method*	*Temp*	*Medium*	*Log of equilibrium constant, remarks*	*Ref*
H^+	gl	30	75% dioxan	K_1 10·54	67S
Cu^{2+}	gl	30	75% dioxan	K_1 9·88±0·07, β_2 19·16	67S

67S F. A. Snavely, D. A. Sweigart, C. H. Yoder, and A. Terzis, *Inorg. Chem.*, 1967, **6**, 1831

$C_{17}H_{16}ON_4$ 4-(4′-Tolylazo)-1-phenyl-5-methylpyrazol-3(2*H*)-one HL 1069.3

Metal	Method	Temp	Medium	Log of equilibrium constant, remarks	Ref
H^+	gl	30	75% dioxan	K_1 10·47	67S
Cu^{2+}	gl	30	75% dioxan	K_1 9·6±0·1, β_2 18·28	67S

67S F. A. Snavely, D. A. Sweigart, C. H. Yoder, and A. Terzis, *Inorg. Chem.*, 1967, **6**, 1831

$C_{17}H_{16}O_2N_4$ 4-(2′-Methoxyphenylazo)-1-phenyl-5-methylpyrazol-3(2*H*)-one HL 1069.4

Metal	Method	Temp	Medium	Log of equilibrium constant, remarks	Ref
H^+	gl	30	75% dioxan	K_1 9·82	67S
Cd^{2+}	gl	30	75% dioxan	K_1 6·6±0·1, β_2 12·26	67S
Co^{2+}	gl	30	75% dioxan	K_1 8·8±0·1, β_2 16·16	67S
Cu^{2+}	gl	30	75% dioxan	K_1 10·3±0·2, β_2 19·01	67S
Ni^{2+}	gl	30	75% dioxan	K_1 9·5±0·1, β_2 17·11	67S
Zn^{2+}	gl	30	75% dioxan	K_1 8·37±0·07, β_2 15·20	67S

67S F. A. Snavely, D. A. Sweigart, C. H. Yoder, and A. Terzis, *Inorg. Chem.*, 1967, **6**, 1831

$C_{17}H_{16}O_2N_4$ 4-(3′-Methoxyphenylazo)-1-phenyl-5-methylpyrazol-3(2*H*)-one HL 1069.5

Metal	Method	Temp	Medium	Log of equilibrium constant, remarks	Ref
H^+	gl	30	75% dioxan	K_1 10·42	67S
Cu^{2+}	gl	30	75% dioxan	K_1 9·4±0·2, β_2 17·90	67S

67S F. A. Snavely, D. A. Sweigart, C. H. Yoder, and A. Terzis, *Inorg. Chem.*, 1967, **6**, 1831

$C_{17}H_{16}O_2N_4$ 4-(4′-Methoxyphenylazo)-1-phenyl-5-methylpyrazol-3(2*H*)-one HL 1069.6

Metal	Method	Temp	Medium	Log of equilibrium constant, remarks	Ref
H^+	gl	30	75% dioxan	K_1 10·61	67S
Cu^{2+}	gl	30	75% dioxan	K_1 9·6±0·3, β_2 17·93	67S

67S F. A. Snavely, D. A. Sweigart, C. H. Yoder, and A. Terzis, *Inorg. Chem.*, 1967, **6**, 1831

1071.1 $C_{17}H_{16}N_4S_2$ 3-Methyl-4-(2′-methylthiophenylazo)-1-phenylpyrazole-5(2*H*)-thione HL

Metal	*Method*	*Temp*	*Medium*	*Log of equilibrium constant, remarks*	*Ref*
H^+	gl	30	75% dioxan	K_1 10·70	64S
Cd^{2+}	gl	30	75% dioxan	K_1 8·36±0·07, 8·94±0·09	64S
Ni^{2+}	gl	30	75% dioxan	K_1 9·5±0·2, 10·3±0·2	64S
Pb^{2+}	gl	30	75% dioxan	K_1 7·8±0·1, K_2 8·6±0·1	64S
Zn^{2+}	gl	30	75% dioxan	K_1 6·6±0·2, K_2 7·7±0·2	64S

64S F. A. Snavely, W. S. Trahanovsky, and F. H. Suydam, *Inorg. Chem.*, 1964, 3, 123

1072.1 $C_{17}H_{20}N_4S$ 1,5-Bis(2,4-dimethylphenyl)-1,2,4,5-tetra-azapentene-3-thione L [bis(2,4-dimethylphenyl)thiocarbazone]

Metal	*Method*	*Temp*	*Medium*	*Log of equilibrium constant, remarks*	*Ref*
H^+	sp	25	0·1 50% dioxan	K_1 6·87	64M
Ni^{2+}	sp	25	0·1 50% dioxan	K_1 6·40	64M
Zn^{2+}	sp	25	0·1 50% dioxan	K_1 4·80	64M

64M K. S. Math, Q. Fernando, and H. Freiser, *Analyt. Chem.*, 1964, **36**, 1762

1073 $C_{17}H_{21}ON$ *NN*-Dimethyl-2-(diphenylmethoxy)ethylamine (benadryl) L

Metal	*Method*	*Temp*	*Medium*	*Log of equilibrium constant, remarks*	*Ref*
H^+	gl	0—45	0·1 (KNO_3)	K_1 (0°) 9·65, (25°) 9·10, (45°) 8·64	64A
Cd^{2+}	pol	0—45	0·1 (KNO_3)	β_2 (0°) 7·86, (25°) 7·28, (45°) 6·89 $\Delta H_{\beta 2}°$ −8·6, $\Delta S_{\beta 2}°$ 4	64A

64A A. C. Andrews and J. K. Romary, *J. Chem. Soc.*, 1964, 405

1074.1 $C_{17}H_{12}O_{10}N_2S_2$ 2-(2′-Carboxyphenylazo)chromotropic acid H_5L

Metal	*Method*	*Temp*	*Medium*	*Log of equilibrium constant, remarks*	*Ref*
H^+	sp	25	0·1 ($NaClO_4$)	K_2 10·17, K_3 3·20	67T
Th^{4+}	sp	25	0·1 ($NaClO_4$)	$K(Th^{4+}+H_2L^{3-} \rightleftharpoons ThHL+H^+)$ 3·8, 5·8 *K* varies with pH.	67T

67T K. Toel, H. Miyata, and T. Harada, *Bull. Chem. Soc. Japan*, 1967, **40**, 1141

$C_{17}H_{13}O_5N_3S$ 1-Amino-2-(2′-carboxyphenylazo)naphthalene-4-sulphonic acid H_2L 1074.2

Metal	Method	Temp	Medium	Log of equilibrium constant, remarks	Ref
H^+	sp	~ 20	?	K_1 3·90	65B
Cu^{2+}	sp	~ 20	?	K_1 7·7	65B

65B B. Budesinsky, *Z. Analyt. Chem.*, 1965, **207**, 241

$C_{17}H_{16}ON_4S$ 3-Methyl-4-(2′-methoxyphenylazo)-1-phenylpyrazole-5(2*H*)-thione HL 1075.1

Metal	Method	Temp	Medium	Log of equilibrium constant, remarks	Ref
H^+	gl	30	75% dioxan	K_1 12·00	64S
Cd^{2+}	gl	30	75% dioxan	K_1 8·78±0·08, K_2 9·9±0·1	64S
Co^{2+}	gl	30	75% dioxan	K_1 11·3±0·1, K_2 10·2±0·3	64S
Cu^{2+}	gl	30	75% dioxan	K_1 >13	64S
Ni^{2+}	gl	30	75% dioxan	K_1 11·56±0·05, K_2 10·25±0·11	64S
Zn^{2+}	gl	30	75% dioxan	K_1 8·2±0·1, K_2 9·2±0·2	64S

64S F. A. Snavely, W. S. Trahanovsky, and F. H. Suydam, *Inorg. Chem.*, 1964, **3**, 123

$C_{17}N_{16}ON_4S$ 5-Methyl-4-(2′-methylthiophenylazo)-1-phenylpyrazol-3(2*H*)-one HL 1075.2

Metal	Method	Temp	Medium	Log of equilibrium constant, remarks	Ref
H^+	gl	30	75% dioxan	K_1 10·64	67S
Cd^{2+}	gl	30	75% dioxan	K_1 6·39±0·07, β_2 12·56	67S
Co^{2+}	gl	30	75% dioxan	K_1 8·7±0·1, β_2 16·02	67S
Cu^{2+}	gl	30	75% dioxan	K_1 >14	67S
Ni^{2+}	gl	30	75% dioxan	K_1 9·91±0·04, β_2 18·65	67S
Zn^{2+}	gl	30	75% dioxan	K_1 7·08±0·01, β_2 13·39	67S

67S F. A. Snavely, D. A. Sweigart, C. H. Yoder, and A. Terzis, *Inorg. Chem.*, 1967, **6**, 1831

$C_{17}H_{16}ON_4S$ 5-Methyl-4-(3′-Methylthiophenylazo)-1-phenylpyrazol-3(2*H*)-one HL 1075.3

Metal	Method	Temp	Medium	Log of equilibrium constant, remarks	Ref
H^+	gl	30	75% dioxan	K_1 10·35	67S
Cu^{2+}	gl	30	75% dioxan	K_1 9·57±0·06, β_2 18·96	67S

67S F. A. Snavely, D. A. Sweigart, C. H. Yoder, and A. Terzis, *Inorg. Chem.*, 1967, **6**, 1831

1078.1 $C_{18}H_{12}O_6$ 2,5-Dihydroxy-3,6-diphenoxy-1,4-benzoquinone H_2L

Metal	*Method*	*Temp*	*Medium*	*Log of equilibrium constant, remarks*	*Ref*
H^+	sp	25	0·5 NaCl	K_1 3·00±0·07, K_2 2·0±0·1	67B
Ge^{IV}	sp	25	0·5 NaCl	$K[Ge(OH)_4 + 2HL^- \rightleftharpoons Ge(OH)_2L_2^{2-}]$ 8·8±0·1	67B

67B A. Beauchamp and R. Benoit, *Bull. Soc. chim. France*, 1967, 672

1079 $C_{18}H_{12}N_2$ 5-Phenyl-1,10-phenanthroline L

Metal	*Method*	*Temp*	*Medium*	*Log of equilibrium constant, remarks*	*Ref*
H^+	gl	25	0·3 (KNO_3) 50% dioxan	K_1 4·03±0·03	61J
Cu^+				See Cu^{2+} below.	
Cu^{2+}	Pt	25	0·3 (KNO_3) 50% dioxan	K_1 4·05 $K(CuL_2^{2+} + Cu^+ \rightleftharpoons CuL_2^+ + Cu^{2+})$ 0·68	61J

61J B. R. James and R. J. P. Williams, *J. Chem. Soc.*, 1961, 2007

1079.1 $C_{18}H_{12}N_2$ 2,2′-Biquinolyl L

Metal	*Method*	*Temp*	*Medium*	*Log of equilibrium constant, remarks*	*Ref*
H^+	gl	25	0·3 (K_2SO_4)	K_1 3·10±0·10	61J
Cu^+	Pt	25	0·3 (K_2SO_4)	β_2 ~ 16·5	61J
Cu^{2+}	Pt	25	0·3 (K_2SO_4)	K_1 4·27, K_2 3·46	61J

61J B. R. James and R. J. P. Williams, *J. Chem. Soc.*, 1961, 2007

1079.2 $C_{18}H_{12}N_6$ 2,4,6-Tri(2′-pyridyl)-1,3,5-triazine L

Metal	*Method*	*Temp*	*Medium*	*Log of equilibrium constant, remarks*	*Ref*
H^+	gl	25	0·1 KCl	K_1 3·10	66B
Fe^{2+}	sp	25	~ 0·05—0·07	β_2 10·24±0·01	66B

66B E. B. Buchanan, jun., D. Crichton and J. R. Bacon, *Talanta*, 1966, 13, 903

$C_{18}H_{15}As$ Triphenylarsine L 1079.3

Metal	Method	Temp	Medium	Log of equilibrium constant, remarks	Ref
Ag^+	sp	25	55·6—75·4% methanol	K_1 (55·6%) 5·81, (75·4%) 5·70	66O
Hg^{2+}	sp	25	0·10 $NaClO_4$ 74·5% methanol	$K(HgCl_2+L \rightleftharpoons HgClL^+ + Cl^-)$ 1·11 $K(HgClL^+ + L \rightleftharpoons HgL_2^{2+} + Cl^-)$ 0·83	66O

66O D. C. Olson and J. Bjerrum, *Acta Chem. Scand.*, 1966, 20, 143

$C_{18}H_{18}O_2$ 3-Phenyl-1-(2′,4′,6′-trimethylphenyl)-propane-1,3-dione HL 1079.4 (mesitoylbenzoylmethane)

Metal	Method	Temp	Medium	Log of equilibrium constant, remarks	Ref
H^+	gl	30	75% dioxan	K_1 13·11	65U
Be^{2+}	gl	30	75% dioxan	K_1 11·79, K_2 11·12	65U
Cu^{2+}	gl	30	75% dioxan	K_1 12·47, K_2 11·61	65U
Ni^{2+}	gl	30	75% dioxan	K_1 10·32, K_2 9·37	65U
Zn^{2+}	gl	30	75% dioxan	K_1 10·07, K_2 9·13	65U

65U E. Uhlemann and E. Frank, *Z. anorg. Chem.*, 1965, 340, 319

$C_{18}H_{11}O_2N$ 2-(2′-Quinolyl)indane-1,3-dione HL 1079.5

Metal	Method	Temp	Medium	Log of equilibrium constant, remarks	Ref
H^+	gl	30	75% dioxan	K_1 14·90	64C
Be^{2+}	gl	30	75% dioxan	K_1 11·78±0·02	64C
Co^{2+}	gl	30	75% dioxan	K_1 9·8±0·2	64C
Cu^{2+}	gl	30	75% dioxan	K_1 11·62±0·02, K_2 4·75±0·02	64C
Mg^{2+}	gl	30	75% dioxan	K_1 7·33±0·02	64C
Mn^{2+}	gl	30	75% dioxan	K_1 9·31±0·02	64C
UO_2^{2+}	gl	30	75% dioxan	K_1 12·95±0·02, K_2 12·07±0·02	64C
Zn^{2+}	gl	30	75% dioxan	K_1 10·53±0·02	64C

64C J. R. Cook and D. F. Martin, *J. Inorg. Nuclear Chem.*, 1964, 26, 571

1079.6 $C_{18}H_{14}O_3N_2$ 2-(2′,4′-Dihydroxyphenylazo)-4-phenylphenol H_3L

Metal	Method	Temp	Medium	Log of equilibrium constant, remarks	Ref
H^+	sp	?	0·100 (KCl)	K_1 11·4, K_2 8·1, K_3 6·7	60D
Ca^{2+}	sp	?	0·100 (KCl)	K_1 1·68	60D
Mg^{2+}	sp	?	0·100 (KCl)	K_1 3·68	60D

60D H. Diehl and J. Ellingboe, *Analyt. Chem.*, 1960, 32, 1120

1081.1 $C_{18}H_{18}ON_4$ 4-(2′-Ethylphenylazo)-5-methyl-1-phenylpyrazol-3(2*H*)-one HL

Metal	Method	Temp	Medium	Log of equilibrium constant, remarks	Ref
H^+	gl	30	75% dioxan	K_1 10·62	67S
Cd^{2+}	gl	30	75% dioxan	K_1 5·03±0·07, β_2 10·33	67S
Co^{2+}	gl	30	75% dioxan	K_1 5·98±0·04, β_2 12·01	67S
Cu^{2+}	gl	30	75% dioxan	K_1 9·29±0·05, β_2 17·83	67S
Ni^{2+}	gl	30	75% dioxan	K_1 6·48±0·04, β_2 12·45	67S

67S F. A. Snavely, D. A. Sweigart, C. H. Yoder, and A. Terzis, *Inorg. Chem.*, 1967, **6**, 1831

1081.2 $C_{18}H_{18}ON_4$ 4-(2′,6′-Dimethylphenylazo)-5-methyl-1-phenylpyrazol-3(2*H*)-one HL

Metal	Method	Temp	Medium	Log of equilibrium constant, remarks	Ref
H^+	gl	30	75% dioxan	K_1 10·72	67S
Cu^{2+}	gl	30	75% dioxan	K_1 10·80±0·04, β_2 20·54	67S

67S F. A. Snavely, D. A. Sweigart, C. H. Yoder, and A. Terzis, *Inorg. Chem.*, 1967, **6**, 1831

1081.3 $C_{18}H_{19}ON_5$ 4-(2-Dimethylaminophenylazo)-3-methyl-1-phenylpyrazol-5(2*H*)-thione HL

Metal	Method	Temp	Medium	Log of equilibrium constant, remarks	Ref
H^+	gl	30	75% dioxan	K_1 12·43	63S
Cd^{2+}	gl	30	75% dioxan	K_1 8·95, K_2 9·34	63S
Co^{2+}	gl	30	75% dioxan	K_1 11·02, K_2 9·88	63S
Cu^{2+}	gl	30	75% dioxan	K_1 >13·0, K_2 7·6	63S
Mn^{2+}	gl	30	75% dioxan	β_2 ~ 14	63S
Ni^{2+}	gl	30	75% dioxan	K_1 12·37, K_2 10·34	63S
Pb^{2+}	gl	30	75% dioxan	K_1 ~ 8·9, K_2 ~ 8·3	63S
Zn^{2+}	gl	30	75% dioxan	K_1 10·14, K_2 8·88	63S

63S F. A. Snavely, C. H. Yoder, and F. H. Suydam, *Inorg. Chem.*, 1963, **2**, 708

$C_{18}H_{20}O_4N_2$ 1,2-Bis(3′-methoxysalicylideneamino)ethane H_2L 1081.4

Metal	*Method*	*Temp*	*Medium*	*Log of equilibrium constant, remarks*	*Ref*
H^+	sp	20	0·3	K_1 10·9, K_2 7·8, K_3 4·4, K_4 3·3, Theorell-Steinhagen buffer	66S
Ni^{2+}	dis	20	0·3 (acetate)	K (?) 5·55	66S
UO_2^{2+}	dis	20	0·3 (acetate)	K_1 19·6	66S

66S J. Stronski, A. Zielinski, A. Saneotus, Z. Stasicka, and B. Budesinsky, *Z. analyt. Chem.*, 1966, **222**, 14

$C_{18}H_{20}O_6N_2$ Ethylenedinitrilo-*NN*′-bis(2′-hydroxyphenyl)-*NN*′-diacetic acid H_4L 1082

Metal	*Method*	*Temp*	*Medium*	*Log of equilibrium constant, remarks*	*Ref*
H^+	sp	20	0·1 KNO_3	K_1 11·85, K_2 10·56	64A
	gl	20	0·1 KNO_3	K_3 8·78, K_4 6·39	64A
	gl	25	0·1 (KNO_3)	K_1 11·68, K_2 10·24, K_3 8·64, K_4 6·32	64S
Fe^{2+}	pol	25	0·1 $NaClO_4$	K_1 14·26	64Sa
Fe^{3+}	sp	20	0·1 KNO_3	K_1 33·91	64A
	sp	25	0·1 ($NaClO_4$)	K_1 33·9 ± 0·1	64S

64A G. Anderegg and F. L'Eplattenier, *Helv. Chim. Acta*, 1964, **47**, 1067
64S K. H. Schroder, *Nature*, 1964, **202**, 898
64Sa K. H. Schroder, *Acta Chem. Scand.*, 1964, **18**, 596

$C_{18}H_{22}O_4N_4$ Ethylenedinitrilo-*NN*′-bis(2′-pyridylmethyl)-*NN*′-diacetic acid H_2L 1083

Errata. 1964, p. 720, Table 1083. *Delete* reference 57L and *substitute* 65L. R. G. Lacoste C. V. Christoffers, and A. E. Martell, *J. Amer. Chem. Soc.*, 1965, **87**, 2385.
Under Medium *for* 0·1 *read* 0·1 (KNO_3).

$C_{18}H_{30}O_{12}N_4$ Triethylenetetraminehexa-acetic acid (TTHA) H_6L 1085

Metal	*Method*	*Temp*	*Medium*	*Log of equilibrium constant, remarks*	*Ref*
H^+	gl	30	0·1 (KCl)	K_1 10·33, K_2 9·35, K_3 5·98, K_4 4·00, K_5 2·52, K_6 2·46	65S, 66S
	gl	25	0·10 (KNO_3)	K_1 10·19, K_2 9·40, K_3 6·16, K_4 4·16, K_5 2·95, K_6 2·42	66E, 67B, 67Ba, 68C, 68W
	gl	25	0·1 KNO_3	K_1 10·82, K_2 9·67, K_3 6·26, K_4 4·08, K_5 2·64	65K
Ag^+	Ag	25	0·1 (KNO_3)	K_1 8·67, $K(AgL^{5-} + H^+ \rightleftharpoons AgHL^{4-})$ 9·11, $K(2Ag^+ + L^{6-} \rightleftharpoons Ag_2L^{4-})$ 5·22	68W
Ca^{2+}	gl	25	0·10 (KNO_3)	$K(CaHL^{3-} + H^+ \rightleftharpoons CaH_2L^{2-})$ 4·87, $K(CaL^{4-} + H^+ \rightleftharpoons CaHL^{3-})$ 8·53, $K(2Ca^{2+} + L^{6-} \rightleftharpoons Ca_2L^{2-})$ 14·2	67B

continued overleaf

1085 $C_{18}H_{30}O_{12}N_4$ (contd)

Metal	*Method*	*Temp*	*Medium*	*Log of equilibrium constant, remarks*	*Ref*
Cd^{2+}	pol	25	0·1 $NaClO_4$	$K(Cd^{2+}+H_2L^{4-} \rightleftharpoons CdH_2L^{2-})$ 10·36 $K(2Cd^{2+}+HL^{5-} \rightleftharpoons Cd_2HL^{-})$ 0·24	65C
Co^{2+}	gl	25	0·10 (KNO_3)	K_1 20·6±0·1 $K(CoH_3L^{-}+H^{+} \rightleftharpoons CoH_4L)$ 1·57±0·05 $K(CoH_2L^{2-}+H^{+} \rightleftharpoons CoH_3L^{-})$ 2·63±0·03 $K(CoHL^{3-}+H^{+} \rightleftharpoons CoH_2L^{2-})$ 4·03±0·01 $K(CoL^{4-}+H^{+} \rightleftharpoons CoHL^{3-})$ 7·97±0·01	67B
Cu^{2+}	gl	?	0·1 KNO_3	K_1 15·4 $K(CuL^{4-}+H^{+} \rightleftharpoons CuHL^{3-})$ 8·06 $K(2Cu^{2+}+L^{6-} \rightleftharpoons Cu_2L^{2-})$ 24·5	65K
	gl	25	0·10 (KNO_3)	K_1 20·3±0·1 $K(CuH_3L^{-}+H^{+} \rightleftharpoons CuH_4L)$ 2·04±0·05 $K(CuH_2L^{2-}+H^{+} \rightleftharpoons CuH_3L^{-})$ 2·86±0·03 $K(CuHL^{3-}+H^{+} \rightleftharpoons CuH_2L^{2-})$ 4·05±0·01 $K(CuL^{4-}+H^{+} \rightleftharpoons CuHL^{3-})$ 7·96±0·01	67B
	gl	25	0·1 (KNO_3)	$K(Cu_2L^{2-}+2A^{4-} \rightleftharpoons Cu_2LA_2^{10-})$ 14·8 where H_4A = 1,2-dihydroxybenzene-3,5-disulphonic acid, $K(Cu_2L^{2-}+A^{2-} \rightleftharpoons Cu_2LA^{4-})$ 7·2, $K(Cu_2LA^{4-}+A^{2-} \rightleftharpoons Cu_2LA_2^{6-})$ 2·75 where H_2A = 8-hydroxyquinoline-5-sulphonic acid	67B
Fe^{2+}	red	25	0·1 $NaClO_4$	K_1 17·1±0·2, K_2 10·2 $K(FeL^{4-}+H^{+} \rightleftharpoons FeHL^{3-})$ 8·67 $K(FeHL^{3-}+H^{+} \rightleftharpoons FeH_2L^{2-})$ 3·25 $K(FeH_2L^{2-}+H^{+} \rightleftharpoons FeH_3L^{-})$ <2 $K(FeL^{4-}+OH^{-} \rightleftharpoons FeLOH^{5-})$ 4·98 $K[FeLOH^{5-}+OH^{-} \rightleftharpoons FeL(OH)_2^{6-}]$ 4·19 $K(Fe_2L^{2-}+OH^{-} \rightleftharpoons Fe_2LOH^{3-})$ 5·27 $K[Fe_2LOH^{3-}+OH^{-} \rightleftharpoons Fe_2L(OH)_2^{4-}]$ 5·18	65S
Fe^{3+}	red	25	0·1 $NaClO_4$	K_1 29·4±0·2 $K(FeL^{3-}+H^{+} \rightleftharpoons FeHL^{2-})$ 7·51 $K(FeHL^{2-}+H^{+} \rightleftharpoons FeH_2L^{-})$ 2·60 $K(FeL^{3-}+OH^{-} \rightleftharpoons FeLOH^{4-})$ 4·20 $K[FeLOH^{4-}+OH^{-} \rightleftharpoons FeL(OH)_2^{5-}]$ 3·50 $K[Fe_2LOH^{-}+OH^{-} \rightleftharpoons Fe_2L(OH)_2^{2-}]$ 2·9	65S
	gl	25	0·1 (KNO_3)	$K[Fe_2(OH)_2L^{2-}+2HA^{3-} \rightleftharpoons Fe_2LA_2^{8-}]$ 5·05, $K[Fe_2(OH)_2L^{2-}+2H^{+} \rightleftharpoons Fe_2L]$ 6·6, $K(Fe_2L+2A^{4-} \rightleftharpoons Fe_2LA_2^{8-})$ 23·5 where H_4A = 1,2-dihydroxybenzene-3,5-disulphonic acid, $K(Fe_2L+2A^{2-} \rightleftharpoons Fe_2LA_2^{4-})$ 18·10 where H_2A = 8-hydroxyquinoline-5-sulphonic acid	67B
Hf^{4+}	ix	?	0·5 $HClO_4$	K_1 19·08 $K[Hf(OH)^{3+}+H_6L \rightleftharpoons HfH_2L+3H^{+}]$ 3·90±0·03	66E
			1—2 $HClO_4$	$K(Hf^{4+}+H_6L \rightleftharpoons HfH_2L+4H^{+})$ (1) 3·5±0·1, (2) 2·84±0·04	
Hg^{2+}	gl	25	0·1 $NaClO_4$	K_1 25·27±0·08, K_2 8·4 $K(HgL^{4-}+H^{+} \rightleftharpoons HgHL^{3-})$ 6·55 $K(HgHL^{3-}+H^{+} \rightleftharpoons HgH_2L^{2-})$ 3·30 $K(Hg_2L^{2-}+OH^{-} \rightleftharpoons Hg_2LOH^{3-})$ 6·1 $K[Hg_2LOH^{3-}+OH^{-} \rightleftharpoons Hg_2L(OH)_2^{4-}]$ 5·9	66S

co[ntd]

Metal	Method	Temp	Medium	Log of equilibrium constant, remarks	Ref
Mg^{2+}	gl	25	0·10 (KNO_3)	$K(MgHL^{3-}+H^+ \rightleftharpoons MgH_2L^{2-})$ 4·65	67B
				$K(MgL^{4-}+H^+ \rightleftharpoons MgHL^{3-})$ 9·31	
				$K(2Mg^{2+}+L^{6-} \rightleftharpoons Mg_2L^{2-})$ 13·9	
Ni^{2+}	gl	25	0·10 (KNO_3)	K_1 18·8±0·1	67B
				$K(NiH_3L^-+H^+ \rightleftharpoons NiH_4L)$ 1·15±0·05	
				$K(NiH_2L^{2-}+H^+ \rightleftharpoons NiH_3L^-)$ 2·74±0·03	
				$K(NiHL^{3-}+H^+ \rightleftharpoons NiH_2L^{2-})$ 4·86±0·01	
				$K(NiL^{4-}+H^+ \rightleftharpoons NiHL^{3-})$ 9·11±0·01	
U^{4+}	gl	25	0·10 (KCl)	$K(UL^{2-}+H^+ \rightleftharpoons ULH^-)$ 2·28±0·03	68C
Zr^{4+}	ix	?	0·5 $HClO_4$	K_1 19·74	66E
				$K[Zr(OH)^{3+}+H_6L \rightleftharpoons ZrH_2L+3H^+]$ 4·76±0·02	
			1—2 $HClO_4$	K_1 (1) 9·74	66E
				$K(Zr^{4+}+H_6L \rightleftharpoons ZrH_2L+4H^+)$ (1) 4·08±0·03, (2) 2·9±0·1	

Errata, 1964, p. 721, Table 1085. *for* reference 64B *read* 65B, T. A. Bohigian jun and A. E. Martell, *Inorg. Chem.* 1965, **4,** 1264.

Under Ca^+ seventh line, *for* 3·01 *read* 4·00.

Under Mg^{2+} seventh line, *for* 3·10 *read* 5·32.

65C G. Conradi, M. Kopanica, and J. Koryta, *Coll. Czech. Chem. Comm.*, 1965, 30, 2029
65K K. S. Klausen, G. O. Kalland, and E. Jacobsen, *Analyt. Chim. Acta*, 1965, **33,** 67
65S K. H. Schroder, *Acta Chem. Scand.*, 1965, **19,** 1797
66E A. N. Ermakov, I. N. Marov, and N. B. Kalinichenko, *Russ. J. Inorg. Chem.*, 1966, **11,** 1404 (2614)
66S K. H. Schroder, *Acta Chem. Scand.*, 1966, **20,** 881
67B T. A. Bohigian jun. and A. E. Martell, *J. Amer. Chem. Soc.*, 1967, **89,** 832
67Ba T. A. Bohigian jun. and A. E. Martell, *J. Inorg. Nuclear Chem.*, 1967, **29,** 453
68C G. H. Carey and A. E. Martell, *J. Amer. Chem. Soc.*, 1968, **90,** 32
68W H. Wikberg and A. Ringbom, *Suomen Kem.*, 1968, **B41,** 177

$C_{18}H_{14}O_{11}N_2S_2$ 2-[2′-(Carboxyhydroxymethyl)phenylazo]chromotropic acid H_5L 1086.1

Metal	Method	Temp	Medium	Log of equilibrium constant, remarks	Ref
H^+	sp	25	0·1 $NaClO_4$	K_2 9·55, K_3 3·10	67M
Th^{4+}	sp	25	0·1	$K(Th^{4+}+H_3L^{2-} \rightleftharpoons ThHL+2H^+)$ 3·9, 5·9, 7·7 *K* varies with pH.	67M

67M H. Miyata, *Bull. Chem. Soc. Japan*, 1967, **40,** 2815

$C_{18}H_{14}O_{11}N_2S_2$ 2-[2′-(Carboxymethoxy)phenylazo]chromotropic acid H_5L 1086.2

Metal	Method	Temp	Medium	Log of equilibrium constant, remarks	Ref
H^+	sp	25	0·1 ($NaClO_4$)	K_2 9·76, K_3 2·99	67M
Th^{4+}	sp	25	0·1 ($NaClO_4$)	$K(Th^{4+}+H_3L^{2-} \rightleftharpoons ThHL+2H^+)$ 3·6, 5·2, 7·6 *K* varies with pH.	67M

67M H. Miyata, *Bull. Chem. Soc. Japan*, 1967, **40,** 1875

1086.3 $C_{18}H_{15}O_2NFe$ 1-(Ferrocenyl)-3-(2′-pyridyl)propane-1,3-dione (ferrocenecarbonyl-2-picolinoylmethane) HL

Metal	*Method*	*Temp*	*Medium*	*Log of equilibrium constant, remarks*	*Ref*
H^+				H^+ constants not given.	
Cu^{2+}	gl	25	0·01 75% dioxan	K_1 13·27, K_2 12·37	67W
Ni^{2+}	gl	25	0·01 75% dioxan	K_1 12·18, K_2 11·43	67W
Zn^{2+}	gl	25	0·01 75% dioxan	K_1 11·38, K_2 10·37	67W

67W L. Wolf, H. Hennig, and I. P. Sereda, *Russ. J. Inorg. Chem.*, 1967, **12**, 231 (450)

1086.4 $C_{18}H_{15}O_2NFe$ 1-(Ferrocenyl)-3-(3′-pyridyl)propane-1,3-dione (ferrocenecarbonylnicotinoylmethane) HL

Metal	*Method*	*Temp*	*Medium*	*Log of equilibrium constant, remarks*	*Ref*
H^+				H^+ constants not given.	
Cu^{2+}	gl	25	0·002—0·01 75%dioxan	K_1 (0·002) 12·80, (0·01) 12·91, K_2 (0·002) 12·81, (0·01) 11·54	67W
Ni^{2+}	gl	25	0·002—0·01 75% dioxan	K_1 (0·002) 11·26, (0·01) 11·34, K_2 (0·002) 11·17, (0·01) 11·16	67W
Zn^{2+}	gl	25	0·002—0·01 75% dioxan	K_1 (0·002) 10·50, (0·01) 10·44, K_2 (0·002) 10·14, (0·01) 9·63	67W

67W L. Wolf, H. Hennig, and I. P. Sereda, *Russ. J. Inorg. Chem.*, 1967, **12**, 231 (450)

1086.5 $C_{18}H_{15}O_2NFe$ 1-(Ferrocenyl)-3-(4′-pyridyl)propane-1,3-dione (ferrocenecarbonylisonicotinoylmethane) HL

Metal	*Method*	*Temp*	*Medium*	*Log of equilibrium constant, remarks*	*Ref*
H^+				H^+ constants not given.	
Cu^{2+}	gl	25	0·002—0·01 75% dioxan	K_1 (0·002) 12·37, (0·01) 12·55, K_2 (0·002) 12·14, (0·01) 11·08	67W
Ni^{2+}	gl	25	0·002—0·01 75% dioxan	K_1 (0·002) 11·12, (0·01) 11·11, K_2 (0·002) 10·78, (0·01) 10·76	67W
Zn^{2+}	gl	25	0·002—0·01 75% dioxan	K_1 (0·002) 10·05, (0·01) 9·97, K_2 (0·002) 9·75, (0·10) 9·48	67W

67W L. Wolf, H. Hennig, and I. P. Sereda, *Russ. J. Inorg. Chem.*, 1967, **12**, 231 (450)

$C_{18}H_{15}O_3N_3S$ Perhydro-1,3-bis(2′-tolyl-2-thiopyrimidine-2,5,6,-trione 5-oxime (di-2-tolylthiovioluric acid) HL 1086.6

Metal	*Method*	*Temp*	*Medium*	*Log of equilibrium constant, remarks*	*Ref*
H^+	gl	?	50% dioxan	K_1 5·35	60S
Fe^{2+}	sp	?	0·2 KNO_3 33% acetone	β_2 5·60, buffer	60S

60S R. P. Singh and N. R. Banerjee, *J. Indian Chem. Soc.*, 1960, **37**, 713

$C_{18}H_{15}O_3N_3S$ Perhydro-1,3-bis(3′-tolyl)-2-thiopyrimidine-2,5,6-trione 5-oxime (di-3-tolylthiovioluric acid) HL 1086.7

Metal	*Method*	*Temp*	*Medium*	*Log of equilibrium constant, remarks*	*Ref*
H^+	gl	?	50% dioxan	K_1 5·90	60S
Fe^{2+}	sp	?	0·2 KNO_3 33% acetone	β_2 6·08, buffer	60S

60S R. P. Singh and N. R. Banerjee, *J. Indian Chem. Soc.* 1960, **37**, 713

$C_{18}H_{15}O_3N_3S$ Perhydro-1,3-bis(4′-tolyl)-2-thiopyrimidine-2,5,6-trione 5-oxime (di-4-tolylthiovioluric acid) HL 1086.8

Metal	*Method*	*Temp*	*Medium*	*Log of equilibrium constant, remarks*	*Ref*
H^+	gl	?	50% dioxan	K_1 5·60	60S
Fe^{2+}	sp	?	0·2 KNO_3 33% acetone	β_2 5·80, buffer	60S

60S R. P. Singh and N. R. Banerjee, *J. Indian Chem. Soc.*, 1960, **37**, 713

$C_{18}H_{15}O_3SP$ 3-(Diphenylphosphino)benzenesulphonic acid HL 1087

Metal	*Method*	*Temp*	*Medium*	*Log of equilibrium constant, remarks*	*Ref*
H^+	sp	25	*I* $HClO_4$	K_1 0·13±0·50 *I*	62W
Bi^{3+}	Bi(Hg)	25	1 HNO_3	K_1 3·7, β_6 21·8	62W
Cu^+	Cu(Hg)	25	1·0 (LiCl)	K_1 5·76, K_2 5·45, K_3 4·91, K_4 3·80	68G
Hg^{2+}	Hg	25	1 KNO_3	K_1 14·3, β_2 24·6, β_3 29·7, β_4 33·0	62S

Erratum. 1964, p. 721, Table 1087. *For p*-(Diphenylphosphino) *etc, read m*-(Diphenylphosphino) *etc.*

62S B. Salvesen and J. Bjerrum, *Acta Chem. Scand.*, 1962, **16**, 735
62W G. Wright and J. Bjerrum, *Acta Chem. Scand.*, 1962, **16**, 1262
68G R. George and J. Bjerrum, *Acta Chem. Scand.*, 1968, **22**, 497

1087.1 $C_{18}H_{15}O_8N_6SAs$ 4-(4′-Sulphophenylazo)anilinoazo-4-nitrobenzene-2-arsonic acid (sulpharsazen) H_3L

Metal	*Method*	*Temp*	*Medium*	*Log of equilibrium constant, remarks*	*Ref*
H^+	sp, gl	20	0·08 (KCl) 4% ethanol	K_1 11·7, K_2 8·7, K_3 5·2	65P
Cd^{2+}	gl, sp	20	0·08 (KCl) 4% ethanol	K_1 9·8 $K(CdL^{2-}+H^+ \rightleftharpoons CdHL^-)$ 8·8	65P
Ni^{2+}	gl, sp	20	0·08 (KCl) 4% ethanol	K_1 8·1 $K(NiL^{2-}+H^+ \rightleftharpoons NiHL^-)$ 8·93	65P
Pb^{2+}	gl, sp	20	0·08 (KCl) 4% ethanol	K_1 16·5 $K(PbL^{2-}+H^+ \rightleftharpoons PbHL^-)$ 5·7	65P
Zn^{2+}	sp, gl	20	0·08 (KCl) 4% ethanol	K_1 10·8 $K(ZnL^{2-}+H^+ \rightleftharpoons ZnHL^-)$ 7·75	65P

65P M. Z. Partashnikova and I. G. Shafran, *J. Analyt. Chem.* (*U.S.S.R.*), 1965, **20**, 288 (313)

1089.1 $C_{19}H_{21}N_5$ 2,6-Bis[*N*-(2′-pyridylmethyl)aminomethyl]pyridine L

Metal	*Method*	*Temp*	*Medium*	*Log of equilibrium constant, remarks*	*Ref*
H^+	gl	25	0·1 (KCl)	K_1 7·55, K_2 6·86	68G
Co^{2+}	gl	25	0·1 (KCl)	K_1 14·8	68G
Cu^{2+}	gl	25	0·1 (KCl)	K_1 18·4	68G
Zn^{2+}	gl	25	0·1 (KCl)	K_1 12·0	68G

68G D. W. Gruenwedel, *Inorg. Chem.*, 1968, **7**, 495

1090 $C_{19}H_{14}O_7S$ 3,3′,4′-Trihydroxyfuchsone-2″-sulphonic acid (Pyrocatechol Violet) H_4L

Metal	*Method*	*Temp*	*Medium*	*Log of equilibrium constant, remarks*	*Ref*
H^+				H^+ constants not given.	
VO^{2+}	sp	25	~0	K (?) 9	67M

67M S. P. Mushran, O. Prakash, and J. N. Awasthi, *Analyt. Chem.*, 1967, **39**, 1307

[illegible]$H_{18}O_4N_4$ 4-[2′-(2″-Carboxyethoxy)phenylazo]-3-methyl-1-phenylpyrazole-5(2*H*)-thione H_2L 1090.1

Metal	*Method*	*Temp*	*Medium*	*Log of equilibrium constant, remarks*	*Ref*
H^+	gl	30	~0 75% dioxan	K_1 10·78, K_2 7·66	65S
Ba^{2+}	gl	30	~0 75% dioxan	K_1 4·07±0·08	65S
Ca^{2+}	gl	30	~0 75% dioxan	K_1 6·5±0·2	65S
Cd^{2+}	gl	30	~0 75% dioxan	K_1 8·8±0·2	65S
Co^{2+}	gl	30	~0 75% dioxan	K_1 10·67±0·02	65S
Cu^{2+}	gl	30	~0 75% dioxan	K_1 14·80±0·05	65S
Mg^{2+}	gl	30	~0 75% dioxan	K_1 4·6±0·2	65S
Mn^{2+}	gl	30	~0 75% dioxan	K_1 7·80±0·08	65S
Ni^{2+}	gl	30	~0 75% dioxan	K_1 11·58±0·07	65S
Pb^{2+}	gl	30	~0 75% dioxan	K_1 9·9±0·1	65S
Sr^{2+}	gl	30	~0 75% dioxan	K_1 4·9±0·1	65S
UO_2^{2+}	gl	30	~0 75% dioxan	K_1 12·1±0·3	65S
Zn^{2+}	gl	30	~0 75% dioxan	K_1 10·80±0·04	65S

65S F. A. Snavely, W. Magen, and D. Kozart, *J. Inorg. Nuclear Chem.*, 1965, 27, 679

$C_{19}H_{20}O_2N_2$ 4-Butyl-1,2-diphenylpyrazolidine-3,5-dione (butazolidine) L 1091.1

Metal	*Method*	*Temp*	*Medium*	*Log of equilibrium constant, remarks*	*Ref*
H^+	gl	20	0·1 (KCl) 50% ethanol	K_1 4·89	57W
Ca^{2+}	gl	20	0·1 KCl 50% ethanol	K_1 1·35	57W
Mg^{2+}	gl	20	0·1 (KCl) 50% ethanol	K_1 1·21	57W
Zn^{2+}	gl	20	0·1 (KCl) 50% ethanol	K_1 1·28	57W

57W K. Wallenfels and H. Sund, *Biochem. Z.*, 1957, 329, 41

1091.2 $C_{19}H_{13}O_4N_3S$ 1-Hydroxy-2-(8′-quinolylazo)naphthalene-4-sulphonic acid H_2L

Metal	Method	Temp	Medium	Log of equilibrium constant, remarks	Ref
H^+	sp	18—22	0·1 ($NaClO_4$) 50% methanol	K_1 10·35, K_2 1·79	67A
Co^{2+}	gl	25	0·1 ($NaClO_4$) 50% methanol	K_1 10·5	67A
Fe^{2+}	gl	25	0·1 ($NaClO_4$) 50% methanol	K_2 9·4	67A
Mn^{2+}	gl	25	0·1 ($NaClO_4$) 50% methanol	K_1 8·6, K_2 7·0	67A
Ni^{2+}	gl	25	0·1 ($NaClO_4$) 50% methanol	K_2 9·3	67A
Zn^{2+}	gl	25	0·1 ($NaClO_4$) 50% methanol	K_1 10·6, K_2 7·2	67A

67A R. G. Anderson and G. Nickless, *Talanta*, 1967, **14**, 1221

1091.3 $C_{19}H_{18}O_3N_4S$ 4-[2′-(2″-Carboxyethylthio)phenylazo]-3-methyl-1-phenylpyrazole-5(2*H*)-thione

Metal	Method	Temp	Medium	Log of equilibrium constant, remarks	Ref
H^+	gl	30	~ 0 75% dioxan	K_1 11·31, K_2 7·87	65S
Cd^{2+}	gl	30	~ 0 75% dioxan	K_1 9·59 ± 0·09	65S
Co^{2+}	gl	30	~ 0 75% dioxan	K_1 11·41 ± 0·04	65S
Cu^{2+}	gl	30	~ 0 75% dioxan	K_1 > 16	65S
Mg^{2+}	gl	30	~ 0 75% dioxan	K_1 3·4	65S
Mn^{2+}	gl	30	~ 0 75% dioxan	K_1 7·3 ± 0·2	65S
Ni^{2+}	gl	30	~ 0 75% dioxan	K_1 13·09 ± 0·06	65S
Pb^{2+}	gl	30	~ 0 75% dioxan	K_1 9·89 ± 0·09	65S
Zn^{2+}	gl	30	~ 0 75% dioxan	K_1 10·12 ± 0·04	65S

65S F. A. Snavely, W. Magen, and D. Kozart, *J. Inorg Nuclear Chem.*, 1965, **27**, 679

1091.4 $C_{20}H_{14}N_4$ 2,2′: 6′,2″: 6″,2‴-Quaterpyridine L

Metal	Method	Temp	Medium	Log of equilibrium constant, remarks	Ref
H^+	hyp	?	0·01	K_1 4·7, K_2 3·3 (same as terpyridine)	64B
Fe^{2+}	sp	25	0·01 H_2SO_4	K_1 8·28, K_2 6·65 ΔH_1 −7·5, ΔS_1 12, ΔH_2 −13·5, ΔS_2 −15	64B
Fe^{3+}	sp	25	0·02 HNO_3	K_1 11·40 ΔH_1 −9, ΔS_1 22	64B

64B A. Bergh, P. O'D. Offenhartz, P. George, and G. P. Haight jun., *J. Chem. Soc.*, 1964, 1533

$C_{20}H_{15}O_4N_3$ 8-hydroxy-7-(3-Nitroanilinofurfuryl)-quinoline HL 1092.1

Metal	Method	Temp	Medium	Log of equilibrium constant, remarks	Ref
H^+	sol, gl	~ 20 (?)	?	K_1 11·92±0·08	61T
Cu^{2+}	sol	25	?	K_{s2} −27·28, acetate buffer	61T
UO_2^{2+}	sol	25	?	K_s ($UO_2L_2HL+2H^+ \rightleftharpoons UO_2^{2+}+3HL$) −30·81 acetate buffer	61T

61T E. P. Trailina, V. V. Zelentsov, I. A. Savich, and V. I. Spitsyn, *Russ. J. Inorg. Chem.*, 1961, **6**, 1047 (2048)

$C_{20}H_{16}O_2N_2$ 1,2-Bis(salicylideneamino)benzene H_2L 1092.2

Metal	Method	Temp	Medium	Log of equilibrium constant, remarks	Ref
H^+	sp	20	0·3	K_1 ~ 12, K_2 8·1, K_3 6·2, K_4 5·4, Theorell–Steinhagen buffer	66S
Ni^{2+}	dis	20	0·3 (acetate)	K (?) 5·60	66S
UO_2^{2+}	dis	20	0·3 (acetate)	K_1 20·9	66S

66S T. Stronski A. Zielinski, A. Samotus, Z. Stasicka, and B. Budesinsky; *Z. analyt. Chem.*, 1966, **222**, 14

$C_{20}H_{24}O_6N_2$ Ethylenedinitrito-*NN'*-bis(2-hydroxybenzyl)-*NN'*-diacetic acid H_4L 1092.3

Metal	Method	Temp	Medium	Log of equilibrium constant, remarks	Ref
H^+	gl	25	0·10 (KNO_3)	K_1 12·46, K_2 11·00, K_3 8·32, K_4 4·64	67L
Ca^{2+}	gl	25	0·10 (KNO_3)	K_1 9·29, $K(Ca^{2+}+HL^{3-} \rightleftharpoons CaHL^-)$ 5·52 $K(Ca^{2+}+H_2L^{2-} \rightleftharpoons CaH_2L)$ 2·02	67L
Cd^{2+}	gl	25	0·10 (KNO_3)	K_1 17·52, $K(Cd^{2+}+HL^{3-} \rightleftharpoons CdHL^-)$ 13·17 $K(Cd^{2+}+H_2L^{2-} \rightleftharpoons CdH_2L)$ 8·85	67L
Co^{2+}	gl	25	0·10 (KNO_3)	K_1 19·89, $K(Co^{2+}+HL^{3-} \rightleftharpoons CoHL^-)$ 15·20 $K(Co^{2+}+H_2L^{2-} \rightleftharpoons CoH_2L)$ 9·76	67L
Cu^{2+}	gl	25	0·10 (KNO_3)	K_1 21·38, $K(Cu^{2+}+HL^{3-} \rightleftharpoons CuHL^-)$ 17·55 $K(Cu^{2+}+H_2L^{2-} \rightleftharpoons CuH_2L)$ 11·73	67L
Fe^{3+}	gl	25	0·10 (KNO_3)	K_1 39·68	67L
Lu^{3+}	gl	25	0·10 (KNO_3)	K_1 20·38 $K(Lu^{3+}+H_2L^{2-} \rightleftharpoons LuH_2L^+)$ 7·06	67L
Mg^{2+}	gl	25	0·10 (KNO_3)	K_1 10·51, $K(Mg^{2+}+HL^{3-} \rightleftharpoons MgHL^-)$ 6·20 $K(Mg^{2+}+H_2L^{2-} \rightleftharpoons MgH_2L)$ 2·21	67L
Mn^{2+}	gl	25	0·10 (KNO_3)	K_1 14·78, $K(Mn^{2+}+HL^{3-} \rightleftharpoons MnHL^-)$ 9·98 $K(Mn^{2+}+H_2L^{2-} \rightleftharpoons MnH_2L)$ 5·56	67L
Ni^{2+}	gl	25	0·10 (KNO_3)	K_1 19·31, $K(Ni^{2+}+HL^{3-} \rightleftharpoons NiHL^-)$ 15·36 $K(Ni^{2+}+H_2L^{2-} \rightleftharpoons NiH_2L)$ 10·81	67L

continued overleaf

1092.3 $C_{20}H_{24}O_6N_2$ (contd)

Metal	*Method*	*Temp*	*Medium*	*Log of equilibrium constant, remarks*	*Ref*
Pb^{2+}	gl	25	0·10 (KNO_3)	K_1 18·24, $K(Pb^{2+}+HL^{3-} \rightleftharpoons PbHL^-)$ 14·76 $K(Pb^{2+}+H_2L^{2-} \rightleftharpoons PbH_2L)$ 10·38	67L
Zn^{2+}	gl	25	0·10 (KNO_3)	K_1 18·37, $K(Zn^{2+}+HL^{3-} \rightleftharpoons ZnHL^-)$ 14·18 $K(Zn^{2+}+H_2L^{2-} \rightleftharpoons ZnH_2L)$ 9·17	67L

67L F. L'Eplattenier, I. Murase, and A. E. Martell, *J. Amer. Chem. Soc.*, 1967, **89,** 837

1094 $C_{20}H_{13}O_7N_3S$ 3-Hydroxy-4-(1′-hydroxy-2′-naphthylazo)-8-nitronaphthalene-1-sulphonic acid (Eriochrome Black T) H_3L

Metal	*Method*	*Temp*	*Medium*	*Log of equilibrium constant, remarks*	*Ref*
H^+	sp	18—20	0·3	K_1 11·31, K_2 6·80	68K
Cd^{2+}	sp	20 (?)	0·3 $NaClO_4$	K_1 12·74	68K
Co^{2+}	sp	20	0·3	K_1 20·0	67K
Cu^{2+}	sp	20	0·3 ($NaClO_4$)	K_1 21·38	67Ka
Pb^{2+}	sp	20 (?)	0·3 $NaClO_4$	K_1 13·19	68K
Zn^{2+}	sp	20	0·3 ($NaClO_4$)	K_1 12·31 $K(Zn^{2+}+L^{3-}+A^{3-} \rightleftharpoons ZnLA^{4-})$ 15·15 $K(ZnA^-+L^{3-} \rightleftharpoons ZnLA^{4-})$ 2·83 where H_3A = nitrilotriacetic acid	67Ka

67K M. Kodama, *Bull. Chem. Soc. Japan*, 1967, **40,** 2575
67Ka M. Kodama and H. Ebine, *Bull. Chem. Soc. Japan*, 1967, **40,** 1857
68K M. Kodama and C. Sasaki, *Bull. Chem. Soc. Japan*, 1968, **41,** 127

1100.1 $C_{21}H_{24}O_2$ 1,3-Bis(2′,4′,6′-trimethylphenyl)propane-1,3-dione (dimesitoylmethane) HL

Metal	*Method*	*Temp*	*Medium*	*Log of equilibrium constant, remarks*	*Ref*
H^+	gl	30	75% dioxan	K_1 12·47	65U
Be^{2+}	gl	30	75% dioxan	K_1 10·66, K_2 9·74	65U
Cu^{2+}	gl	30	75% dioxan	K_1 11·80, K_2 10·99	65U
Ni^{2+}	gl	30	75% dioxan	K_1 9·77, K_2 8·61	65U
Zn^{2+}	gl	30	75% dioxan	K_1 9·51, K_2 8·58	65U

65U E. Uhlemann and E. Frank, *Z. anorg. Chem.*, 1965, **340,** 319

$C_{21}H_{16}N_4S$ 1,5-Di-1-naphthyl-1,2,4,5-tetra-azapenten-3-thione HL 1100.2 (di-1-naphthylthiocarbazone)

Metal	*Method*	*Temp*	*Medium*	*Log of equilibrium constant, remarks*	*Ref*
H^+	sp	25	0·1 50% dioxan	K_1 5·23	64M
Ni^{2+}	sp	25	0·1 50% dioxan	K_1 5·38	64M
Zn^{2+}	sp	25	0·1 50% dioxan	K_1 4·21	64M

64M K. S. Math, Q. Fernando, and H. Freiser, *Analyt. Chem.*, 1964, **36**, 1762

$C_{21}H_{16}N_4S$ 1,5-Di-2′-naphthyl-1,2,4,5-tetra-azapenten-3-thione HL 1100.3 (di-2-naphthylthiocarbazone)

Metal	*Method*	*Temp*	*Medium*	*Log of equilibrium constant, remarks*	*Ref*
H^+	sp	25	0·1 50% dioxan	K_1 5·76	64M
Ni^{2+}	sp	25	0·1 50% dioxan	K_1 6·14	64M
Zn^{2+}	sp	25	0·1 50% dioxan	K_1 5·91	64M

64M K. S. Math, Q. Fernando, and H. Freiser, *Analyt. Chem.*, 1964, **36**, 1762

$H_{19}O_8N_3S$ *N*-[1′-Hydroxy-4′-(4″-sulphophenylazo)-2′-naphthylmethyl]iminodiacetic acid H_4L 1101.1 (methyl napthol orange)

Metal	*Method*	*Temp*	*Medium*	*Log of equilibrium constant, remarks*	*Ref*
H^+	sp	~ 20	0·1 ($NaNO_3$)	K_1 10·50, K_2 4·50, K_3 2·40	63B
Sm^{3+}	sp	~ 20	0·2 ($NaNO_3$)	β_2 20·15	63B
Ti^{4+}	sp	~ 20	0·2 ($NaNO_3$)	β_2 22·96	63B
UO_2^{2+}	sp	~ 20	0·2 ($NaNO_3$)	β_2 12·41	63B

63B B. Budesinsky, *Z. analyt. Chem.*, 1963. **195**, 324

$C_{21}H_{27}O_{14}N_7P_2$ Diphosphopyridine nucleotide (DPN) H_3L 1101.2

Metal	*Method*	*Temp*	*Medium*	*Log of equilibrium constant, remarks*	*Ref*
H^+	gl	20	0·1 (KCl)	K_1 3·67	57W
	gl	25	0·1 KCl	K_1 3·76	58W
Cu^{2+}	gl	20	0·1 (KCl)	K_1 2·16	57W
Zn^{2+}	gl	20	0·1 (KCl)	K_1 1·39	57W
	gl	25	0·1 KCl	K_1 2·06	58W

57W K. Wallenfels and H. Sund, *Biochem. Z.*, 1957, **329**, 41
58W G. Weitzel and T. Speer, *Z. physiol. Chem.*, 1958, **313**, 212

1102 $C_{22}H_{14}O_9$ 4′,4″-Dihydroxyfuchsone-3,3′,3″-tricarboxylic acid (aluminon) H_5L

Metal	*Method*	*Temp*	*Medium*	*Log of equilibrium constant, remarks*	*Ref*
H^+				H^+ constants not given.	
Cu^{2+}	sp	25	?	$K[Cu^{2+}+2HL^{4-} \rightleftharpoons Cu(HL)_2{}^{6-}]$ (?) 8·6±0·2	64S
Dy^{3+}	sp	25	?	$K(Dy^{3+}+HL^{4-} \rightleftharpoons DyHL^-)$ (?) 4·9±0·1	67S
Er^{3+}	sp	25	?	$K(Er^{3+}+HL^{4-} \rightleftharpoons ErHL^-)$ (?) 5·1±0·2	67S
Eu^{3+}	sp	25	?	$K(Eu^{3+}+HL^{4-} \rightleftharpoons EuHL^-)$ (?) 4·6±0·2	67S
Fe^{3+}	sp	25	?	$K(Fe^{3+}+HL^{4-} \rightleftharpoons FeHL^-)$ (?) 4·5	65S
Gd^{3+}	sp	25	?	$K(Gd^{3+}+HL^{4-} \rightleftharpoons GdHL^-)$ (?) 4·8±0·2	67S
Ho^{3+}	sp	25	?	$K(Ho^{3+}+HL^{4-} \rightleftharpoons HoHL^-)$ (?) 5·0±0·4	67S
La^{3+}	sp	25	?	$K(La^{3+}+HL^{4-} \rightleftharpoons LaHL^-)$ (?) 4·3±0·1	66M
Lu^{3+}	sp	25	?	$K(Lu^{3+}+HL^{4-} \rightleftharpoons LuHL^-)$ (?) 5·3±0·4	67S
Nd^{3+}	sp	25	?	$K(Nd^{3+}+HL^{4-} \rightleftharpoons NdHL^-)$ (?) 4·4±0·1	67S
Pr^{3+}	sp	25	?	$K(Pr^{3+}+HL^{4-} \rightleftharpoons PrHL^-)$ (?) 4·2±0·1	67S
Sc^{3+}	sp	25	?	$K(Sc^{3+}+HL^{4-} \rightleftharpoons ScHL^-)$ (?) 4·5±0·2	66M
Sm^{3+}	sp	25	?	$K(Sm^{3+}+HL^{4-} \rightleftharpoons SmHL^-)$ (?) 4·5±0·1	67S
Tb^{3+}	sp	25	?	$K(Tb^{3+}+HL^{4-} \rightleftharpoons TbHL^-)$ (?) 4·8±0·2	67S
Tm^{3+}	sp	25	?	$K(Tm^{3+}+HL^{4-} \rightleftharpoons TmHL^-)$ (?) 5·1±0·1	67S
$UO_2{}^{2+}$	sp	25	?	$K(UO_2{}^{2+}+HL^{4-} \rightleftharpoons UO_2HL^{2-})$ (?) 4·5	65S
Y^{3+}	sp	25	?	$K(Y^{3+}+HL^{4-} \rightleftharpoons YHL^-)$ (?) 4·5±0·2	66M
Yb^{3+}	sp	25	?	$K(Yb^{3+}+HL^{4-} \rightleftharpoons YbHL^-)$ (?) 5·2±0·2	67S

64S S. P. Sangal, *Vijnana Parishad Anusandhan Patrik*, 1964, **7**, 109
65S S. P. Sangal, *Chim. analyt.*, 1965, **47**, 288, 662
66M K. N. Munshi, S. P. Sangal, and A. K. Dey, *J. Indian Chem. Soc.*, 1966, **43**, 115
67S S. P. Sangal, *J. prakt. Chem.*, 1967, **36**, 126

1103.1 $C_{22}H_{16}N_2$ 4,4′-Diphenyl-2,2′-bipyridyl L

Metal	*Method*	*Temp*	*Medium*	*Log of equilibrium constant, remarks*	*Ref*
H^+	gl	25	0·3 (KNO_3) 50% dioxan	K_1 3·25±0·05	61J
Cu^+, Cu^{2+}	Pt	25	0·3 (KNO_3) 50% dioxan	$K(CuL_2{}^{2+}+Cu^+ \rightleftharpoons CuL_2{}^++Cu^{2+})$ 0·30	61J

61J B. R. James and R. J. P. Williams, *J. Chem. Soc.*, 1961, 2007

1103.2 $C_{22}H_{20}O_{13}$ 5-Carboxy-1,3,4,6-tetrahydroxy-8-methyl-2-(1′-oxo-2′,3′,4′,5′-tetrahydroxyhexyl)-anthraquinone (carminic acid)* H_5L

Metal	*Method*	*Temp*	*Medium*	*Log of equilibrium constant, remarks*	*Ref*
Mo^{VI}	sp	19—25	?	$K(MoO_4{}^{2-}+H_5L \rightleftharpoons MoO_3H_3L^{2-})$ (?) 3·8	66K
W^{VI}	sp	19—25	?	$K(WO_4{}^{2-}+H_5L \rightleftharpoons WO_3H_3L^{2-})$ (?) 5·5, acetate buffer	66K

* Geometric configuration of hydroxyl groups not given

66K G. F. Kirkbright, T. S. West, and C. Woodward, *Talanta*, 1966, **13**, 1637

$C_{22}H_{16}O_2N_2$ 2′-Hydroxy-1-(5′-phenyl-phenylazo)-2-naphthol H_2L 1103.3

Metal	Method	Temp	Medium	Log of equilibrium constant, remarks	Ref
H^+	sp	?	0·100 (KCl)	K_1 11·8, K_2 8·0	60D
Ca^{2+}	sp	?	0·100 (KCl)	K_1 1·75	60D
Mg^{2+}	sp	?	0·100 (KCl)	K_1 4·29	60D

60D H. Diehl and J. Ellingboe, *Analyt. Chem.*, 1960, **32**, 1120

$C_{22}H_{17}O_3N_3$ 8-Hydroxy-7-(3′-nitroanilinobenzyl)-quinoline HL 1103.4

Metal	Method	Temp	Medium	Log of equilibrium constant, remarks	Ref
H^+	sol, gl	~ 20 (?)	?	K_1 13·50 ± 0·02	61T
Cu^{2+}	sol	25	?	K_{s2} −25·43, acetate buffer	61T
Ni^{2+}	sol	25	?	K_{s2} −25·00, acetate buffer	61T
UO_2^{2+}	sol	25	?	$K_s(UO_2L_2HL + 2H^+ \rightleftharpoons UO_2^{2+} + 3HL)$ −30·04 acetate buffer	61T

61T E. P. Trailina, V. V. Zelentsov, I. A. Savich, and V. I. Spitsyn, *Russ. J. Inorg. Chem.*, 1961, **6**, 1047 (2048)

$C_{22}H_{24}O_8N_2$ (1,2-Diphenylethylenedinitrilo)tetra-acetic acid) H_4L 1103.5

Metal	Method	Temp	Medium	Log of equilibrium constant, remarks	Ref
H^+	gl	25	0·1 (KCl)	K_1 9·91 ± 0·01, K_2 5·42 ± 0·03, K_3 3·73 ± 0·02, K_4 2·18 ± 0·06	67O
Ba^{2+}	gl	25	0·1 (KCl)	K_1 9·11 ± 0·01	67O
Ca^{2+}	gl	25	0·1 (KCl)	K_1 12·11 ± 0·01	67O
Co^{2+}	gl	25	0·1 (KCl)	K_1 17·9 ± 0·2	67O
Mg^{2+}	gl	25	0·1 (KCl)	K_1 10·40 ± 0·03	67O
Sr^{2+}	gl	25	0·1 (KCl)	K_1 10·12 ± 0·01	67O
Zn^{2+}	gl	25	0·1 (KCl)	K_1 17·8 ± 0·2	67O

67O N. Okaku, K. Toyoda, Y. Moriguchi, and K. Ueno, *Bull. Chem. Soc. Japan*, 1967, **40**, 2326

$C_{22}H_{37}O_{14}N_5$ Tetraethylenepenta-aminehepta-acetic acid (TPHA) H_7L 1105.1

Metal	Method	Temp	Medium	Log of equilibrium constant, remarks	Ref
H^+	gl	25	0·1 (KNO_3)	K_1 9·95, K_2 8·85, K_3 5·56, K_4 3·82, K_5 2·79	68M
Dy^{3+}	gl	25	0·1 (KNO_3)	K_1 20·72, $K(Dy^{3+} + HL^{6-} \rightleftharpoons DyHL^{3-})$ 14·13 $K(Dy^{3+} + L^{7-} + OH^- \rightleftharpoons DyOHL^{5-})$ 5·22	68M
Eu^{3+}	gl	25	0·1 (KNO_3)	K_1 20·72, $K(Eu^{3+} + HL^{6-} \rightleftharpoons EuHL^{3-})$ 14·51 $K(Eu^{3+} + L^{7-} + OH^- \rightleftharpoons EuOHL^{5-})$ 5·23	68M
Gd^{3+}	gl	25	0·1 (KNO_3)	K_1 20·41, $K(Gd^{3+} + HL^{6-} \rightleftharpoons GdHL^{3-})$ 14·11 $K(Gd^{3+} + L^{7-} + OH^- \rightleftharpoons GdLOH^{5-})$ 5·34	68M

continued overleaf

1105.1 $C_{22}H_{37}O_{14}N_5$ (contd)

Metal	Method	Temp	Medium	Log of equilibrium constant, remarks	Ref
Hf^{4+}	ix	?	0·2 $HClO_4$	$K[Hf(OH)^{3+} + H_7L \rightleftharpoons HfH_4L^+ + 2H^+]$ 6·11 ± 0·07	66E
			0·5 $HClO_4$	$K[Hf(OH)^{3+} + H_7L \rightleftharpoons HfH_4L^+ + 2H^+]$ 5·72 ± 0·03	
			1—2 $HClO_4$	$K(Hf^{4+} + H_7L \rightleftharpoons HfH_4L^+ + 3H^+)$ (1) 5·18 ± 0·06, (2) 4·23 ± 0·03	
Ho^{3+}	gl	25	0·1 (KNO_3)	K_1 20·35, $K(Ho^{3+} + HL^{6-} \rightleftharpoons HoHL^{3-})$ 13·97 $K(Ho^{3+} + L^{7-} + OH^- \rightleftharpoons HoOHL^{5-})$ 5·21	68M
La^{3+}	gl	25	0·1 (KNO_3)	K_1 19·10, $K(La^{3+} + HL^{6-} \rightleftharpoons LaHL^{3-})$ 13·12 $K(La^{3+} + L^{7-} + OH^- \rightleftharpoons LaLOH^{5-})$ 4·97	68M
Lu^{3+}	gl	25	0·1 (KNO_3)	K_1 19·14, $K(Lu^{3+} + HL^{6-} \rightleftharpoons LuHL^{3-})$ 13·26 $K(Lu^{3+} + L^{7-} + OH^- \rightleftharpoons LuOHL^{5-})$ 5·11	68M
Nd^{3+}	gl	25	0·1 (KNO_3)	K_1 20·18, $K(Nd^{3+} + HL^{6-} \rightleftharpoons NdHL^{3-})$ 14·10 $K(Nd^{3+} + L^{7-} + OH^- \rightleftharpoons NdOHL^{5-})$ 5·34	68M
Pr^{3+}	gl	25	0·1 (KNO_3)	K_1 19·64, $K(Pr^{3+} + HL^{6-} \rightleftharpoons PrHL^{3-})$ 13·46 $K(Pr^{3+} + L^{7-} + OH^- \rightleftharpoons PrOHL^{5-})$ 5·27	68M
Sm^{3+}	gl	25	0·1 (KNO_3)	K_1 20·50, $K(Sm^{3+} + HL^{6-} \rightleftharpoons SmHL^{3-})$ 14·43 $K(Sm^{3+} + L^{7-} + OH^- \rightleftharpoons SmOHL^{5-})$ 5·29	68M
Tb^{3+}	gl	25	0·1 (KNO_3)	K_1 20·68, $K(Tb^{3+} + HL^{6-} \rightleftharpoons TbHL^{3-})$ 14·47 $K(Tb^{3+} + L^{7-} + OH^- \rightleftharpoons TbOHL^{5-})$ 5·25	68M
Tm^{3+}	gl	25	0·1 (KNO_3)	K_1 20·11, $K(Tm^{3+} + HL^{6-} \rightleftharpoons TmHL^{3-})$ 13·98 $K(Tm^{3+} + L^{7-} + OH^- \rightleftharpoons TmOHL^{5-})$ 5·15	68M
Yb^{3+}	gl	25	0·1 (KNO_3)	K_1 19·75, $K(Yb^{3+} + HL^{6-} \rightleftharpoons YbHL^{3-})$ 13·80 $K(Yb^{3+} + L^{7-} + OH^- \rightleftharpoons YbOHL^{5-})$ 5·23	68M
Zr^{4+}	ix	?	0·5 $HClO_4$	$K[Zr(OH)^{3+} + H_7L \rightleftharpoons ZrH_4L^+ + 2H^+]$ 6·08 ± 0·04	66E
			1—2 $HClO_4$	$K(Zr^{4+} + H_7L \rightleftharpoons ZrH_4L^+ + 3H^+)$ (1) 5·45 ± 0·05, (2) 4·51 ± 0·04	

66E A. N. Ermakov, I. N. Marov, and N. B. Kalinichenko, *Russ. J. Inorg. Chem.*, 1966, **11**, 1404 (2614)
68M S. Misumi, *Nippon Kagaku Zasshi*, 1968, **89**, 723

1105.2 $C_{22}H_{16}O_{14}N_4S_4$ 2,7-Bis(2′-sulphophenylazo)chromotropic acid (Sulphonazo III) H_6L

Metal	Method	Temp	Medium	Log of equilibrium constant, remarks	Ref
H^+	sp	18—22	0·2 (KNO_3)	K_1 14·5, K_2 11·7, K_3 2·9, K_4 2·3, K_5 1·9, K_6 0·9, K_7 0·3, K_8 −2·0	65B
Ba^{2+}	sp	18—22	0·2 (KNO_3)	$K(Ba^{2+} + 2H^+ + L^{6-} \rightleftharpoons BaH_2L^{2-})$ 25·9	65B

65B B. Budesinsky and D. Vrzalova *Z. analyt. Chem.*, 1965, **210**, 161

$C_{22}H_{14}O_{16}N_4Cl_2S_4$ 2,7-Bis(5′-chloro-2′-hydroxy-3′-sulpho-phenylazo)chromotropic acid H_8L 1106.1
(Chlorosulphophenol S)

Metal	Method	Temp	Medium	Log of equilibrium constant, remarks	Ref
H^+	sp	~ 20	0·2 (KNO_3)	K_1 14·5, K_2 11·9, K_3 9·7, K_4 7·1, K_5 2·5, K_6 1·3, K_7 0·5, K_8 −0·5, K_9 −2·5, K_{10} −3·6	65B
Nb^{III}	sp	~ 20	0·2 (KNO_3)	$K(NbO_2^- + 6H^+ + L^{8-} \rightleftharpoons NbO_2H_6L^{3-})$ 53·0	65B

65B B. Budesinsky and B. Savvin, *Z. analyt. Chem.*, 1965, **214,** 189

$C_{22}H_{17}O_{11}N_4S_2As$ 2-(2′-Arsenophenylazo)-7-(phenylazo)chromotropic acid H_6L 1106.2
(Monoarsenazo III)

Metal	Method	Temp	Medium	Log of equilibrium constant, remarks	Ref
H^+	sp	20	?	K_1 12·46, K_2 8·39, K_3 6·64	65B
La^{3+}	sp	20	?	$K(La^{3+} + 2H^+ + 2L^{6-} \rightleftharpoons LaH_2L_2^{7-})$ 50·8	65B

65B B. Budesinsky, *Z. analyt. Chem.*, 1965, **207,** 105

$C_{22}H_{18}O_{14}N_4S_2P_2$ 2,7-Bis(2′-phosphonophenylazo)chromotropic acid H_8L 1106.3
(Phosphonazo III)

Metal	Method	Temp	Medium	Log of equilibrium constant, remarks	Ref
H^+	sp	25	0·2 (KNO_3)	K_1 14·6, K_2 11·3, K_3 9·6, K_4 7·2, K_5 4·5, K_6 1·7, K_7 0·6, K_8 0·3, K_9 −0·4, K_{10} −2·0	67B
La^{3+}	sp	25	0·2 (KNO_3)	$K(La^{3+} + 10H^+ + 2L^{8-} \rightleftharpoons LaH_{10}L_2^{3-})$ 104·8±0·1	67B
UO_2^{2+}	sp	25	0·2 (KNO_3)	$K(UO_2^{2+} + 10H^+ + 2L^{8-} \rightleftharpoons UO_2H_{10}L_2^{4-})$ 106·8±0·1	67B

67B B. Budesinsky, K. Maas, and A. Betdikova, *Coll. Czech. Chem. Comm.*, 1967, **32,** 1528

$C_{22}H_{18}O_{14}N_4S_2As_2$ 2,7-Bis(2′-arsonophenylazo)chromotropic acid (Arsenazo III) H_8L 1106.4

Metal	Method	Temp	Medium	Log of equilibrium constant, remarks	Ref
H^+	sp	~ 20	0·2 $NaNO_3$	K_1 12·33, K_2 7·48, K_3 5·35, K_4 2·41, K_5 2·41	63B
Dy^{3+}	sp	~ 20	0·2 $NaNO_3$	$K(2Dy^{3+} + 2L^{8-} \rightleftharpoons Dy_2L_2^{10-})$ 83·0±0·4	63B
Gd^{3+}	sp	~ 20	0·2 $NaNO_3$	$K(2Gd^{3+} + 2L^{8-} \rightleftharpoons Gd_2L_2^{10-})$ 80·5±0·2	63B
La^{3+}	sp	~ 20	0·2 $NaNO_3$	$K(2La^{3+} + 2L^{8-} \rightleftharpoons La_2L_2^{10-})$ 81·2±0·4	63B
	sp	~ 20	0·2 ($NaNO_3$)	$K(2La^{3+} + 8H^+ + 2L^{8-} \rightleftharpoons La_2H_8L_2^{2-})$ 83·5	64B
				$K(2La^{3+} + 4H^+ + L^{8-} \rightleftharpoons La_2H_4L^{2+})$ 42·5	
Sm^{3+}	sp	~ 20	0·2 $NaNO_3$	$K(2Sm^{3+} + 2L^{8-} \rightleftharpoons Sm_2L_2^{10-})$ 82·1±0·3	63B
Yb^{3+}	sp	~ 20	0·2 $NaNO_3$	$K(2Yb^{3+} + 2L^{8-} \rightleftharpoons Yb_2L_2^{10-})$ 81·9±0·2	63B
Zr^{4+}	sp	~ 20	3—6 $HClO_4$	$K(2Zr^{4+} + 18H^+ + 2L^{8-} \rightleftharpoons Zr_2H_{18}L_2^{10+})$ 87·2	64Ba

63B B. Budesinsky, *Coll. Czech. Chem. Comm.*, 1963, **28,** 2902
64B B. Budesinsky, *Z. analyt. Chem.*, 1964, **202,** 96
64Ba B. Budesinsky, *Z. analyt. Chem.*, 1964, **206,** 401

1106.5 $C_{22}H_{16}O_{14}N_4S_2P_2Cl_2$ 2,7-Bis(4′-chloro-2′-phosphonophenylazo)chromotropic acid (chlorophosphonazo III) **H_8L**

Metal	*Method*	*Temp*	*Medium*	*Log of equilibrium constant, remarks*	*Ref*
H^+	sp	25	0·2 (KNO_3)	K_1 14·6, K_2 11·1, K_3 9·4, K_4 7·0, K_5 4·2, K_6 1·5, K_7 0·6, K_8 0·3, K_9 −0·5, K_{10} −2·1	67B
Ba^{2+}	sp	25	0·2 (KNO_3)	$K(Ba^{2+}+6H^++2L^{8-} \rightleftharpoons BaH_6L_2^{8-})$ 82·5±0·2	67B
Ca^{2+}	sp	25	0·2 (KNO_3)	$K(Ca^{2+}+8H^++2L^{8-} \rightleftharpoons CaH_8L_2^{6-})$ 94·0±0·5	67B
Mg^{2+}	sp	25	0·2 (KNO_3)	$K(Mg^{2+}+4H^++L^{8-} \rightleftharpoons MgH_4L^{2-})$ 47·4±0·5	67B
Sr^{2+}	sp	25	0·2 (KNO_3)	$K(Sr^{2+}+8H^++2L^{8-} \rightleftharpoons SrH_8L_2^{6-})$ 95·6±0·9	67B
UO_2^{2+}	sp	25	0·2 (KNO_3)	$K(UO_2^{2+}+12H^++2L^{8-} \rightleftharpoons UO_2H_{12}L_2^{2-})$ 47·7±0·9	67B

67B B. Budesinsky, K. Maas, and A. Besdekova, *Coll. Czech. Chem. Comm.*, 1967, **32**, 1528

1106.6 $C_{22}H_{17}O_{14}N_4ClS_2P_2$ 2-(4′-Chloro-2′-phosphonophenylazo)-7-(2″-phosphonophenylazo)chromotropic acid **H_8L**

Metal	*Method*	*Temp*	*Medium*	*Log of equilibrium constant, remarks*	*Ref*
H^+	sp	25	0·2 (KNO_3)	K_1 14·6, K_2 11·2, K_3 9·5, K_4 8·1, K_5 4·3, K_6 1·6, K_7 0·6, K_8 0·3, K_9 −0·4, K_{10} −2·0	67B
La^{3+}	sp	25	0·2 (KNO_3)	$K(La^{3+}+10H^++2L^{8-} \rightleftharpoons LaH_{10}L_2^{3-})$ 104·4±0·1	67B
UO_2^{2+}	sp	25	0·2 (KNO_3)	$K(UO_2^{2+}+10H^++2L_2^{8-} \rightleftharpoons UO_2H_{10}L_2^{4-})$ 103·0±0·3	67B

67B B. Budesinsky, K. Maas, and A Besdekova, *Coll. Czech. Chem. Comm.*, 1967, **32**, 1528

1107 $C_{24}H_{16}N_2$ 4,7-Diphenyl-1,10-phenanthroline L

Errata. 1964, p. 727, Table 1107. *For* formula $C_{23}H_{16}N_2$ *read* $\mathbf{C_{24}H_{16}N_2}$.
This Table has been renumbered **1109.1** to correct the ligand sequence.

1107.1 $C_{23}H_{18}O_9S$ 4′-Hydroxy-3,3′-dimethyl-2″-sulphofuchsone-5,5′-dicarboxylic acid (Eriochrome Cyanine R) **H_4L**

Metal	*Method*	*Temp*	*Medium*	*Log of equilibrium constant, remarks*	*Ref*
H^+	sp	?	?	K_1 11·83±0·04, K_2 5·74±0·04, K_3 1·83±0·02	59S, 65L
	sp	18—22	0·1 ($NaClO_4$)	K_1 11·85±0·01, K_2 5·47±0·05, K_3 2·3±0·01, K_4 −4·9 (?)	67S
Be^{2+}	sp	18—22	0·1 ($NaClO_4$)	$K(Be^{2+}+H_2L^{2-} \rightleftharpoons BeHL^-+H^+)$ 0·02 $K(2Be^{2+}+2L^{4-} \rightleftharpoons Be_2L_2^{4-})$ 28·3	67S
Fe^{3+}	sp	17—23	0·1 KCl	K_1 17·9 $K(2Fe^{3+}+L^{4-} \rightleftharpoons Fe_2L^{2+})$ 22·5 $K(2Fe^{3+}+2L^{4-} \rightleftharpoons Fe_2L_2^{2-})$ 37·9	65L

59S V. Suk and V. Miketukova, *Coll. Czech. Chem. Comm.*, 1959, **24**, 3629
65L F. J. Langmyhr and T. Stumpe, *Analyt. Chim. Acta*, 1965, **32**, 535
67S L. Sommer and V. Kuban, *Coll. Czech. Chem. Comm.*, 1967, **32**, 4355

$C_{23}H_{18}O_3N_2$ 7-(4′-Carboxyphenylaminobenzyl)-8-hydroxyquinoline H_2L 1107.2

Metal	Method	Temp	Medium	Log of equilibrium constant, remarks	Ref
H^+	sol, gl	~ 20 (?)	?	K_2 7·48±0·03	61T
Cu^{2+}	sol	25	?	K_{s2} −17·50, acetate buffer	61T
Ni^{2+}	sol	25	?	K_{s2} −16·54, acetate buffer	61T
UO_2^{2+}	sol	25	?	$K_s[UO_2(HL)_2H_2L+2H^+ \rightleftharpoons UO_2^{2+}+3H_2L]$ −22·98 acetate buffer	61T

61T E. P. Trailina, V. V. Zelentsov, I. A. Savich, and V. I. Spitsyn, *Russ. J. Inorg. Chem.*, 1961, **6**, 1047 (2048)

$C_{23}H_{24}O_2N_4$ 4,4′-Methylenebis[2,3-dimethyl-1-phenylpyrazol-5(2*H*)-one] (diantipyrinylmethane) L 1107.3

Metal	Method	Temp	Medium	Log of equilibrium constant, remarks	Ref
Ti^{4+}	sp	20	~ 1	β_3 7·89	62B
Zr^{4+}	sp	?	0·1—1 HCl	β_2 (0·1) 11·5±0·1, (1) 11·8±0·1	63B

62B A. K. Babko and M. M. Tananaiko, *Russ. J. Inorg. Chem.*, 1962, **7**, 286 (562)
63B A. K. Babko and M. I. Shtokalo, *Russ. J. Inorg. Chem.*, 1963, **8**, 564 (1088)

$C_{23}H_{24}ON_2$ 1-[Bis(4′-dimethylaminophenyl)methylene]-2-oxobenzene (2-hydroxy Malachite Green) L 1108

Metal	Method	Temp	Medium	Log of equilibrium constant, remarks	Ref
H^+	sp	20	0·5	K_1 8·57	62C
				$K(L \rightleftharpoons A)$ −2·42, $K(HL^+ \rightleftharpoons HA^+)$ 0·82 where A = hydrated ligand, $C_{23}H_{26}O_2N_2$, with H^+ constants K_1 9·21, K_2 5·37, K_3 4·48	
B^{III}	sp	20	0·5 KCl	$K[B(OH)_4^- + L \rightleftharpoons B(OH)_2(H_{-2}L)^-]$ 4·38	62C
				$K[B(OH)_4^- + HL^+ \rightleftharpoons B(OH)_2(H_{-1}L)]$ 3·60	

Erratum. 1964, p. 727, Table 1108. *Delete* entire entry.

62C R. Cigen, *Acta Chem. Scand.*, 1962, **16**, 192, 1271

$C_{23}H_{16}O_9Cl_2S$ 2″,6″-Dichloro-4′-hydroxy-3,3′dimethyl-3″-sulphofuchsone-5,5′-dicarboxylic acid (Chrome Azurol S) 1109 H_4L

Metal	Method	Temp	Medium	Log of equilibrium constant, remarks	Ref
H^+	sp	20	0·1	K_1 12·21, K_2 4·92, K_3 2·28	62A
	sp	20	0·1 KCl	K_1 11·81±0·03, K_2 4·71±0·03, K_3 2·25±0·05	63L, 66Sa
	sp	18—22	0·1 ($NaClO_4$)	K_1 11·79±0·02, K_2 4·88±0·02, K_3 2·37±0·01, K_4 −4·8	67Sd
	sp	25	0·1 ($NaClO_4$)	K_1 11·75±0·05, K_2 4·88±0·05, K_3 2·25±0·10,	68B
	pH	25	var	K_4 −1·2±0·4	68B

continued overleaf

1109 $C_{23}H_{16}O_9Cl_2S$ (contd)

Metal	*Method*	*Temp*	*Medium*	*Log of equilibrium constant, remarks*	*Ref*
Be^{2+}	sp	20	0·1	K(?) 6·2±0·2	62A
	sp	30	0·1 ($NaClO_4$)	K(?) 4·4, K(?) 4·7	63S
	sp	18—22	0·1 ($NaClO_4$)	$K(Be^{2+}+H_2L^{2-} \rightleftharpoons BeHL^{-}+H^{+})$ 0·05	67Sd
				$K(2Be^{2+}+2L^{4-} \rightleftharpoons Be_2L_2^{4-})$ 26·8	
	sp	25	0·1 ($NaClO_4$)	$K(Be^{2+}+HL^{3-} \rightleftharpoons BeHL^{-})$ 4·66±0·08	68B
				$K(2Be^{2+}+L^{4-} \rightleftharpoons Be_2L)$ 15·8±0·1	
Cu^{2+}	sp	30	0·1 ($NaClO_4$)	K(?) 4·1±0·1	64S
	gl	25	0·1	$K(Cu^{2+}+HL^{3-} \rightleftharpoons CuHL^{-})$ 4·02±0·05	66Sa
				$K(2Cu^{2+}+L^{4-} \rightleftharpoons Cu_2L)$ 13·7±0·1	
Dy^{3+}	sp	25	?	K(?) 4·2±0·1	67S
Eu^{3+}	sp	25	?	K(?) 4·2±0·1	67S
Fe^{3+}	sp	25	0·1	K(?) 4·8±0·1	62S
	sp	20	0·1 KCl	K_1 15·6	63L
				$K(2Fe^{3+}+2L^{4-} \rightleftharpoons Fe_2L_2^{2-})$ 36·2	
				$K(2Fe^{3+}+L^{4-} \rightleftharpoons Fe_2L^{2+})$ 20·2	
Ga^{3+}	sp	25	?	K(?) 4·5±0·1	66D
Gd^{3+}	sp	25	?	K(?) 4·4±0·2	67S
Ho^{3+}	sp	25	?	K(?) 4·3±0·2	67S
In^{3+}	sp	25	?	K(?) 4·4±0·1	64M
La^{3+}	sp	25	?	K(?) 4·8±0·1	67Sc
Lu^{3+}	sp J	25	?	K(?) 10·4	67Sb
Ni^{2+}	sp	25	?	K(?) 9·3±0·2	67Sa
Pd^{2+}	sp	25	?	K(?) 4·8±0·1	63Sb
	sp	25	?	K(?) 5·1±0·2	66S
Sc^{3+}	sp	25	?	K(?) 11·98	67I
	sp	25	?	K(?) 5·5±0·1	67Sc
Sm^{3+}	sp	25	?	K(?) 4·8±0·2	67S
Tb^{3+}	sp	25	?	K(?) 4·7±0·2	67S
Th^{4+}	sp	30	0·15 (NH_4NO_3)	K(?) 4·2±0·1	63Sa
VO^{2+}	sp	25	?	K(?) 4·6	67Se
Y^{3+}	sp	25	?	K(?) 4·3±0·2	67Sc

Erratum. 1964, p. 727, Table 1109. *For* formula $C_{23}H_{15}O_9SCl_2$ *read* $\mathbf{C_{23}H_{16}O_9SCl_2}$.

62A L. P. Adamovich, V. Morgul-Meshkova, and B. V. Yutsis, *J. Analyt. Chem. (U.S.S.R.)*, 1962, **17**, **673** **(678)**

62S R. L. Seth and A. K. Dey, *J. Indian Chem. Soc.*, 1962, **39**, 773

63L E. J. Langmyhr and K. S. Klausen, *Analyt. Chim. Acta*, 1963, **29**, 149

63S S. C. Srivastava and A. K. Dey, *J. Inorg. Nuclear Chem.*, 1963, **25**, 217

63Sa S. C. Srivastava, S. N. Sinha, and A. K. Dey, *Bull. Chem. Soc. Japan*, 1963, **36**, 268

63Sb S. P. Sangal and A. K. Dey, *J. Indian Chem. Soc.*, 1963, **40**, 464

64M K. N. Munshi and A. K. Dey, *J. Indian Chem. Soc.*, 1964, **41**, 340

64S S. C. Srivastava, S. N. Sinha, and A. K. Dey, *Mikrochim. Acta*, 1964, 605

66D C. D. Dwivedi, K. N. Munshi, and A. K. Dey, *J. Indian Chem. Soc.*, 1966, **43**, 111

66S S. P. Sangal, *J. prakt. Chem.*, 1966, **31**, 68

66Sa A. Semb and F. J. Langmyhr, *Analyt. Chim. Acta*, 1966, **35**, 286

67I R. Ishida and N. Hasegawa, *Bull. Chem. Soc. Japan*, 1967, **40**, 1153

67S S. P. Sangal, *J. prakt. Chem.*, 1967, **36**, 126

67Sa P. Sanyal and S. P. Mushran, *Chim. Analyt.*, 1967, **49**, 231

67Sb P. Spacu and S. Plostinaru, *Rev. Roumaine Chim.*, 1967, **12**, 383

67Sc S. N. Sinha, S. P. Sangal, and A. K. Dey, *J. Indian Chem. Soc.*, 1967, **44**, 203

67Sd L. Sommer and V. Kuban, *Coll. Czech. Chem. Comm.*, 1967, **32**, 4355

67Se P. Sanyal, S. P. Sangal, and S. P. Mushran, *Bull. Chem. Soc. Japan*, 1967, **40**, 217

68B W. G. Baldwin and D. R. Stranks, *Austral. J. Chem.*, 1968, **21**, 603

$C_{24}H_{16}N_2$ 4,7-Diphenyl-1,10-phenanthroline L 1109.1

Metal	Method	Temp	Medium	Log of equilibrium constant, remarks	Ref
H^+	gl	25	0·3 (KNO_3) 50% dioxan	K_1 4·30±0·05	61J
Cu^+				See Cu^{2+} below.	
Cu^{2+}	Pt	25	0·3 (KNO_3) 50% dioxan	K_1 ~ 5·7, K_3 3·75 $K(CuL_2^{2+} + Cu^+ \rightleftharpoons CuL_2^+ + Cu^{2+})$ −0·35	61J

Erratum. 1964, p. 727, Table 1107. This Table was numbered 1107 in the 1964 edition, and has been renumbered to correct the ligand sequence.

61J B. R. James and R. J. P. Williams, *J. Chem. Soc.*, 1961, 2007

$C_{24}H_{51}N$ Trioctylamine L 1109.2

Metal	Method	Temp	Medium	Log of equilibrium constant, remarks	Ref
UO_2^{2+}	dis	?	2·5 HNO_3	$K[UO_2^{2+} + 2NO_3^- + LHNO_3 \rightleftharpoons UO_2HL(NO_3)_3]$ (Org. = CCl_4) 0·31±0·02, (Org. = 2-xylene) 0·46±0·01	60S

60S V. B. Shevchenko, V. S. Shmidt, and E. A. Nenarokomov, *Russ. J. Inorg. Chem.*, 1960, 5, 1140 (2354)

$C_{24}H_{51}OP$ Trioctylphosphine oxide L 1109.3

Metal	Method	Temp	Medium	Log of equilibrium constant, remarks	Ref
Zn^{2+}				See 2-thenoyltrifluoroacetone (691) for mixed ligands.	

$C_{24}H_{20}O_{14}N_4S_4$ 2,7-Bis(4′-methyl-2′-sulphophenylazo)chromotropic acid (dimethyl sulphonazo III) H_6L 1109.4

Metal	Method	Temp	Medium	Log of equilibrium constant, remarks	Ref
H^+	sp	25	0·2 (KNO_3)	K_1 14·7, K_2 11·8, K_3 5·7, K_4 4·5, K_5 3·4, K_6 2·4, K_7 2·0, K_8 −1·5	67B
Ba^{2+}	sp	25	0·2 (KNO_3)	K_1 4·24	67B
Sr^{2+}	sp	25	0·2 (KNO_3)	K_1 3·20	67B

67B B. Budesinsky, D. Vrzalova, and A. Beztekova, *Acta Chim. Acad. Sci. Hung.*, 1967, 52, 37

1109.5 $C_{24}H_{18}O_{18}N_4S_2As_2$ 2,7-Bis(2′-arsono-5′-carboxyphenylazo)chromotropic acid (dicarboxyarsenazo III) **H_{10}**

Metal	Method	Temp	Medium	Log of equilibrium constant, remarks	Ref
H^+	sp	~ 20	0·2 (KNO_3)	K_1 14·7, K_2 11·5, K_3 9·0, K_4 6·9, K_5 6·9, K_6 5·0, K_7 3·3, K_8 2·8, K_9 −0·9, K_{10} −1·7, K_{11} −2·5, K_{12} −5·7	65B
La^{3+}	sp	~ 20	0·2 (KNO_3)	$K[La^{3+}+12H^++2L^{10-} \rightleftharpoons La(H_6L)_2{}^{5-}]$ 96·0	65B
Lu^{3+}	sp	~ 20	0·2 (KNO_3)	$K[Lu^{3+}+12H^++2L^{10-} \rightleftharpoons Lu(H_6L)_2{}^{5-}]$ 96·8	65B
Y^{3+}	sp	~ 20	0·2 (KNO_3)	$K[Y^{3+}+12H^++2L^{10-} \rightleftharpoons Y(H_6L)_2{}^{5-}]$ 96·4	65B

65B B. Budesinsky and K. Maas, *Z. analyt. Chem.*, 1965, **210**, 263

1109.6 $C_{24}H_{20}O_{14}N_4Cl_2S_2P_2$ 2,7-Bis(4′-chloro-5′-methyl-2′-phosphonophenylazo)chromotropic acid (methylchlorophosphonazo III) **H_8L**

Metal	Method	Temp	Medium	Log of equilibrium constant, remarks	Ref
H^+	sp	25	0·2 (KNO_3)	K_1 14·6, K_2 11·2, K_3 9·6, K_4 7·2, K_5 4·3, K_6 1·6, K_7 0·6, K_8 0·3, K_9 −0·4, K_{10} −1·9	67B
Ba^{2+}	sp	25	0·2 (KNO_3)	$K(Ba^{2+}+6H^++2L^{8-} \rightleftharpoons BaH_6L^{8-})$ 109·4±0·1	67B
Ca^{2+}	sp	25	0·2 (KNO_3)	$K(Ca^{2+}+8H^++2L^{8-} \rightleftharpoons CaH_8L_2{}^{6-})$ 95·9±0·9	67B
Mg^{2+}	sp	25	0·2 (KNO_3)	$K(Mg^{2+}+4H^++L^{8-} \rightleftharpoons MgH_4L^{2-})$ 47·7±0·9	67B
Sr^{2+}	sp	25	0·2 (KNO_3)	$K(Sr^{2+}+8H^++2L^{8-} \rightleftharpoons SrH_8L_2{}^{6-})$ 96·4±0·7	67B
$UO_2{}^{2+}$	sp	25	0·2 (KNO_3)	$K(UO_2{}^{2+}+12H^++2L^{8-} \rightleftharpoons UO_2H_{12}L_2{}^{2-})$ 108·7±0·2	67B
Yb^{3+}	sp	25	0·2 (KNO_3)	$K(Yb^{3+}+12H^++2L^{8-} \rightleftharpoons YbH_{12}L_2{}^-)$ 105·8±0·1	67B

67B B. Budesinsky, K. Maas, and A. Besdekova, *Coll. Czech. Chem. Comm.*, 1967, **32**, 1528

1109.7 $C_{25}H_{20}O_9$ 4′,3″-Dihydroxy-3,3′,4″-trimethylfuchsone-5,5′,5″-tricarboxylic acid (Chromoxane Violet R) **H_5L**

Metal	Method	Temp	Medium	Log of equilibrium constant, remarks	Ref
H^+	sp	?	0·1	K_1 13·3, K_2 11·3, K_3 4·09±0·02, K_4 3·08±0·03	67L
Al^{3+}	sp	?	0·1	K_1 10·42	67L
Fe^{3+}	sp	?	0·1	K_1 12·53	67L

67L N. F. Lisenko and I. S. Mustafin, *J. Analyt. Chem. (U.S.S.R.)*, 1967, **22**, 20 (25)

1110 $C_{25}H_{48}O_8N_6$ Desferri-ferrioxamin B H_3L

Errata. 1964, p. 728, Table 1110. Under Ni^{2+} first line, equation should *read* $K(Ni^{2+}+HL^{2-} \rightleftharpoons NiHL)$. Reference 63S, *for* A. Schwarzenbach *read* K. Schwarzenbach.

$C_{26}H_{24}N_2$ 4,4′-Bis(phenylethyl)-2,2′-bipyridyl L 1110.1

Metal	Method	Temp	Medium	Log of equilibrium constant, remarks	Ref
H^+	gl	25	0·3 (KNO_3) 50% dioxan	K_1 3·98 ± 0·05	61J
Cu^+, Cu^{2+}	Pt	25	0·3 (KNO_3) 50% dioxan	$K(CuL_2^{2+} + Cu^+ \rightleftharpoons CuL_2^+ + Cu^{2+})$ −0·10	61J

61J B. R. James and R. J. P. Williams, *J. Chem. Soc.*, 1961, 2007

$C_{26}H_{18}O_9N_4S_2$ 1-[3′-(2″-Hydroxy-1″-naphthylazo)-2′-hydroxy-5′-sulphophenylazo]-2-hydroxy-naphthalene-6-sulphonic acid (alizarin acid black SN) 1110.2 H_5L

Metal	Method	Temp	Medium	Log of equilibrium constant, remarks	Ref
H^+	sp	25	0·1 ($NaNO_3$)	K_1 > 14, K_2 12·80, K_3 5·79	62R
Ca^{2+}	sp	25	0·1 ($NaCO_3$)	$K(2Ca^{2+} + HL^{4-} \rightleftharpoons Ca_2L^- + H^+)$ −1·1	62R
				$K(2Ca^{2+} + 2HL^{4-} \rightleftharpoons Ca_2L_2^{6-} + 2H^+)$ −4·5	
				$K(Ca^{2+} + HL^{4-} \rightleftharpoons CaHL^{2-})$ ~ 6—7	
				$K(Ca_2L_2^{6-} + 2Ca^{2+} \rightleftharpoons 2Ca_2L^-)$ 2·3	

62R G. Ross, D. A. Aikens, and C. N. Reilley, *Analyt. Chem.*, 1962, **34**, 1766

$C_{27}H_{30}O_{16}$ 3,3′,4′,5,7-Pentahydroxy-2-phenyl-2-benzopyran-4-one 3-rutinoside (rutin) H_4L 1110.3

Metal	Method	Temp	Medium	Log of equilibrium constant, remarks	Ref
H^+				H^+ constants not given.	
UO_2^{2+}	sp	20	0·5 KNO_3	K (?) 9·35	63D

63D B. Dev and B. D. Jain, *J. Indian Chem. Soc.*, 1963, **40**, 269

$H_{28}O_9N_2S$ 4′-Hydroxy-3,3′-dimethyl-5,5′-bis[*N*-(carboxymethyl)aminomethyl]fuchsone-2″-sulphonic acid (glycine cresol red) H_4L 1114.1

Metal	Method	Temp	Medium	Log of equilibrium constant, remarks	Ref
H^+	sp	?	0·2 ($NaNO_3$)	K_1 12·43, K_2 10·77, K_3 7·12, K_4 4·92, K_5 2·46	63B
Cu^{2+}	sp	?	0·2 ($NaNO_3$)	$K(2Cu^{2+} + 2L^{4-} \rightleftharpoons Cu_2L_2^{4-})$ 54·9 ± 0·2	63B

63B B. Budesinsky and J. Gurovic, *Coll. Czech. Chem. Comm.*, 1963, **28**, 1154

1116 $C_{28}H_{15}O_4N$ 1,1′-Iminodianthraquinone (1,1′-dianthrimide) L

Metal	*Method*	*Temp*	*Medium*	*Log of equilibrium constant, remarks*	*Ref*
Ge^{IV}	sp	?	93·2% H_2SO_4	$K(HGeO_2^+ + HL^+ \rightleftharpoons HGeO_2HL^{2+})$ (?) 2·35	68L
Se^{IV}	sp	?	96% H_2SO_4	$K[2HSeO_2^+ + HL^+ \rightleftharpoons (HSeO_2)_2HL^{3+}]$ 8·75	66D
Te^{IV}	sp	?	93·2% H_2SO_4	$K(HTeO_2^+ + HL^+ \rightleftharpoons HTeO_2HL^{2+})$ (?) 2·36	68L

66D I. Dahl and F. J. Langmyhr, *Analyt. Chim. Acta*, 1966, **35,** 24
68L F. J. Langmyhr and G. Norheim, *Analyt. Chim. Acta*, 1968, **41,** 341

1116.1 $C_{28}H_{15}O_4N$ 1,2′-Iminodianthraquinone (1,2′-dianthrimide) L

Metal	*Method*	*Temp*	*Medium*	*Log of equilibrium constant, remarks*	*Ref*
Se^{IV}	sp	?	96·0% H_2SO_4	$K(HSeO_2^+ + HL^+ \rightleftharpoons HSeO_2L^{2+})$ 4·40	66M

66M J. A. Myhrstad and F. J. Langmyhr, *Acta Chem. Scand.*, 1966, **20,** 2897

1116.2 $C_{28}H_{15}O_4N$ 2,2′-Iminodianthraquinone (2,2′-dianthrimide) L

Metal	*Method*	*Temp*	*Medium*	*Log of equilibrium constant, remarks*	*Ref*
Se^{IV}	sp	?	95·5% H_2SO_4	$K(HSeO_2^+ + HL^+ \rightleftharpoons HSeO_2HL^{2+})$ 3·89 $K[2HSeO_2^+ + 2HL^+ \rightleftharpoons (HSeO_2)_2(HL)_2^{4+}]$ 12·50	66D

66D I. Dahl and F. J. Langmyhr, *Analyt. Chim. Acta*, 1966, **35,** 24

1117.1 $C_{30}H_{18}O_{21}N_6S_6$ Cyclo-tris-7-(1-azo-8-hydroxynaphthalene-3,6-disulphonic acid) (Calcichrome) H_9L

Metal	*Method*	*Temp*	*Medium*	*Log of equilibrium constant, remarks*	*Ref*
H^+	sp	~ 20	0·2 ($NaNO_3$)	K_1 11·5, K_2 7·10	65B
Ca^{2+}	sp	~ 20	0·2 ($NaNO_3$)	$K(Ca^{2+} + 2H^+ + L^{9-} \rightleftharpoons CaH_2L^{5-})$ 26·45	65B

65B A. Besdekova and B. Budesinsky, *Coll. Czech. Chem. Comm.*, 1965, **30,** 811

1117.2 $C_{31}H_{32}O_{13}N_2S$ 5,5′-Bis-*NN*-bis(carboxymethyl)aminomethyl-4′-hydroxy-3,3′-dimethylfuchsone-2″-sulphonic acid (Xylenol Orange) H_6L

Metal	*Method*	*Temp*	*Medium*	*Log of equilibrium constant, remarks*	*Ref*
H^+	sp	~ 20 (?)	0·2 KNO_3	K_1 12·28±0·06, K_2 10·46±0·05, K_3 6·37±0·05, K_6 −1·09±0·07, K_7 −1·74±0·05	60R, 62B, 63B, 63Ba, 63O, 66D, 66K, 67K, 67S

contin

$C_{31}H_{32}O_{13}N_2S$ (contd) 1117.2

Metal	*Method*	*Temp*	*Medium*	*Log of equilibrium constant, remarks*	*Ref*
H^+ (*contd*)	gl	~ 20 (?)	0·2 KNO_3	K_3 6·46 ± 0·07, K_4 3·23 ± 0·05, K_5 2·58 ± 0·05	60R, 62B, 63B, 63Ba, 63O, 66D, 66K, 67K, 67S
Be^{2+}	sp	25	0·1 ($NaClO_4$)	*K* (?) 3·92	65O
Bi^{3+}	sp	16—20	0·1 (?) HNO_3	*K* (?) 7·0	60O
	sp	~ 20 (?)	0·2 $NaNO_3$	$K(2Bi^{3+} + 2L^{6-} \rightleftharpoons Bi_2L_2^{6-})$ 75·6	63Ba
Cd^{2+}	sp	25	0·3 KNO_3	$K(Cd^{2+} + HL^{5-} \rightleftharpoons CdL^{4-} + H^+)$ 3·78	64O
Ce^{3+}	sp	25	?	*K* (?) 5·5, acetate buffer	62T
Dy^{3+}	sp	~ 20	0·2 ($NaNO_3$)	$K(2Dy^{3+} + 2L^{6-} \rightleftharpoons Dy_2L_2^{6-})$ 47·6	63B
Fe^{3+}	sp	?	0·05 $HClO_4$	*K* (?) 5·7	60C
	sp	~ 20 (?)	0·2 ($NaNO_3$)	$K(2Fe^{3+} + L^{6-} \rightleftharpoons Fe_2L)$ 39·80	62B
Ga^{3+}	sp	?	0·01	$K(Ga^{3+} + H_5L^- \rightleftharpoons GaH_5L^{2+})$ 7·0 ± 0·1 $K[GaH_5L^{2+} + H_5L^- \rightleftharpoons Ga(H_5L)_2{}^+]$ 4·5 ± 0·2	66D
	sp	25	?	*K* (?) 4·8 ± 0·1	66Da
	sp	20	0·2 (NaCl)	$K(Ga^{3+} + H_2L^{4-} \rightleftharpoons GaH_2L^-)$ 13·36 ± 0·08	67K
Gd^{3+}	sp	~ 20	0·2 ($NaNO_3$)	$K(2Gd^{3+} + 2L^{6-} \rightleftharpoons Gd_2L_2^{6-})$ 43·1	63B
Hf^{4+}	sp J	?	0·3 $HClO_4$	*K* (?) 6·51	60C
In^{3+}	sp	25	?	*K* (?) 5·0 ± 0·1	66Da
La^{3+}	sp	25	?	*K* (?) 5·8, acetate buffer	62T
	sp	20	0·2 $NaClO_4$	$K(La^{3+} + HL^{5-} \rightleftharpoons LaHL^{2-})$ 11·67 ± 0·09	66K
Lu^{3+}	sp	20	0·2 $NaClO_4$	$K(Lu^{3+} + HL^{5-} \rightleftharpoons LuHL^{2-})$ 14·09 ± 0·07	66K
	sp	20	0·2 (NaCl)	$K(Lu^{3+} + H_2L^{4-} \rightleftharpoons LuH_2L^-)$ 9·94 ± 0·07	67K
Nb^V	sp	25	?	*K* (?) 4·7 ± 0·1	67A
Nd^{3+}	sp	25	?	*K* (?) 6·0, acetate buffer	62T
	sp	25	0·1	$K(Nd^{3+} + H_2L^{4-} \rightleftharpoons NdH_2L^-)$ 6·8	67S
Pd^{2+}	sp	25	?	*K* (?) 10·3	63O
Sc^{3+}	sp	~ 20 (?)	0·2 $NaNO_3$	$K(2Sc^{3+} + 2L^{6-} \rightleftharpoons Sc_2L_2^{6-})$ 61·2	63Ba
	sp	18—20	0·1 ($NaClO_4$)	*K* (?) 5·95	64K
	sp	20	0·2 $NaClO_4$	$K(Sc^{3+} + HL^{5-} \rightleftharpoons ScHL^{2-})$ 18·82 ± 0·06 $K(Sc^{3+} + H_2L^{4-} \rightleftharpoons ScH_2L^-)$ 12·00	66K
Sm^{3+}	sp	~ 20	0·2 ($NaNO_3$)	$K(2Sm^{3+} + 2L^{6-} \rightleftharpoons Sm_2L_2^{6-})$ 47·0	63B
Tb^{4+}	sp	~ 20 (?)	0·2 $NaNO_3$	$K(2Th^{4+} + 2L^{6-} \rightleftharpoons Th_2L_2^{4-})$ 52·5	63Ba
Ti^{IV}	sp	~ 20 (?)	0·2 $NaNO_3$	$K(2Ti^{4+} + 2L^{6-} \rightleftharpoons Ti_2L_2^{4-})$ 57·8	63Ba
	sp	25	~ 0·05 $HClO_4$ 0·5 ($NaClO_4$)	$K(TiO^{2+} + H_6L + H_2O_2 \rightleftharpoons TiH_6LH_2O_2{}^{2+})$ 37·68 $K(TiO^{2+} + H_6L \rightleftharpoons TiOH_5L^+ + H^+)$ 3·46	63O
Tl^{3+}	sp	25	?	*K* (?) 4·8 ± 0·1	66Da
$UO_2{}^{2+}$	sp	~ 20 (?)	0·2 $NaNO_3$	$K[2UO_2{}^{2+} + 2L^{6-} \rightleftharpoons (UO_2)_2L_2{}^{8-}]$ 38·57	62B
	sp	25 (?)	?	*K* (?) 11·46	63O
V^V	sp	~ 20 (?)	0·2 $NaNO_3$	$K[2VO_2{}^+ + 2L^{6-} \rightleftharpoons (VO_2)_2L_2{}^{10-}]$ 63·1	63Ba
	sp	25 (?)	?	*K* (?) 6·45	63O
	sp	25	?	*K* (?) 4·5 ± 0·1	67A
Y^{3+}	sp	25	?	*K* (?) 5·5, acetate buffer	62T
	sp	20	0·2 $NaClO_4$	$K(Y^{3+} + HL^{5-} \rightleftharpoons YHL^{2-})$ 12·81 ± 0·08	66K
Yb^{3+}	sp	~ 20	0·2 ($NaNO_3$)	$K(2Yb^{3+} + 2L^{6-} \rightleftharpoons Yb_2L_2^{6-})$ 45·7	63B
Zn^{2+}	sp	20	0·2 (NaCl)	$K(Zn^{2+} + H_2L^{4-} \rightleftharpoons ZnH_2L^{2-})$ 6·02 ± 0·10	67K
Zr^{4+}	sp	?	0·8 $HClO_4$	*K* (?) 7·60	59C
	sp	~ 20 (?)	0·2 $NaNO_3$	$K(2Zr^{4+} + 2L^{6-} \rightleftharpoons Zr_2L_2^{4-})$ 31·0	63Ba

59C K. L. Cheng, *Talanta*, 1959, **2**, 266
60C K. L. Cheng, *Talanta*, 1960, **3**, 81, 147
60O H. Onishi and N. Ishiwatari, *Bull. Chem. Soc. Japan*, 1960, **33**, 1581

continued overleaf

1117.2 $C_{31}H_{32}O_{13}N_2S$ (contd)

60R B. Rehak and J. Korbl, *Coll. Czech. Chem. Comm.*, 1960, **25**, 797
62B B. Budesinsky, *Z. Analyt. Chem.*, 1962, **188**, 266
62T K. Tonosaki and M. Otomo, *Bull. Chem. Soc. Japan*, 1962, **35**, 1683
63B B. Budesinsky and A. Besdekova, *Z. analyt. Chem.*, 1963, **196**, 17
63Ba B. Budesinsky, *Coll. Czech. Chem. Comm.*, 1963, **28**, 1858
63O M. Otomo, *Bull. Chem. Soc. Japan*, 1963, **36**, 137, 140, 889, 1341, 1577
64K O. V. Kon'kova, *J. Analyt. Chem. (U.S.S.R.)*, 1964, **19**, 63 (73)
64O M. Otomo, *Bull. Chem. Soc. Japan*, 1964, **37**, 504
65O M. Otomo, *Bull. Chem. Soc. Japan*, 1965, **38**, 730
66D R. Doicheva, S. Popova, and E. Mitropolitska, *Talanta*, 1966, **13**, 1345
66Da C. D. Dwivedi, K. N. Munshi, and A. K. Dey, *J. Indian Chem. Soc.*, 1966, **43**, 301
66K V. B. Kumok and V. S. Serebrennikov, *Russ. J. Inorg. Chem.*, 1966, **11**, 47 (90)
67A B. V. Agarwala and A. K. Dey, *J. Indian Chem. Soc.*, 1967, **44**, 691
67K V. N. Kumok, *Russ. J. Inorg. Chem.*, 1967, **12**, 1593 (3010)
67S Z. A. Sheka and E. I. Sinyavskaya, *Russ. J. Inorg. Chem.*, 1967, **12**, 340 (650)

1117.3 $C_{32}H_{22}O_6N_2S_2$ 3,3′-Dimethylene-4,4′-diphenyl-2,2′-biquinolyldisulphonic acid* H_2L

Metal	*Method*	*Temp*	*Medium*	*Log of equilibrium constant, remarks*	*Ref*
H^+	gl	25	0·1	K_1 4·89	67U
Cu^{2+}	gl	25	0·1	K_1 4·75, K_2 5·05	67U
Ni^{2+}	gl	25	0·1	K_1 4·12	67U

* Position of sulphonic acid groups not given.

67U E. Uhlemann and U. Hammerschick, *Z. anorg. Chem.*, 1967, **352**, 53

1117.4 $C_{32}H_{30}O_{16}N_6S_2$ 2,7-Bis[2′-*NN*-bis(carboxymethyl)aminomethylphenylazo]chromotropic acid(amino methylazo III) H_8L

Metal	*Method*	*Temp*	*Medium*	*Log of equilibrium constant, remarks*	*Ref*
H^+	sp	~ 20	0·2 (KNO_3)	K_1 14·1, K_2 13·3, K_3 12·2, K_4 10·3, K_5 9·3, K_6 6·3, K_7 4·9, K_8 3·8, K_9 2·9, K_{10} 1·3, K_{11} −1·0, K_{12} −3·3	65B
Cu^{2+}	sp	~ 20	0·2 (KNO_3)	$K[Cu^{2+}+15H^{+}+3L^{8-} \rightleftharpoons Cu(H_5L)_3^{7-}]$ 195·1±0·3 $K[Cu^{2+}+3H^{+}+3L^{8-} \rightleftharpoons Cu(HL)_3^{19-}]$ 72·2±0·1	65B

65B B. Budesinsky and K. Maas, *Z. analyt. Chem.*, 1965, **214**, 325

1119.1 $C_{33}H_{57}O_9N_3$ See Table 1123.4

1119.2 $C_{33-36}H_{57-63}O_9N_3$ See Table 1113.3

1119.3 $C_{33}H_{24}O_6N_2S_2$ 3,3′-Trimethylene-4,4′-diphenyl-2,2′-biquinolyl-disulphonic acid* H_2L

Metal	*Method*	*Temp*	*Medium*	*Log of equilibrium constant, remarks*	*Ref*
H^+	gl	25	0·1	K_1 4·27	67U
Cu^{2+}	gl	25	0·1	K_1 3·4	67U

* Position of sulphonic acid groups not given.

67U E. Uhlemann and U. Hammerschick, *Z. anorg. Chem.*, 1967, **352**, 53

$C_{34}H_{38}O_6N_4$ Hematoporphyrin IX H_4L 1120

Metal	Method	Temp	Medium	Log of equilibrium constant, remarks	Ref
Cu^{2+}	sp	25	acetic acid	$K(Cu^{2+}+H_6L^{2+} \rightleftharpoons CuH_2L+4H^+)$ (?) 5·26	66B
Zn^{2+}	sp	25	0·1 $NaClO_4$	$K(Zn^{2+}+H_4L \rightleftharpoons ZnH_2L+2H^+)$ (?) −6·89	66B

66B D. A. Brisbin and R. J. Balahura, *Canad. J. Chem.*, 1966, **44**, 2157

$C_{36}H_{60}O_{12}N_4$ See Table 1120.1 1120.1

$C_{44}O_{13}N_2S$ 3,3′-Bis[*NN*-di(carboxymethyl)aminomethyl]-2,2′-dimethyl-5,5′-di(1″-methylethyl)-4′-hydroxyfuchson-2″-sulphonic acid (methyl thymol blue) 1120.2 H_6L

Metal	Method	Temp	Medium	Log of equilibrium constant, remarks	Ref
H^+	gl	?	0·2 ($NaNO_3$)	K_1 13·4, K_2 11·15, K_3 7·4, K_4 3·8, K_5 3·3, K_6 3·0	65T
Al^{3+}	sp	20—24	1 ($NaClO_4$)	$K(2Al^{3+}+2H_4L^{2-} \rightleftharpoons AlH_3L+AlH_4L^++H^+)$(?) 3·88,	67L
				$K[2Al^{3+}+4H_4L^{2-} \rightleftharpoons Al(H_3L)^{3-}+Al(H_4L)_2^-+2H^+]$(?) 6·02	
Fe^{3+}	sp	?	0·1 ($NaClO_4$)	$K(Fe^{3+}+H_2L^{4-} \rightleftharpoons FeH_2L^-)$ 20·56±0·07	68K
				$K(Fe^{3+}+2H^++L^{6-} \rightleftharpoons FeH_2L^-)$ 43·29±0·09	
				$K[FeH_2L^-+H_4L^{2-} \rightleftharpoons Fe(H_3L)_2^{3-}]$ 6·66±0·05	
Ga^{3+}	sp	20—24	1 ($NaClO_4$)	$K(2Ga^{3+}+2H_4L^{2-} \rightleftharpoons GaH_3L+GaH_4L^++H^+)$(?) 4·35,	67L
				$K[2Ga^{3+}+4H_4L^{2-} \rightleftharpoons Ga(H_3L)_2^{3-}+Ga(H_4L)_2^-+2H^+]$(?) 7·58	
La^{3+}	sp	?	0·2 $NaNO_3$	$K(2La^{3+}+2L^{6-} \rightleftharpoons La_2L_2^{6-})$ 35·8	63B
				$K[2La^{3+}+2L^{6-}+2OH^- \rightleftharpoons La_2(OH)_2L_2^{8-}]$ 23·2	
	sp	18—22	0·4 (KCl)	$K(La^{3+}+H_4L^{2-} \rightleftharpoons LaH_2L^-+2H^+)$ −5·85	67M
				$K(2La^{3+}+H_4L^{2-} \rightleftharpoons La_2H_2L^{2+}+2H^+)$ −6·20	
Y^{3+}	gl, sp	?	0·2 ($NaNO_3$)	$K(2Y^{3+}+2L^{6-}+2H^+ \rightleftharpoons Y_2H_2L_2^{4-})$ 50·4±0·6	65T
				$K(Y_2HL_2^{5-}+H^+ \rightleftharpoons Y_2H_2L_2^{4-})$ 8·0	
				$K(Y_2L_2^{6-}+H^+ \rightleftharpoons Y_2HL_2^{5-})$ 9·5	

63B B. Budesinsky and E. Antonescu, *Coll. Czech. Chem. Comm.*, 1963, **28**, 3264
65T G. S. Tereshin, A. R. Rubinshein, and I. V Tananaev, *J. Analyt. Chem.* (*U.S.S.R.*), 1965, **20**, 1138 (1082)
67L N. D. Lukomskaya, T. V. Mal'kova, and K. B. Yatsimirskii, *Russ. J. Inorg. Chem.*, 1967, **12**, 1299 (2462)
67M T. V. Mal'kova, N. A. Fateeva, and Y. B. Yatsimirskii, *Russ. J. Inorg. Chem.*, 1967, **12**, 481 (915)
68K B. Karadalov, D. Kantcheva, and P. Nenova, *Talanta*, 1968, **15**, 525

$H_{44}O_{12}N_2$ 3,3′-Bis[*NN*-bis(carboxymethyl)aminomethyl]-4′-hydroxy-2,2′-dimethyl-5,5′-di(2″propyl)-fuchsone-2″-carboxylic acid (thymolphthalcomplexon) 1120.3 H_6L

Metal	Method	Temp	Medium	Log of equilibrium constant, remarks	Ref
H^+	sp	18—22	0·2 $NaNO_3$	K_1 12·25, K_2 7·35	65B
Ca^{2+}	sp	18—22	0·2 $NaNO_3$	$K(2Ca^{2+}+2L^{6-} \rightleftharpoons Ca_2L_2^{8-})$ 42·74	65B

65B A. Besdekova and B. Budesinsky, *Coll. Czech. Chem. Comm.*, 1965, **30**, 818

1120.4 $C_{40}H_{64}O_{12}$ See Table 1126.1

1120.5 $C_{41}H_{66}O_{12}$ See Table 1124.3

1120.3 $C_xH_yO_8P$ Phosphatidic acid H_2L
(x = 37−45, y = 2x−1 for saturated alkyls)

Metal	*Method*	*Temp*	*Medium*	*Log of equilibrium constant, remarks*	*Ref*
H^+	gl	24—25	0·1	K_1 8·0, K_2 3·0	66A
Ca^{2+}	oth	24	0·03 NaCl 0·05 $(HOCH_2)_3CNH_2$	K_1 4·14±0·03, light scattering	65A
	gl	24	0·1	K_1 4·2±0·1	66A
K^+	gl	24	0·1	K_1 0·9±0·1	66A
Li^+	gl	24	0·1	K_1 1·3±0·1	66A
Mg^{2+}	gl	24	0·1	K_1 4·1±0·1	66A
Na^+	gl	24	0·1	K_1 1·1±0·1	66A

65A M. B. Abramson, R. Katzman, and R. Curi, *J. Colloid. Interface Sci.*, 1965, **20**, 777
66A M. B. Abramson, R. Katzman, H. Gregor, and R. Curci, *Biochemistry*, 1966, **5**, 2207

1120.6 $C_xH_yO_{19}P_3$ Triphosphoinositide H_5L
(x = ?, y = 2x−1 for saturated alkyls)

Metal	*Method*	*Temp*	*Medium*	*Log of equilibrium constant, remarks*	*Ref*
H^+	gl	20	0·1 $(C_2H_7)_4NI$	K_1 8·45±0·05, K_2 6·38±0·05	65H
Ca^{2+}	gl	20	0·1 $(C_3H_7)_4NI$	K_1 5·0±0·1 $K(Ca^{2+}+HL^{4-} \rightleftharpoons CaHL^{2-})$ 3·8±0·1	65H
Mg^{2+}	gl	20	0·1 $(C_3H_7)_4NI$	K_1 5·1±0·3 $K(Mg^{2+}+HL^{4-} \rightleftharpoons MgHL^{2-})$ 3·8±0·1	65H
Ni^{2+}	gl	20	0·1 $(C_3H_7)_4NI$	K_1 6·3±0·5 $K(Ni^{2+}+HL^{4-} \rightleftharpoons NiHL^{2-})$ 4·2±0·3	65H

65H H. S. Henderson and J. G. Fullington, *Biochemistry*, 1965, **4**, 1599

1120.8 $C_xH_yO_{10}NP$ Phosphatidylserine H_2L
(x = ?, y = 2x−2 for saturated alkyls)

Metal	*Method*	*Temp*	*Medium*	*Log of equilibrium constant, remarks*	*Ref*
H^+	gl	20	0·1 $(C_3H_7)_4NI$	K_1 9·93±0·05, K_2 4·42±0·05	65H
Ca^{2+}	gl	20	0·1 $(C_3H_7)_4NI$	K_1 4·1±0·1 $K(Ca^{2+}+HL^- \rightleftharpoons CaHL^+)$ 4·0±0·1	65H
Mg^{2+}	gl	20	0·1 $(C_3H_7)_4NI$	K_1 4·3±0·2 $K(Mg^{2+}+HL^- \rightleftharpoons MgHL^+)$ 3·8±0·2	65H
Ni^{2+}	gl	20	0·1 $(C_3H_7)_4NI$	K_1 7·9±0·1 $K(Ni^{2+}+HL^- \rightleftharpoons NiHL^+)$ 4·6±0·1	65H

65H H. S. Henderson and J. G. Fullington, *Biochemistry*, 1965, **4**, 1599

PART II B—ORGANIC LIGANDS

MACROMOLECULES

TABLES

Macromolecule Albumin 1121

Metal	*Method*	*Temp*	*Medium*	*Log of equilibrium constant, remarks*	*Ref*
H^+	H	25	0·15	*K* (carboxyl) 4·00 ($\bar{\nu}_T = 106$) *K* (imidazole) 6·10 ($\bar{\nu}_T = 17$) *K* (α-amino) 8·00 ($\bar{\nu}_T = 9$) *K* (ε-amino) 9·40 ($\bar{\nu}_T = 58$) *K* (sulphydryl) 9·60 ($\bar{\nu}_T = 0{\cdot}7$) (human) $K = e^{2Z_p\omega}\left[\frac{\bar{\nu}_H}{(\bar{\nu}_T-\bar{\nu}_H)a_{H^+}}\right]$ Z_p = charge on protein molecule ω = constant = 0·0303 for human albumin at 25°, μ = 0·15. (assumed same for bovine albumin) $\bar{\nu}_H$ = no. of protonated sites per molecule $\bar{\nu}_T$ = total no. of donor sites per molecule a_H = activity of H^+ (?)	50T, (52T)
	oth	0	0·15	*K* (imidazole) 6·56, calculated from 50T (human) *K* defined as above (50T) except that $\bar{\nu}_T = 16$	52G
$CdCl^+$	pol	25	0·15 (KCl)	*K* (imidazole) 2·8 (bovine) $K = e^{2Z_pZ_m\omega}\left\{\frac{\bar{\nu}_M}{(\bar{\nu}_T-\bar{\nu}_M-\bar{\nu}_H)[M^{n+}]}\right\}$ Z_m = charge of metal ion (+1 for $CdCl^+$) $\bar{\nu}_M$ = no. of sites covered by metal ion per molecule. Other terms have same meaning as above (50T).	52T
Cu^{2+}	pol	25	0·15 (KNO_3)	*K* (imidazole) 3·7 (bovine) *K* defined above (52T, $CdCl^+$)	52T
Mn^{2+}	nmr, epr	24	0·2 $(CH_3)_4NCl$	*K′* 4·43 ± 0·09 (bovine) (first Mn^{2+} bound) *K′* 3·52 ± 0·09 (bovine) (2nd—6th Mn^{2+} bound) $K'_1 = \frac{[M_b]}{[M^{n+}][L_F]}$ $[M_b]$ = concentration of bound Mn^{2+} ions in solution $[L_F]$ = concentration of free binding sites in solution = $\bar{\nu}_T L_T - M_b$ $\bar{\nu}_T$ = no. of binding sites per molecule L_T = total protein concentration (each binding site considered to consist of two imidazole groups)	63M
Pb^{2+}	pol	25	0·15 (KNO_3)	*K* (imidazole) < 2·3 (bovine) *K* defined above (52T, $CdCl^+$)	52T
Tl^+	pol	25	0·15 KCl	*K* (imidazole) < 0 (bovine) *K* defined above (52T, $CdCl^+$)	52T

continued overleaf

1121 Albumin (contd)

Metal	Method	Temp	Medium	Log of equilibrium constant, remarks	Ref
Zn^{2+}	oth	0	0·15 ($NaNO_3$)	K (imidazole) 2·8±0·1 (human) K (imidazole) 2·7 (guanidinated human) K (imidazole) 2·8 (diazo-esterified human) dialysis $K = \dfrac{\bar{\nu}_M[H^+]K_H e^{2\omega Z_p}}{[Zn^{2+}]\,\bar{\nu}_H L_T}$ $K_H = K$ defined above (H^+, 50T). Other terms as defined above, (Determined $\bar{\nu}_T \simeq 15$, one imidazole bound per Zn^{2+})	52G
	pol	25	0·15 (KCl)	K (imidazole) 2·9 (bovine) K defined above (52T, $CdCl^+$)	52T
	oth	0	0·15 $NaNO_3$	K (imidazole) 2·9±0·1 (mercapto human) dialysis K defined above (52G, Zn^{2+})	54G

Errata. 1964, page 731, Table 1121. *Insert* 38 in the third column *and* 0·16 (NaCl) in the fourth column. *Change* K_1 to †K' (defined above under Mn^{2+}, 63M).

50T C. Tanford, *J. Amer. Chem. Soc.*, 1950, **72**, 441
52G F. R. N. Gurd and D. S. Goodman, *J. Amer. Chem. Soc.*, 1952, **74**, 670
52T C. J. Tanford, *J. Amer. Chem. Soc.*, 1952, **74**, 211
54G F. R. N. Gurd, *J. Phys. Chem.*, 1954, **58**, 788
63M A. S. Mildvan and M. Cohn, *Biochemistry*, 1963, **2**, 910

1122 Macromolecule Carboxypeptidase A H_2L

Erratum. 1964, p. 731, Table 1122. *Insert* in fifth column of each entry—dialysis.

1123 Macromolecule Casein

Metal	Method	Temp	Medium	Log of equilibrium constant, remarks	Ref
Ca^{2+}	oth	5	0·2 NaOH +40g casein/litre	†K 2·73, ultracentrifuge	42C
			0·2 NaOH +40g casein/litre +0·15 NaCl	†K' 2·36, ultracentrifuge	
	oth	38	0·16 NaCl	K_{11} 3·43, K_{12} 3·11, K_{13} 2·90, K_{14} 2·74, K_{15} 2·61, K_{16} 2·49, K_{17} 2·38, K_{18} 2·28, K_{19} 2·18, $K_{1,10}$ 2·08, $K_{1,11}$ 1·97, $K_{1,12}$ 1·85, $K_{1,13}$ 1·72, $K_{1,14}$ 1·56, $K_{1,15}$ 1·36, $K_{1,16}$ 1·04 Calculated from data of E. G. Weir and A. B. Hastings, *J. Biol. Chem.*, 1936, **114**, 397 by $K_x = {}^{\dagger}K^1 + \log \dfrac{16-(x-1)}{x}$; $^{\dagger}K' = 10^{2\cdot23}$	46K

†K' defined in albumin (1121, Mn^{2+}, 63M).

Errata. 1964, page 731, Table 1123. *Insert* 38 in the third column *and* 0·16 (NaCl) in the fourth column. *Change* K_1 to †K^1

42C A. Chanutin, S. Ludewig, and A. V. Masket, *J. Biol Chem.*, 1942, **143**, 737
46K I. M. Klotz, *Arch. Biochem. Biophys.*, 1946, **9**, 109

Macromolecule Cephalin 1123.1

Metal	Method	Temp	Medium	Log of equilibrium constant, remarks	Ref
Ca^{2+}	oth	25	0·16 NaCl	†K' 3·10±0·03, Frog heart (cephalin from beef brain)	43D

†K' defined in albumin (1121, Mn^{2+}, 63M).

43D N. Drinker and H. H. Zinsser, *J. Biol. Chem.*, 1943, **148**, 187

Macromolecule Deoxyribonucleic acid (DNA) 1123.2

Metal	Method	Temp	Medium	Log of equilibrium constant, remarks	Ref
Ag^{+}	Ag	23	0·1 ($NaClO_4$)	K 6·49 ($\omega = -0{\cdot}95$) (*M. lysodeikticus*) K 6·34 ($\omega = -0{\cdot}86$) (*E. coli*) K 6·33 ($\omega = -0{\cdot}61, -0{\cdot}52$) (calf thymus) K 6·62 ($\omega = -1{\cdot}04$) (denatured calf thymus) K for purine binding. $K = \dfrac{\upsilon_b \bar{\nu}_T e^{\omega \upsilon_b}}{(\bar{\nu}' - \upsilon_b \bar{\nu}_T)[M^{n+}]}$ υ_b = moles bound metal ion per mole of DNA bases = 0 to 0·2 $\frac{1}{2}\bar{\nu}_T$ = number of base pairs per molecule of DNA. $\bar{\nu}'$ = number of purine sites per molecule of DNA. ω = interaction constant, varies with each determination. K 5·4 ($\omega = -0{\cdot}99$) (*M. lysodeikticus*) K 4·8 ($\omega = -0{\cdot}82$) (*E. coli*) K 4·8 ($\omega = -0{\cdot}76, -0{\cdot}78$) (calf thymus) K 5·1 ($\omega = -0{\cdot}80$) (denatured calf thymus) K for binding by bases other than purine, terms defined above except υ_b = 0·2 to 0·5. $K = \dfrac{(\upsilon_b \bar{\nu}_T - \bar{\nu}')e^{\omega \upsilon_b}}{(\bar{\nu}' - \upsilon_b)[M^{n+}]}$	62Y
Ca^{2+}	oth	25	0·15—0·2 NaCl	†K'(0·15) 2·10, (0·2) 1·92, dialysis (calf thymus)	57W
	ix	25	0·15 NaCl	†K' 2·21±0·03 (calf thymus)	57W
Cu^{2+}	epr	19—25	0·02 $NaClO_4$	†K' 4·10 (calf thymus), †K' 9·9 (denatured calf thymus)	65R
	kin	5—25	?	†K' (5°) 7·3±0·1, (15°) 7·2±0·1, (25°) 7·2±0·1 $\Delta H \sim 0$, $\Delta S \sim 33$ (calf thymus)	66Y
	sp	25	?	†K' 4·1 (calf thymus)	66Y
	sp	23—27	0·005 $NaNO_3$	†K' 4·0 (salmon sperm) †K' 5·0 (heat-denatured salmon sperm) †K' 3·95 (calf thymus)	67B
	oth	23—27	0·005 $NaNO_3$	†K' 4·3, gel filtration (salmon sperm)	67B
Fe^{3+}	kin	25	~ 0	†K' 5·11±0·06 (rat intestine and liver) $\Delta H \sim 0$. $\Delta S \sim 23$	65K

continued overleaf

1123.2 Deoxyribonucleic acid (contd)

Metal	Method	Temp	Medium	Log of equilibrium constant, remarks	Ref
Mg^{2+}	oth	25	0·15—0·2 NaCl	†K′ (0·15) 2·10, (0·2) 1·92, dialysis (calf thymus)	57W
	oth	5	0·2 NaCl	†K_1 2·45, dialysis (calf thymus)	58Z
	sp	?	0·002 NaCl	*K 5·3 (decreases greatly with increasing Na^+ concentration) (calf thymus)	59S
Mn^{2+}	ix	25	0·15 NaCl	†K′ 2·44 (calf thymus)	57W
UO_2^{2+}	sp	20—25 (?)	0·15 NaCl	*K 6·9 (salmon sperm)	61Z

†K′ defined in albumin (1121, Mn^{2+}, 63M).
*K not clearly defined

57W J. S. Wiberg and W. F. Neuman, *Arch. Biochem. Biophys.*, 1957, **72**, 66
58Z G. Zubay and P. Doty, *Biochim. Biophys. Acta*, 1958, **29**, 47
59S J. Shack and B. S. Bynum, *Nature*, 1959, **184**, 635
61Z C. R. Zobel and M. Beer, *J. Biophys. Biochem. Cytol.*, 1961, **10**, 335
62Y T. Yamane and N. Davidson, *Biochim. Biophys. Acta*, 1962, **55**, 609
65K E. E. Kriss and K. B. Yatsimirskii, *Russ. J. Inorg. Chem.*, 1965, **10**, 1326 (2436)
65R C. Ropars and R. Viovy, *J. Chim. phys.*, 1965, **62**, 408
66Y K. B. Yatsimirskii, E. E. Kriss, and T. I. Akhrameeva, *Proc. Acad. Sci. (U.S.S.R.)*, 1966, **168**, 578 (840)
67B S. E. Bryan and E. Frieden, *Biochemistry*, 1967, **6**, 2728

1123.3 Macromolecule 'Enniatin A'*

[For correct sequence this Table should be re-numbered 111

Metal	Method	Temp	Medium	Log of equilibrium constant, remarks	Ref
K^+	gl	20	1 KI methanol	K_1 3·08±0·04	68W

*Mixture of ~40% Enniatin A, ~15% Enniatin B, the remainder intermediate between Enniatin A and Enniatir

68W H. K. Wipf, L. A. R. Pioda, Z. Stefanac, and W. Simon, *Helv. Chim. Acta*, 1968, **51**, 377

1123.4 $C_{33}H_{57}O_9N_3$,L; 'Enniatin B'

[For correct sequence this Table should be re-numbered 111

Metal	Method	Temp	Medium	Log of equilibrium constant, remarks	Ref
K^+	gl	20	1 KI methanol	K_1 2·93±0·06	68W
Na^+	gl	20	1 NaI methanol	K_1 2·38±0·10	68W

68W H. K. Wipf, L. A. R. Pioda, Z. Stefanac, and W. Simon, *Helv. Chim. Acta*, 1968, **51**, 377

Macromolecule Enolase 1123.5

Metal	Method	Temp	Medium	Log of equilibrium constant, remarks	Ref
Mn^{2+}	nmr, epr	20	0·5 KCl 0·05 *Tris*. HCl	†*K′* 5·0±0·1 (yeast)	63C

†*K′* defined in albumin (1121, Mn^{2+}, 63M)

63C M. Cohn, *Biochemistry*, 1963, **2**, 623

Macromolecule Gelatin 1123.6

Metal	Method	Temp	Medium	Log of equilibrium constant, remarks	Ref
H^+	gl	25	0·2 (KNO_3)	*K* (carboxyl) 4·25 ($\bar{\nu}_T$ = 113), *K* (imidazole) 6·5 ($\bar{\nu}_T$ = 5), *K* (α-amino) 7·5 ($\bar{\nu}_T$ = 3), *K* (ε-amino) 10 ($\bar{\nu}_T$ = 36) (photographic grade bone gelatin) *K* defined as in albumin (1121, 50T) Contribution of 44 guanidine groups considered negligible at pH used. (Molecular weight = 100,000)	66L
Ag^+	pol	25	0·2 (KNO_3)	*K* (carboxyl) 2·17 (pH 6—7) *K* (imidazole) ($\bar{\nu}_M$ = 1) 5·02, ($\bar{\nu}_M$ = 2) 4·85 ($\bar{\nu}_M$ = 3) 4·66, ($\bar{\nu}_M$ = 5) 3·67 (pH 6—7) *K* (α-amino) ($\bar{\nu}_M$ = 1) 7·07, ($\bar{\nu}_M$ = 2) 5·19, ($\bar{\nu}_M$ = 3) 4·23, ($\bar{\nu}_M$ = 5) 3·37 (pH 6—7), ($\bar{\nu}_M$ = 1) 5·73, ($\bar{\nu}_M$ = 2) 5·14, ($\bar{\nu}_M$ = 3) 4·37, ($\bar{\nu}_M$ = 5) 3·37 (pH 8—9) *K* (ε-amino) ($\bar{\nu}_M$ = 1) 5·60, ($\bar{\nu}_M$ = 2) 4·84, ($\bar{\nu}_M$ = 3) 5·25, ($\bar{\nu}_M$ = 5) 5·31 (pH 8—9) *K* defined as in albumin (1121, $CdCl^+$, 52T)	66L
Cd^{2+}	pol	25	0·1	†*K* (imidazole) 3·03	63Mb
	gl	25	0·15 (KCl)	†*K* (carboxyl) 1·76	65M
	oth	25	0·15 (KCl)	†*K* (carbonyl) 1·96, dialysis	65M
Cu^{2+}	pol	20	0·15	†*K* (carboxyl) 2·10	63M
	gl	25	0·15 (KCl)	†*K* (carboxyl) 2·10, *K* (imidazole) 3·28	65M
	oth	25	0·15 (KCl)	†*K* (carboxyl) 2·18, *K* (imidazole) 3·40, dialysis	65M
Pb^{2+}	pol	30	0·15 (KNO_3)	†*K* (carboxyl) 1·87	63Ma
Zn^{2+}	gl	25	0·15 (KCl)	†*K* (carboxyl) 1·87, *K* (imidazole) 2·74	65M
	oth	25	0·15 (KCl)	†*K* (carboxyl) 1·87, *K* (imidazole) 2·91, dialysis	65M

†$K = \dfrac{\bar{\nu}_M}{(\bar{\nu}_T - \bar{\nu}_H' - \bar{\nu}_M)[M^{n+}]}$. Terms have same meaning as in albumin (1121) except that $\bar{\nu}_T$ = 84 for carboxyl, 3·8 for imidazole, 23 for amino and 43 for guanidinium. (Molecular weight = 75,000) (transfusion gelatin).

63M W. V. Malik and Salahuddin, *J. Electroanalyt. Chem.*, 1963, **5**, 147
63Ma W. U. Malik and M. Muzaffaruddin, *J. Electroanalyt. Chem.*, 1963, **6**, 214
63Mb W. U. Malik and Salahuddin, *Nature*, 1963, **200**, 1204
65M W. U. Malik and M. Muzaffaruddin, *Austral. J. Chem.*, 1965, **18**, 1397
66L P. Lanza and I. Mazzei, *J. Electroanalyt. Chem.*, 1966, **12**, 320

1124 Macromolecule Globulin

Metal	*Method*	*Temp*	*Medium*	*Log of equilibrium constant, remarks*	*Ref*
Ca^{2+}	oth	25	0·16 NaCl	$^{\dagger}K'$ 3·0±0·1 (pseudoglobulin) $^{\dagger}K'$ 2·7±0·1 (euglobulin P_1) $^{\dagger}K'$ 2·0±0·1 (euglobulin P_2) $^{\dagger}K'$ 3·1±0·1 (euglobulin P_3). Frog heart (horse serum)	39D

$^{\dagger}K'_{1m}$ defined in albumin (1121, Mn^{2+}, 63M).

Errata. 1964, page 731, Table 1124. *Insert* 38 in the third column *and* 0·16 (NaCl) in the fourth column.

Change K_1 to $^{\dagger}K'$

39D N. Drinker, A. A. Green, and A. B. Hastings, *J. Biol. Chem.*, 1939, **131**, 641

1124.1 Macromolecule Glutamate dehydrogenase

Metal	*Method*	*Temp*	*Medium*	*Log of equilibrium constant, remarks*	*Ref*
Zn^{2+}	sp	20	0·1	$^{*}K$ 7·0, sodium phosphate buffer, (Beef liver). (Equivalent weight = 15,000 per Zn binding site)	65S

$$^{*}K = \frac{[E]\,[Zn],}{[EZn^{2+}]}$$ where E = enzyme and EZn = enzyme – Zn (II) complex (all binding sites assumed equivalent)

65S H. Sund, *Acta Chem. Scand.*, 1965, **19**, 390

1124.2 Macromolecule Insulin

Metal	*Method*	*Temp*	*Medium*	*Log of equilibrium constant, remarks*	*Ref*
H^+	Ag–AgCl	25	0·075	K (α-carboxyl) 3·6±0·3 ($\bar{\nu}_T = 4$), K (β,γ-carboxyl) 4·7±0·1 ($\bar{\nu}_T = 8·5$), K (imidazole) 6·4 ($\bar{\nu}_T = 4$), K (α-amino) 7·4 ($\bar{\nu}_T = 4$), K (ε-amino and phenolic) 9·6±0·1 ($\bar{\nu}_T = 10$), K (guanidyl) 11·9±0·2 ($\bar{\nu}_T = 2$) K defined in albumin (1121, 50T) (Molecular weight = 12,000)	54T
Zn^{2+}	Ag–AgCl	25	0·075	K (imidazole) 6·0 (bovine) K defined in albumin (1121, $CdCl^+$, 52T) (Zn^{2+} combines with two imidazole groups)	54T

54T C. Tanford and J. Epstein, *J. Amer. Chem. Soc.*, 1954, **76**, 2170

$C_{41}H_{66}O_{12}$, L; Monactin [For correct sequence this Table should be re-numbered 1120.5] 1124.3

Metal	Method	Temp	Medium	Log of equilibrium constant, remarks	Ref
K^+	oth	30	0·1 KSCN methanol	K_1 5·5±0·2, osmotic vapour pressure	67P
Na^+	oth	30	0·1 NaSCN methanol	K_1 3·15±0·04, osmotic vapour pressure	67P

67P L. A. R. Pioda, H. A. Wachter, R. E. Dohner, and W. Simon, *Helv. Chim. Acta*, 1967, **50**, 1373

$C_{40}H_{64}O_{12}$, L; Nonactin [For correct sequence this Table should be re-numbered 1120.4] 1126.1

Metal	Method	Temp	Medium	Log of equilibrium constant, remarks	Ref
K^+	oth	30	0·1 KSCN methanol	K_1 3·80±0·07, osmotic vapour pressure	67P
	Ag	20	0·1 KSCN methanol	K_1 3·59±0·08	68W
Na^+	oth	30	0·1 NaSCN methanol	K_1 2·20±0·08, osmotic vapour pressure	67P

67P L. A. R. Pioda, H. A. Wachter, R. E. Dohner, and W. Simon, *Helv. Chim. Acta*, 1967, **50**, 1373
68W H. K. Wipf, L. A. R. Pioda, Z. Stefanac, and W. Simon, *Helv. Chim. Acta*, 1968, **51**, 377

Macromolecule Alkaline phosphatase 1126.2

Metal	Method	Temp	Medium	Log of equilibrium constant, remarks	Ref
Zn^{2+}	oth	25	1 NaCl	K_1 10·22, $K(ZnL+Zn \rightleftharpoons Zn_2L)$ 7·66, (*E. coli*) enzymatic activity	66C

66C S. R. Cohen and I. B. Wilson, *Biochemistry*, 1966, **5**, 904

Macromolecule Polyacrylic acid (linear) 1127

Metal	Method	Temp	Medium	Log of equilibrium constant, remarks	Ref
H^+	gl	25	0—3 KCl	†K' (0) 6·17 (n = 2·0), (0·1) 5·11 (n = 1·68), (1) 4·70 (n = 1·54), (2) 4·55 (n = 1·44), (3) 4·60 (n = 1·44)	55G
			0·2—3 $NaNO_3$	†K' (0·2) 4·91 (n = 1·69), (1) 4·48, (2) 4·30 (n = 1·39), (3) 4·23	
			1—3 KNO_3	†K (1) 4·67, (2) 4·61, (3) 4·57(0·01N PAA) n is given by the relation, $-\log[H^+] = pK + n \log \frac{\alpha}{1-\alpha}$, where α is the fraction of protonated groups of macromolecule neutralized.	
	gl	10—40	1 KCl	ΔH 1·6, ΔS −16	55L

continued overleaf

1127 Polyacrylic acid (linear) (contd)

Metal	Method	Temp	Medium	Log of equilibrium constant, remarks	Ref
Ca^{2+}	gl	25	1 KCl	†K′ 2·0(0·06N PAA)	55Gb
Co^{2+}	gl	25	1 KCl	†K′ 2·6(0·06N PAA)	55Gb
Cu^{2+}	gl	25	0—2 $NaNO_3$	†K′ (0) 9·96, (0·2) 7·48, (2) 6·98(0·01N PAA)	55G
				*K′ (0) −2·38, (0·2) −2·34, (2) −1·98	
			1 KCl	†K′ 5·84(0·01N PAA), 5·78(0·06N PAA), 6·3(0·1N PAA)	
				*K′ −3·56(0·01N PAA), −3·64(0·06N PAA), −3·4(0·1N PAA)	
	gl	10—40	1 KCl	ΔH 0·5, ΔS 28	55L
	oth	25	0—2 $NaNO_3$	‡*K′ (0) −4·38, (0·2) −4·34, (2) −3·98(0·01N PAA)	57Z
			1 KCl	‡*K'_{1m} −5·56(0·01N PAA), −4·86(0·06N PAA), −4·4(0·1N PAA) calculated from data of 55G.	
Li^{+}	gl	25 (?)	0·2 LiCl	†K′ 0·28±0·02	57G
Mg^{2+}	gl	25	1 KCl	†K′ 1·8(0·06N PAA)	55Gb
Mn^{2+}	gl	25	1 KCl	†K′ 3·36(0·06N PAA)	55Gb
Zn^{2+}	gl	25	1 KCl	†K′ 3·32(0·06N PAA)	55Gb

†K′ defined in albumin (1121, Mn^{2+}, 63M). L_F = free unprotonated sites.

$$*K' = \frac{[M_b][H^+]^2}{[M^{n+}][L_F]^2}$$

terms defined in albumin (1121, Mn^{2+}, 63M). (Metal ion bound to two carboxyl groups.) L_F = free protonated sites.

$$‡*K' = \frac{[M_b][H^+]^2}{[M^{n+}][L_F]}$$

55G H. P. Gregor, L. B. Luttinger, and E. M. Lobl, *J. Phys. Chem.*, 1955, **59**, 34

55Gb H. P. Gregor, L. B. Luttinger, and E. M. Lobl, *J. Phys. Chem.*, 1955, **59**, 990

55L E. M. Lobl, L. B. Luttinger, and H. P. Gregor, *J. Phys. Chem.*, 1955, **59**, 559

57G H. P. Gregor and M. Frederick, *J. Polymer Sci.*, 1957, **23**, 451

57Z G. Zubay, *J. Phys. Chem.*, 1957, **61**, 377

1127.1 Macromolecule Polyacrylic acid (XE-89, cross-linked)

Metal	Method	Temp	Medium	Log of equilibrium constant, remarks	Ref
H^{+}	gl	25	0·2—2 $NaNO_3$	†K′ (0·2) 5·66 (n = 1·8), (2) 4·82 (n = 1·7)	55G
	gl	3—40	2 $NaNO_3$	ΔH 0·2, ΔS 23 n defined in polyacrylic acid (1127 H^+).	55L
Cu^{2+}	gl	25	0·2—2 $NaNO_3$	†K′ (0·2) 7·98, (2) 6·24	55G
				*K′ (0·2) −3·34, (2) −3·40	
	gl	3—40	2 $NaNO_3$	ΔH 0·4, ΔS 33	55L

†K′ defined in albumin (1121, Mn^{2+}, 63M). (Metal ion bound to two carboxyl groups.)
*K′ defined in polyacrylic acid (1127, *K'_{1m}).

55G H. P. Gregor, L. B. Luttinger, and E. M. Lobl, *J. Phys. Chem.*, 1955, **59**, 366

55L E. M. Lobl, L. B. Luttinger, and H. P. Gregor, *J. Phys. Chem.*, 1955, **59**, 559

Macromolecule Polyacrylic acid and 7·5% divinylbenzene copolymer 1127.2

Metal	Method	Temp	Medium	Log of equilibrium constant, remarks	Ref
H^+	gl	25	1·0 $NaNO_3$	†K' (n = 1·89)5·31±0·02 n defined in polyacrylic acid (1127, H^+).	68G
Ca^{2+}	gl	4—50	1·0 $NaNO_3$	‡*K' (25°) ($\bar{n}'$ = 0·1)−8·03, ($\bar{n}'$ = 0·2)−8·71, ($\bar{n}'$ = 0·3)−9·39 $\Delta H°$ ($\bar{n}'$ = 0·1)6·6, ($\bar{n}'$ = 0·2)9·5, $\Delta S°$ ($\bar{n}'$ = 0·1)−14, ($\bar{n}'$ = 0·2)−8	68G
Cu^{2+}	gl	4—50	1·0 $NaNO_3$	‡*K' (25°) ($\bar{n}'$ = 0·3)−4·60, ($\bar{n}'$ = 0·4)−4·90, ($\bar{n}'$=0·5)−5·19, ($\bar{n}'$ = 0·6)−5·49 $\Delta H°$ ($\bar{n}'$ = 0·3)5·0±0·2, ($\bar{n}'$ = 0·4)5·0±0·5, ($\bar{n}'$ = 0·5)5·4±0·4, ($\bar{n}'$ = 0·6)6·0±0·3, $\Delta S°$ ($\bar{n}'$ = 0·3)−4±1, ($\bar{n}'$ = 0·4)−6±2, ($\bar{n}'$ = 0·5) −6±1, ($\bar{n}'$ = 0·6)−5±1	68G
Ni^{2+}	gl	4—50	1·0 $NaNO_3$	‡*K' (25°) ($\bar{n}'$ = 0·2)−7·37, ($\bar{n}'$ = 0·3)−7·71, ($\bar{n}'$ = 0·4)−8·05 $\Delta H°$ ($\bar{n}'$ = 0·2)4·5±0·6, ($\bar{n}'$ = 0·3)5·1±0·9, ($\bar{n}'$ = 0·4)6·5±0·5, $\Delta S°$ ($\bar{n}'$ = 0·2)−19±2, ($\bar{n}'$ = 0·3)−18±3, ($\bar{n}'$ = 0·4)−15±2	68G
Zn^{2+}	gl	4—50	1·0 $NaNO_3$	‡*K' (25°) ($\bar{n}'$ = 0·2)−6·65, ($\bar{n}'$ = 0·3)−6·78, ($\bar{n}'$ = 0·4)−6·98, ($\bar{n}'$ = 0·5)−7·22 $\Delta H°$ ($\bar{n}'$ = 0·2)7·9±1·0, ($\bar{n}'$ = 0·3)8·9±0·7, ($\bar{n}'$ = 0·4)10·4±0·4, ($\bar{n}'$ = 0·5)11·7±0·4, $\Delta S°$ ($\bar{n}'$ = 0·2)−4±3, ($\bar{n}'$ = 0·3)−1±2, ($\bar{n}'$ = 0·4)3±1, ($\bar{n}'$ = 0·5)6±1	68G

†K' defined in albumin (1121, Mn^{2+}, 63M).

‡*K' defined in polyacrylic acid (1127).

$\bar{n}' = \frac{\bar{v}_M L_T}{M_T}$, terms defined in albumin (1121) where sites are taken as adjacent pairs of carboxy groups.

68G R. L. Gustafson and J. A. Lirio, *J. Phys. Chem.*, 1968, **72**, 1502

Macromolecule Poly-*N*-benzyliminodiacetic acid (Dowex A-1) 1127.3

Metal	Method	Temp	Medium	Log of equilibrium constant, remarks	Ref
H^+				Used values for *N*-benzyliminodiacetic acid (?)	68E
Co^{2+}	gl	10—40	0·1	†K' (10°) 7·36, (25°) 7·24, (40°) 7·02	68E
Cu^{2+}	ix	23—27	?	†K' 16·9 †K' ($CuA + L \rightleftharpoons CuAL$) 6·17 where H_2A = iminodiacetic acid 8·72 where HA = glycine 7·76 where H_2A = glutamic acid	64L
	gl	25	0·1	†K' 10·54	68E
Ni^{2+}	ix	23—27	?	†K' 15·61 †K' ($NiA + L \rightleftharpoons NiAL$) 6·79 where H_2A = iminodiacetic acid	64L
	gl	10—40	0·1	†K' (10°) 8·13, (25°) 8·06, (40°) 7·63	68E
Zn^{2+}	gl	10—40	0·1	†K' (10°) 7·41, (25°) 7·37, (40°) 7·19	68E

†K' defined in albumin (1121, Mn^{2+}, 63M).

64L H. Loewenschuss and G. Schmuckler, *Talanta*, 1964, **11**, 1399
68E C. Eger, J. A. Marinsky, and W. M. Anspach, *J. Inorg. Nuclear Chem.*, 1968, **30**, 1911

1127.4 Macromolecule Poly-*N*-ethyleneglycine

Metal	*Method*	*Temp*	*Medium*	*Log of equilibrium constant, remarks*	*Ref*
H^+	gl	25	1	†K' (amino) 8·15 (n = 7·45)	62G
				†K' (carboxyl) 2·2 (n = 1)	
				n defined in polyacrylic acid (1127, H^+).	
Cu^{2+}	gl	25	1	†K' 12·04 (metal ion bound by two glycines)	62G
				†K' 14·92 (metal ion bound by three glycines)	

†K' defined in albumin (1121, Mn^{2+}, 63M).

62G D. H. Gold and H. P. Gregor, *J. Phys. Chem.*, 1962, **66**, 246

1127.5 Macromolecule Polyethylene and maleic anhydride copolymer (1 : 1)

Metal	*Method*	*Temp*	*Medium*	*Log of equilibrium constant, remarks*	*Ref*
H^+	gl	25	→ 0	†K' ($H^+ + L^{2-} \rightleftharpoons HL^-$) 9·60 ($n$ = 1·5)	68F
				†K' ($H^+ + HL^- \rightleftharpoons H_2L$) 4·97 ($n$ = 2·2)	
				where H_2L = concentration of dicarboxylic acid groups.	
				n defined in polyacrylic acid (1127, H^+).	
Cd^{2+}	gl	25	→ 0	†K' 9·33±0·03	68F
Co^{2+}	gl	25	→ 0	†K' 8·63±0·03	68F
Cu^{2+}	gl	25	→ 0	†K' 10·65±0·04	68F
Mn^{2+}	gl	25	→ 0	†K' 8·81±0·02	68F
Ni^{2+}	gl	25	→ 0	†K' 8·58±0·03	68F
Zn^{2+}	gl	25	→ 0	†K' 8·70±0·03	68F

†K' defined in albumin (1121, Mn^{2+}, 63M).

68F B. J. Belber, E. M. Hodnett, and N. Purdie, *J. Phys. Chem.*, 1968, **72**, 2496

1127.6 Macromolecule Polymethacrylic acid

Metal	*Method*	*Temp*	*Medium*	*Log of equilibrium constant, remarks*	*Ref*
Cd^{2+}	gl	20	0—0·1 $NaNO_3$	*K' −4·4	64M
Co^{2+}	gl	20	0—0·1 $NaNO_3$	*K' −5·7	64M
Cu^{2+}	gl	20	0—0·1 $NaNO_3$	*K' −3·6	64M
Mg^{2+}	gl	20	0—0·1 $NaNO_3$	*K' −6·2	64M
Ni	gl	20	0—0·1 $NaNO_3$	*K' −5·6	64M
Zn	gl	20	0—0·1 $NaNO_3$	*K' −5·2	64M

*K' defined in polyacrylic acid (1127).

64M M. Mandel and J. C. Leyte, *J. Polymer Sci.*, 1964, **A2**, 2883

Macromolecule Polymethacrylic acid and 1% divinylbenzene copolymer 1127.7

Metal	*Method*	*Temp*	*Medium*	*Log of equilibrium constant, remarks*	*Ref*
H^+	gl	25	0—2 $NaNO_3$	†K′ (~ 0) 7·59 (n = 2·0), (0·2) 6·10 (n = 2·0), (2) 5·40 (n = 1·8) n defined in polyacrylic acid (1127, H^+).	55G
Cu^{2+}	gl	25	0—2 $NaNO_3$	†K′ (~ 0) 10·48, (0·2) 7·40, (2) 5·40 *K′ (~ 0) −4·70, (0·2) −4·80, (2) −5·40	55G
	sp	25	0—2 $NaNO_3$	†K′ (~ 0) 7·15, (0·2) 5·51, (2) 4·97	55G

†K′ defined in albumin (1121, Mn^{2+}, 63M).
(Metal ions bound to two carboxy groups.)
*K′ defined in polyacrylic acid (1127).

55G H. P. Gregor, L. B. Luttinger, and E. M. Lobl, *J. Phys. Chem.*, 1955, **59**, 366

Macromolecule Polymethacrylic acid and 5% divinylbenzene copolymer 1127.8

Metal	*Method*	*Temp*	*Medium*	*Log of equilibrium constant, remarks*	*Ref*
H^+	gl	25	1·0 $NaNO_3$	†K′ 5·91±0·01 (n = 1·60) n defined in polyacrylic acid (1127, H^+).	65G
Ca^{2+}	gl	5—44	1·0 $NaNO_3$	†K′ (n̄′ = 0·5) (25°) 1·41 $\Delta H_1°$ 3·4, $\Delta S_1°$ 18 ‡*K′ (25°) (n̄′ = 0·1)−9·55, (n̄′ = 0·2)−10·37, (n̄′ = 0·3)−11·06 ΔH° (n̄′ = 0·1)9·9, (n̄′ = 0·2)11·5, ΔS° (n̄′ = 0·1)−11, (n̄′ = 0·2)−9	65G
Cu^{2+}	gl	5—44	1·0 $NaNO_3$	†K′ (n̄′ = 0·5) (25°) 3·45 $\Delta H_1°$ 4·3, $\Delta S_1°$ 30 †K′ [n̄′ = 0·5 (?)] (25°) 3·15, ΔH° 4·7, ΔS° 30 ‡*K′ (25°) (n̄′ = 0·2)−6·17, (n̄′ = 0·3)−6·36, (n̄′ = 0·4)−6·49, (n̄′ = 0·5)−6·59, (n̄′ = 0·6)−6·71, (n̄′ = 0·7)−6·83 ΔH° (n̄′ = 0·2)7·8, (n̄′ = 0·3)8·4, (n̄′ = 0·4)8·9, (n̄′ = 0·5)9·8, (n̄′ = 0·6)10·1, (n̄′ = 0·7)10·4, ΔS° (n̄′ = 0·2)−2, (n̄′ = 0·3)−1, (n̄′ = 0·4)0·1, (n̄′ = 0·5)3, (n̄′ = 0·6)3, (n̄′ = 0·7)4	65G
Fe^{3+}	gl	5—44	1·0 $NaNO_3$	†K′ [n̄′ = 0·33 (?)] (25°) 5·50 ΔH° 2·9, ΔS° 35 ‡*K′ (n̄′ = 0·3) (5°) −3·88, (16°) −3·58, (25°) −3·45, (35°) −3·24, (44°) −3·02 ΔH° 8·7	65G
Ni^{2+}	gl	5—44	1·0 $NaNO_3$	†K′ (n̄′ = 0·5) (25°) 2·03 $\Delta H_1°$ 2·9, $\Delta S_1°$ 19 ‡*K′ (25°) (n̄′ = 0·1)−8·59, (n̄′ = 0·2)−9·07, (n̄′ = 0·3)−9·44, (n̄′ = 0·4)−9·84 ΔH° (n̄′ = 0·1)5·3, (n̄′ = 0·2)7·6, (n̄′ = 0·3)8·3, (n̄′ = 0·4)10·6, ΔS° (n̄′ = 0·1)−22, (n̄′ = 0·2)−16, (n̄′ = 0·3)−16, (n̄′ = 0·4)−9	65G

continued overleaf

1127.8 Macromolecule Polymethacrylic acid (contd)

Metal	*Method*	*Temp*	*Medium*	*Log of equilibrium constant, remarks*	*Ref*
Zn^{2+}	gl	5—44	1·0 $NaNO_3$	†K' ($\bar{n}'$ = 0·5) (25°) 2·69 $\Delta H_1°$ 6·1, $\Delta S_1°$ 33 †K' [$\bar{n}'$ = 0·5 (?)] (25°) 2·30, $\Delta H°$ 6·4, $\Delta S°$ 32 ‡*K' (25°) ($\bar{n}'$ = 0·1)−7·53, ($\bar{n}'$ = 0·2)−7·69, ($\bar{n}'$ = 0·3)−7·94, ($\bar{n}'$ = 0·4)−8·20, ($\bar{n}'$ = 0·5)−8·43, ($\bar{n}'$ = 0·6)−8·63 $\Delta H°$ ($\bar{n}'$ = 0·1)10·4, ($\bar{n}'$ = 0·2)11·5, ($\bar{n}'$ = 0·3)12·4, ($\bar{n}'$ = 0·4)14·2, ($\bar{n}'$ = 0·5)15·9, ($\bar{n}'$ = 0·6)16·1, $\Delta S°$ ($\bar{n}'$ = 0·1)0, ($\bar{n}'$ = 0·2)3, ($\bar{n}'$ = 0·3)5, ($\bar{n}'$ = 0·4)10, ($\bar{n}'$ = 0·5)15, ($\bar{n}'$ = 0·6)15	65G

†K' defined in albumin (1121, Mn^{2+}, 63M).
‡*K' defined in polyacrylic acid (1127).
$\bar{n}'$ defined in polyacrylic acid and 7·5% divinylbenzene copolymer (1127.2).

65G R. L. Gustafson and J. A. Lirio, *J. Phys. Chem.*, 1965, **69**, 2849

1127.9 Macromolecule Polymethacrylic acid and 9% divinylbenzene copolymer

Metal	*Method*	*Temp*	*Medium*	*Log of equilibrium constant, remarks*	*Ref*
H^+	gl	25	0—2 $NaNO_3$	†K' (~ 0) 8·25, (0·2) 6·62, (2) 5·94 n = 2·0 in each case. n defined in polyacrylic acid (1127, H^+).	55G
Cu^{2+}	gl	25	0—2 $NaNO_3$	†K' (~ 0) 11·10, (0·2) 7·48, (2) 5·76 *K' (~ 0) −5·40, (0·2) −5·78, (2) −6·12	55G
	sp	25	0—2 $NaNO_3$	†K' (~ 0) 7·71, (0·2) 5·92, (2) 5·00	55G

†K' defined in albumin (1121, Mn^{2+}, 63M).
(Metal ions bound to two carboxyl groups.)
*K' defined in polyacrylic acid (1127).

55G H. P. Gregor, L. B. Luttinger, and E. M. Lobl, *J. Phys. Chem.*, 1955, **59**, 366

1127.10 Macromolecule Polystyrene (54 mole %) and maleic anhydride copolymer

Metal	*Method*	*Temp*	*Medium*	*Log of equilibrium constant, remarks*	*Ref*
Ba^{2+}	gl	25	1 KNO_3	†K' 1·36	54M
Ca^{2+}	gl	25	1 KNO_3	†K' 2·11	54M
Mg^{2+}	gl	25	1 KNO_3	†K' 1·74	54M
Sr^{2+}	gl	25	1 KNO_3	†K' 1·46	54M

†K' defined in albumin (1121, Mn^{2+}, 63M).
(Metal ion bound to two carboxy groups.)

54M H. Morawetz, A. M. Kotliar, and H. Mark, *J. Phys. Chem.*, 1954, **58**, 619

Macromolecule Polyvinyl alcohol 1127.11

Metal	Method	Temp	Medium	Log of equilibrium constant, remarks	Ref
As^{III}	gl	25	0·1	$^{\dagger}K'[As(OH)_4^- + L \rightleftharpoons As(OH)_2(H_{-2}L)^-]$ −0·15	57R
B^{III}	gl	25	0·1	$^{\dagger}K'[B(OH)_4^- + L \rightleftharpoons B(OH)_2(H_{-2}L)^-]$ 0·26	57R
				$^{\dagger}K'[B(OH)_4^- + 2L \rightleftharpoons B(H_{-2}L)_2^-]$ 0·64	
Te^{V}	gl	25	0·1	$^{\dagger}K'[H_5TeO_6^- + L \rightleftharpoons H_4TeO_4(H_{-2}L)^-]$ 0·00 where L is in terms of monomer concentration	57R

$^{\dagger}K'$ defined in albumin (1121, Mn^{2+}, 63M).

57R G. L. Roy, A. L. Laferriere, and J. O. Edwards, *J. Inorg. Nuclear Chem.*, 1957, **4**, 106

Macromolecule Polyvinylethylether (62 mole %) and maleic anhydride copolymer 1127.12

Metal	Method	Temp	Medium	Log of equilibrium constant, remarks	Ref
Ba^{2+}	gl	25	1 KNO_3	$^{\dagger}K'$ 2·00	54M
Ca^{2+}	gl	25	1 KNO_3	$^{\dagger}K'$ 2·45	54M
Mg^{2+}	gl	25	1 KNO_3	$^{\dagger}K'$ 2·30	54M
Sr^{2+}	gl	25	1 KNO_3	$^{\dagger}K'$ 1·96	54M

$^{\dagger}K'$ defined in albumin (1121, $CdCl^+$, 63M). (Metal ion bound to two carboxy groups.)

54M H. Morawetz, A. M. Kotliar, and H. Mark, *J. Phys. Chem.*, 1954, **58**, 619

Macromolecule Poly-*N*-vinylimidazole 1127.13

Metal	Method	Temp	Medium	Log of equilibrium constant, remarks	Ref
Ag^+	gl	25	1 KNO_3	$^{\dagger}K'$ 8·00 (0·01M PVI) (Ag^+ bound to two imidazole groups)	60G
Cu^{2+}	gl	25	0—1 $NaNO_3$	$^{\dagger}K'$ (~ 0) 10·64, (0·1) 12·76, (1) 14·72 (0·01M PVI), (1) 11·00 (0·1M PVI) (Cu^{2+} bound to four imidazole groups)	60G
	gl	24	1·0 $NaNO_3$	$^{\ddagger}K_1$ 2·02, $^{\ddagger}K_2$ 2·33, $^{\ddagger}K_3$ 3·07, $^{\ddagger}K_4$ 3·38 (0·1M PVI)	65L
	gl	25	1 $NaNO_3$	$^{\ddagger}K_1$ 2·90, $^{\ddagger}K_2$ 3·24, $^{\ddagger}K_3$ 4·06, $^{\ddagger}K_4$ 4·36 (0·01M PVI) (calculated from data of 60G)	65L
Zn^{2+}	gl	24	1·0 $NaNO_3$	$^{\ddagger}K_1$ 1·24, $^{\ddagger}K_2$ 1·66, $^{\ddagger}K_3$ 2·43, $^{\ddagger}K_4$ 2·87 (0·1M PVI)	65L
				$^{\ddagger}K_1$ 1·72, $^{\ddagger}K_2$ 2·65, $^{\ddagger}K_3$ 3·35, $^{\ddagger}K_4$ 3·76 (0·01M PVI) (where K_i refer to i imidazole groups bound per metal ion).	

$^{\dagger}K'$ defined in albumin (1121, Mn^{2+}, 63M).
$^{\ddagger}K_n$ corrected for statistical effect: $^{\ddagger}K_1 = K_1/4$, $^{\ddagger}K_2 = 2K_2/3$, $^{\ddagger}K_3 = 3K_3/2$, $^{\ddagger}K_4 = 4K_4$

60G D. H. Gold and H. P. Gregor, *J. Phys. Chem.*, 1960, **64**, 1461, 1464
65L K.-J. Liu and H. P. Gregor, *J. Phys. Chem.*, 1965, **69**, 1252

1129.1 Macromolecule Procarboxypeptidase

Metal	*Method*	*Temp*	*Medium*	*Log of equilibrium constant, remarks*	*Ref*
H^+	gl	25	?	K_1 8·9	67P
Cd^{2+}	oth	4	1 NaCl	K_1 8·4, dialysis	67P
Co^{2+}	oth	4	1 NaCl	K_1 5·4, dialysis	67P
Cu^{2+}	oth	4	1 NaCl	K_1 8·1, dialysis	67P
Hg^{2+}	oth	4	1 NaCl	K_1 18·3, dialysis	67P
Mn^{2+}	oth	4	1 NaCl	K_1 3·4, dialysis	67P
Ni^{2+}	oth	4	1 NaCl	K_1 5·9, dialysis	67P
Zn^{2+}	oth	4	1 NaCl	K_1 9·0, dialysis	67P

67P R. Piras and B. L. Vallee, *Biochemistry*, 1967, **6**, 348

1129.2 Macromolecule Pyruvate kinase

Metal	*Method*	*Temp*	*Medium*	*Log of equilibrium constant, remarks*	*Ref*
Cs^+	sp	25	0·1 $(CH_3)_4NCl$	†K' 1·36±0·04	66S
K^+	sp	25	0·1 $(CH_3)_4NCl$	†K' 0·92±0·08	66S
Mg^{2+}	sp	25 (?)	0·1 KCl, 0·05 Tris	†K' 3·28±0·03	63S
	nmr, epr	24	0·1 KCl, 0·02 Tris	†K' 3·42±0·04	65M
	kin	24	0·1 KCl, 0·02 Tris	†K' 3·4±0·1	65M
	sp	25	0·1 $(CH_3)_4NCl$	†K' 3·04±0·01	66S
Mn^{2+}	nmr, epr	20	0·5 KCl, 0·05 Tris	†K' 4·2±0·2	63C
	sp	25 (?)	0·1 KCl, 0·05 Tris	†K' 4·16±0·03	63S
	nmr, epr	27	0·1 KCl, 0·02 Tris	†K' 4·1±0·2	65M
	kin	29	0·1 KCl, 0·02 Tris	†K' 4·2±0·2	65M
	sp	25	0·1 $(CH_3)_4NCl$	†K' 4·0±0·1	66S
Na^{2+}	sp	25	0·1 $(CH_3)_4NCl$	†K' 0·68±0·04	66S
NH_4^+	sp	25	0·1 $(CH_3)_4NCl$	†K' 1·33±0·06	66S
Rb^+	sp	25	0·1 $(CH_3)_4NCl$	†K' 1·30±0·05	66S

†K' defined in albumin (1121, Mn^{2+}, 63M). (2 metal ions per catalytic site except for Cs^+ and Rb^+ with 3 and NH_4^+ with 4) (pyruvate kinase from rabbit muscle).

63C M. Cohn, *Biochemistry*, 1963, **2**, 623
63S C. H. Suelter and W. Melander, *J. Biol. Chem.*, 1963, **238**, PC4108.
65M A. S. Mildvan and M. Cohn, *J. Biol. Chem.*, 1965, **240**, 238
66S C. H. Suelter, R. Singleton jun, F. J. Kayne, S. Arrington, J. Glass, and A. S. Mildvan, *Biochemistry*, 1966, **5**, 131

1129.3 Macromolecule Ribonucleic acid (RNA)

Metal	*Method*	*Temp*	*Medium*	*Log of equilibrium constant, remarks*	*Ref*
Ca^{2+}	oth	25	0·15 NaCl	†K' 2·32, dialysis (calf liver)	57W
Mg^{2+}	oth	25	0·20 NaCl	†K' 2·09, dialysis (calf liver)	57W

†K' defined in albumin (1121, Mn^{2+}, 63M).

57W J. S. Wiberg and W. F. Neuman, *Arch. Biochem. Biophys.*, 1957, **72**, 66

$C_{36}H_{60}O_{12}N_4$,L; Valinomycin [For correct sequence this Table should be re-numbered 1120.1]

Metal	*Method*	*Temp*	*Medium*	*Log of equilibrium constant, remarks*	*Ref*
K^+	gl	20	1 KI methanol	$K_1 > 3{\cdot}9$	68W
Na^+	gl	20	1 NaI methanol	K_1 $1{\cdot}1 \pm 0{\cdot}4$	68W

68W H. K. Wipf, L. A. R. Pioda, Z. Stefanac, and W. Simon, *Helv. Chim. Acta*, **1968**, **51**, 377

Corrigenda

following is a list of errors in this Supplement noted by the compilers after this ıme had gone to press.

114 Substitute for Errata:
Errata. 1964, p. 365, Table 114. Under Cd^{2+}, first line, *for* 25 *read* 35; *for* Cr^{3+} *read* Cr^{2+}, under Cr^{2+} *for* K_2 2·92 *read* β_2 2·92; under Fe^{2+}, second line, *delete* entire line; p. 366, under Mn^{2+}, first line, *for* K_1 0·84, K_2 1·22 *read* K_1 1·22, K_2 0·85; p. 367, *for* 43L *read* 34L; reference 59P, *for J. Phys. Chem.*, 1958, **62**, 767 *read J. Chem. Soc.*, 1959, 1710; reference 60S, *delete* reference.

130
Under Ag, line 7, for ΔH_2 read $\Delta H_{\beta 2}$
line 9, for ΔS_2 read $\Delta S_{\beta 2}$

130 Add to Errata:
Under Cu^{2+}, seventh line, *for* K_1 6·51, *read* K_1 8·51

143
Add after 11th line (−62B):
gl 25 0–0·2($NaClO_4$) K_1 (→0) 5·684, (0·03) 5·432±0·007, (0·05) 5·382±0·0005, (0·10) 5·281±0·007, (0·15) 5·204±0·006, (0·20) 5·172±0·002 K_2 (→0) 2·854, (0·03) 2·701±0·007, (0·05) 2·69±0·01, (0·10) 2·654±0·002, (0·15) 2·654±0·002, (0·20) 2·618+0·004 62Ba

e 143 Substitute for Errata:
Errata. 1964, p. 385, Table 143. Under H^+, first line, for 40H, *read* 40H, 61N; under Ba^{2+}, third line, *insert* 25 in third column; under Ca^{2+}, second line, *for* $K(Ba^{2+}+HL^- \rightleftharpoons BaHL^+)$ *read* $K(Ca^{2+}+HL^- \rightleftharpoons CaHL^+)$; p. 386, under Mg^{2+}, fourth line, *for* log $\delta K_1/\delta t$ *read* δ log $K_1/\delta t$.

e 154.1 *For* 1-Methylethylamine *read* 2-Propylamine

e 167.1 The principal name should read:
2-Hydroxypropenoic acid phosphate.

e 167.2 *For* Thiopropanoic acid *read* Propanthioic acid

e 167.3 The name should read:
O-Ethyl dithiocarbonate (Xanthic acid)

es 168 and 170.2 Since β-mercaptopropionic acid and 3-mercaptopropanoic are the same compound, the two tables should be combined under 168 and the following name used:
3-Mercaptopropanoic acid (β-mercaptopropionic acid)

le 170.4 Under Ag^+ change β_2 11·5 to $K[Ag^+ + 3HL \rightleftharpoons Ag(HL)_3{}^+]$ 11·5

les 186.1 and 186.2 "Ene" should be deleted from "methylenephosphoric acid" to conform with nomenclature in subsequent Tables.

le 198 Add:
Errata. 1964, p. 409, Table 198. Under H^+, last line, *for* K_{17} *read* K_{12}.

le 216 "Azolidine", and the parentheses round "pyrrolidine" should be deleted, leaving "Pyrrolidine" as the only name.

le 216.9 Footnote should be changed to:
"Positions of hydroxy-groups not stated."

le 223.1 *For* $C_4H_{12}O_3$ *read* $C_4H_{10}O_3$; therefore this Table should be renumbered 217.2.

le 225 The data in this Table are for (*meso*-2,3-butylenedinitrilo)tetra-acetic acid, and should thus form part of Table 932.2.

le 226 The data in this Table are for (DL-2,3-butylenedinitrilo)tetra-acetic acid, and should thus form part of Table 932.1.

CORRIGENDA

Table 231 *For* 2,2′-Iminobis(ethylamine) *read* 1,4,7-Triazaheptane

Table 249.1 This Table should be renumbered 251.

Table 314.1 *For* $C_5H_9O_2N$ *read* $C_5H_{13}O_2N$; therefore this Table should renumbered 338.

Tables 348 and 349 These should not be included as no metal is involved in the measurements.

Table 362 Under H^+, second line:

for	sp	25	0·5 (NaCl)	K_1 8·37±0·03, K_2 −0·71±0·10	66B
read	gl	25	0·5 (NaCl)	K_1 8·37±0·03	66B
	sp	25	var H_2SO_4	K_2 −0·71±0·10	66B

Table 406 *For* 3,3′-Iminobis(propylamine) *read* 1,5,9-Triazanonane

Table 407 *For* Ethylenebis(2′-aminoethylamine) *read* 1,4,7,10-Tetra-azadecane

Table 447.1 *For* 2-Methylazolidin-5-one-2-carboxylic acid *read* 2-Methyl-5-pyrrolidone-2-carboxylic acid.

Table 526.1 The following data were accidentally omitted from this Table:

H^+ gl 25 0—0·2 ($NaClO_4$) K_1 (→0) 5·96±0·01, (0·03) 5·644±0·004, (0·05) 5·592±0·005, (0·10) 5·50±0·01, (0
5·435±0·005, (0·20) 5·394±0·005
K_2 (→0) 3·018±0·005, (0·03) 2·889±0·006, (0·05) 2·848±0·003, (0·10) 2·810±0
(0·15) 2·790±0·003, (0·20) 2·780±0·008

Table 704.2 *For* $C_9H_8O_4$ *read* $C_9H_{10}O_4$; therefore this Table should be renumbered 710.2 and as a consequence Table
becomes 704.2.

...dex of Inorganic Ligands

Numbers in ordinary type refer to Table numbers in the 1964 Edition, those in *italic* type to Tables in this Supplement.

Index of Organic Ligands Excluding Macromolecules

[Numbers are Table Numbers]

…res in *italic* type refer to Special Publication No. 25, the supplement to the 1964 Tables (Special Publication No. 17). The references to the latter are in normal type figures.

INDEX OF ORGANIC LIGANDS EXCLUDING MACROMOLECULES

INDEX OF ORGANIC LIGANDS EXCLUDING MACROMOLECULES

INDEX OF ORGANIC LIGANDS EXCLUDING MACROMOLECULES

INDEX OF ORGANIC LIGANDS EXCLUDING MACROMOLECULES

INDEX OF ORGANIC LIGANDS EXCLUDING MACROMOLECULES

dex for Macromolecules

.mulative Index of Metals

Numbers refer to Table numbers, unless given in parentheses; then they refer to pages
Numbers in ordinary type refer to the 1964 Editions. Numbers in *Italic* type refer to Supplement No. 1, 1970.

Functional Group Index for the Organic Tables in the 1964 edition and the 1970 Supplement

For ease of reference, the compounds have been arranged in alphabetical order.

A

B

C

D

E

F

G

H

I

K

M

N

O

P

R

S

T

U